Lisa Benitez

TEXTBOOK of
MEDICAL PHYSIOLOGY

ninth edition

TEXTBOOK of
MEDICAL PHYSIOLOGY

ARTHUR C. GUYTON, M.D.

Professor Emeritus
Department of Physiology and Biophysics
University of Mississippi Medical Center
Jackson, Mississippi

JOHN E. HALL, Ph.D.

Professor and Chairman
Department of Physiology and Biophysics
University of Mississippi Medical Center
Jackson, Mississippi

W.B. SAUNDERS COMPANY

A Division of Harcourt Brace & Company

PHILADELPHIA – LONDON – TORONTO – MONTREAL – SYDNEY – TOKYO

W.B. SAUNDERS COMPANY
A Division of Harcourt Brace & Company

The Curtis Center
Independence Square West
Philadelphia, Pennsylvania 19106

Library of Congress Cataloging-in-Publication Data

Guyton, Arthur C.
   Textbook of medical physiology / Arthur C. Guyton, John E. Hall.—9th ed.

      p.   cm.

   Includes bibliographical references and index.
   ISBN 0–7216–5944–6
   1. Human physiology.   2. Physiology, Pathological.   I. Hall, John
E. (John Edward).   II. Title.
   [DNLM: 1. Physiology.   QT 104 G992t 1996]

   QP34.5G9   1996
   612—dc20

   DNLM/DLC   94–40510

Cover illustration is *OPUS 1972*, enamel on steel, by Virgil Cantini, with permission of the artist and Mansfield State College, Mansfield, Pennsylvania.

                                                                ISBN   0–7216–5944–6

TEXTBOOK OF MEDICAL PHYSIOLOGY   International Edition   0–7216–6773–2

Copyright © 1996, 1991, 1986, 1981, 1976, 1971, 1966, 1961, 1956 by W.B. Saunders Company.

Copyright renewed 1994, 1984 by Arthur C. Guyton.

All rights reserved. No part of this publication may be reproduced or transmitted in any form or by any means, electronic or mechanical, including photocopy, recording, or any information storage and retrieval system, without permission in writing from the publisher.

Printed in the United States of America.

Last digit is the print number: 9   8   7   6   5   4   3   2   1

To

My Father
For the Uncompromising Principles that Guided His Life

My Mother
For Leading Her Children into Intellectual Pursuits

My Wife
For Her Magnificent Devotion to Her Family

My Children
For Making Everything Worthwhile

A.C.G.

To

My Teachers
For Showing Me the Excitement and Joy of Physiology

My Family
*For Their Abundant Support, For Their Patience and Understanding
and for Their Love.*

J.E.H.

Preface

Each time this *Textbook of Medical Physiology* is revised, we keep wishing that some day physiology could become a completely mature subject without change from year to year. However, this always proves to be far from the truth. Physiology is a vast discipline, and only now are we beginning to make inroads into many of its fundamental secrets. Most importantly, within the past dozen years many new techniques for learning cellular and molecular physiology have become available. Therefore, more and more can we present physiological principles in molecular and physical scientific terms rather than merely as a series of separate biological phenomena. This change we all welcome, but it also makes revision of almost every section of each chapter a necessity.

To help in this massive job of revision, a new author, John Hall, Ph.D., has joined in the preparation of this 9th edition of the *Textbook of Medical Physiology*. The two of us, Dr. Guyton and Dr. Hall, have worked very closely together during the past 20 years. Furthermore, we have written many articles together, so that it has been possible to give this new revision a unified organization that will continue its special usefulness to the student, yet at the same time to maintain a book that is comprehensive enough that students will wish to carry it with them in all later life as a basis for a professional career. As can be expected, Dr. Hall has brought new insights and new bodies of knowledge that have helped immensely in achieving these goals.

The beauty of physiology is that it integrates the individual functions of all the body's different cells and organs into a functional whole, the human or animal body. Indeed, life in the human being relies upon this total function, not on functions of the single parts in isolation from the others.

This brings us to an entirely different subject: how are the separate organs and systems themselves controlled so that no one system overfunctions while others fail to provide their share? Fortunately, the body is endowed with a vast network of feedback controls that achieve the necessary balances, without which we would not be able to live. Physiologists call this high level of internal bodily control *homeostasis*. In disease conditions, more often than not the functional balances become seriously disturbed—that is, homeostasis becomes very poor. When the disturbance is too great, the whole body can no longer live. Therefore, one of the principal goals of any medical physiology text is to emphasize the effectiveness and beauty of the body's homeostatic mechanisms as well as to discuss their aberrant function in disease.

Another goal of each new edition of this text has been to make it as accurate as possible. To help attain this, suggestions and critiques from many physiologists, students, and clinicians throughout the United States and other parts of the world have been sought and used in checking the factual accuracy of the text, as well as its appropriate balance. Yet, even so, because of the likelihood of error in sorting through thousands of bits of information, we wish to issue still a further invitation —in fact, much more than merely an invitation, actually a request—to all readers to send along notices of error or inaccuracy. Indeed, physiologists understand how important feedback is to proper function of the human body; so, too, is feedback equally important for progressive development of a textbook of medical physiology. We hope also that those many persons who have helped already will accept our sincerest thanks for their efforts.

A word of explanation is needed about two features of the text—first, the references, and, second, the two print sizes. The references have been chosen primarily for their presentation of physiological principles and for the quality of their own bibliographies. Use of these references, as well as cross-references from them, can give the student almost complete coverage of the entire field of physiology.

The print is set in two sizes. The material presented in small print is of several different kinds: first, anatomical, chemical, and other information that is needed for the immediate discussion but that most students will learn in more detail in other courses; second, information that is of special importance to certain fields of clinical medicine though not necessarily important to understanding the body's basic physiology; and third, information that will be of value to those students who wish to pursue a subject more deeply than does the average medical student.

In contrast, the material in large print constitutes the fundamental physiological information that students will require in their medical studies and that they will not obtain in other courses. Those teachers who would like to present a limited course of physiology can direct student study primarily to the large type.

Again, we wish to express our deepest appreciation to many others who have helped in the preparation of this book. We are particularly grateful to Ivadelle Osberg Heidke, Susie Zuller, and Gwendolyn Robbins for their excellent secretarial services, to Tomika Mita, Michael Schenk, Cathy Garrity, Angela Gardner, and Myriam Kirkman for their superb work and helpfulness on the illustrations, and to Ms. Kim Kist, Lorraine B. Kilmer, Frank Polizzano, Edna Dick, and Matt Andrews of the staff of the W.B. Saunders Co. for continued editorial and production excellence.

ARTHUR C. GUYTON AND JOHN E. HALL

Contents

UNIT II

MEMBRANE PHYSIOLOGY, NERVE, AND MUSCLE

UNIT III

THE HEART

Chapter 11

The Normal Electrocardiogram . . . .

Chapter 12

Electrocardiographic Interpretation of Cardiac Muscle and Coronary
Abnormalities: Vectorial Analysis . . . .

UNIT IV

THE CIRCULATION

UNIT V

THE KIDNEYS AND BODY FLUIDS

Chapter 26

*Urine Formation by the Kidneys: I. Glomerular Filtration, Renal Blood
Flow, and Their Control*   315

Chapter 29

*Integration of Renal Mechanisms for Control of Blood Volume and
Extracellular Fluid Volume; and Renal Regulation of Potassium,
Calcium, Phosphate, and Magnesium.........................* 367

Chapter 30

Regulation of Acid-Base Balance................................. 385

UNIT VI

BLOOD CELLS, IMMUNITY, AND BLOOD CLOTTING

Chapter 32

Red Blood Cells, Anemia, and Polycythemia 425

Chapter 33

Resistance of the Body to Infection: I. Leukocytes, Granulocytes, the Monocyte-Macrophage System, and Inflammation 435

UNIT VII

RESPIRATION

Chapter 37

Pulmonary Ventilation 477

UNIT X

THE NERVOUS SYSTEM: B. THE SPECIAL SENSES

Chapter 49

The Eye: I. Optics of Vision

Chapter 50

The Eye: II. Receptor and Neural Function of the Retina

UNIT XI

THE NERVOUS SYSTEM: C. MOTOR AND INTEGRATIVE NEUROPHYSIOLOGY

Chapter 59

States of Brain Activity—Sleep; Brain Waves; Epilepsy; Psychoses 761

Chapter 60

The Autonomic Nervous System; The Adrenal Medulla 769

UNIT XIV

ENDOCRINOLOGY AND REPRODUCTION

Chapter 81

Female Physiology Before Pregnancy; and the Female Hormones. 1017

Chapter 82

Pregnancy and Lactation. 1033

Chapter 83
Fetal and Neonatal Physiology 1047

UNIT XV

SPORTS PHYSIOLOGY

Chapter 84
Sports Physiology 1059

INTRODUCTION TO PHYSIOLOGY: THE CELL AND GENERAL PHYSIOLOGY

UNIT I

1. Functional Organization of the Human Body and Control of the "Internal Environment"

2. The Cell and Its Function

3. Genetic Control of Protein Synthesis, Cell Function, and Cell Reproduction

Functional Organization of the Human Body and Control of the "Internal Environment"

CHAPTER 1

The goal of physiology is to explain the physical and chemical factors that are responsible for the origin, development, and progression of life. Each type of life, from the very simple virus up to the largest tree or to the complicated human being, has its own functional characteristics. Therefore, the vast field of physiology can be divided into *viral physiology, bacterial physiology, cellular physiology, plant physiology, human physiology,* and many more subdivisions.

HUMAN PHYSIOLOGY. In human physiology, we are concerned with the specific characteristics and mechanisms of the human body that make it a living being. The very fact that we remain alive is almost beyond our own control, for hunger makes us seek food and fear makes us seek refuge. Sensations of cold make us provide warmth, and other forces cause us to seek fellowship and to reproduce. Thus, the human being is actually an automaton, and the fact that we are sensing, feeling, and knowledgeable beings is part of this automatic sequence of life; these special attributes allow us to exist under widely varying conditions that otherwise would make life impossible.

CELLS AS THE LIVING UNITS OF THE BODY

The basic living unit of the body is the cell, and each organ is an aggregate of many different cells held together by intercellular supporting structures. Each type of cell is specially adapted to perform one or a few particular functions. For instance, the red blood cells, 25 trillion in all, transport oxygen from the lungs to the tissues. Although this type of cell is perhaps the most abundant of all, there are perhaps another 75 trillion cells. The entire body, then, contains about 100 trillion cells.

Although the many cells of the body often differ markedly from each other, all of them have certain basic characteristics that are alike. For instance, in all cells, oxygen combines with the breakdown products of carbohydrate, fat, or protein to release the energy required for cell function. Furthermore, the general mechanisms for changing nutrients into energy are basically the same in all cells, and all the cells also deliver the end products of their chemical reactions into the surrounding fluids.

Almost all cells also have the ability to reproduce, and when cells of a particular type are destroyed from one cause or another, the remaining cells of this type often generate new cells until the supply is replenished.

EXTRACELLULAR FLUID—THE INTERNAL ENVIRONMENT

About 56 per cent of the adult human body is fluid. Although most of this fluid is inside the cells and is called *intracellular fluid*, about one third is in the spaces outside the cells and is called *extracellular fluid*. This extracellular fluid is in constant motion throughout the body. It is rapidly transported in the circulating blood and then mixed between the blood and the tissue fluids by diffusion through the capillary walls.

In the extracellular fluid are the ions and nutrients

needed by the cells for maintenance of cellular life. Therefore, all cells live in essentially the same environment, the extracellular fluid, for which reason the extracellular fluid is called the *internal environment* of the body, or the *milieu intérieur,* a term introduced more than a hundred years ago by the great 19th-century French physiologist Claude Bernard.

Cells are capable of living, growing, and performing their special functions so long as the proper concentrations of oxygen, glucose, different ions, amino acids, fatty substances, and other constituents are available in this internal environment.

DIFFERENCES BETWEEN EXTRACELLULAR AND INTRACELLULAR FLUIDS. The extracellular fluid contains large amounts of *sodium, chloride,* and *bicarbonate ions,* plus nutrients for the cells, such as *oxygen, glucose, fatty acids,* and *amino acids.* It also contains carbon dioxide that is being transported from the cells to the lungs to be excreted, plus other cellular products that are being transported to the kidneys for excretion.

The intracellular fluid differs significantly from the extracellular fluid; particularly, it contains large amounts of *potassium, magnesium,* and *phosphate ions* instead of the sodium and chloride ions found in the extracellular fluid. Special mechanisms for transporting ions through the cell membranes maintain these differences. These mechanisms are discussed in Chapter 4.

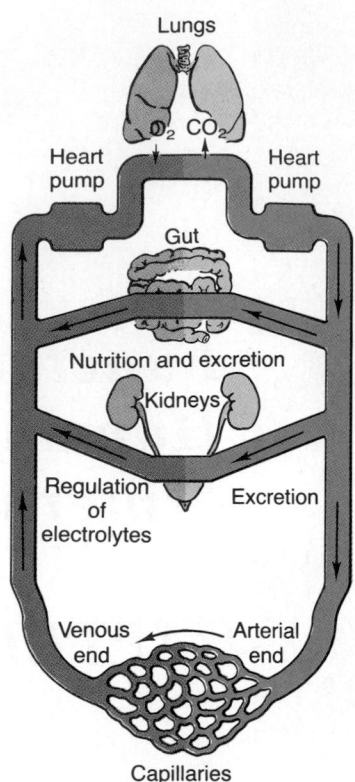

Figure 1–1. General organization of the circulatory system.

"HOMEOSTATIC" MECHANISMS OF THE MAJOR FUNCTIONAL SYSTEMS

Homeostasis

The term *homeostasis* is used by physiologists to mean *maintenance of static or constant conditions in the internal environment.* Essentially all of the organs and tissues of the body perform functions that help to maintain these constant conditions. For instance, the lungs provide oxygen to the extracellular fluid to replenish continually the oxygen that is being used by the cells, the kidneys maintain constant ion concentrations, and the gastrointestinal system provides nutrients.

A large segment of this text is concerned with the manner in which each organ or tissue contributes to homeostasis. To begin this discussion, in this chapter, the different functional systems of the body and their contributions to homeostasis are outlined briefly; then we outline briefly the basic theory of the control systems that cause the functional systems to operate in harmony with one another.

Extracellular Fluid Transport System— The Circulatory System

Extracellular fluid is transported through all parts of the body in two stages. The first stage entails movement of blood around and around the circulatory system, and the second, movement of fluid between the blood capillaries and the cells. Figure 1–1 shows the overall circulation of blood. All the blood in the circulation traverses the entire circuit of the circulation an average of once each minute when the body is at rest and as many as six times each minute when a person becomes extremely active.

As blood passes through the capillaries, continual exchange of extracellular fluid also occurs between the plasma portion of the blood and the interstitial fluid that fills the spaces between the cells, the *intercellular spaces.* This process is shown in Figure 1–2. Note that the capillaries are porous so that large amounts of fluid and its dissolved constituents can *diffuse* back and forth between the blood and the tissue spaces, as shown by the arrows. This process of diffusion is caused by kinetic motion of the molecules in both the plasma and the interstitial fluid. That is, the fluid and dissolved molecules are continually moving and bouncing in all directions within the fluid itself and also through the pores and through the tissue spaces. Few cells are located more than 50 micrometers from a capillary, which ensures diffusion of almost any substance from the capillary to the cell within a few seconds. Thus, the extracellular fluid everywhere in the body, both that of the plasma and that in interstitial spaces, is continually being mixed, thereby maintaining almost complete homogeneity throughout the body.

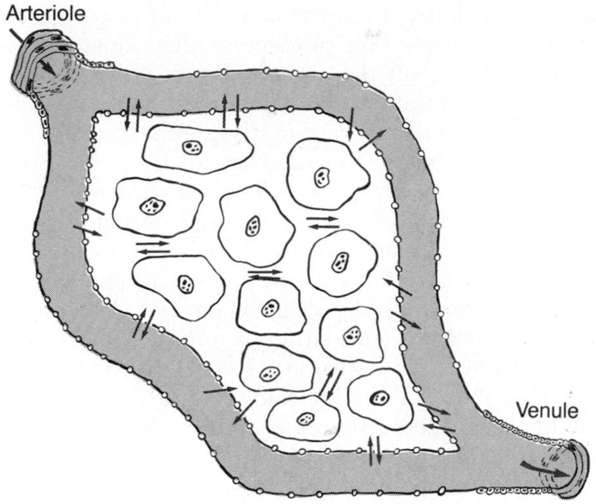

Arteriole

Figure 1–2. Diffusion of fluids through the capillary walls and through the interstitial spaces.

Venule

Origin of Nutrients in the Extracellular Fluid

RESPIRATORY SYSTEM. Figure 1–1 shows that each time the blood passes through the body, it also flows through the lungs. The blood picks up oxygen in the alveoli, thus acquiring the *oxygen* needed by the cells. The membrane between the alveoli and the lumen of the pulmonary capillaries is only 0.4 to 2.0 micrometers thick, and oxygen diffuses by molecular motion through this membrane into the blood in the same manner that water and ions diffuse through the tissue capillaries.

GASTROINTESTINAL TRACT. A large portion of the blood pumped by the heart also passes through the walls of the gastrointestinal organs. Here different dissolved nutrients, including *carbohydrates, fatty acids,* and *amino acids,* are absorbed from the ingested food into the extracellular fluid.

LIVER AND OTHER ORGANS THAT PERFORM PRIMARILY METABOLIC FUNCTIONS. Not all substances absorbed from the gastrointestinal tract can be used in their absorbed form by the cells. The liver changes the chemical compositions of many of these substances to more useable forms, and other tissues of the body— fat cells, gastrointestinal mucosa, kidneys, and endocrine glands—help to modify the absorbed substances or store them until they are needed.

MUSCULOSKELETAL SYSTEM. Sometimes the question is asked, How does the musculoskeletal system fit into the homeostatic functions of the body? The answer is obvious and simple: Were it not for this system, the body could not move to the appropriate place at the appropriate time to obtain the foods required for nutrition. The musculoskeletal system also provides motility for protection against adverse surroundings, without which the entire body, and along with it all the homeostatic mechanisms, could be destroyed instantaneously.

Removal of Metabolic End Products

REMOVAL OF CARBON DIOXIDE BY THE LUNGS. At the same time that blood picks up oxygen in the lungs, *carbon dioxide* is released from the blood into the alveoli, and the respiratory movement of air into and out of the alveoli carries the carbon dioxide to the atmosphere. Carbon dioxide is the most abundant of all the end products of metabolism.

KIDNEYS. Passage of the blood through the kidneys removes most of the other substances besides carbon dioxide from the plasma that are not needed by the cells. These substances include especially different end products of cellular metabolism, such as urea and uric acid; they include as well excesses of ions and water from the food that might have accumulated in the extracellular fluid. The kidneys perform their function by first filtering large quantities of plasma through the glomeruli into the tubules and then reabsorbing into the blood those substances needed by the body, such as glucose, amino acids, appropriate amounts of water, and many of the ions. Most of the substances not needed by the body, especially the metabolic end products such as urea, are poorly reabsorbed and instead pass on through the renal tubules into the urine.

Regulation of Body Functions

NERVOUS SYSTEM. The nervous system is composed of three major parts: the *sensory input portion,* the *central nervous system* (or *integrative portion*), and the *motor output portion.* Sensory receptors detect the state of the body or the state of the surroundings. For instance, receptors present everywhere in the skin apprise one every time an object touches the skin at any point. The eyes are sensory organs that give one a visual image of the surrounding area. The ears also are sensory organs. The central nervous system is composed of the brain and spinal cord. The brain can store information, generate thoughts, create ambition, and determine reactions the body performs in response to the sensations. Appropriate signals are then transmitted through the motor output portion of the nervous system to carry out one's desires.

A large segment of the nervous system is called the *autonomic system.* It operates at a subconscious level and controls many functions of the internal organs, including the level of pumping activity by the heart, movements of the gastrointestinal tract, and glandular secretion.

HORMONAL SYSTEM OF REGULATION. Located in the body are eight major endocrine glands that secrete chemical substances called *hormones.* Hormones are transported in the extracellular fluid to all parts of the body to help regulate cellular function. For instance, thyroid hormone increases the rates of most chemical reactions in all cells. In this way, thyroid hormone helps to set the tempo of bodily activity. Insulin controls glucose metabolism; adrenocortical hormones control sodium ion, potassium ion, and protein metabolism; and parathyroid hormone controls bone calcium and phosphate. Thus, the hormones are a system of regulation that complements the nervous system. The nervous system regulates mainly muscular and secretory activities of the body, whereas the hormonal system regulates mainly metabolic functions.

Reproduction

Sometimes reproduction is not considered a homeostatic function. It does, however, help to maintain static conditions by generating new beings to take the place of those that are dying. This perhaps sounds like a permissive usage of the term homeostasis, but it does illustrate that, in the final analysis, essentially all body structures are so organized that they help maintain the automaticity and continuity of life.

CONTROL SYSTEMS OF THE BODY

The human body has literally thousands of control systems in it. The most intricate of these are the genetic control systems that operate in all cells to control intracellular function as well as all extracellular functions. This subject is discussed in Chapter 3. Many other control systems operate within the organs to control functions of the individual parts of the organs; others operate throughout the entire body to control the interrelations between the organs. For instance, the respiratory system, operating in association with the nervous system, regulates the concentration of carbon dioxide in the extracellular fluid. Also, the liver and pancreas regulate the concentration of glucose in the extracellular fluid. The kidneys regulate concentrations of hydrogen, sodium, potassium, phosphate, and other ions in the extracellular fluid.

Examples of Control Mechanisms

REGULATION OF OXYGEN AND CARBON DIOXIDE CONCENTRATIONS IN THE EXTRACELLULAR FLUID. Because oxygen is one of the major substances required for chemical reactions in the cells, it is fortunate that the body has a special control mechanism to maintain an almost exact and constant oxygen concentration in the extracellular fluid. This mechanism depends principally on the chemical characteristics of *hemoglobin,* which is present in all red blood cells. Hemoglobin combines with oxygen as the blood passes through the lungs. Then, as the blood passes through the tissue capillaries, hemoglobin, because of its own strong chemical affinity for oxygen, does not release the oxygen into the tissue fluid if too much oxygen is already there. If the oxygen concentration is too low, however, sufficient oxygen is released to re-establish adequate tissue oxygen concentration. Thus, the regulation of oxygen concentration in the tissues is vested principally in the chemical characteristics of hemoglobin itself. This regulation is called the *oxygen-buffering function of hemoglobin.*

Carbon dioxide concentration in the extracellular fluid is regulated in quite a different way. Carbon dioxide is a major end product of the oxidative reactions in cells. If all the carbon dioxide formed in the cells should continue to accumulate in the tissue fluids, the mass action of the carbon dioxide itself would soon halt all the energy-giving reactions of the cells. Fortunately, a high carbon dioxide concentration in the blood *excites the respiratory center,* causing a person to breathe rapidly and deeply. This increases the expiration of carbon dioxide and, therefore, its removal from the blood and the extracellular fluid. This process continues until the concentration returns to normal.

REGULATION OF ARTERIAL PRESSURE. Several systems contribute to the regulation of arterial pressure. One of these, the *baroreceptor system,* is a simple and excellent example of a control mechanism. In the walls of the bifurcation region of the carotid arteries in the neck as well as the arch of the aorta are many nerve receptors, called *baroreceptors,* which are stimulated by stretch of the arterial wall. When the arterial pressure becomes great, the baroreceptors send barrages of impulses to the medulla of the brain. Here the impulses inhibit the *vasomotor center,* which in turn decreases the number of impulses transmitted through the sympathetic nervous system to the heart and blood vessels. Lack of these impulses causes diminished pumping activity by the heart and increased ease of blood flow through the peripheral vessels, both of which lower the arterial pressure back toward normal.

Conversely, a fall in arterial pressure relaxes the stretch receptors, allowing the vasomotor center to become more active than usual, thereby causing the arterial pressure to rise back toward normal.

Normal Ranges of Important Extracellular Fluid Constituents

Table 1–1 lists the more important constituents and physical characteristics of extracellular fluid along with their normal values, normal ranges, and maximum limits without causing death for short periods of time. Note the narrowness of the normal range for each one. Values outside these ranges usually are the cause or the result of illness.

Even more important are the limits beyond which abnormalities can cause death. For instance, an increase in the body temperature of only 10° to 12°F (6° to 7°C) above normal can lead to a vicious circle of increasing cellular metabolism that literally destroys the cells. Note also the narrow range for the acid-base balance of the body, with a normal pH value of 7.4 and lethal values only about 0.5 on either side of the normal value. Another important factor is potassium ion because whenever its concentration falls to less than one third of normal, a person is likely to be paralyzed as a result of the nerves' inability to carry nerve signals. Alternatively, if ever the potassium ion concentration rises to two or more times normal, the heart muscle is likely to be severely depressed. Also, when the calcium ion concentration falls below about one half of normal, a person is likely to experience

Table 1–1 SOME IMPORTANT CONSTITUENTS AND PHYSICAL CHARACTERISTICS OF THE EXTRACELLULAR FLUID, THE NORMAL RANGE OF CONTROL, AND THE APPROXIMATE NONLETHAL LIMITS FOR SHORT PERIODS

	Normal Value	Normal Range	Approximate Nonlethal Limits	Units
Oxygen	40	35–45	10–1000	mm Hg
Carbon dioxide	40	35–45	5–80	mm Hg
Sodium ion	142	138–146	115–175	mmol/L
Potassium ion	4.2	3.8–5.0	1.5–9.0	mmol/L
Calcium ion	1.2	1.0–1.4	0.5–2.0	mmol/L
Chloride ion	108	103–112	70–130	mmol/L
Bicarbonate ion	28	24–32	8–45	mmol/L
Glucose	85	75–95	20–1500	mg/dl
Body temperature	98.4 (37.0)	98–98.8 (37.0)	65–110 (18.3–43.3)	°F (°C)
Acid-base	7.4	7.3–7.5	6.9–8.0	pH

tetanic contraction of muscles throughout the body because of spontaneous generation of nerve impulses in the peripheral nerves. When the glucose concentration falls below one half of normal, a person frequently develops extreme mental irritability and sometimes even convulsions.

Thus, consideration of these examples should give one an extreme appreciation of the value and even necessity of the vast numbers of control systems that keep the body operating in health; in absence of any one of these controls, serious illness or death can result.

Characteristics of Control Systems

The aforementioned examples of homeostatic control mechanisms are only a few of the many hundreds to thousands in the body, all of which have certain characteristics in common. They are explained in the following pages.

Negative Feedback Nature of Most Control Systems

Most control systems of the body act by *negative feedback,* which can best be explained by reviewing some of the homeostatic control systems mentioned earlier. In the regulation of carbon dioxide concentration, a high concentration of carbon dioxide in the extracellular fluid increases pulmonary ventilation. This in turn decreases carbon dioxide concentration because the lungs then excrete greater amounts of carbon dioxide out of the body. In other words, the high concentration causes a decreased concentration, which is *negative* to the initiating stimulus. Conversely, if the carbon dioxide concentration falls too low, this causes a feedback increase in the concentration. This response also is negative to the initiating stimulus.

In the arterial pressure–regulating mechanisms, a high pressure causes a series of reactions that promote a lowered pressure, or a low pressure causes a series of reactions that promote an elevated pressure. In both instances, these effects are negative with respect to the initiating stimulus.

Therefore, in general, if some factor becomes excessive or deficient, a control system initiates *negative feedback,* which consists of a series of changes that return the factor toward a certain mean value, thus maintaining homeostatis.

GAIN OF A CONTROL SYSTEM. The degree of effectiveness with which a control system maintains constant conditions is determined by the *gain* of the negative feedback. For instance, let us assume that a large volume of blood is transfused into a person whose baroreceptor pressure control system is not functioning, and the arterial pressure rises from the normal level of 100 mm Hg up to 175 mm Hg. Then, assume the same volume of blood is injected into the same person when the baroreceptor system is functioning, and this time the pressure rises only 25 mm Hg. Thus, the feedback control system has caused a "correction" of −50 mm Hg—that is, from 175 mm Hg to 125 mm Hg. There remains an increase in pressure of +25 mm Hg, called the "error," which means that the control system is not 100 per cent effective in preventing change. The gain of the system is then calculated by the following formula:

$$\text{Gain} = \frac{\text{Correction}}{\text{Error}}$$

Thus, in the baroreceptor system example, the correction is −50 mm Hg and the error still persisting is +25 mm Hg. Therefore, the gain of the person's baroreceptor system for control of arterial pressure is −50 divided by +25, or −2. That is, an extraneous factor that increases or decreases the arterial pressure does so only one third as much as would occur if this control system were not present.

The gains of some other physiological control systems are much greater than that of the baroreceptor system. For instance, the gain of the system controlling body temperature is about −33. Therefore, one can see that the temperature control system is much more effective than the baroreceptor pressure control system.

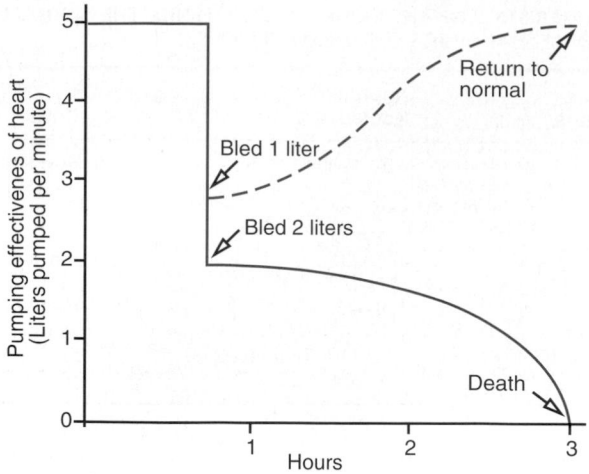

Figure 1–3. Death caused by positive feedback when 2 liters of blood are removed from the circulation.

Positive Feedback: This Sometimes Causes Vicious Circles and Death

One might ask the question, Why do essentially all control systems of the body operate by negative feedback rather than positive feedback? If you consider the nature of positive feedback, you will immediately see that positive feedback does not lead to stability but to instability and often death.

Figure 1–3 shows an instance in which death can ensue from positive feedback. This figure depicts the pumping effectiveness of the heart, showing that the heart of a normal human being pumps about 5 liters of blood per minute. If the person is suddenly bled 2 liters, the amount of blood in the body is decreased to such a low level that not enough is available for the heart to pump effectively. As a result, the arterial pressure falls and the flow of blood to the heart muscle through the coronary vessels diminishes. This results in weakening of the heart, further diminished pumping, further decrease in coronary blood flow, and still more weakness of the heart; the cycle repeats itself again and again until death occurs. Note that each cycle in the feedback results in further weakening of the heart. In other words, the initiating stimulus causes more of the same, which is *positive feedback*.

Positive feedback is better known as a vicious circle, but a mild degree of positive feedback can be overcome by the negative feedback control mechanisms of the body, and a vicious circle fails to develop. For instance, if the person in the aforementioned example were bled only 1 liter instead of 2 liters, the normal negative feedback mechanisms for controlling cardiac output and arterial pressure would overbalance the positive feedback and the person would recover, as shown by the dashed curve of Figure 1–3.

POSITIVE FEEDBACK CAN SOMETIMES BE USEFUL. In rare instances, the body has learned to use positive feedback to its advantage.

Blood clotting is an example of a valuable use of positive feedback. When a blood vessel is ruptured and a clot begins to form, multiple enzymes called *clotting factors* are activated within the clot itself. Some of these enzymes act on other, yet unactivated enzymes of the immediately adjacent blood, activating them and causing still more clot. This process continues until the hole in the vessel is plugged and bleeding no longer occurs. On occasion, this mechanism can itself get out of hand and cause the formation of unwanted clots. In fact, this is what initiates most acute heart attacks, which are caused by a clot beginning on an atherosclerotic plaque in a coronary artery and then growing until the artery is blocked.

Childbirth is another instance in which positive feedback plays a role. When uterine contractions become strong enough for the baby's head to begin pushing through the cervix, stretch of the cervix sends signals through the uterine muscle back to the body of the uterus, causing even more powerful contractions. Thus, the uterine contractions stretch the cervix, and the cervical stretch causes more contractions. When this process becomes powerful enough, the baby is born. If it is not powerful enough, the contractions usually die out, and a few days pass before they begin again.

Finally, another important use of positive feedback is for the generation of nerve signals. That is, when the membrane of a nerve fiber is stimulated, this causes slight leakage of sodium ions through sodium channels in the nerve membrane to the fiber's interior. The sodium ions entering the fiber then change the membrane potential, which in turn causes more opening of channels, more change of potential, still more opening of channels, and so forth. Thus, from a slight beginning, there is an explosion of sodium leakage that creates the nerve action potential. This action potential in turn excites the nerve fiber still further along its length, with the process continuing until the nerve signal goes all the way to all ends of the nerve fiber.

We shall learn that in each case in which positive feedback is useful, the positive feedback itself is part of an overall negative feedback process. For instance, in the case of blood clotting, the positive feedback clotting process is a negative feedback process for maintenance of normal blood volume. And the positive feedback that causes nerve signals allows the nerves to participate in literally thousands of negative feedback nervous control systems.

Some More Complex Types of Control Systems—Adaptive Control System

Later in this text, when we study the nervous system, we shall see that this system contains a morass of interconnected control mechanisms. Some are simple feedback systems similar to those we have already discussed. Many are not. For instance, some movements of the body occur so rapidly that there is not enough time for nerve signals to travel from the peripheral parts of the body all the way to the brain and then back to the periphery again in time to control the movements. Therefore, the brain uses a principle called *feed-forward control* to cause the required muscle contractions. Then, sensory nerve signals from the moving parts apprise the brain in retrospect whether or not the appropriate movement as envisaged by the brain actually has been performed correctly. If not, the brain corrects the feed-forward signals that it sends to the muscles the *next* time the movement is required. Then, once again, if still further correction needs to be made, this, too, will be done for subsequent movements. This is called *adaptive control.* Adaptive control, in a sense, is delayed negative feedback.

Thus, one can see how complex some of the feedback control systems of the body can be. A person's life literally depends on all these. Therefore, a major share of this text is devoted to discussing these life-giving mechanisms.

SUMMARY—AUTOMATICITY OF THE BODY

The purpose of this chapter has been to point out, first, the overall organization of the body and, second, the means by which the different parts of the body operate in harmony. To summarize, the body is actually a *social order of about 100 trillion cells* organized into different functional structures, some of which are called *organs.* Each functional structure provides its share in the maintenance of homeostatic conditions in the extracellular fluid, which is called the *internal environment.* As long as normal conditions are maintained in the internal environment, the cells of the body continue to live and function properly. Thus, each cell benefits from homeostasis, and in turn, each cell contributes its share toward the maintenance of homeostasis. This reciprocal interplay provides continuous automaticity of the body until one or more functional systems lose their ability to contribute their share of function. When this happens, all the cells of the body suffer. Extreme dysfunction leads to death, whereas moderate dysfunction leads to sickness.

REFERENCES

Adolph, E. F.: Physiological adaptations: Hypertrophies and superfunctions. Am. Sci., 60:608, 1972.

Adolph, E. F.: Physiological integrations in action. The Physiologist, 25:(Suppl.) 1, 1982.

Bernard, C.: Lectures on the Phenomena of Life Common to Animals and Plants. Springfield, Ill., Charles C Thomas, 1974.

Brown, J. H. U. (ed.): Engineering Principles in Physiology. Vols. 1 and 2. New York, Academic Press, 1973.

Bryant, P. J., and Simpson, P.: Intrinsic and extrinsic control of growth in developing organs. Q. Rev. Biol., 59:387, 1984.

Burattini, R., and Borgdorff, P.: Closed-loop baroreflex control of total peripheral resistance in the cat: Identification of gains by aid of a model. Cardiovasc. Res., 18:715, 1984.

Cannon, W. B.: The Wisdom of the Body. New York, W. W. Norton & Co., 1932.

Celis, J. E.: Cell biology. San Diego, Academic Press, 1994.

Cox, D. R., et al.: Assessing mapping progress in the human genome project. Science, 265:2031, 1994.

Frisancho, A. R.: Human Adaptation. St. Louis, C. V. Mosby Co., 1979.

Garland, T. Jr., and Carter, P. A.: Evolutionary physiology. Annu. Rev. Physiol., 56:579, 1994.

Gelehrter, T. D., and Collins, F. S.: Principles of Medical Genetics. Baltimore, Williams & Wilkins, 1995.

Guyton, A. C.: Arterial Pressure and Hypertension. Philadelphia, W. B. Saunders Co., 1980.

Guyton, A. C., et al.: Dynamics and Control of the Body Fluids. Philadelphia, W. B. Saunders Co., 1975.

Huffaker, C. B. (ed.): Biological Control. New York, Plenum Press, 1974.

Jones, R. W.: Principles of Biological Regulation: An Introduction to Feedback Systems. New York, Academic Press, 1973.

Klevecz, R. R., et al.: Cellular clocks and oscillators. Int. Rev. Cytol., 86:97, 1984.

Marmarelis, V. Z.: Advanced Methods of Physiological System Modeling. New York, Plenum Publishing Corp., 1994.

McPhee, S. J., et al.: Pathophysiology of Disease: An Introduction to Clinical Medicine. Norwalk, Conn., Appleton & Lange, 1994.

Milhorn, H. T.: The Application of Control Theory to Physiological Systems. Philadelphia, W. B. Saunders Co., 1966.

Moore, K. L., and Persaud, T. V. N.: The Developing Human: Clinically Oriented Embryology. Philadelphia, W. B. Saunders Co., 1993.

Moore, K. L., et al.: Color Atlas of Clinical Embryology. Philadelphia, W. B. Saunders Co., 1994.

Nora, J. J., and Fraser, F. Cl.: Medical Genetics: Principles and Practice. Baltimore, Williams & Wilkins, 1994.

Piva, F., et al.: Regulation of hypothalamic and pituitary function: Long, short, and ultrashort feedback loops. In DeGroot, L. J., et al. (eds.): Endocrinology. Vol. 1. New York, Grune & Stratton, 1979, p. 21.

Randall, J. E., Microcomputers and Physiological Simulation. Reading, Mass., Addison-Wesley Publishing Co., 1980.

Reeve, E. B., and Guyton, A. C.: Physical Bases of Circulatory Transport: Regulation and Exchange. Philadelphia, W. B. Saunders Co., 1967.

Reichel, W.: Clinical Aspects of Aging. Baltimore, Williams & Wilkins, 1995.

Sadler, T. W.: Langman's Medical Embryology. Baltimore, Williams & Wilkins, 1995.

Shenolikar, S., and Nairn, A. C.: Model Systems in Signal Transduction. New York, Raven Press, 1993.

Sweetser, W.: Human Life (Aging and Old Age). New York, Arno Press, 1979.

Thompson, R. F.: The neurobiology of learning and memory. Science, 233:941, 1986.

Toates, F. M.: Control Theory in Biology and Experimental Psychology. London, Hutchinson Education Ltd., 1975.

Yates, F. E. (ed.): Self-Organizing Systems. New York, Plenum Publishing Corp., 1987.

The Cell and Its Function

CHAPTER 2

Each of the 100 trillion or more cells in a human being is a living structure that can survive indefinitely and, in most instances, even reproduce itself, provided its surrounding fluids contain appropriate nutrients. To understand the function of organs and other structures of the body, it is essential that we first understand the basic organization of the cell and the functions of its component parts.

ORGANIZATION OF THE CELL

A typical cell, as seen by the light microscope, is shown in Figure 2–1. Its two major parts are the *nucleus* and the *cytoplasm*. The nucleus is separated from the cytoplasm by a *nuclear membrane,* and the cytoplasm is separated from the surrounding fluids by a *cell membrane.*

The different substances that make up the cell are collectively called *protoplasm.* Protoplasm is composed mainly of five basic substances: water, electrolytes, proteins, lipids, and carbohydrates.

WATER. The principal fluid medium of the cell is water, which is present in most cells besides fat cells in a concentration of between 70 and 85 per cent. Many cellular chemicals are dissolved in the water, whereas others are suspended in particulate or membranous form. Chemical reactions take place among the dissolved chemicals or at the surface boundaries between the suspended particles or membranes and the water.

ELECTROLYTES. The most important electrolytes in the cell are *potassium, magnesium, phosphate, sulfate, bicarbonate,* and small quantities of *sodium, chloride,* and *calcium.* They are discussed in much greater detail in Chapter 4, which considers the interrelations between the intracellular and extracellular fluids.

The electrolytes provide inorganic chemicals for cellular reactions. Also, they are necessary for operation of some of the cellular control mechanisms. For instance, electrolytes acting at the cell membrane allow transmission of electrochemical impulses in nerve and muscle fibers, and the concentrations of certain intracellular electrolytes determine and control the activity of different enzymatically catalyzed reactions that are necessary for cellular metabolism.

PROTEINS. Next to water, the most abundant substance in most cells is proteins, which normally constitute 10 to 20 per cent of the cell mass. These can be divided into two types, *structural proteins* and *globular proteins,* which are mainly *enzymes.*

To get an idea of what is meant by *structural proteins,* one need only note that leather is composed principally of structural proteins and that hair is almost entirely a structural protein. Proteins of this type are present in the cell mainly in the form of long thin filaments that themselves are polymers of many protein molecules. The most prominent use of such intracellular filaments is to provide the contractile mechanism of all muscles. Filaments, however, are also organized into microtubules that provide the "cytoskeletons" of such organelles as cilia, nerve axons, and the mitotic spindles of mitosing cells. Extracellularly, fibrillar proteins are found especially in the collagen and elastin fibers of connective tissue, blood vessels, tendons, ligaments, and so forth.

The *globular proteins* are an entirely different type of protein, usually composed of individual protein

11

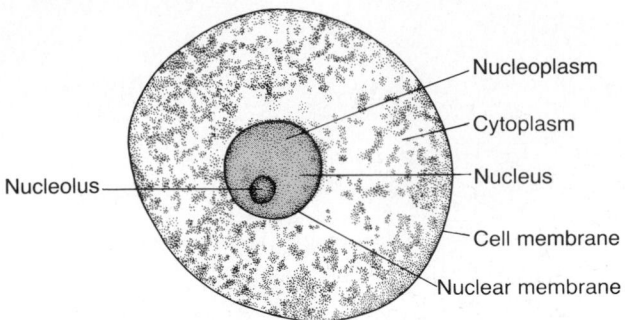

Figure 2–1. Structure of the cell as seen with the light microscope.

molecules or, at most, combinations of a few molecules in a globular form rather than a fibrillar form. These proteins are mainly the enzymes of the cell and, in contrast to the fibrillar proteins, are often soluble in the cell fluid or integral parts of or adherent to membranous structures inside the cell. The enzymes come into direct contact with other substances inside the cell and catalyze chemical reactions. For instance, the chemical reactions that split glucose into its component parts and then combine these with oxygen to form carbon dioxide and water, while at the same time providing energy for cellular function, are catalyzed by a series of protein enzymes.

LIPIDS. Lipids are several types of substances that are grouped together because of their common property of being soluble in fat solvents. The most important lipids in most cells are *phospholipids* and *cholesterol,* which together constitute about 2 per cent of the total cell mass. The special importance of phospholipids and cholesterol is that they are mainly insoluble in water and, therefore, are used to form the cell membrane as well as intracellular membranous barriers that separate the different cell compartments.

In addition to phospholipids and cholesterol, some cells contain large quantities of *triglycerides,* also called *neutral fat.* In the so-called fat cells, triglycerides often account for as much as 95 per cent of the cell mass. The fat stored in these cells represents the body's main storehouse of energy-giving nutrient that can later be dissoluted and used for energy wherever in the body it is needed.

CARBOHYDRATES. Carbohydrates have little structural function in the cell except as part of glycoprotein molecules, but they play a major role in nutrition of the cell. Most human cells do not maintain large stores of carbohydrates, usually averaging about 1 per cent of their total mass but increasing to as much as 3 per cent in muscle cells and, occasionally, 6 per cent in liver cells. Carbohydrate in the form of dissolved glucose is always present in the surrounding extracellular fluid so that it is readily available to the cell. A small amount of carbohydrate is virtually always stored in the cells in the form of *glycogen,* which is an insoluble polymer of glucose and can be used rapidly to supply the cells' energy needs.

PHYSICAL STRUCTURE OF THE CELL

The cell is not merely a bag of fluid, enzymes, and chemicals; it also contains highly organized physical structures, many of which are called *organelles.* The physical nature of each structure is equally as important to the function of the cell as the cell's chemical constituents. For instance, without one of the organelles, the *mitochondria,* more than 95 per cent of the cell's energy supply would cease immediately. Some principal organelles or structures of the cell are shown in Figure 2–2, including the *cell membrane, nuclear membrane, endoplasmic reticulum, Golgi apparatus, mitochondria, lysosomes,* and *centrioles.*

Membranous Structures of the Cell

Essentially all organelles of the cell are covered by membranes composed primarily of lipids and proteins. These membranes include the *cell membrane, nuclear membrane, membrane of the endoplasmic reticulum,* and *membranes of the mitochondria, lysosomes,* and *Golgi apparatus.*

The lipids of the membranes provide a barrier that prevents free movement of water and water-soluble substances from one cell compartment to the other. However, protein molecules in the membrane often penetrate all the way through the membrane, thus providing specialized pathways for passage of specific substances through the membrane. Also, many other membrane proteins are enzymes that catalyze a multitude of different chemical reactions, which are the subjects of numerous discussions in this and subsequent chapters.

Cell Membrane

The cell membrane, which envelops the cell, is a thin, elastic structure only 7.5 to 10 nanometers thick. It is composed almost entirely of proteins and lipids. The approximate composition is proteins, 55 per cent; phospholipids, 25 per cent; cholesterol, 13 per cent; other lipids, 4 per cent; and carbohydrates, 3 per cent.

LIPID BARRIER OF THE CELL MEMBRANE PREVENTS WATER PENETRATION. Figure 2–3 shows the cell membrane. Its basic structure is a *lipid bilayer,* which is a thin film of lipids only 2 molecules thick that is continuous over the entire cell surface. Interspersed in this lipid film are large globular protein molecules.

The basic structure of the lipid bilayer is composed of phospholipid molecules. One part of each phospholipid molecule is soluble in water, that is, *hydrophilic.* The other part is soluble only in fats, that is, *hydrophobic.* The phosphate portion of the phospholipid is hydrophilic, and the fatty acid portion is hydrophobic.

Because the hydrophobic portions of the phospholipid molecules are repelled by water but are mutually

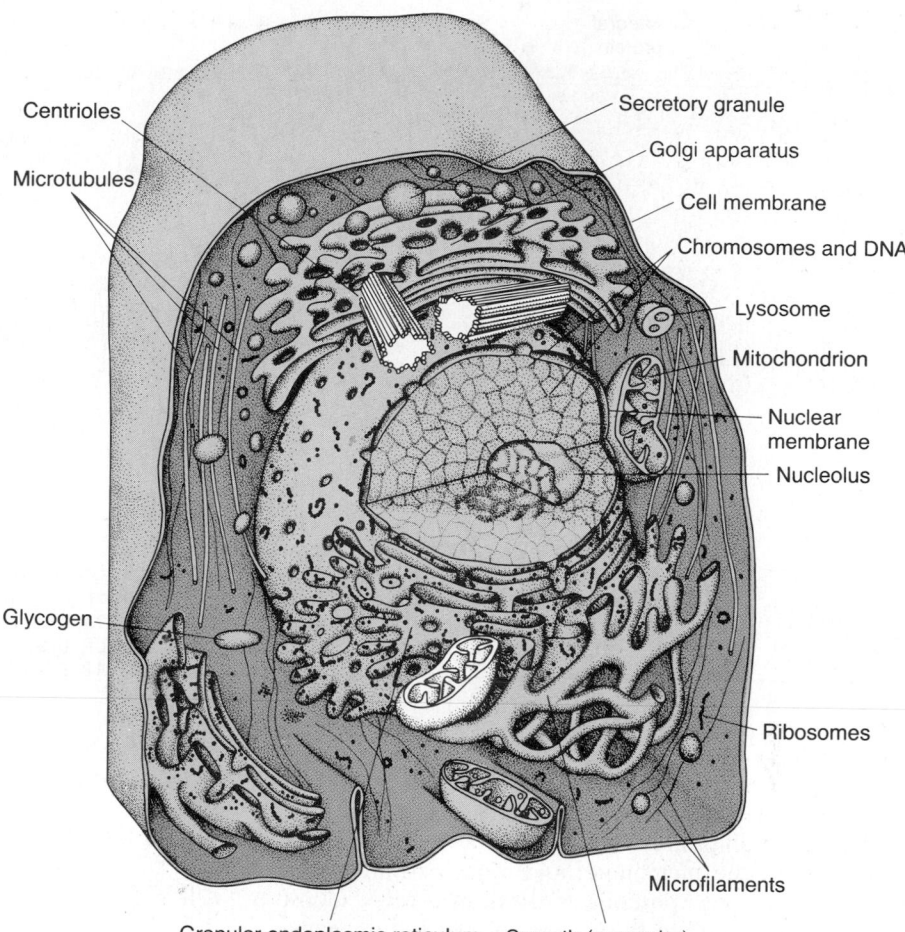

Figure 2–2. Reconstruction of a typical cell, showing the internal organelles in the cytoplasm and in the nucleus.

attracted to each other, they have a natural tendency to line up side by side in the center of the membrane, as shown in Figure 2–3. The hydrophilic phosphate portions cover the two surfaces in contact with the surrounding water.

The membrane lipid bilayer is a major barrier impermeable to the usual water-soluble substances, such as ions, glucose, and urea. On the other hand, fat-soluble substances, such as oxygen, carbon dioxide, and alcohol, can penetrate this portion of the membrane with ease.

A special feature of the lipid bilayer is that it is a *fluid* and not a solid. Therefore, portions of the membrane can literally flow from one point to another along the surface of the membrane. Proteins or other substances dissolved or floating in the lipid bilayer diffuse to all areas of the cell membrane.

The cholesterol molecules in the membrane are also lipid in nature because their steroid nucleus is highly fat soluble. These molecules, in a sense, are dissolved in the phospholipid bilayer of the membrane. They mainly help to determine the degree of permeability of the bilayer to water-soluble constituents of the body fluids. The cholesterol controls much of the fluidity of the membrane as well.

CELL MEMBRANE PROTEINS. Figure 2–3 shows globular masses floating in the lipid bilayer. These are membrane proteins, most of which are *glycoproteins.* Two types of proteins occur: the *integral proteins* that protrude all the way through the membrane and the *peripheral proteins* that are attached only to the surface of the membrane and do not penetrate.

Many of the integral proteins provide structural *channels* (or *pores*) through which water-soluble substances, especially ions, can diffuse between the extracellular and intracellular fluid. These protein channels also have selective properties that cause preferential diffusion of some substances more than others.

Others of the integral proteins act as *carrier proteins* for transporting substances in the direction opposite to their natural direction of diffusion, which is called "active transport." Still others act as *enzymes.*

The peripheral proteins occur either entirely or almost entirely on the inside of the membrane, and they normally are attached to one of the integral proteins. These peripheral proteins function almost entirely as enzymes or as other types of controllers of intracellular function.

MEMBRANE CARBOHYDRATES—THE CELL "GLYCO-CALYX." The membrane carbohydrates occur almost

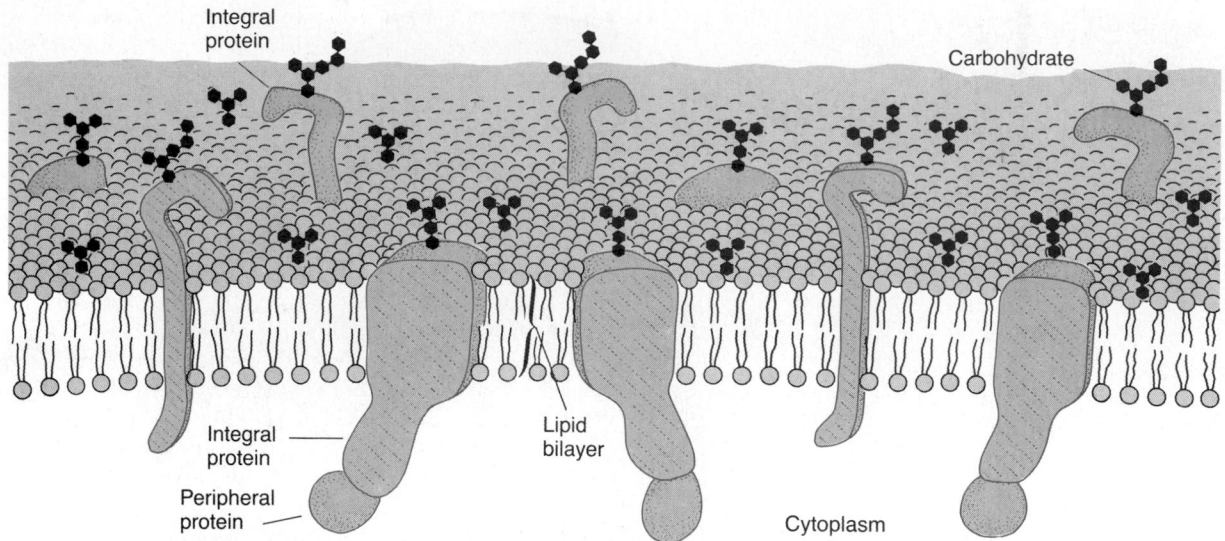

Figure 2–3. Structure of the cell membrane, showing that it is composed mainly from a lipid bilayer of phospholipid molecules, but with large numbers of protein molecules protruding through the layer. Also, carbohydrate moieties are attached to the protein molecules on the outside of the membrane and additional protein molecules on the inside. (From Lodish and Rothman: The assembly of cell membranes. *Sci. Am.*, *240*:48, 1979. ©1979 by Scientific American, Inc. All rights reserved.)

invariably in combination with proteins and lipids in the form of *glycoproteins* and *glycolipids.* In fact, most of the integral proteins are glycoproteins, and about one tenth of the membrane lipid molecules are glycolipids. The "glyco" portions of these molecules almost invariably protrude to the outside of the cell, dangling outward from the cell surface. Many other carbohydrate compounds, called *proteoglycans,* which are mainly carbohydrate substances bound together by small protein cores, often are loosely attached to the outer surface of the cell as well. Thus, the entire outside surface of the cell often has a loose carbohydrate coat called the *glycocalyx.*

The carbohydrate moieties attached to the outer surface of the cell have several important functions: (1) Many of them are negatively charged, which gives most cells an overall negative surface charge that repels other negative objects. (2) The glycocalyx of some cells attaches to the glycocalyx of other cells, thus attaching the cells to one another as well. (3) Many of the carbohydrates act as *receptor substances* for binding hormones like insulin and in doing so activate attached internal proteins that then activate a cascade of intracellular enzymes. (4) Some enter into immune reactions, as we discuss in Chapter 34.

Cytoplasm and Its Organelles

The cytoplasm is filled with both minute and large dispersed particles and organelles, ranging in size from a few nanometers to many micrometers. The clear fluid portion of the cytoplasm in which the particles are dispersed is called *cytosol;* this contains mainly dissolved proteins, electrolytes, and glucose and minute quantities of lipid compounds.

The portion of the cytoplasm immediately beneath the cell membrane frequently contains an interwoven mat of microfilaments composed mainly of actin fibrillae. These provide a semisolid gel-like support for the cell membrane. This zone of the cytoplasm is called the *cortex,* or *ectoplasm.* The cytoplasm between the cortex and the nuclear membrane is more liquefied and is called the *endoplasm.*

Dispersed in the cytoplasm are neutral fat globules, glycogen granules, ribosomes, secretory vesicles, and five especially important organelles: the *endoplasmic reticulum,* the *Golgi apparatus, mitochondria, lysosomes,* and *peroxisomes.*

Endoplasmic Reticulum

Figure 2–2 shows in the cytoplasm a network of tubular and flat vesicular structures called the *endoplasmic reticulum.* The tubules and vesicles all interconnect with one another. Also, their walls are constructed of lipid bilayer membranes that contain large amounts of proteins, similar to the cell membrane. The total surface area of this structure in some cells —the liver cells, for instance—can be as much as 30 to 40 times as great as the cell membrane area.

The detailed structure of a small portion of endoplasmic reticulum is shown in Figure 2–4. The space inside the tubules and vesicles is filled with *endoplasmic matrix,* a fluid medium that is different from the fluid outside the endoplasmic reticulum. Electron micrographs show that the space inside the endoplasmic reticulum is connected with the space between the two membranes of the double nuclear membrane.

Substances formed in some parts of the cell enter the space of the endoplasmic reticulum and are then

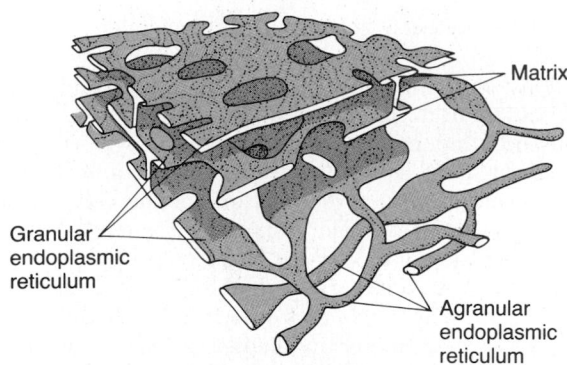

Figure 2–4. Structure of the endoplasmic reticulum. (Modified from De Robertis, Saez, and De Robertis: Cell Biology. 6th ed. Philadelphia, W. B. Saunders Company, 1975.)

conducted to other parts of the cell. Also, the vast surface area of the reticulum and multiple enzyme systems attached to its membranes provide the machinery for a major share of the metabolic functions of the cell.

RIBOSOMES AND THE GRANULAR ENDOPLASMIC RETICULUM. Attached to the outer surfaces of many parts of the endoplasmic reticulum are large numbers of small granular particles called *ribosomes.* Where these are present, the reticulum frequently is called the *granular endoplasmic reticulum.* The ribosomes are composed of a mixture of ribonucleic acid (RNA) and proteins, and they function in the synthesis of protein in the cells, as discussed later in this chapter and in Chapter 3.

AGRANULAR ENDOPLASMIC RETICULUM. Part of the endoplasmic reticulum has no attached ribosomes. This part is called the *agranular,* or *smooth, endoplasmic reticulum.* The agranular reticulum functions in the synthesis of lipid substances and in many other enzymatic processes of the cell.

Golgi Apparatus

The Golgi apparatus, shown in Figure 2–5, is closely related to the endoplasmic reticulum. It has membranes similar to those of the agranular endoplasmic reticulum. It usually is composed of four or more stacked layers of thin, flat enclosed vesicles lying near the nucleus. This apparatus is prominent in secretory cells; in these, it is located on the side of the cell from which the secretory substances are extruded.

The Golgi apparatus functions in association with the endoplasmic reticulum. As shown in Figure 2–5, small "transport vesicles," also called endoplasmic reticulum vesicles or simply *ER vesicles,* continually pinch off from the endoplasmic reticulum and shortly thereafter fuse with the Golgi apparatus. In this way, substances entrapped in the ER vesicles are transported from the endoplasmic reticulum to the Golgi apparatus. The transported substances are then pro-

cessed in the Golgi apparatus to form *lysosomes, secretory vesicles,* or other cytoplasmic components that are discussed later in the chapter.

Lysosomes

Lysosomes, shown in Figure 2–2, are vesicular organelles formed by the Golgi apparatus that then become dispersed throughout the cytoplasm. The lysosomes provide an intracellular digestive system that allows the cell to digest intracellular substances and structures, especially damaged cellular structures, food particles that have been ingested by the cell, and unwanted matter, such as bacteria. The lysosome is quite different from one cell to another, but it usually is 250 to 750 nanometers in diameter. It is surrounded by a typical lipid bilayer membrane and filled with large numbers of small granules 5 to 8 nanometers in diameter, which are protein aggregates of hydrolytic (digestive) enzymes. A hydrolytic enzyme is capable of splitting an organic compound into two or more parts by combining hydrogen from a water molecule with one part of the compound and by combining the hydroxyl portion of the water molecule with the other part of the compound. For instance, protein is hydrolyzed to form amino acids, and glycogen is hydrolyzed to form glucose. About 40 *acid hydrolase enzymes* have been found in lysosomes, and the principal substances that they digest are proteins, carbohydrates, lipids, and derivatives of these.

Ordinarily, the membrane surrounding the lysosome prevents the enclosed hydrolytic enzymes from coming in contact with other substances in the cell and, therefore, prevents their digestive actions. However, many conditions of the cell break the membranes of some of the lysosomes, allowing release of the enzymes. These enzymes then split the organic substances with which they come in contact into small, highly diffusible substances, such as amino acids and glucose. Some of the more specific functions of lysosomes are discussed later in the chapter.

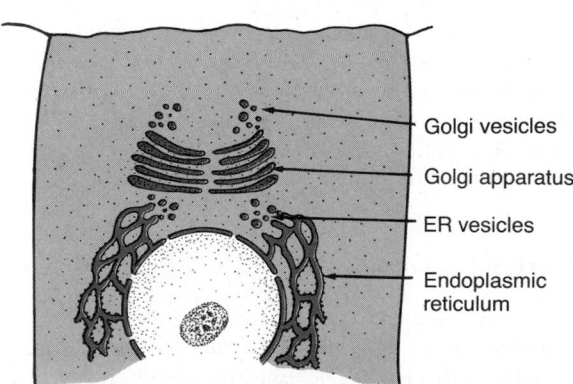

Figure 2–5. A typical Golgi apparatus and its relationship to the endoplasmic reticulum and the nucleus.

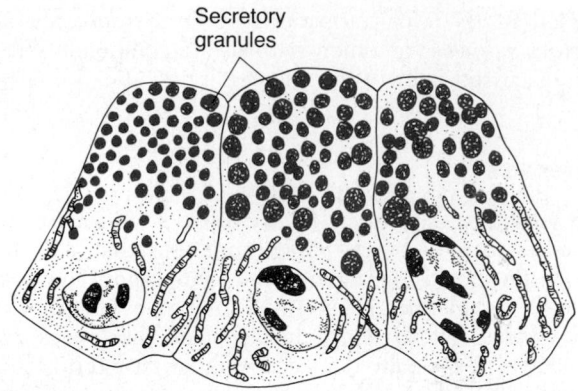

Figure 2–6. Secretory granules in acinar cells of the pancreas.

Peroxisomes

Peroxisomes are similar physically to lysosomes, but they are different in two important ways: First, they are believed to be formed by self-replication (or perhaps by budding off from the smooth endoplasmic reticulum) rather than by the Golgi apparatus. Second, they contain oxidases rather than hydrolases. Several of the oxidases are capable of combining oxygen with hydrogen ions from different intracellular chemicals to form hydrogen peroxide (H_2O_2). The hydrogen peroxide in turn is itself a highly oxidizing substance, and this is used in association with *catalase,* another oxidase enzyme present in large quantities in peroxisomes, to oxidize many substances that might otherwise be poisonous to the cell. For instance, about half the alcohol a person drinks is detoxified by the peroxisomes of the liver cells in this manner.

Secretory Vesicles

One of the important functions of many cells is secretion of special substances. Almost all such secretory substances are formed by the endoplasmic reticulum–Golgi apparatus system and are then released from the Golgi apparatus into the cytoplasm inside storage vesicles called *secretory vesicles* or *secretory granules.* Figure 2–6 shows typical secretory vesicles inside pancreatic acinar cells storing protein proenzymes (enzymes that are not yet activated); the proenzymes are secreted later through the outer cell membrane into the pancreatic duct and thence into the duodenum, where they become activated and perform their digestive functions.

Mitochondria

The mitochondria, shown in Figures 2–2 and 2–7, are called the "powerhouses" of the cell. Without them, the cells would be unable to extract significant amounts of energy from the nutrients and oxygen, and

as a consequence, essentially all cellular functions would cease. Mitochondria are present in basically all portions of the cytoplasm, but the total number per cell varies from less than a hundred up to several thousand, depending on the amount of energy required by each cell. Furthermore, the mitochondria are concentrated in those portions of the cell that are responsible for the major share of its energy metabolism. They are also variable in size and shape; some are only a few hundred nanometers in diameter and globular in shape, whereas others are elongated—as large as 1 micrometer in diameter and as long as 7 micrometers—and still others are branching and filamentous.

The basic structure of the mitochondrion, shown in Figure 2–7, is composed mainly of two lipid bilayer–protein membranes: an *outer membrane* and an *inner membrane.* Many infoldings of the inner membrane form *shelves* onto which oxidative enzymes are attached. In addition, the inner cavity of the mitochondrion is filled with a *matrix* that contains large quantities of dissolved enzymes that are necessary for extracting energy from nutrients. These enzymes operate in association with the oxidative enzymes on the shelves to cause oxidation of the nutrients, thereby forming carbon dioxide and water. The liberated energy is used to synthesize a high-energy substance called *adenosine triphosphate (ATP).* ATP is then transported out of the mitochondrion, and it diffuses throughout the cell to release its energy wherever it is needed for performing cellular functions. The chemical details of ATP formation by the mitochondrion are given in Chapter 67, and some of the basic functions of ATP in the cell are introduced later in this chapter.

Mitochondria are self-replicative, which means that one mitochondrion can form a second one, a third one, and so on, whenever there is need in the cell for increased amounts of ATP. Indeed, the mitochondria contain *deoxyribonucleic acid (DNA)* similar to that found in the nucleus. In Chapter 3 we see that DNA

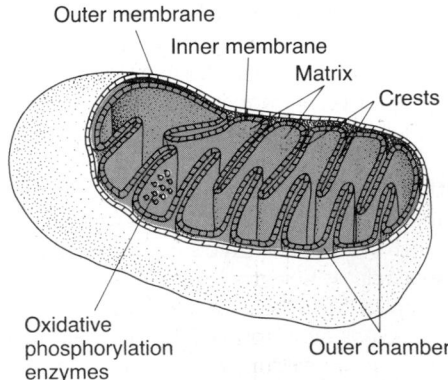

Figure 2–7. Structure of a mitochondrion. (Modified from De Robertis, Saez, and De Robertis: Cell Biology. 6th ed. Philadelphia, W. B. Saunders Company, 1975.)

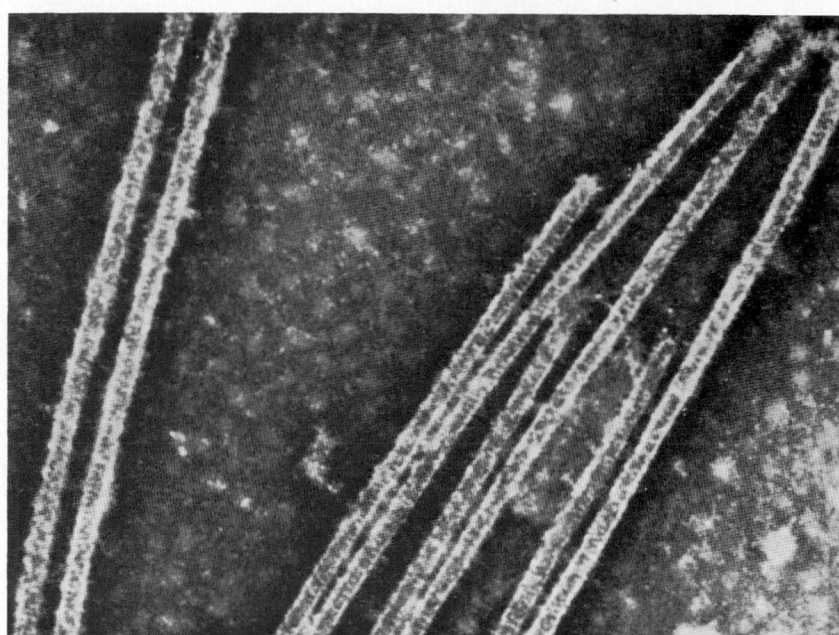

Figure 2–8. Microtubules teased from the flagellum of a sperm. (From Porter: Ciba Foundation Symposium: Principles of Biomolecular Organization. Boston, Little, Brown & Company, 1966.)

is the basic substance of the nucleus that controls replication of the cell and that the DNA of the mitochondrion plays a similar role in the mitochondrion for its own replication.

Filament and Tubular Structures of the Cell

The fibrillar proteins of the cell usually are organized into filaments or tubules. These originate as precursor protein molecules synthesized by ribosomes in the cytoplasm. The precursor molecules polymerize to form *filaments.* We have already pointed out that large numbers of actin filaments frequently occur in the outer zone of the cytoplasm, the zone called the *ectoplasm,* to form an elastic support for the cell membrane. Also, in muscle cells, actin and myosin filaments are organized into a special contractile machine that is the basis of muscle contraction throughout the body, as is discussed in detail in Chapter 6.

A special type of filament composed of polymerized *tubulin* molecules is used in all cells to construct tubular structures, the *microtubules.* Most often these contain 13 tubulin *protofilaments* lying parallel to one another in a circle to form a long hollow cylinder about 25 nanometers in diameter and 1 to many micrometers in length. These cylinders are often arranged in bundles that gives them, en masse, considerable structural strength. However, microtubules are stiff structures that break if bent too severely. Figure 2–8 shows typical microtubules that were teased from the flagellum of a sperm.

Another example of microtubules is the tubular skeletal structure in the center of all cilia that radiates upward from the cell cytoplasm to the tip of the cilium. This structure is shown later in the chapter, in Figure 2–17. Also, the *centrioles* and the *mitotic spindle* of the mitosing cell are both composed of stiff microtubules.

Thus, a primary function of microtubules is to act as a *cytoskeleton,* providing rigid physical structures for certain parts of cells. Also, the cytoplasm often *streams* (flows) in the vicinity of microtubules, which might result from movement of arms that project outward from the microtubules.

Nucleus

The nucleus is the control center of the cell. Briefly, the nucleus contains large quantities of DNA, which we have called *genes* for many years. The genes determine the characteristics of the cell's proteins, including the enzymes of the cytoplasm that control cytoplasmic activities. They also control reproduction; the genes first reproduce themselves, and after this, the cell splits by a special process called *mitosis* to form two daughter cells, each of which receives one of the two sets of genes. All these activities of the nucleus are considered in detail in Chapter 3.

The appearance of the nucleus under the microscope does not give much of a clue to the mechanisms by which it performs its control activities. Figure 2–9 shows the light microscopic appearance of the interphase nucleus (period between mitoses), revealing darkly staining *chromatin material* throughout the nucleoplasm. During mitosis, the chromatin material becomes readily identifiable as the highly structured *chromosomes,* which can then be seen easily with the light microscope, as shown in Chapter 3.

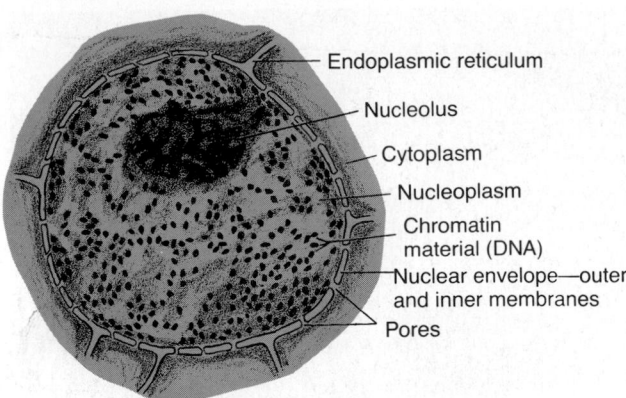

Figure 2–9. Structure of the nucleus.

Endoplasmic reticulum
Nucleolus
Cytoplasm
Nucleoplasm
Chromatin material (DNA)
Nuclear envelope—outer and inner membranes
Pores

Nuclear Membrane

The *nuclear membrane,* also called the *nuclear envelope,* is actually two separate membranes, one layered inside the other. The outer membrane is continuous with the endoplasmic reticulum, and the space between the two nuclear membranes is also continuous with the compartment inside the endoplasmic reticulum, as shown in Figure 2–9.

Both layers of the nuclear membrane are penetrated by several thousand *nuclear pores.* These are large, almost 100 nanometers in diameter. However, large complexes of protein molecules are attached around the edges of the pores so that the central area of the pore is only about 9 nanometers in diameter. Even this size is large enough to allow moderate numbers of molecules up to 44,000 molecular weight to pass through and molecules with molecular weight less than 15,000 to pass extremely rapidly.

Nucleoli

The nuclei of most cells contain one or more lightly staining structures called *nucleoli.* The nucleolus, unlike most of the organelles that we have discussed, does not have a limiting membrane. Instead, it is simply a structure that contains large amounts of RNA and proteins of the types found in ribosomes. The nucleolus becomes considerably enlarged when a cell is actively synthesizing proteins. The genes of five separate chromosome pairs synthesize the ribosomal RNA and then store it in the nucleolus, beginning with a loose fibrillar RNA that later condenses to form granular "subunits" of ribosomes. These in turn are transported through the nuclear membrane pores into the cytoplasm, where they assemble to form the "mature" ribosomes that play an essential role for the formation of proteins, as we discuss more fully in Chapter 3.

COMPARISON OF THE ANIMAL CELL WITH PRECELLULAR FORMS OF LIFE

Many of us think of the cell as the lowest level of life. However, the cell is a very, very complicated

organism, which required many hundreds of million years to develop after the earliest form of life, an organism similar to the present-day *virus,* first appeared on earth. Figure 2–10 shows the relative sizes of the smallest known virus, a large virus, a *rickettsia,* a *bacterium,* and a nucleated cell, demonstrating that the cell has a diameter about 1000 times that of the smallest virus and, therefore, a volume about 1 billion times that of the smallest virus. Correspondingly, the functions and anatomic organization of the cell are also far more complex than those of the virus.

The essential life-giving constituent of the very small virus is a *nucleic acid* embedded in a coat of protein. This nucleic acid is composed of the same basic constituents (DNA or RNA) as found in mammalian cells, and it is capable of reproducing itself if appropriate conditions are available. Thus, the virus is capable of propagating its lineage from generation to generation and, therefore, is a living structure in the same way that the cell and the human being are living structures.

As life evolved, other chemicals besides nucleic acid and simple proteins became integral parts of the organism, and specialized functions began to develop in different parts of the virus. A membrane formed around the virus, and inside the membrane, a fluid matrix appeared. Specialized chemicals developed inside the matrix to perform special functions; many protein enzymes appeared that were capable of catalyzing chemical reactions and, therefore, determining the organism's activities.

In still later stages, particularly in the rickettsial and bacterial stages, *organelles* developed inside the organism, representing physical structures of chemical aggregates that perform functions in a more efficient manner than can be achieved by dispersed chemicals throughout the fluid matrix. Finally, in the nucleated cell, still more complex organelles developed, the most important of which is the *nucleus* itself. The nucleus distinguishes this type of cell from all lower forms of life; this structure provides a control center for all cellular activities, and it provides for exact reproduction of new cells generation after generation, each new

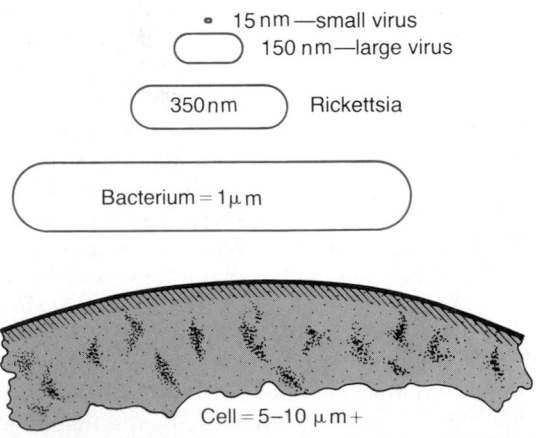

15 nm—small virus
150 nm—large virus
350 nm   Rickettsia
Bacterium = 1 μm
Cell = 5–10 μm+

Figure 2–10. Comparison of sizes of precellular organisms with that of the average cell in the human body.

cell having essentially the same structure as its progenitor.

FUNCTIONAL SYSTEMS OF THE CELL

In the remainder of this chapter, we discuss several representative functional systems of the cell that make it a living organism.

Ingestion by the Cell—Endocytosis

If a cell is to live and grow, it must obtain nutrients and other substances from the surrounding fluids. Most substances pass through the cell membrane by *diffusion* and *active transport*. Diffusion means simply movement through the membrane by random motion of the molecules of the substances, moving either through cell membrane pores or, in the case of lipid-soluble substances, through the lipid matrix of the membrane. Active transport means actual carrying of the substance through the membrane by a protein structure that penetrates all the way through the membrane. These transport mechanisms are so important to cell function that they are presented in detail in Chapter 4. Very large particles enter the cell by a specialized function of the cell membrane called *endocytosis*. The principal forms of endocytosis are *pinocytosis* and *phagocytosis*. Pinocytosis means the ingestion of extremely small vesicles that contain extracellular fluid. Phagocytosis means ingestion of large particles, such as bacteria, cells, and portions of degenerating tissue.

PINOCYTOSIS. Pinocytosis occurs continually at the cell membranes of most cells but especially rapidly in some cells. For instance, it occurs so rapidly in macrophages that about 3 per cent of the total macrophage membrane is engulfed in the form of vesicles each minute. Even so, the pinocytic vesicles are so small—usually only 100 to 200 nanometers in diameter—that most of them can be seen only with the electron microscope.

Pinocytosis is the only means by which most large macromolecules, such as most protein molecules, can enter cells. In fact, the rate at which pinocytic vesicles form usually is enhanced when such macromolecules attach to the cell membrane.

Figure 2–11 demonstrates the successive steps of pinocytosis, showing three molecules of protein attaching to the membrane. These molecules usually attach to specialized *receptors* on the surface of the membrane that are specific for the type of protein that is to be absorbed. The receptors generally are concentrated in small pits on the outer surface of the cell membrane, called *coated pits*. On the inside of the cell membrane beneath these pits is a latticework of a fibrillar protein called *clathrin* as well as other proteins, perhaps including contractile filaments of *actin* and *myosin*. Once the protein molecules have bound with the receptors, the surface properties of the membrane change in such a way that the entire pit invagi-

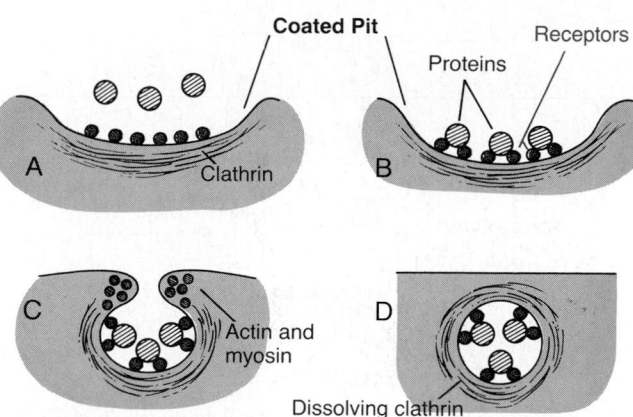

Figure 2–11. Mechanism of pinocytosis.

nates inward and the proteins surrounding the invaginating pit cause its borders to close over the attached proteins as well as over a small amount of extracellular fluid. Immediately thereafter, the invaginated portion of the membrane breaks away from the surface of the cell, forming a *pinocytic vesicle* inside the cytoplasm of the cell.

What causes the cell membrane to go through the necessary contortions for forming pinocytic vesicles remains mainly a mystery. This process requires energy from within the cell; this is supplied by ATP, a high-energy substance that is discussed later in the chapter. Also, it requires the presence of calcium ions in the extracellular fluid, which probably react with contractile protein filaments beneath the coated pits to provide the force for pinching the vesicles away from the cell membrane.

PHAGOCYTOSIS. Phagocytosis occurs in much the same way as pinocytosis except that it involves large particles rather than molecules. Only certain cells have the capability of phagocytosis, most notably the tissue macrophages and some of the white blood cells.

Phagocytosis is initiated when proteins or large polysaccharides on the surface of the particle that is to be phagocytized—such as a bacterium, a dead cell, or other tissue debris—bind with receptors on the surface of the phagocyte. In the case of bacteria, each bacterium usually is already attached to a specific antibody, and it is the antibody that attaches to the phagocyte receptors, dragging the bacterium along with it. This intermediation of antibodies is called *opsonization*, which is discussed in Chapters 33 and 34.

Phagocytosis occurs in the following steps:

1. The cell membrane receptors attach to the surface ligands of the particle.
2. The edges of the membrane around the points of attachment evaginate outward within a fraction of a second to surround the entire particle; then, progressively more and more membrane receptors attach to the particle ligands, all this occurring suddenly in a zipperlike manner to form a closed *phagocytic vesicle*.
3. Actin and other contractile fibrils in the cytoplasm surround the phagocytic vesicle and contract around its outer edge, pushing the vesicle to the interior.

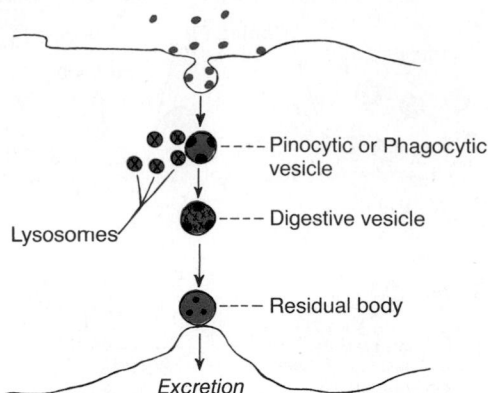

Figure 2–12. Digestion of substances in pinocytic vesicles by enzymes derived from lysosomes.

4. The contractile proteins then pinch the vesicle off, leaving it in the cell interior in the same way that pinocytic vesicles are formed.

Digestion of Pinocytic and Phagocytic Foreign Substances in the Cell— Function of the Lysosomes

Almost immediately after a pinocytic or phagocytic vesicle appears inside a cell, one or more lysosomes become attached to the vesicle and empty their acid hydrolases into the vesicle, as shown in Figure 2–12. Thus, a *digestive vesicle* is formed in which the hydrolases begin hydrolyzing the proteins, carbohydrates, lipids, and other substances in the vesicle. The products of digestion are small molecules of amino acids, glucose, phosphates, and so forth that can then diffuse through the membrane of the vesicle into the cytoplasm. What is left of the digestive vesicle, called the *residual body,* represents the undigestible substances. In most instances, this is finally excreted through the cell membrane by a process called *exocytosis,* which is essentially the opposite of endocytosis.

Thus, the lysosomes may be called the *digestive organs* of the cells.

REGRESSION OF TISSUES AND AUTOLYSIS OF CELLS. Tissues of the body often regress to smaller size. For instance, this occurs in the uterus after pregnancy, in muscles during long periods of inactivity, and in mammary glands at the end of lactation. Lysosomes are responsible for much of this regression. The mechanism by which lack of activity in a tissue causes the lysosomes to increase their activity is unknown.

Another special role of the lysosomes is the removal of damaged cells or damaged portions of cells from tissues—cells damaged by heat, cold, trauma, chemicals, or any other factor. Damage to the cell causes lysosomes to rupture. The released hydrolases begin immediately to digest the surrounding organic substances. If the damage is slight, only a portion of the cell is removed, followed by repair of the cell. If the damage is severe, the entire cell is digested, a process called *autolysis.* In this way, the cell is completely removed, and a new cell of the same type ordinarily is formed by mitotic reproduction of an adjacent cell to take the place of the old one.

The lysosomes also contain bactericidal agents that can kill phagocytized bacteria before they can cause cellular damage. These agents include *lysozyme* that dissolves the bacterial cell membrane, *lysoferrin* that binds iron and other metals that are essential for bacterial growth, and acid at a pH of about 5.0 that activates the hydrolases and inactivates some of the bacterial metabolic systems.

In certain genetic disorders of the body, some of the usual digestive enzymes are missing from the lysosomes, especially enzymes that are required to digest lipid aggregates or glycogen granules. In such instances, extreme quantities of lipids or glycogen often accumulate in the cells of many organs, especially the liver, and lead to early death.

Synthesis and Formation of Cellular Structures by the Endoplasmic Reticulum and the Golgi Apparatus

Specific Functions of the Endoplasmic Reticulum

The extensiveness of the endoplasmic reticulum and the Golgi apparatus, especially in secretory cells, has already been emphasized. These structures are formed primarily of lipid bilayer membranes similar to the cell membrane, and their walls are literally loaded with protein enzymes that catalyze the synthesis of many of the substances required by the cell.

Most of the synthesis begins in the endoplasmic reticulum, but most of the products that are formed are then passed on to the Golgi apparatus, where they are further processed before release into the cytoplasm. But first, let us note the specific products that are synthesized in the specific portions of the endoplasmic reticulum and the Golgi apparatus.

PROTEINS ARE FORMED BY THE GRANULAR ENDOPLASMIC RETICULUM. The granular endoplasmic reticulum is characterized by large numbers of ribosomes attached to the outer surfaces of the reticulum membrane. As we discuss in Chapter 3, protein molecules are synthesized within the structures of the ribosomes. Furthermore, the ribosomes extrude many of the synthesized protein molecules not into the cytosol but instead through the wall of the endoplasmic reticulum to the interior of the endoplasmic vesicles and tubules, called the *endoplasmic matrix.*

Almost as rapidly as the protein molecules enter the endoplasmic matrix, enzymes in the endoplasmic reticulum wall cause rapid changes in these molecules. First, almost all of them are immediately glycosylated, that is, conjugated with carbohydrate moieties, to form *glycoproteins.* Therefore, essentially all the endoplasmic proteins are glycoproteins, in contrast to the proteins that are formed by the ribosomes in the cytosol, which are mainly free proteins. Second, the proteins are cross-linked, folded, and often shortened in chain length to form more compact molecules.

SYNTHESIS OF LIPIDS BY THE ENDOPLASMIC RETICULUM, ESPECIALLY BY THE SMOOTH ENDOPLASMIC RETICULUM. The endoplasmic reticulum also synthesizes lipids, especially phospholipids and cholesterol. These are rapidly incorporated into the lipid bilayer of the endoplasmic reticulum itself, thus causing the endoplasmic reticulum to grow continually. This occurs mainly in the smooth portion of the endoplasmic reticulum.

To keep the endoplasmic reticulum from growing beyond the limits of the cell, small vesicles called *endoplasmic reticulum vesicles* (ER vesicles) or *transport vesicles* continually break away from the smooth reticulum; we see later that most of these vesicles migrate rapidly to the Golgi apparatus.

OTHER FUNCTIONS OF THE ENDOPLASMIC RETICULUM. Other significant functions of the endoplasmic reticulum, especially the smooth reticulum, include the following:

1. It provides the enzymes that control glycogen breakdown when glycogen is to be used for energy.
2. It provides a vast number of enzymes that are capable of detoxifying substances that are damaging to the cell, such as drugs. It achieves detoxification by coagulation, oxidation, hydrolysis, conjugation with glycuronic acid, or in other ways.

Specific Functions of the Golgi Apparatus

SYNTHETIC FUNCTIONS OF THE GOLGI APPARATUS. Although the major function of the Golgi apparatus is to process substances already formed in the endoplasmic reticulum, it also has the capability of synthesizing certain carbohydrates that cannot be formed in the endoplasmic reticulum. This is especially true of sialic acid and galactose. In addition, the Golgi apparatus can cause the formation of large saccharide polymers bound with only small amounts of protein; the most important of these are hyaluronic acid and chondroitin sulfate. A few of the many functions of hyaluronic acid and chondroitin sulfate in the body are as follows: (1) They are the major components of proteoglycans secreted in mucus and other glandular secretions. (2) They are the major components of the ground substance in the interstitial spaces, acting as a filler between collagen fibers and cells. (3) They are principal components of the organic matrix in both cartilage and bone.

PROCESSING OF ENDOPLASMIC SECRETIONS BY THE GOLGI APPARATUS—FORMATION OF VESICLES. Figure 2–13 summarizes the major functions of the endoplasmic reticulum and Golgi apparatus. As substances are formed in the endoplasmic reticulum, especially the proteins, they are transported through the tubules toward the portions of the smooth endoplasmic reticulum that lie nearest the Golgi apparatus. At this point, small *transport vesicles* composed of small envelopes of smooth endoplasmic reticulum continually break away and diffuse to the *deepest layer* of the Golgi apparatus. Inside these vesicles are the synthesized proteins and other products from the endoplasmic reticulum.

The transport vesicles instantly fuse with the Golgi apparatus and empty their contained substances into the vesicular spaces of the Golgi apparatus. Here, ad-

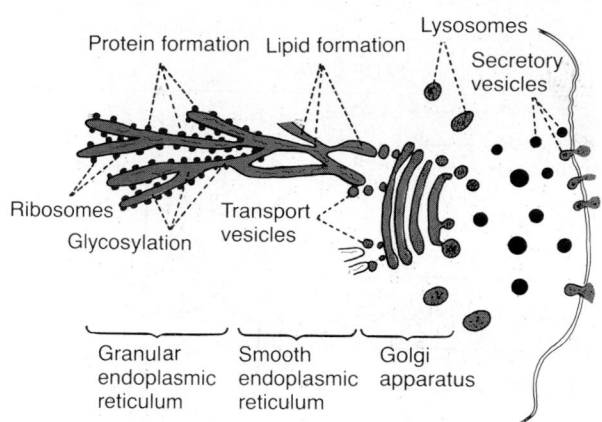

Figure 2–13. Formation of proteins, lipids, and cellular vesicles by the endoplasmic reticulum and Golgi apparatus.

ditional carbohydrate moieties are added to the secretions. Also, a most important function of the Golgi apparatus is to compact the endoplasmic reticular secretions into highly concentrated packets. As the secretions pass toward the outermost layers of the Golgi apparatus, the compaction and processing proceed. Finally, both small and large vesicles continually break away from the Golgi apparatus, carrying with them the compacted secretory substances, and they in turn diffuse throughout the cell.

To give one an idea of the timing of these processes: When a glandular cell is bathed in radioactive amino acids, newly formed radioactive protein molecules can be detected in the granular endoplasmic reticulum within 3 to 5 minutes. Within 20 minutes, the newly formed proteins are present in the Golgi apparatus, and within 1 to 2 hours, radioactive proteins are secreted from the surface of the cell.

TYPES OF VESICLES FORMED BY THE GOLGI APPARATUS—SECRETORY VESICLES AND LYSOSOMES. In a highly secretory cell, the vesicles that are formed by the Golgi apparatus are mainly *secretory vesicles*, containing especially the protein substances that are to be secreted through the surface of the cell membrane. These vesicles diffuse to the cell membrane and then fuse with it and empty their substances to the exterior by the mechanism called *exocytosis*, which is essentially the opposite of endocytosis. Exocytosis, in most cases, is stimulated by entry of calcium ions into the cell; the calcium ions interact with the vesicular membrane in some way not understood to cause its fusion with the cell membrane, then followed by the exocytosis.

On the other hand, some of the vesicles are destined for intracellular use. For instance, specialized portions of the Golgi apparatus form the *lysosomes* that have already been discussed. The membranes of these specialized portions contain chemical receptors that cause acid hydrolase enzymes to attach. In this way, these enzymes are concentrated and then released from the Golgi apparatus in the form of the lysosomal vesicles.

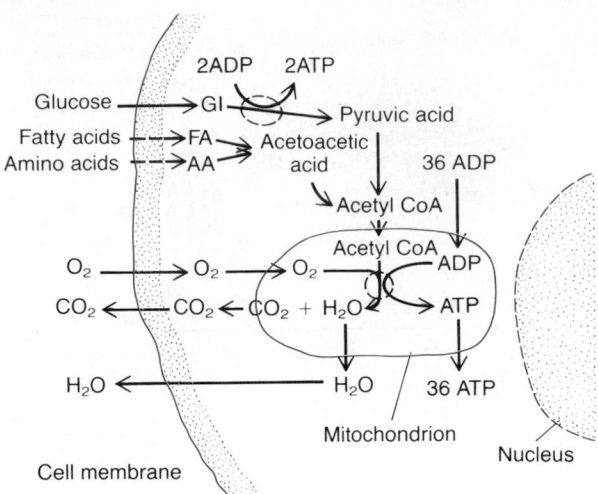

Figure 2–14. Formation of adenosine triphosphate (ATP) in the cell, showing that most of the ATP is formed in the mitochondria.

Use of Intracellular Vesicles to Replenish Cellular Membranes. Many of the vesicles formed by the Golgi apparatus finally fuse with the cell membrane or with the membranes of other intracellular structures, such as the mitochondria and even the endoplasmic reticulum. This increases the expanse of these membranes and thereby replenishes the membranes as they are destroyed. For instance, the cell membrane loses much of its substance every time it forms a phagocytic or pinocytic vesicle, and it is vesicles from the Golgi apparatus that continually replenish the cell membrane.

In summary, the membranous system of the endoplasmic reticulum and the Golgi apparatus represents a highly metabolic organ capable of forming both new cellular structures and secretory substances to be extruded from the cell.

Extraction of Energy from Nutrients— Function of the Mitochondria

The principal substances from which cells extract energy are, on the one hand, oxygen and, on the other hand, one or more of the foodstuffs that react with the oxygen—carbohydrates, fats, and proteins. In the human body, essentially all carbohydrates are converted into *glucose* by the digestive tract and liver before they reach the cell. Similarly, the proteins are converted into *amino acids* and the fats into *fatty acids*. Figure 2–14 shows oxygen and the foodstuffs— glucose, fatty acids, and amino acids—all entering the cell. Inside the cell, the foodstuffs react chemically with the oxygen under the influence of various enzymes that control the rates of the reactions and then also channel the energy that is released in the proper direction. The details of all these digestive and metabolic functions are given in Chapters 62 through 72.

Almost all these oxidative reactions occur inside the mitochondria, and the energy that is released is used to form mainly the very high energy compound *Adenosine triphosphate (ATP)*. Then, the ATP, not the original foodstuffs themselves, is used throughout the cell to energize almost all the intracellular metabolic reactions.

Functional Characteristics of ATP

The formula for ATP is:

$$NH_2$$

Adenine

Phosphate

Ribose

Adenosine triphosphate

ATP is a nucleotide composed of the nitrogenous base *adenine*, the pentose sugar *ribose*, and three *phosphate radicals*. The last two phosphate radicals are connected with the remainder of the molecule by so-called *high-energy phosphate bonds*, which are represented in the formula above by the symbol ~. Each of these bonds contains about 12,000 calories of energy per mole of ATP under the physical conditions of the body (7300 calories under standard conditions), which is much greater than the energy stored in the average chemical bond of other organic compounds, thus giving rise to the term "high-energy" bond. Furthermore, the high-energy phosphate bond is very labile so that it can be split instantly on demand whenever energy is required to promote other cellular reactions.

When ATP releases its energy, a phosphoric acid radical is split away, and *adenosine diphosphate (ADP)* is formed. Then, energy derived from the cellular nutrients causes the ADP and phosphoric acid to recombine to form new ATP, the entire process continuing over and over again. For these reasons, ATP has been called the *energy currency* of the cell because it can be spent and remade again and again, usually having a turnover time of only a few minutes at most.

CHEMICAL PROCESSES IN THE FORMATION OF ATP —ROLE OF THE MITOCHONDRIA. On entry into the cells, glucose is subjected to enzymes in the cytoplasm that convert it into *pyruvic acid* (a process called *glycolysis*). A small amount of ADP is changed into ATP by energy released during this conversion, but this amount accounts for less than 5 per cent of the overall energy metabolism of the cell.

By far the major portion of the ATP formed in the cell is formed in the mitochondria. The pyruvic acid derived from carbohydrates, fatty acids from lipids, and amino acids from proteins are all eventually converted into the compound *acetyl-CoA* in the matrix of the mitochondrion. This substance in turn is further dissoluted—for the purpose of extracting its energy—by another series of enzymes in the mitochondrion matrix, undergoing dissolution in a sequence of chemical reactions called the *citric acid cycle,* or *Krebs cycle.* These chemical reactions are explained in detail in Chapter 67.

In the citric acid cycle, acetyl-CoA is split into its component parts, *hydrogen atoms* and *carbon dioxide.* The carbon dioxide in turn diffuses out of the mitochondria and eventually out of the cell.

The hydrogen atoms, on the other hand, are highly reactive, and they eventually combine with oxygen that has also diffused into the mitochondria. This releases a tremendous amount of energy, which is used by the mitochondria to convert large amounts of ADP to ATP. The processes of these reactions are complex, requiring the participation of large numbers of protein enzymes that are integral parts of the mitochondrial *membranous shelves* that protrude into the mitochondrial matrix. The initial event is removal of an electron from the hydrogen atom, thus converting it to a hydrogen ion. The terminal event is movement of the hydrogen ions through large globular proteins called *ATP synthetase* that protrude like knobs through the membranes of the mitochondrial shelves. Finally, ATP synthetase is an enzyme that uses the energy from the movement of the hydrogen ions to cause conversion of ADP to ATP, while at the same time the hydrogen ions combine with oxygen to form water. The newly formed ATP is transported out of the mitochondria into all parts of the cell cytoplasm and nucleoplasm, where its energy is used to energize cell functions.

This overall process for the formation of ATP is called the *chemiosmotic mechanism* of ATP formation. The chemical and physical details of this mechanism are presented in Chapter 67, and many of the metabolic functions of ATP in the body are presented in Chapters 67 through 71.

USES OF ATP FOR CELLULAR FUNCTION. ATP is used to promote three major categories of cellular functions: (1) *membrane transport,* (2) *synthesis of chemical compounds* throughout the cell, and (3) *mechanical work.* These uses of ATP are illustrated by examples in Figure 2–15: (1) to supply energy for the transport of sodium through the cell membrane, (2) to promote protein synthesis by the ribosomes, and (3) to supply the energy needed during muscle contraction.

In addition to membrane transport of sodium, energy from ATP is required for membrane transport of potassium ions, calcium ions, magnesium ions, phosphate ions, chloride ions, urate ions, hydrogen ions, and still many other ions and various organic substances. Membrane transport is so important to cell function that some cells, the renal tubular cells for instance, use as much as 80 per cent of the ATP formed in the cells for this purpose alone.

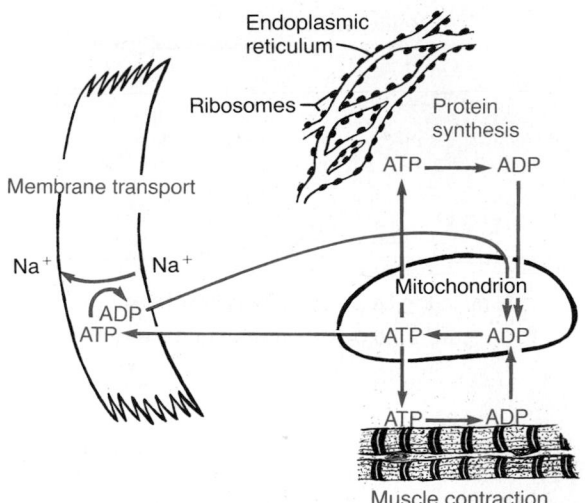

Figure 2–15. Use of adenosine triphosphate to provide energy for three major cellular functions: (1) membrane transport, (2) protein synthesis, and (3) muscle contraction.

In addition to synthesizing proteins, cells synthesize phospholipids, cholesterol, purines, pyrimidines, and a great host of other substances. Synthesis of almost any chemical compound requires energy. For instance, a single protein molecule might be composed of as many as several thousand amino acids attached to one another by peptide linkages; the formation of each of these linkages requires the breakdown of four high-energy bonds; thus, many thousand ATP molecules (or comparable guanosine triphosphate molecules) must release their energy as each protein molecule is formed. Indeed, some cells use as much as 75 per cent of all the ATP formed in the cell simply to synthesize new chemical compounds, especially protein molecules; this is particularly true during the growth phase of cells.

The final major use of ATP is to supply energy for special cells to perform mechanical work. We see in Chapter 6 that each contraction of a muscle fiber requires expenditure of tremendous quantities of ATP. Other cells perform mechanical work in additional ways, especially by *ciliary* and *ameboid motion,* which are described later in this chapter. The source of energy for all these types of mechanical work is ATP.

In summary, therefore, ATP is always available to release its energy rapidly and almost explosively wherever in the cell it is needed. To replace the ATP used by the cell, other, much slower chemical reactions break down carbohydrates, fats, and proteins and use the energy derived from these to form new ATP. More than 95 per cent of this ATP is formed in the mitochondria, which accounts for the name given to the mitochondria, the "powerhouses" of the cell.

Ameboid Locomotion of Cells

By far the most important type of cell movement that occurs in the body is that of the specialized muscle cells in skeletal, cardiac, and smooth muscle, which constitute almost 50 per cent of the entire body mass. The spe-

Movement of Cell

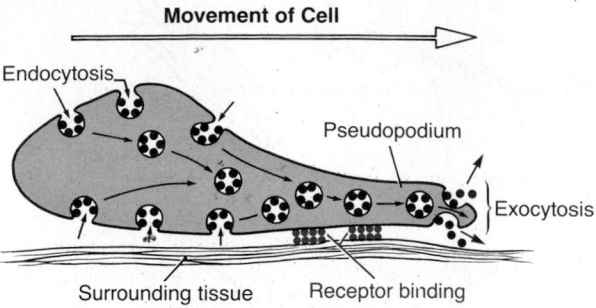

Figure 2–16. Ameboid motion by a cell.

cialized functions of these cells are discussed in Chapters 6 through 9. Two other types of movement—*ameboid locomotion* and *ciliary movement*—occur in other cells.

Ameboid locomotion means movement of an entire cell in relation to its surroundings, such as the movement of white blood cells through tissues. It receives its name from the fact that amebae move in this manner and have provided an excellent tool for studying the phenomenon. Typically, ameboid locomotion begins with protrusion of a *pseudopodium* from one end of the cell. The pseudopodium projects far out away from the cell body and then attaches to the new tissue area, and finally, the remainder of the cell moves toward the pseudopodium. Figure 2–16 demonstrates this process, showing an elongated cell, the right-hand end of which is a protruding pseudopodium. The membrane of this end of the cell is continually moving forward, and the membrane at the left-hand end of the cell is continually following along as the cell moves.

MECHANISM OF AMEBOID LOCOMOTION. Figure 2–16 shows the general principle of ameboid motion. Basically, it results from continual exocytosis that forms new cell membrane at the leading edge of the pseudopodium and continual endocytosis of the membrane in mid and rear portions of the cell. Also, two other effects are essential for forward movement of the cell. The first effect is attachment of the pseudopodium to the surrounding tissues so that it becomes fixed in its leading position, while the remainder of the cell body is pulled forward toward the point of attachment. This attachment is effected by receptor proteins that line the insides of the exocytotic vesicles. When the vesicles become part of the pseudopodial membrane, they open so that their insides evert to the outside and the receptors now protrude to the outside and contact ligands in the surrounding tissues. An especially important ligand is a protein called *fibronectin* bound to the collagen fibers of the tissue.

At the opposite end of the cell, the endocytotic activity pulls the receptors away from their ligands to form endocytotic vesicles. Then, inside the cell, these vesicles stream toward the pseudopodial end of the cell, where they are used to form still new membrane for the pseudopodium.

The second essential for locomotion is to provide the energy required to pull the cell body in the direction of the pseudopodium. Recent experiments suggest the following as an answer to this.

In the cytoplasm of all cells is a moderate to large amount of the protein *actin*. Much of the actin is in the form of single molecules that do not provide any motive power; however, when these polymerize together to form a filamentous network, the network contracts when it binds with still another protein, an actin-binding protein such as *myosin*. The whole process is energized by the high-energy compound ATP. This is what happens in the pseudopodium of a moving cell, where such a network of actin filaments forms anew inside the growing pseudopodium. Contraction also occurs in the ectoplasm of the cell body, where a pre-existing actin network is already present beneath the cell membrane. The contraction in the pseudopodium pulls the cell body toward the pseudopodium, whereas contraction of the ectoplasm of the cell body squeezes the body and its contents toward the pseudopodium.

TYPES OF CELLS THAT EXHIBIT AMEBOID LOCOMOTION. The most common cells that exhibit ameboid locomotion in the human body are the *white blood cells* moving out of the blood into the tissues in the form of *tissue macrophages* or *microphages*. Many other types of cells can move by ameboid locomotion under certain circumstances. For instance, fibroblasts move into a damaged area to help repair the damage, and even some of the germinal cells of the skin, though ordinarily completely sessile cells, move toward a cut area to repair the rent. Finally, cell locomotion is especially important in the development of the embryo and fetus after fertilization of the ovum because embryonic cells often migrate long distances from the primordial sites of origin to new areas during the development of special structures.

CONTROL OF AMEBOID LOCOMOTION—CHEMOTAXIS. The most important factor that usually initiates ameboid locomotion is the process called *chemotaxis*. This results from the appearance of certain chemical substances in the tissues. The chemical substance that causes chemotaxis to occur is called a *chemotactic substance*. Most cells that exhibit ameboid locomotion move toward the source of the chemotactic substance—that is, from an area of lower concentration toward an area of higher concentration—which is called *positive chemotaxis*. Some cells move away from the source, which is called *negative chemotaxis*.

But how does chemotaxis control the direction of ameboid locomotion? Although the answer is not certain, it is known that the side of the cell most exposed to the chemotactic substance develops membrane changes that affect pseudopodial protrusion.

Cilia and Ciliary Movements

A second type of cellular motion, *ciliary movement*, is a whiplike movement of cilia on the surfaces of cells. This occurs in only two places in the human body: on the inside surfaces of the respiratory airways and on the inside surfaces of the uterine tubes (fallopian tubes) of the reproductive tract. In the nasal cavity and lower respiratory airways, the whiplike motion of the cilia causes a layer of mucus to move at a rate of about 1 cm/min toward the pharynx, in this way continually clearing these passageways of mucus and any particles that have become entrapped in the mucus. In the uterine tubes, the cilia cause slow movement of fluid from the ostium of the uterine tube toward the uterine cavity; this movement of fluid transports the ovum from the ovary to the uterus.

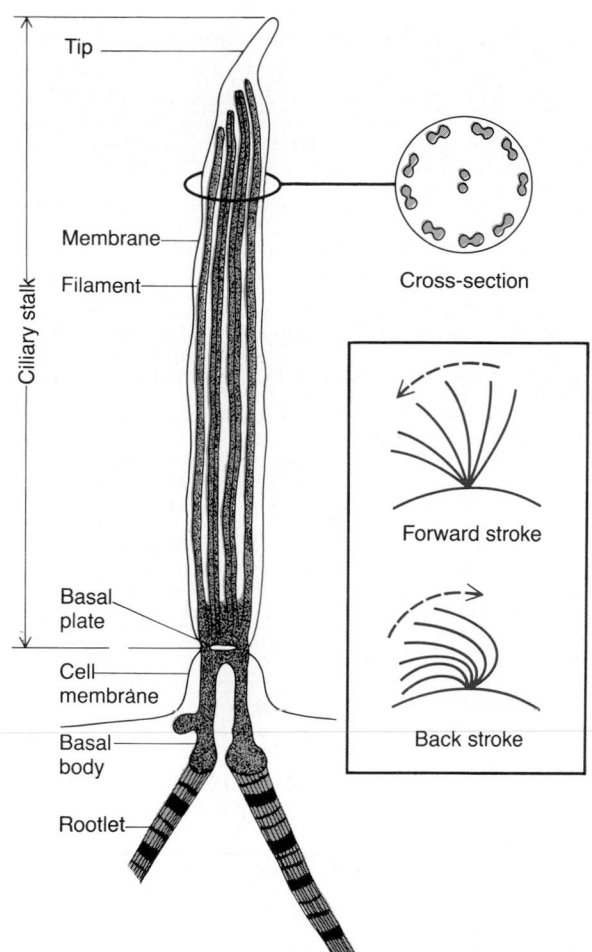

Figure 2–17. Structure and function of the cilium. (Modified from Satir: Cilia. *Sci. Am., 204*:108, 1961. ©1961 by Scientific American, Inc. All rights reserved.)

As shown in Figure 2–17, a cilium has the appearance of a sharp-pointed curved hair that projects 2 to 4 micrometers from the surface of the cell. Many cilia project from each single cell—for instance, as many as 200 cilia on the surface of each epithelial cell in the respiratory tract. The cilium is covered by an outcropping of the cell membrane, and it is supported by 11 microtubules, 9 double tubules located around the periphery of the cilium, and 2 single tubules down the center, as demonstrated in the cross section shown in the figure. Each cilium is an outgrowth of a structure that lies immediately beneath the cell membrane, called the *basal body* of the cilium.

The *flagellum of a sperm* is similar to a cilium; in fact, it has much the same type of structure and same type of contractile mechanism. The flagellum, however, is much longer and moves in quasi-sinusoidal waves instead of with whiplike movements.

In the inset of Figure 2–17, movement of the cilium is shown. The cilium moves forward with a sudden, rapid stroke, 10 to 20 times per second, bending sharply where it projects from the surface of the cell. Then it moves backward slowly in a whiplike manner. The rapid forward-thrusting, whiplike movement pushes the fluid lying adjacent to the cell in the direction that the cilium

moves; then the slow, dragging movement in the other direction has almost no effect on fluid movement. As a result, fluid is continually propelled in the direction of the fast-forward stroke. Because most ciliated cells have large numbers of cilia on their surfaces and because all the cilia are oriented in the same direction, this is an effective means for moving fluids from one part of the surface to another.

MECHANISM OF CILIARY MOVEMENT. Although all aspects of ciliary movement are not clear, we do know the following: First, the nine double tubules and the two single tubules are all linked to one another by a complex of protein cross-linkages; this total complex of tubules and cross-linkages is called the *axoneme*. Second, even after removal of the membrane and destruction of other elements of the cilium besides the axoneme, the cilium can still beat under appropriate conditions. Third, there are two necessary conditions for continued beating of the axoneme after removal of the other structures of the cilium: (1) the presence of ATP and (2) appropriate ionic conditions, including especially appropriate concentrations of magnesium and calcium. Fourth, during forward motion of the cilium, the double tubules on the front edge of the cilium slide outward toward the tip of the cilium while those on the back edge remain in place. Fifth, multiple protein arms, composed of the protein *dynein* that has ATPase activity, project from each double tubule toward an adjacent double tubule.

Given this basic information, it has been determined that the release of energy from ATP in contact with the ATPase dynein arms causes the heads of these arms to "crawl" along the surface of the adjacent double tubule. If the front tubules crawl outward while the back tubules remain stationary, this will cause bending.

The way in which cilia contraction is controlled is not understood. The cilia of some genetically abnormal cells do not have the two central single tubules, and these cilia fail to beat. Therefore, it is presumed that some signal, perhaps an electrochemical signal, is transmitted along these two central tubules to activate the dynein arms.

REFERENCES

Bershadsky, A. D., and Vasiliev, J. M.: Cytoskeleton. New York, Plenum Publishing Corp., 1988.
Bettger, W. J., and McKeehan, W. L.: Mechanisms of cellular nutrition. Physiol. Rev., 66:1, 1986.
Bohr, D. F.: Cell membrane in hypertension. News in Physiol. Sci., 4:85, 1989.
Brown, B. L., and Dobson, P. R. M.: Cell Signalling: Biology and Medicine of Signal Transduction. New York, Raven Press, 1993.
Celis, J. E.: Cell Biology. San Diego, Academic Press, 1994.
Cereijido, M., et al.: Tight junction: Barrier between higher organisms and environment. New Physiol. Sci., 4:72, 1989.
Chien, S. (ed.): Molecular Biology in Physiology. New York, Raven Press.
Conaway, R. C., and Conaway, J. W.: Transcription: Mechanisms and Regulation. New York, Raven Press, 1994.
Cormack, D. H.: Essential Histology. Philadelphia, J. B. Lippincott Co., 1993.
Cristofalo, V. J., and Pignolo, R. J.: Replicative senescence of human fibroblast-like cells in culture. Physiol. Rev., 73:617, 1993.
Damjanov, I.: Color Atlas of Histopathology. Baltimore, Williams & Wilkins, 1995.
Davis, L. G., et al.: Basic Methods in Molecular Biology. 2nd Ed. Norwalk, Conn., Appleton & Lange, 1994.
DeVita, V. T., et al.: Cancer: Principles and Practice of Oncology. 4th Ed. Philadelphia, J. B. Lippincott Co., 1993.
DeMello, W. C. (ed.): Cell-to-Cell Communication. New York, Plenum Publishing Corp., 1987.
DeRobertis, E. D. P., and DeRobertis, E. M. F., Jr.: Cell and Molecular Biology. 8th Ed. Philadelphia, Lea & Febiger, 1987.

Disalvo, E. A., and Simon, S. A.: Permeability and Stability of Lipid Bilayers. Boca Raton, Fla., CRC Press, Inc., 1994.

Fantes, P., and Brooks, R.: The Cell Cycle: A Practical Approach. New York, Oxford University Press, 1994.

Fawcett, D. W.: Bloom & Fawcett: A Textbook of Histology. 11th Ed. Philadelphia, W. B. Saunders Co., 1986.

Fawcett, D. W.: The Cell. 2nd Ed. Philadelphia, W. B. Saunders Co., 1981.

Fitzpatrick, T. B., et al.: Dermatology in General Medicine. Hightstown, NJ, McGraw-Hill, 1993.

Gartner, L. P., and Hiatt, J. L.: Color Atlas of Histology. Baltimore, Williams & Wilkins, 1994.

Gartner, L. P., et al.: Cell Biology and Histology. Baltimore, Williams & Wilkins, 1993.

Goodman, S. R.: Medical Cell Biology. Philadelphia, J. B. Lippincott Co., 1994.

Green, H., and Barrandon, Y.: Cultured epidermal cells and their use in the generation of epidermis. News in Physiol. Sci., 3:53, 1988.

Hoffman, E. K., and Simonsen, L. O.: Membrane mechanisms in volume and pH regulation in vertebrate cells. Physiol. Rev., 69:315, 1989.

Holland, J. F., et al.: Cancer Medicine. Baltimore, Williams & Wilkins, 1993.

Holtzman, E.: Lysosomes. New York, Plenum Publishing Corp., 1989.

Hubbard, A. L., et al.: Biogenesis of endogenous plasma membrane proteins in epithelian cells. Ann. Rev. Physiol., 51:755, 1989.

Kaufman, P. B., et al.: Handbook of Molecular and Cellular Methods in Biology and Medicine. Boca Raton, Fla., CRC Press, Inc., 1995.

Kudlow, J. E., et al. (eds.): Biology of Growth Factors. New York, Plenum Publishing Corp., 1988.

Lane, M. D., et al.: The mitochondrion updated. Science, 234:526, 1986.

Lemasters, J. J., et al. (eds.): Integration of Mitochondrial Function. New York, Plenum Publishing Corp., 1988.

Low, M. G.: Glycosyl-phosphatidylinositol: A versatile anchor for cell surface proteins. FASEB J., 3:1600, 1989.

Machlin, L. J., and Bendich, A.: Free radical tissue damage: Protective role of antioxidant nutrients. FASEB J., 1:441, 1987.

Miller, M., et al.: Nuclear pore complex: Structure, function, and regulation. Physiol. Rev., 71:909, 1991.

Mond, J. J., et al.: Cell Activation: Genetic Approaches. New York, Raven Press, 1991.

Moore, K. L., and Persaud, T. V. N.: The Developing Human: Clinically Oriented Embryology. Philadelphia, W. B. Saunders Co., 1993.

Mumby, M. C., and Walter, G.: Protein serine/threonine phosphatases: Structure, regulation, and functions in cell growth. Physiol. Rev., 73:673, 1993.

Ross, M. H., et al.: Histology: A Text and Atlas. Baltimore, Williams & Wilkins, 1994.

Rodriguez-Boulan, E., and Salas, P. J. I.: External and internal signals for epithelial cell surface polarization. Ann. Rev. Physiol., 51:741, 1989.

Slavik, J.: Fluorescent Probes in Cellular and Molecular Biology. Boca Raton, Fla., CRC Press, Inc., 1994.

Sowers, A. E. (ed.): Cell Fusion. New York, Plenum Publishing Corp., 1987.

Thomas, K. A.: Fibroblast growth factors. FASEB J., 1:434, 1987.

van der Laarse, W. J., et al.: Energetics at the single cell level. News in Physiol. Sci., 4:91, 1989.

Wade, J. B.: Role of membrane fusion in hormonal regulation of epithelial transport. Ann. Rev. Physiol., 48:213, 1986.

White, S. H.: Vol. 1: Membrane Protein Structure Experimental Approaches. Am. Physiol. Soc., 1994.

Genetic Control of Protein Synthesis, Cell Function, and Cell Reproduction

CHAPTER 3

Virtually everyone knows that the genes control heredity from parents to children, but most people do not realize that the same genes control the reproduction and day-by-day function of all cells. The genes control cell function by determining what substances will be synthesized within the cell—what structures, what enzymes, what chemicals.

Figure 3–1 shows the general schema of genetic control. Each gene, which is a nucleic acid called *deoxyribonucleic acid (DNA)*, automatically controls the formation of another nucleic acid, *ribonucleic acid (RNA)*, which spreads throughout the cell and controls the formation of a specific protein. Because there are about 100,000 genes, it is theoretically possible to form a large number of different cellular proteins.

Some of the cellular proteins are *structural proteins*, which, in association with various lipids and carbohydrates, form the structures of the various intracellular organelles that are discussed in Chapter 2. By far, the majority of the proteins are *enzymes* that catalyze the different chemical reactions in the cells. For instance, enzymes promote all the oxidative reactions that supply energy to the cell, and they promote the synthesis of various chemicals, such as lipids, glycogen, and adenosine triphosphate (ATP).

The Genes

Large numbers of genes are attached end on end in extremely long double-stranded, helical molecules of *DNA* having molecular weights measured in the billions. A very short segment of such a molecule is shown in Figure 3–2. This molecule is composed of several simple chemical compounds bound together in a regular pattern explained in the next few paragraphs.

BASIC BUILDING BLOCKS OF DNA. Figure 3–3 shows the basic chemical compounds involved in the formation of DNA. These include (1) *phosphoric acid,* (2) a sugar called *deoxyribose,* and (3) four nitrogenous *bases* (two purines, *adenine* and *guanine,* and two pyrimidines, *thymine* and *cytosine*). The phosphoric acid and deoxyribose form the two helical strands that are the backbone of the DNA molecule, and the bases lie between the two strands and connect them.

NUCLEOTIDES. The first stage in the formation of DNA is the combination of one molecule of phosphoric acid, one molecule of deoxyribose, and one of the four bases to form a nucleotide. Four separate nucleotides are thus formed, one for each of the four bases: *deoxyadenylic, deoxythmidylic, deoxyguanylic,* and *deoxycytidylic acids.* Figure 3–4 shows the chemical structure of deoxyadenylic acid, and Figure 3–5 shows simple symbols for all the four basic nucleotides that form DNA.

ORGANIZATION OF THE NUCLEOTIDES TO FORM TWO STRANDS OF DNA LOOSELY BOUND TO EACH OTHER. Figure 3–6 shows the manner in which multiple numbers of nucleotides are bound together to form two strands of DNA, one shown above in the figure and one below, and the two strands loosely bonded with each other. Note that the backbone of each DNA strand is composed of alternating phosphoric acid and deoxyribose molecules. The purine and pyrimidine bases are attached to the sides of the deoxyribose molecules, and it is by means of loose bonds (the dashed lines) between these purine and pyrimidine bases that the two respective DNA strands are held together. But note carefully that

1. The purine base *adenine* of one strand always bonds with the pyrimidine base *thymine* of the other strand, and

2. The purine base *guanine* always bonds with the pyrimidine base *cytosine.*

Thus, in Figure 3–6, the sequence of complementary pairs of bases is CG, CG, GC, TA, CG, TA, GC, AT, and AT. The bases are bound together by very loose *hydrogen bonding,*

Gene (DNA)

↓

RNA formation

↓

Protein formation

↙   ↘

Cell structure   Cell enzymes

↘   ↙

Cell function

Figure 3–1. General schema by which the genes control cell function.

represented in the figure by dashed lines. Because of the looseness of these bonds, the two strands can pull apart with ease, and they do so many times during the course of their function in the cell.

Now, to put the DNA of Figure 3–6 into its proper physical perspective, one needs merely to pick up the two ends and twist them into a helix. Ten pairs of nucleotides are present in each full turn of the helix in the DNA molecule, as shown in Figure 3–2.

Genetic Code

The importance of DNA lies in its ability to control the formation of other substances in the cell. It does this by means of the so-called *genetic code*. When the two strands of a DNA molecule are split apart, this exposes the purine and pyrimidine bases projecting to the side of each strand, as shown by the top strand in Figure 3–7. It is these projecting bases that form the code.

The genetic code consists of successive "triplets" of bases —that is, each three successive bases is a code word. The successive triplets will eventually control the sequence of amino acids in a protein molecule synthesized in the cell. Note in Figure 3–6 that each of the two strands of the DNA molecule carries its own genetic code. For instance, the top strand, reading from left to right, has the genetic code GGC, AGA, CTT, the triplets being separated from one another by the arrows. As we follow this genetic code through Figures 3–7 and 3–8, we see that these three respective triplets are

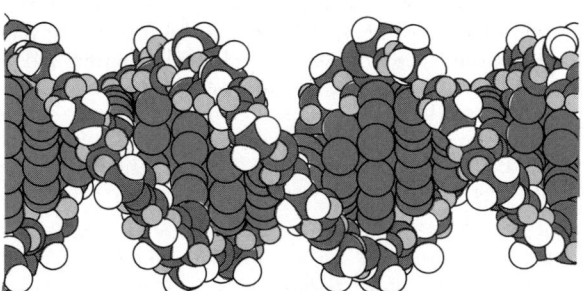

Figure 3–2. The helical, double-stranded structure of the gene. The outside strands are composed of phosphoric acid and the sugar deoxyribose. The internal molecules connecting the two strands of the helix are purine and pyrimidine bases; these determine the "code" of the gene.

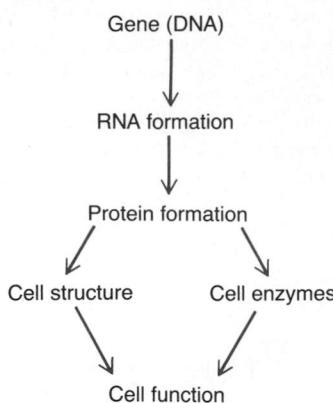

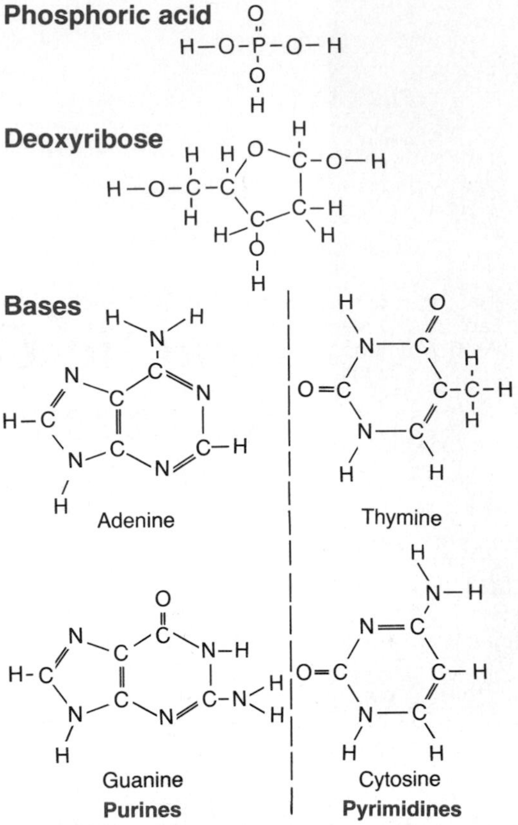

Figure 3–3. The basic building blocks of deoxyribonucleic acid (DNA).

responsible for successive placement of the three amino acids, *proline, serine,* and *glutamic acid,* in a molecule of protein.

THE DNA CODE IS TRANSFERRED TO AN RNA CODE—THE PROCESS OF TRANSCRIPTION

Because almost all DNA is located in the nucleus of the cell and yet most of the functions of the cell are carried out in the cytoplasm, some means must be available for the DNA genes of the nucleus to control the chemical reactions

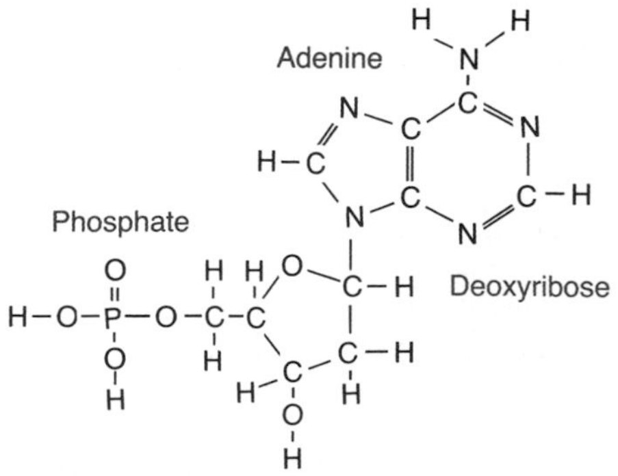

Figure 3–4. Deoxyadenylic acid, one of the nucleotides that make up DNA.

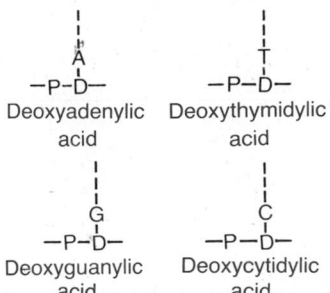

Figure 3–5. The four nucleotides that are combined to form DNA. Each nucleotide contains phosphoric acid (P), deoxyribose (D), plus one of the four nucleotide bases: A, adenine; T, thymine; G, guanine; and C, cytosine.

of the cytoplasm. This is achieved through the intermediary of another type of nucleic acid, *RNA*, the formation of which is controlled by the DNA of the nucleus. In this process, the code is transferred to the RNA, which is called *transcription.* The RNA then diffuses from the nucleus through the nuclear pores into the cytoplasmic compartment, where it controls protein synthesis.

Synthesis of RNA

During synthesis of RNA, the two strands of the DNA molecule separate temporarily; one of these strands is then used as a template for synthesis of the RNA molecules. The code triplets in the DNA cause the formation of *complementary* code triplets (called *codons*) in the RNA; these codons in turn control the sequence of amino acids in a protein to be synthesized later in the cytoplasm. When one strand of DNA is used in this manner to cause the formation of RNA, the opposite strand remains inactive. Each DNA strand in each chromosome is such a long molecule that it carries the code for an average of about 4000 genes.

BASIC BUILDING BLOCKS OF RNA. The basic building blocks of RNA are almost the same as those of DNA except for two differences. First, the sugar deoxyribose is not used in the formation of RNA. In its place is another sugar of slightly different composition, *ribose.* Second, thymine is replaced by another pyrimidine, *uracil.*

FORMATION OF RNA NUCLEOTIDES. The basic building blocks of RNA first form nucleotides exactly as previously described for the synthesis of DNA. Here again, four separate nucleotides are used in the formation of RNA. These nucleotides contain the bases *adenine, guanine, cytosine,* and *uracil.* Note that these are the same as the bases in DNA except for one of them; the uracil in RNA replaces the thymine in DNA.

ACTIVATION OF THE NUCLEOTIDES. The next step in the synthesis of RNA is activation of the nucleotides. This occurs by adding to each nucleotide two phosphate radicals to form triphosphates (shown in Figure 3–7 by the two RNA nucleotides to the far right during RNA chain formation). These last two phosphates are combined with the nucleotide by *high-energy phosphate bonds* derived from the ATP of the cell.

The result of this activation process is that large quantities of energy are made available to each of the nucleotides, and this energy is used in promoting the chemical reactions that add each new RNA nucleotide to the end of the RNA chain.

Assembly of the RNA Molecule from Activated Nucleotides Using the DNA Strand as a Template—The Process of Transcription

Assembly of the RNA molecule is accomplished in the manner shown in Figure 3–7 under the influence of the enzyme *RNA polymerase.* This is a large protein enzyme that has many functional properties necessary for the formation of the RNA molecule. They are as follows:

1. In the DNA strand immediately ahead of the initial gene is a sequence of nucleotides called the *promoter.* The RNA polymerase has an appropriate complementary structure that recognizes this promoter and becomes attached to it. This is an essential step in initiating the formation of the RNA molecule.

2. After the RNA polymerase attaches to the promoter, the polymerase causes unwinding of about two turns of the DNA helix and separation of the unwound portions of the two strands.

3. Then the polymerase moves along the DNA strand, temporarily unwinding and separating the two DNA strands at each stage of its movement. As it moves along, it adds new RNA nucleotides to the end of the newly forming RNA chain by the following steps:

4. First, it causes hydrogen bonds to form between the bases of the DNA strand and bases of RNA nucleotides in the nucleoplasm.

5. Then, one at a time, the RNA polymerase breaks two of the three phosphate radicals away from each of these RNA nucleotides, liberating large amounts of energy from the broken high-energy phosphate bonds; this energy is used to cause covalent linkage of the remaining phosphate on the nucleotide with the ribose on the end of the growing RNA molecule.

6. When the RNA polymerase reaches the end of the DNA gene, it encounters a new sequence of DNA nucleotides called the *chain-terminating sequence;* this causes the polymerase to break away from the DNA strand. Then the polymerase can attach somewhere else on the same or an-

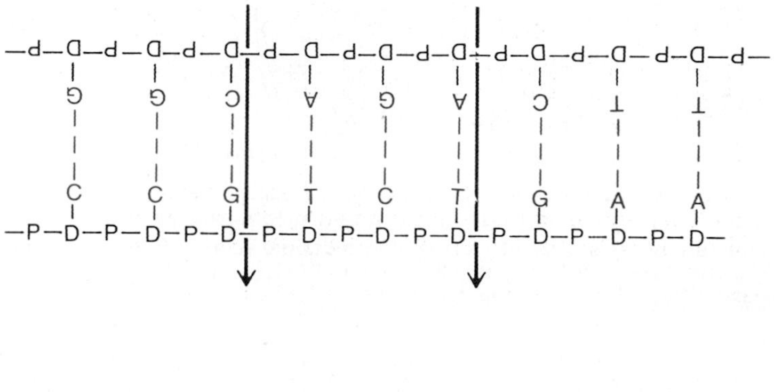

Figure 3–6. Arrangement of deoxyribose nucleotides in a double strand of DNA.

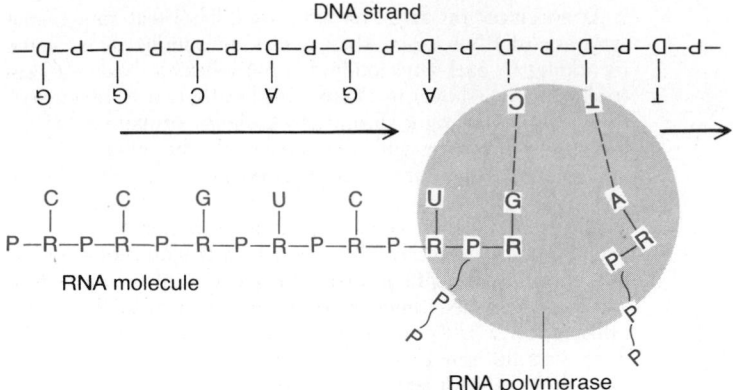

Figure 3–7. Combination of ribose nucleotides with a strand of DNA to form a molecule of ribonucleic acid (RNA) that carries the genetic code from the gene to the cytoplasm. The *RNA polymerase* moves along the DNA strand and builds the RNA molecule.

other DNA strand to be used again and again for forming new RNA chains.

7. As the new RNA strand is formed, its hydrogen bonds with the DNA template break away because the DNA has a high affinity for rebonding with its own complementary DNA strand. Thus, the RNA chain is forced away from the DNA and is released into the nucleoplasm.

It should be remembered again that there are four types of DNA bases and four types of RNA nucleotide bases. Furthermore, these always combine with each other in specific combinations. Therefore, the code that is present in the DNA strand is transmitted in *complementary* form to the RNA molecule. The ribose nucleotide bases always combine with the deoxyribose bases in the following combinations:

DNA base	RNA base
guanine	cytosine
cytosine	guanine
adenine	uracil
thymine	adenine

After the RNA molecules have been released into the nucleoplasm, they still must be processed further before leaving the nucleus to enter the cytoplasm. The reason for this is that the newly transcribed RNA contains many unwanted sequences of RNA nucleotides. Some of these occur at both ends of the RNA strand, and many others occur in the middle of the strand; this unwanted material usually constitutes more than 90 per cent of the total strand. A series of enzymes within the nucleoplasm has the capability of cutting out these unwanted sequences and then resplicing the RNA strand, a process called *RNA splicing*. The RNA is now ready to be used for the formation of proteins.

Three Separate Types of RNA. There are three separate types of RNA, each of which plays an independent and entirely different role in protein formation. They are

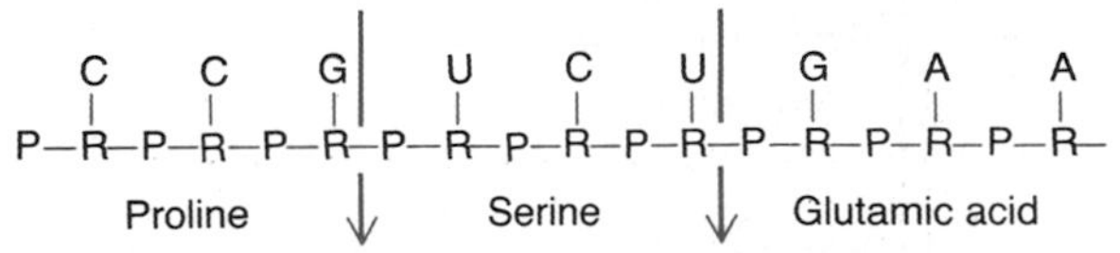

1. *Messenger RNA*, which carries the genetic code to the cytoplasm for controlling the formation of the proteins;
2. *Transfer RNA*, which transports activated amino acids to the ribosomes to be used in assembling the protein molecules; and
3. *Ribosomal RNA*, which, along with about 75 different proteins, forms the *ribosomes,* the physical and chemical structures on which protein molecules are actually assembled.

Messenger RNA—The Codons

Messenger RNA molecules are long, single RNA strands that are suspended in the cytoplasm. These molecules are composed of several hundred to several thousand nucleotides in unpaired strands, and they contain *codons* that are exactly complementary to the code triplets of the DNA genes. Figure 3–8 shows a small segment of a molecule of messenger RNA. Its codons are CCG, UCU, and GAA. These are the codons for the amino acids proline, serine, and glutamic acid. The transcription of these codons from the DNA molecule to the RNA molecule is shown in Figure 3–7.

RNA Codons for the Different Amino Acids. Table 3–1 gives the RNA codons for the 20 common amino acids found in protein molecules. Note that most of the amino acids are represented by more than one codon; also, one codon represents the signal "start manufacturing a protein molecule" and three codons represent "stop manufacturing a protein molecule." In Table 3–1, these two types of codons are designated CI for "chain-initiating" and CT for "chain-terminating."

Transfer RNA—The Anticodons

Another type of RNA that plays an essential role in protein synthesis is called *transfer RNA* because it transfers amino acid molecules to protein molecules as the protein is being synthesized. Each type of transfer RNA combines specifically with 1 of the 20 amino acids that are to be incorporated into proteins. The transfer RNA then acts as a *carrier* to transport its specific type of amino acid to the ribosomes, where protein molecules are forming. In the ribosomes, each specific type of transfer RNA recognizes a particular codon on the messenger RNA, as is described subsequently, and thereby delivers the appropriate amino acid to the appropri-

C   C   G   U   C   U   G   A   A
P—R—P—R—P—R—P—R—P—R—P—R—P—R—P—R—P—R—
Proline ↓ Serine ↓ Glutamic acid

Figure 3–8. Portion of a ribonucleic acid molecule, showing three "codons," CCG, UCU, and GAA, which represent the three amino acids *proline, serine,* and *glutamic acid.*

Table 3-1 RNA CODONS FOR AMINO ACIDS AND FOR START AND STOP

Amino Acid	RNA Codons					
Alanine	GCU	GCC	GCA	GCG		
Arginine	CGU	CGC	CGA	CGG	AGA	AGG
Asparagine	AAU	AAC				
Aspartic acid	GAU	GAC				
Cysteine	UGU	UGC				
Glutamic acid	GAA	GAG				
Glutamine	CAA	CAG				
Glycine	GGU	GGC	GGA	GGG		
Histidine	CAU	CAC				
Isoleucine	AUU	AUC	AUA			
Leucine	CUU	CUC	CUA	CUG	UUA	UUG
Lysine	AAA	AAG				
Methionine	AUG					
Phenylalanine	UUU	UUC				
Proline	CCU	CCC	CCA	CCG		
Serine	UCU	UCC	UCA	UCG	AGC	AGU
Threonine	ACU	ACC	ACA	ACG		
Tryptophan	UGG					
Tyrosine	UAU	UAC				
Valine	GUU	GUC	GUA	GUG		
Start (CI)	AUG					
Stop (CT)	UAA	UAG	UGA			

ate place in the chain of the newly forming protein molecule.

Transfer RNA, which contains only about 80 nucleotides, is a relatively small molecule in comparison with messenger RNA. It is a folded chain of nucleotides with a cloverleaf appearance similar to that shown in Figure 3–9. At one end of the molecule is always an adenylic acid; it is to this that the transported amino acid attaches to a hydroxyl group of the ribose in the adenylic acid. A specific enzyme causes this attachment for each specific type of transfer RNA; this enzyme also determines the type of amino acid that will attach to each respective type of transfer RNA.

Because the function of transfer RNA is to cause attachment of a specific amino acid to a forming protein chain, it is essential that each type of transfer RNA also have specificity for a particular codon in the messenger RNA. The specific code in the transfer RNA that allows it to recognize a specific codon is again a triplet of nucleotide bases and is called an *anticodon*. This is located approximately in the middle of the transfer RNA molecule (at the bottom of the cloverleaf configuration shown in Figure 3–9). During formation of a protein molecule, the anticodon bases combine loosely by hydrogen bonding with the codon bases of the messenger RNA. In this way, the respective amino acids are lined up one after another along the messenger RNA chain, thus establishing the appropriate sequence of amino acids in the protein molecule.

Ribosomal RNA

The third type of RNA in the cell is ribosomal RNA; it constitutes about 60 per cent of the *ribosome*. The remainder of the ribosome is protein, containing about 75 types of proteins that are both structural proteins and enzymes needed in the manufacture of protein molecules.

The ribosome is the physical structure in the cytoplasm on which protein molecules are actually synthesized. However,

it always functions in association with both the other types of RNA as well: transfer RNA transports amino acids to the ribosome for incorporation into the developing protein molecule, whereas messenger RNA provides the information necessary for sequencing the amino acids in proper order for each specific type of protein to be manufactured.

The ribosomes of nucleated cells are composed of two physical subunits, called the *small subunit*, containing 1 RNA molecule and 33 proteins, and the *large subunit*, containing 3 RNAs and more than 40 proteins. Although we have only partial knowledge of the mechanism of protein manufacture in the ribosome, it is known that messenger RNA and transfer RNA first complex with the small subunit. Then the large subunit provides most of the enzymes that promote peptide linkages between the successive amino acids. Thus, the ribosome acts as a manufacturing plant in which the protein molecules are formed.

FORMATION OF RIBOSOMES IN THE NUCLEOLUS. The DNA genes for formation of ribosomal RNA are located in five chromosomal pairs of the nucleus, and each of these chromosomes contains many duplicates of these genes because of the large amount of ribosomal RNA required for cellular function.

As the ribosomal RNA forms, it collects in the *nucleolus*, a specialized structure lying adjacent to the chromosomes. When large amounts of ribosomal RNA are being synthesized, as occurs in cells that manufacture large amounts of protein, the nucleolus is a large structure, whereas in cells that synthesize little protein, the nucleolus may not be seen. The ribosomal RNA is specially processed in the nucleolus, where it binds with "ribosomal proteins" to form granular condensation products that are primordial subunits of the ribosomes. These subunits are then released from the nucleolus and transported through the large pores of the nuclear envelope to almost all parts of the cytoplasm. After the subunits enter the cytoplasm, they are assembled to form mature, functional ribosomes. Therefore, proteins are not formed in the nucleus because the nucleus does not contain mature ribosomes.

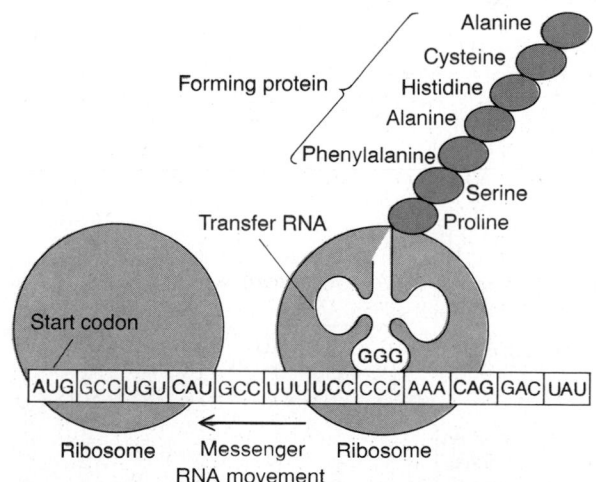

Figure 3–9. A messenger RNA strand is moving through two ribosomes. As each "codon" passes through, an amino acid is added to the growing protein chain, which is shown in the right-hand ribosome. The transfer RNA molecule determines which of the 20 amino acids will be added at each stage of protein formation.

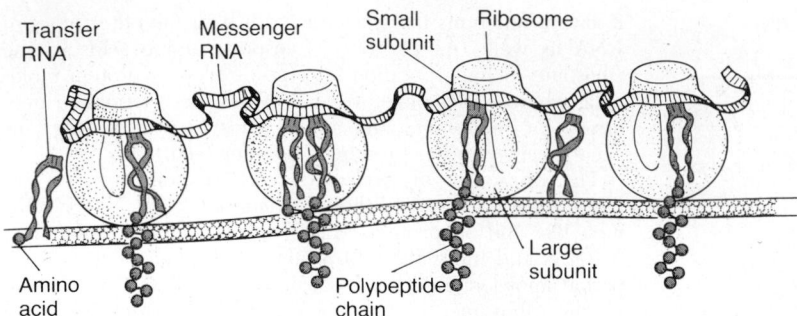

Figure 3–10. Physical structure of the ribosomes as well as their functional relation to messenger RNA, transfer RNA, and the endoplasmic reticulum during the formation of protein molecules. (From Bloom and Fawcett: A Textbook of Histology. 10th ed. Philadelphia, W. B. Saunders Company, 1975.)

Formation of Proteins on the Ribosomes— The Process of Translation

When a molecule of messenger RNA comes in contact with a ribosome, it travels through the ribosome, beginning at a predetermined end of the RNA molecule specified by an appropriate sequence of RNA bases. Then, as shown in Figure 3–9, while the messenger RNA travels through the ribosome, a protein molecule is formed—a process called *translation.* Thus, the ribosome reads the codons of the messenger RNA in much the same way that a tape is "read" as it passes through the playback head of a tape recorder. Then, when a "stop" (or "chain-terminating") codon slips past the ribosome, the end of a protein molecule is signaled and the protein molecule is freed into the cytoplasm.

POLYRIBOSOMES. A single messenger RNA molecule can form protein molecules in several ribosomes at the same time because the RNA strand can pass to a successive ribosome as it leaves the first, as shown in Figure 3–9. The protein molecules are in different stages of development in each ribosome. As a result, clusters of ribosomes frequently occur, 3 to 10 ribosomes being attached to a single messenger RNA at the same time. These clusters are called *polyribosomes.*

It is especially important to note that a messenger RNA can cause the formation of a protein molecule in any ribosome, that there is no specificity of ribosomes for given types of protein. The ribosome is simply the structure in which the chemical reactions take place.

MANY RIBOSOMES ATTACH TO THE ENDOPLASMIC RETICULUM. In Chapter 2, it was noted that many ribosomes become attached to the endoplasmic reticulum. This occurs because the initial ends of many forming protein molecules have amino acid sequences that immediately attach to specific receptor sites on the endoplasmic reticulum; this causes these molecules to penetrate the reticulum wall and enter the endoplasmic reticulum matrix. This occurs while the protein molecule is still being formed by the ribosome, which pulls the ribosome against the endoplasmic reticulum, thereby giving the granular appearance to those portions of the reticulum where proteins are being formed and entering the matrix of the reticulum.

Figure 3–10 shows the functional relation of messenger RNA to the ribosomes and the manner in which the ribosomes attach to the membrane of the endoplasmic reticulum. Note the process of translation occurring in several ribosomes at the same time in response to the same strand of messenger RNA. Note also the newly forming polypeptide (protein) chains passing through the endoplasmic reticulum membrane into the endoplasmic matrix.

Yet it should be noted that except in glandular cells where large amounts of protein-containing secretory vesicles are formed, most proteins synthesized by the ribosomes are released directly into the cytosol. They are the enzymes and internal structural proteins of the cell.

CHEMICAL STEPS IN PROTEIN SYNTHESIS. Some of the chemical events that occur in synthesis of a protein molecule are shown in Figure 3–11. This figure shows representative

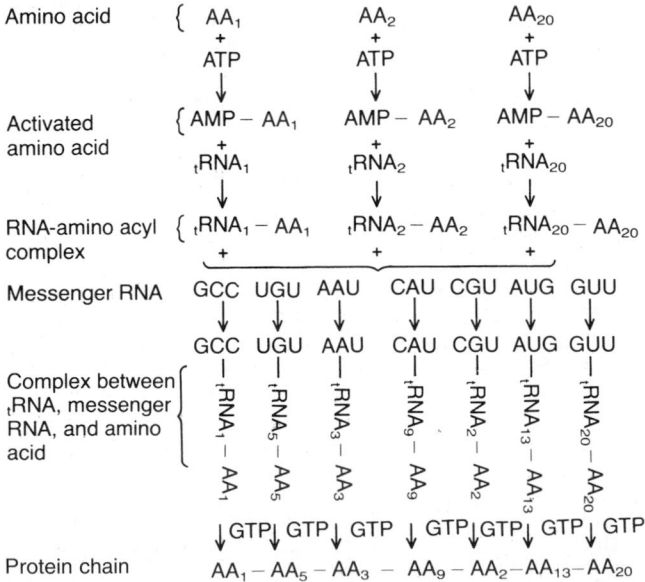

Figure 3–11. Chemical events in the formation of a protein molecule.

reactions for three separate amino acids, AA_1, AA_2, and AA_{20}. The stages of the reactions are the following: (1) Each amino acid is *activated* by a chemical process in which ATP combines with the amino acid to form an *adenosine monophosphate complex with the amino acid*, giving up two high-energy phosphate bonds in the process. (2) The activated amino acid, having an excess of energy, then *combines with its specific transfer RNA to form an amino acid–tRNA complex* and, at the same time, releases the adenosine monophosphate. (3) The transfer RNA carrying the amino acid complex then comes in contact with the messenger RNA molecule in the ribosome, where the anticodon of the transfer RNA attaches temporarily to its specific codon of the messenger RNA, thus lining up the amino acids in appropriate sequence to form a protein molecule. Then, under the influence of the enzyme *peptidyl transferase*, one of the proteins in the ribosome, *peptide bonds* are formed between the successive amino acids, thus adding progressively to the protein chain. These chemical events require the energy from two additional high-energy phosphate bonds, making a total of four high-energy bonds used for each amino acid added to the protein chain. Thus, the synthesis of proteins is one of the most energy-consuming processes of the cell.

Peptide Linkage. The successive amino acids in the protein chain combine with one another according to the typical reaction:

$$\begin{array}{c} \overset{NH_2}{\underset{|}{R-C}}\overset{O}{\underset{||}{-C}}-OH + H-\overset{H}{\underset{|}{N}}-\overset{R}{\underset{|}{C}}-COOH \longrightarrow \\[2em] \overset{NH_2}{\underset{|}{R-C}}\overset{O}{\underset{||}{-C}}-\overset{H}{\underset{|}{N}}-\overset{R}{\underset{|}{C}}-COOH + H_2O \end{array}$$

In this chemical reaction, a hydroxyl radical (OH^-) is removed from the COOH portion of the first amino acid, while a hydrogen (H^+) of the NH_2 portion of the other amino acid is removed. These combine to form water, and the two reactive sites left on the two successive amino acids bond with each other, resulting in a single molecule. This process is called *peptide linkage*. Then, as each additional amino acid is added, a new peptide linkage is formed.

SYNTHESIS OF OTHER SUBSTANCES IN THE CELL

Many thousand protein enzymes formed in the manner just described control essentially all the other chemical reactions that take place in cells. These enzymes promote synthesis of lipids, glycogen, purines, pyrimidines, and hundreds of other substances. We discuss many of these synthetic processes in relation to carbohydrate, lipid, and protein metabolism in Chapters 67 through 69. It is by means of all these substances that the many functions of the cells are performed.

CONTROL OF GENETIC FUNCTION AND BIOCHEMICAL ACTIVITY IN CELLS

From our discussion thus far, it is clear that the genes control both the physical and the chemical functions of the cells. However, the activation of the genes themselves must be controlled as well; otherwise, some parts of the cell might overgrow or some chemical reactions might overact until they should kill the cell. Each cell has powerful internal feedback control mechanisms that keep the various functional operations of the cell in step with one another. For each gene or for each small group of genes (100,000 genes in all), there is at least one such feedback mechanism.

The smallest viruses have contributed immensely to our understanding of cellular function because they are simple enough for biologists to work out almost the exact details of function, molecule by molecule. At a slightly higher level, the bacteria have also been valuable, especially the bacterium *Escherichia coli*, which is immensely abundant in feces. Most of what we discuss in the next few pages has been learned from studies in these lower forms of life. The nucleated cell is so complex that we are only now beginning to understand the more complex control mechanisms invented by this higher form of life. To illustrate the difference in complexity of the nucleated cell (called a *eukaryote*) and the non-nucleated cell (called a *prokaryote*), the eukaryote of the human being has almost 1000 times as much DNA as each *E. coli* bacterium.

There are basically two methods by which the biochemical activities in the cell are controlled. One of these is called *genetic regulation*, in which the activities of the genes themselves are controlled, and the other *enzyme regulation*, in which the degree of activity of the already formed enzymes in the cell is controlled.

Genetic Regulation

THE OPERON OF THE CELL AND ITS CONTROL OF BIOCHEMICAL SYNTHESIS—FUNCTION OF THE PROMOTER. Synthesis of a cellular biochemical product usually requires a series of reactions, and each of these reactions is catalyzed by a special protein enzyme. Formation of all the enzymes needed for the synthetic process often is controlled by a sequence of genes located in series one after the other on the same chromosomal DNA strand. This area of the DNA strand is called an *operon*, and the genes responsible for forming the respective enzymes are called *structural genes*. In Figure 3–12, three respective structural genes are shown in an operon, and it is demonstrated that they control the formation of three respective enzymes that in turn cause synthesis of a specific intracellular product.

Now note in the figure the segment on the DNA strand called the *promoter*. This is a series of nucleotides that has specific affinity for RNA polymerase, as already discussed. The polymerase must bind with this promoter before it can begin traveling along the DNA strand to synthesize RNA. Therefore, the promoter is the essential element in activating the operon.

CONTROL OF THE OPERON BY A REPRESSOR PROTEIN— THE REPRESSOR OPERATOR. Also note in Figure 3–12 an additional band of nucleotides lying in the middle of the promoter. This area is called a *repressor operator* because a "regulatory" protein can bind here and prevent attachment of RNA polymerase to the promoter, thereby blocking transcription of the genes in the operon. Such a regulatory protein is called a *repressor protein*. Each regulatory repressor protein generally exists in two allosteric forms, one that can bind with the operator and repress transcription and the other that does not bind. That is, one of its forms might be a

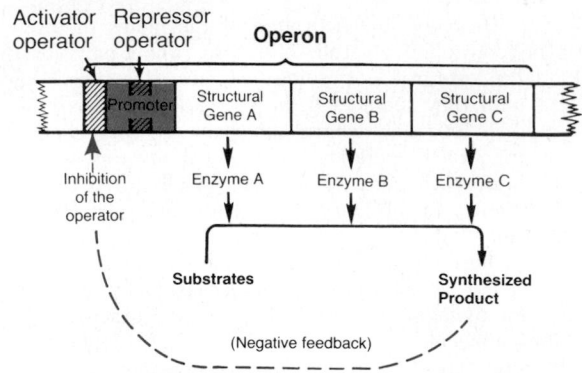

Figure 3–12. Function of the *operon* to control synthesis of a nonprotein intracellular product, such as an intracellular metabolic chemical. Note that the synthesized product exerts a negative feedback to inhibit function of the operon, in this way automatically controlling the concentration of the product itself.

straight protein molecule and the other the same molecule bent at an angle in the middle. Only one of these forms can repress the operator. In turn, various nonprotein substances in the cell, such as some of the cell metabolites, can bind with the repressor protein to change its state. A substance that causes it to change so that it will bind with the operator and stop transcription is called a *repressor substance* or *inhibitor substance.* On the other hand, a substance that changes the repressor protein so that it will break its bond with the operator is called an *activator substance* or *inducer substance* because it activates, or induces, the transcription process by removing the repressor protein.

To illustrate the control of gene transcription by a repressor protein, let us give an example. The saccharide lactose usually is not available to *E. coli* cells as a food substrate. Therefore, the bacterium normally will not synthesize the required enzymes for metabolic use of lactose. However, when lactose does become available, it induces an allosteric conformational change in a repressor protein, causing it to break away from the promotor of the operon that transcribes for the necessary metabolic enzymes. Then, within minutes, RNA polymerase binds with the promoter and travels along the operon, forming the appropriate enzymes to cause breakdown of the lactose. Then, as the lactose begins to disappear from within the cell, the rate of synthesis of the enzymes decreases back to that level only required for the amount of lactose that is available. Thus, one can see the logic of having such regulatory systems in the cell.

CONTROL OF THE OPERON BY AN "ACTIVATOR PROTEIN" —THE "ACTIVATOR OPERATOR." Now note in Figure 3–12 another operator, called the *activator operator,* that lies adjacent to but ahead of the promoter. When a regulatory protein binds to this operator, it helps to attract the RNA polymerase to the promoter, in this way activating the operon. Therefore, a regulatory protein of this type is called an *activator protein.* The operon can be activated or inhibited through the activator operator in ways exactly opposite to control by the repressor operator.

NEGATIVE FEEDBACK CONTROL OF THE OPERON. Finally, note in Figure 3–12 that the presence of a critical amount of a synthesized product in the cell can cause negative feedback inhibition of the operon that is responsible for its synthesis. It can do this either by causing a regulatory repressor protein to bind at the repressor operator or by causing a regulatory activator protein to break its bond with the activator operator. In either case, the operon becomes inhibited. Therefore, once the required synthesized product has become abundant enough, the operon becomes dormant. On the other hand, when the synthesized product becomes degraded in the cell and its concentration falls, the operon once again becomes active. In this way, the concentration of the product is automatically controlled.

OTHER MECHANISMS FOR CONTROL OF TRANSCRIPTION BY THE OPERON. Variations in the basic mechanism for control of the operon have been discovered with rapidity within the past decade. Without giving details, let us list some of them:

1. An operon frequently is controlled by a *regulatory gene* located elsewhere in the genetic complex of the nucleus. That is, the regulatory gene causes the formation of a regulatory protein that in turn acts either as an activator or as a repressor substance to control the operon.

2. Occasionally, many different operons are controlled at the same time by the same regulatory protein. In some instances, the same regulatory protein functions as an activator for one operon and as a repressor for another operon. When multiple operons are controlled simultaneously in this manner, all the operons that function together are called a *regulon.*

3. Some operons are controlled not at the starting point of transcription on the DNA strand but instead farther along the strand. Sometimes the control is not even at the DNA strand itself but during the processing of the RNA molecules in the nucleus before they are released into the cytoplasm; or, rarely, control might occur at the level of protein formation in the cytoplasm during RNA translation by the ribosomes.

4. In eukaryotes, the nuclear DNA is packaged in specific structural units, the *chromosomes.* And within each chromosome, the DNA is wound around small proteins called *histones,* which in turn are held tightly together in a compacted state by still other proteins. As long as the DNA is in this compacted state, it cannot function to form RNA. However, multiple control mechanisms are beginning to be discovered that can cause selected areas of the chromosomes to become decompacted one part at a time so that RNA transcription can occur. Even then, some specific "transcriptor factor" controls the actual rate of transcription by each separate operon. Thus, in the eukaryote, still higher orders of control are used for establishing proper cell function. In addition, signals from outside the cell, such as some of the hormones, can activate specific chromosomal areas and specific transcription factors, thus controlling the chemical machinery for each cell function.

Because there are as many as 100,000 different genes in each human cell, the large number of ways in which genetic activity can be controlled is not surprising. The genetic control systems are especially important for controlling the intracellular concentrations of amino acids, amino acid derivatives, and the intermediate substrates and products of carbohydrate, lipid, and protein metabolism.

Control of Intracellular Function by Enzyme Regulation

In addition to control of cell function by genetic regulation, some cell activities are controlled by intracellular inhibitors or activators that act directly on specific intracellular enzymes. Thus, enzyme regulation represents a second category of mechanisms by which cellular biochemical functions can be controlled.

ENZYME INHIBITION. Some of the chemical substances formed in the cell have a direct feedback effect in inhibiting the specific enzyme systems that synthesize them. Almost always the synthesized product acts on the first enzyme in a sequence, rather than on the subsequent enzymes, usually binding directly with the enzyme and causing an allosteric conformational change that inactivates it. One can readily recognize the importance of inactivating this first enzyme: It prevents buildup of intermediary products that will not be used.

This process of enzyme inhibition is another example of negative feedback control; it is responsible for controlling the intracellular concentrations of some of the amino acids, purines, pyrimidines, vitamins, and other substances.

ENZYME ACTIVATION. Enzymes that are normally inactive often can be activated when needed. An example of this occurs when most of the ATP has been depleted in a cell. In this case, a considerable amount of cyclic adenosine monophosphate (cAMP) begins to be formed as a breakdown product of the ATP; the presence of this cAMP in turn immediately activates the glycogen-splitting enzyme phosphorylase, liberating glucose molecules that are rapidly metabolized and their energy used for replenishment of the ATP stores. Thus, the cAMP acts as an enzyme activator for the enzyme phosphorylase and thereby helps control intracellular ATP concentration.

Another interesting instance of both enzyme inhibition and enzyme activation occurs in the formation of the purines and pyrimidines. These substances are needed by the cell in about equal quantities for formation of DNA and RNA. When purines are formed, they *inhibit* the enzymes that are required for formation of additional purines. However, they *activate* the enzymes for formation of pyrimidines. Conversely, the pyrimidines inhibit their own enzymes but activate the purine enzymes. In this way, there is continual cross-feed between the synthesizing systems for these two substances, resulting in almost exactly equal amounts of the two substances in the cells at all times.

SUMMARY. In summary, there are two principal methods by which cells control proper proportions and proper quantities of different cellular constituents: (1) the mechanism of genetic regulation and (2) the mechanism of enzyme regulation. The genes can be either activated or inhibited, and likewise, the enzyme systems can be either activated or inhibited. These regulatory mechanisms most often function as feedback control systems that continually monitor the cell's biochemical composition and make corrections as needed. But on occasion, substances from without the cell (especially some of the hormones that are discussed at many places in this text) also control the intracellular biochemical reactions by activating or inhibiting one or more of the intracellular control systems.

THE DNA-GENETIC SYSTEM ALSO CONTROLS CELL REPRODUCTION

Cell reproduction is another example of the pervading, ubiquitous role that the DNA-genetic system plays in all life processes. The genes and their regulatory mechanisms determine the growth characteristics of the cells and also when or whether these cells divide to form new cells. In this way, the all-important genetic system controls each stage in the development of the human being from the single-cell fertilized ovum to the whole functioning body. Thus, if there is any central theme to life, it is the DNA-genetic system.

LIFE CYCLE OF THE CELL. The life cycle of a cell is the period from cell reproduction to the next reproduction. When mammalian cells *are not inhibited and are reproducing as rapidly as they can*, this life cycle lasts for 10 to 30 hours. It is terminated by a series of distinct physical events called *mitosis* that cause division of the cell into two new daughter cells. The events of mitosis are shown in Figure 3–13 and are described later. The actual stage of mitosis, however, lasts for only about 30 minutes, so that more than 95 per cent of the life cycle even of rapidly reproducing cells is represented by the interval between mitosis, called *interphase*.

Except in special conditions of rapid cellular reproduction, inhibitory factors almost always slow or stop the uninhibited life cycle of the cell. Therefore, different cells of the body, in actual fact, have life cycle periods that vary from as little as 10 hours for highly stimulated bone marrow cells to an entire lifetime of the human body for nerve cells.

Cell Reproduction Begins With Replication of the DNA

As is true of almost all other important events in the cell, reproduction begins in the nucleus itself. The first step is *replication (duplication) of all DNA in the chromosomes.* Only after this has occurred can mitosis take place.

The DNA begins to be duplicated some 5 to 10 hours

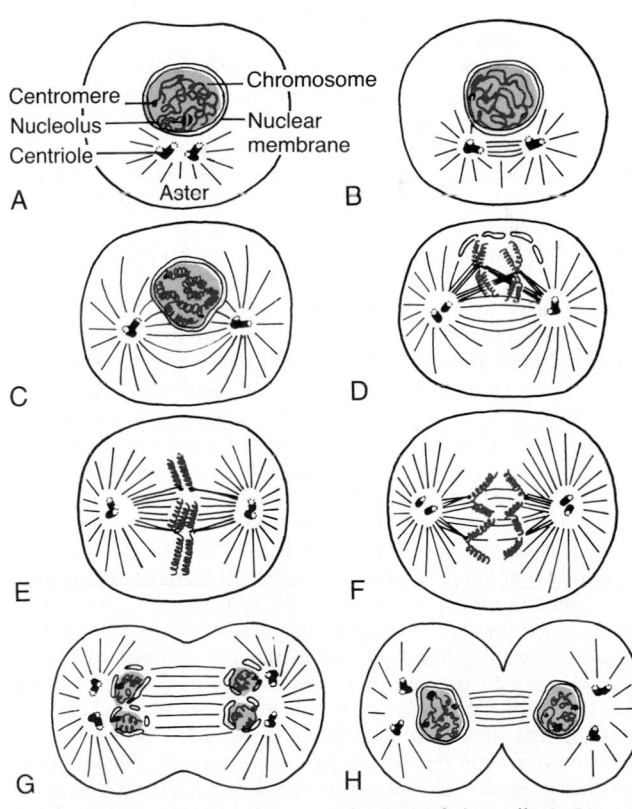

Figure 3–13. Stages in the reproduction of the cell. *A, B,* and *C,* prophase; *D,* prometaphase; *E,* metaphase; *F,* anaphase; *G* and *H,* telophase. (Redrawn from Mazia: How cells divide. *Sci. Am., 205:*102, 1961. © by Scientific American, Inc. All rights reserved.)

before mitosis, and this is completed in 4 to 8 hours. The DNA is duplicated only once, so that the net result is two exact *replicates* of all DNA. These replicates in turn become the DNA in the two new daughter cells that will be formed at mitosis. After replication of the DNA, there is another period of 1 to 2 hours before mitosis begins abruptly. Even during this period, preliminary changes are beginning to take place that will lead to the mitotic process.

CHEMICAL AND PHYSICAL EVENTS OF DNA REPLICATION. DNA is replicated in much the same way that RNA is transcribed by DNA except for a few important differences:

1. Both strands of the DNA in each chromosome are replicated, not simply one of them.

2. Both entire strands of the DNA helix are replicated from end to end rather than small portions of them, as occurs in the transcription of RNA by the genes.

3. The principal enzymes for replicating DNA are a complex of several enzymes called *DNA polymerase,* which is comparable to RNA polymerase. It attaches to and moves along the DNA template strand, while another enzyme, *DNA ligase,* causes bonding of the successive nucleotides to one another, using high-energy phosphate bonds to energize these attachments.

4. Formation of each new DNA strand occurs simultaneously in hundreds of segments along each of the two strands of the helix until the entire strand is replicated. Then the ends of the subunits are joined together by the DNA ligase enzyme.

5. Each newly formed strand of DNA remains attached by loose hydrogen bonding to the original DNA strand that is used as its template. Therefore, two new DNA helixes are formed that are exact duplicates of each other and that are still coiled together.

6. Because the DNA helixes in each chromosome are about 6 centimeters in length and have millions of turns in each helix, it would be impossible for the two newly formed DNA helixes to uncoil from each other were it not for some special mechanism. This is achieved by enzymes that periodically cut each helix along its entire length, rotate each segment enough to cause separation, and then resplice the helix. Thus, the two new helixes become uncoiled.

DNA REPAIR AND DNA "PROOFREADING." During the hour or so between DNA replication and the beginning of mitosis, there is a period of very active repair and "proofreading" of the DNA strands. That is, wherever inappropriate DNA nucleotides have been matched up with the nucleotides of the original template strand, special enzymes cut out the defective areas and replace these with the appropriate complementary nucleotides. This is achieved by the same DNA polymerases and DNA ligase that are used in replication. This repair process is referred to as *DNA proofreading.*

Because of repair and proofreading, the transcription process only rarely makes a mistake. But when a mistake is made, this is called a *mutation;* it in turn causes the formation of some abnormal protein in the cell rather than forming a needed protein, often leading to abnormal cellular function and sometimes even to cell death. Yet, when one realizes that there are 100,000 or more genes in the human genome and that the period from one human generation to another is about 30 years, one would still expect as many as 10 or more mutations in the passage of the genome from parent to child. As still a further protection, however, each human genome is represented by two separate sets of chromosomes with almost identical genes. Therefore, one func-

tional gene of each pair almost always is still available to the child despite mutations.

Chromosomes and Their Replication

The DNA helixes of the nucleus are each packaged as a single chromosome. The human cell contains 46 chromosomes arranged in 23 pairs. Most of the genes in the two chromosomes of each pair are identical or almost identical with each other, so that it is usually stated that the different genes also exist in pairs, although occasionally this is not the case.

Aside from the DNA in the chromosome, there is also a large amount of protein in the chromosome, composed mainly of many small molecules of electropositively charged *histones.* The histones are organized into vast numbers of small, bobbinlike cores. Small segments of each DNA helix are coiled sequentially around one core after another. During mitosis, the successive cores are packed one against the other, thus allowing the tremendously long DNA molecule, having a linear length of 6 centimeters and a molecular weight of about 60 billion, to be packaged in a coiled and folded arrangement of the *mitotic chromosome* that is only a few micrometers in length, 1/10,000 of the stretched-out length of the DNA itself.

The histone cores, as already pointed out, play an important role in the regulation of DNA activity because as long as the DNA is packaged tightly, it cannot function as a template either for the formation of RNA or for the replication of new DNA. Furthermore, some of the regulatory proteins have been shown to *decondense* the histone packaging of the DNA and to allow small segments at a time to form RNA.

Several nonhistone proteins are also major components of chromosomes, functioning both as chromosomal structural proteins and, in connection with the genetic regulatory machinery, as activators, inhibitors, and enzymes.

Replication of the chromosomes in their entirety occurs during the next few minutes after replication of the DNA helixes; the new DNA helixes collect new protein molecules as needed. The two newly formed chromosomes remain temporarily attached to each other (until the time for mitosis) at a point called the *centromere* located near their center. These duplicated but still attached chromosomes are called *chromatids.*

Mitosis

The actual process by which the cell splits into two new cells is called *mitosis.* Once each chromosome has been replicated to form the two chromatids, mitosis follows automatically within 1 or 2 hours.

MITOTIC APPARATUS: FUNCTION OF THE CENTRIOLES. One of the first events of mitosis takes place in the cytoplasm, occurring during the latter part of interphase in or around the small structures called *centrioles.* As shown in Figure 3–13, two pairs of centrioles lie close to each other near one pole of the nucleus. (These centrioles, like the DNA and chromosomes, had also been replicated during interphase, usually shortly before the replication of the DNA.) Each centriole is a small cylindrical body about 0.4 micrometer long and about 0.15 micrometer in diameter, consisting mainly of nine parallel tubular structures arranged in the form of a cylinder. The two centrioles of each pair lie at right angles to each other. Each pair of centrioles, along with attached pericentriolar material, is called a *centrosome.*

Shortly before mitosis is to take place, the two pairs of

centrioles begin to move apart from each other. This is caused by successive polymerization of protein microtubules growing between the respective centriole pairs and actually pushing them apart. At the same time, other microtubules grow radially away from each of the centriole pairs, forming a spiny star, called the *aster,* in each end of the cell. Some of the spines penetrate the nucleus and play a role in separating the two sets of chromatids during mitosis. The complex of microtubules extending between the two centriole pairs is called the *spindle,* and the entire set of microtubules plus the two pairs of centrioles is called the *mitotic apparatus.*

PROPHASE. The first stage of mitosis, called *prophase,* is shown in Figure 3–13A, B, and C. While the spindle is forming, the chromosomes of the nucleus, which in interphase consist of loosely coiled strands, become condensed into well-defined chromosomes.

PROMETAPHASE. During this stage (Figure 3–13D), the growing microtubular spines of the aster puncture and fragment the nuclear envelope. At the same time, multiple microtubules from the aster attach to the chromatids at the centromeres where the paired chromatids are still bound to each other; the tubules then pull one chromatid of each pair toward one cellular pole and its partner toward the opposite pole.

METAPHASE. During metaphase (Figure 3–13E), the two asters of the mitotic apparatus are pushed farther apart. This is believed to occur because the microtubular spines from the two asters, where they interdigitate with each other to form the mitotic spindle, literally push each other away. There is reason to believe that minute contractile protein molecules called "motor molecules," perhaps composed of the muscle protein *actin,* extend between the respective spines and, using a stepping action as in muscle, actively slide the spines in a reverse direction along each other. Simultaneously the chromatids are pulled tightly by their attached microtubules to the very center of the cell, lining up to form the *equatorial plate* of the mitotic spindle.

ANAPHASE. During this phase (Fig. 3–13F), the two chromatids of each chromosome are pulled apart at the centromere. All 46 pairs of chromatids are separated, forming two separate sets of 46 *daughter chromosomes.* One of these sets is pulled toward one mitotic aster and the other toward the other aster as the two respective poles of the dividing cell are pushed apart.

TELOPHASE. In telophase (Fig. 3–13G and H), the two sets of daughter chromosomes are now pulled completely apart. Then the mitotic apparatus dissolutes, and a new nuclear membrane develops around each set of chromosomes. This membrane is formed from portions of the endoplasmic reticulum that are already present in the cytoplasm. Shortly thereafter, the cell pinches in two midway between the two nuclei. This is caused by a contractile ring of *microfilaments* composed of *actin* and probably *myosin,* the two contractile proteins of muscle, forming at the juncture of the newly developing cells and pinching them off from each other.

Control of Cell Growth and Reproduction

We all know that certain cells grow and reproduce all the time, such as the blood-forming cells of the bone marrow, the germinal layers of the skin, and the epithelium of the gut. Many other cells, however, such as smooth muscle cells, may not reproduce for many years. A few cells, such as the neurons and most striated muscle cells, do not reproduce during the entire life of the person, except during the original period of fetal life.

In certain tissues, an insufficiency of some types of cells causes these to grow and reproduce rapidly until appropriate numbers of them are again available. For instance, seven eighths of the liver can be removed surgically, and the cells of the remaining one eighth will grow and divide until the liver mass returns almost to normal. The same occurs for almost all glandular cells, cells of the bone marrow, subcutaneous tissue, intestinal epithelium, and almost any other tissue except highly differentiated cells, such as nerve and muscle cells.

We know little about the mechanisms that maintain proper numbers of the different types of cells in the body. However, experiments have shown at least three ways in which growth can be controlled. First, growth often is controlled by *growth factors* that come from other parts of the body. Some of these circulate in the blood, but others originate in adjacent tissues. For instance, the epithelial cells of some glands, such as the pancreas, will fail to grow without a growth factor from the sublying connective tissue of the gland. Second, most normal cells will stop growing when they have run out of space for growth. This occurs when cells are grown in tissue culture; the cells grow until they contact a solid object and then growth stops. Third, cells grown in tissue culture often stop growing when minute amounts of their own secretions are allowed to collect in the culture medium. This, too, could provide a means for negative feedback control of growth.

REGULATION OF CELL SIZE. Cell size is determined almost entirely by the amount of functioning DNA in the nucleus. If replication of the DNA does not occur, the cell grows to a certain size and thereafter remains at that size. On the other hand, it is possible, by use of the chemical *colchicine,* to prevent formation of the mitotic spindle and therefore to prevent mitosis, even though replication of the DNA continues. In this event, the nucleus contains far greater quantities of DNA than it normally does and the cell grows proportionately larger. It is assumed that this results simply from increased production of RNA and cell proteins, which in turn cause the cell to grow larger.

CELL DIFFERENTIATION

A special characteristic of cell growth and cell division is *cell differentiation,* which means changes in physical and functional properties of cells as they proliferate in the embryo to form the different bodily structures.

The earliest and simplest theory for explaining differentiation was that the genetic composition of the nucleus undergoes changes during successive generations of cells in such a way that one daughter cell inherits a different set of genes from that of the other daughter cell.

This theory has now been disproved in many ways but is especially well illustrated by the following simple experiment. The nucleus from an intestinal mucosal cell of a frog, when surgically implanted into a frog ovum from which the original nucleus has been removed, often causes the formation of a normal frog. This demonstrates that even the intestinal mucosal cell, which is a well-differentiated cell, still carries all the necessary genetic information for development of all structures required in the frog's body.

Therefore, it has become clear that differentiation results not from loss of genes but from selective repression of different genetic operons. In fact, electron micrographs suggest that some segments of DNA helixes wound around histone cores become so condensed that they will no longer uncoil to form RNA molecules. One suggestion for the cause of this effect is the following: It has been supposed that the cellular genome begins at a certain stage of cell differentiation to produce a regulatory *protein* that forever thereafter represses a select group of genes. Therefore, the repressed genes will never function again. Regardless of the mechanism, most mature cells of the human being produce about 8000 to 10,000 proteins rather than the potential 100,000 or more if all genes were active.

Embryological experiments show also that certain cells in an embryo control the differentiation of adjacent cells. For instance, the *primordial chordamesoderm* is called the *primary organizer* of the embryo because it forms a focus around which the rest of the embryo develops. It differentiates into a *mesodermal axis* that contains segmentally arranged *somites* and, as a result of *inductions* in the surrounding tissues, causes formation of essentially all the organs of the body.

Another instance of induction occurs when the developing eye vesicles come in contact with the ectoderm of the head and cause it to thicken into a lens plate that folds inward to form the lens of the eye. Therefore, a large share of the embryo develops as a result of such inductions, one part of the body affecting another part, and this part affecting still other parts.

Thus, although our understanding of cell differentiation is still hazy, we know many control mechanisms by which differentiation *could* occur.

CANCER

Cancer is caused in all or almost all instances by *mutation* or *abnormal activation* of cellular genes that control cell growth and cell mitosis. The abnormal genes are called *oncogenes.* As many as 100 different oncogenes have been discovered. Also present in all cells are *antioncogenes,* which suppress the activation of specific oncogenes. Therefore, loss or inactivation of antioncogenes can allow activation of oncogenes and lead to cancer.

Only a minute fraction of the cells that mutate in the body ever lead to cancer. There are several reasons for this.

First, most mutated cells have less survival capability than normal cells and therefore simply die.

Second, only a few of the mutated cells that do survive become cancerous because even most mutated cells still have the normal feedback controls that prevent excessive growth.

Third, those cells that are potentially cancerous are often, if not usually, destroyed by the body's immune system before they grow into a cancer. This occurs in the following way: Most mutated cells form abnormal proteins within their cell bodies because of their altered genes, and these proteins then stimulate the body's immune system, causing it to form antibodies or sensitized lymphocytes against the cancerous cells, in this way destroying them. In support of this is the fact that in people whose immune systems have been suppressed, such as those who are taking immunosuppressant drugs after transplantation of a kidney or a heart, the probability of developing a cancer is multiplied as much as fivefold.

Fourth, usually several different activated oncogenes are required all at the same time to cause a cancer. For instance, one such gene might promote rapid reproduction of a cell line, but no cancer occurs because there is not a simultaneous mutant gene required to form the needed blood vessels.

But what is it that causes the altered genes? When one realizes that many trillions of new cells are formed each year in the human being, this question should probably better be asked in the following form: Why is it that all of us do not develop literally millions or billions of mutant cancerous cells? The answer is the incredible precision with which DNA chromosomal strands are replicated in each cell before mitosis takes place and also because the proofreading process cuts and repairs any abnormal DNA strand before the mitotic process is allowed to proceed. Yet, despite all these inherited cellular precautions, probably one newly formed cell in every few million still has significant mutant characteristics.

Thus, chance alone is all that is required for mutations to take place, so we may suppose that a large number of cancers are merely the result of an unlucky occurrence.

Yet the probability of mutations can be increased manyfold when a person is exposed to certain chemical, physical, or biological factors. Some of these are the following.

1. It is well known that *ionizing radiation,* such as x-rays, gamma rays, and particle radiations from radioactive substances, and even ultraviolet light can predispose to cancer. Ions formed in tissue cells under the influence of such radiation are highly reactive and can rupture DNA strands, thus causing many mutations.

2. *Chemical substances* of certain types also have a high propensity for causing mutations. Historically, it was long ago discovered that various aniline dye derivatives are likely to cause cancer, so that workers in chemical plants producing such substances, if unprotected, have a special predisposition to cancer. Chemical substances that can cause mutation are called *carcinogens.* The carcinogens that cause by far the greatest number of deaths in our present-day society are those in cigarette smoke. They cause about one quarter of all cancer deaths.

3. *Physical irritants* can also lead to cancer, such as continued abrasion of the linings of the intestinal tract by some types of food. The damage to the tissues leads to rapid mitotic replacement of the cells. The more rapid the mitosis, the greater the chance for mutation.

4. In many families, there is a strong *hereditary tendency to cancer.* This results from the fact that most cancers require not one mutation but two or more mutations before cancer occurs. In those families that are particularly predisposed to cancer, it is presumed that one or more of the genes are already mutated in the inherited genome. Therefore, far fewer additional mutations must take place in such a person before a cancer begins to grow.

5. In laboratory animals, certain types of viruses can cause some kinds of cancer, including leukemia. This occasionally results by either of two ways. First, in the case of DNA viruses, the DNA strand of the virus can insert itself directly into one of the chromosomes and thereby cause the mutation that leads to cancer. In the case of RNA viruses, some of these carry with them an enzyme called *reverse transcriptase* that causes DNA to be transcribed from the RNA. Then the transcribed DNA inserts itself into the animal cell genome, thus leading to cancer.

INVASIVE CHARACTERISTIC OF THE CANCER CELL. The major differences between the cancer cell and the normal cell are: (1) The cancer cell does not respect usual cellular growth limits; the reason for this is that the cells presumably do not require all the same growth factors that are necessary to cause growth of normal cells. (2) Cancer cells are far less adhesive to one another than are normal cells. Therefore, they have a tendency to wander through the tissues, to enter the blood stream, and to be transported all through the body, where they form nidi for numerous new cancerous growths. (3) Some cancers also produce *angiogenic factors* that cause many new blood vessels to grow into the cancer, thus supplying the nutrients required for cancer growth.

WHY DO CANCER CELLS KILL? The answer to this question usually is simple. Cancer tissue competes with normal tissues for nutrients. Because cancer cells continue to proliferate indefinitely, their number multiplying day by day, one can readily understand that the cancer cells will soon demand essentially all the nutrition available to the body or to an essential part of the body. As a result, normal tissues gradually suffer nutritive death.

REFERENCES

Bownes, M.: Differentiation of Cells. New York, Methuen, Inc., 1985.

Busch, H. (ed.): The Cell Nucleus, Nuclear Particles. New York, Academic Press, 1981.

Conaway, R. C., and Conaway, J. W.: Transcription: Mechanisms and Regulation. New York, Raven Press, 1994.

Dice, J. F.: Cellular and molecular mechanisms of aging. Physiol. Rev., 73:149, 1993.

DeVita, V. T., et al.: Cancer: Principles and Practice of Oncology. 4th Ed. Philadelphia, J. B. Lippincott Co., 1993.

DiMauro, S., and Wallace, D. C.: Mitochondrial DNA in Human Pathology. New York, Raven Press, 1993.

Edwards, R. G.: Fetal Tissue Transplants in Medicine. New York, Cambridge University Press, 1992.

Edwards, R. G.: Preconception and Preimplantation Diagnosis of Human Genetic Disease. New York, Cambridge University Press, 1993.

Emery, A. E. H., and Rimoin, D. L.: Principles and Practice of Medical Genetics. 2nd Ed. New York, Churchill Livingstone, 1990.

Fantes, P., and Brooks, R.: The Cell Cycle: A Practical Approach. New York, Oxford University Press, 1994.

Gelehrter, T. D., and Collins, F. S.: Principles of Medical Genetics. Baltimore, Williams & Wilkins, 1995.

Haseltine, W. A., and Wong-Staal, F.: Genetic Structure and Regulation of HIV. New York, Raven Press, 1991.

Haust, M. D., et al.: Genetic Metabolic Diseases. Farmington, Conn., S. Karger Publishers, Inc., 1993.

Hawkins, J.: Gene Structure and Expression. 2nd Ed. New York, Cambridge University Press, 1991.

Hunt, T. (ed.): DNA Makes RNA Makes Protein. New York, Elsevier Science Publishing Co., 1983.

Joyner, A. L.: Gene Targeting: A Practical Approach. New York, Oxford University Press, 1993.

Klevecz, R. R., et al.: Cellular clocks and oscillators. Int. Rev. Cytol., 86:97, 1984.

Kneale, G. G.: DNA-Protein Interactions: Principles and Protocols. Totowa, NJ, Humana Press, Inc., 1994.

Knoppers, B. M., and Chadwick R.: The human genome project: Under an international ethical microscope. Science, 265:2035, 1994.

Lander, E. S., and Schork, N. J.: Genetic dissection of complex traits. Science, 265:2037, 1994.

Larsen, W. J.: Human Embryology. New York, Churchill Livingstone, 1993.

Lindsten, J., and Pettersson, U.: Etiology of Human Disease at the DNA Level. New York, Raven Press, 1991.

Lusis, A. J., et al.: Molecular Genetics of Coronary Artery Disease. Farmington, Conn., S. Karger Publishers, Inc., 1992.

Macara, I. G.: Oncogenes and cellular signal transduction. Physiol. Rev., 69:797, 1989.

McHugh, P. R., and McKusick, V. A.: Genes, Brain, and Behavior. New York, Raven Press, 1991.

Mohri, H., et al.: Biology of the Germ Line. Farmington, Conn., S. Karger Publishers, Inc., 1993.

Mond, J. J., et al.: Cell Activation: Genetic Approaches. New York, Raven Press, 1991.

Moore, K. L., et al.: Color Atlas of Clinical Embryology. Philadelphia, W. B. Saunders Co., 1994.

Murray, J. C., et al.: A comprehensive human linkage map with centimorgan density. Science, 265:2049, 1994.

Nora, J. J., and Fraser, F. Cl.: Medical Genetics: Principles and Practice. Baltimore, Williams & Wilkins, 1994.

Sadler, T. W.: Langman's Medical Embryology. Baltimore, Williams & Wilkins, 1995.

Sara, V. R., et al.: Growth Factors—From Genes to Clinical Application. New York, Raven Press, 1990.

Schleif, R.: DNA looping. Science, 240:127, 1988.

Schopf, J. W.: The evolution of the earliest cells. Sci. Am., 239:110, 1978.

Scriver, C. R., et al.: The Metabolic and Molecular Basis of Inherited Disease. Hightstown, NJ, McGraw-Hill, 1994.

Scriver, C. R., et al. (eds.): The Metabolic Basis of Inherited Disease. New York, McGraw-Hill Book Co., 1989.

Seashore, M. R.: Genetics in Clinical Medicine and Primary Care. Norwalk, Conn., Appleton & Lange, 1995.

Shenolikar, S., and Nairn, A. C.: Model Systems in Signal Transduction. New York, Raven Press, 1993.

Stoffler, G., and Stoffler-Meilicke, M.: Immunoelectron microscopy of ribosomes. Annu. Rev. Biophys. Bioeng., 13:303, 1984.

Strada, S. J., and Hidaka, H.: The Biology of Cyclic Nucleotide Phosphodiesterases. New York, Raven Press, 1992.

Tannock, I. F., and Hill, R. P.: The Basic Science of Oncology. Hightstown, NJ, McGraw-Hill, 1992.

Thompson, M. W., et al.: Thompson and Thompson Genetics in Medicine. Philadelphia, W. B. Saunders Co., 1991.

Varmus, H. E.: Oncogenes and transcriptional control. Science, 238:1337, 1987.

Weiss L., et al.: Interactions of cancer cells with the microvasculature during metastasis. FASEB J., 2:12, 1988.

Yang, S. S., and Warner, H. R. The Underlying Molecular, Cellular and Immunological Factors in Cancer and Aging. New York, Plenum Publishing Corp., 1993.

MEMBRANE PHYSIOLOGY, NERVE, AND MUSCLE

UNIT II

Transport of Ions and Molecules Through the Cell Membrane

CHAPTER *4*

Figure 4–1 gives the approximate compositions of the *extracellular fluid,* which lies outside the cell membranes, and the *intracellular fluid,* inside the cells. Note that the extracellular fluid contains a large amount of *sodium* but only a small amount of *potassium.* Exactly the opposite is true of the intracellular fluid. Also, the extracellular fluid contains a large amount of chloride, whereas the intracellular fluid contains very little. But the concentrations of phosphates, essentially all of which are organic metabolic intermediates, and proteins in the intracellular fluid are considerably greater than those in the extracellular fluid. These many differences are extremely important to the life of the cell. It is the purpose of this chapter to explain how these differences are brought about by the transport mechanisms of the cell membranes.

Lipid Barrier and Transport Proteins of the Cell Membrane

The structure of the cell membrane is discussed in Chapter 2 and shown in Figure 2–3. It consists almost entirely of a *lipid bilayer* with large numbers of protein molecules floating in the lipid, many penetrating all the way through, as shown in Figure 4–2.

The lipid bilayer is not miscible with either the extracellular fluid or the intracellular fluid. Therefore, it constitutes a barrier for the movement of most water molecules and water-soluble substances between the extracellular and intracellular fluid compartments. However, as demonstrated by the left-hand arrow of Figure 4–2, a few substances can penetrate this bilayer and either enter the cell or leave it, passing directly through the lipid substance itself; this is true mainly for lipid-soluble substances, as we see later.

The protein molecules, on the other hand, have entirely different properties for transporting substances through the membrane. Their molecular structures interrupt the continuity of the lipid bilayer and therefore constitute an alternate pathway through the cell membrane. Most of these penetrating proteins, therefore, are *transport proteins.* Different proteins function differently. Some have watery spaces all the way through the molecule and allow free movement of certain ions or molecules; they are called *channel proteins.* Others, called *carrier proteins,* bind with substances that are to be transported, and conformational changes in the protein molecules then move the substances through the interstices of the molecules to the other side of the membrane. Both the channel proteins and the carrier proteins are often highly selective in the type or types of molecules or ions that are allowed to cross the membrane.

DIFFUSION VERSUS ACTIVE TRANSPORT. Transport through the cell membrane, either directly through the lipid bilayer or through the proteins, occurs by one of two basic processes, *diffusion* (which is also called passive transport) or *active transport.* Although there are many variations of these basic mechanisms, as we see later in this chapter, diffusion means random molecular movement of substances molecule by molecule either through intermolecular spaces in the membrane or in combination with a carrier protein. The energy that causes diffusion is the energy of the normal kinetic motion of matter. By contrast, active transport means movement of ions or other substances across the membrane in combination with a carrier protein

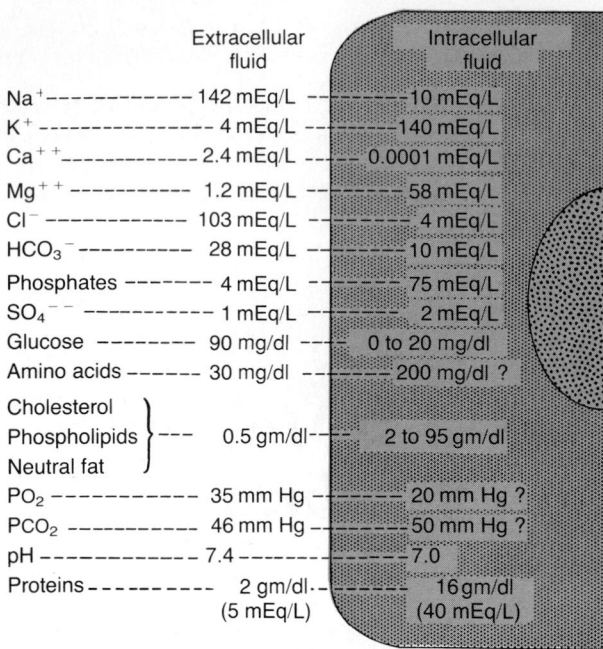

	Extracellular fluid	Intracellular fluid
Na$^+$	142 mEq/L	10 mEq/L
K$^+$	4 mEq/L	140 mEq/L
Ca^{++}	2.4 mEq/L	0.0001 mEq/L
Mg^{++}	1.2 mEq/L	58 mEq/L
Cl$^-$	103 mEq/L	4 mEq/L
HCO$_3^-$	28 mEq/L	10 mEq/L
Phosphates	4 mEq/L	75 mEq/L
SO$_4^{--}$	1 mEq/L	2 mEq/L
Glucose	90 mg/dl	0 to 20 mg/dl
Amino acids	30 mg/dl	200 mg/dl ?
Cholesterol Phospholipids Neutral fat	0.5 gm/dl	2 to 95 gm/dl
PO$_2$	35 mm Hg	20 mm Hg ?
PCO$_2$	46 mm Hg	50 mm Hg ?
pH	7.4	7.0
Proteins	2 gm/dl (5 mEq/L)	16 gm/dl (40 mEq/L)

Figure 4–1. Chemical compositions of extracellular and intracellular fluids.

but additionally *against an energy gradient,* such as from a low concentration state to a high concentration state, a process that requires an additional source of energy besides kinetic energy to cause the movement. Let us explain in more detail the basic physics and physical chemistry of these two processes.

DIFFUSION

All molecules and ions in the body fluids, including both water molecules and dissolved substances, are in constant motion, each particle moving its own separate way. Motion of these particles is what physicists call heat—the greater the motion, the higher the temperature—and the motion never ceases under any conditions except at absolute zero temperature. When

a moving molecule, A, approaches a stationary molecule, B, the electrostatic and internuclear forces of molecule A repel molecule B, transferring some of the energy of motion to molecule B. Consequently, molecule B gains kinetic energy of motion, whereas molecule A slows down, losing some of its kinetic energy. Thus, as shown in Figure 4–3, a single molecule in solution bounces among the other molecules first in one direction, then another, then another, and so forth, bouncing randomly millions of times each second.

This continual movement of molecules among one another in liquids, or in gases, is called *diffusion.* Ions diffuse in the same manner as whole molecules, and even suspended colloid particles diffuse in a similar manner, except that they diffuse far less rapidly than molecular substances because of their large size.

Diffusion Through the Cell Membrane

Diffusion through the cell membrane is divided into two subtypes called *simple diffusion* and *facilitated diffusion.* Simple diffusion means that molecular kinetic movement of molecules or ions occurs through a membrane opening or through intermolecular spaces without the necessity of binding with carrier proteins in the membrane. The rate of diffusion is determined by the amount of substance available, by the velocity of kinetic motion, and by the number of openings in the cell membrane through which the molecules or ions can move. On the other hand, facilitated diffusion requires the interaction of a carrier protein with the molecules or ions. The carrier protein aids passage of the molecules or ions through the membrane, probably by binding chemically with them and shuttling them through the membrane in this form.

Simple diffusion can occur through the cell membrane by two pathways: (1) through the interstices of the lipid bilayer, especially if the diffusing substance is lipid-soluble, and (2) through watery channels in some of the transport proteins, as shown to the left in Figure 4–2.

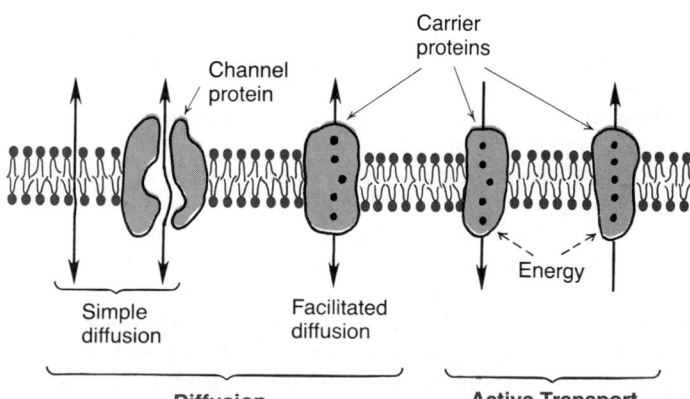

Figure 4–2. Transport pathways through the cell membrane and the basic mechanisms of transport.

Figure 4–3. Diffusion of a fluid molecule during a billionth of a second.

DIFFUSION OF LIPID-SOLUBLE SUBSTANCES THROUGH THE LIPID BILAYER. In experimental studies, the lipids of cells have been separated from the proteins and then reconstituted as artificial membranes consisting of a lipid bilayer but without any transport proteins. Using such an artificial membrane, the transport properties of the lipid bilayer by itself have been determined.

One of the most important factors that determines how rapidly a substance will move through the lipid bilayer is the *lipid solubility* of the substance. For instance, the lipid solubilities of oxygen, nitrogen, carbon dioxide, and alcohols are high, so that all these can dissolve directly in the lipid bilayer and diffuse through the cell membrane in the same manner that diffusion occurs in a watery solution. For obvious reasons, the rate of diffusion of these substances through the membrane is directly proportional to their lipid solubility. Especially large amounts of oxygen can be transported in this way; therefore, oxygen is delivered to the interior of the cell almost as though the cell membrane did not exist.

DIFFUSION OF WATER AND OTHER LIPID-INSOLUBLE MOLECULES THROUGH PROTEIN CHANNELS. Even though water is highly insoluble in the membrane lipids, it readily penetrates the cell membrane, virtually all of it passing through protein channels. The rapidity with which water molecules can penetrate most cell membranes is astounding. As an example, the total amount of water that diffuses in each direction through the red cell membrane during each second is about 100 times as great as the volume of the red cell itself.

Other lipid-insoluble molecules can pass through the protein pore channels in the same way as water molecules if they are small enough. As they become larger, their penetration falls off rapidly. For instance, the diameter of the urea molecule is only 20 per cent greater than that of water. Yet its penetration through the cell membrane pores is about a thousand times less than that of water. Even so, remembering the astonishing rate of water penetration, this amount of penetration still allows rapid transport of urea through the cell membrane.

Diffusion Through Protein Channels and "Gating" of These Channels

The protein channels are believed to be watery pathways through the interstices of the protein molecules. In fact, computerized three-dimensional reconstructions of some of these proteins have demonstrated actual tube-shaped channels from the extracellular to the intracellular ends. Therefore, substances can diffuse by simple diffusion directly through these channels from one side of the membrane to the other. The protein channels are distinguished by two important characteristics: (1) They are often selectively permeable to certain substances. (2) Many of the channels can be opened or closed by *gates*.

SELECTIVE PERMEABILITY OF DIFFERENT PROTEIN CHANNELS. Most protein channels are highly selective for the transport of one or more specific ions or molecules. This results from the characteristics of the channel itself, such as its diameter, its shape, and the nature of the electrical charges along its inside surfaces. To give an example, one of the most important of the protein channels, the so-called *sodium channels,* calculate to be only 0.3 by 0.5 nanometer in size, but more important, the inner surfaces of these channels are *strongly negatively charged,* as shown by the negative signs inside the channel proteins in the top panel of Figure 4–4. These strong negative charges are postulated especially to pull small *dehydrated* sodium ions into these channels, actually pulling the sodium ions away from their hydrating water molecules. Once in the channel, the sodium ions then diffuse in either direction according to the usual laws of diffusion.

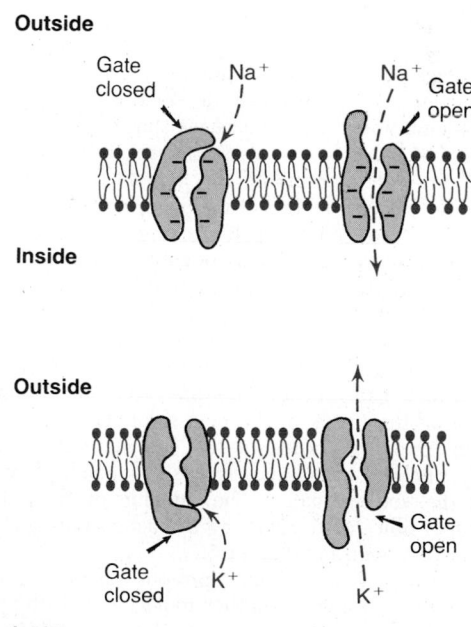

Figure 4–4. Transport of sodium and potassium ions through protein channels. Also shown are conformational changes of the channel protein molecules that open or close the "gates" guarding the channels.

Thus, the sodium channel is specifically selective for the passage of sodium ions.

On the other hand, another set of protein channels is selective for potassium transport, shown in the lower panel of Figure 4–4. These channels calculate to be slightly smaller than the sodium channels, only 0.3 by 0.3 nanometer, but *they are not negatively charged.* Therefore, no strong attractive force is pulling ions into the channels, and the ions are not pulled away from the water molecules that hydrate them. The hydrated form of the potassium ion is considerably smaller than the hydrated form of sodium because the sodium ion has one whole orbital set of electrons less than the potassium ion, which allows the sodium nucleus to attract far more water molecules than can the potassium. Therefore, the smaller hydrated potassium ions can pass easily through this smaller channel, whereas sodium ions are mainly rejected, thus once again providing selective permeability for a specific ion.

GATING OF PROTEIN CHANNELS. Gating of protein channels provides a means for controlling the permeability of the channels. This is shown in both the upper and the lower panels of Figure 4–4 for the sodium and the potassium ion. It is believed that the gates are actual gatelike extensions of the transport protein molecule, which can close over the opening of the channel or can be lifted away from the opening by a conformational change in the shape of the protein molecule itself. In the case of the sodium channels, this gate opens and closes at the end of the channel on the outside of the cell membrane, whereas for the potassium channels, it opens and closes at the inside end of the channel.

The opening and closing of gates are controlled in two principal ways:

1. *Voltage gating.* In this instance, the molecular conformation of the gate responds to the electrical potential across the cell membrane. For instance, as shown in the top panel of Figure 4–4, when there is a strong negative charge on the inside of the cell membrane, the outside sodium gates remain tightly closed; on the other hand, when the inside of the membrane loses its negative charge, these gates open suddenly and allow tremendous quantities of sodium to pass inward through the sodium pores (until still another set of gates at the inside ends of the channels close, as explained in Chapter 5). This is the basic cause of action potentials in nerves that are responsible for nerve signals. Now, look at the potassium gates in the bottom panel of Figure 4–4. These are on the inside of the potassium channels, and they, too, open when the inside of the cell membrane becomes positively charged, but this response is much slower than that for the sodium gates. The opening of these gates is partly responsible for terminating the action potential. These events are discussed in Chapter 5.

2. *Chemical gating.* Some protein channel gates are opened by the binding of another molecule with the protein; this causes a conformational change in the protein molecule that opens or closes the gate. This is called *chemical gating.* One of the most important instances of chemical gating is the effect of acetylcholine on the so-called *acetylcholine channel.* This opens the gate of this channel, providing a negatively charged pore about 0.65 nanometer in diameter that allows all uncharged molecules as well as positive ions smaller than this diameter to pass through. This gate is exceedingly important in the transmission of signals from one nerve cell to another (see Chapter 45) and from nerve cells to muscle cells (see Chapter 7).

Open-State, Closed-State of Gated Channels. Figure 4–5A shows an especially interesting characteristic of voltage-gated channels. This figure shows two recordings of electrical current flowing through a single sodium channel when there was an approximate 25-millivolt potential gradient across the membrane. Note that the channel conducts current either all or none. That is, the gate of the channel snaps open and then snaps closed, each snapping event occurring within a few millionths of a second. This demonstrates the rapidity with which conformational changes can occur during the opening and closing of the protein molecular gates. At one voltage potential the channel may remain closed all the time or almost all the time, whereas at another voltage level it may remain open either all or most of the time. At in-between voltages, as shown in the figure, the gates tend to snap open and closed intermittently, giving an average current flow somewhere between the minimum and the maximum.

Patch-Clamp Method for Recording Ion Current Flow Through Single Channels. One might wonder how it is technically possible to record ion current flow through single channels as shown in Figure 4–5A. This has been achieved by using the "patch-clamp" method shown in Figure 4–5B. Very simply, a micropipette, having a tip diameter of only 1 or 2 micrometers, is abutted against the outside of a cell membrane. Then suction is applied inside the pipette to pull the membrane slightly into the tip of the pipette. This creates a seal where the edges of the pipette touch the cell membrane. The result is a minute "patch" at the tip of the pipette through which current flow can be recorded.

Alternatively, as shown to the right in Figure 4–5B, the small cell membrane patch at the end of the pipette can be torn away from the cell. The pipette with its sealed patch is then inserted into a free solution. This allows the concentration of the ions both inside the micropipette and in the outside solution to be altered as desired. Also, the voltage between the two sides of the membrane can be set at will—that is, "clamped" to a given voltage.

It has been possible to make such patches small enough that one often finds only a single channel protein in the membrane patch that is being studied. By varying the concentrations of different ions and the voltage across the membrane, one can determine the transport characteristics of the channel as well as its gating properties.

Facilitated Diffusion

Facilitated diffusion is also called *carrier-mediated diffusion* because a substance transported in this manner usually cannot pass through the membrane without a specific carrier protein helping it. That is, the carrier

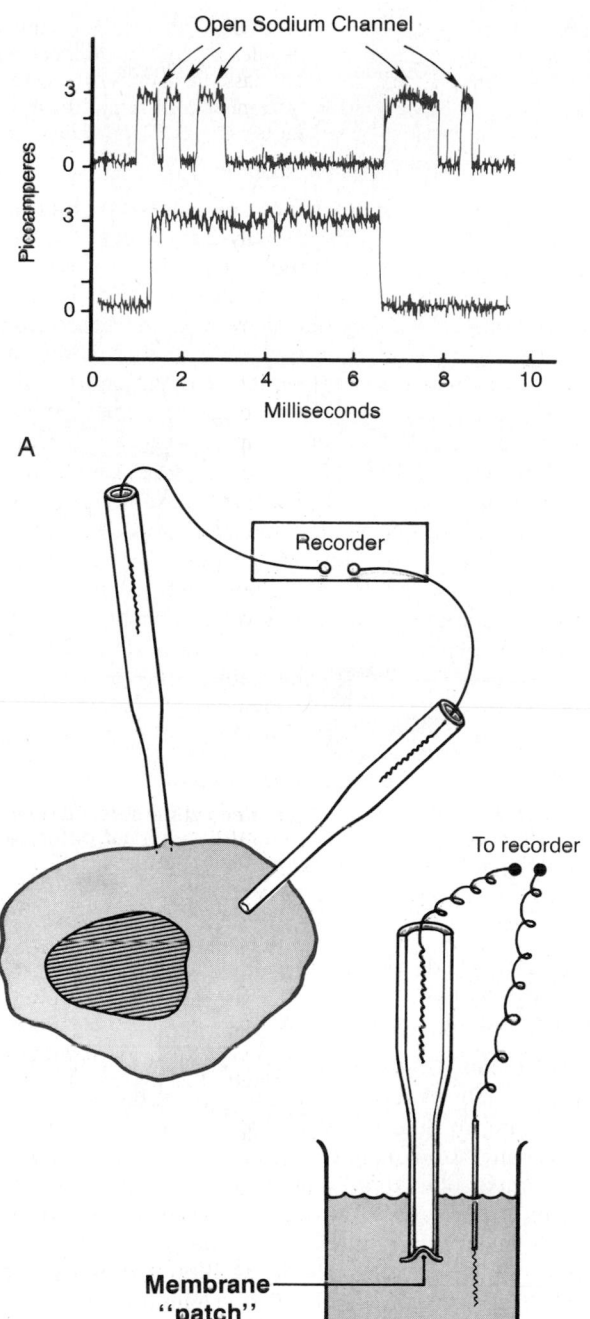

A

B

Figure 4–5. *A*, Record of current flow through a single voltage-gated sodium channel, demonstrating the all-or-none principle for opening of the channel. *B*, The "patch-clamp" method for recording current flow through a single protein channel. To the left, recording is performed from a "patch" of a living cell membrane. To the right, recording is from a membrane patch that has been torn away from the cell.

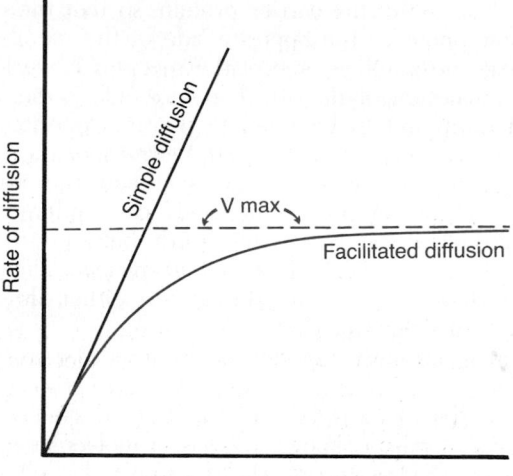

Figure 4–6. Effect of concentration of a substance on rate of diffusion through a membrane in which there is simple diffusion and a membrane in which there is facilitated diffusion. This shows that facilitated diffusion approaches a maximum rate called the V_{max}.

tion of the diffusing substance, in facilitated diffusion, the rate of diffusion approaches a maximum, called V_{max}, as the concentration of the substance increases. This difference between simple diffusion and facilitated diffusion is demonstrated in Figure 4–6, showing that as the concentration of the substance increases, the rate of simple diffusion continues to increase proportionately but showing also the limitation of facilitated diffusion to the V_{max} level.

What is it that limits the rate of facilitated diffusion? A probable answer is the mechanism illustrated in Figure 4–7. This figure shows a carrier protein with a channel large enough to transport a specific molecule part way through the membrane. It also shows a binding "receptor" on the protein carrier. The molecule to be transported enters the channel and becomes bound. Then, in a fraction of a second, a conformational

facilitates the diffusion of the substance to the other side.

Facilitated diffusion differs from simple diffusion through an open channel in the following important way: Although the rate of diffusion through an open channel increases proportionately with the concentra-

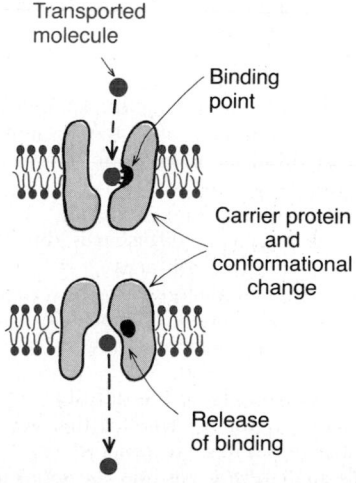

Figure 4–7. Postulated mechanism for facilitated diffusion.

change occurs in the carrier protein, so that the channel now opens to the opposite side of the membrane. Because the binding force of the receptor is weak, the thermal motion of the attached molecule causes it to break away and to be released on the opposite side. The rate at which molecules can be transported by this mechanism can never be greater than the rate at which the carrier protein molecule can undergo conformational change back and forth between its two states. Note specifically that this mechanism allows the transported molecule to "diffuse" in either direction through the membrane.

Among the most important substances that cross cell membranes by facilitated diffusion are *glucose* and most of the *amino acids*. In the case of glucose, the carrier molecule is known to have a molecular weight of about 45,000; it can also transport several other monosaccharides that have structures similar to that of glucose, including especially galactose. Also, insulin can increase the rate of facilitated diffusion of glucose as much as 10-fold to 20-fold. This is the principal mechanism by which insulin controls glucose use in the body, as we discuss in Chapter 78.

Factors That Affect Net Rate of Diffusion

By now it is evident that many substances can diffuse either through the lipid bilayer of the cell membrane or through protein channels. However, please understand clearly that substances that diffuse in one direction can also diffuse in the opposite direction. Therefore, what is usually important is the *net* rate of diffusion of a substance in the desired direction. This net rate is determined by the following factors:

EFFECT OF MEMBRANE PERMEABILITY ON DIFFUSION RATE. The permeability of a membrane for a given substance, designated by the letter P, is expressed as the *net* rate of diffusion of the substance through each unit area of the membrane for a unit concentration difference between the two sides of the membrane (when there are no electrical or pressure differences). The different factors that affect cell membrane permeability are

1.   Thickness of the membrane—the greater the thickness, the less the rate of diffusion.
2.   Lipid solubility—the greater the solubility of the substance in the cell membrane lipids, the greater the quantity of substances that dissolves in the membrane and, therefore, that can pass through.
3.   Number of protein channels through which the substance can pass—the rate of diffusion is directly related to the number of channels per unit area.
4.   The temperature—the greater the temperature, the greater is the thermal motion of molecules and ions in a solution, so that diffusion increases directly in proportion to temperature.
5.   The molecular weight of the diffusing substance—this has a complex effect; the velocity of thermal motion of a dissolved substance is inversely proportional to the square root of its molecular weight. Also, as the molecular diameter approaches the diameter of a channel, the resistance in-

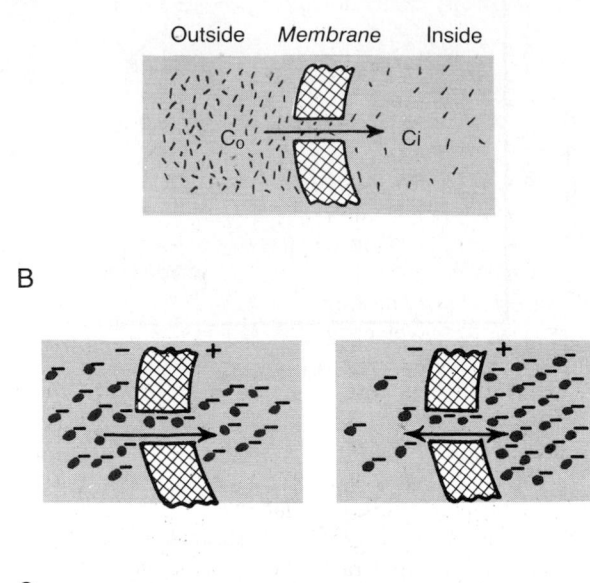

Figure 4–8. Effect of *(A)* concentration difference, *(B)* electrical difference, and *(C)* pressure difference on net diffusion of molecules and ions through a cell membrane.

creases tremendously, so that a membrane is frequently hundreds to millions of times as permeable to very small molecules as to very large ones.

Diffusion Coefficient. Another factor that affects the total rate of diffusion is the area of the membrane. Therefore, to determine the total permeability of a cell membrane, of a capillary, or of any other complete membrane, one must simply multiply its permeability (which expresses the movement of substances through a *unit* membrane area) by the *total* area of the membrane. This total permeability is often expressed as the *diffusion coefficient* for the entire membrane:

$$D = P \times A,$$

where D is the diffusion coefficient, P is the permeability, and A is the total area.

EFFECT OF CONCENTRATION DIFFERENCE ON DIFFUSION THROUGH A MEMBRANE. Figure 4–8A shows a cell membrane with a substance in high concentration on the outside and low concentration on the inside. The rate at which the substance diffuses *inward* is proportional to the concentration of molecules on the *outside* because this concentration determines how many molecules strike the outside of the channels each second. On the other hand, the rate at which molecules diffuse *outward* is proportional to their concentration *inside* the membrane. Therefore, the rate of net diffusion into the cell is proportional to the concentration on the outside *minus* the concentration on the inside or

$$\text{Net diffusion} \propto D\,(C_o - C_i),$$

in which C_o is the concentration on the outside, C_i is the concentration on the inside, and D is the diffusion coefficient of the cell membrane for the substance.

EFFECT OF AN ELECTRICAL POTENTIAL ON THE DIFFUSION OF IONS. If an electrical potential is applied across the membrane, as shown in Figure 4–8B, because of their electrical charges, ions will move through the membrane even though no concentration difference exists to cause their movement. Thus, in the left panel of Figure 4–8B, the concentrations of negative ions are the same on both sides of the membrane, but a positive charge has been applied to the right side of the membrane and a negative charge to the left, creating an electrical gradient across the membrane. The positive charge attracts the negative ions, whereas the negative charge repels them. Therefore, net diffusion occurs from left to right. After much time, large quantities of negative ions will have moved to the right (if we neglect, for the time being, the disturbing effects of the positive ions of the solution), creating the condition shown in the right panel of Figure 4–8B, in which a concentration difference of the same ions has developed in the direction opposite to the electrical potential difference. The concentration difference is now tending to move the ions to the left, whereas the electrical difference is tending to move them to the right. When the concentration difference rises high enough, the two effects balance each other. At normal body temperature (37°C), the electrical difference that will balance a given concentration difference of *univalent* ions—such as sodium (Na^+) ions—can be determined from the following formula, called the *Nernst equation:*

$$\text{EMF (in millivolts)} = \pm 61 \log \frac{C_1}{C_2},$$

in which EMF is the electromotive force (voltage) between side 1 and side 2 of the membrane, C_1 is the concentration on side 1, and C_2 is the concentration on side 2. The polarity of the voltage on side 1 in the equation above is + for negative ions and − for positive ions. This relation is extremely important in understanding the transmission of nerve impulses, for which reason it is discussed in even greater detail in Chapter 5.

EFFECT OF A PRESSURE DIFFERENCE. At times, considerable pressure difference develops between the two sides of a membrane. This occurs, for instance, at the capillary membrane, which has a pressure about 20 mm Hg greater inside the capillary than outside. Pressure actually means the sum of all the forces of the different molecules striking a unit surface area at a given instant. Therefore, when the pressure is higher

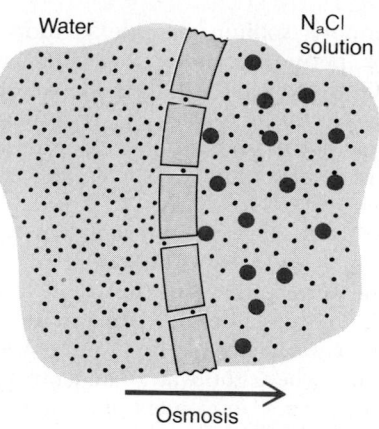

Figure 4–9. Osmosis at a cell membrane when a sodium chloride solution is placed on one side of the membrane and water on the other side.

on one side of a membrane than on the other, this means that the sum of all the forces of the molecules striking the channels on that side of the membrane is greater than on the other side. In most instances, this is caused by greater numbers of molecules striking the membrane per second on one side than on the other side. The result is that increased amounts of energy are available to cause net movement of molecules from the high pressure side toward the low pressure side. This effect is demonstrated in Figure 4–8C, which shows a piston developing high pressure on one side of a cell membrane, thereby causing more molecules to strike the membrane and, therefore, net diffusion to the other side.

Osmosis Across Selectively Permeable Membranes—Net Diffusion of Water

By far the most abundant substance to diffuse through the cell membrane is water. It should be recalled that enough water ordinarily diffuses in each direction through the red cell membrane per second to equal about *100 times the volume of the cell itself.* Yet, *normally,* the amount that diffuses in the two directions is so precisely balanced that not even the slightest *net* movement of water occurs. Therefore, the volume of the cell remains constant. However, under certain conditions, a *concentration difference for water* can develop across a membrane, just as concentration differences for other substances can occur. When this happens, net movement of water does occur across the cell membrane, causing the cell to either swell or shrink, depending on the direction of the net movement. This process of net movement of water caused by a concentration difference of water is called *osmosis.*

To give an example of osmosis, let us assume the conditions shown in Figure 4–9, with pure water on one side of the cell membrane and a solution of sodium chloride on the other side. Water molecules pass through the cell membrane with ease, whereas sodium and chloride ions pass through only with extreme difficulty. Therefore, sodium chloride solution is actually a mixture of permeant water molecules and nonpermeant sodium and chloride ions, and the membrane is said to be *selectively permeable* (or "semipermeable") to water but not to sodium and chloride ions. Yet the presence of the sodium and chloride has displaced some of the water molecules on the side of the membrane where these ions are present and, therefore, has reduced the concentration of water molecules to less than that of pure water. As a result, in the example of Figure 4–9, more water molecules strike the channels on the left side, where there is pure water, than on the right side, where the water concentration has been reduced. Thus, net movement of water occurs from left to right—that is, osmosis occurs from the pure water into the sodium chloride solution.

Osmotic Pressure

If in Figure 4–9 pressure were applied to the sodium chloride solution, osmosis of water into this solution would be slowed, stopped, or even reversed. The amount of pressure required exactly to stop osmosis is called the *osmotic pressure* of the sodium chloride solution.

The principle of a pressure difference opposing osmosis is demonstrated in Figure 4–10, which shows a selectively permeable membrane separating two columns of fluid, one containing water and the other containing a solution of water and any solute that will not penetrate the membrane. Osmosis of water from chamber B into chamber A causes the levels of the fluid columns to become farther and farther apart, until eventually a pressure difference between the two sides of the membrane is developed that is great enough to oppose the osmotic effect. The pressure difference across the membrane at this point is the

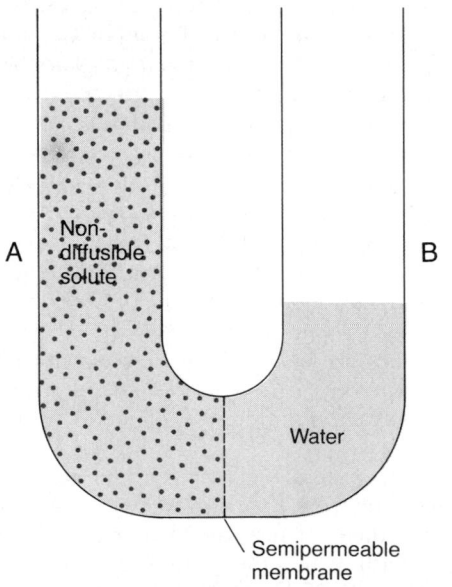

Figure 4–10. Demonstration of osmotic pressure on the two sides of a semipermeable membrane.

osmotic pressure of the solution that contains the nondiffusible solute.

IMPORTANCE OF NUMBERS OF OSMOTIC PARTICLES (OR OF MOLAR CONCENTRATION) IN DETERMINING OSMOTIC PRESSURE. The osmotic pressure exerted by particles in a solution, whether they be molecules or ions, is determined by the *numbers* of particles per unit volume of fluid and not the mass of the particles. The reason for this is that each particle in a solution, regardless of its mass, exerts, on the average, the same amount of pressure against the membrane. That is, all particles are bouncing among one another with, on the average, equal energy. If some particles have greater kinetic energy of movement than others, their impact with the low-energy particles will impart part of their energy to these, thus decreasing the energy level of the high-energy particles and increasing the energy level of the others until the energy level of all of them, averaged over a period of time, is the same. Therefore, the large particles, which have greater mass (m) than the small particles move at lower velocities (v), whereas the small particles move at higher velocities in such a way that their average kinetic energies (k), determined by the equation

$$k = \frac{mv^2}{2},$$

become equal to one another. Therefore, on the average, the kinetic energy of each molecule or ion that strikes a membrane is about the same regardless of its molecular size. Consequently, the factor that determines the osmotic pressure of a solution is the concentration of the solution in terms of numbers of particles (which is the same as its *molar concentration* if it is a nondissociated molecule) and not in terms of mass of the solute.

"OSMOLALITY"—THE OSMOLE. Because the amount of osmotic pressure exerted by a solute is proportional to the concentration of the solute in numbers of molecules or ions, expressing the solute concentration in terms of mass is of no value in determining osmotic pressure. To express the concentration in terms of numbers of particles, the unit called the *osmole* is used in place of grams.

One osmole is 1 gram molecular weight of undissociated solute. Thus, 180 grams of glucose, which is 1 gram molecular weight of glucose, is equal to 1 osmole of glucose because glucose does not dissociate. On the other hand, if the solute dissociates into two ions, 1 gram molecular weight of the solute equals 2 osmoles because the number of osmotically active particles is now twice as great as is the case in the undissociated solute. Therefore, 1 gram molecular weight of sodium chloride, 58.5 grams, is equal to 2 osmoles.

A solution that has *1 osmole of solute dissolved in each kilogram of water* is said to have an *osmolality of 1 osmole per kilogram,* and a solution that has ⅟₁₀₀₀ osmole dissolved per kilogram has an osmolality of 1 milliosmole per kilogram. The normal osmolality of the extracellular and intracellular fluids is about 300 milliosmoles per kilogram.

RELATION OF OSMOLALITY TO OSMOTIC PRESSURE. At normal body temperature, 37°C, a concentration of 1 osmole per liter will cause *19,300 mm Hg* osmotic pressure in the solution. Likewise, 1 milliosmole per liter concentration is equivalent to *19.3 mm Hg* osmotic pressure. Multiplying this value times the 300 milliosmolar concentration of the body fluids gives a total calculated osmotic pressure of these fluids of 5790 mm Hg. The measured value for this, however, averages only about 5500 mm Hg. The reason for this difference is that many of the ions in the body fluids, such as the sodium and the chloride ions, are highly attracted to one another; consequently, they cannot move totally unrestrained in the fluids and create their full osmotic pressure potential. Therefore, on the average, the actual osmotic pressure of the body fluids is about 0.93 times the calculated value.

THE TERM "OSMOLARITY." Because of the difficulty of measuring kilograms of water in a solution, which is required to determine "osmolality," one usually uses another term, "osmolarity," which is the osmolar concentration expressed as *osmoles per liter of solution* rather than osmoles per kilogram of water. Although, strictly speaking, it is osmoles per kilogram of water (osmolality) that determines the osmotic pressure, nevertheless, for dilute solutions, such as those in the body, the quantitative differences between osmolarity and osmolality are less than 1 per cent. Because it is far more practical to measure osmolarity than osmolality, this is the usual practice in almost all physiological studies.

ACTIVE TRANSPORT

From the discussion thus far, it is evident that *no net amount of a substance can diffuse against an "electrochemical gradient,"* which is the sum of all the diffusion forces acting at the membrane—the forces caused by concentration difference, electrical difference, and pressure difference. That is, it is often said that substances cannot diffuse "uphill."

Yet, at times, a large concentration of a substance is required in the intracellular fluid even though the extracellular fluid contains only a minute concentration. This is true, for instance, for potassium ions. Conversely, it is important to keep the concentrations of other ions very low inside the cell even though their concentrations in the extracellular fluid are great. This is especially true for sodium ions. Neither of these two effects could occur by simple diffusion because simple diffusion equilibrates the concentrations on the two sides of the membrane. Instead, some energy source must cause movement of potassium ions "uphill" to the inside of cells and cause movement of sodium ions also "uphill" but in this instance to the outside of the cell. When a cell membrane moves molecules or ions uphill against a concentration gradient (or uphill against an electrical or pressure gradient), the process is called *active transport*.

Among the different substances that are actively transported through at least some cell membranes are sodium ions, potassium ions, calcium ions, iron ions, hydrogen ions, chloride ions, iodide ions, urate ions, several different sugars, and most of the amino acids.

PRIMARY ACTIVE TRANSPORT AND SECONDARY ACTIVE TRANSPORT. Active transport is divided into two types according to the source of the energy used to cause the transport. They are called *primary active transport* and *secondary active transport*. In primary active transport, the energy is derived directly from the breakdown of adenosine triphosphate (ATP) or some other high-energy phosphate compound. In secondary active transport, the energy is derived secondarily from energy that has been stored in the form of ionic concentration differences between the two sides of a membrane, created in the first place by primary active transport. In both instances, transport depends on *carrier proteins* that penetrate through the membrane, the same as is true for facilitated diffusion. However, in active transport, the carrier protein functions differently from the carrier in facilitated diffusion because it is capable of imparting energy to the transported substance to move it against an electrochemical gradient. Let us give some examples of primary active transport and secondary active transport and explain their principles of function more fully.

Primary Active Transport

Sodium-Potassium Pump

Among the substances that are transported by primary active transport are sodium, potassium, calcium, hydrogen, chloride, and a few other ions. However, not all of these substances are transported by the membranes of all cells. Furthermore, some of the pumps function at intracellular membranes rather than (or in addition to) the surface membrane of the cell, such as at the membrane of the muscle sarcoplasmic reticulum or at one of the two membranes of the mitochondria. They all operate by essentially the same basic mechanism.

The active transport mechanism that has been studied in greatest detail is the *sodium-potassium* (Na^+-K^+) *pump*, a transport process that pumps sodium ions outward through the cell membrane and at the same time pumps potassium ions from the outside to the inside. This pump is present in all cells of the body, and it is responsible for maintaining the sodium and potassium concentration differences across the cell membrane as well as for establishing a negative electrical potential inside the cells. Indeed, we see in Chapter 5 that this pump is the basis of nerve function to transmit nerve signals throughout the nervous system.

Figure 4–11 shows the basic physical components of the Na^+-K^+ pump. The *carrier protein* is a complex of two separate globular proteins, a larger one called the α subunit with a molecular weight of about 100,000 and a smaller one called the β subunit with a molecu-

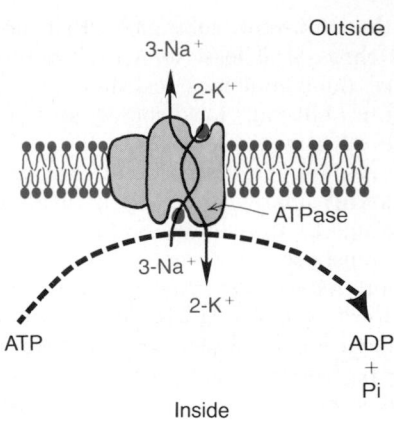

Figure 4–11. Postulated mechanism of the sodium-potassium pump.

lar weight about 55,000. Although the function of the smaller protein is not known, other than that it might anchor the protein complex in the lipid membrane, the larger protein has three specific features that are important for function of the pump:

1.   It has three *receptor sites for binding sodium ions* on the portion of the protein that protrudes to the interior of the cell.
2.   It has two *receptor sites for potassium ions* on the outside.
3.   The inside portion of this protein adjacent or near to the sodium binding sites has ATPase activity.

Now to put the pump into perspective: When three sodium ions bind on the inside of the carrier protein, the ATPase function of the protein becomes activated. This then cleaves one molecule of ATP, splitting it to adenosine diphosphate and liberating a high-energy phosphate bond of energy. This energy is then believed to cause a conformational change in the protein carrier molecule, extruding the sodium ions to the outside and the potassium ions to the inside. The precise mechanism of the conformational change of the carrier is unknown.

IMPORTANCE OF THE NA⁺-K⁺ PUMP IN CONTROLLING CELL VOLUME. One of the most important functions of the Na⁺-K⁺ pump is to control the volume of the cells. Without function of this pump, most cells of the body would swell until they burst. The mechanism for controlling the volume is the following: Inside the cell are large numbers of proteins and other organic compounds that cannot escape from the cell. Most of these are negatively charged and therefore collect around them large numbers of positive ions as well. All these substances then tend to cause osmosis of water to the interior of the cell; unless this is checked, the cell will swell indefinitely until it bursts. The normal mechanism for preventing this is the Na⁺-K⁺ pump. Note again that this device pumps three Na⁺ ions to the outside of the cell for every two K⁺ ions pumped to the interior. Also, the membrane is far less perme-

able to sodium ions than to potassium ions, so that once the sodium ions are on the outside, they have a strong tendency to stay there. Thus, this represents a continual net loss of ions out of the cell, which initiates an opposite osmotic tendency to move water out of the cell. Furthermore, when the cell begins to swell, this automatically activates the Na⁺-K⁺ pump, moving still more ions to the exterior and carrying water with them. Therefore, the Na⁺-K⁺ pump performs a continual surveillance role in maintaining normal cell volume.

ELECTROGENIC NATURE OF THE NA⁺-K⁺ PUMP. The fact that the Na⁺-K⁺ pump moves three Na⁺ ions to the exterior for every two K⁺ ions to the interior means that a net of one positive charge is moved from the interior of the cell to the exterior for each cycle of the pump. This creates positivity outside the cell but leaves a deficit of positive ions inside the cell; that is, it causes negativity on the inside. Therefore, the Na⁺-K⁺ pump is said to be *electrogenic* because it creates an electrical potential across the cell membrane as it pumps.

As discussed in Chapter 5, the Na⁺-K⁺ pump creates still more electrical potential across the membrane for other reasons as well. This is a basic requirement in nerves and muscles to transmit nerve and muscle signals.

Primary Active Transport of Calcium

Another important primary active transport mechanism is the calcium pump. Calcium ions are normally maintained at extremely low concentration in the intracellular cytosol of virtually all cells in the body, at a concentration about 10,000 times less than that in the extracellular fluid. This is achieved mainly by two primary active transport calcium pumps. One is in the cell membrane and pumps calcium to the outside of the cell. The other pumps calcium ions into one or more of the internal vesicular organelles of the cell, such as into the sarcoplasmic reticulum of muscle cells and into the mitochondria in all cells. In each of these instances, the carrier protein penetrates the membrane from side to side and also serves as an ATPase, having the same capability to cleave ATP as the ATPase sodium carrier protein. The difference is that this protein has a highly specific binding site for calcium instead of sodium.

Primary Active Transport of Hydrogen Ions

At two places in the body are important primary active transport systems for hydrogen ions. They are (1) in the gastric glands of the stomach and (2) in the late distal tubules and cortical collecting ducts of the kidneys. In the gastric glands, the deep-lying *parietal cells* have the most potent primary active mechanism for transporting hydrogen ions of any part of the body. This is the basis for secreting hydrochloric acid in the stomach digestive secretions. At the secretory side of the parietal cells, the hydrogen ion concentration can be increased as much as a millionfold and then released in association with chloride ions in the form of hydrochloric acid.

In the renal tubules are special *intercalated cells* in the late distal tubules and cortical collecting ducts that also transport hydrogen ions by primary active transport. In this case, large amounts of hydrogen ions are secreted into the tubules to eliminate them from the body for the purpose of controlling the blood hydrogen ion concentration. The hydrogen ions can be secreted against a concentration gradient of about 900-fold.

At many other points in the body, hydrogen ions are transported by *secondary* active transport, but in these instances, they usually are transported against far less concentration gradients, such as 4 to 1 up to 10 to 1.

Saturation of Primary Active Transport

Active transport saturates in the same way that facilitated diffusion saturates, as shown in Figure 4–6. When the concentration of the substance to be transported is small, the rate of transport increases approximately in proportion to the increase in concentration. At large concentrations, transport approaches a maximum, which is called the V_{max}, as is also true for facilitated diffusion. The saturation is caused by limitation of the rates at which the chemical reactions of binding, release, and carrier conformational changes can occur.

Energetics of Primary Active Transport

The amount of energy required to transport a substance actively through a membrane (aside from energy lost as heat in the chemical reactions) is determined by the degree that the substance is concentrated during transport. Compared with the energy required to concentrate a substance 10-fold, to concentrate it 100-fold requires twice as much energy and to concentrate it 1000-fold requires three times as much. In other words, the energy required is proportional to the *logarithm* of the degree that the substance is concentrated, as expressed by the following formula:

$$\text{Energy (in calories per osmole)} = 1400 \log \frac{C_1}{C_2}$$

That is, in terms of calories, the amount of energy required to concentrate 1 osmole of substance 10-fold is about 1400 calories, or 100-fold, 2800 calories. One can see that the energy expenditure for concentrating substances in cells or for removing substances from cells against a concentration gradient can be tremendous. Some cells, such as those lining the renal tubules as well as many glandular cells, expend as much as 90 per cent of their energy for this purpose alone.

Secondary Active Transport— Co-Transport and Counter-Transport

When sodium ions are transported out of cells by primary active transport, a large concentration gradient of sodium usually develops—very high concentration outside the cell and very low concentration inside. This gradient represents a storehouse of energy because the excess sodium outside the cell membrane is always attempting to diffuse to the interior. Under the appropriate conditions, this diffusion energy of sodium can literally pull other substances along with the sodium through the cell membrane. This phenomenon is called *co-transport;* it is one form of secondary active transport.

For sodium to pull another substance along with it, a coupling mechanism is required. This is achieved by means of still another carrier protein in the cell membrane. The carrier in this instance serves as an attachment point for both the sodium ion and the substance to be co-transported. Once they both are attached, a conformational change occurs in the carrier protein, and the energy gradient of the sodium ion causes both the sodium ion and the other substance to be transported together to the interior of the cell.

In *counter-transport,* sodium ions again attempt to diffuse to the interior of the cell because of their large concentration gradient. However, this time, the substance to be transported is on the inside of the cell and must be transported to the outside. Therefore, the sodium ion binds to the carrier protein where it projects through the exterior surface of the membrane, whereas the substance to be counter-transported binds to the interior projection of the carrier protein. Once both have bound, a conformational change occurs again, with the energy of the sodium ion moving to the interior causing the other substance to move to the exterior.

Sodium Co-Transport of Glucose and Amino Acids

Glucose and many amino acids are transported into most cells against large concentration gradients; the mechanism of this is entirely by the co-transport mechanism shown in Figure 4–12. Note that the transport carrier protein has two binding sites on its exterior side, one for sodium and one for glucose. Also, the concentration of sodium ions is very high on the outside and very low inside, which provides the energy for the transport. A special property of the transport protein is that the conformational change to allow sodium movement to the interior will not occur

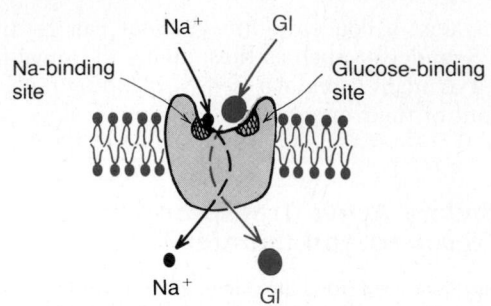

Figure 4–12. Postulated mechanism for sodium co-transport of glucose.

until a glucose molecule also attaches. But when they are both attached, the conformational change takes place automatically, and both the sodium and the glucose are transported to the inside of the cell at the same time. Hence, this is a *sodium-glucose co-transport* mechanism.

Sodium co-transport of the amino acids occurs in the same manner as for glucose, except that it uses a different set of transport proteins. Five *amino acid transport proteins* have been identified, each of which is responsible for transporting one subset of amino acids with specific molecular characteristics.

Sodium co-transport of glucose and amino acids occurs especially in the epithelial cells of the intestinal tract and renal tubules to aid in the absorption of these substances into the blood, as we discuss in later chapters.

OTHER IMPORTANT CO-TRANSPORT MECHANISMS. Two other important co-transport mechanisms are (1) a *sodium-potassium–two chloride* co-transporter that allows two chloride ions to be carried into cells along with one sodium and one potassium ion, all moving in the same direction, and (2) a *potassium-chloride co-transporter* that allows potassium and chloride ions to be transported together from inside cells to the exterior. Still other co-transport mechanisms into at least some cells include co-transport of iodine ions, iron ions, and urate ions.

Sodium Counter-Transport of Calcium and Hydrogen Ions

Two especially important counter-transport mechanisms are *sodium-calcium counter-transport* and *sodium-hydrogen counter-transport*. Calcium counter-transport occurs in all or almost all cell membranes, with sodium ions moving to the interior and calcium ions, to the exterior, both bound to the same transport protein in a counter-transport mode. This is in addition to primary active transport of calcium that occurs in some cells.

Sodium-hydrogen counter-transport occurs in several tissues. An especially important example is in the proximal tubules of the kidneys, where sodium ions move from the lumen of the tubule to the interior of the tubular cells, whereas hydrogen ions are counter-

transported into the lumen. This mechanism is not nearly so powerful for concentrating hydrogen ions as is the primary active transport of hydrogen that occurs in some of the more distal renal tubules, but it can transport such large numbers of hydrogen ions that it is nevertheless a key to hydrogen ion control in the body fluids, as discussed in detail in Chapter 30.

OTHER IMPORTANT COUNTER-TRANSPORT MECHANISMS. Other counter-transport mechanisms include cation exchanges of calcium or sodium ions on one side of the membrane for magnesium or potassium ions on the other side and anion exchanges of chloride ions moving in one direction for bicarbonate ions or sulfate ions moving in the other direction.

Active Transport Through Cellular Sheets

At many places in the body, substances must be transported all the way through a cellular sheet instead of simply through the cell membrane. Transport of this type occurs through the intestinal epithelium, the epithelium of the renal tubules, the epithelium of all exocrine glands, the epithelium of the gallbladder, the membrane of the choroid plexus of the brain, and many other membranes.

The basic mechanism for transport of a substance through a cellular sheet is (1) to provide *active transport* through the cell membrane *on one side* of the cell and then (2) to provide for either *simple diffusion* or *facilitated diffusion* through the membrane *on the opposite side* of the cell.

Figure 4–13 shows one of the mechanisms for transport of sodium ions through the epithelium of the intestines, gallbladder, and renal tubules. This figure shows that the epithelial cells are connected together at the luminal pole by means of tight junctions that mainly block sodium ion diffusion through the spaces between the cells. However, the *brush border* on the luminal surfaces of the cells is permeable to both sodium ions and water. Therefore, sodium and water diffuse readily to the interior of the cell. Then, at the

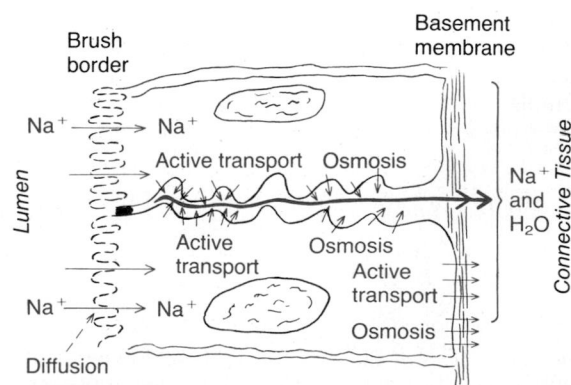

Figure 4–13. Basic mechanism of active transport all the way through a layer of cells.

basal and lateral membranes of the cells, the sodium ions are actively transported into the extracellular fluid. This creates a high sodium ion concentration gradient across these membranes, which in turn causes osmosis of water as well. Thus, the active transport of sodium ions at the basolateral sides of the epithelial cells results in the transport not only of sodium ion but of water also.

Furthermore, any other substances that might be linked by primary active transport or co-transport with the sodium can be transported in this same way. For instance, the positive charges of the sodium ions often pull negatively charged chloride ions along with the sodium. Likewise, when glucose (or amino acids) is co-transported along with sodium through the luminal surface of the cell, the concentration of the glucose increases inside the cell. Then, this glucose is transported by facilitated diffusion through the basolateral borders of the cell, finally entering the extracellular fluid along with the sodium ions, chloride ions, and water.

These are the mechanisms by which almost all the nutrients, ions, and other substances are absorbed into the blood from the intestine; they are also the way in which the same substances are reabsorbed from the glomerular filtrate by the renal tubules.

Throughout this text are numerous examples of each type of transport discussed in this chapter as well as still other variations.

REFERENCES

Agnew, W. S.: Voltage-regulated sodium channel molecules. Annu. Rev. Physiol., 46:517, 1984.

Aidley, D. J.: The Physiology of Excitable Cells, 3rd Ed. New York, Cambridge University Press, 1990.

Almers, W., and Stirling, C.: Distribution of transport proteins over animal membranes. J. Membr. Biol., 77:169, 1984.

Andreoli, T. E., et al. (eds.): Physiology of Membrane Disorders. 2nd Ed. New York, Plenum Publishing Corp., 1986.

Beller, G. A.: Clinical Nuclear Cardiology. Philadelphia, W. B. Saunders Co., 1994.

Biggio, G., and Costa, E. (eds.): Chloride Channels and Their Modulation by Neurotransmitters and Drugs. New York, Raven Press, 1988.

Bishop, M. L., et al.: Clinical Chemistry: Principles, Procedures, Correlations. 2nd Ed. Philadelphia, J. B. Lippincott, 1992.

Büsselberg, D., et al.: Mammalian voltage-activated calcium channel currents are blocked by Pb^{2+}, Zn^{2+}, and Al^{3+}. J. Neurophysiol., 71:1491, 1994.

Byrne, J. H., and Schultz, S. G.: An Introduction to Membrane Transport and Bioelectricity. New York, Raven Press, 1994.

DeLisa, J. A., et al.: Manual of Nerve Conduction Velocity and Clinical Neurophysiology. New York, Raven Press, 1994.

Dempster, F. N., and Brainerd, C. J.: Interference and Inhibition in Cognition. San Diego, Academic Press, 1994.

Dinno, M. A., and Armstrong, W. M. (eds.): Membrane Biophysics III: Biological Transport. New York, Alan R. Liss, Inc., 1988.

DiPolo, R., and Beauge, L.: The calcium pump and sodium-calcium exchange in squid axons. Annu. Rev. Physiol., 45:313, 1983.

Donowitz, M., and Welsh, M. J.: Ca^{2+} and cyclic AMP in regulation of intestinal Na, K, and Cl transport. Annu. Rev. Physiol., 48:135, 1986.

Ellis, D.: Na-Ca exchange in cardiac tissues. Adv. Myocardiol., 5:295, 1985.

Finkelstein, A., et al.: Osmotic swelling of vesicles. Annu. Rev. Physiol., 48:135, 1986.

Forgac, M.: Structure and function of vacuolar class of ATP-driven proton pumps. Physiol. Rev., 69:765, 1989.

Gadsby, D. C.: The Na/K pump of cardiac cells. Annu. Rev. Biophys. Bioeng., 13:373, 1984.

Haas, M.: Properties and diversity of Na-K-Cl cotransporters. Annu. Rev. Physiol., 51:443, 1989.

Haynes, D. H., and Mandveno, A.: Computer modeling of Ca^{2+} pump function of Ca^{2+}-Mg^{2+}-ATPase of sarcoplasmic reticulum. Physiol. Rev., 67:244, 1987.

Hidalgo, C. (ed.): Physical Properties of Biological Membranes and Their Functional Implications. New York, Plenum Publishing Corp., 1988.

Higashida, H., et al.: Molecular Basis of Ion Channels and Receptors Involved in Nerve Excitation, Synaptic Transmission, and Muscle. New York Academy of Sciences, 1994.

Hinrichsen, R. D.: CA^{2+}-Dependent K+ Channels. Boca Raton, Fla., CRC Press, Inc., 1993.

Hoffmann, E. K., and Simonsen, L. O.: Membrane mechanisms in volume and pH regulation in vertebrate cells. Physiol. Rev., 69:315, 1989.

Horisberger, J-D.: The Na,K-Pump: Structure-function Relationship. Boca Raton, Fla., CRC Press, Inc., 1994.

Huguenard, J., and McCormick, D.: Electrophysiology of the Neuron: A Companion to Shepherd's Neurobiology: An Interactive Tutorial. New York, Oxford University Press, 1994.

Keja, J. A., and Kits, K. S.: Single-channel properties of high- and low-voltage-activated calcium channels in rat pituitary melanotropic cells. J. Neurophysiol., 71:840, 1994.

Kim, C. H., and Tedeschi, H. (eds.): Advances in Membrane Biochemistry and Bioenergetics. New York, Plenum Publishing Corp., 1987.

Kostyuk, P. G.: Calcium Ions in Nerve Cell Function, New York, Oxford University Press, 1992.

Lang, F.: Ion Transport in the Regulation of Cell Proliferation in Cellular Physiology and Biochemistry. Farmington, Conn., S. Karger Publishers, Inc., 1992.

Lang, F.: The Molecules of Transport. Farmington, Conn., S. Karger Publishers, Inc., 1993.

Latorre, R., et al.: Varieties of calcium-activated potassium channels. Annu. Rev. Physiol., 51:385, 1989.

Lauger, P.: Dynamics of ion transport systems in membranes. Physiol. Rev., 67:1296, 1987.

Levitan, I. B., and Kaczmarek, L. K.: The Neuron: Cell and Molecular Biology. New York, Oxford University Press, 1991.

McKay, M. C., et al.: Opening of large-conductance calcium-activated potassium channels by the substituted benzimidazolone NS004. J. Neurophysiol., 71:1873, 1994.

Melandri, B. A., et al.: Bioelectrochemistry IV: Nerve Muscle Function—Bioelectrochemistry, Mechanisms, Bioenergetics, and Control. New York, Plenum Publishing Corp., 1994.

Miller, C.: Integral membrane channels: Studies in model membranes. Physiol. Rev., 63:1209, 1983.

Narahashi, T.: Ion Channels. New York, Plenum Publishing Corp., 1988.

Reichert, H.: Introduction to Neurobiology. New York, Oxford University Press, 1993.

Rudel, R., and Lehmann-Horn, F.: Membrane changes in cells from myotonia patients. Physiol. Rev., 65:310, 1985.

Sakmann, B., and Neher, E.: Patch clamp techniques for studying ionic channels in excitable membranes. Annu. Rev. Physiol., 46:455, 1984.

Salkoff, L. B., and Tonouye, M. A.: Genetics of ion channels. Physiol. Rev., 66:301, 1986.

Schatzmann, H. J.: The calcium pump of the surface membrane and of the sarcoplasmic reticulum. Annu. Rev. Physiol., 51:473, 1989.

Schatzmann, H. J.: The red cell calcium pump. Annu. Rev. Physiol., 445:303, 1983.

Schultz, S. G.: A cellular model for active sodium absorption by mammalian colon. Annu. Rev. Physiol., 46:435, 1984.

Schwartz, G. J., and Al-Awqati, Q.: Regulation of transepithelia H+ transport by exocytosis and endocytosis. Annu. Rev. Physiol., 48:153, 1986.

Shepherd, G. M.: Foundations of the Neuron Doctrine. New York, Oxford University Press, 1991.

Shepherd, G. M.: Neurobiology. New York, Oxford University Press, 1994.

Siegel, G. J.: Basic Neurochemistry: Molecular, Cellular, and Medical Aspects. 5th Ed. New York, Raven Press, 1994.

Stamatoyannopoulos, G., et al.: The Molecular Basis of Blood Diseases. Philadelphia, W. B. Saunders Co., 1994.

Stein, W. D. (ed.): The Ion Pumps: Structure, Function, and Regulation. New York, Alan R. Liss, Inc., 1988.

Strange, P. G.: Brain Biochemistry and Brain Disorders. New York, Oxford University Press, 1993.

Verkman, A. S.: Water Channels. Boca Raton, Fla., CRC Press, Inc., 1993.

Wallis, D. I.: Electrophysiology: A Practical Approach. New York, Oxford University Press, 1993.

Waxman, S. G., et al.: The Axon: Structure, Function, and Pathophysiology. New York, Oxford University Press, 1995.

White, S. H.: Membrane Protein Structure: Experimental Approaches. New York, Oxford University Press, 1994.

Membrane Potentials and Action Potentials

CHAPTER 5

Electrical potentials exist across the membranes of essentially all cells of the body. In addition, some cells, such as nerve and muscle cells, are "excitable"—that is, capable of self-generation of electrochemical impulses at their membranes. In most instances, these impulses can be used to transmit signals along the membranes. In still other types of cells, such as glandular cells, macrophages, and ciliated cells, other types of changes in membrane potentials probably play significant roles in controlling many of the cell's functions. The present discussion is concerned with membrane potentials that are generated both at rest and during action by nerve and muscle cells.

BASIC PHYSICS OF MEMBRANE POTENTIALS

Membrane Potentials Caused by Diffusion

Figure 5–1 shows a nerve fiber when there is no active transport of sodium or potassium ions. In Figure 5–1A, the potassium concentration is great inside the membrane, whereas that outside is very low. Let us assume that the membrane in this instance is permeable to the potassium ions but not to any other ions. Because of the large potassium concentration gradient from the inside toward the outside, there is a strong tendency for potassium ions to diffuse outward. As they do so, they carry positive charges to the outside, thus creating a state of electropositivity outside the membrane and electronegativity on the inside because of the negative anions that remain behind and do not diffuse outward along with the potassium. This new potential difference, positive outside and negative inside, repels the positively charged potassium ions that are diffusing outward back in the opposite direction, from the outside toward the inside. Within a millisecond or so, the potential change becomes great enough to block further net diffusion to the exterior despite the high potassium ion concentration gradient. In the normal large mammalian nerve fiber, *the potential difference required is about 94 millivolts, with negativity inside the fiber membrane.*

Figure 5–1B shows the same phenomenon as that in Figure 5–1A but this time with a high concentration of sodium ions outside the membrane and a low sodium concentration inside. These ions are also positively charged, and this time the membrane is highly permeable to the sodium ions but impermeable to all other ions. Diffusion of the sodium ions to the inside creates a membrane potential now of opposite polarity, with negativity outside and positivity inside. Again, the membrane potential rises high enough within milliseconds to block further net diffusion of sodium ions to the inside; however, this time, for the large mammalian nerve fiber, *the potential is about 61 millivolts and with positivity inside the fiber.*

Thus, in both parts of Figure 5–1, we see that a concentration difference of ions across a selectively permeable membrane can, under appropriate conditions, cause the creation of a membrane potential. In later sections of this chapter, we see that many of the rapid changes in membrane potentials observed during the course of nerve and muscle impulse transmission result from the occurrence of rapidly changing diffusion potentials of this nature.

57

Diffusion Potentials

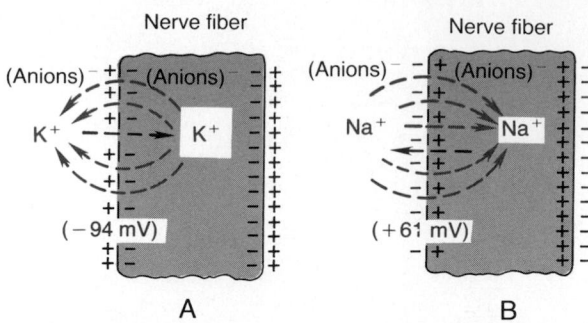

Figure 5–1. *A,* Establishment of a diffusion potential across a cell membrane, caused by potassium ions diffusing from inside the cell to the outside through a membrane that is selectively permeable only to potassium. *B,* Establishment of a diffusion potential when the membrane is permeable only to sodium ions. Note that the internal membrane potential is negative when potassium ions diffuse and positive when sodium ions diffuse because of opposite concentration gradients of these two ions.

RELATION OF THE DIFFUSION POTENTIAL TO THE CONCENTRATION DIFFERENCE—THE NERNST EQUATION.

The potential level across the membrane that prevents net diffusion of an ion in either direction through the membrane is called the *Nernst potential* for that ion, a term that was introduced in the previous chapter. The magnitude of this potential is determined by the *ratio* of the ion concentrations on the two sides of the membrane—the greater this ratio, the greater the tendency for the ions to diffuse in one direction, and therefore the greater is the Nernst potential. The following equation, called the *Nernst equation*, can be used to calculate the Nernst potential for any univalent ion at normal body temperature of 98.6°F (37°C):

$$\text{EMF (millivolts)} = \pm 61 \log \frac{\text{Conc. inside}}{\text{Conc. outside}}$$

When using this formula, it is usually assumed that the potential outside the membrane always remains at exactly zero potential, and the Nernst potential that is calculated is the potential inside the membrane. Also, the sign of the potential is positive (+) if the ion under consideration is a negative ion and negative (−) if it is a positive ion.

Thus, when the concentration of a positive ion (potassium ions, for instance) on the inside is 10 times that on the outside, the log of 10 is 1, so that the Nernst potential calculates to be −61 millivolts inside the membrane.

Calculation of the Diffusion Potential When the Membrane is Permeable to Several Different Ions

When a membrane is permeable to several different ions, the diffusion potential that develops depends on three factors: (1) the polarity of the electrical charge of

each ion, (2) the permeability of the membrane (*P*) to each ion, and (3) the concentrations (*C*) of the respective ions on the inside (*i*) and outside (*o*) of the membrane. Thus, the following formula, called the *Goldman equation*, or the *Goldman-Hodgkin-Katz equation*, gives the calculated membrane potential on the *inside* of the membrane when two univalent positive ions, sodium (Na^+) and potassium (K^+), and one univalent negative ion, chloride (Cl^-), are involved.

$$\text{EMF (millivolts)} = -61 \cdot \log \frac{C_{Na^+_i}P_{Na^+} + C_{K^+_i}P_{K^+} + C_{Cl^-_o}P_{Cl^-}}{C_{Na^+_o}P_{Na^+} + C_{K^+_o}P_{K^+} + C_{Cl^-_i}P_{Cl^-}}$$

Now, let us study the importance and the meaning of this equation. First, sodium, potassium, and chloride ions are the ions most importantly involved in the development of membrane potentials in nerve and muscle fibers as well as in the neuronal cells in the central nervous system. The concentration gradient of each of these ions across the membrane helps determine the voltage of the membrane potential.

Second, the degree of importance of each of the ions in determining the voltage is proportional to the membrane permeability for that particular ion. Thus, if the membrane is impermeable to both potassium and chloride ions, the membrane potential becomes entirely dominated by the concentration gradient of sodium ions alone, and the resulting potential will be equal to the Nernst potential for sodium. The same principle holds for each of the other two ions if the membrane should become selectively permeable for either one of them alone.

Third, a positive ion concentration gradient from *inside* the membrane *to the outside* causes electronegativity inside the membrane. The reason for this is that positive ions diffuse to the outside when their concentration is higher inside than outside. This carries positive charges to the outside but leaves the nondiffusible negative anions on the inside. The opposite effect occurs when there is a negative ion gradient. That is, a chloride ion gradient from the *outside to the inside* causes negativity inside the cell because negatively charged chloride ions then diffuse to the inside, while leaving the nondiffusible positive ions on the outside.

Fourth, we see later that the permeabilities of the sodium and potassium channels undergo rapid changes during conduction of the nerve impulse, whereas the permeability of the chloride channels does not change greatly during this process. Therefore, the changes in the sodium and potassium permeabilities are primarily responsible for signal transmission in the nerves, which is the subject of most of the remainder of this chapter.

Measuring the Membrane Potential

The method of measuring the membrane potential is simple in theory but often difficult in practice because of the small sizes of most of the fibers. Figure 5–2 shows a small pipette filled with a strong electrolyte solution (KCl) that is impaled through the cell membrane to the interior of the fiber. Then another electrode, called the "indifferent electrode," is placed in the extracellular fluid, and the potential difference between the inside and outside of the fiber is measured using an appropriate voltmeter. This voltmeter is a highly sophisticated electronic apparatus that is capable of measuring very small voltages despite extremely high resistance to electrical flow through the tip of the micropipette, which has a diameter usually less than 1 micrometer and a resistance often as great as 1 billion ohms. For recording rapid *changes* in the membrane potential during the transmission of nerve impulses, the microelectrode is connected to an oscilloscope, as explained later in the chapter.

The Cell Membrane as an Electrical Capacitor

In each of the figures shown thus far, the negative and positive ionic charges that cause the membrane potential have been shown to be lined up against the membrane, and we have not spoken of the arrangement of the charges elsewhere in the fluids, either inside the nerve fiber or on the outside in the extracellular fluid. However, Figure 5–3 demonstrates this, showing that everywhere except adjacent to the surfaces of the cell membrane itself the negative and positive charges are equal. This is called the principle of *electrical neutrality;* that is, for every positive ion there is a negative ion nearby to neutralize it; otherwise, electrical potentials of billions of volts would appear within the fluids.

When positive ions are pumped to the outside of the membrane, they line up along the outside of the membrane, and on the inside, the negative ions that have been left behind line up. This creates a *dipole layer* of positive and negative charges between the outside and inside of the membrane, but equal numbers of negative and positive charges are left everywhere else within the fluids. This is the same effect that occurs when the plates of an electrical capacitor become electrically charged—that is, lining up of negative and positive charges on the opposite sides of the dielectric membrane of the capacitor. Therefore, the lipid bilayer of the cell membrane actually functions as a *dielectric* of a cell membrane capacitor, much as mica, paper, and Mylar function as dielectrics in electrical capacitors.

Because of the extreme thinness of the cell membrane (only 7 nanometers), its capacitance is tremendous for its area—about *1 microfarad per square centimeter.*

The lower part of Figure 5–3 shows the electrical potential that will be recorded at each point in or near

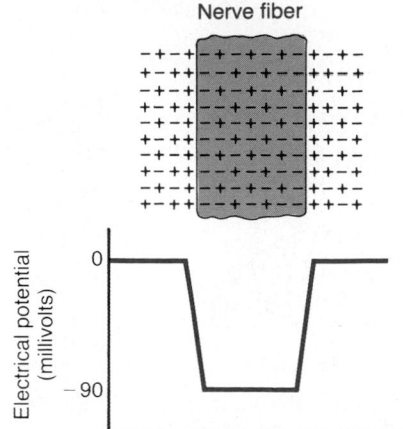

Figure 5–3. Distribution of positively and negatively charged ions in the extracellular fluid surrounding a nerve fiber and in the fluid inside the fiber; note the dipolar alignment of negative charges along the inside surface of the membrane and positive charges along the outside surface. In the lower panel are displayed the abrupt changes in membrane potential that occur at the membranes on the two sides of the fiber.

the nerve fiber membrane, beginning at the left side of the figure and passing to the right. As long as the electrode is outside the nerve membrane, the potential that is recorded is zero, which is the potential of the extracellular fluid. Then, as the recording electrode passes through the electrical dipole layer at the cell membrane, the potential decreases abruptly to − 90 millivolts. Then, moving across the center of the fiber, the potential remains at a steady level but reverses back to zero the instant it passes through the opposite side of the membrane.

The fact that the nerve membrane functions as a capacitor has one especially important point of significance: To create a negative potential inside the membrane, only enough positive ions must be transported outward to develop the electrical dipole layer at the membrane itself. All the remaining ions inside the nerve fiber can still be both positive and negative ions. Therefore, an incredibly small number of ions need to be transferred through the membrane to establish the normal potential of − 90 millivolts inside the nerve fiber— only about 1/5,000,000 to 1/100,000,000 of the total positive charges inside the fiber need be so transferred. Also, an equally small number of positive ions moving from outside to the inside of the fiber can reverse the potential from − 90 millivolts to as much as + 35 millivolts within as little as 1/10,000 of a second. The rapid shifting of ions in this manner causes the nerve signals that we discuss in subsequent sections of this chapter.

RESTING MEMBRANE POTENTIAL OF NERVES

The membrane potential of large nerve fibers when they are not transmitting nerve signals is about − 90 millivolts. That is, the potential *inside the fiber* is 90 millivolts more negative than the potential in the extracellular fluid on the outside of the fiber. In the next

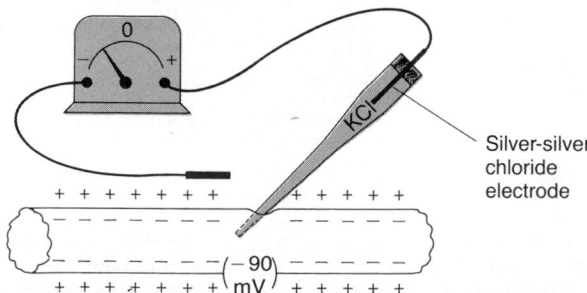

Figure 5–2. Measurement of the membrane potential of the nerve fiber using a microelectrode.

Outside

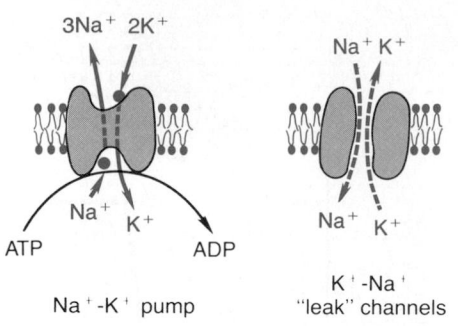

Figure 5–4. Functional characteristics of the Na⁺-K⁺ pump and the potassium-sodium "leak" channels.

few paragraphs, we explain all the factors that determine the level of this potential, but before doing so, we must describe the transport properties of the resting nerve membrane for sodium and potassium.

ACTIVE TRANSPORT OF SODIUM AND POTASSIUM IONS THROUGH THE MEMBRANE—THE SODIUM-POTASSIUM PUMP. First, let us recall from the discussions of Chapter 4 that all cell membranes of the body have a powerful sodium-potassium pump that continually pumps sodium to the outside of the fiber and potassium to the inside, as illustrated on the left-hand side in Figure 5–4. Further, let us remember that this is an *electrogenic pump* because more positive charges are pumped to the outside than to the inside (three Na⁺ ions to the outside for each two K⁺ ions to the inside), leaving a net deficit of positive ions on the inside; this causes a negative charge inside the cell membrane.

The sodium-potassium pump also causes large concentration gradients for sodium and potassium across the resting nerve membrane. These gradients are the following:

Na⁺ (outside):	142 mEq/L
Na⁺ (inside):	14 mEq/L
K⁺ (outside):	4 mEq/L
K⁺ (inside):	140 mEq/L

The ratios of these two respective ions from the inside to the outside are

$$Na^+{}_{inside}/Na^+{}_{outside} = 0.1$$

$$K^+{}_{inside}/K^+{}_{outside} = 35.0$$

LEAKAGE OF POTASSIUM AND SODIUM THROUGH THE NERVE MEMBRANE. To the right in Figure 5–4 is shown a channel protein in the cell membrane through which potassium and sodium ions can leak, called a *potassium-sodium "leak" channel.* The emphasis is on potassium leakage because, on the average, the channels are far more permeable to potassium than to sodium, normally about 100 times as permeable. We

see later that this differential in permeability is exceedingly important in determining the level of the normal resting membrane potential.

Origin of the Normal Resting Membrane Potential

Figure 5–5 shows the important factors in the establishment of the normal resting membrane potential of − 90 millivolts. They are as follows.

CONTRIBUTION OF THE POTASSIUM DIFFUSION POTENTIAL. In Figure 5–5A, we make the assumption that the only movement of ions through the membrane is the diffusion of potassium ions, as demonstrated by the open channels between the potassium inside the membrane and the outside. Because of the high ratio

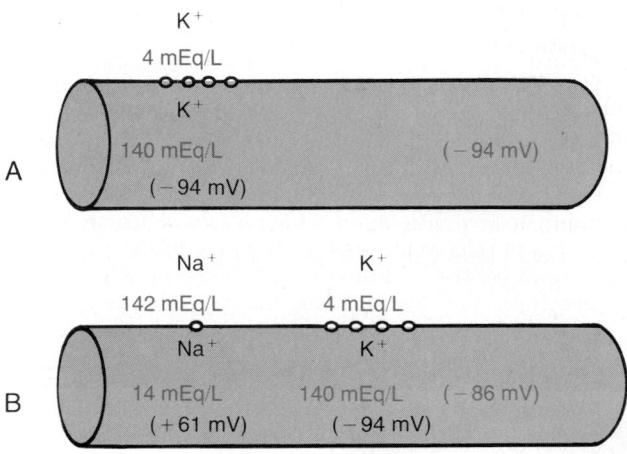

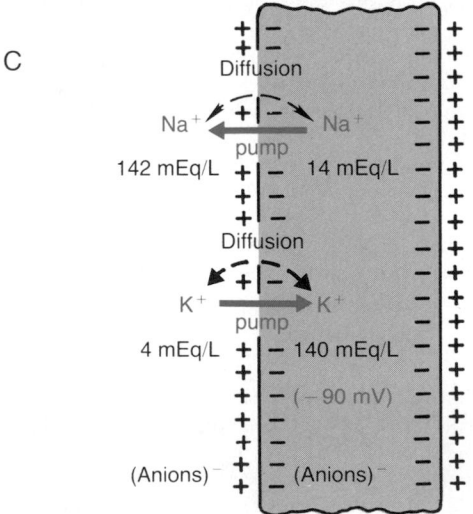

Figure 5–5. Establishment of resting membrane potentials in nerve fibers under three conditions: *A,* when the membrane potential is caused entirely by potassium diffusion alone; *B,* when the membrane potential is caused by diffusion of both sodium and potassium ions; and *C,* when the membrane potential is caused by diffusion of both sodium and potassium ions plus pumping of both these ions by the Na⁺-K⁺ pump.

of potassium ions inside to outside, 35 to 1, the Nernst potential corresponding to this ratio is − 94 millivolts because the logarithm of 35 is 1.54, and this times − 61 millivolts, is − 94 millivolts. Therefore, if potassium ions were the only factor causing the resting potential, this resting potential *inside the fiber* would also be equal to − 94 millivolts, as shown in the figure.

CONTRIBUTION OF SODIUM DIFFUSION THROUGH THE NERVE MEMBRANE. Figure 5–5B shows the addition of slight permeability of the nerve membrane to sodium ions, caused by the minute diffusion of sodium ions through the K⁺-Na⁺ leak channels. The ratio of sodium ions from inside to outside the membrane is 0.1, and this gives a calculated Nernst potential for the inside of the membrane of + 61 millivolts. But also shown in Figure 5–5B is the Nernst potential for potassium diffusion of − 94 millivolts. How do these interact with each other, and what will be the summated potential? This can be answered by using the Goldman equation described earlier. Intuitively, one can see that if the membrane is highly permeable to potassium but only slightly permeable to sodium, it is logical that the diffusion of potassium will contribute far more to the membrane potential than will the diffusion of sodium. In the normal nerve fiber, the permeability of the membrane to potassium is about 100 times as great as to sodium. Using this value in the Goldman equation gives an internal membrane potential of − 86 millivolts, which is near to the potassium potential shown to the right in the figure.

CONTRIBUTION OF THE NA⁺-K⁺ PUMP. In Figure 5–5C, an additional contribution of the Na⁺-K⁺ pump is shown. In this figure, there is continuous pumping of three sodium ions to the outside for each two potassium ions pumped to the inside of the membrane. The fact that more sodium ions are being pumped to the outside than potassium to the inside causes a continual loss of positive charges from inside the membrane; this creates an additional degree of negativity (about − 4 millivolts additional) on the inside beyond that which can be accounted for by diffusion alone. Therefore, as shown in Figure 5–5C, the net membrane potential with all these factors operative at the same time is − 90 millivolts.

In summary, the diffusion potentials alone caused by potassium and sodium diffusion would give a membrane potential of about − 86 millivolts, almost all of this being determined by potassium diffusion. Then, an additional − 4 millivolts is contributed to the membrane potential by the electrogenic Na⁺-K⁺ pump, giving a net resting membrane potential of − 90 millivolts.

The resting membrane potential in large skeletal muscle fibers is about the same as that in large nerve fibers, also − 90 millivolts. However, in both small nerve fibers and small muscle fibers—smooth muscle, for instance—as well as in many of the neurons of the central nervous system, the membrane potential is often as little as − 40 to − 60 millivolts instead of − 90 millivolts.

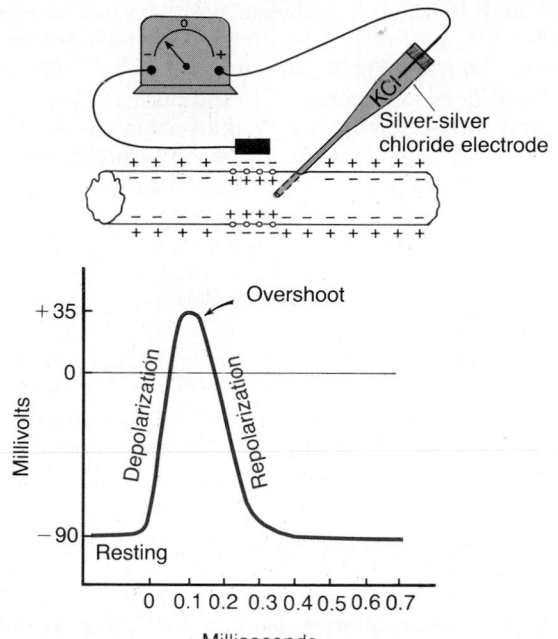

Figure 5–6. Typical action potential recorded by the method shown in the upper panel of the figure.

NERVE ACTION POTENTIAL

Nerve signals are transmitted by *action potentials,* which are rapid changes in the membrane potential. Each action potential begins with a sudden change from the normal resting negative potential to a positive membrane potential and then ends with an almost equally rapid change back to the negative potential. To conduct a nerve signal, the action potential moves along the nerve fiber until it comes to the fiber's end. The upper panel of Figure 5–6 shows the disturbances that occur at the membrane during the action potential, with transfer of positive charges to the interior of the fiber at its onset and return of positive charges to the exterior at its end. The lower panel graphically shows the successive changes in the membrane potential over a few 10,000ths of a second, illustrating the explosive onset of the action potential and the almost equally as rapid recovery.

The successive stages of the action potential are as follows.

RESTING STAGE. This is the resting membrane potential before the action potential occurs. The membrane is said to be "polarized" during this stage because of the large negative membrane potential that is present.

DEPOLARIZATION STAGE. At this time, the membrane suddenly becomes permeable to sodium ions, allowing tremendous numbers of positively charged sodium ions to flow to the interior of the axon. The normal "polarized" state of − 90 millivolts is lost, with the potential rising rapidly in the positive direction. This is called *depolarization.* In large nerve fibers, the membrane potential "overshoots" beyond the zero level and becomes somewhat positive, but in some smaller fibers as well as many central nervous system neurons, the potential merely approaches the zero level and does not overshoot to the positive state.

REPOLARIZATION STAGE. Within a few 10,000ths of a second after the membrane becomes highly permeable to sodium ions, the sodium channels begin to close and the potassium channels open more than they normally do. Then, rapid diffusion of potassium ions to the exterior re-establishes the normal negative resting membrane potential. This is called *repolarization* of the membrane.

To explain more fully the factors that cause both depolarization and repolarization, we need now to describe the special characteristics of yet two other types of transport channels through the nerve membrane: the voltage-gated sodium and potassium channels.

Voltage-Gated Sodium and Potassium Channels

The necessary actor in causing both depolarization and repolarization of the nerve membrane during the action potential is the *voltage-gated sodium channel.* The *voltage-gated potassium channel* also plays an important role in increasing the rapidity of repolarization of the membrane. *These two voltage-gated channels are in addition to the Na^+-K^+ pump and the Na^+-K^+ leak channels.*

Voltage-Gated Sodium Channel—Activation and Inactivation of the Channel

The upper panel of Figure 5–7 shows the voltage-gated sodium channel in three separate states. This channel has two *gates,* one near the outside of the channel called the *activation gate* and another near the inside called the *inactivation gate.* To the left is shown (schematically) the state of these two gates in the normal resting membrane when the membrane potential is − 90 millivolts. In this state, the activation gate is closed, which prevents any entry of sodium ions to the interior of the fiber through these sodium channels.

ACTIVATION OF THE SODIUM CHANNEL. When the membrane potential becomes less negative than during the resting state, rising from − 90 millivolts toward zero, it finally reaches a voltage, usually somewhere between − 70 and − 50 millivolts, that causes a sudden conformational change in the activation gate, flipping it to the open position. This is called the *activated state;* during this state, sodium ions can literally pour inward through the channel, increasing the sodium permeability of the membrane as much as 500- to 5000-fold.

INACTIVATION OF THE SODIUM CHANNEL. To the far right in the upper panel of Figure 5–7 is shown a third state of the sodium channel. The same increase in voltage that opens the activation gate also closes the inactivation gate. The inactivation gate, however, closes a few 10,000ths of a second after the activation gate opens. That is, the conformational change that flips the

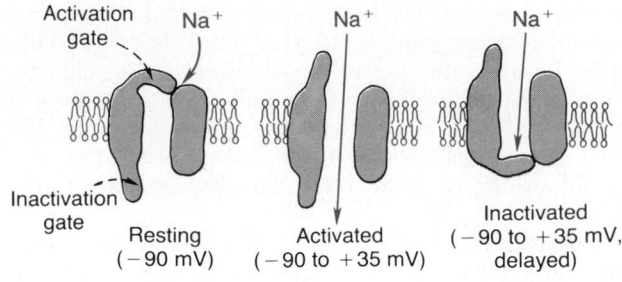

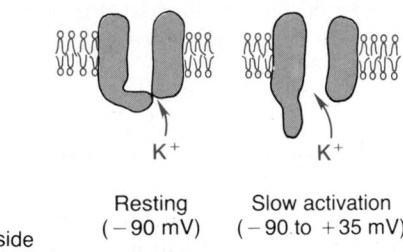

Figure 5–7. Characteristics of the voltage-gated sodium and potassium channels, showing both activation and inactivation of the sodium channels but activation of the potassium channels only when the membrane potential is changed from the normal resting negative value to a positive value.

inactivation gate to the closed state is a slow process, whereas the conformational change that opens the activation gate is a rapid process. Therefore, after the sodium channel has remained open for a few 10,000ths of a second, it closes and sodium ions can no longer pour to the inside of the membrane. At this point, the membrane potential begins to recover back toward the resting membrane state, which is the repolarization process.

An important characteristic of the sodium channel inactivation process is that *the inactivation gate will not re-open until the membrane potential returns either to or nearly to the original resting membrane potential level.* Therefore, it is not possible for the sodium channels to open again without the nerve fiber first repolarizing.

Voltage-Gated Potassium Channel and Its Activation

The lower panel of Figure 5–7 shows the voltage-gated potassium channel in two states: during the resting state and toward the end of the action potential. During the resting state, the gate of the potassium channel is closed, as illustrated to the left in the figure, and potassium ions are prevented from passing through this channel to the exterior. When the membrane potential rises from − 90 millivolts toward zero, this voltage change causes a slow conformational opening of the gate and allows increased potassium diffusion outward through the channel. Because of the slowness of opening of these potassium channels, they

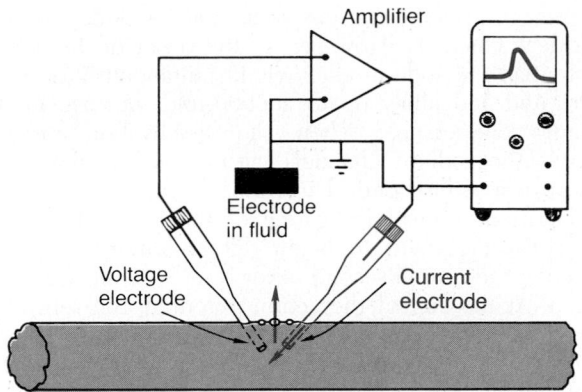

Figure 5–8. "Voltage clamp" method for studying flow of ions through specific channels.

mainly open just at the same time that the sodium channels are beginning to close because of inactivation. Thus, the decrease in sodium entry to the cell and simultaneous increase in potassium exit from the cell greatly speed the repolarization, leading to full recovery of the resting membrane potential within a few 10,000ths of a second.

RESEARCH METHOD FOR MEASURING THE EFFECT OF VOLTAGE ON OPENING AND CLOSING OF THE VOLTAGE-GATED CHANNELS—THE "VOLTAGE CLAMP."

The original research that led to our quantitative understanding of the sodium and potassium channels was so ingenious that it led to Nobel prizes for the scientists responsible, Hodgkin and Huxley. The essence of these studies is shown in Figures 5–8 and 5–9.

Figure 5–8 shows the experimental apparatus called the *voltage clamp*, which is used to measure the flow of ions through the different channels. In using this apparatus, two electrodes are inserted into the nerve fiber. One of these is for the purpose of measuring the voltage of the membrane potential. The other is to conduct electrical current into or out of the nerve fiber. This apparatus is used in the following way: The investigator decides what voltage he or she wants to establish inside the nerve fiber. He or she then adjusts the electronic portion of the apparatus to the desired voltage, and this automatically injects either positive or negative electricity through the current electrode at whatever rate is required to hold the voltage, as measured by the voltage electrode, at the level set by the operator. For instance, when the membrane potential is suddenly increased by this voltage clamp from − 90 millivolts to zero, the voltage-gated sodium and potassium channels open, and sodium and potassium ions begin to pour through the channels. To counterbalance the effect of these ion movements on the desired setting of the intracellular voltage, electrical current is injected automatically through the current electrode of the voltage clamp to maintain the intracellular voltage at the required steady zero level. To achieve this, the current injected must be equal to but of opposite polarity to the net current flow through the membrane channels. To measure how much current flow is occurring at each instant, the current electrode is connected to an oscilloscope that records the current flow, as demonstrated on the screen of the oscilloscope in the figure. Finally, the investigator

adjusts the concentrations of the ions to other than normal levels both inside and outside the nerve fiber and repeats the study. This can be done easily when using large nerve fibers removed from some crustaceans, especially the giant squid axon that in some cases is as large as 1 millimeter in diameter. When sodium is the only permeant ion in the solutions inside and outside the squid axon, the voltage clamp measures current flow only through the sodium channels. When potassium is the only permeant ion, current flow only through the potassium channels is measured.

Another means for studying the flow of ions through an individual type of channel is to block one type of channel at a time. For instance, the sodium channels can be blocked by a toxin called *tetrodotoxin* by applying this toxin to the outside of the cell membrane where the sodium activation gates are located. Conversely, *tetraethylammonium ion* blocks the potassium pores when it is applied to the interior of the nerve fiber.

Figure 5–9 shows the typical changes in conductance of the voltage-gated sodium and potassium channels when the membrane potential is suddenly changed by use of the voltage clamp from − 90 millivolts to + 10 millivolts and 2 milliseconds later back again to − 90 millivolts. Note the sudden opening of the sodium channels (the activation stage) within a small fraction of a millisecond after the membrane potential is increased to the positive value. However, during approximately the next millisecond or so, the sodium channels automatically close (the inactivation stage).

Now, note the opening (activation) of the potassium channels. These open slowly and reach the full open state only after the sodium channels have already become almost completely closed. Furthermore, once the potassium channels open, they remain open for the entire duration of the positive membrane potential and do not close again until after the membrane potential is decreased back to a very negative value.

Finally, let us recall that the voltage-gated channels normally flip to the open state or the closed state suddenly, which was shown in Figure 4–5. Therefore, how is it that the curves in Figure 5–9 are so smooth? The answer is that these curves represent the flow of sodium

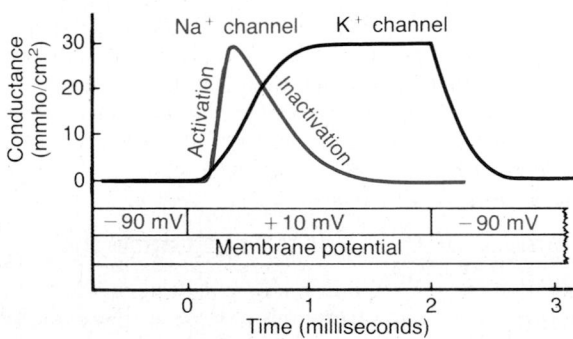

Figure 5–9. Typical changes in conductance of sodium and potassium ion channels when the membrane potential is suddenly increased from the normal resting value of − 90 millivolts to a positive value of + 10 millivolts for 2 milliseconds. This figure shows that the sodium channels open (activate) and then close (inactivate) before the end of the 2 milliseconds, whereas the potassium channels only open (activate), and the rate of opening is much slower than for the sodium channels.

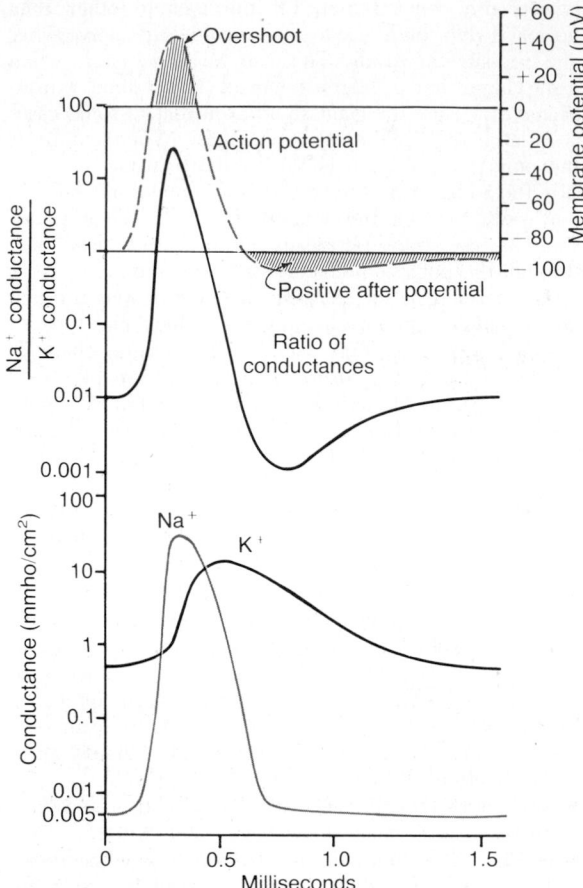

Figure 5–10. Changes in sodium and potassium conductances during the course of the action potential. Note that sodium conductance increases several thousandfold during the early stages of the action potential, whereas potassium conductance increases only about 30-fold during the latter stages of the action potential and for a short period thereafter. (Curves constructed from data in Hodgkin and Huxley papers but transposed from squid axon to apply to the membrane potentials of large mammalian nerve fibers.)

and potassium ions through literally thousands or millions of channels at the same time. Some open at one voltage level, others at another voltage, and so forth. Likewise, some become inactivated at different points in the cycle from the others. Thus, the illustrated curves represent summated current flows through the many channels.

Summary of the Events That Cause the Action Potential

Figure 5–10 shows in summary form the sequential events that occur during and shortly after the action potential. They are as follows.

At the bottom of the figure are shown the changes in membrane conductances for sodium and potassium ions. During the resting state, before the action potential begins, the conductance for potassium ions is shown to be 50 to 100 times as great as the conductance for sodium ions. This is caused by much greater leakage of potassium ions than sodium ions through the leak channels. However, at the onset of the action potential, the sodium channels instantaneously become activated and allow up to a 5000-fold increase in sodium conductance. Then the inactivation process closes the sodium channels within another few fractions of a millisecond. The onset of the action potential also causes voltage gating of the potassium channels, causing them to begin opening more slowly, a fraction of a millisecond after the sodium channels open. At the end of the action potential, the return of the membrane potential to the negative state causes the potassium channels to close back to their original status, but again only after a delay.

In the middle portion of Figure 5–10 is shown the ratio of sodium conductance to potassium conductance at each instant during the action potential, and above this is shown the action potential itself. During the early portion of the action potential, the ratio of sodium to potassium conductance increases more than a thousandfold. Therefore, far more sodium ions now flow to the interior of the fiber than do potassium ions to the exterior. This is what causes the membrane potential to become positive. Then the sodium channels begin to close and, at the same time, the potassium channels open, so that the ratio of conductance now shifts far in favor of high potassium conductance but low sodium conductance. This allows extremely rapid loss of potassium ions to the exterior, whereas essentially no sodium ions flow to the interior. Consequently, the action potential quickly returns to its baseline level.

"Positive" Afterpotential

Note also in Figure 5–10 that the membrane potential becomes even more negative than the original resting membrane potential for a few milliseconds after the action potential is over. Strangely enough, this is called the *"positive" afterpotential,* which is a misnomer because the positive afterpotential is even more negative than the resting potential. The reason for calling it "positive" is that, historically, the first potential measurements were made on the outside of the nerve fiber membrane rather than the inside, and when measured on the outside, this potential causes a positive record on the recording meter rather than a negative one.

The cause of the positive afterpotential is mainly that many potassium channels remain open for several milliseconds after repolarization of the membrane is complete. This allows excess potassium ions to diffuse out of the nerve fiber, leaving an extra deficit of positive ions on the inside, which means more negativity.

Roles of Other Ions During the Action Potential

Thus far, we have considered only the roles of sodium and potassium ions in the generation of the action potential. At least two other types of ions must be considered. They are as follows.

IMPERMEANT NEGATIVELY CHARGED IONS (ANIONS) INSIDE THE AXON. Inside the axon are many negatively charged ions that cannot go through the membrane channels. They include the anions of protein molecules, many organic phosphate compounds, sulfate compounds, and so forth. Because these ions cannot leave the interior of the axon, any deficit of positive ions inside the membrane leaves an excess of these impermeant negative anions. Therefore, these impermeant negative ions are responsible for the negative charge inside the fiber when there is a deficit of the positively charged potassium ions and other positive ions.

CALCIUM IONS. The cell membranes of almost all, if not all, cells of the body have a calcium pump similar to the sodium pump, and calcium serves along with (or instead of) sodium in some cells to cause the action potential. Like the sodium pump, the calcium pump pumps calcium ions from the interior to the exterior of the cell membrane (or into the endoplasmic reticulum), creating a calcium ion gradient of about 10,000-fold. This leaves an internal concentration of calcium ions of about 10^{-7} molar in contrast to an external concentration of about 10^{-3} molar.

In addition, there are voltage-gated calcium channels. These channels are slightly permeable to sodium ions as well as to calcium ions; when they open, both calcium and sodium ions flow to the interior of the fiber. Therefore, these channels are also called $Ca^{++}\text{-}Na^+$ *channels.* The calcium channels are slow to become activated, requiring 10 to 20 times as long for activation as the sodium channels. Therefore, they are also called *slow channels,* in contrast to the sodium channels that are called *fast channels.*

Calcium channels are numerous in both cardiac muscle and smooth muscle. In fact, in some types of smooth muscle, the fast sodium channels are hardly present, so the action potentials then are caused almost entirely by activation of the slow calcium channels.

Increased Permeability of the Sodium Channels When There Is a Deficit of Calcium Ions. The concentration of calcium ions in the extracellular fluid also has a profound effect on the voltage level at which the sodium channels become activated. When there is a deficit of calcium ions, the sodium channels become activated (opened) by very little change of the membrane potential from its normal very negative resting level to a less negative level. Therefore, the nerve fiber becomes highly excitable, sometimes discharging repetitively without provocation rather than remaining in the resting state. In fact, the calcium ion concentration needs to fall only 50 per cent below normal before spontaneous discharge occurs in many peripheral nerves, often causing muscle "tetany" that can be lethal because of tetanic contraction of the respiratory muscles.

The probable way in which calcium ions affect the sodium channels is the following: These ions appear to bind to the exterior surfaces of the sodium channel protein molecule. The positive charges of these calcium ions in turn alter the electrical state of the channel protein itself, in this way increasing the voltage level required to open the gate.

Initiation of the Action Potential

Up to this point, we have explained the changing sodium and potassium permeabilities of the membrane as well as the development of the action potential itself, but we have not explained what initiates the action potential. The answer to this, as follows, is quite simple.

A POSITIVE-FEEDBACK VICIOUS CIRCLE OPENS THE SODIUM CHANNELS. First, as long as the membrane of the nerve fiber remains undisturbed, no action potential occurs in the normal nerve. However, if any event causes enough initial rise in the membrane potential from −90 millivolts up toward the zero level, the rising voltage itself will cause many voltage-gated sodium channels to begin opening. This allows rapid inflow of sodium ions, which causes still further rise of the membrane potential, thus opening still more voltage-gated sodium channels and allowing more streaming of sodium ions to the interior of the fiber. This process is a positive-feedback vicious circle that, once the feedback is strong enough, will continue until all the voltage-gated sodium channels have become activated (opened). Then, within another fraction of a millisecond, the rising membrane potential causes beginning closure of the sodium channels as well as opening of potassium channels, and the action potential soon terminates.

THRESHOLD FOR INITIATION OF THE ACTION POTENTIAL. An action potential will not occur until the initial rise in membrane potential is great enough to create the vicious circle described in the preceding paragraph. This occurs when the number of Na^+ ions entering the fiber becomes greater than the number of K^+ ions leaving the fiber. A sudden rise in membrane potential of 15 to 30 millivolts usually is required. Therefore, a sudden increase in the membrane potential in a large nerve fiber from −90 millivolts up to about −65 millivolts usually causes the explosive development of the action potential. This level of −65 millivolts is said to be the *threshold* for stimulation.

ACCOMMODATION OF THE MEMBRANE—FAILURE TO FIRE DESPITE RISING VOLTAGE. If the membrane potential rises slowly—over many milliseconds instead of a fraction of a millisecond—the slow inactivating gates of the sodium channels will have time to close at the same time that the activating gates are opening. Consequently, the opening of the activating gates will not be as effective in increasing the flow of sodium ions as it is normally. Therefore, a slow increase in the internal voltage of a nerve fiber either requires a higher threshold voltage than normal to cause firing or at times prevents firing entirely, even with a voltage rise all the way to zero or even to positive voltage. This phenomenon is called *accommodation* of the membrane to the stimulus.

PROPAGATION OF THE ACTION POTENTIAL

In the preceding paragraphs we discussed the action potential as it occurs at one spot on the membrane. However, an action potential elicited at any one point on an excitable membrane usually excites adjacent portions of the membrane, resulting in propagation of the action potential. The mechanism of this is demon-

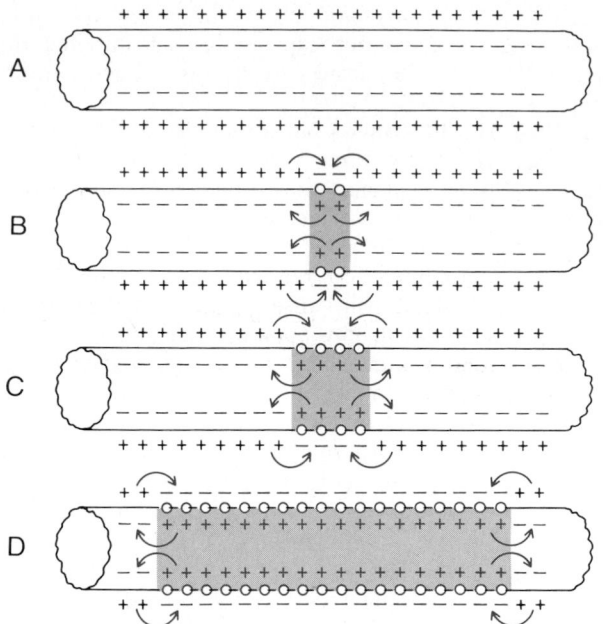

Figure 5–11. Propagation of action potentials in both directions along a conductive fiber.

strated in Figure 5–11. Figure 5–11A shows a normal resting nerve fiber, and Figure 5–11B shows a nerve fiber that has been excited in its midportion—that is, the midportion has suddenly developed increased permeability to sodium. The arrows show a "local circuit" of current flow between the depolarized areas of the membrane and the adjacent resting membrane areas; positive electrical charges carried by the inward diffusing sodium ions flow inward through the depolarized membrane and then for several millimeters along the core of the axon. These positive charges increase the voltage for a distance of 1 to 3 millimeters inside large myelinated fibers to above the threshold voltage value for initiating an action potential. Therefore, the sodium channels in these new areas immediately open, and as shown in Figure 5–11C and D, the explosive action potential spreads. Then these newly depolarized areas produce local circuits of current flow still further along the membrane, causing progressively more and more depolarization. Thus, the depolarization process travels along the entire extent of the fiber. The transmission of the depolarization process along a nerve or muscle fiber is called a *nerve* or *muscle impulse*.

DIRECTION OF PROPAGATION. As demonstrated in Figure 5–11, an excitable membrane has no single direction of propagation, but the action potential will travel in both directions away from the stimulus—and even along all branches of a nerve fiber—until the entire membrane has become depolarized.

ALL-OR-NOTHING PRINCIPLE. Once an action potential has been elicited at any point on the membrane of a normal fiber, the depolarization process will travel over the entire membrane if conditions are right, or it might not travel at all if conditions are not right. This is called the all-or-nothing principle, and it applies to all normal excitable tissues. Occasionally, the action potential will reach a point on the membrane at which it does not generate sufficient voltage to stimulate the next area of the membrane. When this occurs, the spread of depolarization stops. Therefore, for continued propagation of an impulse to occur, the ratio of action potential to threshold for excitation must at all times be greater than 1. This is called the *safety factor* for propagation.

RE-ESTABLISHING SODIUM AND POTASSIUM IONIC GRADIENTS AFTER ACTION POTENTIALS— IMPORTANCE OF ENERGY METABOLISM

The transmission of each impulse along the nerve fiber reduces infinitesimally the concentration differences of sodium and potassium between the inside and outside of the membrane because of diffusion of sodium ions to the inside during depolarization and diffusion of potassium ions to the outside during repolarization. For a single action potential, this effect is so minute that it cannot be measured. Indeed, 100,000 to 50 million impulses can be transmitted by nerve fibers, the number depending on the size of the fiber and several other factors, before the concentration differences have run down to the point that action potential conduction ceases. Even so, with time it becomes necessary to re-establish the sodium and potassium membrane concentration differences. This is achieved by the action of the Na^+-K^+ pump in the same way as that described earlier in the chapter for original establishment of the resting potential. That is, the sodium ions that have diffused to the interior of the cell during the action potentials and the potassium ions that have diffused to the exterior are returned to their original state by the Na^+-K^+ pump. Because this pump requires energy for operation, the process of "recharging" the nerve fiber is an active metabolic one, using energy derived from the adenosine triphosphate (ATP) energy system of the cell. Figure 5–12

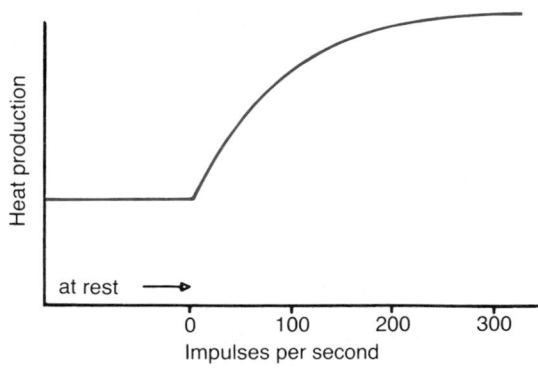

Figure 5–12. Heat production in a nerve fiber at rest and at progressively increasing rates of stimulation.

shows that the nerve fiber produces excess heat, which is a measure of its energy expenditure, when the impulse frequency increases.

A special feature of the sodium-potassium ATPase pump is that its degree of activity is strongly stimulated when excess sodium ions accumulate inside the cell membrane. In fact, the pumping activity increases approximately in proportion to the third power of the sodium concentration. That is, as the internal sodium concentration rises from 10 to 20 mEq/L, the activity of the pump does not merely double but increases about eightfold. Therefore, it can easily be understood how the "recharging" process of the nerve fiber can rapidly be set into motion whenever the concentration differences of sodium and potassium ions across the membrane begin to "run down."

PLATEAU IN SOME ACTION POTENTIALS

In some instances, the excitable membrane does not repolarize immediately after depolarization; instead, the potential remains on a plateau near the peak of the spike for many milliseconds before repolarization begins. Such a plateau is shown in Figure 5–13; one can readily see that the plateau greatly prolongs the period of depolarization. This type of action potential occurs in heart muscle fibers, where the plateau lasts for as long as 2/10 to 3/10 second and causes contraction of the heart muscle to last for this same long period.

The cause of the plateau is a combination of several factors. First, as discussed earlier, in heart muscle, two types of channel enter into the depolarization process: (1) the usual voltage-activated sodium channels, called *fast channels,* and (2) voltage-activated calcium channels, which are slow to open and therefore are called *slow channels.* These channels allow diffusion mainly of calcium ions but of some sodium ions as well. Opening of the fast channels causes the spike portion

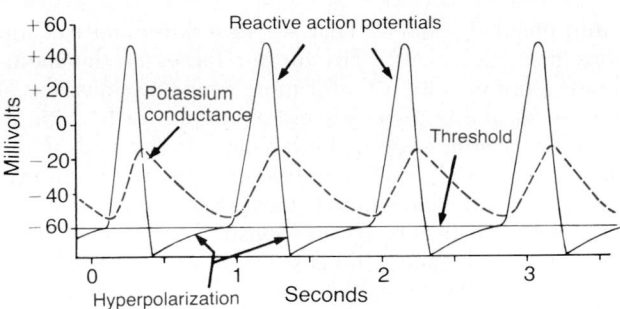

Figure 5–14. Rhythmic action potentials similar to those recorded in the rhythmical control center of the heart. Note their relationship to potassium conductance and to the state of hyperpolarization.

of the action potential, whereas the slow but prolonged opening of the slow channels is mainly responsible for the plateau portion of the action potential.

A second factor sometimes partly responsible for the plateau is that the voltage-gated potassium channels are even slower than usual to open, often not opening greatly until the very end of the plateau. This delays the return of the membrane potential toward the resting value. But then this opening of the potassium channels at the same time that the slow channels begin to close causes rapid return of the action potential from its plateau level back to the negative resting level, accounting for the rapid downslope at the end of the action potential.

RHYTHMICITY OF CERTAIN EXCITABLE TISSUES—REPETITIVE DISCHARGE

Repetitive self-induced discharges, or *rhythmicity,* occur normally in the heart, in most smooth muscle, and in many of the neurons of the central nervous system. These rhythmical discharges cause the rhythmical beat of the heart, peristalsis, and such neuronal events as the rhythmical control of breathing.

Also, almost all other excitable tissues can discharge repetitively if the threshold for stimulation is reduced low enough. For instance, even large nerve fibers and skeletal muscle fibers, which normally are highly stable, discharge repetitively when they are placed in a solution that contains the drug veratrine or when the calcium ion concentration falls below a critical value, both of which increase sodium permeability of the membrane.

Re-Excitation Process Necessary for Spontaneous Rhythmicity. For spontaneous rhythmicity to occur, the membrane, even in its natural state, must already be permeable enough to sodium ions (or to calcium and sodium ions through the slow calcium channels) to allow automatic membrane depolarization. Thus, Figure 5–14 shows that the "resting" membrane potential is only − 60 to − 70 millivolts. This is not enough negative voltage to keep the sodium and cal-

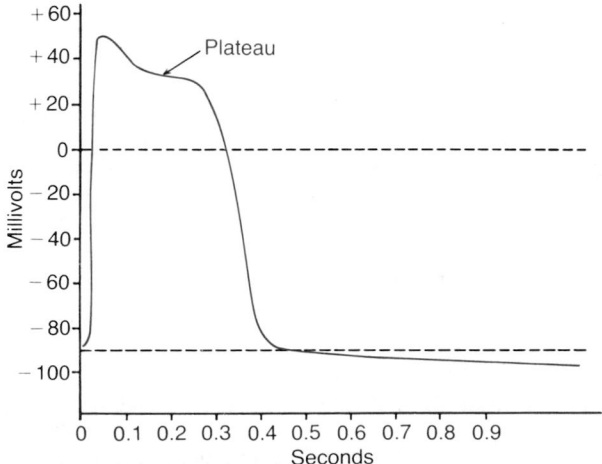

Figure 5–13. Action potential from a Purkinje fiber of the heart, showing a "plateau."

cium channels closed. That is, (1) sodium and calcium ions flow inward; (2) this further increases the membrane permeability; (3) still more ions flow inward; (4) the permeability increases more, and so forth, until an action potential is generated. Then, at the end of the action potential, the membrane repolarizes. Shortly thereafter, depolarization begins again, and a new action potential occurs spontaneously. This cycle continues again and again and causes self-induced rhythmical excitation of the excitable tissue.

Why does the membrane not depolarize immediately after it has become repolarized rather than delaying for nearly a second before the onset of the next action potential? The answer to this can be found by observing the curve labeled "potassium conductance" in Figure 5–14, which shows that toward the end of all action potentials and continuing for a short period thereafter, the membrane becomes excessively permeable to potassium. The excessive outflow of potassium ions carries tremendous numbers of positive charges to the outside of the membrane, creating inside the fiber considerably more negativity than would otherwise occur. This continues for a short period after the preceding action potential is over, thus drawing the membrane potential nearer to the potassium Nernst potential. This is a state called *hyperpolarization,* which is also shown in Figure 5–14. As long as this state exists, re-excitation will not occur; the excess potassium conductance (and the state of hyperpolarization) gradually disappears, as shown in the figure, thereby allowing the membrane potential again to increase until it reaches the *threshold* for excitation; then, suddenly, a new action potential results, the process occurring again and again.

SPECIAL ASPECTS OF SIGNAL TRANSMISSION IN NERVE TRUNKS

Myelinated and Unmyelinated Nerve Fibers. Figure 5–15 shows a cross section of a typical small nerve trunk, revealing many large nerve fibers that comprise most of the cross-sectional area. However, look carefully and you will see many more very small fibers lying between the large ones. The large fibers are *myelinated,* and the small ones are *unmyelinated.* The average nerve trunk contains about twice as many unmyelinated fibers as myelinated fibers.

Figure 5–16 shows a typical myelinated fiber. The central core of the fiber is the *axon,* and the membrane of the axon is the actual *conductive membrane* for conducting the action potential. The axon is filled in its center with *axoplasm,* which is a viscid intracellular fluid. Surrounding the axon is a *myelin sheath* that is often thicker than the axon itself, and about once every 1 to 3 millimeters along the length of the axon the myelin sheath is interrupted by a *node of Ranvier.*

The myelin sheath is deposited around the axon by Schwann cells in the following manner: The membrane of a Schwann cell first envelops the axon. Then the cell rotates around the axon many times, laying down multiple layers of cellular membrane containing the lipid substance *sphingomyelin.* This substance is an excellent insulator that decreases the ion flow through the membrane about 5000-fold and decreases the membrane capacitance as much as 50-fold. At the juncture between each two successive Schwann cells along the axon, a small uninsulated area only 2 to 3 *micro*meters in length remains where ions can still flow with ease between the extracellular fluid and the axon. This area is the node of Ranvier.

"Saltatory" Conduction in Myelinated Fibers from Node to Node. Even though ions cannot flow significantly through the thick myelin sheaths of myelinated nerves, they can flow with considerable ease through the nodes of Ranvier. Therefore, action potentials can occur *only at the nodes.* Yet the action potentials are conducted from node to node, as shown in Figure 5–17; this is called *saltatory conduction.* That is, electrical current flows through the surrounding extracellular fluids outside the myelin sheath as well as through the axoplasm from node to node, exciting successive nodes one after another. Thus, the nerve impulse jumps down the fiber, which is the origin of the term "saltatory."

Saltatory conduction is of value for two reasons. First, by causing the depolarization process to jump long intervals along the axis of the nerve fiber, this mechanism increases the velocity of nerve transmission in myelinated fibers as much as 5-fold to 50-fold. Second, saltatory conduction conserves energy for the axon because only the nodes depolarize, allowing perhaps a hundred times smaller loss of ions than would otherwise be necessary and therefore requiring little metabolism for re-establishing the sodium and potassium concentration differences across the membrane after a series of nerve impulses.

Still another feature of saltatory conduction in large myelinated fibers is the following: The excellent insulation afforded by the myelin membrane and the 50-fold decrease in membrane capacitance allow repolarization to occur with little transfer of ions. Therefore, at the end of the action potential, when the sodium channels begin to close, repolarization occurs so rapidly that the potassium channels usually have not yet begun to open significantly. Therefore, conduction of the nerve impulse in the myelinated nerve fiber is accomplished almost entirely by the sequential changes in the voltage-gated sodium channels, with little contribution by the potassium channels.

Velocity of Conduction in Nerve Fibers

The velocity of conduction in nerve fibers varies from as little as 0.25 m/sec in very small unmyelinated fibers to as high as 100 m/sec (the length of a football field in 1 second) in very large myelinated fibers. The velocity increases approximately with the fiber diameter in myelinated nerve fibers and approximately with the square root of fiber diameter in unmyelinated fibers.

EXCITATION–THE PROCESS OF ELICITING THE ACTION POTENTIAL

Basically, any factor that causes sodium ions to begin to diffuse inward through the membrane in sufficient numbers will set off the automatic regenerative opening

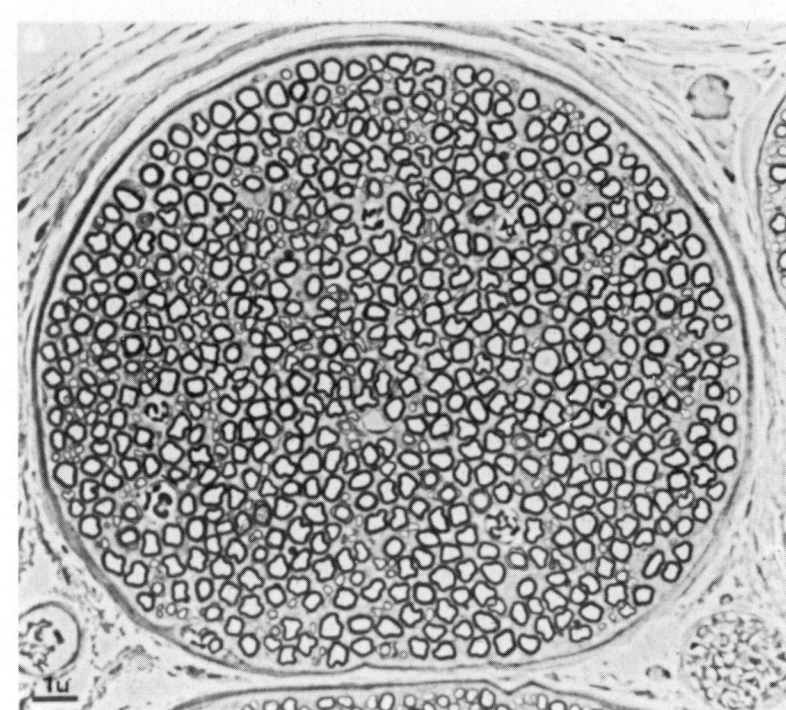

Figure 5–15. Cross section of a small nerve trunk containing myelinated and unmyelinated fibers.

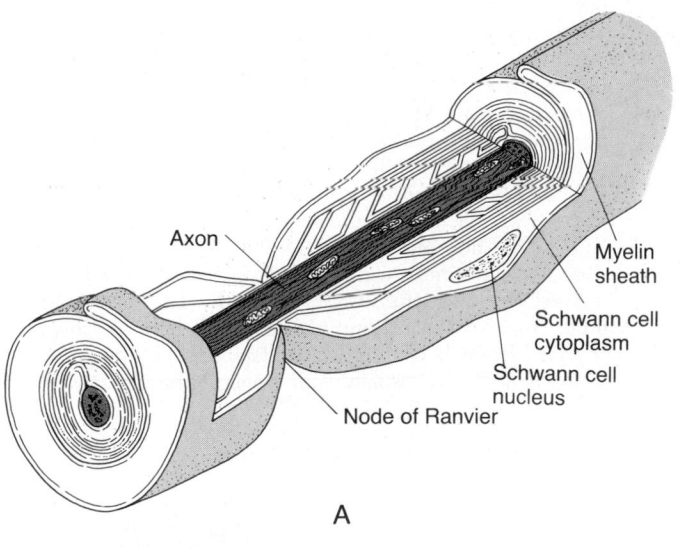

Axon

Myelin sheath

Schwann cell cytoplasm

Schwann cell nucleus

Node of Ranvier

A

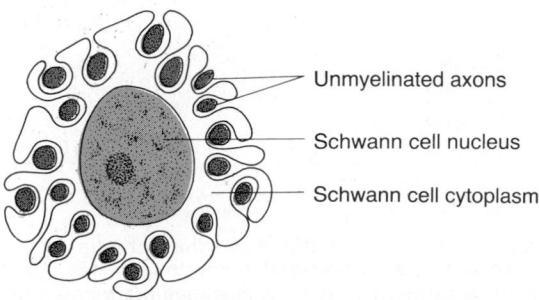

Unmyelinated axons

Schwann cell nucleus

Schwann cell cytoplasm

B

Figure 5–16. Function of the Schwann cell to insulate nerve fibers. *A,* The wrapping of a Schwann cell membrane around a large axon to form the myelin sheath of the myelinated nerve fiber. (Modified form Leeson and Leeson: Histology. Philadelphia, W. B. Saunders Company, 1979.) *B,* Partial wrapping of the membrane and cytoplasm of a Schwann cell around multiple unmyelinated nerve fibers (shown in cross section).

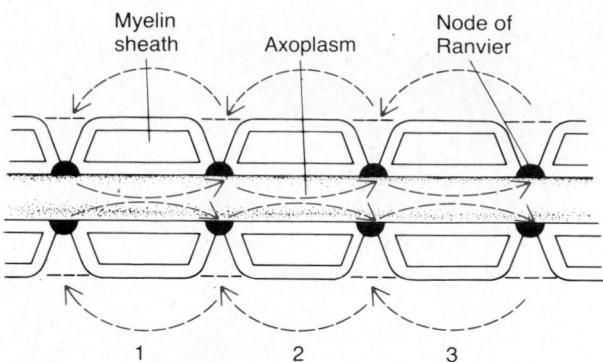

Myelin
sheath   Axoplasm   Node of
                                  Ranvier

1   2   3

Figure 5–17. Saltatory conduction along a myelinated axon.

of the sodium channels. This can result from simple *mechanical* disturbance of the membrane, *chemical* effects on the membrane, or passage of *electricity* through the membrane. All these are used at different points in the body to elicit nerve or muscle action potentials: mechanical pressure to excite sensory nerve endings in the skin, chemical neurotransmitters to transmit signals from one neuron to the next in the brain, and electrical current to transmit signals between muscle cells in the heart and intestine. For the purpose of understanding the excitation process, let us begin by discussing the principles of electrical stimulation.

EXCITATION OF A NERVE FIBER BY A NEGATIVELY CHARGED METAL ELECTRODE. The usual means for exciting a nerve or muscle in the experimental laboratory is to apply electricity at the nerve or muscle surface through two small electrodes, one of which is negatively charged and the other positively charged. When this is done, one finds that the excitable membrane becomes stimulated at the negative electrode.

The cause of these effects is the following: Remember that the action potential is initiated by the opening of voltage-gated sodium channels. Furthermore, these channels are opened by a decrease in the electrical voltage across the membrane. That is, the negative current from the electrode decreases the voltage on the outside of the membrane to a negative value nearer to the voltage of the negative membrane potential inside the fiber. This decreases the electrical voltage across the membrane and allows opening of the sodium channels, thus resulting in an action potential. Conversely, at the anode, the injection of positive charges on the outside of the nerve membrane heightens the voltage difference across the membrane rather than lessening it. This causes a state of "hyperpolarization," which actually decreases the excitability of the fiber rather than causing an action potential.

THRESHOLD FOR EXCITATION AND "ACUTE LOCAL POTENTIALS." A weak electrical stimulus may not be able to excite a fiber. However, when the voltage of the stimulus is increased, there comes a point at which excitation does take place. Figure 5–18 shows the effects of successively applied stimuli of progressing strength. A very weak stimulus at point A causes the membrane potential to change from −90 to −85 millivolts, but this is not sufficient change for the automatic regenerative processes of the action potential to develop. At point B, the stimulus is greater, but here again, the intensity still is not enough. The stimulus does disturb the membrane potential locally for as long

as 1 millisecond or more after both of these weak stimuli. These local potential changes are called *acute local potentials,* and when they fail to elicit an action potential, they are called *acute subthreshold potentials.*

At point C in Figure 5–18, the stimulus is even stronger. Now the local potential has reached barely the level to elicit an action potential, called the threshold level, but this occurs only after a short "latent period." At point D, the stimulus is still stronger, the acute local potential is also stronger, and the action potential occurs after less of a latent period.

Thus, this figure shows that even a very weak stimulus causes a local potential change at the membrane, but the intensity of the local potential must rise to a *threshold level* before the action potential is set off.

"Refractory Period" During Which New Stimuli Cannot Be Elicited

A new action potential cannot occur in an excitable fiber as long as the membrane is still depolarized from the preceding action potential. The reason for this is that shortly after the action potential is initiated, the sodium channels (or calcium channels, or both) become inactivated, and any amount of excitatory signal applied to these channels at this point will not open the inactivation gates. The only condition that will re-open them is for the membrane potential to return either to or almost to the original resting membrane potential level. Then, within another small fraction of a second, the inactivation gates of the channels open and a new action potential can then be initiated.

The period during which a second action potential cannot be elicited, even with a strong stimulus, is called the *absolute refractory period.* This period for large myelinated nerve fibers is about 1/2500 second. Therefore, one can readily calculate that such a fiber can carry a maximum of about 2500 impulses per second.

After the absolute refractory period is a *relative refractory period,* lasting about one quarter to one half as long as the absolute period. During this time, stronger than normal stimuli can excite the fiber. The cause of this relative refractoriness is twofold: (1) During this time, some of the sodium channels still have not been reversed from their inactivation state, and (2) the potassium channels are usually wide open at this time, caus-

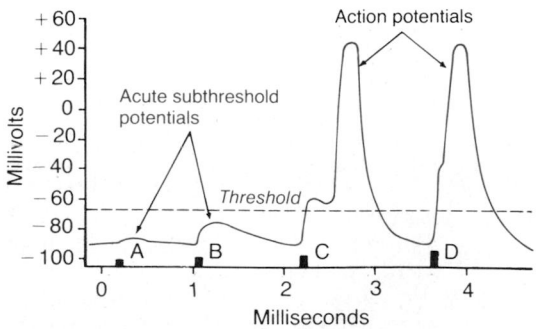

Figure 5–18. Effect of stimuli on the potential of the excitable membrane, showing the development of "acute subthreshold potentials" when the stimuli are below the threshold value required for eliciting an action potential.

ing greatly excess flow of positive potassium ion charges to the outside of the fiber, thus opposing much of the stimulating signal.

Inhibition of Excitability—"Stabilizers" and Local Anesthetics

In contrast to the factors that increase nerve excitability, still others, called *membrane-stabilizing factors, can decrease excitability. For instance, a high extracellular fluid calcium ion concentration* decreases the membrane permeability to sodium ions and simultaneously reduces its excitability. Therefore, calcium ions are said to be a "stabilizer."

LOCAL ANESTHETICS. Among the most important stabilizers are the many substances used clinically as local anesthetics, including *procaine* and *tetracaine*. Most of these act directly on the activation gates of the sodium channels, making it much more difficult for these gates to open and thereby reducing the membrane excitability. When the excitability has been reduced so low that the ratio of *action potential strength to excitability threshold* (called the "safety factor") is reduced below 1.0, a nerve impulse fails to pass through the anesthetized area.

RECORDING MEMBRANE POTENTIALS AND ACTION POTENTIALS

CATHODE RAY OSCILLOSCOPE. Earlier in this chapter, we noted that the membrane potential changes occur rapidly throughout the course of an action potential. Indeed, most of the action potential complex of large nerve fibers takes place in less than ¹⁄₁₀₀₀ second. In some figures of this chapter, an electrical meter has been shown recording these potential changes. However, it must be understood that any meter capable of recording most action potentials must be capable of responding extremely rapidly. For practical purposes, the only common type of meter that is capable of responding accurately to the rapid membrane potential changes is the cathode ray oscilloscope.

Figure 5–19 shows the basic components of a cath-

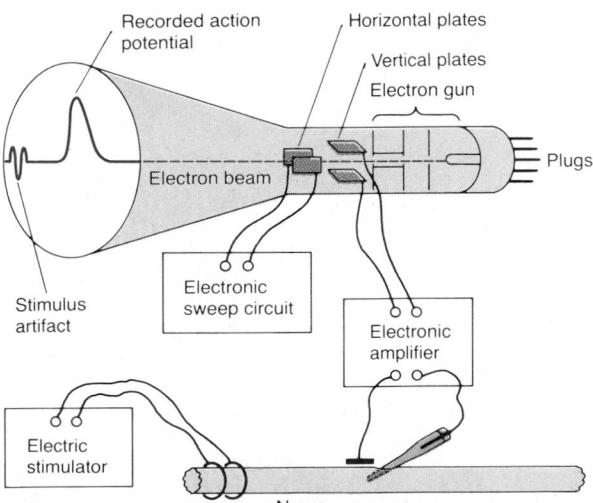

Figure 5–19. Cathode ray oscilloscope for recording transient action potentials.

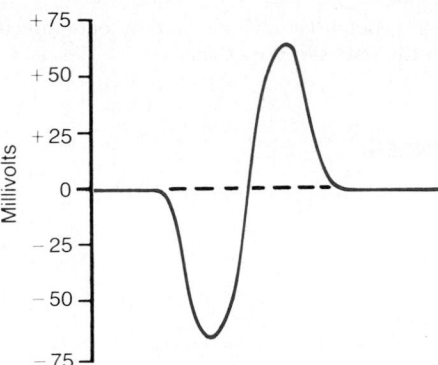

Figure 5–20. Recording of a biphasic action potential.

ode ray oscilloscope. The cathode ray tube itself is composed basically of an *electron gun* and a *fluorescent surface* against which electrons are fired. Where the electrons hit the surface, the fluorescent material glows. If the electron beam is moved across the surface, the spot of glowing light also moves and draws a fluorescent line on the screen.

In addition to the electron gun and fluorescent surface, the cathode ray tube is provided with two sets of electrically charged plates, one set positioned on the two sides of the electron beam and the other set positioned above and below. Appropriate electronic control circuits change the voltages on these plates so that the electron beam can be bent up or down in response to electrical signals coming from recording electrodes on nerves. The beam of electrons is also swept horizontally across the screen at a constant rate. This gives the record shown on the face of the cathode ray tube, giving a time base horizontally and the voltage changes at the nerve electrodes shown vertically. Note at the left end of the record a small *stimulus artifact* caused by the electrical stimulus used to elicit the action potential. Then further to the right is the recorded action potential itself.

RECORDING THE MONOPHASIC ACTION POTENTIAL. Throughout this chapter, "monophasic" action potentials have been shown in the different diagrams. To record these, a micropipette electrode, shown in Figure 5–2, is inserted into the interior of the fiber. Then, as the action potential spreads down the fiber, the changes in the potential inside the fiber are recorded, as shown in Figures 5–6, 5–10, and 5–13.

RECORDING A BIPHASIC ACTION POTENTIAL. When one wants to record impulses from a whole nerve trunk, it is not feasible to place electrodes inside the nerve fibers. Therefore, the usual method of recording is to place two electrodes on the outside of fibers. The record that is obtained is then usually biphasic for the following reasons: When an action potential moving down the nerve fiber reaches the first electrode, it becomes charged negatively, whereas the second electrode is still unaffected. This causes the oscilloscope to record in the negative direction. Then, as the action potential proceeds still farther down the nerve, there comes a point when the membrane beneath the first electrode becomes repolarized, whereas the second electrode is negative, and the oscilloscope records in the opposite direction. Thus, a graphic record approximating that shown in Figure 5–20 is recorded by the oscilloscope, showing a potential change first in one direction and then in the opposite direction.

REFERENCES

Aidley, D. J.: The Physiology of Excitable Cells. 3rd Ed. New York, Cambridge University Press, 1990.

Alvarez-Leefmans, F. J., et al.: Transmembrane ion movements elicited by sodium pump inhibition in *helix aspersa* neurons. J. Neurophysiol., 71: 1787, 1994.

Andersen, O. S., and Koeppe, R. E. II: Molecular determinants of channel function. Physiol. Rev., 72:(Suppl.) S89, 1992.

Armstrong, C. M.: Sodium channels and gating currents. Physiol. Rev., 61:644, 1981.

Armstrong, C. M.: Voltage-dependent ion channels and their gating. Physiol. Rev., 72:(Suppl.) S5, 1992.

Auerbach, A., and Sachs, F.: Patch clamp studies of single ionic channels. Annu. Rev. Biophys. Bioeng. 13:269, 1984.

Bkaily, G.: Membrane Physiopathology. Hingham, Mass., Kluwer Academic Publishers, 1994.

Boulton, A. A., et al.: Patch Clamp Techniques and Protocols. Totowa, NJ, Humana Press, Inc., 1995.

Boulton, A. A., et al.: Voltammetric Methods in Brain Tissue. Totowa, NJ, Humana Press, Inc., 1995.

Bretag, A. H.: Muscle chloride channels. Physiol. Rev., 67:618, 1987.

Byrne, J. H., and Schultz, S. G.: An Introduction to Membrane Transport and Bioelectricity, New York, Raven Press, 1988.

Byrne, J. H., and Schultz, S. G.: An Introduction to Membrane Transport and Bioelectricity: Foundations of General Physiology and Electrochemical Signaling. New York, Raven Press, 1994.

Catterall, W. A.: Cellular and molecular biology of voltage-gated sodium channels. Physiol. Rev., 72:(Suppl.) S15, 1992.

Cole, K. S.: Electrodiffusion models for the membrane of squid giant axon. Physiol. Rev., 45:340, 1965.

DeLisa, J. A., et al.: Manual of Nerve Conduction Velocity and Clinical Neurophysiology. New York, Raven Press, 1994.

DeWeer, P., et al.: Voltage dependence of the Na-K pump. Annu. Rev. Physiol., 50:225, 1988.

Gardner, D.: A time integral of membrane currents. Physiol. Rev., 72:(Suppl.) S1, 1992.

Heckman, C. J., et al.: Reduction in postsynaptic inhibition during maintained electrical stimulation of different nerves in the cat hindlimb. J. Neurophysiol., 71:2281, 1994.

Higashida, H., et al.: Molecular Basis of Ion Channels and Receptors Involved in Nerve Excitation, Synaptic Transmission, and Muscle. New York Academy of Sciences, 1994.

Hille, B.: Gating in sodium channels of nerve. Annu. Rev. Physiol., 38:139, 1976.

Hodgkin, A. L.: The Conduction of the Nervous Impulse. Springfield, Ill., Charles C Thomas, 1963.

Hodgkin, A. L., and Huxley, A. F.: Movement of sodium and potassium ions during nervous activity. Cold Spr. Harb. Symp. Quant. Biol., 17:43, 1952.

Hodgkin, A. L., and Huxley, A. F.: Quantitative description of membrane current and its application to conduction and excitation in nerve. J. Physiol. (Lond.), 117:500, 1952.

Honmou, O., et al.: Delayed depolarization and slow sodium currents in cutaneous afferents. J. Neurophysiol., 71:1627, 1994.

Huguenard, J., and McCormick, D.: Electrophysiology of the Neuron: A Companion to Shepherd's Neurobiology: An Interactive Tutorial. New York, Oxford University Press, 1994.

Kostyuk, P. G.: Calcium Ions in Nerve Cell Function. New York, Oxford University Press, 1992.

Krueger, B. K.: Toward an understanding of structure and function of ion channels. FASEB J., 3:1906, 1989.

Lang, F.: The Molecules of Transport. Farmington, Conn., S. Karger Publishers, Inc., 1993.

Melandri, B. A., et al.: Bioelectrochemistry IV: Nerve Muscle Function— Bioelectrochemistry, Mechanisms, Bioenergetics, and Control. New York, Plenum Publishing Corp., 1994.

Miller, R. J.: Multiple calcium channels and neuronal function. Science, 235:46, 1987.

Narahashi, T.: Ion Channels. New York, Plenum Publishing Corp., 1988.

Pallotta, B. S., and Wagoner, P. K.: Voltage-dependent potassium channels since Hodgkin and Huxley. Physiol. Rev., 72:(Suppl.) S49, 1992.

Patlak, J.: Molecular kinetics of voltage-dependent Na+ channels. Physiol. Rev., 71:1047, 1991.

Pongs, O.: Molecular biology of voltage-dependent potassium channels. Physiol. Rev., 72:(Suppl.) S69, 1992.

Pusch, M., and Jentsch, T. J.: Molecular physiology of voltage-gated chloride channels. Physiol. Rev., 74:813, 1994.

Reichert, H.: Introduction to Neurobiology. New York, Oxford University Press, 1993.

Ross, W. N.: Changes in intracellular calcium during neuron activity. Annu. Rev. Physiol., 51:491, 1989.

Schwarz, W.: The Molecules of Transport. Farmington, Conn., S. Karger Publishers, Inc., 1994.

Shepherd, G. M.: Foundations of the Neuron Doctrine. New York, Oxford University Press, 1991.

Shepherd, G. M.: Neurobiology. New York, Oxford University Press, 1994.

Terzis, J. K., and Smith, K. L.: The Peripheral Nerve: Structure, Function, and Reconstruction. New York, Raven Press, 1990.

Wallis, D. I.: Electrophysiology: A Practical Approach. New York, Oxford University Press, 1993.

Waxman, S. G., et al.: The Axon: Structure, Function, and Pathophysiology. New York, Oxford University Press, 1995.

Yoshimura, Y., and Tsumoto, T.: Dependence of LTP induction on postsynaptic depolarization: A perforated patch-clamp study in visual cortical slices of young rats. J. Neurophysiol., 71:1638, 1994.

Contraction of Skeletal Muscle

*C*HAPTER 6

About 40 per cent of the body is skeletal muscle, and perhaps another 10 per cent is smooth and cardiac muscle. Many of the same principles of contraction apply to all these different types of muscle, but in this chapter, the function of skeletal muscle is considered mainly; the specialized functions of smooth muscle are discussed in Chapter 8 and cardiac muscle, in Chapter 9.

PHYSIOLOGIC ANATOMY OF SKELETAL MUSCLE

Skeletal Muscle Fiber

Figure 6–1 shows the organization of skeletal muscle, demonstrating that all skeletal muscles are made of numerous fibers ranging from 10 to 80 micrometers in diameter. Each of these fibers in turn is made up of successively smaller subunits, also shown in Figure 6–1 and described in subsequent paragraphs.

In most muscles, the fibers extend the entire length of the muscle; except for about 2 per cent of the fibers, each is innervated by only one nerve ending, located near the middle of the fiber.

SARCOLEMMA. The sarcolemma is the cell membrane of the muscle fiber. The sarcolemma consists of a true cell membrane, called the *plasma membrane,* and an outer coat made up of a thin layer of polysaccharide material that contains numerous thin collagen fibrils. At the end of the muscle fiber, this surface layer of the sarcolemma fuses with a tendon fiber, and the tendon fibers in turn collect into bundles to form the muscle tendons and thence insert into the bones.

MYOFIBRILS; ACTIN AND MYOSIN FILAMENTS. Each muscle fiber contains several hundred to several thousand *myofibrils,* which are demonstrated by the many small open dots in the cross-sectional view of Figure 6–1C. Each myofibril (Figure 6–1D and E) in turn has, lying side by side, about 1500 *myosin filaments* and 3000 *actin filaments,* which are large polymerized protein molecules that are responsible for muscle contraction. These can be seen in longitudinal view in the electron micrograph of Figure 6–2 and are represented diagrammatically in Figure 6–1, parts E through L. The thick filaments in the diagrams are *myosin* and the thin filaments are *actin.*

Note that the myosin and actin filaments partially interdigitate and thus cause the myofibrils to have alternate light and dark bands. The light bands contain only actin filaments and are called *I bands* because they are *isotropic* to polarized light. The dark bands contain the myosin filaments as well as the ends of the actin filaments where they overlap the myosin and are called *A bands* because they are *anisotropic* to polarized light. Note also the small projections from the sides of the myosin filaments. These are *cross-bridges.* They protrude from the surfaces of the myosin filaments along the entire extent of the filament except in the very center. Interaction between these cross-bridges and the actin filaments causes contraction.

Figure 6–1E also shows that the ends of the actin filaments are attached to a so-called Z *disc.* From this disc, these filaments extend in both directions to interdigitate with the myosin filaments. The Z disc, which itself is composed of filamentous proteins different from the actin and myosin filaments, passes crosswise across the myofibril and also crosswise from myofibril

Skeletal Muscle

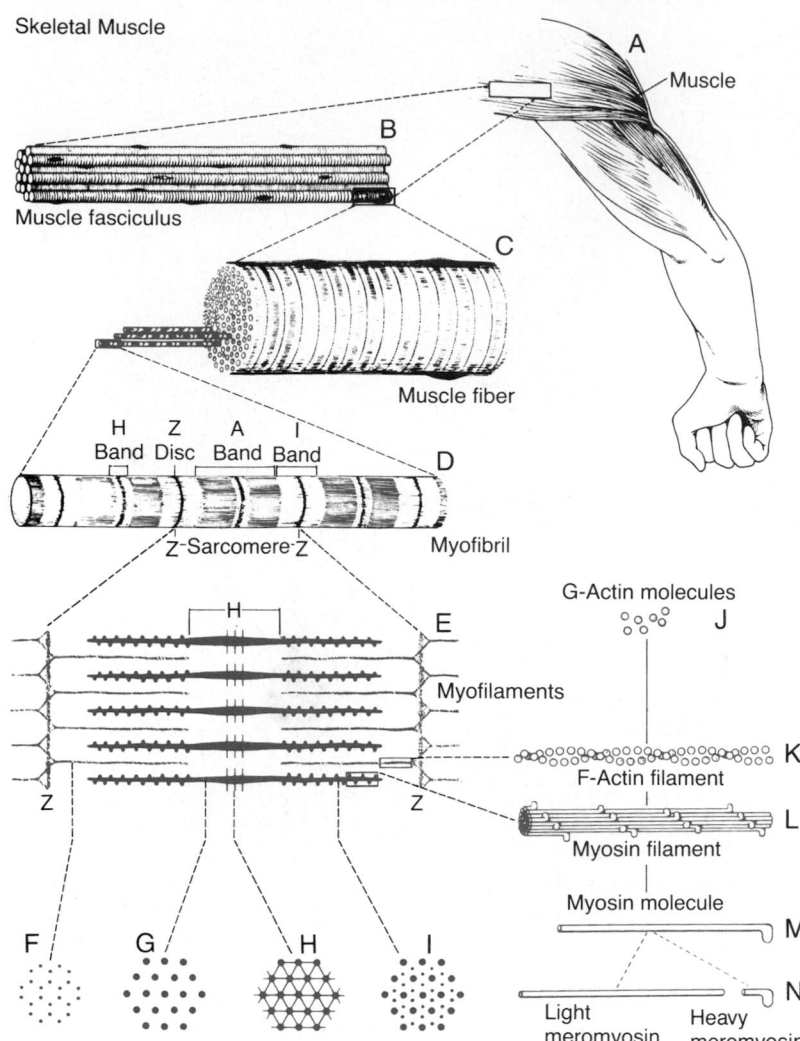

Figure 6–1. Organization of skeletal muscle, from the gross to the molecular level. *F, G, H,* and *I* are cross sections at the levels indicated. (Drawing by Sylvia Colard Keene. Modified from Fawcett: Bloom and Fawcett: A Textbook of Histology. Philadelphia, W. B. Saunders Company, 1986.)

to myofibril, attaching the myofibrils to one another all the way across the muscle fiber. Therefore, the entire muscle fiber has light and dark bands, as do the individual myofibrils. These bands give skeletal and cardiac muscle their striated appearance.

The portion of a myofibril (or of the whole muscle fiber) that lies between two successive Z discs is called a *sarcomere*. When the muscle fiber is at its normal, fully stretched resting length, the length of the sarcomere is about 2 micrometers. At this length, the actin filaments overlap the myosin filaments and are just beginning to overlap one another. We see later that at this length, the sarcomere also is capable of generating its greatest force of contraction.

SARCOPLASM. The myofibrils are suspended inside the muscle fiber in a matrix called *sarcoplasm,* which is composed of usual intracellular constituents. The fluid of the sarcoplasm contains large quantities of potassium, magnesium, phosphate, and protein enzymes. Also present are tremendous numbers of *mitochondria* that lie between and parallel to the myofibrils, a condition that is indicative of the great need of the contracting myofibrils for large amounts of adenosine triphosphate (ATP) formed by the mitochondria.

SARCOPLASMIC RETICULUM. Also in the sarcoplasm is an extensive endoplasmic reticulum, which in the muscle fiber is called the *sarcoplasmic reticulum.* This reticulum has a special organization that is extremely important in the control of muscle contraction, which is discussed in Chapter 7. The electron micrograph of Figure 6–3 shows the arrangement of this sarcoplasmic reticulum and how extensive it can be. The more rapidly contracting types of muscle have especially extensive sarcoplasmic reticula, indicating that this structure is important in causing rapid muscle contraction, as is also discussed later.

GENERAL MECHANISM OF MUSCLE CONTRACTION

The initiation and execution of muscle contraction occurs in the following sequential steps.

1. An action potential travels along a motor nerve to its endings on muscle fibers.

2. At each ending, the nerve secretes a small amount of the neurotransmitter substance *acetylcholine.*

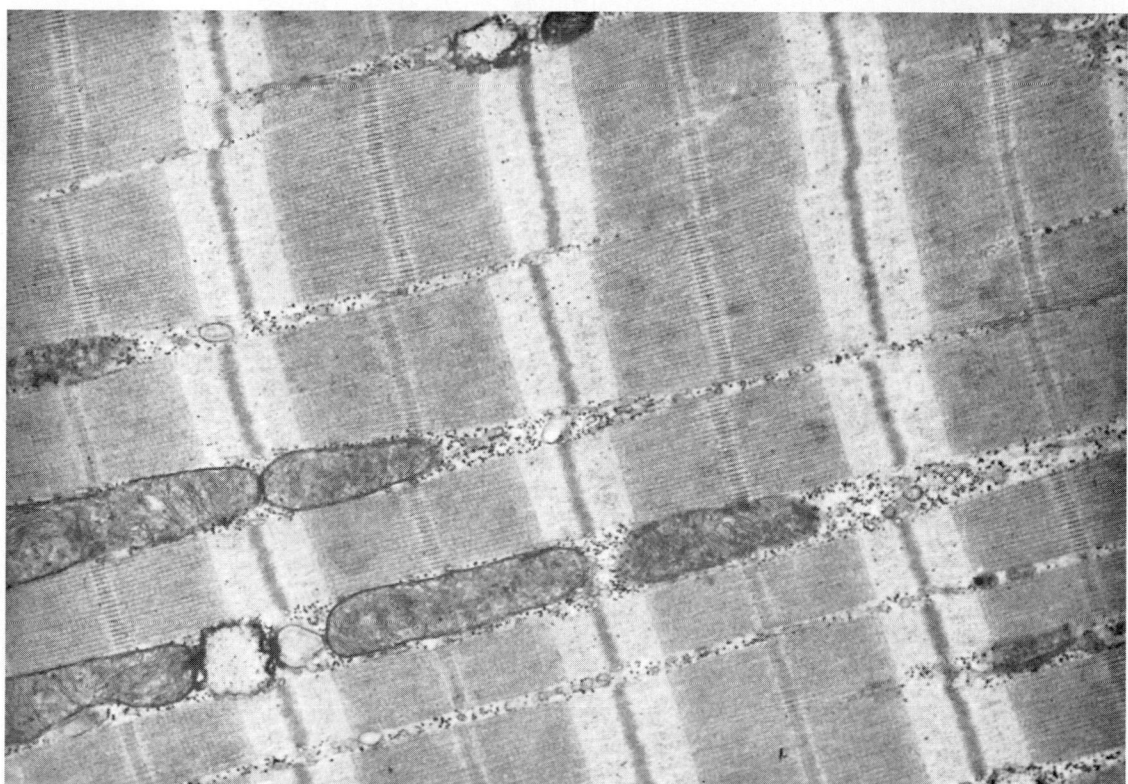

Figure 6–2. Electron micrograph of muscle myofibrils showing the detailed organization of actin and myosin filaments. Note the mitochondria lying between the myofibrils. (From Fawcett: The Cell. Philadelphia, W. B. Saunders Company, 1981.)

3. The acetylcholine acts on a local area of the muscle fiber membrane to open multiple acetylcholine-gated channels through protein molecules in the muscle fiber membrane.

4. Opening of the acetylcholine channels allows large quantities of sodium ions to flow to the interior of the mus-

cle fiber membrane at the point of the nerve terminal. This initiates an action potential in the muscle fiber.

5. The action potential travels along the muscle fiber membrane in the same way that action potentials travel along nerve membranes.

6. The action potential depolarizes the muscle fiber

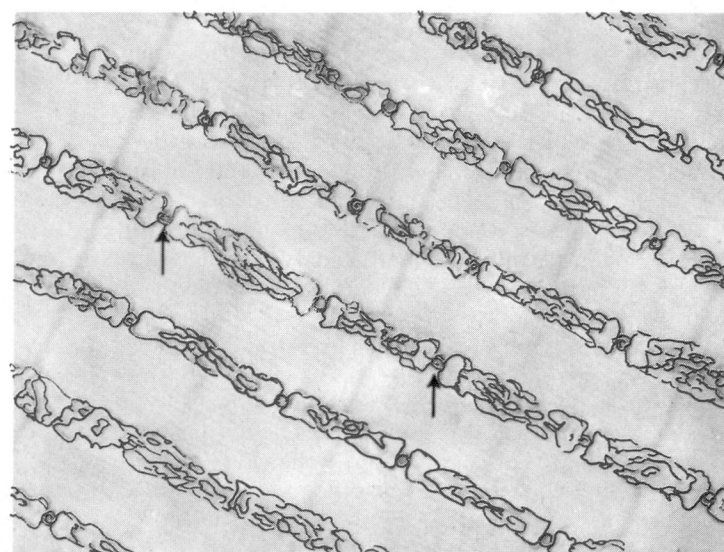

Figure 6–3. Sarcoplasmic reticulum surrounding the myofibrils showing the longitudinal system paralleling the myofibrils. Also shown in cross section are the T tubules *(arrows)* that lead to the exterior of the fiber membrane and that contain extracellular fluid. (From Fawcett: The Cell. Philadelphia, W. B. Saunders Company, 1981.)

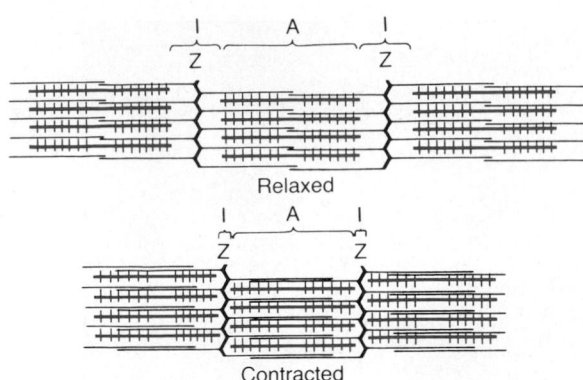

Figure 6–4. Relaxed and contracted states of a myofibril showing sliding of the actin filaments *(black)* into the spaces between the myosin filaments *(red).*

membrane and also travels deeply within the muscle fiber, where it causes the sarcoplasmic reticulum to release into the myofibrils large quantities of calcium ions that have been stored within the reticulum.

7. The calcium ions initiate attractive forces between the actin and myosin filaments, causing them to slide together, which is the contractile process.

8. After a fraction of a second, the calcium ions are pumped back into the sarcoplasmic reticulum, where they remain stored until a new muscle action potential comes along; this removal of the calcium ions from the myofibrils causes muscle contraction to cease.

We now describe the machinery of the contractile process but return to the details of muscle excitation in Chapter 7.

MOLECULAR MECHANISM OF MUSCLE CONTRACTION

SLIDING MECHANISM OF CONTRACTION. Figure 6–4 demonstrates the basic mechanism of muscle contraction. It shows the relaxed state of a sarcomere (top) and the contracted state (bottom). In the relaxed state, the ends of the actin filaments derived from two successive Z discs barely begin to overlap one another, while at the same time lying adjacent to the myosin filaments. On the other hand, in the contracted state, these actin filaments have been pulled inward among the myosin filaments, so that they now overlap one another to a major extent. Also, the Z discs have been pulled by the actin filaments up to the ends of the myosin filaments. Indeed, during intense contraction, the actin filaments can be pulled together so tightly that the ends of the myosin filaments buckle. Thus, muscle contraction occurs by a *sliding filament mechanism.*

But what causes the actin filaments to slide inward among the myosin filaments? This is caused by mechanical forces generated by the interaction of the cross-bridges of the myosin filaments with the actin filaments, as we discuss in the following sections. Under resting conditions, these forces are inhibited, but when an action potential travels over the muscle

fiber membrane, this causes the sarcoplasmic reticulum to release large quantities of calcium ions that rapidly penetrate the myofibrils. These calcium ions in turn activate the forces between the myosin and actin filaments, and contraction begins. But energy is also needed for the contractile process to proceed. This energy is derived from the high-energy bonds of ATP, which is degraded to adenosine diphosphate (ADP) to liberate the energy required.

In the next few sections, we describe what is known about the details of these molecular processes of contraction. To begin this discussion, however, we must characterize in detail the myosin and actin filaments.

Molecular Characteristics of the Contractile Filaments

MYOSIN FILAMENT. The myosin filament is composed of multiple myosin molecules, each having a molecular weight of about 480,000. Figure 6–5A shows an individual molecule; section B shows the organization of the molecules to form a myosin filament as well as interaction of this filament on one side with the ends of two actin filaments.

The *myosin molecule* is composed of six polypeptide chains, two *heavy chains* each with a molecular weight of about 200,000 and four *light chains* with molecular weights of about 20,000 each. The two heavy chains wrap spirally around each other to form a double helix. One end of each of these chains is folded into a globular polypeptide structure called the myosin *head.* Thus, there are two free heads lying side by side at one end of the double helix myosin molecule; the elongated portion of the coiled helix is called the *tail.* The four light chains are also parts of the myosin

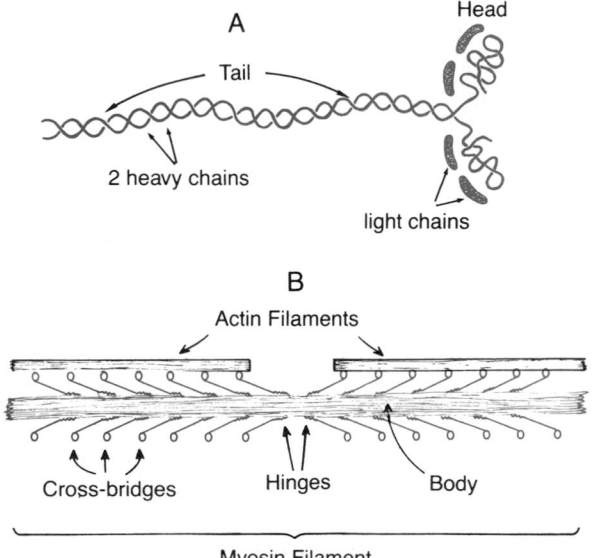

Figure 6–5. *A,* Myosin molecule. *B,* Combination of many myosin molecules to form a myosin filament. Also shown are the cross-bridges and the interaction between the heads of the cross-bridges and adjacent actin filaments.

heads, two to each head. These light chains help control the function of the head during muscle contraction.

The *myosin filament* is made up of 200 or more individual myosin molecules. The central portion of one of these filaments is shown in Figure 6–5B, displaying the tails of the myosin molecules bundled together to form the *body* of the filament, while many heads of the molecules hang outward to the sides of the body. Also, part of the helix portion of each myosin molecule extends to the side along with the head, thus providing an *arm* that extends the head outward from the body, as shown in the figure. The protruding arms and heads together are called *cross-bridges.* Each cross-bridge is believed to be flexible at two points called *hinges,* one where the arm leaves the body of the myosin filament and the other where the two heads attach to the arm. The hinged arms allow the heads either to be extended far outward from the body of the myosin filament or to be brought close to the body. The hinged heads are believed to participate in the actual contraction process, as we discuss in the following sections.

The total length of each myosin filament is uniform, almost exactly 1.6 micrometers. Note, however, that there are no cross-bridge heads in the very center of the myosin filament for a distance of about 0.2 micrometer because the hinged arms extend toward both ends of the myosin filament away from the center; therefore, in the center, there are only tails of the myosin molecules and no heads.

Now, to complete the picture, the myosin filament itself is twisted so that each successive set of cross-bridges is axially displaced from the previous set by 120 degrees. This ensures that the cross-bridges extend in all directions around the filament.

ATPase Activity of the Myosin Head. Another feature of the myosin head that is essential for muscle contraction is that it functions as an ATPase enzyme. As we see later, this property allows the head to cleave ATP and to use the energy derived from the ATP's high-energy phosphate bond to energize the contraction process.

ACTIN FILAMENT. The actin filament is also complex. It is composed of three protein components: *actin, tropomyosin,* and *troponin.*

The backbone of the actin filament is a double-stranded F-actin protein molecule, represented by the two lighter-colored strands in Figure 6–6. The two strands are wound in a helix in the same manner as the myosin molecule but with a complete revolution every 70 nanometers.

Each strand of the double F-actin helix is composed of polymerized G-actin molecules, each having a molecular weight of about 42,000. There are about 13 of these molecules in each revolution of each strand of helix. Attached to each one of the G-actin molecules is one molecule of ADP. It is believed that these ADP molecules are the active sites on the actin filaments with which the cross-bridges of the myosin filaments interact to cause muscle contraction. The active sites

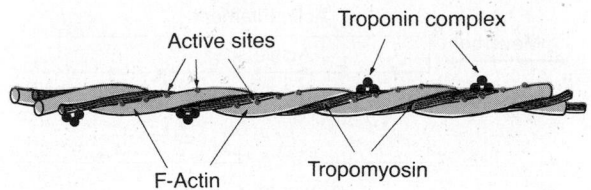

Figure 6–6. Actin filament, composed of two helical strands of F-actin and tropomyosin molecules that fit loosely in the grooves between the actin strands. Attached to one end of each tropomyosin molecule is a troponin complex that initiates contraction.

on the two F-actin strands of the double helix are staggered, giving one active site on the overall actin filament about every 2.7 nanometers.

Each actin filament is about 1 micrometer long. The bases of the actin filaments are inserted strongly into the Z discs, whereas the other ends protrude in both directions into the adjacent sarcomeres to lie in the spaces between the myosin molecules, as shown in Figure 6–4.

Tropomyosin Molecules. The actin filament also contains another protein, *tropomyosin.* Each molecule of tropomyosin has a molecular weight of 70,000 and a length of 40 nanometers. These molecules are connected loosely with the F-actin strands, wrapped spirally around the sides of the F-actin helix. In the resting state, the tropomyosin molecules are believed to lie on top of the active sites of the actin strands, so that attraction cannot occur between the actin and myosin filaments to cause contraction. Each tropomyosin molecule covers about seven of these active sites.

Troponin and Its Role in Muscle Contraction. Attached near one end of each tropomyosin molecule is still another protein molecule called *troponin.* This is actually a complex of three loosely bound protein subunits, each of which plays a specific role in the control of muscle contraction. One of the subunits (troponin I) has a strong affinity for actin, another (troponin T) for tropomyosin, and a third (troponin C) for calcium ions. This complex is believed to attach the tropomyosin to the actin. The strong affinity of the troponin for calcium ions is believed to initiate the contraction process, as is explained in the next section.

Interaction of Myosin, Actin Filaments, and Calcium Ions to Cause Contraction

INHIBITION OF THE ACTIN FILAMENT BY THE TROPONIN-TROPOMYOSIN COMPLEX; ACTIVATION BY CALCIUM IONS. A pure actin filament without the presence of the troponin-tropomyosin complex binds instantly and strongly with the heads of the myosin molecules in the presence of magnesium ions and ATP, both of which are normally abundant in the myofibril. If the troponin-tropomyosin complex is added to the actin filament, this binding does not take place. Therefore, it is believed that the active sites on the normal actin filament of the relaxed muscle are inhibited or physically covered by the troponin-tropomyosin complex.

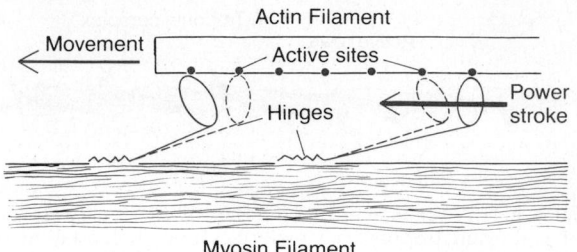

Figure 6–7. "Walk-along" mechanism for contraction of the muscle.

Consequently, the sites cannot attach to the heads of the myosin filaments to cause contraction. Before contraction can take place, the inhibitory effect of the troponin-tropomyosin complex must itself be inhibited.

This brings us to the role of the calcium ions. In the presence of large amounts of calcium ions, the inhibitory effect of the troponin-tropomyosin on the actin filaments is itself inhibited. The mechanism of this is not known, but one suggestion is the following: When calcium ions combine with troponin C, each molecule of which can bind strongly with up to four calcium ions, the troponin complex supposedly undergoes a conformational change that in some way tugs on the tropomyosin molecule and supposedly moves it deeper into the groove between the two actin strands. This "uncovers" the active sites of the actin, thus allowing contraction to proceed. Although this is a hypothetical mechanism, it does emphasize that the normal relation between the tropomyosin-troponin complex and actin is altered by calcium ions, producing a new condition that leads to contraction.

INTERACTION BETWEEN THE "ACTIVATED" ACTIN FILAMENT AND THE MYOSIN CROSS-BRIDGES—THE "WALK-ALONG" THEORY OF CONTRACTION. As soon as the actin filament becomes activated by the calcium ions, the heads of the cross-bridges from the myosin filaments become attracted to the active sites of the actin filament, and this, in some way, causes contraction to occur. Although the precise manner by which this interaction between the cross-bridges and the actin causes contraction is still partly theoretical, a suggested hypothesis for which considerable evidence exists is the "walk-along" theory (or *"ratchet" theory*) *of contraction.*

Figure 6–7 demonstrates the postulated walk-along mechanism for contraction. This figure shows the heads of two cross-bridges attaching to and disengaging from the active sites of an actin filament. It is postulated that when the head attaches to an active site, this attachment simultaneously causes profound changes in the intramolecular forces between the head and arm of the cross-bridge. The new alignment of forces causes the head to tilt toward the arm and to drag the actin filament along with it. This tilt of the head is called the *power stroke.* Then, immediately after tilting, the head automatically breaks away from the active site. Next, the head returns to its normal

perpendicular direction. In this position, it combines with a new active site farther down along the actin filament; then the head tilts again to cause a new power stroke, and the actin filament moves another step. Thus, the heads of the cross-bridges bend back and forth and step by step walk along the actin filament, pulling the ends of the actin filaments toward the center of the myosin filament.

Each one of the cross-bridges is believed to operate independently of all others, each attaching and pulling in a continuous but random cycle. Therefore, the greater the number of cross-bridges in contact with the actin filament at any given time, the greater, theoretically, is the force of contraction.

ATP AS THE SOURCE OF ENERGY FOR CONTRACTION —CHEMICAL EVENTS IN THE MOTION OF THE MYOSIN HEADS. When a muscle contracts, work is performed and energy is required. Large amounts of ATP are cleaved to form ADP during the contraction process. Furthermore, the greater the amount of work performed by the muscle, the greater the amount of ATP that is cleaved, which is called the *Fenn effect.* The following is the sequence of events that is believed to be the means by which this occurs:

1. Before contraction begins, the heads of the cross-bridges bind with ATP. The ATPase activity of the myosin head immediately cleaves the ATP but leaves the cleavage products, ADP plus Pi, bound to the head. In this state, the conformation of the head is such that it extends perpendicularly toward the actin filament but is not yet attached to the actin.

2. Next, when the troponin-tropomyosin complex binds with calcium ions, active sites on the actin filament are uncovered, and the myosin heads then bind with these, as shown in Figure 6–7.

3. The bond between the head of the cross-bridge and the active site of the actin filament causes a conformational change in the head, prompting the head to tilt toward the arm of the cross-bridge. This provides the *power stroke* for pulling the actin filament. The energy that activates the power stroke is the energy already stored, like a "cocked" spring, by the conformational change in the head when the ATP molecule had earlier been cleaved.

4. Once the head of the cross-bridge is tilted, this allows release of the ADP and Pi that were previously attached to the head; at the site of release of the ADP, a new molecule of ATP binds. This binding in turn causes detachment of the head from the actin.

5. After the head has detached from the actin, a new molecule of ATP is cleaved to begin the next cycle leading to the power stroke. That is, the energy again "cocks" the head back to its perpendicular condition ready to begin the new power stroke cycle.

6. Then, when the cocked head with its stored energy derived from the cleaved ATP binds with a new active site on the actin filament, it becomes uncocked and once again provides the power stroke.

Thus, the process proceeds again and again until the actin filament pulls the Z membrane up against the ends of the myosin filaments or until the load on the muscle becomes too great for further pulling to occur.

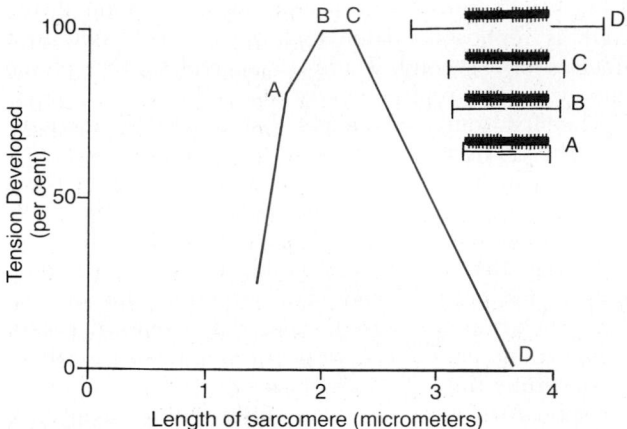

Figure 6–8. Length-tension diagram for a single sarcomere showing maximum strength of contraction when the sarcomere is 2.0 to 2.2 micrometers in length. At the upper right are shown the relative positions of the actin and myosin filaments at different sarcomere lengths from point A to point D. (Modified from Gordon, Huxley, and Julian: The length-tension diagram of single vertebrate striated muscle fibers. *J. Physiol., 171:28P,* 1964.)

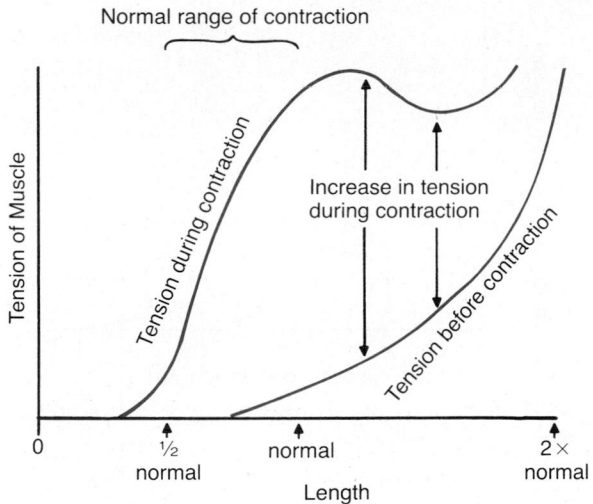

Figure 6–9. Relation of muscle length to force of contraction.

Degree of Actin and Myosin Filament Overlap—Effect on Tension Developed by the Contracting Muscle

Figure 6–8 shows the effect of sarcomere length and of myosin-actin filament overlap on the active tension developed by a contracting muscle fiber. To the right are shown different degrees of overlap of the myosin and actin filaments at different sarcomere lengths. At point D on the diagram, the actin filament has pulled all the way out to the end of the myosin filament with no overlap. At this point, the tension developed by the activated muscle is zero. Then, as the sarcomere shortens and the actin filament begins to overlap the myosin filament, the tension increases progressively until the sarcomere length decreases to about 2.2 micrometers. At this point, the actin filament has already overlapped all the cross-bridges of the myosin filament but has not yet reached the center of the myosin filament. On further shortening, the sarcomere maintains full tension until point B at a sarcomere length of about 2.0 micrometers. At this point, the ends of the two actin filaments begin to overlap each other, in addition to overlapping the myosin filaments. As the sarcomere length falls from 2 micrometers down to about 1.65 micrometers, at point A, the strength of contraction decreases. At this point, the two Z discs of the sarcomere abut the ends of the myosin filaments. Then, as contraction proceeds to still shorter sarcomere lengths, the ends of the myosin filaments are crumpled and, as shown in the figure, the strength of contraction decreases precipitously.

This diagram demonstrates that maximum contraction occurs when there is maximum overlap between the actin filaments and the cross-bridges of the myosin filaments, and it supports the idea that the greater the number of cross-bridges pulling the actin filaments, the greater the strength of contraction.

EFFECT OF MUSCLE LENGTH ON FORCE OF CONTRACTION IN THE INTACT MUSCLE. The upper curve of Figure 6–9 is similar to that in Figure 6–8, but this illustrates the intact, whole muscle rather than a single muscle fiber. The whole muscle has a large amount of connective tissue in it; also, the sarcomeres in different parts of the muscle do not necessarily contract in unison. Therefore, the curve has somewhat different dimensions from those shown for the individual muscle fiber, but it exhibits the same form.

Note in Figure 6–9 that when the muscle is at its normal resting length, which is at a sarcomere length of about 2 micrometers, it contracts with maximum force of contraction. If the muscle is stretched to much greater than normal length before contraction, a large amount of *resting tension* develops in the muscle even before contraction takes place; this tension results from the elastic forces of the connective tissue, sarcolemma, blood vessels, nerves, and so forth. However, the *increase* in tension during contraction, called *active tension,* decreases as the muscle is stretched much beyond its normal length—that is, to a sarcomere length greater than about 2.2 micrometers. This is demonstrated by the decrease in the arrow length in the figure at greater than normal length.

Relation of Velocity of Contraction to Load

A muscle contracts extremely rapidly when it contracts against no load—to a state of full contraction in about 0.1 second for the average muscle. When loads are applied, the velocity of contraction becomes progressively less as the load increases, as shown in Figure 6–10. When the load increases to equal the maximum force that the muscle can exert, the velocity of contraction becomes zero and no contraction results, despite activation of the muscle fiber.

This decreasing velocity with load is caused by the

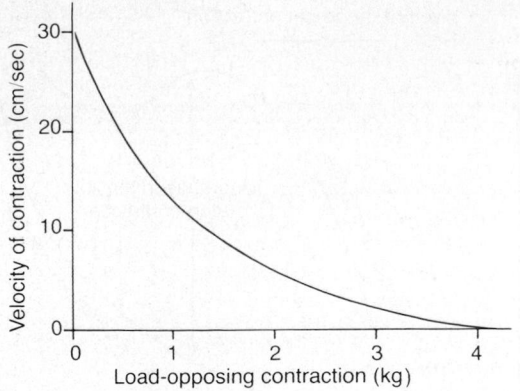

Figure 6–10. Relation of load to velocity of contraction in a skeletal muscle 8 centimeters long.

fact that a load on a contracting muscle is a reverse force that opposes the contractile force caused by muscle contraction. Therefore, the net force that is available to cause velocity of shortening is correspondingly reduced.

ENERGETICS OF MUSCLE CONTRACTION

Work Output During Muscle Contraction

When a muscle contracts against a load, it performs *work*. This means that *energy* is transferred from the muscle to the external load, for example, to lift an object to a greater height or to overcome resistance to movement.

In mathematical terms, work is defined by the following equation:

$$W = L \times D$$

in which W is the work output, L is the load, and D is the distance of movement against the load. The energy required to perform the work is derived from the chemical reactions in the muscle cells during contraction, as we describe in the following sections.

Sources of Energy for Muscle Contraction

We have already seen that muscle contraction depends on energy supplied by ATP. Most of this energy is required to actuate the walk-along mechanism by which the cross-bridges pull the actin filaments, but small amounts are required for (1) pumping calcium from the sarcoplasm into the sarcoplasmic reticulum after the contraction is over and (2) pumping sodium and potassium ions through the muscle fiber membrane to maintain an appropriate ionic environment for the propagation of action potentials.

The concentration of ATP present in the muscle fiber, about 4 millimolar, is sufficient to maintain full contraction for only 1 to 2 seconds at most. After the

ATP is split into ADP, as described in Chapter 2, the ADP is rephosphorylated to form new ATP within a fraction of a second. There are several sources of the energy for this rephosphorylation.

The first source of energy that is used to reconstitute the ATP is the substance *phosphocreatine*, which carries a high-energy phosphate bond similar to those of ATP. The high-energy phosphate bond of the phosphocreatine has a slightly higher amount of free energy than that of the ATP bond, as we discuss more fully in Chapters 67 and 72. Therefore, phosphocreatine is instantly cleaved, and the released energy causes bonding of a new phosphate ion to ADP to reconstitute the ATP. However, the total amount of phosphocreatine is also very little—only about five times as great as the ATP. Therefore, the combined energy of both the stored ATP and the phosphocreatine in the muscle is still capable of causing maximal muscle contraction for only 5 to 8 seconds.

The next important source of energy, which is used to reconstitute both ATP and phosphocreatine, is *glycogen* previously stored in the muscle cells. Rapid enzymatic breakdown of the glycogen to pyruvic acid and lactic acid liberates energy that is used to convert ADP to ATP, and the ATP can then be used directly to energize muscle contraction or to re-form the stores of phosphocreatine. The importance of this "glycolysis" mechanism is twofold. First, the glycolytic reactions occur even in the absence of oxygen, so that muscle contraction can be sustained for a short time when oxygen is not available. Second, the rate of formation of ATP by the glycolytic process is about two and one half times as rapid as ATP formation when the cellular foodstuffs react with oxygen. Unfortunately, so many end products of glycolysis accumulate in the muscle cells that glycolysis alone can sustain maximum muscle contraction for only about 1 minute.

The final source of energy is *oxidative metabolism*. This means the combining of oxygen with the various cellular foodstuffs to liberate ATP. More than 95 per cent of all energy used by the muscles for sustained, long-term contraction is derived from this source. The foodstuffs that are consumed are carbohydrates, fats, and protein. For extremely long-term maximal muscle activity—over a period of many hours—by far the greatest proportion of energy comes from fats, but for periods lasting 2 to 4 hours, as much as one half of the energy can come from stored glycogen before the glycogen is depleted.

The detailed mechanisms of these energetic processes are discussed in Chapters 67 through 72. In addition, the importance of the different mechanisms of energy release in different sports is discussed in Chapter 84 on sports physiology.

EFFICIENCY OF MUSCLE CONTRACTION. The efficiency of an engine or a motor is calculated as the percentage of energy input that is converted into work instead of heat. The percentage of the input energy to the muscle (the chemical energy in the nutrients) that can be converted into work, even under the best of conditions, is less than 25 per cent, the remainder be-

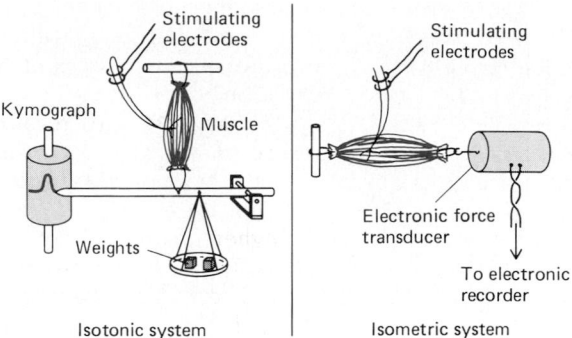

Figure 6–11. Isotonic and isometric recording systems.

coming heat. The reason for this low efficiency is that about one half of the energy in the foodstuffs is lost during the formation of ATP, and even then only 40 to 45 per cent of the energy in the ATP itself can later be converted into work.

Maximum efficiency can be realized only when the muscle contracts at a moderate velocity. If the muscle contracts slowly or without any movement, large amounts of *maintenance heat* are released during contraction even though little or no work is being performed, thereby decreasing the efficiency. On the other hand, if contraction is too rapid, large proportions of the energy are used to overcome the viscous friction within the muscle itself, and this, too, reduces the efficiency of contraction. Ordinarily, maximum efficiency is developed when the velocity of contraction is about 30 per cent of maximum.

CHARACTERISTICS OF WHOLE MUSCLE CONTRACTION

Many features of muscle contraction can be especially well demonstrated by eliciting single *muscle twitches.* This can be accomplished by instantaneous electrical excitation of the nerve to a muscle or by passing a short electrical stimulus through the muscle itself, giving rise to a single, sudden contraction lasting for a fraction of a second.

Isometric Versus Isotonic Contraction. Muscle contraction is said to be *isometric* when the muscle does not shorten during contraction and *isotonic* when it does shorten and with the tension on the muscle remaining constant. Systems for recording the two types of muscle contraction are shown in Figure 6–11.

In the isometric system, the muscle contracts against a force transducer without decreasing the muscle length, as exhibited to the right in Figure 6–11. In the isotonic system, the muscle shortens against a fixed load; this is exhibited to the left in the figure, showing a muscle lifting a pan of weights. The characteristics of isotonic contraction depend on the load against which the muscle contracts as well as on the inertia of the load. On the other hand, the isometric system records strictly changes in force of muscle contraction itself. Therefore, the isometric system is most often used when comparing the functional characteristics of different muscle types.

Series Elastic Component of Muscle Contraction. When muscle fibers contract against a load, those portions of the muscle that do not contract—the tendons, the sarcolemmal ends of the muscle fibers where they attach to the tendons, and perhaps even the hinged arms of the cross-bridges—stretch slightly as the tension increases. Consequently, the contractile part of the muscle must shorten an extra 3 to 5 per cent to make up for the stretch of these elements. The elements of the muscle that stretch during contraction are called the *series elastic component* of the muscle.

Characteristics of Isometric Twitches Recorded from Different Muscles

The body has many sizes of skeletal muscles—from the very small stapedius muscle in the middle ear with only a few millimeters length and a millimeter or so diameter up to the very large quadriceps muscle, a half million times as large as the stapedius. Furthermore, the fibers may be as small as 10 micrometers in diameter or as large as 80 micrometers. Finally, the energetics of muscle contraction vary considerably from one muscle to another. Therefore, it is no wonder that the characteristics of muscle contraction differ among muscles.

Figure 6–12 shows isometric contractions of three types of skeletal muscles: an ocular muscle, which has a duration of *isometric* contraction of less than ¼₀ second; the gastrocnemius muscle, which has a duration of contraction of about ¹⁄₁₅ second; and the soleus muscle, which has a duration of contraction of about ⅕ second. It is interesting that these durations of contractions are adapted to the function of each of the respective muscles. Ocular movements must be extremely rapid to maintain fixation of the eyes on specific objects, and the gastrocnemius muscle must contract moderately rapidly to provide sufficient velocity of limb movement for running and jumping, whereas the soleus muscle is concerned principally with slow contraction for continual support of the body against gravity.

Fast Versus Slow Muscle Fibers. As we discuss more fully in Chapter 84 on sports physiology, every muscle of the body is composed of a mixture of so-called *fast* and *slow* muscle fibers, with still other fibers

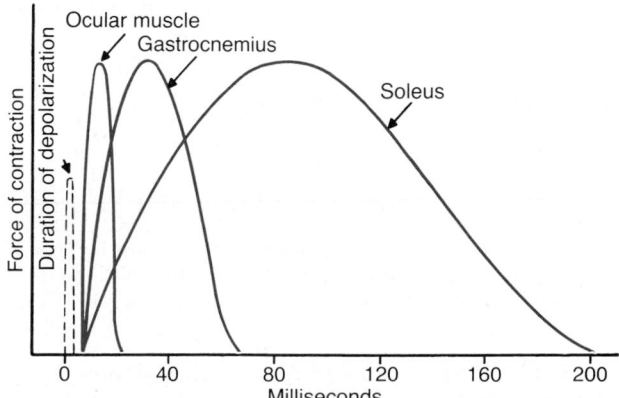

Figure 6–12. Duration of isometric contractions of different types of mammalian muscles, showing also a latent period between the action potential (depolarization) and muscle contraction.

gradated between these two extremes. The muscles that react rapidly are composed mainly of the fast fibers with only small numbers of the slow variety. Conversely, the muscles that respond slowly but with prolonged contraction are composed mainly of slow fibers. The differences between these two types of fibers are the following:

Fast fibers: (1) Much larger fibers for great strength of contraction. (2) Extensive sarcoplasmic reticulum for rapid release of calcium ions to initiate contraction. (3) Large amounts of glycolytic enzymes for rapid release of energy by the glycolytic process. (4) Less extensive blood supply because oxidative metabolism is of secondary importance. (5) Fewer mitochondria, also because oxidative metabolism is secondary.

Slow fibers: (1) Smaller fibers. (2) Also innervated by smaller nerve fibers. (3) More extensive blood vessel system and capillaries to supply extra amounts of oxygen. (4) Greatly increased numbers of mitochondria, also to support high levels of oxidative metabolism. (5) Fibers contain large amounts of myoglobin, an iron-containing protein similar to hemoglobin in red blood cells. Myoglobin combines with oxygen and stores it until needed, and it greatly speeds oxygen transport to the mitochondria. The myoglobin gives the slow muscle a reddish appearance and the name *red muscle*, whereas the deficit of the red myoglobin in fast muscle gives the name *white muscle*.

From these descriptions, one can see that the fast fibers are adapted for rapid and powerful muscle contractions, such as for jumping and short-distance powerful running. The slow fibers are adapted for prolonged, continued muscle activity, such as support of the body against gravity and long-continuing athletic events—for example, marathon races.

Mechanics of Skeletal Muscle Contraction

Motor Unit

Each motoneuron that leaves the spinal cord innervates many different muscle fibers, the number depending on the type of muscle. All the muscle fibers innervated by a single motor nerve fiber are called a *motor unit*. In general, small muscles that react rapidly and whose control must be exact have few muscle fibers (as few as two to three in some of the laryngeal muscles) in each motor unit. On the other hand, the large muscles that do not require very fine control, such as the soleus muscle, may have several hundred muscle fibers in a motor unit. An average figure for all the muscles of the body is questionable, but a good guess would be about 100 muscle fibers to the motor unit.

The muscle fibers in each motor unit are not bunched together in a muscle but instead overlap other motor units in microbundles of 3 to 15 fibers. This interdigitation allows the separate motor units to contract in support of one another rather than entirely as individual segments.

Muscle Contractions of Different Force—Force Summation

Summation means the adding together of individual twitch contractions to increase the intensity of overall muscle contraction. Summation occurs in two ways: (1)

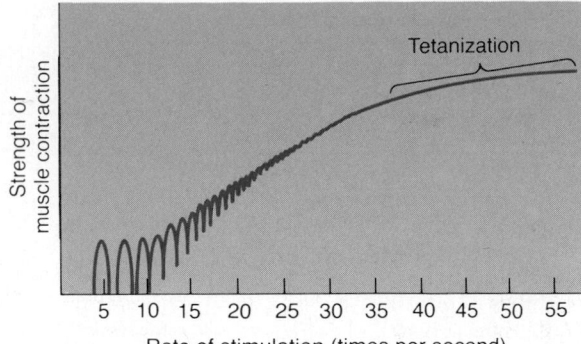

Figure 6–13. Frequency summation and tetanization.

by increasing the number of motor units contracting simultaneously, which is called *multiple fiber summation,* and (2) by increasing the frequency of contraction, which is called *frequency summation* and can lead to *tetanization.*

MULTIPLE FIBER SUMMATION. When the central nervous system sends a weak signal to contract a muscle, the motor units in the muscle that contain the smallest and fewest muscle fibers are stimulated in preference to the larger motor units. Then, as the strength of the signal increases, larger and larger motor units begin to be excited as well, with the largest motor units often having as much as 50 times as much contractile force as the smallest units. This is called the *size principle.* It is important because it allows the gradations of muscle force during weak contraction to occur in small steps, whereas the steps become progressively greater when large amounts of force are required. The cause of this size principle is that the smaller motor units are driven by small motor nerve fibers, and the small motoneurons in the spinal cord are far more excitable than the larger ones, so that they naturally are excited first.

Another important feature of multiple fiber summation is that the different motor units are driven asynchronously by the spinal cord, so that contraction alternates among motor units one after the other, thus providing smooth contraction even at low frequencies of nerve signals.

FREQUENCY SUMMATION AND TETANIZATION. Figure 6–13 shows the principles of frequency summation and tetanization. To the left are displayed individual twitch contractions occurring one after another at low frequency of stimulation. Then, as the frequency increases, there comes a point when each new contraction occurs before the preceding one is over. As a result, the second contraction is added partially to the first, so that the total strength of contraction rises progressively with increasing frequency. When the frequency reaches a critical level, the successive contractions are so rapid that they literally fuse together, and the contraction appears to be completely smooth and continuous, as shown in the figure. This is called *tetanization.* At still slightly higher frequency, the strength of contraction reaches its maximum, so that additional increase in frequency beyond that point has no further effect in increasing contractile force. This occurs because enough calcium ions are then maintained in the muscle sarcoplasm even between action potentials, so that the full contractile state is sustained without allowing relaxation between the action potentials.

MAXIMUM STRENGTH OF CONTRACTION. The maximum strength of tetanic contraction of a muscle operating at a normal muscle length averages between 3 and 4 kilograms per square centimeter of muscle, or 50 pounds per square inch. Because a quadriceps muscle can at times have as much as 16 square inches of muscle belly, as much as 800 pounds of tension may at times be applied to the patellar tendon. One can readily understand, therefore, how it is possible for muscles sometimes to pull their tendons out of the insertions in bones.

CHANGES IN MUSCLE STRENGTH AT THE ONSET OF CONTRACTION—THE STAIRCASE EFFECT (TREPPE). When a muscle begins to contract after a long period of rest, its initial strength of contraction may be as little as one half its strength 10 to 50 muscle twitches later. That is, the strength of contraction increases to a plateau, a phenomenon called the *staircase effect* or *treppe.*

Although all the possible causes of the staircase effect are not yet known, it is believed primarily to be caused by increase of calcium ions in the cytosol because of release of more and more ions from the sarcoplasmic reticulum with each muscle action potential and failure to recapture the ions immediately.

Skeletal Muscle Tone

Even when muscles are at rest, a certain amount of tautness usually remains. This is called *muscle tone.* Because skeletal muscle fibers do not contract without an action potential to stimulate the fibers (except in certain pathological conditions), skeletal muscle tone results entirely from a low rate of nerve impulses coming from the spinal cord. These in turn are controlled partly by impulses transmitted from the brain to the appropriate anterior motoneurons and partly by impulses that originate in *muscle spindles* located in the muscle itself. Both of these are discussed in relation to muscle spindle and spinal cord function in Chapter 54.

Muscle Fatigue

Prolonged and strong contraction of a muscle leads to the well-known state of muscle fatigue. Studies in athletes have shown that muscle fatigue increases in almost direct proportion to the rate of depletion of muscle glycogen. Therefore, most fatigue probably results simply from inability of the contractile and metabolic processes of the muscle fibers to continue supplying the same work output. However, experiments have also shown that transmission of the nerve signal through the neuromuscular junction, which is discussed in Chapter 7, diminishes after prolonged muscle activity, thus further diminishing muscle contraction.

Interruption of blood flow through a contracting muscle leads to almost complete muscle fatigue in 1 or more minutes because of the loss of nutrient supply, especially loss of oxygen.

Lever Systems of the Body

Muscles operate by applying tension to their points of insertion into bones, and the bones in turn form various types of lever systems. Figure 6–14 shows the lever

Figure 6–14. Lever system activated by the biceps muscle.

system activated by the biceps muscle to lift the forearm. If we assume that a large biceps muscle has a cross-sectional area of 6 square inches, the maximum force of contraction would be about 300 pounds. When the forearm is at right angles with the upper arm, the tendon attachment of the biceps is about 2 inches anterior to the fulcrum at the elbow and the total length of the forearm lever is about 14 inches. Therefore, the amount of lifting power that the biceps would have at the hand would be only one seventh of the 300 pounds force, or about 43 pounds. When the arm is fully extended, the attachment of the biceps is much less than 2 inches anterior to the fulcrum and the force with which the forearm can be brought forward is much less than 43 pounds.

In short, an analysis of the lever systems of the body depends on (1) a discrete knowledge of the point of muscle insertion and (2) its distance from the fulcrum of the lever as well as (3) the length of the lever arm and (4) the position of the lever. Many types of movement are required in the body, some of which need great strength and others large distances of movement. For this reason, there are all types of muscle; some are long and contract a long distance, and some are short but have large cross-sectional areas and therefore can provide extreme strengths of contraction over short distances. The study of different types of muscles, lever systems, and their movements is called *kinesiology* and is an important phase of human physioanatomy.

"Positioning" of a Body Part by Contraction of Antagonist Muscles on Opposite Sides of a Joint —"Coactivation" of Antagonist Muscles

Virtually all body movements are caused by simultaneous contraction of antagonist muscles on opposite sides of joints. This is called *coactivation* of the antagonist muscles, and it is controlled by the motor mechanisms of the brain and spinal cord.

The position of each separate part of the body, such as an arm or a leg, is determined by the relative degrees of contraction of the antagonist sets of muscles. For instance, let us assume that an arm or a leg is to be placed in a midrange position. To achieve this, antagonist muscles are excited about equally. Remember that an elongated muscle contracts with more force than a shortened muscle, which was demonstrated in Figure 6–9, showing maximum strength of contraction at full muscle length and almost no strength of contraction at half normal length. Therefore, the longer antagonist muscle contracts with far greater force than the shorter muscle. As the arm or leg moves toward its midposition, the strength of the longer muscle decreases, whereas the strength of the shorter muscle increases until the two strengths equal each other. At this point, movement of the arm or leg stops. Thus, by varying the ratios of the degree of activation of the antagonist muscles, the nervous system directs the positioning of the arm or leg.

We learn in Chapter 54 that the motor nervous system has additional important mechanisms to compensate for different muscle loads when directing this positioning process.

Remodeling of Muscle to Match Function

All the muscles of the body are continually being remodeled to match the functions that are required of them. Their diameters are altered, their lengths are altered, their strengths are altered, their vascular supplies are altered, and even the types of muscle fibers are altered at least slightly. This remodeling process is often quite rapid, within a few weeks. Indeed, experiments have shown that even normally, the muscle contractile proteins can be totally replaced in as little as 2 weeks.

Muscle Hypertrophy and Muscle Atrophy

When the total mass of a muscle enlarges, this is called *muscle hypertrophy*. When it decreases, the process is called *muscle atrophy*.

Virtually all muscle hypertrophy results from increase in the number of actin and myosin filaments in each muscle fiber, thus causing enlargement of the individual muscle fibers, which is called simply *fiber hypertrophy*. This usually occurs in response to contraction of a muscle at maximal or almost maximal force. Hypertrophy occurs to a much greater extent when the muscle is simultaneously stretched during the contractile process. Only a few such strong contractions each day are required to cause almost maximum hypertrophy within 6 to 10 weeks.

The manner in which forceful contraction leads to hypertrophy is not known. It is known, however, that the rate of synthesis of muscle contractile proteins is far greater during developing hypertrophy than their rate of decay, leading to the progressively greater numbers of both actin and myosin filaments in the myofibrils. In turn, the myofibrils themselves split within each muscle fiber to form new myofibrils. Thus, it is mainly this great increase in numbers of additional myofibrils that causes muscle fibers to hypertrophy.

Along with the increasing numbers of myofibrils, the enzyme systems that provide energy also increase. This is especially true of the enzymes for glycolysis, allowing a rapid supply of energy during short-term forceful muscle contraction.

When a muscle remains unused for a long period, the rate of decay of the contractile proteins as well as the numbers of myofibrils occur more rapidly than the rate of replacement. Therefore, muscle atrophy occurs.

ADJUSTMENT OF MUSCLE LENGTH. Another type of hypertrophy occurs when muscles are stretched to a greater-than-normal length. This causes new sarcomeres to be added at the ends of the muscle fibers where they attach to the tendons. In fact, new sarcomeres can be added as rapidly as several per minute, illustrating the rapidity of this type of hypertrophy.

Conversely, when a muscle remains shortened continually to less than its normal length, sarcomeres at the ends of the muscle fibers disappear approximately equally as rapidly. It is by these processes that muscles are continually remolded to have the appropriate length for proper muscle contraction.

HYPERPLASIA OF MUSCLE FIBERS. Under rare conditions of extreme muscle force generation, the actual numbers of muscle fibers have been observed to increase, but by only a few percentage points, in addition to the fiber hypertrophy process. This increase in fiber numbers is called *fiber hyperplasia*. When it does occur, the mechanism is linear splitting of previously enlarged fibers.

Effects of Muscle Denervation

When a muscle loses its nerve supply, it no longer receives the contractile signals that are required to maintain normal muscle size. Therefore, atrophy begins almost immediately. After about 2 months, degenerative changes also begin to appear in the muscle fibers themselves. If the nerve supply grows back to the muscle, full return of function usually occurs in about 3 months, but from that time onward, the capability of functional return becomes less and less, with no return of function after 1 to 2 years.

In the final stage of denervation atrophy, most of the muscle fibers are destroyed and replaced by fibrous and fatty tissue. The fibers that do remain are composed of a long cell membrane with a line-up of muscle cell nuclei but with no contractile properties and with no capability of regenerating myofibrils if a nerve regrows.

The fibrous tissue that replaces the muscle fibers during denervation atrophy has a tendency to continue shortening for many months, which is called *contracture*. Therefore, one of the most important problems in the practice of physical therapy is to keep atrophying muscles from developing debilitating and disfiguring contracture. This is achieved by daily stretching of the muscles or use of appliances that keep the muscles stretched during the atrophying process.

RECOVERY OF MUSCLE CONTRACTION IN POLIOMYELITIS: DEVELOPMENT OF MACROMOTOR UNITS. When some nerve fibers to a muscle are destroyed, as commonly occurs in poliomyelitis, the remaining nerve fibers sprout forth new axons to form many new branches that then innervate many of the paralyzed muscle fibers. This causes large motor units called *macromotor units*, which contain as many as five times the normal number of muscle fibers for each motoneuron in the spinal cord. This decreases the fineness of control that one has over the muscles but allows the muscles to regain strength.

Rigor Mortis

Several hours after death, all the muscles of the body go into a state of *contracture* called rigor mortis; that is, the muscle contracts and becomes rigid even without action potentials. This rigidity is caused by loss of all the ATP, which is required to cause separation of the cross-bridges from the actin filaments during the relaxation process. The muscles remain in rigor until the muscle proteins are destroyed, which usually results from autolysis caused by enzymes released from the lysosomes some 15 to 25 hours later, the process occurring more rapidly at higher temperature.

REFERENCES

Bkaily, G.: Membrane Physiopathology. Hingham, Mass., Kluwer Academic Publishers, 1994.

Engel, A. G., and Franzini-Armstrong, C.: Myology. 2nd Ed. Blue Ridge Summit, Pa., McGraw-Hill, 1994.

Fitts, R. H.: Cellular mechanisms of muscle fatigue. Physiol. Rev., 74:49, 1994.

Franzini-Armstrong, C., and Jorgensen, A. O.: Structure and development of E-C coupling units in skeletal muscle. Annu. Rev. Physiol., 56:509, 1994.

Fu, W.-M., and Huang, F.-L.: L-type CA^{2+} channel is involved in the regulation of spontaneous transmitter release at developing neuromuscular synapses. Neuroscience, 58:131, 1994.

Hanson, S. M., and McGinnis, M. E.: Regeneration of rat sciatic nerves in silicone tubes: Characterization of the response to low intensity d. c. stimulation. Neuroscience, 58:41, 1994.

Haynes, D. H., and Mandveno, A.: Computer modeling of Ca^{2+}-Mg^{2+}-ATPase of sarcoplasmic reticulum. Physiol. Rev., 67:244, 1987.

Hertling and Kessler: Management of Common Musculoskeletal Disorders: Physical Therapy Principles and Methods. Philadelphia, J. B. Lippincott, 1990.

Hinrichsen, R. D.: Ca2-dependent K Channels. Boca Raton, Fla., CRC Press, Inc., 1995.

Hochachka, P. W.: Muscles as Molecular and Metabolic Machines. Boca Raton, Fla., CRC Press, Inc., 1994.

Huang, C. L.-H.: Intramembrane Charge Movements in Striated Muscle. New York, Oxford University Press, 1993.

Hudicka, O., et al.: Angiogenesis in skeletal and cardiac muscle. Physiol. Rev., 72:369, 1992.

Huxley, A. F.: Muscular Contraction. Annu. Rev. Physiol., 50:1, 1988.

Huxley, A. F., and Gordon, A. M.: Striation patterns in active and passive shortening of muscle. Nature (Lond.) 193:280, 1962.

Huxley, H. E., and Faruqi, A. R.: Time-resolved x-ray diffraction studies on vertebrate striated muscle. Annu. Rev. Biophys. Bioeng., 12:381, 1983.

Janmey, P. A.: Phosphoinositides and calcium as regulators of cellular actin assembly and disassembly. Annu. Rev. Physiol., 56:169, 1994.

Johnson, E. W. (ed.): Practical Electromyography. Baltimore, Williams & Wilkins, 1988.

Kakulas, Byron A., et al.: Duchenne Muscular Dystrophy: Animal Models and Genetic Manipulation. New York, Raven Press, 1992.

Kelly, A. M., and Blau, H. M.: Neuromuscular Development and Disease. New York, Raven Press, 1992.

Keynes, R. D.: Nerve and Muscle. 2nd Ed. New York, Cambridge University Press, 1992.

Leach, D., and Schamhardt, H. C.: Animal Locomotion. Farmington, Conn., S. Karger Publishers, Inc., 1993.

Lieber, R. L.: Skeletal Muscle Structure and Function. Baltimore, Williams & Wilkins, 1992.

Lieber, R. L., et al.: *In vivo* measurement of human wrist extensor muscle sarcomere length changes. J. Neurophysiol., 71:874, 1994.

LeVeau, B. F.: Williams & Lissner's Biomechanics of Human Motion. Philadelphia, W. B. Saunders Co., 1992.

Levitan, I. B.: Modulation of ion channels by protein phosphorylation and dephosphorylation. Annu. Rev. Physiol., 56:193, 1994.

Meissner, G.: Ryanodine receptor/Ca^{2+} release channels and their regulation by endogenous effectors. Annu. Rev. Physiol., 56:485, 1994.

Melandri, B. A., et al.: Bioelectrochemistry IV: Nerve Muscle Function—Bioelectrochemistry, Mechanisms, Bioenergetics, and Control. New York, Plenum Publishing Corp., 1994.

Oho, S. J.: Clinical Electromyography: Nerve Conduction Studies. Baltimore, Williams & Wilkins, 1993.

Partridge, L. D., and Benton, L. A.: Muscle, the motor. In Brooks, V. B. (ed.): Handbook of Physiology. Sec. 1, Vol. II. Bethesda, American Physiological Society, 1981, p. 43.

Penn, A. S., et al.: Myasthenia Gravis and Related Disorders: Experimental and Clinical Aspects. New York Academy of Sciences, 1993.

Pollack, G. H.: The cross-bridge theory. Physiol. Rev., 63:1049, 1983.

Pusch, M., and Jentsch, T. J.: Molecular physiology of voltage-gated chloride channels. Physiol. Rev., 74:813, 1994.

Ríos, E., and Pizarro, G.: Voltage sensor of excitation-contraction coupling in skeletal muscle. Physiol. Rev., 71:849, 1994.

Schneider, M. F.: Control of calcium release in functioning skeletal muscle fibers. Annu. Rev. Physiol., 56:463, 1994.

Scott, J. J. A., et al.: Correlating resting discharge with small signal sensitivity and discharge variability in primary endings of cat soleus muscle spindles. J. Neurophysiol., 71:309, 1994.

Sealfon, S. C.: Receptor Molecular Biology. San Diego, Academic Press, 1995.

Seil, F. J.: Neural Injury and Regeneration. New York, Raven Press, 1993.

Silverman, D. G.: Neuromuscular Blockade. Philadelphia, J. B. Lippincott, 1994.

Simmons, R. M.: Muscular Contraction. New York, Cambridge University Press, 1992.

Van der Kloot, W., and Molgó, J.: Quantal acetylcholine release at the vertebrate neuromuscular junction. Physiol. Rev., 74:899, 1994.

Walsh, E. G.: Muscles, Masses and Motion: The Physiology of Normality, Hypotonicity, Spasticity and Rigidity. New York, Cambridge University Press, 1993.

Wasserman, K., et al.: Principles of Exercise Testing and Interpretation. Baltimore, Williams & Wilkins, 1994.

Weinstein, H., and Mehler, E. L.: Ca^{2+}-binding and structural dynamics in the functions of calmodulin. Annu. Rev. Physiol., 56:213, 1994.

Windhorst, U., et al.: Signal transmission from motor axons to group Ia muscle spindle afferents: Frequency responses and second-order non-linearities. Neuroscience, 59:149, 1994.

Excitation of Skeletal Muscle: A. Neuromuscular Transmission and B. Excitation-Contraction Coupling

CHAPTER 7

TRANSMISSION OF IMPULSES FROM NERVES TO SKELETAL MUSCLE FIBERS: NEUROMUSCULAR JUNCTION

The skeletal muscle fibers are innervated by large, myelinated nerve fibers that originate in the large motoneurons of the anterior horns of the spinal cord. As pointed out in Chapter 6, each nerve fiber normally branches many times and stimulates from three to several hundred skeletal muscle fibers. The nerve ending makes a junction, called the *neuromuscular junction,* with the muscle fiber near the fiber's midpoint, and the action potential in the fiber travels in both directions toward the muscle fiber ends. With the exception of about 2 per cent of the muscle fibers, there is only one such junction per muscle fiber.

PHYSIOLOGICAL ANATOMY OF THE NEUROMUSCULAR JUNCTION—MOTOR END PLATE. Figure 7–1A and B show the neuromuscular junction between a large, myelinated nerve fiber and a skeletal muscle fiber. The nerve fiber branches at its end to form a complex of branching nerve *terminals,* which invaginate into the muscle fiber but lie outside the muscle fiber plasma membrane. The entire structure is called the *motor end plate.* It is covered by one or more Schwann cells that insulate it from the surrounding fluids.

Figure 7–1C shows an electron micrographic sketch of the junction between a single-branch axon terminal and the muscle fiber membrane. The invagination of the membrane is called the *synaptic gutter* or *synaptic trough,* and the space between the terminal and the fiber membrane is called the *synaptic cleft.* This space is 20 to 30 nanometers thick and is occupied by a thin layer of spongy reticular fibers called the *basal lamina* through which diffuses extracellular fluid. At the bottom of the gutter are numerous smaller *folds* of the muscle membrane called *subneural clefts,* which greatly increase the surface area at which the synaptic transmitter can act.

In the axon terminal are many mitochondria that supply energy mainly for synthesis of the excitatory transmitter *acetylcholine.* The acetylcholine in turn excites the muscle fiber. Acetylcholine is synthesized in the cytoplasm of the terminal but is rapidly absorbed into many small synaptic vesicles, about 300,000 of which are normally in the terminals of a single end plate. Attached to the matrix of the basal lamina are large quantities of the enzyme *acetylcholinesterase,* which is capable of destroying acetylcholine.

Secretion of Acetylcholine by the Nerve Terminals

When a nerve impulse reaches the neuromuscular junction, about 125 vesicles of acetylcholine are released from the terminals into the synaptic space. Some of the details of this mechanism can be seen in Figure 7–2, which shows an expanded view of a synaptic trough with the neural membrane above and the muscle membrane and its subneural clefts below.

On the inside surface of the neural membrane are linear *dense bars,* shown in cross section in Figure 7–2. To each side of each dense bar are protein particles that penetrate the membrane, believed to be volt-

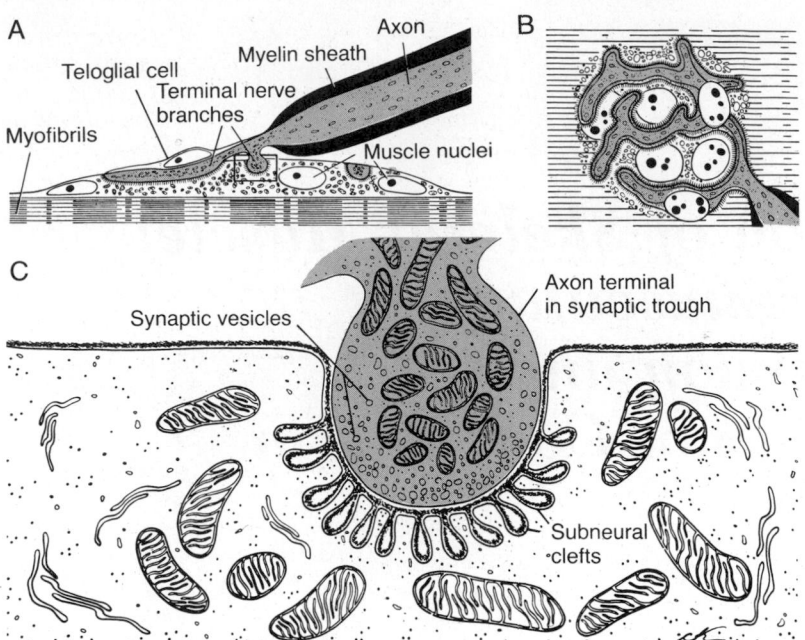

A

Teloglial cell
Terminal nerve branches
Myofibrils
Axon
Myelin sheath
Muscle nuclei

B

C

Synaptic vesicles
Axon terminal in synaptic trough
Subneural clefts

Figure 7–1. Different views of the motor end plate. *A*, Longitudinal section through the end plate. *B*, Surface view of the end plate. *C*, Electron micrographic appearance of the contact point between one of the axon terminals and the muscle fiber membrane, representing the rectangular area shown in *A*. (From Fawcett, as modified from R. Couteaux: Bloom and Fawcett: A Textbook of Histology. Philadelphia, W. B. Saunders Company, 1986.)

age-gated calcium channels. When the action potential spreads over the terminal, these channels open and allow large quantities of calcium to diffuse to the interior of the terminal. The calcium ions in turn are believed to exert an attractive influence on the acetylcholine vesicles, drawing them to the neural membrane adjacent to the dense bars. Some of the vesicles fuse with the neural membrane and empty their acetylcholine into the synaptic space by the process of *exocytosis*.

Although some of the aforementioned details are speculative, it is known that the effective stimulus for causing acetylcholine release from the vesicles is entry of calcium ions. Furthermore, the vesicles are emptied through the membrane adjacent to the dense bars.

EFFECT OF ACETYLCHOLINE ON THE POSTSYNAPTIC MEMBRANE TO OPEN ACETYLCHOLINE-GATED ION CHANNELS. Figure 7–2 shows many acetylcholine receptors in the muscle membrane; these are actually acetylcholine-gated ion channels, located almost entirely near the mouths of the subneural clefts lying immediately below the dense bar areas, where the acetylcholine vesicles empty into the synaptic space.

Each receptor is a large protein complex that has a total molecular weight of 275,000. The complex is composed of five subunit proteins, two *alpha* proteins and one each of *beta, delta,* and *gamma* proteins. These penetrate all the way through the membrane lying side by side in a circle to form a tubular channel. The channel remains constricted until two acetylcholine molecules attach respectively to the two *alpha* subunit proteins. This causes a conformational change that opens the channel, as shown in Figure 7–3; the upper channel in the figure is closed, while the bottom one has been opened by attachment of acetylcholine molecules.

The opened acetylcholine channel has a diameter of about 0.65 nanometer, which is large enough to

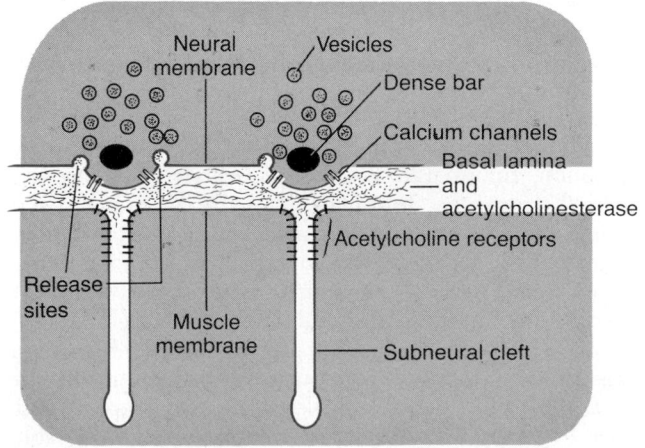

Neural membrane
Vesicles
Dense bar
Calcium channels
Basal lamina and acetylcholinesterase
Acetylcholine receptors
Release sites
Muscle membrane
Subneural cleft

Figure 7–2. Release of acetylcholine from synaptic vesicles at the neural membrane of the neuromuscular junction. Note the proximity of the release sites to the acetylcholine receptors at the mouths of the subneural clefts.

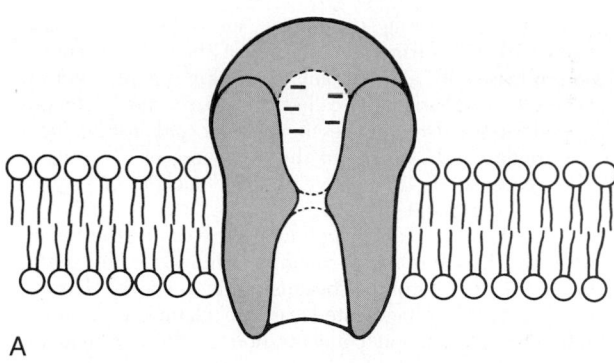

A

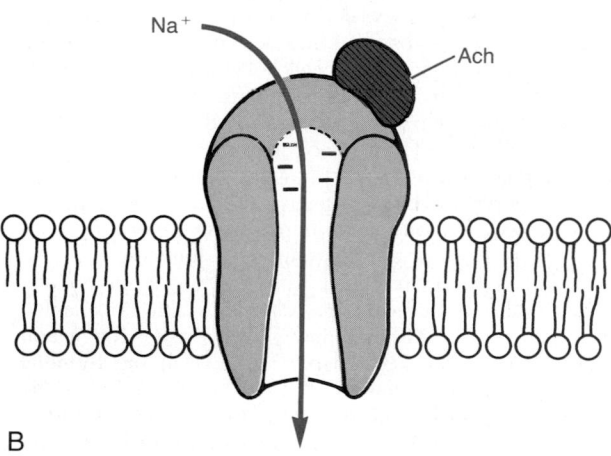

B

Figure 7–3. Acetylcholine channel. *A*, Closed state. *B*, After acetylcholine has become attached and a conformational change has opened the channel, allowing excess sodium to enter the muscle fiber and excite contraction. Note the negative charges at the channel mouth that prevent passage of negative ions.

allow all the important positive ions—sodium (Na⁺), potassium (K⁺), and calcium (Ca⁺⁺)—to move easily through the opening. On the other hand, negative ions, such as chloride ions, do not pass through because of strong negative charges in the mouth of the channel.

In practice, far more sodium ions flow through the acetylcholine channels than any other ions for two reasons. First, there are only two positive ions in great enough concentration to matter greatly, sodium ions in the extracellular fluid and potassium ions in the intracellular fluid. Second, the very negative potential on the inside of the muscle membrane, −80 to −90 millivolts, pulls the positively charged sodium ions to the inside of the fiber, while at the same time preventing the efflux of the potassium ions when they attempt to pass outward.

Therefore, as shown in the lower panel of Figure

7–3, the principal effect of opening the acetylcholine-gated channels is to allow large numbers of sodium ions to pour to the inside of the fiber, carrying with them large numbers of positive charges. This creates a local potential change at the muscle fiber membrane called the *end plate potential*. In turn, this end plate potential initiates an action potential at the muscle membrane and thus causes muscle contraction.

DESTRUCTION OF THE RELEASED ACETYLCHOLINE BY ACETYLCHOLINESTERASE. The acetylcholine, once released into the synaptic space, continues to activate the acetylcholine receptors as long as it persists in the space. However, it is rapidly removed by two means: (1) Most of the acetylcholine is destroyed by the enzyme *acetylcholinesterase* that is attached mainly to the *basal lamina*, the spongy layer of fine connective tissue that fills the synaptic space between the presynaptic terminal and the postsynaptic muscle membrane. (2) A small amount diffuses out of the synaptic space and is then no longer available to act on the muscle fiber membrane.

The short period that the acetylcholine remains in the synaptic space—a few milliseconds at most—is almost always sufficient to excite the muscle fiber. Then the rapid removal of the acetylcholine prevents muscle re-excitation after the fiber has recovered from the first action potential.

END PLATE POTENTIAL AND EXCITATION OF THE SKELETAL MUSCLE FIBER. The sudden insurgence of sodium ions into the muscle fiber when the acetylcholine channels open causes the internal membrane potential in the *local area of the end plate* to increase in the positive direction as much as 50 to 75 millivolts, creating a *local potential* called the *end plate potential*. If one recalls from Chapter 5 that a sudden increase in membrane potential of more than 20 to 30 millivolts is normally sufficient to initiate the positive feedback effect of sodium channel activation, one can understand that the end plate potential created by the acetylcholine stimulation is normally far greater than enough to initiate an action potential in the muscle fiber.

Figure 7–4 shows the principle of an end plate potential initiating the action potential. In this figure are shown three separate end plate potentials. End plate potentials A and C are too weak to elicit an action potential, but they do give the weak local end plate potentials recorded in the figure. In contrast, end plate potential B is much stronger and causes sodium channels to open so that the self-regenerative effect of more and more sodium ions flowing to the interior of the fiber initiates an action potential. The weakness of the end plate potential at point A was caused by poisoning the muscle fiber with *curare*, a drug that blocks the gating action of acetylcholine on the acetylcholine channels by competing with the acetylcholine for the acetylcholine receptor site. The weakness of the end plate potential at point C resulted from the effect of *botulinum toxin*, a bacterial poison that decreases the release of acetylcholine by the nerve terminals.

SAFETY FACTOR FOR TRANSMISSION AT THE NEURO-MUSCULAR JUNCTION; FATIGUE OF THE JUNCTION. Or-

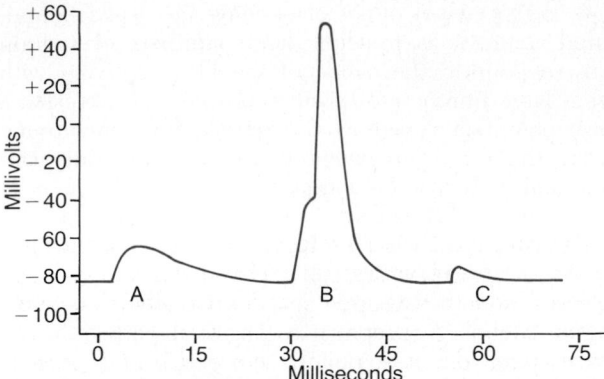

Figure 7–4. End plate potentials. *A,* Weakened end plate potential recorded in a curarized muscle, too weak to elicit an action potential; *B,* normal end plate potential eliciting a muscle action potential; and *C,* weakened end plate potential caused by botulinum toxin that decreases end plate release of acetylcholine, again too weak to elicit a muscle action potential.

dinarily, each impulse that arrives at the neuromuscular junction causes about three times as much end plate potential as that required to stimulate the muscle fiber. Therefore, the normal neuromuscular junction is said to have a high *safety factor.* However, artificial stimulation of the nerve fiber at rates greater than 100 times per second for several minutes often diminishes the number of vesicles of acetylcholine released with each impulse so much that impulses then fail to pass into the muscle fiber. This is called *fatigue* of the neuromuscular junction, and it is analogous to fatigue of the synapse in the central nervous system. Under normal functioning conditions, fatigue of the neuromuscular junction occurs rarely and even then only at the most exhausting levels of muscle activity.

Molecular Biology of Acetylcholine Formation and Release

Because the neuromuscular junction is large enough to be easily studied, it is one of the few synapses of the nervous system at which most of the details of chemical transmission have been worked out. The formation and release of acetylcholine at this junction occurs in the following stages.

1.   Small vesicles, about 40 nanometers in size, are formed by the Golgi apparatus in the cell body of the motoneuron in the spinal cord. These vesicles are then transported by "streaming" of the axoplasm through the core of the axon from the central cell body in the spinal cord all the way to the neuromuscular junction at the tips of the nerve fibers. About 300,000 of these small vesicles collect in the nerve terminals of a single skeletal muscle end plate.

2.   Acetylcholine is synthesized in the cytosol of the terminal nerve fibers but is then transported through the membranes of the vesicles to their interior, where it is stored in highly concentrated form, with about 10,000 molecules of acetylcholine in each vesicle.

3.   Under resting conditions, an occasional vesicle fuses with the surface membrane of the nerve terminal and releases its acetylcholine into the synaptic gutter. When this occurs, a so-called *miniature end plate* potential, about 0.4 millivolt in intensity and lasting for a few milliseconds, occurs in the local area of the muscle fiber because of the action of this "packet" of 10,000 acetylcholine molecules.

4.   When an action potential arrives at the nerve terminal, it opens many calcium channels in the membrane of the terminal because this terminal has an abundance of voltage-gated calcium channels. As a result, the calcium ion concentration in the terminal increases about 100-fold, which in turn increases the rate of fusion of the acetylcholine vesicles with the terminal membrane about 10,000-fold. As each vesicle fuses, its outer surface ruptures through the cell membrane, thus causing *exocytosis* of acetylcholine into the synaptic cleft. About 125 vesicles usually rupture with each action potential. The acetylcholine is then split by acetylcholinesterase into acetate ion and choline, and the choline is actively reabsorbed into the neural terminal to be reused in forming new acetylcholine. This sequence of events occurs in 5 to 10 milliseconds.

5.   After each vesicle has released its acetylcholine, the membrane of the vesicle becomes part of the cell membrane. However, the number of vesicles available in the nerve ending is sufficient to allow transmission of only a few thousand nerve impulses. Therefore, for continued function of the neuromuscular junction, the vesicles need to be retrieved from the nerve membrane. Retrieval is achieved by *endocytosis,* which is explained in Chapter 2. Within a few seconds after the action potential is over, "coated pits" appear on the terminal nerve membrane, caused by contractile proteins of the cytosol, especially the protein *clathrin,* attaching underneath the membrane in the areas of the original vesicles. Within about 20 seconds, the proteins contract and cause the pits to break away to the interior of the membrane, thus forming new vesicles. Within another few seconds, acetylcholine is transported to the interior of these vesicles, and they are then ready for a new cycle of acetylcholine release.

Drugs That Affect Transmission at the Neuromuscular Junction

DRUGS THAT STIMULATE THE MUSCLE FIBER BY ACETYLCHOLINE-LIKE ACTION. Many compounds, including *methacholine, carbachol,* and *nicotine,* have the same effect on the muscle fiber as does acetylcholine. The difference between these drugs and acetylcholine is that they are not destroyed by cholinesterase or are destroyed slowly, so that when once applied to the muscle fiber, the action persists for many minutes to several hours. These drugs work by causing localized areas of depolarization at the motor end plate, where the acetylcholine receptors are located. Then, every time the muscle fiber becomes repolarized elsewhere, these depolarized areas, by virtue of their leaking ions, cause new action potentials, thereby causing a state of spasm.

DRUGS THAT BLOCK TRANSMISSION AT THE NEUROMUSCULAR JUNCTION. A group of drugs known as the *curariform drugs* can prevent passage of impulses from the end plate into the muscle. Thus, D-tubocurarine affects the membrane by competing with acetylcholine for the acetylcholine receptor sites, so that the acetylcholine generated by the end plate cannot increase the permeability of the muscle membrane acetylcholine channels sufficiently to initiate an action potential.

DRUGS THAT STIMULATE THE NEUROMUSCULAR JUNCTION BY INACTIVATING ACETYLCHOLINESTERASE. Three particularly well-known drugs, *neostigmine, physostigmine,* and *diisopropyl fluorophosphate,* inactivate acetylcholinesterase so that the cholinesterase normally in the synapses will not hydrolyze the acetylcholine released at the end plate. As a result, acetylcholine increases in quantity with successive nerve impulses so that large amounts of acetylcholine can accumulate and then repetitively stimulate the muscle fiber. This causes *muscle spasm* when even a few nerve impulses reach the muscle; it can cause death due to laryngeal spasm, which smothers the person.

Neostigmine and physostigmine combine with acetylcholinesterase to inactivate the acetylcholinesterase for up to several hours, after which they are displaced from the acetylcholinesterase so that it once again becomes active. On the other hand, diisopropyl fluorophosphate, which has military potential as a powerful "nerve" gas poison, inactivates acetylcholinesterase for weeks, which makes this a particularly lethal poison.

Myasthenia Gravis

Myasthenia gravis, which occurs in about 1 of every 20,000 persons, causes paralysis because of the inability of the neuromuscular junctions to transmit signals from the nerve fibers to the muscle fibers. Pathologically, antibodies that attack the acetylcholine-gated transport proteins have been demonstrated in the blood of most patients with myasthenia gravis. Therefore, it is believed that myasthenia gravis is an autoimmune disease in which patients have developed antibodies against their own acetylcholine-activated ion channels.

Regardless of the cause, the end plate potentials that occur in the muscle fibers are too weak to adequately stimulate the muscle fibers. If the disease is intense enough, the patient dies of paralysis—in particular, paralysis of the respiratory muscles. The disease usually can be ameliorated by administering *neostigmine* or some other anticholinesterase drug. This allows far more acetylcholine to accumulate in the synaptic cleft. Within minutes, some of these paralyzed people can begin to function almost normally.

MUSCLE ACTION POTENTIAL

Almost everything discussed in Chapter 5 regarding initiation and conduction of action potentials in nerve fibers applies equally well to skeletal muscle fibers except for quantitative differences. Some of the quantitative aspects of muscle potentials are the following:

1. Resting membrane potential: about -80 to -90 millivolts in skeletal fibers—the same as in large myelinated nerve fibers.

2. Duration of action potential: 1 to 5 milliseconds in skeletal muscle—about five times as long as large myelinated nerves.

3. Velocity of conduction: 3 to 5 m/sec—about ⅟₁₈ the velocity of conduction in the large myelinated nerve fibers that excite skeletal muscle.

Spread of the Action Potential to the Interior of the Muscle Fiber by Way of the Transverse Tubule System

The skeletal muscle fiber is so large that action potentials spreading along its surface membrane cause almost no current flow deep within the fiber. To cause contraction, these electrical currents must penetrate to the vicinity of all the separate myofibrils. This is achieved by transmission of the action potentials along *transverse tubules* (T tubules) that penetrate all the way through the muscle fiber from one side to the other. The T tubule action potentials in turn cause the sarcoplasmic reticulum to release calcium ions in the immediate vicinity of all the myofibrils, and these calcium ions then cause contraction. This overall process is called *excitation-contraction* coupling.

EXCITATION-CONTRACTION COUPLING

Transverse Tubule–Sarcoplasmic Reticulum System

Figure 7–5 shows myofibrils surrounded by the T tubule–sarcoplasmic reticulum system. The T tubules are very small and run transverse to the myofibrils. They begin at the cell membrane and penetrate all the way from one side of the muscle fiber to the opposite side. Not shown in the figure is the fact that these tubules branch among themselves so that they form entire *planes* of T tubules interlacing among all the separate myofibrils. Also, *where the T tubules originate from the cell membrane, they are open to the exterior.* Therefore, they communicate with the fluid surrounding the muscle fiber and contain extracellular fluid in their lumens. In other words, the T tubules are internal extensions of the cell membrane. Therefore, when an action potential spreads over a muscle fiber membrane, it spreads along the T tubules to the deep interior of the muscle fiber as well. The action potential currents surrounding these T tubules then elicit the muscle contraction.

Figure 7–5 shows the *sarcoplasmic reticulum* as well, in red. This is composed of two major parts: (1) long *longitudinal tubules* that run parallel to the myofibrils and terminate in (2) large chambers called *terminal cisternae;* these cisternae abut the T tubules. When the muscle fiber is sectioned longitudinally and electron micrographs are made, one sees this abutting of the cisternae against the tubule, which gives the appearance of a *triad* with a small central tubule and a large cisterna on either side. This is shown in Figure 7–5 as well as in the electron micrograph of Figure 6–3.

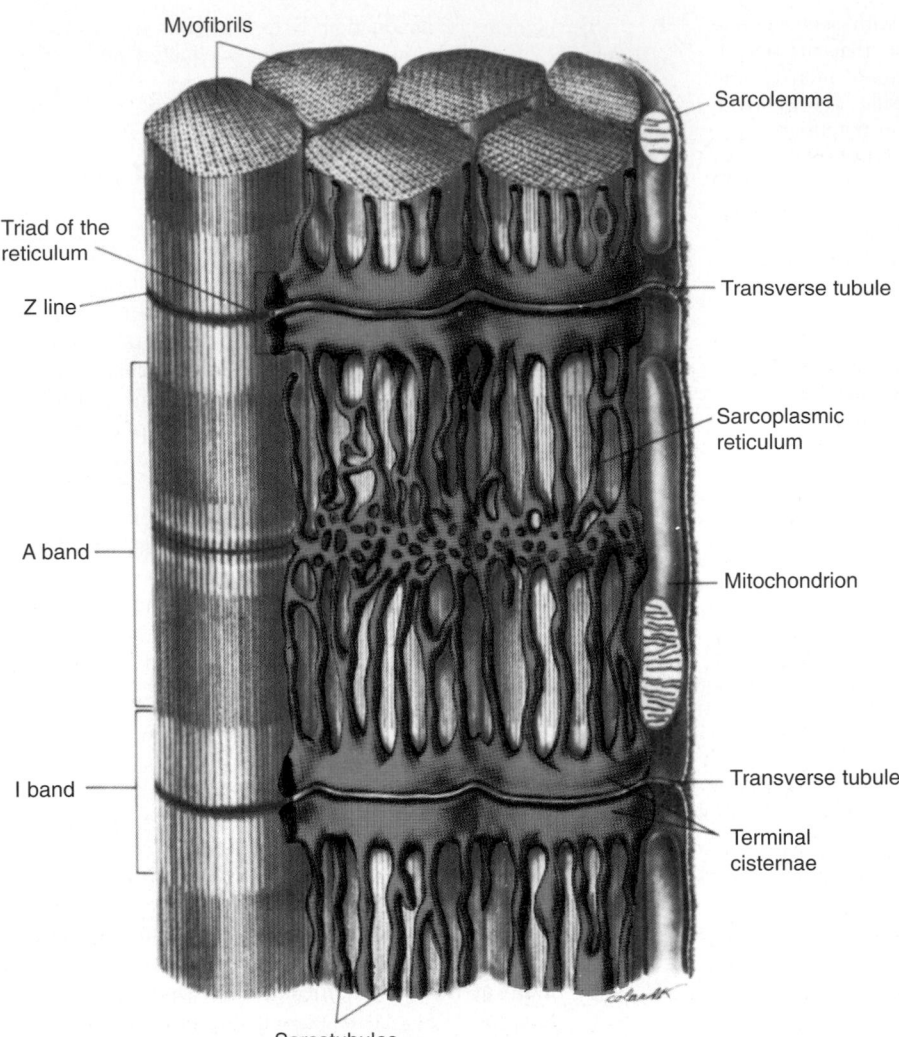

Myofibrils

Triad of the reticulum

Z line

A band

I band

Sarcolemma

Transverse tubule

Sarcoplasmic reticulum

Mitochondrion

Transverse tubule

Terminal cisternae

Sarcotubules

Figure 7–5. Transverse (T) tubule–sarcoplasmic reticulum system. Note the *longitudinal tubules* that terminate in large *cisternae.* The cisternae in turn abut the T tubules. Note also that the T tubules communicate with the outside of the cell membrane. This illustration was drawn from frog muscle, which has one T tubule per sarcomere, located at the Z line. A similar arrangement is found in mammalian heart muscle, but mammalian skeletal muscle has two T tubules per sarcomere, located at the A-I junctions. (From Fawcett: Bloom and Fawcett: A Textbook of Histology. Philadelphia, W. B. Saunders Company, 1986. Modified after Peachey: *J. Cell Biol. 25:*209, 1965. Drawn by Sylvia Colard Keene.)

In the muscle of lower animals, such as the frog, there is a single T tubule network for each sarcomere, located at the level of the Z disc, as shown in Figure 7–5. Cardiac muscle also has this type of T tubule system. In mammalian skeletal muscle there are two T tubule networks for each sarcomere located near the two ends of the myosin filaments, which are the points where the actual mechanical forces of muscle contraction are created. Thus, mammalian skeletal muscle is optimally organized for rapid excitation of muscle contraction.

Release of Calcium Ions by the Sarcoplasmic Reticulum

One of the special features of the sarcoplasmic reticulum is that within its vesicular tubules it contains calcium ions in high concentration, and many of these ions are released when an action potential occurs in the adjacent T tubule.

Figure 7–6 shows that the action potential of the T tubule causes current flow through the tips of the cisternae that abut the T tubule. At these points, each cisterna projects *junctional feet* that attach to the

membrane of the T tubule, presumably facilitating passage of some signal from the T tubule to the cisterna. This signal possibly is electrical current of the action potential itself. There are also reasons to believe that it could be some chemical or mechanical signal. Whatever the signal, it causes rapid opening of large numbers of calcium channels through the membranes of the cisternae and their attached longitudinal tubules of the sarcoplasmic reticulum. These channels remain open for a few milliseconds; during this time, the calcium ions responsible for muscle contraction are released into the sarcoplasm surrounding the myofibrils.

The calcium ions that are thus released diffuse to the adjacent myofibrils, where they bind strongly with troponin C, as discussed in Chapter 6, and this in turn elicits the muscle contraction.

CALCIUM PUMP FOR REMOVING CALCIUM IONS FROM THE SARCOPLASMIC FLUID. Once the calcium ions have been released from the sarcoplasmic tubules and have diffused to the myofibrils, muscle contraction will continue as long as the calcium ions remain in high concentration in the myofibrillar fluid. However, a continually active calcium pump located in the walls of the sarcoplasmic reticulum pumps calcium ions

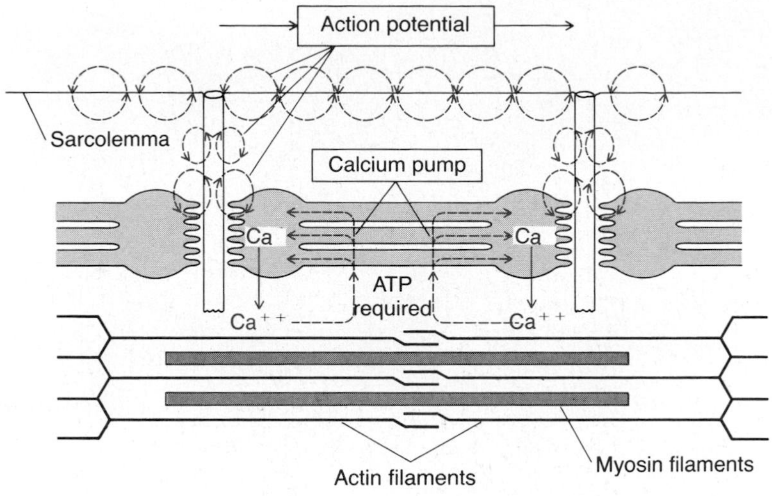

Figure 7–6. Excitation-contraction coupling in the muscle showing an action potential that causes release of calcium ions from the sarcoplasmic reticulum and then re-uptake of the calcium ions by a calcium pump.

away from the myofibrils back into the sarcoplasmic tubules. This pump can concentrate the calcium ions about 10,000-fold inside the tubules. In addition, inside the reticulum is a protein called *calsequestrin* that can bind more than 40 times as much calcium as that in the ionic state, thus providing another 40-fold increase in the storage of calcium. Thus, this massive transfer of calcium into the sarcoplasmic reticulum causes virtual total depletion of calcium ions in the myofibrillar fluid. Therefore, except immediately after an action potential, the calcium ion concentration in the myofibrils is kept at an extremely low level.

Excitatory "Pulse" of Calcium Ions. The normal concentration (less than 10^{-7} molar) of calcium ions in the cytosol that bathes the myofibrils is too little to elicit contraction. Therefore, in the resting state, the troponin-tropomyosin complex keeps the actin filaments inhibited and maintains a relaxed state of the muscle.

On the other hand, full excitation of the T tubule–sarcoplasmic reticulum system causes enough release of calcium ions to increase the concentration in the myofibrillar fluid to as high as 2×10^{-4} molar concentration, which is about 10 times the level required to cause maximum muscle contraction (about 2×10^{-5} molar). Immediately thereafter, the calcium pump depletes the calcium ions again. The total duration of this calcium "pulse" in the usual skeletal muscle fiber lasts about 1/20 of a second, although it may last several times as long as this in some fibers and several times less in others. (In heart muscle, the calcium pulse lasts about 1/3 second because of the long duration of the cardiac action potential.) During this calcium pulse, muscle contraction occurs. If the contraction is to continue without interruption for longer intervals, a series of such pulses must be initiated by a continuous series of repetitive action potentials, as discussed in Chapter 6.

REFERENCES

See references for Chapters 5 and 6.

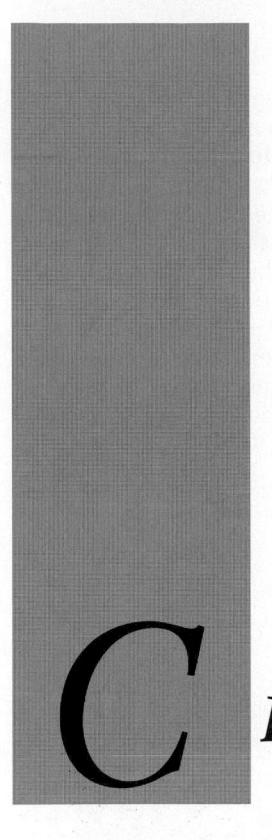

Contraction and Excitation of Smooth Muscle

CHAPTER 8

CONTRACTION OF SMOOTH MUSCLE

In Chapters 6 and 7, the discussion has been concerned with skeletal muscle. We now turn to smooth muscle, which is composed of far smaller fibers—usually 2 to 5 micrometers in diameter and only 20 to 500 micrometers in length—in contrast to the skeletal muscle fibers, which arc as much as 20 times as large in diameter and thousands of times as long. Many of the principles of contraction apply to smooth muscle the same as to skeletal muscle. Most important, essentially the same attractive forces between myosin and actin filaments cause contraction in smooth muscle as in skeletal muscle, but the internal physical arrangement of smooth muscle fibers is entirely different, as we see subsequently.

Types of Smooth Muscle

The smooth muscle of each organ is distinctive from that of most other organs in several ways: physical dimensions, organization into bundles or sheets, response to different types of stimuli, characteristics of innervation, and function. Yet, for the sake of simplicity, smooth muscle can generally be divided into two major types, which are shown in Figure 8–1: *multi-unit smooth muscle* and *unitary* (or *single-unit*) *smooth muscle.*

MULTI-UNIT SMOOTH MUSCLE. This type of smooth muscle is composed of discrete smooth muscle fibers. Each fiber operates independently of the others and often is innervated by a single nerve ending, as occurs for skeletal muscle fibers. Furthermore, the outer sur-

faces of these fibers, like those of skeletal muscle fibers, are covered by a thin layer of basement membrane–like substance, a mixture of fine collagen and glycoprotein fibrillae that helps insulate the separate fibers from one another.

The most important characteristic of multi-unit smooth muscle fibers is that each fiber can contract independently of the others, and their control is exerted mainly by nerve signals. This is in contrast to a major share of the control of visceral smooth muscle by non-nervous stimuli. An additional characteristic is that they seldom exhibit spontaneous contractions.

Some examples of multi-unit smooth muscle found in the body are the smooth muscle fibers of the ciliary muscle of the eye, the iris of the eye, the nictitating membrane that covers the eyes in some lower animals, and the piloerector muscles that cause erection of the hairs when stimulated by the sympathetic nervous system.

UNITARY SMOOTH MUSCLE. The term "unitary" is confusing because it does not mean single muscle fibers. Instead, it means a whole mass of hundreds to millions of muscle fibers that contract together as a single unit. The fibers usually are aggregated into sheets or bundles, and their cell membranes are adherent to one another at multiple points so that force generated in one muscle fiber can be transmitted to the next. In addition, the cell membranes are joined by many *gap junctions* through which ions can flow freely from one cell to the next so that action potentials or simple ion flow can travel from one fiber to the next and cause the muscle fibers to contract together. This type of smooth muscle is also known as *syncytial*

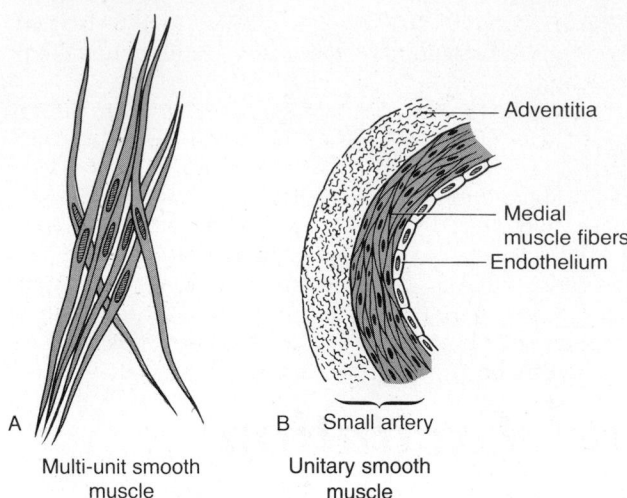

Figure 8–1. Multi-unit and unitary smooth muscle.

smooth muscle because of its interconnections among fibers. Because such muscle is found in the walls of most viscera of the body—including the gut, bile ducts, ureters, uterus, and many blood vessels—it is also called *visceral smooth muscle.*

Contractile Process in Smooth Muscle

Chemical Basis for Smooth Muscle Contraction

Smooth muscle contains both *actin* and *myosin filaments,* having chemical characteristics similar to those of the actin and myosin filaments in skeletal muscle. It does not contain the normal troponin complex that is requisite in the control of skeletal muscle contraction, so that the mechanism for control of contraction is different. This is discussed in detail in a subsequent section of this chapter.

Chemical studies have shown that actin and myosin derived from smooth muscle interact with each other in much the same way that this occurs for actin and myosin derived from skeletal muscle. Furthermore, the contractile process is activated by calcium ions, and adenosine triphosphate (ATP) is degraded to adenosine diphosphate (ADP) to provide the energy for contraction.

On the other hand, there are major differences between the physical organization of smooth muscle and that of skeletal muscle as well as differences in excitation-contraction coupling, control of the contractile process by calcium ions, duration of contraction, and amount of energy required for the contractile process.

Physical Basis For Smooth Muscle Contraction

Smooth muscle does not have the same striated arrangement of the actin and myosin filaments as that found in skeletal muscle. For a long time, it was im-

possible to discern even in electron micrographs any specific organization in the smooth muscle cell that could account for contraction. However, recent special electron micrographic techniques suggest the physical organization exhibited in Figure 8–2. This figure shows large numbers of actin filaments attached to so-called *dense bodies.* Some of these bodies are attached to the cell membrane. Others are dispersed inside the cell and held in place by a scaffold of structural proteins linking one dense body to another. Note in Figure 8–2 that some of the membrane dense bodies of adjacent cells are also bonded together by intercellular protein bridges. It is mainly through these bonds that the force of contraction is transmitted from one cell to the next.

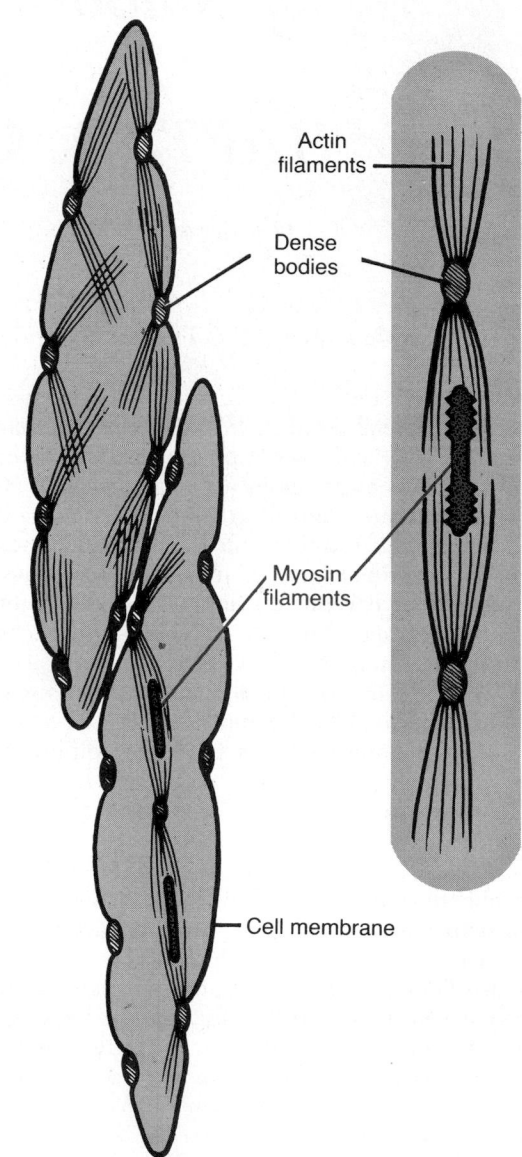

Figure 8–2. Physical structure of smooth muscle. The upper left-hand fiber shows actin filaments radiating from dense bodies. The lower fiber as well as the right-hand insert demonstrate the relation of myosin filaments to the actin filaments.

Interspersed among the many actin filaments in the muscle fiber are a few myosin filaments. These have a diameter more than twice as great as that of the actin filaments. When seen in electron micrographic cross section, one usually finds about 15 times as many actin filaments as myosin filaments. Part of this difference is caused by the fact that the ratio of actin filament length to myosin filament length in smooth muscle is much greater than that in skeletal muscle. Therefore, the probability of seeing excess actin filaments is increased. Nevertheless, one is impressed by the relative sparsity of myosin filaments with respect to actin filaments.

To the right in Figure 8–2 is the postulated structure of individual contractile units within smooth muscle cells, showing large numbers of actin filaments radiating from two dense bodies; these filaments overlap a single myosin filament located midway between the dense bodies. This contractile unit is similar to the contractile unit of skeletal muscle but without the regularity of the skeletal muscle structure; in fact, the dense bodies of smooth muscle serve the same role as the Z discs in skeletal muscle.

Comparison of Smooth Muscle Contraction with Skeletal Muscle Contraction

Although most skeletal muscles contract and relax rapidly, most smooth muscle contraction is prolonged tonic contraction, sometimes lasting hours or even days. Therefore, it is to be expected that both the physical and the chemical characteristics of smooth muscle versus skeletal muscle contraction would differ. The following are some of the differences.

SLOW CYCLING OF THE CROSS-BRIDGES. The rapidity of cycling of the cross-bridges in smooth muscle—that is, their attachment to actin, then release from the actin, and reattachment for the next cycle—is much, much slower in smooth muscle than in skeletal muscle, in fact as little as $\frac{1}{10}$ to $\frac{1}{300}$ the frequency in skeletal muscle. Yet the *fraction of time* that the cross-bridges remain attached to the actin filaments, which is the major factor that determines the force of contraction, is believed to be greatly increased in smooth muscle. A possible reason for the slow cycling is that the cross-bridge heads have far less ATPase activity than that in skeletal muscle, so that degradation of the ATP that energizes the movements of the heads is greatly reduced, with corresponding slowing of the rate of cycling.

ENERGY REQUIRED TO SUSTAIN SMOOTH MUSCLE CONTRACTION. Only $\frac{1}{10}$ to $\frac{1}{300}$ as much energy is required to sustain the same tension of contraction in smooth muscle as in skeletal muscle. This, too, is believed to result because of the slow attachment cycling of the cross-bridges and because only one molecule of ATP is required for each cycle, regardless of its duration.

This economy of energy utilization by smooth muscle is exceedingly important to the overall energy economy of the body because organs, such as the intestines, urinary bladder, gallbladder, and other viscera, must maintain tonic muscle contraction almost indefinitely.

SLOWNESS OF ONSET OF CONTRACTION AND RELAXATION OF SMOOTH MUSCLE. A typical smooth muscle tissue begins to contract 50 to 100 milliseconds after it is excited, reaches full contraction about ½ second later, and then declines in contractile force in another 1 to 2 seconds, giving a total contraction time of 1 to 3 seconds. This is about 30 times as long as a single contraction of an average skeletal muscle. Because of the many types of smooth muscle, contraction of some types can be as short as 0.2 second or as long as 30 seconds.

The slow onset of contraction in smooth muscle as well as the prolonged contraction is probably caused by the slowness of attachment and detachment of the cross-bridges. In addition, the initiation of contraction in response to calcium ions, called the excitation-contraction coupling mechanism, is much slower than in skeletal muscle, as we discuss later.

FORCE OF MUSCLE CONTRACTION. Despite the relatively few myosin filaments in smooth muscle and despite the slow cycling time of the cross-bridges, the maximum force of contraction of smooth muscle is often even greater than that of skeletal muscle—as great as 4 to 6 kg/cm^2 cross-sectional area for smooth muscle in comparison with 3 to 4 kilograms for skeletal muscle. This great force of attraction is postulated to result from the prolonged period of attachment of the myosin cross-bridges to the actin filaments.

PERCENTAGE OF SHORTENING OF SMOOTH MUSCLE DURING CONTRACTION. Another characteristic of smooth muscle that makes it different from skeletal muscle is its ability to shorten a far greater percentage of its length than can skeletal muscle while maintaining almost full force of contraction. Skeletal muscle has a useful distance of contraction of only about one quarter to one third its stretched length, whereas smooth muscle can often contract quite effectively more than two thirds its stretched length. This allows smooth muscle to perform especially important functions in the hollow viscera, allowing the gut, bladder, blood vessels, and other internal bodily structures to change their lumen diameters from very large down to almost zero.

Why this difference between smooth muscle and skeletal muscle? The answer to this is not completely known, but there appear to be two probable reasons. First, it is likely that some contractile units of smooth muscle have optimal overlapping of their actin and myosin filaments at one length of the muscle and others at other lengths, rather than all of these being synchronized together, as usually occurs in skeletal muscle. Therefore, a greater distance of contraction can be achieved. Second, the actin filaments in smooth muscle are much longer than those in skeletal muscle. Therefore, these filaments can be pulled along the myosin filaments for a much greater distance during smooth muscle contraction than occurs in skeletal muscle contraction.

"LATCH" MECHANISM FOR PROLONGED HOLDING CONTRACTIONS OF SMOOTH MUSCLE. Once smooth muscle has developed full contraction, the degree of activation of the muscle can usually be reduced to far less than the initial level and yet the muscle will maintain its full force of contraction. Furthermore, the energy consumed to maintain contraction is often miniscule, sometimes as little as 1/300 the energy required for comparable sustained skeletal muscle contraction. This is called the "latch" mechanism. The same effect occurs to a slight extent in skeletal muscle but to an extent many times less than that in smooth muscle.

The importance of the latch mechanism is that it can maintain prolonged tonic contraction in smooth muscle for hours with little use of energy. Also, little excitatory signal is required from nerve fibers or hormonal sources.

Stress-Relaxation of Smooth Muscle. Another important characteristic of smooth muscle, especially of the visceral type of smooth muscle in many hollow organs, is its ability to return nearly to its original *force* of contraction seconds or minutes after it has been elongated or shortened. For example, a sudden increase in fluid volume in the urinary bladder, thus stretching the smooth muscle in the bladder wall, causes an immediate large increase in pressure in the bladder. However, during the next 15 seconds to a minute or so, despite continued stretch of the bladder wall, the pressure returns almost exactly back to the original level. Then, when the volume is increased by another step, the same effect occurs again. When the volume is suddenly decreased, the pressure falls very low at first but then returns in another few seconds or minutes back to or near the original level. This phenomenon is called *stress-relaxation* and *reverse stress-relaxation.* Its importance is that it allows a hollow organ to maintain about the same amount of pressure inside its lumen regardless of the length of the muscle fibers.

The stress-relaxation phenomenon is probably closely related to the latch phenomenon. When the muscle is first stretched, the latch phenomenon resists change in length. Yet, with successive recycling of the myosin heads during the succeeding seconds to minutes, the heads release and attach further along the actin filaments. Therefore, the length of the muscle eventually changes, whereas the tension in the muscle returns nearly to its original value because the number of myosin cross-bridges causing contractile force is almost the same as had been true previously.

Regulation of Contraction by Calcium Ions

As is true for skeletal muscle, the initiating event in most smooth muscle contraction is an increase in intracellular calcium ions. This increase can be caused by nerve stimulation of the smooth muscle fiber, hormonal stimulation, stretch of the fiber, or even changes in the chemical environment of the fiber.

Yet smooth muscle does not contain troponin, the regulatory protein that is activated by calcium ions to cause skeletal muscle contraction. Instead, smooth muscle contraction is activated by an entirely different mechanism, as follows.

COMBINATION OF CALCIUM IONS WITH CALMODULIN—ACTIVATION OF MYOSIN KINASE AND PHOSPHORYLATION OF THE MYOSIN HEAD. In place of troponin, smooth muscle cells contain a large amount of another regulatory protein called *calmodulin.* Although this protein is similar to troponin, in that it reacts with four calcium ions, it is different in the manner in which it initiates the contraction. Calmodulin does this by activating the myosin cross-bridges. This activation and subsequent contraction occurs in the following sequence.

1. The calcium ions bind with calmodulin.
2. The calmodulin-calcium combination then joins with and activates myosin kinase, a phosphorylating enzyme.
3. One of the light chains of each myosin head, called the *regulatory chain*, becomes phosphorylated in response to the myosin kinase. When this chain is not phosphorylated, the attachment-detachment cycling of the head with the actin filament does not occur. But when the regulatory chain is phosphorylated, the head has the capability of binding with the actin filament and proceeding through the entire cycling process, the same as occurs for skeletal muscle, thus causing muscle contraction.

CESSATION OF CONTRACTION—ROLE OF MYOSIN PHOSPHATASE. When the calcium ion concentration falls below a critical level, the aforementioned processes automatically reverse except for the phosphorylation of the myosin head. Reversal of this requires another enzyme, *myosin phosphatase,* located in the fluids of the smooth muscle cell, which splits the phosphate from the regulatory light chain. Then the cycling stops and the contraction ceases. The time required for relaxation of muscle contraction, therefore, is determined to a great extent by the amount of active myosin phosphatase in the cell.

A Possible Mechanism for Regulation of the Latch Phenomenon

Because of the importance of the latch phenomenon in smooth muscle and because this phenomenon allows long-term maintenance of tone in many smooth muscle organs, many attempts have been made to explain it. Among many of the mechanisms that have been postulated, one of the simplest is the following.

When the myosin kinase and myosin phosphatase enzymes are both strongly activated, the cycling frequency of the myosin heads and the velocity of contraction are great. Then, as the activation of the enzymes decreases, the cycling frequency decreases, but at the same time, the lower activation of the enzymes causes the myosin heads to remain attached to the actin filament for a longer and longer proportion of the cycling period. Therefore, the number of heads attached to the actin filament at any given time remains large. Because the number of heads attached to the actin determines the static force of contraction, tension is maintained, or "latched"; yet little energy is used by the muscle because ATP is not degraded to ADP, except on the rare occasion when a head detaches.

NEURAL AND HORMONAL CONTROL OF SMOOTH MUSCLE CONTRACTION

Although skeletal muscle is activated exclusively by the nervous system, smooth muscle can be stimulated to contract by multiple types of signals: by nervous signals, by hormonal stimulation, by stretch of the muscle, and in several other ways. The principal reason for the difference is that the smooth muscle membrane contains many types of receptor proteins that can initiate the contractile process. Still other receptor proteins inhibit smooth muscle contraction, which is another difference from skeletal muscle. Therefore, in this section, we discuss, first, neural control of smooth muscle contraction, followed by hormonal control and other means of control.

Neuromuscular Junctions of Smooth Muscle

PHYSIOLOGIC ANATOMY OF SMOOTH MUSCLE NEUROMUSCULAR JUNCTIONS. Neuromuscular junctions of the highly structured type found on skeletal muscle fibers do not occur in smooth muscle. Instead, the *autonomic nerve fibers* that innervate smooth muscle generally branch diffusely on top of a sheet of muscle fibers, as shown in Figure 8–3. In most instances, these fibers do not make direct contact with the smooth muscle fibers but instead form so-called *diffuse junctions* that secrete their transmitter substance into the matrix coating of the smooth muscle a few nanometers to a few micrometers away from the muscle cells; the transmitter substance then diffuses to the cells. Furthermore, where there are many layers of muscle cells, the nerve fibers often innervate only the outer layer, and the muscle excitation then travels from this outer layer to the inner layers by action

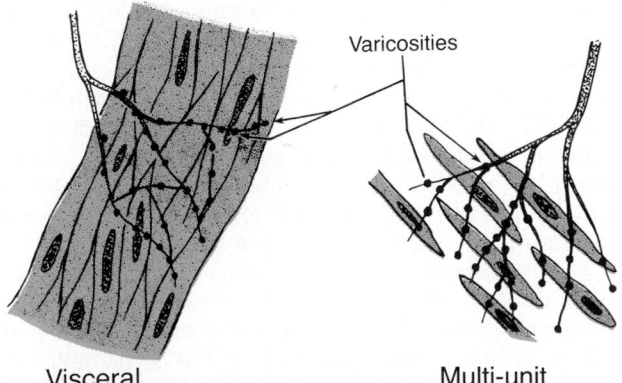

Visceral   Multi-unit

Figure 8–3. Innervation of smooth muscle.

potential conduction in the muscle mass or by subsequent diffusion of the transmitter substance.

The axons that innervate smooth muscle fibers also do not have typical branching end feet of the type in the motor end plate on skeletal muscle fibers. Instead, most of the fine terminal axons have multiple *varicosities* distributed along their axes. At these points the Schwann cells are interrupted so that transmitter substance can be secreted through the walls of the varicosities. In the varicosities are vesicles similar to those in the skeletal muscle end plate that contain transmitter substance. In contrast to the vesicles of skeletal muscle junctions that contain only acetylcholine, the vesicles of the autonomic nerve fiber endings contain *acetylcholine* in some fibers and *norepinephrine* in others—and perhaps, rarely, still other substances.

In a few instances, particularly in the multi-unit type of smooth muscle, the varicosities lie directly on the muscle fiber membrane separated from this membrane by as little as 20 to 30 nanometers—the same width as the synaptic cleft that occurs in the skeletal muscle junction. They are called *contact junctions* and function in much the same way as the skeletal muscle neuromuscular junction; the latent period of contraction of these smooth muscle fibers is considerably shorter than that of fibers stimulated by the diffuse junctions.

EXCITATORY AND INHIBITORY TRANSMITTER SUBSTANCES AT THE SMOOTH MUSCLE NEUROMUSCULAR JUNCTION. Transmitter substances secreted by the autonomic nerves innervating smooth muscle are *acetylcholine* and *norepinephrine,* but they are never secreted by the same nerve fibers. Acetylcholine is an excitatory transmitter substance for smooth muscle fibers in some organs but an inhibitory substance for smooth muscle in other organs. When acetylcholine excites a muscle fiber, norepinephrine ordinarily inhibits it. Conversely, when acetylcholine inhibits a fiber, norepinephrine usually excites it.

But why these different responses? The answer is that both acetylcholine and norepinephrine excite or inhibit smooth muscle by first binding with a *receptor protein* on the surface of the muscle cell membrane. This receptor controls the opening or closing of ion channels or controls some other means for activating or inhibiting the smooth muscle fiber. Furthermore, some of the receptor proteins are *excitatory receptors,* whereas others are *inhibitory receptors.* Thus, the type of receptor determines whether the smooth muscle is inhibited or excited and which of the two transmitters, acetylcholine or norepinephrine, is effective in causing the excitation or inhibition. These receptors are discussed in more detail in Chapter 60 in relation to the function of the autonomic nervous system.

Membrane Potentials and Action Potentials in Smooth Muscle

MEMBRANE POTENTIALS IN SMOOTH MUSCLE. The quantitative value of the membrane potential of smooth muscle is variable from one type of smooth muscle to another and depends on the momentary condition of the muscle. In the normal resting state, the membrane potential is usually about -50 to -60 millivolts, or about 30 millivolts less negative than in skeletal muscle.

ACTION POTENTIALS IN UNITARY SMOOTH MUSCLE. Action potentials occur in unitary smooth muscle, such as visceral muscle, in the same way that they occur in skeletal muscle. They do not normally occur in many, if not most, multi-unit types of smooth muscle, as discussed in a subsequent section.

The action potentials of visceral smooth muscle occur in two forms: (1) spike potentials and (2) action potentials with plateaus.

Spike Potentials. Typical spike action potentials, such as those seen in skeletal muscle, occur in most types of unitary smooth muscle. The duration of this type of action potential is 10 to 50 milliseconds, as shown in Figure 8–4A and B. Such action potentials can be elicited in many ways, for example, by electrical stimulation, by the action of hormones on the smooth muscle, by the action of transmitter substances from nerve fibers, by stretch, or as a result of spontaneous generation in the muscle fiber itself, as discussed subsequently.

Action Potentials with Plateaus. Figure 8–4C shows an action potential with a plateau. The onset of this action potential is similar to that of the typical spike potential. However, instead of rapid repolarization of the muscle fiber membrane, the repolarization is delayed for several hundred to as much as 1000 milliseconds. The importance of the plateau is that it can account for the prolonged periods of contraction that occur in some types of smooth muscle, such as the ureter, the uterus under some conditions, and some types of vascular smooth muscle. (Also, this is the type of action potential seen in cardiac muscle fibers that have a prolonged period of contraction, as we discuss in Chapters 9 and 10.)

Importance of Calcium Channels in Generating the Smooth Muscle Action Potential. The smooth muscle cell membrane has far more voltage-gated calcium channels than does skeletal muscle but few voltage-gated sodium channels. Therefore, sodium participates little in the generation of the action potential in most smooth muscle. Instead, the flow of calcium ions to the interior of the fiber is mainly responsible for the action potential. This occurs in the same self-regenerative way as occurs for the sodium channels in nerve fibers and in skeletal muscle fibers. However, calcium channels open many times more slowly than do sodium channels, but they also remain open much longer. This accounts in large measure for the slow action potentials of smooth muscle fibers.

Another important feature of calcium entry into the cells during the action potential is that the same calcium acts directly on the smooth muscle contractile mechanism to cause contraction, as discussed earlier. Thus, the calcium performs two tasks at once.

SLOW WAVE POTENTIALS IN UNITARY SMOOTH MUSCLE AND SPONTANEOUS GENERATION OF ACTION POTENTIALS. Some smooth muscle is self-excitatory. That is, action potentials arise within the smooth muscle itself without an extrinsic stimulus. This is often associated with a basic *slow wave rhythm* of the membrane potential, especially in intestinal wall smooth muscle. A typical slow wave of this type in the visceral smooth muscle of the gut is shown in Figure 8–4B. The slow wave itself is not an action potential. That is, it is not a self-regenerative process that spreads progressively over the membranes of the muscle fibers. Instead, it is a local property of the smooth muscle fibers that make up the muscle mass.

The cause of the slow wave rhythm is unknown; one suggestion is that the slow waves are caused by waxing and waning of the pumping of sodium ions outward through the muscle fiber membrane; the membrane potential becomes more negative when sodium is pumped rapidly and less negative when the sodium pump becomes less active. Another suggestion is that the conductances of the ion channels increase and decrease rhythmically.

The importance of the slow waves is that they can initiate action potentials. The slow waves themselves cannot cause muscle contraction, but when the potential of the slow wave rises above the level of about -35 millivolts (the approximate threshold for eliciting action potentials in most visceral smooth muscle), an action potential develops and spreads over the muscle

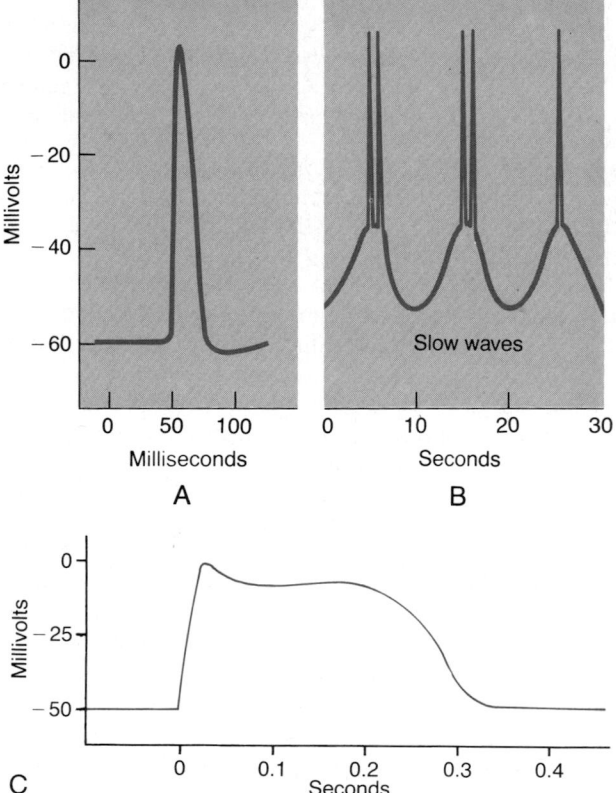

Figure 8–4. *A*, Typical smooth muscle action potential (spike potential) elicited by an external stimulus. *B*, Repetitive spike potentials elicited by slow rhythmical electrical waves that occur spontaneously in the smooth muscle of the intestinal wall. *C*, Action potential with a plateau recorded from a smooth muscle fiber of the uterus.

mass, and then contraction does occur. Figure 8–4B demonstrates this effect, showing that at each peak of the slow wave, one or more action potentials occur. This effect can promote a series of rhythmical contractions of the smooth muscle mass. Therefore, the slow waves are called *pacemaker waves.* In Chapter 62, we see that this type of activity controls the rhythmical contractions of the gut.

EXCITATION OF VISCERAL SMOOTH MUSCLE BY STRETCH. When visceral (unitary) smooth muscle is stretched sufficiently, spontaneous action potentials are usually generated. They result from a combination of the normal slow wave potentials plus a decrease in the negativity of the membrane potential caused by the stretch itself. This response to stretch allows the gut wall, when excessively stretched, to contract automatically and therefore to resist the stretch. For instance, when the gut is overfilled by intestinal contents, a local automatic contraction often sets up a peristaltic wave that moves the contents away from the overfilled intestine.

Depolarization of Multi-Unit Smooth Muscle Without Action Potentials

The smooth muscle fibers of multi-unit smooth muscle (such as the muscle of the iris of the eye or the piloerector muscle) normally contract mainly in response to nerve stimuli. The nerve endings secrete acetylcholine in the case of some multi-unit smooth muscles and norepinephrine in the case of others. In both instances, these transmitter substances cause depolarization of the smooth muscle membrane, and this in turn elicits the contraction. Action potentials usually do not develop. The reason for this is that the fibers are too small to generate an action potential. (When action potentials are elicited in visceral unitary smooth muscle, 30 to 40 smooth muscle fibers must depolarize simultaneously before a self-propagating action potential ensues.) Yet, even without an action potential in the multi-unit smooth muscle fibers, the local depolarization, called the junctional potential, caused by the nerve transmitter substance itself spreads "electrotonically" over the entire fiber and is all that is needed to cause the muscle contraction.

Smooth Muscle Contraction Without Action Potentials—Effect of Local Tissue Factors and Hormones

Probably at least half of all smooth muscle contraction is initiated not by action potentials but by stimulatory factors acting directly on the smooth muscle contractile machinery. The two types of non-nervous and nonaction potential stimulating factors most often involved are (1) local tissue factors and (2) various hormones.

SMOOTH MUSCLE CONTRACTION IN RESPONSE TO LOCAL TISSUE FACTORS. In Chapter 17, we discuss the control of contraction of the arterioles, meta-arterioles, and precapillary sphincters. The smaller of these vessels have little or no nervous supply. Yet the smooth muscle is highly contractile, responding rapidly to changes in local conditions in the surrounding interstitial fluid. In this way, a powerful local feedback control system controls the blood flow to the local tissue area. Some of the specific control factors are as follows:

1. Lack of oxygen in the local tissues causes smooth muscle relaxation and, therefore, vasodilation.
2. Excess carbon dioxide causes vasodilatation.
3. Increased hydrogen ion concentration also causes increased vasodilatation.

Such factors as adenosine, lactic acid, increased potassium ions, diminished calcium ion concentration, and decreased body temperature also cause local vasodilatation.

EFFECTS OF HORMONES ON SMOOTH MUSCLE CONTRACTION. Most of the circulating hormones in the body affect smooth muscle contraction to some degree, and some have profound effects. Some of the more important blood-borne hormones that affect contraction are *norepinephrine, epinephrine, acetylcholine, angiotensin, vasopressin, oxytocin, serotonin,* and *histamine.*

A hormone causes contraction of smooth muscle when the muscle cell membrane contains *hormone-gated excitatory receptors* for the respective hormone. Conversely, the hormone causes inhibition if the membrane contains *inhibitory receptors* rather than excitatory receptors.

MECHANISM OF SMOOTH MUSCLE EXCITATION OR INHIBITION BY HORMONES OR LOCAL TISSUE FACTORS. Some hormone receptors in the smooth muscle membrane open sodium or calcium ion channels and depolarize the membrane the same as after nerve stimulation. Sometimes action potentials result, or rhythmical action potentials that are already occurring may be enhanced. In many instances, depolarization occurs without action potentials; this depolarization is associated with calcium ion entry into the cell that promotes contraction.

Activation of other membrane receptors inhibits contraction. This is achieved by closing sodium and calcium channels to prevent the entry of these positive ions or by opening potassium channels to allow positive potassium ions to flow to the exterior, in both instances increasing the degree of negativity inside the muscle cell, a state called *hyperpolarization.*

Sometimes contraction or inhibition is initiated by hormones without causing any change in the membrane potential. In these instances, the hormone may activate a membrane receptor that does not open any ion channels but instead causes an internal change in the muscle fiber, such as release of calcium ions from the sarcoplasmic reticulum; the calcium then induces contraction. Or, to inhibit contraction, other receptor mechanisms are known to activate the enzyme *adenylate cyclase* or *guanylate cyclase* in the cell membrane; a portion of the enzyme protrudes to the interior of the cell and causes the formation of *cyclic adenosine monophosphate (cAMP)* or *cyclic guanosine monophosphate (cGMP),* so-called *second messengers.* The cAMP or cGMP in turn has many effects, one of which is to change the degree of phosphorylation of several enzymes that indirectly inhibit contraction. Especially, the pump that pumps calcium ions from the sarcoplasm into the sarcoplasmic reticulum is activated as well as the cell membrane pump that pumps calcium ions out of the cell itself; these effects reduce the intracellular calcium ion concentration, thereby inhibiting contraction.

It is not known how most other local tissue factors besides the hormones—such as oxygen lack, excess carbon dioxide, or changes in hydrogen ion concentration—either excite or inhibit smooth muscle contraction. Possible mechanisms include changes in the cell membrane potential, changes in the permeability of the membrane to calcium ions, changes in the intracellular contractile machinery, or some combination.

Source of Calcium Ions That Cause Contraction: (1) Through the Cell Membrane and (2) From the Sarcoplasmic Reticulum

Although the contractile process in smooth muscle, as in skeletal muscle, is activated by calcium ions, the source of the calcium ions differs considerably in smooth muscle; the difference is that the sarcoplasmic reticulum, from which virtually all the calcium ions are derived in skeletal muscle contraction, is only rudimentary in most smooth muscle. Instead, in most types of smooth muscle, almost all the calcium ions that cause contraction enter the muscle cell from the extracellular fluid at the time of the action potential or other stimulus. There is a reasonably high concentration of calcium ions in the extracellular fluid, greater than 10^{-3} molar in comparison with less than 10^{-7} molar in the cell sarcoplasm, and as pointed out earlier, the smooth muscle action potential is caused mainly by influx of calcium ions into the muscle cell. Because the smooth muscle fibers are extremely small (in contrast to the skeletal muscle fibers), these calcium ions can diffuse to all parts of the smooth muscle and elicit the contractile process. The time required for this diffusion to occur is usually 200 to 300 milliseconds and is called the *latent period* before the contraction begins; this latent period is some 50 times as great as that for skeletal muscle contraction.

Additional calcium can enter the smooth muscle fiber by way of *hormone-activated calcium channels;* these, too, cause contraction. Usually, the opening of these channels does not cause an action potential and sometimes not much change in the resting membrane potential because enough potassium ions move to the exterior to maintain an almost normal membrane potential. Even so, contraction continues as long as these calcium channels remain open because it is calcium ions, not a change in membrane potential, that cause contraction. This is one means by which contraction is achieved in smooth muscle without significant change in cell membrane potential.

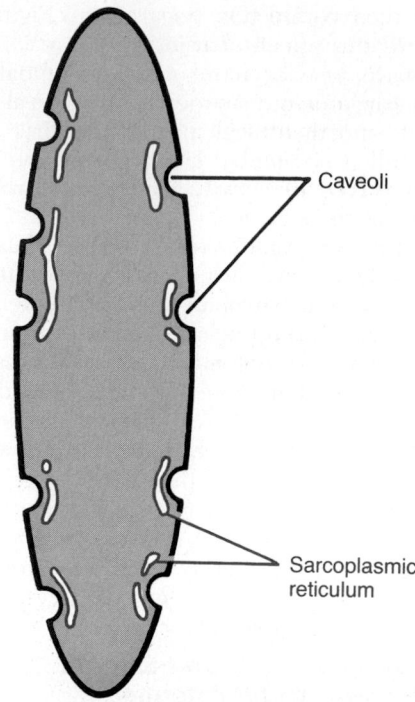

Figure 8–5. Sarcoplasmic tubules in a smooth muscle fiber showing their relation to invaginations in the cell membrane called *caveoli.*

ROLE OF THE SARCOPLASMIC RETICULUM. Some smooth muscle contains a moderately developed sarcoplasmic reticulum. Figure 8–5 shows an example, demonstrating separate sarcoplasmic tubules that lie near the cell membrane. Small invaginations of the membrane, called *caveoli*, abut against the surfaces of these tubules. The caveoli are believed to represent a rudimentary analog of the transverse tubule system of skeletal muscle. When an action potential is transmitted into the caveoli, this seems to excite calcium ion release from the abutting sarcoplasmic tubules in the same way that action potentials in skeletal muscle transverse tubules cause release of calcium ions from the longitudinal sarcoplasmic tubules.

In general, the more extensive the sarcoplasmic reticulum in the smooth muscle fiber, the more rapidly it contracts, presumably because calcium entry through the cell membrane is much slower than internal release of calcium ions from the sarcoplasmic reticulum.

EFFECT OF EXTRACELLULAR CALCIUM ION CONCENTRATION ON SMOOTH MUSCLE CONTRACTION. Although the extracellular fluid calcium ion concentration has almost no effect on the force of contraction of skeletal muscle, this is not true for most smooth muscle. When the extracellular fluid calcium ion concentration falls to a low level, smooth muscle contraction usually almost ceases. In fact, after several minutes of being immersed in a low-calcium medium, even the sarcoplasmic reticulum of smooth muscle fibers loses its calcium supply. Therefore, the force of contraction of smooth muscle is highly dependent on the extracellular fluid calcium ion concentration. We will see in Chapter 9 that this is also true for cardiac muscle.

CALCIUM PUMP. To cause relaxation of the smooth muscle contraction, it is necessary to remove the calcium ions from the intracellular fluids surrounding the actin and myosin filaments. This removal is achieved by calcium pumps that pump the calcium ions out of the smooth muscle fiber back into the extracellular fluid or pump the calcium ions into the sarcoplasmic reticulum. These pumps are slow-acting in comparison with the fast-acting sarcoplasmic reticulum pump in skeletal muscle. Therefore, the duration of smooth muscle contraction is often in the order of seconds rather than hundredths to tenths of a second, as occurs for skeletal muscle.

REFERENCES

Bennett, M. R.: Development of neuromuscular synapses. Physiol. Rev., 63:915, 1983.

Blaustein, M. P., and Hamlyn, J. M.: Sodium transport inhibition, cell calcium, and hypertension. The natriuretic hormone/Na$^+$-Ca^{2+} exchange/hypertension hypothesis. Am. J. Med., 77:45, 1984.

Bohr, D. F., and Webb, R. C.: Vascular smooth muscle function and its changes in hypertension. Am. J. Med., 77:3, 1984.

Bolton, T. B.: Mechanisms of action of transmitters and other substances on smooth muscle. Physiol. Rev., 59:606, 1979.

Borgstrom, P., et al.: An evaluation of the metabolic interaction with myogenic vascular reactivity during blood flow autoregulation. Acta Physiol. Scand., 122:275, 1984.

Bulbring, E. (ed.): Smooth Muscle. An Assessment of Current Knowledge. Austin, University of Texas Press, 1981.

Butler, T. M., and Siegman, M. J.: High-energy phosphate metabolism in vascular smooth muscle. Annu. Rev. Physiol., 47:629, 1985.

Campbell, J. H., and Campbell, G. R.: Endothelial cell influences on vascular smooth muscle phenotype. Annu. Rev. Physiol., 48:295, 1986.

Dowben, R. M. (ed.): Cell and Muscle Motility. New York, Plenum Publishing Corp., 1983.

Furchgott, R. F.: The role of endothelium in the responses of vascular smooth muscle to drugs. Annu. Rev. Pharmacol. Toxicol., 24:175, 1984.

Gabella, G.: Structural apparatus for force transmission in smooth muscle. Physiol. Rev., 64:455, 1984.

Hai, C.-M., and Murphy, R. A.: Ca^{2+}, crossbridge phosphorylation, and contraction. Annu. Rev. Physiol., 51:285, 1989.

Hartshorne, D. J., and Gorecka, A.: Biochemistry of the contractile proteins of smooth muscle. In Bohr, D. F., et al. (eds.): Handbook of Physiology. Sec. 2, Vol. II. Baltimore, Williams & Wilkins, 1980, p. 83.

Hartshorne, D. J., and Siemankowski, R. R.: Regulation of smooth muscle actomyosin. Annu. Rev. Physiol., 43:519, 1981.

Hess, G. P., et al.: Acetylcholine receptor–controlled ion translocation: Chemical kinetic investigations of the mechanism. Annu. Rev. Biophys. Bioeng., 12:443, 1983.

Hirst, G. D. S., and Edwards, F. R.: Sympathetic neuroeffector transmission in arteries and arterioles. Physiol. Rev., 69:546, 1989.

Hochachka, P. W.: Muscles as Molecular and Metabolic Machines. Boca Raton, Fla., CRC Press, Inc., 1994.

Homsher, E.: Muscle enthalpy production and its relationship to actomyosin ATPase. Annu. Rev. Physiol., 49:673, 1987.

Johansson, B., and Somlyo, A. P.: Electrophysiology and excitation-contraction coupling. In Bohr, D. F., et al. (eds.): Handbook of Physiology. Sec. 2, Vol. II. Baltimore, Williams & Wilkins, 1980, p. 301.

Kamm, K. E., and Stull, J. T.: Regulation of smooth muscle contractile elements by second messengers. Annu. Rev. Physiol., 51:299, 1989.

Keynes, R. D.: Nerve and Muscle. 2nd Ed. New York, Cambridge University Press, 1992.

Kito, S., et al. (eds.): Neuroreceptors and Signal Transduction. New York, Plenum Publishing Corp., 1988.

Lambert, J. J., et al.: Drug-induced modification of ionic conductance at the neuromuscular junction. Annu. Rev. Pharmacol. Toxicol., 23:505, 1983.

Lowenstein, W. R.: Junctional intercellular communication: The cell-to-cell membrane channel. Physiol. Rev., 61:829, 1981.

McDonald, T. F., et al.: Regulation and modulation of calcium channels in cardiac, skeletal, and smooth muscle cells. Physiol. Rev., 74:365, 1994.

McKinney, M., and Richelson, E.: The coupling of neuronal muscarinic receptor to responses. Annu. Rev. Pharmacol. Toxicol., 24:121, 1984.

Morgan, D. L., and Proske, U.: Vertebrate slow muscle: Its structure, pattern of innervation, and mechanical properties. Physiol. Rev. 64:103, 1984.

Murphy, R. A.: Muscle cells of hollow organs. News Physiol. Sci., 3:124, 1988.

O'Donnell, M. E., and Owen, N. E.: Regulation of ion pumps and carriers in vascular smooth muscle. Physiol. Rev., 74:683, 1994.

Paul, R. J.: Smooth muscle energetics. Annu. Rev. Physiol., 51:331, 1989.

Purves, D., and Lichtman, J. W.: Specific connections between nerve cells. Annu. Rev. Physiol., 45:553, 1983.

Pusch, M., and Jentsch, T. J.: Molecular physiology of voltage-gated chloride channels. Physiol. Rev., 74:813, 1994.

Putney, J. W., Jr., et al.: How do inositol phosphates regulate calcium signaling? FASEB J., 3:1899, 1989.

Rasmussen, H., et al.: Protein kinase C in the regulation of smooth muscle contraction. FASEB J., 1:177, 1987.

Rosenthal, W., et al.: Control of voltage-dependent Ca^{2+} channels by G protein-coupled receptors. FASEB J., 2:2784, 1988.

Rowland, L. P., et al. (eds.): Molecular Genetics in Diseases of Brain, Nerve, and Muscle. New York, Oxford University Press, 1989.

Schneider, M. F.: Membrane charge movement and depolarization-contraction coupling. Annu. Rev. Physiol., 43:507, 1981.

Seidel, C. L., and Schildmeyer, L. A.: Vascular smooth muscle adaptation to increased load. Annu. Rev. Physiol., 49:489, 1987.

Somlyo, A. P.: Ultrastructure of vascular smooth muscle. In Bohr, D. F., et al. (eds.): Handbook of Physiology. Sec. 2, Vol. II. Baltimore, Williams & Wilkins, 1980, p. 33.

Spray, D. C., and Bennett, M. V. L.: Physiology and pharmacology of gap junctions. Annu. Rev. Physiol., 47:281, 1985.

van Breemen, C., and Saida, K.: Cellular mechanisms regulating [Ca^{2+}]$_i$ smooth muscle. Annu. Rev. Physiol., 51:315, 1989.

Vanhoutte, P. M.: Calcium-entry blockers, vascular smooth muscle and systemic hypertension. Am. J. Cardiol., 55:17B, 1985.

(See also Chapters 5 and 6.)

THE HEART
UNIT III

Heart Muscle; The Heart as a Pump

CHAPTER 9

With this chapter we begin discussion of the heart and circulatory system. The heart, shown in Figure 9–1, is actually two separate pumps: a *right heart* that pumps the blood through the lungs and a *left heart* that pumps the blood through the peripheral organs. In turn, each of these hearts is a pulsatile two-chamber pump composed of an *atrium* and a *ventricle*. The atrium functions principally as a weak primer pump for the ventricle, helping to move the blood into the ventricle. The ventricle in turn supplies the main force that propels the blood through either the pulmonary or the peripheral circulation.

Special mechanisms in the heart provide cardiac rhythmicity and transmit action potentials throughout the heart muscle to cause the heart's rhythmical beat. This rhythmical control system is explained in Chapter 10. In this chapter, we explain how the heart operates as a pump, beginning with the special features of heart muscle itself.

PHYSIOLOGY OF CARDIAC MUSCLE

The heart is composed of three major types of cardiac muscle: atrial muscle, ventricular muscle, and specialized excitatory and conductive muscle fibers. The atrial and ventricular types of muscle contract in much the same way as skeletal muscle except that the duration of contraction is much longer. On the other hand, the specialized excitatory and conductive fibers contract only feebly because they contain few contractile fibrils; instead, they exhibit rhythmicity and varying rates of conduction, providing an excitatory system for the heart.

Physiologic Anatomy of Cardiac Muscle

Figure 9–2 shows a typical histological picture of cardiac muscle, demonstrating the cardiac muscle fibers arranged in a latticework, the fibers dividing, then recombining, and then spreading again. One notes immediately from this figure that cardiac muscle is *striated* in the same manner as typical skeletal muscle. Furthermore, cardiac muscle has typical myofibrils that contain *actin* and *myosin filaments* almost identical to those found in skeletal muscle, and these filaments interdigitate and slide along one another during contraction in the same manner as occurs in skeletal muscle. (See Chapter 6.) In other ways, cardiac muscle is quite different from skeletal muscle, as we shall see.

CARDIAC MUSCLE AS A SYNCYTIUM. The dark areas crossing the cardiac muscle fibers in Figure 9–2 are called *intercalated discs;* they are actually cell membranes that separate individual cardiac muscle cells from one another. That is, cardiac muscle fibers are made up of many individual cells connected in series with one another. Yet electrical resistance through the intercalated disc is only ¼₀₀ the resistance through the outside membrane of the cardiac muscle fiber because the cell membranes fuse with one another in such a way that they form permeable "communicating" junctions (gap junctions) that allow relatively free diffusion of ions. Therefore, from a functional point of view, ions move with ease along the longitudinal axes of the cardiac muscle fibers, so that action potentials travel from one cardiac muscle cell to another, past the intercalated discs, with only slight hindrance. Thus, cardiac muscle is a *syncytium* of many heart muscle cells, in which the cardiac cells are so interconnected that

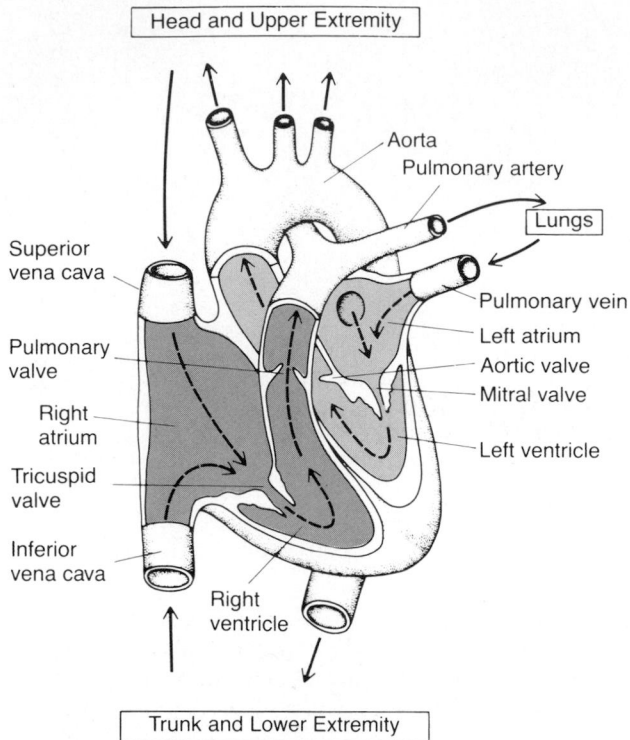

Figure 9–1. Structure of the heart and course of blood flow through the heart chambers.

when one of these cells becomes excited, the action potential spreads to all of them, spreading from cell to cell as well as throughout the latticework interconnections.

The heart is actually composed of two syncytiums: the *atrial syncytium* that constitutes the walls of the two atria and the *ventricular syncytium* that constitutes the walls of the two ventricles. The atria are separated from the ventricles by fibrous tissue that surrounds the valvular openings between the atria and ventricles. Normally, action potentials can be conducted from the atrial syncytium into the ventricular syncytium only by way of a specialized conductive sys-

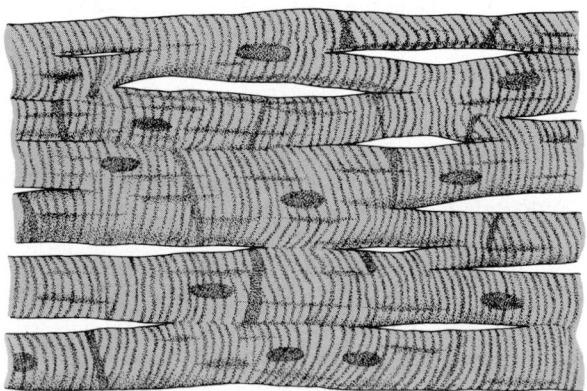

Figure 9–2. "Syncytial," interconnecting nature of cardiac muscle.

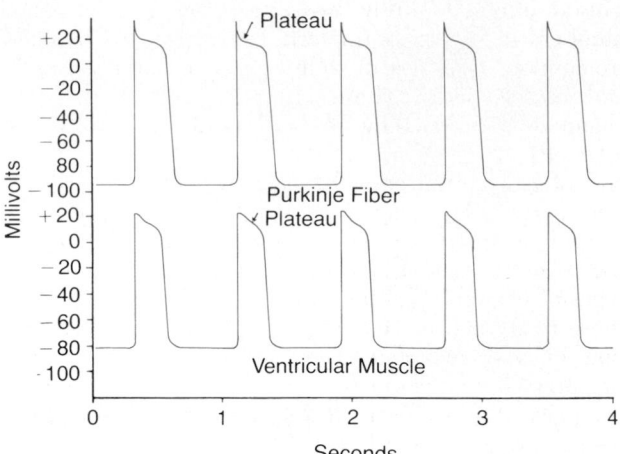

Figure 9–3. Rhythmical action potentials from a Purkinje fiber and from a ventricular muscle fiber recorded by means of microelectrodes.

tem, the *atrioventricular (A-V) bundle,* a bundle of conductive fibers several millimeters in diameter that is discussed in detail in Chapter 10. This division of the muscle mass of the heart into two functional syncytiums allows the atria to contract a short time ahead of ventricular contraction, which is important for the effectiveness of heart pumping.

Action Potentials in Cardiac Muscle

The *resting membrane potential* of normal cardiac muscle is about − 85 to − 95 millivolts and about − 90 to − 100 millivolts in the specialized conductive fibers, the Purkinje fibers, which are discussed in Chapter 10.

The *action potential* recorded in ventricular muscle, shown by the bottom record of Figure 9–3, is 105 millivolts, which means that the membrane potential rises from its normally very negative value to a slightly positive value of about + 20 millivolts. The positive portion is called the *overshoot potential.* After the initial *spike,* the membrane remains depolarized for about 0.2 second in atrial muscle and about 0.3 second in ventricular muscle, exhibiting a *plateau* as shown in Figure 9–3, followed at the end of the plateau by abrupt repolarization. The presence of this plateau in the action potential causes muscle contraction to last 3 to 15 times as long in cardiac muscle as in skeletal muscle.

WHY THE LONG ACTION POTENTIAL AND THE PLATEAU? At this point, we must ask the question: Why is the action potential of cardiac muscle so long, and why does it have a plateau, whereas that of skeletal muscle does not? The basic biophysical answers to these questions are presented in Chapter 5, but they merit summarizing again.

At least two major differences between the membrane properties of cardiac and skeletal muscle account for the prolonged action potential and the plateau in cardiac muscle.

First, the action potential of skeletal muscle is

caused almost entirely by sudden opening of large numbers of so-called *fast sodium channels* that allow tremendous numbers of sodium ions to enter the skeletal muscle fiber. These channels are called "fast" channels because they remain open for only a few 10,000ths of a second and then abruptly close. At the end of this closure, repolarization occurs, and the action potential is over within another 10,000th of a second or so. In cardiac muscle, on the other hand, the action potential is caused by the opening of two types of channels: (1) the same *fast sodium channels* as those in skeletal muscle and (2) another entire population of so-called *slow calcium channels*, also called *calcium-sodium channels.* This second population of channels differs from the fast sodium channels in being slower to open; but more important, they remain open for several tenths of a second. During this time, a large quantity of both calcium and sodium ions flows through these channels to the interior of the cardiac muscle fiber, and this maintains a prolonged period of depolarization, causing the plateau in the action potential. Furthermore, the calcium ions that enter the muscle during this action potential play an important role in helping excite the muscle contractile process, which is another difference between cardiac muscle and skeletal muscle, as we discuss later in this chapter.

The second major functional difference between cardiac muscle and skeletal muscle that helps account for both the prolonged action potential and its plateau is this: Immediately after the onset of the action potential, the permeability of the cardiac muscle membrane for potassium *decreases* about fivefold, an effect that does not occur in skeletal muscle. This decreased potassium permeability may be caused by the excess calcium influx through the calcium channels just noted. Regardless of the cause, the decreased potassium permeability greatly decreases the outflux of potassium ions during the action potential plateau and thereby prevents early return of the potential to its resting level. When the slow calcium-sodium channels do close at the end of 0.2 to 0.3 second and the influx of calcium and sodium ions ceases, the membrane permeability for potassium increases rapidly; this rapid loss of potassium from the fiber returns the membrane potential to its resting level, thus ending the action potential.

VELOCITY OF CONDUCTION IN CARDIAC MUSCLE. The velocity of conduction of the action potential in both atrial and ventricular muscle fibers is about 0.3 to 0.5 m/sec, or about 1/250 the velocity in very large nerve fibers and about 1/10 the velocity in skeletal muscle fibers. The velocity of conduction in the specialized conductive system—the Purkinje fibers—varies from 0.02 to 4 m/sec in different parts of the system, which allows rapid conduction of the excitatory signal in the heart, as explained in Chapter 10.

REFRACTORY PERIOD OF CARDIAC MUSCLE. Cardiac muscle, like all excitable tissue, is refractory to restimulation during the action potential. Therefore, the refractory period of the heart is the interval of time, as shown to the left in Figure 9–4, during which a nor-

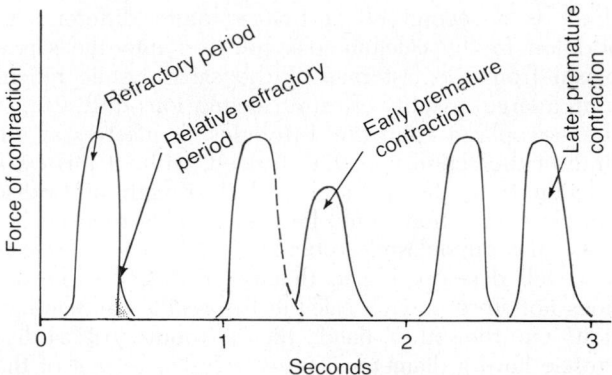

Figure 9–4. Contraction of the heart showing the durations of the refractory period and the relative refractory period, the effect of an early premature contraction, and the effect of a later premature contraction. Note that the premature contractions do not cause wave summation, as occurs in skeletal muscle.

mal cardiac impulse cannot re-excite an already excited area of cardiac muscle. The normal refractory period of the ventricle is 0.25 to 0.3 second, which is about the duration of the action potential. There is an additional *relative refractory period* of about 0.05 second during which the muscle is more difficult than normal to excite but nevertheless can be excited, as demonstrated by the early premature contraction in the second example of Figure 9–4.

The refractory period of atrial muscle is much shorter than that for the ventricles (about 0.15 second), and the relative refractory period is another 0.03 second. Therefore, the rhythmical rate of contraction of the atria can be faster than that of the ventricles.

Excitation-Contraction Coupling—Function of Calcium Ions and the Transverse Tubules

The term "excitation-contraction coupling" means the mechanism by which the action potential causes the myofibrils of muscle to contract. This is discussed for skeletal muscle in Chapter 7. Once again, there are differences in this mechanism in cardiac muscle that have important effects on the characteristics of cardiac muscle contraction.

As is true for skeletal muscle, when an action potential passes over the cardiac muscle membrane, the action potential also spreads to the interior of the cardiac muscle fiber along the membranes of the transverse (T) tubules. The T tubule action potentials in turn act on the membranes of the longitudinal sarcoplasmic tubules to cause instantaneous release of calcium ions into the muscle sarcoplasm from the sarcoplasmic reticulum. In another few thousandths of a second, these calcium ions diffuse into the myofibrils and catalyze the chemical reactions that promote sliding of the actin and myosin filaments along one another; this in turn produces the muscle contraction.

Thus far, this mechanism of excitation-contraction coupling is the same as that for skeletal muscle, but there is a second effect that is quite different. In addition to the calcium ions released into the sarcoplasm from the cisternae of the sarcoplasmic reticulum, a large quantity of extra calcium ions diffuses into the sarcoplasm from the T tubules themselves at the time of the action potential. Indeed, without this extra calcium from the T tubules, the strength of cardiac muscle contraction would be considerably reduced because the sarcoplasmic reticulum of cardiac muscle is less well developed than that of skeletal muscle and does not store enough calcium to provide full contraction. On the other hand, the T tubules of cardiac muscle have a diameter 5 times as great as that of the skeletal muscle tubules, which means a volume 25 times as great. Also, inside the T tubules is a large quantity of mucopolysaccharides that are electronegatively charged and bind an abundant store of even more calcium ions, keeping them always available for diffusion to the interior of the cardiac muscle fiber when the T tubule action potential occurs.

The strength of contraction of cardiac muscle depends to a great extent on the concentration of calcium ions in the extracellular fluids. The reason for this is that the ends of the T tubules open directly to the outside of the cardiac muscle fibers, allowing the same extracellular fluid that is in the cardiac muscle interstitium to percolate through the T tubules as well. Consequently, the quantity of calcium ions in the T tubule system—that is, the availability of calcium ions to cause cardiac muscle contraction—depends to a great extent on the extracellular fluid calcium ion concentration.

By way of contrast, the strength of skeletal muscle contraction is hardly affected by the extracellular fluid calcium concentration because its contraction is caused almost entirely by calcium ions released from the sarcoplasmic reticulum inside the skeletal muscle fiber itself.

At the end of the plateau of the cardiac action potential, the influx of calcium ions to the interior of the muscle fiber is suddenly cut off, and the calcium ions in the sarcoplasm are rapidly pumped back into both the sarcoplasmic reticulum and the T tubules. As a result, the contraction ceases until a new action potential occurs.

DURATION OF CONTRACTION. Cardiac muscle begins to contract a few milliseconds after the action potential begins and continues to contract until a few milliseconds after the action potential ends. Therefore, the duration of contraction of cardiac muscle is mainly a function of the duration of the action potential—about 0.2 second in atrial muscle and 0.3 second in ventricular muscle.

Effect of Heart Rate on Duration of Contraction. When the heart rate increases, the duration of each total cycle of the heart, including both the contraction phase and the relaxation phase, decreases. The duration of the action potential and the period of contraction (systole) also decrease but not by as great a percentage as does the relaxation phase (diastole). At a normal heart rate of 72 beats per minute, the period of contraction is about 0.40 of the entire cycle. At three times

normal heart rate, this period is about 0.65 of the entire cycle, which means that the heart beating at a very fast rate sometimes does not remain relaxed long enough to allow complete filling of the cardiac chambers before the next contraction.

THE CARDIAC CYCLE

The cardiac events that occur from the beginning of one heartbeat to the beginning of the next are called the *cardiac cycle.* Each cycle is initiated by spontaneous generation of an action potential in the sinus node, as explained in Chapter 10. This node is located in the superior lateral wall of the right atrium near the opening of the superior vena cava, and the action potential travels rapidly through both atria and thence through the A-V bundle into the ventricles. Because of a special arrangement of the conducting system from the atria into the ventricles, there is a delay of more than $\frac{1}{10}$ second between passage of the cardiac impulse from the atria into the ventricles. This allows the atria to contract ahead of the ventricles, thereby pumping blood into the ventricles before the strong ventricular contraction. Thus, the atria act as *primer pumps* for the ventricles, and the ventricles then provide the major source of power for moving blood through the vascular system.

Systole and Diastole

The cardiac cycle consists of a period of relaxation called *diastole,* during which the heart fills with blood, followed by a period of contraction called *systole.*

Figure 9–5 shows the different events during the cardiac cycle. The top three curves show the pressure changes in the aorta, left ventricle, and left atrium, respectively. The fourth curve depicts the changes in ventricular volume, the fifth the electrocardiogram, and the sixth a phonocardiogram, which is a recording of the sounds produced by the heart—mainly by the heart valves—as it pumps. It is especially important that the reader study in detail the diagram of this figure and understand the causes of all the events shown.

Relationship of the Electrocardiogram to the Cardiac Cycle

The electrocardiogram in Figure 9–5 shows the *P, Q, R, S,* and *T waves,* which are discussed in Chapters 11, 12, and 13. They are electrical voltages generated by the heart and recorded by the electrocardiograph from the surface of the body. The *P wave* is caused by the *spread of depolarization* through the atria, and this is followed by atrial contraction, which causes a slight rise in the atrial pressure curve immediately after the P wave. About 0.16 second after the onset of the P wave, the *QRS waves* appear as a result of depolarization of the ventricles, which initiates contraction of the ventricles and causes the ventricular pressure to begin

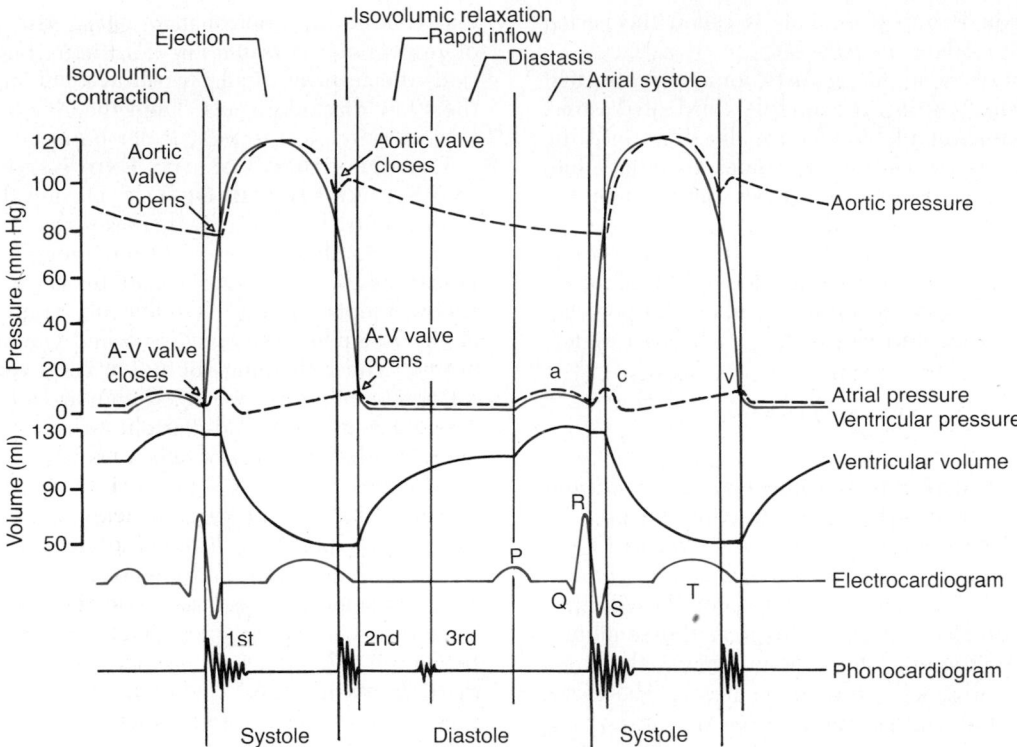

Figure 9–5. Events of the cardiac cycle for left ventricular function showing changes in left atrial pressure, left ventricular pressure, aortic pressure, ventricular volume, the electrocardiogram, and the phonocardiogram.

rising, as also shown in the figure. Therefore, the QRS complex begins slightly before the onset of ventricular systole.

Finally, one observes the *ventricular T wave* in the electrocardiogram. This represents the stage of repolarization of the ventricles at which time the ventricular muscle fibers begin to relax. Therefore, the T wave occurs slightly before the end of ventricular contraction.

Function of the Atria as Primer Pumps

Blood normally flows continually from the great veins into the atria; about 75 per cent of the blood flows directly through the atria into the ventricles even before the atria contract. Then, atrial contraction usually causes an additional 25 per cent filling of the ventricles. Therefore, the atria simply function as primer pumps that increase the ventricular pumping effectiveness as much as 25 per cent. Yet the heart can continue to operate satisfactorily under most conditions even without this extra 25 per cent effectiveness because it normally has the capability of pumping 300 to 400 per cent more blood than is required by the body. Therefore, when the atria fail to function, the difference is unlikely to be noticed unless a person exercises; then acute signs of heart failure occasionally develop, especially shortness of breath.

PRESSURE CHANGES IN THE ATRIA—THE A, C, AND V WAVES. In the atrial pressure curve of Figure 9–5,

three major pressure elevations, called the *a, c,* and *v atrial pressure waves,* can be noted.

The *a wave* is caused by atrial contraction. Ordinarily, the *right* atrial pressure rises 4 to 6 mm Hg during atrial contraction, whereas the *left* atrial pressure rises about 7 to 8 mm Hg.

The *c wave* occurs when the ventricles begin to contract; it is caused partly by slight backflow of blood into the atria at the onset of ventricular contraction but probably mainly by bulging of the A-V valves backward toward the atria because of increasing pressure in the ventricles.

The *v wave* occurs toward the end of ventricular contraction; it results from slow flow of blood into the atria from the veins while the A-V valves are closed during ventricular contraction. Then, when ventricular contraction is over, the A-V valves open, allowing this blood to flow rapidly into the ventricles and causing the v wave to disappear.

Function of the Ventricles as Pumps

FILLING OF THE VENTRICLES. During ventricular systole, large amounts of blood accumulate in the atria because of the closed A-V valves. Therefore, just as soon as systole is over and the ventricular pressures fall again to their low diastolic values, the moderately increased pressures in the atria immediately push the A-V valves open and allow blood to flow rapidly into the ventricles, as shown by the rise of the *ventricular* *volume curve* in Figure 9–5. This is called the *period of rapid filling of the ventricles.*

The period of rapid filling lasts for about the first third of diastole. During the middle third of diastole, only a small amount of blood normally flows into the ventricles; this is blood that continues to empty into the atria from the veins and passes on through the atria directly into the ventricles.

During the last third of diastole, the atria contract and give an additional thrust to the inflow of blood into the ventricles; this accounts for about 25 per cent of the filling of the ventricles during each heart cycle.

EMPTYING OF THE VENTRICLES DURING SYSTOLE

Period of Isovolumic (Isometric) Contraction. Immediately after ventricular contraction begins, the ventricular pressure abruptly rises, as shown in Figure 9–5, causing the A-V valves to close. Then an additional 0.02 to 0.03 second is required for the ventricle to build up sufficient pressure to push the semilunar (aortic and pulmonary) valves open against the pressures in the aorta and pulmonary artery. Therefore, during this period, contraction is occurring in the ventricles, but there is no emptying. This period is called the period of *isovolumic* or *isometric contraction,* meaning by these terms that tension is increasing in the muscle but no shortening of the muscle fibers is occurring. (This is not strictly true because there is apex-to-base shortening and circumferential elongation.)

Period of Ejection. When the left ventricular pressure rises slightly above 80 mm Hg (and the right ventricular pressure slightly above 8 mm Hg), the ventricular pressures now push the semilunar valves open. Immediately, blood begins to pour out of the ventricles, with about 70 per cent of the emptying occurring during the first third of the period of ejection and the remaining 30 per cent during the next two thirds. Therefore, the first third is called the *period of rapid ejection* and the last two thirds, the *period of slow ejection.*

For a peculiar reason, the ventricular pressure falls to a value slightly *below* that in the aorta during the period of slow ejection, despite the fact that some blood is still leaving the left ventricle. The reason is that the blood flowing out of the ventricle has built up momentum. As this momentum decreases during the latter part of systole, the kinetic energy of the momentum is converted into pressure in the aorta, which makes the arterial pressure slightly greater than the pressure inside the ventricle.

PERIOD OF ISOVOLUMIC (ISOMETRIC) RELAXATION. At the end of systole, ventricular relaxation begins suddenly, allowing the intraventricular pressures to fall rapidly. The elevated pressures in the distended large arteries immediately push blood back toward the ventricles, which snaps the aortic and pulmonary valves closed. For another 0.03 to 0.06 second, the ventricular muscle continues to relax, even though the ventricular volume does not change, giving rise to the period of *isovolumic* or *isometric relaxation.* During this period, the intraventricular pressures fall rapidly back to their low diastolic levels. Then the A-V valves open to begin a new cycle of ventricular pumping.

END-DIASTOLIC VOLUME, END-SYSTOLIC VOLUME, AND STROKE VOLUME OUTPUT. During diastole, filling of the ventricles normally increases the volume of each ventricle to about 110 to 120 milliliters. This volume is known as the *end-diastolic volume.* Then, as the ventricles empty during systole, the volume decreases about 70 milliliters, which is called the *stroke volume output.* The remaining volume in each ventricle, about 40 to 50 milliliters, is called the *end-systolic volume.* The fraction of the end-diastolic volume that is ejected is called the *ejection fraction*—usually equal to about 60 per cent.

When the heart contracts strongly, the end-systolic volume can fall to as little as 10 to 20 milliliters. On the other hand, when large amounts of blood flow into the ventricles during diastole, their end-diastolic volumes can become as great as 150 to 180 milliliters in the normal heart. And by both increasing the end-diastolic volume and decreasing the end-systolic volume, the stroke volume output can at times be increased to about double normal.

Function of the Valves

ATRIOVENTRICULAR VALVES. The *A-V valves* (the *tricuspid* and the *mitral* valves) prevent backflow of blood from the ventricles to the atria during systole, and the *semilunar valves* (the *aortic* and *pulmonary* valves) prevent backflow from the aorta and pulmonary arteries into the ventricles during diastole. All these valves, which are shown in Figure 9–6, close and open *passively.* That is, they close when a backward pressure gradient pushes blood backward, and they open when a forward pressure gradient forces blood in the

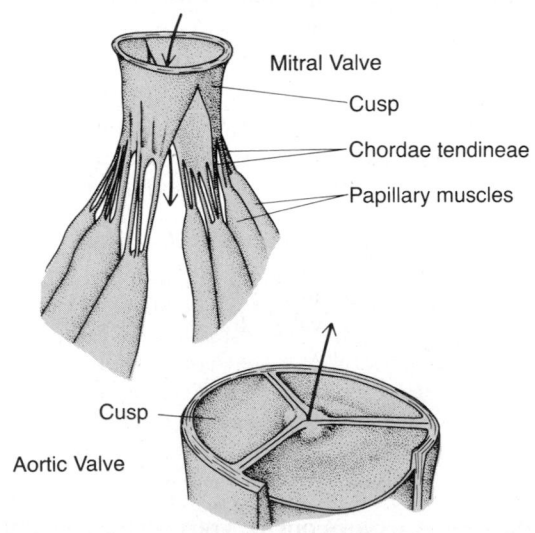

Figure 9–6. Mitral and aortic valves.

forward direction. For anatomical reasons, the thin, filmy A-V valves require almost no backflow to cause closure, whereas the much heavier semilunar valves require rather strong backflow for a few milliseconds.

Function of the Papillary Muscles. Figure 9–6 also shows the papillary muscles that attach to the vanes of the A-V valves by the *chordae tendineae.* The papillary muscles contract when the ventricular walls contract, but contrary to what might be expected, they *do not* help the valves to close. Instead, they pull the vanes of the valves inward toward the ventricles to prevent their bulging too far backward toward the atria during ventricular contraction. If a chorda tendinea becomes ruptured or if one of the papillary muscles becomes paralyzed, the valve bulges far backward, sometimes so far that it leaks severely and results in severe or even lethal cardiac incapacity.

AORTIC AND PULMONARY VALVES. The aortic and pulmonary semilunar valves function quite differently from the A-V valves. First, the high pressures in the arteries at the end of systole cause the semilunar valves to snap to the closed position in comparison with a much softer closure of the A-V valves. Second, because of smaller openings, the velocity of blood ejection through the aortic and pulmonary valves is far greater than that through the much larger A-V valves. Also, because of the rapid closure and rapid ejection, the edges of the semilunar valves are subjected to much greater mechanical abrasion than are the A-V valves. And, finally, the A-V valves are supported by the chordae tendineae, which is not true for the semilunar valves. It is obvious from the anatomy of the aortic and pulmonary valves, as shown in Figure 9–6, that they are well adapted to withstand the extra physical stresses.

The Aortic Pressure Curve

When the left ventricle contracts, the ventricular pressure rises rapidly until the aortic valve opens. Then the pressure in the ventricle rises much less thereafter, as shown in Figure 9–5, because blood immediately flows out of the ventricle into the aorta.

The entry of blood into the arteries causes the walls of these arteries to stretch and the pressure to rise. Then, at the end of systole, after the left ventricle stops ejecting blood and the aortic valve closes, the elastic recoil of the arteries maintains a high pressure in the arteries even during diastole.

A so-called *incisura* occurs in the aortic pressure curve when the aortic valve closes. This is caused by a short period of backward flow of blood immediately before closure of the valve, followed then by sudden cessation of the backflow.

After the aortic valve has closed, pressure in the aorta falls slowly throughout diastole because blood stored in the distended elastic arteries flows continually through the peripheral vessels back to the veins. Before the ventricle contracts again, the aortic pressure usually falls to about 80 mm Hg (diastolic pressure), which is two thirds the maximal pressure of 120 mm Hg (systolic pressure) that occurs in the aorta during ventricular contraction.

The pressure curve in the pulmonary artery is similar to that in the aorta except that the pressures are only about one sixth as great, as discussed in Chapter 14.

Relationship of the Heart Sounds to Heart Pumping

When listening to the heart with a stethoscope, one does not hear the opening of the valves because this is a relatively slowly developing process that makes no noise. However, when the valves close, the vanes of the valves and the surrounding fluids vibrate under the influence of the sudden pressure differentials that develop, giving off sound that travels in all directions through the chest.

When the ventricles contract, one first hears a sound that is caused by closure of the A-V valves. The vibration is low in pitch and relatively long continued and is known as the *first heart sound.* When the aortic and pulmonary valves close at the end of systole, one hears a relatively rapid snap because these valves close rapidly, and the surroundings vibrate for only a short period. This sound is known as the *second heart sound.*

Occasionally, one can hear an *atrial sound* when the atria beat because of vibrations associated with the flow of blood into the ventricles. Also, a *third heart sound* sometimes occurs at about the end of the first third of diastole, believed to be caused by blood flowing with a rumbling motion into the almost-filled ventricles. The precise causes of the heart sounds are discussed more fully in Chapter 23, in relation to auscultation.

Work Output of the Heart

STROKE WORK OUTPUT AND MINUTE WORK OUTPUT. The *stroke work output* of the heart is the amount of energy that the heart converts to work during each heartbeat while pumping blood into the arteries. *Minute work output* is the total amount of energy converted in 1 minute; this is equal to the stroke work output times the heart rate per minute.

Work output of the heart is in two forms. First, by far the major proportion is used to move the blood from the low-pressure veins to the high-pressure arteries. This is called *volume-pressure work* or *external work.* Second, a minor proportion of the energy is used to accelerate the blood to its velocity of ejection through the aortic and pulmonary valves. This is the *kinetic energy of blood flow* component of the work output.

External Work (Volume-Pressure Work). The work performed by the left ventricle to raise the pressure of the blood during each heartbeat (the left ventricular external work output) is equal to the *stroke volume* times (*left ventricular mean ejection pressure minus mean left ventricular input pressure during diastolic filling*). When pressure is expressed in dynes per square centimeter and stroke volume in milliliters, the external work output is in ergs.

Right ventricular external work output is normally about one sixth the work output of the left ventricle because of the difference in systolic pressure against which the two ventricles must pump.

Kinetic Energy of Blood Flow. The additional work output of each ventricle required to create kinetic energy of blood flow is proportional to the mass of blood ejected times the square of velocity of ejection. That is,

$$\text{Kinetic energy} = \frac{mv^2}{2}$$

When the mass is expressed in *grams* of blood ejected and the velocity in *centimeters per second*, the work output is in *ergs*.

Ordinarily, the work output of the left ventricle required to create kinetic energy of blood flow is only about 1 per cent of the total work output of the ventricle and therefore is ignored in the calculation of the total stroke work output. In certain abnormal conditions, such as aortic stenosis in which the blood flows with great velocity through the stenosed valve, more than 50 per cent of the total work output may be required to create kinetic energy of blood flow.

Graphical Analysis of Ventricular Pumping

Figure 9–7 shows a graphical diagram that is especially useful in explaining the pumping mechanics of the left ventricle. The most important components of the diagram are the two heavy black curves labeled "diastolic pressure" and "systolic pressure." These curves are volume-pressure curves.

The diastolic pressure curve is determined by filling the heart with progressively greater quantities of blood and then measuring the diastolic pressure immediately before ventricular contraction occurs, which is the *end-diastolic pressure* of the ventricle.

The systolic pressure curve is determined by preventing any outflow of blood from the heart and measuring the maximum systolic pressure that is achieved during ventricular contraction at each volume of filling.

Until the volume of the ventricle rises above about

150 milliliters, the diastolic pressure does not increase greatly. Therefore, up to this volume, blood can flow easily into the ventricle from the atrium. Above 150 milliliters, the diastolic pressure does increase rapidly, partly because of fibrous tissue in the heart that will stretch no more and partly because the pericardium that surrounds the heart becomes stretched nearly to its limit. During ventricular contraction, the systolic pressure increases rapidly at progressively greater ventricular volumes but reaches a maximum at a ventricular volume of 150 to 170 milliliters. Then, as the volume increases still further, the systolic pressure actually decreases under some conditions, as demonstrated by the falling systolic pressure curve, because at these great volumes, the actin and myosin filaments of the cardiac muscle fibers are pulled apart enough for the strength of cardiac fiber contraction to become less than optimal.

Note especially in the figure that the maximum systolic pressure for the normal unstimulated left ventricle is between 250 and 300 mm Hg, but this varies widely with strength of the heart. For the normal right ventricle, it is between 60 and 80 mm Hg.

"Volume-Pressure Diagram" During the Cardiac Cycle; Cardiac Work Output. The red curves in Figure 9–7 form a loop called the *volume-pressure diagram* of the cardiac cycle for the left ventricle. It is divided into four phases.

Phase I: *Period of filling.* This phase in the volume-pressure diagram begins at a ventricular volume of about 45 milliliters and a diastolic pressure near 0 mm Hg. Forty-five milliliters is the amount of blood that remains in the ventricle after the previous heartbeat and is called the *end-systolic volume.* As venous blood flows into the ventricle from the left atrium, the ventricular volume normally increases to about 115 milliliters, called the *end-diastolic volume,* an increase of 70 milliliters. Therefore, the volume-pressure diagram during phase I extends along the line labeled "I," with the volume increasing to 115 milliliters and the diastolic pressure rising to about 5 mm Hg.

Phase II: *Period of isovolumic contraction.* During isovolumic contraction, the volume of the ventricle does not change because all valves are closed. However, the pressure inside the ventricle rises to equal the pressure in the aorta, a pressure value of about 80 mm Hg, as depicted by the line labeled "II."

Phase III: *Period of ejection.* During ejection, the systolic pressure rises even higher because of still more contraction of the heart. At the same time, the volume of the ventricle decreases because the aortic valve opens and blood now flows out of the ventricle into the aorta. Therefore, the curve labeled "III" traces the changes in volume and systolic pressure during this period of ejection.

Phase IV: *Period of isovolumic relaxation.* At the end of the period of ejection, the aortic valve closes, and the ventricular pressure falls back to the diastolic pressure level. The line labeled "IV" traces this decrease in intraventricular pressure without any change in volume. Thus, the ventricle returns to its starting point, with about 45 milliliters of blood left in the ventricle and at an atrial pressure usually close to 0 mm Hg.

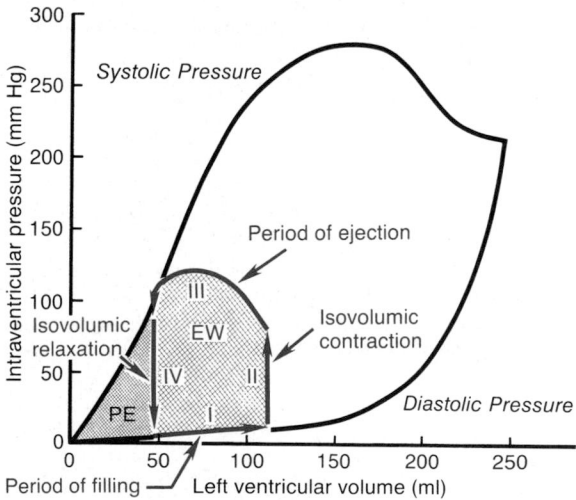

Figure 9–7. Relationship between left ventricular volume and intraventricular pressure during diastole and systole. Also shown by the heavy red lines is the "volume-pressure diagram" that demonstrates the changes in intraventricular volume and pressure during the cardiac cycle.

WORK OUTPUT CALCULATED FROM THE VOLUME-PRESSURE DIAGRAM. Readers well trained in the basic principles of physics should recognize that the area subtended by this volume-pressure diagram (the right-hand portion of the shaded area, labeled EW) is representative of the *net external work output* of the ventricle during its contraction cycle. In experimental studies of cardiac contraction, this diagram is used for calculating cardiac work output.

When the heart pumps large quantities of blood, the work diagram becomes much larger. That is, it extends far to the right because the ventricle now fills with more blood during diastole, it rises much higher because the ventricle contracts with greater pressure, and it usually extends further to the left because the ventricle contracts to a smaller volume—especially if the ventricle is stimulated to increased activity by the sympathetic nervous system.

CONCEPTS OF PRELOAD AND AFTERLOAD. In assessing the contractile properties of muscle, it is important to specify the degree of tension on the muscle when it begins to contract, which is called the *preload,* and to specify the load against which the muscle exerts its contractile force, which is called the *afterload.*

For cardiac contraction, the preload is usually considered to be the end-diastolic pressure when the ventricle has become filled.

The afterload of the ventricle is the pressure in the artery leading from the ventricle. In Figure 9–7, this corresponds to the systolic pressure described by the Phase III curve of the volume-pressure diagram. (Sometimes the afterload is loosely considered to be the resistance in the circulation rather than the pressure.)

The importance of the concepts of preload and afterload is that in many abnormal functional states of the heart or circulation, the pressure during filling of the ventricle (the preload), the arterial pressure against which the ventricle must contract (the afterload), or both are severely altered from the normal.

Chemical Energy for Cardiac Contraction: Oxygen Utilization by the Heart

Heart muscle, like skeletal muscle, uses chemical energy to provide the work of contraction. This energy is derived mainly from oxidative metabolism of fatty acids and, to a lesser extent, other nutrients, especially lactate and glucose. Therefore, the rate of oxygen consumption by the heart is an excellent measure of the chemical energy liberated while the heart performs its work. The different chemical reactions that liberate this energy are discussed in Chapters 67 and 68.

Experimental studies on isolated hearts have shown that the oxygen consumption of the heart, and therefore the chemical energy expended during contraction, is directly related to the total shaded area of Figure 9–7. This shaded portion consists of the *external work,* EW, as explained earlier, and an additional portion called the *potential energy,* labeled PE. The potential energy represents additional work output that could be accomplished by contraction of the ventricle if the ventricle should empty completely all the blood in its chamber with each contraction.

It has also been found experimentally that oxygen consumption is nearly proportional to the *tension* that occurs in the heart muscle during contraction *times the*

duration of time that the contraction persists, called the *tension-time index.* Because tension is high when the systolic pressure is high, correspondingly more oxygen is used. Also, much more chemical energy is expended even at normal systolic pressures when the ventricle is abnormally dilated because the heart muscle tension during contraction is proportional to pressure *times* the diameter of the ventricle. This is especially important in heart failure because the ventricle is then dilated, and paradoxically, the amount of chemical energy required for a given amount of work output must be greater than ever even though the heart is already failing.

EFFICIENCY OF CARDIAC CONTRACTION. During muscle contraction, most of the chemical energy is converted into heat and a much smaller portion into work output. The ratio of work output to chemical energy expenditure is called the efficiency of cardiac contraction, or simply *efficiency of the heart.* The maximum efficiency of the normal heart is between 20 and 25 per cent. In heart failure, this may fall to as low as 5 to 10 per cent.

REGULATION OF HEART PUMPING

When a person is at rest, the heart pumps only 4 to 6 liters of blood each minute. During severe exercise, the heart may be required to pump four to seven times this amount. This section discusses the means by which the heart can adapt to such extreme increases in cardiac output.

The basic means by which the volume pumped by the heart is regulated are (1) intrinsic cardiac regulation of pumping in response to changes in volume of blood flowing into the heart and (2) control of the heart by the autonomic nervous system.

Intrinsic Regulation of Heart Pumping— The Frank-Starling Mechanism

In Chapter 20, we see that the amount of blood pumped by the heart each minute is determined by the rate of blood flow into the heart from the veins, which is called *venous return.* That is, each peripheral tissue of the body controls its own blood flow, and the total of all the local blood flow through all the peripheral tissues returns by way of the veins to the right atrium. The heart in turn automatically pumps this incoming blood into the systemic arteries, so that it can flow around the circuit again.

This intrinsic ability of the heart to adapt to changing volumes of inflowing blood is called the *Frank-Starling mechanism of the heart,* in honor of Frank and Starling, two great physiologists of nearly a century ago. Basically, the Frank-Starling mechanism means that the greater the heart muscle is stretched during filling, the greater will be the force of contraction and the greater will be the quantity of blood pumped into the aorta. Or another way to express this is: *Within physiologically limits, the heart pumps all the blood that comes to it without allowing excessive damming of blood in the veins.*

WHAT IS THE EXPLANATION OF THE FRANK-STARLING MECHANISM? When an extra amount of blood flows into the ventricles, the cardiac muscle itself is stretched to a greater length. This in turn causes the muscle to contract with increased force because the actin and myosin filaments are then brought to a more nearly optimal degree of interdigitation for force generation. Therefore, the ventricle, because of its increased pumping, automatically pumps the extra blood into the arteries. This ability of stretched muscle, up to an optimal length, to contract with increased force is characteristic of all striated muscle, as explained in Chapter 6, not simply of cardiac muscle.

In addition to the important effect of stretching the heart muscle, still another factor increases heart pumping when its volume is increased. Stretch of the right atrial wall directly increases the heart rate by 10 to 20 per cent; this, too, helps increase the amount of blood pumped each minute, although its contribution is much less than that of the Frank-Starling mechanism.

LACK OF EFFECT ON CARDIAC OUTPUT OF CHANGES IN ARTERIAL PRESSURE LOAD. One of the most important consequences of the Frank-Starling mechanism of the heart is that, within reasonable limits, changes in the arterial pressure against which the heart pumps have almost no effect on the rate at which blood is pumped by the heart each minute (the cardiac output). This effect is shown in Figure 9–8, which is a curve extrapolated to the human from data in dogs in which the arterial pressure was progressively changed by constricting the aorta, while the cardiac output was measured simultaneously. The significance of this effect is the following: Regardless of the arterial pressure load up to a reasonable limit, the important factor that determines the amount of blood pumped by the heart is still the rate of entry of blood into the heart.

Ventricular Function Curves

One of the best ways to express the functional ability of the ventricles to pump blood is by ventricular

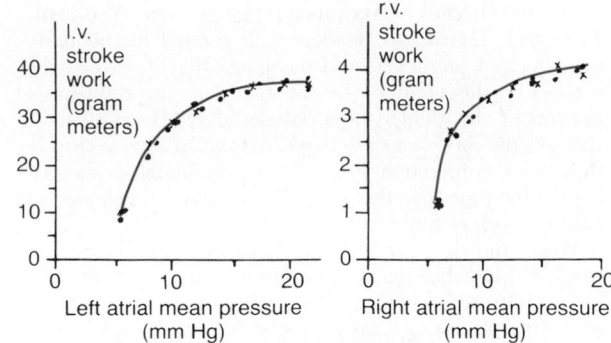

Figure 9–9. Left and right ventricular function curves in a dog depicting ventricular stroke work output as a function of left and right mean atrial pressures. (Curves reconstructed from data in Sarnoff: *Physiol. Rev. 35*:107, 1955.)

function curves, as shown in Figures 9–9 and 9–10. Figure 9–9 shows a type of ventricular function curve called the *stroke work output curve*. Note that as the atrial pressures increase, the stroke work outputs also increase until they reach the limit of the heart's ability.

Figure 9–10 shows another type of ventricular function curve called the *minute ventricular output curve*. These two curves represent function of the two ventricles of the human heart based on data extrapolated from lower animals. As each atrial pressure rises, the respective ventricular volume output per minute also increases.

Thus, ventricular function curves are another way of expressing the Frank-Starling mechanism of the heart. That is, as the ventricles fill to higher atrial pressures, the ventricular volume and strength of cardiac contraction increase, causing the heart to pump increased quantities of blood into the arteries.

Control of the Heart by the Sympathetic and Parasympathetic Nerves

The pumping effectiveness of the heart is highly controlled by the *sympathetic* and *parasympathetic* (vagus) nerves, which abundantly supply the heart, as

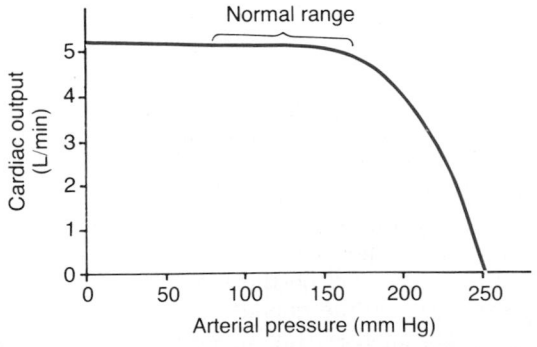

Figure 9–8. Constancy of cardiac output even in the face of wide changes in arterial pressure up to a pressure level of 160 mm Hg. Only when the arterial pressure rises above the normal operating pressure range does the pressure load cause the heart output to fall.

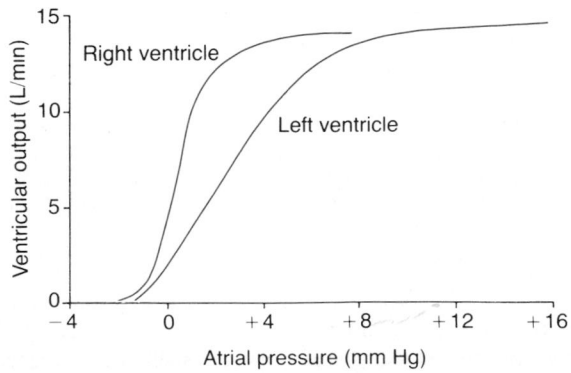

Figure 9–10. Approximate normal right and left ventricular output curves for the normal resting human heart as extrapolated from data obtained in dogs.

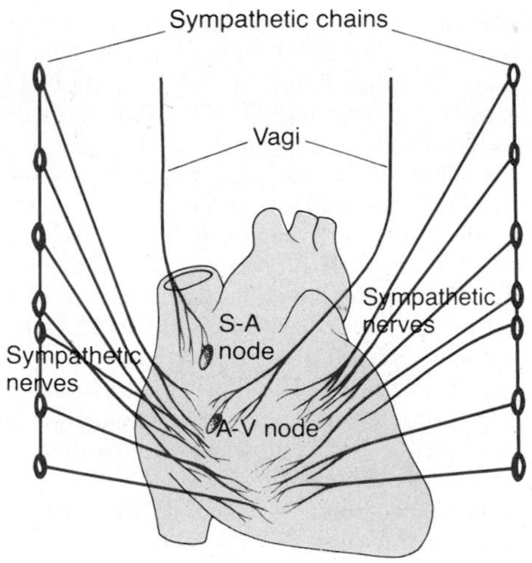

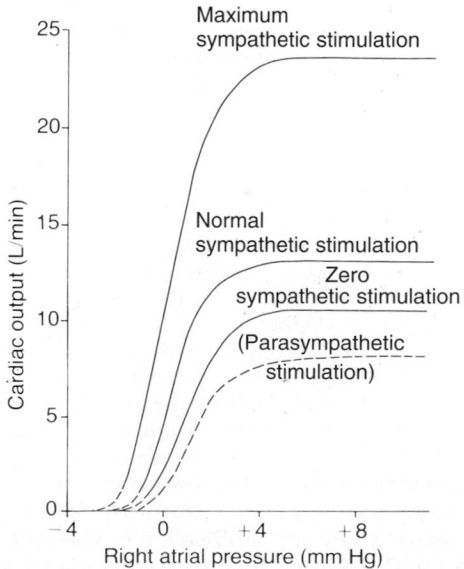

Figure 9–11. Cardiac nerves.

shown in Figure 9–11. The amount of blood pumped by the heart each minute, the *cardiac output,* can often be increased more than 100 per cent by sympathetic stimulation. By contrast, it can be decreased to as low as zero or almost zero by vagal (parasympathetic) stimulation.

EXCITATION OF THE HEART BY THE SYMPATHETIC NERVES. Strong sympathetic stimulation can increase the heart rate in adult humans to 180 to 200 and, rarely, even 250 beats per minute in young people. Also, sympathetic stimulation increases the force with which the heart muscle contracts, therefore increasing the volume of blood pumped and increasing the ejection pressure. Thus, sympathetic stimulation can often increase the cardiac output as much as twofold to threefold, in addition to the increased output that might be caused by the Frank-Starling mechanism already discussed.

On the other hand, inhibition of the sympathetic nervous system can be used to decrease cardiac pumping to a moderate extent in the following way: Under normal conditions, the sympathetic nerve fibers to the heart discharge continuously at a slow rate that maintains pumping at about 30 per cent above that with no sympathetic stimulation. Therefore, when the activity of the sympathetic nervous system is depressed below normal, this decreases both the heart rate and the strength of ventricular contraction, thereby decreasing the level of cardiac pumping to as much as 30 per cent below normal.

PARASYMPATHETIC (VAGAL) STIMULATION OF THE HEART. Strong vagal stimulation of the heart can stop the heartbeat for a few seconds, but then the heart usually "escapes" and beats at a rate of 20 to 40 beats per minute thereafter. In addition, strong vagal stimulation can decrease the strength of heart contraction by 20 to 30 per cent. This decrease is not greater because the vagal fibers are distributed mainly to the atria but not much to the ventricles where the power contraction of the heart occurs. Nevertheless, the great decrease in heart rate combined with a slight decrease in heart contraction can decrease ventricular pumping

50 or more per cent, especially so when the heart is working under great workload.

EFFECT OF SYMPATHETIC OR PARASYMPATHETIC STIMULATION ON THE CARDIAC FUNCTION CURVE. Figure 9–12 shows four cardiac function curves. They are much the same as the ventricular function curves of Figure 9–10. However, they represent function of the entire heart rather than of a single ventricle; they show the relation between the right atrial pressure at the input of the heart and cardiac output into the aorta.

The curves of Figure 9–12 demonstrate that at any given right atrial pressure, the cardiac output increases with increasing sympathetic stimulation and decreases with increasing parasympathetic stimulation. The changes in output caused by nerve stimulation result from *changes in heart rate* and *changes in contractile strength of the heart* because both of these affect cardiac output.

Effect of Heart Rate on Function of the Heart as a Pump

In general, the more times the heart beats per minute, the more blood it can pump, but there are important limitations to this effect. For instance, once the heart rate rises above a critical level, the heart strength itself decreases, presumably because of overuse of metabolic substrates in the cardiac muscle. In addition, the period of diastole between the contractions becomes so reduced that blood does not have time to flow adequately from the atria into the ventricles. For these reasons, when the heart rate is increased artificially by *electrical stimulation,* the normal large animal heart has its peak ability to pump large quantities of blood at a heart rate between 100 and 150 beats per minute. On the other hand, when its rate is increased by *sympathetic stimulation,* it reaches its peak ability

Figure 9–12. Effect on the cardiac output curve of different degrees of sympathetic and parasympathetic stimulation.

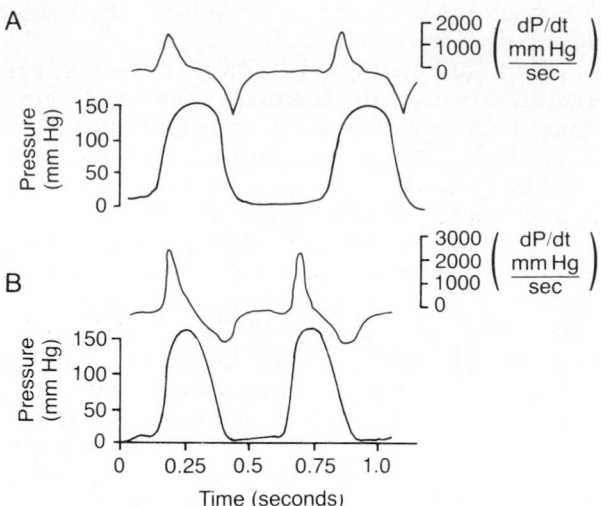

Figure 9–13. Simultaneous recordings of ventricular pressure and dP/dt. *A* shows results from a normal heart, and *B* shows results from a heart stimulated by isoproterenol. (Modified from Mason et al., in Sodeman and Sodeman (eds.): Pathologic Physiology. 6th Ed. Philadelphia, W.B. Saunders Company, 1979.)

to pump blood at a heart rate between 170 and 220 beats per minute. The reason for this difference is that sympathetic stimulation not only increases the heart rate but also increases heart strength. At the same time, it decreases the duration of systolic contraction and allows more time for filling during diastole.

Assessment of Cardiac Contractility

Although it is easy to determine the heart rate by simply timing the pulse, it has always been difficult to determine the strength of contraction of the heart, commonly called *cardiac contractility*. The change in contractility often is exactly opposite to the change in heart rate. Indeed, this effect occurs almost invariably in heart-debilitating diseases.

One of the ways in which cardiac contractility can be determined with great precision is to record one or more of the cardiac function curves. This can be done easily only in laboratory animals. Therefore, many physiologists and clinicians have searched for methods to assess the cardiac contractility in a simple way in the human being. One of these methods is to determine the so-called dP/dt.

DP/DT AS A MEASURE OF CARDIAC CONTRACTILITY. dP/dt means the *rate of change of the ventricular pressure* with respect to time. The dP/dt record is generated by an electronic computer that *differentiates* the measured ventricular pressure wave, thus giving a record of the *rate of change of the ventricular pressure*. Figure 9–13 shows two recordings of the ventricular pressure wave as well as simultaneous recordings (in color) of the dP/dt. In the upper part of the figure, the heart was beating normally, and in the lower part, the heart had been stimulated by isoproterenol, a drug that has essentially the same effect on the heart as sympathetic stimulation.

Note in the upper record that at the same time the

ventricular pressure is increasing at its most rapid rate, the recording of the dP/dt record reaches its greatest height. On the other hand, at the time the ventricular pressure is falling most rapidly, the dP/dt record reaches its lowest level. When the ventricular pressure is neither rising nor falling, the dP/dt record is at zero value.

Experimental studies have shown that the rate of rise of ventricular pressure, the dP/dt, in general correlates well with the strength of contraction of the ventricle. This effect is demonstrated by a comparison of the dP/dt record in the upper part of Figure 9–13 with that in the lower part, showing a much higher peak dP/dt in the lower record when the heart contractility was increased by isoproterenol. Thus, the *peak* dP/dt is often used as a means for comparing the contractilities of hearts in different functional states.

The quantitative value for peak dP/dt is also affected by factors that are not related to cardiac contractility. For instance, the value is increased by both increased input pressure to the left ventricle (the end-diastolic ventricular pressure), which is the *preload* of the ventricle, and the pressure in the aorta against which the heart is pumping the blood, called the *afterload*. Therefore, it is often difficult to use dP/dt as a measure of contractility in comparing hearts from one person to another because one of these factors may differ. For this reason, other quantitative measures have been used in attempts to assess cardiac contractility. One of these has been to use dP/dt divided by the instantaneous pressure in the ventricle, or (dP/dt)/P.

Effect of Potassium and Calcium Ions on Heart Function

In the discussion of membrane potentials in Chapter 5, it is pointed out that potassium ions have a marked effect on membrane potentials and action potentials, and in Chapter 6, it is noted that calcium ions play an especially important role in activating the muscle contractile process. Therefore, it is to be expected that the concentrations of these two ions in the extracellular fluids also has important effects on cardiac pumping.

EFFECT OF POTASSIUM IONS. Excess potassium in the extracellular fluids causes the heart to become extremely dilated and flaccid and slows the heart rate. Large quantities can also block conduction of the cardiac impulse from the atria to the ventricles through the A-V bundle. Elevation of potassium concentration to only 8 to 12 mEq/liter—two to three times the normal value—can cause such weakness of the heart and abnormal rhythm that this can cause death.

These effects are caused partially by the fact that a high potassium concentration in the extracellular fluids decreases the resting membrane potential in the cardiac muscle fibers, as explained in Chapter 5. As the membrane potential decreases, the intensity of the action potential also decreases, which makes the contraction of the heart progressively weaker.

EFFECT OF CALCIUM IONS. An excess of calcium ions causes effects almost exactly opposite to those of potassium ions, causing the heart to go into spastic contraction. This is caused by the direct effect of calcium ions in exciting the cardiac contractile process, as explained earlier in the chapter. Conversely, a deficiency of calcium ions causes cardiac flaccidity, similar to the effect of high potassium. Because the calcium ion levels in the blood are normally regulated within narrow ranges, these cardiac effects of abnormal calcium concentrations seldom are of clinical concern.

Effect of Temperature on the Heart

Increased temperature, as occurs when one has fever, causes greatly increased heart rate, sometimes to as great as double normal. Decreased temperature causes greatly decreased heart rate, falling to as low as a few beats per minute when a person is near death from hypothermia in the range of 60° to 70°F (15.5° to 21.1°C). These effects presumably result from the fact that heat causes increased permeability of the muscle membrane to the ions, resulting in acceleration of the self-excitation process.

Contractile strength of the heart is often enhanced temporarily by a moderate increase in temperature, but prolonged elevation of the temperature exhausts the metabolic systems of the heart and causes weakness.

REFERENCES

Beller, G. A.: Clinical Nuclear Cardiology. Philadelphia, W. B. Saunders Co., 1994.

Brady, A. J.: Mechanical properties of isolated cardiac myocytes. Physiol. Rev., 71:413, 1991.

Cowley, A. W., Jr., and Guyton, A. C.: Heart rate as a determinant of cardiac output in dogs with arteriovenous fistula. Am. J. Cardiol., 28:321, 1971.

Craig, M., et al.: Diagnostic Medical Sonography: Echocardiography, Vol. II. Philadelphia, J. B. Lippincott, 1991.

Driedzic, W. R., and Gesser, H.: Energy metabolism and contractility in ectothermic vertebrate hearts: Hypoxia, acidosis, and low temperature. Physiol Rev., 74:221, 1004.

Fedida, D., et al.: α_1-Adrenoceptors in myocardium: Functional aspects and transmembrane signaling mechanisms. Physiol. Rev., 73:469, 1993.

FitzGerald, P. G.: Gap junction heterogeneity in liver, heart, and lens. News Physiol. Sci., 3:206, 1988.

Fozzard, H. A., et al.: The Heart and Cardiovascular System: Scientific Foundations. New York, Raven Press, 1991.

Gadsby, D. C.: The Na/K pump of cardiac cells. Annu. Rev. Biophys. Bioeng., 13:373, 1984.

Gevers, W.: Protein metabolism of the heart. J. Mol. Cell. Cardiol., 16:3, 1984.

Guyton, A. C.: Determination of cardiac output by equating venous return curves with cardiac response curves. Physiol. Rev., 35:123, 1955.

Guyton, A. C., et al.: Circulatory Physiology: Cardiac Output and Its Regulation. 2nd Ed. Philadelphia, W. B. Saunders Co., 1973.

Hudicka, O., et al.: Angiogenesis in skeletal and cardiac muscle. Physiol. Rev., 72:369, 1992.

Iskandrian, A. S.: Myocardial Viability: Detection and Clinical Relevance. Hingham, Mass., Kluwer Academic Publishers, 1994.

Jacobus, W. E.: Respiratory control and the integration of heart high-energy phosphate metabolism by mitochondrial creating kinase. Annu. Rev. Physiol., 47:707, 1985.

Kelley, W. N.: Essentials of Internal Medicine. 2nd Ed. Philadelphia, J. B. Lippincott, 1994.

Manfredi, J. P., and Holmes, E. W.: Purine salvage pathways in myocardium. Annu. Rev. Physiol., 47:691, 1985.

Mela-Riker, L. M., and Bukoski, R. D.: Regulation of mitochondrial activity in cardiac cells. Annu. Rev. Physiol., 47:645, 1985.

Morgan, H. E., et al.: Biochemical mechanisms of cardiac hypertrophy. Annu. Rev. Physiol., 49:533, 1987.

Nagano, M., et al.: The Adaptive Heart. New York, Raven Press, 1994.

Nozawa, T., et al.: Relation between oxygen consumption and pressure-volume area of in situ dog heart. Am. J. Physiol., 253:H31, 1987.

Page, E., and Shibata, Y.: Permeable junctions between cardiac cells. Annu. Rev. Physiol., 43:431, 1981.

Rovetto, M. J.: Myocardial nucleotide transport. Annu. Rev. Physiol., 47:605, 1985.

Ruegg, J. C.: Dependence of cardiac contractility on myofibrillar calcium sensitivity. News Physiol. Sci., 2:179, 1987.

Sarnoff, S. J.: Myocardial contractility as described by ventricular function curves. Physiol. Rev., 35:107, 1955.

Schwartz, J.-C., et al.: Histaminergic transmission in the mammalian brain. Physiol. Rev., 71:1, 1991.

Sperelakis, N.: Physiology and Pathophysiology of the Heart, 3rd Ed. Hingham, Mass., Kluwer Academic Publishers, 1994.

Starling, E. H.: The Linacre Lecture on the Law of the Heart. London, Longmans Green & Co., 1918.

Suga, H., et al.: Critical evaluation of left ventricular systolic pressure volume area as predictor of oxygen consumption rate. Jpn. J. Physiol., 30:907, 1980.

Suga, H., et al.: Prospective prediction of O_2 consumption from pressure-volume area in dog hearts. Am. J. Physiol., 252:H1258, 1987.

Sugden, P. H.: The effects of hormonal factors on cardiac protein turnover. Adv. Myocardiol., 5:105, 1985.

Sugimoto, T., et al.: Effect of maximal work load on cardiac function. Jpn. Heart J., 14:146, 1973.

Sugimoto, T., et al.: Quantitative effect of low coronary pressure on left ventricular performance. Jpn. Heart J., 9:46, 1968.

Swynghedauw, B.: Developmental and functional adaptation of contractile proteins in cardiac and skeletal muscles. Physiol. Rev., 66:710, 1986.

Winegrad, S.: Calcium release from cardiac sarcoplasmic reticulum. Annu. Rev. Physiol., 44:451, 1982.

Winegrad, S.: Regulation of cardiac contractile proteins. Correlations between physiology and biochemistry. Circ. Res., 55:565, 1984.

(See also Chapter 10.)

Rhythmical Excitation of the Heart

CHAPTER 10

The heart is endowed with a specialized system for (1) generating rhythmical impulses to cause rhythmical contraction of the heart muscle and (2) conducting these impulses rapidly throughout the heart. When this system functions normally, the atria contract about one sixth of a second ahead of ventricular contraction, which allows extra filling of the ventricles before they pump the blood through the lungs and peripheral circulation. Another special importance of the system is that it allows all portions of the ventricles to contract almost simultaneously, which is essential for effective pressure generation in the ventricular chambers.

This rhythmical and conductive system of the heart is susceptible to damage by heart disease, especially by ischemia of the heart tissues resulting from poor coronary blood flow. The consequence is often a bizarre heart rhythm or abnormal sequence of contraction of the heart chambers, and the pumping effectiveness of the heart is often affected severely, even to the extent of causing death.

SPECIALIZED EXCITATORY AND CONDUCTIVE SYSTEM OF THE HEART

Figure 10–1 shows the specialized excitatory and conductive system of the heart that controls cardiac contractions. The figure shows (A) the *sinus node* (also called *sinoatrial* or *S-A node*), in which the normal rhythmical impulse is generated; (B) the *internodal pathways* that conduct the impulse from the sinus node to the atrioventricular (A-V) node; (C) the *A-V node*, in which the impulse from the atria is delayed before passing into the ventricles; (D) the *A-V bundle*, which conducts the impulse from the atria into the ventricles; and (E) the *left* and *right bundles of Purkinje fibers*, which conduct the cardiac impulse to all parts of the ventricles.

Sinus Node

The sinus node is a small, flattened, ellipsoid strip of specialized muscle about 3 millimeters wide, 15 millimeters long, and 1 millimeter thick; it is located in the superior lateral wall of the right atrium immediately below and slightly lateral to the opening of the superior vena cava. The fibers of this node have almost no contractile filaments and are each 3 to 5 micrometers in diameter, in contrast to a diameter of 10 to 15 micrometers for the surrounding atrial muscle fibers. The sinus fibers connect directly with the atrial fibers, so that any action potential that begins in the sinus node spreads immediately into the atria.

Automatic Electrical Rhythmicity of the Sinus Fibers

Many cardiac fibers have the capability of *self-excitation,* a process that can cause automatic rhythmical discharge and contraction. This is especially true of the fibers of the heart's specialized conducting system; the portion of this system that displays self-excitation to the greatest extent is the fibers of the sinus node. For this reason, the sinus node ordinarily controls the rate of beat of the entire heart, as discussed in detail later

121

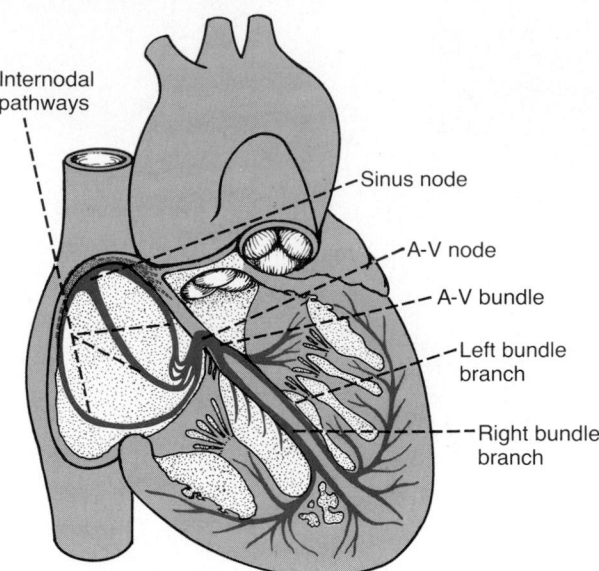

Figure 10–1. Sinus node and the Purkinje system of the heart, showing also the A-V node, atrial internodal pathways, and ventricular bundle branches.

in this chapter. First, let us describe this automatic rhythmicity.

MECHANISM OF SINUS NODAL RHYTHMICITY. Figure 10–2 shows action potentials recorded from a sinus nodal fiber for three heartbeats and, by comparison, a single ventricular muscle fiber action potential. Note that the potential of the sinus nodal fiber between discharges has a negativity of only −55 to −60 millivolts in comparison with −85 to −90 millivolts for the ventricular muscle fiber. The cause of this reduced negativity is that the cell membranes of the sinus fibers are naturally leaky to sodium ions.

Before attempting to explain the rhythmicity of the sinus nodal fibers, first recall from the discussions of Chapters 5 and 9 that in cardiac muscle, three types of membrane ion channels play important roles in causing the voltage changes of the action potential. They are (1) *fast sodium channels,* (2) *slow calcium-sodium channels,* and (3) *potassium channels.* Opening of the fast sodium channels for a few 10,000ths of a second is responsible for the rapid spikelike onset of the action potential observed in ventricular muscle because of rapid influx of positive sodium ions to the interior of the fiber. Then the plateau of the ventricular action potential is caused primarily by slower opening of the slow calcium-sodium channels, which lasts for a few tenths of a second. Finally, increased opening of the potassium channels and diffusion of large amounts of positive potassium ions out of the fiber return the membrane potential to its resting level.

But there is a difference in the function of these channels in the sinus nodal fiber because of the much lesser negativity of the "resting" potential—only −55 millivolts. At this level of negativity, the fast sodium channels have mainly become "inactivated," which means that they have become blocked. The cause of

this is that any time the membrane potential remains less negative than about −60 millivolts for more than a few milliseconds, the inactivation gates on the inside of the cell membrane that close the fast sodium channels become closed and remain so. Therefore, only the slow calcium-sodium channels can open (that is, can become "activated") and thereby cause the action potential. As a result, the action potential is slower to develop than that of the ventricular muscle, and it also recovers with a slow decrement of the potential rather than the abrupt recovery that occurs for the ventricular fiber.

Self-Excitation of Sinus Nodal Fibers. Because of the high sodium ion concentration in the extracellular fluid as well as the negative electrical charge inside the resting sinus nodal fibers, the positive sodium ions outside the fibers even normally tend to leak to the inside. Furthermore, the resting nodal fibers have a moderate number of channels that are already open to the sodium ions. Therefore, influx of positively charged sodium ions causes a rising membrane potential. Thus, as shown in Figure 10–2, the "resting" potential gradually rises between each two heartbeats. When it reaches a *threshold voltage* of about −40 millivolts, the calcium-sodium channels become activated, leading to rapid entry of both calcium and sodium ions, thus causing the action potential. Therefore, basically, the inherent leakiness of the sinus nodal fibers to sodium ions causes their self-excitation.

Why does this leakiness to sodium ions not cause the sinus nodal fibers to remain depolarized all the time? The answer is that two events occurring during the course of the action potential prevent this. First, the calcium-sodium channels become inactivated (that is, they close) within about 100 to 150 milliseconds after opening, and second, at about the same time, greatly increased numbers of potassium channels open. Therefore, the influx of calcium and sodium ions through the calcium-sodium channels ceases, while at the same time large quantities of positive potassium ions diffuse *out* of the fiber, thus terminating the action potential. Furthermore, the potassium channels

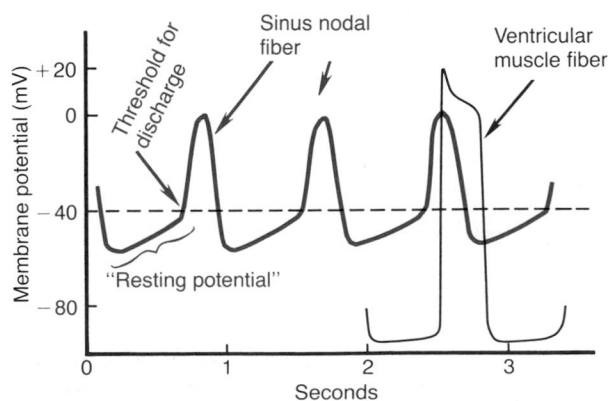

Figure 10–2. Rhythmical discharge of a sinus nodal fiber. Also, the sinus nodal action potential is compared with that of a ventricular muscle fiber.

remain open for another few tenths of a second, carrying a great excess of positive potassium charges out of the cell, which temporarily causes considerable *excess negativity* inside the fiber; this is called *hyperpolarization*. This hyperpolarization initially carries the "resting" membrane potential down to about −55 to −60 millivolts at the termination of the action potential.

Last, we must explain why this new state of hyperpolarization is not maintained forever. The reason is that during the next few tenths of a second after the action potential is over, progressively more and more of the potassium channels begin to close. Now the inward-leaking sodium ions once again overbalance the outward flux of potassium ions, which causes the "resting" potential to drift upward once more, finally reaching the threshold level for discharge at a potential of about −40 millivolts. Then the entire process begins again: self-excitation, recovery from the action potential, hyperpolarization after the action potential is over, drift of the "resting" potential again to threshold, then re-excitation again to elicit another cycle. This process continues indefinitely throughout a person's life.

Internodal Pathways and Transmission of the Cardiac Impulse Through the Atria

The ends of the sinus nodal fibers fuse with the surrounding atrial muscle fibers, and action potentials originating in the sinus node travel outward into these fibers. In this way, the action potential spreads through the entire atrial muscle mass and, eventually, to the A-V node. The velocity of conduction in the atrial muscle is about 0.3 m/sec. Conduction is somewhat more rapid in several small bundles of atrial muscle fibers. One of these, called the *anterior interatrial band*, passes through the anterior walls of the atria to the left atrium and conducts the cardiac impulse at a velocity of about 1 m/sec. In addition, three other small bundles curve through the atrial walls and terminate in the A-V node, also conducting the cardiac impulse at this rapid velocity. These three small bundles, shown in Figure 10–1, are called, respectively, the *anterior, middle,* and *posterior internodal pathways*. The cause of the more rapid velocity of conduction in these bundles is the presence of a number of specialized conduction fibers mixed with the atrial muscle. These fibers are similar to the rapidly conducting Purkinje fibers of the ventricles, which are discussed subsequently.

A-V Node, and Delay in Impulse Conduction from the Atria to the Ventricles

The conductive system is organized so that the cardiac impulse will not travel from the atria into the ventricles too rapidly; this delay allows time for the atria to empty their contents into the ventricles before ventricular contraction begins. It is primarily the A-V

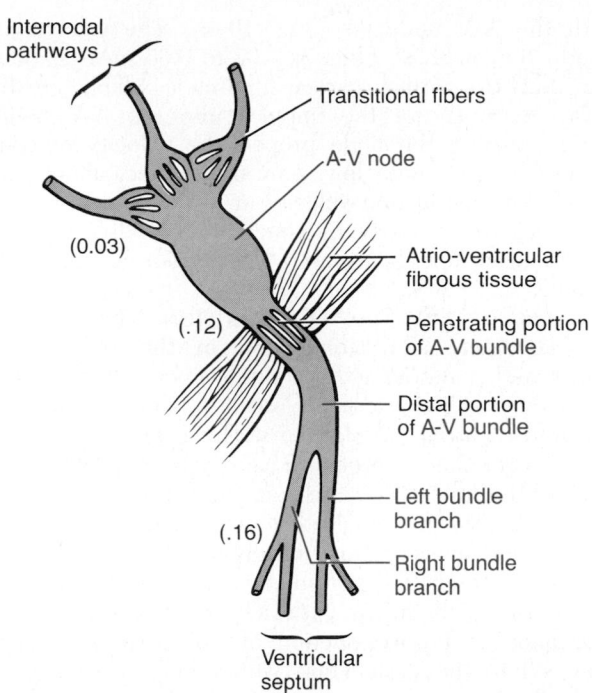

Figure 10–3. Organization of the A-V node. The numbers represent the interval of time from the origin of the impulse in the sinus node. The values have been extrapolated to the human being.

node and its adjacent conductive fibers that delay this transmission of the cardiac impulse from the atria into the ventricles.

The A-V node is located in the posterior septal wall of the right atrium immediately behind the tricuspid valve and adjacent to the opening of the coronary sinus, as shown in Figure 10–1. Figure 10–3 shows diagrammatically the different parts of this node and its connections with the atrial internodal pathway fibers and the A-V bundle. The figure also shows the approximate intervals of time in fractions of a second between the genesis of the cardiac impulse in the sinus node and its appearance at different points in the A-V nodal system. Note that the impulse, after traveling through the internodal pathway, reaches the A-V node about 0.03 second after its origin in the sinus node. Then there is a further delay of 0.09 second in the A-V node itself before the impulse enters the *penetrating portion of the A-V bundle*, where is passes into the ventricles. A final delay of another 0.04 second occurs mainly in this penetrating A-V bundle, which is composed of multiple small fascicles passing through the fibrous tissue separating the atria from the ventricles.

Thus, the total delay in the A-V nodal and A-V bundle system is about 0.13 second, in addition to the initial conduction delay of 0.03 second from the sinus node to the A-V node, making a total delay of 0.16 second. About a quarter of the time lapse occurs in the *transitional fibers*, which are very small fibers that connect the fibers of the atrial internodal pathways with the A-V node (see Fig. 10–3). The velocity of conduction in these fibers is 0.02 to 0.05 m/sec (about 1/12 that in normal cardiac muscle), which greatly delays entrance of the impulse into the A-V node. After entering the node proper, the velocity of conduction is still quite low, only 0.05 m/sec, about one eighth the conduction velocity in normal cardiac muscle. This low velocity of conduction is also approximately true for the penetrating portion of the A-V bundle.

CAUSE OF THE SLOW CONDUCTION. The cause of the extremely slow conduction in the transitional, nodal, and penetrating A-V bundle fibers is partly that their sizes are considerably smaller than the sizes of the normal atrial muscle fibers. However, most of the slow conduction is probably caused by two other entirely different factors. First, all these fibers have resting membrane potentials that are much less negative than the normal resting potential of other cardiac muscle. Second, few gap junctions connect the successive muscle cells in the pathway, so that there is great resistance to the conduction of excitatory ions from one cell to the next. Thus, with both low voltage to drive the ions and great resistance to the movement of the ions, it is easy to see why each succeeding cell is slow to be excited.

Transmission in the Purkinje System

The *Purkinje fibers* lead from the A-V node through the A-V bundle into the ventricles. Except for the initial portion of these fibers where they penetrate the A-V fibrous barrier, they have functional characteristics that are quite the opposite of those of the A-V nodal fibers. They are very large fibers, even larger than the normal ventricular muscle fibers, and they transmit action potentials at a velocity of 1.5 to 4.0 m/sec, a velocity about 6 times that in the usual cardiac muscle and 150 times that in some of the A-V transitional fibers. This allows almost immediate transmission of the cardiac impulse throughout the entire ventricular system.

The rapid transmission of action potentials by Purkinje fibers is believed to be caused by a high level of permeability of the gap junctions at the intercalated discs between the successive cardiac cells that make up the Purkinje fibers. Therefore, ions are transmitted easily from one cell to the next, thus enhancing the velocity of transmission. The Purkinje fibers also have few myofibrils, which means that they barely contract during the course of impulse transmission.

ONE-WAY CONDUCTION THROUGH THE A-V BUNDLE. A special characteristic of the A-V bundle is the inability, except in abnormal states, of action potentials to travel backward in the bundle from the ventricles to the atria. This prevents re-entry of cardiac impulses by this route from the ventricles to the atria, allowing only forward conduction from the atria to the ventricles.

Furthermore, it should be recalled that everywhere except at the A-V bundle, the atrial muscle is separated from the ventricular muscle by a continuous fibrous barrier, a portion of which is shown in Figure 10–3. This barrier normally acts as an insulator to prevent passage of the cardiac impulse between the atria and the ventricles through any other route besides forward conduction through the A-V bundle itself. (In rare instances, an abnormal muscle bridge does penetrate the fibrous barrier elsewhere besides at the A-V bundle. Under such conditions, the cardiac impulse can re-enter the atria from the ventricles and cause a serious cardiac arrhythmia.)

DISTRIBUTION OF THE PURKINJE FIBERS IN THE VENTRICLES. After penetrating the fibrous tissue, between the atrial and ventricular muscle, the distal portion of the A-V bundle passes downward in the ventricular septum for 5 to 15 millimeters toward the apex of the heart, as shown in Figures 10–1 and 10–3. Then the bundle divides into *left* and *right bundle branches* that lie beneath the endocardium of the two respective sides of the septum. Each branch spreads downward to the apex of the ventricle, progressively dividing into smaller branches that course around each ventricular chamber and back toward the base of the heart. The terminal Purkinje fibers penetrate about one third of the way into the muscle mass and then become continuous with the cardiac muscle fibers.

From the time the cardiac impulse enters the bundle branches in the ventricular septum until it reaches the terminations of the Purkinje fibers, the total time that elapses averages only 0.03 second; therefore, once the cardiac impulse enters the Purkinje system, it spreads almost immediately to the entire endocardial surface of the ventricular muscle.

Transmission of the Cardiac Impulse in the Ventricular Muscle

Once the impulse reaches the ends of the Purkinje fibers, it is transmitted through the ventricular muscle mass by the ventricular muscle fibers themselves. The velocity of transmission is now only 0.3 to 0.5 m/sec, one sixth that in the Purkinje fibers.

The cardiac muscle wraps around the heart in a double spiral with fibrous septa between the spiraling layers; therefore, the cardiac impulse does not necessarily travel directly outward toward the surface of the heart but instead angulates toward the surface along the directions of the spirals. Because of this, transmission from the endocardial surface to the epicardial surface of the ventricle requires as much as another 0.03 second, approximately equal to the time required for transmission through the entire ventricular portion of the Purkinje system. Thus, the total time for transmission of the cardiac impulse from the initial bundle branches to the last of the ventricular muscle fibers in the normal heart is about 0.06 second.

Summary of the Spread of the Cardiac Impulse Through the Heart

Figure 10–4 shows in summary form the transmission of the cardiac impulse through the human heart. The numbers on the figure represent the intervals of

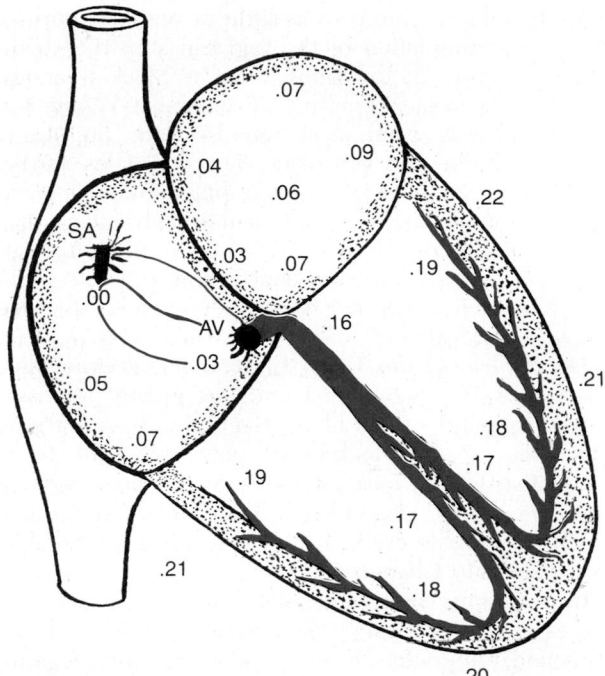

Figure 10–4. Transmission of the cardiac impulse through the heart showing the time of appearance (in fractions of a second) of the impulse in different parts of the heart.

time, in hundredths of a second, that lapse between the origin of the cardiac impulse in the sinus node and its appearance at each respective point in the heart. Note that the impulse spreads at moderate velocity through the atria but is delayed more than 0.1 second in the A-V nodal region before appearing in the ventricular septal A-V bundle. Once it has entered this bundle, it spreads rapidly through the Purkinje fibers to the entire endocardial surfaces of the ventricles. Then the impulse once again spreads slowly through the ventricular muscle to the epicardial surfaces.

It is extremely important that the reader learn in detail the course of the cardiac impulse through the heart and the times of its appearance in each separate part of the heart because a thorough quantitative knowledge of this process is essential to the understanding of electrocardiography, which is discussed in Chapters 11 through 13.

CONTROL OF EXCITATION AND CONDUCTION IN THE HEART

The Sinus Node as the Pacemaker of the Heart

In the discussion thus far of the genesis and transmission of the cardiac impulse through the heart, we have noted that the impulse normally arises in the sinus node. This often is not the case under abnormal conditions because other parts of the heart can exhibit rhythmical contraction in the same way that the sinus

nodal fibers can; this is particularly true of the A-V nodal and Purkinje fibers.

The A-V nodal fibers, when not stimulated from some outside source, discharge at an intrinsic rhythmical rate of 40 to 60 times per minute, and the Purkinje fibers discharge at a rate somewhere between 15 and 40 times per minute. These rates are in contrast to the normal rate of the sinus node of 70 to 80 times per minute.

Therefore, the question that we must ask is, Why does the sinus node control the heart's rhythmicity rather than the A-V node or the Purkinje fibers? The answer to this derives from the fact that the discharge rate of the sinus node is considerably faster than that of either the A-V node or the Purkinje fibers. Each time the sinus node discharges, its impulse is conducted into both the A-V node and the Purkinje fibers, discharging their excitable membranes. Then these tissues as well as the sinus node all recover from the action potential and become hyperpolarized nearly at the same time. But the sinus node loses its hyperpolarization and discharges again much more rapidly than does either of the other two. Thus, the sinus node emits a new impulse before either the A-V node or the Purkinje fibers can reach their own threshold for self-excitation. The new impulse from the sinus node again discharges both the A-V node and the Purkinje fibers. This process continues on and on, the sinus node always exciting these other potentially self-excitatory tissues before self-excitation can occur.

Thus, the sinus node controls the beat of the heart because its rate of rhythmical discharge is greater than that of any other part of the heart. Therefore, the sinus node is the normal *pacemaker* of the heart.

ABNORMAL PACEMAKERS—ECTOPIC PACEMAKER. Occasionally some other part of the heart develops a rhythmical discharge rate that is more rapid than that of the sinus node. For instance, this often occurs in the A-V node or in the Purkinje fibers when one of these functions abnormally. In either of these cases, the pacemaker of the heart shifts from the sinus node to the A-V node or to the excited Purkinje fibers. Under rarer conditions, a point in the atrial or ventricular muscle develops excessive excitability and becomes the pacemaker.

A pacemaker elsewhere than the sinus node is called an *ectopic pacemaker*. An ectopic pacemaker causes an abnormal sequence of contraction of the different parts of the heart and can cause great debility of heart pumping.

Another cause of shift of the pacemaker is blockage of transmission of the impulses from the sinus node to the other parts of the heart. The new pacemaker occurs most frequently at the A-V node or in the penetrating portion of the A-V bundle on the way to the ventricles.

When A-V block occurs—that is, when the cardiac impulse fails to pass from the atria into the ventricles through the A-V nodal and bundle system—the atria continue to beat at the normal rate of rhythm of the sinus node, while a new pacemaker develops in the Purkinje system of the ventricles and drives the ventricular muscle at a new rate somewhere between 15 and 40 beats per minute. After a sudden block, the Purkinje system does not begin to emit its rhythmical impulses until 5 to 30 seconds later because, before the blockage, the Purkinje fibers had been "overdriven" by the rapid sinus impulses and, consequently, are in a suppressed state. During these 5 to 30 seconds, the ventricles fail to pump blood, and the person faints after the first 4 to 5 seconds because of lack of blood flow to the brain. This delayed pickup of the heartbeat is called *Stokes-Adams syndrome*. If the delay period is too long, it can lead to death.

Role of the Purkinje System in Causing Synchronous Contraction of the Ventricular Muscle

It is clear from our description of the Purkinje system that the cardiac impulse arrives at almost all portions of the ventricles within a narrow span of time, normally exciting the first ventricular muscle fiber only 0.06 second ahead of excitation of the last ventricular muscle fiber. This causes all portions of the ventricular muscle in both ventricles to begin contracting at almost the same time. Effective pumping by the two ventricular chambers requires this synchronous type of contraction. If the cardiac impulse traveled through the ventricular muscle slowly, much of the ventricular mass would contract before contraction of the remainder, in which case the overall pumping effect would be greatly depressed. Indeed, in some types of cardiac debilities, several of which are discussed in Chapters 12 and 13, slow transmission does occur, and the pumping effectiveness of the ventricles is decreased perhaps 20 to 30 per cent.

Control of Heart Rhythmicity and Conduction by the Cardiac Nerves: Sympathetic and Parasympathetic Nerves

The heart is supplied with both sympathetic and parasympathetic nerves, as shown in Figure 9–11 of the previous chapter. The parasympathetic nerves (the vagi) are distributed mainly to the sinus and A-V nodes, to a lesser extent to the muscle of the two atria, and even less to the ventricular muscle. The sympathetic nerves, on the other hand, are distributed to all parts of the heart, with a strong representation to the ventricular muscle as well as to all the other areas.

EFFECT OF PARASYMPATHETIC (VAGAL) STIMULATION TO SLOW OR EVEN BLOCK CARDIAC RHYTHM AND CONDUCTION—VENTRICULAR ESCAPE. Stimulation of the parasympathetic nerves to the heart (the vagi) causes the hormone *acetylcholine* to be released at the vagal endings. This hormone has two major effects on the heart. First, it decreases the rate of rhythm of the sinus node and, second, it decreases the excitability of the A-V junctional fibers between the atrial musculature and the A-V node, thereby slowing transmission of the cardiac impulse into the ventricles. Weak to moderate vagal stimulation will slow the rate

of heart pumping often to as little as one-half normal. But strong stimulation of the vagi can stop the rhythmical excitation of the sinus node or block transmission of the cardiac impulse through the A-V junction. In either case, rhythmical impulses are no longer transmitted into the ventricles. The ventricles usually stop beating for 5 to 20 seconds, but then some point in the Purkinje fibers, usually in the ventricular septal portion of the A-V bundle, develops a rhythm of its own and causes ventricular contraction at a rate of 15 to 40 beats per minute. This phenomenon is called *ventricular escape*.

Mechanism of the Vagal Effects. The acetylcholine released at the vagal nerve endings greatly increases the permeability of the fiber membranes to potassium, which allows rapid leakage of potassium out of the conductive fibers. This causes increased negativity inside the fibers, an effect called *hyperpolarization*, which makes this excitable tissue much less excitable, as explained in Chapter 5.

In the sinus node, the state of hyperpolarization decreases the "resting" membrane potential of the sinus nodal fibers to a level considerably more negative than the normal value, to -65 to -75 millivolts rather than the normal level of -55 to -60 millivolts. Therefore, the drift of the resting membrane potential caused by sodium leakage requires much longer to reach the threshold potential for excitation. This greatly slows the rate of rhythmicity of these nodal fibers. And if the vagal stimulation is strong enough, it is possible to stop the rhythmical self-excitation of this node.

In the A-V node, the state of hyperpolarization makes it difficult for the minute junctional fibers, which can generate only small quantities of current during the action potential, to excite the nodal fibers. Therefore, the *safety factor* for transmission of the cardiac impulse through the junctional fibers and into the nodal fibers decreases. A moderate decrease simply delays conduction of the impulse, but a decrease in safety factor below unity (which means so low that the action potential of a fiber cannot cause an action potential in the successive portion of the fiber) blocks conduction.

EFFECT OF SYMPATHETIC STIMULATION ON CARDIAC RHYTHM AND CONDUCTION. Sympathetic stimulation causes essentially the opposite effects on the heart to those caused by vagal stimulation as follows: First, it increases the rate of sinus nodal discharge. Second, it increases the rate of conduction as well as the level of excitability in all portions of the heart. Third, it increases greatly the force of contraction of all the cardiac musculature, both atrial and ventricular, as discussed in Chapter 9.

In short, sympathetic stimulation increases the overall activity of the heart. Maximal stimulation can almost triple the rate of heartbeat and increase the strength of a heart contraction as much as twofold.

Mechanism of the Sympathetic Effect. Stimulation of the sympathetic nerves releases the hormone *norepinephrine* at the sympathetic nerve endings. The precise mechanism by which this hormone acts on cardiac muscle fibers is somewhat doubtful, but the belief is that it increases the permeability of the fiber membrane to sodium and calcium. In the sinus node, an increase of sodium permeability causes a more positive resting potential and an increased rate of upward drift of the membrane potential to the threshold level for self-excitation, both of which accelerate the onset of self-excitation and, therefore, increase the heart rate.

In the A-V node, increased sodium permeability makes it easier for the action potential to excite each succeeding portion of the conducting fiber, thereby decreasing the conduction time from the atria to the ventricles.

The increase in permeability to calcium ions is at least partially responsible for the increase in contractile strength of the cardiac muscle under the influence of sympathetic stimulation because calcium ions play a powerful role in exciting the contractile process of the myofibrils.

REFERENCES

Armstrong, C. M.: Voltage-dependent ion channels and their gating. Physiol. Rev., 72:(Suppl.) S5, 1992.

Beller, G. A.: Clinical Nuclear Cardiology. Philadelphia, W. B. Saunders Co., 1994.

Brutsaert, D. L., and Paulus, W. J.: Contraction and relaxation of the heart as muscle and pump. In Guyton, A. C., and Young, D. B. (eds.): International Review of Physiology: Cardiovascular Physiology III. Vol. 18. Baltimore, University Park Press, 1979, p. 1.

Catalano, J. T.: Guide to ECG Analysis. Philadelphia, J. B. Lippincott, 1993.

Craig, M., et al.: Diagnostic Medical Sonography: Echocardiography, Vol. II. Philadelphia, J. B. Lippincott, 1991.

DiFrancesco, D., and Noble, D.: A model of cardiac electrical activity incorporating ionic pumps and concentration changes. Phil. Trans. R. Soc. Lond. (Biol.), 307:353, 1985.

Ellis, D.: Na-Ca exchange in cardiac tissues. Adv. Myocardiol., 5:295, 1985.

Farah, A. E., et al.: Positive inotropic agents. Annu. Rev. Pharmacol. Toxicol., 24:275, 1985

Fedida, D., et al.: α_1-Adrenoceptors in myocardium: Functional aspects and transmembrane signaling mechanisms. Physiol. Rev., 73:469, 1993.

Fozzard, H. A., et al.: The Heart and Cardiovascular System: Scientific Foundations. New York, Raven Press, 1986.

Gardner, D.: A time integral of membrane currents. Physiol. Rev., 72:(Suppl.) S1, 1992.

Gilmour, R. F., Jr., and Zipes, D. P.: Slow inward current and cardiac arrhythmias. Am. J. Cardiol. 55:89B, 1985.

Gravanis, M. B. (ed.): Cardiovascular Pathophysiology. New York, McGraw-Hill Book Co., 1987.

Hainsworth, R.: Reflexes from the heart. Physiol. Rev., 71:617, 1991.

Huff, J., et al.: ECG Workout: Exercises in Arrhythmia Interpretation. 2nd Ed. Philadelphia, J. B. Lippincott, 1993.

Huizinga, J. D.: Pacemaker Activity and Intercellular Communication. Boca Raton, Fla., CRC Press, Inc., 1995.

Irisawa, H., et al.: Cardiac pacemaking in the sinoatrial node. Physiol. Rev., 73:197, 1993.

Josephson, M. E.: Clinical Cardiac Electrophysiology. Baltimore, Williams & Wilkins, 1993.

Kamino, K.: Optical approaches to ontogeny of electrical activity and related functional organization during early heart development. Physiol. Rev., 71:53, 1991.

Kastor, J. A.: Arrhythmias. Philadelphia, W. B. Saunders Co., 1994.

Katz, A. M.: Physiology of the Heart. New York, Raven Press, 1992.

Latorre, R., et al.: K^+ channels gated by voltage and ions. Annu. Rev. Physiol., 46:485, 1984.

Levy, M. N., et al.: Neural regulation of the heart beat. Annu. Rev. Physiol., 43:443, 1981.

Loewenstein, W. R.: Junctional intercellular communication: The cell-to-cell membrane channel. Physiol. Rev., 61:829, 1981.

Lynch, C. III: Cardiac Electrophysiology: Perioperative Considerations. Philadelphia, J. B. Lippincott, 1994.

Mazzanti, M., and DeFelice, L. J.: K channel kinetics during the spontaneous heart beat in embryonic chick ventricle cells. Biophys. J., 54:1139, 1988.

Mazzanti, M., and DeFelice, L. J.: Na channel kinetics during the spontaneous heart beat in embryonic chick ventricle cells. Biophys. J., 52:95, 1987.

Mazzanti, M., and DeFelice, L. J.: Regulation of the Na-conducting Ca channel during the cardiac action potential. Biophys. J., 51:115, 1987.

McDonald, T. F., et al.: Regulation and modulation of calcium channels in cardiac, skeletal, and smooth muscle cells. Physiol. Rev., 74:365, 1994.

Meijler, F. L.: Atrioventricular conduction versus heart size from mouse to whale. J. Am. Coll. Cardiol., 5:363, 1985.

Meijler, F. L., and Janse, M. J.: Morphology and electrophysiology of the mammalian atrioventricular node. Physiol. Rev., 68:608, 1988.

Pape, H.-C.: Specific bradycardic agents block the hyperpolarization-activated cation current in central neurons. Neuroscience, 59:368, 1994.

Reuter, H.: Ion channels in cardiac cell membranes. Annu. Rev. Physiol., 44:473, 1984.

Sheridan, J. D., and Atkinson, M. M.: Physiological roles of permeable junctions: Some possibilities. Annu. Rev. Physiol., 47:337, 1985.

Sperelakis, N.: Hormonal and neurotransmitter regulation of Ca^{++} influx through voltage-dependent slow channels in cardiac muscle membrane. Membr. Biochem., 5:131, 1984.

Surawicz, B.: Electrophysiologic Basis of ECG and Cardiac Arrhythmias. Baltimore, Williams & Wilkins, 1995.

Verrier, R. L., and Lown, B.: Behavioral stress and cardiac arrhythmias. Annu. Rev. Physiol., 46:155, 1984.

Zipes, D. P., and Jalife, J.: Cardiac Electrophysiology: From Cell to Bedside. Philadelphia, W. B. Saunders Co., 1990.

(See also Chapter 9.)

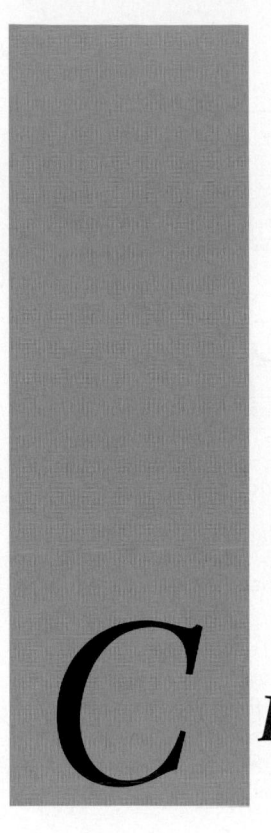

The Normal Electrocardiogram

*C*HAPTER *11*

As the cardiac impulse passes through the heart, electrical currents spread into the tissues surrounding the heart, and a small proportion of these spreads all the way to the surface of the body. If electrodes are placed on the skin on opposite sides of the heart, electrical potentials generated by these currents can be recorded; the recording is known as an *electrocardiogram*. A normal electrocardiogram for two beats of the heart is shown in Figure 11–1.

CHARACTERISTICS OF THE NORMAL ELECTROCARDIOGRAM

The normal electrocardiogram (Fig. 11–1) is composed of a P wave, a QRS complex, and a T wave. The QRS complex is often three separate waves, the Q wave, the R wave, and the S wave, but not always.

The P wave is caused by electrical potentials generated as the atria depolarize before contraction. The QRS complex is caused by potentials generated when the ventricles depolarize before contraction, that is, as the depolarization wave spreads through the ventricles. Therefore, both the P wave and the components of the QRS complex are *depolarization waves*.

The T wave is caused by potentials generated as the ventricles recover from the state of depolarization. This process normally occurs in ventricular muscle 0.25 to 0.35 second after depolarization, and this wave is known as a *repolarization wave*.

Thus, the electrocardiogram is composed of both depolarization and repolarization waves. The principles of depolarization and repolarization are discussed in Chapter 5. The distinction between depolarization waves and repolarization waves is so important in electrocardiography that further clarification is needed.

Depolarization Waves Versus Repolarization Waves

Figure 11–2 shows a muscle fiber in four stages of depolarization and repolarization. During depolarization, the normal negative potential inside the fiber is lost and the membrane potential reverses; that is, it becomes slightly positive inside and negative outside.

In Figure 11–2A, depolarization, demonstrated by the red positive charges inside and negative charges outside, is traveling from left to right, and the first half of the fiber has already depolarized, whereas the remaining half is still polarized. Therefore, the left electrode on the fiber is in an area of negativity where it touches the outside of the fiber, and the right electrode is in an area of positivity; this causes the meter to record positively. To the right of the muscle fiber is shown a record of the potential between the electrodes as recorded by a high-speed recording meter. Note that when depolarization has reached this halfway mark in Figure 11–2A, the record has risen to a maximum positive value.

In Figure 11–2B, depolarization has extended over the entire muscle fiber, and the recording to the right has returned to the zero baseline because both electrodes are now in areas of equal negativity. The completed wave is a *depolarization wave* because it results from spread of depolarization along the entire extent of the muscle fiber.

Figure 11–2C shows repolarization in the muscle fiber, with positivity returning to the outside of the fiber; this has proceeded halfway along the fiber from left to right. At this point, the left electrode is in an area of positivity, and the right electrode is in an area of negativity. This is opposite to the polarity in Figure 11–2A. Consequently, the recording, as shown to the right, becomes negative.

In Figure 11–2D, the muscle fiber has completely repolarized, and both electrodes are now in areas of positivity, so that no potential is recorded between them. Thus, in the

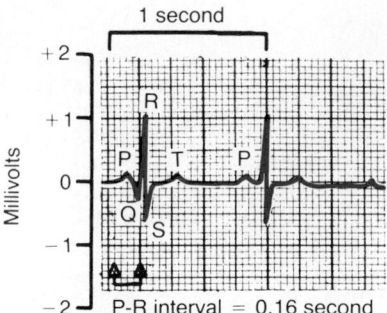

Figure 11–1. Normal electrocardiogram.

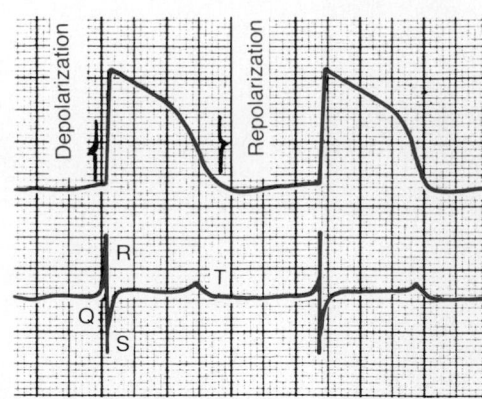

Figure 11–3. *Above,* Monophasic action potential from a ventricular muscle fiber during normal cardiac function showing rapid depolarization and then repolarization occurring slowly during the plateau stage but rapidly toward the end. *Below,* Electrocardiogram recorded simultaneously.

recording to the right, the potential returns once more to the zero level. This completed negative way is a *repolarization wave* because it results from spread of repolarization over the muscle fiber.

RELATION OF THE MONOPHASIC ACTION POTENTIAL OF VENTRICULAR MUSCLE TO THE QRS AND T WAVES. The monophasic action potential of ventricular muscle, discussed in Chapter 10, normally lasts between 0.25 and 0.35 second. The top part of Figure 11–3 shows a monophasic action potential recorded from a microelectrode inserted to the inside of a single ventricular muscle fiber. The upsweep of this action potential is caused by *depolarization,* and the return of the potential to the baseline is caused by *repolarization.*

Note below in the figure the simultaneous recording of the electrocardiogram from this same ventricle, which shows the QRS wave appearing at the beginning of the monophasic action potential and the T wave appearing at the end. Note especially that *no potential is recorded in the electrocardiogram when the ventricular muscle is either completely polar-*

ized or completely depolarized. Only when the muscle is partly polarized and partly depolarized does the current flow from one part of the ventricles to another part and therefore also flow to the surface of the body to cause the electrocardiogram.

Relationship of Atrial and Ventricular Contraction to the Waves of the Electrocardiogram

Before contraction of muscle can occur, depolarization must spread through the muscle to initiate the chemical processes of contraction. Therefore, the P wave occurs at the *beginning of contraction of the atria,* and the QRS wave occurs at the *beginning of contraction of the ventricles.* The ventricles remain contracted until a few milliseconds after repolarization has occurred, that is, until after the end of the T wave.

The ventricular repolarization wave is the T wave of the normal electrocardiogram. Ordinarily, ventricular muscle begins to repolarize in some fibers about 0.20 second after the beginning of the depolarization wave but in many others not until as long as 0.35 second. Thus, the process of repolarization extends over a long period, about 0.15 second. For this reason, the T wave in the normal electrocardiogram is often a prolonged wave, but the voltage of the T wave is considerably less than the voltage of the QRS complex, partly because of its prolonged length.

The atria repolarize about 0.15 to 0.20 second after the P wave. This occurs just at the instant when the QRS wave is being recorded in the electrocardiogram. Therefore, the atrial repolarization wave, known as the *atrial T wave,* is usually obscured by the much larger QRS wave. For this reason, an atrial T wave seldom is observed in the electrocardiogram.

Voltage and Time Calibration of the Electrocardiogram

All recordings of electrocardiograms are made with appropriate calibration lines on the recording paper. Either these calibration lines are already ruled on the paper, as is the

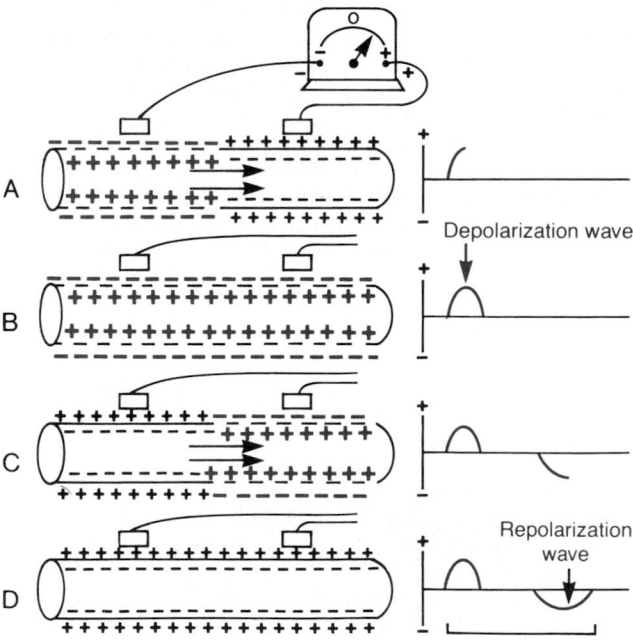

Figure 11–2. Recording the *depolarization wave* (*A* and *B*) and the *repolarization wave* (*C* and *D*) from a cardiac muscle fiber.

case when a pen recorder is used, or they are recorded on the paper at the same time that the electrocardiogram is recorded, which is the case with the photographic types of electrocardiographs.

As shown in Figure 11–1, the horizontal calibration lines are arranged so that 10 small divisions upward or downward in the standard electrocardiogram represent 1 millivolt with positivity in the upward direction and negativity in the downward direction.

The vertical lines on the electrocardiogram are time calibration lines. Each inch in the horizontal direction is 1 second, and each inch in turn is usually broken into five segments by dark vertical lines; the intervals between these dark lines represent 0.20 second. These intervals are then broken into five smaller intervals by thin lines, and each of these represents 0.04 second.

Normal Voltages in the Electrocardiogram. The voltages of the waves in the normal electrocardiogram depend on the manner in which the electrodes are applied to the surface of the body and how close the electrodes are to the heart. When one electrode is placed directly over the heart and a second electrode is placed elsewhere on the body, the voltage of the QRS complex may be as great as 3 to 4 millivolts. Even this voltage is small in comparison with the monophasic action potential of 110 millivolts recorded directly at the heart muscle membrane. When electrocardiograms are recorded from electrodes on the two arms or on one arm and one leg, the voltage of the QRS complex usually is about 1 millivolt from the top of the R wave to the bottom of the S wave; the voltage of the P wave, between 0.1 and 0.3 millivolt; and that of the T wave, between 0.2 and 0.3 millivolt.

P-Q OR P-R INTERVAL. The duration of time between the beginning of the P wave and the beginning of the QRS wave is the interval between the beginning of contraction of the atria and the beginning of contraction of the ventricles. This period is called the P-Q interval. The normal P-Q interval is about 0.16 second. Sometimes this interval is called the P-R interval because the Q wave is frequently absent.

Q-T INTERVAL. Contraction of the ventricle lasts almost from the beginning of the Q wave to the end of the T wave. This interval is called the Q-T interval and ordinarily is about 0.35 second.

RATE OF HEARTBEAT AS DETERMINED FROM THE ELECTROCARDIOGRAM. The rate of heartbeat can be determined easily from an electrocardiogram because the heart rate is the reciprocal of the time interval between two successive beats. If the interval between two beats as determined from the time calibration lines is 1 second, the heart rate is 60 beats per minute. The normal interval between two successive QRS complexes is about 0.83 second. This is a heart rate of 60/0.83 times per minute, or 72 beats per minute.

METHODS FOR RECORDING ELECTROCARDIOGRAMS

Sometimes the electrical currents generated by the cardiac muscle during each beat of the heart change electrical potentials and polarities on the respective sides of the heart in less than 0.01 second. Therefore, it is essential that any apparatus for recording electrocardiograms be capable of responding rapidly to these changes in potentials. In general, two types of recording apparatuses are used for this purpose.

Pen Recorder

Most modern clinical electrocardiographs use a direct pen writing recorder that writes the electrocardiogram with a pen directly on a moving sheet of paper. Sometimes the pen is a thin tube connected at one end to an inkwell, and its recording end is connected to a powerful electromagnet system that is capable of moving the pen back and forth at high speed. As the paper moves forward, the pen records the electrocardiogram. The movement of the pen in turn is controlled by means of appropriate electronic amplifiers connected to electrocardiographic electrodes on the patient.

Other pen recording systems use special paper that does not require ink in the recording stylus. One such paper turns black when it is exposed to heat; the stylus itself is made very hot by electrical current flowing through its tip. Another type turns black when electrical current flows from the tip of the stylus through the paper to an electrode at its back. This leaves a black line everywhere on the paper that the stylus touches.

FLOW OF CURRENT AROUND THE HEART DURING THE CARDIAC CYCLE

Recording Electrical Potentials from a Partially Depolarized Mass of Syncytial Cardiac Muscle

Figure 11–4 shows a syncytial mass of cardiac muscle that has been stimulated at its centralmost point. Before stimulation, all the exteriors of the muscle cells had been positive and the interiors negative. For reasons presented in Chapter 5 in the discussion of membrane potentials, as soon as an area of the cardiac syncytium becomes depolarized, negative charges leak to the outsides of the depolarized muscle fibers, making this surface area electronegative, as represented by the negative signs in Figure 11–4, with respect to the remaining surface of the heart, which is still polarized in the normal manner, as represented by the positive signs. Therefore, a meter connected with its negative terminal on the area of depolarization and its positive terminal on one of the still-polarized areas, as shown to the right in the figure, records positively.

Two other possible electrode placements and meter readings are also demonstrated in Figure 11–4. *These should be studied carefully, and the reader should be able to explain the causes of the respective meter readings.* Because the

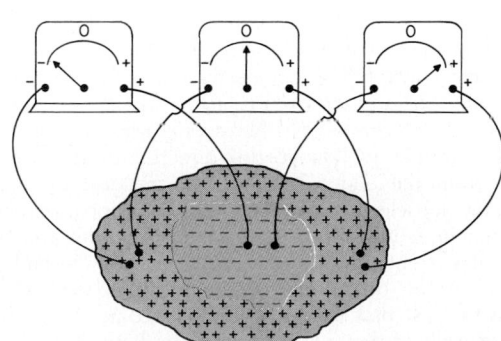

Figure 11–4. Instantaneous potentials developed on the surface of a cardiac muscle mass that has been depolarized in its center.

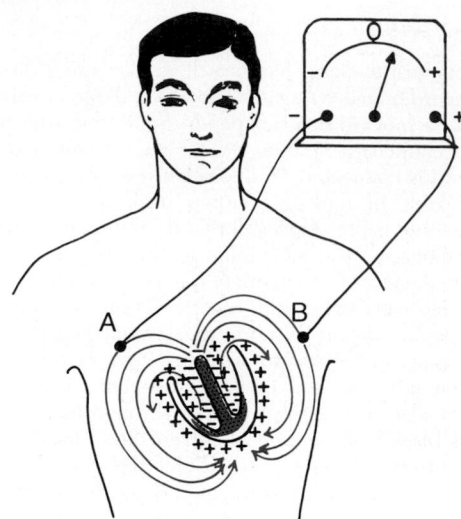

Figure 11–5. Flow of current in the chest around partially depolarized ventricles.

depolarization spreads in all directions through the heart, the potential differences shown in the figure last for only a few milliseconds, and the actual voltage measurements can be accomplished only with a high-speed recording apparatus.

Flow of Electrical Currents in the Chest Around the Heart

Figure 11–5 shows the ventricular muscle lying within the chest. Even the lungs, although mostly filled with air, conduct electricity to a surprising extent, and fluids of the other tissues surrounding the heart conduct electricity even more easily. Therefore, the heart is actually suspended in a conductive medium. When one portion of the ventricles becomes electronegative with respect to the remainder, electrical current flows from the depolarized area to the polarized area in large circuitous routes, as noted in the figure.

It should be recalled from the discussion of the Purkinje system in Chapter 10 that the cardiac impulse first arrives in the ventricles in the septum and shortly thereafter reaches the endocardial surfaces of the remainder of the ventricles, as shown by the colored areas and the negative signs in Figure 11–5. This provides electronegativity on the insides of the ventricles and electropositivity on the outer walls of the ventricles, and current flows through the fluids surrounding the ventricles along elliptical paths, as demonstrated by the curving arrows in the figure. If one algebraically averages all the lines of current flow (the elliptical lines), one finds that the average current flow occurs *with negativity toward the base of the heart* and *with positivity toward the apex*. During most of the remainder of the depolarization process, the current continues to flow in this direction as depolarization spreads from the endocardial surface outward through the ventricular muscle. Immediately before depolarization has completed its course through the ventricles, the average direction of current flow reverses for about $\frac{1}{100}$ second, flowing then in the direction from the apex toward the base because the last part of the heart to become depolarized is the outer walls of the ventricles near the base of the heart.

Thus, in the normal heart, current flows from negative to positive primarily in the direction from the base toward the apex during almost the entire cycle of depolarization except

at the very end. Therefore, if a meter is connected to the surface of the body, as shown in Figure 11–5, the electrode nearer the base will be negative, whereas the electrode nearer the apex will be positive, and the recording meter will show a positive recording in the electrocardiogram.

ELECTROCARDIOGRAPHIC LEADS

Three Bipolar Limb Leads

Figure 11–6 shows electrical connections between the patient's limbs and the electrocardiograph for recording electrocardiograms from the so-called *standard bipolar limb leads*. The term "bipolar" means that the electrocardiogram is recorded from *two* electrodes on the body, in this case on the limbs. Thus, a "lead" is not a single wire connecting from the body but a combination of two wires and their electrodes to make a complete circuit with the electrocardiograph. The electrocardiograph in each instance is represented by an electrical meter in the diagram, although the actual electrocardiograph is a high-speed recording meter with a moving paper.

LEAD I. In recording limb lead I, the *negative terminal of the electrocardiograph is connected to the right arm and the positive terminal, to the left arm.* Therefore, when the point on the chest where the right arm connects to the chest is electronegative with respect to the point where the left arm connects, the electrocardiograph records positively, that is,

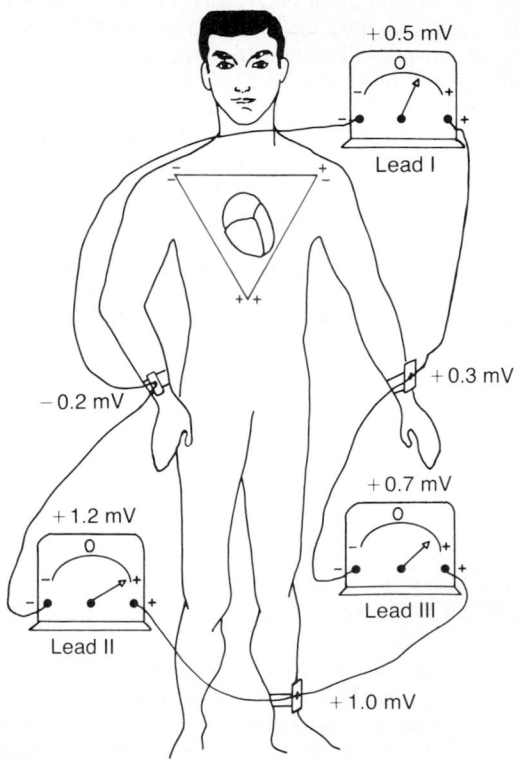

Figure 11–6. Conventional arrangement of electrodes for recording the standard electrocardiographic leads. Einthoven's triangle is superimposed on the chest.

above the zero voltage line in the electrocardiogram. When the opposite is true, the electrocardiograph records below the line.

Lead II. In recording limb lead II, the *negative terminal of the electrocardiograph is connected to the right arm* and the *positive terminal, to the left leg.* Therefore, when the right arm is negative with respect to the left leg, the electrocardiograph records positively.

Lead III. In recording limb lead III, the *negative terminal of the electrocardiograph is connected to the left arm* and the *positive terminal to the left leg.* This means that the electrocardiograph records positively when the left arm is negative with respect to the left leg.

Einthoven's Triangle. In Figure 11–6, a triangle, called *Einthoven's triangle,* is drawn around the area of the heart. This is a diagrammatic means of illustrating that the two arms and the left leg form apices of a triangle surrounding the heart. The two apices at the upper part of the triangle represent the points at which the two arms connect electrically with the fluids around the heart, and the lower apex is the point at which the left leg connects with the fluids.

Einthoven's Law. Einthoven's law states that if the electrical potentials of any two of the three bipolar limb electrocardiographic leads are known at any given instant, the third one can be determined mathematically from the first two by simply summing the first two (but note that the positive and negative signs of the different leads must be observed when making this summation).

For instance, let us assume that momentarily, as noted in Figure 11–6, the right arm is 0.2 millivolt negative with respect to the average potential in the body, the left arm is 0.3 millivolt positive, and the left leg is 1.0 millivolt positive. Observing the meters in the figure, it can be seen that lead I records a positive potential of 0.5 millivolt because this is the difference between the −0.2 millivolt on the right arm and the +0.3 millivolt on the left arm. Similarly, lead III records a positive potential of 0.7 millivolt, and lead II records a positive potential of 1.2 millivolts because these are the instantaneous potential differences between the respective pairs of limbs.

Now, note that *the sum of the voltages in leads I and III equals the voltage in lead II.* That is, 0.5 plus 0.7 equals 1.2. Mathematically, this principle, called Einthoven's law, holds true at any given instant while the electrocardiogram is being recorded.

Normal Electrocardiograms Recorded from the Three Standard Bipolar Limb Leads. Figure 11–7 shows recordings of the electrocardiogram in leads I, II, and III. It is obvious from this figure that the electrocardiograms in these three leads are similar to one another because they all record positive P waves and positive T waves, and the major portion of the QRS complex is also positive in each electrocardiogram.

On analysis of the three electrocardiograms, it can be shown with careful measurements that at any given instant, the sum of the potentials in leads I and III equals the potential in lead II, thus illustrating the validity of Einthoven's law.

Because the recordings from all the bipolar limb leads are similar to one another, it does not matter greatly which lead is recorded when one wants to diagnose the different cardiac arrhythmias because diagnosis of arrhythmias depends mainly on the time relations between the different waves of the cardiac cycle. On the other hand, when one wants to diagnose damage in the ventricular or atrial muscle or in the

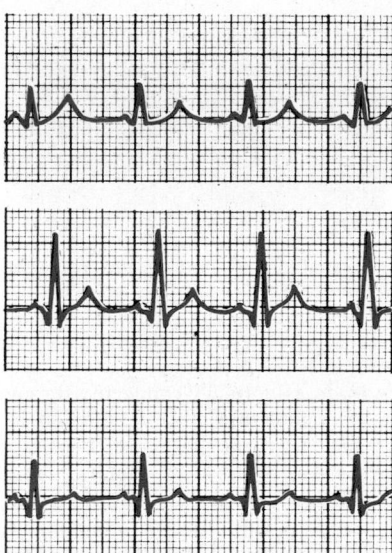

Figure 11–7. Normal electrocardiograms recorded from the three standard electrocardiographic leads.

conducting system, it does matter greatly which leads are recorded because abnormalities of the cardiac muscle change the patterns of the electrocardiograms markedly in some leads and yet may not affect other leads.

Electrocardiographic interpretation of these two types of conditions—cardiac myopathies and cardiac arrhythmias—is discussed separately in Chapters 12 and 13.

Chest Leads (Precordial Leads)

Often electrocardiograms are recorded with an electrode placed on the anterior surface of the chest over the heart at one of the six separate red points shown in Figure 11–8. This electrode is connected to the positive terminal of the electrocardiograph, and the negative electrode, called the

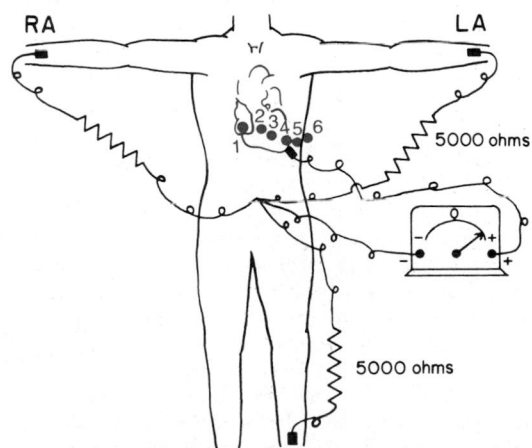

Figure 11–8. Connections of the body with the electrocardiograph for recording chest leads.

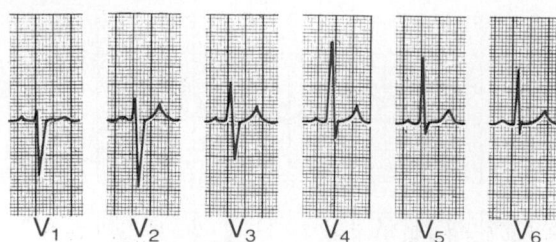

Figure 11–9. Normal electrocardiograms recorded from the six standard chest leads.

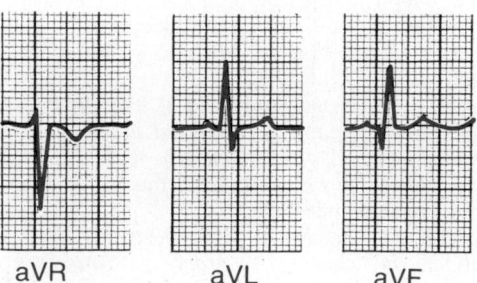

Figure 11–10. Normal electrocardiograms recorded from the three augmented unipolar limb leads.

indifferent electrode, is normally connected through electrical resistances to the right arm, left arm, and left leg all at the same time, as also shown in the figure. Usually six standard chest leads are recorded from the anterior chest wall, the chest electrode being placed respectively at the six points shown in the diagram. The different recordings recorded by the method demonstrated in Figure 11–8 are known as leads V_1, V_2, V_3, V_4, V_5, and V_6.

Figure 11–9 shows the electrocardiograms of the normal heart as recorded from these six standard chest leads. Because the heart surfaces are close to the chest wall, each chest lead records mainly the electrical potential of the cardiac musculature immediately beneath the electrode. Therefore, relatively minute abnormalities in the ventricles, particularly in the anterior ventricular wall, frequently cause marked changes in the electrocardiograms recorded from chest leads.

In leads V_1 and V_2, the QRS recordings of the normal heart are mainly negative because, as shown in Figure 11–8, the chest electrode in these leads is nearer the base of the heart than the apex, which is the direction of electronegativity during most of the ventricular depolarization process. On the other hand, the QRS complexes in leads V_4, V_5, and V_6 are mainly positive because the chest electrode in these leads is nearer the apex, which is the direction of electropositivity during most of depolarization.

Augmented Unipolar Limb Leads

Another system of leads in wide use is the augmented unipolar limb lead. In this type of recording, two of the limbs are connected through electrical resistances to the negative terminal of the electrocardiograph, and the third limb is connected to the positive terminal. When the positive terminal is on the right arm, the lead is known as the aV_R lead; when on the left arm, the aV_L lead; and when on the left leg, the aV_F lead.

Normal recordings of the augmented unipolar limb leads are shown in Figure 11–10. They are all similar to the standard limb lead recordings except that the recording from the aV_R lead is inverted.

REFERENCES

See references for Chapter 13.

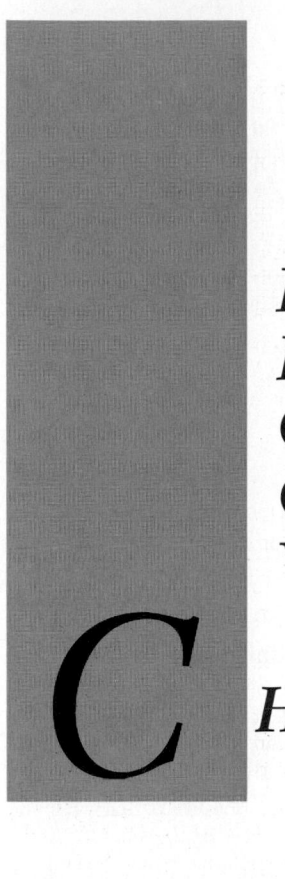

Electrocardiographic Interpretation of Cardiac Muscle and Coronary Abnormalities: Vectorial Analysis

CHAPTER 12

From the discussion in Chapter 10 of impulse transmission through the heart, it is obvious that any change in the pattern of this transmission can cause abnormal electrical potentials around the heart and, consequently, alter the shapes of the waves in the electrocardiogram. For this reason, almost all serious abnormalities of the heart muscle can be detected by analyzing the contours of the different waves in the different electrocardiographic leads.

PRINCIPLES OF VECTORIAL ANALYSIS OF ELECTROCARDIOGRAMS

Use of Vectors to Represent Electrical Potentials

Before it is possible to understand how cardiac abnormalities affect the contours of waves in the electrocardiogram, one must become familiar with the concept of vectors and vectorial analysis as applied to electrical potentials in and around the heart.

Several times in Chapter 11 it is pointed out that heart current flows in a particular direction at a given instant in the cardiac cycle. A vector is an arrow that points in the direction of the electrical potential generated by the current flow *with the arrowhead in the positive direction.* Also, by convention, the length of the arrow is drawn *proportional to the voltage of the potential.*

"RESULTANT" VECTOR IN THE HEART AT ANY GIVEN INSTANT. Figure 12–1 shows, by way of the shaded area and the negative signs, depolarization of the ventricular septum and parts of the lateral endocardial walls of the two ventricles. Electrical current flows between these depolarized areas inside the heart and the nondepolarized areas on the outside of the heart as indicated by the elliptical arrows. Current also flows inside the heart chambers directly from

the depolarized areas toward the polarized areas. Even though a small amount of current flows upward inside the heart, a considerably greater quantity flows downward on the outside of the ventricles toward the apex. Therefore, the summated vector of the generated potential at this particular instant, called the *instantaneous mean vector,* is drawn through the center of the ventricles in a direction from the base of the heart toward the apex. Furthermore, because these currents are considerable in quantity, the potential is large, so that the vector is relatively long.

Denoting the Direction of a Vector in Terms of Degrees

When a vector is horizontal and directed toward the subject's left side, the vector is said to extend in the direction of 0 degrees, as shown in Figure 12–2. From this zero reference point, the scale of vectors rotates clockwise; when the vector extends from above downward, it has a direction of + 90 degrees; when it extends from the subject's left to the right, it has a direction of + 180 degrees; and when it extends upward, it has a direction of − 90 or + 270 degrees.

In a normal heart, the average direction of the vector of the heart during spread of the depolarization wave through the ventricles, called the *mean QRS vector,* is about + 59 degrees, which is shown by vector A drawn through the center of Figure 12–2 in the + 59-degree direction. This means that during most of the depolarization wave, the apex of the heart remains positive with respect to the base of the heart, as discussed later in the chapter.

Axis of Each of the Standard Bipolar Leads and the Unipolar Limb Leads

In Chapter 11, the three standard bipolar and the three unipolar limb leads are described. Each lead is actually a pair of electrodes connected to the body on opposite sides of

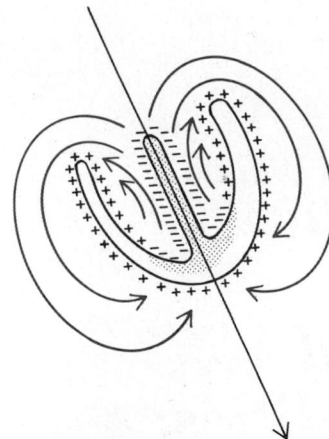

Figure 12–1. Mean vector through the partially depolarized heart.

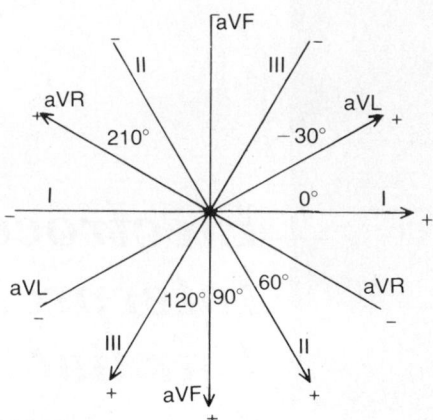

Figure 12–3. Axes of the three bipolar and three unipolar leads.

the heart, and the direction from the negative to the positive electrode is called the axis *of the lead.* Lead I is recorded from two electrodes placed respectively on the two arms. Because the electrodes lie in the horizontal direction with the positive electrode to the left, the axis of lead I is 0 degrees.

In recording lead II, electrodes are placed on the right arm and left leg. The right arm connects to the torso in the upper right-hand corner and the left leg connects in the lower left-hand corner. Therefore, the direction of this lead is about 60 degrees.

By a similar analysis, it can be seen that lead III has an axis of about 120 degrees; lead aV$_R$, 210 degrees; aV$_F$, 90 degrees; and aV$_L$, −30 degrees. The directions of the axes of all these leads are shown in the diagram of Figure 12–3, which is known as the *hexagonal reference system.* The polarities of the electrodes are shown by the plus and minus signs. *The reader must learn these axes and their polarities, particularly for the bipolar limb leads I, II, and III, to understand the remainder of this chapter.*

Vectorial Analysis of Potentials Recorded in Different Leads

Now that we have discussed, first, the conventions for representing potentials across the heart by means of vectors and, second, the axes of the leads, it becomes possible to put them together to determine the potential that will be recorded in each lead for a given vector in the heart.

Figure 12–4 shows a partially depolarized heart; vector A represents the instantaneous mean direction of current flow in the heart and its potential. In this instance, the direction of the potential is 55 degrees, and the voltage of the potential will be assumed to be 2 millivolts. Through the base of vector A is drawn the axis of lead I in the 0-degree direction. To determine how much of the voltage in vector A will be recorded in lead I, a line perpendicular to the axis of lead I is drawn from the tip of vector A to the lead I axis, and a so-called *projected vector (B)* is drawn along the axis. The head of this vector points toward the positive end of the lead I axis, which means that the record momentarily being recorded in the electrocardiogram of lead I will be positive. The voltage recorded will be equal to the length of B divided by the length of A times 2 millivolts, or about 1 millivolt.

Figure 12–5 shows another example of vectorial analysis. In this example, vector A represents the electrical potential at a given instant during ventricular depolarization in another heart, in which the left side of the heart becomes depolar-

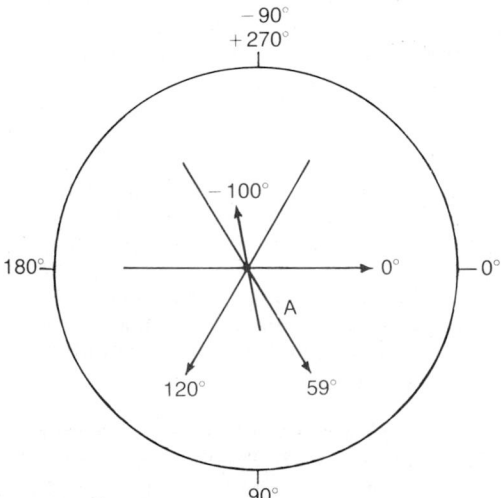

Figure 12–2. Vectors drawn to represent directions of potentials for several different Hearts.

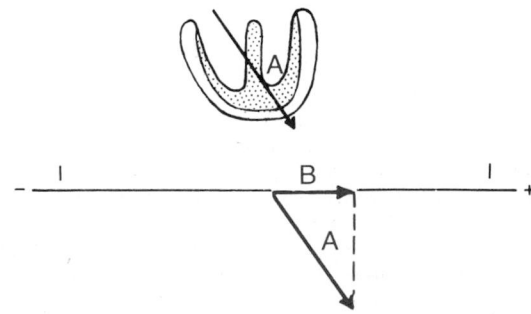

Figure 12–4. Determination of a projected vector B along the axis of lead I when vector A represents the instantaneous potential in the ventricles.

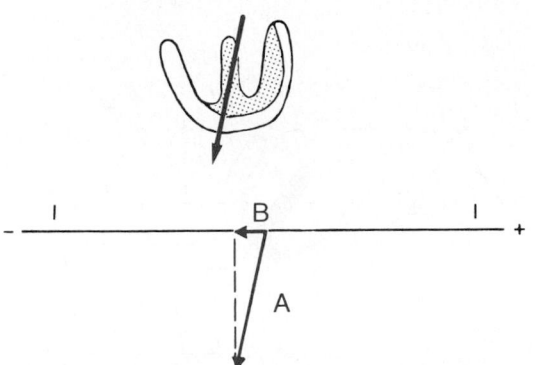

Figure 12–5. Determination of the projected vector B along the axis of lead I when vector A represents the instantaneous potential in the ventricles.

ized more rapidly than the right. In this instance, the vector has a direction of 100 degrees, and the voltage is again 2 millivolts. To determine the potential actually recorded in lead I, we draw a perpendicular line to the lead I axis and find the projected vector *B*. Vector *B* is very short and this time in the negative direction, indicating that at this particular instant, the recording in lead I will be negative (below the zero line in the electrocardiogram), and the voltage recorded will be slight. This figure demonstrates that *when the vector in the heart is in a direction almost perpendicular to the axis of the lead, the voltage recorded in the electrocardiogram of this lead is very low. On the other hand, when the heart vector has almost exactly the same axis as the lead, essentially the entire voltage of the vector will be recorded.*

Vectorial Analysis of Potentials in the Three Standard Bipolar Limb Leads. In Figure 12–6, vector *A* depicts the instantaneous electrical potential of a partially depolarized ventricle. To determine the potential recorded at this instant in the electrocardiogram of each one of the three standard bipolar limb leads, perpendicular lines (the dashed lines) are drawn from the tip of vector *A* to all the lines representing the different leads, as shown in the figure. The projected vector *B* depicts the potential recorded at that instant in lead I, projected vector *C* depicts the potential in lead II, and projected vector *D* depicts the potential in lead III. In each of these, the record in the electrocardiogram is positive—that is, above the zero line—because the pro-

jected vectors point in the positive directions along the axes of all the leads. The potential in lead I is about one half that of the actual potential in the heart as depicted by vector *A*; in lead II, it is almost equal to that in the heart; and in lead III, it is about one third that in the heart.

An identical analysis can be used to determine potentials recorded in augmented limb leads except that the respective axes of these leads (see Fig. 12–3) are used in place of the standard bipolar limb lead axes used in Figure 12–6.

VECTORIAL ANALYSIS OF THE NORMAL ELECTROCARDIOGRAM

Vectors That Occur During Depolarization of the Ventricles—The QRS Complex

When the cardiac impulse enters the ventricles through the atrioventricular bundle, the first part of the ventricles to become depolarized is the left endocardial surface of the septum. This depolarization spreads rapidly to involve both endocardial surfaces of the septum, as demonstrated by the shaded portion of the ventricle in Figure 12–7A. Then the depolarization spreads along the endocardial surfaces of the two ventricles, as shown in Figure 12–7B and C. Finally, it spreads through the ventricular muscle to the outside of the heart, as shown progressively in Figure 12–7C, D, and E.

At each stage of depolarization of the ventricles in Figure 12–7, parts A to E, the instantaneous electrical potential is represented by a vector superimposed on the ventricle in each figure. Each of these vectors is analyzed by the method described in the preceding section to determine the voltages that will be recorded at each instant in each of the three standard electrocardiographic leads. To the right in each stage of the figure is shown the progressive development of the QRS complex. *Keep in mind that a positive vector in a lead will cause the recording in the electrocardiogram to be above the zero line, whereas a negative vector will cause the recording to be below the zero line.*

Before proceeding with further consideration of vectorial analysis, it is essential that this analysis of the successive normal vectors presented in Figure 12–7 be understood. Each of these analyses should be studied in detail by the procedure given previously. A short summary of this sequence is the following.

In Figure 12–7A, the ventricular muscle has just begun to be depolarized, representing an instant about 0.01 second after the onset of depolarization. At this time, the vector is short because only a small portion of the ventricles—the septum—is depolarized. Therefore, all electrocardiographic voltages are low, as recorded to the right of the ventricular muscle for each of the leads. The voltage in lead II is greater than the voltages in leads I and III because the heart vector extends mainly in the same direction as the axis of lead II.

In Figure 12–7B, which represents about 0.02 second after onset of depolarization, the heart vector is long because much of the ventricles has now become depolarized. Therefore, the voltages in all electrocardiographic leads have increased.

In Figure 12–7C, about 0.035 second after onset of depolarization, the heart vector is becoming shorter and the recorded electrocardiographic voltages are less because the outside of the apex of the heart is now electronegative, neutralizing much of the positivity on the other epicardial surfaces of the heart. Also, the axis of the vector is beginning

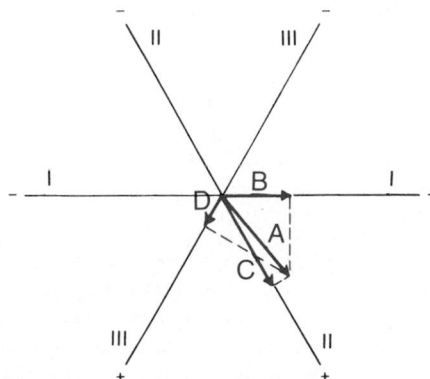

Figure 12–6. Determination of projected vectors in leads I, II, and III when vector A represents the instantaneous potential in the ventricles.

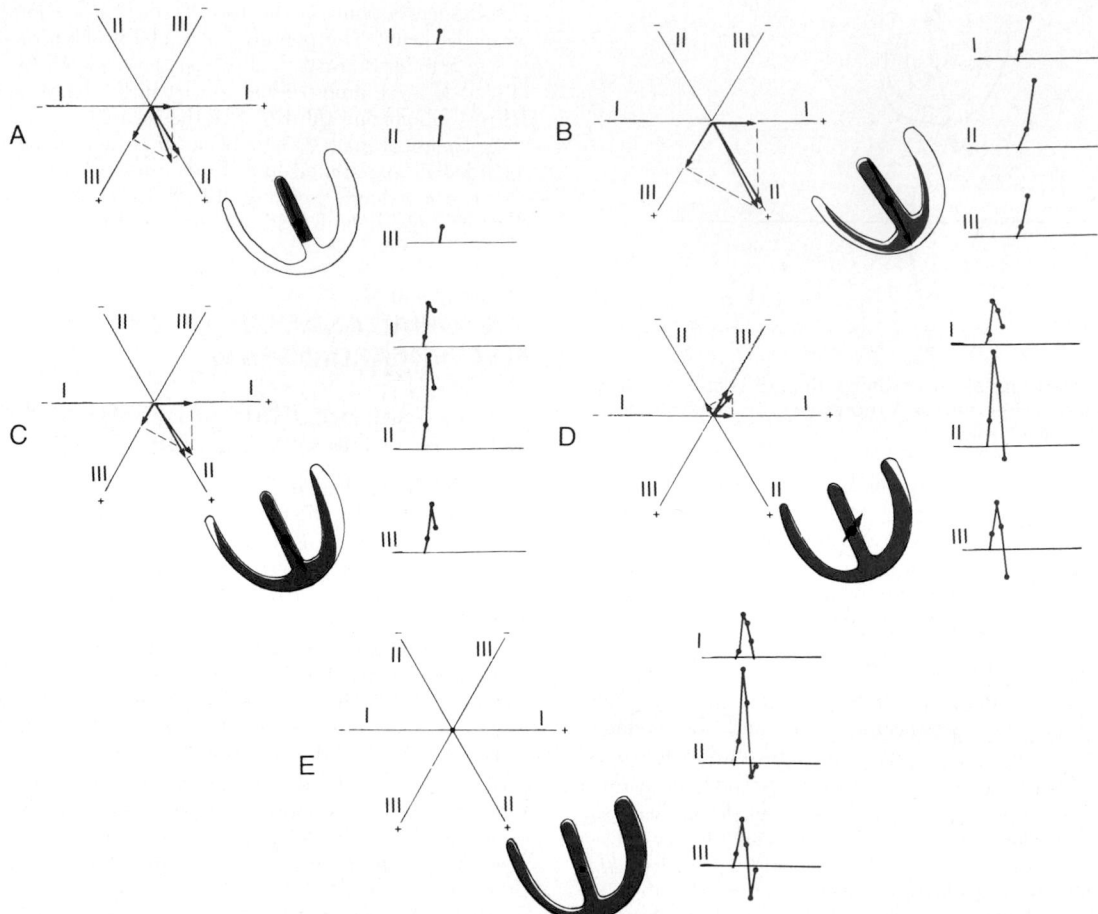

Figure 12–7. Colored areas of the ventricles are depolarized (−); white areas are still polarized (+). *A,* The ventricular vectors and QRS complexes 0.01 second after onset of ventricular depolarization. *B,* 0.02 second after onset of depolarization. *C,* 0.035 second after onset of depolarization. *D,* 0.05 second after onset of depolarization. *E,* After depolarization of the ventricles is complete, 0.06 second after onset.

to shift toward the left side of the chest because the left ventricle is slightly slower to depolarize than the right. Therefore, the ratio of the voltage in lead I to that in lead III is increasing.

In Figure 12–7D, about 0.05 second after onset of depolarization, the heart vector points toward the base of the left ventricle, and it is short because only a minute portion of the ventricular muscle is still polarized positive. Because of the direction of the vector at this time, the voltages recorded in leads II and III are both negative—that is, below the line—whereas the voltage of lead I is still positive.

In Figure 12–7E, about 0.06 second after onset of depolarization, the entire ventricular muscle mass is depolarized, so that no current flows around the heart and no electrical potential is generated. The vector becomes zero, and the voltages in all leads become zero.

Thus, the QRS complexes are completed in the three standard bipolar limb leads.

Sometimes the QRS complex has a slight negative depression at its beginning in one or more of the leads, which is not shown in Figure 12–7; this is the Q wave. When it occurs, it is caused by initial depolarization of the left side of the septum before the right side, which creates a weak vector from left to right for a fraction of a second before the

usual base-to-apex vector occurs. The major positive deflection shown in Figure 12–7 is the R wave, and the final negative deflection is the S wave.

The Electrocardiogram During Repolarization—The T Wave

Once the ventricular muscle has become depolarized, about 0.15 second elapses before sufficient repolarization begins for it to be observed in the electrocardiogram; then repolarization proceeds throughout the ventricular muscle until it is complete at about 0.35 second after the onset of the QRS complex. This process of repolarization causes the T wave in the electrocardiogram.

Because the septum and other endocardial areas of the ventricular muscle depolarize first, it seems logical that these areas should repolarize first as well, but this is not the usual case because the septum and the other endocardial areas have a longer period of contraction and therefore are slower to repolarize than most of the external surfaces of the heart. Therefore, *the greatest portion of ventricular muscle to repolarize first is that located over the entire outer surface of the ventricles and especially near the apex of the heart.* The

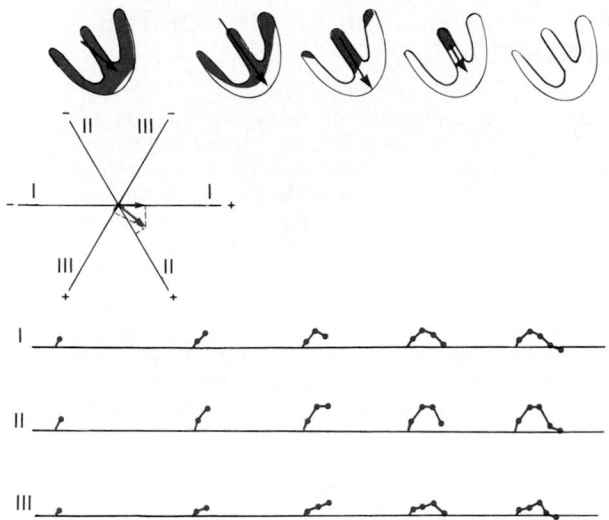

Figure 12–8. Generation of the T wave during repolarization of the ventricles, showing a vectorial analysis of the first stage of repolarization. The total time from the beginning of the T wave to the end is about 0.15 second.

endocardial areas, on the other hand, normally repolarize last. The reason for this abnormal sequence of repolarization is believed to be that the high pressure inside the ventricles during contraction greatly reduces coronary blood flow to the endocardium, thereby slowing the repolarization process in the endocardial areas.

Because the outer, apical surfaces of the ventricles repolarize before the inner, basal surfaces, the positive end of the heart vector during repolarization is toward the apex of the heart. Thus, *the predominant direction of the vector through the heart during repolarization of the ventricles is from base to apex, which is also the predominant direction of the vector during depolarization. As a result, the T wave in the three normal bipolar limb leads is positive, which is also the polarity of most of the normal QRS complex.*

In Figure 12–8, five stages of repolarization of the ventricles are denoted by the progressive increase of the white areas—the repolarized areas. At each stage, the vector extends from the base toward the apex until it disappears in the last stage. At first, the vector is relatively small because the area of repolarization is small. Later, the vector becomes stronger and stronger because of greater degrees of repolarization. Finally, the vector becomes weaker again because the areas of depolarization still persisting become so slight that the total quantity of current flow begins to decrease. These changes demonstrate that the vector is greatest when about half the heart is in the polarized state and about half is depolarized.

The changes in the electrocardiograms of the three standard limb leads during repolarization are noted under each of the ventricles, depicting the progressive stages of repolarization. Over about 0.15 second, the period required for the whole process to take place, the T wave of the electrocardiogram is generated.

Depolarization of the Atria—The P Wave

Depolarization of the atria begins in the sinus node and spreads in all directions over the atria. Therefore, the point of original electronegativity in the atria is about at the point of entry of the superior vena cava where the sinus node lies, and the direction of the electrical potential in the atrium at the beginning of depolarization is in the direction noted in Figure 12–9. Furthermore, the vector remains generally in this direction throughout the process of normal atrial depolarization.

Thus, the vector of current flow during depolarization in the atria points in almost the same direction as that in the ventricles. And because this direction is in the directions of the axes of the three standard bipolar limb leads I, II, and III, the electrocardiograms recorded from the atria during depolarization usually are positive in all three of these leads, as shown in Figure 12–9. The record of atrial depolarization is known as the P wave.

REPOLARIZATION OF THE ATRIA—THE ATRIAL T WAVE. Spread of the depolarization wave through the atrial muscle is *much slower than in the ventricles.* Therefore, the musculature around the sinus node becomes depolarized a long time before the musculature in distal parts of the atria. Because of this, *the area in the atria that becomes repolarized first is the sinus nodal region, the area that had originally become depolarized first,* which is an entirely different situation from that in the ventricles. Thus, when repolarization begins, the region around the sinus node becomes positive with respect to the rest of the atria. Therefore, the atrial repolarization vector is *backward to the vector of depolarization.* (Note again that this is opposite to the effect that occurs in the ventricles.) Therefore, as noted to the right in Figure 12–9, the so-called atrial T wave follows about 0.15 second after the atrial P wave, but this T wave is on the opposite side of the zero reference line from the P wave; that is, it is normally negative rather than positive in the three standard bipolar limb leads. In the normal electrocardiogram, this T wave appears at about the same time that the QRS complex of the ventricle appears. Therefore, it is almost always totally obscured by the larger QRS complex, although in some abnormal states, it does appear in the recorded electrocardiogram.

The Vectorcardiogram

It has been noted in the preceding discussions that the vector of current flow through the heart changes rapidly as the impulse spreads through the myocardium. It changes in

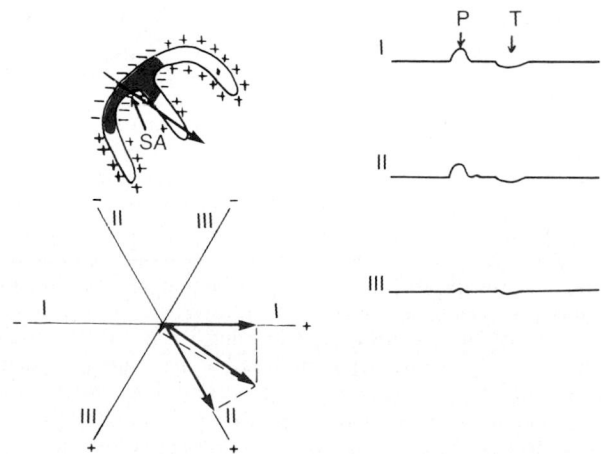

Figure 12–9. Depolarization of the atria and generation of the P wave, showing the vector through the atria and the resultant vectors in the three standard leads. At right are the atrial P and T waves.

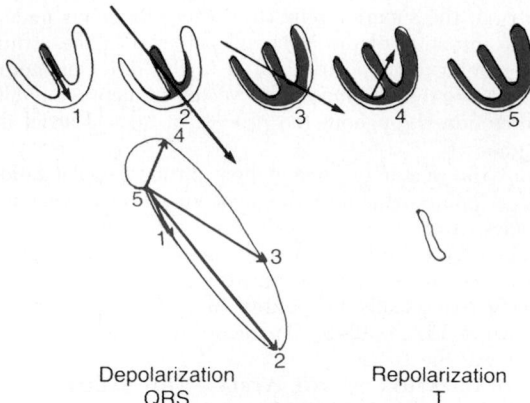

Depolarization   Repolarization
QRS   T

Figure 12–10. QRS and T vectorcardiograms.

two aspects: First, the vector increases and decreases in length because of the increasing and decreasing voltage of the vector. Second, the vector changes its direction because of changes in the average direction of the electrical potential of the heart. The so-called *vectorcardiogram* depicts these changes in the vectors at the different times during the cardiac cycle, as shown in Figure 12–10.

In the vectorcardiogram of Figure 12–10, point 5 is the *zero reference point*, and this point is the negative end of all the vectors. While the heart is quiescent, the positive end of the vector also remains at the zero point because there is no electrical potential. However, just as soon as current begins to flow through the heart at the beginning of ventricular depolarization, the positive end of the vector leaves the zero reference point.

When the septum first becomes depolarized, the vector extends downward toward the apex of the heart, but it is relatively weak, thus generating the first portion of the vectorcardiogram, as shown by the positive end of vector 1. As more of the heart becomes depolarized, the vector becomes stronger and stronger, usually swinging slightly to one side. Thus, vector 2 of Figure 12–10 represents the state of depolarization of the heart about 0.02 second after vector 1. After another 0.02 second, vector 3 represents the potential of the heart, and vector 4 occurs in still another 0.01 second. Finally, the heart becomes totally depolarized, and the vector becomes zero once again, as shown at point 5.

The elliptical figure generated by the positive ends of the vectors is called the *QRS vectorcardiogram.*

Vectorcardiograms can be recorded instantaneously on an oscilloscope by connecting body surface electrodes from above and below the heart to the vertical plates of the oscilloscope and connecting surface electrodes from each side of the heart to the horizontal plates. When the vector changes, the spot of light on the oscilloscope follows the course of the positive end of the changing vector, thus inscribing the vectorcardiogram on the oscilloscopic screen.

"T" Vectorcardiogram. Changing vectors in the heart occur not only during depolarization but also during repolarization because vectors depicting current flow around the ventricles reappear during repolarization. Therefore, a second, smaller vectorcardiogram—the *T vectorcardiogram*—is inscribed during repolarization of the muscle mass; this is depicted to the right in Figure 12–10. Also, a still smaller *P vectorcardiogram* is inscribed during atrial depolarization.

THE MEAN ELECTRICAL AXIS OF THE VENTRICULAR QRS

The vectorcardiogram of the ventricular depolarization wave (the QRS vectorcardiogram) shown in Figure 12–10 is that of a normal heart. Note from this vectorcardiogram that the preponderant direction of the vectors of the ventricles is normally toward the apex of the heart; that is, during most of the cycle of ventricular depolarization, the direction of the electrical potential (negative to positive) is from the base of the ventricles toward the apex. This preponderant direction of the potential during depolarization is called the *mean electrical axis of the ventricles* or the *mean QRS vector.* The mean electrical axis of the normal ventricles is 59 degrees. In certain pathological conditions of the heart, this direction is changed markedly—sometimes even to opposite poles of the heart.

Determining the Electrical Axis from Standard Lead Electrocardiograms

Clinically, the electrical axis of the heart is usually determined from the standard bipolar limb lead electrocardiograms rather than from the vectorcardiogram. Figure 12–11 shows a method for doing this. After recording the standard leads, one determines the maximum potential and polarity of the recording in two of the leads. In lead I of the figure, the recording is positive, and in lead III, the recording is mainly positive but negative during part of the cycle. If any part of the recording is negative, *this negative potential is subtracted from the positive potential* to determine the net potential for that lead, as shown by the arrows to the right of the QRS complexes of leads I and III. (To be even more accurate, some cardiologists subtract the *area* of the negative wave from the *area* of the positive wave.) After subtracting the negative portion of the QRS wave in lead III from the positive portion, each net potential is plotted on the axes of the respective leads, with the base of the potential at the point of intersection of the axes, as shown in Figure 12–11.

If the net potential of lead I is positive, it is plotted in a positive direction along the line depicting lead I. On the other hand, if this potential is negative, it is plotted in a negative direction. Also, for lead III, the net potential is placed with its base at the point of intersection, and, if positive, it is plotted in the positive direction along the line

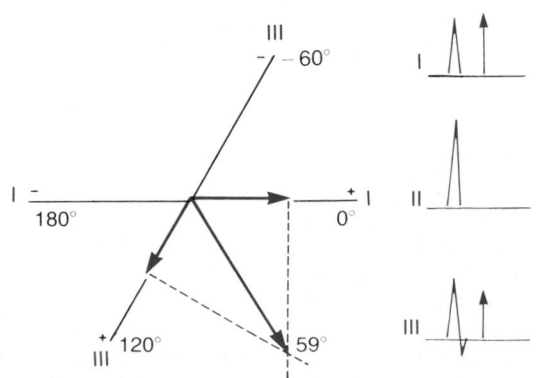

Figure 12–11. Plotting the mean electrical axis of the heart from two electrocardiographic leads.

depicting lead III. If it is negative, it is plotted in the negative direction.

To determine the actual vector of the ventricular mean electrical potential, one draws perpendicular lines from the apices of the two net potentials of leads I and III, respectively. The point of intersection of these two perpendicular lines represents, by vectorial analysis, the apex of the actual mean QRS vector in the ventricles, and the point of intersection of the two lead axes represents the negative end of the actual vector. Therefore, the *mean QRS vector* is drawn between these two points. The approximate average potential generated by the ventricles during depolarization is represented by the length of the vector, and the mean electrical axis is represented by the direction of the vector. Thus, the orientation of the mean electrical axis of the normal ventricles, as determined in Figure 12–11, is 59 degrees.

Abnormal Ventricular Conditions That Cause Axis Deviation

Although the mean electrical axis of the ventricles averages about 59 degrees, this axis can swing to the left even in the normal heart to about 20 degrees or to the right to about 100 degrees. The causes of the normal variations are mainly anatomical differences in the Purkinje system distribution or in the musculature itself of different hearts. Yet a number of conditions can cause axis deviation even beyond these normal limits, as follows.

CHANGE IN THE POSITION OF THE HEART. If the heart itself is angulated to the left, the mean electrical axis of the heart will also shift to the left. Such shift occurs (1) during expiration, (2) when a person lies down because the abdominal contents press upward against the diaphragm, and (3) quite frequently in stocky, fat people whose diaphragms normally press upward against the heart all the time.

Likewise, angulation of the heart to the right causes the mean electrical axis of the ventricles to shift to the right. This condition occurs (1) during inspiration, (2) when a person stands up, and (3) normally in tall, lanky people whose hearts hang downward.

HYPERTROPHY OF ONE VENTRICLE. When one ventricle greatly hypertrophies, the axis of the heart shifts toward the hypertrophied ventricle for two reasons. First, far greater quantity of muscle exists on the hypertrophied side of the heart than on the other side, and this allows excess generation of electrical potential on that side. Second, more time is required for the depolarization wave to travel through the hypertrophied ventricle than through the normal ventricle. Consequently, the normal ventricle becomes depolarized— that is, becomes negative—considerably in advance of the hypertrophied ventricle, and this causes a strong vector from the normal side of the heart toward the hypertrophied side, which remains positively charged. Thus, the axis deviates toward the hypertrophied ventricle.

Vectorial Analysis of Left Axis Deviation Resulting from Hypertrophy of the Left Ventricle. Figure 12–12 shows the three standard bipolar limb leads of an electrocardiogram in which analysis shows left axis deviation with the mean electrical axis pointing in the − 15-degree direction. This is a typical electrocardiogram resulting from increased muscle mass of the left ventricle. In this instance, the axis deviation was caused by *hypertension* (high blood pressure), which caused the left ventricle to hypertrophy to pump blood against the elevated systemic arterial pressure. A similar picture of left axis deviation occurs when the left ventri-

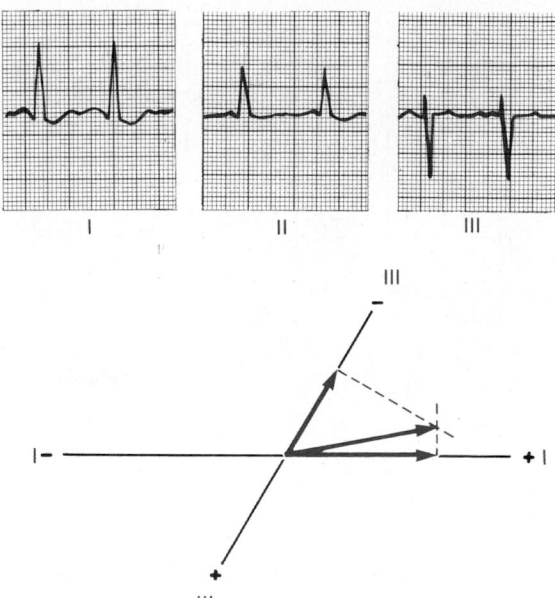

Figure 12–12. Left axis deviation in hypertensive heart disease. Note the slightly prolonged QRS complex as well.

cle hypertrophies as a result of *aortic valvular stenosis, aortic valvular regurgitation,* or any of a number of *congenital heart conditions* in which the left ventricle enlarges while the right ventricle remains relatively normal in size.

Vectorial Analysis of Right Axis Deviation Resulting from Hypertrophy of the Right Ventricle. The electrocardiogram of Figure 12–13 shows intense right axis deviation with an electrical axis of about 170 degrees, which is 111 degrees to the right of the normal 59-degree mean electrical axis of the ventricles. The right axis deviation demonstrated in this figure was caused by hypertrophy of the right ventricle as a result of *pulmonary valve stenosis.* Right axis deviation may also occur in other congenital heart conditions that cause hypertrophy of the right ventricle, such as *tetralogy of Fallot* and *interventricular septal defect.* Also, hypertrophy of the right ventricle as a result of *increased pulmonary vascular resistance* can cause right axis deviation.

BUNDLE BRANCH BLOCK CAUSES AXIS DEVIATION. Ordinarily, the two lateral walls of the ventricles depolarize at almost the same time because both the left and the right bundle branches of the Purkinje system transmit the cardiac impulse to the endocardial surfaces of the two ventricular walls at almost the same instant. As a result, the potentials generated by the two ventricles almost neutralize each other. If one of the major bundle branches is blocked, the cardiac impulse spreads through the normal ventricle long before it spreads through the other. Therefore, depolarization of the two ventricles does not occur even nearly simultaneously, and the depolarization potentials do not neutralize each other. As a result, axis deviation occurs as follows.

Vectorial Analysis of Left Axis Deviation in Left Bundle Branch Block. When the left bundle branch is blocked, cardiac depolarization spreads through the right ventricle two to three times as rapidly as through the left ventricle. Consequently, much of the left ventricle remains polarized (positive) for a long time after the right ventricle has become totally depolarized (negative). Thus, the right ventricle be-

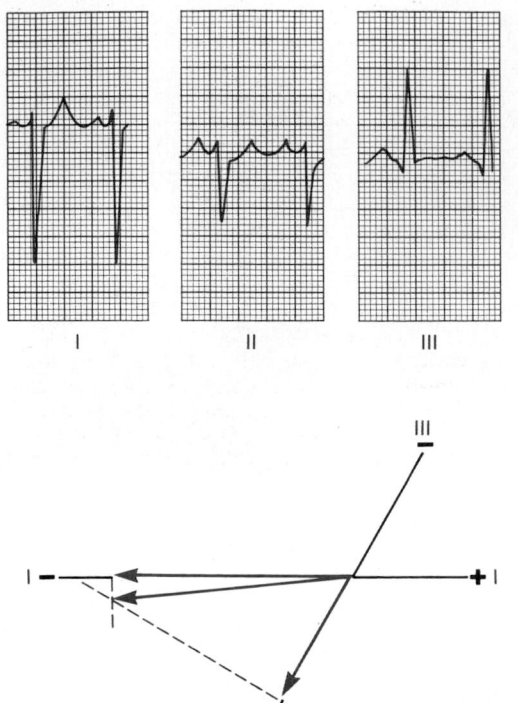

Figure 12–13. High-voltage electrocardiogram in pulmonary valve stenosis with right ventricular hypertrophy. Intense right axis deviation and a slightly prolonged QRS complex are also seen.

comes electronegative while the left ventricle remains electropositive during most of the depolarization process, and a strong vector projects from the right ventricle toward the left ventricle. In other words, there is intense left axis deviation because the positive end of the vector points toward the left ventricle. This is demonstrated in Figure 12–14, which shows typical left axis deviation resulting from left bundle branch block. Note that the axis is about −50 degrees.

Because of slowness of impulse conduction when the Purkinje system is blocked, in addition to the axis deviation, the duration of the QRS complex is also greatly prolonged, which one can see by observing the excessive widths of the QRS waves in Figure 12–14. This is discussed in greater detail later in the chapter. This prolonged QRS complex differentiates this condition from axis deviation caused by hypertrophy.

Vectorial Analysis of Right Axis Deviation in Right Bundle Branch Block. When the right bundle branch is blocked, the left ventricle depolarizes far more rapidly than the right ventricle, so that the left becomes electronegative and the right remains electropositive. Therefore, a strong vector develops with its negative end toward the left ventricle and its positive end toward the right ventricle. In other words, intense right axis deviation occurs.

Right axis deviation caused by right bundle branch block is demonstrated and its vector analyzed in Figure 12–15, which shows an axis of about 105 degrees and a prolonged QRS complex because of slow conduction.

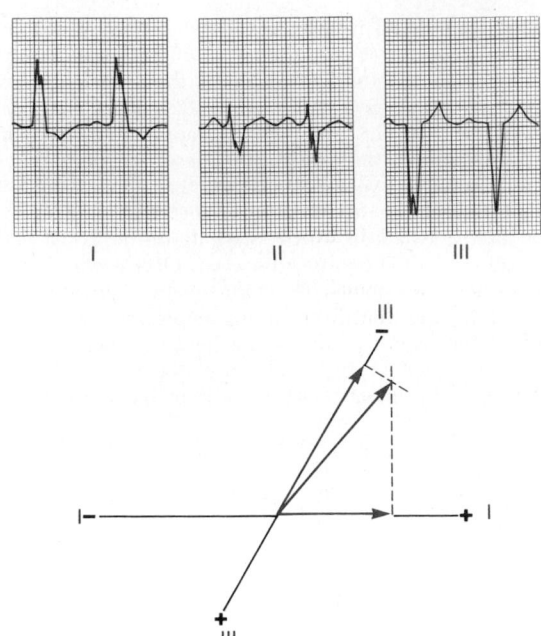

Figure 12–14. Left axis deviation caused by left bundle branch block. Note also the greatly prolonged QRS complex.

CONDITIONS THAT CAUSE ABNORMAL VOLTAGES OF THE QRS COMPLEX

Increased Voltage in the Standard Bipolar Limb Leads

Normally, the voltages in the three standard bipolar limb leads, as measured from the peak of the R wave to the bottom of the S wave, vary between 0.5 and 2.0 millivolts, with lead III usually recording the lowest voltage and lead II the highest. However, these relations are not invariably true even in the normal heart. In general, when the sum of the voltages of all the QRS complexes of the three standard leads is greater than 4 millivolts, one considers that the patient has a high-voltage electrocardiogram.

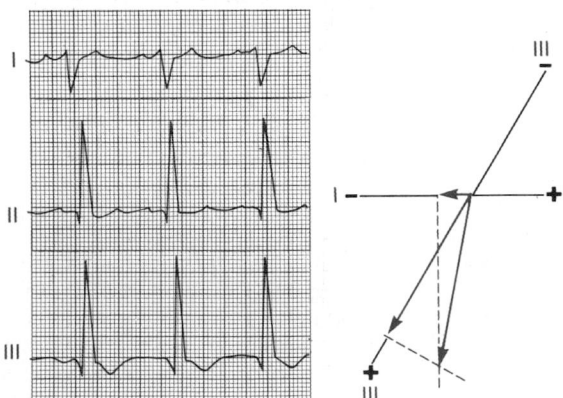

Figure 12–15. Right axis deviation caused by right bundle branch block. Note the greatly prolonged QRS complex as well.

The cause of high-voltage QRS complexes is most often increased muscle mass of the heart, which ordinarily results from *hypertrophy of the muscle* in response to excessive load on one part of the heart or the other. For instance, the right ventricle hypertrophies when it must pump blood through a stenotic pulmonary valve, and the left ventricle hypertrophies when a person has high blood pressure. The increased quantity of muscle causes generation of increased quantities of electricity around the heart. As a result, the electrical potentials recorded in the electrocardiographic leads are considerably greater than normal, as shown in Figures 12–12 and 12–13.

Decreased Voltage of the Electrocardiogram

There are three major causes of decreased voltage in the electrocardiogram. They are, first, abnormalities of the cardiac muscle itself that prevent generation of large quantities of current; second, abnormal conditions around the heart, so that current cannot be conducted from the heart to the surface of the body with ease; and, third, rotation of the apex of the heart to point toward the anterior chest wall, so that the electrical current of the heart flows mainly anteroposteriorly in the chest rather than in the frontal plane of the body, which decreases the voltages in the limb leads.

DECREASED VOLTAGE CAUSED BY CARDIAC MYOPATHIES. One of the most usual causes of decreased voltage of the QRS complex is a series of *old myocardial infarctions* with resultant *diminished muscle mass.* This also causes the depolarization wave to move through the ventricles slowly and prevents major portions of the heart from becoming massively depolarized all at once. Consequently, this condition causes moderate prolongation of the QRS complex along with the decreased voltage. Figure 12–16 shows the typical low-voltage electrocardiogram with prolongation of the QRS complex that one often finds after multiple small infarctions of the heart that have resulted in local blocks and loss of muscle mass throughout the ventricles.

DECREASED VOLTAGE CAUSED BY CONDITIONS SURROUNDING THE HEART. One of the most important causes of decreased voltage in the electrocardiographic leads is *fluid in the pericardium.* Because extracellular fluid conducts electrical currents with great ease, a large portion of electricity flowing out of the heart is conducted from one part of the heart to another through the pericardial effusion. Thus, this effusion effectively "short-circuits" the electrical potentials generated by the heart, thus decreasing the electrocardiographic voltages that reach the outside surfaces of the body. *Pleural effusion,* to a lesser extent, can also "short" the electricity around the heart so that the voltages at the surface of the body and in the electrocardiograms are decreased.

Pulmonary emphysema can decrease the electrocardiographic potentials but by a different method from that of pericardial effusion. In pulmonary emphysema, conduction of electrical current through the lungs is considerably depressed because of the excessive quantity of air in the lungs. Also, the chest cavity enlarges, and the lungs tend to envelop the heart to a greater extent than normally. Therefore, the lungs act as an insulator to prevent spread of electrical voltage from the heart to the surface of the body, and this results in decreased electrocardiographic potentials in the various leads.

PROLONGED AND BIZARRE PATTERNS OF THE QRS COMPLEX

Prolonged QRS Complex as a Result of Cardiac Hypertrophy or Dilatation

The QRS complex lasts as long as depolarization continues to spread through the ventricles—that is, as long as part of the ventricles is depolarized and part is still polarized. Therefore, the cause of a prolonged QRS complex is always *prolonged conduction* of the impulse through the ventricles. Such prolongation often occurs when one or both ventricles are hypertrophied or dilated, owing to the longer pathway that the impulse must then travel. The normal QRS complex lasts 0.06 to 0.08 second, whereas in hypertrophy or dilatation of the left or right ventricle, the QRS complex may be prolonged to 0.09 to 0.12 second.

Prolonged QRS Complex Resulting from Purkinje System Blocks

When the Purkinje fibers are blocked, the cardiac impulse must be conducted by the ventricular muscle instead of by way of the Purkinje system. This decreases the velocity of impulse conduction to about one-third to one-fourth normal. Therefore, if complete block of one of the bundle branches occurs, the duration of the QRS complex is usually increased to 0.14 second or greater.

In general, a QRS complex is considered to be abnormally long when it lasts more than 0.09 second, and when it lasts more than 0.12 second, the prolongation is almost certain to be caused by pathological block of the conduction system somewhere in the ventricles, as shown by the electrocardiograms for bundle branch block in Figures 12–14 and 12–15.

Conditions That Cause Bizarre QRS Complexes

Bizarre patterns of the QRS complex are most frequently caused by two conditions: first, destruction of cardiac muscle in various areas throughout the ventricular system with replacement of this muscle by scar tissue and, second, local blocks in the conduction of impulses by the Purkinje system.

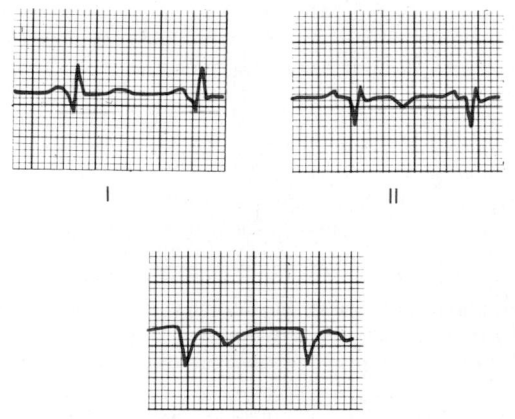

I

II

III

Figure 12–16. Low-voltage electrocardiogram with evidence of local damage throughout the ventricles caused by previous myocardial infarction.

Sometimes local blocks occur at multiple points in the ventricles. As a result, cardiac impulse conduction becomes irregular, causing rapid shifts in voltages and axis deviations. This often causes double or even triple peaks in some of the electrocardiographic leads, such as those shown in Figure 12–14.

CURRENT OF INJURY

Many different cardiac abnormalities, especially those that damage the heart muscle itself, often cause part of the heart to remain partially or totally *depolarized all the time*. When this occurs, current flows between the pathologically depolarized and the normally polarized areas. This is called a *current of injury*. Note especially that *the injured part of the heart is negative because this part is depolarized and emits negative charges into the surrounding fluids, whereas the remainder of the heart is positive.*

Some of the abnormalities that can cause current of injury are (1) *mechanical trauma,* which makes the membranes remain so permeable that full repolarization cannot take place; (2) *infectious processes* that damage the muscle membranes; and (3) *ischemia of local areas of muscle caused by coronary occlusion,* which is by far the most common cause of current of injury in the heart. During ischemia, not enough nutrients from the coronary blood supply are available to the heart muscle to maintain normal function.

Effect of Current of Injury on the QRS Complex

In Figure 12–17, a shaded area in the base of the left ventricle is newly infarcted (loss of coronary blood flow). Therefore, during the T-P interval—that is, when the normal ventricular muscle is polarized—abnormal *negative* current flows from the infarcted area in the base of the left ventricle toward the rest of the ventricles. The vector of this "current of injury," as shown in the first heart in the figure, is in a direction of about 125 degrees, with the base of the vector, the *negative end,* toward the injured muscle. As

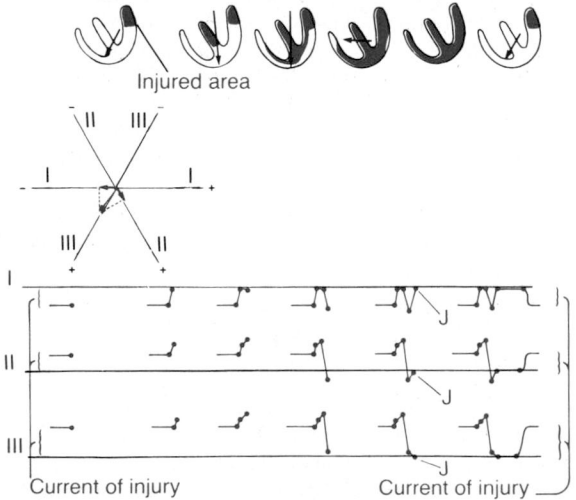

Injured area

Figure 12–17. Effect of a current of injury on the electrocardiogram.

shown in the lower portions of the figure, even before the QRS complex begins, *this vector causes an initial record in lead I below the zero potential line* because the projected vector of the current of injury in lead I points toward the negative end of the lead I axis. In lead II, the record is above the line because the projected vector points toward the positive terminal of lead II. In lead III, the projected vector of the current flow is in the same direction as the polarity of lead III, so that the record is positive. Furthermore, because the vector of the current of injury lies almost exactly along the axis of lead III, the voltage of the current of injury in lead III is much greater than that in either of the other two records.

As the heart then proceeds through its normal process of depolarization, the septum first becomes depolarized, and the depolarization spreads down to the apex and back toward the bases of the ventricles. The last portion of the ventricles to become totally depolarized is the base of the right ventricle because the base of the left ventricle is already totally and permanently depolarized. By vectorial analysis, as shown in the figure, the electrocardiogram generated by the depolarization wave traveling through the ventricles may be constructed graphically, as demonstrated in Figure 12–17.

When the heart becomes totally depolarized at the end of the depolarization process, as noted by the next to last stage in Figure 12–17, all of the ventricular muscle is in a negative state. Therefore, at this instant in the electrocardiogram, no current flows around the musculature of the ventricles because now both the injured heart muscle and the contracting muscle are depolarized.

As repolarization then takes place, all of the heart finally repolarizes except the area of permanent depolarization in the injured base of the left ventricle. Thus, repolarization causes a return of the current of injury in each lead, as noted at the far right in Figure 12–17.

The J Point—The Zero Reference Potential for Analyzing Current of Injury

One would think that the electrocardiograph machines for recording electrocardiograms could determine when no current is flowing around the heart. However, many stray currents exist in the body, such as currents resulting from "skin potentials" and from differences in ionic concentrations in different parts of the body. Therefore, when two electrodes are connected between the arms or between an arm and a leg, these stray currents make it impossible for one to predetermine the exact zero reference level in the electrocardiogram. For these reasons, the following procedure must be used to determine the zero potential level: First, one notes *that exact point at which the wave of depolarization just completes its passage through the heart,* which occurs at the end of the QRS complex. At exactly this point, all parts of the ventricles are depolarized, including both the damaged parts and the normal parts, so that no current is flowing around the heart. Even the current of injury disappears at this point. Therefore, the potential of the electrocardiogram at this instant is at zero voltage. This point is known as the "J point" in the electrocardiogram, as shown in Figure 12–18.

For analysis of the electrical axis of the injury potential caused by a current of injury, a horizontal line is drawn through the electrocardiogram at the level of the J point, and this horizontal line is the zero potential level in the electrocardiogram from which all potentials caused by currents of injury must be measured.

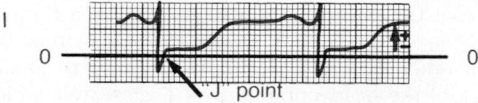

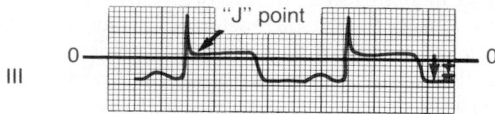

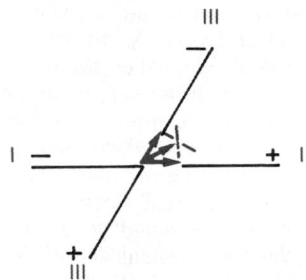

Figure 12–18. J point as the zero reference potential of the electrocardiogram. Also, method for plotting the axis of a current of injury is shown *(lower)*.

USE OF THE J POINT IN PLOTTING THE AXIS OF AN INJURY POTENTIAL. Figure 12–18 shows electrocardiograms recorded from leads I and III, both of which show currents of injury. In other words, the J point of each of these two electrocardiograms is not on the same line as the T-P segment. A horizontal line has been drawn through the J point to represent the zero voltage level in each of the two recordings. The voltage of the current of injury in each lead is the difference between the level of the T-P segment of the electrocardiogram (which is between heartbeats when there is a current of injury) and the zero voltage potential level, as shown by the arrows. In lead I, the recorded voltage caused by the current of injury is above the zero potential level and is, therefore, positive. On the other hand, in lead III, the T-P segment is below the zero voltage level; therefore, the voltages of the current of injury in lead III is negative.

At the bottom in Figure 12–18, the voltages of the current of injury in leads I and III are plotted on the coordinates of these leads, and the resultant vector of the injury potential for the whole ventricular mass is determined by the method already described. In this instance, the vector of the current of injury extends from the right side of the ventricles toward the left and slightly upward, with an axis of about −30 degrees.

If one places the vector of the current of injury directly over the ventricles, *the negative end of the vector points toward the permanently depolarized, "injured" area of the ventricles.* In the instance shown in Figure 12–18, the injured area would be in the lateral wall of the right ventricle.

PHENOMENON OF S-T SEGMENT SHIFT. The portion of the electrocardiogram that occurs between the end of the QRS complex and the beginning of the T wave is called the *S-T segment.* The J point lies at the beginning of this segment. Therefore, every time a current of injury occurs in

one of the electrocardiographic leads, one also finds that the S-T segment and T-P segments of the electrocardiogram are not at the same voltage levels in the record. Actually, it is the T-P segment and not the S-T segment that is shifted away from the zero axis. Most people, however, are conditioned to consider the T-P segment of the electrocardiogram as the reference level rather than the J point. Therefore, when a current of injury is evident in an electrocardiogram, it appears that the S-T segment is shifted from its normal level in the electrocardiogram, and this is called an *S-T segment shift.* When one sees an S-T segment shift in an electrocardiogram, one knows immediately that the electrocardiogram shows the characteristics of a current of injury. In fact, most electrocardiographers do not speak of current of injury but simply speak of S-T segment shift, which means the same thing.

Coronary Ischemia as a Cause of Current of Injury

Insufficient blood flow to the cardiac muscle depresses the metabolism of the muscle for three reasons: (1) oxygen lack, (2) excess accumulation of carbon dioxide, and (3) lack of sufficient nutrients. Consequently, repolarization of the membranes cannot occur in areas of severe myocardial ischemia. Often the heart muscle does not die because the blood flow is sufficient to maintain life of the muscle even though it is not sufficient to cause repolarization of the membranes. As long as this state exists, a current of injury continues to flow during the diastolic portion of each heartbeat.

Extreme ischemia of the cardiac muscle occurs after coronary occlusion, and a strong current of injury flows from the infarcted area of the ventricles during the T-P interval between heartbeats, as shown in Figures 12–19 and 12–20.

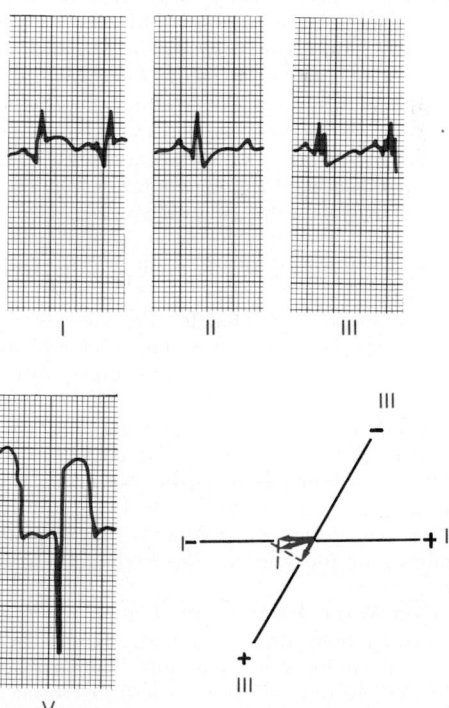

Figure 12–19. Current of injury in acute anterior wall infarction. Note the intense current of injury in lead V_2.

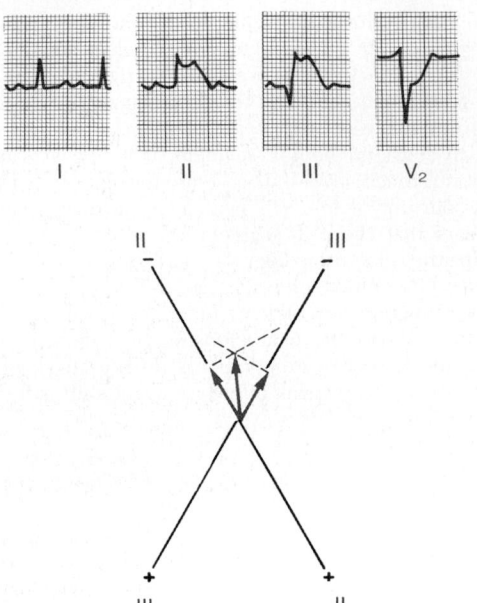

Figure 12–20. Current of injury in acute posterior wall, apical infarction.

Therefore, one of the most important diagnostic features of electrocardiograms recorded after acute coronary thrombosis is the current of injury.

ACUTE ANTERIOR WALL INFARCTION. Figure 12–19 shows the electrocardiogram in the three standard bipolar limb leads and in one chest lead recorded from a patient with acute anterior wall cardiac infarction. The most important diagnostic feature of this electrocardiogram is the intense current of injury of the chest lead. If one draws a zero potential line through the J point of this electrocardiogram, a strong *negative* injury potential during the T-P interval is found, which means that the chest electrode over the front of the heart is in an area of strongly negative potential. In other words, the negative end of the injury potential vector is against the chest wall. This means that the current of injury is emanating from the anterior wall of the ventricles, which diagnoses this condition as anterior wall infarction.

Analyzing the currents of injury in leads I and III, one finds a negative potential caused by the current of injury in lead I and a positive potential for the current of injury in lead III. This means that the resultant vector of the current of injury in the heart is about +150 degrees, with the negative end of the vector pointing toward the left ventricle and the positive end pointing toward the right ventricle. Thus, in this particular electrocardiogram, the current of injury appears to be coming mainly from the left ventricle as well as from the anterior wall of the heart. Therefore, one would suspect that this anterior wall infarction is probably caused by thrombosis of the anterior descending limb of the left coronary artery.

POSTERIOR WALL INFARCTION. Figure 12–20 shows the three standard bipolar limb leads and one chest lead from a patient with posterior wall infarction. The major diagnostic feature of this electrocardiogram is also in the chest lead. If a zero potential reference line is drawn through the J point of this lead, it is readily apparent that during the T-P interval, the potential of the current of injury is positive. This

means that the positive end of the vector is at the chest wall and the negative end (injured end) is away from the chest wall. In other words, the current of injury is coming from the back of the heart opposite to the chest wall, which is the reason this type of electrocardiogram is the basis for diagnosing posterior wall infarction.

If one analyzes the currents of injury in leads II and III of Figure 12–20, it is readily apparent that the injury potential is negative in both leads. By vectorial analysis, as shown in the figure, one finds that the vector of the injury potential is about −95 degrees, with the negative end of the vector pointing downward and the positive end pointing upward. Thus, because the infarct, as indicated by the chest lead, is on the posterior wall of the heart and, as indicated by the currents of injury in leads II and III, is in the apical portion of the heart, one would suspect that this infarct is close to the apex on the posterior wall of the left ventricle.

INFARCTION IN OTHER PARTS OF THE HEART. By the same procedures as those demonstrated in the preceding two discussions of anterior and posterior wall infarctions, it is possible to determine the locus of any infarcted area emitting a current of injury regardless of which part of the heart is involved. In making such vectorial analyses, it must be remembered that *the positive end of the injury potential vector points toward the normal cardiac muscle and the negative end points toward the abnormal portion of the heart that is emitting the current of injury.*

RECOVERY FROM ACUTE CORONARY THROMBOSIS. Figure 12–21 shows a V_3 chest lead from a patient with acute posterior infarction, demonstrating the change in the electrocardiogram of this lead from the day of the attack to 1 week later, then 3 weeks later, and, finally, 1 year later. From this electrocardiogram, it can be seen that the current of injury is strong immediately after the acute attack (T-P segment displaced positively from the J point and S-T segment), but after about 1 week, the current of injury has diminished considerably, and after 3 weeks, it is gone. After that, the electrocardiogram does not change greatly during the next year. This is the usual recovery pattern after acute cardiac infarction of moderate degree when the collateral coronary blood flow is sufficient to re-establish appropriate nutrition to most of the infarcted area.

On the other hand, in some patients with coronary infarction, the infarcted area never redevelops an adequate coronary blood supply; some of the heart muscle dies, and relative coronary insufficiency persists in this area of the heart indefinitely. If the muscle does not die and become replaced by scar tissue, it continually emits a current of injury as long

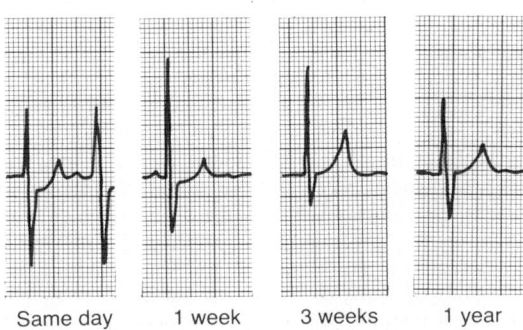

Same day   1 week   3 weeks   1 year

Figure 12–21. Recovery of the myocardium after moderate posterior wall infarction, demonstrating disappearance of the current of injury (lead V_3).

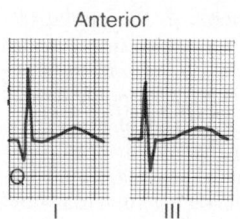 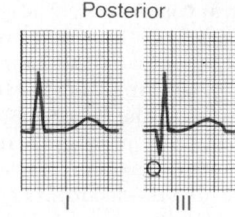

Figure 12–22. Electrocardiograms of previous anterior and posterior wall infarctions, showing the Q wave in lead I in the anterior wall infarction and the Q wave in lead III in the posterior wall infarction.

as the relative ischemia exists, particularly during bouts of exercise when the heart is overloaded.

Old Recovered Myocardial Infarction. Figure 12–22 shows leads I and III after anterior infarction and posterior infarction as these leads appear about a year after the acute episode. These are what might be called the "ideal" configurations of the QRS complex in these types of recovered myocardial infarction. Usually a Q wave develops at the beginning of the QRS complex in lead I in anterior infarction because of loss of muscle mass in the anterior wall of the left ventricle, whereas in posterior infarction, a Q wave develops at the beginning of the QRS complex in lead III because of loss of muscle in the posterior apical part of the ventricle.

These configurations are certainly not those found in all cases of old anterior and posterior cardiac infarction. Local loss of muscle and local areas of conduction block can cause the following abnormalities of the QRS complex: bizarre patterns (the prominent Q waves, for instance), decreased voltage, and prolongation.

Current of Injury in Angina Pectoris. "Angina pectoris" means pain from the heart felt in the pectoral regions of the upper chest. This pain usually radiates into the neck and down the left arm. The pain is typically caused by *relative* ischemia of the heart. No pain is usually felt as long as the person is quiet, but as soon as the person overworks the heart, the pain appears.

A current of injury often occurs during an attack of severe angina pectoris, because the relative coronary insufficiency then becomes great enough to prevent adequate repolarization of the membranes in some areas of the heart during diastole.

ABNORMALITIES IN THE T WAVE

Earlier in the chapter, it was pointed out that the T wave is normally positive in all the standard bipolar limb leads and that this is caused by repolarization of the apex and outer surfaces of the ventricles ahead of the endocardial surfaces. This direction in which repolarization occurs in the heart is backward to the direction in which depolarization takes place. (If the basic principles of the upright T wave in the standard leads have not been understood by now, the reader should become familiar with the earlier, more detailed discussion of this before proceeding to the next few sections.)

The T wave becomes abnormal when the normal sequence of repolarization does not occur. Several factors can change this sequence of repolarization.

Effect of Slow Conduction of the Depolarization Wave on the T Wave

Referring back to Figure 12–14, note that the QRS complex is considerably prolonged. The reason for this prolongation is delayed conduction in the left ventricle as a result of left bundle branch block. The left ventricle becomes depolarized about 0.08 second after depolarization of the right ventricle, which gives a strong mean QRS vector *to the left*. The refractory periods of the right and left ventricular muscle masses are not greatly different from each other. Therefore, the right ventricle begins to repolarize long before the left ventricle; this causes positivity in the right ventricle and negativity in the left ventricle. In other words, the mean axis of the T wave is deviated *to the right*, which is opposite the mean electrical axis of the QRS complex in the same electrocardiogram. Thus, when conduction of the depolarization impulse through the ventricles is greatly delayed, the T wave is almost always of opposite polarity to that of the QRS complex.

In Figure 12–15 and in several figures in Chapter 13, conduction also does not occur through the Purkinje system. As a result, the rate of conduction is greatly slowed, and in each instance, the T wave is of opposite polarity to that of the QRS complex, whether the condition causing this delayed conduction happens to be left bundle branch block, right bundle branch block, premature ventricular contraction, or otherwise.

Prolonged Depolarization in Portions of the Ventricular Muscle as a Cause of Abnormalities in the T Wave

If the apex of the ventricles should have an abnormally long period of depolarization, that is, a prolonged action potential, repolarization of the ventricles would not begin at the apex as it normally does. Instead, the base of the ventricles would repolarize ahead of the apex and the vector of repolarization would point from the apex toward the base of the heart, opposite to the usual vector of repolarization. Consequently, the T wave in all three standard leads would be negative rather than the usual positive. Thus, the simple fact that the apical muscle of the heart has a prolonged period of depolarization is sufficient to cause marked changes in the T wave, even to the extent of changing the entire polarity, as shown in Figure 12–23.

Mild ischemia is by far the most common cause of increased duration of depolarization of cardiac muscle, and when the ischemia occurs in only one area of the heart, the depolarization period of this area increases out of proportion to that in other portions. As a result, definite changes in the T wave may take place. The ischemia may result from chronic, progressive coronary occlusion, acute coronary oc-

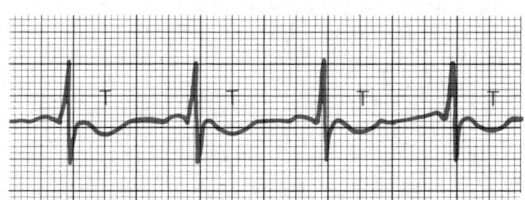

Figure 12–23. Inverted T wave resulting from mild ischemia of the apex of the ventricles.

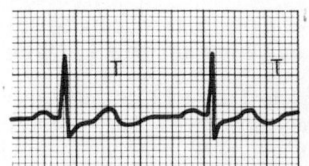

Figure 12–24. Biphasic T wave caused by digitalis toxicity.

clusion, or relative coronary insufficiency that occurs during exercise.

One means for detecting mild coronary insufficiency is to have the patient exercise and to record the electrocardiogram immediately thereafter, noting whether or not changes occur in the T waves. The changes in the T waves need not be specific because any change in the T wave in any lead—inversion, for instance, or a biphasic wave—is often evidence enough that some portion of the ventricular muscle has increased its period of depolarization out of proportion to the rest of the heart, and this is probably caused by relative coronary insufficiency.

Effect of Digitalis on the T Wave. As discussed in Chapter 22, digitalis is a drug that can be used during relative coronary insufficiency to increase the strength of cardiac muscle contraction. Digitalis also increases the period of depolarization of cardiac muscle by about the same proportion in all or most of the ventricular muscle. When overdosages of digitalis are given, however, the depolarization period of one part of the heart may be increased out of proportion to that of other parts. As a result, nonspecific changes, such as T-wave inversion or biphasic T waves, may occur in one or more of the electrocardiographic leads. A biphasic T wave caused by excessive administration of digitalis is shown in Figure 12–24. There is a slight amount of current of injury, too. This probably results from continuous depolarization of part of the ventricular muscle.

Changes in the T wave during digitalis administration are the earliest signs of digitalis toxicity. If still more digitalis is given to the patient, strong currents of injury may develop. Also, digitalis can block conduction of the cardiac impulse to various portions of the heart, so that various arrhythmias can result. It is desirable clinically to keep the effects of digitalis from going beyond the stage of mild T-wave abnormalities. Therefore, the electrocardiograph is used routinely in following digitalized patients.

REFERENCES

See references for Chapter 13.

Cardiac Arrhythmias and Their Electrocardiographic Interpretation

CHAPTER 13

Some of the most distressing types of heart malfunction occur not as a result of abnormal heart muscle but because of an abnormal rhythm of the heart. For instance, sometimes the beat of the atria is uncoordinated with the beat of the ventricles, so that the atria no longer function as primers for the ventricles.

The purpose of this chapter is to discuss the common cardiac arrhythmias and their effects on heart pumping as well as their diagnosis by electrocardiography. The causes of the cardiac arrhythmias are usually one or a combination of the following abnormalities in the rhythmicity-conduction system of the heart:

1. Abnormal rhythmicity of the pacemaker
2. Shift of the pacemaker from the sinus node to other parts of the heart
3. Blocks at different points in the transmission of the impulse through the heart
4. Abnormal pathways of impulse transmission through the heart
5. Spontaneous generation of abnormal impulses in almost any part of the heart

ABNORMAL SINUS RHYTHMS

Tachycardia

The term "tachycardia" means fast heart rate, usually defined as faster than 100 beats per minute. An electrocardiogram recorded from a patient with tachycardia is shown in Figure 13–1. This electrocardiogram is normal except that the rate of heartbeat, as determined from the time intervals between QRS complexes, is about 150 per minute instead of the normal 72 per minute.

The general causes of tachycardia are *increased body tem-* *perature, stimulation of the heart by the sympathetic nerves,* and *toxic conditions of the heart.*

The rate of the heart increases about 10 beats per minute for each degree Fahrenheit (18 beats per degree Celsius) increase in body temperature up to a body temperature of about 105°F (40.5°C); beyond this the heart rate may decrease, owing to progressive weakening of the heart muscle as a result of the fever. Fever causes tachycardia because increased temperature increases the rate of metabolism of the sinus node, which in turn directly increases its excitability and rate of rhythm.

Many factors can cause the sympathetic nervous system to excite the heart, as we discuss at multiple points in this text. For instance, when a patient loses blood and passes into a state of shock or semishock, reflex stimulation of the heart increases its rate to 150 to 180 beats per minute. Also, simple weakening of the myocardium usually increases the heart rate because the weakened heart does not pump blood into the arterial tree to a normal extent, and this elicits sympathetic reflexes to increase the heart rate.

Bradycardia

The term "bradycardia" means a slow heart rate, usually defined as less than 60 beats per minute. Bradycardia is shown by the electrocardiogram in Figure 13–2.

BRADYCARDIA IN ATHLETES. The athlete's heart is considerably stronger than that of a normal person, a fact that allows the athlete's heart to pump a greater stroke volume output per beat. The excessive quantities of blood pumped into the arterial tree with each beat initiate feedback circulatory reflexes or other effects to cause the bradycardia when the athlete is at rest.

VAGAL STIMULATION AS A CAUSE OF BRADYCARDIA. Any circulatory reflex that stimulates the vagus nerve can cause

149

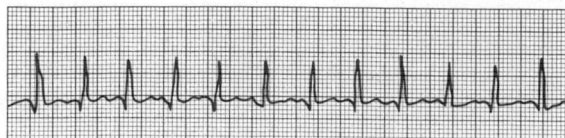

Figure 13–1. Sinus tachycardia (lead I).

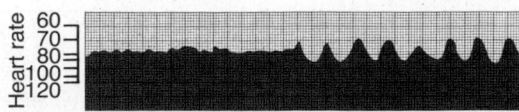

Figure 13–3. Sinus arrhythmia as detected by a cardiotachometer. To the left is the recording taken when the subject was breathing normally; to the right, when breathing deeply.

the heart rate to decrease considerably because of the inhibitory effect that parasympathetic stimulation of acetylcholine has on heart function. Perhaps the most striking example of this occurs in patients with the *carotid sinus syndrome.* In these patients, an arteriosclerotic process in the carotid sinus region of the carotid artery causes excessive sensitivity of the pressure receptors (baroreceptors) located in the arterial wall. As a result, mild pressure on the neck elicits a strong baroreceptor reflex, causing intense vagal stimulation of the heart and extreme bradycardia. Indeed, sometimes this reflex is so powerful that it stops the heart.

Sinus Arrhythmia

Figure 13–3 shows a *cardiotachometer* recording of the heart rate during normal and deep respiration. A cardiotachometer is an instrument that records by the height of successive spikes the duration of the interval between each two QRS complexes in the electrocardiogram. Note from this record that the heart rate increases and decreases about 5 per cent during the various phases of the quiet respiratory cycle. During deep respiration, as shown to the right in Figure 13–3, the heart rate even normally increases and decreases with each respiratory cycle by as much as 30 per cent.

Sinus arrhythmia can result from any one of many circulatory reflexes that alter the strength of the sympathetic and parasympathetic nerve signals to the sinus node. In the respiratory type of sinus arrhythmia shown in Figure 13–3, this results mainly from "spillover" of signals from the medullary respiratory center into the vasomotor center during the inspiratory and expiratory cycles of respiration. The spillover signals cause alternate increase and decrease in the number of impulses transmitted to the heart through the sympathetic and vagus nerves.

ABNORMAL RHYTHMS THAT RESULT FROM IMPULSE CONDUCTION BLOCK

Sinoatrial Block

In rare instances, the impulse from the sinus node is blocked before it enters the atrial muscle. This phenomenon is demonstrated in Figure 13–4, which shows the sudden cessation of P waves with resultant standstill of the atrium.

However, the ventricle picks up a new rhythm, the impulse usually originating in the atrioventricular (A-V) node, so that the ventricular QRS-T complex is slowed but not otherwise altered.

Atrioventricular Block

The only means by which impulses can ordinarily pass from the atria into the ventricles is through the *A-V bundle,* also known as the *bundle of His.* The different conditions that can either decrease the rate of conduction of the impulse through this bundle or block the impulse are as follows:

1. *Ischemia of the A-V nodal or A-V bundle fibers* often delays or blocks conduction from the atria to the ventricles. Coronary insufficiency can cause ischemia of the A-V node and bundle in the same manner that it can cause ischemia of the myocardium.

2. *Compression of the A-V bundle* by scar tissue or by calcified portions of the heart can depress or block conduction from the atria to the ventricles.

3. *Inflammation of the A-V node or A-V bundle* can depress conductivity between the atria and the ventricles. Inflammation results frequently from different types of myocarditis, such as occur in diphtheria and rheumatic fever.

4. *Extreme stimulation of the heart by the vagus nerves* in rare instances blocks impulse conduction through the A-V node. Such vagal excitation occasionally results from strong stimulation of the baroreceptors in people with the *carotid sinus syndrome,* discussed earlier in relation to bradycardia.

INCOMPLETE HEART BLOCK

Prolonged P-R (or P-Q) Interval—First Degree Block.
The normal lapse of time between the *beginning* of the P wave and the *beginning* of the QRS complex is about 0.16 second when the heart is beating at a normal rate. This P-R interval usually decreases in length with faster heartbeat and increases with slower heartbeat. In general, when the P-R interval increases above a value of about 0.20 second in a heart beating at normal rate, the P-R interval is said to be prolonged and the patient is said to have *first degree incomplete heart block.* Figure 13–5 shows an electrocardiogram with a prolonged P-R interval, the interval in this instance being about 0.30 second. Thus, first degree block is defined

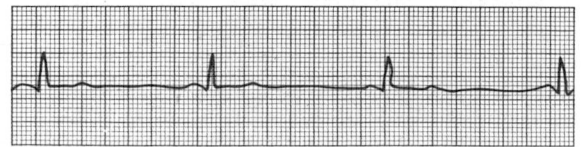

Figure 13–2. Sinus bradycardia (lead III).

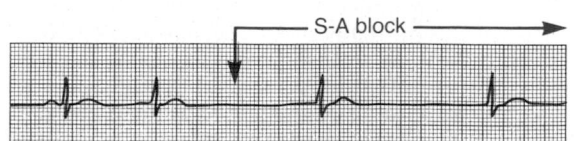

Figure 13–4. Sinoatrial nodal block with A-V nodal rhythm (lead III).

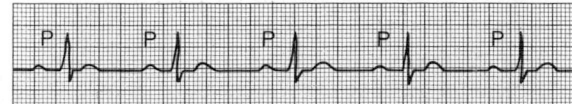

Figure 13–5. Prolonged P-R interval caused by first degree block (lead II).

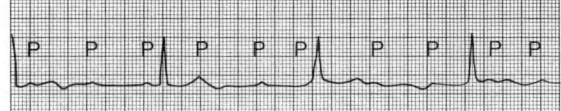

Figure 13–7. Complete A-V block (lead II).

as a delay of conduction from the atria to the ventricles but not actual blockage of conduction.

The P-R interval seldom increases above 0.35 to 0.45 second because by that time, the conduction through the A-V node and bundle is depressed to such an extent that conduction stops entirely. Thus, when a patient's P-R interval is approaching these limits, additional slight increase in the severity of the condition will block impulse conduction rather than simply delay it further.

One of the means for determining the severity of some heart diseases—acute rheumatic fever, for instance—is to measure the P-R interval.

Second Degree Block. When conduction through the A-V junction is slowed until the P-R interval is 0.25 to 0.45 second, sometimes the action potentials traveling through the A-V node are strong enough to pass on through the A-V node and at other times are not strong enough. In this instance, the atria beat at a faster rate than the ventricles, and it is said that there are "dropped beats" of the ventricles. This condition is called *second degree incomplete heart block.*

Figure 13–6 shows P-R intervals of 0.30 second as well as one dropped beat as a result of failure of conduction from the atria to the ventricles.

At times, every other beat of the ventricles is dropped, so that a "2:1 rhythm" develops in the heart, with the atria beating twice for every single beat of the ventricles. Sometimes other rhythms such as 3:2 or 3:1, also develop.

Complete A-V Block (Third Degree Block). When the condition causing poor conduction in the A-V node or A-V bundle becomes severe, complete block of the impulse from the atria into the ventricles occurs. In this instance, the P waves become dissociated from the QRS-T complexes, as shown in Figure 13–7. Note that the rate of rhythm of the atria in this electrocardiogram is about 100 beats per minute, whereas the rate of ventricular beat is less than 40 per minute. Furthermore, there is no relation between the rhythm of the P waves and that of the QRS-T complexes because the ventricles have "escaped" from control by the atria, and they are beating at their own natural rate.

Stokes-Adams Syndrome—Ventricular Escape. In some patients with A-V block, the total block comes and goes; that is, impulses are conducted from the atria into the ventricles for a period of time, and then suddenly no impulses are transmitted. The duration of total block may be a few seconds, a few minutes, or a few hours, or it may be weeks or even longer before conduction returns. In particular, this condition occurs in hearts with borderline ischemia of the conductive system.

Immediately after A-V conduction is first blocked, the ventricles stop contracting for 5 to 30 seconds because of the phenomenon called *overdrive suppression,* which means that ventricular excitability has been suppressed because the ventricles have been driven by the atria at a rate greater than their natural rate or rhythm. Finally, some part of the Purkinje system beyond the block, usually in the distal part of the A-V node beyond the blocked point in the node or in the A-V bundle, begins discharging rhythmically at a rate of 15 to 40 times per minute and acting as the pacemaker of the ventricles. This is called *ventricular escape.*

Because the brain cannot remain active for more than 4 to 5 seconds without blood supply, most patients faint a few seconds after complete block occurs because the heart does not pump any blood for 5 to 30 seconds until the ventricles "escape." After escape, however, the slowly beating ventricles usually pump enough blood to allow rapid recovery from the faint and then to sustain a person. These periodic fainting spells are known as the Stokes-Adams syndrome.

Occasionally the interval of ventricular standstill at the onset of complete block is so long that it becomes detrimental to the patient's health or even causes death. Consequently, most of these patients are provided with an *artificial pacemaker,* a small battery-operated electrical stimulator planted beneath the skin, the electrodes from which are usually connected to the right ventricle. The pacemaker provides continued rhythmical impulses that take control of the ventricles. The batteries are replaced about once every 5 years.

Incomplete Intraventricular Block— Electrical Alternans

Most of the same factors that can cause A-V block can also block impulse conduction in peripheral portions of the ventricular Purkinje system. At times, *incomplete* intraventricular block occurs, so that the impulse sometimes fails to be transmitted to part of the heart during some heart cycles and not during others. The QRS complex may be considerably abnormal during the cycles of partial block. Figure 13–8 shows the condition known as *electrical alternans,* which results from partial intraventricular block every other heartbeat. This electrocardiogram also shows tachycardia, which is

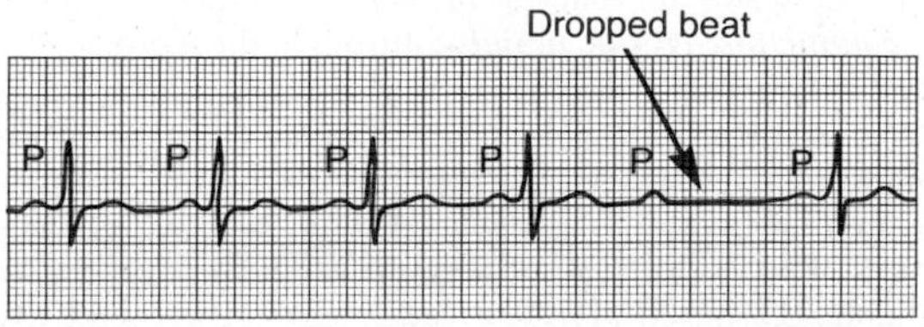

Dropped beat

Figure 13–6. Second degree incomplete A-V block (lead V₃).

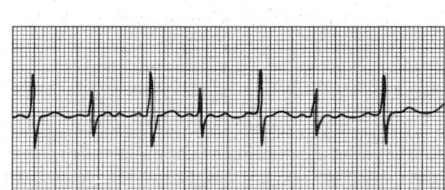

Figure 13–8. Partial intraventricular block—"electrical alternans" (lead III).

probably the reason the block has occurred because when the rate of the heart is rapid, it may be impossible for portions of the Purkinje system to recover from the refractory period quickly enough to respond during each succeeding heartbeat. Also, many conditions that depress the heart, such as ischemia, myocarditis, and digitalis toxicity, can cause incomplete intraventricular block with resultant electrical alternans.

PREMATURE CONTRACTIONS

A premature contraction is a contraction of the heart before the time that normal contraction would have been expected. This condition is also called *extrasystole, premature beat,* or *ectopic beat.*

CAUSES OF PREMATURE CONTRACTIONS. Most premature contractions result from *ectopic foci* in the heart, which emit abnormal impulses at odd times during the cardiac rhythm. Among the possible causes of ectopic foci are (1) local areas of ischemia; (2) small calcified plaques at different points in the heart, which press against the adjacent cardiac muscle so that some of the fibers are irritated; and (3) toxic irritation of the A-V node, Purkinje system, or myocardium caused by drugs, nicotine, or caffeine. Mechanical initiation of premature contractions is also frequent during cardiac catheterization, large numbers of premature contractions often occurring when the catheter enters the right ventricle and presses against the endocardium.

Premature Atrial Contractions

Figure 13–9 shows a single premature atrial contraction. The P wave of this beat occurs too soon in the heart cycle, and the P-R interval is shortened, indicating that the ectopic origin of the beat is near the A-V node. Also, the interval between the premature contraction and the next succeeding contraction is slightly prolonged, which is called a *compensatory pause.* One of the reasons for this is that the premature contraction originated in the atrium some distance from the sinus node, and the impulse had to travel through a considerable amount of atrial muscle before it discharged the sinus node. Consequently, the sinus node discharged late in the premature cycle, and this made the succeeding sinus node discharge also late in appearing.

Premature atrial contractions occur frequently in healthy people and indeed are often found in athletes whose hearts are certain to be in healthy condition. Mild toxic conditions resulting from such factors as excess smoking, lack of sleep, ingestion of too much coffee, alcoholism, and use of various drugs can also initiate such contractions.

PULSE DEFICIT. When the heart contracts ahead of schedule, the ventricles will not have filled with blood normally and the stroke volume output during that contraction is depressed or almost absent. Therefore, the pulse wave

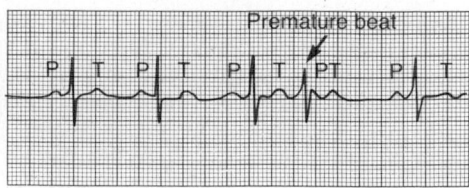

Figure 13–10. A-V nodal premature contraction (lead III).

passing to the periphery after a premature contraction may be so weak that it cannot be felt in the radial artery. Thus, a deficit in the number of pulses felt in the radial pulse occurs in relation to the number of contractions of the heart.

BIGEMINAL PULSE. At times, every other beat of the heart may be a premature contraction. This causes the patient to have a bigeminal pulse—that is, two pulses close together, then a longer diastolic interval, then two again, and so on.

A-V Nodal or A-V Bundle Premature Contractions

Figure 13–10 shows a premature contraction that originates in the A-V node or in the A-V bundle. The P wave is missing from the record of the premature contraction. Instead, the P wave is superimposed onto the QRS-T complex of the premature contraction because the cardiac impulse travels backward into the atria at the same time that it travels forward into the ventricles; this P wave distorts the QRS-T complex, but the P wave itself cannot be discerned as such.

In general, A-V nodal premature contractions have the same significance and causes as atrial premature contractions.

Premature Ventricular Contractions

The electrocardiogram of Figure 13–11 shows a series of premature ventricular contractions (PVCs) alternating with normal contractions. PVCs cause specific effects in the electrocardiogram, as follows:

1. The QRS complex is usually considerably prolonged. The reason is that the impulse is conducted mainly through the slowly conducting muscle of the ventricle rather than through the Purkinje system.

2. The QRS complex has a high voltage for the following reasons: when the normal impulse passes through the heart, it passes through both ventricles about simultaneously; consequently, in the normal heart the depolarization waves of the two sides of the heart partially neutralize each other. When a PVC occurs, the impulse travels in only one direction, so that there is no such neutralization effect, and one entire side of the heart is depolarized ahead of the other, whereas the other entire side is still polarized; this causes intense electrical potentials.

3. After almost all PVCs, the T wave has a potential polarity opposite to that of the QRS complex because the *slow conduction of the impulse* through the cardiac muscle causes the area first depolarized also to repolarize first. As a result, the direction of current flow in the heart during repolarization is opposite to that during depolarization, and the potential of the T wave is reversed to that of the QRS complex. This is not true of the normal T wave, as explained in Chapter 11.

Some PVCs are relatively benign in their origin and result

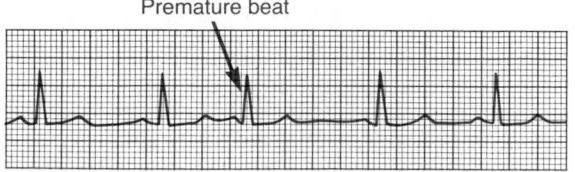

Figure 13–9. Atrial premature contraction (lead I).

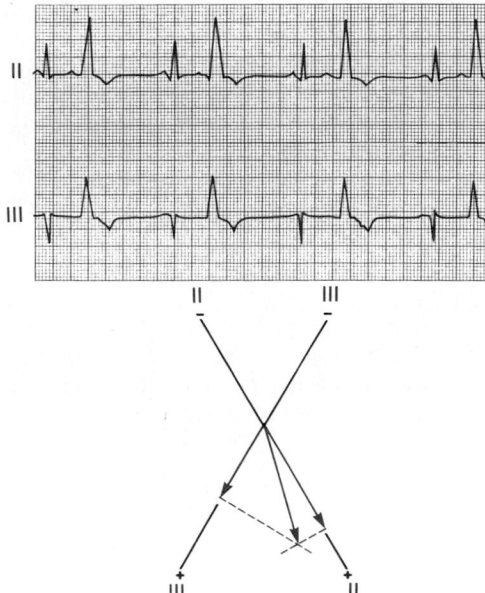

Figure 13–11. Premature ventricular contractions (PVCs) demonstrated by the large abnormal QRS-T complexes (leads II and III). Axis of the premature contractions is plotted in accord with the principles of vectorial analysis explained in Chapter 12; this shows the origin of the PVC to be near the base of the ventricles.

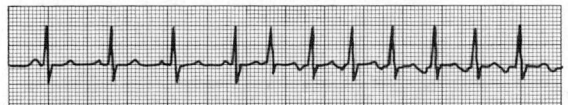

Figure 13–12. Atrial paroxysmal tachycardia—onset in middle of record (lead I).

from factors such as cigarettes, coffee, lack of sleep, various mild toxic states, and even emotional irritability. On the other hand, many other PVCs result from stray impulses or re-entrant signals that originate around the borders of infarcted or ischemic areas of the heart. Therefore, the presence of such PVCs is not to be taken lightly. Statistics show that people with significant numbers of PVCs have a much higher than normal chance of developing spontaneous lethal ventricular fibrillation, presumably initiated by one of the PVCs. This is especially true when the PVCs occur during the vulnerable period for causing fibrillation, just at the end of the T wave when the ventricles are coming out of refractoriness, as explained later in the chapter.

VECTOR ANALYSIS OF THE ORIGIN OF AN ECTOPIC PREMATURE VENTRICULAR CONTRACTION. In Chapter 12, the principles of vectorial analysis are explained. Applying these principles, one can determine from the electrocardiogram in Figure 13–11 the point of origin of the PVC as follows: Note that the potentials of the premature contractions in leads II and III are both strongly positive. Plotting these potentials on the axes of leads II and III and solving by vectorial analysis for the mean QRS vector in the heart, one finds that the vector of this premature contraction has its negative end (origin) at the base of the heart and its positive end toward the apex. Thus, the first portion of the heart to become depolarized during this premature contraction is near the base of the ventricles, which therefore is the locus of the ectopic focus.

PAROXYSMAL TACHYCARDIA

Abnormalities in any portion of the heart, including the atria, the Purkinje system, or the ventricles, can cause rapid rhythmical discharge of impulses that spread in all directions

throughout the heart. This is believed to be caused most frequently by re-entrant pathways that set up local repeated self–re-excitation. Because of the rapid rhythm in the irritable focus, this focus becomes the pacemaker of the heart.

The term "paroxysmal" means that the heart rate usually becomes rapid in paroxysms, with the paroxysms beginning suddenly and lasting for a few seconds, a few minutes, a few hours, or much longer. Then the paroxysms usually end as suddenly as they had begun, with the pacemaker of the heart instantly shifting back to the sinus node.

Paroxysmal tachycardia often can be stopped by eliciting a vagal reflex. A strange type of vagal reflex elicited for this purpose is one that occurs when painful pressure is applied to the eyes. Also, sometimes pressure on the carotid sinuses can elicit enough of a vagal reflex to stop the tachycardia. Various drugs may also be used. Two drugs frequently used are quinidine and lidocaine, both of which depress the normal increase in sodium permeability of the cardiac muscle membrane during the generation of the action potential, thereby often blocking the rhythmical discharge of the focal region that is causing the paroxysmal attack.

Atrial Paroxysmal Tachycardia

Figure 13–12 demonstrates in the middle of the record a sudden increase in rate of heartbeat from about 95 to about 150 beats per minute. On close observation of the electrocardiogram, it can be seen that an inverted P wave occurs before each of the QRS-T complexes during the paroxysm of rapid heartbeat, and this P wave is partially superimposed on the normal T wave of the preceding beat. This indicates that the origin of this paroxysmal tachycardia is in the atrium, but because the P wave is abnormal, the origin is not near the sinus node.

A-V NODAL PAROXYSMAL TACHYCARDIA. Paroxysmal tachycardia often results from an aberrant rhythm that involves the A-V node. This usually causes almost normal QRS-T complexes but missing or obscured P waves.

Atrial or A-V nodal paroxysmal tachycardia, both of which are called *supraventricular tachycardias,* usually occurs in young, otherwise healthy people, and such people generally grow out of the predisposition to the tachycardias after adolescence. In general, supraventricular tachycardia frightens a person tremendously and may cause weakness during the paroxysms but only seldom does permanent harm come from the attacks.

Ventricular Paroxysmal Tachycardia

Figure 13–13 shows a typical short paroxysm of ventricular tachycardia. The electrocardiogram of ventricular paroxysmal tachycardia has the appearance of a series of ventricular premature beats occurring one after another without any normal beats interspersed.

Ventricular paroxysmal tachycardia is usually a serious condition for two reasons. First, this type of tachycardia usually does not occur unless considerable ischemic damage

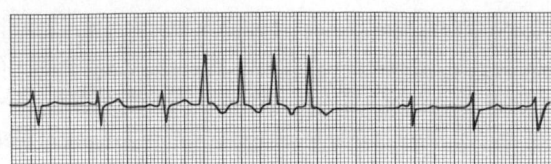

Figure 13–13. Ventricular paroxysmal tachycardia (lead III).

is present in the ventricles. Second, ventricular tachycardia frequently initiates the lethal condition of ventricular fibrillation because of rapid repeated stimulation of the ventricular muscle, as we discuss in the next section.

Sometimes digitalis intoxication causes irritable foci that lead to ventricular tachycardia. On the other hand, quinidine, which increases the refractory period of cardiac muscle as well as its threshold for excitation, may be used to block irritable foci causing ventricular tachycardia.

VENTRICULAR FIBRILLATION

The most serious of all cardiac arrhythmias is *ventricular fibrillation,* which, if not treated instantly, is almost invariably fatal.

Ventricular fibrillation results from cardiac impulses that have gone berserk within the ventricular muscle mass, stimulating first one portion of the ventricular muscle, then another portion, then another, and eventually feeding back onto itself to re-excite the same ventricular muscle over and over—never stopping. When this happens, many small portions of the ventricular muscle will be contracting at the same time, while equally as many other portions will be relaxing. Thus, there is never a coordinate contraction of all the ventricular muscle at once, which is required for a pumping cycle of the heart. Therefore, despite massive flow of stimulatory signals throughout the ventricles, the ventricular chambers neither enlarge nor contract but remain in an indeterminate stage of partial contraction, pumping either no blood or negligible amounts. Therefore, after fibrillation begins, unconsciousness occurs within 4 to 5 seconds for lack of blood flow to the brain, and irretrievable death of tissues begins to occur throughout the body within a few minutes.

Multiple factors can spark the beginning of ventricular fibrillation—with a normal heartbeat one second and a second later the ventricles in fibrillation. Especially likely to initiate fibrillation is (1) sudden electrical shock of the heart or (2) ischemia of the heart muscle, of its specialized conducting system, or both. In either instance, an instantaneous pattern of re-entry signals can be established, so that the contractile impulses travel around and around through the heart muscle. This phenomenon is also called a *circus movement.*

Phenomenon of Re-entry—Circus Movements as the Basis for Ventricular Fibrillation

When the *normal* cardiac impulse has traveled through the extent of the ventricles, it then has no place else to go because all the ventricular muscle is at that time refractory and cannot conduct the impulse further. Therefore, that im-

pulse dies, and the heart awaits a new action potential to begin in the sinus node.

Under some circumstances, this normal sequence of events does not occur. Therefore, let us explain more fully the background conditions that can initiate re-entry and lead to the circus movements of ventricular fibrillation.

Figure 13–14 shows several small cardiac muscle strips cut in the form of circles. If such a strip is stimulated at the 12 o'clock position *so that the impulse travels in only one direction,* the impulse spreads progressively around the circle until it returns to the 12 o'clock position. If the originally stimulated muscle fibers are still in a refractory state, the impulse then dies out because refractory muscle cannot transmit a second impulse. There are three different conditions that can cause this impulse to continue to travel around the circle, that is, to cause re-entry of the impulse into muscle that has already been excited.

First, if the *pathway around the circle is long,* by the time the impulse returns to the 12 o'clock position, the originally stimulated muscle will no longer be refractory and the impulse will continue around the circle again and again.

Second, if the length of the pathway remains constant but the *velocity of conduction becomes decreased* enough, an increased interval of time will elapse before the impulse returns to the 12 o'clock position. By this time, the originally stimulated muscle might be out of the refractory state and the impulse can continue around the circle again and again.

Third, *the refractory period of the muscle might become greatly shortened.* In this case, the impulse could also continue around and around the circle.

All of these conditions occur in different pathological states of the human heart as follows: (1) A long pathway typically occurs in dilated hearts. (2) Decreased rate of conduction frequently results from blockage of the Purkinje system, ischemia of the muscle, high blood potassium levels, or many other factors. (3) A shortened refractory period commonly occurs in response to various drugs, such as epinephrine, or after repetitive electrical stimulation. Thus, in many cardiac disturbances, re-entry can cause abnormal patterns of cardiac contraction or abnormal cardiac rhythms that ignore the pace-setting effects of the sinus node.

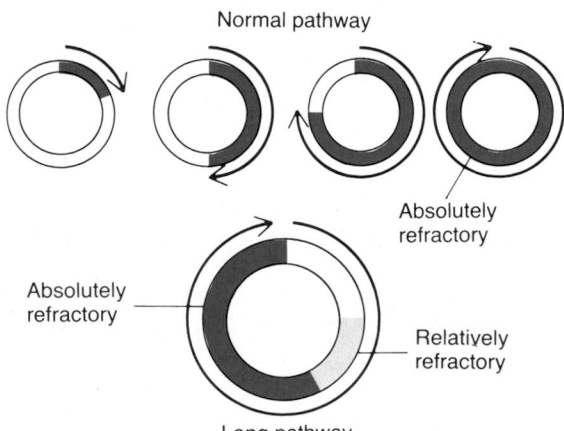

Figure 13–14. Circus movement showing annihilation of the impulse in the short pathway and continued propagation of the impulse in the long pathway.

Chain Reaction Mechanism of Fibrillation

In ventricular fibrillation, one sees many separate and small contractile waves spreading at the same time in different directions over the cardiac muscle. The re-entrant impulses in fibrillation are not simply a single impulse moving in a circle, as shown in Figure 13–14. Instead, they have degenerated into a series of multiple wave fronts that have the appearance of a chain reaction. One of the best ways to explain this process in fibrillation is to describe the initiation of fibrillation by electric shock caused by 60-cycle alternating electric current.

FIBRILLATION CAUSED BY 60-CYCLE ALTERNATING CURRENT. At a central point in the ventricles of heart A in Figure 13–15, a 60-cycle electrical stimulus is applied through a stimulating electrode. The first cycle of the electrical stimulus causes a depolarization wave to spread in all directions, leaving all the muscle beneath the electrode in a refractory state. After about 0.25 second, part of this muscle begins to come out of the refractory state. Some portions of the muscle come out of refractoriness before other portions. This state of events is depicted in heart A by many lighter patches, which represent excitable cardiac muscle, and dark patches, which represent still refractory muscle. New stimuli from the electrode can now cause impulses to travel in certain directions through the heart but not in all directions. Thus, in heart A, certain impulses travel for short distances until they reach refractory areas of the heart and then are blocked. Other impulses pass between the refractory areas and continue to travel in the excitable patches of muscle. Now, several events transpire in rapid succession, all occurring simultaneously and eventuating in a state of fibrillation.

First, block of the impulses in some directions but successful transmission in other directions creates one of the necessary conditions for a re-entrant signal to develop—that is, *transmission of some of the depolarization waves around the heart in only one direction.* As a result, these waves do not run into waves traveling in the opposite direction and therefore do not annihilate themselves on the opposite side of the heart but can continue around and around the ventricles.

Second, the rapid stimulation of the heart causes two changes in the cardiac muscle itself, both of which predis-

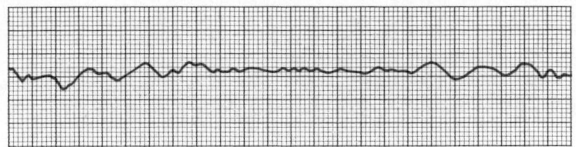

Figure 13–16. Ventricular fibrillation (lead II).

pose to circus movement: (1) the *velocity of conduction through the heart decreases,* which allows a longer interval for the impulses to travel around the heart. (2) The *refractory period of the muscle is shortened,* allowing re-entry of the impulse into previously excited heart muscle within a much shorter period than normally.

Third, one of the most important features of fibrillation is the *division of impulses,* as demonstrated in heart A. When a depolarization wave reaches a refractory area in the heart, it travels to both sides around the area. Thus, a single impulse becomes two impulses. Then, when each of these reaches another refractory area, it, too, divides to form two more impulses. In this way, many new wave fronts are continually being formed in the heart by a progressive *chain reaction* until, finally, there are many small depolarization waves traveling in many directions at the same time. Furthermore, this irregular pattern of impulse travel causes a *circuitous route for the impulses to travel, greatly lengthening the conductive pathway, which is one of the conditions that sustains the fibrillation.* It also results in a continual irregular pattern of patchy refractory areas in the heart. One can readily see that a vicious cycle has been initiated: More and more impulses are formed, these cause more and more patches of refractory muscle, and the refractory patches cause more and more division of the impulses. Therefore, any time a single area of cardiac muscle comes out of refractoriness, an impulse is close at hand to re-enter the area.

Heart B in Figure 13–15 demonstrates the final state that develops in fibrillation. Here one can see many impulses traveling in all directions, some dividing and increasing the number of impulses, whereas others are blocked by refractory areas.

Vulnerable Period for Causing Ventriculation Fibrillation. The period during the heart cycle when there are likely to be simultaneous areas of refractoriness and nonrefractoriness in the heart muscle is at the very moment that the heart is recovering from the previous cardiac cycle—that is, just at the end of the cardiac contraction. Therefore, this instant in the cycle is said to be the *vulnerable period* of the ventricles for fibrillation. In fact, a single electric shock during this vulnerable period frequently can lead to the odd pattern of impulses spreading unidirectionally around refractory areas of muscle, thus initiating fibrillation.

Electrocardiogram in Ventricular Fibrillation

In ventricular fibrillation, the electrocardiogram is bizarre, as seen in Figure 13–16, and ordinarily shows no tendency toward a regular rhythm of any type. In the early phases of ventricular fibrillation, relatively large masses of muscle contract simultaneously, and this causes coarse, irregular waves in the electrocardiogram. After only a few seconds, the coarse contractions of the ventricles disappear and the electrocardiogram changes into a new pattern of low-voltage, very irregular waves. Thus, no repetitive electrocardiographic pattern can be ascribed to ventricular fibrillation except that

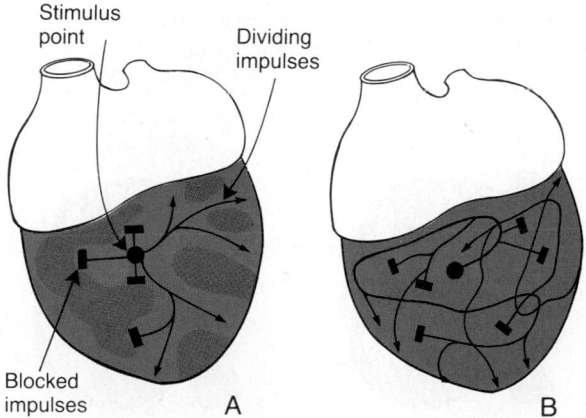

Figure 13–15. *A,* Initiation of fibrillation in a heart when patches of refractory musculature are present. *B,* Continued propagation of *fibrillatory impulses* in the fibrillating ventricle.

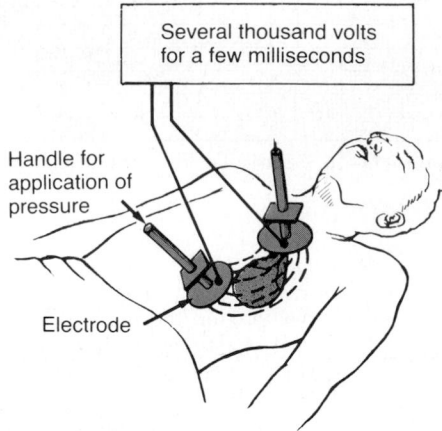

Several thousand volts
for a few milliseconds

Handle for
application of
pressure

Electrode

Figure 13–17. Application of electric current to the chest to stop ventricular fibrillation.

the electrical potentials constantly and spasmodically change because the currents in the heart flow first in one direction and then in another and seldom repeat any specific cycle.

The voltages of the waves in the electrocardiogram in ventricular fibrillation are usually about 0.5 millivolt when ventricular fibrillation first begins, but they decay rapidly so that after 20 to 30 seconds, they are usually only 0.2 to 0.3 millivolt. Minute voltages of 0.1 millivolt or less may be recorded for 10 minutes or longer after ventricular fibrillation begins. As already pointed out, ventricular fibrillation is lethal unless it is stopped by some heroic therapy, such as immediate electroshock through the heart, as explained in the next section.

Electroshock Defibrillation of the Ventricles

Although a weak alternating current almost invariably throws the ventricles into fibrillation, a strong electric current passed through the ventricles for a short interval can stop fibrillation by throwing all the ventricular muscle into refractoriness simultaneously. This is accomplished by passing intense current through electrodes placed on two sides of the heart. The current penetrates most of the fibers of the ventricles, thus stimulating essentially all parts of the ventricles simultaneously and causing them to become refractory. All impulses stop and the heart then remains quiescent for 3 to 5 seconds, after which it begins to beat again, usually with the sinus node or some other part of the heart becoming the pacemaker. However, the same re-entrant focus that had originally thrown the ventricles into fibrillation often is still present, in which case fibrillation may begin again immediately.

When electrodes are applied directly to the two sides or the heart, fibrillation can usually be stopped with 110 volts of 60-cycle alternating current applied for 0.1 second or 1000 volts of direct current applied for a few thousandths of a second. When applied through the chest wall, as shown in Figure 13–17, the usual procedure is to charge a large electric capacitor up to several thousand volts and then to cause the capacitor to discharge in a few thousandths of a second through the electrodes and the heart. In our laboratory, the heart of a single anesthetized dog was defibrillated 130 times through the chest wall, and the animal remained in perfectly normal condition.

Hand Pumping of the Heart (Cardiopulmonary Resuscitation) as an Aid to Defibrillation

Unless defibrillated within 1 minute after fibrillation begins, the heart is usually too weak to be revived by defibrillation alone because of lack of nutrition from coronary blood flow. However, it is still possible to revive the heart by preliminarily pumping it by hand (intermittent squeezing) and then defibrillating it later. In this way, small quantities of blood are delivered into the aorta and a renewed coronary blood supply develops. Then, after a few minutes, electrical defibrillation often becomes possible. Indeed, fibrillating hearts have been pumped by hand for as long as 90 minutes before achieving successful defibrillation.

A technique of pumping the heart without opening the chest consists of intermittent, powerful thrusts of pressure on the chest wall along with artificial respiration. This is called *cardiopulmonary resuscitation*, or simply CPR.

Lack of blood flow to the brain for more than 5 to 8 minutes usually results in permanent mental impairment or even destruction of the brain. Even though the heart should be revived, the person might die from the effects of brain damage or live with permanent mental impairment.

ATRIAL FIBRILLATION

Remember that except for the connection through the A-V bundle, the atrial muscle mass is separated from the ventricular muscle mass, insulated from each other by fibrous tissue. Therefore, ventricular fibrillation often occurs without atrial fibrillation. Likewise, fibrillation often occurs in the atria without ventricular fibrillation; this is shown to the right in Figure 13–18.

The mechanism of atrial fibrillation is identical with that of ventricular fibrillation except that the process occurs only in the atrial muscle mass instead of the ventricular mass. A frequent cause of atrial fibrillation is atrial enlargement resulting from heart valve lesions that prevent the atria from emptying adequately into the ventricles or from ventricular failure with excess damming of blood in the atria. The dilated atrial walls provide the ideal conditions of a long conductive pathway as well as slow conduction, both of which predispose to atrial fibrillation.

Pumping Characteristics of the Atria During Atrial Fibrillation. For the same reasons that the ventricles will not pump blood during ventricular fibrillation, neither do the atria pump blood in atrial fibrillation. Therefore, the atria become useless as primer pumps for the ventricles.

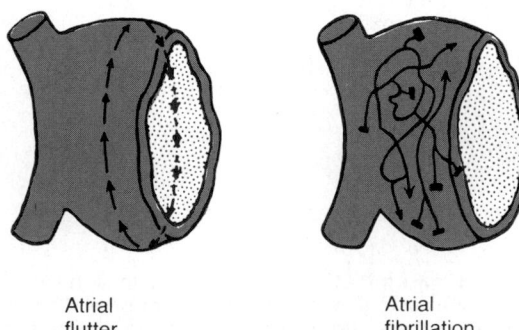

Atrial
flutter

Atrial
fibrillation

Figure 13–18. Pathways of impulses in atrial flutter and atrial fibrillation.

Even so, blood flows passively through the atria into the ventricles, and the efficiency of ventricular pumping is decreased only 20 to 30 per cent. Therefore, in contrast to the lethality of ventricular fibrillation, a person can live for months or even years with atrial fibrillation, although at a reduced efficiency of overall heart pumping.

Electrocardiogram in Atrial Fibrillation. Figure 13–19 shows the electrocardiogram during atrial fibrillation. Numerous small depolarization waves spread in all directions through the atria during atrial fibrillation. Because the waves are weak and many of them are of opposite polarity at any given time, they usually almost completely neutralize one another. Therefore, in the electrocardiogram, one can see either no P waves from the atria or a fine, high-frequency, very low voltage wavy record. On the other hand, the QRS-T complexes are normal unless there is some pathology of the ventricles, but their timing is irregular for the following reasons.

IRREGULARITY OF THE VENTRICULAR RHYTHM DURING ATRIAL FIBRILLATION. When the atria are fibrillating, impulses arrive from the atrial muscle at the A-V node rapidly but also irregularly. Because the A-V node will not pass a second impulse for about 0.35 second after a previous one, at least 0.35 second must elapse between one ventricular contraction and the next, and an additional but variable interval of 0 to 0.6 second occurs before one of the irregular fibrillatory impulses happens to arrive at the A-V node. Thus, the interval between successive ventricular contractions varies from a minimum of about 0.35 second to a maximum of about 0.95 second, causing a very irregular heartbeat. In fact, this irregularity, demonstrated by the variable spacing of the heartbeats in the electrocardiogram of Figure 13–19, is one of the clinical findings used to diagnose the condition. Also, because of the rapid rate of the fibrillatory impulses in the atria, the ventricle is usually driven at a fast heart rate, usually between 125 and 150 beats per minute.

ELECTROSHOCK TREATMENT OF ATRIAL FIBRILLATION. In the same manner that ventricular fibrillation can be converted back to a normal rhythm by electroshock, so also can atrial fibrillation be converted by electroshock. The procedure is essentially the same as for ventricular conversion—passage of a single strong electric shock through the atria—which throws the entire heart into refractoriness for a few seconds; a normal rhythm usually will follow if the heart is capable of this.

Atrial Flutter

Atrial flutter is another condition caused by a circus movement in the atria. It is different from atrial fibrillation in that the electrical signal travels as a single large wave front always in one direction around and around the atrial muscle mass. As shown to the left in Figure 13–18, this wave front usually travels from top to bottom to top around the openings of the superior and inferior venae cavae.

Atrial flutter causes a rapid rate of contraction of the atria,

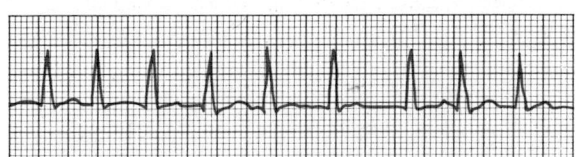

Figure 13–19. Atrial fibrillation (lead I).

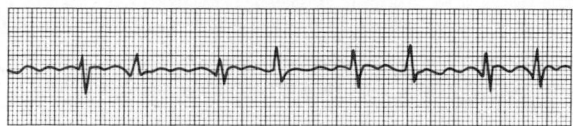

Figure 13–20. Atrial flutter—2 : 1 and 3 : 1 rhythm (lead I).

usually between 200 and 350 beats per minute. However, because one side of the atria is contracting while the other side is relaxing, the amount of blood pumped by the atria is slight. Furthermore, the signals reach the A-V node too rapidly for all of them to be passed into the ventricles because the refractory period of the A-V node and A-V bundle is too long to pass more than a fraction of the atrial signals. Therefore, there are usually two to three beats of the atria for every single beat of the ventricles.

Figure 13–20 shows a typical electrocardiogram in atrial flutter. The P waves are strong because of contraction of semicoordinate masses of muscle. However, note in the record that a QRS-T complex follows an atrial P wave only once for every two to three beats of the atria, giving a 2 : 1 and a 3 : 1 rhythm.

CARDIAC ARREST

A final serious abnormality of the cardiac rhythmicity-conduction system is cardiac arrest. This results from cessation of all rhythmical impulses of the heart. That is, no spontaneous rhythm remains.

Cardiac arrest is especially likely to occur during deep anesthesia when many patients develop severe hypoxia because of inadequate respiration. The hypoxia prevents the muscle fibers and conductive fibers from maintaining normal electrolyte concentration differentials across their membranes, and their excitability may be so affected that the automatic rhythmicity disappears.

In most instances of cardiac arrest, cardiopulmonary resuscitation is quite successful in re-establishing a normal heart rhythm. In some patients, severe myocardial disease leads to permanent or semipermanent cardiac arrest, which can cause immediate death. In many cases, rhythmical electrical impulses from an implanted electronic cardiac pacemaker have been used successfully to keep patients alive for years.

REFERENCES

Catalano, J. T.: Guide to ECG Analysis. Philadelphia, J. B. Lippincott., 1993.
Catterall, W. A.: Cellular and molecular biology of voltage-gated sodium channels. Physiol. Rev., 72:(Suppl.) S15, 1992.
Chou, T.-C.: Electrocardiography in Clinical Practice. Philadelphia, W. B. Saunders Co., 1991.
Craig, M., et al.: Diagnostic Medical Sonography: Echocardiography, Vol. II. Philadelphia, J. B. Lippincott, 1991.
Davis, D.: How to Quickly and Accurately Master ECG Interpretation. Philadelphia, J. B. Lippincott, 1991.
Estes, N. A. M. III, et al.: Implantable Cardioverter-Defibrillators. New York, Marcel Dekker, Inc., 1994.
Falk, R. H., and Podrid, P. J.: Atrial Fibrillation: Mechanisms and Management. New York, Raven Press, 1992.
Fisch, C.: Electrocardiography of Arrhythmias. Baltimore, Williams & Wilkins, 1990.
Guyton, A. C., and Crowell, J. W.: A stereovectorcardiograph. J. Lab. Clin. Med., 40:726, 1952.

Huff, J., et al.: ECG Workout: Exercises in Arrhythmia Interpretation. Philadelphia, J. B. Lippincott, 1993.

Hurst, J. W., et al. (eds.): The Heart. 7th Ed. New York, McGraw-Hill, 1990.

Irisawa, H.: Comparative physiology of the cardiac pacemaker mechanism. Physiol. Rev., 58:461, 1984.

Josephson, M. E.: Clinical Cardiac Electrophysiology. Baltimore, Williams & Wilkins, 1993.

Kastor, J. A.: Arrhythmias. Philadelphia, W. B. Saunders Co., 1994.

Katz, A. M.: Physiology of the Heart. New York, Raven Press, 1992.

Lynch, C. III: Cardiac Electrophysiology: Perioperative Considerations. Philadelphia, J. B. Lippincott, 1994.

Marriott, H. J. L., and Conover, M. B.: Advanced Concepts in Arrhythmias. 2nd Ed. St. Louis, C. V. Mosby., 1989.

Obeid, A. I.: Echocardiography in Clinical Practice. Philadelphia, J. B. Lippincott, 1992.

Oka, Y., and Goldiner, P. L.: Transesophageal Echocardiography. Philadelphia, J. B. Lippincott, 1992.

Pallotta, B. S., and Wagoner, P. K.: Voltage-dependent potassium channels since Hodgkin and Huxley. Physiol. Rev., 72:(Suppl.) S49, 1992.

Pongs, O.: Molecular biology of voltage-dependent potassium channels. Physiol. Rev., 72:(Suppl.) S69, 1992.

Saksena, S., and Goldschlager, N.: Electrical Therapy for Cardiac Arrhythmias: Pacing, Antitachycardia Devices, Catheter Ablation. Philadelphia, W. B. Saunders Co., 1990.

Surawicz, B.: Electrophysiologic Basis of ECG and Cardiac Arrhythmias. Baltimore, Williams & Wilkins, 1995.

Wagner, G. S.: Marriott's Practical Electrocardiography. Baltimore, Williams & Wilkins, 1994.

Zipes, D. P., and Jalife, J.: Cardiac Electrophysiology: From Cell to Bedside. Philadelphia, W. B. Saunders Co., 1990.

THE CIRCULATION

UNIT IV

Overview of the Circulation; Medical Physics of Pressure, Flow, and Resistance

CHAPTER 14

The function of the circulation is to service the needs of the tissues—to transport nutrients to the tissues, to transport waste products away, to conduct hormones from one part of the body to another, and, in general, to maintain an appropriate environment in all the tissue fluids for optimal survival and function of the cells.

Sometimes it is difficult to understand how blood flow is controlled in relation to the tissue's needs and to understand how the heart and circulation are controlled to provide the necessary cardiac output and arterial pressure. Also, what are the mechanisms for controlling blood volume, and how does this relate to all the other functions of the circulation? These are some of the questions that we propose to answer in this section on the circulation.

PHYSICAL CHARACTERISTICS OF THE CIRCULATION

The circulation, shown in Figure 14–1, is divided into the *systemic circulation* and the *pulmonary circulation*. Because the systemic circulation supplies all the tissues of the body except the lungs with blood flow, it is also called the *greater circulation* or *peripheral circulation*.

Although the vascular system in each separate tissue of the body has its own special characteristics, some general principles of vascular function apply in all parts of the system. It is the purpose of this chapter to discuss these general principles.

FUNCTIONAL PARTS OF THE CIRCULATION. Before attempting to discuss the details of function in the circulation, it is important to understand the overall role of each of its parts.

The function of the *arteries* is to transport blood *under high pressure* to the tissues. For this reason, the arteries have strong vascular walls, and blood flows rapidly in the arteries.

The *arterioles* are the last small branches of the arterial system, and they act as *control valves* through which blood is released into the capillaries. The arteriole has a strong muscular wall that is capable of closing the arteriole completely or allowing it to be dilated severalfold, thus having the capability of vastly altering blood flow to the capillaries in response to the needs of the tissues.

The function of the *capillaries* is to exchange fluid, nutrients, electrolytes, hormones, and other substances between the blood and the interstitial fluid. For this role, the capillary walls are very thin and permeable to small molecular substances.

The *venules* collect blood from the capillaries; they gradually coalesce into progressively larger veins.

The *veins* function as conduits for transport of blood from the tissues back to the heart, but equally important, they serve as a major reservoir of blood. Because the pressure in the venous system is very low, the venous walls are thin. Even so, they are muscular, and this allows them to contract or expand and thereby act as a controllable reservoir for extra blood, either a small or a large amount, depending on the needs of the body.

VOLUMES OF BLOOD IN THE DIFFERENT PARTS OF THE CIRCULATION. By far the greater proportion of the blood in the circulation is contained in the sys-

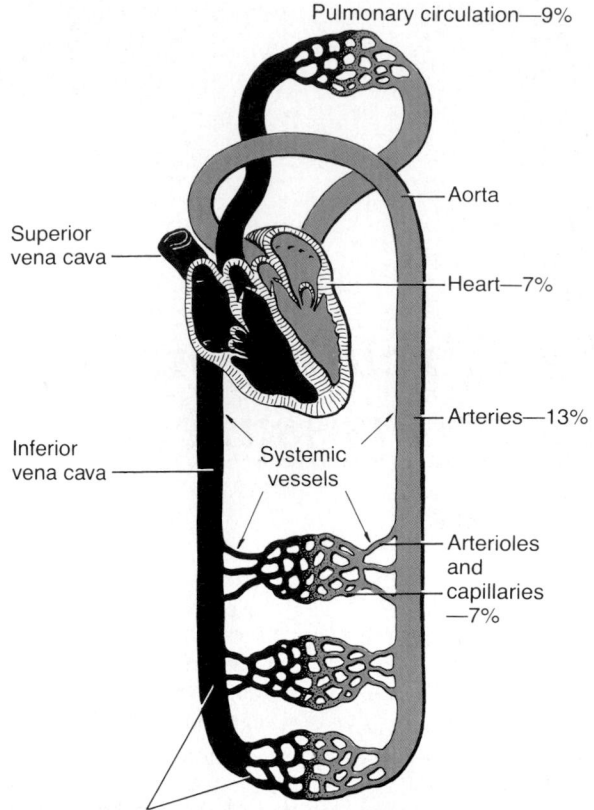

Pulmonary circulation—9%

Aorta

Superior
vena cava

Heart—7%

Inferior
vena cava

Systemic
vessels

Arteries—13%

Arterioles
and
capillaries
—7%

Veins, venules, and venous sinuses—64%

Figure 14–1. Distribution of blood volume in the different portions of the circulatory system.

temic veins. Figure 14–1 shows this, demonstrating that about 84 per cent of the entire blood volume of the body is in the systemic circulation, with 64 per cent in the veins, 13 per cent in the arteries, and 7 per cent in the systemic arterioles and capillaries. The heart contains 7 per cent of the blood and the pulmonary vessels, 9 per cent. Most surprising is the low blood volume in the capillaries of the systemic circulation. Yet it is here that the most important function of the systemic circulation occurs, diffusion of the substances back and forth between the blood and the tissues. This function is so important that it is discussed in detail in Chapter 16.

CROSS-SECTIONAL AREAS AND VELOCITIES OF BLOOD FLOW. If all the systemic vessels of each type were put side by side, their approximate total cross-sectional areas would be as follows:

	cm^2
Aorta	2.5
Small arteries	20
Arterioles	40
Capillaries	2500
Venules	250
Small veins	80
Venae cavae	8

Note particularly the much larger cross-sectional areas of the veins than of the arteries, averaging about four times those of the corresponding arteries. This explains the large storage of blood in the venous system in comparison with that in the arterial system.

Because the same volume of blood flows through each segment of the circulation each minute, the velocity of blood flow is inversely proportional to its cross-sectional area. Thus, under resting conditions, the velocity averages about 33 cm/sec in the aorta but $\frac{1}{1000}$ of this in the capillaries, or about 0.3 mm/sec. However, because the capillaries have a typical length of only 0.3 to 1 millimeter, the blood remains in the capillaries for only 1 to 3 seconds. This is a surprising fact because all diffusion that takes place through the capillary walls must occur in this exceedingly short time.

PRESSURES IN THE VARIOUS PORTIONS OF THE CIRCULATION. Because the heart pumps blood continually into the aorta, the pressure in the aorta is high, averaging about 100 mm Hg. Also, because the pumping by the heart is pulsatile, the arterial pressure fluctuates between a *systolic level* of 120 mm Hg and a *diastolic level* of 80 mm Hg, as shown in Figure 14–2. As the blood flows through the systemic circulation, its pressure falls progressively to about 0 mm Hg by the time it reaches the termination of the venae cavae in the right atrium of the heart.

The pressure in the systemic capillaries varies from as high as 35 mm Hg near the arteriolar ends to as low as 10 mm Hg near the venous ends, but their average "functional" pressure in most vascular beds is about 17 mm Hg, a pressure low enough that little of the plasma leaks out of the porous capillaries even though nutrients can diffuse easily to the tissue cells.

Note to the far right in Figure 14–2 the respective pressures in the different parts of the pulmonary circulation. In the pulmonary arteries, the pressure is pulsatile, just as in the aorta, but the pressure level is far less, at a systolic pressure of about 25 mm Hg and a diastolic pressure of 8 mm Hg, with a mean pulmonary arterial pressure of only 16 mm Hg. The pulmonary capillary pressure averages only 7 mm Hg. Yet the total blood flow through the lungs each minute is the same as that through the systemic circulation. The low pressures of the pulmonary system are in accord with the needs of the lungs because all that is required is to expose the blood in the pulmonary capillaries to the oxygen and other gases in the pulmonary alveoli, and the distances that the blood must flow before returning to the heart are short.

BASIC THEORY OF CIRCULATORY FUNCTION

Although the details of circulatory function are complex, there are three basic principles that underlie all functions of the system.

THE BLOOD FLOW TO EACH TISSUE OF THE BODY IS ALMOST ALWAYS PRECISELY CONTROLLED IN RELATION TO THE TISSUE NEEDS. When tissues are active,

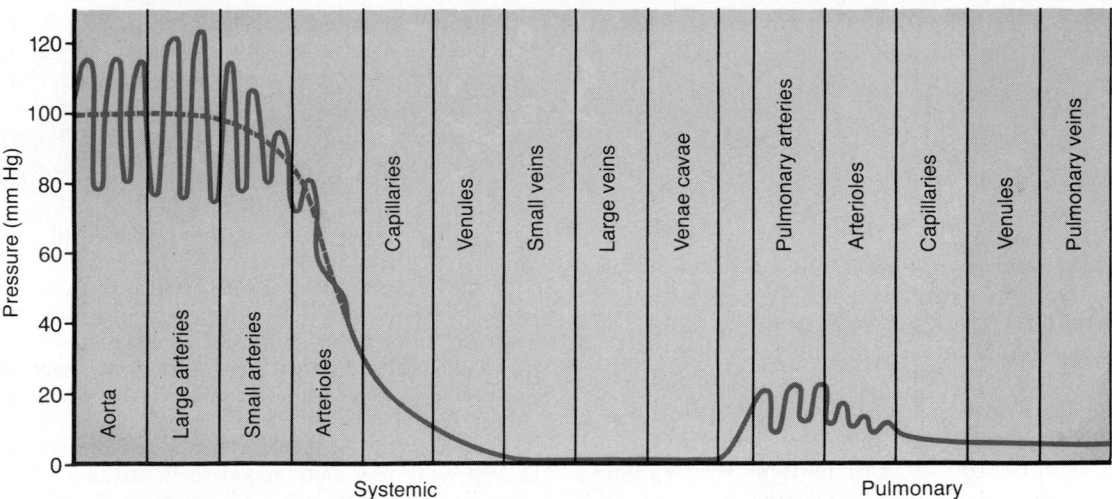

Figure 14–2. Blood pressures in the different portions of the circulatory system.

they need much more blood flow than when at rest—occasionally 20 to 30 times the resting level. Yet the heart normally cannot increase its cardiac output more than four to seven times. Therefore, it is not possible simply to increase the blood flow everywhere in the body when a particular tissue demands increased flow. Instead, the microvessels of each tissue continuously monitor the tissue needs, such as the availability of oxygen and other nutrients and the accumulation of carbon dioxide and other tissue waste products, and these in turn control the local blood flow precisely to the level required for the tissue activity. Also, nervous control of the circulation provides additional specific attributes to tissue blood flow control.

THE CARDIAC OUTPUT IS CONTROLLED MAINLY BY THE SUM OF ALL THE LOCAL TISSUE FLOWS. When blood flows through a tissue, it immediately returns by way of the veins to the heart. The heart responds to this increased inflow of blood by pumping almost all of it immediately back into the arteries from whence it came. In this sense, the heart acts as an automaton, responding to the demands of the tissues. The heart, however, is not perfect in its response. Therefore, it often needs help in the form of special nerve signals to make it pump the required amounts of blood flow.

IN GENERAL, THE ARTERIAL PRESSURE IS CONTROLLED INDEPENDENTLY OF EITHER LOCAL BLOOD FLOW CONTROL OR CARDIAC OUTPUT CONTROL. The circulatory system is provided with an extensive system for controlling the arterial pressure. For instance, if at any time the pressure falls significantly below its normal mean level of about 100 mm Hg, a barrage of nervous reflexes within seconds elicits a series of circulatory changes to raise the pressure back to normal, including increased force of heart pumping, contraction of the large venous reservoirs to provide more blood for the heart, and generalized constriction of most of the arterioles throughout the body, so that more blood will accumulate in the arterial tree. Then, over more prolonged periods, hours and days, the kid-

neys play an additional major role in pressure control both by secreting pressure-controlling hormones and by regulating the blood volume.

The importance of pressure control is that it prevents changes in blood flow in one area of the body from significantly affecting flow elsewhere in the body because the common pressure head for both areas is not allowed to change greatly.

Thus, in summary, the needs of the local tissues are served by the circulation. In the remainder of this chapter, we begin to discuss the basic details of the management of blood flow and the control of cardiac output and arterial pressure.

INTERRELATIONSHIPS AMONG PRESSURE, FLOW, AND RESISTANCE

Flow through a blood vessel is determined by two factors: (1) the *pressure difference* between the two ends of the vessel (also frequently called "pressure gradient"), which is the force that pushes the blood through the vessel, and (2) the impediment to blood flow through the vessel, which is called vascular *resistance*. Figure 14–3 demonstrates these relations, showing a blood vessel segment located anywhere in the circulatory system.

P_1 represents the pressure at the origin of the vessel; at the other end the pressure is P_2. Resistance to flow (R) occurs as a result of friction all along the

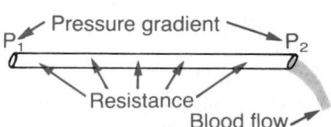

Figure 14–3. Relations among pressure, resistance, and blood flow.

inside of the vessel. The flow through the vessel can be calculated by the following formula, which is called *Ohm's law:*

$$Q = \frac{\Delta P}{R}, \qquad (1)$$

in which Q is blood flow, ΔP is the pressure difference $(P_1 - P_2)$ between the two ends of the vessel, and R is the resistance. This formula states, in effect, that the blood flow is directly proportional to the pressure difference but inversely proportional to the resistance.

Note that it is the *difference* in pressure between the two ends of the vessel, not the absolute pressure in the vessel, that determines the rate of flow. For instance, if the pressure at both ends of the segment were 100 mm Hg and yet no difference existed between the two ends, there would be no flow despite the presence of 100 mm Hg pressure.

Ohm's law expresses the most important of all the relations that the reader needs to understand to comprehend the hemodynamics of the circulation. Because of the extreme importance of this formula, the reader should also become familiar with its other algebraic forms:

$$\Delta P = Q \times R \qquad (2)$$

$$R = \frac{\Delta P}{Q} \qquad (3)$$

Blood Flow

Blood flow means simply the quantity of blood that passes a given point in the circulation in a given period. Ordinarily, blood flow is expressed in *milliliters* or *liters per minute,* but it can be expressed in milliliters per second or in any other unit of flow.

The overall blood flow in the circulation of an adult at rest is about 5000 ml/min. This is called the *cardiac output* because it is the amount of blood pumped by the heart in a unit period.

METHODS FOR MEASURING BLOOD FLOW. Many mechanical and mechanoelectrical devices can be inserted in series with a blood vessel or, in some instances, applied to the outside of the vessel to measure flow. They are called *flowmeters.*

Electromagnetic Flowmeter. One of the most important devices for measuring blood flow *without opening the vessel* is the electromagnetic flowmeter, the principles of which are shown in Figure 14–4. Figure 14–4A shows generation of electromotive force in a wire that is moved rapidly in the cross-wise direction through a magnetic field. This is the well-known principle for production of electricity by the electric generator. Figure 14–4B shows that the same principle applies for generation of electromotive force in blood when it moves through a magnetic field. In this case, a blood vessel is placed between the poles of a strong magnet, and electrodes are placed on the two sides of the vessel perpendicular to the magnetic lines of force. When blood flows through the vessel, electrical voltage pro-

portional to the rate of flow is generated between the two electrodes, and this is recorded using an appropriate meter or electronic recording apparatus. Figure 14–4C shows an actual probe that is placed on a large blood vessel to record its blood flow. The probe contains both the strong magnet and the electrodes.

A special advantage of the electromagnetic flowmeter is that it can record changes in flow that occur in less than 0.01 second, allowing accurate recording of pulsatile changes in flow as well as steady flow.

ULTRASONIC DOPPLER FLOWMETER. Another type of flowmeter that can be applied to the outside of the vessel and that has many of the same advantages as the electromagnetic flowmeter is the ultrasonic Doppler flowmeter, shown in Figure 14–5. A minute piezoelectric crystal is mounted in the wall of the device. This crystal, when energized with an appropriate electronic apparatus, transmits sound at a frequency of several million cycles per second downstream along the flowing blood. A portion of the sound is reflected by the red blood cells in the flowing blood, so that reflected sound waves travel backward from the cells toward the crystal. These reflected waves have a lower frequency than the transmitted wave because the red cells are moving away from the transmitter crystal. This is called the Doppler effect. (It is the same effect that one experiences when a train approaches and passes by while blowing the whistle. Once the whistle has passed by the person, the pitch of the sound from the whistle suddenly becomes much lower than when the train is approaching.) The transmitted wave is intermittently cut off, and the reflected wave is received back onto the crystal and then amplified greatly by the electronic apparatus. Another portion of the apparatus determines the frequency difference between the transmitted wave and the reflected wave, thus also determining the velocity of blood flow.

Like the electromagnetic flowmeter, the ultrasonic Doppler flowmeter is capable of recording rapid, pulsatile changes in flow as well as steady flow.

Laminar Flow of Blood in Vessels

When blood flows at a steady rate through a long, smooth vessel, it flows in *streamlines*, with each layer of blood remaining the same distance from the wall. Also, the central portion of the blood stays in the center of the vessel. This type of flow is called *laminar flow* or *streamline flow*, and it is opposite to *turbulent flow*, which is blood flowing in all directions in the vessel and continually mixing within the vessel, as discussed subsequently.

PARABOLIC VELOCITY PROFILE DURING LAMINAR FLOW. When laminar flow occurs, the velocity of flow in the center of the vessel is far greater than that toward the outer edges. This is demonstrated by the experiment shown in Figure 14–6. In vessel A are two fluids, the one to the left colored by a dye and the one to the right, a clear fluid, but there is no flow in the vessel. Then the fluids are made to flow; a parabolic interface develops between the two fluids, as shown 1 second later in vessel B, showing that the portion of fluid adjacent to the vessel wall has hardly moved, the portion slightly away from the wall has moved a small distance, and the portion in the center of the vessel has moved a long distance. This effect is called the parabolic profile for the velocity of blood flow.

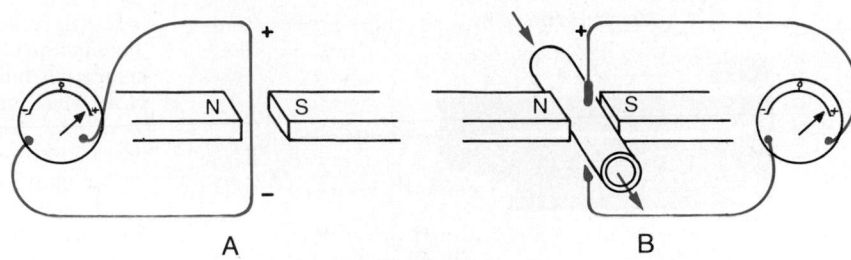

Figure 14–4. Flowmeter of the electromagnetic type, showing *A*, generation of an electromotive force in a wire as it passes through an electromagnetic field; *B*, generation of an electromotive force in electrodes on a blood vessel when the vessel is placed in a strong magnetic field and blood flows through the vessel; and *C*, a modern electromagnetic flowmeter probe for chronic implantation around blood vessels.

The cause of the parabolic profile is the following: The fluid molecules touching the wall hardly move because of adherence to the vessel wall. The next layer of molecules slips over these, the third layer over the second, the fourth layer over the third, and so forth. Therefore, the fluid in the middle of the vessel can move rapidly because many layers of slipping molecules exist between the middle of the vessel and the vessel wall, and each layer toward the center flows progressively more rapidly than the previous layer.

TURBULENT FLOW OF BLOOD UNDER SOME CONDITIONS. When the rate of blood flow becomes too great, when it passes an obstruction in a vessel, when it makes a sharp turn, or when it passes over a rough surface, the flow may become *turbulent* rather than streamline. Turbulent flow means that the blood flows crosswise in the vessel as well as along the vessel, usually forming whorls in the blood called *eddy currents*. They are similar to the whirlpools that one sees in a rapidly flowing river at a point of obstruction.

When eddy currents are present, blood flows with much greater resistance than when the flow is streamline because the eddies add tremendously to the overall friction of flow in the vessel.

The tendency for turbulent flow increases in direct proportion to the velocity of blood flow, in direct proportion to the diameter of the blood vessel, and inversely proportional to the viscosity of the blood divided by its density in accordance with the following equation:

$$\mathrm{Re} = \frac{\mathrm{v} \cdot \mathrm{d}}{\dfrac{\eta}{\rho}}, \qquad (4)$$

in which Re is *Reynolds' number* and is the measure of the tendency for turbulence to occur, v is the velocity of blood flow (in centimeters per second), d is diameter, η is the viscosity (in poises), and ρ is the density. When Reynolds' number rises above 200 to 400, turbulent flow occurs at some branches of vessels but dies out along the smooth portions of the vessels. When Reynolds' number rises above about 2000, turbulence usually occurs even in a straight, smooth vessel. Reynolds' number very often rises to 2000 in some large arteries; as a result, there is almost always some turbulence of flow in some arteries, such as at the root of the aorta and at the major arterial branches.

Blood Pressure

STANDARD UNITS OF PRESSURE. Blood pressure is almost always measured in *millimeters of mercury (mm Hg)* because the mercury manometer (shown in

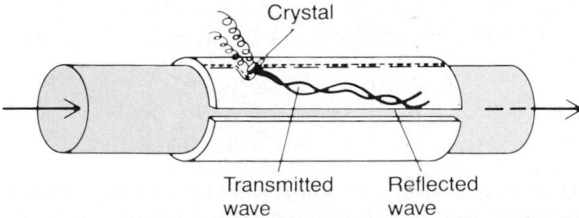

Figure 14–5. Ultrasonic Doppler flowmeter.

Crystal

Transmitted wave   Reflected wave

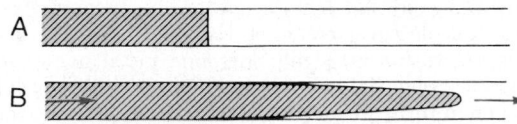

Figure 14–6. Experiment demonstrating parabolic blood flow with much faster flow in the center of a vessel. *A*, Two fluids before flow begins; *B*, the same fluids 1 second after flow begins.

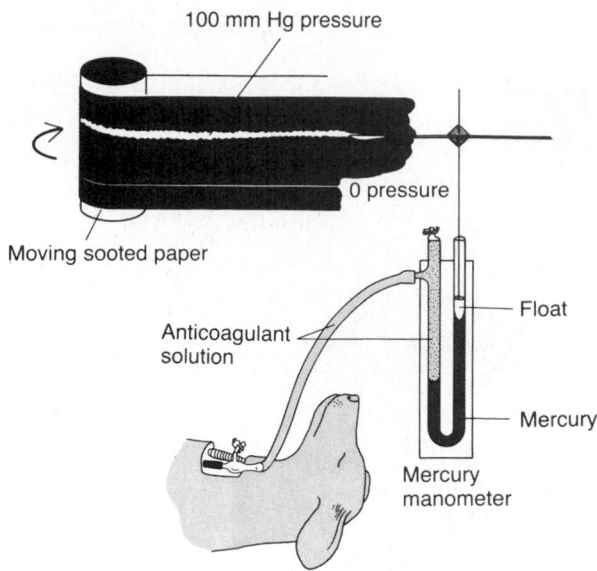

Figure 14–7. Recording arterial pressure with a mercury manometer, a method that has been used in the manner shown above for recording pressure throughout the history of physiology.

Figure 14–7) has been used since antiquity as the standard reference for measuring blood pressure. Actually, blood pressure means the *force exerted by the blood against any unit area of the vessel wall.* When one says that the pressure in a vessel is 50 mm Hg, one means that the force exerted is sufficient to push a column of mercury up to a level 50 mm high. If the pressure is 100 mm Hg, it will push the column of mercury up to 100 millimeters.

Occasionally, pressure is measured in *centimeters of water* (cm H₂O). A pressure of 10 cm H₂O means a pressure sufficient to raise a column of water to a height of 10 centimeters. *One millimeter of mercury equals 1.36 cm H₂O* because the specific gravity of mercury is 13.6 times that of water, and 1 centimeter is 10 times as great as 1 millimeter.

High-Fidelity Methods for Measuring Blood Pressure. The mercury in the mercury manometer has so much *inertia* that it cannot rise and fall rapidly. For this reason, the mercury manometer, although excellent for recording steady pressures, cannot respond to pressure changes that occur more rapidly than about one cycle every 2 to 3 seconds. Whenever it is desired to record rapidly changing pressures, some other type of pressure recorder is needed. Figure 14–8 demonstrates the basic principles of three electronic pressure *transducers* commonly used for converting pressure into electrical signals and then recording the pressure on a high-speed electrical recorder. Each of these transducers uses a very thin, highly stretched metal membrane that forms one wall of the fluid chamber. The fluid chamber in turn is connected through a needle or catheter with the vessel in which the pressure is to be measured. When the pressure is high, the membrane bulges slightly, and when it is low, it returns toward its resting position.

In Figure 14–8A, a simple metal plate is placed a few thousandths of an inch above the membrane. When the membrane bulges, the membrane comes closer to the plate, which increases the electrical capacitance between these two, and this change in capacitance can be recorded by an appropriate electronic system.

In Figure 14–8B, a small iron slug rests on the membrane, and this can be displaced upward into a coil. Movement of the iron changes the *inductance* of the coil, and this, too, can be recorded electronically.

Finally, in Figure 14–8C, a very thin, stretched resistance wire is connected to the membrane. When this wire is greatly stretched, its resistance increases; when it is less stretched, its resistance decreases. These changes also can be recorded by means of an electronic system.

With some of these high-fidelity types of recording systems, pressure cycles up to 500 cycles per second have been recorded accurately. In common use are recorders capable of registering pressure changes that occur as rapidly as 20 to 100 cycles per second, in the manner shown on the recording paper in Figure 14–8.

Resistance to Blood Flow

Units of Resistance. Resistance is the impediment to blood flow in a vessel, but it cannot be measured by any direct means. Instead, resistance must be calculated from measurements of blood flow and pressure difference in the vessel. If the pressure difference between two points in a vessel is 1 mm Hg and the flow is 1 ml/sec, the resistance is said to be 1 *peripheral resistance unit,* usually abbreviated *PRU.*

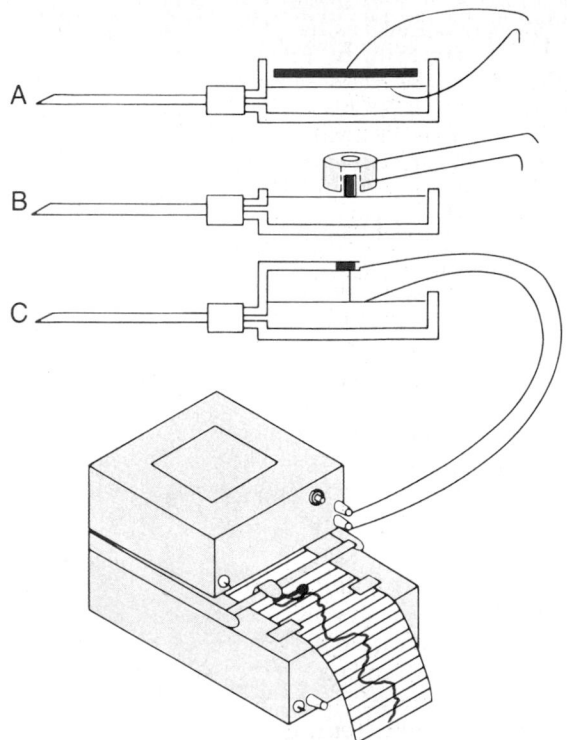

Figure 14–8. Principles of three types of electronic transducers for recording rapidly changing blood pressures (explained in the text).

Expression of Resistance in CGS Units. Occasionally, a basic physical unit called the CGS (centimeters, grams, seconds) unit is used to express resistance. This unit is *dyne seconds/centimeters⁵*. Resistance in these units can be calculated by the following formula:

$$R \left(in \ \frac{dyne \ sec}{cm^5} \right) = \frac{1333 \times mm \ Hg}{ml/sec} \qquad (5)$$

TOTAL PERIPHERAL RESISTANCE AND TOTAL PULMONARY RESISTANCE. The rate of blood flow through the circulatory system when a person is at rest is close to 100 ml/sec, and the pressure difference from the systemic arteries to the systemic veins is about 100 mm Hg. Therefore, in round figures, the resistance of the entire systemic circulation, called the *total peripheral resistance,* is about 100/100 or 1 PRU. In some conditions in which all the blood vessels throughout the body become strongly constricted, the total peripheral resistance rises to as high as 4 PRU, and when the vessels become greatly dilated, it can fall to as little as 0.2 PRU.

In the pulmonary system, the mean right arterial pressure averages 16 mm Hg and the mean left atrial pressure averages 2 mm Hg, giving a net pressure difference of 14 mm. Therefore, in round figures, the *total pulmonary resistance* at rest calculates to be about 0.14 PRU.

"CONDUCTANCE" OF BLOOD IN A VESSEL AND ITS RELATION TO RESISTANCE. Conductance is a measure of the blood flow through a vessel for a given pressure difference. This is generally expressed in terms of milliliters per second per millimeter of mercury pressure, but it can also be expressed in terms of liters per second per millimeter of mercury or in any other units of blood flow and pressure.

It is immediately evident that conductance is the reciprocal of resistance in accord with the following equation:

$$Conductance = \frac{1}{Resistance} \qquad (6)$$

VERY SLIGHT CHANGES IN DIAMETER OF A VESSEL CHANGE ITS CONDUCTANCE TREMENDOUSLY! Slight changes in the diameter of a vessel cause tremendous changes in its ability to conduct blood when the blood flow is streamline. This is demonstrated forcefully by the experiment in Figure 14–9A, which shows three vessels with relative diameters of 1, 2, and 4 but with the same pressure difference of 100 mm Hg between the two ends of the vessels. Although the diameters of these vessels increase only 4-fold, the respective flows are 1, 16, and 256 ml/mm, which is a 256-fold increase in flow. Thus, the conductance of the vessel increases in proportion to the *fourth power of the diameter,* in accord with the following formula:

$$Conductance \propto Diameter^4 \qquad (7)$$

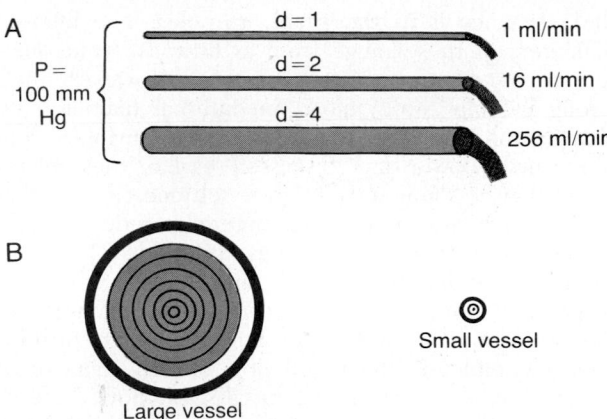

Figure 14–9. *A,* Demonstration of the effect of vessel diameter on blood flow. *B,* Concentric rings of blood flowing at different velocities; the farther away from the vessel wall, the faster the flow.

POISEUILLE'S LAW. The cause of this great increase in conductance when the diameter increases can be explained by referring to Figure 14–9B, which shows cross sections of a large and a small vessel. The concentric rings inside the vessels indicate that the velocity of flow in each ring is different from that in the other rings because of *laminar* flow, which is discussed earlier in the chapter. That is, the blood in the ring touching the wall of the vessel is hardly flowing because of its adherence to the vascular endothelium. The next ring of blood slips past the first ring and, therefore, flows more rapidly. The third, fourth, fifth, and sixth rings likewise flow at progressively increasing velocities. Thus, the blood that is near the wall of the vessel flows extremely slowly, whereas that in the middle of the vessel flows extremely rapidly.

In the small vessel, essentially all the blood is near the wall, so that the extremely rapidly flowing central stream of blood simply does not exist.

By integrating the velocities of all the concentric rings of flowing blood and multiplying them by the areas of the rings, one can derive the following formula, known as Poiseuille's law:

$$Q = \frac{\pi \Delta P r^4}{8 \, \eta l}, \qquad (8)$$

in which Q is the rate of blood flow, ΔP is the pressure difference between the ends of the vessel, r is the radius of the vessel, l is the length of the vessel, and η is the viscosity of the blood.

Note particularly in this equation that the rate of blood flow is directly proportional to the *fourth power of the radius* of the vessel, which demonstrates once again that the diameter of a blood vessel (which is equal to twice the radius) plays by far the greatest role of all factors in determining the rate of blood flow through a vessel.

IMPORTANCE OF THE VESSEL DIAMETER "FOURTH POWER LAW" IN DETERMINING ARTERIOLAR RESISTANCE. In the systemic circulation, about two thirds of the resistance is in the small arterioles. The internal diameters of these range from as little as 4 micrometers to as great as 25 micrometers. However, their strong vascular walls allow the internal diameters to change tremendously, often as much as fourfold. From the fourth power law discussed earlier that relates blood flow to diameter of the vessel, one can see that a 4-fold increase in vessel diameter theoretically could increase the flow as much as 256-fold. Thus, this fourth power law makes it possible for the arterioles, responding with only small changes in diameter to nervous signals or local tissue signals, either to turn off almost completely the blood flow to the tissue or at other times to cause a vast increase in flow. Indeed, ranges of blood flow of more than a hundredfold in small tissue areas have been recorded between the limits of maximum arteriolar constriction and maximum arteriolar dilatation.

Effect of Blood Hematocrit and Viscosity on Vascular Resistance and Blood Flow

Note especially that one of the important factors in Poiseuille's law is the viscosity of the blood. The greater the viscosity, the less the flow in a vessel if all other factors are constant. Furthermore, *the viscosity of normal blood is about three times as great as the viscosity of water.*

But what makes the blood so viscous? It is mainly the large numbers of suspended red cells in the blood, each of which exerts frictional drag against adjacent cells and against the wall of the blood vessel.

HEMATOCRIT. The percentage of the blood that is cells is called the hematocrit. Thus, if a person has a hematocrit of 40, 40 per cent of the blood volume is cells and the remainder is plasma. The hematocrit of men averages about 42, whereas that of women averages about 38. These values vary tremendously, depending on whether or not the person has anemia, the degree of bodily activity, and the altitude at which the person resides. These effects are discussed in relation to the red blood cells and their oxygen transport function in Chapter 32.

Hematocrit is determined by centrifuging blood in a calibrated tube, such as that shown in Figure 14–10. The calibration allows direct reading of the percentage of cells.

EFFECT OF HEMATOCRIT ON BLOOD VISCOSITY. The greater the percentage of cells in the blood—that is, the greater the hematocrit—the more friction there is between successive layers of blood, and this friction determines viscosity. Therefore, the viscosity of blood increases drastically as the hematocrit increases, as shown in Figure 14–11. If we consider the viscosity of whole blood at normal hematocrit to be about 3, this means that three times as much pressure is required to force whole blood as to force water through the same tube. When the hematocrit rises to 60 or 70, which it often does in polycythemia, the blood viscosity

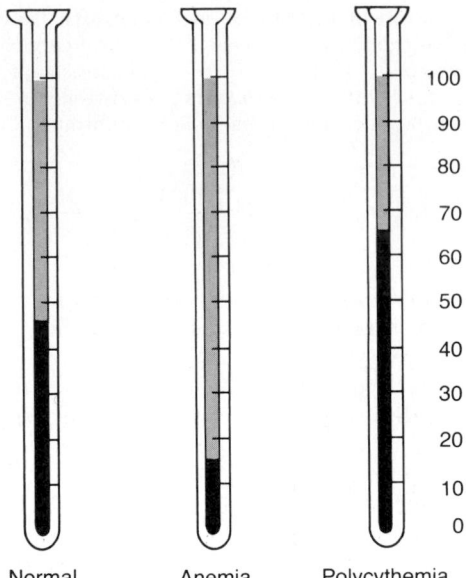

Figure 14–10. Hematocrits in a normal person and in patients with anemia and polycythemia.

can become as great as 10 times that of water and its flow through blood vessels is greatly retarded.

Another factor that affects blood viscosity is the concentration and types of proteins in the plasma, but these effects are so much less important than the effect of hematocrit that they are not significant considerations in most hemodynamic studies. The viscosity of blood plasma is about 1.5 times that of water.

Blood Viscosity in the Microcirculation. Because most resistance in the circulatory system occurs in the very small blood vessels, it is especially important to know how blood viscosity affects blood flow in these minute vessels. At least three additional factors besides hematocrit and plasma proteins affect blood viscosity in these vessels.

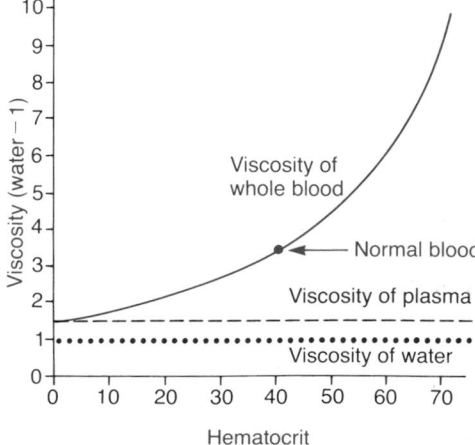

Figure 14–11. Effect of hematocrit on viscosity.

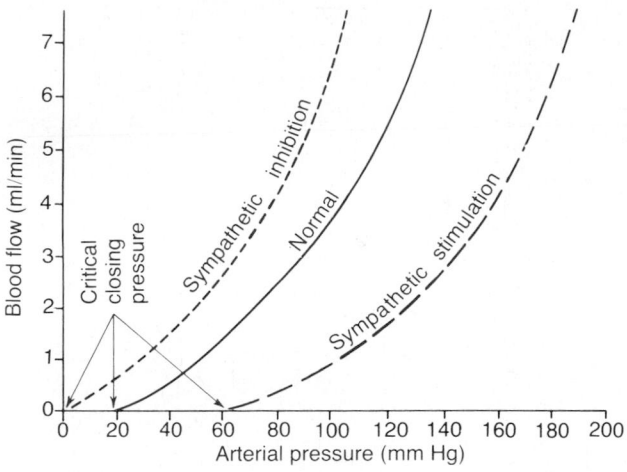

Figure 14–12. Effect of arterial pressure on blood flow through a blood vessel at different degrees of vascular tone caused by increased or decreased sympathetic stimulation of the vessels.

1. Blood flow in minute tubes exhibits far less viscous effect than it does in large vessels. This is called the *Fahraeus-Lindqvist effect.* It begins to appear when the vessel diameter falls below about 1.5 millimeters. In tubes as small as capillaries, the viscosity of whole blood is as little as one half that in large vessels. The Fahraeus-Lindqvist effect is caused by alignment of the red cells as they pass through the vessels. That is, the red cells, instead of moving randomly, line up and move through the vessels as a single plug, thus eliminating the viscous resistance that occurs internally in the blood itself. The Fahraeus-Lindqvist effect, however, is probably more than offset by the next two effects under most conditions.

2. The viscosity of blood increases tremendously as its velocity of flow decreases. Because the velocity of blood flow in the small vessels is extremely slow, often less than 1 mm/sec, blood viscosity can increase as much as 10-fold from this factor alone. This effect is partly caused by adherence of the slowly moving red cells to one another (formation of rouleaux and larger aggregates) as well as to the vessel walls.

3. Cells often become stuck at constrictions in small blood vessels; this happens especially in capillaries where the nuclei of endothelial cells protrude into the capillary lumen. When this occurs, blood flow can become blocked for a fraction of a second or for much longer periods, thus giving an apparent effect of greatly increased viscosity.

Because of these special effects that occur in the minute vessels of the circulatory system, it has been impossible to derive an exact mathematical relation that describes how the hematocrit affects viscosity in the minute vessels, which is the place in the circulatory system where viscosity plays its most important role. Because some of these effects decrease viscosity and others increase viscosity, it is usually assumed that the overall viscous effects in the small vessels are about equivalent to those that occur in the larger vessels.

Effects of Pressure on Vascular Resistance and Tissue Blood Flow

From the discussion thus far, one would expect that an increase in arterial pressure would cause a proportionate increase in blood flow through the various tissues of the body. However, the effect of pressure on blood flow is far greater than one would expect, as shown in Figure 14–12. The reason for this is that an increase in arterial pressure not only increases the force that tends to push blood through the vessels but also distends the vessels at the same time, which decreases their resistance. Thus, increased pressure increases the flow in two ways, and for most tissues, blood flow at 100 mm Hg arterial pressure is usually about four to six times as great as the blood flow at 50 mm Hg.

Note also in Figure 14–12 the large changes in blood flow that can be caused by either increased or decreased sympathetic stimulation of the peripheral blood vessels. Thus, as shown in the figure, inhibition of sympathetic stimulation greatly dilates the vessels and can increase the blood flow twofold or more. Conversely, strong sympathetic stimulation can constrict the vessels so much that blood flow can be decreased to as low as zero for short periods despite high arterial pressure.

REFERENCES

See references for Chapter 15.

Vascular Distensibility, and Functions of the Arterial and Venous Systems

CHAPTER 15

VASCULAR DISTENSIBILITY

A valuable characteristic of the vascular system is that all blood vessels are distensible. We have seen one example of this in Chapter 14, when the pressure in the arterioles is increased; this dilates the arterioles and therefore decreases their resistance. The result is increased blood flow not only because of increased pressure but also because of decreased resistance, usually giving at least twice as much flow increase as one might expect.

Vascular distensibility also plays other important roles in circulatory function. For example, the distensible nature of the arteries allows them to accommodate the pulsatile output of the heart and to average out the pressure pulsations. This provides an almost completely smooth, continuous flow of blood through the very small blood vessels of the tissues.

The most distensible by far of all the vessels are the veins. Even slight increases in pressure cause the veins to store 0.5 to 1.0 liter of extra blood. Therefore, the veins provide a reservoir function for storing large quantities of blood that can be called into use whenever required elsewhere in the circulation.

UNITS OF VASCULAR DISTENSIBILITY. Vascular distensibility is normally expressed as the fractional increase in volume for each millimeter of mercury rise in pressure in accordance with the following formula:

$$\text{Vascular distensibility} = \frac{\text{Increase in volume}}{\text{Increase in pressure} \times \text{Original volume}} \quad (1)$$

That is, if 1 mm Hg causes a vessel that originally contained 10 ml of blood to increase its volume by 1 ml, the distensibility would be 0.1 per mm Hg, or 10 per cent per mm Hg.

Difference in Distensibility of the Arteries and the Veins. Anatomically, the walls of arteries are far stronger than those of veins. Consequently, the veins, on the average, are about 8 times as distensible as the arteries. That is, a given rise in pressure causes about 8 times as much extra blood to fill a vein as to fill an artery of comparable size.

In the pulmonary circulation, the veins are similar to those of the systemic circulation. However, the pulmonary arteries normally operate under pressures about one sixth those in the systemic arterial system, and they have distensibilities about one half those of veins, rather than one eighth as is true of the systemic arteries.

Vascular Compliance (or Capacitance)

In hemodynamic studies, it usually is much more important to know the *total quantity of blood* that can be stored in a given portion of the circulation for each millimeter of mercury pressure rise than to know the distensibility of the individual vessels. This value is called the *compliance* or *capacitance* of the respective vascular bed. That is,

$$\text{Vascular compliance} = \frac{\text{Increase in volume}}{\text{Increase in pressure}} \quad (2)$$

Compliance and distensibility are quite different. A

171

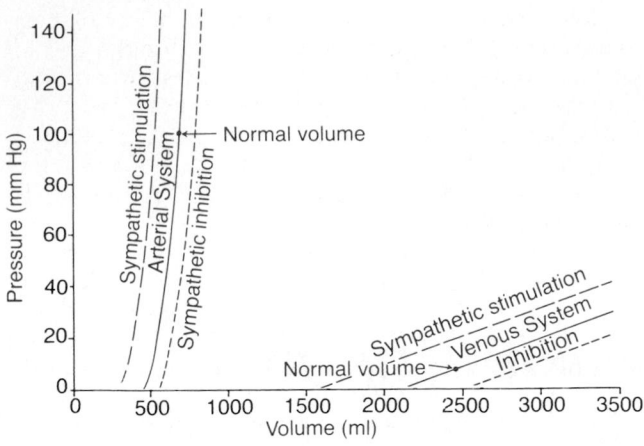

Figure 15–1. Volume-pressure curves of the systemic arterial and venous systems, showing also the effects of sympathetic stimulation and sympathetic inhibition.

highly distensible vessel that has a slight volume may have far less compliance than a much less distensible vessel that has a large volume because *compliance is equal to distensibility times volume.*

The compliance of a vein is about 24 times that of its corresponding artery because it is about 8 times as distensible and it has a volume about 3 times as great (8 × 3 = 24).

Volume-Pressure Curves of the Arterial and Venous Circulations

A convenient method for expressing the relation of pressure to volume in a vessel or in any portion of the circulation is the so-called *volume-pressure curve* (also called the *pressure-volume curve*). The two solid curves of Figure 15–1 represent, respectively, the volume-pressure curves of the normal systemic arterial and venous systems, showing that when the arterial system of the average adult person, including the larger arteries, small arteries, and arterioles, is filled with about 750 milliliters of blood, the mean arterial pressure is 100 mm Hg, but when filled with only 500 milliliters, the pressure falls to zero.

In the entire systemic venous system, on the other hand, the volume of blood normally ranges from 2500 to 3500 milliliters, and tremendous changes in this volume are required to change the venous pressure only a few millimeters of mercury.

EFFECT OF SYMPATHETIC STIMULATION OR SYMPATHETIC INHIBITION ON THE VOLUME-PRESSURE RELATIONS OF THE ARTERIAL AND VENOUS SYSTEMS. Also shown in Figure 15–1 are the effects of sympathetic stimulation and sympathetic inhibition on the volume-pressure curves. It is evident that the increase in vascular smooth muscle tone caused by sympathetic stimulation increases the pressure at each volume of the arteries or veins, whereas sympathetic inhibition decreases the pressure at each volume. Control of the vessels in this manner by the sympathetics is a valuable means for diminishing the dimensions of one segment of the circulation, thus transferring blood to

other segments. For instance, an increase in vascular tone throughout the systemic circulation often causes large volumes of blood to shift into the heart, which is a principal method that the body uses to increase heart pumping.

Sympathetic control of vascular capacity is also especially important during hemorrhage. Enhancement of the sympathetic tone of the vessels, especially of the veins, reduces the vessel sizes so that the circulation continues to operate almost normally even when as much as 25 per cent of the total blood volume has been lost.

Delayed Compliance (Stress-Relaxation) of Vessels

The term "delayed compliance" means that a vessel exposed to increased volume will at first exhibit a large increase in pressure, but delayed stretch of the vessel wall will allow the pressure to return back toward normal over a period of minutes to hours. This effect is shown in Figure 15–2. In this figure, the pressure is being recorded in a small segment of a vein that is occluded at both ends. An extra volume of blood is suddenly injected until the pressure rises from 5 to 12 mm Hg. Even though none of the blood is removed after it is injected, the pressure begins to fall immediately and approaches about 9 mm Hg after several minutes. In other words, the volume of blood injected caused immediate *elastic* distention of the vein, but then the smooth muscle fibers of the vein began to "creep" to longer lengths, and their tensions correspondingly decreased. This effect is a characteristic of all smooth muscle tissue and is called *stress-relaxation,* which is explained in Chapter 8.

After the delayed increase in compliance has taken place in the experiment shown in Figure 15–2, the extra blood volume is suddenly removed, and the pressure immediately falls to a very low value. Subsequently, the smooth muscle fibers begin to readjust their tensions back to their initial values, and after a number of minutes, the normal vascular pressure of 5 mm Hg returns.

Delayed compliance is a valuable mechanism by which the circulation can accommodate much extra

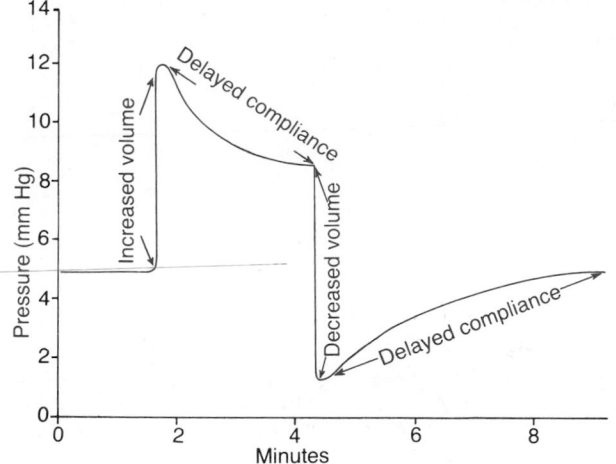

Figure 15–2. Effect on the intravascular pressure of injecting a small volume of blood into a venous segment, demonstrating the principles of delayed compliance.

blood when necessary, such as after too large a transfusion. Also, delayed compliance in the reverse direction is one of the ways in which the circulation automatically adjusts itself over a period of minutes or hours to diminished blood volume after serious hemorrhage.

ARTERIAL PRESSURE PULSATIONS

With each beat of the heart a new surge of blood fills the arteries. Were it not for the distensibility of the arterial system, blood flow through the tissues would occur only during cardiac systole, and no blood flow would occur during diastole. The combination of distensibility of the arteries and their resistance to blood flow reduces the pressure pulsations almost to no pulsations by the time the blood reaches the capillaries; therefore, tissue blood flow usually is continuous rather than occurring in pulses.

A typical record of the *pressure pulsations* at the root of the aorta is shown in Figure 15–3. In the

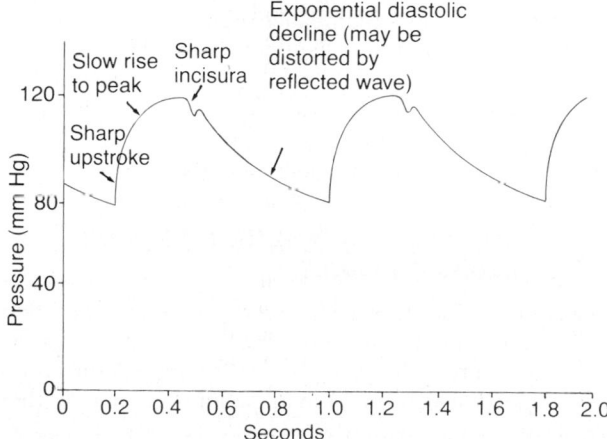

Figure 15–3. Normal pressure pulse contour recorded from the ascending aorta. (From Opdyke: *Fed. Proc. 11*:734, 1952.)

normal young adult, the pressure at the height of each pulse, the *systolic pressure,* is about 120 mm Hg and at its lowest point, the *diastolic pressure,* it is about 80 mm Hg. The difference between these two pressures, about 40 mm Hg, is called the *pulse pressure.*

Two major factors affect the pulse pressure: (1) the *stroke volume output* of the heart and (2) the *compliance (total distensibility)* of the arterial tree. A third, less important factor is the character of ejection from the heart during systole.

In general, the greater the stroke volume output, the greater is the amount of blood that must be accommodated in the arterial tree with each heartbeat, and, therefore, the greater is the pressure rise and fall during systole and diastole, thus causing a greater pulse pressure.

On the other hand, the less the compliance of the arterial system, the greater will be the rise in pressure for a given stroke volume of blood pumped into the arteries. For instance, as demonstrated by the middle top curve in Figure 15–4, sometimes the pulse pressure rises to as much as twice that of normal in old age because the arteries become hardened with *arteriosclerosis* and therefore are noncompliant.

In effect, then, the pulse pressure is determined approximately by the *ratio of stroke volume output to compliance of the arterial tree.* Any condition of the circulation that affects either of these two factors will also affect the pulse pressure.

Abnormal Pressure Pulse Contours

Some conditions of the circulation also cause abnormal contours to the pressure pulse wave in addition to altering the pulse pressure. Especially distinctive among these are aortic stenosis, patent ductus arteriosus, and aortic regurgitation, each of which is also shown in Figure 15–4.

In *aortic stenosis,* the pressure pulse is greatly decreased because of diminished blood flow outward through the stenotic aortic valve.

In *patent ductus arteriosus,* one half or more of the blood pumped into the aorta by the left ventricle flows immediately through the wide open ductus arteriosus into the pulmonary artery, thus allowing the diastolic pressure to fall very low before the next heartbeat.

In *aortic regurgitation,* the aortic valve is absent. Therefore, after each heartbeat, the blood that has just been pumped into the aorta flows immediately back into the left ventricle. As a result, the aortic pressure can fall all the way to zero between heartbeats. Also, there is no incisura in the pulse contour because there is no aortic valve to close.

Transmission of Pressure Pulses to the Peripheral Arteries

When the heart ejects blood into the aorta during systole, at first only the proximal portion of the aorta becomes distended because the inertia of the blood

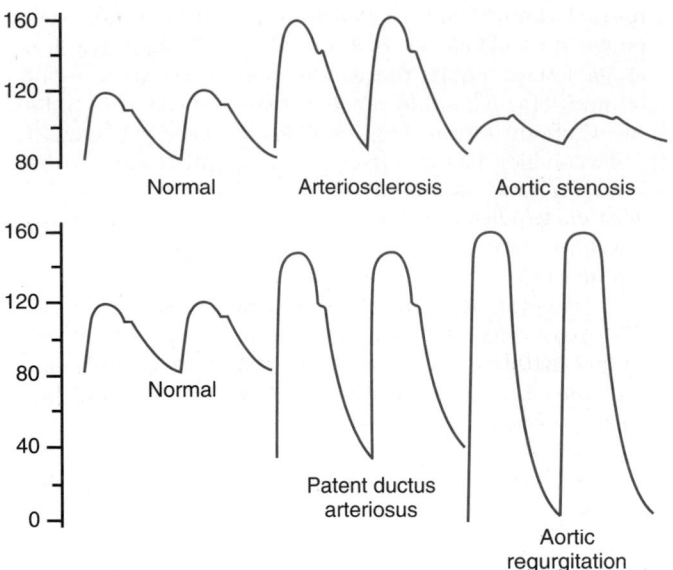

Figure 15–4. Pressure pulse contours in arteriosclerosis, aortic stenosis, patent ductus arteriosus, and aortic regurgitation.

prevents sudden movement all the way to the periphery. However, the rising pressure in the central aorta rapidly overcomes this inertia and the wave front of distention spreads further and further along the aorta, as shown in Figure 15–5. This is called *transmission of the pressure pulse* in the arteries.

The velocity of pressure pulse transmission in the normal aorta is 3 to 5 m/sec; in the large arterial branches, 7 to 10 m/sec; and in the small arteries, 15 to 35 m/sec. In general, the greater the compliance of each vascular segment, the slower is the velocity, which explains the slow transmission in the aorta and the much faster transmission in the much less compliant small distal arteries.

It should also be recognized that in the aorta, the velocity of transmission of the pressure pulse is 15 or more times the velocity of blood flow because the pressure pulse is simply a moving wave of *pressure* that involves little forward movement of blood volume.

DAMPING OF THE PRESSURE PULSES IN THE SMALLER ARTERIES, ARTERIOLES, AND CAPILLARIES. Figure 15–6 shows typical changes in the contours of the pressure pulse as the pulse travels into the peripheral vessels. Note especially in the three lower curves that the intensity of pulsation becomes progressively less in the smaller arteries, the arterioles, and, especially, the capillaries. In fact, only when the aortic pulsations are extremely large or when the arterioles are greatly dilated can pulsations be observed in the capillaries.

This progressive diminishment of the pulsations in the periphery is called *damping* of the pressure pulses. The cause of this is twofold: (1) the resistance to blood movement in the vessels and (2) the compliance of the vessels. The resistance damps the pulsations because a small amount of blood must flow forward at the pulse wave front to distend the next segment of the vessel; the greater the resistance, the more difficult it is for this to occur. The compliance damps the pulsations because the more compliant a vessel, the more the quantity of blood flow at the pulse wave front must be to cause the rise in pressure. Therefore, in effect, *the degree of damping is almost directly proportional to the product of resistance times compliance.*

Clinical Methods for Measuring Systolic and Diastolic Pressures

It is impossible to use the various pressure recorders that require needle insertion into an artery, as described earlier in the chapter, for making routine pressure measurements in human patients, although they are used on occasion when special studies are necessary. Instead, the clinician determines systolic and dia-

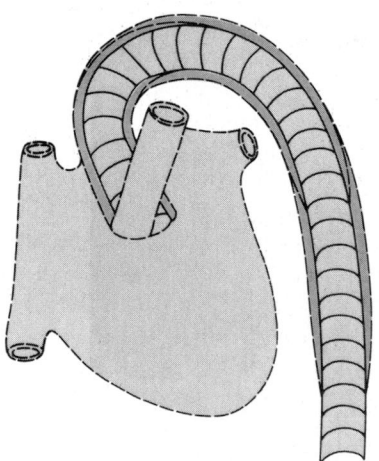

Figure 15–5. Progressive stages in the transmission of the pressure pulse along the aorta.

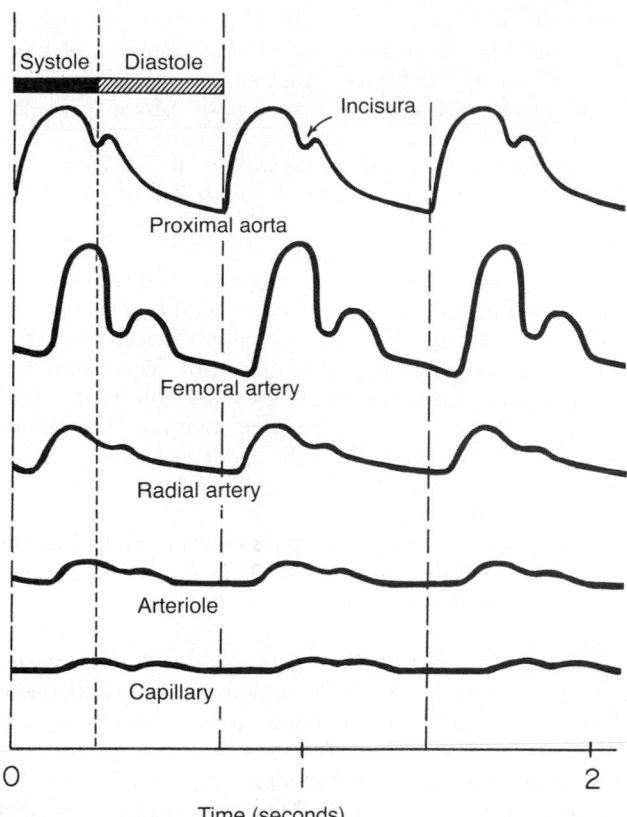

Figure 15–6. Changes in the pulse pressure contour as the pulse wave travels toward the smaller vessels.

stolic pressures by indirect means, usually by the auscultatory method.

AUSCULTATORY METHOD. Figure 15–7 shows the auscultatory method for determining systolic and diastolic arterial pressures. A stethoscope is placed over the antecubital artery and a blood pressure cuff is

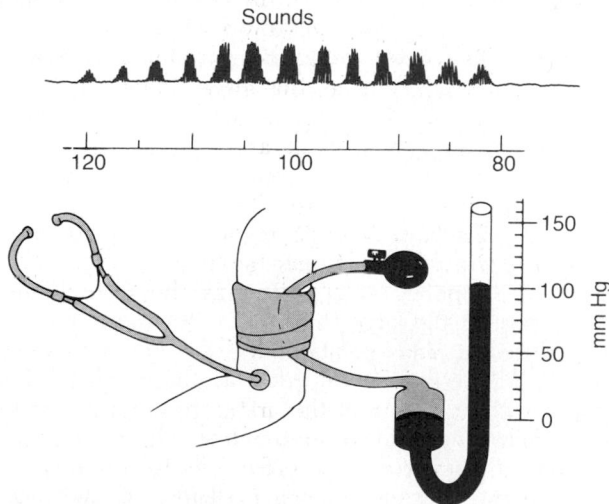

Figure 15–7. Auscultatory method of measuring systolic and diastolic arterial pressures.

inflated around the upper arm. As long as the cuff compresses the arm with so little pressure that the artery remains distended with blood, no sounds are heard by the stethoscope, despite the fact that the blood within the artery is pulsating. When the cuff pressure is great enough to close the artery during part of the arterial pressure cycle, a sound is then heard with each pulsation. These sounds are called *Korotkoff sounds.*

The exact cause of Korotkoff sounds is still debated, but they are believed to be caused mainly by blood jetting through the partly occluded vessel. The jet causes turbulence in the open vessel beyond the cuff, and this sets up the vibrations heard through the stethoscope.

In determining blood pressure by the auscultatory method, the pressure in the cuff is first elevated well above arterial systolic pressure. As long as this pressure is higher than systolic pressure, the brachial artery remains collapsed and no blood jets into the lower artery during any part of the pressure cycle. Therefore, no Korotkoff sounds are heard in the lower artery. But then the cuff pressure is gradually reduced. Just as soon as the pressure in the cuff falls below systolic pressure, blood slips through the artery beneath the cuff during the peak of systolic pressure, and one begins to hear *tapping* sounds in the antecubital artery in synchrony with the heartbeat. As soon as these sounds are heard, the pressure level indicated by the manometer connected to the cuff is about equal to the systolic pressure.

As the pressure in the cuff is lowered still more, the Korotkoff sounds change in quality, having less of the tapping quality and more of a rhythmical, harsher quality. Then, finally, when the pressure in the cuff falls to equal diastolic pressure, the artery no longer closes during diastole, which means that the basic factor causing the sounds (the jetting of blood through a squeezed artery) is no longer present. Therefore, the sounds suddenly change to a muffled quality and then usually disappear entirely after another 5- to 10-millimeter drop in cuff pressure. One notes the manometer pressure when the Korotkoff sounds change to the muffled quality, and this pressure is about equal to the diastolic pressure.

The auscultatory method for determining systolic and diastolic pressures is not entirely accurate, but it usually gives values within 10 per cent of those determined by direct measurement from the arteries.

NORMAL ARTERIAL PRESSURES AS MEASURED BY THE AUSCULTATORY METHOD. Figure 15–8 shows the approximate normal systolic and diastolic pressures at different ages. The progressive increase in pressure with age results from the effects of aging on the blood pressure control mechanisms. We see in Chapter 19 that the kidneys primarily are responsible for this long-term regulation of arterial pressure; it is well known that the kidneys do indeed exhibit definitive changes with age, especially after the age of 50.

The exceptional rise in *systolic* pressure beyond the

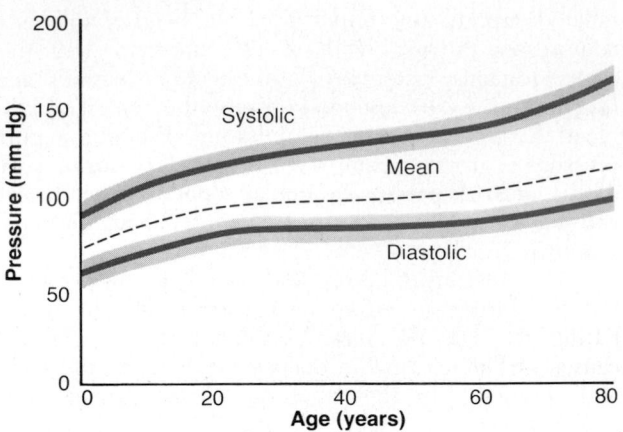

Figure 15–8. Changes in systolic, diastolic, and mean arterial pressures with age. The shaded areas show the approximate normal range.

age of 60 results from hardening of the arteries, which itself is an end-stage result of *atherosclerosis.* This causes a bounding systolic pressure and considerable increase in the pulse pressure, as already explained.

MEAN ARTERIAL PRESSURE. The mean arterial pressure is the average of all the pressures measured millisecond by millisecond over a period of time. It is not equal to the average of systolic and diastolic pressure because the pressure remains nearer to diastolic pressure than to systolic pressure during the greater part of the cardiac cycle. Therefore, the mean arterial pressure is determined about 60 per cent by the diastolic pressure and 40 per cent by the systolic pressure. Note in Figure 15–8 that the mean pressure at all ages is nearer to the diastolic pressure than to the systolic pressure, especially so at older ages.

VEINS AND THEIR FUNCTIONS

For years, the veins have been considered to be nothing more than passageways for flow of blood into the heart, but it is rapidly becoming apparent that they perform many special functions that are necessary to the operation of the circulation. Especially important, they are capable of constricting and enlarging and thereby storing either small or large quantities of blood and making this blood available when it is required by the remainder of the circulation. They also can propel blood forward by means of a so-called venous pump, and they even help to regulate cardiac output, an exceedingly important function that is described in Chapter 20.

Venous Pressures—Right Atrial Pressure (Central Venous Pressure) and Peripheral Venous Pressures

To understand the various functions of the veins, it is first necessary to know something about the pressures in the veins and how they are regulated. Blood from all the systemic veins flows into the right atrium; therefore, the pressure in the right atrium is called the *central venous pressure.* Anything that affects right atrial pressure usually affects venous pressure everywhere in the body.

Right atrial pressure is regulated by a balance between the ability of the heart to pump blood out of the right atrium and *the tendency for blood to flow from the peripheral vessels back into the right atrium.*

If the heart is pumping strongly, the right atrial pressure decreases. On the other hand, weakness of the heart elevates the right atrial pressure. Likewise, any effect that causes rapid inflow of blood into the right atrium from the veins elevates the right atrial pressure. Some of the factors that increase this venous return (and increase the right atrial pressure) are (1) increased blood volume, (2) increased large vessel tone throughout the body with resultant increased peripheral venous pressures, and (3) dilatation of the arterioles, which decreases the peripheral resistance and allows rapid flow of blood from the arteries to the veins.

The same factors that regulate right atrial pressure also enter into the regulation of cardiac output because the amount of blood pumped by the heart depends both on the ability of the heart to pump and the tendency for blood to flow into the heart from the peripheral vessels. Therefore, we discuss the regulation of right atrial pressure in much more depth in Chapter 20 in connection with the regulation of cardiac output.

The *normal right atrial pressure* is about 0 mm Hg, which is about equal to the atmospheric pressure around the body. It can rise to 20 to 30 mm Hg under very abnormal conditions, such as serious heart failure or after massive transfusion of blood, which will cause excessive quantities of blood to attempt to flow into the heart from the peripheral vessels.

The lower limit to the right atrial pressure is usually about −3 to −5 mm Hg, which is the pressure in the chest cavity that surrounds the heart. The right atrial pressure aproaches these low values when the heart pumps with exceptional vigor or when the flow of blood into the heart from the peripheral vessels is greatly depressed, such as after severe hemorrhage.

Venous Resistance and Peripheral Venous Pressure

Large veins have so little resistance to blood flow *when they are distended* that the resistance then is of almost no importance. However, as shown in Figure 15–9, most of the large veins that enter the thorax are compressed at many points by the surrounding tissues, so that blood flow is impeded at these points. For instance, the veins from the arms are compressed by their sharp angulation over the first rib. Second, the pressure in the neck veins often falls so low that the atmospheric pressure on the outside of the neck causes them to collapse. Finally, veins coursing through the abdomen are often compressed by different organs and by the intra-abdominal pressure, so that they usually are at least partially collapsed to an ovoid or a slitlike state. For these reasons, the *large veins do usually offer considerable resistance to blood flow,* and because of this, the pressure in the more peripheral small veins is usually 4 to 7 mm Hg greater than the right atrial pressure.

EFFECT OF HIGH RIGHT ATRIAL PRESSURE ON PERIPHERAL VENOUS PRESSURE. When the right atrial pressure rises above its normal value of 0 mm Hg, blood begins to back up in the large veins and to open them up. The pressures in the peripheral veins do not rise until all the collapsed points between the peripheral veins and the large central veins have opened up. This usually occurs when the right atrial pressure rises to about +4 to 6 mm Hg. Then, as the right atrial pressure rises still further, the additional increase in pressure is reflected by a corresponding rise in peripheral venous pressure. Because the heart must be greatly weakened to cause a rise in right atrial pressure to 4 to 6 mm Hg, one often finds that the peripheral venous pressure is not elevated in the early stages of heart failure.

EFFECT OF ABDOMINAL PRESSURE ON VENOUS PRESSURES OF THE LEG. The normal pressure in the peritoneal cavity of a recumbent person averages about 6 mm Hg, but at times, it can rise to 15 to 30 mm Hg as a result of pregnancy, large tumors, or excessive fluid (called ascites) in the peritoneal cavity. When this happens, the pressure in the veins of the legs must rise *above* the abdominal pressure before the abdominal veins will open and allow the blood to flow from the legs to the heart. Thus, if the intra-abdominal pressure is 20 mm Hg, the lowest possible pressure in the femoral veins is 20 mm Hg.

Effect of Hydrostatic Pressure on Venous Pressure

In any body of water, the pressure at the surface of the water is equal to atmospheric pressure, but the pressure rises 1 mm Hg for each 13.6-millimeter dis-

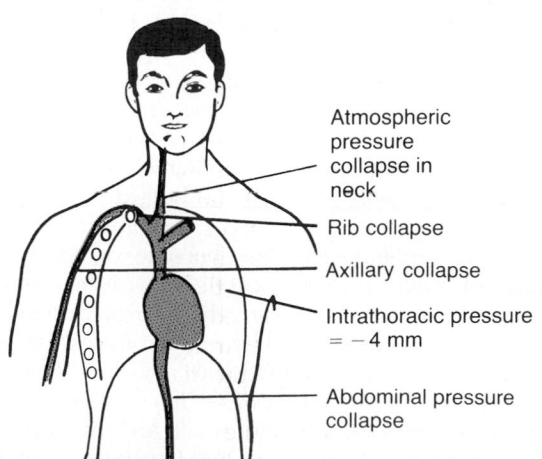

Figure 15–9. Factors that tend to collapse the veins entering the thorax.

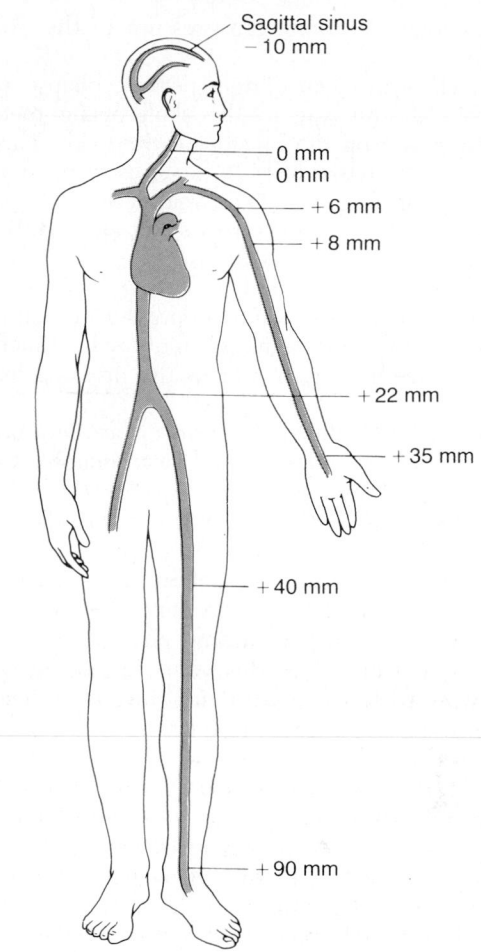

Figure 15–10. Effect of hydrostatic pressure on the venous pressures throughout the body in the standing person.

tance below the surface. This pressure results from the weight of the water and therefore is called *hydrostatic pressure.*

Hydrostatic pressure also occurs in the vascular system of the human being because of the weight of the blood in the vessels, as shown in Figure 15–10. When a person is standing, the pressure in the right atrium remains about 0 mm Hg because the heart pumps into the arteries any excess blood that attempts to accumulate at this point. However, in an adult *who is standing absolutely still,* the pressure in the veins of the feet is about +90 mm Hg simply because of the hydrostatic weight of the blood in the veins between the heart and the feet. The venous pressures at other levels of the body lie proportionately between 0 and 90 mm Hg.

In the arm veins, the pressure at the level of the top rib is usually about +6 mm Hg because of compression of the subclavian vein as it passes over this rib. The hydrostatic pressure down the length of the arm is then determined by the distance below the level of this rib. Thus, if the hydrostatic difference between the level of the rib and the hand is 29 mm Hg, this hydrostatic pressure is added to the 6 mm Hg pressure caused by compression of the vein as it crosses the rib,

In the figure labels for Figure 15-9: Atmospheric pressure collapse in neck; Rib collapse; Axillary collapse; Intrathoracic pressure = –4 mm; Abdominal pressure collapse.

In the figure labels for Figure 15-10: Sagittal sinus –10 mm; 0 mm; 0 mm; +6 mm; +8 mm; +22 mm; +35 mm; +40 mm; +90 mm.

making a total of 35 mm Hg pressure in the veins of the hand.

The neck veins of an upright person collapse almost completely all the way to the skull owing to atmospheric pressure on the outside of the neck. This collapse causes the pressure in these veins to remain zero along their entire extent. The reason for this is that any tendency for the pressure to rise above this level opens the veins and allows the pressure to fall back to zero because of increased flow of the blood. On the other hand, any tendency for the pressure to fall below this level collapses the veins still more, which increases their resistance and again returns the pressure back to zero.

The veins inside the skull, however, are in a noncollapsible chamber, and they will not collapse. Consequently, *negative pressure can exist in the dural sinuses of the head;* in the standing position, the venous pressure in the sagittal sinus is about − 10 mm Hg because of the hydrostatic "suction" between the top of the skull and the base of the skull. Therefore, if the sagittal sinus is opened during surgery, air can be sucked immediately into this vein; the air may even pass downward to cause air embolism in the heart, so that the heart valves will not function satisfactorily, and death can ensue.

EFFECT OF THE HYDROSTATIC FACTOR ON ARTERIAL AND OTHER PRESSURES. The hydrostatic factor also affects the pressures in the peripheral arteries and capillaries in addition to its effects in the veins. For instance, a standing person who has an arterial pressure of 100 mm Hg at the level of the heart has an arterial pressure in the feet of about 190 mm Hg. Therefore, any time one states that the arterial pressure is 100 mm Hg, it generally means that this is the pressure only at the hydrostatic level of the heart but not necessarily elsewhere in the arterial vessels.

Venous Valves and the "Venous Pump": Their Effects on Venous Pressure

Were it not for valves in the veins, the hydrostatic pressure effect would cause the venous pressure in the feet always to be about + 90 mm Hg in a standing adult. However, every time one moves the legs, one tightens the muscles and compresses the veins either in the muscles or adjacent to them, and this squeezes the blood out of the veins. The valves in the veins, shown in Figure 15–11, are arranged so that the direction of blood flow can be only toward the heart. Consequently, every time a person moves the legs or even tenses the muscles, a certain amount of blood is propelled toward the heart and the pressure in the veins is lowered. This pumping system is known as the "venous pump" or "muscle pump," and it is efficient enough that under ordinary circumstances, the venous pressure in the feet of a walking adult remains close to or less than 25 mm Hg.

If a human being stands perfectly still, the venous pump does not work, and the venous pressures in the lower part of the leg will rise to the full hydrostatic

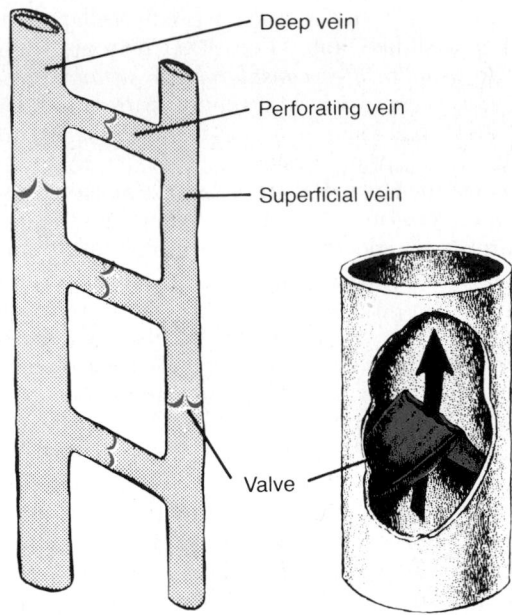

Figure 15–11. Venous valves of the leg.

value of 90 mm Hg in about 30 seconds. The pressures in the capillaries also increase greatly, causing fluid to leak from the circulatory system into the tissue spaces. As a result, the legs swell, and the blood volume diminishes. Indeed, 10 to 20 percent of the blood volume can be lost from the circulatory system within the first 15 minutes of standing absolutely still, as often occurs when a soldier is made to stand at attention.

VENOUS VALVE INCOMPETENCE AND VARICOSE VEINS. The valves of the venous system frequently become "incompetent" or sometimes are even destroyed. This is especially true when the veins have been overstretched by excess venous pressure lasting weeks or months, as occurs in pregnancy or when one stands most of the time. Stretching the veins increases their cross-sectional areas, but the leaflets of the valves do not increase in size. Therefore, the leaflets of the valves no longer close completely. When this develops, the pressure in the veins of the legs increases still more owing to failure of the venous pump; this further increases the sizes of the veins and finally destroys the function of the valves entirely. Thus, the person develops "varicose veins," which are characterized by large, bulbous protrusions of the veins beneath the skin of the entire leg and particularly the lower leg. The venous and capillary pressures become very high, and leakage of fluid from the capillaries causes constant edema in the legs whenever these people stand for more than a few minutes. The edema in turn prevents adequate diffusion of nutritional materials from the capillaries to the muscle and skin cells, so that the muscles become painful and weak and the skin frequently becomes gangrenous and ulcerates. The best treatment for such a condition is continual elevation of the legs to a level at least as high as the heart, but

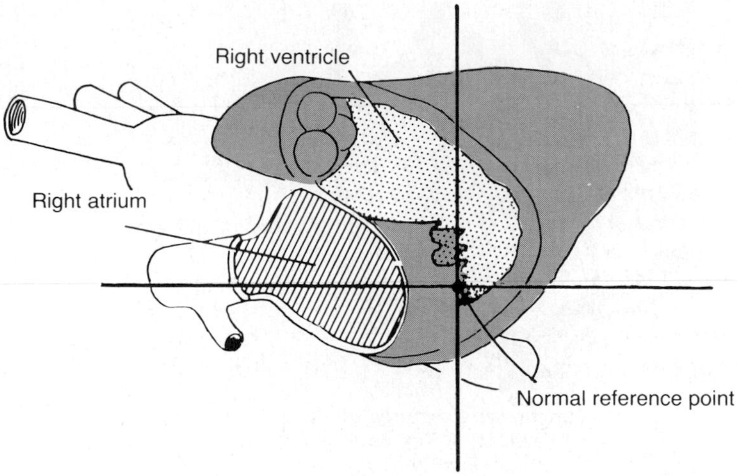

Figure 15–12. Location of the reference point for pressure measurement at the tricuspid valve.

tight binders on the legs are also of considerable aid in preventing the edema and its sequelae.

CLINICAL ESTIMATION OF VENOUS PRESSURE. The venous pressure can often be estimated by simply observing the degree of distention of the peripheral veins—especially the neck veins. For instance, in the sitting position, the neck veins are never distended in the normal quietly resting person. However, when the right atrial pressure becomes increased to as much as 10 mm Hg, the lower veins of the neck begin to protrude, and at 15 mm Hg, essentially all the veins in the neck become distended.

Direct Measurement of Venous Pressure and Right Atrial Pressure. Venous pressure can also be measured with ease by inserting a syringe needle directly into a vein and connecting it to a pressure recorder.

The only means by which *right atrial pressure* can be measured accurately is by inserting a catheter through the veins into the right atrium. Pressures measured through such central venous catheters are used almost routinely in some hospitalized cardiac patients to provide constant assessment of heart pumping ability.

Pressure Reference Level for Measuring Venous and Other Circulatory Pressures

In discussions up to this point, we have often spoken of right atrial pressure as being 0 mm Hg and arterial pressure as being 100 mm Hg, but we have not stated the hydrostatic level in the circulatory system to which this pressure is referred. There is one point in the circulatory system at which hydrostatic pressure factors caused by changes in body position of a normal person usually do not affect the pressure measurement by more than 1 to 2 mm Hg. This is at the level of the tricuspid valve, as shown by the crossed axes in Figure 15–12. Therefore, all pressure measurements discussed in this text are referred to this level, which is called the *reference level for pressure measurement.*

The reason for lack of hydrostatic effects at the tricuspid valve is that the heart automatically prevents significant hydrostatic changes in pressure at this point in the following way.

If the pressure at the tricuspid valve rises slightly above normal, the right ventricle fills to a greater extent than usual, causing the heart to pump blood more rapidly and therefore to decrease the pressure at the tricuspid valve back toward the normal mean value. On the other hand, if the pressure falls, the right ventricle fails to fill adequately, its pumping decreases, and blood dams up in the venous system until the tricuspid pressure again rises to a normal value. In other words, *the heart acts as a feedback regulator of pressure* at the tricuspid valve.

When a human is lying on his or her back, the tricuspid valve is located almost exactly 60 per cent of the chest thickness in front of the back. Therefore, this is the *zero pressure reference level.*

Blood Reservoir Function of the Veins

As pointed out in Chapter 14, more than 60 per cent of all the blood in the circulatory system is usually in the veins. For this reason and because the veins are so compliant, it is said that the venous system serves as a *blood reservoir* for the circulation.

When blood is lost from the body and the arterial pressure begins to fall, pressure reflexes are elicited from the carotid sinuses and other pressure-sensitive areas of the circulation, as discussed in Chapter 18; these in turn send sympathetic nerve signals to the veins, causing them to constrict, and this takes up much of the slack in the circulatory system caused by the lost blood. Indeed, even after as much as 20 per cent of the total blood volume has been lost, the circulatory system often functions almost normally because of this variable reservoir system of the veins.

SPECIFIC BLOOD RESERVOIRS. Certain portions of the circulatory system are so extensive and so compliant that they are called specific "blood reservoirs." These include (1) the *spleen,* which can sometimes decrease in size sufficiently to release as much as 100 milliliters of blood into other areas of the circulation; (2) the *liver,* the sinuses of which can release several hundred milliliters of blood into the remainder of the circulation; (3) the *large abdominal veins,* which

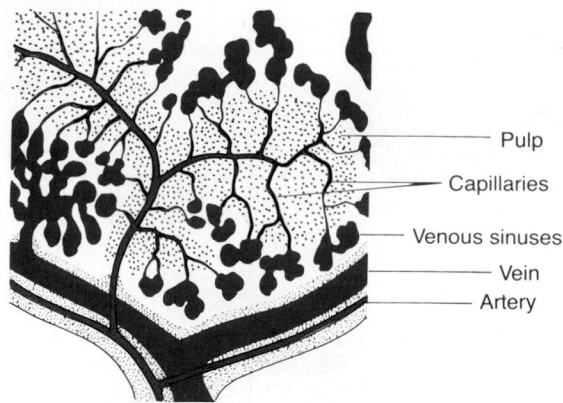

Figure 15-13. Functional structures of the spleen. (Modified from Bloom and Fawcett: A Textbook of Histology. 10th Ed. Philadelphia, W. B. Saunders Company, 1975.)

can contribute as much as 300 milliliters; and (4) the *venous plexus beneath the skin,* which can also contribute several hundred milliliters. The *heart* and the *lungs,* although not parts of the systemic venous reservoir system, must also be considered blood reservoirs. The heart, for instance, becomes reduced in size during sympathetic stimulation and in this way can contribute some 50 to 100 milliliters of blood, and the lungs can contribute another 100 to 200 milliliters when the pulmonary pressures fall to low values.

The Spleen as a Reservoir for Storing Red Blood Cells. Figure 15–13 shows that the spleen has two separate areas for storing blood: the *venous sinuses* and the *pulp.* Small vessels flow directly into the venous sinuses, and the sinuses can swell the same as any other part of the venous system and store whole blood.

In the splenic pulp, the capillaries are so permeable that whole blood oozes through the capillary walls into a trabecular mesh forming the *red pulp.* The red cells are trapped by the trabeculae, whereas the plasma returns into the venous sinuses and then into the general circulation. As a consequence, the red pulp of the splenic pulp is a *special reservoir that contains large quantities of extra red blood cells* that are expelled into the general circulation when the sympathetic nervous system is excited and contracts the spleen or its vessels. In lower animals, this extra storage of red blood cells is much greater than in humans, but even in the human, possibly as much as 50 milliliters of concentrated red blood cells can be released into the circulation, raising the hematocrit 1 to 2 per cent.

In other areas of the splenic pulp are islands of white blood cells, which collectively are called the *white pulp.* Here lymphoid cells are manufactured similar to those manufactured in the lymph nodes. They are part of the body's immune system, which is described in Chapter 34.

Blood-Cleansing Function of the Spleen— Removal of Old Cells. Blood passing through the splenic pulp before it enters the sinuses undergoes thorough squeezing. Therefore, it is to be expected that fragile red blood cells would not withstand the trauma. For this reason, many of the red blood cells destroyed in the body have their final demise in the spleen. After the cells rupture, the released hemoglobin and the cell stroma are ingested by the reticuloendothelial cells of the spleen.

Reticuloendothelial Cells of the Spleen. The pulp of the spleen contains many large phagocytic reticuloendothelial cells, and the venous sinuses are lined with similar cells. These cells act as a cleansing system for the blood, acting in concert with a similar system in the venous sinuses of the liver. When the blood is invaded by infectious agents, the reticuloendothelial cells of the spleen rapidly remove debris, bacteria, parasites, and so forth. Also, in many infectious processes, the spleen enlarges in the same manner that lymph glands enlarge and then performs its cleansing function even more adequately.

REFERENCES

Antonaccio, M. J.: Cardiovascular Pharmacology. New York, Raven Press, 1990.

Aukland, K., and Reed, R. K.: Interstitial-lymphatic mechanisms in the control of extracellular fluid volume. Physiol. Rev., 73:1–78, 1993.

Braunwald, E.: Heart Disease: A Textbook of Cardiovascular Medicine. Philadelphia, W. B. Saunders Co., 1992.

Burg, F. D., et al.: Basic Sciences in Medicine: A Concept-based Approach. Philadelphia, J. B. Lippincott, 1995.

Chien, S.: Red cell deformability and its relevance to blood flow. Annu. Rev. Physiol., 49:177, 1987.

Chien, S., et al.: Blood flow in small tubes. In Renkin, E. M., and Michel, C. C. (eds.).: Handbook of Physiology. Sec. 2, Vol. IV. Bethesda, Md., American Physiological Society, 1984, p. 217.

Dean, R. H., et al.: Current Diagnosis and Treatment in Vascular Surgery. Redding, Mass., Appleton & Lange, 1995.

Dobrin, P. B.: Vascular mechanics. In Shepard, J. T., and Abboud, F. M. (eds.): Handbook of Physiology. Sec. 2, Vol. III. Bethesda, Md., American Physiological Society, 1983, p. 65.

Donald, D. E.: Splanchnic circulation. In Shepherd, J. T., and Abboud, F. M. (eds.): Handbook of Physiology. Sec. 2, Vol. III. Bethesda, Md., American Physiological Society, 1983, p. 219.

Fozzard, H. A., et al.: The Heart and Cardiovascular System: Scientific Foundations. New York, Raven Press, 1986.

Friedewald, V. E., Jr., and Crossen, C.: Vascular Anastomosis. New York, Scientific American Science and Medicine, September/October, 1994.

Fronek, A. (ed.): Noninvasive Diagnostics in Vascular Disease. New York, McGraw-Hill Book Co., 1989.

Goerke, J., and Mines, A. H.: Cardiovascular Physiology. New York, Raven Press, 1988.

Goodman, A. H., et al.: A television method for measuring capillary red cell velocities. J. Appl. Physiol., 37:126, 1974.

Gow, B. S.: Circulation correlates: Vascular impedance, resistance, and capacity. In Bohr, D. F., et al. (eds.): Handbook of Physiology. Sec. 2, Vol. II. Baltimore, Williams & Wilkins, 1980, p. 353.

Green, H. D.: Circulation: Physical principles. In Glasser, O. (ed.): Medical Physics. Chicago, Year Book Medical Publishers, 1944.

Gross, J. F., and Popel, A. (eds): Mathematics of Microcirculation Phenomena. New York, Raven Press, 1980.

Guyton, A. C.: Arterial Pressure and Hypertension. Philadelphia, W. B. Saunders Co., 1980.

Guyton, A. C.: Peripheral circulation. Annu. Rev. Physiol., 21:239, 1959.

Guyton, A. C.: The venous system and its role in the circulation. Mod. Conc. Cardiov. Dis., 17:483, 1958.

Guyton, A. C., and Greganti, F. P.: A physiologic reference point for measuring circulation pressures in the dog—particularly venous pressure. Am. J. Physiol., 185:137, 1956.

Guyton, A. C., and Jones, C. E.: Central venous pressure: Physiological significance and clinical implications. Am. Heart J., 86:431, 1973.

Guyton, A. C., et al.: Cardiac Output and Its Regulation. Philadelphia, W. B. Saunders Co., 1973.

Guyton, A. C., et al.: Evidence for tissue oxygen demand as the major factor causing autoregulation. Circ. Res., 14:60, 1964.

Guyton, A. C., et al.: Pressure-volume curves of the entire arterial and venous systems in the living animal. Am. J. Physiol., 184:253, 1956.

Guyton, J. R.: Mechanical control of smooth muscle growth. In Seidel, C. L., and Weisbrodt, N. W.: Hypertrophic Response in Smooth Muscle. Boca Raton, Fla., CRC Press, 1987, p. 121.

Hirakawa, S., et al.: The role of alpha and beta adrenergic receptors in constriction and dilation of the systemic capacitance vessels: A study with measurements of the mean circulatory pressure in dogs. Jpn. Circ. J., 48:620, 1984.

Hwang, N. H. C., et al.: Advances in Cardiovascular Engineering. New York, Plenum Publishing Corp., 1992.

Katz, A. M.: Physiology of the Heart. New York, Raven Press, 1992.

Krupski, W. C.: Review of Vascular Surgery. Philadelphia, W. B. Saunders Co., 1994.

Lassen, N. A., et al.: Indicator methods for measurement of organ and tissue blood flow. In Shepherd, J. T., and Abboud, F. M. (eds.): Handbook of Physiology. Sec. 2, Vol. III. Bethesda, Md., American Physiological Society, 1983, p. 21.

Mellander, S.: Systemic circulation: Local control. Annu. Rev. Physiol., 32:313, 1970.

Nichols, W. W., and O'Rourke, M. F.: McDonald's Blood Flow in Arteries: Theoretic, Experimental and Clinical Principles. Baltimore, Williams & Wilkins, 1990.

Piene, H.: Pulmonary arterial impedance and right ventricular function. Physiol. Rev., 66:606, 1986.

Rothe, C. F.: Reflex control of veins and vascular capacitance. Physiol. Rev., 63:1281, 1983.

Rothe, C. F.: Venous system: Physiology of the capacitance vessels. In Shepherd, J. T., and Abboud, F. M. (eds.): Handbook of Physiology. Sec. 2, Vol. III. Bethesda, Md., American Physiological Society, 1983, p. 397.

Rowan: Physics and the Circulation. Philadelphia, Heyden & Sons, Inc., 1981.

Ruderman, N., et al.: Hyperglycemia, Diabetes and Vascular Disease. New York, Oxford University Press, 1992.

Schneck, D. J., and Vawter, D. L. (eds.): Biofluid Mechanics. New York, Plenum Press, 1980.

Smith, J. J., and Kampine, J. P.: Circulatory Physiology—The Essentials. Baltimore, Williams & Wilkins, 1990.

Strandness, D. E., and van Breda A.: Vascular diseases: Surgical and interventional therapy. New York, Churchill Livingstone, 1994.

Vanhoutte, P. M.: Vasodilation: Vascular Smooth Muscle, Peptides, Autonomic Nerves, and Endothelium. New York, Raven Press, 1988.

Vanhoutte, P. M., et al.: Local modulation of the adrenergic neuroeffector interaction in the blood vessel wall. Physiol. Rev., 61:151, 1981.

Veith, F. J., et al.: Vascular Surgery: Principles and Practice. Hightstown, NJ, McGraw-Hill, 1994.

Wiedeman, M. P.: Dimensions of blood vessels from distributing artery to collecting vein. Circ. Res., 12:375, 1963.

Willerson, J. T., and Cohn, J. N.: Cardiovascular Medicine. New York, Churchill Livingstone, 1994.

Ziegler, M. G.: Postural hypotension. Annu. Rev. Med., 31:239, 1980.

The Microcirculation and the Lymphatic System: Capillary Fluid Exchange, Interstitial Fluid, and Lymph Flow

CHAPTER 16

In the microcirculation, the most purposeful function of the circulation occurs: transport of nutrients to the tissues and removal of cellular excreta. The small arterioles control the blood flow to each tissue area, and local conditions in the tissues themselves control the diameters of the arterioles in turn. Thus, each tissue, in most instances, controls its own blood flow in relation to its needs, a subject that is discussed in detail in Chapter 17.

The capillaries are extremely thin structures with walls of a single layer of highly permeable endothelial cells. Here interchange of nutrients and cellular excreta occurs between the tissues and the circulating blood. About 10 billion capillaries with a total surface area estimated to be 500 to 700 square meters (about the surface area of a football field) provide this function for the whole body. Indeed, it is rare that any single functional cell of the body is more than 20 to 30 micrometers away from a capillary.

The purpose of this chapter is to discuss the transfer of substances between the blood and interstitial fluid and especially to discuss the factors that affect the transfer of fluid volume through the capillary walls between the circulating blood and the interstitial fluid.

Structure of the Microcirculation and Capillary System

The microcirculation of each organ is specifically organized to serve that organ's special needs. In general, each nutrient artery entering an organ branches six to eight times before the arteries become small enough to be called arterioles, which generally have internal diameters less than 20 micrometers. Then the arterioles themselves branch two to five times, reaching diameters of 5 to 9 micrometers at their ends where they supply blood to the capillaries.

Figure 16–1 shows the structure of a representative capillary bed as seen in the mesentery, showing that blood enters the capillaries through an *arteriole* and leaves by way of a *venule*. Blood from the arteriole passes into a series of *metarterioles,* which are called by some physiologists *terminal arterioles* and which have a structure midway between that of arterioles and capillaries. After leaving the metarteriole, the blood enters the *capillaries,* some of which are large and are called *preferential channels* and others of which are small and are *true capillaries*. After passing through the capillaries, the blood enters the venule and returns to the general circulation.

The arterioles are highly muscular, and their diameters can change manyfold. The metarterioles (the terminal arterioles) do not have a continuous muscular coat, but smooth muscle fibers encircle the vessel at intermittent points, as shown in Figure 16–1 by the large black dots to the sides of the metarteriole.

At the point where the true capillaries originate from the metarterioles, a smooth muscle fiber usually encircles the capillary. This is called the *precapillary sphincter*. This sphincter can open and close the entrance to the capillary.

The venules are considerably larger than the arterioles and have a much weaker muscular coat. Yet it must be remembered that the pressure in the venules is much less than that in the arterioles, so that the venules can still contract considerably despite the weak muscle.

This typical arrangement of the capillary bed is not found in all parts of the body; however, some similar arrangement serves the same purposes. Most important, the metarterioles (and the precapillary sphincters when they also exist) are in close contact with the tissues they

183

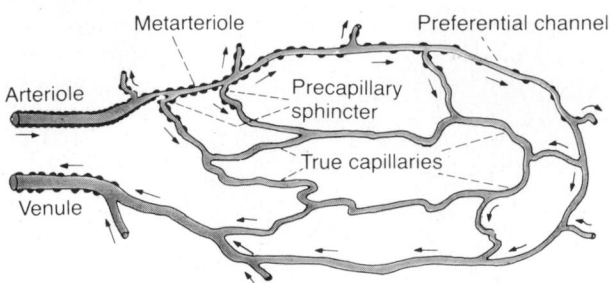

Figure 16–1. Structure of the mesenteric capillary bed. (From Zweifach: Factors Regulating Blood Pressure. New York, Josiah Macy, Jr., Foundation, 1950.)

serve. Therefore, the local conditions of the tissues—the concentrations of nutrients, end products of metabolism, hydrogen ions, and so forth—can cause direct effects on these in controlling the local blood flow in each minute tissue area.

STRUCTURE OF THE CAPILLARY WALL. Figure 16–2 shows the ultramicroscopic structure of a typical capillary wall as found in most organs of the body, especially in the muscles and connective tissue. Note that the wall is composed of a unicellular layer of endothelial cells and surrounded by a basement membrane on the outside. The total thickness of the wall is about 0.5 micrometer.

The diameter of the capillary is 4 to 9 micrometers, barely large enough for red blood cells and other blood cells to squeeze through.

"PORES" IN THE CAPILLARY MEMBRANE. Studying Figure 16–2, one sees two minute passageways connecting the interior of the capillary with the exterior. One of these is the *intercellular cleft,* which is the thin slit that lies between adjacent endothelial cells. Each of these clefts is interrupted periodically by short ridges of protein attachments that hold the endothelial cells together, but each ridge in turn is broken after a short distance, so that in between them fluid can percolate freely through the cleft. The cleft normally has a uniform spacing with a width of about 6 to 7 nanometers (60 to 70 angstroms), slightly smaller than the diameter of an albumin protein molecule.

Because the intercellular clefts are located only at the edges of the endothelial cells, they usually represent no more than 1/1000 of the total surface area of the capillary. Nevertheless, the rate of thermal motion of water molecules as well as most other water-soluble ions and small solutes is so rapid that all of these diffuse with ease between the interior and exterior of the capillaries through these "slit-pores," the intercellular clefts.

Also present in the endothelial cells are many minute *plasmalemmal vesicles.* These form at one surface of the cell by imbibing small packets of plasma or extracellular fluid. They can then move slowly through the endothelial cell. It has also been postulated that some of these vesicles coalesce to form vesicular channels all the way through the membrane, which is also

demonstrated in Figure 16–2. However, careful measurements in laboratory animals have probably proved that these vesicular forms of transport are quantitatively of little importance.

SPECIAL TYPES OF "PORES" OCCUR IN THE CAPILLARIES OF CERTAIN ORGANS. The "pores" in the capillaries of some organs have special characteristics to meet the peculiar needs of the organs. Some of these characteristics are as follows:

1. In the *brain,* the junctions between the capillary endothelial cells are mainly "tight" junctions that allow only very small molecules to pass into the brain tissues. This is called the *blood-brain barrier* and is discussed further in Chapter 61.

2. In the *liver,* the opposite is true. The clefts between the capillary endothelial cells are wide open, so that almost all dissolved substances of the plasma, including the plasma proteins, can pass from the blood into the liver tissues. The pores of the intestinal membranes are midway between those of the muscles and those of the liver.

3. In the *glomerular tufts of the kidney,* numerous small oval windows called *fenestrae* penetrate directly through the middle of the endothelial cells, so that tremendous amounts of substances can be filtered through the glomeruli without having to pass through the clefts between the endothelial cells.

FLOW OF BLOOD IN THE CAPILLARIES—VASOMOTION

Blood usually does not flow continuously through the capillaries. Instead, it flows intermittently, turning on and off every few seconds or minutes. The cause of this intermittency is the phenomenon called *vasomotion,* which means intermittent contraction of the metarterioles and precapillary sphincters.

REGULATION OF VASOMOTION. The most important factor found thus far to affect the degree of opening and closing of the metarterioles and precapillary

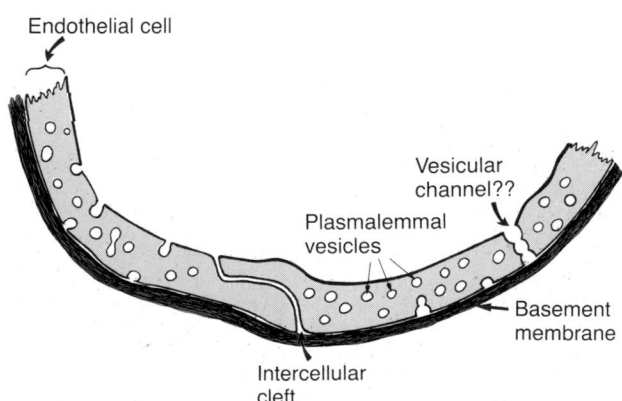

Figure 16–2. Structure of the capillary wall. Note especially the *intercellular cleft* at the junction between adjacent endothelial cells; it is believed that most water-soluble substances diffuse through the capillary membrane along this cleft.

sphincters is the concentration of *oxygen* in the tissues. When the rate of oxygen usage is great, the intermittent periods of blood flow occur more often and the duration of each period of flow lasts longer, thereby allowing the blood to carry increased quantities of oxygen (as well as other nutrients) to the tissues. This effect, along with multiple other factors that control local tissue blood flow, is discussed in Chapter 17.

Average Function of the Capillary System

Despite the fact that blood flow through each capillary is intermittent, so many capillaries are present in the tissues that their overall function becomes averaged. That is, there is an *average rate of blood flow* through each tissue capillary bed, an *average capillary pressure* within the capillaries, and an *average rate of transfer of substances* between the blood of the capillaries and the surrounding interstitial fluid. In the remainder of this chapter, we are concerned with these averages, although one must remember that the average functions are, in reality, the functions of literally billions of individual capillaries, each operating intermittently in response to the local conditions in the tissues.

EXCHANGE OF NUTRIENTS AND OTHER SUBSTANCES BETWEEN THE BLOOD AND INTERSTITIAL FLUID

Diffusion Through the Capillary Membrane

By far the most important means by which substances are transferred between the plasma and the interstitial fluid is by *diffusion*. Figure 16–3 demonstrates this process, showing that as the blood traverses the capillary, tremendous numbers of water molecules and dissolved particles diffuse back and forth through the capillary wall, providing continual mixing between the interstitial fluid and the plasma. Diffusion results from thermal motion of the water molecules and the dissolved substances in the fluid, the different particles moving first in one direction and then another, moving randomly in every direction.

Lipid-Soluble Substances Can Diffuse Directly Through the Cell Walls of the Capillary Endothelium. If a substance is lipid-soluble, it can diffuse directly through the cell membranes of the capillary without having to go through the pores. Such substances include especially oxygen and carbon dioxide. Because these substances can permeate all areas of the capillary membrane, their rates of transport through the capillary membrane are many times the rates for most lipid-insoluble substances, such as sodium ions and glucose.

Water-Soluble Substances Diffuse Only Through Intercellular "Pores" in the Capillary Membrane. Many substances needed by the tissues are soluble in water but cannot pass through the lipid membranes of the endothelial cells; such substances include water molecules themselves, sodium ions, chloride ions, and glucose. Despite the fact that not more than $\frac{1}{1000}$ of the surface area of the capillaries is represented by the intercellular clefts between the endothelial cells, the velocity of thermal molecular motion in the clefts is so great that even this small area is sufficient to allow tremendous diffusion of water and water-soluble substances through these cleft-pores. To give one an idea of the rapidity with which these substances diffuse, *the rate at which water molecules diffuse through the capillary membrane is about 80 times as great as the rate at which plasma itself flows linearly along the capillary.* That is, the water of the plasma is exchanged with the water of the interstitial fluid 80 times before the plasma can go the entire distance through the capillary.

Effect of Molecular Size on Passage Through the Pores. The width of the capillary intercellular cleft-pores, 6 to 7 nanometers, is about 20 times the diameter of the water molecule, which is the smallest molecule that normally passes through the capillary pores. On the other hand, the diameters of plasma protein molecules are slightly greater than the width of the pores. Other substances, such as sodium ions, chloride ions, glucose, and urea, have intermediate diameters. Therefore, the permeability of the capillary pores for different substances varies according to their molecular diameters.

Table 16–1 gives the relative permeabilities of the capillary pores in muscle for substances commonly encountered by the capillary membrane, demonstrating, for instance, that the permeability for glucose molecules is 0.6 times that for water molecules, whereas the permeability for albumin molecules is very, very slight.

A word of caution must be issued at this point. The capillaries in different tissues have extreme differences

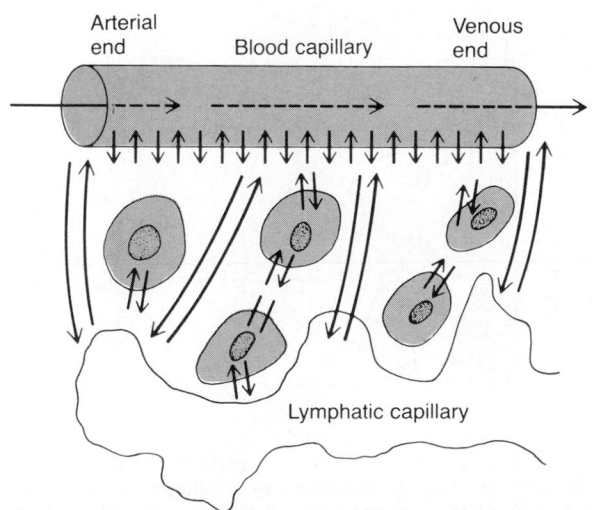

Figure 16–3. Diffusion of fluid molecules and dissolved substances between the capillary and interstitial fluid spaces.

Table 16–1 RELATIVE PERMEABILITY OF MUSCLE CAPILLARY PORES TO DIFFERENT-SIZED MOLECULES

Substance	Molecular Weight	Permeability
Water	18	1.00
NaCl	58.5	0.96
Urea	60	0.8
Glucose	180	0.6
Sucrose	342	0.4
Inulin	5000	0.2
Myoglobin	17,600	0.03
Hemoglobin	68,000	0.01
Albumin	69,000	.001

Modified from Pappenheimer, *Physiol. Rev., 33:*387, 1953.

in their permeabilities. For instance, the membrane of the liver capillary sinusoids is so permeable that even plasma proteins pass freely through these walls almost as easily as water and other substances. Also, the permeability of the renal glomerular membrane for water and electrolytes is about 500 times the permeability of the muscle capillaries, but the glomerular and muscle permeabilities for protein are about the same. When we study these different organs later in this text, it should become clear why some tissues require greater degrees of capillary permeability than others—the liver, for instance, to transfer tremendous amounts of nutrients between the blood and the liver parenchymal cells and the kidneys to allow filtration of large quantities of fluid for the formation of urine.

EFFECT OF CONCENTRATION DIFFERENCE ON NET RATE OF DIFFUSION THROUGH THE CAPILLARY MEMBRANE. The "net" rate of diffusion of a substance through any membrane is proportional to the *concentration difference* between the two sides of the membrane. That is, the greater the difference between the concentrations of any given substance on the two sides of the capillary membrane, the greater will be the net movement of the substance in one direction through the membrane. Thus, the concentration of oxygen in the blood is normally greater than that in the interstitial fluid. Therefore, large quantities of oxygen normally move from the blood toward the tissues. Conversely, the concentration of carbon dioxide is greater in the tissues than in the blood, which causes carbon dioxide to move into the blood and to be carried away from the tissues.

The rates of diffusion through the capillary membranes of most nutritionally important substances are so great that only slight concentration differences suffice to cause more than adequate transport between the plasma and interstitial fluid. For instance, the concentration of oxygen in the interstitial fluid immediately outside the capillary is probably no more than 1 per cent less than the concentration in the plasma of the blood, and yet this 1 per cent difference causes enough oxygen to move from the blood into the interstitial spaces to provide all the oxygen required for tissue metabolism.

THE INTERSTITIUM AND INTERSTITIAL FLUID

About one sixth of the body consists of spaces between cells, which collectively are called the *interstitium.* The fluid in these spaces is the *interstitial fluid.*

The structure of the interstitium is shown in Figure 16–4. It has two major types of solid structures: (1) collagen fiber bundles and (2) proteoglycan filaments. The collagen fiber bundles extend long distances in the interstitium. They are extremely strong and therefore provide most of the tensional strength of the tissues. The proteoglycan filaments, on the other hand, are extremely thin, coiled molecules composed of about 98 per cent hyaluronic acid and 2 per cent protein. These molecules are so thin that they can never be seen with a light microscope and are difficult to demonstrate even with the electron microscope. Nevertheless, they form a mat of very fine reticular filaments aptly described as a "brush pile."

"GEL" IN THE INTERSTITIUM. The fluid in the interstitium is derived by filtration and diffusion from the capillaries. It contains almost the same constituents as plasma except for much lower concentrations of proteins because proteins do not pass outward through the walls of the capillaries with ease. The interstitial fluid is mainly entrapped in the minute spaces among the proteoglycan filaments. This combination of the proteoglycan filaments and the fluid entrapped within them has the characteristics of a *gel* and therefore is called the *tissue gel.*

Because of the large number of proteoglycan filaments, it is difficult for fluid to *flow* easily through the tissue gel. Instead, it mainly *diffuses* through the gel; that is, it moves molecule by molecule from one place to another by kinetic motion rather than by large numbers of molecules moving together.

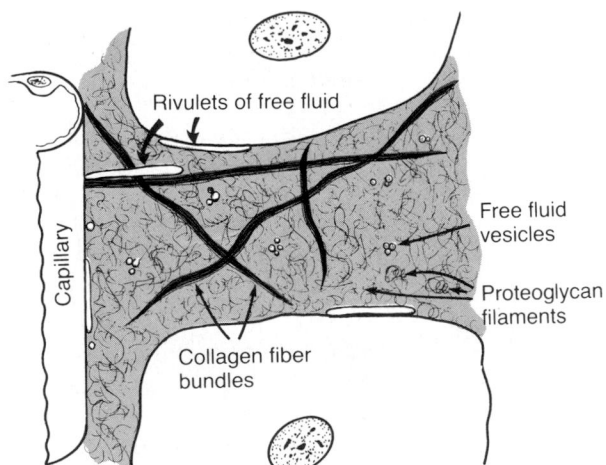

Figure 16–4. Structure of the interstitium. Proteoglycan filaments fill the spaces between the collagen fiber bundles. Free fluid vesicles are seen and small amounts of free fluid in the form of rivulets.

Diffusion through the gel occurs about 95 to 99 per cent as rapidly as it does through free fluid. For the short distances between the capillaries and the tissue cells, this diffusion allows rapid transport through the interstitium not only of water molecules but also electrolytes, nutrients, cellular excreta, oxygen, carbon dioxide, and so forth.

"FREE" FLUID IN THE INTERSTITIUM. Although almost all the fluid in the interstitium is normally entrapped within the tissue gel, occasionally small *rivulets of "free" fluid* and small *free fluid vesicles* are also present, which means fluid that is free of the proteoglycan molecules and therefore can flow freely. When a dye is injected into the circulating blood, it can often be seen to flow through the interstitium in the small rivulets, usually coursing along the surfaces of collagen fibers or surfaces of cells. The amount of "free" fluid present in *normal* tissues is slight, usually much less than 1 per cent. On the other hand, when the tissues develop *edema, these small pockets and rivulets of free fluid expand tremendously* until one half or more of the fluid becomes freely flowing fluid independent of the proteoglycan filaments.

THE PROTEINS IN THE PLASMA AND INTERSTITIAL FLUID MAINLY DETERMINE THE PLASMA AND INTERSTITIAL FLUID VOLUMES

The pressure in the capillaries tends to force fluid and its dissolved substances through the capillary pores into the interstitial spaces. In contrast, osmotic pressure caused by the plasma proteins (called *colloid* osmotic pressure) tends to cause fluid movement by osmosis from the interstitial spaces into the blood; this osmotic pressure prevents significant loss of fluid volume from the blood into the interstitial spaces. Also important is the lymphatic system, which returns back to the circulation the small amounts of protein that do leak into the interstitial spaces. In the remainder of this chapter, we discuss how these effects control the respective volumes of the plasma and the interstitial fluid.

FOUR PRIMARY FORCES THAT DETERMINE FLUID MOVEMENT THROUGH THE CAPILLARY MEMBRANE. Figure 16–5 shows the four primary forces that determine whether fluid will move out of the blood into the interstitial fluid or in the opposite direction; called "Starling forces" in honor of the physiologist who first demonstrated their importance, they are:

1. The *capillary pressure* (Pc), which tends to force fluid outward through the capillary membrane.
2. The *interstitial fluid pressure* (Pif), which tends to force fluid inward through the capillary membrane when Pif is positive but outward when Pif is negative.
3. The *plasma colloid osmotic pressure* (Πp), which tends to cause osmosis of fluid inward through the capillary membrane.

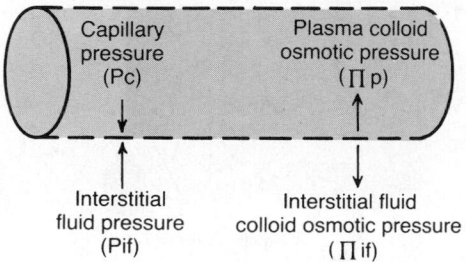

Figure 16–5. Forces operative at the capillary membrane tending to move fluid either outward or inward through the membrane pores.

4. The *interstitial fluid colloid osmotic pressure* (Πif), which tends to cause osmosis of fluid outward through the capillary membrane.

Let us discuss each of these in detail.

Capillary Pressure

Two experimental methods have been used to estimate the capillary pressure: (1) *direct cannulation of the capillaries,* which has given an average mean capillary pressure of about 25 mm Hg, and (2) *indirect functional measurement of the capillary pressure,* which has given a capillary pressure averaging about 17 mm Hg.

MICROPIPETTE METHOD FOR MEASURING CAPILLARY PRESSURE. To measure pressure in a capillary by cannulation, a microscopic glass pipette is thrust directly into the capillary, and the pressure is measured by an appropriate micromanometer system. Using this method, capillary pressures have been measured in capillaries of exposed tissues of lower animals and in large capillary loops of the eponychium at the base of the fingernail in humans. These measurements have given pressures of 30 to 40 mm Hg in the arterial ends of the capillaries, 10 to 15 mm Hg in the venous ends, and about 25 mm Hg in the middle.

ISOGRAVIMETRIC METHOD FOR INDIRECTLY MEASURING "FUNCTIONAL" CAPILLARY PRESSURE. Figure 16–6 demonstrates an *isogravimetric* method for indirectly estimating capillary pressure. This figure shows a section of gut held by one arm of a gravimetric balance. Blood is perfused through the gut. When the arterial pressure is decreased, the resulting decrease in capillary pressure allows the osmotic pressure of the plasma proteins to cause absorption of fluid out of the gut wall and makes the weight of the gut decrease. This immediately causes displacement of the balance arm. To prevent this weight decrease, the venous pressure is raised an amount sufficient to overcome the effect of decreasing the arterial pressure. In other words, the capillary pressure is kept constant while decreasing the arterial pressure but raising the venous pressure.

In the lower part of the figure, the changes in arterial and venous pressures that exactly nullify all weight

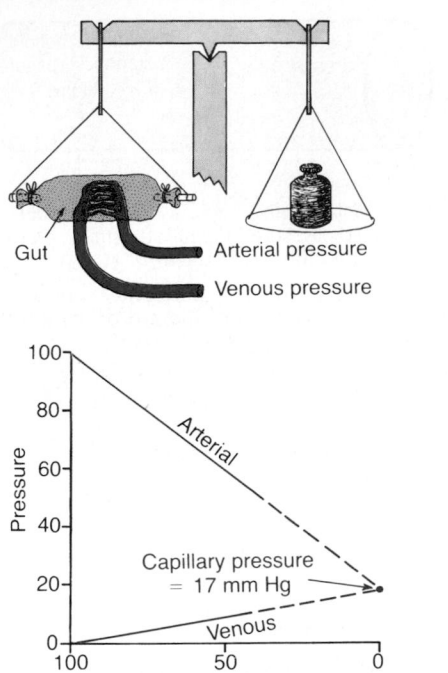

Figure 16-6. Isogravimetric method for measuring capillary pressure.

changes are shown. The arterial and venous lines meet each other at a value of 17 mm Hg. Therefore, the capillary pressure must have remained at this same level of 17 mm Hg throughout these maneuvers, or otherwise filtration or absorption of fluid through the capillary walls would have occurred. Thus, in a roundabout way, the "functional" capillary pressure is measured to be about 17 mm Hg.

WHY IS THE FUNCTIONAL CAPILLARY PRESSURE MUCH LOWER THAN CAPILLARY PRESSURE MEASURED BY THE MICROPIPETTE METHOD? It is clear that the aforementioned two methods do not give the same capillary pressure. However, the isogravimetric method determines the capillary pressure that exactly balances all the other forces tending to move fluid into or out of the capillaries. Because such a balance of forces is the normal state, the average functional capillary pressure must be close to the pressure measured by the isogravimetric method. Therefore, one is justified in believing that the true functional capillary pressure averages about 17 mm Hg.

It is easy to explain why the cannulation methods give higher pressure values. The most important reason is that these measurements are usually made in capillaries whose arterial ends are open and blood is actively flowing into the capillary. However, it should be recalled from the earlier discussion of capillary vasomotion that the metarterioles and precapillary sphincters are normally closed during a large part of the vasomotion cycle. When closed, the pressure in the

capillaries beyond the closures should be almost equal to the pressure at the venous ends of the capillaries, about 10 mm Hg. Therefore, when averaged over a period of time, one would expect the *functional* mean capillary pressure to be much nearer to the pressure in the venous ends of the capillaries than to the pressure in the arterial ends.

There are two other reasons why the functional capillary pressure is less than the values measured by cannulation. One of these is that there are far more venous capillaries than arterial capillaries. Second, the venous capillaries are several times as permeable as the arterial capillaries. Both of these effects further weight the functional capillary pressure to a lower value.

Interstitial Fluid Pressure

As is true for the measurement of capillary pressure, there are several methods for measuring interstitial fluid pressure, and each of these gives slightly different values but usually values that are a few millimeters of mercury less than atmospheric pressure, that is, values called *negative interstitial fluid pressure*. The methods most widely used have been (1) direct cannulation of the tissues with a micropipette, (2) measurement of the pressure from implanted perforated capsules, and (3) measurement of the pressure from a cotton wick inserted into the tissue.

MEASUREMENT OF INTERSTITIAL FLUID PRESSURE USING THE MICROPIPETTE. The same type of micropipette used for measuring capillary pressure can also be used in some tissues for measuring interstitial fluid pressure. The tip of the micropipette is about 1 micrometer in diameter, but even this is 20 or more times larger than the sizes of the spaces between the proteoglycan filaments of the interstitium. Therefore, the pressure that is measured is probably the pressure in a free fluid pocket.

The first pressures measured using the micropipette method ranged from − 1 to + 2 mm Hg but were usually slightly positive. With experience in making such measurements, the more recent pressures have averaged about − 2 mm Hg, giving average pressure values in *loose* tissues that are slightly less than atmospheric pressure.

MEASUREMENT OF INTERSTITIAL FREE FLUID PRESSURE IN IMPLANTED PERFORATED HOLLOW CAPSULES. Figure 16–7 shows an indirect method for measuring interstitial fluid pressure that may be explained as follows: A small hollow plastic capsule perforated by up to a hundred small holes is implanted in the tissues, and the surgical wound is allowed to heal for about 1 month. At the end of that time, tissue will have grown inward through the holes to line the inner surface of the sphere. Furthermore, the cavity is filled with fluid that flows freely through the perforations back and forth between the fluid in the interstitial spaces and the fluid in the cavity. Therefore, the pressure in the cavity should equal the free fluid pressure in the interstitial fluid spaces. A needle is inserted through the skin and

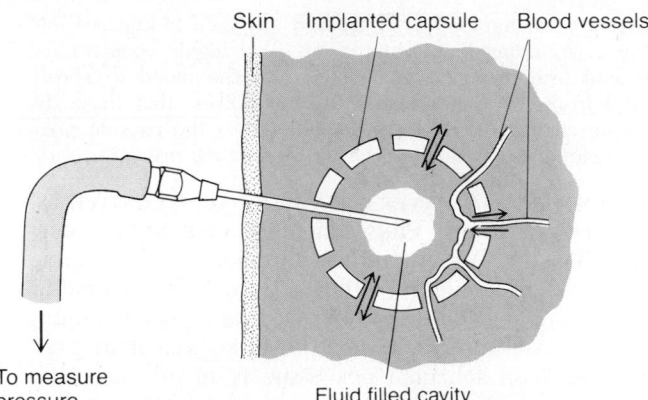

Figure 16–7. Perforated capsule method for measuring interstitial fluid pressure.

through one of the perforations to the interior of the cavity, and the pressure is measured by use of an appropriate manometer.

Interstitial free fluid pressure measured by this method when using 2-cm diameter capsules in normal *loose* subcutaneous tissue averages about − 6 mm Hg, but with smaller capsules, the values are not greatly different from the − 2 mm Hg measured by the micropipette.

MEASUREMENT OF INTERSTITIAL FREE FLUID PRESSURE BY MEANS OF A COTTON WICK. Still another method is to insert into a tissue a small Teflon tube with about eight cotton fibers protruding from its end. The cotton fibers form a "wick" that makes excellent contact with the tissue fluids and transmits interstitial fluid pressure into the Teflon tube; the pressure can then be measured from the tube by usual manometric means. Pressures measured by this technique in *loose* subcutaneous tissue have also been negative, usually measuring − 1 to − 3 mm Hg.

Interstitial Fluid Pressures in Tightly Encased Tissues

Some tissues of the body are surrounded by tight encasements, such as the cranial vault around the brain, the strong fibrous capsule around the kidney, the fibrous sheaths around the muscles, and the sclera around the eye. In most of these, regardless of the method used for measurement, the interstitial fluid pressures are usually positive. However, these interstitial fluid pressures almost invariably are still less than the pressures exerted on the outsides of the tissues by their encasements. For instance, the cerebrospinal fluid pressure surrounding the brain of an animal lying on its side averages about + 10 mm Hg, whereas the brain interstitial fluid pressure averages about + 4 to + 6 mm Hg. In the kidneys, the capsular pressure surrounding the kidney averages about + 13 mm Hg, whereas the reported interstitial fluid pressures have averaged about + 6 mm Hg.

Thus, if one remembers that the pressure exerted on the skin is atmospheric pressure, which is considered to be zero pressure, one might formulate a general rule that the normal interstitial fluid pressure is usually several millimeters of mercury negative with respect to the pressure that surrounds each tissue.

Is the True Interstitial Fluid Pressure in Loose Subcutaneous Tissue Subatmospheric?

The concept that the interstitial fluid pressure is subatmospheric in many tissues of the body began with clinical observations that could not be explained by the previously held concept that interstitial fluid pressure was always positive. Some of the pertinent observations are the following.

1. When a skin graft is placed on a concave surface of the body, such as in an eye socket after removal of the eye, before the skin becomes attached to the sublying socket, fluid tends to collect underneath the graft. Also, the skin attempts to shorten, which tries to pull it away from the concavity. Nevertheless, some negative force underneath the skin causes absorption of the fluid and usually will literally pull the skin back into the concavity.

2. Less than 1 mm Hg of positive pressure is required to inject tremendous volumes of fluid into loose subcutaneous tissues, such as beneath the lower eyelid, in the axillary space, and in the scrotum. Amounts of fluid calculated to be more than 100 times the amount of fluid normally in the interstitial space, when injected into these areas, will cause no more than about 2 mm Hg of positive pressure. The importance of these observations is that they show that such tissues do not have strong fibers that can prevent the accumulation of fluid. Therefore, some other mechanism must be available to prevent such fluid accumulation.

3. In most natural cavities of the body where there is free fluid in dynamic equilibrium with the surrounding interstitial fluids, the pressures that have been measured have been negative. Some of these are the following:

Intrapleural space: − 8 mm Hg
Joint synovial spaces: − 4 to − 6 mm Hg
Epidural space: − 4 to − 6 mm Hg

4. The implanted capsule method for measuring the interstitial fluid pressure can be used to record dynamic changes in this pressure. The changes are approximately those that one would calculate to occur when (1) the arterial

pressure is increased or decreased, (2) fluid is injected into the surrounding tissue spaces, or (3) a highly concentrated colloid osmotic agent is injected into the blood to absorb fluid from the tissue spaces. It is not likely that these dynamic changes could be measured unless the capsule pressure closely approximated the true interstitial pressure.

SUMMARY—AN AVERAGE VALUE FOR NEGATIVE INTERSTITIAL FLUID PRESSURE IN LOOSE SUBCUTANEOUS TISSUE. Although the aforementioned different methods give slightly different values for interstitial fluid pressure, there is now a general belief among most physiologists that the true interstitial fluid pressure in *loose* subcutaneous tissue is slightly less than atmospheric pressure. A pressure value that many are beginning to accept is an average value of about − 3 mm Hg. However, in all tissues with tight fibrous or fascial coverings that hold the tissues tightly together, such as the kidneys, the pressures underneath these constrictive coverings are usually more positive.

Pumping by the Lymphatic System Is the Basic Cause of the Negative Pressure

The lymphatic system is discussed later in the chapter, but we need to understand here the basic role that this system plays in determining interstitial fluid pressure. The lymphatic system is a "scavenger" system that removes excess fluid, protein molecules, debris, and other matter from the tissue spaces. When fluid enters the terminal lymphatic capillaries, any movement of the tissue propels the lymph forward through the lymphatic system, eventually emptying back into the circulation. In this way, any time any free fluid accumulates in the tissues, it is simply pumped away as a consequence of tissue movement. When the amount of fluid leaking from the blood capillaries is slight, as is true for most tissues, research evidence suggests that motion of the tissues and lymphatic capillaries can actually pump a slight intermittent negative pressure that gives an average negativity in the loose tissues (that is, slightly below atmospheric pressure). The details of this lymphatic pumping system are discussed later in the chapter.

Plasma Colloid Osmotic Pressure

PROTEINS IN THE PLASMA CAUSE COLLOID OSMOTIC PRESSURE. The proteins are the only dissolved substances in the plasma and interstitial fluid that do not diffuse readily through the capillary membrane. Furthermore, when small quantities of protein do diffuse into the interstitial fluid, much of these are soon removed from the interstitial spaces by way of the lymph vessels. Therefore, the concentration of protein in the plasma averages about three times as much as that in most interstitial fluid, 7.3 gm/dl in the plasma versus 2 to 3 gm/dl in the interstitial fluid.

In the basic discussion of osmotic pressure in Chap-

ter 4, it is pointed out that only those molecules or ions that fail to pass through the pores of a semipermeable membrane exert osmotic pressure. Because the proteins are the only dissolved constituents that do not readily penetrate the pores of the capillary membrane, it is the dissolved proteins of the plasma and interstitial fluids that are responsible for the osmotic pressure at the capillary membrane. To distinguish this osmotic pressure from that which occurs at the cell membrane, it is called either *colloid osmotic pressure* or *oncotic pressure*. The term "colloid" osmotic pressure is derived from the fact that a protein solution resembles a colloidal solution despite the fact that it is actually a true molecular solution. (The osmotic pressure that results at the cell membrane is called *total osmotic pressure* to distinguish it from the colloid osmotic pressure because essentially all dissolved substances of the body fluids exert osmotic pressure at the cell membrane. This is not true at the capillary membrane because of the large sizes of the capillary pores.)

Donnan Equilibrium Effect on the Colloid Osmotic Pressure. The so-called *Donnan equilibrium effect* causes the colloid osmotic pressure of the plasma to be about 50 per cent greater than that caused by the proteins alone. This results from the fact that the proteins are negative ions, and to balance these negative ions, a large number of positively charged ions (cations), mainly sodium ions, are held near to, but usually not bound to, the electronegative charges of the proteins. These extra cations, therefore, increase the number of osmotically active substances wherever the proteins occur, and they increase the osmotic pressure. Even more important, however (for mathematical reasons that cannot be explained here), the Donnan equilibrium effect becomes progressively more significant the higher the concentration of proteins. This means, as shown in Figure 16–8, that the initial few grams of protein in each 100 milliliters of plasma or interstitial fluid have much less colloid osmotic effect than do the next few grams.

NORMAL VALUES FOR PLASMA COLLOID OSMOTIC PRESSURE. The colloid osmotic pressure of normal human plasma averages about 28 mm Hg; 19 mm of this is caused by the dissolved protein and 9 mm by the cations held in the plasma by the Donnan effect of the proteins, as just discussed.

DIMINISHMENT OF THE COLLOID OSMOTIC PRESSURE CAUSED BY PROTEIN MOLECULES LEAKING THROUGH THE CAPILLARY PORES—THE "REFLECTION COEFFICIENT." Only when the protein molecules are unable to pass through the capillary pores will they cause osmotic pressure. In other words, those molecules that do not pass through when they come to a pore are said to be "reflected" from the pore rather than filtered through it; osmotic pressure is created as a product of this reflection process. When all the molecules are reflected, the pores are said to have a "reflection coefficient" for the protein molecules of 1.0. When none are reflected but instead pass through the pores, the reflection coefficient is said to be 0.0.

One can easily see that when all the protein mole-

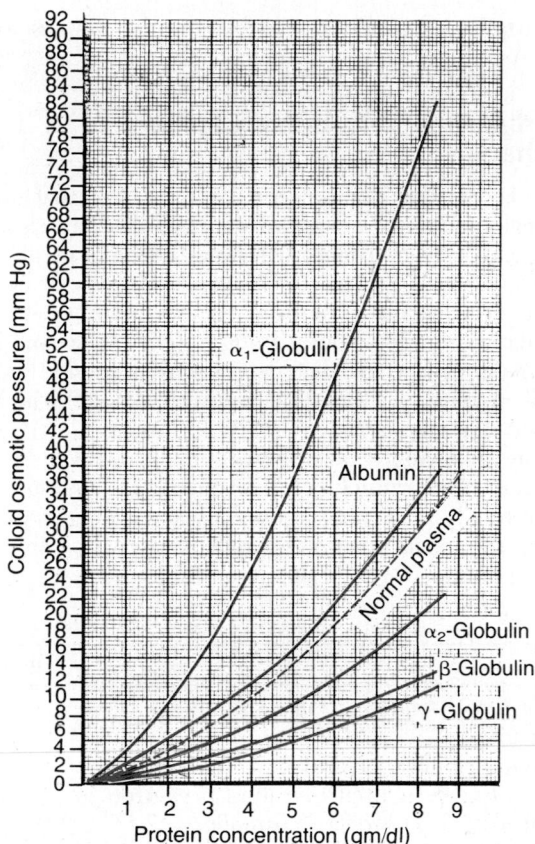

Figure 16–8. Osmotic pressure of five fractions of the plasma proteins at different concentrations. Also, the dashed line shows the osmotic pressure of normal plasma proteins, which are a mixture of the others. (Modified from Ott: *Klin. Wschr.,* *34:*1079, 1956.)

	gm/dl	Πp *(mm Hg)*
Albumin	4.5	21.8
Globulins	2.5	6.0
Fibrinogen	0.3	0.2
Total	7.3	28.0

Thus, about 80 per cent of the total colloid osmotic pressure of the plasma results from the albumin fraction, 20 per cent from the globulins, and almost none from the fibrinogen. Therefore, from the point of view of capillary dynamics, it is mainly albumin that is important.

Figure 16–8 demonstrates graphically the colloid osmotic pressures exerted by different concentrations of albumin and four fractions of globulins.

Interstitial Fluid Colloid Osmotic Pressure

Although the size of the usual capillary pore is smaller than the molecular sizes of the plasma proteins, this is not true of all the pores. Therefore, small amounts of plasma proteins do leak through the pores into the interstitial spaces.

The total quantity of protein in the entire 12 liters of interstitial fluid of the body is actually greater than the total quantity of protein in the plasma itself, but because this volume is four times the volume of plasma, the average protein *concentration* of the interstitial fluid is usually about 40 per cent of that in plasma, or about 3 gm/dl. Referring to the diagram in Figure 16–8, one finds that the average colloid osmotic pressure for this concentration of proteins in the interstitial fluids is about 8 mm Hg when the reflection coefficient of the capillary membranes is 1.0.

Exchange of Fluid Volume Through the Capillary Membrane

Now that the different factors affecting fluid movement through the capillary membrane have been discussed, it is possible to put all these together to see how normal capillaries maintain normal fluid volume distribution between the plasma and the interstitial fluid.

The average capillary pressure at the arterial ends of the capillaries is 15 to 25 mm Hg greater than at the venous ends. Because of this difference, fluid "filters" out of the capillaries at their arterial ends, and at their venous ends, fluid is reabsorbed back into the capillaries. Thus, a small amount of fluid actually "flows" through the tissues from the arterial ends of the capillaries to the venous ends. The dynamics of this flow are as follows.

ANALYSIS OF THE FORCES CAUSING FILTRATION AT THE *ARTERIAL END* OF THE CAPILLARY. The approximate average forces operative at the *arterial end* of the capillary that cause movement through the capillary membrane are shown as follows:

cules are reflected, they all exert full osmotic pressure, but when only one half are reflected—a "reflection coefficient" of 0.5—only half of the protein molecules then exert osmotic pressure. Thus, for the brain capillaries, the reflection coefficient is almost exactly 1, and for muscle capillaries, it approaches 1; therefore, either all or almost all of the plasma colloid osmotic force is exerted. On the other hand, in the liver sinusoids, the reflection coefficient approaches 0, and the plasma proteins there exert almost no osmotic pressure.

EFFECT OF THE DIFFERENT PLASMA PROTEINS ON COLLOID OSMOTIC PRESSURE. The plasma proteins are a mixture of proteins that contains albumin, with an average molecular weight of 69,000; globulins, 140,000; and fibrinogen, 400,000. Thus, 1 gram of globulin contains only half as many molecules as 1 gram of albumin, and 1 gram of fibrinogen contains only one sixth as many molecules as 1 gram of albumin. It should be recalled from the discussion of osmotic pressure in Chapter 4 that the osmotic pressure is determined by the *number of molecules* dissolved in a fluid rather than by the mass of these molecules. Therefore, when corrected for number of molecules rather than mass, the following chart gives both the relative mass concentrations of the different types of proteins in normal plasma and their respective contributions to the total plasma colloid osmotic pressure.

	mm Hg
Forces tending to move fluid outward:	
Capillary pressure	30
Negative interstitial free fluid pressure	3
Interstitial fluid colloid osmotic pressure	8
TOTAL OUTWARD FORCE	41
Forces tending to move fluid inward:	
Plasma colloid osmotic pressure	28
TOTAL INWARD FORCE	28
Summation of forces:	
Outward	41
Inward	28
NET OUTWARD FORCE	13

Thus, the summation of forces at the arterial end of the capillary shows a net *filtration pressure* of 13 mm Hg, tending to move fluid in the outward direction.

This 13 mm Hg filtration pressure causes, on the average, about 0.5 per cent of the plasma in the flowing blood to filter out of the arterial end of the capillaries into the interstitial spaces.

ANALYSIS OF REABSORPTION AT THE VENOUS END OF THE CAPILLARY. The low pressure at the venous end of the capillary changes the balance of forces in favor of absorption as follows.

	mm Hg
Forces tending to move fluid inward:	
Plasma colloid osmotic pressure	28
TOTAL INWARD FORCE	28
Forces tending to move fluid outward:	
Capillary pressure	10
Negative interstitial free fluid pressure	3
Interstitial fluid colloid osmotic pressure	8
TOTAL OUTWARD FORCE	21
Summation of forces:	
Inward	28
Outward	21
NET OUTWARD FORCE	7

Thus, the force that causes fluid to move into the capillary, 28 mm Hg, is greater than that opposing reabsorption, 21 mm Hg. The difference, 7 mm Hg, is the *reabsorption pressure* at the venous ends of the capillaries. This reabsorption pressure is considerably less than the filtration pressure, but remember that the venous capillaries are more numerous and more permeable than the arterial capillaries, so that less pressure is required to cause the inward movement of fluid.

The reabsorption pressure causes about nine tenths of the fluid that has filtered out of the arterial ends of the capillaries to be reabsorbed at the venous ends. The remainder flows into the lymph vessels.

Starling Equilibrium for Capillary Exchange

E. H. Starling pointed out a century ago that under normal conditions, a state of near equilibrium exists at the capillary membrane, whereby the amount of fluid filtering outward from some capillaries equals almost exactly the quantity of fluid that is returned to the circulation by absorption through other capillaries. The slight disequilibrium that does occur accounts for the small amount of fluid that is eventually returned by way of the lymphatics.

The following chart shows the principles of the Starling equilibrium. For this chart, the pressures in the arterial and venous capillaries are averaged to calculate the mean *functional* capillary pressure. This calculates to be 17.3 mm Hg.

	mm Hg
Mean forces tending to move fluid outward:	
Mean capillary pressure	17.3
Negative interstitial free fluid pressure	3.0
Interstitial fluid colloid osmotic pressure	8.0
TOTAL OUTWARD FORCE	28.3
Mean force tending to move fluid inward:	
Plasma colloid osmotic pressure	28.0
TOTAL INWARD FORCE	28.0
Summation of mean forces:	
Outward	28.3
Inward	28.0
NET OUTWARD FORCE	0.3

Thus, for the total capillary circulation, we find a near-equilibrium between the total outward forces, 28.3 mm Hg, and the total inward force, 28.0 mm Hg. This slight imbalance of forces, 0.3 mm Hg, causes slightly more filtration of fluid into the interstitial spaces than reabsorption. This slight excess of filtration is called the *net filtration,* and it is what causes fluid to return to the circulation through the lymphatics. The normal rate of net filtration in the entire body is only about 2 ml/min.

FILTRATION COEFFICIENT. In the previous example, an average net imbalance of forces at the capillary membranes of 0.3 mm Hg causes a net rate of fluid filtration in the entire body of 2 ml/min. Expressing this for each millimeter of mercury imbalance, one finds a net filtration rate of 6.67 milliliters of fluid per minute per millimeter of mercury for the entire body. This expression is the *filtration coefficient.*

The filtration coefficient can also be expressed for different parts of the body in terms of the rate of filtration per minute per millimeter of mercury per

100 grams of tissue. On this basis, the filtration coefficient of the average tissue is about 0.01 ml/min/mm Hg/100 gm of tissue. Because of extreme differences in permeabilities of the capillary systems in different tissues, this coefficient varies more than a hundredfold among the different tissues. It is very small in both brain and muscle, moderately great in subcutaneous tissue, large in the intestine, and extreme in the liver and the glomerulus of the kidney where the pores are either numerous or wide open. By the same token, the permeation of proteins through the capillary membranes varies greatly as well. The concentration of protein in the interstitial fluid of muscles is about 1.5 gm/dl, in subcutaneous tissue 2 gm/dl, in intestine 4 gm/dl, and in liver 6 gm/dl.

Effect of Abnormal Imbalance of Forces at the Capillary Membrane

If the mean capillary pressure rises above 17 mm Hg, the net force tending to cause filtration of fluid into the tissue spaces rises. Thus, a 20-mm Hg rise in mean capillary pressure causes an increase in the net filtration pressure from 0.3 mm Hg to 20.3 mm Hg, which results in 68 times as much net filtration of fluid into the interstitial spaces as normally occurs. To prevent accumulation of excess fluid in the spaces would require 68 times the normal flow of fluid into the lymphatic system, an amount that is usually 2 to 3 times too much for the lymphatics to carry away. As a result, fluid begins to accumulate in the interstitial spaces and edema results.

Conversely, if the capillary pressure falls very low, net reabsorption of fluid into the capillaries occurs instead of net filtration, and the blood volume increases at the expense of the interstitial fluid volume. The effects of these imbalances at the capillary membrane are discussed in Chapter 25 in relation to the development of different kinds of edema.

THE LYMPHATIC SYSTEM

The lymphatic system represents an accessory route by which fluid can flow from the interstitial spaces into the blood. And, most important, the lymphatics can carry proteins and large particulate matter away from the tissue spaces, neither of which can be removed by absorption directly into the blood capillary. This removal of proteins from the interstitial spaces is an essential function, without which we would die within about 24 hours.

Lymph Channels of the Body

Almost all tissues of the body have lymphatic channels that drain excess fluid directly from the interstitial spaces. The exceptions include the superficial portions of the skin, the central nervous system, deeper portions of peripheral nerves, the endomysium of muscles, and the bones. Even these tissues have minute interstitial channels called *prelymphatics* through which interstitial fluid can flow; this fluid eventually flows either into lymphatic vessels or, in the case of the brain, into the cerebrospinal fluid and thence directly back into the blood.

Essentially all the lymph from the lower part of the body flows up the *thoracic duct* and empties into the venous system at the juncture of the *left* internal jugular vein and subclavian vein, as shown in Figure 16–9.

Lymph from the left side of the head, the left arm, and parts of the chest region also enters the thoracic duct before it empties into the veins. Lymph from the right side of the neck and head, the right arm, and parts of the thorax enters the *right lymph duct*, which then empties into the venous system at the juncture of the *right* subclavian vein and internal jugular vein.

TERMINAL LYMPHATIC CAPILLARIES AND THEIR PERMEABILITY. Most of the fluid filtering from the arterial capillaries flows among the cells and is finally reabsorbed back into the *venous ends* of the *blood capillaries;* but on the average, probably about *one tenth* of the fluid enters the *lymphatic capillaries* instead and returns to the blood through the lymphatic system rather than through the venous capillaries. The total quantity of this lymph is normally only 2 to 3 liters each day.

The minute quantity of fluid that returns to the circulation by way of the lymphatics is extremely important because substances of high molecular weight, such as proteins, cannot be reabsorbed in any other way. Yet they can enter the lymphatic capillaries almost unimpeded. The reason for this is a special structure of the lymphatic capillaries, demonstrated in Figure 16–10. This figure shows the endothelial cells of the capillary attached by *anchoring filaments* to the surrounding connective tissue. At the junctions of adjacent endothelial cells, the edge of one endothelial cell usually overlaps the edge of the adjacent cell in such a way that the overlapping edge is free to flap inward, thus forming a minute valve that opens to the interior of the capillary. Interstitial fluid, along with its suspended particles, can push the valve open and flow directly into the lymphatic capillary. But this fluid has difficulty leaving the capillary once it has entered because any backflow will close the flap valve. Thus, the lymphatics have valves at the very tips of the terminal lymphatic capillaries as well as valves along their larger vessels up to the point where they empty into the blood circulation.

Formation of Lymph

Lymph is derived from interstitial fluid that flows into the lymphatics. Therefore, lymph as it first flows from each tissue has almost the same composition as the interstitial fluid.

The protein concentration in the interstitial fluid of most tissues averages about 2 gm/dl, and the protein concentration of lymph flowing from these tissues is near this value. On the other hand, lymph formed in

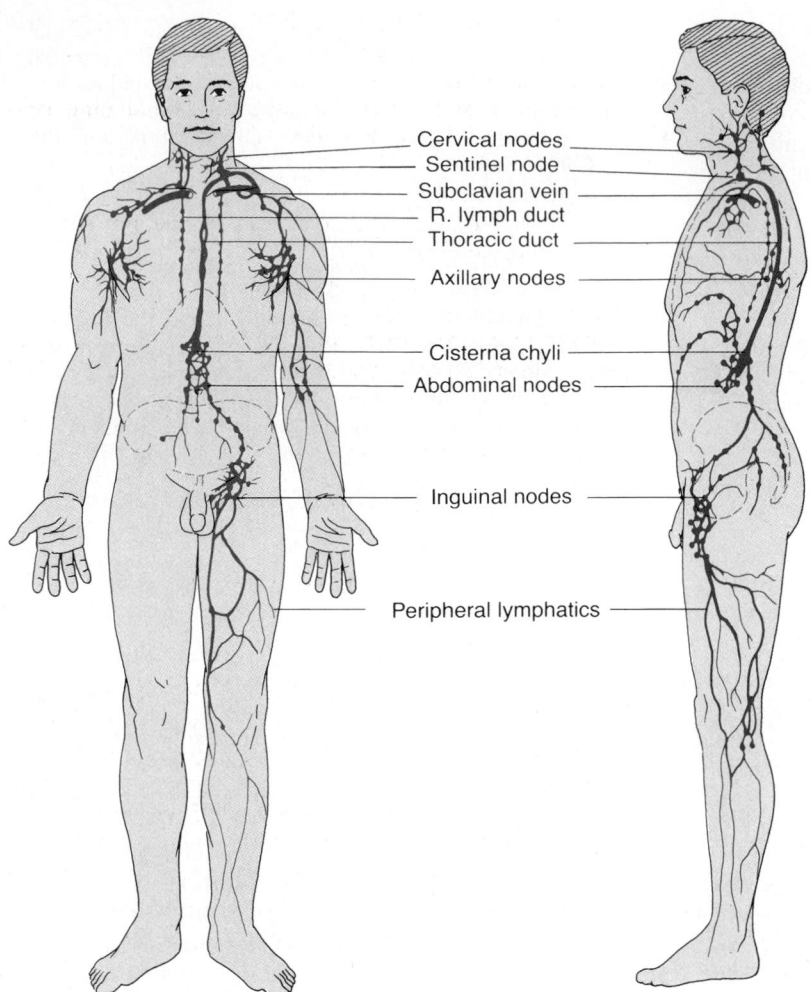

Cervical nodes
Sentinel node
Subclavian vein
R. lymph duct
Thoracic duct
Axillary nodes
Cisterna chyli
Abdominal nodes
Inguinal nodes
Peripheral lymphatics

Figure 16–9. Lymphatic system.

the liver has a protein concentration as high as 6 gm/dl, and lymph formed in the intestines has a protein concentration as high as 3 to 4 gm/dl. Because about two thirds of all lymph normally is derived from the liver and intestines, the thoracic lymph, which is a mixture of lymph from all areas of the body, usually has a protein concentration of 3 to 5 gm/dl.

The lymphatic system is also one of the major routes for absorption of nutrients from the gastrointestinal tract, being responsible principally for the absorption of fats, as discussed in Chapter 65. Indeed, after a fatty meal, thoracic duct lymph sometimes contains as much as 1 to 2 per cent fat.

Finally, even large particles, such as bacteria, can push their way between the endothelial cells of the lymphatic capillaries and in this way enter the lymph. As the lymph passes through the lymph nodes, these particles are removed and destroyed, as discussed in Chapter 33.

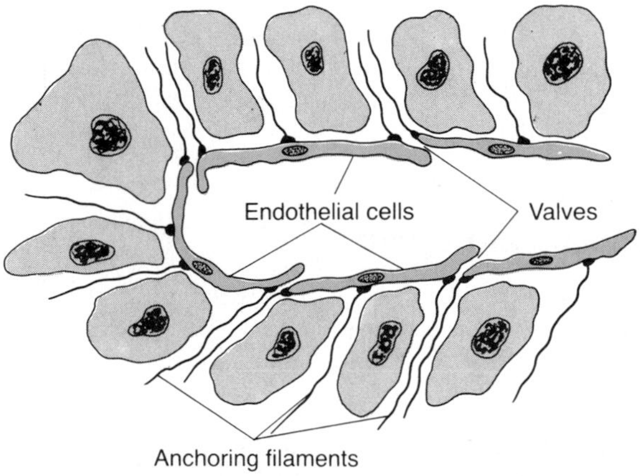

Rate of Lymph Flow

About 100 milliliters of lymph flows through the *thoracic duct* of a resting human per hour, and perhaps another 20 milliliters flows into the circulation each hour through other channels, making a total estimated lymph flow of about 120 ml/hr, between 2 and 3 liters per day.

EFFECT OF INTERSTITIAL FLUID PRESSURE ON LYMPH FLOW. Figure 16–11 shows the effect of dif-

Endothelial cells   Valves

Anchoring filaments

Figure 16–10. Special structure of the lymphatic capillaries that permits passage of substances of high molecular weight into the lymph.

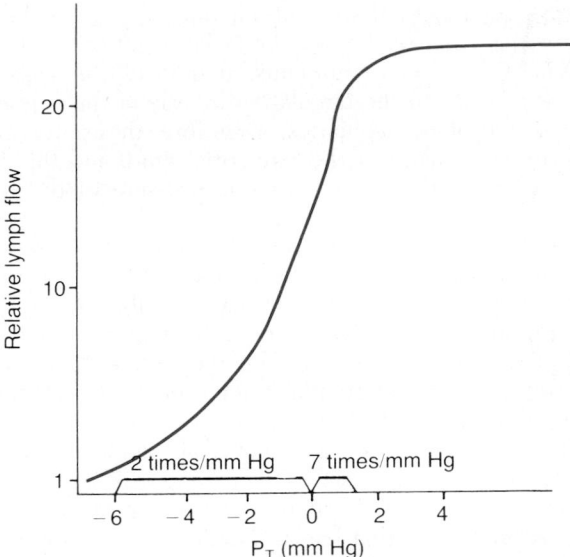

Figure 16–11. Relation between interstitial fluid pressure and lymph flow. Note that lymph flow reaches a maximum as the interstitial pressure rises slightly above atmospheric pressure (0 mm Hg). (Courtesy of Drs. Harry Gibson and Aubrey Taylor.)

ferent levels of interstitial fluid pressure on lymph flow as measured in dog legs. Note that the lymph flow is slight at interstitial fluid pressures more negative than −6 mm Hg. Then, as the pressure rises up to values slightly greater than 0 mm Hg (atmospheric pressure), the flow increases more than 20-fold. Therefore, any factor that increases interstitial fluid pressure will normally also increase the lymph flow. Such factors include the following:

Elevated capillary pressure
Decreased plasma colloid osmotic pressure
Increased interstitial fluid protein
Increased permeability of the capillaries

All of these cause the balance of fluid exchange at the blood capillary membrane to favor fluid movement into the interstitium, thus increasing interstitial fluid volume, interstitial fluid pressure, and lymph flow all at the same time.

However, note that when the interstitial fluid pressure becomes 1 or 2 millimeters greater than atmospheric pressure (0 mm Hg), lymph flow fails to rise further at still higher pressures. This probably results from the fact that the increasing tissue pressure not only increases entry of fluid into the lymphatic capillaries, but also compresses the outside surfaces of the larger lymphatics, thus impeding lymph flow. At these higher pressures, these two factors appear to balance each other almost exactly. (A similar phenomenon is observed in the flow of air from the alveoli of the lungs during expiration; it is called the "maximum expiratory rate," the basic principles of which are explained in detail in Chapter 37.)

THE LYMPHATIC PUMP INCREASES LYMPH FLOW. Valves exist in all lymph channels; typical valves are shown in Figure 16–12 in *collecting lymphatics* into which the lymphatic capillaries empty. In the large lymphatics, valves exist every few millimeters and in the smaller lymphatics, much closer than this.

Intrinsic Pumping by the Collecting Lymphatics and Larger Lymph Vessels. Motion pictures of exposed lymph vessels, both in animals and in humans, show that when a collecting lymphatic or larger lymph vessel becomes stretched with fluid, the smooth muscle in the wall of the vessel automatically contracts. Furthermore, each segment of the lymph vessel between successive valves functions as a separate automatic pump. That is, filling of a segment causes it to contract, and the fluid is pumped through the valve into the next lymphatic segment. This fills the subsequent segment, and a few seconds later it, too, contracts, the process continuing all along the lymph vessel until the fluid is finally emptied. In a very large lymph vessel such as the thoracic duct, this lymphatic pump can generate pressures of 25 to 50 mm Hg if the outflow from the vessel becomes blocked.

Pumping Caused by External Intermittent Compression of the Lymphatics. In addition to the pumping caused by intrinsic contraction of the lymph vessel walls, any external factor that intermittently

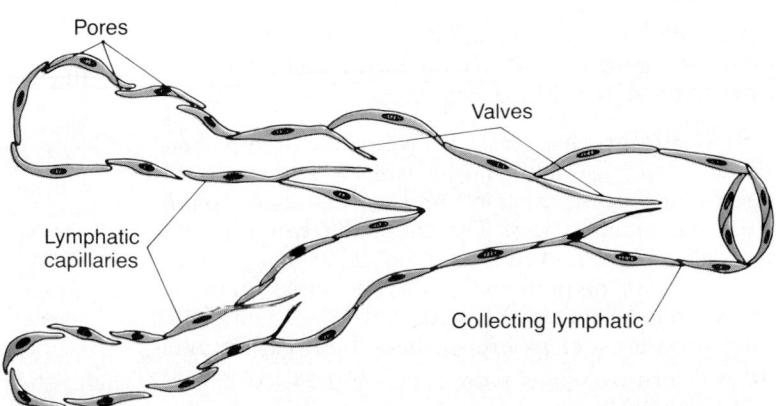

Figure 16–12. Structure of lymphatic capillaries and a collecting lymphatic, showing also the lymphatic valves.

compresses the lymph vessel can cause pumping. In order of their importance, such factors are as follows:

Contraction of surrounding muscles of the body
Movement of the parts of the body
Arterial pulsations
Compression of the tissues by objects outside the body

The lymphatic pump becomes very active during exercise, often increasing lymph flow 10- to 30-fold. On the other hand, during periods of rest, lymph flow is sluggish.

LYMPHATIC CAPILLARY PUMP. Some physiologists believe that in a peculiar way, the terminal lymphatic capillary is capable of pumping lymph, in addition to the lymphatic pump of the larger lymph vessels. As explained earlier in the chapter, the walls of the lymphatic capillaries are tightly adherent to the surrounding tissue cells by means of their anchoring filaments. Therefore, each time excess fluid enters the tissue and causes the tissue to swell, the anchoring filaments pull the lymphatic capillary open, and fluid flows into the capillary through the junctions between the endothelial cells. Then, when the tissue is compressed, the pressure inside the capillary increases and causes the overlapping edges of the endothelial cells to close like valves. Therefore, the pressure pushes the lymph forward into the collecting lymphatic instead of backward through the cell junctions.

The lymphatic capillary endothelial cells contain contractile actomyosin filaments. In some animal tissues (the bat's wing, for example) they have been observed to cause rhythmical contraction of the lymphatic capillaries in the same way that many of the small blood and lymphatic vessels contract rhythmically. Therefore, it is possible that at least part of lymph pumping results from lymph capillary endothelial cell contraction in addition to contraction of the larger muscular lymphatics.

SUMMARY OF FACTORS THAT DETERMINE LYMPH FLOW. From the previous discussion, one can see that the two primary factors that determine lymph flow are (1) the interstitial fluid pressure and (2) the activity of the lymphatic pump. Therefore, one can state that, roughly, *the rate of lymph flow is determined by the product of interstitial fluid pressure and the activity of the lymphatic pump.*

Role of the Lymphatic System in Controlling Interstitial Fluid Protein Concentration, Interstitial Fluid Volume, and Interstitial Fluid Pressure

It is already clear that the lymphatic system functions as an "overflow mechanism" to return to the circulation excess proteins and excess fluid volume from the tissue spaces. Therefore, the lymphatic system also plays a central role in controlling (1) the concentration of proteins in the interstitial fluids, (2) the volume of interstitial fluid, and (3) the interstitial fluid pressure. Let us explain how these factors interact.

First, remember that small amounts of proteins leak continuously out of the blood capillaries into the interstitium. Only minute amounts, if any, of the leaked proteins return to the circulation by way of the venous ends of the blood capillaries. Therefore, these proteins tend to accumulate in the interstitial fluid, and this in turn increases the colloid osmotic pressure of the interstitial fluids.

Second, the increasing colloid osmotic pressure in the interstitial fluid shifts the balance of forces at the blood capillary membranes in favor of fluid filtration into the interstitium. Therefore, fluid is pulled osmotically by these proteins into the interstitium, thus increasing both the interstitial fluid volume and the interstitial fluid pressure.

Third, the increasing interstitial fluid pressure greatly increases the rate of lymph flow, as explained earlier. This in turn carries away the excess volume and excess protein that has accumulated in the spaces.

Thus, once the interstitial fluid protein concentration reaches a certain level and causes a comparable increase in interstitial fluid volume and interstitial fluid pressure, the return of protein and fluid by way of the lymphatic system becomes great enough to balance exactly the rate of leakage of these from the blood capillaries. Therefore, the quantitative values of all these factors reach a steady state; they will remain balanced at these levels until something changes the rate of leakage of proteins and fluid from the blood capillaries.

Significance of Negative Interstitial Fluid Pressure as a Means for Holding the Body Tissues Together

Traditionally, it has been assumed that the different tissues of the body are held together entirely by connective tissue fibers. However, at many places in the body, connective tissue fibers are absent. This occurs particularly at points where tissues slide over one another, such as the skin sliding over the back of the hand or over the face. Yet even at these places, the tissues are held together by the negative interstitial fluid pressure, which is actually a partial vacuum. When the tissues lose their negative pressure, fluid accumulates in the spaces and the condition known as *edema* occurs, which is discussed in Chapter 25.

REFERENCES

Altenkamper, H., et al.: A Color Atlas of Venous Disease. Philadelphia, J. B. Lippincott, 1994.

Arendshorst, W. J., and Gottschalk, C. W.: Glomerular ultrafiltration dynamics: Historical perspective. Am. J. Physiol., 248:F163, 1985.

Aukland, K., and Reed, R. K.: Interstitial-lymphatic mechanisms in the control of extracellular fluid volume. Physiol. Rev., 73:1, 1993.

Bassingthwaighte, J. B., and Sparks, H. V., Jr.: Indicator dilution estimation of capillary endothelial transport. Annu. Rev. Physiol., 48:321, 1986.

Bishop, M. L., et al.: Clinical Chemistry: Principles, Procedures, Correlations. 2nd Ed. Philadelphia, J. B. Lippincott, 1992.

Brace, R. A., and Guyton, A. C.: Effect of hindlimb isolation procedure on isogravimetric capillary pressure and transcapillary fluid dynamics in dogs. Circ. Res., 38:192, 1976.

Brace, R. A., and Guyton, A. C.: Interaction of transcapillary Starling forces in the isolated dog forelimb. Am. J. Physiol., 233:H136, 1977.

Bundgaard, M.: Transport pathways in capillaries—in search of pores. Annu. Rev. Physiol., 42:325, 1980.

Chien, S. (ed.): Vascular Endothelium in Health and Disease. New York, Plenum Publishing Corp., 1988.

Comerota, A. J.: Thrombolytic Therapy in Vascular Disease. Philadelphia, J. B. Lippincott, 1994.

Comper, W. D., and Laurent, T. C.: Physiological function of connective tissue polysaccharides. Physiol. Rev. 58:255, 1978.

Crone, C.: The Malpighi lecture. From "Porositates carnis" to cellular microcirculation. Int. J. Microcir. Clin. Exp., 6:101, 1987.

Crone, C., and Levitt, D. G.: Capillary permeability to small solutes. In Renkin, E. M., and Michel, C. C. (eds.): Handbook of Physiology. Sec. 2, Vol. IV. Bethesda, Md., American Physiological Society, 1984, p. 411.

Duling, B. R., and Desjardins, C.: Capillary hematocrit—What does it mean? News Physiol. Sci., 2:66, 1987.

Folkman, J., and Klagsbrun, M.: Angiogenic factors. Science, 235:442, 1987.

Galioto, G. B.: Tonsils: A Clinically Oriented Update. Farmington, CT, S. Karger Publishers, Inc., 1992.

Gross, J. F., and Popel, A. (eds.): Mathematics of Microcirculation Phenomena. New York, Raven Press, 1980.

Guyton, A. C.: Concept of negative interstitial pressure based on pressures in implanted perforated capsules. Circ. Res., 12:399, 1963.

Guyton, A. C.: Interstitial fluid pressure: II. Pressure-volume curves of interstitial space. Circ. Res., 16:452, 1965.

Guyton, A. C., et al.: Circulatory Physiology II. Dynamics and Control of the Body Fluids. Philadelphia, W. B. Saunders Co., 1975.

Guyton, A. C., et al.: Interstitial fluid pressure. Physiol. Rev., 51:527, 1971.

Guyton, A. C., et al.: Interstitial fluid pressure: III. Its effect on resistance to tissue fluid mobility. Circ. Res., 19:412, 1966.

Guyton, A. C., et al.: Interstitial fluid pressure: IV. Its effect on fluid movement through the capillary wall. Circ. Res., 19:1022, 1966.

Halperin, M. L., and Goldstein, M. B.: Fluid, Electrolyte, and Acid-Base Physiology: A Problem Based Approach. Philadelphia, W. B. Saunders Co., 1994.

Intaglietta, M.: Transcapillary exchange of fluid in single microvessels. In Kaley, G., and Altura, B. M. (eds.): Microcirculation. Vol. 1. Baltimore, University Park Press, 1977, p. 197.

Ioachim, H. L.: Lymph Node Pathology. Philadelphia, J. B. Lippincott, 1994.

Landis, E. M.: Capillary pressure and capillary permeability. Physiol. Rev., 14:404, 1934.

Landis, E. M., and Pappenheimer, J. R.: Exchange of substances through the capillary walls. In Hamilton, W. F. (ed.): Handbook of Physiology. Sec. 2, Vol. 2, Baltimore, Williams & Wilkins, 1963, p. 961.

Meyer, J. L.: The Lymphatic System and Cancer: Mechanisms and Clinical Management. Farmington, CT, S. Karger Publishers, Inc., 1994.

Michel, C. C.: Fluid movements through capillary walls. In Renkin, E. M., and Michel, C. C. (eds.): Handbook of Physiology. Sec. 2, Vol. IV. Bethesda, Md., American Physiological Society, 1984, p. 375.

Nicoll, P. A., and Taylor, A. E.: Lymph formation and flow. Annu. Rev. Physiol., 39:73, 1977.

Pappenheimer, J. R.: Contributions to microvascular research of Jean Leonard Marie Poiseuille. In Renkin, E. M., and Michel, C. C. (eds.): Handbook of Physiology. Sec. 2, Vol. IV. Bethesda, Md., American Physiological Society, 1984, p. 1.

Pappenheimer, J. R.: Passage of molecules through capillary walls. Physiol. Rev., 33:387, 1953.

Pappenheimer, J. R., et al.: Intestinal Absorption and Excretion of Octapeptides Composed of D amino acids. Proc. Natl. Acad. Sci. USA, 91:1942, 1994.

Rhodin, J. A. G.: Architecture of the vessel wall. In Bohr, D. F., et al. (eds.): Handbook of Physiology. Sec. 2, Vol. II. Baltimore, Williams & Wilkins, 1980, p. 1.

Rothschild, M. A., et al.: Albumin synthesis. In Javitt, N. B. (ed.): International Review of Physiology: Liver and Biliary Tract Physiology I. Vol. 21. Baltimore, University Park Press, 1980, p. 249.

Samuelsson, B., et al.: Prostaglandins and Related Compounds. New York, Raven Press, 1991.

Schmid-Schönbein, G. W.: The lymphatic transport mechanisms. Bioeng. Sci. News, 17:51, 1993.

Shepro, D., and D'Amore, P. A.: Physiology and biochemistry of the vascular wall endothelium. In Renkin, E. M., and Michel, C. C. (eds.): Handbook of Physiology. Sec. 2, Vol. IV. Bethesda, Md., American Physiological Society, 1984, p. 103.

Simionescu, N.: Cellular aspects of transcapillary exchange. Physiol. Rev., 63:1536, 1983.

Simionescu, M., and Simionescu, N.: Ultrastructure of the microvascular wall: Functional correlations. In Renkin, E. M., and Michel, C. C. (eds.): Handbook of Physiology. Sec. 2, Vol. IV. Bethesda, Md., American Physiological Society, 1984, p. 41.

Smith, J. J., and Kampine, J. P.: Circulatory Physiology—The Essentials. Baltimore, Williams & Wilkins, 1990.

Stromme, S. B., et al.: Interstitial fluid pressure in terrestrial and semiterrestrial animals. J. Appl. Physiol., 27:123, 1969.

Taylor, A. E., and Granger, D. N.: Exchange of macromolecules across the microcirculation. In Renkin, E. M., and Michel, C. C. (eds.): Handbook of Physiology. Sec. 2, Vol. IV. Bethesda, Md., American Physiological Society, 1984, p. 467.

Taylor, A. E., and Townsley, M. I.: Evaluation of the Starling fluid flux equation. News Physiol. Sci., 2:48, 1987.

Vanhoutte, P. M.: Vasodilation: Vascular Smooth Muscle, Peptides, Autonomic Nerves, and Endothelium. New York, Raven Press, 1988.

Xzerlip, H. M., and Goldfarb, St.: Workshops in Fluid and Electrolyte Disorders. New York, Churchill Livingstone, 1993.

Local Control of Blood Flow by the Tissues, and Humoral Regulation

CHAPTER 17

LOCAL CONTROL OF BLOOD FLOW IN RESPONSE TO TISSUE NEED

One of the most fundamental principles of circulatory function is the ability of each tissue to control its own local blood flow in proportion to its metabolic needs. Furthermore, as the need for blood flow changes, the flow follows the change.

What are some of the specific needs of the tissues for blood flow? The answer to this is manyfold, including the following:

1. Delivery of oxygen to the tissues
2. Delivery of other nutrients, such as glucose, amino acids, fatty acids, and so forth
3. Removal of carbon dioxide from the tissues
4. Removal of hydrogen ions from the tissues
5. Maintenance of proper concentrations of other ions in the tissues
6. Transport of various hormones and other specific substances to the different tissues

Also, certain organs have special requirements. For instance, blood flow to the skin determines heat loss from the body and in this way helps control the body temperature. Delivery of adequate quantities of blood plasma to the kidneys allows the kidneys to excrete the waste products of the body.

We shall see later that most of these factors exert extreme degrees of control over local blood flow.

VARIATIONS IN BLOOD FLOW IN DIFFERENT TIS-SUES AND ORGANS. In general, the greater the metabolism in an organ, the greater its blood flow. Note, for instance, in Table 17–1, the very large blood flows in the various glandular organs—for example, several hundred milliliters per minute per 100 grams of thyroid or adrenal gland tissue and a blood flow of 95 ml/min/100 gm of liver.

Also note the extremely large blood flow through the kidneys, 360 ml/min/100 gm. This extreme amount of flow is required for the kidneys to perform their function of cleansing the blood of waste products.

On the other hand, most surprising is the low blood flow to the resting muscles of the body, even though they constitute between 30 and 40 per cent of the total body mass. In the resting state, the metabolic activity of the muscles is very low, and so also is the blood flow, only 4 ml/min/100 gm. Yet, during heavy exercise, muscle metabolic activity can increase more than 60-fold and the blood flow as much as 20-fold, rising to as high as 80 ml/min/100 gm of muscle.

IMPORTANCE OF BLOOD FLOW CONTROL BY THE LOCAL TISSUES. One might ask the simple question: Why not simply allow a very large blood flow all the time through every tissue of the body, always enough to supply the tissue's needs whether the activity of the tissue is little or great? The answer to this is equally simple: To do this would require many times the amount of blood flow that the heart can pump.

Experiments have shown that the blood flow to each tissue is usually regulated at the minimal level that will supply its requirements, no more, no less. For instance, in tissues for which the most important requirement is delivery of oxygen, the blood flow is always controlled at that level only slightly more than required to maintain full tissue oxygenation but no more than this. By controlling the local blood flow in

Table 17–1 BLOOD FLOW TO DIFFERENT ORGANS AND
TISSUES UNDER BASAL CONDITIONS

	Per cent	Ml/min	Ml/min/ 100 gm
Brain	14	700	50
Heart	4	200	70
Bronchi	2	100	25
Kidneys	22	1100	360
Liver	27	1350	95
Portal	(21)	(1050)	
Arterial	(6)	(300)	
Muscle (inactive state)	15	750	4
Bone	5	250	3
Skin (cool weather)	6	300	3
Thyroid gland	1	50	160
Adrenal glands	0.5	25	300
Other tissues	3.5	175	1.3
Total	100.0	5000	—

Based mainly on data compiled by Dr. L. A. Sapirstein.

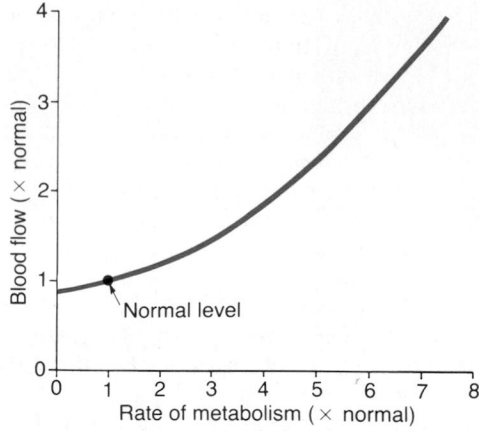

Figure 17–1. Effect of increasing rate of metabolism on tissue blood flow.

such an exact way, the tissues never suffer from nutritional deficiency, and yet the workload on the heart is kept at a minimum.

MECHANISMS OF BLOOD FLOW CONTROL

Local blood flow control can be divided into two phases: (1) acute control and (2) long-term control. Acute control is achieved by rapid changes in local constriction of the arterioles, metarterioles, and precapillary sphincters, occurring within seconds to minutes to provide a rapid means for maintaining appropriate local tissue blood flow. Long-term control, on the other hand, means slow changes in flow over a period of days, weeks, or even months. In general, the long-term changes provide far better control of the flow in proportion to the needs of the tissues. These changes come about as a result of an increase or decrease in the physical sizes and numbers of actual blood vessels supplying the tissues.

Acute Control of Local Blood Flow

EFFECT OF TISSUE METABOLISM ON LOCAL BLOOD FLOW. Figure 17–1 shows the approximate quantitative acute effect on blood flow of increasing the rate of metabolism in a local tissue, such as muscle. Note that an increase in metabolism up to eight times normal increases the blood flow acutely about fourfold. The increase in flow at first is less than the increase in metabolism. However, once the metabolism rises high enough to remove most of the nutrients from the blood, further increase in metabolism can occur only with a concomitant increase in blood flow to supply the required nutrients.

LOCAL BLOOD FLOW REGULATION WHEN OXYGEN AVAILABILITY CHANGES. One of the most necessary of the nutrients is oxygen. Whenever the availability of oxygen to the tissues decreases, such as at high altitude, in pneumonia, in carbon monoxide poisoning (which poisons the ability of hemoglobin to transport oxygen), or in cyanide poisoning (which poisons the ability of the tissues to use oxygen), the blood flow through the tissues increases markedly. Figure 17–2 shows that as the arterial oxygen saturation falls to about 25 per cent of normal, the blood flow through an isolated leg increases about threefold; that is, the blood flow increases almost enough, but not quite, to make up for the decreased amount of oxygen in the blood, thus automatically almost maintaining a constant supply of oxygen to the tissues. Cyanide poisoning of local tissue areas can cause a local blood flow increase as much as sevenfold, thus demonstrating the extreme effect of oxygen deficiency in increasing blood flow.

There are two basic theories for the regulation of local blood flow when either the rate of tissue metabolism changes or the availability of oxygen changes. They are (1) the *vasodilator theory* and (2) the *oxygen demand theory*.

Vasodilator Theory for Local Blood Flow Regulation—Possible Special Role of Adenosine. According to this theory, the greater the rate of me-

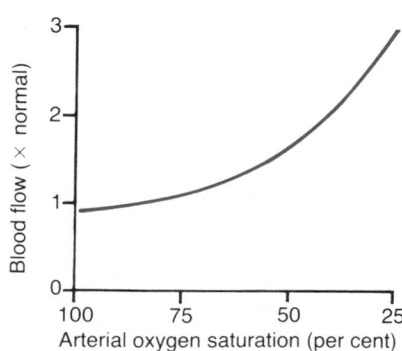

Figure 17–2. Effect of arterial oxygen saturation on blood flow through an isolated dog leg.

tabolism or the less the availability of oxygen or some other nutrients to a tissue, the greater becomes the rate of formation of a *vasodilator substance*. The vasodilator substance then is believed to diffuse back to the precapillary sphincters, metarterioles, and arterioles to cause dilatation. Some of the different vasodilator substances that have been suggested are *adenosine, carbon dioxide, lactic acid, adenosine phosphate compounds, histamine, potassium ions,* and *hydrogen ions.*

Most of the vasodilator theories assume that the vasodilator substance is released from the tissue mainly in response to oxygen deficiency. For instance, experiments have shown that decreased availability of oxygen can cause both adenosine and lactic acid to be released from the tissues; these substances can cause intense vasodilation and therefore could be responsible, or partially responsible, for the local blood flow regulation.

Some physiologists have suggested that the substance *adenosine* is by far the most important of the local vasodilators for controlling local blood flow. For instance, minute quantities of adenosine are released from heart muscle cells whenever coronary blood flow becomes too little, and it is believed that this causes local vasodilation in the heart and thereby returns the blood flow back toward normal. Also, whenever the heart becomes overly active and the heart's metabolism increases, this, too, causes excessive utilization of oxygen, followed by decreased oxygen concentration in the heart muscle with consequent degradation of adenosine triphosphate (ATP), which increases the formation of adenosine. And it is believed that some of this adenosine leaks out of the cells to cause coronary vasodilation, providing increased coronary blood flow to supply the nutrient demands of the active heart.

Although the research evidence is less clear, some physiologists have also suggested that the same adenosine mechanism might also be the most important controller of blood flow in skeletal muscle and many other tissues as well as in the heart.

The problem with the different vasodilator theories of local blood flow regulation has been the following: It has been difficult to prove that sufficient quantities of any single vasodilator substance are indeed formed in the tissues to cause all the measured increase in blood flow either in states of increased tissue metabolic demand or in decreased tissue oxygen. On the other hand, perhaps a combination of all the different vasodilators could increase the blood flow sufficiently.

Oxygen Demand Theory for Local Blood Flow Control. Although the vasodilator theory is accepted by most physiologists, several critical facts have made a few physiologists favor still another theory, which can be called either the oxygen demand theory or, more accurately, the *nutrient demand theory* (because probably other nutrients besides oxygen are involved). Oxygen (and other nutrients as well) is required to maintain vascular muscle contraction. Therefore, in the absence of an adequate supply of oxygen and other

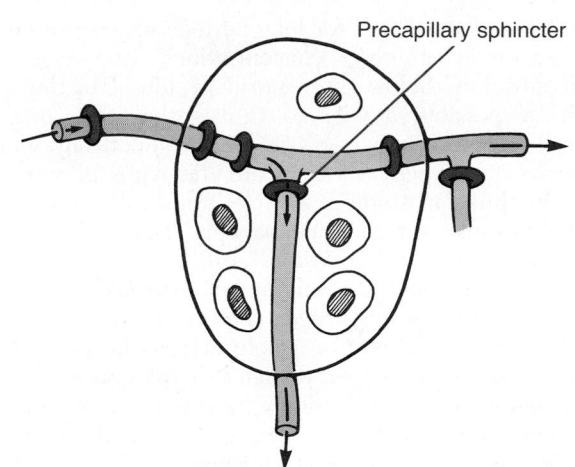

Figure 17–3. Diagram of a tissue unit area for explanation of local feedback control of blood flow.

nutrients, it is reasonable to believe that the blood vessels would naturally dilate. Also, increased utilization of oxygen in the tissues as a result of increased metabolism would theoretically decrease the availability of oxygen to the local blood vessels, and this, too, would cause local vasodilation.

A mechanism by which the oxygen demand theory could operate is shown in Figure 17–3. This figure shows what might be called a "tissue unit," consisting of a metarteriole, a single capillary, and its surrounding tissue. At the origin of the capillary is a *precapillary sphincter,* and around the metarteriole are several other smooth muscle fibers. Observing under a microscope a thin tissue, such as a bat's wing, one sees that the precapillary sphincters are normally either completely open or completely closed, and the degree of constriction of the metarteriole also varies. The number of precapillary sphincters that are open at any given time is about proportional to the requirements of the tissue for nutrition. In addition, the precapillary sphincters and metarterioles often open and close cyclically several times per minute, with the duration of the open phases being about proportional to the metabolic needs of the tissues. The cyclic opening and closing is called *vasomotion.*

Now, let us explain how oxygen concentration in the local tissue could regulate blood flow through the area. Because smooth muscle requires oxygen (or other nutrients or both) to remain contracted, one might assume that the strength of contraction of the sphincters would increase with an increase in oxygen concentration. Consequently, when the oxygen concentration in the tissue rises above a certain level, the precapillary and metarteriole sphincters presumably would close and remain closed until the tissue cells consume the excess oxygen. When the oxygen concentration falls low enough, the sphincters would open once more to begin the cycle again.

The evidence against the oxygen demand theory is that under some conditions, vascular smooth muscle can remain contracted for long periods in the presence of extremely minute concentrations of oxygen—concentrations below those normally found in the tissues. A possible answer to this is that the smooth muscle in the microvessels might be genetically more sensitive to oxygen lack than are the types of smooth muscle thus far studied, muscle removed from large blood vessels, not the microvessels. Indeed, even in very small, isolated, perfused arteries (with internal diameters of about 0.5 millimeter), *marked vasodilation does occur at low oxygen concentrations.*

Thus, on the basis of available data, either a vasodilator theory or an oxygen demand theory could explain local blood flow regulation in response to the metabolic needs of the tissues. Perhaps the truth lies in a combination of the two mechanisms.

POSSIBLE ROLE OF OTHER NUTRIENTS BESIDES OXYGEN IN THE CONTROL OF LOCAL BLOOD FLOW. Under special conditions, it has been shown that lack of glucose in the perfusing blood for longer than a few minutes can cause local tissue vasodilation. Also, it is possible that this same effect occurs when other nutrients, such as amino acids or fatty acids, are deficient, although this has not been studied adequately. In addition, vasodilation occurs in the vitamin deficiency disease *beriberi,* in which the patient usually has deficiencies of the vitamin B substances thiamine, niacin, and riboflavin. In this disease, the peripheral vascular blood flow all over the body can increase twofold to threefold. Because these vitamins are all concerned with the oxidative phosphorylation mechanism for generating ATP in the local tissues, one could suspect that deficiency of these vitamins leads to diminished smooth muscle contractile ability and therefore to the local vasodilation.

Special Examples of "Metabolic" Control of Local Blood Flow

The mechanisms that we have described thus far for local blood flow control are called "metabolic mechanisms" because all of them function in response to the metabolic needs of the tissues. Two additional special instances of metabolic control of local blood flow are *reactive hyperemia* and *active hyperemia.*

REACTIVE HYPEREMIA. When the blood supply to a tissue is blocked for a few seconds to several hours and then is unblocked, the flow through the tissue usually increases to four to seven times normal; the increased flow will continue for a few seconds if the block has lasted only a few seconds but sometimes for as long as many hours if the blood flow has been stopped for an hour or more. This phenomenon is called *reactive hyperemia.* Reactive hyperemia is almost certainly another manifestation of the local "metabolic" blood flow regulation mechanism; that is, lack of flow sets into motion all of those factors that cause vasodilation. After short periods of vascular occlusion, the extra blood flow during the reactive hyperemia phase lasts long enough to repay almost exactly the tissue oxygen deficit that has accrued during the period of occlusion. This mechanism emphasizes the close connection between local blood flow regulation and delivery of oxygen and other nutrients to the tissues.

ACTIVE HYPEREMIA. When any tissue becomes highly active, such as an exercising muscle, a gastrointestinal gland during a hypersecretory period, or even the brain during rapid mental activity, the rate of blood flow through the tissue increases. Here again, by simply applying the basic principles of local blood flow control, one can easily understand this so-called *active hyperemia.* The increase in local metabolism causes the cells to devour the tissue fluid nutrients extremely rapidly and also to release large quantities of vasodilator substances. The result is to dilate the local blood vessels and, therefore, to increase local blood flow. In this way, the active tissue receives the additional nutrients required to sustain its new level of function.

As pointed out earlier, active hyperemia in skeletal muscle can increase the local blood flow as much as 20-fold during intense exercise.

"Autoregulation" of Blood Flow When the Arterial Pressure Changes from Normal— "Metabolic" Versus "Myogenic" Mechanisms

In any tissue of the body, an acute increase in arterial pressure will cause an immediate rise in blood flow. Within less than a minute, the blood flow in most tissues returns most of the way back toward the normal level. This return of flow back toward normal is called "autoregulation of blood flow." After autoregulation has occurred, the local blood flow in most body tissues will be related to arterial pressure approximately in accord with the solid curve labeled "acute" in Figure 17–4. Note that between an arterial pressure of about 70 mm Hg and 175 mm Hg, the blood flow increases only to 30 per cent.

For almost a century, two views have been proposed to explain the acute autoregulation mechanism. They have been called (1) the metabolic theory and (2) the myogenic theory.

The *metabolic theory* can easily be understood by applying the basic principles of local blood flow regulation discussed in earlier sections. Thus, when the arterial pressure becomes too great, the excess flow provides too much oxygen and too many other nutrients to the tissues, and these nutrients then cause the blood vessels to constrict and the flow to return nearly to normal despite the increased pressure.

The *myogenic theory,* on the other hand, suggests that still another mechanism not related to tissue metabolism explains the phenomenon of autoregulation. This theory is based on the observation that sudden stretch of small blood vessels will cause the smooth muscle of the vessel wall to contract. Therefore, it is

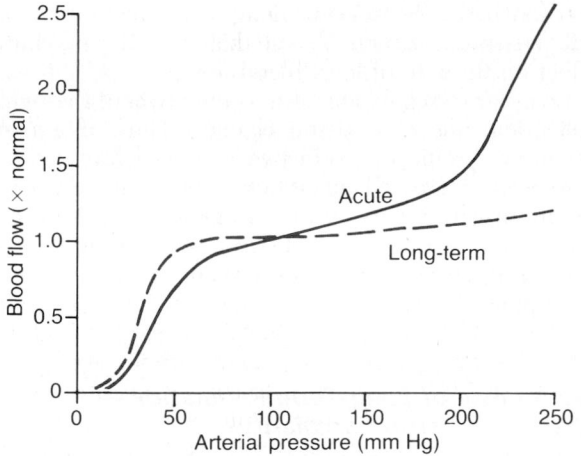

Figure 17–4. Effect of increasing arterial pressure on blood flow through a muscle. The solid curve shows the effect if the arterial pressure is raised over a period of a few minutes. The dashed curve shows the effect if the arterial pressure is raised extremely slowly over a period of many weeks.

believed that when high arterial pressure stretches the vessel, this in turn causes vascular constriction and reduces the blood flow nearly back to normal. Conversely, at low pressures, the degree of stretch of the vessel is less, so that the smooth muscle relaxes and allows increased flow.

Certainly, some experiments are difficult to explain without invoking the myogenic theory of autoregulation. Yet it is doubtful that myogenic autoregulation is a powerful functional mechanism in most of the body for a clear reason. A strong myogenic mechanism everywhere in the body would rapidly lead to death in the following way: An increase in pressure would stretch the blood vessels, which would then cause vasoconstriction. The increased constriction would increase the peripheral resistance and raise the blood pressure still more. This secondary increase in pressure would cause another cycle of stretch followed by still more vasoconstriction and then still more increase in pressure. Thus, a vicious circle would ensue, and if this should occur strongly in all the body at once, the arterial pressure would rise suddenly to such a high level that the heart would fail. Even so, in certain isolated tissues as well as in many intermediate-size arteries, it is difficult to explain some of the phenomena observed without invoking the myogenic theory—but only for certain local tissues, not for the whole body at once, which would be deadly.

It has been suggested especially that the myogenic mechanism protects the capillaries from excessively high blood pressures. That is, if the pressure in the small arteries and arterioles rises too high, these vessels would simply constrict within seconds and prevent this high pressure from being transmitted into the capillaries, which are so weak that excessive pressure could rupture them.

Special Mechanisms of Blood Flow Control in Specific Tissues

Although the general mechanisms for local blood flow control discussed thus far are present in almost all tissues of the body, distinctly different mechanisms operate in a few special areas. They are all discussed in relation to specific organs, but two notable ones are as follows.

1. In the *kidneys,* blood flow control is vested mainly in a mechanism called *tubuloglomerular feedback,* in which the composition of the fluid in the early distal tubule is detected by a tubular epithelial structure called the *macula densa* located where this tubule abuts against the afferent arteriole at the *juxtaglomerular apparatus.* When too much fluid filters from the blood through the glomerulus into the tubular system, appropriate feedback signals from the macula densa cause changes in constriction of both the afferent and efferent arterioles, in this way reducing both renal blood flow and glomerular filtration rate back to or near to normal. The details of this mechanism are discussed in Chapter 26.

2. In the *brain,* in addition to control of blood flow by tissue oxygen concentration, the concentrations of carbon dioxide and hydrogen ions play prominent roles. An increase of either of these dilates the cerebral vessels and allows rapid washout of the excess carbon dioxide or hydrogen ions. This is important because the excitatory function of the brain itself is highly dependent on exact control of both the carbon dioxide concentration and the hydrogen ion concentration. This special mechanism for cerebral blood flow control is presented in Chapter 61.

Mechanism for Dilating the Large Upstream Arteries When Microvascular Blood Flow Increases—The Endothelial-Derived Relaxing Factor (Nitric Oxide)

The local mechanisms for controlling tissue blood flow can dilate only the very small microvessels located in the immediate tissue itself because local feedback caused by vasodilator substances or oxygen deficiency can reach only these vessels, not the intermediate and larger arteries back upstream. Yet, when blood flow through the microvascular portion of the circulation increases, this entrains secondarily another mechanism that does dilate the larger arteries as well. This mechanism is the following.

The endothelial cells lining the arterioles and small arteries synthesize several substances that, when released, can affect the degree of contraction of the arterial wall. The most important of these is a vasodilator substance called *endothelial-derived relaxing factor,* which is composed principally, if not entirely, of nitric oxide that has a half-life in the blood of only 6 seconds. Rapid flow of blood through the arteries causes "shear-stress" on the endothelial cells because of viscous drag of the blood against the vascular walls. This stress contorts the endothelial cells in the direction of flow and causes greatly increased release of nitric oxide. The nitric oxide then relaxes the arterial wall, causing it to dilate.

This is a fortunate mechanism because it causes a secondary increase in the dimensions of the larger blood vessels whenever the microvascular blood flow increases. Without such a response, the effectiveness of local blood flow control would be greatly compromised and often totally ineffectual because much of the resistance to blood flow is in the upstream arterioles and small arteries.

Multiple other stimuli also can cause nitric oxide to be released from the endothelium. These stimuli include acetylcholine, bradykinin, ATP, and others. The nitric oxide in turn causes the local blood vessel to dilate. For instance, when acetylcholine is secreted by autonomic nerve endings, it first causes nitric oxide release from the endothelial cells; then the nitric oxide dilates the local blood vessel. If the endothelial cells have been destroyed or rendered nonfunctional, no dilatation will occur in response to the autonomic stimulation.

(Nitric oxide has recently been found to have multiple other physiological functions, some of which are discussed at other points in this text. Among them are the following: (1) It functions as a powerful toxin released by white blood cells to kill bacteria and tumor cells. (2) It functions as a messenger molecule in several discrete areas of the brain to transmit nerve signals from one neuron to the next. (3) It functions as a vasodilator transmitter substance at the parasympathetic nerve endings in the penis, in this way causing penile erection. (4) It possibly even functions in some of the memory and thought mechanisms of the brain.)

Long-Term Blood Flow Regulation

Thus far, most of the mechanisms for local blood flow regulation that we have discussed act within a few seconds to a few minutes after the local tissue conditions have changed. Yet, even after full function of these acute mechanisms, the blood flow usually is adjusted only about three quarters of the way to the exact requirement of the tissues. For instance, when the arterial pressure is suddenly increased from 100 to 150 mm Hg, the blood flow increases almost instantaneously about 100 per cent. Then, within 30 seconds to 2 minutes, the flow decreases back to about 15 per cent above the original control value. This illustrates the rapidity of the acute type of local regulation, but at the same time, it demonstrates that the regulation is still incomplete because there remains an excess 15 per cent increase in blood flow.

However, over a period of hours, days, and weeks, a long-term type of local blood flow regulation develops in addition to the acute regulation. This long-term regulation gives far more complete regulation than the acute mechanism. For instance, in the example above, if the arterial pressure remains at 150 mm Hg indefinitely, within a few weeks the blood flow through the tissues will gradually reapproach almost exactly the normal value. Figure 17–4 shows by the dashed curve the extreme effectiveness of this long-term local blood flow regulation. Note that once the long-term regula-

tion has had time to occur, long-term changes in arterial pressure between 50 and 250 mm Hg have little effect on the rate of local blood flow.

Long-term regulation also occurs when the metabolic demands of a tissue change. Thus, if a tissue becomes chronically overactive and therefore requires chronically increased quantities of nutrients, the blood supply usually increases within a few weeks almost to match the needs of the tissue—that is, unless the circulatory system has become pathological or too old to respond.

Mechanism of Long-Term Regulation— Change in Tissue Vascularity

The mechanism of long-term local blood flow regulation is a change in the degree of vascularity of the tissues. That is, if the arterial pressure falls to 60 mm Hg and remains at this level for many weeks, the physical structural sizes of the vessels in the tissue increase, and under some conditions, even the number of vessels increases; on the other hand, if the pressure rises to a very high level, the number and sizes of vessels decrease. Likewise, if the metabolism in a given tissue is increased for a prolonged period, vascularity increases; if the metabolism is decreased, vascularity decreases.

Thus, there is reconstruction of the tissue vasculature to meet the needs of the tissues. This reconstruction occurs rapidly (within days) in extremely young animals. It also occurs rapidly in new growth of tissue, such as in scar tissue and cancerous tissue; on the other hand, it occurs slowly in old, well-established tissues. Therefore, the time required for long-term regulation to take place may be only a few days in the neonate or as long as months or even years in the elderly person. Furthermore, the final degree of response is much greater in younger tissues than in older, so that in the neonate, the vascularity will adjust to match almost exactly the needs of the tissue for blood flow, whereas in older tissues, vascularity frequently lags far behind the needs of the tissues.

ROLE OF OXYGEN IN LONG-TERM REGULATION. Oxygen is important not only for acute control of local blood flow but also for long-term control. One effect of this is to increase the vascularity in the tissues of many animals that live at high altitudes, where the atmospheric oxygen is low. A second effect is that fetal chicks hatched in low oxygen have up to twice as much vascular conductivity as is normally true. The effect is also dramatically demonstrated in premature human neonates who are put into oxygen tents for therapeutic purposes. The excess oxygen causes almost immediate cessation of new vascular growth in the retina of the eye and even causes degeneration of some of the capillaries that have already formed. Then when the infants are taken out of the oxygen, there is explosive overgrowth of new vessels to make up for the sudden decrease in available oxygen; indeed, there is often so much overgrowth that the vessels grow into the eye's vitreous humor and eventually cause blindness. (This condition is called *retrolental fibroplasia*.)

Growth of New Vessels—Angiogenesis and Angiogenic Factors

The term "angiogenesis" means growth of new blood vessels. Angiogenesis occurs mainly in response to the presence of angiogenic factors released from (1) ischemic tissues, (2) tissues that are growing rapidly, or (3) tissues that have excessively high metabolic rates.

A dozen or more such angiogenic factors have been found, almost all of which are small peptides. Three of those that have been best characterized are *endothelial cell growth factor, fibroblast growth factor,* and *angiogenin,* each of which has been isolated either from tumors or from other tissues that have inadequate blood supply. Presumably, it is the deficiency of tissue oxygen, other nutrients, or both that leads to the formation of the angiogenic factors.

Essentially all the angiogenic factors promote new vessel growth in the same manner. They cause new vessels to sprout from either small venules or, occasionally, capillaries. The first step is dissolution of the basement membrane of the endothelial cells at the point of sprouting. This is followed by rapid reproduction of new endothelial cells that then stream out of the vessel wall in extended cords directed toward the source of the angiogenic factor. The cells in each cord continue to divide and eventually fold over into a tube. Next, the tube connects with another tube budding from another donor vessel and forms a capillary loop through which blood begins to flow. If the flow is great enough, smooth muscle cells eventually invade the wall, so that some of the new vessels eventually grow to be small arterioles or perhaps even larger arteries. Thus, angiogenesis explains the manner in which metabolic factors in local tissues can cause the growth of new vessels.

Certain other substances, such as some steroid hormones, have exactly the opposite effect on blood vessels, occasionally even causing dissolution of vascular cells and disappearance of vessels. Therefore, blood vessels can increase when needed or at other times disappear.

Development of Collateral Circulation— A Phenomenon of Long-Term Local Blood Flow Regulation

When an artery or a vein is blocked, a new vascular channel usually develops around the blockage and allows at least partial resupply of blood to the affected tissue. The first stage in this process is dilatation of vascular loops around the point of blockage that already connect the vessel above the blockage to the vessel below. This dilatation occurs within the first minute or two, indicating that it is simply a metabolic relaxation of the muscle fibers of the small vessels involved. After this initial opening of these collateral vessels, the blood flow usually will still be less than one quarter that needed to supply the tissue needs. Further opening occurs within the ensuing hours, so that within 1 day, as much as half the tissue needs may be met and within a few days often all the tissue needs. The collateral vessels continue to grow in size for many months thereafter, almost always forming multiple small collateral channels rather than one single large vessel. Under resting conditions, the blood flow usually returns to normal, but the new channels seldom become large enough to supply maximum blood flow during strenuous tissue activity.

Thus, the development of collateral vessels follows the usual principles of both acute and long-term local blood flow control, the acute control being rapid metabolic dilatation, followed by growth and enlargement of the vessels manyfold over a period of weeks and months.

The most important example of the development of collateral blood vessels occurs after thrombosis of one of the coronary arteries. Almost all people by the age of 60 will have had at least one of the smaller coronary vessels to close. Yet most people will not know that this has happened because collaterals will have developed rapidly enough to prevent myocardial damage. It is in those instances in which thrombosis occurs too rapidly for simultaneous development of collaterals that serious heart attacks occur.

HUMORAL REGULATION OF THE CIRCULATION

Humoral regulation of the circulation means regulation by substances secreted or absorbed into the body fluids, such as hormones and ions. Some of these substances are formed by special glands and then transported in the blood throughout the entire body. Others are formed in local tissue areas and cause only local circulatory effects. Among the most important of the humoral factors that affect circulatory function are the following.

Vasoconstrictor Agents

NOREPINEPHRINE AND EPINEPHRINE. Norepinephrine is an especially powerful vasoconstrictor hormone; epinephrine is less so and, in some instances, even causes mild vasodilation, which occasionally occurs in the heart to dilate the coronary arteries during increased heart activity. When the sympathetic nervous system is stimulated in most or all parts of the body during stress or exercise, the sympathetic nerve endings in the individual tissues release norepinephrine that excites the heart, veins, and arterioles. The sympathetic nerves to the adrenal medullae also cause these glands to secrete both norepinephrine and epinephrine into the blood. These hormones then circulate to all areas of the body and cause almost the same excitatory effects on the circulation as direct sympathetic stimulation, thus providing a dual system of control.

ANGIOTENSIN. Angiotensin is one of the most powerful vasoconstrictor substances known. As little as *one millionth* of a gram can increase the arterial pressure of a human 50 mm Hg or more.

The effect of angiotensin is to powerfully constrict the small arterioles. If this occurs in an isolated tissue area, the blood flow to the area can be severely depressed. However, the real importance of angiotensin in the blood is that it normally acts simultaneously on all the arterioles of the body to increase the *total* peripheral resistance, thereby increasing the arterial pressure. Because of this, plus several renal and adrenocortical stimulatory effects of angiotensin, this hormone plays an integral role in the regulation of the arterial pressure, as discussed in detail in Chapter 19.

VASOPRESSIN. Vasopressin, also called *antidiuretic hormone,* is even more powerful than angiotensin as a vasoconstrictor, thus making it perhaps the body's most potent constrictor substance. It is formed in the hypothalamus (see Chapter 75) but transported down the center of nerve axons to the posterior pituitary gland, where it is eventually secreted into the blood.

It is clear that vasopressin could have intense effects on circulatory function. Yet, normally, only minute amounts of vasopressin are secreted, so that most physiologists have thought that vasopressin plays little role in vascular control. On the other hand, experiments have now shown that the concentration of circulating vasopressin during severe hemorrhage can rise high enough to increase the arterial pressure as much as 60 mm Hg, and in many instances, this can, by itself, bring the arterial pressure almost back up to normal.

Also, vasopressin has an *all-important* function in controlling water reabsorption in the renal tubules, discussed in Chapter 28, and therefore to help control body fluid volume. That is why this hormone is also called antidiuretic hormone.

ENDOTHELIN—A POWERFUL VASOCONSTRICTOR IN DAMAGED BLOOD VESSELS. Still another vasoconstrictor substance that ranks along with angiotensin and vasopressin in its vasoconstrictor capability is a large peptide (21 amino acids) called *endothelin,* which requires only nanogram quantities to cause powerful vasoconstriction. This substance is present in the endothelial cells of all or most blood vessels. The usual stimulus for release is damage to the endothelium, such as caused by crushing of the tissues or by injecting a traumatizing chemical into the blood vessel. After severe blood vessel damage, it is probably the release of local endothelin and subsequent vasoconstriction that prevents extensive bleeding from arteries as large as 5 millimeters in diameter that have been broken wide open by crushing injury.

A special function of endothelin might be the constriction of the umbilical artery of a neonate immediately after birth. Also, chronic infusion of endothelin into arteries in most parts of the body will cause prolonged vasoconstriction. It is especially potent in constricting coronary, renal, mesenteric, and cerebral arteries, but it is questionable whether this is a normal function of endothelin except in response to tissue damage.

Vasodilator Agents

BRADYKININ. Several substances called *kinins* that can cause powerful vasodilation are formed in the blood and tissue fluids of some organs. One of these substances is *bradykinin.*

The kinins are small polypeptides that are split away by proteolytic enzymes from alpha$_2$-globulins in the plasma or tissue fluids. A proteolytic enzyme of particular importance is *kallikrein,* which is present in the blood and tissue fluids in an inactive form. Kallikrein is activated by maceration of the blood, tissue inflammation, and other similar chemical and physical effects on the blood or tissues. As kallikrein becomes activated, it acts immediately on the alpha$_2$-globulin to release a kinin called *kallidin* that is then converted by tissue enzymes into *bradykinin.* Once formed, the bradykinin persists for only a few minutes because it is inactivated by the enzyme *carboxypeptidase* or by *converting enzyme,* an enzyme that also plays an essential role in activating angiotensin, as discussed in Chapter 19. The activated kallikrein enzyme is destroyed by a kallikrein inhibitor also present in the body fluids.

Bradykinin causes powerful *arteriolar dilatation* and *increased capillary permeability.* For instance, the injection of 1 *microgram* of bradykinin into the brachial artery of a person increases the blood flow through the arm as much as sixfold, and even smaller amounts injected locally into tissues can cause marked edema because of the increase in capillary pore size.

There is reason to believe that kinins play special roles in regulating blood flow and capillary leakage of fluids in inflamed tissues. It is also believed that bradykinin plays a role in regulating blood flow in the skin as well as in the salivary and gastrointestinal glands.

SEROTONIN. Serotonin (5-hydroxytryptamine) is present in large concentrations in the chromaffin tissue of the intestine and other abdominal structures. Also, it is present in high concentration in the platelets. Serotonin can have either a vasodilator or a vasoconstrictor effect, depending on the condition or the area of the circulation. Even though these effects can be powerful, most functions of serotonin in regulation of the circulation are unknown. Occasionally, tumors composed of chromaffin tissue develop, called *carcinoid tumors.* They secrete tremendous quantities of serotonin and cause mottled areas of vasodilation in the skin; but the very fact that these tremendous quantities of serotonin do not drastically disturb most aspects of circulatory function makes it doubtful that serotonin plays a widespread general role in circulatory regulation.

HISTAMINE. Histamine is released in essentially every tissue of the body when it becomes damaged or inflamed or is the subject of an allergic reaction. Most of the histamine is derived from mast cells in the damaged tissues and from basophils in the blood.

Histamine has a powerful vasodilator effect on the arterioles and, like bradykinin, has the ability to greatly increase capillary porosity, allowing leakage of both fluid and plasma protein into the tissues. In many pathological conditions, the intense arteriolar dilatation and increased capillary porosity produced by histamine cause tremendous quantities of fluid to leak out of the circulation into the tissues, inducing edema. The local vasodilatory and edema-producing effects of histamine are especially prominent in allergic reactions and are discussed in Chapter 34.

PROSTAGLANDINS. Almost every tissue of the body contains small to moderate amounts of several chemically related substances called prostaglandins. These substances have important intracellular effects, but in addition, some of them are released into the local tissue fluids and the circulating blood under both physiological and pathological conditions. Although some of the prostaglandins cause vasoconstriction, most of the more important ones seem to be mainly vasodilator agents. Thus far, no specific pattern of function of the prostaglandins in circulatory control has been found. However, their widespread prevalence in the tissues and their myriad effects on the circulation make them ideal candidates for special roles in circulatory control in local vascular areas. For this reason, these substances are under intensive research investigation, although unequivocal conclusions about their roles in circulatory regulation are evasive.

Effects of Ions and Other Chemical Factors on Vascular Control

Many different ions and other chemical factors can either dilate or constrict local blood vessels. Most of them have little function in the overall *regulation* of the circulation, but their specific effects can be listed as follows.

An increase in *calcium ion* concentration causes vasoconstriction. This results from the general effect of calcium to stimulate smooth muscle contraction, as discussed in Chapter 8.

An increase in *potassium ion* concentration causes vasodilation. This results from the ability of potassium ions to inhibit smooth muscle contraction.

An increase in *magnesium ion* concentration causes powerful vasodilation because magnesium ions inhibit smooth muscle generally.

Increased *sodium ion* concentration causes mild arteriolar dilatation. This results mainly from an increase in osmolality of the fluids rather than from a specific effect of sodium ion itself. *Increased osmolality* of the blood caused by increased quantities of *glucose* or other nonvasoactive substances also causes mild arteriolar dilatation. Decreased osmolality causes mild arteriolar constriction.

The only anions to have significant effects on blood vessels are *acetate* and *citrate*, both of which cause mild degrees of vasodilation.

An *increase in hydrogen ion* concentration (decrease in pH) causes dilatation of the arterioles. A slight *decrease in hydrogen ion* concentration causes arteriolar constriction, but an intense decrease causes dilatation.

An increase in carbon dioxide concentration causes moderate vasodilation in most tissues and marked vasodilation in the brain. However, carbon dioxide, acting on the brain vasomotor center, has an extremely powerful indirect effect, transmitted through the sympathetic nervous vasoconstrictor system, to cause widespread vasoconstriction throughout the body.

REFERENCES

Allegra, C., et al.: Vasomotion and Flowmotion. Farmington, CT, S. Karger Publishers, Inc., 1993.

Antonaccio, M. J., (ed.): Cardiovascular Pharmacology. New York, Raven Press, 1984.

Banchero, N.: Cardiovascular responses to chronic hypoxia. Annu. Rev. Physiol., 49:465, 1987.

Brenner, B. M., and Laragh, J. H.: Advances in Atrial Peptide Research (American Society of Hypertension Symposium Series, Vol. 2). New York, Raven Press, 1988.

Buckley, J. P., and Ferrario, C. M.: Brain Peptides and Catecholamines in Cardiovascular Regulation. New York, Raven Press, 1987.

Chien, S. (ed.): Vascular Endothelium in Health and Disease. New York, Plenum Publishing Corp., 1988.

Cowley, A. W., and Guyton, A. C.: Quantification of intermediate steps in the renin-angiotensin-vasoconstrictor feedback loop in the dog. Circ. Res., 30:557, 1972.

Cowley, A. W., Jr., et al.: Vasopressin: Cellular and Integrative Functions. New York, Raven Press, 1988.

Duling, B.: Oxygen, metabolism, and microcirculatory regulation. In Kaley, G., and Altura, B. M. (eds.): Microcirculation. Vol. II. Baltimore, University Park Press, 1977.

Duling, B. R., and Klitzman, B.: Local control of microvascular function: Role in tissue oxygen supply. Annu. Rev. Physiol., 42:373, 1980.

Folkman, J.: Angiogenesis: What makes blood vessels grow? News Physiol. Sci., 1:199, 1986.

Furchgott, R. F., and Vanhoutte, P. M.: Endothelium-derived relaxing and contracting factors. FASEB J., 3:1007, 1989.

Granger, H. J., and Guyton, A. C.: Autoregulation of the total systemic circulation following destruction of the central nervous system in the dog. Circ. Res., 25:379, 1969.

Guyton, A. C.: Integrative hemodynamics. In Sodeman, W. A., Jr., and Sodeman, T. M. (eds.): Pathologic Physiology: Mechanisms of Disease. 6th Ed. Philadelphia, W. B. Saunders Co., 1979, p., 169.

Guyton, A. C., et al.: Cardiac Output and Its Regulation. Philadelphia, W. B. Saunders Co., 1973.

Guyton, A. C., et al.: Circulation: Overall regulation. Annu. Rev. Physiol., 34:13, 1972.

Haddy, F. J.: The role of a humoral Na^+, K^+-ATPase inhibitor in regulating precapillary vessel tone. J. Cardiovasc. Pharmacol., 6:(Suppl. 2)S439, 1984.

Harris, P. D.: Movement of oxygen in skeletal muscle. News Physiol. Sci., 1:147, 1986.

Hudlicka, O.: Development of microcirculation: Capillary growth and adaptation. In Renkin, E. M., and Michel, C. C. (eds.): Handbook of Physiology. Sec. 2, Vol. IV. Bethesda, Md., American Physiological Society, 1984, p. 165.

Hudlicka, O., et al.: Angiogenesis in skeletal and cardiac muscle. Physiol. Rev., 72:369, 1992.

Johnson, P. C.: The myogenic response. In Bohr, D. F., et al. (eds.): Handbook of Physiology. Sec. 2. Vol. II. Baltimore, Williams & Wilkins, 1980, p. 409.

Kreye, V. A.: Direct vasodilators with unknown modes of action: The nitrocompounds and hydralazine. J. Cardiovasc. Pharmacol., 6:(Suppl. 4)S646, 1984.

Manfredi, J. P., and Holmes, E. W.: Purine salvage pathways in myocardium. Annu. Rev. Physiol., 47:691, 1985.

Maragoudakis, M. E., et al.: Angiogenesis in Health and Disease. New York, Plenum Publishing Corp., 1992.

Moschella, S. L., and Hurley, H. J.: Dermatology. Philadelphia, W. B. Saunders Co., 1992.

Opie, L. H.: Ace inhibitors: Almost too good to be true. New York, Scientific American Science and Medicine, July/August, 1994.

Pearson, J. D., and Gordon, J. L.: Nucleotide metabolism by endothelium. Annu. Rev. Physiol., 47:617, 1985.

Peroutka, S. J.: 5-Hydroxytryptamine receptor subtypes. Annu. Rev. Neurosci., 11:45, 1988.

Renkin, E. M.: Control of microcirculation and blood-tissue exchange. In Renkin, E. M., and Michel, C. C. (eds.): Handbook of Physiology. Sec. 2, Vol. IV. Bethesda, Md., American Physiological Society, 1984, p. 627.

Rosenthal, D. L., and Guyton, A. C.: Hemodynamics of collateral vasodilatation following femoral artery occlusion in anesthetized dogs. Circ. Res., 23:239, 1968.

Rubanyi, G. M.: Endothelin. New York, Oxford University Press, 1992.

Samuelsson, B., et al.: Prostaglandins and Related Compounds. New York, Raven Press, 1991.

Sarrel, P. M., et al.: Estrogen actions in arteries, bone, and brain. New York, Scientific American Science and Medicine, July/August, 1994.

Schmid-Schonbein, G. W.: Granulocyte: Friend and foe. News Physiol. Sci., 3:144, 1988.

Seidel, C. L., and Schildmeyer, L. A.: Vascular smooth muscle adaptation to increased load. Annu. Rev. Physiol., 49:489, 1987.

Smith, J. J., and Kampine, J. P.: Circulatory Physiology—The Essentials. Baltimore, Williams & Wilkins, 1990.

Smith, W. L.: Prostaglandin biosynthesis and its compartmentation in vascular smooth muscle and endothelial cells. Annu. Rev. Physiol., 48:251, 1986.

Steranka, L. R., et al.: Antagonists of B_2 bradykinin receptors. FASEB J., 3:2019, 1989.

Vallee, B. L., et al.: Tumor derived angiogenesis factors from rat Walker 256 carcinoma: An experimental investigation and review. Experientia, 41:1, 1985.

Vanhoutte, P. M.: Endothelium and the control of vascular tissue. News Physiol. Sci., 2:18, 1987.

Vanhoutte, P. M.: Vasodilation: Vascular Smooth Muscle, Peptides, Autonomic Nerves, and Endothelium. New York, Raven Press, 1988.

Vanhoutte, P. M., et al.: Local modulation of the adrenergic neuroeffector interaction in the blood vessel wall. Physiol. Rev., 61:151, 1981.

Williams, M.: Adenosine and Adenosine Receptors. Totowa, NJ, Humana Press, Inc., 1990.

Wyse, G. D., et al.: Responses of human peripheral arteries to agonists: Similarities and differences. Fed. Proc., 44:331, 1985.

Nervous Regulation of the Circulation, and Rapid Control of Arterial Pressure

CHAPTER 18

NERVOUS REGULATION OF THE CIRCULATION

As discussed in Chapter 17, nervous control normally has little to do with adjustment of blood flow tissue by tissue—this is the function of the local tissue blood flow control. Instead, nervous control mainly affects more global functions, such as redistributing the blood flow to different areas of the body, increasing the pumping activity of the heart, and, especially, providing rapid control of arterial pressure.

The means by which the nervous system controls the circulation is almost entirely through the autonomic nervous system. The total function of this system is presented in Chapter 60. Its specific anatomical and functional characteristics relating to circulatory control require special attention here as well.

Autonomic Nervous System

By far the most important part of the autonomic nervous system for regulation of the circulation is the *sympathetic nervous system.* The *parasympathetic nervous system* is also important in contributing to regulation of heart function, as we see later in the chapter.

SYMPATHETIC NERVOUS SYSTEM. Figure 18–1 shows the anatomy of sympathetic nervous control of the circulation. Sympathetic vasomotor nerve fibers leave the spinal cord through all the thoracic and the first one to two lumbar spinal nerves. They pass into the sympathetic chain and thence by two routes to the circulation: (1) through specific *sympathetic nerves* that innervate mainly the vasculature of the internal

viscera and the heart and (2) through the *spinal nerves* that innervate mainly the vasculature of the peripheral areas. The precise pathways of these fibers in the spinal cord and in the sympathetic chains are discussed in Chapter 60.

Sympathetic Innervation of the Blood Vessels. Figure 18–2 shows the distribution of sympathetic nerve fibers to the blood vessels, demonstrating that all the vessels except the capillaries, precapillary sphincters, and most of the metarterioles are innervated.

The innervation of the *small arteries* and *arterioles* allows sympathetic stimulation to increase the *resistance* and thereby to *decrease* the rate of blood flow through the tissues.

The innervation of large vessels, particularly of the *veins,* makes it possible for sympathetic stimulation to *decrease* the volume of these vessels and thereby to alter the volume of the peripheral circulatory system. This can translocate blood into the heart and thereby play a major role in the regulation of cardiovascular function, as we see later in this and subsequent chapters.

Sympathetic Nerve Fibers to the Heart. In addition to sympathetic nerve fibers supplying the blood vessels, sympathetic fibers also go to the heart, as discussed in Chapter 9. It should be recalled that sympathetic stimulation markedly increases the activity of the heart, both increasing the heart rate and enhancing its strength of pumping.

PARASYMPATHETIC CONTROL OF HEART FUNCTION, ESPECIALLY HEART RATE. Although the parasympathetic nervous system is exceedingly important for

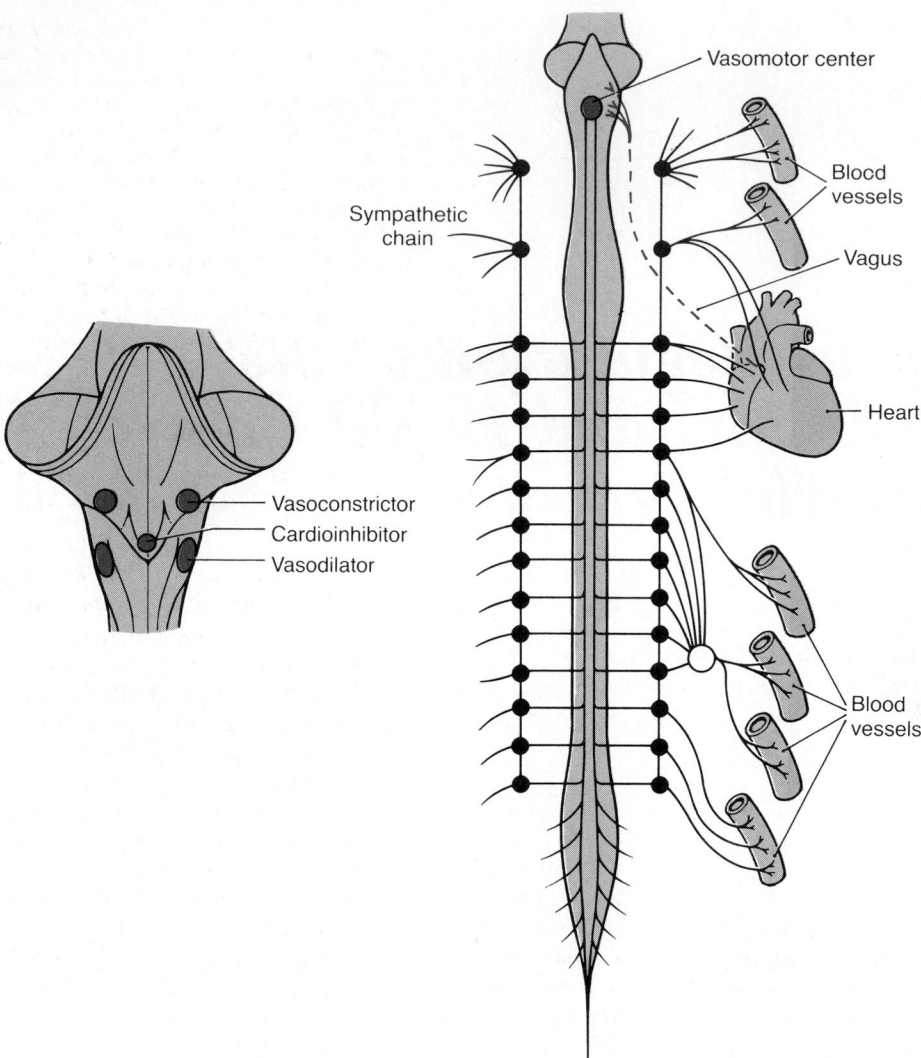

Figure 18–1. Anatomy of sympathetic nervous control of the circulation.

many other autonomic functions of the body, it plays only a minor role in regulation of the circulation. Its only really important circulatory effect is its control of heart rate by way of parasympathetic fibers carried to

the heart in the *vagus nerves,* shown in Figure 18–1 by the dashed colored nerve from the medulla directly to the heart.

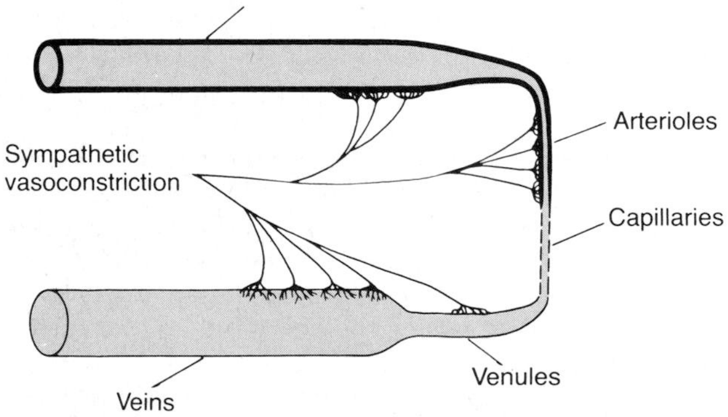

The effects of parasympathetic stimulation on heart function are discussed in detail in Chapter 9. Principally, parasympathetic stimulation causes a marked *decrease* in heart rate and a slight decrease in heart muscle contractility.

Sympathetic Vasoconstrictor System and Its Control by the Central Nervous System

The sympathetic nerves carry tremendous numbers of *vasoconstrictor fibers* and only a few vasodilator fibers. The vasoconstrictor fibers are distributed to essentially all segments of the circulation. This distribution is greater in some tissues than in others. It is especially powerful in the kidneys, gut, spleen, and skin but less potent in skeletal muscle and the brain.

VASOMOTOR CENTER AND ITS CONTROL OF THE VASOCONSTRICTOR SYSTEM. Located bilaterally mainly

Figure 18–2. Sympathetic innervation of the systemic circulation.

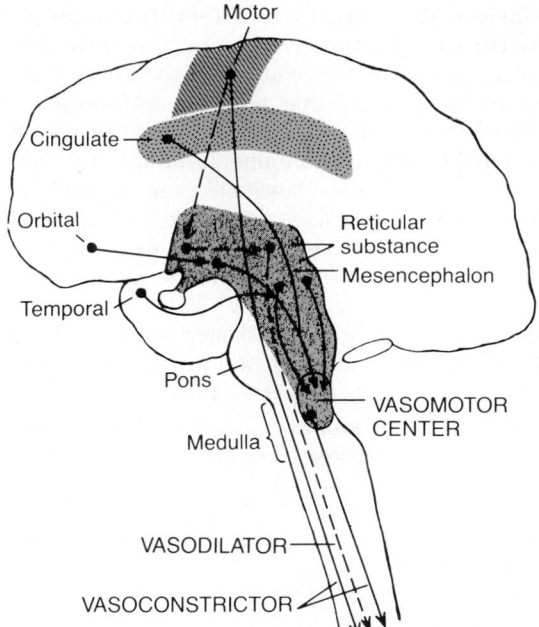

Figure 18–3. Areas of the brain that play important roles in the nervous regulation of the circulation. The dashed lines represent inhibitory pathways.

in the reticular substance of the medulla and lower third of the pons, shown in Figures 18–1 and 18–3, is an area called the *vasomotor center.* The center transmits parasympathetic impulses through the vagus nerves to the heart and transmits sympathetic impulses through the cord and peripheral sympathetic nerves to all or almost all the blood vessels of the body.

Although the total organization of the vasomotor center is still unclear, experiments have made it possible to identify certain important areas in the center, as follows:

1. A *vasoconstrictor area,* called area "C-1," located bilaterally in the anterolateral portions of the upper medulla. The neurons in this area secrete *norepinephrine;* their fibers are distributed throughout the cord, where they excite the vasoconstrictor neurons of the sympathetic nervous system.

2. A *vasodilator area,* called area "A-1," located bilaterally in the anterolateral portions of the lower half of the medulla. The fibers from these neurons project upward to the vasoconstrictor area (C-1) and inhibit the vasoconstrictor activity of that area, thus causing vasodilation.

3. A *sensory area,* area "A-2," located bilaterally in the *tractus solitarius* in the posterolateral portions of the medulla and lower pons. The neurons of this area receive sensory nerve signals mainly from the vagus and glossopharyngeal nerves, and the output signals from this sensory area then help to control the activities of both the vasoconstrictor and vasodilator areas, thus providing "reflex" control of many circulatory functions. An example is the baroreceptor reflex for controlling arterial pressure, which we describe later in the chapter.

CONTINUOUS PARTIAL CONSTRICTION OF THE BLOOD VESSELS CAUSED BY SYMPATHETIC VASOCONSTRICTOR TONE. Under normal conditions, the vasoconstrictor area of the vasomotor center transmits signals continuously to the sympathetic vasoconstrictor nerve fibers over the entire body, causing continuous slow firing of these fibers at a rate of about one half to two impulses per second. This continual firing is called *sympathetic vasoconstrictor tone.* These impulses maintain a partial state of contraction in the blood vessels called *vasomotor tone.*

Figure 18–4 demonstrates the significance of vasoconstrictor tone. In the experiment of this figure, total spinal anesthesia was administered to an animal, which blocked all transmission of sympathetic nerve impulses from the central nervous system to the periphery. As a result, the arterial pressure fell from 100 to 50 mm Hg, demonstrating the effect of loss of vasoconstrictor tone throughout the body. A few minutes later a small amount of the hormone norepinephrine was injected intravenously—norepinephrine is the

Figure 18–4. Effect of total spinal anesthesia on the arterial pressure, showing a marked fall in pressure resulting from loss of vasomotor tone.

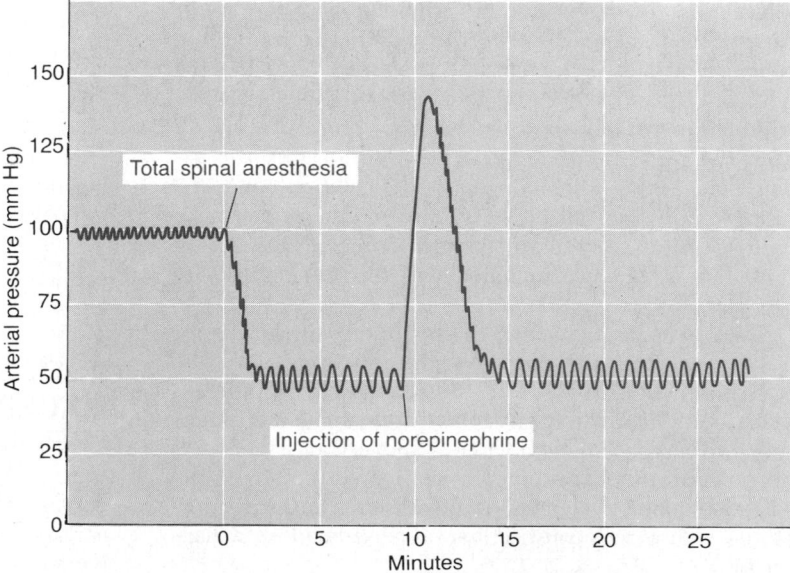

hormonal substance secreted at the endings of sympathetic vasoconstrictor nerve fibers throughout the body. As this hormone was transported in the blood to all the blood vessels, the vessels once again became constricted, and the arterial pressure rose to a level even greater than normal for 1 or 2 minutes until the norepinephrine was itself destroyed.

CONTROL OF HEART ACTIVITY BY THE VASOMOTOR CENTER. At the same time the vasomotor center is controlling the degree of vascular constriction, it is controlling heart activity. The lateral portions of the vasomotor center transmit excitatory impulses through the sympathetic nerve fibers to the heart to increase heart rate and contractility, whereas the medial portion of the vasomotor center, which lies in immediate apposition to the *dorsal motor nucleus of the vagus nerves,* transmits impulses through the vagus nerves to the heart to decrease heart rate. Therefore, the vasomotor center can either increase or decrease heart activity. This ordinarily increases at the same time that vasoconstriction occurs throughout the body and ordinarily decreases at the same time that vasoconstriction is inhibited.

CONTROL OF THE VASOMOTOR CENTER BY HIGHER NERVOUS CENTERS. Large numbers of areas throughout the *reticular substance* of the *pons, mesencephalon,* and *diencephalon* can either excite or inhibit the vasomotor center. This reticular substance is shown in Figure 18–3 by the diffuse shaded area. In general, the more lateral and superior portions of the reticular substance cause excitation, whereas the more medial and inferior portions cause inhibition.

The *hypothalamus* plays a special role in the control of the vasoconstrictor system because it can exert either powerful excitatory or inhibitory effects on the vasomotor center. The *posterolateral portions* of the hypothalamus cause mainly excitation, whereas the *anterior part* can cause mild excitation or inhibition, depending on the precise part of the anterior hypothalamus stimulated.

Many parts of the *cerebral cortex* can also excite or inhibit the vasomotor center. Stimulation of the *motor cortex,* for instance, excites the vasomotor center because of impulses transmitted downward into the hypothalamus and thence to the vasomotor center. Also, stimulation of the *anterior temporal lobe,* the *orbital areas of the frontal cortex,* the *anterior part of the cingulate gyrus,* the *amygdala,* the *septum,* and the *hippocampus* can all either excite or inhibit the vasomotor center, depending on the precise portion of these areas that is stimulated and on the intensity of the stimulus.

Thus, widespread areas of the brain can have profound effects on the cardiovascular function.

NOREPINEPHRINE—THE SYMPATHETIC VASOCONSTRICTOR TRANSMITTER SUBSTANCE. The substance secreted at the endings of the vasoconstrictor nerves is norepinephrine. Norepinephrine acts directly on the so-called alpha receptors of the vascular smooth muscle to cause vasoconstriction, as discussed in Chapter 60.

ADRENAL MEDULLAE AND THEIR RELATION TO THE SYMPATHETIC VASOCONSTRICTOR SYSTEM. Sympathetic impulses are transmitted to the adrenal medullae at the same time that they are transmitted to all the blood vessels. They cause the medullae to secrete both epinephrine and norepinephrine into the circulating blood. These two hormones are carried in the blood stream to all parts of the body, where they act directly on the blood vessels usually to cause vasoconstriction, but sometimes the epinephrine causes vasodilation because it has a potent "beta" receptor stimulatory effect, which often dilates vessels in certain tissues of the body, as discussed in Chapter 60.

Sympathetic Vasodilator System and Its Control by the Central Nervous System

The sympathetic nerves to skeletal muscles carry sympathetic vasodilator fibers as well as constrictor fibers. In lower animals, such as the cat, these fibers release *acetylcholine,* not norepinephrine, at their endings, although in primates, the vasodilator effect might be caused by epinephrine exciting the beta receptors in the muscle vasculature.

The pathways for central nervous system control of the vasodilator system are shown by the dashed lines in Figure 18–3. The principal area of the brain controlling this system is the *anterior hypothalamus.*

IMPORTANCE OF THE SYMPATHETIC VASODILATOR SYSTEM. It is doubtful that the sympathetic vasodilator system plays an important role in the control of the circulation in the human because complete block of the sympathetic nerves to the muscles hardly affects the ability of the muscles to control their own blood flow in response to their needs. Yet it is possible that at the onset of exercise, the sympathetic vasodilator system causes initial vasodilation in the skeletal muscles to allow an *anticipatory increase in blood flow* even before the muscles require increased nutrients.

EMOTIONAL FAINTING—VASOVAGAL SYNCOPY. A particularly interesting vasodilatory reaction occurs in people who experience intense emotional disturbances that cause fainting. The muscle vasodilator system becomes powerfully activated, and at the same time, the vagal cardioinhibitory center transmits strong signals to the heart to slow markedly the heart rate. The arterial pressure falls instantly, which also reduces blood flow to the brain and causes the person to lose consciousness. This overall effect is called *vasovagal syncopy.* Emotional fainting begins with disturbing thoughts in the cerebral cortex. The pathway probably then goes to the vasodilatory center of the anterior hypothalamus, then to the vagal centers of the medulla, and finally through the cord to the vasodilator nerves of the muscles.

ROLE OF THE NERVOUS SYSTEM FOR RAPID CONTROL OF ARTERIAL PRESSURE

One of the most important functions of nervous control of the circulation is its capability to cause rapid increases in arterial pressure. For this purpose, the entire vasoconstrictor and cardioaccelerator functions of the sympathetic nervous system are stimulated as a unit. At the same time, there is reciprocal inhibition of the parasympathetic vagal inhibitory signals to the heart. In consequence, three major changes occur simultaneously, each of which helps to increase the arterial pressure. They are as follows:

1. *Almost all arterioles of the body are constricted.* This greatly increases the total peripheral resistance, impeding the run-off of blood from the arteries and thereby increasing the arterial pressure.

2. *The veins especially but the other large vessels of the circulation as well are strongly constricted.* This displaces blood out of the large peripheral blood vessels toward the heart, thus increasing the volume of blood in the heart chambers. This then causes the heart to beat with far greater force and therefore to pump increased quantities of blood. This, too, increases the arterial pressure.

3. Finally, *the heart itself is directly stimulated by the autonomic nervous system, further enhancing cardiac pumping.* Much of this is caused by an increase in the heart rate, sometimes to as great as three times normal. In addition, sympathetic nervous signals have a direct effect to increase the contractile force of the heart muscle, this, too, increasing the capability of the heart to pump larger volumes of blood. Therefore, under strong sympathetic stimulation, the heart can pump for several minutes two to three times as much blood as under normal conditions. This contributes still more to the rise in arterial pressure.

RAPIDITY OF NERVOUS CONTROL OF ARTERIAL PRESSURE. An especially important characteristic of nervous control of arterial pressure is its rapidity of response, beginning within seconds and often increasing the pressure to two times normal within 5 to 10 seconds. Conversely, sudden inhibition of nervous stimulation can decrease the arterial pressure to as little as one half normal within 10 to 40 seconds. Therefore, nervous control of arterial pressure is by far the most rapid of all our mechanisms for pressure control.

Increase in Arterial Pressure During Muscle Exercise and Other Types of Stress

An important example of the ability of the nervous system to increase the arterial pressure is increased pressure during muscle exercise. During heavy exercise, the muscles require greatly increased blood flow. Part of this increase results from local vasodilation of the muscle vasculature caused by increased metabolism of the muscle cells, as explained in Chapter 17. Additional increase results from simultaneous elevation of arterial pressure during the exercise. In most heavy exercise, the arterial pressure rises about 30 to 40 per cent, which will increase blood flow by about an additional twofold.

The increase in arterial pressure during exercise is believed to result mainly from the following effect: At the same time that the motor areas of the nervous system become activated to cause exercise, most of the reticular activating system of the brain stem is also activated, which includes greatly increased stimulation of the vasoconstrictor and cardioacceleratory areas of the vasomotor center. These raise the arterial pressure instantaneously to keep pace with the increase in muscle activity.

In many other types of stress besides muscle exercise, a similar rise in pressure can also take place. For instance, during extreme fright, the arterial pressure often rises to as high as double normal within a few seconds. This is called the *alarm reaction,* and it provides a head of pressure that can immediately supply blood to any or all muscles of the body that might want to respond instantly to cause flight from danger.

Reflex Mechanisms for Maintaining Normal Arterial Pressure

Aside from the exercise and stress functions of the autonomic nervous system to raise the arterial pressure, there are multiple subconscious special nervous control mechanisms that operate all the time to maintain the arterial pressure at or near its normal operating level. Almost all of them are *negative feedback reflex mechanisms,* which we explain in the next sections.

Arterial Baroreceptor Control System—Baroreceptor Reflexes

By far the best known of the nervous mechanisms for arterial pressure control is the *baroreceptor reflex.* Basically, this reflex is initiated by stretch receptors, called either *baroreceptors* or *pressoreceptors,* which are located in the walls of several of the large systemic arteries. A rise in pressure stretches the baroreceptors and causes them to transmit signals into the central nervous system, and "feedback" signals are then sent back through the autonomic nervous system to the circulation to reduce arterial pressure downward toward the normal level.

PHYSIOLOGIC ANATOMY OF THE BARORECEPTORS AND THEIR INNERVATION. Baroreceptors are spray-type nerve endings that lie in the walls of the arteries; they are stimulated when stretched. A few baroreceptors are located in the wall of almost every large artery of the thoracic and neck regions; but, as shown in Figure 18–5, baroreceptors are extremely abundant in (1) the wall of each internal carotid artery slightly above the carotid bifurcation, an area known as the *carotid sinus,* and (2) the wall of the aortic arch.

Figure 18–5 also shows that signals are transmitted from each carotid sinus through the very small *Hering's nerve* to the glossopharyngeal nerve and then to the *tractus solitarius* in the medullary area of the brain stem. Signals from the arch of the aorta are transmitted through the vagus nerves also into the same area of the medulla.

RESPONSE OF THE BARORECEPTORS TO PRESSURE. Figure 18–6 shows the effect of different arterial pressures on the rate of impulse transmission in a Hering's

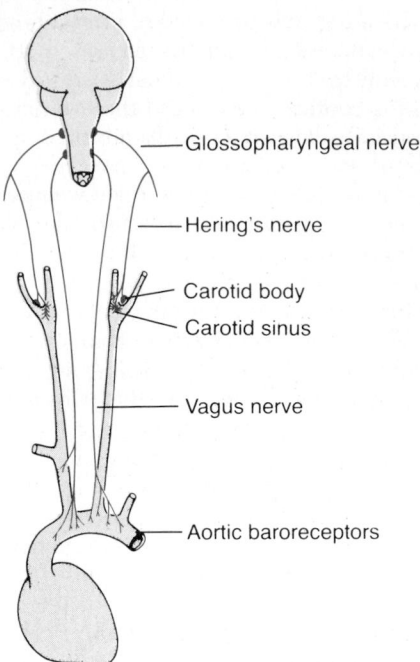

Figure 18–5. The baroreceptor system for controlling arterial pressure.

nerve. Note that the carotid sinus baroreceptors are not stimulated by pressures between 0 and 60 mm Hg, but above 60 mm Hg, they respond progressively more rapidly and reach a maximum at about 180 mm Hg. The responses of the aortic baroreceptors are similar to those of the carotid receptors except that they operate, in general, at pressure levels about 30 mm Hg higher.

Note especially that in the normal operating range of arterial pressure, around 100 mm Hg, even a slight change in pressure causes a strong change in auto-

nomic reflexes to readjust the arterial pressure back toward normal. Thus, the baroreceptor feedback mechanism functions most effectively in the pressure range where it is most needed.

The baroreceptors respond extremely rapidly to changes in arterial pressure; in fact, the rate of impulse firing increases during systole and decreases again during diastole. Furthermore, the baroreceptors *respond much more to a rapidly changing pressure* than to a stationary pressure. That is, if the mean arterial pressure is 150 mm Hg but at that moment is rising rapidly, the rate of impulse transmission may be as much as twice that when the pressure is stationary at 150 mm Hg. On the other hand, if the pressure is falling, the rate might be as little as one quarter that for the stationary pressure.

REFLEX INITIATED BY THE BARORECEPTORS. After the baroreceptor signals have entered the tractus solitarius of the medulla, secondary signals eventually *inhibit the vasoconstrictor center* of the medulla and *excite the vagal center.* The net effects are (1) *vasodilation* of the veins and arterioles throughout the peripheral circulatory system and (2) *decreased heart rate* and *strength of heart contraction.* Therefore, excitation of the baroreceptors by pressure in the arteries reflexly *causes the arterial pressure to decrease* because of both a decrease in peripheral resistance and a decrease in cardiac output. Conversely, low pressure has opposite effects, reflexly causing the pressure to rise back toward normal.

Figure 18–7 shows a typical reflex change in arterial pressure caused by occluding the common carotid arteries. This reduces the carotid sinus pressure; as a result, the baroreceptors become inactive and lose their inhibitory effect on the vasomotor center. The vasomotor center then becomes much more active than usual, causing the arterial pressure to rise and to remain elevated during the 10 minutes that the carotids are occluded. Removal of the occlusion allows the pressure to fall immediately to slightly below nor-

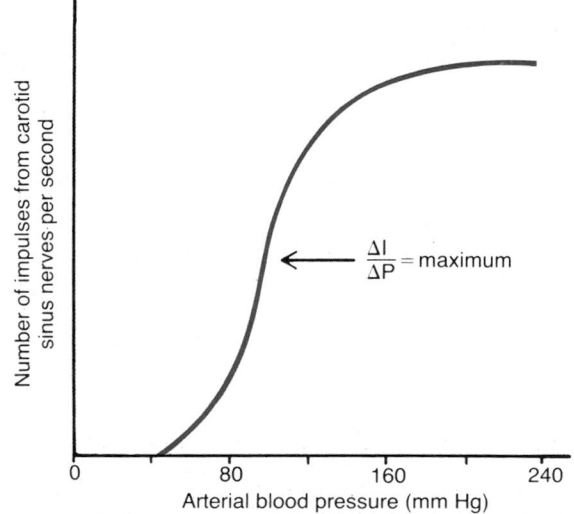

Figure 18–6. Response of the baroreceptors at different levels of arterial pressure.

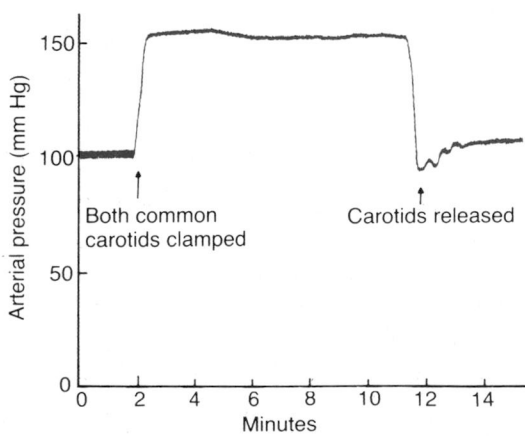

Figure 18–7. Typical carotid sinus reflex effect on arterial pressure caused by clamping both common carotids (after the two vagus nerves have been cut).

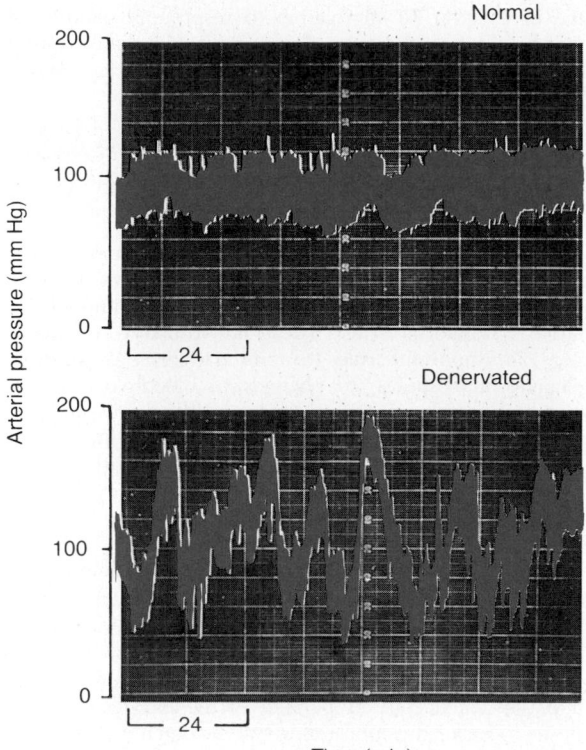

Figure 18–8. Two-hour records of arterial pressure in a normal dog *(above)* and in the same dog *(below)* several weeks after the baroreceptors had been denervated. (From Cowley, Liard, and Guyton: *Circ. Res., 32:*564, 1973. By permission of the American Heart Association, Inc.)

mal as a momentary overcompensation and then to return to normal in another minute or so.

Function of the Baroreceptors During Changes in Body Posture. The ability of the baroreceptors to maintain relatively constant arterial pressure is important when a person stands up after having been lying down. Immediately on standing, the arterial pressure in the head and upper part of the body tends to fall, and marked reduction of this pressure can cause loss of consciousness. The falling pressure at the baroreceptors elicits an immediate reflex, resulting in strong sympathetic discharge throughout the body, and this minimizes the decrease in pressure in the head and upper body.

"Buffer" Function of the Baroreceptor Control System. Because the baroreceptor system opposes either increases or decreases in arterial pressure, it is called a *pressure buffer system,* and the nerves from the baroreceptors are called *buffer nerves.*

Figure 18–8 shows the importance of this buffer function of the baroreceptors. The upper record in this figure shows an arterial pressure recording for 2 hours from a normal dog and the lower record shows an arterial pressure recording from a dog whose baroreceptor nerves from both the carotid sinuses and the aorta have been removed. Note the extreme variability of pressure in the denervated dog caused by simple

events of the day, such as lying down, standing, excitement, eating, defecation, and noises.

Figure 18–9 shows the frequency distributions of the arterial pressures recorded for a 24-hour day in both the normal dog and the denervated dog. Note that when the baroreceptors were normal, the arterial pressure remained throughout the day within a narrow range between 85 and 115 mm Hg—indeed, during most of the day at almost exactly 100 mm Hg. On the other hand, after denervation of the baroreceptors, the frequency distribution curve became the broad, low curve of the figure, showing that the pressure range increased 2.5-fold, frequently falling to as low as 50 mm Hg or rising to over 160 mm Hg. Thus, one can see the extreme variability of pressure in the absence of the arterial baroreceptor system.

In summary, a primary purpose of the arterial baroreceptor system is to reduce the daily variation in arterial pressure to about one half to one third that which would occur were the baroreceptor system not present.

UNIMPORTANCE OF THE BARORECEPTOR SYSTEM FOR LONG-TERM REGULATION OF ARTERIAL PRESSURE—"RESETTING" OF THE BARORECEPTORS. The baroreceptor control system is probably of little or no importance in long-term regulation of arterial pressure for a simple reason: The baroreceptors themselves reset in 1 to 2 days to whatever pressure level they are exposed. That is, if the pressure rises from the normal value of 100 mm Hg to 160 mm Hg, extreme numbers of baroreceptor impulses are at first transmitted. During the next few seconds, the rate of firing diminishes considerably; then it diminishes much more slowly during the next 1 to 2 days, at the end of which time

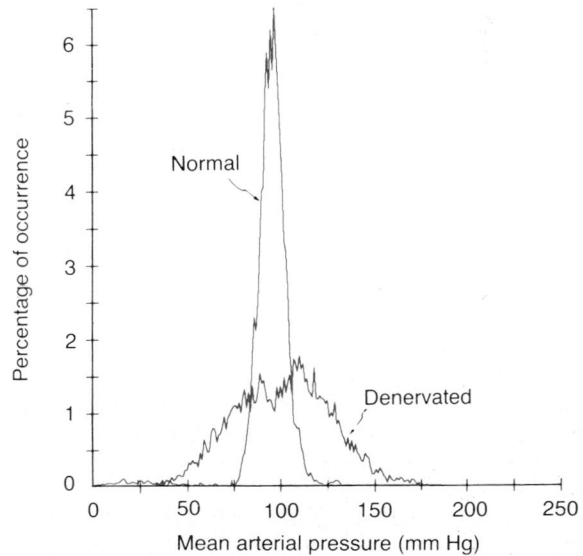

Figure 18–9. Frequency distribution curves of the arterial pressure for a 24-hour period in a normal dog and in the same dog several weeks after the baroreceptors had been denervated. (From Cowley, Liard, and Guyton: *Circ. Res., 32:*564, 1973. By permission of the American Heart Association, Inc.)

the rate will have returned essentially to the normal level despite the fact that the arterial pressure now remains 160 mm Hg. Conversely, when the arterial pressure falls to a very low level, the baroreceptors at first transmit no impulses, but gradually, over about a day, the rate of baroreceptor firing returns to the original control level.

This "resetting" of the baroreceptors prevents the baroreceptor reflex from functioning as a control system for arterial pressure changes that last longer than a few days at a time. In fact, referring again to Figures 18–8 and 18–9, one can see that the average arterial pressure over any prolonged period is almost exactly the same whether the baroreceptors are present or not. This demonstrates the *unimportance of the baroreceptor system for long-term regulation of the arterial pressure* even though it is a potent mechanism for preventing the rapid changes of arterial pressure that occur moment by moment or hour by hour. Prolonged regulation of arterial pressure requires other control systems, principally the renal–body fluid–pressure control system (along with its associated hormonal mechanisms), discussed in Chapter 19.

Control of Arterial Pressure by the Carotid and Aortic Chemoreceptors—Effect of Oxygen Lack on Arterial Pressure

Closely associated with the baroreceptor pressure control system is a chemoreceptor reflex that operates in much the same way as the baroreceptor reflex except that chemoreceptors, instead of stretch receptors, initiate the response.

The *chemoreceptors* are chemosensitive cells sensitive to oxygen lack, carbon dioxide excess, or hydrogen ion excess. They are located in several small organs 1 to 2 millimeters in size: two *carotid bodies*, one of which lies in the bifurcation of each common carotid artery, and several *aortic bodies* adjacent to the aorta. The chemoreceptors excite nerve fibers that, along with the baroreceptor fibers, pass through Hering's nerves and the vagus nerves into the vasomotor center.

Each carotid or aortic body is supplied with an abundant blood flow through a small nutrient artery, so that the chemoreceptors are always in close contact with the arterial blood. Whenever the arterial pressure falls below a critical level, the chemoreceptors become stimulated because of diminished blood flow to the bodies and therefore diminished availability of oxygen as well as excess buildup of carbon dioxide and hydrogen ions that are not removed by the slow flow of blood.

The signals transmitted from the chemoreceptors into the vasomotor center *excite* the vasomotor center, and this elevates the arterial pressure. This reflex helps to return the arterial pressure back toward the normal level whenever it falls too low.

The chemoreceptor reflex is not a powerful arterial pressure controller in the normal arterial pressure range because the chemoreceptors themselves are not stimulated strongly by pressure changes until the arterial pressure falls below 80 mm Hg. Therefore, it is at the lower pressures that this reflex becomes especially important to help prevent still further fall in pressure.

The chemoreceptors are discussed in much more de-

tail in Chapter 41 in relation to respiratory control, in which they play a far more important role than in pressure control.

Atrial and Pulmonary Artery Reflexes That Help Regulate Arterial Pressure and Other Circulatory Factors

Both the atria and the pulmonary arteries have stretch receptors, called *low-pressure receptors,* in their walls similar to the baroreceptor stretch receptors of the large systemic arteries. These low-pressure receptors play an important role to minimize arterial pressure changes in response to changes in blood volume. To give an example, if 300 milliliters of blood are suddenly infused into a dog with all the receptors intact, the arterial pressure will rise only 15 mm Hg. With the *arterial* baroreceptors denervated, the pressure will rise 40 mm Hg. If the *low-pressure receptors* are also denervated, the pressure will rise 100 mm Hg.

Thus, one can see that even though the low-pressure receptors in the pulmonary artery and the atria cannot detect the systemic arterial pressure, they do detect simultaneous increases in pressure in the low-pressure areas of the circulation caused by an increase in volume, and they elicit reflexes parallel to the baroreceptor reflexes to make the total reflex system much more potent for control of arterial pressure.

ATRIAL REFLEXES TO THE KIDNEYS—THE VOLUME REFLEX. Stretch of the atria also causes reflex dilatation of the afferent arterioles in the kidneys, which is the same reflex effect that occurs from the low-pressure receptors to other peripheral arterioles but is unusually potent in the kidneys. Also, signals are transmitted simultaneously to the hypothalamus to decrease the secretion of antidiuretic hormone, thereby indirectly affecting kidney function. The decreased afferent arteriolar resistance causes the glomerular capillary pressure to rise, with resultant increase in filtration of fluid into the kidney tubules. The diminution of antidiuretic hormone diminishes the reabsorption of water from the tubules. The combination of these two effects causes rapid loss of fluid into the urine, which serves as a powerful means to return the blood volume back toward normal. (We also see in Chapter 19 that atrial stretch elicits a hormonal effect on the kidneys—release of *atrial natriuretic peptide*—that adds still further to the rapid loss of fluid in the urine and return of blood volume toward normal.)

All these mechanisms that tend to return the blood volume back toward normal after a volume overload act indirectly as pressure controllers as well as volume controllers because excess volume drives the heart to greater cardiac output and leads, therefore, to greater arterial pressure. This volume reflex mechanism is discussed again in Chapter 29, along with mechanisms of blood volume control.

ATRIAL REFLEX CONTROL OF HEART RATE (THE BAINBRIDGE REFLEX). An increase in atrial pressure also causes an increase in heart rate, sometimes increasing the heart rate as much as 75 per cent. A small part of this increase is caused by a direct effect of the increased atrial volume to stretch the sinus node; it is pointed out in Chapter 10 that such direct stretch can increase the heart rate as much as 15 per cent. An additional 40 to 60 per cent increase in rate is caused by a reflex called the *Bainbridge reflex*. The stretch receptors of the atria that elicit the Bainbridge reflex transmit their afferent signals through the vagus nerves to the medulla of the brain. Then, efferent signals are transmitted back through both the vagal and the sympathetic nerves to increase the heart's rate and, presumably, strength of contraction. Thus, this reflex helps prevent damming of blood in the veins, atria, and pulmonary circulation. This reflex has a different purpose from that of controlling arterial pressure.

Central Nervous System Ischemic Response— Control of Arterial Pressure by the Brain's Vasomotor Center in Response to Diminished Brain Blood Flow

Most nervous control of blood pressure is achieved by reflexes that originate in the baroreceptors, the chemoreceptors, and the low-pressure receptors, all of which are located in the peripheral circulation outside the brain. When blood flow to the vasomotor center in the lower brain stem becomes decreased enough to cause nutritional deficiency, that is, to cause *cerebral ischemia,* the neurons in the vasomotor center itself respond directly to the ischemia and become strongly excited. When this occurs, the systemic arterial pressure often rises to a level as high as the heart can possibly pump. This effect is believed to be caused by failure of the slowly flowing blood to carry carbon dioxide away from the vasomotor center; the local concentration of carbon dioxide then increases greatly and has an extremely potent effect in stimulating the sympathetic nervous control areas in the brain's medulla. It is possible that other factors, such as the buildup of lactic acid and other acidic substances, also contribute to the marked stimulation of the vasomotor center and to the elevation in pressure. This arterial pressure elevation in response to cerebral ischemia is known as the *central nervous system ischemic response,* or simply *CNS ischemic response.*

The magnitude of the ischemic effect on vasomotor activity is tremendous; it can elevate the mean arterial pressure for as long as 10 minutes sometimes to as high as 250 mm Hg. *The degree of sympathetic vasoconstriction caused by intense cerebral ischemia is often so great that some of the peripheral vessels become totally or almost totally occluded.* The kidneys, for instance, often entirely cease their production of urine because of arteriolar constriction in response to the sympathetic discharge. Therefore, *the CNS ischemic response is one of the most powerful of all the activators of the sympathetic vasoconstrictor system.*

IMPORTANCE OF THE CNS ISCHEMIC RESPONSE AS A REGULATOR OF ARTERIAL PRESSURE. Despite the powerful nature of the CNS ischemic response, it does not become significant until the arterial pressure falls far below normal, down to 60 mm Hg and below, reaching its greatest degree of stimulation at a pressure of 15 to 20 mm Hg. Therefore, it is not one of the usual mechanisms for regulating normal arterial pressure. Instead, it operates principally as an *emergency arterial pressure control system that acts rapidly and powerfully to prevent further decrease in arterial pressure whenever blood flow to the brain decreases dangerously close to the lethal level.* It is sometimes called the "last ditch stand" pressure control mechanism.

CUSHING REACTION. The so-called Cushing reaction is a special type of CNS ischemic response that results from increased pressure in the cranial vault. For instance, when the cerebrospinal fluid pressure rises to equal the arterial pressure, it compresses the arteries in the brain and cuts off the blood supply to the brain. This initiates a CNS ischemic response, which causes the arterial pressure to rise. When the arterial pressure has risen to a level higher than the cerebrospinal fluid pressure, blood flows once again into the vessels of the brain to relieve the ischemia. Ordinarily, the blood pressure comes to a new equilibrium level slightly higher than the cerebrospinal fluid pressure, thus allowing blood to continue flowing to the brain. A typical Cushing reaction is shown in Figure 18–10, caused in this instance by pumping fluid under pressure into the cranial vault around the brain.

The Cushing reaction helps protect the vital centers of the brain from loss of nutrition if ever the cerebrospinal fluid pressure rises high enough to compress the cerebral arteries.

DEPRESSANT EFFECT OF PROLONGED ISCHEMIA OF THE VASOMOTOR CENTER. If cerebral ischemia becomes so severe that maximum rise in mean arterial pressure still cannot relieve the ischemia, the neuronal cells begin to suffer metabolically, and within 3 to 10 minutes, they become inactive (they can die if this continues for more than 20 to 60 minutes). The arterial pressure falls to about 40 to 50 mm Hg, which is the level to which the pressure falls when the vasomotor center loses all its tonic vasoconstrictor activity. Therefore, it is fortunate that the ischemic response is powerful, so that arterial pressure can usually rise high enough to correct brain ischemia before it causes nutritional depression and death of the neuronal cells.

SPECIAL FEATURES OF NERVOUS CONTROL OF ARTERIAL PRESSURE

Role of the Skeletal Nerves and Skeletal Muscles in Increasing Cardiac Output and Arterial Pressure

Although most nervous control of the circulation is effected through the autonomic nervous system, at least two conditions in which the skeletal nerves and muscles also play major roles in circulatory responses are the following.

ABDOMINAL COMPRESSION REFLEX. When a baroreceptor or chemoreceptor reflex is elicited, or whenever almost any other factor stimulates the sympathetic vasoconstrictor system, nerve signals are also transmitted simultaneously through skeletal nerves to skeletal muscles of the body, particularly the abdominal muscles. This increases the basal tone of these muscles, which compresses all the venous reservoirs of the abdomen, helping to translocate blood out of the abdominal vascular reservoirs toward the heart. As a result, increased quantities of blood are made available to the heart to be pumped. This overall response is called the *abdominal compression reflex.* The resulting effect on the circulation is the same as that caused by sympathetic vasoconstrictor impulses when they constrict the veins—an increase in both cardiac output and arterial pressure.

The abdominal compression reflex is probably much

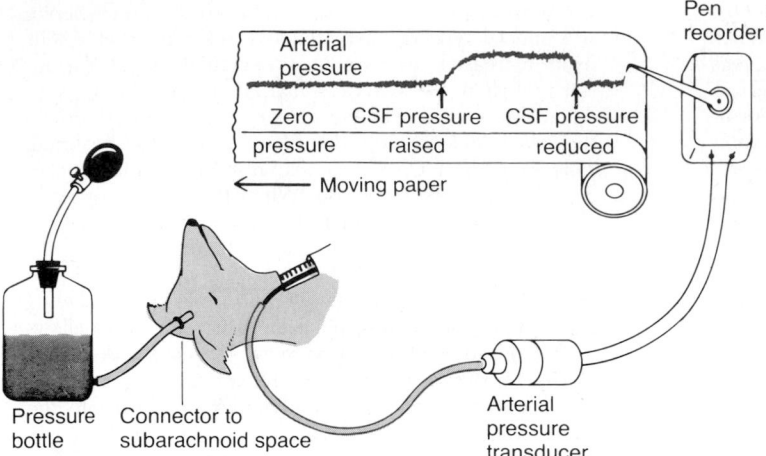

Figure 18–10. Cushing reaction, showing a rise in arterial pressure resulting from increased cerebrospinal fluid (CSF) pressure.

more important than has been realized in the past because it is well known that people whose skeletal muscles have been paralyzed are considerably more prone to hypotensive episodes than are people with normal skeletal muscles.

INCREASED CARDIAC OUTPUT AND ARTERIAL PRESSURE CAUSED BY SKELETAL MUSCLE CONTRACTION DURING EXERCISE. When the skeletal muscles contract during exercise, they compress blood vessels throughout the body. Even anticipation of exercise tightens the muscles, thereby compressing the vessels. The resulting effect is to translocate large quantities of blood from the peripheral vessels into the heart and lungs and, therefore, to increase the cardiac output. This is an essential effect in helping to cause the fivefold to sixfold increase in cardiac output that sometimes occurs in severe exercise. The increase in cardiac output in turn is an essential ingredient in increasing the arterial pressure during exercise, an increase usually in the range of 20 to 60 per cent.

Respiratory Waves in the Arterial Pressure

With each cycle of respiration, the arterial pressure usually rises and falls 4 to 6 mm Hg in a wavelike manner, giving rise to so-called *respiratory waves* in the arterial pressure. The waves result from several different effects, some of which are reflex in nature, as follows:

1. Many impulses that arise in the respiratory center of the medulla "spill over" into the vasomotor center with each respiratory cycle.

2. Every time a person inspires, the pressure in the thoracic cavity becomes more negative than usual, causing the blood vessels in the chest to expand. This reduces the quantity of blood returning to the left side of the heart and thereby momentarily decreases the cardiac output and arterial pressure.

3. The pressure changes caused in the thoracic vessels by respiration can excite vascular and atrial stretch receptors.

Although it is difficult to analyze the exact relation of all these factors in causing the respiratory pressure waves, the net result during normal respiration is usually an increase in arterial pressure during the early part of expiration and a fall in pressure during the remainder of the respiratory cycle. During deep respiration, the blood pressure can rise and fall as much as 20 mm Hg with each respiratory cycle.

Arterial Pressure "Vasomotor" Waves— Oscillation of the Pressure Reflex Control Systems

Often in recording arterial pressure in an animal, in addition to the small pressure waves caused by the respiration, some much larger waves are noted—as great as 10 to 40 mm Hg at times—that rise and fall more slowly than the respiratory waves. The duration of each cycle varies from 26 seconds in the anesthetized dog to 7 to 10 seconds in the human. These waves are called *vasomotor waves* or "Mayer waves." Such records are demonstrated in Figure 18–11, showing the cyclical rise and fall in arterial pressure.

The cause of vasomotor waves is usually oscillation of one or more nervous pressure control mechanisms, some of which are the following.

OSCILLATION OF THE BARORECEPTOR AND CHEMORECEPTOR REFLEXES. The vasomotor waves of Figure 18–11B are the common type of vasomotor waves that are seen almost daily in experimental pressure recordings, although usually much less intense than shown in the figure. They are caused mainly by oscillation of the

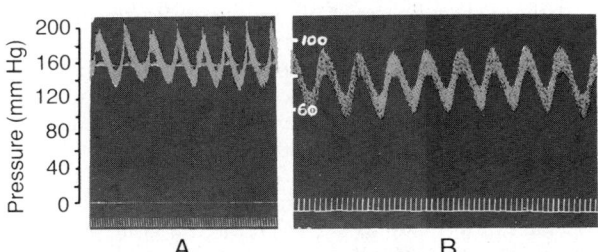

Figure 18–11. *A,* Vasomotor waves caused by oscillation of the CNS ischemic response. *B,* Vasomotor waves caused by baroreceptor reflex oscillation.

baroreceptor reflex. That is, a high pressure excites the baroreceptors; this then inhibits the sympathetic nervous system and lowers the pressure a few seconds later. The decreased pressure reduces the baroreceptor stimulation and allows the vasomotor center to become active once again, elevating the pressure to a high value, but again the response is not instantaneous; it is delayed until a few seconds later. This high pressure then initiates another cycle, and the oscillation continues on and on.

The *chemoreceptor reflex* can also oscillate to give the same type of waves. This reflex usually oscillates simultaneously with the baroreceptor reflex. In fact, it probably plays the major role in causing vasomotor waves when the arterial pressure is in the range of 40 to 80 mm Hg because in this range, chemoreceptor control of the circulation becomes powerful, whereas baroreceptor control becomes weaker.

OSCILLATION OF THE CNS ISCHEMIC RESPONSE. The record in Figure 18–11A resulted from oscillation of the CNS ischemic pressure control mechanism. In this experiment, the cerebrospinal fluid pressure was raised to 160 mm Hg, which compressed the cerebral vessels and initiated a CNS ischemic response. When the arterial pressure rose above 160 mm Hg, the ischemia was relieved and the sympathetic nervous system became inactive. As a result, the arterial pressure fell rapidly back to a lower value, causing medullary ischemia once again. The ischemia then initiated another rise in pressure. Again the ischemia was relieved and the pressure fell. This repeated itself cyclically as long as the cerebrospinal fluid pressure remained elevated.

Thus, any reflex pressure control mechanism can oscillate if the intensity of "feedback" is strong enough and there is a delay between excitation of the pressure receptor and the subsequent pressure response. The vasomotor waves are of considerable theoretical importance because they show that the nervous reflexes that control arterial pressure obey identically the same principles as those applicable to mechanical and electrical control systems. For instance, if the feedback "gain" is too great in the guiding mechanism of an automatic pilot for an airplane and there is also delay in response of the guiding mechanism, the plane will oscillate from side to side instead of following a straight course.

REFERENCES

Andresen, M. C., and Kunze, D. L.: Nucleus tractus solitarius—gateway to neural circulatory control. Annu. Rev. Physiol., 56:93, 1994.

Armour, J. A.: Neurocardiology. New York, Oxford University Press, 1994.

Bishop, V. S., Malliani, A., and Thoren, P.: Cardiac mechanoreceptors. In Sheperd, J. T., and Abboud, F. M. (eds.): Handbook of Physiology. Sec. 2, Vol. III. Bethesda, Md., American Physiological Society, 1983, p. 497.

Blix, A. S., and Folkow, B.: Cardiovascular adjustments to diving in mammals and birds. In Sheperd, J. T., and Abboud, F. M. (eds.): Handbook of Physiology. Sec. 2, Vol. III. Bethesda, Md., American Physiological Society, 1983, p. 917.

Bohr, D. F., and Webb, R. C.: Vascular smooth muscle function and its changes in hypertension. Am. J. Med., 1984.

Brown, A. J., et al.: Cardiovascular and renal responses to chronic vasopressin infusion. Am. J. Physiol. 250:H584, 1986.

Buckley, J. P., et al. (eds.): Brain Peptides and Catecholamines in Cardiovascular Regulation. New York, Raven Press, 1987.

Calaresu, F. R., and Yardley, C. P.: Medullary basal sympathetic tone. Annu. Rev. Physiol., 50:511, 1988.

Coleman, T. G., et al.: Angiotensin and the hemodynamics of chronic salt deprivation. Am. J. Physiol., 229:167, 1975.

Coleridge, H. M., and Coleridge, J. C. G.: Cardiovascular afferents involved in regulation of peripheral vessels. Annu. Rev. Physiol., 42:413, 1980.

Conway, J.: Hemodynamic aspects of essential hypertension in humans. Physiol. Rev., 64:617, 1984.

Cowley, A. W., Jr.: Long-term control of arterial blood pressure. Physiol. Rev., 72:231, 1992.

Cowley, A. W., Jr., and Guyton, A. C.: Baroreceptor reflex contribution in angiotensin II–induced hypertension. Circulation, 50:61, 1974.

Cowley, A. W., Jr., et al.: Interaction of vasopressin and the baroreceptor reflex system in the regulation of arterial pressure in the dog. Circ. Res., 34:505, 1974.

Cruickshank, J. M., and Prichard, B. N. C.: Beta-Blockers in Clinical Practice. New York, Churchill Livingstone, 1994.

Cushing, H.: Concerning a definite regulatory mechanism of the vasomotor center which controls blood pressure during cerebral compression. Bull. Johns Hopkins Hosp., 12:290, 1901.

Dampney, R. A.: Functional organization of central pathways regulating the cardiovascular system. Physiol. Rev., 74:323, 1994.

Dampney, R. A., et al.: Identification of cardiovascular cell groups in the brain stem. Clin. Exp. Hypertens., 6:205, 1984.

Edwards, C. R., and Padfield, P. L.: Angiotensin-converting enzyme inhibitors: Past, present, and bright future. Lancet, 1:30, 1985.

Folkow, B.: Physiological aspects of primary hypertension. Physiol. Rev., 62:347, 1982.

Goldstein, D. S.: Stress, Catecholamines, and Cardiovascular Disease. New York, Oxford University Press, 1995.

Gonzalez, C., et al.: Carotid body chemoreceptors: from natural stimuli to sensory discharges. Physiol. Rev., 74:829, 1994.

Guyton, A. C.: Arterial Pressure and Hypertension. Philadelphia, W. B. Saunders Co., 1980.

Guyton, A. C.: Acute hypertension in dogs with cerebral ischemia. Am. J. Physiol., 154:45, 1948.

Guyton, A. C., and Satterfield, J. H.: Vasomotor waves possibly resulting from CNS ischemic reflex oscillation. Am. J. Physiol., 170:601, 1952.

Guyton, A. C., et al.: Method for studying competence of the body's blood pressure regulatory mechanisms and effect of pressoreceptor denervation. Am. J. Physiol., 164:360, 1951.

Guyton, A. C., et al.: Synthesis of endocrine control in hypertension. Clin. Sci. Molec. Med., 51:319, 1976.

Hainsworth, R.: Reflexes from the heart. Physiol. Rev., 71:617, 1991.

Hall, J. E., and Guyton, A. C.: Changes in renal hemodynamics and renin release caused by increased plasma oncotic pressure. Am. J. Physiol., 231:1550, 1976.

Herd, J. A.: Cardiovascular response to stress in man. Annu. Rev. Physiol., 46:177, 1984.

Irisawa, H., et al.: Cardiac pacemaking in the sinoatrial node. Physiol. Rev., 73:197, 1993.

Herd, J. A.: Cardiovascular response to stress. Physiol. Rev., 71:305, 1991.

Kotchen, T. A., and Frohlich, E. D.: Advances in hypertension 1993. Philadelphia, J. B. Lippincott, 1993.

Krieger, E. M.: Time course of baroreceptor resetting in acute hypertension. Am. J. Physiol., 218:484, 1970.

Lake, C. L.: Clinical Monitoring for Anesthesia and Critical Care. Philadelphia, W. B. Saunders Co., 1994.

Lisney, S. J. W., and Bharali, L. A. M.: The axon reflex: An outdated idea or a valid hypothesis? News Physiol. Sci., 4:45, 1989.

Ludbrook, J.: Reflex control of blood pressure during exercise. Annu. Rev. Physiol., 45:155, 1983.

Mancia, G., and Mark, A. L.: Arterial baroreflexes in humans. In Shepherd, J. T., and Abboud, F. M. (eds.): Handbook of Physiology. Sec. 2, Vol. III. Bethesda, Md., American Physiological Society, 1983, p. 755.

Marshall, J. M.: Peripheral chemoreceptors and cardiovascular regulation. Physiol. Rev., 74:543, 1994.

Mathias, C. J., and Frankel, H. L.: Cardiovascular control in spinal man. Annu. Rev. Physiol., 50:577, 1988.

Opie, L. H. (ed.): Calcium Antagonists and Cardiovascular Disease. New York, Raven Press, 1984.

Persson, P. B., et al.: Cardiopulmonary-arterial baroreceptor interaction in control of blood pressure. News Physiol. Sci., 4:56, 1989.

Randall, W. C. (ed.): Nervous Control of Cardiovascular Function. New York, Oxford University Press, 1984.

Reid, I. A., et al.: The renin-angiotensin system. Annu. Rev. Physiol., 40:377, 1978.

Reid, J. L., and Rubin, P. C.: Peptides and central neural regulation of the circulation. Physiol. Rev., 67:725, 1987.

Rowell, L. B.: Reflex control of regional circulations in humans. J. Autonom. Nerv. Syst., 11:101, 1984.

Sagawa, K.: Baroreflex control of systemic arterial pressure and vascular bed. In Sheperd, J. T., and Abboud, F. M. (eds.): Handbook of Physiology. Sec. 2, Vol. III. Bethesda, Md., American Physiological Society, 1983, p. 453.

Share, L.: Role of vasopressin in cardiovascular regulation. Physiol. Rev., 68:1246, 1988.

Singh, B. N., et al.: Cardiovascular Pharmacology and Therapeutics. New York, Churchill Livingstone, 1994.

Stephenson, R. B.: Modification of reflex regulation of blood pressure by behavior. Annu. Rev. Physiol., 46:133, 1984.

Stiles, G. L., et al.: β-Adrenergic receptors: Biochemical mechanisms of physiological regulation. Physiol., Rev. 64:661, 1984.

Tilkian, A. G., and Conover, M. B.: Understanding Heart Sounds and Murmurs. Philadelphia, W. B. Saunders Co., 1993.

Uther, J. B., and Guyton, A. C.: Cardiovascular regulation following changes in central nervous perfusion pressure in the unanesthetized rabbit. Austral. J. Exp. Biol. Med. Sci., 51:295, 1973.

van Giersbergen, P. L. M., et al.: Involvement of neurotransmitters in the nucleus tractus solitarii in cardiovascular regulation. Physiol. Rev., 72:789, 1992.

Vanhoutte, P. M.: Vasodilation: Vascular Smooth Muscle, Peptides, Autonomic Nerves, and Endothelium. New York, Raven Press, 1988.

Yates, B. J., and Miller, A. D.: Properties of sympathetic reflexes elicited by natural vestibular stimulation: implications for cardiovascular control. J. Neurophysiol., 71:2087, 1994.

Youmans, W. B., et al.: The Abdominal Compression Reaction. Baltimore, Williams & Wilkins. 1963.

Dominant Role of the Kidneys in Long-Term Regulation of Arterial Pressure and in Hypertension: The Integrated System for Pressure Control

C HAPTER 19

Although we see in Chapter 18 that the nervous system has powerful capabilities for rapid, short-term control of arterial pressure, when the arterial pressure changes slowly over many hours or days, the nervous mechanisms gradually lose all or almost all of their ability to oppose the changes. Therefore, what is it that sets the long-term, week-after-week or month-after-month arterial pressure level? We see in this chapter that the kidneys play the dominant role in this control.

RENAL–BODY FLUID SYSTEM FOR ARTERIAL PRESSURE CONTROL

The renal–body fluid system for arterial pressure control is a simple one: When the body contains too much extracellular fluid, the arterial pressure rises. The rising pressure in turn has a direct effect to cause the kidneys to excrete the excess extracellular fluid, thus returning the pressure back toward normal.

In the phylogenetic history of animal development, this renal–body fluid system for pressure control is a primitive one. It is fully operative in one of the lowest of vertebrates, the hagfish. This animal has a low arterial pressure, only 8 to 14 mm Hg, and its pressure increases almost directly in proportion to its blood volume. The hagfish continually drinks sea water, which is absorbed into its blood, increasing the blood volume as well as the pressure. However, when the pressure rises too high, the kidney simply excretes the excess volume into the urine and relieves the pressure. At low pressure, the kidney excretes far less fluid than

is ingested. Therefore, the volume and the pressure build up again to the normal level.

Throughout the ages, this primitive mechanism of pressure control has survived almost exactly as it functions in the hagfish; in the human, kidney output of water and salt is just as sensitive to pressure changes as in the hagfish, if not more so. Indeed, an increase in arterial pressure in the human of only a few millimeters of mercury can double the renal output of water, which is called *pressure diuresis*, as well as the output of salt, which is called *pressure natriuresis*.

In the human, the renal–body fluid system for arterial pressure control, just as in the hagfish, is the fundamental basis of long-term arterial pressure control. However, through the stages of evolution, multiple refinements have been added in the human being to make this system much more exact in its control. An especially important refinement, as we see, has been the addition of the renin-angiotensin mechanism.

Quantitation of Pressure Diuresis as a Basis for Arterial Pressure Control

Figure 19–1 shows the approximate average effect of different arterial pressures on urine volume output, demonstrating markedly increased output of volume as the pressure rises, which is the phenomenon of *pressure diuresis*. The curve in this figure is called a *renal output curve* or *a renal function curve*. At an arterial pressure of 50 mm Hg, the urine output is essentially zero. At 100 mm Hg, it is normal and at 200 mm Hg, about six to eight times normal. Furthermore, not only does increasing the pressure increase the urine volume output but also there is about equal an effect on so-

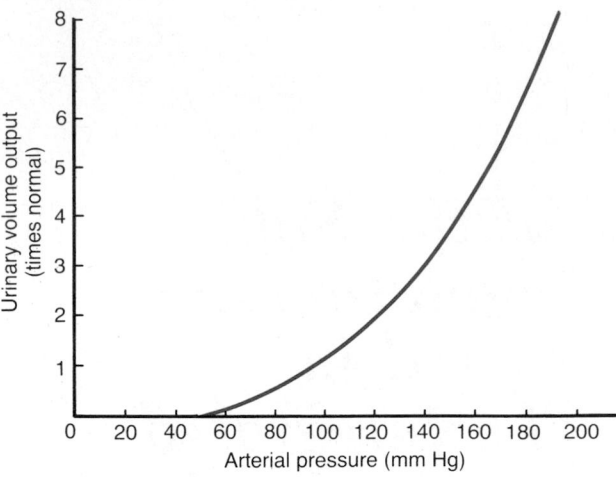

Figure 19–1. Typical renal output curve measured in a perfused isolated kidney, showing pressure diuresis when the arterial pressure rises above normal.

dium output, which is the phenomenon of *pressure natriuresis.*

AN EXPERIMENT DEMONSTRATING THE RENAL–BODY FLUID SYSTEM FOR ARTERIAL PRESSURE CONTROL. Figure 19–2 shows an experiment in dogs in which all the nervous reflex mechanisms for blood pressure control were blocked and the arterial pressure was then suddenly elevated by infusing about 400 milliliters of blood. Note the instantaneous increase in

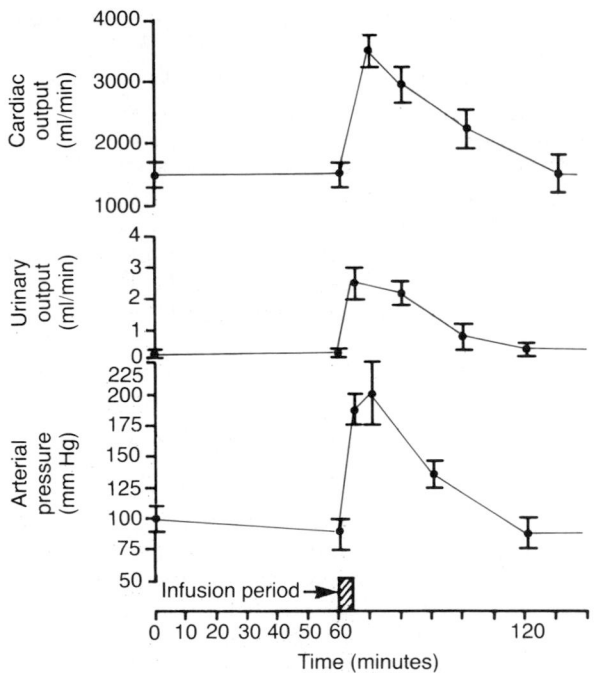

Figure 19–2. Increase in cardiac output, arterial pressure, and urine output caused by increased blood volume in animals whose nervous pressure control mechanisms had been blocked. This figure shows the return of arterial pressure to normal after about an hour of fluid loss into the urine. (Courtesy of Dr. William Dobbs.)

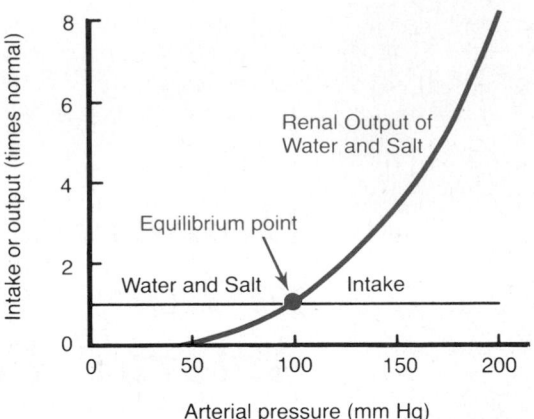

Figure 19–3. Analysis of arterial pressure regulation by equating the renal output curve with the salt and water intake curve. The equilibrium point describes the level to which the arterial pressure will be regulated. (That portion of the salt and water intake that is lost from the body through nonrenal routes is ignored in this and other, similar figures of this chapter.)

cardiac output to about double its normal level and the increase in mean arterial pressure to 205 mm Hg, 115 mm Hg above its resting level. Shown by the middle curve is the effect of this increased arterial pressure on urine output, which increased 12-fold. Along with this tremendous loss of fluid, both the cardiac output and the arterial pressure returned to normal during the subsequent hour. Thus, one sees the extreme capability of the kidneys to eliminate fluid volume from the body and in so doing return the arterial pressure back to normal.

GRAPHICAL ANALYSIS OF PRESSURE CONTROL BY THE RENAL–BODY FLUID MECHANISM, DEMONSTRATING AN "INFINITE GAIN" FEATURE. Figure 19–3 shows a graphical method that can be used for analyzing arterial pressure control by the renal–body fluid system. This analysis is based on two separate curves that intersect each other: (1) the renal output curve for water and salt, which is the same renal output curve as that shown in Figure 19–1, and (2) the curve (or line) that represents the net water and salt intake, that is, the intake minus the amount of water and salt lost from the body in ways other than through the kidneys.

Over a long period, the water and salt output must equal the intake. Furthermore, the only place on the graph in Figure 19–3 at which the output equals the intake is where the two curves intersect, which is called the *equilibrium point*. Now, let us see what will happen if the arterial pressure becomes some value that is different from that at the equilibrium point.

First, assume that the arterial pressure rises to 150 mm Hg. At this level, the graph shows that renal output of water and salt is about three times as great as the intake. Therefore, the body loses fluid, the blood volume decreases, and the arterial pressure decreases. Furthermore, this "negative balance" of fluid will not cease until the pressure falls *all the way* back exactly to the equilibrium point. Indeed, even when the arterial pressure is only 1 mm Hg greater than the equilibrium point, there will still be more loss of water and salt than intake, so that the pressure will continue to fall that last 1 mm Hg *until the pressure returns exactly to the equilibrium point.*

Now, let us see what will happen if the arterial pressure falls below the equilibrium point. This time the intake of water and salt is greater than the output. Therefore, the body fluid volume increases, the blood volume increases, and the arterial pressure rises until once again it returns *exactly* to the equilibrium point.

This return of the arterial pressure *always exactly back to the equilibrium point* is the *infinite gain principle* for control of arterial pressure by the renal–body fluid mechanism.

Two Determinants of the Long-Term Arterial Pressure Level. In Figure 19–3, one can also see that two basic long-term factors determine the long-term arterial pressure level. This can be explained as follows.

As long as the two curves representing (1) renal output of salt and water and (2) salt and water intake remain exactly as they are shown in Figure 19–3, the long-term mean arterial pressure level will always readjust exactly to 100 mm Hg, which is the pressure level depicted by the equilibrium point of this figure. Furthermore, there are only two ways in which the pressure of this equilibrium point can be changed from the 100 mm Hg level. One of these is by shifting the renal output curve for salt and water and the other is by changing the level of the water and salt intake curve. Therefore, expressed simply, the two primary determinants of the long-term arterial pressure level are as follows:

1. The *degree of shift of the renal output curve* for water and salt
2. The *level of the water and salt intake line*

The operation of these two determinants in the control of arterial pressure is shown in Figure 19–4. In Figure 19–4A, some abnormality of the kidney has caused the renal output curve to shift 50 mm Hg in the high-pressure direction (to the right). Note that the equilibrium point has also shifted to 50 mm Hg higher than normal. Therefore, one can state that if the renal output curve shifts to a new pressure level, so also will the arterial pressure follow to this new pressure level within a few days' time.

Figure 19–4B shows how a change in the level of salt and water intake also can change the arterial pressure. In this case, the intake level has increased fourfold and the equilibrium point has shifted to a pressure level of 160 mm Hg, 60 mm Hg above the normal level. Conversely, a decrease in the intake level would reduce the arterial pressure.

Therefore, it is *impossible* to change the long-term mean arterial pressure level to a new value without changing one or both of the two basic determinants of the long-term arterial pressure level, either the level of salt and water intake or the degree of shift of the renal function curve along the pressure axis. However, if

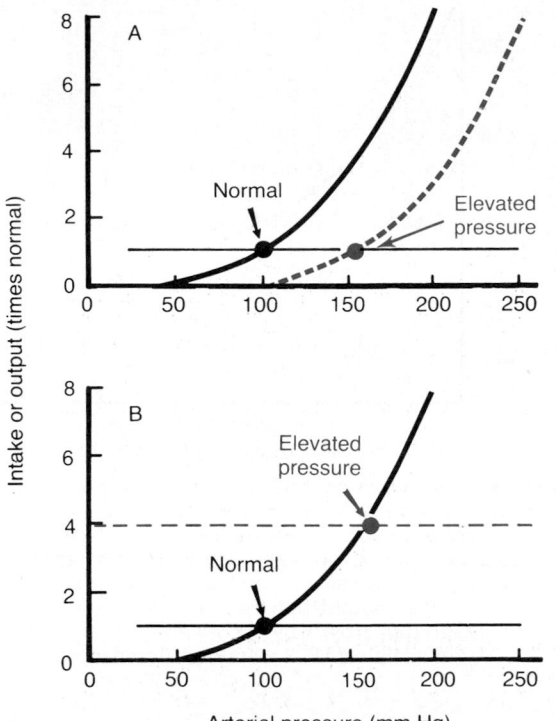

Figure 19–4. Two ways in which the arterial pressure can be increased: *A*, by shifting the renal output curve in the right-hand direction toward a higher pressure level and *B*, by increasing the intake level of salt and water.

either of these is changed, one finds the arterial pressure thereafter to be regulated at a new pressure level, at the pressure level at which these two new curves intersect.

Failure of Increased Total Peripheral Resistance to Elevate the Long-Term Level of Arterial Pressure If Fluid Intake and Renal Function Do Not Change

Now is the chance for the reader to see whether or not he or she really understands the renal–body fluid mechanism for arterial pressure control. Recalling the basic equation for arterial pressure, *arterial pressure* equals *cardiac output* times *total peripheral resistance*, it is clear that an increase in total peripheral resistance should elevate the arterial pressure. Indeed, when the total peripheral resistance is acutely increased, the arterial pressure does rise immediately. Yet if the kidneys continue to function normally, the acute rise in arterial pressure is not maintained. Instead the arterial pressure returns to normal within a day or so. Why?

The answer to this is as follows: Increasing the resistance in the blood vessels everywhere else in the body *besides in the kidneys* does not change the equilibrium point for blood pressure control as dictated by the kidneys (see again Figs. 19–3 and 19–4). Therefore, the kidneys immediately begin to respond to the high arterial pressure with pressure diuresis and pressure natriuresis. Within hours or days, large amounts

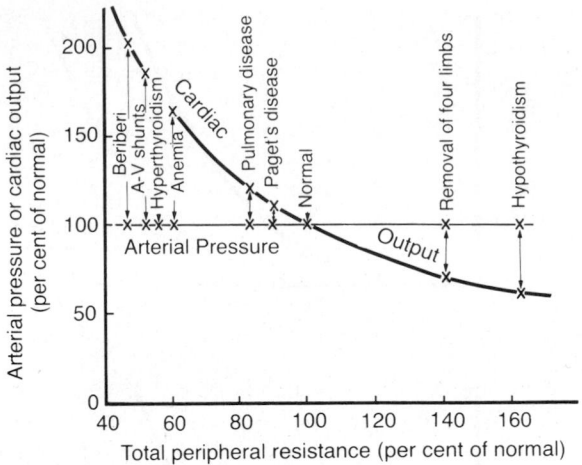

Figure 19–5. Relations of total peripheral resistance to the long-term levels of arterial pressure and cardiac output in different clinical abnormalities. In these conditions, the kidneys were functioning normally. Note that the changes in total peripheral resistance caused equal and opposite changes in cardiac output but had no effect on the arterial pressure. (Reprinted from Guyton: Arterial Pressure and Hypertension, Philadelphia, W. B. Saunders Company, 1980.)

of salt and water are lost from the body, and this continues until the arterial pressure returns to the pressure level of the equilibrium point.

As proof of this principle, that changes in total peripheral resistance will not affect the long-term level of arterial pressure, carefully study Figure 19–5. This figure shows the approximate cardiac outputs and the arterial pressures in different clinical conditions in which the *long-term* total peripheral resistance is either much less than or much greater than normal, but kidney excretion of salt and water is normal or nearly normal. Note in all the different conditions that the arterial pressure is also exactly normal.

(A word of caution! Many times when the total peripheral resistance increases, *this increases the intrarenal vascular resistance at the same time,* which alters the function of the kidney and can cause hypertension by shifting the renal function curve to a higher pressure level, in the manner shown in Figure 19–4A. We see an example of this later in this chapter when we discuss hypertension caused by vasoconstrictor mechanisms. But *it is the increase in renal resistance* that is the culprit, *not the increased total peripheral resistance*—an important distinction!)

How Increased Fluid Volume Elevates the Arterial Pressure—The Role of Autoregulation

The overall mechanism by which increased extracellular volume elevates arterial pressure is given in the schema of Figure 19–6. The sequential events are (1) increased extracellular fluid volume (2) increases the

blood volume, which (3) increases the mean circulatory filling pressure, which (4) increases the venous return of blood to the heart, which (5) increases the cardiac output, which (6) increases the arterial pressure.

Note especially in this schema the two ways in which an increase in cardiac output can increase the arterial pressure. One of these is (1) the direct effect of increased cardiac output in increasing the pressure, and the other is (2) an indirect effect resulting from tissue autoregulation of blood flow. The second effect can be explained as follows.

Referring back to Chapter 17, let us recall that whenever an excess amount of blood flows through a tissue, the local vasculature constricts and decreases the blood flow back toward normal. This phenomenon is called "autoregulation," which means simply regulation of blood flow by the tissue itself. When increased blood volume increases the cardiac output, the blood flow increases in all the tissues of the body, so that this autoregulation mechanism constricts the blood vessels all over the body. This in turn increases the total peripheral resistance.

Finally, because arterial pressure is equal to cardiac output times total peripheral resistance, the secondary increase in total peripheral resistance that results from the autoregulation mechanism helps greatly in increasing the arterial pressure, often accounting for a rise from the normal mean arterial pressure of 100 mm Hg up to 150 mm Hg but with a cardiac output increase of only 5 to 10 per cent. This slight increase in cardiac output is usually unmeasurable.

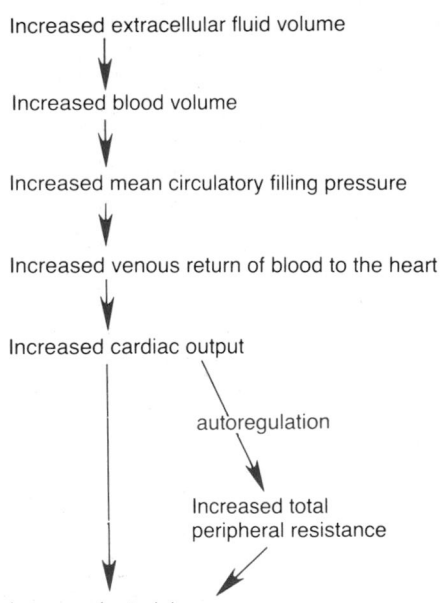

Figure 19–6. Sequential steps by which increased extracellular fluid volume increases the arterial pressure. Note especially that increased cardiac output has both a direct effect of increasing arterial pressure and an indirect effect of increasing the total peripheral resistance.

Importance of Salt in the Renal–Body Fluid Schema for Arterial Pressure Regulation

Although the discussions thus far have emphasized the importance of volume in the regulation of arterial pressure, experimental studies have shown that an increase in salt intake is far more likely to elevate the arterial pressure than is an increase in water intake. The reason for this is that water is normally excreted by the kidneys almost as rapidly as it is ingested, but salt is not excreted so easily. As it accumulates in the body, salt indirectly increases the extracellular fluid volume for two basic reasons:

1. When there is excess salt in the body, the osmolality of the body fluids increases, and this in turn *stimulates the thirst center*, making the person drink extra amounts of water to dilute the extracellular salt to a normal concentration. This increases the extracellular fluid volume.

2. The increase in osmolality in the extracellular fluid also stimulates the hypothalamic–posterior pituitary gland secretory mechanism to secrete increased quantities of *antidiuretic hormone.* This is discussed fully in Chapter 28. The antidiuretic hormone in turn causes the kidneys to reabsorb greatly increased quantities of water from the renal tubular fluid before it is excreted as urine, thereby diminishing the volume of urine while increasing the extracellular fluid volume.

Thus, for these important reasons, the amount of salt that accumulates in the body is the main determinant of the extracellular fluid volume. Because only small increases in extracellular fluid and blood volume can often increase the arterial pressure greatly, the accumulation of even a small amount of extra salt in the body can lead to considerable elevation of the arterial pressure.

Hypertension (High Blood Pressure): This Is Often Caused by Excessive Extracellular Fluid Volume

When a person is said to have hypertension (or "high blood pressure"), it is meant that his or her mean arterial pressure is greater than the upper range of accepted normality. A mean arterial pressure greater than 110 mm Hg (normal is about 90 mm Hg) under resting conditions usually is considered to be hypertensive; this level occurs when the diastolic blood pressure is greater than 90 mm Hg and the systolic pressure greater than about 135 to 140 mm Hg. In severe hypertension, the mean arterial pressure can rise to 150 to 170 mm Hg, with diastolic pressures as high as 130 mm Hg and systolic arterial pressures occasionally as high as 250 mm Hg.

Even moderate elevation of the arterial pressure leads to shortened life expectancy; at high pressures—mean arterial pressures 50 per cent or more above normal—a person can expect to live no more than a few more years. The lethal effects of hypertension are caused mainly in three ways: (1) Excess workload on the heart leads to early development of congestive heart disease, coronary heart disease, or both, often causing death as a result of a heart attack. (2) The high pressure frequently ruptures a major blood vessel in the brain, followed by death of major portions of the brain; this is a *cerebral infarct.* Clinically it is called a "stroke." Depending on what part of the brain is involved, a stroke can cause paralysis, dementia, blindness, or multiple other serious brain disorders. (3) High pressure almost always causes multiple hemorrhages in the kidneys, producing many areas of renal destruction and, eventually, kidney failure, uremia, and death.

The lessons learned from one type of hypertension called "volume-loading hypertension" have been crucial in understanding the role of the renal–body fluid volume mechanism for arterial pressure regulation. Volume-loading hypertension means hypertension caused by excess accumulation of extracellular fluid in the body, some examples of which follow.

EXPERIMENTAL VOLUME-LOADING HYPERTENSION CAUSED BY REDUCED RENAL MASS ALONG WITH SIMULTANEOUS INCREASE IN SALT INTAKE. Figure 19–7 shows a typical experiment demonstrating volume-loading hypertension in a group of dogs with 70 per cent of the renal mass removed. At the first circled point on the curve, the two poles of one of the kidneys were removed, and at the second circled point, the entire opposite kidney was removed, leaving the animals with only 30 per cent of the normal renal mass. Note that removal of this amount of mass increased the arterial pressure an average of only 6 mm Hg. Then, the dogs were required to drink salt solution instead of normal water. Because salt solution fails to quench the thirst, the dogs drank two to four times the normal amounts of volume, and within a few days, their average arterial pressure rose to about 40 mm Hg above normal. After 2 weeks, the dogs were allowed to drink tap water again instead of salt solution; the pressure returned to normal within 2 days. Finally, at the end of the experiment, the dogs were required to drink salt solution again, and this time the pressure rose much more rapidly to an even higher level because the dogs had already learned to tolerate the salt solution and therefore drank much more. Thus, this experiment demonstrates volume-loading hypertension.

If the reader considers again the two basic determinants of long-term arterial pressure regulation, he or she can immediately understand why hypertension occurred in the volume-loading experiment of Figure 19–7. First, reduction of the kidney mass to 30 per cent of normal greatly reduced the ability of the kidneys to excrete salt. Therefore, to continue excreting salt at the same level as before, the arterial pressure had to increase, thus shifting the renal function curve to a higher than normal arterial pressure level. Second, the level of salt and water intake was increased to five to six times normal. Thus, both of the primary determinants of long-term arterial pressure regulation were

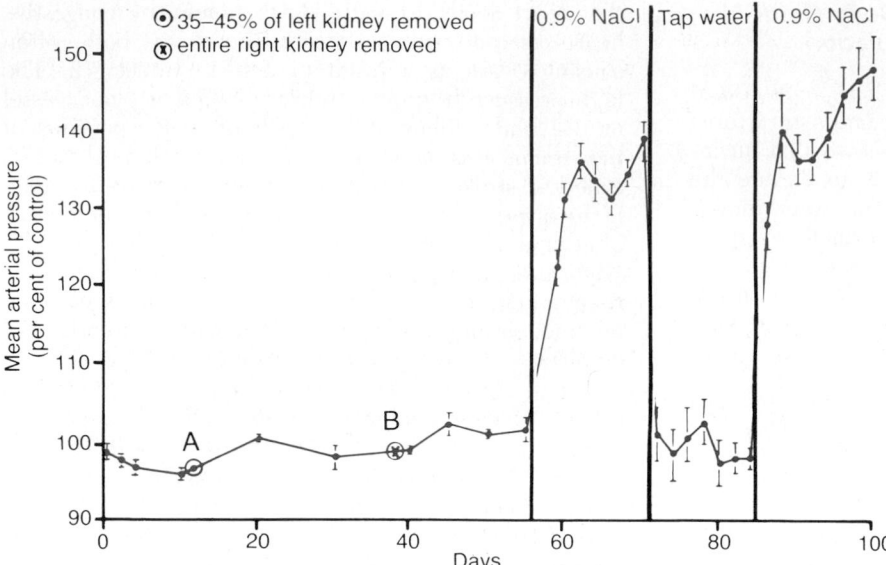

Figure 19–7. Effect on arterial pressure of drinking 0.9 per cent saline solution instead of water in four dogs with 70 per cent of their renal tissue removed. (From Langston, Guyton, Douglas, and Dorsett: *Circ. Res., 12:*508, 1963. By permission of The American Heart Association, Inc.)

increased, giving a combination that led to the large increase in mean arterial pressure.

SEQUENTIAL CHANGES IN CIRCULATORY FUNCTION DURING THE DEVELOPMENT OF VOLUME-LOADING HYPERTENSION. It is especially instructive to study the sequential changes in circulatory function during the progressive development of volume-loading hypertension. Figure 19–8 shows these sequential changes. A week or so before the point labeled "0" days, the kidney mass had already been decreased to only 30 per cent of normal. Then, at this point, the intake of salt and water was increased to about six times normal. The acute effect was to increase extracellular fluid volume, blood volume, and cardiac output to 20 to 40

per cent above normal. Simultaneously, the arterial pressure began to rise but not nearly so much at first as did the fluid volumes and cardiac output. The reason for this slower rise in pressure can be discerned by studying the total peripheral resistance curve, which shows an initial *decrease* in total peripheral resistance. This decrease was caused by the baroreceptor mechanism, discussed in Chapter 18, which tried to prevent the rise in pressure. However, after a few days, the baroreceptors adapted (reset) and no longer opposed the rise in pressure. At this time, the arterial pressure had risen almost to its full height because of the increase in cardiac output, even though the total peripheral resistance was still almost at the normal level.

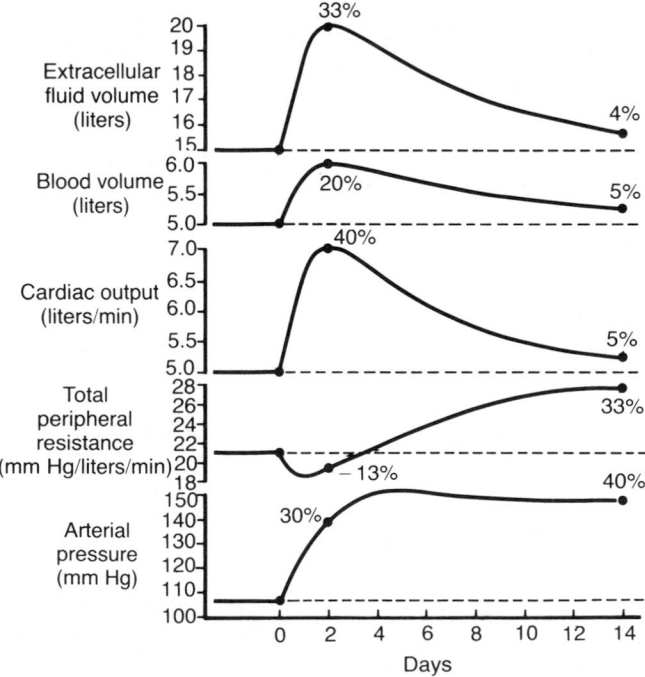

Figure 19–8. Progressive changes in important circulatory system variables during the first few weeks of *volume-loading hypertension.* Note especially the initial rise in cardiac output as the basic cause of the hypertension. Subsequently, the autoregulation mechanism returns the cardiac output almost to normal while at the same time causing a *secondary increase in total peripheral resistance.* (Modified from Guyton: Arterial Pressure and Hypertension. Philadelphia, W. B. Saunders Company, 1980.)

After these early acute changes in the circulatory variables had occurred, more prolonged secondary changes occurred during the next few days and weeks. Especially important was a *progressive increase in total peripheral resistance,* while at the same time *the cardiac output decreased almost all the way back to normal.* Multiple experiments have now shown that these changes were almost certainly caused mainly by *the long-term blood flow autoregulation* mechanism that is discussed in detail in Chapter 17 as well as earlier in this chapter. That is, after the cardiac output had risen to a high level and had initiated the hypertension, the excess blood flow through the tissues then caused progressive constriction of the local arterioles, thus returning the local blood flows and the sum of these, the cardiac output, almost all the way back to normal but simultaneously causing a *secondary increase in total peripheral resistance.*

Note also that the extracellular fluid volume and blood volume returned almost all the way back to normal along with the decrease in cardiac output. This resulted from two factors: First, the increase in arteriolar resistance decreased the capillary pressure, which allowed the fluid in the tissue spaces to be absorbed back into the blood. Second, the elevated arterial pressure now caused the kidneys to excrete the excess volume of fluid that had initially accumulated in the body.

Lastly, let us take stock of the final state of the circulation several weeks after the initial onset of volume-loading. We find the following effects:

1. Hypertension
2. Marked increase in total peripheral resistance
3. Almost complete return of the extracellular fluid volume, blood volume, and cardiac output back to normal

Therefore, we can divide volume-loading hypertension into two separate sequential stages: The first stage results from increased fluid volume causing increased cardiac output. This increase in cardiac output causes the hypertension. The second stage in volume-loading hypertension is characterized by high blood pressure and high total peripheral resistance but return of the cardiac output so near to normal that the usual measuring techniques frequently cannot detect an abnormally elevated output.

The increased total peripheral resistance in volume-loading hypertension occurs after *the hypertension has developed and, therefore, is* secondary *to the hypertension rather than being the cause of the hypertension.*

Volume-Loading Hypertension in Patients Who Have No Kidneys But Are Being Maintained on an Artificial Kidney

When a patient is maintained on an artificial kidney, it is especially important to keep the body fluid volume at a normal level—that is, to remove the appropriate amount of water and salt each time the patient is dialyzed. If this is not done and the extracellular fluid volume is allowed to increase, hypertension will develop in almost exactly the same way as that shown in Figure 19–8. That is, the cardiac output increases at first and causes hypertension. Then the autoregulation mechanism returns the cardiac output back toward normal while causing a secondary increase in total peripheral resistance. Therefore, in the end, the hypertension is a high peripheral resistance type of hypertension.

Hypertension Caused by Primary Aldosteronism

Another type of volume-loading hypertension is caused by excess aldosterone in the body or, occasionally, by excesses of other types of steroids. A small tumor in one of the adrenal glands occasionally secretes large quantities of aldosterone, which is the condition called "primary aldosteronism." As discussed in Chapter 29, aldosterone increases the rate of reabsorption of salt and water by the tubules of the kidneys, thereby reducing the loss of these in the urine while causing an increase in the extracellular fluid volume. Consequently, hypertension occurs; when salt intake is increased at the same time, the hypertension becomes greater. Furthermore, if the condition persists for months or years, the excess aldosterone seems to cause pathological changes in the kidneys that make the kidneys then retain still more salt and water in addition to that caused directly by the aldosterone. Therefore, the hypertension often finally becomes severe.

Here again, in the early stages of this type of hypertension, the cardiac output is increased, but in the later stages, the cardiac output generally returns almost entirely to normal, whereas the total peripheral resistance becomes secondarily elevated.

RENIN-ANGIOTENSIN SYSTEM: ITS ROLE IN PRESSURE CONTROL AND HYPERTENSION

Aside from the capability to control arterial pressure through changes in extracellular fluid volume, the kidneys have another powerful mechanism for controlling pressure. It is the renin-angiotensin system.

Renin is a small protein enzyme released by the kidneys when the arterial pressure falls too low. In turn, it raises the arterial pressure in several ways, thus helping to correct the initial fall in pressure.

Components of the Renin-Angiotensin System

Figure 19–9 shows the functional steps by which the renin-angiotensin system helps to regulate arterial pressure.

Renin is synthesized and stored in an inactive form called *prorenin* in the *juxtaglomerular cells (JG cells)* of the kidneys. The JG cells are modified smooth

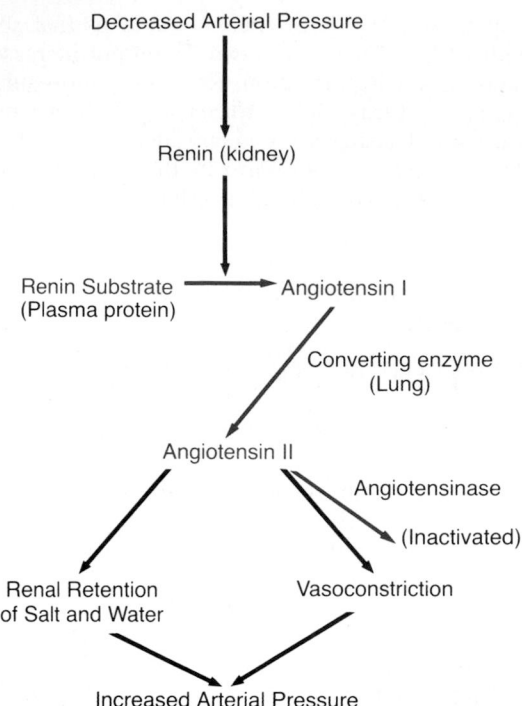

Decreased Arterial Pressure

Renin (kidney)

Renin Substrate
(Plasma protein) → Angiotensin I

Converting enzyme
(Lung)

Angiotensin II

Angiotensinase

(Inactivated)

Renal Retention
of Salt and Water   Vasoconstriction

Increased Arterial Pressure

Figure 19–9. Renin-angiotensin-vasoconstrictor mechanism for arterial pressure control.

muscle cells located in the walls of the afferent arterioles immediately proximal to the glomeruli. When the arterial pressure falls, intrinsic reactions in the kidneys themselves cause many of the prorenin molecules in the JG cells to split and release *renin.* Most of the renin enters the blood and leaves the kidneys to circulate throughout the entire blood stream, although a small amount remains in the local fluids of the kidney and initiates several intrarenal functions.

Renin itself is an enzyme, not a vasoactive substance. Instead, as shown in the schema of Figure 19–9, it acts enzymatically on another plasma protein, a globulin called *renin substrate* (or *angiotensinogen*), to release a 10–amino acid peptide, *angiotensin* I. Angiotensin I has mild vasoconstrictor properties but not enough to cause significant functional changes in circulatory function. The renin persists in the blood for 30 minutes to 1 hour and continues to cause formation of angiotensin I during this entire time.

Within a few seconds after formation of the angiotensin I, two additional amino acids are split from it to form the 8–amino acid peptide *angiotensin II.* This conversion occurs almost entirely during the few seconds while the blood flows through the small vessels of the lungs, catalyzed by the enzyme *converting enzyme* that is present in the endothelium of the lung vessels.

Angiotensin II is an extremely powerful vasoconstrictor, and it has other effects as well that affect the circulation. It persists in the blood only for 1 or 2 minutes because it is rapidly inactivated by multiple blood and tissue enzymes collectively called *angiotensinase.*

During its persistence in the blood, angiotensin II has two principal effects that can elevate arterial pressure. The first of these, *vasoconstriction,* occurs rapidly. Vasoconstriction occurs intensely in the arterioles and to considerably less extent in the veins. Constriction of the arterioles increases the peripheral resistance, thereby raising the arterial pressure, as demonstrated at the bottom of the schema in Figure 19–9. Also, the mild constriction of the veins promotes increased venous return of blood to the heart, thereby helping the heart pump against the increasing pressure.

The second principal means by which angiotensin increases the arterial pressure is to act on the kidneys to *decrease the excretion of both salt and water.* This slowly increases the extracellular fluid volume, which then increases the arterial pressure over a period of hours and days. This long-term effect, acting through the extracellular fluid volume mechanism, is even more powerful than the acute vasoconstrictor mechanism in eventually returning the arterial pressure back to normal.

Rapidity and Intensity of the Vasoconstrictor Pressure Response to the Renin-Angiotensin System

Figure 19–10 shows a typical experiment demonstrating the effect of hemorrhage on the arterial pressure under two separate conditions: (1) with the renin-angiotensin system functioning and (2) without the system functioning (the system was interrupted by a renin-blocking antibody). Note that after the hemorrhage, which caused an acute fall of the arterial pressure to 50 mm Hg, the arterial pressure rose back to 83 mm Hg when the renin-angiotensin system was functional. On the other hand, it rose to only 60 mm Hg when the renin-angiotensin system was blocked. This shows that the renin-angiotensin system is powerful enough to return the arterial pressure at least halfway back to normal after severe hemorrhage.

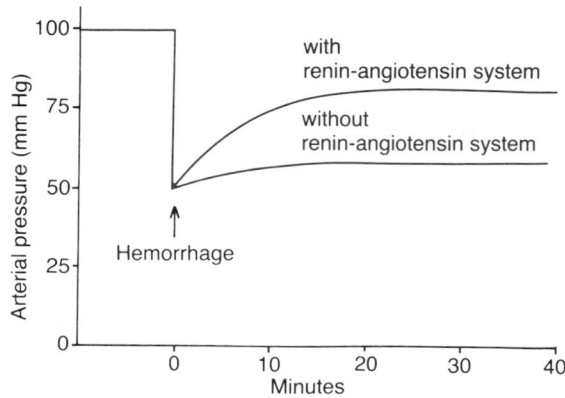

with
renin-angiotensin system

without
renin-angiotensin system

Hemorrhage

Figure 19–10. Pressure-compensating effect of the renin-angiotensin-vasoconstrictor system after severe hemorrhage. (Drawn from experiments by Dr. Royce Brough.)

Therefore, sometimes it can be of lifesaving service to the body, especially in circulatory shock.

Note that the renin-angiotensin vasoconstrictor system requires about 20 minutes to become fully active. Therefore, it is far slower to act than are the nervous reflexes and the sympathetic norepinephrine-epinephrine system.

Effect of Angiotensin to Cause Renal Retention of Salt and Water—An Especially Important Means for Long-Term Control of Arterial Pressure

Angiotensin causes the kidneys to retain both salt and water in two ways:

1. Angiotensin *acts directly on the kidneys to cause salt and water retention.*
2. Angiotensin *causes the adrenal glands to secrete aldosterone, and the aldosterone in turn increases salt and water reabsorption by the kidney tubules.*

The arterial pressure must rise to a considerably increased level to overcome these two fluid-retaining effects of angiotensin. For this reason, whenever excess amounts of angiotensin circulate in the blood, the entire long-term renal–body fluid mechanism for arterial pressure control automatically becomes set to a higher than normal arterial pressure level.

MECHANISMS OF THE DIRECT RENAL EFFECTS OF ANGIOTENSIN TO CAUSE RENAL RETENTION OF SALT AND WATER. Angiotensin has several intrarenal effects that make the kidneys retain salt and water. Probably the most important is to constrict the renal blood vessels, thereby diminishing blood flow through the kidneys. As a result, less fluid filters through the glomeruli into the tubules. Also, the slow flow of blood in the peritubular capillaries reduces their pressure, which allows rapid osmotic reabsorption of fluid from the tubules. Thus, for both of these reasons, less urine is excreted. In addition, angiotensin has a moderate effect on the tubular cells themselves to increase tubular reabsorption of sodium and water. The total result of all these effects is significant, sometimes decreasing urine output fourfold to sixfold.

STIMULATION OF ALDOSTERONE SECRETION BY ANGIOTENSIN AND THE EFFECT OF ALDOSTERONE IN INCREASING SALT AND WATER RETENTION BY THE KIDNEYS. Angiotensin is also one of the most powerful controllers of aldosterone secretion, as we discuss in relation to body fluid regulation in Chapter 29 and in relation to adrenal gland function in Chapter 77. Therefore, when the renin-angiotensin system becomes activated, the rate of aldosterone secretion usually increases at the same time. One of the most important functions of aldosterone is to cause marked increase in sodium reabsorption by the kidney tubules, thus increasing the extracellular fluid sodium. This then causes water retention, as already explained, thus increasing the extracellular fluid volume and leading

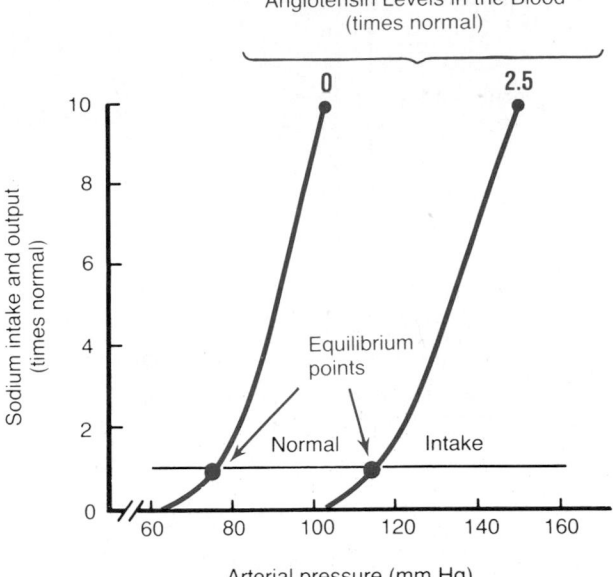

Figure 19–11. Effect of two angiotensin levels on the renal output curve, showing regulation of the arterial pressure at an equilibrium point of 75 mm Hg when the angiotensin level is low and at 115 mm Hg when the angiotensin level is high.

secondarily to still more long-term elevation of the arterial pressure.

Both the direct effect of angiotensin on the kidney and its effect acting through aldosterone are important in long-term arterial pressure control. However, research in our own laboratory has suggested that the direct effect of angiotensin on the kidneys is perhaps three or more times as potent as the indirect effect acting through aldosterone—even though the indirect effect is the one most widely known.

QUANTITATIVE ANALYSIS OF ARTERIAL PRESSURE CHANGES CAUSED BY ANGIOTENSIN. Figure 19–11 shows a quantitative analysis of the effect of angiotensin in arterial pressure control. This figure shows two renal output curves as well as the normal level of sodium intake. The left-hand renal output curve is about that measured in dogs whose renin-angiotensin system had been blocked by the drug captopril (which blocks the conversion of angiotensin I to angiotensin II). The right-hand curve was measured in dogs infused continuously with angiotensin II at a level about 2.5 times the normal rate of angiotensin formation in the blood. Note the shift of the renal output curve toward higher pressure levels under the influence of angiotensin. This shift is caused by both the direct effect of angiotensin on the kidney and the indirect effect acting through aldosterone secretion, as explained previously.

Finally, note the two equilibrium points, one for zero angiotensin at an arterial pressure level of 75 mm Hg and one for elevated angiotensin at a pressure level of 115 mm Hg.

Therefore, the effects of angiotensin on renal retention of water and salt can have a powerful effect in causing chronic elevation of the arterial pressure.

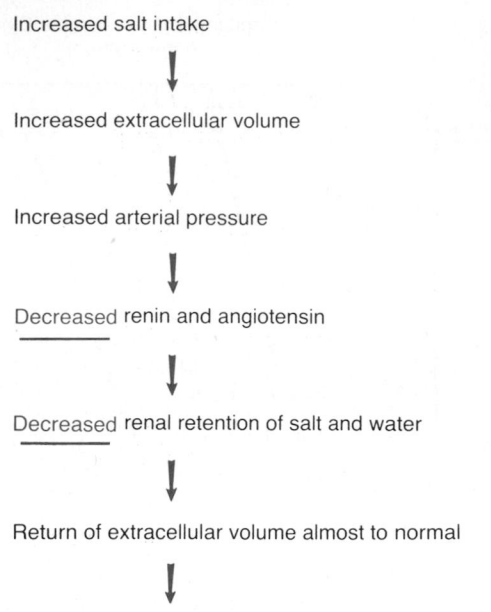

Increased salt intake

⬇

Increased extracellular volume

⬇

Increased arterial pressure

⬇

Decreased renin and angiotensin

⬇

Decreased renal retention of salt and water

⬇

Return of extracellular volume almost to normal

⬇

Return of arterial pressure almost to normal

Figure 19–12. Sequential events by which increased salt intake tends to increase the arterial pressure. However, the schema also shows a feedback decrease in activity of the renin-angiotensin system that returns the arterial pressure back toward the normal level.

Role of the Renin-Angiotensin System in Maintaining a Normal Arterial Pressure Despite Wide Variations in Salt Intake

Probably the most important function of the renin-angiotensin system is to allow a person to eat either a very small or a very large amount of salt without causing great changes in either extracellular fluid volume or arterial pressure. This function is explained by the schema in Figure 19–12, which shows that the direct effect of increased salt intake is to elevate the extracellular fluid volume and this in turn to elevate the arterial pressure. Then, the increased arterial pressure causes increased blood flow through the kidneys, which reduces the rate of secretion of renin to a much lower level; this leads sequentially to decreased renal retention of salt and water, return of the extracellular fluid volume almost to normal, and, finally, return of the arterial pressure also almost to normal. Thus, the renin-angiotensin system is an automatic feedback mechanism that helps maintain the arterial pressure at or near the normal level even when the salt intake is increased. When the salt intake is decreased below normal, exactly the opposite effects take place.

To emphasize the importance of this effect of the renin-angiotensin system, when the system functions normally, the arterial pressure rises no more than 4 to 6 mm Hg in response to as much as a 50-fold increase in salt intake. On the other hand, when the renin-angiotensin system is blocked, the same increase in salt intake will cause the pressure to rise about 10 times as much, sometimes rising 50 to 60 mm Hg.

GRAPHICAL ANALYSIS OF THE COMBINED ROLE OF SODIUM INTAKE AND OF THE RENIN-ANGIOTENSIN SYSTEM IN ARTERIAL PRESSURE CONTROL—THE SODIUM-LOADING RENAL FUNCTION CURVE. Figure 19–13 presents a graphical analysis of the automatic renin-angiotensin feedback mechanism for preventing excessive rise in arterial pressure when sodium intake increases. It shows four levels of salt intake, as demonstrated by the four horizontal lines. To the right of each line is shown the approximate level of circulating angiotensin that one finds in the blood at each of these salt intake levels. Note that at very low salt intake, the concentration of angiotensin in the blood increases to about 2.5 times normal. On the other hand, when the sodium intake is 50 times greater, the level of circulating angiotensin becomes essentially zero.

Now, observe the four renal output curves (the heavy black curves). These are the renal output curves caused by the four levels of angiotensin that occur respectively at the different levels of salt intake. Thus, when the level of angiotensin in the blood is high, at a level 2.5 times normal, it is the renal output curve to the far right that depicts the relation between arterial pressure and sodium output. Conversely, when the circulating level of angiotensin is zero, the far left renal output curve depicts this relation.

Now, finally, match the appropriate levels of sodium intake with the appropriate renal output curves at the different levels of angiotensin and you will be able to determine the equilibrium point for four pairs of intake and output curves. These equilibrium points are represented by the heavy red points on the graph.

As a final maneuver, draw a heavy curve connecting the four equilibrium points, as shown in red in the figure. This curve represents the relation between long-term sodium intake and long-term level of arterial pressure, called the *salt-loading renal function curve*. Note that despite an approximate 50-fold change in sodium intake, from ⅕ normal to 10 times normal, as depicted in the figure, the arterial pressure change is only

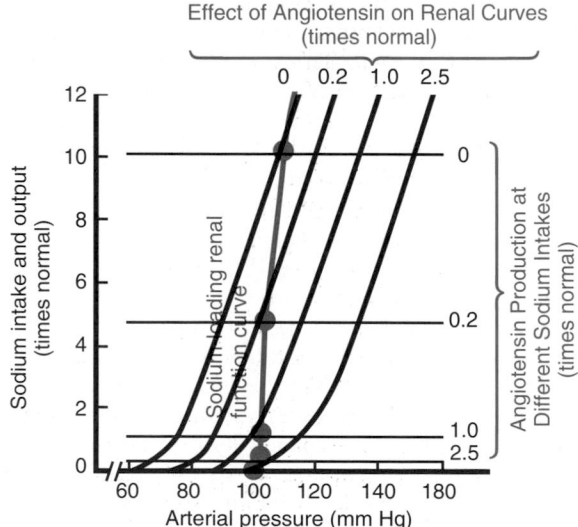

Figure 19–13. Effect of different levels of salt intake on the arterial pressure. The lighter curves are renal output curves. The dark red curve is the sodium-loading renal function curve for a person with normal kidneys.

4 mm Hg. These results have also been observed in animal experiments when the sodium intake has been changed as much as 50-fold. Furthermore, the same experiment performed in humans has shown that changing the sodium intake from $\frac{1}{15}$ normal to 10 times normal—a total range of 150-fold—changes the arterial pressure only about 17 mm Hg.

Types of Hypertension in Which Angiotensin Is Involved: Hypertension Caused by a Renin-Secreting Tumor or by Infusion of Angiotensin II

Occasionally a tumor of the JG cells occurs that secretes tremendous quantities of renin; in turn, equally large quantities of angiotensin II are formed. In all patients in whom this has occurred, severe hypertension has developed. Also, when large amounts of angiotensin are infused continuously for days or weeks on end into animals, similar severe hypertension develops.

We have already noted that angiotensin can increase the arterial pressure in two ways:

1. By constricting the arterioles throughout the entire body, thereby increasing the total peripheral resistance and arterial pressure; this effect occurs within seconds after one begins to infuse angiotensin.
2. By causing the kidneys to retain water and salt; over a period of days, this, too, will cause hypertension.

Which of these causes the chronic hypertension? To answer this, let us suppose that angiotensin had no effect on the renal output curve—that is, the angiotensin did not affect the normal capability of the kidneys to excrete salt and water. In such a theoretical instance, the marked initial rise in arterial pressure caused by angiotensin-induced peripheral vessel constriction would cause intense pressure diuresis and pressure natriuresis because the kidneys would still be normally responsive to the rising arterial pressure. Consequently, the blood volume would decrease rapidly and the arterial pressure would eventually return *all the way* to normal. The reason it would return all the way to normal is that it is only at the normal arterial pressure that urine output would decrease enough to come into balance with the salt and water intake. Therefore, the vasoconstriction in the nonrenal areas of the body, without affecting the kidneys, could not sustain a long-term increase in arterial pressure. Instead, to sustain the increased pressure, it is essential that the renal function curve be shifted to a high pressure level. Thus, once again, we see that the long-term level of arterial pressure can be changed to a new level only by changing one of the two determinants of arterial pressure—in this case, shift of the renal function curve to a high pressure level.

To review, the two determinants of arterial pressure regulation are (1) the level of intake of salt and water (2) the degree of shift of the renal function curve along the pressure axis. In angiotensin-induced hypertension, the angiotensin causes the renal function curve to shift toward higher pressure levels, thus shift-

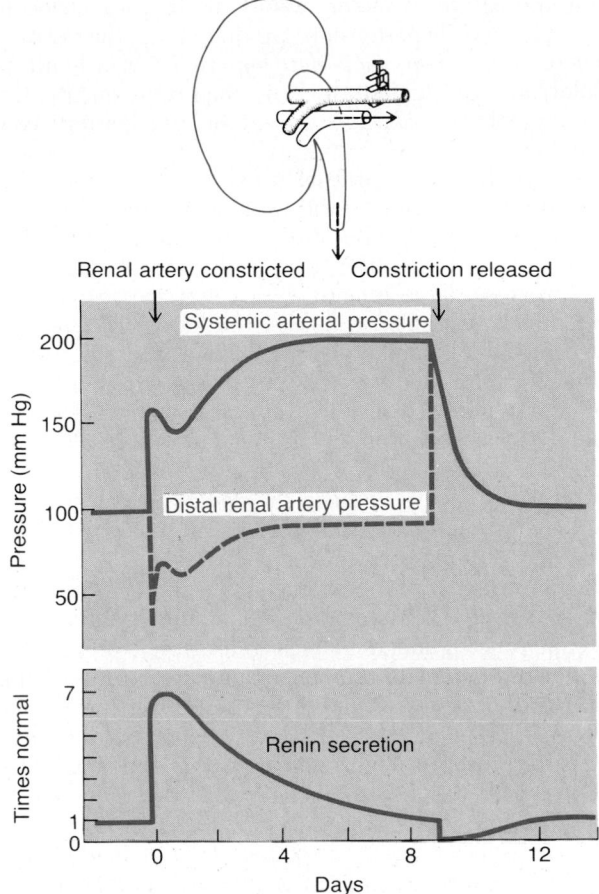

Figure 19–14. Effect of placing a constricting clamp on the renal artery of one kidney after the other kidney has been removed. Note the changes in systemic arterial pressure, renal artery pressure distal to the clamp, and rate of renin secretion. The resulting hypertension is called "one-kidney" Goldblatt hypertension.

ing the equilibrium point for long-term arterial pressure control to a higher level as well. This factor is crucial in maintaining the pressure at a high level even though the initial rise in pressure is caused entirely or almost entirely by the acute effect of angiotensin to increase the total peripheral resistance.

Goldblatt Hypertension

"ONE-KIDNEY" GOLDBLATT HYPERTENSION. When one kidney is removed and a constrictor is placed on the renal artery of the remaining kidney, as shown in Figure 19–14, the immediate effect is greatly reduced pressure in the renal artery beyond the constrictor, as demonstrated by the dashed curve in the figure. Then, within a few minutes, the systemic arterial pressure begins to rise and continues to rise for several days. The pressure usually rises rapidly for the first hour or so, and this is followed by a slower rise over a period of several days to a much higher pressure level. When the systemic arterial pressure reaches its new stable pressure level, the renal arterial pressure will have returned all the way or almost all the way back to normal. The hypertension produced in this way is called *"one-kidney" Goldblatt hypertension* in honor of Goldblatt, who first studied the important quantitative features of hypertension caused by renal artery constriction.

The early rise in arterial pressure in Goldblatt hypertension is caused by the renin-angiotensin vasoconstrictor mechanism. Because of the poor blood flow through the kidney after acute reduction of renal arterial pressure, large quantities of renin are secreted by the kidney, as demonstrated by the lowermost curve in Figure 19–14, and this causes angiotensin to be formed in the blood, as described earlier in the chapter; the angiotensin in turn raises the arterial pressure acutely. The secretion of renin rises to a peak in an hour or so but returns all the way back to normal within 5 to 7 days because the renal arterial pressure by that time has also risen back to normal, so that the kidney is no longer ischemic.

The second rise in arterial pressure is caused by fluid retention; within 5 to 7 days, the fluid volume has increased enough to raise the arterial pressure to its new sustained level. The quantitative value of this sustained pressure level is determined by the degree of constriction of the renal artery. That is, the aortic pressure must rise high enough so that the renal arterial pressure distal to the constrictor will be enough to cause normal urine output.

Note especially that one-kidney Goldblatt hypertension has two phases. The first phase is a vasoconstrictor type of hypertension caused by angiotensin, but this is transient. The second stage is a volume-loading type of hypertension. It is often difficult to tell that this second stage is true volume-loading hypertension because neither the blood volume nor the cardiac output is significantly elevated. Instead, the total peripheral resistance is increased. To understand this, recall that in pure volume-loading hypertension, blood volume and cardiac output are elevated only the first few days during the onset; after the first few days, volume-loading hypertension becomes a high-resistance type of hypertension exactly as seen in the second stage of one-kidney Goldblatt hypertension. The transient increases in blood volume and cardiac output that are normally seen in pure volume-loading hypertension are obscured by the angiotensin vasoconstriction that occurs in Goldblatt hypertension during these early days.

"Two-Kidney" Goldblatt Hypertension. Hypertension also often results when the artery to one kidney is constricted, whereas the artery to the other kidney is normal. This hypertension results from the following mechanism: The constricted kidney retains salt and water because of decreased renal arterial pressure in this kidney. Also, the "normal" kidney retains salt and water because of renin produced by the ischemic kidney. This renin causes the formation of angiotensin, which circulates to the opposite kidney and causes it also to retain salt and water. Thus, both

kidneys, but for different reasons, become salt and water retainers. Consequently, hypertension develops.

Hypertension Caused by Diseased Kidneys That Secrete Renin Chronically. Often patchy areas of one or both kidneys are diseased and become ischemic because of local vascular constrictions, whereas other areas of the kidneys are normal. When this occurs, the almost identical effects occur as in the two-kidney type of Goldblatt hypertension. That is, the patchy ischemic kidney tissue secretes renin, and this in turn, acting through the formation of angiotensin II, causes the remaining kidney mass also to retain salt and water. Indeed, one of the most common causes of renal hypertension is such patchy ischemic kidney disease.

Other Types of Hypertension Caused by Combinations of Volume-Loading and Vasoconstriction

Hypertension in the Upper Part of the Body Caused by Coarctation of the Aorta

One out of every few thousand babies is born with a constricted aorta at a point beyond the arterial branches to the head and arms but proximal to the kidneys, a condition called *coarctation of the aorta.* When this occurs, much of the blood flow to the lower body is carried by collateral blood vessels in the body wall, with much vascular resistance between the upper aorta and the lower aorta. As a consequence, the arterial pressure in the upper part of the body averages 55 per cent higher than that in the lower body.

The mechanism of this upper-body hypertension is almost identical to that of one-kidney Goldblatt hypertension. That is, when a constrictor is placed on the aorta above the renal arteries, the blood pressure in both kidneys at first falls, renin is secreted, angiotensin is formed, and acute hypertension occurs in the upper body because of the vasoconstrictor effects of the angiotensin. Within a few days, salt and water retention occurs so that the arterial pressure in the lower body at the level of the kidneys rises approximately to normal, whereas high pressure persists in the upper body. Now, the kidneys are no longer ischemic, so that the secretion of renin and formation of angiotensin return to normal. Likewise, in coarctation of the aorta, the arterial pressure in the lower body is usually almost normal, whereas the pressure in the upper body is far higher than normal.

Role of Autoregulation in the Hypertension of Coarctation. A significant feature of hypertension caused by aortic coarctation is that the blood flow in the arms, where the pressure averages 55 per cent above normal, is almost exactly normal. Also, the blood flow in the legs, where the pressure is not elevated, is almost exactly normal. How could this be, with the pressure in the upper body 55 per cent greater than that in the lower body? The answer is not that there are differences in vasoconstrictor substances in the bloods of the upper and lower body because the same blood flows to both areas. Likewise, the nervous system innervates both areas of the circulation similarly, so that there is no reason to believe that there is a difference in nervous control of the blood vessels. The only reasonable answer is that long-term autoregulation develops so completely in aortic coarctation that the local blood flow mechanisms have compensated almost 100 per cent for the differences in pressure. The result is that in both the high-pressure area and the low-pressure area, the local blood flow is controlled in accord with the needs of the tissues and not in accord with the level of the pressure.

One of the reasons these observations are so important is that they demonstrate how complete the long-term autoregulation process can be. These observations also show how autoregulation can convert an initial high cardiac output hypertension into a high total peripheral resistance hypertension with normal cardiac output.

Hypertension in Toxemia of Pregnancy

During pregnancy, many patients develop hypertension, which is one of the manifestations of the syndrome called *toxemia of pregnancy*. The principal pathological abnormality that causes this hypertension is believed to be thickening of the glomerular membranes (perhaps caused by an autoimmune process), which reduces the rate of fluid filtration from the glomeruli into the renal tubules. For obvious reasons, the arterial pressure level required to cause normal formation of urine is elevated, and the long-term level of arterial pressure becomes correspondingly elevated. These patients are especially prone to hypertension when they eat large quantities of salt.

Neurogenic Hypertension

Acute hypertension can be caused by strong *stimulation of the sympathetic nervous system*. For instance, when a person becomes excited for any reason or at times during states of anxiety, the sympathetic system becomes excessively stimulated, peripheral vasoconstriction occurs everywhere in the body, and acute hypertension ensues.

ACUTE NEUROGENIC HYPERTENSION CAUSED BY SECTIONING THE BARORECEPTOR NERVES. Another type of acute neurogenic hypertension occurs when the nerves leading from the baroreceptors are cut or when the tractus solitarius is destroyed in each side of the medulla oblongata (these are the areas where the nerves from the baroreceptors connect in the brain stem). The sudden cessation of normal nerve signals from the baroreceptors has the same effect on the nervous pressure control mechanisms as a sudden reduction of the arterial pressure in the aorta and the carotid arteries. That is, the vasomotor center suddenly becomes extremely active and the mean arterial pressure rises from 100 mm Hg to as high as 160 mm Hg. The pressure returns to normal within about 2 days because the response of the vasomotor center to the absent baroreceptor signal fades away, which is called central "resetting" of the baroreceptor pressure control mechanism. Therefore, the neurogenic hypertension caused by sectioning the baroreceptor nerves is only an acute type of hypertension, not a chronic type.

POSSIBLE NEUROGENIC MECHANISMS IN CHRONIC HYPERTENSION. Despite the well-documented specific instances of *acute* hypertension discussed in the previous paragraph, it has been difficult to prove that abnormalities of the nervous system can cause chronic hypertension. Even so, most clinicians have a distinct clinical belief that prolonged nervous tension can maintain prolonged increase in sympathetic stimulation of the blood vessels and kidneys and, therefore, can lead to chronic elevation of the arterial pressure. Research studies thus far in hypertensive patients have failed to give positive proof of continuous excess sympathetic activity.

Probably more likely as a neurogenic mechanism of chronic hypertension is the following: Frequently, the sympathetic nervous system becomes extremely activated for short periods. During such periods, renal blood flow is often decreased to extremely low levels because of severe renovascular vasoconstriction. Experiments have demonstrated that infusing excessive amounts of norepinephrine, the sympathetic transmitter substance, directly into the renal arteries can sometimes cause permanent damage to the kidneys that can possibly lead to a prolonged renal type of hypertension. Thus, it is possible that episodes of extreme sympathetic activity, which are likely to occur in anyone's life, could, under appropriate conditions, lead to renal organic changes that in turn become the basis for permanent hypertension.

Spontaneous Hereditary Hypertension In Lower Animals

Spontaneous hereditary hypertension has been observed in a number of strains of lower animals, including four strains of rats, at least one strain of rabbits, and at least one strain of dogs. In the strain of rats that has been studied to the greatest extent, the Okamoto strain, there is evidence that in the early development of hypertension, the sympathetic nervous system is considerably more active than in normal rats. However, in the late stages of this type of hypertension, two structural changes have been observed in the nephrons of the kidneys: (1) increased preglomerular renal arterial resistance and (2) decreased permeability of the glomerular membranes. These structural changes could easily be the basis for the long-term continuance of the hypertension. In two other strains of hypertensive rats, changes in renal function also have been observed.

Essential Hypertension

About 90 to 95 per cent of all people who have hypertension are said to have "essential hypertension." This term means simply that *the hypertension is of unknown origin*. However, in most patients with essential hypertension, there is a strong hereditary tendency, the same as occurs in strains of hypertensive lower animals.

Some of the characteristics of severe essential hypertension are the following:

1. The mean arterial pressure is increased 40 to 60 per cent.

2. In the later, more severe stages of essential hypertension, the renal blood flow is decreased to about one half normal.

3. The resistance to blood flow through the kidneys is increased twofold to fourfold.

4. Despite the great decrease in renal blood flow, the glomerular filtration rate is often near normal. The reason for this is that the high arterial pressure in hypertension still causes adequate filtration of fluid through the glomeruli into the renal tubules.

5. The cardiac output is about normal.

6. The total peripheral resistance is increased about 40 to 60 per cent, about the same amount that the arterial pressure is increased.

Finally, the most important finding in people with essential hypertension is the following:

7. The kidneys will not excrete adequate amounts of salt and water unless the arterial pressure is high. In other words, if the mean arterial pressure in the essential hypertensive person is 150 mm Hg, reduction of the arterial pressure artificially to the normal value of 100 mm Hg (but without otherwise altering renal function except for the decreased pressure) will cause almost total anuria and the person will retain salt and water until the pressure rises back to the elevated value of 150 mm Hg.

The reason for this failure of the kidneys of essential hypertensive people to excrete salt and water at normal pressure levels is unknown. However, the significant vascular changes in the kidneys suggest that the basic kidney abnormality is these vascular changes.

GRAPHICAL ANALYSIS OF ARTERIAL PRESSURE CONTROL IN ESSENTIAL HYPERTENSION. Figure 19–15 is a graphical analysis of essential hypertension. The curves of this figure are sodium-loading renal function curves of the type shown for the normal kidney by the heavy red curve in Figure 19–13. The reason for using this curve is the following: It is easy to determine the sodium-loading type of curve by simply increasing the level of sodium intake to a new level every few days, then waiting for the renal output of sodium to come into balance with the intake, and at the same time recording the changes in arterial pressure. When this procedure is used in essential hypertensive patients, the two curves to the right in Figure 19–15 are recorded in two groups of essential hypertensive patients, one called (1) *nonsalt-sensitive* patients and the other (2) *salt-sensitive*. Note in both instances that the curves are shifted to the right, to a much higher pressure level than for normal people. Now, let us plot on this same graph (1) a normal level of salt intake and (2) a high level of salt intake representing 3.5 times the normal intake. In the case of the person with nonsalt-sensitive essential hypertension, the arterial pressure does not increase significantly when changing from normal salt intake to high salt intake. On the other hand, in those patients who have salt-sensitive essential hypertension, the high salt intake exacerbates the hypertension.

The reason for the difference between nonsalt-sensitive essential hypertension and salt-sensitive hypertension is presumably structural or functional differences in the kidneys of these two types of hypertensive patients.

TREATMENT OF ESSENTIAL HYPERTENSION. Essen-

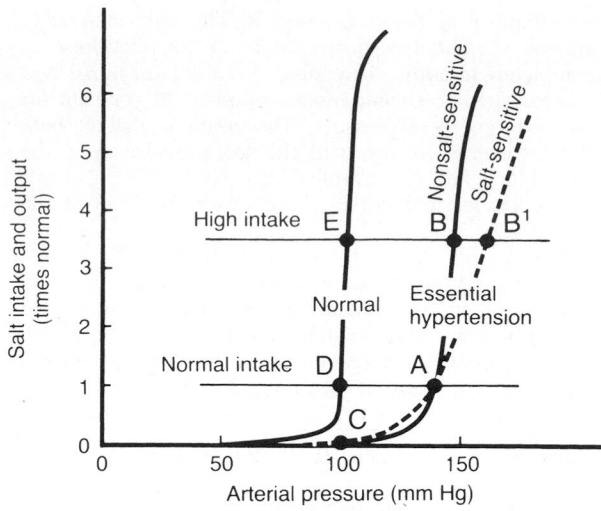

Figure 19–15. Analysis of arterial pressure regulation in non-salt-sensitive essential hypertension and salt-sensitive essential hypertension. (From Guyton et al.: *Annu. Rev. Med. 31*:15, 1980.)

tial hypertension can be treated by two types of drugs: (1) drugs that increase renal blood flow or (2) drugs that decrease tubular reabsorption of salt and water.

Those drugs that increase renal blood flow are the various *vasodilator drugs.* Different ones act in one of the following ways: (1) by inhibiting sympathetic nervous signals to the kidneys or by blocking the action of the sympathetic transmitter substance on the renal vasculature, (2) by directly paralyzing the smooth muscle of the renal vasculature, or (3) by blocking the action of the renin-angiotensin system on the renal vasculature or renal tubules.

Those drugs that reduce reabsorption of salt and water by the tubules include especially drugs that block active transport of sodium through the tubular wall; this blockage in turn prevents the reabsorption of water as well, as explained earlier in the chapter. Such drugs are called *natriuretics* or *diuretics;* they are discussed in greater detail in Chapter 31.

SUMMARY OF THE INTEGRATED, MULTIFACETED SYSTEM FOR ARTERIAL PRESSURE REGULATION

By now, it is clear that arterial pressure is not regulated by a single pressure controlling system but instead by several interrelated systems each of which performs a specific function. For instance, when a person bleeds severely so that the pressure falls suddenly, two problems confront the pressure control system. The first is survival, to return the arterial pressure immediately to a high enough level that the person can live through the acute episode. The second is to return the blood volume eventually to its normal

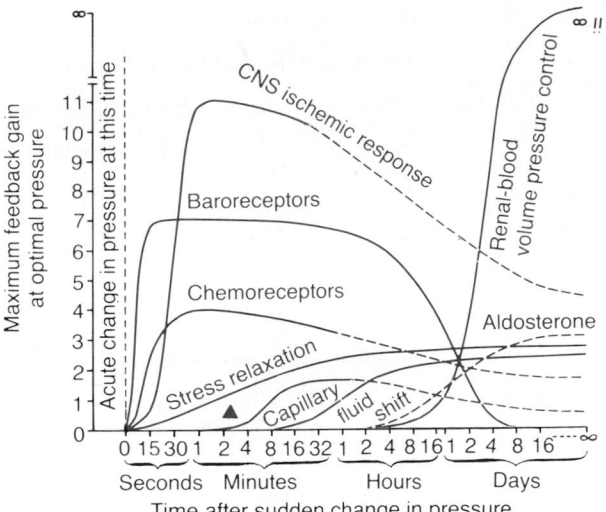

Seconds   Minutes   Hours   Days

Time after sudden change in pressure

▲ Renin-
angiotensin-vasoconstriction

Figure 19–16. Approximate potency of various arterial pressure control mechanisms at different time intervals after the onset of a disturbance to the arterial pressure. Note especially the infinite gain of the renal–body fluid pressure control mechanism that occurs after a few weeks' time. (Reprinted from Guyton: Arterial Pressure and Hypertension. Philadelphia, W. B. Saunders Company, 1980.)

level, so that the circulatory system can re-establish full normality, including return of the arterial pressure all the way back to its normal value, not merely back to a pressure level required for survival.

In Chapter 18, we saw that the first line of defense against acute changes in arterial pressure is the nervous control system. In this chapter, we have emphasized the role of the kidneys in long-term control of arterial pressure. However, there are a few other pieces to the puzzle. Figure 19–16 helps to put these together.

Figure 19–16 shows the approximate immediate (seconds) and long-term (hours and days) control responses, expressed as feedback gain, of eight arterial pressure control mechanisms. These mechanisms can be divided into three groups: (1) those that react rapidly, within seconds or minutes; (2) those that respond over an intermediate time period, minutes or hours; and (3) those that provide long-term arterial pressure regulation, days, months, and years. Let us see how they fit together as a total, integrated system for pressure control.

RAPIDLY ACTING PRESSURE CONTROL MECHANISMS, ACTING WITHIN SECONDS OR MINUTES. The rapidly acting pressure control mechanisms are almost entirely acute nervous reflexes or other nervous responses. Note in Figure 19–16 the three mechanisms that show responses within seconds. They are (1) the baroreceptor feedback mechanism, (2) the central nervous system ischemic mechanism, and (3) the chemoreceptor mechanism. These mechanisms not only

begin to react within seconds, but they are also powerful. After any acute fall in pressure, as might be caused by severe hemorrhage, the nervous mechanisms combine (a) to cause constriction of the veins and provide transfer of blood into the heart, (b) to cause increased heart rate and contractility of the heart and provide greater pumping capacity by the heart, and (c) to cause constriction of the arterioles to impede the flow of the blood out of the arteries; all these effects occur almost instantly to raise the arterial pressure back into a survival range.

When the pressure suddenly rises too high, as might occur in response to a drug or overadministration of a blood transfusion, the same mechanisms operate in the reverse direction, again returning the pressure back toward the normal range.

INTERMEDIATE TIME PERIOD PRESSURE CONTROL MECHANISMS. Several pressure control mechanisms exhibit significant responses only after a few minutes following an acute arterial pressure change. Three of these, shown in Figure 19–16, are (1) the renin-angiotensin vasoconstrictor mechanism, (2) stress-relaxation of the vasculature, and (3) shift of fluid through the capillary walls in and out of the circulation to readjust the blood volume as needed.

We have already described at length the role of the renin-angiotensin vasoconstrictor system to provide a semi-acute means for increasing the arterial pressure when this is needed. The *stress-relaxation mechanism* is demonstrated by the following example: When the pressure in the blood vessels becomes too high, they become stretched and keep on stretching more and more for minutes or hours; as a result, the pressure in the vessels falls toward normal. This continuing stretch of the vessels, called stress-relaxation, can serve as an intermediate-term pressure "buffer."

The *capillary fluid shift mechanism* means simply that any time the capillary pressure falls too low, fluid is absorbed by osmosis from the tissue into the circulation, thus building up the blood volume and increasing the pressure in the circulation. Conversely, when the capillary pressure rises too high, fluid is lost out of the circulation into the tissues, thus reducing the blood volume as well as all the pressures throughout the circulation.

These three intermediate mechanisms become mostly activated within 30 minutes to several hours. Their effect can last for long periods, days if necessary. During this time, the nervous mechanisms usually fatigue and become less and less effective, which explains the importance of these intermediate pressure control measures.

LONG-TERM MECHANISMS FOR ARTERIAL PRESSURE REGULATION. The goal of this chapter has been to explain the role of the kidneys in long-term control of arterial pressure. To the far right in Figure 19–16 is shown the renal–blood volume pressure control mechanism (which is the same as the renal–body fluid pressure control mechanism), demonstrating that it takes a few hours to show significant response. Yet it eventually develops a feedback gain for control of arterial pressure equal to infinity. This means that this mechanism can eventually return the arterial pressure *all the way* back, not merely part way back, to that pressure level that will provide normal output of salt and water by the kidneys. By now, the reader should be familiar with this concept, which has been the major point of this chapter.

It must also be remembered that many factors can affect the pressure-regulating level of the renal–body fluid mechanism. One of these, shown in Figure 19–16, is aldosterone. Within a few hours after a sudden change in the arterial pressure, the effect of the hormone aldosterone on the circulation changes, and this plays an important role in modifying the pressure control characteristics of the renal–body fluid mechanism. Still another factor especially important in the daily control of the arterial pressure is the interaction of the renin-angiotensin system with the aldosterone and renal fluid mechanisms. For instance, a person's salt intake varies tremendously from one day to another. We have seen in this chapter that the salt intake can decrease to as little as 1/10 normal or can rise to 10 to 15 times normal and yet the regulated level of the mean arterial pressure will change only a few millimeters of mercury if the renin-angiotensin-aldosterone system is fully operative, as explained earlier in this chapter. *Without the renin-angiotensin-aldosterone system, the effect on arterial pressure of excessive salt intake is 10 times as great.*

Thus, arterial pressure control begins with the lifesaving measures of the nervous pressure controls, then continues with the sustaining characteristics of the intermediate pressure controls, and, finally, is stabilized at the long-term pressure level by the renal–body fluid mechanism. This long-term mechanism in turn has multiple interactions with the renin-angiotensin-aldosterone system, the nervous system, and several other factors that provide special control capabilities for special purposes.

REFERENCES

Beauchamp, G. K.: The human preference for excess salt. Am. Scientist, 75:27, 1987.

Bennett, W. M., and McCarron, D. A.: Pharmacology and Management of Hypertension. New York, Churchill Livingstone, 1994.

Biglieri, E. G., and Melby, J. C.: Endocrine Hypertension. New York, Raven Press, 1990.

Borgstrom, P., et al.: An evaluation of the metabolic interaction with myogenic vascular reactivity during blood flow autoregulation. Acta Physiol. Scand., 122:275, 1984.

Brunzel, N. A.: Fundamentals of Urine and Body Fluid Analysis. Philadelphia, W. B. Saunders Co., 1994.

Coleman, T. G., and Guyton, A. C.: The pressor role or angiotensin in salt deprivation and renal hypertension. Clin. Sci., 48:458, 1975.

Coleman, T. G., and Guyton, A. C.: Hypertension caused by salt loading in the dog. III. Onset transients of cardiac output and other circulatory variables. Circ. Res., 25:153, 1969.

Coleman, T. G., et al.: Experimental hypertension and the long-term control of arterial pressure. In MTP International Review of Science: Physiology. Baltimore, University Park Press, 1974, Vol. 1, p. 259.

Coleman, T. G., et al.: Regulation of arterial pressure in the anephric state. Circulation, 42:509, 1970.

Conway, J.: Hemodynamic aspects of essential hypertension in humans. Physiol. Rev., 64:617, 1984.

Cowley, A. W. Jr.: Long-term control of arterial blood pressure. Physiol. Rev., 72:231, 1992.

Cowley, A. W., Jr., et al.: Vasopressin: Cellular and Integrative Functions. New York, Raven Press, 1988.

Cowley, A. W., Jr., et al.: Open-loop analysis of the renin-angiotensin system in the dog. Circ. Res., 28:568, 1971.

Cruickshank, J. M., and Prichard, B. N. C.: Beta-Blockers in Clinical Practice. New York, Churchill Livingstone, 1994.

DeClue, J. W., et al.: Subpressor angiotensin infusion, renal sodium handling, and salt-induced hypertension in the dog. Circ. Res., 43(4):503, 1978.

De Santo, N. G., and Alon, U.: Pediatric Hypertension. Farmington, CT, S. Karger Publishers, Inc., 1992.

Folkman, J.: Angiogenesis: What makes blood vessels grow? News Physiol. Sci., 1:199, 1986.

Folkow, B.: Physiological aspects of primary hypertension. Physiol. Rev., 62:347, 1982.

Glorioso, N., et al.: Renovascular Hypertension. New York, Raven Press, 1994.

Granger, H. J., and Guyton, A. C.: Autoregulation of the total systemic circulation following destruction of the central nervous system in the dog. Circ. Res., 25:379, 1969.

Guyton, A. C.: Arterial Pressure and Hypertension. Philadelphia, W. B. Saunders Co., 1980.

Guyton, A. C.: Acute hypertension in dogs with cerebral ischemia. Am. J. Physiol., 154:45, 1948.

Guyton, A. C., and Coleman, T. G.: Quantitative analysis of the pathophysiology of hypertension. Circ. Res., 24:1, 1969.

Guyton, A. C., and Coleman, T. G.: Long-term regulation of the circulation: Interrelationships with body fluid volumes. In Physical Bases of Circulatory Transport Regulation and Exchange. Philadelphia, W. B. Saunders Co., 1967.

Guyton, A. C., et al.: Salt balance and long-term blood pressure control. Annu. Rev. Med., 31:15, 1980.

Guyton, A. C., et al.: Integration and control of circulatory function. Int. Rev. Physiol., 9:341, 1976.

Guyton, A. C., et al.: A systems analysis approach to understanding long-range arterial blood pressure control and hypertension. Circ. Res., 35:159, 1974.

Guyton, A. C., et al.: Arterial pressure regulation: Overriding dominance of the kidneys in long-term regulation and in hypertension. Am. J. Med., 52:584, 1972.

Hall, J. E., et al.: Control of arterial pressure and renal function during glucocorticoid excess in dogs. Hypertension, 2:139, 1980.

Hall, J. E., et al.: Renal hemodynamics in acute and chronic angiotensin II hypertension. Am. J. Physiol., 235(3):F174, 1978.

Holstein-Rathlou, N.-H., and Marsh, D. J.: Renal blood flow regulation and arterial pressure fluctuations: a case study in nonlinear dynamics. Physiol. Rev., 74:637, 1994.

Kaplan, N. M.: Clinical Hypertension. Baltimore, Williams & Wilkins, 1994.

Kaplan, N. M., et al.: The Kidney in Hypertension. New York, Raven Press, 1987.

Kelley, W. N.: Textbook of Internal Medicine. Philadelphia, J. B. Lippincott, 1991.

Kotchen, T. A., and Frohlich, E. D.: Advances in Hypertension: 1993. Philadelphia, J. B. Lippincott, 1993.

Krakoff, L. R.: Management of the Hypertensive Patient. New York, Churchill Livingstone, 1994.

Laragh, J. H., and Brenner, B. M.: Hypertension: Pathophysiology, Diagnosis, and Management. New York, Raven Press, 1994.

Leonetti, G., and Cuspidi, C.: Hypertension in the Elderly. Hingham, MA, Kluwer Academic Publishers, 1994.

Lohmeier, T. E., et al.: Failure of chronic aldosterone infusion to increase arterial pressure in dogs with angiotensin-induced hypertension. Circ. Res., 43(3):381, 1978.

Manning, R. D., Jr., et al.: Essential role of mean circulatory filling pressure in salt-induced hypertension. Am. J. Physiol., 236:R40, 1979.

Norman, R. A., Jr., et al.: Renal function curves in normotensive and spontaneously hypertensive rats. Am. J. Physiol., 234:R98, 1978.

Önen, K., and Diehm, C.: Beta-Blockers in the Treatment of Hypertension in the 90's: Experience with Tertatolol. Farmington, CT, S. Karger Publishers, Inc., 1993.

Opie, L. H.: Ace inhibitors: Almost too good to be true. New York, Scientific American Science and Medicine, July/August, 1994.

Paul, M., et al.: Transgenic rats: new experimental models for the study of candidate genes in hypertension research. Annu. Rev. Phyiol., 56:811, 1994.

Postnov, Y. V., and Orlov, S. N.: Ion transport across plasma membrane in primary hypertension. Physiol. Rev., 65:904, 1985.

Reid, J. L., and Rubin, P. C.: Peptides and central neural regulation of the circulation. Physiol. Rev., 67:725, 1987.

Rosenthal, J., and Stumpe, K. O.: Converting Enzyme Inhibitors in Treatment of Hypertension. Farmington, CT, S. Karger Publishers, Inc., 1993.

Saavedra, J. M., and Timmermans, P. B.: Angiotensin Receptors. New York, Plenum Publishing Corp., 1994.

Seldin, D. W., and Giebisch, G.: The Kidney: Physiology and Pathophysiology. New York, Raven Press, 1992.

Singh, B. N., et al.: Cardiovascular Pharmacology and Therapeutics. New York, Churchill Livingstone, 1993.

Starr, J. M., and Whalley, L. J.: ACE Inhibitors: Central Actions. New York, Raven Press, 1994.

Tierney, L. M., Jr., et al.: Current Medical Diagnosis and Treatment, 1995, Redding, Ma., Appleton and Lange, 1994.

Vallee, B. L., et al.: Tumor-derived angiogenesis factors from rat Walker 256 carcinoma: An experimental investigation and review. Experientia, 41:1, 1985.

Young, D. B., and Guyton, A. C.: Steady state aldosterone dose-response relationships. Circ. Res., 40(2):138, 1977.

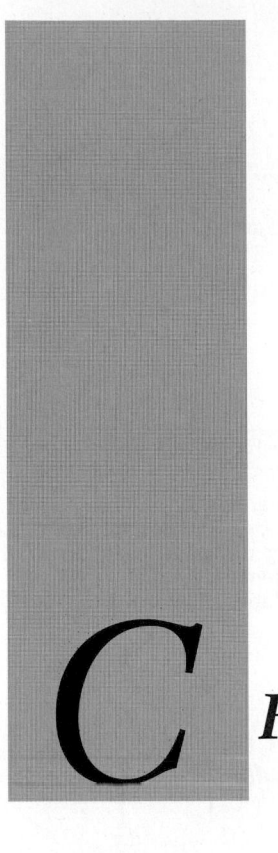

Cardiac Output, Venous Return, and Their Regulation

CHAPTER 20

Cardiac output is the quantity of blood pumped into the aorta each minute by the heart. This is also the quantity of blood that flows through the circulation and is responsible for transporting substances to and from the tissues. Therefore, cardiac output is perhaps the most important factor that we have to consider in relation to the circulation.

Venous return is the quantity of blood flowing from the veins into the right atrium each minute. The venous return and the cardiac output must be equal to each other except for a few heartbeats at a time when blood might be temporarily being stored in or removed from the heart and lungs.

Normal Values for Cardiac Output at Rest and During Activity

Cardiac output varies widely with the level of activity of the body. Therefore, such factors as the level of body metabolism, whether the person is exercising, age, and size of the body as well as a number of other factors can influence the cardiac output.

For young, healthy men, in whom the greatest number of cardiac output measurements have been made, the resting cardiac output averages about 5.6 liters/min. For women, this value is 10 to 20 per cent less. When one considers the factor of age as well—because with increasing age, body activity diminishes—the average cardiac output for the adult, in round numbers, is often stated to be almost exactly 5 liters/min.

CARDIAC INDEX. Because the cardiac output changes markedly with body size, it has been important to find some means by which the cardiac outputs of different-sized people can be compared with one another. Experiments have shown that the cardiac output increases approximately in proportion to the surface area of the body. Therefore, it is frequently stated in terms of the *cardiac index,* which is the *cardiac output per square meter of body surface area.* The normal human weighing 70 kilograms has a body surface area of about 1.7 square meters, which means that the normal average cardiac index for adults is about 3 liters/min/m² of body surface area.

EFFECT OF AGE ON CARDIAC OUTPUT. Figure 20–1 shows the cardiac output, expressed as cardiac index, at different ages. Rising rapidly to a level greater than 4 liters/min/m² at age 10 years, the cardiac index declines to about 2.4 liters/min at age 80 years. We see later in the chapter that the cardiac output is regulated throughout life almost directly in proportion to the overall bodily metabolic activity. Therefore, the declining cardiac index is indicative of declining activity with age.

CONTROL OF CARDIAC OUTPUT BY VENOUS RETURN—ROLE OF THE FRANK-STARLING MECHANISM OF THE HEART

When one states that cardiac output is controlled by venous return, one means that it is not the heart itself that is the primary controller of cardiac output. Instead, it is the various factors of the peripheral circulation that affect the flow of blood into the heart from the veins, called *venous return,* that are the primary controllers.

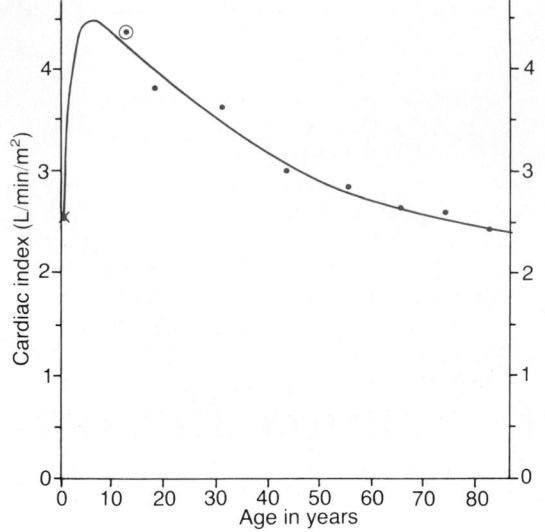

Figure 20–1. Cardiac index at different ages. (From Guyton, Jones, and Coleman: Circulatory Physiology: Cardiac Output and Its Regulation. Philadelphia, W. B. Saunders Company, 1973.)

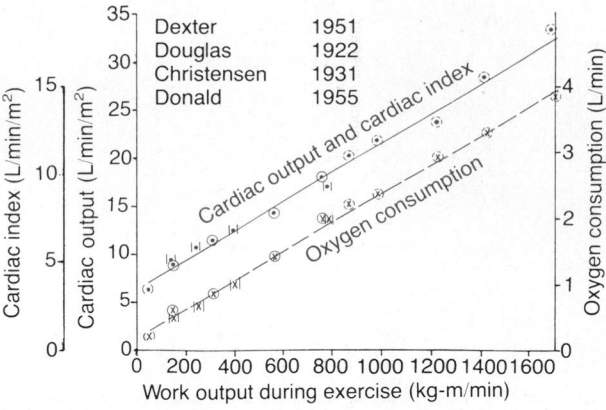

Figure 20–2. Relations between cardiac output and work output (solid curve) and between oxygen consumption and work output (dashed curve) during different levels of exercise. (From Guyton, Jones, and Coleman: Circulatory Physiology: Cardiac Output and Its Regulation. Philadelphia, W. B. Saunders Company, 1973.)

The main reason peripheral factors are usually more important in controlling cardiac output is that the heart has a built-in mechanism that normally allows it to pump automatically whatever amount of blood flows into the right atrium from the veins. This mechanism, called the *Frank-Starling law of the heart,* is discussed in Chapter 9. Basically, this law states that when increased quantities of blood flow into the heart, this stretches the walls of the heart chambers. As a result of the stretch, the cardiac muscle contracts with increased force and, within limits, empties the expanded chambers as much as ever. Therefore, all the extra blood that flows into the heart is automatically pumped without delay into the aorta and flows again through the circulation.

Another important factor, discussed in Chapter 10, is the effect on the heart rate caused by stretching the heart. Stretch of the sinus node in the wall of the right atrium has a direct effect on the rhythmicity of the node itself to increase the heart rate 10 to 15 per cent. In addition, the stretched right atrium initiates a nervous reflex called the *Bainbridge reflex,* passing first to the vasomotor center of the brain and then back to the heart by way of the sympathetic nerves and vagi, also to increase the heart rate. The increase in rate then helps to pump the extra blood.

Therefore, under most normal unstressful conditions, the cardiac output is controlled almost entirely by peripheral factors that determine the venous return. We shall see later in the chapter that when the returning blood is more than the heart can pump, the heart does then limit cardiac output.

Cardiac Output Regulation Is the Sum of All Local Blood Flow Regulations — Body Metabolism Regulates Most Local Blood Flow

The venous return to the heart is the sum of all the local blood flows from the individual segments of the peripheral circulation. Therefore, it follows that cardiac output regulation is the sum of all the local blood flow regulations. The mechanisms of local blood flow regulation are discussed in Chapter 17. In most tissues, blood flow increases mainly in proportion to each tissue's metabolism. For instance, the blood flow almost always increases when tissue oxygen consumption increases, which is shown in Figure 20–2 during different levels of exercise. Note that at each increasing level of exercise, both the oxygen consumption and the cardiac output increase in a parallel manner.

Therefore, the cardiac output is determined principally by the sum of the various factors throughout the body that control local blood flow. All the local blood flows summate to form the venous return, and the heart automatically pumps this returning blood back into the arteries to flow around the system again.

EFFECT OF TOTAL PERIPHERAL RESISTANCE ON THE LONG-TERM CARDIAC OUTPUT LEVEL. Figure 20–3 is the same as Figure 19–5. It is repeated here to illustrate an extremely important principle in cardiac output control: When the arterial pressure is controlled normally, the long-term cardiac output level varies exactly reciprocally to the changes in total peripheral resistance. Note in the figure that when the total peripheral resistance is exactly normal (at the 100 per cent mark in the figure), the cardiac output is also normal. Then, when the total peripheral resistance increases, the cardiac output falls; conversely, when the total peripheral resistance decreases, the cardiac output increases. One can easily understand this by re-

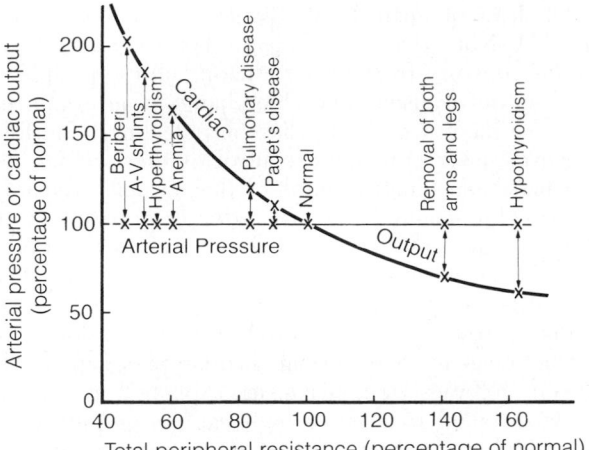

Figure 20–3. Chronic effect of different total peripheral resistances on the cardiac output, showing the reciprocal relation between total peripheral resistance and cardiac output. (Reprinted from Guyton: Arterial Pressure and Hypertension. Philadelphia, W. B. Saunders Company, 1980.)

considering one of the forms of Ohm's law, as expressed in Chapter 14:

$$\text{Cardiac Output} = \frac{\text{Arterial Pressure}}{\text{Total Peripheral Resistance}}$$

The meaning of this formula and also of Figure 20–3 is simply the following: Any time the long-term level of total peripheral resistance changes, the cardiac output will change in exactly the opposite direction.

The Heart Has Limits for the Cardiac Output That It Can Achieve—This Causes a Plateau Level in the Cardiac Output Curve

There are definite limits to the amount of blood that the heart can pump, which are best expressed in the form of cardiac output curves.

Figure 20–4 demonstrates by the heavy black curve the normal cardiac output curve, showing the cardiac output per minute at each level of right atrial pressure. This is one type of *cardiac function curve*, which is discussed in Chapter 9 and shown in Figure 9–12. Note that the plateau level of this normal cardiac output curve is about 13 liters/min, 2.5 times the normal cardiac output of about 5 liters/min. This means that the normal heart, functioning without any excess nervous stimulation, can pump an amount of venous return up to about 2.5 times the normal venous return before the heart becomes a limiting factor in the control of cardaic output.

Shown in red in Figure 20–4 are several other cardiac output curves for hearts that are not pumping normally. The uppermost curves are for *hypereffective* hearts that are pumping better than normal. The lowermost curves are for *hypoeffective hearts* that are pumping at levels below normal.

The Hypereffective Heart and Factors That Can Cause Hypereffectivity

Only two types of factors usually can make the heart a better pump than normal. They are (1) nervous stimulation and (2) hypertrophy of the heart muscle.

EFFECT OF NERVOUS EXCITATION TO INCREASE HEART PUMPING. In Chapter 9, we see that a combination of sympathetic *stimulation* and parasympathetic *inhibition* does two things to increase the pumping effectiveness of the heart: (1) greatly increase the heart rate—sometimes, in young people, to 180 to 200 beats/min—and (2) increase the strength of heart contraction, which is called increased "contractility," to twice its normal strength. Combining these two effects, maximal nervous excitation of the heart can raise the plateau level of the cardiac output curve to almost twice the plateau of the normal curve, as shown by the 25-liter level of the uppermost curve in Figure 20–4.

INCREASED PUMPING EFFECTIVENESS CAUSED BY HEART HYPERTROPHY. A heart that is subjected to increased workload, but not so much excess load that it damages the heart, causes the heart muscle to increase in mass and contractile strength in the same way that heavy exercise causes skeletal muscles to hypertrophy. For instance, it is common for the hearts of marathon runners to be increased in mass 50 to 75 per cent. This increases the plateau level of the cardiac output curve, sometimes 50 to 100 per cent, and therefore allows the heart to pump much greater than usual amounts of cardiac output.

When one combines nervous excitation of the heart

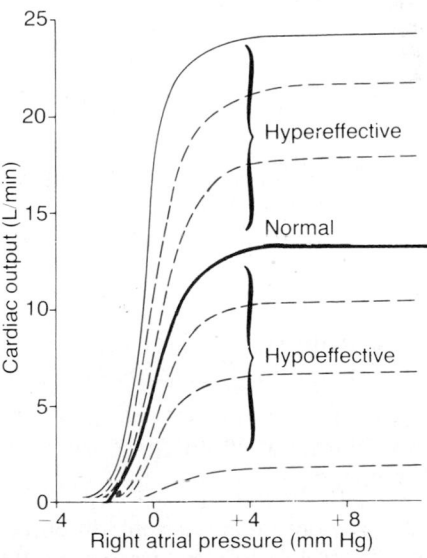

Figure 20–4. Cardiac output curves for the normal heart and for hypoeffective and hypereffective hearts. (From Guyton, Jones, and Coleman: Circulatory Physiology: Cardiac Output and Its Regulation. Philadelphia, W. B. Saunders, 1973.)

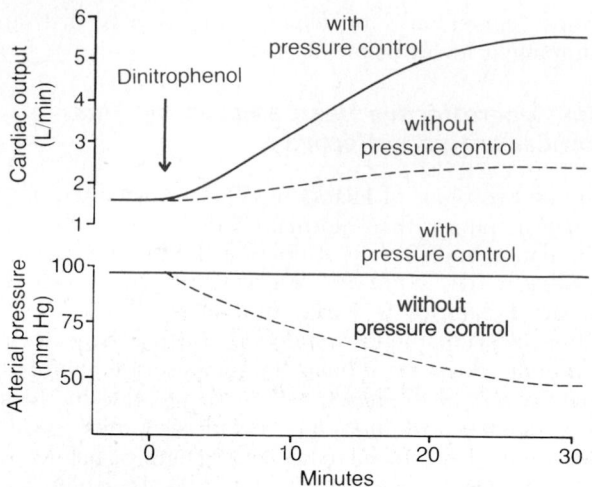

Figure 20–5. Experiment in a dog to demonstrate the importance of nervous control of arterial pressure as a prerequisite for cardiac output control. Note that with pressure control, the metabolic stimulant dinitrophenol increases cardiac output; without pressure control, the arterial pressure falls and the cardiac output rises very little. (Drawn from experiments by Dr. M. Banet.)

and hypertrophy, as occurs in marathon runners, the effect can allow the heart to pump 30 to 40 liters/min; this increased level of pumping is one of the most important factors in determining the runner's running time.

Factors That Cause a Hypoeffective Heart

Any factor that decreases the heart's ability to pump blood causes hypoeffectivity. Some of the factors are the following:

Inhibition of nervous excitation of the heart
Pathological factors that cause abnormal rhythm or rate of heartbeat
Valvular heart disease
Increased arterial pressure against which the heart must pump, such as in hypertension
Congenital heart disease
Myocarditis
Cardiac anoxia
Diphtheritic or other types of myocardial damage or toxicity

What Is the Role of the Nervous System in Controlling Cardiac Output?

Importance of the Nervous System to Maintain the Arterial Pressure When the Cardiac Output Increases

Figure 20–5 shows an important difference in cardiac output control with and without a functioning autonomic nervous system. The solid curves demonstrate the effect in the normal animal of intense dilatation of the peripheral blood vessels caused by administering the drug dinitrophenol, which increased the

metabolism of virtually all tissues of the body about fourfold. Note that with nervous control to keep the arterial pressure from falling, dilating all the peripheral blood vessels caused almost no change in arterial pressure but increased the cardiac output almost fourfold. However, after autonomic control of the nervous system had been blocked, none of the normal circulatory reflexes for maintaining the arterial pressure could function, and vasodilation of the vessels with dinitrophenol (dashed curves) then caused a profound fall in arterial pressure to about one half normal, whereas the cardiac output rose only 1.6-fold instead of 4-fold.

Maintenance of a normal arterial pressure by the nervous reflexes, by mechanisms explained in Chapter 18, is essential to achieve high cardiac outputs when the peripheral tissues dilate their vessels to increase their local blood flow.

EFFECT OF THE NERVOUS SYSTEM TO INCREASE THE ARTERIAL PRESSURE DURING EXERCISE. During exercise, the nervous system not only prevents the arterial pressure from falling but also *increases* the pressure. The cause of this is mainly the following: The same brain activity that sends signals to the peripheral muscles to cause exercise also sends simultaneous signals into the autonomic centers to excite circulatory activity, causing venous constriction, increased heart rate, and increased contractility of the heart; all these changes acting together increase the arterial pressure even above normal, which in turn forces more blood flow through the tissues. This increased pressure often allows the cardiac output to increase during exercise an extra 50 to 100 per cent.

In summary, when the local tissue vessels dilate and attempt to increase the cardiac output above normal, the nervous system plays an exceedingly important role in preventing the arterial pressure from falling to disastrously low levels. In fact, in exercise, the nervous system goes even further, providing additional signals to raise the arterial pressure above normal, which serves to increase the cardiac output an extra 50 to 100 per cent.

PATHOLOGICALLY HIGH AND PATHOLOGICALLY LOW CARDIAC OUTPUTS

In healthy human beings, the cardiac outputs are surprisingly constant from one person to another. However, multiple clinical abnormalities can cause either high or low cardiac outputs. Some of the more important of these are shown in Figure 20–6.

High Cardiac Output Is Almost Always Caused by Reduced Total Peripheral Resistance

The left side of Figure 20–6 identifies conditions that commonly cause cardiac outputs higher than normal. One of the distinguishing features of these conditions is that *they all result from chronically reduced total peripheral resistance.* None of them result from excessive

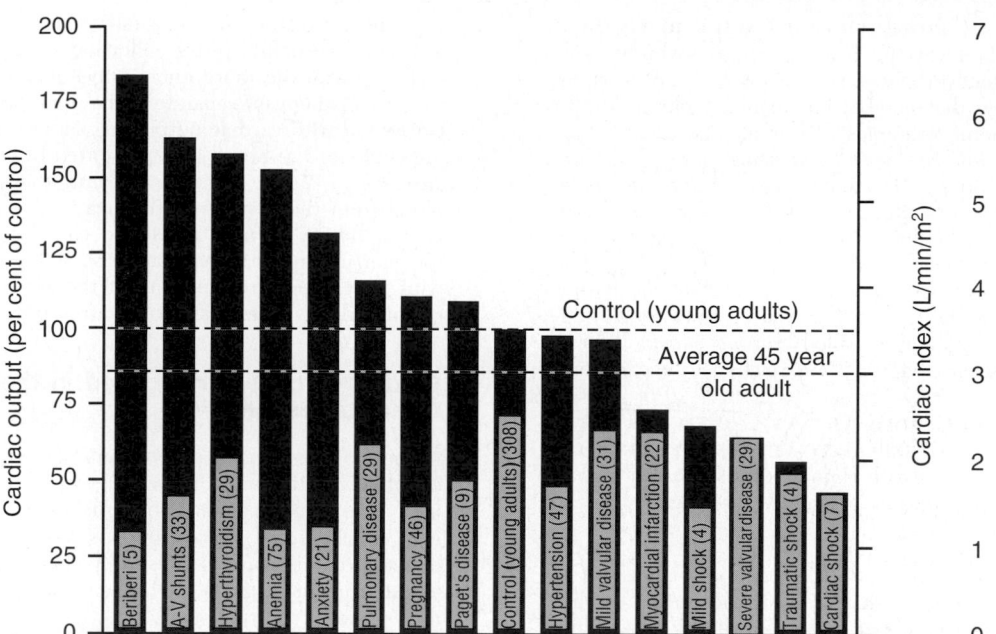

Figure 20–6. Cardiac output in different pathological conditions. The numbers in parentheses indicate number of patients studied in each condition. (From Guyton, Jones, and Coleman: Circulatory Physiology: Cardiac Output and Its Regulation. Philadelphia, W. B. Saunders Company, 1973.)

excitation of the heart itself, which we explain subsequently. For the present, let us look at some of the peripheral factors that can decrease the peripheral resistance and at the same time increase the cardiac output to above normal.

1. *Beriberi.* This disease is caused by insufficient quantity of the vitamin *thiamine* in the diet. Lack of this vitamin causes diminished ability of the tissues to use cellular nutrients, and the local tissue blood flow mechanisms in turn cause marked peripheral vasodilation. Sometimes the total peripheral resistance decreases to as little as one-half normal. Consequently, the long-term level of cardiac output also often increases to twice normal.

2. *Arteriovenous fistula (shunt).* Earlier in the chapter, we pointed out that whenever a fistula (also called a shunt) occurs between a major artery and a major vein, tremendous amounts of blood flow directly from the artery into the vein. This, too, greatly decreases the total peripheral resistance and, likewise, increases the venous return and cardiac output.

3. *Hyperthyroidism.* In hyperthyroidism, the metabolism of all the tissues of the body becomes greatly increased. Oxygen usage increases and vasodilator products are released from the tissues. Therefore, the total peripheral resistance decreases markedly because of the local tissue blood flow control reactions throughout the body; in consequence, the cardiac output often increases to 40 to 80 per cent above normal.

4. *Anemia.* In anemia, two peripheral effects greatly decrease the total peripheral resistance. One of these is reduced viscosity of the blood, resulting from the decreased concentration of red blood cells. The other is diminished delivery of oxygen to the tissues because of the decreased hemoglobin; lack of oxygen delivery

causes local vasodilation. As a consequence, the cardiac output increases greatly.

Any other factor that decreases the total peripheral resistance chronically also increases the cardiac output.

FAILURE OF INCREASED CARDIAC PUMPING TO CAUSE PROLONGED INCREASE OF THE CARDIAC OUTPUT. If the heart is suddenly stimulated excessively, the cardiac output often increases acutely 20 to 30 per cent. This small increase is maintained no longer than a few minutes, even though the heart continues to be strongly stimulated. There are two reasons for this: (1) Excess blood flow through the tissues causes automatic vasoconstriction of the blood vessels because of the autoregulation mechanism discussed in previous chapters, and this reduces the venous return and cardiac output back toward normal. (2) The slightly increased arterial pressure that results after acute cardiac stimulation raises the capillary pressure, and fluid filters out of the capillaries into the tissues. This decreases the blood volume and also decreases the venous return back toward normal. Over a period of hours and days, the increased pressure also causes the kidneys to lose blood fluid volume until the arterial pressure and cardiac output return to normal.

Thus, all the known conditions that cause *chronic* elevation of the cardiac output result from decreased total peripheral resistance and not increased cardiac activity.

Low Cardiac Output

Figure 20–6 shows to the far right several conditions that cause abnormally low cardiac output. They fall into two categories: (1) those abnormalities that cause the pumping effectiveness of the heart to fall too low and (2) those that cause venous return to fall too low.

DECREASED CARDIAC OUTPUT CAUSED BY CARDIAC FACTORS. Whenever the heart becomes severely damaged from whatever cause, its limiting level of pumping may fall below that needed for adequate blood flow to the tissues. Some examples of this include *severe myocardial infarction, severe valvular heart disease, myocarditis, cardiac tamponade,* and certain *cardiac metabolic derangements.* The effect of several of these is shown on the right in Figure 20–6, demonstrating the low cardiac outputs that result.

When the cardiac output falls so low that the tissues throughout the body begin to suffer nutritional deficiency, the condition is called *cardiac shock.* This is discussed fully in Chapter 22 in relation to cardiac failure.

DECREASE IN CARDIAC OUTPUT CAUSED BY PERIPHERAL FACTORS—DECREASED VENOUS RETURN. Any factor that interferes with venous return also can lead to decreased cardiac output. Some of these factors are the following:

1. *Decreased blood volume.* By far, the most common peripheral factor that leads to decreased cardiac output is decreased blood volume, resulting most often from hemorrhage. It is clear why this condition decreases the cardiac output: Loss of blood decreases the filling of the vascular system to such a low level that there is not enough blood in the peripheral vessels to create the pressure required to push blood back to the heart.

2. *Acute venous dilatation.* On some occasions, the peripheral veins become acutely vasodilated. This results most often when the sympathetic nervous system suddenly becomes inactive. For instance, fainting often results from sudden loss of sympathetic nervous system activity, which causes the peripheral capacitative vessels, especially the veins, to dilate markedly. This decreases the filling pressure of the vascular system because the blood volume can no longer create adequate pressure in the now flaccid peripheral blood vessels. As a result, the blood "pools" in the vessels and will not return to the heart.

3. *Obstruction of the large veins.* On rare occasions, the large veins leading into the heart become obstructed, so that the blood in the peripheral vessels cannot flow back into the heart. Consequently, the cardiac output falls markedly.

Regardless of the cause of low cardiac output, whether it be a peripheral factor or a cardiac factor, if ever the cardiac output falls below that level required for adequate nutrition of the tissues, the person is said to suffer *circulatory shock.* This condition can be lethal within a few minutes to a few hours. Circulatory shock is such an important clinical problem that it is discussed in detail in Chapter 24.

A MORE QUANTITATIVE ANALYSIS OF CARDIAC OUTPUT REGULATION

Our discussion of cardiac output regulation thus far is adequate for understanding the factors that control cardiac output in most simple conditions. However, to understand cardiac output regulation in especially stressful situations, such as the extremes of exercise, cardiac fail-

ure, and circulatory shock, a more complex quantitative analysis is presented in the following sections.

To perform the more quantitative analysis, it is necessary to distinguish separately the two primary factors concerned with cardiac output regulation: (1) the pumping ability of the heart, as represented by *cardiac output curves,* and (2) the peripheral factors that affect flow of blood from the veins into the heart, as represented by *venous return curves.* Then one can put these curves together in a quantitative way to show how they interact with each other to determine at the same time cardiac output, venous return, and right atrial pressure.

Cardiac Output Curves Used in the Quantitative Analysis

Some of the cardiac output curves used to depict the quantitative effects of different levels of heart effectiveness have already been shown in Figure 20–4. However, an additional set of curves is required to show the effect on cardiac output caused by changing the external pressure on the outside of the heart, as explained in the next section.

EFFECT OF EXTERNAL PRESSURE OUTSIDE THE HEART ON CARDIAC OUTPUT CURVES. Figure 20–7 shows the effect of changes in external cardiac pressure on the cardiac output curve. The normal external pressure is the normal intrapleural pressure (the pressure in the chest cavity), which is − 4 mm Hg. Note in the figure that a rise in intrapleural pressure to − 2 mm Hg shifts the entire curve to the right by the same amount. This shift occurs because an extra 2 mm Hg right atrial pressure is now required to overcome the increased pressure on the outside of the heart. Likewise, an increase in intrapleural pressure to + 2 mm Hg requires a 6-mm Hg increase in right atrial pressure, which shifts the entire curve 6 mm Hg to the right.

Some of the factors that can alter the intrapleural pressure and thereby shift the cardiac output curve are the following:

1. *Cyclic changes during normal respiration,* which can be ± 2 mm Hg in normal expiration versus inspiration and ± 50 mm Hg in strenuous breathing

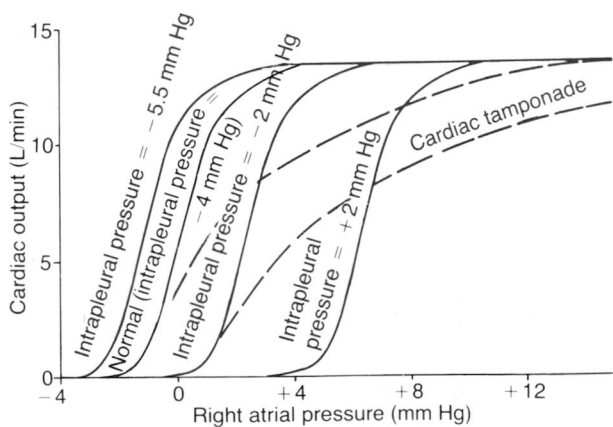

Figure 20–7. Cardiac output curves at different levels of intrapleural pressure and at different degrees of cardiac tamponade. (From Guyton, Jones, and Coleman: Circulatory Physiology: Cardiac Output and Its Regulation. Philadelphia, W. B. Saunders Company, 1973.)

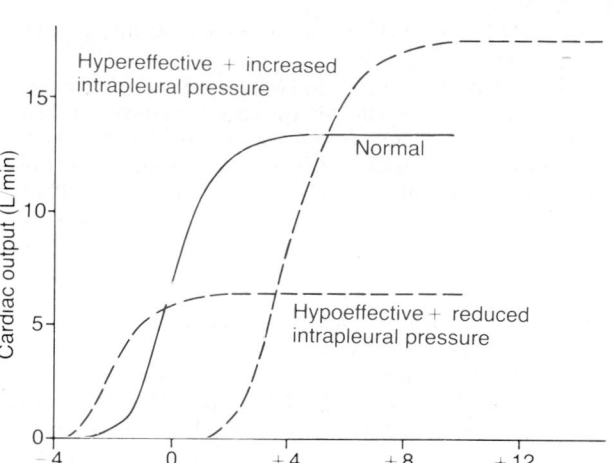

Figure 20–8. Combinations of two major patterns of cardiac output curves, showing the effect of alterations in both extracardiac pressure and effectiveness of the heart as a pump. (From Guyton, Jones, and Coleman: Circulatory Physiology: Cardiac Output and Its Regulation. Philadelphia, W. B. Saunders Company, 1973.)

2. *Breathing against a negative pressure,* which shifts the curve to a more negative atrial pressure (to the left)

3. *Positive pressure breathing,* which shifts the curve to the right

4. *Opening the thoracic cage,* which increases the intrapleural pressure to 0 mm Hg and shifts the curve to the right 4 mm Hg

5. *Cardiac tamponade,* which means the accumulation of a large quantity of fluid in the pericardial cavity around the heart with resultant increase in external cardiac pressure and shifting of the curve to the right. Note in Figure 20–7 that cardiac tamponade shifts the upper parts of the curves further to the right than the lower parts because the external cardiac pressure rises to higher values as the chambers of the heart fill to increased volumes during high cardiac output.

COMBINATIONS OF DIFFERENT PATTERNS OF CARDIAC OUTPUT CURVES. Figure 20–8 shows that the cardiac output curve can change as a result of simultaneous changes in both external cardiac pressure and effectiveness of the heart as a pump. Thus, knowing what is

happening to the external pressure as well as the capability of the heart as a pump, one can express the momentary ability of the heart to pump blood by a single cardiac output curve.

Venous Return Curves

There remains the entire systemic circulation that must be considered before a total analysis of cardiac regulation can be made. To analyze the function of the systemic circulation, we remove the heart and lungs from the circulation of an animal and replace them with a pump and artificial oxygenator system. Then, different factors, such as changes in blood volume, changes in vascular resistances, and changes in right atrial pressure, are altered to determine how the systemic circulation operates in different circulatory states. In these studies, one finds three principal factors that affect venous return to the heart from the systemic circulation. They are as follows:

1. *Right atrial pressure,* which exerts a backward force on the veins to impede flow of blood into the right atrium

2. Degree of filling of the systemic circulation, measured by the *mean systemic filling pressure,* which forces the systemic blood toward the heart (this is the pressure measured everywhere in the systemic circulation when all flow of blood is stopped—we discuss this in detail later)

3. *Resistance to blood flow* between the peripheral vessels and the right atrium

These factors can all be expressed quantitatively by the *venous return curve,* as we explain in the next sections.

Normal Venous Return Curve

In the same way that the cardiac output curve relates the pumping of blood by the heart to right atrial pressure, the *venous return curve relates venous return also to right atrial pressure,* that is, the flow of blood into the heart from the systemic circulation.

The curve in Figure 20–9 is the normal venous return curve. This curve shows that when heart pumping fails and causes the right atrial pressure to rise, the backward force of the rising atrial pressure on the systemic circulation decreases venous return of blood to

Figure 20–9. Normal venous return curve. The *plateau* is caused by collapse of the large veins entering the chest when the right atrial pressure falls below atmospheric pressure. Note also that venous return becomes zero when the right atrial pressure rises to equal the mean systemic filling pressure.

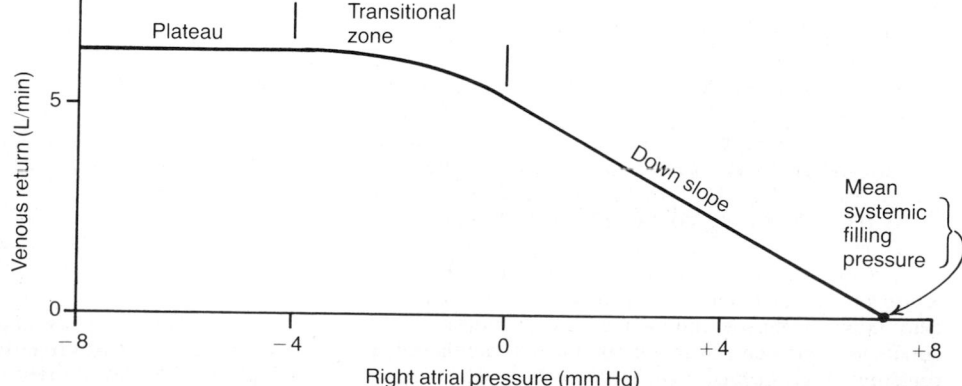

the heart. *If all circulatory reflexes are prevented from acting,* venous return decreases to zero when the right atrial pressure rises to about 7 mm Hg. Such a slight rise in right atrial pressure causes a drastic decrease in venous return because the systemic circulation is a distensible bag, so that any increase in back pressure causes blood to dam up in this bag instead of returning to the heart.

At the same time that the right atrial pressure is rising and causing venous stasis, the pumping by the failing heart also approaches zero, so that the arterial pressure falls to equal the venous pressure. Thus, both the arterial and the venous pressures come to equilibrium when all flow in the systemic circulation ceases at a pressure of 7 mm Hg, which, by definition, is the *mean systemic filling pressure.*

PLATEAU IN THE VENOUS RETURN CURVE AT NEGATIVE ATRIAL PRESSURES CAUSED BY COLLAPSE OF THE VEINS. When the right atrial pressure falls *below* zero—that is, below atmospheric pressure—further increase in venous return ceases rapidly. By the time the right atrial pressure has fallen to about −2 mm Hg, the venous return will have reached a plateau; then it remains at this plateau level even though the right atrial pressure falls to −20 to −50 mm Hg. This plateau is caused by collapse of the veins entering the chest. Negative right atrial pressure sucks the walls of the veins together where they enter the chest, which prevents the negative pressure from sucking blood from the peripheral veins. Instead, the pressure in the veins immediately outside the chest remains almost equal to atmospheric pressure (zero pressure) because that is the pressure of the air pressing through the skin and soft tissues against the outsides of the flaccid veins, causing them to collapse. Therefore, for all practical purposes, the venous pressure where the large veins empty into the chest never falls below 0 mm Hg, despite the fact that the right atrial pressure may fall to very negative values. Consequently, even very negative pressures in the right atrium cannot increase venous return significantly above that which exists at a normal atrial pressure of 0 mm Hg.

Mean Circulatory Filling Pressure and Mean Systemic Filling Pressure, and Their Effect on Venous Return

When the heart pumping is stopped by causing ventricular fibrillation or in any other way, the flow of blood everywhere in the circulation ceases a few seconds later. Without blood flow, the pressures everywhere in the circulation become equal after a minute or so. This equilibrated pressure level is called the *mean circulatory filling pressure.*

EFFECT OF BLOOD VOLUME ON MEAN CIRCULATORY FILLING PRESSURE. The greater the volume of blood in the circulation, the greater is the mean circulatory filling pressure because extra blood volume stresses the walls of the vasculature. The *solid curve* in Figure 20–10 shows the approximate normal effect of different levels of blood volume on the mean circulatory filling pressure. Note that at a blood volume of about 4000 milliliters, the mean circulatory filling pressure is close to zero because this is the "unstressed volume" of the circulation, but at a volume of 5000 milliliters, the filling pressure is the normal value of 7 mm Hg. Similarly, at

still higher volumes, the mean circulatory filling pressure increases almost linearly.

EFFECT OF SYMPATHETIC NERVOUS STIMULATION OF THE CIRCULATION ON MEAN CIRCULATORY FILLING PRESSURE. The two *dashed curves* in Figure 20–10 show the effects, respectively, of high and low levels of sympathetic stimulation on the mean circulatory filling pressure. Strong sympathetic stimulation constricts all the systemic blood vessels as well as the larger pulmonary vessels and even the chambers of the heart. Therefore, the capacity of the system decreases, so that at each level of blood volume, the mean circulatory filling pressure is increased. At normal blood volume, maximal sympathetic stimulation increases the mean circulatory filling pressure from 7 mm Hg to about 2.5 times that value, or about 17 mm Hg. Conversely, complete inhibition of the sympathetic nervous system relaxes the blood vessels as well as the heart, decreasing the mean circulatory filling pressure from the normal value of 7 mm Hg down to about 4 mm Hg.

Before leaving Figure 20–10, note specifically how steep the curves are. This means that even slight changes in blood volume or slight changes in the capacity of the system caused by various levels of sympathetic activity can have large effects on the mean circulatory filling pressure.

MEAN SYSTEMIC FILLING PRESSURE AND ITS RELATION TO MEAN CIRCULATORY FILLING PRESSURE. The *mean systemic filling pressure,* Psf, is slightly different from the mean circulatory filling pressure. It is the pressure measured everywhere in the systemic circulation after blood flow has been stopped by clamping the large blood vessels at the heart, so that the pressures in the systemic circulation can be measured independently from those in the pulmonary circulation. The mean systemic pressure, although almost impossible to measure in the living animal, is the important pressure for deter-

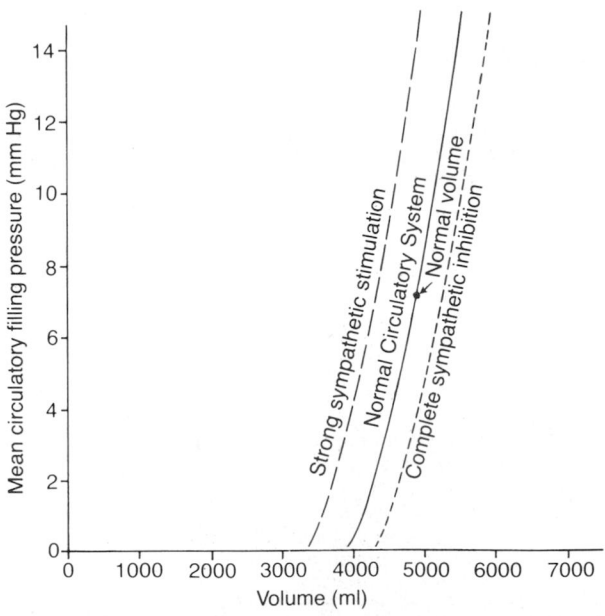

Figure 20–10. Volume-pressure curves of the entire circulation, showing the effect of strong sympathetic stimulation and complete sympathetic inhibition.

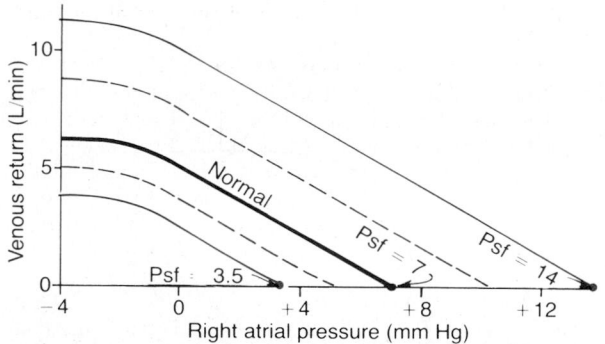

Figure 20–11. Venous return curves, showing the normal curve when the mean systemic filling pressure *(Psf)* is 7 mm Hg and showing the effect of altering the mean systemic filling pressure to either 3.5 or 14 mm Hg. (From Guyton, Jones, and Coleman: Circulatory Physiology: Cardiac Output and Its Regulation, Philadelphia, W. B. Saunders Company, 1973.)

mining venous return. *The mean systemic filling pressure is almost always nearly equal to the mean circulatory filling pressure* because the pulmonary circulation has less than one eighth as much capacitance as the systemic circulation and only about one tenth as much blood volume.

EFFECT ON THE VENOUS RETURN CURVE OF CHANGES IN MEAN SYSTEMIC FILLING PRESSURE. Figure 20–11 shows the effects on the venous return curve caused by increasing or decreasing the mean systemic filling pressure. Note in Figure 20–11 that the normal mean systemic filling pressure is 7 mm Hg. For the uppermost curve in the figure, the mean systemic filling pressure is 14 mm Hg and for the lowermost curve, 3.5 mm Hg. These curves demonstrate that the greater the mean systemic filling pressure, which also means the greater the "tightness" with which the circulatory system is filled with blood, the more the venous return curve shifts *upward* and *to the right.* Conversely, the lower the mean systemic filling pressure, the more the curve shifts *downward* and *to the left.* To express this another way, the greater the system is filled, the easier it is for blood to flow into the heart. And the less the filling, the more difficult it is for blood to flow into the heart.

Pressure Gradient for Venous Return—When This Is Zero, There Is No Venous Return. When the right atrial pressure rises to equal the mean systemic filling pressure, all other pressures in the systemic circulation also come to this same pressure. Therefore, there is no longer any pressure difference between the peripheral vessels and the right atrium. Consequently, there also can no longer be any flow from any peripheral vessels back to the right atrium. However, when the right atrial pressure falls progressively lower than the mean systemic filling pressure, the flow to the heart increases proportionately, as one can see by studying any of the venous return curves in Figure 20–11. That is, *the greater the difference between the mean systemic filling pressure and the right atrial pressure, the greater becomes the venous return.* Therefore, the difference between these two pressures is called the *pressure gradient for venous return.*

Resistance to Venous Return

In the same way that the mean systemic filling pressure represents a pressure pushing the blood in the veins from the periphery toward the heart, there is also resistance to this venous flow of blood. It is called the *resistance to venous return.*

Most of the resistance to venous return occurs in the veins, although some occurs in the arterioles and small arteries as well. Why is venous resistance so important in determining the resistance to venous return? The answer is that when the resistance in the veins increases, blood begins to be dammed up in all parts of the systemic circulation, but the venous pressure rises very little because the veins are highly distensible. Therefore, the rise in pressure upstream from the resistance (that is, in the upstream veins themselves) is not very effective in overcoming the resistance, and the venous return decreases drastically. On the other hand, when the arteriolar and small artery resistances increase, blood accumulates in the arteries, which have a capacitance only ⅟₃₀ as great as that of the veins. Therefore, even slight accumulation of blood in the arteries raises the pressure greatly—30 times as much as in the veins—and this high pressure overcomes much of the increased resistance so that the venous return decreases very little. Mathematically, it turns out that about two thirds of the so-called resistance to venous return is determined by venous resistance and about one third, by the arteriolar and small artery resistance.

Venous return can be calculated by the following formula:

$$VR = \frac{PSF - PRA}{RVR},$$

in which VR is venous return, PSF is mean systemic filling pressure, PRA is right atrial pressure, and RVR is resistance to venous return. In the normal human, the values for these are: venous return equals 5 liters/min, mean systemic filling pressure 7 mm Hg, right atrial pressure 0 mm Hg, and resistance to venous return 1.4 mm Hg per liter of blood flow.

EFFECT OF RESISTANCE TO VENOUS RETURN ON THE VENOUS RETURN CURVE. Figure 20–12 demonstrates the effect of different resistances to venous return on the venous return curve, showing that a *decrease* in this resistance to one-half normal allows twice as much flow of blood and, therefore, *rotates the curve upward* to twice as great a slope. Conversely, an *increase* in resistance to twice normal *rotates the curve downward* to one half as great a slope. However, note that when the right atrial pressure rises to equal the mean systemic filling pressure, venous return becomes zero at all levels of resistance to venous return because when there is no pressure gradient to cause flow of blood, it makes no difference what the resistance is in the circulation; the flow is still zero. Therefore, *the highest level to which the right atrial pressure can rise* regardless of how much the heart might fail is equal to the mean systemic filling pressure.

COMBINATIONS OF VENOUS RETURN CURVE PATTERNS. Figure 20–13 shows the effects on the venous return curve caused by simultaneous changes in mean systemic pressure and resistance to venous return, dem-

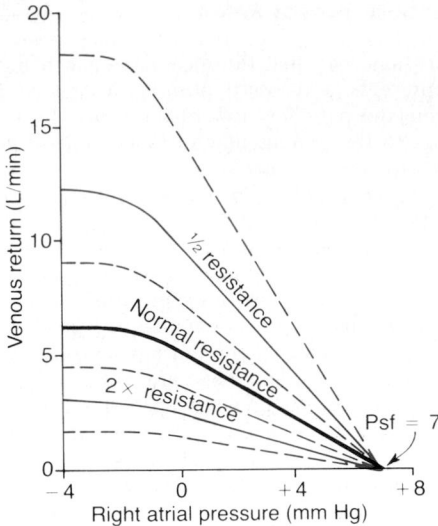

Figure 20–12. Venous return curves, depicting the effect of altering the "resistance to venous return." (From Guyton, Jones, and Coleman: Circulatory Physiology: Cardiac Output and Its Regulation. Philadelphia, W. B. Saunders Company, 1973.)

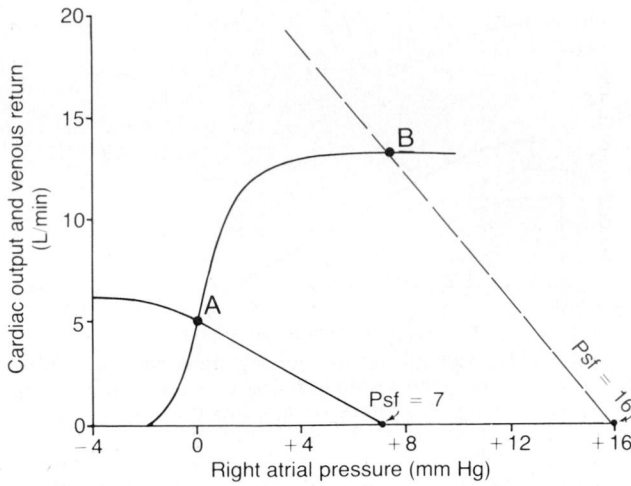

Figure 20–14. Two solid curves demonstrate an analysis of cardiac output and right atrial pressure when the cardiac output and venous return curves are normal. Transfusion of blood equal to 20 per cent of the blood volume causes the venous return curve to become the dashed curve; as a result, the cardiac output and right atrial pressure shift from point A to point B. (Psf = mean systemic filling pressure.)

onstrating that both these factors can operate simultaneously.

Analysis of Cardiac Output and Right Atrial Pressure, Using Simultaneous Cardiac Output and Venous Return Curves

In the complete circulation, the heart and the systemic circulation must operate together. This means that (1) the venous return from the systemic circulation must equal the cardiac output from the heart and (2)

the right atrial pressure is the same for both the heart and the systemic circulation.

Therefore, one can predict the cardiac output and right atrial pressure in the following way: (1) Determine the momentary pumping ability of the heart and depict this in the form of a cardiac output curve; (2) determine the momentary state of flow from the systemic circulation into the heart and depict this in the form of a venous return curve; and (3) "equate" these curves against each other, as shown in Figure 20–14.

The two black curves in the figure depict the *normal cardiac output curve* and the *normal venous return curve*. There is only one point on the graph, point A, at which the venous return equals the cardiac output and at which the right atrial pressure is the same for both the heart and the systemic circulation. Therefore, in the normal circulation, the right atrial pressure, cardiac output, and venous return are all depicted by point A, called the *equilibrium point*.

Effect of Increased Blood Volume on Cardiac Output

A sudden increase in blood volume of about 20 per cent increases the cardiac output to about 2.5 to 3 times normal. An analysis of this effect is shown by the dashed curve of Figure 20–14. Immediately on infusing the large quantity of extra blood, the increased filling of the system causes the mean systemic filling pressure (Psf) to rise to 16 mm Hg, which shifts the venous return curve to the right. At the same time, the increased blood volume distends the blood vessels, thus reducing their resistance and thereby reducing the resistance to venous return, which rotates the curve upward. As a result of these two effects, the venous return curve of Figure 20–14 changes from the solid curve to the dashed curve. This new curve equates with the cardiac output curve at point B, showing that the cardiac

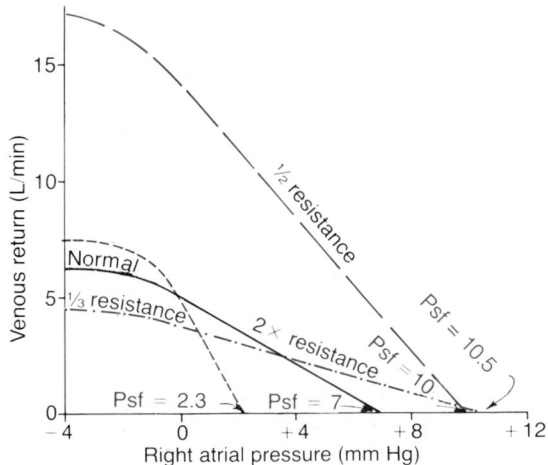

Figure 20–13. Combinations of the major patterns of venous return curves, showing the effects of simultaneous changes in mean systemic filling pressure *(Psf)* and in "resistance to venous return." (From Guyton, Jones, and Coleman: Circulatory Physiology: Cardiac Output and Its Regulation. Philadelphia, W. B. Saunders Company, 1973.)

output increases 2.5 to 3 times and that the right atrial pressure rises to about +8 mm Hg.

COMPENSATORY EFFECTS INITIATED IN RESPONSE TO THE INCREASED BLOOD VOLUME. The increased cardiac output caused by the increased blood volume lasts for only a few minutes because several compensatory effects immediately begin to occur: (1) The increased cardiac output increases the capillary pressure, so that fluid begins to transude out of the capillaries into the tissues, thereby returning the blood volume toward normal. (2) The increased pressure in the veins causes the veins to distend gradually by the mechanism called *stress-relaxation,* especially causing the venous blood reservoirs, such as the liver and spleen, to distend, thus reducing the mean systemic pressure. (3) The excess blood flow through the peripheral tissues causes autoregulatory increase in the peripheral resistance, thus increasing the resistance to venous return. These factors cause the mean systemic filling pressure to return back toward normal and the resistance vessels of the systemic circulation to constrict. Therefore, gradually, over a period of 10 to 40 minutes, the cardiac output returns almost to normal.

Effect of Sympathetic Stimulation on Cardiac Output

Sympathetic stimulation affects both the heart and the systemic circulation: (1) It makes the heart a stronger pump. (2) In the systemic circulation, it increases the mean systemic filling pressure because of contraction of the peripheral vessels—especially the veins—and increases the resistance to venous return. In Figure 20–15, the normal cardiac output and venous return curves are depicted by the dark lines; these equate with each other at point A, which represents a normal venous return and cardiac output of 5 liters/min and a right atrial pressure of 0 mm Hg. Moderate sympathetic stimulation is depicted by the long-dashed curves and maximal sympathetic stimulation, by the dot-dash red curves.

Note that maximal sympathetic stimulation (the red curves) increases the mean systemic filling pressure to 17 mm Hg (depicted by the point at which the venous return curve reaches the zero venous return level) and increases the pumping effectiveness of the heart by nearly 100 per cent. As a result, the cardiac output rises from the normal value at equilibrium point A to about double normal at equilibrium point D—and yet *the right atrial pressure hardly changes.* Thus, different degrees of sympathetic stimulation can increase the cardiac output progressively to about twice normal—at least for short periods until compensatory effects occur.

Effect of Sympathetic Inhibition

The sympathetic nervous system can be blocked by inducing *total spinal anesthesia* or by using some drug, such as *hexamethonium,* that blocks transmission of nerve impulses through the autonomic ganglia. The two lowermost curves in Figure 20–15, the short-dashed curves, show the effect of sympathetic inhibition caused by total spinal anesthesia, demonstrating that (1) the mean systemic filling pressure falls to about 4 mm Hg and (2) the effectiveness of the heart as a pump decreases to about 80 per cent of normal. The cardiac output falls from point A to point B, which is a decrease to about 60 per cent of normal.

Effect of Opening a Large Arteriovenous Fistula

Figure 20–16 shows various stages of circulatory changes that occur after opening a large arteriovenous fistula, that is, a direct opening between a large artery and a large vein.

1. The two curves crossing at point A show the normal condition.
2. The curves crossing at point B show the circulatory condition immediately after opening the large fistula. The principal effects are (1) a sudden and precipitous rotation of the venous return curve upward caused by the large decrease in resistance to venous return when blood is allowed to flow with almost no impediment directly from the large artery into the venous system, bypassing most of the resistance elements of the peripheral circulation, and (2) a slight increase in the level of the cardiac output curve because opening the fistula decreases the peripheral resistance and allows an acute fall in arterial pressure against which the heart must pump. The net result, depicted by point B, is an increase in cardiac output from 5 liters/min to 13 liters/min and an increase in right atrial pressure to about +3 mm Hg.
3. Point C represents the effects about 1 minute later, after the sympathetic nerve reflexes have restored the arterial pressure almost to normal and caused two other effects: (1) an increase in the mean systemic filling pressure because of constriction of all vessels from 7

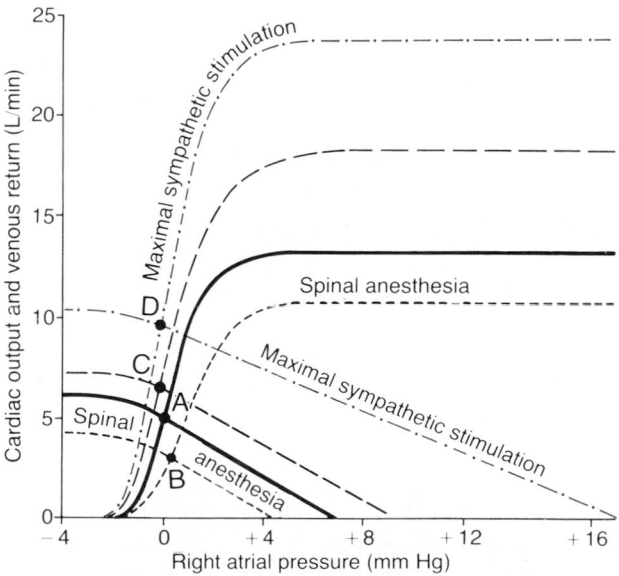

Figure 20–15. Analysis of the effect on cardiac output of (1) moderate sympathetic stimulation—Point C, (2) maximal sympathetic stimulation (colored curves)—Point D, and (3) sympathetic inhibition caused by total spinal anesthesia—Point B. (From Guyton, Jones, and Coleman: Circulatory Physiology: Cardiac Output and Its Regulation. Philadelphia, W. B. Saunders Company, 1973.)

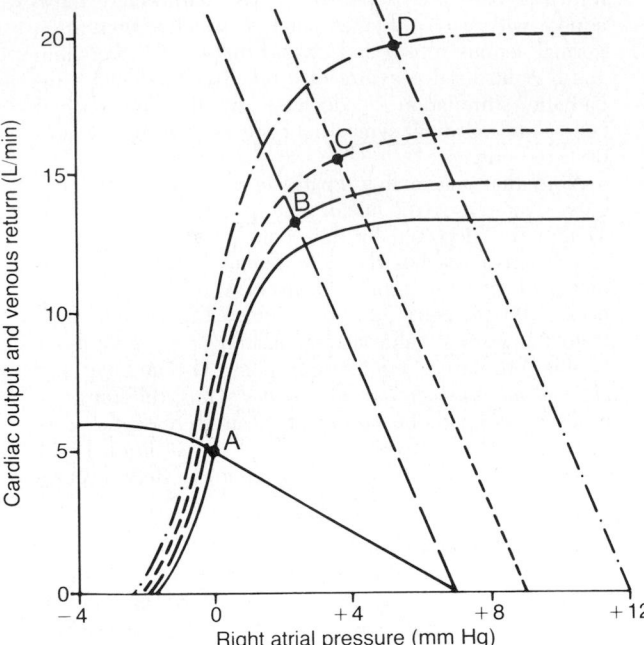

Figure 20–16. Analysis of successive changes in cardiac output and right atrial pressure after a large arteriovenous (A-V) fistula is suddenly opened. The stages of the analysis, as shown by the equilibrium points, are (A) normal conditions, (B) immediately after opening the A-V fistula, (C) 1 minute or so later after the sympathetic reflexes have become active, and (D) several weeks later after the blood volume has increased and the heart has begun to hypertrophy. (From Guyton, Jones, and Coleman: Circulatory Physiology: Cardiac Output and Its Regulation. Philadelphia, W. B. Saunders Company, 1973.)

to 9 mm Hg, thus shifting the venous return curve 2 mm Hg to the right, and (2) further elevation of the cardiac output curve because of sympathetic nervous excitation of the heart. The cardiac output now rises to almost 16 liters/min and the right atrial pressure, to about 4 mm Hg.

4. Point D shows the effect after several more weeks. By this time, the blood volume has increased because the slight reduction in arterial pressure and the sympathetic stimulation have both reduced kidney output of urine. The mean systemic filling pressure has now risen to + 12 mm Hg, shifting the venous return curve another 3 mm Hg to the right. Also, the prolonged increased workload on the heart has caused the heart muscle to hypertrophy, raising the level of the cardiac output curve still further. Therefore, point D shows a cardiac output now of almost 20 liters/min and a right atrial pressure of about 6 mm Hg.

Other Graphical Analyses

In Chapter 21, a graphical analysis of cardiac output regulation during exercise is presented, and in Chapter 22, analyses of cardiac output regulation at various stages of congestive heart failure are shown.

METHODS FOR MEASURING CARDIAC OUTPUT

In animal experiments, one can cannulate the aorta, pulmonary artery, or great veins entering the heart and measure the cardiac output using any type of flowmeter. For instance, an electromagnetic or ultrasonic flowmeter can be placed on the aorta or pulmonary artery to measure cardiac output. In the human, except in rare instances, cardiac output is measured by indirect methods that do not require surgery. Two of the methods com-

monly used are the *oxygen Fick method* and the *indicator dilution method.*

Pulsatile Output of the Heart as Measured by an Electromagnetic or Ultrasonic Flowmeter

Figure 20–17 shows a recording in a dog of blood flow in the root of the aorta made using an electromagnetic flowmeter. It demonstrates that the blood flow rises rapidly to a peak during systole and then, at the end of systole, reverses for a fraction of a second. This reverse flow causes the aortic valve to close. A minute amount of reverse flow continues throughout diastole to supply blood to the coronary vessels.

Measurement of Cardiac Output by the Oxygen Fick Method

The Fick procedure is explained by Figure 20–18. This figure shows that 200 milliliters of oxygen are being absorbed from the lungs into the pulmonary blood each minute. It also shows that the blood entering the right side of the heart has an oxygen concentration of 160 milliliters per liter of blood, whereas that leaving the left side has an oxygen concentration

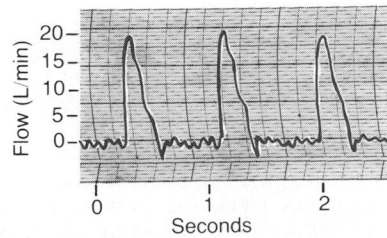

Figure 20–17. Pulsatile blood flow in the root of the aorta recorded by an electromagnetic flowmeter.

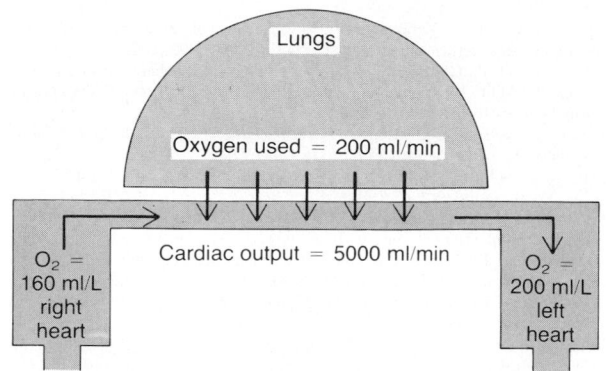

Figure 20–18. Fick principle for determining cardiac output.

of 200 milliliters per liter of blood. From these data, one can calculate that each liter of blood passing through the lungs picks up 40 milliliters of oxygen. Because the total quantity of oxygen absorbed into the blood from the lungs each minute is 200 milliliters, a total of five 1-liter portions of blood must pass through the pulmonary circulation each minute to absorb this amount of oxygen. Therefore, the quantity of blood flowing through the lungs each minute is 5 liters, which is in fact a measure of the cardiac output. Thus, the cardiac output can be calculated by the following formula:

Cardiac output (liters/min) =

$$\frac{O_2 \text{ absorbed per minute by the lungs (ml/min)}}{\text{Arteriovenous } O_2 \text{ difference (ml/liter of blood)}}$$

In applying the Fick procedure, mixed venous blood is usually obtained through a catheter inserted up the brachial vein of the forearm, through the subclavian vein, down to the right atrium, and, finally, into the right ventricle or pulmonary artery. Arterial blood can be obtained from any artery in the body, and the rate of oxygen absorption by the lungs is measured by the

disappearance of oxygen from the respired air, using any type of oxygen meter.

Indicator Dilution Method

In measuring the cardiac output by the indicator dilution method, a small amount of *indicator*, such as a dye, is injected into a large vein or, preferably, the right atrium. This passes rapidly through the right side of the heart, the lungs, and the left side of the heart and, finally, into the arterial system. Then one records the concentration of the dye as it passes through one of the peripheral arteries, giving a curve such as one of the two red curves shown in Figure 20–19. In each of these instances, 5 milligrams of Cardio-Green dye was injected at zero time. In the top recording, none of the dye passed into the arterial tree until about 3 seconds after the injection, but then the arterial concentration of the dye rose rapidly to a maximum in about 6 to 7 seconds. After that, the concentration fell rapidly. Before the concentration reached the zero point, some of the dye had already circulated all the way through some of the peripheral vessels and returned through the heart for a second time. Consequently, the dye concentration in the artery began to rise again. For the purpose of calculation, it is necessary to extrapolate the early downslope of the curve to the zero point, as shown by the dashed portion of the curve. In this way, the *time-concentration curve* of the dye in an artery, without recirculation, can be measured in its first portion and estimated reasonably accurately in its latter portion.

Once the time-concentration curve has been determined, one can then calculate the mean concentration of dye in the arterial blood for the duration of the curve. In the top example of Figure 20–19, this was done by measuring the area under the entire curve and then averaging the concentration of dye for the duration of the curve; one can see from the shaded rectangle straddling the upper curve of the figure that the average concentration of dye was about 0.25 mg/dl of blood and that the duration of this average value was 12 seconds. A total of 5 mg of dye was injected at the beginning of the experiment. For blood carrying only 0.25 mg of dye in each deciliter to carry the entire 5 mg of dye through the heart and lungs in 12 seconds, a total of 20 1-deciliter portions of blood would need to pass through the heart during this time, which would be the same as a cardiac output of 2 liters/12 sec, or 10 liters/min.

We leave it to the reader to calculate the cardiac output from the bottom curve of Figure 20–19.

To summarize, the cardiac output can be determined using the following formula:

Cardiac output (ml/min) =

$$\frac{\text{Milligrams of dye injected} \times 60}{\left(\begin{array}{c}\text{Average concentration of dye}\\ \text{in each milliliter of blood}\\ \text{for the duration of the curve}\end{array}\right) \times \left(\begin{array}{c}\text{Duration of}\\ \text{the curve}\\ \text{in seconds}\end{array}\right)}$$

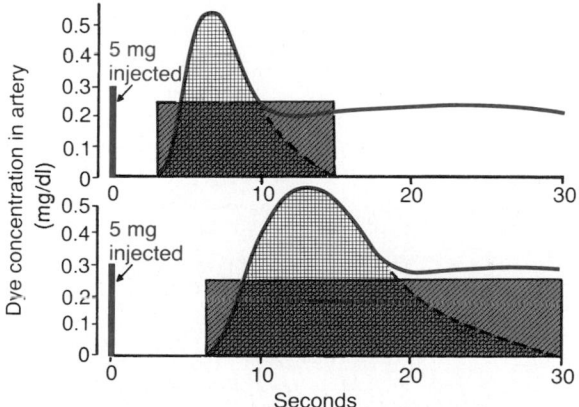

Figure 20–19. Dye concentration curves used to calculate two separate cardiac output levels by the dilution method. (The rectangular areas are the calculated average concentrations of dye in the arterial blood for the durations of the respective curves.)

REFERENCES

Banet, M., and Guyton, A. C.: Effect of body metabolism on cardiac output: Role of the central nervous system. Am. J. Physiol., 220:662, 1971.

Bishop, V. S., and Stone, H. L.: Quantitative description of ventricular output curves in conscious dogs. Circ. Res., 20:581, 1967.

Bishop, V. S., et al.: Cardiac function curves in conscious dogs. Am. J. Physiol., 207:677, 1964.

Brady, A. J.: Mechanical properties of isolated cardiac myocytes. Physiol. Rev., 71:413, 1991.

Braunwald, E.: Heart Disease: A Textbook for Cardiovascular Medicine. Philadelphia, W. B. Saunders Co., 1992.

Bullock, B. L., and Rosendahl, P. P.: Pathophysiology: adaptations and alterations in function. 3rd Ed. Philadelphia, J. B. Lippincott, 1992.

Coleman, T. G., et al.: Control of cardiac output of regional blood flow distribution. Ann. Biomed. Eng., 2:149, 1974.

Dobbs, W. A., Jr., et al.: Relative importance of nervous control of cardiac output and arterial pressure. Am. J. Cardiol., 27:507, 1971.

Donald, D. E., and Shepherd, J. T.: Response to exercise in dogs with cardiac denervation. Am. J. Physiol., 205:393, 1963.

Fermoso, J. D., et al.: Mechanism of decrease in cardiac output caused by opening the chest. Am. J. Physiol., 207:1112, 1964.

Fallon, J. T.: Cardiovascular Pathophysiology. Philadelphia, J. B. Lippincott, 1994.

Figulla, H. R., and Scholz, K. H.: Circulatory Support Devices in Interventional Cardiology. Farmington, CT, S. Karger Publishers, Inc., 1994.

Grodins, F. S.: Integrative cardiovascular physiology: A mathematical synthesis of cardiac and blood vessel hemodynamics. Q. Rev. Biol., 34:93, 1959.

Guyton, A. C.: Essential cardiovascular regulation—the control linkages between bodily needs and circulatory function. In Dickinson, C. J., and Marks, J. (eds.): Developments in Cardiovascular Medicine. Lancaster, England, MTP Press, 1978, p. 265.

Guyton, A. C.: An overall analysis of a cardiovascular regulation. Anesth. Analg., 56:761, 1977.

Guyton, A. C.: Regulation of cardiac output. N. Engl. J. Med., 277:805, 1967.

Guyton, A. C.: Venous return. In Hamilton, W. F. (ed.): Handbook of Physiology. Sec. 2, Vol. 2. Baltimore, Williams & Wilkins, 1963, p. 1099.

Guyton, A. C.: Determination of cardiac output by equating venous return curves with cardiac response curves. Physiol. Rev., 35:123, 1955.

Guyton, A. C., et al: Cardiac Output and Its Regulation. Philadelphia, W. B. Saunders Co., 1973.

Guyton, A. C., et al.: Circulation: Overall regulation. Annu. Rev. Physiol., 34:13, 1972.

Guyton, A. C., et al.: Systems analysis of arterial pressure regulation and hypertension. Ann. Biomed. Eng., 1:254, 1972.

Guyton, A. C., et al.: Autoregulation of the total systemic circulation and its relation to control of cardiac output and arterial pressure. Circ. Res., 28:(Suppl. 1)93, 1971.

Guyton, A. C., et al.: Instantaneous increase in mean circulatory pressure and cardiac output at the onset of muscular activity. Circ. Res., 11:431, 1962.

Guyton, A. C., et al.: Relative importance of venous and arterial resistance in controlling venous return and cardiac output. Am. J. Physiol., 196:1008, 1959.

Guyton, A. C., et al.: Venous return at various right atrial pressures and the normal venous return curve. Am. J. Physiol., 189:609, 1957.

Jones, C. E., et al.: Cardiac output and physiological mechanisms in circulatory shock. In MTP International Review of Science: Physiology. Vol. 1. Baltimore, University Park Press, 1974, p. 233.

Kenner, T. (ed.): Cardiovascular System Dynamics. Models and Measurements. New York, Plenum Publishing Corp., 1982.

Lake, C. L.: Clinical Monitoring for Anesthesia and Critical Care. Philadelphia, W. B. Saunders Co., 1994.

Longhurst, J. C., and Mitchell, J. H.: Reflex control of the circulation by afferents from skeletal muscle. In Guyton, A. C., and Young, D. B. (eds.): International Review of Physiology: Cardiovascular Physiology III. Vol. 18. Baltimore, University Park Press, 1979, p. 125.

Lyons, K. P.: Cardiovascular Nuclear Medicine. East Norwalk, Conn., Appleton & Lange, 1988.

Marcus, M. L., et al.: Cardiac Imaging. Philadelphia, W. B. Saunders Co., 1991.

Mark, A. L., and Mancia, G.: Cardiopulmonary baroreflexes in humans. In Shepherd, J. T., and Abboud, F. M. (eds.): Handbook of Physiology. Sec. 2, Vol. III. Bethesda, Md., American Physiological Society, 1983, p. 795.

Mitchell, J. H., and Wildenthal, K.: Static (isometric) exercise and the heart: Physiological and clinical considerations. Annu. Rev. Med., 25:369, 1974.

Nagano, M., et al.: The Cardiomyopathic Heart. New York, Raven Press, 1994.

Prather, J. W., et al.: Effect of blood volume, mean circulatory pressure and stress relaxation on cardiac output. Am. J. Physiol., 216:467, 1969.

Robertson, J. I. S., and Birkenhager, W. H.: Cardiac Output Measurement. Philadelphia, W. B. Saunders Co., 1991.

Rothe, C. F.: Mean circulatory filling pressure: its meaning and measurement. J. Appl. Physiol., 74:499, 1993.

Rothe, C. F.: Reflex control of veins and vascular capacitance. Physiol. Rev., 63:1281, 1983.

Rothe, C. F.: Venous system: Physiology of the capacitance vessels. In Shepherd, J. T., and Abboud, F. M. (eds.): Handbook of Physiology. Sec. 2, Vol. III. Bethesda, Md., American Physiological Society, 1983, p. 397.

Sarnoff, S., and Mitchell, J. H.: The regulation of the performance of the heart. Am. J. Med., 30:747, 1961.

Sugimoto, T., et al.: Effect of tachycardia on cardiac output during normal and increased venous return. Am. J. Physiol., 211:288, 1966.

Varat, M. A., et al.: Cardiovascular effects of anemia. Am. Heart J., 83:415, 1972.

Willerson, J. T., and Cohn, J. N.: Cardiovascular medicine. New York, Churchill Livingstone, 1994.

Wyngaarden, J. B., et al.: Cecil Textbook of Medicine. Philadelphia, W. B. Saunders Co., 1992.

Muscle Blood Flow and Cardiac Output During Exercise; the Coronary Circulation and Ischemic Heart Disease

CHAPTER 21

In this chapter we consider blood flow to the skeletal muscles and coronary blood flow to the heart. Regulation of both of these is achieved mainly by local control of vascular resistance by tissue metabolic needs. In addition, many related subjects, such as cardiac output control during exercise, the characteristics of heart attacks, and the pain of angina pectoris, are discussed.

BLOOD FLOW IN SKELETAL MUSCLES AND ITS REGULATION DURING EXERCISE

Very strenuous exercise is the most stressful condition that the normal circulatory system faces. This is true because in some conditions the blood flow in muscles can increase more than 20-fold (a greater increase than in any other tissue of the body) and because there is such a large mass of skeletal muscle in the body. The product of these two factors is so great that the total muscle blood flow in the healthy young adult can increase during heavy exercise from the normal level of less than 1 liter/min to as great as 20 liters/min, high enough to increase the cardiac output to five times normal in the nonathlete and in the well-trained athlete to six to seven times normal.

Rate of Blood Flow Through the Muscles

During rest, blood flow through skeletal muscle averages 3 to 4 ml/min/100 gm of muscle. During ex-

treme exercise, this rate can increase 15- to 25-fold, rising to 50 to 80 ml/100 gm of muscle.

INTERMITTENT FLOW DURING MUSCLE CONTRACTION. Figure 21–1 shows a study of blood flow changes in the calf muscles of the human leg during strong rhythmical muscular exercise. Note that the flow increases and decreases with each muscle contraction, decreasing during the contraction phase and increasing between contractions. At the end of the rhythmical contractions, the blood flow remains high for a few seconds longer but then fades toward normal over the next few minutes.

The cause of the lower flow during the muscle contraction phase of exercise is compression of the blood vessels by the contracted muscle. During strong *tetanic* contraction, which causes sustained compression of the blood vessels, the blood flow can be almost stopped.

OPENING OF MUSCLE CAPILLARIES DURING EXERCISE. During rest, some of the muscle capillaries have little or no flowing blood. But during strenuous exercise, all the capillaries open up. This opening up of dormant capillaries also diminishes the distance that oxygen and other nutrients must diffuse from the capillaries to the muscle fibers and probably contributes a twofold to threefold increased surface area through which nutrients can diffuse from the blood.

Control of Blood Flow Through the Skeletal Muscles

LOCAL REGULATION—DECREASED OXYGEN IN MUSCLE GREATLY ENHANCES FLOW. The tremendous increase in muscle blood flow that occurs during skeletal muscle activity is caused primarily by local effects

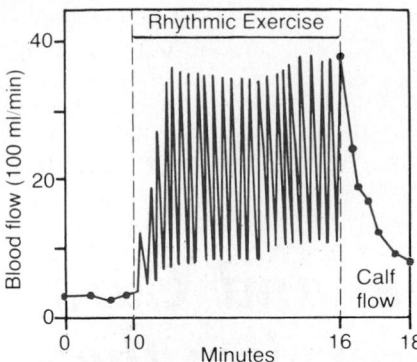

Figure 21–1. Effects of muscle exercise on blood flow in the calf of a leg during strong rhythmical contraction. The blood flow was much less during contraction than between contractions. (From Barcroft and Dornhorst: *J. Physiol., 109*:402, 1949.)

acting directly on the muscle arterioles to cause vasodilation.

This local increase in blood flow during muscle contraction is probably caused by several factors all operating at the same time. One of the most important is reduction of oxygen in the muscle tissues. That is, during muscle activity, the muscle uses oxygen rapidly, thereby decreasing the oxygen concentration in the tissue fluids. This in turn causes vasodilation either because the vessel walls cannot maintain contraction in the absence of oxygen or because oxygen deficiency causes release of vasodilator substances. The vasodilator substance that has been suggested most widely in recent years has been adenosine, but experiments have shown that even large amounts of adenosine infused directly into a muscle artery cannot sustain vasodilation in skeletal muscle for more than about 2 hours. Yet, even after the muscle blood vessels have become insensitive to the vasodilator effects of adenosine, these same vessels will still dilate fully in response to muscle activity.

Other vasodilator factors activated during muscle contraction include potassium ions, acetylcholine, adenosine triphosphate (ATP), lactic acid, and carbon dioxide. We still do not know quantitatively how great a role each of these plays in increasing muscle blood flow during muscle activity; this subject is discussed in more detail in Chapter 17.

Nervous Control of Muscle Blood Flow. In addition to the local tissue regulatory mechanism, the skeletal muscles are provided with sympathetic vasoconstrictor nerves and, in some species of animals, sympathetic vasodilator nerves.

Sympathetic Vasoconstrictor Nerves. The sympathetic vasoconstrictor nerve fibers secrete norepinephrine and, when maximally stimulated, can decrease blood flow through the muscles to perhaps one half to one fourth normal. This represents rather poor vasoconstriction in comparison with that caused by sympathetic nerves in some other areas of the body in which blood flow can be almost completely blocked. Yet even this degree of vasoconstriction is of physiological im-

portance in circulatory shock and during other periods of stress when it is desirable to reduce blood flow through the many muscles of the body.

In addition to the norepinephrine secreted at the sympathetic vasoconstrictor nerve endings, the adrenal medullae secrete large amounts of norepinephrine plus even several times more epinephrine into the circulating blood during strenuous exercise. The circulating norepinephrine acts on the muscle vessels to cause a vasoconstrictor effect similar to that caused by direct sympathetic nerve stimulation. The epinephrine, on the other hand, often has a slight to moderate vasodilator effect because epinephrine excites more of the *beta* receptors of the vessels, which are vasodilator receptors, in contrast to the *alpha* vasoconstrictor receptors excited mainly by the norepinephrine. These receptors are discussed in Chapter 60.

Sympathetic Vasodilator Fibers. In the cat and some other lower animals, there are also sympathetic *vasodilator* fibers that secrete acetylcholine—this hormone in turn causing some vasodilation. Such fibers have not yet been proved to occur in the human. Instead, as noted previously, circulating epinephrine from the adrenal medullae acting at beta receptors in the muscle arterioles sometimes seems to cause mild vasodilation, and this may subserve the same function as the vasodilator system of the lower animals.

Circulatory Readjustments During Exercise

Three major effects occur during exercise that are essential for the circulatory system to supply the tremendous blood flow required by the muscles. They are (1) mass discharge of the sympathetic nervous system throughout the body with consequent stimulatory effects on the circulation, (2) increase in arterial pressure, and (3) increase in cardiac output.

Mass Sympathetic Discharge

At the onset of exercise, signals are transmitted not only from the brain to the muscle to cause muscle contraction but also from the muscle control centers of the brain into the vasomotor center to initiate mass sympathetic discharge. Simultaneously, the parasympathetic signals to the heart are greatly attenuated. Therefore, three major circulatory effects result.

First, the *heart is stimulated* to greatly increased heart rate and increased pumping strength as a result of the sympathetic drive to the heart and release of the heart from the normal parasympathetic inhibition.

Second, most of the arterioles of the peripheral circulation are strongly contracted except the arterioles in the active muscles, which are strongly vasodilated by the local vasodilator effects in the muscles themselves. Thus, the heart is stimulated to supply the increased blood flow required by the muscles, and blood flow through most nonmuscular areas of the body is temporarily reduced, thereby temporarily "lending" their blood supply to the muscles. This effect accounts for as much as 2 liters of extra blood flow to the muscles, which is exceedingly important when one thinks of the wild animal running for its life because even a fractional increase in running speed may make the difference between life and death. Two of the peripheral circulatory systems, the *coronary* and *cerebral systems, are spared this vasoconstrictor effect* because both these circulatory areas have poor vasoconstrictor innervation—fortunately so because both the heart and the brain are as essential to exercise as are the skeletal muscles.

Third, the muscle walls of the *veins and other capacitative areas of the circulation are contracted powerfully*, which greatly *increases the mean systemic filling pressure*. As we learned in Chapter 20, this is one of the most important factors in promoting venous return of blood to the heart and, therefore, in increasing the cardiac output.

A Muscle Reflex That May Further Stimulate the Sympathetic Nervous System. Aside from the sympathetic stimulation caused by direct signals from the brain, reflex signals from the contracting muscles are believed to pass up the spinal cord to the vasomotor center and to excite the sympathetic nerves still further. These signals have been postulated to be initiated by metabolic endproducts acting on small sensory nerve endings in the muscle tissue.

Increase in Arterial Pressure During Exercise—An Important Result of Increased Sympathetic Activity

One of the most important effects of increased sympathetic activity in exercise is to increase the arterial pressure. This results from multiple stimulatory effects, including (1) vasoconstriction of the arterioles and small arteries in most of the tissues of the body besides the active muscles, (2) increased pumping activity by the heart, and (3) a great increase in mean systemic filling pressure caused mainly by venous contraction. These effects working together virtually always increase the arterial pressure during exercise. This increase can be as little as 20 or as great as 80 mm Hg, depending on the conditions under which the exercise is performed. When a person performs exercise under tense conditions but uses only a few muscles, the sympathetic response still occurs everywhere in the body, but vasodilation occurs in only those few muscles that are active. Therefore, the net effect is mainly one of vasoconstriction, often increasing the mean arterial pressure to as high as 170 mm Hg. Such a condition might occur in a person standing on a ladder and nailing with a hammer on the ceiling above. The tenseness of the situation is obvious, and yet the amount of muscle vasodilation is relatively slight.

On the other hand, when a person performs whole-body exercise, such as running or swimming, the increase in arterial pressure is often only 20 to 40 mm Hg. The lack of a tremendous rise in pressure results from the extreme vasodilation that occurs in large masses of muscle.

In rare instances, people are found in whom the sympathetic nervous system is absent, either congenitally or because of surgical removal. When such a person exercises, instead of the arterial pressure rising, the pressure falls—sometimes to as low as one-half normal—and the cardiac output rises only about one third as much as it does normally. Therefore, one can readily understand the major importance of increased sympathetic activity during exercise.

Why Is the Arterial Pressure Rise Important During Exercise? When muscles are stimulated maximally in a laboratory experiment but without allowing the arterial pressure to rise, the blood flow will seldom rise more than about eightfold. Yet we know from studies of marathon runners that the blood flow through the muscles can increase from as little as 1 liter/min for the whole body during rest to at least 20 liters/min during maximal activity. Therefore, it is clear that the muscle blood flow can increase much more than occurs in simple laboratory experiments. What is the difference? The difference is mainly that in normal exercise, the arterial pressure rises at the same time. Let us assume, for instance, that the arterial pressure rises 30 per cent, a common rise during exercise. This 30 per cent increase causes 30 per cent more force to push the blood through the tissue vessels. This is not the most important effect of the increasing pressure because the pressure also dilates the blood vessels and allows about another 100 per cent increase in flow. Thus, a 30 per cent increase over the original eightfold increase in blood flow would increase the flow to 10.4 times normal, and doubling this again would increase the flow to more than 20 times normal.

Importance of the Increase in Cardiac Output in Exercise

Many different physiological effects occurring at the same time during exercise cause the cardiac output to increase along with the increase in degree of exercise. This increase in output in turn is essential to supply the large amounts of oxygen and other nutrients needed by the working muscles. In fact, the ability of the circulatory system to provide increased cardiac output during heavy exercise is equally as important as the strength of the muscles themselves in setting the limit for the performance of muscle work. For instance, those marathon runners who can increase their cardiac outputs the most are generally the same ones who have the record-breaking times.

Graphical Analysis of the Changes in Cardiac Output in Heavy Exercise. Figure 21–2 shows a graphical analysis of the large increase in cardiac output that occurs in heavy exercise. The black cardiac output and venous return curves crossing at point A give the analysis for the normal circulation, whereas the red curves and point B analyze heavy exercise. Note that the great increase in cardiac output requires significant changes in both the cardiac output curve and the venous return curve, as follows.

The increased level of the cardiac output curve is easy to understand. It results almost entirely from sympathetic stimulation of the heart that causes both in-

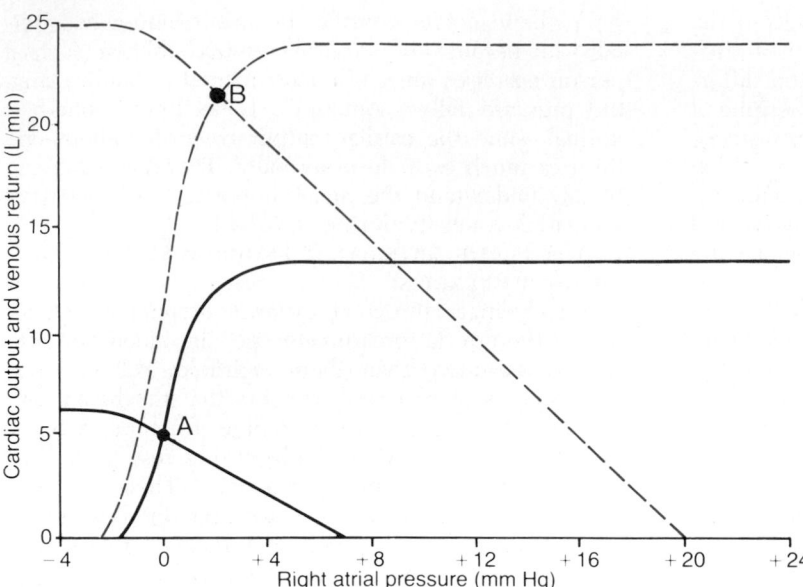

Figure 21–2. Graphical analysis of the changes in cardiac output and right atrial pressure with the onset of strenuous exercise.

creased heart rate, often up to rates of 170 to 190 beats/min, and increased strength of contraction of the heart, often to as much as twice normal. Without the increased level of the output curve the increase in cardiac output would be limited to the plateau level of the normal heart, which would mean a maximum increase of cardiac output of only about 2.5-fold rather than the 4-fold that can commonly be achieved and the 7-fold that can be achieved in some marathon runners.

Now study the venous return curves. If no change occurred from the normal venous return curve, the cardiac output could hardly rise at all in exercise because the upper plateau level of the normal curve is only 6 liters/min. Yet two important changes do occur: (1) The mean systemic filling pressure rises tremendously at the outset of heavy exercise. This results partly from sympathetic stimulation of the veins and other capacitive parts of the circulation. In addition, tensing of the abdominal and other muscles of the body compresses many of the internal vessels, thus providing more compression of the entire capacitative vascular system with greater increase in the mean systemic filling pressure. In maximal exercise, these two effects together can increase the mean systemic filling pressure from a normal level of 7 mm Hg to as high as 30 mm Hg. (2) The slope of the venous return curve rotates upward. This is caused by decreased resistance in virtually all the blood vessels in the active muscle tissue, which causes the resistance to venous return to decrease, thus increasing the slope of the venous return curve. The combination of increased mean systemic filling pressure and decreased resistance to venous return raises the entire level of the venous return curve.

In response to the changes in both the venous return curve and the cardiac output curve, the new equilibrium point in Figure 21–2 for cardiac output and right atrial pressure is now point B, in contrast to the normal level at point A. Note especially that the right atrial pressure has hardly changed, having risen only 1.5 mm Hg. In fact, in a person with a strong heart, the right atrial pressure often falls below normal because of the greatly increased sympathetic stimulation of the heart during exercise.

CORONARY CIRCULATION

About one third of all deaths in the affluent society of the Western world result from coronary artery disease, and almost all elderly people have at least some impairment of the coronary artery circulation. For this reason, the normal and pathological physiology of the coronary circulation is one of the most important subjects in medicine.

Physiologic Anatomy of the Coronary Blood Supply

Figure 21–3 shows the heart with its coronary blood supply. Note that the main coronary arteries lie on the surface of the heart and smaller arteries penetrate from the surface into the cardiac muscle mass. It is almost entirely through these arteries that the heart receives its nutritive blood supply. Only the inner 75 to 100 micrometers of the endocardial surface can obtain significant amounts of nutrition directly from the blood in the

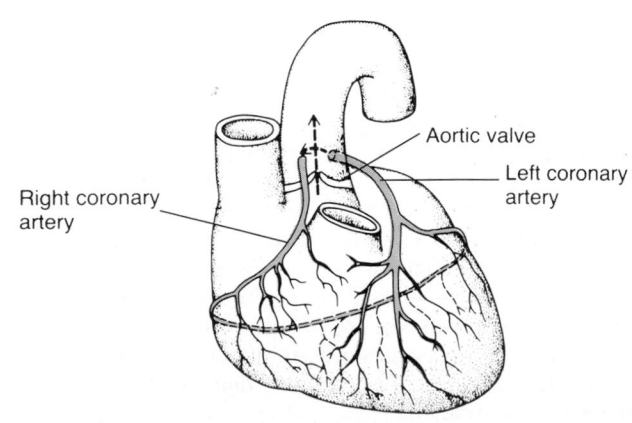

Figure 21–3. Coronary vessels.

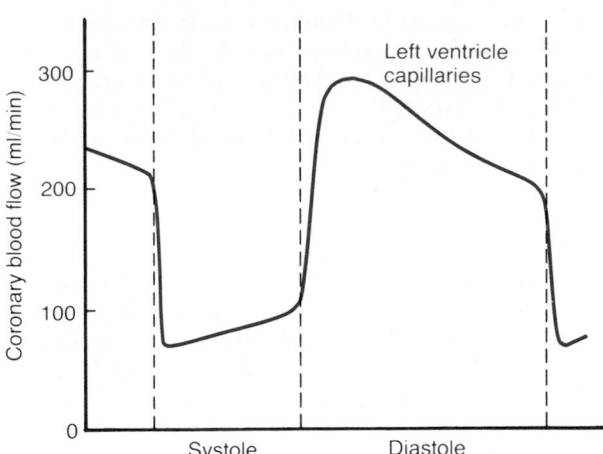

Figure 21–4. Phasic flow of blood through the coronary capillaries of the human left ventricle (as extrapolated from studies in dogs).

cardiac chambers, so that this source of muscle nutrition is minuscule.

The *left coronary artery* supplies mainly the anterior and lateral portions of the left ventricle, whereas the *right coronary artery* supplies most of the right ventricle as well as the posterior part of the left ventricle in 80 to 90 per cent of people.

Most of the venous blood flow from the left ventricle leaves by way of the *coronary sinus*—which is about 75 per cent of the total coronary blood flow—and most of the venous blood from the right ventricle flows through the small *anterior cardiac veins* directly into the right atrium, not by way of the coronary sinus. A small amount of coronary blood flows back into the heart through minute *thebesian veins*, which empty directly into all chambers of the heart.

Normal Coronary Blood Flow

The resting coronary blood flow in the human averages about 225 ml/min, which is about 0.7 to 0.8 milliliter per gram of heart muscle, or 4 to 5 per cent of the total cardiac output.

In strenuous exercise, the heart in the young adult increases its cardiac output fourfold to sevenfold, and it pumps this blood against a higher than normal arterial pressure. Consequently, the work output of the heart under severe conditions may increase sixfold to eightfold. The coronary blood flow increases threefold to fourfold to supply the extra nutrients needed by the heart. This increase is not as much as the increase in workload, which means that the ratio of coronary blood flow to energy expenditure by the heart decreases. The "efficiency" of cardiac utilization of energy increases to make up for this relative deficiency of blood supply.

PHASIC CHANGES IN CORONARY BLOOD FLOW— EFFECT OF CARDIAC MUSCLE COMPRESSION. Figure 21–4 shows the average blood flow *through the nutrient capillaries* of the left ventricular coronary system in milliliters per minute in the human heart during systole and diastole, as extrapolated from experiments

in lower animals. Note from this diagram that the blood flow in the left ventricle falls to a low value during *systole,* which is opposite to the flow in other vascular beds of the body. The reason for this is the strong compression of the left ventricular muscle around the intramuscular vessels during systole.

During *diastole,* the cardiac muscle relaxes and no longer obstructs the blood flow through the left ventricular capillaries, so that blood now flows rapidly during all of diastole.

Blood flow through the coronary capillaries of the right ventricle also undergoes phasic changes during the cardiac cycle, but because the force of contraction of the right ventricle is far less than that of the left ventricle, the inverse phasic changes are only partial in contrast to those in the left ventricle.

EPICARDIAL VERSUS SUBENDOCARDIAL BLOOD FLOW—EFFECT OF INTRAMYOCARDIAL PRESSURE. During cardiac contraction, all the cardiac muscle squeezes toward the centers of the ventricles. That is, the ventricular muscle adjacent to the ventricular chambers (the subendocardial muscle) squeezes the blood in the ventricle, the muscle in the middle layer of the ventricle squeezes both the blood in the ventricle *and the subendocardial muscle,* and the outermost muscle squeezes both the middle and the subendocardial muscle as well as the blood in the ventricle. Therefore, during systole, a tissue pressure gradient develops within the heart muscle itself, with the pressure in the subendocardial muscle almost as great as the pressure inside the ventricle, whereas the pressure in the outer layer of the heart muscle is only slightly greater than atmospheric pressure. The importance of this pressure gradient is that the intramyocardial pressure in the inner layers of the heart muscle is so much greater than in the outer layers that it compresses the subendocardial blood vessels far more than the outer vessels.

Figure 21–5 demonstrates the special arrangement of the coronary vessels at different depths in the heart, showing on the surface of the cardiac muscle the large epicardial coronary arteries that supply the heart. Smaller, intramuscular arteries penetrate the muscle, supplying the needed nutrients en route to the endocardium. Then lying immediately beneath the endocardium is a plexus of subendocardial arteries. During systole, blood flow through the subendocardial plexus of the left ventricle, where the contractile force of the heart muscle is great, falls almost to zero. To compen-

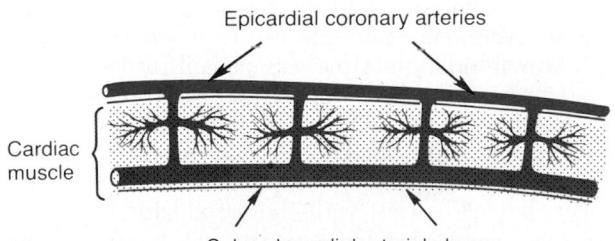

Figure 21–5. Diagram of the epicardial, intramuscular, and subendocardial coronary vasculature.

sate for this almost total lack of flow during systole, the subendocardial arterial plexus is more extensive than the nutrient arteries in the middle and outer layers of the heart. Therefore, during diastole, blood flow in the subendocardial arteries is considerably greater than is blood flow in the outermost arteries. Later in the chapter, we see that this peculiar difference between blood flow in the epicardial and subendocardial arteries plays an important role in certain types of coronary ischemia.

Control of Coronary Blood Flow

Local Metabolism Is the Primary Controller of Coronary Flow

Blood flow through the coronary system is regulated almost entirely by local arterial vasodilation in response to cardiac muscle need for nutrition. This mechanism works equally well when the nerves to the heart are intact or are removed. That is, whenever the vigor of contraction is increased, regardless of cause, the rate of coronary blood flow simultaneously increases. Conversely, decreased activity is accompanied by decreased coronary flow. This local regulation of blood flow is almost identical with that occurring in many other tissues, especially in the skeletal muscles all over the body.

OXYGEN DEMAND AS A MAJOR FACTOR IN LOCAL BLOOD FLOW REGULATION. Blood flow in the coronaries usually is regulated almost exactly in proportion to the need of the cardiac musculature for oxygen. Even in the normal resting state, about 70 per cent of the oxygen in the coronary arterial blood is removed as the blood passes through the heart; and because not much oxygen is left, not much additional oxygen can be supplied to the heart musculature unless the blood flow increases. The blood flow does increase almost in proportion to the metabolic consumption of oxygen by the heart.

Yet the exact means by which increased oxygen consumption causes coronary dilatation has not been determined. It is speculated by many research workers that a decrease in the oxygen concentration in the heart causes vasodilator substances to be released from the muscle cells and that these dilate the arterioles. The substance with the greatest vasodilator propensity is *adenosine*. In the presence of very low concentrations of oxygen in the muscle cells, a large proportion of the cell's ATP degrades to adenosine monophosphate; then small portions of this are further degraded to release adenosine into the tissue fluids of the heart muscle. After the adenosine causes vasodilation, much of it is reabsorbed into the cardiac cells to be reused.

Adenosine is not the only vasodilator product that has been identified. Others include adenosine phosphate compounds, potassium ions, hydrogen ions, carbon dioxide, bradykinin, and, possibly, prostaglandins.

Yet difficulties exist with the vasodilator hypothesis. For one, agents that block or partially block the vasodilator effect of adenosine do not prevent the coronary vasodilation caused by increased heart muscle activity. Second, studies in skeletal muscle have shown that continued infusion of adenosine will maintain vascular dilatation for only 1 to 3 hours, and yet muscle activity will still dilate the local blood vessels even when the adenosine can no longer dilate them.

Yet another theory to explain the coronary artery dilatation should be remembered until proved false: In the absence of adequate amounts of oxygen in the cardiac muscle, not only does the heart muscle itself suffer oxygen deficiency but so do the arteriolar muscle walls. This could easily cause local vasodilation because of lack of the required energy to keep the coronary vessels contracted against the high arterial pressure. But this, too, has its problems because the coronary arteries require only minute amounts of oxygen to maintain full contraction.

Nervous Control of Coronary Blood Flow

Stimulation of the automatic nerves to the heart can affect coronary blood flow both directly and indirectly. The direct effects result from direct action of the nervous transmitter substances, acetylcholine from the vagus nerves and norepinephrine from the sympathetic nerves, on the coronary vessels themselves. The indirect effects result from secondary changes in coronary blood flow caused by increased or decreased activity of the heart.

The indirect effects, which are mostly opposite to the direct effects, play a far more important role in normal control of coronary blood flow. Thus, sympathetic stimulation, which releases norepinephrine, increases both heart rate and heart contractility as well as its rate of metabolism. In turn, the increased activity of the heart sets off local blood flow regulatory mechanisms for dilating the coronary vessels, and the blood flow increases approximately in proportion to the metabolic needs of the heart muscle. In contrast, vagal stimulation, with its release of acetylcholine, slows the heart and has a slight depressive effect on heart contractility. Both these effects decrease cardiac oxygen consumption and, therefore, indirectly constrict the coronaries.

DIRECT EFFECTS OF NERVOUS STIMULI ON THE CORONARY VASCULATURE. The distribution of parasympathetic (vagal) nerve fibers to the ventricular coronary system is so slight that parasympathetic stimulation has only a slight direct effect to *dilate* the coronaries.

There is much more extensive sympathetic innervation of the coronary vessels. In Chapter 60, we see that the sympathetic transmitter substances norepinephrine and epinephrine can have either vascular constrictor or vascular dilator effects, depending on the presence or absence of constrictor or dilator receptors in the blood vessel walls. The constrictor receptors are called *alpha receptors* and the dilator receptors are called *beta receptors*. Both alpha and beta receptors exist in the coronary vessels. In general, the epicardial coronary vessels have a preponderance of alpha receptors, whereas the intramuscular arteries may have a preponderance of beta receptors. Therefore, sympathetic stimulation can, at least theoretically, cause slight overall coronary constriction or dilatation but usually more constriction. Yet, in some people, the alpha vasoconstrictor effects seem to be disproportionately severe, and these people can have vasospastic myocardial ischemia during periods of excess sympathetic drive, often with resultant anginal pain.

It must be pointed out again, however, that metabolic factors—especially myocardial oxygen consumption—are the major controllers of myocardial blood flow. Whenever the direct effects of nervous stimulation alter the coronary blood flow, the metabolic control of coronary flow usually overrides the direct coronary nervous effects within seconds or minutes.

Special Features of Cardiac Muscle Metabolism

The basic principles of cellular metabolism, which are discussed in Chapters 67 through 72, apply to cardiac muscle the same as for other tissues, but there are quantitative differences. Most important, under resting conditions, cardiac muscle normally uses mainly fatty acids for its energy instead of carbohydrates, with about 70 per cent of the normal metabolism being derived from fatty acids. However, as is true of other tissues, under anaerobic or ischemic conditions, cardiac metabolism must call on the anaerobic glycolysis mechanisms for energy. This can supply little extra energy in relation to the large energy requirements of the heart. Also, glycolysis uses tremendous quantities of the blood glucose and at the same time forms large amounts of lactic acid in the cardiac tissue, which is probably one of the causes of cardiac pain in cardiac ischemic conditions, as discussed later in the chapter.

As is true in other tissues, more than 95 per cent of the metabolic energy liberated from the foods is used to form ATP in the mitochondria. This ATP in turn acts as the conveyer of energy for cellular function. In severe coronary ischemia the ATP degrades to adenosine diphosphate, adenosine monophosphate, and adenosine. Because the cell membrane is permeable to adenosine, much of this can be lost from the muscle cells into the circulating blood. This released adenosine is believed to be one of the substances that causes dilatation of the coronary arterioles during coronary hypoxia, as discussed earlier. The loss of adenosine also has a serious cellular consequence. Within as little as 30 minutes of severe coronary ischemia, as occurs after a myocardial infarct or during cardiac arrest, about one half of the adenine base can be lost from the cardiac muscle cells. Furthermore, this loss can be replaced by new synthesis of adenine at a rate of only 2 per cent per hour. Therefore, once a serious bout of ischemia has persisted for 30 or more minutes, relief of the coronary ischemia may be too late to save the lives of the cardiac cells. This almost certainly is one of the major causes of cardiac cellular death after myocardial ischemia and one of the most important causes of cardiac debility in the late stages of circulatory shock, as discussed in Chapter 24.

Ischemic Heart Disease

The most common cause of death in Western culture is ischemic heart disease, which results from insufficient coronary blood flow. About 35 per cent of people in the United States die of this cause. Some deaths occur suddenly as a result of an acute coronary occlusion or of fibrillation of the heart, whereas others occur slowly over a period of weeks to years as a result of progressive weakening of the heart pumping process. In this chapter, we discuss the coronary ischemia problem itself as well as acute coronary occlusion and myocardial infarction. In Chapter 22, we discuss congestive heart failure, the most frequent cause of which is progressive coronary ischemia.

ATHEROSCLEROSIS AS THE CAUSE OF ISCHEMIC HEART DISEASE. The most frequent cause of diminished coronary blood flow is atherosclerosis. The atherosclerotic process is discussed in connection with lipid metabolism in Chapter 68; briefly, this process is the following: In certain people who have a genetic predisposition to atherosclerosis or in people who eat excessive quantities of cholesterol and other fats, large quantities of cholesterol gradually become deposited beneath the endothelium at many points in arteries throughout the body. Later, these areas of deposit are invaded by fibrous tissue and frequently become calcified. The net result is the development of *atherosclerotic plaques* that protrude into the vessel lumens and either block or partially block blood flow.

A common site for development of atherosclerotic plaques is the first few centimeters of the major coronary arteries.

Acute Coronary Occlusion

Acute occlusion of a coronary artery frequently occurs in a person who already has serious underlying atherosclerotic coronary heart disease but almost never occurs in a person with a normal coronary circulation. This condition can result from any one of several effects, two of which are the following.

1. The atherosclerotic plaque can cause a local blood clot called a *thrombus*, which in turn occludes the artery. The thrombus usually occurs where the plaque has broken through the endothelium, thus coming in direct contact with the flowing blood. Because the plaque presents an unsmooth surface to the blood platelets begin to adhere to it, fibrin begins to be deposited, and blood cells become entrapped to form a clot that grows until it occludes the vessel. Or, occasionally, the clot breaks away from its attachment on the atherosclerotic plaque and flows to a more peripheral branch of the coronary arterial tree, where it blocks the artery at that point. A thrombus that flows along the artery in this way and occludes the vessel more distally is called an *embolus*.

2. Many clinicians believe that local spasm of a coronary artery can also cause sudden occlusion. The spasm might result from direct irritation of the smooth muscle of the arterial wall by the edges of the arteriosclerotic plaque, or it might result from nervous reflexes that cause coronary muscle contraction. The spasm may then lead to *secondary* thrombosis of the vessel.

THE LIFESAVING VALUE OF COLLATERAL CIRCULATION IN THE HEART. The degree of damage to the heart caused either by slowly developing atherosclerotic constriction of the coronary arteries or by sudden occlusion is determined to a great extent by the degree of collateral circulation that has already developed or that can develop within a short period after the occlusion.

In a normal heart, almost no communications exist among the larger coronary arteries. But many anastomoses do exist among the smaller arteries sized 20 to 250 micrometers in diameter, as shown in Figure 21–6.

When a sudden occlusion occurs in one of the larger coronary arteries, the small anastomoses dilate within a few seconds. But the blood flow through these minute collaterals is usually less than one half that needed to keep alive the cardiac muscle that they supply; the diameters of the collateral vessels do not enlarge further for the next 8 to 24 hours. But then collateral flow does begin to increase, doubling by the 2nd or 3rd day and often reaching normal or almost normal coronary flow in the previously ischemic muscle within about 1 month. In fact, the flow is capable of increasing even further with increased metabolic loads. Because of these developing collateral channels, many patients recover from various types of coronary occlusion when the area of muscle involved is not too great.

When atherosclerosis constricts the coronary arteries slowly over a period of many years rather than suddenly, collateral vessels can develop at the same time that the atherosclerosis does. Therefore, the person may never experience an acute episode of cardiac dysfunction. Eventually, the sclerotic process develops beyond the limits of even the collateral blood supply to provide the needed blood flow, and sometimes even the collaterals develop atherosclerosis. When this occurs, the heart muscle becomes severely limited in its work output, often so much so that the heart cannot pump even the normally required amounts of blood flow. This is one of the most common causes of cardiac failure and occurs in vast numbers of older people.

Myocardial Infarction

Immediately after an acute coronary occlusion, blood flow ceases in the coronary vessels beyond the occlusion except for small amounts of collateral flow from surrounding vessels. The area of muscle that has either zero flow or so little flow that it cannot sustain cardiac muscle function is said to be *infarcted*. The overall process is called a *myocardial infarction*.

Soon after the onset of the infarction, small amounts of collateral blood seep into the infarcted area, and this, combined with progressive dilatation of the local blood vessels, causes the area to become overfilled with stagnant blood. Simultaneously the muscle fibers use the last vestiges of the oxygen in the blood, causing the hemoglobin to become totally reduced and dark blue. The infarcted area takes on a bluish brown hue and the blood vessels of the area appear to be engorged despite the lack of blood flow. In later stages, the vessel walls become highly permeable and leak fluid, the tissue becomes edematous, and the cardiac muscle cells begin to swell because of diminished cellular metabolism. Within a few hours of almost no blood supply, the cells die.

Cardiac muscle requires about 1.3 milliliters of oxygen per 100 grams of muscle tissue per minute to remain alive. This is in comparison with about 8 milliliters of oxygen per 100 grams delivered to the normal resting left ventricle each minute. Therefore, if there is even 15 to 30 per cent of normal resting coronary blood flow, the muscle will not die. In the central portion of a large infarct, however, where there is almost no collateral blood flow, the muscle does die.

SUBENDOCARDIAL INFARCTION. The subendocardial muscle frequency becomes infarcted even when there is no evidence of infarction in the outer surface portions of the muscle. The reason for this is that the subendocardial muscle even normally has difficulty obtaining adequate blood flow because the blood vessels in the subendocardium are so intensely compressed by systolic contraction of the heart, as explained earlier. Therefore, any condition that compromises blood flow to any area of the heart usually causes damage first in the subendocardial regions, and the damage then spreads outward toward the epicardium.

Causes of Death After Acute Coronary Occlusion

The major causes of death after acute myocardial infarction are (1) decreased cardiac output; (2) damming of blood in the pulmonary or systemic veins with death resulting from edema, especially pulmonary edema; (3) fibrillation of the heart; and, occasionally, (4) rupture of the heart.

DECREASED CARDIAC OUTPUT—CARDIAC SHOCK. When some of the cardiac muscle fibers are not functioning and others are too weak to contract with great force, the overall pumping ability of the affected ventricle is proportionately depressed. Indeed, the overall pumping strength of the heart is often decreased more than one might expect because of a phenomenon called *systolic stretch*, which is shown in Figure 21–7. When the normal portions of the ventricular muscle contract, the ischemic muscle, whether this be dead or simply nonfunctional, instead of contracting is forced outward by the pressure that develops inside the ventricle. Therefore, much of the pumping force of the ventricle is dissipated by bulging of the area of nonfunctional cardiac muscle.

When the heart becomes incapable of contracting with sufficient force to pump enough blood into the arterial tree, cardiac failure and death of the peripheral

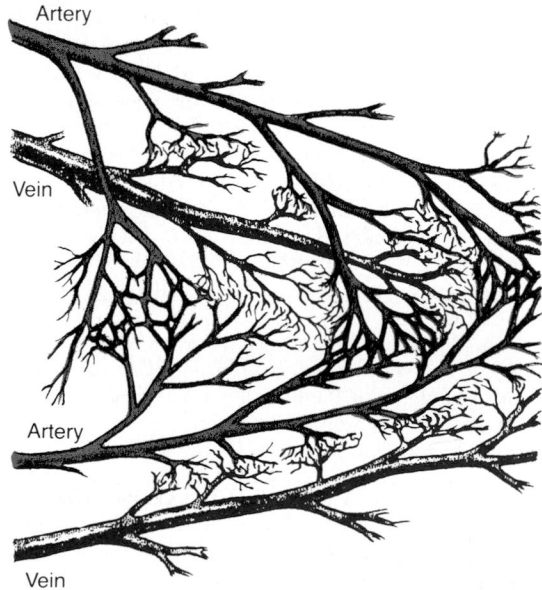

Figure 21–6. Minute anastomoses of the coronary arterial system.

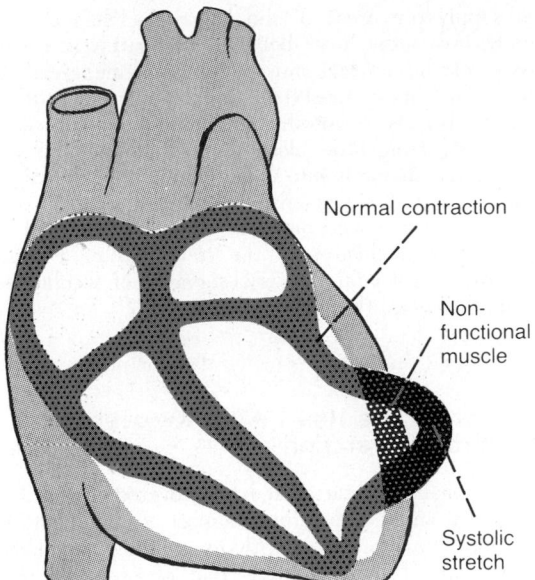

Figure 21–7. Systolic stretch in an area of ischemic cardiac muscle.

Normal contraction

Non-
functional
muscle

Systolic
stretch

tissues ensue as a result of peripheral ischemia. This condition is called *coronary shock, cardiogenic shock, cardiac shock,* or *low cardiac output failure*. It is discussed in Chapter 22. Cardiac shock almost always occurs when more than 40 per cent of the left ventricle is infarcted. Death occurs in about 85 per cent of patients who develop cardiac shock.

DAMMING OF BLOOD IN THE VENOUS SYSTEM. When the heart is not pumping blood forward, it must be damming blood in the blood vessels of the lungs or in the systemic circulation. This increases both the left and the right atrial pressures and leads to increased capillary pressures, particularly in the lungs. These effects often cause little difficulty during the first few hours after the myocardial infarction. Instead, the symptoms develop a few days later for the following reason: The diminished cardiac output leads to diminished blood flow to the kidneys, and for reasons that are discussed in Chapter 22, the kidneys then fail to excrete enough urine. This adds progressively to the blood volume and, therefore, to the congestive symptoms. Consequently, many patients who seemingly are getting along well develop acute pulmonary edema several days after a myocardial infarction and often die within a few hours after appearance of the initial edema symptoms.

FIBRILLATION OF THE VENTRICLES AFTER MYOCARDIAL INFARCTION. Many people who die of coronary occlusion die because of sudden ventricular fibrillation. The tendency to develop fibrillation is especially great after a large infarction, but fibrillation can occur after small occlusions as well. Indeed, some patients with coronary insufficiency die acutely from fibrillation without any infarction.

There are two especially dangerous periods after coronary infarction during which fibrillation is most likely to occur. The first is during the first 10 minutes after the infarction occurs. Then there is a short period of relative safety, followed by a secondary period of cardiac irritability beginning 1 hour or so later and lasting for another few hours. Fibrillation can also occur for many days after the infarct.

At least four factors enter into the tendency for the heart to fibrillate.

1. Acute loss of blood supply to the cardiac muscle causes *rapid depletion of potassium* from the ischemic musculature. This increases the potassium concentration in the extracellular fluids surrounding the cardiac muscle fibers. Experiments in which potassium has been injected into the coronary system have demonstrated that an elevated extracellular potassium concentration increases the irritability of the cardiac musculature and, therefore, its likelihood of fibrillating.

2. Ischemia of the muscle causes an *"injury current,"* which is described in Chapter 12 in relation to electrocardiograms in patients with acute myocardial infarction. The ischemic musculature cannot repolarize its membranes, so that the external surface of this muscle remains negative with respect to the normal cardiac muscle membrane elsewhere in the heart. Therefore, electric current flows from this ischemic area of the heart to the normal area and can elicit abnormal impulses that can cause fibrillation.

3. *Powerful sympathetic reflexes* often develop after massive infarction, principally because the heart does not pump an adequate volume of blood into the arterial tree. The sympathetic stimulation also increases the irritability of the cardiac muscle and thereby predisposes to fibrillation.

4. The myocardial infarction itself often *causes the ventricle to dilate excessively*. This increases the pathway length for impulse conduction in the heart and frequently causes *abnormal conduction pathways around the infarcted area* of the cardiac muscle. Both of these effects predispose to development of circus movements. As discussed in Chapter 13, excess prolongation of the conduction pathway in the ventricles allows an impulse to re-enter muscle that is already recovering from refractoriness, thereby initiating a subsequent cycle of excitation and causing the process to continue on and on.

RUPTURE OF THE INFARCTED AREA. During the first day of an acute infarct, there is little danger of rupture of the ischemic portion of the heart, but a few days after a large infarct occurs, the dead muscle fibers begin to degenerate and the dead tissue to become stretched very thin. If this happens, the dead muscle bulges outward severely with each heart contraction and the systolic stretch becomes greater and greater until finally the heart ruptures. In fact, one of the means used in assessing the progress of a severe myocardial infarction is to record by x-rays whether the degree of systolic stretch is worsening.

When a ventricle does rupture, the loss of blood into the pericardial space causes rapid development of *cardiac tamponade*—that is, compression of the heart from the outside by blood collecting in the pericardial cavity. Because of this compression of the heart, blood cannot flow into the right atrium, and the patient dies of suddenly decreased cardiac output.

Stages of Recovery from Acute Myocardial Infarction

The upper part of Figure 21–8 shows the effects of acute coronary occlusion, on the left, in a patient with a small area of muscle ischemia and, on the right, in a patient with a large area of ischemia. When the area of

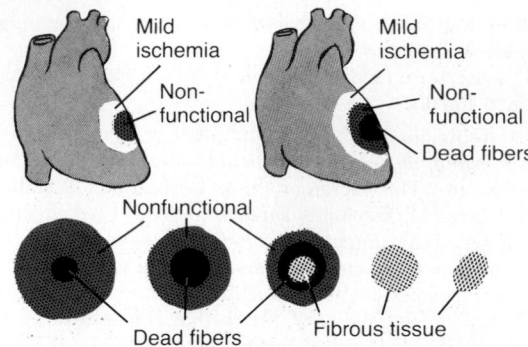

Figure 21–8. *Top,* Small and large areas of coronary ischemia. *Bottom,* Stages of recovery from myocardial infarction.

ischemia is small, little or no death of the muscle cells may occur but part of the muscle often does become temporarily nonfunctional because of inadequate nutrition to support muscle contraction.

When the area of ischemia is large, some of the muscle fibers in the center of the area die rapidly, within 1 to 3 hours in an area of total cessation of coronary blood supply. Immediately around the dead area is a nonfunctional area because of failure of contraction and usually failure of impulse conduction. Then, extending circumferentially around the nonfunctional area is an area that is still contracting but weakly so because of mild ischemia.

REPLACEMENT OF DEAD MUSCLE BY SCAR TISSUE. In the lower part of Figure 21–8., the various stages of recovery after a large myocardial infarction are shown. Shortly after the occlusion, the muscle fibers in the center of the ischemic area die. Then, during the ensuing days, this area of dead fibers grows because many of the marginal fibers finally succumb to the prolonged ischemia. At the same time, owing to the enlargement of the collateral arterial channels growing into the outer rim of the infarcted area, the nonfunctional area of muscle becomes smaller and smaller. After a few days to 3 weeks, most of the nonfunctional area of muscle becomes functional again or dies—one or the other. In the meantime, fibrous tissue begins developing among the dead fibers because ischemia stimulates growth of fibroblasts and promotes development of greater than normal quantities of fibrous tissue. Therefore, the dead muscle tissue is gradually replaced by fibrous tissue. Then, because it is a general property of fibrous tissue to undergo progressive contraction and dissolution, the fibrous scar may grow smaller over a period of several months to a year.

Finally, the normal areas of the heart gradually hypertrophy to compensate at least partially for the lost cardiac musculature. By these means, the heart recovers either partially or almost completely.

VALUE OF REST IN TREATING MYOCARDIAL INFARCTION. The degree of cellular death is determined by the *degree of ischemia × the degree of metabolism* of the heart muscle. When the metabolism of the heart muscle is greatly increased, such as during exercise, in severe emotional strain, or as a result of fatigue, the heart needs increased oxygen and other nutrients for sustaining its life. Furthermore, anastomotic blood vessels that supply blood to ischemic areas of the heart must also

still supply the areas of the heart that they normally supply. When the heart becomes excessively active, the vessels of the normal musculature become greatly dilated. This allows most of the blood flowing into the coronary vessels to flow through the normal muscle tissue, thus leaving little blood to flow through the small anastomotic channels into the ischemic area, so that the ischemic condition worsens. This condition is called the "coronary steal" syndrome. Consequently, one of the most important factors in the treatment of a patient with myocardial infarction is observance of absolute rest during the recovery process.

Function of the Heart After Recovery from Myocardial Infarction

Occasionally, a heart that has recovered from a large myocardial infarction returns almost to full functional capability, but more frequently its pumping capability is permanently decreased below that of a normal heart. This does not mean that the person is necessarily a cardiac invalid or that the resting cardiac output is depressed below normal because the normal person's heart is capable of pumping about 300 per cent more blood per minute than the body requires—that is, a person has a "cardiac reserve" of 300 per cent. Even when the cardiac reserve is reduced to as little as 100 per cent, the person can still perform normal activity of a quiet, restful type but not strenuous exercise that would overload the heart.

Pain in Coronary Disease

Normally, a person cannot "feel" his or her heart, but ischemic cardiac muscle often does exhibit pain sensation. Exactly what causes this pain is not known, but it is believed that ischemia causes the muscle to release acidic substances, such as lactic acid, or other pain-promoting products, such as histamine, kinins, or cellular proteolytic enzymes, that are not removed rapidly enough by the slowly moving blood. The high concentrations of these abnormal products then stimulate the pain endings in the cardiac muscle, and pain impulses are conducted through the sympathetic afferent nerve fibers into the central nervous system.

Angina Pectoris

In most people who develop progressive constriction of their coronary arteries, cardiac pain, called *angina pectoris,* begins to appear whenever the load on the heart becomes too great in relation to the coronary blood flow. This pain is usually felt beneath the upper sternum and is often referred to surface areas of the body, most commonly to the left arm and left shoulder but also frequently to the neck and even to the side of the face or to the opposite arm and shoulder. The reason for this distribution of pain is that the heart originates during embryonic life in the neck, as do the arms. Therefore, both of these structures receive pain nerve fibers from the same spinal cord segments.

Most people who have chronic angina pectoris feel the pain when they exercise and when they experience emotions that increase metabolism of the heart or temporarily constrict the coronary vessels because of sympathetic vasoconstrictor nerve signals. The pain usually lasts for only a few minutes. However, some patients have such severe and lasting ischemia that the pain is present all the time. The pain is frequently described as hot, pressing, and constricting; it is of such quality that it usually makes the patient stop all activity and come to a complete state of rest.

TREATMENT WITH DRUGS. Several vasodilator drugs, when administered during an acute anginal attack, usually give immediate relief from the pain. Commonly used vasodilators are *nitroglycerin* and *other nitrate drugs.*

A second class of drugs that are used for prolonged treatment of angina pectoris is the *beta blockers,* such as propranolol. They block the sympathetic beta receptors, which prevents sympathetic stimulation of heart rate and cardiac metabolism during exercise or emotional episodes. Therefore, therapy with a beta blocker decreases the need of the heart for extra metabolic oxygen during stressful conditions. For obvious reasons, this can greatly reduce the number of anginal attacks as well as their severity.

Surgical Treatment of Coronary Disease

AORTIC-CORONARY BYPASS SURGERY. In many patients with coronary ischemia, the constricted areas of the coronary vessels are located at only a few discrete points and the coronary vessels beyond these points are normal or almost normal. A surgical procedure has been developed in the past 30 years, called *aortic-coronary bypass,* for anastomosing small vein grafts to the aorta and to the sides of the more peripheral coronary vessels. One to five such grafts usually are performed during the operations, each of which supplies a peripheral coronary artery beyond a block. The vein that typically is used for the graft is the long superficial saphenous vein removed from the patient's leg.

Anginal pain is relieved in most patients. Also, in patients whose hearts have not become too severely damaged before the operation, the coronary bypass procedure may provide the patient with a normal survival expectation. On the other hand, if the heart has been severely damaged, the bypass procedure is likely to be of little value.

CORONARY ANGIOPLASTY. During the past 15 years, a procedure has been used to open partially blocked coronary vessels before they become totally occluded. This procedure, called *coronary artery angioplasty,* is the following: A small balloon-tipped catheter, about 1 millimeter in diameter, is passed under radiographic guidance into the coronary system and pushed through the partially occluded artery until the balloon portion of the catheter straddles the partially occluded point. Then the balloon is inflated with several atmospheres of pressure, which stretches the diseased artery almost to the point of bursting. After this procedure is performed, the blood flow through the vessel often increases threefold to fourfold, and more than three quarters of the patients who undergo the procedure are relieved of the coronary ischemic symptoms for at least several years,

although many of the patients still eventually require coronary bypass surgery.

REFERENCES

Ardisson, D., et al.: Drug Evaluation in Angina Pectoris. Hingham, MA, Kluwer Academic Publishers, 1994.

Becker, R. C.: The Modern Era of Coronary Thrombolysis. Hingham, MA, Kluwer Academic Publishers, 1994.

Beller, G. A.: Clinical Nuclear Cardiology. Philadelphia, W. B. Saunders Co., 1994.

Braunwald, E.: Heart Disease: A Textbook of Cardiovascular Medicine. Philadelphia, W. B. Saunders Co., 1992.

Catravas, J. D., et al.: Vascular Endothelium: Physiological Basis of Clinical Problems II. New York, Plenum Publishing Corp., 1993.

Cernaianu, A. C., and DelRossi, A. J.: Cardiac surgery. New York, Plenum Publishing Corp., 1994.

Chilian, W. M., and Marcus, M. L.: Coronary vascular adaptations to myocardial hypertrophy. Annu. Rev. Physiol., 49:477, 1987.

Cohn, P. F.: Silent myocardial ischemia: Present status. Mod. Concepts Cardiovas. Dis., 56:1, 1987.

Crawford, M. H.: Current Diagnosis and Treatment in Cardiology. Redding, MA, Appleton and Lange, 1994.

Dean, R. H., et al.: Current Diagnosis and Treatment in Vascular Surgery. Redding, MA, Appleton and Lange, 1995.

Dembroski, T. M., et al.: Biobehavioral Bases of Coronary Heart Disease. Farmington, CT, S. Karger Publishers, Inc., 1983.

Dietz, H. C. III, and Pyeritz, R. E.: Molecular genetic approaches to the study of human cardiovascular disease. Annu. Rev. Physiol., 56:763, 1994.

Estafanous, F. G., et al.: Cardiac Anesthesia: Principles and Clinical Practice. Philadelphia, J. B. Lippincott, 1994.

Fallon, J. T.: Cardiovascular Pathophysiology: A Problem-Oriented Approach. Philadelphia, J. B. Lippincott, 1994.

Faxon, D. P.: Practical Angioplasty. New York, Raven Press, 1994.

Fuster, V., and Chesebro, J. A.: Symposium on thrombosis and antithrombotic therapy. Am. J. Cardiol., 8:1B, 1986.

Ghista, D. N.: Clinical Cardiac Assessment, Interventions, and Assist-Technology. Farmington, CT, S. Karger Publishers, Inc., 1990.

Gimbrone, Jr., M. A., and Schoen, F. J.: Cardiovascular Pathology: Clinicopathologic Correlations and Pathogenetic Mechanisms. Baltimore, Williams & Wilkins, 1994.

Goldberger, E. (ed.): Treatment of Cardiac Emergencies. St. Louis, C. V. Mosby Co., 1989.

Goldbourt, U., et al.: Genetic Factors in Coronary Heart Disease. Hingham, MA, Kluwer Academic Publishers, 1994.

Gregg, D. E.: Coronary Circulation in Health and Disease. Philadelphia, Lea & Febiger, 1950.

Guyton, R. A., and Daggett, W. M.: The evolution of myocardial infarction: Physiological basis for clinical intervention. Int. Rev. Physiol., 9:305, 1976.

Guyton, A. C., et al.: Cardiac Output and Its Regulation. Philadelphia, W. B. Saunders Co., 1973.

Hurst, J. W., et al. (eds.): Atlas of the Heart. New York, McGraw-Hill Book Co., 1988.

Kedes, L.: Regulation of myocardial adaptation. New York, Scientific American Science and Medicine, July/August, 1994.

Kelley, W. N.: Textbook of Internal Medicine. Philadelphia, J. B. Lippincott, 1991.

Kirklin, J. W., and Barratt-Boyes, B. G.: Cardiac Surgery. New York, Churchill Livingstone, 1993.

Krupski, W. C.: Review of Vascular Surgery. Philadelphia, W. B. Saunders Co., 1994.

Lawrie: Coronary Artery Bypass Surgery. Chicago, Year Book Medical Publishers, 1985.

Lusis, A. J., et al.: Molecular Genetics of Coronary Artery Disease. Farmington, CT, S. Karger Publishers, Inc., 1992.

Miller, A. J.: Diagnosis of Chest Pain. New York, Raven Press, 1988.

Nabel, E. G., and Nabel, G. J.: Complex models for the study of gene function in cardiovascular biology. Annu. Rev. Physiol., 56:741, 1994.

Opie, L. H., et al.: The Heart: Physiology and Metabolism. New York, Raven Press, 1991.

Parmley, W. W., and Chatterjee, K.: Cardiology: Clinical Text in Three Looseleaf Volumes. Philadelphia, J. B. Lippincott, 1994.

Reeder, G. S., and Vliestra, R. E.: Coronary angioplasty: 1986. Mod. Concepts Cardiovas. Dis., 55:49, 1986.

Reiber, J. H. C., and Serruys, P. W.: Progress in Quantitative Coronary Arteriography. Hingham, MA, Kluwer Academic Publishers, 1994.

Rigotti, N. A., et al.: Exercise and coronary heart disease. Annu. Rev. Med., 34:391, 1983.

Roubin, G. S., et al.: Interventional Cardiovascular Medicine. New York, Churchill Livingstone, 1994.

Schlant, R. C., et al.: Hurst's The Heart. Hightstown, NJ, McGraw-Hill, 1994.

Singh, B. N., et al.: Cardiovascular Pharmacology and Therapeutics. New York, Churchill Livingstone, 1993.

Sharma, B., et al.: Intracoronary prostaglandin E_1 plus streptokinase in acute myocardial infarction. Am. J. Cardiol., 58:1161, 1986.

Shepherd, J. T., and Vanhoutte, P. M.: Spasm of the coronary arteries: Causes and consequences (the scientist's viewpoint). Mayo Clin. Proc., 60:33, 1985.

Stone, H. L.: Control of the coronary circulation during exercise. Annu. Rev. Physiol., 45:213, 1983.

Sugimoto, T., et al.: Quantitative effect of low coronary pressure on left ventricular performance. Jpn. Heart J., 9:46, 1968.

Sundberg, C. J.: Exercise and training during graded leg ischaemia in healthy man. Acta Physiol. Scand., 150:Supplement 615, 1994.

Topol, E. J., Textbook of Interventional Cardiology. Philadelphia, W. B. Saunders Co., 1994.

Virmani, R., and Forman, M. B.: Nonatherosclerotic Ischemic Heart Disease. New York, Raven Press, 1989.

Weber, P. C., and Leaf, A.: Atherosclerosis: Cellular Interactions, Growth Factors, and Lipids. New York, Raven Press, 1994.

Weber, P. C., and Leaf, A.: Atherosclerosis: Its Pathogenesis and the Role of Cholesterol. New York, Raven Press, 1991.

Willerson, J. T., and Cohn, J. N.: Cardiovascular Medicine. New York, Churchill Livingstone, 1994.

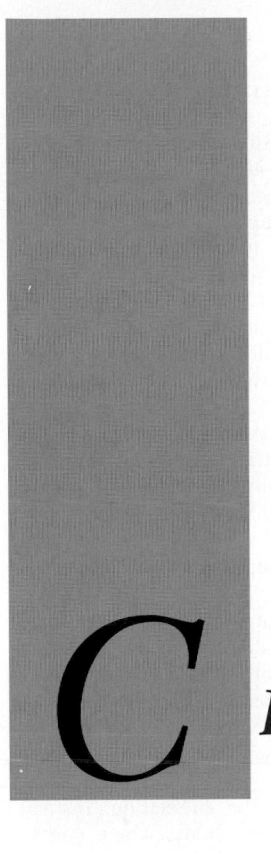

Cardiac Failure

CHAPTER 22

Perhaps the most important ailment that must be treated by the physician is cardiac failure ("heart failure"), which can result from any heart condition that reduces the ability of the heart to pump blood. The cause usually is decreased contractility of the myocardium resulting from diminished coronary blood flow, but failure to pump can also be caused by damage to the heart valves, external pressure around the heart, vitamin B deficiency, primary cardiac muscle disease, or any other abnormality that makes the heart a hypoeffective pump. In this chapter, we discuss primarily cardiac failure caused by ischemic heart disease. In Chapter 23, we discuss valvular and congenital heart disease.

DEFINITION OF CARDIAC FAILURE. The term "cardiac failure" means simply *failure of the heart to pump enough blood to satisfy the needs of the body.* Cardiac failure may be manifest by (1) a decrease in cardiac output and/or (2) a damming of the blood in the veins behind the left or right side of the heart—even though the cardiac output may be normal or, at times, above normal, as we discuss later.

DYNAMICS OF THE CIRCULATION IN CARDIAC FAILURE

Acute Effects of Moderate Cardiac Failure

If a heart suddenly becomes severely damaged, such as by myocardial infarction, the pumping ability of the heart is immediately depressed. As a result, two essential effects occur: (1) reduced cardiac output and (2) damming of blood in the veins, resulting in increased systemic venous pressure. These effects are shown graphically in Figure 22–1. This figure shows, first, a normal cardiac output curve, and point A on this curve is the normal operating point, with a normal cardiac output under resting conditions of 5 liters/min and a right atrial pressure of 0 mm Hg.

Immediately after the heart becomes damaged, the cardiac output curve becomes greatly reduced, falling to the lowest short-dashed curve at the bottom of the graph. Within a few seconds, a new circulatory state is established at point B rather than point A, showing that the cardiac output has fallen to 2 liters/min, about two fifths normal, whereas the right atrial pressure has risen to 4 mm Hg because blood returning to the heart is dammed up in the right atrium. This low cardiac output is still sufficient to sustain life for perhaps a few hours, but it is likely to be associated with fainting. This acute stage usually lasts for only a few seconds because sympathetic reflexes occur immediately that can compensate, to a great extent, for the damaged heart as follows.

COMPENSATION FOR ACUTE CARDIAC FAILURE BY SYMPATHETIC REFLEXES. When the cardiac output falls precariously low, many of the circulatory reflexes discussed in Chapter 18 are immediately activated. The best known of these is the baroreceptor reflex, which is activated by diminished arterial pressure. It is probable that the chemoreceptor reflex, central nervous system ischemic response, and even reflexes that originate in the damaged heart also contribute to the nervous response. But whatever the reflexes might be, the sympathetics become strongly stimulated within a few seconds, and the parasympathetics become reciprocally inhibited at the same time.

Strong sympathetic stimulation has two major effects on the circulation: first, on the heart itself and, second, on the peripheral vasculature. If all the ventricular musculature is diffusely damaged but is still functional, sympathetic stimulation strengthens this damaged musculature. If part of the muscle is nonfunctional and part of it is still normal, the normal muscle is strongly stimulated by sympathetic stimulation, in this way partially compensating for the nonfunctional muscle. Thus, *the heart, one way or another, becomes a stronger pump, often as much as 100 per cent stronger, under the influence of the sympathetic impulses.* This effect is also demonstrated in Figure 22–1, which shows about

265

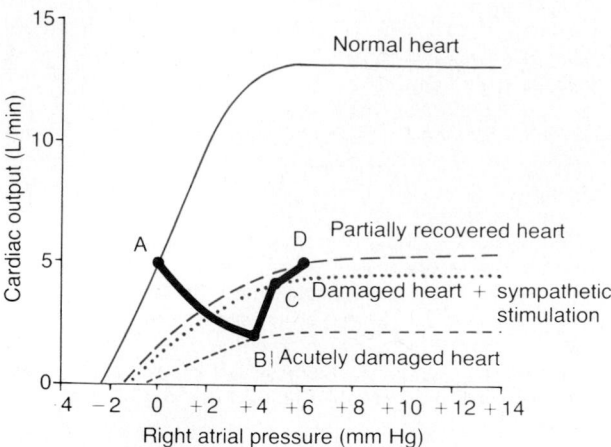

Figure 22–1. Progressive changes in the cardiac output curve after acute myocardial infarction. The cardiac output and right atrial pressure change progressively from point A to point D (illustrated by the heavy line).

two-fold elevation of the cardiac output curve after sympathetic compensation (the dotted curve).

Sympathetic stimulation also increases the tendency for venous return because it increases the tone of most of the blood vessels of the circulation, especially the veins, *raising the mean systemic filling pressure* to 12 to 14 mm Hg, almost 100 per cent above normal. As discussed in Chapter 20, this greatly increases the tendency for blood to flow back to the heart. Therefore, the damaged heart also becomes primed with more inflowing blood than usual and the right atrial pressure rises still further, which helps the heart to pump still larger quantities of blood. Thus, in Figure 22–1, the new circulatory state is depicted by point C, showing a cardiac output of 4.2 liters/min and a right atrial pressure of 5 mm Hg.

The sympathetic reflexes become maximally developed in about 30 seconds. Therefore, a person who has a sudden, moderate heart attack might experience nothing more than cardiac pain and a few seconds of fainting. Shortly thereafter, with the aid of the sympathetic reflex compensations, the cardiac output may return to a level adequate to sustain the person who remains quiet, although the pain might persist.

Chronic Stage of Failure—Fluid Retention Helps to Compensate Cardiac Output

After the first few minutes of an acute heart attack, a prolonged secondary state begins, characterized mainly by two events: (1) retention of fluid by the kidneys and (2) usually progressive recovery of the heart over a period of several weeks to months, as discussed in Chapter 21.

Renal Retention of Fluid and Increase in Blood Volume

A low cardiac output has a profound effect on renal function, sometimes causing anuria when the cardiac output falls to one-half to two-thirds normal. In general, the urine output remains reduced as long as the cardiac output is significantly less than normal, and it usually does not return to normal after an acute heart attack until the cardiac output rises either all the way back to normal or almost to normal.

BENEFICIAL EFFECTS OF MODERATE FLUID RETENTION IN CARDIAC FAILURE. Many cardiologists formerly considered fluid retention always to have a detrimental effect in cardiac failure. It is now known that a moderate increase in body fluid and blood volume is an important factor in helping to compensate for the diminished pumping ability of the heart because it increases the tendency for venous return. The increased blood volume increases the venous return in two ways: First, it increases the mean systemic filling pressure, which *increases the pressure gradient for causing flow of blood toward the heart.* Second, it distends the veins, which *reduces the venous resistance* and thereby allows increased ease of flow of blood to the heart.

If the heart is not too greatly damaged, this increased tendency for venous return can often fully compensate for the heart's diminished pumping ability—so much so in fact that even if the heart's pumping ability is reduced to 40 to 50 per cent of normal, the increased venous return can often cause an entirely normal cardiac output under quiet resting conditions.

When the heart's maximum pumping ability is reduced still more, the blood flow to the kidneys becomes permanently too low for the urine output of salt and water to equal the intake. Therefore, fluid retention begins and continues indefinitely unless major therapeutic procedures are used to prevent this. Furthermore, because the heart is already pumping at its maximum pumping capacity, this excess fluid no longer has a beneficial effect on the circulation. Instead, severe edema develops throughout the body, which can be detrimental in itself and can lead to death.

DETRIMENTAL EFFECTS OF EXCESS FLUID RETENTION IN SEVERE STAGES OF CARDIAC FAILURE. In contrast to the beneficial effects of moderate fluid retention in cardiac failure, in severe failure with extreme excesses of fluid retention, the fluid begins to have serious physiological consequences. They include overstretching of the heart, thus weakening the heart still more; filtration of fluid into the lungs to cause pulmonary edema and consequent deoxygenation of the blood; and development of extensive edema in all the peripheral tissues of the body. These detrimental effects of excessive fluid are discussed in subsequent sections of the chapter.

Recovery of the Myocardium After Myocardial Infarction

After a heart becomes suddenly damaged as a result of myocardial infarction, the natural reparative processes of the body begin immediately to help restore normal cardiac function. Thus, a new collateral blood supply begins to penetrate the peripheral portions of the infarcted area, often causing much of the muscle in the fringe areas to become functional again. Also, the undamaged musculature hypertrophies, in this way offsetting much of the cardiac damage.

The degree of recovery depends on the type of cardiac damage, and it varies from no recovery to almost complete recovery. After myocardial infarction, the heart ordinarily recovers rapidly during the first few days and weeks and achieves most of its final state of recovery within 5 to 7 weeks, although some recovery continues for months.

CARDIAC OUTPUT CURVE AFTER PARTIAL RECOVERY. The long-dashed curve of Figure 22–1 shows function of the partially recovered heart 1 week or so after the acute myocardial infarction. By this time, considerable fluid also has been retained in the body and the tendency for venous return has increased markedly; therefore, the right atrial pressure has risen even more. As a result, the state of the circulation is now changed from point C to point D, which represents a *normal* cardiac output of 5 liters/min but a right atrial pressure elevated to 6 mm Hg.

Because the cardiac output has returned to normal, renal output also will have returned to normal, and no further fluid retention will occur. Therefore, except for the high right atrial pressure represented by point D in this figure, the person now has essentially normal cardiovascular dynamics *as long as he or she remains at rest.*

If the heart recovers to a significant extent and if adequate fluid retention occurs, the sympathetic stimulation gradually abates toward normal for the following reasons: The partial recovery of the heart can do the same thing for the cardiac output curve as sympathetic stimulation, and the increased blood volume can do the same thing for venous return as sympathetic contraction of the veins. Thus, as these two factors develop, the fast pulse rate, cold skin, and pallor resulting from sympathetic stimulation in the acute stage of cardiac failure gradually disappear.

Summary of the Changes That Occur After Acute Cardiac Failure—"Compensated Heart Failure"

To summarize the events discussed in the past few sections describing the dynamics of circulatory changes after an acute, moderate heart attack, we may divide the stages into (1) the instantaneous effect of the cardiac damage; (2) compensation by the sympathetic nervous system, which occurs mainly within the first 30 seconds to 1 minute; and (3) chronic compensations resulting from partial cardiac recovery and renal retention of fluid. All these changes are shown graphically by the heavy line in Figure 22–1. The progression of this line shows the normal state of the circulation (point A), the state a few seconds after the heart attack but before sympathetic reflexes have occurred (point B), the rise in cardiac output toward normal caused by sympathetic stimulation (point C), and final return of the cardiac output to normal after several days to several weeks of cardiac recovery and fluid retention (point D). This final state is called *compensated heart failure.*

COMPENSATED HEART FAILURE. Note especially in Figure 22–1 that the pumping ability of the heart, as depicted by the cardiac output curve, is still depressed to less than one-half normal. This demonstrates that factors that increase the right atrial pressure (principally the increased blood volume caused by retention of fluid) can maintain the cardiac output at a normal level despite continued weakness of the heart. However, one of the results of chronic cardiac weakness is the chronic increase in right atrial pressure; in Figure 22–1,

it is shown to be 6 mm Hg. Many people, especially in old age, have normal resting cardiac outputs but mildly to moderately elevated right atrial pressures because of various degrees of compensated heart failure. These people may not know that they have cardiac damage because the damage more often than not has occurred a little at a time, and the compensation has occurred concurrently with the progressive stages of damage.

When a person is in compensated heart failure, any attempt to perform heavy exercise usually causes immediate return of the symptoms of acute failure because the heart is not able to increase its pumping capacity to the levels required to sustain the exercise. Therefore, it is said that the *cardiac reserve* is reduced in compensated heart failure. This concept of cardiac reserve is discussed more fully later in the chapter.

Dynamics of Severe Cardiac Failure— Decompensated Heart Failure

If the heart becomes severely damaged, no amount of compensation, either by sympathetic nervous reflexes or by fluid retention, can make the weakened heart pump a normal cardiac output. As a consequence, the cardiac output cannot rise to a high enough value to bring about return of normal renal function. Fluid continues to be retained, the person develops more and more edema, and this state of events eventually leads to death. This is called *decompensated heart failure.*

Thus, the main basis of decompensated heart failure is *failure of the heart to pump sufficient blood to make the kidneys excrete daily the necessary amounts of fluid.*

ANALYSIS OF DECOMPENSATED HEART FAILURE. Figure 22–2 shows a greatly depressed cardiac output curve, depicting the function of a heart that has become extremely weakened and cannot be strengthened. Point A on this curve represents the approximate state of the circulation before any compensation has occurred, and point B, the state after sympathetic stimulation has compensated as much as it can but before excess fluid retention has begun. At point B, the cardiac output has risen to 4 liters/min and the right atrial pressure, to 5 mm Hg. The person appears to be in reasonably good condition, but this state will not remain static for the following reason: The cardiac output has not risen quite high enough to cause adequate kidney excretion of fluid; therefore, fluid retention continues unabated and can eventually be the cause of death. These events can be explained quantitatively in the following way.

Note the dashed line at a cardiac output level of 5 liters in Figure 22–2. This is the critical cardiac output level that is

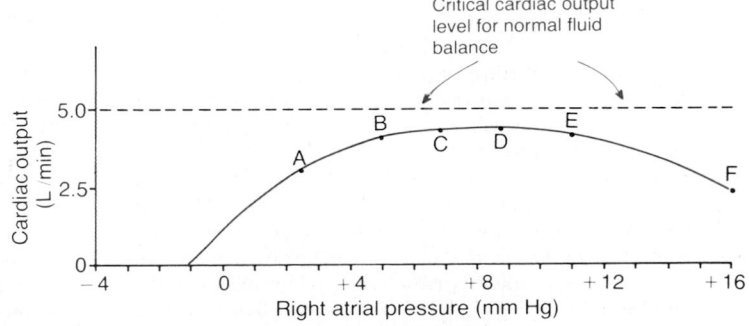

Figure 22–2. Greatly depressed cardiac output curve that indicates decompensated heart disease. Progressive fluid retention raises the right atrial pressure, and the cardiac output progresses from point A to point F.

required to make the kidneys re-establish normal fluid balance—that is, for the output of salt and water to be as great as the intake of these. At any cardiac output below this level, all the fluid-retaining mechanisms discussed in the earlier section remain in play and the body fluid volumes increase progressively. Because of this progressive increase in fluid volume, the mean systemic filling pressure continues to rise; this forces progressively increasing quantities of blood into the right atrium, thus increasing the right atrial pressure. After 1 day or so, the state of the circulation changes in Figure 22–2 from point B to point C—the right atrial pressure rising to 7 mm Hg and the cardiac output, to 4.2 liters/min. Note again that the cardiac output is still not high enough to cause normal renal output of fluid; therefore, fluid continues to be retained, and after another day or so, the right atrial pressure rises to 9 mm Hg and the circulatory state becomes that depicted by point D. Still, the cardiac output is not enough to establish normal fluid balance.

After another few days of fluid retention, the right atrial pressure has risen still further, but by now, the cardiac function curve is beginning to decline toward a lower level. This decline is caused by overstretch of the heart, edema of the heart muscle, and other factors that diminish the heart's pumping performance. It is now clear that further retention of fluid will be more detrimental than beneficial to the circulation. Yet the cardiac output still is not high enough to bring about normal renal function, and fluid retention not only continues but accelerates because of the falling cardiac output. Consequently, within a few days, the state of the circulation has reached point F on the curve, with the cardiac output less than 2.5 liters/min and the right atrial pressure 16 mm Hg. This state now has approached or reached incompatibility with life, and the patient dies.

This state of heart failure in which the failure continues to worsen is called *decompensated heart failure*.

Thus, one can see from this analysis that failure of the cardiac output ever to rise to the critical level required for normal renal function results in (1) progressive retention of more and more fluid, which causes (2) progressive elevation of the mean systemic filling pressure, and (3) progressive elevation of the right atrial pressure until finally the heart is so overstretched or so edematous that it cannot pump even moderate quantities of blood and, therefore, fails completely. Clinically, one detects this serious condition of decompensation principally by the progressive edema, especially edema of the lungs, which leads to bubbling rales in the lungs and dyspnea (air hunger). All clinicians know that lack of appropriate therapy when this state of events occurs leads to rapid death.

TREATMENT OF DECOMPENSATION. The decompensation process can often be stopped by (1) strengthening the heart in any one of several ways, especially by administration of a cardiotonic drug, such as digitalis, so that the heart becomes strong enough to pump adequate quantities of blood required to make the kidneys function normally again, or (2) administering diuretics and reducing water and salt intake, which brings about a balance between fluid intake and output despite the low cardiac output.

Both methods stop the decompensation process by re-establishing normal fluid balance, so that at least as much fluid leaves the body as enters it.

Mechanism of Action of the Cardiotonic Drugs Such as Digitalis. Cardiotonic drugs, such as digitalis, when administered to a person with a normal heart, have little effect on increasing the contractile strength of the cardiac muscle. When administered to a person with a chronically failing heart, the same drugs can sometimes increase the strength of

the failing myocardium 50 to 100 per cent. Therefore, they are the mainstay of therapy in chronic heart failure.

The way in which digitalis and other cardiotonic glycosides strengthen heart contraction is believed to be by increasing the quantity of calcium ions in the muscle fibers. In the failing heart muscle, the sarcoplasmic reticulum fails to accumulate normal quantities of calcium and, therefore, cannot release enough calcium to cause full contraction of the muscle. The effect of digitalis is to depress the calcium pump of the cardiac muscle cell membrane that normally pumps calcium ions out of the muscle cells. This mechanism is required normally to keep the concentration of calcium from rising too high in these cells because calcium ions have a natural tendency to leak inward through the membrane. However, in the case of the failing heart, the extra calcium is needed, and it does increase the muscle contractile force. Therefore, it is usually beneficial to depress the calcium pumping mechanism a moderate amount using digitalis, thus allowing the intracellular calcium level to rise.

UNILATERAL LEFT HEART FAILURE

In the discussions thus far in this chapter, we have considered failure of the heart as a whole. Yet, in a large number of patients, especially those with early acute failure, left-sided failure predominates over right-sided failure, and in rare instances, the right side fails without significant failure of the left side. Therefore, we need especially to discuss the special features of unilateral left heart failure.

When the left side of the heart fails without concomitant failure of the right side, blood continues to be pumped into the lungs with usual right heart vigor, whereas it is not pumped adequately out of the lungs into the systemic circulation. As a result, the *mean pulmonary filling pressure* rises because of the shift of large volumes of blood from the systemic circulation into the pulmonary circulation.

As the volume of blood in the lungs increases, the pulmonary capillary pressure also increases, and if this rises above a value approximately equal to the colloid osmotic pressure of the plasma, about 28 mm Hg, fluid begins to filter out of the capillaries into the interstitial spaces and alveoli, resulting in pulmonary edema.

Thus, among the most important problems of left heart failure are *pulmonary vascular congestion* and *pulmonary edema*. In severe, acute left heart failure, pulmonary edema occasionally occurs so rapidly that it causes death by suffocation in 20 to 30 minutes, which we discuss more fully later in the chapter.

"HIGH-OUTPUT CARDIAC FAILURE"— THIS CAN OCCUR EVEN IN A NORMAL HEART THAT IS OVERLOADED

A condition called "high-output cardiac failure" sometimes occurs in people whose cardiac output is much higher than normal but who also have signs of cardiac failure because the left or right or both atrial pressures are much too high, and there usually is accumulation of edema. The true problem is often not failure of the heart's pumping ability but *overloading of the heart with too much venous return*. Most commonly this is caused by some circulatory abnormality that drastically decreases the total peripheral resistance, such as the following:

1. *Arteriovenous fistula,* in which blood is shunted directly from the arteries to the veins. This greatly increases the venous return and frequently causes congestive symptoms similar to those of typical heart failure, even though the heart is normal and is pumping a tremendously increased cardiac output.

2. *Beriberi heart disease,* which is caused by lack of the B vitamins, especially thiamine. This causes poor metabolic utilization of nutrients throughout the body. In turn, the tissue metabolic mechanisms that control local blood flow dilate all the local blood vessels, profoundly reducing the total peripheral resistance and often doubling the venous return. The heart is usually weakened as well; yet, even with the weakness of the heart, the cardiac output is still as much as twice normal. Therefore, severe congestive symptoms occur both because of the excess venous return and because of the weakened heart, so that there is *both excess loading of the heart and true heart failure.*

3. *Thyrotoxicosis,* which, because of its stimulatory effect on body metabolism, causes generalized systemic vasodilation. Here again, the thyrotoxicosis frequently weakens the heart as well, so that there is often both overloading of the heart and true heart failure, occurring simultaneously.

LOW-OUTPUT CARDIAC FAILURE— CARDIOGENIC SHOCK

In many instances after acute heart attacks and often also after prolonged periods of slow progressive cardiac deterioration, the heart becomes incapable of pumping even the minimal amount of blood flow required to keep the body alive. Consequently, all the body tissues begin to suffer and even to deteriorate, often leading to death within a few hours to a few days. The picture then is one of circulatory shock, as explained in Chapter 24. Even the cardiovascular system suffers from lack of nutrition, and it, too, along with the remainder of the body, deteriorates, thus hastening death. This circulatory shock syndrome caused by inadequate cardiac pumping is called *cardiogenic shock* or simply *cardiac shock.* Sometimes it is called the *power failure syndrome.* Once a person develops cardiogenic shock, the survival rate, even with the best of therapy, is usually less than 15 per cent.

VICIOUS CIRCLE OF CARDIAC DETERIORATION IN CARDIOGENIC SHOCK. The discussion of circulatory shock in Chapter 24 emphasizes the tendency for the heart to become progressively more damaged when its coronary blood supply is reduced during the course of shock. That is, the low arterial pressure that occurs during shock reduces the coronary supply, which makes the heart still weaker, which makes the shock still worse, the process eventually becoming a vicious circle of cardiac deterioration. In cardiogenic shock caused by myocardial infarction, this problem is greatly compounded by the already existing coronary blockage. For instance, in a normal heart, the arterial pressure usually must be reduced below about 45 mm Hg before cardiac deterioration sets in. However, in a heart that already has a blocked major coronary vessel, deterioration sets in when the arterial pressure falls to 80 to 90 mm Hg. In other words, even the minutest amount of fall in arterial pressure can set off a vicious circle of cardiac deterioration after myocardial infarction. For this reason, in treating myocardial infarction, it is extremely important to prevent even short periods of hypotension.

PHYSIOLOGY OF TREATMENT. Often a patient dies of cardiogenic shock before the various compensatory processes can return the cardiac output to a life-sustaining level. Therefore, treatment of this condition is one of the most important problems in the management of acute heart attacks. Immediate digitalization of the heart is often used for strengthening the heart if the remaining functioning ventricular muscle shows signs of deterioration. Also, infusion of whole blood, plasma, or a blood pressure–raising drug is used to sustain the arterial pressure. If the arterial pressure can be raised high enough, the coronary blood flow may be elevated to a high enough value to prevent the vicious circle of deterioration until appropriate compensatory mechanisms in the body can correct the shock.

Some success has also been achieved in saving the lives of patients with cardiogenic shock using one of the following procedures: (1) surgically removing the clot in the coronary artery, often in combination with coronary bypass graft, or (2) catheterizing the blocked coronary artery and infusing either *streptokinase* or *tissue-type plasminogen activator* enzymes that cause dissolution of the clot. The results occasionally are astounding when the procedure is instituted within the 1st hour but of little, if any, benefit after 3 hours.

EDEMA IN PATIENTS WITH CARDIAC FAILURE

INABILITY OF ACUTE CARDIAC FAILURE TO CAUSE PERIPHERAL EDEMA. Although acute left heart failure can cause terrific congestion of the lungs, with rapid development of pulmonary edema and even death within minutes to hours, *acute* heart failure of any type does not cause immediate development of peripheral edema. This can best be explained by referring to Figure 22–3. When a previously normal heart acutely fails as a pump, the aortic pressure falls and the right atrial pressure rises. As the cardiac output approaches zero, these two pressures approach each other at an equilibrium value of about 13 mm Hg. Capillary pressure must also fall from its normal value of 17 mm Hg to the equilibrium pressure of 13 mm Hg. Thus, *severe acute cardiac failure causes a fall in peripheral capillary pressure rather than a rise.* Animal experiments as well as experience

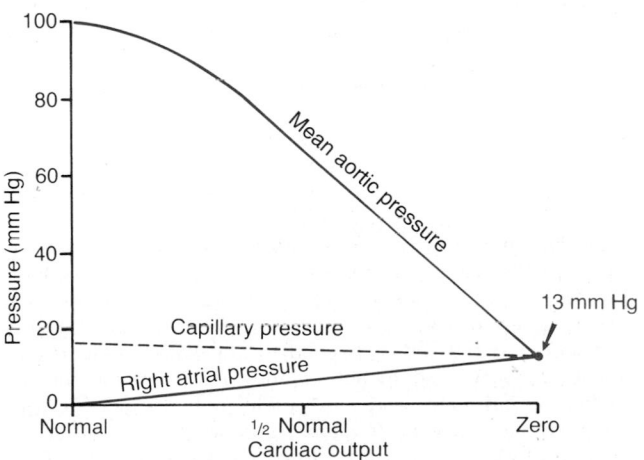

Figure 22–3. Progressive changes in mean aortic pressure, capillary pressure, and right atrial pressure as the cardiac output falls from normal to zero.

in humans show that acute cardiac failure does not cause immediate development of peripheral edema.

Fluid Retention by the Kidneys—The Cause of Peripheral Edema in Heart Failure

After the 1st day or so of total heart failure or of right-sided heart failure, peripheral edema does begin to occur but principally *because of fluid retention by the kidneys*. The retention of fluid increases the mean systemic filling pressure, resulting in increased tendency for blood to return to the heart. This elevates the right atrial pressure to a still higher value and returns the arterial pressure back toward normal. Therefore, *the capillary pressure also rises markedly*, thus causing loss of fluid into the tissues and development of severe edema.

There are three known causes of the reduced renal output of urine during cardiac failure, all of which are perhaps equally important but in different ways.

1. Decreased Glomerular Filtration. A decrease in cardiac output has a tendency to reduce the glomerular pressure in the kidneys because of (1) *reduced arterial pressure* and (2) *intense sympathetic constriction of the afferent arterioles of the kidney*. As a consequence, except in the mildest degrees of heart failure, the glomerular filtration rate becomes less than normal. It is clear from the discussion of kidney function in Chapters 26 through 29 that *even a slight decrease in glomerular filtration often markedly decreases urine output*. When the cardiac output falls to about one-half normal, this can result in almost complete anuria.

2. Activation of the Renin-Angiotensin System and Increased Reabsorption of Water and Salt by the Renal Tubules. The reduced blood flow to the kidneys causes marked increase in renin output, and this in turn causes the formation of angiotensin by the mechanism described in Chapter 19. The angiotensin has a direct effect on the arterioles of the kidneys to further decrease the blood flow through the kidneys. This especially reduces the peritubular capillary pressure, thus promoting greatly increased reabsorption of both water and salt from the renal tubules. Therefore, the net loss of water and salt into the urine is greatly decreased, so that the quantities of salt and water in the body fluids increase.

3. Increased Aldosterone Secretion. In the chronic stage of heart failure, large quantities of aldosterone are secreted by the adrenal cortex. This results mainly from the effect of angiotensin in stimulating aldosterone secretion. Also, some of the increase in aldosterone secretion might result from increased plasma potassium because excess potassium is one of the most powerful stimuli known for aldosterone secretion, and the potassium concentration rises in response to reduced renal function in cardiac failure. The elevated aldosterone levels further increase the reabsorption of sodium from the renal tubules, and this in turn leads to a secondary increase in water reabsorption for two reasons. First, as the sodium is reabsorbed, it reduces the osmotic pressure in the tubules while increasing the osmotic pressure in the renal interstitial fluids; these changes promote osmosis of water into the blood. Second, the absorbed sodium and anions to go with it, mainly chloride ions, increase the osmotic concentration of the extracellular fluid everywhere in the body, and this elicits *antidiuretic hormone* secretion by the hypothalamic–posterior pituitary system, which is discussed in Chapter 29. The antidiuretic hormone then promotes a still greater increase in tubular reabsorption of water.

Role of Atrial Natriuretic Factor to Delay the Onset of Cardiac Decompensation. *Atrial natriuretic factor (ANF)* is a hormone released by the atrial walls of the heart when they become stretched. Because heart failure almost always causes excessive increase in both the right and the left atrial pressures that stretch the atrial walls, the circulating levels of ANF in the blood increase five- to tenfold in severe heart failure. The ANF in turn has a direct effect on the kidneys to greatly increase their excretion of salt and water. Therefore, ANF plays a natural role to help prevent the extreme congestive symptoms of cardiac failure. The renal effects of ANF are discussed in Chapter 29.

Acute Pulmonary Edema in Chronic Heart Failure—A Lethal Vicious Circle

A frequent cause of death in heart failure is acute pulmonary edema occurring in patients who have had chronic heart failure for a long time. When this occurs in a person without new cardiac damage, it usually is set off by some temporary overload of the heart, such as might result from a bout of heavy exercise, some emotional experience, or even a severe cold. This acute pulmonary edema is believed to result from the following vicious circle:

1. A temporarily increased load on the already weak left ventricle results from increased venous return from the peripheral circulation. Because of the limited pumping capacity of the left heart, blood begins to dam up in the lungs.
2. The increased blood in the lungs elevates the pulmonary capillary pressure, and a small amount of fluid begins to transude into the lung tissues and alveoli.
3. The increased fluid in the lungs diminishes the degree of oxygenation of the blood.
4. The decreased oxygen in the blood further weakens the heart and causes peripheral vasodilatation.
5. The peripheral vasodilatation increases venous return from the peripheral circulation still more.
6. The increased venous return further increases the damming of the blood in the lungs, leading to still more transudation of fluid, more arterial oxygen desaturation, more venous return, and so forth. Thus, a vicious circle has been established.

Once this vicious circle has proceeded beyond a certain critical point, it will continue until death of the patient unless heroic therapeutic measures are used. The types of heroic therapeutic measures that can reverse the process and save the patient's life include the following:

1. Putting tourniquets on both arms and legs to sequester much of the blood in the veins and, therefore, decrease the workload on the left side of the heart
2. Bleeding the patient
3. Giving a rapidly acting diuretic, such as furosemide, to cause rapid loss of fluid from the body
4. Giving the patient pure oxygen to breathe to reverse the blood desaturation, heart deterioration, and peripheral vasodilatation
5. Giving the patient a rapidly acting cardiotonic drug, such as ouabain, to strengthen the heart

This vicious circle of acute pulmonary edema can proceed so rapidly that death can occur within 20 minutes to 1 hour. Therefore, any procedure that is to be successful must be instituted immediately.

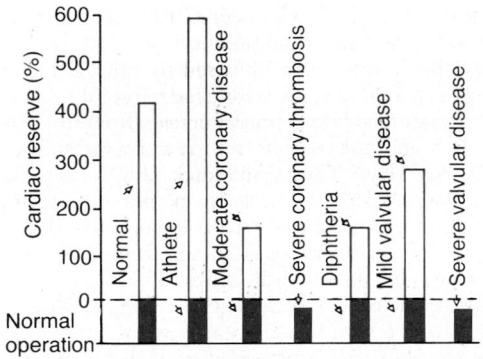

Figure 22–4. Cardiac reserve in different conditions, showing less than zero reserve for two of the conditions.

CARDIAC RESERVE

The maximum percentage that the cardiac output can increase above normal is called the *cardiac reserve.* Thus, in the healthy young adult, the cardiac reserve is 300 to 400 per cent. In the athletically trained person, it is occasionally 500 to 600 per cent, whereas in heart failure, there is no reserve. As an example of normal reserve, during severe exercise, the cardiac output of the healthy young adult can rise to about five times normal; this is an increase above normal of 400 per cent—that is, a cardiac reserve of 400 per cent.

Any factor that prevents the heart from pumping blood satisfactorily decreases the cardiac reserve. This can result from ischemic heart disease, primary myocardial disease, vitamin deficiency, damage to the myocardium, valvular heart disease, and many other factors, some of which are shown in Figure 22–4.

DIAGNOSIS OF LOW CARDIAC RESERVE—EXERCISE TEST. So long as people with low cardiac reserve remain in a state of rest, they usually will not know that they have heart disease. A diagnosis of low cardiac reserve usually can be easily made by requiring a person to exercise either on a treadmill or by walking up and down steps. The increased load on the heart rapidly uses up the small amount of reserve that is available, and the cardiac output fails to rise high enough to sustain the body's new level of activity. The acute effects are as follows:

1. Immediate and sometimes extreme shortness of breath (dyspnea) resulting from the heart's not pumping sufficient blood to the tissues, thereby causing tissue ischemia and creating a sensation of air hunger
2. Extreme muscle fatigue resulting from muscle ischemia, thus limiting the person's ability to continue with the exercise
3. Excessive increase in heart rate because the nervous reflexes to the heart overreact in an attempt to overcome the inadequate cardiac output

Exercise tests are part of the armamentarium of the cardiologist. These tests take the place of cardiac output measurements that cannot be made with ease in most clinical settings.

APPENDIX

Quantitative Graphical Method for Analysis of Cardiac Failure

Although it is possible to understand most of the general principles of cardiac failure using mainly qualitative logic, as we have done thus far in this chapter, one can grasp the importance of the different factors in cardiac failure with far greater depth by using quantitative approaches. One such approach is the graphical method for analysis of cardiac output regulation introduced in Chapter 20. In the remaining sections of this chapter, we analyze several aspects of cardiac failure, using this graphical technique.

Graphical Analysis of Acute Heart Failure and Chronic Compensation

Figure 22–5 shows cardiac output and venous return curves for different states of the heart and of the peripheral circulation. The two *solid curves* are (1) the normal cardiac output curve and (2) the normal venous return curve. As pointed out in Chapter 20, there is only one point on each of these two curves at which the circulatory system can operate so long as these two curves remain normal. This point is where the two curves cross at point A. Therefore, the normal state of the circulation is a cardiac output and venous return of 5 liters/min and a right atrial pressure of 0 mm Hg.

EFFECT OF ACUTE HEART ATTACK. During the first few seconds after a moderately severe heart attack, the cardiac output curve falls to the lowermost *long-dashed curve.* During these few seconds, the venous return curve still has not changed because the peripheral circulatory system is still operating normally. Therefore, the new state of the circulation is depicted by *point B,* where the new cardiac output curve crosses the normal venous return curve. Thus, the right atrial pressure rises to 4 mm Hg, whereas the cardiac output falls to 2 liters/min.

EFFECT OF SYMPATHETIC REFLEXES. Within the next 30 seconds, the sympathetic reflexes become very active. They affect both curves, raising them to the *short-dashed curves.* Sympathetic stimulation can increase the plateau level of the cardiac output curve 100 per cent. It can also increase the mean systemic filling pressure (depicted by the point where the venous return curve crosses the zero venous return axis) by several millimeters of mercury—in this figure from a normal value of 7 mm Hg to 10 mm Hg. This increase in

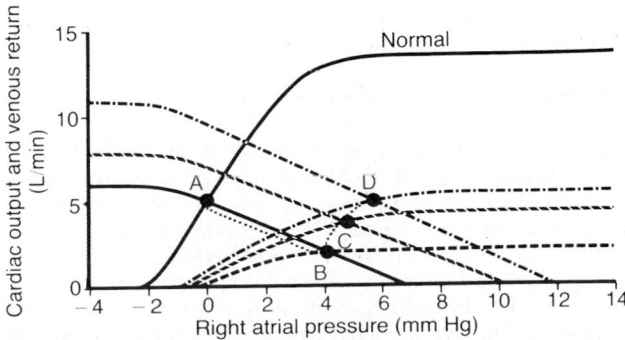

Figure 22–5. Progressive changes in cardiac output and right atrial pressure during different stages of cardiac failure.

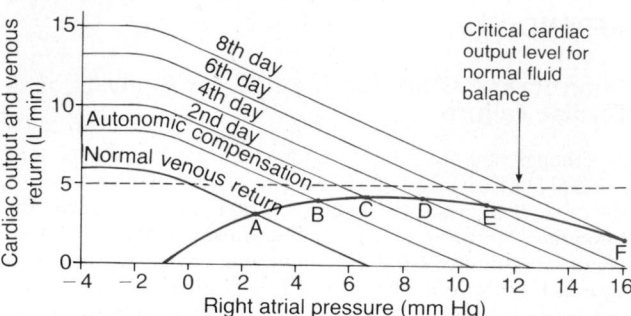

Figure 22–6. Graphical analysis of decompensated heart disease, showing progressive shift of the venous return curve to the right as a result of fluid retention.

mean systemic filling pressure shifts the entire venous return curve to the right and upward. The new cardiac output and venous return curves now equilibrate at point C, that is, at a right atrial pressure of 5 mm Hg and a cardiac output of 4 liters/min.

COMPENSATION DURING THE NEXT FEW DAYS. During the ensuing week, the cardiac output and venous return curves rise to the *dot-dashed curves* because of (1) some recovery of the heart and (2) renal retention of salt and water, which raises the mean systemic filling pressure still further—this time up to 12 mm Hg. The two new curves, the dot-dashed curves, now equilibrate at *point D.* Thus, the cardiac output has now returned to normal. The right atrial pressure, however, has risen still further to 6 mm Hg. Because the cardiac output is now normal, renal output is also normal, so that a new state of equilibrated fluid balance has been achieved. The circulatory system will continue to function at point D and remain stable, with a normal cardiac output and an elevated right atrial pressure, until some additional extrinsic factor changes either the cardiac output curve or the venous return curve.

Using this technique for analysis, one can see especially the importance of moderate fluid retention and how it eventually leads to a new stable state of the circulation; one can also see the interrelation between the mean systemic filling pressure and cardiac pumping at various degrees of cardiac failure.

Note that the events described in Figure 22–5 are the same as those presented in Figure 22–1, but in Figure 22–5, they are presented in a more quantitative manner.

Graphical Analysis of "Decompensated" Cardiac Failure

The cardiac output curve in Figure 22–6 is the same as that shown in Figure 22–2, a very low curve that, in this case, has already reached a degree of recovery as great as this heart can achieve. In this figure, we have added venous return curves that occur during successive days after the acute fall of the cardiac output curve to this low level. At *point A,* the curve at time zero equates with the normal venous return curve to give a cardiac output of about 3 liters/min. However, stimulation of the sympathetic nervous system, caused by this low cardiac output, increases the mean systemic filling pressure within 30 seconds from 7 to 10.5 mm Hg and shifts the venous return curve upward and to the right to give the curve labeled "autonomic compensation." Thus, the new venous return curve equates with the cardiac output curve at *point B.* The cardiac output has been

improved to a level of 4 liters/min but at the expense of an additional rise in right atrial pressure to 5 mm Hg.

The cardiac output of 4 liters/min is still too low to cause the kidneys to function normally. Therefore, fluid is retained, the mean systemic filling pressure rises from 10.5 to almost 13 mm Hg, and the venous return curve becomes that labeled "second day." This equilibrates with the cardiac output curve at *point C,* and the cardiac output rises to 4.2 liters/min and the right atrial pressure rises to 7 mm Hg.

During the succeeding days, the cardiac output never rises high enough to re-establish normal renal function. Fluid continues to be retained, the mean systemic filling pressure continues to rise, the venous return curve continues to shift to the right, and the equilibrium point between the venous return curve and the cardiac output curve shifts progressively to *point D,* to *point E,* and, finally, to *point F.* The equilibration process is now on the downslope of the cardiac output curve, so that further retention of fluid causes only more severe cardiac edema and a detrimental effect on cardiac output. The condition accelerates downhill until death occurs.

Thus, "decompensation" results from the fact that the cardiac output curve never rises to the critical level of 5 liters/min required to re-establish normal balance between fluid input and output.

TREATMENT OF DECOMPENSATED HEART DISEASE WITH DIGITALIS. Let us assume that the stage of decompensation has already reached point E in Figure 22–6, and let us proceed to the same point E in Figure 22–7. At this point, digitalis is given to strengthen the heart and to raise the cardiac output curve to the level of the heavy curve in Figure 22–7. This does not immediately change the venous return curve. Therefore, the new cardiac output curve equates with the venous return curve at *point G.* The cardiac output is now 5.7 liters/min, a value greater than the critical level of 5 liters required for normal fluid balance. Therefore, the kidneys eliminate far more fluid than normally, causing *diuresis,* a well-known therapeutic effect of digitalis.

The progressive loss of fluid over a period of several days reduces the mean systemic filling pressure back down to 11.5 mm Hg, and the new venous return curve becomes the heavy curve labeled "several days later." This curve equates with the cardiac output curve of the digitalized heart at *point H,* at an output of 5 liters/min and a right atrial pressure of 4.6 mm Hg. This cardiac output is precisely that required for normal fluid balance. Therefore, no additional fluid will be lost and none will be gained. Consequently, the circulatory system has now stabilized, or in other words, the decompen-

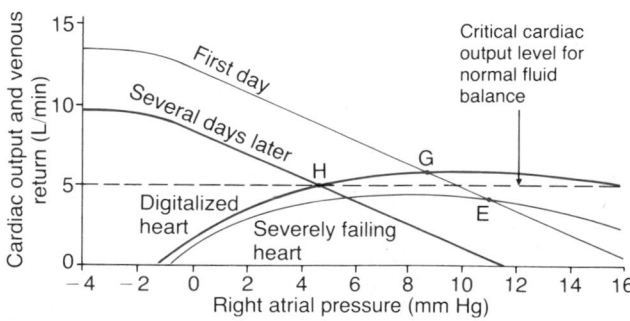

Figure 22–7. Treatment of decompensated heart disease, showing the effect of digitalis in elevating the cardiac output curve, this in turn causing increased urine output and progressive shift of the venous return curve to the left.

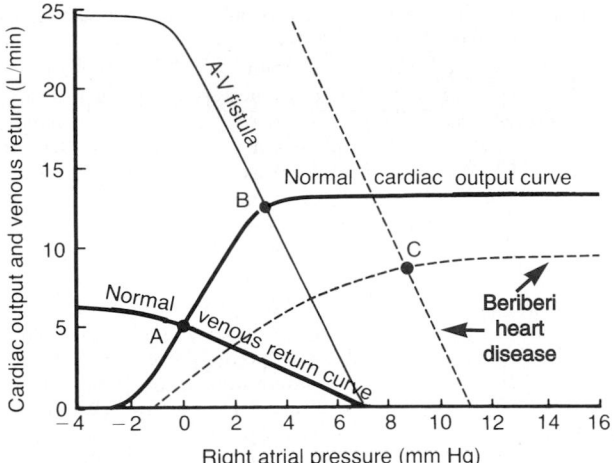

Figure 22–8. Graphical analysis of two types of conditions that can cause high output cardiac failure, A-V fistula, and beriberi heart disease.

BERIBERI. The dashed lines in Figure 22–8 show the approximate changes in the cardiac output and venous return curves caused by beriberi. The decreased level of the cardiac output curve is caused by weakening of the heart because of the avitaminosis (mainly lack of thiamine) that causes the beriberi syndrome. The weakening of the heart has decreased the blood flow to the kidneys. Therefore, the kidneys have retained a large amount of fluid, which in turn has increased the mean systemic filling pressure (represented by the point where the venous return curve intersects with the zero cardiac output level) from the normal value of 7 mm Hg up to 11 mm Hg. This has shifted the venous return curve to the right. Finally, the venous return curve has rotated upward from the normal value because the avitaminosis has dilated the peripheral blood vessels, as explained in Chapter 17.

The two dashed curves intersect with each other at *point C,* which describes the circulatory condition in beriberi, with a right atrial pressure in this instance of 9 mm Hg and a cardiac output about 65 per cent above normal; this high cardiac output occurs despite the weak heart, as demonstrated by the depressed plateau level of the cardiac output curve.

sation of the heart failure has been "compensated." To state this another way, the final steady-state condition of the circulation is defined by the crossing point of three curves: the cardiac output curve, the venous return curve, and the critical level for normal fluid balance. The compensatory mechanisms automatically stabilize the circulation when all three curves cross at the same point.

Graphical Analysis of High-Output Cardiac Failure

Figure 22–8 gives an analysis of two types of high-output cardiac failure. One of these is caused by an arteriovenous fistula that overloads the heart because of excessive venous return, even though the pumping capability of the heart is not depressed. The other is caused by beriberi, in which the venous return is greatly increased because of diminished systemic vascular resistance, but at the same time, the pumping capability of the heart is depressed.

ARTERIOVENOUS FISTULA. The heavy black curves of the figure depict the normal cardiac output and normal venous return curves. These equate with each other at *point A,* which depicts a normal cardiac output of 5 liters/min and a normal right atrial pressure of 0 mm Hg.

Now let us assume that the systemic resistance (the total peripheral resistance) becomes greatly decreased because of opening a large arteriovenous fistula (a direct opening between a large artery and a large vein). The venous return curve rotates upward to give the curve labeled "A-V fistula." This venous return curve equates with the cardiac output curve at *point B,* with a cardiac output of 12.5 liters/min and a right atrial pressure of 3 mm Hg. Thus, the cardiac output has become greatly elevated, the right atrial pressure is slightly elevated, and there are mild signs of peripheral congestion. If the person attempts to exercise, he or she will have little cardiac reserve because the heart is already being used almost to maximum capacity to pump the extra blood through the arteriovenous fistula. This condition resembles a failure condition and is called "high-output failure," but in reality, the heart is overloaded by excess venous return.

REFERENCES

Baumgartner, W. A., et al.: Heart and Heart-Lung Transplantation. Philadelphia, W. B. Saunders Co., 1990.
Braunwald, E.: Heart Disease: A Textbook for Cardiovascular Medicine. Philadelphia, W. B. Saunders Co., 1992.
Braunwald, E., et al.: Mechanism of Contractility of the Normal and Failing Heart. Boston, Little, Brown, 1968.
Civetta, J. M.: Critical Care. Philadelphia, J. B. Lippincott, 1992.
Crawford, M. H.: Current Diagnosis and Treatment in Cardiology. Redding, MA, Appleton and Lange, 1994.
Delgado, J. N., and Reemers, W. A.: Wilson and Gisvold's Textbook of Organic Medicinal and Pharmaceutical Chemistry. Philadelphia, J. B. Lippincott, 1991.
Fallon, J. T.: Cardiovascular Pathophysiology: A Problem-Oriented Approach. Philadelphia, J. B. Lippincott, 1994.
Fein, F. S., and Sonnenblick, E. H.: Diabetic cardiomyopathy. Prog. Cardiovasc. Dis., 27:255, 1985.
Flye, M. W.: Atlas of Organ Transplantation. Philadelphia, W. B. Saunders Co., 1994.
Folkow, B., and Svanborg, A.: Physiology of cardiovascular aging. Physiol. Rev., 73:725, 1993.
Francis, G. S.: Neurohumoral mechanisms involved in congestive heart failure. Am. J. Cardiol., 55:15A, 1985.
Gilbert, I., and Henning, R. J.: Adenocarcinoma of the lung presenting with pericardial tamponade: Report of a case and review of the literature. Heart-Lung, 14:83, 1985.
Gimbrone, Jr., M. A., and Schoen, F. J.: Cardiovascular Pathology: Clinicopathologic Correlations and Pathogenetic Mechanisms. Baltimore, Williams & Wilkins, 1994.
Goldberger, E.: Treatment of Cardiac Emergencies. St. Louis, C. V. Mosby Co., 1989.
Guyton, A. C.: The systemic venous system in cardiac failure. J. Chronic Dis., 9:465, 1959.
Guyton, A. C., et al.: Cardiac Output and Its Regulation. Philadelphia, W. B. Saunders Co., 1973.
Kellermann, J. J., and Braunwald, E.: Silent Myocardial Ischemia: A Critical Appraisal. Farmington, CT, S. Karger Publishers, Inc., 1990.
Kelly, W. N.: Essentials of Internal Medicine. 2nd Ed. Philadelphia, J. B. Lippincott, 1994.
Lakatta, E. G.: Cardiovascular regulatory mechanisms in advanced age. Physiol. Rev., 73:197, 1993.
McManus: Cardiovascular Complications of Exercise. Chicago, Year Book Medical Publishers, 1985.
Miller, A. J.: Diagnosis of Chest Pain. New York, Raven Press, 1988.
Nagano, M., et al.: The Cardiomyopathic Heart. New York, Raven Press, 1994.
Orrego, F.: Calcium and the mechanism of action of digitalis. Gen. Pharmacol., 15:273, 1984.
Parmley, W. W., and Chatterjee, K.: Cardiology: Clinical Text in Three Looseleaf Volumes. Philadelphia, J. B. Lippincott, 1994.

Rakel, R. E.: Conn's Current Therapy. Philadelphia, W. B. Saunders Co., 1994.

Rubin, E., and Farber, J. L.: Pathology. Philadelphia, J. B. Lippincott, 1994.

Samuel, P., et al.: The role of diet in the etiology and treatment of atherosclerosis. Annu. Rev. Med., 34:179, 1983.

Schlant, R. C., et al.: Hurst's The Heart. Hightstown, NJ, McGraw-Hill, 1994.

Shoemaker, W. C., et al.: Society of Critical Care Medicine, The Textbook of Critical Care. 2nd Ed. Philadelphia, W. B. Saunders Co., 1988.

Tintinalli, J. E., et al.: Emergency Medicine: A Comprehensive Study Guide. 2nd Ed. New York, McGraw-Hill, 1988.

Wasserman, K., et al.: Principles of Exercise Testing and Interpretation. Philadelphia, Lea & Febiger, 1987.

Wyngaarden, J. B., et al.: Cecil Textbook of Medicine. Philadelphia, W. B. Saunders Co., 1992.

(See also Chapter 21.)

Heart Sounds; Dynamics of Valvular and Congenital Heart Defects

CHAPTER 23

HEART SOUNDS

The function of the heart valves is discussed in Chapter 9, wherein it is pointed out that *closure* of the valves is associated with audible sounds. No sounds occur when the valves open. The purpose of the first section of this chapter is to discuss the factors that cause the sounds in the heart under normal and abnormal conditions.

Normal Heart Sounds

Listening with a stethoscope to a normal heart, one hears a sound usually described as "lub, dub, lub, dub . . ." The "lub" is associated with closure of the atrioventricular (A-V) valves at the beginning of systole and the "dub," with closure of the semilunar (aortic and pulmonary) valves at the end of systole. The "lub" sound is called the *first heart sound* and the "dub," the *second heart sound* because the normal cycle of the heart is considered to start with the beginning of systole when the A-V valves close.

CAUSES OF THE FIRST AND SECOND HEART SOUNDS. The earliest explanation for the cause of the heart sounds was that the slapping together of the valve leaflets sets up vibrations, but this has been shown to cause little, if any, of the sound because of the cushioning effect of the blood. Instead, the cause is *vibration of the taut valves immediately after closure* along with *vibration of the adjacent blood, walls of the heart, and major vessels around the heart.* That is, in generating the first heart sound, contraction of the ventricles causes sudden backflow of blood against the A-V valves (the tricuspid and mitral valves), causing them to bulge toward the atria until the chordae tendineae abruptly stop the backbulging. The elastic tautness of the valves then causes the backsurging blood to bounce forward again into each respective ventricle. This sets the blood and the ventricular walls as well as the taut valves into vibration and causes vibrating turbulence in the blood. The vibrations then travel through the adjacent tissues to the chest wall, where they can be heard as sound by the stethoscope.

The second heart sound results from sudden closure of the semilunar valves. When the semilunar valves close, they bulge backward toward the ventricles and their elastic stretch recoils the blood back into the arteries, which causes a short period of reverberation of blood back and forth between the walls of the arteries and the semilunar valves as well as between the valves and the ventricular walls. The vibrations set up in the arterial walls are then transmitted along the arteries. When the vibrations of the vessels or ventricles come into contact with a "sounding board," such as the chest wall, they create sound that can be heard.

DURATION AND PITCH OF THE FIRST AND SECOND HEART SOUNDS. The duration of each of the heart sounds is slightly more than 0.10 second—the first sound about 0.14 second and the second about 0.11 second. The reason for the shorter second sound is that the semilunar valves are more taut than the A-V valves, so that they vibrate for a shorter period than do the A-V valves.

The audible range of frequencies (pitch) in the first and second heart sounds, as shown in Figure 23–1, begins at the lowest frequency the ear can detect and

275

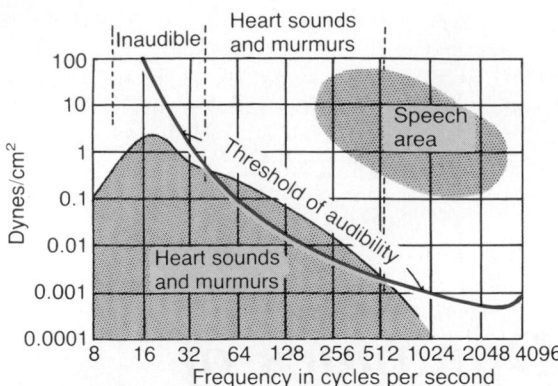

Figure 23–1. Amplitude of different frequency vibrations in the heart sounds and heart murmurs in relation to the threshold of audibility, showing that the range of sounds that can be heard is between 40 and 500 cycles/sec. (Modified from Butterworth, Chassin, and McGrath: Cardiac Auscultation. New York, Grune and Stratton.)

goes up to about 500 cycles/sec. When special electronic apparatus is used to record these sounds, by far the largest proportion of the recorded sound is at frequencies below the audible range, going down to 3 to 4 cycles/sec and peaking at about 20 cycles/sec. For this reason, some heart sounds can be recorded electronically in phonocardiograms when they cannot be heard with a stethoscope.

The second heart sound normally has a higher frequency than the first heart sound for two reasons: (1) the tautness of the semilunar valves in comparison with much less taut A-V valves and (2) the greater elasticity coefficient of the arteries that provide the principal vibrating chambers for the second sound in comparison with the much looser ventricular chambers that provide the vibrating system of the first heart sound. The clinician uses these sound differences to distinguish special characteristics of the two respective sounds.

THIRD HEART SOUND. Occasionally a weak, rumbling third heart sound is heard at the beginning of the *middle third of diastole.* A logical but unproved explanation of this sound is oscillation of blood back and forth between the walls of the ventricles initiated by inrushing blood from the atria. This is analogous to running water from a faucet into a sack, the inrushing water reverberating back and forth between the walls of the sack to cause vibrations in the walls. The reason the third heart sound does not occur until the middle third of diastole is believed to be that in the early part of diastole, the heart is not filled sufficiently to create even the small amount of elastic tension in the ventricles necessary for reverberation. The frequency of this sound is usually so low that the ear cannot hear the sound; yet it can often be recorded in the phonocardiogram.

ATRIAL HEART SOUND (FOURTH HEART SOUND). An atrial heart sound can be recorded in some people in the phonocardiogram, but it can almost never be heard with a stethoscope because of its low frequency—

usually 20 cycles/sec or less. This sound occurs when the atria contract, and presumably, it is caused by inrush of blood into the ventricles, which initiates vibrations similar to those of the third heart sound.

Areas for Auscultation of Normal Heart Sounds

Listening to the sounds of the body, usually with the aid of a stethoscope, is called *auscultation.* Figure 23–2 shows the areas of the chest wall from which the different valvular sounds can best be distinguished. The sounds from all the valves can be heard from all these areas, although the sound from the designated valve area is as loud for that particular valve, *relative to the other valve sounds,* as it ever will be. The cardiologist distinguishes the sounds from the different valves by a process of elimination; that is, he or she moves the stethoscope from one area to another, noting the loudness of the sounds in different areas and gradually picking out the sound components from each valve.

The areas for listening to the different heart sounds are not directly over the valves themselves. The aortic area is upward along the aorta because of sound transmission up the aorta, the pulmonic area is upward along the pulmonary artery, the tricuspid area is over the right ventricle, and the mitral area is over the apex of the heart, which is the portion of the left ventricle nearest the surface of the chest because the heart is rotated so that most of the left ventricle lies hidden behind the right ventricle. In other words, the sounds caused by the A-V valves are transmitted to the chest wall through each respective ventricle, and the sounds from the semilunar valves are transmitted especially along the great vessels leading from the heart.

Phonocardiogram

If a microphone specially designed to detect low-frequency sound is placed on the chest, the heart sounds can be amplified and recorded by a high-speed recording apparatus. The recording is called a *phono-*

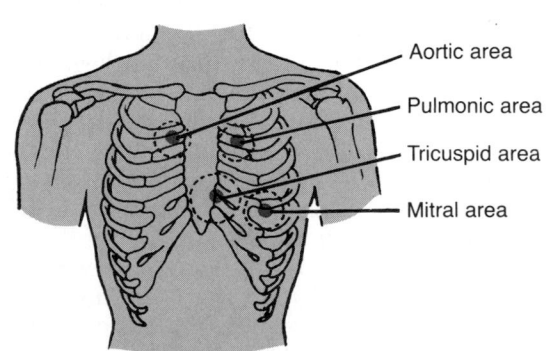

Figure 23–2. Chest areas from which each valve sound is best heard.

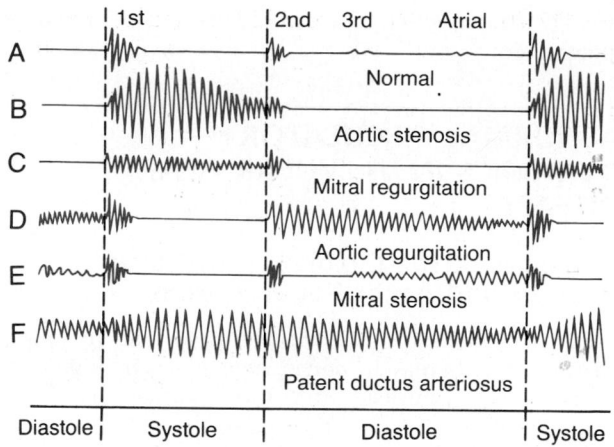

Figure 23-3. Phonocardiograms from normal and abnormal hearts.

cardiogram, and the heart sounds appear as waves, as shown schematically in Figure 23-3. Record A is a recording of normal heart sounds, showing the vibrations of the first, second, and third heart sounds and even the atrial sound. Note specifically that the third and atrial heart sounds are each a very low rumble. The third heart sound can be recorded in only one third to one half of all people, and the atrial heart sound can be recorded in perhaps one fourth of all people.

Valvular Lesions

Rheumatic Valvular Lesions

By far the greatest number of valvular lesions results from rheumatic fever. Rheumatic fever is an autoimmune disease in which the heart valves are likely to be damaged or destroyed. It is usually initiated by streptococcal toxin in the following manner.

The sequence of events almost always begins with a preliminary streptococcal infection caused specifically by group A hemolytic streptococci, such as a sore throat, scarlet fever, or middle ear infection. The streptococci release several different proteins against which antibodies are formed, the most important of which seems to be a protein called the "M" antigen. The antibodies then react not only with this M antigen but also against many different tissues of the body, often causing severe immunological damage. These reactions continue to take place as long as the antibodies persist in the blood—6 months or more.

Rheumatic fever causes damage in many parts of the body but especially in certain susceptible areas, such as the heart valves. The degree of heart valve damage is directly correlated with the titer and the persistence of these antibodies. The principles of immunity that relate to this type of reaction are discussed in Chapter

34, and it is noted in Chapter 31 also that acute glomerular nephritis has a similar basis.

In rheumatic fever, large hemorrhagic, fibrinous, bulbous lesions grow along the inflamed edges of the heart valves. Because the mitral valve receives more trauma during valvular action than any of the other valves, it is the one most often seriously damaged, and the aortic valve is second most frequently damaged. The right heart valves, the tricuspid and pulmonary valves, are usually affected much less severely, probably because the stresses that act on these valves are slight compared with those that act on the left heart valves.

SCARRING OF THE VALVES. The lesions of acute rheumatic fever frequently occur on adjacent valve leaflets simultaneously, so that the edges of the leaflets become stuck together. Then, weeks, months, or years later, the lesions become scar tissue, permanently fusing portions of adjacent valve leaflets. Also, the free edges of the leaflets, which are normally filmy and free-flapping, become solid, scarred masses.

A valve in which the leaflets adhere to one another so extensively that blood cannot flow through satisfactorily is said to be *stenosed*. On the other hand, when the valve edges are so destroyed by scar tissue that they cannot close as the ventricles contract, *regurgitation,* or backflow, of blood occurs when the valve should be closed. Stenosis usually does not occur without the coexistence of at least some degree of regurgitation, and vice versa.

OTHER CAUSES OF VALVULAR LESIONS. Stenosis or lack of one or more leaflets of a valve frequently occurs as a congenital defect. Complete lack of leaflets is rare, although congenital stenosis is common, as is discussed later in this chapter.

Heart Murmurs Caused by Valvular Lesions

As shown by the phonocardiograms in Figure 23-3, many abnormal heart sounds, known as "heart murmurs," occur when there are abnormalities of the valves, as follows.

MURMUR OF AORTIC STENOSIS. In aortic stenosis, blood is ejected from the left ventricle through only a small opening of the aortic valve. Because of the resistance to ejection, sometimes the pressure in the left ventricle rises as high as 300 mm Hg, whereas the pressure in the aorta is still normal. Thus, a nozzle effect is created *during systole,* with blood jetting at tremendous velocity through the small opening of the valve. This causes *severe turbulence* of the blood in the root of the aorta. The turbulent blood impinging against the aortic walls causes intense vibration, and a loud murmur is transmitted throughout the upper aorta and even into the larger arteries of the neck. This sound is harsh and in severe stenosis occasionally is so loud that it can be heard several feet away from the patient. Also, the sound vibrations can often be felt with the hand on the upper chest and lower neck, a phenomenon known as a "thrill."

MURMUR OF AORTIC REGURGITATION. In aortic regurgitation, no sound is heard during systole, but *during diastole,* blood flows backward from the aorta into the left ventricle, causing a "blowing" murmur of relatively high pitch and with a swishing quality heard maximally over the left ventricle. This murmur results from *turbulence* of blood jetting backward into the blood already in the left ventricle.

MURMUR OF MITRAL REGURGITATION. In mitral regurgitation, blood flows backward through the mitral valve into the left atrium *during systole.* This also causes a high-frequency "blowing," swishing sound similar to that of aortic regurgitation, transmitted most strongly into the left atrium. However, the left atrium is so deep within the chest that it is difficult to hear this sound directly over the atrium. As a result, the sound of mitral regurgitation is transmitted to the chest wall mainly through the left ventricle instead, and it is usually best heard at the apex of the heart.

MURMUR OF MITRAL STENOSIS. In mitral stenosis, blood passes with difficulty through the stenosed mitral valve from the left atrium into the left ventricle, and because the pressure in the left atrium seldom rises above 30 mm Hg except for short periods, a great pressure differential forcing blood from the left atrium into the left ventricle never develops. Consequently, the abnormal sounds heard in mitral stenosis are usually weak and of very low frequency, so that most of the sound spectrum is below the low-frequency end of human hearing.

During the early part of diastole, the ventricle has so little blood in it and its walls are so flabby that blood does not reverberate back and forth between the walls of the ventricle. For this reason, even in severe mitral stenosis, no murmur might be heard during the first third of diastole. Then, after the first third, the ventricle is stretched enough for blood to reverberate, and a low rumbling murmur often begins. In mild stenosis, the murmur lasts only during the first half of the middle third of diastole, but in severe stenosis, it can begin earlier and persist for the remainder of diastole. Because of the extremely low frequency of the sound vibrations, one can sometimes feel these, called a "thrill," over the apex of the heart even when the sound cannot be heard through the stethoscope.

PHONOCARDIOGRAMS OF VALVULAR MURMURS. Phonocardiograms B, C, D, and E of Figure 23–3 show, respectively, idealized records obtained from patients with aortic stenosis, mitral regurgitation, aortic regurgitation, and mitral stenosis. It is obvious from these phonocardiograms that the aortic stenotic lesion causes the loudest of all the murmurs and the mitral stenotic lesion, the weakest. The phonocardiograms show how the intensity of the murmurs varies during different portions of systole and diastole, and the relative timing of each murmur is also evident. Note especially that the murmurs of aortic stenosis and mitral regurgitation occur only during systole, whereas the murmurs of aortic regurgitation and mitral stenosis occur only during diastole. If the reader does not understand this timing, a moment's pause should be taken until it is understood.

ABNORMAL CIRCULATORY DYNAMICS IN VALVULAR HEART DISEASE

Dynamics of the Circulation in Aortic Stenosis and Aortic Regurgitation

In *aortic stenosis,* the left ventricle fails to empty adequately, whereas in *aortic regurgitation,* blood returns to the ventricle after the ventricle has been emptied. Therefore, in either case, the *net stroke volume output* of the heart is reduced.

Several important compensations take place that can ameliorate the severity of the circulatory defects. Some of these compensations are the following.

HYPERTROPHY OF THE LEFT VENTRICLE. In both aortic stenosis and aortic regurgitation, the left ventricular musculature hypertrophies because of the increased ventricular workload. In regurgitation, the ventricular chamber also enlarges to hold all the regurgitant blood. Sometimes the left ventricle muscle mass increases fourfold to fivefold, creating a tremendously large left side of the heart. When the aortic valve is seriously stenosed, this hypertrophied muscle allows the left ventricle to develop as much as 400 mm Hg intraventricular pressure during occasional periods of peak activity; even at rest, the pressure differential across the stenotic valve is sometimes 100 mm Hg.

In severe aortic regurgitation, sometimes the hypertrophied muscle allows the left ventricle to pump a stroke volume output as high as 250 milliliters, although as much as three fourths of this blood returns to the ventricle during diastole.

INCREASE IN BLOOD VOLUME. Another effect that helps to compensate for the diminished net pumping by the left ventricle is increased blood volume. This results from an initial slight decrease in arterial pressure plus peripheral circulatory reflexes that this decrease induces, both of which diminish renal output of urine until the blood volume increases and the mean arterial pressure returns to normal. Also, red cell mass eventually increases because of a slight degree of tissue hypoxia.

The increase in blood volume tends to increase venous return to the heart. This in turn increases the ventricular end-diastolic volume and end-diastolic pressure, causing the left ventricle to pump with the extra power required to overcome the abnormal pumping dynamics.

Eventual Failure of the Left Ventricle, and Development of Pulmonary Edema

In the early stages of aortic stenosis or aortic regurgitation, the intrinsic ability of the left ventricle to adapt to increasing loads prevents significant abnormalities in circulatory function in the person at rest other than increased work output required of the left ventricle. Therefore, very considerable degrees of aortic stenosis or aortic regurgitation often occur before the person knows that he or she has serious heart disease, such as a resting left ventricular systolic pressure as high as 200 mm Hg in aortic stenosis or a left ventricular stroke volume output as high as double normal in aortic regurgitation.

Beyond critical stages in these two aortic valve lesions, the left ventricle finally cannot keep up with the work demand. As a consequence, the left ventricle dilates and cardiac output begins to fall; blood simultaneously dams up in the left atrium and lungs behind the failing left ventricle. The left atrial pressure rises progressively, and at pressures above 30 to 40 mm Hg mean left atrial pressure, edema appears in the lungs, as discussed in detail in Chapter 38.

MYOCARDIAL ISCHEMIA IN AORTIC VALVULAR DISEASE. Because of the high intraventricular pressure during systole in aortic stenosis, little blood flows through the coronary arteries during systole. Therefore, extra blood flow is required during diastole to make up the difference. The hypertrophied muscle of the left ventricle also frequently has a relatively deficient coronary vasculature. Finally, the intraventricular pressure sometimes remains high during diastole, thereby compressing the inner layers of the heart and diminishing coronary blood flow. For all these reasons, the patient frequently experiences severe degrees of coronary ischemia and angina.

In aortic regurgitation, the problem is further compounded because the diastolic aortic pressure frequently falls very low as the aortic blood regurgitates back into the ventricle. Because most of the left ventricular coronary blood flow occurs during diastole, this low pressure can be particularly detrimental to the flow. The effect is especially serious for the subendocardial myocardium in which almost zero flow occurs during systole. Therefore, again, coronary ischemia occurs, with concomitant anginal pain and even death of the subendocardial cardiac muscle fibers.

Dynamics of Mitral Stenosis and Mitral Regurgitation

In mitral stenosis, blood flow from the left atrium into the left ventricle is impeded, and in mitral regurgitation, much of the blood that has flowed into the left ventricle during diastole leaks back into the left atrium during systole rather than being pumped into the aorta. Both of these conditions reduce the net movement of blood from the left atrium into the left ventricle.

PULMONARY EDEMA IN MITRAL VALVULAR DISEASE. The buildup of blood in the left atrium causes progressive increase in left atrial pressure, and this eventually can result in the development of serious pulmonary edema. Ordinarily, lethal edema does not occur until the mean left atrial pressure rises above 30 mm Hg and sometimes to as high as 40 mm Hg because the lung lymphatic vasculature enlarges manyfold and can carry fluid away from the lung tissues extremely rapidly.

ENLARGED LEFT ATRIUM AND ATRIAL FIBRILLATION. The high left atrial pressure also causes progressive enlargement of the left atrium, which increases the distance that the cardiac impulse must travel in the atrial wall. This pathway eventually becomes so long that it predisposes to the development of circus movements. Therefore, in late stages of mitral valvular disease, especially mitral stenosis, atrial fibrillation usually occurs. This further reduces the pumping effectiveness of the heart and, therefore, causes still further cardiac debility.

COMPENSATIONS IN MITRAL VALVULAR DISEASE. As also occurs in aortic valvular disease and in many types of congenital heart disease, the blood volume increases in mitral valvular disease, principally because of diminished excretion of fluid by the kidneys. This increases venous return to the heart, thereby helping to overcome the effect of the cardiac debility. Cardiac output does not fall more than minimally until the late stages of mitral valvular disease, even though the left atrial pressure is rising.

As the left atrial and pulmonary capillary pressures rise, blood begins to dam up in the lungs, and the incipient edema of the lungs causes intense pulmonary arteriolar constriction; these two effects together then increase systolic pulmonary arterial pressure sometimes to as high as 60 mm Hg. This in turn causes hypertrophy of the right side of the heart, which partially compensates for its increased workload.

Circulatory Dynamics During Exercise in Patients with Valvular Lesions

During exercise, large quantities of venous blood are returned to the heart from the peripheral circulation. Therefore, all the dynamic abnormalities that occur in the different types of valvular heart disease become tremendously exacerbated. Even in mild valvular heart disease, in which the symptoms may be unrecognizable at rest, severe symptoms often develop during heavy exercise. For instance, in patients with aortic valvular lesions, exercise can cause acute left ventricular failure followed by *acute pulmonary edema*. Also, in patients with mitral disease, exercise can cause so much damming of blood in the lungs that serious or even lethal pulmonary edema can ensue within 10 minutes or more.

Even in some mild to moderate cases of valvular disease, the patient's *cardiac reserve* diminishes in proportion to the severity of the valvular dysfunction. That is, the cardiac output does not increase as much as it should during exercise. Therefore, the muscles of the body fatigue rapidly.

ABNORMAL CIRCULATORY DYNAMICS IN CONGENITAL HEART DEFECTS

Occasionally, the heart or its associated blood vessels are malformed during fetal life; the defect is called a *congenital anomaly.* There are three major types of congenital anomalies of the heart and its associated vessels: (1) *stenosis* of the channel of blood flow at some point in the heart or in a closely allied major vessel, (2) an abnormality that allows blood to flow backward from the left side of the heart or aorta to the right side of the heart or pulmonary artery, thus failing to flow through the systemic circulation—this is called a *left-to-right shunt,* and (3) an abnormality that allows blood to flow directly from the right side of the heart into the left side of the heart, thus failing to flow through the lungs—this is called a *right-to-left shunt.*

The effects of the different stenotic lesions are easily understood. For instance, *congenital aortic valve stenosis* causes the same dynamic effects as aortic valve stenosis caused by valvular lesions. Another type of congenital stenosis is *coarctation of the aorta,* which causes the arterial pressure in the upper part of the body (above the level of the coarctation) usually to be about 55 per cent greater than the pressure in the lower body because of the great resistance to blood flow to the lower body, as discussed in Chapter 19.

Now, let us consider two other common types of congenital heart defects: *patent ductus arteriosus,* a left-to-right shunt, and *tetralogy of Fallot,* a right-to-left shunt.

Patent Ductus Arteriosus— A Left-to-Right Shunt

During fetal life, the lungs are collapsed and the elastic compression of the lungs that keeps the alveoli collapsed also keeps the blood vessels collapsed. Therefore, the resistance to blood flow through the lungs is so great that the pulmonary arterial pressure is high in the fetus. Because of low resistance to blood flow from the aorta through the large vessels of the placenta, the pressure in the aorta is lower than in the pulmonary artery. This causes almost all the pulmonary arterial blood to flow through a special artery present in the fetus that connects the pulmonary artery with the aorta (Fig. 23–4), called the *ductus arteriosus,* thus bypassing the lungs. This allows immediate recirculation of the blood through the systemic arteries of the fetus without the blood going through the lungs. This lack of blood flow through the lungs is not detrimental to the fetus because the blood is oxygenated by the placenta of the mother.

CLOSURE OF THE DUCTUS AFTER BIRTH. As soon as a baby is born, the lungs inflate; not only do the alveoli fill with air but also the resistance to blood flow through the pulmonary vascular tree decreases tremendously, allowing pulmonary arterial pressure to fall. Simultaneously, the aortic pressure rises because of sudden cessation of blood flow through the placenta. Thus, the pressure in the pulmonary artery falls, whereas that in the aorta rises. As a result, forward blood flow through the ductus ceases suddenly at birth, and, in fact, blood begins to flow backward through the ductus from the aorta into the pulmonary artery. This new state of backward blood flow causes the ductus arteriosus to become occluded within a few hours to a few days in most babies, so that blood flow through the ductus does not persist. The ductus is believed to close because the aortic blood now flowing through the ductus has about twice as high an oxygen concentration as the pulmonary artery blood that had been flowing in the ductus during fetal life. The oxygen in turn constricts the ductus muscle. This is discussed further in Chapter 83.

In at least some babies with persistent ductus, the failure of the ductus to close seems to result from excessive dilation caused by prostaglandins in the ductus wall. The administration of indomethacin, which blocks the prostaglandin dilation effect, causes the ductus to close in many of these cases.

In about 1 of every 5500 babies, the ductus never closes, causing the condition known as *patent ductus arteriosus,* which is shown in Figure 23–4.

DYNAMICS OF THE CIRCULATION WITH A PERSISTENT PATENT DUCTUS. During the early months of an infant's life, a patent ductus usually does not cause severely abnormal function. As the child grows older, the differential between the pressure in the aorta and that in the pulmonary artery progressively increases, with corresponding increase in the backward flow of blood from the aorta to the pulmonary artery. Also, the diameter of the partially closed ductus often increases with time, making the condition worse.

Recirculation Through the Lungs. In the older child with a patent ductus, one half to two thirds of the aortic blood flows backward through the ductus into the pulmonary artery, then through the lungs, and finally back into the left ventricle, passing through the lungs and left side of the heart two or more times for every one time that it passes through the systemic circulation.

These people do not show cyanosis until late in life, when the heart fails or the lungs become congested. Indeed, early in life, the arterial blood is often better oxygenated than normally because of the extra times of passage through the lungs. Yet, because of the tremendous accessory flow of blood around and around through the lungs and left side of the heart, the output of the left ventricle in patent ductus arteriosus is often two to three times normal.

Diminished Cardiac and Respiratory Reserve. The major effects of patent ductus arteriosus on the patient are low cardiac and respiratory reserve. The left ventricle is already pumping about two or more times the normal cardiac output, and the maximum that it can possibly pump even after hypertrophy of the heart has occurred is about four to seven times normal. Therefore, during exercise, the net blood flow through the remainder of the body can never increase

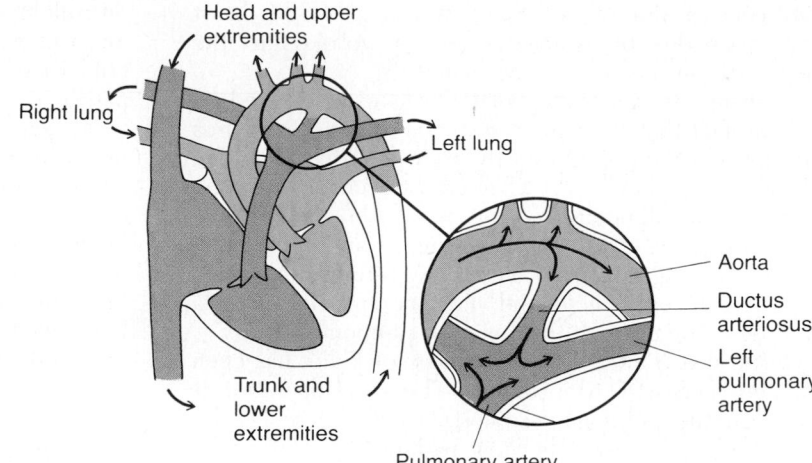

Figure 23–4. Patent ductus arteriosus, showing the degree of blood oxygenation in the different parts of the circulation.

to the levels required for strenuous activity. With even moderately strenuous exercise, the person is likely to become weak and may even faint from momentary heart failure.

Also, the high pressures in the pulmonary vessels caused by excess flow through the lungs often lead to pulmonary congestion and pulmonary edema.

As a result of the increased load on the heart and especially because the pulmonary congestion becomes progressively more severe with age, most patients with uncorrected patent ductus die between ages 20 and 40 years.

MACHINERY MURMUR. In the infant with patent ductus arteriosus, occasionally no abnormal heart sounds are heard because the quantity of reversed blood flow may be insufficient. As the baby grows older, reaching ages 1 to 3 years, a harsh, blowing murmur begins to be heard in the pulmonic area of the chest. This sound is much more intense during systole when the aortic pressure is high and much less intense during diastole when the aortic pressure falls very low, so that the murmur waxes and wanes with each beat of the heart, creating the so-called "machinery murmur." The idealized phonocardiogram of this murmur is shown in Figure 23–3F.

SURGICAL TREATMENT. Surgical treatment of patent ductus arteriosus is extremely simple; all one needs to do is to ligate the patent ductus or divide it and sew the two ends.

Tetralogy of Fallot—A Right-to-Left Shunt

Tetralogy of Fallot is shown in Figure 23–5. It is the most common cause of the "blue baby" in which most of the blood fails to flow through the lungs, and therefore, the aortic blood is mainly still unoxygenated venous blood. In this condition, four abnormalities of the heart occur simultaneously.

1. The aorta originates from the right ventricle rather than the left, or it overrides a hole in the septum, as shown in the figure, receiving blood from both ventricles.

2. The pulmonary artery is stenosed, so that much less than normal amounts of blood pass from the right ventricle into the lungs; instead, the blood passes into the aorta.

3. Blood from the left ventricle flows either through a ventricular septal hole into the right ventricle and then into the aorta or directly into the aorta, overriding this hole.

4. Because the right side of the heart must pump large quantities of blood against the high pressure in the aorta, its musculature is highly developed, causing an enlarged right ventricle.

ABNORMAL CIRCULATORY DYNAMICS. It is readily apparent that the major physiological difficulty caused by tetralogy of Fallot is the shunting of blood past the lungs without its becoming oxygenated. As much as 75

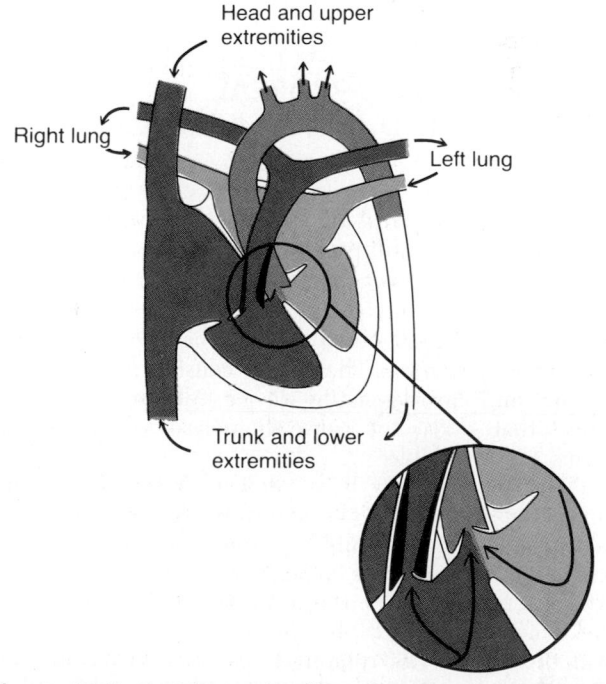

Figure 23–5. Tetralogy of Fallot, showing the degree of blood oxygenation in the different parts of the circulation.

per cent of the venous blood returning to the heart may pass directly from the right ventricle into the aorta without becoming oxygenated.

A diagnosis of tetralogy of Fallot is usually based on (1) the fact that the baby is *cyanotic* (blue), (2) records of high systolic pressure in the right ventricle recorded through a catheter, (3) characteristic changes in the radiological silhouette of the heart showing an enlarged right ventricle, and (4) angiograms (x-ray pictures) showing abnormal blood flow through the interventricular septal hole and into the overriding aorta and less flow through the stenosed pulmonary artery.

SURGICAL TREATMENT. Tetralogy of Fallot has been treated successfully by surgery. The usual operation is to open the pulmonary stenosis, close the septal defect, and reconstruct the flow pathway into the aorta. When surgery is successful, the average life expectancy increases from only 3 to 4 years to 50 or more years.

Causes of Congenital Anomalies

One of the most common causes of congenital heart defects is a virus infection in the mother during the first trimester of pregnancy when the fetal heart is being formed. Defects are particularly prone to develop when the mother contracts German measles at this time—so often indeed that obstetricians advise termination of pregnancy if German measles occurs in the first trimester. Some congenital defects of the heart are hereditary because the same defect has been known to occur in identical twins as well as in succeeding generations. Children of patients surgically treated for congenital heart disease have about 10 times as much chance of having congenital heart disease as do other children. Congenital defects of the heart are also frequently associated with other congenital defects of the body.

USE OF EXTRACORPOREAL CIRCULATION DURING CARDIAC SURGERY

It is almost impossible to repair intracardiac defects while the heart is still pumping. Therefore, many types of artificial *heart-lung machines* have been developed to take the place of the heart and lungs during the course of operation. Such a system is called an *extracorporeal circulation*. The system consists principally of a pump and an oxygenating device. Almost any type of pump that does not cause hemolysis of the blood seems to be suitable.

The principles that have been used for oxygenating blood are (1) bubbling oxygen through the blood and then removing the bubbles from the blood before passing it back into the patient, (2) dripping the blood downward over the surfaces of large areas of plastic sheet in the presence of oxygen, (3) passing the blood over the surfaces of rotating discs, and (4) passing the blood between thin membranes or through thin tubes that are porous to oxygen and carbon dioxide.

The different systems have been fraught with many difficulties, including hemolysis of the blood, development of small clots in the blood, likelihood of small bubbles of oxygen or small emboli of antifoam agent passing into the arteries of the patient, necessity for large quantities of blood to prime the entire system, failure to exchange adequate quantities of oxygen, and necessity to use heparin in the system to prevent blood coagulation, the heparin also interfering with adequate hemostasis during the surgical procedure. Yet, despite these difficulties, in the hands of experts, patients can be kept on artificial heart-lung machines for many hours while operations are performed on the inside of the heart.

HYPERTROPHY OF THE HEART IN VALVULAR AND CONGENITAL HEART DISEASE

Hypertrophy of cardiac muscle is one of the most important mechanisms by which the heart adapts to increased workloads, whether these loads be caused by increased pressure against which the heart muscle must contract or by increased blood volume that must be pumped. Some investigators believe that the increased tension in the muscle causes the hypertrophy; others believe that the increased metabolic rate of the muscle is the primary stimulus. Regardless of which of these is correct, one can calculate about how much hypertrophy will occur in each chamber of the heart by multiplying ventricular output times the pressure against which the ventricle must work, with emphasis on the pressure. Thus, hypertrophy occurs in most types of valvular and congenital disease, sometimes causing heart weights as great as 800 grams instead of the normal 300 grams.

DETRIMENTAL EFFECTS OF THE LATE STAGES OF HYPERTROPHY. Although physiological hypertrophy of heart muscle is usually beneficial to cardiac function, sometimes extreme degrees of hypertrophy lead to failure. One of the reasons for this is that the coronary vasculature typically does not increase to the same extent as the mass of muscle. A second reason is that fibrosis often develops in the muscle, especially in the subendocardial muscle where the coronary blood flow is poor anyway, with fibrous tissue replacing degenerating muscle fibers. Because of the sometimes disproportionate increase in muscle mass relative to coronary flow, relative ischemia may develop as the muscle hypertrophies, and coronary insufficiency easily ensues. Therefore, anginal pain is a frequent accompaniment of many valvular and congenital heart diseases. Also, heart failure ensues easily and usually enters a vicious circle of rapid progression and death once subendocardial fibrosis begins.

REFERENCES

Bodnar, E., and Frater, R.: Replacement Cardiac Valves. Hightstown, NJ, McGraw-Hill, 1992.

Braunwald, E.: Heart Disease: A Textbook for Cardiovascular Medicine. Philadelphia, W. B. Saunders Co., 1992.

Castaneda, A. R., et al.: Cardiac Surgery of the Neonate and Infant. Philadelphia, W. B. Saunders Co., 1994.

Cernaianu, A. C., and DelRossi, A. J.: Cardiac Surgery. New York, Plenum Publishing Corp., 1994.

DePace, N. L., et al.: Acute severe mitral regurgitation. Pathophysiology, clinical recognition, and management. Am. J. Med., 78:293, 1985.

Dickinson, C. J.: The aetiology of clubbing and hypertrophic osteoarthropathy. Eur. J. Clin. Invest., 23:330, 1993.

Dietz, H. C. III, and Pyeritz, R. E.: Molecular genetic approaches to the study of human cardiovascular disease. Annu. Rev. Physiol., 56:763, 1994.

Dunn, J. M.: Cardiac Valve Disease in Children. New York, Elsevier Science Publishing Co., 1988.

Edmunds, L. H., Jr., et al.: Atlas of Cardiothoracic Surgery. Philadelphia, Lea & Febiger, 1989.

Emmanouilides, G. C., et al.: Moss & Adams' Heart Disease in Infants, Children, and Adolescents. Baltimore, Williams & Wilkins Co., 1994.

Erickson, B.: Heart Sounds and Murmurs: A Practical Guide. St. Louis, C. V. Mosby Co., 1987.

Fallon, J. T.: Cardiovascular Pathophysiology. Philadelphia, J. B. Lippincott, 1994.

Feigenbaum, H.: Echocardiography. Baltimore, Williams & Wilkins, 1994.

Fishman, M. C., et al.: Medicine. Philadelphia, J. B. Lippincott, 1991.

Gravlee, G. P., et al.: Cardiopulmonary Bypass: Principles and Practice. Baltimore, Williams & Wilkins, 1993.

Grossman, W. (ed.): Cardiac Catheterization and Angiography. 3rd ed. Philadelphia, Lea & Febiger, 1986.

Hallman, G. L., et al.: Surgical Treatment of Congenital Heart Disease. 3rd Ed. Philadelphia, Lea & Febiger, 1987.

Higgins, C. B., et al.: Congenital Heart Disease: Echocardiography and Magnetic Resonance Imaging. New York, Raven Press, 1990.

Hurst, J. W., and Alpert, J. S.: Diagnostic Atlas of the Heart. New York, Raven Press, 1994.

Kirklin, J. W., and Barratt-Boyes, B. G.: Cardiac Surgery, 2nd Ed. New York, Churchill Livingstone, 1993.

Lake, C. L.: Clinical Monitoring for Anesthesia and Critical Care. Philadelphia, W. B. Saunders Co., 1994.

Long, W. A.: Fetal and Neonatal Cardiology. Philadelphia, W. B. Saunders Co., 1990.

Lyons, K. P.: Cardiovascular Nuclear Medicine. East Norwalk, Conn., Appleton & Lange, 1988.

Marcus, M. L., et al.: Cardiac Imaging. Philadelphia, W. B. Saunders Co., 1991.

Montgomery, W. H., and Atkins, J. A.: Decision Making in Emergency Cardiology. St. Louis, C. V. Mosby Co., 1989.

Nabel, E. G., and Nabel, G. J.: Complex models for the study of gene function in cardiovascular biology. Annu. Rev. Physiol., 56:741, 1994.

Parmley, W. W., et al.: Cardiology: Clinical Text in Three Looseleaf Volumes. Philadelphia, J. B. Lippincott, 1995.

Perloff, J. K.: The Clinical Recognition of Congenital Heart Disease. Philadelphia, W. B. Saunders Co., 1994.

Roubin, G. S., et al.: Interventional Cardiovascular Medicine. New York, Churchill Livingstone, 1994.

Rutherford, R. B.: Vascular Surgery. 3rd Ed. Philadelphia, W. B. Saunders Co., 1989.

Stark, J., and de Leval, M.: Surgery for Congenital Heart Defects. Philadelphia, W. B. Saunders Co., 1993.

Taussig, H.: Congenital Malformations of the Heart Vol. 1: General Considerations. 2nd Ed. Vol. 2: Specific Malformations. 2nd Ed. Cambridge, Mass., Harvard University Press, 1960.

Topol, E. J., Textbook of Interventional Cardiology. Philadelphia, W. B. Saunders Co., 1994.

Weyman, A. E.: Principles and Practice of Echocardiography. Baltimore, Williams & Wilkins, 1993.

(See also Chapters 21 and 22.)

Circulatory Shock and Physiology of Its Treatment

CHAPTER 24

Circulatory shock means generalized inadequacy of blood flow throughout the body to the extent that the body tissues are damaged because of too little flow, especially too little delivery of oxygen and other nutrients to the tissue cells. Even the cardiovascular system itself—the heart musculature, walls of the blood vessels, vasomotor system, and other circulatory parts—begins to deteriorate, so that the shock becomes progressively worse.

PHYSIOLOGICAL CAUSES OF SHOCK

Circulatory Shock Caused by Decreased Cardiac Output

Shock usually results from inadequate cardiac output. Therefore, any factor that reduces the cardiac output will probably lead to circulatory shock. Two types of factors can severely reduce the cardiac output.

1. *Cardiac abnormalities that decrease the ability of the heart to pump blood.* They include especially (1) *myocardial infarction* but also (2) *toxic states of the heart*, (3) *severe heart valve dysfunction*, (4) heart *arrhythmias*, and other conditions. The circulatory shock that results from diminished cardiac pumping ability is called *cardiogenic shock*. This is discussed in detail in Chapter 22, where it is pointed out that about 85 per cent of people who develop cardiogenic shock do not survive.

2. *Factors that decrease the venous return.* The most common cause of decreased venous return is (1) *diminished blood volume*, but venous return also can be reduced as a result of (2) decreased *vascular tone*, especially of the venous blood reservoirs, or (3) *obstruction to blood flow* at some point in the circulation, especially in the venous return pathway to the heart.

Circulatory Shock That Occurs Without Diminished Cardiac Output

Occasionally, the cardiac output is normal or even greater than normal and yet the person is in circulatory shock. This can result from (1) *excessive metabolism of the body, so that even a normal cardiac output is inadequate,* or (2) *abnormal tissue perfusion patterns, so that most of the cardiac output is passing through blood vessels besides those that are supplying the local tissues with nutrition.* These conditions are present most frequently in the type of shock called *septic shock,* which is also called "blood poisoning."

The specific causes of shock are discussed later in the chapter. For the time being, it is important to note that all of them lead to *inadequate delivery of nutrients to all the tissues or at least to critical tissues and inadequate removal of cellular waste products from the tissues.*

What Happens to the Arterial Pressure in Circulatory Shock?

In the minds of many physicians, the arterial pressure level is the principal measure of the adequacy of circulatory function. However, the arterial pressure often can be seriously misleading because many times a person may be in severe shock and still have almost a normal pressure because powerful nervous reflexes keep the pressure from falling. At other times the arterial pressure can fall to one-half normal, but the person still has normal tissue perfusion and is not in shock.

In most types of shock, especially that caused by severe blood loss, the arterial blood pressure does decrease at the same time the cardiac output decreases, although usually not as much as the decrease in output.

Tissue Deterioration Is the End Stage of Circulatory Shock, Whatever the Cause

Once circulatory shock reaches a critical state of severity, regardless of its initiating cause, *the shock itself breeds more shock.* That is, the inadequate blood flow causes the circulatory system to begin to deteriorate. This in turn causes even more decrease in cardiac output, and a vicious circle ensues, with progressively increasing circulatory shock, still less adequate tissue perfusion, still more shock, and so forth until death. It is with this late stage of the circulatory shock that we are especially concerned because appropriate physiological treatment can often reverse the rapid slide to oblivion.

Stages of Shock

Because the characteristics of circulatory shock change at different degrees of severity, shock is divided into the following three major stages:

1. A *nonprogressive stage* (sometimes called the *compensated stage*), from which the normal circulatory compensatory mechanisms will eventually cause full recovery without help from outside therapy
2. A *progressive stage,* in which the shock becomes steadily worse until death
3. An *irreversible stage,* in which the shock has progressed to such an extent that all forms of known therapy are inadequate to save the person's life, even though, for the moment, the person is still alive.

Now, let us discuss the stages of circulatory shock caused by decreased blood volume, which illustrate the basic principles. Then we can consider the special characteristics of other initiating causes of shock.

SHOCK CAUSED BY HYPOVOLEMIA— HEMORRHAGIC SHOCK

Hypovolemia means diminished blood volume. Hemorrhage is the most common cause of hypovolemic shock.

Hemorrhage *decreases the filling pressure of the circulation* and, as a consequence, decreases venous return. As a result, the cardiac output falls below normal and shock ensues. All degrees of shock can result from hemorrhage, from the mildest diminishment of cardiac output to almost complete cessation of output, depending on the amount of blood loss.

Relationship of Bleeding Volume to Cardiac Output and Arterial Pressure

Figure 24–1 shows the approximate effects on both cardiac output and arterial pressure of removing blood from the circulatory system over a period of about 30 minutes. About 10 per cent of the total blood volume can be removed with almost no effect on either arterial pressure or cardiac output, but greater blood loss usually diminishes the cardiac output first and later the pressure, both of these falling to zero when about 35 to 45 per cent of the total blood volume has been removed.

SYMPATHETIC REFLEX COMPENSATIONS IN SHOCK— THEIR SPECIAL VALUE TO MAINTAIN ARTERIAL PRESSURE. The decrease in arterial pressure and also decreases in pressures in the low-pressure areas of the thorax after hemor-

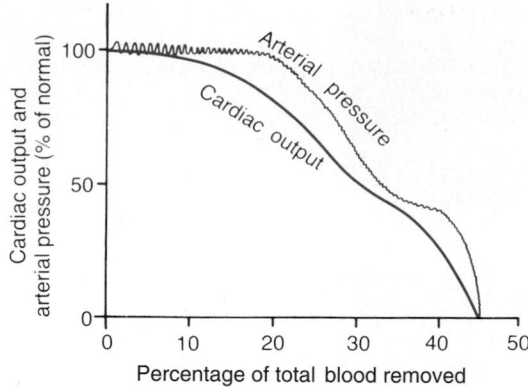

Figure 24–1. Effect of hemorrhage on cardiac output and arterial pressure.

rhage initiate powerful sympathetic reflexes (initiated mainly by the baroreceptors and the low-pressure vascular stretch receptors). These reflexes stimulate the sympathetic vasoconstrictor system throughout the body, resulting in three important effects: (1) The arterioles constrict in most parts of the body, thereby increasing the total peripheral resistance. (2) The veins and venous reservoirs constrict, thereby helping to maintain adequate venous return despite diminished blood volume. (3) Heart activity increases markedly, sometimes increasing the heart rate from the normal value of 72 beats/min to 170 to 200 beats/min.

Value of the Reflexes. In the absence of the sympathetic reflexes, only 15 to 20 per cent of the blood volume can be removed over a period of 30 minutes before a person dies; this is in contrast to a 30 to 40 per cent loss of blood volume that a person can sustain when the reflexes are intact. Therefore, the reflexes extend the amount of blood loss that can occur without causing death to about twice that which is possible in their absence.

Greater Effect of the Reflexes in Maintaining Arterial Pressure Than in Maintaining Cardiac Output. Referring again to Figure 24–1, note that the arterial pressure is maintained at or near normal levels in the hemorrhaging person longer than is the cardiac output. The reason for this is that the sympathetic reflexes are geared more for maintaining arterial pressure than for maintaining output. They increase the arterial pressure mainly by increasing the total peripheral resistance, which has no beneficial effect on cardiac output; however, the sympathetic constriction of the veins is important to keep venous return and cardiac output from falling too much, in addition to their role in maintaining the arterial pressure.

Especially interesting is the second plateau in the arterial pressure curve of Figure 24–1. This results from activation of the central nervous system ischemic response, which causes extreme stimulation of the sympathetic nervous system, as discussed in Chapter 18, when the arterial pressure falls below 50 mm Hg. This effect of the central nervous system ischemic response can be called the "last-ditch stand" of the sympathetic reflexes in their attempt to keep the arterial pressure from falling too low.

Protection of Coronary and Cerebral Blood Flow by the Reflexes. A special value of the maintenance of normal arterial pressure even in the face of decreasing cardiac output is protection of blood flow through the coronary and cerebral circulatory systems. The sympathetic stimulation does not cause significant constriction of either the cerebral or the cardiac vessels. In addition, in both these vascular beds, local autoregulation is excellent, which prevents moderate decreases in arterial pressure from significantly affecting their blood flows. Therefore, blood flow through the heart and brain is maintained essentially at normal levels as long as the arterial pressure does not fall below about 70 mm Hg, despite the fact that blood flow in many other areas of the body might be decreased to as little as one-quarter normal by this time because of vasospasm.

Nonprogressive and Progressive Hemorrhagic Shock

Figure 24–2 shows an experiment that we performed in dogs to demonstrate the effects of different degrees of hemorrhage on the subsequent course of arterial pressure. The dogs were bled rapidly until their arterial pressures fell to different levels. The dogs whose pressures fell immediately to no lower than 45 mm Hg (groups I, II, and III) all eventually recovered; the recovery occurred rapidly if the pressure fell only slightly (group I) but occurred slowly if it fell almost to the 45-mm Hg level (group III). When the arterial pressure fell below 45 mm Hg (groups IV, V, and VI), all the dogs died, although many of them hovered between life and death for many hours before the circulatory system began to deteriorate.

This experiment demonstrates that the circulatory system can recover as long as the degree of hemorrhage is no greater than a certain critical amount. Crossing this critical amount by even a few milliliters of blood loss makes the eventual difference between life and death. Thus, hemorrhage beyond a certain critical level causes shock to become *progressive.* That is, *the shock itself causes still more shock,* the condition becoming a vicious circle that leads eventually to deterioration of the circulation and to death.

Nonprogressive Shock—Compensated Shock

If shock is not severe enough to cause its own progression, the person eventually recovers. Therefore, shock of this lesser degree is called *nonprogressive shock.* It is also called *compensated shock,* meaning that the sympathetic reflexes and other factors compensate enough to prevent further deterioration of the circulation.

The factors that cause a person to recover from moderate degrees of shock are all the negative feedback control mechanisms of the circulation that attempt to return cardiac output and arterial pressure back to normal levels. They include the following:

1. *Baroreceptor reflexes,* which elicit powerful sympathetic stimulation of the circulation
2. *Central nervous system ischemic response,* which elicits even more powerful sympathetic stimulation throughout the body but is not activated significantly until the arterial pressure falls below 50 mm Hg
3. *Reverse stress-relaxation of the circulatory system,* which causes the blood vessels to contract down around the diminished blood volume, so that the blood volume that is available will more adequately fill the circulation
4. *Formation of angiotensin,* which constricts the peripheral arteries and causes increased conservation of water and salt by the kidneys, both of which help prevent progression of the shock
5. *Formation of vasopressin (antidiuretic hormone),* which constricts the peripheral arteries and veins and greatly increases water retention by the kidneys
6. *Compensatory mechanisms that return the blood volume back toward normal,* including absorption of large quantities of fluid from the intestinal tract, absorption of fluid into the blood capillaries from the interstitial spaces of the body, conservation of water and salt by the kidneys, and increased thirst and increased appetite for salt, which make the person drink water and eat salty foods if able

The sympathetic reflexes provide immediate help toward bringing about recovery because they become maximally activated within 30 seconds after hemorrhage. The angiotensin and vasopressin mechanisms as well as the reverse stress-relaxation that causes contraction of the blood vessels and venous reservoirs all require 10 minutes to 1 hour to respond completely, but nevertheless, they aid greatly in increasing the arterial pressure or increasing the circulatory filling pressure and thereby increasing the return of blood to the heart. Finally, the readjustment of blood volume by absorption of fluid from the interstitial spaces and the intestinal tract as well as the ingestion and absorption of additional quantities of fluid and salt may require from 1 to 48 hours, but recovery eventually takes place, provided the shock does not become severe enough to enter the progressive stage.

"Progressive Shock" Is Caused by a Vicious Circle of Cardiovascular Deterioration

Once shock has become severe enough, the structures of the circulatory system begin to deteriorate and various types of positive feedback develop that can cause a vicious circle of progressively decreasing cardiac output. This is called "progressive shock." Figure 24–3 shows some of these positive feedbacks that further depress the cardiac output in shock. Some of the more important of these feedbacks are the following.

CARDIAC DEPRESSION. When the arterial pressure falls low enough, *coronary blood flow decreases below that required for adequate nutrition of the myocardium.* This weakens the heart and, thereby, decreases the cardiac output still more. Thus, a positive feedback cycle has developed whereby the shock becomes more and more severe.

Figure 24–4 shows cardiac output curves extrapolated to the human heart from experiments in dogs, demonstrating progressive deterioration of the heart at different times after the onset of shock. A dog was bled until the arterial pressure fell to 30 mm Hg, and the pressure was held at this level by further bleeding or retransfusion of blood as required. Note that there was little deterioration of the heart during the first 2 hours, but by 4 hours, the heart had deteriorated about 40

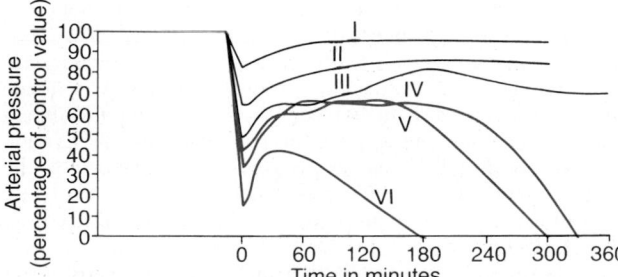

Figure 24–2. Course of arterial pressure in dogs after different degrees of acute hemorrhage. Each curve represents the average results from six dogs.

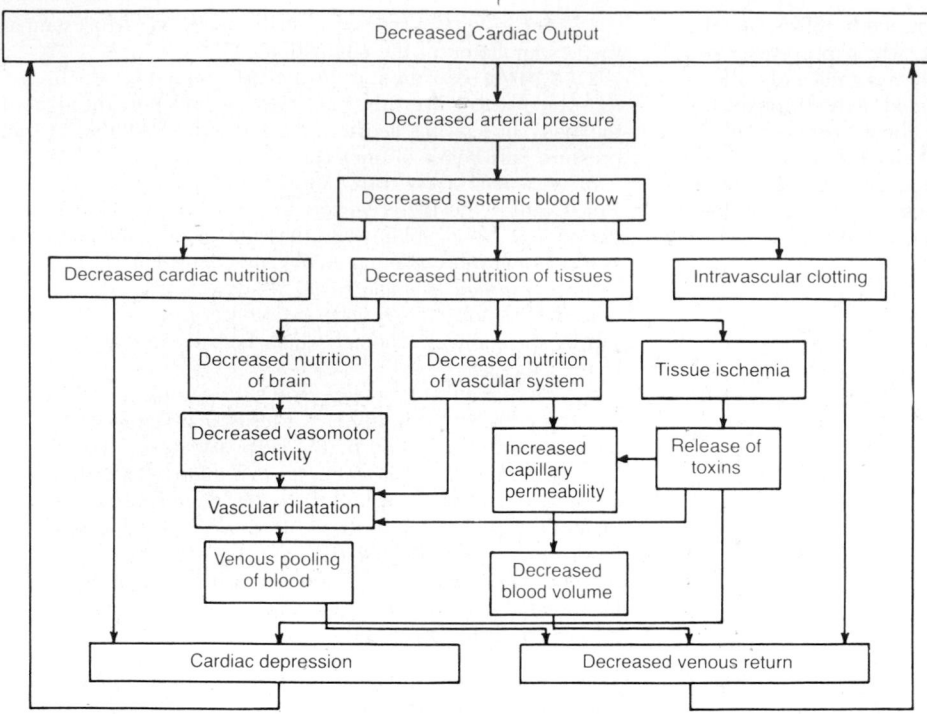

Figure 24–3. Different types of feedback that can lead to progression of shock.

per cent; then, rapidly, during the last hour of the experiment, the heart deteriorated almost completely.

Thus, one of the important features of progressive shock, whether it be hemorrhagic in origin or caused in any other way, is eventual progressive deterioration of the heart. In the early stages of shock, this plays a very little role in the condition of the person, partly because deterioration of the heart is not severe during the 1st hour or so of shock but mainly because the heart has tremendous reserve capability that makes it normally capable of pumping 300 to 400 per

cent more blood than is required by the body for adequate nutrition. In the latest stages of shock, however, deterioration of the heart is probably the most important factor in the progression of the shock.

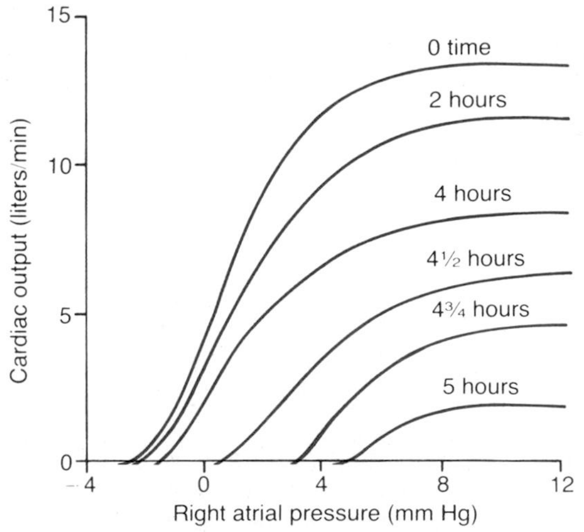

VASOMOTOR FAILURE. In the early stages of shock, various circulatory reflexes cause intense activity of the sympathetic nervous system. This, as discussed previously, helps delay depression of the cardiac output and especially helps prevent decreased arterial pressure. However, there comes a point at which diminished blood flow to the vasomotor center so depresses the center that it becomes progressively less active and finally totally inactive. For instance, complete circulatory arrest to the brain causes, during the first 4 to 8 minutes, the most intense of all sympathetic discharges, but by the end of 10 to 15 minutes, the vasomotor center becomes so depressed that no further evidence of sympathetic discharge can be demonstrated. The vasomotor center usually does not fail in the early stages of shock—only in the late stages.

BLOCKAGE OF THE MINUTE VESSELS—SLUDGED BLOOD. In time, blockage occurs in many of the minute vessels in the circulatory system, and this causes the shock to progress. The initiating cause of this blockage is sluggish blood flow in the microvessels. Because tissue metabolism continues despite the low flow, large amounts of acid, both carbonic acid and lactic acid, continue to empty into the local blood vessels and increase greatly the local acidity of the blood. This acid, plus other deterioration products from the ischemic tissues, causes local blood agglutination or minute blood clots, leading to minute plugs in the small vessels. Even if the vessels do not become plugged, the tendency for the cells to stick to one another makes it more difficult for blood to flow through the microvasculature, giving rise to the term *sludged blood.*

INCREASED CAPILLARY PERMEABILITY. After many hours of capillary hypoxia and lack of other nutrients, the permeability of the capillaries gradually increases and large quanti-

Figure 24–4. Cardiac output curves of the heart at different times after hemorrhagic shock begins. (These curves are extrapolated to the human heart from data obtained in dog experiments by Dr. J. W. Crowell.)

ties of fluid begin to transude into the tissues. This further decreases the blood volume with resultant further decrease in cardiac output, making the shock still more severe.

Capillary hypoxia does not cause increased capillary permeability until the late stages of prolonged shock.

RELEASE OF TOXINS BY ISCHEMIC TISSUES. Throughout the history of research in the field of shock, it has been suggested that shock causes tissues to release toxic substances, such as histamine, serotonin, and tissue enzymes, that cause further deterioration of the circulatory system. Quantitative studies have especially proved the significance of at least one toxin, *endotoxin*, in many types of shock.

Endotoxin. Endotoxin is released from the bodies of dead gram-negative bacteria in the intestines. Diminished blood flow to the intestines causes enhanced formation and absorption of this toxic substance. The circulating toxin then causes greatly increased cellular metabolism, despite the inadequate nutrition of the cells, and has a specific effect to cause cardiac depression. Although endotoxin can play a major role in some types of shock, especially septic shock, discussed later in the chapter, how much endotoxin is released during hemorrhagic shock and whether or not it is an especially important factor in the progression of this type of shock are unclear.

GENERALIZED CELLULAR DETERIORATION. As shock becomes severe, many signs of generalized cellular deterioration occur throughout the body. One organ especially affected is the *liver*. This occurs mainly because of lack of enough nutrients to support the normally high rate of metabolism in liver cells but also partly because of the extreme vascular exposure of the liver cells to any toxic or other abnormal metabolic factor in shock.

Among the damaging cellular effects that are known to occur in most body tissues are the following:

1. Active transport of sodium and potassium through the cell membrane is greatly diminished. As a result, sodium and chloride accumulate in the cells and potassium is lost from the cells. In addition, the cells begin to swell.
2. Mitochondrial activity in the liver cells as well as in many other tissues of the body becomes severely depressed.
3. Lysosomes begin to split open in widespread tissue areas, with intracellular release of hydrolases that cause further intracellular deterioration.
4. Cellular metabolism of nutrients, such as glucose, eventually becomes greatly depressed in the last stages of shock. The activities of some hormones are depressed as well, including as much as 200-fold depression in the action of insulin.

All these effects contribute to further deterioration of many organs of the body, including especially (1) the *liver*, with depression of its many metabolic and detoxification functions; (2) the *lungs*, with eventual development of pulmonary edema and poor ability to oxygenate the blood; and (3) the *heart*, thereby further depressing the contractility of the heart.

Tissue Necrosis in Severe Shock—Patchy Areas of Necrosis Occur Because of Patchy Blood Flows in Different Organs. Not all cells of the body are equally damaged by shock because some tissues have better blood supplies than others. For instance, the cells adjacent to the arterial ends of capillaries receive better nutrition than the cells adjacent to the venous ends of the same capillaries. Therefore, one would expect more nutritive deficiency around the venous ends of capillaries than elsewhere. This is

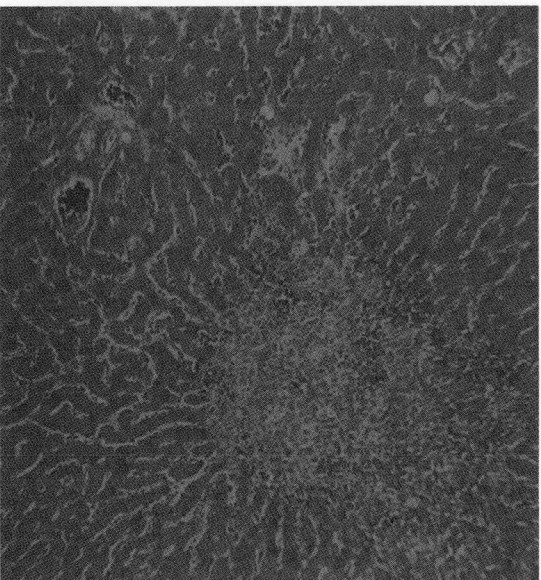

Figure 24–5. Necrosis of the central portion of a liver lobule in severe shock. (Courtesy of Dr. J. W. Crowell.)

precisely the effect that Crowell found in studying tissue areas in many parts of the body. For instance, Figure 24–5 shows necrosis in the center of a liver lobule, the portion of the lobule that is last to be bathed by the blood as it passes through the liver sinusoids.

Similar punctate lesions occur in heart muscle, although here a definite repetitive pattern, such as occurs in the liver, cannot be demonstrated. Nevertheless, the cardiac lesions probably play an important role in leading to the final irreversible stage of shock. Deteriorative lesions also occur in the kidneys, especially in the epithelium of the kidney tubules, leading to kidney failure and subsequent uremic death several days later. Deterioration of the lungs also often leads to respiratory distress and death several days later—called the *shock lung syndrome.*

ACIDOSIS IN SHOCK. Most of the metabolic derangements that occur in shocked tissue can lead to acidosis. Especially important is the poor delivery of oxygen to the tissues, which greatly diminishes oxidative metabolism of the foodstuffs. When this occurs, the cells obtain their energy by the anaerobic process of glycolysis, which leads to tremendous quantities of *excess lactic acid* in the blood. In addition, the poor blood flow through the tissues prevents normal removal of carbon dioxide; the carbon dioxide reacts locally in the cells with water to form high concentrations of intracellular carbonic acid; this in turn reacts with the various tissue buffers to form still other intracellular acidic substances.

Thus, another deteriorative effect of shock is both generalized and local tissue acidosis, leading to still further progression of the shock itself.

Positive Feedback Deterioration of Tissues in Shock and the Vicious Circle of Progressive Shock

All the factors just discussed that can lead to further progression of shock are types of *positive feedback.* That is, each increase in the degree of shock causes a further increase in the shock.

However, positive feedback does not necessarily lead to a vicious circle. Whether or not a vicious circle develops depends on the intensity of the positive feedback. In mild degrees of shock, the negative feedback mechanisms—the sympathetic reflexes, reverse stress-relaxation mechanism of the blood reservoirs, absorption of fluid into the blood from the interstitial spaces, and others—can easily overcome the positive feedback influences and, therefore, cause recovery. In severe degrees of shock, the positive feedback mechanisms become more and more powerful, thus leading to such rapid deterioration of the circulation that all the negative feedback systems of circulatory control acting together cannot return the cardiac output to normal.

Considering once again the principles of positive feedback and vicious circles discussed in Chapter 1, one can readily understand why there is a critical cardiac output level above which a person in shock recovers and below which the person enters a vicious circle of circulatory deterioration and death.

Irreversible Shock

After shock has progressed to a certain stage, transfusion or any other type of therapy becomes incapable of saving the person's life. The person is then said to be in the *irreversible stage of shock*. Ironically, even in this irreversible stage, therapy can, on rare occasions, still return the arterial pressure and even the cardiac output to normal or near normal for short periods, but the circulatory system nevertheless continues to deteriorate, and death ensues in another few minutes to few hours.

Figure 24–6 demonstrates this effect, showing that transfusion during the irreversible stage can sometimes cause the cardiac output (as well as the arterial pressure) to return to normal. However, the cardiac output soon begins to fall again, and subsequent transfusions have less and less effect. Something has changed in the cells of the heart, other tissues, or both that may not necessarily affect the *immediate* ability of the heart to pump blood but over a long period does depress this ability and results in death. Now the question remains: What factor or factors lead to the eventual total deterioration of circulatory function?

The answer to this question seems to be, simply, that beyond a certain point, so much tissue damage has occurred, so many destructive enzymes have been released into the body fluids, so much acidosis has developed, and so many other destructive factors are now in progress that even a normal cardiac output cannot reverse the continuing deterioration. Therefore, in severe shock, a stage is eventually

reached beyond which the person is destined to die even though vigorous therapy can still return the cardiac output to normal for short periods.

DEPLETION OF CELLULAR HIGH-ENERGY PHOSPHATE RESERVES IN IRREVERSIBLE SHOCK. The high-energy phosphate reserves in the tissues of the body, especially in the liver and the heart, are greatly diminished in severe degrees of shock. Essentially all the creatine phosphate is degraded, and almost all the adenosine triphosphate has been degraded to adenosine diphosphate, adenosine monophosphate, and, eventually, adenosine. Then much of this adenosine diffuses out of the cells into the circulating blood and is converted into uric acid, a substance that cannot re-enter the cells to reconstitute the adenosine phosphate system. New adenosine can be synthesized at a rate of only about 2 per cent of the normal cellular amount an hour, meaning that once depleted, the high-energy phosphate stores of the cells are difficult to replenish. One of the most devastating end results of deterioration in shock, and one that is perhaps the most significant in the development of the final state of irreversibility, is this cellular depletion of the high-energy compounds.

Hypovolemic Shock Caused by Plasma Loss

Loss of plasma from the circulatory system, even without the loss of whole blood, can be severe enough to reduce the total blood volume markedly, in this way causing typical hypovolemic shock similar in almost all details to that caused by hemorrhage. Severe plasma loss occurs in the following conditions.

1. *Intestinal obstruction* is often a cause of severely reduced plasma volume. The distention of the intestine in obstruction partly blocks venous blood flow in the intestinal circulation, which increases intestinal capillary pressure. This in turn causes fluid to leak from the capillaries into the intestinal walls and intestinal lumen. Also, the lost fluid has a high content of protein, thereby reducing the total plasma protein as well as the plasma volume.

2. In patients who have *severe burns* or other denuding conditions of the skin, so much plasma almost always is lost through the denuded areas that the plasma volume becomes markedly reduced.

The hypovolemic shock that results from plasma loss has almost the same characteristics as the shock caused by hemorrhage, except for one additional complicating factor—the blood viscosity increases greatly as a result of plasma loss, and this exacerbates the sluggishness of blood flow.

Loss of fluid from all fluid compartments of the body is called *dehydration;* this, too, can reduce the blood volume and cause hypovolemic shock similar to that resulting from hemorrhage. Some of the causes of this type of shock are (1) excessive sweating, (2) fluid loss in severe diarrhea or vomiting, (3) excess loss of fluid by nephrotic kidneys, (4) inadequate intake of fluid and electrolytes, and (5) destruction of the adrenal cortices, with consequent failure of the kidneys to reabsorb sodium, chloride, and water because of the absence of the hormone aldosterone.

Hypovolemic Shock Caused by Trauma

One of the most common causes of circulatory shock is trauma to the body. Often the shock results simply from hemorrhage caused by the trauma, but it can also occur even

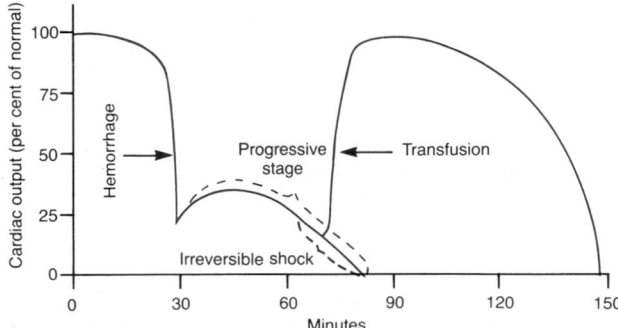

Figure 24–6. Failure of transfusion to prevent death in irreversible shock.

without hemorrhage because contusion of the body can damage the capillaries sufficiently to allow excessive loss of plasma into the tissues. This results in greatly reduced plasma volume with resultant hypovolemic shock.

The *pain* associated with serious trauma can be an additional aggravating factor in traumatic shock because pain sometimes seems to inhibit the vasomotor center, thereby inhibiting sympathetic signals to the circulation. This in turn increases the vascular capacitance and reduces the venous return. Various attempts have also been made to implicate toxic factors released by the traumatized tissues as one of the causes of shock after trauma. However, cross-transfusion experiments have failed to show significant toxic elements.

In summary, traumatic shock seems to result mainly from hypovolemia, although there may also be a moderate degree of concomitant neurogenic shock caused by the pain.

NEUROGENIC SHOCK—INCREASED VASCULAR CAPACITY

Shock occasionally results without any loss of blood volume. Instead, the *vascular capacity* increases so much that even the normal amount of blood becomes incapable of adequately filling the circulatory system. One of the major causes of this is *sudden loss of vasomotor tone* throughout the body, causing especially massive dilatation of the veins. The resulting condition is known as *neurogenic shock*.

The relation of vascular capacity to blood volume is discussed in Chapter 15, where it is pointed out that either an increase in vascular capacity or a decrease in blood volume *reduces the mean systemic filling pressure,* which in turn reduces the venous return to the heart. Diminished venous return caused by vascular dilatation is called "venous pooling" of blood.

CAUSES OF NEUROGENIC SHOCK. Some of the factors that can cause loss of vasomotor tone include the following:

1. *Deep general anesthesia* often depresses the vasomotor center enough to cause vasomotor collapse, with resulting neurogenic shock.
2. *Spinal anesthesia*, especially when this extends all the way up the spinal cord, blocks the sympathetic outflow from the nervous system and can be a potent cause of neurogenic shock.
3. *Brain damage* is often a cause of vasomotor collapse. Many patients who have had brain concussion or contusion of the basal regions of the brain develop profound neurogenic shock. Also, even though the first few minutes of medullary ischemia cause extreme vasomotor activity, prolonged ischemia (lasting longer than 5 to 10 minutes) can cause inactivation of the vasomotor neurons and development of severe neurogenic shock.

VASOVAGAL SYNCOPE—EMOTIONAL FAINTING. The circulatory collapse that results from "emotional" fainting usually is not caused by vasomotor failure but by strong emotional activation of parasympathetic signals to slow the heart and also activation of inverse sympathetic signals to dilate the peripheral vasculature, thereby decreasing cardiac output and arterial pressure. The fainting that results from an emotional disturbance is called *vasovagal syncope* to differentiate it from other types of fainting that result from other neurogenic causes of reduced cardiac output.

ANAPHYLACTIC SHOCK

"Anaphylaxis" is an allergic condition in which the cardiac output and arterial pressure often fall drastically. This is discussed in Chapter 34. It results primarily from an antigen–antibody reaction that takes place immediately after an antigen to which the person is sensitive has entered the circulation. One of the principal effects is to cause the *basophils* in the blood and *mast* cells in the pericapillary tissues to release *histamine* or a *histamine-like substance*. The histamine in turn causes (1) an increase in vascular capacity because of venous dilatation, (2) dilatation of the arterioles with resultant greatly reduced arterial pressure, and (3) greatly increased capillary permeability with rapid loss of fluid and protein into the tissue spaces. The net effect is a great reduction in venous return and often such serious shock that the person dies within minutes.

Furthermore, intravenous injection of large amounts of histamine causes "histamine shock," which has characteristics almost identical to those of anaphylactic shock, although usually less severe.

SEPTIC SHOCK

A condition that was formerly known by the popular name of "blood poisoning" is now called by most clinicians *septic shock*. This simply means widely disseminated bacterial infection to many areas of the body, with the infection being borne through the blood from one tissue to another and causing extensive damage. There are many varieties of septic shock because of the many types of bacterial infection that can cause it and because infection in one part of the body produces different effects from those caused by infection elsewhere in the body.

Septic shock is extremely important to the clinician because this type of shock, more frequently than any other kind of shock besides cardiogenic shock, causes patient death in the modern hospital.

Some of the typical causes of septic shock include the following:

1. Peritonitis caused by spread of infection from the uterus and fallopian tubes, frequently resulting from instrumental abortion performed under unsterile conditions
2. Peritonitis resulting from rupture of the gut, sometimes caused by intestinal disease and sometimes by wounds
3. Generalized infection resulting from spread of a simple skin infection, such as streptococcal or staphylococcal infection
4. Generalized gangrenous infection resulting specifically from gas gangrene bacilli, spreading first through the tissues themselves and finally by way of the blood to the internal organs, especially the liver
5. Infection spreading into the blood from the kidney or urinary tract, often caused by colon bacilli

SPECIAL FEATURES OF SEPTIC SHOCK. Because of the multiple types of septic shock, it is difficult to categorize this condition. Some features are as follows:

1. High fever
2. Often marked vasodilation throughout the body, especially in the infected tissues
3. High cardiac output in perhaps one half of patients caused by vasodilation in the infected tissues and by high metabolic rate and vasodilation elsewhere in the body, re-

sulting from bacterial toxin stimulation of cellular metabolism and from the high body temperature

4. Sludging of the blood, presumably caused by red cell agglutination in response to degenerating tissues

5. Development of microclots in widespread areas of the body, a condition called *disseminated intravascular coagulation*. Also, this causes the clotting factors to be used up so that hemorrhages occur into many other tissues, especially the gut wall and the intestinal tract.

In the early stages of septic shock, the patient usually does not have signs of circulatory collapse but only signs of the bacterial infection. As the infection becomes more severe, the circulatory system usually becomes involved either directly or as a secondary result of toxins from the bacteria. *There finally comes a point at which deterioration of the circulation becomes progressive in the same way that progression occurs in all other types of shock. The end stages of septic shock are not greatly different from the end stages of hemorrhagic shock,* even though the initiating factors are markedly different in the two conditions.

ENDOTOXIN SHOCK. A special type of septic shock is endotoxin shock. It frequently occurs when a large segment of the gut becomes strangulated and loses most of its blood supply. The gut rapidly becomes gangrenous, and the bacteria in the gut multiply rapidly. Most of these bacteria are so-called gram-negative bacteria, mainly colon bacilli, that contain a toxin called *endotoxin*. Another condition that also frequently causes colon bacilli septicemia is extension of urinary tract infections into the blood.

On entering the circulation, endotoxin causes an effect similar to that of anaphylaxis, often resulting in severe shock. Further compounding the circulatory depression is a direct effect of endotoxin on the heart to decrease myocardial contractility.

STILL OTHER EFFECTS OF SHOCK ON THE BODY

DECREASED TISSUE METABOLISM AND CELLULAR DETERIORATION IN HYPOVOLEMIC SHOCK. In hypovolemic shock, the decreased cardiac output reduces the delivery of oxygen and other nutrients to the tissues. This in turn reduces the metabolism of virtually all cells of the body, leading to multiple types of cellular damage, including (1) decreased ability of the mitochondria to synthesize adenosine triphosphate, (2) decreased ability of the cell's membrane pump to keep the sodium concentration low and the potassium concentration high inside the cell, (3) depressed processing of nutrients by the cell's metabolic machinery, and (4) eventual rupture of many lysosomes, releasing digestive enyzmes within the cells that cause intracellular destruction and even death of cells. All these effects lead to still greater cellular deterioration and further decrease in total bodily metabolism. Most people can continue to live for only a few hours if the cardiac output falls to one-third normal.

MUSCLE WEAKNESS. One of the earliest symptoms of shock is severe muscle weakness that is also associated with profound and rapid fatigue whenever patients attempt to use their muscles. This results from the diminished supply of nutrients—especially oxygen—to the muscles.

DECREASED BODY TEMPERATURE. Because of the depressed metabolism in shock, the amount of heat liberated in the body is reduced (except in septic shock, in which the infection may cause an opposite effect). As a result, the body

temperature decreases if the body is exposed to even the slightest cold.

DEPRESSED MENTAL FUNCTION. In the early stages of shock, a person is usually conscious, although signs of mental haziness may be noted. As the shock progresses, the person falls into a state of stupor, and in the last stages of shock, even the subconscious mental functions, including vasomotor control and respiration, fail.

A person who recovers from shock usually exhibits no permanent impairment of mental functions.

REDUCED RENAL FUNCTION AND RENAL DETERIORATION. Even the slight decreases in cardiac output and arterial pressure that occur in the early stages of shock greatly diminish or even abolish urine output, mainly because the glomerular pressure falls below the critical level required for filtration. Other renal effects are discussed in Chapters 26 through 29. The retention of fluid by the kidneys is usually helpful in shock because it aids in preventing further diminishment of blood volume.

In the late stages of shock, the renal tubular epithelial cells deteriorate extremely rapidly because even normally they have a high metabolism and require large amounts of nutrients. The result is severe *tubular necrosis* with tubular cell death and sloughing and blockage of the tubules. Even though a person might survive the initial stages of the shock itself, the kidney damage often causes renal shutdown, with death occurring 1 week or so later because of uremia.

PHYSIOLOGY OF TREATMENT IN SHOCK

Replacement Therapy

BLOOD AND PLASMA TRANSFUSION. If a person is in shock caused by hemorrhage, the best possible therapy is usually transfusion of whole blood. If the shock is caused by plasma loss, the best therapy is administration of plasma; when dehydration is the cause, administration of the appropriate electrolytic solution can correct the shock.

Whole blood is not always available, such as under battlefield conditions. Plasma can usually substitute adequately for whole blood because it increases the blood volume and restores normal hemodynamics. Plasma cannot restore a normal hematocrit, but the human can usually stand a decrease in hematocrit to about one-third normal before serious consequences result if the cardiac output is adequate. Therefore, in acute conditions, it is reasonable to use plasma in place of whole blood for treatment of hemorrhagic and most other types of hypovolemic shock.

Sometimes plasma also is unavailable. In these instances, various *plasma substitutes* have been developed that perform almost exactly the same hemodynamic functions as plasma. One of these is dextran solution.

DEXTRAN SOLUTION AS A PLASMA SUBSTITUTE. The principal requirement of a truly effective plasma substitute is that it remain in the circulatory system—that is, not filter through the capillary pores into the tissue spaces. In addition, the solution must be nontoxic and contain appropriate electrolytes to prevent derangement of the extracellular fluid electrolytes on administration. To remain in the circulation, the plasma substitute must contain some substance that has a large enough molecular size to exert colloid osmotic pressure.

One of the most satisfactory substances developed thus far for this purpose is dextran, a large polysaccharide polymer of glucose. Certain bacteria secrete dextran as a by-product of their growth, and commercial dextran is manufactured by a bacterial culture procedure. By varying the growth conditions of the bacteria, the molecular weight of the dextran can be controlled to the desired value. Dextrans of appropriate molecular size do not pass through the capillary pores and, therefore, can replace plasma proteins as colloid osmotic agents.

Few toxic reactions have been observed when using dextran to provide colloid osmotic pressure; therefore, solutions of this substance have proved to be a satisfactory substitute for plasma in much fluid replacement therapy.

Treatment of Shock with Sympathomimetic Drugs—Sometimes Useful, Sometimes Not

A *sympathomimetic drug* is a drug that mimics sympathetic stimulation. These drugs include norepinephrine, epinephrine, and a large number of long-acting drugs that have the same effects as epinephrine and norepinephrine.

In two types of shock, sympathomimetic drugs have proved to be especially beneficial. The first of these is *neurogenic shock*, in which the sympathetic nervous system is severely depressed. Administering a sympathomimetic drug takes the place of the diminished sympathetic activity and can often restore full circulatory function.

The second type of shock in which sympathomimetic drugs are valuable is *anaphylactic shock*, in which histamine plays a prominent role. The sympathomimetic drugs have a vasoconstrictor effect that opposes the vasodilating effect of histamine. Therefore, either norepinephrine or another sympathomimetic drug is often lifesaving.

Sympathomimetic drugs have not proved to be valuable in hemorrhagic shock. The reason is that in this type of shock, the sympathetic nervous system has almost always already become maximally activated by the circulatory reflexes, and so much norepinephrine and epinephrine are circulating in the blood that the sympathomimetic drugs have essentially no additional beneficial effect.

Other Therapy

TREATMENT BY THE HEAD-DOWN POSITION. When the pressure falls too low in most types of shock, especially hemorrhagic and neurogenic shock, placing the patient with the head as much as 12 inches lower than the feet helps tremendously in promoting venous return and thereby increasing cardiac output. This head-down position is the first essential in the treatment of many types of shock.

OXYGEN THERAPY. Because the major deleterious effect of most types of shock is too little delivery of oxygen to the tissues, giving the patient oxygen to breathe can be of benefit in some instances. This frequently is of far less value than one might expect because the problem usually is not inadequate oxygenation of the blood in the lungs but inadequate transport of the blood after it is oxygenated.

TREATMENT WITH GLUCOCORTICOIDS (ADRENAL CORTEX HORMONES THAT CONTROL GLUCOSE METABOLISM). Glucocorticoids are frequently given to patients in severe shock for several reasons: (1) Experiments have shown empirically that glucocorticoids frequently increase the strength of the heart in the late stages of shock; (2) glucocorticoids stabilize the lysosomal membranes and prevent release of lysosomal enzymes into the cytoplasm of the cells, thus preventing deterioration from this source; and (3) glucocorticoids might aid in the metabolism of glucose by the severely damaged cells.

CIRCULATORY ARREST

A condition closely allied to circulatory shock is circulatory arrest, in which all blood flow stops. This occurs frequently on the surgical operating table as a result of *cardiac arrest* or *ventricular fibrillation*.

Ventricular fibrillation can usually be stopped by strong electroshock of the heart, the basic principles of which were described in Chapter 13.

Cardiac arrest often results from too little oxygen in the anesthetic gaseous mixture or from a depressant effect of the anesthesia itself. A normal cardiac rhythm can usually be restored by removing the anesthetic and then applying cardiopulmonary resuscitation procedures for a few minutes while supplying the patient's lungs with adequate quantities of ventilatory oxygen.

Effect of Circulatory Arrest on the Brain

A special problem in circulatory arrest is to prevent detrimental effects in the brain as a result of the arrest. In general, more than about 5 minutes of circulatory arrest causes at least some degree of permanent brain damage in more than one half of patients. Circulatory arrest for as long as 10 minutes almost universally destroys most, if not all, of the mental powers.

For many years, it was taught that this detrimental effect on the brain was caused by the acute cerebral hypoxia that occurs during the circulatory arrest. However, experiments have shown that if blood clots are prevented from occurring in the blood vessels of the brain, this will also prevent the rapid deterioration of the brain during circulatory arrest. For instance, in one animal experiment performed by J. W. Crowell, all the animal's blood was removed from the blood vessels at the beginning of circulatory arrest and then replaced at the end of the circulatory arrest, so that no intravascular blood clotting could occur. In this experiment, the brain was able to withstand up to 30 minutes of circulatory arrest without permanent brain damage. Also, administration of heparin or streptokinase before cardiac arrest was shown to increase the survivability of the brain up to two to four times as long as is usually the case. Therefore, it is likely that the severe brain damage that occurs after circulatory arrest results mainly from permanent blockage of many small or even large blood vessels by blood clots, thus causing prolonged ischemia and eventual death of the neurons.

REFERENCES

Achauer, B. M.: Management of the Burned Patient. East Norwalk, Conn., Appleton & Lange, 1987.

Bernton, E. W., et al.: Opioids and neuropeptides: Mechanisms in circulatory shock. Fed. Proc., 44:190, 1985.

Bongard, F. S., and Sue, D. Y.: Current Critical Care Diagnosis and Treatment. Redding, MA, Appleton and Lange, 1994.

Braunwald, E. (ed.): Heart Disease. A Textbook of Cardiovascular Medicine. Philadelphia, W. B. Saunders Co., 1988.

Burtis, C. A., and Ashwood, E. R.: Tietz Textbook of Clinical Chemistry. Philadelphia, W. B. Saunders Co., 1994.

Carlson, R. W., and Geheb, M. A.: Principles & Practice of Medical Intensive Care. Philadelphia, W. B. Saunders Co., 1993.

Civetta, J. M.: Critical Care. Philadelphia, J. B. Lippincott, 1992.

Crowell, J. W., and Guyton, A. C.: Evidence favoring a cardiac mechanism in irreversible hemorrhagic shock. Am. J. Physiol., 201:893, 1961.

Crowell, J. W., and Smith, E. E.: Oxygen deficit and irreversible hemorrhagic shock. Am. J. Physiol., 206:313, 1964.

Eisenberg, M., et al.: Emergency Medical Therapy. Philadelphia, W. B. Saunders Co., 1994.

Fallon, J. T.: Cardiovascular Pathophysiology: A Problem-Oriented Approach. Philadelphia, J. B. Lippincott, 1994.

Feliciano, D. V., et al.: Trauma. 3rd Ed. Redding, MA, Appleton and Lange, 1995.

Geller, E. R.: Shock and Resuscitation. Hightstown, NJ, McGraw-Hill, 1993.

Guyton, A. C., and Crowell, J. W.: Dynamics of the heart in shock. Fed. Proc., 20:51, 1961.

Guyton, A. C., et al.: Cardiac Output and Its Regulation. Philadelphia, W. B. Saunders Co., 1973.

Harwood-Nuss, A. L., et al.: The Clinical Practice of Emergency Medicine. Philadelphia, J. B. Lippincott, 1991.

Heffernan, J. J., et al.: Clinical Problems in Acute Care Medicine. Philadelphia, W. B. Saunders Co., 1989.

Jones, C. E., et al.: A cause-effect relationship between oxygen deficit and irreversible hemorrhagic shock. Surgery, 127:93, 1968.

Lake, C. L.: Clinical Monitoring for Anesthesia and Critical Care. Philadelphia, W. B. Saunders Co., 1994.

Luce, J. M., and Pierson, D. J.: Critical Care Medicine. Philadelphia, W. B. Saunders Co., 1988.

Pousada, L., et al.: Emergency Medicine. Baltimore, Williams & Wilkins, 1994.

Rothe, C. F., et al.: Control of total vascular resistance in hemorrhagic shock in the dog. Circ. Res., 12:667, 1963.

Salzman, S. K., and Faden, A. I.: The Neurobiology of Central Nervous System Trauma. New York, Oxford University Press, 1994.

Schmid-Schönbein, H., and Teitel, P. (eds.): Basic Aspects of Blood Trauma. Hingham, Mass., Kluwer Boston, 1979.

Shoemaker, W. C., et al. (eds.): Society of Critical Care Medicine, The Textbook of Critical Care. Philadelphia, W. B. Saunders Co., 1988.

Sibbald, W. J. (ed.): Synopsis of Critical Care. Baltimore, Williams & Wilkins, 1988.

Smith, E. E., and Crowell, J. W.: Effect of hemorrhagic hypotension on oxygen consumption of dogs. Am. J. Physiol., 207:647, 1964.

Ward, K. M., et al.: Clinical Laboratory Instrumentation and Automation: Principles, Applications and Selection. Philadelphia, W. B. Saunders Co., 1994.

Weil, M. H., et al.: Diagnosis and Treatment of Shock. Baltimore, Williams & Wilkins, 1994.

Winkelman, J. L.: Essentials of Basic Life Support. Minneapolis, Burgess, 1981.

Worthley, L. I. G.: Synopsis of Intensive Care Medicine. New York, Churchill Livingstone, 1994.

Yu, B. P.: Cellular defenses against damage from reactive oxygen species. Physiol. Rev., 74:139, 1994.

THE KIDNEYS AND BODY FLUIDS

UNIT V

The Body Fluid Compartments: Extracellular and Intracellular Fluids; Interstitial Fluid and Edema

CHAPTER 25

The maintenance of a relatively constant volume and a stable composition of the body fluids is essential for homeostasis, as discussed in Chapter 1. Some of the most important problems in clinical medicine arise because of abnormalities in the control systems that maintain this constancy of the body fluids. In this chapter and in the following chapters on the kidneys, we discuss the overall regulation of body fluid volume, regulation of the constituents of the extracellular fluid, regulation of acid-base balance, and control of fluid exchange between extracellular and intracellular compartments.

FLUID: INTAKE AND OUTPUT MUST BE BALANCED DURING STEADY-STATE CONDITIONS

The total amount of body fluid volume and the total amounts of solutes as well as their concentrations are relatively constant during steady-state conditions, as required for homeostasis. This constancy is remarkable because there is continuous exchange of fluid and solutes with the external environment as well as within the different compartments of the body. For example, there is a highly variable fluid intake that must be carefully matched by equal output from the body to prevent body fluid volumes from increasing or decreasing.

Daily Intake of Water

Water is added to the body by two major sources: (1) it is ingested in the form of liquids or water in the food, which together normally add about 2100 ml/day to the body fluids, and (2) it is synthesized in the body as a result of oxidation of carbohydrates, adding about 200 ml/day. This provides a total water intake of about 2300 ml/day (Table 25–1). Intake of fluid is highly variable among different people and even within the same person on different days, depending on climate, habits, and level of physical activity.

Daily Loss of Body Water

INSENSIBLE FLUID LOSS. A variable intake of water must be carefully matched to the daily fluid losses from the body. Some of the fluid losses cannot be precisely regulated. For example, there is a continuous loss of fluid by evaporation from the respiratory tract and diffusion through the skin, which together account for about 700 ml/day of fluid loss under normal conditions. This is termed *insensible water loss* because we are not consciously aware of it, even though it occurs continually in all living humans.

The insensible water loss through the skin occurs independently of sweating and is present even in people who are born without sweat glands; the average water loss by diffusion through the skin is about 300 to 400 ml/day. This loss is minimized by the cholesterol-filled cornified layer of the skin, which provides a barrier against excessive loss by diffusion. When the cornified layer becomes denuded, as occurs with extensive burns, the rate of evaporation can increase as much as 10-fold, to 3 to 5 liters/day. For this reason, burn victims must be given large amounts of fluid, usually intravenously, to balance fluid loss.

Insensible fluid loss through the respiratory tract

Table 25–1 DAILY INTAKE AND OUTPUT OF WATER (in ml/day)

	Normal	Prolonged Heavy Exercise
Intake		
Fluids ingested	2100	?
From metabolism	200	200
Total intake	2300	?
Output		
Insensible—Skin	350	350
Insensible—Lungs	350	650
Sweat	100	5000
Feces	100	100
Urine	1400	500
Total output	2300	6600

averages about 300 to 400 ml/day. As air enters the respiratory tract, it becomes saturated with moisture, to a vapor pressure of about 47 mm Hg, before it is expelled. Because the vapor pressure of the inspired air is usually less than 47 mm Hg, water is continuously lost through the lungs with respiration. In cold weather, the atmospheric vapor pressure decreases to nearly 0, causing an even greater loss of fluid from the lungs as the temperature decreases. This explains the dry feeling in the respiratory passages in cold weather.

FLUID LOSS IN SWEAT. The amount of fluid lost by sweating is highly variable, depending on physical activity and environmental temperature. The volume of sweat normally is only about 100 ml/day, but in very hot weather or during heavy exercise, water loss in sweat occasionally increases to 1 to 2 liters/hour. This would rapidly deplete the body fluids if intake were not also increased owing to activation of the thirst mechanism discussed in Chapter 29.

WATER LOSS IN FECES. Only a small amount of water (100 ml/day) normally is lost in the feces. This can increase to several liters a day in people with severe diarrhea. For this reason, severe diarrhea can be life-threatening if not corrected within a few days.

WATER LOSS BY THE KIDNEYS. The remaining water loss from the body occurs in the urine excreted by the kidneys. There are multiple mechanisms that control the rate of urine excretion. In fact, the most important means by which the body maintains a balance between fluid intake and output as well as balance between intake and output of most electrolytes in the body is by controlling the rates at which the kidneys excrete these substances. For example, urine volume can be as low as 0.5 liter/day in a dehydrated person or as high as 20 liters/day in a person who has been drinking tremendous amounts of fluid.

This extreme variability is also true for most of the electrolytes of the body, such as sodium, chloride, and potassium. In some people, sodium intake may be as low as 20 mEq/day, whereas in others, sodium intake may be 300 to 500 mEq/day. The kidneys are faced with the tasks of adjusting their excretion rates of water and electrolytes to match precisely the intakes of

these substances as well as compensating for excessive losses of fluids and electrolytes that occur in certain disease states. In Chapters 26 through 30, we discuss the mechanisms that allow the kidneys to perform these remarkable tasks.

BODY FLUID COMPARTMENTS

The total body fluid is distributed among two major compartments: the *extracellular fluid* and the *intracellular fluid* (Fig. 25–1). The extracellular fluid in turn is divided into the interstitial fluid and the blood plasma.

There is another small compartment of fluid that is referred to as *transcellular fluid*. This compartment includes fluid in the synovial, peritoneal, pericardial, and intraocular spaces as well as the cerebrospinal fluid; it is usually considered to be a specialized type of extracellular fluid, although in some cases, its composition may differ markedly from that of the plasma or interstitial fluid. All the transcellular fluids together constitute about 1 to 2 liters.

In a normal 70-kilogram adult human, the total body

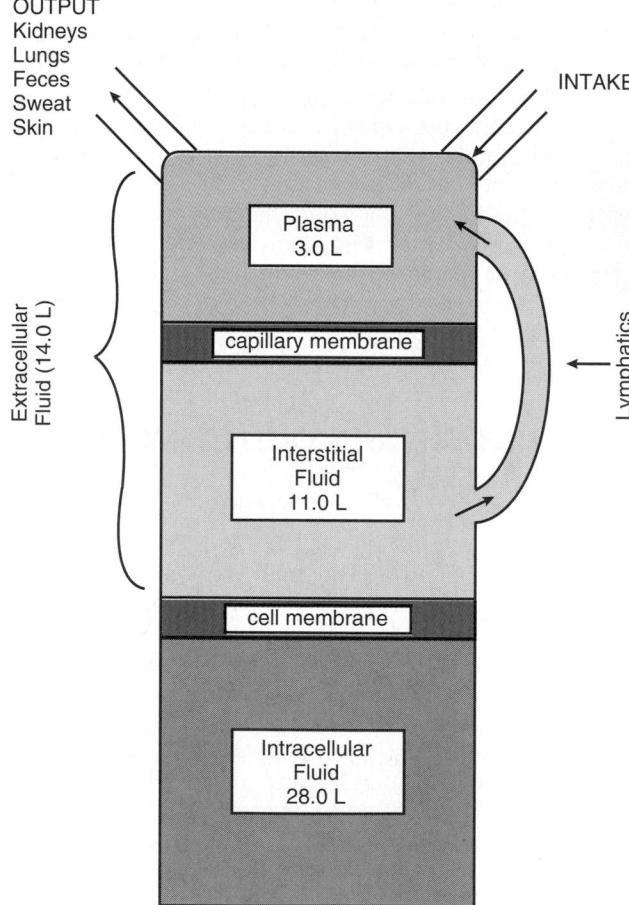

Figure 25–1. Summary of body fluid regulation, including the major body fluid compartments and the membranes that separate these compartments. The values shown are for an "average" 70-kilogram man.

water averages about 60 per cent of the body weight, or about 42 liters. This percentage can change, depending on age, sex, and degree of obesity. As a person grows older, the percentage of total body weight that is fluid gradually decreases. This is due in part to the fact that aging is usually associated with an increased percentage of the body weight that is fat, which in turn decreases the percentage of water in the body. Because women normally have more body fat than men, they contain slightly less water than men in proportion to their body weight. Therefore, when we discuss the "average" body fluid compartments, we should realize that variations exist, depending on age, sex, and degree of obesity.

INTRACELLULAR FLUID COMPARTMENT

About 28 of the 42 liters of fluid in the body are inside the 75 trillion cells and are collectively called the *intracellular fluid*. Thus, the intracellular fluid constitutes about 40 per cent of the total body weight in an "average" man.

The fluid of each cell contains its individual mixture of different constituents, but the concentrations of these substances are reasonably similar from one cell to another. In fact, the composition of cell fluids is remarkably similar even in different animals, ranging from the most primitive micro-organisms to humans. For this reason, the intracellular fluid of all the different cells together is considered to be one large fluid compartment.

EXTRACELLULAR FLUID COMPARTMENT

All the fluids outside the cells are collectively called the *extracellular fluid*. Together these fluids account for about 20 per cent of the body weight, or about 14 liters in a normal 70-kilogram adult. The two largest compartments of the extracellular fluid are the *interstitial fluid*, which makes up about three fourths of the extracellular fluid, and the *plasma*, which makes up almost one fourth of the extracellular fluid, or about 3 liters. The plasma is the noncellular part of the blood and communicates continuously with the interstitial fluid through the pores of the capillary membranes. These pores are highly permeable to almost all solutes in the extracellular fluid except the proteins. Therefore, the extracellular fluids are constantly mixing, so that the plasma and interstitial fluids have about the same composition except for proteins, which have a higher concentration in the plasma.

BLOOD VOLUME

Blood contains both extracellular fluid (the fluid in plasma) and intracellular fluid (the fluid in the red blood cells). However, blood is considered to be a separate fluid compartment because it is contained in a chamber of its own, the circulatory system. The blood volume is especially important to the control of cardiovascular dynamics.

The average blood volume of normal adults is about 8 per cent of body weight, or about 5 liters. On the average, about 60 per cent of the blood is plasma and 40 per cent is red blood cells, but these values can vary considerably in different people, depending on sex, weight, and other factors.

HEMATOCRIT (PACKED RED CELL VOLUME). The hematocrit is the fraction of the blood composed of red blood cells, as determined by centrifuging blood in a "hematocrit tube" until the cells become tightly packed in the bottom of the tube. It is impossible to completely pack the red cells together; therefore, about 3 to 4 per cent of the plasma remains entrapped among the cells, and the true hematocrit is only about 96 per cent of the measured hematocrit.

In normal men, the measured hematocrit is about 0.40, and in normal women, it is about 0.36. In severe *anemia*, the hematocrit may fall as low as 0.10, a value that is barely sufficient to sustain life. On the other hand, there are some conditions in which there is excessive production of red blood cells, resulting in *polycythemia*. In these conditions, the hematocrit can rise to 0.65.

CONSTITUENTS OF EXTRACELLULAR AND INTRACELLULAR FLUIDS

Comparisons of the compositions of the extracellular fluid, including the plasma, interstitial fluid, and intracellular fluid are shown in Figures 25–2 and 25–3 and in Table 25–2.

Ionic Compositions of Plasma and Interstitial Fluid Are Similar

Because the plasma and interstitial fluids are separated only by highly permeable capillary membranes, their ionic compositions are similar. The most important difference between these two compartments is the higher concentration of protein in the plasma; the capillaries have a low permeability to the plasma proteins and, therefore, leak only small amounts of proteins into the interstitial spaces in most tissues.

Because of the *Donnan effect,* the concentration of positively charged ions (cations) is slightly greater (about 2 per cent) in the plasma than in the interstitial fluid; this effect is the following: the plasma proteins have a net negative charge and, therefore, tend to bind cations, such as sodium and potassium ions, thus holding extra amounts of these cations in the plasma along with the plasma proteins. On the other hand, negatively charged ions (anions) tend to have a slightly higher concentration in the interstitial fluid, compared with the plasma, because the negative charges of the plasma proteins repel the negatively charged anions.

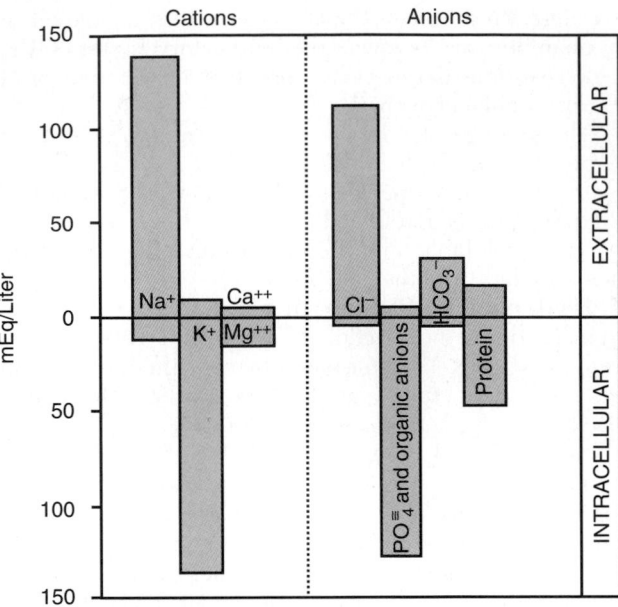

Figure 25–2. Major cations and anions of the intracellular and extracellular fluids.

chloride ions, reasonably large amounts of bicarbonate ions, but only small quantities of potassium, calcium, magnesium, phosphate, and organic acid ions.

The composition of extracellular fluid is carefully regulated by various mechanisms but especially by the kidneys, as discussed later. This allows the cells to remain continually bathed in a fluid that contains the proper concentration of electrolytes and nutrients for optimal cell function.

Important Constituents of the Intracellular Fluid

The intracellular fluid is separated from the extracellular fluid by a selective cell membrane that is highly permeable to water but not to most of the electrolytes in the body. The cell membrane maintains a fluid composition inside the cells that is similar among different cells of the body.

In contrast to the extracellular fluid, the intracellular fluid contains only small quantities of sodium and chloride ions and almost no calcium ions. Instead, it contains large amounts of potassium and phosphate ions plus moderate quantities of magnesium and sulfate ions, all of which have low concentrations in the extracellular fluid. Also, cells contain large amounts of protein, almost four times as much as in the plasma.

For practical purposes, however, the concentrations of ions in the interstitial fluid and plasma are considered to be about equal.

Referring again to Figure 25–2, one can see that the extracellular fluid, including the plasma and the interstitial fluid, contains large amounts of sodium and

MEASUREMENT OF FLUID VOLUMES IN THE DIFFERENT BODY FLUID COMPARTMENTS; THE INDICATOR-DILUTION PRINCIPLE

The volume of a fluid compartment in the body can be measured by placing a substance in the compartment, allowing it to disperse evenly throughout the compartment's fluid, and then analyzing the extent to which the substance has become diluted. Figure 25–4 shows this "indicator-dilution" method for measuring the volume of a fluid compartment, which is based on the principle of conservation of mass. This means that the total mass of a substance after dispersion in the fluid compartment will be the same as the total mass injected into the compartment.

In the example shown in Figure 25–4, a small amount of dye or other substance contained in syringe A is injected into a chamber, and the substance is allowed to disperse throughout the chamber until it becomes mixed in equal concentrations in all areas, as shown in chamber B. Then a sample of fluid containing the dispersed substance is removed and the concentration of the substance is analyzed chemically, photoelectrically, or by other means. If none of the substance leaks out of the compartment, then the total mass of substance in the compartment (Volume B × Concentration of B) will equal the total mass of the substance injected (Volume A × Concentration of A). By simple rearrangement of the equation, one can

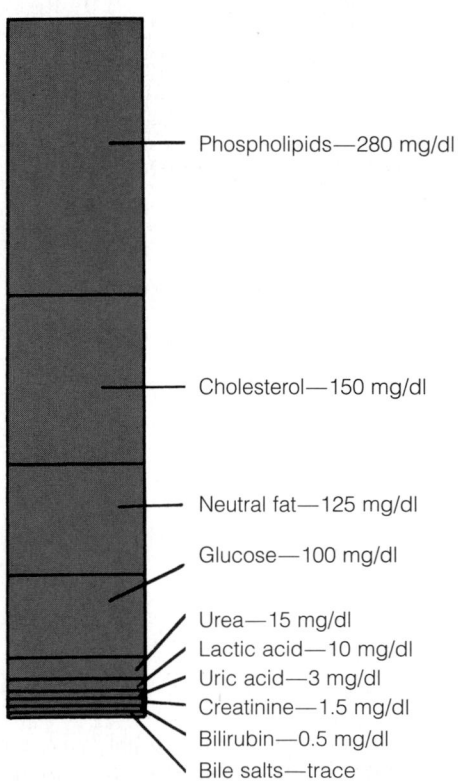

Phospholipids—280 mg/dl

Cholesterol—150 mg/dl

Neutral fat—125 mg/dl

Glucose—100 mg/dl

Urea—15 mg/dl
Lactic acid—10 mg/dl
Uric acid—3 mg/dl
Creatinine—1.5 mg/dl
Bilirubin—0.5 mg/dl
Bile salts—trace

Figure 25–3. Non-electrolytes of the plasma.

Table 25–2 OSMOLAR SUBSTANCES IN EXTRACELLULAR AND INTRACELLULAR FLUIDS

	Plasma (mOsm/liter of H_2O)	Interstitial (mOsm/liter of H_2O)	Intracellular (mOsm/liter of H_2O)
Na^+	142	139	14
K^+	4.2	4.0	140
Ca^{++}	1.3	1.2	0
Mg^+	0.8	0.7	20
$Cl-$	108	108	4
HCO_3^-	24	28.3	10
$HPO_4^-, H2PO_4^-$	2	2	11
SO_4^-	0.5	0.5	1
Phosphocreatine			45
Carnosine			14
Amino acids	2	2	8
Creatine	0.2	0.2	9
Lactate	1.2	1.2	1.5
Adenosine triphosphate			5
Hexose monophosphate			3.7
Glucose	5.6	5.6	
Protein	1.2	0.2	4
Urea	4	4	4
Others	4.8	3.9	10
Total mOsm/liter	301.8	300.8	301.2
Corrected osmolar activity (mOsm/liter)	282.0	281.0	281.0
Total osmotic pressure at 37° C (mm Hg)	5443	5423	5423

calculate the unknown volume of chamber B as:

$$\text{Volume B} = \frac{\text{Volume A} \times \text{Concentration A}}{\text{Concentration B}}$$

Note that all one needs to know for this calculation is (1) the total amount of substance injected into the chamber (the numerator of the equation) and (2) the

Indicator Mass A = Volume A × Concentration A

Indicator Mass A = Indicator Mass B

Indicator Mass B = Volume B × Concentration B

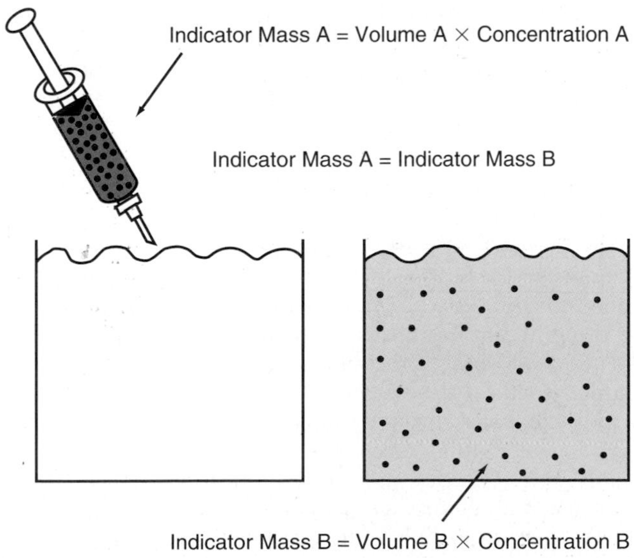

Figure 25–4. Indicator-dilution method for measuring fluid volumes.

concentration of the fluid in the chamber after the substance has been dispersed (the denominator). In this example, if 1 milliliter of a solution containing 10 mg/ml of dye is dispersed into chamber B and the final concentration in chamber B is 0.01 milligram for each milliliter of fluid, then the unknown volume of chamber B can be calculated as follows:

$$\text{Volume B} = \frac{1 \text{ ml} \times 10 \text{ mg/ml}}{.01 \text{ mg/ml}} = 1000 \text{ ml}$$

This method can be used to measure the volume of virtually any compartment in the body as long as (1) the indicator disperses evenly throughout the compartment, (2) the indicator disperses only in the compartment that is being measured, and (3) the indicator is not metabolized or excreted. Several substances can be used to measure the volume of each of the different body fluids.

Determination of Volumes of Specific Body Fluid Compartments

MEASUREMENT OF TOTAL BODY WATER. Radioactive water (tritium, ^3H_2O) or heavy water (deuterium, ^2H_2O) can be used to measure total body water. These forms of water mix with the total body water within a few hours after being injected into the blood, and the dilution principle can be used to calculate total body water.

Another substance that has been used to measure total body water is *antipyrine*, which is very lipid-soluble and can rapidly penetrate cell membranes and distribute itself uniformly throughout the intracellular and extracellular compartments.

MEASUREMENT OF EXTRACELLULAR FLUID VOLUME. The volume of extracellular fluid can be estimated using any of several substances that disperse in the plasma and interstitial fluid but do not readily permeate the cell membrane. They include radioactive sodium, radioactive chloride, radioactive iothalamate, thiosulfate ion, and inulin. When any one of these substances is injected into the blood, it usually disperses almost completely throughout the extracellular fluids within 30 to 60 minutes. Some of these substances, however, such as radioactive sodium, may diffuse into the cell in small amounts. Therefore, one frequently speaks of the *sodium space,* or the *inulin space,* instead of calling the measurement the true extracellular fluid volume.

CALCULATION OF INTRACELLULAR VOLUME. The intracellular volume cannot be measured directly. However, it can be calculated as:

Intracellular volume = Total body water

− Extracellular volume

MEASUREMENT OF PLASMA VOLUME. To measure plasma volume, a substance must be used that does not readily penetrate capillary membranes but remains in the vascular system after vascular injection. One of the most commonly used substances for measuring plasma volume is serum albumin labeled with radioactive iodine (^{125}I-albumin). Also, dyes that avidly bind to the plasma proteins, such as *Evans blue dye* (also called *T-1824*), can be used to measure plasma volume.

CALCULATION OF INTERSTITIAL FLUID VOLUME. Interstitial fluid volume cannot be measured directly, but it can be calculated as:

Interstitial fluid volume = Extracellular fluid volume

− Plasma volume

MEASUREMENT OF BLOOD VOLUME. If one measures plasma volume using the methods described above, blood volume can also be calculated if one knows the *hematocrit,* which is the fraction of the total blood volume composed of cells, using the following equation:

$$\text{Total blood volume} = \frac{\text{Plasma volume}}{1 - \text{Hematocrit}}$$

For example, if plasma volume is 3.0 liters and hematocrit is 0.40, then total blood volume would be calculated as

$$\frac{3 \text{ liters}}{1.0 - 0.4} = 5.0 \text{ liters}$$

Another way to measure blood volume is to inject into the circulation red blood cells that have been labeled with radioactive material. After these mix in the circulation, the radioactivity of a mixed blood sample can be measured and the total blood volume can be calculated using the dilution principle. A substance frequently used to label the red blood cells is radioactive chromium (^{51}Cr), which binds tightly with the red blood cells.

REGULATION OF FLUID EXCHANGE AND OSMOTIC EQUILIBRIA BETWEEN INTRACELLULAR AND EXTRACELLULAR FLUID

A frequent problem in the treatment of seriously ill patients is the difficulty of maintaining adequate fluids in one or both of the intracellular and extracellular compartments. As discussed in Chapter 16 and later in this chapter, the relative amounts of extracellular fluid distributed between the plasma and interstitial spaces are determined mainly by the balance of hydrostatic and colloid osmotic forces across the capillary membrane. The distribution of fluid between intracellular and extracellular compartments, on the other hand, is determined mainly by the osmotic effect of the smaller solutes—especially sodium, chloride, and other electrolytes—acting across the cell membrane. The reason for this is that the cell membranes are highly permeable to water but relatively impermeable even to small ions such as sodium and chloride. Therefore, water moves across the cell membrane rapidly, so that the intracellular fluid remains isotonic with the extracellular fluid.

In the next section, we discuss the interrelations between intracellular and extracellular fluid volumes and the osmotic factors that can cause shifts of fluid between these two compartments.

BASIC PRINCIPLES OF OSMOSIS AND OSMOTIC PRESSURE

The basic principles of osmosis and osmotic pressure are presented in Chapter 4. Therefore, we review here only the most important aspects of these principles as they apply to volume regulation.

Osmosis is the net diffusion of water from a region of high water concentration to one that has a lower water concentration. When a solute is added to pure water, this reduces the concentration of water in the mixture. Thus, the higher the solute concentration in a solution, the lower the water concentration. Furthermore, water diffuses from a region of low solute concentration (high water concentration) to one that has a high solute concentration (low water concentration).

Because the cell membrane is relatively impermeable to most solutes but highly permeable to water, whenever there is a higher concentration of solute on one side of the cell membrane, water diffuses across the membrane toward the region of higher solute concentration. Thus, if a solute such as sodium chloride is added to the extracellular fluid, water rapidly diffuses from the cells through the cell membranes into the extracellular fluid until the water concentration on both sides of the membrane becomes equal. Conversely, if a solute such as sodium chloride is removed from the extracellular fluid, thereby raising the water concentration, water diffuses from the extracellular fluid through the cell membranes and into the cells.

The rate of diffusion of water is called the *rate of osmosis*.

RELATION BETWEEN MOLES AND OSMOLES. Because the water concentration of a solution depends on the number of solute particles in the solution, a concentration term is needed to describe the total concentration of solute particles, regardless of their exact composition. The total number of particles in a solution is measured in terms of "*osmoles.*" One osmole (osm) is equal to 1 mole (mol; 6.02×10^{23}) of solute particles. Therefore, a solution containing 1 mole of glucose in each liter has a concentration of 1 osm/liter. If a molecule dissociates into two ions (giving two particles), such as sodium chloride ionizing to give chloride and sodium ions, then a solution containing 1 mol/liter will have an osmotic concentration of 2 osm/liter. Likewise, a solution that contains 1 mole of a molecule that dissociates into three ions, such as sodium sulfate (Na_2SO_4), will contain 3 osm/liter. Thus, the term osmole refers to the number of osmotically active particles in a solution rather than to the molar concentration.

In general, the osmole is too large a unit for expressing osmotic activity of solutes in the body fluids. Therefore, the term *milliosmole* (mOsm), which equals 1/1000 osmole, is commonly used.

OSMOLALITY AND OSMOLARITY. The osmolal concentration of solution is called *osmolality* when the concentration is expressed as *osmoles per kilogram of water;* it is called *osmolarity* when it is expressed as *osmoles per liter of solution.* In dilute solutions such as the body fluids, these two terms can be used almost synonymously because the differences are small. In most cases, it is easier to express body fluid quantities in liters rather than in kilograms of water. Therefore, most of the calculations used clinically and the calculations expressed in the next several chapters are based on osmolarities rather than osmolalities.

OSMOTIC PRESSURE. Osmosis of water molecules across a selectively permeable membrane can be opposed by applying a pressure in the direction opposite that of the osmosis. The precise amount of pressure required to prevent the osmosis is called the *osmotic pressure.* It is important to realize that osmotic pressure is not the pressure that causes net diffusion of water through a membrane. Instead, it is equal to the amount of pressure that must be applied to prevent the net diffusion of water through the membrane. Osmotic pressure, therefore, is an indirect measurement of the water and solute concentrations of a solution. The higher the osmotic pressure of a solution, the lower the water concentration but the higher the solute concentration of the solution.

RELATION BETWEEN OSMOTIC PRESSURE AND OSMOLARITY. The osmotic pressure of a solution is directly proportional to the concentration of osmotically active particles in that solution. This is true regardless of whether the solute is a large molecule or a small molecule. For example, one molecule of albumin with a molecular weight of 70,000 has the same osmotic effect as one molecule of glucose with a molecular weight of 180. On the other hand, one molecule of sodium chloride has two osmotically active particles, Na^+ and Cl^-, and, therefore, has twice the osmotic effect of either an albumin molecule or a glucose molecule. Thus, the osmotic pressure of a solution is proportional to its osmolarity, a measure of the concentration of solute particles.

Expressed mathematically, according to van't Hoff's law, osmotic pressure (π) can be calculated as:

$$\pi = CRT,$$

where C is the concentration of solutes in osmoles per liter, R is the ideal gas constant, and T is the absolute temperature in degrees kelvin ($273° +$ centigrade°). If π is expressed in millimeters of mercury (mm Hg), the unit of pressure commonly used for biological fluids, and T is normal body temperature ($273° + 37° = 310°$ kelvin), the value of π calculates to be about 19,300 mm Hg for a solution having a concentration of 1.0 osm/liter. This means that for a concentration of 1.0 *mOsm*/liter, π is equal to 19.3 mm Hg. Thus, for each milliosmole concentration gradient across the cell membrane, 19.3 mm Hg osmotic pressure is exerted.

CALCULATION OF THE OSMOTIC PRESSURE OF A SOLUTION. Using van't Hoff's law, one can calculate the potential osmotic pressure of a solution, assuming that the cell membrane is impermeable to the solute. For example, the osmotic pressure of a 0.9 per cent sodium chloride solution is calculated as follows: A 0.9 per cent solution means that there are 0.9 gram of sodium chloride per 100 milliliters of solution, or 9 gm/liter. Because the molecular weight of sodium chloride is 58.5 gm/mol, the molarity of the solution is 9 gm/liter divided by 58.5 gm/mol, or about 0.154 mol/liter. Because each molecule of sodium chloride is equal to 2 osmoles, the osmolarity of the solution is .154 × 2, or .308 osm/liter. Therefore, the osmolarity of this solution is 308 mOsm/liter. The potential osmotic pressure of this solution would, therefore, be 308 mOsm/liter × 19.3 mm Hg/mOsm/liter, or 5944 mm Hg.

This calculation is only an approximation because sodium and chloride ions do not behave entirely as independent particles in solution because of interionic attraction between these ions. One can correct for these deviations from the predictions of van't Hoff's law by using a correction factor called the *osmotic coefficient.* For sodium chloride, the osmotic coefficient is about 0.93. Therefore, the actual osmolarity of a 0.9 per cent sodium chloride solution is 308 × 0.93, or about 286 mOsm/liter. For practical reasons, sometimes the osmotic coefficients of different solutes are neglected in determining the osmolarity and osmotic pressures of physiological solutions.

OSMOLARITY OF THE BODY FLUIDS. Turning back to Table 25–2, note the approximate osmolarity of the various osmotically active substances in plasma, interstitial fluid, and extracellular fluid. Note that about 80 per cent of the total osmolarity of the interstitial fluid and plasma is due to sodium and chloride ions, whereas for intracellular fluid, almost one half of the osmolality is due to potassium ions and the remainder is divided among many other intracellular substances.

As shown at the bottom of Table 25–2, the total osmolarity of each of the three compartments is about 300 mOsm/liter, with the plasma being about 1 mOsm/liter greater than that of the interstitial and intracellular fluids. The slight difference between plasma and interstitial fluid is caused by the osmotic effects of the plasma proteins, which maintain about 20 mm Hg greater pressure in the capillaries than in the surrounding interstitial spaces, as discussed in Chapter 16.

CORRECTED OSMOLAR ACTIVITY OF THE BODY FLUIDS. At the bottom of Table 25–2 are shown *corrected osmolar activities* of plasma, interstitial fluid, and intracellular fluid. The reason for these corrections is that molecules and ions in solution exert interionic and intermolecular attraction or repulsion from one solute molecule to the next, and these two effects can cause, respectively, a decrease or an increase in the osmotic "activity" of the dissolved substance.

TOTAL OSMOTIC PRESSURE EXERTED BY THE BODY FLUIDS. Table 25–2 also shows the total osmotic pressure in millimeters of mercury that would be exerted by each of the different fluids if it were placed on one side of the cell membrane with pure water on the other side. Note that this total pressure averages about 5443 mm Hg for plasma, which is 19.3 times the corrected osmolarity of 282 mOsm/liter for plasma.

OSMOTIC EQUILIBRIUM IS MAINTAINED BETWEEN INTRACELLULAR AND EXTRACELLULAR FLUIDS

Large osmotic pressures can develop across the cell membrane with relatively small changes in the concentration of solutes in the extracellular fluid. As discussed above, for each milliosmole concentration gradient of an *impermeant solute* (one that will not permeate the cell membrane), about 19.3 mm Hg osmotic pressure is exerted across the cell membrane. If the cell membrane is exposed to pure water and the osmolarity of intracellular fluid is 280 mOsm/liter, the potential osmotic pressure that can develop across the cell membrane is more than 5400 mm Hg. This demonstrates the large force that can act to move water across the cell membrane when the intracellular and extracellular fluids are not in osmotic equilibrium. As a result of these forces, relatively small changes in concentration of impermeant solutes in the extracellular fluid can cause tremendous changes in cell volume.

ISOTONIC, HYPOTONIC, AND HYPERTONIC FLUIDS. The effects of different concentrations of impermeant solutes in the extracellular fluid on cell volume are shown in Figure 25–5. If a cell is placed in a solution of impermeant solutes having an osmolarity of

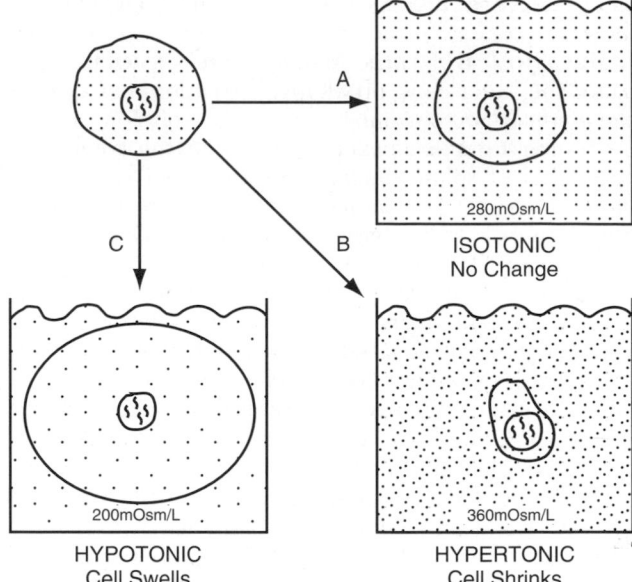

Figure 25–5. Effects of isotonic (*A*), hypertonic (*B*), and hypotonic (*C*) solutions on cell volume.

280 mOsm/liter, the cells will not shrink or swell because the water concentrations in the intracellular and extracellular fluids are equal and the solutes cannot enter or leave the cell. Such a solution is said to be *isotonic* because it neither shrinks nor swells the cells. Examples of isotonic solutions are a 0.9 per cent solution of sodium chloride or a 5 per cent glucose solution. These solutions are important in clinical medicine because they can be infused into the blood without the danger of upsetting osmotic equilibrium between the intracellular and extracellular fluids.

If a cell is placed into a solution that has a lower concentration of impermeant solutes (less than 280 mOsm/liter), water will diffuse into the cell causing it to swell; water will continue to diffuse into the cell, diluting the intracellular fluid while also concentrating the extracellular fluid until both solutions have about the same osmolarity. Solutions of sodium chloride with a concentration of less than 0.9 per cent are *hypotonic* and cause cells to swell.

If a cell is placed in a solution having a higher concentration of impermeant solutes, water will flow out of the cell into the extracellular fluid, concentrating the intracellular fluid and diluting the extracellular fluid. In this case, the cell will shrink until the two concentrations become equal. A solution that causes cells to shrink is said to be *hypertonic;* sodium chloride solutions of greater than 0.9 per cent are hypertonic.

ISOSMOTIC, HYPEROSMOTIC, AND HYPO-OSMOTIC FLUIDS. The terms isotonic, hypotonic, and hypertonic refer to whether or not solutions will cause a change in cell volume. The tonicity of solutions depends on the concentrations of impermeant solutes. Some solutes, however, can permeate the cell membrane. Solutions with an osmolarity the same as the cell are called *isosmotic*, regardless of whether the solute can penetrate the cell membrane.

The terms *hyperosmotic* and *hypo-osmotic* refer to solutions that have a higher osmolarity or lower osmolarity, respectively, compared with the normal extracellular fluid, without regard for whether the solute permeates the cell membrane.

Highly permeating substances, such as urea, can cause transient shifts in fluid volumes between the intracellular and extracellular fluids, but given enough time, the concentrations of these substances eventually become equal in the two compartments and have little effect on intracellular volume under steady-state conditions.

OSMOTIC EQUILIBRIUM BETWEEN INTRACELLULAR AND EXTRACELLULAR FLUIDS IS RAPIDLY ATTAINED. The transfer of fluid across the cell membrane occurs so rapidly that any differences in osmolarities between these two compartments are usually corrected within seconds or, at the most minutes. This rapid movement of water across the cell membrane does not mean that complete equilibrium occurs between the intracellular and extracellular compartments throughout the whole body within the same short period. The reason for this is that fluid usually enters the body through the gut and must be transported by the blood to all tissues before complete osmotic equilibrium can occur. In normal people, it may take as long as 30 minutes to achieve osmotic equilibrium everywhere in the body after drinking water.

VOLUMES AND OSMOLALITIES OF EXTRACELLULAR AND INTRACELLULAR FLUID IN ABNORMAL STATES

Abnormalities of the composition and volumes of body fluids are among the most common and important of clinical problems and are of concern for almost all seriously ill, hospitalized patients. Understanding and treating these disorders require knowledge of fluid shifts between intracellular and extracellular compartments before and after therapy.

Some of the different factors that can cause extracellular and intracellular volumes to change markedly are ingestion of water, dehydration, intravenous infusions of different types of solutions, loss of large amounts of fluid from the gastrointestinal tract, and loss of abnormal amounts of fluid by sweating or through the kidneys.

One can calculate both the changes in intracellular and extracellular fluid volumes and the types of therapy that must be instituted if the following basic principles are kept in mind:

1. *Water moves rapidly across cell membranes;* therefore, the osmolarities of intracellular and extracellular fluids remain almost exactly equal to each other except for a few minutes after a change in one of the compartments.

2. *The cell membrane is almost completely impermeable to many solutes;* therefore, the number of osmoles in the extracellular or intracellular fluid remains constant unless solutes are added to or lost from the extracellular compartment.

With these basic principles in mind, we can analyze the effects of different abnormal fluid conditions on extracellular and intracellular fluid volumes and osmolarities.

Calculation of Water Deficit in Dehydration

Suppose you have a 70-kilogram comatose patient in whom the thirst mechanism is inoperative, and after analysis of his plasma, you find a plasma osmolarity of 320 mOsm/liter. How much water should be administered to restore plasma osmolality to 280 mOsm/liter, assuming that fluid and solutes are not excreted? Also, after this fluid is administered and osmotic equilibrium has occurred, what would be the final intracellular and extracellular fluid volumes?

To answer these questions, we can first approximate the patient's intracellular and extracellular fluid volumes and the total milliosmoles of solute in each compartment. If we assume that extracellular fluid volume is 20 per cent of body weight and intracellular fluid volume is 40 per cent of body weight, we can calculate the volumes and total milliosmoles in the intracellular and extracellular compartments. Extracellular fluid volume is about 14 liters and intracellular fluid volume is about 28 liters (this is not quite correct because the patient is dehydrated). The total quantity of milliosmoles in each of these compartments is calculated by multiplying the volume times the concentration, which is 320 mOsm/liter in each compartment. This gives us the initial conditions of the patient.

STEP 1. INITIAL CONDITIONS

	Volume (Liters)	Concentration (mOsm/liter)	Total mOsm
Extracellular fluid	14	320	4,480
Intracellular fluid	28	320	8,960
Total body fluid	42	320	13,440

The second step is to determine the body fluid volumes that would be necessary to achieve a concentration of 280 mOsm/liter, assuming that solutes and water are not lost from the body. To do this, we can assume that the total milliosmoles in the body remain constant and that the total milliosmoles in each compartment remain constant. After the proper amount of fluid has been added and osmotic equilibrium has occurred, all the fluid compartments will have the same concentration, 280 mOsm/liter. Therefore, the volume of each compartment can be calculated by dividing the total milliosmoles in each compartment by the concentration (280 mOsm/liter) in each compartment. For the total body fluids, the volume is calculated as 13,440 mOsm/280 mOsm/liter, which yields a volume of 48 liters. Intracellular volume is calculated by dividing 8960 milliosmoles by 280 mOsm/liter to yield a volume of 32 liters. Extracellular volume is calculated by dividing 4480 milliosmoles by 280 mOsm/liter to yield a volume of 16 liters.

STEP 2

	Volume (Liters)	Concentration (mOsm/liter)	Total mOsm
Extracellular fluid	16	280	4,480
Intracellular fluid	32	280	8,960
Total body fluid	48	280	13,440

The last step of this problem is to calculate the amount of fluid that must be added to achieve these conditions. Because the total volume required to yield a plasma osmolarity of 280 mOsm/liter is 48 liters, the total volume that must be added is 6 liters (48 − 42 liters). Of the 6 liters that must be added to achieve these final conditions, 2 liters stayed in the extracellular space and 4 liters diffused into the cells.

In making this approximation, we should realize that some of the fluid and solutes would be excreted during

the course of administering the water. Also, one would not administer all of the 6 liters of water immediately but would titrate the patient to achieve the desired concentration of 280 mOsm/liter. This type of analysis, however, provides a quantitative approach to correcting the abnormalities of fluid volumes in this patient caused by simple dehydration.

Effect of Adding Saline Solution to the Extracellular Fluid

If an *isotonic* saline solution is added to the extracellular fluid compartment, the osmolarity of the extracellular fluid does not change; therefore, no osmosis occurs through the cell membranes. The only effect is an increase in extracellular fluid volume (Fig. 25–6A). The sodium and chloride largely remain in the extracellular fluid because the cell membrane behaves as though it were virtually impermeable to the sodium chloride.

If a *hypertonic* solution is added to the extracellular fluid, the extracellular osmolarity increases and causes osmosis of water out of the cells into the extracellular compartment (Fig. 25–6B). Again, almost all the added sodium chloride remains in the extracellular compartment and fluid diffuses from the cells into the extracellular space to achieve osmotic equilibrium. The net effect is an increase in extracellular volume (greater than the volume of fluid added), a decrease in intracellular volume, and a rise in osmolarity in both compartments.

If a *hypotonic* solution is added to the extracellular fluid, the osmolarity of the extracellular fluid decreases

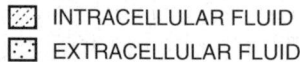

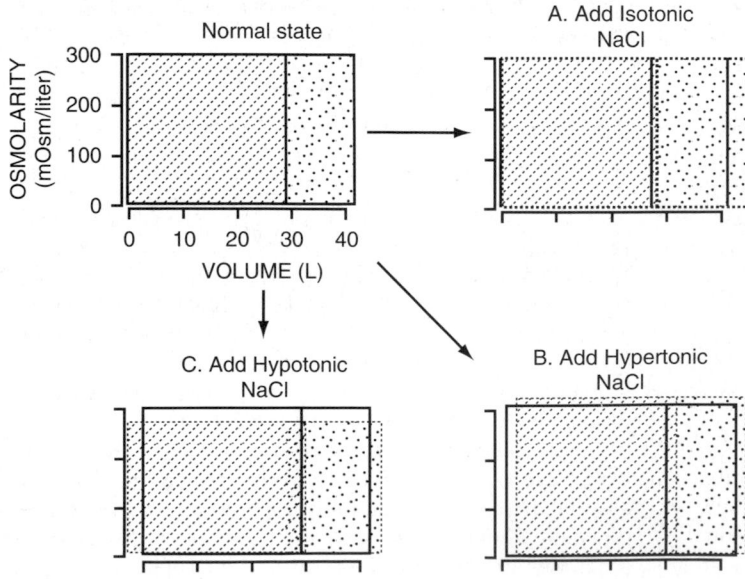

Figure 25–6. Effect of adding isotonic, hypertonic, and hypotonic solutions to the extracellular fluid after osmotic equilibrium. The normal state is indicated by the solid lines and the shifts from normal are shown by the dashed lines. The volumes of intracellular and extracellular fluid compartments are shown in the abscissa of each diagram, and the osmolarities of these compartments are shown on the ordinates.

and some of the extracellular water diffuses into the cells until the intracellular and extracellular compartments have the same osmolarity (Fig. 25–6C). Both the intracellular and the extracellular volumes are increased by addition of hypotonic fluid, although the intracellular volume increases to a greater extent.

CALCULATION OF FLUID SHIFTS AND OSMOLARITIES AFTER INFUSION OF HYPERTONIC SALINE. As an additional example, we can calculate the sequential effects on extracellular and intracellular fluid volumes and osmolarities if we infuse 2 liters of a hypertonic 2.9 per cent sodium chloride solution into the extracellular fluid compartment of a 70-kilogram patient with an initial plasma osmolarity of 280 mOsm/liter.

The first step is to calculate the initial conditions, including the volumes, concentrations, and total milliosmoles in each compartment. Assuming that extracellular fluid volume is 20 per cent of body weight and intracellular fluid volume is 40 per cent of body weight, the following volumes and concentrations can be calculated.

STEP 1. INITIAL CONDITIONS

	Volume (Liters)	Concentration (mOsm/liter)	Total mOsm
Extracellular fluid	14	280	3,920
Intracellular fluid	28	280	7,840
Total body fluid	42	280	11,760

Next, we calculate the total milliosmoles added to the extracellular fluid in 2 liters of 2.9 per cent sodium chloride. A 2.9 per cent solution means that there are 2.9 gm/100 ml, or 29 grams of sodium chloride per liter. Because the molecular weight of sodium chloride is about 58 gm/mol, this means that there is about 0.5 mole of sodium chloride per liter of solution. For 2 liters of solution, this would be 1.0 mole of sodium chloride. Because 1 mole of sodium chloride is about equal to 2 osmoles (because sodium chloride has two osmotically active particles per mole), the net effect of adding 2 liters of this solution is to add 2000 milliosmoles of sodium chloride to the extracellular fluid.

In Step 2, we calculate the instantaneous effect of adding 2000 milliosmoles of sodium chloride to the extracellular fluid along with 2 liters of volume. Instantaneously, there would be no change in the *intracellular fluid* concentration or volume and there would be no osmotic equilibrium. In the *extracellular fluid*, however, there would be an additional 2000 milliosmoles of total solute, yielding a total of 5920 milliosmoles. Because the extracellular compartment now has 16 liters of volume, the concentration can be calculated by dividing 5920 milliosmoles by 16 liters to yield a concentration of 370 mOsm/liter. The following values would occur instantly after adding the solution.

STEP 2. INSTANTANEOUS EFFECT OF ADDING 2 LITERS OF 2.9 PER CENT SODIUM CHLORIDE

	Volume (Liters)	Concentration (mOsm/liter)	Total mOsm
Extracellular fluid	16	370	5,920
Intracellular fluid	28	280	7,840
Total body fluid	44	No equilibrium	13,760

In the third step, we calculate the volumes and concentrations that would occur after osmotic equilibrium. In this case, the concentrations in the intracellular and extracellular fluid compartments would be equal and can be calculated by dividing the total milliosmoles in the body, 13,760, by the total volume, which is now 44 liters. This yields a concentration of 312.7 mOsm/liter. Therefore, all the body fluid compartments will have this same concentration after osmotic equilibrium. If we assume that no solute or water has been lost from the body and that there is no movement of sodium chloride into or out of the cells, we then calculate the volumes of the intracellular and extracellular compartments: the intracellular fluid volume is calculated by dividing the total milliosmoles in the intracellular fluid (7840) by the concentration (312.7 mOsm/liter) to yield a volume of 25.1 liters. Extracellular fluid volume is calculated by dividing the total milliosmoles in extracellular fluid (5920) by the concentration (312.7 mOsm/liter) to yield a volume of 18.9 liters. These calculations, again, are based on the assumption that the sodium chloride added to the extracellular fluid remains there and does not move into the cells.

STEP 3. EFFECT OF ADDING 2 LITERS OF 2.9 PER CENT SODIUM CHLORIDE AFTER OSMOTIC EQUILIBRIUM

	Volume (Liters)	Concentration (mOsm/liter)	Total mOsm
Extracellular fluid	18.9	312.7	5,920
Intracellular fluid	25.1	312.7	7,840
Total body fluid	44.0	312.7	13,760

Thus, one can see from this example that adding 2 liters of a hypertonic sodium chloride solution causes a 4.9-liter increase in extracellular fluid volume while decreasing intracellular fluid volume by 2.9 liters.

This method of calculating changes in intracellular and extracellular fluid volumes and osmolarities can be applied to virtually any clinical problem of fluid volume regulation. The reader should be familiar with such calculations because an understanding of the mathematical aspects of osmotic equilibria between intracellular and extracellular fluid compartments is essential for understanding almost all fluid abnormalities of the body and their treatment.

GLUCOSE AND OTHER SOLUTIONS ADMINISTERED FOR NUTRITIVE PURPOSES

Many types of solutions are administered intravenously to provide nutrition to people who cannot otherwise take adequate amounts of nutrition. Glucose solutions are widely used, and amino acid and homogenized fat solutions are used to a lesser extent. When these solutions are administered, their concentrations of osmotically active substances are usually adjusted nearly to isotonicity or they are given slowly enough that they do not upset the osmotic equilibria of the body fluids. After the glucose or other nutrients are metabolized, an excess of water often remains, especially if additional fluid is ingested. Ordinarily, the kidneys excrete this in the form of a very dilute urine. The net result is, therefore, addition only of the nutrients to the body.

CLINICAL ABNORMALITIES OF FLUID VOLUME REGULATION: HYPONATREMIA AND HYPERNATREMIA

The primary measurement that is readily available to the clinician in evaluating a patient's fluid status often is the plasma sodium concentration. Plasma osmolarity is not routinely measured, but because sodium and its associated anions (mainly chloride) account for more than 90 per cent of the solute in the extracellular fluid, plasma sodium concentration is a reasonable indicator of plasma osmolarity under many conditions. When plasma sodium concentration is reduced below normal (below about 142 mEq/L), a person is said to have *hyponatremia*. When plasma sodium concentration is elevated above normal, a person is said to have *hypernatremia*.

Causes of Hyponatremia: Excess Water or Loss of Sodium

Decreased plasma sodium concentration can result from loss of sodium chloride from the extracellular fluid or addition of excess water to the extracellular fluid. A primary loss of sodium chloride usually results in *hypoosmotic dehydration* and is associated with decreased extracellular fluid volume. Conditions that can cause hyponatremia owing to loss of sodium chloride include excessive sweating, diarrhea, and vomiting. Overuse of diuretics that inhibit the ability of the kidneys to conserve sodium and certain types of sodium-wasting kidney diseases can also cause modest degrees of hyponatremia. Finally, Addison's disease, which results from decreased secretion of the hormone aldosterone, impairs the ability of the kidneys to reabsorb sodium and can cause a modest degree of hyponatremia.

Hyponatremia can also be associated with excess water retention, which dilutes the sodium in the extracellular fluid, a condition that is referred to as *hypoosmotic overhydration*. For example, excessive secretion of antidiuretic hormone, which causes the kidney tubules to reabsorb more water, can lead to hyponatremia and overhydration.

Causes of Hypernatremia: Water Loss or Excess Sodium

Increased plasma sodium concentration, which also causes increased osmolarity, can be due either to loss of water from the extracellular fluid, which concentrates the sodium ions, or to an excess of sodium in the extracellular fluid. When there is primary loss of water from the extracellular fluid, this results in *hyperosmotic dehydration*. This condition can occur from an inability to secrete antidiuretic hormone, which is needed for the kidneys to conserve water. As a result of lack of antidiuretic hormone, the kidneys excrete large amounts of dilute urine (a disorder referred to as *diabetes insipidus*), causing dehydration and increased concentration of sodium chloride in the extracellular fluid. In certain types of renal diseases, the kidneys cannot respond to antidiuretic hormone, also causing a type of "*nephrogenic*" *diabetes insipidus*. A more common cause of hypernatremia associated with decreased extracellular fluid volume is dehydration caused by a water intake that is less than the water lost from the body, as can occur with sweating during heavy exercise.

Hypernatremia can also occur as a result of excessive sodium chloride added to the extracellular fluid. This often results in *hyperosmotic overhydration* because excess extracellular sodium chloride is usually associated with at least some degree of water retention by the kidneys as well. For example, excessive secretion of the sodium-retaining hormone aldosterone can cause a mild degree of hypernatremia and overhydration. The reason that the hypernatremia is not more severe is that increased aldosterone secretion causes the kidneys to reabsorb greater amounts of water as well as sodium.

Thus, in analyzing abnormalities of plasma sodium concentration and deciding on proper therapy, one should first determine whether the abnormality is caused by a primary loss or gain of sodium or a primary loss or gain of water.

EDEMA: EXCESS FLUID IN THE TISSUES

Edema refers to the presence of excess fluid in the body tissues. In most instances, edema occurs mainly in the extracellular fluid compartment, but it can involve intracellular fluids as well.

INTRACELLULAR EDEMA

Two conditions are especially prone to cause intracellular swelling: (1) depression of the metabolic systems of the tissues and (2) lack of adequate nutrition to the cells. For example, when blood flow to a tissue is decreased, the delivery of oxygen and nutrients is reduced; if the blood flow becomes too low to maintain normal tissue metabolism, the cell membrane ionic pumps become depressed. When this occurs, sodium ions that even normally leak into the interior of the cell can no longer be pumped out of the cells, and the excess sodium ions inside the cells cause osmosis of water into the cells. Sometimes this can increase intracellular volume of a tissue area—even of an entire ischemic leg, for example—to two to three times normal. When this occurs, it is usually a prelude to death of the tissue.

Intracellular edema can also occur in inflamed tissues; inflammation usually has a direct effect on the cell membranes to increase their permeability, allowing sodium and other ions to diffuse into the interior of the cell with subsequent osmosis of water into the cells.

EXTRACELLULAR EDEMA

Extracellular fluid edema occurs when there is excess fluid accumulation in the extracellular spaces. There are two general causes of extracellular edema: (1) abnormal leakage of fluid from the plasma to the interstitial spaces across the capillaries and (2) failure of the lymphatics to return fluid from the interstitium back into the blood. The most common clinical cause of interstitial fluid accumulation is excessive capillary fluid filtration, as discussed later.

Factors That Can Increase Capillary Filtration

To understand the causes of excessive capillary filtration, it is useful to review the determinants of capillary filtration discussed in Chapter 16. Mathematically, capillary filtration rate can be expressed as:

$$\text{Filtration} = K_f \times (P_c - P_{if} - \pi_c + \pi_{if}),$$

where K_f is the capillary filtration coefficient (the product of the permeability and surface area of the capillaries), P_{if} is the interstitial fluid hydrostatic pressure, π_c is the capillary plasma colloid osmotic pressure, and π_{if} is the interstitial fluid colloid osmotic pressure. From this equation, one can see *that any one of the following changes can increase the capillary filtration rate:* increased capillary filtration coefficient, increased capillary hydrostatic pressure, or decreased plasma colloid osmotic pressure.

Lymphatic Blockage Causes Edema

When lymphatic blockage occurs, edema can become especially severe because plasma proteins that leak into the interstitium have no way to be removed. The rise in protein concentration raises the colloid osmotic pressure of the interstitial fluid, which draws even more fluid out of the capillaries.

Blockage of lymph flow can be especially severe with infections of the lymph nodes, such as occurs with infection by filaria nematodes. Also, blockage of the lymph vessels can occur in certain types of cancer or after surgery in which lymph vessels are removed or obstructed. For example, after radical mastectomy, large numbers of lymph vessels have been removed, impairing removal of fluid from the breast and arm areas and causing edema and swelling of the tissue spaces. A few lymph vessels eventually regrow after this type of surgery, so that the interstitial edema is usually temporary.

Summary of Causes of Extracellular Edema

A large number of conditions can cause fluid accumulation in the interstitial spaces by abnormal leakage of fluid from the capillaries or by preventing the lymphatics from returning fluid from the interstitium back to the circulation. The following is a list of different conditions that can cause extracellular edema by these two types of abnormalities:

I. Increased Capillary Pressure
   A. Excessive kidney retention of salt and water
      1. Acute or chronic kidney failure
      2. Mineralocorticoid excess
   B. High venous pressure
      1. Heart failure
      2. Venous obstruction
      3. Failure of venous pumps
        (a) Paralysis of muscles
        (b) Immobilized parts of body
        (c) Failure of venous valves
   C. Decreased arteriolar resistance
      1. Excessive body heat
      2. Insufficiency of sympathetic nervous system
      3. Vasodilator drugs
II. Decreased Plasma Proteins
   A. Loss of proteins in urine (nephrotic syndrome)
   B. Loss of protein from denuded skin areas
      1. Burns
      2. Wounds
   C. Failure to produce proteins
      1. Liver disease
      2. Serious protein or caloric malnutrition
III. Increased Capillary Permeability
   A. Immune reactions that cause release of histamine and other immune products
   B. Toxins
   C. Bacterial infections
   D. Vitamin deficiency, especially vitamin C
   E. Prolong ischemia
   F. Burns
IV. Blockage of Lymph Return
   A. Cancer
   B. Infections (e.g., filaria nematodes)
   C. Surgery
   D. Congenital absence or abnormality of lymphatic vessels

From this outline, one can see that there are three major factors that cause increased capillary filtration of fluid and protein into the interstitium: (1) increased capillary hydrostatic pressure, (2) decreased plasma colloid osmotic pressure, and (3) increased capillary permeability, which causes leakage of proteins and fluid through the pores of the capillaries.

EDEMA CAUSED BY HEART FAILURE. One of the most serious and most common causes of edema is heart failure. In heart failure, the heart fails to pump blood normally from the veins into the arteries; this raises venous pressure and capillary pressure, causing more capillary filtration. In addition, the arterial pressure tends to fall, causing decreased excretion of salt and water by the kidneys, which increases blood volume and further raises capillary hydrostatic pressure to cause still more edema. Also, diminished blood flow to the kidneys stimulates secretion of renin, causing increased formation of angiotensin II and increased secretion of aldosterone, both of which cause additional salt and water retention by the kidneys. Thus, in untreated heart failure, all these factors acting together cause serious generalized extracellular edema.

In patients with left-sided heart failure but without significant failure of the right side of the heart, blood is pumped into the lungs normally by the right side of the heart but cannot escape easily from the pulmonary veins to the left side of the heart because this part of the heart has been greatly weakened. Consequently, all the pulmonary vascular pressures, including pulmonary capillary pressure, rise far above normal, causing serious and life-threatening pulmonary edema. When untreated, fluid accumulation in the lungs can rapidly progress, causing death within a few hours.

Edema Caused by Kidney Retention of Salt and Water. As discussed earlier, most sodium chloride added to the blood remains in the extracellular compartment and only small amounts enter the cells. Therefore, in kidney diseases that compromise urinary excretion of salt and water, large amounts of sodium chloride and water are added to the extracellular fluid. Most of this salt and water leak from the blood into the interstitial spaces, but some remains in the blood. The main effects of this are to cause (1) widespread increases in interstitial fluid volume (extracellular edema) and (2) hypertension because of the increase in blood volume, as explained in Chapter 19. As an example, children who develop acute glomerulonephritis, in which the renal glomeruli are injured by inflammation and, therefore, fail to filter adequate amounts of fluid, also develop serious extracellular fluid edema in the entire body; along with the edema, these children usually develop severe hypertension.

Edema Caused by Decreased Plasma Proteins. A reduction in plasma concentration of proteins either because of failure to produce normal amounts of proteins or because of leakage of proteins from the plasma causes the plasma colloid osmotic pressure to fall. This in turn leads to increased capillary filtration throughout the body as well as extracellular edema.

One of the most importance causes of decreased plasma protein concentration is loss of proteins in the urine in certain kidney diseases, a condition referred to as *nephrotic syndrome*. Multiple types of renal diseases can damage the membranes of the renal glomeruli, causing the membranes to become leaky to the plasma proteins and often allowing large quantities of these proteins to pass into the urine. When this loss exceeds the ability of the body to synthesize proteins, a reduction in plasma protein concentration occurs. Serious generalized edema occurs when the plasma protein concentration falls below 2.5 gm/100 ml.

Cirrhosis of the liver is another condition that causes a reduction in plasma protein concentration. Cirrhosis means development of large amounts of fibrous tissue throughout the liver parenchymal cells. One result is failure of these cells to produce sufficient plasma proteins, thus leading to decreased plasma colloid osmotic pressure and the generalized edema that goes with this condition.

Another way that liver cirrhosis causes edema is that the fibrosis compresses the abdominal portal venous drainage vessels as they pass through the liver before emptying back into the general circulation. Blockage of this portal venous outflow raises capillary hydrostatic pressure throughout the gastrointestinal area and further increases filtration of fluid out of the plasma into the intra-abdominal areas. When this occurs, the combined effects of decreased plasma protein concentration and high portal capillary pressures cause transudation of large amounts of fluid and protein into the abdominal cavity, a condition referred to as *ascites.*

SAFETY FACTORS THAT NORMALLY PREVENT EDEMA

Even though many abnormalities can cause edema, usually the abnormality must be severe before edema develops. The reason for this is that three major safety factors prevent fluid accumulation in the interstitial spaces: (1) low compliance of the interstitium when interstitial fluid pressure is in the negative pressure range, (2) the ability of lymph flow to increase 10- to 50-fold, and (3) the washdown of interstitial fluid protein concentration, which reduces interstitial fluid colloid osmotic pressure as capillary filtration increases.

Safety Factor Caused by Low Compliance of the Interstitium in the Negative Pressure Range

In Chapter 16, we note that interstitial fluid hydrostatic pressure in most loose subcutaneous tissues of the body is slightly less than atmospheric pressure, averaging about -3 mm Hg. This slight suction in the tissues helps to hold the tissues together. Figure 25–7 shows the approximate relations between different levels of interstitial fluid pressure and interstitial fluid volume, as extrapolated to the human being from animal studies. At an interstitial fluid pressure of -3 mm Hg, the interstitial fluid volume is about 12 liters. Note also in Figure 25–7 that as long as the interstitial fluid pressure is in the negative range, small changes in interstitial fluid volume are associated with relatively large changes in interstitial fluid hydrostatic pressure. Therefore, in the negative pressure range, the *compliance* of the tissues, defined as the change in volume per millimeter of mercury pressure change, is low.

How does the low compliance of the tissues in the negative pressure range act as a safety factor against edema? To answer this question, recall the determinants of capillary filtration discussed previously. When interstitial fluid hydrostatic pressure increases, this increased pressure tends to oppose further capillary filtration. Therefore, as long as the interstitial fluid hydrostatic pressure is in the negative pressure range, small increases in interstitial fluid volume cause rela-

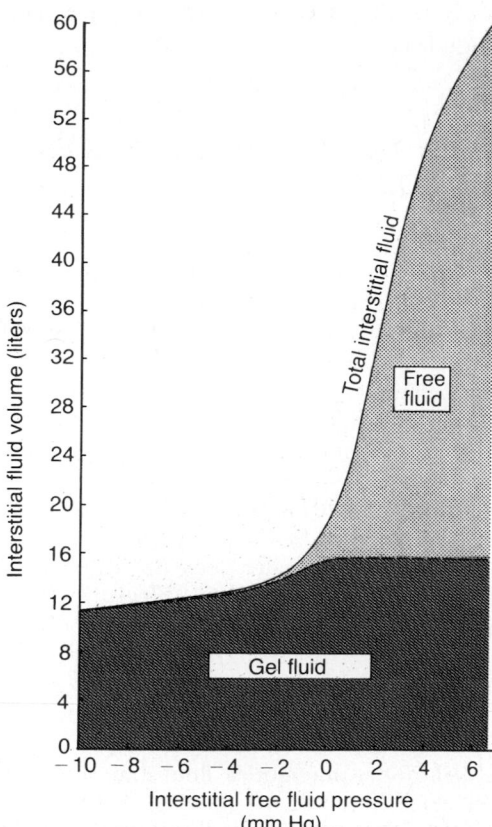

Figure 25–7. Relation between interstitial fluid hydrostatic pressure and interstitial fluid volumes, including total volume, free fluid volume, and gel fluid volume, for loose tissues such as skin. Note that significant amounts of free fluid occur only when the interstitial fluid pressure becomes positive. (Modified from Guyton, Granger, and Taylor: *Physiol. Rev.* 51: 527, 1971).

tively large increases in interstitial fluid hydrostatic pressure, opposing further filtration of fluid into the tissues.

Because the normal interstitial fluid hydrostatic pressure is − 3 mm Hg, the interstitial fluid hydrostatic pressure must increase by about 3 mm Hg before large amounts of fluid will begin to accumulate in the tissues. Therefore, the safety factor against edema for this effect is a change of interstitial fluid pressure of about 3 mm Hg.

Once interstitial fluid pressure rises above 0 mm Hg, the compliance of the tissues increases markedly, allowing large amounts of fluid to accumulate in the tissues with relatively small additional increases in interstitial fluid hydrostatic pressure. Thus, in the positive tissue pressure range, this safety factor against edema is lost because of the large increase in compliance of the tissues that occurs.

IMPORTANCE OF INTERSTITIAL GEL IN PREVENTING FLUID ACCUMULATION IN THE INTERSTITIUM. Note from Figure 25–7 that in normal tissues with negative interstitial fluid pressure, virtually all the fluid in the interstitium is in gel form. That is, the fluid is bound in a proteoglycan meshwork so that there are virtually no "free" fluid spaces larger than a few hundredths of a micrometer in diameter. The importance of the gel is that it prevents fluid from *flowing* easily through the tissues because of impediment from the "brush pile" of trillions of proteoglycan filaments. Also, when the interstitial fluid pressure falls to very negative values, the gel does not contract greatly because the meshwork of proteoglycan filaments offers an elastic resistance to compression. Therefore, in the negative fluid pressure range, the interstitial fluid volume does not change greatly, regardless of whether the degree of suction is only a few millimeters of mercury negative pressure or 10 to 20 mm Hg negative pressure. In other words, the compliance of the tissues is very low in the negative pressure range.

By contrast, when interstitial fluid pressure rises to the positive pressure range, there is a tremendous accumulation of *free fluid* in the tissues. In this pressure range, the tissues are compliant, allowing large amounts of fluid to accumulate with relatively small additional increases in interstitial fluid hydrostatic pressure. Most of the extra fluid that accumulates is "free fluid" because it pushes the brush pile of proteoglycan filaments apart. Therefore, the fluid can now flow freely through the tissue spaces because it is not in the gel form. When this occurs, the edema is said to be *pitting edema* because one can press the thumb against the tissue area and push the fluid out of the area. When the thumb is removed, a pit is left in the skin for a few seconds until the fluid flows back from the surrounding tissues.

This type of edema is distinguished from *nonpitting edema*, which occurs when the tissue cells swell instead of the interstitium swelling or when the fluid in the interstitium becomes clotted with fibrinogen so that it cannot move freely within the tissue spaces.

IMPORTANCE OF THE PROTEOGLYCAN FILAMENTS AS A "SPACER" FOR THE CELLS AND IN PREVENTING RAPID FLOW OF FLUID IN THE TISSUES. The proteoglycan filaments, along with much larger collagen fibrils in the interstitial spaces, act as a "spacer" between the cells. Nutrients and ions do not diffuse readily through cell membranes; therefore, without adequate spacing between the cells, these nutrients, electrolytes, and cell waste products could not be rapidly exchanged between the blood capillaries and cells located at a distance from one another.

The proteoglycan filaments also prevent fluid from flowing too easily through the tissue spaces. If it were not for the proteoglycan filaments, the simple act of a person standing up would cause large amounts of interstitial fluid to flow from the upper body to the lower body. When too much fluid accumulates in the interstitium, as occurs in edema, this extra fluid creates large channels that allow the fluid to flow readily through the interstitium. Therefore, when severe edema occurs in the legs, some of the edema fluid often can be decreased by simply elevating the legs.

Even though fluid does not *flow* easily through the tissues in the presence of the compacted proteoglycan filaments, different substances within the fluid can *diffuse* through the tissues at least 95 per cent as easily as they normally diffuse. Therefore, the usual diffusion of nutrients to the cells and removal of waste products from the cells is not compromised by the proteoglycan filaments of the interstitium.

Increased Lymph Flow as a Safety Factor Against Edema

A major function of the lymphatic system is to return to the circulation the fluid and proteins filtered from the capillaries into the interstitium. Without this continuous return of the filtered proteins and fluid to the blood, the plasma volume would be rapidly depleted and interstitial edema would occur simultaneously.

The lymphatics act as a safety factor against edema because lymph flow can increase 10- to 50-fold when fluid begins to accumulate in the tissues. This allows the lymphatics to carry away large amounts of fluid and proteins in response to increased capillary filtration, preventing the interstitial pressure from rising into the positive pressure range. The safety factor caused by increased lymph flow has been calculated to be about 7 mm Hg.

"Washdown" of the Interstitial Fluid Protein as a Safety Factor Against Edema

As increased amounts of fluid are filtered into the interstitium, the interstitial fluid pressure increases, causing increased lymph flow. In most tissues, the protein concentration of the interstitium decreases as lymph flow is increased because larger amounts of protein are carried away than can be filtered out of the capillaries; the reason for this is that the capillaries are relatively impermeable to proteins, compared with the lymph vessels. Therefore, the proteins are "washed out" of the interstitial fluid as lymph flow increases.

Because the interstitial fluid colloid osmotic pressure caused by the proteins tends to draw fluid out of the capillaries, decreasing the interstitial fluid proteins lowers the net filtration force across the capillaries and tends to prevent further accumulation of fluid. The safety factor from this effect has been calculated to be about 7 mm Hg.

Summary of Safety Factors That Prevent Edema

Putting together all the safety factors against edema we find the following:

1. The safety factor caused by low tissue compliance in the negative pressure range is about 3 mm Hg.
2. The safety factor caused by increased lymph flow is about 7 mm Hg.
3. The safety factor caused by washdown of proteins from the interstitial spaces is 7 mm Hg.

Therefore, the total safety factor against edema is about 17 mm Hg. This means that the capillary pressure in a peripheral tissue could theoretically rise by 17 mm Hg, or approximately double the normal value, before significant edema would occur.

FLUIDS IN THE POTENTIAL SPACES OF THE BODY

Perhaps the best way to describe a "potential space" is to list some examples: the pleural cavity, pericardial cavity, peritoneal cavity, and synovial cavities, including both the joint cavities and the bursae. Virtually all these potential spaces have surfaces that almost touch each other with only a thin layer of fluid in between, and the surfaces slide over each other. To facilitate the sliding, a viscous proteinaceous fluid lubricates the surfaces.

Fluid Is Exchanged Between the Capillaries and the Potential Spaces. The surface membrane of a potential space usually does not offer significant resistance to the passage of fluids, electrolytes, or even proteins, which all move back and forth between the space and the interstitial fluid in the surrounding tissue with relative ease. Therefore, each potential space is in reality a large tissue space. Consequently, fluid in the capillaries adjacent to the potential space diffuses not only into the interstitial fluid but also into the potential space.

Lymphatic Vessels Drain Protein from the Potential Spaces. Proteins collect in the potential spaces because of leakage out of the capillaries, similar to the collection of protein in the interstitial spaces throughout the body. The protein must be removed through lymphatics or other channels and returned to the circulation. Each potential space is either directly or indirectly connected with lymph vessels. In some cases, such as the pleural cavity and peritoneal cavity, large lymph vessels arise directly from the cavity itself.

Edema Fluid in the Potential Spaces Is Called "Effusion." When edema occurs in the subcutaneous tissues adjacent to the potential space, edema fluid usually collects in the potential space as well and this fluid is called *effusion*. Thus, lymph blockage or any one of the multiple abnormalities that can cause excessive capillary filtration can cause effusion in the same way that interstitial edema is caused. The abdominal cavity is especially prone to collect effusion fluid, and in this instance, the effusion is called *ascites*. In serious cases, 20 liters or more of ascitic fluid can accumulate.

The other potential spaces, such as the pleural cavity, pericardial cavity, and joint spaces, can become seriously swollen when there is generalized edema. Also, injury or local infection in any one of the cavities often blocks the lymph drainage, causing isolated swelling in any one of the cavities.

The dynamics of fluid exchange in the pleural cavity are discussed in detail in Chapter 38. These dynamics are mainly representative of all the other potential spaces as well. It is especially interesting that the fluid pressure in most or all of the potential spaces in the nonedematous state is *negative* in the same way that this pressure is negative (subatmospheric) in loose subcutaneous tissue. For instance, the interstitial fluid hydrostatic pressure is about -7 to -8 mm Hg in the pleural cavity, -3 to -5 mm Hg in the joint spaces, and -5 to -6 mm Hg in the pericardial cavity.

REFERENCES

Adair, T. A., and Guyton, A. C.: Modification of lymph by lymph nodes III. Effect of increased lymph hydrostatic pressure. Am. J. Physiol., 249:H777, 1985.

Adair, T. A., and Montani, J. P.: Dynamics of lymph formation and its modification. In Olszewski, W. L. (ed.): Lymph Stasis: Pathophysiology Diagnosis and Treatment. Boca Raton, FL, CRC Press, 1991.

Adoph, E. F.: Physiology of Man in the Desert. New York, Inerscience Publishers, 1947.

Andreoli, T. E. (ed.): Membrane Physiology. New York, Plenum Press, 1980.

Aukland, K., and Reed, R. K.: Interstitial-lymphatic mechanisms in the control of extracellular fluid volume. Physiol. Rev., 73:1, 1993.

Aukland, K.: Why don't our feet swell in the upright position? News Physiolog. Sci., 9:214, 1994.

Badr, K., and Ichikawa, I.: Physical and biological properties of body fluid electrolytes. In Ichikawa, I. (ed.): Pediatric Textbook of Fluids and Electrolytes. Baltimore, Williams & Wilkins, 1990.

Berl, T.: Treating hyponatremia. Kidney Int., 37:1006, 1990.

Cogan, M. G.: Fluid and Electrolytes—Physiology and Pathophysiology. Norwalk, CT, Appleton & Lange, 1991.

Dawson, D. C.: Water transport—principles and perspectives. In Seldin, D. W., and Giebisch, G. (eds.): The Kidney—Physiology and Pathophysiology. New York, Raven Press, 1992.

DeWeer, P.: Cellular sodium-potassium transport. In Seldin, D. W., and Giebisch, G. (eds.): The Kidney: Physiology and Pathophysiology, 2nd Ed., New York, Raven Press, 1992.

Guyton, A. C., et al.: Circulatory Physiology II; Dynamics and Control of Body Fluids. Philadelphia, W. B. Saunders Co., 1975.

Guyton, A. C., et al.: Interstitial fluid pressure. Physiol. Rev., 51:527, 1971.

Levy, M.: Hepatorenal syndrome. Kidney Int., 43:737, 1993.

McKnight, D. C., et al.: Physiological and pathophysiological responses to changes in extracellular osmolality. In Seldin, D. W., and Giebisch, G. (eds.): The Kidney—Physiology and Pathophysiology. New York, Raven Press, 1992.

Michel, C. C.: Fluid movements through capillary walls. In Renkin, E. M., and Michel, C. C. (eds.): Handbook of Physiology, Section 2, The Cardiovascular System, Vol. IV. Baltimore, Williams & Wilkins, 1984.

Narins, R. G. (ed.): Maxwell & Kleeman's Clinical Disorders of Fluid and Electrolyte Metabolism, 5th Ed. New York, McGraw-Hill, 1994.

Rose, B. D.: Clinical physiology of acid-base and electrolyte disorders. New York, McGraw-Hill, 1994.

Rose, B. D.: New approach to disturbances in the plasma sodium concentration. Am. J. Med., 81:1033, 1986.

Schrier, R. W.: Body fluid regulation in health and disease: A unifying hypothesis. Ann. Intern. Med. 113:155, 1990.

Schultz, S. G.: Basic Principles of Membrane Transport, Cambridge, Cambridge University Press, 1980.

Smith, K.: Fluids and Electrolytes: A Conceptual Approach, New York, Churchill Livingstone, 1980.

Spring, K. R., and Hoffmann, E. K.: Cellular volume control. In Seldin, D. W., and Giebisch, G. (eds.): The Kidney—Physiology and Pathophysiology, New York, Raven Press, 1992.

Taylor, A. E.: Capillary fluid filtration, Starling forces and lymph flow. Circ. Res., 49:557–575, 1981.

Taylor, A. E., and Granger, D. N.: Exchange of macromolecules across the microcirculation. In Renkin, E. M., and Michel, C. C. (eds.): Handbook of Physiology. Section 2. The Cardiovascular System, Vol. IV. Baltimore, Williams & Wilkins, 1984.

Verbalis, J. G.: Hyponatremia: answered and unanswered questions. Am. J. Kidney Dis., 18:546, 1991.

Urine Formation by the Kidneys: I. Glomerular Filtration, Renal Blood Flow, and Their Control

CHAPTER 26

MULTIPLE FUNCTIONS OF THE KIDNEYS IN HOMEOSTASIS

Most people are familiar with one important function of the kidneys—to rid the body of waste materials that are either ingested or produced by metabolism. A second function that is especially critical is to control the volume and composition of the body fluids. For water and virtually all electrolytes in the body, balance between intake (due to ingestion or metabolic production) and output (due to excretion or metabolic consumption) is maintained in large part by the kidneys. This regulatory function of the kidneys maintains the stable environment of the cells necessary for them to perform their various activities.

The kidneys perform their most important functions by filtering the plasma and removing substances from the filtrate at a variable rate, depending on the needs of the body. Ultimately, the kidneys "clear" unwanted substances from the filtrate (and therefore from the blood) by excreting them in the urine while returning substances that are needed back to the blood.

Although this chapter and the next few chapters focus mainly on the control of renal excretion, it is important to recognize that the kidneys serve multiple functions, including the following:

Regulation of water and electrolyte balances
Regulation of body fluid osmolality and electrolyte concentrations
Regulation of acid-base balance
Excretion of metabolic waste products and foreign chemicals
Regulation of arterial pressure

Secretion of hormones
Gluconeogenesis

REGULATION OF WATER AND ELECTROLYTE BALANCES. For maintenance of homeostasis, excretion of water and electrolytes must be precisely matched to intake. If intake exceeds excretion, the amount of that substance in the body will increase. If intake is less than excretion, the amount of that substance in the body will decrease.

Intakes of water and many electrolytes usually are governed mainly by a person's eating and drinking habits, necessitating the kidneys to adjust their excretion rates to match intakes of the various substances. Figure 26–1 shows the response of the kidneys to a sudden 10-fold increase in sodium intake from a low level of 30 mEq/day to a high level of 300 mEq/day. Within 2 to 3 days after raising sodium intake, renal excretion also increases to about 300 mEq/day, so that a balance between intake and output is re-established. However, during the 2 to 3 days of renal adaptation to the high sodium intake, there is a modest accumulation of sodium that raises extracellular fluid volume slightly and triggers hormonal changes and other compensatory responses that signal the kidneys to increase their sodium excretion.

The capacity of the kidneys to alter their excretion of sodium in response to changes in sodium intake is enormous. Experimental studies have shown that in normal people, sodium intake can be increased to 1500 mEq/day (more than 10 times normal) or decreased to 10 mEq/day (less than 1/10 normal) with relatively modest changes in extracellular fluid volume or plasma sodium concentration. This is also true for

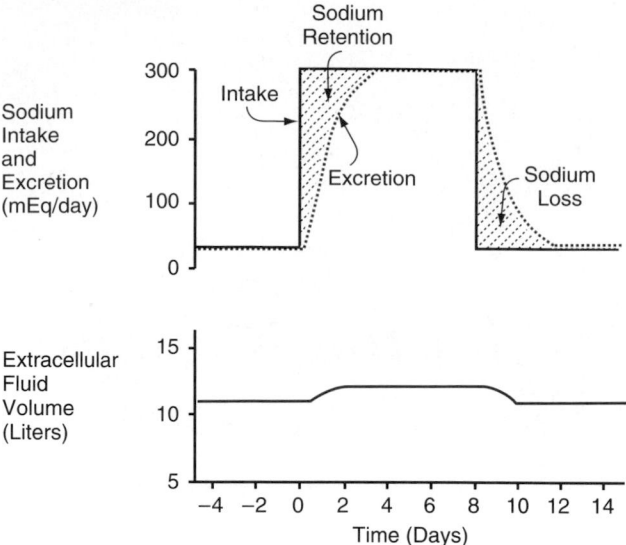

Figure 26-1. Effect of increasing sodium intake 10-fold (from 30 to 300 mEq/day) on urinary sodium excretion and extracellular fluid volume. The shaded areas represent the net sodium retention or the net sodium loss, determined from the difference between sodium intake and sodium excretion.

water and for most other electrolytes, such as chloride, potassium, calcium, hydrogen, magnesium, and phosphate ions. In the next few chapters, we discuss the specific mechanisms that permit the kidneys to perform these amazing feats of homeostasis.

EXCRETION OF METABOLIC WASTE PRODUCTS AND FOREIGN CHEMICALS. The kidneys are the primary means for eliminating waste products of metabolism that are no longer needed by the body. These products include *urea* (from the metabolism of amino acids), *creatinine* (from muscle creatine), *uric acid* (from nucleic acids), the *end products of hemoglobin breakdown* (such as bilirubin), and *metabolites of various hormones.* As with the electrolytes, these waste products must be eliminated from the body as rapidly as they are produced. The kidneys also eliminate most toxins and other foreign substances that are either produced by the body or ingested, such as pesticides, drugs, and food additives.

REGULATION OF ARTERIAL PRESSURE. As discussed in Chapter 19, the kidneys play a dominant role in long-term regulation of arterial pressure by excreting variable amounts of sodium and water. In addition, the kidneys contribute to short-term arterial pressure regulation by secreting vasoactive factors or substances, such as renin, that lead to formation of vasoactive products (for example, angiotensin II).

REGULATION OF ACID-BASE BALANCE. The kidneys contribute to acid-base regulation, along with the lungs and body fluid buffers, by excreting acids and by regulating the body fluid buffer stores. The kidneys are the only means for eliminating from the body certain types of acids generated by metabolism of proteins, such as sulfuric and phosphoric acid.

REGULATION OF ERYTHROCYTE PRODUCTION. The kidneys secrete *erythropoietin,* which stimulates the production of red blood cells, as discussed in Chapter 32. One important stimulus for erythropoietin secretion by the kidneys is hypoxia. In the normal person, the kidneys account for almost all the erythropoietin secreted into the circulation. In people with severe kidney disease or who have had their kidneys removed and been placed on hemodialysis, severe anemia develops as a result of decreased erythropoietin production.

REGULATION OF 1,25-DIHYDROXY VITAMIN D₃ PRODUCTION. The kidneys produce the active form of vitamin D, 1,25-dihydroxy vitamin D_3, by hydroxylating this vitamin at the number "1" position. As discussed in Chapter 79, vitamin D plays an important role in calcium and phosphate regulation.

GLUCOSE SYNTHESIS. The kidneys synthesize glucose from amino acids and other precursors during prolonged fasting, a process referred to as *gluconeogenesis.* The kidneys' capacity to add glucose to the blood during prolonged periods of fasting rivals that of the liver.

With chronic kidney disease or acute failure of the kidneys, these homeostatic functions are disrupted and severe abnormalities of body fluid volumes and composition rapidly occur. With complete renal failure, enough accumulation in the body of potassium, acids, fluid, and other substances occurs within a few days to cause death, unless clinical interventions such as hemodialysis are initiated to restore, at least partially, the body fluid and electrolyte balances.

PHYSIOLOGICAL ANATOMY OF THE KIDNEYS

General Organization of the Kidneys and Urinary Tract

The two kidneys lie on the posterior wall of the abdomen, outside the peritoneal cavity (Fig. 26–2). Each kidney of the adult human weighs about 150 grams and is about the size of a clenched fist. The medial side of each kidney contains an indented region called the *hilum* through which pass the renal artery and vein, lymphatics, nerve supply, and ureter that carries the final urine from the kidney to the bladder, where it is stored until emptied. If the kidney is bisected from top to bottom, the two major regions that can be visualized are the outer *cortex* and the inner region referred to as the *medulla.* The medulla of the kidney is divided into multiple cone-shaped masses of tissue called *renal pyramids.* The base of each pyramid originates at the border between the cortex and medulla and terminates in the *papilla,* which projects into the space of the *renal pelvis,* a funnel-shaped continuation of the upper end of the ureter. The outer border of the pelvis is divided into open-ended pouches called *major calices* that extend downward and divide into

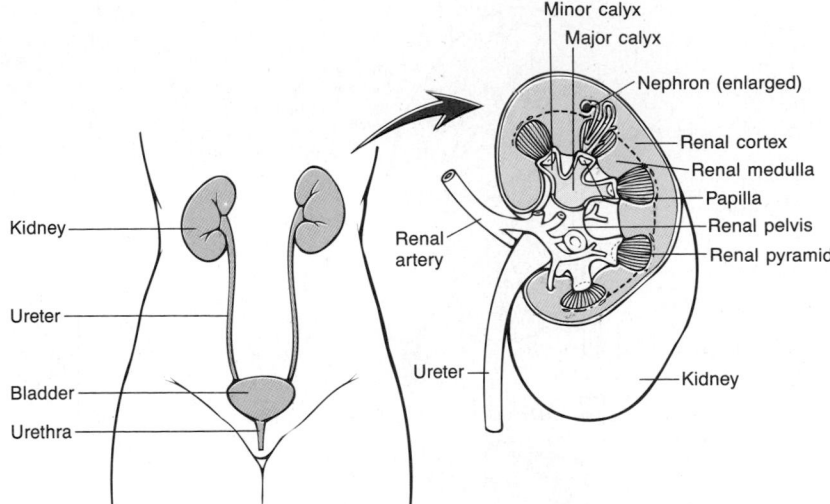

Figure 26-2. General organization of the kidneys and the urinary system.

minor calices, which collect urine from the tubules of each *papilla.* The walls of the calices, pelvis, and ureter contain contractile elements that propel the urine toward the *bladder,* where urine is stored until it is emptied by *micturition,* discussed later.

Renal Blood Supply

Blood flow to the two kidneys is normally 21 per cent of the cardiac output, or about 1200 ml/min. The renal artery enters the kidney through the hilum along with the ureter and renal vein and then branches progressively to form the *interlobar arteries, arcuate arteries, interlobular arteries* (also called radial arteries), and *afferent* arterioles, which lead to the *glomerular capillaries* in the glomeruli where large amounts of fluid and solutes (except the plasma proteins) are filtered to begin urine formation (Fig. 26–3). The distal ends of the capillaries of each glomerulus coalesce to form the *efferent arteriole,* which leads to a second capillary network, the *peritubular capillaries,* that surround the renal tubules.

The renal circulation is unique in that it has two capillary beds, the glomerular and peritubular capillaries, which are arranged in series and separated by the efferent arterioles that help to regulate the hydrostatic pressures in both sets of capillaries. High hydrostatic pressure in the glomerular capillaries (about 60 mm Hg) causes a rapid fluid filtration, whereas a much lower hydrostatic pressure in the peritubular capillaries (about 13 mm Hg) permits rapid fluid reabsorption. By adjusting the resistances of the afferent and efferent arterioles, the kidneys can regulate the hydrostatic pressures in both the glomerular and the peritubular capillaries, thereby changing the rate of glomerular filtration and/or tubular reabsorption in response to body homeostatic demands.

The peritubular capillaries empty into the vessels of the venous system, which run parallel to the arteriolar vessels and progressively form the *interlobular vein,*

arcuate vein, interlobar vein, and *renal vein,* which leaves the kidney beside the renal artery and ureter.

The Nephron Is the Functional Unit of the Kidney

Each kidney in the human is made up of about 1 million *nephrons,* each capable of forming urine. The kidney cannot regenerate new nephrons. Therefore, with renal injury, disease, or normal aging, there is a gradual decrease in nephron number. After age 40, the number of functioning nephrons usually decreases about 10 per cent every 10 years; thus, at age 80, many people have 40 per cent fewer functioning nephrons than they did at age 40. This loss is not life-threatening because adaptive changes in the remaining nephrons allow them to excrete the proper amounts of water, electrolytes, and waste products, as discussed in Chapter 31.

Each nephron has two major components: (1) a *glomerulus* (glomerular capillaries) through which large amounts of fluid are filtered from the blood and (2) a long *tubule* in which the filtered fluid is converted into urine on its way to the pelvis of the kidney (refer again to Fig. 26–3).

The glomerulus is composed of a network of branching and anastomosing glomerular capillaries that, compared with other capillary networks, have a high hydrostatic pressure (about 60 mm Hg). The glomerular capillaries are covered by epithelial cells, and the total glomerulus is encased in *Bowman's capsule.* Fluid filtered from the glomerular capillaries flows into Bowman's capsule and then into the *proximal tubule,* which lies in the cortex of the kidney (Fig. 26–4).

From the proximal tubule, fluid flows into the *loop of Henle* that dips into the renal medulla. Each loop consists of a *descending and an ascending limb.* The walls of the descending limb and the lower end of the ascending limb are very thin and, therefore, are called the *thin segment of the loop of Henle.* After the as-

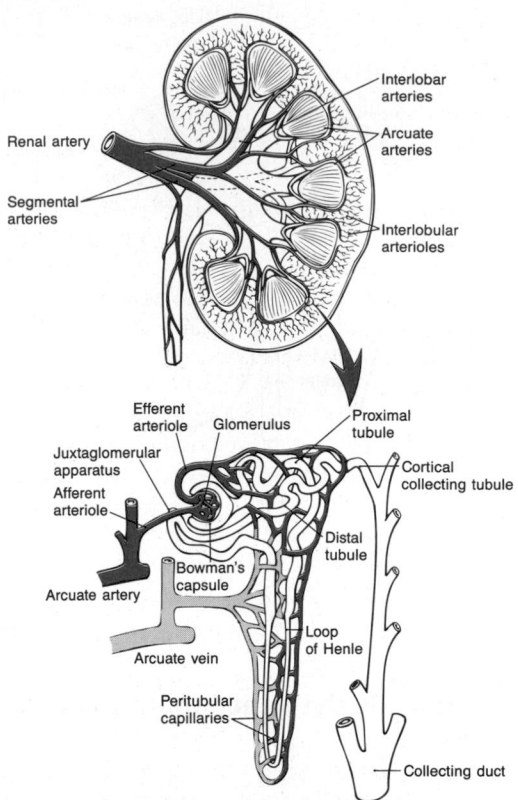

Figure 26-3. Section of the human kidney showing the major vessels that supply the blood flow to the kidney and schematic of the microcirculation of each nephron.

cending limb of the loop has returned part of the way back to the cortex, its wall becomes thick like that of other portions of the tubular system and is, therefore, referred to as the *thick segment of the ascending limb*.

At the end of the thick ascending limb is a short segment, which is actually a plaque in its wall, known

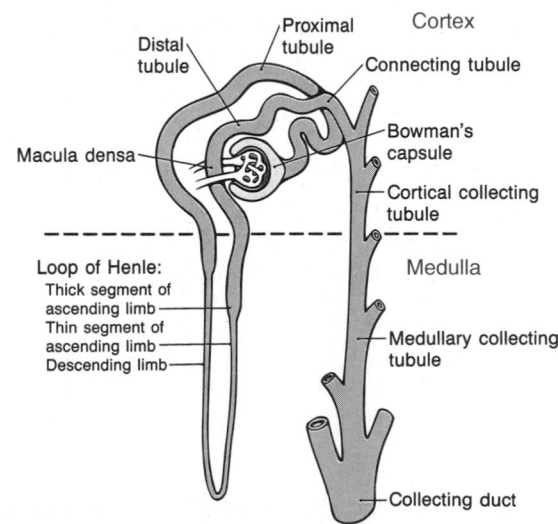

Figure 26-4. Basic tubular segments of the nephron. The relative lengths of the different tubular segments are not drawn to scale.

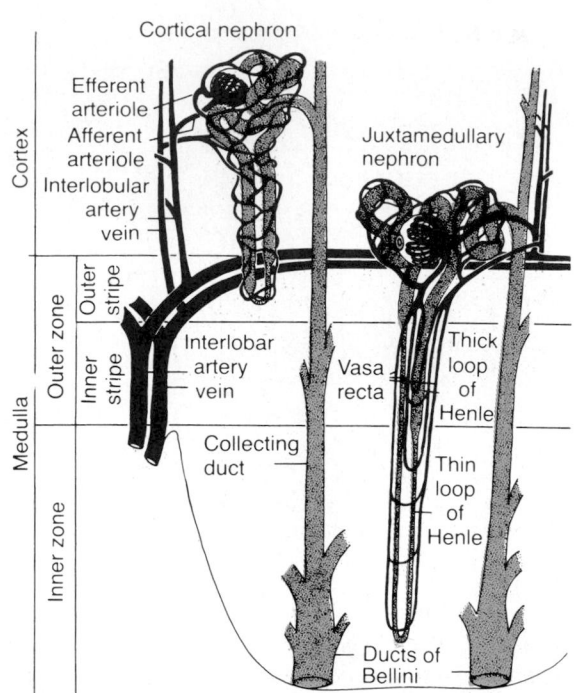

Figure 26-5. Schematic of relations between blood vessels and tubular structures and differences between cortical and juxtamedullary nephrons. (From Pitts: *Physiology of the Kidney and Body Fluids.* Chicago, Yearbook Medical Publishers, 1974.)

as the *macula densa*. As we discuss later, the macula densa plays an important role in controlling nephron function. Beyond the macula densa, fluid enters the *distal tubule,* which, like the proximal tubule, lies in the renal cortex. This is followed by the *connecting tubule* and the *cortical collecting tubule,* which lead to the *cortical collecting duct.* The initial parts of 8 to 10 cortical collecting ducts join to form a single larger collecting duct that runs downward into the medulla and becomes the *medullary collecting duct.* The collecting ducts merge to form progressively larger ducts that eventually empty into the renal pelvis through the tips of the *renal papillae.* In each kidney, there are about 250 of the very large collecting ducts, each of which collects urine from about 4000 nephrons.

REGIONAL DIFFERENCES IN NEPHRON STRUCTURE: CORTICAL AND JUXTAMEDULLARY NEPHRONS. Although each nephron has all the components described above, there are some differences, depending on how deep the nephrons lie within the kidney mass. Those nephrons that have glomeruli located in the outer cortex are called *cortical nephrons;* they have short loops of Henle that penetrate only a short distance into the medulla (Fig. 26–5).

About 20 to 30 per cent of the nephrons have glomeruli that lie deep in the renal cortex near the medulla and are called *juxtamedullary nephrons.* These nephrons have long loops of Henle that dip deeply into the medulla, in some cases all the way to the tips of the renal papillae.

Basic Mechanisms of Renal Excretion

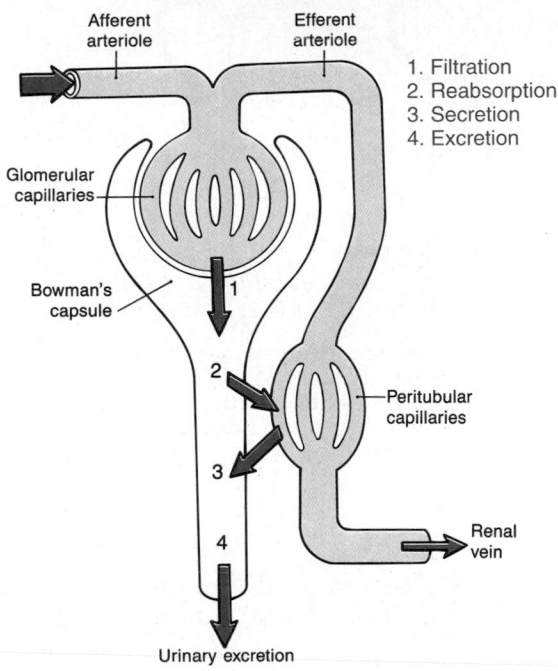

1. Filtration
2. Reabsorption
3. Secretion
4. Excretion

Excretion = Filtration − Reabsorption + Secretion

Figure 26-6. Basic kidney processes that determine the composition of the urine. Urinary excretion rate of a substance is equal to the rate at which the substance is filtered minus its reabsorption rate plus the rate at which it is secreted from the peritubular capillary blood into the tubules.

The vascular structures supplying the juxtamedullary nephrons also differ from those of cortical nephrons. For the cortical nephrons, the entire tubular system is surrounded by an extensive network of peritubular capillaries. For the juxtamedullary nephrons, long efferent arterioles extend from the glomeruli down into the outer medulla and then divide into specialized peritubular capillaries called *vasa recta* that extend downward into the medulla, lying side by side with the loops of Henle. Like the loops of Henle, the vasa recta return toward the cortex and empty into the cortical veins. This specialized network of capillaries in the medulla plays an essential role in the formation of a concentrated urine.

URINE FORMATION RESULTS FROM GLOMERULAR FILTRATION, TUBULAR REABSORPTION, AND TUBULAR SECRETION

The rates at which different substances are excreted in the urine represent the sum of three renal processes, shown in Figure 26–6: (1) glomerular filtration, (2) reabsorption of substances from the renal tubules into the blood, and (3) secretion of substances from the blood into the renal tubules. Expressed mathematically,

Urinary excretion rate = Filtration rate
    − Reabsorption rate + Secretion rate

Urine formation begins with filtration from the glomerulary capillaries into Bowman's capsule of a large amount of fluid that is virtually free of protein. Most substances in the plasma, except for proteins, are freely filtered so that their concentrations in the glomerular filtrate in Bowman's capsule are almost the same as in the plasma. As filtered fluid leaves Bowman's capsule and passes through the tubules, it is modified by reabsorption of water and specific solutes back into the blood or by secretion of other substances from the peritubular capillaries into the tubules.

Figure 26–7 shows the renal handling of four hypothetical substances. The substance shown in Panel A is freely filtered by the glomerular capillaries but is neither reabsorbed nor secreted. Therefore, its excretion rate is equal to the rate at which it was filtered. Certain waste products in the body, such as creatinine, are handled by the kidneys in this manner, allowing excretion of essentially all that is filtered.

In Panel B, the substance is freely filtered but is also partly reabsorbed from the tubules back into the blood. Therefore, the rate of urinary excretion is less than the rate of filtration at the glomerular capillaries. In this case, the excretion rate is calculated as the filtration rate minus the reabsorption rate. This is typical for many of the electrolytes of the body.

In Panel C, the substance is freely filtered at the glomerular capillaries but is not excreted into the urine because all the filtered substance is reabsorbed from the tubules back into the blood. This pattern occurs for some of the nutritional substances in the blood, such as amino acids and glucose, allowing them to be conserved in the body fluids.

The substance in Panel D is freely filtered at the glomerular capillaries and is not reabsorbed, but additional quantities of this substance are secreted from the peritubular capillary blood into the renal tubules. This allows the substance to be rapidly cleared from the blood and excreted in large amounts in the urine. The excretion rate in this case is calculated as filtration rate plus tubular secretion rate.

For each substance in the plasma, a particular combination of filtration, reabsorption, and secretion occurs. The rate at which the substance is excreted in the urine depends on the relative rates of these three basic renal processes.

Filtration, Reabsorption, and Secretion of Different Substances

In general, tubular reabsorption is quantitatively more important than tubular secretion in the formation of urine, but secretion plays an important role in determining the amounts of potassium and hydrogen

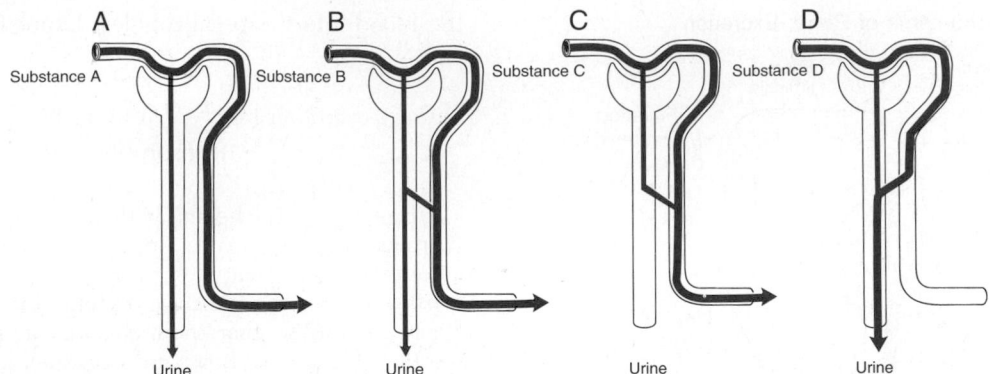

Figure 26-7. Renal handling of four hypothetical substances. The substance in Panel A is freely filtered but not reabsorbed. The substance in Panel B is freely filtered but part of the filtered load is reabsorbed back in the blood. The substance in Panel C is freely filtered but is not excreted in the urine because all the filtered substance is reabsorbed from the tubules into the blood. The substance in Panel D is freely filtered and is not reabsorbed but is secreted from the peritubular capillary blood into the renal tubules.

ions and a few other substances that are excreted in the urine. Most substances that must be cleared from the blood, especially the end products of metabolism such as urea, creatinine, uric acid, and urates, are poorly reabsorbed and are, therefore, excreted in large amounts in the urine. Certain foreign substances and chemicals are also poorly reabsorbed but, in addition, are secreted from the blood into the tubules, so that their excretion rates are high. On the other hand, electrolytes, such as sodium ions, chloride ions, and bicarbonate ions, are highly reabsorbed, so that only small amounts appear in the urine. Certain nutritional substances, such as amino acids and glucose, are completely reabsorbed from the tubules and do not appear in the urine even though large amounts are filtered by the glomerular capillaries.

Each of the processes—glomerular filtration, tubular reabsorption, and tubular secretion—is regulated according to the needs of the body. For example, when there is excess sodium in the body, the rate at which sodium is filtered increases and a smaller fraction of the filtered sodium is reabsorbed, resulting in increased urinary excretion of sodium.

For most substances, the rates of filtration and reabsorption are extremely large relative to the rates of excretion. Therefore, subtle adjustments of filtration or reabsorption can lead to relatively large changes in renal excretion. For example, an increase in glomerular filtration rate (GFR) of only 10 per cent (from 180 to 198 liters/day) would raise urine volume 13-fold (from 1.5 to 19.5 liters/day) if tubular reabsorption remained constant. In reality, changes in glomerular filtration and tubular reabsorption usually act in a coordinated manner to produce the necessary changes in renal excretion.

WHY ARE LARGE AMOUNTS OF SOLUTES FILTERED AND THEN REABSORBED BY THE KIDNEYS? One might question the wisdom of filtering such large amounts of water and solutes and then reabsorbing most of these substances. One advantage of a high GFR is that this

allows the kidneys to rapidly remove waste products from the body that depend primarily on glomerular filtration for their excretion. Most waste products are poorly reabsorbed by the tubules and, therefore, depend on a high GFR for effective removal from the body.

A second advantage of a high GFR is that it allows all the body fluids to be filtered and processed by the kidney many times each day. Because the entire plasma volume is only about 3 liters, whereas the GFR is about 180 liters/day, this means that the entire plasma can be filtered and processed about 60 times each day. This high GFR allows the kidneys to precisely and rapidly control the volume and composition of the body fluids.

GLOMERULAR FILTRATION—THE FIRST STEP IN URINE FORMATION

Composition of the Glomerular Filtrate

Urine formation begins with filtration of large amounts of fluid through the glomerular capillaries into Bowman's capsule. Like most capillaries, the glomerular capillaries are relatively impermeable to proteins, so that the filtered fluid (called the glomerular filtrate) is essentially protein-free and devoid of cellular elements, including red blood cells. The concentrations of other constituents of the plasma, including salts and organic molecules that are not bound to the plasma proteins, such as glucose and amino acids, are similar in the plasma and the glomerular filtrate. Exceptions to this generalization include a few low-molecular-weight substances, such as calcium and fatty acids, that are not freely filtered because they are partially bound to the plasma proteins. Almost one half of the plasma calcium and most of the plasma fatty acids are bound to proteins, and these bound portions are not filtered from the glomerular capillaries.

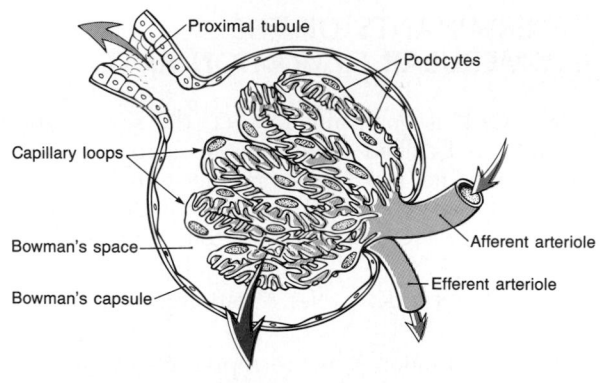

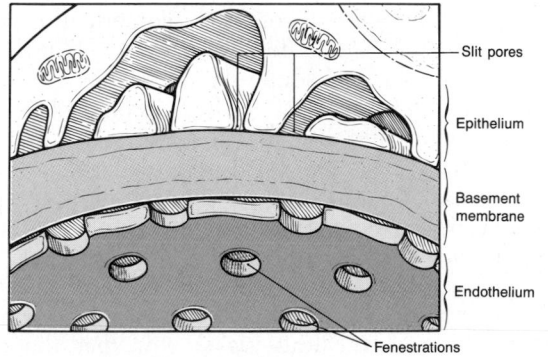

Figure 26–8. *A,* Basic ultrastructure of the glomerular capillaries. *B,* Cross section of the glomerular capillary membrane and its major components: (1) capillary endothelium, (2) basement membrane, and (3) epithelium (podocytes).

GFR Is About 20 Per Cent of the Renal Plasma Flow

As in other capillaries, the GFR is determined by (1) the balance of hydrostatic and colloid osmotic forces acting across the capillary membrane and (2) the capillary filtration coefficient (K_f), the product of the permeability and filtering surface area of the capillaries. The glomerular capillaries have a much higher rate of filtration than most other capillaries because of a high glomerular hydrostatic pressure and a large K_f. In the normal adult human, the GFR averages 125 ml/min, or 180 liters/day. The fraction of the renal plasma flow that is filtered (the filtration fraction) averages about 0.2; this means that about 20 per cent of the plasma flowing through the kidney is filtered by the glomerulary capillaries. The filtration fraction is calculated as follows:

$$\text{Filtration fraction} = \text{GFR/Renal plasma flow.}$$

Glomerular Capillary Membrane

The glomerular capillary membrane is similar to that of other capillaries, except that it has three (instead of the usual two) major layers: (1) *the endothelium* of the capillary, (2) *a basement membrane,* and (3) a layer of *epithelial cells (podocytes)* surrounding the outer surface of the capillary basement membrane (Fig. 26–8).

Together, these layers make up the filtration barrier, which, despite the three layers, filters several hundred times as much water and solutes as the usual capillary membrane. Even with this high rate of filtration, the glomerular capillary membrane normally prevents the filtration of plasma proteins.

The high filtration rate across the glomerular capillary membrane is due partly to its special characteristics. The capillary *endothelium* is perforated by thousands of small holes called *fenestrae,* similar to the fenestrated capillaries found in the liver. Because the fenestrations are relatively large, the endothelium does not act as a major barrier for plasma proteins.

Surrounding the endothelium is the *basement membrane,* which consists of a meshwork of collagen and proteoglycan fibrillae that have large spaces through which can filter large amounts of water and small solutes. The basement membrane effectively prevents filtration of plasma proteins, in part because of strong negative electrical charges associated with the proteoglycans.

The final part of the glomerular membrane is a layer of epithelial cells that line the outer surface of the glomerulus. These cells are not continuous but have long foot-like processes (podocytes) that encircle the outer surface of the capillaries (refer again to Fig. 26–8). The foot processes are separated by gaps called *slit-pores* through which the glomerular filtrate moves. Although the epithelial cells may provide some restriction to filtration, the *primary restriction point for plasma proteins appears to be the basement membrane.*

THE FILTERABILITY OF SOLUTES IS DETERMINED BY THEIR SIZE AND ELECTRICAL CHARGE. The glomerular capillary membrane is thicker than most other capillaries, but it is also much more porous and, therefore, filters fluid at a high rate. Despite the high filtration rate, the glomerular filtration barrier is selective in determining which molecules will filter, based on their size and electrical charge. Table 26–1 lists the effect of molecular size on filterability of different molecules. A filterability of 1.0 means that the substance is as freely filtered as water; a filterability of 0.75 means that the substance is filtered only 75 per cent as rapidly as water. Note that electrolytes such as sodium and small organic compounds such as glucose are freely filtered. As the molecular weight of the mole-

Table 26–1 FILTERABILITY OF SUBSTANCES BY GLOMERULAR CAPILLARIES DECREASES WITH INCREASING MOLECULAR WEIGHT

Substance	Molecular Weight	Filterability
Water	18	1.0
Sodium	23	1.0
Glucose	180	1.0
Inulin	5,500	1.0
Myoglobin	17,000	0.75
Albumin	69,000	0.005

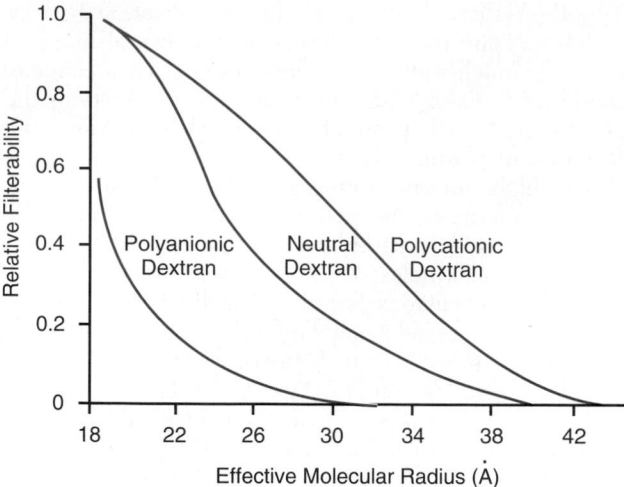

Figure 26-9. Effects of size and electrical charge of dextran on its filterability by the glomerular capillaries. A value of 1.0 indicates that the substance is filtered as freely as water, whereas a value of 0 indicates that it is not filtered. Dextrans are polysaccharides that can be manufactured as neutral molecules, or with negative or positive charges, and with varying molecular weights.

cule approaches that of albumin, the filterability rapidly decreases to low levels, approaching zero.

The filterability of a substance is also determined by charge of the molecule. In general, *negatively charged large molecules are filtered less easily than positively charged molecules of equal molecular size.* The molecular diameter of the plasma protein albumin is only about 6 nanometers, whereas the pores of the glomerular membrane are thought to be about 8 nanometers (80 angstroms). Albumin is restricted from filtration, however, because of its negative charge and the electrostatic repulsion exerted by negative charges of the basement membrane proteoglycans.

Figure 26–9 shows how electrical charge affects the filtration of different molecular weight dextrans by the glomerulus. Dextrans are polysaccharides that can be manufactured as neutral molecules or with negative or positive charges. Note that for any given molecular radius, positively charged molecules are filtered much more readily than negatively charged molecules. Neutral dextrans are also filtered more readily than negatively charged dextrans of equal molecular weight. The reason for these differences in filterability is that the negative charges of the basement membrane provide an important means for restricting plasma proteins, which normally have a net negative charge. In certain kidney diseases, the negative charges on the basement membrane are lost even before there are noticeable changes in kidney histology, a condition that is referred to as *minimal change nephropathy;* as a result of this loss of negative changes on the basement membranes, some of the lower-molecular-weight proteins, especially albumin, are filtered and appear in the urine, a condition known as *proteinuria* or *albuminuria.*

DETERMINANTS OF THE GLOMERULAR FILTRATION RATE

The GFR is determined by (1) the sum of the hydrostatic and colloid osmotic forces across the glomerular membrane, which gives *the net filtration pressure,* and (2) K_f. Expressed mathematically, the GFR equals the product of K_f and the net filtration pressure:

$$GFR = K_f \times \text{Net filtration pressure.}$$

The net filtration pressure represents the sum of the hydrostatic and colloid osmotic forces that either favor or oppose filtration across the glomerular capillaries (Fig. 26–10). These forces include (1) hydrostatic pressure inside the glomerular capillaries (glomerular hydrostatic pressure, P_G), which promotes filtration; (2) the hydrostatic pressure in Bowman's capsule (P_B) outside the capillaries, which opposes filtration; (3) the colloid osmotic pressure of the glomerular capillary plasma proteins (π_G), which opposes filtration; and (4) the colloid osmotic pressure of the proteins in Bowman's capsule (π_B), which promotes filtration. (Under normal conditions, the concentration of protein in the glomerular filtrate is so low that the colloid osmotic pressure of the Bowman's capsule fluid is considered to be zero.)

Although the normal values for the determinants of GFR have not been measured directly in humans, they have been estimated in animals such as dogs and rats. Based on the results in animals, the approximate normal forces favoring and opposing glomerular filtration in humans are believed to be as follows (see Fig. 26–10):

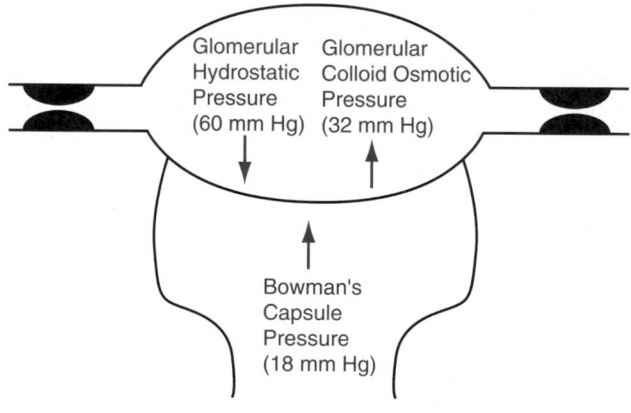

Net Filtration Pressure = Glomerular Hydrostatic Pressure − Bowman's Capsule Pressure − Glomerular Oncotic Pressure
(10 mm Hg)   (60 mm Hg)   (18 mm Hg)   (32 mm Hg)

Figure 26-10. Summary of forces causing filtration by the glomerular capillaries. The values shown are estimates for normal humans.

Forces Favoring Filtration (mm Hg)

Glomerular hydrostatic pressure	60
Bowman's capsule colloid osmotic pressure	0

Forces Opposing Filtration (mm Hg)

Bowman's capsule hydrostatic pressure	18
Glomerular capillary colloid osmotic pressure	32

$$\text{Net filtration pressure} = 60 - 18 - 32$$

$$= +10 \text{ mm Hg}$$

The GFR can, therefore, be expressed as:

$$GFR = K_f \times (P_G - P_B - \pi_G + \pi_B).$$

Some of these values can change markedly under different physiological conditions, whereas others are altered mainly in disease states, as discussed below.

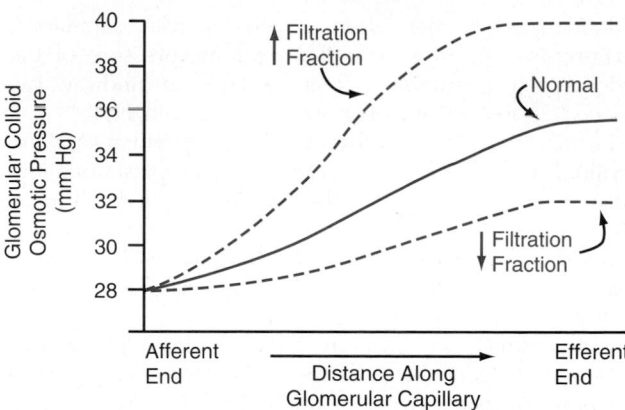

Figure 26-11. Increase in colloid osmotic pressure in plasma flowing through the glomerular capillary. Normally, about one fifth of the fluid in the glomerular capillaries filters into Bowman's capsule, thereby concentrating the plasma proteins that are not filtered. Increases in filtration fraction (GFR/renal plasma flow) increase the rate at which the plasma colloid osmotic pressure rises along the glomerular capillary; decreases in filtration fraction have the opposite effect.

Increased Glomerular Capillary Filtration Coefficient (K_f) Increases GFR

The K_f is a measure of the product of the hydraulic conductivity and surface area of the glomerular capillaries. The K_f cannot be measured directly, but it is estimated experimentally by dividing the rate of glomerular filtration by net filtration pressure:

$$K_f = GFR/\text{Net filtration pressure.}$$

Because total GFR for both kidneys is about 125 ml/min and the net filtration pressure is 10 mm Hg, the normal K_f is calculated to be about 12.5 ml/min/mm Hg of filtration pressure. When K_f is expressed per 100 grams of kidney weight, it averages about 4.2 ml/min/mm Hg per 100 grams of kidney weight, a value about 400 times as high as the K_f of most other capillary systems of the body; the average K_f of all the other tissues in the body is only about 0.01 ml/min/mm Hg per 100 grams. This high K_f for the glomerular capillaries contributes tremendously to their rapid rate of fluid filtration.

Theoretically, increased K_f raises GFR, whereas decreased K_f reduces GFR. However, changes in K_f probably do not provide a primary mechanism for the normal day-to-day regulation of GFR. But some diseases lower K_f by reducing the number of functional glomerular capillaries (thereby reducing the surface area for filtration) or by increasing the thickness of the glomerular capillary membrane and reducing its hydraulic conductivity. For example, chronic, uncontrolled hypertension or diabetes mellitus gradually reduces K_f by increasing the thickness of the glomerular capillary basement membrane and, eventually, by damaging the capillaries so severely that there is a severe or even complete loss of capillary function.

Increased Bowman's Capsule Hydrostatic Pressure Decreases GFR

Direct measurements, using micropipettes, of hydrostatic pressure in Bowman's capsule and at different points in the proximal tubule suggest that a reasonable estimate for Bowman's capsule pressure in humans is about 18 mm Hg under normal conditions. Increasing the hydrostatic pressure in Bowman's capsule reduces GFR, whereas decreasing this pressure raises GFR. However, changes in Bowman's capsule pressure normally do not serve as a primary means for regulating GFR.

In certain pathological states associated with obstruction of the urinary tract, Bowman's capsule pressure can increase markedly, causing serious reduction of GFR. For example, precipitation of calcium or of uric acid may lead to "stones" that lodge in the urinary tract, often in the ureter, thereby obstructing outflow of the urinary tract and raising Bowman's capsule pressure. This reduces GFR and eventually can damage or even destroy the kidney unless the obstruction is relieved.

Increased Glomerular Capillary Colloid Osmotic Pressure Decreases GFR

As blood passes from the afferent arteriole through the glomerular capillaries to the efferent arterioles, the plasma protein concentration increases about 20 per cent (Fig. 26–11). The reason for this is that about one fifth of the fluid in the capillaries filters into Bowman's capsule, thereby concentrating the glomerular plasma proteins that are not filtered. Assuming that the normal colloid osmotic pressure of plasma entering the glomerular capillaries is 28 mm Hg, this value normally rises to about 36 mm Hg by the time the blood reaches the efferent end of the capillaries. Therefore, the average colloid osmotic pressure of the glomerular capillary plasma proteins is midway between 28 and 36 mm Hg, or about 32 mm Hg.

Thus, two factors influence the glomerular capillary colloid osmotic pressure: (1) the arterial plasma colloid osmotic pressure and (2) the fraction of plasma filtered by the glomerular capillaries (filtration fraction). Increasing the arterial plasma colloid osmotic pressure raises the glomerular capillary colloid osmotic pressure, which in turn decreases GFR.

Increasing the filtration fraction also concentrates the plasma proteins and raises the glomerular colloid osmotic pressure (refer again to Fig. 26–11). Because filtration fraction is defined as GFR/renal plasma flow, the filtration fraction can be increased either by raising GFR or by reducing the renal plasma flow. For example, a reduction in renal plasma flow with no initial change in GFR would tend to increase filtration fraction, which would raise the glomerular capillary colloid osmotic pressure and tend to reduce GFR. For this reason, changes in renal blood flow can influence GFR independently of changes in the glomerular hydrostatic pressure. With increasing renal blood flow, a lower fraction of the plasma is initially filtered out of the glomerular capillaries, causing a slower rise in the glomerular capillary colloid osmotic pressure and less inhibitory effect on GFR. *Consequently, even with a constant glomerular hydrostatic pressure, a greater rate of blood flow into the glomerulus tends to increase GFR and a lower rate of blood flow into the glomerulus tends to decrease GFR.*

Increased Glomerular Capillary Hydrostatic Pressure Increases GFR

The glomerular capillary hydrostatic pressure in humans has not been measured directly but has been estimated to be about 60 mm Hg under normal conditions. Changes in glomerular hydrostatic pressure serve as the primary means for physiological regulation of GFR. Increases in glomerular hydrostatic pressure raise GFR, whereas decreases in glomerular hydrostatic pressure reduce GFR.

Glomerular hydrostatic pressure is determined by three variables, each of which is under physiological control: (1) *arterial pressure,* (2) *afferent arteriolar resistance,* and (3) *efferent arteriolar resistance.* Increased arterial pressure tends to raise glomerular hydrostatic pressure and, therefore, to increase GFR. (However, as discussed below, this effect is buffered by autoregulatory mechanisms that maintain a relatively constant glomerular pressure as blood pressure fluctuates.)

Increased resistance of afferent arterioles reduces glomerular hydrostatic pressure and decreases GFR. Conversely, dilation of the afferent arterioles increases both the glomerular hydrostatic pressure and GFR (Fig. 26–12).

Constriction of the efferent arterioles increases the resistance to outflow from the glomerular capillaries.

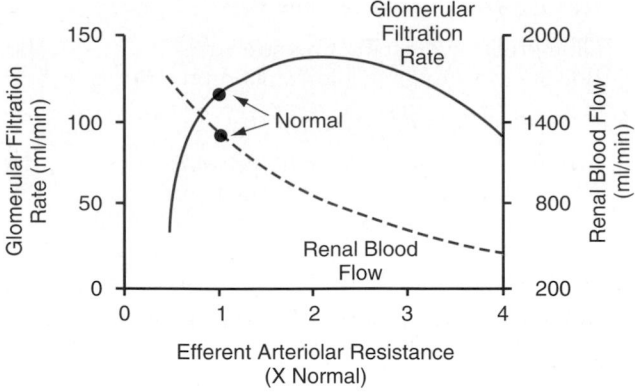

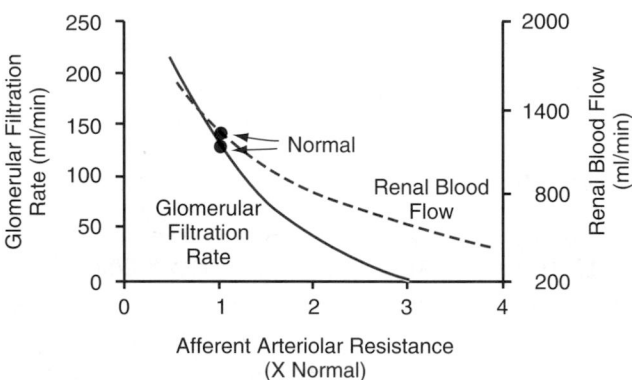

Figure 26-12. Effect of change in afferent arteriolar resistance or efferent arteriolar resistance on glomerular filtration rate and renal blood flow.

This raises the glomerular hydrostatic pressure, and as long as the increase in efferent resistance does not reduce renal blood flow too much, GFR increases slightly (see Fig. 26–12). However, because efferent arteriolar constriction also reduces renal blood flow, the filtration fraction and glomerular colloid osmotic pressure increase as efferent arteriolar resistance increases. Therefore, if the constriction of efferent arterioles is severe (more than about a threefold increase in efferent arteriolar resistance), the rise in colloid osmotic pressure exceeds the increase in glomerular capillary hydrostatic pressure caused by efferent arteriolar constriction. When this occurs, the *net force* for filtration actually decreases, causing a reduction in GFR.

Thus, efferent arteriolar constriction has a biphasic effect on GFR. At moderate levels of constriction there is a slight increase in GFR, but with severe constriction, there is a decrease in GFR. The primary cause of the eventual decrease in GFR is as follows: as efferent constriction becomes severe and as plasma protein concentration increases, there is a rapid, nonlinear increase in colloid osmotic pressure caused by the Donnan effect; the higher the protein concentration, the more rapidly the colloid osmotic pressure rises because of the interaction of ions bound to the

Table 26-2 SUMMARY OF DETERMINANTS OF GFR

$$\uparrow K_f \rightarrow \uparrow GFR$$
$$\uparrow P_B \rightarrow \downarrow GFR$$
$$\uparrow \pi_G \rightarrow \downarrow GFR$$
$$\bullet \quad \uparrow \pi_A \rightarrow \uparrow \pi_G$$
$$\bullet \quad \uparrow FF \rightarrow \uparrow \pi_G$$
$$\uparrow P_G \rightarrow \uparrow GFR$$
$$\bullet \quad \downarrow R_A \rightarrow \uparrow P_G$$
$$\bullet \quad \uparrow R_E \rightarrow \uparrow P_G$$

K_f, glomerular capillary filtration coefficient; P_G, glomerular capillary hydrostatic pressure; P_B, Bowman's capsule hydrostatic pressure; R_A, afferent arteriolar resistance; R_E, efferent arteriolar resistance; π_G, glomerular capillary colloid osmotic pressure; π_A, arterial colloid osmotic pressure; FF, filtration fraction.

plasma proteins, which also exert an osmotic effect, as discussed in Chapter 16.

To summarize, constriction of afferent arterioles always reduces GFR. However, the effect of efferent arteriolar constriction depends on the severity of the constriction; modest efferent constriction raises GFR, but severe efferent constriction (more than a threefold increase in resistance) tends to reduce GFR.

Table 26-2 summarizes the effect of increasing or decreasing the different determinants of GFR.

RENAL BLOOD FLOW

In a normal 70-kilogram man, the combined blood flow through both kidneys is about 1200 ml/min, or about 21 per cent of the cardiac output. Considering the fact that the two kidneys constitute only about 0.4 per cent of the total body weight, one can readily see that they receive an extremely high blood flow compared with other organs.

As with other tissues, blood flow supplies the kidneys with nutrients and removes waste products. However, the high flow to the kidneys greatly exceeds this need. The purpose of this additional flow is to supply enough plasma for the high rates of glomerular filtration that are necessary for precise regulation of body fluid volumes and solute concentrations. As might be expected, the mechanisms that regulate renal blood flow are closely linked to the control of GFR and the excretory functions of the kidneys.

Determinants of Renal Blood Flow

Renal blood flow is determined by the pressure gradient across the renal vasculature (the difference between renal artery and renal vein hydrostatic pressures), divided by the total renal vascular resistance:

Renal blood flow =

$$\frac{(\text{Renal artery pressure} - \text{Renal vein pressure})}{\text{Total renal vascular resistance}}$$

Renal artery pressure is about equal to systemic arterial pressure and renal vein pressure averages about 3 to 4 mm Hg under most conditions. As in other vascular beds, the total vascular resistance through the kidneys is determined by the sum of the resistances in the individual vasculature segments, including the arteries, arterioles, capillaries, and veins (Table 26-3). Most of the renal vascular resistance resides in three major segments: the interlobular arteries, afferent arterioles, and efferent arterioles. Resistance of these vessels is controlled by the sympathetic nervous system, various hormones, and local internal renal control mechanisms, as discussed below. An increase in the resistance of any of the vascular segments of the kidneys tends to reduce the renal blood flow, whereas a decrease in vascular resistance increases renal blood flow if renal artery and renal vein pressures remain constant.

Although changes in arterial pressure have some influence on renal blood flow, the kidneys have effective mechanisms for maintaining renal blood flow and GFR relatively constant over an arterial pressure range between 80 and 170 mm Hg, a process called *autoregulation*. This capacity for autoregulation occurs through mechanisms that are completely intrinsic to the kidneys, as we discuss in detail later in this chapter.

Blood Flow in the Vasa Recta of the Renal Medulla Is Very Low Compared with Flow in the Renal Cortex

The outer part of the kidney, the renal cortex, receives most of the kidney's blood flow because blood flow in the renal medulla accounts for only 1 to 2 per cent of the total renal blood flow. Flow to the renal medulla is supplied by a specialized portion of the peritubular capillary system called the *vasa recta*. These vessels descend into the medulla in parallel with the loops of Henle and then loop back along with the loops of Henle and return to the cortex before emptying into the venous system. As discussed in Chapter 28, the vasa recta play an important role in allowing the kidneys to form a concentrated urine.

Table 26-3 APPROXIMATE PRESSURES AND RESISTANCES IN THE CIRCULATION OF A NORMAL KIDNEY

| Vessel | Pressure in Vessel | | % of Total Renal Vascular Resistance |
	Beginning	*End*	
Renal artery	100	100	~0
Interlobar, arcuate, and interlobular arteries	~100	85	~16%
Afferent arteriole	85	60	~26%
Glomerular capillaries	60	59	~1%
Efferent arteriole	59	18	~43%
Peritubular capillaries	18	8	~10%
Interlobar, interlobular, and arcuate veins	8	4	~4%
Renal vein	4	~4	~0%

Table 26–4 HORMONES AND AUTACOIDS THAT INFLUENCE GFR

Hormone or Autacoid	Effect on GFR
Norepinephrine	↓
Epinephrine	↓
Endothelin	↓
Angiotensin II	←→ (prevents ↓)
Endothelial-derived nitric oxide	↑
Prostaglandins	↑

PHYSIOLOGICAL CONTROL OF GLOMERULAR FILTRATION AND RENAL BLOOD FLOW

The determinants of GFR that are most variable and subject to physiological control include the glomerular hydrostatic pressure and the glomerular capillary colloid osmotic pressure. These variables in turn are influenced by the sympathetic nervous system, hormones and autacoids (vasoactive substances that are released in the kidneys and act locally), and other feedback controls that are intrinsic to the kidneys.

Sympathetic Nervous System Activation Decreases GFR

Essentially all the blood vessels of the kidneys, including both the afferent and the efferent arterioles, are richly innervated by sympathetic nerve fibers. Strong activation of the renal sympathetic nerves can constrict the renal arterioles and decrease renal blood flow and GFR. Moderate or mild sympathetic stimulation has little influence on renal blood flow and GFR. For example, reflex activation of the sympathetic nervous system resulting from moderate decreases in pressure at the carotid sinus baroreceptors or cardiopulmonary receptors has little influence on renal blood flow or GFR. Moreover, because the baroreceptors adapt within minutes or hours to sustained changes in arterial pressure, it is unlikely that these reflex mechanisms have an important role in long-term control of renal blood flow and GFR. The renal sympathetic nerves seem to be most important in reducing GFR during severe, acute disturbances, lasting for a few minutes to a few hours, such as those elicited by the defense reaction, brain ischemia, or severe hemorrhage. In the normal resting person, there appears to be little sympathetic tone to the kidneys.

Hormonal and Autacoid Control of Renal Circulation

There are several hormones and autacoids that can also influence GFR and renal blood flow, as summarized in Table 26–4.

NOREPINEPHRINE, EPINEPHRINE, AND ENDOTHELIN CONSTRICT RENAL BLOOD VESSELS AND DECREASE GFR. Hormones that constrict afferent and efferent arterioles, causing reductions in GFR and renal blood flow, include *norepinephrine* and *epinephrine* released from the adrenal medulla. In general, blood levels of these hormones parallel the activity of the sympathetic nervous system; thus, norepinephrine and epinephrine have little influence on renal hemodynamics except under extreme conditions, such as severe hemorrhage.

Another vasoconstrictor, *endothelin,* is a peptide that can be released by damaged vascular endothelial cells of the kidneys as well as other tissues. The physiological role of this autacoid is not completely understood. However, endothelin may contribute to hemostasis (minimizing blood loss) when a blood vessel is severed, which damages the endothelium and releases this powerful vasoconstrictor. Researchers have also found that plasma endothelin levels are increased in certain disease states associated with vascular injury, such as toxemia of pregnancy, acute renal failure, and chronic uremia; whether endothelin contributes to renal vasoconstriction and decreased GFR in these pathophysiological conditions is unknown.

ANGIOTENSIN II CONSTRICTS EFFERENT ARTERIOLES. A powerful renal vasoconstrictor, *angiotensin II,* can be considered as a circulating hormone as well as a locally produced autacoid because it is formed in the kidneys as well as in the systemic circulation. Because angiotensin II preferentially constricts efferent arterioles, increased angiotensin II levels raise glomerular hydrostatic pressure while reducing renal blood flow. It should be kept in mind that increased angiotensin II formation usually occurs in circumstances associated with decreased arterial pressure or volume depletion, which tend to decrease GFR. In these circumstances, the increased level of angiotensin II, by constricting efferent arterioles, helps to *prevent* decreases in glomerular hydrostatic pressure and GFR; at the same time, though, the reduction in renal blood flow caused by efferent arteriolar constriction contributes to decreased flow through the peritubular capillaries, which in turn causes increased reabsorption of sodium and water, as discussed in Chapter 27.

Thus, increased angiotensin II levels that occur with a low sodium diet or volume depletion help to preserve GFR and to maintain a normal excretion of metabolic waste products, such as urea and creatinine, that depend on glomerular filtration for their excretion; at the same time, the angiotensin II–induced constriction of efferent arterioles causes increased reabsorption of sodium and water, which helps to restore blood volume and blood pressure. This effect of angiotensin II in helping to "autoregulate" GFR is discussed further.

ENDOTHELIAL-DERIVED NITRIC OXIDE DECREASES RENAL VASCULAR RESISTANCE AND INCREASES GFR. An autacoid that decreases renal vascular resistance and is released by the vascular endothelium throughout the body is *endothelial-derived nitric oxide.* A basal level of nitric oxide production appears to be important for preventing excessive vasoconstriction of the kidneys and allowing them to excrete normal amounts of sodium and water. Administration of drugs that inhibit the formation of nitric oxide increases renal vascular resistance and decreases GFR and urinary sodium excretion, eventually causing high blood pressure. This finding has led researchers to speculate that impaired nitric oxide production may contribute to renal vasoconstriction and increased blood pressure in some hypertensive patients.

OTHER RENAL VASODILATORS THAT INCREASE GFR. Hormones and autacoids that cause vasodilation and increased renal blood flow and GFR include the prostaglandins (PGE_2 and PGI_2) and bradykinin. These substances are discussed in Chapter 17. Although these vasodilators do not appear to be of major importance in regulating renal blood flow or GFR in normal conditions, they may dampen the renal vasoconstrictor effects of the sympathetic nerves on angiotensin II, especially their effects on the afferent arterioles. By opposing vasoconstriction of afferent arterioles, the prostaglandins may help to prevent excessive reductions in GFR and renal blood flow. For example, the administration of nonsteroidal anti-inflammatory agents, such as aspirin, that inhibit prostaglandin synthesis may cause significant reductions in GFR under stressful conditions, such as volume depletion or after surgery.

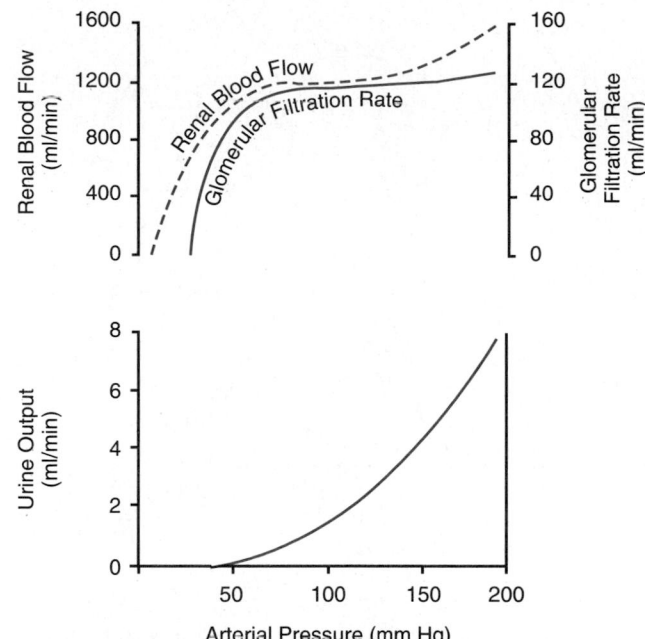

Figure 26-13. Autoregulation of renal blood flow and glomerular filtration rate but lack of autoregulation of urine flow during changes in renal arterial pressure.

AUTOREGULATION OF GFR AND RENAL BLOOD FLOW

Feedback mechanisms intrinsic to the kidneys normally keep the renal blood flow and GFR relatively constant, despite marked changes in arterial blood pressure. These mechanisms still function in blood-perfused kidneys that have been removed from the body, independent of systemic influences. This relative constancy of GFR and renal blood flow is referred to as *autoregulation* (Fig. 26–13).

The primary function of blood flow autoregulation in most other tissues besides the kidneys is to maintain delivery of oxygen and other nutrients to the tissues at a normal level and to remove the waste products of metabolism, despite changes in the arterial pressure. In the kidneys, the normal blood flow is much higher than required for these functions. The major function of autoregulation in the kidneys is to maintain a relatively constant GFR and to allow precise control of renal excretion of water and solutes.

The GFR normally remains autoregulated (that is, remains relatively constant) throughout the day, despite considerable arterial pressure fluctuations that occur during a person's usual activities. For instance, a decrease in arterial pressure to as low as 75 mm Hg or an increase to as high as 160 mm Hg changes GFR only a few percentage points. In general, renal blood flow is autoregulated in parallel with GFR, but GFR is

more efficiently autoregulated under certain conditions.

Importance of GFR Autoregulation in Preventing Extreme Changes in Renal Excretion

The autoregulatory mechanisms of the kidney are not 100 per cent perfect, but they do prevent potentially large changes in GFR and renal excretion of water and solutes that would otherwise occur with changes in blood pressure. One can understand the quantitative importance of autoregulation by considering the relative magnitudes of glomerular filtration, tubular reabsorption, and renal excretion and the changes in renal excretion that would occur without autoregulatory mechanisms. Normally, GFR is about 180 liters/day and tubular reabsorption is 178.5 liters/day, leaving 1.5 liters/day of fluid to be excreted in the urine. In the absence of autoregulation, a relatively small increase in blood pressure (from 100 to 125 mm Hg) would cause a similar 25 per cent increase in GFR (from about 180 to 225 liters/day). If tubular reabsorption remained constant at 178.5 liters/day, this would increase the urine flow to 46.5 liters/day (the difference between GFR and tubular reabsorption), making a total increase in urine of more than 30-fold! Because the total plasma volume is only about 3 liters, such a change would quickly deplete the blood volume.

But in reality, such a change in arterial pressure exerts much less of an effect on urine volume for two reasons: (1) renal autoregulation prevents large changes in GFR that would otherwise occur and

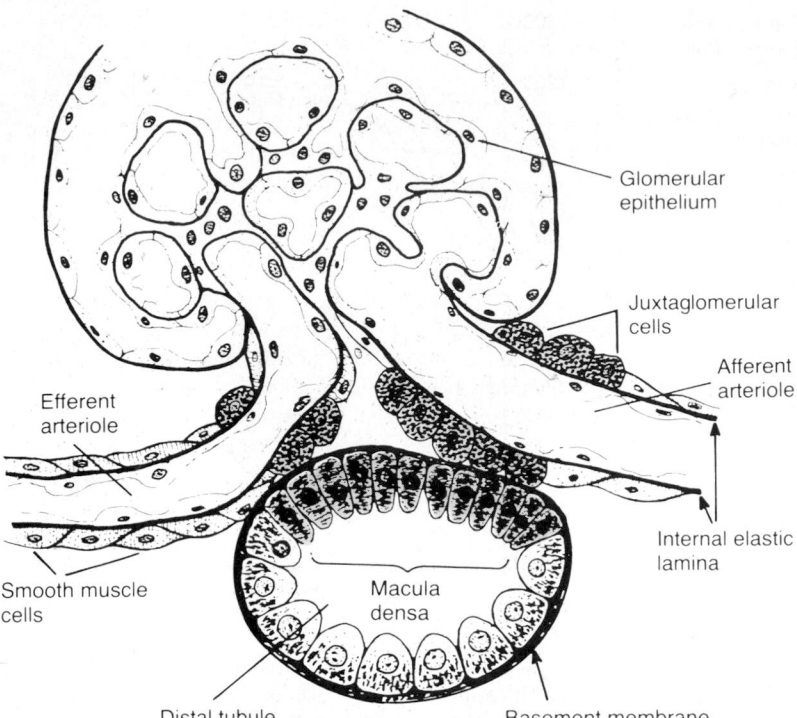

Glomerular
epithelium

Juxtaglomerular
cells

Afferent
arteriole

Efferent
arteriole

Internal elastic
lamina

Smooth muscle
cells

Macula
densa

Distal tubule

Basement membrane

Figure 26-14. Structure of the juxtaglomerular apparatus, demonstrating its possible feedback role in the control of nephron function.

(2) there are additional adaptive mechanisms in the renal tubules that allow them to increase their reabsorption rate when GFR rises, a phenomenon referred to as *glomerulotubular balance* (discussed in Chapter 27). Yet, even with these special control mechanisms, changes in arterial pressure still have significant effects on renal excretion of water and sodium; this is referred to as *pressure diuresis* or *pressure natriuresis,* and it is crucial in regulation of body fluid volumes and arterial pressure, as discussed in Chapter 29.

Role of Tubuloglomerular Feedback in Autoregulation of GFR

To perform the function of autoregulation, the kidneys have a feedback mechanism that links changes in sodium chloride concentration at the macula densa with the control of renal arteriolar resistance. This feedback helps to ensure a relatively constant delivery of sodium chloride to the distal tubule and helps to prevent spurious fluctuations in renal excretion that would otherwise occur. In many circumstances, this feedback autoregulates renal blood flow and GFR in parallel. However, because this mechanism is specifically directed toward stabilizing GFR, there are instances in which GFR is autoregulated at the expense of changes in renal blood flow, as discussed below.

The tubuloglomerular feedback mechanism has two components that act together to control GFR: (1) an afferent arteriolar feedback mechanism and (2) an efferent arteriolar feedback mechanism. These feedback mechanisms depend on special anatomical arrangements of the *juxtaglomerular complex* (Fig. 26–14).

The juxtaglomerular complex consists of *macula*

densa cells in the initial portion of the distal tubule and *juxtaglomerular cells* in the walls of the afferent and efferent arterioles. The macula densa is a specialized group of epithelial cells in the distal tubules that comes in close contact with the afferent and efferent arterioles. The macula densa cells contain Golgi apparati, which are intracellular secretory organelles, directed toward the arterioles, suggesting that these cells may be secreting a substance toward the arterioles.

DECREASED MACULA DENSA SODIUM CHLORIDE CAUSES DILATION OF AFFERENT ARTERIOLES AND INCREASED RENIN RELEASE. The macula densa cells sense changes in volume delivery to the distal tubule by way of signals that are not completely understood. Experimental studies suggest that decreased GFR may slow the flow rate in the loop of Henle, causing an increase in reabsorption of sodium and chloride ions in the ascending loop of Henle and thereby reducing the concentration of sodium chloride at the macula densa cells. This decrease in sodium chloride concentration in turn initiates a signal from the macula densa that has two effects (Fig. 26–15): (1) it decreases resistance of the afferent arterioles, which raises glomerular hydrostatic pressure and helps to return GFR toward normal, and (2) it increases renin release from the juxtaglomerular cells of the afferent and efferent arterioles, which are the major storage sites for renin. Renin released from these cells then functions as an enzyme to increase the formation of angiotensin I, which is converted to angiotensin II. Finally, the angiotensin II constricts the efferent arterioles, thereby increasing glomerular hydrostatic pressure and returning GFR toward normal.

These two components of the tubuloglomerular

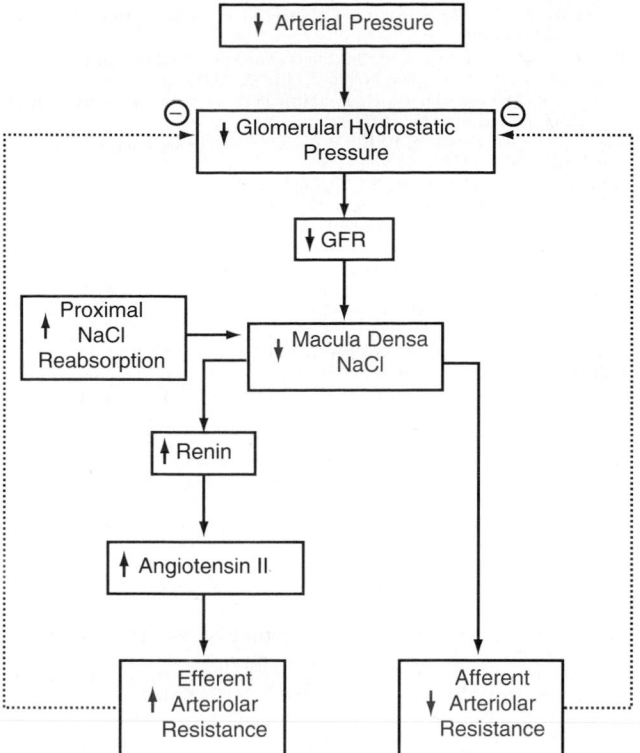

Figure 26-15. Macula densa feedback mechanism for autoregulation of glomerular hydrostatic pressure and glomerular filtration rate during decreased renal arterial pressure.

feedback mechanism, operating together by way of the special anatomical structure of the juxtaglomerular apparatus, provide feedback signals to both the afferent and the efferent arterioles for efficient autoregulation of GFR during changes in arterial pressure. When both of these mechanisms are functioning together, the GFR changes only a few percentage points even with large fluctuations in arterial pressure between the limits of 75 and 160 mm Hg.

BLOCKADE OF ANGIOTENSIN II FORMATION FURTHER REDUCES GFR DURING RENAL HYPOPERFUSION. As discussed above, a preferential constrictor action of angiotensin II on efferent arterioles helps to prevent serious reductions in glomerular hydrostatic pressure and GFR when renal perfusion pressure falls below normal. The administration of drugs that block the formation of angiotensin II (angiotensin converting enzyme inhibitors) or that block the action of angiotensin II (angiotensin II antagonists) causes greater reductions in GFR than usual when the renal arterial pressure falls below normal. Therefore, an important complication of using these drugs to treat patients who have hypertension because of renal artery stenosis (partial blockage of the renal artery) is a severe decrease in GFR that can actually cause acute renal failure. Nevertheless, angiotensin II blocking drugs can be useful therapeutic agents in many patients with hypertension, congestive heart failure, and other conditions as long as they are monitored to ensure that severe decreases in GFR do not occur.

Myogenic Autoregulation of Renal Blood Flow and GFR

A second mechanism that contributes to the maintenance of a relatively constant renal blood flow and GFR is the ability of individual blood vessels to resist stretching during increased arterial pressure, a phenomenon referred to as the *myogenic mechanism.* Studies of individual blood vessels (especially small arterioles) throughout the body have shown that they respond to increased wall tension or wall stretch by contraction of the vascular smooth muscle. Stretch of the vascular wall allows increased movement of calcium ions from the extracellular fluid into the cells, causing them to contract through the mechanisms discussed in Chapter 8. This contraction serves to prevent overdistention of the vessel and at the same time, by raising vascular resistance, helps to prevent excessive increases in renal blood flow and GFR when arterial pressure increases.

Although the myogenic mechanism probably operates in most arterioles throughout the body, its importance in renal blood flow and GFR autoregulation has been questioned by some physiologists because this pressure-sensitive mechanism has no means of directly detecting changes in renal blood flow or GFR per se.

Other Factors That Increase Renal Blood Flow and GFR: High Protein Intake and Increased Blood Glucose

Although renal blood flow and GFR are relatively stable under most conditions, there are circumstances in which these variables change significantly. For example, *a high protein intake is known to increase both renal blood flow and GFR.* With a chronic high-protein diet, such as occurs with diets that contain large amounts of meat, the increase in GFR and renal blood flow is due partly to growth of the kidneys. However, GFR and renal blood flow increase 20 to 30 per cent within 1 or 2 hours after a person eats a high-protein meat meal.

The exact mechanisms by which this occurs are still not completely understood, but one possible explanation is the following: A high-protein meal increases release of amino acids into the blood, which are reabsorbed in the proximal tubule. Because amino acids and sodium are reabsorbed together by the proximal tubules, increased amino acid reabsorption also stimulates sodium reabsorption in the proximal tubules. This decreases sodium delivery to the macula densa, which in turn elicits a tubuloglomerular feedback–mediated decrease in resistance of the afferent arterioles, as discussed above. The decreased afferent arteriolar resistance then raises renal blood flow and GFR. This increased GFR allows sodium excretion to be maintained at a nearly normal level while increasing the excretion of the waste products of protein metabolism, such as urea.

A similar mechanism may also explain the marked increases in renal blood flow and GFR that occur with large increases in blood glucose levels in uncontrolled diabetes mellitus. Because glucose, like some of the amino acids, is also reabsorbed along with sodium in the proximal tubule, increased glucose delivery to the tubules causes them to reabsorb excess sodium along with glucose. This in turn decreases delivery of sodium chloride to the macula densa, activating a tubuloglomerular feedback–mediated dilation of the afferent arterioles and subsequent increases in renal blood flow and GFR.

These examples demonstrate that renal blood flow and GFR per se are not the primary variables controlled by the tubuloglomerular feedback mechanism. The main purpose of this feedback is to ensure a constant delivery of sodium chloride to the distal tubule, where final processing of the urine takes place. Thus, disturbances that tend to increase reabsorption of sodium chloride at tubular sites before the macula densa would tend to elicit increased renal blood flow and GFR, which would then help to return distal sodium chloride delivery toward normal so that normal rates of sodium and water excretion can be maintained (refer again to Fig. 26–15).

An opposite sequence of events occurs when proximal tubular reabsorption is reduced. For example, with damaged proximal tubules, which can occur as a result of heavy metals or large doses of drugs such as tetracyclines, there is decreased ability of the tubules to reabsorb sodium chloride. As a consequence, large amounts of sodium chloride are delivered to the distal tubule, which, without appropriate compensations, would quickly cause excessive volume depletion. One of the important compensatory responses appears to be a tubuloglomerular feedback–mediated renal vasoconstriction that occurs in response to the increased sodium chloride delivery to the macula densa in these circumstances. These examples again demonstrate the importance of this feedback mechanism in making sure that the distal tubule receives the proper rate of delivery of sodium chloride, other tubular fluid solutes, and tubular fluid volume so that appropriate amounts of these substances are excreted in the urine.

REFERENCES

Arendshorst, W. J., and Navar, L. G.: Renal circulation and glomerular hemodynamics. In Schirer, R. W., and Gottschalk, C. W. (eds.): Diseases of the Kidney. 5th Ed. Boston, Little, Brown, 1993.

Beeuwkes, R., III: The vascular organization of the kidney. Annu. Rev. Physiol., 42:531, 1980.

Braam, B., et al.: Relevance of the tubuloglomerular feedback mechanism in pathophysiology. J. Am. Soc. Nephrol., 4:1275, 1993.

Brenner, B. M., and Humes, H. D.: Mechanics of glomerular ultrafiltration. N. Engl. J. Med., 148:277, 1977.

Davis, J. O., and Freeman, R. H.: Mechanisms regulating renin release. Physiol. Rev., 56:1, 1976.

Dworkin, L. D., and Brenner, B. M.: Biophysical basis of glomerular filtration. In Seldin, D. W., and Giebisch, G. (eds.): The Kidney: Physiology and Pathophysiology, 2nd Ed. New York, Raven Press, 1992.

Hall, J. E., and Brands, M. W.: Intrarenal actions of angiotensin II: Physiology and pathophysiology. In McGregor, G. A., and Sever, P. (eds.): Current Advances in ACE Inhibition. London, Churchill Livingstone, 1993.

Hall, J. E., and Brands, M. W.: Intrarenal and circulating angiotensin II and renal function. In Robertson, J. I. S., and Nicholls, M. G. (eds.): The Renin-Angiotensin System. London, Gower Medical Publishing, 1993.

Hall, J. E., and Brands, M. W.: The renin-angiotensin-aldosterone system: Renal mechanisms and circulatory hemostasis. In Seldin, D. W., and Giebisch, G. (eds.): The Kidney: Physiology and Pathophysiology, 2nd Ed. New York, Raven Press, 1992, p. 1455–1504.

Hura, C., and Stein, J. H.: Renal blood flow. In Windhager, E. E. (ed.): Handbook of Physiology, Section 8, Renal Physiology. New York, Oxford University Press, 1992.

Kriz, W., and Kaissling, B.: Structural organization of the mammalian kidney. In Seldin, D. W., and Giebisch, G. (eds.): The Kidney: Physiology and Pathophysiology. New York, Raven Press, 1992.

Navar, L. G.: Renal autoregulation: Perspectives from whole kidney and single nephron studies. Am. J. Physiol., 234:F357, 1978.

Persson, A. E. G., and Boberg, U.: The Juxtaglomerular Apparatus. Amsterdam, Elsevier, 1988.

Schnermann, J., et al.: Tubuloglomerular feedback control of renal vascular resistance. In Windhager, E. E. (ed.): Handbook of Physiology, Section 8, Renal Physiology. New York, Oxford University Press, 1992.

Smith, H. W.: The Kidney: Structure and Function in Health and Disease. New York, Oxford University Press, 1951.

Steinhausen, M., et al.: Glomerular blood flow. Kidney Int., 38:769, 1990.

Tisher, C. C., and Madsen, K. M.: Anatomy of the Kidney. In Brenner, B. M., and Rector, F. C., Jr. (eds.): The Kidney, 4th Ed. Philadelphia, W. B. Saunders Co., 1991.

Ulfendahl, H. R., and Wolgast, M.: Renal circulation and lymphatics. In Seldin, D. W., and Giebisch, G. (eds.): The Kidney: Physiology and Pathophysiology. New York, Raven Press, 1992.

Vander, A. J.: Renal Physiology, 4th Ed. New York, McGraw-Hill, 1991.

Wesson, L. G., Jr.: Physiology of the Human Kidney. New York, Grune & Stratton, 1969.

Urine Formation by the Kidneys: II. Tubular Processing of the Glomerular Filtrate

CHAPTER 27

REABSORPTION AND SECRETION BY THE RENAL TUBULES

As the glomerular filtrate enters the renal tubules, it flows sequentially through the successive parts of the tubule—the *proximal tubule,* the *loop of Henle,* the *distal tubule,* the *collecting tubule,* and, finally, the *collecting duct*—before it is excreted as urine. Along this course, some substances are selectively reabsorbed from the tubules back into the blood, whereas others are secreted from the blood into the tubular lumen. Eventually, the urine that is formed and all the substances in the urine represent the sum of three basic renal processes—glomerular filtration, tubular reabsorption, and tubular secretion—as follows:

Urinary excretion = Glomerular filtration

        − Tubular reabsorption + Tubular secretion

For many substances, reabsorption plays a much more important role than does secretion in determining the final urinary excretion rate. However, secretion accounts for a significant amount of potassium ions, hydrogen ions, and a few other substances that appear in the urine.

Tubular Reabsorption Is Selective and Quantitatively Large

Table 27–1 shows the renal handling of several substances that are all freely filtered in the kidneys and reabsorbed at variable rates.

The rate at which each of these substances is filtered is calculated as

Filtration = Glomerular filtration rate

        × Plasma concentration

This calculation assumes that the substance is freely filtered and not bound to plasma proteins. For example, if plasma glucose concentration is 1 gm/liter, the amount of glucose filtered each day is about 180 liters/day × 1 gm/liter, or 180 gm/day. Because virtually none of the filtered glucose is normally excreted, the rate of glucose reabsorption is also 180 gm/day.

From Table 27–1, two things are immediately apparent. First, *the processes of glomerular filtration and tubular reabsorption are quantitatively very large relative to urinary excretion for many substances.* This means that a small change in glomerular filtration or tubular reabsorption can potentially cause a relatively large change in urinary excretion. For example, a 10 per cent decrease in tubular reabsorption, from 178.5 to 160.7 liters/day, would increase urine volume from 1.5 to 19.3 liters/day (almost a 13-fold increase) if the glomerular filtration rate (GFR) remained constant. However, in reality, changes in tubular reabsorption and glomerular filtration are closely coordinated, so that large fluctuations in urinary excretion are avoided.

Second, unlike glomerular filtration, which is relatively nonselective (that is, essentially all solutes in the plasma are filtered except the plasma proteins or substances bound to them), *tubular reabsorption is highly*

331

Table 27-1 FILTRATION, REABSORPTION, AND EXCRETION RATES OF DIFFERENT SUBSTANCES BY THE KIDNEYS

	Amount Filtered	Amount Reabsorbed	Amount Excreted	% of Filtered Load Reabsorbed
Glucose (gm/day)	180	180	0	100
Bicarbonate (mEq/day)	4,320	4,318	2	>99.9
Sodium (mEq/day)	25,560	25,410	150	99.4
Chloride (mEq/day)	19,440	19,260	180	99.1
Urea (gm/day)	46.8	23.4	23.4	50
Creatinine (gm/day)	1.8	0	1.8	0

selective. Some substances, such as glucose and amino acids, are almost completely reabsorbed from the tubules, so that the urinary excretion rate is essentially zero. Many of the ions in the plasma, such as sodium, chloride, and bicarbonate, are also highly reabsorbed, but their rates of reabsorption and urinary excretion are variable, depending on the needs of the body. Certain waste products, such as urea and creatinine, on the other hand, are poorly reabsorbed from the tubules and excreted in relatively large amounts. Therefore, by controlling the rate at which they reabsorb different substances, the kidneys regulate the excretion of solutes independently of one another, a capability that is essential for precise control of the composition of body fluids. In this chapter, we discuss the mechanisms that allow the kidneys to selectively reabsorb or secrete different substances at variable rates.

Active Transport

Active transport can move a solute against an electrochemical gradient and requires energy derived from metabolism. Transport that is coupled directly to an energy source, such as the hydrolysis of adenosine triphosphate (ATP), is termed *primary active transport.* A good example of this is the sodium-potassium ATPase pump that functions throughout most parts of the renal tubule. Transport that is coupled *indirectly* to an energy source, such as that due to an ion gradient, is referred to as *secondary active transport.* Reabsorption of glucose by the renal tubule is an example of secondary active transport. Although solutes can be reabsorbed by active and/or passive mechanisms by the tubule, water is always reabsorbed by a passive (nonactive) physical mechanism called *osmosis,* which means water diffusion from a region of low

TUBULAR REABSORPTION INCLUDES PASSIVE AND ACTIVE MECHANISMS

For a substance to be reabsorbed, it must first be transported (a) across the tubular epithelial membranes into the renal interstitial fluid and then (b) through the peritubular capillary membrane back into the blood (Fig. 27–1). Thus, reabsorption of water and solutes includes a series of transport steps. Reabsorption across the tubular epithelium into the interstitial fluid includes active or passive transport by way of the same basic mechanisms discussed in Chapter 4 for transport across other membranes of the body. For instance, water and solutes can be transported either through the cell membranes themselves (*transcellular route*) or through the junctional spaces between the cells (*paracellular route*). Then, after absorption across the tubular epithelial cells into the interstitial fluid, water and solutes are transported the rest of the way through the peritubular capillary walls into the blood by *ultrafiltration (bulk flow)* that is mediated by hydrostatic and colloid osmotic forces. The peritubular capillaries behave very much like the venous ends of most other capillaries because there is a net reabsorptive force that moves the fluid and solutes from the interstitium into the blood.

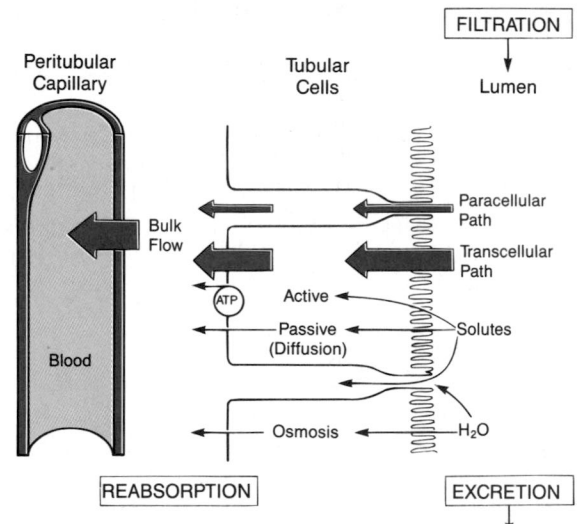

Figure 27–1. Reabsorption of filtered water and solutes from the tubular lumen across the tubular epithelial cells, through the renal interstitium, and back into the blood. Solutes are transported through the cells (transcellular route) by passive diffusion or active transport or between the cells (paracellular route) by diffusion. Water is transported through the cells and between the tubular cells by osmosis. Transport of water and solutes from the interstitial fluid into the peritubular capillaries occurs by ultrafiltration (bulk flow). ATP, adenosine triphosphate.

solute concentration (high water concentration) to one of high solute concentration (low water concentration).

SOLUTES CAN BE TRANSPORTED THROUGH EPITHELIAL CELLS OR BETWEEN CELLS. Renal tubular cells, like other epithelial cells, are held together by *tight junctions.* Lateral intercellular spaces lie behind the tight junctions and separate the epithelial cells of the tubule. Solutes can be reabsorbed or secreted across the cells by way of the *transcellular pathway* or between the cells by moving across the tight junctions and intercellular spaces, by way of the *paracellular pathway.* Sodium is a substance that moves through both routes, although most of the sodium is transported through the transcellular pathway. In some nephron segments, especially the proximal tubule, water is also reabsorbed across the paracellular pathway, and substances dissolved in the water, especially potassium, magnesium, and chloride ions, are carried with the reabsorbed fluid between the cells.

PRIMARY ACTIVE TRANSPORT THROUGH THE TUBULAR MEMBRANE IS LINKED TO HYDROLYSIS OF ATP. *The special importance of primary active transport is that it can move solutes against an electrochemical gradient.* The energy for this active transport comes from the hydrolysis of ATP by way of membrane-bound ATPase; the ATPase is also a component of the carrier mechanism that binds and moves solutes across the cell membranes. The primary active transporters that are known include *sodium-potassium ATPase, hydrogen ATPase, hydrogen-potassium ATPase,* and *calcium ATPase.*

A good example of a primary active transport system is the reabsorption of sodium ions across the proximal tubular membrane, as shown in Figure 27–2. On the basolateral sides of the tubular epithelial cell, the cell membrane has an extensive sodium-potassium ATPase system that hydrolyses ATP and uses the released energy to transport sodium ions out of the cell into the interstitium. At the same time, potassium is transported from the interstitium to the inside of the cell. The operation of this ion pump maintains low intracellular sodium and high intracellular potassium concentrations and creates a net negative charge of about − 70 millivolts within the cell. This pumping of sodium out of the cell across the *basolateral* membrane of the cell favors passive diffusion of sodium across the *luminal* membrane of the cell, from the tubular lumen into the cell, for two reasons: (1) There is a concentration gradient favoring sodium diffusion into the cell because intracellular sodium concentration is low (12 mEq/L) and tubular fluid sodium concentration is high (140 mEq/L). (2) The negative − 70-millivolt intracellular potential attracts the positive sodium ions from the tubular lumen into the cell.

Active reabsorption of sodium by sodium-potassium ATPase occurs in most parts of the tubule. In certain parts of the nephron, there are additional provisions for moving large amounts of sodium into the cell. In the proximal tubule, there is an extensive brush border on the luminal side of the membrane (the side that

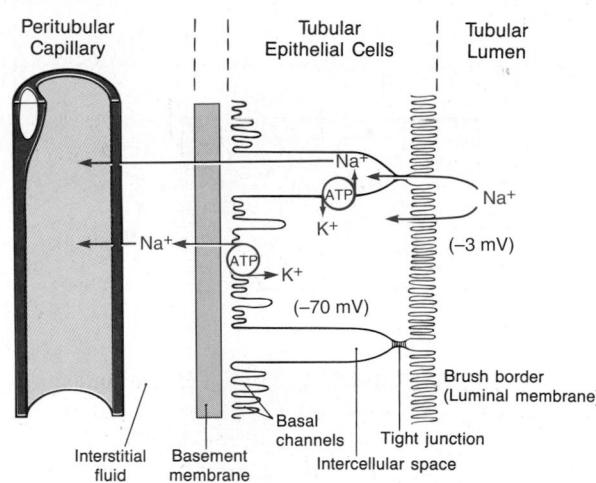

Figure 27–2. Basic mechanism for active transport of sodium through the tubular epithelial cell. The sodium-potassium pump transports sodium from the interior of the cell across the basolateral membrane, creating a low intracellular sodium concentration and a negative intracellular electrical potential. The low intracellular sodium concentration and the negative electrical potential cause sodium ions to diffuse from the tubular lumen into the cell through the brush border.

faces the tubular lumen) that multiplies the surface area about 20-fold. There are also sodium carrier proteins that bind sodium ions on the luminal surface of the membrane and release them inside the cell, providing *facilitated diffusion* of sodium through the membrane into the cell. These sodium carrier proteins are also important for secondary active transport of other substances, such as glucose and amino acids, as discussed later.

Thus, the net reabsorption of sodium ions from the tubular lumen back into the blood involves at least three steps:

1. Sodium is transported across the basolateral membrane against an electrochemical gradient by the sodium-potassium ATPase pump.

2. Sodium diffuses across the luminal membrane (also called the apical membrane) into the cell down an electrochemical gradient established by the sodium-potassium ATPase pump on the basolateral side of the membrane.

3. Sodium, water, and other substances are reabsorbed from the interstitial fluid into the peritubular capillaries by ultrafiltration, a passive process driven by the hydrostatic and colloid osmotic pressure gradients.

SECONDARY ACTIVE REABSORPTION THROUGH THE TUBULAR MEMBRANE. In secondary active transport, two or more substances interact with a specific membrane protein (a carrier molecule) and are co-transported together across the membrane. As one of the substances (for instance, sodium) diffuses down its electrochemical gradient, the energy released is used to drive another substance (for instance, glucose) against its electrochemical gradient. Thus, secondary active transport does not require energy directly from

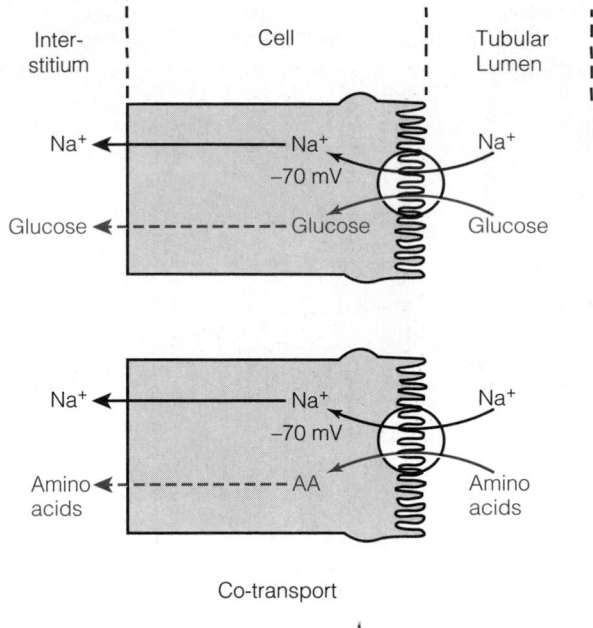

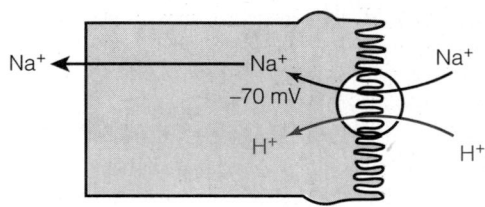

Co-transport

Counter Transport

Figure 27–3. Mechanisms of secondary active transport. The upper two cells show the *co-transport* of glucose or amino acids along with sodium ions through the brush border of the tubular epithelial cells, followed by facilitated diffusion through the basolateral membranes. The third cell shows the *counter-transport* of hydrogen ions from the interior of the cell across the brush border member and into the tubular lumen; movement of sodium ions into the cell, down an electrochemical gradient established by the sodium-potassium pump on the basolateral membrane, provides the energy for transport of the hydrogen ions from inside the cell into the tubular lumen.

ATP or from other high-energy phosphate sources. Rather, the direct source of the energy is that liberated by the simultaneous facilitated diffusion of another transported substance down its own electrochemical gradient.

Figure 27–3 shows secondary active transport of glucose and amino acids in the proximal tubule. In both instances, a specific carrier protein in the brush border combines with a sodium ion and an amino acid or a glucose molecule at the same time. These transport mechanisms are so efficient that they remove vir-

tually all the glucose and amino acids from the tubular lumen. After entry into the cell, glucose and amino acids exit across the basolateral membranes by facilitated diffusion, driven by the high glucose and amino acid concentrations in the cell.

Although transport of glucose against a chemical gradient does not directly use ATP, the reabsorption of glucose depends on energy expended by the primary active sodium-potassium ATPase pump in the basolateral membrane. Because of the activity of this pump, an electrochemical gradient for facilitated diffusion of sodium across the luminal membrane is maintained, and it is this downhill diffusion of sodium to the interior of the cell that provides the energy for the simultaneous uphill transport of glucose across the luminal membrane. Thus, this reabsorption of glucose is referred to as "secondary active transport" because glucose itself is reabsorbed uphill against a chemical gradient, but it is "secondary" to primary active transport of sodium.

Another important point to remember is that a substance is said to undergo "active" transport when at least one of the steps in the reabsorption involves primary or secondary active transport, even though other steps in the reabsorption process may be passive. For glucose reabsorption, secondary active transport occurs at the luminal membrane, but passive facilitated diffusion occurs at the basolateral membrane, and passive uptake by bulk flow occurs at the peritubular capillaries.

SECONDARY ACTIVE SECRETION INTO THE TUBULES. A few substances are secreted into the tubules by secondary active transport. This often involves *counter-transport* of the substance with sodium ions. In counter-transport, the energy liberated from the downhill movement of one of the substances (for example, sodium ions) enables uphill movement of a second substance in the opposite direction.

One example of counter-transport, shown in Figure 27–3, is the active secretion of hydrogen ions coupled to sodium reabsorption in the luminal membrane of the proximal tubule. In this case, sodium entry into the cell is coupled with hydrogen extrusion from the cell by sodium-hydrogen counter-transport. This transport is mediated by a specific protein in the brush border of the luminal membrane. As sodium is carried to the interior of the cell, hydrogen ions are forced outward in the opposite direction into the tubular lumen. The basic principles of primary and secondary active transport are discussed in additional detail in Chapter 4.

PINOCYTOSIS—AN ACTIVE TRANSPORT MECHANISM FOR REABSORPTION OF PROTEINS. Some parts of the tubule, especially the proximal tubule, reabsorb large molecules such as proteins by *pinocytosis*. In this process, the protein attaches to the brush border of the luminal membrane, and this portion of the membrane then invaginates to the interior of the cell until it is completely pinched off and a vesicle is formed containing the protein. Once inside the cell, the protein is digested into its constituent amino acids, which are reabsorbed through the basolateral membrane into the

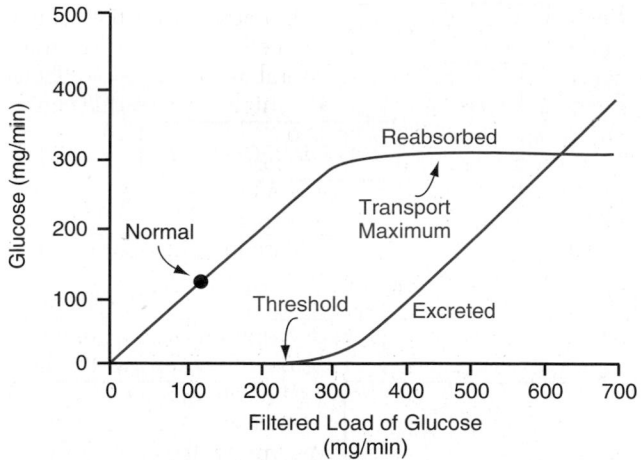

Figure 27–4. Relations among the filtered load of glucose, the rate of glucose reabsorption by the renal tubules, and the rate of glucose excretion in the urine. The *transport maximum* is the maximum rate at which glucose can be reabsorbed from the tubules. The *threshold* for glucose refers to the filtered load of glucose at which glucose first begins to appear in the urine.

interstitial fluid. Because pinocytosis requires energy, it is considered a form of active transport.

TRANSPORT MAXIMUM FOR SUBSTANCES THAT ARE ACTIVELY REABSORBED. For most substances that are actively reabsorbed or secreted, there is a limit to the rate at which the solute can be transported, often referred to as the *transport maximum*. This limit is due to saturation of the specific transport systems involved when the amount of solute delivered to the tubule (referred to as *tubular load*) exceeds the capacity of the carrier proteins and specific enzymes involved in the transport process.

The glucose transport system in the proximal tubule is a good example. Normally, measurable glucose does not appear in the urine because essentially all the filtered glucose is reabsorbed in the proximal tubule. However, when the filtered load exceeds the capability of the tubules to reabsorb glucose, urinary excretion of glucose does occur. In the adult human, the transport maximum for glucose averages about 320 mg/min, whereas the filtered load of glucose is only about 125 mg/ml (GFR × plasma glucose = 125 ml/min × 1 mg/ml). With large increases in GFR and/or plasma glucose concentration that increase the filtered load of glucose above 320 mg/min, the excess glucose filtered is not reabsorbed but passes into the urine.

Figure 27–4 shows the relation between tubular load of glucose, tubular transport maximum for glucose, and rate of glucose loss in the urine. Note that when the tubular load is at its normal level, 125 mg/min, there is no loss of glucose in the urine. However, when the tubular load rises above 220 mg/min, a small amount of glucose begins to appear in the urine, a point that is termed the *threshold* for glucose. *Note that this appearance of glucose in the urine (at the threshold) occurs before the transport maximum is reached.* One reason for the difference between

threshold and transport maximum is that not all nephrons have the same transport maximum for glucose, and some of the nephrons excrete glucose before others have reached their transport maximum. *The overall transport maximum for the kidneys is reached when all nephrons have reached their maximal capacity to reabsorb glucose.*

The plasma glucose of a *normal* person almost never becomes high enough to cause excretion of glucose in the urine. However, in uncontrolled *diabetes mellitus,* plasma glucose may rise to high levels, causing the filtered load of glucose to exceed the transport maximum and resulting in urinary glucose excretion. Some of the important transport maximums for substances *actively reabsorbed* by the tubules are as follows:

Substance	Transport Maximum
Glucose	320 ng/min
Phosphate	0.10 mM/min
Sulfate	0.06 mM/min
Amino acids	1.5 mM/min
Urate	15 mg/min
Lactate	75 mg/min
Plasma protein	30 mg/min

TRANSPORT MAXIMUMS FOR SUBSTANCES THAT ARE ACTIVELY SECRETED. Substances that are *actively secreted* also exhibit transport maximums as follows:

Substance	Transport Maximum
Creatinine	16 mg/min
Paraminohippuric acid	80 mg/min

SUBSTANCES THAT ARE TRANSPORTED BUT DO NOT EXHIBIT A TRANSPORT MAXIMUM. The reason that actively transported solutes often exhibit a transport maximum is that the transport carrier system becomes saturated as tubular load increases. *Substances that are passively reabsorbed do not demonstrate a transport maximum* because their rate of transport is determined by other factors, such as (1) the electrochemical gradient for diffusion of the substance across the membrane, (2) the permeability of the membrane for the substance, and (3) the time that the fluid containing the substance remains within the tubule. Transport of this type is referred to as *gradient-time transport* because the rate of transport depends on the electrochemical gradient and the time that the substance is in the tubule, which in turn depends on the tubular flow rate.

Some actively transported substances also have characteristics of gradient-time transport. An example is sodium reabsorption in the proximal tubule. The main reason that sodium transport in the proximal tubule does not exhibit a transport maximum is that other factors limit the reabsorption rate besides the maximum rate of active transport. For example, in the proximal tubules, the maximum transport capacity of the basolateral sodium-potassium ATPase pump is usually far greater than the actual rate of net sodium reabsorption. One of the reasons for this is that a significant amount of sodium transported out of the cell leaks back into the tubular lumen through the epithelial tight junctions. The rate at which this back-leak occurs depends on several factors, including (a) the permeability of the tight junctions and (b) the interstitial physical forces, which determine the rate of bulk flow reabsorption from the interstitial fluid into the peritubular capillaries. Therefore, sodium transport in the proximal tubules obeys mainly gradient-time transport principles rather than tubular maximum transport characteristics. This means that the greater the concentration of sodium in the proximal tubules, the greater its reabsorption rate. Also, the slower the flow rate of tubular fluid, the greater the percentage of sodium that can be reabsorbed from the proximal tubules.

In the more distal parts of the nephron, the epithelial cells have much tighter junctions and transport much smaller amounts of sodium. In these segments, sodium reabsorption exhibits a transport maximum similar to that for other actively transported substances. Furthermore, this transport maximum can be increased in response to certain hormones, such as *aldosterone.*

Passive Water Reabsorption by Osmosis Is Coupled Mainly to Sodium Reabsorption

When solutes are transported out of the tubule by either primary or secondary active transport, their concentrations tend to decrease inside the tubule while increasing in the renal interstitium. This creates a concentration difference that causes osmosis of water in the same direction that the solutes are transported, from the tubular lumen to the renal interstitium. Some parts of the renal tubule, especially the proximal tubule, are highly permeable to water, so that reabsorption of water occurs so rapidly that there is only a small concentration gradient for solutes across the tubular membrane.

A large part of the osmotic flow of water occurs through the so-called *tight junctions* between the epithelial cells as well as through the cells themselves. The reason for this, as already discussed, is that the junctions between the cells are not as tight as their name would imply, and they allow significant diffusion of water and other small ions. This is especially true in the proximal tubules, which have a high permeability for water and a smaller but significant permeability to most ions, such as sodium, chloride, potassium, calcium, and magnesium.

As water moves across the tight junctions by osmosis, it can also carry with it some of the solutes, a process referred to as *solvent drag.* And because the reabsorption of water, organic solutes, and ions is coupled to sodium reabsorption, changes in sodium reabsorption significantly influence the reabsorption of water and many other solutes.

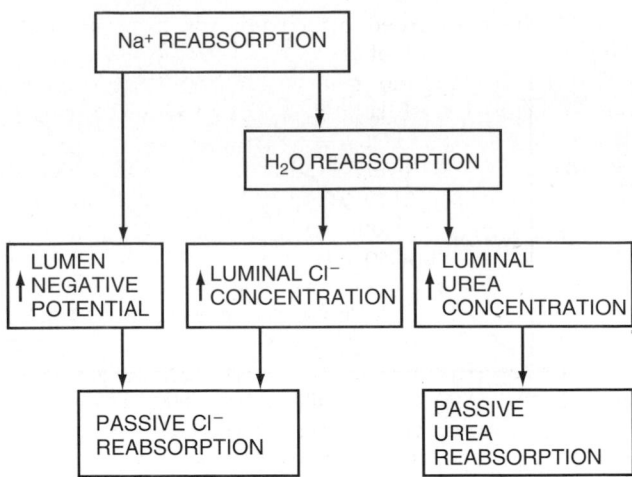

Figure 27–5. Mechanisms by which water, chloride, and urea reabsorption are coupled to sodium reabsorption.

In the more distal parts of the nephron, beginning in the loop of Henle and extending through the collecting tubule, the tight junctions become far less permeable to water and solutes and the epithelial cells also have a greatly decreased membrane surface area. Therefore, water cannot move easily across the tubular membrane by osmosis. However, antidiuretic hormone (ADH) greatly increases the water permeability in the distal and collecting tubule, as discussed later.

Thus, water movement across the tubular epithelium can occur only if the membrane is permeable to water, no matter how large the osmotic gradient. In the proximal tubule, the water permeability is always high, and water is reabsorbed as rapidly as the solutes. In the ascending loop of Henle, water permeability is always low, so that almost no water is reabsorbed, despite a large osmotic gradient. Water permeability in the last parts of the tubules—the distal tubule, collecting tubule, and collecting ducts—can be high or low, depending on the presence or absence of ADH.

Reabsorption of Chloride, Urea, and Other Solutes by Passive Diffusion

When sodium is reabsorbed through the tubular epithelial cell, negative ions such as chloride are transported along with the sodium because of electrical potentials. That is, the transport of positively charged sodium ions out of the lumen leaves the inside of the lumen negatively charged, compared with the interstitial fluid. This causes chloride ions to diffuse *passively* through the *paracellular pathway* (that is, between the cells). Additional reabsorption of chloride ions occurs because of a chloride concentration gradient that develops when water is reabsorbed from the tubule by osmosis, thereby concentrating the chloride ions in the tubular lumen (Fig. 27–5). Thus, the active reabsorption of sodium is closely coupled to the passive reabsorption of chloride by way of an electrical potential and a chloride concentration gradient. Chloride ions can also be reabsorbed by secondary active transport. The most important of the secondary active transport processes for chloride reabsorption involves co-transport of chloride with sodium across the luminal membrane.

Urea is also passively reabsorbed from the tubule but to a much lesser extent than chloride ions. As water is reabsorbed from the tubules (by osmosis coupled to sodium reabsorption), urea concentration in the tubular lumen increases (see Fig. 27–5). This creates a concentration gradient favoring the reabsorption of urea. However, urea does not permeate the tubule nearly so much as water. Therefore, about one half of the urea that is filtered by the glomerular capillaries is passively reabsorbed from the tubules. The remainder of the urea passes into the urine, allowing the kidneys to excrete large amounts of this waste product of metabolism.

Another waste product of metabolism, creatinine, is an even larger molecule than urea and is essentially impermeant to the tubular membrane. Therefore, almost none of the creatinine that is filtered is reabsorbed, so that virtually all the creatinine filtered by the glomerulus is excreted in the urine.

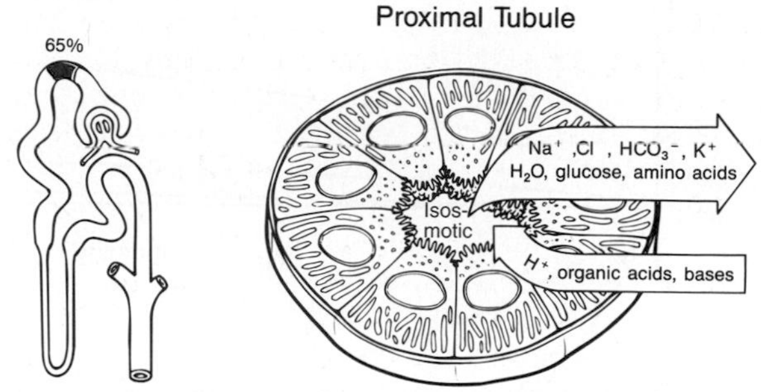

REABSORPTION AND SECRETION ALONG DIFFERENT PARTS OF THE NEPHRON

In the previous sections, we discussed the basic principles by which water and solutes are transported across the tubular membrane. With these generalizations in mind, we can now discuss the different characteristics of the individual tubular segments that enable them to perform their specific excretory functions. Only the tubular transport functions that are quantitatively most important are discussed, especially as they relate to the reabsorption of sodium, chloride, and water. In subsequent chapters, we discuss the reabsorption and secretion of other specific substances in different parts of the tubular system.

Proximal Tubular Reabsorption

Normally, about 65 per cent of the filtered load of sodium and water and a slightly lower percentage of filtered chloride are reabsorbed by the proximal tubule before the filtrate reaches the loops of Henle. These percentages can be increased or decreased in different physiological conditions, as discussed later.

THE PROXIMAL TUBULES HAVE A HIGH CAPACITY FOR ACTIVE AND PASSIVE REABSORPTION. The high capacity of the proximal tubule for reabsorption results from its special cellular characteristics, as shown in Figure 27–6. The proximal tubule epithelial cells are highly metabolic and have large numbers of mitochondria to support potent active transport processes. In addition, the proximal tubular cells have an extensive brush border on the luminal (apical) side of the membrane as well as an extensive labyrinth of intercellular and basal channels, all of which together provide an extensive membrane surface area on the luminal and basolateral sides of the epithelium for rapid transport of sodium ions and other substances.

Proximal Tubule

Figure 27–6. Characteristics of the proximal tubule, showing the cellular ultrastructure and the primary transport characteristics. The proximal tubule reabsorbs about 65 per cent of the filtered sodium, chloride, bicarbonate, and potassium and essentially all the filtered glucose and amino acids. The proximal tubule also secretes organic acids, bases, and hydrogen ions into the tubular lumen.

The extensive membrane surface of the epithelial brush border is also loaded with protein carrier molecules that transport a large fraction of the sodium ions across the luminal membrane linked by way of the *co-transport* mechanism with multiple organic nutrients such as amino acids and glucose. The remainder of the sodium is transported from the tubular lumen into the cell by *counter-transport* mechanisms, which reabsorb sodium while secreting other substances into the tubular lumen, especially hydrogen ions. As discussed in Chapter 30, the secretion of hydrogen ions into the tubular lumen is an important step in the removal of bicarbonate ions from the tubule (by combining H^+ with the HCO_3^- to form H_2CO_3, which then dissociates into H_2O and CO_2).

Although the sodium-potassium ATPase pump provides the major force for reabsorption of sodium, chloride, and water throughout the proximal tubule, there are some differences in the mechanisms by which sodium and chloride are transported through the luminal side of the early and late portions of the proximal tubular membrane. In the first half of the proximal tubule, sodium is reabsorbed by co-transport along with glucose, amino acids, and other solutes. But in the second half of the proximal tubule, little glucose and amino acids remain to be reabsorbed. Instead, sodium is now reabsorbed mainly with chloride ions. The second half of the proximal tubule has a relatively high concentration of chloride (around 140 mEq/L) compared with the early proximal tubule (about 105 mEq/L) because when sodium is reabsorbed, it preferentially carries with it glucose, bicarbonate, and organic ions in the early proximal tubule, leaving behind a solution that has a higher concentration of chloride. In the second half of the proximal tubule, the higher chloride concentration favors the diffusion of

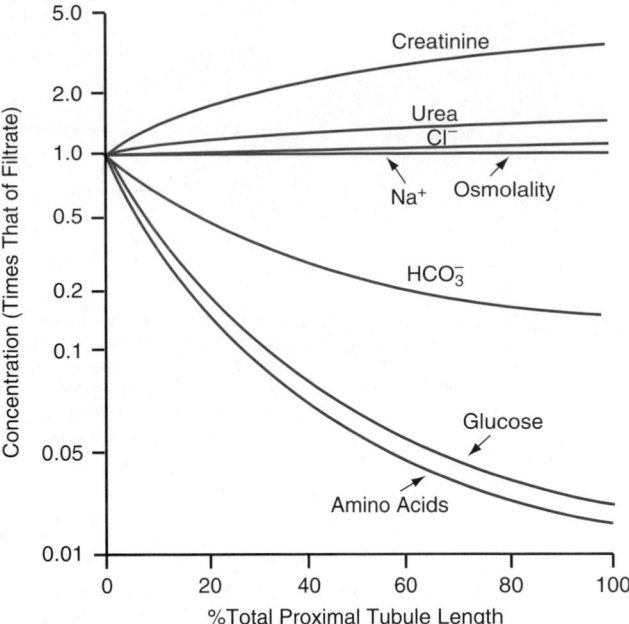

Figure 27–7. Changes in concentrations of different substances in tubular fluid along the proximal convoluted tubule relative to the concentrations of these substances in the glomerular filtrate. A value of 1.0 indicates that the concentration of the substance in the tubular fluid is the same as the concentration in the glomerular filtrate. Values below 1.0 indicate that the substance is reabsorbed more avidly than water, whereas values above 1.0 indicate that the substance is reabsorbed to a lesser extent than water.

this ion from the tubule lumen through the intercellular junctions into the renal interstitial fluid.

CONCENTRATIONS OF SOLUTES ALONG THE PROXIMAL TUBULE. Figure 27–7 summarizes the changes in concentrations of various solutes along the proximal tubule. Although the *amount* of sodium in the tubular fluid decreases markedly along the proximal tubule, the *concentration* of sodium (and the total osmolarity) remains relatively constant because water permeability of the proximal tubules is so great that water reabsorption keeps pace with sodium reabsorption. Certain organic solutes, such as glucose, amino acids, and bicarbonate, are much more avidly reabsorbed than water, so that their concentrations decrease markedly along the length of the proximal tubule. Other organic solutes that are less permeant and not actively reabsorbed, such as urea, increase their concentration along the proximal tubule. The total solute concentration, as reflected by osmolarity, remains essentially the same all along the proximal tubule because of the extremely high permeability of this part of the nephron to water.

SECRETION OF ORGANIC ACIDS AND BASES BY THE PROXIMAL TUBULE. The proximal tubule is also an important site for secretion of organic acids and bases such as *bile salts, oxalate, urate, and catecholamines.* Many of these substances are the end products of metabolism and must be rapidly removed from the body. The *secretion* of these substances into the proxi-

mal tubule plus *filtration* into the proximal tubule by the glomerular capillaries and the almost total lack of reabsorption in any portion of the tubular system, all combined, contribute to rapid excretion in the urine.

In addition to the waste products of metabolism, the kidneys secrete many potentially harmful drugs or toxins directly through the tubular cells into the tubules and rapidly clear these substances from the blood. In the case of certain drugs, such as penicillin and salicylates, the rapid clearance by the kidneys creates a problem in maintaining a therapeutically effective drug concentration.

Another compound that is rapidly secreted by the proximal tubule is para-aminohippuric acid (PAH). PAH is secreted so rapidly that the normal person can clear about 90 per cent of the PAH from the plasma flowing through the kidneys and excrete it in the urine. For this reason, the rate of PAH clearance can be used as an index of the renal plasma flow, as discussed later.

Solute and Water Transport in the Loop of Henle

The loop of Henle consists of three functionally distinct segments: *the descending thin segment, the ascending thin segment, and the thick ascending segment.* The thin descending and thin ascending segments, as their names imply, have thin epithelial membranes with no brush borders, few mitochondria, and minimal levels of metabolic activity (Fig. 27–8).

The descending part of the thin segment is highly permeable to water and moderately permeable to most solutes, including urea and sodium. The function of this nephron segment is mainly to allow simple diffusion of substances through its walls. About 20 per cent of the filtered water is reabsorbed in the loop of Henle, and almost all of this occurs in the descending thin limb because the ascending limb, including both the thin and the thick portions, is virtually impermeable to water, a characteristic that is important for concentrating the urine.

The thick segment of the loop of Henle, which begins about half way up the ascending limb, has thick epithelial cells that have high metabolic activity and are capable of active reabsorption of sodium, chloride, and potassium (see Fig. 27–8). About 25 per cent of the filtered loads of sodium, chloride, and potassium are reabsorbed in the loop of Henle, mostly in the thick ascending limb. Considerable amounts of other ions, such as calcium, bicarbonate, and magnesium, are also reabsorbed in the thick ascending loop of Henle. The thin segment of the ascending limb has a much lower reabsorptive capacity than the thick segment, and the descending thin limb does not reabsorb significant amounts of any of these solutes.

An important component of solute reabsorption in the thick ascending limb is the sodium-potassium ATPase pump in the epithelial cell basolateral membranes. As in the proximal tubule, the reabsorption of other solutes in the thick segment of the ascending

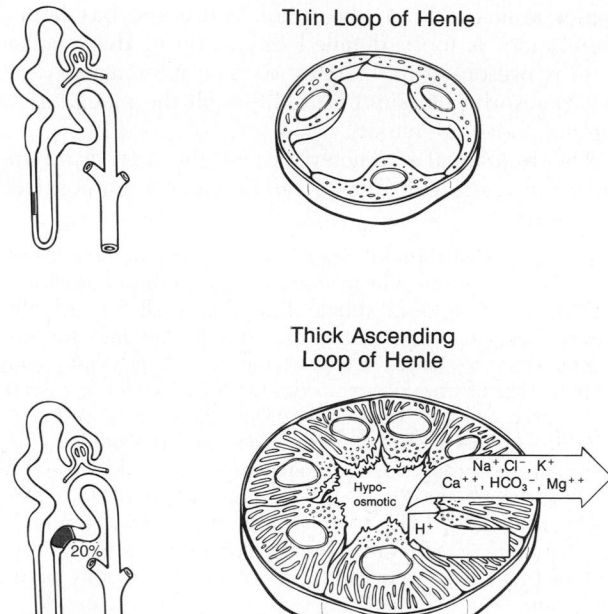

Thin Loop of Henle

**Thick Ascending
Loop of Henle**

20%

Figure 27–8. Characteristics of the thin descending loop of Henle *(top)* and the thick ascending segment of the loop of Henle *(bottom).* The descending part of the thin segment of the loop of Henle is highly permeable to water and moderately permeable to most solutes but has few mitochondria and little or no active reabsorption. The thick ascending limb of the loop of Henle reabsorbs about 25 per cent of the filtered loads of sodium, chloride, and potassium as well as large amounts of calcium, bicarbonate, and magnesium. This segment also secretes hydrogen ions into the tubular lumen.

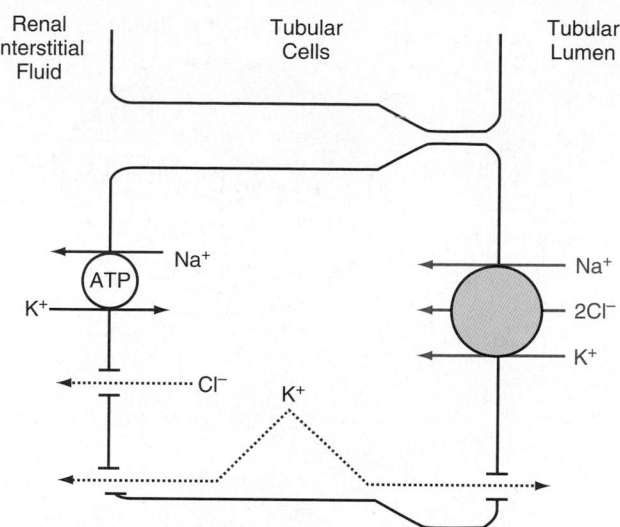

Figure 27–9. Mechanisms of sodium, chloride, and potassium transport in the thick ascending loop of Henle. The sodium-potassium ATPase pump in the basolateral cell membrane maintains a low intracellular sodium concentration and a negative electrical potential in the cell. The 1-sodium, 2-chloride, 1-potassium co-transporter in the luminal membrane transports these three ions from the tubular lumen into the cell, using the potential energy released by diffusion of sodium down an electrochemical gradient into the cells.

loop of Henle is closely linked to the reabsorptive capability of the sodium-potassium ATPase pump, which maintains a low intracellular sodium concentration. The low intracellular sodium concentration in turn provides a favorable gradient for movement of sodium from the tubular fluid into the cell. *In the thick ascending loop, movement of sodium across the luminal membrane is mediated primarily by a 1-sodium, 2-chloride, 1-potassium cotransporter* (Fig. 27–9). This cotransport protein carrier in the luminal membrane uses the potential energy released by downhill diffusion of sodium into the cell to drive the reabsorption of potassium into the cell against a concentration gradient.

The thick ascending limb also has a sodium-hydrogen counter-transport mechanism in its luminal cell membrane that mediates sodium reabsorption and hydrogen secretion in this segment.

Because the thick segment of the ascending loop of Henle is virtually impermeable to water, most of the water delivered to this segment remains in the tubule, despite reabsorption of large amounts of solute. Thus, the tubular fluid in the ascending limb becomes very dilute as it flows toward the distal tubule, a feature that is important in allowing the kidneys to dilute or concentrate the urine under different conditions, as we discuss much more fully later.

Distal Tubule

The thick segment of the ascending limb of the loop of Henle empties into the *distal tubule.* The very first portion of the distal tubule forms part of the *juxtaglomerular complex* that provides feedback control of GFR and blood flow in this same nephron. The next early part of the distal tubule is highly convoluted and has many of the same reabsorptive characteristics of the thick segment of the ascending limb of the loop of Henle. That is, it avidly reabsorbs most of the ions, including sodium, potassium, and chloride, but is virtually impermeable to water and urea. For this reason, it is referred to as the *diluting segment* because it also dilutes the tubular fluid.

Late Distal Tubule and Cortical Collecting Tubule

The second half of the distal tubule and the subsequent cortical collecting tubule have similar functional characteristics. Anatomically, they are composed of two distinct cell types, the *principal cells and the intercalated cells* (Fig. 27–10). The principal cells reabsorb sodium and water from the lumen and secrete potassium ions into the lumen. The intercalated cells reabsorb potassium ions and secrete hydrogen ions into the tubular lumen.

THE PRINCIPAL CELLS REABSORB SODIUM AND SECRETE POTASSIUM. Sodium *reabsorption* and potassium *secretion* by the principal cells depend on the activity of a sodium-potassium ATPase pump in each cell's basolateral membrane. This pump maintains a

Early Distal Tubule

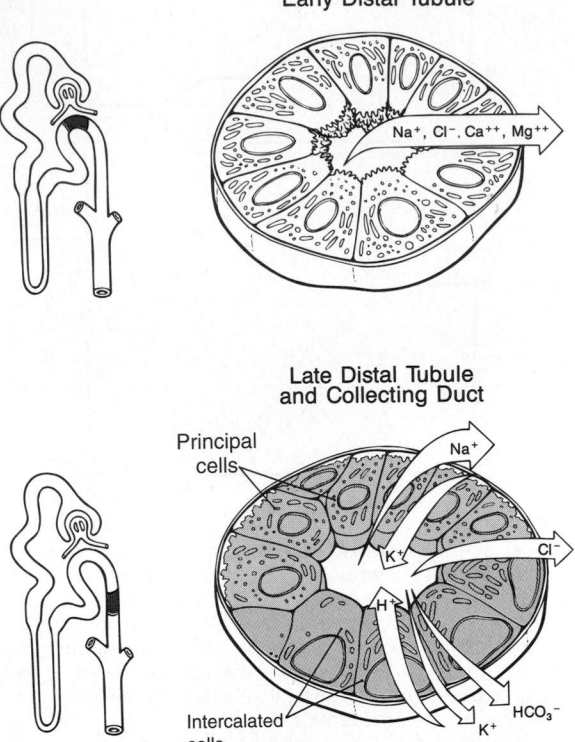

**Late Distal Tubule
and Collecting Duct**

Principal
cells

Intercalated
cells

Figure 27–10. Cellular ultrastructure and transport characteristics of the early distal tubules and the late distal tubules and collecting tubules. The early distal tubule has many of the same characteristics as the thick ascending loop of Henle and reabsorbs sodium, chloride, calcium, and magnesium but is virtually impermeable to water and urea. The late distal tubules and cortical collecting tubules are composed of two distinct cell types, the *principal cells* and the *intercalated cells.* The principal cells reabsorb sodium from the lumen and secrete potassium ions into the lumen. The intercalated cells reabsorb potassium ions and bicarbonate ions from the lumen and secrete hydrogen ions into the lumen. The reabsorption of water from this segment of the nephron is controlled by the concentration of antidiuretic hormone.

low sodium concentration inside the cell and, therefore, favors sodium diffusion into the cell through special channels. The secretion of potassium by these cells from the blood into the tubular lumen involves two steps: (1) Potassium enters the cell because of the sodium-potassium ATPase pump, which maintains a high intracellular potassium concentration. (2) Once in the cell, potassium diffuses down its concentration gradient across the luminal membrane into the tubular fluid.

THE INTERCALATED CELLS AVIDLY SECRETE HYDROGEN AND REABSORB BICARBONATE AND POTASSIUM IONS. Hydrogen ion secretion by the intercalated cells is mediated by a hydrogen-ATPase transport mechanism. Hydrogen is generated in this cell by the action of carbonic anhydrase on water and carbon dioxide to form carbonic acid, which then dissociates into hydrogen ions and bicarbonate ions. The hydrogen ions are then secreted into the tubular lumen, and for each hydrogen ion secreted, a bicarbonate ion be-

comes available for reabsorption across the basolateral membrane. A more detailed discussion of this mechanism is presented in Chapter 30. The intercalated cells also reabsorb potassium ions, although the mechanisms are not well understood.

The functional characteristics of the *late distal tubule* and *cortical collecting tubule* can be summarized as follows:

1. The tubular membranes of both segments are almost completely impermeable to urea, similar to the diluting segment of the early distal tubule; thus, almost all the urea that enters these segments passes on through and into the collecting duct to be excreted in the urine, although some reabsorption of urea occurs in the medullary collecting ducts.

2. Both the late distal tubule and the cortical collecting tubule segments reabsorb sodium ions, and the rate of reabsorption is controlled by hormones, especially aldosterone. At the same time, these segments secrete potassium ions from the peritubular capillary blood into the tubular lumen, a process that is also controlled by aldosterone and by other factors such as the concentration of potassium ions in the body fluids.

3. The *intercalated cells* of these nephron segments avidly secrete hydrogen ions by an active hydrogen-ATPase mechanism. This process is different from the secondary active secretion of hydrogen ions by the proximal tubule because it is capable of secreting hydrogen ions against a large concentration gradient, as much as 1000 to 1. This is in contrast to the relatively small gradient (4- to 10-fold) for hydrogen ions that can be achieved by secondary active secretion in the proximal tubule. Thus, the intercalated cells play a key role in acid-base regulation of the body fluids.

4. The permeability of the late distal tubule and cortical collecting duct to water is controlled by the concentration of *ADH,* which is also called *vasopressin.* With high levels of ADH, these tubular segments are permeable to water, but in the absence of ADH, they are virtually impermeable to water. This special characteristic provides an important mechanism for controlling the degree of dilution or concentration of the urine.

Medullary Collecting Duct

Although the medullary collecting ducts reabsorb less than 10 per cent of the filtered water and sodium, they are the final site for processing the urine and, therefore, play an extremely important role in determining the final urine output of water and solutes.

The epithelial cells of the collecting ducts are nearly cuboidal in shape with smooth surfaces and relatively few mitochondria. Special characteristics of this tubular segment are as follows:

1. The permeability of the medullary collecting duct to water is controlled by the level of ADH. With high levels of ADH, water is avidly reabsorbed into the medullary interstitium, thereby reducing the urine volume while at the same time concentrating most of the solutes in the urine.

2. Unlike the cortical collecting tubule, the medullary collecting duct is permeable to urea. Therefore, some of the tubular urea is reabsorbed into the medullary interstitium, helping to raise the osmolality in this region of the kidneys and contributing to the kidneys' overall ability to form a concentrated urine.

3. The medullary collecting duct is capable of secreting

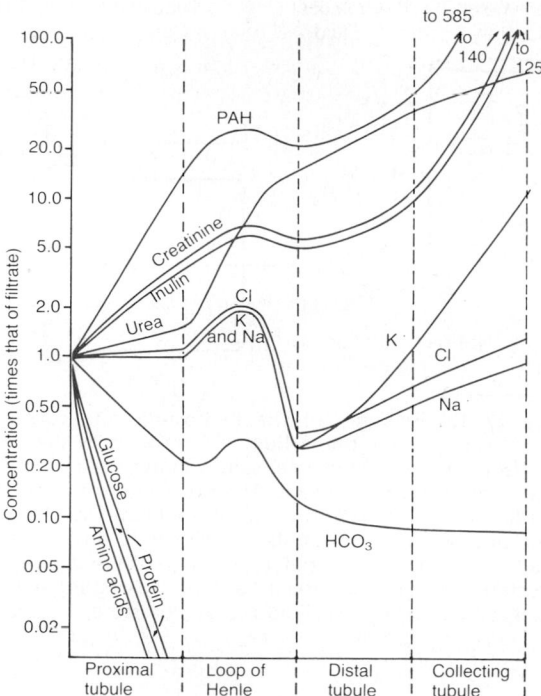

Figure 27–11. Changes in average concentrations of different substances at different points in the tubular system relative to the concentration of that substance in the glomerular filtrate. A value of 1.0 indicates that the concentration of the substance in the tubular fluid is the same as the concentration of that substance in the glomerular filtrate. Values below 1.0 indicate that the substance is reabsorbed more avidly than water, whereas values above 1.0 indicate that the substance is reabsorbed to a lesser extent than water.

hydrogen ions against a large concentration gradient, as also occurs in the cortical collecting tubule. Thus, the medullary collecting duct also plays a key role in regulating acid-base balance.

Summary of Concentrations of Different Solutes in the Different Tubular Segments

Whether or not a solute will become concentrated in a tubular fluid is determined by the relative degree of reabsorption of that solute versus the reabsorption of water. If a greater percentage of water is reabsorbed, the substance becomes more concentrated. If a greater percentage of the solute is reabsorbed, the substance becomes more diluted.

Figure 27–11 shows the degree of concentration of the most important substances in the different tubular segments. All the values in this figure are relative concentrations, with each substance considered to have a normal concentration of 1.0 in the glomerular filtrate, as shown at the left-hand side of the figure. As the filtrate moves along the tubular system, the concentration rises to progressively greater than 1.0 if more water is reabsorbed than solutes or the concentration becomes progressively less than 1.0 if more solute is

reabsorbed than water. Also, if a substance is secreted by the tubular epithelium into the tubule, this, too, will increase its concentration in the tubular fluid.

The substances represented at the top of Figure 27–11, such as creatinine and urea, become highly concentrated in the urine. In general, these substances are not needed by the body, and the kidneys have become adapted to reabsorb them only slightly or not at all or even to secrete them into the tubules, thereby excreting especially great quantities into the urine.

On the other hand, the substances, represented toward the bottom of the figure, such as glucose and amino acids, are all strongly reabsorbed; these are all substances that the body needs to conserve and almost none of them are lost in the urine.

REGULATION OF TUBULAR REABSORPTION

Because it is essential to maintain a precise balance between tubular reabsorption and glomerular filtration, there are multiple nervous, hormonal, and local control mechanisms that regulate tubular reabsorption, just as there are for control of glomerular filtration. An important feature of tubular reabsorption is that reabsorption of some solutes can be regulated independently of others, especially through hormonal control mechanisms.

Glomerulotubular Balance—The Ability of the Tubules to Increase Reabsorption Rate in Response to Increased Tubular Load

One of the most basic mechanisms for controlling tubular reabsorption rate is the intrinsic ability of the tubules to increase their reabsorption rate in response to increased tubular load (increased tubular inflow). This phenomenon is referred to as *glomerulotubular balance.* For example, if GFR is increased from 125 to 150 ml/min, the absolute rate of proximal tubular reabsorption also increases from about 81 ml/min (65 per cent of GFR) to about 97.5 ml/min (65 per cent of GFR). Thus, glomerulotubular balance refers to the fact that the total rate of reabsorption increases as the filtered load increases, even though the percentage of GFR reabsorbed in the proximal tubular remains relatively constant at about 65 per cent.

Some degree of glomerulotubular balance also occurs in other tubular segments, especially the loop of Henle. The precise mechanisms responsible for this are not well understood but may be due partly to changes in physical forces in the tubule and surrounding renal interstitium, as discussed later. It is clear that the mechanisms for glomerulotubular balance can occur independently of hormones and can be demonstrated in completely isolated kidneys or even in a completely isolated proximal tubular segment.

The importance of glomerulotubular balance is that

it helps to prevent overloading of the distal tubular segments when GFR increases. Glomerulotubular balance acts as a second line of defense to buffer the effects of spontaneous changes in GFR on urine output. (The first line of defense discussed above includes the renal autoregulatory mechanisms, especially tubuloglomerular feedback, that help to prevent changes in GFR.) Working together, the autoregulatory and glomerulotubular balance mechanisms prevent large changes in fluid flow in the distal tubules when the arterial pressure changes or when there are other disturbances that would otherwise wreak havoc with the maintenance of sodium and volume homeostasis.

Peritubular Capillary and Renal Interstitial Fluid Physical Forces

Hydrostatic and colloid osmotic forces govern the rate of reabsorption across the peritubular capillaries, just as these physical forces control filtration in the glomerular capillaries. Changes in peritubular capillary reabsorption can in turn influence the hydrostatic and colloid osmotic pressures of the renal interstitium and, ultimately, reabsorption of water and solutes from the renal tubules.

NORMAL VALUES FOR PHYSICAL FORCES AND REABSORPTION RATE. As the glomerular filtrate passes through the renal tubules, normally more than 99 per cent of the water and most of the solutes are reabsorbed. Fluid and electrolytes are reabsorbed from the tubules into the renal interstitium and from there into the peritubular capillaries. The normal rate of peritubular capillary reabsorption is about 124 ml/min.

Reabsorption across the peritubular capillaries can be calculated as:

$$\text{Reabsorption} = K_f \times \text{Net reabsorptive force}$$

The net reabsorptive force represents the sum of the hydrostatic and colloid osmotic forces that either favor or oppose reabsorption across the peritubular capillaries. These forces include (1) hydrostatic pressure inside the peritubular capillaries (peritubular hydrostatic pressure [P_c]), which opposes reabsorption; (2) hydrostatic pressure in the renal interstitium (P_{if}) outside the capillaries, which favors reabsorption; (3) colloid osmotic pressure of the peritubular capillary plasma proteins (π_c), which favors reabsorption; and (4) colloid osmotic pressure of the proteins in the renal interstitium (π_{if}), which opposes reabsorption.

Figure 27–12 shows the approximate normal forces that favor and oppose peritubular reabsorption. Because the normal peritubular capillary pressure averages about 13 mm Hg and renal interstitial fluid hydrostatic pressure averages 6 mm Hg, there is a positive hydrostatic pressure gradient from the peritubular capillary to the interstitial fluid of about 7 mm Hg that opposes fluid reabsorption. This is more than counterbalanced by the colloid osmotic pressures that favor reabsorption. That is, the plasma colloid osmotic pressure, which favors reabsorption, is about

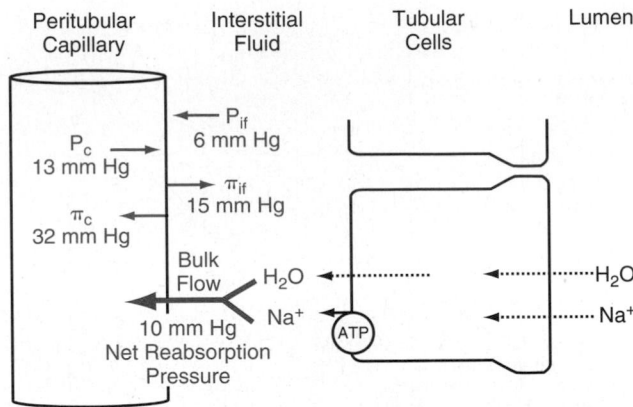

Figure 27–12. Summary of the hydrostatic and colloid osmotic forces that determine fluid reabsorption by the peritubular capillaries. The numerical values shown are estimates of the normal values for humans. The net reabsorptive pressure is normally about 10 mm Hg, causing fluid and solutes to be reabsorbed into the peritubular capillaries as they are transported across the renal tubular cells. P_c, peritubular capillary hydrostatic pressure; P_{if}, interstitial fluid hydrostatic pressure; π_c, peritubular capillary colloid osmotic pressure; π_{if}, interstitial fluid colloid osmotic pressure.

32 mm Hg and the colloid osmotic pressure of the interstitium, which opposes reabsorption, is 15 mm Hg, causing a net colloid osmotic force of about 17 mm Hg favoring reabsorption. Therefore, subtracting the net hydrostatic forces that oppose reabsorption (7 mm Hg) from the net colloid osmotic forces that favor reabsorption (17 mm Hg) gives a net reabsorptive force of about 10 mm Hg. This is a high value, similar to that found in the glomerular capillaries, but in the opposite direction.

The other factor that contributes to the high rate of fluid reabsorption in the peritubular capillaries is a large filtration coefficient (K_f) because of the high hydraulic conductivity and large surface area of the capillaries. Because the reabsorption rate is normally about 124 ml/min and net reabsorption pressure is 10 mm Hg, K_f normally is about 12.4 ml/min/mm Hg of net reabsorption pressure.

REGULATION OF PERITUBULAR CAPILLARY PHYSICAL FORCES. The two determinants of peritubular capillary reabsorption that are directly influenced by renal hemodynamic changes are the hydrostatic and colloid osmotic pressures of the peritubular capillaries.

The *peritubular capillary hydrostatic pressure* is influenced by (1) the *arterial pressure;* increases in arterial pressure tend to raise peritubular capillary hydrostatic pressure and decrease reabsorption rate. This effect is buffered to some extent by autoregulatory mechanisms that maintain relatively constant renal blood flow as well as relatively constant hydrostatic pressures in the renal blood vessels; and (2) *resistances of the afferent and efferent arterioles;* increase in resistance of either the afferent or the efferent arterioles reduces peritubular capillary hydrostatic pressure. Although constriction of the efferent arterioles increases

Table 27–2 FACTORS THAT CAN INFLUENCE PERITUBULAR CAPILLARY REABSORPTION

↑ P_c → ↓ Reabsorption
• ↓ R_A → ↑ P_c
• ↓ R_E → ↑ P_c
• ↑ Arterial Pressure → ↑ P_c
↑ π_c → ↑ Reabsorption
• ↑ π_A → ↑ π_c
• ↑ FF → ↑ π_c
↑ K_f → ↑ Reabsorption

P_c, peritubular capillary hydrostatic pressure; R_A and R_E, afferent and efferent arteriolar resistances, respectively; π_c, peritubular capillary colloid osmotic pressure; π_A, arterial plasma colloid osmotic pressure; FF, filtration fraction; K_f, peritubular capillary filtration coefficient.

glomerular capillary hydrostatic pressure, it lowers peritubular capillary hydrostatic pressure.

The second major determinant of peritubular capillary reabsorption is the *colloid osmotic pressure* of the plasma in these capillaries; raising the colloid osmotic pressure increases peritubular capillary reabsorption. *The colloid osmotic pressure of peritubular capillaries is determined by* (1) the *systemic plasma colloid osmotic pressure;* increasing the plasma protein concentration of systemic blood tends to raise peritubular capillary colloid osmotic pressure, thereby increasing reabsorption; and (2) *the filtration fraction;* the higher the filtration fraction, the greater the fraction of plasma filtered through the glomerulus and, consequently, the more concentrated the protein becomes in the plasma that remains behind. Thus, increasing the filtration fraction also tends to increase peritubular capillary reabsorption rate. Because filtration fraction is defined as the ratio of GFR/renal plasma flow, increased filtration fraction can occur as a result of increased GFR or decreased renal plasma flow. Some renal vasoconstrictors, such as angiotensin II, decrease the renal plasma flow and increase filtration fraction, as discussed later.

Changes in the peritubular K_f can also influence reabsorption rate because K_f is a measure of the permeability and surface area of the capillaries. Increases in K_f raise reabsorption, whereas decreases in K_f lower peritubular capillary reabsorption. K_f remains relatively constant in most physiological conditions.

Table 27–2 summarizes the factors that can influence the peritubular capillary reabsorption rate.

Renal Interstitial Hydrostatic and Colloid Osmotic Pressures. Ultimately, changes in peritubular capillary physical forces influence tubular reabsorption by changing the physical forces in the renal interstitium surrounding the tubules. For example, a decrease in the reabsorptive force across the peritubular capillary membranes, caused by either increased peritubular hydrostatic pressure or decreased peritubular capillary colloid osmotic pressure, reduces the uptake of fluid and solutes from the interstitium into the peritubular capillaries. This in turn raises renal interstitial fluid hydrostatic pressure and decreases in-

terstitial fluid colloid osmotic pressure because of dilution of the proteins in the renal interstitium. These changes in the physical forces of the renal interstitium then decrease the net reabsorption of fluid from the renal tubules into the interstitium, especially in the proximal tubules.

The mechanisms by which changes in interstitial fluid hydrostatic and colloid osmotic pressures influence tubular reabsorption can be understood by examining the pathways through which solute and water are reabsorbed (Fig. 27–13). Once the solutes enter the intercellular channels or renal interstitium by active transport or passive diffusion, water is drawn from the tubular lumen into the interstitium by osmosis. And once the water and solutes are in the interstitial spaces, they can either be swept up into the peritubu-

Normal

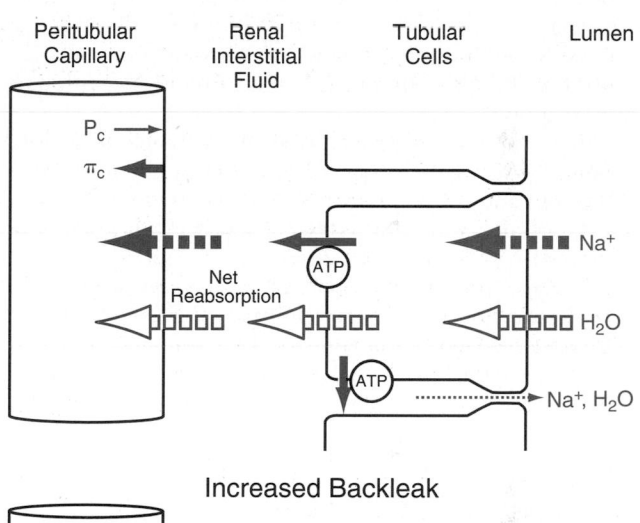

Increased Backleak

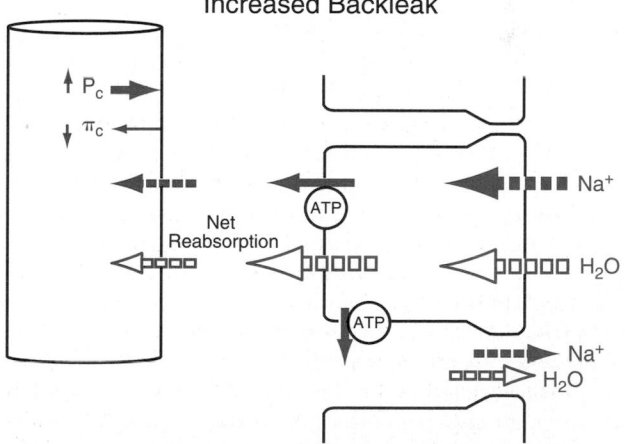

Figure 27–13. Proximal tubular and peritubular capillary reabsorption under normal conditions *(top)* and during decreased peritubular capillary reabsorption *(bottom)* caused by either increasing peritubular capillary hydrostatic pressure (P_c) or decreasing peritubular capillary colloid osmotic pressure (π_c). Reduced peritubular capillary reabsorption in turn decreases the net reabsorption of sodium and water by increasing the amounts of sodium and water that leak back into the tubular lumen through the tight junctions of the tubular epithelial cells, especially in the proximal tubule.

lar capillaries or diffuse back through the epithelial junctions into the tubular lumen. The so-called tight junctions between the epithelial cells of the proximal tubule are actually quite leaky, so that considerable amounts of sodium can diffuse in both directions through these junctions. With the normal high rate of peritubular capillary reabsorption, the net movement of water and solutes is toward the peritubular capillaries with little backleak into the lumen of the tubule. On the other hand, during reductions in peritubular capillary reabsorption, there is increased interstitial fluid hydrostatic pressure and a tendency for greater amounts of solute and water to backleak into the tubular lumen, thereby reducing the rate of net reabsorption (refer again to Fig. 27–13).

The opposite is true when there is increased peritubular capillary reabsorption, above the normal level. An initial increase in reabsorption by the peritubular capillaries tends to reduce interstitial fluid hydrostatic pressure and raise interstitial fluid colloid osmotic pressure. Both of these forces favor movement of fluid and solutes out of the tubular lumen and into the interstitium; therefore, backleak of water and solutes into the tubular lumen is reduced and net tubular reabsorption increases.

Thus, through changes in the hydrostatic and colloid osmotic pressures of the renal interstitium, the uptake of water and solutes by the peritubular capillaries is closely matched to the net reabsorption of water and solutes from the tubular lumen into the interstitium. Therefore, in general, *forces that increase peritubular capillary reabsorption also increase reabsorption from the renal tubules. Conversely, hemodynamic changes that inhibit peritubular capillary reabsorption also inhibit tubular reabsorption of water and solutes.*

Effect of Arterial Pressure on Urine Output—The Pressure-Natriuresis and Pressure-Diuresis Mechanisms

Even small increases in arterial pressure often cause marked increases in urinary excretion of sodium and water, phenomena that are referred to as *pressure natriuresis* and *pressure diuresis.* Because of the autoregulatory mechanisms described in Chapter 26, increasing the arterial pressure between the limits of 75 and 160 mm Hg usually has only a small effect on renal blood flow and GFR. The slight increase in GFR that does occur contributes in part to the effect of increased arterial pressure on urine output. When GFR autoregulation is impaired, as often occurs in kidney disease, increases in arterial pressure cause much larger increases in GFR and, therefore, have an even greater effect on sodium and water excretion.

A second effect of increased renal arterial pressure that raises urine output is that it decreases the percentage of the filtered load of sodium and water that is reabsorbed by the tubules. The mechanisms responsible for this effect result in part from a slight increase in peritubular capillary hydrostatic pressure, especially in the vasa recta of the renal medulla, and subsequent

increase in the renal interstitial fluid hydrostatic pressure. As discussed earlier, an increase in the renal interstitial fluid hydrostatic pressure enhances backleak of sodium into the tubular lumen, thereby reducing the net reabsorption of sodium and water and further increasing the rate of urine output when renal arterial pressure rises.

A third factor that contributes to the pressure-natriuresis and pressure-diuresis mechanisms is reduced angiotensin II formation. Angiotensin II itself increases sodium reabsorption by the tubules; it also stimulates aldosterone secretion, which further increases sodium reabsorption. Therefore, decreased angiotensin II formation contributes to the decreased tubular sodium reabsorption that occurs when arterial pressure is increased.

Hormonal Control of Tubular Reabsorption

Precise regulation of body fluid volumes and solute concentrations requires the kidneys to excrete different solutes and water at variable rates, sometimes independently of one another. For example, when potassium intake is increased, the kidneys must excrete more potassium while maintaining normal excretion of sodium and other electrolytes. Likewise, when sodium intake is changed, the kidneys must appropriately adjust urinary sodium excretion without allowing major changes in excretion of other electrolytes and water. Several hormones in the body provide this specificity of tubular reabsorption for different electrolytes and water. Table 27–3 summarizes some of the most important hormones for regulating tubular reabsorption, their principal sites of action on the renal tubule, and their effects on solute and water excretion. Some of these hormones are discussed in more detail in Chapters 28 and 29, but we briefly review their renal tubular actions in the next few paragraphs.

ALDOSTERONE INCREASES SODIUM REABSORPTION AND INCREASES POTASSIUM SECRETION. Aldosterone, secreted by the zona glomerulosa cells of the adrenal cortex, is an important regulator of sodium reabsorption and potassium secretion by the renal tubules. *The primary site of aldosterone action is on the principal cells of the cortical collecting tubule.* The mechanism by which aldosterone increases sodium reabsorption while at the same time increasing potassium secretion is by stimulating the sodium-potassium ATPase pump on the basolateral side of the cortical collecting tubule membrane. Aldosterone also increases the sodium permeability of the luminal side of the membrane.

In the absence of aldosterone, as occurs with adrenal destruction or malfunction (Addison's disease), there is marked loss of sodium from the body and accumulation of potassium. Conversely, excess aldosterone secretion, as occurs in patients with adrenal tumors (Conn's syndrome), is associated with sodium retention and potassium depletion. Although day-to-day regulation of sodium balance can be maintained as long as minimal levels of aldosterone are present, the

Table 27–3 HORMONES THAT REGULATE TUBULAR REABSORPTION

Hormone	Site of Action	Effects
Aldosterone	Distal tubule/collecting duct	↑ NaCl, H_2O reabsorption, ↑ K^+ secretion
Angiotensin II	Proximal tubule	↑ NaCl, H_2O reabsorption, ↑ H^+ secretion
Antidiuretic hormone	Distal tubule/collecting duct	↑ H_2O reabsorption
Atrial natriuretic peptide	Distal tubule/collecting duct	↓ NaCl reabsorption
Parathyroid hormone	Proximal tubules, thick ascending loop of Henle/distal tubules	↓ $PO_4^=$ reabsorption, ↑ Ca^{++} reabsorption

inability to appropriately adjust aldosterone secretion greatly impairs the regulation of renal potassium excretion and potassium concentration of the body fluids. Thus, aldosterone is even more important as a regulator of potassium concentration than it is for sodium concentration.

ANGIOTENSIN II INCREASES SODIUM AND WATER REABSORPTION. Angiotensin II is perhaps the body's most powerful sodium-retaining hormone. As discussed in Chapter 19, angiotensin II formation increases in circumstances associated with low blood pressure and/or low extracellular fluid volume, such as during hemorrhage or loss of salt and water from the body fluids. The increased formation of angiotensin II helps to return blood pressure and extracellular volume toward normal by increasing sodium and water reabsorption from the renal tubules through three main effects:

1. *Angiotensin II stimulates aldosterone secretion,* which in turn increases sodium reabsorption.

2. *Angiotensin II constricts the efferent arterioles,* which has two effects on peritubular capillary dynamics that raise sodium and water reabsorption. First, efferent arteriolar constriction reduces peritubular capillary hydrostatic pressure, which increases net tubular reabsorption, especially from the proximal tubules. Second, efferent arteriolar constriction, by reducing renal blood flow, raises filtration fraction in the glomerulus and increases the concentration of proteins and the colloid osmotic pressure in the peritubular capillaries; this increases the reabsorptive force at the peritubular capillaries and raises tubular reabsorption of sodium and water.

3. *Angiotensin II directly stimulates sodium reabsorption, especially in the proximal tubules.* One of the direct effects of angiotensin II is to stimulate the sodium-potassium ATPase pump on the tubular epithelial cell basolateral membrane. A second effect is to stimulate sodium-hydrogen exchange in the luminal membrane, especially in the proximal tubule. Thus, angiotensin II stimulates sodium transport across both the luminal and the basolateral surfaces of the epithelial cell membrane in the proximal tubule.

These multiple actions of angiotensin II cause marked sodium retention by the kidneys when angiotensin II levels are increased.

ADH INCREASES WATER REABSORPTION. The most important renal action of ADH is to increase the water permeability of the distal tubule, collecting tubule, and collecting duct epithelia. This effect helps the body to conserve water in circumstances such as dehydration. In the absence of ADH, the permeability of the distal

tubules and collecting ducts to water is low, causing the kidneys to excrete large amounts of dilute urine. Thus, the actions of ADH play a key role in controlling the degree of dilution or concentration of the urine, as discussed further in Chapter 28.

ATRIAL NATRIURETIC PEPTIDE DECREASES SODIUM AND WATER REABSORPTION. Specific cells of the cardiac atria, when distended because of plasma volume expansion, secrete a peptide called *atrial natriuretic peptide.* Increased levels of this peptide in turn inhibit the reabsorption of sodium and water by the renal tubules, especially in the collecting ducts. This decreased sodium and water reabsorption increases urinary excretion which helps to return blood volume back toward normal.

PARATHYROID HORMONE INCREASES CALCIUM REABSORPTION. Parathyroid hormone is one of the most important calcium-regulating hormones in the body. Its principal action in the kidneys is to increase tubular reabsorption of calcium, especially in the thick ascending limb of the loop of Henle and in the distal tubule. Parathyroid hormone also has other actions, including inhibition of phosphate reabsorption by the proximal tubule and stimulation of magnesium reabsorption by the loop of Henle, as discussed in Chapter 29.

Sympathetic Nervous System Activation Increases Sodium Reabsorption

Activation of the sympathetic nervous system can decrease sodium and water excretion by constricting both the afferent and the efferent arterioles, thereby reducing GFR. Sympathetic activation also increases sodium reabsorption in the proximal tubule and the thick ascending limb of the loop of Henle. And finally, sympathetic nervous system stimulation increases renin release and angiotensin II formation, which adds to the overall effect to increase tubular reabsorption and decrease renal excretion of sodium.

USE OF CLEARANCE METHODS TO QUANTIFY KIDNEY FUNCTION

The rates at which different substances are "cleared" from the plasma provide a useful way of quantitating the effectiveness with which the kidneys excrete various substances. By definition, the *renal clearance of a substance is the volume of plasma that is completely cleared of the substance by the kidneys per unit time.* This is somewhat of an abstract concept because there is no single volume of plasma that is *completely* cleared of a substance. However, this concept provides a useful way of quantifying the excretory function of the kidneys and, as discussed below, can be used to quantify the rate at which blood flows through the kidneys as well as the basic functions of the kidney: glomerular filtration, tubular reabsorption, and tubular secretion.

To illustrate the clearance principle, consider the following example: If the plasma passing through the kidneys contains 1 milligram of a substance in each milliliter and 1 milligram of this substance is also excreted into the urine each minute, then 1 ml/min of the plasma is "cleared" of the substance. Thus, clearance refers to the volume of plasma that would be necessary to supply the amount of substance excreted in the urine per unit time. Stated mathematically,

$$C_s \times P_s = U_s \times V,$$

where C_s is the clearance rate of a substance s, P_s is the plasma concentration of the substance, V is the urine flow rate, and U_s is the urine concentration of that substance. Rearranging this equation, clearance can be expressed as:

$$C_s = \frac{U_s \times V}{P_s}$$

Thus, renal clearance of a substance is calculated from the urinary excretion rate ($U_s \times V$) of that substance divided by its plasma concentration.

Inulin Clearance Can Be Used to Estimate GFR

If a substance existed that was freely filtered (filtered as freely as water) and if it was not reabsorbed or secreted by the renal tubules, then the rate at which that substance was excreted in the urine ($U_s \times V$) would be equal to the rate at which the substance was filtered by the kidneys ($GFR \times P_s$). Thus,

$$GFR \times P_s = U_s \times V$$

The GFR, therefore, can be calculated as the clearance of the substance as follows:

$$GFR = \frac{U_s \times V}{P_s} = C_s$$

A substance that fits these criteria is inulin, a polysaccharide molecule with a molecular weight of about 5200. Inulin, which is not produced in the body, is found in the roots of certain plants and must be administered intravenously into a patient to measure GFR.

Figure 27–14 shows the renal handling of inulin. In

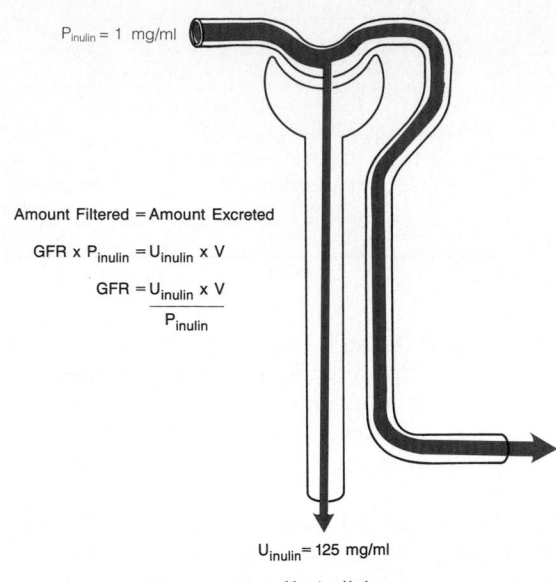

Measurement of Glomerular
Filtration Rate

$P_{inulin} = 1$ mg/ml

Amount Filtered = Amount Excreted

$GFR \times P_{inulin} = U_{inulin} \times V$

$GFR = \dfrac{U_{inulin} \times V}{P_{inulin}}$

$U_{inulin} = 125$ mg/ml

$V = 1$ ml/min

Figure 27–14. Measurement of glomerular filtration rate from the renal clearance of inulin. Inulin is freely filtered by the glomerular capillaries but is not reabsorbed by the renal tubules. Pinulin, plasma inulin concentration; Uinulin, urine inulin concentration; V, urine flow rate.

this example, the plasma concentration is 1 mg/ml, urine concentration is 125 mg/ml, and urine flow rate is 1 ml/min. Therefore, 125 mg/min of inulin passes into the urine. Then, inulin clearance is calculated as the urine excretion rate of inulin divided by the plasma concentration, which yields a value of 125 ml/min. Thus, 125 milliliters of plasma flowing through the kidneys must be filtered to deliver the inulin that appears in the urine.

Inulin is not the only substance that can be used for determining GFR. Other substances that have been used clinically to estimate GFR include *creatinine* and *radioactive iothalamate.* Because creatinine is a by-product of skeletal muscle metabolism, it is present in the plasma at a relatively constant concentration and does not require intravenous infusion into the patient. For this reason, creatinine clearance is perhaps the most widely used method of estimating GFR clinically. However, creatinine is not a perfect marker for GFR because a small amount of it is secreted by the tubules, so that the amount of creatinine excreted in the urine slightly exceeds the amount filtered. There is normally a slight error in measuring plasma creatinine that leads to an overestimate of the plasma concentration, and fortuitously, these two errors tend to cancel each other. Therefore, the clearance of creatinine provides a reasonable estimate of GFR.

PAH Clearance Can Be Used to Estimate Renal Plasma Flow

Theoretically, if a substance is *completely* cleared from the plasma, the clearance rate of that substance is equal to the total renal plasma flow. In other words, the

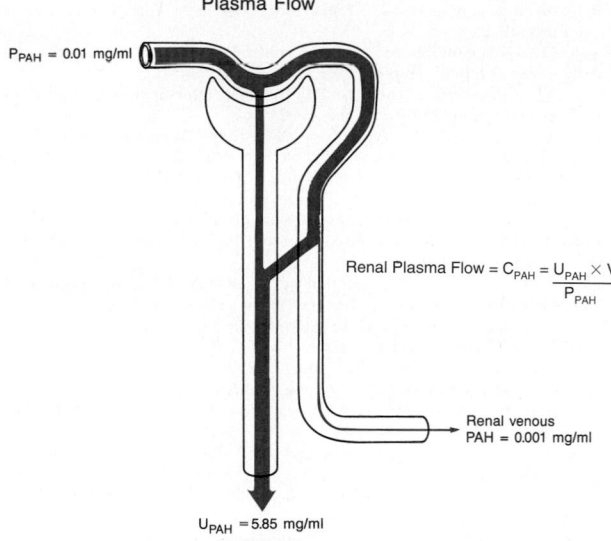

Figure 27–15. Measurement of renal plasma flow from the renal clearance of para-aminohippuric acid (PAH). PAH is freely filtered by the glomerular capillaries and is also secreted from the peritubular capillary blood into the tubular lumen. The amount of PAH in the plasma of the renal artery is about equal to the amount of PAH excreted in the urine. Therefore, the renal plasma flow can be calculated from the clearance of PAH (C_{PAH}). To be more accurate, one can correct for the percentage of PAH that is still in the blood when it leaves the kidneys. P_{PAH}, arterial plasma PAH concentration; U_{PAH}, urine PAH concentration; V, urine flow rate.

amount of the substance delivered to the kidneys in the blood (renal plasma flow $\times$ P_s) would be equal to the amount excreted in the urine ($U_s \times V$). Thus, renal plasma flow (RPF) could be calculated as:

$$RPF = \frac{U_s \times V}{P_s} = C_s$$

Because the GFR is only about 20 per cent of the total plasma flow, a substance that is completely cleared from the plasma must be excreted by tubular secretion as well as glomerular filtration (Fig. 27–15). There is no known substance that is *completely* cleared by the kidneys. One substance, however, PAH, is about 90 per cent cleared from the plasma. Therefore, the clearance of PAH can be used as an approximation of renal plasma flow. To be more accurate, one can correct for the percentage of PAH that is still in the blood when it leaves the kidneys. The percentage of substance removed from the blood is known as the *extraction ratio of PAH* and averages about 90 per cent in normal kidneys. In diseased kidneys, this extraction ratio may be reduced because of inability of damaged tubules to secrete PAH into the tubular fluid.

The calculation of renal plasma flow can be demonstrated by the following example: Assume that the plasma concentration of PAH is 0.01 mg/ml, urine concentration is 5.85 mg/ml, and urine flow rate is 1 ml/min. PAH clearance can be calculated from the rate of urinary PAH excretion (5.85 mg/ml $\times$ 1 ml/min) di-

vided by the plasma PAH concentration (0.01 mg/ml). Thus, clearance of PAH calculates to be 585 ml/min.

If the extraction ratio for PAH is 90 per cent, the actual renal plasma flow can be calculated by dividing 585 ml/min by 0.9, yielding a value of 650 ml/min. Thus, total renal plasma flow can be calculated as:

Total Renal Plasma Flow

= Clearance of PAH/Extraction Ratio of PAH

The extraction ratio (E_{PAH}) is calculated as the difference between the renal arterial (P_{PAH}) and renal venous (V_{PAH}) PAH concentration, divided by the renal arterial PAH concentration:

$$E_{PAH} = \frac{P_{PAH} - V_{PAH}}{P_{PAH}}$$

One can calculate the total blood flow through the kidneys from the total renal plasma flow and hematocrit (the percentage of red blood cells in the blood). If the hematocrit is 0.45 and the total renal plasma flow is 650 ml/min, the total blood flow through both kidneys is 650/(1 − 0.45), or 1182 ml/min.

Filtration Fraction Is Calculated from GFR Divided by Renal Plasma Flow

To calculate the filtration fraction, which is the fraction of plasma that filters through the glomerular membrane, one must first know the renal plasma flow (PAH clearance) and the GFR (inulin clearance). If renal plasma flow is 650 ml/min and GFR is 125 ml/min, the filtration fraction (FF) is calculated as:

$$FF = GFR/RPF = 125/650 = 0.19$$

Calculation of Tubular Reabsorption or Secretion from Renal Clearances

If the rates of glomerular filtration and renal excretion of a substance are known, one can calculate whether there is a net reabsorption or a net secretion of that substance by the renal tubules. For example, if the rate of excretion of the substance ($U_s \times V$) is less than the filtered load of the substance (GFR $\times$ P_s), then some of the substance must have been reabsorbed from the renal tubules. Conversely, if the excretion rate of the substance is greater than its filtered load, then the rate of which it appears in the urine represents the sum of the rate of glomerular filtration plus tubular secretion.

The following example demonstrates the calculation of tubular reabsorption. Assume the following laboratory values for a patient were obtained:

Urine flow rate = 1 ml/min;
Urine concentration of sodium (U_{Na}) = 70 mEq/L = 70 μEq/ml
Plasma sodium concentration = 140 mEq/L = 140 μEq/ml
GFR (inulin clearance) = 100 ml/min

In this example, the filtered sodium load is GFR $\times$ P_{Na}, or 100 ml/min $\times$ 140 μEq/ml = 14,000 μEq/min.

Urinary sodium excretion ($U_{Na} \times$ urine flow rate) is 70 μEq/min. Therefore, tubular reabsorption of sodium is the difference between the filtered load and urinary excretion, or 14,000 μEq/min − 70 μEq/min = 13,930 μEq/min.

COMPARISONS OF INULIN CLEARANCE WITH CLEARANCES OF DIFFERENT SOLUTES. The following generalizations can be made by comparing the clearance of a substance with the clearance of inulin, a measure of GFR: (1) if the clearance rate of the substance equals that of inulin, the substance is only filtered and not reabsorbed or secreted; (2) if the clearance rate of a substance is less than inulin clearance, the substance must have been reabsorbed by the nephron tubules; and (3) if the clearance rate of a substance is greater than that of inulin, the substance must be secreted by the nephron tubules. Listed below are the approximate clearance rates for some of the substances normally handled by the kidneys:

Substance	Clearance Rate (ml/min)
Glucose	0
Sodium	0.9
Chloride	1.3
Potassium	12.0
Phosphate	25.0
Inulin	125
Creatinine	140

REFERENCES

Andreoli, T. E., et al. (eds.): Physiology of Membrane Disorders, 2nd Ed. New York, Plenum Publishing Corp., 1986.

Aukland, K., et al.: Renal cortical interstitium and fluid absorption by peritubular capillaries. Am. J. Physiol., 266:F175, 1994.

Brenner, B. M., and Rector, F. C., Jr. (eds.): The Kidney, 4th Ed. Philadelphia, W. B. Saunders Co., 1991.

Breyer, M. D., and Ando, Y.: Hormonal signaling and regulation of salt and water transport in the collecting duct. Ann. Rev. Physiol., 56:711, 1994.

Burckhardt, G., and Greger, R.: Principles of electrolyte transport across plasma membranes of renal tubular cells. In Windhager, E. E. (ed.): Handbook of Physiology, Section 8, Renal Physiology. New York, Oxford University Press, 1992.

Cogan, M. G.: Angiotensin II: A potent controller of sodium transport in the early proximal tubule. Hypertension, 15:451, 1990.

Granger, J. P.: Pressure natriuresis: Role of renal interstitial hydrostatic pressure. Hypertension, 19 (Suppl. I):I, 1992.

Hall, J. E., and Brands, M. W.: Intrarenal and circulating angiotensin II and renal function. In Robertson, J. I. S. and Nicholls, M. G. (eds.): The Renin-Angiotensin System. London, Gower Medical Publishing, 1993.

Hall, J. E., and Brands, M. W.: The renin-angiotensin-aldosterone system: Renal mechanisms and circulatory hemostasis. In Seldin, D. W., and Giebisch, G. (eds.): The Kidney: Physiology and Pathophysiology, 2nd Ed. New York, Raven Press, 1992.

Hall, J. E.: Control of sodium excretion by angiotensin: Intrarenal mechanisms and blood pressure regulation. Am. J. Physiol., 250:R960, 1986.

Kinne, R. K. H.: Selectivity and direction: Plasma membranes in renal transport. Am. J. Physiol., 260:F153, 1991.

Knox, F. G., and Granger, J. P.: Control of sodium excretion: The kidney produces under pressure. News Physiological Sciences, 2:26, 1982.

Koeppen, B. M., and Stanton, B. A.: Renal Physiology. St. Louis, Mosby-Year Book, 1992.

Lohmeier, T. L., et al.: Angiotensin and ANP during chronically controlled increments in atrial pressure. Am. J. Physiol., 266:R989, 1994.

Marver, D. (ed.): Corticosteroids and the kidney. Semin. Nephrol., 10:311, 1990.

Reeves, W. B., and Andreoli, T. E.: Renal epithelial chloride channels. Ann. Rev. Physiol., 54:29, 1992.

Roman, R. J., and Zou, A.-P.: Influence of the renal medullary circulation on the control of sodium excretion. Am. J. Physiol., 245:R963, 1993.

Schafer, J. A., and Williams, J. C., Jr.: Transport of metabolic substrates by the proximal nephron. Ann. Rev. Physiol., 47:103, 1985.

Schafer, J. A., et al.: Mechanisms of fluid transport across renal tubules. In Windhager, E. E. (ed.): Handbook of Physiology, Section 8, Renal Physiology. New York, Oxford University Press, 1992.

Schafer, J. A.: Salt and water homeostasis—is it just a matter of good bookkeeping? J. Am. Soc. Nephrol., 4:1929, 1994.

Schuster, V. L., and Seldon, D. W.: Renal clearance. In Seldin, D. W., and Giebisch, G. (eds.): The Kidney: Physiology and Pathophysiology, Vol. I. New York, Raven Press, 1992.

Seldin, D. W., and Giebisch, G. (eds.): The Regulation of Sodium and Chloride Balance. New York, Raven Press, 1990.

Smith, H. W.: The Kidney: Structure and Function in Health and Disease. New York, Oxford University Press, 1951.

Vander, A. J.: Renal Physiology, 4th Ed. New York, McGraw-Hill, 1991.

Wilcox, C. S., et al.: Glomerular-tubular balance and proximal regulation. In Seldin, D. W., and Giebisch, G. (eds.): The Kidney: Physiology and Pathophysiology, 2nd Ed. New York, Raven Press, 1992.

Zeidel, M. L.: Hormonal regulation of inner medullary collecting duct sodium transport. Am. J. Physiol., 26:F159, 1993.

Regulation of Extracellular Fluid Osmolarity and Sodium Concentration

CHAPTER 28

For the cells of the body to function properly, they must be bathed in extracellular fluid with a relatively constant concentration of electrolytes and other solutes. The total concentration of solutes in the extracellular fluid—and therefore the osmolarity—is determined by the amount of solute divided by the volume of the extracellular fluid. Thus, to a large extent, extracellular fluid sodium concentration and osmolarity are regulated by the amount of extracellular water. The body water in turn is controlled by (1) fluid intake, which is regulated by factors that determine thirst, and (2) renal excretion of water, which is controlled by multiple factors that influence glomerular filtration and tubular reabsorption.

In this chapter, we discuss specifically (1) the mechanisms that cause the kidneys to eliminate excess water by excreting a dilute urine; (2) the mechanisms that cause the kidneys to conserve water by excreting a concentrated urine; (3) the renal feedback mechanisms that control the extracellular fluid sodium concentration and osmolarity; and (4) the thirst and salt appetite mechanisms that determine the intakes of water and salt, which also help to control extracellular fluid volume, osmolarity, and sodium concentration.

THE KIDNEY EXCRETES EXCESS WATER BY FORMING A DILUTE URINE

The normal kidney has tremendous capability to vary the relative proportions of solutes and water in the urine in response to various challenges. When there is excess water in the body and body fluid osmolarity is reduced, the kidney can excrete urine with an osmolarity as low as 50 mOsm/liter, a concentration that is only about one sixth the osmolarity of normal extracellular fluid. Conversely, when there is a deficit of water and extracellular fluid osmolarity is high, the kidney can excrete urine with a concentration of about 1200 to 1400 mOsm/liter. Equally important, the kidney can excrete a large volume of dilute urine or a small volume of concentrated urine without major changes in rates of excretion of solutes such as sodium and potassium. This ability to regulate water excretion independently of solute excretion is necessary for survival, especially when fluid intake is limited.

Antidiuretic Hormone Controls Urine Concentration

There is a powerful feedback system for regulating plasma osmolarity and sodium concentration that operates by altering renal excretion of water independently of the rate of solute excretion. A primary effector of this feedback is *antidiuretic hormone (ADH)*, also called *vasopressin.*

When osmolarity of the body fluids increases above normal (that is, the solutes in the body fluids become too concentrated), the posterior pituitary gland secretes more ADH, which increases the permeability of the distal tubules and collecting ducts to water. This allows large amounts of water to be reabsorbed and decreases urine volume but does not markedly alter the rate of renal excretion of the solutes.

When there is excess water in the body and extracellular fluid osmolarity is reduced, the secretion of

349

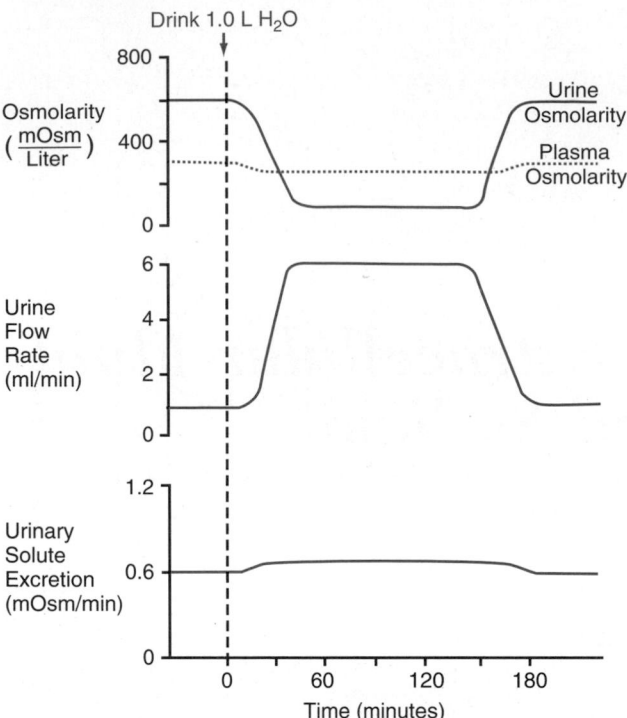

Figure 28–1. Water diuresis in a human after ingestion of 1 liter of water. Note that after water ingestion, urine volume increases and urine osmolarity decreases, causing the excretion of a large volume of dilute urine; however, the total amount of solute excreted by the kidneys remains relatively constant. These responses of the kidneys prevent plasma osmolarity from decreasing markedly during excess water ingestion.

ADH by the posterior pituitary becomes decreased, thereby reducing the permeability of the distal tubule and collecting ducts to water, which causes large amounts of dilute urine to be excreted. Thus, the presence or absence of ADH determines, to a large extent, whether the kidney excretes a dilute or a concentrated urine.

Renal Mechanisms for Excreting a Dilute Urine

When there is a large excess of water in the body, the kidney can excrete as much as 20 liters/day of dilute urine, with a concentration as low as 50 mOsm/ liter. The kidney performs this impressive feat by continuing to reabsorb solutes while failing to reabsorb large amounts of water in the distal parts of the nephron, including the late distal tubule and the collecting ducts. Figure 28–1 shows the approximate renal responses in a human after ingestion of 1 liter of water. Note that urine volume increases to about six times normal within 45 minutes after the water has been drunk. However, the total amount of solute excreted remains relatively constant because the urine formed becomes very dilute and urine osmolarity decreases from 600 to about 100 mOsm/liter. Thus, after ingestion of excess water, the kidney rids the body of

the excess water but does not excrete excess amounts of solutes.

When the glomerular filtrate is initially formed, its osmolarity is about the same as plasma (300 mOsm/ liter). To excrete excess water, it is necessary to dilute the filtrate as it passes along the tubule. This is achieved by reabsorbing solutes to a greater extent than water, as shown in Figure 28–2, but this occurs only in certain segments of the tubular system as follows.

As fluid flows through the proximal tubule, solutes and water are reabsorbed in equal proportions, so that little change in osmolarity occurs; that is, the proximal tubule fluid remains isosmotic to the plasma, with an osmolarity of about 300 mOsm/liter. As fluid passes down the descending loop of Henle, water is reabsorbed by osmosis and the tubular fluid reaches equilibrium with the surrounding interstitial fluid of the renal medulla, which is very hypertonic—about four times the osmolarity of the original glomerular filtrate. Therefore, the tubular fluid becomes more concentrated as it flows into the inner medulla.

In the ascending limb of the loop of Henle, especially the thick segment, sodium, potassium, and chloride are avidly reabsorbed. However, this portion of the tubular segment is impermeable to water, even in the presence of large amounts of ADH. Therefore, the tubular fluid becomes more dilute as it flows up the ascending loop of Henle into the early distal tubule, with the osmolarity decreasing progressively to about 100 mOsm/liter by the time the fluid enters the early distal tubular segment. *Thus, regardless of whether ADH is present or absent, fluid leaving the early distal tubular segment is hypo-osmotic, with an osmolarity of only about one third the osmolarity of plasma.*

As the dilute fluid in the early distal tubule passes into the late distal convoluted tubule, cortical collecting duct, and collecting duct, there is additional reabsorption of sodium chloride. In the absence of ADH, this portion of the tubule is also impermeable to water, and the additional reabsorption of solutes causes the tubular fluid to become even more dilute, decreasing its osmolarity to as low as 50 mOsm/liter. The failure to reabsorb water and the continued reabsorption of solutes leads to a large volume of dilute urine.

To summarize, the mechanism for forming a dilute urine is to continue reabsorbing solutes from the distal segments of the tubular system while failing to reabsorb water. Fluid leaving the ascending loop of Henle and early distal tubule is always dilute, regardless of the level of ADH. In the absence of ADH, the urine is further diluted in the late distal tubule and collecting ducts, and a large volume of dilute urine is excreted.

THE KIDNEY CONSERVES WATER BY EXCRETING A CONCENTRATED URINE

The ability of the kidney to form a urine that is more concentrated than plasma is essential for survival of mammals that live on land, including humans.

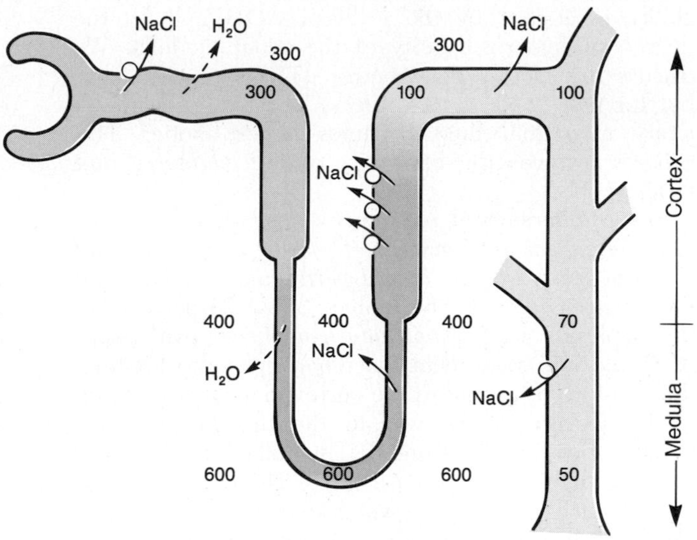

Figure 28–2. Formation of a dilute urine when ADH levels are very low. Note that in the ascending loop of Henle, the tubular fluid becomes very dilute. In the distal tubules and collecting tubules, the tubular fluid is further diluted by the reabsorption of sodium chloride and the failure to reabsorb water when ADH levels are very low. The failure to reabsorb water and continued reabsorption of solutes lead to a large volume of dilute urine. (Numerical values are in milliosmoles per liter.)

Water is continuously lost from the body through various routes, including the lungs by evaporation into the expired air, the gastrointestinal tract by way of the feces, the skin through evaporation and perspiration, and the kidneys through the excretion of urine. Fluid intake is required to match this loss, but the ability of the kidney to form a small volume of concentrated urine minimizes the intake of fluid required to maintain homeostasis, a function that is especially important when water is in short supply.

When there is a water deficit in the body, the kidney forms a concentrated urine by continuing to excrete solutes while increasing water reabsorption and decreasing the volume of urine formed. The human kidney can produce a maximal urine concentration of 1200 to 1400 mOsm/liter, four to five times the osmolarity of plasma. Some desert animals, such as the Australian hopping mouse, can concentrate urine to as high as 10,000 mOsm/liter. This ability allows this mouse to survive in the desert without drinking water; sufficient water can be obtained through the food ingested and water produced in the body by metabolism of the food. Animals adapted to aquatic environments, such as the beaver, have minimal urine concentrating ability; they can concentrate the urine to only about 500 mOsm/liter.

Obligatory Urine Volume

The maximal concentrating ability of the kidney dictates how much urine volume must be excreted each day to rid the body of waste products of metabolism and ions that are ingested. A normal 70-kilogram human must excrete about 600 milliosmoles of solute each day. If maximal urine concentrating ability is 1200 mOsm/liter, the *minimal* volume of urine that must be excreted, called the *obligatory urine volume*, can be calculated as:

$$\frac{600 \text{ mOsm/day}}{1200 \text{ mOsm/L}} = 0.5 \text{ L/day}$$

This minimal loss of volume in the urine contributes to dehydration, along with water loss from the skin, respiratory tract, and the gastrointestinal tract, when water is not available to drink.

The limited ability of the human kidney to concentrate the urine to a maximal concentration of 1200 mOsm/liter explains why severe dehydration occurs if one attempts to drink seawater. The concentration of salt in the oceans averages about 3 per cent sodium chloride, with an osmolarity between 2000 and 2400 mOsm/liter. Drinking 1 liter of seawater with a concentration of 2400 mOsm/liter would provide a total solute intake of 2400 milliosmoles. If maximal urine concentrating ability is 1200 mOsm/liter, the amount of urine volume needed to excrete 2400 milliosmoles would be 2400 milliosmoles divided by 1200 mOsm/liter, or 2 liters. Thus, for every liter of seawater drunk, 2 liters of urine volume would be required to rid the body of solutes ingested. This would result in a net fluid loss of 1 liter for every liter of seawater drunk, explaining the rapid dehydration that occurs in shipwreck victims who drink seawater. On the other hand, a shipwreck victim's pet Australian hopping mouse could drink with impunity all the seawater it wanted.

Requirements for Excreting a Concentrated Urine—High ADH Levels and Hyperosmotic Renal Medulla

The basic requirements for forming a concentrated urine are (1) a high level of ADH, which increases the permeability of the distal tubules and collecting ducts to water, thereby allowing these tubular segments to avidly reabsorb water, and (2) a high osmolarity of the renal medullary interstitial fluid, which provides the osmotic gradient necessary for water reabsorption to occur in the presence of high levels of ADH.

The renal medullary interstitium surrounding the collecting ducts normally is very hyperosmotic, so that when ADH levels are high, water moves through the tubular membrane by osmosis into the renal interstitium; from there it is carried away by the vasa recta back into the blood. Thus, the urine concentrating ability is limited by the level of ADH and by the degree of hyperosmolarity of the renal medulla. We discuss the factors that control ADH secretion later, but for now, what is the process by which renal medullary interstitial fluid becomes hyperosmotic? This process involves the operation of the *countercurrent mechanism.*

The countercurrent mechanism depends on the special anatomical arrangement of the loops of Henle and the vasa recta, the specialized peritubular capillaries of the renal medulla. In the human, about 25 per cent of the nephrons are *juxtamedullary nephrons,* with loops of Henle and vasa recta that go deeply into the medulla before returning to the cortex. Some of the loops of Henle dip all the way to the tips of the renal papillae that project from the medulla into the renal pelvis. Paralleling the long loops of Henle are the vasa recta, which also loop down into the medulla before returning to the renal cortex. And finally, the collecting ducts, which carry urine through the hyperosmotic renal medulla before it is excreted, also play a critical role in the countercurrent mechanism.

The Countercurrent Mechanism Produces a Hyperosmotic Renal Medullary Interstitium

The osmolarity of interstitial fluid in almost all parts of the body is about 300 mOsm/liter, which is similar to the plasma osmolarity. (As discussed in Chapter 25, the *corrected osmolar activity,* which accounts for intermolecular attraction and repulsion, is about 282 mOsm/liter.) The osmolarity of the interstitial fluid in the medulla of the kidney is much higher, increasing progressively to about 1200 mOsm/liter (in some cases to as high as 1400 mOsm/liter) in the pelvic tip of the medulla. This means that the renal medullary interstitium has accumulated solutes in great excess of water. Once the high solute concentration in the medulla is achieved, it is maintained by a balanced inflow and outflow of solutes and water in the medulla.

The major factors that contribute to the buildup of solute concentration into the renal medulla are as follows:

1. Active transport of sodium ions and co-transport of potassium, chloride, and other ions out of the thick portion of the ascending limb of the loop of Henle into the medullary interstitium
2. Active transport of ions from the collecting ducts into the medullary interstitium
3. Passive diffusion of large amounts of urea from the inner medullary collecting ducts into the medullary interstitium
4. Diffusion of only small amounts of water from the medullary tubules into the medullary interstitium, far less than the reabsorption of solutes into the medullary interstitium

SPECIAL CHARACTERISTICS OF THE LOOP OF HENLE THAT CAUSE SOLUTES TO BE TRAPPED IN THE RENAL MEDULLA. The transport characteristics of the loop of

Table 28–1 SUMMARY OF TUBULE CHARACTERISTICS— URINE CONCENTRATION

	Active NaCl Transport	Permeability		
		H₂O	NaCl	Urea
Thin descending limb	0	+ + + + +	+	+
Thin ascending limb	0	0	+	+
Thick ascending limb	+ + + + +	0	0	0
Distal tubule	+	+ ADH	0	0
Cortical collecting tubule	+	+ ADH	0	0
Inner medullary collecting duct	+	+ ADH	0	+ + +

Henle are summarized in Table 28–1, along with the characteristics of the distal tubule, cortical collecting tubules, and inner medullary collecting ducts.

The most important cause of the high medullary osmolarity is active transport of sodium and co-transport of potassium, chloride, and other ions from the thick ascending loop of Henle into the interstitium. This pump is capable of establishing about a 200-milliosmole concentration gradient between the tubular lumen and the interstitial fluid. Because the thick ascending limb is virtually impermeable to water, the solutes pumped out are not followed by osmotic flow of water into the interstitium. Thus, the active transport of sodium and other ions out of the thick ascending loop adds solutes in excess of water to the renal medullary interstitium. There is some passive reabsorption of sodium chloride from the thin ascending limb of Henle's loop, which is also impermeable to water, adding further to the high solute concentration of the renal medullary interstitium.

The descending limb of Henle's loop, in contrast to the ascending limb, is very permeable to water, and the tubular fluid osmolarity quickly becomes equal to the renal medullary osmolarity. Therefore, water diffuses out of the descending limb of Henle's loop into the interstitium, and the tubular fluid osmolarity gradually rises as it flows toward the tip of the loop of Henle.

STEPS INVOLVED IN CAUSING THE HYPEROSMOTIC RENAL MEDULLARY INTERSTITIUM. With these characteristics of the loop of Henle in mind, let us now discuss how the renal medulla becomes hyperosmotic. First, assume that the loop of Henle is filled with fluid with a concentration of 300 mOsm/liter, the same as that leaving the proximal tubule (Fig. 28–3, step 1). Next, the active pump of the *thick ascending limb* on the loop of Henle is turned on, reducing the concentration inside the tubule and raising the interstitial concentration; this pump establishes a 200-mOsm/liter concentration gradient between the tubular fluid and the interstitial fluid (step 2). The limit to the gradient is about 200 mOsm/liter because paracellular diffusion of ions back into the tubule eventually counterbalances transport of ions out of the lumen when the 200-mOsm/liter concentration gradient is achieved.

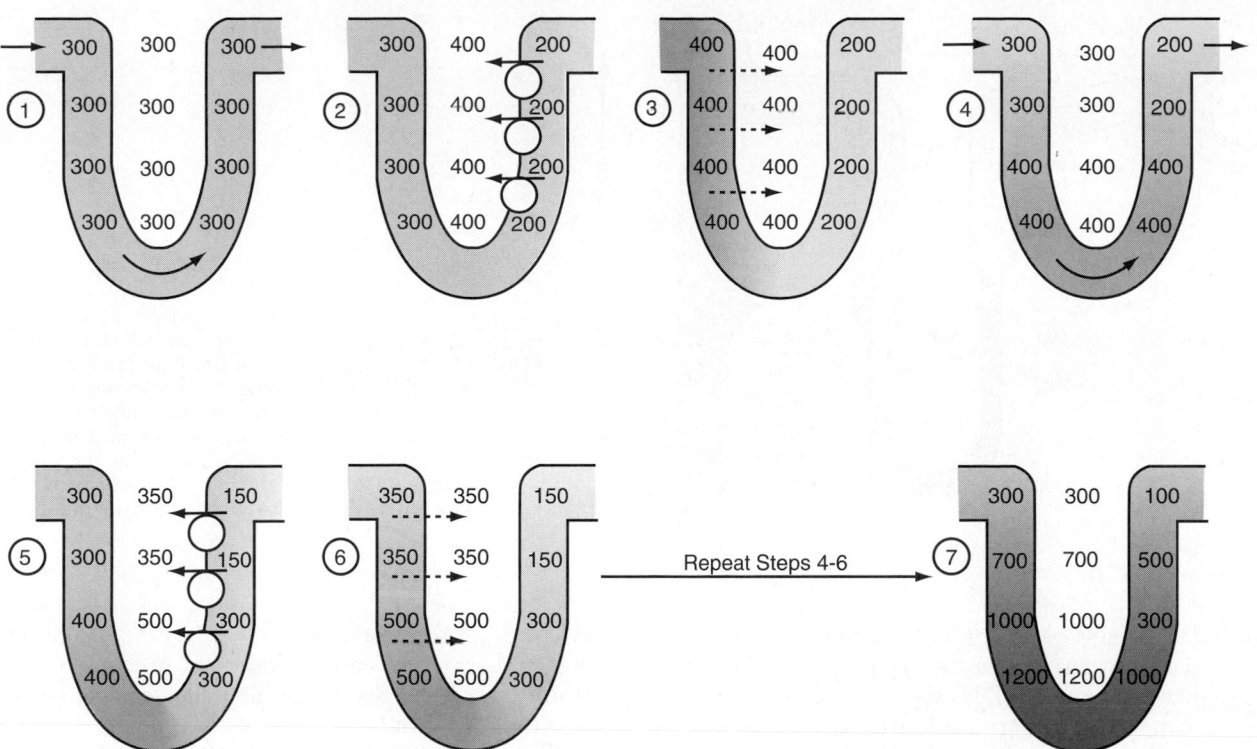

Figure 28–3. Countercurrent multiplier system in the loop of Henle for producing a hyperosmotic renal medulla (Numerical values are in milliosmoles per liter.)

The third step is that the tubular fluid in the *descending limb of the loop of Henle* and the interstitial fluid quickly reach osmotic equilibrium because of osmosis of water out of the descending limb. The interstitial osmolarity is maintained at 400 mOsm/liter because of continued transport of ions out of the thick ascending loop of Henle. Thus, by itself, the active transport of sodium chloride out of the thick ascending limb is capable of establishing only a 200-mOsm/liter concentration gradient, much less than that achieved by the countercurrent system.

The fourth step is additional flow of fluid into the loop of Henle from the proximal tubule, which causes the hyperosmotic fluid previously formed in the descending limb to flow into the ascending limb. Once this fluid is in the ascending limb, additional ions are pumped into the interstitium, with water remaining behind, until a 200-mOsm/liter osmotic gradient is established, with the interstitial fluid osmolarity rising to 500 mOsm/liter (step 5). Then once again, the fluid in the descending limb reaches equilibrium with the hyperosmotic medullary interstitial fluid (step 6), and as the hyperosmotic tubular fluid from the descending limb of the loop of Henle flows into the ascending limb, still more solute is continuously pumped out of the tubules and deposited into the medullary interstitium.

These steps are repeated over and over, with the net effect of adding more and more solute to the medulla in excess of water; with sufficient time, *this process gradually traps solutes in the medulla and multiplies the concentration gradient established by the active pumping of ions out of the thick ascending loop of Henle, eventually raising the interstitial third osmolarity to 1200 to 1400 mOsm/liter.*

Thus, the repetitive reabsorption of sodium chloride by the thick ascending loop of Henle and continued inflow of new sodium chloride from the proximal tubule into the loop of Henle is called the *countercurrent multiplier.* The sodium chloride reabsorbed from the ascending loop of Henle keeps adding to the newly arrived sodium chloride, thus "multiplying" its concentration in the medullary interstitium.

Role of the Distal Tubule and Collecting Ducts in Excreting a Concentrated Urine

When the tubular fluid leaves the loop of Henle and flows into the distal convoluted tubule in the renal cortex, the fluid is dilute, with a osmolarity of only about 100 mOsm/liter (Fig. 28–4). The early distal tubule further dilutes the tubular fluid because this segment, like the ascending loop of Henle, actively transports sodium chloride out of the tubule but is relatively impermeable to water.

As fluid flows into the cortical collecting tubule, the amount of water reabsorbed is critically dependent on the plasma concentration of ADH. In the absence of ADH, this segment is almost impermeable to water

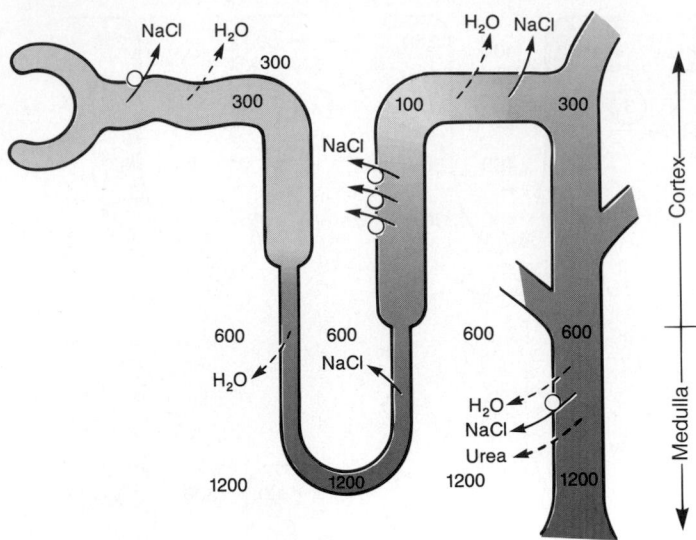

Figure 28–4. Formation of a concentrated urine when ADH levels are high. Note that the fluid leaving the loop of Henle is dilute but becomes concentrated as water is absorbed from the distal tubules and collecting tubules. With high ADH levels, the osmolarity of the urine is about the same as the osmolarity of the renal medullary interstitial fluid in the papilla, which is about 1200 mOsm/liter. (Numerical values are in milliosmoles per liter.)

and fails to reabsorb water but continues to reabsorb solutes and further dilutes the urine. When there is a high concentration of ADH, the cortical collecting tubule becomes highly permeable to water, so that large amounts of water are now reabsorbed from the tubule into the cortex interstitium, where it is swept away by the rapidly flowing peritubular capillaries. *The fact that these large amounts of water are reabsorbed into the cortex, rather than into the renal medulla, helps to preserve the high medullary interstitial fluid osmolarity.*

As the tubular fluid flows along the medullary collecting ducts, there is further water reabsorption from the tubular fluid into the interstitium, but the total amount of water is relatively small compared with that added to the cortex interstitium. The reabsorbed water is quickly carried away by the vasa recta into the venous blood. When high levels of ADH are present, the collecting ducts become permeable to water, so that the fluid at the end of the collecting ducts has essentially the same osmolarity as the interstitial fluid of the renal medulla—about 1200 mOsm/liter (see Fig. 28–3). Thus, by reabsorbing as much water as possible, the kidneys form a highly concentrated urine, excreting normal amounts of solutes in the urine while adding water back to the extracellular fluid and compensating for deficits of body water.

Urea Contributes to Hyperosmotic Renal Medullary Interstitium and to a Concentrated Urine

Thus far, we have considered only the contribution of sodium chloride to the hyperosmotic renal medullary interstitium. However, urea contributes about 40 per cent of the osmolarity (500 mOsm/liter) of the renal medullary interstitium when the kidney is forming a maximally concentrated urine. Unlike sodium chloride, urea is passively reabsorbed from the tubule.

When there is a water deficit and blood concentrations of ADH are high, large amounts of urea are passively reabsorbed from the inner medullary collecting ducts into the interstitium.

The mechanism for reabsorption of urea into the renal medulla is as follows: As water flows up the ascending loop of Henle and into the distal and cortical collecting tubules, little urea is reabsorbed because these segments are impermeable to urea (see Table 28–1). In the presence of high concentrations of ADH, water is reabsorbed rapidly from the cortical collecting tubule and the urea concentration increases rapidly because urea is not very permeant in this part of the tubule. Then, as the tubular fluid flows into the inner medullary collecting ducts, still further water reabsorption takes place, causing an even higher concentration of urea in the fluid. This high concentration of urea in the tubular fluid of the inner medullary collecting duct causes large amounts of urea to diffuse out of the tubule into the renal interstitium because this segment is highly permeable to urea, and ADH increases this permeability even more. The simultaneous movement of water and urea out of the inner medullary collecting ducts maintains a high concentration of urea in the tubular fluid and, eventually, in the urine, even though urea is being reabsorbed.

The fundamental role of urea in contributing to urine concentrating ability is evidenced by the fact that people who are fed a high-protein diet, yielding large amounts of urea as a nitrogenous "waste" product, can concentrate their urine much better than people whose protein intake and urea production are low. Malnutrition is associated with a low urea concentration in the medullary interstitium and considerable impairment of urine concentrating ability.

RECIRCULATION OF UREA FROM THE COLLECTING DUCT TO THE LOOP OF HENLE CONTRIBUTES TO HYPEROSMOTIC RENAL MEDULLA. A normal person excretes about 40 to 60 per cent of the filtered load of urea. In general, the rate of urea excretion is deter-

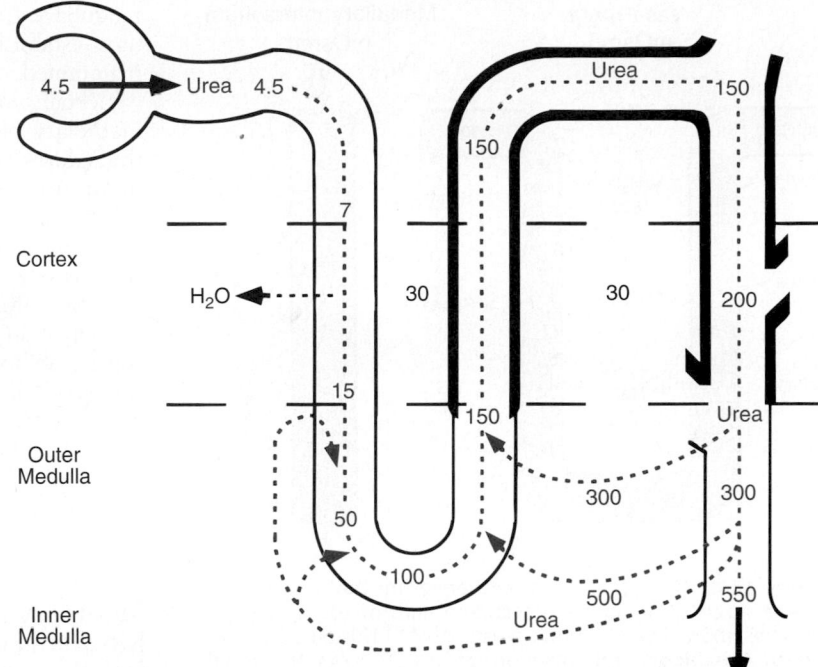

Figure 28–5. Recirculation of urea absorbed from the collecting duct into the interstitial fluid. This urea then passes into the loop of Henle, through the distal tubule, and, finally, back into the collecting duct. The recirculation of urea helps to trap urea in the renal medulla and contributes to the hyperosmolarity of the renal medulla. (Numerical values are milliosmolarities of urea during antidiuresis, when large amounts of ADH are present.)

mined mainly by two factors: (1) the concentration of urea in the plasma and (2) the glomerular filtration rate (GFR). In patients with renal disease who have large reductions of GFR, the plasma urea concentration increases markedly, returning the filtered urea load and urea excretion rate to the normal level (equal to the rate of urea production), despite the reduced GFR.

As urea enters the proximal tubule, a small amount of urea reabsorption takes place, but even so, the tubular fluid urea concentration increases because urea is not nearly as permeant as water. The concentration of urea continues to rise as the tubular fluid flows into the thin segments of the loop of Henle, partly because of water reabsorption out of the descending loop of Henle but also because of some diffusion of urea into the thin loop of Henle from the medullary interstitium (Fig. 28–5).

The thick limb of the loop of Henle, the distal tubule, and the cortical collecting tubule are all relatively impermeable to urea. When the kidney is forming a concentrated urine and high levels of ADH are present, the reabsorption of water from the distal tubule and cortical collecting tubule further raises the tubular fluid concentration of urea. And as this urea flows into the inner medullary collecting duct, the high tubular fluid concentration of urea causes urea to diffuse into the medullary interstitium. A moderate share of the urea that moves into the medullary interstitium eventually diffuses into the thin loop of Henle, so that it passes upward through the ascending loop of Henle, the distal tubule, through the cortical collecting tubule, and back down into the medullary collecting duct again. In this way, urea can recirculate through these

terminal parts of the tubular system several times before it is excreted. Each time around the circuit contributes to a higher concentration of urea.

This urea recirculation provides an additional mechanism for forming a hyperosmotic renal medulla. Because urea is the most abundant of the waste products that must be excreted by the kidneys, this mechanism for concentrating urea before it is excreted is essential to the economy of the body fluid when water is in short supply.

Countercurrent Exchange in the Vasa Recta Preserves Hyperosmolarity of the Renal Medulla

Blood flow must be provided to the renal medulla to supply the metabolic needs of the cells in this part of the kidney. Without a special medullary blood flow system, the solutes pumped into the renal medulla by the countercurrent multiplier system would be rapidly dissipated.

There are two special features of the renal medullary blood flow that contribute to the preservation of the high solute concentrations:

1. *The medullary blood flow is low,* accounting for only 1 to 2 per cent of the total renal blood flow. This sluggish blood flow is sufficient to supply the metabolic needs of the tissues but helps to minimize solute loss from the medullary interstitium.

2. *The vasa recta serve as countercurrent exchangers,* minimizing washout of solutes from the medullary interstitium.

The countercurrent exchange mechanism operates as

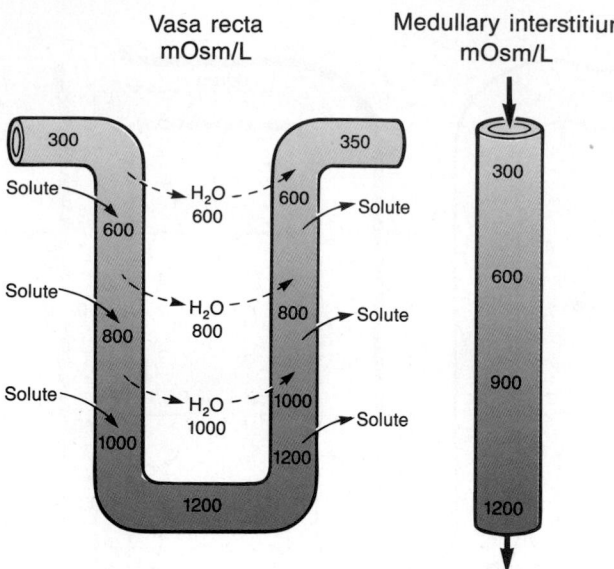

Figure 28–6. Countercurrent exchange in the vasa recta. Plasma flowing down the descending limb of the vasa recta becomes more hyperosmotic because of diffusion of water out of the blood and diffusion of solutes from the renal interstitial fluid into the blood. In the ascending limb of the vasa recta, solutes diffuse back into the interstitial fluid and water diffuses back into the vasa recta. As shown on the right side of the figure, large amounts of solutes would be lost from the renal medulla without the U shape of the vasa recta capillaries (Numerical values are in milliosmoles per liter.)

follows (Fig. 28–6): Blood enters and leaves the medulla by way of the vasa recta at the boundary of the cortex and renal medulla. The vasa recta, like other capillaries, are highly permeable to solutes in the blood, except for the plasma proteins. As blood descends into the medulla toward the papillae, it becomes progressively more concentrated, partly by solute entry from the interstitium and partly by loss of water into the interstitium. By the time the blood reaches the tips of the vasa recta, it has a concentration of about 1200 mOsm/liter, the same as the medullary interstitium. As blood ascends back toward the cortex, it becomes progressively less concentrated as solutes diffuse back out into the medullary interstitium and as water moves into the vasa recta.

Thus, although there is a large amount of fluid and solute exchange across the vasa recta, there is little net dilution of the concentration of the interstitial fluid at each level of the renal medulla because of the U shape of the vasa recta capillaries, which act as countercurrent exchangers. *Thus, the vasa recta do not create the medullary hyperosmolarity but do prevent it from being dissipated.*

The U-shaped structure of the vessels minimizes loss of solute from the interstitium but does not prevent the bulk flow of fluid and solutes into the blood through the usual colloid osmotic and hydrostatic pressures that favor reabsorption in these capillaries. Thus, under steady-state conditions, the vasa recta carry away only as much solute and water as absorbed from the

medullary tubules, and the high concentration of solutes established by the countercurrent mechanism is maintained.

Certain vasodilators can markedly increase renal medullary blood flow, thereby "washing out" some of the solutes from the renal medulla and reducing maximum urine concentrating ability. Large increases in arterial pressure can also increase the blood flow of the renal medulla to a greater extent than in other regions of the kidney and tend to wash out the hyperosmotic interstitium, thereby reducing urine concentrating ability. As discussed above, maximum concentrating ability of the kidney is determined not only by the level of ADH but also by the osmolarity of the renal medulla interstitial fluid. Even with maximal levels of ADH, urine concentrating ability will be reduced if medullary blood flow increases enough to reduce the hyperosmolarity in the renal medulla.

Summary of Urine Concentrating Mechanism and Changes in Osmolarity in Different Segments of the Tubules

The changes in osmolarity and volume of the tubular fluid as it passes through the different parts of the nephron are shown in Figure 28–7.

PROXIMAL TUBULE. About 65 per cent of the filtered electrolytes are reabsorbed in the proximal tubule. However, the tubular membranes are highly permeable to water, so that whenever solutes are reabsorbed, water also diffuses through the tubular membrane by osmosis. Therefore, the osmolarity of the fluid remains about the same as the glomerular filtrate, 300 mOsm/liter.

DESCENDING LOOP OF HENLE. As fluid flows down the descending loop of Henle, water is absorbed into the medulla. The descending limb is highly permeable to water but much less permeable to sodium chloride and urea. Therefore, the osmolarity of the fluid into the descending loop gradually increases until it is equal to that of the surrounding interstitial fluid, which is about 1200 mOsm/liter when the blood concentration of ADH is high. When a dilute urine is being formed, owing to low ADH concentrations, the medullary interstitial osmolarity is less than 1200 mOsm/liter; consequently, the descending loop tubular fluid osmolarity also becomes less concentrated. This is due partly to the fact that less urea is absorbed into the medullary interstitium from the collecting ducts when ADH levels are low and the kidney is forming a large volume of dilute urine.

THIN ASCENDING LOOP OF HENLE. The thin ascending limb is essentially impermeable to water but is more permeable to sodium chloride. Because of the high concentration of sodium chloride in the tubular fluid, owing to water removal from the descending loop of Henle, there is some passive diffusion of sodium chloride from the thin ascending limb into the medullary interstitium. Thus, the tubular fluid becomes more dilute as the sodium chloride diffuses out

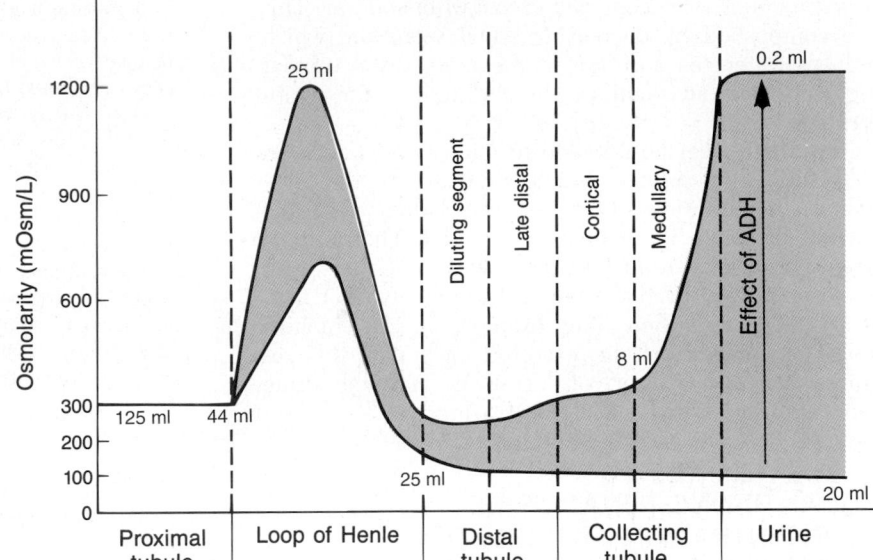

Figure 28–7. Changes in osmolarity of the tubular fluid as it passes through the different tubular segments in the presence of high levels of ADH and in the absence of ADH. (Numerical values indicate the approximate volumes in milliliters per minute or osmolarities in milliosmoles per liter of fluid flowing along the different tubular segments.)

of the tubule and water remains in the tubule. Some of the urea absorbed into the medullary interstitium from the collecting ducts also diffuses into the ascending limb, thereby returning the urea to the tubular system and helping to prevent its washout from the renal medulla. This *urea recycling* is an additional mechanism that contributes to the hyperosmotic renal medulla.

THICK ASCENDING LOOP OF HENLE. The thick part of the ascending loop of Henle is also virtually impermeable to water, but large amounts of sodium, chloride, potassium, and other ions are actively transported from the tubule into the medullary interstitium. Therefore, fluid in the thick ascending limb of the loop of Henle becomes very dilute, falling to a concentration of about 100 mOsm/liter.

EARLY DISTAL TUBULE. The early distal tubule has properties similar to those of the thick ascending loop of Henle, so that further dilution of the tubular fluid occurs as solutes are reabsorbed while water remains in the tubule.

LATE DISTAL TUBULE AND CORTICAL COLLECTING TUBULES. In the late distal tubule and cortical collecting tubules, the osmolarity of the fluid depends on the presence or absence of ADH. With high levels of ADH, these tubules are highly permeable to water, thus causing large amounts of water to be reabsorbed. Urea, however, is not very permeant in this part of the nephron, resulting in increased urea concentration as water is reabsorbed. This allows most of the urea delivered to the distal tubule and collecting tubule to pass into the inner medullary collecting ducts, from which it is eventually reabsorbed or excreted in the urine. In the absence of ADH, little water is reabsorbed in the late distal tubule and cortical collecting tubule; therefore, osmolarity decreases even further because of continued active reabsorption of ions from these segments.

INNER MEDULLARY COLLECTING DUCTS. The concentration of fluid in the inner medullary collecting ducts also depends on (a) ADH and (b) the osmolarity of the medullary interstitium established by the countercurrent mechanism. In the presence of large amounts of ADH, these ducts are highly permeable to water, and water diffuses from the tubule into the interstitium until osmotic equilibrium is reached, with the tubular fluid having about the same concentration as the renal medullary interstitium (1200 mOsm/liter). Thus, a very concentrated but small volume of urine is produced when ADH levels are high. And because water reabsorption increases urea concentration in the tubular fluid, plus the fact that the inner medullary collecting ducts are highly permeable to urea, much of the highly concentrated urea in the ducts diffuses out of the tubular lumen into the medullary interstitium. This absorption of the urea into the renal medulla but not the renal cortex contributes to the high osmolarity of the medullary interstitium and the high concentrating ability of the kidney.

There are several important points to consider that may not be obvious from this discussion. First, although sodium chloride is one of the principal solutes that contributes to the hyperosmolarity of the medullary interstitium, *the kidney can, when needed, excrete a highly concentrated urine that contains little sodium chloride.* The hyperosmolarity of the urine in these circumstances is due to high concentrations of other solutes, especially of waste products such as urea and creatinine. One condition in which this occurs is dehydration accompanied by low sodium intake. As discussed in Chapter 29, low sodium intake stimulates formation of the hormones angiotensin II and aldosterone, which together cause avid sodium reabsorption from the tubules while leaving the urea and other solutes to maintain the highly concentrated urine.

Second, *large quantities of dilute urine can be ex-*

creted without increasing the excretion of sodium. This is accomplished by decreasing ADH secretion, which reduces water reabsorption in the more distal tubular segments without significantly altering sodium reabsorption.

And finally, we should keep in mind that there is an *obligatory urine volume,* which is dictated by the maximum concentrating ability of the kidney and the amount of solute that must be excreted. Therefore, if large amounts of solute must be excreted, they must be accompanied by the minimal amount of water necessary to excrete them. For example, if 1200 milliosmoles of solute must be excreted each day, this requires *at least* 1 liter of urine if maximal urine concentrating ability is 1200 mOsm/liter.

QUANTIFYING RENAL URINE CONCENTRATION AND DILUTION: "FREE WATER" AND OSMOLAR CLEARANCES

The process of concentrating or diluting the urine requires the kidneys to excrete water and solutes somewhat independently. When the urine is dilute, water is excreted in excess of solutes. Conversely, when the urine is concentrated, solutes are excreted in excess of water.

The total clearance of solutes from the blood can be expressed as the *osmolar clearance* (C_{osm}); this measures the volume of plasma cleared of solutes each minute, in the same way that clearance of a single substance is calculated:

$$C_{osm} = \frac{U_{osm} \times V}{P_{osm}},$$

where U_{osm} is the urine osmolarity, V is the urine flow rate, and P_{osm} is the plasma osmolarity. For example, if plasma osmolarity is 300 mOsm/liter, urine osmolarity is 600 mOsm/liter, and urine flow rate is 1 ml/min (0.001 liter/min), the rate of osmolar excretion is 0.6 mOsm/min (600 mOsm/liter × 0.001 liter/min) and osmolar clearance is 0.6 mOsm/min divided by 300 mOsm/liter, or 0.002 liter/min (2.0 ml/min). This means that 2 milliliters of plasma are being cleared of solute each minute.

THE RELATIVE RATES AT WHICH SOLUTES AND WATER ARE EXCRETED CAN BE ASSESSED USING THE CONCEPT OF "FREE-WATER CLEARANCE." *Free-water clearance* (C_{H_2O}) is calculated as the difference between water excretion (urine flow rate) and osmolar clearance:

$$C_{H_2O} = V - C_{osm} = V - \frac{(U_{osm} \times V)}{(P_{osm})}$$

Thus, the rate of free-water clearance represents the rate at which solute-free water is excreted by the kidneys. *When free-water clearance is positive, excess water is being excreted by the kidneys; when free-water clearance is negative, excess solutes are being removed from the blood by the kidneys and water is being conserved.*

Using the example discussed above, if urine flow rate is 1 ml/min and osmolar clearance is 2 ml/min, free-water clearance would be − 1 ml/min. This means that instead of water being cleared from the kidneys in excess of solutes, the kidneys are actually returning water back to the systemic circulation, as occurs during water deficits. Thus, whenever urine osmolarity is greater than plasma osmolarity, free-water clearance will be negative, indicating water conservation.

When the kidneys are forming a dilute urine (that is, urine osmolarity is less than plasma osmolarity), free-water clearance will be a positive value, denoting that water is being removed from the plasma by the kidneys in excess of solutes. Thus, water free of solutes, called "free water," is being lost from the body and the plasma is being concentrated when free-water clearance is positive.

DISORDERS OF URINARY CONCENTRATING ABILITY

An impairment in the ability of the kidneys to concentrate or dilute the urine appropriately can occur with one or more of the following abnormalities:

1. *Inappropriate secretion of ADH.* Either too much or too little ADH secretion results in abnormal fluid handling by the kidneys.

2. *Impairment of the countercurrent mechanism.* A hyperosmotic medullary interstitium is required for maximal urine concentrating ability. No matter how much ADH is present, maximal urine concentration is limited by the degree of hyperosmolarity of the medullary interstitium.

3. *Inability of the distal tubule, collecting tubule, and collecting ducts to respond* to ADH

FAILURE TO PRODUCE ADH: "CENTRAL" DIABETES INSIPIDUS. An inability to produce or release ADH from the posterior pituitary can be caused by head injuries or infections or it can be congenital. Because the distal tubular segments cannot reabsorb water in the absence of ADH, this condition, called *"central" diabetes insipidus,* results in the formation of a large volume of dilute urine, with urine volumes that can exceed 15 liters/day. The thirst mechanisms, discussed later in this chapter, are activated when excessive water is lost from the body; therefore, as long as the person drinks enough water, large decreases in body fluid water do not occur. The primary abnormality observed clinically in people with this condition is the large volume of dilute urine. However, if water intake is restricted, as can occur in a hospital setting when fluid intake is restricted or the patient is unconscious (for example, because of a head injury), severe dehydration can rapidly occur.

INABILITY OF THE KIDNEYS TO RESPOND TO ADH: "NEPHROGENIC" DIABETES INSIPIDUS. There are circumstances in which normal or elevated levels of ADH are present but the distal tubular segments cannot respond appropriately. This condition is referred to as *"nephrogenic" diabetes insipidus* because the abnormality resides in the kidneys. This abnormality can be due to either failure of the countercurrent mechanism to form a hyperosmotic renal medullary interstitium or failure of the distal and collecting tubules and collecting ducts to respond to ADH. In either case, large volumes of dilute urine are formed, which tends to cause dehydration unless fluid intake is increased by the same amount as urine volume is increased.

Many types of renal diseases can impair the concentrating mechanism, especially those that damage the renal medulla. Also, impairment of the function of the loop of Henle, as occurs with diuretics that inhibit electrolyte reabsorption by this segment, can compromise urine concentrating ability. And certain drugs, such as lithium (used to treat manic-depressive disorders) and tetracyclines (used as antibiotics), can impair the ability of the distal nephron segments to respond to ADH.

Thus, the major causes of impaired urinary concentrating ability include those factors that (a) impair the release of ADH or (b) impair the ability of the kidneys to respond to ADH by preventing the effect of ADH on the tubule or by interfering with the formation of a hyperosmotic renal medullary interstitium.

CONTROL OF EXTRACELLULAR FLUID OSMOLARITY AND SODIUM CONCENTRATION

The regulation of extracellular fluid osmolarity and sodium concentration are closely linked because sodium is the most abundant ion in the extracellular compartment. Plasma sodium concentration is normally regulated within close limits of 140 ± 5 mEq/liter, with an average concentration of about 142 mEq/liter. Osmolarity averages about 300 mOsm/liter and seldom changes more than ± 2 to 3 per cent. As discussed in Chapter 25, these variables must be precisely controlled because they determine the distribution of fluid between the intracellular and extracellular compartments.

Estimating Plasma Osmolarity from Plasma Sodium Concentration

In most clinical laboratories, plasma osmolarity is not routinely measured. However, because sodium and its associated anions account for about 94 per cent of the solute in the extracellular compartment, plasma osmolarity (P_{osm}) can be roughly approximated as:

$$P_{osm} = 2.1 \times \text{Plasma sodium concentration.}$$

For instance, with a plasma sodium concentration of 142 mEq/liter, the plasma osmolarity would be estimated from the formula above to be 298.2 mOsm/liter. To be more exact, especially in conditions associated with renal disease, the contribution of two other solutes, glucose and urea, should be included. Such estimates of plasma osmolarity are usually accurate within a few percentage points of those measured directly.

Normally, sodium ions and associated anions (primarily bicarbonate and chloride) represent about 94 per cent of the extracellular osmoles, with glucose and urea contributing about 3 to 5 per cent of the total osmolarity. However, because urea easily permeates

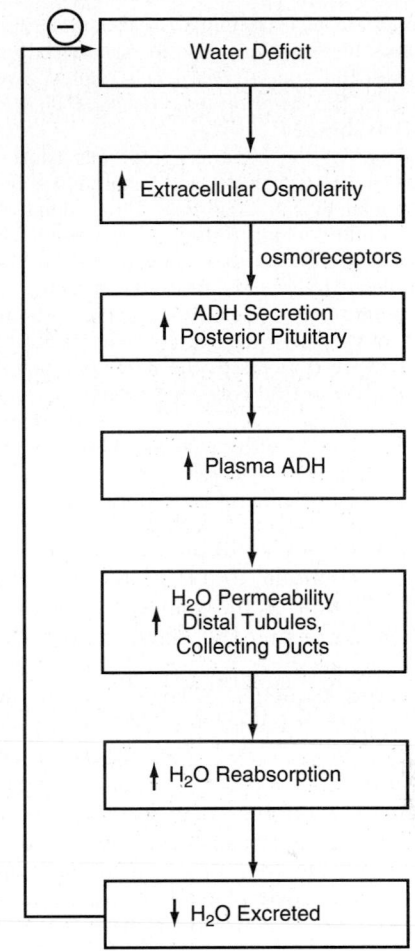

Figure 28–8. Osmoreceptor-ADH feedback mechanism for regulating extracellular fluid osmolarity in response to a water deficit.

most cell membranes, it exerts little *effective* osmotic pressure under steady-state conditions. Therefore, the sodium ions in the extracellular fluid and associated anions are the principal determinants of fluid movement across the cell membrane. Consequently, we can discuss the control of osmolarity and control of sodium ion concentration at the same time.

Although multiple mechanisms control the amount of sodium and water excretion by the kidneys, *two primary systems are especially involved in regulating the concentration of sodium and osmolarity of extracellular fluid:* (1) *the osmoreceptor-ADH system and* (2) *the thirst mechanism.*

Osmoreceptor-ADH Feedback System

Figure 28–8 shows the basic components of the osmoreceptor-ADH feedback system for control of extracellular fluid sodium concentration and osmolarity. When osmolarity (plasma sodium concentration) increases above normal because of water deficit, for example, this feedback system operates as follows:

1. An increase in extracellular fluid osmolarity (which in practical terms means an increase in plasma sodium concentration) causes the special nerve cells called *osmoreceptor cells,* located in the *anterior hypothalamus* near the supraoptic nuclei, to shrink.

2. Shrinkage of the osmoreceptor cells causes them to fire, sending nerve signals to additional nerve cells in the supraoptic nuclei, which then relay these signals down the stalk of the pituitary gland to the posterior pituitary.

3. These action potentials conducted to the posterior pituitary stimulate the release of ADH, which is stored in secretory granules (or vesicles) in the nerve endings.

4. ADH enters the blood stream and is transported to the kidneys, where it increases the water permeability of the late distal tubules, cortical collecting tubules, and inner medullary collecting ducts.

5. The increased water permeability in the distal nephron segments causes increased water reabsorption and excretion of a small volume of concentrated urine.

Thus, water is conserved in the body while sodium and other solutes continue to be excreted in the urine. This causes dilution of the solutes in the extracellular fluid, thereby correcting the initial excessively concentrated extracellular fluid.

The opposite sequence of events occurs when the extracellular fluid becomes too dilute (hypo-osmotic). For example, with excess water ingestion and a decrease in extracellular fluid osmolarity, less ADH is formed, the renal tubules decrease their permeability for water, less water is reabsorbed, and a large volume of dilute urine is formed. This in turn concentrates the body fluids and returns plasma osmolarity toward normal.

ADH Synthesis in Supraoptic and Paraventricular Nuclei of the Hypothalamus and ADH Release from the Posterior Pituitary

Figure 28–9 shows the neuroanatomy of the hypothalamus and the pituitary gland, where ADH is synthesized and released. The hypothalamus contains two types of *magnocellular (large) neurons that synthesize ADH in the supraoptic and paraventricular nuclei of the hypothalamus,* about five sixths in the supraoptic nuclei and about one sixth in the paraventricular nuclei. Both of these nuclei have axonal extensions to the posterior pituitary. Once ADH is synthesized, it is transported down the axons of the neurons to their tips, terminating in the posterior pituitary gland. When the supraoptic and paraventricular nuclei are stimulated by changes in osmolarity or other factors, nerve impulses pass down these nerve endings, changing their membrane permeability and increasing calcium entry. ADH stored in the secretory granules (also called vesicles) of the nerve endings is released in response to increased calcium entry. The released ADH is then carried away in the capillary blood of the posterior pituitary into the systemic circulation.

Secretion of ADH in response to an osmotic stimulus is rapid, so that plasma ADH levels can increase

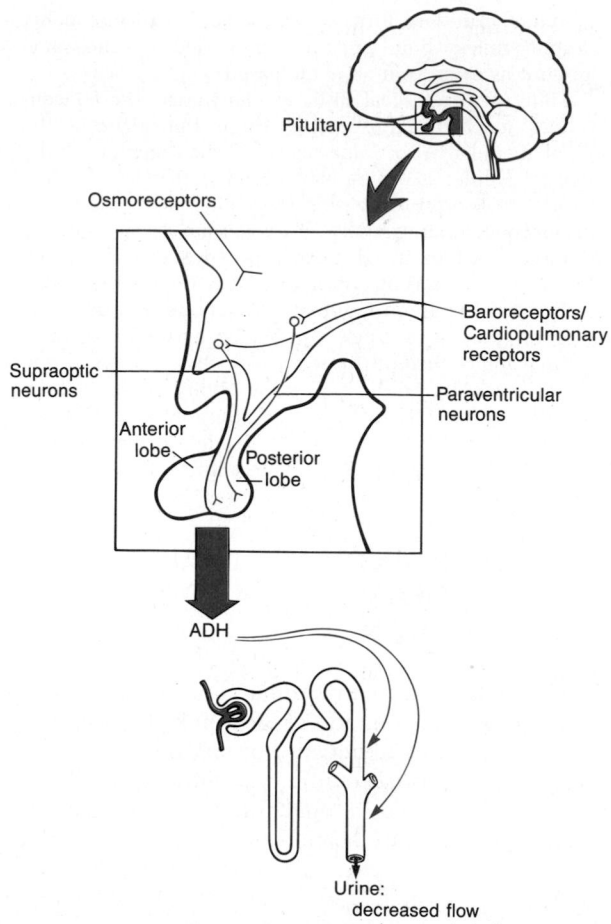

Figure 28–9. Neuroanatomy of the hypothalamus, where ADH is synthesized, and the posterior pituitary gland, where ADH is released.

severalfold within minutes, thereby providing a rapid means for altering renal excretion of water.

A second neuronal area important in controlling osmolarity and ADH secretion is a broad area located along the *anteroventral region of the third ventricle,* called the *AV3V region.* At the upper part of this region is a structure called the *subfornical organ,* and at the inferior part is another structure called the *organum vasculosum* of the *lamina terminalis.* Between these two organs is the *median preoptic nucleus,* which has multiple nervous connections with the two organs as well as with the supraoptic nuclei and the blood pressure control centers in the medulla of the brain. Lesions of the AV3V region cause multiple deficits in the control of ADH secretion, thirst, sodium appetite, and blood pressure control. Electrical stimulation of this region or stimulation by angiotensin II can alter ADH secretion, thirst, and sodium appetite.

In the vicinity of the AV3V region and the supraoptic nuclei are neuronal cells that are excited by small increases in extracellular fluid osmolarity; hence, the term *osmoreceptors* has been used to describe these neurons. These cells send nerve signals to the supraoptic nuclei to control their firing and secretion of ADH. It is also likely that they induce thirst in response to increased extracellular fluid osmolarity.

Both the subfornical organ and the organum vasculosum of the lamina terminalis have vascular supplies that lack the typical blood-brain barrier present elsewhere in the brain that impedes the diffusion of most ions from the blood into the brain tissue. This makes it possible for ions and other solutes to cross between the blood and the local interstitial fluid in this region. As a result, the osmoreceptors rapidly respond to changes in osmolarity of the extracellular fluid, exerting powerful control over the secretion of ADH and over thirst, as discussed below.

Cardiovascular Reflex Stimulation of ADH Release by Decreased Arterial Pressure and/or Decreased Blood Volume

ADH release is also controlled by cardiovascular reflexes in response to decreases in blood pressure and/or blood volume, including *(1) the arterial baroreceptor reflex and (2) the cardiopulmonary reflexes,* both of which are discussed in Chapter 18. These reflex pathways originate in high-pressure regions of the circulation, such as the aortic arch and carotid sinus, and in the low-pressure regions, especially in the cardiac atria. Afferent stimuli are carried by the vagus and glossopharyngeal nerves with synapses in the nuclei of the tractus solitarius. Projections from these nuclei relay signals to the hypothalamic nuclei that control ADH synthesis and secretion.

Thus, in addition to increased osmolarity, two other stimuli increase ADH secretion: (1) decreased arterial pressure and (2) decreased blood volume. Whenever blood pressure and blood volume are reduced, such as occurs during hemorrhage, increased ADH secretion causes increased fluid reabsorption by the kidneys, helping to restore blood pressure and blood volume toward normal.

Quantitative Importance of Cardiovascular Reflexes and Osmolarity in Stimulating ADH Secretion

As shown in Figure 28-10, either a decrease in effective blood volume or an increase in extracellular fluid osmolarity stimulates ADH secretion. However, ADH is considerably more sensitive to small changes in osmolarity than to similar changes in blood volume. For example, a change in plasma osmolarity of only 1 per cent is sufficient to increase ADH levels. By contrast, after blood loss, plasma ADH levels do not change appreciably until blood volume is reduced by about 10 per cent. With further decreases in blood volume, ADH levels rapidly increase. Thus, with severe decreases in blood volume, the cardiovascular reflexes play a major role in stimulating ADH secretion. The usual day-to-day regulation of ADH secretion

during simple dehydration, however, is effected mainly by changes in plasma osmolarity.

Other Stimuli for ADH Secretion

ADH secretion can also be increased or decreased by other stimuli to the central nervous system as well as by various drugs and hormones, as shown in Table 28-2. For example, *nausea* is a potent stimulus for ADH release, which may increase to as much as 100 times normal after vomiting. Also, drugs such as *nicotine* and *morphine* stimulate ADH release, whereas *some drugs, such as alcohol, inhibit ADH release.* The marked diuresis that occurs after ingestion of alcohol is due in part to inhibition of ADH release.

The detailed mechanisms by which ADH secretion is stimulated or inhibited by many drugs are poorly understood, but recognition of the effects of these drugs can be important in understanding the causes of disordered sodium and volume regulation in some clinical situations.

ROLE OF THIRST IN CONTROLLING EXTRACELLULAR FLUID OSMOLARITY AND SODIUM CONCENTRATION

The kidneys minimize fluid loss during water deficits through the osmoreceptor-ADH feedback system. In addition, fluid intake is necessary to counterbalance

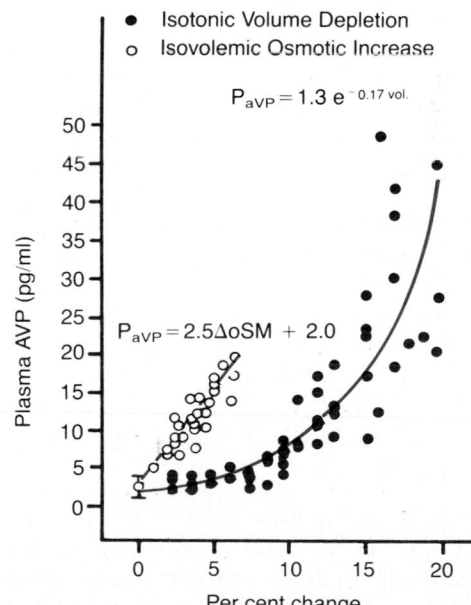

Figure 28-10. The effect of increased plasma osmolarity or decreased blood volume on the level of plasma ADH, also called arginine vasopressin (AVP). (From Dunn et al: *J. Clin. Invest.,* 52:3212, 1973. By copyright permission of the American Society of Clinical Investigation.)

Table 28-2 REGULATION OF ADH SECRETION

Increase ADH	Decrease ADH
↑ Plasma osmolality	↓ Plasma osmolality
↓ Blood volume	↑ Blood volume
↓ Blood pressure	↑ Blood pressure
Nausea	
Hypoxia	
Drugs:	Drugs:
Morphine	Alcohol
Nicotine	Clonidine (antihypertensive drug)
Cyclophosphamide	Haloperidol (dopamine blocker)

whatever fluid loss does occur through sweating and breathing and through the gastrointestinal tract. Fluid intake is regulated by the thirst mechanism, which, together with the osmoreceptor-ADH mechanism, maintains precise control of extracellular fluid osmolarity and sodium concentration.

Many of the same stimuli involved in controlling ADH secretion also increase thirst, which is defined as the conscious desire for water.

Central Nervous System Centers for Thirst

Referring again to Figure 28–9, the same area along the anteroventral wall of the third ventricle that promotes ADH release also stimulates thirst. Located anterolaterally in the preoptic nucleus is another small area that, when stimulated electrically, causes immediate drinking that continues as long as the stimulation lasts. All these areas together are called the *thirst center*.

The neurons of the thirst center respond to injections of hypertonic salt solutions by stimulating drinking behavior. These cells almost certainly function as osmoreceptors to activate the thirst mechanism, in the same way that the osmoreceptors stimulate ADH release.

Increased osmolarity of the cerebrospinal fluid in the third ventricle has essentially the same effect to promote drinking. It is likely that the *organum vasculosum of the lamina terminalis*, which lies immediately beneath the ventricular surface at the inferior end of the AV3V region, is intimately involved in mediating this response.

Stimuli for Thirst

Table 28–3 summarizes some of the known stimuli for thirst. One of the most important is *increased extracellular fluid osmolarity, which causes intracellular dehydration in the thirst centers*, thereby stimulating the sensation of thirst. The value of this response is obvious: it helps to dilute extracellular fluids and returns osmolarity toward normal.

Decreases in extracellular fluid volume and arterial pressure also stimulate thirst by a pathway that is in-

dependent of the one stimulated by increased plasma osmolarity. Thus, blood volume loss by hemorrhage stimulates thirst even though there might be no change in plasma osmolarity. This probably occurs because of neural input from cardiopulmonary and systemic arterial baroreceptors in the circulation.

A third important stimulus for thirst is angiotensin II. Studies in animals have shown that angiotensin II acts on the subfornical organ and on the organum vasculosum of the lamina terminalis, areas that are outside the blood-brain barrier, thus allowing peptides such as angiotensin II to diffuse into the tissues of these brain areas. Because angiotensin II is also stimulated by factors associated with hypovolemia and low blood pressure, its effect on thirst helps to restore blood volume and blood pressure toward normal, along with the other actions of angiotensin II on the kidneys to decrease fluid excretion.

Still other factors can influence water intake. *Dryness of the mouth and mucous membranes of the esophagus* can elicit the sensation of thirst. As a result, a thirsty person may receive relief from thirst almost immediately after drinking water even though the water has not been absorbed from the gastrointestinal tract and has not yet had an effect on extracellular fluid osmolarity. And in animals that have an esophageal opening to the exterior, so that water is never absorbed into the blood, partial relief of thirst occurs after drinking, although the relief is only temporary. Also, gastrointestinal distention may partially alleviate thirst; for instance, simple inflation of a balloon in the stomach can often relieve thirst. However, relief of thirst sensations through gastrointestinal or pharyngeal mechanisms is short-lived; the desire to drink is completely satisfied only when plasma osmolarity and/or blood volume returns to normal.

The ability of animals and humans to "meter" fluid intake is important because it prevents overhydration. After a person drinks water, 30 to 60 minutes may be required for the water to be reabsorbed and distributed throughout the body. If the thirst sensation were not temporarily relieved after drinking water, the person would continue to drink more and more, eventually leading to overhydration and excess dilution of the body fluids. Experimental studies have repeatedly shown that thirsty animals never drink more than the amount of water needed to relieve the state of dehydration. It is remarkable that animals drink almost exactly the amount necessary to return plasma osmolarity and volume to normal.

Table 28-3 CONTROL OF THIRST

Increase Thirst	Decrease Thirst
↑ Osmolality	↓ Osmolality
↓ Blood volume	↑ Blood volume
↓ Blood pressure	↑ Blood pressure
↑ Angiotensin II	↓ Angiotensin II
Dryness of mouth	Gastric distention

Threshold for Osmolar Stimulus of Drinking

The kidneys must continually excrete at least some fluid, even in a dehydrated person, to rid the body of excess solutes that are ingested or produced by metabolism. Water is also lost by evaporation from the lungs and the gastrointestinal tract and by evaporation and sweating from the skin. Therefore, there is always a tendency for dehydration, with resultant increased extracellular fluid sodium concentration and osmolarity. When the sodium concentration increases only about 2 mEq/liter above normal, the thirst mechanism is activated, causing a desire to drink water. This is called the *threshold for drinking*. Thus, even small increases in plasma osmolarity are normally followed by water intake, which restores extracellular fluid osmolarity and volume toward normal. In this way, the extracellular fluid osmolarity and sodium concentration are precisely controlled.

Integrated Responses of Osmoreceptor-ADH and Thirst Mechanisms in Controlling Extracellular Fluid Osmolarity and Sodium Concentration

In the normal person, the osmoreceptor-ADH and thirst mechanisms work in parallel to precisely regulate extracellular fluid osmolarity and sodium concentration, despite the constant challenges of dehydration. Even with additional challenges, such as high salt intake, these feedback systems are able to keep plasma osmolarity reasonably constant. Figure 28–11 shows

that an increase in sodium intake to as high as six times normal has almost no effect on plasma sodium concentration as long as the ADH and thirst mechanisms are both functioning normally.

When either the ADH or the thirst mechanism fails, the other ordinarily can still control extracellular osmolarity and sodium concentration with reasonable effectiveness, so long as there is enough fluid intake to balance the daily obligatory urine volume and water losses caused by respiration, sweating, or gastrointestinal losses. However, if both the ADH and thirst mechanisms fail simultaneously, neither sodium concentration nor osmolarity can be adequately controlled; thus, when sodium intake is increased after blocking the total ADH-thirst system, relatively large changes in plasma sodium concentration do occur. In the absence of the ADH-thirst mechanisms, no other feedback mechanism is capable of increasing water ingestion or conserving water by the kidneys during changes in sodium intake.

Role of Angiotensin II and Aldosterone in Controlling Extracellular Fluid Osmolarity and Sodium Concentration

As discussed in Chapter 27, both angiotensin II and aldosterone play an important role in regulating sodium reabsorption by the renal tubules. When sodium intake is low, increased levels of these hormones stimulate sodium reabsorption by the kidneys and, therefore, prevent large sodium losses, even though sodium intake may be reduced to as low as 10 per cent of normal. Conversely, with high sodium intake, decreased formation of these hormones permits the kidneys to excrete large amounts of sodium.

Because of the importance of angiotensin II and aldosterone in regulating sodium excretion by the kidneys, one might mistakenly infer that they also play an important role in regulating extracellular fluid sodium concentration. Although these hormones increase the *amount* of sodium in the extracellular fluid, they also increase the amount of water in the extracellular fluid by increasing reabsorption of water along with the sodium. Therefore, angiotensin II and aldosterone have little effect on sodium *concentration*, except under extreme conditions.

This relative unimportance of aldosterone in regulating sodium concentration of the extracellular fluid is shown by the experiment of Figure 28–12. This figure shows the effect on plasma sodium concentration of changing sodium intake more than sixfold under two conditions: (1) under normal conditions and (2) after the aldosterone feedback system has been blocked by removing the adrenal glands and infusing the animals with aldosterone at a constant rate so that plasma levels could not change upward or downward. Note that when sodium intake was increased sixfold, plasma concentration changed only about 1 to 2 per cent in either case. This indicates that even without a functional aldosterone feedback system, plasma sodium concentration can be well regulated. The same type of experiment has been conducted after blocking angiotensin II formation, with the same result.

There are two primary reasons why changes in angio-

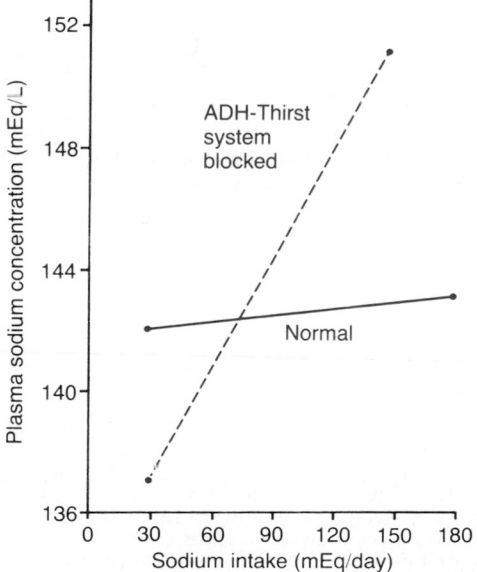

Figure 28–11. Effect of large changes in sodium intake on extracellular fluid sodium concentration in dogs (1) under normal conditions and (2) after the ADH and thirst feedback systems had been blocked. Note that control of extracellular fluid sodium concentration is poor in the absence of these feedback systems. (Courtesy of Dr. David B. Young.)

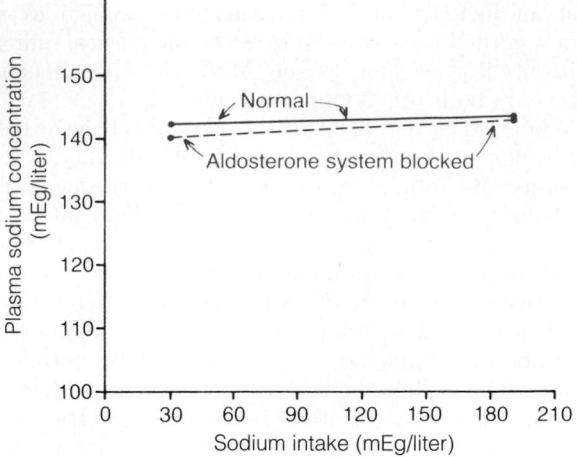

Figure 28–12. Effect of large changes in sodium intake on extracellular fluid sodium concentration in dogs (1) under normal conditions and (2) after the aldosterone feedback system had been blocked. Note that sodium concentration is maintained relatively constant over this wide range of sodium intakes, with or without aldosterone feedback control. (Courtesy of Dr. David B. Young.)

tensin II and aldosterone do not have a major effect on plasma sodium concentration. First, as discussed above, angiotensin II and aldosterone increase both sodium and water reabsorption by the renal tubules, leading to increases in extracellular fluid volume and sodium *quantity* but little change in sodium *concentration.* Second, as long as the ADH-thirst mechanism is functional, any tendency toward increased plasma sodium concentration is compensated for by increased water intake or increased plasma ADH secretion, which tends to dilute the extracellular fluid back toward normal. The ADH-thirst system far overshadows the angiotensin II and aldosterone systems for regulating sodium concentration under normal conditions. Even in patients with *primary aldosteronism,* who have extremely high levels of aldosterone, the plasma sodium concentration still increases only about 2 to 3 mEq/liter above normal.

Under extreme conditions, caused by complete loss of aldosterone secretion because of adrenalectomy or in patients with Addison's disease (severely impaired secretion or total lack of aldosterone), there is tremendous loss of sodium by the kidneys, which then can lead to reductions in plasma sodium concentration. One of the reasons for this is that with large losses of sodium, there will eventually develop severe volume depletion and decreased blood pressure, which can activate the thirst mechanism through the cardiovascular reflexes. This now leads to a further dilution of the plasma sodium concentration, even though the increased water intake helps to minimize the decrease in body fluid volumes under these conditions.

Thus, there are extreme situations in which plasma sodium concentration may change significantly, even with a functional ADH-thirst mechanism. Even so, the ADH-thirst mechanism is by far the most powerful feedback system in the body for controlling extracellular fluid osmolarity and sodium concentration.

SALT-APPETITE MECHANISM FOR CONTROLLING EXTRACELLULAR FLUID SODIUM CONCENTRATION AND VOLUME

Maintenance of normal extracellular fluid volume and sodium concentration requires a balance between sodium excretion and sodium intake. In modern civilizations, sodium intake is almost always greater than necessary for homeostasis. In fact, the average sodium intake for individuals in industrialized cultures eating processed foods usually ranges between 100 and 200 mEq/day, even though humans can survive and function normally on 10 to 20 mEq/day. Thus, most people eat far more sodium than is necessary for homeostasis, and there is evidence that our normally high sodium intake may contribute to certain cardiovascular diseases, such as hypertension.

Salt appetite is due in part to the fact that animals and humans like salt and eat it regardless of whether they are salt-deficient. There is also a regulatory component to salt appetite in which there is a behavioral drive to obtain salt when there is sodium deficiency in the body. This is particularly important in herbivores, which naturally eat a low-sodium diet, but salt craving may also be important in humans who have extreme deficiency of sodium, such as occurs in Addison's disease. In this instance, there is deficiency of aldosterone secretion, which causes excessive loss of sodium in the urine and leads to (a) decreased extracellular fluid volume and (b) decreased sodium concentration; both of these changes elicit the desire for salt.

In general, the two primary stimuli that are believed to excite salt appetite are (1) decreased extracellular fluid sodium concentration and (2) decreased blood volume or blood pressure, associated with circulatory insufficiency. These are the same major stimuli that elicit thirst.

The neuronal mechanism for salt appetite is analogous to that of the thirst mechanism. Some of the same neuronal centers in the AV3V region of the brain seem to be involved because lesions in this region frequently affect simultaneously both thirst and salt appetite in animals. Also, circulatory reflexes elicited by low blood pressure or decreased blood volume affect both thirst and salt appetite at the same time.

REFERENCES

Bankir, L., and Kriz, W.: Adaptation of the kidney to protein intake and to urine concentrating ability, similar consequences in health and CRF. Kidney Int., 47:7, 1995.

Cowley, A. W., Jr., et al.: Vasopressin-Cellular and Integrative Functions. New York, Raven Press, 1988.

DuBose, T. D., Jr., et al.: Ammonium transport in the kidney: New physiological concepts and their clinical application. J. Am. Soc. Nephrol., 1:1193, 1991.

Fitzsimmons, J. T.: Physiology and pathophysiology of thirst and sodium appetite. In Seldin, D. W., and Giebisch, G. (eds.): The Kidney: Physiology and Pathophysiology, 2nd Ed. New York, Raven Press, 1992.

Hall, J. E.: The kidney. In Ingram, D., and Bloch, R. F. (eds.): Mathematical Methods in Medicine. New York, John Wiley & Sons, 1984.

Harris, H. W. Jr., et al.: Current understanding of the biology and molecular structure of the antidiuretic hormone–stimulated water transport pathway. J. Clin. Invest., 88:1, 1991.

Jamison, R. L., and Gerhig, J. J.: Urinary concentration and dilution: Physiology. In Windhager, E. E. (ed.): Handbook of Physiology, Section 8, Renal Physiology, Vol. II. New York, Oxford University Press, 1992.

Knepper, M. A., and Rector, F. C., Jr.: Urinary Concentration and Dilution. In Brenner, B. M., and Rector, F. C., Jr. (eds.): The Kidney, 4th Ed. Philadelphia, W. B. Saunders Co., 1991.

Kokko, J.: The role of the collecting duct in urinary concentration. Kidney Int., 31:606, 1987.

Marsh, D. J., and Knepper, M. A.: Renal handling of urea. In Windhager, E. E. (ed.): Handbook of Physiology, Section 8, Renal Physiology, Vol. II. New York, Oxford University Press, 1992.

Robertson, G. L.: Physiology of ADH secretion. Kidney Int., (Suppl. 21):S-20, 1987.

Roy, D. R., et al.: Countercurrent mechanism and its regulation. In Seldin, D. W., and Giebisch, G. (eds.): The Kidney: Physiology and Pathophysiology, New York, Raven Press, 1992.

Vander, A. T.: Renal Physiology, 4th Ed. New York, McGraw Hill, Inc., 1991.

Wolff, S. D., and Balaban, R. S.: Regulation of the predominant renal medullary organic solutes. Ann. Rev. Physiol., 52:727, 1990.

Young, D. B., et al.: Control of extracellular sodium concentration by antidiuretic hormone-thirst feedback mechanism. Am. J. Physiol., 232:R145, 1977.

Young, D. B., et al.: Effectiveness of the aldosterone-sodium and -potassium feedback control system. Am. J. Physiol., 231:945, 1976.

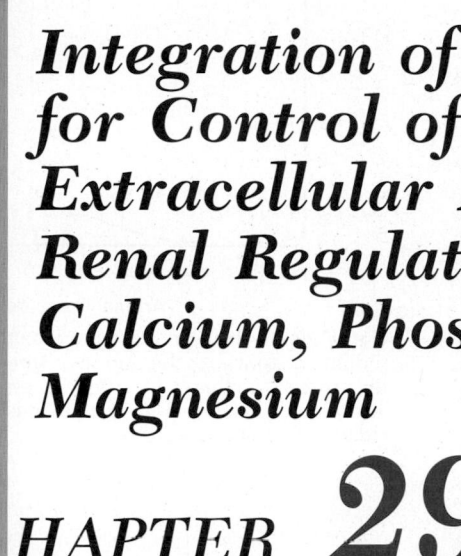

Integration of Renal Mechanisms for Control of Blood Volume and Extracellular Fluid Volume; and Renal Regulation of Potassium, Calcium, Phosphate, and Magnesium

*C*HAPTER *29*

Extracellular fluid volume is determined mainly by the balance between intake and output of water and salt. In most cases, salt and fluid intakes are dictated mainly by a person's habits rather than by physiological control mechanisms. Therefore, the burden of extracellular volume regulation is usually placed on the kidneys, which must adapt their excretion of salt and water to match intake of salt and water under steady-state conditions.

In discussing the regulation of extracellular fluid volume, we also consider the factors that regulate the amount of sodium chloride in the extracellular fluid, because changes in sodium chloride content of extracellular fluid usually cause parallel changes in extracellular fluid volume, provided the antidiuretic hormone (ADH)-thirst mechanisms are also operative. When the ADH-thirst mechanisms are functioning normally, a change in the amount of sodium chloride in the extracellular fluid is matched by a similar change in the amount of extracellular water, so that osmolality and sodium concentration are maintained relatively constant.

In this chapter, we discuss how the various factors that control salt and water excretion by the kidneys are coordinated to control extracellular fluid volume and blood volume under normal conditions as well as during pathophysiological states.

CONTROL MECHANISMS FOR REGULATING SODIUM AND WATER EXCRETION

An important fact that should be remembered in considering the overall control of sodium excretion—

or excretion of any electrolyte for that matter—is that under steady-state conditions, excretion by the kidneys is determined by intake. To maintain life, a person must over the long term excrete almost precisely the amount of sodium ingested. Therefore, even with disturbances that cause major changes in kidney function, balance between intake and output of sodium usually is restored within a few days.

If disturbances of kidney function are not too severe, sodium balance may be achieved mainly by intrarenal adjustments with minimal changes in extracellular fluid volume or other systemic adjustments. But when perturbations to the kidneys are severe and intrarenal compensations are exhausted, systemic adjustments must be invoked, such as changes in blood pressure, changes in circulating hormones, and alterations of sympathetic nervous system activity. These adjustments are costly in terms of overall homeostasis because they cause other changes throughout the body that may, in the long run, be damaging. These compensations, however, are necessary because a sustained imbalance between sodium intake and excretion would quickly lead to fluid accumulation or fluid loss and cause cardiovascular collapse within a few days. Thus, one can view the systemic adjustments that occur in response to abnormalities of kidney function as a necessary trade-off that brings sodium excretion back in balance with intake.

Sodium Excretion Is Controlled by Altering Glomerular Filtration or Tubular Sodium Reabsorption Rates

The two variables that influence sodium and water excretion are the rate of filtration and the rate of reabsorption:

Excretion = Glomerular filtration
$$- \text{Tubular reabsorption}$$

Glomerular filtration rate (GFR) normally is about 180 liters/day, tubular reabsorption is 178.5 liters/day, and urine excretion is 1.5 liters/day. Thus, small changes in GFR or tubular reabsorption potentially can cause large changes in renal excretion. For example, a 5 per cent increase in GFR (to 189 liters/day) would cause a 9 liter/day increase in urine volume, if tubular compensations did not occur; this would quickly cause catastrophic changes in body fluid volumes. Similarly, small changes in tubular reabsorption, in the absence of compensatory adjustments of GFR, can also lead to dramatic changes in urine volume and sodium excretion. Tubular reabsorption and GFR usually are regulated precisely, so that excretion by the kidneys can be exactly matched to intake of water and electrolytes.

Even with disturbances that alter GFR or tubular reabsorption, changes in urinary excretion are minimized by various buffering mechanisms. For example, if a kidney becomes greatly vasodilated and GFR increases (as can occur with certain drugs or high fevers), this raises sodium chloride delivery to the tubules, which in turn leads to at least two intrarenal compensations: (1) increased tubular reabsorption of much of the extra sodium chloride filtered, called *glomerulotubular balance,* and (2) *macula densa feedback,* in which increased sodium chloride delivery to the distal tubule causes afferent arteriolar constriction and return of GFR toward normal. Likewise, abnormalities of tubular reabsorption in the proximal tubule or loop of Henle are partially compensated for by these same intrarenal feedbacks.

Because neither of these two mechanisms operates perfectly to restore distal sodium chloride delivery all the way back to normal, changes in either GFR or tubular reabsorption can lead to significant changes in urine sodium and water excretion. When this happens, other feedback mechanisms come into play, such as changes in blood pressure and changes in various hormones, that eventually return sodium excretion to equal sodium intake. In the next few sections, we review how these mechanisms operate together to control sodium and water balance and in doing this also to control extracellular fluid volume. We should keep in mind, however, that all these feedback mechanisms control renal excretion of sodium and water by altering either GFR or tubular reabsorption.

IMPORTANCE OF PRESSURE NATRIURESIS AND PRESSURE DIURESIS IN MAINTAINING BODY SODIUM AND FLUID BALANCE

Perhaps the most powerful mechanism for control of blood volume and extracellular fluid volume as well as for the maintenance of sodium and fluid balance is the

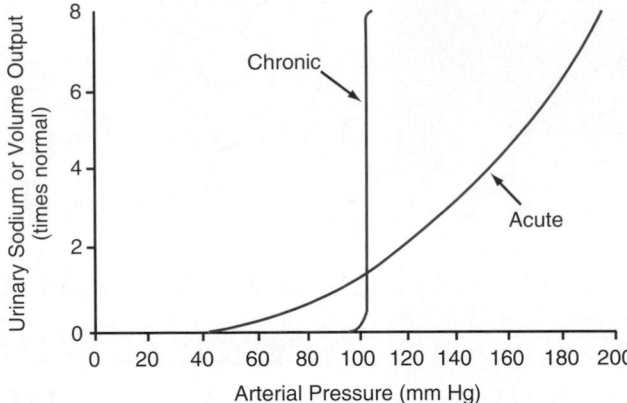

Figure 29–1. Acute and chronic effects of arterial pressure on sodium output by the kidneys (pressure natriuresis). Note that chronic increases in arterial pressure cause much greater increases in sodium output than measured during acute increases in arterial pressure.

effect of blood pressure on sodium and water excretion—called the *pressure natriuresis* and *pressure diuresis* mechanisms. As discussed in Chapter 19, this feedback between the kidneys and the circulatory system also plays a dominant role in long-term blood pressure regulation.

Pressure diuresis refers to the effect of increased blood pressure to raise urinary volume excretion, whereas pressure natriuresis refers to the rise in sodium excretion that occurs with elevated blood pressure. Because pressure diuresis and natriuresis usually occur in parallel, we refer to these mechanisms simply as "pressure natriuresis" in the following discussion.

Figure 29–1 shows the effect of arterial pressure on urinary sodium output. Note that acute increases in blood pressure of 30 to 50 mm Hg cause a twofold to threefold increase in urinary sodium output. This effect is independent of changes in activity of the sympathetic nervous system or of various hormones, such as angiotensin II, ADH, or aldosterone, because pressure natriuresis can be demonstrated in an isolated kidney that has been removed from the influence of these factors. With chronic increases in blood pressure, the effectiveness of pressure natriuresis is greatly enhanced because the increased blood pressure also, after a short time delay, suppresses renin release and, therefore, decreases formation of angiotensin II and aldosterone. As discussed previously, decreased levels of angiotensin II and aldosterone inhibit renal tubular reabsorption of sodium, thereby amplifying the direct effects of increased blood pressure to raise sodium and water excretion.

Pressure Natriuresis and Diuresis Are Key Components of a Renal–Body Fluid Feedback for Regulating Body Fluid Volumes and Arterial Pressure

The effect of increased blood pressure to raise urine output is part of a powerful feedback system that operates to maintain balance between fluid intake and

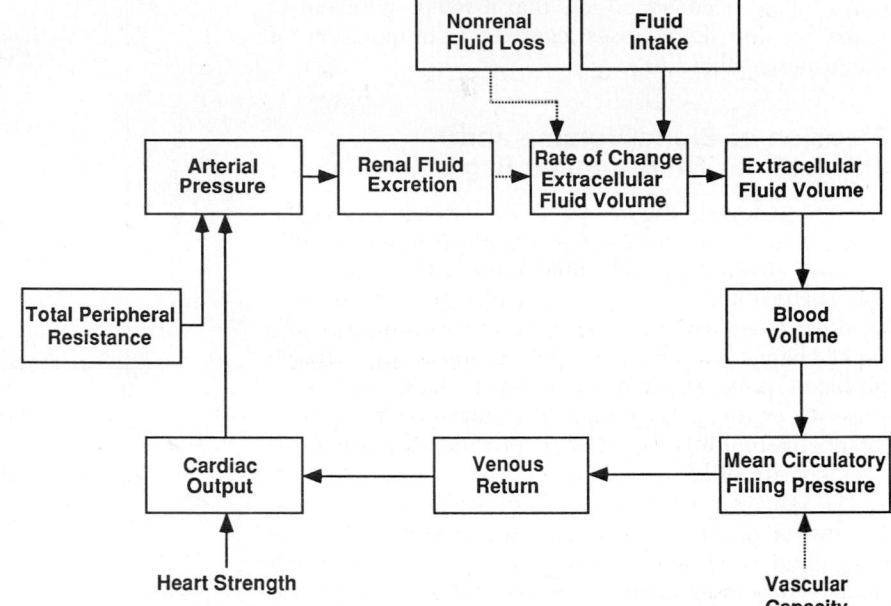

Figure 29–2. Basic renal–body fluid feedback mechanism for control of blood volume, extracellular fluid volume, and arterial pressure. Solid lines indicate positive effects and dashed lines indicate negative effects.

output, as shown in Figure 29–2. This is essentially the same mechanism that is discussed in Chapter 19 for arterial pressure control. As pointed out in that chapter, the extracellular fluid volume, blood volume, cardiac output, arterial pressure, and urine output are all controlled at the same time as separate parts of this basic feedback mechanism.

During changes in sodium and fluid intake, this feedback mechanism helps to maintain fluid balance and to minimize changes in blood volume, extracellular fluid volume, and arterial pressure as follows:

1. An increase in fluid intake (assuming that sodium accompanies the fluid intake) above the level of urine output causes a temporary accumulation of fluid in the body.
2. As long as fluid intake exceeds urine output, fluid accumulates in the blood and interstitial spaces, causing parallel increases in blood volume and extracellular fluid volume. As discussed later, the actual increases in these variables are small because of the effectiveness of this feedback.
3. An increase in blood volume raises mean circulatory filling pressure.
4. An increase in mean circulatory filling pressure raises the pressure gradient for venous return.
5. An increased pressure gradient for venous return elevates cardiac output.
6. An increased cardiac output raises arterial pressure.
7. An increased arterial pressure increases urine output by way of pressure diuresis. The steepness of the normal pressure natriuresis relation indicates that only a slight increase in blood pressure is required to raise urinary excretion severalfold.
8. The increased fluid excretion balances the increased intake, and further accumulation of fluid is prevented.

Thus, the renal–body fluid feedback mechanism operates to prevent continuous accumulation of salt and water in the body during increased salt and water intake. As long as kidney function is normal and the pressure diuresis mechanism is operating effectively,

large changes in salt and water intake can be accommodated with only slight changes in blood volume, extracellular fluid volume, cardiac output, and arterial pressure.

The opposite sequence of events occurs when fluid intake falls below normal. In this case, there is a tendency for decreased blood volume and extracellular fluid volume as well as reduced arterial pressure. Even a small decrease in blood pressure causes a large decrease in urine output, thereby again allowing fluid balance to be maintained with minimal changes in blood pressure, blood volume, or extracellular fluid volume. The effectiveness of this mechanism in preventing large changes in blood volume is demonstrated in Figure 29–3, which shows that changes in blood volume are almost imperceptible despite large variations in daily intake of water and electrolytes, except

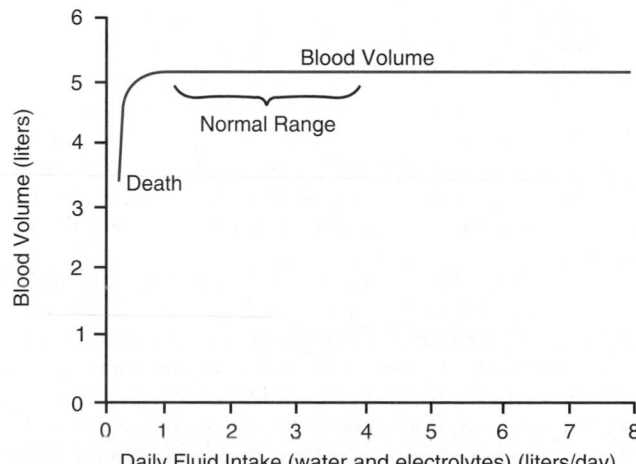

Figure 29–3. Approximate effect of changes in daily fluid intake on blood volume. Note that blood volume remains relatively constant in the normal range of daily fluid intakes.

when intake becomes so low that it is not sufficient to make up for fluid losses caused by evaporation or other inescapable losses.

Precision of Blood Volume and Extracellular Fluid Volume Regulation

From studying Figure 29–2, one can see why the blood volume remains almost exactly constant despite extreme changes in daily fluid intake. The reason for this is the following: (1) a slight change in blood volume causes a marked change in cardiac output, (2) a slight change in cardiac output causes a large change in blood pressure, and (3) a slight change in blood pressure causes a large change in urine output. These factors all multiply together to provide effective feedback control of blood volume.

The same sequence of events occurs whenever there is a loss of whole blood because of hemorrhage. In this case, fluid is retained by the kidneys, whereas other parallel processes occur to reconstitute the red blood cells and plasma proteins in the blood. If abnormalities of red blood cell volume remain, such as occurs when there is deficiency of erythropoietin or other factors needed to stimulate red blood cell production, the plasma volume will simply make up the difference and the overall blood volume will return essentially to normal despite the low red blood cell mass.

DISTRIBUTION OF EXTRACELLULAR FLUID BETWEEN THE INTERSTITIAL SPACES AND VASCULAR SYSTEM

From Figure 29–2 it is also apparent that blood volume and extracellular fluid volume are usually controlled in parallel with each other. Ingested fluid initially goes into the blood, but it rapidly becomes distributed between the interstitial spaces and the plasma. Therefore, blood volume and extracellular fluid volume usually are controlled simultaneously. There are circumstances, however, in which the distribution of extracellular fluid between the interstitial spaces and blood can vary greatly. As discussed in Chapter 25, *the principal factors that can cause accumulation of fluid in the interstitial spaces include (1) increased capillary pressure, (2) decreased plasma colloid osmotic pressure, and (3) increased permeability of the capillaries.* In all these conditions, an unusually high proportion of the extracellular fluid becomes distributed to the interstitial spaces.

Figure 29–4 shows the normal distribution of fluid between the interstitial spaces and the vascular system and the distribution that occurs in edema states. When small amounts of fluid accumulate in the blood as a result of either too much fluid intake or a decrease in renal output of fluid, about 20 to 30 per cent of it stays in the blood and increases the blood volume. The remainder is distributed to the interstitial spaces. When the extracellular fluid volume rises more than

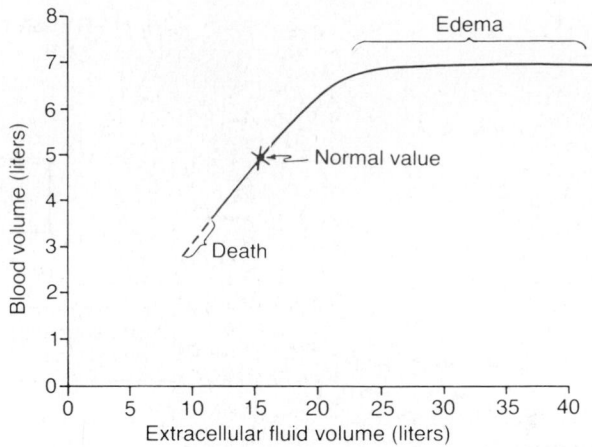

Figure 29–4. Approximate relation between extracellular fluid volume and blood volume, showing a nearly linear relation in the normal range but also showing the failure of blood volume to continue rising when the extracellular fluid volume becomes excessive. When this occurs, the additional extracellular fluid volume resides in the interstitial spaces and edema occurs.

30 to 50 per cent above normal, almost all the additional fluid goes into the interstitial spaces and little remains in the blood. This occurs because once the interstitial fluid pressure rises from its normally negative value to become positive, the tissue interstitial spaces become compliant and large amounts of fluid then pour into the tissues without interstitial fluid pressure rising much more. In other words, the safety factor against edema, owing to a rising interstitial fluid pressure that counteracts fluid accumulation in the tissues, is lost once the tissues become highly compliant. Thus, under normal conditions, the interstitial spaces act as an "overflow" reservoir for excess fluid, sometimes increasing in volume 10 to 30 liters. This causes edema, as explained in Chapter 25, but it also acts as an important overflow release valve for the circulation, protecting the cardiovascular system against dangerous overload that could lead to pulmonary edema and cardiac failure.

To summarize, extracellular fluid volume and blood volume are controlled simultaneously, but the quantitative amounts of fluid distribution between the interstitium and the blood depend on the physical properties of the circulation and the interstitial spaces as well as the dynamics of fluid exchange through the capillary membranes.

NERVOUS AND HORMONAL FACTORS INCREASE THE EFFECTIVENESS OF RENAL–BODY FLUID FEEDBACK CONTROL

In Chapter 27, we discuss the nervous and hormonal factors that influence GFR and tubular reabsorption and, therefore, renal excretion of salt and water. In normal people, these nervous and hormonal mechanisms act in concert with the pressure natriuresis and pressure diuresis mechanisms, making them more effective in minimizing the changes in blood volume, extracellular fluid volume, and arterial pressure that occur in response to day-to-day challenges. However, abnormalities of kidney function or of the various nervous and hormonal factors that influence the kidneys can lead to serious changes in blood pressure and body fluid volumes, as discussed below.

Sympathetic Nervous System Control of Renal Excretion: The Arterial Baroreceptor and Low-Pressure Stretch Receptor Reflexes

Because the kidneys receive extensive sympathetic innervation, changes in sympathetic activity can alter renal sodium and water excretion as well as regulation of extracellular fluid volume under some conditions. For example, when blood volume is reduced by hemorrhage, the pressures in the pulmonary blood vessels and other low-pressure regions of the thorax decrease, causing reflex activation of the sympathetic nervous system. This in turn increases renal sympathetic nerve activity, which has several effects to decrease sodium and water excretion: (1) constriction of the renal arterioles, with resultant decreased GFR; (2) increased tubular reabsorption of salt and water; and (3) stimulation of renin release and increased angiotensin II and aldosterone formation, both of which further increase tubular reabsorption. And if the reduction in blood volume is great enough to lower systemic arterial pressure, further activation of the sympathetic nervous system occurs because of decreased stretch of the arterial baroreceptors located in the carotid sinus and aortic arch. All these reflexes together play an important role in the rapid restitution of blood volume that occurs in acute conditions such as hemorrhage. Also, reflex inhibition of renal sympathetic activity may contribute to the rapid elimination of excess fluid in the circulation that occurs acutely after eating a meal that contains large amounts of salt and water.

Role of Angiotensin II in Controlling Renal Excretion

One of the body's most powerful controllers of sodium excretion is angiotensin II. Changes in sodium and fluid intake are associated with reciprocal changes in angiotensin II formation, and this in turn contributes greatly to the maintenance of body sodium and fluid balances. That is, when sodium intake is elevated above normal, renin secretion is decreased, causing decreased angiotensin II formation. Because angiotensin II has several important effects to increase tubular reabsorption of sodium, as explained in Chapter 27, a reduced level of angiotensin II decreases tubular reabsorption of sodium and water, thus increasing the kidneys' excretion of sodium and water. The net result is to minimize the rise in extracellular fluid volume and

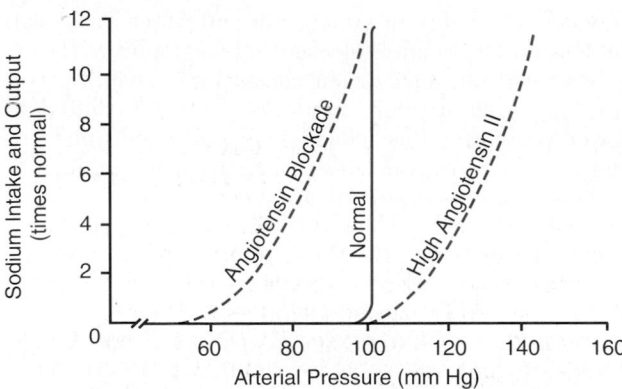

Figure 29–5. Effect of excessive angiotensin II formation and effect of blocking angiotensin II formation on the renal–pressure natriuresis curve. Note that high levels of angiotensin II formation decrease the slope of pressure natriuresis, making blood pressure very sensitive to changes in sodium intake. Blockade of angiotensin II formation shifts pressure natriuresis to lower blood pressures.

arterial pressure that would otherwise occur when sodium intake increases.

Conversely, when sodium intake is reduced below normal, increased levels of angiotensin II cause sodium and water retention and oppose reductions in arterial blood pressure that would otherwise occur. Thus, changes in activity of the renin-angiotensin system act as a powerful amplifier of the pressure natriuresis mechanism for maintaining stable blood pressures and body fluid volumes.

IMPORTANCE OF ANGIOTENSIN II IN INCREASING THE EFFECTIVENESS OF PRESSURE NATRIURESIS. The importance of angiotensin II in making the pressure natriuresis mechanism more effective is shown in Figure 29–5. Note that when the angiotensin control of natriuresis is fully functional, the pressure natriuresis curve is steep (normal curve), indicating that only minor changes in blood pressure are needed to increase sodium excretion when sodium intake is raised. In contrast, when angiotensin levels cannot be decreased in response to increased sodium intake (high angiotensin II curve), as occurs in certain diseases associated with impaired ability to decrease renin secretion, the pressure natriuresis curve is not nearly as steep. Therefore, when sodium intake is raised, much greater increases in arterial pressure are needed to increase sodium excretion and maintain sodium balance. For example, in the normal person, a 10-fold increase in sodium intake causes an increase of only a few millimeters of mercury in arterial pressure, whereas in subjects who cannot suppress angiotensin II formation appropriately in response to excess sodium, the same rise in sodium intake causes blood pressure to rise as much as 50 mm Hg. Thus, the inability to suppress angiotensin II formation when there is excess sodium reduces the slope of pressure natriuresis and makes arterial pressure very salt sensitive, as discussed in Chapter 19.

The use of drugs to block the effects of angiotensin II has proved to be important clinically for improving the kidneys' ability to excrete salt and water. As shown in Figure 29–5, after blockade of angiotensin II formation with an angiotensin converting enzyme inhibitor, the renal–pressure natriuresis curve is shifted to lower pressures; this indicates an enhanced ability of the kidneys to excrete sodium because normal levels of sodium excretion can now be maintained at reduced arterial pressures. This shift of pressure natriuresis provides the basis for the chronic blood pressure–lowering effects in hypertensive patients of the angiotensin converting enzyme inhibitors.

EXCESSIVE ANGIOTENSIN II DOES NOT CAUSE LARGE INCREASES IN EXTRACELLULAR FLUID VOLUME. Although angiotensin II is one of the most powerful sodium- and water-retaining hormones in the body, neither a decrease nor an increase in circulating angiotensin II has a large effect on extracellular fluid volume or blood volume. The reason for this is that with large increases in angiotensin II levels, as occurs with a renin-secreting tumor of the kidney, the high angiotensin II levels initially cause sodium and water retention by the kidneys and a small increase in extracellular fluid volume. This also initiates a rise in arterial pressure, which quickly increases kidney output of sodium and water, thereby overcoming the sodium- and water-retaining effects of the angiotensin II and re-establishing a balance between intake and output of sodium at a higher blood pressure. Conversely, after blockade of angiotensin II formation, as occurs when an angiotensin-converting enzyme inhibitor is administered, there is initial loss of sodium and water, but the fall in blood pressure offsets this effect and sodium excretion is once again restored to normal.

Role of Aldosterone in Controlling Renal Excretion

Aldosterone increases sodium reabsorption, especially in the cortical collecting tubule. The increased sodium reabsorption is also associated with increased water reabsorption and potassium secretion. Therefore, the net effect of aldosterone is to make the kidneys retain sodium and water but to increase potassium excretion in the urine.

The function of aldosterone in regulating sodium balance is closely related to that described above for angiotensin II. That is, with reduction in sodium intake, the increased angiotensin II levels that occur stimulate aldosterone secretion, which in turn contributes to the reduction in urinary sodium excretion and, therefore, to the maintenance of sodium balance. Conversely, with high sodium intake, suppression of aldosterone formation decreases tubular reabsorption, allowing the kidneys to excrete larger amounts of sodium. Thus, changes in aldosterone formation also aid the pressure natriuresis mechanism in maintaining sodium balance during variations in salt intake.

DURING CHRONIC OVERSECRETION OF ALDOSTERONE, THE KIDNEYS "ESCAPE" FROM SODIUM RETENTION AS THE ARTERIAL PRESSURE RISES. Although aldosterone has powerful effects on sodium reabsorption, when there is excessive infusion of aldosterone or excessive formation of aldosterone, as occurs in patients with tumors of the adrenal gland (Conn's syndrome), the increased sodium reabsorption and decreased sodium excretion by the kidneys are transient. After 1 to 3 days of sodium and volume retention, the extracellular fluid volume rises by about 10 to 15 per cent and there is a simultaneous increase in arterial blood pressure. When the arterial pressure rises sufficiently, the kidneys "escape" from the sodium and water retention and thereafter excrete amounts of sodium equal to the daily intake, despite continued presence of high levels of aldosterone.

The primary reason for the escape is the pressure natriuresis and diuresis that occur when the arterial pressure rises. In experiments in which renal artery pressure has been prevented from increasing as systemic arterial pressure rises, pressure natriuresis is prevented and escape does not occur. Thus, as long as pressure natriuresis is functioning normally, excessive aldosterone secretion usually causes no more than a 10 to 15 per cent increase in extracellular fluid volume. (High levels of aldosterone cause marked loss of potassium and decreases in extracellular fluid potassium concentration, as discussed later.)

In patients with adrenal insufficiency who do not secrete enough aldosterone (Addison's disease), there is increased excretion of sodium and water, reduction in extracellular fluid volume, and a tendency toward low blood pressure. In the complete absence of aldosterone, the volume depletion may be severe unless the person is allowed to eat large amounts of salt and drink large amounts of water to balance the increased urine output of salt and water.

Role of ADH in Controlling Renal Excretion

As discussed in Chapter 28, ADH plays an important role in allowing the kidneys to form a small volume of concentrated urine while excreting normal amounts of salt. This effect is especially important during water deprivation, which strongly elevates plasma levels of ADH that in turn increase water reabsorption by the kidneys and help to minimize the decreases in extracellular fluid volume and arterial pressure that would otherwise occur. Water deprivation for 24 to 48 hours normally causes only a small decrease in extracellular fluid volume and arterial pressure. However, if the effects of ADH are blocked with a drug that antagonizes the action of ADH to promote water reabsorption in the distal and collecting tubules, the same period of water deprivation causes a substantial fall in both extracellular fluid volume and arterial pressure.

Conversely, when there is excess extracellular volume, *decreased* ADH levels reduce reabsorption of water by the kidneys, thus helping to rid the body of the excess volume.

EXCESS ADH SECRETION USUALLY CAUSES ONLY SMALL INCREASES IN EXTRACELLULAR FLUID VOLUME BUT LARGE DECREASES IN SODIUM CONCENTRATION. Although ADH is important in regulating extracellular fluid volume, excessive levels of ADH seldom cause large increases in arterial pressure or extracellular fluid volume. Infusion of large amounts of ADH into animals initially causes renal retention of water and a 10 to 15 per cent increase in extracellular fluid volume. As the arterial pressure rises in response to this increased volume, much of the excess volume is excreted because of the pressure diuresis mechanism. After several weeks of ADH infusion, the blood volume and extracellular fluid volume are elevated no more than 5 to 10 per cent and the arterial pressure is also elevated by less than 10 mm Hg. The same is true for patients with *inappropriate ADH syndrome*, in which ADH levels may be elevated severalfold. Thus, high levels of ADH do not cause major increases of either body fluid volume or arterial pressure, although *high ADH levels can cause severe reductions in extracellular sodium ion concentration*. The reason for this is that increased water reabsorption by the kidneys dilutes the extracellular sodium, and at the same time, the small increase in blood pressure that does occur causes loss of sodium from the extracellular fluid in the urine through pressure natriuresis.

In patients who have lost their ability to secrete ADH because of destruction of the supraoptic nuclei, the urine volume may become 5 to 10 times normal. This is almost always compensated for by ingestion of enough water to maintain fluid balance. If free access to water is prevented, the inability to secrete ADH may lead to marked reductions in blood volume and arterial pressure.

Role of Atrial Natriuretic Peptide in Controlling Renal Excretion

Thus far, we have discussed mainly the role of sodium- and water-retaining hormones in controlling extracellular fluid volume. However, some researchers believe that so-called natriuretic hormones may also contribute to volume regulation. One of the most important of the natriuretic hormones is a peptide released by the cardiac atrial muscle fibers, referred to as *atrial natriuretic peptide (ANP)*. The stimulus for release of this peptide appears to be overstretch of the atria, which can result from excess blood volume. Once released by the cardiac atria, ANP enters the circulation and acts on the kidneys to cause small increases in GFR and decreases in sodium reabsorption by the collecting duct. These combined actions of ANP lead to increased excretion of salt and water, which helps to compensate for the excess blood volume.

Changes in ANP levels probably help to minimize changes in blood volume during various disturbances, such as increased salt and water intake. However, excessive production of ANP or even complete lack of ANP does not cause major changes in blood volume because these effects can easily be overcome by small changes in blood pressure, acting through pressure natriuresis. For example, infusions of large amounts of ANP initially raise urine output of salt and water and cause slight decreases in blood volume. In less than 24 hours, this effect is overcome by a slight decrease in blood pressure that returns urine output toward normal, despite continued excess of ANP.

INTEGRATED RESPONSES TO CHANGES IN SODIUM INTAKE

The integration of the different control systems that regulate sodium and fluid excretion under normal conditions can be summarized by examining the homeostatic responses to progressive increases in dietary sodium intake. As discussed previously, the kidneys have an amazing capability to match their excretion of salt and water to intakes that can range from as low as one tenth of normal to as high as 10 times normal.

As sodium intake is increased, sodium output initially lags slightly behind intake. The time delay results in a small increase in the cumulative sodium balance, which causes a slight increase in extracellular fluid volume. It is mainly this small increase in extracellular fluid volume that triggers various mechanisms in the body to increase sodium excretion. One of the first mechanisms to be triggered is a low-pressure receptor reflex that originates from the stretch receptors of the right atrium and the pulmonary blood vessels. Signals from the stretch receptors go to the brain stem and there inhibit sympathetic nerve activity to the kidneys to decrease tubular sodium reabsorption. This mechanism is most important in the first few hours—or perhaps the first day—after a large increase in salt and water intake.

A second response also occurs: expansion of extracellular fluid volume leads to a small increase in arterial pressure, which by itself increases sodium excretion through pressure natriuresis.

Then a third response occurs: the increased arterial pressure also suppresses angiotensin II formation, which in turn decreases tubular sodium reabsorption by eliminating the normal effect of angiotensin II to increased sodium reabsorption. Also, the reduced angiotensin II decreases aldosterone secretion, thus further reducing tubular sodium reabsorption.

Finally, expansion of extracellular fluid volume may also stimulate natriuretic systems, especially ANP, which contribute still further to the increased sodium excretion.

Thus, the combined activation of natriuretic systems and suppression of sodium- and water-retaining systems leads to an increase in sodium excretion when sodium intake is increased. The opposite changes take place when sodium intake is reduced below normal levels.

CONDITIONS THAT CAUSE LARGE INCREASES IN BLOOD VOLUME AND EXTRACELLULAR FLUID VOLUME

Despite the powerful regulatory mechanisms that maintain blood volume and extracellular fluid volume reasonably constant, there are abnormal conditions that can cause large increases in both of these variables. Almost all these conditions result from circulatory abnormalities as follows.

Increased Blood Volume and Extracellular Fluid Volume Caused by Heart Diseases

In congestive heart failure, blood volume may increase 15 to 20 per cent and extracellular fluid volume sometimes increases by 200 per cent or more. The reason for this can be understood by re-examination of Figure 29–2. Initially, heart failure reduces cardiac output and, consequently, decreases arterial pressure. This in turn activates multiple sodium-retaining systems, especially the renin-angiotensin, aldosterone, and sympathetic nervous systems. In addition, the low blood pressure itself causes the kidneys to retain salt and water. Therefore, the kidneys retain volume in an attempt to return the arterial pressure and cardiac output toward normal. Indeed, if the heart failure is not too severe, the rise in blood volume can often return cardiac output and arterial pressure virtually all the way to normal, and sodium excretion will eventually increase back to normal, although there will remain excess extracellular fluid volume and blood volume to keep the weakened heart pumping adequately. However, if the heart is greatly weakened, arterial pressure will not be able to increase enough to restore urine output to normal. When this occurs, the kidneys continue to retain volume until the person develops severe circulatory congestion and eventually dies of edema, especially pulmonary edema.

Therefore, in myocardial failure, heart valvular disease, and congenital abnormalities of the heart, an important circulatory compensation is an increase in blood volume, which helps to return cardiac output and blood pressure to normal. This allows even the weakened heart to pump a life-sustaining level of cardiac output.

Increased Blood Volume Caused by Increased Capacity of the Circulation

Any condition that increases vascular capacity will also cause the blood volume to increase to fill this extra capacity. An increase in vascular capacity initially reduces mean circulatory filling pressure, which leads to decreased cardiac output and decreased arterial pressure. The fall in pressure causes salt and water retention by the kidneys until the blood volume increases sufficiently to fill the extra capacity. For example, in pregnancy, the increased vascular capacity of the uterus, placenta, and other enlarged organs of the woman's body regularly increases the blood volume 15 to 25 per cent. Similarly, in patients who have large varicose veins of the legs, which in rare instances may hold up to an extra liter of blood, the blood volume simply increases to fill the extra vascular capacity. In these cases, salt and water are retained by the kidneys until the total vascular bed is filled enough to raise blood pressure to the level required to balance renal output of fluid with daily intake of fluid.

CONDITIONS THAT CAUSE LARGE INCREASES IN EXTRACELLULAR FLUID VOLUME BUT WITH NORMAL BLOOD VOLUME

There are several conditions in which extracellular fluid volume becomes markedly increased but blood volume remains normal or even slightly reduced. These conditions are usually initiated by leakage of fluid and protein into the interstitium, which tends to decrease the blood volume. The kidneys' response to these conditions is similar to the response after hemorrhage. That is, the kidneys retain salt and water in an attempt to restore blood volume toward normal.

Nephrotic Syndrome—Loss of Plasma Proteins in the Urine and Sodium Retention by the Kidneys

The general mechanisms that lead to extracellular edema are reviewed in Chapter 25. One of the most important clinical causes of edema is the so-called *nephrotic syndrome*. In nephrotic syndrome, the glomerular capillaries leak large amounts of protein into the filtrate and the urine because of an increased permeability of the glomerulus. Thirty to 50 grams of plasma protein can be lost in the urine each day, sometimes causing the plasma protein concentration to fall to less than one-third normal. As a consequence of the decreased plasma protein concentration, the plasma colloid osmotic pressure falls to low levels. This causes the capillaries all over the body to filter large amounts of fluid into the various tissues, which in turn causes edema and decreases the plasma volume.

Renal sodium retention in nephrotic syndrome occurs through multiple mechanisms activated by leakage of protein and fluid from the plasma into the interstitial fluid, including activation of various sodium-retaining systems such as the renin-angiotensin system, aldosterone, and possibly the sympathetic nervous system. The kidneys continue to retain sodium and water until plasma volume is restored nearly to normal. However, because of the large amount of sodium and water retention, the plasma protein concentration becomes further diluted, causing still more fluid to leak into the tissues of the body. The net result is massive fluid retention by the kidneys until tremendous extracellular edema occurs unless treatment is instituted to restore the plasma proteins.

Liver Cirrhosis—Decreased Synthesis of Plasma Proteins by the Liver and Sodium Retention by the Kidneys

A similar sequence of events occurs in cirrhosis of the liver as in nephrotic syndrome, except that in liver cirrhosis, the reduction in plasma protein concentration results from destruction of the liver cells, thus reducing the ability of the liver to synthesize enough plasma proteins. Cirrhosis is also associated with large amounts of fibrous tissue in the liver structure, which greatly impedes the flow of portal blood through the liver. This in turn raises capillary pressure throughout the portal vascular bed, which also contributes to the leakage of fluid and proteins into the peritoneal cavity, a condition called *ascites*. Once fluid and protein are lost from the circulation, the renal responses are similar to those observed in other conditions associated with decreased plasma volume. That is, the kidneys continue to retain salt and water until plasma volume and arterial pressure are restored to normal.

REGULATION OF POTASSIUM EXCRETION AND POTASSIUM CONCENTRATION IN THE EXTRACELLULAR FLUID

Extracellular fluid potassium concentration normally is regulated precisely at about 4.2 mEq/liter, seldom rising or falling more than ± 0.3 mEq/liter. This precise control is necessary because many of the cell functions are sensitive to changes in extracellular fluid potassium concentration. For instance, an increase in plasma potassium concentration of only 4 mEq/liter can cause cardiac arrhythmias and higher concentrations can lead to cardiac arrest of fibrillation.

A special difficulty in regulating extracellular potassium concentration is the fact that about 95 per cent of the total body potassium is contained in the cells and only 2 per cent in the extracellular fluid (Fig. 29–6). For a 70-kilogram adult, who has about 28 liters of intracellular fluid (40 per cent of body weight) and 14 liters of extracellular fluid (20 per cent of body weight), about 3920 milliequivalents of potassium are inside the cells and only about 59 milliequivalents are in the extracellular fluid. Also, the potassium contained in a single meal is often as high as 50 milliequivalents, and the daily intake usually ranges between 50 and 200 mEq/day; therefore, failure to rapidly rid the extracellular fluid of the ingested potassium could cause life-threatening *hyperkalemia* (increased plasma potassium concentration). Likewise, a small loss of potassium from the extracellular fluid could cause severe *hypokalemia* (low plasma potassium concentration) in the absence of rapid and appropriate compensatory responses.

Maintenance of potassium balance depends primarily on excretion by the kidneys because the amount excreted in the feces is only about 5 to 10 per cent of the potassium intake. Thus, the maintenance of normal

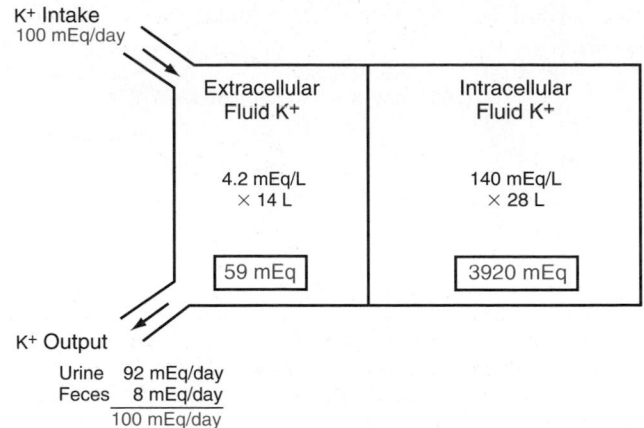

Figure 29–6. Normal potassium intake, distribution of potassium in the body fluids, and potassium output from the body.

potassium balance requires the kidneys to adjust their potassium excretion rapidly and precisely to wide variations in intake, as is also true for most other electrolytes.

Overview of Renal Potassium Excretion

Potassium excretion is determined by the sum of three renal processes: (1) the rate of potassium filtration (GFR multiplied by the plasma potassium concentration), (2) the rate of potassium reabsorption by the tubules, and (3) the rate of potassium secretion by the tubules. The normal rate of potassium filtration is about 756 mEq/day (GFR, 180 liters/day multiplied by plasma potassium, 4.2 mEq/liter); this rate of filtration is usually relatively constant because of the autoregulatory mechanisms for GFR discussed previously and the precision with which plasma potassium concentration is regulated. Severe decreases in GFR in certain renal diseases can cause serious potassium accumulation and hyperkalemia.

Figure 29–7 summarizes the tubular handling of potassium under normal conditions. About 65 per cent of the filtered potassium is reabsorbed in the proximal tubule. Another 25 to 30 per cent of the filtered potassium is reabsorbed in the loop of Henle, especially in the thick ascending part where potassium is actively co-transported along with sodium and chloride. In both the proximal tubule and the loop of Henle, a relatively constant fraction of the filtered potassium load is reabsorbed. Abnormalities of potassium reabsorption in these segments can influence potassium excretion, but most of the day-to-day variation of potassium excretion is not due to changes in reabsorption in the proximal tubule or loop of Henle.

MOST OF THE DAILY VARIATION IN POTASSIUM EXCRETION IS CAUSED BY CHANGES IN POTASSIUM SECRETION IN THE DISTAL AND COLLECTING TUBULES. The most important sites for regulating potassium ex-

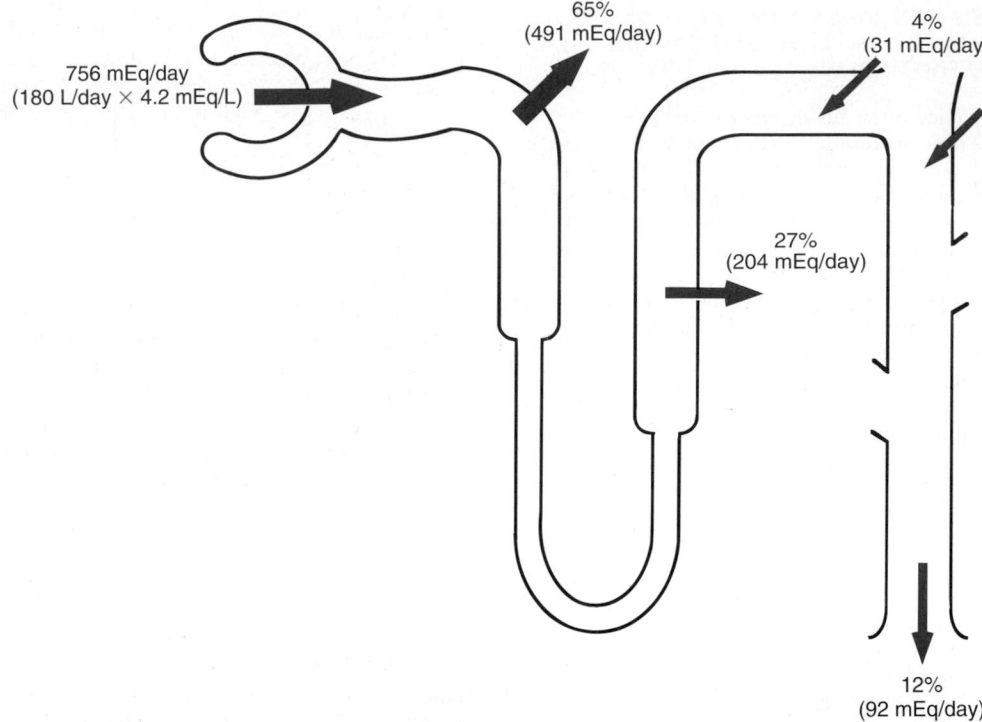

Figure 29–7. Renal tubular sites of potassium reabsorption and secretion. Potassium is reabsorbed in the proximal tubule and in the ascending loop of Henle, so that only about 8 per cent of the filtered load is delivered to the distal tubule. Secretion of potassium into the distal tubule and collecting ducts adds to the amount delivered, so that the daily excretion is about 12 per cent of the potassium filtered at the glomerular capillaries. The percentages indicate how much of the filtered load is reabsorbed or secreted into the different tubular segments.

cretion are the distal tubules and cortical collecting tubules. In these tubular segments, potassium can at times be reabsorbed or at other times be secreted, depending on the needs of the body. With a normal potassium intake of 100 mEq/day, the kidneys must excrete about 92 mEq/day (the remaining 8 mEq is lost in the feces). About one third (31 mEq/day) of this amount of potassium is secreted into the distal tubule and collecting ducts.

With high potassium intakes, the required extra excretion of potassium is achieved almost entirely by increasing the secretion of potassium into the distal tubules and collecting tubules. In fact, with extremely high potassium diets, the rate of potassium excretion can exceed the amount of potassium in the glomerular filtrate, indicating a powerful mechanism for secreting potassium in the distal segments of the nephron.

When potassium intake is reduced below normal, the secretion rate of potassium in the distal tubule and collecting tubules decreases, causing a reduction in urinary potassium secretion. With extreme reductions in potassium intake, there is net reabsorption of potassium in the distal segments of the nephron and potassium excretion can fall to 1 per cent of the potassium in the glomerular filtrate (to less than 10 mEq/day).

With potassium intakes below this level, severe hypokalemia can develop.

Thus, most of the day-to-day regulation of potassium excretion occurs in the late distal tubule and cortical collecting tubules, where potassium can be either reabsorbed or secreted, depending on the needs of the body. In the next section, we consider the basic mechanisms of potassium secretion and the factors that regulate this process.

Potassium Secretion in the Late Distal Tubules and Cortical Collecting Tubules

The cells in the late distal tubules and cortical collecting tubules that secrete potassium are called *principal cells* and make up about 90 per cent of the epithelial cells in these regions. Figure 29–8 shows the basic cellular mechanisms of potassium secretion by the principal cells.

Secretion of potassium from the blood into the tubular lumen is a two-step process, beginning with uptake from the interstitium into the cell by the sodium-potassium adenosine triphosphatase (ATPase) pump in the basolateral membrane of the cell; this pump moves sodium out of the cell into the interstitium and at the same time moves potassium to the interior of the cell.

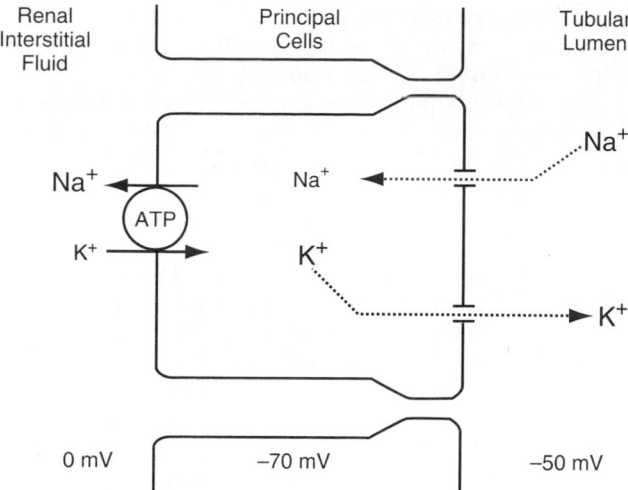

Figure 29–8. Mechanisms of potassium secretion and sodium reabsorption by the principal cells of the distal tubule and collecting tubules.

The second step of the process is passive diffusion of potassium from the interior of the cell into the tubular fluid. The sodium-potassium ATPase pump creates a high intracellular potassium concentration, which provides the driving force for passive diffusion of potassium from the cell into the tubular lumen. The luminal membrane of the principal cells is highly permeable to potassium. One reason for this high permeability is that there are special channels that are specifically permeable to potassium ions, thus allowing these ions to diffuse across the membrane.

CONTROL OF POTASSIUM SECRETION BY THE PRINCIPAL CELLS. The primary factors that control potassium secretion by the principal cells of the late distal tubule and cortical collecting tubule are (1) the activity of the sodium-potassium ATPase pump, (2) the electrochemical gradient for potassium secretion from the blood to the tubular lumen, and (3) the permeability of the luminal membrane for potassium. These three determinants of potassium secretion are in turn regulated by the factors discussed below.

THE INTERCALATED CELLS CAN REABSORB POTASSIUM DURING POTASSIUM DEPLETION. In circumstances associated with severe potassium depletion, there is a cessation of potassium secretion and actually net reabsorption of potassium in the late distal tubules and collecting ducts. This reabsorption occurs through the *intercalated cells;* although this reabsorptive process is not completely understood, one mechanism believed to contribute is a hydrogen potassium ATPase transport mechanism located in the luminal membrane. This transporter reabsorbs potassium in exchange for hydrogen secreted into the tubular lumen and the potassium then diffuses through the basolateral membrane of the cell into the blood. This transporter is necessary to allow potassium reabsorption during extracellular fluid potassium depletion, but under normal conditions, it plays a small role in controlling the excretion of potassium.

Summary of Factors That Regulate Potassium Secretion: Plasma Potassium Concentration, Aldosterone, Tubular Flow Rate, and Hydrogen Ion

Because normal regulation of potassium excretion occurs mainly as a result of changes in potassium secretion by the principal cells of the distal tubule and collecting tubules, we discuss the primary factors that influence secretion by these cells. The most important factors that *stimulate* potassium secretion by the principal cells include (1) increased extracellular fluid potassium concentration, (2) increased aldosterone, and (3) increased tubular flow rate.

One factor that *decreases* potassium secretion is increased hydrogen ion concentration (acidosis).

INCREASED EXTRACELLULAR FLUID POTASSIUM CONCENTRATION STIMULATES POTASSIUM SECRETION. The rate of potassium secretion in the late distal tubules and cortical collecting tubules is directly stimulated by increased extracellular fluid potassium concentration, leading to increases in potassium excretion, as shown in Figure 29–9. This effect is especially pronounced when extracellular fluid potassium concentration rises above about 4.1 mEq/liter, slightly less than the normal concentration. This effect, therefore, serves as one of the most important mechanisms for increasing potassium secretion and regulating extracellular fluid potassium ion concentration.

There are three mechanisms by which increased extracellular fluid potassium concentration raises potassium secretion: (1) Increased extracellular fluid potassium concentration stimulates the sodium-potassium

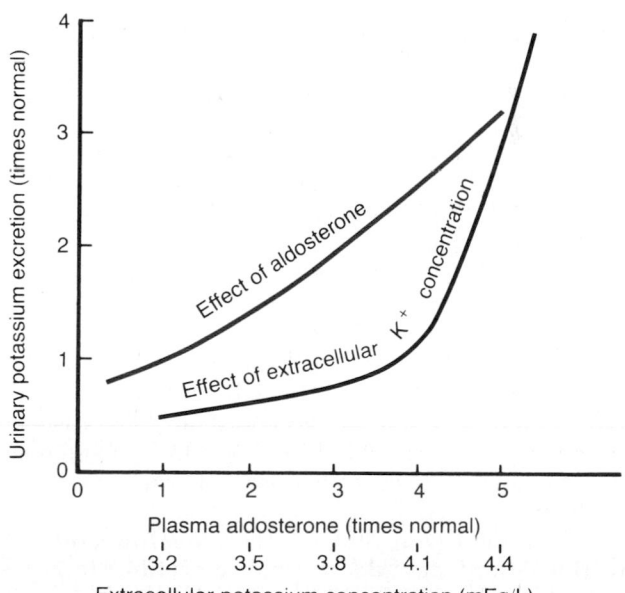

Figure 29–9. Effect of (1) plasma aldosterone concentration and (2) extracellular potassium ion concentration on the rate of urinary potassium excretion. These factors stimulate potassium secretion by the principal cells of the cortical collecting tubules. (Drawn from data in Young: *Am. J. Physiol.* 244: F28, 1983.)

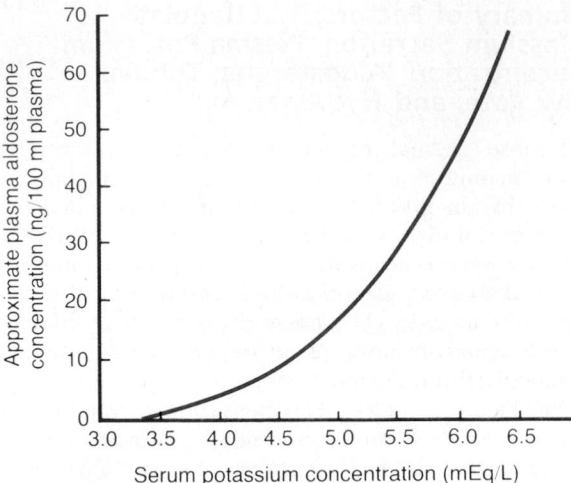

Figure 29–10. Effect of extracellular fluid potassium ion concentration on plasma aldosterone concentration. Note that small changes in potassium concentration cause large changes in aldosterone concentration.

ATPase pump, thereby increasing potassium uptake across the basolateral membrane. This in turn increases intracellular potassium ion concentration, causing potassium to diffuse across the luminal membrane into the tubule. (2) Increased extracellular potassium concentration increases the potassium gradient from the renal interstitial fluid to the interior of the epithelial cell; this reduces backleakage of potassium ions from inside the cells through the basolateral membrane. (3) Increased potassium concentration stimulates aldosterone secretion by the adrenal cortex, which further stimulates potassium secretion, as discussed next.

ALDOSTERONE STIMULATES POTASSIUM SECRETION. In Chapter 28, we discuss the fact that aldosterone stimulates active reabsorption of sodium ions by the principal cells of the late distal tubules and collecting ducts. This effect is mediated through a sodium-potassium ATPase pump that transports sodium outward through the basolateral membrane of the cell and into the blood at the same time that it pumps potassium into the cell. Thus, aldosterone also has a powerful effect to control the rate at which the principal cells secrete potassium.

A second effect of aldosterone is to increase the permeability of the luminal membrane for potassium, further adding to the effectiveness of aldosterone in stimulating potassium secretion. Therefore, aldosterone has a powerful effect to increase potassium excretion, as shown in Figure 29–9.

Increased Extracellular Potassium Ion Concentration Stimulates Aldosterone Secretion. In negative feedback control systems, the factor that is controlled usually has a feedback effect on the controller. In the case of the aldosterone-potassium control system, the rate of aldosterone secretion by the adrenal gland is controlled strongly by extracellular fluid potassium ion concentration. Figure 29–10 shows that an increase in plasma potassium concentration of about 3

mEq/liter can increase plasma aldosterone concentration from nearly 0 to as high as 60 ng/100 ml, a concentration almost 10 times normal.

The effect of potassium ion concentration to stimulate aldosterone secretion is part of a powerful feedback system for regulating potassium excretion, as shown in Figure 29–11. In this feedback system, an increase in plasma potassium concentration stimulates aldosterone secretion and, therefore, increases the blood level of aldosterone (Block 1). The increase in blood aldosterone then causes a marked increase in potassium excretion by the kidneys (Block 2). The increased potassium excretion then reduces the extracellular fluid potassium concentration back toward normal (Blocks 3 and 4). Thus, this feedback mechanism acts synergistically with the direct effect of increased extracellular potassium concentration to elevate potassium excretion when potassium intake is raised (Fig. 29–12).

Blockade of the Aldosterone Feedback System Greatly Impairs Control of Potassium Concentration. In the absence of aldosterone secretion, as occurs in patients with Addison's disease, renal secretion of potassium is impaired, thus causing extracellular fluid potassium concentration to rise to dangerously high levels. Conversely, with excess aldosterone secretion (primary aldosteronism), potassium secretion becomes greatly increased, causing potassium loss by the kidneys and thus leading to hypokalemia.

The special quantitative importance of the aldosterone feedback system in controlling potassium concentration is shown in Figure 29–13. In this experiment, potassium intake was increased almost sevenfold in dogs under two conditions: (1) under normal conditions and (2) after the aldosterone feedback system had been blocked by removing the adrenal glands and placing the animals on a fixed rate of aldosterone infusion so that plasma aldosterone concentration could neither increase nor decrease.

Note that in the normal animals, a sevenfold increase in potassium intake caused only a slight increase in potassium concentration, from 4.2 to 4.3 mEq/liter. Thus, when the aldosterone feedback system is func-

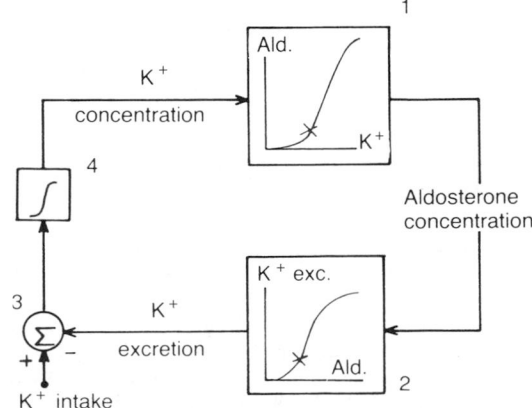

Figure 29–11. Basic feedback mechanism for control of extracellular fluid potassium concentration by aldosterone.

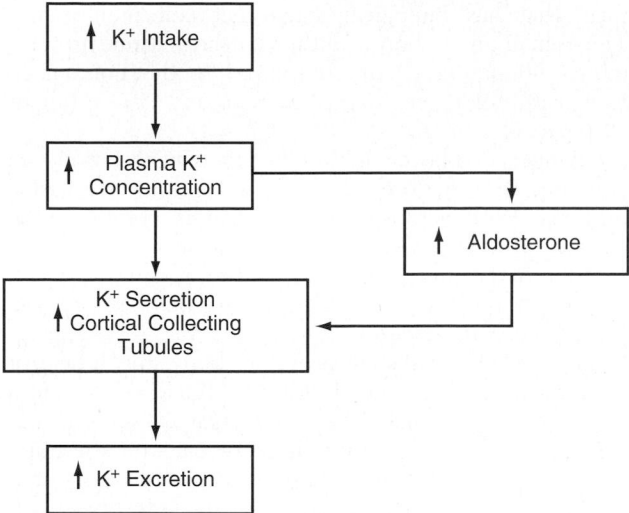

Figure 29–12. Primary mechanisms by which high potassium intake raises potassium excretion. Note that increased plasma potassium concentration directly raises potassium secretion by the cortical collecting tubules and indirectly increases potassium secretion by raising plasma aldosterone concentration.

tioning normally, potassium concentration is precisely controlled, despite large changes in potassium intake.

When the aldosterone feedback system was blocked, the same increases in potassium intake caused a much larger increase in potassium concentration, from 3.8 to almost 4.7 mEq/liter. Thus, control of potassium con-

centration is greatly impaired when the aldosterone feedback system is blocked. A similar impairment of potassium regulation is observed in humans with poorly functioning aldosterone feedback systems, such as occurs in patients with either primary aldosteronism (too much aldosterone) or Addison's disease (too little aldosterone).

INCREASED DISTAL TUBULAR FLOW RATE STIMULATES POTASSIUM SECRETION. A rise in distal tubular flow rate, as occurs with volume expansion, high sodium intake, or diuretic drug treatment, stimulates potassium secretion. Conversely, a decrease in distal tubular flow rate, as caused by sodium depletion, reduces potassium secretion.

The mechanism for the effect of high-volume flow rate is as follows: When potassium is secreted into the tubular fluid, the luminal concentration of potassium increases, thereby reducing the driving force for potassium diffusion across the luminal membrane. But with increased tubular flow rate, the secreted potassium is continuously flushed down the tubule, so that the rise in tubular potassium concentration becomes minimized. Therefore, net potassium secretion is stimulated by increased tubular flow rate.

The effect of increased tubular flow rate is especially important in helping to preserve normal potassium excretion during changes in sodium intake. For example, with a high sodium intake, there is decreased aldosterone secretion, which by itself would tend to decrease the rate of potassium secretion and, therefore, reduce urinary excretion of potassium. The high distal tubular flow rate that occurs with a high sodium intake tends to increase potassium secretion (Fig. 29–14), as discussed in the previous paragraph. Therefore, the two effects of high sodium intake, decreased aldosterone secretion and the high tubular flow rate, counterbalance each other, so that there is little change in potassium excretion. And likewise, with a low sodium intake, there is also little change in potassium excretion because of the counterbalancing effects of increased aldosterone secretion and decreased tubular flow rate on potassium secretion.

ACUTE ACIDOSIS DECREASES POTASSIUM SECRETION. Acute increases in hydrogen ion concentration of the extracellular fluid (acidosis) reduce potassium secretion, whereas decreased hydrogen ion concentration (alkalosis) increases potassium secretion. The primary mechanism by which increased hydrogen ion concentration inhibits potassium secretion is by reducing the activity of the sodium-potassium ATPase pump. This in turn decreases intracellular potassium concentration and subsequent passive diffusion of potassium across the luminal membrane into the tubule.

With more prolonged acidosis, lasting over a period of several days, there is an increase in urinary potassium excretion. The mechanism for this effect is due in part to an effect of chronic acidosis to inhibit proximal tubular sodium chloride and water reabsorption, which increases distal volume delivery, thereby stimulating the secretion of potassium. This effect overrides the inhibitory effect of hydrogen ions on the sodium-potassium

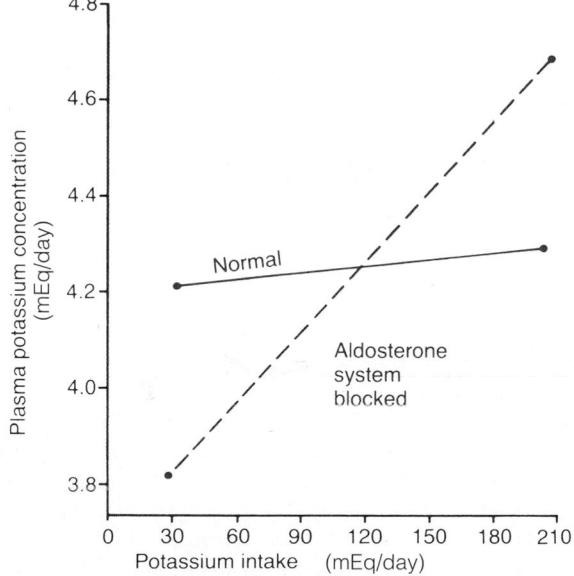

Figure 29–13. Effect of large changes in potassium intake on extracellular fluid potassium concentration (1) under normal conditions (solid line) and (2) after the aldosterone feedback had been blocked (dashed line). Note that after blockade of the aldosterone system, regulation of potassium concentration was greatly impaired. (Courtesy of Dr. David B. Young.)

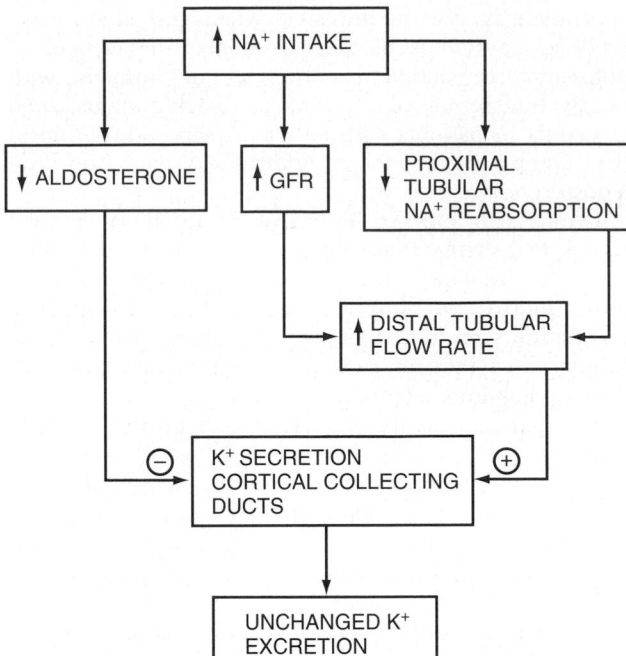

Figure 29–14. Effect of high sodium intake on renal excretion of potassium. Note that a high-sodium diet decreases plasma aldosterone, which tends to decrease potassium secretion by the cortical collecting tubules. However, the high-sodium diet simultaneously increases fluid delivery to the cortical collecting duct, which tends to increase potassium secretion. The opposing effects of a high-sodium diet counterbalance each other, so that there is little change in potassium excretion.

ATPase pump. *Thus, chronic acidosis leads to a loss of potassium, whereas acute acidosis leads to decreased potassium excretion.*

CONTROL OF RENAL CALCIUM EXCRETION AND EXTRACELLULAR CALCIUM ION CONCENTRATION

The mechanisms for regulating of calcium ion concentration are discussed in detail in Chapter 79, along with the endocrinology of the calcium-regulating hormones, parathyroid hormone (PTH) and calcitonin. Therefore, calcium ion regulation is discussed only briefly in this chapter.

Extracellular fluid calcium ion concentration normally remains tightly controlled within a few percentage points of its normal level, 2.4 mEq/liter. When calcium ion concentration falls to low levels (*hypocalcemia*), the excitability of nerve and muscle cells increases markedly and can in extreme cases result in *hypocalcemic tetany*. This is characterized by spastic skeletal muscle contractions. *Hypercalcemia* (increased calcium concentration) depresses neuromuscular excitability and can lead to cardiac arrhythmias.

About 50 per cent of the total calcium in the plasma (5.0 mEq/liter) exists in the ionized form, which is the

form that has biological activity at cell membranes. The remainder is bound either to the plasma proteins (about 40 per cent) or complexed in the nonionized form with anions such as phosphate and citrate (about 10 per cent).

Changes in plasma hydrogen ion concentration can influence the degree of calcium binding to plasma proteins. With acidosis, less calcium is bound to the plasma proteins. Conversely, in alkalosis, a greater amount of calcium is bound to the plasma proteins. Therefore, *patients with alkalosis are more susceptible to hypocalcemic tetany.*

As with other substances in the body, the intake of calcium must be balanced with the net loss of calcium over the long term. Unlike ions such as sodium and chloride, however, a large share of calcium excretion occurs in the feces. Only about 10 per cent of ingested calcium normally is absorbed from the intestinal tract, and the remainder is excreted in the feces. Under certain conditions, fecal calcium excretion can exceed calcium ingestion because calcium can also be secreted into the intestinal lumen. Therefore, the gastrointestinal tract and the regulatory mechanisms that influence intestinal calcium absorption and secretion play a major role in calcium homeostasis, as discussed in Chapter 79.

Almost all the calcium in the body (99 per cent) is stored in the bone, with only about 1 per cent in the intracellular fluid and 0.1 per cent in the extracellular fluid. The bone, therefore, acts as a large reservoir for storing calcium and as a source of calcium when extracellular fluid calcium concentration tends to decrease. *One of the most important regulators of bone uptake and release of calcium is PTH.* When extracellular fluid calcium concentration falls below normal, the parathyroid glands are directly stimulated by the low calcium levels to promote increased secretion of PTH. This hormone then acts directly on the bones to increase the resorption of bone salts (release of salts from the bones) and, therefore, to release large amounts of calcium into the extracellular fluid, thereby returning calcium levels back toward normal. When calcium ion concentration is elevated, PTH secretion decreases, so that almost no bone resorption now occurs; instead, excess calcium now is deposited in the bones because of new bone formation. Thus, the day-to-day regulation of calcium ion concentration is mediated in large part by the effect of PTH on bone resorption.

The bones, however, do not have an inexhaustible supply of calcium. Therefore, over the long term, the intake of calcium must be balanced with calcium excretion by the gastrointestinal tract and the kidneys. The most important regulator of calcium reabsorption at both of these sites is PTH. *Thus, PTH regulates plasma calcium concentration through three main effects: (1) by stimulating bone resorption; (2) by stimulating activation of vitamin D, which then increases intestinal reabsorption of calcium; and (3) by directly increasing renal tubular calcium reabsorption* (Fig. 29–15). The control of gastrointestinal calcium reab-

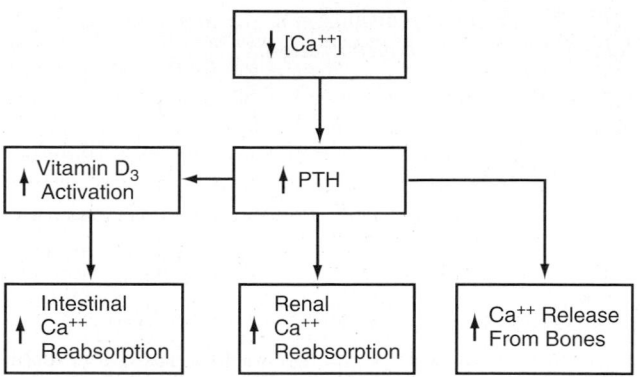

Figure 29–15. Compensatory responses to decreased plasma ionized calcium concentration mediated by parathyroid hormone and vitamin D.

sorption and calcium exchange in the bone are discussed elsewhere, and the remainder of this section focuses on the mechanisms that control renal calcium excretion.

Control of Calcium Excretion by the Kidneys

Because calcium is both filtered and reabsorbed in the kidneys but not secreted, the rate of renal calcium excretion is calculated as:

Renal calcium excretion
$$= \text{Calcium filtered} - \text{Calcium reabsorbed}$$

Only about 50 per cent of the plasma calcium is ionized, with the remainder being bound to the plasma proteins. Therefore, only about 50 per cent of the plasma calcium can be filtered at the glomerulus. Normally, about 99 per cent of the filtered calcium is reabsorbed by the tubules, with only about 1 per cent of the filtered calcium being excreted. About 65 per cent of the filtered calcium is reabsorbed in the proximal tubule, 25 to 30 per cent is reabsorbed in the loop of Henle, and 4 to 9 per cent is reabsorbed in the distal tubule and collecting tubules. This pattern of reabsorption is similar to that for sodium.

As is true with the other ions, calcium excretion is adjusted to meet the body's needs. With an increase in calcium intake, there is also increased renal calcium excretion, although much of the increase of calcium intake is eliminated in the feces. With calcium depletion, calcium excretion by the kidneys decreases as a result of enhanced tubular reabsorption.

One of the primary controllers of renal tubular calcium reabsorption is PTH. With increased levels of PTH, there is increased calcium reabsorption in the thick ascending loop of Henle and distal tubule, which reduces urinary excretion of calcium. Conversely, reduction of PTH promotes calcium excretion by decreasing reabsorption in the loop of Henle and distal tubules.

In the proximal tubule, calcium reabsorption usually parallels sodium and water reabsorption. Therefore, in instances of extracellular volume expansion or increased arterial pressure—both of which decrease proximal sodium and water reabsorption—there is also reduction in calcium reabsorption and, consequently, increased urinary excretion of calcium. Conversely, with extracellular volume contraction or decreased blood pressure, calcium excretion decreases primarily because of increased proximal tubular reabsorption.

Another factor that influences calcium reabsorption is the plasma concentration of phosphate. An increase in plasma phosphate stimulates PTH, which increases calcium reabsorption by the renal tubules, thereby reducing calcium excretion. The opposite occurs with reduction in plasma phosphate concentration.

Calcium reabsorption is also stimulated by metabolic acidosis and inhibited by metabolic alkalosis. Most of the effect of hydrogen ion concentration on calcium excretion results from changes in calcium reabsorption in the distal tubule.

A summary of the factors that are known to influence calcium reabsorption by the renal tubules is shown in Table 29–1.

REGULATION OF RENAL PHOSPHATE EXCRETION

Phosphate excretion by the kidneys is controlled primarily by an overflow mechanism that can be explained as follows: The renal tubules have a normal transport maximum for reabsorbing phosphate of about 0.1 mM/min. When less than this amount of phosphate is present in the glomerular filtrate, essentially *all* the filtered phosphate is reabsorbed. When more than this is present, the *excess* is excreted. Therefore, phosphate normally begins to spill into the urine when its concentration in the extracellular fluid rises above a threshold of about 0.8 mM/liter, which gives a tubular load of phosphate of about 0.1 mM/min, assuming a GFR of 125 ml/min. Because most people ingest large quantities of phosphate in milk products and meat, the concentration of phosphate is usually maintained above 1 mM/liter, a level at which there is continual excretion of phosphate into the urine.

Table 29–1 FACTORS THAT ALTER RENAL TUBULAR CALCIUM REABSORPTION.

↑ Calcium Reabsorption	↓ Calcium Reabsorption
↑ PTH	↓ PTH
↓ Extracellular fluid volume	↑ Extracellular fluid volume
↓ Blood pressure	↑ Blood pressure
↑ Plasma phosphate	↓ Plasma phosphate
Metabolic acidosis	Metabolic alkalosis
Vitamin D$_3$	

Changes in tubular phosphate reabsorption can also influence phosphate excretion. For instance, a diet low in phosphate can, over time, increase the reabsorptive transport maximum for phosphate, thereby reducing the tendency for phosphate to spill over into the urine.

PTH can play a significant role in regulating phosphate concentration through two effects: (1) PTH promotes bone resorption, thereby dumping large amounts of phosphate ions into the extracellular fluid from the bone salts, and (2) PTH decreases the transport maximum for phosphate by the renal tubules, so that a greater proportion of the tubular phosphate is lost in the urine. *Thus, whenever plasma PTH is increased, tubular phosphate reabsorption is decreased and more phosphate is excreted.* These interrelations between the phosphate, PTH, and calcium are discussed in more detail in Chapter 79.

CONTROL OF RENAL MAGNESIUM EXCRETION AND EXTRACELLULAR MAGNESIUM ION CONCENTRATION

More than one half of the body's magnesium is stored in the bones; most of the rest resides within the cells, with less than 1 per cent located in the extracellular fluid. Although the total plasma magnesium concentration is about 1.8 mEq/liter, more than one half of this is bound to plasma proteins. Therefore, the free ionized concentration of magnesium is only about 0.8 mEq/liter.

The normal daily intake of magnesium is about 250 to 300 mg/day, but only about one half of this intake is absorbed by the gastrointestinal tract. To maintain magnesium balance, the kidneys must excrete this absorbed magnesium, about one half the daily intake of magnesium or 125 to 150 mg/day. The kidneys normally excrete about 10 to 15 per cent of the magnesium in the glomerular filtrate.

Renal excretion of magnesium can increase markedly during magnesium excess or decrease to almost nil during magnesium depletion. Because magnesium is involved in many biochemical processes in the body, including activation of many enzymes, its concentration must be closely regulated.

Regulation of magnesium excretion is achieved mainly by changing tubular reabsorption. The proximal tubule usually reabsorbs only about 25 per cent of the filtered magnesium. The primary site of reabsorption is the loop of Henle, where about 65 per cent of the filtered load of magnesium is reabsorbed. Only a small amount (usually less than 5 per cent) of the filtered magnesium is reabsorbed in the distal and collecting tubules.

The mechanisms that regulate magnesium excretion are not well understood, but *the following disturbances lead to increased magnesium excretion: (1) increased extracellular fluid magnesium concentration, (2) extra-cellular volume expansion, and (3) increased extracellular fluid calcium concentration.*

REFERENCES

Agus, A. S., et al.: Disorders of calcium and magnesium homeostasis. Am. J. Med., 82:34, 1987.
Alberola, A., et al.: Renal hemodynamic effects of angiotensin II (AII): Interactions with endothelium derived nitric oxide. Am. J. Physiol., 267:R1472, 1994.
Andersson, D., and Rundgren, M.: Thirst and its disorders. Annu. Rev. Med., 33:231, 1982.
Ballerman, B. J. and Zeidel, M. L.: Atrial natriuretic hormone. In Seldin, D. W. and Giebisch, G. (eds.): The Kidney: Physiology and Pathophysiology, 2nd Ed. New York, Raven Press, 1992.
Berndt, T. J. and Knox, F. G.: Renal regulation of phosphate excretion. In Seldin, D. W. and Giebisch, G. (eds.): The Kidney: Physiology and Pathophysiology, 2nd Ed. New York, Raven Press, 1992.
Brands, M. W., et al.: Chronic converting enzyme inhibition improves cardiac output and fluid balance during heart failure. Am. J. Physiol., 264:R414, 1993.
Brenner, B. M. and Rector, F. C., Jr.: The Kidney, 4th Ed. Phildelphia, W. B. Saunders Co., 1991.
Brenner, B. M., et al.: Diverse biological actions of atrial natriuretic peptide. Physiol. Rev., 70:665, 1990.
Carretero, O. A. and Scicli, A. G.: The kallikrein-kinin system as a regulator of cardiovascular and renal function. In Laragh, J. H. and Brenner, B. M. (eds): Hypertension: Pathophysiology, Diagnosis and Management, 2nd Ed. Raven Press Ltd., New York, 1995.
Conrad, K. P., and Dunn, M. J.: Renal prostaglandins and other eicosanoids. In Windhager, E. E. (ed.): Handbook of Physiology, Section 8, Renal Physiology. New York, Oxford University Press, 1992.
Cowley, A. W., Jr.: Long-term control of arterial pressure. Physiol. Rev., 72:231, 1992.
deWardener, H. E., and Clarkson, E. M.: Concept of natriuretic hormone. Physiol. Rev., 65:658, 1985.
DiBona, G. F.: Role of renal nerves in edema formation. News Physiolog. Sci., 9:183, 1994.
Dunn, M. J. (ed.): Prostaglandins and the Kidney. Biochemistry, Physiology, Pharmacology, and Clinical Applications. New York, Plenum Publishing Corp., 1983.
Fitzsimmons, J. T.: Physiology and pathophysiology of thirst and sodium appetite. In Seldin, D. W. and Giebisch, G. (eds.): The Kidney. Physiology and Pathophysiology, 2nd Ed. New York, Raven Press, 1992.
Granger, J. P.: Pressure natriuresis: role of renal interstitial hydrostatic pressure. Hypertension 19 (Suppl. I):I9–I17, 1992.
Gross, M., and Kumar, R.: Vitamin D endocrine system and calcium and phosphorous homeostasis. In Windhager, E. E. (ed.): Handbook of Physiology, Section 8, Renal Physiology. New York, Oxford University Press, 1992.
Guyton, A. C., et al.: Dynamics and Control of the Body Fluids. Philadelphia, W. B. Saunders Co., 1975.
Guyton, A. C.: Abnormal renal function and autoregulation in essential hypertension. Hypertension 18 (Suppl. III):III49, 1991.
Guyton, A. C.: Blood pressure control—Special role of the kidneys and body fluids. Science, 252:1813, 1991.
Guyton, A. C.: Kidneys and fluids in pressure regulation: Small volume but large pressure changes. Hypertension, 19 (Suppl. I):I2, 1992.
Hall, J. E., and Brands, M. W.: Intrarenal and circulating angiotensin II and renal function. In Robertson, J. I. S. and Nicholls, M. G. (eds.): The Renin-Angiotensin System, Vol. I. London, Gower Medical Publishing, 1993.
Hall, J. E., et al.: Control of sodium excretion and arterial pressure by intrarenal mechanisms and the renin-angiotensin system. In Laragh, J. H. and Brenner, B. M. (eds.): Hypertension: Pathophysiology, Diagnosis, and Management. New York, Raven Press, 1994.
Hall, J. E., et al.: Abnormal pressure natriuresis: A cause or a consequence of hypertension? Hypertension 15:547, 1990.
Hall, J. E., et al.: Role of vasopressin and pressure diuresis in regulation of arterial pressure and body fluid volumes. In Cowley, A. W., Jr., et al. (eds.): Vasopressin, Cellular and Integrative Functions. New York, Raven Press, 1988.
Hall, J. E.: Control of sodium excretion by angiotensin: Intrarenal mechanisms and blood pressure regulation. Am. J. Physiol., 250:R960, 1986.
Hall, J. E.: Role of angiotensin and aldosterone in volume homeostasis. In Brenner, B. M., and Stein, J. H. (eds.): Body Fluid Homeostasis. Contemporary Issues in Nephrology, Vol. 17. New York, Churchill Livingston, 1986.
Honig, A.: Salt and water metabolism in acute high-altitude hypoxia: Role of peripheral arterial chemoreceptors. News Physiol. Sci., 4:109, 1989.

King, A. J.: Endothelins: Multifunctional peptides with potent vasoactive properties. In: Laragh, J. H. and Brenner, B. M. (eds): Hypertension: Pathophysiology, Diagnosis and Management, 2nd Ed. New York, Raven Press, 1995.

Knox, F. G., and Granger, J. P.: Control of sodium excretion: An integrative approach. In Windhager, E. E. (ed.): Handbook of Physiology, Section 8, Renal Physiology. New York, Oxford University Press, 1992.

Laragh, J. H., and Brenner, B. M.: Hypertension: Pathophysiology, Diagnosis and Management, 2nd Ed. Raven Press Ltd., New York, 1995.

Laragh, J. H., and Sealey, J. E.: Renin-angiotensin-aldosterone system and the renal regulation of sodium, potassium, and blood pressure homeostasis. In Windhager, E. E. (ed.): Handbook of Physiology, Section 8, Renal Physiology, Vol. II. New York, Oxford University Press, 1992.

Lohmeier, T. E., and Yang, H. M.: Preservation of renal function by angiotensin during chronic adrenergic stimulation. Hypertension, 17:278, 1991.

Manning, R. D., and Hu, L.: Nitric oxide regulates renal hemodynamics and urinary sodium excretion in dogs. Hypertension, 23:619, 1994.

Manning, R. E.: Effects of hypoproteinemia on blood volume and arterial pressure in volume-loaded dogs. Am. J. Physiol., 159:H1317, 1990.

Robertson, G. L., and Berl, T.: Pathophysiology of water metabolism. In Brenner, B. M., and Rector, F. C. Jr. (eds.): The Kidney, 4th Ed. Philadelphia, W. B. Saunders Co., 1986.

Robertson, G. L.: Regulation of vasopressin secretion. In Seldin, D. W., and Giebisch, G. (eds.): The Kidney: Physiology and Pathophysiology, 2nd Ed. New York, Raven Press, 1992.

Rossier, B. C., and Palmer, L. G.: Mechanisms of aldosterone action on sodium and potassium transport. In Seldin, D. W., and Giebisch, G. (eds.): The Kidney: Physiology and Pathophysiology, 2nd Ed. New York, Raven Press, 1992.

Schrier, R. W.: Pathogenesis of sodium and water retention in high-output and low output cardiac failure, nephrotic syndrome, cirrhosis, and pregnancy. N. Engl. J. Med., 319:106, 1988.

Seldin, D. W., and Giebisch, G. (eds.): The Regulation of Potassium Balance. New York, Raven Press, 1990.

Stanton, B. A., and Giebisch, G. H.: Renal potassium transport. In Windhager, F. E. (ed.): Handbook of Physiology, Section 8, Renal Physiology. New York, Oxford University Press, 1992.

Stanton, B. A.: Renal potassium transport: morphological and functional adaptations. Am. J. Physiol., 257:R989, 1989.

Suki, W. N., and Rouse, D.: Renal transport of calcium, magnesium. In Brenner B. M., and Rector, F. C., Jr. (eds.): The Kidney, 4th Ed. Philadelphia, W. B. Saunders Co., 1991.

Sutton, R. L., and Dirks, J. H.: Disturbances of calcium and magnesium metabolism. In Brenner, B. M., and Rector, F. C. Jr. (eds.): The Kidney, 4th Ed. Philadelphia, W. B. Saunders Co., 1986.

Tannen, R. L.: Disorders of potassium balance. In Brenner, B. M. and Rector, F. C. Jr. (eds.): The Kidney, 4th Ed. Philadelphia, W. B. Saunders Co., 1986.

Wilcox, C. S., et al.: Glomerular-tubular balance and proximal regulation. In Seldin, D. W., and Giebisch, G. (eds.): The Kidney: Physiology and Pathophysiology, 2nd Ed. New York, Raven Press, 1992.

Wilkins, F. C., Jr., et al.: Systemic hemodynamics and renal function during long-term pathophysiological increases in circulating endothelin. Am. J. Physiol., 268:R375, R381, 1995.

Young, D. B.: Analysis of long-term potassium regulation. Endocr. Rev., 6:24, 1985.

Young, D. B.: Potassium homeostasis and blood-pressure and sodium-volume regulation. In Laragh, J. H., and Brenner, B. M. (eds.): Hypertension: Pathophysiology, Diagnosis and Management. New York. Raven Press, 1990.

Young, D. B.: Quantitative analysis of aldosterone's role in potassium regulation. Am. J. Physiol., 255:F811, 1988.

Regulation of Acid-Base Balance

CHAPTER 30

Regulation of hydrogen ion balance is similar in some ways to the regulation of other ions in the body. For instance, to achieve homeostasis, there must be a balance between the intake or production of hydrogen ions and the net removal of hydrogen ions from the body. And, as is true for other ions, the kidneys play a key role in regulating hydrogen ion removal. However, precise control of extracellular fluid hydrogen ion concentration involves much more than simple elimination of hydrogen ions by the kidneys. There are also multiple acid-base buffering mechanisms involving the blood, cells, and lungs that are essential in maintaining normal hydrogen ion concentrations in both the extracellular and the intracellular fluid.

In this chapter, the various mechanisms that contribute to the regulation of hydrogen ion concentration are discussed, with special emphasis on the control of renal hydrogen ion secretion and renal reabsorption, production, and excretion of bicarbonate ions, one of the key components of acid-base control systems in the different body fluids.

Hydrogen Ion Concentration Is Precisely Regulated

Precise hydrogen ion regulation is essential because the activities of almost all enzyme systems in the body are influenced by hydrogen ion concentration. Therefore, changes in hydrogen concentration alter virtually all cell and body functions.

The hydrogen ion concentration of the body fluids normally is kept at a low level, compared with other ions. For example, the concentration of sodium in extracellular fluid (142 mEq/liter) is about 3.5 million times as great as the normal concentration of hydrogen ion, which averages only .00004 mEq/liter. Equally important, the normal variation in hydrogen ion concentration in extracellular fluid is only about one millionth as great as the normal variation in sodium ion concentration. Thus, the precision with which hydrogen ion is regulated emphasizes its importance to the various cell functions.

Acids and Bases—Their Definitions and Meanings

A hydrogen ion is a single free proton released from a hydrogen atom. Molecules containing hydrogen atoms that can release hydrogen ions in solutions are referred to as *acids*. An example is hydrochloric acid (HCl), which ionizes in water to form hydrogen ions (H^+) and chloride ions (Cl^-). Likewise, carbonic acid (H_2CO_3) ionizes in water to form H^+ and bicarbonate ions (HCO_3^-).

A *base* is an ion or a molecule that can accept a hydrogen ion. For example, the bicarbonate ion, HCO_3^-, is a base because it can combine with a hydrogen ion to form H_2CO_3. Likewise, $HPO_4^=$ is a base because it can accept a hydrogen ion to form $H_2PO_4^-$. The proteins in the body also function as bases because some of the amino acids that make up proteins have net negative charges that readily accept hydrogen ions. The protein hemoglobin in the red blood cells and proteins in the other cells of the body are among the most important of the body's bases.

The term "base" is often used synonymously with the term "alkali." An alkali is a molecule formed by combination of one or more of the alkaline metals—sodium, potassium, lithium, and so forth—with a highly basic ion such as a hydroxyl ion (OH^-). The basic portion of these molecules reacts quickly with hydrogen ions to remove them from solution and are, therefore, typical bases. For similar reasons, the term "alkalosis" refers to excess removal of hydrogen ions from the body fluids, in contrast to the excess addition of hydrogen ions, which is referred to as "acidosis."

STRONG AND WEAK ACIDS AND BASES. A strong acid is one that rapidly dissociates and releases especially large amounts of H^+ in solution. An example is HCl. Weak acids have less tendency to dissociate their ions and, therefore, release H^+ with less vigor. An example is H_2CO_3. A strong base is one that reacts rapidly and strongly with H^+ and, therefore, quickly removes these from a solution. A typical example is OH^-, which reacts with H^+ to form water (H_2O). A typical weak base is HCO_3^- because it binds with H^+ much more weakly than does OH^-. Most of the acids and bases in the extracellular fluid that are concerned with normal acid-base regulation are weak acids and bases. The most important ones that we discuss in detail are H_2CO_3 and bicarbonate base.

NORMAL HYDROGEN ION CONCENTRATION AND pH OF BODY FLUIDS AND CHANGES THAT OCCUR IN ACIDOSIS AND ALKALOSIS. As discussed above, the blood hydrogen ion concentration normally is maintained within tight limits around a normal value of about .00004 mEq/liter (40 nEq/liter). Normal variations are only about 3 to 5 nEq/liter, but under extreme conditions, the hydrogen ion concentration can vary from as low as 10 nEq/liter to as high as 160 nEq/liter without causing death.

Because hydrogen ion concentration normally is low and because these small numbers are cumbersome, it is customary to express hydrogen ion concentration on a logarithm scale, using pH units. pH is related to the actual hydrogen ion concentration by the following formula (hydrogen ion concentration $[H^+]$ is expressed in *equivalents* per liter):

$$pH = \log \frac{1}{[H^+]} = -\log [H^+]$$

For example, the normal $[H^+]$ is 40 mEq/liter (.00000004 Eq/liter). Therefore, the normal pH is:

$$pH = -\log [.00000004]$$
$$pH = 7.4$$

From this formula, one can see that pH is inversely related to the hydrogen ion concentration; therefore, a low pH corresponds to a high hydrogen ion concentration and a high pH corresponds to a low hydrogen concentration.

The normal pH of arterial blood is 7.4, whereas the pH of venous blood and interstitial fluids is about 7.35 because of the extra amounts of carbon dioxide (CO_2) released from the tissues to form H_2CO_3 in these fluids (Table 30–1). Because the normal pH of arterial blood is 7.4, a person is considered to have acidosis when the pH falls below this value and to have alkalosis when the pH rises above 7.4. The lower limit of pH at which a person can live more than a few hours is about 6.8, and the upper limit is about 8.0.

Intracellular pH usually is slightly lower than plasma pH because the metabolism of the cells produces acid,

Table 30–1 pH AND H^+ CONCENTRATION OF BODY FLUIDS

	H^+ Concentration mEq/L	pH
Extracellular fluid		
Arterial blood	4.0×10^{-5}	7.40
Venous blood	4.5×10^{-5}	7.35
Interstitial fluid	4.5×10^{-5}	7.35
Intracellular fluid	1×10^{-3} to 4×10^{-5}	6.0 to 7.4
Urine	3×10^{-2} to 1×10^{-5}	4.5 to 8.0
Gastric HCl	160	0.8

especially H_2CO_3. Depending on the type of cells, the pH of intracellular fluid has been estimated to range between 6.0 and 7.4. Hypoxia of the tissues and poor blood flow to the tissues can cause acid accumulation and, therefore, can decrease intracellular pH.

The pH of urine can range from 4.5 to 8.0, depending on the acid-base status of the extracellular fluid. As discussed later, the kidneys play a major role in correcting abnormalities of extracellular fluid hydrogen ion concentration by excreting acids or bases at variable rates.

An extreme example of an acidic body fluid is the HCl secreted into the stomach by the oxyntic (parietal cells) of the stomach mucosa, as discussed in Chapter 64. The hydrogen ion concentration in these cells is about 4 million times greater than the hydrogen concentration in blood, with a pH of 0.8.

In the remainder of this chapter, we discuss the regulation of extracellular fluid hydrogen ion concentration.

DEFENSES AGAINST CHANGES IN HYDROGEN ION CONCENTRATION: BUFFERS, LUNGS, AND KIDNEYS

There are three primary systems that regulate the hydrogen ion concentration in the body fluids to prevent acidosis or alkalosis: (1) *the chemical acid-base buffer systems of the body fluids,* which immediately combine with acid or base to prevent excessive changes in hydrogen ion concentration; (2) *the respiratory center,* which regulates the removal of CO_2 (and, therefore, H_2CO_3) from the extracellular fluid; and (3) *the kidneys,* which can excrete either acid or alkaline urine, thereby readjusting the extracellular fluid hydrogen ion concentration toward normal during acidosis or alkalosis.

When there is a change in hydrogen ion concentration, the *buffer systems* of the body fluids react within a fraction of a second to minimize these changes. Buffer systems do not eliminate hydrogen ions from the body or add them to the body but only keep them tied up until balance can be re-established. The second line of defense, the *respiratory system,* also acts within a few minutes to eliminate CO_2 and, therefore, H_2CO_3 from the body. These first two lines of defense keep the hydrogen ion concentration from changing too much until the more slowly responding third line of defense, the *kidneys*, can eliminate the excess acid or base from the body. Although the kidneys are relatively slow to respond, compared with the other defenses, they are over a period of hours to several days by far the most powerful of the acid-base regulatory systems.

BUFFERING OF HYDROGEN IONS IN THE BODY FLUIDS

A buffer is any substance that can reversibly bind hydrogen ions. The general form of the buffering reaction is:

$$Buffer + H^+ \rightleftarrows H\ Buffer$$

In this example, a free H^+ combines with the buffer to form a weak acid (H Buffer) that can either remain as an unassociated molecule or dissociate back to Buffer and H^+. When the hydrogen ion concentration increases, the reaction is forced to the right and more hydrogen ions bind to the buffer, as long as available buffer is present. Conversely, when the hydrogen ion concentration decreases, the reaction shifts toward the left and hydrogen ions are released from the buffer. In this way, changes in hydrogen ion concentration are minimized.

The importance of the body fluid buffers can be quickly realized if one considers the low concentration of hydrogen ion in the body fluids and the relatively large amounts of acids produced by the body each day. For example, about 80 milliequivalents of hydrogen is either ingested or produced each day by metabolism, whereas the hydrogen ion concentration of the body fluids normally is only about .00004 mEq/liter. Without buffering, the daily production and ingestion of acids would cause huge changes in body fluid hydrogen ion concentration.

The action of acid-base buffers can perhaps best be explained by considering the buffer system that is quantitatively the most important in the extracellular fluid—the bicarbonate buffer system.

THE BICARBONATE BUFFER SYSTEM

The bicarbonate buffer system consists of a water solution that contains two ingredients: (1) a weak acid, H_2CO_3, and (2) a bicarbonate salt, such as $NaHCO_3$.

H_2CO_3 is formed in the body by the reaction of CO_2 with H_2O:

$$CO_2 + H_2O \xrightleftharpoons[]{\text{carbonic anhydrase}} H_2CO_3$$

This reaction is slow, and exceedingly small amounts of H_2CO_3 are formed unless the enzyme *carbonic anhydrase* is present. This enzyme is especially abundant in the walls of the lung alveoli, where CO_2 is released; carbonic anhydrase is also present in the epithelial

cells of the renal tubules, where CO_2 reacts with H_2O to form H_2CO_3.

H_2CO_3 ionizes weakly to form small amounts of H^+ and HCO_3^-:

$$H_2CO_3 \rightleftarrows H^+ + HCO_3^-$$

The second component of the system, bicarbonate salt, occurs predominantly as sodium bicarbonate ($NaHCO_3$) in the extracellular fluid. $NaHCO_3$ ionizes almost completely to form bicarbonate ions (HCO_3^-) and sodium ions (Na^+) as follows:

$$Na\ HCO_3 \rightleftarrows Na^+ + HCO_3^-.$$

Now, putting the entire system together, we have the following:

$$CO_2 + H_2O \rightleftarrows H_2CO_3 \rightleftarrows H^+ + \underbrace{HCO_3^-}_{+\ Na^+}.$$

Because of the weak dissociation of H_2CO_3, the H^+ concentration is extremely small.

When a strong acid such as HCl is added to the bicarbonate buffer solution, the increased hydrogen ions released from the acid ($HCl \rightarrow H^+ + Cl^-$) are buffered by HCO_3^-:

$$\uparrow H^+ + HCO_3^- \rightarrow H_2CO_3 \rightarrow CO_2 + H_2O$$

As a result, more H_2CO_3 is formed, causing increased CO_2 and H_2O production. From these reactions, one can see that the hydrogen ions from the strong acid, HCl, react with HCO_3^- to form the very weak acid H_2CO_3, which in turn forms CO_2 and H_2O. The excess CO_2 greatly stimulates respiration, which eliminates the CO_2 from the extracellular fluid.

The opposite reactions take place when a strong base, such as sodium hydroxide (NaOH), is added to the bicarbonate buffer solution:

$$NaOH + H_2CO_3 \rightarrow NaHCO_3 + H_2O$$

In this case, the hydroxyl ion (OH^-) from the NaOH combines with H_2CO_3 to form additional HCO_3^-. Thus, the weak base $NaHCO_3$ replaces the strong base NaOH. At the same time, the concentration of H_2CO_3 decreases (because it reacts with NaOH), causing more CO_2 to combine with H_2O to replace the H_2CO_3.

$$CO_2 + H_2O \rightarrow H_2CO_3 \rightarrow \uparrow HCO_3^- + H^+$$
$$+ \qquad\qquad +$$
$$NaOH \qquad\qquad Na$$

The net result, therefore, is a tendency for the CO_2 levels in the blood to decrease; but the decreased CO_2 in the blood inhibits respiration and decreases the rate of CO_2 expiration. The rise in blood HCO_3^- that occurs is compensated for by increased renal excretion of HCO_3^-.

Quantitative Dynamics of the Bicarbonate Buffer System

All acids, including H_2CO_3, are ionized to some extent. From mass balance considerations, the concentrations of hydrogen ions and bicarbonate ions are proportional to the concentration of H_2CO_3.

$$H_2CO_3 \rightleftharpoons H^+ + HCO_3^-$$

For any acid, the concentration of the acid relative to its dissociated ions is defined by the *dissociation constant K′*:

$$K' = \frac{H^+ \times HCO_3^-}{H_2CO_3} \qquad (1)$$

This equation indicates that in an H_2CO_3 solution, the amount of free hydrogen ions is equal to:

$$H^+ = K' \times \frac{H_2CO_3}{HCO_3^-}. \qquad (2)$$

The concentration of undissociated H_2CO_3 cannot be measured in solution because it rapidly dissociates into CO_2 and H_2O or to H^+ and HCO_3^-. However, the CO_2 dissolved in the blood is directly proportional to the amount of undissociated H_2CO_3. Therefore, equation 2 can be rewritten as

$$H^+ = K \times \frac{CO_2}{HCO_3^-} \qquad (3)$$

The dissociation constant (K) for equation 3 is only about ¼₀₀ of the dissociation constant (K′) of equation 2 because the proportionality ratio between H_2CO_3 and CO_2 is 1 to 400.

Equation 3 is written in terms of the total amount of CO_2 dissolved in solution. However, most clinical laboratories measure the blood CO_2 tension (P_{CO_2}) rather than the actual amount of CO_2. Fortunately, the amount of CO_2 in the blood is a linear function of P_{CO_2} times the solubility coefficient for CO_2; under physiological conditions, the solubility coefficient for CO_2 is 0.03 mmol/mm Hg at body temperature. This means that 0.03 millimole of H_2CO_3 is present in the blood for each millimeter of mercury P_{CO_2} measured. Therefore, equation 3 can be rewritten as

$$H^+ = K \times \frac{(0.03 \times P_{CO_2})}{HCO_3^-} \qquad (4)$$

HENDERSON-HASSELBALCH EQUATION. As discussed above, it is customary to express hydrogen ion concentration in pH units rather than in actual concentrations. Recall that pH is defined as pH = $-\log H^+$.

The dissociation constant can be expressed in a similar manner:

$$pK = -\log K$$

Therefore, we can express the hydrogen ion concentration in equation 4 in pH units by taking the negative logarithm of that equation, which yields

$$-\log H^+ = -\log pK - \log \frac{(0.03 \times P_{CO_2})}{HCO_3^-}. \qquad (5)$$

Therefore,

$$pH = pK - \log \frac{(0.03 \times P_{CO_2})}{HCO_3^-}. \qquad (6)$$

Rather than work with a negative logarithm, we can change the sign of the logarithm and invert the numerator and denominator in the last term, using the law of logarithms to yield

$$pH = pK + \log \frac{HCO_3^-}{(0.03 \times P_{CO_2})}. \qquad (7)$$

For the bicarbonate buffer system, the pK is 6.1, and equation 7 can be written as

$$pH = 6.1 + \log \frac{HCO_3^-}{0.03 \times P_{CO_2}}. \qquad (8)$$

Equation 8 is the Henderson-Hasselbalch equation, and with it one can calculate the pH of a solution if the molar concentration of bicarbonate ion and the P_{CO_2} are known.

From the Henderson-Hasselbalch equation, it is apparent that an increase in bicarbonate ion concentration causes the pH to rise, shifting the acid-base balance toward alkalosis. And an increase in P_{CO_2} causes the pH to decrease, shifting the acid-base balance toward acidosis.

The Henderson-Hasselbalch equation, in addition to defining the determinants of normal pH regulation and acid-base balance in the extracellular fluid, provides insight into the physiological control of acid and base composition of the extracellular fluid. As discussed later, *the bicarbonate concentration is regulated mainly by the kidneys, whereas the P_{CO_2} in extracellular fluid is controlled by the rate of respiration.* By increasing the rate of respiration, the lungs remove CO_2 from the plasma, and by decreasing respiration, P_{CO_2} is elevated. Normal physiological acid-base homeostasis results from the coordinated efforts of both of these organs, the lungs and the kidneys, and acid-base disorders occur when one or both of these control mechanisms are impaired, thus altering either the bicarbonate concentration or the P_{CO_2} of extracellular fluid.

When disturbances of acid-base balance result from a primary change in extracellular fluid bicarbonate concentration, they are referred to as *metabolic* acid-base disorders. Therefore, acidosis caused by a primary decrease in bicarbonate concentration is termed *metabolic acidosis*, whereas alkalosis caused by a primary increase in bicarbonate concentration is called *metabolic alkalosis*. Acidosis caused by an increase in P_{CO_2} is called

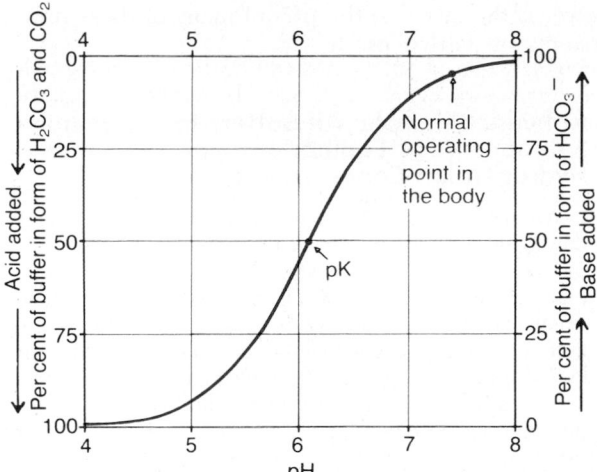

Figure 30–1. Titration curve for bicarbonate buffer system showing the pH of extracellular fluid when the percentages of buffer in the form of HCO_3^- and CO_2 (or H_2CO_3) are altered.

respiratory acidosis, whereas alkalosis caused by a decrease in P_{CO_2} is termed *respiratory alkalosis*.

BICARBONATE BUFFER SYSTEM TITRATION CURVE. Figure 30–1 shows the changes in pH of the extracellular fluid when the ratio of HCO_3^- to CO_2 in extracellular fluid is altered. When the concentrations of these two components are equal, the right-hand portion of equation 8 becomes the log of 1, which is equal to 0. Therefore, when the two components of the buffer system are equal, the pH of the solution is the same as the pK (6.1) of the bicarbonate buffer system. When base is added to the system, part of the dissolved CO_2 is converted into HCO_3^-, causing an increase in the ratio of HCO_3^- to CO_2 and increasing the pH, as is evident from the Henderson-Hasselbalch equation. When acid is added, it is buffered by HCO_3^-, which is then converted into dissolved CO_2, decreasing the ratio of HCO_3^- to CO_2 and decreasing the pH of the extracellular fluid.

"Buffer Power" Is Determined by the Amount and Relative Concentrations of the Buffer Components. From the titration curve in Figure 30–1, several points are apparent. First, the pH of the system is the same as the pK when each of the components (HCO_3^- and CO_2) constitutes 50 per cent of the total concentration of the buffer system. Second, the buffer system is most effective in the central part of the curve, where the pH is near the pK of the system. This means that the change in pH for any given amount of acid or base added to the system is least when the pH is near the pK of the system. The buffer system is still reasonably effective for 1.0 pH unit on either side of the pK, which for the bicarbonate buffer system extends from a pH of about 5.1 to 7.1 units. Beyond these limits, the buffering power rapidly diminishes. And when all the CO_2 has been converted into HCO_3^- or when all the HCO_3^- has been converted into CO_2, the system has no more buffering power.

In addition to the relative concentrations of the buffers, the absolute concentration of the buffers is an important factor in determining the buffer power of a system. With low concentrations of the buffers, only a small amount of acid or base added to the solution changes the pH considerably.

THE BICARBONATE BUFFER SYSTEM IS THE MOST IMPORTANT EXTRACELLULAR BUFFER. From the titration curve shown in Figure 30–1, one would not expect the bicarbonate buffer system to be powerful for two reasons: (1) The pH of the extracellular fluid is about 7.4, whereas the pK of the bicarbonate buffer system is 6.1. This means that there is about 20 times as much of the bicarbonate buffer system in the form of HCO_3^- as in the form of dissolved CO_2. For this reason, this system operates on the portion of the buffering curve where the slope is low and the buffering power is poor. (2) The concentrations of the two elements of the bicarbonate system, CO_2 and HCO_3^-, are not great.

Despite these characteristics, the bicarbonate buffer system is the most powerful extracellular buffer in the body. This apparent paradox is due mainly to the fact that the two elements of the buffer system, HCO_3^- and CO_2, are regulated, respectively, by the kidneys and the lungs, as discussed below. As a result of this regulation, the pH of the extracellular fluid can be precisely controlled by the relative rate of removal and addition of HCO_3^- by the kidneys and the rates of removal of CO_2 by the lungs.

The Phosphate Buffer System and Its Importance as an Intracellular and Renal Tubular Fluid Buffer

Although the phosphate buffer system is not of major importance as an extracellular fluid buffer, it plays a major role in buffering renal tubular fluid and intracellular fluids.

The main elements of the phosphate buffer system are $H_2PO_4^-$ and $HPO_4^=$. When a strong acid such as HCl is added to a mixture of these two substances, the hydrogen is accepted by the base $HPO_4^=$ and converted to $H_2PO_4^-$:

$$HCl + Na_2HPO_4 \rightarrow NaH_2PO_4 + NaCl$$

The result of this reaction is that the strong acid, HCl, is replaced by an additional amount of a weak acid, NaH_2PO_4, and the decrease in pH is minimized.

When a strong base, such as NaOH, is added to the buffer system, the OH^- is buffered by the $H_2PO_4^-$ to form additional amounts of $HPO_4^= + H_2O$:

$$NaOH + NaH_2PO_4 \rightarrow Na_2HPO_4 + H_2O$$

In this case, a strong base, NaOH, is traded for a weak base, Na_2HPO_4, causing only a slight increase in the pH.

The phosphate buffer system has a pK of 6.8, which is not far from the normal pH of 7.4 in the body fluids; this allows the system to operate near its maximum buffering power. However, its concentration in the extracellular fluid is low, only about 8 per cent of the concentration of the bicarbonate buffer. Therefore, the total buffering power of the phosphate system in the extracellular fluid is much less than that of the bicarbonate buffering system.

In contrast to its rather insignificant role as an extracellular buffer, *the phosphate buffer is especially important in the tubular fluids of the kidneys* for two reasons: (1) phosphate usually becomes greatly concentrated in the tubules, thereby increasing the buffering power of the phosphate system, and (2) the tubular fluid usually has a considerably lower pH than extracellular fluid, bringing the operating range of the buffer closer to the pK (6.8) of the system.

The phosphate buffer system is also important in buffering intracellular fluids because the concentration of phosphate in these fluids is many times that in the extracellular fluids. Also, the pH of intracellular fluids is lower than that of extracellular fluid and therefore usually closer to the pK of the phosphate buffer system, compared with the extracellular fluid.

Proteins Are Important Intracellular Buffers

Proteins are among the most plentiful buffers in the body because of their high concentrations, especially within the cells.

The pH of the cells, although slightly lower than in the extracellular fluid, nevertheless changes approximately in proportion to extracellular fluid pH changes. There is a slight amount of diffusion of hydrogen and bicarbonate ions through the cell membrane, although these ions require several hours to come to equilibrium with the extracellular fluid, except for rapid equilibrium that occurs in the red blood cells. CO_2, however, can rapidly diffuse through all the cell membranes. *This diffusion of the elements of the bicarbonate buffer system causes the pH in intracellular fluids to change when there are changes in extracellular pH.* For this reason, the buffer systems within the cells help to prevent changes in pH of extracellular fluids but may take several hours to become maximally effective.

In the red blood cell, hemoglobin is an important buffer as follows:

$$H^+ + Hb \leftrightarrows HHb$$

Experimental studies have shown that 60 to 70 per cent of the total chemical buffering of the body fluids is inside the cells, and most of this results from the intracellular proteins. However, except for the red blood cells, the slowness of movement of hydrogen ions and bicarbonate ions through the cell membranes often delays for several hours the maximum ability of the intracellular proteins to buffer extracellular acid-base abnormalities.

Another factor besides the high concentration of proteins in the cells that contributes to their buffering power is the fact that the pKs of many of these protein systems are fairly close to 7.4.

Isohydric Principle: All Buffers in a Common Solution Are in Equilibrium with the Same Hydrogen Ion Concentration

We have been discussing buffer systems as though they operated individually in the body fluids. However, they all work together because hydrogen ions are common to the reactions of all the systems. Therefore, whenever there is a change in hydrogen ion concentration in the extracellular fluid, the balance of all the buffer system changes at the same time. This phenomenon is called the *isohydric principle* and is illustrated by the following formula:

$$H^+ = K_1 \times \frac{HA_1}{A_1} = K_2 \times \frac{HA_2}{A_2} = K_3 \times \frac{HA_3}{A_3}$$

K_1, K_2, K_3, are the dissociation constants of three respective acids, HA_1, HA_2, HA_3, and A_1, A_2, A_3 are the concentrations of the free negative ions that constitute the bases of the three buffer systems.

An implication of this principle is that any condition that changes the balance of one of the buffer systems also changes the balance of all the others because the buffer systems actually buffer one another by shifting hydrogen ions back and forth from one to the other.

RESPIRATORY REGULATION OF ACID-BASE BALANCE

The second line of defense against acid-base disturbances is control of extracellular fluid CO_2 concentration by the lungs. In discussing the Henderson-Hasselbalch equation, we noted that an increase in P_{CO_2} of extracellular fluid decreases the pH, whereas a decrease in P_{CO_2} raises the pH. Therefore, by adjusting the P_{CO_2} either up or down, the lungs can effectively regulate the hydrogen ion concentration of the extracellular fluid. An increase in ventilation eliminates CO_2 from extracellular fluid, which, by mass action, reduces the hydrogen ion concentration. Conversely, decreased ventilation increases CO_2, thus also increasing hydrogen ion concentration in the extracellular fluid.

Pulmonary Expiration of CO_2 Balances Metabolic Formation of CO_2

CO_2 is formed continually in the body by intracellular metabolic processes. After it is formed, it diffuses from the cells into the interstitial fluids and blood, and the flowing blood transports it to the lungs, where it diffuses into the alveoli and then is transferred to the atmosphere by pulmonary ventilation. On the average, about 1.2 mol/liter of dissolved CO_2 normally is in the extracellular fluids, corresponding to a P_{CO_2} of 40 mm Hg.

If the rate of metabolic formation of CO_2 increases, the P_{CO_2} of the extracellular fluid is likewise increased. Conversely, decreased metabolic rate lowers the P_{CO_2}. If the rate of pulmonary ventilation is increased, CO_2 is blown off from the lungs and the P_{CO_2} in the extracellular fluid decreases. Therefore, changes in either pulmonary ventilation or the rate of CO_2 formation by the tissues can change the extracellular fluid P_{CO_2}.

Increasing Alveolar Ventilation Decreases Extracellular Fluid Hydrogen Ion Concentration and Raises pH

If the metabolic formation of CO_2 remains constant, then the only other factor that affects P_{CO_2} in extracellular fluid is the rate of alveolar ventilation. The higher the alveolar ventilation, the lower the P_{CO_2}, and conversely, the lower the alveolar ventilation rate, the higher the P_{CO_2}. As discussed previously, when CO_2 concentration increases, the H_2CO_3 concentration and hydrogen ion concentration also increase, thereby lowering extracellular fluid pH.

Figure 30–2 shows the approximate changes in pH in the blood that are caused by increasing or decreasing the rate of alveolar ventilation. Note that increasing alveolar ventilation to about twice normal raises the pH of the extracellular fluids by about 0.23. If the pH of the body fluids is 7.40 with normal alveolar ventilation, doubling the ventilation rate raises the pH to about 7.63. Conversely, a decrease in alveolar ventilation to one-fourth normal reduces the pH by 0.45. That is, if at a normal alveolar ventilation the pH is 7.4, reducing the ventilation to one-fourth normal reduces the pH to 6.95. Because the alveolar ventilation rate can change markedly, from as low as 0 to as high as 15 times normal, one can easily understand how much the pH of the body fluids can be changed by the respiratory system.

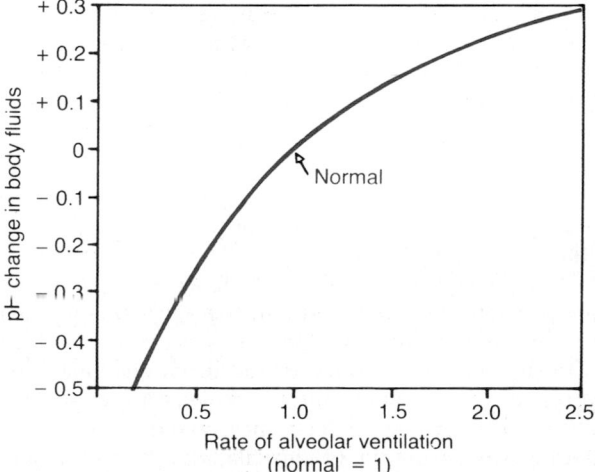

Figure 30–2. Change in extracellular fluid pH caused by increased or decreased rate of alveolar ventilation, expressed as times normal.

Increased Hydrogen Ion Concentration Stimulates Alveolar Ventilation

Not only does the alveolar ventilation rate influence hydrogen ion concentration by changing the P_{CO_2} of the body fluids, but the hydrogen ion concentration in turn affects the rate of alveolar ventilation. Thus, Figure 30–3 shows that alveolar ventilation rate increases four to five times the normal rate as pH decreases from the normal value of 7.4 to the strongly acidic value of 7.0. Conversely, when plasma pH rises above 7.4, this causes a decrease in ventilation rate. As one can see from the graph, the change in ventilation rate per unit pH change is much greater at reduced levels of pH (corresponding to elevated hydrogen ion concentration) compared with increased levels of pH (corresponding to decreased hydrogen ion concentration). The reason for this is that as alveolar ventilation rate decreases, owing to an increase in pH (decreased hydrogen ion concentration), the amount of oxygen added to the blood decreases and the partial pressure of oxygen (P_{O_2}) in the blood also decreases, which has an effect to stimulate ventilation rate. Therefore, the respiratory compensation for an increase in pH is not nearly as effective as the response to a marked reduction in pH.

FEEDBACK CONTROL OF HYDROGEN ION CONCENTRATION BY THE RESPIRATORY SYSTEM. Because increased hydrogen ion concentration stimulates respiration and because increased alveolar ventilation in turn decreases the hydrogen ion concentration, the respiratory system acts as a typical negative feedback controller of hydrogen ion concentration:

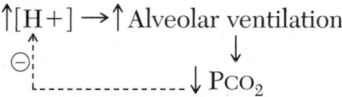

That is, whenever the hydrogen ion concentration increases above normal, the respiratory system is stimulated and alveolar ventilation increases. This decreases the P_{CO_2} in extracellular fluids and reduces hydrogen ion concentration back toward normal. Conversely, if hydrogen ion concentration falls below normal, the respiratory center becomes depressed, alveolar ventilation decreases, and hydrogen ion concentration increases back toward normal.

EFFICIENCY OF RESPIRATORY CONTROL OF HYDROGEN ION CONCENTRATION. Respiratory control cannot return the hydrogen ion concentration all the way back to normal when some disturbance outside the respiratory system has altered pH. Ordinarily, the respiratory mechanism for controlling hydrogen ion concentration has an effectiveness between 50 and 75 per cent, corresponding to a *feedback gain* of 1 to 3. That is, if the hydrogen ion concentration is suddenly increased by adding acid to the extracellular fluid and pH falls from 7.4 to 7.0, the respiratory system can return the pH to a value of about 7.2 to 7.3. This response occurs within 3 to 12 minutes.

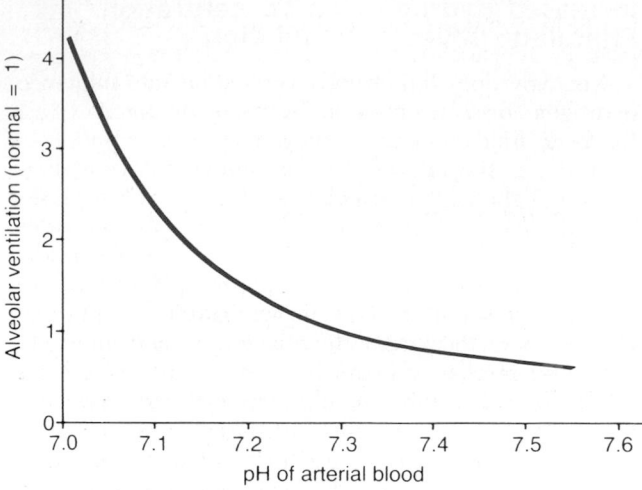

Figure 30–3. Effect of blood pH on the rate of alveolar ventilation. (Constructed from data obtained by Gray: *Pulmonary Ventilation and Its Regulation.* Springfield, IL, Charles C Thomas.)

BUFFERING POWER OF THE RESPIRATORY SYSTEM. *Respiratory regulation of acid-base balance is a physiological type of buffer system* because it acts rapidly and keeps the hydrogen ion concentration from changing too much until the much more slowly responding kidneys can eliminate the imbalance. In general, the overall buffering power of the respiratory system is one to two times as great as the buffering power of all other chemical buffers in the extracellular fluid combined. That is, one to two times as much acid or base can normally be buffered by this mechanism as by the chemical buffers.

We have discussed thus far the role of the *normal* respiratory mechanism as a means of buffering changes in hydrogen ion concentration. However, *abnormalities of respiration* can also cause changes in hydrogen ion concentration. For example, an impairment of lung function, such as severe emphysema, decreases the ability of the lungs to eliminate CO_2; this in turn causes a buildup of CO_2 in the extracellular fluid and a tendency toward *respiratory acidosis.* Also, the ability to respond to metabolic acidosis is impaired because the compensatory reductions in P_{CO_2} that would normally occur by means of increased ventilation are blunted. In these circumstances, the kidneys represent the sole remaining physiological mechanism for returning pH toward normal after the initial chemical buffering in the extracellular fluid has occurred.

RENAL CONTROL OF ACID-BASE BALANCE

The kidneys control acid-base balance by excreting either an acidic or a basic urine. Excreting an acidic urine reduces the amount of acid in extracellular fluid, whereas excreting a basic urine removes base from the extracellular fluids.

The overall mechanism by which the kidneys excrete acidic or basic urine is as follows: Large numbers of bicarbonate ions are filtered continuously into the tubules, and if they are excreted into the urine, this removes base from the blood. On the other hand, large numbers of hydrogen ions are also secreted into the tubular lumen by the tubular epithelial cells, thus removing acid from the blood. If more hydrogen ions are secreted than bicarbonate ions are filtered, there will be a net loss of acid from the extracellular fluids. Conversely, if more bicarbonate is filtered than hydrogen is secreted, there will be a net loss of base.

As discussed previously, the body produces each day about 80 milliequivalents of nonvolatile acids, mainly from the metabolism of proteins. These acids are called *nonvolatile* because they are not H_2CO_3 and, therefore, cannot be excreted by the lungs. The primary mechanism for removal of these acids from the body is by renal excretion. The kidneys must also prevent the loss of bicarbonate in the urine, a task that is quantitatively more important than the excretion of nonvolatile acids. Each day the kidneys filter about 4320 milliequivalents of bicarbonate (180 liters/day × 24 mEq/liter), and under normal conditions, almost all of this is reabsorbed from the tubules, thereby conserving the primary buffer system of the extracellular fluids.

As discussed later, the reabsorption of bicarbonate and the excretion of hydrogen ions are both accomplished through the process of hydrogen ion secretion by the tubule. Because the bicarbonate ion must react with a secreted hydrogen ion to form H_2CO_3 before it can be reabsorbed, 4320 milliequivalents of hydrogen ions must be secreted each day just to reabsorb the filtered bicarbonate. Then an additional 80 milliequivalents of hydrogen ions must be secreted to rid the body of the nonvolatile acids produced each day, for a total of 4400 milliequivalents of hydrogen ions secreted into the tubular fluid each day.

When there is a reduction in the extracellular fluid hydrogen ion concentration (alkalosis), the kidneys fail to reabsorb all the filtered bicarbonate, thereby increasing the excretion of bicarbonate. Because bicarbonate ions normally buffer hydrogen in the extracellular fluid, this loss of a bicarbonate is the same as adding a hydrogen ion to the extracellular fluid. Therefore, in alkalosis, the removal of bicarbonate ions raises the extracellular fluid hydrogen ion concentration back toward normal.

In acidosis, the kidneys do not excrete bicarbonate into the urine but reabsorb all the filtered bicarbonate and produce new bicarbonate, which is added back to the extracellular fluid. This reduces the extracellular fluid hydrogen ion concentration back toward normal.

Thus, the kidneys regulate extracellular fluid hydrogen ion concentrations through three basic mechanisms: (1) secretion of hydrogen ions, (2) reabsorption of filtered bicarbonate ions, and (3) production of new bicarbonate ions. All of these processes are accomplished through the same basic mechanism as discussed in the next few sections.

SECRETION OF HYDROGEN IONS AND REABSORPTION OF BICARBONATE IONS BY THE RENAL TUBULE

Hydrogen ion secretion and bicarbonate reabsorption occur in virtually all parts of the tubules except the descending and ascending thin limbs of the loop of Henle. Figure 30–4 summarizes bicarbonate reabsorption along the tubule. Keep in mind that for each bicarbonate reabsorbed, there must be a hydrogen ion secreted. About 80 to 90 per cent of the bicarbonate reabsorption (and hydrogen ion secretion) occurs in the proximal tubule, so that only a small amount of bicarbonate flows into the distal tubules and collecting ducts. In the thick ascending loop of Henle, another 10 per cent of the filtered bicarbonate is reabsorbed

and the remainder of the reabsorption takes place in the distal tubule and collecting duct. As discussed previously, the mechanism by which bicarbonate is reabsorbed also involves tubular secretion of hydrogen ions, but there are some differences in the way that different tubular segments accomplish this task.

Hydrogen Ions Are Secreted by Secondary Active Transport in the Early Tubular Segments

The epithelial cells of the proximal tubule, the thick segment of the ascending loop of Henle, and the distal tubule all secrete hydrogen ions into the tubular fluid by sodium-hydrogen counter-transport, as shown in Figure 30–5. This secondary active secretion of hydrogen ions is coupled with the transport of sodium into the cell at the luminal membrane, and the energy for hydrogen ion secretion against a concentration gradient is derived from the sodium gradient favoring sodium movement into the cell. This gradient is established by the sodium-potassium adenosine triphosphatase (ATPase) pump in the basolateral membrane. More than 90 per cent of the bicarbonate is reabsorbed in this manner, requiring about 3900 milliequivalents of hydrogen to be secreted each day by the tubules. This mechanism, however, does not establish a very high hydrogen ion concentration in the tubular fluid; the tubular fluid becomes very acidic only in the latter parts of the tubular system.

Figure 30–5 shows how the process of hydrogen ion secretion achieves bicarbonate reabsorption. The secretory process begins when CO_2 either diffuses into the tubular cells or is formed by metabolism in the

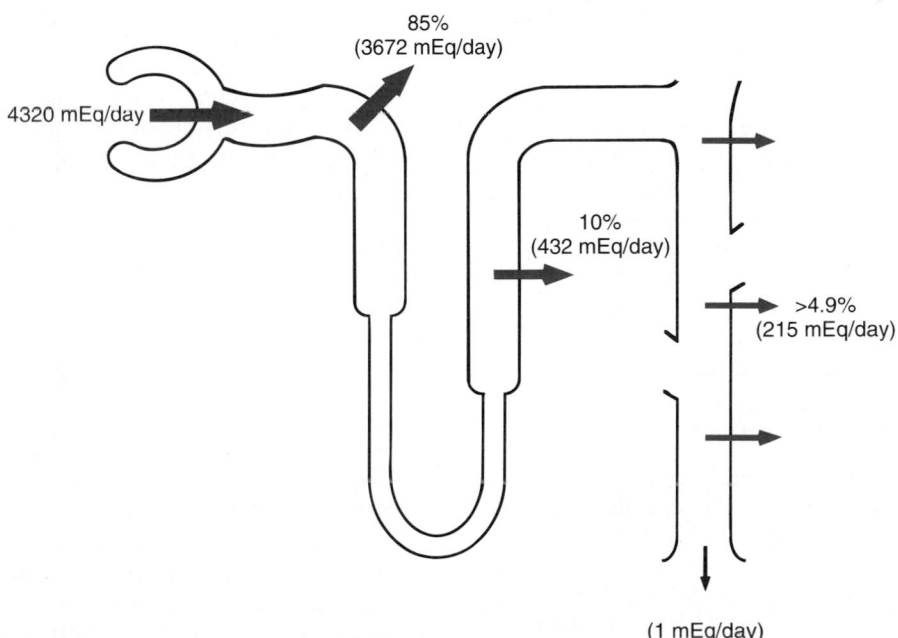

Figure 30–4. Reabsorption of bicarbonate in different segments of the renal tubule. The percentages of the filtered load of bicarbonate absorbed by the various tubular segments are shown as well as the number of milliequivalents reabsorbed per day under normal conditions.

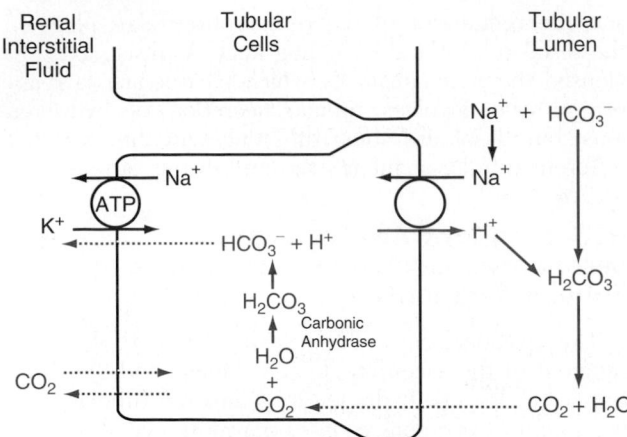

Figure 30–5. Cellular mechanisms for (1) active secretion of hydrogen ions into the renal tubule; (2) tubular reabsorption of bicarbonate by combination with hydrogen ions to form carbonic acid, which dissociates to form carbon dioxide and water; and (3) sodium ion reabsorption in exchange for the hydrogen ions secreted. This pattern of hydrogen ion secretion occurs in the proximal tubule.

tubular epithelial cells. CO_2, under the influence of the enzyme *carbonic anhydrase*, combines with H_2O to form H_2CO_3, which dissociates into HCO_3^- and H^+. The hydrogen ions are secreted from the cell into the tubular lumen by sodium-hydrogen counter-transport. That is, when sodium moves from the lumen of the tubule to the interior of the cell, it first combines with a carrier protein in the luminal border of the cell membrane; at the same time, a hydrogen ion in the interior of the cells combines with the carrier protein. The sodium moves into the cell down a concentration gradient that has been established by the sodium-potassium ATPase pump in the basolateral membrane. The gradient for sodium movement into the cell then provides the energy for moving hydrogen ions in the opposite direction from the interior of the cell to the tubular lumen.

The bicarbonate ion generated in the cell (when hydrogen ion dissociates from H_2CO_3) then moves downhill across the basolateral membrane into the renal interstitial fluid and the peritubular capillary blood. The net result is that for every hydrogen ion secreted into the tubular lumen, a bicarbonate ion enters the blood.

Filtered Bicarbonate Ions Are Reabsorbed by Interaction with Hydrogen Ions in the Tubules

Bicarbonate ions do not readily permeate the luminal membranes of the renal tubular cells; therefore, bicarbonate ions that are filtered by the glomerulus cannot be directly reabsorbed. Instead, bicarbonate is reabsorbed by a special process in which it first combines with hydrogen ions to form H_2CO_3, which eventually becomes CO_2 and H_2O, as shown in Figure 30–5.

This reabsorption of bicarbonate ions is initiated by a reaction in the tubules between bicarbonate ions

filtered at the glomerulus and hydrogen ions secreted by the tubular cells. The H_2CO_3 formed then dissociates into CO_2 and H_2O. The CO_2 can move easily across the tubular membrane; therefore, it instantly diffuses into the tubular cell, where it recombines with H_2O, under the influence of carbonic anhydrase, to generate a new H_2CO_3 molecule. This H_2CO_3 in turn dissociates to form bicarbonate ion and hydrogen ion; the bicarbonate ion then diffuses through the basolateral membrane into the interstitial fluid and is taken up into the peritubular capillary blood. *Thus, each time a hydrogen ion is formed in the tubular epithelial cells, a bicarbonate ion is also formed and released back into the blood.* The net effect of these reactions is a "reabsorption" of bicarbonate ions from the tubules, although the bicarbonate ions that actually enter the extracellular fluid are not the same ones that are filtered into the tubules.

BICARBONATE IONS ARE "TITRATED" AGAINST HYDROGEN IONS IN THE TUBULES. Under normal conditions, the rate of tubular hydrogen ion secretion is about 4400 mEq/day and the rate of filtration by bicarbonate ions is about 4320 mEq/day. Thus, the quantities of these two ions entering the tubules are almost equal, and they combine with each other to form CO_2 and H_2O. Therefore, it is said that bicarbonate ions and hydrogen ions normally "titrate" each other in the tubules.

The titration process is not quite exact because there is usually a slight excess of hydrogen ions in the tubules to be excreted in the urine. These excess hydrogen ions (about 80 mEq/day) rid the body of nonvolatile acids produced by metabolism. As discussed below, most of these hydrogen ions are not excreted as free hydrogen ions but rather in combination with other urinary buffers, especially phosphate and ammonia.

When there is an excess of bicarbonate ions over hydrogen ions in the urine, as occurs in metabolic alkalosis, the excess bicarbonate ions cannot be reabsorbed; therefore, the excess bicarbonate ions are left in the tubules and eventually excreted into the urine, which helps to correct the metabolic alkalosis.

In acidosis, there is an excess amount of hydrogen ions compared with bicarbonate ions, causing complete reabsorption of the bicarbonate, and the excess hydrogen ions pass into the urine. These excess hydrogen ions are buffered in the tubules by phosphate and ammonia and eventually excreted as salts. Thus, the basic mechanism by which the kidneys correct either acidosis or alkalosis is incomplete titration of hydrogen ions against bicarbonate ions, leaving one or the other of these to pass into the urine and, therefore, be removed from the extracellular fluid.

Primary Active Secretion of Hydrogen Ions in the Intercalated Cells of Late Distal Tubules and Collecting Ducts

Beginning in the late distal tubules and continuing through the remainder of the tubular system, the tubular epithelium secretes hydrogen ions by *primary*

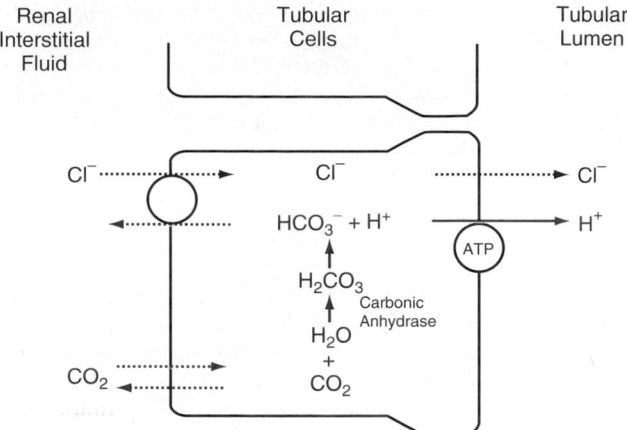

Figure 30–6. Primary active secretion of hydrogen ions through the luminal membrane of the epithelial cells of the distal and collecting tubules. Note that one bicarbonate is absorbed for each hydrogen ion secreted and a chloride ion is passively secreted along with the hydrogen ion. This pattern of hydrogen ion secretion occurs in the intercalated cells of the late distal tubules and collecting tubules.

active transport. The characteristics of this transport are different from those discussed for the proximal tubule and loop of Henle.

The mechanism for primary active hydrogen ion secretion is shown in Figure 30–6. It occurs at the luminal membrane of the tubular cell, where hydrogen ions are transported directly by a specific protein, a hydrogen-transporting ATPase. The energy required for pumping the hydrogen ions is derived from the breakdown of ATP to adenosine diphosphate.

Primary active secretion of hydrogen ions occurs in a special type of cell called the *intercalated cells* of the late distal tubules and in the collecting ducts. Hydrogen secretion in these cells is accomplished in two steps: (1) the dissolved CO_2 in this cell combines with H_2O to form H_2CO_3 and (2) the H_2CO_3 then dissociates into bicarbonate ions that are reabsorbed into the blood plus hydrogen ions that are secreted into the tubule by means of the hydrogen-ATPase mechanism. For each hydrogen ion secreted, a bicarbonate is reabsorbed, similar to the process in the proximal tubules. The main difference is that hydrogen moves across the luminal membrane by an active H^+ pump instead of by counter-transport, as occurs in the early parts of the nephron.

Although the secretion of hydrogen ions in the late distal tubule and collecting ducts accounts for only about 5 per cent of the total hydrogen ions secreted, this mechanism is important in forming a maximally acidic urine. In the proximal tubules, hydrogen ion concentration can be increased only about threefold to fourfold, although large *amounts* of hydrogen ion are secreted by this nephron segment. On the other hand, hydrogen ion concentration can be increased as much as 900-fold in the collecting ducts. This decreases the pH of the tubular fluid to about 4.5, which is the lower limit of pH that can be achieved in normal kidneys.

COMBINATION OF EXCESS HYDROGEN IONS WITH PHOSPHATE AND AMMONIA BUFFERS IN THE TUBULE—A MECHANISM FOR GENERATING NEW BICARBONATE IONS

When hydrogen ions are secreted in excess of the bicarbonate filtered into the tubular fluid, only a small part of the excess hydrogen ions can be excreted in the ionic form (H^+) in the urine. The reason for this is that the minimal urine pH is about 4.5, corresponding to a hydrogen ion concentration of $10^{-4.5}$ mEq/liter, or 0.03 mEq/liter. Thus, for each liter of urine formed, a maximum of only about 0.03 milliequivalent of free hydrogen ion can be excreted. To excrete the 80 milliequivalents of nonvolatile acid formed by metabolism each day, about 2667 liters of urine would have to be excreted if the hydrogen ions remained free in solution.

The excretion of large amounts of hydrogen ions (on occasion as much as 500 mEq/day) in the urine is accomplished primarily by combining the hydrogen ions with buffers in the tubular fluid. The most important buffers are phosphate buffer and ammonia buffer. There are other weak buffer systems, such as urate and citrate, that are much less important.

When hydrogen ions are titrated in the tubular fluid with bicarbonate, this results in the reabsorption of a bicarbonate ion for each hydrogen ion secreted, as discussed above. But when there are excess hydrogen ions in the urine, they combine with buffers other than bicarbonate, and this results in the generation of new bicarbonate ions that can also enter the blood. Thus, when there are excess hydrogen ions in the extracellular fluid, the kidneys not only reabsorb all the filtered bicarbonate but also generate new bicarbonate, thereby helping to replenish the bicarbonate lost from the extracellular fluid in acidosis. In the next two sections, the mechanisms by which phosphate and ammonia buffers contribute to the generation of new bicarbonate are discussed.

The Phosphate Buffer System Carries Excess Hydrogen Ions into the Urine and Generates New Bicarbonate

The phosphate buffer system is composed of $HPO_4^=$ and $H_2PO_4^-$. Both become concentrated in the tubular fluid because of their relatively poor reabsorption and because of the reabsorption of water from the tubular fluid. Therefore, although phosphate is not an important extracellular fluid buffer, it is much more effective as a buffer in the tubular fluid.

Another factor that makes phosphate important as a tubular buffer is the fact that the pK of this system is about 6.8. Under normal conditions, the urine is slightly acidic and the urine pH is near the pK of the phosphate buffer system. Therefore, in the tubules, the phosphate buffer system normally functions near its most effective range of pH.

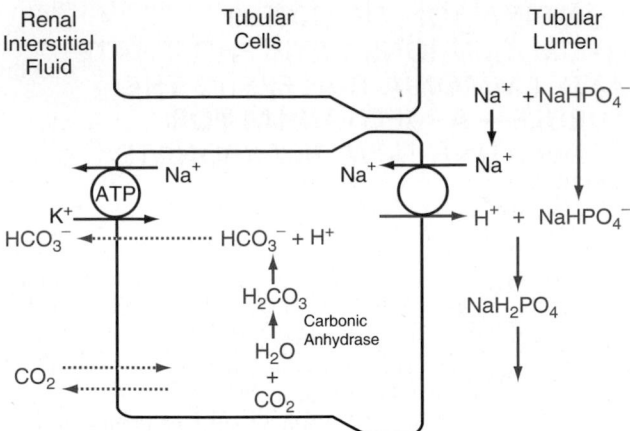

Figure 30–7. Buffering of secreted hydrogen ions by filtered phosphate ($NaHPO_4^-$). Note that a new bicarbonate is returned to the blood for each $NaHPO_4^-$ that reacts with a secreted hydrogen ion.

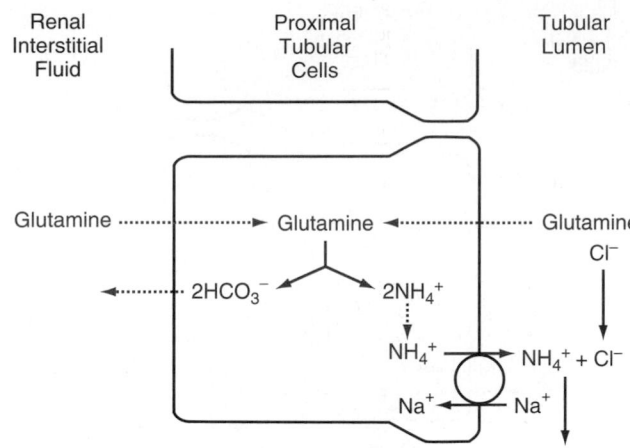

Figure 30–8. Production and secretion of ammonium (NH_4^+) by proximal tubular cells. Glutamine is metabolized in the cell, yielding NH_4^+ and bicarbonate. The ammonium ion (NH_4^+) is actively secreted into the lumen by means of a sodium-NH_4^+ pump. For each glutamine molecule metabolized, two NH_4^+ are produced and secreted and two HCO_3^- are returned to the blood.

Figure 30–7 shows the sequence of events by which hydrogen ions are excreted in combination with phosphate buffer and the mechanism by which new bicarbonate is added to the blood. The process of hydrogen ion secretion into the tubules is the same as described earlier. As long as there are excess bicarbonate ions in the tubular fluid, most of the secreted hydrogen ions combine with bicarbonate ions. However, once all the bicarbonate has been reabsorbed and is no longer available to bind with hydrogen ions, any excess hydrogen ions can combine with $HPO_4^=$ and other tubular buffers. After the hydrogen ion combines with $HPO_4^=$ to form $H_2PO_4^-$, it can be excreted as a sodium salt (NaH_2PO_4), carrying with it the excess hydrogen.

There is one important difference in this sequence of hydrogen ion excretion from that discussed above. In this case, the bicarbonate ion that is generated in the tubular cell and that enters the peritubular blood represents a net gain of bicarbonate by the blood, rather than merely a replacement of filtered bicarbonate. *Therefore, whenever a hydrogen ion secreted into the tubular lumen combines with a buffer other than bicarbonate, the net effect is addition of a new bicarbonate ion to the blood.* This demonstrates one of the mechanisms by which the kidneys are able to replenish the extracellular fluid stores of bicarbonate.

Under normal conditions, much of the filtered phosphate is reabsorbed, and only about 30 to 40 mEq/day is available for buffering hydrogen ions. Therefore, much of the buffering of excess hydrogen ions in the tubular fluid in acidosis occurs through the ammonia buffer system.

Excretion of Excess Hydrogen Ions and Generation of New Bicarbonate by the Ammonia Buffer System

A second special buffer system in the tubular fluid that is even more important quantitatively than the phosphate buffer system is composed of ammonia (NH_3) and the ammonium ion (NH_4^+). Ammonium ion is synthesized from glutamine, which is actively transported into the epithelial cells of the proximal tubules, thick ascending limb of the loop of Henle, and distal tubules (Fig. 30–8). Once inside the cell, each molecule of glutamine is metabolized to form two NH_4^+ and two HCO_3^- ions. The NH_4^+ is secreted into the tubular lumen by a counter-transport mechanism in exchange for sodium, which is reabsorbed. The HCO_3^- moves across the basolateral membrane along with the reabsorbed sodium ion (Na^+) into the interstitial fluid and is taken up by the peritubular capillaries. Thus, for each molecule of glutamine metabolized in the proximal tubules, two NH_4^+ ions are secreted into the urine and two HCO_3^- ions are reabsorbed into the blood. *The HCO_3^- generated by this process constitutes new bicarbonate.*

In the collecting tubules, the addition of NH_4^+ ions to the tubular fluids occurs through a different mechanism (Fig. 30–9). Here, hydrogen ion is secreted by the tubular membrane into the lumen, where it combines with ammonia (NH_3) to form NH_4^+, which is then excreted. The collecting ducts are permeable to NH_3, which can easily diffuse into the tubular lumen. However, the luminal membrane of this part of the tubules is much less permeable to NH_4^+; therefore, once the hydrogen ion has reacted with NH_3 to form NH_4^+, the NH_4^+ is trapped in the tubular lumen and eliminated in the urine. *For each NH_4^+ excreted, a new HCO_3^- is generated and added to the blood.*

Chronic Acidosis Increases NH_4^+ Excretion. One of the most important features of the renal ammonium-ammonia buffer system is that it is subject to physiological control. An increase in extracellular fluid hydrogen ion concentration stimulates renal glutamine metabolism and, therefore, increases the formation of NH_4^+ and new bicarbonate to be used in hydrogen

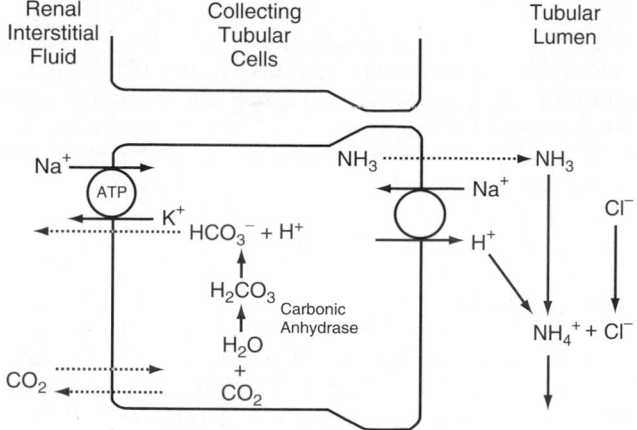

Figure 30–9. Buffering of hydrogen ion secretion by ammonia (NH_3) in the collecting tubules. Ammonia diffuses into the tubular lumen, where it reacts with secreted hydrogen ions to form NH_4^+, which is then excreted. For each NH_4^+ excreted, a new HCO_3^- is formed in the tubular cells and returned to the blood.

ion buffering; a decrease in hydrogen ion concentration has the opposite effect.

Under *normal conditions*, the amount of hydrogen ion eliminated by the ammonia buffer system accounts for about 50 per cent of the acid excreted and 50 per cent of the new bicarbonate generated by the kidneys. However, with *chronic acidosis*, the rate of NH_4^+ excretion increases markedly to as much as 500 mEq/day. *Therefore, with chronic acidosis, the dominant mechanism by which acid is eliminated is excretion of NH_4^+.* This also provides the dominant mechanism for generating new bicarbonate during chronic acidosis.

QUANTIFYING RENAL ACID-BASE EXCRETION

Based on the principles discussed above, we can quantitate the kidneys' net excretion of acid or net bicarbonate addition or elimination from the blood as follows.

Bicarbonate excretion is calculated as the urine flow rate multiplied by urinary bicarbonate concentration. This number indicates how rapidly the kidneys are removing bicarbonate ions from the blood (which is the same as adding a hydrogen ion to the blood). In alkalosis, the loss of bicarbonate ions (addition of hydrogen ions to the blood) helps to return the plasma pH toward normal.

The amount of new bicarbonate contributed to the blood at any given time is equal to the amount of hydrogen ions secreted that end up in the tubular lumen with non-bicarbonate urinary buffers. As discussed above, the primary sources of non-bicarbonate urinary buffers are NH_4^+ and phosphate. Therefore, the amount of bicarbonate added to the blood (and hydrogen ion excreted by NH_4^+) is calculated by measuring NH_4^+ excretion (urine flow rate multiplied by urinary NH_4^+ concentration).

The rest of the non-bicarbonate, non-NH_4^+ buffer excreted in the urine is measured by determining a value known as *titratable acid*. The amount of titratable acid in the urine is measured by titrating the urine with a strong base, such as NaOH, to a pH of 7.4, the pH of normal plasma and the pH of the glomerular filtrate. This titration reverses the events that occurred in the tubular lumen when the tubular fluid was titrated by excreted hydrogen ions. Therefore, the number of milliequivalents of NaOH required to return the urinary pH to 7.4 equals the number of milliequivalents of H^+ added to the tubular fluid that combined with phosphate and other organic buffers. The titratable acid measurement does not include hydrogen ions in association with NH_4^+ because the pK of the ammonia-ammonium reaction is 9.2, and titration with NaOH to a pH of 7.4 does not remove the hydrogen ions from NH_4^+.

Thus, the *net acid excretion* by the kidneys can be assessed as:

$$\text{Net acid excretion} = NH_4^+ \text{ excretion} \\ + \text{Urinary titratable acid} - \text{Bicarbonate excretion}$$

The reason we subtract bicarbonate excretion is that loss of bicarbonate ions is the same as adding hydrogen ions to the blood. To maintain acid-base balance, the net acid excretion must equal the nonvolatile acid production in the body. In acidosis, the net acid excretion increases markedly, especially because of increased NH_4^+ excretion, thereby removing acid from the blood. The net acid excretion also equals the rate of net bicarbonate addition to the blood. *Therefore, in acidosis, there is a net addition of bicarbonate back to the blood as more NH_4^+ and urinary titratable acid are excreted.*

In alkalosis, titratable acid and NH_4^+ excretion drops to 0, whereas HCO_3^- excretion increases. *Therefore, in alkalosis, there is a negative net acid secretion.* This means that there is a net loss of bicarbonate from the blood (which is the same as adding hydrogen ion to the blood) and that no new bicarbonate is generated by the kidneys.

Regulation of Renal Tubular Hydrogen Ion Secretion

As discussed above, hydrogen ion secretion by the tubular epithelium is necessary for both bicarbonate reabsorption and generation of new bicarbonate associated with titratable acid formation. Therefore, the rate of hydrogen ion secretion must be carefully regulated if the kidneys are to perform effectively their functions in acid-base homeostasis. Under normal conditions, the kidney tubules must secrete at least enough hydrogen ion to reabsorb almost all the bicarbonate that is filtered, and there must be enough hydrogen ions left over to be excreted as titratable acid or NH_4^+ to rid the body of the nonvolatile acids produced each day from metabolism.

In alkalosis, tubular secretion of hydrogen ion must

be reduced to a level that is too low to achieve complete bicarbonate reabsorption, enabling the kidneys to increase bicarbonate excretion. In this condition, titratable acid and ammonia are not excreted because there are no excess hydrogen ions available to combine with non-bicarbonate buffers; therefore, there is no new bicarbonate added to the urine in alkalosis. During acidosis, on the other hand, the tubular hydrogen ion secretion must be increased sufficiently to reabsorb all the filtered bicarbonate and still have enough hydrogen ions left over to excrete large amounts of NH_4 and titratable acid, thereby contributing large amounts of new bicarbonate ions to the blood.

The most important stimuli for increasing hydrogen ion secretion by the tubules in acidosis are (1) an increase in P_{CO_2} of the extracellular fluid and (2) an increase in hydrogen ion concentration of the extracellular fluid (decreased pH).

The tubular cells respond directly to an increase in P_{CO_2} of the blood, as occurs in respiratory acidosis, with an increase in the rate of hydrogen ion secretion as follows: The increased P_{CO_2} raises the P_{CO_2} of the tubular cells, causing increased formation of H^+ in the tubular cells, which in turn stimulates the secretion of hydrogen ions. The second factor that stimulates hydrogen ion secretion is an increase in extracellular fluid hydrogen ion concentration (decreased pH).

A special factor that can increase hydrogen ion secretion under some pathophysiological conditions is excessive aldosterone secretion. Aldosterone stimulates the secretion of hydrogen ions by the intercalated cells of the collecting duct. Therefore, oversecretion of aldosterone, as occurs in Conn's syndrome, can cause excessive secretion of hydrogen ion into the tubular fluid and, consequently, increased amounts of bicarbonate added back to the blood. This in turn usually causes alkalosis in patients with excessive aldosterone secretion.

In alkalosis, there is reduction of hydrogen ion secretion. This can occur as a result of a decreased extracellular P_{CO_2}, as occurs in respiratory alkalosis, or it can occur as a result of a decrease in hydrogen ion concentration per se, as occurs in both respiratory and metabolic alkalosis.

RENAL CORRECTION OF ACIDOSIS— INCREASED EXCRETION OF HYDROGEN IONS AND ADDITION OF BICARBONATE IONS TO THE EXTRACELLULAR FLUID

Now that we have described the mechanisms by which the kidneys secrete hydrogen ions and reabsorb bicarbonate ions, we can explain how the kidneys readjust the pH of the extracellular fluids when it becomes abnormal.

Referring to equation 8, the Henderson-Hasselbalch equation, we can see that acidosis occurs when the ratio of HCO_3^- to CO_2 in the extracellular fluid de-

creases, thereby causing a decrease in pH. If this ratio decreases because of a fall in HCO_3^-, the acidosis is referred to as *metabolic acidosis*. If the pH falls because of an increase in P_{CO_2}, the acidosis is referred to as *respiratory acidosis*.

Regardless of whether the acidosis is respiratory or metabolic, both conditions cause a decrease in the ratio of bicarbonate to hydrogen ions in the renal tubular fluid. As a result, there is an excess of hydrogen ions in the renal tubules, causing complete reabsorption of bicarbonate ions and still leaving additional hydrogen ions available to combine with the urinary buffers, NH_4^+ and $HPO_4^=$. Thus, in acidosis, the kidneys reabsorb all the filtered bicarbonate and contribute new bicarbonate through the formation of NH_4^+ and titratable acid.

In metabolic acidosis, an excess of hydrogen ions over bicarbonate ions occurs in the tubular fluid primarily because of decreased filtration of bicarbonate ions. This decreased filtration of bicarbonate ions is caused mainly by a decrease in the extracellular fluid concentration of bicarbonate, an effect that occurs in metabolic acidosis, as discussed earlier. *In respiratory acidosis, the excess hydrogen ions in the tubular fluid is due mainly to the rise in extracellular fluid P_{CO_2}, which stimulates hydrogen ion secretion.*

As discussed previously, with chronic acidosis, regardless of whether it is respiratory or metabolically mediated, there is an increase in the production of NH_4^+, which further contributes to the excretion of hydrogen ions and the addition of new bicarbonate ions to the extracellular fluid. With severe chronic acidosis, as much as 500 mEq/day of hydrogen ions can be excreted in the urine, mainly in the form of NH_4^+; this in turn contributes up to 500 mEq/day of new bicarbonate that is added to the blood.

Thus, with chronic acidosis, the increased secretion of hydrogen ions by the tubules helps to eliminate excess hydrogen ions from the body and increases the quantity of bicarbonate ions in the extracellular fluid. This increases the bicarbonate part of the bicarbonate buffer system, which, in accordance with the Henderson-Hasselbalch equation, helps to raise the extracellular pH and corrects the acidosis. If the acidosis is metabolically mediated, additional compensation by the lungs causes a reduction in P_{CO_2}, also helping to correct the acidosis.

Table 30–2 summarizes the characteristics associated with respiratory and metabolic acidosis as well as with respiratory and metabolic alkalosis, which are discussed in the next section. Note that *in respiratory acidosis, there is a reduction in pH, an increase in extracellular fluid hydrogen ion concentration, and an increase in P_{CO_2}, which is the initial cause of the acidosis. The compensatory response is an increase in plasma HCO_3^-, caused by addition of new bicarbonate to the extracellular fluid by the kidneys.* The rise in HCO_3^- helps to offset the increase in P_{CO_2}, thereby returning the plasma pH toward normal.

In *metabolic acidosis*, there is also a decrease in pH and a rise in extracellular fluid hydrogen ion concen-

Table 30–2 CHARACTERISTICS OF PRIMARY ACID-BASE DISTURBANCES

	pH	H⁺	PCO₂	HCO₃⁻
	pH	H^+	P_{CO_2}	HCO_3^-
Normal	7.4	40 nEq/L	40 mm Hg	24 mEq/L
Respiratory acidosis	↓	↑	⇑	↑
Respiratory alkalosis	↑	↓	⇓	↓
Metabolic acidosis	↓	↑	↓	⇓
Metabolic alkalosis	↑	↓	↑	⇑

The primary event is indicated by the double arrows (⇑ or ⇓). Note that respiratory acid-base disorders are initiated by an increase or a decrease in P_{CO_2}, whereas metabolic disorders are initiated by an increase or decrease in HCO_3^-.

tration. However, in this case, the primary abnormality is a decrease in plasma HCO_3^-. *The primary compensations include increased ventilation rate, which reduces P_{CO_2}, and renal compensation, which, by adding new bicarbonate to the extracellular fluid, helps to minimize the initial fall in extracellular HCO_3^- concentration.*

RENAL CORRECTION OF ALKALOSIS— DECREASED TUBULAR SECRETION OF HYDROGEN IONS AND INCREASED EXCRETION OF BICARBONATE IONS

The compensatory responses to alkalosis are basically opposite to those that occur in acidosis. In alkalosis, the ratio of HCO_3^- to CO_2 in the extracellular fluid increases, causing a rise in pH (a decrease in hydrogen ion concentration), as is evident from the Henderson-Hasselbalch equation. Regardless of whether the alkalosis is caused by metabolic or respiratory abnormalities, there is still an increase in the ratio of bicarbonate ions to hydrogen ions in the renal tubular fluid. The net effect of this is an excess of bicarbonate ions that cannot be reabsorbed from the tubules and are, therefore, excreted in the urine. Thus, in alkalosis, bicarbonate is removed from the extracellular fluid by renal excretion, which has the same effect as adding a hydrogen ion to the extracellular fluid. This helps to return the hydrogen ion concentration and pH back toward normal.

Table 30–2 also shows the overall characteristics of respiratory and metabolic alkalosis. In *respiratory alkalosis*, there is an increase in extracellular fluid pH and a decrease in hydrogen ion concentration. *The cause of the alkalosis is a decrease in plasma P_{CO_2}, caused by hyperventilation.* The reduction in P_{CO_2} then leads to a decrease in the rate of hydrogen ion secretion by the renal tubules. The decrease in hydrogen ion secretion reduces the amount of hydrogen ions in the renal tubular fluid. Consequently, there are not enough hydrogen ions to react with all the HCO_3^- that is filtered. Therefore, the HCO_3^- that cannot react with hydrogen ions is not reabsorbed and is ex-

creted in the urine. This results in a decrease in plasma HCO_3^- concentration and correction of the alkalosis. *Therefore, the compensatory response to a primary reduction in P_{CO_2} in respiratory alkalosis is a reduction in plasma bicarbonate concentration, caused by increased renal excretion of bicarbonate.*

In *metabolic alkalosis*, there is also an increase in plasma pH and a decrease in hydrogen ion concentration. *The cause of metabolic alkalosis, however, is a rise in the extracellular fluid bicarbonate ion concentration.* This is partly compensated for by a reduction in respiration rate, which increases P_{CO_2} and helps to return the extracellular fluid pH toward normal. In addition, the increase in bicarbonate concentration in the extracellular fluid leads to an increase in the filtered load of bicarbonate, which in turn causes an excess of bicarbonate ions over hydrogen ions secreted in the renal tubular fluid. The excess bicarbonate ions in the tubular fluid fail to be reabsorbed because they do not have hydrogen ions to react with, and therefore they are excreted in the urine. In *metabolic alkalosis, the primary compensations are decreased ventilation, which raises P_{CO_2}, and increased renal bicarbonate ion excretion, which helps to compensate for the initial rise in extracellular fluid bicarbonate ion concentration.*

CLINICAL CAUSES OF ACID-BASE DISORDERS

Respiratory Acidosis Is Caused by Decreased Ventilation and Increased PCO₂

From the previous discussion, it is obvious that any factor that decreases the rate of pulmonary ventilation also increases the P_{CO_2} of extracellular fluid. This causes an increase in H_2CO_3 and hydrogen ion concentration, thus resulting in acidosis. Because the acidosis is caused by an abnormality in respiration, it is called *respiratory acidosis*.

Respiratory acidosis occurs frequently from pathological conditions that damage the respiratory centers or that decrease the ability of the lungs to eliminate CO_2. For example, damage to the respiratory center in the medulla oblongata can lead to respiratory acidosis. Also, obstruction of the passageways of the respiratory tract, pneumonia, or decreased pulmonary membrane surface area as well as any factor that interferes with the exchange of the gases between the blood and the alveolar air can cause respiratory acidosis.

In respiratory acidosis, the compensatory responses available are (1) the buffers of the body fluids and (2) the kidneys, which require several days to compensate for the disorder.

Respiratory Alkalosis Results from Increased Ventilation and Decreased PCO₂

Respiratory alkalosis is caused by overventilation by the lungs. Rarely does this occur because of physical pathological conditions. However, a psychoneurosis can occasionally cause overbreathing to the extent that a person becomes alkalotic. Also, a physiological type of respiratory alkalosis occurs when a person ascends to high altitude. The low oxygen content of the air stimulates respiration, which causes excess loss of CO_2 and development of mild respiratory alkalosis. Again, the major means for compensation are the chemical buffers of the body fluids and the ability of the kidneys to increase bicarbonate excretion.

Metabolic Acidosis Results from Decreased Extracellular Fluid Bicarbonate Concentration

The term *metabolic acidosis* refers to all other types of acidosis besides those caused by excess CO_2 in the body fluids. Metabolic acidosis can result from several general causes: (1) failure of the kidneys to excrete metabolic acids normally formed in the body, (2) formation of excess quantities of metabolic acids in the body, (3) addition of metabolic acids to the body by ingestion or infusion of acids, and (4) loss of base from the body fluids, which has the same effect as adding an acid to the body fluids. Some specific conditions that cause metabolic acidosis are the following.

RENAL TUBULAR ACIDOSIS. This type of acidosis results from a defect in renal excretion of hydrogen ion or in reabsorption of bicarbonate, or both. These disorders are generally of two types: (1) impairment of renal tubular bicarbonate reabsorption, causing the loss of bicarbonate in the urine, or (2) inability of the renal tubular hydrogen secretory mechanism to establish a normal acidic urine, causing the excretion of an alkaline urine. In these cases, inadequate amounts of titratable acid and NH_4^+ are excreted, so that there is net accumulation of acid in the body fluids.

DIARRHEA. Severe diarrhea is probably the most frequent cause of metabolic acidosis. The cause of this acidosis is the loss of large amounts of sodium bicarbonate into the feces. The gastrointestinal secretions normally contain large amounts of bicarbonate, and diarrhea results in the loss of these bicarbonate ions from the body, which has the same effect as losing large amounts of bicarbonate in the urine. This form of metabolic acidosis is particularly serious and can be a cause of death, especially in young children.

VOMITING. Vomiting of gastric contents alone would cause loss of acid and a tendency toward alkalosis because the stomach secretions are highly acidic. However, vomiting of large amounts of the contents from deeper in the gastrointestinal tract, which sometimes occurs, causes loss of bicarbonate and results in metabolic acidosis in the same way that diarrhea causes acidosis.

DIABETES MELLITUS. Diabetes mellitus is caused by lack of insulin secretion by the pancreas, which in turn prevents the normal use of glucose for metabolism. Instead, some of the fats are split into acetoacetic acid, and this is metabolized by the tissues for energy in place of glucose. With severe diabetes mellitus, blood acetoacetic acid levels can rise very high, thus causing severe metabolic acidosis. In an attempt to compensate for this acidosis, large amounts of acid are excreted in the urine, sometimes as much as 500 mmol/day.

INGESTION OF ACIDS. Rarely are large amounts of acids ingested in normal foods. However, severe metabolic acidosis occasionally can result from poisoning as a result of the ingestion of certain acidic poisons. Some of these include acetylsalicylics (aspirin) and methyl alcohol (which when metabolized forms formic acid).

CHRONIC RENAL FAILURE. When kidney function declines markedly, there is a buildup of the anions of weak acids in the body fluids that are not being excreted by the kidneys. In addition, the decreased glomerular filtration rate reduces the excretion of phosphates and NH_4^+, which reduces the amount of bicarbonate added back to the body fluids. Thus, chronic renal failure can be associated with severe metabolic acidosis.

Metabolic Alkalosis Is Caused by Increased Extracellular Fluid Bicarbonate Concentration

When there is excess retention of bicarbonate or loss of hydrogen ion from the body, this results in metabolic alkalosis. Metabolic alkalosis is not nearly as common as metabolic acidosis, but some of the causes of metabolic alkalosis are as follows.

ALKALOSIS CAUSED BY ADMINISTRATION OF DIURETICS (EXCEPT THE CARBONIC ANHYDRASE INHIBITORS). All diuretics cause increased flow of fluid along the tubules, usually causing increased flow in the distal and collecting tubules. This in turn leads to increased reabsorption of sodium ions from these parts of the nephrons. Because the sodium reabsorption here is coupled with hydrogen ion secretion, the enhanced sodium reabsorption also leads to an increase in hydrogen ion secretion and an increase in bicarbonate reabsorption. These changes in turn lead to the development of alkalosis, characterized by increased extracellular fluid bicarbonate concentration.

EXCESS ALDOSTERONE CAUSES METABOLIC ALKALOSIS. When large amounts of aldosterone are secreted by the adrenal glands, a mild metabolic alkalosis develops. As discussed previously, aldosterone promotes extensive reabsorption of sodium ions from the distal and collecting tubules and at the same time stimulates the secretion of hydrogen ions by the intercalated cells of the collecting tubules. This increased secretion of hydrogen ions leads to increased excretion of hydrogen ions by the kidneys and, therefore, metabolic alkalosis.

VOMITING OF GASTRIC CONTENTS CAUSES METABOLIC ALKALOSIS. Vomiting of the gastric contents alone, without vomiting the lower gastrointestinal contents, causes loss of the HCl secreted by the stomach mucosa. The net result is a loss of acid from the extracellular fluid and development of metabolic alkalosis. This type of alkalosis occurs especially in neonates who have pyloric obstruction caused by hypertrophied pyloric sphincter muscles.

METABOLIC ALKALOSIS CAUSED BY INGESTION OF ALKALINE DRUGS. One of the common causes of metabolic alkalosis is ingestion of alkaline drugs, such as sodium bicarbonate, for the treatment of gastritis or peptic ulcer.

TREATMENT OF ACIDOSIS OR ALKALOSIS

The best treatment for acidosis or alkalosis is to correct the condition that has caused the abnormality. This often can be difficult, especially in chronic disease states that cause impaired lung function or kidney failure. In these circumstances, various agents can be used to neutralize the excess acid or base in the extracellular fluid.

To neutralize excess acid, large amounts of *sodium bicarbonate* can be ingested by mouth. The sodium bicarbonate is absorbed from the gastrointestinal tract into the blood and increases the bicarbonate portion of the bicarbonate buffer system, thereby increasing pH toward normal. Sodium bicarbonate can also be infused intravenously, but because of the potentially dangerous physiological effects of such treatment, other substances are often used instead, such as *sodium lactate* and *sodium gluconate*. The lactate and gluconate portions of the molecules are metabolized in the body, leaving the sodium in the extracellular fluids in the form of sodium bicarbonate and thereby increasing the pH of the fluids toward normal.

For the treatment of alkalosis, ammonium chloride can be administered by mouth. When the ammonium chloride is absorbed into the blood, the ammonia portion is converted by the liver into urea. This reaction liberates HCl, which immediately reacts with the buffers of the body fluids to shift the hydrogen ion concentration in the acidic direction. Ammonium chloride occasionally is infused intravenously, but the ammonium ion is also highly toxic, and this procedure can be dangerous. Another substance used occasionally is *lysine monohydrochloride*.

CLINICAL MEASUREMENTS AND ANALYSIS OF ACID-BASE DISORDERS

Appropriate therapy of acid-base disorders requires proper diagnosis. For the simple acid-base disorders described above, one can make a diagnosis from an analysis of three measurements from an arterial blood sample: the pH, the plasma bicarbonate concentration, and the P_{CO_2}.

The diagnosis of simple acid-base disorders involves several steps, as shown in Figure 30–10. By examining the pH, one can determine whether the disorder is acidosis or alkalosis. A pH less than 7.4 indicates acidosis, whereas a pH greater than 7.4 indicates alkalosis.

The second step is to examine the plasma P_{CO_2} and bicarbonate concentration. The normal value for P_{CO_2} is 40 mm Hg and for bicarbonate, 24 mEq/liter. If the disorder has been characterized as acidosis and the plasma P_{CO_2} is increased, then there must be a respiratory component to the acidosis. After renal compensation, the plasma bicarbonate concentration in respiratory acidosis would tend to increase above normal. *Therefore, the expected values for a simple respiratory acidosis would be reduced plasma pH, increased P_{CO_2}, and increased plasma bicarbonate concentration after partial renal compensation.*

For metabolic acidosis, there would also be a decrease in plasma pH. However, with metabolic acidosis, the primary abnormality is a decrease in plasma bicarbonate concentration. Therefore, if a low pH is associated with a low bicarbonate concentration, there must be a metabolic component to the acidosis. In simple metabolic acidosis, the P_{CO_2} is reduced because of partial respiratory compensation, in contrast to respiratory acidosis in which P_{CO_2} is increased. *Therefore, in metabolic acidosis, one would expect a low pH, a low plasma bicarbonate concentration, and a reduction in P_{CO_2} after partial respiratory compensation.*

The procedures for categorizing the types of alkalosis involve the same basic steps. First, alkalosis implies that there is an increase in plasma pH. If the increase in pH is associated with decreased P_{CO_2}, then there must be a respiratory component to the alkalosis. If, on the other hand, the rise in pH is associated with increased HCO_3^-, there must be a metabolic component to the alkalosis. *Therefore, with simple respiratory alkalosis, one would expect to find the following plasma values relative to normal: increased pH, decreased P_{CO_2}, and decreased plasma HCO_3^- concentration. In simple metabolic alkalosis, one would expect to find increased pH, increased plasma HCO_3^-, and increased P_{CO_2}.*

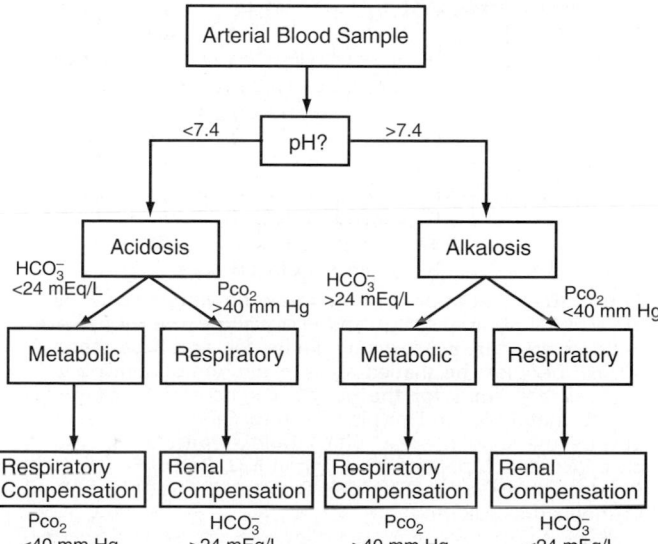

Figure 30–10. Analysis of simple acid-base disorders. If the compensatory responses are markedly different than those shown at the bottom of the figure, one should expect a mixed acid-base disorder.

Complex Acid-Base Disorders and the Use of the Acid-Base Nomogram for Diagnosis

In some instances, acid-base disorders are not accompanied by proper compensatory responses. When this occurs, the abnormality is referred to as a *mixed acid-base disorder*. This means that there are two or more underlying causes for the acid-base disturbance. For example, a patient with low pH would be categorized as acidotic. If the disorder was metabolically mediated, this would also be accompanied by a low plasma bicarbonate concentration and, after appropriate respiratory compensation, a low P_{CO_2}. However, if the low plasma pH and low bicarbonate concentration are associated with elevated P_{CO_2}, one would suspect that there may also be a respiratory component to the acidosis as well as a metabolic component. Therefore, this disorder would be categorized as a mixed acidosis. This could occur, for example, in a patient with acute bicarbonate loss from the gastrointestinal tract because of diarrhea who also has emphysema and underlying respiratory acidosis in addition to the metabolic acidosis caused by the diarrhea.

A convenient way to diagnose acid-base disorders is to use an acid-base nomogram, as shown in Figure 30–11. This diagram can be used to determine the type of acidosis or alkalosis a person has as well as its severity. In this acid-base diagram, pH, bicarbonate concentration, and P_{CO_2} values intersect according to the Henderson-Hasselbalch equation. The central open circle shows normal values and the deviations that can still be considered within the normal range. The shaded areas of the diagram show the 95 per cent confidence limits for the normal compensations to simple metabolic and respiratory disorders.

When using this diagram, one must assume that sufficient time has elapsed for a full compensatory response, which is 6 to 12 hours for the ventilatory compensations in primary metabolic disorders and 3 to 5 days for the metabolic compensations to primary respiratory disorders. If a value is within the shaded area, this suggests that there is a simple acid-base disturbance. On the other hand, if the values for pH, bicarbonate, or P_{CO_2} lie outside the shaded area, this suggests that there may be a mixed acid-base disorder.

It is important to recognize that finding an acid-base value within the shaded area does not *always* mean that there is a simple acid-base disorder. With this reservation in mind, the acid-base diagrams can be used as a quick means of determining the specific type and severity of an acid-base disorder.

For example, assume that the arterial plasma from a patient yields the following values: pH 7.30, plasma bicarbonate concentration 12.0 mEq/liter, and plasma P_{CO_2} 25 mm Hg. With these values, one can look at the diagram and find that this represents a simple metabolic acidosis, with appropriate respiratory compensation that reduces the P_{CO_2} from its normal value of 40 mm Hg to 25 mm Hg.

A second example would be a patient with the following values: pH 7.15, plasma bicarbonate concentration 17 mEq/liter, and plasma P_{CO_2} 50 mm Hg. In this example, the patient is acidotic, and there appears to be a metabolic component because the plasma bicarbonate concentration is lower than the normal value of 24 mEq/liter. However, the respiratory compensation that would normally reduce P_{CO_2} is absent, and P_{CO_2} is slightly increased above the normal value of 40 mm Hg. This is consistent with a mixed acid-base disturbance with metabolic acidosis as well as a respiratory component to the acidosis.

The acid-base diagram serves as a quick way to assess the type and severity of disorders that may be contributing to abnormal pH, P_{CO_2}, and plasma bicarbonate concentrations. In a clinical setting, the patient's history and other physical findings also provide important clues concerning causes and treatment of the acid-base disorders.

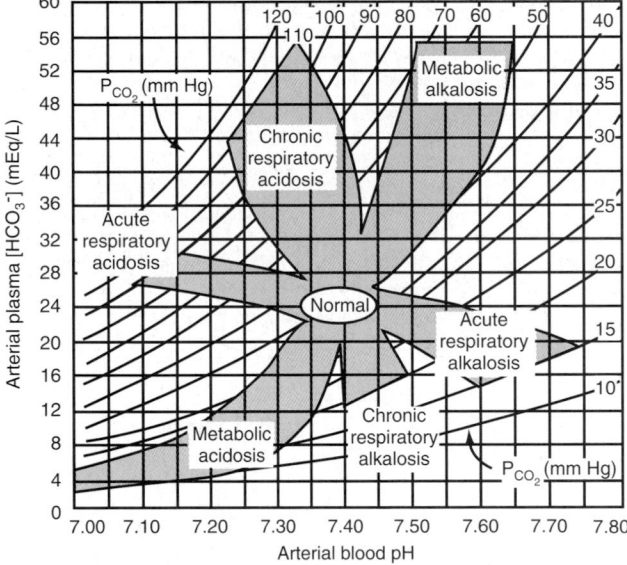

Figure 30–11. Acid-base nomogram showing arterial blood pH, arterial plasma HCO_3^-, and P_{CO_2} values. The central open circle shows the approximate limits for acid-base status in normal people. The shaded areas in the nomogram show the approximate limits for the normal compensations caused by simple metabolic and respiratory disorders. For values lying outside the shaded areas, one should suspect a mixed acid-base disorder. (Adapted from Cogan MG, Rector FC, Jr.: *Acid-Base Disorders in the Kidney,* 3rd ed. Philadelphia: W. B. Saunders Co., 1986.)

REFERENCES

Alpern, R. J., and Rector, F. C., Jr.: Renal acidification: Cellular mechanisms of tubular transport and regulation. In Windhager, E. E. (ed.): Handbook of Physiology, Section 8, Renal Physiology. New York, Oxford University Press, 1992.

Alpern, R. J.: Cell mechanisms of proximal tubule acidification. Physiol. Rev., 70:79, 1990.

Galla, J. H., et al.: Adaptations to chloride depletion alkalosis. Am. J. Physiol., 261:R771, 1991.

Good, D. W.: Ammonium transport by the thick ascending limb of Henle's loop. Ann. Rev. Physiol., 56:623, 1994.

Good, D. W.: Regulation of bicarbonate and ammonium reabsorption on the thick ascending limbs of the rat. Kidney Int., 40(Suppl. 33): S36, 1991.

Halperin, M. L., and Goldstein, M. B.: Fluid, Electrolyte and Acid-Base Physiology, 2nd Ed. Philadelphia, W. B. Saunders Co., 1994.

Halperin, M. L., and Jungas, R. L.: The metabolic production and renal disposal of hydrogen ions: an examination of the biochemical processes. Kidney Int., 24:709, 1983.

Narins, R., and Emmett, M.: Simple and mixed acid-base disorders: a practical approach. Medicine, 59:161, 1980.

Rose, B. D.: Clinical physiology of acid-base and electrolyte disorders. McGraw-Hill, Inc., New York, 1994.

Schuster, V. L.: Function and regulation of collecting duct intercalated cells. Annu. Rev. Physiol., 55:267, 1993.

Schwartz, J. H.: Renal acid-base transport: the regulatory role of the inner medullary collecting duct. Kidney Int., 47:333, 1995.

Tannen, R. L.: Control of acid secretion by the kidney. Annu. Rev. Med., 31:35, 1980.

Vander, A. T.: Renal Physiology, 4th Ed. New York, McGraw-Hill, 1991.

Weinman, E. J., and Shenolikan S.: Regulation of the renal brush border membrane Na^+-H^+ exchanger. Annu. Rev. Physiol., 55:289, 1993.

Micturition, Diuretics, and Kidney Diseases

CHAPTER *31*

MICTURITION

Micturition is the process by which the urinary bladder empties when it becomes filled. This involves two main steps: (1) the bladder fills progressively until the tension in its walls rises above a threshold level, which then elicits the second step; (2) a nervous reflex called the *micturition reflex* occurs that empties the bladder or, if this fails, at least causes a conscious desire to urinate. Although the micturition reflex is an autonomic spinal cord reflex, it can also be inhibited or facilitated by centers in the cerebral cortex or brain stem.

PHYSIOLOGIC ANATOMY AND NERVOUS CONNECTIONS OF THE BLADDER

The urinary bladder, shown in Figure 31–1, is a smooth muscle chamber composed of two main parts: (1) *the body,* which is the major part of the bladder in which urine collects, and (2) *the neck,* which is a funnel-shaped extension of the body, passing inferiorly and anteriorly into the urogenital triangle and connecting with the urethra. The lower part of the bladder neck is also called the *posterior urethra* because of its relation to the urethra.

The smooth muscle of the bladder is called the *detrusor muscle.* Its muscle fibers extend in all directions and, when contracted, can increase the pressure in the bladder to 40 to 60 mm Hg. Thus, contraction of the detrusor muscle is a major step for emptying the bladder. Smooth muscle cells of the detrusor muscle fuse with one another so that low-resistance electrical pathways exist from one muscle cell to the other. Therefore, an action potential can spread throughout the detrusor muscle, from one muscle cell to the next, to cause contraction of the entire bladder at once.

On the posterior wall of the bladder, lying immediately above the bladder neck, is a small triangular area called the *trigone.* At the lowermost apex of the trigone is the opening of the bladder through the bladder neck into the *posterior urethra,* and the two ureters enter the bladder at the uppermost angles of the trigone. The trigone can be identified by the fact that its *mucosa,* the inner lining of the bladder, is smooth, in contrast to the remaining bladder mucosa, which is folded to form *rugae.* Each ureter, as it enters the bladder, courses obliquely through the detrusor muscle and then passes another 1 to 2 centimeters beneath the bladder mucosa before emptying into the bladder.

The bladder neck (posterior urethra) is 2 to 3 centimeters long, and its wall is composed of detrusor muscle interlaced with a large amount of elastic tissue. The muscle in this area is called the *internal sphincter.* Its natural tone normally keeps the bladder neck and posterior urethra empty of urine and, therefore, prevents emptying of the bladder until the pressure in the main part of the bladder rises above a critical threshold.

Beyond the posterior urethra, the urethra passes through the *urogenital diaphragm,* which contains a layer of muscle called the *external sphincter* of the bladder. This muscle is a voluntary skeletal muscle in contrast to the muscle of the bladder body and bladder neck, which contain only smooth muscle. The external sphincter muscle is under voluntary control of the nervous system and can be used to consciously prevent urination even when involuntary controls are attempting to empty the bladder.

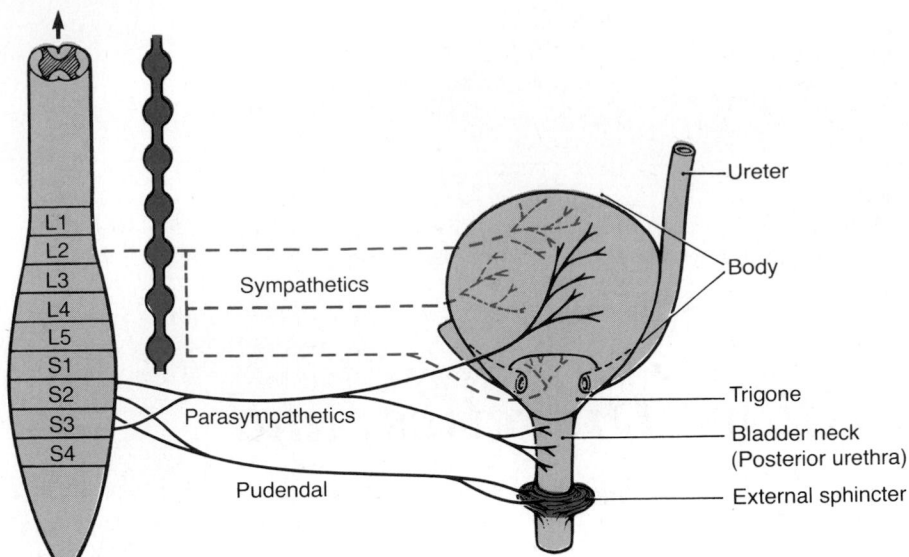

Figure 31–1. Urinary bladder and its innervation.

Innervation of the Bladder

The principal nerve supply of the bladder is by way of the *pelvic nerves*, which connect with the spinal cord through the *sacral plexus*, mainly connecting with cord segments S-2 and S-3. Coursing through the pelvic nerves are both *sensory nerve fibers* and *motor nerve fibers*. The sensory fibers detect the degree of stretch in the bladder wall. Stretch signals from the posterior urethra are especially strong and are mainly responsible for initiating the reflexes that cause bladder emptying.

The motor nerves transmitted in the pelvic nerves are *parasympathetic fibers*. These terminate on ganglion cells located in the wall of the bladder. Short postganglionic nerves then innervate the detrusor muscle.

In addition to the pelvic nerves, two other types of innervation are important in bladder function. Most important are the *skeletal motor fibers* transmitted through the *pudendal nerve* to the external bladder sphincter. These are *somatic nerve fibers* that innervate and control the voluntary skeletal muscle of the sphincter. Also, the bladder receives *sympathetic innervation* from the sympathetic chain through the *hypogastric nerves*, connecting mainly with the L-2 segment of the spinal cord. These sympathetic fibers probably stimulate mainly the blood vessels and have little to do with bladder contraction. Some sensory nerve fibers also pass by way of the sympathetic nerves and may be important in the sensations of fullness and, in some instances, pain.

TRANSPORT OF URINE FROM THE KIDNEY THROUGH THE URETERS AND INTO THE BLADDER

Urine that is expelled from the bladder has essentially the same composition as fluid flowing out of the collecting ducts; there are no significant changes in the composition of urine as it flows through the renal calices and ureters to the bladder.

Urine flowing from the collecting ducts into the renal calices stretches the renal calices and increases their inherent pacemaker activity, which in turn initiates peristaltic contractions that spread to the renal pelvis and then downward along the length of the ureter, thereby forcing urine from the renal pelvis toward the bladder. The walls of the ureters contain smooth muscle and are innervated by both sympathetic and parasympathetic nerves as well as an intramural plexus of neurons and nerve fibers that extends along the entire length of the ureters. As with other visceral smooth muscle, *peristaltic contractions in the ureter are enhanced by parasympathetic stimulation and inhibited by sympathetic stimulation.*

The ureters enter the bladder through the *detrusor muscle* in the trigone region of the bladder, as shown in Figure 31–1. Normally, the ureters course obliquely for several centimeters through the bladder wall. The normal tone of the detrusor muscle in the bladder wall tends to compress the ureter, thereby preventing backflow of urine from the bladder when pressure builds up in the bladder during micturition or bladder compression. Each peristaltic wave along the ureter increases the pressure within the ureter so that the region passing through the bladder wall opens and allows urine to flow into the bladder.

In some people, the distance that the ureter courses through the bladder wall is less than normal, so that contraction of the bladder during micturition does not always lead to complete occlusion of the ureter. As a result, some of the urine in the bladder is propelled backward into the ureter, a condition called *vesicoureteral reflux*. Such reflux can lead to enlargement of the ureters and, if severe, can increase the pressure in the renal calices and structures of the renal medulla, causing damage to these regions.

PAIN SENSATIONS IN THE URETERS, AND THE URETERORENAL REFLEX. The ureters are well supplied with pain nerve fibers. When a ureter becomes blocked (for example, by a ureteral stone), intense

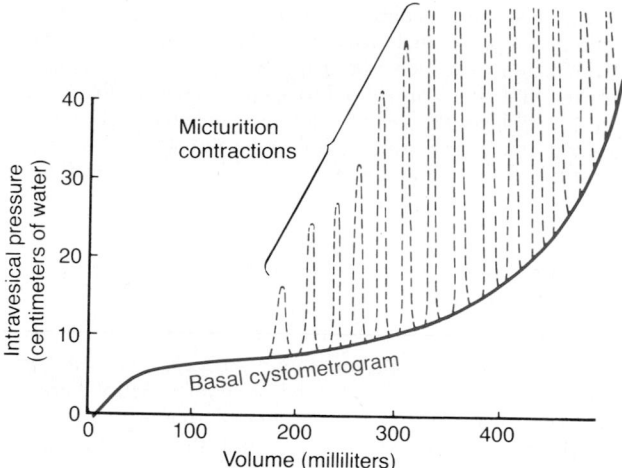

Figure 31–2. Normal cystometrogram, showing also acute pressure waves (dashed spikes) caused by micturition reflexes.

reflex constriction occurs associated with severe pain. Also, the pain impulses cause a sympathetic reflex back to the kidney to constrict the renal arterioles, thereby decreasing urine output from the kidney. This effect is called the *ureterorenal reflex* and is important for preventing excessive flow of fluid into the pelvis of the kidney with a blocked ureter.

FILLING OF THE BLADDER AND BLADDER WALL TONE; THE CYSTOMETROGRAM

Figure 31–2 shows the approximate changes in intravesicular pressure as the bladder fills with urine. When there is no urine in the bladder, the intravesicular pressure is about 0, but by the time 30 to 50 milliliters of urine has collected, the pressure rises to 5 to 10 centimeters of water. Additional urine—20 to 300 milliliters —can collect with only a small additional rise in pressure; this constant level of pressure is caused by intrinsic tone of the bladder wall itself. Beyond 300 to 400 milliliters, collection of more urine in the bladder causes the pressure to rise rapidly.

Superimposed on the tonic pressure changes during filling of the bladder are periodic acute increases in pressure that last from a few seconds to more than a minute. The pressure peaks may rise only a few centimeters of water or may rise to more than 100 centimeters of water. These pressure peaks are called *micturition waves* in the cystometrogram caused by the micturition reflex.

MICTURITION REFLEX

Referring again to Figure 31–2, one can see that as the bladder fills, many superimposed *micturition contractions* begin to appear, as shown by the dashed spikes. They are the result of a stretch reflex initiated by *sensory stretch receptors* in the bladder wall, espe-

cially by the receptors in the posterior urethra when this area begins to fill with urine at the higher bladder pressures. Sensory signals from the bladder stretch receptors are conducted to the sacral segments of the cord through the *pelvic nerves* and then reflexively back again to the bladder through the *parasympathetic nerve fibers* by way of these same nerves.

When the bladder is only partially filled, these micturition contractions usually relax spontaneously after a fraction of a minute, the detrusor muscles stop contracting, and pressure falls back to the baseline. As the bladder continues to fill, the micturition reflexes become more frequent and cause greater contractions of the detrusor muscle.

Once a micturition reflex begins, it is "self-regenerative." That is, initial contraction of the bladder further activates the stretch receptors to cause still further increase in sensory impulses to the bladder and posterior urethra, which causes further increase in reflex contraction of the bladder; thus, the cycle is repeated again and again until the bladder has reached a strong degree of contraction. Then, after a few seconds to more than a minute, the self-regenerative reflex begins to fatigue and the regenerative cycle of the micturition reflex ceases, permitting the bladder to relax.

Thus, the micturition reflex is a single complete cycle of (1) progressive and rapid increase of pressure, (2) a period of sustained pressure, and (3) return of the pressure to the basal tone of the bladder. Once a micturition reflex has occurred but has not succeeded in emptying the bladder, the nervous elements of this reflex usually remain in an inhibited state for a few minutes to 1 hour or more before another micturition reflex occurs. As the bladder becomes more and more filled, micturition reflexes occur more and more often and more and more powerfully.

Once the micturition reflex becomes powerful enough, it causes another reflex, which passes through the *pudendal nerves* to the *external sphincter* to inhibit it. If this inhibition is more potent in the brain than the voluntary constrictor signals to the external sphincter, urination will occur. If not, urination will not occur until the bladder fills still further and the micturition reflex becomes more powerful.

Facilitation or Inhibition of Micturition by the Brain

The micturition reflex is a completely autonomic spinal cord reflex, but it can be inhibited or facilitated by centers in the brain. These centers include (1) strong *facilitory* and *inhibitory centers in the brain stem, located mainly in the pons,* and (2) several *centers located in the cerebral cortex* that are mainly inhibitory but can become excitatory.

The micturition reflex is the basic cause of micturition, but the higher centers normally exert final control of micturition as follows:

1. The higher centers keep the micturition reflex partially inhibited except when micturition is desired.
2. The higher centers can prevent micturition, even if

the micturition reflex does occur, by continual tonic contraction of the external bladder sphincter until a convenient time presents itself.

3. When it is time to urinate, the cortical centers can facilitate the sacral micturition centers to help initiate a micturition reflex and at the same time inhibit the external urinary sphincter so that urination can occur.

Voluntary urination is usually initiated in the following way: First, a person voluntarily contracts his or her abdominal muscles, which increases the pressure in the bladder and allows extra urine to enter the bladder neck and posterior urethra under pressure, thus stretching their walls. This stimulates the stretch receptors, which excites the micturition reflex and simultaneously inhibits the external urethral sphincter. Ordinarily, all the urine will be emptied, with rarely more than 5 to 10 milliliters left in the bladder.

ABNORMALITIES OF MICTURITION

THE ATONIC BLADDER CAUSED BY DESTRUCTION OF SENSORY NERVE FIBERS. Micturition reflex contraction cannot occur if the sensory nerve fibers from the bladder to the spinal cord are destroyed, thereby preventing transmission of stretch signals from the bladder. When this happens, a person loses bladder control, despite intact efferent fibers from the cord to the bladder and despite intact neurogenic connections within the brain. Instead of emptying periodically, the bladder fills to capacity and overflows a few drops at a time through the urethra. This is called *overflow incontinence*.

A common cause of atonic bladder is crush injury to the sacral region of the spinal cord. Certain diseases can also cause damage to the dorsal root nerve fibers that enter the spinal cord. For example, syphilis can cause constrictive fibrosis around the dorsal root nerve fibers, destroying these fibers. This condition is called *tabes dorsalis,* and the resulting bladder condition is called *tabetic bladder.*

THE AUTOMATIC BLADDER CAUSED BY SPINAL CORD DAMAGE ABOVE THE SACRAL REGION. If the spinal cord is damaged above the sacral region but the sacral cord segments are still intact, typical micturition reflexes can still occur. However, they are no longer controlled by the brain. During the first few days to several weeks after the damage to the cord has occurred, the micturition reflexes are suppressed because of the state of "spinal shock" caused by the sudden loss of facilitory impulses from the brain stem and cerebrum. However, if the bladder is emptied periodically by catheterization to prevent bladder injury caused by overstretching of the bladder, the excitability of the micturition reflex gradually increases until typical micturition reflexes return; then, periodic (but unannounced) bladder emptying occurs.

Some patients can still control urination in this condition by stimulating the skin (scratching or tickling) in the genital region, which sometimes elicits a micturition reflex.

THE UNINHIBITED NEUROGENIC BLADDER CAUSED BY LACK OF INHIBITORY SIGNALS FROM THE BRAIN. Another abnormality of micturition is the so-called uninhibited neurogenic bladder, which results in frequent and relatively uncontrolled micturition. This condition derives from partial damage in the spinal cord or the brain stem that interrupts most of the inhibitory signals. Therefore, facilitory impulses passing continually down the cord keep the sacral centers so excitable that even a small quantity of urine will elicit an uncontrollable micturition reflex, and thereby promote frequent urination.

DIURETICS AND THEIR MECHANISMS OF ACTION

A diuretic is a substance that increases the rate of urine volume output, as the name implies. Most diuretics also increase urinary excretion of solutes, especially sodium and chloride. In fact, most diuretics that are used clinically act by decreasing the rate of sodium reabsorption from the tubules, which in turn causes *natriuresis* (increased sodium output) and this in turn causes *diuresis* (increased water output). That is, increased water output, in most cases, occurs secondary to inhibition of tubular sodium reabsorption because sodium remaining in the tubules acts osmotically to decrease water reabsorption. Because the renal tubular reabsorption of many solutes, such as potassium, chloride, magnesium, and calcium, is also influenced secondarily by sodium reabsorption, many diuretics raise renal output of these solutes as well.

The most common clinical use of diuretics is to reduce extracellular fluid volume, especially in diseases associated with edema and hypertension. As discussed in Chapter 25, loss of sodium from the body mainly decreases extracellular fluid volume; therefore, diuretics are most often administered in clinical conditions in which extracellular fluid volume is expanded.

Some diuretics can increase urine output more than 20-fold within a few minutes after they are administered. However, the effect of most diuretics on renal output of salt and water subsides within a few days (Fig. 31–3). This is due to activation of other compensatory mechanisms initiated by decreased extracellular fluid volume. For example, a decrease in extracellular fluid volume often reduces arterial pressure and glomerular filtration rate (GFR) and increases renin secretion and angiotensin II formation; all these responses, together, eventually override the chronic effects of the diuretic on urine output. Thus, in the steady state, urine output becomes equal to intake but only after reduction in arterial pressure and extracellular fluid volume has occurred, which relieves the hypertension or edema that prompted the use of diuretics in the first place.

A large number of diuretics available for clinical use have different mechanisms of action and, therefore, inhibit tubular reabsorption at different sites along the renal nephron. The general classes of diuretics and their mechanisms of action are shown in Table 31–1.

Osmotic Diuretics Decrease Water Reabsorption by Increasing Osmotic Pressure of Tubular Fluid

Injection into the blood stream of substances that are not easily reabsorbed by the renal tubules—such as urea, mannitol, and sucrose—causes a marked increase in the concentration of osmotically active molecules or

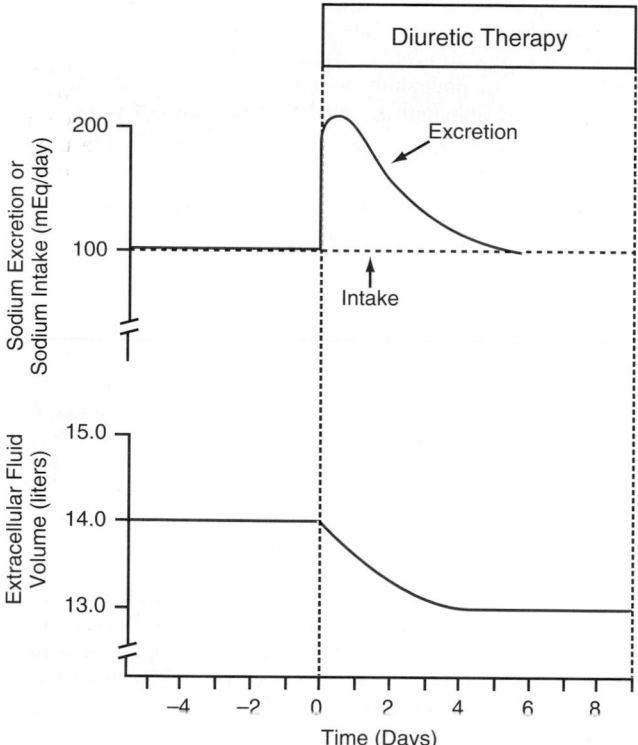

Figure 31–3. Sodium excretion and extracellular fluid volume during diuretic administration. The immediate increase in sodium excretion is accompanied by a decrease in extracellular fluid volume. If sodium intake is held constant, compensatory mechanisms will eventually return sodium excretion to equal sodium intake, thus re-establishing sodium balance.

ions in the tubules. The osmotic pressure of these solutes then greatly reduces water reabsorption, flushing large amounts of tubular fluid on into the urine.

Large volumes of urine are also formed in certain disease states associated with excess solutes that fail to be reabsorbed from the tubular fluid. For example, when the blood glucose concentration rises to high levels in diabetes mellitus, the increased filtered load of glucose into the tubules exceeds their capacity to reabsorb glucose (that is, exceeds their *transport maximum* for glucose). Above a glucose concentration of 250 mg/dl plasma, little of the extra glucose is reabsorbed by

the tubules; instead, the excess glucose remains in the tubules, acts as an osmotic diuretic, and causes rapid loss of fluid into the urine. In fact, the name "diabetes" refers to this high level of urine output. In patients with diabetes mellitus, the high urine output is balanced by a high level of fluid intake owing to activation of the thirst mechanism.

"Loop" Diuretics Decrease Active Sodium-Chloride-Potassium Reabsorption in the Thick Ascending Loop of Henle

Furosemide, ethacrynic acid, and *bumetanide* are powerful diuretics that decrease active reabsorption in the thick ascending limb of the loop of Henle by blocking the 1-sodium, 2-chloride, 1-potassium co-transporter located in the luminal membrane of the epithelial cells. These diuretics are among the most powerful of the clinically used diuretics.

By blocking active sodium-chloride-potassium co-transport in the luminal membrane of the loop of Henle, the loop diuretics raise urine output of sodium, chloride, potassium, and other electrolytes as well as water for several reasons: (1) They greatly increase the quantities of solutes delivered to the distal parts of the nephrons, and these act as osmotic agents to prevent water reabsorption as well. (2) They disrupt the countercurrent multiplier system by decreasing absorption of ions from the loop of Henle into the medullary interstitium, thereby decreasing the osmolarity of the medullary interstitial fluid. Because of this effect, the loop diuretics impair the ability of the kidneys to either concentrate or dilute the urine. *Urinary dilution* is impaired because the inhibition of sodium and chloride reabsorption in the loop of Henle causes more of these ions to be excreted along with increased water excretion. *Urinary concentration,* on the other hand, is impaired because the renal medullary interstitial fluid concentration of these ions, and therefore renal medullary osmolarity, is reduced. Consequently, reabsorption of fluid from the collecting ducts is decreased, so that the maximal concentrating ability of the kidneys is also greatly reduced. In addition, decreased renal medullary interstitial fluid osmolarity reduces absorption of water from the descending loop of Henle. Because of these multiple effects, 20 to 30 per cent of the glomerular filtrate may be delivered into the urine, causing, under acute conditions, urine output to be as great as 25 times normal for a period of at least a few minutes.

Table 31–1 CLASSES OF DIURETICS, THEIR MECHANISMS OF ACTION, AND TUBULAR SITES OF ACTION

Class of Diuretic	Mechanism of Action	Tubular Site of Action
Osmotic diuretics	Inhibit water and solute reabsorption by increasing osmolarity of tubular fluid	Mainly proximal tubules
Loop diuretics	Inhibit Na-K-Cl co-transport in luminal membrane	Thick ascending loop of Henle
Thiazide diuretics	Inhibit Na-Cl co-transport in luminal membrane	Early distal tubules
Carbonic anhydrase inhibitors	Inhibit H^+ secretion and HCO_3^- reabsorption, which reduces Na^+ reabsorption	Proximal tubules
Competitive inhibitors of aldosterone	Inhibit action of aldosterone on tubular receptor, decrease Na^+ reabsorption, and decrease K^+ secretion	Collecting tubules
Sodium channel blockers	Block entry of Na^+ into Na^+ channels of luminal membrane, decrease Na^+ reabsorption, and decrease K^+ secretion	Collecting tubules

Thiazide Diuretics Inhibit Sodium-Chloride Reabsorption in the Early Distal Tubule

The thiazide derivatives, such as *chlorothiazide,* act mainly on the early distal tubules to block the sodium-chloride co-transporter in the luminal membrane of the tubular cells. Under favorable conditions, these agents cause 5 to 10 per cent of the glomerular filtrate to pass into the urine. This is about the same amount of sodium normally reabsorbed by the distal tubules.

Carbonic Anhydrase Inhibitors Block Sodium-Bicarbonate Reabsorption in the Proximal Tubules

Acetazolamide (Diamox) inhibits the enzyme *carbonic anhydrase,* which is critical for the reabsorption of bicarbonate in the proximal tubule, as discussed in Chapter 30. Carbonic anhydrase is abundant in the proximal tubule, the primary site of action of carbonic anhydrase inhibitors. Some carbonic anhydrase is also present in other tubular cells, such as in the intercalated cells of the collecting tubule.

Bicarbonate ions are not permeable to the luminal membrane of the tubule and, therefore, are not reabsorbed directly from the tubular lumen. Instead, bicarbonate in the tubular fluid reacts with hydrogen ions secreted into the tubular fluid to form carbonic acid (H_2CO_3), which dissociates to form carbon dioxide (CO_2) and water (H_2O). The CO_2 diffuses into the tubular cell, where it can recombine with H_2O to form H_2CO_3 if carbonic anhydrase is present. The H_2CO_3 dissociates in the cell to form HCO_3^- (which can then diffuse across the basolateral membrane and be absorbed into the blood) and H^+. Therefore, when carbonic anhydrase is blocked, bicarbonate cannot be reabsorbed from the tubular fluid.

Because hydrogen ion secretion and bicarbonate reabsorption in the proximal tubules are coupled to sodium reabsorption, through the sodium-hydrogen ion counter-transport mechanism in the luminal membrane, decreasing bicarbonate reabsorption also reduces sodium reabsorption. The blockage of sodium and bicarbonate reabsorption from the tubular fluid causes these ions to remain in the tubules and act as an osmotic diuretic. Predictably, a disadvantage of the carbonic anhydrase inhibitors is that they cause some degree of acidosis because of the excessive loss of bicarbonate ions in the urine.

Competitive Inhibitors of Aldosterone Decrease Sodium Reabsorption from and Potassium Secretion into the Cortical Collecting Tubule

Spironolactone and similar substances compete with aldosterone for receptor sites in the cortical collecting tubular epithelial cells and, therefore, can decrease the reabsorption of sodium and secretion of potassium in this tubular segment. As a consequence, sodium remains in the tubules and acts as an osmotic diuretic, causing increased excretion of water as well as sodium. Because these drugs also block the effect of aldosterone to promote potassium secretion in the tubules, they decrease the excretion of potassium; in some instances, this causes extracellular fluid potassium concentration to in-crease excessively. For this reason, spironolactone and other aldosterone inhibitors are referred to as *"potassium-sparing"* *diuretics.* Many of the other diuretics cause loss of potassium in the urine, in contrast to the aldosterone antagonists, which "spare" the loss of potassium.

Diuretics That Block Sodium Channels in the Collecting Tubules Decrease Sodium Reabsorption

Amiloride and *triamterene* also inhibit sodium reabsorption and potassium secretion in the collecting tubules, similar to the effects of spironolactone. However, at the cellular level, these drugs act directly to block the entry of sodium into the sodium channels of the luminal membrane of the collecting tubule epithelial cells. Because of this decreased sodium entry into the epithelial cell, there is also decreased sodium transport across the cell's basolateral membrane and, therefore, decreased activity of the sodium-potassium adenosine triphosphatase pump. This decreased activity in turn reduces the transport of potassium into the cell and ultimately decreases the secretion of potassium into the tubular fluid. For this reason, the sodium channel blockers are also potassium-sparing diuretics and decrease the urinary excretion rate of potassium.

KIDNEY DISEASES

Diseases of the kidney are among the most important causes of death and disability in many countries throughout the world. For example, in 1994, more than 15 million people in the United States were estimated to have kidney disease, which appears to be a major cause of lost work time.

Several kidney diseases can be lumped into two main categories: (1) *acute renal failure,* in which the kidneys abruptly stop working entirely or almost entirely but may eventually recover nearly normal function, and (2) *chronic renal failure,* in which there is progressive loss of function of more and more nephrons that gradually decreases overall kidney function. Within these two general categories, there are vast arrays of specific kidney diseases that can affect the kidney blood vessels, the glomeruli, the tubules, the renal interstitium, and parts of the urinary tract outside the kidney, including the ureters and bladder. In this chapter, we discuss specific physiological abnormalities that occur in only a few of the most important types of kidney diseases.

ACUTE RENAL FAILURE

The causes of acute renal failure can be divided into three main categories:

1. Acute renal failure resulting from decreased blood supply to the kidneys; this condition is often referred to as *prerenal acute renal failure* to reflect the fact that the abnormality occurs *before* the kidney. This can be a consequence

of (a) heart failure with reduced cardiac output and low blood pressure or (b) conditions associated with diminished blood volume and low blood pressure, such as severe hemorrhage.

2. *Intrarenal acute renal failure* resulting from abnormalities within the kidney itself, including those that affect the blood vessels, glomeruli, or tubules.

3. *Post-renal acute renal failure,* meaning obstruction of the urinary collecting system anywhere from the calices to the outflow from the bladder; the most important causes of obstruction of the urinary tract outside the kidney are kidney stones, caused by precipitation of calcium, urate, or cystine.

Prerenal Acute Renal Failure Caused by Decreased Blood Flow to the Kidney

The kidneys normally receive an abundant blood supply of about 1200 ml/min, or about 20 to 25 per cent of the cardiac output. The main value of this high blood flow to the kidneys is to provide enough plasma for the high rates of glomerular filtration needed for effective regulation of body fluid volumes and solute concentrations. Therefore, decreased renal blood flow is usually accompanied by decreased GFR and decreased urine output of water and solutes. Consequently, conditions that acutely diminish blood flow to the kidneys usually cause *oliguria,* which refers to diminished urine output below the level of intake of water and solutes. This causes accumulation of water and solutes in the body fluids. If renal blood flow is markedly reduced, total cessation of urine output can occur, a condition referred to as *anuria.*

As long as renal blood flow does not fall below about 20 per cent of normal, acute renal failure can usually be reversed if the cause of the ischemia is corrected before damage to the renal cells has occurred. Unlike some tissues, the kidney can endure a relatively large reduction in renal blood flow before actual damage to the renal cells occurs. The reason for this is that as renal blood flow is reduced, the GFR and amount of sodium chloride filtered by the glomeruli (as well as the filtration rate of water and other electrolytes) are reduced. This decreases the amount of sodium chloride that must be reabsorbed by the tubules, which uses most of the energy and oxygen consumed by the normal kidney. Therefore, as renal blood flow and GFR fall, the requirement for renal oxygen consumption is also reduced. As the GFR approaches zero, oxygen consumption of the kidney approaches the rate that is required to keep the renal tubular cells alive even when they are not reabsorbing sodium. When blood flow is reduced below this basal requirement, which is usually less than 20 per cent of the total renal blood flow, the renal cells start to become hypoxic and further decreases in renal blood flow, if prolonged, will cause damage, or even death, of the renal cells, especially the tubular epithelial cells. If the cause of prerenal acute renal failure is not corrected and ischemia of the kidney persists longer than a few hours, this type of renal failure can evolve into "intrarenal acute renal failure," as discussed below.

Acute reduction of renal blood flow is a common cause of acute renal failure in hospitalized patients. Table 31-2 shows some of the common causes of decreased renal blood flow and prerenal acute renal failure.

Intrarenal Acute Renal Failure Caused by Abnormalities Within the Kidney

Abnormalities that originate within the kidney and that abruptly diminish urine output fall into the general category of *intrarenal acute renal failure.* This category of acute renal failure can be further divided into (a) conditions that injure the glomerular capillaries or other small renal vessels, (b) conditions that damage the renal tubular epithelium, and (c) conditions that cause damage to the renal interstitium. This type of classification refers to the primary site of injury, but because the renal vasculature and tubular system are functionally interdependent, damage to the renal blood vessels can lead to tubular damage, and primary tubular damage can lead to damage of the renal blood vessels. Causes of intrarenal acute renal failure are listed in Table 31-3.

ACUTE RENAL FAILURE CAUSED BY GLOMERULONEPHRITIS. Acute glomerulonephritis is a type of *intrarenal* acute renal failure usually caused by an abnormal immune reaction that damages the glomeruli. In about 95 per cent of the patients with this disease, damage to the glomeruli occurs 1 to 3 weeks after an infection elsewhere in the body, usually caused by certain types of group A beta steptococci. The infection may have been a streptococcal sore throat, streptococcal tonsillitis, or even streptococcal infection of the skin. It is not the infection itself that damages the kidneys. Instead, as antibodies develop during the succeeding few weeks against the streptococcal antigen, the antibodies and antigen react with each other to form an insoluble immune complex that becomes entrapped in the glomeruli, especially in the basement membrane portion of the glomeruli.

Once the immune complex has deposited in the glomeruli, many of the cells of the glomeruli begin to proliferate but mainly the epithelial cells and mesangial cells that lie between the endothelium and the epithelium. In addition, large numbers of white blood cells become entrapped in the glomeruli. Many of the glomeruli become blocked by this inflammatory reaction, and those that are not blocked usually become excessively permeable, allowing both protein and red blood cells to leak from the blood of the glomerular capillaries into the glomerular filtrate. In severe cases, either total or almost complete renal shutdown occurs.

Table 31-2 SOME CAUSES OF PRERENAL ACUTE RENAL FAILURE

1. **Intravascular Volume Depletion**
   Hemorrhage (due to trauma, surgery, postpartum, gastrointestinal)
   Diarrhea or vomiting
   Burns
2. **Cardiac Failure**
   Myocardial infarction
   Valvular damage
3. **Primary Renal Hemodynamic Abnormalities**
   Renal artery stenosis, embolism, or thrombosis of renal artery or vein
   Excessive blockade of prostaglandin synthesis (aspirin)
4. **Peripheral Vasodilation and Resultant Hypotension**
   Anaphylactic shock
   Anesthesia
   Sepsis, severe infections

Table 31–3 SOME CAUSES OF INTRARENAL ACUTE
RENAL FAILURE

1. **Small Vessel and/or Glomerular Injury**
   Vasculitis (polyarteritis nodosa)
   Cholesterol emboli
   Malignant hypertension
   Acute glomerulonephritis
2. **Tubular Epithelial Injury (Tubular Necrosis)**
   Acute tubular necrosis due to ischemia
   Acute tubular necrosis due to toxins (heavy metals,
      ethylene glycol, insecticides, poison mushrooms,
      carbon tetrachloride)
3. **Renal Interstitial Injury**
   Acute pyelonephritis
   Acute allergic interstitial nephritis

The acute inflammation of the glomeruli usually subsides in about 2 weeks, and in most patients, the kidneys return to almost normal function within the next few weeks to few months. Sometimes, however, many of the glomeruli are destroyed beyond repair, and in a small percentage of patients, progressive renal deterioration continues indefinitely, leading to progressive *chronic renal failure,* as described in a subsequent section of this chapter.

Tubular Necrosis as a Cause of Acute Renal Failure. Another cause of intrarenal acute renal failure is *tubular necrosis,* which means destruction of epithelial cells in the tubules. Some common causes of tubular necrosis are (1) severe ischemia and inadequate supply of oxygen and nutrients to the tubular epithelial cells and (2) poisons, toxins, or medications that destroy the tubular epithelial cells.

Acute Tubular Necrosis Caused by Severe Renal Ischemia. Severe ischemia of the kidney can result from circulatory shock or any other disturbance that severely impairs the blood supply to the kidney. If the ischemia is severe enough to seriously impair the delivery of nutrients and oxygen to the renal tubular epithelial cells and if the insult is prolonged, damage or eventual destruction of the epithelial cells can occur. When this happens, tubular cells "slough off" and plug many of the nephrons, so that there is no urine output from the blocked nephrons; the affected nephrons often fail to excrete urine even when renal blood flow is restored to normal, as long as the tubules remain plugged.

The most common causes of ischemic damage to the tubular epithelium are the prerenal causes of acute renal failure associated with circulatory shock, as discussed earlier in this chapter.

Acute Tubular Necrosis Caused by Toxins or Medications. There is a long list of renal poisons and medications that can damage the tubular epithelium and cause acute renal failure. Some of these are *carbon tetrachloride, heavy metals* (such as mercury and lead), *ethylene glycol* (which is a major component in antifreeze), *various insecticides,* and various *medications* (such as tetracyclines), which are used as antibiotics, and cis-platinum, which is used in treating certain cancers. Each of these substances has a specific toxic action on the renal tubular epithelial cells, causing death of many of them. As a result, the epithelial cells slough away from the basement membrane and plug the tubules. In some instances, the basement membrane also is destroyed. If the basement membrane remains intact,

new tubular epithelial cells can grow along the surface of the membrane, so that the tubule repairs itself within 10 to 20 days.

Postrenal Acute Renal Failure Caused by Abnormalities of the Lower Urinary Tract

Multiple abnormalities in the lower urinary tract can block or partially block urine flow and therefore lead to acute renal failure even when the kidneys' blood supply and other functions are initially normal. If the urine output of only one kidney is diminished, no major change in body fluid composition will occur because the contralateral kidney can increase its urine output sufficiently to maintain relatively normal levels of extracellular electrolytes and solutes as well as normal extracellular fluid volume. With this type of renal failure, normal kidney function can be restored if the basic cause of the problem is corrected within a few hours. But chronic obstruction of the urinary tract, lasting for several days or weeks, can lead to irreversible kidney damage.

Some of the causes of postrenal acute failure include (1) bilateral obstruction of the ureters or renal pelvises caused by large stones or blood clots, (2) bladder obstruction, and (3) obstruction of the urethra.

Physiological Effects of Acute Renal Failure

When acute renal failure is moderate, the major physiological effect is retention in the blood and extracellular fluid of water, waste products of metabolism, and electrolytes. This can lead to water and salt overload, which in turn can lead to edema and hypertension. Excessive retention of potassium, however, often provides an even more serious threat to patients with acute renal failure because increases in plasma potassium concentration (hyperkalemia) to more than about 8 mEq/liter (only twice normal) can be fatal. Because the kidneys are also unable to excrete sufficient hydrogen ions, patients with acute renal failure develop metabolic acidosis, which in itself can be lethal or can aggravate the hyperkalemia.

In the most severe cases of acute renal failure, complete anuria occurs. The patient will die within 8 to 14 days unless kidney function is restored or unless an artificial kidney is used to rid the body of the excessive retained water, electrolytes, and waste products of metabolism. Other effects of diminished urine output as well as treatment with an artificial kidney are discussed in the next section in relation to chronic renal failure.

CHRONIC RENAL FAILURE: AN IRREVERSIBLE DECREASE IN THE NUMBER OF FUNCTIONAL NEPHRONS

Chronic renal failure results from irreversible loss of large numbers of functioning nephrons. Serious clinical symptoms often do not occur until the number of functional nephrons falls to at least 70 per cent below

Table 31–4 SOME CAUSES OF CHRONIC RENAL FAILURE

1. **Immunological disorders**
   Glomerulonephritis
   Polyarteritis nodosa
   Lupus erythematosus
2. **Metabolic disorders**
   Diabetes mellitus
   Amyloidosis
3. **Renal vascular disorders**
   Atherosclerosis
   Nephrosclerosis
4. **Infections**
   Pyelonephritis
   Tuberculosis
5. **Primary tubular disorders**
   Nephrotoxins (analgesics, heavy metals)
6. **Urinary tract obstruction**
   Renal calculi
   Hypertrophy of prostate
   Urethral constriction
7. **Congenital disorders**
   Polycystic disease
   Congenital absence of kidney tissue (renal hypoplasia)

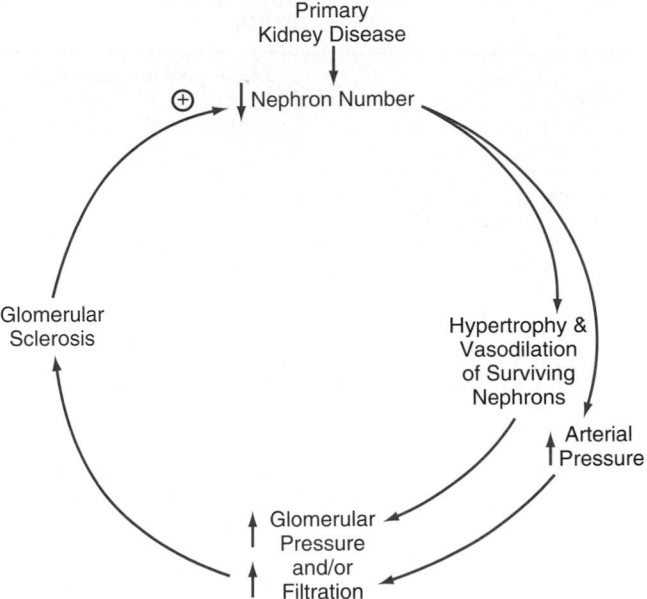

Figure 31–4. Vicious circle that can occur with primary kidney disease. Loss of nephrons because of disease may increase pressure and flow in the surviving glomerular capillaries, which in turn may eventually injure these "normal" capillaries as well, thus causing progressive sclerosis and eventual loss of these glomeruli.

normal. In fact, relatively normal blood concentrations of most electrolytes and normal body fluid volumes can still be maintained until the number of functioning nephrons decreases below 20 to 30 per cent of normal.

Table 31–4 gives some of the most important causes of chronic renal failure. In general, chronic renal failure, like acute renal failure, can occur because of disorders of the blood vessels, glomeruli, tubules, renal interstitium, and lower urinary tract.

Despite the wide variety of diseases that can lead to chronic renal failure, the end result is essentially the same—a decrease in the number of functional nephrons.

Vicious Circle of Chronic Renal Failure Leading to End-Stage Renal Failure

In many cases, an initial insult to the kidney leads to progressive deterioration of kidney function and further loss of nephrons to a point at which the person must be placed on dialysis treatment or transplanted with a functional kidney to survive. This condition is referred to as *end-stage renal failure.*

Studies in laboratory animals have shown that surgical removal of large portions of the kidney initially causes adaptive changes in the remaining nephrons that lead to increased single-nephron blood flow, increased GFR, and increased urine output in the surviving nephrons. The exact mechanisms responsible for these changes are not well understood but involve hypertrophy (growth of the various structures of the surviving nephrons) as well as functional changes that decrease vascular resistance and tubular reabsorption in the surviving nephrons. These adaptive changes permit a person to excrete normal amounts of water

and solutes even with a kidney mass reduced to 20 to 30 per cent of normal. Over a period of several years, the renal functional changes may lead to further injury of the remaining nephrons, particularly injury to the glomeruli of these nephrons.

The cause of this additional injury is not known, but some investigators believe that it may be related in part to increased pressure or stretch of the remaining glomeruli that occurs as a result of functional vasodilation or increased blood pressure; the chronic increase in pressure and stretch of the small arterioles and glomeruli is believed eventually to cause sclerosis of these vessels (replacement of normal tissue with connective tissue). These sclerotic lesions can eventually obliterate the glomerulus, leading to further reduction in kidney function, further adaptive changes in the remaining nephrons, and a slowly progressing vicious circle that eventually terminates in end-stage renal failure (Fig. 31–4). The only proven method of slowing down this progressive loss of kidney function is to lower arterial pressure and glomerular hydrostatic pressure, especially by using drugs such as the angiotensin-converting enzyme inhibitors, which block the formation of angiotensin II.

Table 31–5 gives the most common causes of end-stage renal failure. In the early 1980s, *glomerulonephritis* in all of its various forms was believed to be the most common initiating cause of end-stage renal failure. In recent years, *diabetes mellitus* and *hypertension* have become recognized as the leading causes of end-stage renal failure.

Table 31–5 MOST COMMON CAUSES OF END-STAGE RENAL DISEASE (ESRD)

Cause	Percentage of Total ESRD Patients
Diabetes mellitus	28
Hypertension	25
Glomerulonephritis	21
Polycystic kidney disease	4
Other/Unknown	22

Injury of the Renal Vasculature as a Cause of Chronic Renal Failure

Many types of vascular lesions can lead to renal ischemia and death of kidney tissue. The most common of these are (1) *atherosclerosis* of the larger renal arteries, with progressive sclerotic constriction of the vessels; (2) *fibromuscular hyperplasia* of one or more of the large arteries, which also causes occlusion of the vessels; and (3) *nephrosclerosis,* a common condition caused by sclerotic lesions of the smaller arteries, arterioles, and glomeruli.

Atherosclerotic or hyperplastic lesions of the large arteries frequently affect one kidney more than the other and, therefore, cause unilaterally diminished kidney function. As discussed in Chapter 19, hypertension often occurs when the artery of one kidney is constricted while the artery of the other kidney is still normal, a condition analogous to "two-kidney Goldblatt hypertension."

Benign nephrosclerosis, the most common form of kidney disease, is seen to at least some extent in about 70 per cent of postmortem examinations in people who die after the age of 60. This type of vascular lesion occurs in the smaller interlobular arteries and in the afferent arterioles of the kidney. It is believed to begin with leakage of plasma through the intimal membrane of these vessels. This causes fibrinoid deposits to develop in the medial layers of these vessels, followed by progressive thickening of the vessel wall that eventually constricts the vessels and, in some cases, occludes them. Because there is essentially no collateral circulation among the smaller renal arteries, occlusion of one or more of these causes destruction of a comparable number of nephrons. Therefore, much of the kidney tissue becomes replaced by small amounts of fibrous tissue. When sclerosis occurs in the glomeruli, the glomerular injury is referred to as *glomerulosclerosis.*

Nephrosclerosis and glomerulosclerosis occur to some extent in most people after the fourth decade of life, causing about a 10 per cent decrease in the number of functional nephrons each 10 years after age 40 (Fig. 31–5). This loss of glomeruli and overall nephron function is reflected by progressive decrease in both renal blood flow and GFR. Even in "normal" people, kidney plasma flow and GFR decrease by 40 to 50 per cent by age 80.

The frequency and severity of nephrosclerosis and glomerulosclerosis are greatly increased by concurrent *hypertension* and/or *diabetes mellitus.* In fact, diabetes mellitus and hypertension are the two most important causes of end-stage renal failure, as discussed previously. Thus, benign nephrosclerosis in association with

severe hypertension can lead to a rapidly progressing *malignant nephrosclerosis.* The characteristic histological features of malignant nephrosclerosis include large amounts of fibrinoid deposits in the arterioles and progressive thickening of the vessels, with severe ischemia occurring in the affected nephrons. For unknown reasons, the incidences of malignant nephrosclerosis and severe glomerulosclerosis are significantly higher in African-Americans than in Caucasians of similar ages who have similar degrees of severity of hypertension and/or diabetes.

Injury to the Glomeruli as a Cause of Chronic Renal Failure—Glomerulonephritis

Chronic glomerulonephritis is caused by any one of several diseases that cause inflammation and damage to the capillary loops in the glomeruli of the kidneys. In contrast to the acute form of this disease, chronic glomerulonephritis is a slowly progressive disease that often leads to irreversible renal failure. It may be a primary kidney disease, following acute glomerulonephritis, or it may be secondary to systemic diseases, such as *lupus erythematosus.*

In most cases, chronic glomerulonephritis begins with accumulation of precipitated antigen-antibody complexes in the glomerular membrane. In contrast to acute glomerulonephritis, streptococcal infections account for only a small percentage of patients with the chronic form of glomerulonephritis. The result of the antigen-antibody complex accumulation in the glomerular membranes is inflammation, progressive thickening of the membranes, and eventual invasion of the glomeruli by fibrous tissue. In the later stages of the disease, the glomerular capillary filtration coefficient becomes greatly reduced because of decreased numbers of filtering capillaries in the glomerular tufts and because of thickened glomerular membranes. In the final stages of the disease, many glomeruli are replaced by fibrous tissue and are, therefore, unable to filter fluid.

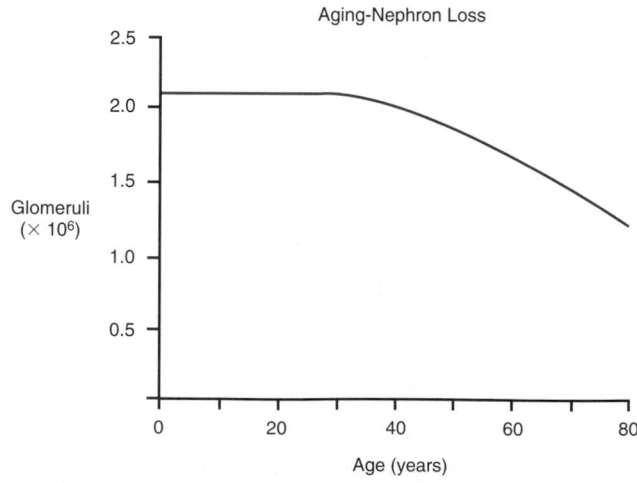

Figure 31–5. Effect of aging on the number of functional glomeruli.

Injury of the Renal Interstitium as a Cause of Chronic Renal Failure—Pyelonephritis

Primary or secondary disease of the renal interstitium is referred to as *interstitial nephritis*. In general, this can result from vascular, glomerular, or tubular damage that destroys individual nephrons, or it can involve primary damage to the renal interstitium by poisons, drugs, and bacterial infections.

Renal interstitial injury caused by bacterial infection is called *pyelonephritis*. The infection can result from different types of bacteria but especially from *Escherichia coli* that originate from fecal contamination of the urinary tract. These bacteria reach the kidneys either by way of the blood stream or, more commonly, by ascension from the lower urinary tract by way of the ureters to the kidneys.

Although the normal bladder is able to clear bacteria readily, there are two general clinical conditions that may interfere with the normal flushing of bacteria from the bladder: (1) the inability of the bladder to empty completely, leaving residual urine in the bladder, and (2) the existence of obstruction of urine outflow. With impaired ability to flush the bacteria from the bladder, the bacteria multiply and the bladder becomes inflamed, a condition that is termed *cystitis*. Once cystitis has occurred, it may remain localized without ascending to the kidney, or in some people, bacteria may reach the renal pelvis because of a pathological condition in which urine is propelled up one or both of the ureters during micturition. This condition is called *vesicoureteral reflux* and is due to the failure of the bladder wall to occlude the ureter during micturition; as a result, some of the urine is propelled upward toward the kidney, carrying with it bacteria that can reach the renal pelvis and renal medulla, where they can initiate the infection and inflammation associated with pyelonephritis.

Because pyelonephritis begins in the renal medulla, it usually affects the function of the medulla more than it affects the cortex, at least in the initial stages. Because one of the primary functions of the medulla is to provide the countercurrent mechanism for concentrating urine, patients with pyelonephritis frequently have marked impairment of their ability to concentrate the urine.

With long-standing pyelonephritis, invasion of the kidneys by bacteria causes not only damage to the renal medulla interstitium but also results in progressive damage of renal tubules, glomeruli, and other structures throughout the kidney. Consequently, large parts of functional renal tissue are lost and chronic renal failure can develop.

Nephrotic Syndrome—Excretion of Protein in the Urine Because of Increased Glomerular Permeability

Many patients with kidney disease develop the so-called *nephrotic syndrome,* which is characterized by loss of large quantities of plasma proteins into the urine. In some instances, this occurs without evidence of other major abnormalities of kidney function, but more often it is associated with some degree of renal failure.

The cause of the protein loss in the urine is increased permeability of the glomerular membrane. Therefore, any disease that increases the permeability of this membrane can cause the nephrotic syndrome. Such diseases include (1) *chronic glomerulonephritis,* which affects primarily the glomeruli and often causes greatly increased permeability of the glomerular membrane; (2) *amyloidosis,* which results from deposition of an abnormal proteinoid substance in the walls of the blood vessels and seriously damages the basement membrane of the glomeruli; and (3) *minimal change nephrotic syndrome,* which is associated with no major abnormality in the glomerular capillary membrane that can be detected with light microscopy. As discussed in Chapter 26, minimal change nephropathy has been found to be associated with loss of the negative charges that are normally present in the glomerular capillary basement membrane. Immunological studies have also shown abnormal immune reactions in some cases, suggesting that the loss of the negative charges may have resulted from antibody attack on the membrane.

Loss of normal negative charges in the basement membrane of the glomerular capillaries allows proteins, especially albumin, to pass through the glomerular membrane with ease because the negative charges in the basement membrane normally repel the negatively charged plasma proteins.

Minimal change nephropathy can occur in adults, but more frequently it occurs in children between the ages of 2 and 6 years. Increased permeability of the glomerular capillary membrane occasionally allows as much as 40 grams of plasma protein loss into the urine each day, which is an extreme amount for a young child. Therefore, the child's plasma protein concentration often falls below 2 gm/dl, and the colloid osmotic pressure falls from a normal value of 28 to 6 to 8 mm Hg. As a consequence of this low colloid osmotic pressure in the plasma, large amounts of fluid leak from the capillaries all over the body into most of the tissues, causing a severe edema, as discussed in Chapter 25.

Abnormal Nephron Function in Chronic Renal Failure

LOSS OF FUNCTIONAL NEPHRONS REQUIRES THE SURVIVING NEPHRONS TO EXCRETE MORE WATER AND SOLUTES. The kidney normally filters about 180 liters of fluid each day at the glomerular capillaries and then transforms this filtrate into urine as the fluid flows along successive nephron segments, so that the final urine volume and composition are carefully matched to the water and solute load imposed by the daily intake of fluid and food. Therefore, it would be reasonable to suspect that decreasing the number of functional nephrons, which would reduce the GFR, would also cause major decreases in renal excretion of water and solutes. Yet patients who have lost as much as 70 per cent of their nephrons are able to excrete normal amounts of water and electrolytes without serious accumulation of any of these in the body fluids. Further reduction in the number of nephrons, however, leads to electrolyte and fluid retention, and death usually ensues when the number of nephrons falls below 5 to 10 per cent of normal.

In contrast to the electrolytes, many of the waste products of metabolism, such as urea and creatinine, accumulate almost in proportion to the number of nephrons that have been destroyed. The reason for this is that substances such as creatinine and urea depend largely on glomerular filtration for their excretion, and they are not reabsorbed as avidly as the electrolytes. Creatinine, for example, is not reabsorbed, so that the excretion rate is equal to the rate at which it is filtered:

$$\text{Creatinine filtration rate} = \text{GFR} \times \text{Plasma creatinine}$$
$$\text{concentration} = \text{Creatinine excretion rate}$$

Therefore, if GFR decreases, the creatinine excretion rate also transiently decreases, causing accumulation of creatinine in the body fluids and raising creatinine concentration until the excretion rate of creatinine returns to normal—the same rate at which creatinine is produced in the body (Fig. 31–6). Thus, under steady-state conditions, the creatinine excretion rate equals the rate of creatinine production, despite reductions in GFR; however, this normal rate of creatinine excretion occurs at the expense of elevated plasma creatinine concentration, as shown in curve A of Figure 31–7.

Some solutes, such as phosphate, urate, and hydro-

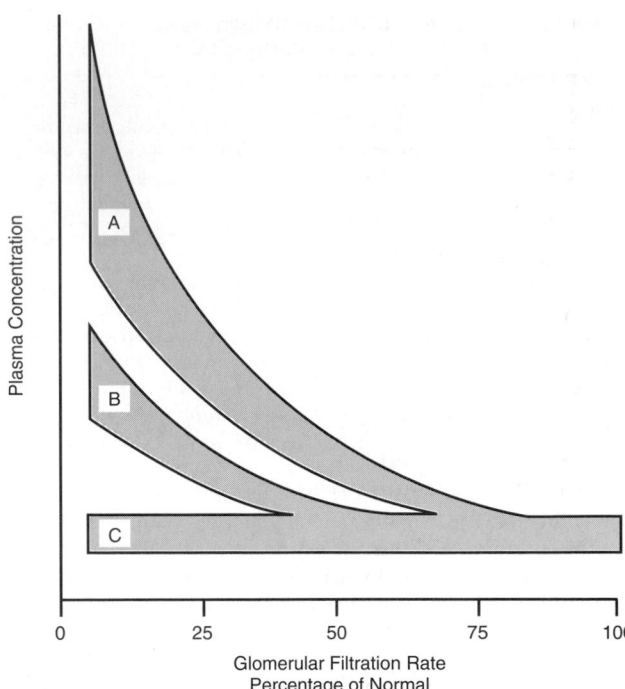

Figure 31–7. Representative patterns of adaptation for different types of solutes in chronic renal failure. Curve A shows the approximate changes in plasma concentrations of solutes such as creatinine and urea that are filtered and poorly reabsorbed. Curve B shows the approximate concentrations for solutes such as phosphate and urate. Curve C shows the approximate concentrations for solutes such as sodium and chloride.

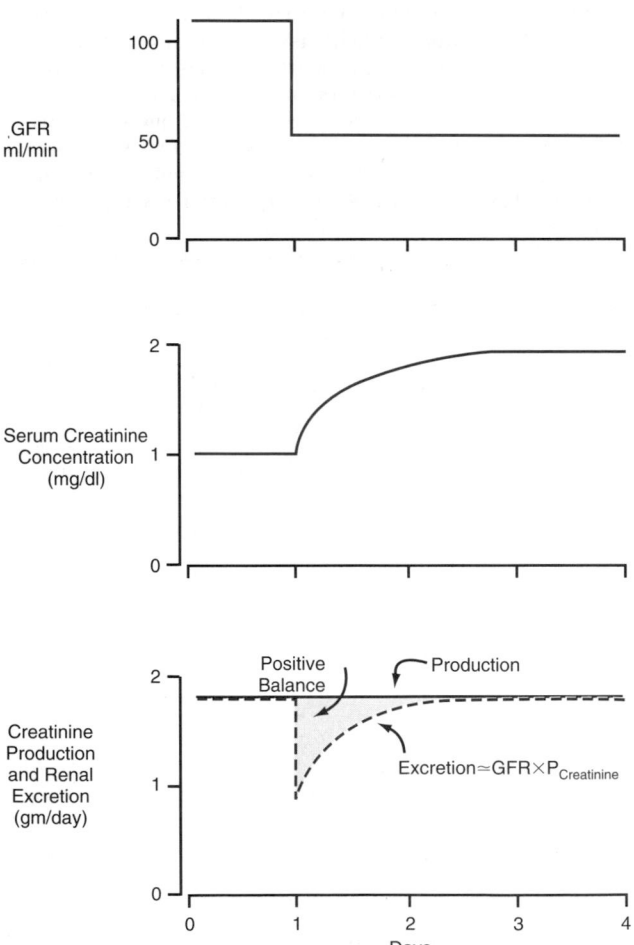

Figure 31–6. Effect of reducing glomerular filtration rate by 50 per cent on serum creatinine concentration and on creatinine excretion rate when the production rate of creatinine remains constant.

gen ions, are often maintained near the normal range until GFR falls below 20 to 30 per cent of normal. Thereafter, the plasma concentrations of these substances rise but not in proportion to the fall in GFR, as shown in curve B of Figure 31–7. Maintenance of relatively constant plasma concentrations of these solutes as GFR declines is accomplished by excreting progressively larger fractions of the amounts of these solutes that are filtered at the glomerular capillaries; this occurs by decreasing the rate of tubular reabsorption or, in some instances, by increasing tubular secretion rates.

In the case of sodium and chloride ions, their plasma concentrations are maintained virtually constant even with severe decreases in GFR (curve C of Fig. 31–7). This is accomplished by greatly decreasing tubular reabsorption of these electrolytes. For example, with a 75 per cent loss of functional nephrons, each surviving nephron must excrete four times as much sodium and four times as much volume as under normal conditions (Table 31–6).

Part of this adaptation occurs because of increased blood flow and increased GFR in each of the surviving nephrons, owing to hypertrophy of the blood vessels and glomeruli, as well as functional changes that cause the blood vessels to vasodilate. Even with large decreases in the total GFR, normal rates of renal excretion can still be maintained by decreasing the rate at which the tubules reabsorb water and solutes.

Table 31-6 TOTAL KIDNEY EXCRETION AND EXCRETION PER NEPHRON IN RENAL FAILURE

	Normal	75% Loss of Nephrons
Number of nephrons	2,000,000	500,000
Total GFR (ml/min)	125	40
Single nephron GFR (nl/min)	62.5	80
Volume excreted for all nephrons (ml/min)	1.5	1.5
Volume excreted per nephron (nl/min)	.75	3.0

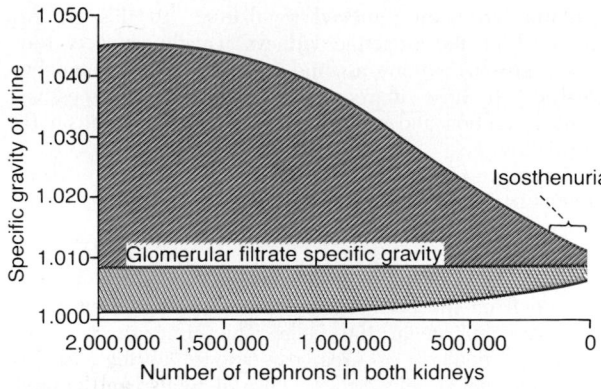

Figure 31-8. Development of isosthenuria in patients with decreased numbers of functional nephrons.

Isosthenuria—Inability of the Kidney to Concentrate or Dilute the Urine. One important effect of the rapid rates of tubular flow that occur in the remaining nephrons of diseased kidneys is that the renal tubules lose their ability to concentrate or dilute the urine. The concentrating ability of the kidney is impaired mainly because (1) the rapid flow of tubular fluid through the collecting ducts prevents adequate water reabsorption and (2) the rapid flow through both the loop of Henle and the collecting ducts prevents the countercurrent mechanism from operating effectively to concentrate the medullary interstitial fluid solutes. Therefore, as progressively more nephrons are destroyed, the maximum concentrating ability of the kidney declines, and urine osmolarity and specific gravity (a measure of the total solute concentration) approach the osmolarity and specific gravity of the glomerular filtrate, as shown in Figure 31-8.

The diluting mechanism in the kidney is also impaired when the number of nephrons decreases because the rapid flushing of fluid through the loops of Henle and the high load of solutes such as urea cause a relatively high solute concentration in the tubular fluid of this part of the nephron. As a consequence, the diluting capacity of the kidney is impaired and the minimal urine osmolality and specific gravity approach those of the glomerular filtrate.

Because the concentrating mechanism becomes impaired to a greater extent than the diluting mechanism in chronic renal failure, an important clinical renal functional test is to determine how well the kidneys can concentrate urine when the person is dehydrated for 12 or more hours.

Effects of Renal Failure on the Body Fluids—Uremia

The effect of renal failure on the body fluids depends on (1) the water and food intake and (2) the degree of impairment of renal function. Assuming that the person with complete renal failure continues to ingest the same amounts of water and food, the concentrations of different substances in the extracellular fluid are approximately those shown in Figure 31-9. Important effects include (1) *generalized edema* resulting from water and salt retention, (2) *acidosis* resulting from failure of the kidneys to rid the body of normal acidic products, (3) *high concentration of the non-protein nitrogens*—especially urea, creatinine, and uric acid—resulting

from failure of the body to excrete the metabolic end products of proteins, and (4) *high concentrations of other substances* excreted by the kidney, including *phenols, sulfates, phosphates, potassium,* and *guanidine bases.* This total condition is called *uremia* because of the high concentration of urea in the body fluids.

Water Retention and Development of Edema in Renal Failure. If water intake is restricted immediately after acute renal failure begins, the total body fluid content may become only slightly increased. If fluid intake is not limited and the patient drinks in response to the normal thirst mechanisms, the body fluids begin to increase immediately and rapidly.

With chronic partial kidney failure, accumulation of fluid may not be severe, as long as salt and fluid intake are not excessive, until kidney function falls to 30 per cent of normal or lower. The reason for this, as discussed previously, is that the surviving nephrons excrete larger amounts of salt and water. Even the small fluid retention that does occur, along with increased secretion of renin and angiotensin II that usually occurs in ischemic kidney disease, often causes severe hypertension in chronic renal failure. In patients with kidney function so reduced as to require dialysis to preserve life, almost all develop hypertension. In most of these patients, severe reduction of salt intake or removal of extracellular fluid by dialysis can control the hypertension. The remaining patients continue to have hypertension even after excess

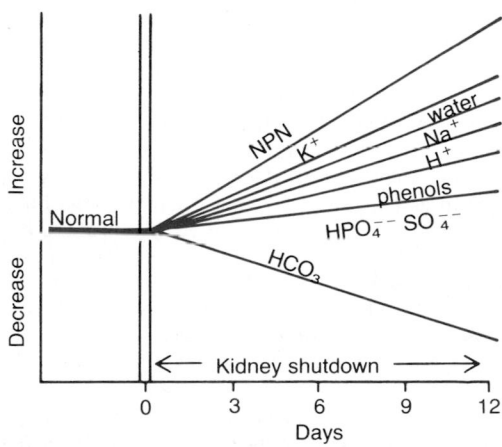

Figure 31-9. Effect of kidney shutdown on extracellular fluid constituents.

sodium has been removed by dialysis. In this group, removal of the ischemic kidneys usually corrects the hypertension (as long as fluid retention is prevented by dialysis) because it removes the source of excessive renin secretion and subsequent increased angiotensin II formation.

UREMIA—INCREASE IN UREA AND OTHER NON-PROTEIN NITROGENS (AZOTEMIA). The non-protein nitrogens include urea, uric acid, creatinine, and a few less important compounds. These, in general, are the end products of protein metabolism and must be removed from the body to ensure continued normal protein metabolism in the cells. The concentrations of these, particularly of urea, can rise to as high as 10 times normal during 1 to 2 weeks of total renal failure. With chronic renal failure, the concentrations rise approximately in proportion to the degree of reduction in functional nephrons. For this reason, measuring the concentrations of these substances, especially of urea and creatinine, provides an important means for assessing the degree of renal failure.

ACIDOSIS IN RENAL FAILURE. Each day the body normally produces about 50 to 80 millimoles more metabolic acid than metabolic alkali. Therefore, when the kidneys fail to function, acid accumulates in the body fluids. The buffers of the body fluids normally can buffer 500 to 1000 millimoles of acid without lethal increases in extracellular fluid hydrogen ion concentration, and the phosphate compounds in the bones can buffer an additional few thousand millimoles of hydrogen ion. However, when this buffering power is used up, the blood pH eventually falls drastically and the patient will become comatose and die as the pH falls below about 6.8.

ANEMIA IN CHRONIC RENAL FAILURE CAUSED BY DECREASED ERYTHROPOIETIN SECRETION. Patients with severe chronic renal failure almost always develop *anemia.* The most important cause of this is decreased renal secretion of *erythropoietin,* which stimulates the bone marrow to produce red blood cells. If the kidneys are seriously damaged, they are unable to form adequate quantities of erythropoietin, which leads to diminished red blood cell production and consequent anemia.

OSTEOMALACIA IN CHRONIC RENAL FAILURE CAUSED BY DECREASED PRODUCTION OF ACTIVE VITAMIN D AND BY PHOSPHATE RETENTION BY THE KIDNEYS. Prolonged renal failure also causes *osteomalacia,* a condition in which the bones are partially absorbed and, therefore, become greatly weakened. An important cause of this condition is the following: vitamin D must be converted by a two-stage process, first in the liver and then in the kidneys, into 1,25-dihydroxycholecalciferol before it is able to promote calcium absorption from the intestine. Therefore, serious damage to the kidney greatly reduces the blood concentration of *active* vitamin D, which in turn decreases intestinal absorption of calcium and the availability of calcium to the bones.

Another important cause of demineralization of the skeleton in chronic renal failure is the rise in serum phosphate concentration that occurs as a result of decreased GFR. This rise in serum phosphate causes increased binding of phosphate with calcium in the plasma, thus decreasing the plasma serum *ionized* calcium concentration, which in turn stimulates *parathyroid hormone* secretion. This secondary hyperparathyroidism then stimulates release of calcium from bones, causing further demineralization of the bones.

Hypertension and Kidney Disease

As discussed earlier in this chapter, hypertension can exacerbate injury to the glomeruli and blood vessels of the kidneys and is, therefore, a major cause of end-stage renal failure. Conversely, abnormalities of kidney function can cause hypertension, as discussed in detail in Chapter 19. Thus, the relation between hypertension and kidney disease can, in some instances, propagate a vicious cycle: primary kidney damage leads to increased blood pressure, which in turn causes further damage to the kidneys, further increases in blood pressure, and so forth, until end-stage renal failure develops.

Not all types of kidney disease cause hypertension because damage to certain portions of the kidney will cause uremia without hypertension. Nevertheless, some types of renal damage are particularly prone to cause hypertension. A classification of kidney disease relative to hypertensive or nonhypertensive effects is the following.

RENAL LESIONS THAT REDUCE THE ABILITY OF THE KIDNEYS TO EXCRETE SODIUM AND WATER PROMOTE HYPERTENSION. Renal lesions that decrease the ability of the kidneys to excrete sodium and water almost invariably cause hypertension. Therefore, lesions that either *decrease* GFR or *increase* tubular reabsorption usually lead to hypertension of varying degrees. Some specific types of renal abnormalities that can cause hypertension are as follows:

1. *Increased renal vascular resistance,* which reduces renal blood flow and GFR. An example is hypertension caused by renal artery stenosis.
2. *Decreased glomerular capillary filtration coefficient,* which reduces GFR. An example of this is chronic glomerulonephritis, which causes inflammation and thickening of the glomerular capillary membranes, thereby reducing the glomerular capillary filtration coefficient.
3. *Excessive tubular sodium reabsorption.* An example is hypertension caused by excessive aldosterone secretion, which increases sodium reabsorption mainly in the cortical collecting tubules.

Once hypertension has developed, renal excretion of sodium and water returns to normal because the high arterial pressure causes pressure natriuresis and pressure diuresis, so that intake and output of sodium and water become balanced once again. Even when there are large increases in renal vascular resistance or decreases in glomerular capillary coefficient, the GFR may still return to nearly normal levels after the arterial blood pressure rises. Likewise, when tubular reabsorption is increased, as occurs with excessive aldosterone secretion, urinary excretion rate is initially reduced but then returns to normal as arterial pressure rises. Thus, after hypertension develops, there may be no sign of impaired excretion of sodium and water other than the hypertension. As explained in Chapter 19, normal excretion of sodium and water at an elevated arterial pressure means that pressure natriuresis and pressure diuresis have been reset to a higher arterial pressure.

HYPERTENSION CAUSED BY PATCHY RENAL DAMAGE AND INCREASED RENAL SECRETION OF RENIN. If one part of the kidney is ischemic and the remainder is not ischemic, such as occurs when one renal artery is severely constricted, the ischemic renal tissue secretes large quantities of renin. This secretion leads to the formation of angiotensin II, which can cause hypertension. The most likely sequence of events in causing this hypertension, as discussed in Chapter 19, is (1) the ischemic kidney tissue itself excretes less than normal amounts of water and salt; (2) the renin secreted by the ischemic kidney, and subsequent increased angiotensin II formation, affects the nonischemic kidney tissue, causing it also to retain salt and water; and (3) excess salt and water cause hypertension in the usual manner.

A similar type of hypertension can result when patchy areas of one or both kidneys become ischemic as a result of arteriosclerosis or vascular injury in patchy portions of the kidneys. When this occurs, the ischemic nephrons excrete less salt and water but secrete greater amounts of renin, which causes increased angiotensin II formation. The high levels of angiotensin II then impair the capability of the surrounding otherwise normal nephrons to excrete sodium and water. As a result, hypertension develops, which restores the overall excretion of sodium and water by the kidney, so that balance between intake and output of salt and water is maintained but at the expense of high blood pressure.

Kidney Diseases That Cause Loss of Entire Nephrons Lead to Renal Failure But May Not Cause Hypertension. Loss of large numbers of whole nephrons, such as occurs with the loss of one kidney and part of another kidney, almost always leads to renal failure if the amount of kidney tissue lost is great enough. If the remaining nephrons are normal and the salt intake is not excessive, this condition might not cause clinically significant hypertension because even a slight rise in blood pressure will raise the GFR and decrease tubular sodium reabsorption sufficiently to promote enough water and salt excretion in the urine, even with the few nephrons that remain intact. On the other hand, a patient with this type of abnormality may become severely hypertensive if additional stresses are imposed, such as eating a large amount of salt. In this case, the kidneys simply cannot clear adequate quantities of salt with the small number of functioning nephrons that remain.

SPECIFIC TUBULAR DISORDERS

In Chapter 27, it is pointed out that specific transport mechanisms are responsible for transporting different individual substances across the tubular epithelial membranes. In Chapter 3, it is also pointed out that each cellular enzyme and each carrier protein is formed in response to a respective gene in the nucleus. If any required gene happens to be absent or abnormal, the tubules may be deficient in one of the appropriate carrier proteins or one of the required enzymes needed for solute transport by the renal tubular epithelial cells. For this reason, many hereditary tubular disorders occur for the transport of individual substances or groups of substances through the tubular membrane. In addition, damage to the tubular epithelial membrane by toxins or ischemia can cause important renal tubular disorders.

Renal Glycosuria—Failure of the Kidneys to Reabsorb Glucose. In this condition, the blood glucose concentration may be normal, but the transport mechanism for tubular reabsorption of glucose is greatly limited or absent. Consequently, despite a normal blood glucose level, large amounts of glucose pass into the urine each day. Because diabetes mellitus is also associated with the presence of glucose in the urine, renal glycosuria, which is a relatively benign condition, must be ruled out before making a diagnosis of diabetes mellitus.

Aminoaciduria—Failure of the Kidneys to Reabsorb Amino Acids. Some amino acids share mutual transport systems for reabsorption, whereas other amino acids have distinct transport systems of their own. Rarely, a condition called *generalized aminoaciduria* results from deficient reabsorption of all amino acids; more frequently, deficiencies of specific carrier systems may result in (1) *essential cystinuria*, in which large amounts of cystine fail to be reabsorbed and often crystallize in the urine to form renal stones; (2) *simple glycinuria*, in which glycine fails to be reabsorbed; or (3) *beta-amino-isobutyric aciduria*, which occurs in about 5 per cent of all people but apparently has no major clinical significance.

Renal Hypophosphatemia—Failure of the Kidneys to Reabsorb Phosphate. In renal hypophosphatemia, the renal tubules fail to reabsorb large enough quantities of phosphate ions when the phosphate concentration of the body fluids falls very low. This condition usually does not cause serious immediate abnormalities because phosphate concentration of the extracellular fluids can vary widely without causing major cellular dysfunction. Over a long period, a low phosphate level causes diminished calcification of the bones, thus causing the person to develop rickets. Furthermore, this type of rickets is refractory to vitamin D therapy, in contrast to rapid response of the usual type of rickets, as discussed in Chapter 79.

Renal Tubular Acidosis—Failure of the Tubules to Secrete Hydrogen Ions. In this condition, the renal tubules are unable to secrete adequate amounts of hydrogen ions. As a result, large amounts of sodium bicarbonate are continually lost in the urine. This causes a continued state of metabolic acidosis in the blood and extracellular fluid, as discussed in Chapter 30. This type of renal abnormality can be caused by hereditary disorders or it can occur as a result of widespread injury to the renal tubules.

Nephrogenic Diabetes Insipidus—Failure of the Kidneys to Respond to Antidiuretic Hormone. Occasionally, the renal tubules do not respond to antidiuretic hormone, causing large quantities of dilute urine to be excreted. As long as the person is supplied with plenty of water, this condition seldom causes severe difficulty. However, when adequate quantities of water are not available, the person rapidly becomes dehydrated.

Fanconi Syndrome—A Generalized Reabsorptive Defect of the Renal Tubules. Fanconi syndrome is usually associated with increased urinary excretion of virtually all amino acids, glucose, and phosphate. In severe cases, other manifestations are also observed, such as (1) failure to reabsorb sodium bicarbonate, which results in metabolic acidosis; (2) increased excretion of potassium and sometimes calcium; and (3) nephrogenic diabetes insipidus.

There are multiple causes of Fanconi syndrome, which results from a generalized inability of the renal tubular cells to transport various substances. Some of these causes include (1) hereditary defects in cell transport mechanisms, (2) toxins or drugs that injure the renal tubular epithelial cells, and (3) injury to the renal tubular cells as a result of ischemia. The proximal tubular cells are especially affected in Fanconi syndrome caused by tubular injury because these cells reabsorb and secrete many of the drugs and toxins that can cause damage.

TREATMENT OF RENAL FAILURE BY DIALYSIS WITH AN ARTIFICIAL KIDNEY

Severe loss of kidney function, either acutely or chronically, is a threat to life and requires removal of toxic waste products and restoration of body fluid volume and composition toward normal. This can be accomplished by dialysis with an artificial kidney. In certain types of acute renal failure, an artificial kidney may be used to tide the patient over until the kidneys resume their function. If the loss of kidney function is irreversible, it is necessary to perform dialysis chronically to maintain life. Thousands of people with irreversible renal failure or even total kidney removal are being maintained for 15 to 20 years by dialysis with artificial kidneys. Because dialysis cannot maintain completely normal body fluid composition and cannot replace all the multiple functions performed by the kidney, the health of patients maintained on artificial kidneys usually remains significantly impaired. A better treatment for permanent loss of kidney function is to restore functional kidney tissue by means of a kidney transplant.

BASIC PRINCIPLES OF DIALYSIS. The basic principle of the artificial kidney is to pass blood through minute blood channels bounded by a thin membrane. On the other side of the membrane is a *dialyzing fluid* into which unwanted substances in the blood pass by diffusion.

Figure 31–10 shows the components of one type of artificial kidney in which blood flows continually between two thin membranes of cellophane; outside the membrane is a dialyzing fluid. The cellophane is porous enough to allow the constituents of the plasma, except the plasma proteins, to diffuse in both directions—from plasma into the dialyzing fluid or from the dialyzing fluid back into the plasma. If the concentration of a substance is greater in the plasma than in the dialyzing fluid, there will be *net* transfer of the substance from the plasma into the dialyzing fluid.

The rate of movement of solute across the dialyzing membrane depends on (1) the concentration gradient of the solute between the two solutions, (2) the permeability of the membrane to the solute, (3) the surface area of the membrane, and (4) the length of time that the blood and fluid remain in contact with the membrane.

Thus, the maximum rate of solute transfer occurs initially when the concentration gradient is greatest (when dialysis is begun) and slows down as the concentration gradient is dissipated. In a flowing system, as is the case with "hemodialysis," in which blood and dialysate fluid flow through the artificial kidney, the dissipation of the concentration gradient can be reduced and diffusion of solute across the membrane can be optimized by increasing the flow rate of either or both the blood and the dialyzing fluid.

In normal operation of the artificial kidney, blood flows continually or intermittently back into the vein. The total amount of blood in the artificial kidney at any

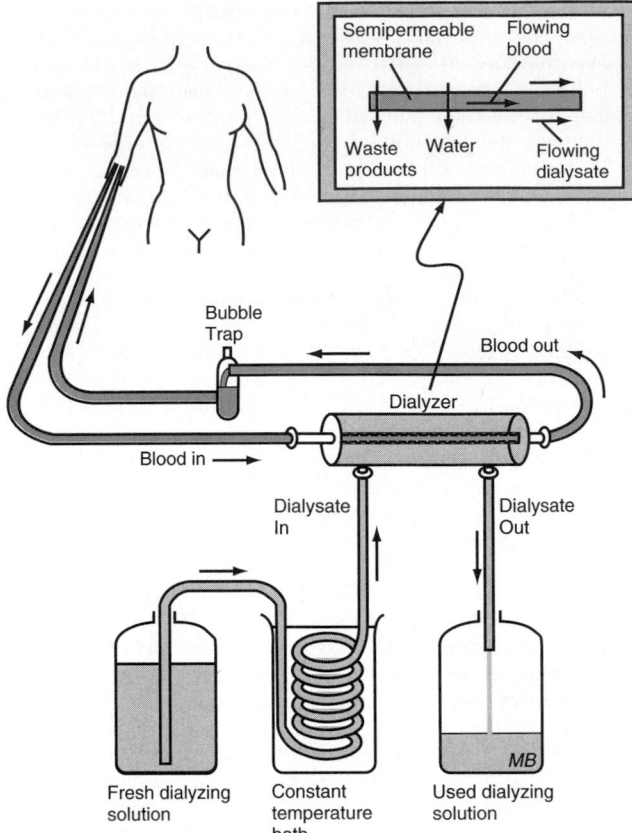

Figure 31–10. Principles of dialysis with an artificial kidney.

one time is usually less than 500 milliliters, the rate of flow may be several hundred milliliters per minute, and the total diffusion surface area is between 0.6 and 2.5 square meters. To prevent coagulation of the blood in the artificial kidney, a small amount of heparin is infused into the blood as it enters the artificial kidney.

In addition to diffusion of solutes, mass transfer of solutes and water can be produced by applying a hydrostatic pressure to force the fluid and solutes by the process of filtration across the membranes of the dialyzer; such filtration is called "bulk flow."

DIALYZING FLUID. Table 31–7 compares the constituents in a typical dialyzing fluid with those in normal plasma and uremic plasma. Note that the concentrations of the ions and other substances in dialyzing fluid are not the same as the concentrations in normal plasma or in uremic plasma. Instead, they are adjusted to levels that are needed to cause appropriate movement of water and solutes through the membrane during the dialysis.

Note that there is no phosphate, urea, urate, sulfate, or creatinine in the dialyzing fluid; however, these are present in high concentrations in the uremic blood. Therefore, when the uremic patient is dialyzed, these substances are lost in large quantities into the dialyzing fluid.

The effectiveness of the artificial kidney can be expressed in terms of the amount of plasma that is cleared of different substances each minute, which, as discussed in Chapter 27, is the primary means for expressing the functional effectiveness of the kidneys themselves to rid

Table 31–7 COMPARISON OF DIALYZING FLUID WITH NORMAL AND UREMIC PLASMA

Constituent	Normal Plasma	Dialyzing Fluid	Uremic Plasma
Electrolytes (mEq/liter)			
Na^+	142	133	142
K^+	5	1.0	7
Ca^{++}	3	3.0	2
Mg^{++}	1.5	1.5	1.5
Cl^-	107	105	107
HCO_3^-	24	35.7	14
$Lactate^-$	1.2	1.2	1.2
$HPO_4^=$	3	0	9
$Urate^-$	0.3	0	2
$Sulfate^=$	0.5	0	3
Nonelectrolytes (mg/dl)			
Glucose	100	125	100
Urea	26	0	200
Creatinine	1	0	6

the body of unwanted substances. Most artificial kidneys can clear urea from the plasma at a rate of 100 to 225 ml/min, which shows that at least for the excretion of urea, the artificial kidney can function about twice as rapidly as two normal kidneys together, whose urea clearance is only 70 ml/min. Yet the artificial kidney is used for only 4 to 6 hours per day, three times a week. Therefore, the overall plasma clearance is still considerably limited when the artificial kidney replaces the normal kidneys. Also, it is important to keep in mind that the artificial kidney cannot replace some of the other functions of the kidneys, such as secretion of erythropoietin, which is necessary for red blood cell production.

REFERENCES

Adler, S., et al.: Diabetic nephropathy: Pathogenesis and treatment. Annu. Rev. Med., 44:303, 1993.

Appel, G. B., et al.: Renal vascular complications of systemic lupus erythematosus. J. Am. Soc. Nephrol., 4:1499, 1994.

Beck, F., et al.: Pathophysiology and pathobiochemistry of acute renal failure. In Seldin, D. W. and Giebisch, G. (eds.): The Kidney: Physiology and Pathophysiology. New York, Raven Press, 1992.

Bonventre, J. V.: Mechanisms of ischemic acute renal failure. Kidney Int., 43:1160, 1993.

Brater, D. C.: Resistance to diuretics: Mechanisms and clinical implications. Adv. Nephrol., 22:349, 1993.

Brenner, B. M., and Lazarus, J. M. (eds.): Acute Renal Failure. Philadelphia, W. B. Saunders Co., 1983.

Brenner, B. M., et al.: Clinical Nephrology. Philadelphia, W. B. Saunders Co., 1987.

Brenner, B. M., et al.: Glomeruli and blood pressure—less of one, more of the other? Am. J. Hypertens., 1:335, 1988.

Brenner, B. M.: Nephron adaptation to renal injury or ablation. Am. J. Physiol., 249:F324, 1985.

Capasso, G., et al.: Acidification in mammalian cortical distal tubule. Kidney Int., 45:1543, 1994.

D'Agati, V.: The many masks of focal segmental glomerulosclerosis. Kidney Int., 46:1223, 1994.

Depner, T. A.: Assessing adequacy of hemodialysis: Urea modeling. Kidney Int., 45:1522, 1994.

Dirks, J. H., and Sulton, R. A. L. (eds.): Diuretics: Physiology, Pharmacology and Clinical Use. Philadelphia, W. B. Saunders Co., 1986.

Fine, L. G., and Norman, J. T.: Renal growth responses to acute and chronic injury: Routes to therapeutic intervention. J. Am. Soc. Nephrol., 2:S206, 1992.

Guyton, A. C., et al.: The dominant role of the kidneys in long-term arterial pressure regulation in normal and hypertensive states. In Laragh, J. H., and Brenner, B. M. (eds.): Hypertension: Pathophysiology, Diagnosis and Management. New York, Raven Press, 1994.

Hall, J. E., and Granger, J. P.: Role of sodium and fluid excretion in hypertension. In Swales, J. D. (ed.): Textbook of Hypertension. Oxford, Blackwell Scientific Publishing, 1994.

Hall, J. E., et al.: Abnormal pressure natriuresis: A cause or a consequence of hypertension? Hypertension 15:547, 1990.

Hall, J. E.: Renal and cardiovascular mechanisms of hypertension in obesity. Hypertension, 23:381, 1994.

Hall, J. E.: Renal function in 1-kidney, 1-clip hypertension and low-renin essential hypertension. Am. J. Hypertens., 4(Suppl.): 523s, 1991.

Hammerman, M. R., et al.: Role of growth factors in regulation of renal growth. Annu. Rev. Physiol. 55:305, 1993.

Hester, R. L., et al.: Non-invasive determination of recirculation in the patient on dialysis. ASAIO J., 38:M190, 1992.

Humphreys, M. H.: Mechanisms and management of nephrotic syndrome. Kidney Int., 43:1160, 1993.

Johnson, R. J.: The glomerular response to injury: progression or resolution? Kidney Int., 45:1769, 1994.

Keane, W. F., et al.: Progression of renal disease. Kidney Int., 45(Suppl. 45) S-1, 1994.

Koeppen, B. M., and Stanton, B. A.: Renal Physiology. St. Louis, Mosby-Year Book, 1992.

Kurtzman, N. A.: Disorders of distal acidification. Kidney Int., 38:720, 1990.

Leaf, A., and Cotran, R. S.: Renal Pathophysiology, 2nd Ed. New York, Oxford University Press, 1980.

Massry, S. G., and Glassock, R. J.: Textbook of Nephrology, 2nd Ed. Baltimore, Williams & Wilkins, 1988.

Rose, B.: Diuretics. Kidney Int., 39:336, 1991.

Schrier, R. W., and Gottschalk, C. W. (eds.): Diseases of the Kidney, 5th Ed. Boston, Little, Brown, 1992.

Seidman, E. J. (ed.): Current Urologic Therapy. Philadelphia, W. B. Saunders Co., 1994.

Seldin, D. W., and Giebisch, G. (eds.): The Regulation of Acid-Base Balance. New York, Raven Press, 1990.

BLOOD CELLS, IMMUNITY, AND BLOOD CLOTTING

UNIT VI

Red Blood Cells, Anemia, and Polycythemia

CHAPTER 32

With this chapter we begin a discussion of the blood cells and other cells closely related to those of the blood: the cells of the macrophage system and the lymphatic system. We first present the functions of red blood cells, which are the most abundant of all the cells of the body and are necessary for delivery of oxygen to the tissues.

RED BLOOD CELLS

The major function of red blood cells, also known as erythrocytes, is to transport *hemoglobin,* which in turn carries oxygen from the lungs to the tissues. In some lower animals, hemoglobin circulates as free protein in the plasma, not enclosed in red blood cells. When it is free in the plasma of the human being, about 3 per cent of it leaks through the capillary membrane into the tissue spaces or through the glomerular membrane of the kidney into the glomerular filtrate each time the blood passes through the capillaries. Therefore, for hemoglobin to remain in the blood stream, it must exist inside red blood cells.

The red blood cells have other functions besides transport of hemoglobin. For instance, they contain a large quantity of *carbonic anhydrase,* which catalyzes the reaction between carbon dioxide and water, increasing the rate of this reversible reaction several thousandfold. The rapidity of this reaction makes it possible for the water in blood to react with large quantities of carbon dioxide and thereby transport it from the tissues to the lungs in the form of the bicarbonate ion (HCO_3^-). Also, the hemoglobin in the cells is an excellent *acid-base buffer* (as is true of most proteins), so that the red blood cells are responsible for most of the buffering power of whole blood.

SHAPE AND SIZE OF RED BLOOD CELLS. Normal red blood cells, shown in Figure 32–3, are biconcave discs having a mean diameter of about 7.8 micrometers and a thickness at the thickest point of 2.5 micrometers and in the center of 1 micrometer or less. The average volume of the red blood cell is 90 to 95 cubic micrometers.

The shapes of red blood cells can change remarkably as the cells pass through capillaries. Actually, the red blood cell is a "bag" that can be deformed into almost any shape. Furthermore, because the normal cell has a great excess of cell membrane for the quantity of material inside, deformation does not stretch the membrane greatly and, consequently, does not rupture the cell, as would be the case with many other cells.

CONCENTRATION OF RED BLOOD CELLS IN THE BLOOD. In normal men, the average number of red blood cells per cubic millimeter is 5,200,000 ($\pm$ 300,000) and in normal women, 4,700,000 ($\pm$ 300,000). The altitude at which the person lives affects the number of red blood cells he or she has. This is discussed later.

QUANTITY OF HEMOGLOBIN IN THE CELLS. Red blood cells have the ability to concentrate hemoglobin in the cell fluid up to about 34 gm/dl of cells. The concentration never rises above this value because this is a metabolic limit of the cell's hemoglobin-forming mechanism. Furthermore, in normal people, the percentage of hemoglobin is almost always near the maximum in each cell. However, when hemoglobin formation is deficient in the bone marrow, the percentage of hemoglobin in the cells may fall considerably below this value, and the volume of the red cell may decrease as well because of diminished hemoglobin to fill the cell.

When the hematocrit (the percentage of the blood that is cells—normally 40 to 45 per cent) and the quantity of hemoglobin in each respective cell are normal, the whole blood of men contains an average of 16 grams of hemoglobin per

425

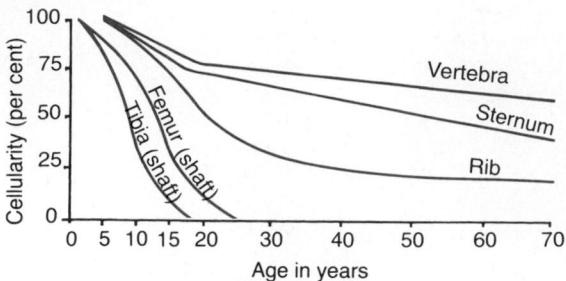

Figure 32–1. Relative rates of red blood cell production in the different bones at different ages.

deciliter and of women, an average of 14 gm/dl. As discussed in connection with the transport of oxygen in Chapter 40, each gram of pure hemoglobin is capable of combining with about 1.39 milliliters of oxygen. Therefore, in normal man, more than 21 milliliters of oxygen can be carried in combination with hemoglobin in each deciliter of blood, and in normal woman, 19 milliliters of oxygen can be carried.

Production of Red Blood Cells

AREAS OF THE BODY THAT PRODUCE RED BLOOD CELLS. In the early few weeks of embryonic life, primitive, nucleated red blood cells are produced in the *yolk sac.* During the middle trimester of gestation, the *liver* is the main organ for production of red blood cells, although reasonable numbers of red blood cells are also produced by the *spleen* and *lymph nodes.* Then, during the last month or so of gestation and after birth, red blood cells are produced exclusively by the *bone marrow.*

As demonstrated in Figure 32–1, the bone marrow of essentially all bones produces red blood cells until a person is 5 years old; but the marrow of the long bones, except for the proximal portions of the humeri and tibiae, becomes quite fatty and produces no more red blood cells after about age 20 years. Beyond this age, most red cells are produced in the marrow of the membranous bones, such as the vertebrae, sternum, ribs, and ilia. Even in these bones, the marrow becomes less productive as age increases.

Genesis of Blood Cells

PLURIPOTENTIAL HEMOPOIETIC STEM CELLS, GROWTH INDUCERS, AND DIFFERENTIATION INDUCERS. In the bone marrow are cells called *pluripotential hemopoietic stem cells,* from which all the cells in the circulating blood are derived. Figure 32–2 shows the successive divisions of the pluripotential cells to form the different peripheral blood cells. As these cells reproduce, continuing throughout the life of the person, a portion of them remains exactly like the original pluripotential cells and is retained in the bone marrow to maintain a supply of these, although their numbers do diminish with age. The larger portion of the reproduced stem cells, however, differentiates to form the other cells shown to the right in Figure 32–2. The early offspring still cannot be recognized as different from the pluripotential stem cells, even though they have already become committed to a particular line of cells and are called *committed stem cells.*

The different committed stem cells, when grown in cul-

ture, will produce colonies of specific types of blood cells. A committed stem cell that produces erythrocytes is called a *colony-forming unit–erythrocyte,* and the abbreviation CFU-E is used to designate this type of stem cell. Likewise, colony-forming units that form granulocytes and monocytes have the designation CFU-GM, and so forth.

Growth and reproduction of the different stem cells are controlled by multiple proteins called *growth inducers.* Four major growth inducers have been described, each having different characteristics. One of these, *interleukin-3,* promotes growth and reproduction of virtually all the different types of stem cells, whereas the others induce growth of only specific types of committed stem cells.

The growth inducers promote growth but not differentiation of the cells. This is the function of still another set of proteins, called *differentiation inducers.* Each of these causes one type of stem cell to differentiate one or more steps toward a final type of adult blood cell.

Formation of the growth inducers and differentiation inducers is itself controlled by factors outside the bone marrow. For instance, in the case of red blood cells, exposure of the body to low oxygen for a long period results in growth induction, differentiation, and production of greatly increased numbers of erythrocytes, as we shall discuss later in the chapter. In the case of some of the white blood cells, infectious diseases cause growth, differentiation, and eventual formation of specific types of white blood cells that are needed to combat the infection.

Stages of Differentiation of Red Blood Cells

The first cell that can be identified as belonging to the red blood cell series is the *proerythroblast,* shown in Figure 32–3. Under appropriate stimulation, large numbers of these cells are formed from the CFU-E stem cells.

Once the proerythroblast has been formed, it divides several more times, eventually forming many mature red blood cells. The first-generation cells are called *basophil erythroblasts* because they stain with basic dyes; the cell at this time has accumulated very little hemoglobin. In the succeeding generations, as shown in Figure 32–3, the cells become filled with hemoglobin to a concentration of about 34 per cent, the nucleus condenses to a small size, and its final remnant is extruded from the cell. At the same time, the endoplasmic reticulum is reabsorbed. The cell at this stage is called a *reticulocyte* because it still contains a small amount of basophilic material, consisting of remnants of the Golgi apparatus, mitochondria, and a few other cytoplasmic organelles. During this reticulocyte stage, the cells pass from the bone marrow into the blood capillaries by diapedesis (squeezing through the pores of the capillary membrane).

The remaining basophilic material in the reticulocyte normally disappears within 1 to 2 days, and the cell is then the *mature erythrocyte.* Because of the short life of the reticulocytes, their concentration among all the red cells of the blood is normally slightly less than 1 per cent.

Regulation of Red Blood Cell Production—Role of Erythropoietin

The total mass of red blood cells in the circulatory system is regulated within narrow limits, so that an adequate number of red cells is always available to provide sufficient tissue oxygenation and yet so that the cells do not become so concentrated that they impede blood flow. What we know

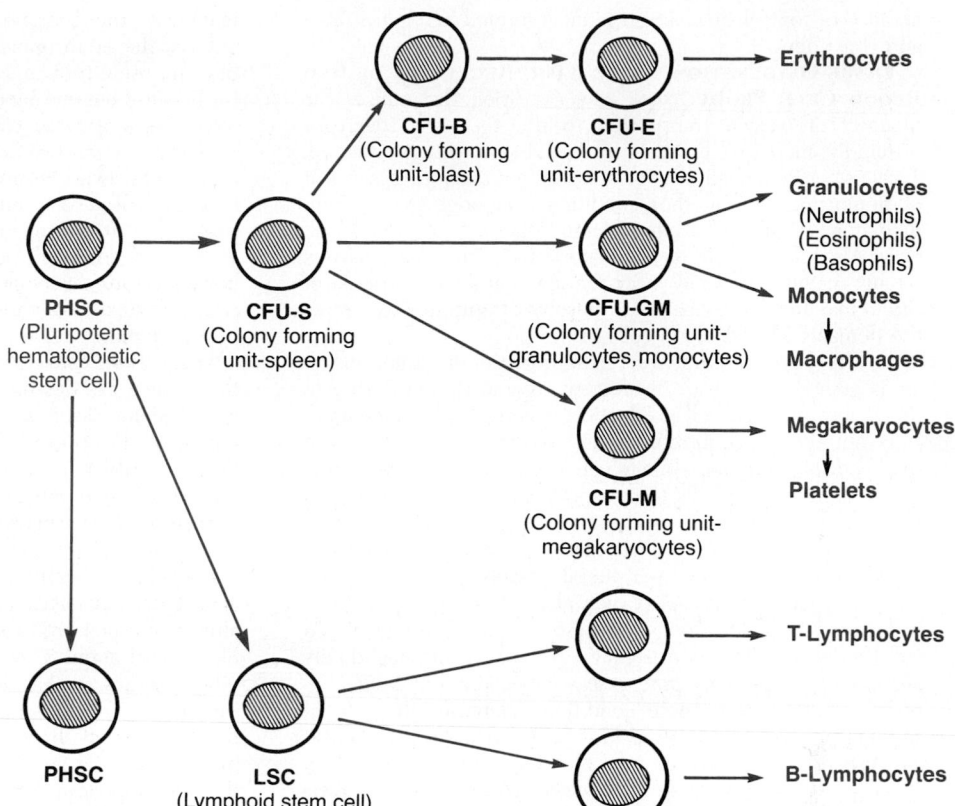

Figure 32–2. Formation of the multiple different peripheral blood cells from the original pluripotential hematopoietic stem cell (PHSC) in the bone marrow.

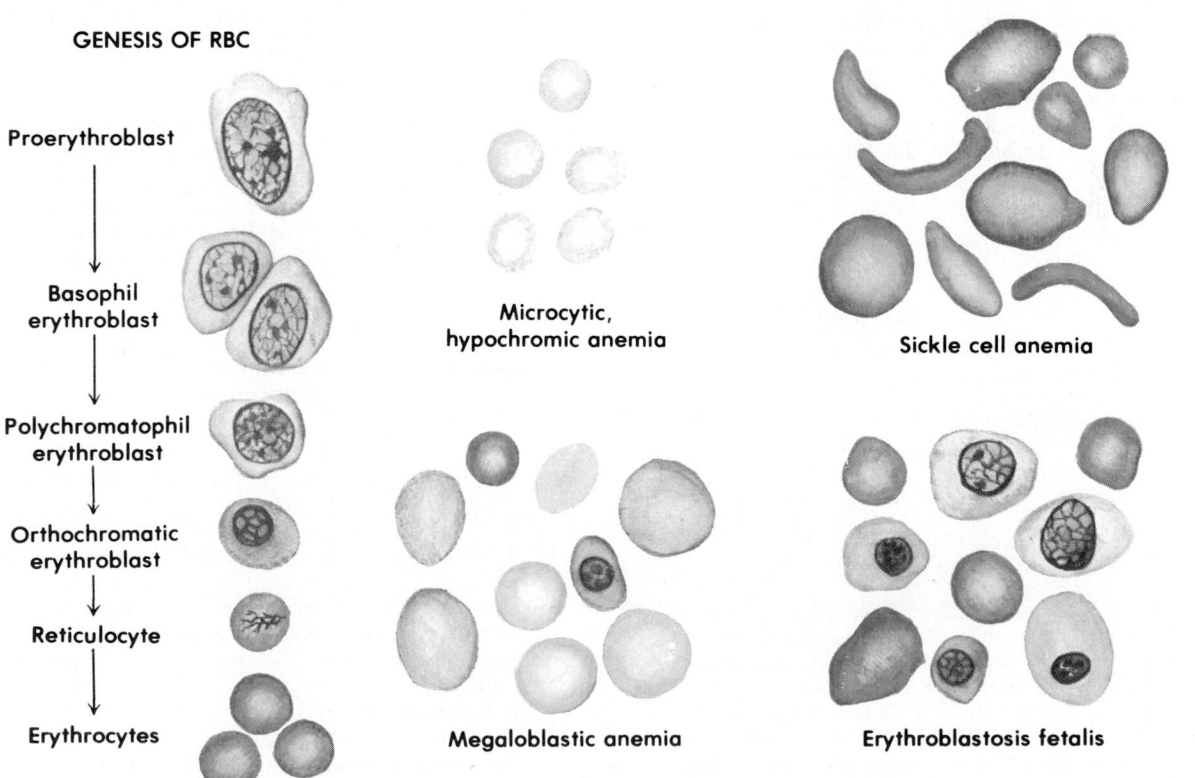

Figure 32–3. Genesis of red blood cells and red blood cells in different types of anemias.

about this control mechanism is diagrammed in Figure 32–4 and is as follows.

TISSUE OXYGENATION AS THE BASIC REGULATOR OF RED BLOOD CELL PRODUCTION. Any condition that causes the quantity of oxygen transported to the tissues to decrease ordinarily increases the rate of red blood cell production. Thus, when a person becomes extremely *anemic* as a result of hemorrhage or another condition, the bone marrow immediately begins to produce large quantities of red blood cells. Also, destruction of major portions of the bone marrow by any means, especially x-ray therapy, causes hyperplasia of the remaining bone marrow, thereby attempting to supply the demand for red blood cells in the body.

At very *high altitudes,* where the quantity of oxygen in the air is greatly decreased, insufficient oxygen is transported to the tissues and red cell production is considerably increased. It is not the concentration of red blood cells in the blood that controls the rate of red cell production but the functional ability of the cells to transport oxygen to the tissues in relation to the tissue demand for oxygen.

Various diseases of the circulation that cause decreased blood flow through the peripheral vessels, and particularly those that cause failure of oxygen absorption by the blood as it passes through the lungs, can also increase the rate of red cell production. This is especially apparent in prolonged *cardiac failure* and many *lung diseases* because the tissue hypoxia resulting from these conditions increases the rate of red cell production, with resultant increase in the hematocrit and usually increase in the total blood volume.

ERYTHROPOIETIN, ITS FUNCTION TO STIMULATE RED CELL PRODUCTION, AND ITS FORMATION IN RESPONSE TO HYPOXIA. The principal factor that stimulates red blood cell production is a circulating hormone called *erythropoietin,* a glycoprotein with a molecular weight of about 34,000. In the absence of erythropoietin, hypoxia has little or no effect in stimulating red blood cell production. On the other hand, when the erythropoietin system is functional, hypoxia causes marked increase in erythropoietin production, and the erythropoietin in turn enhances red blood cell production until the hypoxia is relieved.

Role of the Kidneys in the Formation of Erythropoietin. In the normal person, about 90 per cent of all erythropoietin is formed in the kidneys; the remainder is formed mainly in the liver. It is not known exactly where in the kidneys the erythropoietin is formed. One likely possibility is that the renal tubular epithelial cells secrete the erythropoietin because anemic blood is unable to deliver enough oxygen from the peritubular capillaries to the highly oxygen-consuming tubular cells, thus stimulating erythropoietin production.

At times, hypoxia in other parts of the body but not in the kidneys will also stimulate erythropoietin secretion, which suggests that there might be some nonrenal sensor that sends an additional signal to the kidneys to produce this hormone. In particular, both norepinephrine and epinephrine and several of the prostaglandins stimulate erythropoietin production.

When both kidneys are removed from a person or when the kidneys are destroyed by renal disease, the person invariably becomes very anemic because the 10 per cent of the normal erythropoietin formed in other tissues (mainly in the liver) is sufficient to cause only one third to one half as much red blood cell formation as needed by the body.

Effect of Erythropoietin on Erythrogenesis. On placing an animal or a person in an atmosphere of low oxygen, erythropoietin begins to be formed within minutes to hours, and it reaches maximum production within 24 hours. Yet almost no new red blood cells appear in the circulating blood until about 5 days later. From this fact, as well as still other studies, it has been determined that the important effect of erythropoietin is to stimulate the production of proerythroblasts from hemopoietic stem cells in the bone marrow. In addition, once the proerythroblasts are formed, the erythropoietin causes these cells also to pass more rapidly through the different erythroblastic stages than normally, further speeding up the production of new cells. The rapid production of cells continues as long as the person remains in the low oxygen state or until enough red blood cells are produced to carry adequate amounts of oxygen to the tissues despite the low oxygen; at this time, the rate of erythropoietin production decreases to a level that will maintain the required number of red cells but not an excess.

In the absence of erythropoietin, few red blood cells are formed by the bone marrow. At the other extreme, when large quantities of erythropoietin are formed and if there is plenty of iron available and other required nutrients, the rate of red blood cell production can rise to perhaps ten or more times normal. Therefore, the erythropoietin control mechanism for red blood cell production is a powerful one.

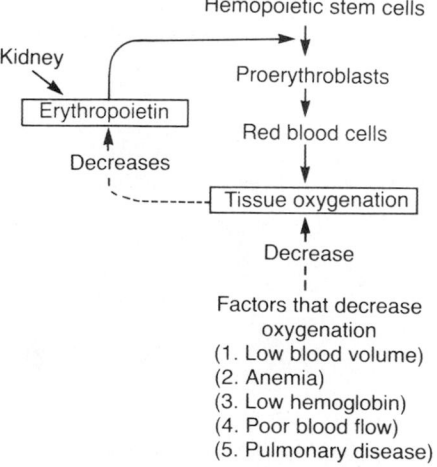

Figure 32–4. Function of the erythropoietin mechanism to increase the production of red blood cells when various factors decrease tissue oxygenation.

Maturation of Red Blood Cells—Requirement for Vitamin B₁₂ (Cyanocobalamin) and Folic Acid

Because of the continuing need to replenish red blood cells, the cells of the bone marrow are among the most rapidly growing and reproducing cells of the entire body. Therefore, as would be expected, their maturation and rate of production are affected greatly by a person's nutritional status.

Especially important for final maturation of the red blood cells are the two vitamins, *vitamin B₁₂* and *folic acid.* Both of these are essential for the synthesis of DNA because each in a different way is required for the formation of thymidine triphosphate, one of the essential building blocks of DNA. Therefore, lack of either vitamin B₁₂ or folic acid causes diminished DNA and, consequently, failure of nuclear maturation and division. Furthermore, the erythroblastic cells of the bone marrow, in addition to failing to proliferate rapidly, produce mainly larger than normal red cells called *macro-*

cytes, and the cell has a flimsy membrane and is often irregular, large, and oval instead of the usual biconcave disc. These poorly formed cells, after entering the circulating blood, are capable of carrying oxygen normally, but their fragility causes them to have a short life, one-half to one-third normal. Therefore, it is said that vitamin B_{12} or folic acid deficiency causes *maturation failure* in the process of erythropoiesis.

The cause of the abnormal cells seems to be as follows: The inability of the cells to synthesize adequate quantities of DNA leads to slow reproduction of the cells but does not prevent excess formation of RNA by the DNA in those cells that do succeed in being produced. Therefore, the quantity of RNA in each cell becomes greater than normal, leading to excess production of cytoplasmic hemoglobin and other constituents, which causes the cells to enlarge. Yet, because of possible abnormalities of all the cell's DNA, the structural components of the cell membrane and cytoskeleton are also malformed, which leads to abnormal cell shapes and especially greatly increased cell membrane fragility.

Maturation Failure Caused by Poor Absorption of Vitamin B_{12}—Pernicious Anemia. A common cause of maturation failure is failure to absorb vitamin B_{12} from the gastrointestinal tract. This often occurs in *pernicious anemia,* in which the basic abnormality is an *atrophic gastric mucosa* that fails to secrete normal gastric secretions. The parietal cells of the gastric glands secrete a glycoprotein called *intrinsic factor,* which combines with vitamin B_{12} of the food and makes the B_{12} available for absorption by the gut. It does this in the following way: (1) The intrinsic factor binds tightly with the vitamin B_{12}. In this bound state, the B_{12} is protected from digestion by the gastrointestinal enzymes. (2) Still in the bound state, the intrinsic factor binds to specific receptor sites on the brush border membranes of the mucosal cells in the ileum. (3) Vitamin B_{12} is transported into the blood during the next few hours by the process of pinocytosis, carrying the intrinsic factor and the vitamin together through the membrane.

Lack of intrinsic factor, therefore, causes loss of much of the vitamin because of both digestive enzyme action in the gut and failure of its absorption.

Once vitamin B_{12} has been absorbed from the gastrointestinal tract, it is stored in large quantities in the liver and then released slowly as needed to the bone marrow and other tissues of the body. The minimum amount of vitamin B_{12} required each day to maintain normal red cell maturation is only 1 to 3 micrograms, and the normal store in the liver and other body tissues is about 1000 times this amount. Therefore, 3 to 4 years of defective B_{12} absorption are required to cause maturation failure anemia.

Failure of Maturation Caused by Deficiency of Folic Acid (Pteroylglutamic Acid). Folic acid is a normal constituent of green vegetables, some fruits, liver, and other meats. However, it is easily destroyed during cooking. Also, people with gastrointestinal absorption abnormalities, such as the frequently occurring small intestinal disease called *sprue,* often have serious difficulty in absorbing both folic acid and vitamin B_{12}. Therefore, in many instances of maturation failure, the cause is deficiency of absorption of both folic acid and vitamin B_{12}.

Formation of Hemoglobin

Synthesis of hemoglobin begins in the proerythroblasts and continues slightly even into the reticulocyte stage because when the reticulocytes leave the bone marrow and

I.   2 succinyl-CoA + 2 glycine ⟶ (pyrrole)

II.   4 pyrrole ⟶ protoporphyrin IX
III. protoporphyrin IX + Fe^{++} ⟶ heme
IV.   heme + polypeptide ⟶ hemoglobin chain (α or β)
V.   2 α chains + 2 β chains ⟶ hemoglobin A

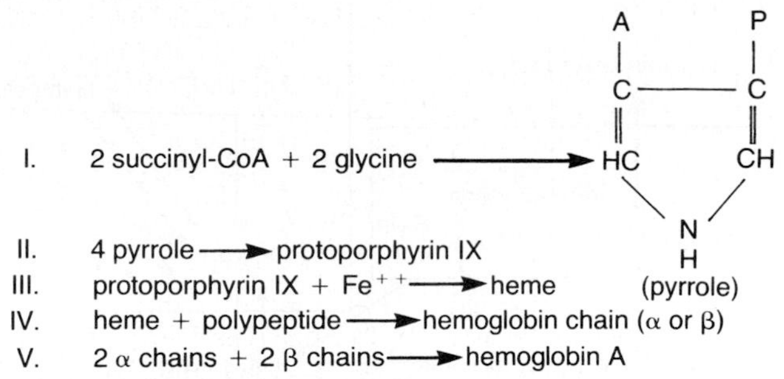

Figure 32–5. Formation of hemoglobin.

pass into the blood stream, they continue to form minute quantities of hemoglobin for another day or so.

Figure 32–5 shows the basic chemical steps in the formation of hemoglobin. First, succinyl-CoA, formed in the Krebs' cycle as explained in Chapter 67, binds with glycine to form a pyrrole molecule. In turn, four pyrroles combine to form protoporphyrin IX, which then combines with iron to form the *heme* molecule. Finally, each heme molecule combines with a long polypeptide chain, called a *globin,* synthesized by the ribosomes, forming a subunit of hemoglobin called a *hemoglobin chain* (Fig. 32–6). Each of these chains has a molecular weight of about 16,000; four of them in turn bind together loosely to form the whole hemoglobin molecule.

There are several slight variations in different subunit hemoglobin chains, depending on the amino acid composition of the polypeptide portion. The different types of chains are designated *alpha chains, beta chains, gamma chains,* and *delta chains.* The most common form of hemoglobin in the adult human being, *hemoglobin A,* is a combination of *two alpha chains* and *two beta chains.*

Because each chain has a heme prosthetic group, there are 4 iron atoms in each hemoglobin molecule; each of these

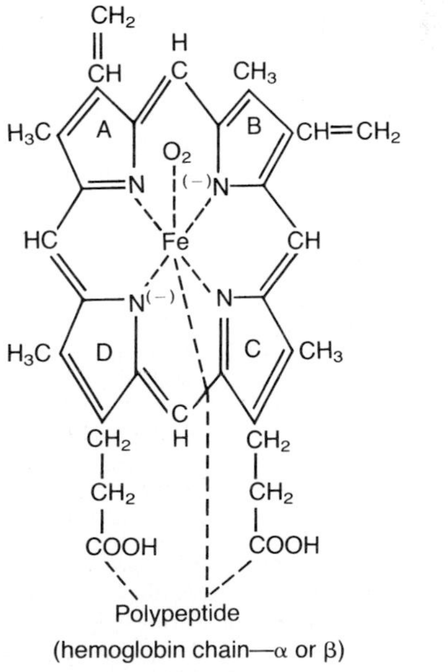

Figure 32–6. Basic structure of the hemoglobin molecule, showing one of the four heme chains that bind together to form the hemoglobin molecule.

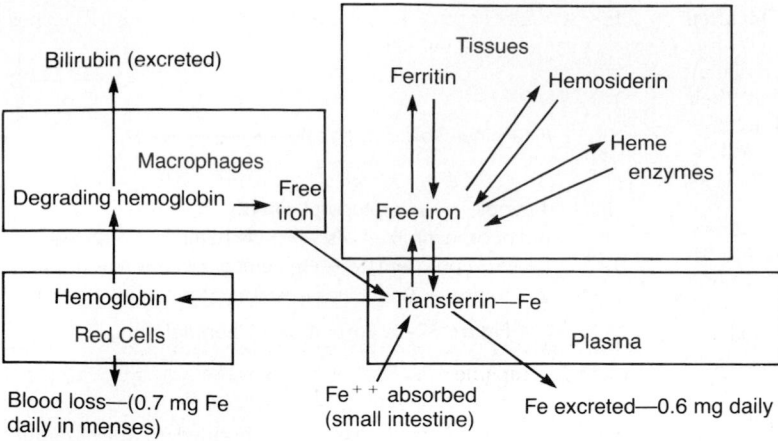

Figure 32–7. Iron transport and metabolism.

can bind with 1 molecule of oxygen, making a total of 4 molecules of oxygen (or 8 oxygen atoms) that can be transported by each hemoglobin molecule. Hemoglobin A has a molecular weight of 64,458.

The nature of the hemoglobin chains determines the binding affinity of the hemoglobin for oxygen. Abnormalities of the chains can alter the physical characteristics of the hemoglobin molecule as well. For instance, in *sickle cell anemia,* the amino acid valine is substituted for glutamic acid at one point in each of the two beta chains. When this type of hemoglobin is exposed to low oxygen, it forms elongated crystals inside the red blood cells that are sometimes 15 micrometers in length. These make it almost impossible for the cells to pass through many small capillaries, and the spiked ends of the crystals are likely to rupture the cell membranes, thus leading to sickle cell anemia.

COMBINATION OF HEMOGLOBIN WITH OXYGEN. The most important feature of the hemoglobin molecule is its ability to combine loosely and reversibly with oxygen. This ability is discussed in detail in Chapter 40 in relation to respiration because the primary function of hemoglobin in the body depends on its ability to combine with oxygen in the lungs and then to release this oxygen readily in the tissue capillaries where the gaseous tension of oxygen is much lower than in the lungs.

Oxygen *does not* combine with the two positive bonds of the iron in the hemoglobin molecule. Instead, it binds loosely with one of the so-called coordination bonds of the iron atom. This is an extremely loose bond so that the combination is easily reversible. Furthermore, the oxygen does not become ionic oxygen but is carried as molecular oxygen, composed of two oxygen atoms, to the tissues where, because of the loose, readily reversible combination, it is released into the tissue fluids in the form of dissolved molecular oxygen, rather than ionic oxygen.

Iron Metabolism

Because iron is important for formation of hemoglobin, myoglobin, and other substances, such as the cytochromes, cytochrome oxidase, peroxidase, and catalase, it is essential to understand the means by which iron is utilized in the body.

The total quantity of iron in the body averages 4 to 5 grams, about 65 per cent of which is in the form of hemoglobin. About 4 per cent is in the form of myoglobin, 1 per cent is in the form of the various heme compounds that promote intracellular oxidation, 0.1 per cent is combined with the protein transferrin in the blood plasma, and 15 to 30 per cent is stored mainly in the reticuloendothelial system and liver parenchymal cells, principally in the form of ferritin.

TRANSPORT AND STORAGE OF IRON. Transport, storage, and metabolism of iron in the body are shown in Figure 32–7 and may be explained as follows: When iron is absorbed from the small intestine, it immediately combines in the blood plasma with a beta globulin, *apotransferrin,* to form *transferrin,* which is then transported in the plasma. The iron is loosely combined with the globulin molecule and, consequently, can be released to any of the tissue cells at any point in the body. Excess iron in the blood is deposited in all cells of the body *but especially* in the liver hepatocytes and less in the reticuloendothelial cells of the bone marrow. In the cell cytoplasm, it combines mainly with a protein, *apoferritin,* to form *ferritin.* Apoferritin has a molecular weight of about 460,000, and varying quantities of iron can combine in clusters of iron radicals with this large molecule; therefore, ferritin may contain only a small amount of iron or a large amount. This iron stored as ferritin is called *storage iron.*

Smaller quantities of the iron in the storage pool are stored in an extremely insoluble form called *hemosiderin.* This is especially true when the total quantity of iron in the body is more than the apoferritin storage pool can accommodate. Hemosiderin forms especially large clusters in the cells and, consequently, can be stained and observed microscopically as large particles in tissue slices by histological techniques. Ferritin can also be stained, but the ferritin particles are so small and dispersed that they usually can be seen only with the electron microscope.

When the quantity of iron in the plasma falls very low, iron is removed from ferritin quite easily but much less easily from hemosiderin. The iron is then transported in the form of transferrin in the plasma to the portions of the body where it is needed.

A unique characteristic of the transferrin molecule is that it binds strongly with receptors in the cell membranes of erythroblasts in the bone marrow. Then, along with its bound iron, it is ingested into the erythroblasts by endocytosis. There the transferrin delivers the iron directly to the mitochondria, where heme is synthesized. In people who do not have adequate quantities of transferrin in their blood, failure to transport iron to the erythroblasts in this manner can cause severe hypochromic anemia—that is, decreased numbers of red cells that contain less hemoglobin than normal.

When red blood cells have lived their life span and are destroyed, the hemoglobin released from the cells is ingested by the cells of the monocyte-macrophage system. There free iron is liberated, and it is then mainly stored in the ferritin pool or reused for formation of new hemoglobin.

DAILY LOSS OF IRON. A man excretes about 1 milligram of iron each day, mainly into the feces. Additional quantities of iron are lost whenever bleeding occurs. For a woman, the menstrual loss of blood brings the iron loss to an average value of about 2 mg/day.

Absorption of Iron from the Gastrointestinal Tract

Iron is absorbed from all parts of the small intestine, mostly by the following mechanism. The liver secretes moderate amounts of *apotransferrin* into the bile that flows through the bile duct into the duodenum. In the small intestine, the apotransferrin binds with free iron and with some iron compounds such as hemoglobin and myoglobin from meat, two of the most important sources of iron in the diet. This combination is called *transferrin.* It in turn is attracted to and binds with receptors in the membranes of the intestinal epithelial cells. Then, by pinocytosis, the transferrin molecule, carrying with it its iron store, is absorbed into the epithelial cells and later is released on the blood side of these cells in the form of *plasma transferrin.*

The rate of iron absorption is extremely slow, with a maximum rate of only a few milligrams per day. This means that when tremendous quantities of iron are present in the food, only small proportions of this can be absorbed.

REGULATION OF TOTAL BODY IRON BY CONTROLLING RATE OF ABSORPTION. When the body has become saturated with iron so that essentially all the apoferritin in the iron storage areas is already combined with iron, the rate of absorption of iron from the intestinal tract becomes greatly decreased. On the other hand, when the iron stores have been depleted of iron, the rate of absorption can become accelerated probably to five or more times as great as when the iron stores are saturated. Thus, the total body iron is regulated to a great extent by altering the rate of absorption.

Feedback Mechanisms for Regulating Iron Absorption. Two mechanisms that play at least some role in regulating iron absorption are the following: (1) When essentially all the apoferritin in the body has become saturated with iron, it becomes difficult for transferrin to release iron to the tissues. As a consequence, the transferrin, which is normally only one-third saturated with iron, now becomes almost fully bound with iron, so that the transferrin accepts almost no new iron from the mucosal cells. Then, as a final stage of this process, the buildup of excess iron in the mucosal cells themselves depresses active absorption of iron from the intestinal lumen. (2) When the body already has excess stores of iron, the liver decreases its rate of formation of apotransferrin, thus reducing the concentration of this iron-transporting molecule in the plasma and the bile. Therefore, less iron is then absorbed by the intestinal apotransferrin mechanism, and less iron can also be transported away from the intestinal epithelial cells in the plasma by plasma transferrin.

Yet, despite these feedback control mechanisms for regulating iron absorption, when a person eats extremely large amounts of iron compounds, excess iron does enter the blood and can lead to massive deposition of hemosiderin in the reticuloendothelial cells throughout the body. At times, this can be very damaging.

DESTRUCTION OF RED BLOOD CELLS

When red blood cells are delivered from the bone marrow into the circulatory system, they normally circulate an average of 120 days before being destroyed. Even though mature red cells do not have a nucleus, mitochondria, or endoplasmic reticulum, they nevertheless have cytoplasmic enzymes that are capable of metabolizing glucose and forming small amounts of adenosine triphosphate and, especially, the reduced form of nicotinamide–adenine dinucleotide phosphate (NADPH). The NADPH in turn serves the red cell in several important ways: (1) maintaining the pliability of the cell membrane; (2) maintaining membrane transport of ions; (3) keeping the iron of the cell's hemoglobin in the ferrous form, rather than the ferric form (which causes the formation of methemoglobin that will not carry oxygen); and (4) preventing oxidation of the proteins in the red cell. These metabolic systems of the red cell become progressively less active with time, and the cells become more and more fragile, presumably because their life processes wear out.

Once the red cell membrane becomes fragile, the cell ruptures during passage through some tight spot of the circulation. Many of the red cells fragment in the spleen, where they squeeze through the red pulp of the spleen. Here the spaces between the structural trabeculae of the red pulp, through which most of the cells must pass, are only 3 micrometers wide, in comparison with the 8-micrometer diameter of the red cell. When the spleen is removed, the number of abnormal red cells and old cells circulating in the blood increases considerably.

DESTRUCTION OF HEMOGLOBIN. The hemoglobin released from the cells when they burst is phagocytized almost immediately by macrophages in many parts of the body but especially in the liver (the Kupffer cells), spleen, and bone marrow. During the next few hours to days, the macrophages release the iron from the hemoglobin back into the blood to be carried by transferrin either to the bone marrow for production of new red blood cells or to the liver and other tissues for storage in the form of ferritin. The porphyrin portion of the hemoglobin molecule is converted by the macrophages, through a series of stages, into the bile pigment *bilirubin,* which is released into the blood and later secreted by the liver into the bile; this is discussed in relation to liver function in Chapter 70.

THE ANEMIAS

Anemia means a deficiency of red blood cells, which can be caused by either too rapid loss or too slow production of red blood cells. Some types of anemia and their physiological causes are the following:

BLOOD LOSS ANEMIA. After rapid hemorrhage, the body replaces the plasma within 1 to 3 days, but this leaves a low concentration of red blood cells. If a second hemorrhage does not occur, the red blood cell concentration usually returns to normal within 3 to 6 weeks.

In chronic blood loss, a person frequently cannot absorb enough iron from the intestines to form hemoglobin as rapidly as it is lost. Red cells are then produced with too little hemoglobin inside them, giving rise to *microcytic hypochromic anemia*, which is shown in Figure 32–3.

APLASTIC ANEMIA. *Bone marrow aplasia* means lack of a functioning bone marrow. For instance, a person exposed to gamma ray radiation from a nuclear bomb blast is likely to sustain complete destruction of bone marrow, followed in a few weeks by lethal anemia. Likewise, excessive x-ray treatment, certain industrial chemicals, and even drugs to which the person might be sensitive can cause the same effect.

MEGALOBLASTIC ANEMIA. From the earlier discussion of vitamin B_{12}, folic acid, and intrinsic factor from the stomach mucosa, one can readily understand that loss of any one of these factors can lead to slow reproduction of the erythroblasts in the bone marrow. As a result, these grow too large, with odd shapes, and are called *megaloblasts*. Thus, atrophy of the stomach mucosa, as occurs in *pernicious anemia*, or loss of the entire stomach as the result of total gastrectomy can lead to megaloblastic anemia. Also, patients who have intestinal sprue, in which folic acid, B_{12}, and other vitamin B compounds are poorly absorbed, often develop megaloblastic anemia. Because the erythroblasts cannot proliferate rapidly enough to form normal numbers of red blood cells, the cells that are formed are mostly oversized, of bizarre shapes, and have fragile membranes. These cells rupture easily, leaving the person in dire need of an adequate number of red cells.

HEMOLYTIC ANEMIA. Different abnormalities of the red blood cells, many of which are hereditarily acquired, make the cells fragile, so that they rupture easily as they go through the capillaries, especially through the spleen. Even though the number of red blood cells formed is normal, or even much greater than normal in some hemolytic diseases, the red cell life span is so short that serious anemia results. Some of these types of anemia are the following:

In *hereditary spherocytosis*, the red cells are very small and *spherical*, rather than being biconcave discs. These cells cannot be compressed because they do not have the normal loose, baglike cell membrane structure of the biconcave discs. On passing through the splenic pulp, they are easily ruptured by even slight compression.

In *sickle cell anemia*, which is present in 0.3 to 1.0 per cent of West African and American blacks, the cells contain an abnormal type of hemoglobin called *hemoglobin S*, caused by abnormal beta chains of the hemoglobin molecule, as explained earlier in the chapter. When this hemoglobin is exposed to low concentrations of oxygen, it precipitates into long crystals inside the red blood cell. These crystals elongate the cell and give it the appearance of being a sickle, rather than a biconcave disc. The precipitated hemoglobin also damages the cell membrane, so that the cells become highly fragile, leading to serious anemia. Such patients frequently go into a vicious circle called a sickle cell disease "crisis," in which low oxygen tension in the tissues causes sickling, which causes ruptured red cells, this in turn causing still further decrease in oxygen tension and still more sickling and red cell destruction. Once the process starts, it progresses rapidly, leading to serious decrease in red blood cell mass within a few hours and, often, death.

In *erythroblastosis fetalis*, Rh-positive red blood cells in the fetus are attacked by antibodies from an Rh-negative mother. These antibodies make the cells fragile, leading to rapid rupture and causing the child to be born with serious anemia. This is discussed in Chapter 35 in relation to the Rh factor of blood. The extremely rapid formation of new red cells to make up for the destroyed cells that occurs in erythroblastosis fetalis causes a large number of early *blast* forms of red cells to be released into the blood.

Hemolysis also occasionally results from transfusion reactions, malaria, and reactions to certain drugs.

EFFECTS OF ANEMIA ON THE CIRCULATORY SYSTEM

The viscosity of the blood, which is discussed in Chapter 14, depends almost entirely on the concentration of red blood cells. In severe anemia, the blood viscosity may fall to as low as 1.5 times that of water rather than the normal value of about 3. This decreases the resistance to blood flow in the peripheral vessels so that far greater than normal quantities of blood flow through the tissues and then return to the heart. Moreover, hypoxia resulting from diminished transport of oxygen by the blood causes the peripheral tissue vessels to dilate, allowing still further increase in return of blood to the heart, increasing the cardiac output to a still higher level. Thus, one of the major effects of anemia is greatly *increased workload on the heart*.

The increased cardiac output in anemia partially offsets many of the effects of anemia because even though each unit quantity of blood carries only small quantities of oxygen, the rate of blood flow may be increased enough so that almost normal quantities of oxygen are delivered to the tissues. However, when a person with anemia begins to exercise, the heart is not capable of pumping much greater quantities of blood than it is already pumping. Consequently, during exercise, which greatly increases the tissue demand for oxygen, extreme tissue hypoxia results and acute cardiac failure often ensues.

POLYCYTHEMIA

SECONDARY POLYCYTHEMIA. Whenever the tissues become hypoxic because of too little oxygen in the atmosphere, such as at high altitudes, or because of failure of delivery of oxygen to the tissues, as occurs in cardiac failure, the blood-forming organs automatically produce large quantities of red blood cells. This condition is called *secondary polycythemia*, and the red cell count commonly rises to 6 to 7 million/mm^3.

A common type of secondary polycythemia, called *physiologic polycythemia*, occurs in natives who live at altitudes of 14,000 to 17,000 feet. The blood count is generally 6 to 7 million/mm^3; this is associated with the ability of these people to perform high levels of continuous work even in a rarefied atmosphere.

POLYCYTHEMIA VERA (ERYTHREMIA). In addition to those people who have physiologic polycythemia, others have a condition known as *polycythemia vera*, in which the red blood cell count may be 7 to 8 million and the hematocrit 60 to 70 per cent. Polycythemia vera is caused by a gene aberration that occurs in the hemocytoblastic cell line that produces the blood cells. The blast cells no longer stop producing red cells when too many cells are already present. This causes excess production of red blood cells in the same manner that a tumor of a breast causes excess production of a specific type of breast cell. It usually causes excess production of white blood cells and platelets as well.

In polycythemia vera, not only does the hematocrit increase but the total blood volume also increases, rarely to almost twice normal. As a result, the entire vascular system becomes intensely engorged. In addition, many of the capillaries become plugged by the viscous blood because the viscosity of the blood in polycythemia vera sometimes increases from the normal of 3 times the viscosity of water to 10 times that of water.

Effect of Polycythemia on the Circulatory System

Because of the greatly increased viscosity of the blood in polycythemia, the flow of blood through the vessels is often sluggish. In accordance with the factors that regulate the return of blood to the heart, as discussed in Chapter 20, increasing the viscosity tends to *decrease* the rate of venous return to the heart. On the other hand, the blood volume is greatly increased in polycythemia, which tends to *increase* the venous return. Actually, the cardiac output in polycythemia is not far from normal because these two factors more or less neutralize each other.

The arterial pressure is normal in most people with polycythemia, although in about one third of them the pressure is elevated. This means that the blood pressure–regulating mechanisms can usually offset the tendency for increased blood viscosity to increase peripheral resistance and, thereby, to increase arterial pressure. Beyond certain limits, these regulations fail.

The color of the skin depends to a great extent on the quantity of blood in the subpapillary venous plexus. In polycythemia vera, the quantity of blood in this plexus is greatly increased. Furthermore, because the blood passes sluggishly through the skin capillaries before entering the venous plexus, a larger than normal quantity of hemoglobin is deoxygenated before the blood enters the plexus. The blue color of this deoxygenated hemoglobin masks the red color of the oxygenated hemoglobin. Therefore, a person with polycythemia vera ordinarily has a ruddy complexion with a bluish (cyanotic) tint to the skin.

REFERENCES

Bauer, C., and Kurtz, A.: Erythropoietin production in the kidney. News Physiol. Sci., 2:69, 1987.

Bauer, C., and Kurtz, A.: Oxygen sensing in the kidney and its relation to erythropoietin production. Annu. Rev. Physiol., 51:845, 1989.

Brown, B. A.: Hematology: Principles and Procedures. Baltimore, Williams & Wilkins, 1993.

Burtis, C. A., and Ashwood, E. R.: Tietz Textbook of Clinical Chemistry. Philadelphia, W. B. Saunders Co., 1994.

Charlton, R. W., and Bothwell, T. H.: Iron absorption. Annu. Rev. Med., 34:55, 1983.

Chasis, J. A., and Shohet, S. B.: Red cell biochemical anatomy and membrane properties. Annu. Rev. Physiol., 49:237, 1987.

Clark, M. R.: Senescence of red blood cells: Progress and problems. Physiol. Rev. 68:503, 1988.

Ersley, A. H., and Gabuzda, T. G.: Basic Pathophysiology of Blood, 3rd Ed. Philadelphia, W. B. Saunders Co., 1985.

Fomon, S. J., and Zlotkin, S.: Nutritional Anemias. New York, Raven Press, 1992.

Furth, R. van: Hemopoietic Growth Factors and Mononuclear Phagocytes. Farmington, CT, S. Karger Publishers, Inc., 1993.

Gale, R. P., et al.: Blood Stem Cell Transplants. New York, Cambridge University Press, 1994.

Golde, D. W., and Gasson, J. C.: Hormones that stimulate the growth of blood cells. Sci. Am., July 1988, p. 62.

Grignani, F., et al.: Genotypic, Phenotypic, and Functional Aspects of Haematopoiesis. New York, Raven Press, 1987.

Handin, R. I., et al.: Blood: Principles and Practice of Hematology, Philadelphia, J. B. Lippincott, 1994.

Hansen, R. M., et al.: Failure to suspect and diagnose thalassemic syndromes. Interpretation of RBC indices by the nonhematologist. Arch. Intern. Med., 145:93, 1985.

Harris, J. R.: Erythroid Cells. New York, Plenum Publishing Corp., 1990.

Hoffman, R., et al.: Hematology: Basic Principles and Practice. New York, Churchill Livingstone, 1994.

Huebers, H. A., and Finch, C. A.: The physiology of transferrin and transferrin receptors. Physiol. Rev., 67:520, 1987.

Jelkmann, W.: Erythropoietin: structure, control of production, and function. Physiol. Rev., 72:449, 1992.

Johnson, G. R.: Haemopoietic multipotential stem cells in culture. Clin. Haematol., 13:309, 1984.

Johnstone, R. M., and Teng, K.: Membrane remodelling during reticulocyte maturation. News Physiol. Sci., 4:37, 1989.

Knowles, D. M.: Neoplastic Hematopathology. Baltimore, Williams & Wilkins, 1992.

Lee, G. R., et al.: Wintrobe's Clinical Hematology. Baltimore, Williams & Wilkins, 1993.

Mackinney, A. A. Jr.: The Pathophysiology of Blood. New York, John Wiley & Sons, 1984.

McKenzie, S., et al.: Textbook of Hematology. Baltimore, Williams & Wilkins, 1994.

Magallon Martinez, M.: Inside Hemophilia: Milestones in Hemophilia and von Willebrand's Disease in the Last 25 Years. Farmington, CT, S. Karger Publishers, Inc., 1992.

Miller, D. R., et al. (eds.): Blood Diseases of Infancy and Childhood. St. Louis, C. V. Mosby Co., 1989.

Nikinmaa, M.: Membrane transport and control of hemoglobin-oxygen affinity in nucleated erythrocytes. Physiol. Rev., 72:301, 1992.

Ogawa, M., et al.: Differentiation and proliferation kinetics of hemopoietic stem cells in culture. Prog. Clin. Biol. Res., 148:35, 1984.

Petz, L. D.: Red cell transfusion problems in immunohematologic disease. Annu. Rev. Med., 39:355, 1988.

Plowman, P. N.: Hematology and Immunology. New York, Elsevier Science Publishing, 1987.

Rodack, B. F.: Diagnostic Hematology. Philadelphia, W. B. Saunders Co., 1994.

Rodgers, G. P., et al.: Sickle cell anemia. New York, Scientific American Science and Medicine, September/October, 1994.

Rossi, E. C., et al.: Principles of Transfusion Medicine. Baltimore, Williams & Wilkins, 1990.

Scharff, O., and Foder, B.: Regulation of cytosolic calcium in blood cells. Physiol. Rev., 73:547, 1993.

Schmid-Schonbein, H., et al.: Biology of red cells: Non-nucleated erythrocytes as fluid drop-like cell fragments. Int. J. Microcirc. Clin. Exp., 3:161, 1984.

Stamatoyannopoulos, G., et al.: The Molecular Basis of Blood Diseases. Philadelphia, W. B. Saunders Co., 1994.

Ward, K. M., et al.: Clinical Laboratory Instrumentation and Automation: Principles, Applications and Selection. Philadelphia, W. B. Saunders Co., 1994.

Weatherall, D. J. (ed.): Advances in Red Blood Cell Biology. New York, Raven Press, 1982.

Williams, W. J., et al. (eds.): Hematology, 4th Ed. New York, McGraw-Hill Book Co., 1989.

Resistance of the Body to Infection: I. Leukocytes, Granulocytes, the Monocyte-Macrophage System, and Inflammation

CHAPTER 33

Our bodies are at all times exposed to bacteria, viruses, fungi, and parasites, all of which occur even normally and to varying degrees in the skin, the mouth, the respiratory passageways, the intestinal tract, the lining membranes of the eyes, and even the urinary tract. Many of these agents are capable of causing serious disease if they invade the deeper tissues. In addition, we are exposed intermittently to other highly infectious bacteria and viruses besides those that are normally present, and these can cause lethal diseases, such as pneumonia, streptococcal infection, and typhoid fever.

Our bodies have a special system for combating the different infectious and toxic agents. This is composed of the blood leukocytes (the white blood cells) and tissue cells derived from the leukocytes. These cells all work together in two ways to prevent disease: (1) by actually destroying invading agents by phagocytosis and (2) by forming antibodies and sensitized lymphocytes, one or both of which may destroy or inactivate the invader. This chapter is concerned with the first of these methods, and Chapter 34 is concerned with the second.

LEUKOCYTES (WHITE BLOOD CELLS)

The leukocytes are the *mobile units* of the body's protective system. They are formed partially in the bone marrow (the *granulocytes* and *monocytes* and a few *lymphocytes*) and partially in the lymph tissue (*lymphocytes* and *plasma cells*). After formation, they are transported in the blood to the different parts of the body where they are to be used. The real value of the white blood cells is that most of them are specifically transported to areas of serious infection and inflammation, thereby providing a rapid and potent defense against any infectious agent that might be present. As we see later, the granulocytes and monocytes have a special capability to "seek out and destroy" a foreign invader.

General Characteristics of Leukocytes

TYPES OF WHITE BLOOD CELLS. Six types of white blood cells are normally found in the blood. They are *polymorphonuclear neutrophils, polymorphonuclear eosinophils, polymorphonuclear basophils, monocytes, lymphocytes,* and, occasionally, *plasma cells*. In addition, there are large numbers of *platelets,* which are fragments of a seventh type of white cell found in the bone marrow, the *megakaryocyte*. The first three types of cells, the polymorphonuclear cells, all have a granular appearance, as shown in cell numbers 7, 10, and 12 in Figure 33–1, for which reason they are called *granulocytes,* or, in clinical terminology, "polys" because of the multiple nuclei.

The granulocytes and the monocytes protect the body against invading organisms mainly by ingesting them—that is, by *phagocytosis*. The lymphocytes and plasma cells function mainly in connection with the immune system; this is discussed in Chapter 34. Finally, the function of platelets is mainly to activate the blood-clotting mechanism, which is discussed in Chapter 36.

CONCENTRATIONS OF THE DIFFERENT WHITE BLOOD CELLS IN THE BLOOD. The adult human being has about 7000 white blood cells per microliter of blood. The normal percentages of the different types of white blood cells are approximately the following:

Polymorphonuclear neutrophils	62.0%
Polymorphonuclear eosinophils	2.3%
Polymorphonuclear basophils	0.4%
Monocytes	5.3%
Lymphocytes	30.0%

The number of platelets, which are only cell fragments, in each microliter of blood is normally about 300,000.

435

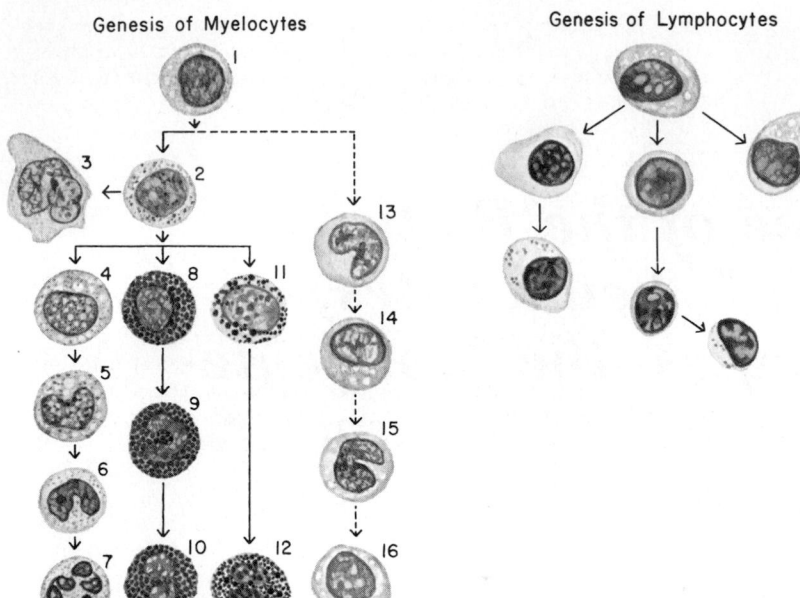

Genesis of Myelocytes

Genesis of Lymphocytes

Figure 33–1. Genesis of the white blood cells. The different cells of the myelogenous series are 1, myeloblast; 2, promyelocyte; 3, mega-karyocyte; 4, neutrophil myelocyte; 5, young neutrophil metamyelocyte; 6, "band" neutro-phil metamyelocyte; 7, polymorphonuclear neutrophil; 8, eosinophil myelocyte; 9, eosi-nophil metamyelocyte; 10, polymorphonu-clear eosinophil; 11, basophil myelocyte; 12, polymorphonuclear basophil; 13–16, stages of monocyte formation.

Genesis of the Leukocytes

The early differentiation of the pluripotential hemopoietic stem cell into the different types of committed stem cells is shown in Figure 32–2 in the previous chapter. Aside from those cells committed to the formation of red blood cells, two major lineages of *white blood cells* are formed, the myelocytic and the lymphocytic lineages. To the left in Figure 33–1 is shown the *myelocytic lineage* beginning with the *myeloblast,* and to the right is shown the *lymphocytic lineage* beginning with the *lymphoblast.*

The granulocytes and monocytes are formed only in the bone marrow. Lymphocytes and plasma cells are produced mainly in the various lymphogenous organs, including the lymph glands, spleen, thymus, tonsils, and various pockets of lymphoid tissue elsewhere in the body, especially in the bone marrow and in so-called Peyer's patches underneath the epithelium in the gut wall.

The white blood cells formed in the bone marrow, especially the granulocytes, are stored within the marrow until they are needed in the circulatory system. Then, when the need arises, various factors that are discussed later cause them to be released. Normally, about three times as many granulocytes are stored in the marrow as circulate in the entire blood. This represents about a six-day supply of granulocytes.

The lymphocytes are mostly stored in the various areas of lymphoid tissue except for the small number of lymphocytes that are temporarily being transported in the blood.

As shown in Figure 33–1, megakaryocytes (cell #3) are also formed in the bone marrow and are part of the myelog-enous group of bone marrow cells. These megakaryocytes fragment in the bone marrow, the small fragments known as *platelets* or *thrombocytes* passing then into the blood.

Life Span of the White Blood Cells

The main reason white blood cells are present in the blood is to be transported from the bone marrow or lym-

phoid tissue to the areas of the body where they are needed. The life of the granulocytes once released from the bone marrow is normally 4 to 8 hours circulating in the blood and another 4 to 5 days in the tissues. In times of serious tissue infection, this total life span is often shortened to only a few hours because the granulocytes then proceed rapidly to the infected area, perform their functions, and in the process are themselves destroyed.

The monocytes also have a short transit time, 10 to 20 hours, in the blood before wandering through the capillary membranes into the tissues. Once in the tissues they swell to much larger sizes to become *tissue macrophages,* and in this form, they can live for months or even years unless they are destroyed by performing phagocytic function. These tissue macrophages form the basis of the tissue macrophage system, discussed in much greater detail later, that provides a continuing defense in the tissues against infection.

Lymphocytes enter the circulatory system continually along with the drainage of lymph from the lymph nodes and other lymphoid tissue. Then, after a few hours, they pass back into the tissues by diapedesis and then re-enter the lymph and return either to lymphoid tissue or to the blood again and again; thus, there is continual circulation of the lymphocytes through the body. The lymphocytes have life spans of weeks, months, or even years, but this depends on the body's need for these cells.

The platelets in the blood are replaced about once every 10 days; in other words, about 30,000 platelets are formed each day for each microliter of blood.

DEFENSIVE PROPERTIES OF (1) NEUTROPHILS AND (2) MACROPHAGES

It is mainly the neutrophils and macrophages that attack and destroy invading bacteria, viruses, and other injurious agents. The neutrophils are mature cells that can attack and

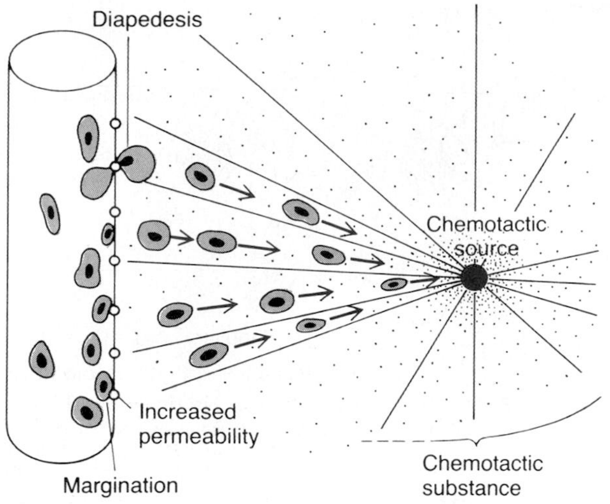

Figure 33–2. Movement of neutrophils by *chemotaxis* toward an area of tissue damage.

destroy bacteria and viruses even in the circulating blood. On the other hand, the macrophages begin life as blood monocytes that are immature cells while still in the blood and have little ability to fight infectious agents. However, once they enter the tissues, they begin to swell, sometimes increasing their diameters as much as fivefold, to as great as 80 micrometers, a size that can barely be seen with the naked eye. Also, extremely large numbers of lysosomes develop in the cytoplasm, giving the cytoplasm the appearance of a bag filled with granules. These cells are now called *macrophages,* and they are extremely capable of combating disease agents.

WHITE BLOOD CELLS ENTER THE TISSUE SPACES BY DIAPEDESIS. Neutrophils and monocytes can squeeze through the pores of the blood vessels by diapedesis. That is, even though a pore is much smaller than the size of the cell, a small portion of the cell slides through the pore at a time, the portion sliding through being momentarily constricted to the size of the pore, as shown in Figure 33–2.

WHITE BLOOD CELLS MOVE THROUGH TISSUE SPACES BY AMEBOID MOTION. Both neutrophils and macrophages move through the tissues by ameboid motion, described in Chapter 2. Some of the cells can move at velocities as great as 40 μm/min, several times their own length each minute.

WHITE BLOOD CELLS ARE ATTRACTED TOWARD INFLAMED TISSUE AREAS BY CHEMOTAXIS. Many different chemical substances in the tissues cause both neutrophils and macrophages to move toward the source of the chemical. This phenomenon, shown in Figure 33–2, is known as *chemotaxis.* When a tissue becomes inflamed, at least a dozen different products are formed that can cause chemotaxis toward the inflamed area. They include (1) some of the bacterial toxins, (2) degenerative products of the inflamed tissues themselves, (3) several reaction products of the "complement complex" (discussed in Chapter 34) that is activated in inflamed tissues, and (4) several reaction products caused by plasma clotting in the inflamed area, as well as other substances.

As shown in Figure 33–2, chemotaxis depends on a concentration gradient of the chemotactic substance. The concentration is greatest near the source, which causes the directional movement of the white cells. Chemotaxis is

effective up to 100 micrometers away from an inflamed tissue; because almost no tissue area is more than 50 micrometers away from a capillary, the chemotactic signal can easily move vast hordes of white cells from the capillaries into the inflamed area.

Phagocytosis

The most important function of the neutrophils and macrophages is phagocytosis, which means cellular ingestion of the offending agent.

Phagocytes must be selective of the material that is phagocytized; otherwise, some of the normal cells and structures of the body would be ingested. Whether or not phagocytosis will occur depends especially on three selective procedures.

First, most natural structures in the tissues have smooth surfaces, which resist phagocytosis. But if the surface is rough, the likelihood of phagocytosis is increased.

Second, most natural substances of the body have protective protein coats that repel the phagocytes. On the other hand, dead tissues and most foreign particles frequently have no protective coats, which also makes them subject to phagocytosis.

Third, the body has a specific means of recognizing certain foreign materials. This is a function of the immune system that is described in Chapter 34. The immune system develops antibodies against infectious agents like bacteria. The antibodies then adhere to the bacterial membranes and thereby make the bacteria especially susceptible to phagocytosis. To do this, the antibody molecule also combines with the C3 product of the *complement* cascade, which is an additional part of the immune system that is discussed in Chapter 34. The C3 molecules than attach themselves to receptors on the phagocyte membrane, thus initiating phagocytosis. The entire process is called *opsonization.*

Phagocytosis by Neutrophils. The neutrophils entering the tissues are already mature cells that can immediately begin phagocytosis. On approaching a particle to be phagocytized, the neutrophil first attaches itself to the particle and then projects pseudopodia in all directions around the particle. The pseudopodia meet each other on the opposite side and fuse. This creates an enclosed chamber that contains the phagocytized particle. Then the chamber invaginates to the inside of the cytoplasmic cavity and breaks away from the outer cell membrane to form a free-floating *phagocytic vesicle* (also called a *phagosome*) inside the cytoplasm.

A neutrophil can usually phagocytize 5 to 20 bacteria before the neutrophil itself becomes inactivated and dies.

Phagocytosis by Macrophages. Macrophages, when activated by the immune system as described in Chapter 34, are much more powerful phagocytes than the neutrophils, often capable of phagocytizing as many as 100 bacteria. They also have the ability to engulf much larger particles, even whole red blood cells, or, occasionally, even malarial parasites, whereas neutrophils are not capable of phagocytizing particles much larger than bacteria. Also, macrophages, after digesting particles, can extrude the residual products and often survive many more months.

ONCE PHAGOCYTIZED, MOST PARTICLES ARE DIGESTED BY INTRACELLULAR ENZYMES. Once a foreign particle has been phagocytized, lysosomes and other cytoplasmic granules immediately come in contact with the phagocytic vesicle, and their membranes fuse with those of the vesicle, thereby dumping many digestive enzymes and bactericidal agents into the vesicle. Thus, the phagocytic vesicle now becomes a *digestive vesicle,* and digestion of the phagocytized particle begins immediately.

Neutrophils and macrophages both have an abundance of lysosomes filled with *proteolytic enzymes* especially geared for digesting bacteria and other foreign protein matter. The lysosomes of macrophages (but not of neutrophils) also contain large amounts of *lipases,* which digest the thick lipid membranes possessed by some bacteria.

BOTH NEUTROPHILS AND MACROPHAGES CAN KILL BACTERIA. In addition to digestion of ingested bacteria in the phagosomes, neutrophils and macrophages contain bactericidal agents that kill most bacteria even when the lysosomal enzymes fail to digest them. This is especially important because some bacteria have protective coats or other factors that prevent their destruction by the digestive enzymes. Much of the killing effect results from several powerful *oxidizing agents* formed by enzymes in the membrane of the phagosome or by the special organelle called the *peroxisome.* These oxidizing agents include large quantities of *superoxide* (O_2^-), *hydrogen peroxide* (H_2O_2), and *hydroxyl ions* ($-OH^-$), all of which are lethal to most bacteria even in small quantities. Also, one of the lysosomal enzymes, myeloperoxidase, catalyzes the reaction between H_2O_2 and chloride ions to form hypochlorite, which is exceedingly bactericidal.

Some bacteria, however, notably the tubercle bacillus, have coats that are resistant to lysosomal digestion and at the same time also secrete substances that resist even the killing effects of the neutrophils and macrophages. These bacteria often are responsible for many of the chronic diseases.

MONOCYTE-MACROPHAGE SYSTEM (RETICULOENDOTHELIAL SYSTEM)

In the paragraphs above we describe the macrophages mainly as mobile cells that are capable of wandering through the tissues. A large portion of the monocytes, however, on entering the tissues and after becoming macrophages, become attached to the tissues and remain attached for months or even years until they are called on to perform specific protective functions. They have the same capabilities as the mobile macrophages to phagocytize large quantities of bacteria, viruses, necrotic tissue, or other foreign particles in the tissue. When appropriately stimulated, they can break away from their attachments and once again become mobile macrophages that respond to chemotaxis and all the other stimuli related to the inflammatory process. Thus, the body has a widespread "monocyte-macrophage system" in virtually all areas.

The combination of monocytes, mobile macrophages, fixed tissue macrophages, and a few specialized endothelial cells in the bone marrow, spleen, and lymph nodes is also called the *reticuloendothelial system.* However, all or almost all these cells originate from monocytic stem cells; therefore, the reticuloendothelial system is almost synonymous with the monocyte-macrophage system. Yet because the term "reticuloendothelial system" is much better known in medical literature than the term "monocyte-macrophage system," it should be remembered as a generalized phagocytic system located in all tissues but especially so in those tissue areas where large quantities of particles, toxins, and other unwanted substances must be destroyed.

TISSUE MACROPHAGES IN THE SKIN AND SUBCUTANEOUS TISSUES (HISTIOCYTES). Although the skin is normally impregnable to infectious agents, this no longer holds true

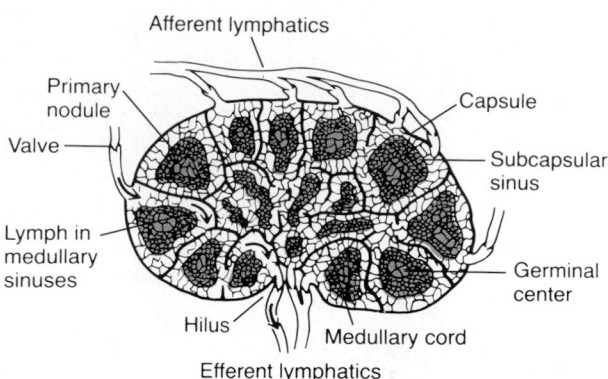

Figure 33–3. Functional diagram of a lymph node. (Redrawn from Ham: Histology. Philadelphia, J. B. Lippincott Company, 1971.)

when the skin is broken. When infection does begin in the subcutaneous tissues and local inflammation ensues, the tissue macrophages can divide in situ and form still more macrophages. Then they perform the usual functions of attacking and destroying the infectious agents as described earlier.

MACROPHAGES OF THE LYMPH NODES. Essentially no particulate matter, such as bacteria, that enters the tissues can be absorbed directly through the capillary membranes into the blood. Instead, if the particles are not destroyed locally in the tissues, they enter the lymph and flow through the lymphatic vessels to the lymph nodes located intermittently along the course of the lymphatic. The foreign particles are trapped there in a meshwork of sinuses lined by *tissue macrophages.*

Figure 33–3 demonstrates the general organization of the lymph node, showing lymph entering through the lymph node capsule by way of *afferent lymphatics,* then flowing through the node to the *medullary sinuses,* and finally passing out of the *hilus* into the *efferent lymphatics.* Large numbers of macrophages line the sinuses, and if any particles enter the sinuses, the macrophages phagocytize them and prevent general dissemination throughout the body.

ALVEOLAR MACROPHAGES IN THE LUNGS. Another route by which invading organisms frequently enter the body is through the lungs. Large numbers of tissue macrophages are present as integral components of the alveolar walls. They can phagocytize particles that become entrapped in the alveoli. If the particles are digestible, the macrophages can also digest them and release the digestive products into the lymph. If the particle is not digestible, the macrophages often form a "giant cell" capsule around the particle until such time—if ever—that it can be slowly dissolved. Such capsules are frequently formed around tubercle bacilli, silica dust particles, and even carbon particles.

MACROPHAGES (KUPFFER CELLS) IN THE LIVER SINUSES. Still another favorite route by which bacteria invade the body is through the gastrointestinal tract. Large numbers of bacteria constantly pass through the gastrointestinal mucosa into the portal blood. Before this blood enters the general circulation, it must pass through the sinuses of the liver; these sinuses are lined with tissue macrophages called *Kupffer cells,* shown in Figure 33–4. These cells form such an effective particulate filtration system that almost none of the bacteria from the gastrointestinal tract succeeds in passing from the portal blood into the general systemic circulation. Indeed, motion pictures of phagocytosis by Kupffer

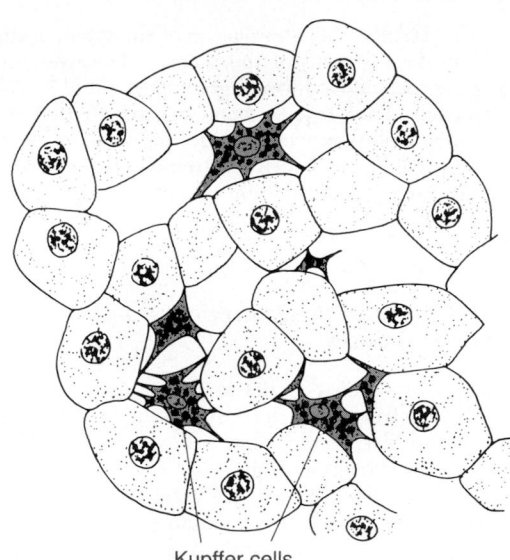

Kupffer cells

Figure 33–4. Kupffer cells lining the liver sinusoids, showing phagocytosis of India ink particles. (Redrawn from Copenhaver *et al.*: Bailey's Textbook of Histology. Baltimore, Williams & Wilkins, 1969.)

cells have demonstrated phagocytosis of a single bacterium in less than one-hundredth of a second.

MACROPHAGES OF THE SPLEEN AND BONE MARROW. If an invading organism does succeed in entering the general circulation, there remain other lines of defense by the tissue macrophage system, especially by macrophages of the spleen and bone marrow. In both these tissues, macrophages have become entrapped by the reticular meshwork of the two organs, and when foreign particles come in contact with these, they are phagocytized.

The spleen is similar to the lymph nodes, except that blood, instead of lymph, flows through the tissue spaces of the spleen. Figure 33–5 shows the spleen's general structure, demonstrating a small peripheral segment. Note that a small artery penetrates from the splenic capsule into the *splenic pulp* and terminates in small capillaries. The capillaries are highly porous, allowing whole blood to pass out of the capillaries into the *cords of the red pulp.* The blood then gradually *squeezes* through the trabecular meshwork of the cords and eventually returns to the circulation through the endothelial walls of the *venous sinuses.* The trabeculae of the red pulp are lined with vast numbers of macrophages, and the venous sinuses are also lined with macrophages. This peculiar passage of blood through the cords of the red pulp provides an exceptional means for phagocytosis of unwanted debris in the blood, including especially old and abnormal red blood cells.

INFLAMMATION AND FUNCTION OF NEUTROPHILS AND MACROPHAGES

Inflammation

When tissue injury occurs, whether caused by bacteria, trauma, chemicals, heat, or any other phenomenon, multiple substances that cause dramatic secondary changes in the tissues are released by the injured tissues. The entire complex of tissue changes is called *inflammation.*

Inflammation is characterized by (1) vasodilatation of the local blood vessels with consequent excess local blood flow, (2) increased permeability of the capillaries with leakage of large quantities of fluid into the interstitial spaces, (3) often clotting of the fluid in the interstitial spaces because of excessive amounts of fibrinogen and other proteins leaking from the capillaries, (4) migration of large numbers of granulocytes and monocytes into the tissue, and (5) swelling of the tissue cells. Some of the many tissue products that cause these reactions are *histamine, bradykinin, serotonin, prostaglandins,* several different *reaction products of the complement system* (which are described in Chapter 34), *reaction products of the blood-clotting system,* and multiple *hormonal substances called lymphokines* that are released by sensitized T cells (part of the immune system, as also discussed in Chapter 34). Several of these substances strongly activate the macrophage system, and within a few hours, the macrophages begin to devour the destroyed tissues; at times, the macrophages also further injure the still-living tissue cells.

"WALLING-OFF" EFFECT OF INFLAMMATION. One of the first results of inflammation is to "wall off" the area of injury from the remaining tissues. The tissue spaces and the lymphatics in the inflamed area are blocked by fibrinogen clots so that fluid barely flows through the spaces. This walling-off process delays the spread of bacteria or toxic products.

The intensity of the inflammatory process is usually proportional to the degree of tissue injury. For instance, staphylococci invading the tissues liberate extremely lethal cellular toxins. As a result, inflammation develops rapidly—indeed, much more rapidly than the staphylococci themselves can multiply and spread. Therefore, local staphylococcal infection is characteristically walled off rapidly and prevented from spreading through the body. On the other hand, streptococci do not cause such intense local tissue destruction. Therefore, the walling-off process develops slowly while the streptococci reproduce and migrate. As a result, streptococci often have a far greater tendency to spread through the body and cause death than do staphylococci, even though staphylococci are far more destructive to the tissues.

Macrophage and Neutrophil Response During Inflammation

THE TISSUE MACROPHAGE IS A FIRST LINE OF DEFENSE AGAINST INFECTION. Within minutes after inflammation begins, the macrophages already present in the tissues, whether histiocytes in the subcutaneous tissues, alveolar

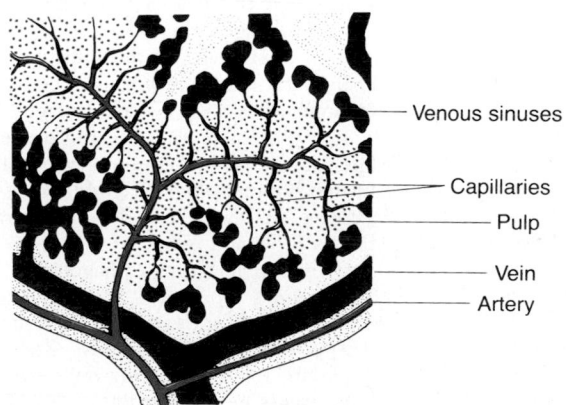

— Venous sinuses

— Capillaries
— Pulp

— Vein
— Artery

Figure 33–5. Functional structures of the spleen. (Modified from Bloom and Fawcett: Textbook of Histology. Philadelphia, W. B. Saunders Company, 1975.)

macrophages in the lungs, microglia in the brain, or others, immediately begin their phagocytic actions. When activated by the products of infection and inflammation, the first effect is rapid enlargement of each of these cells. Next, many of the previously sessile macrophages break loose from their attachments and become mobile, forming the first line of defense against infection during the first hour or so. The numbers of these early mobilizable macrophages often are not great.

NEUTROPHIL INVASION OF THE INFLAMED AREA IS A SECOND LINE OF DEFENSE. Within the first hour or so after inflammation begins, large numbers of neutrophils begin to invade the inflamed area from the blood. This is caused by products from the inflamed tissues that initiate the following reactions: (1) They alter the inside surface of the capillary endothelium, causing neutrophils to stick to the capillary walls in the inflamed area. This effect is called *margination* and is shown in Figure 33–2. (2). They cause the endothelial cells of the capillaries and small venules to separate easily, allowing openings large enough for neutrophils to pass by *diapedesis* into the tissue spaces. (3) Other products of the inflammation cause *chemotaxis* of the neutrophils toward the injured tissues, as explained in an earlier section.

Thus, within several hours after tissue damage begins, the area becomes well supplied with neutrophils. Because the blood neutrophils are already mature cells, they are ready to begin immediately their scavenger functions for killing bacteria and removing foreign matter.

Acute Increase in Neutrophils in the Blood—"Neutrophilia." Also within a few hours after the onset of acute, severe inflammation, the number of neutrophils in the blood sometimes increases fourfold to fivefold—from a normal of 4000 to 5000 to 15,000 to 25,000 neutrophils per microliter. This is called *neutrophilia*, which means an increase in the number of neutrophils in the blood. Neutrophilia is caused by products of inflammation that enter the blood stream, then are transported to the bone marrow, and there act on the marrow capillaries and on stored neutrophils to mobilize these immediately into the circulating blood. This makes more neutrophils available to the inflamed tissue area.

A SECOND MACROPHAGE INVASION OF THE INFLAMED TISSUE IS A THIRD LINE OF DEFENSE. Along with the invasion of neutrophils, monocytes from the blood enter the inflamed tissue and enlarge to become macrophages. However, the number of monocytes in the circulating blood is low; also, the storage pool of monocytes in the bone marrow is much less than that of neutrophils. Therefore, the buildup of macrophages in the inflamed tissue area is much slower than that of neutrophils, requiring several days to become effective. Furthermore, even after invading the inflamed tissue, monocytes are still immature cells, requiring 8 hours or more to swell to much larger sizes and develop tremendous quantities of lysosomes, only then acquiring the full capacity for phagocytosis. Yet after several days to several weeks, the macrophages finally come to dominate the phagocytic cells of the inflamed area because of greatly increased bone marrow production of monocytes, as explained below.

As already pointed out, macrophages can phagocytize far more bacteria (about five times as many) and far larger particles, including even neutrophils themselves and large quantities of necrotic tissue, than can neutrophils. Also, the macrophages play an important role in initiating the development of antibodies, as we discuss in Chapter 34.

INCREASED PRODUCTION OF GRANULOCYTES AND MONOCYTES BY THE BONE MARROW IS A FOURTH LINE OF DEFENSE. The fourth line of defense is greatly increased production of both granulocytes and monocytes by the bone marrow. This results from stimulation of the granulocytic and monocytic progenitor cells of the marrow. However, it takes 3 to 4 days before newly formed granulocytes and monocytes reach the stage of leaving the bone marrow. If the stimulus from the inflamed tissue continues, the bone marrow can continue to produce these cells in tremendous quantities for months and even years, sometimes at rates of production 20 to 50 times normal.

Feedback Control of the Macrophage and Neutrophil Responses

Although more than two dozen factors have been implicated in the control of the macrophage-neutrophil response to inflammation, five of these are believed to play dominant roles. They are shown in Figure 33–6 and consist of (1) *tumor necrosis factor* (TNF), (2) *interleukin-1* (IL-1), (3) *granulocyte-monocyte colony stimulating factor* (GM-CSF), (4) *granulocyte colony stimulating factor* (G-CSF), and (5) *monocyte colony stimulating factor* (M-CSF).

These factors are formed by activated macrophages and T cells in the inflamed tissues and in smaller quantities by other inflamed tissue cells.

The cause of the increased production of granulocytes and monocytes by the bone marrow is mainly the three colony stimulating factors, one of which, GM-CSF, stimulates both granulocyte and monocyte production and the other two, G-CSF and M-CSF, granulocyte and monocyte production, respectively.

This combination of TNF, IL-1, and colony stimulating factors, along with other important factors, provides a powerful feedback mechanism that begins with tissue inflammation and then proceeds to formation of defensive white blood cells and, finally, removal of the cause of the inflammation.

Formation of Pus

When the neutrophils and macrophages engulf large numbers of bacteria and necrotic tissue, essentially all the neutrophils and many, if not most, of the macrophages eventually die. After several days, a cavity is often excavated in the inflamed tissues that contains varying portions of necrotic tissue, dead neutrophils, dead macrophages, and tissue fluid. Such a mixture is commonly known as *pus*. After the infection has been suppressed, the dead cells and necrotic tissue in the pus gradually autolyze over a period of days, and the end products are usually absorbed into the surrounding tissues until most of the evidence of tissue damage is gone.

EOSINOPHILS

The eosinophils normally constitute about 2 per cent of all the blood leukocytes. Eosinophils are weak phagocytes, and they exhibit chemotaxis, but in comparison with the neutrophils, it is doubtful that the eosinophils are of significant importance in protection against the usual types of infection.

On the other hand, eosinophils are often produced in large numbers in people with parasitic infections, and they migrate into tissues diseased by parasites. Although most of the parasites are too large to be phagocytized by the eosinophils or by any other phagocytic cells, nevertheless the eosinophils attach themselves by way of special surface mole-

Inflammation

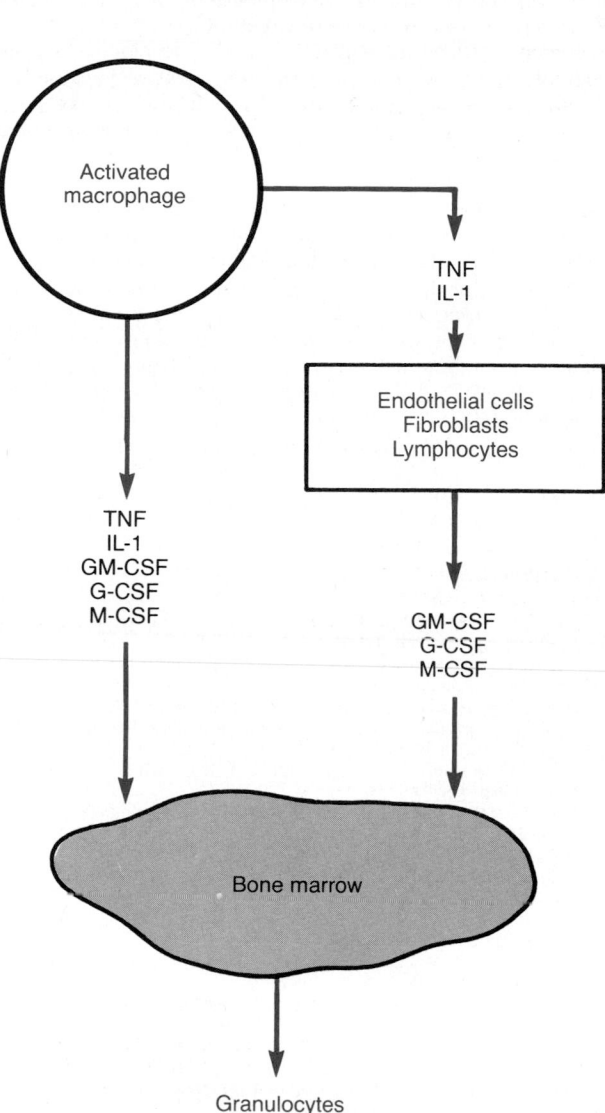

Figure 33–6. Control of bone marrow production of granulocytes and monocyte-macrophages in response to multiple growth factors released from activated macrophages in an inflamed tissue. (TNF, tumor necrosis factor; IL-1, interleukin-1; GM-CSF, granulocyte-monocyte colony stimulating factor; G-CSF, granulocyte colony stimulating factor; M-CSF, monocyte colony stimulating factor.)

cules to the parasites and release substances that kill many of them. For instance, one of the most widespread infections of the world is *schistosomiasis,* a parasitic infection found in as many as one third of the population of some Third World countries; the parasite invades literally any part of the body. Eosinophils attach themselves to the juvenile forms of the parasite and kill many of them. They do so in several ways: (1) by releasing hydrolytic enzymes from their granules, which are modified lysosomes; (2) probably by also releasing highly reactive forms of oxygen that are especially lethal; and (3) by releasing from the granules a highly larvacidal polypeptide called *major basic protein.* In the United States, another parasitic disease that causes eosinophilia is *trichinosis,* which results from invasion of the muscles by the *Trichi-*

nella parasite ("pork worm") after a person eats uncooked pork.

Eosinophils also have a special propensity to collect in tissues in which allergic reactions have occurred, such as in the peribronchial tissues of the lungs in people with asthma and in the skin after allergic skin reactions. This is caused at least partly by the fact that many mast cells and basophils participate in allergic reactions, as we discuss in the next section; these mast cells and basophils release an *eosinophil chemotactic factor* that causes eosinophils to migrate toward the inflamed allergic tissue. The eosinophils are believed to detoxify some of the inflammation-inducing substances released by the mast cells and basophils and probably also to phagocytize and destroy allergen–antibody complexes, thus preventing spread of the local inflammatory process.

BASOPHILS

The basophils in the circulating blood are similar to the large *mast* cells located immediately outside many of the capillaries in the body. Both mast cells and basophils liberate *heparin* into the blood, a substance that can prevent blood coagulation as well as speed the removal of fat particles from the blood after a fatty meal.

The mast cells and basophils also release histamine as well as smaller quantities of bradykinin and serotonin. Indeed, it is mainly the mast cells in inflamed tissues that release these substances during inflammation.

The mast cells and basophils play an exceedingly important role in some types of allergic reactions because the type of antibody that causes allergic reactions, the IgE type (see Chapter 34), has a special propensity to become attached to mast cells and basophils. Then, when the specific antigen subsequently reacts with the antibody, the resulting attachment of the antigen to the antibody causes the mast cell or basophil to rupture and release exceedingly large quantities of histamine, bradykinin, serotonin, heparin, slow-reacting substance of anaphylaxis, and a number of lysosomal enzymes. These in turn cause local vascular and tissue reactions that cause many, if not most, of the allergic manifestations. These reactions are discussed in greater detail in Chapter 34.

LEUKOPENIA

A clinical condition known as *leukopenia* or *agranulocytosis* occasionally occurs in which the bone marrow stops producing white blood cells, leaving the body unprotected against bacteria and other agents that might invade the tissues.

Normally, the human body lives in symbiosis with many bacteria because all the mucous membranes of the body are constantly exposed to large numbers of bacteria. The mouth almost always contains various spirochetal, pneumococcal, and streptococcal bacteria, and these same bacteria are present to a lesser extent in the entire respiratory tract. The gastrointestinal tract is especially loaded with colon bacilli. Furthermore, one can always find bacteria in the eyes, urethra, and vagina. Any decrease in the number of white blood cells immediately allows invasion of the tissues by the bacteria that are already present in the body. Within two days after the bone marrow stops producing white blood cells, ulcers may appear in the mouth and colon, or the person develops some form of severe respiratory infection. Bacteria from the ulcers then rapidly invade the surrounding tissues and the blood. Without treatment, death often ensues in less than a week after acute total leukopenia begins.

Irradiation of the body by gamma rays caused by a nuclear explosion or exposure to drugs and chemicals that contain benzene or anthracene nuclei is likely to cause aplasia of the bone marrow. Indeed, some of the common drugs, such as chloramphenicol (an antibiotic), thiouracil (used to treat thyrotoxicosis), and even the various barbiturate hypnotics, on occasion cause agranulocytosis (or *bone marrow aplasia* in which no cells of any type—red cells included—are produced in the bone marrow), thus setting off the entire infective sequence of this malady.

After irradiation injury to the bone marrow, some stem cells, myeloblasts, and hemocytoblasts usually remain undestroyed in the marrow and are capable of regenerating the bone marrow, provided sufficient time is available. A patient properly treated with antibiotics and other drugs to ward off infection usually develops enough new bone marrow within several weeks to several months for blood cell concentrations to return to normal.

THE LEUKEMIAS

Uncontrolled production of white blood cells is caused by cancerous mutation of a myelogenous or lymphogenous cell. This causes leukemia, which is usually characterized by greatly increased numbers of abnormal white blood cells in the circulating blood.

TYPES OF LEUKEMIA. Leukemias are divided into two general types: the *lymphogenous leukemias* and the *myelogenous leukemias.* The lymphogenous leukemias are caused by cancerous production of lymphoid cells, usually beginning in a lymph node or other lymphogenous tissue and then spreading to other areas of the body. The second type of leukemia, myelogenous leukemia, begins by cancerous production of young myelogenous cells in the bone marrow and then spreads throughout the body, so that white blood cells are produced in many extramedullary organs, especially in the lymph nodes, spleen, and liver.

In myelogenous leukemia, the cancerous process occasionally produces partially differentiated cells, resulting in what might be called *neutrophilic leukemia, eosinophilic leukemia, basophilic leukemia,* or *monocytic leukemia.* More frequently, however, the leukemia cells are bizarre and undifferentiated and not identical to any of the normal white blood cells. Usually the more undifferentiated the cells, the more *acute* is the leukemia, often leading to death within a few months if untreated. With some of the more differentiated cells, the process can be quite *chronic,* sometimes developing slowly over 10 to 20 years.

Leukemic cells, especially the very undifferentiated cells, are usually nonfunctional in providing the usual protection against infection associated with white blood cells.

Effects of Leukemia on the Body

The first effect of leukemia is metastatic growth of leukemic cells in abnormal areas of the body. Leukemic cells of the bone marrow may reproduce so greatly that they invade the surrounding bone, causing pain and, eventually, a tendency to easy bone fracture. Almost all leukemias spread to the spleen, lymph nodes, liver, and other especially vascular regions, regardless of whether the origin of the leukemia is in the bone marrow or the lymph nodes. In each of these

areas, the rapidly growing cells invade the surrounding tissues, utilizing the metabolic elements of these tissues and, consequently, causing tissue destruction.

Common effects in leukemia are the development of infections, severe anemia, and bleeding tendency caused by thrombocytopenia (lack of platelets). These effects result mainly from displacement of the normal bone marrow and lymphoid cells by the nonfunctional leukemic cells.

Finally, perhaps the most important effect of leukemia on the body is the excessive use of metabolic substrates by the growing cancerous cells. The leukemic tissues reproduce new cells so rapidly that tremendous demands are made on the body fluids for foodstuffs, especially the amino acids and vitamins. Consequently, the energy of the patient is greatly depleted, and the excessive utilization of amino acids causes rapid deterioration of the normal protein tissues of the body. Thus, while the leukemic tissues grow, the other tissues become debilitated. After metabolic starvation has continued long enough, it alone is sufficient to cause death.

REFERENCES

Askonas, B. A., and Bancroft, B. J.: Interaction of African trypanosomes with the immune system. Phil. Trans. R. Soc. Lond. (Biol.), 307:41, 1984.

Barrett, K. E., and Metcalfe, D. D.: Mast cell heterogeneity: Evidence and implications. J. Clin. Immunol., 4:253, 1984.

Besemer, J., et al.: Colony stimulating factors. Triangle, 25:177, 1986.

Bocci, V.: What are the roles of interferons in physiological conditions? News Physiol. Sci., 3:201, 1988.

Brown, B. A.: Hematology: Principles and Procedures. Baltimore, Williams & Wilkins, 1993.

Cacciola, E., et al.: Hemopoietic Growth Factors, Oncogenes and Cytokines in Clinical Hematology. Farmington, CT, S. Karger Publishers, Inc., 1994.

Dinarello, C. A.: Biology of interleukin 1. FASEB J., 2:108, 1988.

Feuerstein, G., and Hallenbeck, J. M.: Leukotrienes in health and disease. FASEB J., 1:186, 1987.

Flandrin, G., et al.: An Atlas of Leukemias: Cytology, Histology and Cytogenetics. Philadelphia, J. B. Lippincott Co., 1994.

Ford-Hutchinson, A. W.: Leukotrienes: Their formation and role as inflammatory mediators. Fed. Proc., 44:25, 1985.

Forman, H. J., and Thomas, M. J.: Oxidant production and bactericidal activity in phagocytes. Annu. Rev. Physiol., 48:669, 1986.

Franceschi, C., et al.: Aging and Cellular Defense Mechanisms. New York, New York Academy of Sciences, 1992.

Gale, R. P., et al.: Blood Stem Cell Transplants. New York, Cambridge University Press, 1994.

Galli, S. J., et al.: Basophils and mast cells: Morphologic insights into their biology, secretory patterns, and function. Prog. Allergy, 34:1, 1984.

Gallin, E. K.: Ion channels in leukocytes. Physiol. Rev. 71:775, 1991.

Gallin, J. I., et al.: Inflammation: Basic Principles and Clinical Correlates. New York, Raven Press, 1992.

Graziano, F. M., and Lemanske, R. F., Jr. (eds.): Clinical Immunology. Baltimore, Williams & Wilkins, 1988.

Greene, W. C.: AIDS and the immune system. Sci. Am., Sept., p. 98, 1993.

Handin, R. I., et al.: Blood: Principles and Practice of Hematology. Philadelphia, J. B. Lippincott, 1994.

Harris, J. R.: Lymphocytes and Granulocytes. New York, Plenum Publishing Corp., 1991.

Hoeprich, P. D., et al.: Infectious Diseases. Philadelphia, J. B. Lippincott, 1994.

Hogg, J. C.: Neutrophil kinetics and lung injury. Physiol. Rev., 67:1249, 1987.

Holleb, A. I. (ed.): Interferon: A Current Perspective. In Ca-A Cancer Journal for Clinicians, Vol. 38: No. 5. New York, American Cancer Society by H & W Publishing, 1988.

Horton, M. A.: Macrophages and Related Cells. New York, Plenum Publishing Corp., 1993.

Ioachim, H. L.: Lymph Node Pathology. Philadelphia, J. B. Lippincott, 1994.

Knowles, D. M.: Neoplastic Hematopathology. Baltimore, Williams & Wilkins, 1992.

Lee, G. R., et al.: Wintrobe's Clinical Hematology. Baltimore, Williams & Wilkins, 1993.

Mackinney, A. A., Jr.: The Pathophysiology of Blood. New York, John Wiley & Sons, 1984.

McKenzie, S., et al.: Textbook of Hematology. Baltimore, Williams & Wilkins, 1994.

Mandell, G. L., et al.: Principles and Practice of Infectious Diseases. 4th Ed. New York, Churchill Livingstone, 1994.

Metcalf, J. A., et al.: Laboratory Manual of Neutrophil Function. New York, Raven Press, 1986.

Michna, H.: The Human Macrophage System: Activity and Functional Morphology. Farmington, CT, S. Karger Publishers, Inc., 1988.

Miller, D. R., et al. (eds.): Blood Disease of Infancy and Childhood. St. Louis, C. V. Mosby Co., 1989.

Ogawa, M., et al.: Differentiation and proliferative kinetics of hemopoietic stem cells in culture. Prog. Clin. Biol. Res., 148:35, 1984.

Olsen, E. G., and Spry, C. J.: Relation between eosinophilia and endomyocardial disease. Prog. Cardiovasc. Dis., 21:241, 1985.

Omann, G. M., et al.: Signal transduction and cytoskeletal activation in the neutrophil. Physiol. Rev., 67:285, 1987.

Pardoll, D. M., et al.: The unfolding story of T cell receptor c. FASEB J., 1:103, 1987.

Paul, W. E.: Infectious diseases and the immune system. Sci. Am., Sept., p. 90, 1993.

Rodack, B. F.: Diagnostic Hematology. Philadelphia, W. B. Saunders Co., 1994.

Rubin, R. H., and Young, L. S.: Clinical Approach to Infection in the Compromised Host. 3rd Ed. New York, Plenum Publishing Corp., 1994.

Sacher, R. A., et al.: Cellular and Humoral Immunotherapy and Apheresis. Farmington, CT, S. Karger Publishers, Inc., 1991.

Sachs, L.: The molecular control of blood cell development. Science, 238:1374, 1987.

Samuelsson, B., et al.: Leukotrienes and lipoxins: Structures, biosynthesis, and biological effects. Science, 237:1171, 1987.

Schmid-Schonbein, G. W.: Granulocyte: Friend and foe. News Physiol. Sci., 3:144, 1988.

Sigal, L. H., and Ron, Y.: Immunology and Inflammation. Hightstown, NJ, McGraw-Hill, 1994.

Staub, N. C.: Pulmonary intravascular macrophages. Annu. Rev. Physiol., 56:47, 1994.

Thelen, M., et al.: Neutrophil signal transduction and activation of the respiratory burst. Physiol. Rev., 73:797, 1993.

Thomas, E. D.: Application of Basic Science to Hematopoiesis and Treatment of Disease. New York, Raven Press, 1993.

Thompson, R. B.: A Short Textbook of Haematology. Baltimore, Urban & Schwarzenberg, 1984.

Vane, J., and Botting, R.: Inflammation and the mechanism of action of anti-inflammatory drugs. FASEB J., 1:89, 1987.

Warnke, R. A., and Link, M. P.: Identification and significance of cell markers in leukemia and lymphoma. Annu. Rev. Med., 34:117, 1983.

Williams, W. J., et al. (eds.): Hematology, 4th Ed. New York, McGraw-Hill Book Co., 1990.

Young, J. D.: Killing of target cells by lymphocytes: A mechanistic view. Physiol. Rev., 69:250, 1989.

Resistance of the Body to Infection: II. Immunity and Allergy

CHAPTER 34

INNATE IMMUNITY

The human body has the ability to resist almost all types of organisms or toxins that tend to damage the tissues and organs. This capacity is called *immunity*. Much of immunity is *acquired immunity* that does not develop until after the body is first attacked by a bacterial disease or a toxin, often requiring weeks or months to develop. An additional portion of immunity results from general processes, rather than from processes directed at specific disease organisms. This is called *innate immunity*. It includes the following:

1. Phagocytosis of bacteria and other invaders by white blood cells and cells of the tissue macrophage system, as described in Chapter 33
2. Destruction by the acid secretions of the stomach and the digestive enzymes of organisms swallowed into the stomach
3. Resistance of the skin to invasion by organisms
4. Presence in the blood of certain chemical compounds that attach to foreign organisms or toxins and destroy them. Some of these compounds are (1) *lysozyme*, a mucolytic polysaccharide that attacks bacteria and causes them to dissolute; (2) *basic polypeptides*, which react with and inactivate certain types of gram-positive bacteria; (3) the *complement complex* that is described later, a system of about 20 proteins that can be activated in various ways to destroy bacteria; and (4) *natural killer lymphocytes* that can recognize and destroy foreign cells, tumor cells, and even some infected cells.

This innate immunity makes the human body resistant to such diseases as some paralytic viral infections of animals, hog cholera, cattle plague, and distemper—a viral disease that kills a large percentage of dogs that become afflicted with it. On the other hand, many lower animals are resistant or even immune to many human diseases, such as poliomye-

litis, mumps, human cholera, measles, and syphilis, which are destructive or even lethal to human beings.

ACQUIRED IMMUNITY

In addition to its innate immunity, the human body has the ability to develop extremely powerful specific immunity against individual invading agents such as lethal bacteria, viruses, toxins, and even foreign tissues from other animals. This is called *acquired immunity*. Acquired immunity is caused by a special immune system that forms antibodies and activated lymphocytes that attack and destroy the specific organisms or toxins. It is with this immune mechanism and some of its associated reactions—especially the allergies—that the remainder of this chapter is concerned.

Acquired immunity can often bestow extreme protection. For instance, certain toxins, such as the paralytic toxin of botulinum or the tetanizing toxin of tetanus, can be protected against in doses as high as 100,000 times the amount that would be lethal without immunity. This is the reason the process known as "vaccination" is so important in protecting human beings against disease and against toxins, as explained in the course of this chapter.

Basic Types of Acquired Immunity

Two basic but closely allied types of acquired immunity occur in the body. In one of them the body develops circulating *antibodies*, which are globulin molecules in the blood that are capable of attacking the invading agent. This type of immunity is called *humoral immunity* or *B-cell immunity* (because B lymphocytes produce the antibodies). The second type of acquired immunity is achieved through the formation of large numbers of *activated lymphocytes* that are specifically designed to destroy the foreign agent. This type of

445

immunity is called *cell-mediated immunity* or *T-cell immunity* (because the activated lymphocytes are T lymphocytes).

We shall see shortly that both the antibodies and the activated lymphocytes are formed in the lymphoid tissue of the body. First, let us discuss the initiation of the immune process by *antigens*.

Both Types of Acquired Immunity Are Initiated by Antigens

Because acquired immunity does not occur until after first invasion by a foreign organism or toxin, it is clear that the body must have some mechanism for recognizing the initial invasion. Each toxin or each type of organism almost always contains one or more specific chemical compounds in its makeup that are different from all other compounds. In general, they are proteins or large polysaccharides, and it is they that initiate the acquired immunity. These substances are called *antigens*.

For a substance to be antigenic, it usually must have a high molecular weight, 8000 or greater. Furthermore, the process of antigenicity usually depends on regularly recurring molecular groups, called *epitopes*, on the surface of the large molecule, which explains why proteins and large polysaccharides are almost always antigenic because they both have this type of stereochemical characteristic.

HAPTENS. Although substances with molecular weights of less than 8000 only seldom act as antigens, immunity can nevertheless be developed against substances of low molecular weight in a special way, as follows: If the low-molecular-weight compound, which is called a *hapten,* first combines with a substance that is antigenic, such as a protein, then the combination will elicit an immune response. The antibodies or activated lymphocytes that develop against the combination will then often react separately against either the protein or the hapten. Therefore, on second exposure to the hapten, some of the antibodies or lymphocytes will react with it before it can spread through the body and cause damage.

The haptens that elicit immune responses of this type are usually low-molecular-weight drugs, chemical constituents in dust, breakdown products of dandruff from animals, degenerative products of scaling skin, various industrial chemicals, the toxin of poison ivy, and so forth.

Lymphocytes Are the Basis of Acquired Immunity

Acquired immunity is the product of the body's lymphocyte system. In people who have a genetic lack of lymphocytes or whose lymphocytes have been destroyed by radiation or chemicals, no acquired immunity can develop. And within days after birth, such a person dies of fulminating bacterial infection unless treated by heroic measures. Therefore, it is clear that the lymphocytes are essential to survival of the human being.

The lymphocytes are located most extensively in the *lymph nodes,* but they are also found in special lymphoid tissues, such as the *spleen, submucosal areas of the gastrointestinal tract,* and the *bone marrow.* The lymphoid tissue is distributed advantageously in the body to intercept the invading organisms or toxins before they can spread too widely. In most instances, the invading agent first enters the tissue fluids and then is carried by way of lymph vessels to the lymph node or other lymphoid tissue. For instance, the lymphoid tissue of the gastrointestinal tract is exposed immediately to antigens invading through the gut. The lym-

phoid tissue of the throat and pharynx (the tonsils and adenoids) is well located to intercept antigens that enter by way of the upper respiratory tract. The lymphoid tissue in the lymph nodes is exposed to antigens that invade the peripheral tissues of the body. And, finally, the lymphoid tissue of the spleen and bone marrow plays the specific role of intercepting antigenic agents that have succeeded in reaching the circulating blood.

TWO TYPES OF LYMPHOCYTES PROMOTE, RESPECTIVELY, CELL-MEDIATED IMMUNITY AND HUMORAL IMMUNITY— THE T AND THE B LYMPHOCYTES. Although most of the lymphocytes in normal lymphoid tissue look alike when studied under the microscope, these cells are distinctly divided into two major populations. One of the populations, the T lymphocytes, is responsible for forming the activated lymphocytes that provide cell-mediated immunity, and the other population, the B lymphocytes, is responsible for forming the antibodies that provide humoral immunity.

Both types of lymphocytes are derived originally in the embryo from *pluripotent hemopoietic stem cells* that differentiate and form lymphocytes. The lymphocytes that are formed eventually end up in the lymphoid tissue, but before doing so, they are further differentiated or "preprocessed" in the following ways.

The lymphocytes that are destined to eventually form activated T lymphocytes first migrate to and are preprocessed in the *thymus* gland, which is why they are called "T" *lymphocytes.* They are responsible for cell-mediated immunity.

The other population of lymphocytes—the B lymphocytes that are destined to form antibodies—are preprocessed in the liver during midfetal life and in the bone marrow in late fetal life and after birth. This population of cells was first discovered in birds in which the preprocessing occurs in the *bursa of Fabricius,* a structure not found in mammals. For this reason, these lymphocytes are called "B" lymphocytes, and they are responsible for humoral immunity.

Figure 34–1 shows the two lymphocyte systems for the formation, respectively, of the activated T lymphocytes and the antibodies.

Preprocessing of the T and B Lymphocytes

Although all the lymphocytes of the body originate from the lymphocytic committed stem cells of the embryo, these stem cells themselves are incapable of forming either activated T lymphocytes or antibodies. Before they can do so, they must be further differentiated in appropriate processing areas in the thymus or in the B-cell processing areas.

THE THYMUS GLAND PREPROCESSES THE T LYMPHOCYTES. The T lymphocytes, after their origination in the bone marrow, first migrate to the thymus gland. Here they divide rapidly and at the same time develop extreme diversity for reacting against different specific antigens. That is, one of the thymic lymphocytes develops specific reactivity against one antigen. Then the next lymphocyte develops specificity against another antigen. This continues until there are different thymic lymphocytes with specific reactivities against literally millions of different antigens. These different types of processed T lymphocytes now leave the thymus and spread throughout the body to lodge in lymphoid tissue everywhere.

The thymus also makes certain that any T lymphocytes leaving the thymus will not react against proteins or other antigens that are present in the body's own tissues; otherwise, the T lymphocytes would be lethal to the body in only a few days. The thymus selects which T lymphocytes will be

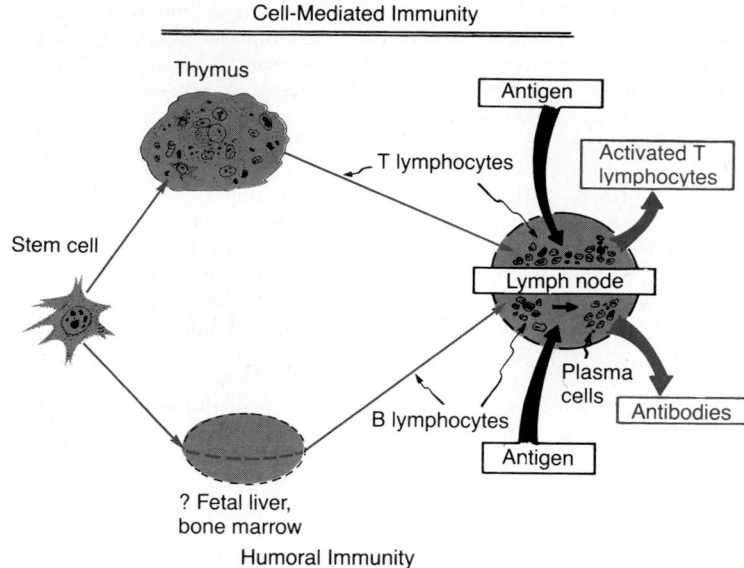

Figure 34–1. Formation of antibodies and sensitized lymphocytes by a lymph node in response to antigens. This figure also shows the origin of *thymic* (T) and *bursal* (B) lymphocytes that are responsible for the cell-mediated and humoral immune processes of the lymph nodes.

released by first mixing them with virtually all the specific "self-antigens" from the body's own tissues. If a T lymphocyte reacts, it is destroyed and phagocytized instead of being released, which is what happens to as many as 90 per cent of the cells. Thus, the only cells that are finally released are those that are nonreactive against the body's own antigens—instead, only against antigens from an outside source, such as from a bacterium, a toxin, or even a transplanted tissue from another person.

Most of the preprocessing of the T lymphocytes in the thymus occurs shortly before birth of a baby and for a few months after birth. Beyond this period, removal of the thymus gland diminishes (but does not eliminate) the T-lymphocytic immune system. However, removal of the thymus several months before birth can prevent development of all cell-mediated immunity. Because this cellular type of immunity is mainly responsible for rejection of transplanted organs, such as hearts and kidneys, one can transplant organs with little likelihood of rejection if the thymus is removed from an animal a reasonable period before birth.

The Liver and Bone Marrow Preprocess the B Lymphocytes. Much less is known about the details for preprocessing B lymphocytes than those for preprocessing T lymphocytes. In the human being, B lymphocytes are known to be preprocessed in the liver during midfetal life and in the bone marrow during late fetal life and after birth.

B lymphocytes are different from T lymphocytes in two ways: First, instead of the whole cell developing reactivity against the antigen, as occurs for the T lymphocytes, the B lymphocytes actively secrete antibodies that are the reactive agents. These agents are large protein molecules that are capable of combining with and destroying the antigenic agent, which is explained elsewhere in this chapter and in Chapter 33. Second, the B lymphocytes have even greater diversity than the T lymphocytes, thus forming many, many millions—and perhaps even billions—of types of B-lymphocyte antibodies with different specific reactivities.

After preprocessing, the B lymphocytes, like the T lymphocytes, migrate to the lymphoid tissue throughout the body where they lodge near to but slightly removed from the T-lymphocyte areas.

T Lymphocytes and B Lymphocyte Antibodies React Highly Specifically Against Specific Antigens—Role of Lymphocyte Clones

When a specific antigen comes in contact with the T and B lymphocytes in the lymphoid tissue, certain of the T lymphocytes become activated to form activated T cells and certain of the B lymphocytes form antibodies. The activated T cells and antibodies in turn react highly specifically against the particular type of antigen that initiated their development. The mechanism of this specificity is the following:

Millions of Specific Lymphocytes Are Preformed in the Lymphoid Tissue. There are millions of different types of preformed B lymphocytes and preformed T lymphocytes that are capable of forming the highly specific antibodies or T cells when stimulated by the appropriate antigens. Each one of these preformed lymphocytes is capable of forming only one type of antibody or one type of T cell with a single type of specificity. And only the specific type of antigen with which it can react can activate it. Once the specific lymphocyte is activated by its antigen, it reproduces wildly, forming tremendous numbers of duplicate lymphocytes. If it is a B lymphocyte, its progeny will eventually secrete antibodies that then circulate throughout the body. If it is a T lymphocyte, its progeny are sensitized T cells that are released into the lymph and carried to the blood and then circulated through all the tissue fluids and back into the lymph, sometimes circulating around and around in this circuit for months or years.

All the different lymphocytes that are capable of forming one specificity of antibody or T cell are called a *clone of lymphocytes.* That is, the lymphocytes in each clone are alike and are derived originally from one or a few early lymphocytes of its specific type.

Origin of the Many Clones of Lymphocytes

Only several hundred to a thousand genes code for the different types of antibodies and T lymphocytes. At first, it was a mystery how it was possible for so few genes to code for the millions of different specificities of antibody molecules or T cells that can be produced by the lymphoid tissue, especially when one considers that a single gene is usually necessary for the formation of each different type of protein. This mystery has now been solved. The whole gene for forming each type of T cell or B cell is never present in the original stem cells from which the functional immune cells are formed. Instead, there are only multiple "gene segments"—actually hundreds of such segments—but not whole genes. During preprocessing of the respective T- and B-cell lymphocytes, these gene segments become mixed up with one another in random combinations, in this way finally forming the whole genes. Because of the several hundred types of gene segments, as well as the millions of different orders in which the segments can be arranged in single cells, one can understand the millions or even billions of different cell gene types that can occur. For each functional T or B lymphocyte that is finally formed, the gene structure codes for only a single antigen specificity. These mature cells then become the highly specific T and B cells that spread to and populate the lymphoid tissue.

Mechanism for Activating a Clone of Lymphocytes

Each clone of lymphocytes is responsive to only a single type of antigen (or to similar antigens that have almost exactly the same stereochemical characteristics). The reason for this is the following: In the case of the B lymphocytes, each one has on the surface of its cell membrane about 100,000 antibody molecules that will react highly specifically with only the one specific type of antigen. Therefore, when the appropriate antigen comes along, it immediately attaches to the cell membrane; this leads to the activation process, which we describe in more detail subsequently. In the case of the T lymphocytes, molecules similar to antibodies, called *surface receptor proteins* (or T-cell markers), are on the surface of the T-cell membrane, and these, too, are highly specific for the one specified activating antigen.

ROLE OF MACROPHAGES IN THE ACTIVATION PROCESS. Aside from the lymphocytes in lymphoid tissue, literally millions of macrophages are present in the same tissue. These line the sinusoids of the lymph nodes, spleen, and other lymphoid tissue, and they lie in apposition to many of the lymph node lymphocytes. Most invading organisms are first phagocytized and partially digested by the macrophages, and the antigenic products are liberated into the macrophage cytosol. The macrophages then pass these antigens by cell-to-cell contact directly to the lymphocytes, thus leading to activation of the specified clones. The macrophages in addition secrete a special activating substance that promotes the growth and reproduction of the specific lymphocytes. The substance is called *interleukin-1*.

ROLE OF THE T CELLS IN THE ACTIVATION OF THE B LYMPHOCYTES. Most antigens activate both T lymphocytes and B lymphocytes at the same time. Some of the T cells that are formed, called *helper cells,* in turn secrete specific substances (collectively called *lymphokines*) that further activate the B lymphocytes. Indeed, without the help of these T cells, the quantity of antibodies formed by the B lymphocytes is usually slight. We discuss this cooperative relationship between helper cells and B cells again after we have an opportunity to describe the mechanisms of the T-cell system of immunity.

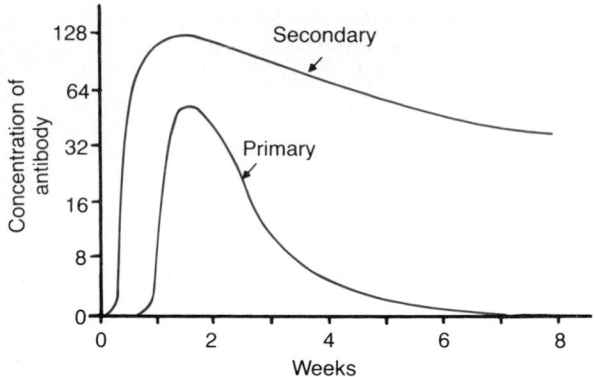

Figure 34–2. Time course of the antibody response in the circulating blood to a *primary* injection of antigen and to a *secondary* injection several months later.

Specific Attributes of the B-Lymphocyte System—Humoral Immunity and the Antibodies

FORMATION OF ANTIBODIES BY PLASMA CELLS. Before exposure to a specific antigen, the clones of B lymphocytes remain dormant in the lymphoid tissue. On entry of a foreign antigen, the macrophages in the lymphoid tissue phagocytize the antigen and then present it to the adjacent B lymphocytes. In addition, the antigen is presented to T cells at the same time, and activated "helper" T cells then also contribute to the activation of the B lymphocytes, as we discuss more fully later. Those B lymphocytes specific for the antigen immediately enlarge and take on the appearance of *lymphoblasts*. Some of the lymphoblasts further differentiate to form *plasmablasts*, which are the precursors of *plasma cells*. In these cells, the cytoplasm expands and the rough endoplasmic reticulum vastly proliferates. They then begin to divide at a rate of about once every 10 hours for about nine divisions, giving in 4 days a total population of about 500 cells for each original plasmablast. The mature plasma cell then produces gamma globulin antibodies at an extremely rapid rate—about 2000 molecules per second for each plasma cell. The antibodies are secreted into the lymph and carried to the circulating blood. This process continues for several days or weeks until final exhaustion and death of the plasma cells.

FORMATION OF "MEMORY" CELLS—DIFFERENCE BETWEEN PRIMARY RESPONSE AND SECONDARY RESPONSE. Some of the lymphoblasts formed by activation of a clone of B lymphocytes do not go on to form plasma cells but instead form moderate numbers of new B lymphocytes similar to those of the original clone. In other words, the B-cell population of the specifically activated clone becomes greatly enhanced. And the new B lymphocytes are added to the original lymphocytes of the clone. They also circulate throughout the body to populate all the lymphoid tissue, but immunologically, they remain dormant until activated once again by a new quantity of the same antigen. These lymphocytes are called *memory cells*. Subsequent exposure to the same antigen will then cause a much more rapid and much more potent antibody response because there are many more memory cells than there were original B lymphocytes of the specific clone.

Figure 34–2 shows the differences between the *primary*

response for forming antibodies that occurs on first exposure to a specific antigen and the *secondary response* that occurs after a second exposure to the same antigen. Note the delay in the appearance of the primary response, its weak potency, and its short life. The secondary response, by contrast, begins rapidly after exposure to the antigen (often within hours), is far more potent, and forms antibodies for many months, rather than for only a few weeks.

The increased potency and duration of the secondary response explains why *vaccination* is usually accomplished by injecting antigen in multiple doses with periods of several weeks or several months between injections.

Nature of the Antibodies

The antibodies are gamma globulins called *immunoglobulins,* and they have molecular weights between 160,000 and 970,000. They usually constitute about 20 per cent of all the plasma proteins.

All the immunoglobulins are composed of combinations of *light* and *heavy polypeptide chains;* most are a combination of 2 light and 2 heavy chains, as shown in Figure 34–3. Some of the immunoglobulins, though, have combinations of as many as 10 heavy and 10 light chains, which gives rise to the high-molecular-weight immunoglobulins. Yet in all immunoglobulins, each heavy chain is paralleled by a light chain at one of its ends, thus forming a heavy–light pair, and there are always at least 2 and as many as 10 such pairs in each immunoglobulin molecule.

Figure 34–3 shows by the shaded area a designated end of each light and heavy chain, called the *variable portion;* the remainder of each chain is called the *constant portion.* The variable portion is different for each specificity of antibody, and it is this portion that attaches specifically to a particular type of antigen. The constant portion of the antibody determines other properties of the antibody, establishing such factors as diffusivity of the antibody in the tissues, adherence of the antibody to specific structures within the tissues, attachment to the *complement complex,* the ease with which the antibodies pass through membranes, and other biological properties of the antibody.

SPECIFICITY OF ANTIBODIES. Each antibody is specific for a particular antigen; this is caused by its unique structural organization of amino acids in the variable portions of both the light and heavy chains. The amino acid organization has a different steric shape for each antigen specificity, so that when an antigen comes in contact with it, multiple prosthetic groups of the antigen fit as a mirror image with those of the antibody, thus allowing rapid bonding between the antibody and the antigen. The bonding is noncovalent, but when the antibody is highly specific, there are so many bonding sites that the antibody–antigen coupling is nevertheless exceedingly strong, held together by (1) hydrophobic bonding, (2) hydrogen bonding, (3) ionic attractions, and (4) van der Waals forces. It also obeys the thermodynamic mass action law:

$$K_a = \frac{\text{concentration of bound antibody–antigen}}{\text{concentration of antibody} \times \text{concentration of antigen}}$$

K_a is called the *affinity constant* and is a measure of how tightly the antibody binds with the antigen.

Note, especially, in Figure 34–3 that there are two variable sites on the illustrated antibody for attachment of antigens, making this type of antibody *bivalent.* A small proportion of the antibodies, which consist of combinations of up to 10 light and 10 heavy chains, have as many as 10 *binding* sites.

CLASSES OF ANTIBODIES. There are five general classes of antibodies, respectively named *IgM, IgG, IgA, IgD,* and *IgE.* Ig stands for immunoglobulin, and the other five respective letters designate the respective classes.

For the purpose of our present limited discussion, two of these classes of antibodies are of particular importance: IgG, which is a bivalent antibody and constitutes about 75 per cent of the antibodies of the normal person, and IgE, which constitutes only a small percentage of the antibodies but is especially involved in allergy. The IgM class is also interesting because a large share of the antibodies formed during the primary response are of this type. These antibodies have 10 binding sites that make them exceedingly effective in protecting the body against invaders, even though there are not many IgM antibodies.

Mechanisms of Action of Antibodies

Antibodies act mainly in two ways to protect the body against invading agents: (1) by direct attack on the invader and (2) by activation of the complement system that then has multiple means of its own for destroying the invader.

DIRECT ACTION OF ANTIBODIES ON INVADING AGENTS. Figure 34–4 shows antibodies (designated by the Y-shaped bars) reacting with antigens (designated by the shaded dumbbells). Because of the bivalent nature of the antibodies and the multiple antigen sites on most invading agents, the antibodies can inactivate the invading agent in one of several ways, as follows:

1. *Agglutination,* in which multiple large particles with antigens on their surfaces, such as bacteria or red cells, are bound together into a clump
2. *Precipitation,* in which the molecular complex of soluble antigen (such as tetanus toxin) and antibody becomes so large that it is rendered insoluble and precipitates
3. *Neutralization,* in which the antibodies cover the toxic sites of the antigenic agent
4. *Lysis,* in which some potent antibodies are occasionally capable of directly attacking membranes of cellular agents and thereby causing rupture of the cell

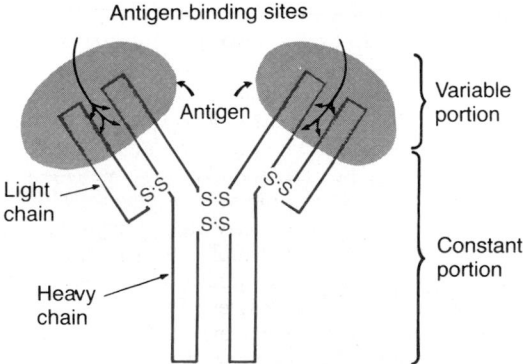

Antigen-binding sites

Light chain

Heavy chain

Antigen

S-S

Variable portion

Constant portion

Figure 34–3. Structure of the typical IgG antibody, showing it to be composed of two heavy polypeptide chains and two light polypeptide chains. The antigen binds at two different sites on the variable portions of the chains.

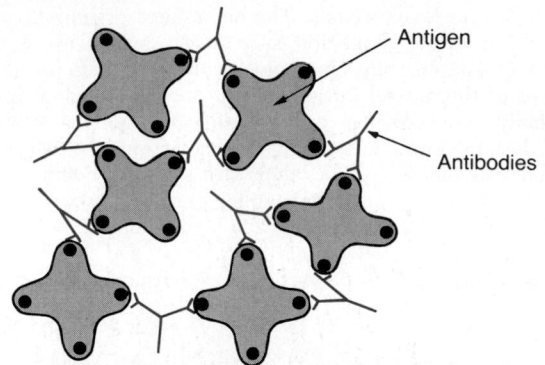

Figure 34–4. Binding of antigen molecules to one another by bivalent antibodies.

These direct actions of antibodies attacking the antigenic invaders probably, under normal conditions, are not strong enough to play a major role in protecting the body against the invader. Most of the protection comes through the *amplifying* effects of the complement system described below.

Complement System for Antibody Action

"Complement" is a collective term that describes a system of about 20 proteins, many of which are enzyme precursors. The principal actors in this system are 11 proteins designated C1 through C9, B, and D, shown in Figure 34–5. All these are present normally among the plasma proteins as well as among the plasma proteins that leak out of the capillaries into the tissue spaces. The enzyme precursors are normally inactive, but they can be activated in two ways: (1) the *classical pathway* and (2) the *alternate pathway*.

CLASSICAL PATHWAY. The classical pathway is activated by an antigen–antibody reaction. That is, when an antibody binds with an antigen, a specific reactive site on the "constant" portion of the antibody becomes uncovered, or activated, and this in turn binds directly with the C1 molecule of the complement system, setting into motion a "cascade" of sequential reactions, shown in Figure 34–5, beginning with activation of the proenzyme C1 itself. Only a few antigen–antibody combinations are required to activate many molecules in the first stage of the complement system. The C1 enzymes that are formed *then activate successively increasing quantities of enzymes* in the later stages of the system, so that from a small beginning, an extremely large "amplified" reaction occurs. Multiple end products are formed, as shown in the figure, and several of these cause important effects that help to prevent damage by the invading organism or toxin. Among the more important effects are the following:

1. *Opsonization* and *phagocytosis.* One of the products of the complement cascade, C3b, strongly activates phagocytosis by both neutrophils and macrophages, causing them to engulf the bacteria to which the antigen–antibody complexes are attached. This process is called *opsonization.* It often enhances the number of bacteria that can be destroyed by many hundredfold.

2. *Lysis.* One of the most important of all the products of the complement cascade is the *lytic complex,* which is a combination of multiple complement factors and is designated C5b6789. This has a direct effect of rupturing the cell membranes of bacteria or other invading organisms.

3. *Agglutination.* The complement products also change the surfaces of the invading organisms, causing them to adhere to one another, thus promoting agglutination.

4. *Neutralization of viruses.* The complement enzymes and other complement products can attack the structures of some viruses and thereby render them nonvirulent.

5. *Chemotaxis.* Fragment C5a causes chemotaxis of neutrophils and macrophages, thus causing large numbers of these phagocytes to migrate into the local region of the antigenic agent.

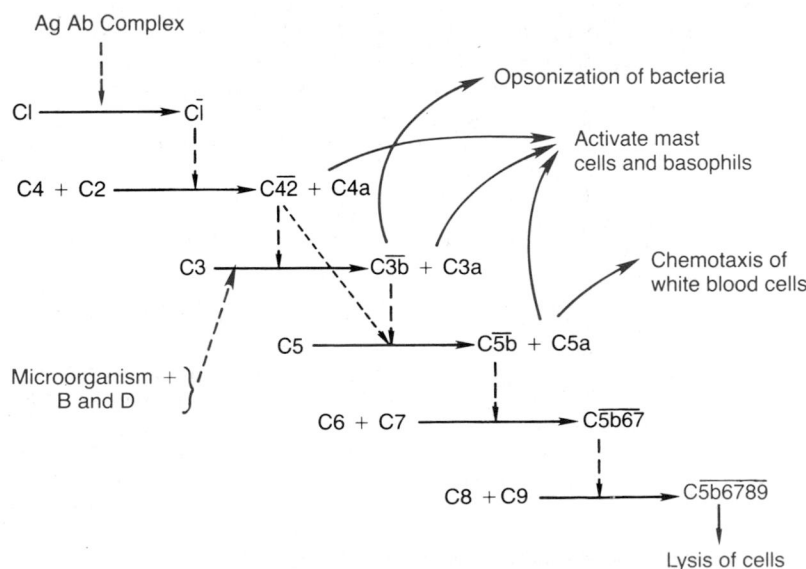

Figure 34–5. Cascade of reactions during activation of the classical pathway of complement. (Modified from Alexander and Good: Fundamentals of Clinical Immunology. Philadelphia, W. B. Saunders Company, 1977.)

6. *Activation of mast cells and basophils.* Fragments C3a, C4a, and C5a all activate mast cells and basophils, causing them to release histamine, heparin, and several other substances into the local fluids. These substances in turn cause increased local blood flow, increased leakage of fluid and plasma protein into the tissue, and other local tissue reactions that help inactivate or immobilize the antigenic agent. The same factors play a major role in inflammation, which is discussed in Chapter 33, and allergy, as we discuss later.

7. *Inflammatory effects.* In addition to inflammatory effects caused by activation of the mast cells and basophils, several other complement products contribute to local inflammation. These products cause the already increased blood flow to increase still further, the capillary leakage of proteins to be increased, and the proteins to coagulate in the tissue spaces, thus preventing movement of the invading organism through the tissues.

ALTERNATE PATHWAY. Sometimes the complement system is activated without the intermediation of an antigen–antibody reaction. This occurs especially in response to large polysaccharide molecules in the cell membranes of some invading micro-organisms. These substances react with complement factors B and D, forming an activation product that activates factor C3, setting off the remainder of the complement cascade beyond the C3 level. Thus, essentially all the same final products of the system are formed as in the classical pathway, and these cause the same effects as those just listed to protect the body against the invader.

Because the alternate pathway does not involve an antigen–antibody reaction, it is one of the first lines of defense against invading micro-organisms, capable of functioning even before a person becomes immunized against the organism.

Special Attributes of the T-Lymphocyte System—Activated T Cells and "Cell-Mediated Immunity"

RELEASE OF ACTIVATED T CELLS FROM LYMPHOID TISSUE AND FORMATION OF MEMORY CELLS. On exposure to the proper antigens, as presented by adjacent macrophages, the T lymphocytes of the specific lymphoid tissue clone proliferate and release large numbers of activated T cells in ways that parallel antibody release by the activated B cells. The principal difference is that instead of releasing antibodies, whole activated T cells are formed and released into the lymph. They then pass into the circulation and are distributed throughout the body, passing through the capillary walls into the tissue spaces, back into the lymph and blood once again, and circulating again and again throughout the body, sometimes lasting for months or even years.

Also, *T-lymphocyte memory cells* are formed in the same way that B memory cells are formed in the antibody system. That is, when a clone of T lymphocytes is activated by an antigen, many of the newly formed lymphocytes are preserved in the lymphoid tissue to become additional T lymphocytes of that specific clone; in fact, these memory cells even spread throughout the lymphoid tissue of the entire body. Therefore, on subsequent exposure to the same antigen, the release of activated T cells occurs far more rapidly and much more powerfully than in the first response.

ANTIGEN RECEPTORS ON THE T LYMPHOCYTES. Antigens bind with *receptor molecules* on the surfaces of T cells in the same way that they bind with antibodies. These receptor molecules are composed of a variable unit similar to the

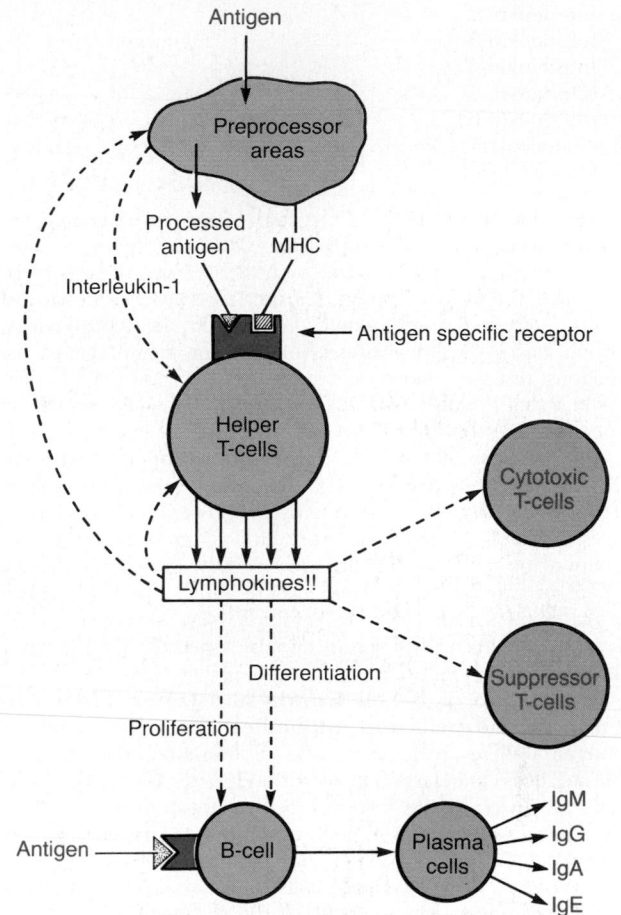

Figure 34–6. Regulation of the immune system, emphasizing the pivotal role of the helper T cells.

variable portion of the humoral antibody, but its stem section is firmly bound to the cell membrane. There are as many as 100,000 receptor sites on a single T cell.

Several Types of T Cells and Their Different Functions

It has become clear that there are multiple types of T cells. They are classified into three major groups: (1) *helper T cells*, (2) *cytotoxic T cells*, and (3) *suppressor T cells*. The functions of each of these are quite distinct.

Helper T Cells—Their Role in Overall Regulation of Immunity

The helper T cells are by far the most numerous of the T cells, usually constituting more than three quarters of all of them. As their name implies, they *help* in the functions of the immune system and in many ways. In fact, they serve as the major regulator of virtually all immune functions, as shown in Figure 34–6. They do this by forming a series of protein mediators, called *lymphokines*, that act on other cells of the immune system as well as on bone marrow cells. Among the important lymphokines secreted by the helper T cells are the following:

Interleukin-2
Interleukin-3
Interleukin-4
Interleukin-5
Interleukin-6
Granulocyte-monocyte colony stimulating factor
Interferon-γ

SPECIFIC REGULATORY FUNCTIONS OF THE LYMPHOKINES.
In the absence of the lymphokines from the helper T cells, the remainder of the immune system is almost paralyzed. In fact, it is the helper T cells that are inactivated or destroyed by the acquired immunodeficiency syndrome (AIDS) virus, which leaves the body almost totally unprotected against infectious disease, therefore leading to the now well-known rapid lethal effects of AIDS. Some of the specific regulatory functions are the following.

Stimulation of Growth and Proliferation of Cytotoxic T Cells and Suppressor T Cells. In the absence of helper T cells, the clones for producing cytotoxic T cells and suppressor T cells are activated very little by most antigens. The lymphokine interleukin-2 has an especially strong stimulatory effect in causing growth and proliferation of both cytotoxic and suppressor T cells. In addition, several of the other lymphokines have less potent effects, especially interleukin-4 and interleukin-5.

Stimulation of B-Cell Growth and Differentiation to Form Plasma Cells and Antibodies. The direct actions of antigen to cause B-cell growth, proliferation, formation of plasma cells, and secretion of antibodies are also slight without the "help" of the helper T cells. Almost all the interleukins participate in the B-cell response, but especially interleukins 4, 5, and 6. In fact, these three interleukins have such potent effects on the B cells that they have been called *B-cell stimulating factors* or *B-cell growth factors.*

Activation of the Macrophage System. The lymphokines also affect the macrophages. First, they slow or stop the migration of the macrophages after they have been chemotactically attracted into the inflamed tissue area, thus causing great accumulation of macrophages. Second, they activate the macrophages to cause far more efficient phagocytosis, allowing them to attack and destroy greatly increasing numbers of invading organisms.

Feedback, Stimulatory Effect on the Helper Cells Themselves. Some of the lymphokines, especially interleukin-2, have a direct positive feedback effect in stimulating activation of the helper T cells themselves. This acts as an amplifier in further enhancing the helper cell response as well as the entire immune response to an invading antigen.

Cytotoxic T Cells

The cytotoxic T cell is a direct-attack cell that is capable of killing micro-organisms and, at times, even some of the body's own cells. For this reason, these cells are called *killer cells.* The receptor proteins on the surfaces of the cytotoxic cells cause them to bind tightly to those organisms or cells that contain their binding-specific antigen. Then, they kill the attacked cell in the manner shown in Figure 34–7. After binding, the cytotoxic T cell secretes *hole-forming proteins,* called *perforins,* that literally punch large round holes in the membrane of the attacked cell. Then fluid flows rapidly into the cell from the interstitial space. In addition, the cytotoxic ... leases cytotoxic substances directly into the attacked ... immediately, the attacked cell becomes greatly ... usually dissolves shortly thereafter.

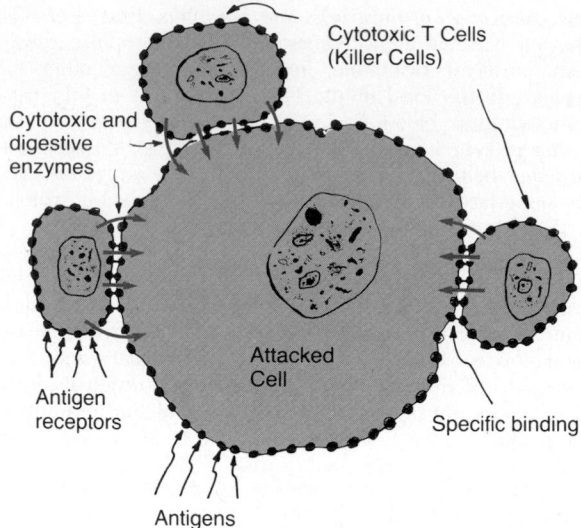

Figure 34–7. Direct destruction of an invading cell by sensitized lymphocytes.

Especially important is that these cytotoxic killer cells can pull away from the victim cells after they have punched holes and delivered cytotoxic substances and then move on to kill many more cells. Indeed, even after destruction of all the invaders, many of these cells persist for months thereafter in the tissues.

Some of the cytotoxic T cells are especially lethal to tissue cells that have been invaded by viruses because many virus particles become entrapped in the membranes of these cells and attract the T cells in response to the viral antigenicity. The cytotoxic cells also play an important role in destroying cancer cells, heart transplant cells, or other types of cells that are foreign to the person's own body.

Suppressor T Cells

Much less is known about the suppressor T cells than about the others, but they are capable of suppressing the functions of both cytotoxic and helper T cells. It is believed that these suppressor functions serve the purpose of regulating the activities of the other cells, keeping them from causing excessive immune reactions that might be severely damaging to the body. For this reason, the suppressor cells are, along with the helper T cells, classified as *regulatory T cells.* One scenario for the function of the regulatory suppressor T cells is the following: The helper T cells activate the suppressor T cells; these cells in turn act as a negative feedback controller of the helper T cells; and this automatically sets the level of activity of the helper T-cell system. It is also probable that the suppressor T cells play an important role in limiting the ability of the immune system to attack a person's own body tissues, called *immune tolerance,* as we discuss in the next section.

Tolerance of the Acquired Immunity System to One's Own Tissues—Role of Preprocessing in the Thymus and Bone Marrow

If a person should become immune to his or her own tissues, the process of acquired immunity would destroy the individual's own body. The immune mechanism normally "recognizes" a person's own tissues as being distinctive from bacteria or viruses, and the person's immunity system forms few antibodies or activated T cells against the person's own antigens. This phenomenon is known as *self-tolerance* to the body's own tissues.

MOST TOLERANCE RESULTS FROM CLONE SELECTION DURING PREPROCESSING. It is believed that most of the phenomenon of tolerance develops during the preprocessing of the T lymphocytes in the thymus and the B lymphocytes in the B-lymphocyte preprocessing—in the human being in the bone marrow. The reason for this belief is that injecting a strong antigen into a fetus at the time the lymphocytes are being preprocessed in these two areas prevents the development of clones of lymphocytes in the lymphoid tissue that are specific for the injected antigen. Also, experiments have shown that specific immature lymphocytes in the thymus, when exposed to a strong antigen, become lymphoblastic, proliferate considerably, and then combine with the stimulating antigen—an effect that is believed to cause the cells themselves to be destroyed by the thymic epithelial cells before they can migrate to and colonize the lymphoid tissue.

Therefore, it is believed that during the preprocessing of lymphocytes in the thymus and the bone marrow, all or most of those clones of lymphocytes that are specific for the body's own tissues are self-destroyed because of their continual exposure to the body's antigens.

ROLE OF THE SUPPRESSOR T CELLS IN THE DEVELOPMENT OF TOLERANCE. The suppressor T cells are probably responsible for still another type of "self-tolerance." For instance, sometimes an autoimmune reaction occurs acutely against one of the body's tissues but after a few days or a few weeks disappears even though autoimmune antibodies persist in the circulating plasma. What has happened is that the number of suppressor T cells specifically sensitized to the offending self-antigen has greatly increased. It is believed that these suppressor T cells function to counteract the effects of the autoimmune antibodies as well as the sensitized helper cells and sensitized cytotoxic T cells, thus blocking the immune attack on the tissue. However, this is not entirely clear.

FAILURE OF THE TOLERANCE MECHANISM CAUSES AUTOIMMUNE DISEASES. Sometimes people lose some of their immune tolerance to their own tissues. This occurs to a greater extent the older a person becomes. It usually occurs after destruction of some of the body's tissues, which releases considerable quantities of "self-antigens" that circulate in the body and presumably cause acquired immunity in the form of either activated T cells or antibodies.

Several specific diseases that result from autoimmunity include (1) *rheumatic fever*, in which the body becomes immunized against tissues in the joints and heart, especially the heart valves, after exposure to a specific type of streptococcal toxin that has an epitope in its molecular structure similar to the structure of some of the body's own self-antigens; (2) one type of *glomerulonephritis*, in which the person becomes immunized against the basement membranes of glomeruli; (3) *myasthenia gravis*, in which immunity develops against the acetylcholine receptor proteins of the neuromuscular junction, causing paralysis; and (4) *lupus erythematosus*, in which the person becomes immunized against many different body tissues at the same time, a disease that causes extensive damage and often rapid death.

Vaccination

Vaccination has been used for many years to produce acquired immunity against specific diseases. A person can be vaccinated by injecting dead organisms that are no longer capable of causing disease but that still have their chemical antigens. This type of vaccination is used to protect against typhoid fever, whooping cough, diphtheria, and many other types of bacterial diseases. Also, immunity can be achieved against toxins that have been treated with chemicals so that their toxic nature has been destroyed even though their antigens for causing immunity are still intact. This procedure is used in vaccinating against tetanus, botulism, and other, similar toxic diseases. And, finally, a person can be vaccinated by infection of the person with live organisms that have been "attenuated." That is, these organisms either have been grown in special culture media or have been passed through a series of animals until they have mutated enough that they will not cause disease but do still carry the specific antigens. This procedure is used to protect against poliomyelitis, yellow fever, measles, smallpox, and many other viral diseases.

Passive Immunity

Thus far, all the acquired immunity we have discussed has been *active immunity*. That is, the person's body develops either antibodies or activated T cells in response to invasion of the body by a foreign antigen. However, temporary immunity can be achieved in a person without injecting any antigen. This is done by infusing antibodies, activated T cells, or both obtained from the blood of someone else or from some other animal that has been actively immunized against the antigen. The antibodies last for 2 to 3 weeks, and during that time, the person is protected against the invading disease. Activated T cells last for a few weeks if transfused from another person and for a few hours to a few days if transfused from an animal. Such transfusion of antibodies or lymphocytes to confer immunity is called *passive immunity*.

ALLERGY AND HYPERSENSITIVITY

An important undesirable side effect of immunity is the development, under some conditions, of allergy or other types of immune hypersensitivity. There are several types of allergy and other hypersensitivities, some of which occur only in people who have a specific allergic tendency.

Allergy Caused by Activated T Cells: Delayed-Reaction Allergy

This type of allergy can cause skin eruptions in response to certain drugs or chemicals, particularly some cosmetics and household chemicals, to which one's skin is often exposed. Another example of such allergic hypersensitivity is the skin eruption caused by exposure to poison ivy.

Delayed-reaction allergy is caused by activated T cells and not by antibodies. In the case of poison ivy, the toxin of poison ivy in itself does not cause much harm to the tissues. However, on repeated exposure, it does cause the formation of activated helper and cytotoxic T cells. Then, after subsequent exposure to the poison ivy toxin, within a day or so the activated T cells diffuse from the circulating blood in sufficient numbers into the skin to respond to the poison ivy toxin and elicit a cell-mediated type of immune reaction. Remembering that this type of immunity can cause release of many toxic substances from the activated T cells as well as extensive invasion of the tissues by macrophages and their subsequent effects, one can well understand that the eventual result of some delayed-reaction allergies can be serious tissue damage. The damage normally occurs in the tissue area where the instigating antigen is present, such as the skin in the case of poison ivy, or the lungs to cause lung edema and asthmatic attacks in the case of some air-borne antigens.

Allergies in the So-Called Allergic Person with Excess IgE Antibodies

Some people have an "allergic" tendency. Their allergies are called *atopic allergies* because they are caused by a nonordinary response of the immune system. The allergic tendency is genetically passed on from parent to child and is characterized by the presence of large quantities of IgE *antibodies*. These antibodies are called *reagins* or *sensitizing antibodies* to distinguish them from the more common IgG antibodies. When an *allergen* (defined as an antigen that reacts specifically with a specific type of IgE reagin antibody) enters the body, an allergen–reagin reaction takes place and a subsequent allergic reaction takes place.

A special characteristic of the IgE antibodies (the reagins) is a strong propensity to attach to mast cells and basophils. Indeed, a single mast cell or basophil can bind as many as half a million molecules of IgE antibodies. Then, when an antigen (an allergen) that has multiple binding sites binds with several of the IgE antibodies attached to a mast cell or basophil, this causes an immediate change in the membrane of the cell, perhaps resulting from a simple physical effect of the antibody molecules being pulled together by the antigen. At any rate, many of the mast cells and basophils rupture; others release their granules without rupturing and secrete additional substances not already preformed in the granules. Some of the many substances that are either released immediately or secreted shortly thereafter include *histamine, slow-reacting substance of anaphylaxis* (which is a mixture of toxic leukotrienes), *eosinophil chemotactic substance*, a *protease*, a *neutrophil chemotactic substance, heparin*, and *platelet activating factors*. These substances cause such phenomena as dilatation of the local blood vessels, attraction of eosinophils and neutrophils to the reactive site, damage to the local tissues by the protease, increased permeability of the capillaries and loss of fluid into the tissues, and contraction of local smooth muscle cells. Therefore, a number of different types of abnormal tissue responses can occur, depending on the type of tissue in which the allergen–reagin reaction occurs. Among the different types of allergic reactions caused in this manner are the following:

ANAPHYLAXIS. When a specific allergen is injected directly into the circulation, it can react in widespread areas of the body with the basophils of the blood and the mast cells located immediately outside the small blood vessels if these have been sensitized by attachment of IgE reagins. Therefore, a widespread allergic reaction occurs throughout the vascular system and in closely associated tissues. This is called *anaphylaxis*. The *histamine* released into the circulation causes body-wide vasodilation as well as increased permeability of the capillaries with resultant marked loss of plasma from the circulation. Many people who experience this reaction die of circulatory shock within a few minutes unless they are treated with epinephrine to oppose the effects of the histamine. But also released from the cells is a mixture of leukotrienes, which are also called *slow-reacting substance of anaphylaxis*. These leukotrienes cause spasm of the smooth muscle of the bronchioles, eliciting an asthma-like attack and sometimes causing death by suffocation.

URTICARIA. Urticaria results from antigen entering specific skin areas and causing localized anaphylactoid reactions. *Histamine* released locally causes (1) vasodilatation that induces an immediate *red flare* and (2) increased local permeability of the capillaries that leads to local circumscribed areas of swelling of the skin in another few minutes. The swellings are commonly called *hives*. Administration of antihistamine drugs to a person before exposure will prevent the hives.

HAY FEVER. In hay fever, the allergen–reagin reaction occurs in the nose. *Histamine* released in response to the reaction causes local vascular dilatation, with resultant increased capillary pressure, as well as increased capillary permeability. Both of these effects cause rapid fluid leakage into the tissues of the nose, and the nasal linings become swollen and secretory. Here again, use of antihistamine drugs can prevent this swelling reaction. Other products of the allergen–reagin reaction still cause irritation of the nose, still eliciting the typical sneezing syndrome despite drug therapy.

ASTHMA. Asthma often occurs in the "allergic" type of person. In these, the allergen–reagin reaction occurs in the bronchioles of the lungs. Here, the most important product released from the mast cells seems to be the *slow-reacting substance of anaphylaxis*, which causes spasm of the bronchiolar smooth muscle. Consequently, the person has difficulty breathing until the reactive products of the allergic reaction have been removed. Administration of antihistaminics has little effect on the course of asthma because histamine does not appear to be the major factor eliciting the asthmatic reaction.

REFERENCES

Ada, G. L., and Griffin, P. D.: Vaccines for Fertility Regulation: The Assessment of Their Safety and Efficacy. New York, Cambridge University Press, 1991.

Bierman, C. W., and Pearlman, D. S.: Allergic Diseases from Infancy to Adulthood, 2nd Ed. Philadelphia, W. B. Saunders Co., 1988.

Blalock, J. E.: Neuroimmunoendocrinology. Farmington, CT, S. Karger Publishers, Inc., 1992.

Broder, S., et al.: Textbook of AIDS Medicine. Baltimore, Williams & Wilkins, 1994.

Bryant, N. J.: Laboratory Immunology and Serology. Philadelphia, W. B. Saunders Co., 1992.

Collegium Internationale Allergologicum: Chemical Mediators and Cellular Interactions in Clinical Immunology. Farmington, CT, S. Karger Publishers, Inc., 1992.

Colvin, R. B., et al.: Diagnostic Immunopathology. New York, Raven Press, 1994.

Cotman, C. W., et al. (eds.): The Neuro-Immune-Endocrine Connection. New York, Raven Press, 1987.

Cruse, J. M., et al.: Clinical and Molecular Aspects of Autoimmune Diseases. Farmington, CT, S. Karger Publishers, Inc., 1992.

DeVita, V. T., Jr., et al.: AIDS: Etiology, Diagnosis, Treatment and Prevention. 3rd Ed. Philadelphia, J. B. Lippincott, 1992.

Evans, A. S. (ed.): Viral Infections of Humans, 3rd Ed. New York, Plenum Publishing Corp., 1989.

Fahey, J. L., et al.: AIDS/HIV Reference Guide for Medical Professionals. Baltimore, Williams & Wilkins, 1995.

Granstein, R. D.: Mechanisms of Immune Regulation. Farmington, CT, S. Karger Publishers, Inc., 1994.

Graziano, F. M., and Lemanske, R. F. Jr. (eds.): Clinical Immunology. Baltimore, Williams & Wilkins, 1988.

Greene, W. C.: AIDS and the immune system. Sci. Am., Sept., p. 98, 1993.

Heberman, R. B.: Adoptive therapy for cancer with interleukin-2-activated killer cells. Cancer Bull., 39:6, 1987.

Hoeprich, P. D., et al.: Infectious Diseases. 5th Ed. Philadelphia, J. B. Lippincott, 1994.

Janeway, C. A., Jr.: How the immune system recognizes invaders. Sci. Am., Sept., p. 72, 1993.

Kelley, W. N., et al.: Textbook of Rheumatology. Philadelphia, W. B. Saunders Co., 1993.

Kiyono, H., and McGhee, J. R.: Mucosal Immunology: Intraepithelial Lymphocytes. New York, Raven Press, 1994.

Kiyono, H., et al.: Molecular Aspects of Immune Response and Infectious Diseases. New York, Raven Press, 1990.

Kouttab, N. M., and Maizel, A. L.: Interleukin-2—Its role in the regulation of T-cell proliferation. Cancer Bull., 39:51, 1987.

Lahita, R. G.: Systemic Lupus Erythematosus. 2nd Ed. New York, Churchill Livingstone, 1992.

Lessof, M. H., et al.: Allergy: Immunological and Clinical Aspects, 2nd Ed. Baltimore, Williams & Wilkins, 1988.

Levinson, W. E., and Jawetz, E.: Medical Microbiology and Immunology: Examination and Board Review. 3rd Ed. Norwalk, CT, Appleton and Lange, 1994.

Lichtenstein, L. M.: Allergy and the immune system. Sci. Am., Sept., p. 116, 1993.

Lockey, R. F., and Bukantz, S. C.: Principles of Immunology and Allergy. Philadelphia, W. B. Saunders Co., 1987.

Mandell, G. L., et al.: Principles and Practice of Infectious Diseases. 4th Ed. New York, Churchill Livingstone, 1994.

Marrack, P. and Kappler, J. W.: How the immune system recognizes the body. Sci. Am., Sept., p. 80, 1993.

Mier, J. W.: Therapeutic uses of recombinant interleukin-2 in patients with cancer. Cancer Bull., 39:19, 1987.

Miyajima, A., et al.: Coordinate regulation of immune and inflammatory responses by T cell-derived lymphokines. FASEB J., 2:2462, 1988.

Nance, S. T.: Alloimmunity: 1993 and Beyond. Farmington, CT, S. Karger Publishers, Inc., 1993.

Nance, S. T.: Clinical and Basic Science Aspects of Immunohematology. Farmington, CT, S. Karger Publishers, Inc., 1991.

Norrby, E.: Immunochemistry of AIDS. Farmington, CT, S. Karger Publishers, Inc., 1993.

Nossal, G. J. V.: Immunologic tolerance: Collaboration between antigen and lymphokines. Science, 245:147, 1989.

Paige, C. J., and Wu, G. E.: The B cell repertoire. FASEB J., 3:1818, 1989.

Patterson, R., et al.: Allergic Diseases: Diagnosis and Management. Philadelphia, J. B. Lippincott, 1993.

Paul, W. E.: Fundamental Immunology. New York, Raven Press, 1994.

Plotkin, S. A., and Mortimer, E. J. Jr.: Vaccines. Philadelphia, W. B. Saunders Co., 1994.

Reynolds, H. Y.: Immunologic system in the respiratory tract. Physiol. Rev., 71:1117, 1991.

Ritter, M. A., and Ladyman, H. M.: Monoclonal Antibodies: Production, Engineering and Clinical Application. New York, Cambridge University Press, 1994.

Samelson, L. E.: Lymphocyte Activation. Farmington, CT, S. Karger Publishers, Inc., 1994.

Sedlacek, H.-H., et al.: Antibodies as Carriers of Cytotoxicity. Farmington, CT, S. Karger Publishers, Inc., 1992.

Sell, S., and Berkower, I.: Immunology, Immunopathology, and Immunity. Redding, MA, Appleton & Lange, 1995.

Shelhamer, J., et al. Respiratory Disease in the Immunosuppressed Host. Philadelphia, J. B. Lippincott, 1991.

Silverman, D. G., et al.: Review of Clinical Anesthesia Biology. Philadelphia, J. B. Lippincott, 1994.

Smith, K. A.: Interleukin-2: inception, impact, and implications. Science, 240:1169, 1988.

Steinman, L.: Autoimmune disease. Sci. Am., Sept., p. 106, 1993.

Stites, D. P., et al.: Basic and Clinical Immunology, 8th Ed. Redding, MA, Appleton & Lange, 1994.

Streilein, J. W.: Immune regulation and the eye: A dangerous compromise. FASEB J., 1:199, 1987.

Turgeon, M. L.: Fundamentals of Immunohematology. Baltimore, Williams & Wilkins, 1995.

Unanue, E. R., and Allen, P. M.: The basis for the immunoregulatory role of macrophages and other accessory cells. Science, 236:551, 1987.

Weiner, D. B.: Acquired Immune Deficiency Syndrome. Farmington, CT, S. Karger Publishers, Inc., 1992.

Weissman, I. L., and Cooper, M. D.: How the immune system develops. Sci. Am., Sept., p. 64, 1993.

Wigzell, H.: The immune system as a therapeutic agent. Sci. Am., Sept., p. 126, 1993.

Blood Groups; Transfusion; Tissue and Organ Transplantation

C HAPTER 35

ANTIGENICITY CAUSES IMMUNE REACTIONS OF BLOOD

When blood transfusions from one person to another were first attempted, the transfusions were successful only in some instances. Often immediate or delayed agglutination and hemolysis of the red blood cells occurred, resulting in typical transfusion reactions that occasionally led to death. Soon it was discovered that the bloods of different people usually have different antigenic and immune properties, so that antibodies in the plasma of one blood react with antigens on the surfaces of the red cells of another blood. For this reason, it is easy for blood from a donor to be mismatched with that of a recipient. If proper precautions are taken, one can determine ahead of time whether or not appropriate antibodies and antigens are present in the donor and recipient bloods to cause a reaction.

MULTIPLICITY OF ANTIGENS IN THE BLOOD CELLS. At least 30 commonly occurring antigens and hundreds of other, rare antigens, each of which can at times cause antigen–antibody reactions, have been found in human blood cells, especially on the surfaces of the cell membranes. Most of them are weak and therefore of importance principally for studying the inheritance of genes to establish parentage and so forth. Two particular groups of antigens are more likely than the others to cause blood transfusion reactions. They are the *O-A-B* system of antigens and the *Rh* system.

O-A-B BLOOD GROUPS

A and B Antigens—Agglutinogens

Two antigens—type A and type B—occur on the surfaces of the red blood cells in a large proportion of the population. It is these antigens (also called agglutinogens because they often cause blood cell agglutination) that cause blood transfusion reactions. Because of the way these agglutinogens are inherited, people may have neither of them on their cells, they may have one, or they may have both simultaneously.

MAJOR O-A-B BLOOD TYPES. In transfusing blood from one person to another, the bloods of donors and recipients are normally classified into four major O-A-B types, as shown in Table 35–1, depending on the presence or absence of the two agglutinogens, the A and B agglutinogens. When neither A nor B agglutinogen is present, the blood group is *type O*. When only type A agglutinogen is present, the blood is *type A*. When only type B agglutinogen is present, the blood is *type B*. When both A and B agglutinogens are present, the blood is *type AB*.

Relative Frequencies of the Different Blood Types. The prevalence of the different blood types among caucasoids is approximately as follows:

O	47%
A	41%
B	9%
AB	3%

It is obvious from these percentages that the O and A genes occur frequently, whereas the B gene is infrequent.

GENETIC DETERMINATION OF THE AGGLUTINOGENS. Two genes, one on each of two paired chromosomes, determine the O-A-B blood groups. These two genes are allelomorphic genes that can be any one of three types but only one type on each chromosome: type O, type A, or type B. The type O gene is either functionless or almost functionless, so that it causes no significant type O agglutinogen on the cells. On the other hand, the type A and type B genes do cause strong agglutinogens on the cells.

The six possible combinations of genes, as shown in Table 35–1, are OO, OA, OB, AA, BB, and AB. These combina-

457

Table 35–1 BLOOD GROUPS WITH THEIR GENOTYPES AND THEIR CONSTITUENT AGGLUTINOGENS AND AGGLUTININS

Genotypes	Blood Types	Agglutinogens	Agglutinins
OO	O	—	Anti-A and Anti-B
OA or AA	A	A	Anti-B
OB or BB	B	B	Anti-A
AB	AB	A and B	—

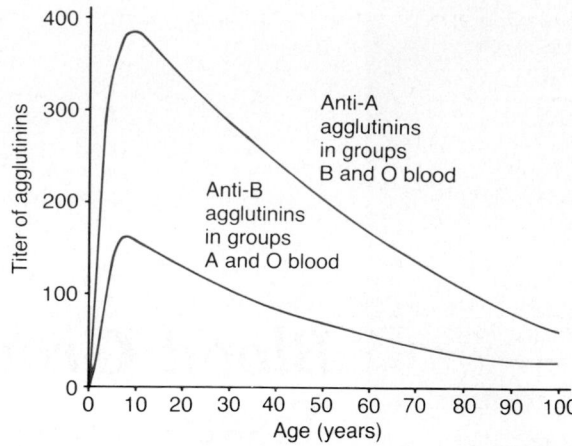

Figure 35–1. Average titers of anti-A and anti-B agglutinins in the blood of people in group B and group A at different ages.

tions of genes are known as the *genotypes,* and each person is one of the six genotypes.

One can observe from the table that a person with genotype OO produces no agglutinogens, and therefore, the blood type is O. A person with genotype OA or AA produces type A agglutinogens and, therefore, has blood type A. Genotypes OB and BB give type B blood and genotype AB gives type AB blood.

Agglutinins

When type A agglutinogen *is not present* in a person's red blood cells, antibodies known as anti-A *agglutinins* develop in the plasma. Also, when type B agglutinogen *is not present* in the red blood cells, antibodies known as anti-B *agglutinins* develop in the plasma.

Thus, referring once again to Table 35–1, note that group O blood, although containing no agglutinogens, does contain both *anti-A* and *anti-B agglutinins;* group A blood contains type A agglutinogens and *anti-B agglutinins;* and group B blood contains type B agglutinogens and *anti-A agglutinins.* Finally, group AB blood contains both A and B agglutinogens but no agglutinins.

TITER OF THE AGGLUTININS AT DIFFERENT AGES. Immediately after birth, the quantity of agglutinins in the plasma is almost zero. Two to eight months after birth, an infant begins to produce agglutinins—anti-A agglutinins when type A agglutinogens are not present in the cells and anti-B agglutinins when type B agglutinogens are not in the cells. Figure 35–1 shows the changing titer of the anti-A and anti-B agglutinins at different ages. A maximum titer is usually reached at 8 to 10 years of age, and this gradually declines throughout the remaining years of life.

ORIGIN OF AGGLUTININS IN THE PLASMA. The agglutinins are gamma globulins, as are other antibodies, and they are produced by the same cells that produce antibodies to any other antigens. Most of them are IgM and IgG immunoglobulin molecules.

But why are these agglutinins produced in people who do not have the respective agglutinogens in their red blood cells? The answer to this is that small amounts of group A and B antigens enter the body in the food, in bacteria, and in other ways, and these substances initiate the development of the anti-A or anti-B agglutinins. For instance, infusion of group A antigen into a recipient having a non-A blood type causes a typical immune response with formation of greater quantities of anti-A agglutinins than ever. Also, the neonate has few, if any, agglutinins, showing that agglutinin formation occurs almost entirely after birth.

Agglutination Process in Transfusion Reactions

When bloods are mismatched so that anti-A or anti-B plasma agglutinins are mixed with red blood cells that contain A or B agglutinogens, respectively, the red cells agglutinate by the following process: The agglutinins attach themselves to the red blood cells. Because the agglutinins have 2 binding sites (IgG type) or 10 sites (IgM type), a single agglutinin can attach to two or more different red blood cells at the same time, thereby causing the cells to adhere to each other. This causes the cells to clump together, which is the process of agglutination. Then these clumps plug small blood vessels throughout the circulatory system. During the ensuing few hours to a few days, either physical distortion of the cells or attack by phagocytic white blood cells destroys the agglutinated cells, releasing hemoglobin into the plasma, which is called "hemolysis" of the red blood cells.

ACUTE HEMOLYSIS OCCURS IN SOME TRANSFUSION REACTIONS. Sometimes, when recipient and donor bloods are mismatched, immediate hemolysis of red cells occurs in the circulating blood. In this case, the antibodies cause lysis of the red blood cells by activating the complement system. This in turn releases proteolytic enzymes (the *lytic complex*) that rupture the cell membranes, as described in Chapter 34.

Immediate intravascular hemolysis is far less common than agglutination followed by *delayed* hemolysis because not only does there have to be a high titer of antibodies for this to occur but also a different type of antibody seems to be required, mainly the IgM antibodies; these antibodies are called *hemolysins.*

Blood Typing

Before giving a transfusion, it is necessary to determine the blood type of the recipient and the blood type of the donor blood so that the bloods can be appropriately matched. This is called *blood typing,* and it is performed in the following way: The red blood cells are first diluted with saline. One portion is then mixed with anti-A agglutinin and another portion is mixed with anti-B agglutinin. After several minutes, the mixture is observed under a microscope. If

Table 35–2 BLOOD TYPING SHOWING AGGLUTINATION OF CELLS OF THE DIFFERENT BLOOD TYPES WITH ANTI-A AND ANTI-B AGGLUTININS

Red Blood Cell Types	Sera	
	Anti-A	Anti-B
O	–	–
A	+	–
B	–	+
AB	+	+

the red blood cells have become clumped—that is, "agglutinated"—one knows that an antibody–antigen reaction has resulted.

Table 35–2 lists the presence (+) or absence (−) of agglutination with each of the four types of blood. Type O red blood cells have no agglutinogens and, therefore, do not react with either the anti-A or the anti-B serum. Type A blood has A agglutinogens and, therefore, agglutinates with anti-A agglutinins. Type B blood has B agglutinogens and agglutinates with anti-B serum. Type AB blood has both A and B agglutinogens and agglutinates with both types of serum.

Rh BLOOD TYPES

Along with the O-A-B blood group system, the Rh system is important in the transfusion of blood. The major difference between the O-A-B system and the Rh system is the following: In the O-A-B system, the agglutinins responsible for causing transfusion reactions develop spontaneously, whereas in the Rh system, spontaneous agglutinins almost never occur. Instead, the person must first be massively exposed to an Rh antigen, usually by transfusion of blood or by a mother's having a baby who has the antigen, before enough agglutinins to cause a significant transfusion reaction will develop.

Rh Antigens—"Rh-Positive" and "Rh-Negative" People. There are six common types of Rh antigens, each of which is called an Rh factor. These types are designated C, D, E, c, d, and e. A person who has a C antigen does not have the c antigen, but the person missing the C antigen always has the c antigen. The same is true for the D-d and E-e antigens. Also, because of the manner of inheritance of these factors, each person has one of each of the three pairs of antigens.

The type D antigen is widely prevalent in the population and considerably more antigenic than the other Rh antigens. Therefore, anyone who has this type of antigen is said to be *Rh positive*, whereas a person who does not have type D antigen is said to be *Rh negative*. However, it must be noted that even in Rh-negative people, some of the other Rh antigens can still cause transfusion reactions, although they usually are much milder.

About 85 per cent of all white people are Rh positive and 15 per cent, Rh negative. In American blacks, the percentage of Rh positives is about 95, whereas in African blacks, it is virtually 100 per cent.

Rh Immune Response

Formation of Anti-Rh Agglutinins. When red blood cells containing Rh factor or even protein breakdown products of such cells are injected into a person whose blood does not contain the same factor—that is, into the Rh-negative person—anti-Rh agglutinins develop slowly, the maximum concentration of agglutinins occurring about 2 to 4 months later. This immune response occurs to a much greater extent in some people than in others. On multiple exposure to the Rh factor, the Rh-negative person eventually becomes strongly "sensitized" to Rh factor.

Characteristics of Rh Transfusion Reactions. If an Rh-negative person has never before been exposed to Rh-positive blood, transfusion of Rh-positive blood into that person causes no immediate reaction. However, in some of these people, anti-Rh antibodies develop in sufficient quantities during the next 2 to 4 weeks to cause agglutination of the transfused cells that are still circulating in the blood. These cells are then hemolyzed by the tissue macrophage system. Thus, a delayed transfusion reaction occurs, although it is usually mild. On subsequent transfusion of Rh-positive blood into the same person, who is now immunized against the Rh factor, the transfusion reaction is greatly enhanced and can be as severe as the transfusion reactions caused by type A or B blood.

Erythroblastosis Fetalis ("Hemolytic Disease of the Newborn")

Erythroblastosis fetalis is a disease of the fetus and neonate characterized by agglutination and phagocytosis of the red blood cells. In most instances of erythroblastosis fetalis, the mother is Rh negative and the father is Rh positive. The baby has inherited the Rh-positive antigen from the father, and the mother has developed anti-Rh agglutinins from exposure to the baby's Rh antigen. In turn, the mother's agglutinins diffuse through the placenta into the fetus to cause red blood cell agglutination.

Prevalence of the Disease. An Rh-negative mother having her first Rh-positive child usually does not develop sufficient anti-Rh agglutinins to cause any harm. However, about 3 per cent of second Rh-positive babies exhibit some signs of erythroblastosis fetalis; about 10 per cent of the third babies exhibit the disease; and the incidence rises progressively with subsequent pregnancies.

Effect of the Mother's Antibodies on the Fetus. After anti-Rh anitbodies have formed in the mother, they diffuse slowly through the placental membrane into the fetus's blood. There they cause agglutination of the fetus's blood. The agglutinated red blood cells subsequently hemolyze, releasing hemoglobin into the blood. The macrophages then convert the hemoglobin into bilirubin, which causes the skin to yellow (jaundice). The antibodies can also attack and damage other cells of the body.

Clinical Picture of Erythroblastosis. The jaundiced, erythroblastotic neonate is usually anemic at birth, and the anti-Rh agglutinins from the mother usually circulate in the infant's blood for 1 to 2 months after birth, destroying more and more red blood cells.

The hemopoietic tissues of the infant attempt to replace the hemolyzed red blood cells. The liver and spleen become greatly enlarged and produce red blood cells in the same manner that they normally do during the middle of gestation. Because of the rapid production of cells, many early forms, including many nucleated blastic forms, are emptied into the circulatory system, and it is because of the presence of these that the disease is called erythroblastosis fetalis.

Although the severe anemia of erythroblastosis fetalis is usually the cause of death, many children who barely survive the anemia exhibit permanent mental impairment or damage to motor areas of the brain because of precipitation of bilirubin in the neuronal cells, causing their destruction, a condition called *kernicterus.*

TREATMENT OF THE ERYTHROBLASTOTIC NEONATE. The usual treatment for erythroblastosis fetalis is to replace the neonate's blood with Rh-negative blood. About 400 milliliters of Rh-negative blood is infused over a period of 1.5 or more hours while the neonate's own Rh-positive blood is being removed. This procedure may be repeated several times during the first few weeks of life, mainly to keep the bilirubin level low and thereby prevent kernicterus. By the time these transfused Rh-negative cells are replaced with the infants's own Rh-positive cells, a process that requires 6 or more weeks, the anti-Rh agglutinins that had come from the mother will have been destroyed.

Transfusion Reactions Resulting from Mismatched Blood Types

If donor blood of one blood type is transfused to a recipient of another blood type, a transfusion reaction is likely to occur in which the red blood cells *of the donor blood* are agglutinated. It is rare that the transfused blood causes agglutination *of the recipient's cells,* for the following reason: The plasma portion of the donor blood immediately becomes dilated by all the plasma of the recipient, thereby decreasing the titer of the infused agglutinins to a level too low to cause agglutination. On the other hand, the infused blood does not dilute the agglutinins in the recipient's plasma to a major extent. Therefore, the recipient's agglutinins can still agglutinate the donor cells.

As explained earlier, all transfusion reactions eventually cause either immediate hemolysis resulting from hemolysins or later hemolysis resulting from phagocytosis of agglutinated cells. The hemoglobin released from the red cells is then converted by the phagocytes into bilirubin and later excreted into the bile by the liver, as discussed in Chapter 70. The concentration of bilirubin in the body fluids often rises high enough to cause *jaundice*—that is, the person's tissues become colored with yellow bile pigment. But if liver function is normal, jaundice usually does not appear unless more than 400 milliliters of blood is hemolyzed in less than a day.

ACUTE KIDNEY SHUTDOWN AFTER TRANSFUSION REACTIONS. One of the most lethal effects of transfusion reactions is kidney shutdown, which can begin within a few minutes to a few hours and continue until the person dies of renal failure.

The kidney shutdown seems to result from three causes: First, the antigen–antibody reaction of the transfusion reaction releases toxic substances from the hemolyzing blood that cause powerful renal vasoconstriction. Second, the loss of circulating red cells along with production of toxic substances from the hemolyzed cells and from the immune reaction often causes circulatory shock. The arterial blood pressure falls very low and the renal blood flow and urine output decrease. Third, if the total amount of free hemoglobin in the circulating blood is greater than that quantity that can bind with haptoglobin (a plasma protein that can bind small amounts of hemoglobin), much of the excess leaks through the glomerular membranes into the kidney tubules. If this amount is still slight, it can be reabsorbed through the tubular epithelium into the blood and will cause no harm; if it is great, then only a small percentage is reabsorbed. Yet water continues to be reabsorbed, causing the tubular hemoglobin concentration to rise so high that it precipitates and blocks many of the tubules; this is especially true if the urine is acidic. Thus, renal vasoconstriction, circulatory shock, and tubular blockage all add together to cause acute renal shutdown. If the shutdown is complete and fails to open up, the patient dies within a week to 12 days, as explained in Chapter 31, unless treated with the artificial kidney.

TRANSPLANTATION OF TISSUES AND ORGANS

Many of the different antigens of red blood cells that cause transfusion reactions are present in other cells of the body as well, and there are also many other antigens. Consequently, any foreign cells transplanted into a recipient can produce immune responses and immune reactions. In other words, most recipients are just as able to resist invasion by foreign tissue cells as to resist invasion by foreign bacteria or red cells.

AUTOGRAFTS, ISOGRAFTS, ALLOGRAFTS, AND XENOGRAFTS. A transplant of a tissue or whole organ from one part of the same animal to another part is called an *autograft;* from one identical twin to another, an *isograft;* from one human being to another or from any animal to another animal of the same species, an *allograft;* and from a lower animal to a human being or from an animal of one species to one of another species, a *xenograft.*

TRANSPLANTATION OF CELLULAR TISSUES. In the case of autografts and isografts, cells in the transplant contain virtually the same types of antigens and will almost always live indefinitely if an adequate blood supply is provided; in the case of allografts and xenografts, immune reactions almost always occur, causing death of all the cells in the graft within 1 to 5 weeks after transplantation unless some specific therapy is used to prevent the immune reaction. The cells of allografts usually persist longer than those of xenografts because the antigenic makeup of the allograft is 15 to 50 times nearer to that of the recipient's tissues than is true for the xenograft. When the tissues are properly "typed" and are similar between donor and recipient for their cellular antigens, successful long-term allografts will occur but usually only if simultaneous drug therapy is used to minimize the immune reaction.

Some of the different cellular tissues and organs that have been transplanted either experimentally or for therapeutic benefit from one person to another are skin, kidney, heart, liver, glandular tissue, bone marrow, and lung. Many kidney allografts have been successful for 5 to 15 years and allograft liver and heart transplants, for 1 to 15 years.

Attempts to Overcome the Immune Reaction in Transplanted Tissue

Because of the extreme potential importance of transplanting certain tissues and organs, serious attempts have been made to prevent the antigen–antibody reactions associated with transplantation. The following specific procedures have met with certain degrees of clinical or experimental success.

Tissue Typing—The HLA Complex of Antigens

The most important antigens for causing graft rejection are a complex called the HLA antigens. Only six of these antigens are present on the cell surfaces of any one person, but there are about 150 different types to choose from. Therefore, this represents more than a trillion possible combinations. Consequently, it is virtually impossible for two persons, except in the case of identical twins, to have the same six HLA antigens. Development of significant immunity against any one of these antigens can cause graft rejection.

The HLA antigens occur on the white blood cells as well as on the tissue cells. Therefore, tissue typing for these antigens is done on the membranes of lymphocytes that have been separated from the person's blood. The lymphocytes are mixed with appropriate antisera and complement; after appropriate incubation, the cells are tested for membrane damage, usually by testing the rate of uptake by the lymphocytic cells of a supravital dye.

Some of the HLA antigens are not severely antigenic, for which reason a precise match of some of the antigens between donor and recipient is not essential to allow allograft acceptance. Therefore, by obtaining the best possible match between donor and recipient, the grafting procedure has become far less hazardous. The best success has been with tissue-type matches between siblings and between parent and child. The match in identical twins is exact, so that transplants between identical twins are almost never rejected because of immune reactions.

Prevention of Graft Rejection by Suppressing the Immune System

If the immune system were completely suppressed, graft rejection would not occur. In fact, in an occasional person who has serious depression of the immune system, grafts can be successful without the use of therapy to prevent rejection. In the normal person, even with the best possible tissue typing, allografts seldom resist rejection for more than a few weeks without the use of some therapeutic procedure to suppress the immune system. Furthermore, because the T cells are mainly the portion of the immune system important for killing the cells of the grafts, their suppression is much more important than suppression of plasma antibodies. Some of the therapeutic agents that have been used for this purpose include the following:

1. *Glucocorticoid hormones isolated from the adrenal cortex glands (or drugs with glucocorticoid activity)*, which suppress the growth of all lymphoid tissue and, therefore, decrease the formation of both antibodies and T cells

2. *Various drugs that have a toxic effect on the lymphoid system* and, therefore, block the formation of antibodies and T cells, especially the drug azathioprine (Imuran)

3. *Cyclosporine*, which has a specific inhibitory effect on the formation of helper T cells and, therefore, is especially efficacious in blocking the T-cell rejection reaction. This has proved to be the most valuable of all the drugs because it does not depress some other portions of the immune system.

Use of these agents leaves the person unprotected from infectious disease; therefore, sometimes bacterial and viral infections become rampant. In addition, the incidence of cancer is several times as great in an immunosuppressed person, presumably because the immune system is important in destroying many early cancer cells before they can begin to proliferate.

To summarize, transplantation of living tissues in human beings until the present has had limited but important success. On the other hand, when someone succeeds in blocking the immune response of the recipient to a donor organ without at the same time destroying the recipient's specific immunity for disease, the story will change overnight.

REFERENCES

Baumgartner, W. A., and Reitz, B. A.: Heart and Heart-Lung Transplantation. Philadelphia, W. B. Saunders Co., 1989.

Beutler, E., et al.: Williams' Hematology. Hightstown, NJ, McGraw-Hill, 1994.

Boren, T., and Falk, P.: Helicobacter pylori binds to blood group antigens. New York, Scientific American Science and Medicine, September/October, 1994.

Bryant, N. J.: Laboratory Immunology and Serology. Philadelphia, W. B. Saunders Co., 1992.

Coleman, R. W., et al.: Hemostasis and thrombosis: Basic principles and clinical practice. 3rd Ed. Philadelphia, J. B. Lippincott, 1994.

Demetris, A. J., et al.: Pathology of hepatic transplantation: A review of 62 adult allograft recipients immunosuppressed with a cyclosporin/steroid regimen. Am. J. Pathol., 118:151, 1985.

Doherty, P. C., et al.: Immunological surveillance of tumors in the context of major histocompatibility complex restriction of T cell formation. Adv. Cancer Res., 42:1, 1984.

Fowler, M. B., and Schroeder, J. S.: Current status of cardiac transplantation. Mod. Conc. Cardio. Dis., 55:37, 1986.

Hackel, E., and AuBuchon, J. P.: Advances in Transplantation. Farmington, CT, S. Karger Publishers, Inc., 1993.

Hancock, W. W.: Analysis of intragraft effector mechanisms associated with human renal allograft rejection: Immunohistological studies with monoclonal antibodies. Immunol. Rev., 77:61, 1984.

Handin, R. I., et al.: Blood: Principles and Practice of Hematology. Philadelphia, J. B. Lippincott, 1994.

Hillman, R. S.: Hematology in Clinical Practice. Hightstown, NJ, McGraw-Hill, 1994.

Hobbs, J. R. (ed.): Thymic Factor Therapy. New York, Raven Press, 1985.

Kahan, B. D., et al.: Clinical and experimental studies with cyclosporine in renal transplantation. Surgery, 97:125, 1985.

Lee, G. R., et al.: Wintrobe's Clinical Hematology. Baltimore, Williams & Wilkins, 1993.

McKenzie, S., et al.: Textbook of Hematology. Baltimore, Williams & Wilkins, 1994.

Messmer, K., et al.: Microcirculation in Organ Transplantation. Farmington, CT, S. Karger Publishers, Inc., 1994.

Miller, D. R., et al. (eds.): Blood Diseases of Infancy and Childhood. St. Louis, C. V. Mosby Co., 1989.

Morris, P. J.: Kidney Transplantation, 3rd Ed. Philadelphia, W. B. Saunders Co., 1988.

Pisciotto, P. T.: Blood Transfusion Therapy. Farmington, CT, S. Karger Publishers, Inc., 1993.

Rodack, B. F.: Diagnostic Hematology. Philadelphia, W. B. Saunders Co., 1994.

Rodger, J. C., and Drake, B. L.: The enigma of the fetal graft. Am. Sci., 75:51, 1987.

Rossi, E. C., et al.: Principles of Transfusion Medicine. Baltimore, Williams & Wilkins, 1990.

Rudmann, S. V.: Blood Banking: Concepts and Applications. Philadelphia, W. B. Saunders Co., 1994.

Sacher, R. A., and AuBuchon, J. P.: Marrow Transplantation: Practical and Technical Aspects of Stem Cell Reconstitution. Farmington, CT, S. Karger Publishers, Inc., 1992.

Singal, D. P.: Immunological Effects of Blood Transfusion. Boca Raton, FL, CRC Press, Inc., 1994.

Stamatoyannopoulos, G., et al.: The Molecular Basis of Blood Diseases. Philadelphia, W. B. Saunders Co., 1994.

Strom, T. B.: Toward more selective therapies to block graft rejection. AFR Nephrol. Letter, 4:13, 1987.

Ward, K. M., et al.: Clinical Laboratory Instrumentation and Automation: Principles, Applications and Selection. Philadelphia, W. B. Saunders Co., 1994.

Williams, W. J., et al. (eds.): Hematology, 4th Ed. New York, McGraw-Hill Book Co., 1990.

Zmijewski, C. M.: Human leukocyte antigen matching in renal transplantation: Review and current status. J. Surg. Res., 38:66, 1985.

Hemostasis and Blood Coagulation

CHAPTER 36

EVENTS IN HEMOSTASIS

The term *hemostasis* means prevention of blood loss. Whenever a vessel is severed or ruptured, hemostasis is achieved by several mechanisms, including (1) vascular spasm, (2) formation of a platelet plug, (3) formation of a blood clot as a result of blood coagulation, and (4) eventual growth of fibrous tissue into the blood clot to close the hole in the vessel permanently.

Vascular Constriction

Immediately after a blood vessel is cut or ruptured, the stimulus of the trauma to the vessel causes the wall of the vessel to contract; this instantaneously reduces the flow of blood from the vessel rupture. The contraction results from nervous reflexes, local myogenic spasm, and local humoral factors from the traumatized tissues and blood platelets. The nervous reflexes are initiated by pain or other impulses that originate from the traumatized vessel or from nearby tissues. Most of the vasoconstriction probably results from local myogenic contraction of the blood vessels initiated by direct damage to the vascular wall. For the smaller vessels, the platelets are responsible for much of the vasoconstriction by releasing the vasoconstrictor substance *thromboxane A₂*. The more a vessel is traumatized, the greater the degree of spasm; this means that a sharply cut blood vessel usually bleeds much more than does a vessel ruptured by crushing. This local vascular spasm can last for many minutes or even hours, during which time the ensuing processes of platelet plugging and blood coagulation can take place.

The value of vascular spasm as a means of hemostasis is demonstrated by the fact that people whose legs have been severed by crushing types of trauma sometimes have such intense spasm in vessels as large as the anterior tibial artery that there is not lethal loss of blood.

Formation of the Platelet Plug

If the rent in the blood vessel is very small—and many very small vascular holes do develop throughout the body each day—it is often sealed by a *platelet plug*, rather than by a blood clot. To understand this, it is important that we first discuss the nature of platelets themselves.

Physical and Chemical Characteristics of Platelets

Platelets are minute round or oval discs 2 to 4 micrometers in diameter. They are formed in the bone marrow from *megakaryocytes*, which are extremely large cells of the hemopoietic series in the bone marrow that fragment into platelets either in the bone marrow or soon after entering the blood, especially as they try to squeeze through the pulmonary capillaries. The normal concentration of platelets in the blood is between 150,000 and 300,000 per microliter.

Platelets have many functional characteristics of whole cells, even though they do not have nuclei and cannot reproduce. In their cytoplasm are such active factors as (1) *actin* and *myosin molecules*, similar to those found in muscle cells, as well as still another contractile protein, *thrombosthenin*, that can cause the platelets to contract; (2) residuals of both the *endoplasmic reticulum* and the *Golgi apparatus* that synthesize various enzymes and store large quantities of calcium ions; (3) mitochondria and enzyme systems that are capable of forming *adenosine triphosphate* and *adenosine diphosphate (ADP)*; (4) enzyme systems that synthesize *prostaglandins*, which are local hormones that cause many types of vascular and other local tissue reactions; (5) an important protein called *fibrin-stabilizing factor*, which we discuss later in relation to blood coagulation; and (6) a *growth factor* that can cause vascular endothelial cells, vascular smooth muscle cells, and fibroblasts to multiply and grow, thus causing cellular growth that helps repair damaged vascular walls.

The cell membrane of the platelets is also important. On

463

its surface is a coat of *glycoproteins* that causes it to avoid adherence to normal endothelium and yet to adhere to *injured* areas of the vessel wall, especially to injured endothelial cells and even more so to any exposed collagen from deeper in the vessel wall. In addition, the membrane contains large amounts of *phospholipids* that play several activating roles at multiple points in the blood-clotting process, as we discuss later.

Thus, the platelet is an active structure. It has a half-life in the blood of 8 to 12 days, at the end of which time its life processes run out. Then it is eliminated from the circulation mainly by the tissue macrophage system; more than one half the platelets are removed by the macrophages in the spleen, where the blood passes through a latticework of tight trabeculae.

Mechanism of the Platelet Plug

Platelet repair of vascular openings is based on several important functions of the platelet itself: When platelets come in contact with a damaged vascular surface, such as the collagen fibers in the vascular wall or even damaged endothelial cells, they immediately change their characteristics drastically. They begin to swell; they assume irregular forms with numerous irradiating pseudopods protruding from their surfaces; their contractile proteins contract forcefully and cause the release of granules that contain multiple active factors; they become sticky so that they stick to the collagen fibers; they secrete large quantities of *ADP;* and their enzymes form *thromboxane A₂,* which is also secreted into the blood. The ADP and thromboxane in turn act on nearby platelets to activate them as well, and the stickiness of these additional platelets causes them to adhere to the originally activated platelets. Therefore, at the site of any rent in a vessel, the damaged vascular wall or extravascular tissues elicit a vicious circle of activation of successively increasing numbers of platelets that themselves attract more and more additional platelets, thus forming a *platelet plug.* This is at first a fairly loose plug, but it is usually successful in blocking the blood loss if the vascular opening is small. Then, during the subsequent process of blood coagulation, to be described in subsequent paragraphs, *fibrin threads* form that attach to the platelets, thus constructing a tight and unyielding plug.

IMPORTANCE OF THE PLATELET METHOD FOR CLOSING VASCULAR HOLES. If the rent in a vessel is small, the platelet plug by itself can stop blood loss, but if there is a large hole, a blood clot in addition to the platelet plug is required to stop the bleeding.

The platelet-plugging mechanism is extremely important for closing the minute ruptures in very small blood vessels that occur hundreds of times daily. Indeed, multiple small holes through the endothelial cells themselves are often closed by platelets actually fusing with the endothelial cells to form additional endothelial cell membrane. A person who has few platelets develops each day literally hundreds of small hemorrhagic areas under the skin and throughout the internal tissues, but this does not occur in the normal person.

Blood Coagulation in the Ruptured Vessel

The third mechanism for hemostasis is formation of the blood clot. The clot begins to develop in 15 to 20 seconds if the trauma of the vascular wall has been severe and in 1 to 2 minutes if the trauma has been minor. Activator substances both from the traumatized vascular wall and from platelets

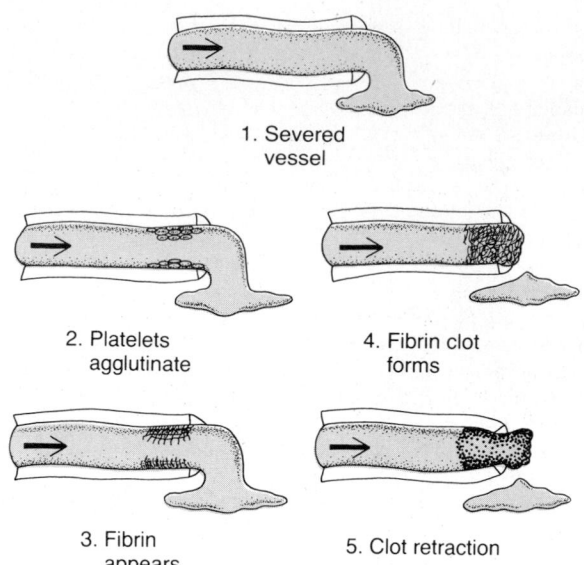

Figure 36–1. Clotting process in the traumatized blood vessel. (Modified from W. H. Seegers: Hemostatic Agents. Springfield, Ill., Charles C Thomas, 1948.)

1. Severed vessel
2. Platelets agglutinate
3. Fibrin appears
4. Fibrin clot forms
5. Clot retraction occurs

and blood proteins adhering to the traumatized vascular wall initiate the clotting process. The physical events of this process are shown in Figure 36–1 and the chemical events are discussed in detail later in the chapter.

Within 3 to 6 minutes after rupture of a vessel, if the vessel opening is not too large, the entire opening or broken end of the vessel is filled with clot. After 20 minutes to an hour, the clot retracts; this closes the vessel still further. Platelets play an important role in this clot retraction, as is also discussed later.

Fibrous Organization or Dissolution of the Blood Clot

Once a blood clot has formed, it can follow one of two courses: (1) It can become invaded by fibroblasts, which subsequently form connective tissue all through the clot, or (2) it can dissolve. The usual course for a clot that forms in a small hole of a vessel wall is invasion by fibroblasts, beginning within a few hours after the clot is formed (which is promoted at least partially by the *growth factor* secreted by platelets). This continues to complete organization of the clot into fibrous tissue within about 1 to 2 weeks. On the other hand, when additional blood coagulates to form a larger clot, such as blood that has leaked into tissues, special substances within the clot itself usually become activated, and these then function as enzymes to dissolve the clot, as discussed later in the chapter.

MECHANISM OF BLOOD COAGULATION

BASIC THEORY. More than 50 important substances that affect blood coagulation have been found in the blood and tissues, some that promote coagulation, called *procoagulants,* and others that inhibit coagulation, called *anticoagulants.* Whether or not the blood will coagulate depends on the balance between these two groups of substances. The anticoagulants normally predominate, and the blood does not coagulate; but when a vessel is ruptured, procoagulants in the area of damage become "activated" and override the anticoagulants, and then a clot does develop.

General Mechanism. All research workers in the field of blood coagulation agree that clotting takes place in three essential steps: (1) In response to rupture of the vessel or damage to the blood itself, a complex cascade of chemical reactions occurs in the blood involving more than a dozen blood coagulation factors. The net result is formation of a complex of activated substances collectively called *prothrombin activator*. (2) The prothrombin activator catalyzes the conversion of *prothrombin* into *thrombin*. (3) The thrombin acts as an enzyme to convert *fibrinogen* into *fibrin fibers* that enmesh platelets, blood cells, and plasma to form the clot.

Let us discuss first the mechanism by which the blood clot itself is formed, beginning with the conversion of prothrombin to thrombin; then we will come back to the initiating stages in the clotting process by which prothrombin activator is formed.

Conversion of Prothrombin to Thrombin

After prothrombin activator has been formed as a result of rupture of the blood vessel or as a result of damage to special activator substances in the blood, the prothrombin activator then, in the presence of sufficient amounts of ionic Ca^{++}, causes conversion of prothrombin to thrombin (Fig. 36–2). The thrombin in turn causes polymerization of fibrinogen molecules into fibrin fibers within another 10 to 15 seconds. Thus, the rate-limiting factor in causing blood coagulation is usually the formation of prothrombin activator and not the subsequent reactions beyond that point because these terminal steps normally occur rapidly to form the clot itself.

Platelets also play an important role in the conversion of prothrombin to thrombin because much of the prothrombin first attaches to prothrombin receptors on the platelets that have already bound to the damaged tissue. Then this binding accelerates the formation of thrombin from the prothrombin, occurring exactly in the tissue where the clot is needed.

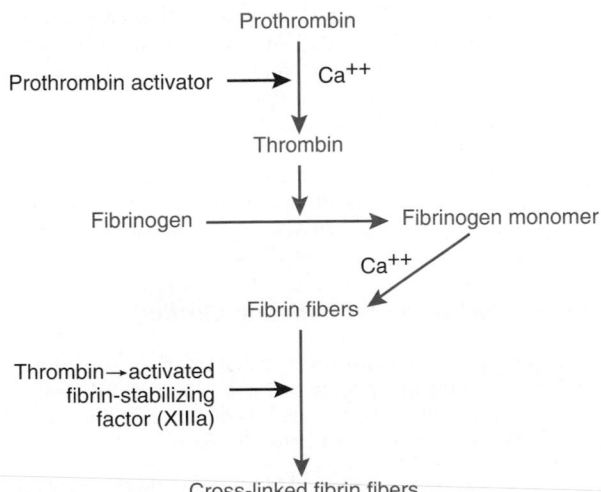

Figure 36–2. Schema for conversion of prothrombin to thrombin and polymerization of fibrinogen to form fibrin fibers.

PROTHROMBIN AND THROMBIN. Prothrombin is a plasma protein, an alpha$_2$-globulin, having a molecular weight of 68,700. It is present in normal plasma in a concentration of about 15 mg/dl. It is an unstable protein that can split easily into smaller compounds, one of which is *thrombin*, which has a molecular weight of 33,700, almost exactly one half that of prothrombin.

Prothrombin is formed continually by the liver, and it is continually being used throughout the body for blood clotting. If the liver fails to produce prothrombin, its concentration in the plasma falls too low to provide normal blood coagulation within one to several days.

Vitamin K is required by the liver for normal formation of prothrombin as well as four other clotting factors that we discuss later. Therefore, either lack of vitamin K or the presence of liver disease that prevents normal prothrombin formation can decrease the prothrombin level so low that a bleeding tendency results.

Conversion of Fibrinogen to Fibrin— Formation of the Clot

FIBRINOGEN. Fibrinogen is a high-molecular-weight protein (340,000) that occurs in the plasma in quantities of 100 to 700 mg/dl. Fibrinogen is formed in the liver, and liver disease occasionally decreases the concentration of circulating fibrinogen, as it does the concentration of prothrombin, as pointed out earlier.

Because of its large molecular size, little fibrinogen normally leaks into the interstitial fluids; and because it is one of the essential factors in the coagulation process, interstitial fluids ordinarily coagulate poorly, if at all. Yet when the permeability of the capillaries becomes pathologically increased, fibrinogen does then leak into the tissue fluids in sufficient quantities to allow clotting of these fluids in much the same way that plasma and whole blood clot.

ACTION OF THROMBIN ON FIBRINOGEN TO FORM FIBRIN. Thrombin is a protein *enzyme* with proteolytic capabilities. It acts on fibrinogen to remove four low-molecular-weight peptides from each molecule of fibrinogen, forming a molecule of *fibrin monomer* that has the automatic capability of polymerizing with other fibrin monomer molecules. Therefore, many fibrin monomer molecules polymerize within seconds into *long fibrin fibers* that form the *reticulum* of the clot.

In the early stages of this polymerization, the fibrin monomer molecules are held together by weak noncovalent hydrogen bonding, and the newly forming fibers also are not cross-linked with one another; therefore, the resultant clot is weak and can be broken apart with ease. Still another process occurs during the next few minutes that greatly strengthens the fibrin reticulum. This involves a substance called *fibrin-stabilizing factor* that is normally present in small amounts in the plasma globulins but it is also released from platelets entrapped in the clot. Before fibrin-stabilizing factor can have an effect on the fibrin fibers, it must itself be activated. The same thrombin that causes fibrin formation also activates the fibrin-stabilizing factor. Then this activated substance operates as an enzyme to cause *covalent bonds* between the fibrin monomer molecules as well as multiple cross-linkages between the adjacent fibrin fibers, thus adding tremendously to the three-dimensional strength of the fibrin meshwork.

THE BLOOD CLOT. The clot is composed of a meshwork of fibrin fibers running in all directions and entrapping blood cells, platelets, and plasma. The fibrin fibers also adhere to damaged surfaces of blood vessels; therefore, the blood clot becomes adherent to any vascular opening and thereby prevents blood loss.

CLOT RETRACTION—SERUM. Within a few minutes after a clot is formed, it begins to contract and usually expresses most of the fluid from the clot within 20 to 60 minutes. The fluid expressed is called *serum* because all its fibrinogen and most of the other clotting factors have been removed; in this way, serum differs from plasma. Serum cannot clot because of lack of these factors.

Platelets are necessary for clot retraction to occur. Therefore, failure of clot retraction is an indication that the number of platelets in the circulating blood is low. Electron micrographs of platelets in blood clots show that they become attached to the fibrin fibers in such a way that they actually bond different fibers together. Furthermore, platelets entrapped in the clot continue to release procoagulant substances, one of which is fibrin-stabilizing factor, which causes more and more cross-linking bonds between the adjacent fibrin fibers. In addition, the platelets themselves contribute directly to clot contraction by activating platelet thrombosthenin, actin, and myosin molecules, which are contractile proteins in the platelets and cause strong contraction of the platelet spicules attached to the fibrin. This also helps compress the fibrin meshwork into a smaller mass. The contraction is activated or accelerated by thrombin as well as by calcium ions released from the calcium stores in the mitochondria, endoplasmic reticulum, and Golgi apparatus of the platelets.

As the clot retracts, the edges of the broken blood vessel are pulled together, thus possibly or probably contributing to the ultimate state of hemostasis.

Vicious Circle of Clot Formation

Once a blood clot has started to develop, it normally extends within minutes into the surrounding blood. That is, the clot itself initiates a vicious circle (positive feedback) to promote more clotting. One of the most important causes of this is the fact that the proteolytic action of thrombin allows it to act on many of the other blood-clotting factors in addition to fibrinogen. For instance, thrombin has a direct proteolytic effect on prothrombin itself, tending to convert this into still more thrombin, and it acts on some of the blood-clotting factors responsible for the the formation of prothrombin activator. (These effects, discussed in subsequent paragraphs, include acceleration of the actions of Factors VIII, IX, X, XI, and XII and aggregation of platelets.) Once a critical amount of thrombin is formed, a vicious circle develops that causes still more blood clotting and more thrombin to be formed; thus, the blood clot continues to grow until something stops its growth.

Initiation of Coagulation: Formation of Prothrombin Activator

Now that we have discussed the clotting process itself, we must turn to the more complex mechanisms that initiate the clotting in the first place. These mechanisms can be set into play by trauma to the vascular wall and adjacent tissues, trauma to the blood, or contact of the blood with damaged endothelial cells or with collagen and other tissue elements outside the blood vessel endothelium. In each instance, they lead to the formation of *prothrombin activator*, which then causes prothrombin conversion to thrombin and all the subsequent clotting steps.

Prothrombin activator is generally considered to be

Table 36–1 CLOTTING FACTORS IN THE BLOOD AND THEIR SYNONYMS

Clotting Factor	Synonyms
Fibrinogen	Factor I
Prothrombin	Factor II
Tissue factor	Factor III; tissue thromboplastin
Calcium	Factor IV
Factor V	Proaccelerin; labile factor; Ac-globulin (Ac-G)
Factor VII	Serum prothrombin conversion accelerator (SPCA); proconvertin; stable factor
Factor VIII	Antihemophilic factor (AHF); antihemophilic globulin (AHG); antihemophilic factor A
Factor IX	Plasma thromboplastin component (PTC); Christmas factor; antihemophilic factor B
Factor X	Stuart factor; Stuart-Prower factor
Factor XI	Plasma thromboplastin antecedent (PTA); antihemophilic factor C
Factor XII	Hageman factor
Factor XIII	Fibrin-stabilizing factor
Prekallikrein	Fletcher factor
High-molecular-weight kininogen	Fitzgerald factor; HMWK
Platelets	

formed in two ways, although, in reality, the two ways interact constantly with each other: (1) by the *extrinsic pathway* that begins with trauma to the vascular wall and surrounding tissues and (2) by the *intrinsic pathway* that begins in the blood itself.

In both the extrinsic and the intrinsic pathway a series of different plasma proteins, especially beta-globulins, play major roles. Along with the other factors already discussed that enter into the clotting process, they are called *blood-clotting factors*, and for the most part, they are *inactive* forms of proteolytic enzymes. When converted to the active forms, their enzymatic actions cause the successive, cascading reactions of the clotting process.

Most of the clotting factors, which are listed in Table 36–1, are designated by Roman numerals. When one wants to indicate the activated form of the factor, he or she adds a small letter "a" after the Roman numeral, such as Factor VIIIa to indicate the activated state of Factor VIII.

Extrinsic Mechanism for Initiating Clotting

The extrinsic mechanism for initiating the formation of prothrombin activator begins with a traumatized vascular wall or extravascular tissues and occurs according to the following steps, as shown in Figure 36–3.

1. *Release of tissue factor.* Traumatized tissue releases a complex of several factors called *tissue factor* or *tissue thromboplastin*. This is composed especially of *phospholipids* from the membranes of the tissues and a *lipoprotein complex* containing an important *proteolytic enzyme*.

Extrinsic Pathway

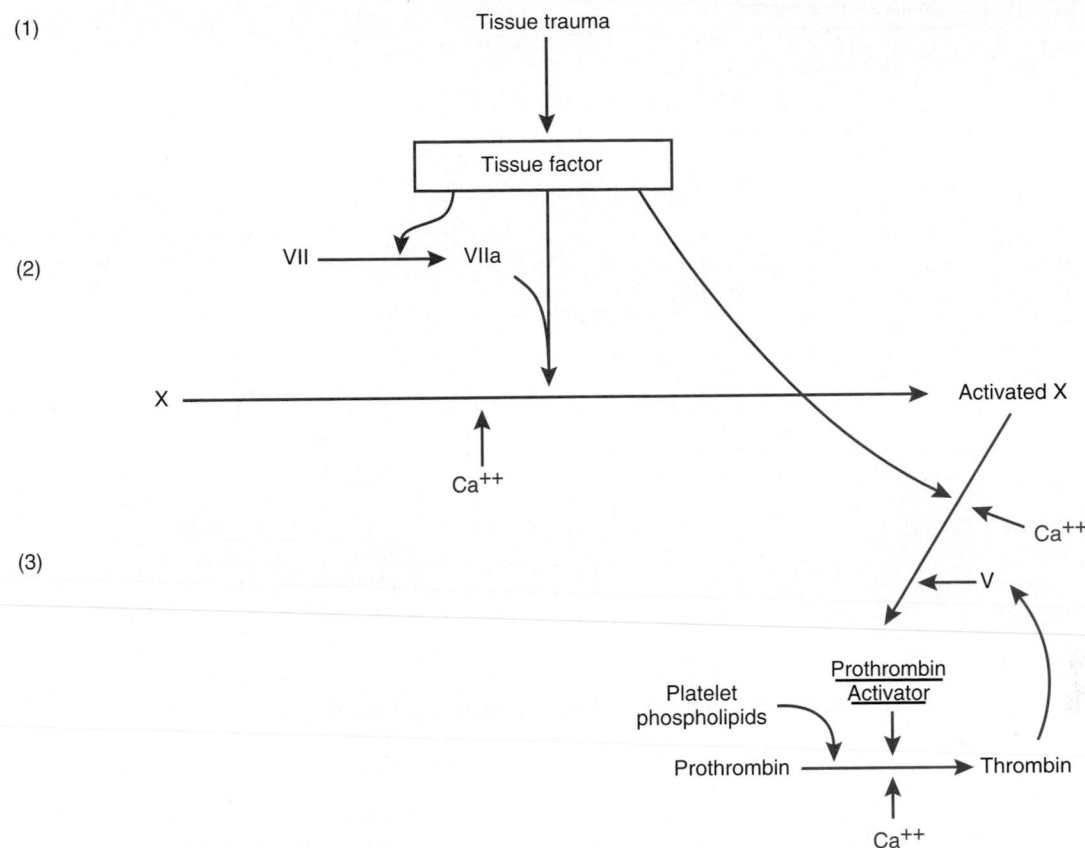

Figure 36–3. Extrinsic pathway for initiating blood clotting.

2. *Activation of Factor X—role of Factor VII and tissue factor.* The lipoprotein complex of tissue factor further complexes with blood coagulation Factor VII and, in the presence of calcium ions, acts enzymatically on Factor X to form *activated Factor X.*

3. *Effect of activated Factor X to form prothrombin activator—role of Factor V.* The activated Factor X combines immediately with tissue phospholipids that are part of tissue factor or with additional phospholipids released from platelets as well as with Factor V to form the complex called *prothrombin activator.* Within a few seconds, this splits prothrombin to form thrombin, and the clotting process proceeds as already explained. At first, the Factor V in the prothrombin activator complex is inactive, but once clotting begins and thrombin begins to form, the proteolytic action of thrombin activates Factor V. This then becomes an additional strong accelerator of prothrombin activation. Thus, in the final prothrombin activator complex, activated Factor X is the actual protease that causes splitting of prothrombin to thrombin, activated Factor V greatly accelerates this protease activity, and phospholipids act as a vehicle that further accelerates the process. Note especially the *positive feedback* effect of thrombin, acting through Factor V, in accelerating the entire process once it begins.

Intrinsic Mechanism for Initiating Clotting

The second mechanism for initiating the formation of prothrombin activator, and therefore for initiating clotting, begins with trauma to the blood itself or exposure of the blood to collagen in a traumatized blood vessel wall and then continues through the following series of cascading reactions shown in Figure 36–4.

1. *Blood trauma causes activation of Factor XII and release of platelet phospholipids.* Trauma to the blood or exposure of the blood to vascular wall collagen alters two important clotting factors in the blood: Factor XII and the platelets. When Factor XII is disturbed, such as by coming into contact with collagen or with a wettable surface such as glass, it takes on a new configuration that converts it into a proteolytic enzyme called "activated Factor XII." Simultaneously, the blood trauma also damages the platelets because of adherence to either collagen or a wettable surface (or by damage in other ways), and this releases platelet phospholipid that contains the lipoprotein called platelet factor 3, which also plays a role in subsequent clotting reactions.

2. *Activation of Factor XI.* The activated Factor XII acts enzymatically on Factor XI to activate this as well, which is the second step in the intrinsic pathway. This reaction also requires *HMW (high-molecular-weight) kininogen* and is accelerated by prekallikrein.

3. *Activation of Factor IX by activated Factor XI.* The activated Factor XI then acts enzymatically on Factor IX to activate this factor also.

4. *Activation of Factor X—role of Factor VIII.* The activated Factor IX, acting in concert with activated Factor VIII and with the platelet phospholipids and factor 3 from the

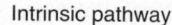

Intrinsic pathway

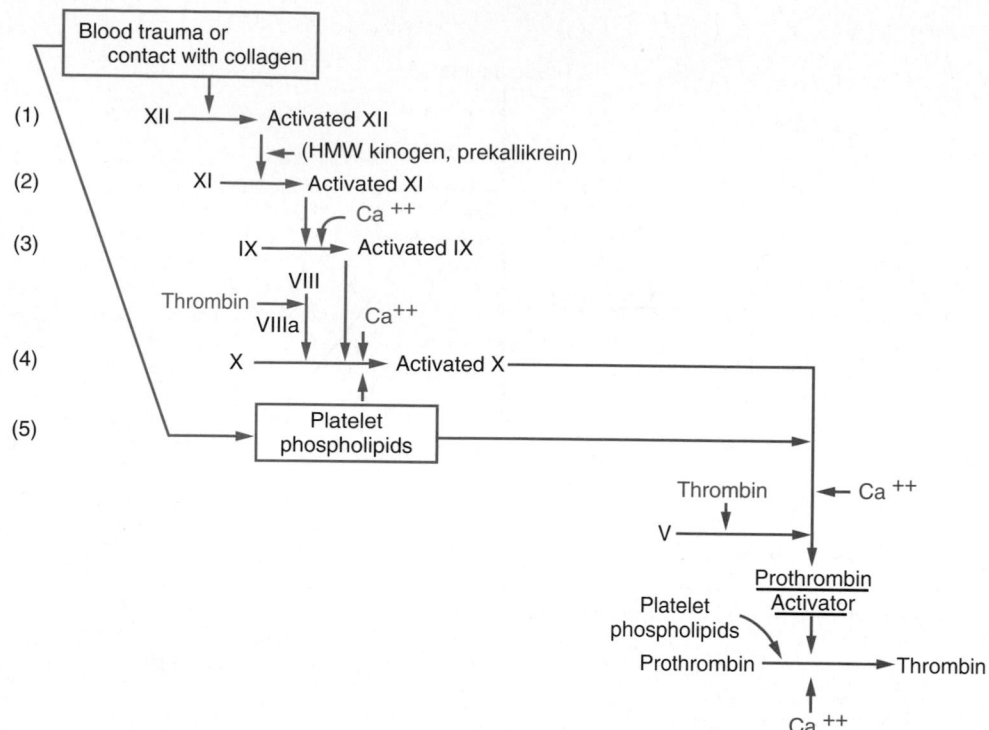

Figure 36–4. Intrinsic pathway for initiating blood clotting.

traumatized platelets, activates Factor X. It is clear that when either Factor VIII or platelets are in short supply, this step is deficient. Factor VIII is the factor that is missing in the person who has classic *hemophilia,* for which reason it is called *antihemophilic factor.* Platelets are the clotting factor that is lacking in the bleeding disease called *thrombocytopenia.*

5. *Action of activated Factor X to form prothrombin activator—role of Factor V.* This step in intrinsic pathway is the same as the last step in the extrinsic pathway. That is, activated Factor X combines with Factor V and platelet or tissue phospholipids to form the complex called *prothrombin activator.* The prothrombin activator in turn initiates within seconds the cleavage of prothrombin to form thrombin, thereby setting into motion the final clotting process, as described earlier.

Role of Calcium Ions in the Intrinsic and Extrinsic Pathways

Except for the first two steps in the intrinsic pathway, calcium ions are required for promotion or acceleration of all the reactions. Therefore, in the absence of calcium ions, blood clotting does not occur.

In the living body, the calcium ion concentration seldom falls low enough to significantly affect the kinetics of blood clotting. On the other hand, when blood is removed from a person, it can be prevented from clotting by reducing the calcium ion concentration below the threshold level for clotting, either by deionizing the calcium by reacting it with substances such as *citrate ion* or by precipitating the calcium with substances such as *oxalate ion.*

Interaction Between the Extrinsic and Intrinsic Pathways—Summary of Blood-Clotting Initiation

It is clear from the schemas of the intrinsic and extrinsic systems above that after blood vessels rupture, clotting is initiated by both pathways simultaneously. Tissue factor initiates the extrinsic pathway, whereas contact of Factor XII and the platelets with collagen in the vascular wall initiates the intrinsic pathway.

An especially important difference between the extrinsic and intrinsic pathways is that *the extrinsic pathway* can be explosive in nature; once initiated, its speed of occurrence is limited only by the amount of tissue factor released from the traumatized tissues and by the quantities of Factors X, VII, and V in the blood. With severe tissue trauma, clotting can occur in as little as 15 seconds. The intrinsic pathway is much slower to proceed, usually requiring 1 to 6 minutes to cause clotting.

Prevention of Blood Clotting in the Normal Vascular System—The Intravascular Anticoagulants

ENDOTHELIAL SURFACE FACTORS. Probably the most important factors for preventing clotting in the normal vascular system are (1) the *smoothness* of the endothelium, which prevents contact activation of the intrinsic clotting system; (2) a layer of *glycocalyx,* a mucopolysaccharide adsorbed to the inner surface of the endothelium, which repels the clotting factors and platelets, thereby preventing activation of clotting; and (3) a protein bound with the endothelial membrane, *thrombomodulin*, which binds thrombin. The binding of thrombomodulin with thrombin not only slows the clotting process by removing thrombin but the thrombomodulin–thrombin complex also activates a plasma protein, *protein C*, that acts as an anticoagulant by *inactivating* activated Factors V and VIII.

When the endothelial wall is damaged, its smoothness and its glycocalyx–thrombomodulin layer are both lost, which activates both Factor XII and the platelets, thus setting off the intrinsic pathway of clotting. If Factor XII and platelets come in contact with the subendothelial collagen, the activation is even more powerful.

ANTITHROMBIN ACTION OF FIBRIN AND ANTITHROMBIN III. Among the most important anticoagulants in the blood itself are those that remove thrombin from the blood. The most powerful of these are (1) the *fibrin fibers* that themselves are formed during the process of clotting and (2) an alpha-globulin called *antithrombin III* or *antithrombin-heparin cofactor.*

While a clot is forming, about 85 to 90 per cent of the thrombin formed from the prothrombin becomes adsorbed to the fibrin fibers as they develop. This helps prevent the spread of thrombin into the remaining blood and, therefore, prevents excessive spread of the clot.

The thrombin that does not adsorb to the fibrin fibers soon combines with antithrombin III, which blocks the effect of the thrombin on the fibrinogen and then inactivates the bound thrombin during the next 12 to 20 minutes.

HEPARIN. Heparin is another powerful anticoagulant. Yet its concentration in the blood is normally slight, so that only under limited physiological conditions does it have significant anticoagulant effects. On the other hand, it is widely used in medical practice to prevent intravascular clotting.

The heparin molecule is a highly negatively charged conjugated polysaccharide. By itself, it has little or no anticoagulant property, but when it combines with antithrombin III, this increases a hundredfold to a thousandfold the effectiveness of antithrombin III in removing thrombin and thus acts as an anticoagulant. Therefore, in the presence of excess heparin, the removal of thrombin from the circulating blood by antithrombin III is almost instantaneous.

The complex of heparin and antithrombin III removes several other activated coagulation factors in addition to thrombin, further enhancing the effectiveness of anticoagulation. The others include activated Factors XII, XI, IX, and X.

Heparin is produced by many different cells of the human body, but especially large quantities are formed by the basophilic *mast cells* located in the pericapillary connective tissue throughout the body. These cells continually secrete small quantities of heparin that diffuse into the circulatory system. Also, the *basophil cells* of the blood, which are functionally almost identical with the mast cells, release small quantities of heparin into the plasma.

Mast cells are abundant in the tissue surrounding the capillaries of the lungs and to a lesser extent in the capillaries of the liver. It is easy to understand why large quantities of heparin might be needed in these areas because the capillaries of the lungs and liver receive many embolic clots formed in the slowly flowing venous blood; sufficient formation of heparin prevents further growth of the clots.

ALPHA₂-MACROGLOBULIN. *Alpha₂-macroglobulin* is a large globulin molecule having a molecular weight of 360,000. It is similar to antithrombin–heparin complex in that it combines with the proteolytic coagulation factors. However, its activity is not accelerated by heparin. Its function is mainly to act as a binding agent for several of the coagulation factors and prevent their proteolytic actions until they can be destroyed

in various ways but not by the alpha₂-macroglobulin itself. In this way, it probably plays an important role even normally in preventing blood clotting.

Lysis of Blood Clots—Plasmin

The plasma proteins contain a euglobulin called *plasminogen* or *profibrinolysin*, which, when activated, becomes a substance called *plasmin* or *fibrinolysin*. Plasmin is a proteolytic enzyme that resembles trypsin, the most important proteolytic digestive enzyme of pancreatic secretion. It digests the fibrin fibers as well as other substances in the surrounding blood, such as fibrinogen, Factor V, Factor VIII, prothrombin, and Factor XII. Therefore, whenever plasmin is formed in a blood clot, it can cause lysis of the clot and destruction of many of the clotting factors, thereby sometimes even causing hypocoagulability of the blood.

ACTIVATION OF PLASMINOGEN TO FORM PLASMIN: THEN LYSIS OF CLOTS. When a clot is formed, a large amount of plasminogen is trapped in the clot along with other plasma proteins. This will not become plasmin or cause lysis of the clot until it is activated. The injured tissues and vascular endothelium very slowly release a powerful activator called *tissue plasminogen activator* (t-PA) that a day or so later, after the clot has stopped the bleeding, eventually converts plasminogen to plasmin and removes the clot. In fact, many small blood vessels in which the blood flow has been blocked by clots are reopened by this mechanism.

A PLASMIN INHIBITOR, ALPHA₂-ANTIPLASMIN. Plasmin not only destroys fibrin fibers but also functions as a proteolytic enzyme to digest fibrinogen and a number of other clotting factors. Small amounts of plasmin are formed in the blood all the time, which could seriously impede the activation of the clotting system were it not for the fact that the blood also contains another factor, *alpha₂-antiplasmin*, that binds with plasmin and inhibits it. Therefore, the rate of plasmin formation must rise above a certain critical level before it becomes effective.

SIGNIFICANCE OF THE PLASMIN SYSTEM. The lysis of blood clots allows slow clearing (over a period of several days) of extraneous clotted blood in the tissues and sometimes allows reopening of clotted vessels. An especially important function of the plasmin system is to remove minute clots from the millions of tiny peripheral vessels that eventually would become occluded were there no way to clean them.

CONDITIONS THAT CAUSE EXCESSIVE BLEEDING IN HUMAN BEINGS

Excessive bleeding can result from deficiency of any one of the many blood-clotting factors. Three particular types of bleeding tendencies that have been studied to the greatest extent are discussed: (1) bleeding caused by vitamin K deficiency, (2) hemophilia, and (3) thrombocytopenia (platelet deficiency).

Decreased Prothrombin, Factor VII, Factor IX, and Factor X Caused by Vitamin K Deficiency

With few exceptions, almost all the blood-clotting factors are formed by the liver. Therefore, diseases of the liver such as *hepatitis, cirrhosis,* and *acute yellow atrophy* all can sometimes depress the clotting system so greatly that the patient develops a severe tendency to bleed.

Another cause of depressed formation of clotting factors by the liver is vitamin K deficiency. Vitamin K is necessary for formation of five of the important clotting factors, *prothrombin, Factor VII, Factor IX, Factor X*, and *protein C*. In the absence of vitamin K, insufficiency of these coagulation factors can lead to a serious bleeding tendency.

Vitamin K is continually synthesized in the intestinal tract by bacteria, so that vitamin K deficiency seldom occurs in the normal person because of its absence from the diet, except in neonates before they establish their intestinal bacterial flora. In gastrointestinal disease, vitamin K deficiency often occurs as a result of poor absorption of fats from the gastrointestinal tract because vitamin K is fat-soluble and ordinarily is absorbed into the blood along with the fats.

One of the most prevalent causes of vitamin K deficiency is failure of the liver to secrete bile into the gastrointestinal tract (which occurs either as a result of obstruction of the bile ducts or as a result of liver disease) because lack of bile prevents adequate fat digestion and absorption and, therefore, depresses vitamin K absorption as well. Thus, liver disease often causes decreased production of prothrombin and the other factors both because of poor vitamin K absorption and because of the diseased liver cells. Because of this, vitamin K is injected into all patients with liver disease or obstructed bile ducts before the performance of any surgical procedure. Ordinarily, if vitamin K is given to a deficient patient 4 to 8 hours before operation and the liver parenchymal cells are at least one-half normal in function, sufficient clotting factors will be produced to prevent excessive bleeding during the operation.

Hemophilia

Hemophilia is a bleeding tendency that occurs almost exclusively in males. In 85 per cent of cases, it is caused by *deficiency of Factor VIII;* this type of hemophilia is called *hemophilia A* or *classic hemophilia.* About 1 of every 10,000 males in the United States has classic hemophilia. In the other 15 per cent of the so-called hemophilia patients, the bleeding tendency is caused by deficiency of Factor IX. Both of these factors are transmitted genetically by way of the female chromosome as a recessive trait. Therefore, almost never will a woman have hemophilia because at least one of her two X chromosomes will have the appropriate genes. If one of her X chromosomes is deficient, she will be a *hemophilia carrier*, transmitting the disease to half her male offspring and transmitting the carrier state to half her female offspring.

The bleeding trait in hemophilia can have various degrees of severity, depending on the severity of the genetic deficiency. Bleeding usually does not occur except after trauma, but the degree of trauma required to cause severe and prolonged bleeding may be so mild that it is hardly noticeable. And bleeding can often last for weeks after extraction of a tooth.

Factor VIII is composed of two components, a large component with a molecular weight in the millions and a smaller component with a molecular weight of about 230,000. The smaller component is most important in the intrinsic pathway for clotting, and deficiency of this part of Factor VIII causes classic hemophilia. Another bleeding disease with somewhat different characteristics, called von Willebrand's disease, results from loss of the large component.

When a person with classic hemophilia develops severe, prolonged bleeding, almost the only therapy that is truly effective is injection of purified Factor VIII. The cost of Factor VIII is high, and its availability is limited because it can be gathered only from human blood and only in extremely small quantities. A genetically engineered Factor VIII should be available for human use soon.

Thrombocytopenia

Thrombocytopenia means the presence of a very low quantity of platelets in the circulatory system. People with thrombocytopenia have a tendency to bleed, as do hemophiliacs, except that the bleeding is usually from many small venules or capillaries, rather than from larger vessels, as in hemophilia. As a result, small punctate hemorrhages occur throughout all the body tissues. The skin of such a person displays many small, purplish blotches, giving the disease the name *thrombocytopenic purpura.* As stated earlier, platelets are especially important for repair of minute breaks in capillaries and other small vessels.

Ordinarily, bleeding does not occur until the number of platelets in the blood falls below 50,000 per microliter, rather than the normal 150,000 to 300,000. Levels as low as 10,000 per microliter are frequently lethal.

Even without making specific platelet counts on the blood, sometimes one can suggest the existence of thrombocytopenia by noting whether or not a clot of the person's blood retracts because, as pointed out earlier, clot retraction is normally dependent on release of multiple additional coagulation factors from the large numbers of platelets entrapped in the fibrin mesh of the clot.

Most people with thrombocytopenia have the disease known as *idiopathic thrombocytopenia,* which means "thrombocytopenia of unknown cause." In most of these people, it has been discovered that specific antibodies have formed and react against the platelets themselves to destroy them. Occasionally these have developed because of transfusions from other people, but they usually result from development of autoimmunity to the person's own platelets, the cause of which is not known.

Relief from bleeding for 1 to 4 days can often be effected in the patient with thrombocytopenia by giving *fresh whole blood transfusions* that contain large amounts of platelets. Also, *splenectomy* is usually very helpful, sometimes effecting almost a complete cure because the spleen removes large numbers of platelets, particularly damaged ones, from the blood.

THROMBOEMBOLIC CONDITIONS IN THE HUMAN BEING

THROMBI AND EMBOLI. An abnormal clot that develops in a blood vessel is called a *thrombus.* Once a clot has developed, continued flow of blood past the clot is likely to break it away from its attachment to flow along with the blood; such freely flowing clots are known as *emboli.* Emboli generally do not stop flowing until they come to a narrow point in the circulatory system. Thus, emboli that originate in large arteries or in the left side of the heart eventually plug either smaller systemic arteries or aterioles in the brain, kidneys, or elsewhere. Emboli that originate in the venous system and in the right side of the heart flow into the vessels of the lung to cause pulmonary arterial embolism.

CAUSES OF THROMBOEMBOLIC CONDITIONS. The causes of thromboembolic conditions in the human being are usually twofold: (1) Any *roughened endothelial surface of a vessel*—as may be caused by arteriosclerosis, infection, or trauma—is likely to initiate the clotting process. (2) Blood often clots *when it flows very slowly* through blood vessels because small quantities of thrombin and other procoagulants are always being formed. These are generally removed from the blood by the macrophage system, mainly the Kupffer cells of the liver. If the blood is flowing too slowly, the concentrations of the procoagulants in local areas often rise high enough to initiate clotting, but when the blood flows rapidly, they are rapidly mixed with large quantities of blood and are removed during passage through the liver.

USE OF t-PA OR STREPTOKINASE IN TREATING INTRAVASCULAR CLOTS. Genetically engineered t-PA is available. When delivered directly to a thrombosed area through a catheter, it is effective in activating plasminogen to plasmin, which in turn dissolves intravascular clots. For instance, if used within the first hour or so after thrombotic occlusion of a coronary artery, the heart is often spared serious damage.

Another substance that similarly activates plasminogen to plasmin is *streptokinase*. This is produced by some types of β-hemolytic streptococci and can be used with about equal success as t-PA in dissolving the clots of the coronary artery.

Femoral Thrombosis and Massive Pulmonary Embolism

Because clotting almost always occurs when blood flow is blocked for many hours in any vessel of the body, the immobility of patients confined to bed plus the practice of propping the knees with underlying pillows often cause intravascular clotting because of blood stasis in one or more of the leg veins for hours at a time. Then the clot grows, mainly in the direction of the slowly moving blood, sometimes growing the entire length of the leg veins and occasionally even up into the common iliac vein and inferior vena cava. Then, about 1 time out of every 10, a large part of the clot disengages from its attachments to the vessel wall and flows freely with the venous blood into the right side of the heart and then into the pulmonary arteries to cause massive blockage of the pulmonary arteries, called *massive pulmonary embolism*. If the clot is large enough to occlude both the pulmonary arteries, immediate death ensues. If only one pulmonary artery or a smaller branch is blocked, death may not occur, or the embolism may lead to death a few hours to several days later because of further growth of the clot within the pulmonary vessels. But, again, t-PA therapy can be a lifesaver.

Disseminated Intravascular Coagulation

Occasionally the clotting mechanism becomes activated in widespread areas of the circulation, giving rise to the condition called *disseminated intravascular coagulation*. This often results from the presence of large amounts of traumatized or dying tissue in the body that releases great quantities of tissue factor into the blood. Frequently, the clots are small but numerous, and they plug a large share of the small peripheral blood vessels. This occurs especially in septicemic shock, in which either circulating bacteria or bacterial toxins—especially *endotoxins*—activate the clotting mechanisms. The plugging of the small peripheral vessels greatly diminishes the delivery of oxygen and other nutrients to the tissues—a situation that exacerbates the shock picture. It is partly for this reason that septicemic shock is lethal in 85 per cent or more of patients.

A peculiar effect of disseminated intravascular coagulation is that the patient frequently begins to bleed. The reason for this is that so many of the clotting factors are removed by the widespread clotting that too few procoagulants remain to allow normal hemostasis of the remaining blood.

ANTICOAGULANTS FOR CLINICAL USE

In some thromboembolic conditions, it is desirable to delay the coagulation process. Various anticoagulants have been developed for this purpose. The ones most useful clinically are heparin and the coumarins.

Heparin as an Intravenous Anticoagulant

Commercial heparin is extracted from several different animal tissues and prepared in almost pure form. Injection of relatively small quantities, about 0.5 to 1 milligram per kilogram of body weight, causes the blood-clotting time to increase from a normal of about 6 minutes to 30 or more minutes. Furthermore, this change in clotting time occurs instantaneously, thereby immediately preventing or slowing further development of the thromboembolic condition.

The action of heparin lasts about 1.5 to 4 hours. The injected heparin is destroyed by an enzyme in the blood known as *heparinase*.

In the treatment of a patient with heparin, sometimes too much heparin is given, and *serious* bleeding crises occur. In these instances, *protamine* acts specifically as an antiheparin, and the clotting mechanism can be reverted to normal by administering this substance. Protamine combines electrostatically with heparin and inactivates it because it carries strong positive electrical charges, whereas heparin carries strong negative charges.

Coumarins as Anticoagulants

When a coumarin, such as *warfarin*, is given to a patient, the plasma levels of prothrombin and Factors VII, IX, and X, all formed by the liver, begin to fall, indicating that warfarin has a potent depressant effect on liver formation of these compounds. Warfarin causes this effect by competing with vitamin K for reactive sites in the enzymatic processes for formation of prothrombin and the other three clotting factors, thereby blocking the action of vitamin K.

After administration of an effective dose of warfarin, the coagulant activity of the blood decreases to about 50 per cent of normal by the end of 12 hours and to about 20 per cent of normal by the end of 24 hours. In other words, the coagulation process is not blocked immediately but must await the consumption of the prothrombin and other factors already present in the plasma. Normal coagulation returns 1 to 3 days after discontinuing therapy.

Prevention of Blood Coagulation Outside the Body

Although blood removed from the body and held in a glass test tube normally clots in about 6 minutes, blood collected in *siliconized containers* often does not clot for 1 hour or more. The reason for this delay is that preparing the surfaces of the containers with silicone prevents contact activation of platelets and Factor XII, which are the two effects that initiate the intrinsic clotting mechansim. On the other hand, untreated glass containers allow contact activation and rapid development of clots.

Heparin can be used for preventing coagulation of blood outside the body as well as in the body. Heparin is especially used in all surgical procedures in which the blood must be passed through a heart-lung machine or artificial kidney and then back into the person.

Various substances that *decrease the concentration of calcium ions* in the blood can also be used for preventing blood coagulation outside the body. For instance, soluble *oxalate* compounds mixed in very small quantity with a sample of blood cause precipitation of calcium oxalate from the plasma and thereby decrease the ionic calcium levels so much that blood coagulation is blocked.

A second calcium-deionizing agent used for preventing coagulation is *sodium, ammonium,* or *potassium citrate.* The citrate ion combines with calcium in the blood to cause an un-ionized calcium compound, and the lack of ionic calcium prevents coagulation. Citrate anticoagulants have an important advantage over the oxalate anticoagulants because oxalate is toxic to the body, whereas moderate quantities of citrate can be injected intravenously. After injection, the citrate ion is removed from the blood within a few minutes by the liver and is polymerized into glucose or metabolized directly for energy. Consequently, 500 milliliters of blood that has been rendered incoagulable by citrate can ordinarily be injected into a recipient within a few minutes without dire consequences. If the liver is damaged or if large quantities of citrated blood or plasma are given too rapidly, the citrate ion may not be removed quickly enough, and the citrate can then greatly depress the level of calcium ion in the blood, which can result in tetany and convulsive death.

BLOOD COAGULATION TESTS

Bleeding Time

When a sharp knife is used to pierce the tip of the finger or lobe of the ear, bleeding ordinarily lasts for 1 to 6 minutes. The time depends largely on the depth of the wound and the degree of hyperemia in the finger at the time of the test. Lack of several of the clotting factors can prolong the bleeding time, but this is especially prolonged by lack of platelets.

Clotting Time

Many methods have been devised for determining clotting times. The one most widely used is to collect blood in a chemically clean glass test tube and then to tip the tube back and forth about every 30 seconds until the blood has clotted. By this method, the normal clotting time is about 6 to 10 minutes.

Procedures using multiple test tubes have been devised for determining clotting time more accurately. Clotting times are also dependent on the condition of the glass itself and even on the size of the tube, which makes a high degree of standardization necessary to obtain accurate results. A typical condition that causes a prolonged clotting time is hemophilia, but a deficiency of any of the factors in the intrinsic pathway for clotting can be the culprit.

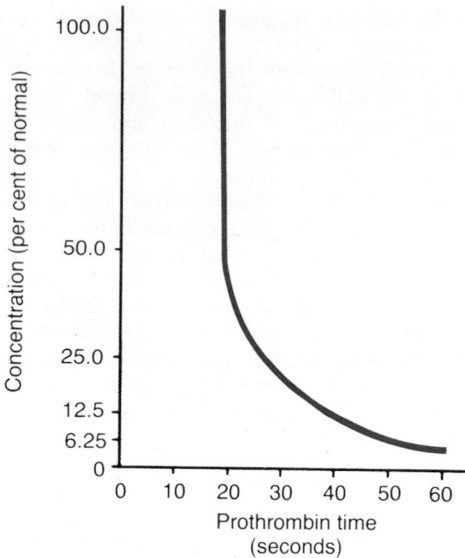

Figure 36–5. Relation of prothrombin concentration in the blood to the prothrombin time.

Prothrombin Time

The prothrombin time gives an indication of the total quantity of prothrombin in the blood. Figure 36–5 shows the relation of prothrombin concentration to prothrombin time. The means for determining prothrombin time is the following:

Blood removed from the patient is immediately oxalated so that none of the prothrombin can change into thrombin. Later, a large excess of calcium ion and tissue factor is suddenly mixed with the oxalated blood. The calcium nullifies the effect of the oxalate and the tissue factor activates the prothrombin-to-thrombin reaction by means of the extrinsic clotting pathway. The time required for coagulation to take place is known as the *prothrombin time.* The normal prothrombin time is about 12 seconds, but this depends to a certain extent on the exact procedure used. In each laboratory, a curve relating prothrombin concentration to prothrombin time, such as that shown in Figure 36–5, is ordinarily drawn for the method used so that the prothrombin in the blood can be quantitated.

Tests similar to that for prothrombin time have been devised to determine the quantities of other clotting factors in the body. In each of these tests, excesses of calcium ions and all the other factors besides the one being tested are added to oxalated blood all at once and then the time of coagulation is determined in the same manner as the usual prothrombin time. If the factor is deficient, the time is considerably prolonged.

REFERENCES

Bloom, A. L., et al.: Haemostasis and Thrombosis. New York, Churchill Livingstone, 1994.

Carpentier, A.: Low molecular weight (LMW) heparin derivatives in experimental extra-corporeal circulation (ECC). Haemostasis, 14:325, 1984.

Chavin, S. I.: Factor VIII: Structure and function in blood clotting. Am. J. Hematol., 16:297, 1984.

Claeson, G., et al.: The Design of Synthetic Inhibitors of Thrombin. New York, Plenum Publishing Corp., 1993.

Coller, B. S. (ed.): Progress in Hemostasis and Thrombosis, Vol. 9. Philadelphia, W. B. Saunders Co., 1989.

Colman, R. W., et al.: Hemostasis and Thrombosis: Basic Principles and Clinical Practice. Philadelphia, J. B. Lippincott, 1994.

Comerota, A. J.: Thrombolytic Therapy in Vascular Disease. Philadelphia, J. B. Lippincott, 1994.

Dvorak, A. M.: Basophil and Mast Cell Degranulation and Recovery. New York, Plenum Publishing Corp., 1991.

Esmon, C. T.: The regulation of natural anticoagulant pathways. Science, 235:1348, 1987.

Haber, E., et al.: Innovative approaches to plasminogen activator therapy. Science, 243:51, 1989.

Handin, R. I., et al.: Blood: Principles and Practice of Hematology. Philadelphia, J. B. Lippincott, 1994.

Harris, J. R.: Megakaryocytes, Platelets, Macrophages, and Eosinophils. New York, Plenum Publishing Corp., 1991.

Hassan, H. J., et al.: Intragenic Factor IX restriction site polymorphism in hemophilia B variants. Blood, 65:441, 1985.

Hathaway, W. E. and Goodnight, S. H., Jr.: Disorders of Hemostasis and Thrombosis. Hightstown, NJ, McGraw-Hill, 1993.

Hilgartner, M. W. and Pochedly, C.: Hemophilia in the Child and Adult, 3rd Ed. New York, Raven Press, 1989.

Holmsen, H. Nucleotide metabolism of platelets. Annu. Rev. Physiol., 47:677, 1985.

Kane, K. K.: Fibronolysis: A review. Ann. Clin. Lab. Sci., 14:443, 1984.

Kurtz, S. R. and Brubaker, D. B.: Clinical Decisions in Platelet Therapy. Farmington, CT, S. Karger Publishers, Inc., 1992.

Lee, G. R., et al.: Wintrobe's Clinical Hematology. Baltimore, Williams & Wilkins, 1993.

Mann, K. G.: The biochemistry of coagulation. Clin. Lab. Med., 4:207, 1984.

Miller, D. R., et al. (eds.): Blood Diseases of Infancy and Childhood. St. Louis, C. V. Mosby Co., 1989.

Monasterio, J., et al.: 13th International Congress on Thrombosis: Plenary Lectures. Farmington, CT, S. Karger Publishers, Inc., 1994.

Renck, H.: Bleeding and Thrombotic Disorders in Surgical Patients. East Norwalk, Conn., Appleton & Lange, 1988.

Rodack, B. F.: Diagnostic Hematology. Philadelphia, W. B. Saunders Co., 1994.

Rodgers, G. M.: Hemostatic properties of normal and perturbed vascular cells. FASEB J., 2:116, 1988.

Rudmann, S. V.: Blood Banking: Concepts and Applications. Philadelphia, W. B. Saunders Co., 1994.

Sherry, S.: Fibrinolysis, Thrombosis and Hemostasis. Baltimore, Williams & Wilkins, 1992.

Siess, W.: Molecular mechanisms of platelet activation. Physiol. Rev., 69:58, 1989.

Stamatoyannopoulos, G., et al.: The Molecular Basis of Blood Diseases. Philadelphia, W. B. Saunders Co., 1994.

Williams, W. J., et al.: Hematology. New York, McGraw-Hill Book Co., 1990.

RESPIRATION

UNIT VII

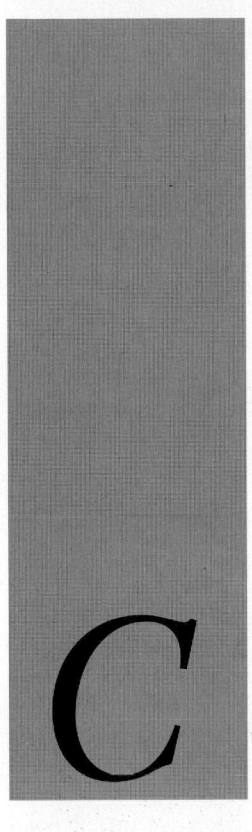

Pulmonary Ventilation

CHAPTER 37

The goals of respiration are to provide oxygen to the tissues and to remove carbon dioxide. To achieve these goals, respiration can be divided into four major functional events: (1) pulmonary ventilation, which means the inflow and outflow of air between the atmosphere and the lung alveoli; (2) diffusion of oxygen and carbon dioxide between the alveoli and the blood; (3) transport of oxygen and carbon dioxide in the blood and body fluids to and from the cells; and (4) regulation of ventilation and other facets of respiration. This chapter is a discussion of pulmonary ventilation and the subsequent five chapters cover the other respiratory functions as well as the physiology of special respiratory problems.

MECHANICS OF PULMONARY VENTILATION

Muscles That Cause Lung Expansion and Contraction

The lungs can be expanded and contracted in two ways: (1) by downward and upward movement of the diaphragm to lengthen or shorten the chest cavity and (2) by elevation and depression of the ribs to increase and decrease the anteroposterior diameter of the chest cavity. Figure 37–1 shows these two methods.

Normal quiet breathing is accomplished almost entirely by the first of the two methods above, that is, by movement of the diaphragm. During inspiration, contraction of the diaphragm pulls the lower surfaces of the lungs downward. Then, during expiration, the dia-

phragm simply relaxes, and the *elastic recoil* of the lungs, chest wall, and abdominal structures compresses the lungs. During heavy breathing, however, the elastic forces are not powerful enough to cause the necessary rapid expiration, so that the extra required force is achieved mainly by contraction of the *abdominal muscles*, which pushes the abdominal contents upward against the bottom of the diaphragm.

The second method for expanding the lungs is to raise the rib cage. This expands the lungs because in the natural resting position, the ribs slant downward, as shown on the left side of Figure 37–1, thus allowing the sternum to fall backward toward the vertebral column. But when the rib cage is elevated, the ribs project almost directly forward so that the sternum now also moves forward, away from the spine, making the anteroposterior thickness of the chest about 20 per cent greater during maximum inspiration than during expiration. Therefore, the muscles that elevate the chest cage are classified as muscles of inspiration, and the muscles that depress the chest cage are classified as muscles of expiration. The most important muscles that raise the rib cage are the *external intercostals*, but others that help are (1) the *sternocleidomastoid* muscles, which lift upward on the sternum; (2) the *anterior serrati*, which lift many of the ribs; and (3) the *scaleni*, which lift the first two ribs.

The muscles that pull the rib cage downward during expiration are (1) the *abdominal recti,* which have the powerful effect of pulling downward on the lower ribs at the same time that they and the other abdominal muscles compress the abdominal contents upward toward the diaphragm, and (2) the *internal intercostals.*

477

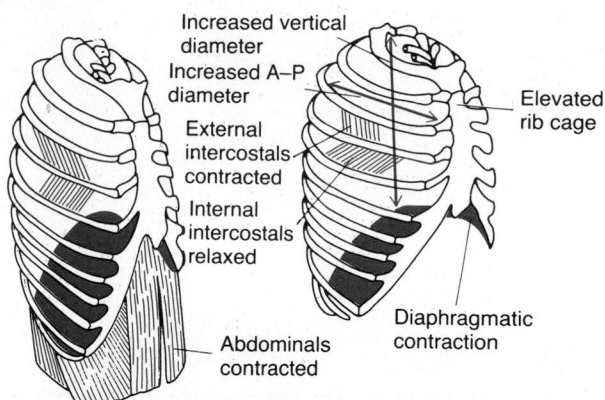

Increased vertical diameter
Increased A–P diameter
External intercostals contracted
Internal intercostals relaxed
Elevated rib cage
Diaphragmatic contraction
Abdominals contracted

Figure 37–1. Expansion and contraction of the thoracic cage during expiration and inspiration, demonstrating especially diaphragmatic contraction, elevation of the rib cage, and function of the intercostals.

Figure 37–1 also shows the mechanism by which the external and internal intercostals act to cause inspiration and expiration. To the left, the ribs during expiration are angled downward and the external intercostals are elongated forward and downward. As they contract, they pull the upper ribs forward in relation to the lower ribs, and this causes leverage on the ribs to raise them upward, thereby causing inspiration. The internal intercostals function exactly oppositely, functioning as expiratory muscles, because they angle between the ribs in the opposite direction and cause opposite leverage.

Movement of Air In and Out of the Lungs—and the Pressures That Cause It

The lung is an elastic structure that will collapse like a balloon and expel all its air through the trachea whenever there is no force to keep it inflated. Also, there are no attachments between the lung and the walls of the chest cage except where it is suspended at its hilum from the mediastinum. Instead, the lung literally floats in the thoracic cavity, surrounded by a thin layer of *pleural fluid* that lubricates the movements of the lungs within the cavity. Furthermore, continual suction of excess fluid into lymphatic channels maintains a slight suction between the visceral surface of the lung pleura and the parietal pleural surface of the thoracic cavity. Therefore, the two lungs are held to the thoracic wall as if glued there, except that they can slide freely, well lubricated, as the chest expands and contracts.

Pleural Pressure and Its Changes During Respiration

Pleural pressure is the pressure of the fluid in the narrow space between the lung pleura and the chest wall pleura. As noted above, this is normally a slight suction, which means a slightly negative pressure. The normal pleural pressure at the beginning of inspiration is about − 5 centimeters of water, which is the amount

of suction that is required to hold the lungs open to their resting level. Then, during normal inspiration, the expansion of the chest cage pulls the surface of the lungs with still greater force and creates a still more negative pressure down to an average of about − 7.5 centimeters of water.

These relations between pleural pressure and changing lung volume are demonstrated in Figure 37–2, showing in the bottom panel the increasing negativity of the pleural pressure from − 5 to − 7.5 during inspiration and in the top panel an increase in lung volume of 0.5 liter. Then, during expiration, the events are essentially reversed.

Alveolar Pressure

Alveolar pressure is the pressure inside the lung alveoli. When the glottis is open and no air is flowing into or out of the lungs, the pressures in all parts of the respiratory tree, all the way to the alveoli, are all equal to atmospheric pressure, which is considered to be 0 centimeter water pressure. To cause inward flow of air during inspiration, the pressure in the alveoli must fall to a value slightly below atmospheric pressure. The second dark line of Figure 37–2 demonstrates in normal inspiration a decrease in alveolar pressure to about − 1 centimeter of water. This slight negative pressure is enough to move about 0.5 liter of air into the lungs in the 2 seconds required for inspiration.

During expiration, opposite changes occur: The alveolar pressure rises to about + 1 centimeter of water, and this forces the 0.5 liter of inspired air out of the lungs during the 2 to 3 seconds of expiration.

Transpulmonary Pressure. Finally, note in Figure 37–2 the pressure difference between the alveolar pressure and the pleural pressure. This is called the *transpulmonary pressure.* It is the pressure difference between the alveoli and the outer surfaces of the lungs, and it is a measure of the elastic forces in the lungs that tend to collapse the lungs at each point of expansion, called the *recoil pressure.*

Compliance of the Lungs

The extent to which the lungs expand for each unit increase in transpulmonary pressure is called their *compliance.* The normal total compliance of both lungs together in the average adult human being averages about 200 ml/cm of water pressure, but this varies approximately in proportion to the person's non-fat weight. That is, every time the transpulmonary pressure increases by 1 centimeter of water, the lungs expand 200 milliliters.

Compliance Diagram of the Lungs. Figure 37–3 is a diagram relating lung volume changes to changes in transpulmonary pressure. Note that the relation is different for inspiration and expiration. Each curve is recorded by changing the transpulmonary pressure in small steps and allowing the lung volume to come to a steady level between successive steps. The two curves

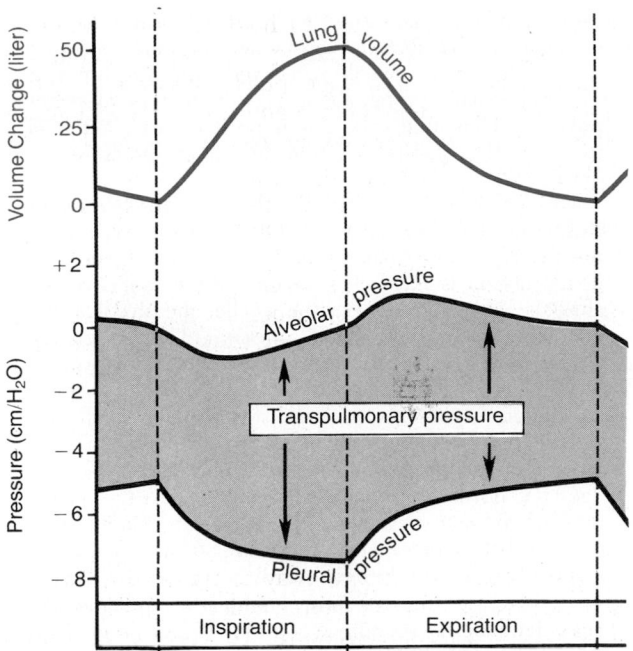

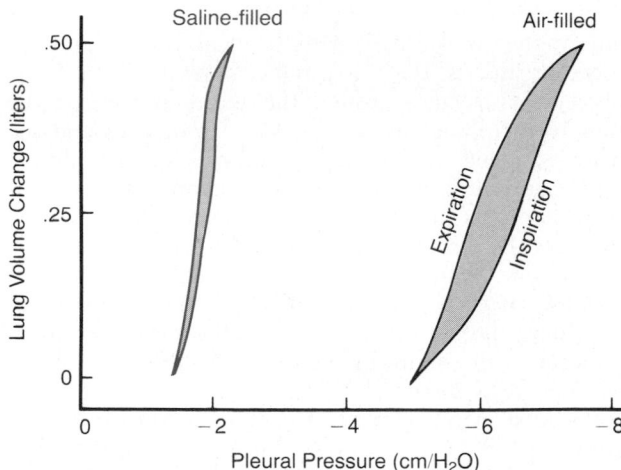

Figure 37–4. Comparison of the compliance diagrams of air-filled and saline-filled lungs.

Figure 37–2. Changes in lung volume, alveolar pressure, pleural pressure, and transpulmonary pressure during normal breathing.

are called the *inspiratory compliance curve* and the *expiratory compliance curve,* and the entire diagram is called the *compliance diagram of the lungs.*

The characteristics of the compliance diagram are determined by the elastic forces of the lungs. These can be divided into two parts: (1) the *elastic forces of the lung tissue* itself and (2) the *elastic force caused by surface tension of the fluid that lines the inside walls of the alveoli* and other lung air spaces.

The elastic forces of the lung tissue are determined

mainly by the elastin and collagen fibers interwoven among the lung parenchyma. In the deflated lungs, these fibers are in an elastically contracted and kinked state, and then, when the lungs are expanded, the fibers become stretched and unkinked, thereby elongating but still exerting elastic force to return to their natural state.

The elastic forces caused by surface tension are much more complex. However, surface tension accounts for about two thirds of the total elastic forces in the normal lungs. The significance of surface tension is shown in Figure 37–4, in which are compared the compliance of the lungs when filled with air and that when filled with saline solution. When filled with air, there is an interface between the alveolar lining fluid and the air in the alveoli. In the case of the saline solution–filled lung, there is no air–fluid interface, and therefore, the surface tension effect is not present— only tissue elastic forces are operative in the saline solution–filled lung.

Note that the transpulmonary pressures required to expand the air-filled lungs are about three times as great as the pressures required to expand the saline solution–filled lungs. Thus, one can conclude that the tissue elastic forces tending to cause collapse of the air-filled lung represent only about one third of the total lung elasticity, whereas the surface tension forces represent about two thirds.

The surface tension elastic forces of the lungs also change tremendously when the substance called "surfactant" is not present in the alveolar fluid. Therefore, let us discuss surfactant and its relation to the surface tension forces.

"Surfactant," Surface Tension, and Collapse of the Lungs

PRINCIPLE OF SURFACE TENSION. When water forms a surface with air, the water molecules on the surface of the water have an extra strong attraction for

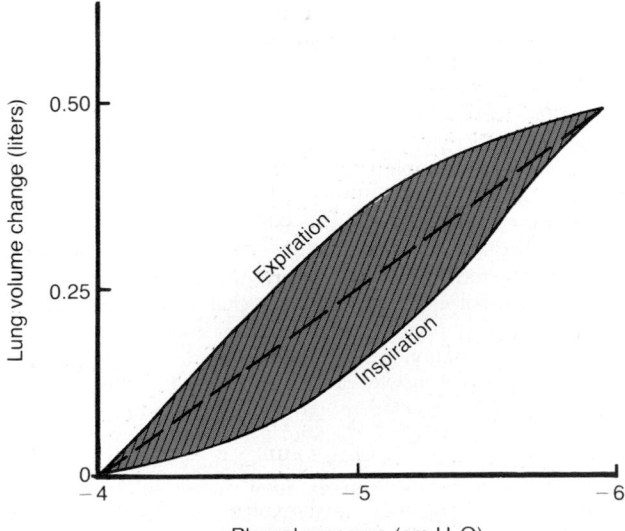

Figure 37–3. Compliance diagram in a normal person. This diagram shows the compliance of the lungs alone.

one another. As a result, the water surface is always attempting to contract. This is what holds raindrops together; that is, there is a tight contractile membrane of water molecules around the entire surface of the raindrop. Now let us reverse these principles and see what happens on the inner surfaces of the alveoli. Here, the water surface is also attempting to contract. This attempts to force the air out of the alveoli through the bronchi, and in doing so, it causes the alveoli (and other air spaces in the lungs) to attempt to collapse. Because this occurs in all the air spaces of the lungs, the net effect is to cause an elastic contractile force of the entire lungs, which is called the *surface tension elastic force.*

"SURFACTANT" AND ITS EFFECT ON SURFACE TENSION. Surfactant is a *surface active agent,* which means that when it spreads over the surface of a fluid, it greatly reduces the surface tension. It is secreted by special surfactant-secreting epithelial cells that constitute about 10 per cent of the surface area of the alveoli. These cells are granular, containing lipid inclusions; they are called *type II alveolar epithelial cells.*

Surfactant is a complex mixture of several phospholipids, proteins, and ions. The most important components are the phospholipid *dipalmitoylphosphatidylcholine, surfactant apoproteins,* and *calcium ions.* The dipalmitoylphosphatidylcholine, along with several less important phospholipids, is responsible for reducing the surface tension. They do not dissolve in the fluid; instead, they spread over its surface because one portion of each phospholipid molecule is hydrophilic and dissolves in the water lining the alveoli, whereas the lipid portion of the molecule is hydrophobic and oriented toward the air, forming a lipid hydrophobic surface exposed to the air. This surface has from one-twelfth to one-half the surface tension of a pure water surface; the exact surface tension depends on the concentration and orientation of the surfactant molecules on the surface. The importance of the surfactant apoproteins and the calcium ions is that in their absence, the dipalmitoylphosphatidylcholine spreads so slowly over the fluid surface that it cannot function effectively.

In quantitative terms, the surface tension of different watery fluids is about the following: pure water, 72 dynes/cm; normal fluids lining the alveoli but without surfactant, 50 dynes/cm; fluids lining the alveoli with surfactant included, between 5 and 30 dynes/cm.

COLLAPSE PRESSURE OF OCCLUDED ALVEOLI CAUSED BY SURFACE TENSION. If the air passages leading from the air spaces of the lungs are blocked, the surface tension tending to cause collapse of the spaces will create a positive pressure in the alveoli, attempting to push the air out. The amount of pressure generated in this way in a spherical air space can be calculated from the following formula:

$$\text{Pressure} = \frac{2 \times \text{Surface Tension}}{\text{Radius}}$$

For the average-sized alveolus with a radius of about 100 micrometers and lined with normal surfactant, this calculates to be about 4 centimeters of water pressure (3 mm Hg). If the alveoli were lined with pure water, it would calculate to be about 18 centimeters of water pressure. Thus, one sees how important surfactant is in reducing the amount of transpulmonary pressure required to keep the lungs expanded.

EFFECT OF SIZE OF THE ALVEOLI ON THE COLLAPSE PRESSURE CAUSED BY SURFACE TENSION. Note from the formula above that the collapse pressure generated in the alveoli is *inversely* affected by the radius of the alveolus, which means that the smaller the alveolus, the greater becomes the collapse pressure. Thus, when the alveoli have one-half the normal radius, only 50 instead of 100 micrometers, the collapse pressures noted above are doubled. This is specifically significant in small premature babies, many of whom have alveoli with radii less than one-quarter normal. Furthermore, surfactant does not normally begin to be secreted into the alveoli until between the 6th and 7th month of gestation and, in some babies, even later than that, so that many premature babies have little or no surfactant in the alveoli. Therefore, the lungs of these babies have extreme collapse tendencies, sometimes 30 mm Hg or more, and this causes the condition called *respiratory distress syndrome of the newborn.* It is fatal if not treated with strong measures, especially properly applied continuous positive pressure breathing.

ROLE OF SURFACTANT, "INTERDEPENDENCE," AND LUNG FIBROUS TISSUE IN "STABILIZING" THE SIZES OF THE ALVEOLI. Now, let us see what would happen if many of the alveoli in the lung were very small and others very large. The collapse tendency of the smaller alveoli would be far greater than that of the larger ones. Therefore, the smaller alveoli theoretically would tend to collapse, decreasing their volume in the lungs, and this loss of volume in a portion of the lungs would cause expansion of the larger alveoli. As the smaller alveoli became still smaller, their collapse tendency would become even greater, whereas the collapse tendency of the expanding, larger alveoli would become less. Therefore, theoretically, all the smaller alveoli would collapse totally, and this would pull the larger alveoli to still larger sizes. This phenomenon is called *instability of the alveoli.*

In practice, this phenomenon of instability of the alveoli does not occur in the normal lung, although it does occur under special conditions, such as when there is little surfactant in the alveolar fluid and the lung volume is decreased at the same time. There are several reasons why instability does not occur in the normal lung. One of these is the phenomenon called *interdependence* between the adjacent alveoli, alveolar ducts, and other air spaces. That is, each of these spaces splints the others in such a way that a large alveolus usually cannot exist adjacent to a small alveolus because they share common septal walls. This is the interdependence phenomenon.

A second reason why instability does not occur is that the lung is constructed of about 50,000 functional units, each containing one or a few alveolar ducts and their associated alveoli. All these are surrounded by fibrous septa that penetrate from the lung surface into the lung parenchyma. This fibrous tissue acts as an additional splint.

Finally, one must not forget the role of surfactant in opposing instability. It does this in two ways: First, it reduces the total amount of surface tension, and this allows the interdependence phenomenon and the fibrous tissue to overcome the surface tension effects. Second, as an alveolus becomes smaller, the surfactant molecules on the alveolar surface become squeezed together, increasing their concentration, and this reduces the surface tension still further. Thus, the smaller the alveolus becomes, the less becomes its surface tension, which opposes the extra tendency of smaller alveoli to collapse. Conversely, as alveoli become larger, the surface concentration of surfactant decreases, and the surface tension becomes greater, and this opposes further enlargement of these already large alveoli.

Effect of the Thoracic Cage on Lung Expansibility

Thus far, we have discussed the expansibility of the lungs alone without considering the thoracic cage. The thoracic cage has its own elastic and viscous characteristics, similar to those of the lungs; even if the lungs were not present in the thorax, muscular effort would be required to expand the thoracic cage.

Compliance of the Thorax and the Lungs Together

The compliance of the entire pulmonary system (the lungs and thoracic cage together) is measured while expanding the lungs of a totally relaxed or paralyzed person. To do this, air is forced into the lungs a little at a time while recording the lung pressures and volumes. To inflate this total pulmonary system, almost twice as much pressure is needed as to inflate the same lungs after removal from the chest cage. Therefore, the compliance of the combined lung-thorax system is almost exactly one-half that of the lungs alone —110 milliliters of volume per centimeter of water for the combined system, compared with 200 ml/cm for the lungs alone. Furthermore, when the lungs are expanded to high volumes or compressed to low volumes, the limitations of the chest become extreme; when near these limits, the compliance of the combined lung-thorax system can be as little as one-fifth that of the lungs alone.

"Work" of Breathing

We have already pointed out that during normal quiet respiration, almost all respiratory muscle contraction occurs only during inspiration, whereas expiration is almost entirely a passive process caused by elastic recoil of the lung and chest cage structures. Thus, the respiratory muscles normally perform "work" only to cause inspiration and not to cause expiration.

The work of inspiration can be divided into three fractions: (1) that required to expand the lungs against the lung and chest elastic forces, called *compliance work* or *elastic work;* (2) that required to overcome the viscosity of the lung and chest wall structures, called *tissue resistance work;* and (3) that required to overcome airway resistance during the movement of air into the lungs, called *airway resistance work.* These three types

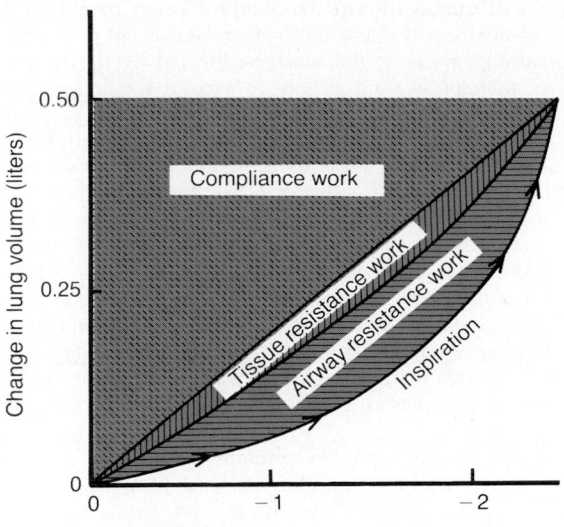

Figure 37–5. Graphic representation of the three types of work accomplished during inspiration: (1) compliance work, (2) tissue resistance work, and (3) airway resistance work.

of work are demonstrated graphically by the three shaded areas in Figure 37–5. In this diagram, the curve labeled "inspiration" shows the progressive *change in pleural pressure* and *lung volume* during inspiration. Also, the total shaded area of the figure represents the total work performed on the lungs by the inspiratory muscles during the act of inspiration. The shaded area in turn is divided into three segments that represent the three types of work performed during inspiration. They can be explained as follows.

COMPLIANCE WORK. The dotted area represents the compliance work that is required to expand the lungs against the elastic forces. This can be calculated by multiplying the volume of expansion times the average pressure required to cause the expansion. This is equal to the area represented by the dots. That is,

$$\text{Compliance work} = \frac{\Delta V \cdot \Delta P}{2},$$

where ΔV is the increase in volume and ΔP is the increase in intrapleural pressure.

TISSUE RESISTANCE WORK. The area represented by the vertical bars is proportional to the work expenditure required to overcome the viscosity of the lungs.

AIRWAY RESISTANCE WORK. Finally, the area in Figure 37–5 marked with horizontal bars represents the work required to overcome the resistance to airflow through the respiratory passageways.

ADDITIONAL WORK REQUIRED TO EXPAND AND CONTRACT THE THORACIC CAGE. The work of breathing calculated in Figure 37–5 applies only to the lungs, not the thoracic cage. However, we have seen that the compliance of the total lung-thorax system is only slightly more than one-half that of the lungs alone. Therefore, almost twice as much energy is required for normal expansion and contraction of the total lung-thorax system as for expansion of the lungs alone.

COMPARISON OF THE DIFFERENT TYPES OF WORK. It is clear from Figure 37–5 that during normal quiet breathing, most of the work performed by the respiratory muscles is used simply to expand the lungs. Normally, only a small percentage of the total work is used to overcome tissue resistance (tissue viscosity) and somewhat more is used to overcome airway resistance. On the other hand, during heavy breathing, when air must flow through the respiratory passageways at high velocity, the greater proportion of the work is used to overcome airway resistance.

In pulmonary disease, all three types of work are frequently vastly increased. Compliance work and tissue resistance work are especially increased by diseases that cause fibrosis of the lungs, and airway resistance work is especially increased by diseases that obstruct the airways.

During normal quiet respiration, no muscle "work" is performed during expiration because expiration results from elastic recoil of the lungs and chest. In heavy breathing or when airway resistance and tissue resistance are great, expiratory work does occur and sometimes becomes even greater than inspiratory work. This is especially true in asthma, which often increases airway resistance manyfold during expiration but much less so during inspiration for reasons that are explained later.

ENERGY REQUIRED FOR RESPIRATION. During normal quiet respiration, only 3 to 5 per cent of the total energy expended by the body is required to energize the pulmonary ventilatory process. During heavy exercise, the amount of energy required can increase as much as 50-fold, especially if the person has any degree of increased airway resistance or decreased pulmonary compliance. Therefore, one of the major limitations in the intensity of exercise that a person can perform is the person's ability to provide enough muscle energy for the respiratory process alone.

PULMONARY VOLUMES AND CAPACITIES

Recording Changes in Pulmonary Volume— Spirometry

A simple method for studying pulmonary ventilation is to record the volume movement of air into and out of the lungs, a process called *spirometry*. A typical, basic spirometer is shown in Figure 37–6. It consists of a drum inverted over a chamber of water, with the drum counterbalanced by a weight. In the drum is a breathing gas, usually air or oxygen; a tube connects the mouth with the gas chamber. When one breathes in and out of the chamber, the drum rises and falls and an appropriate recording is made on a moving sheet of paper.

Figure 37–7 shows a spirogram indicating changes in lung volume under different conditions of breathing. For ease in describing the events of pulmonary ventilation, the air in the lungs has been subdivided on this diagram into four *volumes* and four *capacities*, which are the following.

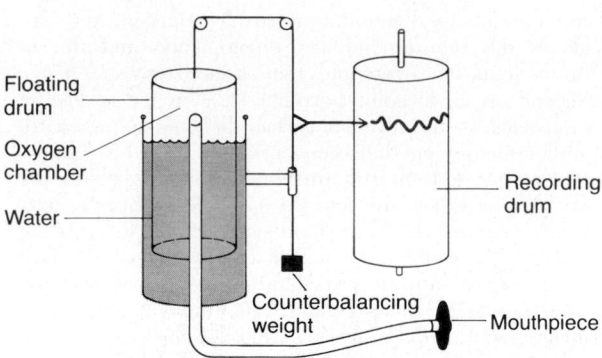

Figure 37–6. Spirometer.

Pulmonary "Volumes"

To the left in Figure 37–7 are listed four pulmonary lung "volumes," which, when added together, equal the maximum volume to which the lungs can be expanded. The significance of each of these volumes is the following:

1. The *tidal volume* is the volume of air inspired or expired with each normal breath; it amounts to about 500 milliliters in the average *young man*.

2. The *inspiratory reserve volume* is the extra volume of air that can be inspired over and above the normal tidal volume; it is usually equal to about 3000 milliliters.

3. The *expiratory reserve volume* is the extra amount of air that can be expired by forceful expiration after the end of a normal tidal expiration; this normally amounts to about 1100 milliliters.

4. The *residual volume* is the volume of air remaining in the lungs after the most forceful expiration. This volume averages about 1200 milliliters.

Pulmonary "Capacities"

In describing events in the pulmonary cycle, it is sometimes desirable to consider two or more of the volumes above together. Such combinations are called *pulmonary capacities*. To the right in Figure 37–7 are listed the pulmonary capacities, which can be described as follows:

1. The *inspiratory capacity* equals the *tidal volume* plus the *inspiratory reserve volume*. This is the amount of air (about 3500 milliliters) a person can breathe beginning at the normal expiratory level and distending the lungs to the maximum amount.

2. The *functional residual capacity* equals the *expiratory reserve volume* plus the *residual volume*. This is the amount of air that remains in the lungs at the end of normal expiration (about 2300 milliliters).

3. The *vital capacity* equals the *inspiratory reserve volume* plus the *tidal volume* plus the *expiratory reserve vol-*

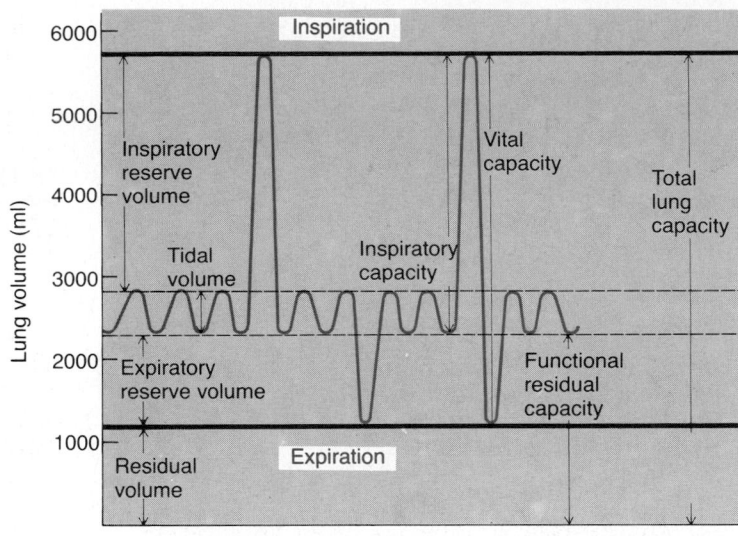

Figure 37–7. Diagram showing respiratory excursions during normal breathing and during maximal inspiration and maximal expiration.

ume. This is the maximum amount of air a person can expel from the lungs after first filling the lungs to their maximum extent and then expiring to the maximum extent (about 4600 milliliters).

4. The *total lung capacity* is the maximum volume to which the lungs can be expanded with the greatest possible inspiratory effort (about 5800 milliliters); it is equal to the vital capacity plus the residual volume.

All pulmonary volumes and capacities are about 20 to 25 per cent less in women than in men, and they are greater in large and athletic people than in small and asthenic people.

Abbreviations and Symbols Used in Pulmonary Function Studies

Spirometry is only one of many measurement procedures that the pulmonary physician uses daily. Furthermore, we shall see in subsequent discussions that many of the measurement procedures depend heavily on mathematical computations. To simplify these calculations as well as the presentation of pulmonary function data, a number of abbreviations and symbols have become standardized. The more important of these are given in Table 37–1. Using these symbols, we present here a few simple algebraic exercises showing some of the interrelations among the pulmonary volumes and capacities; the student should think through and verify

Table 37–1 ABBREVIATIONS AND SYMBOLS FOR PULMONARY FUNCTION

V_T	tidal volume	P_B	atmospheric pressure
FRC	functional residual capacity	Palv	alveolar pressure
ERV	expiratory reserve volume	Ppl	pleural pressure
RV	residual volume	P_{O_2}	partial pressure of oxygen
IC	inspiratory capacity	P_{CO_2}	partial pressure of carbon dioxide
IRV	inspiratory reserve volume	P_{N_2}	partial pressure of nitrogen
TLC	total lung capacity	Pa_{O_2}	partial pressure of oxygen in arterial
VC	vital capacity		blood
Raw	resistance of tracheobronchial tree to flow of air into the lung	Pa_{CO_2}	partial pressure of carbon dioxide in arterial blood
C	compliance	PA_{O_2}	partial pressure of oxygen in alveolar
V_D	volume of dead space gas		gas
V_A	volume of alveolar gas	PA_{CO_2}	partial pressure of carbon dioxide in
$\dot{V}_I$	inspired volume of ventilation per minute		alveolar gas
		PA_{H_2O}	partial pressure of water in alveolar gas
$\dot{V}_E$	expired volume of ventilation per minute	R	respiratory exchange ratio
$\dot{V}_A$	alveolar ventilation per minute	$\dot{Q}$	cardiac output
$\dot{V}_{O_2}$	rate of oxygen uptake per minute	$\dot{Q}s$	shunt flow
$\dot{V}_{CO_2}$	amount of carbon dioxide eliminated per minute	Ca_{O_2}	concentration of oxygen in arterial blood
$\dot{V}_{CO}$	rate of carbon monoxide uptake per minute	$C\overline{V}_{O_2}$	concentration of oxygen in mixed venous blood
DL_{O_2}	diffusing capacity of the lung for oxygen	S_{O_2}	percentage saturation of hemoglobin with oxygen
DL_{CO}	diffusing capacity of the lung for carbon monoxide	Sa_{O_2}	percentage saturation of hemoglobin with oxygen in arterial blood

these interrelations.

$$VC = IRV + V_T + ERV$$
$$VC = IC + ERV$$
$$TLC = VC + RV$$
$$TLC = IC + FRC$$
$$FRC = ERV + RV$$

Determination of Functional Residual Capacity, Residual Volume, and Total Lung Capacity—Helium Dilution Method

The functional residual capacity, which is the volume of air that normally remains in the lungs between breaths, is important to lung function. Its value changes markedly in some types of pulmonary disease, for which reason it is often desirable to measure this capacity. The spirometer cannot be used in a direct way to measure the functional residual capacity because the air in the residual volume of the lungs cannot be expired into the spirometer, and this volume constitutes about one half of the functional residual capacity. To measure functional residual capacity, the spirometer must be used in an indirect manner, usually by means of a helium dilution method, as follows.

A spirometer of known volume is filled with air mixed with helium at a known concentration. Before breathing from the spirometer, the person expires normally. At the end of this expiration, the remaining volume in the lungs is equal to the functional residual capacity. At this point, the subject immediately begins to breathe from the spirometer, and the gases of the spirometer begin to mix with the gases of the lungs. As a result, the helium becomes diluted by the functional residual capacity gases, and the volume of the functional residual capacity can then be calculated from the degree of dilution of the helium with the use of the following formula:

$$FRC = \left(\frac{Ci_{He}}{Cf_{He}} - 1\right) Vi_{Spir} ,$$

in which

FRC is *functional residual capacity*
Ci_{He} is *initial concentration of helium in the spirometer*
Cf_{He} is *final concentration of helium in the spirometer*
Vi_{Spir} is *initial volume of the spirometer*

Once the functional residual capacity has been determined, the residual volume can then be determined by subtracting the expiratory reserve volume from the functional residual capacity. Also, the total lung capacity can be determined by adding the inspiratory capacity to the functional residual capacity. That is,

$$RV = FRC - ERV$$

and

$$TLC = FRC + IC$$

MINUTE RESPIRATORY VOLUME EQUALS RESPIRATORY RATE TIMES TIDAL VOLUME

The *minute respiratory volume* is the total amount of new air moved into the respiratory passages each minute; this is equal to the *tidal volume* times the *respiratory rate*. The normal tidal volume is about 500 milliliters, and the normal respiratory rate is about 12 breaths per minute. Therefore, the *minute respiratory volume averages about 6 liters/min*. A person can live for a short period with a minute respiratory volume as low as 1.5 liters/min and a respiratory rate of two to four breaths per minute.

The respiratory rate occasionally rises to 40 to 50 per minute, and the tidal volume can become as great as the vital capacity, about 4600 milliliters in a young man. This can give a minute respiratory volume greater than 200 liters/min, or more than 30 times normal. Most people cannot sustain more than one half to two thirds these values for longer than 1 minute or so.

ALVEOLAR VENTILATION

The ultimate importance of the pulmonary ventilatory system is to continually renew the air in the gas exchange areas of the lungs where the air is in proximity to the pulmonary blood. These areas include the alveoli, alveolar sacs, alveolar ducts, and respiratory bronchioles. The rate at which new air reaches these areas is called *alveolar ventilation*. Strangely, however, during normal quiet respiration, the volume of air in the tidal air is only enough to fill the respiratory passageways down as far as the terminal bronchioles, with only a small portion of the inspired air actually flowing all the way into the alveoli. Therefore, how does new air move this last, short distance from the terminal bronchioles into the alveoli? The answer: by *diffusion*. Diffusion is caused by the kinetic motion of molecules, each gas molecule moving at high velocity among the other molecules. The velocity of movement of the molecules in the respiratory air is so great and the distances so short from the terminal bronchioles to the alveoli that the gases move this remaining distance in only a fraction of a second.

Dead Space and Its Effect on Alveolar Ventilation

Some of the air a person breathes never reaches the gas exchange areas but instead goes to fill respiratory passages where gas exchange does not occur, such as

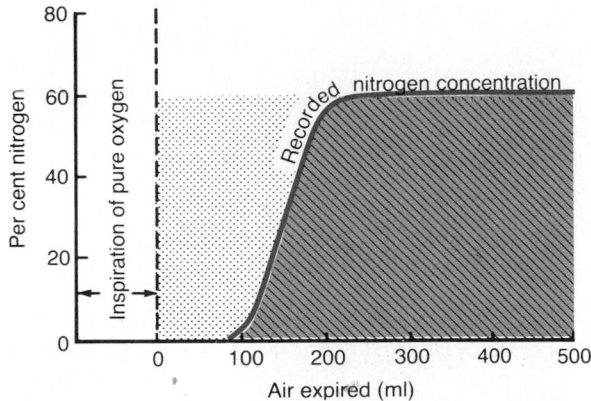

Figure 37–8. Continuous record of the changes in nitrogen concentration in the expired air after a previous inspiration of pure oxygen. This record can be used to calculate dead space.

in the nose, pharynx, and trachea. This air is called *dead space air* because it is not useful for the gas exchange process; the respiratory passages where no gas exchange takes place is called the *dead space*.

On expiration, the air in the dead space is expired first, before any of the air from the alveoli reaches the atmosphere. Therefore, the dead space is equally disadvantageous for removal of the expiratory gases from the lungs.

MEASUREMENT OF THE DEAD SPACE VOLUME. A simple method for measuring dead space volume is demonstrated by the graph in Figure 37–8. In making this measurement, the subject suddenly takes a deep breath of oxygen. This fills the entire dead space with pure oxygen. Some oxygen also mixes with the alveolar air but does not completely replace this air. Then the person expires through a rapidly recording nitrogen meter, which makes the record shown in the figure. The first portion of the expired air comes from the dead space regions of the respiratory passageways where the air has been completely replaced by oxygen. Therefore, in the early part of the record, only oxygen appears and the nitrogen concentration is zero. Then, when alveolar air begins to reach the nitrogen meter, the nitrogen concentration rises rapidly because alveolar air containing large amounts of nitrogen begins to mix with the dead space air. After still more air has been expired, all the dead space air has been washed from the passages and only alveolar air remains. The recorded nitrogen concentration reaches a plateau level equal to its concentration in the alveoli, as shown to the right in the figure. With a little thought, the student can see that the area covered by the dots represents the air that has no nitrogen in it; this area is a measure of the volume of dead space air. For exact quantification, the following equation is used:

$$V_D = \frac{\text{Area of dots} \times V_E}{\text{Area of hatching} + \text{Area of dots}},$$

where V_D is *dead space air* and V_E is the *total volume of expired air.*

Let us assume, for instance, that the area of the dots on the graph is 30 square centimeters, the area of the hatching is 70 square centimeters, and the total volume expired is 500 milliliters. The dead space then would be

$$\frac{30}{30 + 70} \times 500, \text{ or } 150 \text{ ml}$$

NORMAL DEAD SPACE VOLUME. The normal dead space air in a young man is about 150 milliliters. This increases slightly with age.

ANATOMIC VERSUS PHYSIOLOGIC DEAD SPACE. The method just described for measuring the dead space measures the volume of all the space of the respiratory system besides the alveoli and their other closely related gas exchange areas; this space is called the *anatomic dead space*. On occasion, some of the alveoli themselves are nonfunctional or are only partially functional because of absent or poor blood flow through adjacent pulmonary capillaries. Therefore, from a functional point of view, these alveoli must also be considered dead space. When the alveolar dead space is included in the total measurement of dead space, this is then called *physiologic dead space*, in contradistinction to the anatomic dead space. In a normal person, the anatomic and physiologic dead spaces are nearly equal because all alveoli are functional in the normal lung, but in a person with partially functional or nonfunctional alveoli in some parts of the lungs, sometimes the physiologic dead space is as much as 10 times the volume of the anatomic dead space, or 1 to 2 liters. These problems are discussed further in Chapter 39 in relation to pulmonary gaseous exchange and in Chapter 42 in relation to certain pulmonary diseases.

Rate of Alveolar Ventilation

Alveolar ventilation per minute is the total volume of new air entering the alveoli (and other adjacent gas exchange areas) each minute. It is equal to the respiratory rate times the amount of new air that enters the alveoli with each breath:

$$\dot{V}_A = \text{Freq} \cdot (V_T - V_D),$$

where $\dot{V}_A$ is the *volume of alveolar ventilation per minute*, Freq is the *frequency of respiration per minute*, V_T is the *tidal volume*, and V_D is the *dead space volume*.

Thus, with a normal tidal volume of 500 milliliters, a normal dead space of 150 milliliters, and a respiratory rate of 12 breaths per minute, alveolar ventilation equals $12 \times (500 - 150)$, or 4200 ml/min.

Alveolar ventilation is one of the major factors determining the concentrations of oxygen and carbon dioxide in the alveoli. Therefore, almost all discussions

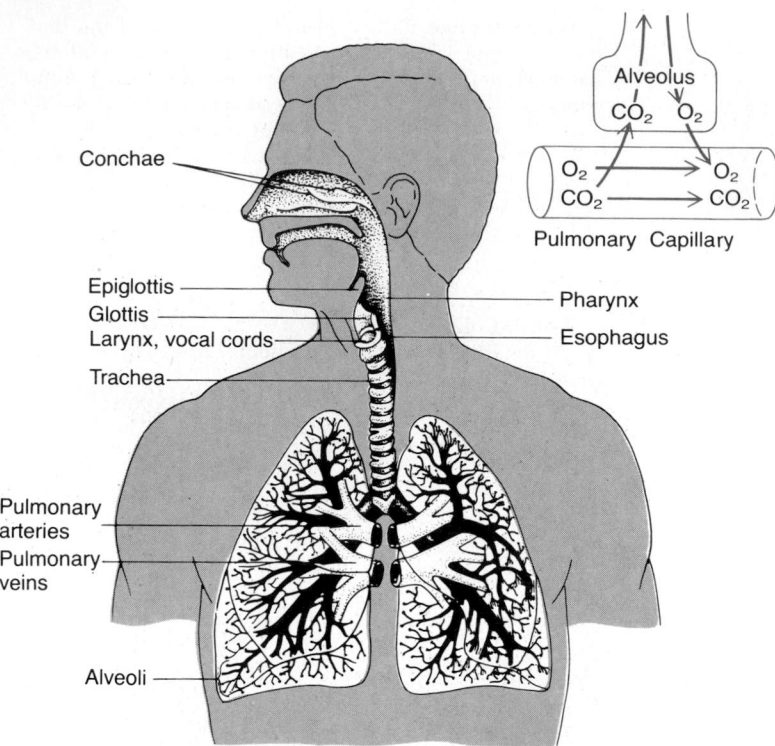

Conchae

Alveolus
CO_2 O_2

Pulmonary Capillary

O_2 O_2
CO_2 CO_2

Epiglottis
Glottis
Larynx, vocal cords
Trachea

Pharynx
Esophagus

Pulmonary arteries
Pulmonary veins

Alveoli

Figure 37–9. Respiratory passages.

of gaseous exchange in the following chapters on the respiratory system emphasize alveolar ventilation.

FUNCTIONS OF THE RESPIRATORY PASSAGEWAYS

Trachea, Bronchi, and Bronchioles

Figure 37–9 shows the respiratory system, demonstrating especially the respiratory passageways. The air is distributed to the lungs by way of the trachea, bronchi, and bronchioles. The *trachea* is called the *first-generation respiratory passageway,* and the two *main right* and *left bronchi* are the *second-generation;* each division thereafter is an additional generation. There are between 20 and 25 generations before the air finally reaches the alveoli.

One of the most important problems in all the respiratory passageways is to keep them open to allow easy passage of the air to and from the alveoli. To keep the trachea from collapsing, multiple cartilage rings extend about five sixths of the way around the trachea. In the walls of the bronchi, less extensive cartilage plates also maintain a reasonable amount of rigidity yet allow sufficient motion for the lungs to expand and contract. These plates become progressively less extensive in the later generations of bronchi and are gone in the bronchioles, which usually have diameters less than 1.5 millimeters. The bronchioles are not prevented from collapsing by rigidity of their walls. Instead, they are expanded by the same transpulmonary pressures that expand the alveoli. That is, as the alveoli enlarge, so also do the bronchioles.

MUSCULAR WALL OF THE BRONCHI AND BRONCHIOLES AND ITS CONTROL. In all areas of the trachea

and bronchi not occupied by cartilage plates, the walls are composed mainly of smooth muscle. Also, the walls of the bronchioles are almost entirely smooth muscle, with the exception of the most terminal bronchiole, called the *respiratory bronchiole,* which has only a few smooth muscle fibers. Many obstructive diseases of the lung result from narrowing of the smaller bronchi and bronchioles, often because of excessive contraction of the smooth muscle itself.

RESISTANCE TO AIR FLOW IN THE BRONCHIAL TREE. Under *normal respiratory conditions,* air flows through the respiratory passageways so easily that less than 1 centimeter of water pressure gradient from the alveoli to the atmosphere is sufficient to cause enough air flow for quiet breathing. The greatest amount of resistance to air flow occurs not in the minute air passages of the bronchioles but in some of the larger bronchi near to the trachea. The reason for this high resistance is that there are relatively few of these larger bronchi in comparison with about 65,000 parallel terminal bronchioles, through each of which only a minute amount of air must pass.

Yet in disease conditions, the smaller bronchioles often do play a far greater role in determining air flow resistance for two reasons: (1) because of their small size, they are easily occluded; (2) because they have a greater percentage of smooth muscle in the walls, they constrict easily.

NERVOUS AND LOCAL CONTROL OF THE BRONCHIOLAR MUSCULATURE—SYMPATHETIC DILATATION OF THE BRONCHIOLES. Direct control of the bronchioles by sympathetic nerve fibers is relatively weak because few of these fibers penetrate to the central portions of the lung. However, the bronchial tree is very much exposed to circulating norepinephrine and epinephrine released into the blood by sympathetic stimulation of the adrenal gland medullae. Both of these hormones, especially epinephrine because of its greater stimulation of *beta receptors,* cause dilatation of the bronchial tree.

PARASYMPATHETIC CONSTRICTION OF THE BRONCHIOLES. A few parasympathetic nerve fibers derived from the vagus nerves also penetrate the lung parenchyma. These nerves secrete acetylcholine and, when activated, cause mild to moderate constriction of the bronchioles. When a disease process such as asthma has already caused some constriction, parasympathetic superimposed nervous stimulation often worsens the condition. When this occurs, administration of drugs that block the effects of acetylcholine, such as *atropine,* can sometimes relax the respiratory passages enough to relieve the obstruction.

Sometimes the parasympathetic nerves are activated by reflexes that originate in the lungs. Most of these begin with irritation of the epithelial membrane of the respiratory passageways themselves, initiated by noxious gases, dust, cigarette smoke, or bronchial infection. Also, a bronchiolar constrictor reflex often occurs when microemboli occlude small pulmonary arteries.

LOCAL FACTORS OFTEN CAUSE BRONCHIOLAR CONSTRICTION. Several substances formed in the lungs themselves are often quite active in causing bronchiolar constriction. Two of the most important of these are *histamine* and *slow reactive substance of anaphylaxis.* Both of them are released in the lung tissues by mast cells during allergic reactions, especially allergic reactions caused by pollen in the air. Therefore, they play key roles in causing the airway obstruction that occurs in allergic asthma; this is especially true of the slow reactive substance of anaphylaxis.

Also, the same irritants that cause parasympathetic vasoconstrictor reflexes of the airways—smoke, dust, sulfur dioxide, and some of the acidic elements in smog—can initiate local, non-nervous reactions that cause obstructive constriction of the airways.

Mucous Coat of the Respiratory Passageways, and Action of Cilia to Clear the Passageways

All the respiratory passages, from the nose to the terminal bronchioles, are kept moist by a layer of mucus that coats the entire surface. The mucus is secreted partly by individual goblet cells in the epithelial lining of the passages and partly by small submucosal glands. In addition to keeping the surfaces moist, the mucus traps small particles out of the inspired air and keeps most of them from ever reaching the alveoli. The mucus itself is removed from the passages in the following manner.

The entire surface of the respiratory passages, both in the nose and in the lower passages down as far as the terminal bronchioles, is lined with ciliated epithelium, with about 200 cilia on each epithelial cell. These cilia beat continually at a rate of 10 to 20 times per second by the mechanism explained in Chapter 2, and the direction of their "power stroke" is always toward the pharynx. That is, the cilia in the lungs beat upward, whereas those in the nose beat downward. This continual beating causes the coat of mucus to flow slowly, at a velocity of about 1 cm/min, toward the pharynx. Then the mucus and its entrapped particles are either swallowed or coughed to the exterior.

Cough Reflex

The bronchi and trachea are so sensitive to light touch that excessive amounts of foreign matter or other cause of irritation initiates the cough reflex. The larynx and carina (the point where the trachea divides into the bronchi) are especially sensitive, and the terminal bronchioles and even the alveoli are sensitive to corrosive chemical stimuli such as sulfur dioxide gas and chlorine. Afferent impulses pass from the respiratory passages mainly through the vagus nerves to the medulla. There, an automatic sequence of events is triggered by the neuronal circuits of the medulla, causing the following effect.

First, about 2.5 liters of air is inspired. Second, the epiglottis closes, and the vocal cords shut tightly to entrap the air within the lungs. Third, the abdominal muscles contract forcefully, pushing against the diaphragm while other expiratory muscles, such as the internal intercostals, also contract forcefully. Consequently, the pressure in the lungs rises to 100 mm Hg or more. Fourth, the vocal cords and the epiglottis suddenly open widely, so that air under pressure in the lungs *explodes* outward. Indeed, sometimes this air is expelled at velocities ranging from 75 to 100 miles per hour. Furthermore, and important, the strong compression of the lungs collapses the bronchi and trachea by causing the noncartilaginous parts of these to invaginate inward, so that the exploding air actually passes through *bronchial* and *tracheal slits.* The rapidly moving air usually carries with it any foreign matter that is present in the bronchi or trachea.

Sneeze Reflex

The sneeze reflex is very much like the cough reflex except that it applies to the nasal passageways instead of the lower respiratory passages. The initiating stimulus of the sneeze reflex is irritation in the nasal passageways, the afferent impulses passing in the fifth nerve to the medulla, where the reflex is triggered. A series of reactions similar to those for the cough reflex takes place; however, the uvula is depressed, so that large amounts of air pass rapidly through the nose, thus helping clear the nasal passages of foreign matter.

Respiratory Functions of the Nose

As air passes through the nose, three distinct functions are performed by the nasal cavities: (1) The *air is warmed* by the extensive surfaces of the conchae and septum, a total area of about 160 square centimeters, shown in Figure 37–9. (2) The *air is almost completely humidified* even before it passes beyond the nose. (3) The *air is filtered.* These functions together are called the *air conditioning function* of the upper respiratory passageways. Ordinarily, the temperature of the inspired air rises to within 1°F of body temperature and within 2 to 3 per cent of full saturation with water vapor before it reaches the trachea. When a person breathes air through a tube directly into the trachea (as through a tracheostomy), the cooling and especially the drying effect in the lower lung can lead to serious lung crusting and infection.

FILTRATION FUNCTION OF THE NOSE. The hairs at the entrance to the nostrils are important for filtering out large particles. Much more important, though, is the removal of particles by *turbulent precipitation.* That is, the air passing through the nasal passageways hits many obstructing vanes: the conchae (also called "turbinates" because they cause turbulence of the air), the septum, and the pharyngeal wall. Each time air hits one of these obstructions it must change its direction of movement; the particles suspended in the air, having far more mass and momentum than air, cannot change their direction of travel as rapidly as can the air. Therefore, they continue forward, striking the surfaces of the obstructions, and are entrapped in the mucous coating and transported by the cilia to the pharynx to be swallowed.

Size of Particles Entrapped in the Respiratory Passages. The nasal turbulence mechanism for removing particles from air is so effective that almost no particles larger than 6 micrometers in diameter enter the lungs through the nose. This size is smaller than the size of red blood cells.

Of the remaining particles, many that are between 1 and 5 micrometers *settle* out in the smaller bronchioles as a result of *gravitational precipitation.* For instance, terminal bronchiolar disease is common in coal miners because of settled dust particles. Some of the still smaller particles (smaller than 1 micrometer in diameter) *diffuse* against the walls of the alveoli and adhere to the alveolar fluid. But many particles smaller than 0.5 micrometer in diameter remain suspended in the alveolar air and are later expelled by expiration. For instance, the particles of cigarette smoke have a particle size of about 0.3 micrometer. Almost none of these particles are precipitated in the respiratory passageways before they reach the alveoli. However, up to one third of them do precipitate in the alveoli by the diffusion process, with the balance remaining suspended and expelled in the expired air.

Many of the particles that become entrapped in the alveoli are removed by *alveolar macrophages,* as explained in Chapter 33, and others are carried away by the lung lymphatics. An excess of particles causes growth of fibrous tissue in the alveolar septa, leading to permanent debility.

Vocalization

Speech involves not only the respiratory system but also (1) specific speech nervous control centers in the cerebral cortex, which are discussed in Chapter 57,

(2) respiratory control centers of the brain, and (3) the articulation and resonance structures of the mouth and nasal cavities. Speech is composed of two mechanical functions: (1) *phonation,* which is achieved by the larynx, and (2) *articulation,* which is achieved by the structures of the mouth.

PHONATION. The larynx is especially adapted to act as a vibrator. The vibrating element is the *vocal folds,* commonly called the *vocal cords.* The vocal folds protrude from the lateral walls of the larynx toward the center of the glottis; they are stretched and positioned by several specific muscles of the larynx itself.

Figure 37–10B shows the vocal folds as they are seen when looking into the glottis with a laryngoscope. During normal breathing, the folds are wide open to allow easy passage of air. During phonation, the folds close together so that passage of air between them will cause vibration. The pitch of the vibration is determined mainly by the degree of stretch of the folds but also by how tightly the folds are approximated to one another and by the mass of their edges.

Figure 37–10A shows a dissected view of the vocal folds after removing the mucous lining. Immediately inside each fold is a strong elastic ligament called the *vocal ligament.* This is attached anteriorly to the large *thyroid cartilage,* which is the cartilage that protrudes from the anterior surface of the neck and is called the "Adam's apple." Posteriorly, the vocal ligament is attached to the *vocal processes* of two *arytenoid cartilages.* Both the thyroid cartilage and the arytenoid cartilages in turn articulate from below with another cartilage not shown in Figure 37–10, the *cricoid cartilage.*

The vocal folds can be stretched by either forward rotation of the thyroid cartilage or posterior rotation of the arytenoid cartilages, activated by muscles from the thyroid cartilage and arytenoid cartilages to the cricoid cartilage. Muscles located within the vocal folds lateral to the vocal ligaments, the thyroarytenoid muscles, can pull the arytenoid cartilages toward the thyroid cartilage and, therefore, loosen the vocal folds. Also, slips of these muscles can change the *shapes and masses of the vocal fold edges,* sharpening them to emit high-pitched sounds and blunting them for the more bass sounds.

Finally, several other sets of small laryngeal muscles lie between the arytenoid cartilages and the cricoid cartilage and can rotate these cartilages inward or outward or pull their bases together or apart to give the various configurations of the vocal folds shown in Figure 37–10B.

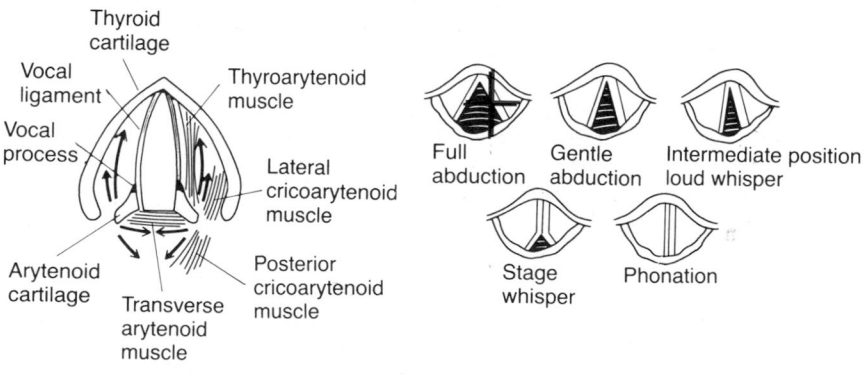

Figure 37–10. Laryngeal function in phonation showing in section B the positions of the vocal folds during different types of phonation. (Modified from Greene: The Voice and Its Disorders. 4th ed. Philadelphia, J. B. Lippincott Company, 1980.)

ARTICULATION AND RESONANCE. The three major organs of articulation are the *lips, tongue,* and *soft palate.* They need not be discussed in detail because we are all familiar with their movements during speech and other vocalizations.

The resonators include the *mouth,* the *nose* and *associated nasal sinuses,* the *pharynx,* and even the *chest cavity.* Again, we are all familiar with the resonating qualities of these structures. For instance, the function of the nasal resonators is demonstrated by the change in the quality of the voice when a person has a severe cold that blocks the air passages to these resonators.

REFERENCES

Agostoni, F.: Thickness and pressure of the pleural liquid. In Fishman, A. P., and Hecht, H. H. (eds.). The Pulmonary Circulation and Interstitial Space, Chicago, University of Chicago Press, 1969, p. 65.

Bailey, B. J., and Biller, H. F.: Surgery of the Larynx. Philadelphia, W. B. Saunders Co., 1985.

Barash, P. G., et al.: Handbook of Clinical Anesthesia Biology. Philadelphia, J. B. Lippincott Co., 1993.

Barnes, P. J.: Airway neuropeptides: Roles in fine tuning and in disease? News Physiol. Sci., 4:116, 1989.

Barnes, P. J.: Modulation of neurotransmission in airways. Physiol. Rev., 72:669, 1992.

Bartlett, D., Jr.: Respiratory functions of the larynx. Physiol. Rev., 69:33, 1989.

Baumgartner, W. A., et al.: Heart and Heart-Lung Transplantation. Philadelphia, W. B. Saunders Co., 1990.

Benumof, J. L.: Anesthesia for Thoracic Surgery. Philadelphia, W. B. Saunders Co., 1994.

Bless, D. M. (ed.): Vocal Fold Physiology. Contemporary Research and Clinical Issues. San Diego, College-Hill Press, 1983.

Bourbon, J. R., and Rieutort, M.: Pulmonary surfactant: biochemistry, physiology, and pathology. News Physiol. Sci., 2:129, 1987.

Brewis, R. A. L., and Geddes, D. M.: Respiratory Medicine. Philadelphia, W. B. Saunders Co., 1990.

Burg, F. D., et al.: Basic Sciences in Medicine: A Concept-based Approach. Philadelphia, J. B. Lippincott, 1995.

Burri, P. H.: Fetal and postnatal development of the lung. Annu. Rev. Physiol., 46:617, 1984.

Coburn, R. F. (ed.): Airway Smooth Muscle in Health and Disease. New York, Plenum Publishing Corp., 1989.

Coté, C. J., et al.: A Practice of Anesthesia for Infants and Children. Philadelphia, W. B. Saunders Co., 1993.

Deslauriers, J., and Lacquet, L. K.: The Pleural Space, Vol. 6. In Delarue, N. C., and Eschapasse, H. (eds.), International Trends in General Thoracic Surgery Series. St. Louis, C. V. Mosby Co., 1989.

Donald, P. J., et al.: The Sinuses. New York, Raven Press, 1994.

Donner, C. F., et al.: From Alveoli Back to Bronchi. Farmington, CT, S. Karger Publishers, Inc., 1992.

Drazen, J. M., et al.: High-frequency ventilation. Physiol. Rev., 64:505, 1984.

Epstein, M.: Renal effects of head-out water immersion in humans: a 15-year update. Physiol. Rev., 72:563, 1992.

Fairbanks, D. N. F.: Snoring and Obstructive Sleep Apnea. New York, Raven Press, 1994.

Fisher, A. B., and Chander, A.: Intracellular processing of surfactant lipids in the lung. Annu. Rev. Physiol., 47:789, 1985.

Fishman, A. P.: Assessment of Pulmonary Function. New York, McGraw-Hill, 1980.

Ford, C. N., and Bless, D. M.: Phonosurgery: Assessment and Surgical Management of Voice Disorders. New York, Raven Press, 1991.

Forster, R. E., II: Introduction to respiratory physiology. Annu. Rev. Physiol., 49:555, 1987.

Fujimura, O.: Vocal Physiology: Voice Production, Mechanisms and Function (Vocal Fold Physiology, Vol. 2). New York, Raven Press, 1988.

Gabella, G.: Innervation of airway smooth muscle: Fine structure. Annu. Rev. Physiol., 49:583, 1987.

George, R. B., et al.: Chest Medicine. Baltimore, Williams & Wilkins Co., 1995.

Gil, J.: Organization of microcirculation in the lung. Annu. Rev. Physiol., 42:177, 1980.

Gil, J.: Histological preservation and ultrastructure of alveolar surfactant. Annu. Rev. Physiol., 47:753, 1985.

Gonzalez, N. C., and Fedde, M. R. (eds.): Oxygen Transfer from Atmosphere to Tissues. New York, Plenum Publishing Corp., 1988.

Green, M., and Moxham, J.: The respiratory muscles. Clin. Sci., 68:1, 1985.

Grippi, M. A.: Pulmonary Pathophysiology: A Problem-oriented Approach. Philadelphia, J. B. Lippincott, 1994.

Harada, R. N., and Repine, J. E.: Pulmonary host defense mechanisms. Chest, 87:247, 1985.

Harding, R.: Function of the larynx in the fetus and newborn. Annu. Rev. Physiol., 46:645, 1984.

Hodgkin, J. E., and Petty, T. L.: Chronic Obstructive Pulmonary Disease: Current Concepts. Philadelphia, W. B. Saunders Co., 1987.

Kaplan, J. A.: Cardiac Anesthesia. Philadelphia, W. B. Saunders Co., 1993.

King, R. J.: Composition and metabolism of the apolipoproteins of pulmonary surfactant. Annu. Rev. Physiol., 47:775, 1985.

Kirby, R. R., and Gravenstein, N.: Clinical Anesthesia Practice. Philadelphia, W. B. Saunders Co., 1994.

Laszlo, G.: Pulmonary Function: A Guide for Clinicians. New York, Cambridge Univ. Press, 1994.

Leff, A. R., and Schumacker, P. T.: Respiratory Physiology: Basics and Applications. Philadelphia, W. B. Saunders Co., 1993.

Levitzky, M. G.: Pumonary Physiology. New York, McGraw-Hill Book Co., 1982.

Lundberg, J. M., and Coburn, R. F.: Polypeptide-containing neurons in airway smooth muscle. Annu. Rev. Physiol., 49:557, 1987.

Mines, A. H.: Respiratory Physiology. New York, Raven Press, 1993.

Minoo, P., and King, R. J.: Epithelial-mesenchymal interactions in lung development. Annu. Rev. Physiol., 56:13, 1994.

Morrow, P. E.: Factors determining hygroscopic aerosol deposition in airways. Physiol. Rev., 66:330, 1986.

Mortola, J. P.: Dynamics of breathing in newborn mammals. Physiol. Rev., 67:187, 1987.

Murray, J. F.: The Normal Lung, 2nd Ed. Philadelphia, W. B. Saunders Co., 1986.

Murray, J. F., and Nadel, J. A.: Textbook of Respiratory Medicine. Philadelphia, W. B. Saunders Co., 1994.

Newman, J. D. (ed.): The Physiological Control of Mammalian Vocalization. New York, Plenum Publishing Corp., 1988.

Nicolosi, L., et al. Terminology of Communication Disorders: Speech-Language-Hearing, 3rd Ed. Baltimore, Williams & Wilkins, 1988.

Notter, R. H., and Finkelstein, J. N.: Pulmonary surfactant: An interdisciplinary approach. J. Appl. Physiol., 57:1613, 1984.

Perelman, R. H., et al.: Developmental aspects of lung lipids. Annu. Rev. Physiol., 47:803, 1985.

Rahn, H., et al.: The pressure-volume diagram of the thorax and lung. Am. J. Physiol., 146:161, 1946.

Robertson, B., et al.: Experimental and Clinical Aspects of Surfactant Replacement. Farmington, CT, S. Karger Publishers, Inc., 1992.

Ryan, U. S.: Metabolic activity of pulmonary endothelium. Annu. Rev. Physiol., 48:263, 1986.

Saldana, M. J.: Pathology of Pulmonary Disease. Philadelphia, J. B. Lippincott, 1994.

Sant'Ambrogio, G., and Mathew, O. P.: Control of upper airway muscles. News Physiol. Sci., 3:167, 1988.

Schneiderman, C. R.: Basic Anatomy and Physiology in Speech and Hearing. San Diego, College-Hill Press, 1984.

Stoelting, R. K.: Pharmacology and Physiology in Anesthetic Practice. Philadelphia, J. B. Lippincott Co., 1991.

Taylor, A. E., et al.: Clinical Respiratory Physiology. Philadelphia, W. B. Saunders Co., 1989.

van Golde, L. M. G., et al.: The pulmonary surfactant system: Biochemical aspects and functional significance. Physiol. Rev., 68:374, 1988.

Weibel, E. R., and Bachofen, H.: Structural design of the alveolar septum and fluid exchange. In Fishman, A. P., and Renkin, E. M. (eds.): Pulmonary Edema. Baltimore, Waverly Press, 1979, p. 1.

Weibel, E. R.: Scaling of structural and functional variables in the respiratory system. Annu. Rev. Physiol., 49:147, 1987.

Welsh, M. J.: Electrolyte transport by airway epithelia. Physiol. Rev., 67:1143, 1987.

West, J. B.: Pulmonary Pathophysiology—The Essentials. Baltimore, Williams & Wilkins Co., 1992.

Wichert, P. von: Lung Surfactant: Basic Research in the Pathogenesis of Lung Disorders. Farmington, CT, S. Karger Publishers, Inc., 1994.

Pulmonary Circulation; Pulmonary Edema; Pleural Fluid

Chapter 38

The quantity of blood flowing through the lungs is essentially equal to that flowing through the systemic circulation. However, certain problems related to distribution of blood flow and other hemodynamics are special to the pulmonary circulation and are especially important to the gas exchange function of the lungs. Therefore, the present discussion is concerned specifically with these special features of the pulmonary circulation.

PHYSIOLOGIC ANATOMY OF THE PULMONARY CIRCULATORY SYSTEM

PULMONARY VESSELS. The pulmonary artery extends only 5 centimeters beyond the apex of the right ventricle and then divides into the right and left main branches, which supply blood to the two respective lungs. The pulmonary artery is also thin, with a wall thickness about twice that of the venae cavae and one-third that of the aorta. The pulmonary arterial branches are all very short. However, all the pulmonary arteries, even the smaller arteries and arterioles, have larger diameters than their counterpart systemic arteries. This, combined with the fact that the vessels are very thin and distensible, gives the pulmonary arterial tree a large compliance, averaging almost 7 ml/mm Hg, which is similar to that of the entire systemic arterial tree. This large compliance allows the pulmonary arteries to accommodate about two thirds of the stroke volume output of the right ventricle.

The pulmonary veins, like the pulmonary arteries, are short, but their distensibility characteristics are similar to those of the veins in the systemic circulation.

BRONCHIAL VESSELS. Blood also flows to the lungs through several bronchial arteries, amounting to about 1 to 2 per cent of the total cardiac output. This bronchial arterial blood is *oxygenated* blood, in contrast to the partially deoxygenated blood in the pulmonary arteries. It supplies the supporting tissues of the lungs, including the connective tissue, septa, and large and small bronchi. After this bronchial arterial blood has passed through the supporting tissues, it empties into the pulmonary veins and *enters the left atrium*, rather than passing back to the right atrium. Therefore, the flow into the left atrium and the left ventricular output are about 1 to 2 per cent greater than the right ventricular output.

LYMPHATICS. Lymphatics extend from all the supportive tissues of the lung, beginning in the connective tissue spaces that surround the terminal bronchioles and coursing to the hilum of the lung and thence mainly into the right lymphatic duct. Particulate matter entering the alveoli is partly removed by way of these channels, and plasma protein leaking from the lung capillaries is also removed from the lung tissues, thereby helping to prevent edema.

PRESSURES IN THE PULMONARY SYSTEM

PRESSURE PULSE CURVE IN THE RIGHT VENTRICLE. The pressure pulse curves of the right ventricle and pulmonary artery are shown in the lower portion of Figure 38–1. These curves are contrasted with the much higher aortic pressure curve shown in the upper portion of the figure. The systolic pressure in the right ventricle of the normal human being averages about

491

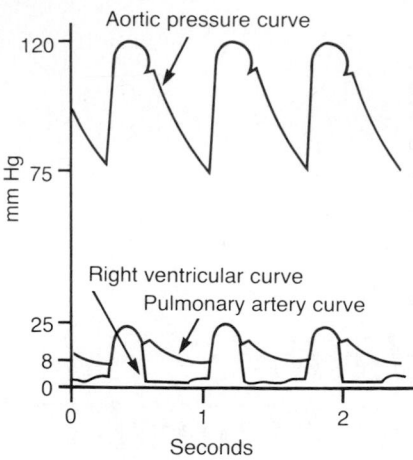

Figure 38–1. Pressure pulse contours in the right ventricle, pulmonary artery, and aorta.

25 mm Hg and the diastolic pressure averages about 0 to 1 mm Hg, values that are only one-fifth those for the left ventricle.

PRESSURES IN THE PULMONARY ARTERY. During *systole*, the pressure in the pulmonary artery is essentially equal to the pressure in the right ventricle, as also shown in Figure 38–1. However, after the pulmonary valve closes at the end of systole, the ventricular pressure falls precipitously, whereas the pulmonary arterial pressure falls more slowly as blood flows through the capillaries of the lungs.

As shown in Figure 38–2, the *systolic pulmonary arterial pressure* averages about 25 mm Hg in the normal human being; the *diastolic pulmonary arterial pressure*, about 8 mm Hg; and the *mean pulmonary arterial pressure*, 15 mm Hg.

PULMONARY CAPILLARY PRESSURE. The mean pulmonary capillary pressure, as diagrammed in Figure 38–2, has been estimated by indirect means to be about 7 mm Hg. The importance of this low capillary pressure is discussed in detail later in the chapter in relation to fluid exchange functions of the capillary.

LEFT ATRIAL AND PULMONARY VENOUS PRESSURES. The mean pressure in the left atrium and the major pulmonary veins averages about 2 mm Hg in the recumbent human being, varying from as low as 1 mm Hg to as high as 5 mm Hg.

It usually is not reasonable to measure the left atrial pressure directly in the normal human being because it is difficult to pass a catheter through the heart chambers into the left atrium. However, the left atrial pressure often can be estimated closely by measuring the so-called *pulmonary wedge pressure*. This is achieved by inserting a catheter through the right side of the heart and then through the pulmonary artery into one of the small branches of the pulmonary artery and, finally, pushing the catheter until it wedges tightly in the artery. The pressure then measured through the catheter, called the "wedge pressure," is about 5 mm Hg. Because all blood flow has been stopped in the

small artery and the blood vessels extending from the artery make almost direct connection through the pulmonary capillaries with the blood in the pulmonary veins, this wedge pressure is usually only 2 to 3 mm Hg greater than the left atrial pressure. Also, when the left atrial pressure rises to high values, so also rises the pulmonary wedge pressure. Therefore, wedge pressure measurements are used frequently for studying changes in left atrial pressure in congestive heart failure.

BLOOD VOLUME OF THE LUNGS

The blood volume of the lungs is about 450 milliliters, about 9 per cent of the total blood volume of the circulatory system. About 70 milliliters of this is in the capillaries, and the remainder is divided about equally between the arteries and the veins.

LUNGS AS A BLOOD RESERVOIR. Under various physiological and pathological conditions, the quantity of blood in the lungs can vary from as little as one-half up to twice normal. For instance, when a person blows out air so hard that high air pressure is built up in the lungs—such as when blowing a trumpet—as much as 250 milliliters of blood can be expelled from the pulmonary circulatory system into the systemic circulation. Also, loss of blood from the systemic circulation by hemorrhage can be partly compensated for by automatic shift of blood from the lungs into the systemic vessels.

Shift of Blood Between the Pulmonary and Systemic Circulatory Systems as a Result of Cardiac Pathology. Failure of the left side of the heart or increased resistance to blood flow through the mitral valve as a result of mitral stenosis or mitral regurgitation causes blood to dam up in the pulmonary circulation, sometimes increasing the pulmonary blood volume as much as 100 per cent and causing increases in the pulmonary vascular pressures.

Because the volume of the systemic circulation is about nine times that of the pulmonary system, a shift of blood from one system to the other affects the

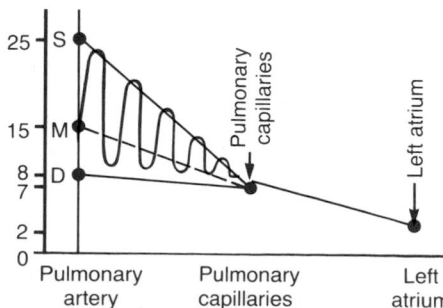

Figure 38–2. Pressures in the different vessels of the lungs.

pulmonary system greatly but usually has only mild systemic effects.

BLOOD FLOW THROUGH THE LUNGS AND ITS DISTRIBUTION

The blood flow through the lungs is essentially equal to the cardiac output. Therefore, the factors that control cardiac output—mainly peripheral factors, as discussed in Chapter 20—also control pulmonary blood flow. Under most conditions, the pulmonary vessels act as passive, distensible tubes that enlarge with increasing pressure and narrow with decreasing pressure. For adequate aeration of the blood, it is important for the blood to be distributed to those segments of the lungs where the alveoli are best oxygenated. This is achieved by the following mechanism.

EFFECT OF DIMINISHED ALVEOLAR OXYGEN ON LOCAL ALVEOLAR BLOOD FLOW—AUTOMATIC CONTROL OF PULMONARY BLOOD FLOW DISTRIBUTION. When the concentration of oxygen in alveoli decreases below normal—especially so when it falls below 70 per cent of normal (below 70 mm Hg Po_2)—the adjacent blood vessels slowly constrict during the ensuing 3 to 10 minutes, the vascular resistance increasing more than fivefold at extremely low oxygen levels. This is *opposite to the effect* normally observed in systemic vessels, which dilate rather than constrict in response to low oxygen. It is believed that the low oxygen concentration causes some yet undiscovered vasoconstrictor substance to be released from the lung tissue, this substance in turn promoting constriction of the small arteries. It has been suggested that this vasoconstrictor might be secreted by the alveolar epithelial cells when they become hypoxic.

This effect of a low oxygen level on pulmonary vascular resistance has an important function: to distribute blood flow where it is most effective. That is, when some of the alveoli are poorly ventilated, so that the oxygen concentration in them becomes low, the local vessels constrict. This in turn causes most of the blood to flow through other areas of the lungs that are better aerated, thus providing an automatic control system for distributing blood flow to the different pulmonary areas in proportion to their degree of ventilation.

PAUCITY OF AUTONOMIC NERVOUS CONTROL OF BLOOD FLOW IN THE LUNGS. Although nerves innervate the lung tissues, it is doubtful that they have a major function in normal control of pulmonary blood flow. Normally, stimulation of the vagal fibers to the lungs causes a slight decrease in pulmonary vascular resistance, and stimulation of the sympathetics causes a slight increase in resistance, both effects perhaps too little to be more than marginally important.

However, research workers have described reflexes in the pulmonary vascular system that might be of clinical importance under certain conditions. For instance, small emboli that occlude the small pulmonary arteries cause a reflex that elicits sympathetic vasoconstriction through-

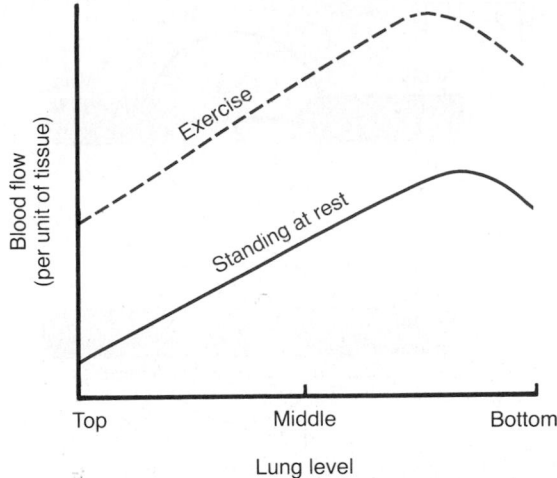

Figure 38–3. Blood flow at different levels in the lung of an upright person both at rest and during exercise. Note that when the person is at rest, the blood flow is very low at the top of the lungs and most of the flow is through the lower lung.

out the lungs, leading to an increase in pulmonary arterial pressure. The significance of this reflex is not certain.

In contrast to the slight effect (almost none) that sympathetic stimulation has on the small resistance vessels of the lung, it does have a great effect in constricting the large pulmonary capacitative vessels, especially the veins. This large-vessel constriction provides a means by which sympathetic stimulation can displace much of the extra blood in the lungs into other segments of the circulation when needed elsewhere to combat low blood pressure.

EFFECT OF HYDROSTATIC PRESSURE GRADIENTS IN THE LUNGS ON REGIONAL PULMONARY BLOOD FLOW

In Chapter 15, it is pointed out that the pressure in the foot of a standing person can be as much as 90 mm Hg greater than the pressure at the level of the heart. This is caused by *hydrostatic pressure*—that is, by the weight of the blood itself. The same effect, but to a lesser degree, occurs in the lungs. In the normal, upright adult, the lowest point in the lungs is about 30 centimeters below the highest point. This represents a 23-mm Hg pressure difference, about 15 mm Hg of which are above the heart and 8 below. That is, the pulmonary arterial pressures in the uppermost portion of the lung of a standing person are about 15 mm Hg less than the pulmonary arterial pressure at the level of the heart, and the pressure in the lowest portion of the lungs is about 8 mm Hg greater. Such pressure differences have profound effects on blood flow through the different areas of the lungs. This is demonstrated by the lower curve in Figure 38–3, which depicts blood flow per unit of lung tissue against hydrostatic level in the lung. Note that in the standing position, at rest, there is little flow in the top of the lung but about five times this much flow in the lower lung. To help explain these differences, one often describes the lung as divided into three zones, as shown in Figure 38–4, in which the patterns of blood flow are quite different. Let us explain these differences.

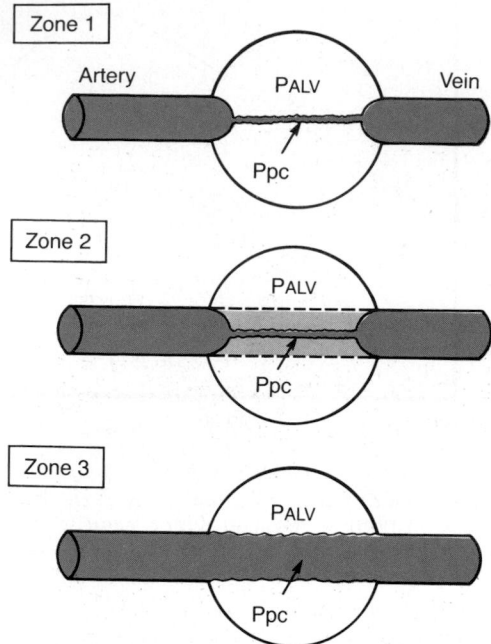

Figure 38-4. Mechanics of blood flow in the three blood flow zones of the lung: *zone 1, no flow* because alveolar pressure is greater than arterial pressure; *zone 2, intermittent flow* because systolic arterial pressure rises higher than alveolar pressure but diastolic pressure falls below alveolar pressure; *zone 3, continuous flow* because arterial pressure remains greater than alveolar pressure at all times.

Zones 1, 2, and 3 of Pulmonary Blood Flow

The capillaries in the alveolar walls are distended by the blood pressure inside them, but simultaneously, they are compressed by the alveolar pressure on their outsides. Therefore, any time the alveolar air pressure becomes greater than the capillary blood pressure, the capillaries close and there is no blood flow. Under different normal and pathological lung conditions, one may find any one of three possible zones of pulmonary blood flow as follows:

Zone 1: No blood flow during any part of the cardiac cycle because the local capillary pressure in that area of the lung never rises higher than the alveolar pressure during any part of the cardiac cycle

Zone 2: Intermittent blood flow only during the pulmonary arterial pressure peaks because the systolic pressure is greater than the alveolar pressure but the diastolic pressure is less than alveolar pressure

Zone 3: Continuous blood flow because the alveolar capillary pressure remains greater than alveolar pressure during the entire cardiac cycle

Normally, the lungs have only zones 2 and 3 blood flow—zone 2 (intermittent flow) in the apices and zone 3 (continuous flow) in all the lower areas. Let us explain. When a person is in the upright position, the pulmonary arterial pressure at the lung apex is about 15 mm Hg less than the pressure at the level of the heart. Therefore, the apical systolic pressure is only 10 mm Hg (25 mm Hg at heart level minus 15 mm Hg hydrostatic pressure difference). This is greater than the zero alveolar pressure, so that blood flows through the pul-

monary apical blood vessels during systole. On the other hand, during diastole, the 8 mm Hg diastolic pressure at the level of the heart is not sufficient to push the blood up the 15-mm Hg hydrostatic pressure gradient required to cause diastolic flow. Therefore, blood flow through the apical part of the lung is intermittent, with flow during systole but cessation of flow during diastole; this is called zone 2 blood flow. Zone 2 blood flow begins in the normal lungs about 10 centimeters above the level of the heart and extends from there to the top of the lungs.

In the lower regions of the lungs, from about 10 centimeters above the level of the heart all the way to the bottom, the pulmonary arterial pressure during both systole and diastole remains greater than the zero alveolar pressure. Therefore, there is continuous flow, which means zone 3 blood flow. Also, when a person is in a lying position, no part of the lung is more than a few centimeters above the level of the heart. Blood flow then in a normal person is entirely zone 3 blood flow, including the apices.

ZONE 1 BLOOD FLOW OCCURS ONLY UNDER ABNORMAL CONDITIONS. Zone 1 blood flow, which is blood flow at no time during the cardiac cycle, occurs when either the pulmonary systolic arterial pressure is too low or the alveolar pressure is too high to allow flow. For instance, if an upright person is breathing against a positive air pressure so that the intra-alveolar pressure is at least 10 mm Hg greater than normal but the pulmonary systolic pressure is normal, then one would expect zone 1 blood flow—no blood flow—at least in the lung apices. Another instance in which zone 1 blood flow occurs is in an upright person whose pulmonary systolic pressure is exceedingly low, as might occur in hypovolemic states.

Effect of Exercise on Blood Flow Through the Different Parts of the Lungs. Referring again to Figure 38–3, one sees that the blood flow in all parts of the lung increases during exercise. The increase in flow in the top of the lung may be 700 to 800 per cent, whereas the increase in the lower part of the lung may be no more than 200 to 300 per cent. The reason for these differences is the considerably higher pulmonary vascular pressures that occur during exercise, which effectively convert the entire lung into a zone 3 pattern of flow.

Effect of Increased Cardiac Output on the Pulmonary Circulation During Heavy Exercise

During heavy exercise, the blood flow through the lungs increases fourfold to sevenfold. This extra flow is accommodated in the lungs in two ways: (1) by increasing the number of open capillaries, sometimes as much as threefold, and (2) by distending all the capillaries and increasing the rate of flow through each capillary more than twofold. In the normal person, these two changes together decrease the pulmonary vascular resistance so much that the pulmonary arterial pressure rises very little even during maximum exercise; this effect is shown in Figure 38–5.

The ability of the lungs to accommodate greatly increased blood flow during exercise conserves the energy of the right side of the heart and prevents a significant rise in pulmonary capillary pressure, therefore preventing development of pulmonary edema during the increased cardiac output.

Function of the Pulmonary Circulation When the Left Atrial Pressure Rises as a Result of Left-Sided Heart Failure

When the left side of the heart fails, blood begins to dam up in the left atrium. As a result, the left atrial pressure can rise on occasion from its normal value of 1 to 5 mm Hg to 40 to 50 mm Hg. The initial atrial pressure rise, up to about 7 mm Hg, has almost no effect on pulmonary circulatory function because this initial rise merely expands the venules and opens up more capillaries so that blood continues to flow with almost equal ease from the pulmonary arteries. Figure 38–6 demonstrates this effect, showing almost no change in pulmonary arterial pressure at the lower degrees of increased left atrial pressures. Because the left atrial pressure in a healthy person almost never rises above +6 mm Hg even during the most strenuous exercise, the changes in left atrial pressure have virtually no effect on pulmonary circulatory function except when the left side of the heart fails.

When the left atrial pressure rises to greater than 7 or 8 mm Hg, then further increases in left atrial pressure cause almost equally great increases in pulmonary arterial pressure, as also shown in Figure 38–6, with concomitant increased load on the right side of the heart.

Also, any increase in left atrial pressure above 7 or 8 mm Hg increases the capillary pressure almost equally as much. When the left atrial pressure has risen above 30 mm Hg, causing similar increases in capillary pressure, pulmonary edema is then likely to develop, as we discuss later in the chapter.

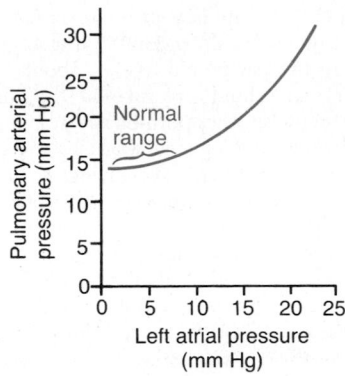

Figure 38–6. Effect of left atrial pressure on pulmonary arterial pressure.

PULMONARY CAPILLARY DYNAMICS

Exchange of gases between the alveolar air and the pulmonary capillary blood is discussed in Chapter 39. However, it is important for us to note here that the alveolar walls are lined with so many capillaries that in most places, the capillaries almost touch one another. Therefore, it has often been said that the capillary blood flows in the alveolar walls as a "sheet," rather than in individual vessels.

PULMONARY CAPILLARY PRESSURE. No direct measurements of pulmonary capillary pressure have been made. However, "isogravimetric" measurement of pulmonary capillary pressure, using a technique described in Chapter 16, has given a value of 7 mm Hg. This is probably nearly correct because the mean left atrial pressure is about 2 mm Hg and the mean pulmonary arterial pressure is only 15 mm Hg, so that the mean pulmonary capillary pressure must lie somewhere between these two values.

LENGTH OF TIME BLOOD STAYS IN THE CAPILLARIES. From histologic study of the total cross-sectional area of all the pulmonary capillaries, it can be calculated that when the cardiac output is normal, blood passes through the pulmonary capillaries in about 0.8 second. Increasing the cardiac output sometimes shortens this time to as little as 0.3 second, but the shortening would be much greater were it not for the fact that additional capillaries, which normally are collapsed, open up to accommodate the increased blood flow. Thus, in less than 1 second, blood passing through the capillaries becomes oxygenated and loses its excess carbon dioxide.

Capillary Exchange of Fluid in the Lungs, and Pulmonary Interstitial Fluid Dynamics

The dynamics of fluid exchange through the lung capillaries are *qualitatively* the same as for peripheral tissues. However, *quantitatively*, there are important differences.

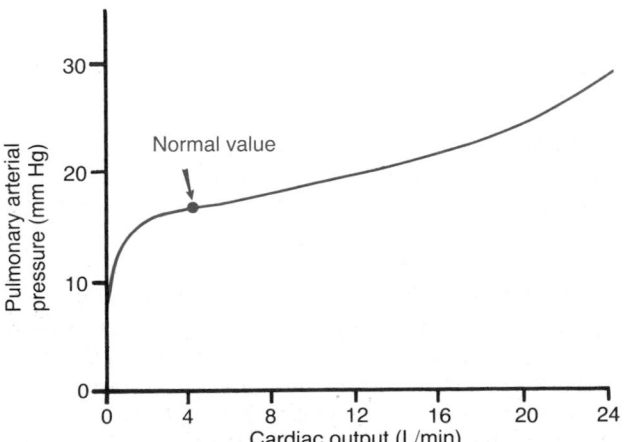

Figure 38–5. Effect on the pulmonary arterial pressure of increasing the cardiac output.

1.   The pulmonary capillary pressure is low, about 7 mm Hg, in comparison with a considerably higher functional capillary pressure in the peripheral tissues, about 17 mm Hg.

2.   The interstitial fluid pressure in the lung is slightly more negative than in the peripheral subcutaneous tissue. (This has been measured in two ways: by a pipette inserted into the pulmonary interstitium, giving a value of about −5 mm Hg, and by measuring the absorption pressure of fluid from the alveoli, giving a value of about −8 mm Hg.)

3.   The pulmonary capillaries are relatively leaky to protein molecules, so that the colloid osmotic pressure of the pulmonary interstitial fluids is about 14 mm Hg in comparison with less than one half this in the peripheral tissues.

4.   The alveolar walls are extremely thin, and the alveolar epithelium covering the alveolar surfaces is so weak that it is ruptured by any positive pressure in the interstitial spaces greater than atmospheric pressure (greater than 0 mm Hg), which allows dumping of fluid from the interstitial spaces into the alveoli.

Now let us see how these quantitative differences affect pulmonary fluid dynamics.

INTERRELATION BETWEEN INTERSTITIAL FLUID PRESSURE AND OTHER PRESSURES IN THE LUNG. Figure 38–7 shows a pulmonary capillary, a pulmonary alveolus, and a lymphatic capillary draining the interstitial space between the capillary and the alveolus. Note the balance of forces at the capillary membrane as follows:

	mm Hg
Forces tending to cause movement of fluid outward from the capillaries and into the pulmonary interstitium:	
Capillary pressure	7
Interstitial fluid colloid osmotic pressure	14
Negative interstitial fluid pressure	8
TOTAL OUTWARD FORCE	29
Forces tending to cause absorption of fluid into the capillaries:	
Plasma colloid osmotic pressure	28
TOTAL INWARD FORCE	28

Thus, the normal outward forces are slightly greater than the inward forces. The *net mean filtration pressure* at the pulmonary capillary membrane can be calculated as follows:

	mm Hg
Total outward force	+29
Total inward force	−28
NET MEAN FILTRATION PRESSURE	+1

This net filtration pressure causes a slight continual flow of fluid from the pulmonary capillaries into the interstitial spaces, and except for a small amount that evaporates in the alveoli, this fluid is pumped back to the circulation through the pulmonary lymphatic system.

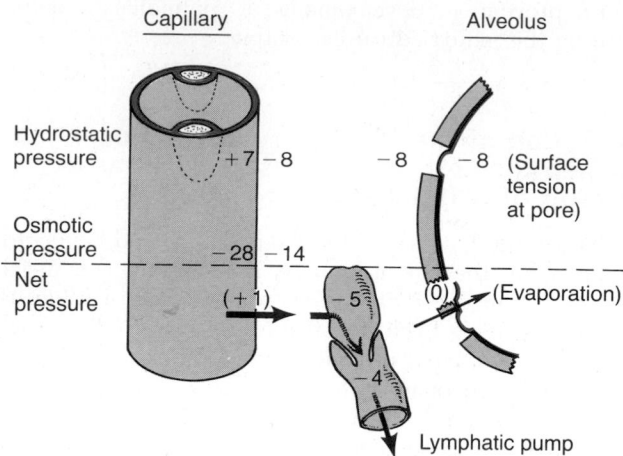

Pressures Causing Fluid Movement

Figure 38–7. Hydrostatic and osmotic forces at the capillary *(left)* and alveolar membrane *(right)* of the lungs. Also shown is a lymphatic *(center)* that pumps fluid from the pulmonary interstitial spaces. (Modified from Guyton, Taylor, and Granger: Dynamics and Control of the Body Fluids. Philadelphia, W. B. Saunders Company, 1975.)

NEGATIVE INTERSTITIAL PRESSURE AND THE MECHANISM FOR KEEPING THE ALVEOLI "DRY." One of the most important problems in lung function is to understand why the alveoli do not fill up with fluid. One's first impulse is to say that the alveolar epithelium keeps fluid from leaking out of the interstitial spaces into the alveoli. This is not true because there are always a small number of openings between the alveolar epithelial cells through which even large protein molecules as well as large quantities of water and electrolytes can pass.

However, if one remembers that the pulmonary capillaries and the pulmonary lymphatic system normally maintain a slight *negative pressure* in the interstitial spaces, then it is clear that whenever extra fluid appears in the alveoli, it will simply be sucked mechanically into the lung interstitium through the small openings between the alveolar epithelial cells. Then the excess fluid is either carried away through the pulmonary lymphatics or absorbed into the pulmonary capillaries. Thus, under normal conditions, the alveoli are kept in a "dry" state except for a small amount of fluid that seeps from the epithelium onto the lining surfaces of the alveoli to keep them moist.

Pulmonary Edema

Pulmonary edema occurs in the same way that edema occurs elsewhere in the body. Any factor that causes the pulmonary interstitial fluid pressure to rise from the negative range into the positive range will cause sudden filling of the pulmonary interstitial spaces and alveoli with large amounts of free fluid.

The most common causes of pulmonary edema are as follows:

1. Left-sided heart failure or mitral valvular disease with consequent great increase in pulmonary capillary pressure and flooding of the interstitial spaces and alveoli.

2. Damage to the pulmonary capillary membrane caused by infections such as pneumonia or by breathing noxious substances such as chlorine gas or sulfur dioxide gas. Each of these causes rapid leakage of both plasma proteins and fluid out of the capillaries.

PULMONARY "INTERSTITIAL FLUID" EDEMA VERSUS PULMONARY "ALVEOLAR" EDEMA. The interstitial fluid volume of the lungs usually cannot increase more than about 50 per cent (representing less than 100 milliliters of fluid) before the alveolar epithelial membranes rupture and fluid begins to pour from the interstitial spaces into the alveoli. The cause of this is simply the slight tensional strength of the pulmonary alveolar epithelium; that is, even the slightest positive pressure in the interstitial fluid spaces—probably even +1 mm Hg—seems to cause immediate rupture of this epithelium.

Therefore, except in the mildest cases of pulmonary edema, edema fluid always enters the alveoli; if this edema becomes severe enough, it can cause death by suffocation.

"PULMONARY EDEMA SAFETY FACTOR." All the factors that tend to prevent edema in the peripheral tissues also tend to prevent edema in the lungs. That is, before positive interstitial fluid pressure can occur and cause edema, all the following factors must be overcome: (1) the normal negativity of the interstitial fluid pressure of the lungs, (2) the lymphatic pumping of fluid out of the interstitial spaces, and (3) the increased osmosis of fluid into the pulmonary capillaries caused by decreased protein in the interstitial fluid when the lymph flow increases.

In experiments in animals, it has been found that the pulmonary capillary pressure normally must rise to a value at least equal to the plasma colloid osmotic pressure before significant pulmonary edema occurs. To give an example, Figure 38–8 shows the effect of different levels of left atrial pressure to increase the rate of edema formation in dogs. Remember that every time the left atrial pressure rises to high values, the pulmonary capillary pressure rises to a level 1 to 2 mm Hg greater than the left atrial pressure. In this figure, as soon as the left atrial pressure rose above 23 mm Hg (with the pulmonary capillary pressure above about 25 mm Hg), fluid began to accumulate in the lungs and increased rapidly, with further increases in pressure. Yet below 25 mm Hg pulmonary capillary pressure, there was no significant increase in pulmonary fluid. The plasma colloid osmotic pressure in dogs is almost equal to this 25 mm Hg critical pressure level. Therefore, in the human being, who normally has a plasma colloid osmotic pressure of 28 mm Hg, one can predict that the pulmonary capillary pressure must rise from the normal level of 7 mm Hg to more than 28 mm Hg to cause pulmonary edema, giving a safety factor against pulmonary edema of about 21 mm Hg.

Safety Factor in Chronic Conditions. When the pulmonary capillary pressure remains elevated chronically (for at least 2 weeks), the lungs become even more resistant to pulmonary edema because the lymph vessels can expand greatly, increasing their capability for carrying fluid away from the interstitial spaces perhaps as much as 10-fold. Therefore, in a patient with chronic mitral stenosis, a pulmonary capillary pressure of 40 to 45 mm Hg has been measured without the development of significant pulmonary edema.

Thus, in chronic pulmonary edema, the safety factor against pulmonary edema can probably rise to 30 to 35 mm Hg, in comparison with a normal value of 21 mm Hg under acute conditions.

RAPIDITY OF DEATH IN ACUTE PULMONARY EDEMA. When the pulmonary capillary pressure does rise even slightly above the safety factor level, lethal pulmonary edema can occur within hours, and even within 20 to 30 minutes if the capillary pressure rises 25 to 30 mm Hg above the safety factor level. Thus, in acute left-sided heart failure, in which the pulmonary capillary pressure occasionally rises to 50 mm Hg, death frequently ensues in less than 30 minutes from acute pulmonary edema.

FLUIDS IN THE PLEURAL CAVITY

When the lungs expand and contract during normal breathing, they slide back and forth within the pleural cavity. To facilitate this, a thin layer of mucoid fluid lies between the parietal and visceral pleurae.

Figure 38–9 shows the dynamics of fluid exchange in the pleural space. Each of the two pleurae is a porous mesenchymal serous membrane through which small amounts of interstitial fluid transude continually into the pleural space. These fluids carry with them tissue proteins, giving the pleural fluid a mucoid characteristic, which is what allows extremely easy slippage of the moving lungs.

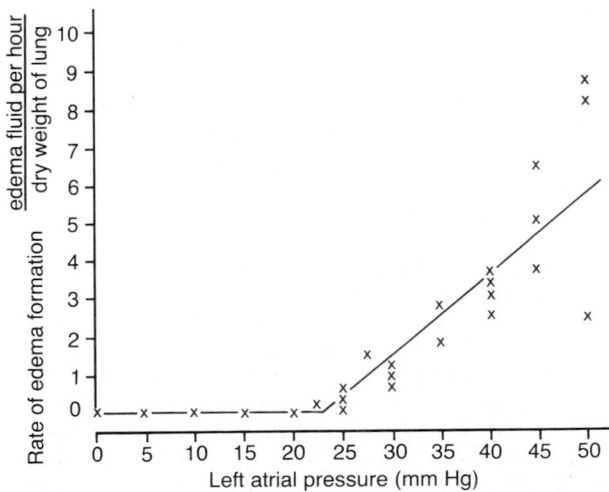

Figure 38–8. Rate of fluid loss into the lung tissues when the left atrial pressure (and also pulmonary capillary pressure) is increased. (From Guyton and Lindsey: *Circ. Res.,* 7:649, 1959. By permission of The American Heart Association, Inc.)

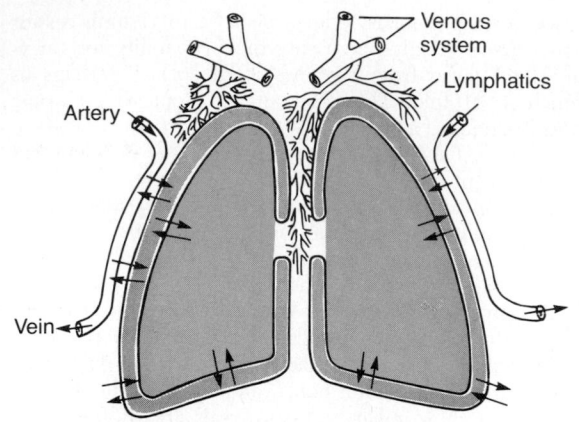

Figure 38–9. Dynamics of fluid exchange in the intrapleural spaces.

The total amount of fluid in each pleural cavity is slight, only a few milliliters. Whenever the quantity becomes more than barely enough to separate the two pleurae, the excess is pumped away by lymphatic vessels opening directly from the pleural cavity into (1) the mediastinum, (2) the superior surface of the diaphragm, and (3) the lateral surfaces of the parietal pleura. Therefore, the *pleural space*—the space between the parietal and visceral pleurae—is called a *potential space* because it normally is so narrow that it is not obviously a physical space.

NEGATIVE PRESSURE IN THE PLEURAL FLUID. Because the recoil tendency of the lungs causes them to try to collapse, a negative force is always required on the outside of the lungs to keep the lungs expanded. This is provided by a negative pressure in the normal pleural space. The basic cause of this negative pressure is the pumping of fluid from the space by the lymphatics (which is also the basis of the negative pressure found in most tissue spaces of the body). Because the normal collapse tendency of the lungs is about − 4 mm Hg (− 5 or − 6 centimeters of water), the pleural fluid pressure must always be at least as negative as − 4 mm Hg to keep the lungs expanded. Actual measurements have shown it usually to be about − 7 mm Hg, which is a few millimeters of mercury more negative than the collapse pressure of the lungs. Thus, the negativity of the pleural fluid keeps the normal lungs pulled tightly against the parietal pleura of the chest cavity except for the extremely thin layer of mucoid fluid that acts as a lubricant.

PLEURAL EFFUSION. Pleural effusion means the collection of large amounts of free fluid in the pleural space. It is analogous to edema fluid in the tissues and can be called edema of the pleural cavity. The possible causes of effusion are the same as the causes of edema in other tissues—discussed in Chapter 25—including (1) blockage of lymphatic drainage from the pleural cavity; (2) cardiac failure, which causes excessively high peripheral and pulmonary capillary pressures, leading to excessive transudation of fluid into the pleural cavity; (3) greatly reduced plasma colloid osmotic pressure, thus also allowing excessive transudation of fluid; and (4) infection or any other cause of inflammation of the pleural surfaces of the pleural cavity, which breaks down the capillary membranes and allows rapid dumping of both plasma proteins and fluid into the cavity.

REFERENCES

Agostoni, E.: Mechanics of the pleural space. Physiol. Rev., 52:57, 1972.

Arakawa, M., et al.: Pulmonary blood volume and pulmonary extravascular water volume in men. Jpn. Circ. J., 49:475, 1985.

Bergofsky, E. H.: Humoral control of the pulmonary circulation. Annu. Rev. Physiol., 42:221, 1980.

Cournand, A.: Some aspects of the pulmonary circulation in normal man and in chronic cardiopulmonary diseases. Circulation, 2:641, 1950.

Culver, B. H., and Butler, J.: Mechanical influences on the pulmonary microcirculation. Annu. Rev. Physiol., 42:187, 1980.

Dawson, C. A.: Role of pulmonary vasomotion in physiology of the lung. Physiol. Rev., 64:544, 1984.

Deslauriers, J., and Lacquet, L. K.: The Pleural Space, Vol. 6. In Delarue, N. C., and Eschapasse, H. (eds.): International Trends in General Thoracic Surgery Series. St. Louis, C. V. Mosby Co., 1989.

Downing, S. E., and Lee, J. C.: Nervous control of the pulmonary circulation. Annu. Rev. Physiol., 42:199, 1980.

Effros, R. M.: Pulmonary microcirculation and exchange. In Renkin, E. M., and Michel, C. C. (eds.): Handbook of Physiology. Sec. 2, Vol. IV. Bethesda, Md., American Physiological Society, 1984, p. 865.

Epstein, M.: Renal effects of head-out water immersion in humans: a 15-year update. Physiol. Rev., 72:563, 1992.

Fishman, A. P., and Pietra, G. G.: Hemodynamic pulmonary edema. In Fishman, A. P. and Renkin, E. M. (eds.): Pulmonary Edema. Baltimore, Waverly Press, 1979, p. 79.

Fishman, A. P., and Renkin, E. M. (eds.): Pulmonary Edema. Baltimore, American Physiological Society, 1979.

Fishman, A. P.: Vasomotor regulation of the pulmonary circulation. Annu. Rev. Physiol., 42:211, 1980.

Fraser, R. S., et al.: Synopsis of Diseases of the Chest. Philadelphia, W. B. Saunders Co., 1994.

Gaar, K. A., Jr., et al.: Effect of lung edema on pulmonary capillary pressure. Am. J. Physiol., 216:1370, 1969.

Gaar, K. A., Jr., et al.: Effects of capillary pressure and plasma protein on development of pulmonary edema. Am. J. Physiol., 213:79, 1967.

Gil, J.: Organization of microcirculation in the lung. Annu. Rev. Physiol., 42:177, 1980.

Grover, R. F., et al.: Pulmonary Circulation. In Shepherd, J. T., and Abboud, F. M. (eds.): Handbook of Physiology, Sec. 2, Vol. III. Bethesda, American Physiological Society, 1983, p. 103.

Grippi, M. A.: Pulmonary Pathophysiology. Philadelphia, J. B. Lippincott Co., 1994.

Guyton, A. C., and Lindsey, A. W.: Effect of elevated left atrial pressure and decreased plasma protein concentration on the development of pulmonary edema. Circ. Res., 7:649, 1959.

Guyton, A. C., et al.: Circulatory Physiology. II. Dynamics and Control of the Body Fluids. Philadelphia, W. B. Saunders Co., 1975.

Guyton, A. C., et al.: Forces governing water movement in the lung. In Fishman, A. P., and Renkin, E. M. (eds.): Pulmonary Edema. Baltimore, Waverly Press, 1979, p. 65.

Hills, B. A.: The pleural interface. Thorax, 40:1, 1985.

Iinuma, J., et al.: Fluid volume balance between pulmonary intravascular space and extravascular space in dogs. Jpn. Circ. J., 50:818, 1986.

Kapoor, A. S., and Laks, H.: Atlas of Heart-lung Transplantation. Hightstown, NJ, McGraw-Hill, 1994.

Laine, G. A., et al.: A new look at pulmonary edema. News Physiol. Sci., 1:150, 1986.

Malik, A. B.: Pulmonary microembolism. Physiol. Rev., 63:1114, 1983.

Mark, A. L., and Mancia, G.: Cardiopulmonary baroreflexes in humans. In Shepherd, J. T., and Abboud, F. M. (eds.): Handbook of Physiology, Sec. 2, Vol. III. Bethesda, American Physiological Society, 1983, p. 795.

Meyer, B. J., et al.: Interstitial fluid pressure V. Negative pressure in the lungs. Circ. Res., 22:263, 1968.

Nijkamp, F. P., et al.: Mechanisms of β-adrenergic receptor regulation in lungs and its implications for physiological responses. Physiol. Rev., 72:323, 1992.

Notter, R. H., and Finkelstein, J. N.: Pulmonary surfactant: An interdisciplinary approach. J. Appl. Physiol., 57:1613, 1984.

Parker, J. C., et al.: Pulmonary interstitial and capillary pressures estimated from intra-alveolar fluid pressures. J. Appl. Physiol., 44(2):267, 1978.

Parker, J. C., et al.: Pulmonary transcapillary exchange and pulmonary edema. In Guyton, A. C., and Young, D. B. (eds.): International Review of Physiology: Cardiovascular Physiology III. Vol. 18. Baltimore, University Park Press, 1979, p. 261.

Piene, H.: Pulmonary arterial impedance and right ventricular function. Physiol. Rev., 66:606, 1986.

Ryan, U. S.: Metabolic activity of pulmonary endothelium. Annu. Rev. Physiol., 48:263, 1986.

Saldana, M. J.: Pathology of Pulmonary Disease. Philadelphia, J. B. Lippincott Co., 1994.

Staub, N. C.: Pulmonary edema due to increased microvascular permeability. Annu. Rev. Med., 32:291, 1981.

Staub, N. C.: Pulmonary edema. Physiol. Rev., 54:678, 1974.

Takaya, T., et al.: Pulmonary vein blood flow velocity waveform. Jpn. Circ. J., 50:405, 1986.

Taylor, A. E., et al.: Na^{24} space, D$_2$O space, and blood volume in isolated dog lung. Am. J. Physiol., 211:66, 1965.

Taylor, A. E., et al.: Permeability of the alveolar membrane to solutes. Circ. Res., 16:353, 1965.

Visscher, M. B., et al.: The physiology and pharmacology of lung edema. Pharm. Rev., 8:389, 1956.

West, J. B.: Pulmonary Pathophysiology: The Essentials. 4th Ed. Baltimore, Williams & Wilkins Co., 1992.

West, J. B.: Respiratory Physiology—The Essentials. 5th Ed. Baltimore, Williams & Wilkins Co., 1994.

Physical Principles of Gas Exchange; Diffusion of Oxygen and Carbon Dioxide Through the Respiratory Membrane

CHAPTER 39

After the alveoli are ventilated with fresh air, the next step in the respiratory process is *diffusion* of oxygen from the alveoli into the pulmonary blood and diffusion of carbon dioxide in the opposite direction. The process of diffusion is simply random molecular motion of molecules intertwining their ways in both directions through the respiratory membrane. However, in respiratory physiology, one is concerned not only with the basic mechanism by which diffusion occurs but also with the *rate* at which it occurs; this is a much more complicated problem, requiring a deeper understanding of the physics of diffusion and gas exchange.

PHYSICS OF GAS DIFFUSION AND GAS PARTIAL PRESSURES

Molecular Basis of Gas Diffusion

All the gases that are of concern in respiratory physiology are simple molecules that are free to move among one another, which is the process called "diffusion." This is also true of the gases dissolved in the fluids and tissues of the body.

For diffusion to occur, there must be a source of energy. This is provided by the kinetic motion of the molecules themselves. That is, except at absolute zero temperature, all molecules of all matter are continually undergoing motion. For free molecules that are not physically attached to others, this means linear movement at high velocity until they strike other molecules. Then they bounce away in new directions and continue again until striking still other molecules. In this way, the molecules move rapidly and randomly among one another.

NET DIFFUSION OF A GAS IN ONE DIRECTION—EFFECT OF A CONCENTRATION GRADIENT. If a gas chamber or a solution has a high concentration of a gas at one end of the chamber and a low concentration at the other end, as shown in Figure 39–1, net diffusion of the gas will occur from the high concentration area toward the low area. The reason for this is obvious: There are far more molecules at end A of the chamber to diffuse toward end B than there are molecules to diffuse in the opposite direction. Therefore, the rates of diffusion in each of the two directions are proportionately different, as demonstrated by the lengths of the two arrows.

Gas Pressures in a Mixture of Gases— Partial Pressures of Individual Gases

Pressure is caused by the constant impact of kinetically moving molecules against a surface. Therefore, the pressure of a gas acting on the surfaces of the respiratory passages and alveoli is proportional to the summated force of impact of all the molecules striking the surface at any given instant. This means that *the total pressure is directly proportional to the concentration of the gas molecules.*

In respiratory physiology, one deals with mixtures of gases, mainly of oxygen, nitrogen, and carbon dioxide. The rate of diffusion of each of these gases is directly proportional to the pressure caused by this gas alone, which is called the *partial pressure* of the gas. Let us explain the concept of partial pressure.

Consider air, which has an approximate composition of 79 per cent nitrogen and 21 per cent oxygen. The total pressure of this mixture at sea level averages

501

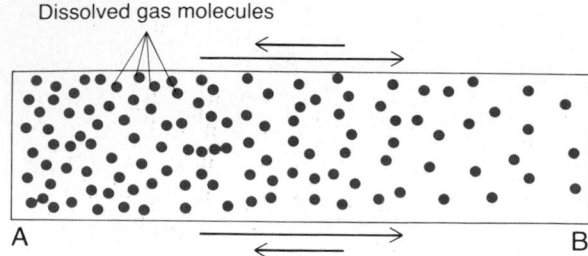

Dissolved gas molecules

A   B

Figure 39–1. Net diffusion of oxygen from one end of a chamber to the other.

760 mm Hg, and it is clear from the description of the molecular basis of pressure above that each gas contributes to the total pressure in direct proportion to its concentration. Therefore, 79 per cent of the 760 mm Hg is caused by nitrogen (about 600 mm Hg) and 21 per cent by oxygen (about 160 mm Hg). Thus, the "partial pressure" of nitrogen in the mixture is 600 mm Hg, and the "partial pressure" of oxygen is 160 mm Hg; the total pressure is 760 mm Hg, the sum of the individual partial pressures.

The partial pressures of the individual gases in a mixture are designated by the symbols P_{O_2}, P_{CO_2}, P_{N_2}, P_{H_2O}, P_{He}, and so forth.

Pressures of Gases Dissolved in Water and Tissues

Gases dissolved in water or the body tissues also exert pressure because the dissolved molecules are moving randomly and have kinetic energy. Furthermore, when the molecules of a gas dissolved in fluid encounter a surface like the membrane of a cell, they exert their own pressure in the same way that a gas in the gas phase exerts its own individual partial pressure. The pressures of the separate dissolved gases are designated similarly as for the partial pressures of the gases in the gas state, that is, P_{O_2}, P_{CO_2}, P_{N_2}, P_{He}.

FACTORS THAT DETERMINE THE PRESSURE OF A GAS DISSOLVED IN A FLUID. The pressure of a gas in a solution is determined not only by its concentration but also by the *solubility coefficient* of the gas. That is, some types of molecules, especially carbon dioxide, are physically or chemically attracted to water molecules, whereas others are repelled. When molecules are attracted, far more of them can then become dissolved without building up excess pressure within the solution. On the other hand, in the case of those that are repelled, excessive pressures will develop for many fewer dissolved molecules. These relations can be expressed by the following formula, which is *Henry's law:*

$$\text{Pressure} = \frac{\text{Concentration of dissolved gas}}{\text{Solubility coefficient}}$$

When pressure is expressed in atmospheres (1 atmosphere pressure equals 760 mm Hg) and concentration is expressed in volume of gas dissolved in each volume of water, the solubility coefficients for important respiratory gases at body temperature are the following:

Oxygen	0.024
Carbon dioxide	0.57
Carbon monoxide	0.018
Nitrogen	0.012
Helium	0.008

From this table, one can see that carbon dioxide is more than 20 times as soluble as oxygen, and oxygen is moderately more soluble than any of the other three gases.

DIFFUSION OF GASES BETWEEN THE GAS PHASE IN THE ALVEOLI AND THE DISSOLVED PHASE IN THE PULMONARY BLOOD. The partial pressure of each gas in the alveolar respiratory gas mixture tends to force molecules of that gas into solution first in the alveolar membrane and then in the blood of the alveolar capillaries. Conversely, the molecules of the same gas that are already dissolved in the blood are bouncing randomly in the fluid of the blood, and some escape back into the alveoli. The rate at which they escape is directly proportional to their partial pressure in the blood. But in which direction will *net diffusion* of the gas occur? The answer to this is that net diffusion is determined by the difference between the two partial pressures. If the partial pressure is greater in the gas phase in the alveoli, as is normally true for oxygen, then more molecules will go into the blood than in the other direction. Alternately, if the pressure of the gas is greater in the blood, which is normally true for carbon dioxide, then net diffusion will occur toward the gas phase in the alveoli.

Vapor Pressure of Water

When air enters the respiratory passageways, water immediately evaporates from the surfaces of these passages and humidifies the air. This results from the fact that water molecules, like the different dissolved gas molecules, are continually escaping from the water surface into the gas phase. The pressure that the water molecules exert to escape through the surface is called the *vapor pressure* of the water. At normal body temperature, 37°C, this vapor pressure is 47 mm Hg. Therefore, once the gas mixture has become fully humidified—that is, once it is in "equilibrium" with the surrounding water—the partial pressure of the water vapor in the gas mixture is also 47 mm Hg. This partial pressure, like the other partial pressures, is designated P_{H_2O}.

The vapor pressure of water depends entirely on the temperature of the water. The greater the temperature, the greater the kinetic activity of the molecules and, therefore, the greater the likelihood that the water molecules will escape from the surface of the water into the gas phase. For instance, the water vapor pressure at 0°C is 5 mm Hg and at 100°C is 760 mm Hg. But the most important value to remember is the vapor pressure at body temperature, 47 mm Hg; this value appears in many of our subsequent discussions.

Diffusion of Gases Through Fluids— Pressure Difference Causes Net Diffusion

Now, let us return to the problem of diffusion. From the discussion above, it is already clear that when the pressure of a gas is greater in one area than in another area, there will be net diffusion from the high-pressure area toward the low-pressure area. For instance, returning to Figure 39–1, one can readily see that the molecules in the area of high pressure, because of their greater number, have a greater statistical chance of moving randomly into the area of low pressure than do molecules attempting to go in the other direction. However, some molecules do bounce randomly from the area of low pressure toward the area of high pressure. Therefore, the *net diffusion* of gas from the area of high pressure to the area of low pressure is equal to the number of molecules bouncing in this forward direction *minus* the number bouncing in the opposite direction, and this in turn is proportional to the gas pressure difference between the two areas, called simply the *pressure difference for diffusion.*

QUANTIFYING THE NET RATE OF DIFFUSION IN FLUIDS. In addition to the pressure difference, several other factors affect the rate of gas diffusion in a fluid. They are (1) the solubility of the gas in the fluid, (2) the cross-sectional area of the fluid, (3) the distance through which the gas must diffuse, (4) the molecular weight of the gas, and (5) the temperature of the fluid. In the body, the last of these factors, the temperature, remains reasonably constant and usually need not be considered.

The greater the solubility of the gas, the greater will be the number of molecules available to diffuse for any given pressure difference. Also, the greater the cross-sectional area of the diffusion pathway, the greater will be the total number of molecules to diffuse. On the other hand, the greater the distance that the molecules must diffuse, the longer will it take the molecules to diffuse the entire distance. Finally, the greater the velocity of kinetic movement of the molecules, which is inversely proportional to the square root of the molecular weight, the greater is the rate of diffusion of the gas. All these factors can be expressed in a single formula, as follows:

$$D \propto \frac{\Delta P \times A \times S}{d \times \sqrt{MW}},$$

in which D is the diffusion rate, ΔP is the pressure difference between the two ends of the diffusion pathway, A is the cross-sectional area of the pathway, S is the solubility of the gas, d is the distance of diffusion, and MW is the molecular weight of the gas.

It is obvious from this formula that the characteristics of the gas itself determine two factors of the formula:

solubility and molecular weight, and together they are proportional to the *diffusion coefficient* of the gas. That is, the diffusion coefficient is proportional to $S/\sqrt{MW}$; also, the relative rates at which different gases at the same pressure levels will diffuse are proportional to their diffusion coefficients. Considering the diffusion coefficient for oxygen to be 1, the *relative* diffusion coefficients for different gases of respiratory importance in the body fluids are as follows:

Oxygen	1.0
Carbon dioxide	20.3
Carbon monoxide	0.81
Nitrogen	0.53
Helium	0.95

Diffusion of Gases Through Tissues

The gases that are of respiratory importance are all highly soluble in lipids and, consequently, are also highly soluble in cell membranes. Because of this, the major limitation to the movement of gases in tissues is the rate at which the gases can diffuse through the tissue water instead of through the cell membranes. Therefore, diffusion of gases through the tissues, including through the respiratory membrane, is almost equal to the diffusion of gases through water, as given in the list above.

COMPOSITION OF ALVEOLAR AIR— ITS RELATION TO ATMOSPHERIC AIR

Alveolar air does not have the same concentrations of gases as atmospheric air by any means, which can readily be seen by comparing the alveolar air composition in column 5 of Table 39–1 with the composition of atmospheric air in column 1. There are several reasons for the differences. First, the alveolar air is only partially replaced by atmospheric air with each breath. Second, oxygen is constantly being absorbed from the alveolar air. Third, carbon dioxide is constantly diffusing from the pulmonary blood into the alveoli. And, fourth, dry atmospheric air that enters the respiratory passages is humidified even before it reaches the alveoli.

HUMIDIFICATION OF THE AIR AS IT ENTERS THE RESPIRATORY PASSAGES. Column 1 of Table 39–1 shows that atmospheric air is composed almost entirely

Table 39–1 PARTIAL PRESSURES OF RESPIRATORY GASES AS THEY ENTER AND LEAVE THE LUNGS (AT SEA LEVEL)

	Atmospheric Air* (mm Hg)		Humidified Air (mm Hg)		Alveolar Air (mm Hg)		Expired Air (mm Hg)	
N_2	597.0	(78.62%)	563.4	(74.09%)	569.0	(74.9%)	566.0	(74.5%)
O_2	159.0	(20.84%)	149.3	(19.67%)	104.0	(13.6%)	120.0	(15.7%)
CO_2	0.3	(0.04%)	0.3	(0.04%)	40.0	(5.3%)	27.0	(3.6%)
H_2O	3.7	(0.50%)	47.0	(6.20%)	47.0	(6.2%)	47.0	(6.2%)
TOTAL	760.0	(100.00%)	760.0	(100.00%)	760.0	(100.0%)	760.0	(100.0%)

* On an average cool, clear day.

of nitrogen and oxygen; it normally contains almost no carbon dioxide and little water vapor. However, as soon as the atmospheric air enters the respiratory passages, it is exposed to the fluids that cover the respiratory surfaces. Even before the air enters the alveoli, it becomes (for all practical purposes) totally humidified.

The partial pressure of water vapor at normal body temperature of 37°C is 47 mm Hg, which, therefore, is the partial pressure of water in the alveolar air. Because the total pressure in the alveoli cannot rise to more than the atmospheric pressure (760 mm Hg), this water vapor simply *dilutes* all the other gases in the inspired air. In column 3 of Table 39–1, it can be seen that humidification of the air dilutes the oxygen partial pressure at sea level from an average of 159 mm Hg in atmospheric air to 149 mm Hg in the humidified air, and it dilutes the nitrogen partial pressure from 597 to 563 mm Hg.

Rate at Which Alveolar Air Is Renewed by Atmospheric Air

In Chapter 37, it is pointed out that the *functional residual capacity* of the lungs, the amount of air remaining in the lungs at the end of normal expiration, measures about 2300 milliliters. Yet only 350 milliliters of new air is brought into the alveoli with each normal respiration, and this same amount of old alveolar air is expired. Therefore, the amount of alveolar air replaced by new atmospheric air with each breath is only one seventh of the total, so that many breaths are required to exchange most of the alveolar air. Figure 39–2 shows this slow rate of renewal of the alveolar air. In the first alveolus of the figure, an excess amount of a gas is in all the alveoli, but note that even at the end of 16 breaths, the excess gas still has not been completely removed from the alveoli.

Figure 39–3 demonstrates graphically the rate at which an excess of a gas in the alveoli is normally removed, showing that with normal alveolar ventilation, about one half the gas is removed in 17 seconds. When a person's rate of alveolar ventilation is only one-half normal, one half the gas is removed in 34 seconds, and when the rate of ventilation is twice normal, one half is removed in about 8 seconds.

IMPORTANCE OF THE SLOW REPLACEMENT OF ALVEOLAR AIR. The slow replacement of alveolar air is of

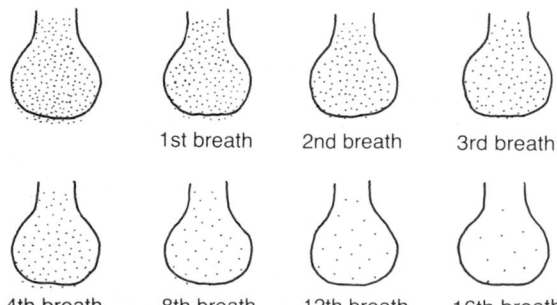

Figure 39–2. Expiration of a gas excess from the alveoli with successive breaths.

1st breath   2nd breath   3rd breath

4th breath   8th breath   12th breath   16th breath

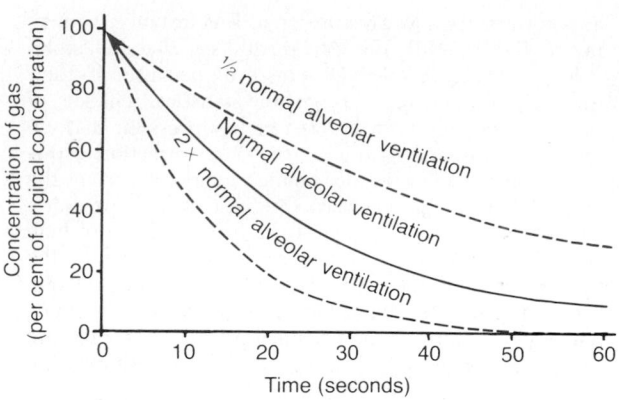

Figure 39–3. Rate of removal of excess gas from the alveoli.

particular importance in preventing sudden changes in gas concentrations in the blood.

This makes the respiratory control mechanism much more stable than it would otherwise be and helps to prevent excessive increases and decreases in tissue oxygenation, tissue carbon dioxide concentration, and tissue pH when respiration is temporarily interrupted.

Oxygen Concentration and Partial Pressure in the Alveoli

Oxygen is continually being absorbed into the blood of the lungs, and new oxygen is continually being breathed into the alveoli from the atmosphere. The more rapidly oxygen is absorbed, the lower becomes its concentration in the alveoli; on the other hand, the more rapidly new oxygen is breathed into the alveoli from the atmosphere, the higher becomes its concentration. Therefore, oxygen concentration in the alveoli, and its partial pressure as well, is controlled by, first, the rate of absorption of oxygen into the blood and, second, the rate of entry of new oxygen into the lungs by the ventilatory process.

Figure 39–4 shows the effect both of alveolar ventilation and of rate of oxygen absorption into the blood on the alveolar partial pressure of oxygen ($P_{A_{O_2}}$). The solid curve represents oxygen absorption at a rate of 250 ml/min and the dotted curve, at 1000 ml/min. At a normal ventilatory rate of 4.2 liters/min and an oxygen consumption of 250 ml/min, the normal operating point in Figure 39–4 is point A. The figure also shows that when 1000 milliliters of oxygen is being absorbed each minute, as occurs during moderate exercise, the rate of alveolar ventilation must increase fourfold to maintain the alveolar P_{O_2} at the normal value of 104 mm Hg.

Another effect shown in Figure 39–4 is that an extremely marked increase in alveolar ventilation can never increase the alveolar P_{O_2} above 149 mm Hg as long as the person is breathing normal atmospheric air at sea level pressure because this is the maximum P_{O_2} of oxygen in humidified air at this pressure. If the person breathes gases that contain partial pressures of oxygen higher than 149 mm Hg, the alveolar P_{O_2} can

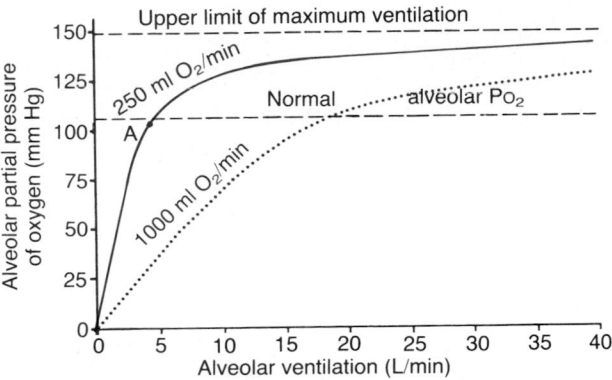

Figure 39–4. Effect on the alveolar P_{O_2} caused by different levels of alveolar ventilation at two rates of oxygen absorption, 250 ml/min and 1000 ml/min, from the alveoli.

approach these higher pressures at high rates of ventilation.

CO₂ Concentration and Partial Pressure in the Alveoli

Carbon dioxide is continually being formed in the body and then discharged into the alveoli; it is continually being removed from the alveoli by ventilation. Figure 39–5 shows the effects on the alveolar P_{CO_2} of both alveolar ventilation and two rates of carbon dioxide excretion. The solid curve represents a normal rate of carbon dioxide excretion of 200 ml/min. At the normal rate of alveolar ventilation of 4.2 liters/min, the operating point for alveolar P_{CO_2} is at point A in Figure 39–5—that is, 40 mm Hg.

Two other facts are also evident from Figure 39–5: First, *the alveolar P_{CO_2} increases directly in proportion to the rate of carbon dioxide excretion*, as represented by the elevation of the dotted curve for 800 milliliters CO_2 excretion per minute. Second, *the alveolar P_{CO_2} decreases in inverse proportion to alveolar ventilation*. Therefore, the concentrations and partial pressures of

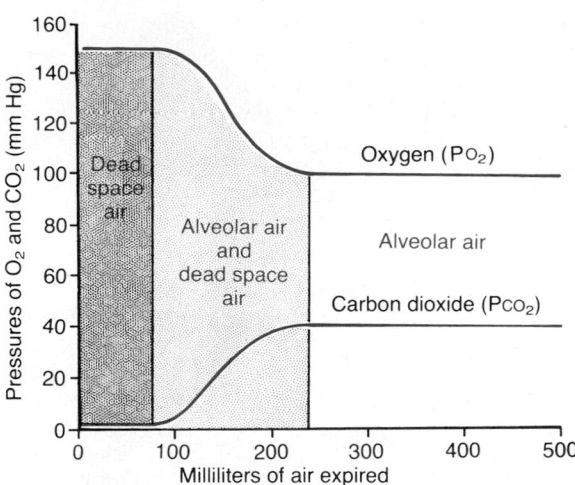

Figure 39–6. Oxygen and carbon dioxide partial pressures in the various portions of normal expired air.

both oxygen and carbon dioxide in the alveoli are determined by the rates of absorption or excretion of the two gases and by the level of alveolar ventilation.

Expired Air

Expired air is a combination of dead space air and alveolar air, and its overall composition is, therefore, determined by, first, the proportion of the expired air that is dead space air and, second, the proportion that is alveolar air. Figure 39–6 shows the progressive changes in oxygen and carbon dioxide partial pressures in the expired air during the course of expiration. The first portion of this air, the dead space air, is typical humidified air, as shown in column 3 of Table 39–1. Then, progressively, more and more alveolar air becomes mixed with the dead space air until all the dead space air has finally been washed out and nothing but alveolar air is expired at the end of expiration. Therefore, the means for collecting alveolar air for study is simply to collect a sample of the last portion of expired air.

Normal expired air, containing both dead space air and alveolar air, has gas concentrations and partial pressures approximately as shown in column 7 of Table 39–1—that is, concentrations between those of alveolar air and humidified atmospheric air.

DIFFUSION OF GASES THROUGH THE RESPIRATORY MEMBRANE

RESPIRATORY UNIT. Figure 39–7 shows the "respiratory unit," which is composed of a *respiratory bronchiole, alveolar ducts, atria,* and *alveoli* (of which there are about 300 million in the two lungs, each alveolus having an average diameter of about 0.2 millimeter). The alveolar walls are extremely thin, and within them is an almost solid network of interconnecting capillaries, shown in Figure 39–8. Indeed, because of the extensiveness of the capillary plexus, the flow of blood in the alveolar wall has been described as a "sheet" of flowing blood. Thus, it is obvious that the alveolar

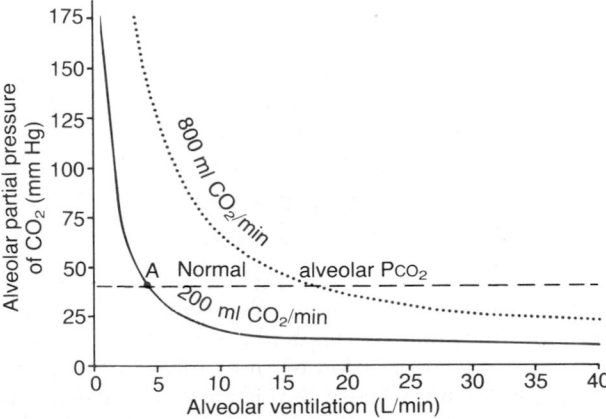

Figure 39–5. Effect on alveolar P_{CO_2} of alveolar ventilation at two rates of carbon dioxide excretion from the blood, 200 ml/min and 800 ml/min.

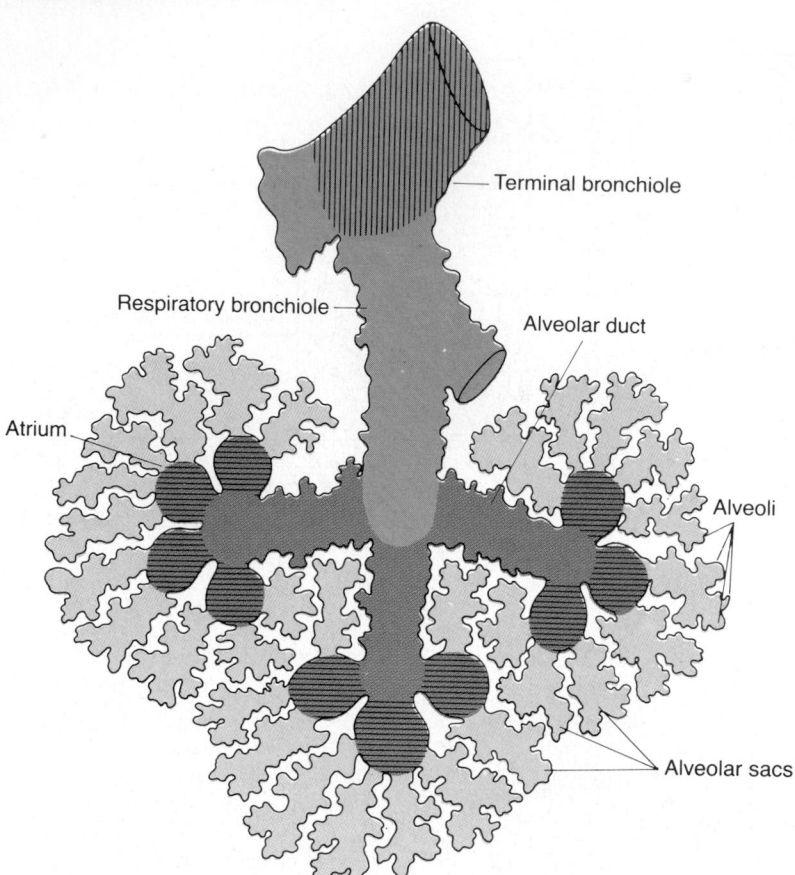

Figure 39–7. Respiratory lobule. (From W. S. Miller: The Lung. Springfield, Ill., Charles C Thomas, 1947.)

gases are in very, very close proximity to the blood of the capillaries. Consequently, gas exchange between the alveolar air and the pulmonary blood occurs through the membranes of all the terminal portions of the lungs, not merely in the alveoli themselves. These membranes are collectively known as the *respiratory membrane,* also called the *pulmonary membrane.*

RESPIRATORY MEMBRANE. Figure 39–9 shows to the left the ultrastructure of the respiratory membrane drawn in cross section and to the right a red blood cell. It also shows the diffusion of oxygen from the alveolus into the red blood cell and diffusion of carbon dioxide in the opposite direction. Note the following different layers of the respiratory membrane:

1. A layer of fluid lining the alveolus and containing surfactant that reduces the surface tension of the alveolar fluid

2. The alveolar epithelium composed of thin epithelial cells

3. An epithelial basement membrane

4. A thin interstitial space between the alveolar epithelium and the capillary membrane

5. A capillary basement membrane that in many places fuses with the epithelial basement membrane

6. The capillary endothelial membrane

Despite the large number of layers, the overall thickness of the respiratory membrane in some areas is as little as 0.2 micrometer, and it averages about 0.6 micrometer except where there are cell nuclei.

From histological studies, it has been estimated that the total surface area of the respiratory membrane is about 70 square meters in the normal adult. This is equivalent to the floor area of a 25-by-30-foot room. The total quantity of blood in the capillaries of the lung at any given instant is 60 to 140 milliliters. Now imagine this small amount of blood spread over the entire surface of a 25-by-30-foot floor, and it is easy to understand the rapidity of respiratory exchange of gases.

The average diameter of the pulmonary capillaries is only about 5 micrometers, which means that red blood cells must squeeze through them. Therefore, the red blood cell membrane usually touches the capillary wall so that oxygen and carbon dioxide need not pass through significant amounts of plasma as they diffuse between the alveolus and the red cell. This, too, increases the rapidity of diffusion.

Factors That Affect the Rate of Gas Diffusion Through the Respiratory Membrane

Referring to the earlier discussion of diffusion through water, one can apply the same principles and same formulas to diffusion of gases through the respiratory membrane. Thus, the factors that determine how rapidly a gas will pass through the membrane are (1) the *thickness of the membrane,* (2) the *surface area of the membrane,* (3) the *diffusion coefficient* of the

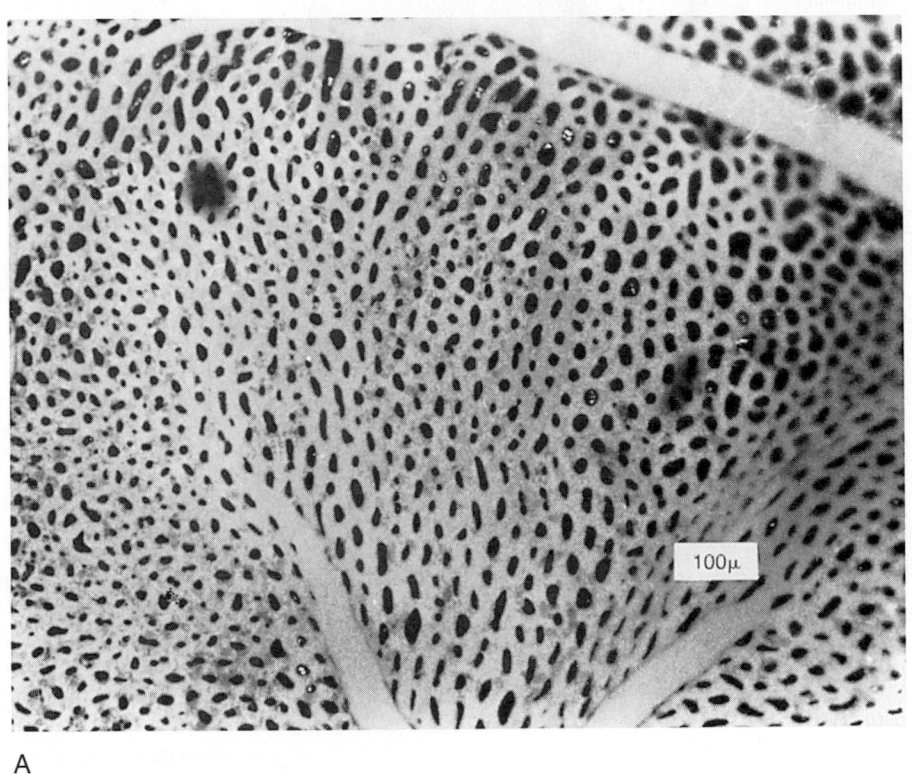

A

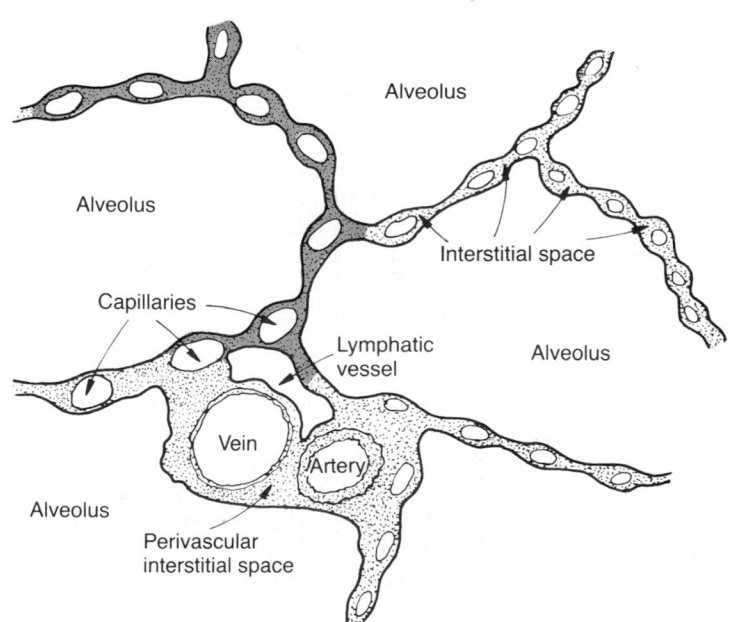

Alveolus

Alveolus

Interstitial space

Capillaries

Lymphatic
vessel

Alveolus

Vein

Artery

Alveolus

Perivascular
interstitial space

B

Figure 39–8. *A*, Surface view of capillaries in the alveolar wall. (From Maloney and Castle: *Resp. Physiol.*, 7:150, 1969. Reproduced by permission of ASP Biological and Medical Press. North-Holland Division.) *B*, Cross-sectional view of alveolar walls and their vascular supply.

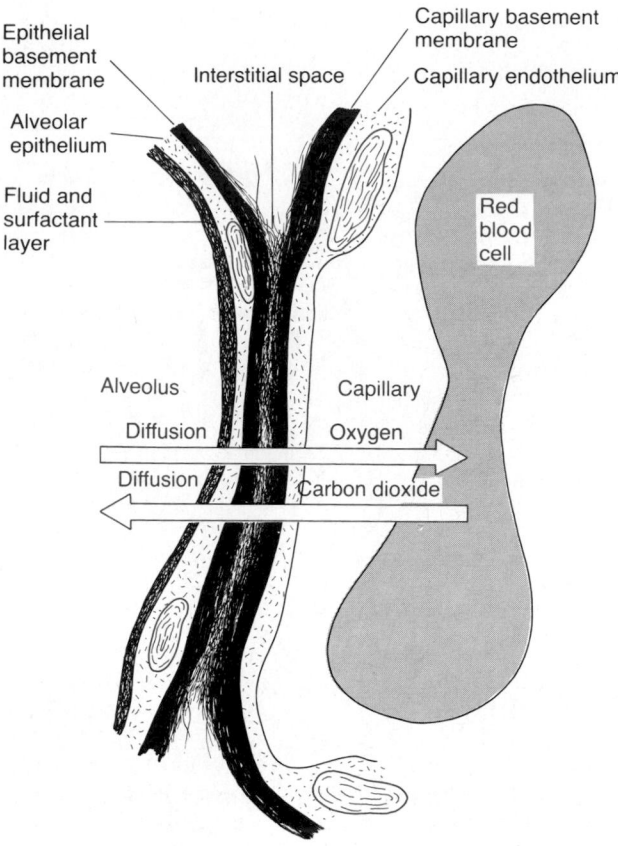

Figure 39–9. Ultrastructure of the respiratory membrane as shown in cross section.

gas in the substance of the membrane, and (4) the *pressure difference* between the two sides of the membrane.

The *thickness of the respiratory membrane* occasionally increases—for instance, as a result of edema fluid in the interstitial space of the membrane and in the alveoli—so that the respiratory gases must then diffuse not only through the membrane but also through this fluid. Also, some pulmonary diseases cause fibrosis of the lungs, which can increase the thickness of some portions of the respiratory membrane. Because the rate of diffusion through the membrane is inversely proportional to the thickness of the membrane, any factor that increases the thickness to more than two to three times normal can interfere significantly with normal respiratory exchange of gases.

The *surface area of the respiratory membrane* can be greatly decreased by many conditions. For instance, removal of an entire lung decreases the total surface area to one-half normal. Also, in *emphysema*, many of the alveoli coalesce, with dissolution of many alveolar walls. Therefore, the new chambers are much larger than the original alveoli, but the total surface area of the respiratory membrane is often decreased as much as fivefold because of loss of the alveolar walls. When the total surface area is decreased to about one-third to one-fourth normal, exchange of gases through the membrane is impeded to a significant degree *even*

under resting conditions. During competitive sports and other strenuous exercise, even the slightest decrease in surface area of the lungs can be a serious detriment to respiratory exchange of gases.

The *diffusion coefficient* for the transfer of each gas through the respiratory membrane depends on its *solubility* in the membrane and, inversely, on the *square root* of its *molecular weight.* The rate of diffusion in the respiratory membrane is almost exactly the same as that in water, for reasons explained earlier. Therefore, for a given pressure difference, carbon dioxide diffuses through the membrane about 20 times as rapidly as oxygen. Oxygen in turn diffuses about twice as rapidly as nitrogen.

The *pressure difference* across the respiratory membrane is the difference between the partial pressure of the gas in the alveoli and the pressure of the gas in the blood. The partial pressure represents a measure of the total number of molecules of a particular gas striking a unit area of the alveolar surface of the membrane in unit time, and the pressure of the gas in the blood represents the number of molecules that attempt to escape from the blood in the opposite direction. Therefore, the difference between these two pressures is a measure of the *net tendency* for the gas to move through the membrane. When the partial pressure of a gas in the alveoli is greater than the pressure of the gas in the blood, as is true for oxygen, net diffusion from the alveoli into the blood occurs; but when the pressure of the gas in the blood is greater than the partial pressure in the alveoli, as is true for carbon dioxide, net diffusion from the blood into the alveoli occurs.

Diffusing Capacity of the Respiratory Membrane

The ability of the respiratory membrane to exchange a gas between the alveoli and the pulmonary blood can be expressed in quantitative terms by its *diffusing capacity,* which is defined as the *volume of a gas that diffuses through the membrane each minute for a pressure difference of 1 mm Hg.*

All the factors discussed above that affect diffusion through the respiratory membrane can affect the diffusing capacity.

DIFFUSING CAPACITY FOR OXYGEN. In the average young man, the *diffusing capacity for oxygen* under resting conditions averages *21 ml/min/mm Hg.* In functional terms, what does this mean? The mean oxygen pressure difference across the respiratory membrane during normal, quiet breathing is about 11 mm Hg. Multiplication of this pressure by the diffusing capacity (11 × 21) gives a total of about 230 milliliters of oxygen diffusing through the respiratory membrane each minute; this is equal to the rate at which the body uses oxygen.

Change in Oxygen-Diffusing Capacity During Exercise. During strenuous exercise or other conditions that greatly increase pulmonary blood flow and alveolar ventilation, the diffusing capacity for oxygen

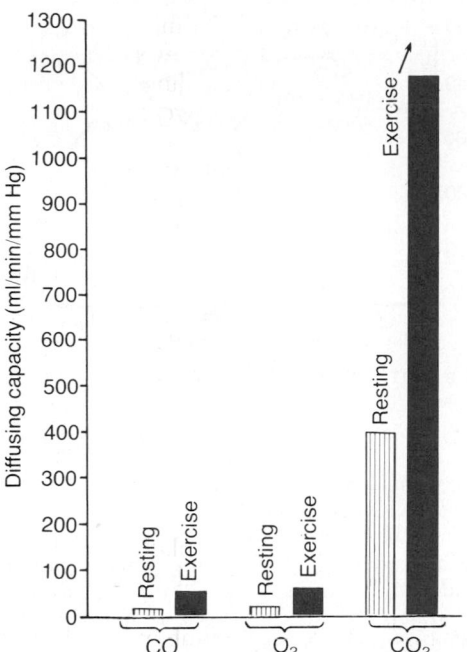

Figure 39–10. Diffusing capacities for carbon monoxide, oxygen, and carbon dioxide in the normal lungs.

increases in young men to a maximum of about 65 ml/min/mm Hg, which is three times the diffusing capacity under resting conditions. This increase is caused by several factors, among which are (1) opening up of a number of previously dormant pulmonary capillaries or extra dilatation of already open capillaries, thereby increasing the surface area of the blood into which the oxygen can diffuse; and (2) a better match between the ventilation of the alveoli and the perfusion of the alveolar capillaries with blood, called the "ventilation-perfusion ratio," which is explained in detail later in this chapter. Therefore, during exercise, the oxygenation of the blood is increased not only by increased alveolar ventilation but also by a greater capacity of the respiratory membrane for transmitting oxygen into the blood.

DIFFUSING CAPACITY FOR CARBON DIOXIDE. The diffusing capacity for carbon dioxide has never been measured because of the following technical difficulty: Carbon dioxide diffuses through the respiratory membrane so rapidly that the average P_{CO_2} in the pulmonary blood is not far different from the P_{CO_2} in the alveoli—the average difference is less than 1 mm Hg —and with the available techniques, this difference is too small to be measured.

Nevertheless, measurements of diffusion of other gases have shown that the diffusing capacity varies directly with the diffusion coefficient of the particular gas. Because the diffusion coefficient of carbon dioxide is 20 times that of oxygen, one would expect a diffusing capacity for carbon dioxide under resting conditions of about 400 to 450 ml/min/mm Hg and during exercise of about 1200 to 1300 ml/min/mm Hg.

Figure 39–10 compares the measured or calculated diffusing capacities of oxygen, carbon dioxide, and carbon monoxide at rest and during exercise, showing the extreme diffusing capacity of carbon dioxide and the effect of exercise on the diffusing capacity of each of these gases.

MEASUREMENT OF DIFFUSING CAPACITY—THE CARBON MONOXIDE METHOD. The oxygen-diffusing capacity can be calculated from measurements of (1) alveolar P_{O_2}, (2) P_{O_2} in the pulmonary capillary blood, and (3) the rate of oxygen uptake by the blood. However, to measure the P_{O_2} in the pulmonary capillary blood is so difficult and so imprecise that it is not practical to measure oxygen-diffusing capacity by such a direct procedure except on an experimental basis.

To obviate the difficulties encountered in measuring oxygen-diffusing capacity directly, physiologists usually measure carbon monoxide–diffusing capacity instead and then calculate the oxygen-diffusing capacity from this. The principle of the carbon monoxide method is the following: A small amount of carbon monoxide is breathed into the alveoli, and the partial pressure of the carbon monoxide in the alveoli is measured from appropriate alveolar air samples. The carbon monoxide pressure in the blood is essentially zero because hemoglobin combines with this gas so rapidly that its pressure never has time to build up. Therefore, the pressure difference of carbon monoxide across the respiratory membrane is equal to its partial pressure in the alveoli. Then, by measuring the volume of carbon monoxide absorbed in a period of time and dividing this by the alveolar carbon monoxide partial pressure, one can determine accurately the carbon monoxide–diffusing capacity.

To convert carbon monoxide–diffusing capacity to oxygen-diffusing capacity, the value is multiplied by a factor of 1.23 because the diffusion coefficient for oxygen is 1.23 times that for carbon monoxide. Thus, the average diffusing capacity for carbon monoxide in young men at rest is 17 ml/min/mm Hg and the diffusing capacity for oxygen is 1.23 times this, or 21 ml/min/mm Hg.

EFFECT OF THE VENTILATION-PERFUSION RATIO ON ALVEOLAR GAS CONCENTRATION

In the early part of this chapter, we learned that two factors determine the P_{O_2} and the P_{CO_2} in the alveoli: (1) the rate of alveolar ventilation and (2) the rate of transfer of oxygen and carbon dioxide through the respiratory membrane. These earlier discussions made the assumption that all the alveoli are ventilated equally and that blood flow through the alveolar capillaries is the same for each alveolus. However, even normally to some extent, and especially in many lung diseases, some areas of the lungs are well ventilated but have almost no blood flow, whereas other areas may have excellent blood flow but little or no ventilation. In either of these conditions, gas exchange through the respiratory membrane is seriously impaired, and the person may suffer severe respiratory distress despite both normal *total* ventilation and normal *total* pulmonary blood flow, but with the ventilation and blood flow going to different parts of the lungs. Therefore, a highly quantitative concept has been developed to help us understand respiratory exchange when there is imbalance between alveolar ventilation and alveolar blood flow. This concept is called the *ventilation-perfusion ratio*.

In quantitative terms, the ventilation-perfusion ratio is expressed as $\dot{V}_A/\dot{Q}$. When $\dot{V}_A$ (alveolar ventilation) is normal for a given alveolus and $\dot{Q}$ (blood flow) is also normal for the same alveolus, then the ventilation-perfusion ratio ($\dot{V}_A/\dot{Q}$) is also said to be normal. However, when the ventilation ($\dot{V}_A$) is zero and yet there is still perfusion ($\dot{Q}$) of the alveolus, the ventilation-perfusion ratio, $\dot{V}_A/\dot{Q}$, is zero. Or, at the other extreme, when there is adequate ventilation ($\dot{V}_A$) but zero perfusion ($\dot{Q}$), then the ratio, $\dot{V}_A/\dot{Q}$, is infinity. At a ratio of either zero or infinity, there is no exchange of gases through the respiratory membrane of the affected alveoli, which explains the importance of this concept. Therefore, let us explain the respiratory consequences of these two extremes.

ALVEOLAR OXYGEN AND CARBON DIOXIDE PARTIAL PRESSURES WHEN $\dot{V}_A/\dot{Q}$ EQUALS ZERO. When $\dot{V}_A/\dot{Q}$ is equal to zero—that is, without any alveolar ventilation— the air in the alveolus comes to equilibrium with the blood oxygen and carbon dioxide because these gases diffuse between the blood and the alveolar air. Because the blood that perfuses the capillaries is venous blood returning to the lungs from the systemic circulation, it is the gases in this blood that come to equilibrium with the alveolar gases. In Chapter 40, we see that the normal venous blood ($\bar{v}$) has a P_{O_2} of 40 mm Hg and a P_{CO_2} of 45 mm Hg. Therefore, these are also the normal partial pressures of these two gases in alveoli that have blood flow but no ventilation.

ALVEOLAR OXYGEN AND CARBON DIOXIDE PARTIAL PRESSURES WHEN $\dot{V}_A/\dot{Q}$ EQUALS INFINITY. The effect on the alveolar gas partial pressures is entirely different when $\dot{V}_A/\dot{Q}$ equals infinity from the effect when $\dot{V}_A/\dot{Q}$ equals zero because now there is no capillary blood flow to carry oxygen away or to bring carbon dioxide to the alveoli. Therefore, instead of the alveolar gases coming to equilibrium with the venous blood, the alveolar air now becomes equal to the humidified inspired air. That is, the air that is inspired loses no oxygen to the blood and gains no carbon dioxide. And because normal inspired and humidified air has a P_{O_2} of 149 mm Hg and a P_{CO_2} of 0 mm Hg, these are the partial pressures of these two gases in the alveoli.

GAS EXCHANGE, AND ALVEOLAR PARTIAL PRESSURES WHEN $\dot{V}_A/\dot{Q}$ IS NORMAL. When there is both normal alveolar ventilation and normal alveolar capillary blood flow (which means normal alveolar perfusion), exchange of oxygen and carbon dioxide through the respiratory membrane is nearly optimal, and alveolar P_{O_2} is normally at a level of 104 mm Hg, which lies between that of the inspired air (149 mm Hg) and that of venous blood (40 mm Hg). Likewise, alveolar P_{CO_2} lies between the two extremes; it is normally 40 mm Hg, in contrast to 45 mm Hg in venous blood and 0 mm Hg in the inspired air. Thus, under normal conditions, the alveolar air P_{O_2} averages 104 mm Hg and the P_{CO_2}, 40 mm Hg.

P_{O_2}-P_{CO_2}, $\dot{V}_A/\dot{Q}$ Diagram

The concepts presented in the sections above can be shown usefully in graphic form, as demonstrated in Figure 39–11, called the P_{O_2}-P_{CO_2}, $\dot{V}_A/\dot{Q}$ diagram. The curve in the diagram represents all possible P_{O_2} and

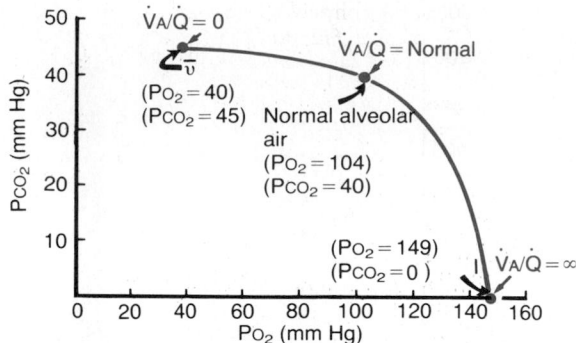

Figure 39–11. Normal P_{O_2}-P_{CO_2}, P_{CO_2}, $\dot{V}_A/\dot{Q}$ diagram.

P_{CO_2} combinations between the limits of $\dot{V}_A/\dot{Q}$ equals zero and $\dot{V}_A/\dot{Q}$ equals infinity when the gas pressures in the venous blood are normal and the person is breathing air at sea level pressure. Thus, point $\bar{v}$ is the plot of P_{O_2} and P_{CO_2} when $\dot{V}_A/\dot{Q}$ equals zero. At this point, the P_{O_2} is 40 mm Hg and the P_{CO_2} is 45 mm Hg, which are the values in venous blood.

At the other end of the curve, when $\dot{V}_A/\dot{Q}$ equals infinity, point I represents inspired air, showing P_{O_2} to be 149 mm Hg while P_{CO_2} is zero.

Also plotted on the curve is the point that represents normal alveolar air when $\dot{V}_A/\dot{Q}$ is normal. At this point, P_{O_2} is 104 mm Hg and P_{CO_2} is 40 mm Hg.

Concept of "Physiologic Shunt" (When $\dot{V}_A/\dot{Q}$ Is Below Normal)

Whenever $\dot{V}_A/\dot{Q}$ is below normal, there is not ventilation enough to provide the oxygen needed to oxygenate fully the blood flowing through the alveolar capillaries. Therefore, a certain fraction of the venous blood passing through the pulmonary capillaries does not become oxygenated. This fraction is called *shunted blood*. Also, some additional blood flows through the bronchial vessels, rather than through the alveolar capillaries, normally about 2 per cent of the cardiac output; this, too, is unoxygenated, shunted blood.

The total quantitative amount of shunted blood per minute is called the *physiologic shunt*. This physiologic shunt is measured in the clinical pulmonary function laboratory by analyzing the concentration of oxygen in both mixed venous blood and arterial blood along with simultaneous measurement of the cardiac output. From these values, the physiologic shunt can be calculated by the following equation:

$$\frac{\dot{Q}_{PS}}{\dot{Q}_T} = \frac{Ci_{O_2} - Ca_{O_2}}{Ci_{O_2} - C\bar{v}_{O_2}},$$

in which

$\dot{Q}_{PS}$ is the physiologic shunt blood flow per minute;
$\dot{Q}_T$ is cardiac output per minute;
Ci_{O_2} is the concentration of oxygen in the arterial blood if there is "ideal" ventilation-perfusion ratio;
Ca_{O_2} is the measured concentration of oxygen in the arterial blood; and
$C\bar{v}_{O_2}$ is the measured concentration of oxygen in the mixed venous blood.

The greater the physiologic shunt, the greater the *amount of blood that fails to be oxygenated* as it passes through the lungs.

Concept of "Physiologic Dead Space" (When VA/Q Is Greater Than Normal)

When ventilation of some of the alveoli is great but the alveolar blood flow is low, there is far more available oxygen in the alveoli than can be transported away from the alveoli by the flowing blood. Thus, the ventilation of these alveoli is said to be *wasted*. In addition, the ventilation of the anatomical dead space areas of the respiratory passageways is wasted. The sum of these two types of wasted ventilation is called the *physiologic dead space*. This is measured in the clinical pulmonary function laboratory by making appropriate blood and expiratory gas measurements and using the following equation, called the Bohr equation:

$$\frac{V_{D_{phys}}}{V_T} = \frac{Pa_{CO_2} - P\overline{E}_{CO_2}}{Pa_{CO_2}},$$

in which

$V_{D_{phys}}$ is the physiologic dead space;

V_T is the tidal volume;

Pa_{CO_2} is the partial pressure of carbon dioxide in the arterial blood; and

$P\overline{E}_{CO_2}$ is the average partial pressure of carbon dioxide in the entire expired air.

When the physiologic dead space is great, much of the *work of ventilation* is wasted effort because so much of the ventilated air never reaches the blood.

Abnormalities of Ventilation-Perfusion Ratio

ABNORMAL VA/QS IN THE UPPER AND LOWER NORMAL LUNG. In a normal person in the upright position, both blood flow and alveolar ventilation are considerably less in the upper part of the lung than in the lower part; however, blood flow is decreased considerably more than is ventilation. Therefore, at the top of the lung, $\dot{V}_A/\dot{Q}$ is as much as 2.5 times as great as the ideal value, which causes a moderate degree of *physiologic dead space* in this area of the lung.

At the other extreme, in the bottom of the lung, there is slightly too little ventilation in relation to blood flow, with $\dot{V}_A/\dot{Q}$ as low as 0.6 times the ideal value. In this area, a small fraction of the blood fails to become normally oxygenated, and this represents a *physiologic shunt*.

In both extremes of the lung, inequalities of ventilation and perfusion decrease the effectiveness of the lung slightly for exchange of oxygen and carbon dioxide. However, during exercise, the blood flow to the upper part of the lung increases markedly, so that far less physiologic dead space occurs and the effectiveness of gas exchange approaches optimum.

ABNORMAL VA/Q IN CHRONIC OBSTRUCTIVE LUNG DISEASE. Most people who smoke for prolonged periods develop various degrees of bronchial obstruction; in a large share of them, this condition eventually becomes so severe that they develop serious alveolar air trapping and resultant emphysema. The emphysema in turn causes many of the alveolar walls to be destroyed. Thus, two abnormalities occur in smokers to cause abnormal $\dot{V}_A/\dot{Q}$. First, because many of the small bronchioles are obstructed, the alveoli beyond the obstructions are unventilated, causing a $\dot{V}_A/\dot{Q}$ that approaches zero. Second, in those areas of the lung where the alveolar walls have been mainly destroyed but there is still alveolar ventilation, most of the ventilation is wasted because of inadequate blood flow to transport the blood gases.

In chronic obstructive lung disease, some areas of the lung exhibit *serious physiologic shunt* and other areas, *serious physiologic dead space*. Both of these conditions tremendously decrease the effectiveness of the lungs as gas exchange organs, sometimes reducing the effectiveness to as little as 1/10 normal. In fact, this is the most prevalent cause of pulmonary disability today.

REFERENCES

Arieff, A. I.: Hypoxia, Metabolic Acidosis and the Circulation. New York, Oxford University Press, 1992.

Bauer, C., et al. (eds.): Biophysics and Physiology of Carbon Dioxide. New York, Springer-Verlag, 1980.

Bidani, A., and Crandall, E. D.: Velocity of CO_2 exchanges in the lungs. Annu. Rev. Physiol., 50:639, 1988.

Brown, D. L.: Risk and Outcome in Anesthesia. Philadelphia, J. B. Lippincott, 1992.

Drazen, J. M., et al.: High-frequency ventilation. Physiol. Rev., 64:505, 1984.

Forster, R. E., and Crandall, E. D.: Pulmonary gas exchange. Annu. Rev. Physiol., 38:69, 1976.

Forster, R. E.: Diffusion of gases. In Fenn, W. O., and Rahn, H. (eds.): Handbook of Physiology. Sec. 3, Vol. 1. Baltimore, Williams & Wilkins, 1964, p. 839.

Forster, R. E.: Interpretation of measurements of pulmonary diffusing capacity. In Fenn, W. O., and Rahn, H. (eds.): Handbook of Physiology. Sec. 3, Vol. 2. Baltimore, Williams & Wilkins, 1965, p. 1435.

Glass, M. L., and Wood, S. C.: Gas exchange and control of breathing in reptiles. Physiol. Rev., 63:232, 1983.

Gonzalcs, N. C., and Fedde, M. R. (eds.): Oxygen Transfer from Atmosphere to Tissues. New York, Plenum Publishing Corp., 1988.

Grippi, M. A.: Pulmonary Pathophysiology: A Problem-oriented Approach. Philadelphia, J. B. Lippincott, 1994.

Guyton, A. C., et al.: An arteriovenous oxygen difference recorder. J. Appl. Physiol., 10:158, 1957.

Huch, A., et al. (eds.): Continuous Transcutaneous Monitoring. New York, Plenum Publishing Corp., 1987.

Hughes, J. M.: Proceedings: Regional differences in gas exchange. Proc. R. Soc. Med., 6:974, 1973.

Johansen, K.: Comparative physiology: Gas exchange and circulation in fishes. Annu. Rev. Physiol., 33:569, 1971.

Jones, N. L.: Blood Gases and Acid-Base Physiology. New York, B. C. Decker, 1980.

Kirby, R. R., and Gravenstein, N.: Clinical Anesthesia Practice. Philadelphia, W. B. Saunders Co., 1994.

Klocke, R. A.: Velocity of CO_2 exchange in blood. Annu. Rev. Physiol., 50:625, 1988.

Lane, E. E., and Walker, J. F.: Clinical Arterial Blood Gas Analysis. St. Louis, C. V. Mosby Co., 1987.

Leff, A. R., and Schumacker, P. T.: Respiratory Physiology: Basics and Applications. Philadelphia, W. B. Saunders Co., 1993.

Levitsky, M. G.: Pulmonary Physiology. New York, McGraw-Hill Book Co., 1982.

Levitzky, M. G., and Hall, S. M. Cardiopulmonary Physiology in Anesthesiology. Hightstown, NJ, McGraw-Hill, 1994.

Malley, W. J.: Clinical Blood Gases: Application and Noninvasive Alternatives. Philadelphia, W. B. Saunders Co., 1990.

Milhorn, H. T., Jr., and Pulley, P. E., Jr.: A theoretical study of pulmonary capillary gas exchange and venous admixture. Biophys. J., 8:337, 1968.

Mochizuki, M., et al. (eds): Oxygen Transport to Tissue X. New York, Plenum Publishing Corp., 1988.

Olson, L. E.: Physiology laboratories quantifying gas exchange in health and disease. The Physiologist, 28:170, 1985.

Otis, A. B.: Quantitative relationships in steady-state gas exchange. In Fenn, W. O., and Rahn, H. (eds.): Handbook of Physiology. Sec. 3, Vol. 1. Baltimore, Williams & Wilkins, 1964, p. 681.

Paiva, M., and Engel, L. A.: Theoretical studies of gas mixing and ventilation distribution in the lung. Physiol. Rev., 67:750, 1987.

Piper, J., and Scheid, P.: Comparative physiology of respiration: Functional analysis of gas exchange organs in vertebrates. Int. Rev. Physiol., 14:219, 1977.

Piper, J., and Scheid, P.: Respiration: Alveolar gas exchange. Annu. Rev. Physiol., 33:131, 1971.

Quintanilha, A. (ed.): Oxygen Radicals in Biological Systems. New York, Plenum Publishing Corp., 1988.

Rahn, H., and Farhi, E. E.: Ventilation, perfusion, and gas exchange—the Va/Q concept. In Fenn, W. O., and Rahn, H. (eds.): Handbook of Physiology. Sec. 3, Vol. 1. Baltimore, Williams & Wilkins, 1964, p. 125.

Rahn, H., and Fenn, W. O.: A Graphical Analysis of Respiratory Gas Exchange. Washington, D.C., American Physiological Society, 1955.

Saldana, M. J.: Pathology of Pulmonary Disease. Philadelphia, J. B. Lippincott, 1994.

Wagner, P. D.: Diffusion and chemical reaction in pulmonary gas exchange. Physiol. Rev., 57:257, 1977.

Wagner, P. D.: Ventilation-perfusion relationships. Annu. Rev. Physiol., 42:235, 1980.

Weibel, E. R.: Morphological basis of alveolar capillary gas exchange. Physiol. Rev., 53:419, 1973.

West, J. B.: Pulmonary gas exchange. Int. Rev. Physiol., 14:83, 1977.

West, J. B.: Pulmonary Pathophysiology: The Essentials. Baltimore, Williams & Wilkins Co., 1992.

West, J. B.: Respiratory Physiology—The Essentials. Baltimore, Williams & Wilkins Co., 1994.

Transport of Oxygen and Carbon Dioxide in the Blood and Body Fluids

*C*HAPTER *40*

Once oxygen has diffused from the alveoli into the pulmonary blood, it is transported principally in combination with hemoglobin to the tissue capillaries, where it is released for use by the cells. The presence of hemoglobin in the red cells of the blood allows the blood to transport 30 to 100 times as much oxygen as could be transported simply in the form of dissolved oxygen in the water of the blood.

In the tissue cells, oxygen reacts with various foodstuffs to form large quantities of carbon dioxide. This in turn enters the tissue capillaries and is transported back to the lungs. Carbon dioxide, like oxygen, also combines with chemical substances in the blood that increase carbon dioxide transport 15- to 20-fold.

The purpose of this chapter is to present both qualitatively and quantitatively the physical and chemical principles of oxygen and carbon dioxide transport in the blood and body fluids.

PRESSURES OF OXYGEN AND CARBON DIOXIDE IN THE LUNGS, BLOOD, AND TISSUES

In the discussions of Chapter 39, it is pointed out that gases can move from one point to another by diffusion and that the cause of this movement is always a pressure difference from the first point to the other. Thus, oxygen diffuses from the alveoli into the pulmonary capillary blood because the oxygen pressure (Po_2) in the alveoli is greater than the Po_2 in the pulmonary blood. Then, in the tissues, a higher Po_2 in the capillary blood causes oxygen to diffuse to the cells.

Conversely, when oxygen is metabolized in the cells to form carbon dioxide, the carbon dioxide pressure (Pco_2) rises to a high value, which causes carbon dioxide to diffuse into the tissue capillaries. Similarly, it diffuses out of the blood into the alveoli because the Pco_2 in the pulmonary capillary blood is greater than that in the alveoli.

Basically, then, the transport of oxygen and carbon dioxide by the blood depends on both diffusion and the movement of blood. We now need to consider quantitatively the factors responsible for these effects.

Uptake of Oxygen by the Pulmonary Blood

The top part of Figure 40–1 shows a pulmonary alveolus adjacent to a pulmonary capillary, demonstrating diffusion of oxygen molecules between the alveolar air and the pulmonary blood. The Po_2 of the gaseous oxygen in the alveolus averages 104 mm Hg, whereas the Po_2 of the venous blood entering the capillary averages only 40 mm Hg because a large amount of oxygen has been removed from this blood as it has passed through the peripheral tissues. Therefore, the *initial* pressure difference that causes oxygen to diffuse into the pulmonary capillary is $104 - 40$, or 64 mm Hg. The curve below the capillary shows the rapid rise in blood Po_2 as the blood passes through the capillary, showing that the Po_2 rises essentially to equal that of the alveolar air by the time the blood has moved a third of the distance through the capillary, becoming almost 104 mm Hg.

UPTAKE OF OXYGEN BY THE PULMONARY BLOOD DURING EXERCISE. During strenuous exercise, a per-

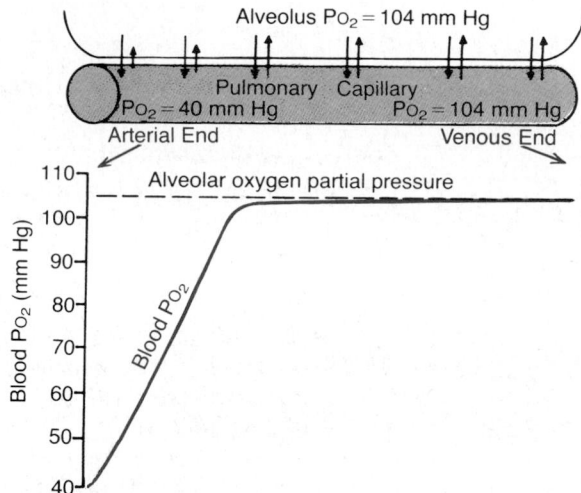

Figure 40–1. Uptake of oxygen by the pulmonary capillary blood. (The curve in this figure was constructed from data in Milhorn and Pulley: *Biophys. J.,* 8:337, 1968.)

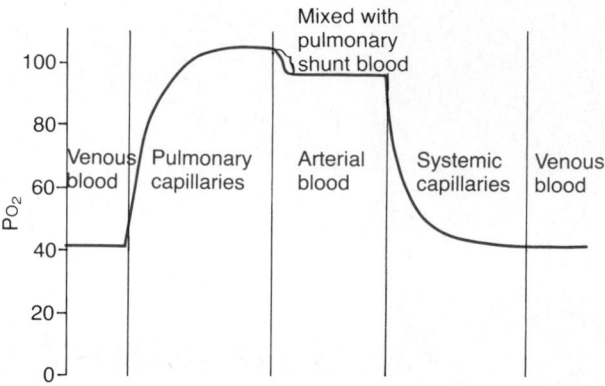

Figure 40–2. Changes in Po_2 in the pulmonary capillary blood, arterial blood, and systemic capillary blood, demonstrating the effect of "venous admixture."

son's body may require as much as 20 times the normal amount of oxygen. Also, because of the increased cardiac output, the time that the blood remains in the capillary may be reduced to less than one-half normal, despite the fact that additional capillaries open up. Oxygenation of the blood could suffer for both of these reasons. Yet because of the great *safety factor* for diffusion of oxygen through the pulmonary membrane, the blood is still *almost saturated* with oxygen when it leaves the pulmonary capillaries. Let us explain this as follows.

First, it is pointed out in Chapter 39 that the diffusing capacity for oxygen increases almost threefold during exercise; this results mainly from increased surface area of capillaries participating in the diffusion but also from more nearly ideal ventilation-perfusion ratio in the upper part of the lungs.

Second, note in Figure 40–1 that during normal pulmonary blood flow, the blood becomes almost saturated with oxygen by the time it has passed through one third of the pulmonary capillary, and little additional oxygen enters the blood during the latter two thirds of its transit. That is, the blood normally stays in the lung capillaries about three times as long as necessary to cause full oxygenation. Therefore, in exercise, even with the shortened time of exposure in the capillaries, the blood still can become fully oxygenated or nearly so.

Transport of Oxygen in the Arterial Blood

About 98 per cent of the blood that enters the left atrium from the lung has passed through the alveolar capillaries and has become oxygenated up to a Po_2 of about 104 mm Hg, as just explained. Another 2 per cent of the blood has passed directly from the aorta through the bronchial circulation, which supplies mainly the deep tissues of the lungs and is not exposed to the pulmonary air. This blood flow represents "shunt" flow, meaning blood that is shunted past the gas exchange areas. On leaving the lungs the Po_2 of the shunt blood is about that of normal venous blood, about 40 mm Hg. This blood combines in the pulmonary veins with the oxygenated blood from the alveolar capillaries; this mixing of the bloods is called *venous admixture of blood,* and it causes the Po_2 of the blood pumped by the left side of the heart into the aorta to fall to about 95 mm Hg. These changes in blood Po_2 at different points in the circulatory system are shown in Figure 40–2.

Diffusion of Oxygen from the Peripheral Capillaries into the Tissue Fluid

When the arterial blood reaches the peripheral tissues, its Po_2 in the capillaries is still 95 mm Hg. On the other hand, as shown in Figure 40–3, the Po_2 in the interstitial fluid that surrounds the tissue cells averages only 40 mm Hg. Thus, there is a tremendous initial pressure difference that causes oxygen to diffuse rapidly from the blood into the tissues, so rapidly that the capillary Po_2 falls almost to equal the 40 mm Hg pressure in the interstitium. Therefore, the Po_2 of the blood leaving the tissue capillaries and entering the veins is also about 40 mm Hg.

EFFECT OF RATE OF BLOOD FLOW ON INTERSTITIAL FLUID Po_2. If the blood flow through a particular tissue becomes increased, greater quantities of oxygen are

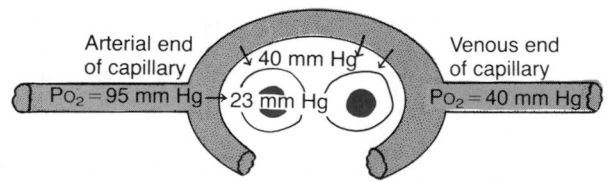

Figure 40–3. Diffusion of oxygen from a tissue capillary to the cells.

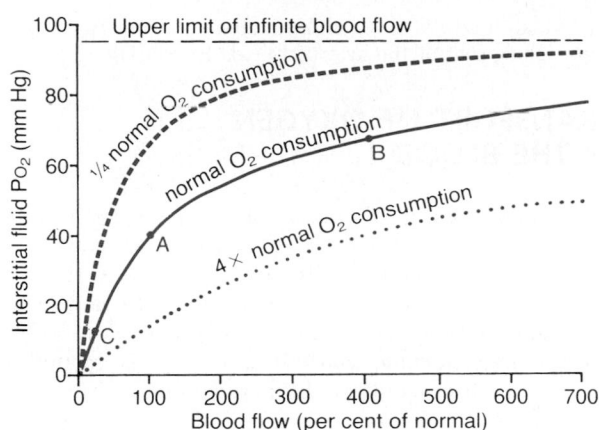

Figure 40–4. Effect of blood flow and rate of oxygen consumption on tissue P_{O_2}.

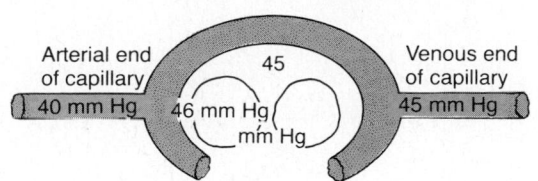

Figure 40–5. Uptake of carbon dioxide by the blood in the capillaries.

transported into the tissue in a given period, and the tissue P_{O_2} becomes correspondingly increased. This effect is shown in Figure 40–4. Note that an increase in flow to 400 per cent of normal increases the P_{O_2} from 40 mm Hg at point A in the figure to 66 mm Hg at point B. However, the upper limit to which the P_{O_2} can rise, even with maximal blood flow, is 95 mm Hg because this is the oxygen pressure in the arterial blood.

EFFECT OF RATE OF TISSUE METABOLISM ON INTERSTITIAL FLUID P_{O_2}. If the cells use more oxygen for metabolism than normally, this tends to reduce the interstitial fluid P_{O_2}. Figure 40–4 also demonstrates this effect, showing reduced interstitial fluid P_{O_2} when the cellular oxygen consumption is increased and increased P_{O_2} when the consumption is decreased.

In summary, tissue P_{O_2} is determined by a balance between (1) the rate of oxygen transport to the tissues in the blood and (2) the rate at which the oxygen is used by the tissues.

Diffusion of Oxygen from the Capillaries to the Cells

Oxygen is always being used by the cells. Therefore, the intracellular P_{O_2} remains lower than the P_{O_2} in the capillaries. Also, in many instances, there is considerable distance between the capillaries and the cells. Therefore, the normal intracellular P_{O_2} ranges from as low as 5 mm Hg to as high as 40 mm Hg, averaging (by direct measurement in lower animals) 23 mm Hg. Because only 1 to 3 mm Hg of oxygen pressure is normally required for full support of the chemical processes that use oxygen in the cell, one can see that even this low cellular P_{O_2} of 23 mm Hg is more than adequate and provides a large safety factor.

Diffusion of Carbon Dioxide from the Tissue Cells into the Tissue Capillaries and from the Pulmonary Capillaries into the Alveoli

When oxygen is used by the cells, most of it becomes carbon dioxide, and this increases the intracellular P_{CO_2}. Therefore, carbon dioxide diffuses from the cells into the tissue capillaries and is then carried by the blood to the lungs. In the lungs, it diffuses from the pulmonary capillaries into the alveoli. Thus, at each point in the gas transport chain, carbon dioxide diffuses in a direction exactly opposite that of the diffusion of oxygen. Yet there is one major difference between the diffusion of carbon dioxide and that of oxygen: *carbon dioxide can diffuse about 20 times as rapidly as oxygen.* Therefore, the pressure differences that cause carbon dioxide diffusion are, in each instance, far less than the pressure differences required to cause oxygen diffusion. These pressures are the following:

1. Intracellular P_{CO_2}, about 46 mm Hg; interstitial P_{CO_2}, about 45 mm Hg; thus, there is only a 1-mm Hg pressure differential, as shown in Figure 40–5.

2. P_{CO_2} of the arterial blood entering the tissues, 40 mm Hg; P_{CO_2} of the venous blood leaving the tissues, about 45 mm Hg; thus, as also shown in Figure 40–5, the tissue capillary blood comes almost exactly to equilibrium with the interstitial P_{CO_2}, also 45 mm Hg.

3. P_{CO_2} of the venous blood entering the pulmonary capillaries, 45 mm Hg; P_{CO_2} of the alveolar air, 40 mm Hg; thus, only a 5-mm Hg pressure difference causes all the required carbon dioxide diffusion out of the pulmonary capillaries into the alveoli. Furthermore, as shown in Figure 40–6, the P_{CO_2} of the pulmonary capillary blood falls almost exactly to equal the alveolar P_{CO_2} of 40 mm Hg before it has passed more than about one-third the distance through the

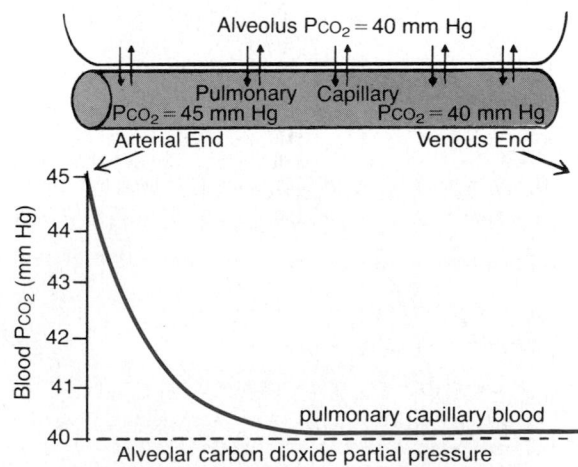

Figure 40–6. Diffusion of carbon dioxide from the pulmonary blood into the alveolus. (This curve was constructed from data in Milhorn and Pulley: *Biophys. J., 8:*337, 1968.)

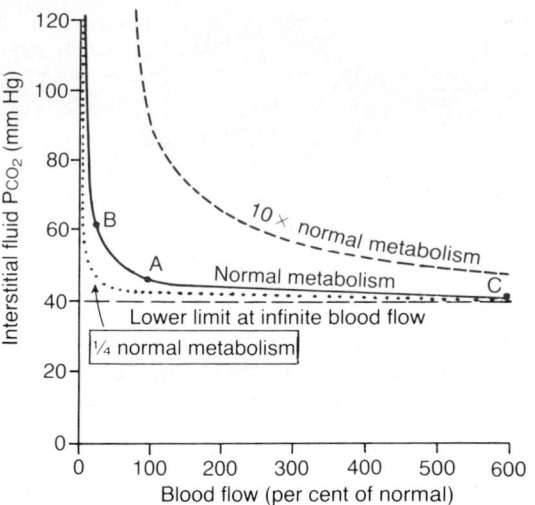

Figure 40–7. Effect of blood flow and metabolic rate on tissue P_{CO_2}.

capillaries. This is the same effect that was observed earlier for oxygen diffusion.

Effect of Tissue Metabolism and Blood Flow on Interstitial P_{CO_2}. Tissue capillary blood flow and tissue metabolism affect the P_{CO_2} in a way exactly opposite the way in which they affect tissue P_{O_2}. Figure 40–7 shows these effects, as follows.

1. A decrease in blood flow from normal, point A, to one-quarter normal, point B, increases the tissue P_{CO_2} from the normal value of 45 mm Hg to the elevated level of 60 mm Hg. Conversely, increasing the blood flow to six times normal, point C, decreases the P_{CO_2} from the normal value of 45 mm Hg to 41 mm Hg, almost down to a level equal to the P_{CO_2} in the arterial blood, 40 mm Hg, entering the tissue capillaries.

2. Note also that a 10-fold increase in metabolic rate greatly elevates the P_{CO_2} at all levels of blood flow, whereas decreasing the metabolism to one-quarter normal causes the

interstitial fluid P_{CO_2} to fall to about 41 mm Hg, closely approaching that of the arterial blood, 40 mm Hg.

TRANSPORT OF OXYGEN IN THE BLOOD

Normally, about 97 per cent of the oxygen transported from the lungs to the tissues is carried in chemical combination with hemoglobin in the red blood cells. The remaining 3 per cent is carried in the dissolved state in the water of the plasma and cells. Thus, *under normal conditions,* oxygen is carried to the tissues almost entirely by hemoglobin.

Reversible Combination of Oxygen with Hemoglobin

The chemistry of hemoglobin is presented in Chapter 32, where it is pointed out that the oxygen molecule combines loosely and reversibly with the heme portion of the hemoglobin. When the P_{O_2} is high, as in the pulmonary capillaries, oxygen binds with the hemoglobin, but when the P_{O_2} is low, as in the tissue capillaries, oxygen is released from the hemoglobin. This is the basis for almost all oxygen transport from the lungs to the tissues.

OXYGEN-HEMOGLOBIN DISSOCIATION CURVE. Figure 40–8 shows the oxygen-hemoglobin dissociation curve, which demonstrates a progressive increase in the percentage of the hemoglobin that is bound with oxygen as the blood P_{O_2} increases, which is called the *per cent saturation of the hemoglobin.* Because the blood in the arteries usually has a P_{O_2} of about 95 mm Hg, one can see from the dissociation curve that the *usual oxygen saturation of arterial blood is about 97 per cent.* On the other hand, in normal venous blood returning from the tissues, the P_{O_2} is about 40 mm Hg and *the saturation of the hemoglobin is about 75 per cent.*

MAXIMUM AMOUNT OF OXYGEN THAT CAN COMBINE WITH THE HEMOGLOBIN OF THE BLOOD. The blood of a normal person contains about 15 grams of

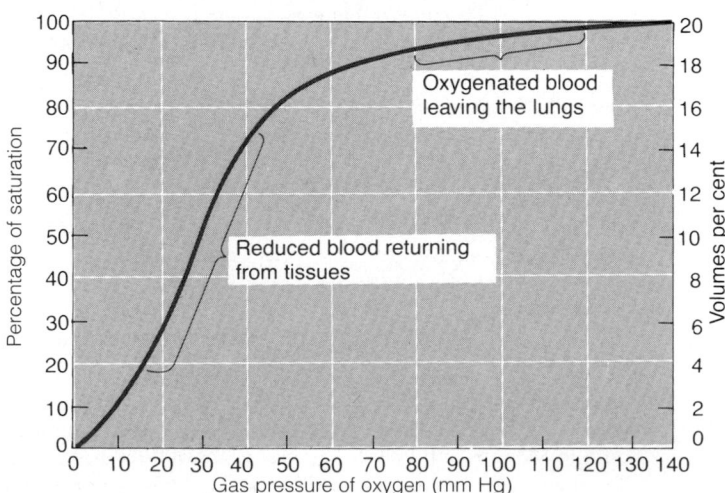

Figure 40–8. Oxygen-hemoglobin dissociation curve.

hemoglobin in each 100 milliliters of blood, and each gram of hemoglobin can bind with a maximum of about 1.34 milliliters of oxygen (1.39 milliliters when the hemoglobin is chemically pure, but this is reduced by impurities such as methemoglobin). Therefore, on the average, the hemoglobin in 100 milliliters of blood can combine with a total of almost exactly 20 milliliters of oxygen when the hemoglobin is 100 per cent saturated. This is usually expressed as 20 *volumes per cent.* The oxygen-hemoglobin dissociation curve for the normal person can also be expressed in terms of volume per cent of oxygen, as shown by the right-most scale in Figure 40–8, instead of per cent saturation of hemoglobin.

AMOUNT OF OXYGEN RELEASED FROM THE HEMO-GLOBIN IN THE TISSUES. The total quantity of oxygen *bound with hemoglobin* in normal arterial blood, which is 97 per cent saturated, is about 19.4 milliliters per 100 milliliters of blood. This is shown in Figure 40–9. On passing through the tissue capillaries, this amount is reduced, on the average, to 14.4 milliliters (Po₂ of 40 mm Hg, 75 per cent saturated hemoglobin). Thus, *under normal conditions, about 5 milliliters of oxygen are transported to the tissues by each 100 milliliters of blood.*

TRANSPORT OF OXYGEN DURING STRENUOUS EXER-CISE. In heavy exercise, the muscle cells use oxygen at a rapid rate, which, in extreme cases, can cause the interstitial fluid Po₂ to fall to as low as 15 mm Hg. At this pressure, only 4.4 milliliters of oxygen remains bound with the hemoglobin in each 100 milliliters of blood, as also shown in Figure 40–9. Thus, 19.4 − 4.4, or 15 milliliters, is the quantity of oxygen then transported by each 100 milliliters of blood. Thus, three times as much oxygen is transported in each volume of blood that passes through the tissues as normally. And when one remembers that the cardiac output can increase to six to seven times normal in well-trained marathon runners, multiplying these two gives a 20-fold increase in oxygen transport to the tissues: this is about the limit that can be achieved. (We see later in the chapter that several other factors facilitate the transport of oxygen in exercise, so that the muscle Po₂ often falls little below normal.)

UTILIZATION COEFFICIENT. The percentage of the blood that gives up its oxygen as it passes through the tissue capillaries is called the *utilization coefficient.* The normal value for this is about 25 per cent, evident from the preceding discussion. During strenuous exercise, the utilization coefficient in the entire body can increase to 75 to 85 per cent. And in local tissue areas where the blood flow is extremely slow or the metabolic rate high, utilization coefficients approaching 100 percent have been recorded—that is, essentially all the oxygen is removed.

Effect of Hemoglobin to "Buffer" the Tissue Oxygen Po₂

Although hemoglobin is necessary for transport of oxygen to the tissues, it performs another major function essential to life. This is its function as a "tissue oxygen buffer" system. That is, the hemoglobin in the blood is mainly responsible for stabilizing the oxygen pressure in the tissues. This can be explained as follows.

ROLE OF HEMOGLOBIN IN MAINTAINING CONSTANT Po₂ IN THE TISSUES. Under basal conditions, the tissues require about 5 milliliters of oxygen from each deciliter of blood passing through the tissue capillaries. Referring again to the oxygen-hemoglobin dissociation curve in Figure 40–9, one can see that for the 5 milliliters of oxygen to be released, the Po₂ must fall to about 40 mm Hg. Therefore, the tissue Po₂ normally will not rise above this 40-mm Hg level because if such should occur, the oxygen needed by the tissues will not be released from the hemoglobin. In this way, the hemoglobin normally sets an upper limit on the gas pressure in the tissues at about 40 mm Hg.

On the other hand, in heavy exercise, extra large amounts of oxygen (as much as 20 times normal!) must be delivered from the hemoglobin to the tissues. But this can be achieved with little further decrease in tissue Po₂—down to a level of 15 to 25 mm Hg—because of the steep slope of the dissociation curve and because of the increase in tissue blood flow caused by the decreased Po₂; that is, a small fall in Po₂ causes large amounts of oxygen to be released.

It can be seen, then, that the hemoglobin in the blood automatically delivers oxygen to the tissues at a pressure that is held rather tightly between about 15 and 40 mm Hg.

WHEN ATMOSPHERIC OXYGEN CONCENTRATION CHANGES MARKEDLY, THE BUFFER EFFECT OF HE-MOGLOBIN STILL MAINTAINS ALMOST CONSTANT TIS-SUE Po₂. The normal Po₂ in the alveoli is about 104 mm Hg, but as one ascends a mountain or ascends in an airplane, the Po₂ can easily fall to less than one-half this amount. Or when one enters areas of compressed air, such as deep in the sea or in pressurized chambers, the Po₂ may rise to 10 times this level. Even so, the tissue Po₂ changes little. Let us explain.

It can be seen from the oxygen-hemoglobin dissociation curve in Figure 40–8 that when the alveolar Po₂ is decreased to as low as 60 mm Hg, the arterial he-

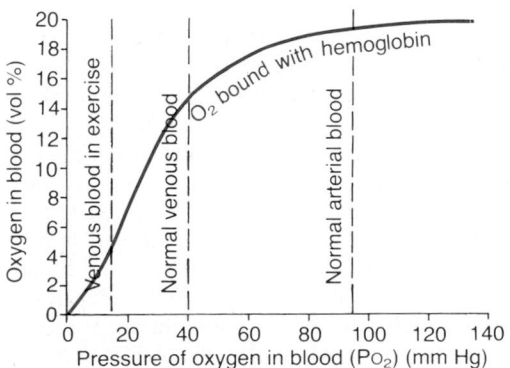

Figure 40–9. Effect of blood Po₂ on the quantity of oxygen bound with hemoglobin in each 100 milliliters of blood.

moglobin is still 89 per cent saturated, only 8 per cent below the normal saturation of 97 per cent. Furthermore, the tissues still remove about 5 milliliters of oxygen from each deciliter of blood passing through the tissues; to remove this oxygen, the P_{O_2} of the venous blood falls to 35 mm Hg, only 5 mm Hg below the normal value. Thus, the tissue P_{O_2} hardly changes, despite the marked fall in alveolar P_{O_2} from 104 to 60 mm Hg.

On the other hand, when the alveolar P_{O_2} rises as high as 500 mm Hg, the maximum oxygen saturation of hemoglobin can never rise above 100 per cent, which is only 3 per cent above the normal level of 97 per cent. Also, a small amount of additional oxygen dissolves in the fluid of the blood, as will be discussed subsequently. Then, when the blood passes through the tissue capillaries, it still loses several milliliters of oxygen to the tissues, which automatically reduces the P_{O_2} of the capillary blood to a value only a few millimeters greater than the normal 40 mm Hg.

Consequently, the level of alveolar oxygen may vary greatly—from 60 to more than 500 mm Hg P_{O_2}—and still the P_{O_2} in the tissue does not vary more than a few millimeters from normal, demonstrating beautifully the tissue oxygen buffer function of the blood hemoglobin.

Factors That Shift the Oxygen-Hemoglobin Dissociation Curve—Their Importance for Oxygen Transport

The oxygen-hemoglobin dissociation curves of Figures 40–8 and 40–9 are for normal, average blood. However, a number of factors can displace the dissociation curve in one direction or the other in the manner shown in Figure 40–10. This figure shows that when the blood becomes slightly acidic, with the pH decreasing from the normal value of 7.4 to 7.2, the oxygen-hemoglobin dissociation curve shifts, on the average, about 15 per cent to the right. On the other

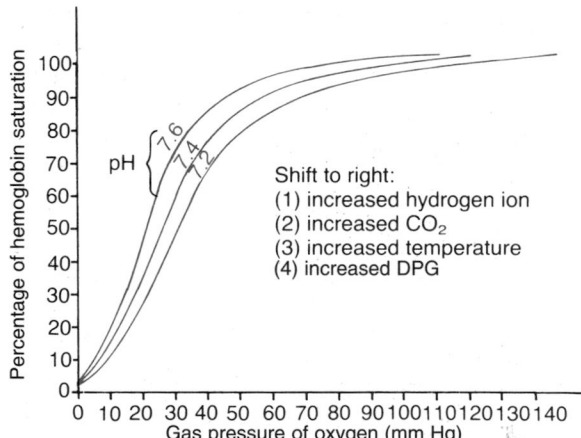

Figure 40–10. Shift of the oxygen-hemoglobin dissociation curve to the right by increases in (1) hydrogen ions, (2) CO_2, (3) temperature, or (4) 2,3-diphosphoglycerate (DPG).

hand, an increase in the pH to 7.6 shifts the curve a similar amount to the left.

In addition to pH changes, several other factors are known to shift the curve. Three of these, all of which shift the curve to the *right,* are (1) increased carbon dioxide concentration, (2) increased blood temperature, and (3) increased 2,3-diphosphoglycerate (DPG), a phosphate compound normally present in the blood but in different concentrations under different conditions.

A condition that shifts the dissociation curve to the *left* is the presence in the blood of large quantities of *fetal hemoglobin,* a type of hemoglobin normally present in the fetus before birth that is different from normal hemoglobin called *adult hemoglobin.* A shift of the curve to the left when fetal hemoglobin is present causes increased oxygen release to the fetal tissues under the hypoxic conditions in which the fetus exists. This is discussed further in Chapter 82.

INCREASED DELIVERY OF OXYGEN TO THE TISSUES WHEN CARBON DIOXIDE AND HYDROGEN IONS SHIFT THE OXYGEN-HEMOGLOBIN DISSOCIATION CURVE— THE BOHR EFFECT. A shift of the oxygen-hemoglobin dissociation curve in response to changes in the blood carbon dioxide and hydrogen ions has a significant effect in enhancing oxygenation of the blood in the lungs and then again in enhancing release of oxygen from the blood in the tissues. This is called the *Bohr effect,* and it can be explained as follows: As the blood passes through the lungs, carbon dioxide diffuses from the blood into the alveoli. This reduces the blood P_{CO_2} and decreases the hydrogen ion concentration because of the resulting decrease in blood carbonic acid. Both these effects shift the oxygen-hemoglobin dissociation curve to the left and upward, as shown in Figure 40–10. Therefore, the quantity of oxygen that binds with the hemoglobin at any given alveolar P_{O_2} becomes considerably increased, thus allowing greater oxygen transport to the tissues. Then when the blood reaches the tissue capillaries, exactly the opposite effects occur. Carbon dioxide entering the blood from the tissues shifts the curve rightward, which displaces oxygen from the hemoglobin and therefore delivers oxygen to the tissues at a higher P_{O_2} than would otherwise occur.

EFFECT OF DPG. The normal DPG in the blood keeps the oxygen-hemoglobin dissociation curve shifted slightly to the right all the time. In addition, in hypoxic conditions that last longer than a few hours, the quantity of DPG in the blood increases considerably, thus shifting the oxygen-hemoglobin dissociation curve even further to the right. This causes oxygen to be released to the tissues at as much as 10 mm Hg higher oxygen pressure than would be the case without this increased DPG. Therefore, under some conditions, this can be an important mechanism for adaptation to hypoxia, especially to hypoxia caused by poor tissue blood flow. However, the presence of the excess DPG also makes it more difficult for the hemoglobin to combine with oxygen in the lungs when the alveolar P_{O_2} is reduced, thereby sometimes causing as much harm as good.

Therefore, the DPG shift of the dissociation curve is beneficial in some conditions but not so in others.

SHIFT OF THE DISSOCIATION CURVE DURING EXERCISE. In exercise, several factors shift the dissociation curve considerably to the right. The exercising muscles release large quantities of carbon dioxide; this, plus several acids released by the exercising muscle, increases the hydrogen ion concentration in the muscle capillary blood. In addition, the temperature of the muscle often rises 2° to 3°C, which can increase the Po_2 level for release of oxygen into the muscle as much as 15 mm Hg. All these factors act together to shift the oxygen-hemoglobin dissociation curve *of the muscle capillary blood* considerably to the right. This right-hand shift of the curve sometimes allows oxygen to be released to the muscle at Po_2s as great as 40 mm Hg (equal to the normal resting value) even when 75 to 85 per cent of the oxygen is being removed from the hemoglobin. Then, in the lungs, the shift occurs in the opposite direction, thus allowing pickup of extra amounts of oxygen from the alveoli.

Metabolic Use of Oxygen by the Cells

EFFECT OF INTRACELLULAR Po_2 ON RATE OF OXYGEN USAGE. Only a minute level of oxygen pressure is required in the cells for normal intracellular chemical reactions to take place. The reason for this is that the respiratory enzyme systems of the cell, which are discussed in Chapter 67, are geared so that when the cellular Po_2 is more than 1 mm Hg, oxygen availability is no longer a limiting factor in the rates of the chemical reactions. Instead, the main limiting factor then is the *concentration of adenosine diphosphate* (ADP) in the cells. This effect is demonstrated in Figure 40–11, which shows the relation between intracellular Po_2 and rate of oxygen usage at different concentrations of

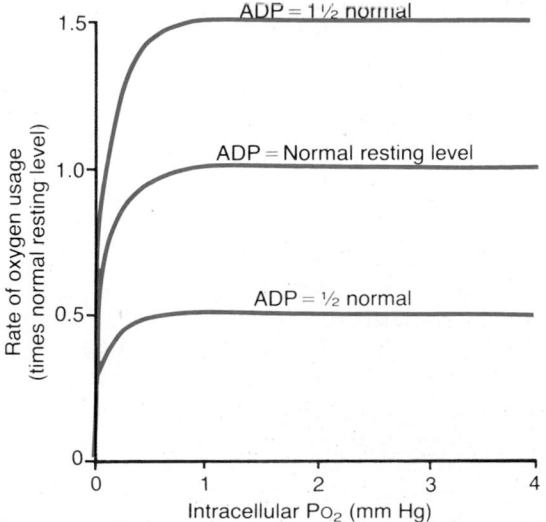

Figure 40–11. Effect of intracellular Po_2 on rate of oxygen usage by the cells. Note that increasing the intracellular concentration of adenosine diphosphate (ADP) increases the rate of oxygen usage.

ADP. Note that whenever the intracellular Po_2 is above 1 mm Hg, the rate of oxygen usage becomes constant for any given concentration of ADP in the cell. On the other hand, when the ADP concentration is altered, the rate of oxygen usage changes in proportion to the change in ADP concentration.

As explained in Chapter 3, when adenosine triphosphate (ATP) is used in the cells to provide energy, it is converted into ADP. The increasing concentration of ADP in turn increases the metabolic usage of both oxygen and the various nutrients that combine with the oxygen to release energy. This energy is used to reconvert the ADP to ATP. Therefore, *under normal operating conditions, the rate of oxygen usage by the cells is controlled ultimately by the rate of energy expenditure within the cells—that is, by the rate at which ADP is formed from ATP.* Only in abnormal hypoxic states does the availability of oxygen become a limiting condition.

EFFECT OF DIFFUSION DISTANCE FROM THE CAPILLARY TO THE CELL ON OXYGEN USAGE. Cells are seldom more than 50 micrometers away from a capillary, and oxygen normally can diffuse readily enough from the capillary to the cell to supply all the required amounts of oxygen for metabolism. However, occasionally, cells are located farther than this from the capillaries, and the rate of oxygen diffusion to these cells is so low that the intracellular Po_2 falls below the critical level of 1 mm Hg required to maintain maximum intracellular metabolism. Thus, under these conditions, oxygen usage by the cells is said to be *diffusion limited* and is no longer determined by the amount of ADP formed in the cells. This almost never occurs except in pathological states.

EFFECT OF BLOOD FLOW ON METABOLIC USE OF OXYGEN. The total amount of oxygen available each minute for use in any given tissue is determined by (1) the quantity of oxygen transported in each deciliter of blood and (2) the rate of blood flow. If the rate of blood flow falls to zero, the amount of available oxygen also falls to zero. Thus, there are times when the rate of blood flow through a tissue can be so low that the tissue Po_2 falls below the critical 1 mm Hg required for maximal intracellular metabolism. Under these conditions, the rate of tissue usage of oxygen is *blood flow limited.* Neither diffusion-limited nor blood flow–limited oxygen states can continue for long because the cells then usually receive less oxygen than is required for continuation of the life of the cells themselves.

Transport of Oxygen in the Dissolved State

At the normal arterial Po_2 of 95 mm Hg, about 0.29 milliliter of oxygen is dissolved in every deciliter of water in the blood. Then when the Po_2 of the blood falls to 40 mm Hg in the tissue capillaries, only 0.12 milliliter of oxygen remains dissolved. In other words, 0.17 milliliter of oxygen is normally transported in the dissolved state to the tissues by each deciliter of blood. This compares with almost 5.0 milliliters transported by

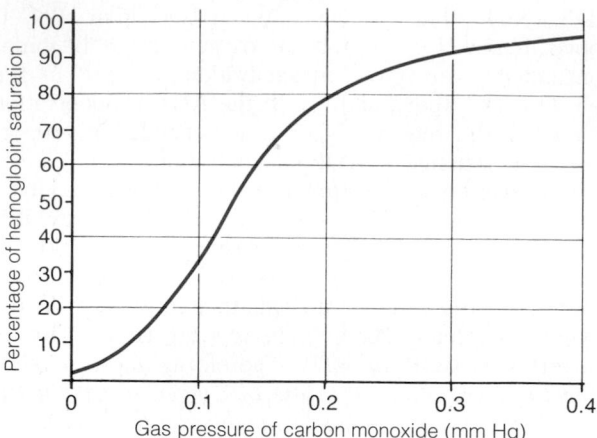

Figure 40–12. Carbon monoxide–hemoglobin dissociation curve. Note the extremely low carbon monoxide pressures at which carbon monoxide combines with hemoglobin.

the hemoglobin. Therefore, the amount of oxygen transported to the tissues in the dissolved state is normally slight, only about 3 per cent of the total, as compared with 97 per cent transported by the hemoglobin. During strenuous exercise, when hemoglobin release of oxygen to the tissues increases another threefold, the relative quantity then transported in the dissolved state falls to as little as 1.5 per cent. If a person breathes oxygen at very, very high alveolar P_{O_2}s, the amount then transported in the dissolved state can become much greater, sometimes so much so that serious excesses of oxygen occur in the tissues and "oxygen poisoning" ensues. This often leads to seizures and even death, as discussed in detail in Chapter 44 in relation to high-pressure breathing, as occurs in deep-sea divers.

Combination of Hemoglobin with Carbon Monoxide—Displacement of Oxygen

Carbon monoxide combines with hemoglobin at the same point on the hemoglobin molecule as does oxygen and, therefore, can displace oxygen from the hemoglobin. Furthermore, it binds with about 250 times as much tenacity as oxygen, which is demonstrated by the carbon monoxide–hemoglobin dissociation curve in Figure 40–12. This curve is almost identical with the oxygen-hemoglobin dissociation curve, except that the pressures of the carbon monoxide, shown on the abscissa, are at a level 1/250 of those in the oxygen-hemoglobin dissociation curve of Figure 40–8. Therefore, a carbon monoxide pressure of only 0.4 mm Hg in the alveoli, 1/250 that of the alveolar oxygen (100 mm Hg P_{O_2}), allows the carbon monoxide to compete equally with the oxygen for combination with the hemoglobin and causes one-half the hemoglobin in the blood to become bound with carbon monoxide instead of with oxygen. Therefore, a carbon monoxide pressure only a little greater than 0.4 mm Hg (about 0.6 mm Hg, or a concentration of about 0.1 per cent in the air) can be lethal.

A patient severely poisoned with carbon monoxide can be advantageously treated by administering pure oxygen because oxygen at high alveolar pressures displaces carbon monoxide from its combination with hemoglobin far more rapidly than can oxygen at the low pressure of normal 20 per cent atmospheric oxygen.

The patient can also be benefited by simultaneous administration of a few per cent carbon dioxide because this strongly stimulates the respiratory center. This increases alveolar ventilation and reduces the alveolar carbon monoxide concentration, which allows increased carbon monoxide release from the blood.

With intensive oxygen and carbon dioxide therapy, carbon monoxide can be removed from the blood perhaps 10 or more times as rapidly as without therapy.

TRANSPORT OF CARBON DIOXIDE IN THE BLOOD

Transport of carbon dioxide by the blood is not nearly so great a problem as transport of oxygen because even in the most abnormal conditions, carbon dioxide can usually be transported in far greater quantities than can oxygen. However, the amount of carbon dioxide in the blood does have much to do with acid-base balance of the body fluids, which is discussed in Chapter 30.

Under normal resting conditions, *an average of 4 milliliters of carbon dioxide is transported from the tissues to the lungs in each deciliter of blood.*

Chemical Forms in Which Carbon Dioxide Is Transported

To begin the process of carbon dioxide transport, carbon dioxide diffuses out of the tissue cells in the dissolved molecular carbon dioxide form (but not to any significant extent in the bicarbonate form because the cell membrane is almost impermeable to bicarbonate ions). On entering the capillary, the carbon dioxide initiates a host of almost instantaneous physical and chemical reactions, shown in Figure 40–13, that are essential for carbon dioxide transport.

TRANSPORT OF CARBON DIOXIDE IN THE DISSOLVED STATE. A small portion of the carbon dioxide is transported in the dissolved state to the lungs. It will

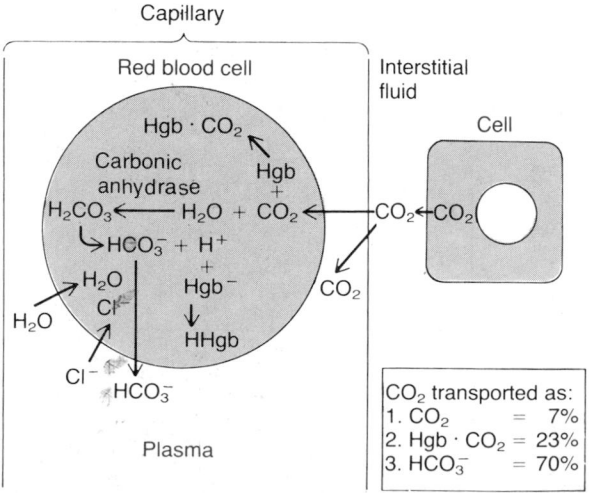

Figure 40–13. Transport of carbon dioxide in the blood.

be recalled that the Pco_2 of venous blood is 45 mm Hg and that of arterial blood is 40 mm Hg. The amount of carbon dioxide dissolved in the fluid of the blood at 45 mm Hg is about 2.7 ml/dl (2.7 volumes per cent). The amount dissolved at 40 mm Hg is about 2.4 milliliters, or a difference of 0.3 milliliter. Therefore, only about 0.3 milliliter of carbon dioxide is transported in the form of dissolved carbon dioxide by each deciliter of blood. This is about 7 per cent of all the carbon dioxide transported.

TRANSPORT OF CARBON DIOXIDE IN THE FORM OF BICARBONATE ION

Reaction of Carbon Dioxide with Water in the Red Blood Cells—Effect of Carbonic Anhydrase.
The dissolved carbon dioxide in the blood reacts with water to form carbonic acid. This reaction would occur much too slowly to be of importance were it not for the fact that inside the red blood cells is an enzyme called *carbonic anhydrase*, which catalyzes the reaction between carbon dioxide and water, accelerating its rate about 5000-fold. Instead of requiring many seconds or minutes to occur, as is true in the plasma, the reaction occurs so rapidly in the red blood cells that it reaches almost complete equilibrium within a fraction of a second. This allows tremendous amounts of carbon dioxide to react with the red cell water even before the blood leaves the tissue capillaries.

Dissociation of the Carbonic Acid into Bicarbonate and Hydrogen Ions.
In another fraction of a second, the carbonic acid formed in the red cells dissociates into *hydrogen* and *bicarbonate ions*. Most of the hydrogen ions then combine with the hemoglobin in the red blood cells because the hemoglobin protein is a powerful acid-base buffer. In turn, many of the bicarbonate ions diffuse from the red cells into the plasma while chloride ions diffuse into the red cells to take their place. This is made possible by the presence of a special *bicarbonate-chloride carrier protein* in the red cell membrane that shuttles these two ions in opposite directions at rapid velocities. Thus, the chloride content of venous red blood cells is greater than that of arterial red cells, a phenomenon called the *chloride shift*.

The reversible combination of carbon dioxide with water in the red blood cells under the influence of carbonic anhydrase accounts for about 70 per cent of the carbon dioxide transported from the tissues to the lungs. Thus, this means of transporting carbon dioxide is by far the most important of all the methods for transport. Indeed, when a carbonic anhydrase inhibitor (acetazolamide) is administered to an animal to block the action of carbonic anhydrase in the red blood cells, carbon dioxide transport from the tissues becomes poor—so poor that the tissue Pco_2 can be made to rise to 80 mm Hg instead of the normal 45 mm Hg.

TRANSPORT OF CARBON DIOXIDE IN COMBINATION WITH HEMOGLOBIN AND PLASMA PROTEINS—CARBAMINOHEMOGLOBIN.
In addition to reacting with water, carbon dioxide reacts directly with amine radicals of the hemoglobin molecules to form the compound *car-*

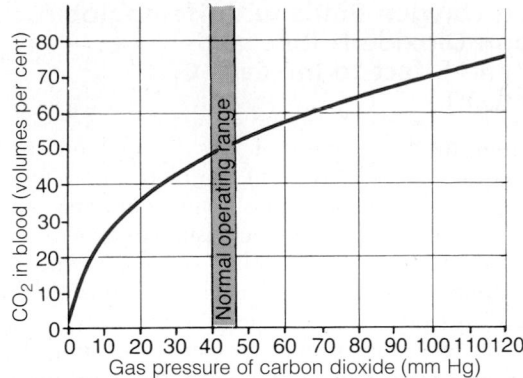

Figure 40–14. Carbon dioxide dissociation curve.

baminohemoglobin (CO_2Hgb). This combination of carbon dioxide with the hemoglobin is a reversible reaction that occurs with a loose bond, so that the carbon dioxide is easily released into the alveoli where the Pco_2 is lower than in the tissue capillaries. A small amount of carbon dioxide also reacts in the same way with the plasma proteins, but this is much less significant because the quantity of these proteins is only one fourth as great as the quantity of hemoglobin.

The *theoretical* quantity of carbon dioxide that can be carried from the tissues to the lungs in the carbamino combination with hemoglobin and plasma proteins is about 30 per cent of the total quantity transported—that is, normally about 1.5 milliliters of carbon dioxide in each deciliter of blood. However, this reaction is much slower than the reaction of carbon dioxide with water inside the red blood cells. It is doubtful that this mechanism provides transport of more than 20 per cent of the total quantity of carbon dioxide.

Carbon Dioxide Dissociation Curve

It is apparent that carbon dioxide can exist in the blood in many forms: (1) as free carbon dioxide and (2) in chemical combinations with water, hemoglobin, and plasma protein. The total quantity of carbon dioxide combined with the blood in all these forms depends on the Pco_2. The curve shown in Figure 40–14 depicts this dependence of total blood carbon dioxide in all its forms on Pco_2; this curve is called the *carbon dioxide dissociation curve*.

Note that the normal blood Pco_2 ranges between the limits of 40 mm Hg in arterial blood and 45 mm Hg in venous blood, which is a narrow range. Note also that the normal concentration of carbon dioxide in the blood in all its different forms is about 50 volumes per cent but that only 4 volumes per cent of this is exchanged during normal transport of carbon dioxide from the tissues to the lungs. That is, the concentration rises to about 52 volumes per cent as the blood passes through the tissues and falls to about 48 volumes per cent as it passes through the lungs.

When Oxygen Binds with Hemoglobin, Carbon Dioxide Is Released—The Haldane Effect to Increase CO₂ Transport

Earlier in the chapter, it was pointed out that an increase in carbon dioxide in the blood will cause oxygen to be displaced from the hemoglobin and that this is an important factor in increasing oxygen transport. The reverse is also true: binding of oxygen with hemoglobin tends to displace carbon dioxide from the blood. Indeed, this effect, called the *Haldane effect,* is quantitatively far more important in promoting carbon dioxide transport than is the Bohr effect in promoting oxygen transport.

The Haldane effect results from the simple fact that combination of oxygen with hemoglobin in the lungs causes the hemoglobin to become a stronger acid. This in turn displaces carbon dioxide from the blood and into the alveoli in two ways: (1) The more highly acidic hemoglobin has less tendency to combine with carbon dioxide to form carbaminohemoglobin, thus displacing much of the carbon dioxide that is present in the carbamino form from the blood. (2) The increased acidity of the hemoglobin also causes it to release an excess of hydrogen ions, and these in turn bind with bicarbonate ions to form carbonic acid; then this dissociates into water and carbon dioxide, and the carbon dioxide is released from the blood into the alveoli.

In the tissue capillaries, the Haldane effect causes increased pickup of carbon dioxide because of oxygen removal from the hemoglobin, and in the lungs, it causes increased release of carbon dioxide because of oxygen pickup by the hemoglobin.

Figure 40–15 demonstrates quantitatively the significance of the Haldane effect on the transport of carbon dioxide from the tissues to the lungs. This figure shows small portions of two carbon dioxide dissociation curves, the solid curve when the P_{O_2} is 100 mm Hg, which is the case in the lungs, and the dashed curve

when the P_{O_2} is 40 mm Hg, which is the case in the tissue capillaries. Point A on the dashed curve shows that the normal P_{CO_2} of 45 mm Hg in the tissues causes 52 volumes per cent of carbon dioxide to combine with the blood. On entering the lungs, the P_{CO_2} falls to 40 mm Hg while the P_{O_2} rises to 100 mm Hg. If the carbon dioxide dissociation curve did not shift because of the Haldane effect, the carbon dioxide content of the blood would fall only to 50 volumes per cent, which would be a loss of only 2 volumes per cent of carbon dioxide. However, the increase in P_{O_2} in the lungs decreases the carbon dioxide dissociation curve from the dashed curve to the solid curve of the figure, so that the carbon dioxide content falls to 48 volumes per cent (point B). This represents an additional 2 volume per cent loss of carbon dioxide. Thus, the Haldane effect approximately doubles the amount of carbon dioxide released from the blood in the lungs and approximately doubles the pickup of carbon dioxide in the tissues.

Change in Blood Acidity During Carbon Dioxide Transport

The carbonic acid formed when carbon dioxide enters the blood in the tissues decreases the blood pH. However, the reaction of this acid with the buffers of the blood prevents the hydrogen ion concentration from rising greatly (and the pH from falling greatly). Ordinarily, arterial blood has a pH of about 7.41, and as the blood acquires carbon dioxide in the tissue capillaries, the pH falls to a venous value of about 7.37. In other words, a pH change of 0.04 unit takes place. The reverse occurs when carbon dioxide is released from the blood in the lungs, the pH rising to the arterial value once again. In exercise or other conditions of high metabolic activity or when the blood flow through the tissues is sluggish, the decrease in pH in the tissue blood (and in the tissues themselves) can be as much as 0.50, thus causing severe tissue acidosis.

RESPIRATORY EXCHANGE RATIO

The discerning student will have noted that normal transport of oxygen from the lungs to the tissues by each deciliter of blood is about 5 milliliters, whereas normal transport of carbon dioxide from the tissues to the lungs is about 4 milliliters. Thus, under normal resting conditions, only about 82 per cent as much carbon dioxide is expired from the lungs as there is oxygen uptake by the lungs. The ratio of carbon dioxide output to oxygen uptake is called the *respiratory exchange ratio* (R). That is,

$$R = \frac{\text{Rate of carbon dioxide output}}{\text{Rate of oxygen uptake}}$$

The value for R changes under different metabolic conditions. When a person is using exclusively carbohydrates for body metabolism, R rises to 1.00. On the other hand, when the person is using exclusively fats for metabolic energy, the R level falls to as low as 0.7. The reason for this difference is that when oxygen is metab-

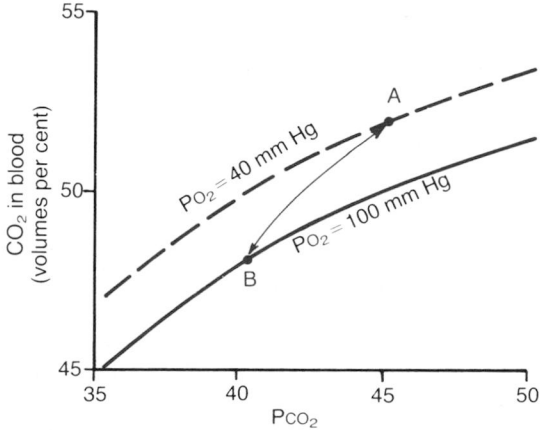

Figure 40–15. Portions of the carbon dioxide dissociation curves when the P_{O_2} is 100 mm Hg and 40 mm Hg, respectively. The arrow represents the Haldane effect on the transport of carbon dioxide, as discussed in the text.

olized with carbohydrates, one molecule of carbon dioxide is formed for each molecule of oxygen consumed; whereas when oxygen reacts with fats, a large share of the oxygen combines with hydrogen atoms from the fats to form water instead of carbon dioxide. In other words, the *respiratory quotient of the chemical reactions* in the tissues when fats are metabolized is about 0.70 instead of 1.00. The tissue respiratory quotient is discussed in Chapter 71.

For a person on a normal diet consuming average amounts of carbohydrates, fats, and proteins, the average value for R is considered to be 0.825.

REFERENCES

Arieff, A. I.: Hypoxia, Metabolic Acidosis and the Circulation. New York, Oxford University Press, 1992.

Bauer, C., et al. (eds.): Biophysics and Physiology of Carbon Dioxide. New York, Springer-Verlag, 1980.

Cherniack, N. S., and Logobardo, G. S.: Oxygen and carbon dioxide gas stores of the body. Physiol. Rev., 50:196, 1970.

Crowell, J. W., and Smith, E. E.: Determinants of the optimal hematocrit. J. Appl. Physiol., 22:501, 1967.

Forster, R. E.: Pulmonary ventilation and blood gas exchange. In Sodeman, W. A., Jr., and Sodeman, W. A. (eds.): Pathologic Physiology: Mechanisms of Disease, 5th Ed. Philadelphia, W. B. Saunders Co., 1974, p. 371.

Gaar, K. A., Jr.: Oxygen transport: A simple model for study and examination. Physiologist, 28:412, 1985.

George, R. B., et al.: Chest Medicine. Baltimore, Williams & Wilkins, 1995.

Gonzalez, N. C., and Fedde, M. R.: Oxygen Transfer from Atmosphere to Tissues. New York, Plenum Publishing Corp., 1988.

Grippi, M. A.: Pulmonary Pathophysiology. Philadelphia, J. B. Lippincott, 1994.

Grodins, F. S., and Yamashiro, S. M.: Optimization of mammalian respiratory gas transport system. Annu. Rev. Biophys. Bioeng., 4:115, 1973.

Haldane, J. S., and Priestley, J. G.: Respiration. New Haven, Yale University Press, 1935.

Jones, C. E., et al.: Determination of mean tissue oxygen tensions by implanted perforated capsules. J. Appl. Physiol., 26:630, 1969.

Jones, N. L.: Blood Gases and Acid-Base Physiology. New York, B. C. Decker, 1980.

Kilmartin, J. V., and Rossi-Bernardi, L.: Interaction of hemoglobin with hydrogen ions, carbon dioxide, and organic phosphates. Physiol. Rev., 53:836, 1973.

Konigsberg, W.: Protein structure and molecular dysfunction: Hemoglobin. In Bondy, P. K., and Rosenberg, L. E. (eds.): Metabolic Control and Disease, 8th Ed. Philadelphia, W. B. Saunders Co., 1980, p. 27.

Lane, E. E., and Walker, J. F.: Clinical Arterial Blood Gas Analysis. St. Louis, C. V. Mosby Co., 1987.

Leff, A. R., and Schumacker, P. T.: Respiratory Physiology: Basics and Applications. Philadelphia, W. B. Saunders Co., 1993.

Levitzky, M. G., and Hall, S. M. Cardiopulmonary Physiology in Anesthesiology. Hightstown, NJ, McGraw-Hill, 1994.

Malley, W. J.: Clinical Blood Gases: Application and Noninvasive Alternatives. Philadelphia, W. B. Saunders Co., 1990.

Michel, C. C.: The transport of oxygen and carbon dioxide by the blood. In MTP International Review of Science: Physiology. Vol. 2. Baltimore, University Park Press, 1974, p. 67.

Minoo, P., and King, R. J.: Epithelial-mesenchymal interactions in lung development. Annu. Rev. Physiol., 56:13, 1994.

Mochizuki, M., et al. (eds.): Oxygen Transport to Tissue X. New York, Plenum Publishing Corp., 1988.

Murray, J. F., and Nadel, J. A.: Textbook of Respiratory Medicine. Philadelphia, W. B. Saunders Co., 1994.

Nikinmaa, M.: Membrane transport and control of hemoglobin-oxygen affinity in nucleated erythrocytes. Physiol. Rev., 72:301, 1992.

Perutz, M. F.: Hemoglobin structure and respiratory transport. Sci. Am., 239(6):92, 1987.

Riggs, A. F.: The Bohr effect. Annu. Rev. Physiol., 50:181, 1988.

Stainsby, W. N., and Barclay, J. K.: Oxygen uptake by striated muscle. Muscle Biol., 1:273, 1972.

Tamoru, M., et al.: In vivo study of tissue oxygen metabolism using optical and nuclear magnetic resonance. Annu. Rev. Physiol., 51:813, 1989.

Wagner, P. D.: Diffusion and chemical reaction in pulmonary gas exchange. Physiol. Rev., 57:257, 1977.

Wagner, P. D.: The oxyhemoglobin dissociation curve and pulmonary gas exchange. Semin. Hematol., 11:405, 1974.

West, J. B.: Pulmonary Pathophysiology: The Essentials. Baltimore, Williams & Wilkins, 1992.

West, J. B.: Respiratory Physiology—The Essentials. Baltimore, Williams & Wilkins, 1994.

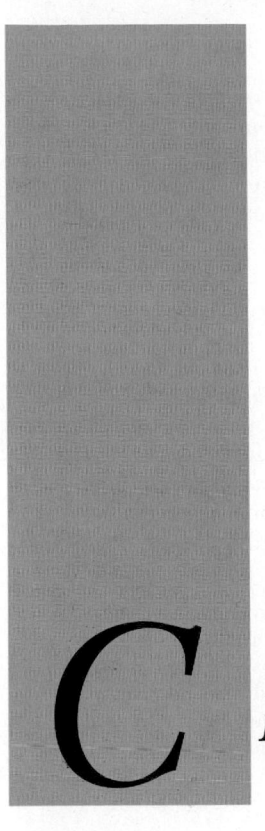

Regulation of Respiration

*C*HAPTER *41*

The nervous system normally adjusts the rate of alveolar ventilation almost exactly to the demands of the body so that the arterial blood oxygen pressure (Po_2) and carbon dioxide pressure (Pco_2) are hardly altered even during moderate to strenuous exercise and most other types of respiratory stress.

This chapter describes the operation of this neurogenic system for regulation of respiration.

RESPIRATORY CENTER

The "respiratory center" is composed of several groups of neurons located *bilaterally* in the medulla oblongata and pons, as shown for one side in Figure 41–1. It is divided into three major collections of neurons: (1) a *dorsal respiratory group*, located in the dorsal portion of the medulla, which mainly causes inspiration; (2) a *ventral respiratory group*, located in the ventrolateral part of the medulla, which can cause either expiration or inspiration, depending on which neurons in the group are stimulated; and (3) the *pneumotaxic center*, located dorsally in the superior portion of the pons, which helps control the rate and pattern of breathing. The dorsal respiratory group of neurons plays the most fundamental role in the control of respiration. Therefore, let us discuss its function first.

Dorsal Respiratory Group of Neurons—Its Inspiratory and Rhythmical Functions

The dorsal respiratory group of neurons extends most of the length of the medulla. All or most of its neurons are located within the *nucleus of the tractus solitarius*, although additional neurons in the adjacent reticular substance of the medulla probably also play important roles in respiratory control. The nucleus of the tractus solitarius is also the sensory termination of both the vagal and the glossopharyngeal nerves, which transmit sensory signals into the respiratory center from the peripheral chemoreceptors, the baroreceptors, and several types of receptors in the lung. All the signals from these peripheral areas help in the control of respiration, as we discuss in subsequent sections of this chapter.

RHYTHMICAL INSPIRATORY DISCHARGES FROM THE DORSAL RESPIRATORY GROUP. The basic rhythm of respiration is generated mainly in the dorsal respiratory group of neurons. Even when all the peripheral nerves entering the medulla are sectioned and the brain stem is transected both above and below the medulla, this group of neurons still emits repetitive bursts of *inspiratory* action potentials. The basic cause of these repetitive discharges, however, is unknown. In primitive animals, neural networks have been found in which activity of one set of neurons excites a second set, which in turn inhibits the first. Then after a period of time the mechanism repeats itself, continuing throughout the life of the animal. Therefore, most respiratory physiologists believe that some similar network of neurons located entirely within the medulla, probably involving not only the dorsal respiratory group but adjacent areas of the medulla as well, is responsible for the basic rhythm of respiration.

INSPIRATORY "RAMP" SIGNAL. The nervous signal that is transmitted to the primary inspiratory muscles such as the diaphragm is not an instantaneous burst of action potentials. Instead, in normal respiration, it

525

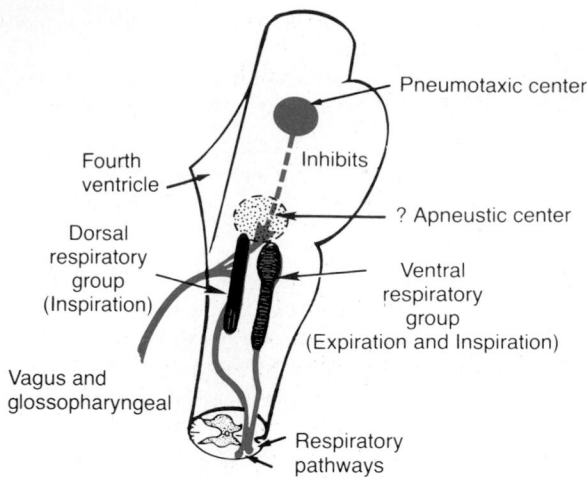

Figure 41–1. Organization of the respiratory center.

begins weakly and increases steadily in a ramp manner for about 2 seconds. It abruptly ceases for approximately the next 3 seconds, which turns off the excitation of the diaphragm and allows the elastic recoil of the chest wall and lungs to cause expiration. Then the inspiratory signal begins again for still another cycle, and again and again, with expiration occurring in between. Thus, the inspiratory signal is a *ramp signal.* The obvious advantage of this is that it causes a steady increase in the volume of the lungs during inspiration, rather than inspiratory gasps.

There are two ways in which the inspiratory ramp is controlled. They are as follows:

1. Control of the rate of increase of the ramp signal, so that during active respiration the ramp increases rapidly and therefore fills the lungs rapidly as well

2. Control of the limiting point at which the ramp suddenly ceases. This is the usual method for controlling the rate of respiration; that is, the earlier the ramp ceases, the shorter the duration of inspiration. For reasons not understood, this also shortens the duration of expiration. Thus, the frequency of respiration is increased.

Pneumotaxic Center Limits the Duration of Inspiration and Increases the Respiratory Rate

The pneumotaxic center, located dorsally in the *nucleus parabrachialis* of the upper pons, transmits signals to the inspiratory area. The primary effect of these is to control the "switch-off" point of the inspiratory ramp, thus controlling the duration of the filling phase of the lung cycle. When the pneumotaxic signal is strong, inspiration might last for as little as 0.5 second, thus filling the lungs only slightly; but when the pneumotaxic signals are weak, inspiration might continue for 5 or more seconds, thus filling the lungs with a great excess of air.

Therefore, the function of the pneumotaxic center is primarily to limit inspiration. This has a secondary effect of increasing the rate of breathing because limita-

tion of inspiration also shortens expiration and the entire period of respiration. A strong pneumotaxic signal can increase the rate of breathing to 30 to 40 breaths per minute, whereas a weak pneumotaxic signal may reduce the rate to only a few breaths per minute.

Ventral Respiratory Group of Neurons Functions in Both Inspiration and Expiration

Located about 5 millimeters anterior and lateral to the dorsal respiratory group of neurons is the ventral respiratory group of neurons, found in the *nucleus ambiguus* rostrally and the *nucleus retroambiguus* caudally. The function of this neuronal group differs from that of the dorsal respiratory group in several important ways.

1. The neurons of the ventral respiratory group remain almost totally *inactive* during normal quiet respiration. Therefore, normal quiet breathing is caused only by repetitive inspiratory signals from the dorsal respiratory group transmitted mainly to the diaphragm, and expiration results from elastic recoil of the lungs and thoracic cage.

2. There is no evidence that the ventral respiratory neurons participate in the basic rhythmical oscillation that controls respiration.

3. When the respiratory drive for increased pulmonary ventilation becomes greater than normal, respiratory signals spill over into the ventral respiratory neurons from the basic oscillating mechanism of the dorsal respiratory area. As a consequence, the ventral respiratory area does then contribute its share to the respiratory drive as well.

4. Electrical stimulation of some of the neurons in the ventral group causes inspiration, whereas stimulation of others causes expiration. Therefore, these neurons contribute to both inspiration and expiration. They are especially important in providing the powerful expiratory signals to the abdominal muscles during expiration. Thus, this area operates more or less as an overdrive mechanism when high levels of pulmonary ventilation are required.

Possibility of an "Apneustic Center" in the Lower Pons

To add confusion to our knowledge about respiratory center function, there is another strange center in the lower part of the pons called the *apneustic center.* Its function can be demonstrated only when the vagus nerves to the medulla have been sectioned and when the connections from the pneumotaxic center have been blocked by transecting the pons in its midregion. Then the apneustic center of the lower pons sends signals to the dorsal respiratory group of neurons that prevent or retard the "switch-off" of the inspiratory ramp signal. Therefore, the lungs become almost completely filled with air, and only occasional short expiratory gasps occur.

The function of the apneustic center is not understood, but it presumably operates in association with the pneumotaxic center to control the depth of inspiration.

Lung Inflation Signals Limit Inspiration—
The Hering-Breuer Inflation Reflex

In addition to the neural mechanisms operating entirely within the brain stem, reflex nerve signals from the lungs help to control respiration. Most important, located in the muscular portions of the walls of the bronchi and bronchioles throughout the lungs are *stretch receptors* that transmit signals through the *vagi* into the dorsal respiratory group of neurons when the lungs become overstretched. These signals affect inspiration in much the same way as signals from the pneumotaxic center; that is, when the lungs become overly inflated, the stretch receptors activate an appropriate feedback response that "switches off" the inspiratory ramp and thus stops further inspiration. This is called the *Hering-Breuer inflation reflex*. This reflex also increases the rate of respiration, the same as is true for signals from the pneumotaxic center.

In human beings, the Hering-Breuer reflex probably is not activated until the tidal volume increases to greater than about 1.5 liters. Therefore, this reflex appears to be mainly a protective mechanism for preventing excess lung inflation rather than an important ingredient in the normal control of ventilation.

Control of Overall Respiratory Center Activity

Up to this point, we have discussed the basic mechanisms for causing inspiration and expiration, but it is also important to know how the intensity of the respiratory control signals is increased or decreased to match the ventilatory needs of the body. For example, during heavy exercise, the rates of oxygen usage and carbon dioxide formation are often increased to as much as 20 times normal, requiring commensurate increases in pulmonary ventilation.

The major purpose of the remainder of this chapter is to discuss this control of ventilation in response to the needs of the body.

CHEMICAL CONTROL OF RESPIRATION

The ultimate goal of respiration is to maintain proper concentrations of oxygen, carbon dioxide, and hydrogen ions in the tissues. It is fortunate, therefore, that respiratory activity is highly responsive to changes in each of these.

Excess carbon dioxide or hydrogen ions mainly stimulate the respiratory center itself, causing greatly increased strength of both the inspiratory and the expiratory signals to the respiratory muscles.

Oxygen, on the other hand, does not have a significant *direct* effect on the respiratory center of the brain in controlling respiration. Instead, it acts almost entirely on peripheral chemoreceptors located in the carotid and aortic bodies, and these in turn transmit

appropriate nervous signals to the respiratory center for control of respiration.

Let us discuss first the stimulation of the respiratory center itself by carbon dioxide and hydrogen ions.

Direct Chemical Control of Respiratory Center Activity by Carbon Dioxide and Hydrogen Ions

CHEMOSENSITIVE AREA OF THE RESPIRATORY CENTER. We have discussed mainly three areas of the respiratory center: the dorsal respiratory group of neurons, the ventral respiratory group, and the pneumotaxic center. It is believed that none of these are affected directly by changes in blood carbon dioxide concentration or hydrogen ion concentration. Instead, an additional neuronal area, a sensitive *chemosensitive area*, shown in Figure 41–2, is located bilaterally lying only one-fifth millimeter beneath the ventral surface of the medulla. This area is highly sensitive to changes in either blood P_{CO_2} or hydrogen ion concentration, and it in turn excites the other portions of the respiratory center.

Response of the Chemosensitive Neurons to Hydrogen Ions—Probably the Primary Stimulus

The sensor neurons in the chemosensitive area are especially excited by hydrogen ions; in fact, it is believed that hydrogen ions are perhaps the only important direct stimulus for these neurons. However, hydrogen ions do not easily cross the blood-brain barrier or the blood–cerebrospinal fluid barrier. For this reason, changes in hydrogen ion concentration in the blood have considerably less effect in stimulating the chemosensitive neurons than do changes in carbon

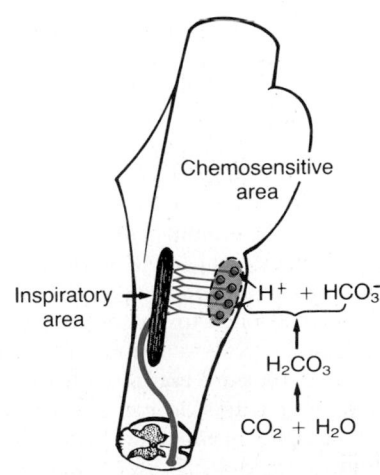

Figure 41–2. Stimulation of the inspiratory area by the *chemosensitive area* located bilaterally in the medulla, lying only a fraction of a millimeter beneath the ventral medullary surface. Note also that hydrogen ions stimulate the chemosensitive area, whereas it is carbon dioxide in the fluid that gives rise to most of the hydrogen ions.

dioxide, even though carbon dioxide is believed to stimulate these neurons secondarily by changing the hydrogen ion concentration, as explained below.

Effect of Blood Carbon Dioxide on Stimulating the Chemosensitive Area

Although carbon dioxide has little direct effect in stimulating the neurons in the chemosensitive area, it does have a potent indirect effect. It does this by reacting with the water of the tissues to form carbonic acid. This in turn dissociates into hydrogen and bicarbonate ions; the hydrogen ions then have a potent direct stimulatory effect. These reactions are shown in Figure 41–2.

Why is it that blood carbon dioxide has a more potent effect in stimulating the chemosensitive neurons than do blood hydrogen ions? The answer is that the blood-brain barrier and the blood–cerebrospinal fluid barrier are both almost completely impermeable to hydrogen ions, whereas carbon dioxide passes through both these barriers almost as if they did not exist. Consequently, whenever the blood P_{CO_2} increases, so also does the P_{CO_2} of both the interstitial fluid of the medulla and the cerebrospinal fluid. In both these fluids, the carbon dioxide immediately reacts with the water to form hydrogen ions. Thus, paradoxically, more hydrogen ions are released into the respiratory chemosensitive sensory area when the blood carbon dioxide concentration increases than when the blood hydrogen ion concentration increases. For this reason, respiratory center activity is affected considerably more by changes in blood carbon dioxide than by changes in blood hydrogen ions, a fact that we subsequently discuss quantitatively.

IMPORTANCE OF CEREBROSPINAL FLUID P_{CO_2} IN STIMULATING THE CHEMORECEPTIVE AREA. Changing the P_{CO_2} in the cerebrospinal fluid that bathes the surface of the brain stem chemoreceptive area excites respiration in the same way that increased P_{CO_2} in the medullary interstitial fluids excites respiration. And, even more important, the excitation by way of the cerebrospinal fluid is several times as rapid. The reason for this is believed to be that the cerebrospinal fluid has little protein acid-base buffer. Therefore, the hydrogen ion concentration increases almost instantly when carbon dioxide enters the cerebrospinal fluid from the extensive arachnoid blood vessels. By contrast, the brain tissues have much more protein buffer, so that the change there in hydrogen ion concentration in response to carbon dioxide is delayed. Consequently, the initial rapid excitation of the respiratory system by carbon dioxide entering the cerebrospinal fluid occurs within seconds, in comparison with 1 minute or more for the stimulation through the brain interstitial fluid.

DECREASED STIMULATORY EFFECT OF CARBON DIOXIDE AFTER THE FIRST 1 TO 2 DAYS. The excitation of the respiratory center by carbon dioxide is great the first few hours but then gradually declines over the next 1 to 2 days, decreasing to about one-fifth

the initial effect. Part of this decline results from renal readjustment of the hydrogen ion concentration back toward normal after the carbon dioxide first increases the hydrogen concentration. The kidneys achieve this by increasing the blood bicarbonate. This binds with the hydrogen ions in the blood and cerebrospinal fluid to reduce their concentration. But even more important, over a period of hours, the bicarbonate ions also slowly diffuse through the blood-brain and blood–cerebrospinal fluid barriers and combine directly with the hydrogen ions around the respiratory neurons as well, thus reducing the hydrogen ions back to normal or near to normal.

Therefore, a change in blood carbon dioxide concentration has a potent *acute* effect on controlling respiratory drive but only a weak *chronic* effect after a few days' adaptation.

Quantitative Effects of Blood P_{CO_2} and Hydrogen Ion Concentration on Alveolar Ventilation

Figure 41–3 shows quantitatively the approximate effects of blood P_{CO_2} and blood pH (which is an inverse logarithmic measure of hydrogen ion concentration) on alveolar ventilation. Note the marked increase in ventilation caused by the increase in P_{CO_2}. But note also the much smaller effect on ventilation of in-

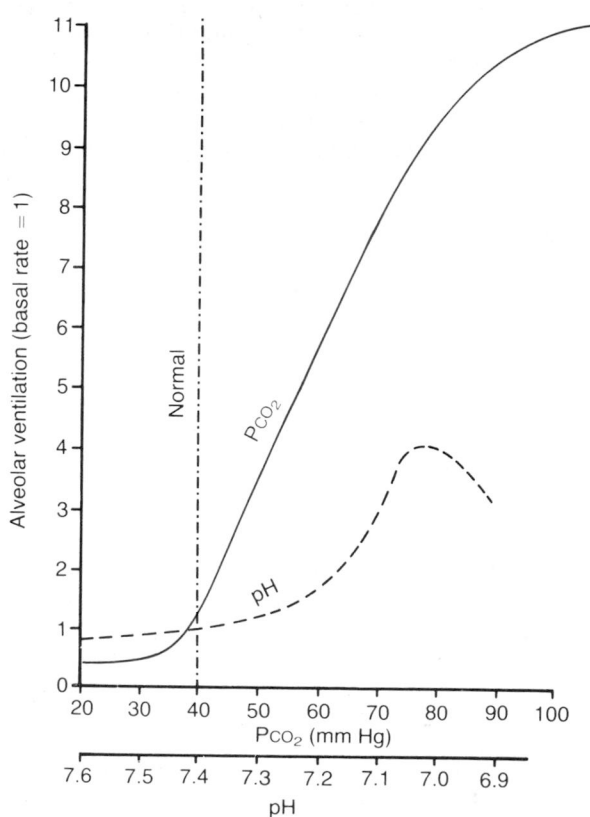

Figure 41–3. Effects of increased arterial P_{CO_2} and decreased arterial pH (increased hydrogen ion concentration) on the rate of alveolar ventilation.

creased hydrogen ion concentration (that is, decreased pH).

Finally, note the great change in alveolar ventilation in the normal *blood* P_{CO_2} range between 35 and 60 mm Hg. This demonstrates the tremendous effect that carbon dioxide changes have in controlling respiration. By contrast, the change in respiration in the normal *blood* pH range between 7.3 and 7.5 is more than 10 times less.

Unimportance of Oxygen for Direct Control of the Respiratory Center

Changes in oxygen concentration have virtually no *direct* effect on the respiratory center itself to alter respiratory drive (although it does have an indirect effect, acting through the peripheral chemoreceptors, as explained in the next section).

We learned in Chapter 40 that the hemoglobin oxygen buffer system delivers almost exactly normal amounts of oxygen to the tissues even when the pulmonary P_{O_2} changes from a value as low as 60 mm Hg up to as high as 1000 mm Hg. Therefore, except under special conditions, proper delivery of oxygen can occur despite changes in lung ventilation ranging from slightly below one-half normal to as high as 20 or more times normal. On the other hand, this is not true for carbon dioxide because both the blood and tissue P_{CO_2} change almost exactly inversely with the rate of pulmonary ventilation; thus, evolution has made carbon dioxide the major controller of respiration, not oxygen.

Yet for those special conditions in which the tissues do get into trouble for lack of oxygen, the body has a special mechanism for respiratory control located in the peripheral chemoreceptors, outside the brain respiratory center; this mechanism responds when the blood oxygen falls too low, mainly below a P_{O_2} of 60 to 70 mm Hg, as explained in the next section.

PERIPHERAL CHEMORECEPTOR SYSTEM FOR CONTROL OF RESPIRATORY ACTIVITY—ROLE OF OXYGEN IN RESPIRATORY CONTROL

In addition to control of respiratory activity by the respiratory center itself, still another mechanism is available for controlling respiration. This is the *peripheral chemoreceptor system,* shown in Figure 41–4. Special nervous chemical receptors, called *chemoreceptors,* are located in several areas outside the brain, and they are especially important for detecting changes in oxygen in the blood, although they also respond to changes in carbon dioxide and hydrogen ion concentrations. The chemoreceptors in turn transmit nervous signals to the respiratory center in the brain to help regulate respiratory activity.

By far the largest number of chemoreceptors are located in the *carotid bodies.* However, a sizable number are in the *aortic bodies,* also shown in Figure

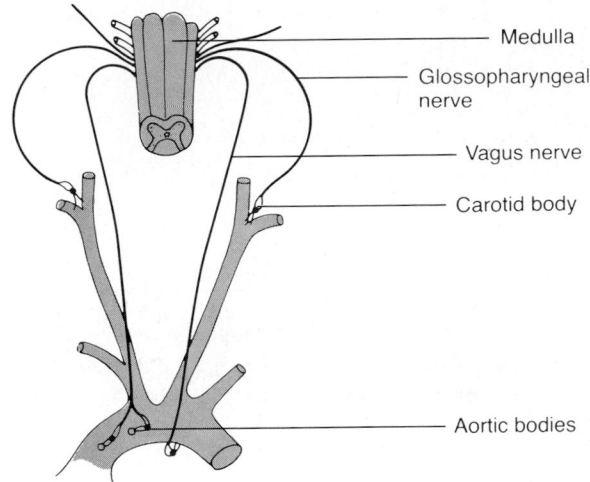

Figure 41–4. Respiratory control by the carotid and aortic bodies.

41–4; a few are located elsewhere in association with other arteries of the thoracic and abdominal regions of the body. The *carotid bodies* are located bilaterally in the bifurcations of the common carotid arteries, and their afferent nerve fibers pass through Hering's nerves to the *glossopharyngeal nerves* and then to the dorsal respiratory area of the medulla. The *aortic bodies* are located along the arch of the aorta; their afferent nerve fibers pass through the *vagi* also to the dorsal respiratory area. Each of these chemoreceptor bodies receives a special blood supply through a minute artery directly from the adjacent arterial trunk. Furthermore, the blood flow through these bodies is extreme, 20 times the weight of the bodies themselves each minute. Therefore, the percentage removal of oxygen from the flowing blood is virtually zero. This means that *the chemoreceptors are exposed at all times to arterial blood,* not venous blood, and their P_{O_2}s are arterial P_{O_2}s.

STIMULATION OF THE CHEMORECEPTORS BY DECREASED ARTERIAL OXYGEN. Changes in arterial oxygen concentration have *no* direct stimulatory effect on the respiratory center itself, but when the oxygen concentration in the arterial blood falls below normal, the chemoreceptors become strongly stimulated. This is demonstrated in Figure 41–5, which shows the effect of different levels of *arterial* P_{O_2} on the rate of nerve impulse transmission from a carotid body. Note that the impulse rate is particularly sensitive to changes in arterial P_{O_2} in the range between 60 and 30 mm Hg, the range in which the arterial hemoglobin saturation with oxygen decreases rapidly.

EFFECT OF CARBON DIOXIDE AND HYDROGEN ION CONCENTRATION ON CHEMORECEPTOR ACTIVITY. An increase in either carbon dioxide concentration or hydrogen ion concentration also excites the chemoreceptors and in this way indirectly increases respiratory activity. However, the direct effects of both these factors in the respiratory center itself are so much more

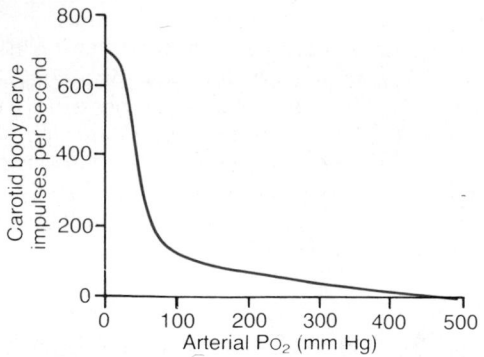

Figure 41–5. Effect of arterial PO_2 on impulse rate from the carotid body of a cat. (Curve drawn from data from several sources, but primarily from Von Euler.)

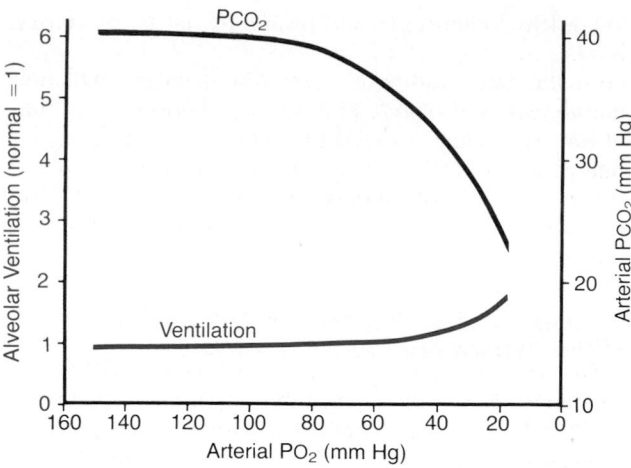

Figure 41–6. The lower curve shows the effect of arterial PO_2 on alveolar ventilation when the ventilation was allowed to decrease the arterial PCO_2. Note the far less stimulation of ventilation by low PO_2 than in Figure 41–7 because the fall in PCO_2 had a depressant effect on ventilation. (Drawn from data in Gray: Pulmonary Ventilation and Its Physiological Regulation. Springfield, Ill., Charles C Thomas.)

powerful than their effects mediated through the chemoreceptors (about seven times as powerful) that for most practical purposes, these indirect effects through the chemoreceptors do not need to be considered. Yet there is one difference between the peripheral and central effects of carbon dioxide: the peripheral stimulation of the chemoreceptors occurs as much as five times as rapidly as central stimulation, so that the peripheral chemoreceptors might increase the rapidity of response to carbon dioxide at the onset of exercise.

Basic Mechanism of Stimulation of the Chemoreceptors by Oxygen Deficiency. The exact means by which low PO_2 excites the nerve endings in the carotid and aortic bodies is still unknown. However, these bodies have multiple highly characteristic glandular-like cells, called *glomus cells*, that synapse directly or indirectly with the nerve endings. For this reason, some investigators have suggested that these cells might function as the chemoreceptors and then in turn stimulate the nerve endings. However, other studies suggest that the nerve endings themselves are directly sensitive to the low PO_2.

Quantitative Effect of Low Arterial PO_2 on Alveolar Ventilation

When a person breathes air that has too little oxygen, this decreases the blood PO_2 and excites the carotid and aortic chemoreceptors, thereby increasing respiration. The effect usually is much less than one would expect because the increased respiration removes carbon dioxide from the lungs and thereby decreases both blood PCO_2 and hydrogen ion concentration. These two changes then severely depress the respiratory center, as discussed earlier, so that the final effect of the chemoreceptors in increasing respiration in response to low PO_2 is mostly counteracted. This is clearly evident in the experimental results demonstrated in Figure 41–6, showing a marked decrease in blood PCO_2 and only a slight rise in pulmonary ventilation in response to a fivefold decrease in arterial blood PO_2.

The effect of low arterial PO_2 on alveolar ventilation

is far greater under some other conditions, two of which are (1) when arterial carbon dioxide and hydrogen ion concentrations remain normal despite increased respiration and (2) breathing of oxygen at low concentrations for many days.

Effect of Low Arterial PO_2 When Arterial Carbon Dioxide and Hydrogen Ion Concentrations Remain Normal

Figure 41–7 shows the effect of low arterial PO_2 on alveolar ventilation when the PCO_2 and the hydrogen

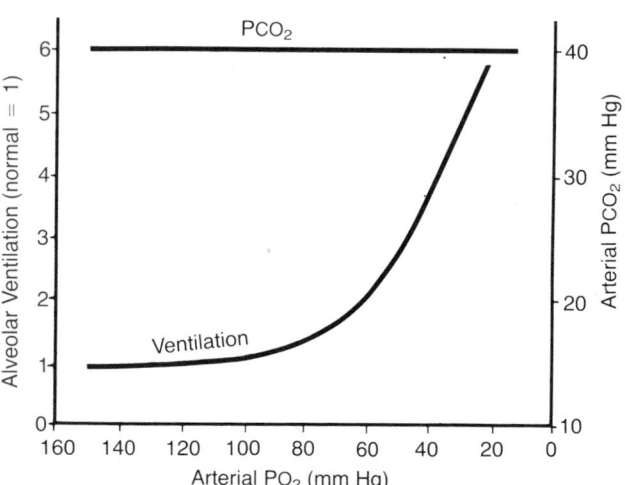

Figure 41–7. The lower curve demonstrates the effect of different levels of arterial PO_2 on alveolar ventilation, showing a six-fold increase in ventilation as the PO_2 falls from the normal level of 100 mm Hg down to 20 mm Hg. The upper curve shows that the arterial PCO_2 was kept at a constant level during the measurements of this study; pH was also kept constant.

ion concentration remain constant at their normal levels. In other words, the ventilatory drive due to both the carbon dioxide and the hydrogen ions does not change; only the ventilatory drive due to the effect of low oxygen on the chemoreceptors is present. The figure shows almost no effect on ventilation as long as the arterial P_{O_2} remains greater than 100 mm Hg. At pressures lower than 100 mm Hg, ventilation approximately doubles when the arterial P_{O_2} falls to 60 mm Hg and can increase as much as fivefold at very low P_{O_2}s. Under these conditions, low arterial P_{O_2} *can* drive the ventilatory process quite strongly.

CONDITIONS IN WHICH VENTILATORY DRIVE BY LOW ARTERIAL P_{O_2} IS NOT BLOCKED BY DECREASES IN P_{CO_2} AND HYDROGEN IONS. In pneumonia, emphysema, or almost any other pulmonary condition that prevents adequate exchange of gases through the pulmonary membrane, too little oxygen is absorbed into the arterial blood, and at the same time the arterial P_{CO_2} and hydrogen ion concentration usually remain near normal or at times are even increased because of poor transport of carbon dioxide through the membrane. In these instances, the ventilatory drive of the low P_{O_2} is not blocked by changes in blood P_{CO_2} and hydrogen ions. The low oxygen is then important in helping to increase respiration. In fact, if one is given a high concentration of oxygen to breathe, one loses the stimulatory drive from the low arterial P_{O_2} and pulmonary ventilation often decreases enough to cause death because of excessive buildup of lethal P_{CO_2} and hydrogen ion concentration.

Also, in some instances of extremely heavy exercise, especially in people with slightly compromised pulmonary systems, the arterial P_{CO_2} and hydrogen ion concentration rise at the same time that the P_{O_2} falls. In such conditions, the decreasing arterial P_{O_2} combines with the increasing carbon dioxide and increasing hydrogen ions to give extremely strong ventilatory drive.

Chronic Breathing of Low Oxygen Stimulates Respiration Many Times As Much—The Phenomenon of Acclimatization

Mountain climbers have found that when they ascend a mountain slowly over a period of days, rather than a period of hours, they can withstand far lower atmospheric oxygen concentrations than when they ascend rapidly. This is called *acclimatization* to the low oxygen. The reason for this is that the respiratory center in the brain stem loses within 2 to 3 days about four fifths of its sensitivity to changes in arterial P_{CO_2} and hydrogen ions. Therefore, the blow-off of carbon dioxide that normally would inhibit respiration now fails to do so, and the low oxygen now can drive the respiratory system to a much higher level of alveolar ventilation than under acute low-oxygen conditions. Instead of a 70 per cent increase in ventilation that might occur on acute exposure to low oxygen, the alveolar ventilation often increases 400 to 500 per cent after 2 to 3 days of low oxygen; this helps immensely

in supplying additional oxygen to the climber. To give a practical example, even experienced mountain climbers can get into difficulty for lack of oxygen if they ascend within 1 day to an altitude of only 18,000 to 20,000 feet. Yet Mount Everest, at an altitude of more than 29,000 feet, has been climbed to its peak without supplemental oxygen; the ascent was made in slow stages that allowed complete acclimatization of the respiratory drive to low P_{O_2}.

Composite Effects of P_{CO_2}, pH, and P_{O_2} on Alveolar Ventilation

Figure 41–8 gives a quick overview of the manner in which the chemical factors P_{O_2}, P_{CO_2}, and pH—all together—affect alveolar ventilation. To understand this diagram, first observe the four solid curves. These curves were recorded at different levels of arterial P_{O_2}: 40 mm Hg, 50 mm Hg, 60 mm Hg, and 100 mm Hg. For each of these curves, the P_{CO_2} was changed from low to high levels. These curves represent the effect of P_{CO_2} on alveolar ventilation. Thus, this "family" of solid curves represents the combined effects of alveolar P_{CO_2} and P_{O_2} on ventilation.

Now observe the dashed curves. The solid curves were measured at a blood pH of 7.4; the dashed curves, at a pH of 7.3. We now have two families of curves representing the combined effects of P_{CO_2} and P_{O_2} on ventilation at two pH values. Still different families of curves would be displaced to the right at higher pHs and displaced to the left at lower pHs.

Thus, using this diagram, one can predict the level of alveolar ventilation for most combinations of alveolar P_{CO_2}, alveolar P_{O_2}, and arterial pH.

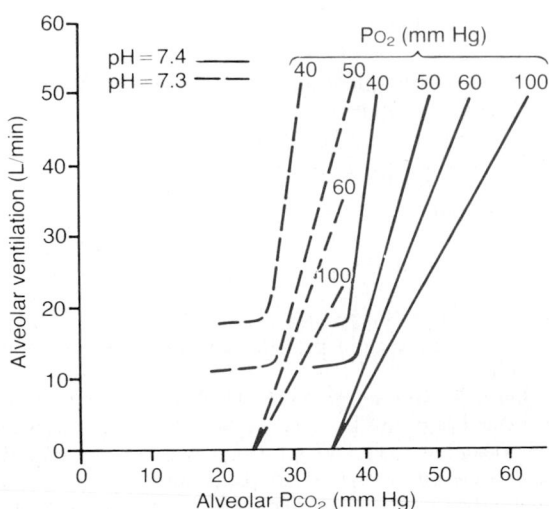

Figure 41–8. Composite diagram showing the interrelated effects of P_{CO_2}, P_{O_2}, and pH on alveolar ventilation. (Drawn from data presented in Cunningham and Lloyd: The Regulation of Human Respiration. Philadelphia, F. A. Davis Company.)

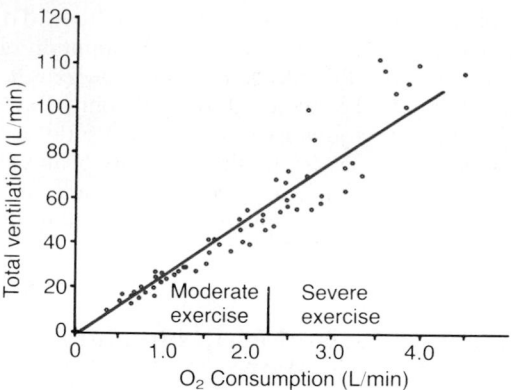

Figure 41–9. Effect of exercise on oxygen consumption and ventilatory rate. (From J. S. Gray: Pulmonary Ventilation and Its Physiological Regulation. Springfield, Ill., Charles C Thomas, 1950.)

REGULATION OF RESPIRATION DURING EXERCISE

In strenuous exercise, oxygen consumption and carbon dioxide formation can increase as much as 20-fold. Yet, in the healthy athlete, alveolar ventilation ordinarily increases almost exactly in step with the increased level of metabolism, as demonstrated by the relation between oxygen consumption and ventilation shown in Figure 41–9. Therefore, the arterial P_{O_2}, P_{CO_2}, and pH remain *almost exactly normal*.

In trying to analyze the factors that cause increased ventilation during exercise, one is tempted immediately to ascribe this to the chemical alterations in the body fluids during exercise, including increase of carbon dioxide, increase of hydrogen ions, and decrease of oxygen. However, this is questionable because measurements of arterial P_{CO_2}, pH, and P_{O_2} show that none of these usually changes significantly, so that none of them becomes abnormal enough to stimulate the respiration.

Therefore, the question must be asked: What is it during exercise that causes the intense ventilation? This question has not been answered fully, but at least two effects seem to be predominantly concerned.

1. The brain, on transmitting impulses to the contracting muscles, is believed to transmit collateral impulses into the brain stem to excite the respiratory center. This is analogous to the stimulatory effect of the higher centers of the brain on the vasomotor center of the brain stem during exercise, causing a rise in arterial pressure as well as an increase in ventilation.

2. During exercise, the body movements, especially of the arms and legs, are believed to increase pulmonary ventilation by exciting joint and muscle proprioceptors that then transmit excitatory impulses to the respiratory center. The reason for believing this is that even passive movements of the arms and legs often increase pulmonary ventilation severalfold, but this will not occur if the sensory nerves from the arms and legs have been blocked.

Still other factors also may be important in increas-

ing pulmonary ventilation during exercise. For instance, some experiments even suggest that hypoxia developing in the muscles during exercise elicits afferent nerve signals to the respiratory center to excite respiration. Also, because the exercising muscles form tremendous amounts of carbon dioxide and use tremendous amounts of oxygen, the P_{CO_2} and P_{O_2} change markedly between the inspiratory cycle of respiration and the expiratory cycle. Some experiments suggest that these wide *variations* in the blood gases stimulate respiration, even though the *mean* values remain almost exactly normal. However, because a large share of the total increase in ventilation begins immediately on the initiation of exercise before the blood chemicals have time to change, most of the increase in respiration probably results from the two neurogenic factors noted above, *stimulatory impulses from the higher centers of the brain and proprioceptive stimulatory reflexes*.

INTERRELATION BETWEEN CHEMICAL FACTORS AND NERVOUS FACTORS IN THE CONTROL OF RESPIRATION DURING EXERCISE. When a person exercises, usually the nervous factors stimulate the respiratory center almost exactly the proper amount to supply the extra oxygen requirements for the exercise and to blow off the extra carbon dioxide. Occasionally, however, the nervous signals are either too strong or too weak in their stimulation of the respiratory center. Then the chemical factors play a significant role in bringing about the final adjustment in respiration required to keep the carbon dioxide and hydrogen ion concentrations of the body fluids as nearly normal as possible. This effect is demonstrated in Figure 41–10, which shows by the lower curve the changes in the alveolar ventilation during a 1-minute period of exercise and, by the upper curve, the changes in P_{CO_2}. Note that at the onset of the exercise, the alveolar ventilation increases instantaneously without an initial increase in

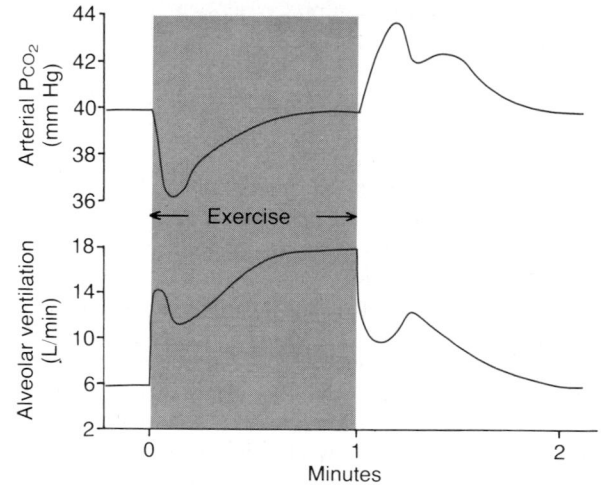

Figure 41–10. Changes in alveolar ventilation and arterial P_{CO_2} during a 1-minute period of exercise and also after termination of the exercise. (Extrapolated to the human being from data in dogs; from Bainton: *J. Appl. Physiol.*, 33:778, 1972.)

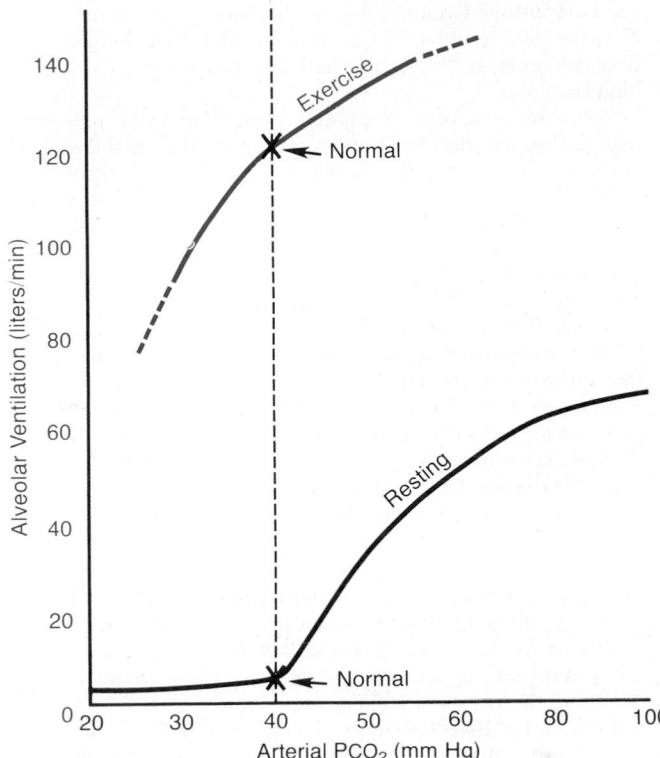

Figure 41–11. Approximate effect of maximum exercise in an athlete to shift the alveolar P_{CO_2}–ventilation response curve to a much higher than normal level. The shift, believed to be caused by neurogenic factors, is almost exactly the right amount to maintain the arterial P_{CO_2} at the normal level of 40 mm Hg both in the resting state and during heavy exercise.

arterial P_{CO_2}. In fact, this onset increase in ventilation is usually great enough that it actually *decreases* arterial P_{CO_2} below normal, as shown in the figure. The reason the ventilation forges ahead of the buildup of blood carbon dioxide is that the brain provides an "anticipatory" stimulation of respiration at the onset of exercise, causing extra alveolar ventilation even before this is needed. However, after about 30 to 40 seconds, the amount of carbon dioxide released into the blood from the active muscles then approximately matches the increased rate of ventilation, and the arterial P_{CO_2} returns essentially to normal, as shown toward the end of the 1-minute period of exercise in the figure.

Figure 41–11 summarizes the control of respiration in still another way, this time more quantitatively. The lower curve of this figure shows the effect of different levels of arterial P_{CO_2} on alveolar ventilation when the body is at rest—that is, not exercising. The upper curve shows the approximate shift of this ventilatory curve caused by the neurogenic drive to the respiratory center that occurs during heavy exercise. The crosses on the two curves show the arterial P_{CO_2}s first in the resting state and then in the exercising state. Note in both instances that the P_{CO_2} is at the normal level of 40 mm Hg. In other words, the neurogenic factor shifts the curve about 20-fold in the upward direction, so that the ventilation almost matches the rate of oxygen consumption and rate of carbon dioxide release, thus keeping the arterial P_{O_2} and P_{CO_2} near to their normal values.

The upper curve of Figure 41–11 also shows that if,

during exercise, the arterial P_{CO_2} does change from the normal value of 40 mm Hg, it has its usual stimulatory effect on ventilation at P_{CO_2}s greater than 40 and its usual depressant effect at P_{CO_2}s less than 40 mm Hg.

POSSIBILITY THAT THE NEUROGENIC FACTOR FOR CONTROL OF VENTILATION DURING EXERCISE IS A LEARNED RESPONSE. Many experiments suggest that the brain's ability to shift the ventilatory response curve during exercise, as shown in Figure 41–11, is at least partly a *learned* response. That is, with repeated exercise, the brain becomes progressively more able to provide the proper amount of brain signal required to keep the blood chemical factors at their normal levels. Also, there is much reason to believe that some of the higher learning centers of the brain are important in this learned neurogenic respiratory control factor—probably even the cerebral cortex. One important reason for believing this is that when the cerebral cortex is anesthetized, the unanesthetized respiratory control system in the brain stem loses its special ability to maintain the arterial blood gases near normal during induced muscle activity.

OTHER FACTORS THAT AFFECT RESPIRATION

VOLUNTARY CONTROL OF RESPIRATION. Thus far, we have discussed the involuntary system for control of respiration. However, we all know that respiration can be controlled voluntarily and that one can hyperventilate or hypoventilate to such an extent that serious derangements in P_{CO_2}, pH, and P_{O_2} can occur in the blood.

Voluntary control of respiration seems not to be mediated through the respiratory center of the medulla. Instead, the nervous pathway for voluntary control passes directly from the cortex and other higher centers downward through the corticospinal tract to the spinal neurons that drive the respiratory muscles.

EFFECT OF IRRITANT RECEPTORS IN THE AIRWAYS. The epithelium of the trachea, bronchi, and bronchioles is supplied with sensory nerve endings called *pulmonary irritant receptors* that are stimulated by some irritants that enter the respiratory airways. They cause coughing and sneezing, as discussed in Chapter 39. They possibly also cause bronchial constriction in such diseases as asthma and emphysema.

FUNCTION OF LUNG "J RECEPTORS." A few sensory nerve endings occur in the alveolar walls in *juxtaposition* to the pulmonary capillaries, from whence comes the name "J receptors." They are stimulated especially when the pulmonary capillaries become engorged with blood or when pulmonary edema occurs in such conditions as congestive heart failure. Although the functional role of the J receptors is not known, their excitation perhaps does give the person a feeling of dyspnea.

EFFECT OF BRAIN EDEMA. The activity of the respiratory center may be depressed or even inactivated by acute brain edema resulting from brain concussion. For instance, the head might be struck against some solid object, following which the damaged brain tissues swell, compressing the cerebral arteries against the cranial vault and thus totally or partially blocking the cerebral blood supply.

Occasionally, respiratory depression resulting from brain edema can be relieved temporarily by intravenous injection of hypertonic solutions such as highly concentrated mannitol solution. These solutions osmotically remove some of the fluids of the brain, thus relieving intracranial pressure and sometimes re-establishing respiration within a few minutes.

ANESTHESIA. Perhaps the most prevalent cause of respiratory depression and respiratory arrest is overdosage with anesthetics or narcotics. For instance, sodium pentobarbital is a poor anesthetic because it depresses the respiratory center considerably more than many other anesthetics, such as halothane. At one time, morphine was used as an anesthetic, but this drug is now used only as an adjunct to anesthetics because it greatly depresses the respiratory center while having much less ability to anesthetize the cerebral cortex.

Periodic Breathing

An abnormality of respiration called *periodic breathing* occurs in a number of disease conditions. The person breathes deeply for a short interval and then breathes slightly or not at all for an additional interval, the cycle repeating itself over and over again.

The most common type of periodic breathing, *Cheyne-Stokes breathing,* is characterized by slowly waxing and waning respiration, occurring over and over again about every 40 to 60 seconds.

BASIC MECHANISM OF CHEYNE-STOKES BREATHING. The basic cause of Cheyne-Stokes breathing is the following: When a person overbreathes, thus blowing off too much carbon dioxide from the pulmonary blood and also increasing the blood oxygen, it takes several seconds before the changed pulmonary blood can be transported to the brain and inhibit the excess ventilation. By this time, the person has already overventilated for an extra few seconds. Therefore, when the respiratory center does eventually respond, it becomes depressed too much because of the overventilation, and now the opposite cycle begins. That is, carbon dioxide builds up and oxygen decreases in the pulmonary blood. Then, again, it takes a few seconds before the brain can respond to the new changes. When the brain does respond, the person breathes hard once again. The cycle repeats itself again and again.

The basic cause of Cheyne-Stokes breathing is present in everyone. However, this mechanism is highly "damped." That is, the fluids of the blood and the respiratory center control areas have large amounts of stored and chemically bound carbon dioxide and oxygen. Normally, the lungs cannot build up enough extra carbon dioxide or depress the oxygen sufficiently in a few seconds to cause the next cycle of the periodic breathing. Yet under two separate conditions the damping factors are overridden, and Cheyne-Stokes breathing does occur.

1. When there is a long delay in the transport of the blood from the lungs to the brain, the gas changes in the blood continue for many more seconds than usual. Under these conditions, the storage capacities of the blood and tissues for the gases is exceeded; then the periodic respiratory drive becomes extreme, and Cheyne-Stokes breathing begins. This type of Cheyne-Stokes breathing often occurs in patients with severe cardiac failure because the left side of the heart is greatly enlarged and the blood flow is slow, thus delaying the transport of blood gases from the lungs to the brain. In fact, in patients with chronic heart failure, Cheyne-Stokes breathing is likely to occur on and off for months.

2. A second cause of Cheyne-Stokes breathing is increased negative feedback gain in the respiratory control areas. This means that a change in blood carbon dioxide or oxygen now causes far greater change in ventilation than normally. For instance, instead of the normal 2- to 3-fold increase in ventilation when the P_{CO_2} rises 3 mm Hg, the same 3 mm Hg might increase the ventilation 10- to 20-fold. The brain feedback tendency for periodic breathing is now strong enough to cause Cheyne-Stokes breathing without extra blood flow delay between the lungs and brain. This type of Cheyne-Stokes breathing occurs mainly in patients with brain damage. The brain damage often turns off the respiratory drive entirely for a few seconds; then an increase in blood carbon dioxide turns it back on with great force. Cheyne-Stokes breathing of this type is frequently a prelude to death.

Typical records of changes in pulmonary and respiratory center P_{CO_2}s during Cheyne-Stokes breathing are shown in Figure 41–12. Note that the P_{CO_2} of the pulmonary blood changes in advance of the P_{CO_2} of the

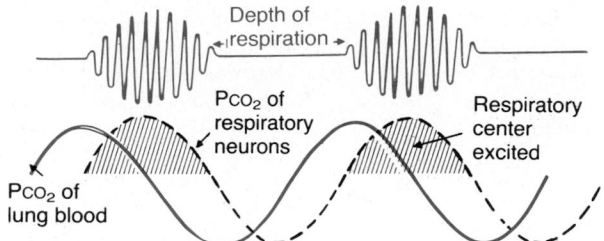

Figure 41-12. Cheyne-Stokes breathing, showing the changing P_{CO_2} in the pulmonary blood *(solid line)* and the delayed changes in P_{CO_2} of the fluids of the respiratory center *(dashed line)*.

respiratory neurons. And the depth of respiration corresponds with the P_{CO_2} in the brain, not with the P_{CO_2} in the pulmonary blood where the ventilation is occurring.

REFERENCES

Acker, H.: P_{O_2} chemoreception in arterial chemoreceptors. Annu. Rev. Physiol., 51:835, 1989.

Barnes, P. J.: Modulation of neurotransmission in airways. Physiol. Rev., 72:669, 1992.

Beckerman, R. C., et al.: Respiratory Control Disorders in Infants and Children. Baltimore, Williams & Wilkins, 1992.

Boynton, B. R., et al.: New Therapies for Neonatal Respiratory Failure: A Physiological Approach. New York, Cambridge University Press, 1994.

Bullock, B. L., and Rosendahl, P. P.: Pathophysiology: Adaptations and Alterations in Function. 3rd Ed. Philadelphia, J. B. Lippincott, 1992.

Cohen, M. I.: Central determinants of respiratory rhythm. Annu. Rev. Physiol., 43:91, 1981.

Coleridge, H. M., and Coleridge, J. C. G.: Pulmonary reflexes: neural mechanisms of pulmonary defense. Annu. Rev. Physiol., 56:69, 1994.

Dempsey, J. A., and Pack, A. I.: Regulation of Breathing. 2nd Ed. New York, Marcel Dekker, Inc., 1994.

Feldman, J. L., and Ellenberger, H. H.: Central coordination of respiratory and cardiovascular control in mammals. Annu. Rev. Physiol., 50:593, 1988.

Gonzalez, C., et al.: Carotid body chemoreceptors: From natural stimuli to sensory discharges. Physiol. Rev., 74:829, 1994.

Grippi, M. A.: Pulmonary Pathophysiology: A Problem-oriented Approach. Philadelphia, J. B. Lippincott, 1994.

Guyton, A. C., et al.: Basic oscillating mechanism of Cheyne-Stokes breathing. Am. J. Physiol., 187:395, 1956.

Haddad, G. G., and Mellins, R. B.: Hypoxia and respiratory control in early life. Annu. Rev. Physiol., 46:629, 1984.

Honig, A.: Salt and water metabolism in acute high-altitude hypoxia: Role of peripheral arterial chemoreceptors. News Physiol. Sci., 4:109, 1989.

Jansen, A. H., and Cherniack, V.: Development of respiratory control. Physiol. Rev., 63:437, 1983.

Karczewski, W. A., et al.: Control of Breathing During Sleep and Anesthesia. New York, Plenum Publishing Corp., 1988.

Lahiri, S., et al.: Response and Adaptation to Hypoxia. New York, Oxford University Press, 1991.

Laszlo, G.: Pulmonary Function: A Guide for Clinicians. New York, Cambridge University Press, 1994.

Leff, A. R., and Schumacker, P. T.: Respiratory Physiology: Basics and Applications. Philadelphia, W. B. Saunders Co., 1993.

Levitzky, M. G., and Hall, S. M. Cardiopulmonary Physiology in Anesthesiology. Hightstown, NJ, McGraw-Hill, 1994.

Milhorn, H. T., Jr., and Guyton, A. C.: An analog computer analysis of Cheyne-Stokes breathing. J. Appl. Physiol., 20:328, 1965.

Milhorn, H. T., Jr., et al.: A mathematical model of the human respiratory control system. Biophys. J., 5:27, 1965.

Mitchell, G. S., et al.: Changes in the V_I-V_{CO_2} relationship during exercise in goats: Role of carotid bodies. J. Appl. Physiol., 57:1894, 1984.

Murray, J. F., Nadel, J. A.: Textbook of Respiratory Medicine. Philadelphia, W. B. Saunders Co., 1994.

Read, D. J., and Henderson-Smart, D. J.: Regulation of breathing in the newborn during different behavioral states. Annu. Rev. Physiol., 46:675, 1984.

Rigatto, H.: Control of ventilation in the newborn. Annu. Rev. Physiol., 46:661, 1984.

Rowell, L. B., and Sheriff, D. D.: Are muscle "chemoreflexes" functionally important? News Physiol. Sci., 3:250, 1988.

Saldana, M. J.: Pathology of Pulmonary Disease. Philadelphia, J. B. Lippincott, 1994.

Schlaefke, M. E. (ed.): Central Neurone Environment and the Control Systems of Breathing and Circulation. New York, Springer-Verlag, 1983.

Sinclair, J. D.: Respiratory drive in hypoxia: Carotid body and other mechanisms compared. News Physiol. Sci., 2:57, 1987.

Von Euler, C., and Lagercrantz, H.: Neurobiology of the Control of Breathing. New York, Raven Press, 1987.

Von Euler, C., and Lagercrantz, H. (eds.): Central Nervous Control Mechanisms in Breathing. New York, Pergamon Press, 1979.

Walker, D. W.: Peripheral and central chemoreceptors in the fetus and newborn. Annu. Rev. Physiol., 46:687, 1984.

West, J. B.: Pulmonary Pathophysiology: The Essentials. Baltimore, Williams & Wilkins, 1992.

West, J. B.: Respiratory Physiology—The Essentials. Baltimore, Williams & Wilkins, 1994.

Whipp, B. J.: Ventilatory control during exercise in humans. Annu. Rev. Physiol., 45:393, 1983.

Respiratory Insufficiency—Pathophysiology, Diagnosis, Oxygen Therapy

CHAPTER 42

The diagnosis and treatment of most respiratory disorders have come to depend heavily on an understanding of the basic physiological principles of respiration and gas exchange. Some diseases of respiration result from inadequate ventilation, whereas others result from abnormalities of diffusion through the pulmonary membrane or transport from the lungs to the tissues. In each of these instances, the therapy is often entirely different, so that it is no longer satisfactory simply to make a diagnosis of "respiratory insufficiency."

Useful Methods for Studying Respiratory Abnormalities

In the previous few chapters, we discuss a number of methods for studying respiratory abnormalities, including measuring vital capacity, tidal air, functional residual capacity, dead space, physiologic shunt, and physiologic dead space. This array of measurements is only part of the armamentarium of the clinical pulmonary physiologist. Some other interesting tools available are described.

Study of Blood Gases and pH

Among the most fundamental of all tests of pulmonary performance are determinations of the blood P_{O_2}, P_{CO_2}, and pH. Indeed, it is often important to make these measurements rapidly as an aid in determining appropriate therapy for acute respiratory distress or acute abnormalities of acid-base balance. Several simple and rapid methods have been developed to make these measurements in only a few minutes, using no more than a few drops of blood. They are the following.

DETERMINATION OF BLOOD pH. Blood pH is measured using a glass pH electrode of the type used in all chemical laboratories. However, the electrodes used for this purpose

are miniaturized, so that no more than a drop or so of blood need be used. The voltage generated by the glass electrode is a direct measure of the pH, and this is generally read directly from a voltmeter scale or it is recorded on a chart.

DETERMINATION OF BLOOD P_{CO_2}. A glass electrode pH meter can be used to determine blood P_{CO_2} in the following way: When a weak solution of sodium bicarbonate is exposed to carbon dioxide, the carbon dioxide dissolves in the solution until an equilibrium state is established. In this equilibrium state, the pH of the solution is a function of the carbon dioxide and bicarbonate ion concentrations in accordance with the Henderson-Hasselbalch equation discussed in Chapter 30; that is,

$$pH = \log \frac{HCO_3^-}{CO_2} + 6.1$$

With appropriate calibration of the pH meter for the concentration of bicarbonate ion used, the carbon dioxide concentration (or P_{CO_2}) can be read directly from the meter.

When this apparatus is used to measure P_{CO_2} in blood, a miniature glass electrode is coated with a thin solution of sodium bicarbonate, and this is separated from the blood by a thin plastic membrane that allows carbon dioxide to diffuse from the blood into the solution. Here again, only a drop or so of blood is required.

DETERMINATION OF BLOOD P_{O_2}. The concentration of oxygen in a fluid can be measured by a technique called *polarography*. Electric current is made to flow between a small negative electrode and the solution. If the voltage of the electrode is more than -0.6 volt different from the voltage of the solution, oxygen will deposit on the electrode. Furthermore, the rate of current flow through the electrode will be directly proportional to the concentration of oxygen (and therefore P_{O_2} as well). In practice, a negative platinum

537

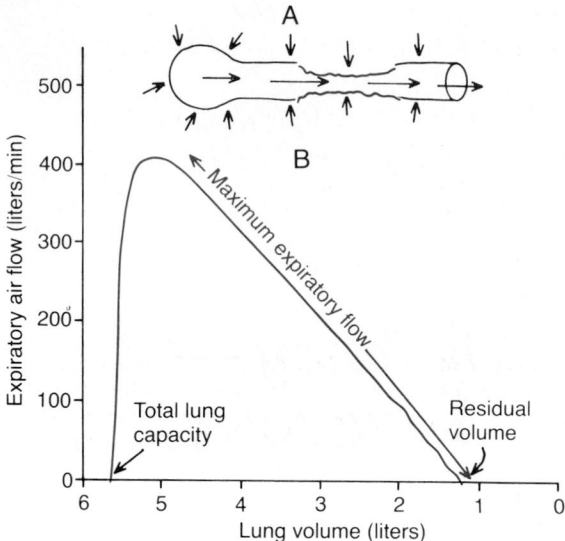

Figure 42–1. *A,* Collapse of the respiratory passageway during maximum expiratory effort, an effect that limits the expiratory flow rate. *B,* Effect of lung volume on the maximum expiratory air flow, showing decreasing maximum expiratory air flow as the lung volume becomes smaller.

electrode with a surface area of about 1 square millimeter is used, and this is separated from the blood by a thin plastic membrane that allows diffusion of oxygen but not diffusion of proteins or other substances that will "poison" the electrode.

Often all three of the measuring devices for pH, P_{CO_2}, and P_{O_2} are built into the same apparatus, and all these measurements can be made within a minute or so using a single, small sample of blood. Thus, changes in the blood gases and pH can be followed almost moment by moment.

Measurement of Maximum Expiratory Flow

In many respiratory diseases, particularly asthma, the resistance to air flow becomes especially great during expiration, sometimes causing tremendous difficulty in breathing. This has led to the concept called the *maximum expiratory flow,* which can be defined as follows: When a person expires with great force, the expiratory air flow reaches a maximum flow beyond which the flow cannot be increased even with greatly increased additional force. This is the maximum expiratory flow. The maximum expiratory flow is much greater when the lungs are filled with a large volume of air than when they are almost empty. These principles can be understood by referring to Figure 42–1.

Figure 42–1A shows the effect of pressure applied to the outsides of the alveoli and the passageways caused by compressing the chest cage. The arrows indicate that the same amount of pressure is applied to the outsides of both the alveoli and the bronchioles. Therefore, not only does this pressure force air from the alveoli into the bronchioles, but it tends also to collapse the bronchioles at the same time, which will oppose the movement of the air to the exterior. Once the bronchioles have become almost completely collapsed, further expiratory force can still greatly increase the alveolar pressure, but it also increases the degree of bronchiolar collapse and the airway resistance by an equal amount, thus preventing further increase in flow. Therefore,

beyond a critical degree of expiratory force, a maximum expiratory flow has been reached.

Figure 42–1B shows the effect of bronchiolar collapse on the maximum expiratory flow. The curve recorded in this section shows the maximum expiratory flow at all levels of lung volume after a healthy person first inhales as much air as possible and then expires with maximum expiratory effort until he or she can expire no more. Note that the person quickly reaches a *maximum expiratory air flow* of more than 400 liters/min. But it does not matter how much additional expiratory effort the person exerts; this is still the maximum flow that he or she can achieve.

Note also that as the lung volume becomes smaller, the maximum expiratory flow also becomes less. The main reason for this is that in the enlarged lung the bronchi and bronchioles are held open partially by way of elastic pull on their outsides by lung structural elements; however, as the lung becomes smaller, these structures are relaxed, so that the bronchi and bronchioles are collapsed more easily by external pressure, thus progressively reducing the maximum expiratory flow as well.

ABNORMALITIES OF THE MAXIMUM EXPIRATORY FLOW-VOLUME CURVE. Figure 42–2 shows once more the normal maximum expiratory flow-volume curve, along with two additional curves recorded in two types of lung diseases: constricted lungs and partial airway obstruction.

Note that the *constricted lungs* have both reduced total lung capacity (TLC) and reduced residual volume (RV). Furthermore, because the lung cannot expand to its normal volume, even with the greatest possible expiratory effort, the maximal expiratory flow cannot rise to equal that of the normal curve. Constricted lung diseases include fibrotic diseases of the lung itself, such as *tuberculosis, silicosis,* and others, and diseases that constrict the chest cage, such as *kyphosis, scoliosis,* and *fibrotic pleurisy.*

In diseases with *airway obstruction,* it is usually much more difficult to expire than to inspire because the closing tendency of the airways is greatly increased by the positive pressure in the chest during expiration. By contrast, the negative pleural pressure of inspiration actually "pulls" the airways open at the same time that it expands the alveoli. Therefore, air tends to enter the lung easily but then become trapped in the lungs. Over a period of months or years, this effect increases both the total lung capacity and the residual volume, as shown by the left-hand red curve in

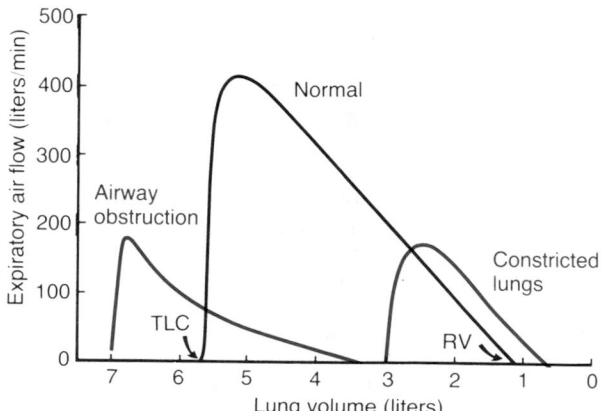

Figure 42–2. Effect of two respiratory abnormalities—constricted lungs and airway obstruction—on the maximum expiratory flow-volume curve.

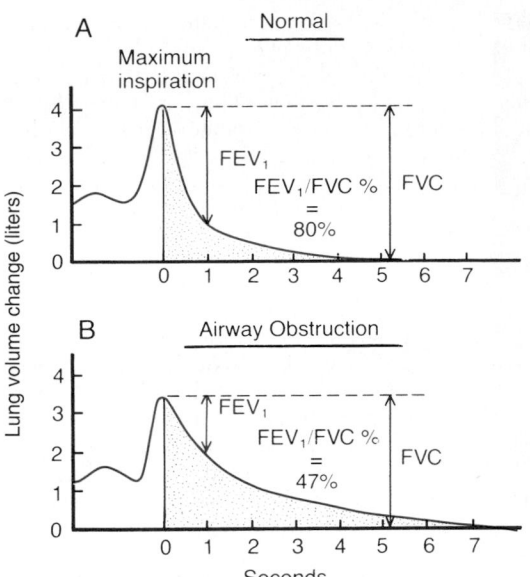

Figure 42–3. Recordings during a forced vital capacity maneuver: *A,* in the normal person and *B,* in the person with airway obstruction. (The "zero" on the volume scale is equal to the residual volume.)

Figure 42–2. Also, because of the obstruction of the airways and because they collapse more easily than normal airways, the maximum expiratory flow is greatly reduced. The classic disease that causes severe airway obstruction is *asthma.* Serious airway obstruction also occurs in some stages of *emphysema.*

Forced Expiratory Vital Capacity and Forced Expiratory Volume

Another exceedingly useful clinical pulmonary test, and one that is also simple, is to make a record on a spirometer of the *forced expiratory vital capacity* (FVC). Such a record is shown in Figure 42–3A for a person with normal lungs and in Figure 42–3B for a person with airway obstruction. In performing the forced expiratory vital capacity maneuver, the person first inspires maximally to the total lung capacity, then exhales into the spirometer with maximum expiratory effort as rapidly and as completely as possible. The downslope of the record as recorded against time represents the *forced vital capacity,* as shown in the figure.

Now, study the difference between the two records for normal lungs and airway obstruction. The total volume changes of the forced vital capacities are nearly equal, indicating only a moderate difference in basic lung volumes in the two persons. On the other hand, there is a major difference in the amounts of air that these people can expire each second, especially during the first second. Therefore, it is customary to compare the recorded forced expiratory volume during the first second (FEV$_1$) with the normal. In the normal person, the percentage of the forced vital capacity that is expired in the first second divided by the total forced vital capacity (FEV$_1$/FVC%) is about 80 per cent. However, note in Figure 42–3B that with airway obstruction, this value was decreased to only 47 per cent. In serious airway obstruction, as often occurs in acute asthma, this can decrease to less than 20 per cent.

PHYSIOLOGICAL PECULIARITIES OF SPECIFIC PULMONARY ABNORMALITIES

Chronic Pulmonary Emphysema

The term *pulmonary emphysema* literally means excess air in the lungs. However, when one speaks of chronic pulmonary emphysema, a complex obstructive and destructive process of the lungs generally is meant, and in most instances it is a consequence of long-term smoking. It results from the following major pathophysiological events in the lungs:

1. *Chronic infection* caused by inhaling smoke or other substances that irritate the bronchi and bronchioles. The principal reason for the chronic infection is that the irritant seriously deranges the normal protective mechanisms of the airways, including partial paralysis of the cilia of the respiratory epithelium by the effects of nicotine, so that mucus cannot be moved easily out of the passageways; stimulation of excess mucus secretion, which further exacerbates the condition; and inhibition of the alveolar macrophages, so that they become less effective in combating infection.

2. The infection, excess mucus, and inflammatory edema of the bronchiolar epithelium together cause *chronic obstruction* of many of the smaller airways.

3. The obstruction of the airways makes it especially difficult to expire, thus causing *entrapment of air in the alveoli* and overstretching them. This, combined with the lung infection, causes *marked destruction of as much as 50 to 80 per cent of the alveolar walls.* Therefore, the final picture of the emphysematous lung is that shown in Figures 42–4 *(top)* and 42–5.

The physiological effects of chronic emphysema are extremely varied, depending on the severity of the disease and the relative degree of bronchiolar obstruction versus lung parenchymal destruction. Among the different abnormalities are the following.

1. The bronchiolar obstruction greatly *increases airway resistance* and results in greatly increased work of breathing. It is especially difficult for the person to move air through the bronchioles during expiration because the compressive force on the outside of the lung not only compresses the alveoli but also compresses the bronchioles, which further increases their resistance during expiration.

2. The marked loss of alveolar walls greatly *decreases the diffusing capacity* of the lung, which reduces the ability of the lungs to oxygenate the blood and remove carbon dioxide.

3. The obstructive process is frequently much worse in some parts of the lungs than in other parts, so that some portions of the lungs are well ventilated, whereas other portions are poorly ventilated. This often causes *extremely abnormal ventilation-perfusion ratios,* with a very low $\dot{V}a/\dot{Q}$ in some parts *(physiologic shunt),* resulting in poor aeration of the blood, and a very high $\dot{V}a/\dot{Q}$ in other parts *(physiologic dead space),* resulting in wasted ventilation, both effects occurring in the same lungs.

4. Loss of large portions of the alveolar walls also decreases the number of pulmonary capillaries through which blood can pass. As a result, the pulmonary vascular resistance increases markedly, causing pulmonary hypertension. This in turn overloads the right side of the heart and frequently causes right-sided heart failure.

Chronic emphysema usually progresses slowly over many years. The person develops hypoxia and hypercapnia because

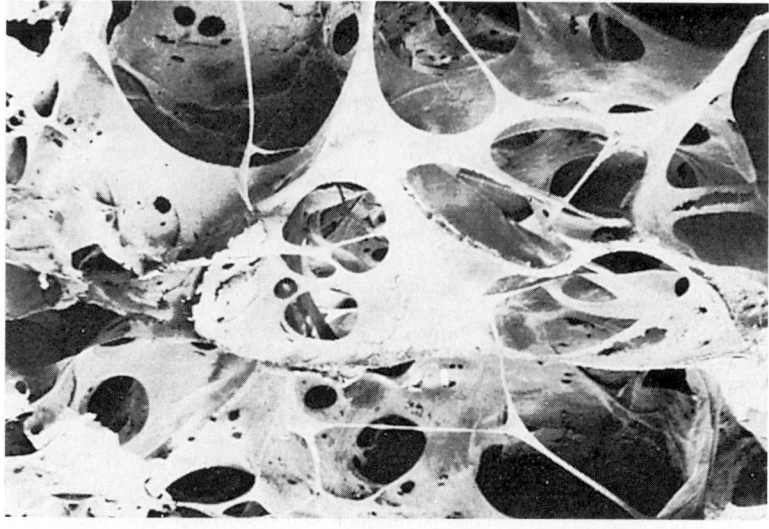

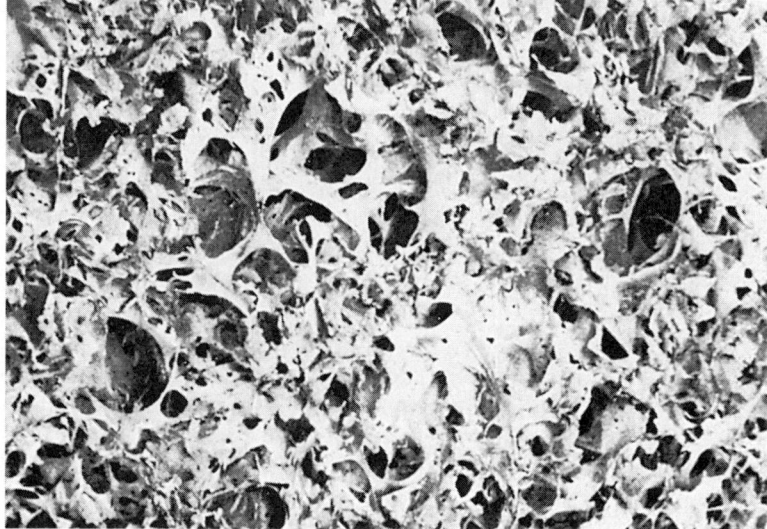

Figure 42–4. Contrast of the emphysematous lung *(above)* with the normal lung *(below)*, showing extensive alveolar destruction. (Reproduced with permission of Patricia Delaney and the Department of Anatomy, The Medical College of Wisconsin.)

of hypoventilation of many alveoli and because of loss of alveolar walls. The net result of all these effects is severe, prolonged, devastating air hunger that can last for years until the hypoxia and hypercapnia cause death—a high penalty to pay for smoking.

Pneumonia

The term *pneumonia* includes any inflammatory condition of the lung in which some or all of the alveoli are filled with fluid and blood cells, as shown in Figure 42–5. A common

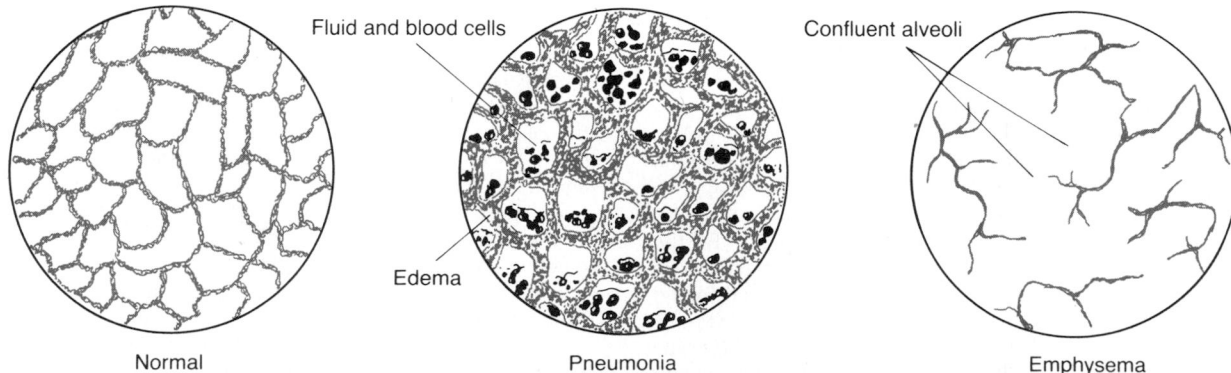

Normal

Fluid and blood cells

Edema

Pneumonia

Confluent alveoli

Emphysema

Figure 42–5. Pulmonary changes in pneumonia and emphysema.

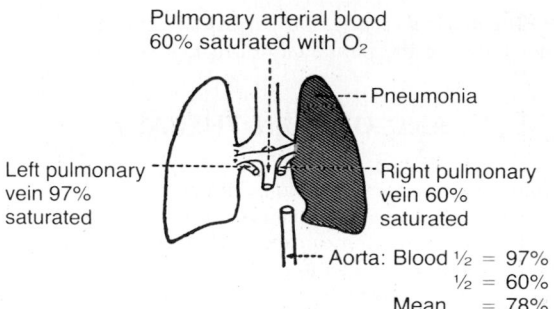

Figure 42-6. Effect of pneumonia on arterial blood oxygen saturation.

type of pneumonia is *bacterial pneumonia*, caused most frequently by pneumococci. This disease begins with infection in the alveoli; the pulmonary membrane becomes inflamed and highly porous so that fluid and even red and white blood cells pass out of the blood into the alveoli. Thus, the infected alveoli become progressively filled with fluid and cells, and the infection spreads by extension of bacteria from alveolus to alveolus. Eventually, large areas of the lungs, sometimes whole lobes or even a whole lung, become "consolidated," which means that they are filled with fluid and cellular debris.

In pneumonia, the pulmonary function of the lungs changes in different stages of the disease. In the early stages, the pneumonia process might well be localized to only one lung, with alveolar ventilation reduced while blood flow through the lung continues normally. This results in two major pulmonary abnormalities: (1) reduction in the total available surface area of the respiratory membrane and (2) decreased ventilation-perfusion ratio. Both these effects cause reduced diffusing capacity, which results in *hypoxemia* (low blood oxygen) and *hypercapnia* (high blood carbon dioxide).

Figure 42-6 shows the effect of the decreased ventilation-perfusion ratio in pneumonia, showing that the blood passing through the aerated lung becomes 97 per cent saturated, whereas that passing through the unaerated lung remains only 60 per cent saturated, causing the mean saturation of the aortic blood to be about 78 per cent, which is far below normal.

Atelectasis

Atelectasis means collapse of the alveoli. It can occur in a localized area of a lung, in an entire lobe, or in an entire lung. Its most common causes are (1) obstruction of the airway and (2) lack of surfactant in the fluids lining the alveoli.

Airway Obstruction. The airway obstruction type of atelectasis usually results from (1) blockage of many small bronchi with mucus or (2) obstruction of a major bronchus by either a large mucous plug or some solid object such as cancer. The air entrapped beyond the block is absorbed within minutes to hours by the blood flowing in the pulmonary capillaries. If the lung tissue is pliable enough, this will lead simply to collapse of the alveoli. However, if the lung is rigid because of fibrotic tissue and cannot collapse, absorption of air from the alveoli creates tremendously negative pressures within the alveoli and pulls fluid out of the pulmonary capillaries into the alveoli, thus causing the alveoli to fill

completely with edema fluid. This almost always is the effect that occurs when an entire lung becomes atelectatic, a condition called *massive collapse* of the lung, because the solidity of the chest wall and the mediastinum allows the lung to decrease only to about one-half normal size, rather than collapsing completely.

The effects on overall pulmonary function caused by *massive collapse* (atelectasis) of an entire lung are shown in Figure 42-7. Collapse of the lung tissue not only occludes the alveoli but almost always also increases the resistance to blood flow through the pulmonary vessels. This resistance increase occurs partially because of the collapse itself, which compresses and folds the vessels as the volume of the lung decreases. In addition, hypoxia in the collapsed alveoli causes additional vasoconstriction, as explained in Chapter 38.

Because of the vascular constriction, blood flow through the atelectatic lung becomes slight. Most of the blood is routed through the ventilated lung and therefore becomes well aerated. In the situation shown in Figure 42-7, five sixths of the blood passes through the aerated lung and only one sixth through the unaerated lung. As a result, the overall ventilation-perfusion ratio is only moderately compromised, so that the aortic blood has only mild oxygen desaturation despite total loss of ventilation in an entire lung.

Lack of Surfactant. The secretion and function of surfactant in the alveoli are discussed in Chapter 37. It is pointed out that the substance surfactant is secreted by special alveolar epithelial cells into the fluids that line the alveoli. This substance decreases the surface tension in the alveoli 2- to 10-fold, which plays a major role in preventing alveolar collapse. However, in a number of conditions, such as in *hyaline membrane disease* (also called *respiratory distress syndrome*), which often occurs in newborn premature babies, the quantity of surfactant secreted by the alveoli is greatly depressed. As a result, the surface tension of the alveolar fluid is increased so much that it causes a serious tendency for the lungs of these babies to collapse or become filled with fluid, as explained in Chapter 37; many of these infants die of suffocation as increasing portions of the lungs become atelectatic.

Asthma

Asthma is characterized by spastic contraction of the smooth muscle in the bronchioles, which causes extremely difficult breathing. It occurs in 3 to 5 per cent of all people at some time in life. The usual cause is hypersensitivity of the bronchioles to foreign substances in the air. In younger patients, under age 30 years, the asthma in about 70 per

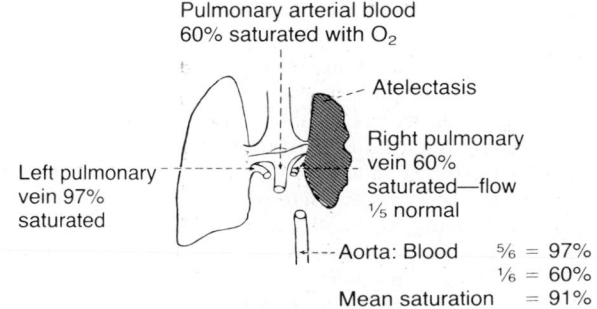

Figure 42-7. Effect of atelectasis on arterial blood oxygen saturation.

cent is caused by allergic hypersensitivity, especially sensitivity to plant pollens. In order people, the cause is almost always hypersensitivity to nonallergic types of irritants in the air, such as irritants in smog.

The allergic reaction that occurs in the allergic type of asthma is believed to occur in the following way: The typically allergic person has a tendency to form abnormally large amounts of IgE antibodies, and these antibodies cause allergic reactions when they react with their specific antigens, as explained in Chapter 34. In asthma, these antibodies are mainly attached to mast cells that lie in the lung interstitium in close association with the bronchioles and small bronchi. When a person breathes in pollen to which he or she is sensitive (that is, to which the person has developed IgE antibodies), the pollen reacts with the mast cell–attached antibodies and causes these cells to release several different substances. Among them are *histamine, slow-reacting substance of anaphylaxis* (which is a mixture of leukotrienes), *eosinophilic chemotactic factor,* and *bradykinin.* The combined effects of all these factors, especially of the slow-reacting substance of anaphylaxis, are to produce (1) localized edema in the walls of the small bronchioles as well as secretion of thick mucus into the bronchiolar lumens and (2) spasm of the bronchiolar smooth muscle. Therefore, the airway resistance increases greatly.

As discussed earlier in this chapter, the bronchiolar diameter becomes more reduced during expiration than during inspiration in asthma because the increased intrapulmonary pressure during expiratory effort compresses the outsides of the bronchioles. Because the bronchioles are already partially occluded, further occlusion resulting from the external pressure creates especially severe obstruction during expiration. The asthmatic person usually can inspire quite adequately but has great difficulty expiring. Clinical measurements show greatly reduced maximum expiratory rate and timed expiratory volume. Also, this results in dyspnea, or "air hunger," which is discussed later in the chapter.

The functional residual capacity and the residual volume of the lung become greatly increased during the asthmatic attack because of the difficulty in expiring air from the lungs. Also, over a period of years, the chest cage becomes permanently enlarged, causing a "barrel chest," and the functional residual capacity and residual volume become permanently increased.

Tuberculosis

In tuberculosis, the tubercle bacilli cause a peculiar tissue reaction in the lungs, including (1) invasion of the infected region by macrophages and (2) walling off of the lesion by fibrous tissue to form the so-called "tubercle." This walling-off process helps to limit further transmission of the tubercle bacilli in the lungs and therefore is part of the protective process against the infection. However, in about 3 per cent of all people who contract tuberculosis, if untreated, the walling-off process fails and tubercle bacilli spread throughout the lungs, often causing extreme destruction of lung tissue with formation of large abscess cavities. Thus, tuberculosis in its late stages causes many areas of fibrosis throughout the lungs and reduces the total amount of functional lung tissue. These effects cause (1) increased "work" on the part of the respiratory muscles to cause pulmonary ventilation and *reduced vital capacity and breathing capacity;* (2) *reduced total respiratory membrane surface area* and *increased thickness of the respiratory membrane,* these causing progressively diminished pulmonary diffusing capacity;

and (3) *abnormal ventilation-perfusion ratio* in the lungs, further reducing the pulmonary diffusing capacity.

HYPOXIA AND OXYGEN THERAPY

Almost any of the conditions discussed in the past few sections of this chapter can cause serious degrees of cellular hypoxia. In some of these, oxygen therapy is of great value; in others, it is of moderate value; in still others, it is of almost no value. Therefore, it is important to understand the different types of hypoxia; then we can readily discuss the physiological principles of oxygen therapy. The following is a descriptive classification of the causes of hypoxia:

1. Inadequate oxygenation of the lungs because of extrinsic reasons
   a. Deficiency of oxygen in atmosphere
   b. Hypoventilation (neuromuscular disorders)
2. Pulmonary disease
   a. Hypoventilation due to increased airway resistance or decreased pulmonary compliance
   b. Uneven alveolar ventilation-perfusion ratio (including increased physiologic dead space and physiologic shunt)
   c. Diminished respiratory membrane diffusion
3. Venous-to-arterial shunts ("right-to-left" cardiac shunts)
4. Inadequate oxygen transport by the blood to the tissues
   a. Anemia or abnormal hemoglobin
   b. General circulatory deficiency
   c. Localized circulatory deficiency (peripheral, cerebral, coronary vessels)
   d. Tissue edema
5. Inadequate tissue capability of using oxygen
   a. Poisoning of cellular enzymes
   b. Diminished cellular metabolic capacity because of toxicity, vitamin deficiency, or other factors

This classification of the types of hypoxia is mainly self-evident from the discussions earlier in the chapter. Only one of the types of hypoxia in the classification needs further elaboration; this is the hypoxia caused by inadequate capability of the cells to use oxygen.

INADEQUATE TISSUE CAPABILITY OF USING OXYGEN. The classic cause of inability of the tissues to use oxygen is *cyanide poisoning,* in which the action of the enzyme cytochrome oxidase is completely blocked by the cyanide—to such an extent that the tissues simply cannot utilize the oxygen even though plenty is available. Also, deficiencies of some oxidative enzymes or of other elements in the tissue oxidative system can lead to this type of hypoxia. A special example occurs in the disease beriberi, in which several important steps in the tissue utilization of oxygen and the formation of carbon dioxide are compromised because of vitamin B deficiency.

EFFECTS OF HYPOXIA ON THE BODY. Hypoxia, if severe enough, can cause death of the cells, but in less severe degrees it results principally in (1) depressed mental activity, sometimes culminating in coma, and (2) reduced work capacity of the muscles. These effects are discussed in Chapter 43 in relation to high-altitude physiology.

Oxygen Therapy in the Different Types of Hypoxia

Oxygen can be administered by (1) placing the patient's head in a "tent" that contains air fortified with oxygen, (2) allowing the patient to breathe either pure oxygen or high concentrations of oxygen from a mask, or (3) administering oxygen through an intranasal tube.

Oxygen therapy is of great value in certain types of hypoxia but of almost zero value in other types. Recalling the basic physiological principles of the different types of hypoxia, one can readily decide when oxygen therapy will be of value and, if so, how valuable.

In *atmospheric hypoxia,* oxygen therapy can correct the depressed oxygen level in the inspired gases and, therefore, provide 100 per cent effective therapy.

In *hypoventilation hypoxia,* a person breathing 100 per cent oxygen can move five times as much oxygen into the alveoli with each breath as when breathing normal air. Therefore, here again oxygen therapy can be extremely beneficial. (However, this provides no benefit for the excess blood carbon dioxide also caused by the hypoventilation.)

In *hypoxia caused by impaired alveolar membrane diffusion,* essentially the same result occurs as in hypoventilation hypoxia because oxygen therapy can increase the P_{O_2} in the lungs from a normal value of about 100 mm Hg to as high as 600 mm Hg. This raises the oxygen diffusion gradient between the alveoli and the blood from a normal value of 60 mm Hg to as high as 560 mm Hg, or an increase of more than 800 per cent. This highly beneficial effect of oxygen therapy in diffusion hypoxia is demonstrated in Figure 42–8, which shows that the pulmonary blood in this patient with pulmonary edema picks up oxygen four times as rapidly as it would with no therapy.

In *hypoxia caused by anemia, abnormal hemoglobin transport of oxygen, circulatory deficiency, or physiologic shunt,* oxygen therapy is of much less value because normal oxygen is already available in the alveoli. The problem instead is that appropriate mechanisms for transporting the oxygen to the tissues are deficient. Even so, a small amount of extra oxygen, between 7 and 30 per cent, can be transported in the dissolved state in the blood when alveolar oxygen is increased to maximum even though the amount transported by the hemoglobin is hardly altered. This small amount of extra oxygen may be the difference between life and death.

In the different types of *hypoxia caused by inadequate tissue use of oxygen,* there is abnormality neither of oxygen pickup by the lungs nor of transport to the tissues. Instead, the tissue metabolic enzyme system is simply incapable of

using the oxygen that is delivered. Therefore, it is doubtful that oxygen therapy is of any measurable benefit.

HYPERCAPNIA

Hypercapnia means excess carbon dioxide in the body fluids.

One might suspect on first thought that any respiratory condition that causes hypoxia would also cause hypercapnia. However, hypercapnia usually occurs in association with hypoxia only when the hypoxia is caused by *hypoventilation* or *circulatory deficiency.* The reasons for this are the following.

Hypoxia caused by *too little oxygen in the air, too little hemoglobin,* or *poisoning of the oxidative enzymes* has to do only with the availability of oxygen or use of oxygen by the tissues. Therefore, it is readily understandable that hypercapnia is *not* a concomitant of these types of hypoxia.

Also, in hypoxia resulting from poor diffusion through the pulmonary membrane or through the tissues, serious hypercapnia usually does not occur at the same time because carbon dioxide diffuses 20 times as rapidly as oxygen. Also, if hypercapnia does begin to occur, this immediately stimulates pulmonary ventilation, which corrects the hypercapnia but not necessarily the hypoxia.

In hypoxia caused by hypoventilation, carbon dioxide transfer between the alveoli and the atmosphere is affected as much as is oxygen transfer. Therefore, hypercapnia always results along with the hypoxia. And in circulatory deficiency, diminished flow of blood decreases the removal of carbon dioxide from the tissues, resulting in tissue hypercapnia. However, the transport capacity of the blood for carbon dioxide is more than three times that for oxygen, so that even here, the tissue hypercapnia is much less than the tissue hypoxia.

When the alveolar P_{CO_2} rises above about 60 to 75 mm Hg, the person by then is breathing about as rapidly and deeply as he can, and "air hunger," also called *dyspnea,* becomes severe. As the P_{CO_2} rises to 80 to 100 mm Hg, the person becomes lethargic and sometimes even semicomatose. Anesthesia and death can result when the P_{CO_2} rises to 120 to 150 mm Hg. Also, at the higher levels of P_{CO_2}, the excess carbon dioxide now begins to depress respiration rather than stimulate it, thus causing a vicious cycle of still more carbon dioxide, further decrease in respiration, then more carbon dioxide, and so forth—culminating rapidly in a respiratory death.

Cyanosis

The term "cyanosis" means blueness of the skin, and its cause is excessive amounts of deoxygenated hemoglobin in the skin blood vessels, especially in the capillaries. This deoxygenated hemoglobin has an intense dark blue–purple color that is transmitted through the skin.

It is *not* the *percentage* deoxygenation of the hemoglobin that causes the bluish hue of the skin but principally the *concentration of deoxygenated hemoglobin without regard to the amount of oxygenated hemoglobin.* The reason for this is that the red color of oxygenated blood is weak in comparison with the dark blue color of deoxygenated blood. Therefore, when the two are mixed together, the oxygenated blood has relatively little coloring effect in comparison with that of the deoxygenated blood.

In general, definite cyanosis appears whenever the *arterial blood* contains more than 5 grams deoxygenated hemoglobin

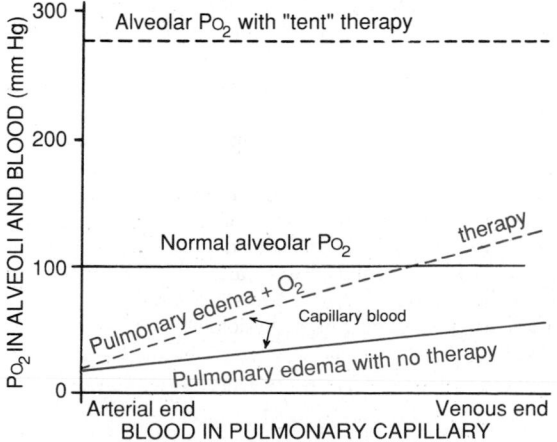

Figure 42–8. Absorption of oxygen into the pulmonary capillary blood in pulmonary edema with and without oxygen therapy.

in each deciliter of blood. A person with *anemia* almost never becomes cyanotic because there is not enough hemoglobin for 5 grams of it to be deoxygenated in the arterial blood. On the other hand, in a person with excess red blood cells, as occurs in *polycythemia vera*, the great excess of available hemoglobin leads frequently to cyanosis even under otherwise normal conditions.

Dyspnea

Dyspnea means mental anguish associated with inability to ventilate enough to satisfy the demand for air. A common synonym is "air hunger."

At least three factors often enter into the development of the sensation of dyspnea. They are (1) abnormality of the respiratory gases in the body fluids, especially hypercapnia and, to much less an extent, hypoxia; (2) the amount of work that must be performed by the respiratory muscles to provide adequate ventilation; and (3) state of mind.

A person becomes very dyspneic especially from excess buildup of carbon dioxide in body fluids. At times, however, the levels of both carbon dioxide and oxygen in the body fluids are normal, but to attain this normality of the respiratory gases, the person has to breathe forcefully. In these instances, the forceful activity of the respiratory muscles frequently gives the person a sensation of severe dyspnea.

Finally, the person's respiratory functions may be normal and still dyspnea may be experienced because of an abnormal state of mind. This is called *neurogenic dyspnea* or *emotional dyspnea*. For instance, almost anyone momentarily thinking about the act of breathing may suddenly start taking breaths a little more deeply than ordinarily because of a feeling of mild dyspnea. This feeling is greatly enhanced in people who have a psychological fear of not being able to receive a sufficient quantity of air, such as on entering small or crowded rooms.

ARTIFICIAL RESPIRATION

RESUSCITATOR. Many types of resuscitators are available, and each has its own characteristic principles of operation. The resuscitator shown in Figure 42–9A consists of a supply of oxygen or air; a mechanism for applying intermittent positive pressure and, with some machines, negative pressure as well; and a mask that fits over the face of the patient or a connector for connecting the equipment to an endotracheal tube. This apparatus forces air through the mask into the lungs of the patient during the positive-pressure cycle and then usually allows the air to flow passively out of the lungs during the remainder of the cycle.

Earlier resuscitators often caused such severe damage to the lungs because of excessive positive pressure that their usage was at one time greatly decried. However, resuscitators now have adjustable positive-pressure limits that are commonly set at 12 to 15 cm H_2O pressure for normal lungs but sometimes much higher for noncompliant lungs.

TANK RESPIRATOR. Figure 42–9B shows the tank respirator with a patient's body inside the tank and the head protruding through a flexible but airtight collar. At the end of the tank opposite the patient's head is a motor-driven leather diaphragm that moves back and forth with sufficient excursion to raise and lower the pressure inside the tank. As the leather diaphragm moves inward, positive pressure develops around the body and causes expiration; as the diaphragm moves outward, negative pressure causes inspiration. Check valves on the respirator control the positive and negative pressure. Ordinarily these pressures are adjusted so that the negative pressure that causes inspiration falls to -10 to -20 cm H_2O and the positive pressure rises to 0 to $+5$ cm H_2O.

EFFECT OF THE RESUSCITATOR AND THE TANK RESPIRATOR ON VENOUS RETURN. When air is forced into the lungs under positive pressure or when the pressure around the patient's body is greatly reduced but with the trachea ex-

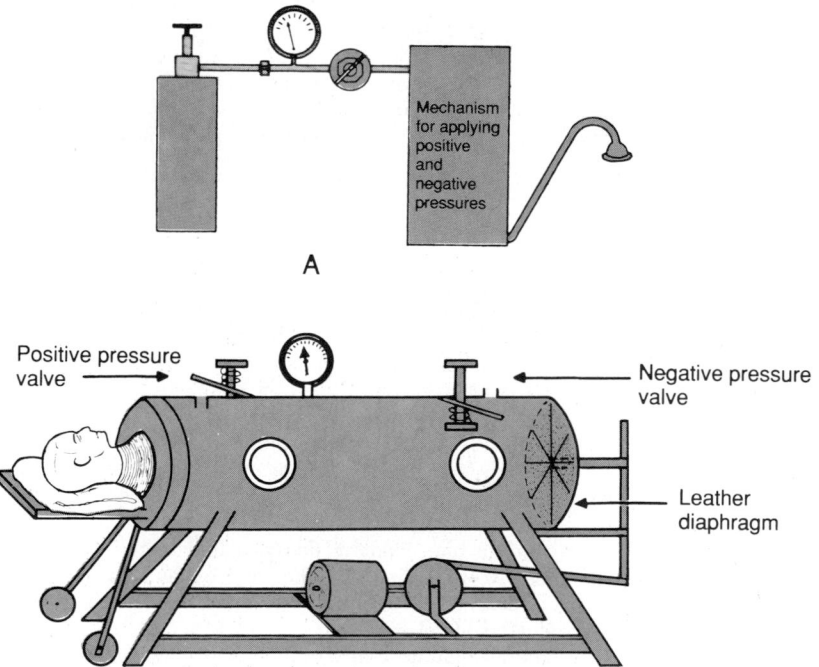

Figure 42–9. *A*, Resuscitator. *B*, Tank respirator.

posed to atmospheric pressure through the nose, as in the case of the tank respirator, the pressure inside the chest cavity is greater than the pressure everywhere else in the body. Therefore, the flow of blood into the chest from the peripheral veins becomes impeded. As a result, use of excessive pressures with either the resuscitator or the tank respirator can reduce the cardiac output—sometimes to lethal levels. For instance, continuous exposure for more than a few minutes to greater than 30 mm Hg positive pressure in the lungs may cause death because of inadequate venous return to the heart.

REFERENCES

Arieff, A. I.: Hypoxia, Metabolic Acidosis and the Circulation. New York, Oxford University Press, 1992.

Barash, P. G., et al.: Handbook of Clinical Anesthesia Biology. Philadelphia, J. B. Lippincott, 1993.

Bates, D. V.: Respiratory Function in Disease, 3rd Ed. Philadelphia, W. B. Saunders Co., 1989.

Baumgartner, W. A., et al.: Heart and Heart-Lung Transplantation. Philadelphia, W. B. Saunders Co., 1990.

Benumof, J. L.: Anesthesia for Thoracic Surgery. Philadelphia, W. B. Saunders Co., 1994.

Berte, J. B.: Critical Care: The Lung, 2nd Ed. East Norwalk, Conn., Appleton & Lange, 1986.

Brewis, R. A. L., and Geddes, D. M.: Respiratory Medicine. Philadelphia, W. B. Saunders Co., 1990.

Casaburi, R., and Petty, T. L.: Principles and Practice of Pulmonary Rehabilitation. Philadelphia, W. B. Saunders Co., 1993.

Chumel, H., et al. Pulmonary Infections and Immunity. New York, Plenum Publishing Corp., 1994.

Crapo, J. D.: Morphologic changes in pulmonary oxygen toxicity. Annu. Rev. Physiol., 49:721, 1987.

Crystal, R. G., and West, J. B.: Lung Injury. New York, Raven Press, 1992.

Crystal, R. G., et al.: The Lung: Scientific Foundations. New York, Raven Press, 1991.

Eisenberg, M., et al: Emergency Medical Therapy. Philadelphia, W. B. Saunders Co., 1984.

Fishman, A. P.: Update: Pulmonary Diseases and Disorders. Hightstown, NJ, McGraw-Hill, 1992.

Fraser, R. S., et al.: Synopsis of Diseases of the Chest. Philadelphia, W. B. Saunders Co., 1994.

Gardner, D. E., et al.: Toxicology of the Lung. New York, Raven Press, 1993.

George, R. B., et al.: Chest Medicine. Baltimore, Williams & Wilkins, 1995.

Grippi, M. A.: Pulmonary Pathophysiology. Philadelphia, J. B. Lippincott, 1994.

Hodgkin, J. E., and Petty, T. L.: Chronic Obstructive Pulmonary Disease: Current Concepts. Philadelphia, W. B. Saunders Co., 1987.

Imlay, J. A., and Linn, S.: DNA damage and oxygen radical toxicity. Science, 240:1302, 1988.

Kelley, W. N.: Essentials of Internal Medicine. 2nd Ed. Philadelphia, J. B. Lippincott, 1994.

Killian, K. J., and Campbell, E. J. M.: Dyspnea and exercise. Annu. Rev. Physiol., 45:465, 1983.

Kirby, R. R., and Gravenstein, N.: Clinical Anesthesia Practice. Philadelphia, W. B. Saunders Co., 1994.

Lahiri, S., et al.: Response and Adaptation to Hypoxia. New York, Oxford University Press, 1991.

Laszlo, G.: Pulmonary Function: A Guide for Clinicians. New York, Cambridge University Press, 1994.

Loughlin, G. M., and Eigen, H.: Respiratory Disease in Children. Baltimore, Williams & Wilkins Co., 1994.

Miller, A. A. (ed.): Pulmonary Function Tests: A Guide for the Student and House Officer. Philadelphia, W. B. Saunders Co., 1987.

Murray, J. F., Nadel, J. A.: Textbook of Respiratory Medicine. Philadelphia, W. B. Saunders Co., 1994.

Niederman, M. S., et al.: Respiratory Infections: A Scientific Basis for Management. Philadelphia, W. B. Saunders Co., 1994.

Niederman, M. S.: Respiratory Infections in the Elderly. New York, Raven Press, 1991.

Pearson, F. G.: Thoracic Surgery. New York, Churchill Livingstone, 1995.

Rakel, R. E.: Conn's Current Therapy. Philadelphia, W. B. Saunders Co., 1994.

Rossman, M. D., and MacGregor, R. R.: Tuberculosis. Hightstown, NJ, McGraw-Hill, 1994.

Rubin, E., and Farber, J. L.: Pathology. Philadelphia, J. B. Lippincott, 1994.

Saldana, M. J.: Pathology of Pulmonary Disease. Philadelphia, J. B. Lippincott, 1994.

Shelhamer, J., et al. Respiratory Disease in the Immunosuppressed Host. Philadelphia, J. B. Lippincott, 1991.

Silverman, D. G., et al.: Review of Clinical Anesthesia Biology. Philadelphia, J. B. Lippincott, 1994.

Smith, L. L.: The response of the lung to foreign compounds that produce free radicals. Annu. Rev. Physiol., 49:681, 1987.

Staub, N. C.: Pulmonary edema. Physiol. Rev., 54:678, 1974.

Stoelting, R. K.: Pharmacology and Physiology in Anesthetic Practice. Philadelphia, J. B. Lippincott, 1991.

Tierney, L. M., Jr., et al. Current Medical Diagnosis and Treatment, 1995, Redding, MA, Appleton & Lange, 1994.

Way, J. L.: Cyanide intoxication and its mechanism of antagonism. Annu. Rev. Pharmacol. Toxicol., 24:451, 1984.

West, J. B.: Pulmonary Pathophysiology—The Essentials. Baltimore, Williams & Wilkins, 1992.

Witek, Jr., T. J. and Schachter, E. N.. Pharmacology and Therapeutics in Respiratory Care. Philadelphia, W. B. Saunders Co., 1994.

AVIATION, SPACE, AND DEEP SEA DIVING PHYSIOLOGY

UNIT VIII

Aviation, High Altitude, and Space Physiology

CHAPTER 43

As we have ascended to higher and higher altitudes in aviation, mountain climbing, and space vehicles, it has become progressively more important to understand the effects of altitude and low gas pressures as well as several other factors—acceleratory forces, weightlessness, and so forth—on the human body. This chapter deals with these problems.

EFFECTS OF LOW OXYGEN PRESSURE ON THE BODY

BAROMETRIC PRESSURES AT DIFFERENT ALTITUDES. Table 43–1 gives the approximate *barometric* and *oxygen pressures* at different altitudes, showing that at sea level, the barometric pressure is 760 mm Hg; at 10,000 feet, only 523 mm Hg; and at 50,000 feet, 87 mm Hg. This decrease in barometric pressure is the basic cause of all the hypoxia problems in high-altitude physiology because as the barometric pressure decreases, the oxygen partial pressure decreases proportionally, remaining at all times slightly less than 21 per cent of the total barometric pressure—at sea level about 159 mm Hg but at 50,000 feet only 18 mm Hg.

Alveolar PO2 at Different Elevations

CARBON DIOXIDE AND WATER VAPOR DECREASE THE ALVEOLAR OXYGEN. Even at high altitudes, carbon dioxide is continually excreted from the pulmonary blood into the alveoli. Also, water vaporizes into the inspired air from the respiratory surfaces. Therefore, these two gases dilute the oxygen in the alveoli, thus reducing the oxygen concentration.

Water vapor pressure in the alveoli remains 47 mm Hg as long as the body temperature is normal, regardless of altitude. In the case of carbon dioxide, during exposure to very

high altitudes, the alveolar P_{CO_2} falls from the sea level value of 40 mm Hg to lower values. In the acclimatized person, who increases his or her ventilation about fivefold, the decrease is to about 7 mm Hg because of increased respiration.

Now let us see how the pressures of these two gases affect the alveolar oxygen. For instance, assume that the barometric pressure falls to 253 mm Hg, which is the measured value at the top of 29,028-foot Mount Everest. Forty-seven millimeters of mercury of this must be water vapor, leaving only 206 mm Hg for all the other gases. In the acclimatized person, 7 mm of the 206 mm Hg must be carbon dioxide, leaving only 199 mm Hg. If there were no use of oxygen by the body, one fifth of this 199 mm Hg would be oxygen and four fifths would be nitrogen; or the P_{O_2} in the alveoli would be 40 mm Hg. Some of this remaining alveolar oxygen would be absorbed into the blood, leaving about 35 mm Hg oxygen pressure in the alveoli. Therefore, at the summit of Mount Everest, only the best of acclimatized people can barely survive when breathing air. But the effect is very different when the person is breathing pure oxygen, as we see in the following discussions.

ALVEOLAR P_{O_2} AT DIFFERENT ALTITUDES. The fifth column of Table 43–1 shows the approximate P_{O_2}s in the alveoli at different altitudes when one is breathing air for both the unacclimatized and the acclimatized person. At sea level, the alveolar P_{O_2} is 104 mm Hg; at 20,000 feet altitude, it falls to about 40 mm Hg in the unacclimatized person but only to 53 mm Hg in the acclimatized. The difference between these two is that the alveolar ventilation increases about five times as much in the acclimatized person, as we discuss later.

SATURATION OF HEMOGLOBIN WITH OXYGEN AT DIFFERENT ALTITUDES. Figure 43–1 shows arterial oxygen saturation at different altitudes while breathing air and while breathing oxygen. Up to an altitude of about 10,000 feet, even when air is breathed, the arterial oxygen saturation

Table 43-1 EFFECTS OF ACUTE EXPOSURE TO LOW ATOMOSPHERIC PRESSURES ON ALVEOLAR GAS CONCENTRATIONS AND ARTERIAL OXYGEN SATURATION*

			Breathing Air			Breathing Pure Oxygen		
Altitude (ft)	Barometric Pressure (mm Hg)	Po$_2$ in Air (mm Hg)	Pco$_2$ in Alveoli (mm Hg)	Po$_2$ in Alveoli (mm Hg)	Arterial Oxygen Saturation (%)	Pco$_2$ in Alevoli (mm Hg)	Po$_2$ in Alveoli (mm Hg)	Arterial Oxygen Saturation (%)
0	760	159	40 (40)	104 (104)	97 (97)	40	673	100
10,000	523	110	36 (23)	67 (77)	90 (92)	40	436	100
20,000	349	73	24 (10)	40 (53)	73 (85)	40	262	100
30,000	226	47	24 (7)	18 (30)	24 (38)	40	139	99
40,000	141	29				36	58	84
50,000	87	18				24	16	15

* Numbers in parentheses are acclimatized values.

remains at least as high as 90 per cent. Above 10,000 feet, the arterial oxygen saturation falls progressively, as shown by the left-hand curve of the figure, until it is only 70 per cent at 20,000 feet and very, very much less at still higher altitudes.

Effect of Breathing Pure Oxygen on Alveolar Po$_2$ at Different Altitudes

When a person breathes pure oxygen instead of air, most of the space in the alveoli formerly occupied by nitrogen now becomes occupied by oxygen. Therefore, at 30,000 feet, the aviator could have an alveolar Po$_2$ as high as 139 mm Hg instead of the 18 mm Hg when breathing air (see Table 43-1).

The second curve of Figure 43-1 shows the arterial oxygen saturation at different altitudes when one is breathing pure oxygen. Note that the saturation remains above 90 per cent until the aviator ascends to about 39,000 feet; then it falls rapidly to about 50 per cent at about 47,000 feet.

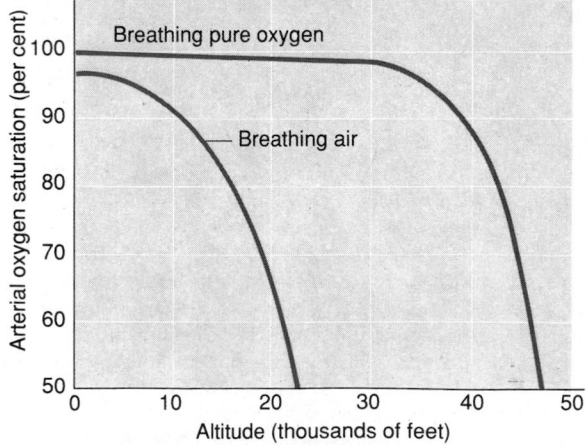

Figure 43-1. Effect of high altitude on arterial oxygen saturation when one is breathing air and when breathing pure oxygen.

The "Ceiling" When Breathing Air and When Breathing Oxygen in an Unpressurized Airplane

Comparing the two arterial oxygen saturation curves in Figure 43-1, one notes that an aviator breathing oxygen can ascend to far higher altitudes than one not breathing oxygen. For instance, the arterial saturation at 47,000 feet when one is breathing oxygen is about 50 per cent and is equivalent to the arterial oxygen saturation at 23,000 feet when one is breathing air. In addition, because an unacclimatized person may remain conscious until the arterial oxygen saturation falls to 50 per cent, for short exposure times, the ceiling for an aviator in an unpressurized airplane when breathing air is about 23,000 feet and when breathing pure oxygen about 47,000 feet, provided the oxygen-supplying equipment operates perfectly.

Acute Effects of Hypoxia

Some of the important acute effects of hypoxia, beginning at an altitude of about 12,000 feet, are drowsiness, lassitude, mental and muscle fatigue, sometimes headache, occasionally nausea, and sometimes euphoria. All of these progress to a stage of twitchings or seizures above 18,000 feet and end, above 23,000 feet in the unacclimatized person, in coma.

One of the most important effects of hypoxia is decreased mental proficiency, which decreases judgment, memory, and the performance of discrete motor movements. For instance, if an unacclimatized aviator stays at 15,000 feet for 1 hour, mental proficiency ordinarily will have fallen to about 50 per cent of normal and after 18 hours at this level to about 20 per cent of normal.

Acclimatization to Low Po$_2$

A person remaining at high altitudes for days, weeks, or years becomes more and more acclimatized to the low Po$_2$, so that it causes fewer deleterious effects on the body and becomes possible for the person to work harder without hypoxic effects or to ascend to still higher altitudes. The principal means by which acclimatization comes about are (1) a great increase in pulmonary ventilation, (2) increased red blood cells, (3) increased diffusing capacity of the lungs, (4) increased vascularity of the tissues, and (5) increased ability of the cells to use oxygen despite the low Po$_2$.

INCREASED PULMONARY VENTILATION. On immediate exposure to very low P_{O_2}, the hypoxic stimulation of the chemoreceptors increases alveolar ventilation to a maximum of about 65 per cent above normal. This is an immediate compensation for the high altitude, and it alone allows the person to rise several thousand feet higher than would be possible without the increased ventilation. Then, if the person remains at a very high altitude for several days, the ventilation gradually increases to an average of about five times normal (400 per cent above normal). The basic cause of this gradual increase is the following.

The immediate 65 per cent increase in pulmonary ventilation on rising to a high altitude blows off large quantities of carbon dioxide, reducing the P_{CO_2} and increasing the pH of the body fluids. Both these changes *inhibit* the respiratory center and thereby *oppose the effect of low P_{O_2} to stimulate the peripheral respiratory chemoreceptors in the carotid and aortic bodies*. During the ensuing 2 to 5 days, this inhibition fades away, allowing the respiratory center now to respond with full force to the chemoreceptor stimuli resulting from hypoxia, and the ventilation increases to about five times normal. The cause of this fading inhibition is believed to be mainly a reduction of bicarbonate ion concentration in the cerebrospinal fluid as well as the brain tissues. This in turn decreases the pH in the fluids surrounding the chemosensitive neurons of the respiratory center, thus increasing the activity of the center. (See Chapter 41 for a more detailed discussion of this effect.)

INCREASE IN RED BLOOD CELLS AND HEMOGLOBIN DURING ACCLIMATIZATION. As stated in Chapter 32, hypoxia is the principal stimulus for causing an increase in red blood cell production. Ordinarily, in full acclimatization to low oxygen, the hematocrit rises from a normal value of 40 to 45 to an average of about 60, with an average increase in hemoglobin concentration from a normal of 15 gm/dl to about 20 gm/dl.

In addition, the blood volume increases, often by 20 to 30 per cent, resulting in a total increase in circulating hemoglobin of 50 or more per cent.

This increase in hemoglobin and blood volume is a slow one, having almost no effect until after 2 weeks, reaching half development in a month or so, and becoming fully developed only after many months.

INCREASED DIFFUSING CAPACITY AFTER ACCLIMATIZATION. It will be recalled that the normal diffusing capacity for oxygen through the pulmonary membrane is about 21 ml/mm Hg/min, and this diffusing capacity can increase as much as threefold during exercise. A similar increase in diffusing capacity occurs at high altitude. Part of the increase probably results from greatly increased pulmonary capillary blood volume, which expands the capillaries and increases the surface through which oxygen can diffuse into the blood. Another part results from an increase in lung volume, which expands the surface area of the alveolar membrane. A final part results from an increase in pulmonary arterial pressure; this forces blood into greater numbers of alveolar capillaries than normally—especially in the upper parts of the lungs, which are poorly perfused under usual conditions.

CIRCULATORY SYSTEM IN ACCLIMATIZATION—INCREASED CAPILLARITY. The cardiac output often increases as much as 30 per cent immediately after a person ascends to high altitude but then decreases back toward normal *as the blood hematocrit increases,* so that the amount of oxygen transported to the tissues remains about normal—that is, unless the altitude becomes so high that severe hypoxia results.

Another circulatory adaptation is an increase in numbers of capillaries in the tissues, which is called *increased capillarity.* This occurs especially in animals born and bred at high altitudes but less so for animals that later in life have become exposed to the altitude. In active tissues exposed to chronic hypoxia, the increase in capillarity is especially marked. For instance, the capillary density in the right ventricular muscle increases many percentage points because of the combined effects of hypoxia and excess work load on the right ventricle caused by pulmonary hypertension at high altitude (the hypertension results from pulmonary vasoconstriction caused by low alveolar oxygen concentration).

CELLULAR ACCLIMATIZATION. In animals native to altitudes of 13,000 to 17,000 feet, mitochondria and some cellular oxidative enzyme systems are slightly more plentiful than in sea-level inhabitants. Therefore, it is presumed that acclimatized human beings as well as these animals can use oxygen more effectively than can their sea-level counterparts, but this is uncertain.

Natural Acclimatization of Natives Living at High Altitudes

Many natives in the Andes and in the Himalayas live at altitudes above 13,000 feet—one group in the Peruvian Andes lives at an altitude of 17,500 feet and works a mine at an altitude of 19,000 feet. Many of these natives are born at these altitudes and live there all their lives. In all aspects of acclimatization, the natives are superior to even the best-acclimatized lowlanders, even though the lowlanders might also have lived at high altitudes for 10 or more years. Acclimatization of the natives begins in infancy. The chest size, especially, is greatly increased, whereas the body size is somewhat decreased, giving a high ratio of ventilatory capacity to body mass. In addition, their hearts, particularly the right side of the heart, which provides a high pulmonary arterial pressure to pump blood through a greatly expanded pulmonary capillary system, are considerably larger than the hearts of lowlanders.

The delivery of oxygen by the blood to the tissues is also highly facilitated in these natives. For instance, Figure 43–2 shows the hemoglobin-oxygen dissociation curves for natives who live at sea level and for their counterparts who live at 15,000 feet. Note that the arterial oxygen P_{O_2} in the natives at high altitude is only 40 mm Hg, but because of the greater quantity of hemoglobin the quantity of oxygen in the arterial blood is greater than in the blood of the natives at the lower altitude. Note also that the venous P_{O_2} in the high-altitude natives is only 15 mm Hg less than the venous P_{O_2} for the lowlanders, despite the low arterial P_{O_2}, indicating that oxygen transport to the tissues is exceedingly effective in the naturally acclimatized high-altitude natives.

Work Capacity at High Altitudes: The Effect of Acclimatization

In addition to the mental depression caused by hypoxia, as discussed earlier, the work capacity of all the muscles is greatly decreased in hypoxia. This includes not only the skeletal muscles but also the cardiac muscle, so that even the maximum level of cardiac output is reduced. In general, the work capacity is reduced in direct proportion to the decrease in maximum rate of oxygen uptake that the body can achieve.

To give an idea of the importance of acclimatization for work capacity, consider this: The work capacities in percent-

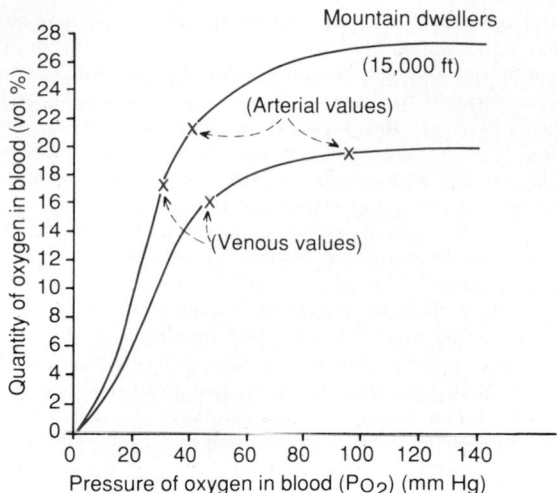

Figure 43–2. Oxygen-dissociation curves for blood of high-altitude residents *(top curve)* and sea-level residents *(lower curve),* showing the respective arterial and venous P_{O_2}s and oxygen contents as recorded in their native surroundings. (From Oxygen-dissociation curves for bloods of high-altitude and sea-level residents. PAHO Scientific Publication No. 140, Life at High Altitudes, 1966.)

age points of normal for unacclimatized and acclimatized people at an altitude of 17,000 feet are as follows:

	Per cent
Unacclimatized	50
Acclimatized for 2 months	68
Native living at 13,200 feet but working at 17,000 feet	87

Thus, naturally acclimatized natives can achieve a daily work output even at this high altitude almost equal to that of a normal person at sea level, but even well-acclimatized lowlanders can almost never achieve this result.

Chronic Mountain Sickness

Occasionally, a person who remains at high altitude too long develops chronic mountain sickness, in which the following effects occur: (1) the red cell mass and hematocrit become exceptionally high, (2) the pulmonary arterial pressure becomes elevated even more than the normal elevation that occurs during acclimatization, (3) the right side of the heart becomes greatly enlarged, (4) the peripheral arterial pressure begins to fall, (5) congestive heart failure ensues, and (6) death often follows unless the person is removed to a lower altitude. The cause of this sequence of events is probably threefold: First, the red cell mass becomes so great that the blood viscosity increases severalfold; this now actually *decreases* tissue blood flow so that oxygen delivery begins to decrease as well. Second, the pulmonary arterioles become exceptionally vasospastic because of the lung hypoxia. This results from the hypoxic vascular constrictor effect that normally operates to divert blood flow from low oxygen to high oxygen alveoli, as explained in Chapter 38. But because all the alveoli are in the low-oxygen state, all the arterioles become constricted, the pulmonary arterial pressure rises excessively, and the right side of the heart

fails. Third, the pulmonary arteriolar spasm diverts much of the blood flow through nonalveolar pulmonary vessels, thus causing an excess of pulmonary shunt blood flow where the blood is not oxygenated; this further compounds the problem. Most of these people recover within days or weeks when they are moved to a lower altitude.

Acute Mountain Sickness and High-Altitude Pulmonary Edema

A small percentage of people who ascend rapidly to high altitudes become acutely sick and can die if not given oxygen or removed to a low altitude. The sickness begins from a few hours up to about 2 days after the ascent. Two events frequently occur.

1. *Acute cerebral edema.* This is believed to result from local vasodilatation of the cerebral blood vessels, caused by the hypoxia. Dilatation of the arterioles increases the capillary pressure, which in turn causes fluid to leak into the cerebral tissues. The cerebral edema can then lead to severe disorientation and other effects related to cerebral dysfunction.

2. *Acute pulmonary edema.* The cause of this is still unknown, but a suggested answer is the following: The severe hypoxia causes the pulmonary arterioles to constrict potently, but the constriction is much greater in some parts of the lungs than in other parts, so that more and more of the pulmonary blood flow is forced through fewer and fewer still unconstricted pulmonary vessels. The postulated result is that the capillary pressure in these areas of the lungs becomes especially high and local edema occurs. Extension of the process to progressively more and more areas of the lungs leads to severe pulmonary dysfunction that can be lethal. Allowing the person to breathe oxygen usually reverses the process within hours.

Some people have extreme pulmonary vascular reactivity to hypoxia, many times the reactivity of the normal person; these people are especially susceptible to acute high-altitude pulmonary edema.

EFFECTS OF ACCELERATORY FORCES ON THE BODY IN AVIATION AND SPACE PHYSIOLOGY

Because of rapid changes in velocity and direction of motion in airplanes and spacecraft, several types of acceleratory forces often affect the body during flight. At the beginning of flight, simple linear acceleration occurs; at the end of flight, deceleration; and every time the vehicle turns, centrifugal acceleration.

Centrifugal Acceleratory Forces

When an airplane makes a turn, the force of centrifugal acceleration is determined by the following relation:

$$f = \frac{mv^2}{r},$$

in which f is the centrifugal acceleratory force, m is the mass of the object, v is the velocity of travel, and r is the radius of curvature of the turn. From this formula, it is obvious that as the velocity increases, the force of centrifugal acceleration increases in proportion to the square of the velocity. It is also obvious that the force of acceleration is directly proportional to the sharpness of the turn (the less the radius).

MEASUREMENT OF ACCELERATORY FORCE—"G." When a man is simply sitting in his seat, the force with which he is pressing against the seat results from the pull of gravity and is equal to his weight. The intensity of this force is said to be + 1 G because it is equal to the pull of gravity. If the force with which he presses against the seat becomes five times his normal weight during pull-out from a dive, the force acting on the seat is + 5 G.

If the airplane goes through an outside loop so that the man is held down by his seat belt, *negative G* is applied to his body; if the force with which he is thrown against his belt is equal to the weight of his body, the negative force is − 1 G.

Effects of Centrifugal Acceleratory Force (Positive G) on the Body

Effects on the Circulatory System. The most important effect of centrifugal acceleration is on the circulatory system because blood is mobile and can be translocated by centrifugal forces.

When the aviator is subjected to *positive G,* the blood is centrifuged toward the lower part of the body. Thus, if the centrifugal acceleratory force is + 5 G and the person is in an immobilized standing position, the hydrostatic pressure in the veins of the feet is five times normal, or about 450 mm Hg; even in the sitting position, this pressure is nearly 300 mm Hg. As the pressure in the vessels of the lower part of the body increases, the vessels passively dilate and a major proportion of the blood from the upper body is translocated into these lower vessels. Because the heart cannot pump unless blood returns to it, the greater the quantity of blood "pooled" in the lower body, the less becomes the cardiac output.

Figure 43–3 shows the changes in systolic and diastolic arterial pressures in the upper body when an acceleratory force of + 3.3 G is suddenly applied to a sitting person. Note that both these pressures fall below 22 mm Hg for the first few seconds after the acceleration begins but then return to a systolic pressure of about 55 mm Hg and a diastolic pressure of 20 mm Hg within another 10 to 15 seconds. This

secondary recovery is caused mainly by activation of the baroreceptor reflexes.

Acceleration greater than 4 to 6 G causes "blackout" of vision within a few seconds and unconsciousness shortly thereafter. If this greater degree of acceleration is continued, the person will die.

Effects on the Vertebrae. Extremely high acceleratory forces for even a fraction of a second can fracture the vertebrae. The degree of positive acceleration that the average person can withstand in the sitting position before vertebral fracture occurs is about 20 G.

NEGATIVE G. The effects of negative G on the body are less dramatic acutely but possibly more damaging permanently than the effects of positive G. An aviator can usually go through outside loops up to negative acceleratory forces of − 4 to − 5 G without harm except for intense momentary hyperemia of the head, although occasionally psychotic disturbances lasting for 15 to 20 minutes occur therafter as a result of brain edema.

Occasionally, negative G forces can be so great and centrifugation of the blood into the head so great that the cerebral blood pressure reaches 300 to 400 mm Hg, sometimes causing small vessels on the surface of the head and in the brain to rupture. However, the vessels inside the cranium show less tendency for rupture than would be expected for the following reason: the cerebrospinal fluid is centrifuged toward the head at the same time that blood is centrifuged toward the cranial vessels, and the greatly increased pressure of this fluid acts as a cushioning buffer on the outside of the brain to prevent vascular rupture.

Because the eyes are not protected by the cranium, intense hyperemia occurs in them during strong negative G. As a result, the eyes often become temporarily blinded with "redout."

PROTECTION OF THE BODY AGAINST CENTRIFUGAL ACCELERATORY FORCES. Specific procedures and apparatus have been developed to protect aviators against the circulatory collapse that occurs during positive G. First, if the aviator tightens the abdominal muscles to an extreme degree and leans forward to compress the abdomen, some of the pooling of blood in the large vessels of the abdomen can be prevented, thereby delaying the onset of blackout. Also, special "anti-G" suits have been devised to prevent pooling of blood in the lower abdomen and legs. The simplest of these applies positive pressure to the legs and abdomen by inflating compression bags as the G increases. Theoretically, a pilot submerged in a tank or suit of water might have little effect on the circulation caused by very high degrees of acceleration, both positive and negative, because the pressures developed in the water and pressing on the outside of the body during centrifugal acceleration would almost exactly balance the forces acting on the body. However, the presence of air in the lungs still allows displacement of the heart, lung tissues, and diaphragm into seriously abnormal positions despite submersion in water. Therefore, even if this procedure were used, the limit of safety would still be less than about 10 G.

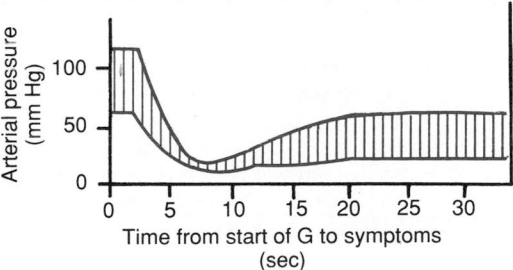

Figure 43–3. Changes in systolic and diastolic arterial pressures after abrupt and continuing exposure to an acceleratory force of + 3.3 G. (Modified from Martin and Henry: *J. Aviation Med., 22:*382, 1951.)

Effects of Linear Acceleratory Forces on the Body

ACCELERATORY FORCES IN SPACE TRAVEL. Unlike an airplane, a spacecraft cannot make rapid turns; therefore, centrifugal acceleration is of little importance except when the spacecraft goes into abnormal gyrations. On the other hand,

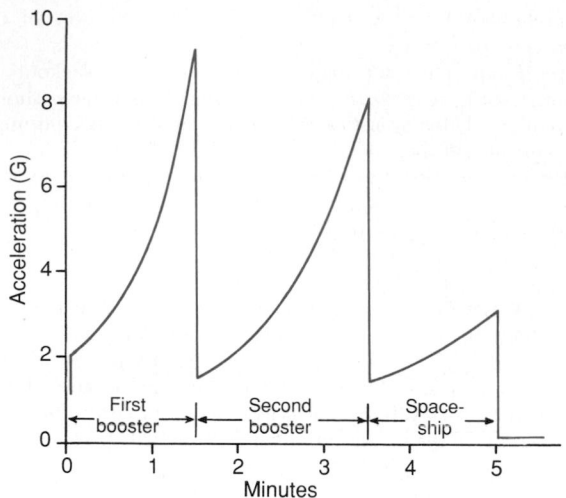

Figure 43–4. Acceleratory forces during the takeoff of a spacecraft.

a distance of 1400 feet, the person will be falling at a "terminal velocity" of 109 to 119 miles per hour (175 feet per second).

If the parachutist has already reached the terminal velocity of fall before opening the parachute, an "opening shock load" of up to 1200 pounds can occur on the parachute shrouds.

The usual-sized parachute slows the fall of the parachutist to about one-ninth the terminal velocity. In other words, the speed of landing is about 20 feet per second, and the force of impact against the earth is ⅟₈₁ the impact force without a parachute. Even so, the force of impact is still great enough to cause considerable damage to the body unless the parachutist is properly trained in landing. Actually, the force of impact with the earth is about the same as that which would be experienced by jumping from a height of about 6 feet. Unless forewarned, the parachutist will be tricked by the senses into striking the earth with extended legs, and this will result in tremendous deceleratory forces along the skeletal axis of the body, resulting in fracture of the pelvis, vertebrae, or leg. Consequently, the trained parachutist strikes the earth with knees bent but muscles taut to cushion the shock of landing.

blast-off acceleration and landing deceleration can be tremendous; both of these are types of linear acceleration.

Figure 43–4 shows a typical profile of the acceleration during blast-off in a three-stage spacecraft, demonstrating that the first-stage booster causes acceleration as high as 9 G and the second-stage booster, as high as 8 G. In the standing position, the human body could not withstand this much acceleration, but in a semireclining position *transverse to the axis of acceleration,* this amount of acceleration can be withstood with ease despite the fact that the acceleratory forces continue for as long as several minutes at a time. Therefore, we see the reason for the reclining seats used by the astronauts.

Problems also occur during deceleration when the spacecraft re-enters the atmosphere. A person traveling at Mach 1 (the speed of sound and of fast airplanes) can be safely decelerated in a distance of about 0.12 mile, whereas a person traveling at a speed of Mach 100 (a speed possible in interplanetary space travel) would require a distance of about 10,000 miles for safe deceleration. The principal reason for this difference is that the total amount of energy that must be dispelled during deceleration is proportional to the *square* of the velocity, which alone increases the required distance about 10,000-fold. But in addition to this, a human being can withstand far less deceleration if the period of deceleration lasts for a long time than for a short time. Therefore, deceleration must be accomplished much more slowly from the high velocities than is necessary at lower velocities.

DECELERATORY FORCES ASSOCIATED WITH PARACHUTE JUMPS. When the parachuting aviator leaves the airplane, the velocity of fall is exactly 0 feet per second at first. However, because of the acceleratory force of gravity, within 1 second, the velocity of fall is 32 feet per second (if there is no air resistance), in 2 seconds it is 64 feet per second, and so on. As the rate of fall increases, the air resistance tending to slow the fall also increases. Finally, the deceleratory force of the air resistance exactly balances the acceleratory force of gravity, so that after falling for about 12 seconds (when near to the surface of the earth, where air density is greatest) and

"ARTIFICIAL CLIMATE" IN THE SEALED SPACECRAFT

Because there is no atmosphere in outer space, an artificial atmosphere and climate must be provided. Most important, the oxygen concentration must remain high enough and the carbon dioxide concentration low enough to prevent suffocation. In some of the earlier space missions, a capsule atmosphere containing pure oxygen at about 260 mm Hg pressure was used, but in the space shuttle, gases about equal to those in normal air are used, with four times as much nitrogen as oxygen and a total pressure of 760 mm Hg. The presence of nitrogen in the mixture greatly diminishes the likelihood of fire and explosion. It also protects against the development of local patches of lung atelectasis that often occur when breathing pure oxygen because oxygen is absorbed rapidly when small bronchi are temporarily blocked by mucous plugs.

For space travel lasting more than several months, it will be impractical to carry along an adequate oxygen supply and enough carbon dioxide absorbent. For this reason, "recycling techniques" have been proposed for use of the same oxygen over and over again. Some recycling processes depend on purely physical procedures, such as distillation and electrolysis of water, to release oxygen. Others depend on biological methods, such as use of algae with their large store of chlorophyll, to generate foodstuffs and at the same time release oxygen from carbon dioxide by photosynthesis. A completely practical system for recycling is yet to be achieved.

WEIGHTLESSNESS IN SPACE

A person in an orbiting satellite or a nonpropelled spacecraft experiences *weightlessness.* That is, the person is not drawn toward the bottom, sides, or top of the spacecraft but simply floats inside its chambers. The cause of this is not failure of gravity to pull on the body because gravity from any nearby heavenly body is still active. However, the gravity is exactly balanced by the centrifugal force of the orbital trajectory acting on both the spacecraft and the person at the same time, so that both are pulled with exactly the same acceleratory forces and in the same direction. For this reason, the person simply is not attracted toward any wall of the spacecraft.

PHYSIOLOGICAL PROBLEMS OF WEIGHTLESSNESS. The physiological problems of weightlessness have not proved to be severe. Most of the problems that do occur appear to be related to three effects of the weightlessness: (1) motion sickness during the first few days of travel, (2) translocation of fluids within the body because of failure of gravity to cause hydrostatic pressures, and (3) diminished physical activity because no strength of muscle contraction is required to oppose the force of gravity.

Almost 50 per cent of astronauts experience motion sickness, with nausea and sometimes vomiting, during the first 2 to 5 days of space travel. This probably results from an unfamiliar pattern of motion signals arriving in the equilibrium centers of the brain but a lack of gravitational signals.

The observed effects of prolonged stay in space are the following: (1) decrease in blood volume, (2) decrease in red cell mass, (3) decrease in muscle strength and work capacity, (4) decrease in maximum cardiac output, and (5) loss of calcium and phosphate from the bones as well as loss of bone mass. Most of these same effects also occur in people who lie in bed for an extended period of time. For this reason, extensive exercise programs are carried out during prolonged space laboratory missions, and most of the effects above are greatly reduced, except for some of the bone loss. In previous space laboratory expeditions in which the exercise program had been less vigorous, the astronauts had severely decreased work capacities for the first few days after returning to earth. They also had a tendency to faint (and still do to some extent) when they stood up during the first day or so after return to gravity because of their diminished blood volume and perhaps diminished responses of the arterial pressure control mechanisms.

The effects of weightlessness that do occur usually reach their maximum within the first few days to weeks after the astronaut enters the space environment and are not progressive thereafter, with one exception: the bone loss continues for many months, presumably because the stimulation of bone deposition requires the extra stresses of gravity. Nevertheless, it appears that with an appropriate exercise program, the physiological effects of weightlessness will not be a serious problem even during prolonged space voyages.

REFERENCES

American Physiological Society: High Altitude and Man. Washington, D.C., American Physiological Society, 1984.

Blomqvist, C. G., and Stone, H. L.: Cardiovascular adjustments to gravitational stress. In Shepard, J. T., and Abboud, F. M. (eds.): Handbook of Physiology. Sec. 2, Vol. III, Bethesda, Md., American Physiological Society, 1983, p. 1025.

Brendel, W. (ed.): High Altitude Physiology and Medicine. I. Physiology of Adaptation. New York, Springer-Verlag, 1982.

Burton, R. R., et al.: Man at high sustained +G acceleration: A review. Aerospace Med., 45:1115, 1974.

Coleridge, H. M., and Coleridge, J. C. G.: Pulmonary reflexes: neural mechanisms of pulmonary defense. Annu. Rev. Physiol., 56:69, 1994.

DeHart, R. L. (ed.): Fundamentals of Aerospace Medicine. Philadelphia, Lea & Febiger, 1985.

Driedzic, W. R., and Gesser, H.: Energy metabolism and contractility in ectothermic vertebrate hearts: hypoxia, acidosis, and low temperature. Physiol. Rev., 74:221, 1994.

Grover, R. F., et al.: High-altitude pulmonary edema. In Fishman, A. P., and Renkin, E. M. (eds.): Pulmonary Edema. Baltimore, Waverly Press, 1979, p. 229.

Honig, A.: Salt and water metabolism in acute high-altitude hypoxia: Role of peripheral arterial chemoreceptors. News Physiol. Sci., 4:109, 1989.

Life Sciences Report, National Aeronautics and Space Administration, Washington, D.C., December 1987.

McCormack, P. D., et al. (eds.): Terrestrial Space Radiation and Its Biological Effects. New York, Plenum Publishing Corp., 1988.

Monge, C., and León-Velarde, F.: Physiological adaptation to high altitude: oxygen transport in mammals and birds. Physiol. Rev., 71:1135, 1991.

Monge, C., et al.: Laying eggs at high altitude. News Physiol. Sci., 3:69, 1988.

Nicogossian, A. E., et al.: Space Physiology and Medicine. Baltimore, Williams & Wilkins, 1994.

Proceedings of the Ninth Annual Meeting of the IUPS Commission on Gravitational Physiology. Physiologist (Suppl.) 31(1), 1988.

Quintanilha, A. (ed.): Reactive Oxygen Species in Chemistry, Biology, and Medicine. New York, Plenum Publishing Corp., 1988.

Reynolds, H. Y.: Immunologic system in the respiratory tract. Physiol. Rev., 71:1117, 1991.

Sloan, A. W.: Man in Extreme Environments. Springfield, Ill., Charles C Thomas, 1979.

Smith, E. E., and Crowell, J. W.: Role of the hematocrit in altitude acclimatization. Aerospace Med., 38:39, 1966.

Smith, E. E., and Crowell, J. W.: Influence of hematocrit ratio on survival of unacclimatized dogs at simulated high altitude. Am. J. Physiol., 205:1172, 1063.

Staub, N. C.: Pulmonary intravascular macrophages. Annu. Rev. Physiol. 50:47, 1994.

Sutton, J. R., et al.: Exercise at altitude. Annu. Rev. Physiol., 45:427, 1983.

Talbot, J. M., and Fisher, K. D.: Space sickness. Physiologist, 27:423, 1984.

West, J. B.: Climbing Mount Everest without oxygen. News Physiol. Sci., 1:25, 1986.

West, J. B.: Human physiology at extreme altitude on Mount Everest. Science, 223:784, 1984.

West, J. B.: Man in space. News Physiol. Sci., 1:198, 1986.

Yu, B. P.: Cellular defenses against damage from reactive oxygen species. Physiol. Rev., 74:139, 1994.

Physiology of Deep Sea Diving and Other Hyperbaric Conditions

CHAPTER 44

When humans descend beneath the sea, the pressure around them increases tremendously. To keep the lungs from collapsing, air must be supplied also under high pressure. This exposes the blood in the lungs to extremely high alveolar gas pressure, a condition called *hyperbarism*. Beyond certain limits, these high pressures can cause tremendous alterations in the body physiology.

RELATIONSHIP OF SEA DEPTH TO PRESSURE. A column of sea water 33 feet deep exerts the same pressure at its bottom as all the atmosphere above the earth. Therefore, a person 33 feet beneath the ocean surface is exposed to a pressure of 2 atmospheres, 1 atmosphere of pressure caused by the air above the water and the second atmosphere by the weight of the water itself. At 66 feet the pressure is 3 atmospheres and so forth, in accord with the table in Figure 44–1.

EFFECT OF DEPTH ON THE VOLUME OF GASES—BOYLE'S LAW. Another important effect of depth is the compression of gases to smaller and smaller volumes. The lower part of Figure 44–1 shows a bell jar at sea level containing 1 liter of air. At 33 feet beneath the sea, where the pressure is 2 atmospheres, the volume has been compressed to only one-half liter and at 8 atmospheres (233 feet) to one-eighth liter. Thus, the volume to which a given quantity of gas is compressed is inversely proportional to the pressure. This is the physical principle called *Boyle's law*, which is extremely important in diving because increased pressures can collapse air chambers of the diver's body, including the lungs, and often cause serious damage.

Many times in this chapter it is necessary to refer to *actual volume* versus *sea-level volume*. For instance, we might speak of an actual volume of 1 liter at a depth of 300 feet; this is the same *quantity* of air as a sea-level volume of 10 liters.

EFFECT OF HIGH PARTIAL PRESSURES OF GASES ON THE BODY

The gases to which a diver breathing air is normally exposed are nitrogen, oxygen, and carbon dioxide, and each of these at times can cause serious physiological effects at high pressures.

NITROGEN NARCOSIS AT HIGH NITROGEN PRESSURES. About four fifths of the air is nitrogen. At sea level pressure, the nitrogen has no known effect on bodily function, but at high pressures it can cause varying degrees of narcosis. When the diver remains beneath the sea for an hour or more and is breathing compressed air, the depth at which the first symptoms of mild narcosis appear is about 120 feet, at which level he begins to exhibit joviality and to lose many of his cares. At 150 to 200 feet, he becomes drowsy. At 200 to 250 feet, his strength wanes considerably, and he often becomes too clumsy to perform the work required of him. Beyond 250 feet (8.5 atmospheres pressure), the diver usually becomes almost useless as a result of nitrogen narcosis if he remains at these depths too long.

Nitrogen narcosis has characteristics similar to those of alcohol intoxication, and for this reason it has frequently been called "raptures of the depths."

The mechanism of the narcotic effect is believed to be the same as that of essentially all the gas anesthetics. That is, nitrogen dissolves freely in the fats of the body, and it is presumed that it, like most other anesthetic gases, dissolves in the membranes of the neurons and, because of its *physi-*

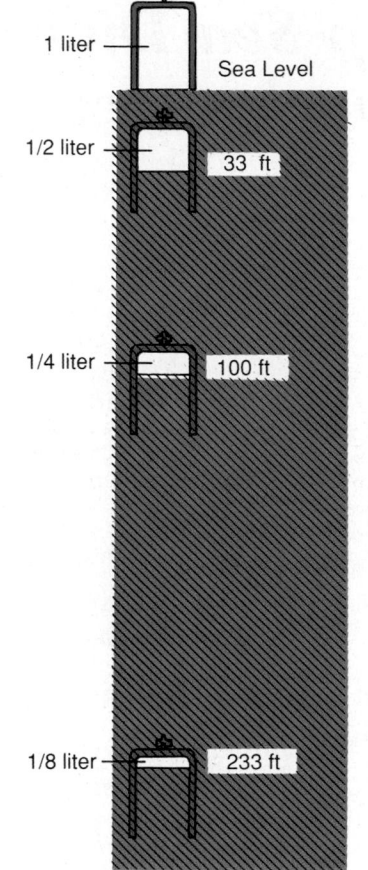

Depth (feet)	Atmosphere (s)
Sea level	1
33	2
66	3
100	4
133	5
166	6
200	7
300	10
400	13
500	16

Figure 44–1. Effect of sea depth on pressure *(top table)* and on gas volumes *(bottom).*

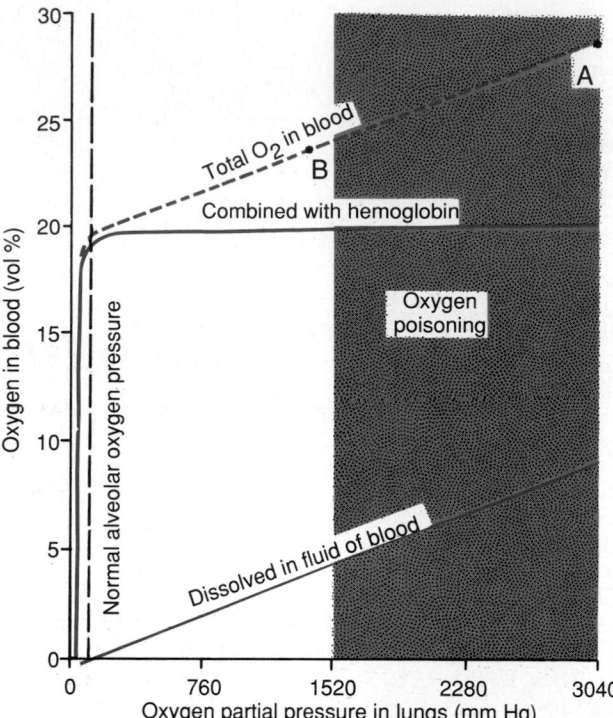

Figure 44–2. Quantity of oxygen dissolved in the fluid of the blood and in combination with hemoglobin at very high Po_2s.

cal effect on altering ionic conductance through the membranes, reduces neuronal excitability.

Oxygen Toxicity at High Pressures

Effect of Extremely High Po_2 on Blood Oxygen Transport. When the Po_2 in the blood rises far above 100 mm Hg, the amount of oxygen dissolved in the water of the blood increases markedly. This is shown in Figure 44–2, which depicts the same oxygen-hemoglobin dissociation curve as that shown in Chapter 40, except that the alveolar Po_2 is now extended to more than 3000 mm Hg. Also depicted is the volume of oxygen dissolved in the fluid of the blood at each Po_2 level. Note that in the normal range of alveolar Po_2, almost none of the total oxygen in the blood is accounted for by dissolved oxygen, but as the pressure rises progressively into the thousands of millimeters of mercury, a large portion of the total oxygen is then dissolved, rather than bound with hemoglobin.

Effect of High Alveolar Po_2 on Tissue Po_2. Let us assume that the Po_2 in the lungs is about 3000 mm Hg (4 atmospheres pressure). Referring to Figure 44–2, one finds that this represents a total oxygen content in each 100 milliliters of blood of about 29 volumes per cent, as demonstrated by point A in the figure—20 volumes per cent bound with hemoglobin and 9 volumes per cent dissolved in the blood water. As this blood passes through the tissue capillaries and the tissues use their normal amount of oxygen, about 5 milliliters from each 100 milliliters of blood, the oxygen content on leaving the tissue capillaries is still 24 volumes per cent (point B in the figure). At this point, the Po_2 is still about 1200 mm Hg, which means that oxygen is delivered to the tissues at this extremely high pressure instead of at the normal value of 40 mm Hg. Thus, once the alveolar Po_2 rises above a critical level, the hemoglobin-oxygen buffer mechanism (discussed in Chapter 40) is no longer capable of keeping the tissue Po_2 in the normal safe range between 20 and 60 mm Hg.

ACUTE OXYGEN POISONING. Because of the extremely high tissue Po_2 that occurs when oxygen is breathed at a very high alveolar oxygen pressure, one can readily understand that this can be detrimental to many of the body's tissues. This is especially true of the brain. In fact, exposure to 4 atmospheres pressure of oxygen (Po_2 = 3040 mm Hg) will cause seizures followed by coma in most people after 30 minutes. The seizures often occur without warning and, for obvious reasons, are likely to be lethal to divers submerged beneath the sea.

Other symptoms encountered in acute oxygen poisoning include *nausea, muscle twitchings, dizziness, disturbances of vision, irritability,* and *disorientation.*

Exercise greatly increases a diver's susceptibility to oxygen toxicity, causing symptoms to appear much earlier and with far greater severity than in the resting person.

Excessive Intracellular Oxidation as the Cause of Nervous System Oxygen Toxicity—"Oxidizing Free Radicals." Molecular oxygen (O_2) has little capability of oxidizing other chemical compounds. Instead, it must first be converted into an "active" form of oxygen. There are several forms of active oxygen; they are called *oxygen free radicals*. One of the most important of these is the superoxide free radical O_2^-, and another is the peroxide radical in the form of hydrogen peroxide. Even when the tissue P_{O_2} is normal at the level of 40 mm Hg, small amounts of free radicals are continually being formed from the dissolved molecular oxygen. However, the tissues also contain multiple enzymes that rapidly remove these free radicals, including especially peroxidases, catalases, and superoxide dismutases. Therefore, so long as the hemoglobin-oxygen buffering mechanism functions properly and maintains a normal tissue P_{O_2}, the oxidizing free radicals are removed so rapidly that they have little or no effect in the tissues.

On the other hand, above a critical alveolar P_{O_2} (above about 2 atmospheres P_{O_2}), the hemoglobin-oxygen buffering mechanism fails, and the tissue P_{O_2} can then rise to hundreds or thousands of millimeters of mercury. Then, the amounts of oxidizing free radicals literally swamp the enzyme systems for removing them, and now they do have serious destructive and even lethal effects on the cells. One of the principal effects is to oxidize the polyunsaturated fatty acids that are essential components of many of the membranous structures of the cells, and another effect is to oxidize some of the cellular enzymes, thus damaging severely the cellular metabolic systems. The nervous tissues are especially susceptible because of their high lipid content. Therefore, most of the acute lethal effects of acute oxygen toxicity are related to brain dysfunction.

Chronic Oxygen Poisoning Causes Pulmonary Disability. A person can be exposed to only 1 atmosphere pressure of oxygen almost indefinitely without developing the *acute* oxygen toxicity of the nervous system just described. However, after only 12 or so hours of 1 atmosphere oxygen exposure, *lung passageway congestion, pulmonary edema*, and *atelectasis* caused by damage to the linings of the bronchi and alveoli begin to develop. The reason for this effect in the lungs and not in the other tissues is that the air spaces of the lungs are directly exposed to the high oxygen pressure, whereas oxygen is delivered to the other tissues at almost normal P_{O_2} because of the hemoglobin-oxygen buffer system, so long as the air P_{O_2} remains less than about 2 atmospheres.

CARBON DIOXIDE TOXICITY AT GREAT DEPTHS IN THE SEA. If the diving gear is properly designed and functions properly, the diver has no problem due to carbon dioxide toxicity because depth alone does not increase the carbon dioxide partial pressure in the alveoli. This is true because depth does not increase the rate of carbon dioxide production in the body; as long as the diver continues to breathe a normal tidal volume, he continues to expire the carbon dioxide as it is formed, maintaining his alveolar carbon dioxide partial pressure at a normal value.

In certain types of diving gear, however, such as the diving helmet and the different types of rebreathing apparatuses, carbon dioxide can frequently build up in the dead space air of the apparatus and be rebreathed by the diver. Up to an alveolar carbon dioxide pressure (P_{CO_2}) of about 80 mm Hg, twice that of normal alveoli, the diver either tolerates or partly tolerates this buildup, his minute respiratory volume increasing up to a maximum of 8- to 11-fold to compensate for the increased carbon dioxide. Beyond the 80-mm Hg level the situation becomes intolerable, and eventually the respiratory center begins to be depressed, rather than excited, because of the negative metabolic effects of the P_{CO_2}. The diver's respiration then begins to fail, rather than to compensate. In addition, the diver develops severe respiratory acidosis; varying degrees of lethargy, narcosis, and finally anesthesia ensue, as discussed in Chapter 42.

Decompression of the Diver After Exposure to High Pressures

When a person breathes air under high pressure for a long time, the amount of nitrogen dissolved in the body fluids becomes great. The reason for this is the following: The blood flowing through the pulmonary capillaries becomes saturated with nitrogen to the same high pressure as that in the breathing mixture. Over several hours, enough nitrogen is carried to all the tissues of the body to saturate the tissues also with dissolved nitrogen. Because nitrogen is not metabolized by the body, it remains dissolved until the nitrogen pressure in the lungs decreases, at which time the nitrogen is then removed by the reverse respiratory process, but this removal takes hours to occur, which is the source of multiple problems collectively called "decompression sickness."

VOLUME OF NITROGEN DISSOLVED IN THE BODY FLUIDS AT DIFFERENT DEPTHS. At sea level, almost 1 liter of nitrogen is dissolved in the entire body. A little less than one half of this is dissolved in the water of the body and a little more than one half in the fat of the body. This is true despite the fact that fat constitutes only 15 per cent of the normal body because nitrogen is five times as soluble in fat as in water.

After the diver has become saturated with nitrogen, the *sea-level volume of nitrogen* dissolved in the body at the different depths is as follows:

Feet	Liters
33	2
100	4
200	7
300	10

Several hours are required for the gas pressures of nitrogen in all the body tissues to come nearly to equilibrium with the gas pressure of nitrogen in the alveoli. The reason for this is that the blood does not flow rapidly enough and the nitrogen does not diffuse rapidly enough to cause instantaneous equilibrium. The nitrogen dissolved in the water of the body comes to almost complete equilibrium in less than 1 hour, but the fat, requiring five times as much transport of nitrogen and having a relatively poor blood supply, reaches equilibrium only after several hours. For this reason, if a person remains at deep levels for only a few minutes, not much nitrogen dissolves in the body fluids and tissues, whereas if the person remains at a deep level for several hours, both the fluids and the tissues become almost saturated with nitrogen.

DECOMPRESSION SICKNESS (SYNONYMS: BENDS, COMPRESSED AIR SICKNESS, CAISSON DISEASE, DIVER'S PARALYSIS, DYSBARISM). If a diver has been beneath the sea long enough that large amounts of nitrogen have dissolved in his body and then he suddenly comes back to the surface of the

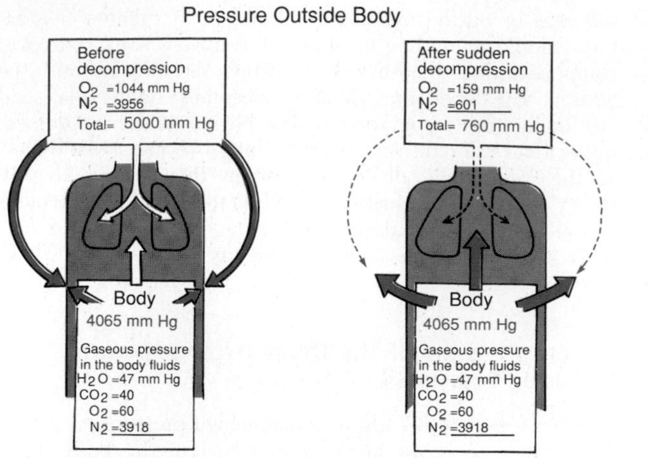

Pressure Outside Body

Figure 44–3. Gaseous pressure both inside and outside the body, showing at left saturation of the body to high gas pressures when breathing air at a total pressure of 5000 mm Hg, and at right the great excess of intrabody pressure that is responsible for bubble formation in the tissues when the body is returned to the normal pressure of 760 mm Hg.

sea, significant quantities of nitrogen bubbles can develop in his body fluids either intracellularly or extracellularly and cause minor or serious damage in almost any area of the body, depending on the number and sizes of bubbles formed; this is "decompression sickness."

The principles underlying bubble formation are shown in Figure 44–3. To the left, the diver's tissues have become equilibrated to a high nitrogen pressure, 3918 mm Hg, dissolving about 6.5 times the normal amount of nitrogen in the tissues. As long as the diver remains deep beneath the sea, the pressure against the outside of his body (5000 mm Hg) compresses all the body tissues sufficiently to keep the gases still dissolved. But when the diver suddenly rises to sea level, the pressure on the outside of his body becomes only 1 atmosphere (760 mm Hg), whereas the pressure inside the body fluids is the sum of the pressures of water vapor, carbon dioxide, oxygen, and nitrogen, or a total of 4065 mm Hg, which is far greater than the pressure on the outside of the body and about 97 per cent of which is caused by the dissolved nitrogen. Therefore, the gases can escape from the dissolved state and form actual bubbles both in the tissues and especially in the blood, where they plug the small blood vessels. The bubbles may not appear for many minutes to hours because sometimes the gases can remain dissolved in the "super-saturated" state for hours before bubbling.

Exercise hastens the formation of bubbles during decompression because of increased agitation of the tissues and fluids. This is an effect analogous to that of shaking an opened bottle of soda pop to release the bubbles.

Symptoms of Decompression Sickness. Most of the symptoms of decompression sickness are caused by gas bubbles blocking blood vessels in the different tissues. At first, only the smallest vessels are blocked by minute bubbles, but as the bubbles coalesce, progressively larger vessels are affected. Tissue ischemia and sometimes tissue death are the result.

In most people with decompression sickness, the symptoms are pain in the joints and muscles of the legs or arms, affecting about 89 per cent of those who develop decompression sickness. The joint pain accounts for the term "bends" that is often applied to this condition.

In 5 to 10 per cent of people with decompression sickness, nervous system symptoms occur, ranging from dizziness in about 5 per cent to paralysis or collapse and unconsciousness in as many as 3 per cent. The paralysis may be temporary, but in some instances, the damage is permanent.

Finally, about 2 per cent of people with decompression sickness develop "the chokes," caused by massive numbers of microbubbles plugging the capillaries of the lungs; this is characterized by serious shortness of breath, often followed by severe pulmonary edema and, occasionally, death.

NITROGEN ELIMINATION FROM THE BODY; DECOMPRESSION TABLES. If a diver is brought to the surface slowly, the dissolved nitrogen is eliminated through the lungs rapidly enough to prevent decompression sickness. About two thirds of the total nitrogen is liberated in 1 hour and about 90 per cent in 6 hours.

Special decompression tables have been prepared by the U.S. Navy that detail procedures for safe decompression. To give the student an idea of the decompression process, a diver who has been breathing air and has been on the bottom for 60 minutes at a depth of 190 feet is decompressed according to the following schedule:

10 minutes at 50 feet depth
17 minutes at 40 feet depth
19 minutes at 30 feet depth
50 minutes at 20 feet depth
84 minutes at 10 feet depth

Thus, for a work period on the bottom of only 1 hour, the total time for decompression is about 3 hours.

TANK DECOMPRESSION, AND TREATMENT OF DECOMPRESSION SICKNESS. Another procedure widely used for decompression of professional divers is to put the diver into a pressurized tank and then to lower the pressure gradually back to normal atmospheric pressure, using essentially the same time schedule as noted above.

Tank decompression is even more important for treating people in whom symptoms of decompression sickness develop minutes or even hours after they have returned to the surface. In this case, the diver is recompressed immediately to a deep level. Then decompression is carried out over a time period several times as long as the usual decompression period.

"SATURATION DIVING" AND USE OF HELIUM-OXYGEN MIXTURES IN DEEP DIVES. When divers must work at very deep levels—between 250 feet and nearly 1000 feet—they frequently live in a large compression tank for days or weeks at a time, remaining compressed at a pressure level near that at which they will be working. This keeps the tissues and fluids of the body saturated with the gases to which they will be exposed while diving. Then, when they return to the same tank after working, there are not significant changes in pressure, and so decompression bubbles do not occur.

In very deep dives, especially during saturation diving, helium is usually used in the gas mixture instead of nitrogen for three reasons: (1) it has only about one-fifth the narcotic effect of nitrogen; (2) only about one half as much volume of helium dissolves in the body tissues as nitrogen, and the volume that does dissolve diffuses out of the tissues several times as rapidly as does nitrogen, thus reducing the problem of decompression sickness; and (3) the low density of helium (one-seventh the density of nitrogen) keeps the airway resistance for breathing at a minimum, which is extremely important because highly compressed nitrogen is so dense that airway resistance can become extreme, sometimes making the work of breathing beyond endurance.

Finally, in very deep dives it is important to reduce the oxygen concentration in the gaseous mixture because otherwise oxygen toxicity would result. For instance, at a depth of 700 feet (22 atmospheres of pressure), a 1 per cent oxygen mixture will provide all the oxygen required by the diver, whereas a 21 per cent mixture of oxygen (the percentage in air) delivers a Po_2 to the lungs of more than 4 atmospheres, a level likely to cause seizures in as little as 30 minutes.

SCUBA DIVING (SELF-CONTAINED UNDERWATER BREATHING APPARATUS)

Before the 1940s, almost all diving was done using a diving helmet connected to a hose through which air was pumped to the diver from the surface. Then, in 1943, Jacques Cousteau developed and popularized the *self-contained underwater breathing apparatus,* popularly known as the SCUBA apparatus. The type of SCUBA apparatus used in more than 99 per cent of all sports and commercial diving is the *open circuit demand system* shown in Figure 44–4. This system consists of the following components: (1) one or more tanks of compressed air or some other breathing mixture, (2) a first-stage "reducing" valve for reducing the pressure from the tanks to a constant low pressure level, (3) a combination inhalation "demand" valve and exhalation valve that allows air to be pulled into the lungs with slight negative pressure of breathing and then to be exhausted into the sea at slight positive pressure, and (4) a mask and tube system with small "dead space."

The demand system operates as follows: The first-stage reducing valve reduces the pressure from the tanks so that the air delivered to the mask has a pressure slightly greater than the surrounding water pressure. The breathing mixture does not flow continually into the mask. Instead, with each inspiration, slight negative pressure in the demand valve on the mask pulls the diaphragm of the demand valve inward, and this automatically releases air from the hose into the mask and lungs. In this way, only the amount of air needed for inhalation enters the mask. Then, on expiration, the air cannot go back into the tank but instead is expired into the sea through the expiration valve.

The most important problem in use of the self-contained underwater breathing apparatus is the time limit that one can remain beneath the surface; for instance, only a few minutes are possible at a 200-foot depth. The reason for this is that tremendous airflow from the tanks is required to wash carbon dioxide out of the lungs—the greater the depth, the greater the airflow in terms of *quantity* of air that is required because the *volumes* have been compressed to small sizes.

SPECIAL PHYSIOLOGICAL PROBLEMS IN SUBMARINES

ESCAPE FROM SUBMARINES. Essentially the same problems as those of deep sea diving are often met in relation to submarines, especially when it is necessary to escape from a submerged submarine. Escape is possible from as deep as 300 feet without using any special type of apparatus. However, proper use of rebreathing devices, especially when using helium, theoretically can allow escape from as deep as 600 feet or perhaps more.

One of the major problems of escape is prevention of air embolism. As the person ascends, the gases in the lungs expand and sometimes rupture a pulmonary vessel, allowing the gases to enter the pulmonary vascular system and cause embolism of the circulation. Therefore, as the person ascends, he must consciously exhale continually.

Exhalation of the expanding gases from the lungs during ascent, even without breathing, is often rapid enough to blow off the accumulating carbon dioxide in the lungs. This keeps the concentration of carbon dioxide from building up in the blood and keeps the person from having the desire to breathe. Therefore, he can hold his breath for an extra long time during ascent.

HEALTH PROBLEMS IN THE SUBMARINE INTERNAL ENVIRONMENT. Except for escape, submarine medicine generally centers around several engineering problems to keep hazards out of the internal environment. In atomic submarines, there exists the problem of radiation hazards, but with appropriate shielding, the amount of radiation received by the crew submerged beneath the sea has been less than the normal radiation received above the surface of the sea from cosmic rays.

Second, poisonous gases on occasion escape into the atmosphere of the submarine and must be controlled early. For instance, during several weeks' submergence, cigarette smoking by the crew can liberate sufficient amounts of carbon monoxide, if it is not removed from the air, to cause carbon monoxide poisoning, and on occasion, even freon gas has been found to diffuse through the tubes in refrigeration systems in sufficient quantity to cause toxicity.

HYPERBARIC OXYGEN THERAPY

It has been learned that the intense oxidizing properties of high-pressure oxygen (*hyperbaric oxygen*) can have valuable therapeutic effects in several important clinical conditions. Therefore, large pressure tanks are now available in many medical centers into which patients can be placed and treated with hyperbaric oxygen. The oxygen is usually administered at Po_2s of 2 to 3 atmospheres of pressure through a mask or intratracheal tube, while the gas around the body

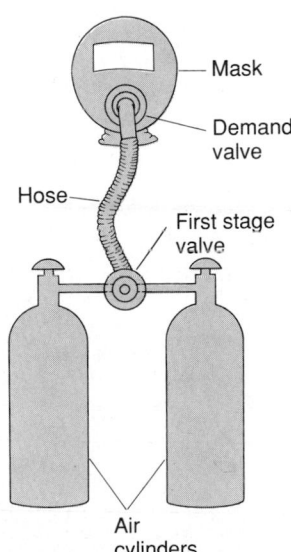

Figure 44–4. Open-circuit demand type of SCUBA apparatus.

is normal air compressed to the same high-pressure level. It is believed that the same oxidizing free radicals responsible for oxygen toxicity are also responsible for the therapeutic benefits. Some of the conditions in which hyperbaric oxygen therapy has been especially beneficial are the following.

Probably the most successful use of hyperbaric oxygen has been in the treatment of *gas gangrene*. The bacteria that cause this condition, *clostridial organisms,* grow best under anaerobic conditions and stop growing at oxygen pressures greater than about 70 mm Hg. Therefore, hyperbaric oxygenation of the tissues can frequently stop the infectious process entirely and thus convert a condition that formerly was almost 100 per cent fatal into one that is cured in most instances of early treatment with hyperbaric therapy.

Results also suggest that hyperbaric oxygenation might have almost as dramatic an effect in curing *leprosy* as in curing gas gangrene—also because of the susceptibility of the leprosy bacillus to destruction by high oxygen pressures.

Other conditions in which hyperbaric oxygen therapy has been either valuable or possibly valuable include decompression sickness, arterial gas embolism, carbon monoxide poisoning, osteomyelitis, and myocardial infarction.

REFERENCES

Bennett, P. B., and Elliott, D. H.: The Physiology and Medicine of Diving. Philadelphia, W. B. Saunders Co., 1993.

Bove, A. A., and Davis, J. C.: Diving Medicine. Philadelphia, W. B. Saunders Co., 1990.

Brauer, R. W.: Problems of exposure to high pressures. News Physiol. Sci., 1:192, 1986.

Burakovsky, V. I.: Hyperbaric Oxygenation and Its Value in Cardiovascular Surgery. Chicago, Imported Publications, 1981.

Cole, T.: Deep duty. Popular Mechanics, November, 1986, p. 71.

Crapo, J. D.: Morphologic changes in pulmonary oxygen toxicity. Annu. Rev. Physiol. 48:721, 1986.

Crystal, R. G., and West, J. B.: Lung Injury. New York, Raven Press, 1992.

Elsner, R., and de Burgh Daly, M.: Coping with asphyxia: Lessons from seals. News Physiol. Sci., 3:65, 1988.

Epstein, M.: Renal effects of head-out water immersion in humans: a 15-year update. Physiol. Rev., 72:563, 1992.

Fisher, A. B., et al.: Oxygen toxicity of the lung: Biochemical aspects. In Fishman, A. P., and Renkin, E. M. (eds.): Pulmonary Edema. Baltimore, Waverly Press, 1979, p. 207.

Fridovich, I., and Freeman, B.: Antioxidant defenses in the lung. Annu. Rev. Physiol., 48:693, 1986.

Fridovich, I.: Superoxide radical: An endogenous toxicant. Annu. Rev. Pharmacol. Toxicol., 23:239, 1983.

Gamarra, J. A.: Decompression Sickness. Hagerstown, Md., Harper & Row, 1974.

Gooden, B. A.: Drowning and the diving reflex in man. Med. J. Aust., 2:583, 1972.

Halsey, M. J.: Effects of high pressure on the central nervous system. Physiol. Rev., 62:1341, 1982.

Hochachka, B. W., and Murphy, B.: Metabolic status during diving and recovery in marine animals. In Robertshaw, D. (ed.): International Review of Physiology: Environmental Physiology III. Vol. 20, Baltimore, University Park Press, 1979, p. 253.

Hochachka, B. W., and Storey, K. B.: Metabolic consequences of diving in animals and man. Science, 187:613, 1975.

Jamieson, D., et al.: The relation of free radical production to hyperoxia. Annu. Rev. Physiol., 48:703, 1986.

Miller, J. W.: Vertical Excursions Breathing Air from Nitrogen-Oxygen or Air Saturation Exposures. Washington, D.C., U.S. Government Printing Office, 1976.

Oxygen Free Radicals and Tissue Damage, Ciba Foundation Symposium. New York, Excerpta Medica, 1979.

Pryor, W. A.: Oxy-radicals and related species. Annu. Rev. Physiol., 48:657, 1986.

Shilling, C. W., and Beckett, M. W. (eds.): Underwater Physiology IV. Bethesda, Md., Federation of American Societies for Experimental Biology, 1978.

Sloan, A. W.: Man in Extreme Environments. Springfield, Ill., Charles C Thomas, 1979.

Smith, L. L.: The response of the lung to foreign compounds that produce free radicals. Annu. Rev. Physiol., 48:681, 1986.

Undersea Medical Society: Glossary of Diving and Hyperbaric Terms. Bethesda, Md., Undersea Medical Society, 1978.

Zimmerman, A. M. (ed.): High Pressure Effects on Cellular Processes. New York, Academic Press, 1970.

THE NERVOUS SYSTEM: A. GENERAL PRINCIPLES AND SENSORY PHYSIOLOGY

UNIT IX

Organization of the Nervous System; Basic Functions of Synapses and Transmitter Substances

*C*HAPTER *45*

The nervous system, along with the endocrine system, provides most of the control functions for the body. In general, the nervous system controls the rapid activities of the body, such as muscle contractions, rapidly changing visceral events, and even the rates of secretion of some endocrine glands. The endocrine system, by contrast, regulates principally the metabolic functions of the body.

The nervous system is unique in the vast complexity of the control actions it can perform. It receives literally millions of bits of information from the different sensory organs and then integrates all these to determine the response to be made by the body. The purpose of this chapter is to present, first, a general outline of the overall mechanisms by which the nervous system performs such functions. Then we will discuss the functions of central nervous system synapses, which are the basic structures that control the passage of signals into, through, and out of the nervous system. In succeeding chapters, we analyze in detail the functions of the individual parts of the nervous system. Before beginning this discussion, the reader should refer to Chapters 5 and 7, which present the principles of membrane potentials and transmission of signals in nerves and through neuromuscular junctions.

GENERAL DESIGN OF THE NERVOUS SYSTEM

Central Nervous System Neuron—The Basic Functional Unit

The central nervous system is composed of more than 100 billion neurons. Figure 45–1 shows a typical neuron of a type found in the cerebral motor cortex. The incoming signals enter the neuron through synapses on the neuronal dendrites or cell body; for different types of neurons, there may be only a few hundred or as many as 200,000 such synaptic connections from the input fibers. On the other hand, the output signal travels by way of a single axon leaving the neuron, but this axon has many separate branches to other parts of the nervous system or peripheral body.

A special feature of most synapses is that the signal normally passes only in the forward direction except under rare conditions. This allows the signals to be conducted in the required directions for performing necessary nervous functions. We shall also see that the neurons are organized into a great multitude of neural networks that determine the functions of the nervous system.

Sensory Division of the Nervous System—Sensory Receptors

Most activities of the nervous system are initiated by sensory experience emanating from *sensory receptors*, whether visual receptors, auditory receptors, tactile receptors on the surface of the body, or other kinds of receptors. This sensory experience can cause an immediate reaction, or its memory can be stored in the brain for minutes, weeks, or years and then can help determine bodily reactions at some future date.

Figure 45–2 shows a portion of the sensory system, the *somatic* portion, which transmits sensory information from the receptors of the entire body surface and from some deep structures. This information enters

565

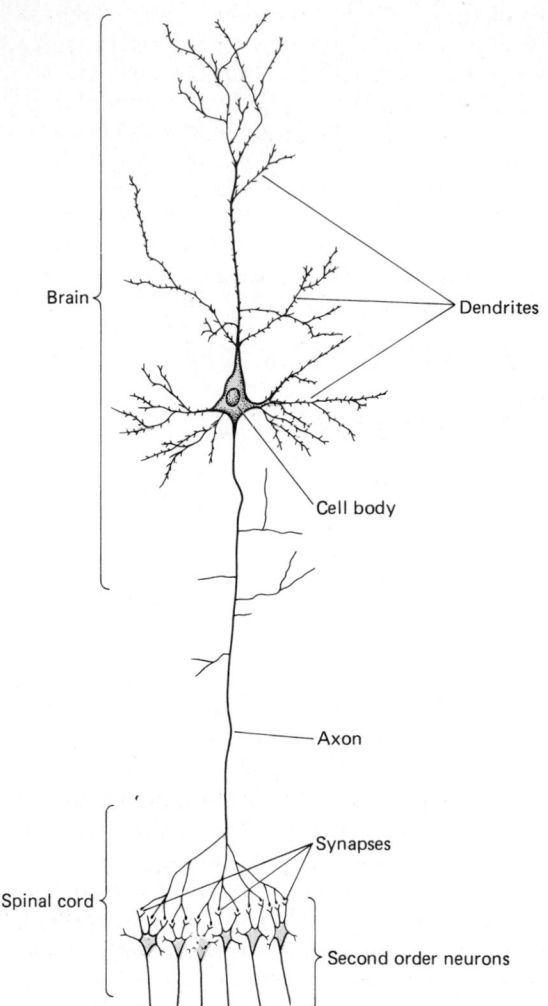

Figure 45–1. Structure of a large neuron of the brain, showing its important functional parts. (From Guyton: Basic Neuroscience: Anatomy and Physiology, Philadelphia, W. B. Saunders Company, 1987.)

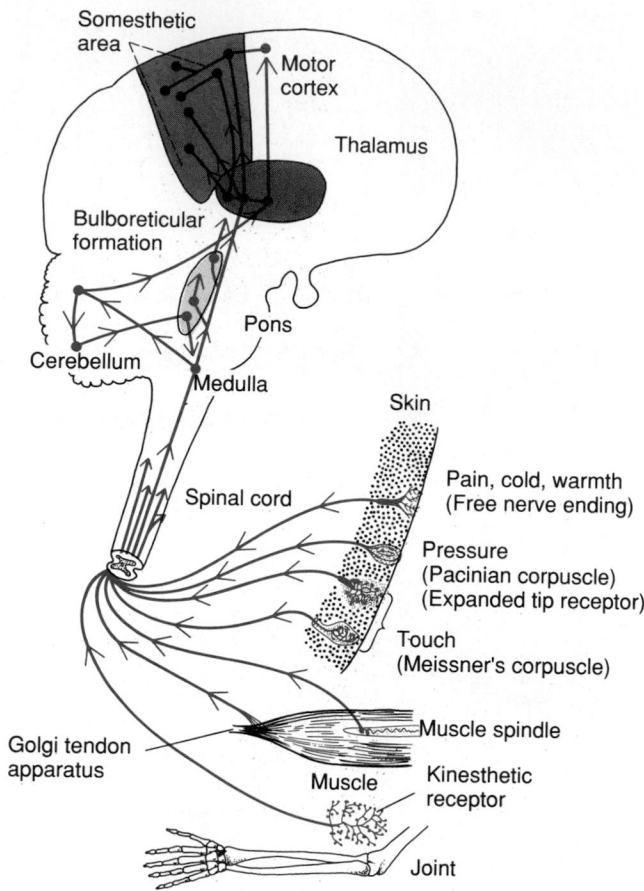

Figure 45–2. Somatic sensory axis of the nervous system.

the central nervous system through the peripheral nerves and is conducted to multiple sensory areas in (1) the spinal cord at all levels; (2) the reticular substance of the medulla, pons, and mesencephalon; (3) the cerebellum; (4) the thalamus; and (5) the somesthetic areas of the cerebral cortex. But in addition to these sensory areas, signals are then relayed to essentially all other parts of the nervous system.

Motor Division—The Effectors

The most important ultimate role of the nervous system is to control the various bodily activities. This is achieved by controlling (1) contraction of skeletal muscles throughout the body, (2) contraction of smooth muscle in the internal organs, and (3) secretion by both exocrine and endocrine glands in many parts of the body. These activities are collectively called *motor functions* of the nervous system, and the muscles and glands are called *effectors* because they perform the functions dictated by the nerve signals.

Figure 45–3 shows the *motor axis* of the nervous system for controlling skeletal muscle contraction. Operating parallel to this axis is another, similar system, called the *autonomic nervous system,* for control of the smooth muscles, glands, and other internal bodily systems; this is discussed in Chapter 60. Note in Figure 45–3 that the skeletal muscles can be controlled from many levels of the central nervous system, including (1) the spinal cord; (2) the reticular substance of the medulla, pons, and mesencephalon; (3) the basal ganglia; (4) the cerebellum; and (5) the motor cortex. Each of these areas plays its own specific role, the lower regions being concerned primarily with the automatic, instantaneous motor responses of the body to sensory stimuli, and the higher regions with deliberate movements controlled by the thought process of the cerebrum.

Processing of Information— "Integrative" Function of the Nervous System

The major function of the nervous system is to process incoming information in such a way that *appropriate* motor responses occur. More than 99 per cent

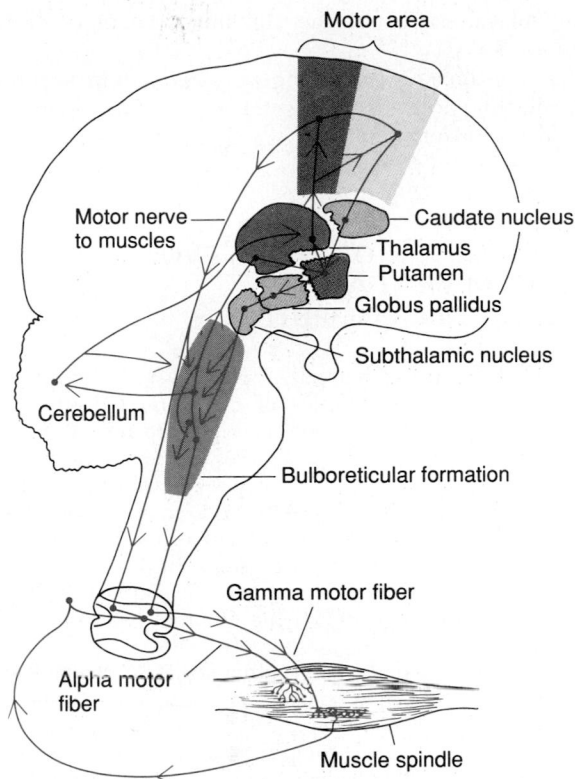

Figure 45–3. Motor axis of the nervous system.

of all sensory information is discarded by the brain as irrelevant and unimportant. For instance, one is ordinarily unaware of the parts of the body that are in contact with clothing as well as of the seat pressure when sitting. Likewise, attention is drawn only to an occasional object in one's field of vision, and even the perpetual noise of our surroundings is usually relegated to the background.

After the important sensory information has been selected, it is then channeled into proper motor regions of the brain to cause the desired responses. This channeling of information is called the *integrative function* of the nervous system. Thus, if a person places a hand on a hot stove, the desired response is to lift the hand. There are other associated responses, too, such as moving the entire body away from the stove and perhaps even shouting with pain. Yet even these responses represent activity by only a small fraction of the total motor system of the body.

ROLE OF SYNAPSES IN PROCESSING INFORMATION. The synapse is the junction point from one neuron to the next and, therefore, is an advantageous site for control of signal transmission. Later in this chapter, we discuss the details of synaptic function. However, it is important to point out here that the synapses determine the directions that the nervous signals spread in the nervous system. Some synapses transmit signals from one neuron to the next with ease, and others transmit signals only with difficulty. Also, facilitatory

and inhibitory signals from other areas in the nervous system can control synaptic transmission, sometimes opening the synapses for transmission and at other times closing them. In addition, some post-synaptic neurons respond with large numbers of output impulses and others respond with only a few. Thus, the synapses perform a selective action, often blocking the weak signals while allowing the strong signals to pass, often selecting and amplifying certain weak signals, and often channeling the signals in many directions, rather than in one direction.

Storage of Information—Memory

Only a small fraction of the important sensory information causes an immediate motor response. Much of the remainder is stored for future control of motor activities and for use in the thinking processes. Most of this information storage occurs in the *cerebral cortex,* but even the basal regions of the brain and perhaps even the spinal cord can store small amounts of information.

The storage of information is the process we call *memory,* and this, too, is a function of the synapses. That is, each time certain types of sensory signals pass through sequences of synapses, these synapses become more capable of transmitting the same signals the next time, which process is called *facilitation.* After the sensory signals have passed through the synapses a large number of times, the synapses become so facilitated that signals generated within the brain itself can also cause transmission of impulses through the same sequences of synapses even though the sensory input has not been excited. This gives the person a perception of experiencing the original sensation, although, in effect, they are only memories of the sensations.

We know much too little about the precise mechanisms by which facilitation of synapses occurs in the memory process, but what is known about this and other details of the memory process are discussed in Chapter 57.

Once memories have been stored in the nervous system, they become part of the processing mechanism. The thought processes of the brain compare new sensory experiences with the stored memories; the memories help to select the important new sensory information and to channel this into appropriate storage areas for future use or into motor areas to cause immediate bodily responses.

MAJOR LEVELS OF CENTRAL NERVOUS SYSTEM FUNCTION

The human nervous system has inherited specific characteristics from each stage of evolutionary development. From this heritage, three major levels of the central nervous system have specific functional attributes: (1) the *spinal cord level,* (2) the *lower brain level,* and (3) the *higher brain* or *cortical level.*

Spinal Cord Level

We often think of the spinal cord as being only a conduit for signals from the periphery of the body to the brain or in the opposite direction from the brain back to the body. This is far from the truth. Even after the spinal cord has been cut in the high neck region, many highly organized spinal cord functions still occur. For instance, neuronal circuits in the cord can cause (1) walking movements, (2) reflexes that withdraw portions of the body from painful objects, (3) reflexes that stiffen the legs to support the body against gravity, and (4) reflexes that control local blood vessels, gastrointestinal movements, and so forth, in addition to many other functions.

In fact, the upper levels of the nervous system often operate not by sending signals directly to the periphery of the body but by sending signals to the control centers of the cord, simply "commanding" the cord centers to perform their functions.

Lower Brain Level

Many, if not most, of what we call subconscious activities of the body are controlled in the lower areas of the brain—in the medulla, pons, mesencephalon, hypothalamus, thalamus, cerebellum, and basal ganglia. Subconscious control of arterial pressure and respiration is achieved mainly in the medulla and pons. Control of equilibrium is a combined function of the older portions of the cerebellum and the reticular substance of the medulla, pons, and mesencephalon. Feeding reflexes, such as salivation in response to the taste of food and the licking of the lips, are controlled by areas in the medulla, pons, mesencephalon, amygdala, and hypothalamus; many emotional patterns, such as anger, excitement, sexual response, reaction to pain, and reaction to pleasure, can occur in animals without a cerebral cortex.

Higher Brain or Cortical Level

After recounting all the nervous system functions that can occur at the cord and lower brain levels, what is left for the cerebral cortex to do? The answer to this is complex, but it begins with the fact that the cerebral cortex is an extremely large memory storehouse. The cortex never functions alone but always in association with the lower centers of the nervous system.

Without the cerebral cortex, the functions of the lower brain centers are often imprecise. The vast storehouse of cortical information usually converts these functions to determinative and precise operations.

Finally, the cerebral cortex is essential for most of our thought processes, but it cannot function by itself in this. In fact, it is the lower brain centers that cause *wakefulness* in the cerebral cortex, thus opening its

bank of memories to the thinking machinery of the brain.

Thus, each portion of the nervous system performs specific functions. But it is the cortex that opens the world up for one's mind.

COMPARISON OF THE NERVOUS SYSTEM WITH AN ELECTRONIC COMPUTER

When electronic computers were first developed in many different laboratories of the world by as many different scientists, it soon became apparent that all these machines have many features in common with the nervous system. First, they all have input circuits that are comparable to the sensory portion of the nervous system and output circuits that are comparable to the motor portion of the nervous system. In the conducting pathways between the inputs and the outputs are the mechanisms for performing the different types of computations.

In simple computers, the output signals are controlled directly by the input signals, operating in a manner similar to that of the simple reflexes of the spinal cord. In the more complex computers, the output is determined both by input signals and by information that has already been stored in memory in the computer, which is analogous to the more complex reflex and processing mechanisms of our higher nervous system. Furthermore, as the computers become even more complex, it is necessary to add still another unit, called the *central processing unit,* which determines the sequence of all operations. This unit is analogous to the mechanism in our brain that allows us to direct our attention first to one thought or sensation or motor activity, then to another, and so forth, until complex sequences of thought or action take place.

Figure 45–4 is a simple block diagram of a modern computer. Even a rapid study of this diagram demonstrates its similarity to the nervous system. The fact that the basic components of the general-purpose computer are analogous to those of the human nervous system demonstrates that the brain is basically a computer that continuously collects sensory information and uses this

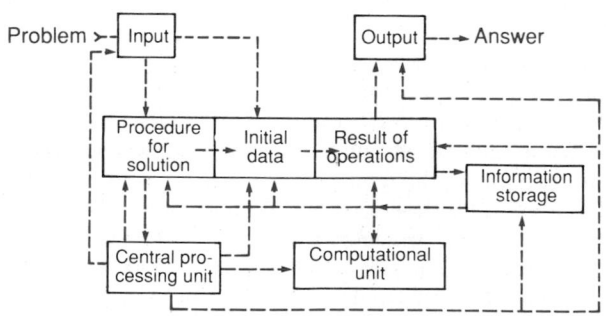

Figure 45–4. Block diagram of a general-purpose electronic computer, showing the basic components and their interrelations.

along with stored information to compute the daily course of bodily activity.

CENTRAL NERVOUS SYSTEM SYNAPSES

Every medical student is aware that information is transmitted in the central nervous system mainly in the form of nerve action potentials, called simply "nerve impulses," through a succession of neurons, one after another. However, it is not immediately apparent that each impulse (1) may be blocked in its transmission from one neuron to the next, (2) may be changed from a single impulse into repetitive impulses, or (3) may be integrated with impulses from other neurons to cause highly intricate patterns of impulses in successive neurons. All these functions can be classified as *synaptic functions of neurons*.

Types of Synapses—Chemical and Electrical

Nerve signals are transmitted from one neuron to the next through interneuronal junctions called *synapses*. There are two types of synapses: (1) the *chemical synapse* and (2) the *electrical synapse*.

Almost all the synapses used for signal transmission in the central nervous system of the human being are *chemical synapses*. In these, the first neuron secretes a chemical substance called a *neurotransmitter* at the synapse, and this transmitter in turn acts on receptor proteins in the membrane of the next neuron to excite the neuron, inhibit it, or modify its sensitivity in some other way. More than 40 transmitter substances have been discovered thus far. Some of the best known are acetylcholine, norepinephrine, histamine, gamma-aminobutyric acid (GABA), glycine, serotonin, and glutamate.

Electrical synapses, on the other hand, are characterized by direct channels that conduct electricity from one cell to the next. Most of these consist of small protein tubular structures called *gap junctions* that allow free movements of ions from the interior of one cell to the next. These junctions are discussed in Chapter 4. Only a few gap junctions have been found in the central nervous system, and their significance in general is not known. On the other hand, it is by way of gap junctions and other, similar junctions that action potentials are transmitted from one smooth muscle fiber to the next in visceral smooth muscle (Chapter 8) and from one cardiac muscle cell to the next in cardiac muscle (Chapter 10).

ONE-WAY CONDUCTION AT CHEMICAL SYNAPSES. Chemical synapses have one exceedingly important characteristic that makes them highly desirable as the form of transmission of nervous system signals: They always transmit the signals in one direction—that is, from the neuron that secretes the transmitter, called the *presynaptic neuron*, to the neuron on which the transmitter acts, called the *postsynaptic neuron*. This is the principle of *one-way conduction* at chemical synapses, and it is quite different from conduction through electrical synapses that can transmit signals in either direction.

Think for a moment about the extreme importance of the one-way conduction mechanism. It allows signals to be directed toward specific goals. Indeed, it is this specific transmission of signals to discrete and highly focused areas both in the nervous system and at the terminals of the peripheral nerves that allows the nervous system to perform its myriad functions of sensation, motor control, memory, and many others.

Physiologic Anatomy of the Synapse

Figure 45–5 shows a typical *anterior motor neuron* in the anterior horn of the spinal cord. It is composed of three major parts: the *soma*, which is the main body of the neuron; a single *axon*, which extends from the soma into a peripheral nerve that leaves the spinal cord; and the *dendrites*, which are great numbers of branching projections of the soma that extend as much as 1 millimeter into the surrounding areas of the cord.

As many as 10,000 or more small knobs called *presynaptic terminals* lie on the surfaces of the dendrites and soma of the motor neuron, about 80 to 95 per cent of them on the dendrites and only 5 to 20 per cent on the soma. These presynaptic terminals are the

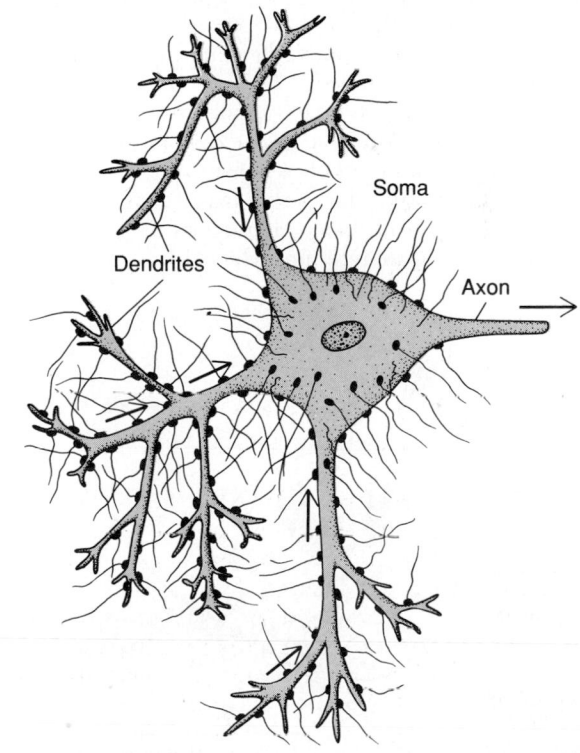

Figure 45–5. Typical motor neuron, showing presynaptic terminals on the neuronal soma and dendrites. Note also the single axon.

ends of nerve fibrils that originate from many other neurons. Later, it will become evident that many of these presynaptic terminals are *excitatory*—that is, they secrete a substance that excites the postsynaptic neuron; many others are *inhibitory*—they secrete a substance that inhibits the postsynaptic neuron.

Neurons in other parts of the cord and brain differ markedly from the anterior motor neuron in (1) the size of the cell body; (2) the length, size, and number of dendrites, ranging in length from almost none to many centimeters; (3) the length and size of the axon; and (4) the number of presynaptic terminals, which may range from only a few to as many as 200,000. These differences make neurons in different parts of the nervous system react differently to incoming signals and, therefore, perform different functions.

PRESYNAPTIC TERMINALS. Electron microscopic studies of the presynaptic terminals show that they have varied anatomical forms, but most resemble small round or oval knobs and, therefore, are called *terminal knobs, boutons, end-feet,* or *synaptic knobs*.

Figure 45–6 shows the basic structure of the presynaptic terminal. It is separated from the postsynaptic neuronal soma by a *synaptic cleft* having a width usually of 200 to 300 angstroms. The terminal has two internal structures important to the excitatory or inhibitory functions of the synapse: the *transmitter vesicles* and the *mitochondria*. The transmitter vesicles contain a *transmitter substance* that, when released into the synaptic cleft, either *excites* or *inhibits* the postsynaptic neuron—excites if the neuronal membrane contains *excitatory receptors*, inhibits if it contains *inhibitory receptors*. The mitochondria provide adenosine triphosphate, which supplies the energy to synthesize new transmitter substance.

When an action potential spreads over a presynaptic terminal, the membrane depolarization causes emptying of a small number of vesicles into the cleft; the released transmitter in turn causes an immediate change in the permeability characteristics of the postsynaptic neuronal membrane, which leads to excitation

or inhibition of the postsynaptic neuron, depending on its receptor characteristics.

Mechanism by Which Action Potentials Cause Transmitter Release at the Presynaptic Terminals—Role of Calcium Ions

The cell membrane covering the presynaptic terminals, which is called the *presynaptic membrane,* contains large numbers of *voltage-gated calcium channels.* This is quite different from the other areas of the nerve fiber, which contain few of these channels. When an action potential depolarizes the terminal, large numbers of calcium ions flow into the terminal through these calcium channels. The quantity of transmitter substance that is released into the synaptic cleft is directly related to the number of calcium ions that enter the terminal. The precise mechanism by which the calcium ions cause this release is not known but is believed to be the following.

When the calcium ions enter the presynaptic terminal, it is believed that they bind with protein molecules on the inner surfaces of the presynaptic membrane, called *release sites.* This in turn causes the transmitter vesicles in the local vicinity to bind also with the membrane and fuse with it and, finally, to open to the exterior by the process called *exocytosis.* A few vesicles usually release their transmitter into the cleft after each single action potential. For those vesicles that store the neurotransmitter acetylcholine, between 2000 and 10,000 molecules of acetylcholine are present in each, and there are enough vesicles in the presynaptic terminal to transmit from a few hundred to more than 10,000 action potentials.

Action of the Transmitter Substance on the Postsynaptic Neuron—The Function of Receptor Proteins

At the synapse, the membrane of the postsynaptic neuron contains large numbers of *receptor proteins,* also shown in Figure 45–6. These receptors have two important components: (1) a *binding component* that protrudes outward from the membrane into the synaptic cleft—here it binds with the neurotransmitter from the presynaptic terminal—and (2) an *ionophore component* that passes all the way through the membrane to the interior of the postsynaptic neuron. The ionophore in turn is one of two types: (1) an *ion channel* that allows passage of specified types of ions through the channel or (2) a *"second messenger" activator* that is not an ion channel but instead protrudes into the cell cytoplasm and activates one or more substances inside the postsynaptic neuron. These substances in turn serve as "second messengers" to change specific cellular functions.

ION CHANNELS. The ion channels in the postsynaptic neuronal membrane are usually of two types: (1) *cation channels* most often allow sodium ions to pass

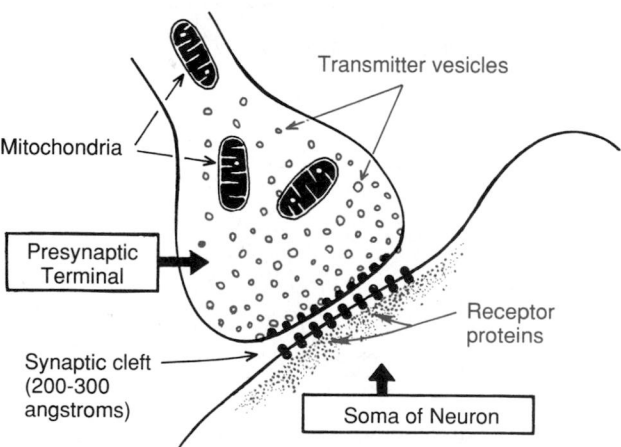

Figure 45–6. Physiologic anatomy of the synapse.

Transmitter vesicles

Mitochondria

Presynaptic Terminal

Synaptic cleft (200-300 angstroms)

Receptor proteins

Soma of Neuron

but sometimes potassium or calcium ions and (2) *anion channels* that allow mainly chloride ions to pass but also minute quantities of other anions.

The cation channels that conduct sodium ions are lined with negative charges. These charges attract the positively charged sodium ions into the channel when the channel diameter increases to a size larger than the hydrated sodium ion. But those same negative charges repel chloride ions and other anions and prevent their passage.

For the anion channels, when the channel diameters become large enough, chloride ions pass into the channels and on through to the opposite side, whereas the sodium, potassium, and calcium cations are blocked mainly because the sizes of their hydrated ions are too large to pass.

We will learn later that opening the sodium channels excites the postsynaptic neuron. Therefore, a transmitter substance that opens sodium channels is called an *excitatory transmitter*. On the other hand, opening chloride channels inhibits the neuron, and transmitter substances that open these are called *inhibitory transmitters*.

When a transmitter substance activates an ion channel, the channel usually opens within a fraction of a millisecond; when the transmitter substance is no longer present, the channel closes equally rapidly. Therefore, the opening and closing of ion channels provide a means for rapid activation or rapid inhibition of postsynaptic neurons.

"Second Messenger" System in the Postsynaptic Neuron. Many functions of the nervous system— for instance, the process of memory—require prolonged changes in neurons for seconds to months after the initial transmitter substance is gone. The ion channels are not suitable for causing prolonged postsynaptic neuronal changes because these channels close within

milliseconds after the transmitter substance is no longer present. In many instances, prolonged neuronal action is achieved by activating a "second messenger" chemical system inside the postsynaptic neuronal cell itself, and then the second messenger causes the prolonged effect.

There are several types of second messenger systems. One of the most prevailing types in neurons uses a group of proteins called *G-proteins*. Figure 45–7 shows in the upper left corner a membrane receptor protein that has been activated by a transmitter substance. A G-protein is attached to the portion of the receptor protein that protrudes to the interior of the cell. The G-protein in turn consists of three components: an alpha (α) component that is the activator portion of the G-protein, and beta (β) and gamma (γ) components that attach the G-protein to the inside of the cell membrane adjacent to the receptor protein. On activation by a nerve impulse, the alpha portion of the G-protein separates from the beta and gamma portions and then is free to move within the cytoplasm of the cell.

Inside the cytoplasm, the separated alpha component performs one or more of multiple functions, depending on the specific characteristic of each type of neuron. Shown in Figure 45–7 are four changes that can occur. They are as follows:

1. *Opening specific ion channels through the postsynaptic cell membrane.* Shown in the upper right of the figure is a potassium channel that is opened in response to the G-protein; this channel often stays open for a prolonged time, in contrast to the rapid closure of directly activated ion channels that do not use the second messenger system.

2. *Activation of cyclic adenosine monophosphate (AMP) or cyclic guanosine monophosphate (GMP) in the neuronal cell.* Recall that either cyclic AMP or cyclic GMP can control highly specific metabolic machinery in the neuron and,

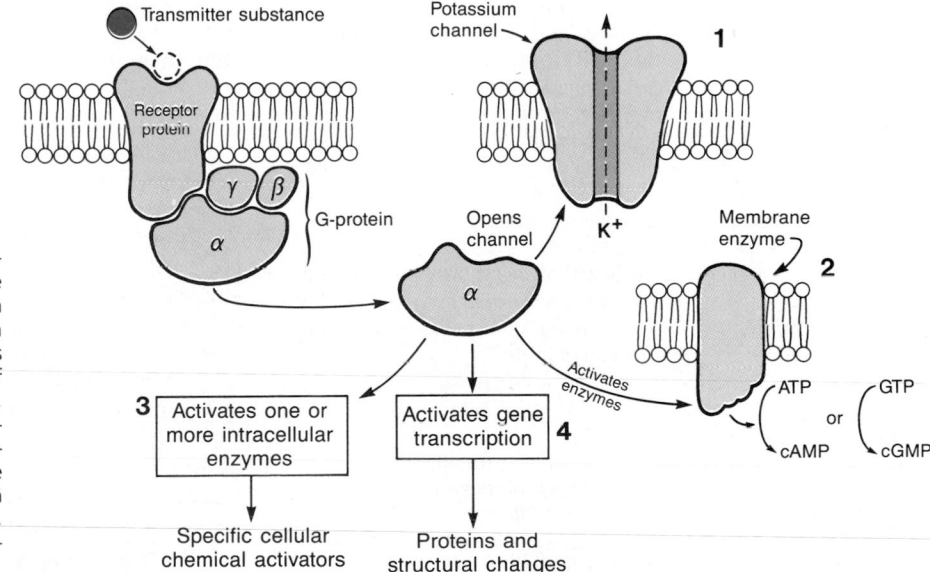

Figure 45–7. "Second messenger" system by which a transmitter substance from an initial neuron can activate a second neuron by first releasing a "G-protein" into the second neuron's cytoplasm. Four subsequent effects of the G-protein are shown, including *1*, opening an ion channel in the membrane of the second neuron; *2*, activating an enzyme system in the neuron's membrane; *3*, activating an intracellular enzyme system; and/or *4*, causing gene transcription in the second neuron.

therefore, can initiate any one of many chemical results, including long-term changes in cell structure itself, thus altering the long-term excitability of the neuron.

3. *Activation of one or more intracellular enzymes.* The G-protein can directly activate one or more intracellular enzymes. In turn the enzymes can cause any one of many specific chemical functions in the cell.

4. *Activation of gene transcription.* This is perhaps the most important of the second messenger systems of the postsynaptic neuron. Gene transcription can cause the formation of new proteins within the neuron, and these can change the metabolic machinery of the cell or its structure. Indeed, it is well known that structural changes of appropriately activated neurons do occur, especially in long-term memory processes.

Therefore, it is clear that activation of second messenger systems within the neuron, whether they be of the G-protein type or of other types, is extremely important for changing the response characteristics of different neuronal pathways. We return to this subject in more detail in Chapter 57 when we discuss memory functions of the nervous system.

Excitatory and Inhibitory Receptors in the Postsynaptic Membrane

Some postsynaptic receptors, when activated, cause excitation of the postsynaptic neuron and others cause inhibition. The importance of having inhibitory as well as excitatory types of receptors is that this gives an additional dimension to nervous function, allowing restraint of nervous action as well as excitation.

The different molecular and membrane mechanisms used by the different receptors to cause excitation or inhibition include the following.

EXCITATION

1. Opening of sodium channels to allow large numbers of positive electrical charges to flow to the interior of the postsynaptic cell. This raises the membrane potential in the positive direction up toward the threshold level for excitation. It is by far the most widely used means of causing excitation.

2. Depressed conduction through chloride or potassium channels, or both. This decreases the diffusion of negatively charged chloride ions to the inside of the postsynaptic neuron or decreases the diffusion of positively charged ions to the outside. In either instance, the effect is to make the internal membrane potential more positive than normally, which is excitatory.

3. Various changes in the internal metabolism of the cell to excite cell activity or, in some instances, increase in the number of excitatory membrane receptors or decrease in the number of inhibitory membrane receptors.

INHIBITION

1. Opening of chloride ion channels through the receptor molecule. This allows rapid diffusion of negatively charged chloride ions from outside the postsynaptic neuron to the inside, thereby carrying negative charges inward and increasing the negativity inside, which is inhibitory.

2. Increase in the conductance of potassium ions through the receptor. This allows positive potassium ions to diffuse to the exterior, which is also inhibitory.

3. Activation of receptor enzymes that inhibit cellular metabolic functions or that increase the number of inhibitory synaptic receptors or decrease the number of excitatory receptors.

Chemical Substances That Function as Synaptic Transmitters

More than 50 chemical substances have been proved or postulated to function as synaptic transmitters. Most of them are listed in Tables 45–1 and 45–2, which give two groups of synaptic transmitters. One comprises small-molecule, rapidly acting transmitters. The other is made up of a large number of neuropeptides of much larger molecular size and much more slowly acting.

The small-molecule, rapidly acting transmitters are the ones that cause most of the acute responses of the nervous system, such as transmission of sensory signals to the brain and motor signals back to the muscles. The neuropeptides, on the other hand, usually cause more prolonged actions, such as long-term changes in numbers of receptors, long-term opening or closure of certain ion channels, and, possibly, even long-term changes in numbers of synapses or sizes of synapses.

Small-Molecule, Rapidly Acting Transmitters

In most cases, the small-molecule types of transmitters are synthesized in the cytosol of the presynaptic terminal and then are absorbed by active transport into the many transmitter vesicles in the terminal. Then, each time an action potential reaches the presynaptic terminal, a few vesicles at a time release their transmitter into the synaptic cleft; this usually occurs within a millisecond or less by the mechanism described earlier. The subsequent action of the small-molecule type of transmitter on the postsynaptic membrane receptors usually also occurs within another millisecond or less. Most often the effect is to activate a receptor protein that increases or decreases conductance through ion

Table 45–1 SMALL-MOLECULE, RAPIDLY ACTING TRANSMITTERS

Class I:
Acetylcholine
Class II: *The Amines*
Norepinephrine
Epinephrine
Dopamine
Serotonin
Histamine
Class III: *Amino Acids*
γ-Aminobutyric acid (GABA)
Glycine
Glutamate
Aspartate
Class IV:
Nitric oxide (NO)

Table 45–2 NEUROPEPTIDE, SLOWLY ACTING TRANSMITTERS

A. *Hypothalamic-releasing hormones*
   Thyrotropin-releasing hormone
   Luteinizing hormone–releasing hormone
   Somatostatin (growth hormone inhibitory factor)
B. *Pituitary peptides*
   ACTH
   β-Endorphin
   α-Melanocyte-stimulating hormone
   Prolactin
   Luteinizing hormone
   Thyrotropin
   Growth hormone
   Vasopressin
   Oxytocin
C. *Peptides that act on gut and brain*
   Leucine enkephalin
   Methionine enkephalin
   Substance P
   Gastrin
   Cholecystokinin
   Vasoactive intestinal polypeptide (VIP)
   Neurotensin
   Insulin
   Glucagon
D. *From other tissues*
   Angiotensin II
   Bradykinin
   Carnosine
   Sleep peptides
   Calcitonin

channels; an example is to increase sodium conductance, which causes excitation, or to increase potassium or chloride conductance, which causes inhibition. Occasionally the small-molecule types of transmitters can stimulate receptor-activated enzymes instead of opening ion channels, thus changing the internal metabolic machinery of the cell.

RECYCLING OF THE SMALL-MOLECULE TYPES OF VESICLES. The vesicles that store and release small-molecule transmitters are continually recycled, that is, used over and over again. After they fuse with the synaptic membrane and open to release their transmitter substance, the vesicle membrane at first simply becomes part of the synaptic membrane. However, within seconds to minutes, the vesicle portion of the membrane invaginates back to the inside of the presynaptic terminal and pinches off to form a new vesicle. It still contains the appropriate transport proteins required for concentrating new transmitter substance inside the vesicle.

Acetylcholine is a typical small-molecule transmitter that obeys the principles of synthesis and release stated above. This transmitter substance is synthesized in the presynaptic terminal from acetyl coenzyme A and choline in the presence of the enzyme *choline acetyltransferase*. Then it is transported into its specific vesicles. When the vesicles later release the acetylcholine into the synaptic cleft, the acetylcholine is rapidly split again to acetate and choline by the enzyme *cholinesterase*, which is bound to the proteoglycan reticulum that fills the space of the synaptic cleft. Then the vesicles are recycled, and choline also is actively transported back into the terminal to be used again for synthesis of new acetylcholine.

CHARACTERISTICS OF SOME OF THE MORE IMPORTANT SMALL-MOLECULE TRANSMITTERS. The most important of the small-molecule transmitters are the following.

Acetylcholine is secreted by neurons in many areas of the brain but specifically by the large pyramidal cells of the motor cortex, by several different neurons in the basal ganglia, by the motor neurons that innervate the skeletal muscles, by the preganglionic neurons of the autonomic nervous system, by the postganglionic neurons of the parasympathetic nervous system, and by some of the postganglionic neurons of the sympathetic nervous system. In most instances, acetylcholine has an excitatory effect; however, it is known to have inhibitory effects at some of the peripheral parasympathetic nerve endings, such as inhibition of the heart by the vagus nerves.

Norepinephrine is secreted by many neurons whose cell bodies are located in the brain stem and hypothalamus. Specifically, norepinephrine-secreting neurons located in the *locus ceruleus* in the pons send nerve fibers to widespread areas of the brain and help control the overall activity and mood of the mind, such as increasing the level of wakefulness. In most of these areas, norepinephrine probably activates excitatory receptors, but in a few areas, it activates inhibitory receptors instead. Norepinephrine is also secreted by most of the postganglionic neurons of the sympathetic nervous system, where it excites some organs but inhibits others.

Dopamine is secreted by neurons that originate in the substantia nigra. The termination of these neurons is mainly in the striatal region of the basal ganglia. The effect of dopamine is usually inhibition.

Glycine is secreted mainly at synapses in the spinal cord. It probably always acts as an inhibitory transmitter.

GABA is secreted by nerve terminals in the spinal cord, cerebellum, basal ganglia, and many areas of the cortex. It is believed always to cause inhibition.

Glutamate is secreted by the presynaptic terminals in many of the sensory pathways as well as in many areas of the cortex. It probably always causes excitation.

Serotonin is secreted by nuclei that originate in the median raphe of the brain stem and project to many brain and spinal cord areas, especially to the dorsal horns of the spinal cord and to the hypothalamus. Serotonin acts as an inhibitor of pain pathways in the cord, and its action in the higher regions of the nervous system is believed to help control the mood of the person, perhaps even to cause sleep.

Nitric oxide is a newly discovered small-molecule transmitter substance. It occurs especially in areas of the brain that are responsible for long-term behavior and for memory. Therefore, this new transmitter system might help to explain behavior and memory functions that thus far have defied understanding. Nitric oxide is different from other small-molecule transmitters in its mechanisms of formation in the presynaptic terminal and in its action on the postsynaptic neuron. It is not preformed and stored in vesicles in the presynaptic terminal as are other transmitters. Instead, it is synthesized almost instantly as needed, and it then diffuses out of the presynaptic terminals over a period of seconds rather than being released in vesicular packets. Next, it diffuses into the immediately adjacent postsynaptic neuron as well as into other postsynaptic neurons nearby. In the postsynaptic neuron, it usually does not greatly alter the membrane potential but instead changes intracellular metabolic functions that modify neuronal excitability for seconds, minutes, or perhaps even longer.

Neuropeptides

The neuropeptides are an entirely different group of transmitters that are synthesized differently and whose actions are usually slow and in other ways quite different from those of the small-molecule transmitters.

The neuropeptides are not synthesized in the cytosol of the presynaptic terminals. Instead, they are synthesized as integral parts of large-protein molecules by the ribosomes in the neuronal cell body. The protein molecules then first enter the endoplasmic reticulum of the cell body and subsequently the Golgi apparatus, where two changes occur: First, the protein is enzymatically split into smaller fragments and thereby releases either the neuropeptide itself or a precursor of it. Second, the Golgi apparatus packages the neuropeptide into minute transmitter vesicles that are released into the cytoplasm. Then the transmitter vesicles are transported all the way to the tips of the nerve fibers by *axonal streaming* of the axon cytoplasm, traveling at the slow rate of only a few centimeters per day. Finally, these vesicles release their transmitter in response to action potentials in the same manner as for small-molecule transmitters. However, the vesicle is autolyzed and is not reused.

Because of this laborious method of forming the neuropeptides, much smaller quantities of them are usually released than for the small-molecule transmitters. This is partly compensated for by the fact that the neuropeptides are generally a thousand or more times as potent as the small-molecule transmitters. Another important characteristic of the neuropeptides is that they usually cause much more prolonged actions. Some of these actions include prolonged closure of calcium pores, prolonged changes in the metabolic machinery of cells, prolonged changes in activation or deactivation of specific genes in the cell nucleus, and prolonged alterations in numbers of excitatory or inhibitory receptors. Some of these effects can last for days or perhaps even months or years. Our knowledge of the functions of the neuropeptides is only beginning to develop.

Usually Only a Single Type of Small-Molecule Transmitter Is Released by Each Type of Neuron

Almost invariably, only a single type of small-molecule transmitter is released by each type of neuron. However, the terminals of the same neuron may also release one or more neuropeptides at the same time. Yet whichever small-molecule transmitter and neuropeptides are released at one terminal of the neuron, these same transmitters will be released at all other terminals of the same neuron, whether they are few in number or many thousand and also wherever they terminate within the nervous system or in peripheral organs.

After a Transmitter Substance Is Released at a Synapse, It Must Be Removed

After a transmitter is released at a nerve ending, it is either destroyed or removed in some other way to prevent continued action forever thereafter. In the case of the neuropeptides, they are removed mainly by diffusion into the surrounding tissues, followed by destruction within a few minutes to several hours by specific or nonspecific enzymes. For the small-molecule, rapidly acting transmitters, removal usually occurs within a few milliseconds. This is achieved in one of the following ways:

1. By *diffusion* of the transmitter out of the synaptic cleft into the surrounding fluids
2. By *enzymatic destruction* within the cleft itself. For instance, in the case of acetylcholine, the enzyme *cholinesterase* is present in the cleft, bound in the proteoglycan matrix that fills the space. Each molecule of this enzyme can split as many as 10 molecules of acetylcholine each millisecond, thus inactivating this transmitter substance. Similar effects occur for other transmitters.
3. By *active transport back into the presynaptic terminal itself* and reuse. This is called *transmitter re-uptake.* It occurs especially prominently at the presynaptic terminals of the sympathetic nervous system for the re-uptake of norepinephrine, as we discuss in Chapter 60.

The degree to which each of these methods of removal is used is different for each type of transmitter and for each type of neuron.

Electrical Events During Neuronal Excitation

The electrical events in neuronal excitation have been studied especially in the large motor neurons of the anterior horns of the spinal cord. Therefore, the events to be described in the next few sections pertain essentially to these neurons. Except for quantitative differences, they apply to most other neurons of the nervous system as well.

RESTING MEMBRANE POTENTIAL OF THE NEURONAL SOMA. Figure 45–8 shows the soma of a motor neuron, indicating the resting membrane potential to

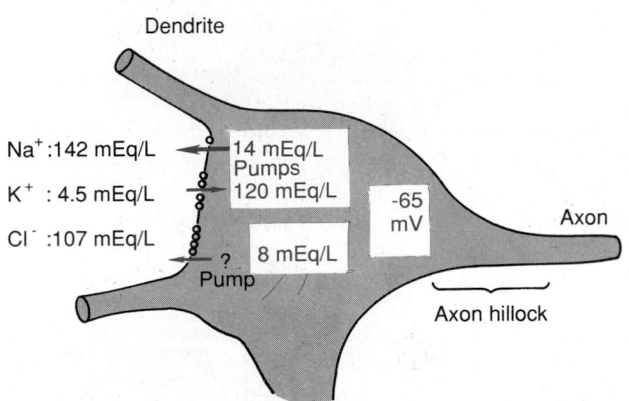

Figure 45–8. Distribution of sodium, potassium, and chloride ions across the neuronal somal membrane; origin of the intrasomal membrane potential.

be about − 65 millivolts. This is somewhat less than the − 90 millivolts found in large peripheral nerve fibers and skeletal muscle fibers; the lower voltage is important because it allows both positive and negative control of the degree of excitability of the neuron. That is, decreasing the voltage to a less negative value makes the membrane of the neuron more excitable, whereas increasing this voltage to a more negative value makes the neuron less excitable. This is the basis for the two modes of function of the neuron—either excitation or inhibition—as explained in detail in the next sections.

Concentration Differences of Ions Across the Neuronal Somal Membrane. Figure 45–8 also shows the concentration differences across the neuronal somal membrane of the three ions that are most important for neuronal function: sodium ions, potassium ions, and chloride ions.

At the top, the sodium ion concentration is shown to be great in the extracellular fluid (142 mEq/liter) but low inside the neuron (14 mEq/liter). This sodium concentration gradient is caused by a strong sodium pump that continually pumps sodium out of the neuron.

The figure also shows that the potassium ion concentration is large inside the neuronal soma (120 mEq/liter) but low in the extracellular fluid (4.5 mEq/liter). It shows that there is also a potassium pump (the other half of the Na$^+$-K$^+$ pump, as described in Chapter 4) that pumps potassium to the interior. However, potassium ions can leak through ion channels in the resting neuronal membrane much more easily than can sodium ions.

Figure 45–8 shows the chloride ion to be of high concentration in the extracellular fluid but low concentration inside the neuron. It also shows that the membrane is quite permeable to chloride ions and that there may be a weak chloride pump. Yet most of the reason for the low concentration of chloride ions inside the neuron is the − 65 millivolts in the neuron. That is, this negative voltage repels the negatively charged

chloride ions, forcing them outward through the pores until the concentration difference is much greater outside the membrane than inside.

Let us recall at this point what we learned in Chapters 4 and 5 about the relation of ionic concentration differences to membrane potentials. It will be recalled that an electrical potential across the membrane can exactly oppose the movement of ions through a membrane, despite concentration differences between the outside and inside of the membrane, if the potential is of proper polarity and magnitude. Such a potential that exactly opposes movement of each type of ion is called the Nernst potential for that ion; the equation for this is the following:

$$\text{EMF(mV)} = \pm 61 \times \log \left(\frac{\text{Concentration inside}}{\text{Concentration outside}} \right),$$

where EMF is the Nernst potential in millivolts on the *inside of the membrane.* The potential will be negative (−) for a positive ion and positive (+) for a negative ion.

Now, let us calculate the Nernst potential that will exactly oppose the movement of each of the three separate ions: sodium, potassium, and chloride.

For the sodium concentration difference shown in Figure 45–8, 142 mEq/liter on the exterior and 14 mEq/liter on the interior, the membrane potential that would exactly oppose sodium ion movement through the sodium channels would be + 61 millivolts. However, the actual membrane potential is − 65 millivolts, not + 61 millivolts. Therefore, net quantities of sodium ions normally diffuse inward through the sodium channels; however, not many sodium ions will diffuse because most of the sodium channels are normally closed. Furthermore, those sodium ions that do diffuse to the interior are normally pumped immediately back to the exterior by the sodium pump.

For potassium ions, the concentration gradient is 120 mEq/liter inside the neuron and 4.5 mEq/liter outside. This gives a Nernst potential of − 86 millivolts inside the neuron, which is more negative than the − 65 that actually exists. Therefore, there is a net tendency for potassium ions to diffuse to the outside of the neuron, but this is opposed by the continual pumping of these potassium ions back to the interior.

Finally, the chloride ion gradient, 107 mEq/liter outside and 8 mEq/liter inside, yields a Nernst potential of − 70 millivolts inside the neuron, which is slightly more negative than the actual value of − 65 millivolts measured. Therefore, chloride ions tend normally to leak to the interior of the neuron, whereas those that do leak are moved back to the exterior, perhaps by an active chloride pump.

Keep these three Nernst potentials in mind and remember the direction in which the different ions tend to diffuse, because this information is important in understanding both excitation and inhibition of the neuron by synaptic activation of receptor ion channels.

Origin of the Resting Membrane Potential of the Neuronal Soma. The immediate cause of the −65 millivolt resting membrane potential of the neuronal soma is the high concentration of potassium ions inside the neuronal cell membrane and its low concentration outside. In the resting state, the membrane is much more permeable to potassium ions than to sodium ions. Therefore, the high potassium concentration inside the membrane, multiplied by the high permeability for potassium, causes large numbers of the positively charged potassium ions to diffuse outward. Because there are large numbers of negatively charged ions still inside the soma that cannot diffuse outward through the membrane—protein ions, phosphate ions, and many others—extrusion of the excess positive potassium ions to the exterior leaves these nondiffusible negative ions inside the cell unbalanced by positive ions. Therefore, the interior of the neuron becomes negatively charged as the result of the potassium diffusion. This principle is discussed in more detail in Chapter 5 in relation to the resting membrane potential of nerve fibers. In addition, as also explained in Chapter 5, the excess pumping of sodium ions outward through the membrane by the sodium-potassium pump is another much longer-term cause of intracellular negativity.

Uniform Distribution of the Potential Inside the Soma. The interior of the neuronal soma contains a highly conductive electrolytic solution, the intracellular fluid of the neuron. Furthermore, the diameter of the neuronal soma is large (from 10 to 80 micrometers in diameter), causing there to be almost no resistance to conduction of electric current from one part of the somal interior to another part. Therefore, any change in potential in any part of the intrasomal fluid causes an almost exactly equal change in potential at all other points inside the soma. This is an important principle because it plays a major role in the "summation" of signals entering the neuron from multiple sources, as we shall see in subsequent sections of this chapter.

EFFECT OF SYNAPTIC EXCITATION ON THE POSTSYNAPTIC MEMBRANE—THE EXCITATORY POSTSYNAPTIC POTENTIAL. Figure 45–9A shows the resting neuron with an unexcited presynaptic terminal resting on its surface. The resting membrane potential everywhere in the soma is −65 millivolts.

Figure 45–9B shows a presynaptic terminal that has secreted a transmitter into the cleft between the terminal and the neuronal somal membrane. This transmitter acts on a membrane excitatory receptor *to increase the membrane's permeability to Na⁺.* Because of the large sodium concentration gradient and large electrical negativity inside the neuron, sodium ions rush to the inside of the membrane.

The rapid influx of the positively charged sodium ions to the interior neutralizes part of the negativity of the resting membrane potential. Thus, in Figure 45–9B, the resting membrane potential has increased from −65 to −45 millivolts. This increase in voltage above the normal resting neuronal potential—that is, to a less negative value—is called the *excitatory post-*

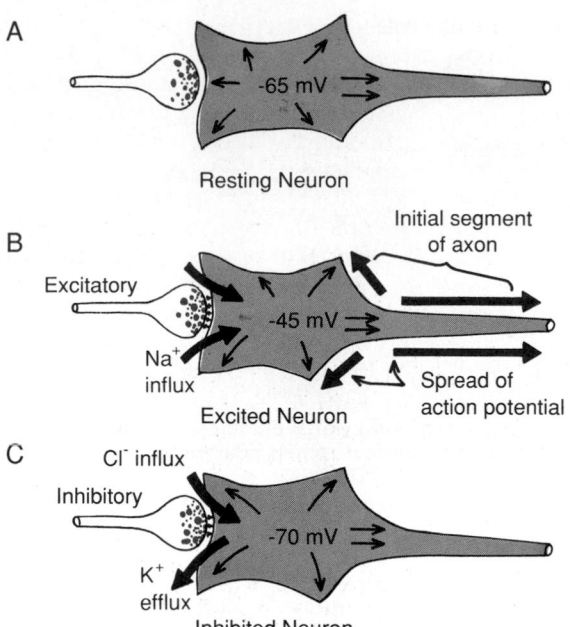

Figure 45–9. Three states of a neuron. *A,* Resting neuron. *B,* Neuron in an excited state, with a more positive—that is, less negative—intraneuronal potential caused by sodium influx. *C,* Neuron in an inhibited state, with a more negative intraneuronal membrane potential caused by potassium ion efflux, chloride ion influx, or both.

synaptic potential (or EPSP) because if this potential rises high enough, it will elicit an action potential in the neuron, thus exciting it. In this case, the EPSP is +20 millivolts (20 millivolts more positive than the initial value).

However, we must issue a word of warning at this point. Discharge of a single presynaptic terminal can never increase the neuronal potential from −65 millivolts all the way up to −45 millivolts. An increase of this magnitude requires the simultaneous discharge of many terminals—about 40 to 80 for the usual anterior motor neuron—at the same time or in rapid succession. This occurs by a process called *summation,* which is discussed in detail in the next sections.

GENERATION OF ACTION POTENTIALS IN THE INITIAL SEGMENT OF THE AXON LEAVING THE NEURON—THRESHOLD FOR EXCITATION. When the excitatory postsynaptic potential rises high enough, there comes a point at which this initiates an action potential in the neuron. However, the action potential does not begin on the somal membrane adjacent to the excitatory synapses. Instead, it begins in the initial segment of the axon leaving the neuronal soma. The main reason for this point of origin of the action potential is that the soma has relatively few voltage-gated sodium channels in its membrane, which makes it difficult for the excitatory postsynaptic potential to open the required number of channels to elicit an action potential. On the other hand, the membrane of the initial segment has seven times as great a concentration of voltage-gated sodium channels and, therefore, can generate an action potential with much greater ease than can the soma. The excitatory postsynaptic potential that will elicit an action potential at the initial segment is between + 10 and + 20 millivolts. This is in contrast to the + 30 millivolts or more required on the soma.

Once the action potential begins, it travels both peripherally along the axon and usually also backward over the soma. In some instances, it travels backward into the dendrites, too, but not into all of them, because they, like the neuronal soma, also have very few voltage-gated sodium channels and therefore frequently cannot generate action potentials at all.

Thus, in Figure 45–9B, the *threshold* for excitation of the neuron is shown to be about − 45 millivolts, which represents an excitatory postsynaptic potential of + 20 millivolts—that is, 20 millivolts more positive than the normal resting neuronal potential of − 65 millivolts.

Electrical Events in Neuronal Inhibition

EFFECT OF INHIBITORY SYNAPSES ON THE POSTSYNAPTIC MEMBRANE—THE INHIBITORY POSTSYNAPTIC POTENTIAL. The inhibitory synapses open mainly the chloride channels, instead of sodium channels, allowing easy passage of chloride ions. Now, to understand how the inhibitory synapses inhibit the postsynaptic neuron, we must recall what we learned about the Nernst potential for chloride ions. We calculated the Nernst potential for chloride ions to be about − 70 millivolts. This potential is more negative than the − 65 millivolts normally present inside the resting neuronal membrane. Therefore, opening the chloride channels will allow negatively charged chloride ions to move to the interior, which will make the membrane potential more negative than normal, and opening potassium channels will allow positively charged potassium ions to move to the exterior, which also will make the membrane potential more negative than usual. This increases the degree of intracellular negativity, which is called *hyperpolarizaton*. It inhibits the neuron because the membrane potential is now further away than ever from the threshold for excitation. Therefore, an increase in negativity beyond the normal resting membrane potential level is called the *inhibitory postsynaptic potential* (IPSP).

Thus, Figure 45–9C shows the effect on the membrane potential caused by activation of inhibitory synapses, allowing chloride influx into the cell or potassium efflux from the cell, with the membrane potential decreasing from its normal value of − 65 millivolts to the more negative value of − 70 millivolts. This membrane potential that is 5 millivolts more negative is the IPSP. Thus, the IPSP in this instance is − 5 millivolts.

THERE IS ANOTHER METHOD FOR INHIBITING NEURONS WITHOUT CAUSING AN IPSP—"SHORT CIRCUITING" OF THE MEMBRANE. Sometimes activation of the inhibitory synapses causes little or no IPSP but nevertheless inhibits the neuron. One good example of this is the following.

In some neurons, the concentration difference across the membrane for chloride ions causes a chloride Nernst potential that is exactly equal to its resting potential. Therefore, when the inhibitory channels open, there is no net flow of chloride ions to cause an inhibitory postsynaptic potential. Yet the chloride ions do diffuse bidirectionally through the wide-open channels many times as rapidly as normally, and this high flux inhibits the neuron in the following way: When excitatory synapses cause sodium ions to flow into the neuron, the wide-open chloride channels cause far less excitatory postsynaptic potential than usual. The reason is that any change in the membrane potential now makes the potential different from the chloride Nernst potential. Therefore, net chloride ion flow through the membrane is no longer zero; instead extra chloride ions flow rapidly through the wide-open chloride channels, and their electronegativity nullifies most of the electropositivity of the sodium-induced excitatory postsynaptic potential. As a result, the amount of excitatory influx of sodium ions required to overcome the chloride flux to cause excitation may be 5 to 20 times normal.

This tendency for the chloride ions to maintain the membrane potential near the resting value when the inhibitory channels are wide open is called "short circuiting" of the membrane, thus making the sodium current caused by excitatory synapses ineffective in exciting the cell.

To express the phenomenon of short circuiting more mathematically, one needs to recall the Goldman equation from Chapter 5. This equation shows that the membrane potential is determined by summation of the tendencies for all the different ions to carry electrical charges through the membrane in the two directions. The membrane potential will approach the Nernst equilibrium potentials for those ions that permeate the membrane to the greatest extent. When the inhibitory channels are wide open, the chloride and potassium ions permeate the membrane greatly. Therefore, when the excitatory channels are opened simultaneously with the inhibitory channels, the summated effect of the inhibitory channels makes it difficult for the opened excitatory channels to raise the neuronal potential up to the threshold value for excitation.

Presynaptic Inhibition

In addition to the inhibition caused by inhibitory synapses operating at the neuronal membrane, which is called *postsynaptic inhibition*, another type of inhibition often occurs in the presynaptic terminals before the signal ever reaches the synapse. This type of inhibition, called *presynaptic inhibition*, occurs in the following way.

In presynaptic inhibition, the inhibition is caused by discharge of inhibitory synapses that lie on the presynaptic terminal nerve fibrils before their endings terminate on the postsynaptic neuron. In most instances, the inhibitory transmitter substance released is GABA.

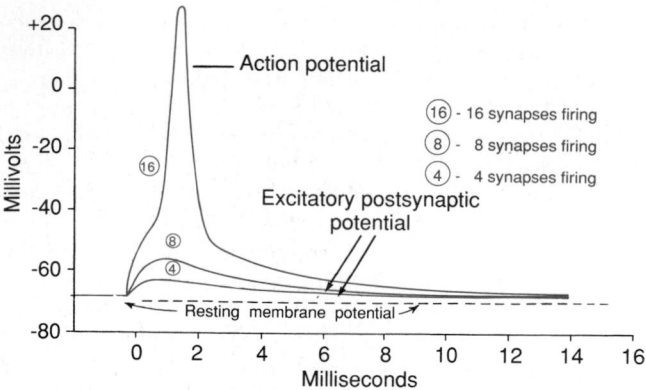

Figure 45–10. Excitatory postsynaptic potentials, showing that simultaneous firing of only a few synapses will not cause sufficient summated potential to elicit an action potential, but the simultaneous firing of many synapses will raise the summated potential to the threshold for excitation and cause a superimposed action potential.

This has the specific effect of opening anion channels, allowing large numbers of chloride ions to diffuse into the terminal fibril. The negative charges of these ions cancel much of the excitatory effect of the positively charged sodium ions that enter the terminal fibrils when an action potential arrives. Therefore, the action potential itself in these terminal fibrils becomes greatly reduced, thus also reducing the degree of excitation of the postsynaptic neuron.

Presynaptic inhibition occurs in many of the sensory pathways in the nervous system. That is, adjacent terminal nerve fibers mutually inhibit one another, which minimizes the sideways spread of signals in sensory tracts. We discuss this phenomenon more fully in subsequent chapters.

Time Course and Summation of Postsynaptic Potentials

Time Course of Postsynaptic Potentials. When a synapse excites the anterior motor neuron, the neuronal membrane remains highly permeable to sodium ions for only 1 to 2 milliseconds. During this time, the sodium ions diffuse rapidly to the interior of the postsynaptic cell to increase its intraneuronal potential, thus creating the *excitatory postsynaptic potential*, shown by the two lower curves of Figure 45–10. This potential then slowly declines over the next 15 milliseconds because this is the time required for the excess positive charges to leak out of the excited neuron and to re-establish the normal resting membrane potential.

Precisely the opposite effect occurs for the inhibitory postsynaptic potential (IPSP). That is, the inhibitory synapse increases the permeability of the membrane to potassium or chloride ions, or both, for 1 to 2 milliseconds, and this decreases the intraneuronal potential to a more negative value than normal, thereby creating the IPSP. This potential also dies away over the next 15 milliseconds.

Other types of transmitter substances acting on other neurons can excite or inhibit for hundreds of milliseconds or even for seconds, minutes, or hours.

This is especially true for the neuropeptide transmitter substances.

Spatial Summation in Neurons—The Threshold for Firing

It has already been pointed out that excitation of a single presynaptic terminal on the surface of a neuron will almost never excite the neuron. The reason for this is that sufficient transmitter substance is released by a single terminal to cause an excitatory postsynaptic potential usually no more than 0.5 to 1 millivolt, instead of the required 10 to 20 millivolts to reach the usual threshold for excitation. However, during excitation in a neuronal pool, many presynaptic terminals are usually stimulated at the same time. Even though these terminals are spread over wide areas of the neuron, their effects can still *summate*. The reason for this is the following: It has already been pointed out that a change in the potential at any single point within the soma will cause the potential to change everywhere in the soma almost exactly equally. This is true because of the very high electrical conductivity within the large neuronal cell body. Therefore, for each excitatory synapse that discharges simultaneously, the intrasomal potential becomes more positive by as much as a fraction of a millivolt. When the excitatory postsynaptic potential becomes great enough, the *threshold for firing* will be reached and an action potential will develop spontaneously in the initial segment of the axon. This effect is demonstrated in Figure 45–10, which shows several excitatory postsynaptic potentials. The bottom postsynaptic potential in the figure was caused by simultaneous stimulation of four synapses; the next higher potential was caused by stimulation of two times as many synapses; finally, a still higher excitatory postsynaptic potential was caused by stimulation of four times as many synapses. In this last instance, an action potential was generated in the initial axon segment.

This effect of summing simultaneous postsynaptic potentials by activating multiple terminals on widely spaced areas of the membrane is called *spatial summation*.

Temporal Summation

Each time a terminal fires, the released transmitter substance opens the membrane channels for a millisecond or so. Because the postsynaptic potential lasts up to 15 milliseconds, a second opening of the same channels can increase the postsynaptic potential to a still greater level; therefore, the more rapid the rate of terminal stimulation, the greater the effective postsynaptic potential. Thus, successive postsynaptic potentials caused by discharges from a single presynaptic terminal, if they occur rapidly enough, can summate in the same way that postsynaptic potentials can summate from widely distributed terminals over the surface of the neuron. This summation is called *temporal summation.*

SIMULTANEOUS SUMMATION OF INHIBITORY AND EXCITATORY POSTSYNAPTIC POTENTIALS. If an inhibitory postsynaptic potential is tending to decrease the membrane potential to a more negative value while an excitatory postsynaptic potential is tending to increase the potential at the same time, these two effects can either completely nullify each other or partially nullify each other. Also, inhibitory "short circuiting" of the membrane potential can nullify much of an excitatory potential. Thus, if a neuron is being excited by an excitatory postsynaptic potential, then an inhibitory signal from another source can easily reduce the postsynaptic potential to less than the threshold value for excitation, thus turning off the activity of the neuron.

Facilitation of Neurons

Often the summated postsynaptic potential is excitatory in nature, but has not risen high enough to reach the threshold for excitation. When this happens, the neuron is said to be *facilitated.* That is, its membrane potential is nearer the threshold for firing than normal, but not yet at the firing level. Nevertheless, another signal entering the neuron from some other source can then excite the neuron very easily. Diffuse signals in the nervous system often facilitate large groups of neurons so that they can respond quickly and easily to signals arriving from second sources.

Special Functions of Dendrites in Exciting Neurons

LARGE SPATIAL FIELD OF EXCITATION OF THE DENDRITES. The dendrites of the anterior motor neurons extend for 500 to 1000 micrometers in all directions from the neuronal soma. Therefore, these dendrites can receive signals from a large spatial area around the motor neuron. This provides vast opportunity for summation of signals from many separate presynaptic nerve fibers.

It is also important that between 80 and 95 per cent of all the presynaptic terminals terminate on the dendrites of the anterior motor neuron, in contrast to only 5 to 20 per cent terminating on the neuronal soma.

Therefore, the preponderant share of the excitation is provided by signals transmitted by way of the dendrites.

MANY DENDRITES CANNOT TRANSMIT ACTION POTENTIALS—BUT THEY CAN TRANSMIT SIGNALS BY ELECTROTONIC CONDUCTION. Many dendrites fail to transmit action potentials because their membranes have relatively few voltage-gated sodium channels, so that their thresholds for excitation are too high for action potentials to occur. Yet they do transmit *electrotonic current* down the dendrites to the soma. Transmission of electrotonic current means the direct spread of current by electrical conduction in the fluids of the dendrites with no generation of action potentials. Stimulation (or inhibition) of the neuron by this current has special characteristics, as follows.

Decrement of Electrotonic Conduction in the Dendrites—Greater Excitatory (or Inhibitory) Effect by Synapses Near the Soma. In Figure 45–11, a number of excitatory and inhibitory synapses are shown stimulating the dendrites of a neuron. On the two dendrites to the left in the figure are shown excitatory effects near the tip ends of the dendrites; note the high levels of the excitatory postsynaptic potentials at these ends—that is, the less negative membrane potentials at these points. However, a large share of the excitatory postsynaptic potential is lost before it reaches the soma. The reason for this is that the dendrites are long and thin, and their membranes are also thin and excessively permeable to potassium and chloride ions, making them "leaky" to electric current. Therefore, before the excitatory potentials can reach the soma, a large share of the potential is lost by leakage through the membrane. This decrease in membrane potential as it spreads electrotonically along

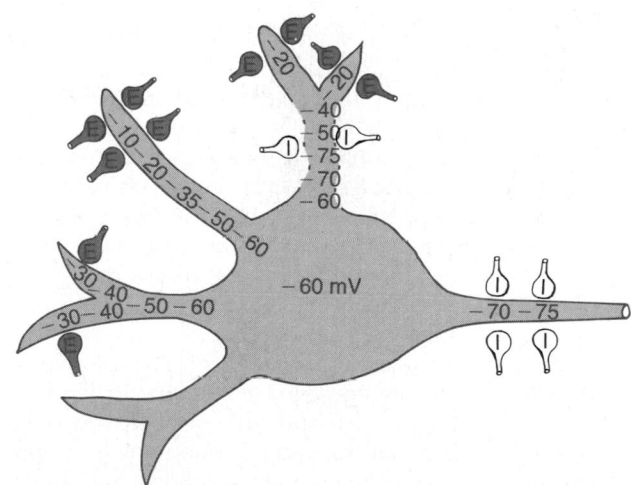

Figure 45–11. Stimulation of a neuron by presynaptic terminals located on dendrites, showing, especially, decremental conduction of excitatory electrotonic potentials in the two dendrites to the left and inhibition of dendritic excitation in the dendrite that is uppermost. A powerful effect of inhibitory synapses at the initial segment of the axon is also shown.

dendrites toward the soma is called *decremental conduction.*

It is also obvious that the nearer the excitatory synapse is to the soma of the neuron, the less will be the total decrement of conduction. Therefore, those synapses that lie near the soma have far more effect to cause excitation or inhibition than those that lie far away from the soma.

SUMMATION OF EXCITATION AND INHIBITION IN DENDRITES. The uppermost dendrite of Figure 45–11 is shown to be stimulated by both excitatory and inhibitory synapses. At the tip of the dendrite is a strong excitatory postsynaptic potential, but nearer to the soma are two inhibitory synapses acting on the same dendrite. These inhibitory synapses provide a hyperpolarizing voltage that completely nullifies the excitatory effect and indeed transmits a small amount of inhibition by electrotonic conduction toward the soma. Thus, dendrites can summate excitatory and inhibitory postsynaptic potentials in the same way that the soma can.

Also shown in the figure are several inhibitory synapses located directly on the axon hillock and initial axon segment. This location provides especially powerful inhibition because it has the direct effect of increasing the threshold for excitation at the very point where the action potential is normally generated.

Relation of State of Excitation of the Neuron to the Rate of Firing

"EXCITATORY STATE." The "excitatory state" of a neuron is defined as the summated degree of excitatory drive to the neuron. If there is a higher degree of excitation than inhibition of the neuron at any given instant, then it is said that there is an *excitatory state.* On the other hand, if there is more inhibition than excitation, then it is said that there is an *inhibitory state.*

When the excitatory state of a neuron rises above the threshold for excitation, then the neuron will fire repetitively as long as the excitatory state remains at this level.

WHAT IS THE MECHANISM FOR TRANSLATING INCREASED EXCITATORY STATE INTO INCREASED FIRING RATE OF A NEURON? One might ask the question: when the excitatory state is above threshold value, why does the neuron not continue to fire all the time at its maximum rate of firing? The basic reason why this is not true is the following.

Every time the initial segment of a postsynaptic neuron fires, not only does the action potential spread into the axon, but it also spreads backward over the surface of the neuronal soma and sometimes even into some of the dendrites. Then, at the end of the action potential, the soma membrane potential returns to a very negative, hyperpolarized state. Therefore, the excitatory state of the neuron must be re-established before still another action potential occurs. And much new excitatory current must enter the neuron at many synapses to re-establish the excitatory state; all this requires time. The greater the number of excitatory

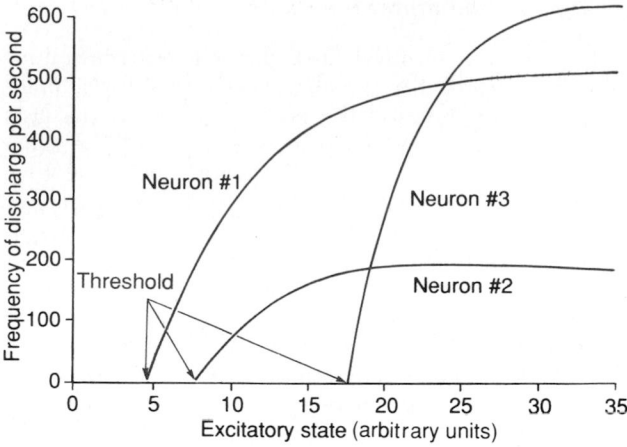

Figure 45–12. Response characteristics of different types of neurons to progressively increasing levels of excitatory state.

synapses being stimulated, the more rapidly will the excitatory state be re-established and a new action potential occur. Therefore, with increased firing of the excitatory synapses, the more rapid will be the rate of firing of the neuron. Yet each type of neuron responds differently from the others, some firing only slowly whereas others burst with firing activity.

Figure 45–12 shows theoretical responses of three types of neurons to varying levels of the excitatory state. Note that neuron 1 has a low threshold for excitation, whereas neuron 3 has a high threshold. But note also that neuron 2 has the lowest maximum frequency of discharge, whereas neuron 3 has the highest maximum frequency.

Some neurons in the central nervous system fire continuously because even the normal excitatory state is above the threshold level. Their frequency of firing can usually be increased still more by further increasing their excitatory state. Or the frequency may be decreased, or firing even be stopped, by superimposing an inhibitory state on the neuron.

Thus, different neurons respond differently, have different thresholds for excitation, and have widely differing maximal frequencies of discharge. With a little imagination, one can readily understand the importance of having neurons with these many types of response characteristics to perform the widely varying functions of the nervous system.

THE RATE OF FIRING OF A NEURON IS DETERMINED BY HOW MUCH ITS EXCITATORY STATE IS ABOVE THRESHOLD. As long as the excitatory state is below threshold, there will be no firing of the postsynaptic neuron. As the excitatory state rises above threshold, the neuron begins to fire. When only slightly above threshold, the rate of firing is slow, only a few impulses per second. But then, as the excitatory state rises still higher, the firing rate of the neuron also increases. Yet there is an upper limit to the rate at which each type of neuron will fire.

Thus, for each type of neuron, there are three characteristics of the firing response to excitation. They are as follows:

1. Each type of neuron has its own specific excitatory threshold at which it begins to fire.
2. Each type of neuron has its own specific rate of increase in firing rate as the excitatory state rises above threshold.
3. Each type of neuron has a characteristic maximum rate of firing when the excitatory state rises far above the threshold level.

SOME SPECIAL CHARACTERISTICS OF SYNAPTIC TRANSMISSION

FATIGUE OF SYNAPTIC TRANSMISSION. When excitatory synapses are repetitively stimulated at a rapid rate, the number of discharges by the postsynaptic neuron is at first very great, but it becomes progressively less in succeeding milliseconds or seconds. This is called *fatigue* of synaptic transmission.

Fatigue is an exceedingly important characteristic of synaptic function because when areas of the nervous system become overexcited, fatigue causes them to lose this excess excitability after a while. For example, fatigue is probably the most important means by which the excess excitability of the brain during an epileptic seizure is finally subdued so that the seizure ceases. Thus, the development of fatigue is a protective mechanism against excess neuronal activity. This is discussed further in the description of reverberating neuronal circuits in Chapter 46.

The mechanism of fatigue is mainly exhaustion of the stores of transmitter substance in the presynaptic terminals, particularly because the excitatory terminals on many, if not most, neurons can store enough excitatory transmitter for only about 10,000 normal synaptic transmissions, so that the transmitter can be exhausted in only a few seconds to a few minutes of rapid stimulation. Part of the fatigue process probably results from two other factors as well: (1) progressive inactivation of many of the postsynaptic membrane receptors and (2) slow buildup of abnormal concentrations of ions inside the *postsynaptic* neuronal cell, which causes an inhibitory effect on the postsynaptic neuron.

POST-TETANIC FACILITATION. When a rapidly repetitive series of impulses stimulates an excitatory synapse for a period of time and then a rest period is allowed, the synapse will often become for a period of seconds or minutes even more responsive to subsequent stimulation than normally. This is called *post-tetanic facilitation.*

Experiments have shown that post-tetanic facilitation is caused mainly by the buildup of excess calcium ions in the *presynaptic* terminals because the calcium pump pumps too slowly to remove all of these immediately after each action potential. These accumulated calcium ions cause more and more vesicular release of transmitter substance, occasionally increasing the release of transmitter to a rate twice that of normal.

The physiological significance of post-tetanic facilitation is still doubtful, and it may have no real significance. However, neurons could possibly store information with this mechanism. Therefore, post-tetanic facilitation might well be a mechanism of "short-term" memory in the central nervous system.

EFFECT OF ACIDOSIS AND ALKALOSIS ON SYNAPTIC TRANSMISSION. The neurons are highly responsive to changes in pH of the surrounding interstitial fluids. *Alkalosis greatly increases neuronal excitability.* For instance, a rise in arterial blood pH from the 7.4 norm to 7.8 to 8.0 often causes cerebral seizures because of increased excitability of the neurons. This can be demonstrated especially well by having a person who is predisposed to epileptic seizures overbreathe. The overbreathing elevates the pH of the blood only momentarily, but even this short interval can precipitate an epileptic attack.

On the other hand, *acidosis greatly depresses neuronal activity;* a fall in pH from 7.4 to below 7.0 usually causes a comatose state. For instance, in severe diabetic or uremic acidosis, coma always develops.

EFFECT OF HYPOXIA ON SYNAPTIC TRANSMISSION. Neuronal excitability is also highly dependent on an adequate supply of oxygen. Cessation of oxygen for only a few seconds can cause complete inexcitability of some neurons. This is seen when the cerebral blood flow is temporarily interrupted, because within 3 to 7 seconds, the person becomes unconscious.

EFFECT OF DRUGS ON SYNAPTIC TRANSMISSION. Many drugs are known to increase the excitability of neurons, and others are known to decrease the excitability. For instance, caffeine, theophylline, and theobromine, which are found in coffee, tea, and cocoa, respectively, all increase neuronal excitability, presumably by reducing the threshold for excitation of the neurons. Also, strychnine is one of the best known of all the agents that increase the excitability of neurons. However, it does not reduce the threshold for excitation of the neurons; instead, it *inhibits the action of some of the inhibitory transmitters* on the neurons, especially the inhibitory effect of glycine in the spinal cord. In consequence, the effects of the excitatory transmitters become overwhelming, and the neurons become so excited that they go into rapidly repetitive discharge, resulting in severe tonic muscle spasms.

Most anesthetics increase the membrane threshold for excitation and thereby decrease synaptic transmission at many points in the nervous system. Because most of the anesthetics are lipid-soluble, it has been reasoned that they might change the physical characteristics of the neuronal membranes, making them less responsive to excitatory agents.

SYNAPTIC DELAY. In transmission of an action potential from a presynaptic neuron to a postsynaptic neuron, a certain amount of time is consumed in the process of (1) discharge of the transmitter substance by the presynaptic terminal, (2) diffusion of the transmitter to the postsynaptic neuronal membrane, (3) action of the transmitter on the membrane receptor, (4) action of the receptor to increase the membrane permeability, and (5) inward diffusion of sodium to raise the excitatory postsynaptic potential to a high enough value to elicit an action potential. The *minimal* period of time required for all these events to take place, even when large numbers of excitatory synapses are stimulated simultaneously, is about 0.5 millisecond. This is called the *synaptic delay*. It is important for the following reason: Neurophysiologists can measure the *minimal* delay time between an input volley of impulses and an output volley and from this can estimate the number of series neurons in the circuit.

REFERENCES

Adams, J. H. and Graham, D. I.: An Introduction to Neuropathology. 2nd Ed. New York, Churchill Livingstone, 1994.

Adams, R. D. and Victor, M. Principles of Neurology. 5th Ed. Blue Ridge Summit, PA, McGraw-Hill, 1993.

Aird, R. B.: Foundations of Modern Neurology: A Century of Progress. New York, Raven Press, 1994.

Aminoff, M.: Neurology and General Medicine: The Neurological Aspects of Medical Disorders. 2nd Ed. New York, Churchill Livingstone, 1994.

Andersen, O. S., and Koeppe, R. E. II: Molecular determinants of channel function. Physiol. Rev., 72:(Suppl.)S89, 1992.

Barr, M. L., and Kiernan, J. A.: The Human Nervous System: An Anatomical Viewpoint. Philadelphia, J. B. Lippincott, 1993.

Baudry, M., et al.: Synaptic Plasticity. Cambridge, MA, The MIT Press, 1993.

Biggio, G., et al.: GABAergic Synaptic Transmission: Molecular, Pharmacological, and Clinical Aspects. New York, Raven Press, 1992.

Burt, A. M.: Textbook of Neuroanatomy. Philadelphia, W. B. Saunders Co., 1993.

Catterall, W. A.: Cellular and molecular biology of voltage-gated sodium channels. Physiol. Rev., 72:(Suppl.)S15, 1992.

Conn, P. M.: Neuroscience in Medicine. Philadelphia, J. B. Lippincott, 1994.

Cooper, J. R., et al.: The Biochemical Basis of Neuropharmacology. New York, Oxford University Press, 1991.

Duckett, S.: Pediatric Neuropathology. Baltimore, Williams & Wilkins, 1994.

Elbert, T., et al.: Chaos and physiology: deterministic chaos in excitable cell assemblies. Physiol. Rev., 74:1, 1994.

Fenichel, G. M.: Clinical Pediatric Neurology: A Signs and Symptoms Approach. Philadelphia, W. B. Saunders Co., 1993.

Finger, S.: Origins of Neuroscience: A History of Explorations into Brain Function. New York, Oxford University Press, 1994.

Frazer, A., et al.: Biological Bases of Brain Function and Disease. New York, Raven Press, 1994.

Griffith, W. H., et al.: Whole-cell and single-channel calcium currents in guinea pig basal forebrain neurons. J. Neurophysiol., 71:2359, 1994.

Hendelman, W. J.: Student's Atlas of Neuroanatomy. Philadelphia, W. B. Saunders Co., 1994.

Huguenard, J. and McCormick, D.: Electrophysiology of the Neuron: A Companion to Shepherd's Neurobiology: An Interactive Tutorial. New York, Oxford University Press, 1994.

Jacobs, B. L. and Azmitia, E. C.: Structure and function of the brain serotonin system. Physiol. Rev., 72:165, 1992.

Jennes, L. and Conn. P. M.: Atlas of the Brain: With Medical Correlations. Philadelphia, J. B. Lippincott, 1994.

Kuno, M.: Synapse: Function, Plasticity, and Neurotrophism. New York, Oxford University Press, 1995.

Langer, S. Z., et al.: Serotonin Receptor Subtypes: Pharmacological Significance and Clinical Implications. Farmington, CT, S. Karger Publishers, Inc., 1992.

Lawrence, A. J. and Jarrott, B.: L-Glutamate as a neurotransmitter at baroreceptor afferents: evidence from *in vivo* microdialysis. Neuroscience, 58:585, 1994.

Le Moal, M. and Simon, H.: Mesocorticolimbic dopaminergic network: functional and regulatory roles. Physiol. Rev., 71:155, 1991.

Lüscher, H.-R. and Clamann, H. P.: Relation between structure and function in information transfer in spinal monosynaptic reflex. Physiol. Rev., 72:71, 1992.

Marcus, E. M., et al.: An Introduction to the Neurosciences. Baltimore, Williams & Wilkins, 1994.

Melandri, B. A., et al.: Bioelectrochemistry IV: Nerve Muscle Function-Bioelectrochemistry, Mechanisms, Bioenergetics, and Control. New York, Plenum Publishing Corp., 1994.

Meldrum, B. S., et al.: Excitatory Amino Acids. New York, Raven Press, 1991.

Menkes, J. H.: Textbook of Child Neurology. Baltimore, Williams & Wilkins, 1995.

Mercuri, N. B., et al.: Effects of anoxia on rat midbrain dopamine neurons. J. Neurophysiol., 71:1165, 1994.

Otsuka, M. and Yoshioka, K.: Neurotransmitter functions of mammalian tachykinins. Physiol. Rev., 73:229, 1993.

Pang, S. F., et al.: Putative Melatonin Receptors in Peripheral Tissues. Farmington, CT, S. Karger Publishers, Inc., 1993.

Parent, A.: Human Neuroanatomy. Baltimore, Williams & Wilkins, 1995.

Pongs, O.: Molecular biology of voltage-dependent potassium channels. Physiol. Rev., 72:(Suppl.) S69, 1992.

Premkumar, L. S. and Gage, P. W.: Potassium channels activated by $GABA_B$ agonists and serotonin in cultured hippocampal neurons. J. Neurophysiol., 71:2570, 1994.

Putney, J. W., Jr.: Inositol Phosphates and Calcium Signalling. New York, Raven Press, 1992.

Rall, W., et al.: Matching dendritic neuron models to experimental data. Physiol. Rev., 72:(Suppl.) S159, 1992.

Reichert, H.: Introduction to Neurobiology. New York, Oxford University Press, 1993.

Rowland, L. P.: Merritt's Textbook of Neurology. Baltimore, Williams & Wilkins, 1995.

Runge, V. M.: Magnetic Resonance Imaging of the Brain. Philadelphia, J. B. Lippincott, 1994.

Schwartz, J. C., et al.: Histaminergic transmission in the mammalian brain. Physiol. Rev., 71:1, 1991.

Sealfon, S. C.: Receptor Molecular Biology. San Diego, CA, Academic Press, 1995.

Seil, F. J.: Neural Injury and Regeneration. New York, Raven Press, 1993.

Shepherd, G. M.: Neurobiology. New York, Oxford University Press, 1994.

Siegel, G. J., et al.: Basic Neurochemistry: Molecular, Cellular, and Medical Aspects. New York, Raven Press, 1994.

Stein, D. G., et al.: Healing the Damaged Brain: The Last Frontiers of Neuroscience. New York, Oxford University Press, 1994.

Strand, F. L., et al.: Neuropeptide hormones as neurotrophic factors. Physiol. Rev., 71:1017, 1991.

Strange, P. G.: Brain Biochemistry and Brain Disorders. New York, Oxford University Press, 1993.

Vincent, S. R.: Nitric Oxide in the Nervous System. San Diego, CA, Academic Press, 1995.

Walton, J.: Brain's Diseases of the Nervous System. New York, Oxford University Press, 1994.

Waxman, S. G. and deGroot, J.: Correlative Neuroanatomy. 22nd Ed. Redding, MA, Appleton & Lange, 1994.

Weiner, W. J. and Goetz, C. G.: Neurology for the Non-Neurologist. Philadelphia, J. B. Lippincott, 1994.

Willard, F. H. and Perl, D. B.: Medical Neuroanatomy: A Problem-Oriented Manual with Annotated Atlas Biology. Philadelphia, J. B. Lippincott, 1993.

Sensory Receptors; Neuronal Circuits for Processing Information

CHAPTER 46

Input to the nervous system is provided by the sensory receptors that detect such sensory stimuli as touch, sound, light, pain, cold, and warmth. The purpose of this chapter is to discuss the basic mechanisms by which these receptors change sensory stimuli into nerve signals and, also, how the information conveyed in the signals is processed in the nervous system.

TYPES OF SENSORY RECEPTORS AND THE SENSORY STIMULI THEY DETECT

Table 46–1 lists and classifies most of the body's sensory receptors. This table shows that there are five types of sensory receptors: (1) *mechanoreceptors,* which detect mechanical deformation of the receptor or of tissues adjacent to the receptor; (2) *thermoreceptors,* which detect changes in temperature, some receptors detecting cold and others warmth; (3) *nociceptors* (pain receptors), which detect damage occurring in the tissues, whether physical damage or chemical damage; (4) *electromagnetic receptors,* which detect light on the retina of the eye; and (5) *chemoreceptors,* which detect taste in the mouth, smell in the nose, oxygen level in the arterial blood, osmolality of the body fluids, carbon dioxide concentration, and perhaps other factors that make up the chemistry of the body.

In this chapter, we discuss the function of a few specific types of receptors, primarily peripheral mechanoreceptors, to illustrate some of the principles by which receptors operate. Other receptors are discussed

in other chapters in relation to the sensory systems that they subserve.

Figure 46–1 shows some of the types of mechanoreceptors found in the skin or in deep tissues of the body, and Table 46–1 gives their respective sensory functions. All these receptors are discussed in the following chapters in relation to the respective sensory systems.

Differential Sensitivity of Receptors

The first question that must be answered is, how do two types of sensory receptors detect different types of sensory stimuli? The answer is, by "differential sensitivities." That is, each type of receptor is highly sensitive to one type of stimulus for which it is designed and yet is almost nonresponsive to normal intensities of the other types of sensory stimuli. Thus, the rods and cones are highly responsive to light but are almost completely nonresponsive to heat, cold, pressure on the eyeballs, and chemical changes in the blood. The osmoreceptors of the supraoptic nuclei in the hypothalamus detect minute changes in the osmolality of the body fluids but have never been known to respond to sound. Finally, pain receptors in the skin are almost never stimulated by usual touch or pressure stimuli but do become highly active the moment tactile stimuli become severe enough to damage the tissues.

Modality of Sensation—The "Labeled Line" Principle

Each of the principal types of sensation that we can experience—pain, touch, sight, sound, and so forth—

583

Table 46-1 CLASSIFICATION OF SENSORY RECEPTORS

Mechanoreceptors
  Skin tactile sensibilities
    (epidermis and dermis)
    Free nerve endings
    Expanded tip endings
      Merkel's discs
      Plus several other variants
    Spray endings
      Ruffini's endings
    Encapsulated endings
      Meissner's corpuscles
      Krause's corpuscles
    Hair end-organs
  Deep tissue sensibilities
    Free nerve endings
    Expanded tip endings
    Spray endings
      Ruffini's endings
    Encapsulated endings
      Pacinian corpuscles
      Plus a few other variants
    Muscle endings
      Muscle spindles
      Golgi tendon receptors
  Hearing
    Sound receptors of cochlea
  Equilibrium
    Vestibular receptors
  Arterial pressure
    Baroreceptors of carotid sinuses and aorta
Thermoreceptors
  Cold
    Cold receptors
  Warmth
    Warm receptors
Nociceptors
  Pain
    Free nerve endings
Electromagnetic Receptors
  Vision
    Rods
    Cones
Chemoreceptors
  Taste
    Receptors of taste buds
  Smell
    Receptors of olfactory epithelium
  Arterial oxygen
    Receptors of aortic and carotid bodies
  Osmolality
    Probably neurons in or near supraoptic nuclei
  Blood CO_2
    Receptors in or on surface of medulla and in aortic
      and carotid bodies
  Blood glucose, amino acids, fatty acids
    Receptors in hypothalamus

is called a *modality* of sensation. Yet despite the fact that we experience these different modalities of sensation, nerve fibers transmit only impulses. Therefore, how is it that different nerve fibers transmit different modalities of sensation?

The answer to this is that each nerve tract terminates at a specific point in the central nervous system, and the type of sensation felt when a nerve fiber is stimulated is determined by the point in the nervous system to which the fiber leads. For instance, if a pain fiber is stimulated, the person perceives pain regardless of what type of stimulus excites the fiber. The stimulus can be electricity, heating the fiber itself, crushing the fiber, or stimulation of the pain nerve ending by damage to the tissue cells. Yet the person still perceives pain. Likewise, if a touch fiber is stimulated by exciting a touch receptor electrically or in any other way, the person perceives touch because touch fibers lead to specific touch areas in the brain. Similarly, fibers from the retina of the eye terminate in the vision areas of the brain, fibers from the ear terminate in the auditory areas of the brain, and temperature fibers terminate in the temperature areas.

This specificity of nerve fibers for transmitting only one modality of sensation is called the *"labeled line" principle*.

TRANSDUCTION OF SENSORY STIMULI INTO NERVE IMPULSES

Local Currents at Nerve Endings—Receptor Potentials

All sensory receptors have one feature in common. Whatever the type of stimulus that excites the receptor, its immediate effect is to change the membrane potential of the receptor. This change in potential is called a *receptor potential*.

MECHANISMS OF RECEPTOR POTENTIALS. Different receptors can be excited in one of several ways to cause receptor potentials: (1) by mechanical deformation of the receptor, which stretches the receptor membrane and opens ion channels; (2) by application

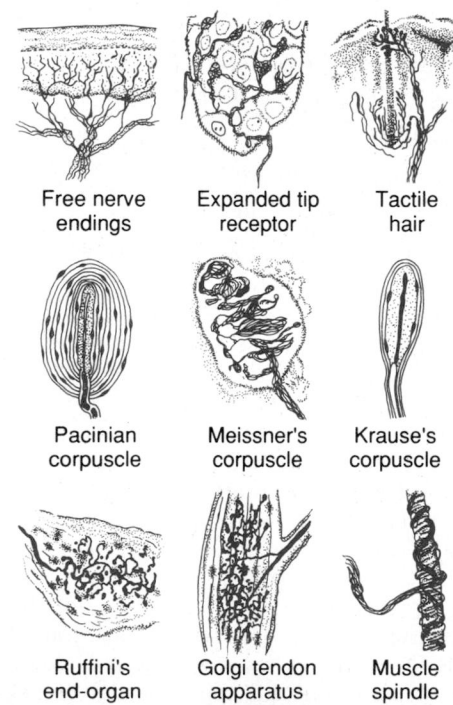

Free nerve endings   Expanded tip receptor   Tactile hair

Pacinian corpuscle   Meissner's corpuscle   Krause's corpuscle

Ruffini's end-organ   Golgi tendon apparatus   Muscle spindle

Figure 46-1. Several types of somatic sensory nerve endings.

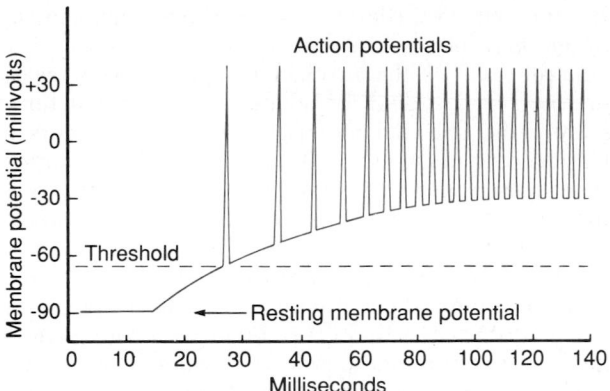

Figure 46–2. Typical relation between receptor potential and action potentials when the receptor potential rises above the threshold level.

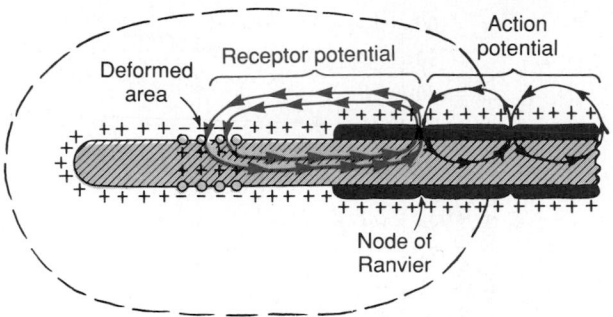

Figure 46–3. Excitation of a sensory nerve fiber by a receptor potential produced in a pacinian corpuscle. (Modified from Loëwenstein: *Ann. N. Y. Acad. Sci.*, *94*:510, 1961.)

of a chemical to the membrane, which also opens ion channels; (3) by change of the temperature of the membrane, which alters the permeability of the membrane; and (4) by the effects of electromagnetic radiation, such as light on the receptor, which either directly or indirectly changes the membrane characteristics and allows ions to flow through membrane channels. It will be recognized that these four means of exciting receptors correspond in general with the different types of known sensory receptors. In all instances, the basic cause of the change in membrane potential is a change in receptor membrane permeability, which allows ions to diffuse more or less readily through the membrane and thereby change the transmembrane potential.

RECEPTOR POTENTIAL AMPLITUDE. The maximum amplitude of most sensory receptor potentials is around 100 millivolts. This is about the same maximum voltage recorded in action potentials and is also the change in voltage when the membrane becomes maximally permeable to sodium ions.

RELATION OF THE RECEPTOR POTENTIAL TO ACTION POTENTIALS. When the receptor potential rises above the *threshold* for eliciting action potentials in the nerve fiber attached to the receptor, then action potentials begin to appear, as illustrated in Figure 46–2. Note also that the more the receptor potential rises above the threshold level, the greater becomes the action potential frequency. Thus, the receptor potential stimulates the sensory nerve fiber in the same way that the excitatory postsynaptic potential in the central nervous system neuron stimulates the neuron's axon.

Receptor Potential of the Pacinian Corpuscle—An Illustrative Example of Receptor Function

The student should at this point restudy the anatomical structure of the pacinian corpuscle shown in Figure 46–1. Note that the corpuscle has a central nerve fiber extending through its core. Surrounding this are multiple concentric capsule layers, so that compression anywhere on the outside of the corpuscle will elongate, indent, or otherwise deform the central fiber.

Now study Figure 46–3, which shows only the central fiber of the pacinian corpuscle after all but one capsule layer have been removed by microdissection. The very tip of the central fiber inside the capsule is unmyelinated, but the fiber does become myelinated shortly before leaving the corpuscle to enter the peripheral sensory nerve.

The figure also shows the mechanism by which a receptor potential is produced in the pacinian corpuscle. Observe the small area of the terminal fiber that has been deformed by compression of the corpuscle, and note that ion channels have opened in the membrane, allowing positively charged sodium ions to diffuse to the interior of the fiber. This creates increased positivity inside the fiber, which is a "receptor potential." The receptor potential in turn induces a *local circuit* of current flow, shown by the red arrows, that spreads along the nerve fiber. At the first node of Ranvier, which itself lies inside the capsule of the pacinian corpuscle, the local current flow depolarizes the fiber membrane at the node, and this then sets off typical action potentials that are transmitted along the nerve fiber toward the central nervous system.

RELATION BETWEEN STIMULUS INTENSITY AND THE RECEPTOR POTENTIAL. Figure 46–4 shows the changing amplitude of the receptor potential caused by progressively stronger mechanical compression applied experimentally to the central core of a pacinian corpuscle. Note that the amplitude increases rapidly at first but then progressively less rapidly at high stimulus strength.

In general, the frequency of repetitive action potentials transmitted from sensory receptors increases approximately in proportion to the increase in receptor potential. Putting this information together with the data in Figure 46–4, one can see that even though a very weak sensory stimulus can usually elicit at least some sensory signal, very intense stimulation of the receptor causes progressively less and less further increase in numbers of action potentials. This is an exceedingly important principle, used by almost all sensory receptors. It allows the receptor to be sensitive to

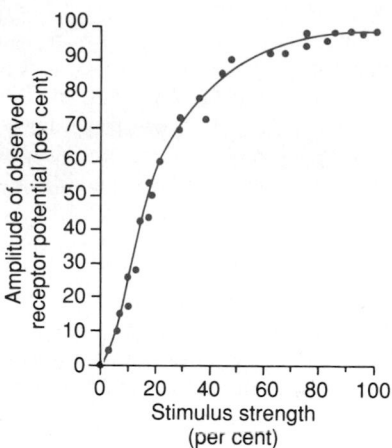

Figure 46–4. Relation of amplitude of receptor potential to strength of a mechanical stimulus applied to a pacinian corpuscle. (From Loëwenstein: *Ann. N. Y. Acad. Sci., 94:*510, 1961.)

weak sensory experience and yet not reach a maximum firing rate until the sensory experience is extreme. This allows the receptor to have an extreme range of response, from very weak to very intense.

Adaptation of Receptors

A special characteristic of all sensory receptors is that they *adapt* either partially or completely to their stimuli after a period of time. That is, when a continuous sensory stimulus is applied, the receptors respond at a high impulse rate at first and then at a progressively slower rate until finally many of them no longer respond.

Figure 46–5 shows typical adaptation of certain types of receptors. Note that the pacinian corpuscle adapts extremely rapidly and hair receptors adapt within a second or so, whereas some joint capsule and muscle spindle receptors adapt slowly.

Furthermore, some sensory receptors adapt to a far greater extent than others. For example, the pacinian corpuscles adapt to "extinction" within a few hundredths of a second, and the receptors at the bases of the hairs adapt to extinction within a second or more. It is probable that all other *mechanoreceptors* also adapt completely eventually, but some require hours or days to do so, for which reason they are called "nonadapting" receptors. The longest measured time for complete adaptation of a mechanoreceptor is about 2 days, which is the total adaptation time for many carotid and aortic baroreceptors.

Some of the nonmechanoreceptors, the chemoreceptors and pain receptors for instance, probably never adapt completely.

MECHANISMS BY WHICH RECEPTORS ADAPT. Adaptation of receptors is an individual property of each type of receptor, in much the same way that development of a receptor potential is an individual property. For instance, in the eye, the rods and cones adapt by

changing the concentrations of their light-sensitive chemicals (which is discussed in Chapter 50).

In the case of the mechanoreceptors, the receptor that has been studied for adaptation in greatest detail is again the pacinian corpuscle. Adaptation occurs in this receptor in two ways. First, the pacinican corpuscle is a viscoelastic structure so that when a distorting force is suddenly applied to one side of the corpuscle, this force is instantly transmitted by the viscous component of the corpuscle directly to the same side of the central nerve fiber, thus eliciting the receptor potential. However, within a few hundredths of a second, the fluid within the corpuscle redistributes, so that the pressure becomes essentially equal all through the corpuscle; this now applies even pressure on all sides of the central nerve fiber, so that the receptor potential is no longer elicited. Thus, the receptor potential appears at the onset of compression but then disappears within a small fraction of a second even though the compression continues.

Then, when the distorting force is removed from the corpuscle, essentially the reverse events occur. The sudden removal of the distortion from one side of the corpuscle allows rapid expansion on that side, and a corresponding distortion of the central core occurs once more. Again, within hundredths of a second, the pressure becomes equalized all through the corpuscle, and the stimulus is lost. Nevertheless, this disturbance of the central core fiber signals the offset of compression in much the same way that it signals the onset of compression.

The second mechanism of adaptation of the pacinian corpuscle, but a much slower one, results from a process called *accommodation*, which occurs in the nerve fiber itself. That is, even if by chance the central core fiber should continue to be distorted, as can be achieved after the capsule has been removed and the core compressed with a stylus, the tip of the nerve fiber itself gradually becomes "accommodated" to the stimulus. This probably results from "inactivation" of the sodium channels in the nerve fiber membrane,

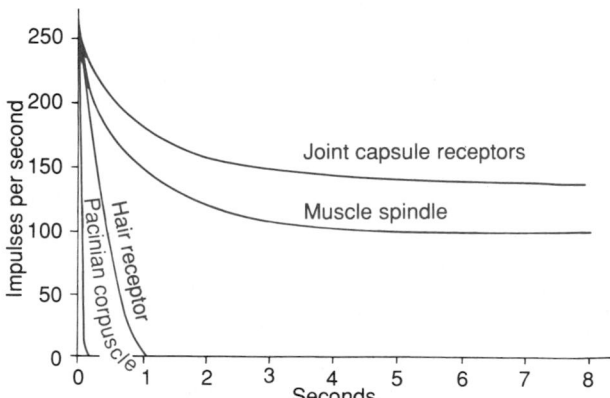

Figure 46–5. Adaptation of different types of receptors, showing rapid adaptation of some receptors and slow adaptation of others.

which means that the sodium current flow itself through the channels in some way causes them gradually to close, as explained in Chapter 5.

Presumably, these same two general mechanisms of adaptation apply also to the other types of mechanoreceptors. That is, part of the adaptation results from readjustments in the structure of the receptor itself, and part results from accommodation in the terminal nerve fibril.

Slowly Adapting Receptors Detect Continuous Stimulus Strength—The "Tonic" Receptors. The slowly adapting receptors continue to transmit impulses to the brain as long as the stimulus is present (or at least for many minutes or hours). Therefore, they keep the brain constantly apprised of the status of the body and its relation to its surroundings. For instance, impulses from the muscle spindles and Golgi tendon apparatus allow the central nervous system to know the status of muscle contraction and the load on the muscle tendon at each instant.

Other types of slowly adapting receptors include the receptors of the macula in the vestibular apparatus, the pain receptors, the baroreceptors of the arterial tree, the chemoreceptors of the carotid and aortic bodies, and some of the tactile receptors, such as Ruffini's endings and Merkel's discs.

Because the slowly adapting receptors can continue to transmit information for many hours, they are called *tonic* receptors. Most of the slowly adapting receptors, especially the mechanoreceptors, will adapt to extinction if the intensity of the stimulus remains absolutely constant for several hours or days. Because of our continually changing bodily state, these receptors almost never completely adapt.

Rapidly Adapting Receptors Detect *Change* in Stimulus Strength—The "Rate Receptors," "Movement Receptors," or "Phasic Receptors." Receptors that adapt rapidly cannot be used to transmit a continuous signal because these receptors are stimulated only when the stimulus strength changes. Yet they react strongly *while a change is actually taking place*. Furthermore, the number of impulses transmitted is directly related to the *rate at which the change takes place*. Therefore, these receptors are called *rate* receptors, *movement* receptors, or *phasic* receptors. Thus, in the case of the pacinian corpuscle, sudden pressure applied to the tissue excites this receptor for a few milliseconds, and then its excitation is over even though the pressure continues. But later, it transmits a signal again when the pressure is released. In other words, the pacinian corpuscle is exceedingly important in apprising the nervous system of rapid tissue deformations, but it is useless for transmitting information about constant conditions in the body.

Importance of the Rate Receptors—Their Predictive Function. If one knows the rate at which some change in bodily status is taking place, one can predict the state of the body a few seconds or even a few minutes later. For instance, the receptors of the semicircular canals in the vestibular apparatus of the ear detect the rate at which the head begins to turn when one runs around a curve. Using this information, a person can predict how much he or she will turn within the next 2 seconds and can adjust the motion of the legs *ahead of time* to keep from losing balance. Likewise, receptors located in or near the joints help detect the rates of movement of the different parts of the body. Therefore, when one is running, information from these receptors allows the nervous system to predict where the feet will be during any precise fraction of a second, and appropriate motor signals can be transmitted to the muscles of the legs to make any necessary anticipatory corrections in position so that the person will not fall. Loss of this predictive function makes it impossible for the person to run.

NERVE FIBERS THAT TRANSMIT DIFFERENT TYPES OF SIGNALS AND THEIR PHYSIOLOGICAL CLASSIFICATION

Some signals need to be transmitted to or from the central nervous system extremely rapidly; otherwise, the information would be useless. An example of this is the sensory signals that apprise the brain of the momentary positions of the legs at each fraction of a second during running. At the other extreme, some types of sensory information, such as that depicting prolonged, aching pain, do not need to be transmitted rapidly, so that slowly conducting fibers will suffice. Nerve fibers come in all sizes between 0.2 and 20 micrometers in diameter—the larger the diameter, the greater the conducting velocity. The range of conducting velocities is between 0.5 and 120 m/sec.

The upper half of Figure 46–6 gives two classifications of nerve fibers that are in general use. One of them is a general classification that includes both sensory and motor fibers, including the autonomic nerve fibers. The other is a classification of sensory nerve fibers that is used primarily by sensory neurophysiologists.

General Classification. In the general classification, the fibers are divided into types A and C, and the type A fibers are further subdivided into α, β, γ, and δ fibers.

Type A fibers are the typical myelinated fibers of spinal nerves. Type C fibers are the small, unmyelinated nerve fibers that conduct impulses at low velocities. The C fibers constitute more than one half the sensory fibers in most peripheral nerves as well as all the postganglionic autonomic fibers.

The sizes, velocities of conduction, and functions of the different nerve fiber types are also given in the figure. Note that a few large fibers can transmit impulses at velocities as great as 120 m/sec, a distance in 1 second that is longer than a football field. On the other hand, the smallest fibers transmit impulses as slowly as 0.5 m/sec, requiring about 2 seconds to go from the big toe to the spinal cord.

Alternate Classification Used by Sensory Physiologists. Certain recording techniques have made it possible to separate the type Aα fibers into two subgroups; yet these same recording techniques cannot dis-

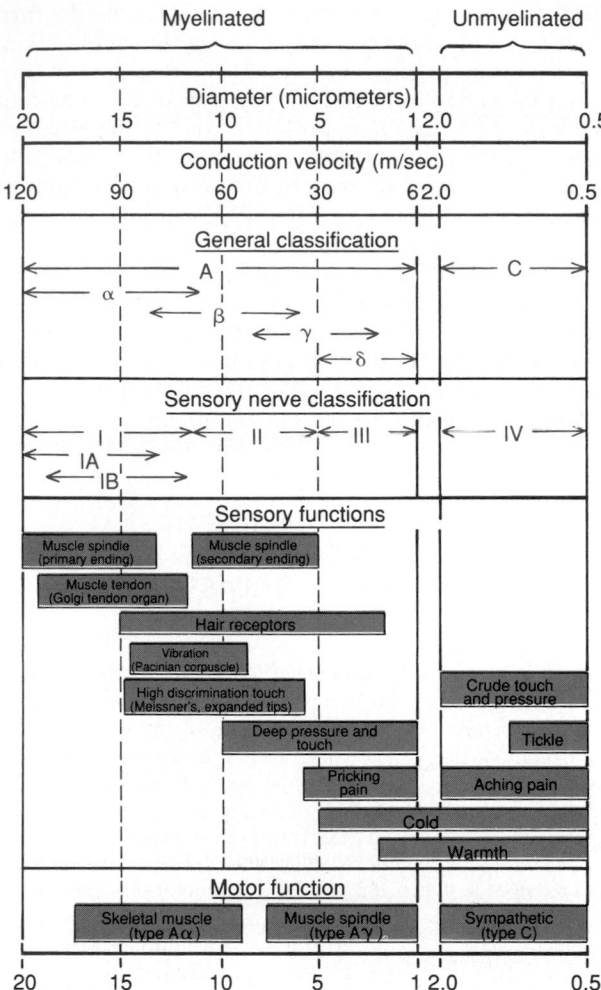

Figure 46–6. Physiological classifications and functions of nerve fibers.

TRANSMISSION OF SIGNALS OF DIFFERENT INTENSITY IN NERVE TRACTS—SPATIAL AND TEMPORAL SUMMATION

One of the characteristics of each signal that always must be conveyed is its intensity, as, for instance, the intensity of pain. These different gradations of intensity can be transmitted either by using increasing numbers of parallel fibers or by sending more impulses along a single fiber. These two mechanisms are called, respectively, spatial summation and temporal summation.

SPATIAL SUMMATION. Figure 46–7 shows the phenomenon of *spatial summation,* whereby increasing signal strength is transmitted by using progressively greater numbers of fibers. This figure shows a section of skin innervated by a large number of parallel pain nerve fibers. Each of these arborizes into hundreds of minute *free nerve endings* that serve as pain receptors. The entire cluster of fibers from one pain fiber frequently covers an area of skin as large as 5 centimeters in diameter. This area is called the *receptor field* of that fiber. The number of endings is large in the center of the field but diminishes toward the periphery. One can also see from the figure that the arborizing fibrils overlap those from other pain fibers. Therefore, a pinprick of the skin usually stimulates endings from many different pain fibers simultaneously. When the pinprick is in the center of the receptive field of a particular pain fiber, the degree of stimulation of that fiber is far greater than when it is in the periphery of the field.

Thus, in the lower part of Figure 46–7 are shown three views of the cross section of the nerve bundle leading from the skin area. To the left is shown the effect of a weak stimulus, with only a single nerve fiber in the middle of the bundle stimulated strongly

tinguish easily between Aβ and Aγ fibers. Therefore, the following classification is frequently used by sensory physiologists:

Group Ia. Fibers from the annulospiral endings of muscle spindles (average about 17 microns in diameter; these are α-type A fibers in the general classification)

Group Ib. Fibers from the Golgi tendon organs (average about 16 micrometers in diameter; these also are α-type A fibers)

Group II. Fibers from most discrete cutaneous tactile receptors and from the flower-spray endings of the muscle spindles (average about 8 micrometers in diameter; these are β- and γ-type A fibers in the general classification)

Group III. Fibers carrying temperature, crude touch, and pricking pain sensations (average about 3 micrometers in diameter; they are δ-type A fibers in the general classification)

Group IV. Unmyelinated fibers carrying pain, itch, temperature, and crude touch sensations (0.5 to 2 micrometers in diameter; they are called type C fibers in the general classification).

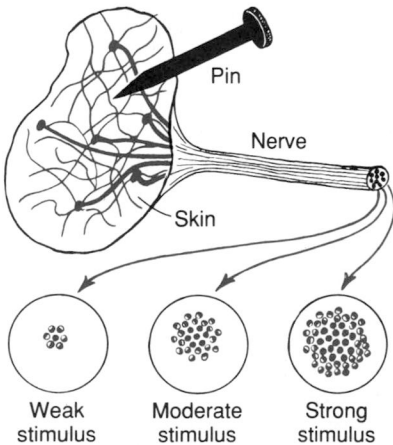

Figure 46–7. Pattern of stimulation of pain fibers in a nerve trunk leading from an area of skin pricked by a pin. This is an example of *spatial summation.*

(represented by the solid fiber), whereas several adjacent fibers are stimulated weakly (half-solid fibers). The other two views of the nerve cross section show the effect, respectively, of a moderate stimulus and a strong stimulus, with progressively more fibers being stimulated. Thus, the stronger signals spread to more and more fibers. This is the phenomenon of spatial summation.

TEMPORAL SUMMATION. A second means for transmitting signals of increasing strength is by increasing the *frequency* of nerve impulses in each fiber, which is called *temporal summation*. Figure 46–8 demonstrates this, showing in the upper part a changing strength of signal and in the lower part the actual impulses transmitted by the nerve fiber.

TRANSMISSION AND PROCESSING OF SIGNALS IN NEURONAL POOLS

The central nervous system comprises literally hundreds or even thousands of neuronal pools, some of which contain few neurons, whereas others have vast numbers. For instance, the entire cerebral cortex could be considered to be a single large neuronal pool. Other neuronal pools include the different basal ganglia, the specific nuclei in the thalamus, cerebellum, mesencephalon, pons, and medulla. Also, the entire dorsal gray matter of the spinal cord could be considered one long pool of neurons and the entire anterior gray matter another long neuronal pool.

Each pool has its own special characteristics of organization that cause it to process signals in its own special way, thus allowing the total consortium of pools to achieve the multitude of functions of the nervous system. Yet despite their differences in function, the pools also have many similar principles of function, described in the following pages.

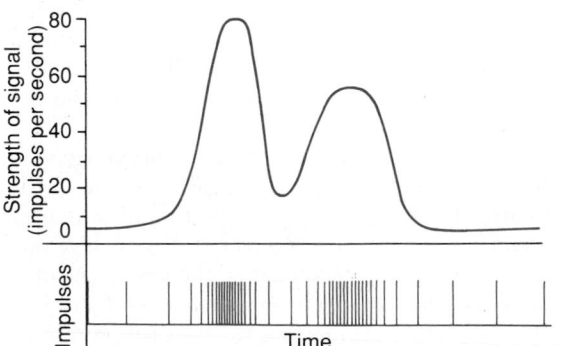

Figure 46–8. Translation of signal strength into a frequency-modulated series of nerve impulses, showing *above* the strength of signal and *below* the separate nerve impulses. This is an example of *temporal summation.*

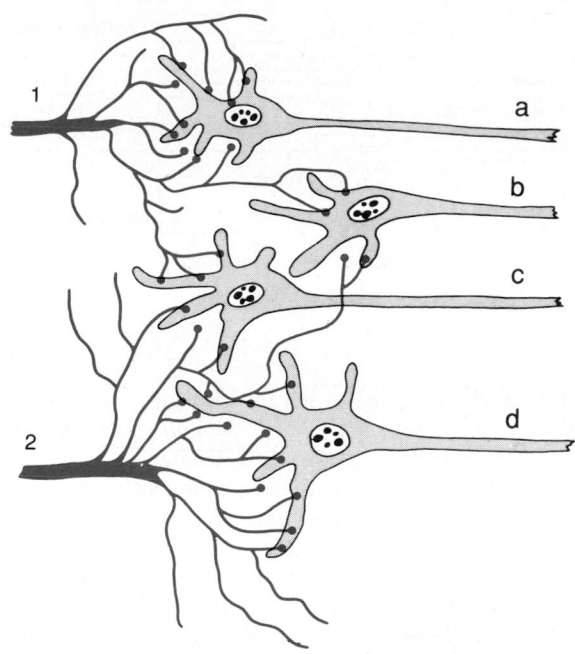

Figure 46–9. Basic organization of a neuronal pool.

Relaying of Signals Through Neuronal Pools

ORGANIZATION OF NEURONS FOR RELAYING SIGNALS. Figure 46–9 is a schematic diagram of several neurons in a neuronal pool, showing "input" fibers to the left and "output" fibers to the right. Each input fiber divides hundreds to thousands of times, providing an average of a thousand or more terminal fibrils that spread over a large area in the pool to synapse with the dendrites or cell bodies of the neurons in the pool. The dendrites usually also arborize and spread for hundreds to thousands of micrometers in the pool. The neuronal area stimulated by each incoming nerve fiber is called its *stimulatory field*. Note in Figure 46–9 that large numbers of the terminals from each input fiber lie on the nearest neuron in its "field," but progressively fewer terminals lie on the neurons farther away.

THRESHOLD AND SUBTHRESHOLD STIMULI—EXCITATION AND FACILITATION. From the discussion of synaptic function in Chapter 45, it will be recalled that discharge of a single excitatory presynaptic terminal almost never causes an action potential in the postsynaptic neuron. Instead, large numbers of input terminals must discharge on the same neuron either simultaneously or in rapid succession to cause excitation. For instance, in Figure 46–9 let us assume that six terminals must discharge simultaneously to excite any one of the neurons. If the student will count the number of terminals on each one of the neurons from each input fiber, he or she will see that input *fiber 1* has more than enough terminals to cause *neuron a* to discharge. Therefore, the stimulus from input fiber 1

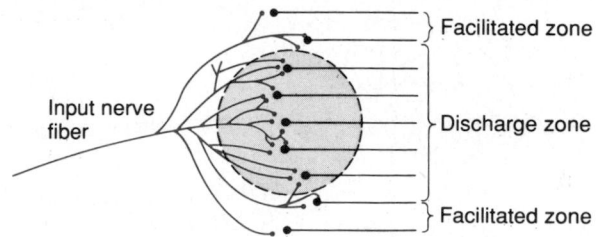

Figure 46–10. "Discharge" and "facilitated" zones of a neuronal pool.

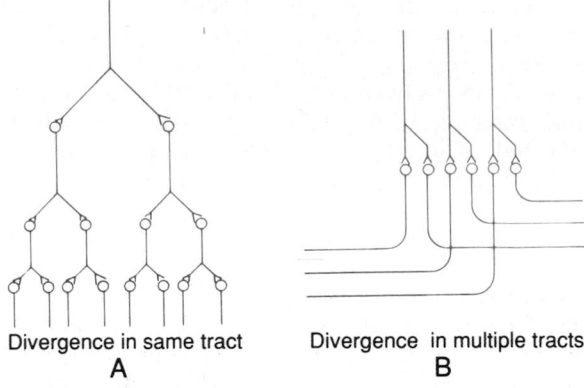

Divergence in same tract
A

Divergence in multiple tracts
B

Figure 46–11. "Divergence" in neuronal pathways. *A,* Divergence within a pathway to cause "amplification" of the signal. *B,* Divergence into multiple tracts to transmit the signal to separate areas.

to this neuron is said to be an *excitatory stimulus;* it is also called a *suprathreshold stimulus* because it is above the threshold required for excitation.

Input fiber 1 also contributes terminals to neurons b and c, but not enough to cause excitation. Nevertheless, discharge of these terminals makes both these neurons more likely to be excited by signals arriving through other incoming nerve fibers. Therefore, the stimulus to these neurons is said to be *subthreshold,* and the neurons are said to be *facilitated.*

Similarly, for input *fiber 2,* the stimulus to *neuron d* is a suprathreshold stimulus, and the stimulus to *neurons b* and *c* is a subthreshold, but facilitating, stimulus.

Figure 46–9 represents a highly condensed version of a neuronal pool because each input nerve fiber usually provides massive numbers of branching terminals to hundreds or thousands of neurons in its distribution "field," as shown in Figure 46–10. In the central portion of the field in this figure, designated by the darkened circle area, all the neurons are stimulated by the incoming fiber. Therefore, this is said to be the *discharge zone* of the incoming fiber, also called the *excited zone* or *liminal zone.* To either side, the neurons are facilitated but not excited, and these areas are called the *facilitated zone,* also called the *subthreshold zone* or *subliminal zone.*

INHIBITION OF A NEURONAL POOL. We must also remember that some incoming fibers inhibit neurons, rather than exciting them. This is the opposite of facilitation, and the entire field of the inhibitory branches is called the *inhibitory zone.* The degree of inhibition in the center of this zone is great because of large numbers of endings in the center; it becomes progressively less toward its edges.

Divergence of Signals Passing Through Neuronal Pools

Often it is important for signals entering a neuronal pool to excite far greater numbers of nerve fibers leaving the pool. This phenomenon is called *divergence.* Two major types of divergence occur and have entirely different purposes.

An *amplifying* type of divergence is shown in Figure 46–11A. This means simply that an input signal spreads to an increasing number of neurons as it passes through successive orders of neurons in its path.

This type of divergence is characteristic of the corticospinal pathway in its control of skeletal muscles, with a single large pyramidal cell in the motor cortex capable, under highly facilitated conditions, of exciting as many as 10,000 muscle fibers.

The second type of divergence, shown in Figure 46–11B, is *divergence into multiple tracts.* In this case, the signal is transmitted in two directions from the pool. For instance, information transmitted in the dorsal columns of the spinal cord takes two courses in the lower part of the brain: (1) into the cerebellum and (2) on through the lower regions of the brain to the thalamus and cerebral cortex. Likewise, in the thalamus, almost all sensory information is relayed both into still deeper structures of the thalamus and at the same time to discrete regions of the cerebral cortex.

Convergence of Signals

"Convergence" means signals from multiple inputs converging to excite a single neuron. Figure 46–12A shows *convergence from a single source.* That is, multiple terminals from an incoming fiber tract terminate on the same neuron. The importance of this is that neurons are almost never excited by an action potential from a single input terminal. But action potentials from multiple input terminals will provide enough "spatial summation" to bring the neuron to the threshold required for discharge.

Convergence can also result from input signals (excitatory or inhibitory) *from multiple sources,* as shown in Figure 46–12B. For instance, the interneurons of the spinal cord receive converging signals from (1) peripheral nerve fibers entering the cord, (2) propriospinal fibers passing from one segment of the cord to another, (3) corticospinal fibers from the cerebral cortex, and (4) several other long pathways descending from the brain into the spinal cord. Then the signals from the interneurons converge on the anterior motor neurons to control muscle function.

Such convergence allows summation of information

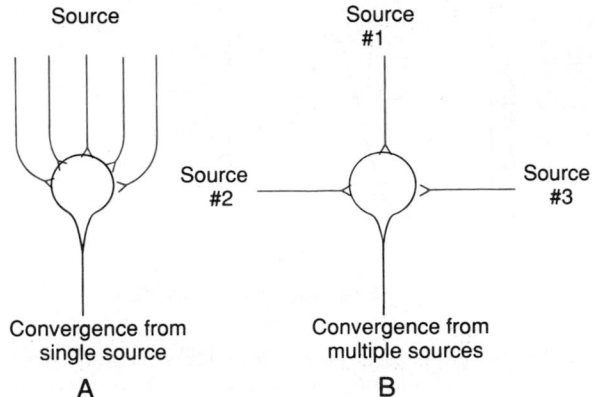

Figure 46-12. "Convergence" of multiple input fibers on a single neuron. *A*, Multiple input fibers from a single source. *B*, Input fibers from multiple sources.

from different sources, and the resulting response is a summated effect of all the different types of information. Therefore, convergence is one of the important means by which the central nervous system correlates, summates, and sorts different types of information.

Neuronal Circuit Causing Both Excitatory and Inhibitory Output Signals

Sometimes an incoming signal to a neuronal pool causes an output excitatory signal going in one direction and at the same time an inhibitory signal going elsewhere. For instance, at the same time that an excitatory signal is transmitted by one set of neurons in the spinal cord to cause forward movement of a leg, an inhibitory signal is transmitted simultaneously through a separate set of neurons to inhibit the muscles on the back of the leg so that they will not oppose the forward movement. This type of circuit is characteristic for controlling all antagonistic pairs of muscles, and it is called the *reciprocal inhibition circuit.*

Figure 46-13 shows the means by which the inhibition is achieved. The input fiber directly excites the excitatory output pathway, but it stimulates an intermediate *inhibitory neuron* (neuron 2), which then secretes a different type of transmitter substance to inhibit the second output pathway from the pool. This

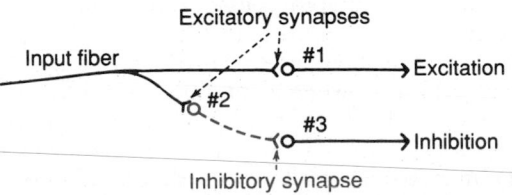

Figure 46-13. Inhibitory circuit. Neuron 2 is an inhibitory neuron.

type of circuit is also important in preventing overactivity in many parts of the brain.

Prolongation of a Signal by a Neuronal Pool—"Afterdischarge"

Thus far, we have considered signals that are merely relayed through neuronal pools. However, in many instances, a signal entering a pool causes a prolonged output discharge, called *afterdischarge*, lasting even after the incoming signal is over for a few milliseconds to as long as many minutes. The most important mechanisms by which afterdischarge occurs are the following.

SYNAPTIC AFTERDISCHARGE. When excitatory synapses discharge on the surfaces of dendrites or the soma of a neuron, a postsynaptic potential develops in the neuron that lasts for many milliseconds, especially so when some of the long-acting synaptic transmitter substances are involved. As long as this potential lasts, it can continue to excite the neuron, causing it to transmit a continuous train of output impulses, as was explained in Chapter 45. Thus, as a result of this synaptic "afterdischarge" mechanism alone, it is possible for a single instantaneous input signal to cause a sustained signal output (a series of repetitive discharges) lasting for many milliseconds.

REVERBERATORY (OSCILLATORY) CIRCUIT AS A CAUSE OF SIGNAL PROLONGATION. One of the most important of all circuits in the entire nervous system is the *reverberatory,* or *oscillatory, circuit.* Such circuits are caused by positive feedback within the neuronal circuit that feeds back to re-excite the input of the same circuit. Consequently, once stimulated, the circuit discharges repetitively for a long time.

Several possible varieties of reverberatory circuits are shown in Figure 46-14, the simplest—in Figure 46-14*A*—involving only a single neuron. In this case, the output neuron simply sends a collateral nerve fiber back to its own dendrites or soma to restimulate itself; therefore, once the neuron discharges, the feedback stimuli could help keep the neuron discharging for a long time thereafter.

Figure 46-14*B* shows a few additional neurons in the feedback circuit, which would give a longer period between the initial discharge and the feedback signal. Figure 46-14*C* shows a still more complex system in which both facilitatory and inhibitory fibers impinge on the reverberating circuit. A facilitatory signal enhances the intensity and frequency of reverberation, whereas an inhibitory signal depresses or stops the reverberation.

Figure 46-14*D* shows that most reverberating pathways are constituted of many parallel fibers, and at each cell station, the terminal fibrils diffuse widely. In such a system, the total reverberating signal can be either weak or strong, depending on how many parallel nerve fibers are momentarily involved in the reverberation.

Characteristics of Signal Prolongation from a Reverberatory Circuit. Figure 46-15 shows output

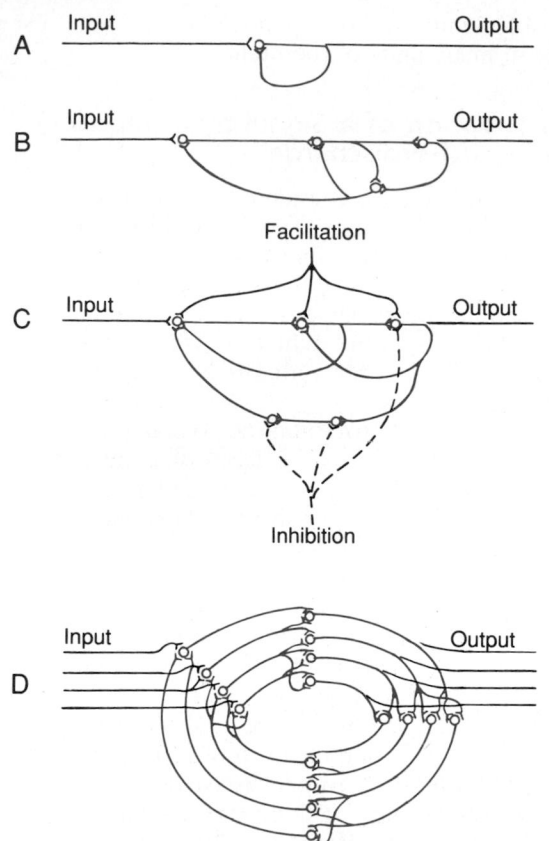

Figure 46–14. Reverberatory circuits of increasing complexity.

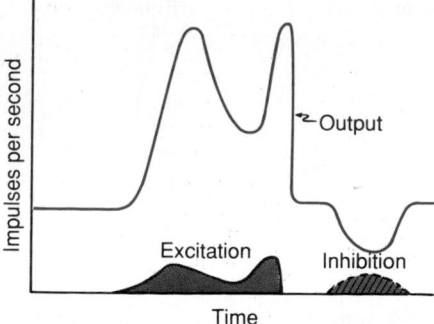

Figure 46–16. Continuous output from either a reverberating circuit or a pool of intrinsically discharging neurons. This figure also shows the effect of excitatory or inhibitory input signals.

signals from a typical reverberatory circuit. The input stimulus need last only 1 millisecond or so, and yet the output can last for many milliseconds or even minutes. The figure demonstrates that the intensity of the output signal usually increases to a high value early in the reverberation and then decreases to a critical point, at which it suddenly ceases entirely. The cause of this sudden cessation of reverberation is fatigue of synaptic junctions in the circuit, because fatigue beyond a certain critical level lowers the stimulation of the next neuron in the circuit below threshold level so that the circuit feedback is suddenly broken. The duration of the signal before cessation can also be controlled by

signals from other parts of the brain that inhibit or facilitate the circuit.

Almost these exact patterns of output signals are recorded from the motor nerves exciting a muscle involved in the flexor reflex after pain stimulation of the foot (as shown in Figure 46–18).

Continuous Signal Output from Neuronal Circuits

Some neuronal circuits emit output signals continuously even without excitatory input signals. At least two mechanisms can cause this effect: (1) continuous intrinsic neuronal discharge and (2) continuous reverberatory signals.

CONTINUOUS DISCHARGE CAUSED BY INTRINSIC NEURONAL EXCITABILITY. Neurons, like other excitable tissues, discharge repetitively if their level of excitatory membrane potential rises above a certain threshold level. The membrane potentials of many neurons even normally are high enough to cause them to emit impulses continually. This occurs especially in large numbers of the neurons of the cerebellum as well as in most of the interneurons of the spinal cord. The rates at which these cells emit impulses can be increased by excitatory signals or decreased by inhibitory signals; the inhibitory signals often can decrease the rate of firing to zero.

CONTINUOUS SIGNALS EMITTED FROM REVERBERATING CIRCUITS AS A MEANS FOR TRANSMITTING INFORMATION. A reverberating circuit that does not fatigue enough to stop reverberation can also be a source of continual impulses. And excitatory impulses entering the reverberating pool can increase the output signal, while inhibition can decrease or even extinguish the signal.

Figure 46–16 shows a continual output signal from a pool of neurons, whether a pool emitting impulses because of intrinsic neuronal excitability or as a result of reverberation. Note that an excitatory input signal greatly increases the output signal, whereas an inhibitory input signal greatly decreases the output. Those

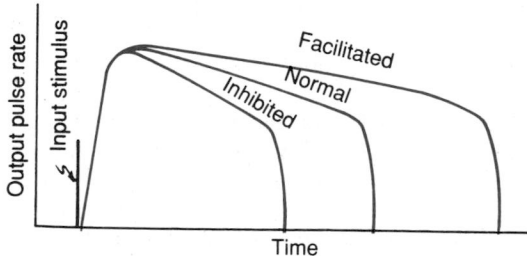

Figure 46–15. Typical pattern of the output signal from a reverberatory circuit after a single input stimulus, showing the effects of facilitation and inhibition.

students who are familiar with radio transmitters will recognize this to be a *carrier wave* type of information transmission. That is, the excitatory and inhibitory control signals are not the *cause* of the output signal, but they do *control* it. Note that this carrier wave system allows *decrease* in signal intensity as well as increase, whereas up to this point, the types of information transmission that we have discussed have been only positive information, rather than negative information. This type of information transmission is used by the autonomic nervous system to control such functions as vascular tone, gut tone, degree of constriction of the iris, heart rate, and others. That is, each of these can be either increased or decreased by the same input nerve fibers—increased when the rate of firing rises above normal and decreased when the firing falls below normal.

Rhythmical Signal Output

Many neuronal circuits emit rhythmical output signals—for instance, the rhythmical respiratory signal originating in the respiratory centers of the medulla and pons. This respiratory rhythmical signal continues throughout life, whereas other rhythmical signals, such as those that cause scratching movements by the hind leg of a dog or the walking movements in an animal, require input stimuli into the respective circuits to initiate the rhythmical signals.

All or almost all rhythmical signals that have been studied experimentally have been found to result from reverberating circuits or a succession of sequential reverberating circuits that feed excitatory or inhibitory signals from one to the next.

Excitatory or inhibitory signals can increase or decrease the rhythmical amplitude of the signal output. Figure 46–17, for instance, shows the rhythmical respiratory signal in the phrenic nerve. However, when the carotid body is stimulated by increasing arterial oxygen deficiency, the frequency and amplitude of the rhythmical signal pattern increase progressively.

INSTABILITY AND STABILITY OF NEURONAL CIRCUITS

Almost every part of the brain connects either directly or indirectly with every other part, and this creates a serious problem. If the first part excites the second, the second the third, the third the fourth, and so on until finally the signal re-excites the first part, it is clear that an excitatory signal entering any part of the brain would set off a continuous cycle of re-excitation of all parts. If this should occur, the brain would be inundated by a mass of uncontrolled reverberating signals—signals that would be transmitting no information but, nevertheless, would be consuming the circuits of the brain so that none of the informational signals could be transmitted. Such an effect occurs in widespread areas of the brain during *epileptic seizures.*

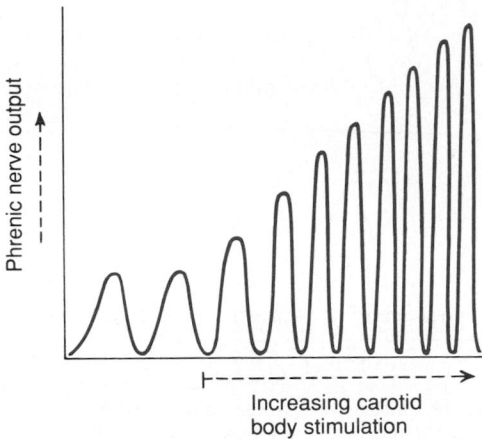

Figure 46–17. The rhythmical output of summated nerve impulses from the respiratory center, showing that progressively increasing stimulation of the carotid body increases both the intensity and the frequency of the phrenic nerve signal to the diaphragm to increase respiration.

How does the central nervous system prevent this from happening all the time? The answer seems to lie in two basic mechanisms that function throughout the central nervous system: (1) inhibitory circuits and (2) fatigue of synapses.

Inhibitory Circuits as a Mechanism for Stabilizing Nervous System Function

Two types of inhibitory circuits in widespread areas of the brain help prevent excessive spread of signals: (1) inhibitory feedback circuits that return from the termini of pathways back to the initial excitatory neurons of the same pathways—these circuits occur in virtually all the sensory nervous pathways; they inhibit either the input neurons or intermediate neurons in the sensory pathway when the termini become overly excited; and (2) some neuronal pools that exert gross inhibitory control over widespread areas of the brain —for instance, many of the basal ganglia exert inhibitory influences throughout the motor control system.

Synaptic Fatigue as a Means of Stabilizing the Nervous System

Synaptic fatigue means simply that synaptic transmission becomes progressively weaker the more prolonged and more intense the period of excitation. Figure 46–18 shows three successive records of a flexor reflex elicited in an animal caused by inflicting pain in the footpad of the paw. Note in each record that the strength of contraction progressively "decrements"— that is, its strength diminishes; this is presumed to be caused by *fatigue* of the synapses in the flexor reflex circuit. Furthermore, the shorter the interval between the successive flexor reflexes, the less is the intensity of the subsequent reflex response. Thus, in most neuronal circuits that are overused, the sensitivities of the circuits become depressed.

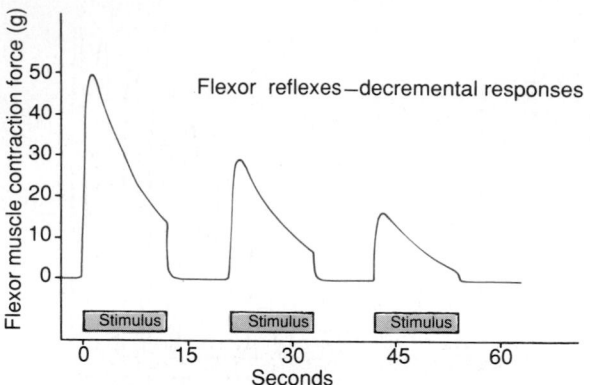

Figure 46–18. Successive flexor reflexes showing fatigue of conduction through the reflex pathway.

AUTOMATIC SHORT-TERM ADJUSTMENT OF PATHWAY SENSITIVITY BY THE FATIGUE MECHANISM. Now let us apply this phenomenon of fatigue to pathways in the brain. Those that are overused usually become fatigued, and so their sensitivies are reduced. On the other hand, those that are underused become rested, and their sensitivities increase. Thus, fatigue and recovery from fatigue constitute an important short-term means of moderating the sensitivities of the different nervous system circuits, helping to keep them operating in a range of sensitivity that allows effective function.

LONG-TERM CHANGES IN SYNAPTIC SENSITIVITY CAUSED BY AUTOMATIC DOWNGRADING OR UPGRADING OF SYNAPTIC RECEPTORS. The long-term sensitivities of synapses can be changed tremendously by downgrading the number of receptor proteins at the synaptic sites when there is overactivity and upgrading the receptors when there is underactivity. The mechanism for this is the following: Receptor proteins are being formed constantly by the endoplasmic reticular–Golgi apparatus system and are constantly being inserted into the receptor neuron synaptic membrane. However, when the synapses are overused so that excesses of transmitter substance combine with the receptor proteins, many of these receptors are inactivated permanently and removed from the synaptic membrane.

It is indeed fortunate that downgrading or upgrading of receptors as well as other control mechanisms for adjusting synaptic sensitivity continually adjust the sensitivity in each circuit to almost the exact level required for proper function. Think for a moment how serious it would be if the sensitivities of only a few of these circuits should be abnormally high; one might then expect almost continual muscle cramps, seizures, psychotic disturbances, hallucinations, mental tension, or other nervous disorders. But the automatic controls normally readjust the sensitivities of the circuits back

to a controllable range of reactivity any time the circuits begin to be too active or too depressed.

REFERENCES

Abeles, M.: Corticonics: Neural Circuits of the Cerebral Cortex. New York, Cambridge University Press, 1991.

Aubin, J.-P.: Neural Networks and Qualitative Physics: A Viability Approach. New York, Cambridge University Press, 1994.

Baudry, M. and Davis J. L.: Long-term Potentiation. Cambridge, MA, The MIT Press, 1994.

Churchland, P. S. and Sejnowski, T. J.: The Computational Brain. Cambridge, MA, The MIT Press, 1992.

Crinella, F. M.: Brain Mechanisms. New York, New York Academy of Sciences, 1994.

Dempster, F. N. and Brainerd, C. J.: Interference and Inhibition in Cognition. San Diego, CA, Academic Press, 1994.

Dietz, V.: Human neuronal control of automatic functional movements: interaction between central programs and afferent input. Physiol. Rev., 72:33, 1992.

Elbert, T., et al.: Chaos and physiology: deterministic chaos in excitable cell assemblies. Physiol. Rev., 74:1, 1994.

Frazer, A., et al.: Biological Bases of Brain Function and Disease. New York, Raven Press, 1994.

Friesen, W. O. and Friesen, J. A.: NeuroDynamix: Neuronal Models for Neurophysiology. New York, Oxford University Press, 1994.

Gardner, D.: The Neurobiology of Neural Networks. Cambridge, MA, The MIT Press, 1993.

Harris-Warrick, R. M., et al.: Dynamic Biological Networks. Cambridge, MA, The MIT Press, 1992.

Honavar, V. and Uhr, L.: Artificial Intelligence and Neural Networks. San Diego, CA, Academic Press, 1994.

Kendall, D. A. and Hill, S. J.: Signal Transduction Protocols. Totowa, NJ, Humana Press, Inc., 1995.

Kitterle, F. L.: Hemispheric Communication. Hillsdale, NJ, Lawrence Erlbaum Assoc. Inc., 1994.

Koch, C. and Davis, J. L.: Large-scale Neuronal Theories of the Brain. Cambridge, MA, The MIT Press, 1994.

Landy, M. S. and Movshon, A.: Computational Models of Visual Processing. Cambridge, MA, The MIT Press, 1991.

Marmarelis, V. Z.: Advanced Methods of Physiological System Modeling. New York, Plenum Publishing Corp., 1994.

McBain, C. J. and Mayer, M. L.: N-methyl-D-aspartic acid receptor structure and function. Physiol. Rev., 74:723, 1994.

Milligan, G.: Signal Transduction: A Practical Approach. New York, Oxford University Press, 1992.

Mountcastle, V. B.: Central nervous mechanisms in mechanoreceptive sensibility. In Darian-Smith, I. (ed.): Handbook of Physiology. Sec. 1, Vol. III. Bethesda, Md., American Physiological Society, 1984, p. 789.

Ono, T., et al.: Brain Mechanisms of Perception and Memory: From Neuron to Behavior. New York, Oxford University Press, 1993.

Peretto, P.: An Introduction to the Modelling of Neural Networks. New York, Cambridge University Press, 1992.

Purves, D.: Neural Activity and the Growth of the Brain. New York, Cambridge University Press, 1994.

Robinson, D. A.: Integrating with neurons. Annu. Rev. Physiol., 12:33, 1989.

Scott, S. A.: Sensory Neurons: Diversity, Development, and Plasticity. New York, Oxford University Press, 1992.

Shapley, R.: Contrast Sensitivity. Cambridge, MA, The MIT Press, 1993.

Shenolikar, S. and Nairn, A. C.: Model Systems in Signal Transduction. New York, Raven Press, 1993.

Stenger, D. A. and McKenna, T. M.: Enabling Technologies for Cultured Neural Networks. San Diego, CA, Academic Press, 1994.

Thatcher, R. W., et al.: Functional Neuroimaging. San Diego, CA, Academic Press, 1994.

Valiant, L. G.: Circuits of the Mind. New York, Oxford University Press, 1994.

Warman, E. N., et al.: Reconstruction of hippocampal CA1 pyramidal cell electrophysiology by computer simulation. J. Neurophysiol., 71:2033, 1994.

Wolpaw, J. R. and Schmidt, J. T.: Activity-driven CNS Changes in Learning and Development. New York, New York Academy of Sciences, 1991.

Yockey, H. P.: Information Theory and Molecular Biology. New York, Cambridge University Press, 1992.

Ziv, I., et al.: Simulator for neural networks and action potentials: description and application. J. Neurophysiol., 71:294, 1994.

Zornetzer, S. F., et al.: An Introduction to Neural and Electronic Networks. 2nd Ed. San Diego, CA, Academic Press, 1994.

Somatic Sensations: I. General Organization; the Tactile and Position Senses

CHAPTER 47

The *somatic senses* are the nervous mechanisms that collect sensory information from the body. These senses are in contradistinction to the *special senses,* which mean specifically vision, hearing, smell, taste, and equilibrium.

Classification of Somatic Senses

The somatic senses can be classified into three physiological types: (1) the *mechanoreceptive somatic senses,* which include both *tactile* and *position* sensations that are stimulated by mechanical displacement of some tissue of the body; (2) the *thermoreceptive senses,* which detect heat and cold; and (3) the *pain sense,* which is activated by any factor that damages the tissues. This chapter deals with the mechanoreceptive tactile and position senses, and Chapter 48 discusses the thermoreceptive and pain senses.

The tactile senses include *touch, pressure, vibration,* and *tickle* senses, and the position senses include *static position* and *rate of movement* senses.

OTHER CLASSIFICATIONS OF SOMATIC SENSATIONS. Somatic sensations are also often grouped together in other classes that are not necessarily mutually exclusive, as follows.

Exteroreceptive sensations are those from the surface of the body. *Proprioceptive sensations* are those having to do with the physical state of the body, including position sensations, tendon and muscle sensations, pressure sensations from the bottom of the feet, and even the sensation of equilibrium, which is generally considered a "special" sensation rather than a somatic sensation.

Visceral sensations are those from the viscera of the body; in using this term, one usually refers specifically to sensations from the internal organs.

The *deep sensations* are those that come from the deep tissues, such as from fasciae, muscles, and bone. These include mainly "deep" pressure, pain, and vibration.

DETECTION AND TRANSMISSION OF TACTILE SENSATIONS

INTERRELATION BETWEEN THE TACTILE SENSATIONS OF TOUCH, PRESSURE, AND VIBRATION. Although touch, pressure, and vibration are frequently classified as separate sensations, they are all detected by the same types of receptors. There are three principal differences among them: (1) touch sensation generally results from stimulation of tactile receptors in the skin or in tissues immediately beneath the skin; (2) pressure sensation generally results from deformation of deeper tissues; (3) vibration sensation results from rapidly repetitive sensory signals, but some of the same types of receptors as those for touch and pressure are used, specifically the rapidly adapting types of receptors.

TACTILE RECEPTORS. At least six entirely different types of tactile receptors are known, but many more similar to these also exist. Some of these receptors were shown in Figure 46–1, and their special characteristics are the following.

First, some *free nerve endings,* which are found everywhere in the skin and in many other tissues, can detect touch and pressure. For instance, even light

595

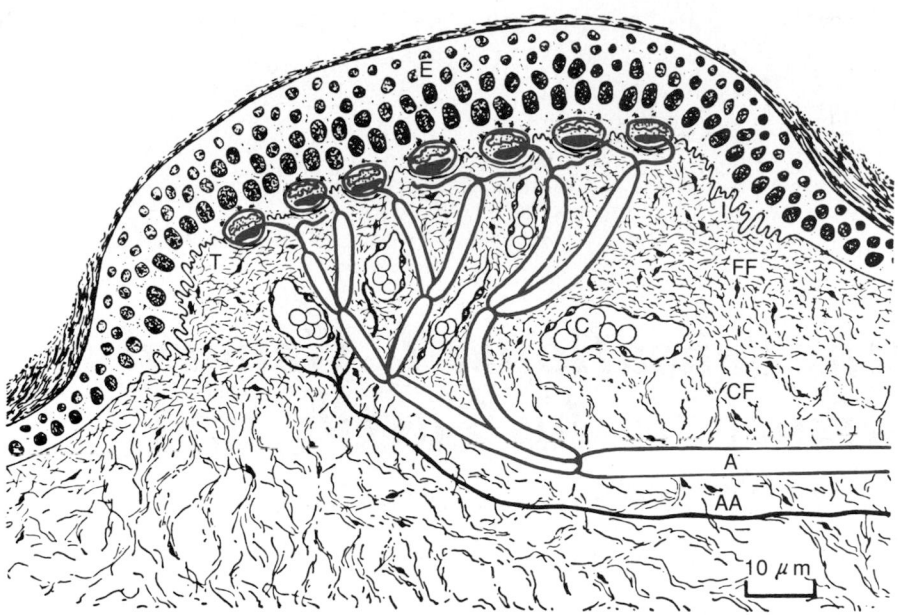

Figure 47–1. Iggo dome receptor. Note the multiple numbers of Merkel's discs innervated by a single large myelinated fiber and abutting tightly the undersurface of the epithelium. (From Iggo and Muir: *J. Physiol., 200:*763, 1969.)

contact with the cornea of the eye, which contains no other type of nerve ending besides free nerve endings, can nevertheless elicit touch and pressure sensations.

Second, a touch receptor illustrated in Figure 46–1 of the previous chapter and of special sensitivity is *Meissner's corpuscle,* an elongated encapsulated nerve ending that excites a large (type Aβ) myelinated sensory nerve fiber. Inside the capsulation are many branching terminal nerve filaments. These receptors are present in the nonhairy parts of the skin (called *glabrous skin*) and are particularly abundant in the fingertips, lips, and other areas of the skin where one's ability to discern spatial characteristics of touch sensations is highly developed. Meissner's corpuscles adapt in a fraction of a second after they are stimulated, which means that they are particularly sensitive to movement of very light objects over the surface of the skin as well as to low-frequency vibration.

Third, the fingertips and other areas that contain large numbers of Meissner's corpuscles also contain large numbers of *expanded tip tactile receptors,* one type of which is *Merkel's discs,* shown in Figure 47–1. The hairy parts of the skin also contain moderate numbers of expanded tip receptors, even though they have almost no Meissner's corpuscles. These receptors differ from Meissner's corpuscles in that they transmit an initially strong but partially adapting signal and then a continuing weaker signal that adapts only slowly. Therefore, they are responsible for giving steady-state signals that allow one to determine continuous touch of objects against the skin. Merkel's discs are often grouped together in a single receptor organ called the *Iggo dome receptor,* which projects upward against the underside of the epithelium of the skin, as also shown in Figure 47–1. This causes the epithelium at this point to protrude outward, thus creating a dome and constituting an extremely sensitive receptor. Also note

that the entire group of Merkel's discs is innervated by a single large type of myelinated nerve fiber (type Aβ). These receptors, along with the Meissner's corpuscles discussed above, play extremely important roles in localizing touch sensations to the specific surface areas of the body and in determining the texture of what is felt.

Fourth, slight movement of any hair on the body stimulates the nerve fiber entwining its base. Thus, each hair and its basal nerve fiber, called the *hair end-organ,* are also a touch receptor. This receptor adapts readily and, therefore, like Meissner's corpuscles, detects mainly movement of objects on the surface of the body or initial contact with the body.

Fifth, located in the deeper layers of the skin and also in deeper tissues are many *Ruffini's end-organs,* which are multibranched, encapsulated endings, as shown in Figure 46–1 in the previous chapter. These endings adapt very little and, therefore, are important for signaling continuous states of deformation of the skin and deeper tissues, such as heavy and continuous touch signals and pressure signals. They are also found in joint capsules and help signal the degree of joint rotation.

Sixth, *pacinian corpuscles,* which are discussed in detail in Chapter 46, lie both immediately beneath the skin and also deep in the fascial tissues of the body. They are stimulated only by rapid movement of the tissues because they adapt in a few hundredths of a second. Therefore, they are particularly important for detecting tissue vibration or other rapid changes in the mechanical state of the tissues.

TRANSMISSION OF TACTILE SENSATIONS IN PERIPHERAL NERVE FIBERS. Almost all the specialized sensory receptors, such as Meissner's corpuscles, Iggo dome receptors, hair receptors, pacinian corpuscles, and Ruffini's endings, transmit their signals in type Aβ

nerve fibers that have transmission velocities ranging from 30 to 70 m/sec. On the other hand, free nerve ending tactile receptors transmit signals mainly by way of the small type Aδ myelinated fibers that conduct at velocities of 5 to 30 m/sec. Some tactile free nerve endings transmit by way of type C unmyelinated fibers at velocities from a fraction of a meter up to 2 m/sec; these send signals into the spinal cord and lower brain stem, probably subserving mainly the sensation of tickle.

The more critical types of sensory signals—those that help to determine precise localization on the skin, minute gradations of intensity, or rapid changes in sensory signal intensity—are all transmitted in more rapidly conducting types of sensory nerve fibers. On the other hand, the cruder types of signals, such as crude pressure, poorly localized touch, and especially tickle, are transmitted by way of much slower, very small nerve fibers that require much less space in the nerve bundle than the faster fibers.

Detection of Vibration

All the different tactile receptors are involved in detection of vibration, although different receptors detect different frequencies of vibration. Pacinian corpuscles can signal vibrations from 30 to 800 cycles per second because they respond extremely rapidly to minute and rapid deformations of the tissues, and they also transmit their signals over type Aβ nerve fibers, which can transmit more than 1000 impulses per second.

Low-frequency vibrations up to 80 cycles per second, on the other hand, stimulate other tactile receptors, especially Meissner's corpuscles, which are less rapidly adapting than pacinian corpuscles.

Tickling and Itch

Neurophysiological studies have demonstrated the existence of very sensitive, rapidly adapting mechanoreceptive free nerve endings that elicit only the tickle and itch sensation. Furthermore, these endings are found almost exclusively in the superficial layers of the skin, which is also the only tissue from which the tickle and itch sensation usually can be elicited. This sensation is transmitted by small type C, unmyelinated fibers similar to those that transmit the aching, slow type of pain.

The purpose of the itch sensation is presumably to call attention to mild surface stimuli such as a flea crawling on the skin or a fly about to bite, and the elicited signals then excite the scratch reflex or other maneuvers that rid the host of the irritant.

Itch can be relieved by scratching if this removes the irritant or if the scratch is strong enough to elicit pain. The pain signals are believed to suppress the itch signals in the cord by lateral inhibition, as described in Chapter 48.

SENSORY PATHWAYS FOR TRANSMISSION OF SOMATIC SIGNALS INTO THE CENTRAL NERVOUS SYSTEM

Almost all sensory information from the somatic segments of the body enters the spinal cord through the dorsal roots of the spinal nerves (with the exception of a few small fibers of questionable importance that enter the ventral roots). However, from the entry point of the cord and then to the brain, the sensory signals are carried through one of two alternate sensory pathways: (1) the *dorsal column–medial lemniscal system* or (2) the *anterolateral system*. These two systems again come together partially at the level of the thalamus.

The dorsal column–medial lemniscal system, as its name implies, carries signals mainly in the *dorsal columns* of the cord and then, after synapsing and crossing to the opposite side in the medulla, upward through the brain stem to the thalamus by way of the *medial lemniscus*. On the other hand, signals of the anterolateral system, after originating in the dorsal horns of the spinal gray matter, cross to the opposite side of the cord and ascend through the anterior and lateral white columns of the cord to terminate at all levels of the brain stem and in the thalamus.

The dorsal column–medial lemniscal system is composed of large, myelinated nerve fibers that transmit signals to the brain at velocities of 30 to 110 m/sec, whereas the anterolateral system is composed of much smaller myelinated fibers (averaging 4 micrometers in diameter) that transmit signals at velocities ranging from a few meters per second up to 40 m/sec.

Another difference between the two systems is that the dorsal column–medial lemniscal system has a high degree of spatial orientation of the nerve fibers with respect to their origin, whereas the anterolateral system has a much smaller degree of spatial orientation.

These differences immediately characterize the types of sensory information that can be transmitted by the two systems. That is, sensory information that must be transmitted rapidly and with temporal and spatial fidelity is transmitted mainly in the dorsal column–medial lemniscal system, whereas that which does not need to be transmitted rapidly or with great spatial fidelity is transmitted mainly in the anterolateral system.

The anterolateral system has a special capability that the dorsal system does not have: the ability to transmit a broad spectrum of sensory modalities—pain, warmth, cold, and crude tactile sensations; most of these are discussed in detail in Chapter 48. The dorsal system is limited to the more discrete types of mechanoreceptive sensations alone.

With this differentiation in mind, we can now list the types of sensations transmitted in the two systems.

Dorsal Column–Medial Lemniscal System

1. Touch sensations requiring a high degree of localization of the stimulus

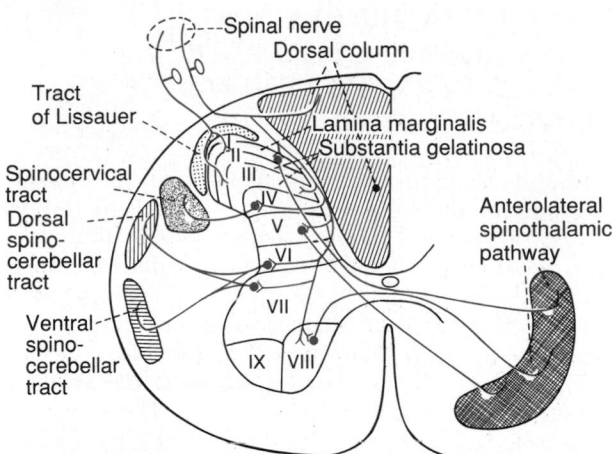

Figure 47–2. Cross section of the spinal cord, showing the anatomical laminae I through IX of the cord gray matter and the ascending sensory tracts (in red) in the white columns of the spinal cord.

2.   Touch sensations requiring transmission of fine gradations of intensity
3.   Phasic sensations, such as vibratory sensations
4.   Sensations that signal movement against the skin
5.   Position sensations
6.   Pressure sensations having to do with fine degrees of judgment of pressure intensity

Anterolateral System

1.   Pain
2.   Thermal sensations, including both warm and cold sensations
3.   Crude touch and pressure sensations capable of only crude localizing ability on the surface of the body
4.   Tickle and itch sensation
5.   Sexual sensations

TRANSMISSION IN THE DORSAL COLUMN–MEDIAL LEMNISCAL SYSTEM

Anatomy of the Dorsal Column–Medial Lemniscal System

On entering the spinal cord from the spinal nerve dorsal roots, the large myelinated fibers from specialized mechanoreceptors divide almost immediately to form a *medial branch* and a *lateral branch,* as shown by the right-hand spinal root fiber in Figure 47–2. The medial branch turns upward in the dorsal column and proceeds by way of the dorsal column pathway all the way to the brain.

The lateral branch enters the dorsal horn of the cord gray matter and then divides many times, synapsing with neurons in almost all parts of the intermediate and anterior portions of the cord gray matter. The neurons that are excited in turn serve three functions: (1) A few of them give off second-order fibers that re-enter the dorsal column to make up about 15 per cent of all the dorsal column fibers, and a few other second-order fibers enter the posterolateral column, forming the

spinocervical tract that rejoins the dorsal column system in the neck and low medulla. (2) Many of the neurons elicit local spinal cord reflexes, which are discussed in Chapter 54. (3) Others give rise to the spinocerebellar tracts, which we discuss in Chapter 56 in relation to the function of the cerebellum.

DORSAL COLUMN–MEDIAL LEMNISCAL PATHWAY. Note in Figure 47–3 that the nerve fibers entering the dorsal columns pass uninterrupted up to the medulla where they synapse in the *dorsal column nuclei* (the *cuneate* and *gracile nuclei*). From here, *second-order neurons* decussate immediately to the opposite side and then continue upward to the thalamus through bilateral brain stem pathways called the *medial lemnisci.* In this pathway through the brain stem, the medial lemniscus is joined by additional fibers from the *sensory nuclei of the trigeminal nerve;* these fibers subserve the same sensory functions for the head that the dorsal column fibers subserve for the body.

In the thalamus, the medial lemniscal fibers from the dorsal columns terminate in the *ventral posterolateral nucleus,* whereas those from the trigeminal nuclei terminate in the *ventral posteromedial nucleus.* These two nuclei, along with the posterior thalamic nuclei, where some fibers from the anterolateral system terminate, are

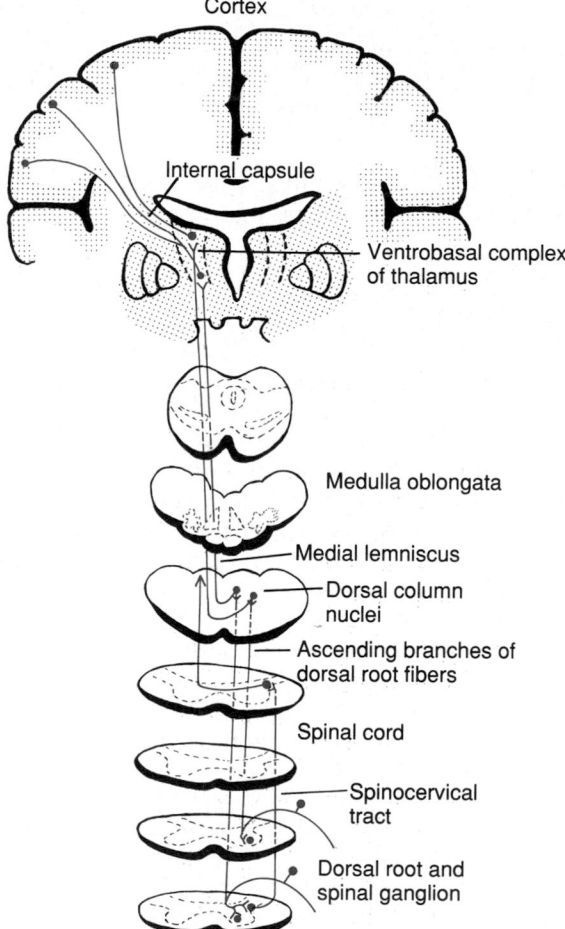

Figure 47–3. The dorsal column and spinocervical pathways for transmitting critical types of tactile signals. (Modified from Ranson and Clark: Anatomy of the Nervous System. Philadelphia, W. B. Saunders Company, 1959.)

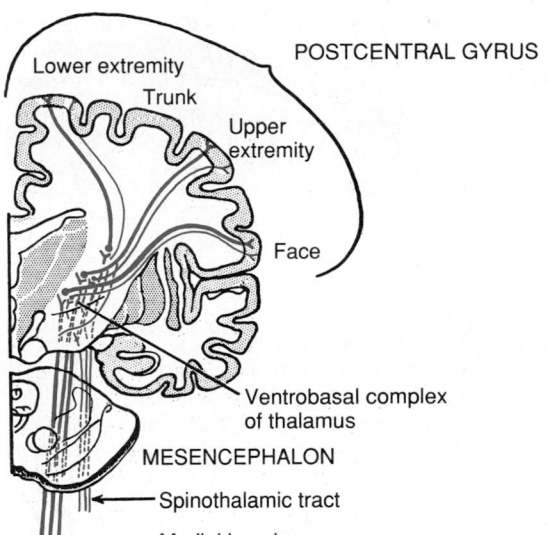

Figure 47–4. Projection of the dorsal column–medial lemniscal system from the thalamus to the somatic sensory cortex. (Modified from Brodal: Neurological Anatomy in Relation to Clinical Medicine. New York, Oxford University Press, 1969.)

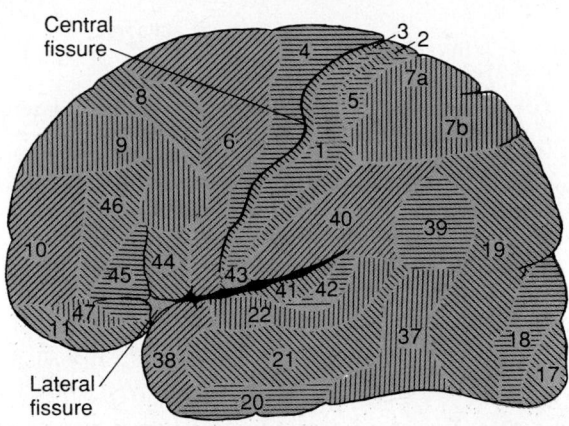

Figure 47–5. Structurally distinct areas, called "Brodmann areas," of the human cerebral cortex. Note specifically areas 1, 2, and 3, which comprise the *primary somatic sensory area I,* and areas 5 and 7, which comprise the *somatic sensory association area.* (From Everett: Functional Neuroanatomy. 5th ed. Philadelphia, Lea & Febiger, 1965. Modified from Brodmann.)

together called the *ventrobasal complex.* From the ventrobasal complex, *third-order nerve fibers* project, as shown in Figure 47–4, mainly to the *postcentral gyrus* of the *cerebral cortex* (see Fig. 47–6), which is called *somatic sensory area I.*

Spatial Orientation of the Nerve Fibers in the Dorsal Column–Medial Lemniscal System

One of the distinguishing features of the dorsal column–medial lemniscal system is a distinct spatial orientation of nerve fibers from the individual parts of the body that is maintained throughout. For instance, in the dorsal columns, the fibers from the lower parts of the body lie toward the center, whereas those that enter the spinal cord at progressively higher segmental levels form successive layers laterally.

In the thalamus, the distinct spatial orientation is still maintained, with the tail end of the body represented by the most lateral portions of the ventrobasal complex and the head and face represented in the medial component of the complex. Because of the crossing of the medial lemnisci in the medulla, the left side of the body is represented in the right side of the thalamus, and the right side of the body is represented in the left side of the thalamus.

Somatic Sensory Cortex

Before discussing the role of the cerebral cortex in somatic sensation, we need to give an orientation to the various areas of the cortex. Figure 47–5 is a map of the human cerebral cortex, showing that it is divided into about 50 distinct areas called *Brodmann areas* based on histological structural differences. The

map itself is important because its numbered areas have come to be used by virtually all neurophysiologists and neurologists in referring to the different functional areas of the human cortex.

Note in the figure the large *central fissure* (also called "central sulcus") that extends horizontally across the brain. In general, sensory signals from all modalities of sensation terminate in the cerebral cortex posterior to the central fissure. In general, the anterior half of the parietal lobe is concerned almost entirely with reception and interpretation of somatic sensory signals and the posterior half, with still higher levels of interpretation. Visual signals terminate in the occipital lobe and auditory signals in the temporal lobe.

On the other hand, the portion of the cortex anterior to the central fissure and constituting the posterior half of the frontal lobe is devoted almost entirely to control of the muscles and body movements. A major share of this motor control is directed by signals from the sensory portions of the cortex, which keep the motor cortex informed about the positions and motions of the different body parts.

SOMATIC SENSORY AREAS I AND II. Figure 47–6 shows two areas in the anterior parietal lobe called *somatic sensory area I* and *somatic sensory area II.* The reason for this division into two areas is that a distinct and separate spatial orientation of the different parts of the body is found in each of these two areas. However, somatic sensory area I is so much more extensive and so much more important than sensory area II that in popular usage, the term "somatic sensory cortex" almost always means area I. Somatic sensory area I has a high degree of localization of the different parts of the body, as shown by the names of virtually all parts of the body in Figure 47–6. By contrast, localization is poor in somatic sensory area II, although roughly, the face is represented anteriorly, the arms centrally, and the legs posteriorly.

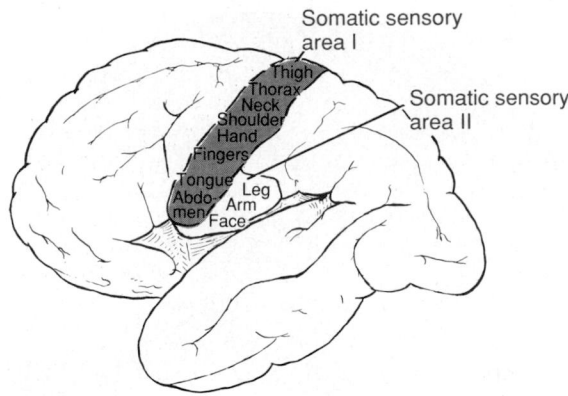

Figure 47–6. Two somatic sensory cortical areas, somatic sensory areas I and II.

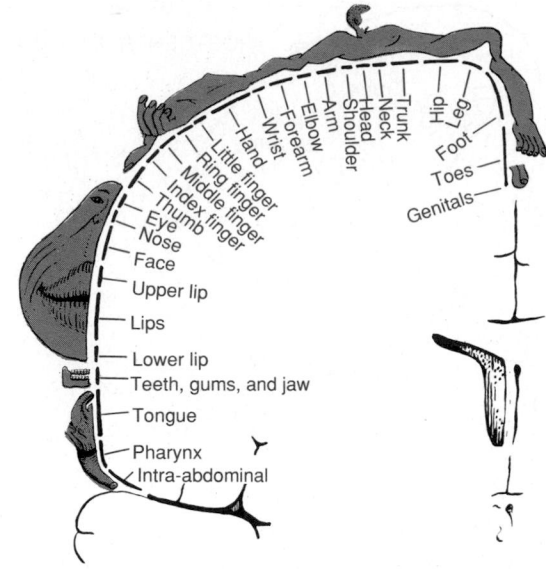

Figure 47–7. Representation of the different areas of the body in the somatic sensory area I of the cortex. (From Penfield and Rasmussen: Cerebral Cortex of Man: A Clinical Study of Localization of Function. New York, Macmillan Company, 1968.)

So little is known about the function of somatic sensory area II that it cannot be discussed intelligently. It is known that some signals enter this area from the brain stem, transmitted upward from both sides of the body. In addition, many signals come secondarily from sensory area I as well as from other sensory areas of the brain, even from the visual and auditory areas.

By contrast, function of somatic sensory area I has been studied in great detail, and most of what we know about somatic sensation seems to be explained by the functions of somatic sensory area I.

SPATIAL ORIENTATION OF SIGNALS FROM DIFFERENT PARTS OF THE BODY IN SOMATIC SENSORY AREA I. Somatic sensory area I lies immediately behind the central fissure, located in the postcentral gyrus of the human cerebral cortex (in Brodmann areas III, I, and II).

A distinct spatial orientation exists in this area for reception of somatic sensory nerve signals from the different areas of the body. Figure 47–7 shows a cross section through the brain at the level of the postcentral gyrus, demonstrating the representations of the different parts of the body in separate regions of somatic sensory area I. Note, however, that each side of the cortex receives sensory information exclusively from the opposite side of the body (with the exception of a very small amount of sensory information from the same side of the face).

Some areas of the body are represented by large areas in the somatic cortex—the lips the greatest of all, followed by the face and thumb—whereas the trunk and lower part of the body are represented by relatively small areas. The sizes of these areas are directly proportional to the number of specialized sensory receptors in each respective peripheral area of the body. For instance, a great number of specialized nerve endings are found in the lips and thumb, whereas only a few are present in the skin of the trunk.

Note also that the head is represented in the most

lateral portion of somatic sensory area I, whereas the lower part of the body is represented medially.

Layers of the Somatic Sensory Cortex and Their Function

The cerebral cortex contains *six* layers of neurons, beginning with layer I next to the surface and extending progressively deeper to layer VI, as shown in Figure 47–8. As would be expected, the neurons in each layer perform functions different from those in other layers. Some of these functions are as follows.

1. The incoming sensory signal excites neuronal layer IV first; then the signal spreads both toward the surface of the cortex and toward the deeper layers.

2. Layers I and II receive a diffuse, nonspecific input from lower brain centers that can facilitate a specific region of the cortex; this system is described in Chapter 57. This input mainly controls the overall level of excitability of the region stimulated.

3. The neurons in layers II and III send axons to other, related portions of the cerebral cortex, including to the opposite side of the brain through the corpus callosum.

4. The neurons in layers V and VI send axons to the deeper parts of the nervous system. Those in layer V are generally larger and project to more distant areas, such as the basal ganglia, brain stem, and spinal cord to provide control signals to these areas. From layer VI, especially large numbers of axons extend to the thalamus, providing feedback signals from the cerebral cortex to the thalamus.

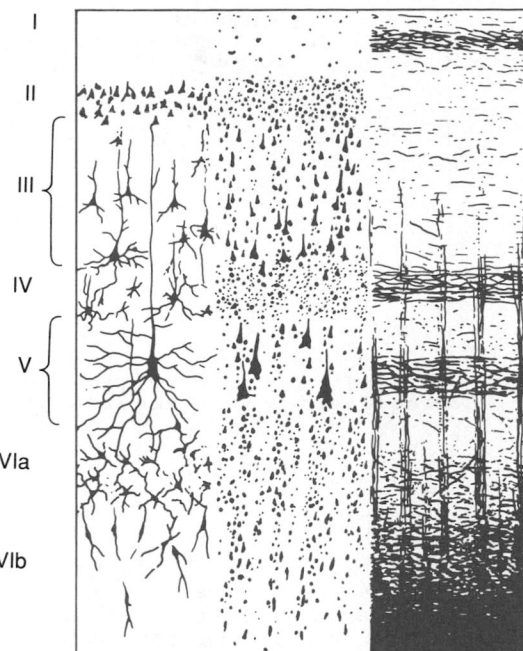

Figure 47–8. Structure of the cerebral cortex, showing *I,* molecular layer; *II,* external granular layer; *III,* layer of pyramidal cells; *IV,* internal granular layer; *V,* large pyramidal cell layer; and *VI,* layer of fusiform or polymorphic cells. (From Ranson and Clark [after Brodmann]: Anatomy of the Nervous System. Philadelphia, W. B. Saunders Company, 1959.)

The Sensory Cortex Contains Vast Numbers of Vertical Columns of Neurons; Each of These Detects a Different Sensory Spot on the Body with a Specific Sensory Modality

Functionally, the neurons of the somatic sensory cortex are arranged in vertical columns extending all the way through the six layers of the cortex, each column having a diameter of 0.3 to 0.5 millimeter and containing perhaps 10,000 neuronal cell bodies. Each of these columns serves a single specific sensory modality, some columns responding to stretch receptors around joints, some to stimulation of tactile hairs, others to discrete localized pressure points on the skin, and so forth. At layer IV, where the signals first enter the cortex, the columns of neurons function almost entirely separately from one another. At other levels of the columns, interactions occur that allow beginning analysis of the meanings of the sensory signals.

In the most anterior portion of the postcentral gyrus, located deep in the central fissure in Brodmann area 3a, a disproportionately large share of the vertical columns respond to muscle, tendon, or joint stretch receptors. Many of the signals from these in turn spread directly to the motor cortex located immediately anterior to the central fissure and help control muscle function. As one proceeds more posteriorly in somatic sensory cortex I, more and more of the vertical columns respond to the slowly adapting cutaneous

receptors, and then still further posteriorly, greater numbers of the columns are sensitive to deep pressure.

In the most posterior portion of somatic sensory area I, about 6 per cent of the vertical columns respond only when a stimulus moves across the skin in a particular direction. Thus, this is a still higher order of interpretation of sensory signals; the process becomes even more complex as the signals spread further backward from somatic sensory area I into the central and posterior parietal cortex, an area called the *somatic association area,* as we discuss subsequently.

Functions of Somatic Sensory Area I

The functional capabilities of different areas of the somatic sensory cortex have been determined by selective recording of electrical signals or selective excision of the different portions. Widespread bilateral excision of somatic sensory area I causes loss of the following types of sensory judgment:

1. The person is unable to localize discretely the different sensations in the different parts of the body. However, he or she can localize these sensations crudely, such as to a particular hand, which indicates that the brain stem, thalamus, or parts of the cerebral cortex not normally considered to be concerned with somatic sensations can perform some degree of localization.
2. He is unable to judge critical degrees of pressure against his body.
3. He is unable to judge the weights of objects.
4. He is unable to judge shapes or forms of objects. This is called *astereognosis.*
5. He is unable to judge texture of materials because this type of judgment depends on highly critical sensations caused by movement of the fingers over the surface to be judged.

Note in the list that nothing has been said about loss of pain and temperature sense. In the absence of somatic sensory area I, the appreciation of these sensory modalities may not be altered significantly in either quality or intensity. But more important, the pain and temperature sensations are poorly localized, indicating that both pain and temperature localization probably depend mainly on the topographical map of the body in somatic sensory area I to localize the source.

Somatic Sensory Association Areas

Brodmann areas 5 and 7 of the cerebral cortex, which are located in the parietal cortex behind somatic sensory area I (Fig. 47–5), play important roles in deciphering the sensory information that enters the somatic sensory areas. Therefore, these areas are called the *somatic sensory association areas.*

Electrical stimulation in the somatic association area can occasionally cause a person to experience a complex somatic sensation, sometimes even the "feeling" of an object such as a knife or a ball. Therefore, it seems clear that the somatic sensory association area combines information from multiple points in the primary somatic sensory area to decipher its meaning. This also fits with the anatomical arrangement of the neuronal tracts that enter the somatic sensory association area because it receives signals from (1) somatic sensory area I, (2) the ventrobasal nuclei of the thalamus, (3) other areas of the thalamus, (4) the visual cortex, and (5) the auditory cortex.

EFFECT OF REMOVING THE SOMATIC ASSOCIATION AREA—AMORPHOSYNTHESIS. When the somatic association area is removed on one side, the person loses the ability to recognize complex objects and complex forms by the process of feeling them on the opposite side of the body. In addition, he or she loses most of the sense of form of his or her own body on the opposite side. Especially interesting, the person is mainly oblivious to the opposite side of the body— that is, forgets that it is there. Therefore, he also often forgets to use the other side for motor functions as well. Likewise, when feeling objects, the person will tend to feel only one side of the object and forget that the other side even exists. This complex sensory deficit is called *amorphosynthesis.*

Overall Characteristics of Signal Transmission and Analysis in the Dorsal Column–Medial Lemniscal System

BASIC NEURONAL CIRCUIT IN THE DORSAL COL-UMN–MEDIAL LEMNISCAL SYSTEM. The lower part of Figure 47–9 shows the basic organization of the neuronal circuit of the dorsal column pathway, demonstrating that at each synaptic stage, divergence occurs. The upper part of the figure shows that a single receptor stimulus on the skin does not cause all the cortical neurons with which that receptor connects to discharge at the same rate. Instead, the cortical neurons that discharge to the greatest extent are those in a central part of the cortical "field" for each respective receptor. Thus, a weak stimulus causes only the centralmost neurons to fire. A stronger stimulus causes still more neurons to fire, but those in the center still discharge at a considerably more rapid rate than do those farther away from the center.

TWO-POINT DISCRIMINATION. A method frequently used to test tactile discrimination is to determine a person's so-called "two-point discriminatory ability." In this test, two needles are pressed lightly against the skin, and the subject determines whether two points of stimulus are felt or one point. On the tips of the fingers, a person can distinguish two separate points even when the needles are as close together as 1 to 2 millimeters. However, on the person's back, the needles must usually be as far apart as 30 to 70 millimeters before two separate points can be detected. The reason for this difference is the different numbers of specialized tactile receptors in the two areas.

Figure 47–10 shows the mechanism by which the dorsal column pathway as well as all other sensory pathways transmit two-point discriminatory informa-

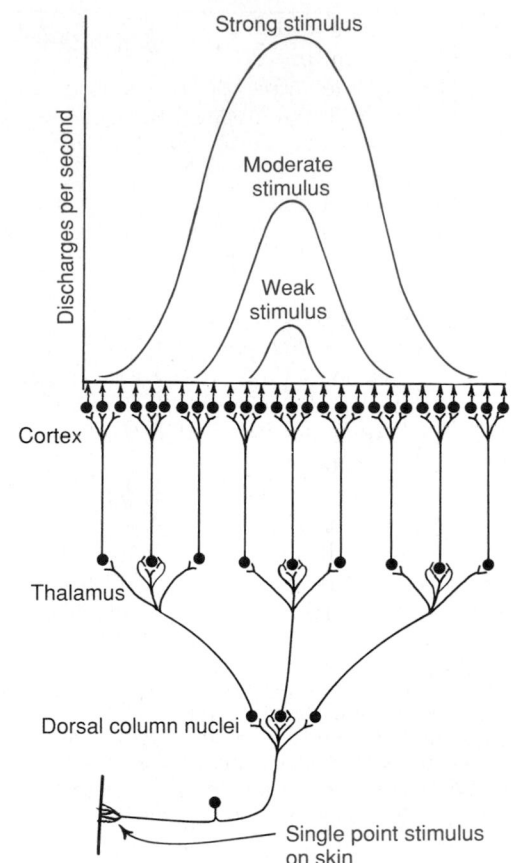

Figure 47–9. Transmission of a pinpoint stimulus signal to the cortex.

tion. This figure shows two adjacent points on the skin that are strongly stimulated as well as the area of the somatic sensory cortex (greatly enlarged) that is excited by signals from the two stimulated points. The solid

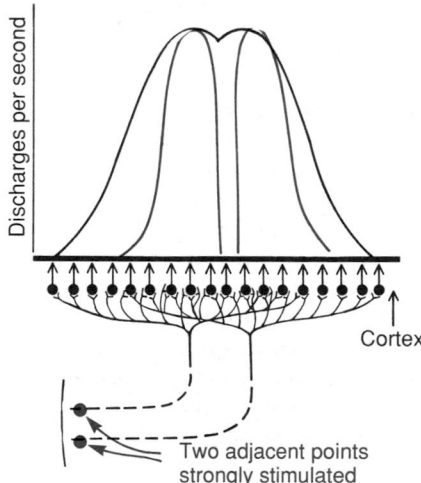

Figure 47–10. Transmission of signals to the cortex from two adjacent pinpoint stimuli. The solid black curve represents the pattern of cortical stimulation without "surround" inhibition, and the two colored curves represent the pattern with "surround" inhibition.

black curve shows the spatial pattern of cortical excitation when both skin points are stimulated simultaneously. Note that the resultant zone of excitation has two separate peaks. These two peaks, separated by a valley, allow the sensory cortex to detect the presence of two stimulatory points, rather than a single point. The capability of the sensorium to distinguish between two points of stimulation is strongly influenced by another mechanism, *lateral inhibition,* as explained in the next section.

EFFECT OF LATERAL INHIBITION IN INCREASING THE DEGREE OF CONTRAST IN THE PERCEIVED SPATIAL PATTERN. As pointed out in Chapter 46, virtually every sensory pathway, when excited, gives rise simultaneously to lateral inhibitory signals; these spread to the sides of the excitatory signal and inhibit adjacent neurons. For instance, consider an excited neuron in a dorsal column nucleus. Aside from the central excitatory signal, short collateral pathways transmit inhibitory signals to the surrounding neurons. These mainly pass through additional interneurons that secrete an inhibitory transmitter.

The importance of *lateral inhibition* is that it blocks lateral spread of the excitatory signals and, therefore, increases the degree of contrast in the sensory pattern perceived in the cerebral cortex.

In the case of the dorsal column system, lateral inhibitory signals occur at each synaptic level—for instance, in the dorsal column nuclei, the ventrobasal nuclei of the thalamus, and the cortex itself. At each of these levels, the lateral inhibition helps to block lateral spread of the excitatory signal. As a result, the peaks of excitation stand out and much of the surrounding diffuse stimulation is blocked. This effect is demonstrated by the two colored curves in Figure 47–10, showing complete separation of the peaks when the intensity of the lateral inhibition, also called *surround inhibition,* is great. This mechanism accentuates the contrast between the areas of peak stimulation and the surrounding areas, thus greatly increasing the contrast or sharpness of the perceived spatial pattern.

TRANSMISSION OF RAPIDLY CHANGING AND REPETITIVE SENSATIONS. The dorsal column system is of particular value for apprising the sensorium of rapidly changing peripheral conditions. Based on recorded action potentials, this system can "follow" changing stimuli that occur in as little as 1/400 second.

Vibratory Sensation. Vibratory signals are rapidly repetitive and can be detected as vibration up to 700 cycles per second. The higher-frequency vibratory signals originate from the pacinian corpuscles, but lower-frequency signals (below about 200 per second) can originate from Meissner's corpuscles as well. These signals are transmitted only in the dorsal column pathway. For this reason, application of vibration with a tuning fork to different peripheral parts of the body is an important tool used by the neurologist for testing the functional integrity of the dorsal columns.

Psychic Interpretation of Sensory Stimulus Intensity

The ultimate goal of most sensory stimulation is to apprise the psyche of the state of the body and its surroundings. Therefore, it is important that we discuss briefly some of the principles related to transmission of sensory *stimulus intensity* to the higher levels of the nervous system.

The first question that comes to mind is, how is it possible for the sensory system to transmit sensory experiences of tremendously varying intensities? For instance, the auditory system can detect the weakest possible whisper but can also discern the meanings of an explosive sound, even though the sound intensities of these two experiences can vary more than 10 billion times; the eyes can see visual images with light intensities that vary as much as a half million times; or the skin can detect pressure differences of 10,000 to 100,000 times.

As a partial explanation of these effects, Figure 46–4 in the previous chapter shows the relation of the receptor potential produced by the pacinian corpuscle to the intensity of the sensory stimulus. At low stimulus intensity, slight changes in intensity increase the potential markedly, whereas at high levels of stimulus intensity, further increases in receptor potential are slight. Thus, the pacinian corpuscle is capable of accurately measuring extremely minute changes in stimulus at low-intensity levels, but at high-intensity levels, the change in stimulus must be much greater to cause the same amount of change in receptor potential.

The transduction mechanism for detecting sound by the cochlea of the ear demonstrates still another method for separating gradations of stimulus intensity. When sound causes vibration at a specific point on the basilar membrane, weak vibration stimulates only those hair cells at the point of maximum vibration. But as the vibration intensity increases, not only do these hair cells become more intensely stimulated but still many more hair cells in each direction farther away from the maximum vibratory point also become stimulated. Thus, signals transmitted over progressively increasing numbers of nerve fibers is another mechanism by which stimulus intensity is transmitted into the central nervous system. This mechanism, plus the direct effect of stimulus intensity on impulse rate in each nerve fiber, as well as several other mechanisms, makes it possible for some sensory systems to operate reasonably faithfully at stimulus intensity levels changing as much as hundreds of thousands to billions of times.

IMPORTANCE OF THE TREMENDOUS INTENSITY RANGE OF SENSORY RECEPTION. Were it not for the tremendous intensity range of sensory reception that we can experience, the various sensory systems would more often than not operate in the wrong range. This is demonstrated by the attempts of most people to adjust the light exposure on a camera without using a light meter. Left to intuitive judgment of light intensity, a person almost always overexposes the film on bright days and greatly underexposes the film at twilight. Yet that person's eyes are capable of discriminating with great detail the visual objects in bright sunlight and at twilight; the camera cannot do this because of the narrow critical range of light intensity required for proper exposure of film.

Judgment of Stimulus Intensity

Physiopsychologists have evolved numerous methods for testing one's judgment of sensory stimulus intensity, but only rarely do the results from the different methods agree with one another. Yet the basic principle of decreasing intensity discrimination as the sensory intensity increases is applicable to virtually all sensory modalities. Two formulations of this principle are widely discussed in the physiopsychology field of sensory interpretation: the *Weber-Fechner principle* and the *power principle.*

WEBER-FECHNER PRINCIPLE—DETECTION OF "RATIO" OF STIMULUS STRENGTH. In the mid-1800s, Weber first and Fechner later proposed the principle that *gradations of stimulus strength are discriminated approximately in proportion to the logarithm of stimulus strength.* That is, a typical test of this principle might show that a person can barely detect a 1-gram increase in weight when already holding 30 grams or a 10-gram increase when already holding 300 grams. Thus, the *ratio* of the change in stimulus strength required for detection of a change remains essentially constant, about 1 to 30, which is what the logarithmic principle means. To express this mathematically,

Interpreted signal strength = log (Stimulus) + Constant

More recently, it has become evident that the Weber-Fechner principle is quantitatively accurate only for the higher intensities of visual, auditory, and cutaneous sensory experience and applies only poorly to most other types of sensory experience.

Yet the Weber-Fechner principle is still a good one to remember because it emphasizes that the greater the background sensory intensity, the greater also must be the additional change in stimulus for the psyche to detect the change.

POWER LAW. Another attempt by physiopsychologists to find a good mathematical relation is the following formula, known as the power law:

Interpreted signal strength = $K \cdot (\text{Stimulus} - k)^y$

In this formula, the exponent y and the constants K and k are different for each type of sensation.

When this power law relation is plotted on a graph using double logarithmic coordinates, as shown in Figure 47–11, and when appropriate quantitative values for the constants y, K, and k are found, a linear relation can be attained between interpreted stimulus strength and actual stimulus strength over a large range for almost any type of sensory perception. However, as shown in the figure, even this power law relation fails to hold satisfactorily at both very low and very high stimulus strengths.

Position Senses

The *position senses* are frequently also called *proprioceptive senses.* They can be divided into two subtypes: (1) *static position sense,* which means conscious perception of the orientation of the different parts of the body with respect to one another, and (2) *rate of*

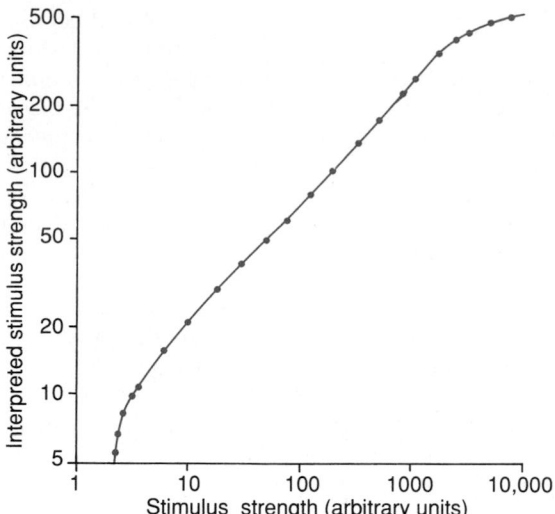

Figure 47–11. Graphical demonstration of the "power law" relation between actual stimulus strength and strength that the psyche interprets it to be. Note that the power law does not hold at either very weak or very strong stimulus strengths.

movement sense, also called *kinesthesia* or *dynamic proprioception.*

POSITION SENSORY RECEPTORS. Knowledge of position, both static and dynamic, depends on knowing the degrees of angulation of all joints in all planes and their rates of change. Therefore, multiple different types of receptors help to determine joint angulation and are used together for position sense. Both skin tactile receptors and deep receptors near the joints are used. In the case of the fingers, where skin receptors are in great abundance, as much as half of position recognition is believed to be detected through the skin receptors. On the other hand, for most of the larger joints of the body, deep receptors are more important.

For determining joint angulation in midranges of motion, the most important receptors are believed to be the *muscle spindles.* They are also exceedingly important in helping to control muscle movement, as we see in Chapter 54. When the angle of a joint is changing, some muscles are being stretched while others are loosened, and the stretch information from the spindles is passed into the computational system of the spinal cord and higher regions of the dorsal column system for deciphering the complex interrelations of joint angulations.

At the extremes of joint angulation, the stretch of the ligaments and deep tissues around the joints is an additional important factor in determining position. Types of sensory endings used for this are the pacinian corpuscles, Ruffini's endings, and receptors similar to the Golgi tendon receptors found in muscle tendons.

The pacinian corpuscles and muscle spindles are especially adapted for detecting rapid rates of change. Therefore, it is likely that these are the receptors most responsible for detecting rate of movement.

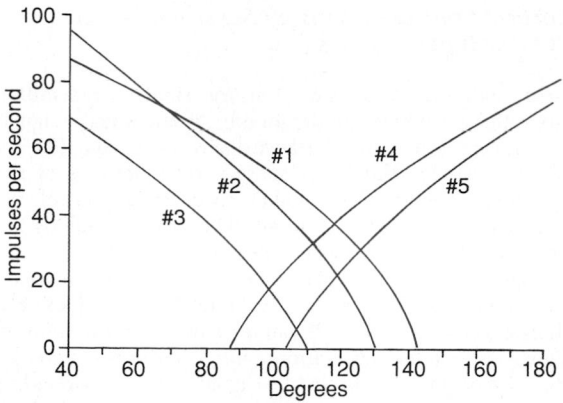

Figure 47–12. Typical responses of five different neurons in the knee joint receptor field of the thalamic ventrobasal complex when the knee joint is moved through its range of motion. (The curves were constructed from data in Mountcastle et al.: *J. Neurophysiol., 26:*807, 1963.)

PROCESSING OF POSITION SENSE INFORMATION IN THE DORSAL COLUMN–MEDIAL LEMNISCAL PATHWAY. Despite the usual faithfulness of transmission of signals from the periphery to the sensory cortex in the dorsal column–medial lemniscal system, there nevertheless seems to be some processing of position sense signals before they reach the cerebral cortex. For instance, referring to Figure 47–12, one sees that the thalamic neurons that respond to joint rotation are of two types: (1) those maximally stimulated when the joint is at full rotation and (2) those maximally stimulated when the joint is at minimal rotation. Furthermore, the intensity of neuronal excitation changes over angles of 40 to 60 degrees of rotation. Yet, for the individual joint receptors, each of these responds only over 20 to 30 degrees of rotation, and the angle of rotation for maximum stimulation is often in the middle range of motion rather than at one or the other extremes of rotation. Thus, the signals from the individual joint receptors have been integrated in the space domain by the time they reach the thalamic neurons, demonstrating some degree of processing of the signals either in the cord or in the thalamus.

TRANSMISSION OF CRUDE TACTILE SENSORY SIGNALS IN THE ANTEROLATERAL PATHWAY

The anterolateral pathway, in contrast to the dorsal column pathway, transmits sensory signals that do not require highly discrete localization of the signal source and that do not require discrimination of fine gradations of intensity. The types of signals transmitted include pain, heat, cold, crude tactile, tickle and itch, and sexual sensations. In Chapter 48, pain and temperature sensations are discussed; this chapter is still concerned principally with transmission of the tactile sensations, but now with the less critical types.

Anatomy of the Anterolateral Pathway

The anterolateral fibers originate mainly in laminae I, IV, V, and VI (see Fig. 47–2) in the dorsal horns where many of the dorsal root sensory nerve fibers terminate after entering the cord (note specifically the terminations of the lateralmost, red-colored nerve fiber entering the spinal cord). Then, as shown in Figure 47–13, the fibers cross in the anterior commissure of the cord to the opposite anterior and lateral white columns, where they turn upward toward the brain. These fibers ascend rather diffusely throughout the anterolateral columns. However, anatomical studies suggest a partial differentiation of this pathway into an anterior division, called the *anterior spinothalamic tract*, and a lateral division, called the *lateral spinothalamic tract*. Also encompassed in the anterolateral pathway are a *spinoreticular pathway* (to the reticular substance of the brain stem) and a *spinotectal tract* (to the tectum of the mesencephalon). However, it has been difficult to make these differentiations using electrical recording techniques.

The upper terminus of the anterolateral pathway is mainly twofold: (1) throughout the *reticular nuclei of*

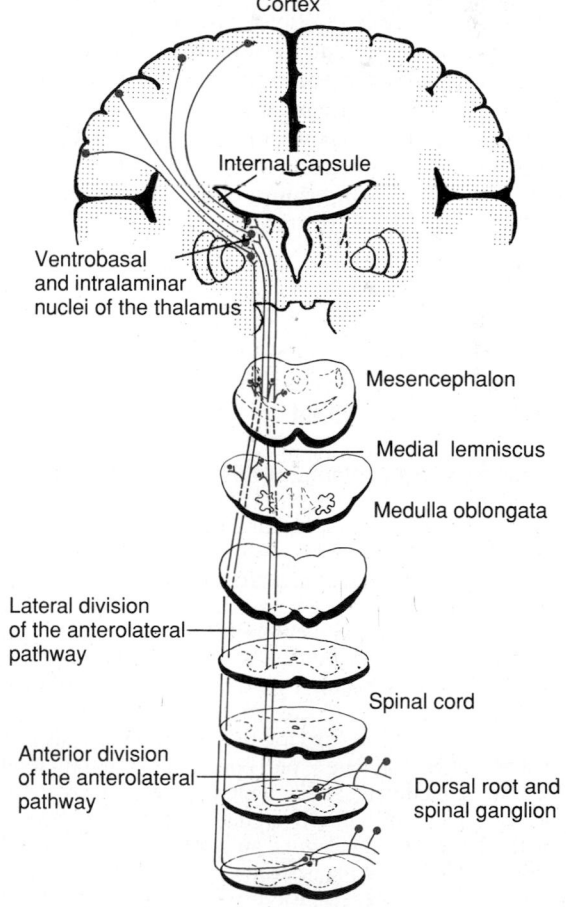

Figure 47–13. Anterior and lateral divisions of the anterolateral pathway.

the brain stem and (2) in two different nuclear complexes of the thalamus, the *ventrobasal complex* and the *intralaminar nuclei*. In general, the tactile signals are transmitted mainly into the ventrobasal complex, terminating in the same *ventral posterior lateral* and *medial* thalmic nuclei where the dorsal column tactile signals terminate. From here, the signals are transmitted to the somatosensory cortex along with the signals from the dorsal columns.

On the other hand, only some of the pain signals project to the ventrobasal complex. Instead, most pain signals enter the reticular nuclei of the brain stem and then are relayed from the brain stem to the intralaminar nuclei of the thalamus, as discussed in greater detail in Chapter 48.

CHARACTERISTICS OF TRANSMISSION IN THE ANTEROLATERAL PATHWAY. In general, the same principles apply to transmission in the anterolateral pathway as in the dorsal column–medial lemniscal system except for the following differences: (1) the velocities of transmission are only one-third to one-half those in the dorsal column–medial lemniscal system, ranging between 8 and 40 m/sec; (2) the degree of spatial localization of signals is poor, especially in the pain pathways; (3) the gradations of intensities are also far less accurate, most of the sensations being recognized in 10 to 20 gradations of strength, rather than as many as 100 gradations for the dorsal column system; and (4) the ability to transmit rapidly repetitive signals is poor.

Thus, it is evident that the anterolateral system is a cruder type of transmission system than the dorsal column–medial lemniscal system. Even so, certain modalities of sensation are transmitted only in this system and not at all in the dorsal column–medial lemniscal system. They are pain, temperature, tickle and itch, and sexual sensations in addition to crude touch and pressure.

SOME SPECIAL ASPECTS OF SOMATIC SENSORY FUNCTION

Function of the Thalamus in Somatic Sensation

When the somatosensory cortex of a human being is destroyed, that person loses most critical tactile sensibilities, but a slight degree of crude tactile sensibility does return. Therefore, it must be assumed that the thalamus (as well as other lower centers) has a slight ability to discriminate tactile sensation even though the thalamus normally functions mainly to relay this type of information to the cortex.

On the other hand, loss of the somatosensory cortex has little effect on one's perception of pain sensation and only a moderate effect on the perception of temperature. Therefore, there is much reason to believe that the brain stem, thalamus, and other associated basal regions of the brain play perhaps the dominant role in discrimination of these sensibilities. It is interesting that these sensibilities appeared very early in the phylogenetic development of animalhood, whereas the critical tactile sensibilities were a late development.

Cortical Control of Sensory Sensitivity—"Corticofugal" Signals

In addition to somatic sensory signals transmitted from the periphery to the brain, "corticofugal" signals are transmitted in the backward direction from the cerebral cortex to the lower sensory relay stations of the thalamus, medulla, and spinal cord; they control the sensitivity of the sensory input. The corticofugal signals are inhibitory, so that when the input intensity becomes too great, the corticofugal signals automatically decrease the transmission in the relay nuclei. This does two things: First, it decreases lateral spread of the sensory signals into adjacent neurons and, therefore, increases the degree of contrast in the signal pattern. Second, it keeps the sensory system operating in a range of sensitivity that is not so low that the signals are ineffectual or so high that the system is swamped beyond its capacity to differentiate sensory patterns.

This principle of corticofugal sensory control is used by all the different sensory systems, not only the somatic system, as explained in subsequent chapters.

Segmental Fields of Sensation—The Dermatomes

Each spinal nerve innervates a "segmental field" of the skin called a *dermatome*. The different dermatomes are shown in Figure 47–14. However, they are shown in the figure as if there were distinct borders between the

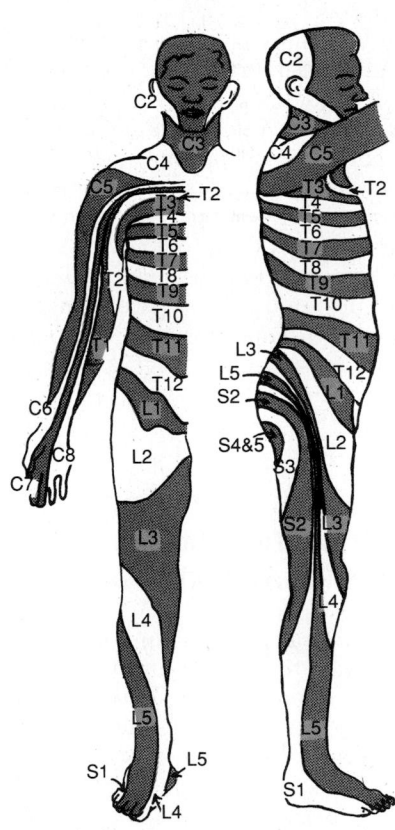

Figure 47–14. Dermatomes. (Modified from Grinker and Sahs: Neurology. Springfield, Ill., Charles C Thomas, 1966.)

adjacent dermatomes, which is far from true because much overlap exists from segment to segment.

The figure shows that the anal region of the body lies in the dermatome of the most distal cord segment: dermatome S-5. In the embryo, this is the tail region and the most distal portion of the body. The legs develop from the lumbar and upper sacral segments (L-2 to S-3), rather than from the distal sacral segments, which is evident from the dermatomal map. One can use a dermatomal map such as that shown in Figure 47–14 to determine the level in the spinal cord at which various cord injuries may have occurred when the peripheral sensations are disturbed.

REFERENCES

Akil, H., and Lewis, J. W. (eds.): Neurotransmitters and Pain Control. New York, S. Karger Publishers, Inc., 1987.

Akil, H., et al.: Endogenous opioids: Etiology and function. Annu. Rev. Neurosci., 7:223, 1984.

American Physiological Society: Sensory Processes. Washington, D.C., American Physiological Society, 1984.

Amit, Z., and Galina, Z. H.: Stress-induced analgesia: Adaptive pain suppression. Physiol. Rev., 68:1091, 1988.

Basbaum, A. I., and Fields, H. L.: Endogenous pain control systems: Brainstem spinal pathways and endorphin circuitry. Annu. Rev. Neurosci., 7:309, 1984.

Bernhard, J. D.: Itch. Hightstown, NJ, McGraw-Hill, 1994.

Berthoz, A.: Multisensory Control of Movement. New York, Oxford University Press, 1993.

Besson, J. M., and Chaouch, A.: Peripheral and spinal mechanisms of nociception. Physiol. Rev., 67:67, 1987.

Bevan, L., et al.: Proprioceptive coordination of movement sequences: discrimination of joint angle versus angular distance. J. Neurophysiol., 71:1862, 1994.

Cohen, B., et al.: Sensing and Controlling Motion: Vestibular and Sensorimotor Function. New York, New York Academy of Sciences, 1992.

Cordo, P., et al.: Proprioceptive coordination of movement sequences: role of velocity and position information. J. Neurophysiol., 71:1848, 1994.

Darian-Smith, I.: The sense of touch: Performance and peripheral neural processes. In Darian-Smith, I. (ed.): Handbook of Physiology. Sec. 1, Vol. III. Bethesda, Md., American Physiological Society, 1984, p. 739.

Darian-Smith, I.: Thermal sensibility. In Darian-Smith I. (ed.): Handbook of Physiology. Sec. 1, Vol. III. Bethesda, Md., American Physiological Society, 1984, p. 879.

Dietz, V.: Human neuronal control of automatic functional movements: interaction between central programs and afferent input. Physiol. Rev., 72:33, 1992.

Dubner, R., and Bennett, G. J.: Spinal and trigeminal mechanisms of nociception. Annu. Rev. Neurosci., 6:381, 1983.

Duckett, S.: Pediatric Neuropathology. Baltimore, Williams & Wilkins, 1994.

Emmers, R.: Somesthetic System of the Rat. New York, Raven Press, 1988.

Fields, H. L. (ed.): Pain: Mechanisms and Management. New York, McGraw-Hill Book Co., 1987.

Finlayson, M. A. J., and Garner, S.: Brain Injury Rehabilitation: Clinical Considerations. Baltimore, Williams & Wilkins, 1994.

Foreman, R. D., and Blair, R. W.: Central organization of sympathetic cardiovascular response to pain. Annu. Rev. Physiol., 50:607, 1988.

Guyton, A. C., and Reeder, R. C.: Pain and contracture in poliomyelitis. Arch. Neurol. Psychiatr., 63:954, 1950.

Hnik, P., et al. (eds.): Mechanoreceptors. Development, Structure and Function. New York, Plenum Publishing Corp., 1988.

Hochberg, J.: Perception. In Darian-Smith, I. (ed.): Handbook of Physiology. Sec. 1, Vol. III. Bethesda, Md., American Physiological Society, 1984, p. 75.

Hyvarinen, J.: Posterior parietal lobe of the primate brain. Physiol. Rev., 62:1060, 1982.

Iggo, A., et al. (eds.): Nociception and Pain. New York, Cambridge University Press, 1986.

Joynt, R. J.: Clinical Neurology: Clinical Text in Four Looseleaf Volumes. Philadelphia, J. B. Lippincott. (Annual updates)

Kruger, L. (ed.): Neural Mechanisms of Pain. New York, Raven Press, 1984.

Lucente, F. E., and Cooper, B. C.: Management of Facial, Head and Neck Pain. Philadelphia, W. B. Saunders Co., 1989.

Lund, J. S. (ed.): Sensory Processing in the Mammalian Brain. New York, Oxford University Press, 1988.

Mountcastle, V. B.: Central nervous mechanisms in mechanoreceptive sensibility. In Darian-Smith, I. (ed.): Handbook of Physiology. Sec. 1, Vol. III. Bethesda, Md., American Physiological Society, 1984, p. 789.

Nakanishi, S.: Substance P precursor and kininogen: Their structures, gene organizations, and regulation. Physiol. Rev., 67:1117, 1987.

Ono, T., et al.: Brain Mechanisms of Perception and Memory: From Neuron to Behavior. New York, Oxford University Press, 1993.

Paintal, A. S.: The visceral sensations—some basic mechanisms. Prog. Brain Res., 67:3, 1986.

Perl, E. R.: Pain and nociception. In Darian-Smith, I. (ed.): Handbook of Physiology. Sec. 1, Vol. III. Bethesda, Md., American Physiological Society, 1984, p. 915.

Price, D. D.: Psychological and Neural Mechanisms of Pain. New York, Raven Press, 1988.

Saper, J. P. (ed.): Controversies and Clinical Variants of Migraine. New York, Pergamon Press, 1987.

Schacter, D. L.: Memory Systems 1994. Cambridge, MA, The MIT Press, 1994.

Scheibel, A. B.: The brain stem reticular core and sensory function: In Darian-Smith, I. (ed.): Handbook of Physiology. Sec. 1, Vol. III. Bethesda, Md., American Physiological Society, 1984, p. 213.

Scott, S. A.: Sensory Neurons: Diversity, Development, and Plasticity. New York, Oxford University Press, 1992.

Tollison, C. D., et al.: Handbook of Chronic Pain Management. Baltimore, Williams & Wilkins, 1988.

Udin, S. B., and Fawcett, J. W.: Formation of topographic maps. Annu. Rev. Neurosci., 11:289, 1988.

Somatic Sensations: II. Pain, Headache, and Thermal Sensations

CHAPTER 48

Many, if not most, ailments of the body cause pain. Furthermore, the ability to diagnose different diseases depends to a great extent on a doctor's knowledge of the different qualities of pain. For these reasons, this chapter is devoted mainly to pain and to the physiological bases of some associated clinical phenomena.

THE PURPOSE OF PAIN. Pain is mainly a protective mechanism for the body; it occurs whenever any tissues are being damaged, and it causes the individual to react to remove the pain stimulus. Even such simple activities as sitting for a long time on the ischia can cause tissue destruction because of lack of blood flow to the skin where the skin is compressed by the weight of the body. When the skin becomes painful as a result of the ischemia, the person normally shifts weight unconsciously. But a person who has lost the pain sense, such as after spinal cord injury, fails to feel the pain and, therefore, fails to shift. This soon results in skin ulceration at the areas of pressure.

TYPES OF PAIN AND THEIR QUALITIES—FAST PAIN AND SLOW PAIN

Pain has been classified into two major types: *fast pain* and *slow pain*. Fast pain is felt within about 0.1 second after a pain stimulus is applied, whereas slow pain begins only after 1 second or more and then increases slowly over many seconds and sometimes even minutes. During the course of this chapter, we shall see that the conduction pathways for these two types of pain are different and that each of them has specific qualities.

Fast pain is also described by many alternate names, such as *sharp pain, pricking pain, acute pain,* and *electric pain.* This type of pain is felt when a needle is stuck into the skin, when the skin is cut with a knife, or when the skin is acutely burned. It is also felt when the skin is subjected to electric shock. Fast, sharp pain is not felt in most of the deeper tissues of the body.

Slow pain also goes by multiple additional names, such as *slow burning pain, aching pain, throbbing pain, nauseous pain,* and *chronic pain.* This type of pain is usually associated with *tissue destruction.* It can lead to prolonged, unbearable suffering. It can occur both in the skin and in almost any deep tissue or organ.

PAIN RECEPTORS AND THEIR STIMULATION

ALL PAIN RECEPTORS ARE FREE NERVE ENDINGS. The pain receptors in the skin and other tissues are all free nerve endings. They are widespread in the superficial layers of the *skin* as well as in certain internal tissues, such as the *periosteum,* the *arterial walls,* the *joint surfaces,* and the *falx* and *tentorium* of the cranial vault. Most other deep tissues are not extensively supplied with pain endings but are sparsely supplied; nevertheless, any widespread tissue damage can still summate to cause the slow-chronic-aching type of pain in these areas.

THREE TYPES OF STIMULI EXCITE PAIN RECEPTORS —MECHANICAL, THERMAL, AND CHEMICAL. Pain can

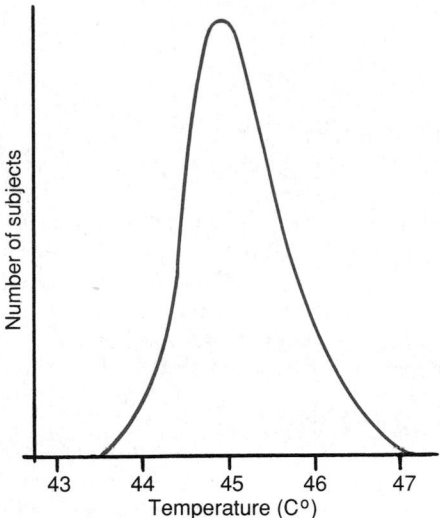

Figure 48–1. Distribution curve obtained from a large number of subjects of the minimal skin temperature that causes pain. (Modified from Hardy: *J. Chronic Dis., 4:*22, 1956.)

be elicited by multiple types of stimuli. They are classified as *mechanical, thermal,* and *chemical pain stimuli.* In general, fast pain is elicited by the mechanical and thermal types of stimuli, whereas slow pain can be elicited by all three types.

Some of the chemicals that excite the chemical type of pain include *bradykinin, serotonin, histamine, potassium ions, acids, acetylcholine,* and *proteolytic enzymes.* In addition, *prostaglandins* and *substance P* enhance the sensitivity of pain endings but do not directly excite them. The chemical substances are especially important in stimulating the slow, suffering type of pain that occurs after tissue injury.

NONADAPTING NATURE OF PAIN RECEPTORS. In contrast to most other sensory receptors of the body, the pain receptors adapt very little and sometimes not at all. In fact, under some conditions, the excitation of the pain fibers becomes progressively greater, especially so for the slow aching nauseous pain, as the pain stimulus continues. This increase in sensitivity of the pain receptors is called *hyperalgesia.*

One can readily understand the importance of this failure of pain receptors to adapt because it allows them to keep the person apprised of a damaging stimulus that causes the pain as long as it persists.

Rate of Tissue Damage as a Cause of Pain

The average person first begins to perceive pain when the skin is heated above 45°C, as shown in Figure 48–1. This is also the temperature at which the tissues begin to be damaged by heat; indeed, the tissues are eventually destroyed if the temperature remains above this level indefinitely. Therefore, it is immediately apparent that pain resulting from heat is closely correlated with the ability of heat to damage the tissues.

Furthermore, the intensity of pain has also been closely correlated with the rate of tissue damage from causes other than heat—bacterial infection, tissue ischemia, tissue contusion, and so forth.

SPECIAL IMPORTANCE OF CHEMICAL PAIN STIMULI DURING TISSUE DAMAGE. Extracts from damaged tissues cause intense pain when injected beneath the normal skin. All the chemicals listed above that excite the chemical pain receptors are found in these extracts. One chemical that seems to be more painful than the others is *bradykinin.* Many research workers have suggested that bradykinin might be the single agent most responsible for causing the tissue damage type of pain. Also, the intensity of the pain felt correlates with the local increase in potassium ion concentration. It should be remembered, too, that proteolytic enzymes can directly attack the nerve endings and excite pain by making their membranes more permeable to ions.

TISSUE ISCHEMIA AS A CAUSE OF PAIN. When blood flow to a tissue is blocked, the tissue becomes very painful within a few minutes. And the greater the rate of metabolism of the tissue, the more rapidly the pain appears. For instance, if a blood pressure cuff is placed around the upper arm and inflated until the arterial blood flow ceases, exercise of the forearm muscles sometimes can cause severe muscle pain within 15 to 20 seconds. In the absence of muscle exercise, the pain may not appear for 3 to 4 minutes.

One of the suggested causes of pain during ischemia is accumulation of large amounts of lactic acid in the tissues, formed as a consequence of the anaerobic metabolism (metabolism without oxygen) that occurs during ischemia. It is also possible that other chemical agents, such as bradykinin and proteolytic enzymes, are formed in the tissues because of cell damage and that these, rather than lactic acid, stimulate the pain nerve endings.

MUSCLE SPASM AS A CAUSE OF PAIN. Muscle spasm is also a common cause of pain, and it is the basis of many clinical pain syndromes. This pain probably results partially from the direct effect of muscle spasm in stimulating mechanosensitive pain receptors. It possibly results also from the indirect effect of muscle spasm to compress the blood vessels and cause ischemia. Also, the spasm increases the rate of metabolism in the muscle tissue at the same time, thus making the relative ischemia even greater, creating ideal conditions for release of chemical pain-inducing substances.

DUAL TRANSMISSION OF PAIN SIGNALS INTO THE CENTRAL NERVOUS SYSTEM

Even though all pain receptors are free nerve endings, these endings use two separate pathways for transmitting pain signals into the central nervous sys-

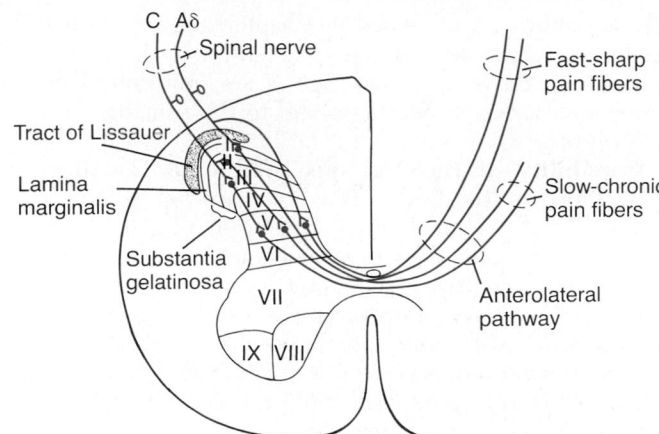

Figure 48–2. Transmission of both acute-sharp and slow-chronic pain signals into and through the spinal cord on the way to the brain stem.

tem. The two pathways at least partially correspond to the two types of pain, a *fast-sharp pain pathway* and a *slow-chronic pain pathway.*

PERIPHERAL PAIN FIBERS—"FAST" AND "SLOW" FIBERS. The fast-sharp pain signals are elicited by either mechanical or thermal pain stimuli; they are transmitted in the peripheral nerves to the spinal cord by small type Aδ fibers at velocities of between 6 and 30 m/sec. On the other hand, the slow-chronic type of pain is specifically elicited by the chemical types of pain stimuli but also at times by persisting mechanical or thermal stimuli; this slow-chronic pain is transmitted by type C fibers at velocities of between 0.5 and 2 m/sec.

Because of this double system of pain innervation, a sudden onset of painful stimulus often gives a "double" pain sensation: a fast-sharp pain that is transmitted to the brain by the Aδ fiber pathway followed a second or so later by a slow pain that is transmitted by the C fiber pathway. The sharp pain apprises the person rapidly of a damaging influence and, therefore, plays an important role in making the person react immediately to remove himself or herself from the stimulus. On the other hand, the slow pain tends to become more and more painful over a period of time. This sensation eventually gives one the intolerable suffering of long-continued pain.

On entering the spinal cord from the dorsal spinal roots, the pain fibers terminate on neurons in the dorsal horns. Here again, there are two systems for processing the pain signals on their way to the brain, as shown in Figures 48–2 and 48–3.

Dual Pain Pathways in the Cord and Brain Stem—The Neospinothalamic Tract and the Paleospinothalamic Tract

On entering the spinal cord, the pain signals take two pathways to the brain, through the *neospinothalamic tract* and through the *paleospinothalamic tract.*

NEOSPINOTHALAMIC TRACT FOR FAST PAIN. The fast type Aδ pain fibers transmit mainly mechanical

and acute thermal pain. They terminate mainly in lamina I (lamina marginalis) of the dorsal horns, as shown in Figure 48–2, and there excite second-order neurons of the neospinothalamic tract. These give rise to long fibers that cross immediately to the opposite side of the cord through the anterior commissure and then pass upward to the brain in the anterolateral columns.

Termination of the Neospinothalamic Tract in the Brain Stem and Thalamus. A few fibers of the neospinothalamic tract terminate in the reticular areas of the brain stem, but most pass all the way to the thalamus, terminating in the *ventrobasal complex* along with the dorsal column–medial lemniscal tract for tac-

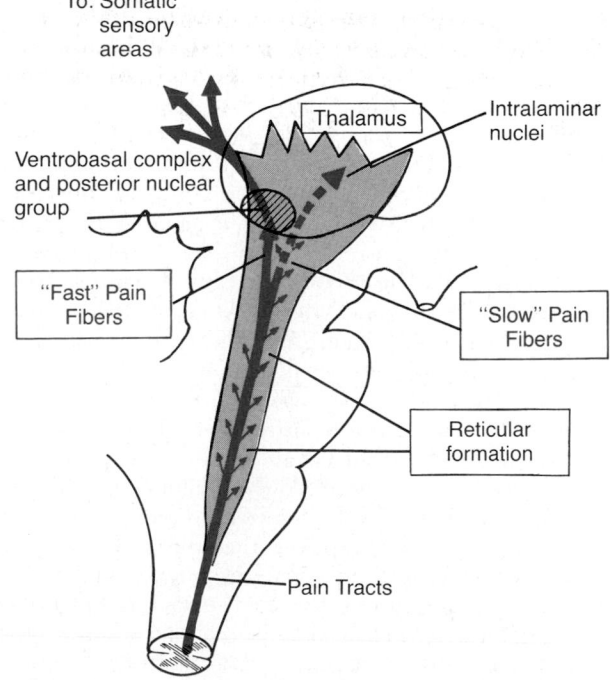

Figure 48–3. Transmission of pain signals into the hindbrain, thalamus, and cortex by way of the fast pricking pain pathway and the slow burning pain pathway.

tile sensations, as discussed in Chapter 47. A few also terminate in the posterior nuclear group of the thalamus. From these areas, the signals are transmitted to other basal areas of the brain and to the somatic sensory cortex.

Capability of the Nervous System to Localize Fast Pain in the Body. The fast-sharp type of pain can be localized much more exactly in the different parts of the body than can slow-chronic pain. However, even fast pain, when only pain receptors are stimulated without simultaneously stimulating tactile receptors, is still quite poorly localized, often only within 10 centimeters or so of the stimulated area. Yet when tactile receptors that excite the dorsal column–medial lemniscal system are also stimulated, the localization can be nearly exact.

Glutamate, the Probable Neurotransmitter of the Type Aδ Fast Pain Fibers. It is believed that glutamate is the neurotransmitter substance secreted in the spinal cord at the type Aδ pain nerve fiber endings. This is one of the most widely used excitatory transmitters in the central nervous system, usually having a period of action lasting for only a few milliseconds.

PALEOSPINOTHALAMIC PATHWAY FOR TRANSMITTING SLOW-CHRONIC PAIN. The paleospinothalamic pathway is a much older system and transmits pain mainly carried in the peripheral slow-chronic type C pain fibers, although it does transmit some signals from type Aδ fibers as well. In this pathway, the peripheral fibers terminate almost entirely in laminas II and III of the dorsal horns, which together are called the *substantia gelatinosa,* as shown by the lateralmost dorsal root fiber in Figure 48–2. Most of the signals then pass through one or more additional short fiber neurons within the dorsal horns themselves before entering laminas V through VIII, also in the dorsal horn. Here the last neuron in the series gives rise to long axons that mostly join the fibers from the fast pathway, passing first through the anterior commissure to the opposite side of the cord and then upward to the brain in the same anterolateral pathway.

Substance P, the Probable Slow-Chronic Neurotransmitter of the Type C Nerve Endings. Research experiments suggest that the type C pain fiber terminals entering the spinal cord might secrete both glutamate transmitter and substance P transmitter. The glutamate transmitter acts instantaneously and lasts for only a few milliseconds. On the other hand, substance P is released much more slowly, building up in concentration over a period of seconds or even minutes. In fact, it has been suggested that the "double" pain sensation one feels after a pinprick might result either partly or entirely from the fact that the glutamate transmitter gives a fast pain sensation, whereas the substance P transmitter gives a more lagging sensation. Regardless of the yet unknown details, it does seem clear that glutamate is the neurotransmitter most involved in transmitting fast pain into the central nervous system, whereas substance P (and other related peptides) are concerned with slow-chronic pain.

Projection of the Paleospinothalamic Pathway, Carrying Slow-Chronic Pain Signals, into the Brain Stem and Thalamus. The slow-chronic paleospinothalamic pathway terminates widely in the brain stem, in the large pink-shaded area shown in Figure 48–3. Only one-tenth to one-fourth of the fibers pass all the way to the thalamus. Instead, they terminate principally in one of three areas: (1) the *reticular nuclei* of the medulla, pons, and mesencephalon; (2) the *tectal area* of the mesencephalon deep to the superior and inferior colliculi; or (3) the *periaqueductal gray region* surrounding the aqueduct of Sylvius. These lower regions of the brain appear to be important in the appreciation of the suffering types of pain because animals with their brains sectioned above the mesencephalon to block any pain signals reaching the cerebrum still evince undeniable evidence of suffering when any part of the body is traumatized.

From the brain stem pain areas, multiple short-fiber neurons relay the pain signals upward into the intralaminar and central lateral nuclei of the thalamus and into certain portions of the hypothalamus and other adjacent regions of the basal brain.

Very Poor Capability of the Nervous System to Localize Pain Transmitted in the Slow-Chronic Pathway. Localization of pain transmitted in the paleospinothalamic pathway is poor. For instance, the slow-chronic type of pain can usually be localized only to a major part of the body such as to one arm or leg but not to a detailed point on the arm or leg. This is in keeping with the multisynaptic, diffuse connectivity of this pathway. It explains why patients often have serious difficulty in localizing the source of some chronic types of pain.

FUNCTION OF THE RETICULAR FORMATION, THALAMUS, AND CEREBRAL CORTEX IN THE APPRECIATION OF PAIN. Complete removal of the somatic sensory areas of the cerebral cortex does not destroy an animal's ability to perceive pain. Therefore, it is likely that pain impulses entering the reticular formation, thalamus, and other lower centers can cause conscious perception of pain. This does not mean that the cerebral cortex has nothing to do with normal pain appreciation; indeed, electrical stimulation of the cortical somatic sensory areas causes a person to perceive mild pain in about 3 per cent of the points stimulated. It is believed that the cortex plays an important role in interpreting the quality of pain even though pain perception might be principally a function of lower centers.

SPECIAL CAPABILITY OF PAIN SIGNALS TO AROUSE NERVOUS EXCITABILITY. Electrical stimulation in the reticular areas of the brain stem and in the intralaminar nuclei of the thalamus, the areas where the slow-suffering type of pain terminates, has a strong arousal effect on nervous activity throughout the brain. In fact, these two areas constitute part of the brain's principal arousal system, which is discussed in Chapter

59. This explains why a person with severe pain is frequently strongly aroused as well as why it is almost impossible for a person to sleep when he or she is subjected to pain.

SURGICAL INTERRUPTION OF PAIN PATHWAYS. Often a person has such severe and intractable pain (often resulting from rapidly spreading cancer) that it is necessary to relieve the pain. To do this, the pain pathway can be destroyed at any one of several points. If the pain is in the lower part of the body, a *cordotomy* in the upper thoracic region often relieves the pain for a few weeks to a few months. To do this, the spinal cord on the side opposite the pain is sectioned almost entirely through its anterolateral ' quadrant, which interrupts the anterolateral sensory pathway.

The cordotomy, however, is not always successful in relieving the pain for two reasons. First, many of the pain fibers from the upper part of the body do not cross to the opposite side of the spinal cord until they have reached the brain, so that the cordotomy does not transect these fibers. Second, pain frequently returns several months later, caused partly by sensitization of other pain pathways that normally are too weak to be effectual (for instance, sparse pathways in the dorsolateral cord) and by fibrous tissue from the cordotomy stimulating still remaining fibers. This new pain is often even more objectionable than the original pain.

Another experimental operative procedure to relieve pain has been to place lesions in the intralaminar nuclei in the thalamus, which often relieve the suffering type of pain while leaving intact one's appreciation of "acute" pain, an important protective mechanism.

PAIN SUPPRESSION ("ANALGESIA") SYSTEM IN THE BRAIN AND SPINAL CORD

The degree to which a person reacts to pain varies tremendously. This results partly from the capability of the brain itself to suppress the input of pain signals to the nervous system by activating a pain control system, called an *analgesia system.*

The analgesia system is shown in Figure 48–4. It consists of three major components (plus other, accessory components): (1) The *periaqueductal gray and periventricular areas* of the mesencephalon and upper pons surrounding the aqueduct of Sylvius and adjacent to portions of the third and fourth ventricles. Neurons from these areas send their signals to (2) the *raphe magnus nucleus,* a thin midline nucleus located in the lower pons and upper medulla, and the *nucleus reticularis paragigantocellularis* located laterally in the medulla. From these nuclei, the signals are transmitted down the dorsolateral columns in the spinal cord to (3) a *pain inhibitory complex located in the dorsal horns of the spinal cord.* At this point, the analgesia signals can block the pain before it is relayed to the brain.

Electrical stimulation either in the periaqueductal gray area or in the raphe magnus nucleus can almost completely suppress many strong pain signals entering by way of the dorsal spinal roots. Also, stimulation of

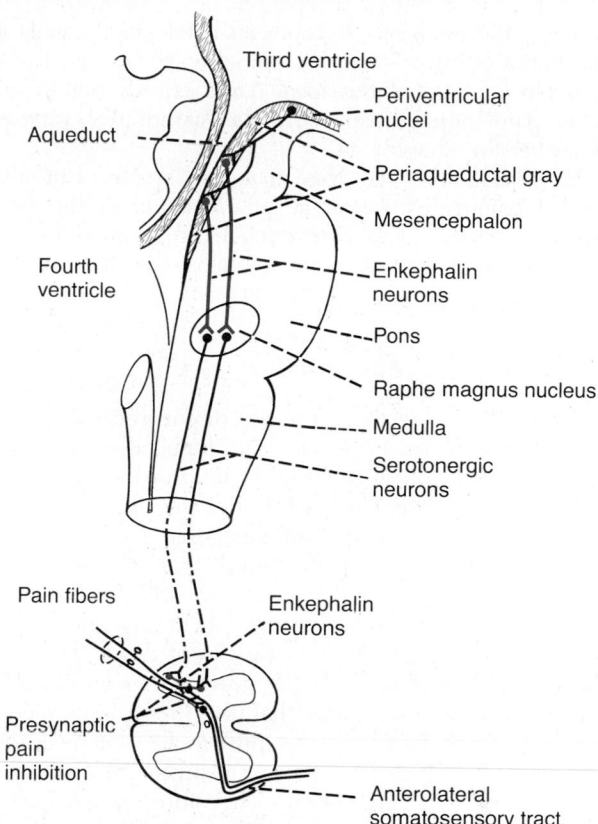

Figure 48–4. Analgesia system of the brain and spinal cord, showing inhibition of incoming pain signals at the cord level and the presence of *enkephalin-secreting neurons* that suppress pain signals in both the cord and the brain stem.

areas at still higher levels of the brain that in turn excite the periaqueductal gray, especially the *periventricular nuclei in the hypothalamus* lying adjacent to the third ventricle and to a lesser extent the *medial forebrain bundle* also in the hypothalamus, can suppress pain, although not quite so much so.

Several transmitter substances are involved in the analgesia system; especially involved are the *enkephalins* and *serotonin.* Many of the nerve fibers derived from both periventricular nuclei and the periaqueductal gray area secrete enkephalin at their endings. Thus, as shown in Figure 48–4, the endings of many of the fibers in the raphe magnus nucleus release enkephalin. The fibers originating in this nucleus but terminating in the dorsal horns of the spinal cord secrete serotonin at their endings. The serotonin in turn causes local cord neurons to secrete enkephalin. Enkephalin is then believed to cause *presynaptic inhibition* and *postsynaptic inhibition* of both incoming type C and type Aδ pain fibers where they synapse in the dorsal horns. It probably achieves the presynaptic inhibition by blocking calcium channels in the membranes of the nerve terminals. Because it is calcium ions that cause release of transmitter at the synapse, such calcium blockage would result in presynaptic inhibition. Furthermore, the blockage appears to last for prolonged periods because after activating the analgesia system, analgesia often lasts for many minutes or even for hours.

Thus, the analgesia system can block pain signals at the initial entry point to the spinal cord. In fact, it can also block many of the local cord reflexes that result from pain signals, especially the withdrawal reflexes described in Chapter 54.

It is probable that this analgesia system can also inhibit pain transmission at other points in the pain pathway, especially in the reticular nuclei in the brain stem and in the intralaminar nuclei of the thalamus.

The Brain's Opiate System—The Endorphins and Enkephalins

More than 25 years ago it was discovered that injection of minute quantities of morphine either into the periventricular nucleus around the third ventricle or into the periaqueductal gray area of the brain stem will cause an extreme degree of analgesia. In subsequent studies, it has now been found that morphine-like agents, mainly the opiates, act at still many other points in the analgesia system, including the dorsal horns of the spinal cord. Because most drugs that alter the excitability of neurons do so by acting on synaptic receptors, it was assumed that the "morphine receptors" of the analgesia system must in fact be receptors for some morphine-like neurotransmitter that is naturally secreted in the brain. Therefore, an extensive search was set into motion for a natural opiate of the brain. About a dozen such opiate-like substances have now been found in different points of the nervous system; all are breakdown products of three large protein molecules: *proopiomelanocortin*, *proenkephalin*, and *prodynorphin*. Furthermore, multiple areas of the brain have been shown to have opiate receptors, especially the areas in the analgesia system. Among the more important of the opiate substances are *β-endorphin*, *met-enkephalin*, *leu-enkephalin*, and *dynorphin*.

The two enkephalins are found most importantly in the brain stem and spinal cord in the portions of the analgesia system described earlier, and *β-endorphin* is present in both the hypothalamus and the pituitary gland. Dynorphin is found in much less quantity mainly in the same areas as the enkephalins.

Thus, although all the fine details of the brain's opiate system are not entirely understood, activation of the analgesia system either by nervous signals entering the periaqueductal gray area and adjacent periventricular areas or by morphine-like drugs can totally or almost totally suppress many pain signals entering through the peripheral nerves.

Inhibition of Pain Transmission by Tactile Sensory Signals

Another important landmark in the saga of pain control was the discovery that stimulation of large type Aβ sensory fibers from the peripheral tactile receptors can depress the transmission of pain signals. This effect presumably results from a type of local lateral inhibition. It explains why such simple maneuvers as rubbing the skin near painful areas is often effective in relieving pain.

And it probably also explains why liniments are often useful in the relief of pain. This mechanism and simultaneous psychogenic excitation of the central analgesia system are probably also the basis of pain relief by acupuncture.

Treatment of Pain by Electrical Stimulation

Several clinical procedures have been developed for suppressing pain by electrical stimulation of large sensory nerve fibers. The stimulating electrodes are placed on selected areas of the skin or, on occasion, implanted over the spinal cord supposedly to stimulate the dorsal sensory columns.

In some patients, electrodes have been placed stereotaxically in the intralaminar nuclei of the thalamus or in the periventricular or periaqueductal area of the diencephalon. The patient can then personally control the degree of stimulation. Dramatic relief has been reported in some instances. Also, the pain relief often lasts for as long as 24 hours after only a few minutes of stimulation.

REFERRED PAIN

Often a person feels pain in a part of his or her body that is considerably remote from the tissues causing the pain. This pain is called *referred pain*. The pain usually is initiated in one of the visceral organs and referred to an area on the body surface. Also, the pain may be referred to an area of the body not exactly coincident with the location of the viscus producing the pain. A knowledge of these different types of referred pain is important in clinical diagnosis because many visceral ailments cause no other clinical signs except referred pain.

MECHANISM OF REFERRED PAIN. Figure 48–5 shows the most likely mechanism by which most pain is referred. In the figure, branches of visceral pain fibers are shown to synapse in the spinal cord with some of the same second-order neurons that receive pain fibers from the skin. When the visceral pain

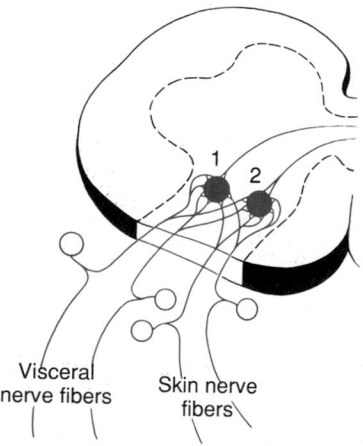

Figure 48–5. Mechanism of referred pain and referred hyperalgesia.

fibers are stimulated, pain signals from the viscera are then conducted through at least some of the same neurons that conduct pain signals from the skin, and the person has the feeling that the sensations originate in the skin itself.

VISCERAL PAIN

In clinical diagnosis, pain from the different viscera of the abdomen and chest is one of the few criteria that can be used for diagnosing visceral inflammation, disease, and other ailments. In general, the viscera have sensory receptors for no other modalities of sensation besides pain. Also, visceral pain differs from surface pain in several important aspects.

One of the most important differences between surface pain and visceral pain is that highly localized types of damage to the viscera seldom cause severe pain. For instance, a surgeon can cut the gut entirely in two in a patient who is awake without causing significant pain. On the other hand, any stimulus that causes *diffuse stimulation of pain nerve endings* throughout a viscus causes pain that can be severe. For instance, ischemia caused by occluding the blood supply to a large area of gut stimulates many diffuse pain fibers at the same time and can result in extreme pain.

Causes of True Visceral Pain

Any stimulus that excites pain nerve endings in diffuse areas of the viscera causes visceral pain. Such stimuli include ischemia of visceral tissue, chemical damage to the surfaces of the viscera, spasm of the smooth muscle in a hollow viscus, distention of a hollow viscus, and stretching of the ligaments.

Essentially, all the true visceral pain that originates in the thoracic and abdominal cavities is transmitted through pain nerve fibers that run in the autonomic nerves, mainly the sympathetic nerves. These fibers are small type C fibers and, therefore, can transmit only the chronic-aching-suffering type of pain.

ISCHEMIA. Ischemia causes visceral pain in the same way that it does in other tissues, presumably because of the formation of acidic metabolic end products or tissue-degenerative products, such as bradykinin, proteolytic enzymes, or others that stimulate the pain nerve endings.

CHEMICAL STIMULI. On occasion, damaging substances leak from the gastrointestinal tract into the peritoneal cavity. For instance, proteolytic acidic gastric juice often leaks through a ruptured gastric or duodenal ulcer. This juice causes widespread digestion of the visceral peritoneum, thus stimulating extremely broad areas of pain fibers. The pain is usually severe.

SPASM OF A HOLLOW VISCUS. Spasm of the gut, the gallbladder, a bile duct, the ureter, or any other hollow viscus can cause pain possibly by mechanical stimulation of the pain endings. Or its cause might be diminished blood flow to the muscle combined with increased metabolic need of the muscle for nutrients. Thus, *relative* ischemia could develop which causes severe pain.

Often pain from a spastic viscus occurs in the form of *cramps*, the pain increasing to a high degree of severity and then subsiding, this process continuing rhythmically once every few minutes. The rhythmical cycles result from rhythmical contraction of smooth muscle. For instance, each time a peristaltic wave travels along an overly excitable spastic gut, a cramp occurs. The cramping type of pain frequently occurs in gastroenteritis, constipation, menstruation, parturition, gallbladder disease, or ureteral obstruction.

OVERDISTENTION OF A HOLLOW VISCUS. Extreme overfilling of a hollow viscus also results in pain, presumably because of overstretch of the tissues themselves. Overdistention can also collapse the blood vessels that encircle the viscus or that pass into its wall, thus perhaps promoting ischemic pain.

Insensitive Viscera

A few visceral areas are almost insensitive to pain of any type. These include the parenchyma of the liver and the alveoli of the lungs. Yet the liver *capsule* is extremely sensitive to both direct trauma and stretch, and the *bile ducts* are also sensitive to pain. In the lungs, even though the alveoli are insensitive, the *bronchi* and the *parietal pleura* are both very sensitive to pain.

Parietal Pain Caused by Visceral Damage

In addition to true visceral pain, pain sensations are transmitted from some viscera through nonvisceral nerve fibers that innervate the parietal peritoneum, pleura, or pericardium.

When a disease affects a viscus, the disease process often spreads to the parietal peritoneum, pleura, or pericardium. These parietal surfaces, like the skin, are supplied with extensive pain innervation from the spinal nerves, not from the sympathetic nerves. Therefore, pain from the parietal wall overlying the viscus is frequently sharp. To emphasize the difference between this pain and true visceral pain: a knife incision through the *parietal* peritoneum is very painful, even though a similar cut through the visceral peritoneum or through a gut wall is not very painful, if painful at all.

Localization of Visceral Pain—The "Visceral" and the "Parietal" Transmission Pathways

Pain from the different viscera is frequently difficult to localize for a number of reasons. First, the brain does not know from firsthand experience that the different internal organs exist, and therefore, any pain that originates internally can be localized only generally. Second, sensations from the abdomen and thorax are transmitted through two pathways to the central nervous system— the *true visceral pathway* and the *parietal pathway.* The true visceral pain is transmitted via sensory fibers in the autonomic nerves (both sympathetic and parasympathetic), and the sensations are *referred* to surface areas of the body often far from the painful organ.

On the other hand, parietal sensations are conducted *directly* into the local spinal nerves from the parietal peritoneum, pleura, or pericardium, and these sensations are usually *localized directly over the painful area.*

LOCALIZATION OF *REFERRED PAIN* TRANSMITTED IN THE VISCERAL PATHWAYS. When visceral pain is referred to the surface of the body, the person generally

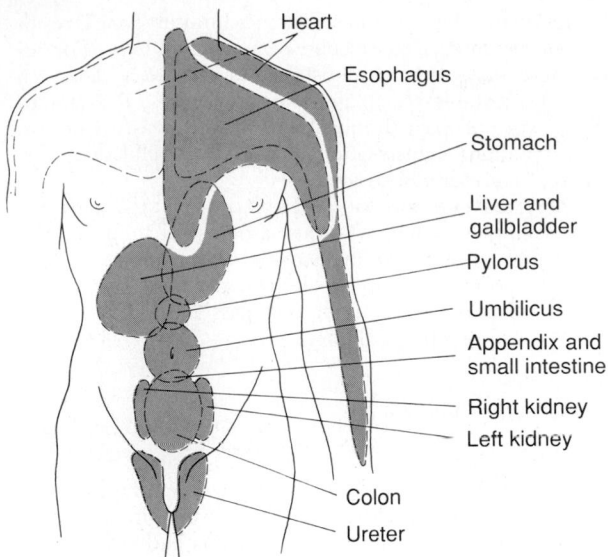

Figure 48–6. Surface areas of referred pain from different visceral organs.

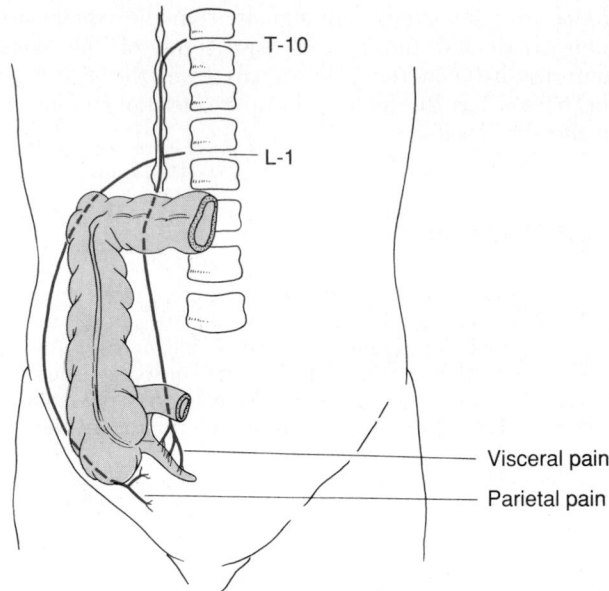

Figure 48–7. Visceral and parietal transmission of pain from the appendix.

localizes it in the dermatomal segment from which the visceral organ originated in the embryo, not necessarily where the visceral organ now lies. For instance, the heart originated in the neck and upper thorax, so that the heart's visceral pain fibers pass up the sympathetic nerves and enter the cord between segments C-3 and T-5. Therefore, as shown in Figure 48–6, pain from the heart is referred to the side of the neck, over the shoulder, over the pectoral muscles, down the arm, and into the substernal area of the chest. These are the areas of the body surface that send their own sensory nerve fibers into the C-3 to T-5 cord segments. Most frequently, the pain is on the left side, rather than on the right because the left side of the heart is much more frequently involved in coronary disease than the right.

The stomach originated approximately from the seventh to the ninth thoracic segments of the embryo. Therefore, stomach pain is referred to the anterior epigastrium above the umbilicus, which is the surface area of the body subserved by the seventh through ninth thoracic segments. And Figure 48–6 shows several other surface areas to which visceral pain is referred from other organs, representing in general the areas in the embryo from which the respective organs originated.

PARIETAL PATHWAY FOR TRANSMISSION OF ABDOMINAL AND THORACIC PAIN. Pain from the viscera is frequently localized to two surface areas of the body at the same time because of the dual transmission of pain through the referred visceral pathway and the direct parietal pathway. Thus, Figure 48–7 shows dual transmission from an inflamed appendix. Pain impulses pass from the appendix through the sympathetic visceral pain fibers into the sympathetic nerves and then into the spinal cord at about T-10 or T-11; this pain is referred to an area around the umbilicus and is of the aching, cramping type. On the other hand, pain impulses also often originate in the parietal peritoneum where the inflamed appendix touches or is adherent to the abdominal wall. These cause pain of the sharp type directly over the irritated peritoneum in the right lower quadrant of the abdomen.

SOME CLINICAL ABNORMALITIES OF PAIN AND OTHER SOMATIC SENSATIONS

Hyperalgesia

A pain pathway sometimes becomes excessively excitable; this gives rise to *hyperalgesia,* which means hypersensitivity to pain. The basic causes of hyperalgesia are (1) excessive sensitivity of the pain receptors themselves, which is called *primary hyperalgesia,* and (2) facilitation of sensory transmission, which is called *secondary hyperalgesia.*

An example of primary hyperalgesia is the extreme sensitivity of sunburned skin, which is believed to result from sensitization of the pain endings by local tissue products of the burn—perhaps histamine, perhaps prostaglandins, perhaps others. Secondary hyperalgesia frequently results from lesions in the spinal cord or the thalamus. Several of these lesions are discussed in subsequent sections.

Thalamic Syndrome

Occasionally the posterolateral branch of the posterior cerebral artery, a small artery that supplies the posteroventral portion of the thalamus, becomes blocked by thrombosis, so that the nuclei of this area of the thalamus degenerate, whereas the medial and anterior nuclei of the thalamus remain intact. The patient suffers a series of abnormalities, as follows: First, loss of almost all sensations from the opposite side of the body occurs because of destruction of the relay nuclei. Second, ataxia (inability to control body movements precisely) may be evident because of loss of position and kinesthetic signals normally relayed through the thalamus to the cortex. Third, after a few weeks to a few months, some sensory perception in the opposite side of the body returns, but strong stimuli are usually necessary to elicit this. When the sensations do occur, they are poorly localized, almost always painful, sometimes lancinating, regardless of the type of stimulus applied to the body. Fourth, the person is likely to perceive many affective sensations of extreme unpleasantness or, rarely, extreme pleasantness; the unpleasant ones often lead to associated emotional tirades.

The medial nuclei of the thalamus are not destroyed by thrombosis of the artery. It is believed that in the thalamic syndrome, these medial nuclei become facilitated and give rise to enhanced sensitivity of the slow-chronic paleospinothalamic pain pathway that transmits suffering types of pain and causes many secondary affective perceptions.

Herpes Zoster (Shingles)

Occasionally a herpes virus infects a dorsal root ganglion. This causes severe pain in the dermatomal segment normally subserved by the ganglion, thus eliciting a segmental type of pain that circles halfway around the body. The disease is called *herpes zoster,* or "shingles," because of the eruption described below.

The cause of the pain is presumably excitation of the neuronal cells in the dorsal root ganglion by the virus infection. Aside from causing pain, the virus is carried by neuronal cytoplasmic flow outward through the peripheral axons to their cutaneous terminals. Here the virus causes a rash that vesiculates within a few days and within another few days crusts over, all of this occurring within the dermatomal area served by the infected dorsal root.

Tic Douloureux

Lancinating pain occurs in some people over one side of the face in the sensory distribution area (or part of the area) of the fifth or ninth nerves; this phenomenon is called *tic douloureux* (or *trigeminal neuralgia* or *glossopharyngeal neuralgia*). The pain feels like sudden electric shocks, and it may appear for only a few seconds at a time or may be almost continuous. Often it is set off by exceedingly sensitive trigger areas on the surface of the face, in the mouth, or in the throat—almost always by a mechanoreceptive stimulus instead of a pain stimulus. For instance, when the patient swallows a bolus of food, as the food touches a tonsil, it might set off a severe lancinating pain in the mandibular portion of the fifth nerve.

The pain of tic douloureux can usually be blocked by cutting the peripheral nerve from the hypersensitive area. The sensory portion of the fifth nerve is often sectioned immediately inside the cranium, where the motor and sensory roots of the fifth nerve are separate from each other, so that the motor portions, which are needed for many of the jaw movements, are spared while the sensory elements are destroyed. This operation leaves the side of the face anesthetic, which in itself may be annoying. Furthermore, sometimes the operation is unsuccessful, indicating that the lesion that causes the pain is in the sensory nucleus in the brain stem and not in the peripheral nerves.

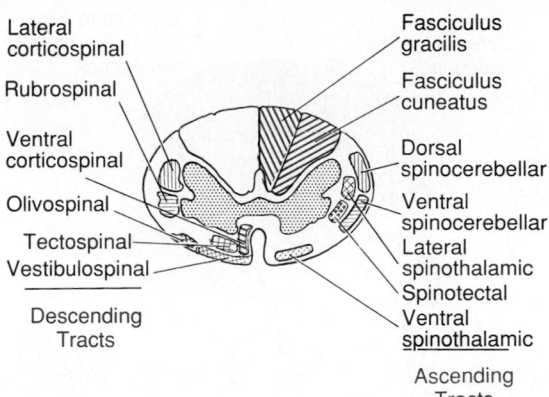

Figure 48–8. Cross section of the spinal cord, showing principal ascending tracts on the right and principal descending tracts on the left.

Brown-Séquard Syndrome

If the spinal cord is transected entirely, all sensations and motor functions distal to the segment of transection are blocked, but if only one half of the spinal cord is transected on a single side, the so-called Brown-Séquard syndrome occurs. The following effects of such a transection occur, and these can be predicted from a knowledge of the cord fiber tracts shown in Figure 48–8: All motor functions are blocked on the side of the transection in all segments below the level of the transection. Yet only some of the modalities of sensation are lost on the transected side, and others are lost on the opposite side. The sensations of pain, heat, and cold—sensations served by the spinothalamic pathway—are lost on the opposite side of the body in all dermatomes two to six segments below the level of the transection. By contrast, the sensations that are transmitted only in the dorsal and dorsolateral columns—kinesthetic and position sensations, vibration sensation, discrete localization, and two-point discrimination—are lost on the side of the transection in all dermatomes below the level of the transection. Discrete, light touch is impaired on the side of the transection because the principal pathway for transmission of light touch, the dorsal columns, is transected. "Crude touch," which is poorly localized, still persists because of transmission in the opposite spinothalamic tract.

HEADACHE

Headaches are referred pain to the surface of the head from the deep structures. Many headaches result from pain stimuli arising inside the cranium, but others result from pain arising outside the cranium, such as from the nasal sinuses.

Headache of Intracranial Origin

PAIN-SENSITIVE AREAS IN THE CRANIAL VAULT. The brain itself is almost totally insensitive to pain. Even

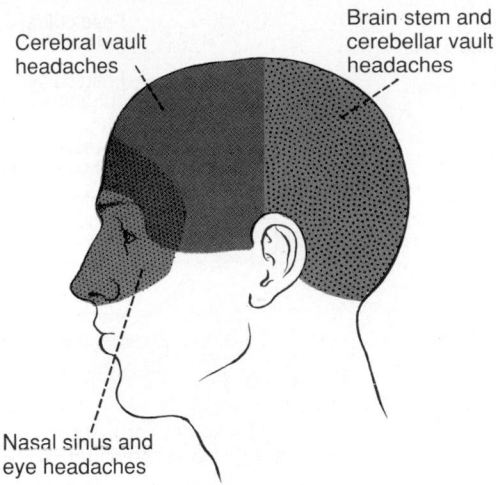

Figure 48–9. Areas of headache resulting from different causes.

cutting or electrically stimulating the sensory areas of the cortex only occasionally causes pain; instead, it causes pins-and-needles types of paresthesias on the area of the body represented by the portion of the sensory cortex stimulated. Therefore, it is likely that much or most of the pain of headache is not caused by damage within the brain itself.

On the other hand, *tugging on the venous sinuses around the brain, damaging the tentorium,* or *stretching the dura at the base of the brain* can all cause intense pain that is recognized as headache. Also, almost any type of traumatizing, crushing, or stretching stimulus to the *blood vessels of the meninges* and, to a lesser extent, of *the brain* can cause headache. A sensitive structure is the middle meningeal artery, and neurosurgeons are careful to anesthetize this artery specifically when performing brain operations under local anesthesia.

AREAS OF THE HEAD TO WHICH INTRACRANIAL HEADACHE IS REFERRED. Stimulation of pain receptors in the intracranial vault above the tentorium, including the upper surface of the tentorium itself, initiates impulses in the cerebral portion of the fifth nerve and, therefore, causes referred headache to the front half of the head in the surface areas supplied by this same portion of the fifth cranial nerve, as shown in Figure 48–9.

On the other hand, pain impulses from beneath the tentorium enter the central nervous system mainly through the glossopharyngeal, vagal, and second cervical nerves, which also supply the scalp behind the ear. Therefore, subtentorial pain stimuli cause "occipital headache" referred to the posterior part of the head, as shown in Figure 48–9.

TYPES OF INTRACRANIAL HEADACHE

Headache of Meningitis. One of the most severe headaches of all is that resulting from meningitis, which causes inflammation of all the meninges, including the sensitive areas of the dura and the sensitive areas around the venous sinuses. Such intense damage as this can cause extreme headache pain referred over the entire head.

Headache Caused by Low Cerebrospinal Fluid Pressure. Removing as little as 20 milliliters of fluid from the spinal canal, particularly if the person remains in an upright position, often causes intense intracranial headache. Removing this quantity of fluid removes the flotation for the brain that is normally provided by the cerebrospinal fluid. Therefore, the weight of the brain stretches and otherwise distorts the various dural surfaces and thereby elicits the pain that causes the headache.

Migraine Headache. Migraine headache is a special type of headache that is thought to result from abnormal vascular phenomena, although the exact mechanism is unknown.

Migraine headaches often begin with various prodromal sensations, such as nausea, loss of vision in part of the field of vision, visual aura, and other types of sensory hallucinations. Ordinarily, the prodromal symptoms begin 30 minutes to 1 hour before the beginning of the headache. Therefore, any theory that explains migraine headache must also explain the prodromal symptoms.

One of the *theories* of the cause of migraine headaches is that prolonged emotion or tension causes reflex vasospasm of some of the arteries of the head, including arteries that supply the brain. The vasospasm theoretically produces ischemia of portions of the brain, and this is responsible for the prodromal symptoms. Then, as a result of the intense ischemia, something happens to the vascular wall, perhaps exhaustion of smooth muscle contraction, to allow it to become flaccid and incapable of maintaining vascular tone for 24 to 48 hours. The blood pressure in the vessels causes them to dilate and pulsate intensely, and it is postulated that the excessive stretching of the walls of the arteries—including some extracranial arteries, such as the temporal artery—causes the actual pain of migraine headaches.

Other *theories* of the cause of migraine headaches include spreading cortical depression, psychological abnormalities, and vasospasm caused by excess local potassium in cerebral extracellular fluid.

ALCOHOLIC HEADACHE. As many people have experienced, a headache usually follows an alcoholic binge. It is most likely that alcohol, because it is toxic to tissues, directly irritates the meninges and causes intracranial pain.

HEADACHE CAUSED BY CONSTIPATION. Constipation causes headache in many people. Because it has been shown that constipation headache can occur in people whose spinal cords have been cut, we know that this headache is not caused by nervous impulses from the colon. Therefore, it possibly results from absorbed toxic products or from changes in the circulatory system resulting from loss of fluid into the gut.

Extracranial Types of Headache

HEADACHE RESULTING FROM MUSCLE SPASM. Emotional tension often causes many of the muscles of the head, including especially those muscles attached to the scalp and the neck muscles attached to the occiput, to become spastic, and it is postulated that this is one of the common causes of headache. The pain of the spastic head muscles supposedly is referred to the overlying areas of the head and gives one the same type of headache as do intracranial lesions.

HEADACHE CAUSED BY IRRITATION OF THE NASAL AND ACCESSORY NASAL STRUCTURES. The mucous membranes of the nose and all the nasal sinuses are sensitive to pain but not intensely so. Nevertheless, infection or other irritative processes in widespread areas of the nasal structures usually summate and cause headache that is referred behind the eyes or, in the case of frontal sinus infection, to the frontal surfaces of the forehead and scalp, as shown in Figure 48–9. Also, pain from the lower sinuses—such as the maxillary sinuses—can be felt in the face.

HEADACHE CAUSED BY EYE DISORDERS. Difficulty in focusing one's eyes clearly may cause excessive contraction of the ciliary muscles in an attempt to gain clear vision. Even though these muscles are extremely small, tonic contraction of them can, it is believed, be the cause of retro-orbital headache. Also, excessive attempts to focus the eyes can result in reflex spasm in various facial and extraocular muscles, which is a possible cause of headache.

A second type of headache that originates in the eyes occurs when the eyes are exposed to excessive irradiation by light rays, especially ultraviolet light. Watching the sun or the arc of an arc-welder for even a few seconds may result in headache that lasts from 24 to 48 hours. The headache sometimes results from "actinic" irritation of the conjunctivae, and the pain is referred to the surface of the head or retro-orbitally. However, focusing intense light from an arc or the sun on the retina can burn the retina, and this could result in headache.

THERMAL SENSATIONS

Thermal Receptors and Their Excitation

The human being can perceive different gradations of cold and heat, progressing from *freezing cold* to *cold* to *cool* to *indifferent* to *warm* to *hot* to *burning hot.*

Thermal gradations are discriminated by at least three types of sensory receptors: the cold receptors, the warmth receptors, and pain receptors. The pain receptors are stimulated only by extreme degrees of heat or cold and, therefore, are responsible, along with the cold and warmth receptors, for "freezing cold" and "burning hot" sensations.

The cold and warmth receptors are located immediately under the skin at discrete but separated points, each having a stimulatory diameter of about 1 millimeter. In most areas of the body, there are 3 to 10 times as many cold receptors as warmth receptors, and the number in different areas of the body varies from 15 to 25 cold points per square centimeter in the lips to 3 to 5 cold points per square centimeter in the finger to less than 1 cold point per square centimeter in some broad surface areas of the trunk. There are correspondingly fewer numbers of warmth points.

Although it is quite certain, on the basis of psychological tests, that there are distinctive warmth nerve endings, they have not been identified histologically. They are presumed to be free nerve endings because warmth signals are transmitted mainly over type C nerve fibers at transmission velocities of only 0.4 to 2 m/sec.

On the other hand, a definitive cold receptor has been identified. It is a special, small type $A\delta$ myelinated nerve ending that branches a number of times,

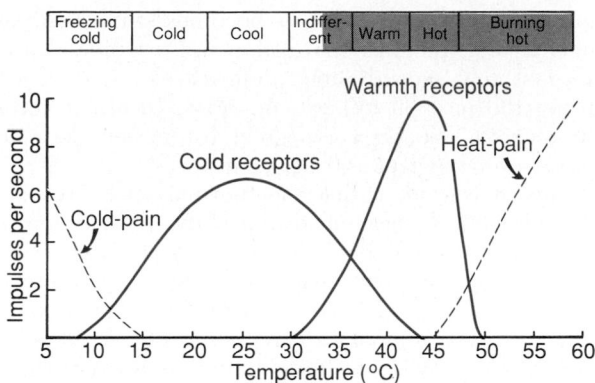

Figure 48–10. Frequencies of discharge of (1) a cold-pain fiber, (2) a cold fiber, (3) a warmth fiber, and (4) a heat-pain fiber. (The responses of these fibers are drawn from original data collected in separate experiments by Zotterman, Hensel, and Kenshalo.)

the tips of which protrude into the bottom surfaces of basal epidermal cells. Signals are transmitted from these receptors via type A delta nerve fibers at velocities of about 20 m/sec. Some cold sensations are believed also to be transmitted in type C nerve fibers, which suggests that some free nerve endings also might function as cold receptors.

STIMULATION OF THERMAL RECEPTORS—SENSATIONS OF COLD, COOL, INDIFFERENT, WARM, AND HOT. Figure 48–10 shows the effects of different temperatures on the responses of four nerve fibers: (1) a pain fiber stimulated by cold, (2) a cold fiber, (3) a warmth fiber, and (4) a pain fiber stimulated by heat. Note especially that these fibers respond differently at different levels of temperature. For instance, in the *very* cold region, only the pain fibers are stimulated (if the skin becomes even colder so that it nearly freezes or actually does freeze, even these fibers cannot be stimulated). As the temperature rises to 10° to 15°C, pain impulses cease, but the cold receptors begin to be stimulated, reaching peak stimulation at about 24°C and fading out slightly above 40°C. Above about 30°C, the warmth receptors become stimulated, but these also fade out at about 49°C. Finally, at around 45°C, pain fibers begin to be stimulated by heat, and paradoxically, some of the cold fibers begin to be stimulated again, possibly because of damage to the cold endings caused by the excessive heat.

One can understand from Figure 48–10, therefore, that a person determines the different gradations of thermal sensations by the relative degrees of stimulation of the different types of endings. One can understand also from this figure why extreme degrees of cold or heat can both be painful and why both these sensations, when intense enough, may give almost the same quality of sensation—that is, freezing cold and burning hot sensations feel almost alike; they are both painful.

STIMULATORY EFFECTS OF RISING AND FALLING TEMPERATURE—ADAPTION OF THERMAL RECEPTORS. When a cold receptor is suddenly subjected to an abrupt fall in temperature, it becomes strongly stimulated at first, but this stimulation fades rapidly during the first few seconds and progressively more slowly during the next 30 minutes or more. In other words, the receptor "adapts" to a great extent but does not appear ever to adapt 100 per cent.

Thus, it is evident that the thermal senses respond markedly to *changes in temperature* in addition to being able to respond to steady states of temperature. This means, therefore, that when the temperature of the skin is actively falling, a person feels much colder than when the temperature remains cold at the same level. Conversely, if the temperature is actively rising, the person feels much warmer than he or she would at the same temperature if it were constant.

The response to changes in temperature explains the extreme degree of heat that one feels on first entering a tub of hot water and the extreme degree of cold felt on going from a heated room to the out-of-doors on a cold day.

Mechanism of Stimulation of the Thermal Receptors

It is believed that the cold and warmth receptors are stimulated by changes in their metabolic rates, these changes resulting from the fact that temperature alters the rates of intracellular chemical reactions more than twofold for each 10°C change. In other words, thermal detection probably results not from direct physical effects of heat or cold on the nerve endings but from chemical stimulation of the endings as modified by the temperature.

SPATIAL SUMMATION OF THERMAL SENSATIONS. Because the number of cold or warm endings in any one surface area of the body is slight, it is difficult to judge gradations of temperature when small areas are stimulated. However, when a large area of the body is stimulated all at once, the thermal signals from the entire area summate. For instance, rapid changes in temperature almost as little as 0.01°C can be detected if this change affects the entire surface of the body simultaneously. On the other hand, temperature changes 100 times this great might not be detected when the skin surface affected is only 1 square centimeter or so in size.

Transmission of Thermal Signals in the Nervous System

In general, thermal signals are transmitted in almost parallel but not exactly the same pathways as pain

signals. On entering the spinal cord, the signals travel for a few segments upward or downward in the *tract of Lissauer* and then terminate mainly in laminae I, II, and III of the dorsal horns—the same as for pain. After a small amount of processing by one or more cord neurons, the signals enter long, ascending thermal fibers that cross to the opposite anterolateral sensory tract and terminate in both (1) the reticular areas of the brain stem and (2) the ventrobasal complex of the thalamus. A few thermal signals are also relayed to the somatic sensory cortex from the ventrobasal complex. Occasionally a neuron in somatic sensory area I has been found by microelectrode studies to be directly responsive to either cold or warm stimuli in specific areas of the skin. Furthermore, it is known that removal of the postcentral gyrus in the human being reduces but does not abolish the ability to distinguish gradations of temperature.

REFERENCES

Aronoff, G. M.: Evaluation and Treatment of Chronic Pain. Baltimore, Williams & Wilkins, 1992.

Bonica, J. J., et al.: The Management of Pain. Baltimore, Williams & Wilkins, 1990.

Casey, K. L.: Pain and Central Nervous System Disease: The Central Pain Syndromes. New York, Raven Press, 1991.

Cervero, F: Sensory innervation of the viscera: peripheral basis of visceral pain. Physiol. Rev., 74:95, 1994.

Dimitrijevic, M. R., and Wall, P. D.: Altered Sensation and Pain. Farmington, CT, S. Karger Publishers, Inc., 1990.

Hamill, R. J., and Rowlingson, J. C.: Handbook of Critical Care Pain Management. Hightstown, NJ, McGraw-Hill, 1994.

Handwerker, H. O., and Kobal, G.: Psycophysiology of experimentally induced pain. Physiol. Rev., 73:639, 1993.

Jacobson, A. L., and Donlon, W. C.: Headache and Facial Pain: Diagnosis and Management. New York, Raven Press, 1990.

Light, A. R.: The Initial Processing of Pain and Its Descending Control: Spinal and Trigeminal Systems. Farmington, CT, S. Karger Publishers, Inc., 1992.

Nashold, B. S., Jr., and Ovelmen-Levitt, J.: Deafferentation Pain Syndromes: Pathophysiology and Treatment. New York, Raven Press, 1991.

Nyhus, L. M., et al.: Abdominal Pain: A Guide to Rapid Diagnosis. Redding, MA, Appleton & Lange, 1994.

Olesen, J., and Schoenen, J.: Tension-Type Headache: Classification, Mechanisms, and Treatment. New York, Raven Press, 1993.

Olesen, J., et al.: The Headaches. New York, Raven Press, 1993.

Olesen, J.: Headache Classification and Epidemiology. New York, Raven Press, 1994.

Olesen, J.: Migraine and Other Headaches: The Vascular Mechanisms. New York, Raven Press, 1991.

Patt, R. B.: Cancer Pain. Philadelphia, J. B. Lippincott, 1993.

Porter, R. W.: Management of Back Pain. 2nd Ed. New York, Churchill Livingstone, 1994.

Rovit, R. L., et al.: Trigeminal Neuralgia. Baltimore, Williams & Wilkins, 1990.

Sicuteri, F., et al.: Pain Versus Man. New York, Raven Press, 1992.

Tollison, C. D.: Handbook of Pain Management. Baltimore, Williams & Wilkins, 1994.

Tollison, C. D.: Headache: Diagnosis and Treatment. Baltimore, Williams & Wilkins, 1993.

Warfield, C. A.: Principles and Practice of Pain Management. Hightstown, NJ, McGraw-Hill, 1993.

Wiener, S. L.: Differential Diagnosis of Acute Pain: By Body Region. Hightstown, NJ, McGraw-Hill, 1993.

THE NERVOUS SYSTEM: B. THE SPECIAL SENSES

UNIT X

The Eye: I. Optics of Vision

CHAPTER 49

PHYSICAL PRINCIPLES OF OPTICS

Before it is possible to understand the optical system of the eye, the student must be thoroughly familiar with the basic physical principles of optics, including the physics of refraction, a knowledge of focusing, depth of focus, and so forth. Therefore, a brief review of these physical principles is first presented and then the optics of the eye is discussed.

Refraction of Light

REFRACTIVE INDEX OF A TRANSPARENT SUBSTANCE. Light rays travel through air at a velocity of about 300,000 km/sec but much slower through transparent solids and liquids. The refractive index of a transparent substance is the *ratio* of the velocity of light in air to the velocity in the substance. The refractive index of air itself is 1.00.

If light travels through a particular type of glass at a velocity of 200,000 km/sec, the refractive index of this glass is 300,000 divided by 200,000, or 1.50.

REFRACTION OF LIGHT RAYS AT AN INTERFACE BETWEEN TWO MEDIA WITH DIFFERENT REFRACTIVE INDICES. When light waves traveling forward in a beam, as shown in the upper part of Figure 49–1, strike an interface that is perpendicular to the beam, the waves enter the second refractive medium without deviating from their course. The only effect that occurs is decreased velocity of transmission and shorter wavelength, as shown in the figure by the shorter distances between wave fronts. On the other hand, as shown in the lower part of the figure, if the light waves pass through an angulated interface, the light waves bend if the refractive indices of the two media are different from each other. In this particular figure, the light waves are leaving air, which has a refractive index of 1.00, and are entering a block of glass having a refractive index of 1.50. When the beam first strikes the angulated interface, the lower edge of the beam enters the glass ahead of the upper edge. The wave front in the upper portion of the beam continues to travel at a velocity of 300,000 km/sec, whereas that which has entered the glass travels at a velocity of 200,000 km/sec. This causes the upper portion of the wave front to move ahead of the lower portion, so that the wave front is no longer vertical but angulated to the right. Because *the direction in which light travels is always perpendicular to the plane of the wave front*, the direction of travel of the light beam now bends downward.

The bending of light rays at an angulated interface is known as *refraction*. Note particularly that the degree of refraction increases as a function of (1) the ratio of the two refractive indices of the two transparent media and (2) the degree of angulation between the interface and the entering wave front.

Application of Refractive Principles to Lenses

THE CONVEX LENS FOCUSES LIGHT RAYS. Figure 49–2 shows parallel light rays entering a convex lens. The light rays passing through the center of the lens strike the lens exactly perpendicular to the lens surface and, therefore, pass through the lens without being refracted. Toward either edge of the lens, however, the light rays strike a progressively more angulated interface. Therefore, the outer rays bend more and more toward the center, which is called *convergence* of the rays. Half the bending occurs when the rays enter the

623

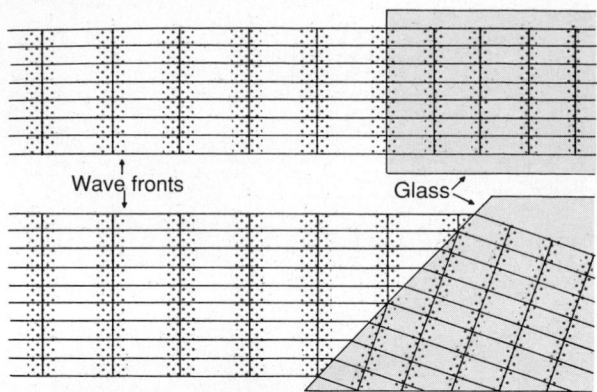

Figure 49–1. Wave fronts entering *(top)* a glass surface perpendicular to the light rays and *(bottom)* a glass surface angulated to the light rays. This figure demonstrates that the distance between waves after they enter the glass is shortened to about two thirds that in air. It also shows that light rays striking an angulated glass surface are bent.

lens and half as they exit from the opposite side. (At this time, the student should pause and analyze why the rays still bend toward the center on leaving the lens.)

Finally, if the lens is ground with exactly the proper curvature, parallel light rays passing through each part of the lens will be bent exactly enough so that all the rays will pass through a single point, which is called the *focal point.*

THE CONCAVE LENS DIVERGES LIGHT RAYS. Figure 49–3 shows the effect of a concave lens on parallel light rays. The rays that enter the very center of the lens strike an interface that is perpendicular to the beam and, therefore, do not refract. The rays at the edge of the lens enter the lens ahead of the rays toward the center. This is opposite to the effect in the convex lens, and it causes the peripheral light rays to *diverge* away from the light rays that pass through the center of the lens.

Thus, the concave lens *diverges* light rays, whereas the convex lens *converges* light rays.

CYLINDRICAL LENSES BEND LIGHT RAYS IN ONLY ONE PLANE—COMPARISON WITH SPHERICAL LENSES. Figure 49–4 shows both a convex *spherical* lens and a convex *cylindrical* lens. Note that the cylindrical lens bends light rays from the two sides of the lens but not from the top or the bottom. That is, bending occurs in one plane but not the other. Thus, parallel light rays are bent to a *focal line.* On the other hand, light rays that

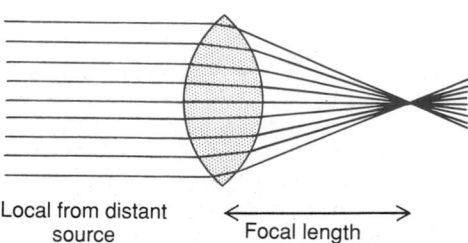

Figure 49–2. Bending of light rays at each surface of a convex spherical lens, showing that parallel light rays are focused to a point focus.

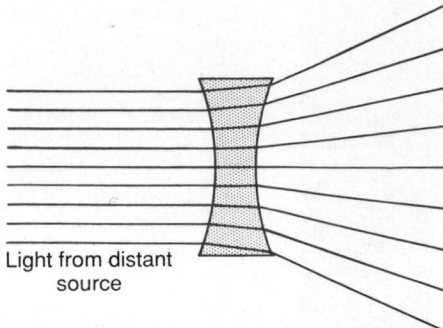

Figure 49–3. Bending of light rays at each surface of a concave spherical lens, showing that parallel light rays are diverged by a concave lens.

pass through the spherical lens are refracted at all edges of the lens (in both planes) toward the central ray, and all the rays come to a *focal point.*

The cylindrical lens is well demonstrated by a test tube full of water. If the test tube is placed in a beam of sunlight and a piece of paper is brought progressively closer to the opposite side of the tube, a certain distance will be found at which the light rays come to a *focal line.* On the other hand, the spherical lens is demonstrated by an ordinary magnifying glass. If such a lens

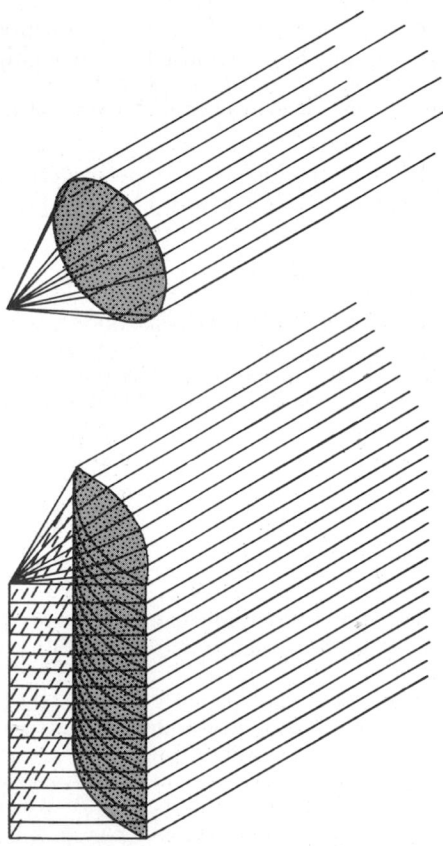

Figure 49–4. *Top:* Point focus of parallel light rays by a spherical convex lens. *Bottom:* Line focus of parallel light rays by a cylindrical convex lens.

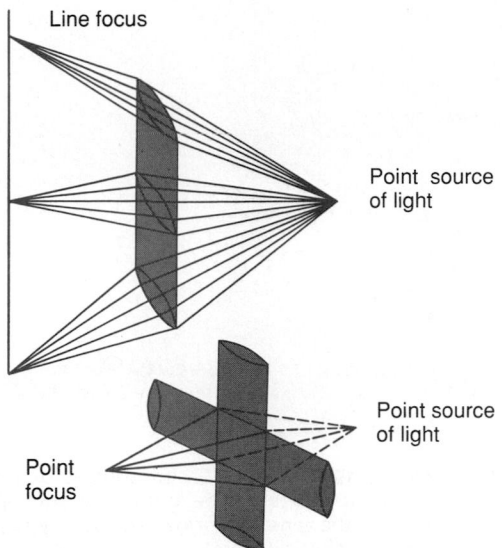

Figure 49–5. *Top:* Focusing of light from a point source to a line focus by a cylindrical lens. *Bottom:* Two cylindrical convex lenses at right angles to each other, demonstrating that one lens converges light rays in one plane and the other lens converges light rays in the plane at right angles. The two lenses combined give the same point focus as that obtained with a single spherical convex lens.

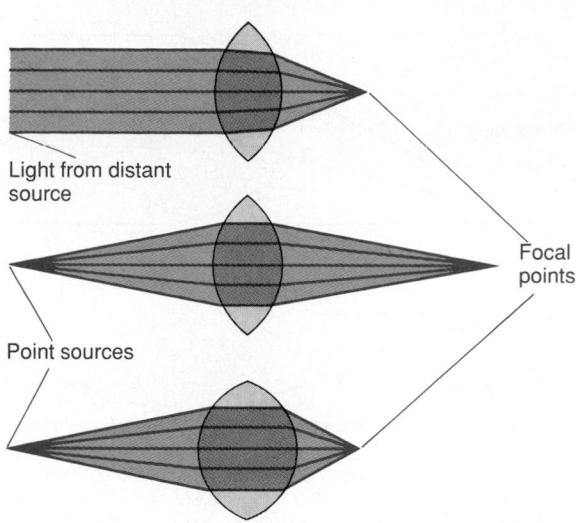

Figure 49–6. The upper two lenses of this figure have the same strength, but the light rays entering the top lens are parallel, whereas those entering the second lens are diverging; the effect of parallel versus diverging rays on the focal distance is shown. The bottom lens has far more refractive power than either of the other two lenses, demonstrating that the stronger the lens, the nearer to the lens is the point focus.

is placed in a beam of sunlight and a piece of paper is brought progressively closer to the lens, the light rays will impinge on a common focal point at an appropriate distance.

Concave cylindrical lenses *diverge* light rays in only one plane in the same manner that *convex* cylindrical lenses *converge* light rays in one plane.

Combination of Two Cylindrical Lenses at Right Angles Equals a Spherical Lens. The lower portion of Figure 49–5 shows two convex cylindrical lenses at right angles to each other. The vertical cylindrical lens causes convergence of the light rays that pass through the two sides of the lens, and the horizontal lens converges the top and bottom rays. Thus, all the light rays come to a single-point focus. In other words, *two cylindrical lenses crossed at right angles to each other perform the same function as one spherical lens of the same refractive power.*

Focal Length of a Lens

The distance beyond a convex lens at which *parallel* rays converge to a common focal point is called the *focal length* of the lens. The diagram at the top of Figure 49–6 demonstrates this focusing of parallel light rays.

In the middle diagram, the light rays that enter the convex lens are not parallel but diverging because the origin of the light is a point source not far away from the lens itself. Because these rays are diverging outward from the point source, it can be seen from the diagram that they do not focus at the same distance away from the lens as do parallel rays. In other words, when rays of light that are already diverging enter a convex lens, the distance of focus on the other side of the lens is

farther from the lens than is the case when the entering rays are parallel to each other.

In the lowest diagram of Figure 49–6 are shown light rays that are diverging toward a convex lens with far greater curvature than that of the upper two lenses of the figure. In this diagram, the distance from the lens at which the light rays come to a focus is exactly the same as that from the lens in the first diagram, in which the lens was less convex but the rays entering it were parallel. This demonstrates that both parallel rays and diverging rays can be focused at the same distance behind a lens, provided the lens changes its convexity.

The relation of focal length of the lens, distance of the point source of light, and distance of focus is expressed by the following formula:

$$\frac{1}{f} = \frac{1}{a} + \frac{1}{b},$$

in which f is the focal length of the lens, a the distance of the point source of light from the lens, and b the distance of focus on the other side of the lens.

Formation of an Image by a Convex Lens

The upper drawing of Figure 49–7 shows a convex lens with two point sources of light to the left. Because light rays pass through the center of a convex lens without being refracted in either direction, the light rays from each point source of light are shown to come to a point focus on the opposite side of the lens *directly in line with the point source and the center of the lens.*

Any object in front of the lens is in reality a mosaic of point sources of light. Some of these points are very bright, some are very weak, and they vary in color. And

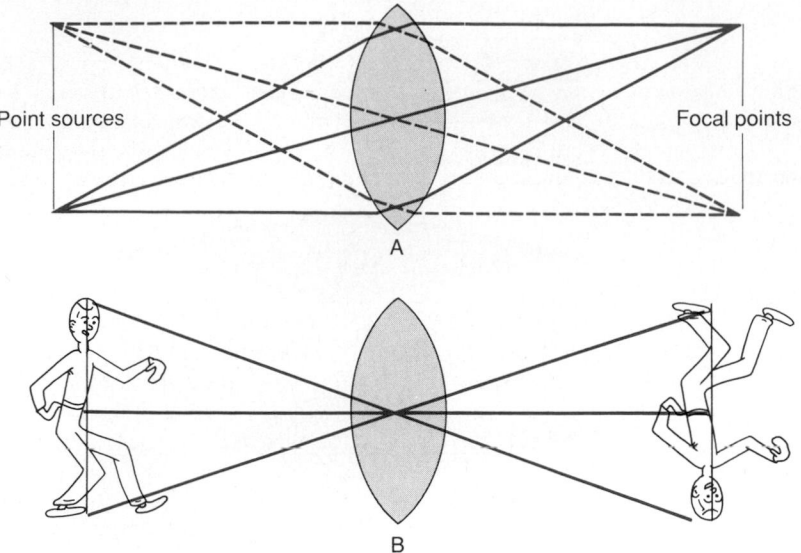

Point sources   Focal points

A

B

Figure 49–7. *A,* Two point sources of light focused at two separate points on the opposite side of the lens. *B,* Formation of an image by a convex spherical lens.

each point source of light on the object comes to a separate point focus on the opposite side of the lens in line with the lens center. If a white sheet of paper is placed at the focus distance from the lens, one can see an image of the object, as demonstrated in the lower portion of Figure 49–7. However, this image is upside down with respect to the original object, and the two lateral sides of the image are reversed. This is the method by which the lens of a camera focuses images on the camera film.

Measurement of the Refractive Power of a Lens—The Diopter

The more a lens bends light rays, the greater is its "refractive power." This refractive power is measured in terms of *diopters.* The refractive power in diopters of a convex lens is equal to 1 meter divided by its focal length. Thus, a spherical lens that converges parallel light rays to a focal point 1 meter beyond the lens has a refractive power of $+1$ diopter, as shown in Figure 49–8. If the lens is capable of bending parallel light rays twice as much as a lens with a power of $+1$ diopter, it is said to have a strength of $+2$ diopters, and the light

rays come to a focal point 0.5 meter beyond the lens. A lens capable of converging parallel light rays to a focal point only 10 centimeters (0.10 meter) beyond the lens has a refractive power of $+10$ diopters.

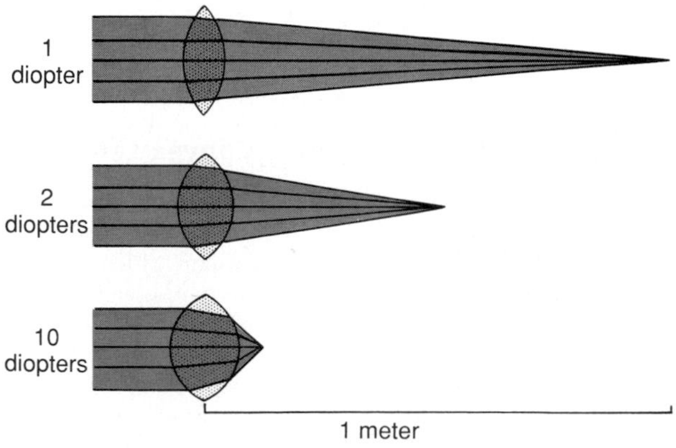

The refractive power of concave lenses cannot be stated in terms of the focal distance beyond the lens because the light rays diverge, rather than focusing to a point. However, if a concave lens diverges light rays at the same rate that a 1-diopter convex lens converges them, the concave lens is said to have a dioptric strength of -1. Likewise, if the concave lens diverges the light rays as much as a $+10$-diopter lens converges them, it is said to have a strength of -10 diopters.

Note particularly that concave lenses "neutralize" the refractive power of convex lenses. Thus, placing a 1-diopter concave lens immediately in front of a 1-diopter convex lens results in a lens system with zero refractive power.

The strengths of cylindrical lenses are computed in the same manner as the strengths of spherical lenses, except that the *axis* of the cylindrical lens must also be stated in addition to its strength. If a cylindrical lens focuses parallel light rays to a line focus 1 meter beyond the lens, it has a strength of $+1$ diopter. On the other hand, if a cylindrical lens of a concave type *diverges* light rays as much as a $+1$-diopter cylindrical lens *converges* them, it has a strength of -1 diopter. If the focused line is horizontal, its axis is said to be 0 degrees. If it is vertical, its axis is 90 degrees.

OPTICS OF THE EYE

The Eye as a Camera

The eye, shown in Figure 49–9, is optically equivalent to the usual photographic camera. It has a lens system, a variable aperture system (the pupil), and a retina that corresponds to the film. The lens system of the eye is composed of four refractive interfaces: (1) the interface between air and the anterior surface of the cornea, (2) the interface between the posterior

1 diopter

2 diopters

10 diopters

1 meter

Figure 49–8. Effect of lens strength on the focal distance.

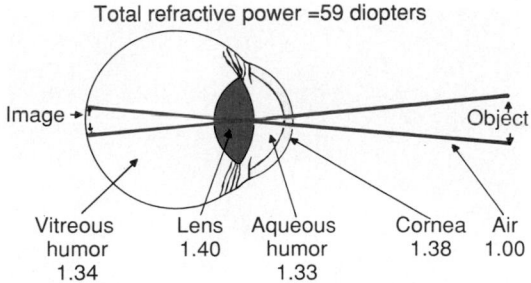

Figure 49–9. The eye as a camera. The numbers are the refractive indices.

surface of the cornea and the aqueous humor, (3) the interface between the aqueous humor and the anterior surface of the crystalline lens of the eye, and (4) the interface between the posterior surface of the lens and the vitreous humor. The refractive index of air is 1; the cornea, 1.38; the aqueous humor, 1.33; the crystalline lens (on the average), 1.40; and the vitreous humor, 1.34.

Reduced Eye. If all the refractive surfaces of the eye are algebraically added together and then considered to be one single lens, the optics of the normal eye may be simplified and represented schematically as a "reduced eye." This is useful in simple calculations. In the reduced eye, a single refractive surface is considered to exist with its central point 17 millimeters in front of the retina and to have a total refractive power of about 59 diopters when the lens is accommodated for distant vision.

Most of the refractive power of the eye is provided not by the crystalline lens but by the anterior surface of the cornea. The principal reason for this is that the refractive index of the cornea is markedly different from that of air.

On the other hand, the total refractive power of the crystalline lens of the eye, as it normally lies in the eye surrounded by fluid on each side, is only 20 diopters, about one-third the total refractive power of the eye's lens system. If this lens were removed from the eye and then surrounded by air, its refractive power would be about six times as great. The reason for this difference is that the fluids surrounding the lens have refractive indices not greatly different from the refractive index of the lens itself, and the smallness of the differences greatly decreases the amount of light refraction at the lens interfaces. But the importance of the crystalline lens is that its curvature can be increased markedly to provide "accommodation," which is discussed later in the chapter.

FORMATION OF AN IMAGE ON THE RETINA. In the same manner that a glass lens can focus an image on a sheet of paper, the lens system of the eye can focus an image on the retina. The image is inverted and reversed with respect to the object. However, the mind perceives objects in the upright position despite the upside-down orientation on the retina because the brain is trained to consider an inverted image as the normal.

Mechanism of Accommodation

The refractive power of the crystalline lens of the eye can be voluntarily increased from 20 diopters to about 34 diopters in young children; this is a total "accommodation" of 14 diopters. To do this, the shape of the lens is changed from that of a moderately convex lens to that of a very convex lens. The mechanism is the following.

In the young person, the lens is composed of a strong elastic capsule filled with viscous, proteinaceous, but transparent fibers. When the lens is in a relaxed state with no tension on its capsule, it assumes an almost spherical shape, owing mainly to the elasticity of the lens capsule. However, as shown in Figure 49–10, about 70 very inelastic ligaments (called *zonules*) attach radially around the lens, pulling the lens edges toward the outer circle of the eyeball. These ligaments are constantly tensed by their attachments to the ciliary body at the anterior border of the choroid and retina. The tension on the ligaments causes the lens to remain relatively flat under normal resting conditions of the eye.

At the insertions of the lens ligaments in the ciliary body is a muscle called the *ciliary muscle*. This muscle has two sets of smooth muscle fibers, the *meridional fibers* and the *circular fibers*. The meridional fibers extend to the corneoscleral junction. When these muscle fibers contract, the peripheral insertions of the lens ligaments are pulled forward and medially toward the cornea, thereby releasing a certain amount of tension on the lens. The circular fibers are arranged circularly all the way around the eye so that when they contract, a sphincter-like action occurs, decreasing the diameter of the circle of ligament attachments; this also allows the ligaments to pull less on the lens capsule.

Thus, contraction of either set of smooth muscle fibers in the ciliary muscle relaxes the ligaments to the lens capsule, and the lens assumes a more spherical shape, like that of a balloon, because of the natural elasticity of its capsule. Therefore, when the ciliary

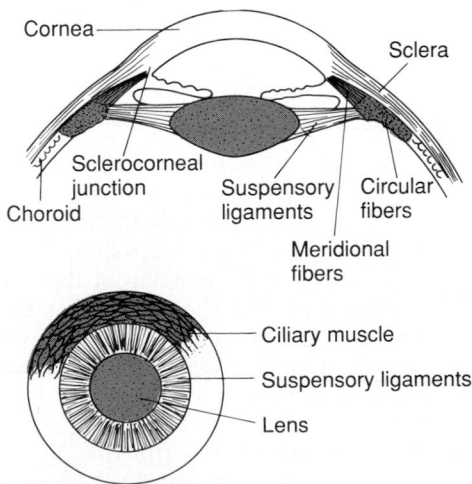

Figure 49–10. Mechanism of accommodation (focusing).

muscle is completely relaxed, the dioptric strength of the lens is as weak as it can become. On the other hand, when the ciliary muscle contracts as strongly as possible, the dioptric strength of the lens becomes maximal.

ACCOMMODATION IS CONTROLLED BY THE PARASYMPATHETIC NERVES. The ciliary muscle is controlled almost entirely by parasympathetic nerve signals transmitted to the eye from the third cranial nerve nucleus in the brain stem, as explained in Chapter 51. Stimulation of the parasympathetic nerves contracts the ciliary muscle, which relaxes the lens ligaments and increases the refractive power. With an increased refractive power, the eye is capable of focusing on objects nearer at hand than when the eye has less refractive power. Consequently, as a distant object moves toward the eye, the number of parasympathetic impulses impinging on the ciliary muscle must be progressively increased for the eye to keep the object constantly in focus. (Sympathetic stimulation has a weak effect in relaxing the ciliary muscle, but this plays almost no role in the normal accommodation mechanism, the neurology of which is discussed in Chapter 51.)

PRESBYOPIA. As a person grows older, the lens grows larger and thicker and becomes far less elastic, partly because of progressive denaturation of the lens proteins. Therefore, the ability of the lens to change shape progressively decreases with age. The power of accommodation decreases from about 14 diopters in the child to less than 2 diopters by ages 45 to 50 and to about 0 at age 70. Thereafter, the lens is almost totally nonaccommodating, a condition known as "presbyopia."

Once a person has reached the state of presbyopia, each eye remains focused permanently at an almost constant distance; this distance depends on the physical characteristics of each person's eyes. The eyes can no longer accommodate for both near and far vision. Therefore, to see clearly both in the distance and nearby, an older person must wear bifocal glasses with the upper segment normally focused for far-seeing and the lower segment focused for near-seeing (that is, for reading).

Pupillary Diameter

A major function of the iris is to increase the amount of light that enters the eye during darkness and to decrease the amount of light that enters the eye in bright light. The reflexes for controlling this mechanism are considered in the discussion of the neurology of the eye in Chapter 51. The amount of light that enters the eye through the pupil is proportional to the *area* of the pupil or to the *square of the diameter* of the pupil. The pupil of the human eye can become as small as about 1.5 millimeters and as large as 8 millimeters in diameter. Therefore, the quantity of light entering the eye can change about 30 times as a result of changes in pupillary aperture.

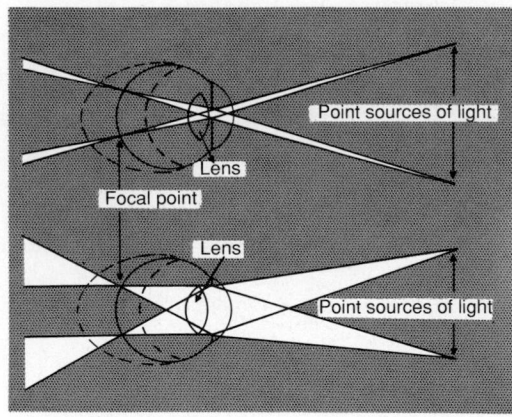

Figure 49–11. Effect of small and large pupillary apertures on the depth of focus.

THE DEPTH OF FOCUS OF THE LENS SYSTEM INCREASES WITH DECREASING PAPILLARY DIAMETER. Figure 49–11 shows two eyes that are exactly alike except for the diameters of the pupillary apertures. In the upper eye, the pupillary aperture is small, and in the lower eye, the aperture is large. In front of each of these two eyes are two small point sources of light, and light from each passes through the pupillary aperture and focuses on the retina. Consequently, in both eyes, the retina sees two spots of light in perfect focus. It is evident from the diagrams, however, that if the retina is moved forward or backward to an out-of-focus position, the size of each spot will not change much in the upper eye, but in the lower eye, the size of each spot will increase greatly, becoming a "blur circle." In other words, the upper lens system has far greater *depth of focus* than the bottom lens system. When a lens system has great depth of focus, the retina can be considerably displaced from the focal plane or the lens strength can change considerably and still the image remains in sharp focus, whereas when a lens system has a shallow depth of focus, moving the retina only slightly away from the focal plane causes extreme blurring.

The greatest possible depth of focus occurs when the pupil is extremely small. The reason for this is that, with a very small aperture, all light rays pass through the center of the lens and the central-most rays are always in focus, as explained earlier.

Errors of Refraction

EMMETROPIA. As shown in Figure 49–12, the eye is considered to be normal, or "emmetropic," if parallel light rays from distant objects are in sharp focus on the retina *when the ciliary muscle is completely relaxed.* This means that the emmetropic eye can see all distant objects clearly, with its ciliary muscle relaxed. However, to focus objects at close range, the eye must contract its ciliary muscle and thereby provide appropriate degrees of accommodation.

HYPEROPIA. Hyperopia, which is also known as "farsightedness," is usually due either to an eyeball that is too short or, occasionally, to a lens system that is too

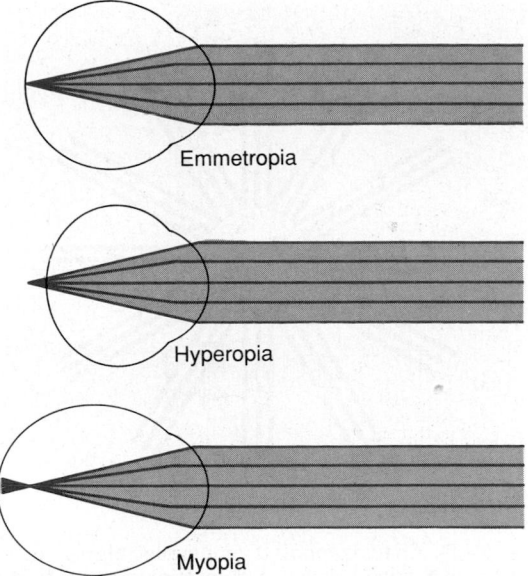

Figure 49-12. Parallel light rays focus on the retina in emmetropia, behind the retina in hyperopia, and in front of the retina in myopia.

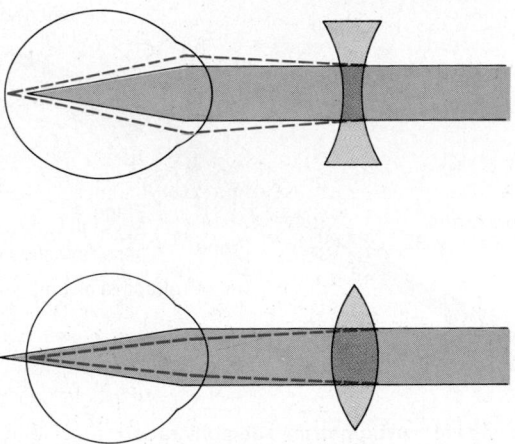

Figure 49-13. Correction of myopia with a concave lens and correction of hyperopia with a convex lens.

weak. In this condition, as seen in the middle panel of Figure 49–12, parallel light rays are not bent sufficiently by the relaxed lens system to come to a focus by the time they reach the retina. To overcome this abnormality, the ciliary muscle must contract to increase the strength of the lens. Therefore, the farsighted person is capable, by using the mechanism of accommodation, of focusing distant objects on the retina. If the person has used only a small amount of strength in the ciliary muscle to accommodate for the distant objects, then he or she still has much accommodative power left, and objects closer and closer to the eye can also be focused sharply until the ciliary muscle has contracted to its limit.

In old age, when the lens becomes presbyopic, the farsighted person often is not able to accommodate his or her lens sufficiently to focus even distant objects, much less to focus near objects.

MYOPIA. In myopia, or "nearsightedness," when the ciliary muscle is completely relaxed, the light rays coming from distant objects are focused in front of the retina, as shown in the lowest panel of Figure 49–12. This is usually due to too long an eyeball, but it can result from too much refractive power in the lens system of the eye.

No mechanism exists by which the eye can decrease the strength of its lens to less than that which exists when the ciliary muscle is completely relaxed. Therefore, the myopic person has no mechanism by which he or she can ever focus distant objects sharply on the retina. However, as an object comes nearer to the person's eye, it finally comes near enough that its image can be focused. Then, when the object comes still closer to the eye, the person can use his or her mechanism of accommodation to keep the image focused clearly. Therefore, a myopic person has a definite limiting "far point" for clear vision.

Correction of Myopia and Hyperopia by Use of Lenses. It will be recalled that light rays passing through a concave lens diverge. Therefore, if the refractive surfaces of the eye have too much refractive power, as in *myopia,* some of this excessive refractive power can be neutralized by placing in front of the eye a concave spherical lens, which will diverge rays. Such correction is demonstrated in the upper diagram of Figure 49–13.

On the other hand, in a person who has *hyperopia*—that is, someone who has too weak a lens system—the abnormal vision can be corrected by adding refractive power with a convex lens in front of the eye. This correction is demonstrated in the lower diagram of Figure 49–13.

One usually determines the strength of the concave or convex lens needed for clear vision by "trial and error"—that is, by trying first a strong lens and then a stronger or weaker lens until the one that gives the best visual acuity is found.

Astigmatism

Astigmatism is a refractive error of the eye that causes the visual image in one plane to focus at a different distance from that of the plane at right angles. This most often results from too great a curvature of the cornea in one of its planes. A lens surface like the side of an egg lying sidewise to the incoming light would be an example of an astigmatic lens. The degree of curvature in the plane through the long axis of the egg is not nearly so great as the degree of curvature in the plane through the short axis.

Because the curvature of the astigmatic lens along one plane is less than the curvature along the other plane, light rays striking the peripheral portions of the lens in one plane are not bent nearly so much as the rays striking the peripheral portions of the other plane. This is demonstrated in Figure 49–14, which shows rays of light emanating from a point source and passing through an oblong, astigmatic lens. The light rays in the vertical plane, indicated by plane BD, are refracted greatly by the astigmatic lens because of the greater curvature in the vertical direction than in the horizontal direction. However, the light rays in the horizontal

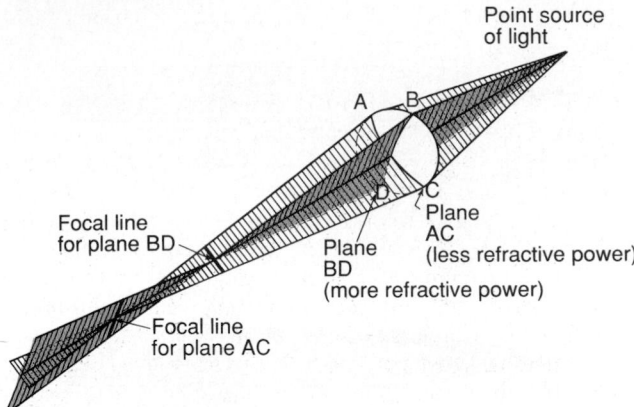

Figure 49–14. Astigmatism, demonstrating that light rays focus at one focal distance in one focal plane and at another focal distance in the plane at right angles.

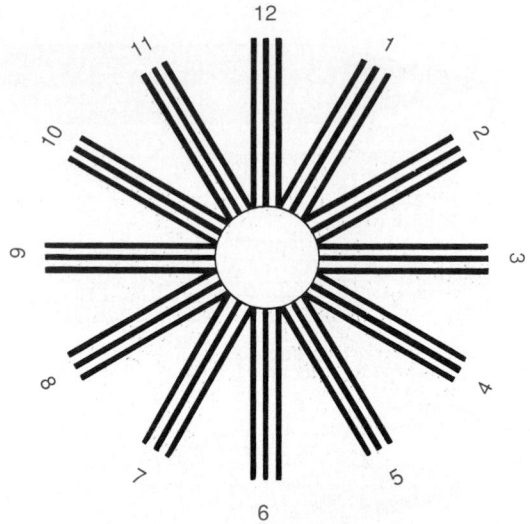

Figure 49–15. Chart composed of parallel black bars at different angular orientations for determining the axis of astigmatism.

plane, indicated by plane AC, are bent not nearly so much as the light rays in the vertical plane. It is obvious, therefore, that the light rays passing through an astigmatic lens do not all come to a common focal point because the light rays passing through one plane focus far in front of those passing through the other plane.

The accommodative power of the eye can never compensate for astigmatism because during accommodation, the curvature of the eye lens changes equally in both planes. In other words, each of the two planes requires a different degree of accommodation to be corrected, so that the two planes are never corrected at the same time without the help of glasses. Thus, vision without the aid of glasses never occurs with a sharp focus in astigmatism.

Correction of Astigmatism with a Cylindrical Lens. One may consider an astigmatic eye as having a lens system made up of two cylindrical lenses of different strengths and placed at right angles to each other. Therefore, to correct for astigmatism, the usual procedure is to find a spherical lens by "trial and error" that corrects the focus in one of the two planes of the astigmatic lens. Then an additional cylindrical lens is used to correct the error in the remaining plane. To do this, both the *axis* and the *strength* of the required cylindrical lens must be determined.

There are several methods for determining the axis of the abnormal cylindrical component of the lens system of an eye. One of these methods is based on the use of parallel black bars of the type shown in Figure 49–15. Some of these parallel bars are vertical, some horizontal, and some at various angles to the vertical and horizontal axes. After placing various spherical lenses in front of the astigmatic eye by trial and error, a strength of lens will usually be found that will cause sharp focus of one set of these parallel bars but will not correct the fuzziness of the set of bars at right angles to the sharp bars. It can be shown from the physical principles of optics discussed earlier in this chapter that the *axis* of the *out-of-focus* cylindrical component of the optical system is parallel to the bars that are fuzzy. Once this axis is found, the examiner tries progressively stronger and weaker positive or negative *cylindrical* lenses, the axes of which are placed parallel to the out-of-focus bars, until the patient sees all the crossed bars with equal

clarity. When this has been accomplished, the examiner directs the optician to grind a special lens combining both the spherical correction and the cylindrical correction at the appropriate axis.

Correction of Optical Abnormalities by Use of Contact Lenses

In recent years, glass or plastic contact lenses have been fitted snugly against the anterior surface of the cornea. These lenses are held in place by a thin layer of tears that fills the space between the contact lens and the anterior eye surface.

A special feature of the contact lens is that it nullifies almost entirely the refraction that normally occurs at the anterior surface of the cornea. The reason for this is that the tears between the contact lens and the cornea have a refractive index almost equal to that of the cornea so that no longer does the anterior surface of the cornea play a significant role in the eye's optical system. Instead, the anterior surface of the contact lens now plays the major role. Thus, the refraction of this contact lens substitutes for the cornea's usual refraction. This is especially important in people whose eye refractive errors are caused by an abnormally shaped cornea, people, for example, who have an odd-shaped, bulging cornea—a condition called *keratoconus*. Without the contact lens, the bulging cornea causes such severe abnormality of vision that almost no glasses can correct the vision satisfactorily; when a contact lens is used, however, the corneal refraction is neutralized and normal refraction by the anterior surface of the contact lens is substituted in its place.

The contact lens has several other advantages as well, including (1) the lens turns with the eye and gives a broader field of clear vision than do usual glasses and (2) the contact lens has little effect on the size of the object that the person sees through the lens; on the other hand, lenses placed several centimeters in front of the eye do affect the size of the image in addition to correcting the focus.

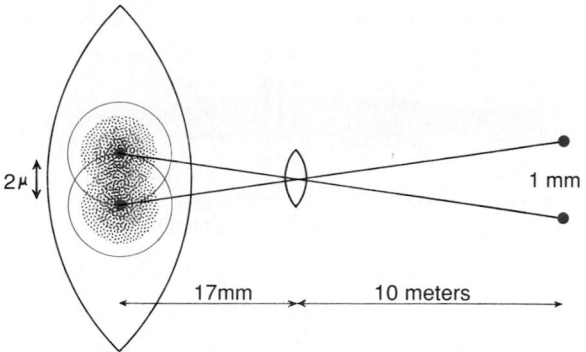

Figure 49–16. Maximum visual acuity for two point sources of light.

Cataracts

Cataracts are an especially common eye abnormality that occurs mainly in older people. A cataract is a cloudy or opaque area or areas in the lens. In the early stage of cataract formation, the proteins in some of the lens fibers become denatured. Later, these same proteins coagulate to form opaque areas in place of the normal transparent protein fibers.

When a cataract has obscured light transmission so greatly that it seriously impairs vision, the condition can be corrected by surgical removal of the lens. When this is done, the eye loses a large portion of its refractive power, which must be replaced by a powerful convex lens in front of the eye; or an artificial lens may be implanted inside the eye in place of the removed lens.

Visual Acuity

Theoretically, light from a distant point source, when focused on the retina, should be infinitely small. However, because the lens system of the eye is not perfect, such a retinal spot ordinarily has a total diameter of about 11 micrometers even with maximal resolution of the optical system. It is brightest in its very center and shades off gradually toward the edges, as shown by the two-point images in Figure 49–16.

The average diameter of cones *in the fovea* of the retina, the central part of the retina where vision is most highly developed, is about 1.5 micrometers, which is one-seventh the diameter of the spot of light. Nevertheless, because the spot of light has a bright center point and shaded edges, a person can normally distinguish two separate points if their centers lie as much as 2 micrometers apart on the retina, which is slightly greater than the width of a foveal cone. This discrimination between points is also shown in Figure 49–16.

The normal visual acuity of the human eye for discriminating between point sources of light is about 25 seconds of arc. That is, when light rays from two separate points strike the eye with an angle of at least 25 seconds between them, they can usually be recognized as two points instead of one. This means that a person with normal acuity looking at two bright pin-point spots of light 10 meters away can barely distinguish the spots as separate entities when they are 1.5 to 2 millimeters apart.

The fovea is less than 0.5 millimeter (less than 500 micrometers) in diameter, which means that maximum visual acuity occurs in less than 2 degrees of the visual field. Outside this foveal area, the visual acuity becomes progressively poorer, decreasing more than 10-fold as the periphery is approached. This is caused by the connection of many rods and cones to the same optic nerve fiber in the nonfoveal, more peripheral parts of the retina, as discussed in Chapter 51.

CLINICAL METHOD FOR STATING VISUAL ACUITY. The test chart for testing eyes usually is placed 20 feet away from the tested person, and if the person can see the letters of the size that he or she should be able to see at 20 feet, he or she is said to have 20/20 vision: that is, normal vision. If the person can see only letters that he or she should be able to see at 200 feet, he or she is said to have 20/200 vision. In other words, the clinical method for expressing visual acuity is to use a mathematical fraction that expresses the ratio of two distances, which is also the ratio of one's visual acuity to that of the normal person.

Determination of Distance of an Object from the Eye—Depth Perception

The visual apparatus normally perceives distance by three major means. This phenomenon is known as *depth perception*. These means are (1) the size of the image of known objects on the retina, (2) the phenomenon of moving parallax, and (3) the phenomenon of stereopsis.

DETERMINATION OF DISTANCE BY SIZES OF RETINAL IMAGES OF KNOWN OBJECTS. If one knows that a person whom one is viewing is 6 feet tall, one can determine how far away the person is simply by the size of the person's image on one's retina. One does not consciously think about the size, but one's brain has learned to calculate automatically from image sizes the distances of objects when the dimensions are known.

DETERMINATION OF DISTANCE BY MOVING PARALLAX. Another important means by which the eyes determine distance is that of moving parallax. If people look off into the distance with their eyes completely still, they perceive no moving parallax, but when they move their head to one side or the other, the images of objects close move rapidly across the retinas while the images of distant objects remain almost completely stationary. For instance, in moving the head 1 inch with an object only 1 inch in front of the eye, the image moves almost all the way across the retinas, whereas the image of an object 200 feet away from the eyes does not move perceptibly. Thus, by this mechanism of moving parallax, one can tell the *relative distances* of different objects even though only one eye is used.

DETERMINATION OF DISTANCE BY STEREOPSIS—BINOCULAR VISION. Another method by which one

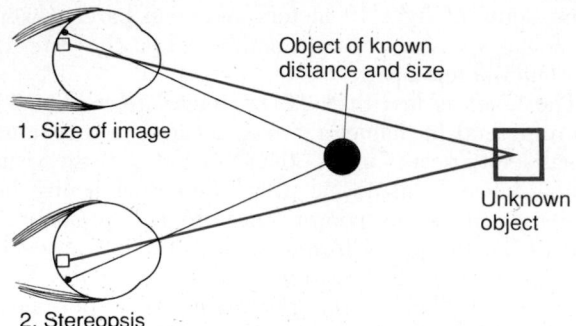

Figure 49–17. Perception of distance (1) by the size of the image on the retina and (2) as a result of stereopsis.

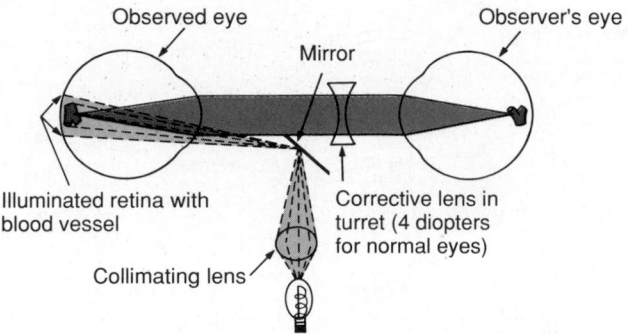

Figure 49–18. Optical system of the ophthalmoscope.

perceives parallax is that of binocular vision. Because one eye is a little more than 2 inches to one side of the other eye, the images on the two retinas are different one from the other—that is, an object that is 1 inch in front of the bridge of the nose forms an image on the left side of the retina of the left eye but on the right side of the retina of the right eye, whereas a small object 20 feet in front of the nose has its image at closely corresponding points in the middle of each retina. This type of parallax is demonstrated in Figure 49–17, which shows the images of a black spot and a square actually reversed on the two retinas because they are at different distances in front of the eyes. This gives a type of parallax that is present all the time when both eyes are being used. It is almost entirely this binocular parallax (or stereopsis) that gives a person with two eyes far greater ability to judge relative distances *when objects are nearby* than a person who has only one eye. However, stereopsis is virtually useless for depth perception at distances beyond 200 feet.

OPHTHALMOSCOPE

The ophthalmoscope is an instrument through which an observer can look into another person's eye and see the retina with clarity. Although the ophthalmoscope appears to be a relatively complicated instrument, its principles are simple. The basic components are shown in Figure 49–18 and may be explained as follows.

If a bright spot of light is on the retina of an *emmetropic eye,* light rays from this spot diverge toward the lens system of the eye, and after passing through the lens system, they are parallel with one another because the retina is located one focal length distance behind the lens. Then, when these parallel rays pass into an emmetropic eye of another person, they focus back again to a point focus on the retina of the second person, because his or her retina is also one focal length distance behind the lens. Therefore, any spot of light on the retina of the observed eye projects to a focal spot on the retina of the observing eye. Thus, if the retina of one person is made to emit light, the image of his or her retina will be focused on the retina of the observer, provided the two eyes are simply looking into each other. These principles apply only to completely emmetropic eyes.

To make an ophthalmoscope, one need only devise a means for illuminating the retina to be examined. Then, the reflected light from that retina can be seen by the observer simply by putting the two eyes close to each other. To illuminate the retina of the observed eye, an angulated mirror or a segment of a prism is placed in front of the observed eye in such a manner, as shown in Figure 49–18, that light from a bulb is reflected into the observed eye. Thus, the retina is illuminated through the pupil, and the observer sees into the subject's pupil by looking over the edge of the mirror or prism or *through* an appropriately designed prism so that the light will not have to enter the pupil at an angle.

It was noted above that these principles apply only to people with completely emmetropic eyes. If the refractive power of either eye is abnormal, it is necessary to correct this refractive power for the observer to see a sharp image of the observed retina. Therefore, the usual ophthalmoscope has a series of lenses mounted on a turret so that the turret can be rotated from one lens to another, and the correction for abnormal refractive power of either or both eyes can be made by selecting a lens of appropriate strength. In normal young adults, when the two eyes come close together, a natural accommodative reflex occurs that causes an approximate + 2-diopter increase in the strength of the lens of each eye. To correct for this, it is necessary that the lens turret be rotated to approximately a − 4-diopter correction.

FLUID SYSTEM OF THE EYE— INTRAOCULAR FLUID

The eye is filled with *intraocular fluid,* which maintains sufficient pressure in the eyeball to keep it distended. Figure 49–19 demonstrates that this fluid can be divided into two portions, the *aqueous humor,* which lies in front and to the sides of the lens, and the fluid of the *vitreous humor,* which lies between the lens and the retina. The aqueous humor is a freely flowing fluid, whereas the vitreous humor, sometimes called the *vitreous body,* is a gelatinous mass held together by a fine fibrillar network composed primarily of greatly elongated proteoglycan molecules. Substances can *diffuse* slowly in the vitreous humor, but there is little *flow* of fluid.

Aqueous humor is continually being formed and reabsorbed. The balance between formation and reab-

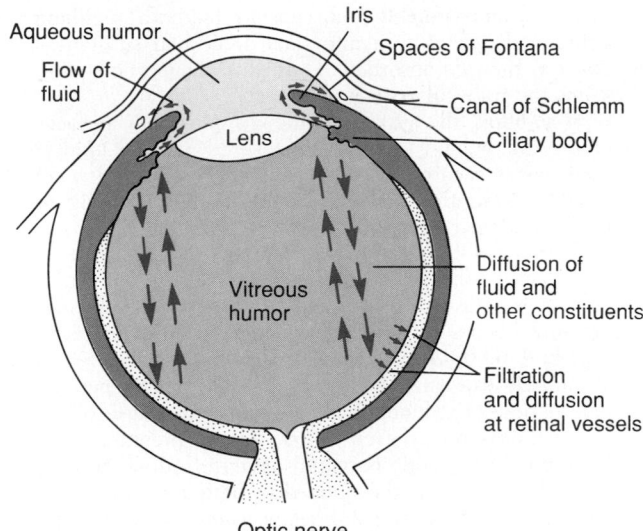

Figure 49–19. Formation and flow of fluid in the eye.

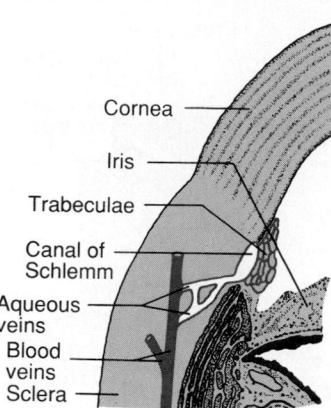

Figure 49–21. Anatomy of the iridocorneal angle, showing the system for outflow of aqueous humor into the conjunctival veins.

sorption of aqueous humor regulates the total volume and pressure of the intraocular fluid.

Formation of Aqueous Humor by the Ciliary Body

Aqueous humor is formed in the eye *at an average rate of 2 to 3 microliters each minute.* Essentially all of it is secreted by the *ciliary processes,* which are linear folds that project from the *ciliary body* into the space behind the iris where the lens ligaments and ciliary muscle also attach to the eyeball. A cross section of these ciliary processes is shown in Figure 49–20, and their relation to the fluid chambers of the eye can be seen in Figure 49–19. Because of their folded architecture, the total surface area of the ciliary processes is about 6 square centimeters in each eye—a large area,

considering the small size of the ciliary body. The surfaces of these processes are covered by highly secretory epithelial cells, and immediately beneath them is a highly vascular area.

Aqueous humor is formed almost entirely as an active secretion of the epithelium lining the ciliary processes. Secretion begins with active transport of sodium ions into the spaces between the epithelial cells. The sodium ions in turn pull chloride and bicarbonate ions along with them to maintain electrical neutrality. Then all these ions together cause osmosis of water from the sublying tissue into the same epithelial intercellular spaces, and the resulting solution washes from the spaces onto the surfaces of the ciliary processes. In addition, several nutrients are transported across the epithelium by active transport or facilitated diffusion; they include amino acids, ascorbic acid, and glucose.

Outflow of Aqueous Humor from the Eye

After aqueous humor is formed by the ciliary processes, it flows, as shown in Figure 49–19, *between the ligaments of the lens* and then *through the pupil into the anterior chamber of the eye.* Here, the fluid flows into the *angle between the cornea and the iris* and then through a meshwork of *trabeculae,* finally entering the *canal of Schlemm,* which empties into extraocular veins. Figure 49–21 demonstrates the anatomical structures at this iridocorneal angle, showing that the spaces between the trabeculae extend all the way from the anterior chamber to the canal of Schlemm. The canal of Schlemm in turn is a thin-walled vein that extends circumferentially all the way around the eye. Its endothelial membrane is so porous that even large protein molecules as well as small particulate matter up to the size of red blood cells can pass from the anterior chamber into the canal of Schlemm. Even though the canal of Schlemm is actually a venous blood vessel, so much aqueous humor normally flows into it that it is filled only with aqueous humor rather than with blood. Also, the small veins that lead from the canal of Schlemm to the larger veins of the eye usually contain only aqueous humor, and they are called *aqueous veins.*

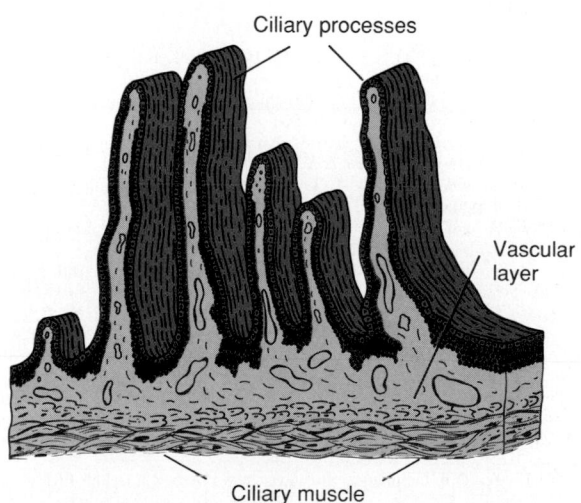

Figure 49–20. Anatomy of the ciliary processes.

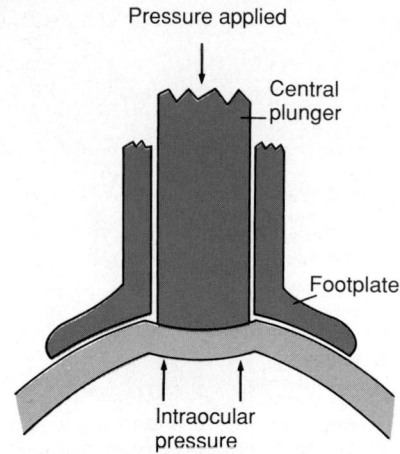

Figure 49–22. Principles of the tonometer.

Intraocular Pressure

The average normal intraocular pressure is about 15 mm Hg, with a range from 12 to 20.

TONOMETRY. Because it is impractical to pass a needle into a patient's eye for measurement of intraocular pressure, this pressure is measured clinically by means of a tonometer, the principle of which is shown in Figure 49–22. The cornea of the eye is anesthetized with a local anesthetic, and the footplate of the tonometer is placed on the cornea. A small force is then applied to a central plunger, causing the part of the cornea beneath the plunger to be displaced inward. The amount of displacement is recorded on the scale of the tonometer, and this in turn is calibrated in terms of intraocular pressure.

REGULATION OF INTRAOCULAR PRESSURE. Intraocular pressure remains constant in the normal eye, normally within about ±2 mm Hg. The level of this pressure is determined mainly by the resistance to outflow of aqueous humor from the anterior chamber into the canal of Schlemm. This outflow resistance results from the meshwork of trabeculae through which the fluid must percolate on its way from the lateral angles of the anterior chamber to the wall of the canal of Schlemm. These trabeculae have minute openings of only 2 to 3 micrometers. The rate of fluid flow into the canal increases markedly as the pressure rises. At about 15 mm Hg in the normal eye, the amount of fluid leaving the eye by way of the canal of Schlemm averages 2.5 μl/min and equals the inflow of fluid from the ciliary body. Therefore, the pressure normally remains at about this level of 15 mm Hg.

CLEANSING OF THE TRABECULAR SPACES AND OF THE INTRAOCULAR FLUID. When large amounts of debris occur in the aqueous humor, as occurs after hemorrhage into the eye or during intraocular infection, the debris is likely to accumulate in the trabecular spaces leading to the canal of Schlemm; this debris can prevent adequate reabsorption of fluid from the anterior chamber, sometimes causing glaucoma, as explained subsequently. However, on the surfaces of the trabecular plates are large numbers of phagocytic cells. Also, immediately outside the canal of Schlemm is a layer of interstitial gel that contains large numbers of reticuloendothelial cells that have an extremely high capacity for both engulfing debris and degrading it into small molecular substances that can then be absorbed. Thus, this phagocytic system keeps the trabecular spaces cleaned.

In addition, the surface of the iris and other surfaces of the eye behind the iris are covered with an epithelium that is capable of phagocytizing proteins and small particles from the aqueous humor, thereby helping to maintain a clear fluid.

GLAUCOMA, A PRINCIPAL CAUSE OF BLINDNESS. Glaucoma is one of the most common causes of blindness. It is a disease of the eye in which the intraocular pressure becomes pathologically high, sometimes rising acutely to 60 to 70 mm Hg. Pressures rising above 20 to 30 mm Hg can cause loss of vision when maintained for long periods. Extremely high pressures can cause blindness within days or even hours. As the pressure rises, the axons of the optic nerve are compressed where they leave the eyeball at the optic disc. This compression is believed to block axonal flow of cytoplasm from the neuronal cell bodies in the retina to the extended optic nerve fibers entering the brain. The result is lack of appropriate nutrition of the fibers, which eventually causes death of the involved neurons. It is possible that compression of the retinal artery, which also enters the eyeball at the optic disc, adds as well to the neuronal damage by reducing nutrition to the retina.

In most cases of glaucoma, the abnormally high pressure results from increased resistance to fluid outflow through the trabecular spaces into the canal of Schlemm at the iridocorneal junction. For instance, in acute eye inflammation, white blood cells and tissue debris can block these trabecular spaces and cause acute increase in intraocular pressure. In chronic conditions, especially in older age, fibrous occlusion of the trabecular spaces appears to be the likely culprit.

Glaucoma can sometimes be treated by placing drops in the eye that contain a drug that diffuses into the eyeball and reduces secretion or increases absorption of aqueous humor. When drug therapy fails, operative techniques to open the spaces of the trabeculae or to make channels directly between the fluid space of the eyeball and the subconjunctival space outside the eyeball can often effectively reduce the pressure.

REFERENCES

Abrams, D.: Duke-Elder's Practice of Refraction. 10th Ed. New York, Churchill Livingstone, 1993.

Albert, D. M., and Jakobiec, F. A.: Principles and Practice of Ophthalmology. Philadelphia, W. B. Saunders Co., 1994.

Apple, D. J., et al.: Intraocular Lenses: Evolution, Design, Complications and Pathology. Baltimore, Williams & Wilkins, 1988.

Bartley, G. B., and Liesegang, T. J.: Essentials of Ophthalmology. Philadelphia, J. B. Lippincott, 1992.

Bennett, E. S., and Henry, V. A.: Clinical Manual of Contact Lenses. Philadelphia, J. B. Lippincott, 1994.

Bentley, P. J.: The crystalline lens of the eye: An optical microcosm. News in Physiol. Sci., 1:195, 1986.

Bill, A.: Circulation in the eye. In Renkin, E. M., and Michel, C. C. (eds.): Handbook of Physiology. Sec. 2, Vol. IV. Bethesda, Md., American Physiological Society, 1984, p. 1001.

Caldwell, D. R.: Cataracts. New York, Raven Press, 1988.

Catalano, R. A., and Nelson, L. B.: Pediatric Ophthalmology: A Text/Atlas. Redding, MA, Appleton & Lange, 1994.

Cavanagh, H. D.: The Cornea: Transactions of the World Congress on the Cornea III. New York, Raven Press, 1988.

Chawla, H. B.: Ophthalmology, 2nd Ed. New York, Churchill Livingstone, 1994.

Collins, R., and Van der Werff, T. J.: Mathematical Models of the Dynamics of the Human Eye. New York, Springer-Verlag, 1980.

Coster, D.: Physics for Ophthalmologists. New York, Churchill Livingstone, 1994.

Davson, H.: Physiology of the Eye. Hightstown, NJ, McGraw-Hill, 1990.

Drance, S. M., et al.: Pharmacology of Glaucoma. Baltimore, Williams & Wilkins, 1992.

Duncan, G., and Jacob, T. J.: Calcium and the physiology of cataract. Ciba Found. Symp., 106:132, 1984.

Elliot, R. H.: A Treatise on Glaucoma. Huntington, NY, R. E. Krieger, 1979.

Eskridge, J. B., et al.: Clinical Procedures in Optometry. Philadelphia, J. B. Lippincott, 1991.

Fischbarg, J., and Lim, J. J.: Fluid and electrolyte transports across corneal endothelium. Curr. Top. Eye Res., 4:201, 1984.

Fraunfelder, F. T., and Roy, H.: Current Ocular Therapy 4. Philadelphia, W. B. Saunders Co., 1994.

Frisén, L.: Clinical Tests of Vision. New York, Raven Press, 1990.

Guyton, D. L.: Sights and Sounds in Ophthalmology: Ocular Motility and Binocular Vision. St. Louis, C. V. Mosby Co., 1989.

Jaffe, N. S.: Cataract Surgery and Its Complications. St. Louis, C. V. Mosby Co., 1983.

Kaufman, H. E., et al.: Corneal and Refractive Surgery. Philadelphia, J. B. Lippincott, 1991.

Kavner, R. S., and Dusky, L.: Total Vision. New York, A & W Publishers, 1980.

Koretz, J. F., and Handelman, G. H.: How the human eye focuses. Sci. Am., July, 1988, p. 92.

Kratz, R. P., et al.: Cataracts. Philadelphia, J. B. Lippincott, 1991.

Kuszak, J. R., et al.: Sutures of the crystalline lens: A review. Scan. Electron Miscrosc. 3:1369, 1984.

Lee, J. R.: Contact Lens Handbook. Philadelphia, W. B. Saunders Co., 1986.

Leydhecker, W., and Krieglstein, G. K. (eds.): Recent Advances in Glaucoma. New York, Springer-Verlag, 1979.

Margo, C. E., et al.: Diagnostic Problems in Clinical Ophthalmology. Philadelphia, W. B. Saunders Co., 1994.

Michaels, D. D.: Basic Refraction Techniques. New York, Raven Press, 1988.

Minckler, D. S., et al.: Glaucoma. Philadelphia, J. B. Lippincott, 1992.

Moses, R. A.: Adler's Physiology of the Eye; Clinical Application. 7th Ed. St. Louis, C. V. Mosby, 1981.

Ostler, H. B., and Ostler, M. W.: Diseases of the External Eye and Adnexa. Baltimore, Williams & Wilkins, 1993.

Piatigorsky, J.: Lens crystallins and their genes: diversity and tissue-specific expression. FASEB J., 3:1933, 1989.

Ritch, R., et al. (eds.): The Glaucomas. St. Louis, C. V. Mosby Co., 1989.

Roth, H. W., and Roth-Wittig, M.: Contact Lenses. Hagerstown, MD, Harper & Row, 1980.

Safir, A. (ed.): Refraction and Clinical Optics. Hagerstown, MD, Harper & Row, 1980.

Scheiman, M., and Wick, B.: Clinical Management of Binocular Vision. Philadelphia, J. B. Lippincott, 1993.

Shields, M. B.: Textbook of Glaucoma. Baltimore, Williams & Wilkins, 1992.

Stenson, S. M.: Contact Lenses: Guide to Selection, Fitting, and Management of Complications. East Norwalk, CT, Appleton & Lange, 1987.

Varma, R., and Spaeth, G. L.: The Optic Nerve in Glaucoma. Philadelphia, J. B. Lippincott, 1992.

Vaughan, D., et al.: General Ophthalmology. 14th Ed. Redding, MA, Appleton & Lange, 1995.

Whitnall, S. E.: The Anatomy of the Human Orbit and Accessory Organs of Vision. Huntington, NY, R. E. Krieger Publishing Co., 1979.

Wiederholt, M.: Ion transport by the cornea. News in Physiol. Sci., 3:97, 1988.

Wills Eye Hospital: The Wills Eye Manual. Philadelphia, J. B. Lippincott, 1993.

Yellott, J. I., Jr., et al.: The beginnings of visual perception: The retinal image and its initial encoding. In Darian-Smith, I. (ed.): Handbook of Physiology. Sec. 1, Vol. III. Bethesda, Md., American Physiological Society, 1984, p. 257.

The Eye: II. Receptor and Neural Function of the Retina

CHAPTER 50

The retina is the light-sensitive portion of the eye that contains the cones, which are responsible for color vision, and the rods, which are mainly responsible for vision in the dark. When the rods and cones are excited, signals are transmitted through successive neurons in the retina itself and, finally, into the optic nerve fibers and cerebral cortex. The purpose of this chapter is to explain specifically the mechanisms by which the rods and cones detect both white and colored light and then convert the visual image into optic nerve signals.

ANATOMY AND FUNCTION OF THE STRUCTURAL ELEMENTS OF THE RETINA

LAYERS OF THE RETINA. Figure 50–1 shows the functional components of the retina arranged in layers from the outside to the inside as follows: (1) pigment layer, (2) layer of rods and cones projecting into the pigment, (3) outer limiting membrane, (4) outer nuclear layer containing the cell bodies of the rods and cones, (5) outer plexiform layer, (6) inner nuclear layer, (7) inner plexiform layer, (8) ganglionic layer, (9) layer of optic nerve fibers, and (10) inner limiting membrane.

After light passes through the lens system of the eye and then through the vitreous humor, it enters the retina from the inside (see Figure 50–1); that is, it passes first through the ganglion cells and then through the plexiform layers, nuclear layer, and limiting membranes before it finally reaches the layer of rods and cones located all the way on the outer side of the retina. This distance is a thickness of several hundred micrometers; visual acuity is decreased by this passage through such nonhomogeneous tissue. However, in the central region of the retina, as is discussed below, the inside layers are pulled aside for prevention of this loss of acuity.

FOVEAL REGION OF THE RETINA AND ITS IMPORTANCE IN ACUTE VISION. A minute area in the center of the retina, shown in Figure 50–2, called the *fovea* and occupying a total area of a little more than 1 square millimeter, is especially capable of acute and detailed vision. The central portion of the fovea, only 0.3 millimeter in diameter, is called the *central fovea;* this area is composed entirely of cones, and its cones also have a special structure that aids their detection of detail in the visual image, especially a long slender body, in contradistinction to much fatter cones located further peripherally in the retina. Also, in this region, the blood vessels, the ganglion cells, the inner nuclear layer of cells, and the plexiform layers are all displaced to one side rather than resting directly on top of the cones. This allows light to pass unimpeded to the cones.

RODS AND CONES. Figure 50–3 is a diagrammatic representation of the essential components of a photoreceptor (either a rod or a cone). As shown in Figure 50–4, the outer segment of the cones is conical in shape. In general, the rods are narrower and longer than the cones, but this is not always the case. In the peripheral portions of the retina, the rods are 2 to 5 micrometers in diameter, whereas the cones are 5 to 8 micrometers in diameter; in the central part of the retina, in the fovea, the cones are slender and have a diameter of only 1.5 micrometers.

To the right in Figure 50–3 are labeled the major functional segments of either a rod or a cone: (1) the *outer segment,* (2) the *inner segment,* (3) the *nucleus,* and (4) the *synaptic body.* In the outer segment, the light-sensitive photochemical is found. In the case of the rods, this is *rhodopsin,* and in the cones, it is one of three "color" photochemicals, usually called simply *color*

OUTSIDE

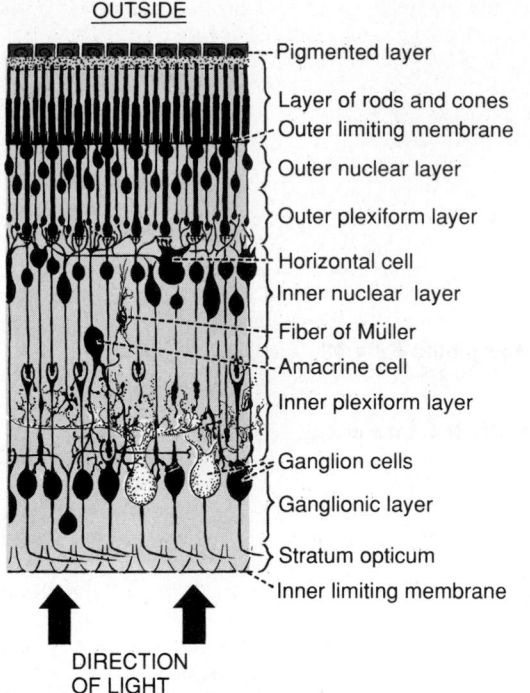

Pigmented layer
Layer of rods and cones
Outer limiting membrane
Outer nuclear layer
Outer plexiform layer
Horizontal cell
Inner nuclear layer
Fiber of Müller
Amacrine cell
Inner plexiform layer
Ganglion cells
Ganglionic layer
Stratum opticum
Inner limiting membrane

DIRECTION
OF LIGHT

Figure 50–1. Plan of the retinal neurons. (Modified from Polyak: The Retina. ©1941 by The University of Chicago. All rights reserved.)

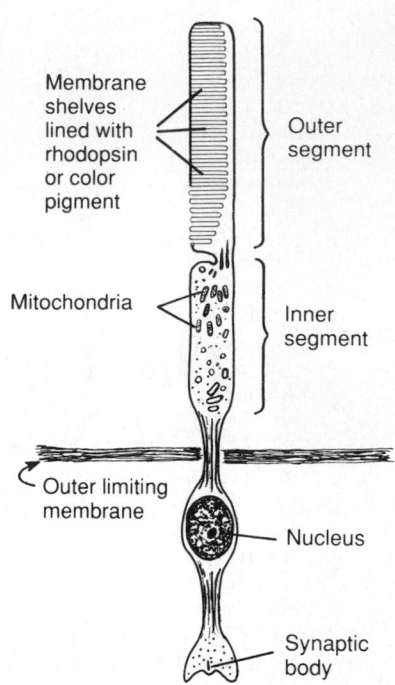

Membrane shelves lined with rhodopsin or color pigment

Outer segment

Mitochondria

Inner segment

Outer limiting membrane

Nucleus

Synaptic body

Figure 50–3. Schematic drawing of the functional parts of the rods and cones.

pigments, that function almost exactly the same as rhodopsin except for differences in spectral sensitivity.

Note in Figures 50–3 and 50–4 the large numbers of discs in both the rods and the cones. In the cones, each of the discs is actually an infolded shelf of cell membrane; in the rods, this is also true near the base of the rod. However, toward the tip of the rod, the discs separate from the membrane and are flat sacs lying totally inside the cell. There are as many as 1000 discs in each rod or cone.

Both rhodopsin and the color pigments are conju-

gated proteins. They are incorporated into the membranes of the discs in the form of transmembrane proteins. The concentrations of these photosensitive pigments in the discs are so great that the pigments themselves constitute about 40 per cent of the entire mass of the outer segment.

The inner segment contains the usual cytoplasm of the cell with the usual cytoplasmic organelles. Particularly important are the mitochondria; we see later that the mitochondria in this segment play the important role in providing the energy for function of the photoreceptors.

The synaptic body is the portion of the rod or cone

Figure 50–2. Photomicrograph of the macula and of the fovea in its center. Note that the inner layers of the retina are pulled to the side to decrease the interference with light transmission. (From Fawcett: Bloom and Fawcett: A Textbook of Histology. 11th ed. Philadelphia, W. B. Saunders Company, 1986; courtesy of H. Mizoguchi.)

that connects with the subsequent neuronal cells, the horizontal and bipolar cells, that represent the next stages in the vision chain.

PIGMENT LAYER OF THE RETINA. The black pigment *melanin* in the pigment layer prevents light reflection throughout the globe of the eyeball; this is extremely important for clear vision. This pigment performs the same function in the eye as the black coloring inside the bellows of a camera. Without it, light rays would be reflected in all directions within the eyeball and would cause diffuse lighting of the retina rather than the normal contrast between dark and light spots required for formation of precise images.

The importance of melanin in the pigment layer and choroid is well illustrated by its absence in *albinos,* people who are hereditarily lacking in melanin pigment in all parts of their bodies. When an albino enters a bright area, light that impinges on the retina is reflected in all directions by the white unpigmented surfaces of the retina and underlying sclera, so that a single discrete spot of light that would normally excite only a few rods or cones is reflected everywhere and excites many of the receptors. Therefore, the visual acuity of albinos, even with the best of optical correction, is seldom better than 20/100 to 20/200.

The pigment layer also stores large quantities of *vitamin A.* This vitamin A is exchanged back and forth through the membranes of the outer segments of the rods and cones, which themselves are embedded in the pigment layers. We shall see later that vitamin A is an important precursor of the photosensitive pigments and

that this interchange of vitamin A is important for adjustment of the light sensitivity of the receptors.

THE BLOOD SUPPLY OF THE RETINA—THE CENTRAL RETINAL ARTERY AND THE CHOROID. The nutrient blood supply for the internal layers of the retina is derived from the central retinal artery, which enters the eyeball along with the optic nerve and then divides to supply the entire inside retinal surface. Thus, to a great extent, the retina has its own blood supply independent of the other structures of the eye.

However, the outermost layer of the retina is adherent to the *choroid,* which is a highly vascular tissue between the retina and the sclera. Also, the outer layers of the retina, especially the outer segments of the rods and cones, depend mainly on diffusion from the choroid vessels for their nutrition, especially for their oxygen.

Retinal Detachment. The neural retina occasionally detaches from the pigment epithelium. In some instances, the cause of such detachment is injury to the eyeball that allows fluid or blood to collect between the retina and the pigment epithelium, but often it is also caused by contracture of fine collagenous fibrils in the vitreous humor, which pull the retina unevenly toward the interior of the globe.

Partly because of diffusion across the detachment gap and partly because of the independent blood supply to the retina through the retinal artery, the detached retina can resist degeneration for days and can become functional once again if surgically replaced in its normal relation with the pigment epithelium. But if not replaced soon, the retina finally is destroyed and is then unable to function even after surgical repair.

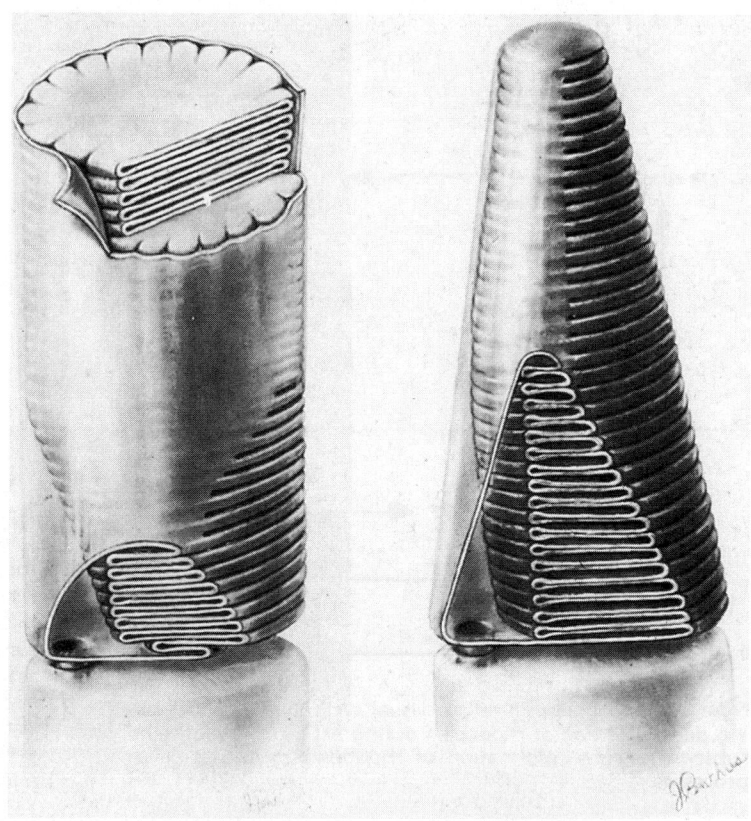

Figure 50–4. Membranous structures of the outer segments of a rod *(left)* and a cone *(right).* (Courtesy of Dr. Richard Young.)

PHOTOCHEMISTRY OF VISION

Both the rods and cones contain chemicals that decompose on exposure to light and, in the process, excite the nerve fibers leading from the eye. The chemical in the *rods* is called *rhodopsin;* the light-sensitive chemicals in the *cones,* called *cone pigments,* have compositions only slightly different from that of rhodopsin.

In this section, we discuss principally the photochemistry of rhodopsin, but we can apply almost exactly the same principles to the cone pigments.

Rhodopsin-Retinal Visual Cycle, and Excitation of the Rods

RHODOPSIN AND ITS DECOMPOSITION BY LIGHT ENERGY. The outer segment of the rod that projects into the pigment layer of the retina has a concentration of about 40 per cent of the light-sensitive pigment called *rhodopsin,* or *visual purple.* This substance is a combination of the protein *scotopsin* and the carotenoid pigment *retinal* (also called "retinene"). Furthermore, the retinal is a particular type called 11-*cis* retinal. This *cis* form of the retinal is important because only this form can bind with scotopsin to synthesize rhodopsin.

When light energy is absorbed by rhodopsin, the rhodopsin begins within trillionths of a second to decompose, as shown at the top of Figure 50–5. The cause of this is photoactivation of electrons in the retinal portion of the rhodopsin, which leads to an instantaneous change (on the order of trillionths of a second) of the *cis* form of retinal into an all-*trans*

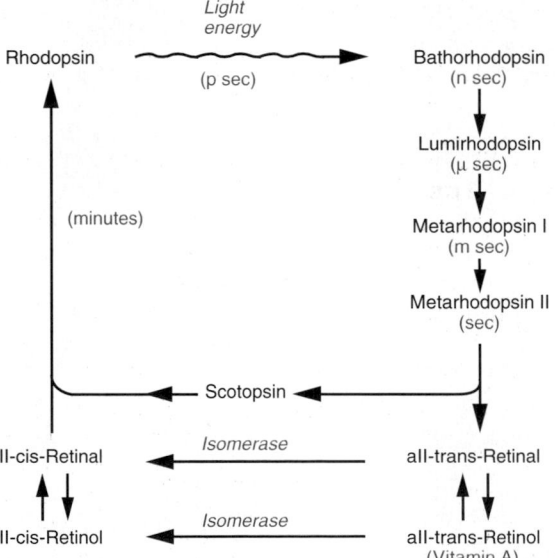

Figure 50–5. Rhodopsin-retinal visual cycle in the rod, showing decomposition of rhodopsin during exposure to light and subsequent slow reformation of rhodopsin by the chemical processes.

form, which still has the same chemical structure as the *cis* form but has a different physical structure—a straight molecule rather than an angulated molecule. Because the three-dimensional orientation of the reactive sites of the all-*trans* retinal no longer fits with the orientation of the reactive sites on the protein scotopsin, it begins to pull away from the scotopsin. The immediate product is *bathorhodopsin,* which is a partially split combination of the all-*trans* retinal and scotopsin. Bathorhodopsin itself is an extremely unstable compound and decays in nanoseconds to *lumirhodopsin.* This then decays in microseconds to *metarhodopsin I,* then in about a millisecond to *metarhodopsin II,* and, finally, much more slowly (in seconds) into the completely split products: *scotopsin* and *all-trans retinal.* It is the metarhodopsin II, also called *activated rhodopsin,* that excites electrical changes in the rods that then transmit the visual image into the central nervous system, as we discuss later.

REFORMATION OF RHODOPSIN. The first stage in reformation of rhodopsin, as shown in Figure 50–5, is to reconvert the all-*trans* retinal into 11-*cis* retinal. This process requires metabolic energy and is catalyzed by the enzyme *retinal isomerase.* Once the 11-*cis* retinal is formed, it automatically recombines with the scotopsin to re-form rhodopsin, which then remains stable until its decomposition is again triggered by absorption of light energy.

ROLE OF VITAMIN A IN THE FORMATION OF RHODOPSIN. Note in Figure 50–5 that there is a second chemical route by which all-*trans* retinal can be converted into 11-*cis* retinal. This is by conversion of the all-*trans* retinal first into *all-trans retinol,* which is one form of vitamin A. Then, the all-*trans* retinol is converted into 11-*cis* retinol under the influence of the enzyme isomerase. And, finally, the 11-*cis* retinol is converted into 11-*cis* retinal that combines with scotopsin to form rhodopsin.

Vitamin A is present both in the cytoplasm of the rods and in the pigment layer of the retina. Therefore, vitamin A is normally always available to form new retinal when needed. On the other hand, when there is excess retinal in the retina, the excess is converted back into vitamin A, thus reducing the amount of light-sensitive pigment in the retina. We shall see later that this interconversion between retinal and vitamin A is especially important in long-term adaptation of the retina to different light intensities.

NIGHT BLINDNESS. Night blindness occurs in any person with severe vitamin A deficiency. The simple reason for this is that not enough vitamin A is then available to form adequate quantities of retinal. Therefore, the amount of rhodopsin that can be formed becomes severely depressed. This condition is called night blindness because the amount of light available at night is then too little to permit adequate vision, although in daylight the cones can still be excited despite reduction of their color pigments as well.

For night blindness to occur, a person usually must remain on a vitamin A–deficient diet for months because large quantities of vitamin A are normally stored in the liver and can be made available to the eyes. Once night blindness does develop, sometimes it can be reversed in less than 1 hour by intravenous injection of vitamin A.

Excitation of the Rod When Rhodopsin Is Activated

THE ROD RECEPTOR POTENTIAL IS HYPERPOLARIZING, NOT DEPOLARIZING. When the rod is exposed to light, the resulting receptor potential is different from the receptor potentials in almost all other sensory receptors. That is, excitation of the rod causes *increased negativity* of the rod membrane potential, which is a state of *hyperpolarization.* This is exactly opposite to the decreased negativity (the process of "depolarization") that occurs in almost all other sensory receptors.

But how does activation of rhodopsin cause hyperpolarization? The answer is that *when rhodopsin decomposes, it decreases the membrane conductance for sodium ions in the outer segment of the rod.* This causes hyperpolarization of the entire rod membrane in the following way.

Figure 50–6 shows movement of sodium ions in a complete electrical circuit through the inner and outer segments of the rod. The inner segment continually pumps sodium from inside the rod to the outside, thereby creating a negative potential on the inside of the entire cell. However, the outer segment of the rod, where the photoreceptor discs are located, is entirely different: here the rod membrane, in the *dark* state, is very leaky to sodium ions. Therefore, sodium ions continually leak back to the inside of the rod and thereby neutralize much of the negativity on the inside of the entire cell. Thus, under normal dark conditions when the rod is not excited, there is a reduced amount of electronegativity inside the membrane of the rod, normally measuring about -40 millivolts rather than the more usual -70 to -80 millivolts found in most sensory receptors.

When the rhodopsin in the outer segment of the rod is exposed to light and begins to decompose, this *decreases* the outer segment conductance of sodium to the interior of the rod, even though sodium ions continue to be pumped out of the inner segment. Thus, more sodium ions now leave the rod than leak back in. Because they are positive ions, their loss from inside the rod creates increased negativity inside the membrane, and the greater the amount of light energy striking the rod, the greater the electronegativity becomes—that is, the greater the degree of *hyperpolarization.* At maximum light intensity, the membrane potential approaches -70 to -80 millivolts, which is near the equilibrium potential for potassium ions across the membrane.

Duration of the Receptor Potential and Logarithmic Relation of the Receptor Potential to Light Intensity. When a sudden pulse of light strikes the retina, the transient hyperpolarization that occurs in rods—that is, the receptor potential that occurs—reaches a peak in about 0.3 second and lasts for more than a second. In cones, these changes occur four times as fast. Therefore, a visual image impinged on the rods of the retina for only 1 millionth of a second nevertheless can cause the sensation of seeing the image sometimes for longer than a second.

Another characteristic of the receptor potential is that it is approximately proportional to the logarithm of the light intensity. This is exceedingly important because it allows the eye to discriminate light intensities through a range many thousand times as great as would be possible otherwise.

MECHANISM BY WHICH RHODOPSIN DECOMPOSITION *DECREASES* MEMBRANE SODIUM CONDUCTANCE—THE EXCITATION "CASCADE." Under optimal conditions, a single photon of light, the smallest possible quantal unit of light energy, can cause a measurable receptor potential in a rod of about 1 millivolt. Only 30 photons of light will cause half saturation of the rod. How can such a small amount of light cause such great excitation? The answer is that the photoreceptors have an extremely sensitive chemical cascade that amplifies the stimulatory effects about a millionfold, as follows:

1. The *photon activates an electron* in the *11-cis retinal* portion of the rhodopsin; this leads to the formation of *metarhodopsin II,* which is the active form of rhodopsin, as already discussed and shown in Figure 50–5.

2. The *activated rhodopsin* functions as an enzyme to activate many molecules of *transducin,* a protein present in

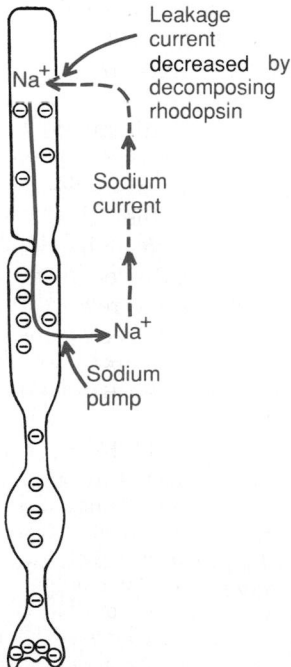

Figure 50–6. Theoretical basis for the generation of a hyperpolarization receptor potential caused by rhodopsin decomposition.

an inactive form in the membranes of the discs and cell membrane of the rod.

3. The *activated transducin* in turn activates many more molecules of *phosphodiesterase*.

4. *Activated phosphodiesterase* is another enzyme; it immediately hydrolyzes many, many molecules of *cyclic* guanosine monophosphate (cGMP), thus destroying it. Before being destroyed, the cGMP had been bound with the sodium channel protein of the rod's outer membrane in a way to "splint" it in the open state. But in light, when phosphodiesterase hydrolyzes the cGMP, this removes the splinting and causes the sodium channels to close. Several hundred channels close for each originally activated molecule of rhodopsin. Because the sodium flux through each of these channels has been extremely rapid, flow of more than a million sodium ions is blocked by the channel closure before the channel opens again. This diminishment of sodium ion flow is what excites the rod, as already discussed.

5. Within about a second, another enzyme, *rhodopsin kinase*, which is always present in the rod, inactivates the activated rhodopsin (the metarhodopsin II) and the entire cascade reverses back to the normal state with open sodium channels.

Thus, the rods have invented an important chemical cascade that amplifies the effect of a single photon of light to cause movement of millions of sodium ions. This explains the extreme sensitivity of the rods under dark conditions.

The cones are about 30 to 300 times less sensitive than the rods, but even this allows color vision in any light greater than very, very dim twilight.

Photochemistry of Color Vision by the Cones

It was pointed out at the outset of this discussion that the photochemicals in the cones have almost exactly the same chemical composition as that of rhodopsin in the rods. The only difference is that the protein portions, the opsins, called *photopsins* in the cones, are different from the scotopsin of the rods. The *retinal* portion of all the visual pigments is exactly the same in the cones as in the rods. The color-sensitive pigments of the cones, therefore, are combinations of retinal and photopsins.

In the discussion of color vision later in the chapter, it will become evident that one of three types of color pigments is present in each of the different cones, thus making the cones selectively sensitive to different colors: blue, green, and red. These color pigments are called, respectively, *blue-sensitive pigment, green-sensitive pigment,* and *red-sensitive pigment.* The absorption characteristics of the pigments in the three types of cones show peak absorbancies at light wavelengths, respectively, of 445, 535, and 570 nanometers. They are also the wavelengths for peak light sensitivity for each type of cone, which begins to explain how the retina differentiates the colors. The approximate absorption curves for these three pigments are shown in Figure 50–7. Also shown is the absorption curve for the rhodopsin of the rods, having a peak at 505 nanometers.

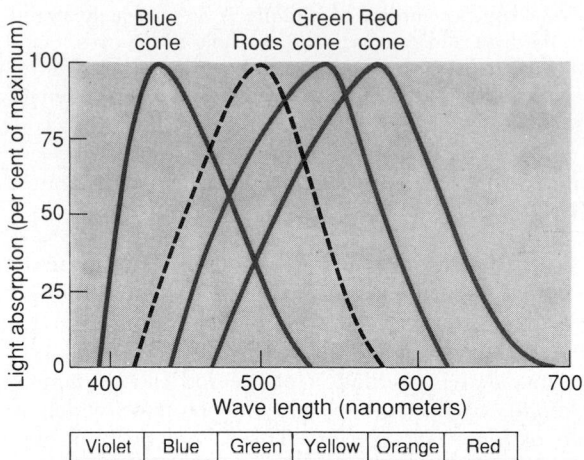

Figure 50–7. Light absorption by the respective pigments of the rods and the three color-receptive cones of the human retina. (Drawn from curves recorded by Marks, Dobelle, and MacNichol, Jr.: *Science, 143:*1181, 1964, and by Brown and Wald: *Science, 144:*45, 1964. Copyright 1964 by the American Association for the Advancement of Science.)

Automatic Regulation of Retinal Sensitivity—Light and Dark Adaptation

LIGHT AND DARK ADAPTATION. If a person has been in bright light for a long time, large proportions of the photochemicals in both the rods and the cones will have been reduced to retinal and opsins. Furthermore, much of the retinal of both the rods and the cones will have also been converted into vitamin A. Because of these two effects, the concentrations of the photosensitive chemicals remaining in the rods and cones are considerably reduced, and the sensitivity of the eye to light is correspondingly reduced. This is called *light adaptation.*

On the other hand, if the person remains in darkness for a long time, the retinal and opsins in the rods and cones are converted back into the light-sensitive pigments. Furthermore, vitamin A is reconverted back into retinal to give still additional light-sensitive pigments, the final limit being determined by the amount of opsins in the rods and cones. This is called *dark adaptation.*

Figure 50–8 shows the course of dark adaptation when a person is exposed to total darkness after having been exposed to bright light for several hours. Note that the sensitivity of the retina is very low on first entering the darkness, but within 1 minute, the sensitivity has already increased 10-fold—that is, the retina can respond to light of 1/10 the previously required intensity. At the end of 20 minutes, the sensitivity has increased about 6000-fold and at the end of 40 minutes, about 25,000-fold.

The resulting curve of Figure 50–8 is called the *dark adaptation curve.* Note, however, the inflection in the curve. The early portion of the curve is caused by adaptation of the cones because all the chemical events of vision, including adaptation, occur about four

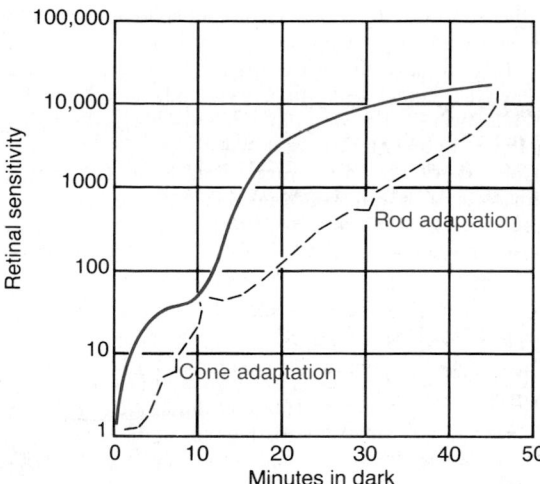

Figure 50–8. Dark adaptation, demonstrating the relation of cone adaptation to rod adaptation.

times as rapidly in cones as in rods. On the other hand, the cones do not achieve anywhere near the same degree of sensitivity change in darkness as the rods. Therefore, despite rapid adaptation, the cones cease adapting after only a few minutes, whereas the slowly adapting rods continue to adapt for many minutes and even hours, their sensitivity increasing tremendously. In addition, still more sensitivity of the rods is caused by convergence of 100 or more rods onto a single ganglion cell in the retina; these rods summate to increase their sensitivity, as discussed later in the chapter.

OTHER MECHANISMS OF LIGHT AND DARK ADAPTATION. In addition to adaptation caused by changes in concentrations of rhodopsin or color photochemicals, the eye has two other mechanisms for light and dark adaptation. The first of these is a *change in pupillary size,* as discussed in Chapter 49. This can cause adaptation of approximately 30-fold within a fraction of a second, because of changes in the amount of light allowed through the pupillary opening.

The other mechanism is *neural adaptation,* involving the neurons in the successive stages of the visual chain in the retina itself and in the brain. That is, when the light intensity first increases, the intensities of the signals transmitted by the bipolar cells, horizontal cells, amacrine cells, and ganglion cells are all intense. However, most of these signals decrease rapidly at different stages of transmission in the neural circuit. Although the degree of this adaptation is only a fewfold rather than the many thousand-fold that occurs during adaptation of the photochemical system, neural adaptation, like pupillary adaptation, occurs in a fraction of a second, in contrast to the many minutes to hours required for full adaptation by the photochemicals.

VALUE OF LIGHT AND DARK ADAPTATION IN VISION. Between the limits of maximal dark adaptation and maximal light adaptation, the eye can change its sensitivity to light as much as 500,000 to 1,000,000 times, the sensitivity automatically adjusting to changes in illumination.

Because the registration of images by the retina requires detection of both dark and light spots in the image, it is essential that the sensitivity of the retina always be adjusted so that the receptors respond to the lighter areas but not to the darker areas. An example of maladjustment of retinal adaptation occurs when a person leaves a movie theater and enters the bright sunlight. Then, even the dark spots in the images seem exceedingly bright, and as a consequence, the entire visual image is bleached, having little contrast between its different parts. This is poor vision, and it remains poor until the retina has adapted sufficiently for the darker areas of the image no longer to stimulate the receptors excessively.

Conversely, when a person enters darkness, the sensitivity of the retina is usually so slight that even the light spots in the image cannot excite the retina. After dark adaptation, the light spots begin to register. As an example of the extremes of light and dark adaptation, the intensity of sunlight is about 10 billion times that of starlight; yet the eye can function both in bright sunlight and, at least to some degree, in starlight.

COLOR VISION

From the preceding sections, we know that different cones are sensitive to different colors of light. This section is a discussion of the mechanisms by which the retina detects the different gradations of color in the visual spectrum.

Tricolor Mechanism of Color Detection

All theories of color vision are based on the well-known observation that the human eye can detect almost all gradations of colors when only red, green, and blue monochromatic lights are appropriately mixed in different combinations.

SPECTRAL SENSITIVITIES OF THE THREE TYPES OF CONES. On the basis of color vision tests, the spectral sensitivities of the three types of cones in human beings have been proved to be essentially the same as the light absorption curves for the three types of pigment found in the respective cones. They appear in Figure 50–7 and are also shown slightly differently in Figure 50–9. These curves can explain most of the phenomena of color vision.

INTERPRETATION OF COLOR IN THE NERVOUS SYSTEM. Referring to Figure 50–9, one can see that an orange monochromatic light with a wavelength of 580 nanometers stimulates the red cones to a stimulus value of about 99 (99 per cent of the peak stimulation at optimum wavelength); it stimulates the green cones to a stimulus value of about 42 but the blue cones not at all. Thus, the ratios of stimulation of the three types of cones in this instance are 99:42:0. The nervous

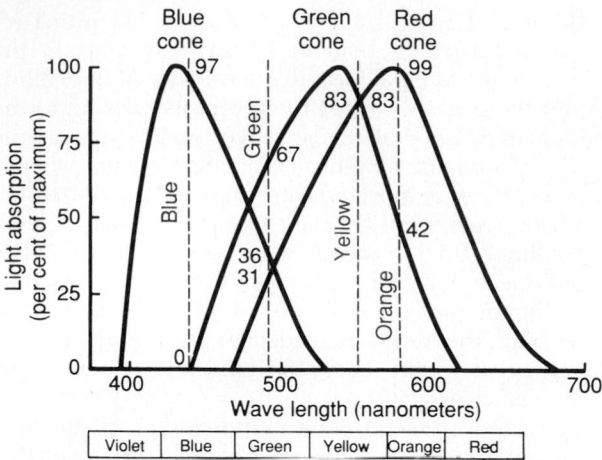

Figure 50–9. Demonstration of the degree of stimulation of the different color-sensitive cones by monochromatic lights of four colors: blue, green, yellow, and orange.

system interprets this set of ratios as the sensation of orange. On the other hand, a monochromatic blue light with a wavelength of 450 nanometers stimulates the red cones to a stimulus value of 0, the green cones cones to a value of 0, and the blue cones to a value of 97. This set of ratios—0:0:97—is interpreted by the nervous system as blue. Likewise, ratios of 83:83:0 are interpreted as yellow and 31:67:36, as green.

Perception of White Light. About equal stimulation of all the red, green, and blue cones gives one the sensation of seeing white. Yet there is no wavelength of light corresponding to white; instead, white is a combination of all the wavelengths of the spectrum. Furthermore, the sensation of white can be achieved by stimulating the retina with a proper combination of only three chosen colors that stimulate the respective types of cones about equally.

CHANGES IN THE COLOR OF AN ILLUMINATING LIGHT DO NOT ALTER THE PERCEIVED COLORS OF A VISUAL SCENE—THE PHENOMENON OF COLOR CONSTANCY. When Edwin Land was developing the Polaroid color camera, he noted that, within limits, changing the color of a light illuminating a scene will alter the hue of a color picture taken by the camera but does not significantly alter the hue of the scene as observed by the human eye. This phenomenon is called *color constancy;* to this day, it has not been completely explained. What is believed to occur is the following.

First, the brain computes from all the colors in the scene the overall hue of the entire vision. This computation is helped when some areas in the picture are known by the person to be white. Using this information of overall hue, the brain then adjusts mathematically for the changed color of the illuminating light. The exact neural mechanism for doing this has not been explained. Yet throughout the primary visual cortex of the brain are irregular peg-shaped blocks of cells, called

by the simple name "blobs," that display electrical signals consistent with color constancy when the illuminating light changes its wavelength. Therefore, it is believed that somewhere in the neighborhood of these blobs is located the computational machinery that allows for this phenomenon of color constancy.

Color constancy is of survival value to the foraging animal when it must distinguish nutritious food from poisonous plants both in bright sunlight and in the rose-colored dawn of day.

Color Blindness

RED-GREEN COLOR BLINDNESS. When a single group of color-receptive cones is missing from the eye, the person is unable to distinguish some colors from others. For instance, one can see in Figure 50–9 that green, yellow, orange, and red colors, which are the colors between the wavelengths of 525 and 675 nanometers, are normally distinguished one from the other by the red and green cones. If either of these two cones is missing, the person no longer can use this mechanism for distinguishing these four colors; the person is especially unable to distinguish red from green and, therefore, is said to have *red-green color blindness.*

The person with loss of red cones is called a *protanope;* his overall visual spectrum is noticeably shortened at the long wavelength end because of lack of the red cones. The color blind person who lacks green cones is called a *deuteranope;* this person has a perfectly normal visual spectral width because red cones are now available to detect the long wavelength red color.

Red-green color blindness is a genetic disease that occurs almost exclusively in males but is transmitted through the female. That is, genes in the female X chromosome code for the respective cones. Yet color blindness almost never occurs in the female because at least one of her two X chromosomes will almost always have a normal gene for each type of cone. But the male has only one X chromosome, so that a missing gene will lead to color blindness in him.

Because the X chromosome in the male is always inherited from the mother, never from the father, color blindness is passed from mother to son, and the mother is said to be a *color blindness* carrier; this is the case for about 8 per cent of all women.

BLUE WEAKNESS. Only rarely are blue cones missing, although sometimes they are under-represented, which is a state also genetically inherited, giving rise to the phenomenon called blue weakness.

COLOR TEST CHARTS. A rapid method for determining color blindness is based on the use of spot-charts such as those shown in Figure 50–10. These charts are arranged with a confusion of spots of several different colors. In the top chart, the normal person will read "74," whereas the red-green color blind person will read "21." In the bottom chart, the normal person will read "42," whereas the red blind "protanope" will read "2," and the green blind "deuteranope" will read "4."

If one will study these charts while at the same time observing the spectral sensitivity curves of the different cones in Figure 50–9, it can be readily understood how excessive emphasis can be placed on spots of certain colors by color blind people.

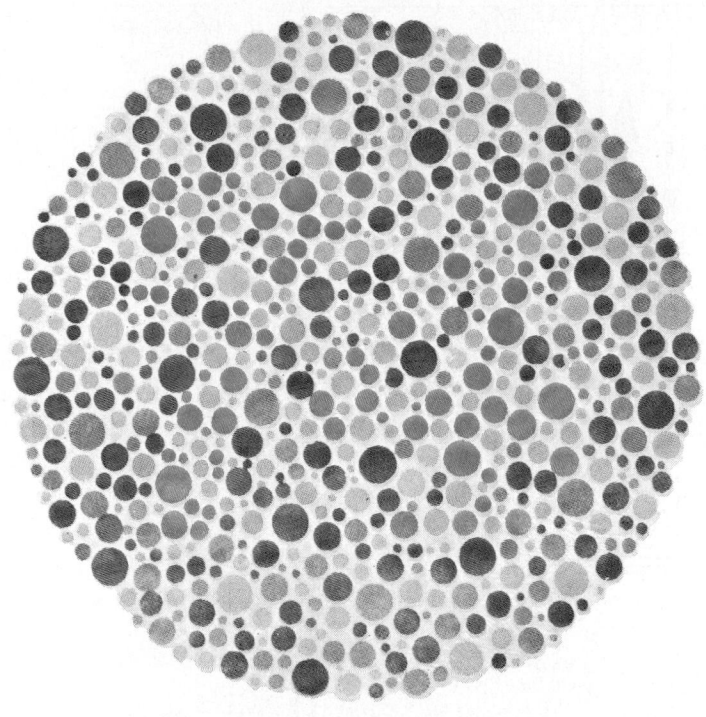

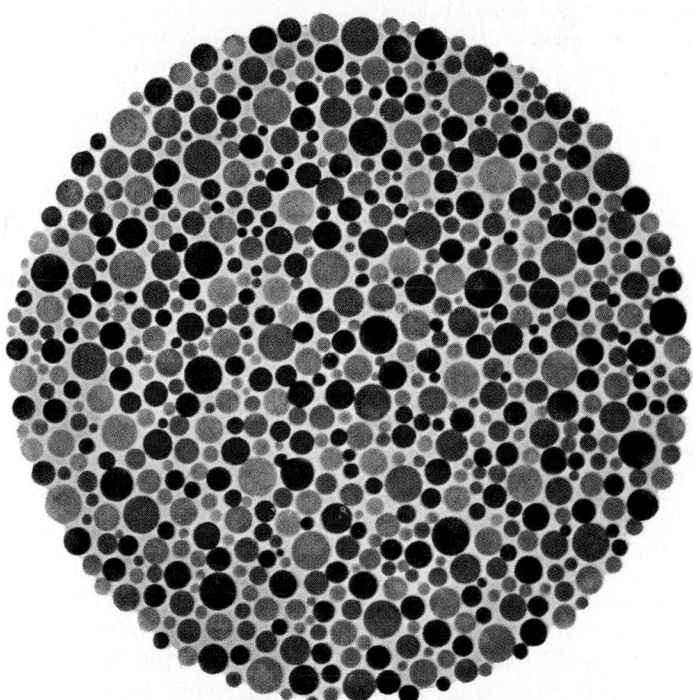

Figure 50–10. Two Ishihara charts. *Upper:* In this chart, the normal person reads "74," whereas the red-green color blind person reads "21." *Lower:* In this chart, the red-blind person (protanope) reads "2," whereas the green-blind person (deuteranope) reads "4." The normal person reads "42." (From Ishihara: Tests for Colour-Blindness. Tokyo, Kanehara and Co.)

NEURAL FUNCTION OF THE RETINA

Neural Circuitry of the Retina

The first figure of this chapter, Figure 50–1, shows the tremendous complexity of neural organization in the retina. To simplify this, Figure 50–11 presents the basic essentials of the retina's neural connections, showing on the left the circuit in the peripheral retina and at the right the circuit in the foveal retina. The different neuronal cell types are as follows:

1. The photoreceptors themselves: the *rods* and *cones*

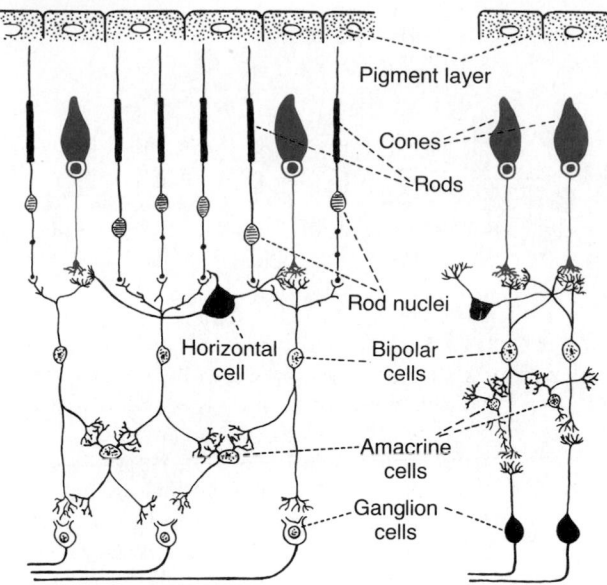

Figure 50–11. Neural organization of the retina: peripheral area to the left, foveal area to the right.

2. The *horizontal cells,* which transmit signals horizontally in the outer plexiform layer from the rods and cones to the bipolar cell dendrites

3. The *bipolar cells,* which transmit signals from the rods, cones, and horizontal cells to the inner plexiform layer, where they synapse with ganglion cells and amacrine cells

4. The *amacrine cells,* which transmit signals in two directions, either directly from bipolar cells to ganglion cells or horizontally within the inner plexiform layer between the axons of the bipolar cells, the dendrites of the ganglion cells, and/or other amacrine cells

5. The *ganglion cells,* which transmit output signals from the retina through the optic nerve into the brain

Still a sixth type of neuronal cell in the retina, not shown in the figure, is the *interplexiform* cell. This cell transmits signals in the retrograde direction from the inner plexiform layer to the outer plexiform layer. These signals are all inhibitory and believed to control the lateral spread of visual signals by the horizontal cells in the outer plexiform layer. Their role possibly is to help control the degree of contrast in the visual image.

THE VISUAL PATHWAY FROM THE CONES TO THE GANGLION CELLS IS DIFFERENT FROM THE ROD PATHWAY. As is true for many of our other sensory systems, the retina has both an old type of vision based on rod vision and a new type of vision based on cone vision. The neurons and nerve fibers that conduct the visual signals for cone vision are considerably larger than those that conduct the visual signals for rod vision, and the signals are conducted to the brain two to five times as rapidly. Also, the circuitries for the two systems are slightly different as follows.

To the far right in Figure 50–11 is shown the visual pathway from the foveal portion of the retina, repre-

senting the new, fast cone system. This shows three neurons in the direct pathway: (1) cones, (2) bipolar cells, and (3) ganglion cells. In addition, horizontal cells transmit inhibitory signals laterally in the outer plexiform layer, and amacrine cells transmit signals laterally in the inner plexiform layer.

To the left in Figure 50–11 is shown the neural connections for the peripheral retina where both rods and cones are present. Three bipolar cells are shown; the middle of these connects only to rods, representing the old visual system. In this case, the output from the bipolar cell passes only to amacrine cells, and these in turn relay the signals to the ganglion cells. Thus, for pure rod vision, there are four neurons in the direct visual pathway: (1) rods, (2) bipolar cells, (3) amacrine cells, and (4) ganglion cells. Also, both horizontal and amacrine cells provide lateral connectivity.

The other two bipolar cells shown in the peripheral retinal circuitry of Figure 50–11 connect with both rods and cones; the outputs of these bipolar cells pass both directly to ganglion cells and also by way of amacrine cells.

NEUROTRANSMITTERS RELEASED BY RETINAL NEURONS. The neurotransmitters used for synaptic transmission in the retina have not all been entirely delineated. However, both the rods and the cones release *glutamate* at their synapses with the bipolar cells. And histological and pharmacological studies have shown there to be many types of amacrine cells secreting at least eight types of transmitter substances, including *gamma-aminobutyric acid, glycine, dopamine, acetylcholine,* and *indolamine,* all of which normally function as inhibitory transmitters. The transmitters of the bipolar, horizontal, and interplexiform cells are unclear, but at least some of the synapses of horizontal cells involve electrical transmission rather than chemical transmission.

TRANSMISSION OF MOST SIGNALS OCCURS IN THE RETINAL NEURONS BY ELECTROTONIC CONDUCTION, NOT BY ACTION POTENTIALS. The only retinal neurons that always transmit visual signals by means of action potentials are the ganglion cells, and they send their signals all the way to the brain. Occasionally action potentials have also been recorded in amacrine cells, although the importance of these action potentials is questionable. Otherwise, all the retinal neurons conduct their visual signals by *electrotonic conduction,* which can be explained as follows.

Electrotonic conduction means direct flow of electric current, not action potentials, in the neuronal cytoplasm from the point of excitation all the way to the output synapses. In fact, even in the rods and cones, conduction from their outer segments where the visual signals are generated to the synaptic bodies is by electrotonic conduction. That is, when hyperpolarization occurs in response to light in the outer segment, almost the same degree of hyperpolarization is conducted by direct electric current flow to the synaptic body, and no action potential occurs. Then, when the transmitter from a rod or cone stimulates a bipolar cell or horizontal cell, once again the signal is transmitted from the input to the output by direct electric current flow, not by action potentials.

The importance of electrotonic conduction is that it allows *graded conduction* of signal strength. Thus, for the rods and cones, the strength of the hyperpolarizing output signal is directly related to the intensity of illumination; the signal is not all-or-none, as would be the case for action potential conduction.

Lateral Inhibition to Enhance Visual Contrast—Function of the Horizontal Cells

The horizontal cells, shown in Figure 50–11, connect laterally between the synaptic bodies of the rods and cones as well as with the dendrites of the bipolar cells. The outputs of the horizontal cells are always inhibitory. Therefore, this lateral connection provides the same phenomenon of lateral inhibition that is important in all other sensory systems, that is, helping to ensure transmission of visual patterns with proper visual contrast into the central nervous system. This phenomenon is demonstrated in Figure 50–12, which shows a minute spot of light focused on the retina. The visual pathway from the central-most area where the light strikes is excited, whereas the area to the side, called the "surround," is inhibited. In other words, instead of the excitatory signal spreading widely in the retina because of the spreading dendritic and axonal trees in the plexiform layers, transmission through the horizontal cells puts a stop to this by providing lateral inhibition in the surrounding area. This is an essential mechanism to allow high visual accuracy in transmitting contrast borders in the visual image. It is probable that some of the amacrine cells provide additional lateral inhibition and further enhancement of visual contrast in the inner plexiform layer of the retina as well.

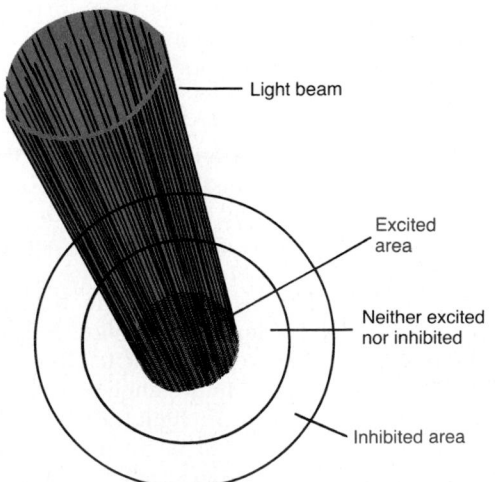

Figure 50–12. Excitation and inhibition of a retinal area caused by a small beam of light, demonstrating the principle of "lateral inhibition."

Excitation of Some Bipolar Cells and Inhibition of Others—The Depolarizing and Hyperpolarizing Bipolar Cells

Two types of bipolar cells provide opposing excitatory and inhibitory signals in the visual pathway, the *depolarizing bipolar cell* and the *hyperpolarizing bipolar cell.* That is, some bipolar cells depolarize when the rods and cones are excited and others hyperpolarize.

There are two possible explanations for this difference in response in the two types of bipolar cells. One explanation is that the two bipolar cells are of entirely different types, one responding to the glutamate neurotransmitter released by the rods and cones by depolarizing and the other responding by hyperpolarizing. The other possibility is that one of the bipolar cells receives direct excitation from the rods and cones, whereas the other receives its signal indirectly through a horizontal cell. Because the horizontal cell is an inhibitory cell, this would reverse the polarity of the electrical response.

Regardless of the mechanism for the two types of bipolar responses, the importance of this phenomenon is that it allows one half of the bipolar cells to transmit positive signals and the other half to transmit negative signals. We shall see later that both positive and negative signals are used in transmitting visual information to the brain.

Another importance of this reciprocal relation between depolarizing and hyperpolarizing bipolar cells is that it provides a second mechanism for lateral inhibition in addition to the horizontal cell mechanism. Because depolarizing and hyperpolarizing bipolar cells lie immediately against each other, this gives a mechanism for separating contrast borders in the visual image even when the border lies exactly between two adjacent photoreceptors. By contrast, the horizontal cell mechanism for lateral inhibition operates over a much greater distance.

Amacrine Cells and Their Functions

About 30 types of amacrine cells have been identified by morphological or histochemical means. The functions of a half dozen types of amacrine cells have been characterized, and all of them are different from one another. It is probable that other amacrine cells have many additional functions yet to be determined.

One type of amacrine cell is part of the direct pathway for rod vision—that is, from rod to bipolar cells to amacrine cells to ganglion cells.

Another type of amacrine cell responds strongly at the onset of a visual signal but the response dies out rapidly. Other amacrine cells respond strongly at the offset of visual signals, but again the response dies quickly. Still other amacrine cells respond both when a light is turned either on or off, signalling simply a change in illumination irrespective of direction.

Still another type of amacrine cell responds to

movement of a spot across the retina in a specific direction; therefore, these amacrine cells are said to be *directional sensitive.*

In a sense, then, amacrine cells are types of interneurons that help in the beginning analysis of visual signals before they ever leave the retina.

Ganglion Cells

CONNECTIVITY OF THE GANGLION CELLS WITH CONES IN THE FOVEA AND WITH RODS AND CONES IN THE PERIPHERAL RETINA. Each retina contains about 100 million rods and 3 million cones; yet the number of ganglion cells is only about 1.6 million. Thus, an average of 60 rods and 2 cones converge on each optic nerve fiber.

Major differences exist between the peripheral retina and the central retina. As one approaches the fovea, fewer rods and cones converge on each optic fiber, and the rods and cones both become more slender. These two effects progressively increase the acuity of vision toward the central retina. And in the very center, in the *central fovea,* there are only slender cones, about 35,000 of them, and no rods. Also, the number of optic nerve fibers leading from this part of the retina is almost equal to the number of cones, as shown to the right in Figure 50–11. This mainly explains the high degree of visual acuity in the central retina in comparison with much poorer acuity peripherally.

Another difference between the peripheral and central portions of the retina is a much greater sensitivity of the peripheral retina to weak light. This results partly from the fact that rods are 30 to 300 times more sensitive to light than are cones, but it is further magnified by the fact that as many as 200 rods converge on the same optic nerve fiber in the more peripheral portions of the retina, so that the signals from the rods summate to give even more intense stimulation of the peripheral ganglion cells.

Three Types of Retinal Ganglion Cells and Their Respective Fields

There are three distinct groups of ganglion cells designated as W, X, and Y cells. Each of these serves a different function.

TRANSMISSION OF ROD VISION BY THE W CELLS. The W cells, constituting about 40 per cent of all the ganglion cells, are small, having a diameter less than 10 micrometers and transmitting signals in their optic nerve fibers at the slow velocity of only 8 m/sec. These ganglion cells receive most of their excitation from rods, transmitted by way of small bipolar cells and amacrine cells. They have broad fields in the retina because their dendrites spread widely in the inner plexiform layer, receiving signals from broad areas.

On the basis of histology as well as physiological experiments, it appears that the W cells are especially sensitive for detecting directional movement anywhere in the field of vision, and they probably also are important for much of our crude rod vision under dark conditions.

TRANSMISSION OF THE VISUAL IMAGE AND COLOR BY THE X CELLS. The most numerous of the ganglion cells are the X cells, representing 55 per cent of the total. They are of medium diameter, between 10 and 15 micrometers, and transmit signals in their optic nerve fibers at about 14 m/sec.

The X cells have small fields because their dendrites do not spread widely in the retina. Because of this, the signals represent rather discrete retinal locations. Therefore, it is through the X cells that the visual image itself is mainly transmitted. Also, because every X cell receives input from at least one cone, X cell transmission is probably responsible for all color vision.

FUNCTION OF THE Y CELLS TO TRANSMIT INSTANTANEOUS CHANGES IN THE VISUAL IMAGE. The Y cells are the largest of all, up to 35 micrometers in diameter, and they transmit their signals to the brain as fast or faster than 50 m/sec. They are the fewest of all the ganglion cells, representing only 5 per cent of the total. Also, they have broad dendritic fields, so that signals are picked up by these cells from widespread retinal areas.

The Y ganglion cells respond like many of the amacrine cells to rapid changes in the visual image, either rapid movement or rapid change in light intensity, sending bursts of signals for only a fraction of a second. These ganglion cells undoubtedly apprise the central nervous system almost instantaneously when an abnormal visual event occurs anywhere in the visual field but without specifying with great accuracy the location of the event other than to give appropriate clues for moving the eyes toward the exciting vision.

Excitation of the Ganglion Cells

SPONTANEOUS, CONTINUOUS ACTION POTENTIALS IN THE GANGLION CELLS. It is from the ganglion cells that the long fibers of the optic nerve lead into the brain. Because of the distance involved, the electrotonic method of conduction is no longer appropriate; therefore, ganglion cells transmit their signals by means of action potentials instead. Furthermore, even when unstimulated, they still transmit continuous impulses at rates varying between 5 and 40 per second, with the nerve fibers from the large ganglion cells firing more rapidly. The visual signals in turn are superimposed onto this background ganglion cell firing.

TRANSMISSION OF CHANGES IN LIGHT INTENSITY— THE ON-OFF RESPONSE. Many ganglion cells are especially excited by *changes* in light intensity. This is demonstrated by the records of nerve impulses in Figure 50–13, showing in the upper panel rapid impulses for a fraction of a second when a light was first turned on, but the rapidity decreased in a fraction of a sec-

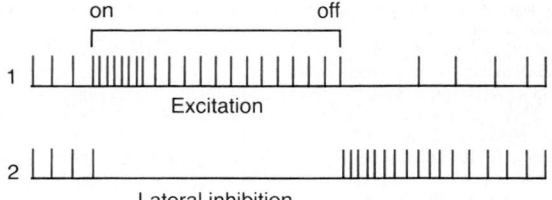

Figure 50–13. Responses of ganglion cells to light in (1) an area excited by a spot of light and (2) an area immediately adjacent to the excited spot; the ganglion cells in this area are inhibited by the mechanism of lateral inhibition. (Modified from Granit: Receptors and Sensory Perception: A Discussion of Aims, Means, and Results of Electrophysiological Research Into the Process of Reception. New Haven, Conn., Yale University Press, 1955.)

ond. The lower tracing is from a ganglion cell located laterally to the spot of light; this cell was markedly inhibited when the light was turned on, because of lateral inhibition. Then, when the light was turned off, the opposite effects occurred. Thus, these records are called "on-off" and "off-on" responses. The opposite directions of these responses to light are caused, respectively, by the depolarizing and hyperpolarizing bipolar cells, and the transient nature of the responses was probably at least partly generated by the amacrine cells, many of which have similar transient responses themselves.

This capability of the eyes to detect *change* in light intensity is strongly developed in the peripheral retina as well as in the central retina. For instance, a minute gnat flying across the peripheral field of vision is instantaneously detected. On the other hand, the same gnat sitting quietly remains below the threshold of visual detection.

Transmission of Signals Depicting Contrasts in the Visual Scene—The Role of Lateral Inhibition

Many of the ganglion cells do not respond to the actual level of illumination of the scene; instead, they respond mainly to contrast borders in the scene. Because it seems that this is the major means by which the pattern of the scene is transmitted to the brain, let us explain how this process occurs.

When flat light is applied to the entire retina—that is, when all the photoreceptors are stimulated equally by the incident light—the contrast type of ganglion cell is neither stimulated nor inhibited. The reason for this is that the signals transmitted *directly* from the photoreceptors through the depolarizing bipolar cells are excitatory, whereas the signals transmitted *laterally* through the hyperpolarizing bipolar cells as well as through the horizontal cells are inhibitory. Thus, the direct excitatory signal through one pathway is likely to

be neutralized by the inhibitory signals through the lateral pathways. One circuit for this is demonstrated in Figure 50–14, which shows three photoreceptors; the central one of these receptors excites a depolarizing bipolar cell. The two receptors on either side are connected to the same bipolar cell through inhibitory horizontal cells that neutralize the direct excitatory signal if these receptors are also stimulated by light.

Now, let us examine what happens when a contrast border occurs in the visual scene. Referring again to Figure 50–14, let us assume that the central photoreceptor is stimulated by a bright spot of light while one of the two lateral receptors is in the dark. The bright spot of light excites the direct pathway through the bipolar cell. Then, in addition, the fact that one of the lateral photoreceptors is in the dark causes one of the horizontal cells to be inhibited. In turn, this cell loses its inhibitory effect on the bipolar cell, and this allows still more excitation of the bipolar cell. Thus, where visual contrasts occur, the signals through the direct and lateral pathways accentuate one another.

In summary, the mechanism of lateral inhibition functions in the eye in the same way that it functions in most other sensory systems—that is, to provide contrast detection and enhancement.

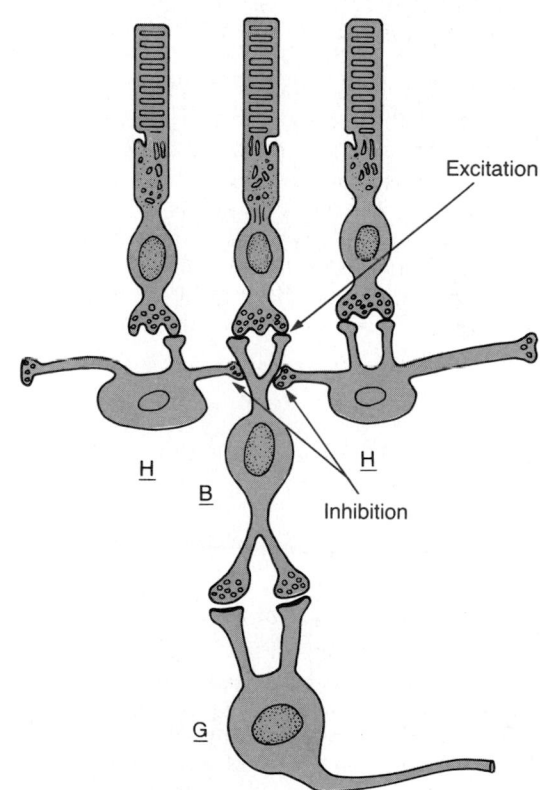

Figure 50–14. Typical arrangement of rods, horizontal cells (H), a bipolar cell (B), and a ganglion cell (G) in the retina, showing excitation at the synapses between the rods and the horizontal cells but inhibition between the horizontal cells and the bipolar cells.

Transmission of Color Signals by the Ganglion Cells

A single ganglion cell may be stimulated by a number of cones or by only a few. When all three types of cones—the red, blue, and green types—stimulate the same ganglion cell, the signal transmitted through the ganglion cell is the same for any color of the spectrum. Therefore, this signal plays no role in the detection of the different colors. Instead, it is a "white" signal.

On the other hand, some of the ganglion cells are excited by only one color type of cone but inhibited by a second type. For instance, this frequently occurs for the red and green cones, red causing excitation and green causing inhibition—or vice versa, with green causing excitation and red, inhibition. The same type of reciprocal effect occurs between blue cones on the one hand and a combination of red and green cones on the other hand, giving a reciprocal excitation inhibition relation between the blue and yellow colors.

The mechanism of this opposing effect of colors is the following: One color-type cone excites the ganglion cell by the direct excitatory route through a depolarizing bipolar cell, whereas the other color type inhibits the ganglion cell by the indirect inhibitory route through a hyperpolarizing bipolar cell.

The importance of these color-contrast mechanisms is that they represent a mechanism by which the retina itself begins to differentiate colors. Thus, each color-contrast type of ganglion cell is excited by one color but inhibited by the "opponent color." Therefore, color analysis begins in the retina and is not entirely a function of the brain.

REFERENCES

Albert, D. M., and Jakobiec, F. A.: Principles and Practice of Ophthalmology. Philadelphia, W. B. Saunders Co., 1994.

Bartley, G. B., Liesegang, T. J.: Essentials of Ophthalmology. Philadelphia, J. B. Lippincott, 1992.

Bloomfield, S. A.: Orientation-sensitive amacrine and ganglion cells in the rabbit retina. J. Neurophysiol., 71:1672, 1994.

Bloom, S. M., and Brucker, A. J.: Laser Surgery of the Posterior Segment. Philadelphia, J. B. Lippincott, 1991.

Bovino, J. A.: Macular Surgery. Redding, MA, Appleton & Lange, 1994.

Daw, N. W., et al.: The function of synaptic transmitters in the retina. Annu. Rev. Neurosci., 12:205, 1989.

DeValois, R. L., and Jacobs, G. H.: Neural mechanisms of color vision. In Darian-Smith, I. (ed.): Handbook of Physiology. Sec. 1, Vol. III. Bethesda, Md., American Physiological Society, 1984, p. 525.

Dowling, J. E., and Dubin, M. W.: The vertebrate retina. In Darian-Smith, I. (ed.): Handbook of Physiology. Sec. 1, Vol. III. Bethesda, Md., American Physiological Society, 1984, p. 317.

Finlay, B. L., and Sengelaub, D. R. (eds.): Development of the Vertebrate Retina. New York, Plenum Publishing Corp., 1989.

Freeman, W. R.: Practical Atlas of Retinal Disease and Therapy. New York, Raven Press, 1993.

Friedlaender, M. H.: Allergy and Immunology of the Eye. New York, Raven Press, 1993.

Hampton, G. R., and Nelsen, P. T.: Age-Related Macular Degeneration: Principles and Practice. New York, Raven Press, 1992.

Hillman, P., et al.: Transduction in invertebrate photoreceptors: Role of pigment bistability. Physiol. Rev., 63:668, 1983.

Huismans, H., et al.: The Photographed Fundus. Baltimore, Williams & Wilkins, 1988.

Iyer, P. V., and Rowland, R.: Ophthalmic Pathology. New York, Churchill Livingstone, 1994.

Liebman, P. A., et al.: The molecular mechanism of visual excitation and its relation to the structure and composition of the rod outer segment. Annu. Rev. Physiol., 49:765, 1987.

Margo, C. E., et al.: Diagnostic Problems in Clinical Ophthalmology. Philadelphia, W. B. Saunders Co., 1994.

Montgomery, G.: Seeing With the Brain. Discover, December 1988, p. 52.

Newsome, D. A.: Retinal Dystrophies and Degenerations. New York, Raven Press, 1988.

Olk, R. J., and Lee, C. M.: Diabetic Retinopathy: Practical Management. Philadelphia, J. B. Lippincott, 1993.

Owen, W. G.: Ionic conductances in rod photoreceptors. Annu. Rev. Physiol., 49:743, 1987.

Peyman, G. A., and Schulman, J. A.: Intravitreal Surgery: Principles and Practice. 2nd Ed. Redding, MA, Appleton & Lange, 1994.

Pugh, E. N., Jr.: The nature and identity of the internal excitational transmitter of vertebrate phototransduction. Annu. Rev. Physiol., 49:715, 1987.

Ryan, S. J., et al. (eds.): Retina. St. Louis, C. V. Mosby Co., 1989.

Smiddy, W. E., et al.: Retinal Surgery and Ocular Trauma. Philadelphia, J. B. Lippincott, 1994.

Sporn, M. B., et al.: The Retinoids: Biology, Chemistry, and Medicine. New York, Raven Press, 1994.

Stillman, A. J.: Current concepts in photoreceptor physiology. Physiologist, 28:122, 1985.

Wassle, H., and Boycott, B. B.: Functional Architecture of the Mammalian Retina. Physiol. Rev., 71:447, 1991.

Wu, S. M.: Synaptic Transmission in the Outer Retina. Annu. Rev. Phyiol., 56:141, 1994.

The Eye: III. Central Neurophysiology of Vision

CHAPTER 51

Visual Pathways

Figure 51–1 shows the principal visual pathways from the two retinas to the *visual cortex*. After nerve impulses leave the retinas, they pass backward through the *optic nerves*. At the *optic chiasm,* all the fibers from the nasal halves of the retinas cross to the opposite side, where they join the fibers from the opposite temporal retinas to form the *optic tracts.* The fibers of each optic tract synapse in the *dorsal lateral geniculate nucleus,* and from here, the *geniculocalcarine fibers* pass by way of the *optic radiation* (or *geniculocalcarine tract*) to the *primary visual cortex* in the calcarine area of the occipital lobe.

In addition, visual fibers pass to older areas of the brain: (1) from the optic tracts to the *suprachiasmatic nucleus of the hypothalamus,* presumably for controlling circadian rhythms; (2) into the *pretectal nuclei,* for eliciting reflex movements of the eyes to focus on objects of importance and for activating the pupillary light reflex; (3) into the *superior colliculus,* for control of rapid directional movements of the two eyes; and (4) into the *ventral lateral geniculate nucleus* of the thalamus and then into surrounding basal regions of the brain, presumably to help control some of the body's behavioral functions.

Thus, the visual pathways can be divided roughly into an *old system* to the midbrain and base of the forebrain and a *new system* for direct transmission into the visual cortex. The new system is responsible in humans for perception of virtually all aspects of visual form, colors, and other conscious vision. On the other hand, in many primitive animals, even visual form is detected by the older system, using the superior colliculus in the same manner that the visual cortex is used in mammals.

Function of the Dorsal Lateral Geniculate Nucleus

The optic nerve fibers of the new visual system all terminate in the *dorsal lateral geniculate nucleus,* located at the dorsal end of the thalamus and also called simply the *lateral geniculate body,* as shown in Figure 51–1. The dorsal lateral geniculate nucleus serves two principal functions: First, it relays visual information from the optic tract to the *visual cortex* by way of the *optic radiation* (also called the *geniculocalcarine tract*). This relay function is accurate, so much so that there is exact point-to-point transmission with a high degree of spatial fidelity all the way from the retina to the visual cortex.

It will be recalled that one half the fibers in each optic tract after passing the optic chiasm are derived from one eye and one half are derived from the other eye, representing corresponding points on the two retinas. However, the signals from the two eyes are kept apart in the dorsal lateral geniculate nucleus. This nucleus is composed of six nuclear layers. Layers II, III, and V (from ventral to dorsal) receive signals from the temporal portion of the ipsilateral retina, whereas layers I, IV, and VI receive signals from the nasal retina of the opposite eye. The respective retinal areas of the two eyes connect with neurons that are approximately superimposed over one another in the paired

651

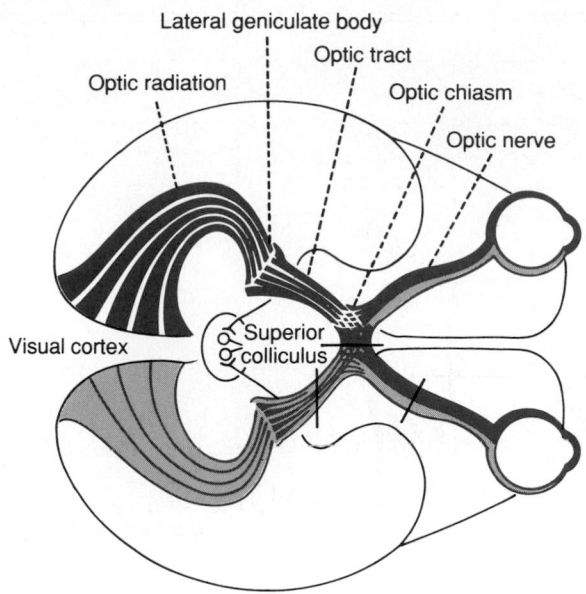

Figure 51–1. Principal visual pathways from the eyes to the visual cortex. (Modified from Polyak: The Retina. © 1941 by The University of Chicago. All rights reserved.)

layers, and similar parallel transmission is preserved all the way back to the visual cortex.

The second major function of the dorsal lateral geniculate nucleus is to "gate" the transmission of signals to the visual cortex—that is, to control how much of the signal is allowed to pass to the cortex. The nucleus receives gating control signals from two major sources: (1) from *corticofugal fibers* returning in a backward direction from the primary visual cortex to the lateral geniculate nucleus and (2) from the *reticular areas of the mesencephalon*. Both of these are inhibitory and, when stimulated, can literally turn off transmission through selected portions of the dorsal lateral geniculate nucleus. Therefore, it is assumed that both of these gating circuits help to control the visual information that is allowed to pass.

Finally, the dorsal lateral geniculate nucleus is divided in another way: (1) Layers I and II are called *magnocellular layers* because they contain large neurons. These receive their input almost entirely from the large type Y retinal ganglion cells. This magnocellular system provides a rapidly conducting pathway to the visual cortex. On the other hand, it is color blind, transmitting only black and white information. Also, its point-to-point transmission is poor because there are not so many Y ganglion cells, and their dendrites spread widely in the retina. (2) Layers III through VI are called *parvocellular layers* because they contain large numbers of small to medium-sized neurons. These neurons receive their input almost entirely from the type X retinal ganglion cells that (a) transmit color and (b) convey accurate point-to-point spatial information but at only a moderate velocity of conduction, rather than at high velocity.

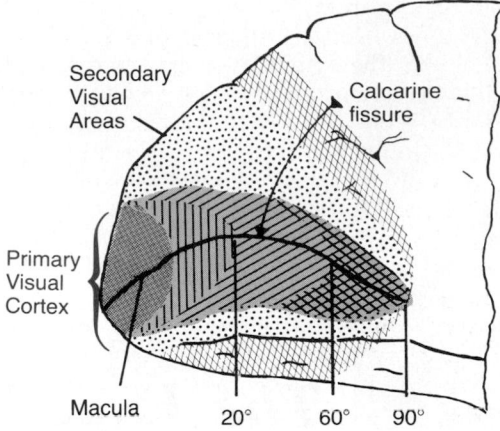

Figure 51–2. Visual cortex in the calcarine fissure area of the medial occipital cortex.

ORGANIZATION AND FUNCTION OF THE VISUAL CORTEX

Figures 51–2 and 51–3 show that the *visual cortex* is located primarily in the occipital lobes. Like the cortical representations of the other sensory systems, the visual cortex is divided into a *primary visual cortex* and *secondary visual areas*.

PRIMARY VISUAL CORTEX. The primary visual cortex (see Fig. 51–2) lies in the *calcarine fissure area*, extending to the *occipital pole* on the medial aspect of each occipital cortex. This area is the terminus of di-

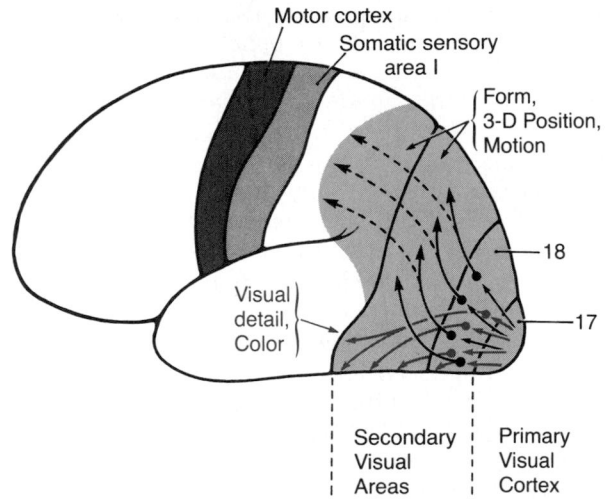

Figure 51–3. Transmission of visual signals from the primary visual cortex into secondary visual areas. Note that the signals representing form, third-dimensional position, and motion are transmitted mainly superiorly into the superior portions of the occipital lobe and the posterior portions of the parietal lobe. By contrast, the signals for visual detail and color are transmitted mainly into the anteroventral portion of the occipital lobe and the ventral portion of the posterior temporal lobe.

rect visual signals from the eyes. Signals from the macular area of the retina terminate near the occipital pole, as shown in the figure, whereas signals from the more peripheral retina terminate in concentric circles anterior to the pole and along the calcarine fissure. The upper portion of the retina is represented superiorly and the lower portion, inferiorly. Note in the figure the especially large area that represents the macula. It is to this region that the retinal fovea transmits its signals. The fovea is responsible for the highest degree of visual acuity. Based on retinal area, the fovea has several hundred times as much representation in the primary visual cortex as do the peripheral portions of the retina.

The primary visual cortex is coextensive with *Brodmann cortical area 17* (see the diagram of the Brodmann areas in Fig. 47–5). It is also called *visual area I* or simply *V-1*. Still another name for the primary visual cortex is *striate cortex* because this area has a grossly striated appearance.

SECONDARY VISUAL AREAS OF THE CORTEX. The secondary visual areas, also called *visual association areas,* lie lateral, anterior, superior, and inferior to the primary visual cortex. Most of these areas fold outward over the lateral surfaces of the occipital cortex, as shown in Figure 51–3. Secondary signals are transmitted to these areas for analysis of visual meanings. For instance, on all sides of the primary visual cortex is *Brodmann area 18* (see Fig. 51–3), which is where virtually all signals from the primary visual cortex pass next. Therefore, Brodmann area 18 is called *visual area II* or simply *V-2.* The other, more distant secondary visual areas have specific designations, V-3, V-4, and so forth, up to more than a dozen areas. The importance of all these areas is that various aspects of the visual image are progressively dissected and analyzed.

Layered Structure of the Primary Visual Cortex

Like almost all other portions of the cerebral cortex, the primary visual cortex has six distinct layers, as shown in Figure 51–4. As is true for the other sensory systems, the geniculocalcarine fibers terminate mainly in layer IV. But this layer, too, is organized in subdivisions. The rapidly conducted signals from the Y retinal ganglion cells terminate in layer IVcα, and from here they are relayed vertically both outward toward the cortical surface and inward toward deeper levels.

The visual signals from the medium-sized optic nerve fibers, derived from the X ganglion cells in the retina, also terminate in layer IV but at points different from the Y signals in layers IVa and IVcβ, the shallowest and deepest portions of layer IV, shown to the right in Figure 51–4. From here, these signals also are transmitted vertically both toward the surface of the cortex and to deeper layers. It is these X ganglion pathways that transmit the accurate point-to-point type of vision as well as color vision.

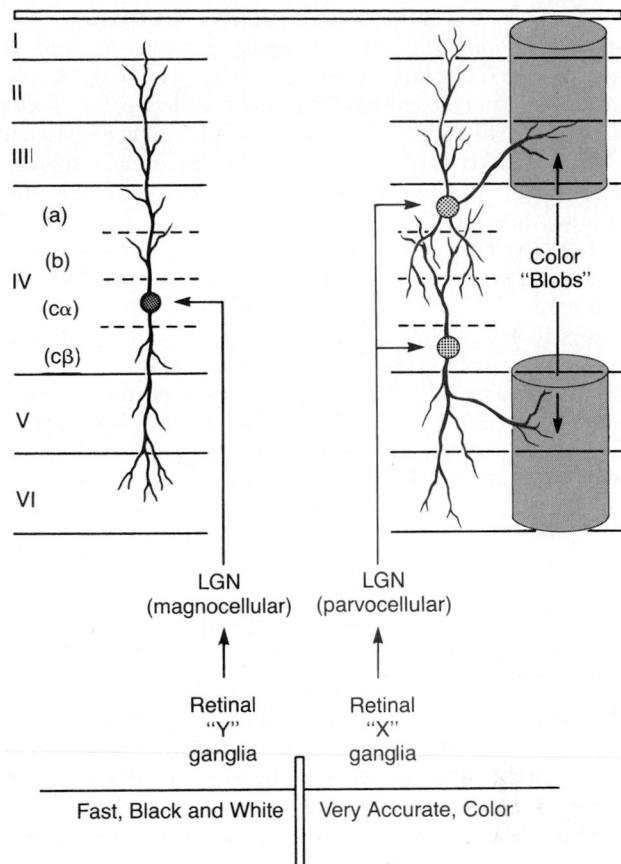

Figure 51–4. Six layers of the primary visual cortex. The connections shown on the left side of the figure originate in the magnocellular layers of the lateral geniculate nucleus (LGN) and transmit rapidly changing black and white visual signals. The pathways to the right originate in the parvocellular layers (layers III through VI) of the lateral geniculate nucleus; they transmit signals that depict accurate detail as well as color. Note especially the small areas of the visual cortex called "color blobs" that are necessary for detection of color.

VERTICAL NEURONAL COLUMNS IN THE VISUAL CORTEX. The visual cortex is organized structurally into several million vertical columns of neuronal cells, each column having a diameter of 30 to 50 micrometers. The same vertical columnar organization is found throughout the cerebral cortex. Each column represents a functional unit. One can calculate roughly that the number of neurons in each of the visual vertical columns is perhaps 1000 or more.

After the optic signals terminate in layer IV, they are further processed as they spread both outward and inward along each vertical column unit. This processing is believed to decipher separate bits of visual information at successive stations along the pathway. The signals that pass outward to layers I, II, and III eventually transmit signals for short distances laterally in the cortex. On the other hand, the signals that pass inward to layers V and VI excite neurons that transmit signals much greater distances.

"Color Blobs" in the Visual Cortex. Interspersed among the primary visual columns as well as among the columns of some of the secondary visual areas are special column-like areas called *color blobs.* They receive lateral signals from the adjacent visual columns and respond specifically to color signals. Therefore, it is presumed that these blobs are the primary areas for deciphering color.

Interaction of Visual Signals from the Separate Eyes. Recall that the visual signals from the two separate eyes are relayed through separate neuronal layers in the lateral geniculate nucleus. And these signals still remain separated from each other when they arrive in layer IV of the primary visual cortex. In fact, layer IV is interlaced with horizontal zebra-like stripes of neuronal columns, each stripe about 0.5 millimeter wide; the signals from one eye enter the columns of every other stripe, alternating with the signals from the second eye. As the signals spread vertically into the other layers of the cortex, this separation is lost because of lateral spread of the visual signals. In the meantime, the cortex deciphers whether the respective areas of the two visual images are "in register" with each other—that is, whether corresponding points on the two retinas fit with each other. In turn, this deciphered information is used to control the movements of the separate eyes so that they will fuse with each other (brought into "register"). The information observed about degree of register of images from the two eyes also allows a person to distinguish distances of objects by the mechanism of stereopsis.

Two Major Pathways for Analysis of Visual Information—(1) The Fast "Position" and "Motion" Pathway; (2) The Accurate Color Pathway

Figure 51–3 shows that after leaving the primary visual cortex, the visual information is analyzed in two major pathways in the secondary visual areas.

1. Analysis of Three-Dimensional Position, Gross Form, and Motion of Objects. One of the analytical pathways, demonstrated in Figure 51–3 by the broad black arrows, analyzes the third-dimension positions of visual objects in the space around the body. From this information, this pathway also analyzes the gross form of the visual scene as well as motion in the scene. In other words, this pathway tells "where" every object is at each instant and whether it is moving. After leaving the primary visual cortex (Brodmann area 17), the signals of this pathway next synapse in visual area II (Brodmann area 18), then flow generally into the posterior midtemporal area, and then upward into the broad occipitoparietal cortex. At the anterior border of this last area, the signals overlap with signals from the posterior somatic association areas that analyze three-dimensional aspects of somatosensory signals. The signals transmitted in this position-form-motion pathway are mainly from the large Y optic nerve fibers of the retinal Y ganglion cells, trans-

mitting rapid signals but depicting only black and white with no color.

2. Analysis of Visual Detail and Color. The red arrows in Figure 51–3, passing from the primary visual cortex (Brodmann area 17) into visual area II (Brodmann area 18) and then into the inferior ventral and medial regions of the occipital and temporal cortex, demonstrate the principal pathway for analysis of visual detail. Separate portions of this pathway specifically dissect out color as well. Therefore, this pathway is concerned with such visual feats as recognizing letters, reading, determining the texture of surfaces, determining detailed colors of objects, and deciphering from all this information "what" the object is and its meaning.

NEURONAL PATTERNS OF STIMULATION DURING ANALYSIS OF THE VISUAL IMAGE

Analysis of Contrasts in the Visual Image. If a person looks at a blank wall, only a few neurons in the primary visual cortex will be stimulated, regardless of whether the illumination of the wall is bright or weak. Therefore, the question must be asked, What does the visual cortex detect? To answer this, let us now place on the wall a large solid cross, as shown to the left in Figure 51–5. To the right is shown the spatial pattern of the majority of the excited neurons in the visual cortex. *Note that the areas of maximum excitation occur along the sharp borders of the visual pattern.* Thus, the visual signal in the primary visual cortex is concerned mainly with the *contrasts* in the visual scene, rather than with the noncontrasting areas. We saw in Chapter 50 that this is true of most of the retinal ganglion cells as well because equally stimulated adjacent retinal receptors mutually inhibit one another. But at any border in the visual scene where there is a change from dark to light or light to dark, mutual inhibition does not occur, and the intensity of stimulation is proportional to the *gradient of contrast—* that is, the greater the sharpness of contrast and the greater the intensity difference between the light and dark areas, the greater the degree of stimulation.

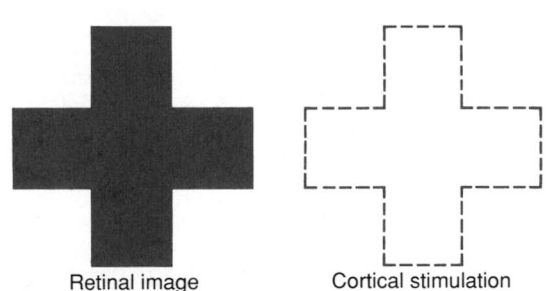

Retinal image   Cortical stimulation

Figure 51–5. Pattern of excitation that occurs in the visual cortex in response to a retinal image of a dark cross.

THE VISUAL CORTEX ALSO DETECTS ORIENTATION OF LINES AND BORDERS—THE "SIMPLE" CELLS. Not only does the visual cortex detect the existence of lines and borders in the different areas of the retinal image, but it also detects the direction of orientation of each line or border—that is, whether it is vertical or horizontal or lies at some degree of inclination. This is believed to result from linear organizations of mutually inhibiting cells that excite second-order neurons when inhibition occurs all along a line of cells where there is a contrast edge. Thus, for each such orientation of a line, a specific neuronal cell is stimulated. And a line oriented in a different direction excites a different cell. These neuronal cells are called *simple cells.* They are found mainly in layer IV of the primary visual cortex.

DETECTION OF LINE ORIENTATION WHEN THE LINE IS DISPLACED LATERALLY OR VERTICALLY IN THE VISUAL FIELD—"COMPLEX" CELLS. As the signal progresses farther away from layer IV, some neurons now respond to lines still oriented in the same direction but not position-specific. That is, the line can be displaced moderate distances laterally or vertically in the field and still the neuron will be stimulated if the line has the same direction. These cells are called *complex cells.*

DETECTION OF LINES OF SPECIFIC LENGTHS, ANGLES, OR OTHER SHAPES. Some neurons in the outer layers of the primary visual columns as well as neurons in some secondary visual areas are stimulated only by lines or borders of specific lengths, or by specific angulated shapes, or by images that have other characteristics. Thus, these neurons detect still higher orders of information from the visual scene.

Thus, as one goes farther into the analytical pathway of the visual cortex, progressively more characteristics of each visual scene are deciphered.

Detection of Color

Color is detected in much the same way that lines are detected: by means of color contrast. For instance, a red area is often contrasted against a green area, or a blue area against a red, or a green area against a yellow. All these colors can also be contrasted against a white area within the visual scene. In fact, it is this contrasting against white that is believed to be mainly responsible for the phenomenon called "color constancy" discussed in Chapter 50; that is, when the color of the illuminating light changes, the color of the "white" changes with the light and appropriate computation in the brain allows red to be interpreted as red even though the illuminating light has changed the color of the "red" entering the eyes.

The mechanism of color contrast analysis depends on the fact that contrasting colors, called "opponent colors," mutually excite specific neuronal cells. It is presumed that the initial details of color contrast are detected by simple cells, whereas more complex contrasts are detected by complex and hypercomplex cells.

Serial Analysis of the Visual Image Versus Parallel Analysis

From the foregoing discussion, it should by now be clear that the visual image is deciphered and analyzed by both serial and parallel pathways. The sequence from simple to complex to hypercomplex cells is serial analysis, with more and more details being deciphered. The transmission of different types of visual information into different brain locations represents parallel processing. The combination of both types of these analyses gives one full interpretation of a visual scene. The highest levels of analysis are still mainly beyond physiological understanding.

Effect of Removing the Primary Visual Cortex

Removal of the primary visual cortex in the human being causes loss of conscious vision. However, psychological studies demonstrate that such people can still react subconsciously to changes in light intensity, to movement in the visual scene, and even to some gross patterns of vision. These reactions include turning the eyes, turning the head, and avoidance. This vision is believed to be subserved by neuronal pathways that pass from the optic tracts mainly into the superior colliculi and other portions of the older visual system.

FIELDS OF VISION; PERIMETRY

The *field of vision* is the visual area seen by an eye at a given instant. The area seen to the nasal side is called the *nasal field of vision,* and the area seen to the lateral side is called the *temporal field of vision.*

To diagnose blindness in specific portions of the retinas, one charts the field of vision for each eye by a process called *perimetry.* This is done by having the subject look with one eye toward a central spot directly in front of the eye. Then a small dot of light or a small object is moved back and forth in all areas of the field of vision, and the person indicates when the spot of light or object can be seen and when it cannot. Thus, the field of vision for the left eye is plotted as shown in Figure 51–6.

In all perimetry charts, a *blind spot* caused by lack of rods and cones in the retina over the *optic disc* is found about 15 degrees lateral to the central point of vision, as shown in the figure.

ABNORMALITIES IN THE FIELDS OF VISION. Occasionally blind spots are found in portions of the field of vision other than the optic disc area. Such blind spots are called *scotomata;* they frequently result from damage to the optic nerve resulting from glaucoma (too much fluid pressure in the eyeball), from allergic reactions in the retina, or from toxic conditions, such as lead poisoning and excessive use of tobacco.

Still another condition that can be diagnosed by perimetry is *retinitis pigmentosa.* In this disease, portions of the retina degenerate, and excessive melanin pigment

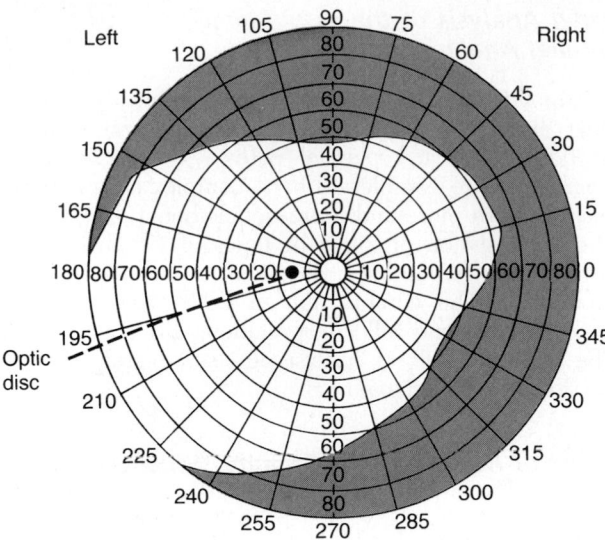

Figure 51–6. Perimetry chart, showing the field of vision for the left eye.

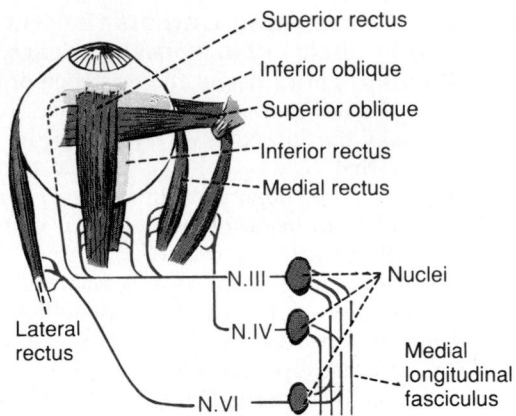

Figure 51–7. Extraocular muscles of the eye and their innervation.

deposits in the degenerated areas. Retinitis pigmentosa usually causes blindness in the peripheral field of vision first and then gradually encroaches on the central areas.

Effect of Lesions in the Optic Pathway on the Fields of Vision. Destruction of an entire *optic nerve* causes blindness of the affected eye. Destruction of the *optic chiasm,* as shown by the longitudinal line across the chiasm in Figure 51–1, prevents the passage of impulses from the nasal halves of the two retinas to the opposite optic tracts. Therefore, the nasal halves of the retinas are both blinded, which means that the person is blind in both temporal fields of vision *because the image of the field of vision is inverted on the retina;* this condition is called *bitemporal hemianopsia.* Such lesions frequently result from tumors of the pituitary gland pressing upward on the optic chiasm.

Interruption of an *optic tract,* which is shown by another line in Figure 51–1, denervates the corresponding half of each retina on the same side as the lesion, and as a result, neither eye can see objects to the opposite side of the head. This condition is known as *homonymous hemianopsia.* Destruction of the *optic radiation* or the *visual cortex* of one side also causes homonymous hemianopsia. A common condition that destroys the visual cortex is thrombosis of the posterior cerebral artery, which often infarcts the occipital cortex except for part of the foveal vision area, thus often sparing central vision.

One can differentiate a lesion in the optic tract from a lesion in the geniculocalcarine tract or visual cortex by determining whether impulses can still be transmitted into the pretectal nuclei to initiate a pupillary light reflex, which is discussed later in the chapter.

EYE MOVEMENTS AND THEIR CONTROL

To make full use of the abilities of the eyes, almost equally as important as the system for interpretation of the visual signals from the eyes is the cerebral control

system for directing the eyes toward the object to be viewed.

MUSCULAR CONTROL OF EYE MOVEMENTS. The eye movements are controlled by three pairs of muscles, shown in Figure 51–7: (1) the *medial* and *lateral recti,* (2) the *superior* and *inferior recti,* and (3) the *superior* and *inferior obliques.* The medial and lateral recti contract reciprocally mainly to move the eyes from side to side. The superior and inferior recti contract reciprocally to move the eyes mainly upward or downward. And the oblique muscles function mainly to rotate the eyeballs to keep the visual fields in the upright position.

NEURAL PATHWAYS FOR CONTROL OF EYE MOVEMENTS. Figure 51–7 also shows the nuclei of the third, fourth, and sixth cranial nerves and their innervation of the ocular muscles. Shown, too, are the interconnections among these three nuclei through the *medial longitudinal fasciculus.* Either by way of this fasciculus or by way of other closely associated pathways, each of the three sets of muscles to each eye is *reciprocally* innervated so that one muscle of the pair relaxes while the other contracts.

Figure 51–8 demonstrates cortical control of the oculomotor apparatus, showing spread of signals from the occipital visual areas through occipitotectal and occipitocollicular tracts into the pretectal and superior colliculus areas of the brain stem. In addition, a frontotectal tract passes from the frontal cortex into the pretectal area. From both the pretectal and the superior colliculus areas, the oculomotor control signals then pass to the nuclei of the oculomotor nerves. Strong signals are also transmitted into the oculomotor system from the vestibular nuclei by way of the medial longitudinal fasciculus.

Fixation Movements of the Eyes

Perhaps the most important movements of the eyes are those that cause the eyes to "fix" on a discrete portion of the field of vision.

Fixation movements are controlled by two neuronal

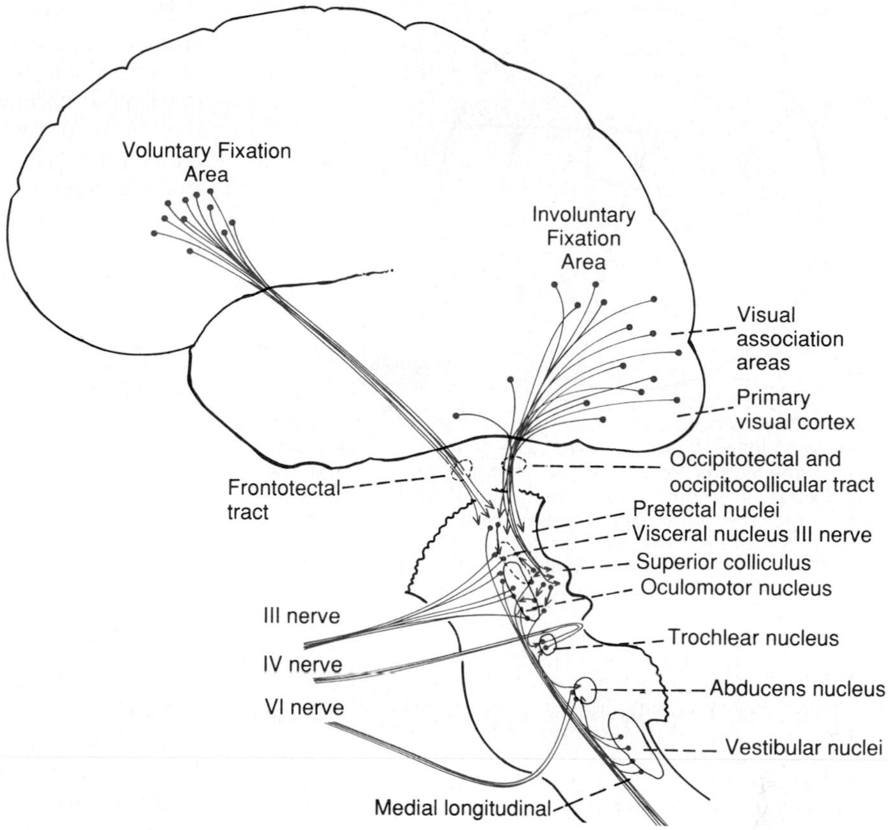

Figure 51–8. Neural pathways for control of conjugate movement of the eyes.

mechanisms. The first of these allows one to move the eyes voluntarily to find the object on which one wants to fix one's vision; this is called the *voluntary fixation mechanism.* The second is an involuntary mechanism that holds the eyes firmly on the object once it has been found; this is called the *involuntary fixation mechanism.*

The voluntary fixation movements are controlled by a small cortical field located bilaterally in the premotor cortical regions of the frontal lobes, as shown in Figure 51–8. Bilateral dysfunction or destruction of these areas makes it difficult or almost impossible for a person to "unlock" the eyes from one point of fixation and then move them to another point. It is usually necessary for the person to blink the eyes or put a hand over the eyes for a short time, which then allows the eyes to be moved.

On the other hand, the fixation mechanism that causes the eyes to "lock" on the object of attention once it is found is controlled by *secondary visual areas of the occipital cortex*—mainly Brodmann area 19, located anterior to visual areas V-1 and V-2 (Brodmann areas 17 and 18). When this area is destroyed bilaterally, an animal has difficulty keeping its eyes directed toward a given fixation point or becomes unable to do so.

To summarize, the posterior eye fields automatically "lock" the eyes on a given spot of the visual field and thereby prevent movement of the image across the retina. To unlock this visual fixation, voluntary impulses must be transmitted from the "voluntary" eye fields located in the frontal areas.

MECHANISM OF THE INVOLUNTARY LOCKING FIXATION—ROLE OF THE SUPERIOR COLLICULI. The involuntary locking type of fixation discussed in the previous section results from a negative feedback mechanism that prevents the object of attention from leaving the foveal portion of the retina. The eyes even normally have three types of continuous but almost imperceptible movements: (1) a *continuous tremor* at a rate of 30 to 80 cycles per second caused by successive contractions of the motor units in the ocular muscles, (2) a *slow drift* of the eyeballs in one direction or another, and (3) sudden *flicking movements* that are controlled by the involuntary fixation mechanism. When a spot of light has become fixed on the foveal region of the retina, the tremorous movements cause the spot to move back and forth at a rapid rate across the cones, and the drifting movements cause it to drift slowly across the cones. Each time the spot of light drifts as far as the edge of the fovea, a sudden reflex reaction occurs, producing a flicking movement that moves the spot away from this edge back toward the foveal center. Thus, an automatic response moves the image back toward the central portion of the fovea. These drifting and flicking motions are demonstrated in Figure 51–9, which shows by the dashed lines the slow drifting across the retina and by the solid lines

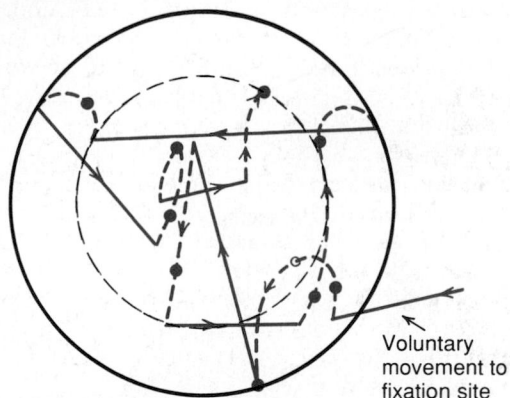

Voluntary movement to fixation site

Figure 51–9. Movements of a spot of light on the fovea, showing sudden "flicking" movements to move the spot back toward the center of the fovea whenever it drifts to the foveal edge. (The dashed lines represent slow drifting movements, and the solid lines represent sudden flicking movements.) (Modified from Whitteridge: Handbook of Physiology. Vol. 2, Sec. 1. Baltimore, Williams & Wilkins, 1960.)

the flicks that keep the image from leaving the foveal region.

This involuntary fixation capability is mostly lost when the superior colliculi are destroyed. After the signals for fixation originate in the visual fixation areas of the occipital cortex, they pass to the superior colliculi, probably from there to reticular areas around the oculomotor nuclei, and then into the motor nuclei themselves.

SACCADIC MOVEMENT OF THE EYES—A MECHANISM OF SUCCESSIVE FIXATION POINTS. When the visual scene is moving continually before the eyes, such as when a person is riding in a car or turning around, the eyes fix on one highlight after another in the visual field, jumping from one to the next at a rate of two to three jumps per second. The jumps are called *saccades,* and the movements are called *opticokinetic movements.* The saccades occur so rapidly that not more than 10 per cent of the total time is spent in moving the eyes, with 90 per cent of the time being allocated to the fixation sites. Also, the brain suppresses the visual image during the saccades, so that one is unconscious of the movements from point to point.

Saccadic Movements During Reading. During the process of reading, a person usually makes several saccadic movements of the eyes for each line. In this case, the visual scene is not moving past the eyes, but the eyes are trained to move by means of several successive saccades across the visual scene to extract the important information. Similar saccades occur when a person observes a painting, except that the saccades occur in upward, sideways, downward, and angulated directions one after another from one highlight of the painting to another, then another, and so forth.

FIXATION ON MOVING OBJECTS—"PURSUIT MOVEMENTS." The eyes can also remain fixed on a moving object, which is called *pursuit movement.* A highly developed cortical mechanism automatically detects the course of movement of an object and then gradually develops a similar course of movement of the eyes. For instance, if an object is moving up and down in a wavelike form at a rate of several times per second, the eyes at first may be unable to fixate on it. However, after a second or so, the eyes begin to jump coarsely in approximately the same wavelike pattern of movement as that of the object. Then, after a few more seconds, the eyes develop progressively smoother movements and finally follow the wave movement almost exactly. This represents a high degree of automatic subconscious computational ability by the visual system for controlling eye movements.

The Superior Colliculi Are Mainly Responsible for Turning the Eyes and Head Toward a Visual Disturbance

Even after the visual cortex has been destroyed, a sudden visual disturbance in a lateral area of the visual field often causes immediate turning of the eyes in that direction. This does not occur if the superior colliculi have also been destroyed. To support this function, the various points of the retina are represented topologically in the superior colliculi in the same way as in the primary visual cortex, although with less accuracy. Even so, the principal direction of a flash of light in a peripheral retinal field is mapped by the colliculi, and secondary signals are then transmitted to the oculomotor nuclei to turn the eyes. To help in this directional movement of the eyes, the superior colliculi also have topological maps of somatic sensations from the body and acoustic signals from the ears.

The optic nerve fibers from the eyes to the colliculi that are responsible for these rapid turning movements are branches from the rapidly conducting Y fibers, with one branch going to the visual cortex and the other going to the superior colliculi. (The superior colliculi and other regions of the brain stem are also strongly supplied with visual signals transmitted in type W optic nerve fibers. These represent the older visual pathway, but their function is unclear.)

In addition to causing the eyes to turn toward the visual disturbance, signals are relayed from the superior colliculi through the *medial longitudinal fasciculus* to other levels of the brain stem to cause turning of the whole head and even of the whole body toward the direction of the disturbance. Other types of disturbances besides visual, such as strong sounds or even stroking of the side of the body, cause similar turning of the eyes, head, and body but only if the superior colliculi are intact. Therefore, the superior colliculi play a global role in orienting the eyes, head, and body with respect to external disturbances whether visual, auditory, or somatic.

Fusion of the Visual Images from the Two Eyes

To make the visual perceptions more meaningful, the visual images in the two eyes normally *fuse* with each other on "corresponding points" of the two retinas.

The visual cortex plays an important role in fusion. It was pointed out earlier in the chapter that corresponding points of the two retinas transmit visual signals to different neuronal layers of the lateral geniculate body, and these signals in turn are relayed to parallel stripes of neurons in the visual cortex. Interactions occur between the stripes of cortical neurons; these interactions cause *interference patterns of excitation* in some of the local neuronal cells when the two visual images are not precisely "in register"—that is, not precisely fused. This excitation presumably provides the signal that is transmitted to the oculomotor apparatus to cause convergence or divergence or rotation of the eyes so that fusion can be re-established. Once the corresponding points of the retinas are precisely in register with each other, the excitation of the specific cells in the visual cortex is greatly diminished or disappears.

Neural Mechanism of Stereopsis for Judging Distances of Visual Objects

In Chapter 49, it is pointed out that because the two eyes are more than 2 inches apart, the images on the two retinas are not exactly the same. That is, the right eye sees a little more of the right-hand side of the object and the left eye, a little more of the left-hand side, and the closer the object, the greater the disparity. Therefore, even when the two eyes are fused with each other, it is still impossible for all corresponding points in the two visual images to be in register at the same time. Furthermore, the nearer the object is to the eyes, the less the degree of register. This degree of nonregister provides the mechanism for *stereopsis*, an important mechanism for judging distances of visual objects up to distances of about 200 feet (60 meters).

The neuronal cellular mechanism for stereopsis is based on the fact that some of the fiber pathways from the retinas to the visual cortex stray 1 to 2 degrees on either side of the central pathway. Therefore, some optic pathways from the two eyes are exactly in register for objects 2 meters away; still another set of pathways is in register for objects 25 meters away. Thus, the distance is determined by which set of pathways interacts with another. This phenomenon is called *depth perception*, which is another name for stereopsis.

Strabismus

Strabismus, also called *squint* or *cross-eyedness*, means lack of fusion of the eyes in one or more of the coordinates described above. The basic types of strabis-

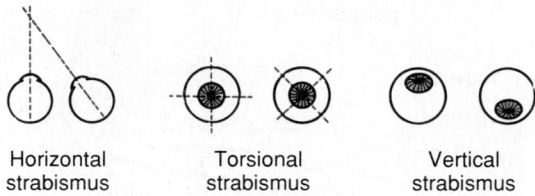

Figure 51–10. Basic types of strabismus.

mus are shown in Figure 51–10: (1) *horizontal strabismus*, (2) *vertical strabismus*, and (3) *torsional strabismus*. Combinations of two or even all three of the different types of strabismus often occur.

Strabismus is often caused by an abnormal "set" of the fusion mechanism of the visual system. That is, in the early efforts of the child to fixate the two eyes on the same object, one of the eyes fixates satisfactorily while the other fails to fixate, or they both fixate satisfactorily but never simultaneously. Soon the patterns of conjugate movements of the eyes become abnormally "set" in the neuronal control pathways themselves so that the eyes never fuse.

SUPPRESSION OF THE VISUAL IMAGE FROM A REPRESSED EYE. In a few patients with strabismus, the eyes alternate in fixing on the object of attention. In other patients, one eye alone is used all the time while the other eye becomes repressed and is never used for precise vision. The visual acuity of the repressed eye develops only slightly, usually remaining 20/400 or less. If the dominant eye then becomes blinded, vision in the repressed eye can develop only to a slight extent in the adult but far more in young children. This demonstrates that visual acuity is highly dependent on proper development of the central synaptic connections from the eyes. In fact, even anatomically, the numbers of neuronal connections diminish in the cortical stripes that receive signals from the repressed eye.

AUTONOMIC CONTROL OF ACCOMMODATION AND PUPILLARY APERTURE

AUTONOMIC NERVES TO THE EYES. The eye is innervated by both parasympathetic and sympathetic nerve fibers, as shown in Figure 51–11. The parasympathetic preganglionic fibers arise in the *Edinger-Westphal nucleus* (the visceral nucleus of the third nerve) and then pass in the *third nerve* to the *ciliary ganglion*, which lies immediately behind the eye. Here the preganglionic fibers synapse with postganglionic parasympathetic neurons that in turn send fibers through the *ciliary nerves* into the eyeball. These nerves excite (a) the ciliary muscle that controls focusing of the eye lens and (b) the sphincter of the iris that constricts the pupil.

The sympathetic innervation of the eye originates in the *intermediolateral horn cells* of the first thoracic segment of the spinal cord. From here, sympathetic fibers enter the sympathetic chain and pass upward to the *superior cervical ganglion*, where they synapse

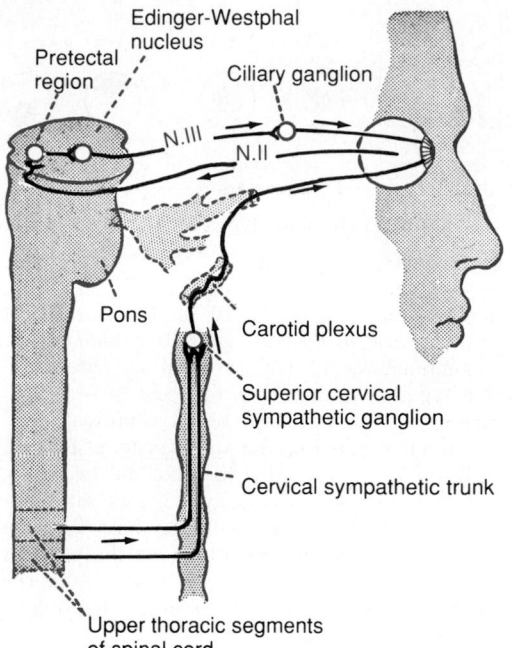

Figure 51–11. Autonomic innervation of the eye, showing also the reflex arc of the light reflex. (Modified from Ranson and Clark: Anatomy of the Nervous System. Philadelphia, W. B. Saunders Company, 1959.)

with postganglionic neurons. Fibers from these spread along the carotid artery and successively smaller arteries until they reach the eye. There the sympathetic fibers innervate the radial fibers of the iris as well as several extraocular muscles around the eye, which are discussed shortly in relation to Horner's syndrome. Also, they supply weak innervation to the ciliary muscle.

Control of Accommodation (Focusing the Eyes)

The accommodation mechanism—that is, the mechanism that focuses the lens system of the eye—is essential for a high degree of visual acuity. Accommodation results from contraction or relaxation of the ciliary muscle, with contraction causing increased strength of the lens system, as explained in Chapter 49, and relaxation causing decreased strength. The question that must be answered now is: How does a person adjust accommodation to keep the eyes in focus all the time?

Accommodation of the lens is regulated by a negative feedback mechanism that automatically adjusts the focal power of the lens for the highest degree of visual acuity. When the eyes have been fixed on some far object and must then suddenly fix on a near object, the lens accommodates for maximum acuity of vision usually within less than 1 second. Although the precise control mechanism that causes this rapid and accurate focusing of the eye is unclear, some of the known features are the following.

First, when the eyes suddenly change the distance

of the fixation point, the lens always changes its strength in the proper direction to achieve a new state of focus. In other words, the lens does not make a mistake and change its strength in the wrong direction in an attempt to find the focus.

Second, different types of clues that can help the lens change its strength in the proper direction include the following: (1) *Chromatic aberration* appears to be important. That is, the red light rays focus slightly posteriorly to the blue light rays because the lens bends blue rays more than red rays. The eyes appear to be able to detect which of these two types of rays is in better focus, and this clue relays information to the accommodation mechanism whether to make the lens stronger or weaker. (2) When the eyes fixate on a near object, they also converge toward each other. The neural mechanisms for *convergence cause a simultaneous signal to strengthen the lens of the eye.* (3) *Because the fovea lies in a hollowed-out depression that is deeper than the remainder of the retina, the clarity of focus in the depth of the fovea versus the clarity of focus on the edges is different.* It has been suggested that this also gives clues as to which way the strength of the lens needs to be changed. (4) It has been found that *the degree of accommodation of the lens oscillates slightly* all the time at a frequency up to twice per second. The visual image becomes clearer when the oscillation of the lens strength is changing in the appropriate direction and becomes poorer when the lens strength is changing in the wrong direction. This could give a rapid clue as to which way the strength of the lens needs to change to provide appropriate focus.

It is presumed that the brain cortical areas that control accommodation closely parallel those that control fixation movements of the eyes, with final integration of the visual signals in Brodmann cortical areas 18 and 19 and transmission of motor signals to the ciliary muscle through the pretectal area in the brain stem and then into the Edinger-Westphal nucleus.

Control of Pupillary Diameter

Stimulation of the parasympathetic nerves excites the pupillary sphincter muscle, thereby decreasing the pupillary aperture; this is called *miosis*. On the other hand, stimulation of the sympathetic nerves excites the radial fibers of the iris and causes pupillary dilatation, called *mydriasis*.

PUPILLARY LIGHT REFLEX. When light is shone into the eyes, the pupils constrict, a reaction called the *pupillary light reflex*. The neuronal pathway for this reflex is demonstrated by the upper two black traces in Figure 51–11. When light impinges on the retina, the resulting impulses pass first through the optic nerves and then to the pretectal nuclei. From here, impulses pass to the *Edinger-Westphal nucleus* and, finally, back through the *parasympathetic nerves* to constrict the sphincter of the iris. In darkness, the reflex becomes inhibited, which results in dilatation of the pupil.

The function of the light reflex is to help the eye

adapt extremely rapidly to changing light conditions, as explained in Chapter 50. The limits of pupillary diameter are about 1.5 millimeters on the small side and 8 millimeters on the large side. Therefore, the range of light and dark adaptation that can be effected by the pupillary reflex is about 30 to 1.

PUPILLARY REFLEXES OR REACTIONS IN CENTRAL NERVOUS SYSTEM DISEASE. Certain central nervous system diseases block the transmission of visual signals from the retinas to the Edinger-Westphal nucleus. Such blocks frequently occur as a result of *central nervous system syphilis, alcoholism, encephalitis,* and so forth. The block usually occurs in the pretectal region of the brain stem, although it can result from destruction of the small afferent fibers in the optic nerves.

The final nerve fibers in the pathway through the pretectal area to the Edinger-Westphal nucleus are of the inhibitory type. Therefore, when their inhibitory effect is lost, the nucleus becomes chronically active, causing the pupils thereafter to remain partially constricted in addition to their failure to respond to light.

Yet the pupils can still constrict some more if the Edinger-Westphal nucleus is stimulated through some other pathway. For instance, when the eyes fixate on a near object, the signals that cause accommodation of the lens and those that cause convergence of the two eyes cause a mild degree of pupillary constriction at the same time. This is called *pupillary reaction to accommodation.* Such a pupil that fails to respond to light but does respond to accommodation and that also is very small (an *Argyll Robertson pupil*) is an important diagnostic sign of central nervous system disease—very often syphilis.

HORNER'S SYNDROME. The sympathetic nerves to the eye are occasionally interrupted, and this interruption frequently occurs in the cervical sympathetic chain. This results in *Horner's syndrome,* which consists of the following effects: First, because of interruption of fibers to the pupillary dilator muscle, the pupil remains persistently constricted to a smaller diameter than that of the pupil of the opposite eye. Second, the superior eyelid droops because this eyelid is normally maintained in an open position during the waking hours partly by contraction of a smooth muscle embedded in the lid and innervated by the sympathetics. Therefore, destruction of the sympathetics makes it impossible to open the superior eyelid nearly as widely as normally. Third, the blood vessels on the corresponding side of the face and head become persistently dilated. And fourth, sweating cannot occur on the side of the face and head affected by Horner's syndrome.

REFERENCES

Albert, D. M., and Jakobiec, F. A.: Principles and Practice of Ophthalmology. Philadelphia, W. B. Saunders Co., 1994.
Andersen, R. A.: Visual and eye movement functions of the posterior parietal cortex. Annu. Rev. Neurosci., 12:377, 1989.
Bahill, A. T., and Hamm, T. M.: Using open-loop experiments to study physiological systems, with examples from the human eye-movement systems. News Physiol. Sci., 4:104, 1989.
Blakemore, C.: Vision: Coding and Efficiency. New York, Cambridge University Press, 1991.
Blasdel, G. G.: Visualization of neuronal activity in monkey striate cortex. Annu. Rev. Physiol., 51:561, 1989.
Buser, P., and Imbert, M.: Vision. Cambridge, MA, The MIT Press, 1992.
Buttner, E. J. (ed.): Neuroanatomy of the Oculomotor System. New York, Elsevier Science Publishing Co., 1984.
Cohen, B., and Bodis-Wollner, I.: Vision and the Brain: The Organization of the Central Visual System. New York, Raven Press, 1990.
DeValois, R. L., and DeValois, K. K.: Spatial Vision. New York, Oxford University Press, 1988.
Farah, M. J., and Ratcliff, G.: The Neuropsychology of High-level Vision. Hillsdale, NJ, Lawrence Erlbaum Assoc. Inc., 1994.
Farah, M. J.: Visual Agnosia. Cambridge, MA, The MIT Press, 1990.
Garraghty, P. E., and Sur, M.: Competitive interactions influencing the development of retinal axonal arbors in cat lateral geniculate nucleus. Physiol. Rev., 73:529, 1993.
Jones, G. M.: The remarkable vestibuloocular reflex. News Physiol. Sci., 2:85, 1987.
Lam, D. M-K., and Bray, G. M.: Regeneration and Plasticity in the Mammalian Visual System. Cambridge, MA, The MIT Press, 1992.
Livingstone, M., and Hubel, D.: Segregation of form, color, movement, and depth: Anatomy, physiology, and perception. Science, 240:740, 1988.
Lund, J. S.: Anatomical organization of Macaque monkey striate visual cortex. Annu. Rev. Neurosci., 11:253, 1988.
Margo, C. E., et al.: Diagnostic Problems in Clinical Ophthalmology. Philadelphia, W. B. Saunders Co., 1994.
Miller, N. R.: Walsh & Hoyt's Clinical Neuro-Ophthalmology. Baltimore, Williams & Wilkins, 1994.
Peters, A., and Jones, E. G. (eds.): Visual Cortex, New York, Plenum Publishing Corp., 1985.
Poggio, G. F., and Poggio, T.: The analysis of stereopsis. Annu. Rev. Neurosci., 7:379, 1984.
Rauschecker, J. P.: Mechanisms of Visual Plasticity: Hebb Synapses, NMDA Receptors, and Beyond. Physiol. Rev., 71:587, 1991.
Reinecke, R. D., and Parks, M. M.: Strabismus, 3rd Ed. East Norwalk, Conn., Appleton & Lange, 1987.
Ringo, J. L., et al.: Eye movements modulate activity in hippocampal, parahippocampal, and inferotemporal neurons. J. Neurophysiol., 71:1285, 1994.
Scheiman, M., and Wick, B.: Clinical Management of Binocular Vision. Philadelphia, J. B. Lippincott, 1993.
Sharpe, J. A., and Barber, H. O.: The Vestibulo-Ocular Reflex and Vertigo. New York, Raven Press, 1993.
Simpson, J. I.: The accessory optic system. Annu. Rev. Neurosci., 7:13, 1984.
Simons, K.: Early Visual Development: Normal and Abnormal. New York, Oxford University Press, 1993.
Smith, A. T., and Snowden, R. J.: Visual Detection of Motion. San Diego, CA, Academic Press, 1994.
Sparks, D. L.: Translation of sensory signals into commands for control of saccadic eye movements: Role of primate superior colliculus. Physiol. Rev., 66:118, 1986.
Spoor, T. C.: Atlas of Optic Nerve Disorders. New York, Raven Press, 1994.
Streri, A.: Seeing, Reaching, Touching: The Relations between Vision and Touch in Infancy. MIT Press, 1994.
Walsh, T. J.: Neuro-Ophthalmology: Clinical Signs and Symptoms. Baltimore, Williams & Wilkins, 1992.
Wills Eye Hospital: The Wills Eye Manual. Philadelphia, J. B. Lippincott, 1993.
Wolfe, J. M. (ed.): The Mind's Eye. New York, W. H. Freeman and Company, 1986.
Wright, K. W., et al.: Strabismus. Philadelphia, J. B. Lippincott, 1991.

The Sense of Hearing

CHAPTER 52

The purpose of this chapter is to describe and explain the mechanisms by which the ear receives sound waves, discriminates their frequencies, and, finally, transmits auditory information into the central nervous system, where its meaning is deciphered.

TYMPANIC MEMBRANE AND THE OSSICULAR SYSTEM

Conduction of Sound from the Tympanic Membrane to the Cochlea

Figure 52–1 shows the *tympanic membrane* (commonly called the *eardrum*) and the *ossicular system*, which conducts sound through the middle ear to the cochlea. The tympanic membrane is cone-shaped. Attached to the center of the tympanic membrane is the *handle* of the *malleus*. At its other end, the malleus is bound to the *incus* by ligaments, so that whenever the malleus moves, the incus moves with it. The opposite end of the incus in turn articulates with the stem of the *stapes*, and the *faceplate* of the stapes lies against the membranous labyrinth in the opening of the oval window, where sound waves are conducted into the inner ear, the *cochlea*.

The ossicles of the middle ear are suspended by ligaments in such a way that the combined malleus and incus act as a single lever having its fulcrum approximately at the border of the tympanic membrane.

The articulation of the incus with the stapes causes the stapes to push forward on the cochlear fluid every time the tympanic membrane and the handle of the malleus move inward and to pull backward on the

fluid every time the malleus moves outward, which promotes inward and outward motion of the faceplate at the oval window.

The handle of the malleus is constantly pulled inward by the *tensor tympani muscle*, which keeps the tympanic membrane tensed. This allows sound vibrations on *any* portion of the tympanic membrane to be transmitted to the malleus, which would not be true if the membrane were lax.

IMPEDANCE MATCHING BY THE OSSICULAR SYSTEM. The amplitude of movement of the stapes faceplate with each sound vibration is only three fourths as much as the amplitude of the handle of the malleus. Therefore, the ossicular lever system does not increase the movement distance of the stapes, as is commonly believed. Instead, the system actually reduces the distance but increases the *force* of movement about 1.3 times. In addition, the surface area of the tympanic membrane is about 55 square millimeters, whereas the surface area of the stapes averages 3.2 square millimeters. This 17-fold difference times the 1.3-fold ratio of the lever system causes about 22 times as much *pressure* to be exerted on the fluid of the cochlea as is exerted by the sound wave against the tympanic membrane. Because fluid has far greater inertia than air, it is easily understood that increased amounts of pressure are needed to cause vibration in the fluid. Therefore, the tympanic membrane and ossicular system provide *impedance matching* between the sound waves in air and the sound vibrations in the fluid of the cochlea. Indeed, the impedance matching is about 50 to 75 per cent of perfect for sound frequencies between 300 and 3000 cycles per second, which allows utilization of most of the energy in the incoming sound waves.

663

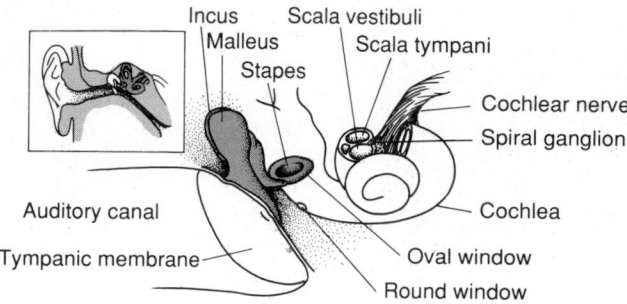

Figure 52–1. Tympanic membrane, ossicular system of the middle ear, and inner ear.

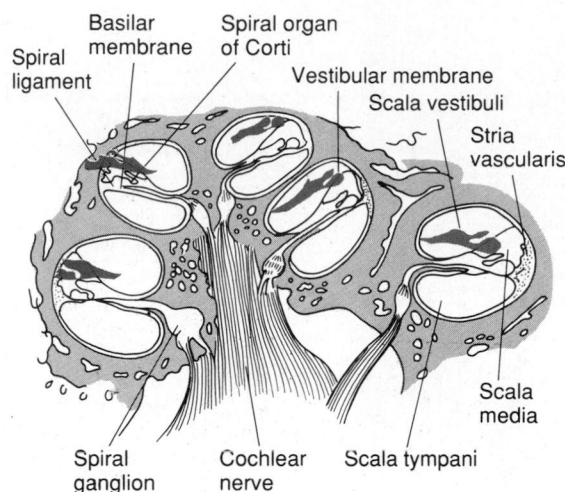

Figure 52–2. Cochlea. (From Goss [ed.]: Gray's Anatomy of the Human Body. Philadelphia, Lea & Febiger.)

In the absence of the ossicular system and tympanum, sound waves can travel directly through the air of the middle ear and enter the cochlea at the oval window. However, the sensitivity for hearing is then 15 to 20 decibels less than for ossicular transmission—equivalent to a decrease from a medium to a barely perceptible voice level.

ATTENUATION OF SOUND BY CONTRACTION OF THE STAPEDIUS AND TENSOR TYMPANI MUSCLES. When loud sounds are transmitted through the ossicular system into the central nervous system, a reflex occurs after a latent period of 40 to 80 milliseconds to cause contraction especially of the *stapedius muscle* and to a lesser extent the *tensor tympani muscle*. The tensor tympani muscle pulls the handle of the malleus inward while the stapedius muscle pulls the stapes outward. These two forces oppose each other and thereby cause the entire ossicular system to develop a high degree of rigidity, thus greatly reducing the ossicular conduction of low-frequency sound, mainly frequencies below 1000 cycles per second.

This *attenuation reflex* can reduce the intensity of lower-frequency sound transmission by 30 to 40 decibels, which is about the same difference as that between a loud voice and the sound of a whisper. The function of this mechanism is probably twofold:

1. To *protect* the cochlea from damaging vibrations caused by excessively loud sound.
2. To *mask* low-frequency sounds in loud environments. This usually removes a major share of the background noise and allows a person to concentrate on sounds above 1000 cycles per second, where most of the pertinent information in voice communication is transmitted.

Another function of the tensor tympani and stapedius muscles is to decrease a person's hearing sensitivity to his or her own speech. This effect is activated by collateral signals transmitted to these muscles at the same time that the brain activates the voice mechanism.

Transmission of Sound Through Bone

Because the inner ear, the *cochlea,* is embedded in a bony cavity in the temporal bone called the *bony labyrinth,* vibrations of the entire skull can cause fluid vibrations in the cochlea itself. Therefore, under appropriate conditions, a tuning fork or an electronic vibrator placed on any bony protuberance of the skull, but especially on the mastoid process, causes the person to hear the sound. The energy available even in loud sound in the air, however, is not sufficient to cause hearing through the bone except when a special electromechanical sound-transmitting device is applied directly to the bone.

THE COCHLEA

Functional Anatomy of the Cochlea

The cochlea is a system of coiled tubes, shown in Figure 52–1 and in cross section in Figures 52–2 and 52–3. It consists of three tubes coiled side by side: (1) the *scala vestibuli*, (2) the *scala media*, and (3) the *scala tympani*. The scala vestibuli and scala media are separated from each other by *Reissner's membrane* (also called the *vestibular membrane*), shown in Figure 52–3; the scala tympani and scala media are separated from each other by the *basilar membrane*. On the surface of the basilar membrane lies the *organ of Corti*, which contains a series of electromechanically sensitive cells, the *hair cells*. They are the receptive end organs that generate nerve impulses in response to sound vibrations.

Figure 52–4 diagrams the functional parts of the uncoiled cochlea for conduction of sound vibrations. First, note that Reissner's membrane is missing from this figure. This membrane is so thin and so easily moved that it does not obstruct the passage of sound vibrations from the scala vestibuli into the scala media. Therefore, so far as the conduction of sound is concerned, the scala vestibuli and scala media are considered to be a single chamber. (The importance of Reissner's membrane is to maintain a special fluid in

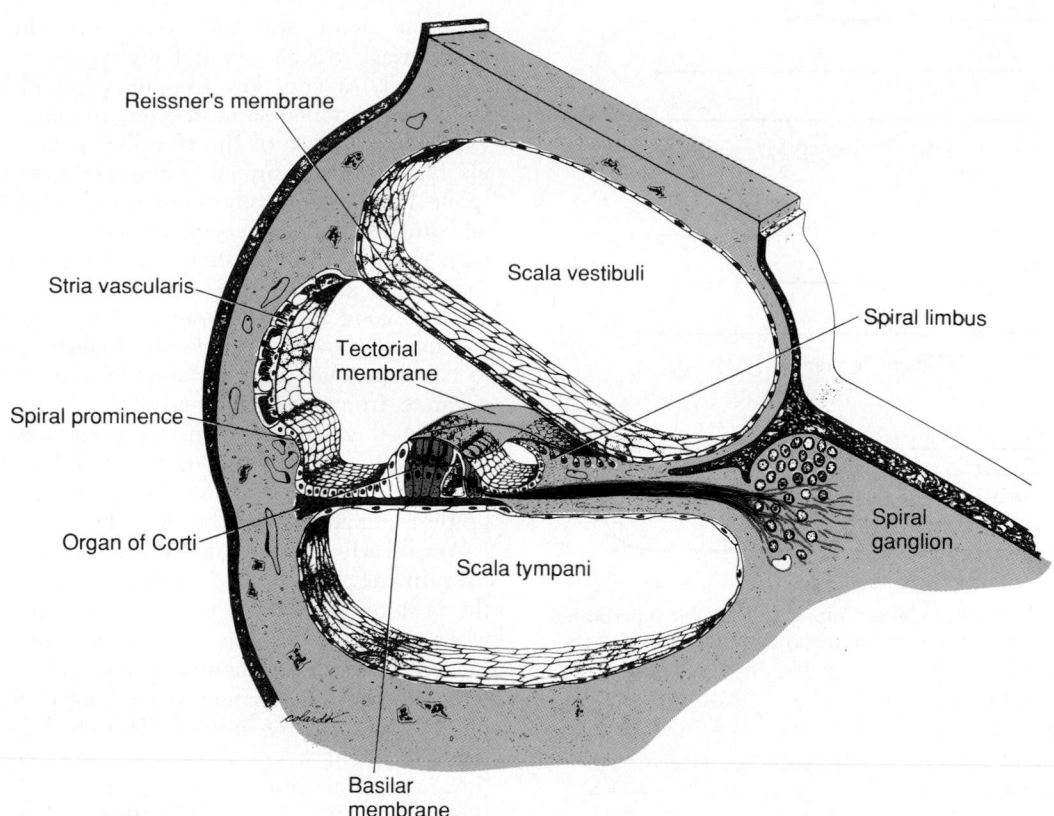

Figure 52–3. Section through one of the turns of the cochlea. (Drawn by Sylvia Colard Keene. From Fawcett: A Textbook of Histology, 11th Ed. Philadelphia, W. B. Saunders Company, 1986.)

the scala media that is required for normal function of the sound-receptive hair cells, as discussed later in the chapter.)

Sound vibrations enter the scala vestibuli from the faceplate of the stapes at the oval window. The faceplate covers this window and is connected with the window's edges by a relatively loose annular ligament so that it can move inward and outward with the sound vibrations. Inward movement causes the fluid to move into the scala vestibuli and scala media, and outward movement causes the fluid to move backward.

BASILAR MEMBRANE AND RESONANCE IN THE COCHLEA. The basilar membrane is a fibrous membrane that separates the scala media from the scala tympani. It contains 20,000 to 30,000 *basilar fibers* that project from the bony center of the cochlea, the *modiolus*, toward the outer wall. These fibers are stiff, elastic, reedlike structures that are fixed at their basal ends in the central bony structure of the cochlea (the modiolus) but not fixed at their distal ends, except that the distal ends are embedded in the loose basilar membrane. Because the fibers are stiff and free at one end, they can vibrate like reeds of a harmonica.

The lengths of the basilar fibers increase progressively as one goes from the base of the cochlea to its apex, from a length of about 0.04 millimeter near the oval and round windows to 0.5 millimeter at the tip of the cochlea, a 12-fold increase in length.

The diameters of the fibers, on the other hand, decrease from the base to the helicotrema, so that their overall stiffness decreases more than 100-fold. As a result, the stiff, short fibers near the oval window of the cochlea vibrate best at a high frequency, whereas the long limber fibers near the tip of the cochlea vibrate best at a low frequency.

Thus, high-frequency resonance of the basilar membrane occurs near the base, where the sound waves enter the cochlea through the oval window; low-frequency resonance occurs near the apex, mainly because of differences in stiffness of the fibers but also because of increased "loading" of the basilar membrane with extra masses of fluid that must vibrate with the membrane at the apex.

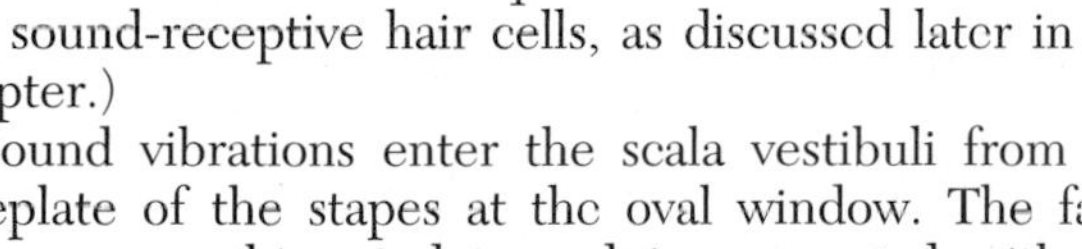

Figure 52–4. Movement of fluid in the cochlea after forward thrust of the stapes.

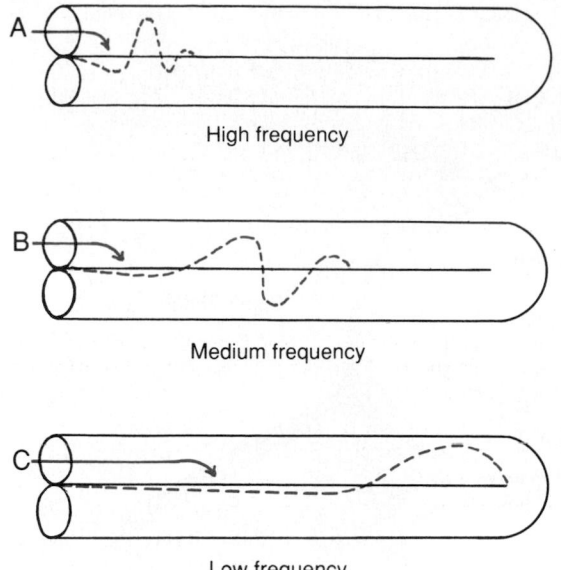

Figure 52–5. "Traveling waves" along the basilar membrane for high-, medium-, and low-frequency sounds.

Transmission of Sound Waves in the Cochlea—The "Traveling Wave"

When the foot of the stapes moves inward against the oval window, the round window must bulge outward because the cochlea is bounded on all sides by bony walls. Therefore, the initial effect of a sound wave entering at the oval window is to cause the basilar membrane at the base of the cochlea to bulge in the direction of the round window. However, the elastic tension that is built up in the basilar fibers as they bend toward the round window initiates a wave that "travels" along the basilar membrane toward the helicotrema, as shown in Figure 52–5. Figure 52–5A shows movement of a high-frequency wave down the basilar membrane; Figure 52–5B, a medium-frequency wave; and Figure 52–5C, a very low frequency wave. Movement of the wave along the basilar membrane is comparable to the movement of a pressure wave along the arterial walls, which is discussed in Chapter 15, and it is also comparable to the wave that travels along the surface of a pond.

PATTERN OF VIBRATION OF THE BASILAR MEMBRANE FOR DIFFERENT SOUND FREQUENCIES. Note in Figure 52–5 the different patterns of transmission for sound waves of different frequencies. Each wave is relatively weak at the outset but becomes strong when it reaches that portion of the basilar membrane that has a natural resonant frequency equal to the respective sound frequency. At this point, the basilar membrane can vibrate back and forth with such ease that the energy in the wave is dissipated. Consequently, the wave dies out at this point and fails to travel the remaining distance along the basilar membrane. Thus, a high-frequency sound wave travels only a short distance along the basilar membrane before it reaches its

resonant point and dies out; a medium-frequency sound wave travels about halfway and then dies out; and finally, a very low frequency sound wave travels the entire distance along the membrane.

Another feature of the traveling wave is that it travels fast along the initial portion of the basilar membrane but progressively more slowly as it goes further and further into the cochlea. The cause of this is the high coefficient of elasticity of the basilar fibers near the stapes and a progressively decreasing coefficient further along the membrane. This rapid initial transmission of the wave allows the high-frequency sounds to travel far enough into the cochlea to spread out and separate from one another on the basilar membrane. Without this, all the high-frequency waves would be bunched together within the first millimeter or so of the basilar membrane, and their frequencies could not be discriminated one from the other.

Amplitude Pattern of Vibration of the Basilar Membrane. The dashed curves of Figure 52–6A show the position of a sound wave on the basilar membrane when the stapes (a) is all the way inward, (b) has moved back to the neutral point, (c) is all the way outward, and (d) has moved back again to the neutral point but is moving inward. The shaded area around these different waves shows the extent of vibration of the basilar membrane during a complete vibratory cycle. This is the *amplitude pattern of vibration* of the basilar membrane for this particular sound frequency.

Figure 52–6B shows the amplitude patterns of vibration for different frequencies, demonstrating that the maximum amplitude for 8000 cycles occurs near

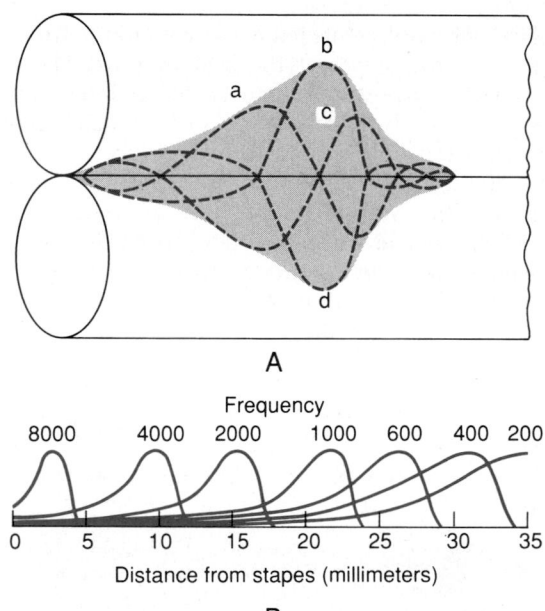

Figure 52–6. *A*, Amplitude pattern of vibration of the basilar membrane for a medium-frequency sound. *B*, Amplitude patterns for sounds of all frequencies between 200 and 8000 per second, showing the points of maximum amplitude (the resonance points) on the basilar membrane for the different frequencies.

the base of the cochlea, whereas that for frequencies less than 200 cycles per second is all the way at the tip of the basilar membrane near the helicotrema where the scala vestibuli opens into the scala tympani.

The principal method by which sound frequencies, especially those above 200 cycles per second, are discriminated from one another is based on the "place" of maximum stimulation of the nerve fibers from the organ of Corti lying on the basilar membrane, as is explained in the next section.

Function of the Organ of Corti

The organ of Corti, in Figures 52–2, 52–3, and 52–7, is the receptor organ that generates nerve impulses in response to vibration of the basilar membrane. Note that the organ of Corti lies on the surface of the basilar fibers and basilar membrane. The actual sensory receptors in the organ of Corti are two types of *hair cells*: a single row of *internal* (or "*inner*") *hair cells*, numbering about 3500 and measuring about 12 micrometers in diameter, and three to four rows of *external* (or "*outer*") *hair cells*, numbering about 12,000 and having diameters of only about 8 micrometers. The bases and sides of the hair cells synapse with a network of cochlear nerve endings. Between 90 and 95 per cent of these endings terminate on the inner hair cells, which emphasizes the special importance of these for the detection of sound. The nerve fibers from these endings lead to the *spiral ganglion of Corti*, which lies in the modiolus (the center) of the cochlea. The spiral ganglion in turn sends axons, a total of about 30,000, into the *cochlear nerve* and then into the central nervous system at the level of the upper medulla. The relation of the organ of Corti to the spiral ganglion and to the cochlear nerve is shown in Figure 52–2.

EXCITATION OF THE HAIR CELLS. Note in Figure 52–7 that minute hairs, or *stereocilia*, project upward from the hair cells and either touch or are embedded in the surface gel coating of the *tectorial membrane*, which lies above the stereocilia in the scala media. These hair cells are similar to the hair cells found in the macula and cristae ampullaris of the vestibular

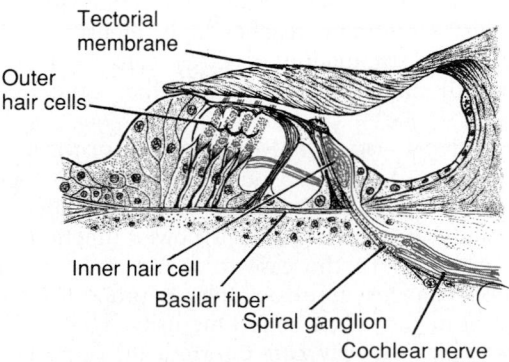

Figure 52–7. Organ of Corti, showing especially the hair cells and the tectorial membrane against the projecting hairs.

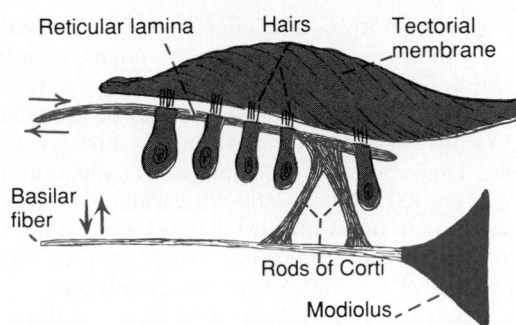

Figure 52–8. Stimulation of the hair cells by the to-and-fro movement of the hairs projecting into the gel coating of the tectorial membrane.

apparatus, which are discussed in Chapter 55. Bending of the hairs in one direction depolarizes the hair cells, and bending them in the opposite direction hyperpolarizes them. This in turn excites the nerve fibers synapsing with their bases.

Figure 52–8 shows the mechanism by which vibration of the basilar membrane excites the hair endings. The upper ends of the hair cells are fixed tightly in a rigid structure composed of a flat plate, called the *reticular lamina*, supported by triangular *rods of Corti*, which in turn are attached tightly to the bases of the basilar fibers. Therefore, the basilar fiber, the rods of Corti, and the reticular lamina all move as a rigid unit.

Upward movement of the basilar fiber rocks the reticular lamina upward and *inward* toward the modiolus. Then, when the basilar membrane moves downward, the reticular lamina rocks downward and *outward*. The inward and outward motion causes the hairs to shear back and forth against the tectorial membrane. Thus, the hair cells are excited whenever the basilar membrane vibrates.

AUDITORY SIGNALS ARE TRANSMITTED MAINLY BY THE INNER HAIR CELLS. Even though there are three to four times as many outer hair cells as inner hair cells, about 90 per cent of the auditory nerve fibers are stimulated by the inner cells rather than by the outer cells. Yet, despite this, if the outer cells are damaged while the inner cells remain fully functional, a large amount of hearing loss occurs. Therefore, a concept has been proposed that the outer hair cells in some way control the sensitivity of the inner hair cells for different sound pitches. In support, a large number of retrograde nerve fibers pass from the brain stem to the vicinity of the outer hair cells, suggesting that a retrograde nervous mechanism for control of the ear's sensitivity to different sound pitches might be effected through the outer hair cells.

HAIR CELL RECEPTOR POTENTIALS AND EXCITATION OF AUDITORY NERVE FIBERS. The stereocilia, the "hairs" protruding from the ends of the hair cells, are stiff structures because each of these has a rigid structural protein framework, as explained for all cilia in Chapter 2. Each hair cell has about 100 stereocilia on its apical border. These stereocilia become progressively longer on the side away from the modiolus, and the tops of the shorter stereocilia are each attached by a thin filament to the side of its adjacent longer stereocilium. Therefore, whenever the cilia are bent in the direction of the longer ones, the tips of the smaller stereocilia are tugged outward from the surface of the hair cell. This causes a mechanical transduction that opens 200 to 300 cation-conducting channels, allowing rapid movement of positively charged potassium ions into the tips of the stereocilia, which in turn causes depolarization of the entire hair cell membrane.

Thus, when the basilar fibers bend toward the scala vestibuli, the hair cells depolarize, and in the opposite direction they hyperpolarize, thereby generating an alternating hair cell receptor potential. This in turn stimulates the cochlear nerve endings that synapse with the bases of the hair cells. It is believed that a rapidly acting neurotransmitter is released by the hair cells at these synapses during depolarization. It is possible that the transmitter substance is glutamate, but this is not certain.

Endocochlear Potential. To explain even more fully the electrical potentials generated by the hair cells, we need to explain another electrical phenomenon called the endocochlear potential: The scala media is filled with a fluid called *endolymph*, in contradistinction to the *perilymph* present in the scala vestibuli and scala tympani. The scala vestibuli and scala tympani communicate directly with the subarachnoid space around the brain, so that the perilymph is almost identical with cerebrospinal fluid. On the other hand, the endolymph that fills the scala media is an entirely different fluid secreted by the *stria vascularis*, a highly vascular area on the outer wall of the scala media. Endolymph contains a high concentration of potassium and a low concentration of sodium, which is exactly opposite to the contents of perilymph.

An electrical potential of about +80 millivolts exists all the time between endolymph and perilymph, with positivity inside the scala media and negativity outside. This is called the *endocochlear potential*, and it is generated by continual transport of positive potassium ions into the scala media by the stria vascularis.

The importance of the endocochlear potential is that the tops of the hair cells project through the reticular lamina and are bathed by the endolymph of the scala media, whereas perilymph bathes the lower bodies of the hair cells. Furthermore, the hair cells have a negative intracellular potential of −70 millivolts with respect to the perilymph but −150 millivolts with respect to the endolymph at their upper surfaces, where the hairs project into the endolymph. It is believed that this high electrical potential at the tips of the stereocilia greatly sensitizes the cell, thereby increasing its ability to respond to the slightest sound.

Determination of Sound Frequency— The "Place" Principle

From earlier discussions in this chapter, it is already apparent that low-frequency sounds cause maximal activation of the basilar membrane near the apex of the cochlea, sounds of high frequency activate the basilar membrane near the base of the cochlea, and interme-diate frequencies activate the membrane at intermediate distances between these two extremes. Furthermore, there is spatial organization of the nerve fibers in the cochlear pathway all the way from the cochlea to the cerebral cortex. And recording of signals in the auditory tracts of the brain stem and in the auditory receptive fields of the cerebral cortex shows that specific neurons are activated by specific sound frequencies. Therefore, the major method used by the nervous system to detect different sound frequencies is to determine the positions along the basilar membrane that are most stimulated. This is called the *place principle* for determination of sound frequency.

Yet, referring again to Figure 52–6, one can see that the distal end of the basilar membrane at the helicotrema is stimulated by all sound frequencies below 200 cycles per second. Therefore, it has been difficult to understand from the place principle how one can differentiate between low sound frequencies, from 200 down to 20. It is postulated that these low frequencies are discriminated mainly by the so-called *volley* or *frequency principle*. That is, low-frequency sounds, from 20 to 2000 cycles per second, can cause volleys of impulses synchronized at the same frequencies, and these impulses are transmitted by the cochlear nerve into the cochlear nuclei. It is believed that the cochlear nuclei then distinguish the different frequencies. In fact, destruction of the entire apical half of the cochlea, which destroys the basilar membrane where all the lower-frequency sounds are normally detected, still does not eliminate the discrimination of low-frequency sound.

Determination of Loudness

Loudness is determined by the auditory system in at least three ways: First, as the sound becomes louder, the amplitude of vibration of the basilar membrane and hair cells also increases, so that the hair cells excite the nerve endings at more rapid rates. Second, as the amplitude of vibration increases, it causes more and more of the hair cells on the fringes of the resonating portion of the basilar membrane to become stimulated, thus causing *spatial summation* of impulses— that is, transmission through many nerve fibers, rather than through a few. Third, outer hair cells do not become stimulated significantly until the vibration of the basilar membrane reaches high intensity, and it is possible that stimulation of these cells in some way apprises the nervous system that the sound is then loud.

DETECTION OF CHANGES IN LOUDNESS—THE POWER LAW. As pointed out in Chapter 46, a person interprets changes in intensity of sensory stimuli approximately in proportion to a power function of the actual intensity. In the case of sound, the interpreted sensation changes approximately in proportion to the cube root of the actual sound intensity. To express this another way, the ear can discriminate differences in sound intensity from the softest whisper to the loudest possible noise, representing *approximately 1 trillion*

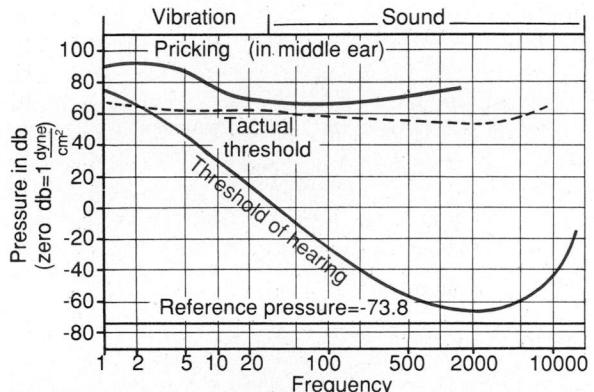

Figure 52–9. Relation of the threshold of hearing and of somesthetic perception (pricking and tactual threshold) to the sound energy level at each sound frequency. (Modified from Stevens and Davis: Hearing. New York, John Wiley & Sons.)

times increase in sound energy or 1 million times increase in amplitude of movement of the basilar membrane. Yet the ear interprets this much difference in sound level as approximately a 10,000-fold change. Thus, the scale of intensity is greatly "compressed" by the sound perception mechanisms of the auditory system. This allows a person to interpret differences in sound intensities over an extremely wide range, a far broader range than would be possible were it not for compression of the intensity scale.

DECIBEL UNIT. Because of the extreme changes in sound intensities that the ear can detect and discriminate, sound intensities are usually expressed in terms of the logarithm of their actual intensities. A 10-fold increase in sound energy is called 1 *bel*, and 0.1 bel is called 1 *decibel*. One decibel represents an actual increase in sound energy of 1.26 times.

Another reason for using the decibel system in expressing changes in loudness is that, in the usual sound intensity range for communication, the ears can barely distinguish an approximately 1-decibel *change* in sound intensity.

THRESHOLD FOR HEARING SOUND AT DIFFERENT FREQUENCIES. Figure 52–9 shows the pressure thresholds at which sounds of different frequencies can barely be heard by the ear. This figure demonstrates that a 3000-cycle-per-second sound can be heard even when its intensity is as low as 70 decibels below 1 dyne/cm² sound pressure level, which is one ten-millionth microwatt per square centimeter. On the other hand, a 100-cycle-per-second sound can be detected only if its intensity is 10,000 times as great as this.

FREQUENCY RANGE OF HEARING. The frequencies of sound that a young person can hear, before aging has occurred in the ears, is stated to be between 20 and 20,000 cycles per second. However, referring again to Figure 52–9, we see that the sound range depends to a great extent on loudness. If the loudness is 60 decibels below 1 dyne/cm² sound pressure level, the sound range is 500 to 5000 cycles per second; only with intense sounds can the complete range of 20 to 20,000 cycles be achieved. In old age, this frequency range falls to 50

to 8000 cycles per second or less, as discussed later in the chapter.

CENTRAL AUDITORY MECHANISMS

Auditory Pathway

Figure 52–10 shows the major auditory pathways. It shows that nerve fibers from the *spiral ganglion of Corti* enter the *dorsal* and *ventral cochlear nuclei* located in the upper part of the medulla. At this point, all the fibers synapse, and second-order neurons pass mainly to the opposite side of the brain stem to terminate in the *superior olivary nucleus*. Some second-order fibers also pass ipsilaterally to the superior olivary nucleus on the same side. From the superior olivary nucleus, the auditory pathway then passes upward through the *lateral lemniscus*; some of the fibers terminate in the *nucleus of the lateral lemniscus*. Many bypass this nucleus and pass on to the inferior colliculus, where all or almost all of them terminate. From here, the pathway passes to the *medial geniculate nucleus*, where all the fibers again synapse. And finally, the pathway proceeds by way of the *auditory radiation* to the *auditory cortex*, located mainly in the superior gyrus of the temporal lobe.

Several points of importance should be noted. First, signals from both ears are transmitted through the pathways of both sides of the brain with slight preponderance of transmission in the contralateral pathway. In at least three places in the brain stem, crossing-over occurs between the two pathways: (1) in the trapezoid body, (2) in the commissure of Probst between the

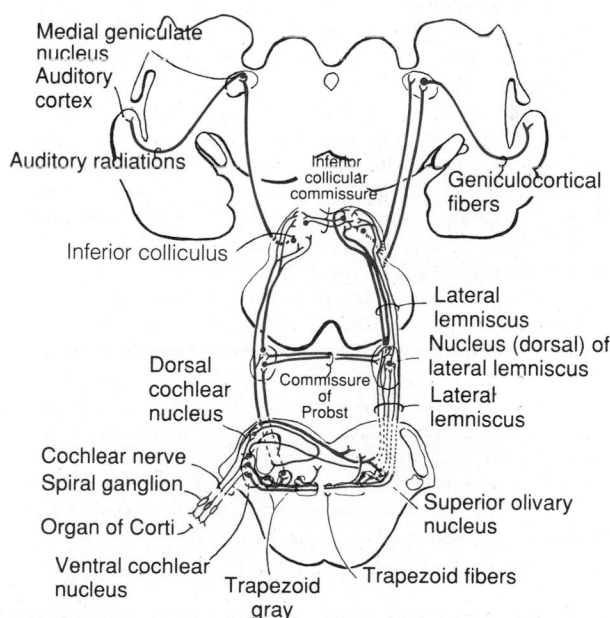

Figure 52–10. Auditory pathway. (Modified from Crosby, E., Humphrey, T., and Lauer, E.: Correlative Anatomy of the Nervous System. New York, Macmillan Publishing Co., 1962. Copyright 1962 by Macmillan Publishing Co. Reprinted with permission.)

two nuclei of the lateral lemnisci, and (3) in the commissure connecting the two inferior colliculi.

Second, many collateral fibers from the auditory tracts pass directly into the *reticular activating system of the brain stem.* This system projects diffusely upward in the brain stem and downward into the spinal cord and activates the entire nervous system in response to a loud sound. Other collaterals go to the *vermis of the cerebellum,* which is also activated instantaneously in the event of a sudden noise.

Third, a high degree of spatial orientation is maintained in the fiber tracts from the cochlea all the way to the cortex. In fact, there are *three* spatial representations of sound frequencies in the cochlear nuclei, *two* representations in the inferior colliculi, *one precise* representation for discrete sound frequencies in the auditory cortex, and *at least five other less precise* representations in the auditory cortex and auditory association areas.

FIRING RATES AT DIFFERENT LEVELS OF THE AUDITORY PATHWAY. Single nerve fibers entering the cochlear nuclei from the auditory nerve can fire at rates up to at least 1000 per second, the rate being determined mainly by the loudness of the sound. At sound frequencies up to 2000 to 4000 cycles per second, the auditory nerve impulses are often synchronized with the sound waves, but they do not necessarily occur with every wave.

In the auditory tracts of the brain stem, the firing is usually no longer synchronized with the sound frequency except at sound frequencies below 200 cycles per second. And above the level of the inferior colliculi, even this synchronization is mainly lost. These findings demonstrate that the sound signals are not transmitted unchanged directly from the ear to the higher levels of the brain; instead, information from the sound signals begins to be dissected from the impulse traffic at levels as low as the cochlear nuclei. We have more to say about this later, especially in relation to perception of direction from which sound comes.

Another significant feature of the auditory pathways is that low rates of impulse firing continue even in the absence of sound, with this occurring from the cochlear nerve fibers all the way to the auditory cortex. When the basilar membrane moves toward the scala vestibuli, the impulse traffic increases; when the basilar membrane moves toward the scala tympani, the impulse traffic decreases. Thus, the presence of this background signal allows information to be transmitted from the basilar membrane when the membrane moves in either direction: positive information in one direction and negative information in the opposite direction. Were it not for the background signal, only the positive half of the information could be transmitted. This type of so-called carrier wave method for transmitting information is used in many parts of the brain, as discussed in several of the succeeding chapters.

Function of the Cerebral Cortex in Hearing

The projection areas of the auditory pathway to the cerebral cortex are shown in Figure 52–11, which

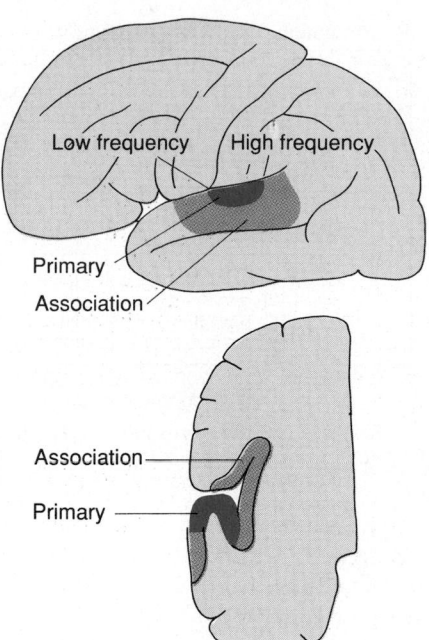

Figure 52–11. Auditory cortex.

demonstrates that the auditory cortex lies principally on the *supratemporal plane of the superior temporal gyrus* but also extends into the *lateral border of the temporal lobe,* over much of the *insular cortex,* and even into the lateral portion of the *parietal operculum.*

Two separate areas are shown in Figure 52–11: the *primary auditory cortex,* shown in dark red, and the *auditory association cortex* (also called the *secondary auditory cortex*), shown in light red. The primary auditory cortex is directly excited by projections from the medial geniculate body, whereas the auditory association areas are excited secondarily by impulses from the primary auditory cortex as well as by projections from thalamic association areas adjacent to the medial geniculate body.

SOUND FREQUENCY PERCEPTION IN THE PRIMARY AUDITORY CORTEX. At least six *tonotopic maps* have been found in the primary auditory cortex and auditory association areas. In each of these maps, high-frequency sounds excite neurons at one end of the map, whereas low-frequency sounds excite the neurons at the opposite end. In most, the low-frequency sounds are located anteriorly, as shown in Figure 52–11, and the high-frequency sounds are located posteriorly. This is not true for all the maps. The question that one must ask is why does the auditory cortex have so many different tonotopic maps? The answer is presumably that each of the separate areas dissects out some specific feature of the sounds. For instance, one of the large maps in the primary auditory cortex almost certainly discriminates the sound frequencies themselves and gives the person the psychic sensation of sound pitches. Another one of the maps probably is used to detect the direction from which the sound comes. And other auditory cortex areas detect special qualities, such as sudden onset of sounds, or perhaps special modulations of sounds, such as noise versus pure frequency sounds, and so forth.

The frequency range to which each individual neuron in the auditory cortex responds is much narrower than that in the cochlear and brain stem relay nuclei. Referring back to Figure 52–6B, we note that the basilar membrane near the base of the cochlea is stimulated by all frequency sounds, and in the cochlear nuclei, this same breadth of sound representation is found. Yet by the time the excitation has reached the cerebral cortex, most sound-responsive neurons respond only to a narrow range of frequencies, rather than a broad range. Therefore, somewhere along the pathway, processing mechanisms "sharpen" the frequency response. It is believed that this sharpening effect is caused mainly by the phenomenon of lateral inhibition, which is discussed in Chapter 46 in relation to mechanisms for transmitting information in nerves. That is, stimulation of the cochlea at one frequency causes inhibition of signals caused by sound frequencies on either side of the stimulated frequency, which results from collateral fibers angling off the primary signal pathway and exerting inhibitory influences on adjacent pathways. The same effect has been demonstrated to be important in sharpening patterns of somesthetic images, visual images, and other types of sensations.

A large share of the neurons in the auditory cortex, especially in the auditory association cortex, do not respond to specific sound frequencies in the ear. It is believed that these neurons "associate" different sound frequencies with one another or associate sound information with information from other sensory areas of the cortex. Indeed, the parietal portion of the auditory association cortex partly overlaps somatic sensory area II, which could provide easy opportunity for association of auditory information with somatic sensory information.

DISCRIMINATION OF SOUND "PATTERNS" BY THE AUDITORY CORTEX. Complete bilateral removal of the auditory cortex does not prevent a cat or monkey from detecting sounds or reacting in a crude manner to the sounds. However, it does greatly reduce or sometimes even abolish its ability to discriminate different sound pitches and especially *patterns of sound*. For instance, an animal that has been trained to recognize a combination or sequence of tones, one following the other in a particular pattern, loses this ability when the auditory cortex is destroyed; furthermore, it cannot relearn this type of response. Therefore, the auditory cortex is important in the discrimination of *tonal* and *sequential sound patterns*.

Destruction of both primary auditory cortices in the human being is said to reduce greatly one's sensitivity for hearing. However, destruction on one side only slightly reduces hearing in the opposite ear but does not cause deafness in the ear because of the many crossover connections from side to side in the auditory neural pathway. This does affect one's ability to localize the source of sound because comparative signals in both cortices are required for this localization function.

Lesions in the human being that affect the auditory association areas but not the primary auditory cortex do not decrease the person's ability to hear and differentiate sound tones and even to interpret at least simple patterns of sound. However, he or she will often be unable to interpret the *meaning* of the sound heard. For instance, lesions in the posterior portion of the superior temporal gyrus, which is called Wernicke's area and is part of the auditory association cortex, often make it impossible for the person to interpret the meanings of words even though he hears them perfectly well and can even repeat them. These functions of the auditory association areas and their relation to the overall intellectual functions of the brain are discussed in detail in Chapter 57.

Determination of the Direction from Which Sound Emanates

A person determines the horizontal direction from which sound emanates by two principal mechanisms: (1) by the time lag between the entry of sound into one ear and into the opposite ear and (2) by the difference between the intensities of the sounds in the two ears. The first mechanism functions best at frequencies below 3000 cycles per second, and the intensity mechanism operates best at higher frequencies because the head acts as a sound barrier at these frequencies. The time lag mechanism discriminates direction much more exactly than the intensity mechanism because the time lag mechanism does not depend on extraneous factors but only on an exact interval of time between two acoustical signals. If a person is looking straight toward the sound, the sound reaches both ears at exactly the same instant, whereas if the right ear is closer to the sound than the left ear, the sound signals from the right ear enter the brain ahead of those from the left ear.

The two mechanisms above cannot tell whether the sound is emanating from in front of the person or behind or from above or below. This discrimination is achieved mainly by the *pinnae* of the two ears. The shape of the pinna changes the *quality* of the sound entering the ear, depending on the direction from which the sound comes. It does this by emphasizing specific sound frequencies from the different directions.

NEURAL MECHANISMS FOR DETECTING SOUND DIRECTION. Destruction of the auditory cortex on both sides of the brain, whether in human beings or lower mammals, causes loss of almost all ability to detect the direction from which sound comes. Yet the mechanism for this detection process begins in the superior olivary nuclei in the brain stem, even though it requires neural pathways all the way from these nuclei to the cortex for interpretation of the signals. The mechanism is believed to be the following.

First, the superior olivary nucleus is divided into two sections: (1) the *medial superior olivary nucleus* and

(2) the *lateral superior olivary nucleus*. The lateral nucleus is concerned with detecting the direction from which the sound is coming by the *difference in intensities of the sound* reaching the two ears, presumably by simply comparing the two intensities and sending an appropriate signal to the auditory cortex to estimate the direction.

The *medial superior olivary nucleus*, on the other hand, has a specific mechanism for *detecting time lag between acoustical signals entering the two ears*. This nucleus contains large numbers of neurons that have two major dendrites, one projecting to the right and the other to the left. The acoustical signal from the right ear impinges on the right dendrite, and the signal from the left ear impinges on the left dendrite. The intensity of excitation of each neuron is highly sensitive to a specific time lag between the two acoustical signals from the two ears. The neurons near one border of the nucleus respond maximally to a short time lag, whereas those near the opposite border respond to a long time lag; those in between respond to intermediate time lags. Thus, a spatial pattern of neuronal stimulation develops in the medial superior olivary nucleus, with sound from directly in front of the head stimulating one set of olivary neurons maximally and sounds from different side angles stimulating other sets of neurons on opposite sides of the straight front neurons. This spatial orientation of signals is then transmitted all the way to the auditory cortex, where sound direction is determined by the locus of the maximally stimulated neurons. It is believed that the signals for determining sound direction are transmitted through a different pathway and excite a different locus in the cerebral cortex from the transmission pathway and termination locus for the tonal patterns of sound.

This mechanism for detection of sound direction indicates again how specific information in sensory signals is dissected out as the signals pass through different levels of neuronal activity. In this case, the "quality" of sound direction is separated from the "quality" of sound tones at the level of the superior olivary nuclei.

Centrifugal Signals from the Central Nervous System to Lower Auditory Centers

Retrograde pathways have been demonstrated at each level of the nervous system from the auditory cortex to the cochlea. The final pathway is mainly from the superior olivary nucleus to the hair cells themselves in the organ of Corti.

These retrograde fibers are inhibitory. Indeed, direct stimulation of discrete points in the olivary nucleus has been shown to inhibit specific areas of the organ of Corti, reducing their sound sensitivities 15 to 20 decibels. One can readily understand how this could allow a person to direct attention to sounds of particular qualities while rejecting sounds of other qualities. This is readily demonstrated when one listens to a single instrument in a symphony orchestra.

HEARING ABNORMALITIES

Types of Deafness

Deafness is usually divided into two types: first, that caused by impairment of the cochlea or auditory nerve, which is usually classed as "nerve deafness," and, second, that caused by impairment of the mechanisms for transmitting sound into the cochlea, which is usually called "conduction deafness." If either the cochlea or the auditory nerve is destroyed, the person is permanently deaf. However, if the cochlea and nerve are still intact but the tympanum-ossicular system has been destroyed or ankylosed ("frozen" in place by fibrosis or calcification), sound waves can still be conducted into the cochlea by means of bone conduction from a sound generator applied to the skull.

AUDIOMETER. To determine the nature of hearing disabilities, the audiometer is used. Simply an earphone connected to an electronic oscillator capable of emitting pure tones ranging from low frequencies to high frequencies, the instrument is calibrated so that the zero intensity level of sound at each frequency is the loudness that can barely be heard by the normal person, based on previous studies of normal people. A calibrated volume control can increase or decrease the loudness of each tone above or below the zero level. If the loudness of a tone must be increased to 30 decibels above normal before it can be heard, the person is said to have a *hearing loss* of 30 decibels for that particular tone.

In performing a hearing test using an audiometer, one tests about 8 to 10 frequencies covering the auditory spectrum, and the hearing loss is determined for each of these frequencies. Then the so-called audiogram is plotted as shown in Figures 52–12 and 52–13, depicting the hearing loss for each of the frequencies in the auditory spectrum.

The audiometer, in addition to being equipped with an earphone for testing air conduction by the ear, is equipped with an electronic vibrator for testing bone conduction from the mastoid process into the cochlea.

Audiogram in Nerve Deafness. In nerve deafness— this term includes damage to the cochlea, the auditory nerve, or the central nervous system circuits from the

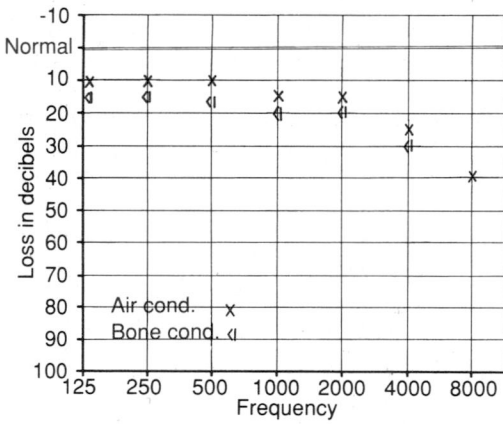

Figure 52–12. Audiogram of the old-age type of nerve deafness.

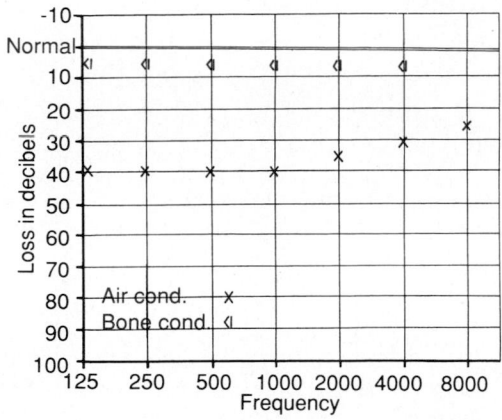

Figure 52–13. Audiogram of deafness resulting from middle ear sclerosis.

ear—the person has decreased or total loss of ability to hear sound as tested by both air conduction and bone conduction. An audiogram depicting partial nerve deafness is shown in Figure 52–12. In this figure, the deafness is mainly for high-frequency sound. Such deafness could be caused by damage to the base of the cochlea. This type of deafness occurs to some extent in almost all older people.

Other patterns of nerve deafness frequently occur as follows: (1) deafness for low-frequency sounds caused by excessive and prolonged exposure to very loud sounds (the rock band or the jet airplane engine) because low-frequency sounds are usually louder and more damaging to the organ of Corti and (2) deafness for all frequencies caused by drug sensitivity of the organ of Corti, especially sensitivity to some antibiotics, such as streptomycin, kanamycin, and chloramphenicol.

Audiogram in Conduction Deafness. A common type of deafness is that caused by fibrosis of the middle ear after repeated infection in the middle ear or fibrosis that occurs in the hereditary disease called *otosclerosis*. In this instance, the sound waves cannot be transmitted easily through the ossicles from the tympanic membrane to the oval window. Figure 52–13 shows an audiogram from a person with "middle ear deafness" of this type. In this case, the bone conduction is essentially normal, but air conduction is greatly depressed at all frequencies, more so at the low frequencies. In some instances of conduction deafness, the faceplate of the stapes becomes "ankylosed" by bony overgrowth to the edges of the oval window. In this case, the person becomes totally deaf for air conduction but can be made to hear again almost normally by removing the stapes and replacing it with a minute Teflon or metal prosthesis that transmits the sound from the incus to the oval window.

REFERENCES

Aitkin, L. M., et al.: Central neural mechanisms of hearing. In Darian-Smith, I. (ed.): Handbook of Physiology. Sec. 1, Vol. III. Bethesda, Md., American Physiological Society, 1984, p. 675.

Altschuler, R. A., et al.: Neurobiology of Hearing: The Central Auditory System. New York, Raven Press, 1991.

Altschuler, R. A., et al.: Neurobiology of Hearing: The Cochlea. New York, Raven Press, 1986.

Bailey, B. J., et al.: Head and Neck Surgery—Otolaryngology. Philadelphia, J. B. Lippincott, 1993.

Benjamin, B., et al.: A Color Atlas of Otorhinolaryngology. Philadelphia, J. B. Lippincott, 1994.

Borg, E., and Counter, S. A.: The middle-ear muscles. Sci. Am., August, 1989.

Bluestone, C. D., and Klein, J. O.: Otitis Media in Infants and Children. Philadelphia, W. B. Saunders Co., 1994.

Buser, P., and Imbert, M.: Audition. Cambridge, MA, The MIT Press, 1992.

Canalis, R. F., and Goodhill, V.: Goodhill's Textbook of Otology. Philadelphia, J. B. Lippincott, 1994.

Dhillon, R. S., and East, C. A.: Ear, Nose and Throat: An Illustrated Colour Text. New York, Churchill Livingstone, 1994.

Eybalin, M.: Neurotransmitters and Neuromodulators of the Mammalian Cochlea. Physiol. Rev., 73:309, 1993.

Fraysse, B., and Deguine, O.: Cochlear Implants: New Perspectives. Farmington, CT, S. Karger Publishers, Inc., 1993.

Fujimura, O.: Vocal Physiology: Voice Production, Mechanisms and Functions. New York, Raven Press, 1988.

Glasscock, M. E. III, et al.: Handbook of Vertigo. New York, Raven Press, 1990.

Glasscock, M., III: Shambaugh's Surgery of the Ear, 4th Ed. Philadelphia, W. B. Saunders Co., 1989.

Grandori, F., et al.: Cochlear Mechanisms and Otoacoustic Emissions. Farmington, CT, S. Karger Publishers, Inc., 1990.

Hawke, M., et al.: Diseases of the Ear: Clinical and Pathologic Aspects. Philadelphia, Lea & Febiger, 1987.

Hudspeth, A. J.: Mechanoelectrical transduction by hair cells in the acousticolateralis sensory system. Annu. Rev. Neurosci., 6:187, 1983.

Hudspeth, A. J.: The cellular basis of hearing: The biophysics of hair cells. Science, 230:745, 1985.

Jahn, A. F., and Santos-Sacchi, J.: Physiology of the Ear. New York, Raven Press, 1988.

Lee, K. J.: Essential Otolaryngology: Head and Neck Surgery. 6th Ed. Redding, MA, Appleton & Lange, 1994.

Lucente, F. E., and Sobol, S. M.: Essentials of Otolaryngology. New York, Raven Press, 1993.

Masterton, R. B., and Imig, T. J.: Neural mechanisms of sound localization. Annu. Rev. Physiol., 46:275, 1984.

Nadol, J. B. Jr., and Schuknecht, H. F.: Surgery of the Ear and Temporal Bone. New York, Raven Press, 1993.

Nager, G. T., and Hyams, V. J.: Pathology of the Ear and Temporal Bone. Baltimore, Williams & Wilkins, 1994.

Nelken, I., and Young, E. D.: Two separate inhibitory mechanisms shape the responses of dorsal cochlear nucleus type IV units to narrowband and wideband stimuli. J. Neurophysiol., 71:2446, 1994.

Patuzzi, R., and Robertson, D.: Tuning in the mammalian cochlea. Physiol. Rev., 68:1009, 1988.

Rhode, W. S.: Cochlear mechanisms. Annu. Rev. Physiol., 46:231, 1984.

Schuknecht, H. F.: Pathology of the Ear. Baltimore, Williams & Wilkins, 1993.

Singh, R. P.: Anatomy of Hearing and Speech. New York, Oxford University Press, 1980.

Sterkers, O., et al.: How are inner ear fluids formed? News Physiol. Sci., 2:176, 1987.

Syka, J., and Masterton, R. B. (eds.): Auditory Pathway. Structure and Function. New York, Plenum Publishing Corp., 1988.

Weiss, T. F.: Relation of receptor potentials of cochlear hair cells to spike discharges of cochlear neurons. Annu. Rev. Physiol., 46:247, 1984.

Wever, E. G., and Lawrence, M.: Physiological Acoustics. Princeton, Princeton University Press, 1954.

The Chemical Senses— Taste and Smell

CHAPTER 53

The senses of taste and smell allow us to separate undesirable or even lethal foods from those that are nutritious. And the sense of smell allows animals to recognize the proximity of other animals, or even individuals among animals. Finally, both senses are strongly tied to primitive emotional and behavioral functions of our nervous systems.

SENSE OF TASTE

Taste is mainly a function of the *taste buds* in the mouth, but it is common experience that one's sense of smell contributes strongly to taste perception. In addition, the texture of food, as detected by tactual senses of the mouth, and the presence of such substances in the food as pepper, which stimulate pain endings, greatly condition the taste experience. The importance of taste lies in the fact that it allows a person to select food in accord with desires and often in accord with the metabolic needs of the tissues for specific nutritive substances.

Primary Sensations of Taste

The identities of the specific chemicals that excite different taste receptors are still incomplete. Even so, psychophysiological and neurophysiological studies have identified at least 13 possible or probable chemical receptors in the taste cells as follows: 2 sodium receptors, 2 potassium receptors, 1 chloride receptor, 1 adenosine receptor, 1 inosine receptor, 2 sweet receptors, 2 bitter receptors, 1 glutamate receptor, and 1 hydrogen ion receptor.

For practical analysis of taste, the receptor capabilities above have been collected into four general categories called the *primary sensations of taste*. They are *sour, salty, sweet,* and *bitter.*

We know that a person can perceive literally hundreds of different tastes. They are all supposed to be combinations of the elementary sensations in the same manner that all the colors we can see are combinations of the three primary colors, as described in Chapter 50.

SOUR TASTE. The sour taste is caused by acids, and the intensity of the taste sensation is approximately proportional to the logarithm of the *hydrogen ion concentration.* That is, the more acidic the acid, the stronger becomes the sensation.

SALTY TASTE. The salty taste is elicited by ionized salts. The quality of the taste varies somewhat from one salt to another because the salts also elicit other taste sensations besides saltiness. The cations of the salts are mainly responsible for the salty taste, but the anions also contribute to a lesser extent.

SWEET TASTE. The sweet taste is not caused by any single class of chemicals. A list of some of the types of chemicals that cause this taste includes sugars, glycols, alcohols, aldehydes, ketones, amides, esters, amino acids, some small proteins, sulfonic acids, halogenated acids, and inorganic salts of lead and beryllium. Note specifically that most of the substances that cause a sweet taste are organic chemicals. It is especially interesting that slight changes in the chemical structure, such as addition of a simple radical, can often change the substance from sweet to bitter.

BITTER TASTE. The bitter taste, like the sweet taste, is not caused by any single type of chemical agent;

675

Table 53–1 RELATIVE TASTE INDICES OF DIFFERENT SUBSTANCES

Sour Substances	Index	Bitter Substances	Index	Sweet Substances	Index	Salty Substances	Index
Hydrochloric acid	1	Quinine	1	Sucrose	1	NaCl	1
Formic acid	1.1	Brucine	11	1-propoxy-2-amino-		NaF	2
Chloracetic acid	0.9	Strychnine	3.1	4-nitrobenzene	5000	CaCl$_2$	1
Acetyllactic acid	0.85	Nicotine	1.3	Saccharin	675	NaBr	0.4
Lactic acid	0.85	Phenylthiourea	0.9	Chloroform	40	NaI	0.35
Tartaric acid	0.7	Caffeine	0.4	Fructose	1.7	LiCl	0.4
Malic acid	0.6	Veratrine	0.2	Alanine	1.3	NH$_4$Cl	2.5
Potassium H tartrate	0.58	Pilocarpine	0.16	Glucose	0.8	KCl	0.6
Acetic acid	0.55	Atropine	0.13	Maltose	0.45		
Citric acid	0.46	Cocaine	0.02	Galactose	0.32		
Carbonic acid	0.06	Morphine	0.02	Lactose	0.3		

(From Derma: *Proc. Oklahoma Acad. Sci.*, 27:9, 1947; and Pfaffman: Handbook of Physiology. Sec. I, Vol. I. Baltimore, Williams & Wilkins, 1959, p. 507.)

here again, the substances that give the bitter taste are almost entirely organic substances. Two particular classes of substances are especially likely to cause bitter taste sensations: (1) long-chain organic substances that contain nitrogen and (2) alkaloids. The alkaloids include many of the drugs used in medicines, such as quinine, caffeine, strychnine, and nicotine.

Some substances that at first taste sweet have a bitter aftertaste. This is true of saccharin, which makes this substance objectionable to some people.

The bitter taste, when it occurs in high intensity, usually causes the person or animal to reject the food. This is undoubtedly an important purposive function of the bitter taste sensation because many deadly toxins found in poisonous plants are alkaloids, which all cause intensely bitter taste.

Threshold for Taste

The threshold for stimulation of the sour taste by hydrochloric acid averages 0.0009 N; for stimulation of the salty taste by sodium chloride, 0.01 M; for the sweet taste by sucrose, 0.01 M; and for the bitter taste by quinine, 0.000008 M. Note especially how much more sensitive is the bitter taste sense than all the others, which would be expected because this sensation provides an important protective function.

Table 53–1 gives the relative taste indices (the reciprocals of the taste thresholds) of different substances. In this table, the intensities of the four primary sensations of taste are referred, respectively, to the intensities of taste of hydrochloric acid, quinine, sucrose, and sodium chloride, each of which is considered to have a taste index of 1.

TASTE BLINDNESS. Many people are taste blind for certain substances, especially for different types of thiourea compounds. A substance used frequently by psychologists for demonstrating taste blindness is *phenylthiocarbamide*, for which about 15 to 30 per cent of all people exhibit taste blindness, the exact percentage depending on the method of testing and the concentration of the substance.

Taste Bud and Its Function

Figure 53–1 shows a taste bud, which has a diameter of about ⅟₃₀ millimeter and a length of about ⅟₁₆ millimeter. The taste bud is composed of about 50 modified epithelial cells, some of which are supporting cells called *sustentacular cells* and others of which are *taste cells*. The taste cells are continually being replaced by mitotic division from the surrounding epithelial cells, so that some are young cells and others are mature cells that lie toward the center of the bud and soon break up and dissolve. The life span of each taste cell is about 10 days in lower mammals but is unknown for humans.

The outer tips of the taste cells are arranged around a minute *taste pore,* shown in Figure 53–1. From the tip of each cell, several *microvilli*, or *taste hairs*, protrude outward into the taste pore to approach the cavity of the mouth. These microvilli provide the receptor surface for taste.

Interwoven among the bodies of the taste cells is a branching terminal network of several *taste nerve fibers* that are stimulated by the taste receptor cells. Some of these fibers invaginate into folds of the taste

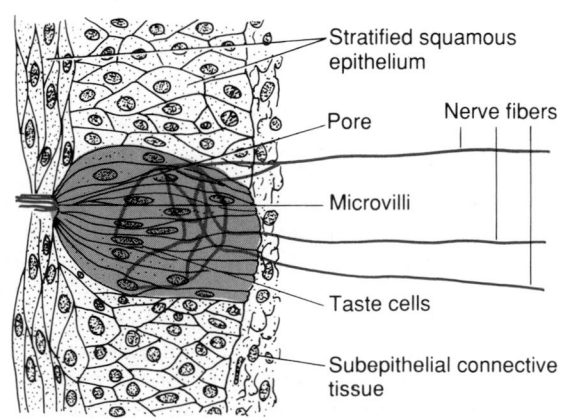

Figure 53–1. Taste bud.

cell membranes. Many vesicles form beneath the cell membrane near the fibers. It is believed that these vesicles contain a neurotransmitter substance that is released through the cell membrane to excite the nerve fiber endings in response to taste stimulation.

LOCATION OF THE TASTE BUDS. The taste buds are found on three types of papillae of the tongue, as follows: (1) A large number of taste buds are on the walls of the troughs that surround the circumvallate papillae, which form a **V** line on the posterior surface of the tongue. (2) Moderate numbers of taste buds are on the fungiform papillae over the flat anterior surface of the tongue. (3) Moderate numbers are on the foliate papillae located in the folds along the lateral surfaces of the tongue. Additional taste buds are located on the palate and a few on the tonsillar pillars, the epiglottis, and even in the proximal esophagus. Adults have 3000 to 10,000 taste buds, and children a few more. Beyond the age of 45 years, many taste buds degenerate, causing the taste sensation to become progressively less critical.

Especially important in relation to taste is the tendency for taste buds subserving particular primary sensations of taste to be located in special areas. The sweet and salty tastes are located *principally* on the tip of the tongue, the sour taste on the two lateral sides of the tongue, and the bitter taste on the posterior tongue and soft palate.

SPECIFICITY OF TASTE BUDS FOR THE PRIMARY TASTE STIMULI. Microelectrode studies from single taste buds show that each taste bud usually *responds to only one of the four primary taste stimuli when the taste substance is in low concentration*. But at high concentration, most buds can be excited by two, three, or even four of the primary taste stimuli as well as by a few other taste stimuli that do not fit into the "primary" categories.

MECHANISM OF STIMULATION OF TASTE BUDS

Receptor Potential. The membrane of the taste cell, like that of other sensory receptor cells, is negatively charged on the inside with respect to the outside. Application of a taste substance to the taste hairs causes partial loss of this negative potential—that is, the taste cell is *depolarized*. The decrease in potential, within a wide range, is approximately proportional to the logarithm of concentration of the stimulating substance. This change in potential in the taste cell is the *receptor potential* for taste.

The mechanism by which most stimulating substances react with the taste villi to initiate the receptor potential is by binding of the taste chemicals to protein receptor molecules that protrude through the villus membrane. This in turn opens ion channels, which allow sodium ions to enter and depolarize the cell. Then the taste chemical is gradually washed away from the taste villus by the saliva, which removes the stimulus. The type of receptor protein in each taste villus determines the type of taste that will elicit responses.

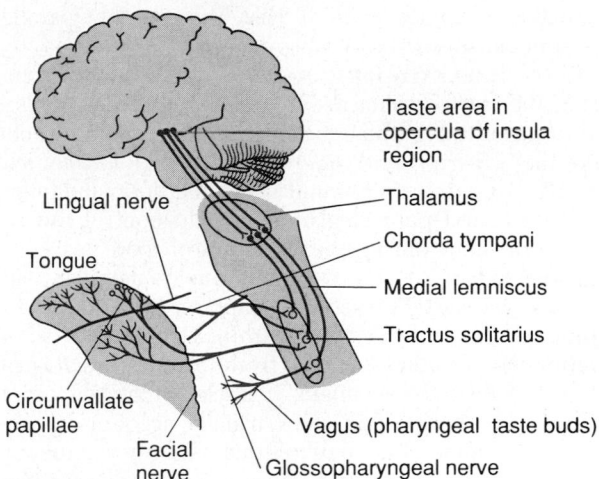

Figure 53–2. Transmission of taste impulses into the central nervous system.

Generation of Nerve Impulses by the Taste Bud. On first application of the taste stimulus, the rate of discharge of the nerve fibers rises to a peak in a small fraction of a second but then adapts within the next 2 seconds back to a lower, steady level. Thus, a strong immediate signal is transmitted by the taste nerve, and a weaker continuous signal is transmitted as long as the taste bud is exposed to the taste stimulus.

Transmission of Taste Signals into the Central Nervous System

Figure 53–2 shows the neuronal pathways for transmission of taste signals from the tongue and pharyngeal region into the central nervous system. Taste impulses from the anterior two thirds of the tongue pass first into the *fifth nerve*, then through the *chorda tympani* into the *facial nerve*, and, finally, into the *tractus solitarius* in the brain stem. Taste sensations from the circumvallate papillae on the back of the tongue and from other posterior regions of the mouth are transmitted through the *glossopharyngeal nerve* also into the *tractus solitarius* but at a slightly lower level. Finally, a few taste signals are transmitted into the *tractus solitarius* from the base of the tongue and other parts of the pharyngeal region by way of the *vagus nerve*.

All taste fibers synapse in the *nuclei of the tractus solitarius* and send second-order neurons to a small area of the *ventral posterior medial nucleus of the thalamus* located slightly medial to the thalamic terminations of the facial regions of the dorsal column–medial lemniscal system. From the thalamus, third-order neurons are transmitted to the *lower tip of the postcentral gyrus in the parietal cortex*, where it curls deep into the sylvian fissure and into the adjacent *opercularinsular area*, also in the sylvian fissure. This lies slightly lateral, ventral, and rostral to the tongue area of somatic area I.

From this description of the taste pathways, it immediately becomes evident that they closely parallel the somatic sensory pathways from the tongue.

TASTE REFLEXES INTEGRATED IN THE BRAIN STEM. From the tractus solitarius, a large number of impulses are transmitted within the brain stem itself directly into the *superior* and *inferior salivatory nuclei,* and these in turn transmit impulses to the submandibular, sublingual, and parotid glands to help control the secretion of saliva during the ingestion of food.

ADAPTATION OF TASTE. Everyone is familiar with the fact that taste sensations adapt rapidly, often with almost complete adaptation within a minute or so of continuous stimulation. Yet, from electrophysiological studies of taste nerve fibers, it is clear that adaptation of the taste buds themselves usually accounts for no more than about one half of this. Therefore, the extreme degree of adaptation that occurs in the sensation of taste almost certainly occurs in the central nervous system itself, although the mechanism and site of this are not known. At any rate, it is a mechanism different from that of most other sensory systems, which adapt mainly at the receptors.

Taste Preference and Control of the Diet

Taste preferences mean simply that an animal will choose certain types of food in preference to others, and the animal automatically uses this to help control the type of diet it eats. Furthermore, its taste preferences often change in accord with the needs of the body for certain specific substances. The following experiments demonstrate this ability of animals to choose food in accord with the needs of their bodies. First, adrenalectomized animals automatically select drinking water with a high concentration of sodium chloride in preference to pure water, and this in many instances is sufficient to supply the needs of the body and prevent death as a result of salt depletion. Second, an animal given injections of excessive amounts of insulin develops a depleted blood sugar, and it automatically chooses the sweetest food from among many samples. Third, parathyroidectomized animals automatically choose drinking water with a high concentration of calcium chloride.

The same phenomena are also observed in many instances of everyday life. For instance, the salt licks of the desert region are known to attract animals from far and wide, and even the human being rejects any food that has an unpleasant affective sensation, which certainly in many instances protects our bodies from undesirable substances.

The phenomenon of taste preference almost certainly results from some mechanism located in the central nervous system and not from a mechanism in the taste receptors themselves, although it is true that the receptors often do become sensitized to the needed nutrient. An important reason for believing taste preference to be mainly a central phenomenon is

that previous experience with unpleasant or pleasant tastes plays a major role in determining one's different taste preferences. For instance, if a person becomes sick soon after eating a particular type of food, the person generally develops a negative taste preference, or *taste aversion,* for that particular food thereafter; the same effect can be demonstrated in animals.

SENSE OF SMELL

Smell is the least understood of our senses. This results partly from the fact that the sense of smell is a subjective phenomenon that cannot be studied with ease in lower animals. Still another complicating problem is that the sense of smell is poorly developed in the human being in comparison with the sense of smell in some lower animals.

Olfactory Membrane

The olfactory membrane, the histology of which is shown in Figure 53–3, lies in the superior part of each nostril, shown in Figure 53–4. Medially, the membrane folds slightly downward over the surface of the superior septum, and laterally, it folds over the superior turbinate and even over a small portion of the upper surface of the middle turbinate. In each nostril, the olfactory membrane has a surface area of about 2.4 square centimeters.

OLFACTORY CELLS. The receptor cells for the smell sensation are the *olfactory cells,* which are actually bipolar nerve cells derived originally from the central nervous system itself. There are about 100 million of these cells in the olfactory epithelium interspersed among *sustentacular cells,* as shown in Figure 53–3. The mucosal end of the olfactory cell forms a knob from which 6 to 12 *olfactory hairs,* or *cilia,* 0.3 mi-

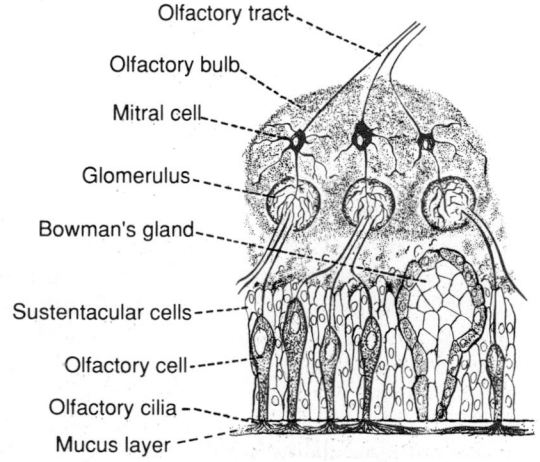

Olfactory tract
Olfactory bulb
Mitral cell
Glomerulus
Bowman's gland
Sustentacular cells
Olfactory cell
Olfactory cilia
Mucus layer

Figure 53–3. Organization of the olfactory membrane.

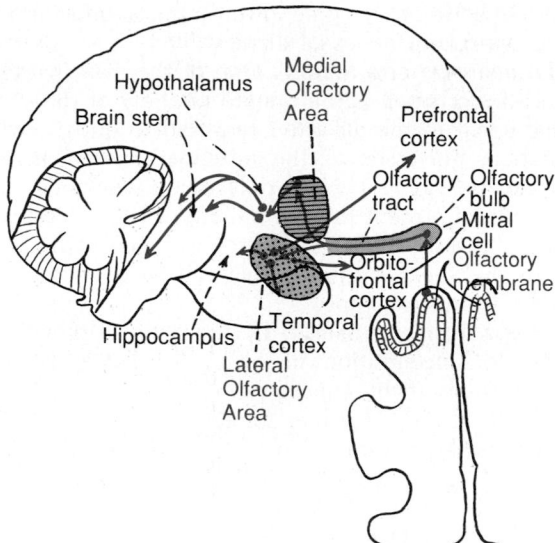

Figure 53-4. Neural connections of the olfactory system.

crometer in diameter and up to 200 micrometers in length, project into the mucus that coats the inner surface of the nasal cavity. These projecting olfactory cilia form a dense mat in the mucus, and it is these cilia that react to odors in the air and then stimulate the olfactory cells, as discussed later. Spaced among the olfactory cells in the olfactory membrane are many small *glands of Bowman* that secrete mucus onto the surface of the olfactory membrane.

Stimulation of the Olfactory Cells

MECHANISM OF EXCITATION OF THE OLFACTORY CELLS. The portion of each olfactory cell that responds to the olfactory chemical stimuli is the cilia. The odorant substance, on coming in contact with the olfactory surface, first diffuses into the mucus that covers the cilia. Then it binds with a *receptor protein* that protrudes through the ciliary membrane. This receptor is a long molecule that threads its way through the membrane seven times, folding inward and outward. The odorant binds with the portion of the receptor that folds to the outside. The inside of the folding receptor, on the other hand, is coupled to a so-called *G-protein,* itself a combination of three subunits. On excitation of the receptor, an *alpha* subunit breaks away from the G-protein and immediately activates adenyl cyclase that is attached to the inside of the ciliary membrane near the receptor cell body. The activated cyclase in turn converts many molecules of intracellular adenosine triphosphate into cyclic adenosine monophosphate (cAMP). Finally, this cAMP activates another nearby membrane protein, a gated sodium ion channel, that allows large numbers of sodium ions to pour into the receptor cell

cytoplasm. The sodium ions contribute positivity to the inside of the cell membrane, thus exciting the olfactory neuron and transmitting action potentials into the central nervous system by way of an olfactory nerve.

The importance of this indirect mechanism for activating olfactory nerves is that it greatly multiplies the excitatory effect of even the weakest odorant. To summarize: (a) Activation of the receptor by the odorant activates the G-protein complex. (b) This in turn activates multiple molecules of adenyl cyclase inside the olfactory cell membrane, which in turn (c) causes formation of still many times more molecules of cAMP. And, finally, (d) the cAMP opens still many more sodium ion channels. Therefore, even the minutest concentration of a specific odorant initiates a cascading effect that opens extremely large numbers of sodium channels. This accounts for the exquisite sensitivity of the olfactory neurons to even the slightest amount of odorant.

In addition to the basic chemical mechanism by which the olfactory cells are stimulated, several physical factors affect the degree of stimulation. First, only volatile substances that can be sniffed into the nostrils can be smelled. Second, the stimulating substance must be at least slightly water-soluble, so that it can pass through the mucus to reach the olfactory cells. And, third, it is helpful to be at least slightly lipid-soluble, presumably because the lipid constituents of the cilium membrane repel non–lipid-soluble odorants from the membrane receptor proteins.

MEMBRANE POTENTIALS AND ACTION POTENTIALS IN OLFACTORY CELLS. The membrane potential of unstimulated olfactory cells, as measured by microelectrodes, averages about -55 millivolts. At this potential, most of the cells generate continuous action potentials at a very slow rate, varying from once every 20 seconds up to two to three per second.

Most odorants cause depolarization of the olfactory cell membrane, decreasing the negative potential in the cell from -55 down to as low as -30 millivolts or even less. Along with this, the number of action potentials increases to about 20 per second, which is a high rate for the minute, fraction-of-a-micrometer olfactory nerve fibers.

A few odorants hyperpolarize the olfactory cell membrane, thus decreasing, instead of increasing, the nerve firing rate.

Over a wide range, the rate of olfactory nerve impulses changes approximately in proportion with the logarithm of the stimulus strength, which demonstrates that the olfactory receptors tend to obey principles of transduction similar to those of other sensory receptors.

ADAPTATION. The olfactory receptors adapt about 50 per cent in the first second or so after stimulation. Thereafter, they adapt very little and very slowly. Yet we all know from our own experience that smell sensations adapt almost to extinction within a minute or so after one enters a strongly odorous atmosphere.

Because this psychological adaptation is far greater than the degree of adaptation of the receptors themselves, it is almost certain that most of the adaptation occurs in the central nervous system, which seems also to be true for the adaptation of taste sensations. A postulated neuronal mechanism for this adaptation is the following: Large numbers of centrifugal nerve fibers pass from the olfactory regions of the brain backward along the olfactory tract and terminate on special inhibitory cells in the olfactory bulb, the *granule cells.* It is postulated that after the onset of an olfactory stimulus, the central nervous system gradually develops a strong feedback inhibition to suppress relaying of the smell signals through the olfactory bulb.

Search for the Primary Sensations of Smell

In past years, most physiologists were convinced that the many smell sensations were subserved by a few rather discrete primary sensations, in the same way that vision and taste are subserved by a select few sensations. Based on psychological studies, one attempt to classify these sensations was the following:

1. Camphoraceous
2. Musky
3. Floral
4. Pepperminty
5. Ethereal
6. Pungent
7. Putrid

It is certain that this list does not represent the true primary sensations of smell. Indeed, multiple clues in recent years, including specific studies on the genes that encode for the receptor proteins, suggest at least 100 primary sensations of smell and perhaps as many as 1000—a marked contrast to only *three* primary sensations of color detected by the eyes and only a few primary sensations of taste detected by the tongue. Further support for the many primary sensations of taste is that people have been found who have *odor blindness* for single substances; such discrete odor blindness has been identified for more than 50 substances. It is presumed that odor blindness for each substance represents a lack of the appropriate receptor protein in olfactory cells for that substance.

AFFECTIVE NATURE OF SMELL. Smell, even more so than taste, has the affective qualities of either pleasantness or unpleasantness. Because of this, smell is probably even more important than taste in the selection of food. Indeed, a person who has previously eaten food that has disagreed with him is often nauseated by even the smell of that same type of food on a second occasion. Other types of odors that have proved to be unpleasant in the past may also provoke a disagreeable feeling; on the other hand, perfume of the right quality can wreak havoc with masculine emo-

tions. In addition, in some lower animals, odors are the primary excitant of sexual drive.

THRESHOLD FOR SMELL. One of the principal characteristics of smell is the minute quantity of the stimulating agent in the air often required to effect a smell sensation. For instance, the substance *methyl mercaptan* can be smelled when only one 25 billionth of a milligram is present in each milliliter of air. Because of this low threshold, this substance is mixed with natural gas to give the gas an odor that can be detected when it leaks from a gas pipe.

GRADATIONS OF SMELL INTENSITIES. Although the threshold concentrations of substances that evoke smell are extremely slight, concentrations only 10 to 50 times above the threshold values for many, if not most, substances evoke maximum intensity of smell. This is in contrast to most other sensory systems of the body, in which the ranges of detection are tremendous—for instance, 500,000 to 1 in the case of the eyes and 1 trillion to 1 in the case of the ears. This perhaps can be explained by the fact that smell is concerned more with detecting the presence or absence of odors than with quantitative detection of their intensities.

Transmission of Smell Signals into the Central Nervous System

The olfactory portions of the brain are among its oldest structures, and much of the remainder of the brain developed around these olfactory beginnings. In fact, part of the brain that originally subserved olfaction later evolved into the basal brain structures that in the human being control emotions and other aspects of behavior; this is the system we call the *limbic system,* discussed in Chapter 58.

TRANSMISSION OF OLFACTORY SIGNALS INTO THE OLFACTORY BULB. The olfactory bulb, which is also called cranial nerve I, is shown in Figure 53–4. Although it looks like a nerve, in reality it is an anterior outgrowth of brain tissue from the base of the brain having a bulbous enlargement, the *olfactory bulb,* at its end that lies over the *cribriform plate* separating the brain cavity from the upper reaches of the nasal cavity. The cribriform plate has multiple small perforations through which an equal number of small nerves pass upward from the olfactory membrane in the nasal cavity to enter the olfactory bulb in the cranial cavity. Figure 53–3 demonstrates the close relation between the *olfactory cells* in the olfactory membrane and the olfactory bulb, showing short axons terminating in multiple globular structures within the olfactory bulb called *glomeruli.* Each bulb has several thousand such glomeruli, each of which is the terminus for about 25,000 axons from olfactory cells. Each glomerulus also is the terminus for dendrites from about 25 large *mitral cells* and about 60 smaller *tufted cells,* the cell bodies of which lie also in the olfactory bulb superior to the glomeruli. These cells in turn send axons through the olfactory tract to transmit the olfactory sensations into the central nervous system.

Research suggests that different glomeruli respond to different odors. Therefore, it is possible that the specific glomeruli that are stimulated are the real clue to the analysis of different odor signals transmitted into the central nervous system.

The Very Old, the Old, and the Newer Olfactory Pathways into the Central Nervous System

The olfactory tract enters the brain at the anterior junction between the mesencephalon and cerebrum; there the tract divides into two pathways, as shown in Figure 53–4, one passing medially into the *medial olfactory area* and the other passing laterally into the *lateral olfactory area*. The medial olfactory area represents a very old olfactory system, whereas the lateral olfactory area is the input to both a less old olfactory system and a newer system.

THE VERY OLD OLFACTORY SYSTEM—THE MEDIAL OLFACTORY AREA. The medial olfactory area consists of a group of nuclei located in the midbasal portions of the brain anterior to the hypothalamus. Most conspicuous are the *septal nuclei,* which are midline nuclei that feed into the hypothalamus and other primitive portions of the brain's limbic system, the system that is concerned with basic behavior (described in Chapter 58).

The importance of this medial olfactory area is best understood by considering what happens in animals when the lateral olfactory areas on both sides of the brain are removed and only the medial system remains. The answer is that this hardly affects the more primitive responses to olfaction, such as licking the lips, salivation, and other feeding responses caused by the smell of food or primitive emotional drives associated with smell. On the other hand, removal of the lateral areas does abolish the more complicated olfactory conditioned reflexes.

THE LESS OLD OLFACTORY SYSTEM—THE LATERAL OLFACTORY AREA. The lateral olfactory area is composed mainly of the *prepyriform* and *pyriform cortex* plus the *cortical portion of the amygdaloid nuclei.* From these areas, signal pathways pass into almost all portions of the limbic system, especially into less primitive portions, such as the hippocampus, which seem to be most important for learning to like or dislike certain foods, depending on experiences with the foods. For instance, it is this lateral olfactory area and its many connections with the limbic behavioral system that cause a person to develop absolute aversion to foods that have previously caused nausea and vomiting.

An important feature of the lateral olfactory area is that many signal pathways from this area feed directly into an older part of cerebral cortex called the paleocortex in the anteromedial portion of the temporal lobe. This is the only area of the entire cerebral cortex where sensory signals pass directly to the cortex without passing through the thalamus.

THE NEWER PATHWAY. Still a newer olfactory pathway has now been found that passes through the thalamus, passing to the dorsomedial thalamic nucleus and then to the lateroposterior quadrant of the orbitofrontal cortex. Based on studies in monkeys, this newer system probably helps especially in the conscious analysis of odor.

SUMMARY. Thus, there appears to be a *very old* olfactory system that subserves the basic olfactory reflexes, an *old* system that provides automatic but learned control of food intake and aversion to toxic and unhealthy foods, and, finally, a *newer* system that is comparable to most of the other cortical sensory systems and is used for conscious perception of olfaction.

CENTRIFUGAL CONTROL OF ACTIVITY IN THE OLFACTORY BULB BY THE CENTRAL NERVOUS SYSTEM. Many nerve fibers that originate in the olfactory portions of the brain pass peripherally in the olfactory tract to the olfactory bulb, that is, "centrifugally" from the brain to the periphery. These terminate on a large number of small *granule cells* located among the mitral and tufted cells in the bulb. These in turn send short, inhibitory *dendrites* to the mitral and tufted cells. It is believed that this inhibitory feedback to the olfactory bulb might be a means of helping to sharpen one's specific capability of distinguishing one odor from another.

ELECTRICAL ACTIVITY IN THE OLFACTORY NERVES AND TRACTS. Electrophysiological studies show that the mitral and tufted cells are continually active. Superimposed on this background are increases or decreases in impulse traffic caused by different odors. Thus, the olfactory stimuli *modulate* the frequency of impulses in the olfactory system and in this way transmit the olfactory information.

REFERENCES

Alberts, J. R.: Producing and interpreting experimental olfactory deficits. Physiol. Behav., 12:657, 1974.

Beauchemin, V., et al.: Quantitative autoradiographic studies of the effects of bilateral olfactory bulbectomy in the rat brain: central- and peripheral-type benzodiazepine receptors. Neuroscience, 58:527, 1994.

Benjamin, B., et al.: A Color Atlas of Otorhinolaryngology. Philadelphia, J. B. Lippincott, 1994.

Berkowicz, D. A., et al.: Evidence for glutamate as the olfactory receptor cell neurotransmitter. J. Neurophysiol., 71:2557, 1994.

Davis, J. L.: Olfaction. Cambridge, MA, The MIT Press, 1991.

Doty, R. L.: Handbook of Olfaction and Gustation. New York, Marcel Dekker, Inc., 1994.

Douek, E.: The Sense of Smell and Its Abnormalities. New York, Churchill Livingstone, 1974.

Farbman, A. I.: Cell Biology of Olfaction. New York, Cambridge University Press, 1991.

Getchell, T. V., et al.: Smell and Taste in Health and Disease. New York, Raven Press, 1991.

Getchell, T. V.: Functional properties of vertebrate olfactory receptor neurons. Physiol. Rev., 66:772, 1986.

Kashara, Y. (ed.): Proceedings of the Seventeenth Japanese Symposium on Taste and Smell. Arlington, VA, IRL Press, 1984.

Libri, V., et al.: A comparison of the muscarinic response and morphological properties of identified cells in the guinea-pig olfactory cortex in vitro. Neuroscience, 59:331, 1994.

Lynch, M. A., et al.: Burket's Oral Medicine. Philadelphia, J. B. Lippincott, 1994.

Margolis, F. L., and Getchell, T. V. (eds.): Molecular Neurobiology of the Olfactory System. New York, Plenum Publishing Corp., 1988.

McBurney, D. H.: Taste and olfaction: Sensory discrimination. In Darian-Smith, I. (ed.): Handbook of Physiology. Sec. 1, Vol. III, Bethesda, Md., American Physiological Society, 1984, p. 1067.

McLaughlin, S., and Margolskee, R.: The sense of taste. Am. Sci., 82:538, 1994.

Moulton, D. G., and Beidler, L. M.: Structure and function in the peripheral olfactory system. Physiol. Rev., 47:1, 1967.

Norgren, R.: Central neural mechanisms of taste. In Darian-Smith, I. (ed.): Handbook of Physiology. Sec. 1, Vol. III. Bethesda, Md., American Physiological Society, 1984, p. 1087.

Oakley, B., and Benjamin, R. M.: Neural mechanisms of taste. Physiol. Rev., 46:173, 1966.

Roper, S. D.: The cell biology of vertebrate taste receptors. Annu. Rev. Neurosci., 12:329, 1989.

Scott, T. R., et al.: Gustatory neural coding in the monkey cortex: the quality of saltiness. J. Neurophysiol., 71:1692, 1994.

Shepherd, G. M.: The olfactory bulb: A simple system in the mammalian brain. In Brookhart, J. M., and Mountcastle, V. B. (eds.): Handbook of Physiology. Sec. 1, Vol. I. Baltimore, Md., Williams & Wilkins, 1977, p. 945.

Takagi, S. F.: The olfactory nervous system of the Old World monkey. Jpn. J. Physiol., 34:51, 1984.

Zotterman, Y.: Olfaction and Taste. New York, Macmillan Co., 1963.

THE NERVOUS SYSTEM: C. MOTOR AND INTEGRATIVE NEUROPHYSIOLOGY

UNIT XI

Motor Functions of the Spinal Cord; the Cord Reflexes

CHAPTER 54

In the discussion of the nervous system thus far, we have considered principally the input of sensory information. In the following chapters, we discuss the origin and output of motor signals—the signals that cause muscle contraction, secretory function, and other motor effects throughout the body.

Sensory information is integrated at all levels of the nervous system and causes appropriate motor responses, beginning in the spinal cord with relatively simple reflexes, extending into the brain stem with still more complicated responses, and finally extending to the cerebrum, where the most complicated responses are controlled.

In this chapter, we discuss the control of muscle function by the spinal cord. The spinal cord is not merely a conduit for sensory signals to the brain or for motor signals from the brain back to the periphery. In fact, without the special neuronal circuits of the cord, even the most complex motor control systems in the brain cannot cause any purposeful muscle movement. To give an example, there is no neuronal circuit anywhere in the brain that causes the specific to-and-fro movement of the legs that is required in walking. Instead, the circuits for these movements are in the cord, and the brain simply sends *command* signals to set into motion the walking process. Thus, under appropriate conditions, a cat or dog with its cord transected in the neck can be made to walk in a crude manner.

Yet let us not belittle the role of the brain as well, because the brain gives the sequential directions to control the cord activities, to promote turning movements when they are required, to lean the body forward during acceleration, to change the movements

from walking to jumping as needed, and to monitor continuously and control equilibrium. All this is done through "analytical" and "command" signals from the brain. But it also requires the many neuronal circuits of the spinal cord that are themselves the objects of the commands. These circuits in turn provide all but a small fraction of the direct control of the muscles.

EXPERIMENTAL PREPARATIONS FOR STUDYING CORD REFLEXES—THE SPINAL ANIMAL AND THE DECEREBRATE ANIMAL. Two types of experimental preparations have been especially useful in studying spinal cord function: (1) the *spinal animal*, in which the spinal cord is transected, frequently in the neck so that most of the cord still remains functional, and (2) the *decerebrate animal*, in which the brain stem is transected in the middle to lower part of the mesencephalon.

Immediately after preparing a *spinal animal*, most spinal cord function is severely depressed below the level of the transection. After a few hours in lower animals and a few days to weeks in monkeys, most of the intrinsic spinal cord functions return nearly to normal and provide a suitable experimental preparation for study.

In the *decerebrate animal*, the brain stem is transected at the middle to lower mesencephalic level, which blocks the normal inhibitory signals to the pontile reticular and vestibular nuclei from the higher control centers of the brain. This causes these nuclei to become tonically active, transmitting facilitatory signals to most of the spinal cord motor control circuits. The result is that the spinal cord motor reflexes become overly excitable and, therefore, easy to activate by even the slightest sensory input signals to the cord. Using this preparation, one can easily study the intrinsic motor functions of the cord itself.

685

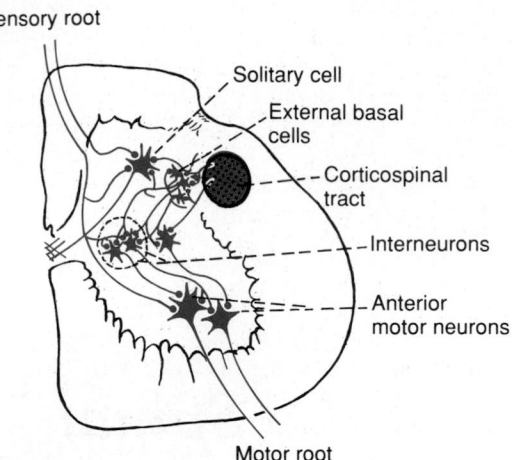

Figure 54–1. Connections of the sensory fibers and corticospinal fibers with the interneurons and anterior motor neurons of the spinal cord.

Organization of the Spinal Cord for Motor Functions

The cord gray matter is the integrative area for the cord reflexes and other motor functions. Figure 54–1 shows the typical organization of the cord gray matter in a single cord segment. Sensory signals enter the cord almost entirely through the sensory (posterior) roots. After entering the cord, every sensory signal travels to two separate destinations. First, one branch of the sensory nerve terminates in the gray matter of the cord and elicits local segmental reflexes and other effects. Second, another branch transmits signals to higher levels of the nervous system—to higher levels in the cord itself, to the brain stem, or even to the cerebral cortex, as described in earlier chapters.

Each segment of the spinal cord between one spinal nerve and the next has several million neurons in its gray matter. Aside from the sensory relay neurons discussed in Chapters 47 and 48, these neurons are of two types, the *anterior motor neurons* and the *interneurons.*

ANTERIOR MOTOR NEURONS. Located in each segment of the anterior horns of the cord gray matter are several thousand neurons that are 50 to 100 per cent larger than most of the others and are called *anterior motor neurons.* They give rise to the nerve fibers that leave the cord by way of the anterior roots and innervate the skeletal muscle fibers. The neurons are of two types, *alpha motor neurons* and *gamma motor neurons.*

Alpha Motor Neurons. The alpha motor neurons give rise to large type A alpha (Aα) nerve fibers averaging 14 micrometers in diameter that innervate the large skeletal muscle fibers, as shown in Figure 54–2. Stimulation of a single nerve fiber excites from as few as three to as many as several hundred skeletal muscle fibers, which are collectively called the *motor unit.* Transmission of nerve impulses into skeletal muscles and their stimulation of the muscles are discussed in Chapters 6 and 7.

Gamma Motor Neurons. In addition to the alpha motor neurons that excite contraction of the skeletal muscle fibers, about one half as many much smaller gamma motor neurons are located in the spinal cord anterior horns along with the alpha motor neurons. These gamma motor neurons transmit impulses through type A gamma (Aγ) fibers, averaging 5 micrometers in diameter, to small, special skeletal muscle fibers called *intrafusal fibers,* also shown in Figure 54–2. These fibers are part of the *muscle spindle,* which is discussed later in the chapter.

INTERNEURONS. Interneurons are present in all areas of the cord gray matter—in the dorsal horns, the anterior horns, and the intermediate areas between these two, as shown in Figure 54–1. These cells are numerous—about 30 times as numerous as the anterior motor neurons. They are small and highly excitable, often exhibiting spontaneous activity and capable of firing as rapidly as 1500 times per second. They have many interconnections one with the other, and many of them directly innervate the anterior motor neurons, as shown in Figure 54–1. The interconnections among the interneurons and anterior motor neurons are responsible for most of the integrative functions of the spinal cord that are discussed in the remainder of this chapter.

Essentially all the different types of neuronal circuits described in Chapter 46 are found in the interneuron pool of cells of the spinal cord, including the *diverging, converging,* and *repetitive-discharge* circuits. In this chapter, we see many applications of these different circuits to the performance of specific reflex acts by the spinal cord.

Only a few incoming sensory signals from the spinal nerves or signals from the brain terminate directly on the anterior motor neurons. Most of these signals are transmitted first through interneurons, where they are appropriately processed. Thus, in Figure 54–1, it is shown that the corticospinal tract terminates almost entirely on interneurons, and it is only after the signals from this tract have been integrated in the interneuron pool with signals from other spinal tracts or from the

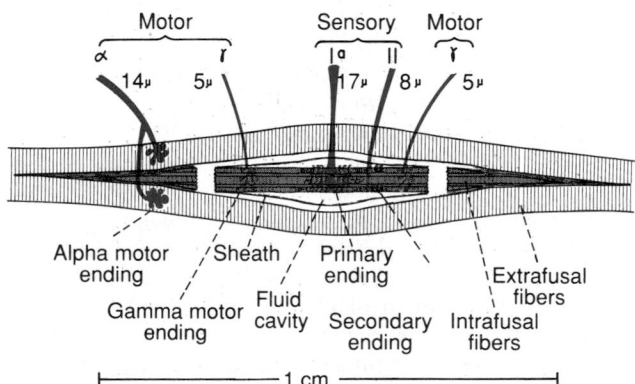

Figure 54–2. Muscle spindle, showing its relation to the large extrafusal skeletal muscle fibers. Note also both the motor and the sensory innervation of the muscle spindle and the extrafusal large muscle fibers.

spinal nerves that they finally impinge on the anterior motor neurons to control muscle function.

RENSHAW CELL INHIBITORY SYSTEM. Located also in the ventral horns of the spinal cord in close association with the motor neurons are a large number of small interneurons called *Renshaw cells*. Almost immediately after the axon leaves the body of the anterior motor neuron, collateral branches from the axon pass to the adjacent Renshaw cells. These in turn are inhibitory cells that transmit inhibitory signals back to the nearby motor neurons. Thus, stimulation of each motor neuron tends to inhibit the surrounding motor neurons, an effect called *recurrent inhibition*. This effect is probably important for the following major reason.

It shows that the motor system uses the principle of lateral inhibition to focus, or sharpen, its signals in the same way that the sensory system uses this principle— that is, to allow unabated transmission of the primary signal while suppressing the tendency for signals to spread to adjacent neurons.

Multisegmental Connections in the Spinal Cord—The Propriospinal Fibers

More than one half of all the nerve fibers that ascend and descend in the spinal cord are *propriospinal fibers*. These fibers run from one segment of the cord to another. In addition, the sensory fibers as they enter the cord from the posterior cord roots bifurcate and branch both up and down the spinal cord, some of the branches transmitting signals only a segment or two, whereas others transmit signals many segments. These ascending and descending fibers of the cord provide pathways for the multisegmental reflexes described later in this chapter, including reflexes that coordinate simultaneous movements in the forelimbs and hindlimbs.

MUSCLE SENSORY RECEPTORS— MUSCLE SPINDLES AND GOLGI TENDON ORGANS—AND THEIR ROLES IN MUSCLE CONTROL

Proper control of muscle function requires not only excitation of the muscle by the anterior motor neurons but also continuous sensory feedback of information from each muscle to the spinal cord, giving the status of the muscle at each instant. That is, what is the length of the muscle, what is its instantaneous tension, and how rapidly is its length or tension changing? To provide this information, the muscles and their tendons are supplied abundantly with two special types of sensory receptors: (1) *muscle spindles*, which are distributed throughout the belly of the muscle and send information to the nervous system about either the muscle length or rate of change of its length, and (2) *Golgi tendon organs*, which are located in the muscle tendons and transmit information about tendon tension or the rate of change of tension.

The signals from these two receptors are either entirely or almost entirely for the purpose of muscle control itself because they operate almost entirely at a subconscious level. Even so, they transmit tremendous amounts of information not only into the spinal cord but also to the cerebellum and even the cerebral cortex, helping each of these portions of the nervous system in its function for controlling muscle contraction.

Receptor Function of the Muscle Spindle

STRUCTURE AND MOTOR INNERVATION OF THE MUSCLE SPINDLE. The physiological organization of the muscle spindle is shown in Figure 54–2. Each spindle is 3 to 10 millimeters long. It is built around 3 to 12 very small *intrafusal muscle fibers* that are pointed at their ends and attached to the glycocalyx of the surrounding large *extrafusal* skeletal muscle fibers. Each intrafusal fiber is a small skeletal muscle fiber. However, the central region of each of these fibers— that is, the area midway between its two ends—has either no or few actin and myosin filaments. Therefore, this central portion does not contract when the ends do. Instead, it functions as a sensory receptor, as we describe later. The end portions that do contract are excited by small *gamma motor nerve fibers* that originate from the small gamma motor neurons in the anterior horns of the spinal cord, as described earlier. These fibers are also called *gamma efferent fibers*, in contradistinction to the large *alpha efferent fibers* that innervate the extrafusal skeletal muscle.

SENSORY INNERVATION OF THE MUSCLE SPINDLE. The receptor portion of the muscle spindle is its central portion, where the intrafusal muscle fibers have no contractile elements. As shown in Figure 54–2 and in more detail in Figure 54–3, sensory fibers originate in this area. They are stimulated by stretching of this midportion of the spindle. One can readily see that the muscle spindle receptor can be excited in two ways:

1. Lengthening the whole muscle will stretch the midportion of the spindle and, therefore, excite the receptor.
2. Even if the length of the entire muscle does not change, contraction of the end portions of the spindle's intrafusal fibers will also stretch the midportions of the fibers and therefore excite the receptor.

Two types of sensory endings are found in the receptor area of the muscle spindle. They are the primary ending and the secondary ending.

Primary Ending. In the very center of the receptor area, a large sensory fiber encircles the central portion of each intrafusal fiber, forming the so-called *primary ending* or *annulospiral ending*. This nerve fiber is a type Ia fiber averaging 17 micrometers in diameter, and it transmits sensory signals to the spinal cord at a velocity of 70 to 120 m/sec, as rapidly as any type of sensory nerve fiber in the entire body.

Secondary Ending. Usually one but sometimes two smaller sensory nerve fibers, type II fibers with an average diameter of 8 micrometers, innervate the receptor region on one side of the primary ending, as shown in Figures 54–2 and 54–3. This sensory ending

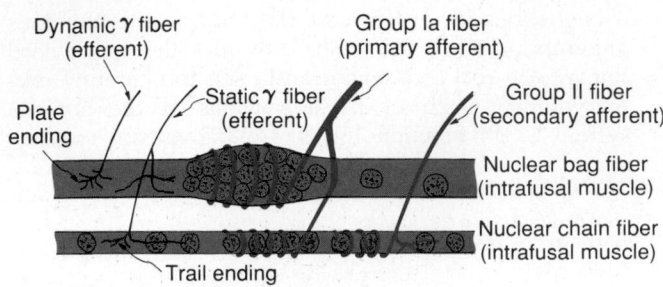

Figure 54–3. Details of nerve connections to the nuclear bag and nuclear chain muscle spindle fibers. (Modified from Stein: *Physiol. Rev.,* 54:225, 1974, and Boyd: *Philios. Trans. R. Soc. Lond. [Biol Sci.],* 245:81, 1962.)

is called the *secondary ending;* it mainly encircles the intrafusal fibers in the same way that the type Ia fiber does.

DIVISION OF THE INTRAFUSAL FIBERS INTO NUCLEAR BAG AND NUCLEAR CHAIN FIBERS—DYNAMIC AND STATIC RESPONSES OF THE MUSCLE SPINDLE. There are also two types of intrafusal fibers: (1) *nuclear bag fibers* (one to three in each spindle), in which a large number of nuclei are congregated in an expanded bag in the central portion of the receptor area, as shown by the top fiber in Figure 54–3, and (2) *nuclear chain fibers* (three to nine), which are about half as large in diameter and half as long as the nuclear bag fibers and have nuclei aligned in a chain throughout the receptor area, as shown by the bottom fiber in the figure. The primary nerve ending (from the 17-micrometer fiber) innervates both the nuclear bag intrafusal fibers *and* the nuclear chain fibers. On the other hand, the secondary ending (from the 8-micrometer fiber) usually innervates only the nuclear chain fibers. These relations are shown in Figure 54–3.

Response of Both the Primary and the Secondary Endings to the Length of the Receptor—The "Static" Response. When the receptor portion of the muscle spindle is stretched *slowly*, the number of impulses transmitted from both the primary and the secondary endings increases almost directly in proportion to the degree of stretching, and the endings continue to transmit these impulses for several minutes. This effect is called the *static response* of the spindle receptor, meaning simply that both the primary and the secondary endings continue to transmit their signals for as long as the receptor itself remains stretched. Because the *nuclear chain* type of intrafusal fiber is innervated by both the primary and the secondary endings, it is believed that these nuclear chain fibers are mainly responsible for the static response.

Response of the Primary Ending (but Not the Secondary Ending) to the Rate of Change of Receptor Length—The "Dynamic" Response. When the length of the spindle receptor increases suddenly, the primary ending (but not the secondary ending) is stimulated especially powerfully, much more powerfully than the stimulus caused by the static response. This excess stimulus of the primary ending is called the *dynamic response,* which means that the primary ending responds extremely actively to a rapid *rate of*

change in spindle length. Even when the length of a spindle receptor increases only a fraction of a micrometer, if this increase occurs in a fraction of a second, the primary receptor transmits tremendous numbers of excess impulses into the Ia fiber but only *while the length is actually increasing.* As soon as the length has stopped increasing, the rate of impulse discharge returns back to the level of the much smaller static response that is still present in the signal.

Conversely, when the spindle receptor shortens, this change momentarily decreases the impulse output from the primary ending; then, as soon as the receptor area has reached its new shortened length, impulses reappear in the Ia fiber within a fraction of a second.

Thus, the primary ending sends extremely strong signals to the spinal cord to apprise it of any change in length of the spindle receptor area.

Because only the primary endings transmit the dynamic response and almost all the nuclear bag intrafusal fibers have only primary endings, it is assumed that the nuclear bag fibers are responsible for the powerful dynamic response.

CONTROL OF THE STATIC AND DYNAMIC RESPONSES BY THE GAMMA MOTOR NERVES. The gamma motor nerves to the muscle spindle can be divided into two types: gamma-dynamic (gamma-d) and gamma-static (gamma-s). The first of these excites mainly the nuclear bag intrafusal fibers and the second, mainly the nuclear chain intrafusal fibers. When the gamma-d fibers excite the nuclear bag fibers, the dynamic response of the muscle spindle becomes tremendously enhanced, whereas the static response is hardly affected. On the other hand, stimulation of the gamma-s fibers, which excite the nuclear chain fibers, enhances the static response while having little influence on the dynamic response. We shall see in subsequent paragraphs that these two types of responses of the muscle spindle are exceedingly important in different types of muscle control.

CONTINUOUS DISCHARGE OF THE MUSCLE SPINDLES UNDER NORMAL CONDITIONS. Normally, particularly when there is a slight amount of gamma nerve excitation, the muscle spindles emit sensory nerve impulses continuously. Stretching the muscle spindles increases the rate of firing, whereas shortening the spindle decreases this rate of firing. Thus, the spindles can send to the spinal cord either *positive signals*—that is, increased numbers of impulses to indicate increasing

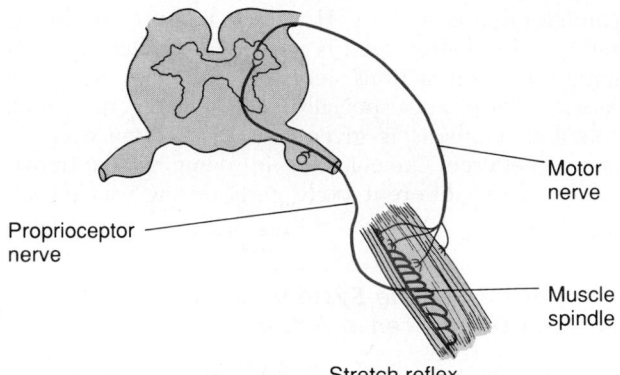

Figure 54-4. Neuronal circuit of the stretch reflex.

stretch of a muscle—or *negative signals*—decreased numbers of impulses below the normal level to indicate that the muscle is being unstretched.

Muscle Stretch Reflex

The simplest manifestation of muscle spindle function is the *muscle stretch reflex* (also called myotatic reflex)—that is, whenever a muscle is stretched, excitation of the spindles causes reflex contraction of the large skeletal muscle fibers of the same muscle and closely allied synergistic muscles.

NEURONAL CIRCUITRY OF THE STRETCH REFLEX. Figure 54–4 demonstrates the basic circuit of the muscle spindle stretch reflex, showing a type Ia nerve fiber originating in a muscle spindle and entering the dorsal root of the spinal cord. Then, in contrast to most other nerve fibers entering the cord, one branch of it passes directly to the anterior horn of the cord gray matter and synapses directly with anterior motor neurons that send nerve fibers mainly back to the same muscle from whence the muscle spindle fiber originated. Thus, this is a *monosynaptic pathway* that allows a reflex signal to return with the shortest possible delay back to the muscle after excitation of the spindle.

Some of the type II fibers from the secondary spindle endings also terminate monosynaptically with the anterior motor neurons. However, most of the type II fibers (as well as many collaterals from the Ia fibers from the primary endings) terminate on multiple interneurons in the cord gray matter, and these in turn transmit more delayed signals to the anterior motor neurons or serve other functions.

DYNAMIC STRETCH REFLEX VERSUS STATIC STRETCH REFLEX. The stretch reflex can be divided into two components: the dynamic stretch reflex and the static stretch reflex. The *dynamic stretch reflex* is elicited by the potent dynamic signal transmitted from the primary endings of the muscle spindles, caused by rapid stretch of the muscle. That is, when a muscle is suddenly stretched, a strong signal is transmitted to the spinal cord, and this causes an instantaneous, strong reflex contraction of the same muscle from which the signal originated. Thus, the reflex functions to oppose sudden changes in the length of the muscle because the muscle contraction opposes the stretch.

The dynamic stretch reflex is over within a fraction of a second after the muscle has been stretched to its new length, but then a weaker *static stretch reflex* continues for a prolonged period thereafter. This reflex is elicited by the continuous static receptor signals transmitted by both the primary and the secondary endings. The importance of the static stretch reflex is that it continues to cause muscle contraction as long as the muscle is maintained at excessive length. The muscle contraction in turn opposes the force that is causing the excess length.

NEGATIVE STRETCH REFLEX. When a muscle is suddenly shortened, exactly opposite effects occur because of decreased nerve impulses from the spindles. If the muscle is already taut, any sudden release of the load on the muscle that allows it to shorten will elicit both dynamic and static reflex *muscle inhibition* rather than reflex excitation. Thus, *this negative stretch reflex* opposes the shortening of the muscle in the same way that the positive stretch reflex opposes lengthening of the muscle. Therefore, one can begin to see that the stretch reflex tends to maintain the status quo for the length of a muscle.

Damping Function of the Dynamic and Static Stretch Reflexes

An especially important function of the stretch reflex is its ability to prevent some types of oscillation or jerkiness of body movements. This is a *damping*, or smoothing, function. An example is the following.

Use of the Damping Mechanism in Smoothing Muscle Contraction. Signals from other parts of the nervous system are often transmitted to a muscle in an unsmooth form, increasing in intensity for a few milliseconds, then decreasing in intensity, then changing to another intensity level, and so forth. When the muscle spindle apparatus is not functioning satisfactorily, the muscle contraction is jerky during the course of such a signal. This effect is demonstrated in Figure 54–5, which shows an experiment in which a sensory nerve signal entering one side of the cord is transmitted to a motor nerve on the other side of the cord to excite a muscle. In curve A, the muscle spindle reflex of the excited muscle is intact. Note that the contraction is relatively smooth even though the sensory nerve is excited at a slow frequency of 8 per second. Curve B, on the other hand, is the same experiment in an animal whose muscle spindle sensory nerves from the muscle had been sectioned 3 months earlier. Note the unsmooth muscle contraction. Thus, curve A demonstrates graphically the ability of the damping mechanism of the muscle spindle to smooth muscle contractions even though the input signals to the muscle motor system may themselves be jerky. This effect can

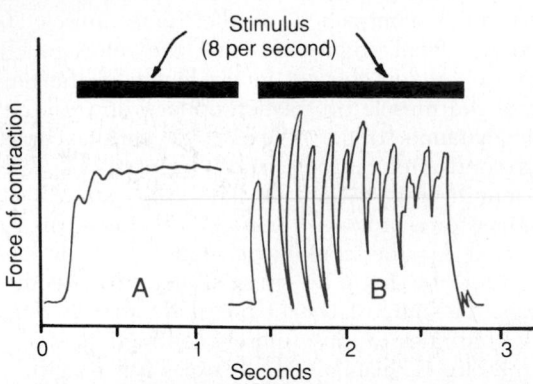

Figure 54–5. Muscle contraction caused by a spinal cord signal under two conditions: *A*, in a normal muscle, and *B*, in a muscle whose muscle spindles had been denervated by section of the posterior roots of the cord 82 days previously. Note the smoothing effect of the muscle spindle reflex in *A*. (Modified from Creed, et al.: Reflex Activity of the Spinal Cord. New York, Oxford University Press, 1932.)

also be called a *signal averaging* function of the muscle spindle reflex.

Role of the Muscle Spindle in Voluntary Motor Activity

To emphasize the importance of the gamma efferent system, one needs to recognize that 31 per cent of all the motor nerve fibers to the muscle are gamma efferent fibers rather than large type A alpha motor fibers. Whenever signals are transmitted from the motor cortex or from any other area of the brain to the alpha motor neurons, in most instances, the gamma motor neurons are stimulated simultaneously, an effect called *coactivation* of the alpha and gamma motor neurons. This causes both the extrafusal and the intrafusal muscle fibers to contract at the same time.

The purpose of contracting the muscle spindle fibers at the same time that the large skeletal muscle fibers contract is twofold: First, it keeps the length of the receptor portion of the muscle spindle from changing and therefore keeps the muscle spindle from opposing the muscle contraction. Second, it maintains proper damping function of the muscle spindle regardless of change in muscle length. For instance, if the muscle spindle should not contract and relax along with the large muscle fibers, the receptor portion of the spindle would sometimes be flail and at other times be overstretched, in neither instance operating under optimal conditions for spindle function.

Brain Areas for Control of the Gamma Motor System

The gamma efferent system is excited by signals from the *bulboreticular facilitatory* region of the brain stem and, secondarily, by impulses transmitted into the bulboreticular area from (a) the *cerebellum*, (b) the *basal ganglia*, and even (c) the *cerebral cortex*. Little is known about the precise mechanisms of control of the gamma efferent system. However, because the bulboreticular facilitatory area is particularly concerned with antigravity contractions and because the antigravity muscles have an especially high density of muscle spindles, emphasis is given to the importance of the gamma efferent mechanism in damping the movements of the different body parts during walking and running.

The Muscle Spindle System Stabilizes Body Position During Tense Action

One of the most important functions of the muscle spindle system is to stabilize body position during tense motor action. To do this, the bulboreticular facilitatory region and its allied areas of the brain stem transmit excitatory signals through the gamma nerve fibers to the intrafusal muscle fibers of the muscle spindles. This shortens the ends of the spindles and stretches the central receptor regions, thus increasing their signal output. However, the spindles on both sides of each joint are activated at the same time. Therefore, reflex excitation of the skeletal muscles on both sides of the joint also increases, producing tight, tense muscles opposing each other at the joint. The net effect is that the position of the joint becomes strongly stabilized, and any force that tends to move the joint from its current position is opposed by the highly sensitized stretch reflex.

Any time a person must perform a muscle function that requires a high degree of delicate and exact positioning, excitation of the appropriate muscle spindles by signals from the bulboreticular facilitatory region of the brain stem stabilizes the positions of the major joints. This aids tremendously in performing the additional detailed voluntary movements (of fingers or other body parts) required for intricate motor procedures.

Clinical Applications of the Stretch Reflex

The stretch reflex is elicited almost every time the clinician performs a physical examination. The purpose is to determine how much background excitation, or "tone," the brain is sending to the spinal cord. This reflex is elicited as follows.

KNEE JERK AND OTHER MUSCLE JERKS. Clinically, a method used to determine the sensitivity of the stretch reflexes is to elicit the knee jerk and other muscle jerks. The knee jerk can be elicited by simply striking the patellar tendon with a reflex hammer; this stretches the quadriceps muscle and initiates a *dynamic stretch reflex* that in turn causes the lower leg to jerk forward. The upper part of Figure 54–6 shows a myogram from the quadriceps muscle recorded during a knee jerk.

Similar reflexes can be obtained from almost any muscle of the body either by striking the tendon of the muscle or by striking the belly of the muscle itself. In other words, sudden stretch of muscle spindles is all that is required to elicit a stretch reflex.

The muscle jerks are used by neurologists to assess the degree of facilitation of spinal cord centers. When

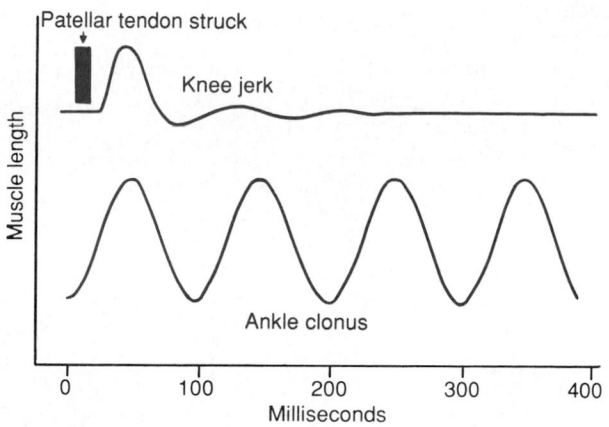

Figure 54–6. Myograms recorded from the quadriceps muscle during elicitation of the knee jerk and from the gastrocnemius muscle during ankle clonus.

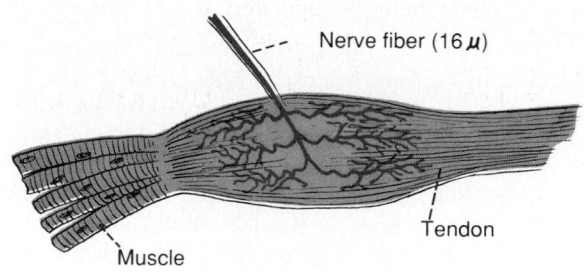

Figure 54–7. Golgi tendon organ.

large numbers of facilitatory impulses are being transmitted from the upper regions of the central nervous system into the cord, the muscle jerks are greatly exacerbated. On the other hand, if the facilitatory impulses are depressed or abrogated, the muscle jerks are considerably weakened or absent. These reflexes are used most frequently in determining the presence or absence of muscle spasticity after lesions in the motor areas of the brain or muscle spasticity in diseases that excite the bulboreticular facilitatory area of the brain stem. Ordinarily, large lesions in the contralateral motor areas of the cerebral cortex, especially those caused by strokes or brain tumors, cause greatly exacerbated muscle jerks.

CLONUS. Under appropriate conditions, the muscle jerks can oscillate, a phenomenon called *clonus* (see lower myogram, Fig. 54–6). Oscillation can be explained particularly well in relation to ankle clonus, as follows.

If a person standing on tiptoes suddenly drops his or her body downward to stretch one of the gastrocnemius muscles, impulses are transmitted from the muscle spindles into the spinal cord. These impulses reflexly excite the stretched muscle, which lifts the body up again. After a fraction of a second, the reflex contraction of the muscle dies out and the body falls again, thus stretching the spindles a second time. Again a dynamic stretch reflex lifts the body, but this, too, dies out after a fraction of a second, and the body falls once more to elicit still a new cycle. In this way, the stretch reflex of the gastrocnemius muscle continues to oscillate, often for long periods; this is clonus.

Clonus ordinarily occurs only if the stretch reflex is highly sensitized by facilitatory impulses from the brain. For instance, in the decerebrate animal, in which the stretch reflexes are highly facilitated, clonus develops readily. Therefore, to determine the degree of facilitation of the spinal cord, neurologists test patients for clonus by suddenly stretching a muscle and keeping a steady stretching force applied to the muscle. If clonus occurs, the degree of facilitation is certain to be high.

Golgi Tendon Reflex

THE GOLGI TENDON ORGAN HELPS TO CONTROL MUSCLE TENSION. The Golgi tendon organ, shown in Figure 54–7, is an encapsulated sensory receptor through which a small bundle of muscle tendon fibers pass. About 10 to 15 muscle fibers are usually connected in series with each Golgi tendon organ, and the organ is stimulated by the tension produced by this small bundle of muscle fibers. Thus, the major difference between the excitation of the Golgi tendon organ and the muscle spindle is that the spindle detects muscle length and changes in muscle length, whereas the tendon organ detects muscle *tension*.

The tendon organ, like the primary receptor of the muscle spindle, has both a *dynamic response* and a *static response*, responding intensely when the muscle tension suddenly increases (the dynamic response) but within a fraction of a second settling down to a lower level of steady-state firing that is almost directly proportional to the muscle tension (the static response). Thus, the Golgi tendon organs provide the nervous system with instantaneous information on the degree of tension in each small segment of each muscle.

TRANSMISSION OF IMPULSES FROM THE TENDON ORGAN INTO THE CENTRAL NERVOUS SYSTEM. Signals from the tendon organ are transmitted through large, rapidly conducting type Ib nerve fibers, averaging 16 micrometers in diameter, only slightly smaller than those from the primary ending of the muscle spindle. These fibers, like those from the primary endings, transmit signals both into local areas of the cord and through long fiber pathways such as the spinocerebellar tracts into the cerebellum and through still other tracts to the cerebral cortex. The local cord signal excites a single *inhibitory* interneuron that in turn inhibits the anterior motor neuron. This local circuit directly inhibits the individual muscle without affecting adjacent muscles. The signals to the brain are discussed in Chapter 56.

Inhibitory Nature of the Tendon Reflex and Its Importance

When the Golgi tendon organs of a muscle are stimulated by increased muscle tension, signals are transmitted into the spinal cord to cause reflex effects in the respective muscle. This reflex is entirely inhibitory. Thus, this reflex provides a *negative feedback* mechanism that prevents the development of too much tension on the muscle.

When tension on the muscle and, therefore, on the tendon becomes extreme, the inhibitory effect from the tendon organ can be so great that it leads to a sudden reaction in the spinal cord and instantaneous relaxation of the entire muscle. This effect is called the *lengthening reaction;* it is possibly or even probably a protective mechanism to prevent tearing of the muscle or avulsion of the tendon from its attachments to the bone. We know, for instance, that direct electrical stimulation of muscles in the laboratory, which cannot be opposed by this negative reflex, can cause such destructive effects.

POSSIBLE ROLE OF THE TENDON REFLEX TO EQUALIZE CONTRACTILE FORCE AMONG THE MUSCLE FIBERS. Another likely function of the Golgi tendon reflex is to equalize the contractile forces of the separate muscle fibers. That is, those fibers that exert excess tension become inhibited by the reflex, whereas those that exert too little tension become more excited because of absence of reflex inhibition. This would spread the muscle load over all the fibers and especially would prevent damage in isolated areas of a muscle where small numbers of fibers might be overloaded.

Function of the Muscle Spindles and Golgi Tendon Organs in Conjunction with Motor Control from Higher Levels of the Brain

Although we have emphasized the function of the muscle spindles and Golgi tendon organs in spinal cord control of motor function, these two sensory organs also apprise the higher motor control centers of instantaneous changes taking place in the muscles. For instance, the dorsal spinocerebellar tracts carry instantaneous information from both the muscle spindles and the Golgi tendon organs directly to the cerebellum at conduction velocities approaching 120 m/sec. Additional pathways transmit similar information into the reticular regions of the brain stem and, to a lesser extent, all the way to the motor areas of the cerebral cortex. We learn in Chapters 55 and 56 that information from these receptors is crucial for feedback control of motor signals that originate in all these areas.

FLEXOR REFLEX AND THE WITHDRAWAL REFLEXES

In the spinal or decerebrate animal, almost any type of cutaneous sensory stimulus on a limb is likely to cause the flexor muscles of the limb to contract, thereby withdrawing the limb from the stimulating object. This is called the *flexor reflex.*

In its classic form, the flexor reflex is elicited most powerfully by stimulation of pain endings, such as by a pinprick or heat, for which reason it is also called a *nociceptive reflex,* or simply *pain reflex.* Stimulation of the touch receptors can also elicit a weaker and less prolonged flexor reflex.

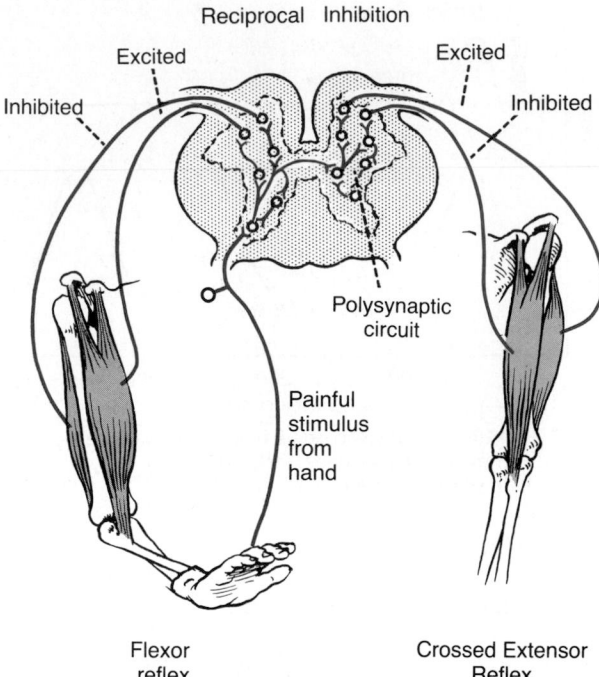

Figure 54–8. Flexor reflex, crossed extensor reflex, and reciprocal inhibition.

If some part of the body besides one of the limbs is painfully stimulated, this part, in a similar manner, will be *withdrawn from the stimulus,* but the reflex may not be confined to flexor muscles even though it is basically the same type of reflex. Therefore, the many patterns of reflexes of this type in the different areas of the body are called the *withdrawal reflexes.*

NEURONAL MECHANISM OF THE FLEXOR REFLEX. The left-hand portion of Figure 54–8 shows the neuronal pathways for the flexor reflex. In this instance, a painful stimulus is applied to the hand; as a result, the flexor muscles of the upper arm become reflexly excited, thus withdrawing the hand from the painful stimulus.

The pathways for eliciting the flexor reflex do not pass directly to the anterior motor neurons but instead pass first into the interneuron pool of neurons and only secondarily to the motor neurons. The shortest possible circuit is a three- or four-neuron arc; most of the signals of the reflex traverse many more neurons than this and involve the following basic types of circuits: (1) diverging circuits to spread the reflex to the necessary muscles for withdrawal; (2) circuits to inhibit the antagonist muscles, called *reciprocal inhibition circuits;* and (3) circuits to cause a prolonged repetitive afterdischarge even after the stimulus is over.

Figure 54–9 shows a typical myogram from a flexor muscle during a flexor reflex. Within a few milliseconds after a pain nerve begins to be stimulated, the flexor response appears. Then, in the next few seconds,

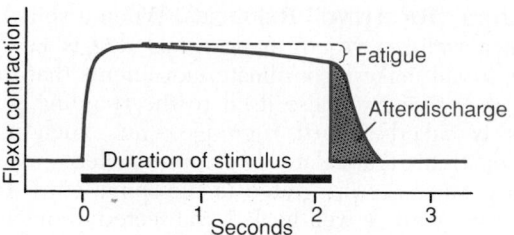

Figure 54–9. Myogram of the flexor reflex, showing rapid onset of the reflex, an interval of fatigue, and, finally, afterdischarge after the stimulus is over.

the reflex begins to *fatigue*, which is characteristic of essentially all the more complex integrative reflexes of the spinal cord. Then, soon after the stimulus is over, the contraction of the muscle begins to return toward the base line, but because of *afterdischarge*, it will not return all the way for many milliseconds. The duration of the afterdischarge depends on the intensity of the sensory stimulus that elicited the reflex; a weak tactile stimulus causes almost no afterdischarge in contrast to an afterdischarge lasting for a second or more after a strong pain stimulus.

The afterdischarge that occurs in the flexor reflex almost certainly results from both types of repetitive-discharge circuits discussed in Chapter 46. Electrophysiological studies indicate that the immediate afterdischarge, lasting for about 6 to 8 milliseconds, results from repetitive firing of the excited interneurons themselves. The prolonged afterdischarge that occurs after strong pain stimuli almost certainly results from recurrent pathways that excite reverberating interneuron circuits, these transmitting impulses to the anterior motor neurons sometimes for several seconds after the incoming sensory signal is over.

Thus, the flexor reflex is appropriately organized to withdraw a pained or otherwise irritated part of the body away from the stimulus. Furthermore, because of the afterdischarge, the reflex can hold the irritated part away from the stimulus for 0.1 to 3 seconds after the irritation is over. During this time, other reflexes and actions of the central nervous system can move the entire body away from the painful stimulus.

PATTERN OF WITHDRAWAL. The pattern of withdrawal that results when the flexor reflex (or the many other types of withdrawal reflexes) is elicited depends on the sensory nerve that is stimulated. Thus, a painful stimulus on the inside of the arm not only elicits a flexor reflex in the arm but also contracts the abductor muscles to pull the arm outward. In other words, the integrative centers of the cord cause those muscles to contract that can most effectively remove the pained part of the body from the object that causes pain. This same principle, called the principle of "local sign," applies to any part of the body but especially to the limbs because they have highly developed flexor reflexes.

CROSSED EXTENSOR REFLEX

About 0.2 to 0.5 second after a stimulus elicits a flexor reflex in one limb, the opposite limb begins to extend. This is called the *crossed extensor reflex.* Extension of the opposite limb can push the entire body away from the object causing the painful stimulus in the withdrawn limb.

NEURONAL MECHANISM OF THE CROSSED EXTENSOR REFLEX. The right-hand portion of Figure 54–8 shows the neuronal circuit responsible for the crossed extensor reflex, demonstrating that signals from the sensory nerves cross to the opposite side of the cord to excite the extensor muscles. Because the crossed extensor reflex usually does not begin until 200 to 500 milliseconds after the initial pain stimulus, it is certain that many interneurons are involved in the circuit between the incoming sensory neuron and the motor neurons of the opposite side of the cord responsible for the crossed extension. Furthermore, after the painful stimulus is removed, the crossed extensor reflex continues for an even longer period of afterdischarge than that for the flexor reflex. Therefore, again, it is presumed that this prolonged afterdischarge results from reverberatory circuits among the internuncial cells.

Figure 54–10 shows a typical myogram recorded from a muscle involved in a crossed extensor reflex. This demonstrates the relatively long latency before the reflex begins and the long afterdischarge period at the end of the stimulus. The prolonged afterdischarge would be of benefit in holding the body away from a painful object until other nervous reactions could cause the body to move away.

RECIPROCAL INHIBITION AND RECIPROCAL INNERVATION

In the foregoing paragraphs, we have pointed out several times that excitation of one group of muscles is usually associated with inhibition of another group. For

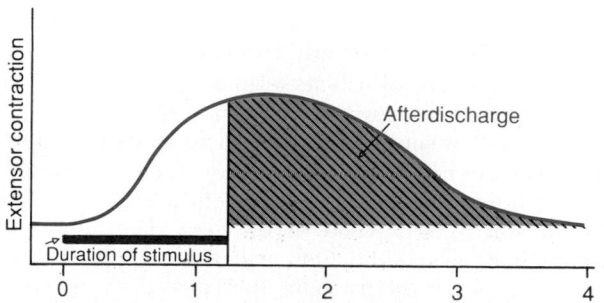

Figure 54–10. Myogram of a crossed extensor reflex, showing slow onset but prolonged afterdischarge.

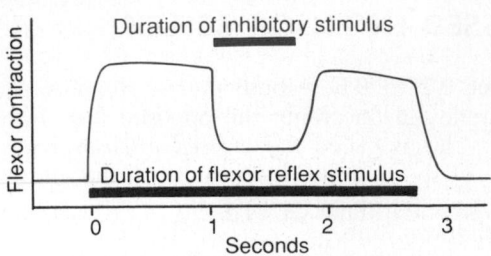

Figure 54–11. Myogram of a flexor reflex, showing reciprocal inhibition caused by an inhibitory stimulus from a stronger flexor reflex in the opposite limb.

instance, when a stretch reflex excites one muscle, it often, at the same time, inhibits the antagonist muscles. This is the phenomenon of *reciprocal inhibition,* and the neuronal circuit that causes this reciprocal relation is called *reciprocal innervation.* Likewise, reciprocal relations often exist between the two sides of the cord, as exemplified by the flexor and extensor reflexes described above.

Figure 54–11 shows a typical example of reciprocal inhibition. In this instance, a moderate but prolonged flexor reflex is elicited from one limb of the body; while this reflex is still being elicited, a stronger flexor reflex is elicited in the opposite limb. This reflex then sends reciprocal inhibitory signals to the first limb and depresses its degree of flexion. Finally, removal of the stronger reflex allows the original reflex to reassume its previous intensity.

REFLEXES OF POSTURE AND LOCOMOTION

Postural and Locomotive Reflexes of the Cord

POSITIVE SUPPORTIVE REACTION. Pressure on the footpad of a decerebrate animal causes the limb to extend against the pressure being applied to the foot. Indeed, this reflex is so strong that an animal whose spinal cord has been transected for several months— that is, after the reflexes have become exacerbated— can often be placed on its feet and the reflex will stiffen the limbs sufficiently to support the weight of the body: the animal will stand in a rigid position. This reflex is called the *positive supportive reaction.*

The positive supportive reaction involves a complex circuit in the interneurons similar to those responsible for the flexor and cross extensor reflexes. The locus of the pressure on the pad of the foot determines the direction in which the limb will extend; pressure on one side causes extension in that direction, an effect called the *magnet reaction.* This helps to keep an animal from falling to that side.

CORD "RIGHTING" REFLEXES. When a spinal cat or even a well-recovered young spinal dog is laid on its side, it will make incoordinate movements that indicate that it is trying to raise itself to the standing position. This is called a *cord righting reflex.* Such a reflex demonstrates that relatively complex reflexes associated with posture are integrated in the spinal cord. Indeed, a puppy with a well-healed transected thoracic cord between the level for the forelimbs and the hindlimbs can right itself from the lying position and even walk on its hindlimbs. And in the case of the opossum with a similar transection of the thoracic cord, the walking movements of the hindlimbs are hardly different from those in the normal opossum—except that the hindlimb movements are not synchronized with those of the forelimbs, as is normally the case.

Stepping and Walking Movements

RHYTHMICAL STEPPING MOVEMENTS OF A SINGLE LIMB. Rhythmical stepping movements are frequently observed in the limbs of spinal animals. Indeed, even when the lumbar portion of the spinal cord is separated from the remainder of the cord and a longitudinal section is made down the center of the cord to block neuronal connections between the two sides of the cord and between the two limbs, each hindlimb can still perform individual stepping functions. Forward flexion of the limb is followed a second or so later by backward extension. Then flexion occurs again, and the cycle is repeated over and over.

This oscillation back and forth between the flexor and extensor muscles can occur even after the sensory nerves have been cut, and it seems to result mainly from mutually reciprocal inhibition circuits that oscillate between the neurons controlling agonist and antagonist muscles within the matrix of the cord itself.

The sensory signals from the footpads and from the position sensors around the joints play a strong role in controlling foot pressure and rate of stepping when the foot is allowed to walk along a surface. In fact, the cord mechanism for control of stepping can be still more complex. For instance, if during the forward thrust of the foot the top of the foot encounters an obstruction, the forward thrust will stop temporarily, but then in rapid sequence, the foot will be lifted higher and proceed forward to be placed over the obstruction. This is the *stumble reflex.* Thus, the cord is an intelligent walking controller.

RECIPROCAL STEPPING OF OPPOSITE LIMBS. If the lumbar spinal cord is not split down its center as noted above, every time stepping occurs in the forward direction in one limb, the opposite limb ordinarily steps backward. This effect results from reciprocal innervation between the two limbs.

DIAGONAL STEPPING OF ALL FOUR LIMBS—THE "MARK TIME" REFLEX. If a well-healed spinal animal with the spinal transection above the forelimb area of the cord is held up from the floor and its legs are allowed to fall downward, as shown in Figure 54–12,

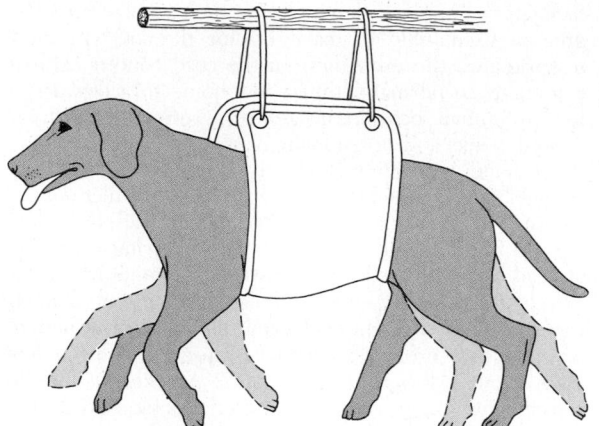

Figure 54–12. Diagonal stepping movements exhibited by a spinal animal.

the stretch on the limbs occasionally elicits stepping reflexes that involve all four limbs. In general, stepping occurs diagonally between the forelimbs and hindlimbs. This diagonal response is another manifestation of reciprocal innervation, this time occurring the entire distance up and down the cord between the forelimbs and hindlimbs. Such a walking pattern is called a *mark time reflex.*

GALLOPING REFLEX. Another type of reflex that occasionally develops in the spinal animal is the galloping reflex, in which both forelimbs move backward in unison while both hindlimbs move forward; this is followed a second or so later by opposite movements of the limbs. This often occurs when stretch or pressure stimuli are applied almost exactly equally to opposite limbs at the same time, whereas unequal stimulation of one side versus the other elicits the diagonal walking reflex. This is in keeping with the normal patterns of walking and of galloping because in walking, only one forelimb and one hindlimb at a time are stimulated, and this would predispose to continued walking. Conversely, when the animal strikes the ground during galloping, both the forelimbs and the hindlimbs on both sides are stimulated about equally; this would predispose to further galloping and, therefore, would continue this pattern of motion in contradistinction to the walking pattern.

SCRATCH REFLEX

An especially important cord reflex in some animals is the scratch reflex, which is initiated by the *itch and tickle sensation.* It involves two functions: (1) a *position sense* that allows the paw to find the exact point of irritation on the surface of the body and (2) a *to-and-fro scratching movement.*

The *to-and-fro movement,* like the stepping movements of locomotion, involves reciprocal innervation circuits that cause oscillation, which can still function even when all the sensory roots from the oscillating limb are sectioned, as is true for basic walking movements.

The *position sense* of the scratch reflex is a highly developed function. Even though a flea might be crawling as far forward as the shoulder of a spinal animal, the hind paw can still find its position even though 19 muscles in the limb must be contracted simultaneously in a precise pattern to bring the paw to the position of the crawling flea. To make the reflex even more complicated, when the flea crosses the midline, the first paw stops scratching and the opposite paw begins the to-and-fro motion and eventually finds the flea.

SPINAL CORD REFLEXES THAT CAUSE MUSCLE SPASM

In human beings, local muscle spasm is often observed. The mechanism of this has not been elucidated to complete satisfaction even in laboratory animals, but it is known that pain stimuli can cause reflex spasm of local muscles. This presumably is the cause of much, if not most, of the muscle spasm observed in localized regions of the human body.

MUSCLE SPASM RESULTING FROM A BROKEN BONE. One type of clinically important spasm occurs in muscles that surround a broken bone. This seems to result from the pain impulses initiated from the broken edges of the bone, which cause the muscles that surround the area to contract powerfully and tonically. Relief of the pain by injection of a local anesthetic relieves the spasm; a deep general anesthetic also relieves the spasm. One of these procedures is often necessary before the spasm can be overcome sufficiently for the two ends of the bone to be set back into appropriate positions.

ABDOMINAL MUSCLE SPASM IN PERITONITIS. Another type of local spasm caused by cord reflexes is the abdominal spasm resulting from irritation of the parietal peritoneum by peritonitis. Here, again, relief of the pain caused by the peritonitis allows the spastic muscle to relax. Almost the same type of spasm often occurs during surgical operations; pain impulses from the parietal peritoneum cause the abdominal muscles to contract extensively and sometimes extrude the intestines through the surgical wound. For this reason, deep anesthesia is usually required for intra-abdominal operations.

MUSCLE CRAMPS. Still another type of local spasm is the typical muscle cramp. Electromyographic studies indicate that the cause of at least some muscle cramps is the following.

Any local irritating factor or metabolic abnormality of a muscle, such as severe cold, lack of blood flow to the muscle, or overexercise of the muscle, can elicit pain or other types of sensory impulses that are transmitted from the muscle to the spinal cord, thus causing reflex muscle contraction. The contraction in turn is believed to stimulate the same sensory receptors still more, which causes the spinal cord to increase the intensity of contraction. Thus, a positive feedback develops so that a small amount of initial irritation causes more and more contraction until a full-blown muscle cramp ensues.

AUTONOMIC REFLEXES IN THE SPINAL CORD

Many types of segmental autonomic reflexes occur in the spinal cord, most of which are discussed in other chapters. Briefly, these include (1) changes in vascular tone resulting from changes in local skin heat (see Chap. 73); (2) sweating, which results from localized heat on the surface of the body (see Chap. 73); (3) intestinointestinal reflexes that control some motor functions of the gut (see Chap. 62); (4) peritoneointestinal reflexes that inhibit gastrointestinal motility in response to peritoneal irritation (see Chap. 66); and (5) evacuation reflexes for emptying the full bladder (see Chap. 31) and the colon (see Chap. 63). In addition, all the segmental reflexes can at times be elicited simultaneously in the form of the so-called mass reflex as follows.

MASS REFLEX. In a spinal animal or human being, sometimes the spinal cord suddenly becomes excessively active, causing massive discharge of large portions of the cord. The usual stimulus that causes this is a strong nociceptive stimulus to the skin or excessive filling of a viscus, such as overdistention of the bladder or the gut. Regardless of the type of stimulus, the resulting reflex, called the *mass reflex*, involves large portions or even all of the cord, and its pattern of reaction is the same. The effects are (1) a major portion of the body goes into strong flexor spasm, (2) the colon and bladder are likely to evacuate, (3) the arterial pressure often rises to maximal values—sometimes to a systolic pressure well over 200 mm Hg—and (4) large areas of the body break out into profuse sweating.

The precise neuronal mechanism of the mass reflex is unknown. Because it lasts for minutes, it presumably results from activation of great masses of reverberating circuits that excite large areas of the cord at once. This is similar to the mechanism of epileptic seizures that also involve reverberating circuits but that occur in the brain instead of the cord.

SPINAL CORD TRANSECTION AND SPINAL SHOCK

When the spinal cord is suddenly transected in the upper neck, essentially all cord functions, including the cord reflexes, immediately become depressed to the point of total silence, a reaction called *spinal shock*. The reason for this is that normal activity of the cord neurons depends to a great extent on continual tonic excitation by discharges of nerve fibers entering the cord from higher centers, particularly discharges transmitted through the reticulospinal tracts, vestibulospinal tracts, and corticospinal tracts.

After a few hours to a few weeks, the spinal neurons gradually regain their excitability. This seems to be a natural characteristic of neurons everywhere in the nervous system—that is, after they lose their source of facilitatory impulses, they increase their own natural degree of excitability to make up for the loss. In most nonprimates, the excitability of the cord centers returns essentially to normal within a few hours to a day or so, but in human beings, the return is often delayed for several weeks and occasionally is never complete; on the other hand, sometimes recovery is excessive, with resultant hyperexcitability of some or all cord functions.

Some of the spinal functions specifically affected during or after spinal shock are the following: (1) The arterial blood pressure falls drastically—sometimes to as low as 40 mm Hg—thus demonstrating that sympathetic activity becomes blocked almost to extinction. The pressure ordinarily returns to normal within a few days, even in the human being. (2) All skeletal muscle reflexes integrated in the spinal cord are blocked during the initial stages of shock. In lower animals, a few hours to a few days are required for these reflexes to return to normal, and in human beings, 2 weeks to several months are usually required. Sometimes, both in animals and in humans, some reflexes eventually become hyperexcitable, particularly if a few facilitatory pathways remain intact between the brain and the cord while the remainder of the spinal cord is transected. The first reflexes to return are the stretch reflexes, followed in order by the progressively more complex reflexes: the flexor reflexes, postural antigravity reflexes, and remnants of stepping reflexes. (3) The sacral reflexes for control of bladder and colon evacuation are suppressed in human beings for the first few weeks after cord transection, but they eventually return. These effects are discussed in Chapters 31 and 66.

REFERENCES

Bannister, C. M., and Tew, B.: Current Concepts in Spina Bifida and Hydrocephalus. New York, Cambridge University Press, 1992.

Berthoz, A.: Multisensory Control of Movement. New York, Oxford University Press, 1993.

Bridwell, K. H., and DeWald, R. L.: The Textbook of Spinal Surgery. Philadelphia, J. B. Lippincott, 1991.

Brooks, V. B.: The Neural Basis of Motor Control. New York, Oxford University Press, 1986.

Burke, R. E.: Motor units: Anatomy, physiology, and functional organization. In Brooks, V. B. (ed.): Handbook of Physiology. Sec. 1, Vol. II. Bethesda, Md., American Physiological Society, 1981, p. 345.

Burt, A. M.: Textbook of Neuroanatomy. Philadelphia, W. B. Saunders Co., 1993.

Cordo, P., and Harnad, S.: Movement Control. New York, Cambridge University Press, 1994.

Creed, R. S., et al.: Reflex Activity of the Spinal Cord. New York, Oxford University Press, 1932.

DeLisa, J. A., et al.: Manual of Nerve Conduction Velocity and Clinical Neurophysiology. New York, Raven Press, 1994.

Dietz, V.: Human neuronal control of automatic functional movements: interaction between central programs and afferent input. Physiol. Rev., 72:33, 1992.

Emonet-Denand, F., et al.: How muscle spindles signal changes of muscle length. News Physiol. Sci., 3:105, 1988.

Hammond, D. L.: New insights regarding organization of spinal cord pain pathways. News Physiol. Sci., 4:98, 1989.

Hasan, A., and Stuart, D. G.: Animal solutions to problems of movement control: The role of proprioceptors. Annu. Rev. Neurosci., 11:199, 1988.

Heckman, C. J.: Computer simulations of the effects of different synaptic input systems on the steady-state input-output structure of the motoneuron pool. J. Neurophysiol., 71:1727, 1994.

Hendelman, W. J.: Student's Atlas of Neuroanatomy. Philadelphia, W. B. Saunders Co., 1994.

Hnik, P., et al. (eds.): Mechanoreceptors. Development, Structure, and Function. New York, Plenum Publishing Corp., 1988.

Houk, J. C.: Control strategies in physiological systems. FASEB J., 2:97, 1988.

Illis, L. S.: Spinal Cord Dysfunction. New York, Oxford University Press, 1988.

Jami, L.: Golgi tendon organs in mammalian skeletal muscle: functional properties and central actions. Physiol. Rev., 72:623, 1992.

Janig, W., and McLachlan, E. M.: Organization of lumbar spinal outflow to distal colon and pelvic organs. Physiol. Rev., 67:1332, 1987.

Kurlan, R.: Treatment of Movement Disorders. Philadelphia, J. B. Lippincott, 1994.

Lüscher, H.-R., and Clamann, H. P.: Relation between structure and function in information transfer in spinal monosynaptic reflex. Physiol. Rev., 72:71, 1992.

Matthews, P. B. C.: Muscle spindles: Their messages and their fusimotor supply. In Brooks, V. B. (ed.): Handbook of Physiology. Sec. 1, Vol. II. Bethesda, Md., American Physiological Society, 1981, p. 189.

Mendell, L. M.: Modifiability of spinal synapses. Physiol. Rev., 64:260, 1984.

Porter, R., and Lemon, R.: Corticospinal Function and Voluntary Movement. New York, Oxford University Press, 1993.

Redman, S. J.: Monosynaptic transmission in the spinal cord. News Physiol. Sci., 1:171, 1986.

Rowell, L. B.: Reflex control of regional circulation in humans. J. Auton. Nerv. Syst., 11:101, 1984.

Rowland, L. P.: Amyotrophic Lateral Sclerosis and Other Motor Neuron Diseases. New York, Raven Press, 1991.

Shenolikar, S., and Nairn, A. C.: Model Systems in Signal Transduction. New York, Raven Press, 1993.

Sherrington, C. S.: The Integrative Action of the Nervous System. New Haven, Conn., Yale University Press, 1911.

Weiner, W. J.: Emergent and Urgent Neurology. Philadelphia, J. B. Lippincott, 1992.

Windhorst, U.: Shaping Static Elbow Torque-angle Relationships by Spinal Cord Circuits: A Theoretical Study. Neuroscience, 59:713, 1994.

Cortical and Brain Stem Control of Motor Function

CHAPTER 55

In this chapter, we discuss control of the body's movements by the cerebral cortex and brain stem. These two neural areas, along with the basal ganglia and cerebellum, which are discussed in Chapter 56, control the complex movements that the human being and other higher animals have developed for their special purposes.

Virtually all "voluntary" movements involve conscious activity in the cerebral cortex. This does not mean that each contraction of each muscle is willed by the cortex itself. Instead, most control by the cortex involves activation at the same time of multiple patterns of function stored in lower brain areas—in the cord, brain stem, basal ganglia, and cerebellum—and these lower centers in turn send most of the specific activating signals to the muscles. For a few types of movements, the cortex does have almost a direct pathway to the anterior motor neurons of the cord, bypassing other motor centers on the way, especially for control of the fine dexterous movements of the fingers and hands. This chapter and Chapter 56 explain the interplay among the different motor areas of the brain and spinal cord that provides this overall synthesis of voluntary motor function.

THE MOTOR CORTEX AND CORTICOSPINAL TRACT

Figure 55–1 shows the functional areas of the cerebral cortex. Anterior to the central sulcus, occupying approximately the posterior one third of the frontal lobes, is the *motor cortex*. Posterior to the central sulcus is the *somatic sensory cortex*, an area discussed in detail in earlier chapters that feeds many signals into the motor cortex for control of motor activities.

The motor cortex itself is further divided into three subareas, each of which has its own topographical representation of muscle groups and specific motor functions of the body: (1) the *primary motor cortex*, (2) the *premotor area,* and (3) the *supplementary motor area.*

Primary Motor Cortex

The primary motor cortex, shown in Figure 55–1, lies in the first convolution of the frontal lobes anterior to the central sulcus. It begins laterally in the sylvian fissure, spreads superiorly to the uppermost portion of the brain, and then dips deep into the longitudinal fissure. This area is the same as area 4 in Brodmann's classification of the brain cortical areas, shown in Figure 47–5.

Figure 55–1 lists the topographical representations of the different muscle areas of the body in the primary motor cortex, beginning with the face and mouth region, near the sylvian fissure; the arm and hand area, in the midportions of the primary motor cortex; the trunk, near the apex of the brain; and the leg and foot areas, in that part of the primary motor cortex that dips into the longitudinal fissure. This topographical organization is demonstrated even more graphically in Figure 55–2, which shows the degrees of representation of the different muscle areas as mapped by Penfield and Rasmussen. This mapping was done by electrically stimulating the different areas of the motor cortex in human beings who were undergoing neurosurgical operations. Note that more than one half of

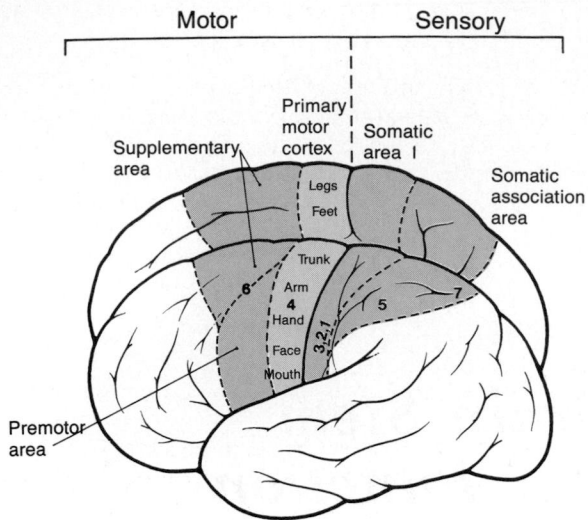

Figure 55–1. Motor and somatosensory functional areas of the cerebral cortex.

the entire primary motor cortex is concerned with controlling the hands and the muscles of speech. Point stimulations in these hand and speech motor areas often cause contraction of a single muscle. But in those areas with lesser degrees of representation, such as the trunk area, electrical stimulation contracts a group of muscles instead.

Premotor Area

The premotor area, also shown in Figure 55–1, lies immediately anterior to the lateral portions of the primary motor cortex, projecting 1 to 3 centimeters ante-

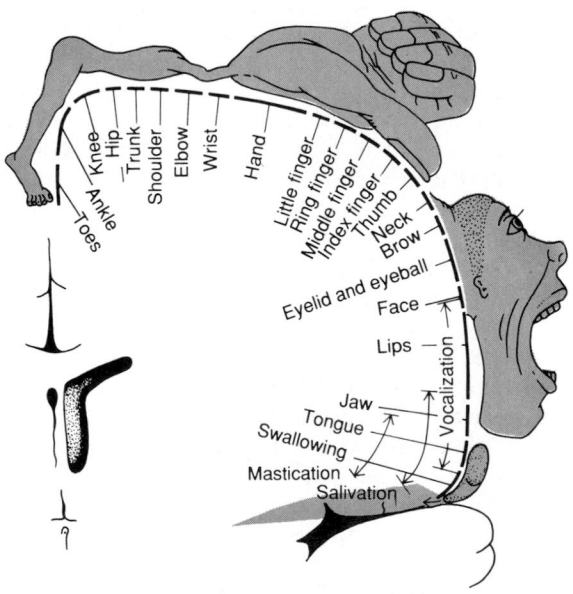

Figure 55–2. Degree of representation of the different muscles of the body in the motor cortex. (From Penfield and Rasmussen: The Cerebral Cortex of Man: A Clinical Study of Localization of Function. New York, Macmillan Co., 1968.)

riorly and extending inferiorly into the sylvian fissure and superiorly about two thirds of the way to the longitudinal fissure; there it abuts the supplementary motor area. The topographical organization of the premotor cortex is roughly the same as that of the primary motor cortex, with the mouth and face areas located most laterally and then in the upward direction the hand, arm, trunk, and leg areas. The premotor area occupies a large share of area 6 in the Brodmann classification of brain topology.

Most nerve signals generated in the premotor area cause patterns of movement that involve groups of muscles that perform specific tasks. For instance, the task may be to position the shoulders and arms so that the hands become properly oriented to perform specific tasks. To achieve these results, the premotor area sends its signals either directly into the primary motor cortex to excite multiple groups of muscles or, more likely, by way of the basal ganglia and then back through the thalamus to the primary motor cortex. Thus, the premotor cortex, basal ganglia, thalamus, and primary motor cortex constitute a complex overall system for control of many of the body's patterns of coordinated muscle activity.

Supplementary Motor Area

The supplementary motor area has still another topographical organization for control of motor function. It lies immediately superior to the premotor area, lying mainly in the longitudinal fissure but extending a few centimeters over the edge onto the superiormost portion of the lateral cortex.

Considerably stronger electrical stimuli are required in the supplementary motor area to cause muscle contraction than in the other motor areas. When contractions are elicited, they are often bilateral rather than unilateral. For instance, stimulation frequently leads to bilateral grasping movements of both hands simultaneously; these movements are perhaps rudiments of the hand functions required for climbing. Also, there may be rotation of the trunk, rotation of the hands, movement of the eyes, or fixation of the shoulders. In general, this area functions in concert with the premotor area to provide attitudinal movements, fixation movements of the different segments of the body, positional movements of the head and eyes, and so forth, as background for the finer motor control of the arms and hands by the premotor area and primary motor cortex.

Some Specialized Areas of Motor Control Found in the Human Motor Cortex

Neurosurgeons have found a few highly specialized motor regions of the human cerebral cortex, located mainly in the premotor areas, as shown in Figure 55–3, that control specific motor functions. These regions have been localized either by electrical stimulation or by noting the loss of motor function when

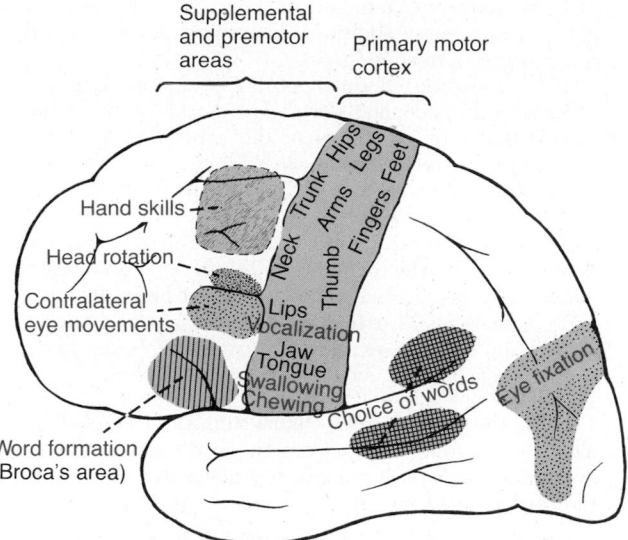

Figure 55–3. Representation of the different muscles of the body in the motor cortex and location of other cortical areas responsible for specific types of motor movements.

destructive lesions have occurred in specific cortical areas. Some of the more important of these are the following.

BROCA'S AREA AND SPEECH. Figure 55–3 shows a premotor area lying immediately anterior, to the primary motor cortex and immediately above the sylvian fissure labeled "word formation." This region is called *Broca's area.* Damage to it does not prevent a person from vocalizing, but it does make it impossible for the person to speak whole words rather than incoordinate utterances or an occasional simple word such as "no" or "yes." A closely associated cortical area also causes appropriate respiratory function, so that respiratory activation of vocal cords can occur simultaneously with the movements of the mouth and tongue during speech. Thus, the premotor activities that are related to Broca's area are highly complex.

"VOLUNTARY" EYE MOVEMENT FIELD. Immediately above Broca's area is a locus for controlling eye movements. Damage to this area prevents a person from voluntarily moving the eyes toward different objects. Instead, the eyes tend to lock on specific objects, an effect controlled by signals from the occipital cortex, as explained in Chapter 51. This frontal area also controls eyelid movements such as blinking.

HEAD ROTATION AREA. Still slightly higher in the motor association area, electrical stimulation will elicit head rotation. This area is closely associated with the eye movement field and is presumably related to directing the head toward different objects.

AREA FOR HAND SKILLS. In the premotor area immediately anterior to the primary motor cortex for the hands and fingers is a region neurosurgeons have called an area for hand skills. That is, when tumors or other lesions cause destruction in this area, the hand movements become incoordinate and nonpurposeful, a condition called *motor apraxia.*

Transmission of Signals from the Motor Cortex to the Muscles

Motor signals are transmitted directly from the cortex to the spinal cord through the *corticospinal tract* and indirectly through multiple accessory pathways that involve the *basal ganglia, cerebellum,* and various *nuclei of the brain stem.* In general, the direct pathways are concerned more with discrete and detailed movements, especially of the distal segments of the limbs, particularly the hands and fingers.

Corticospinal Tract (Pyramidal Tract)

The most important output pathway from the motor cortex is the *corticospinal tract,* also called the *pyramidal tract,* which is shown in Figure 55–4.

The corticospinal tract originates about 30 per cent from the primary motor cortex, 30 per cent from the premotor and supplementary motor areas, and 40 per cent from the somatic sensory areas posterior to the central sulcus. After leaving the cortex, it passes through the posterior limb of the internal capsule (between the caudate nucleus and the putamen of the basal ganglia) and then downward through the brain stem, forming the *pyramids of the medulla.* By far the majority of the pyramidal fibers then cross to the opposite side and descend in the *lateral corticospinal tracts* of the cord, finally terminating principally on the interneurons in the intermediate regions of the cord gray matter but also a few on sensory relay neurons in the dorsal horn, and others directly on the anterior motor neurons that cause muscle contraction.

A few of the fibers do not cross to the opposite side in the medulla but pass ipsilaterally down the cord in the *ventral corticospinal tracts,* but many of these fibers also cross to the opposite side of the cord either in the neck or in the upper thoracic region. These fibers are perhaps concerned with control by the supplementary motor area of bilateral postural movements.

The most impressive fibers in the pyramidal tract are a population of large myelinated fibers with a mean diameter of 16 micrometers. These fibers originate from the *giant pyramidal cells,* also called *Betz cells,* that are found only in the primary motor cortex. These cells are about 60 micrometers in diameter, and their fibers transmit nerve impulses to the spinal cord at a velocity of about 70 m/sec, the most rapid rate of transmission of any signals from the brain to the cord. There are about 34,000 of these large Betz cell fibers in each corticospinal tract. The total number of fibers in each corticospinal tract is more than 1 million, so these large fibers represent only 3 per cent of all of them. The other 97 per cent are mainly fibers smaller than 4 micrometers in diameter that are believed to conduct (a) background tonic signals to the motor areas of the cord or (b) feedback signals from the cortex to control the intensities of various sensory signals going to the brain.

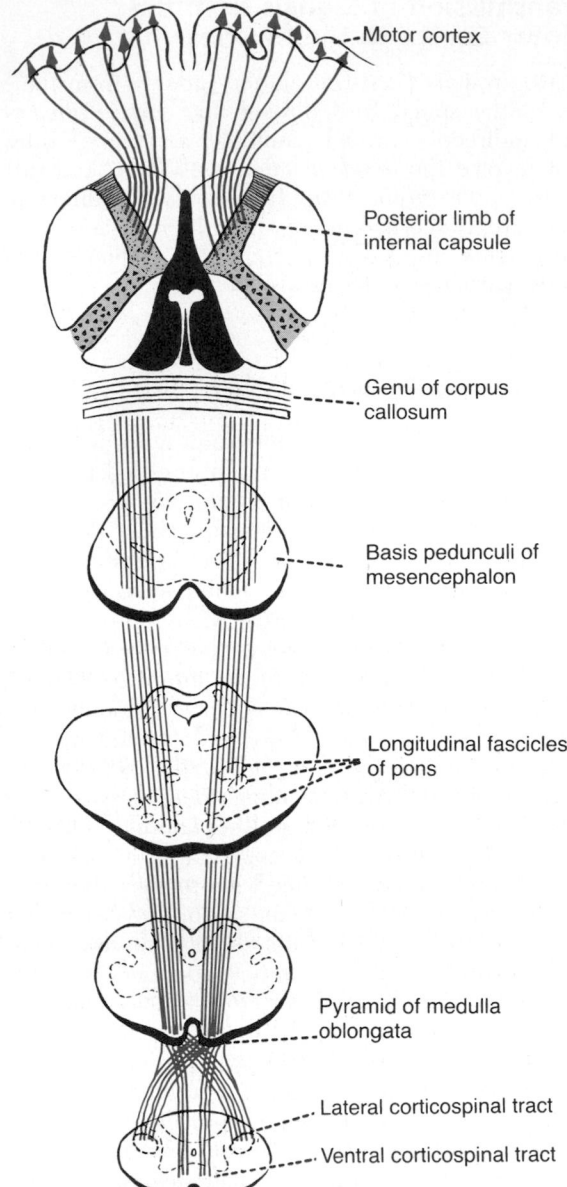

- Motor cortex
- Posterior limb of internal capsule
- Genu of corpus callosum
- Basis pedunculi of mesencephalon
- Longitudinal fascicles of pons
- Pyramid of medulla oblongata
- Lateral corticospinal tract
- Ventral corticospinal tract

Figure 55–4. Pyramidal tract. (Modified from Ranson and Clark: Anatomy of the Nervous System. Philadelphia, W. B. Saunders Co., 1959.)

Other Fiber Pathways from the Motor Cortex

The motor cortex gives rise to large numbers of fibers from the cortex or collaterals from the pyramidal tract that go to deeper regions of the cerebrum and into the brain stem, including the following:

1. The axons from the giant Betz cells send short collaterals back to the cortex itself. It is believed that these collaterals mainly inhibit adjacent regions of the cortex when the Betz cells discharge, thereby "sharpening" the boundaries of the excitatory signal.

2. A large body of fibers passes into the *caudate nucleus* and *putamen*. From here, additional pathways extend through several neurons into the brain stem, as discussed in Chapter 56.

3. A moderate number of fibers pass to the *red nuclei*. From these, additional fibers pass down the cord through the *rubrospinal tract*.

4. A moderate number of fibers deviate into the *reticular substance* and *vestibular nuclei* of the brain stem; from here signals go to the cord by way of the *reticulospinal* and *vestibulospinal tracts,* and others go to the cerebellum by way of the *reticulocerebellar* and *vestibulocerebellar tracts.*

5. A tremendous number of fibers synapse in the pontile nuclei, which give rise to the *pontocerebellar fibers,* carrying signals into the cerebellar hemispheres.

6. Collaterals also terminate in the *inferior olivary nuclei,* and from here, secondary *olivocerebellar fibers* transmit signals to many areas of the cerebellum.

Thus, the basal ganglia, brain stem, and cerebellum all receive strong signals from the corticospinal system every time a signal is transmitted down the spinal cord to cause a motor activity.

Incoming Fiber Pathways to the Motor Cortex

The functions of the motor cortex are controlled mainly by the somatic sensory system but also, to a lesser degree, by the other sensory systems, such as hearing and vision. Once the sensory information is derived from these sources, the motor cortex operates in association with the basal ganglia and cerebellum to excite the appropriate course of motor action. The more important incoming fiber pathways to the motor cortex are as follows:

1. Subcortical fibers from adjacent regions of the cortex, especially from (a) the somatic sensory areas of the parietal cortex, (b) the adjacent areas of the frontal cortex anterior to the motor cortex, and (c) subcortical fibers from the visual and auditory cortices

2. Subcortical fibers that pass through the corpus callosum from the opposite cerebral hemisphere. These fibers connect corresponding areas of the cortices in the two sides of the brain.

3. Somatic sensory fibers that arrive directly from the ventrobasal complex of the thalamus. These transmit mainly cutaneous tactile signals and joint and muscle signals.

4. Tracts from the ventrolateral and ventroanterior nuclei of the thalamus, which in turn receive tracts from the cerebellum and basal ganglia. These tracts provide signals that are necessary for coordination between the functions of the motor cortex, basal ganglia, and cerebellum.

5. Fibers from the intralaminar nuclei of the thalamus. These fibers control the general level of excitability of the motor cortex in the same manner that they also control the general level of excitability of most other regions of the cerebral cortex.

The Red Nucleus Serves as an Alternate Pathway for Transmitting Cortical Signals to the Spinal Cord

The *red nucleus,* located in the mesencephalon, functions in close association with the corticospinal tract. As shown in Figure 55–5, it receives a large

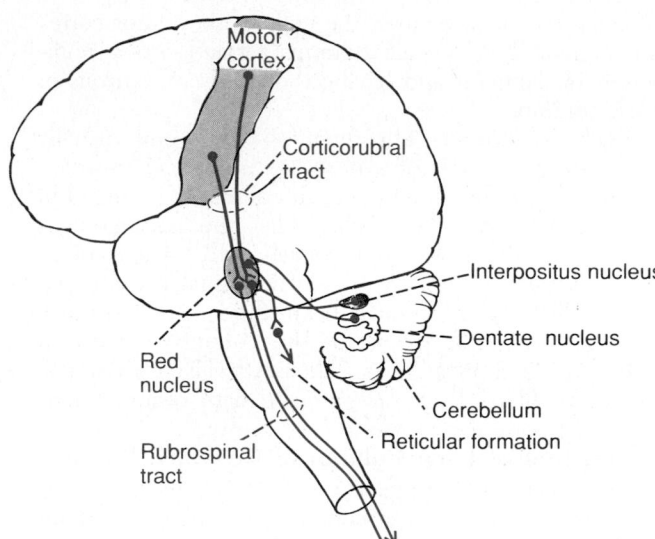

Figure 55–5. Corticorubrospinal pathway for motor control, showing also the relation of this pathway to the cerebellum.

number of direct fibers from the primary motor cortex through the *corticorubral tract* as well as branching fibers from the corticospinal tract as it passes through the mesencephalon. These fibers synapse in the lower portion of the red nucleus, the *magnocellular portion*, that contains large neurons similar to the Betz cells in the motor cortex. These large neurons give rise to the *rubrospinal tract*, which crosses to the opposite side in the lower brain stem and follows a course immediately adjacent and anterior to the corticospinal tract into the lateral columns of the spinal cord. The rubrospinal fibers terminate mostly on the interneurons of the intermediate areas of the cord gray matter along with the corticospinal fibers, but some of the rubrospinal fibers terminate directly on the anterior motor neurons, along with some corticospinal fibers.

The red nucleus also has close connections with the cerebellum, similar to the connections between the motor cortex and the cerebellum.

FUNCTION OF THE CORTICORUBROSPINAL SYSTEM. The magnocellular portion of the red nucleus has a somatographical representation of all the muscles of the body, as is true of the motor cortex. Therefore, stimulation of a single point in this portion of the red nucleus will cause contraction of either a single muscle or a small group of muscles. However, the fineness of representation of the different muscles is far less developed than in the motor cortex. This is especially true in human beings who have a relatively small red nucleus.

The corticorubrospinal pathway serves as an accessory route for transmission of relatively discrete signals from the motor cortex to the spinal cord. When the corticospinal fibers are destroyed without destroying this corticorubrospinal pathway, discrete movements can still occur, except that the movements of the fingers and hands are considerably impaired. Wrist movements are still well developed, which is not true when the corticorubrospinal pathway is also blocked.

Therefore, the pathway through the red nucleus to the spinal cord is associated far more with the corticospinal system than with the other major brain stem motor pathway, the vestibuloreticulospinal system that controls mainly the axial and girdle muscles of the body, as we discuss later in the chapter. Furthermore, the rubrospinal tract lies in the lateral columns of the spinal cord, along with the corticospinal tract, and terminates on the interneurons and motor neurons that control the more distal muscles of the limbs. Therefore, the corticospinal and rubrospinal tracts together are called the *lateral motor system of the cord*, in contradistinction to the vestibuloreticulospinal system that lies mainly medially in the cord and is called the *medial motor system of the cord.*

"Extrapyramidal" System

The term *extrapyramidal motor system* is widely used in clinical circles to denote all those portions of the brain and brain stem that contribute to motor control that are not part of the direct corticospinal-pyramidal system. This includes pathways through the basal ganglia, the reticular formation of the brain stem, the vestibular nuclei, and often the red nuclei. This is such an all-inclusive and diverse group of motor control areas that it is difficult to ascribe specific neurophysiological functions to the extrapyramidal system as a whole. For this reason, the term "extrapyramidal" is beginning to have less usage clinically as well as physiologically.

Excitation of the Spinal Cord by the Primary Motor Cortex and the Red Nucleus

VERTICAL COLUMNAR ARRANGEMENT OF THE NEURONS IN THE MOTOR CORTEX. In Chapters 47 and 51, it is pointed out that the cells in the somatic sensory cortex and visual cortex—and in virtually all other parts of the cortex—are organized in vertical columns of cells. In a like manner, the cells of the motor cortex are organized in vertical columns a fraction of a millimeter in diameter and having thousands of neurons in each column.

Each column of cells functions as a unit, usually stimulating a group of synergistic muscles but sometimes a single muscle. Also, each column is arranged in six distinct layers of cells, like the arrangement throughout almost all the cerebral cortex. The pyramidal cells that give rise to the corticospinal fibers all lie in the fifth layer of cells from the cortical surface, whereas the input signals to the column of cells all enter layers 2 through 4. The sixth layer gives rise mainly to fibers that communicate with other regions of the cerebral cortex itself.

Function of Each Column of Neurons. The neurons of each column operate as an integrative processing system, using information from multiple input sources to determine the output response from the column. In addition, each column can function as an amplifying system to stimulate large numbers of pyramidal fibers to the same muscle or to synergistic muscles simultaneously. This is important because stimulation of a single pyramidal cell can seldom excite a muscle. Instead, 50 to 100 pyramidal cells usually need to be excited simultaneously or in rapid succession to achieve muscle contraction.

Dynamic and Static Signals Transmitted by the Pyramidal Neurons. If a strong signal is sent at first to a muscle to cause initial rapid contraction, then a much weaker signal can maintain the contraction for long periods thereafter. This is the manner in which excitation for causing muscle contractions is usually provided. To do this, each column of cells excites two populations of pyramidal cell neurons, one called *dynamic neurons* and the other *static neurons*. The dynamic neurons are excessively excited for a short period at the beginning of a contraction, causing the initial *development of force*. Then the static neurons fire at a much slower rate, but they continue at this slow rate indefinitely to *maintain the force* of contraction as long as the contraction is required.

The neurons of the red nucleus have similar dynamic and static characteristics, except that more dynamic neurons are in the red nucleus and more static neurons are in the primary motor cortex. This perhaps relates to the fact that the red nucleus is closely allied with the cerebellum, and the cerebellum plays an important role in the rapid initiation of muscle contraction, as explained in Chapter 56.

Somatic Sensory Feedback to the Motor Cortex Helps to Control the Precision of Muscle Contraction

When nerve signals from the motor cortex cause a muscle to contract, somatic sensory signals return from the activated region of the body to the neurons in the motor cortex that are causing the action. Most of these somatic sensory signals arise in the muscle spindles, the tendon organs of the muscle tendons, or in the tactile receptors of the skin overlying the muscles. The somatic signals often cause a positive feedback enhancement of the muscle contraction in the following ways: In the case of the muscle spindles, if the fusimotor muscle fibers in the spindles contract more than the large skeletal muscle itself does, then the central portions of the spindles become stretched and, therefore, are excited; the signals from these spindles stimulate the pyramidal cells in the motor cortex, which further excites the muscle, helping its contraction catch up with the contraction of the spindles. In the case of the tactile receptors, if the muscle contraction causes compression of the skin against an object, such as compression of the fingers around an object that is being grasped, the signals from these receptors cause further excitement of the muscles and, therefore, increase the muscle contraction, such as increasing the tightness of the grasp of the hand.

Stimulation of the Spinal Motor Neurons

Figure 55–6 shows a cross section of a spinal cord segment, demonstrating multiple motor tracts entering the cord from the brain and a representative anterior motor neuron. The corticospinal tract and the rubrospinal tract lie in the dorsal portions of the lateral columns. At most levels of the cord, their fibers terminate mainly on interneurons in the intermediate area of the cord gray matter. However, in the cervical enlargement of the cord where the hands and fingers are represented, large numbers of both corticospinal and rubrospinal fibers terminate directly on the anterior motor neurons, thus allowing a direct route from the brain for activating muscle contraction. This is in keeping with the fact that the primary motor cortex

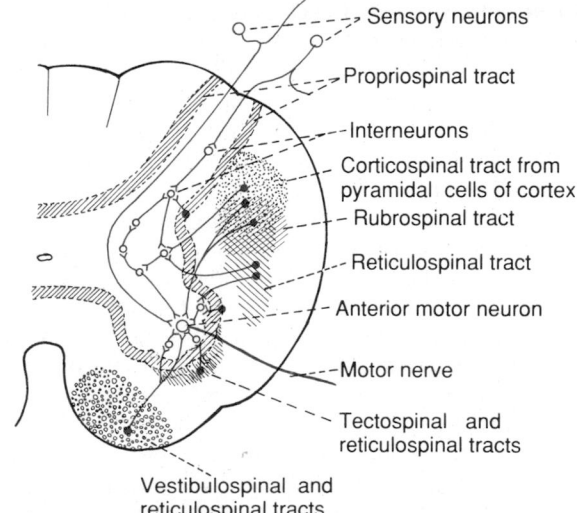

Sensory neurons
Propriospinal tract
Interneurons
Corticospinal tract from pyramidal cells of cortex
Rubrospinal tract
Reticulospinal tract
Anterior motor neuron
Motor nerve
Tectospinal and reticulospinal tracts
Vestibulospinal and reticulospinal tracts

Figure 55–6. Convergence of all the different motor pathways on the anterior motor neurons.

has an extremely high degree of representation for fine control of hand, finger, and thumb actions.

PATTERNS OF MOVEMENT ELICITED BY SPINAL CORD CENTERS. From Chapter 54, recall that the spinal cord can provide specific reflex patterns of movement in response to sensory nerve stimulation. Many of these patterns are also important when the anterior motor neurons are excited by signals from the brain. For instance, the stretch reflex is functional at all times, helping to damp the motor movements initiated from the brain and probably providing at least part of the motive power required to cause the muscle contractions when the intrafusal fibers of the spindles contract more than the large skeletal muscle fibers, thus eliciting reflex "servo-assist" stimulation of the muscle in addition to the direct stimulation by the corticospinal fibers.

Also, when a brain signal excites an agonist muscle, it is not necessary to transmit an inverse signal to relax the antagonist at the same time; this transmission will be achieved by the reciprocal innervation circuit that is always present in the cord for coordinating the functions of antagonistic pairs of muscles.

Finally, other reflex mechanisms, such as withdrawal, stepping and walking, scratching, and postural mechanisms, can be activated by "command" signals from the brain. Thus, simple signals from the brain can initiate many of our normal motor activities, particularly for such functions as walking and the attainment of different postural attitudes of the body.

Effect of Lesions in the Motor Cortex or Corticospinal Pathway—The "Stroke"

The motor cortex or corticospinal pathway can be damaged, especially by the common abnormality called a "stroke." This is caused either by a ruptured blood vessel that hemorrhages into the brain or by thrombosis of one of the major arteries supplying the brain, in either case causing loss of blood supply to the cortex or to the corticospinal tract where it passes through the internal capsule between the caudate nucleus and the putamen. Also, experiments have been performed in animals to remove selectively different parts of the motor cortex.

REMOVAL OF THE PRIMARY MOTOR CORTEX (THE AREA PYRAMIDALIS). Removal of a portion of the primary motor cortex—the area that contains the giant Betz pyramidal cells—in a monkey causes varying degrees of paralysis of the represented muscles. If the sublying caudate nucleus and the adjacent premotor and supplementary motor areas are not damaged, gross postural and limb "fixation" movements can still be performed, but the animal *loses voluntary control of discrete movements of the distal segments of the limbs, especially the hands and fingers.* This does not mean that the hand and finger muscles themselves cannot contract, but that the animal's ability to control the fine movements is gone.

From these results, one can conclude that the area pyramidalis is essential for voluntary initiation of finely controlled movements, especially of the hands and fingers.

MUSCLE SPASTICITY CAUSED BY LESIONS THAT DAMAGE LARGE AREAS ADJACENT TO THE MOTOR CORTEX. Ablation of the primary motor cortex alone causes *hypotonia*, not spasticity, because the primary motor cortex normally exerts a continual tonic stimulatory effect on the motor neurons of the spinal cord; when this is removed, hypotonia results.

On the other hand, most lesions of the motor cortex, especially those caused by a stroke, involve not only the primary motor cortex but also adjacent cortical areas and deeper structures of the cerebrum, especially the basal ganglia. In these instances, muscle spasm almost invariably occurs in the afflicted muscle areas on the opposite side of the body (because the motor pathways cross to the opposite side). This spasm is not caused by loss of the primary motor cortex or blockage of the corticospinal fibers to the cord.

Instead, it is believed that the spasm results mainly from damage to accessory pathways from the nonpyramidal portions of the cortex that normally inhibit the vestibular and reticular brain stem nuclei. When these nuclei lose this inhibition (that is, they are said to be "disinhibited"), they become spontaneously active and cause excessive spastic tone in the involved areas of the body, as we discuss more fully later in the chapter. This is the spasticity that normally accompanies a "stroke" in the human being.

ROLE OF THE BRAIN STEM IN CONTROLLING MOTOR FUNCTION

The brain stem consists of the *medulla, pons,* and *mesencephalon.* In one sense, it is an extension of the spinal cord upward into the cranial cavity because it contains motor and sensory nuclei that perform motor and sensory functions for the face and head regions in the same way that the anterior and posterior gray horns of the spinal cord perform these same functions from the neck down. But in another sense, the brain stem is its own master because it provides many special control functions, such as the following:

1. Control of respiration
2. Control of the cardiovascular system
3. Control of gastrointestinal function
4. Control of many stereotyped movements of the body
5. Control of equilibrium
6. Control of eye movement

Finally, the brain stem serves as a way station for "command signals" from still higher neural centers that "command" the brain stem to initiate or modify specific control functions throughout the body.

In the following sections, we discuss the role of the brain stem in controlling whole-body movement and equilibrium. Especially important are the brain stem's *reticular nuclei* and *vestibular nuclei,* plus the *vestibular apparatus* that sends most of the equilibrium control signals to the vestibular nuclei and, to a lesser extent, the reticular nuclei.

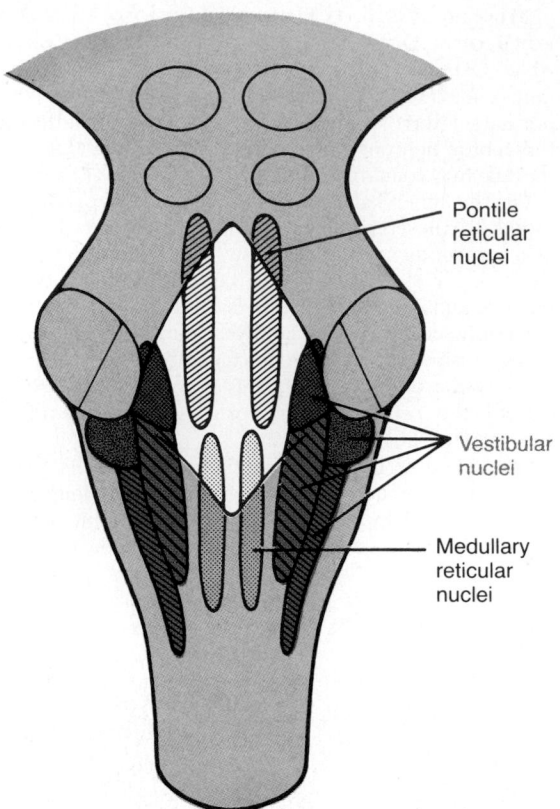

Figure 55–7. Locations of the reticular and vestibular nuclei in the brain stem.

Support of the Body Against Gravity—Roles of the Reticular and Vestibular Nuclei

Excitatory-Inhibitory Antagonism Between Pontine and Medullary Reticular Nuclei

Figure 55–7 shows the locations of the reticular and vestibular nuclei. The recticular nuclei are divided into two major groups: (1) the *pontine reticular nuclei*, located slightly posteriorly and laterally in the pons and extending into the mesencephalon, and (2) the *medullary reticular nuclei*, which extend the entire extent of the medulla, lying ventrally and medially near the midline. These two sets of nuclei function mainly antagonistically to each other, the pontine exciting the antigravity muscles and the medullary inhibiting them. The pontine reticular nuclei transmit excitatory signals downward into the cord through the *pontine* (or *medial*) *reticulospinal tract*, shown in Figure 55–8. The fibers of this pathway terminate on the medial anterior motor neurons that excite the axial muscles of the body that support the body against gravity, that is, the muscles of the vertebral column and the extensor muscles of the limbs.

The pontine reticular nuclei have a high degree of natural excitability. In addition, they receive especially strong excitatory signals from the vestibular nuclei as well as from the deep nuclei of the cerebellum.

Therefore, when the pontine reticular excitatory system is unopposed by the medullary reticular system, it causes powerful excitation of the antigravity muscles throughout the body, so much so that four-legged animals can then be placed in a standing position, supporting the body against gravity without any signals from the higher levels of the brain.

MEDULLARY RETICULAR SYSTEM. The medullary reticular nuclei, on the other hand, transmit *inhibitory* signals to the same antigravity anterior motor neurons by way of a different tract, the *medullary* (or *lateral*) *reticulospinal tract*, also shown in Figure 55–8. The medullary reticular nuclei receive strong input collaterals from (1) the corticospinal tract, (2) the rubrospinal tract, and (3) other motor pathways. These normally activate the medullary reticular inhibitory system to counterbalance the excitatory signals from the pontine reticular system, so that under normal conditions, the body muscles are relaxed.

Yet, under other conditions, signals from higher areas of the brain can "disinhibit" the medullary system when the brain wishes for excitation by the pontine system to cause standing. Or at other times, excitation of the medullary reticular system can inhibit antigravity muscles in certain portions of the body to allow those portions to perform other motor activities, which would be impossible if the antigravity muscles opposed the necessary movements.

Therefore, the excitatory and inhibitory reticular nuclei constitute a controllable system that is manipulated by motor signals from the cortex and elsewhere to provide the necessary muscle contractions for standing against gravity and yet to inhibit appropriate groups of muscles as needed so that other functions can be performed as required.

Role of the Vestibular Nuclei in Exciting the Antigravity Muscles

The vestibular nuclei, shown in Figure 55–7, function in association with the pontine reticular nuclei to excite the antigravity muscles. The *lateral vestibular*

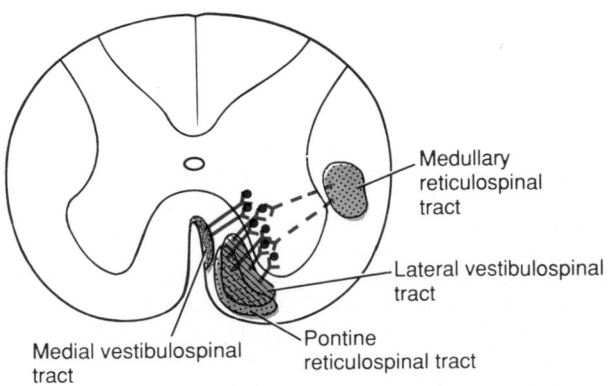

Figure 55–8. Vestibulospinal and reticulospinal tracts descending in the spinal cord to excite *(solid lines)* or inhibit *(dashed lines)* the anterior motor neurons that control the body's axial musculature.

nuclei (the most lateral vestibular area in the figure), especially, transmit strong excitatory signals by way of both the *lateral* and the *medial vestibulospinal tracts* in the anterior column of the spinal cord, as shown in Figure 55–8. In fact, without the support of the vestibular nuclei, the pontine reticular system loses much of its excitation of the axial antigravity muscles. The specific role of the vestibular nuclei, however, is to control selectively the excitatory signals to the different antigravity muscles to maintain equilibrium in response to signals from the vestibular apparatus. We discuss this more fully later in the chapter.

The Decerebrate Animal Develops Spastic Rigidity

When the brain stem is sectioned below the midlevel of the mesencephalon, leaving both the pontine and the medullary reticular systems as well as the vestibular system intact, the animal develops a condition called *decerebrate rigidity*. This rigidity does not occur in all muscles of the body but does occur in the antigravity muscles—the muscles of the neck and trunk and extensors of the legs.

The cause of decerebrate *rigidity* is blockage of the normally strong excitatory input to the medullary reticular nuclei from the cerebral cortex, red nuclei, and basal ganglia. As a result, the medullary reticular inhibitor system becomes nonfunctional, thus allowing full overactivity of the pontine excitatory system and development of rigidity.

A specific characteristic of decerebrate rigidity is that the antigravity muscles exhibit the phenomenon called *spasticity* as well as rigidity. This means that any attempt to change the position of a limb or other part of the body, especially attempts to stretch the muscles suddenly, is resisted by powerful stretch reflexes described in Chapter 54. This occurs because the pontine and vestibular antigravity signals to the cord selectively excite the gamma motor neurons in the spinal cord much more than they excite the alpha motor neurons. This tightens the intrafusal muscle fibers of the muscle spindles, which in turn strongly sensitizes the stretch reflex feedback loop.

We see later that other types of rigidity occur in other neuromotor diseases, especially in lesions of the basal ganglia. In many of these, the rigidity involves all muscles equally without the excessive spastic stretch reflex component.

VESTIBULAR SENSATIONS AND THE MAINTENANCE OF EQUILIBRIUM

Vestibular Apparatus

The vestibular apparatus is the organ that detects sensations of equilibrium. It is composed of a system of bony tubes and chambers in the petrous portion of the temporal bone called the *bony labyrinth* and within this a system of membranous tubes and chambers called the *membranous labyrinth.* The membranous labyrinth is the functional part of the apparatus.

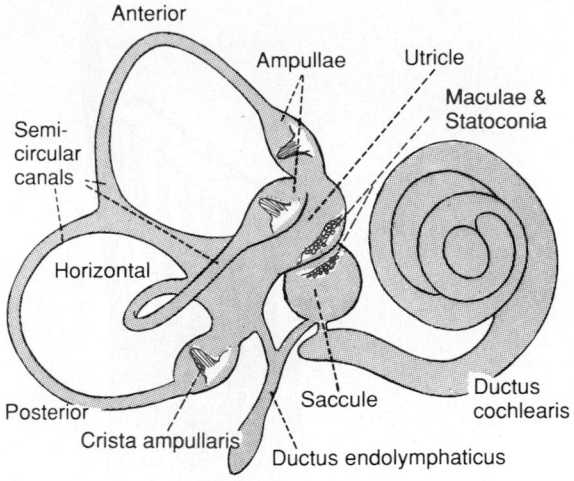

Membranous Labyrinth

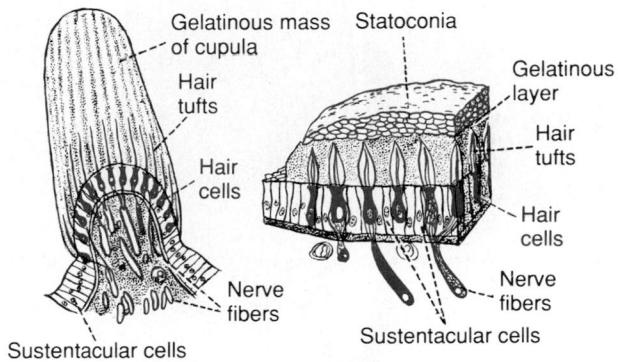

Crista Ampullaris and Macula

Figure 55–9. Membranous labyrinth and organization of the crista ampullaris and the macula. (Modified from Goss: Gray's Anatomy of the Human Body. Philadelphia, Lea & Febiger; modified from Kolmer by Buchanan: Functional Neuroanatomy. Philadelphia, Lea & Febiger.)

The top of Figure 55–9 shows the membranous labyrinth; it is composed mainly of the *cochlea,* three *semicircular ducts,* and two large chambers known as the *utricle* and the *saccule.* The cochlea is the major sensory area for hearing that is discussed in Chapter 52 and has nothing to do with equilibrium. However, the *semicircular ducts,* the *utricle,* and the *saccule* are all integral parts of the equilibrium mechanism.

THE MACULAE—THE SENSORY ORGANS OF THE UTRICLE AND THE SACCULE FOR DETECTING ORIENTATION OF THE HEAD WITH RESPECT TO GRAVITY. Located on the inside surface of each utricle and saccule, as shown in the top diagram of Figure 55–9, is a small sensory area slightly over 2 millimeters in diameter called a *macula.* The macula of the utricle lies mainly in the horizontal plane on the inferior surface of the utricle and plays an important role in determining the orientation of the head with respect to the direction of gravitational force when a person is upright. On the other hand, the macula of the saccule is located mainly

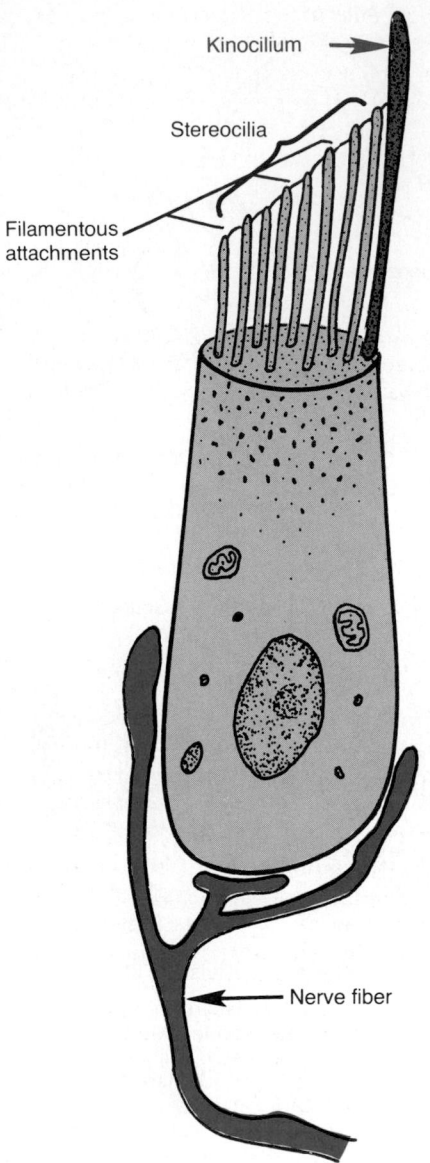

Figure 55–10. Hair cell of the membranous labyrinth of the equilibrium apparatus.

in a vertical plane and, therefore, is important in equilibrium when a person is lying down.

Each macula is covered by a gelatinous layer in which many small calcium carbonate crystals called *statoconia* are imbedded. Also in the macula are thousands of *hair cells*, one of which is shown in Figure 55–10; these project *cilia* up into the gelatinous layer. The bases and sides of the hair cells synapse with sensory endings of the vestibular nerve.

The statoconia have a specific gravity two to three times as great as the specific gravity of the surrounding fluid and tissues. Therefore, the weight of the statoconia will bend the cilia in the direction of gravitational pull.

DIRECTIONAL SENSITIVITY OF THE HAIR CELLS— THE KINOCILIUM. Each hair cell has 50 to 70 small

cilia called *stereocilia*, plus one large cilium, the *kinocilium*, as shown in Figure 55–10. The kinocilium is located always to one side, and the stereocilia become progressively shorter toward the other side of the cell. Minute filamentous attachments, almost invisible even to the electron microscope, connect the tip of each stereocilium to the next longer stereocilium and, finally, to the kinocilium. Because of these attachments, when the stereocilia and kinocilium bend in the direction of the kinocilium, the filamentous attachments tug one after the other on the stereocilia, pulling them in the outward direction from the cell body. This opens several hundred channels in each cilium membrane for conducting positive ions, and large quantities of these positive ions pour into the cell from the surrounding endolymphatic fluid, causing *depolarization*. Conversely, bending the pile of cilia in the opposite direction (away from the kinocilium) reduces the tension on the attachments, and this closes the ion channels, thus causing *hyperpolarization*.

Under normal resting conditions, the nerve fibers leading from the hair cells transmit continuous nerve impulses at rates of about 100 per second. When the cilia are bent toward the kinocilium, the impulse traffic can increase to several hundred per second; conversely, bending the cilia in the opposite direction decreases the impulse traffic, often turning it off completely. Therefore, as the orientation of the head in space changes and the weight of the statoconia (whose specific gravity is two to three times that of the surrounding tissues) bends the cilia, appropriate signals are transmitted to the brain to control equilibrium.

In each macula, the different hair cells are oriented in different directions so that some of them are stimulated when the head bends forward, some when it bends backward, others when it bends to one side, and so forth. Therefore, a different pattern of excitation occurs in the nerve fibers from the macula for each position of the head; it is this "pattern" that apprises the brain of the head's orientation.

SEMICIRCULAR DUCTS. The three semicircular ducts in each vestibular apparatus, known as the *anterior, posterior,* and *lateral (horizontal) semicircular ducts,* are arranged at right angles to one another so that they represent all three planes in space. When the head is bent forward about 30 degrees, the lateral semicircular ducts are located approximately horizontal with respect to the surface of the earth. The anterior ducts are then located in vertical planes that project *forward and 45 degrees outward,* and the posterior ducts are also in vertical planes but project *backward and 45 degrees outward.*

Each semicircular duct has an enlargement at one of its ends called the *ampulla,* and the ducts are filled with a viscous fluid called *endolymph.* Flow of this fluid from one of the ducts into the ampulla excites the sensory organ of the ampulla in the following manner: Figure 55–11 shows in each ampulla a small crest called a *crista ampullaris.* On top of this crista is

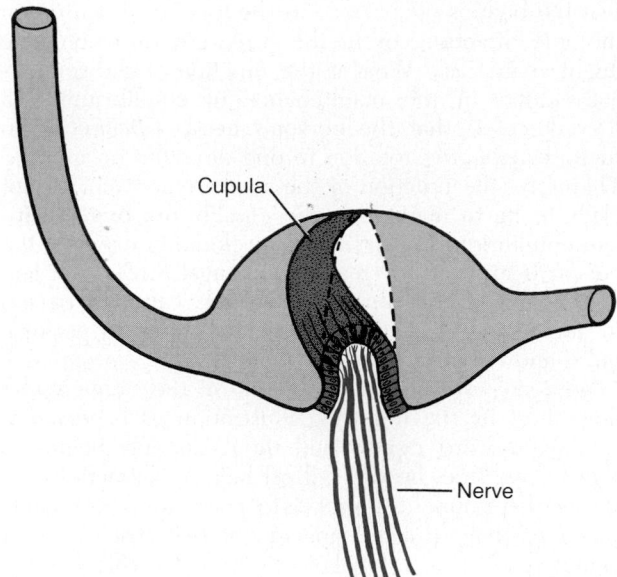

Figure 55–11. Movement of the cupula and its embedded hairs at the onset of rotation.

a gelatinous mass, the *cupula*. When the head begins to rotate in any direction, the inertia of the fluid in one or more of the semicircular ducts will cause the fluid to remain stationary while the semicircular duct rotates with the head. This causes fluid flow from the duct into the ampulla, bending the cupula to one side, as demonstrated by the position of the shaded cupula in Figure 55–11. Rotation of the head in the opposite direction causes the cupula to bend to the opposite side.

Into the cupula are projected hundreds of cilia from hair cells located along the ampullary crest. The kinocilia of all these hair cells is directed toward the same side of the cupula, and bending the cupula in that direction causes depolarization of the hair cells, whereas bending it in the opposite direction hyperpolarizes the cells. From the hair cells, appropriate signals are sent by way of the *vestibular nerve* to apprise the central nervous system of changes in the rate and direction of rotation of the head in the three planes of space.

Function of the Utricle and Saccule in the Maintenance of Static Equilibrium

It is especially important that the different hair cells are oriented in all different directions in the maculae of the utricles and saccules, so that at different positions of the head, different hair cells become stimulated. The "patterns" of stimulation of the different hair cells apprise the nervous system of the position of the head with respect to the pull of gravity. In turn, the vestibular, cerebellar, and reticular motor systems excite the appropriate postural muscles to maintain proper equilibrium.

The vestibular system functions extremely effectively for maintaining equilibrium when the head is in the near-vertical position. Indeed, a person can determine as little as a half-degree of malequilibrium when the body leans from the precise upright position. On the other hand, as the body is leaned further and further from upright, determination of head orientation by the vestibular sense becomes poorer and poorer. In the human being, extreme vestibular sensitivity in the upright position is of major importance for maintenance of precise vertical equilibrium, which is the most essential function of the vestibular apparatus.

DETECTION OF LINEAR ACCELERATION BY THE UTRICLE AND SACCULE MACULAE. When the body is suddenly thrust forward—that is, when the body accelerates—the statoconia, which have greater inertia than the surrounding fluid, fall backward on the hair cell cilia and information of malequilibrium is sent into the nervous centers, causing the person to feel as though he or she were falling backward. This automatically causes the person to lean forward until the anterior shift of the statoconia caused by leaning exactly equals the tendency for the statoconia to fall backward. At this point, the nervous system detects a state of proper equilibrium and therefore leans the body no further forward. Thus, the maculae operate to maintain equilibrium during linear acceleration in exactly the same manner as they operate in static equilibrium.

The maculae *do not* operate for the detection of linear *velocity*. When runners first begin to run, they must lean far forward to keep from falling over backward because of *acceleration*, but once they have achieved running speed, they would not have to lean forward if they were running in a vacuum. When running in air, they lean forward to maintain equilibrium only because of the air resistance against their bodies; in this instance, it is not the maculae that make them lean but pressure of the air acting on pressure end organs in the skin, which initiate the appropriate equilibrium adjustments to prevent falling.

Detection of Head Rotation by the Semicircular Ducts

When the head suddenly *begins* to rotate in any direction (this is called *angular acceleration*), the endolymph in the semicircular ducts, because of its inertia, tends to remain stationary while the semicircular ducts themselves turn. This causes relative fluid flow in the ducts in the direction opposite to the rotation of the head.

Figure 55–12 shows a typical discharge signal from a single hair cell in the crista ampullaris when an animal is rotated for 40 seconds, demonstrating that (1) even when the cupula is in its resting position, the hair cell emits a tonic discharge of about 100 impulses per second; (2) when the animal begins to rotate, the hairs bend to one side and the rate of discharge increases greatly; and (3) with continued rotation, the

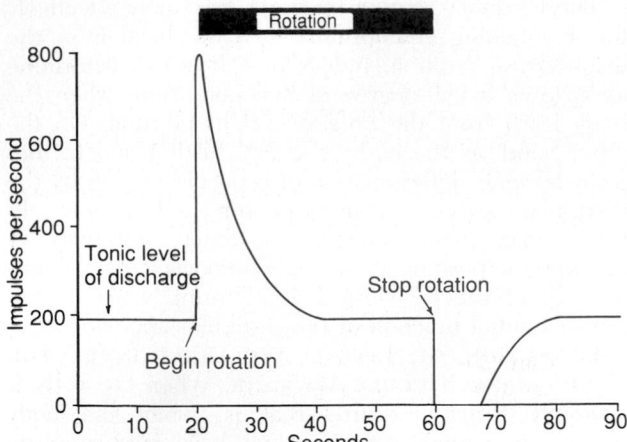

Figure 55–12. Response of a hair cell when a semicircular canal is stimulated first by rotation and then by stopping rotation.

excess discharge of the hair cell gradually subsides back to the resting level in an additional few seconds.

The reason for this adaptation of the receptor is that within a second or more of rotation, back pressure from the bent cupula causes the endolymph to rotate as rapidly as the semicircular canal itself; then, in an additional 5 to 20 seconds, the cupula slowly returns to its resting position in the middle of the ampulla because of its own elastic recoil.

When the rotation suddenly stops, exactly opposite effects take place: the endolymph continues to rotate while the semicircular duct stops. This time the cupula is bent in the opposite direction, causing the hair cell to stop discharging entirely. After another few seconds, the endolymph stops moving and the cupula gradually returns to its resting position, thus allowing the discharge of the hair cell to return to its normal tonic level, as shown to the right in Figure 55–12.

Thus, the semicircular duct transmits a signal of one polarity when the head *begins* to rotate and of opposite polarity when it *stops* rotating. Furthermore, at least some hair cells will always respond to rotation in any plane—horizontal, sagittal, or coronal—because fluid movement always occurs in at least one semicircular duct.

RATE OF ANGULAR ACCELERATION REQUIRED TO STIMULATE THE SEMICIRCULAR DUCTS. The angular acceleration required to stimulate the semicircular ducts in the human being averages about 1 degree per second per second. In other words, when one begins to rotate, the velocity of rotation must be as much as 1 degree per second by the end of the first second, 2 degrees per second by the end of the second second, 3 degrees per second by the end of the third second, and so forth for the person barely to detect that the rate of rotation is increasing.

"PREDICTIVE" FUNCTION OF THE SEMICIRCULAR DUCT SYSTEM IN THE MAINTENANCE OF EQUILIBRIUM. Because the semicircular ducts do not detect

that the body is off balance in the forward direction, in the side direction, or in the backward direction, one might at first ask: What is the function of the semicircular ducts in the maintenance of equilibrium? All they detect is that the person's head is *beginning* to rotate or *stopping* rotation in one direction or another. Therefore, the function of the semicircular ducts is not likely to be to maintain static equilibrium or to maintain equilibrium during linear acceleration or when the person is exposed to steady centrifugal forces. Yet loss of function of the semicircular ducts causes a person to have poor equilibrium when attempting to perform *rapid* and *intricate* body movements.

We can explain the function of the semicircular ducts best by the following illustration. If a person is running forward rapidly and then suddenly begins to turn to one side, he will fall off balance a fraction of a second later unless appropriate corrections are made *ahead of time.* But the macula of the utricle cannot detect that he is off balance until *after* this has occurred. On the other hand, the semicircular ducts will have already detected that the person is turning, and this information can easily apprise the central nervous system of the fact that the person *will* fall off balance within the next fraction of a second or so unless some *anticipatory correction* is made. In other words, the semicircular duct mechanism *predicts ahead of time* that malequilibrium is going to occur even before it does occur and thereby causes the equilibrium centers to make appropriate preventive adjustments. In this way, the person need not fall off balance before he begins to correct the situation.

Removal of the flocculonodular lobes of the cerebellum prevents normal function of the semicircular ducts but has less effect on the function of the macular receptors. It is especially interesting in this connection that the cerebellum serves as a "predictive" organ for most other rapid movements of the body as well as those having to do with equilibrium. These other functions of the cerebellum are discussed in Chapter 56.

Vestibular Postural Reflexes

Sudden changes in the orientation of an animal in space elicit vestibular reflexes that help to maintain equilibrium and posture. For instance, if an animal is suddenly pushed to the right, even before it can fall more than a few degrees, its right legs extend instantaneously. In other words, this mechanism *anticipates* that the animal will be off balance in a few seconds and makes appropriate adjustments to prevent this.

Another type of vestibular postural reflex occurs when the animal suddenly falls forward. When this occurs, the forelegs extend forward, the extensor muscles tighten, and the muscles in the back of the neck stiffen to prevent the animal's head from striking the ground. This reflex is probably also of importance in locomotion because in the case of the galloping horse, the downward thrust of the head automatically provides reflex thrust of the forelimbs to the forward position, preparing for the next gallop.

Vestibular Mechanism for Stabilizing the Eyes

When a person changes direction of movement rapidly or even leans the head sideways, forward, or backward, it would be impossible to maintain a stable image on the retinae of the eyes unless the person had some automatic control mechanism to stabilize the direction of gaze of the eyes. In addition, the eyes would be of little use in detecting an image unless they remained "fixed" on each object long enough to gain a clear image. Each time the head is suddenly rotated, signals from the semicircular ducts cause the eyes to rotate in a direction equal and opposite to the rotation of the head. This results from reflexes transmitted through the *vestibular nuclei* and the *medial longitudinal fasciculus* to the *ocular nuclei* that were described in Chapter 51.

Other Factors Concerned with Equilibrium

NECK PROPRIOCEPTORS. The vestibular apparatus detects the orientation and movements *only of the head.* Therefore, it is essential that the nervous centers also receive appropriate information depicting the orientation of the head with respect to the body. This information is transmitted from the proprioceptors of the neck and body directly to the vestibular and reticular nuclei of the brain stem and indirectly by way of the cerebellum.

Among the most important proprioceptive information needed for maintenance of equilibrium is that transmitted by the *joint receptors of the neck.* When the head is leaned in one direction by bending the neck, impulses from the neck proprioceptors keep the vestibular apparatus from giving the person a sense of malequilibrium. They do this by transmitting signals that exactly oppose the signals transmitted from the vestibular apparatuses. However, *when the entire body* leans in one direction, the impulses from the vestibular apparatuses *are not opposed* by signals from the neck proprioceptors; therefore, the person in this instance does perceive a change in equilibrium status of the entire body.

PROPRIOCEPTIVE AND EXTEROCEPTIVE INFORMATION FROM OTHER PARTS OF THE BODY. Proprioceptive information from other parts of the body besides the neck is also important in the maintenance of equilibrium. For instance, pressure sensations from the footpads can tell one (1) whether weight is distributed equally between the two feet and (2) whether weight is more forward or backward on the feet.

An instance in which exteroceptive information is especially necessary for maintenance of equilibrium occurs when a person is running. The air pressure against the front of the body signals that a force is opposing the body in a direction different from that caused by gravitational pull; as a result, the person leans forward to oppose this.

IMPORTANCE OF VISUAL INFORMATION IN THE MAINTENANCE OF EQUILIBRIUM. After destruction of the vestibular apparatus and even after loss of most proprioceptive information from the body, a person can still use the visual mechanisms reasonably effectively for maintaining equilibrium. Even slight linear or rotational movement of the body instantaneously shifts the visual images on the retina, and this information is relayed to the equilibrium centers. Some people with destruction of the vestibular apparatus have almost normal equilibrium as long as their eyes are open and as long as they perform all motions slowly. But when moving rapidly or when the eyes are closed, equilibrium is immediately lost.

Neuronal Connections of the Vestibular Apparatus with the Central Nervous System

Figure 55–13 shows the central connections of the vestibular nerve. Most of the vestibular nerve fibers end in the vestibular nuclei, which are located approximately at the junction of the medulla and the pons, but some fibers pass without synapsing to the brain stem reticular nuclei and the cerebellar fastigial nuclei, uvula, and flocculonodular lobes. The fibers that end in the vestibular nuclei synapse with second-order neurons that also send fibers into the cerebellum as well as into the vestibulospinal tracts and the medial longitudinal fasciculus and to other areas of the brain stem, particularly the reticular nuclei.

The primary pathway for the equilibrium reflexes begins in the vestibular nerves and passes next to both the vestibular nuclei and the cerebellum. Next, signals are sent into the reticular nuclei of the brain stem as well as down the spinal cord by way of the vestibulospinal and reticulospinal tracts. In turn, the signals to the cord control the interplay between facilitation and inhibition of the antigravity muscles, thus automatically controlling equilibrium.

The *flocculonodular lobes* of the cerebellum seem to be especially concerned with equilibrium signals from the semicircular ducts because destruction of these lobes gives almost exactly the same clinical symptoms as destruction of the semicircular ducts themselves. That is, severe injury to either causes loss of equilibrium during *rapid changes in direction of motion* but does not seriously disturb equilibrium under static conditions.

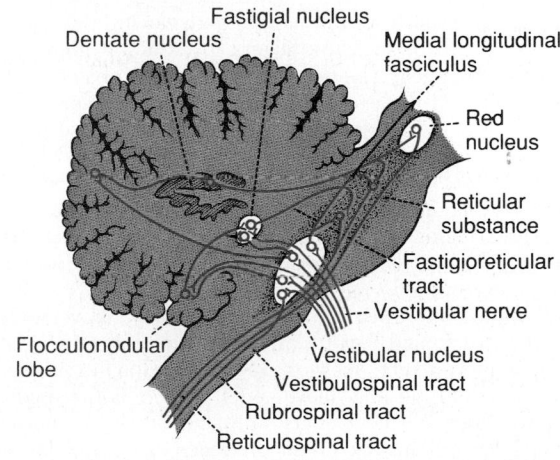

Figure 55–13. Connections of vestibular nerves through the vestibular nuclei (the oval white area) with other areas of the central nervous system.

It is also believed that the *uvula* of the cerebellum plays a similar important role in static equilibrium.

Signals transmitted upward in the brain stem from both the vestibular nuclei and the cerebellum by way of the *medial longitudinal fasciculus* cause corrective movements of the eyes every time the head rotates, so that the eyes can remain fixed on a specific visual object. Signals also pass upward (either through this same tract or through reticular tracts) to the cerebral cortex, probably terminating in a primary cortical center for equilibrium located in the parietal lobe deep in the sylvian fissure, on the opposite side of the fissure from the auditory area of the superior temporal gyrus. These signals apprise the psyche of the equilibrium status of the body.

The *vestibular nuclei* on either side of the brain stem are divided into four subdivisions: (1 and 2) the *superior* and *medial vestibular nuclei,* which receive signals mainly from the semicircular ducts and in turn send large numbers of nerve signals into the *medial longitudinal fasciculus* to cause corrective movements of the eyes as well as signals through the *medial vestibulospinal tract* to cause appropriate movements of the neck and head; (3) the *lateral vestibular nucleus,* which receives its innervation primarily from the utricle and saccule and in turn transmits outflow signals to the spinal cord through the *lateral vestibulospinal tract* to control body movement; and (4) the *inferior vestibular nucleus,* which receives signals from both the semicircular ducts and the utricle and in turn sends signals into both the cerebellum and the reticular formation of the brain stem.

FUNCTIONS OF SPECIFIC BRAIN STEM NUCLEI IN CONTROLLING SUBCONSCIOUS, STEREOTYPED MOVEMENTS

Rarely, a child is born without brain structures above the mesencephalic region, a condition called *anencephaly.* Some of these children have been kept alive for many months. They are able to perform essentially all the functions of feeding, such as suckling, extrusion of unpleasant food from the mouth, and moving the hands to the mouth to suck the fingers. In addition, they can yawn and stretch. They can cry and follow objects with movements of the eyes and head. Also, placing pressure on the upper anterior parts of their legs will cause them to pull to the sitting position.

Therefore, it is clear that many of the stereotyped motor functions of the human being are integrated in the brain stem. Unfortunately, the loci of most of the different motor control systems have not been found, except for the following.

STEREOTYPED BODY MOVEMENTS. Most movements of the trunk and head can be classified into several simple movements, such as forward flexion, extension, rotation, and turning movements of the entire body. These types of movements are controlled by special nuclei located mainly in the mesencephalic and lower diencephalic regions. For instance, *rotational movements* of the head and eyes are controlled by the *interstitial nucleus.* This nucleus lies in the mesencephalon in close

approximation to the *medial longitudinal fasciculus,* through which it transmits a major portion of its control impulses. The *raising movements* of the head and body are controlled by the *prestitial nucleus,* which is located approximately at the juncture of the diencephalon and mesencephalon. On the other hand, the *flexing movements* of the head and body are controlled by the *nucleus precommissuralis* located at the level of the posterior commissure. Finally, *turning movements* of the entire body, which are much more complicated, involve both the pontile and the mesencephalic reticular nuclei.

REFERENCES

Asanuma, H.: The pyramidal tract. In Brooks, V. B. (ed.): Handbook of Physiology. Sec. 1. Vol. II. Bethesda, Md., American Physiological Society, 1981, p. 703.

Berthoz, A., Multisensory Control of Movement. New York, Oxford University Press, 1993.

Brodal, P.: The Central Nervous System: Structure and Function. New York, Oxford University Press, 1992.

Cordo, P., and Harnad, S.: Movement Control. New York, Cambridge University Press, 1994.

Brooks, V. B.: The Neural Basis of Motor Control. New York, Oxford University Press, 1986.

Dutia, M. B.: Mechanisms of head stabilization. News Physiol. Sci., 4:101, 1989.

Evarts, E. V., et al. (eds.): Motor System in Neurobiology. New York, Elsevier Science Publishing Co., 1986.

Evarts, E. V.: Role of motor cortex in voluntary movements in primates. In Brooks, V. B. (ed.): Handbook of Physiology. Sec. 1, Vol. II. Bethesda, Md., American Physiological Society, 1981, p. 1083.

Fournier, E., and Pierrot-Deseilligny, E.: Changes in transmission in some reflex pathways during movement in humans. News Physiol. Sci., 4:29, 1989.

Frazer, A., et al.: Biological Bases of Brain Function and Disease. New York, Raven Press, 1994.

Goldberg, J. M., and Fernandez, C.: The vestibular system. In Darian-Smith, I. (ed.): Handbook of Physiology, Sec. 1, Vol. III. Bethesda, Md., American Physiological Society, 1984, p. 977.

Hasan, Z., and Stuart, D. G.: Animal solutions to problems of movement control: The role of proprioceptors. Annu. Rev. Neurosci., 11:199, 1988.

Jacobs, B. L., and Azmitia, E. C.: Structure and function of the brain serotonin system. Physiol. Rev., 72:165, 1992.

Jami, L.: Golgi tendon organs in mammalian skeletal muscle: functional properties and central actions. Physiol. Rev., 72:623, 1992.

Kuypers, H. B. J. M.: Anatomy of the descending pathways. In Brooks, V. B. (ed.): Handbook of Physiology. Sec. 1, Vol. II. Bethesda, Md., American Physiological Society, 1981, p. 597.

Keifer, J., and Houk, J. C.: Motor function of the cerebellorubrospinal system. Physiol. Rev., 74:509, 1994.

Kurlan, R.: Treatment of Movement Disorders. Philadelphia, J. B. Lippincott, 1994.

Luschei, E. S., and Goldberg, L. J.: Neural mechanisms of mandibular control: Mastication and voluntary biting. In Brooks, V. B. (ed.): Handbook of Physiology. Sec. 1, Vol. II. Bethesda, Md., American Physiological Society, 1981, p. 1237.

Macpherson, J. M.: Changes in a postural strategy with inter-paw distance. J. Neurophysiol., 71:931, 1994.

Narabayashi, H., et al.: Parkinson's Disease: From Basic Research to Treatment. New York, Raven Press, 1993.

Ohye, C., et al.: Motor Thalamus. Farmington, CT, S. Karger Publishers, Inc., 1993.

Oosterveld, W. J. (ed.): Audio-Vestibular System and Facial Nerve. New York, S. Karger, 1977.

Pearson, K.: The control of walking. Sci. Am., 235(6):72, 1976.

Penfield, W., and Rasmussen, T.: The Cerebral Cortex of Man. New York, Macmillan Co., 1950.

Peterson, B. W., and Richmond, F. J. (eds.): Control of Head Movement. New York, Oxford University Press, 1988.

Porter, R., and Lemon, R.: Corticospinal Function and Voluntary Movement. New York, Oxford University Press, 1993.

Poulton, E. C.: Human manual control. In Brooks, V. B. (ed.): Handbook of Physiology. Sec. 1, Vol. II. Bethesda, Md., American Physiological Society, 1981, p. 1337.

Rowland, L. P.: Merritt's Textbook of Neurology. Baltimore, Williams & Wilkins, 1995.

Salcman, M.: Neurologic Emergencies: Recognition and Management. New York, Raven Press, 1990.

Scheibel, A. B.: The brain stem reticular core and sensory function. In Darian-Smith, I. (ed.): Handbook of Physiology. Sec. 1, Vol. III. Bethesda, Md., American Physiological Society, 1984, p. 213.

Schwartz, J.-C., et al.: Histaminergic transmission in the mammalian brain. Physiol. Rev., 71:1, 1991.

Sherrington, C. S.: Decerebrate rigidity and reflex coordination of movements. J. Physiol. (Lond.), 22:319, 1898.

Shimazu, H., and Shinoda, Y.: Vestibular and Brain Stem Control of Eye, Head, and Body Movements. Farmington, CT, S. Karger Publishers, Inc., 1992.

Silverman, A. J.: Magnocellular neurosecretory system. Annu. Rev. Neurosci., 6:357, 1983.

van Giersbergen, P. L. M., et al.: Involvement of neurotransmitters in the nucleus tractus solitarii in cardiovascular regulation. Physiol. Rev., 72:789, 1992.

Van Soest, A. J., et al.: A control strategy for the execution of explosive movements from varying starting positions. J. Neurophysiol., 71:1390, 1994.

Wiesendanger, M., and Miles, T. S.: Ascending pathway of low-threshold muscle afferents to the cerebral cortex and its possible role in motor control. Physiol. Rev., 62:1234, 1982.

The Cerebellum, the Basal Ganglia, and Overall Motor Control

CHAPTER 56

Aside from the cerebral cortical areas for control of muscle activity, two other brain structures are also essential for normal motor function. They are the *cerebellum* and the *basal ganglia*. Yet neither of these two can initiate muscle function by themselves. Instead, *they always function in association with other systems of motor control.*

Basically, the cerebellum plays major roles in the timing of motor activities and in rapid progression from one movement to the next. It also helps to control intensity of muscle contraction when the muscle load changes as well as controlling the necessary instantaneous interplay between agonist and antagonist muscle groups.

The basal ganglia, on the other hand, help to plan and control complex patterns of muscle movement, controlling relative intensities of the sequential movements, directions of movements, and sequencing of multiple successive and parallel movements for achieving specific complicated motor goals.

This chapter explains the basic mechanisms of function of the cerebellum and basal ganglia and discusses what we know about the overall brain mechanisms for achieving the intricate coordination of total motor activity.

THE CEREBELLUM AND ITS MOTOR FUNCTIONS

The cerebellum has long been called a *silent area* of the brain, principally because electrical excitation of this structure does not cause any sensation and rarely causes any motor movement. Removal of the cerebellum, however, does cause movement to become highly abnormal. The cerebellum is especially vital to the control of rapid muscular activities, such as running, typing, playing the piano, and even talking. Loss of this area of the brain can cause almost total incoordination of these activities even though its loss causes paralysis of no muscles.

But how is it that the cerebellum can be so important when it has no direct ability to cause muscle contraction? The answer is that it helps to *sequence the motor activities* and also *monitors and makes corrective adjustments in the body's motor activities so that they will conform to the motor signals directed by the motor cortex and other parts of the brain.* It receives continuously updated information on the desired program of muscle contractions from these motor control areas; it also receives continuous sensory information from the peripheral parts of the body giving the sequential changes in the status of each part of the body—its position, rate of movement, forces acting on it, and so forth. The cerebellum *compares* the actual movements as depicted by the peripheral sensory feedback information with the movements intended by the motor system. If the two do not compare favorably, then instantaneous appropriate corrective signals are transmitted back into the motor system to increase or decrease the levels of activation of the specific muscles.

In addition, the cerebellum aids the cerebral cortex in planning the next sequential movement a fraction of a second in advance while the current movement is still being executed, thus helping one to progress smoothly from one movement to the next. Also, it

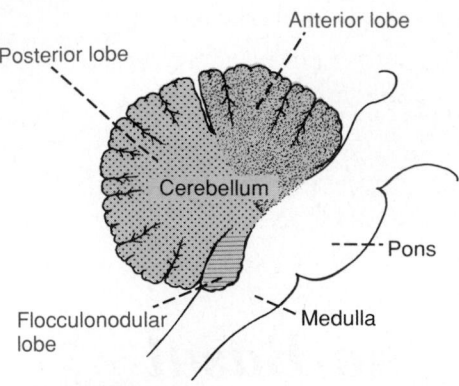

Figure 56–1. Anatomical lobes of the cerebellum as seen from the lateral side.

learns by its mistakes—that is, if a movement does not occur exactly as intended, the cerebellar circuit learns to make a stronger or weaker movement the next time. To do this, changes occur in the excitability of the appropriate cerebellar neurons, thus bringing the subsequent contractions into better correspondence with the intended movements.

Anatomical Functional Areas of the Cerebellum

Anatomically, the cerebellum is divided into three lobes by two deep fissures, as shown in Figures 56–1 and 56–2: (1) the *anterior lobe*, (2) *the posterior lobe*, and (3) the *flocculonodular lobe.* The flocculonodular lobe is the oldest of all portions of the cerebellum; it developed along with (and functions with) the vestibular system in controlling body equilibrium, as discussed in Chapter 55.

LONGITUDINAL FUNCTIONAL DIVISIONS OF THE ANTERIOR AND POSTERIOR LOBES. From a functional point of view, the anterior and posterior lobes are organized not by lobes but along the longitudinal axis, as demonstrated in Figure 56–2, which shows a posterior view of the human cerebellum after the lower end of the posterior cerebellum has been rolled downward from its normally hidden position. Note down the center of the cerebellum a narrow band separated from the remainder of the cerebellum by shallow grooves. This is called the *vermis.* In this area, most cerebellar control functions for the muscle movements of the axial body, neck, shoulders, and hips are located.

To each side of the vermis is a large, laterally protruding *cerebellar hemisphere,* and each of these hemispheres is divided into an *intermediate zone* and a *lateral zone.*

The intermediate zone of the hemisphere is concerned with controlling muscle contractions in the distal portions of the upper and lower limbs, especially the hands and fingers and feet and toes.

The lateral zone of the hemisphere operates at a much more remote level because this area joins with the cerebral cortex in the overall planning of sequential motor movements. Without this lateral zone, most discrete motor activities of the body lose their appropriate timing and sequencing and therefore become incoordinate, as we discuss more fully later.

TOPOGRAPHICAL REPRESENTATION OF THE BODY IN THE VERMIS AND INTERMEDIATE ZONES. In the same manner that the sensory cortex, motor cortex, basal ganglia, red nuclei, and reticular formation all have topographical representations of the different parts of the body, so also is this true for the vermis and intermediate zones of the cerebellum. Figure 56–3 shows two such representations. Note that the axial portions of the body lie in the vermal part of the cerebellum, whereas the limbs and facial regions lie in the intermediate zones. These topographical representations receive afferent nerve signals from all the respective parts of the body as well as from corresponding topographical motor areas in the cortex and brain stem. In turn, they send motor signals into the same respective topographical areas of the motor cortex, red nucleus, and reticular formation.

However, note that the large lateral portions of the cerebellar hemispheres *do not* have topographical representations of the body. These areas of the cerebellum

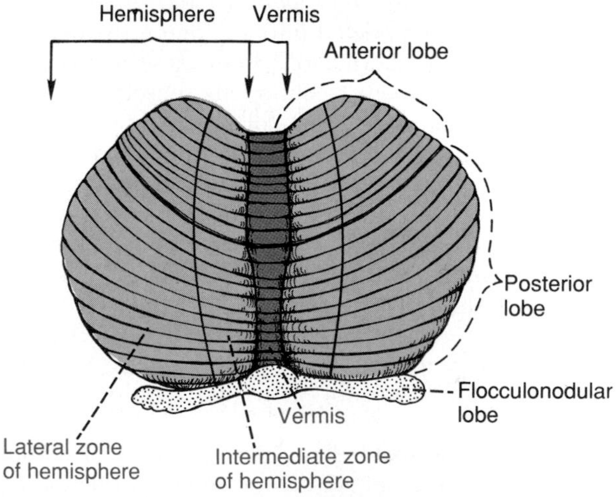

Figure 56–2. Functional parts of the cerebellum as seen from the posteroinferior view, with the inferiormost portion of the cerebellum rolled outward to flatten the surface.

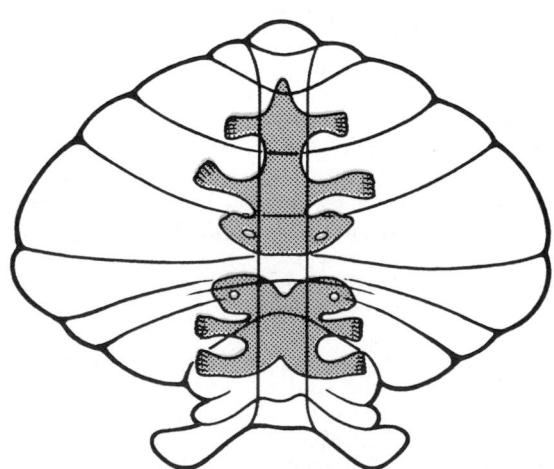

Figure 56–3. Somatosensory projection areas in the cerebellar cortex.

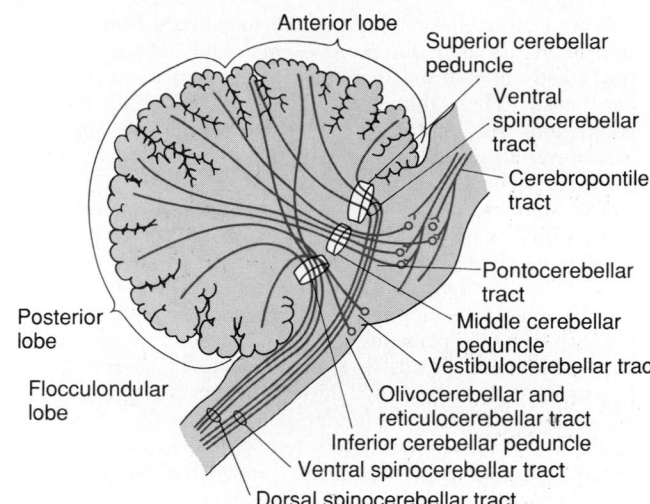

Figure 56–4. Principal afferent tracts to the cerebellum.

receive their input signals almost entirely and exclusively from the cerebral cortex, especially the motor and premotor areas of the frontal cortex and the somatosensory and sensory association areas of the parietal cortex. It is believed that this connectivity with the association areas allows the lateral portions of the cerebellar hemispheres to play important roles in planning and coordinating the body's *rapid* sequential muscular activities as they occur within fractions of a second.

Input Pathways to the Cerebellum

AFFERENT PATHWAYS FROM OTHER PARTS OF THE BRAIN. The basic input pathways to the cerebellum are shown in Figure 56–4. An extensive and important afferent pathway is the *corticopontocerebellar pathway*, which originates in the *motor* and *premotor cortices* and in the *somatosensory cortex* as well and then passes by way of the *pontile nuclei* and *pontocerebellar tracts* mainly to the lateral division of the cerebellar hemisphere on the opposite side.

In addition, important afferent tracts originate in the brain stem; they include (1) an extensive *olivocerebellar tract*, which passes from the *inferior olive* to all parts of the cerebellum and is excited in the olive by fibers from the *motor cortex, basal ganglia*, widespread areas of the *reticular formation,* and *spinal cord;* (2) *vestibulocerebellar fibers*, some of which originate in the vestibular apparatus itself and others from the vestibular nuclei and almost all of which terminate in the *flocculonodular lobe* and *fastigial nucleus* of the cerebellum; and (3) *reticulocerebellar fibers*, which originate in different portions of the reticular formation and terminate mainly in the midline cerebellar areas (the vermis).

AFFERENT PATHWAYS FROM THE PERIPHERY. The cerebellum also receives important sensory signals directly from the peripheral parts of the body mainly through four tracts on each side, two of which are located dorsally in the cord and two ventrally. The two most important of these tracts are shown in Figure 56–5: the *dorsal spinocerebellar tract* and the *ventral spinocerebellar tract* (plus similar tracts from the neck and facial regions). The dorsal tracts enter the cerebellum through the inferior cerebellar peduncle and terminate

in the vermis and intermediate zones of the cerebellum on the same side as their origin. The two ventral tracts enter the same areas of the cerebellum through the superior cerebellar peduncle, but they terminate in both sides of the cerebellum.

The signals transmitted in the dorsal spinocerebellar tracts come mainly from the muscle spindles and to a lesser extent from other somatic receptors throughout the body, such as the Golgi tendon organs, large tactile receptors of the skin, and joint receptors. All these signals apprise the cerebellum of the momentary status of muscle contraction, degree of tension on the muscle tendons, positions and rates of movement of the parts of the body, and forces acting on the surfaces of the body.

On the other hand, the ventral spinocerebellar tracts receive less information from the peripheral receptors. Instead, they are excited mainly by the motor signals

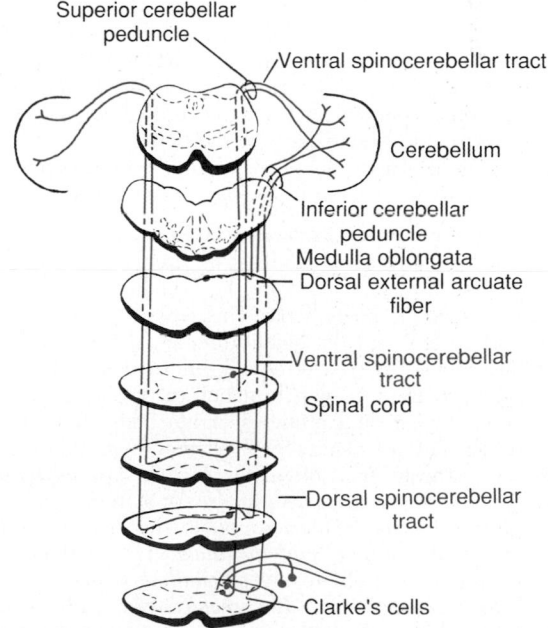

Figure 56–5. Spinocerebellar tracts.

arriving in the anterior horns of the spinal cord from (a) the brain through the corticospinal and rubrospinal tracts and (b) internal motor pattern generators in the cord itself. Thus, this ventral fiber pathway tells the cerebellum what motor signals have arrived at the anterior horns; this feedback is called the *efference copy* of the anterior horn motor drive.

The spinocerebellar pathways can transmit impulses at velocities of up to 120 m/sec, which is the most rapid conduction of any pathway in the central nervous system. This extremely rapid conduction is important for the instantaneous apprisal of the cerebellum of the changes that take place in peripheral muscle actions.

In addition to signals from the spinocerebellar tracts, signals are transmitted into the cerebellum through the spinal dorsal columns to the dorsal column nuclei of the medulla and then relayed from here to the cerebellum. Likewise, signals are transmitted through the *spinoreticular pathway* to the reticular formation of the brain stem and through the *spino-olivary pathway* to the inferior olivary nucleus and then relayed from both these areas to the cerebellum. Thus, the cerebellum continually collects information about the movements and the positions of all parts of the body even though it is operating at a subconscious level.

Output Signals from the Cerebellum

DEEP CEREBELLAR NUCLEI AND THE EFFERENT PATHWAYS. Located deep in the cerebellar mass are three *deep cerebellar nuclei*—the *dentate, interposed,* and *fastigial*. The *vestibular nuclei* in the medulla also function in some respects as if they were deep cerebellar nuclei because of their direct connections with the cortex of the flocculonodular lobe. All the deep cerebellar nuclei receive signals from two sources: (1) the cerebellar cortex and (2) the sensory afferent tracts to the cerebellum. Each time an input signal arrives in the cerebellum, it divides and goes in two directions: (1) directly to one of the deep nuclei and (2) to a corresponding area of the cerebellar cortex overlying the deep nucleus. Then, a short time later, the cerebellar cortex relays its output signals back to the same deep nucleus. Thus, all the input signals that enter the cerebellum eventually end in the deep nuclei. From the deep nuclei, output signals leave the cerebellum and are then distributed to other parts of the brain.

The general plan of the major efferent pathways leading out of the cerebellum is shown in Figure 56–6:

1. A pathway that originates in the *midline structures of the cerebellum* (the *vermis*) and then passes through the *fastigial nuclei* into the *medullary* and *pontile regions of the brain stem*. This circuit functions in close association with the equilibrium apparatus and the vestibular nuclei to control equilibrium and also, in association with the reticular formation of the brain stem, to control the postural attitudes of the body. It is discussed in detail in Chapter 55 in relation to equilibrium.

2. A pathway that originates in the *intermediate zone of the cerebellar hemisphere* and then passes through the *interposed nucleus* (a) to the *ventrolateral and ventroanterior nuclei of the thalamus* and then to the *cerebral cortex*, (b) to several *midline structures of the thalamus* and then to the *basal ganglia*, and (c) to the *red nucleus* and *reticular formation* of the upper portion of the brain stem. This circuit helps to coordi-

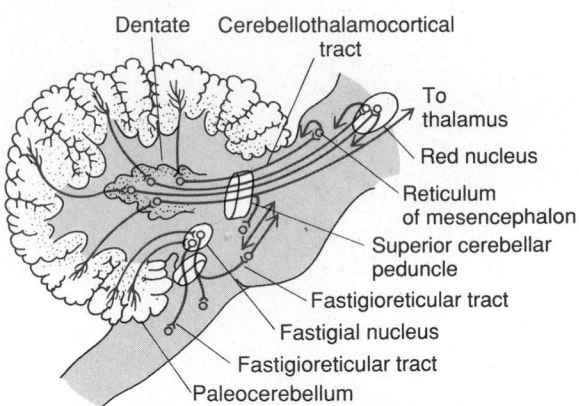

Figure 56–6. Principal efferent tracts from the cerebellum.

nate mainly the reciprocal contractions of agonist and antagonist muscles in the peripheral portions of the limbs, especially in the hands, fingers, and thumbs.

3. A pathway that begins in the *cortex of the lateral zone of the cerebellar hemisphere* and then passes to the *dentate nucleus*, next to the *ventrolateral and ventroanterior nuclei of the thalamus*, and, finally, to the *cerebral cortex*. This pathway plays an important role in helping coordinate sequential motor activities initiated by the cerebral cortex.

Neuronal Circuit of the Cerebellum

The human cerebellar cortex is actually a large folded sheet, about 17 centimeters wide by 120 centimeters long, with the folds lying crosswise, as shown in Figures 56–2 and 56–3. Each fold is called a *folium*. And lying deep beneath the folded mass of cerebellar cortex are the *deep cerebellar nuclei*.

Functional Unit of the Cerebellar Cortex— The Purkinje Cell and the Deep Nuclear Cell

The cerebellum has about 30 million nearly identical functional units, one of which is shown to the left in Figure 56–7. This functional unit centers on a single very large *Purkinje cell*, 30 million of which are in the cerebellar cortex, and on a corresponding *deep nuclear cell*.

To the top and right in Figure 56–7, the three major layers of the cerebellar cortex are shown: the *molecular layer*, *Purkinje cell layer*, and *granular cell layer*. Beneath these cortical layers, in the center of the cerebellar mass, are the deep nuclei.

NEURONAL CIRCUIT OF THE FUNCTIONAL UNIT. Shown in the left half of Figure 56–7 is the neuronal circuit of the functional unit, which is repeated with little variation 30 million times in the cerebellum. The output from the functional unit is from a *deep nuclear cell*. This cell is continually under both excitatory and inhibitory influences. The excitatory influences arise from direct connections with afferent fibers that enter the cerebellum from the brain or the periphery. The

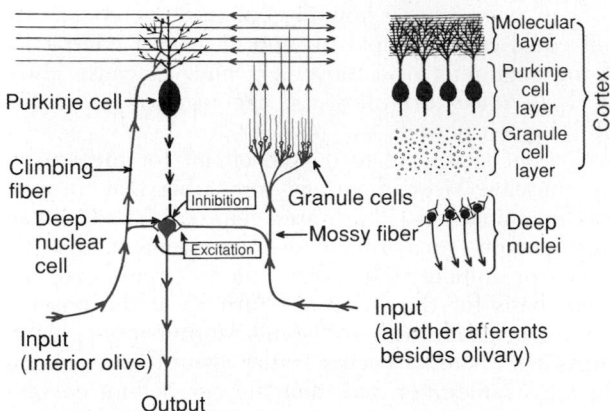

Figure 56–7. The left side of this figure shows the basic neuronal circuit of the cerebellum, with excitatory neurons shown in red and the Purkinje cell (an inhibitory neuron) shown in black. To the right is shown the physical relation of the deep cerebellar nuclei to the cerebellar cortex with its three layers.

inhibitory influence arises entirely from the Purkinje cell in the cortex of the cerebellum.

The afferent inputs to the cerebellum are mainly of two types, one called the *climbing fiber type* and the other called the *mossy fiber type*.

The climbing fibers *all originate from the inferior olive of the medulla.* There is one climbing fiber for about 5 to 10 Purkinje cells. After sending branches to several deep nuclear cells, the climbing fiber projects all the way to the molecular layer of the cerebellar cortex, where it makes about 300 synapses with the soma and dendrites of each Purkinje cell. This climbing fiber is distinguished by the fact that a single impulse in it will always cause a single, prolonged (up to 1 second), and peculiar type of action potential in each Purkinje cell with which it connects, beginning with a strong spike and followed by a trail of weakening secondary spikes. This action potential is called the *complex spike*.

The mossy fibers are all the other fibers that enter the cerebellum from multiple sources: the higher brain, brain stem, and spinal cord. These fibers also send collaterals to excite the deep nuclear cells. Then they proceed to the granular layer of the cortex, where they synapse with hundreds to thousands of *granule cells*. In turn, the granule cells send small axons, less than 1 micrometer in diameter, up to the outer surface of the cerebellar cortex to enter the molecular layer. Here the axons divide into two branches that extend 1 to 2 millimeters in each direction parallel to the folia. There are literally billions of these *parallel nerve fibers* because there are some 500 to 1000 granule cells for every Purkinje cell. It is into this molecular layer that the dendrites of the Purkinje cells project, and 80,000 to 200,000 of these parallel fibers synapse with each Purkinje cell. As these fibers pass along their 1- to 2-millimeter course, each of them contacts about 250 to 500 Purkinje cells.

The mossy fiber input to the Purkinje cell is quite different from the climbing fiber input because their synaptic connections are weak, so that large numbers of mossy fibers must be stimulated simultaneously to excite the Purkinje cell. Furthermore, this activation usually takes the form of a much weaker short-duration action potential called a *simple spike*, rather than the prolonged complex action potential caused by climbing fiber input.

THE PURKINJE CELLS AND DEEP NUCLEAR CELLS FIRE CONTINUOUSLY UNDER NORMAL RESTING CONDITIONS. One characteristic of both Purkinje cells and deep nuclear cells is that normally both of them fire continuously; the Purkinje cell fires at about 50 to 100 action potentials per second and the deep nuclear cells at much higher rates. Furthermore, the output activity of both these cells can be modulated upward or downward.

BALANCE BETWEEN EXCITATION AND INHIBITION OF THE DEEP CEREBELLAR NUCLEI. Referring again to the circuit of Figure 56–7, *one should note that direct stimulation of the deep nuclear cells by both the climbing and the mossy fibers excites them.* By contrast, *the signals arriving from the Purkinje cells inhibit them.* Normally, the balance between these two effects is slightly in favor of excitation, so that the output from the deep nuclear cell remains relatively constant at a moderate level of continuous stimulation. However, in the execution of rapid motor movements, the brain at first greatly increases the excitation. Then a few milliseconds later feedback inhibitory signals from the Purkinje cells occur. In this way, there is first a rapid excitatory signal sent by the deep nuclear cells into the motor pathway of the brain and brain stem to enhance the motor movement, but this is followed within a few milliseconds by an inhibitory signal. This inhibitory signal resembles a "delay-line" negative feedback signal of the type that is effective in providing *damping*. That is, when the motor system is excited, a negative feedback signal occurs after a short delay to stop the muscle movement from overshooting its mark, which would otherwise cause oscillation.

OTHER INHIBITORY CELLS IN THE CEREBELLUM. In addition to the deep nuclear cells, granule cells, and Purkinje cells, three other types of neurons are located in the cerebellum: *basket cells, stellate cells*, and *Golgi cells.* All these are inhibitory cells with short axons. Both the basket cells and the stellate cells are located in the molecular layer of the cortex, lying among and stimulated by the small parallel fibers. These cells in turn send their axons at right angles across the parallel fibers and cause *lateral inhibition* of the adjacent Purkinje cells, thus sharpening the signal in the same manner that lateral inhibition sharpens the contrast of signals in many other areas of the nervous system. The Golgi cells, on the other hand, lie beneath the parallel fibers, although their dendrites are also stimulated by the parallel fibers. Their axons then feed back to inhibit the granule cells. The function of this feedback is to limit the duration of the signal transmitted into the cerebellar cortex from the granule cells. That is, within a short fraction of a second after the granule cells are stimulated, their initial burst of excitation is reduced back to a lower level of excitation by the Golgi cell feedback.

Turn-On/Turn-Off and Turn-Off/Turn-On Output Signals from the Cerebellum

The typical function of the cerebellum is to help provide rapid turn-on signals for agonist muscles and simultaneous reciprocal turn-off signals for the antagonist muscles at the onset of a movement. Then on approaching the termination of the movement, the cerebellum is mainly responsible for timing and executing the turn-off signals to the agonists and turn-on signals to the antagonists. Although the exact details are not fully known, one can speculate from the basic cerebellar circuit of Figure 56–7 how this might work as follows.

First, let us suppose that the turn-on/turn-off pattern of agonist/antagonist contraction at the onset of movement begins with signals from the cerebral cortex. These signals pass through noncerebellar brain stem and cord pathways directly to the agonist muscle to begin the initial contraction. At the same time, parallel signals are sent by way of the pontile mossy fibers into the cerebellum. One branch of each mossy fiber goes directly to deep nuclear cells in the dentate or other deep cerebellar nucleus; this instantly sends an excitatory signal back into the corticospinal motor system, either by way of return signals through the thalamus to the cortex or by way of neuronal circuitry in the brain stem, to support the muscle contraction signal that had already been begun by the cerebral cortex. As a consequence, the turn-on signal, after a few milliseconds, becomes even more powerful than it was at the start because it is now the sum of both the cortical and the cerebellar signals. This is the normal effect when the cerebellum is intact, but in the absence of the cerebellum, the secondary extra supportive signal is missing. This cerebellar support makes the turn-on muscle contraction much stronger than it would be otherwise.

Now, what causes the turn-off signal for the agonist muscles at the termination of the movement? Remember that all mossy fibers have a second branch that transmits signals by way of the granule cells to the cerebellar cortex and eventually to the Purkinje cells, and the Purkinje cells in turn *inhibit* the deep nuclear cells. This pathway passes through some of the smallest nerve fibers known in the nervous system, the parallel fibers of the cerebellar cortical molecular layer that have diameters of only a fraction of a millimeter. Also, the signals from these fibers are weak, so that they require a finite period of time to build up enough excitation in the dendrites of the Purkinje cell to excite it. But once the Purkinje cell is excited, it sends *inhibitory* signals to the same deep nuclear cells that had originally turned on the movement. Therefore, this could turn off the cerebellar excitation of the agonist muscles.

Thus, one can see how the complete cerebellar circuit could cause a rapid turn-on of agonist contraction at the beginning of a movement and yet cause also a *precisely timed* turn-off of the same agonist contraction after a given period.

Now let us speculate on the circuit for the antagonist muscles. Most important, remember that throughout the spinal cord there are reciprocal agonist/antagonist circuits for virtually every movement that the cord can initiate. Therefore, these circuits are the major basis for the antagonist turn-off at the onset of movement and turn-on at its termination, always mirroring whatever occurs in the agonist muscles. But we must remember, too, that the cerebellum contains several other types of inhibitory cells besides the Purkinje cells. The functions of some of these are still to be determined; these, too, could play roles in the initial inhibition of the antagonist muscles at the onset of a movement and their subsequent excitation at the end of a movement.

These mechanisms are still mainly speculation. They are presented here only to illustrate possible ways by which the cerebellum does indeed cause turn-on and turn-off signals in the agonist and antagonist muscles and with controlled timing as well.

The Purkinje Cells "Learn" to Correct Motor Errors—The Role of the Climbing Fibers

The degree to which the cerebellum supports the onset and offset of muscle contractions as well as the timing of the contractions can be learned by the cerebellum itself. Typically, when a person first performs a new motor act, the degree of motor enhancement by the cerebellum at the onset of contraction, the degree of inhibition at the end of contraction, and the timing of these are almost always incorrect for precise performance of the movement. But after the act has been performed many times, these individual events become progressively more precise, sometimes requiring only a few movements before the desired result is achieved but at other times requiring hundreds of movements.

How do these adjustments come about? The exact answer is not known, although it is known that sensitivity levels of cerebellar circuits themselves progressively adapt during the training process. Especially, the long-term sensitivity of the Purkinje cells to respond to the parallel fibers from the granule cells becomes altered. Furthermore, this sensitivity change is brought about by signals from the climbing fibers entering the cerebellum from the inferior olivary complex.

Under resting conditions, the climbing fibers fire about once per second. But each time they do fire, they cause extreme oscillatory depolarization of the entire dendritic tree of the Purkinje cell, lasting for up to a second. During this time, the Purkinje cell fires with one initial strong output spike followed by a series of diminishing spikes. When a person performs a new movement for the first time and feedback signals from the muscle and joint proprioceptors denote that the achieved movement does not match the intended movement, the firing by the climbing fibers changes markedly, either greatly increased or decreased as needed, up to a maximum of about four per second or all the way down to zero. These climbing fiber signals in some way alter the long-term sensitivity of the Purkinje cells to the subsequent signals from the mossy fiber circuit. That is, the greater or lesser the climbing fiber input, the greater becomes the accumulative change in long-term sensitivity to the mossy fiber input. Over a period of time, this change in sensitivity, along with other possible "learning" functions of the cerebellum, is believed to make the timing and other aspects of cerebellar control of movements approach perfection. When this has been achieved, the climbing fibers no longer send their "error" signals to the cerebellum to cause further change.

Finally, we need to answer how the climbing fibers themselves know to alter their own rate of firing when a performed movement is imperfect. What is known about this is that the inferior olivary complex receives full information from the corticospinal tracts as well as from other motor centers in the brain stem detailing the *intent* of each motor movement; it also receives full information from the sensory nerve endings in the muscles and surrounding tissues detailing the movement that actually occurs. Therefore, it is believed that the inferior olivary complex then functions as a *comparator* to test how well the actual performance matches the intended performance. If there is a match, no change in firing of the climbing fibers occurs. But if there is a mismatch, then the climbing fibers are stimulated or inhibited as needed in proportion to the degree of mismatch, thus leading to progressive changes in Purkinje cell sensitivity until no further mismatch occurs—or so the theory goes.

Function of the Cerebellum in Overall Motor Control

The nervous system uses the cerebellum to coordinate motor control functions at three levels, as follows:

1.   The *vestibulocerebellum.* This consists principally of the small flocculonodular cerebellar lobes (that lie under the posterior cerebellum) and adjacent portions of the vermis.
2.   The *spinocerebellum.* This consists of most of the vermis of the posterior and anterior cerebellum plus the adjacent intermediate lobes on both sides of the vermis. It provides the circuitry for coordinating mainly the movements of the distal portions of the limbs, especially of the hands and fingers.
3.   The *cerebrocerebellum.* This consists of the large lateral zones of the cerebellar hemispheres, lateral to the intermediate lobes. It receives virtually all its input from the motor cortex and adjacent premotor and somatosensory cortices of the brain. It transmits its output information in the upward direction back to the brain; it functions in a feedback manner with all the cortical sensorimotor system to plan sequential voluntary body and limb movements, plan-

ning these as much as tenths of a second in advance of the actual movements.

The Vestibulocerebellum—Its Function with the Brain Stem and Spinal Cord to Control Equilibrium and Postural Movements

The vestibulocerebellum originated phylogenetically at about the same time that the vestibular apparatus developed. Furthermore, as discussed in Chapter 55, loss of the flocculonodular lobes and adjacent portions of the vermis of the cerebellum, which constitute the vestibulocerebellum, causes extreme disturbance of equilibrium and postural movements.

We still must ask the question, what role does the vestibulocerebellum play in equilibrium that cannot be provided by the other neuronal machinery of the brain stem? A clue is the fact that in people with vestibulocerebellar dysfunction, equilibrium is far more disturbed *during performance of rapid motions* than during stasis, especially so when these movements involve *changes in direction* of movement of the body that stimulate the semicircular ducts. This suggests that the vestibulocerebellum is especially important in controlling the balance between agonist and antagonist muscle contractions of the spine, hips, and shoulders during *rapid changes* in body positions as required by the vestibular apparatus.

One of the major problems in controlling balance is the time required to transmit position signals and velocity of movement signals from the different parts of the body to the brain. Even when the most rapidly conducting sensory pathways are used, up to 120 m/sec in the spinocerebellar afferent tracts, the delay for transmission from the feet to the brain is still 15 to 20 milliseconds. The feet of a person running rapidly can move as much as 10 inches during this time. Therefore, it is never possible for the return signals from the peripheral parts of the body to reach the brain at the same time that the movements actually occur. How, then, is it possible for the brain to know when to stop a movement to perform the next sequential act, especially when the movements are performed rapidly? The answer is that the signals from the periphery tell the brain not only positions of the different parts of the body but also how rapidly and in what directions they are moving. It is believed to be the function of the vestibulocerebellum then to *calculate* from these rates and directions where the different parts of the body will be during the next few milliseconds. The results of these calculations are the key to the brain's progression to the next sequential movement.

Thus, during the control of equilibrium, it is presumed that the information from the vestibular apparatus is used in a typical feedback control circuit to provide almost instantaneous correction of postural motor signals as necessary for maintaining equilibrium even during extremely rapid motion, including rapidly changing directions of motion. The feedback signals from the peripheral areas of the body help in this process. Their help is mediated mainly through the *cerebellar vermis* that functions in association with the axial and girdle muscles of the body; it is the role of the vestibulocerebellum to help the brain stem vestibular and reticular nuclei compute the required positions of the respective parts of the body at any given time, despite the long delay time from the periphery to the cerebellum.

The Spinocerebellum—Feedback Control of Distal Limb Movements by Way of the Intermediate Cerebellar Cortex and the Interposed Nucleus

As shown in Figure 56–8, the intermediate zone of each cerebellar hemisphere receives two types of information when a movement is performed: (1) direct information from the motor cortex and red nucleus, telling the cerebellum the sequential *intended plan of movement* for the next few fractions of a second, and (2) feedback information from the peripheral parts of the body, especially from the distal parts of the limbs, telling the cerebellum what *actual movements* result. After the intermediate zone of the cerebellum has compared the intended movements with the actual movement, the deep nuclear cells of the interposed

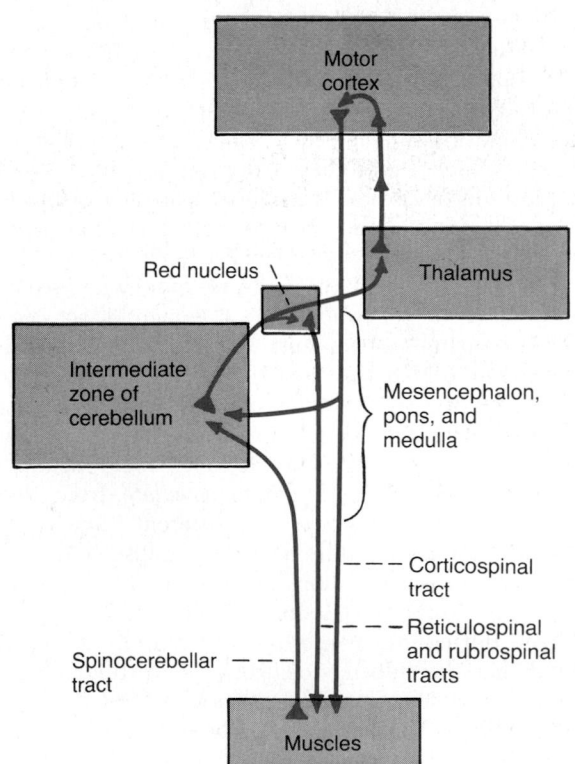

Figure 56–8. Cerebral and cerebellar control of voluntary movements, involving especially the intermediate zone of the cerebellum.

nucleus sends *corrective* output signals (a) back to the *motor cortex* through relay nuclei in the *thalamus* and (b) to the *magnocellular portion* (the lower portion) *of the red nucleus,* which gives rise to the *rubrospinal tract.* The rubrospinal tract in turn joins the corticospinal tract in innervating the lateralmost motor neurons in the anterior horns of the spinal cord gray matter, the neurons that control the distal parts of the limbs, particularly the hands and fingers.

This part of the cerebellar motor control system provides smooth, coordinate movements of the agonist and antagonist muscles of the distal limbs for the performance of acute purposeful patterned movements. The cerebellum seems to compare the "intentions" of the higher levels of the motor control system, as transmitted to the intermediate cerebellar zone through the corticopontocerebellar tract, with the "performance" by the respective parts of the body as transmitted back to the cerebellum from the periphery. In fact, the ventral spinocerebellar tract even transmits back to the cerebellum an "efference" copy of the actual motor control signals that reach the anterior motor neurons, and this, too, is integrated with the signals arriving from the muscle spindles and other proprioceptor sensory organs, transmitted principally in the dorsal spinocerebellar tract. We learned earlier that similar comparator signals also go to the inferior olivary complex; if the signals do not compare favorably, the olivary-Purkinje cell system along with possible other cerebellar learning mechanisms eventually correct the motions until they perform the desired function.

Once the cerebellum has learned its role in each pattern of movement, it provides rapid turn-on of agonist muscle activity at the onset of each movement while inhibiting the antagonist muscles. Then, near the end of the movement, the cerebellar circuit again plays the major role in rapid turn-off of the agonist muscles and turn-on of the antagonist muscles. The point at which the reversal of excitation between agonist and antagonist muscles occurs depends on (1) the rate of movement and (2) the previously learned knowledge of the inertia of the system. The faster the movement and the greater the inertia, the earlier the reversal point must occur in the course of movement to stop the movement at the proper point.

FUNCTION OF THE CEREBELLUM TO PREVENT OVERSHOOT OF MOVEMENTS AND TO "DAMP" MOVEMENTS. Almost all movements of the body are "pendular." For instance, when an arm is moved, momentum develops, and the momentum must be overcome before the movement can be stopped. Because of the momentum, all pendular movements have a tendency to *overshoot.* If overshooting does occur in a person whose cerebellum has been destroyed, the conscious centers of the cerebrum eventually recognize this and initiate a movement in the reverse direction to bring the arm to its intended position. But the arm, by virtue of its momentum, overshoots once more, and appropriate corrective signals must again be instituted. Thus, the arm oscillates back and forth past its intended point for several cycles before it finally fixes on its mark. This effect is called an *action tremor*, or *intention tremor.*

However, if the cerebellum is intact, appropriate learned, subconscious signals stop the movement precisely at the intended point, thereby preventing the overshoot as well as the tremor. This is the basic characteristic of a damping system. All control systems regulating pendular elements that have inertia must have damping circuits built into the mechanisms. In the motor control system of the central nervous system, the cerebellum provides most of this damping function.

CEREBELLAR CONTROL OF BALLISTIC MOVEMENTS. Most rapid movements of the body, such as the movements of the fingers in typing, occur so rapidly that it is not possible to receive feedback information either from the periphery to the cerebellum or from the cerebellum back to the motor cortex before the movements are over. These movements are called *ballistic movements,* meaning that the entire movement is preplanned and set into motion to go a specific distance and then to stop. Another important example is the saccadic movements of the eyes, in which the eyes jump from one position to the next when reading or when looking at successive points along a road as a person is moving in a car.

Much can be understood about the function of the cerebellum by studying the changes that occur in the ballistic movements when the cerebellum is removed. Three major changes occur: (1) the movements are slow to develop and do not have the extra onset surge that the cerebellum usually gives to an agonist movement, (2) the force development is weak, and (3) the movements are slow to turn off, usually allowing the movement to go well beyond the intended mark. Therefore, in the absence of the cerebellar circuit, the motor cortex has to think extra hard to turn ballistic movements on and again has to think hard and take extra time to turn the movement off. Thus, the automatism of ballistic movements is lost.

If one will consider once again the circuitry of the cerebellum as described earlier in the chapter, one will see that it is beautifully organized to perform this biphasic, first excitatory and then delayed inhibitory, function that is required for preplanned rapid movements. One will also see that the time delay circuits of the cerebellar cortex are fundamental to this particular ability of the cerebellum.

The Cerebrocerebellum—Function of the Large Lateral Zone of the Cerebellar Hemisphere to Plan, Sequence, and Time Complex Movements

In human beings, the lateral zones of the two cerebellar hemispheres are highly developed and greatly enlarged, along with the human abilities to plan and perform intricate sequential patterns of movement, es-

pecially with the hands and fingers, and to speak. Yet, strangely enough, the large lateral zones of the cerebellar hemispheres have no direct input of information from the peripheral parts of the body. Also, almost all the communication between these lateral cerebellar areas and the cortex is not with the primary motor cortex itself but with the premotor area and primary and association somatosensory areas. Even so, destruction of the lateral zones of the cerebellar hemispheres along with their deep nuclei, the dentate nuclei, can lead to extreme incoordination of complex purposeful movements of the hands, fingers, and feet and of the speech apparatus. This has been difficult to understand because of lack of direct communication between this part of the cerebellum and the primary motor cortex. However, experimental studies suggest that these portions of the cerebellum are concerned with two other important but indirect aspects of motor control: (1) the planning of sequential movements and (2) the "timing" of the sequential movements.

PLANNING OF SEQUENTIAL MOVEMENTS. The planning of sequential movements seems to be related to the fact that the lateral zones of the hemispheres communicate with the premotor and sensory portions of the cerebral cortex and that there is also two-way communication between these same cortex areas and corresponding areas of the basal ganglia. It seems that the "plan" of the sequential movements is transmitted from the sensory and premotor areas of the cortex to the lateral zones of the cerebellar hemispheres, and two-way traffic between the cerebellum and the cortex is necessary to provide appropriate transition from one movement to the next. An exceedingly interesting observation that supports this view is that many of the neurons in the dentate nuclei display the activity pattern of the next sequential movement that is yet to follow while the present movement is still occurring. Thus, the lateral zones appear to be involved not with what is happening at a given moment but with *what will be happening during the next sequential movement.*

To summarize, one of the most important features of normal motor function is one's ability to progress smoothly from one movement to the next in orderly succession. In the absence of the large lateral zones of the cerebellar hemispheres, this capability is seriously disturbed, especially for rapid movements occurring one after the other within tenths of a second.

TIMING FUNCTION. Another important function of the lateral cerebellar hemispheres is to provide appropriate timing for each succeeding movement. In the absence of the lateral cerebellar zones, one loses the subconscious ability to predict ahead of time how far the different parts of the body will move in a given time. Without this timing capability, the person becomes unable to determine when the next movement should begin. As a result, the succeeding movement may begin too early or, more likely, too late. Therefore, cerebellar lesions cause complex movements (such as those required for writing, running, or even talking) to become incoordinate, lacking in the ability to progress in an orderly sequence from one movement to the next. Such cerebellar lesions are said to cause *failure of smooth progression of movements.*

EXTRAMOTOR PREDICTIVE FUNCTIONS OF THE CEREBROCEREBELLUM. The cerebrocerebellum also plays a role in predicting events other than movements of the body. For instance, the rates of progression of both auditory and visual phenomena can be predicted, and both of these require cerebellar participation. As an example, a person can predict from the changing visual scene how rapidly he or she is approaching an object. A striking experiment that demonstrates the importance of the cerebellum in this ability is the removal of portions of the cerebellum in monkeys. Such a monkey occasionally charges the wall of a corridor and literally bashes its brains out because it is unable to predict when it will reach the wall.

We are only now beginning to learn about these extramotor predictive functions of the cerebellum. It is quite possible that the cerebellum provides a "time base," perhaps using time-delay circuits, against which signals from other parts of the central nervous system can be compared. It is often stated that the cerebellum is especially helpful in interpreting *spatiotemporal relations* in sensory information.

Clinical Abnormalities of the Cerebellum

An important feature of clinical cerebellar abnormalities is that destruction of small portions of the cerebellar *cortex* seldom causes detectable abnormalities in motor function. In fact, several months after as much as one half of the cerebellar cortex has been removed, if the deep cerebellar nuclei are not removed along with the cortex, the motor functions of the animal appear to be almost normal as long as the animal performs all movements slowly. Thus, the remaining portions of the motor control system are capable of compensating tremendously for loss of parts of the cerebellum.

Therefore, to cause serious and continuing dysfunction of the cerebellum, the cerebellar lesion usually must involve one or more of the deep cerebellar nuclei—the *dentate, interposed,* or *fastigial nuclei*—as well as the cerebellar cortex.

DYSMETRIA AND ATAXIA. Two of the most important symptoms of cerebellar disease are dysmetria and ataxia. As pointed out earlier, in the absence of the cerebellum, the subconscious motor control system cannot predict how far movements will go. Therefore, the movements ordinarily overshoot their intended mark, and then the conscious portion of the brain overcompensates in the opposite direction for the succeeding movements. This effect is called *dysmetria,* and it results in incoordinate movements that are called *ataxia.*

Dysmetria and ataxia can also result from lesions in the spinocerebellar tracts because feedback information from the moving parts of the body is essential for accurate control of the movements.

Past Pointing. Past pointing means that in the ab-

sence of the cerebellum, a person ordinarily moves the hand or some other moving part of the body considerably beyond the point of intention. This probably results from the fact that normally the cerebellum provides most of the motor signal that turns off a movement after it has begun; if the cerebellum is not available to do this, the movement ordinarily goes beyond the intended point. Therefore, past pointing is actually a manifestation of dysmetria.

FAILURE OF PROGRESSION

Dysdiadochokinesia. When the motor control system fails to predict where the different parts of the body will be at a given time, it "loses" the parts during rapid motor movements. As a result, the succeeding movement may begin much too early or much too late, so that no orderly "progression of movement" can occur. One can demonstrate this readily by having a patient with cerebellar damage turn one hand upward and downward at a rapid rate. The patient rapidly "loses" all perception of the instantaneous position of the hand during any portion of the movement. As a result, a series of stalled attempted but jumbled movements occurs instead of the normal coordinate upward and downward motions. This is called *dysdiadochokinesia.*

Dysarthria. Another instance in which failure of progression occurs is in talking because the formation of words depends on rapid and orderly succession of individual muscle movements in the larynx, mouth, and respiratory system. Lack of coordination between these and inability to predict either the intensity of the sound or the duration of each successive sound cause jumbled vocalization, with some syllables loud, some weak, some held long, some held for short intervals, and resultant speech that is almost unintelligible. This is called *dysarthria.*

INTENTION TREMOR. When a person who has lost the cerebellum performs a voluntary act, the movements tend to oscillate, especially when they approach the intended mark, first overshooting the mark and then vibrating back and forth several times before settling on the mark. This reaction is called an *intention tremor* or an *action tremor,* and it results from cerebellar overshooting and failure of the cerebellar system to damp the motor movements.

Cerebellar Nystagmus. Cerebellar nystagmus is a tremor of the eyeballs that occurs usually when one attempts to fixate the eyes on a scene to one side of the head. This off-center type of fixation results in rapid, tremulous movements of the eyes rather than a steady fixation, and it is another manifestation of the failure of damping by the cerebellum. It occurs especially when the flocculonodular lobes are damaged; in this instance it is associated with loss of equilibrium because of dysfunction of the pathways through the flocculonodular cerebellum from the semicircular ducts.

HYPOTONIA. Loss of the deep cerebellar nuclei, particularly the dentate and interposed nuclei, causes decreased tone of the peripheral musculature on the side of the lesion, although after several months, the cerebral motor cortex usually compensates for this by an increase in its intrinsic activity. The hypotonia results from loss of cerebellar facilitation of the motor cortex and brain stem motor nuclei by the tonic discharge of the deep cerebellar nuclei.

THE BASAL GANGLIA— THEIR MOTOR FUNCTIONS

The basal ganglia, like the cerebellum, are another accessory motor system that functions not by itself but always in close association with the cerebral cortex and corticospinal motor system. In fact, the basal ganglia receive virtually all their input signals from the cortex itself and in turn return almost all their output signals back to the cortex.

Figure 56–9 shows the anatomical relations of the basal ganglia to the other structures of the brain. These ganglia consist of the *caudate nucleus, putamen, globus pallidus, substantia nigra,* and *subthalamic nucleus.* They are located mainly lateral to the thalamus, occupying a large portion of the deeper regions of both cerebral hemispheres. Note also that almost all the motor and sensory nerve fibers connecting the cerebral cortex and spinal cord pass between the major masses of the basal ganglia, the *caudate nucleus* and the *putamen.* This mass of nerve fibers is called the *internal capsule* of the brain. It is important to our current discussion because of the intimate association between the basal ganglia and the corticospinal system for motor control.

NEURONAL CIRCUITRY OF THE BASAL GANGLIA. The anatomical connections between the basal ganglia and the other elements of motor control are complex, as shown in Figure 56–10. To the left is shown the motor cortex, thalamus, and associated brain stem and cerebellar circuitry. To the right is the major circuitry of the basal ganglia system, showing the tremendous number of interconnections among the basal ganglia themselves plus extensive input and output pathways between the motor regions of the brain and the basal ganglia.

We try in the next few sections to dissect out the major pathways of action among the basal ganglia and other portions of the motor control system, and we attempt to describe their functional attributes. We concentrate especially on two major circuits, the *putamen circuit* and the *caudate circuit.*

Function of the Basal Ganglia in Executing Patterns of Motor Activity—The Putamen Circuit

One of the principal roles of the basal ganglia in motor control is to function in association with the corticospinal system to control complex patterns of motor activity. An example is the writing of letters of the alphabet. When there is serious damage to the basal ganglia, the cortical system of motor control can no longer provide these patterns. Instead, one's writing becomes crude, as if one were learning for the first time how to write.

Other patterns that require the basal ganglia are cutting paper with scissors, hammering nails, shooting basketball through a hoop, passing a football, throwing a baseball, the movements of shoveling dirt, some aspects of vocalization, controlled movements of the eyes, and virtually any other of our skilled movements.

NEURAL PATHWAYS OF THE PUTAMEN CIRCUIT. Figure 56–11 shows the principal pathways through the basal ganglia for executing learned patterns of movement. They begin mainly in the premotor and supplementary motor areas of the motor cortex as well as in the primary somatosensory area of the sensory cortex.

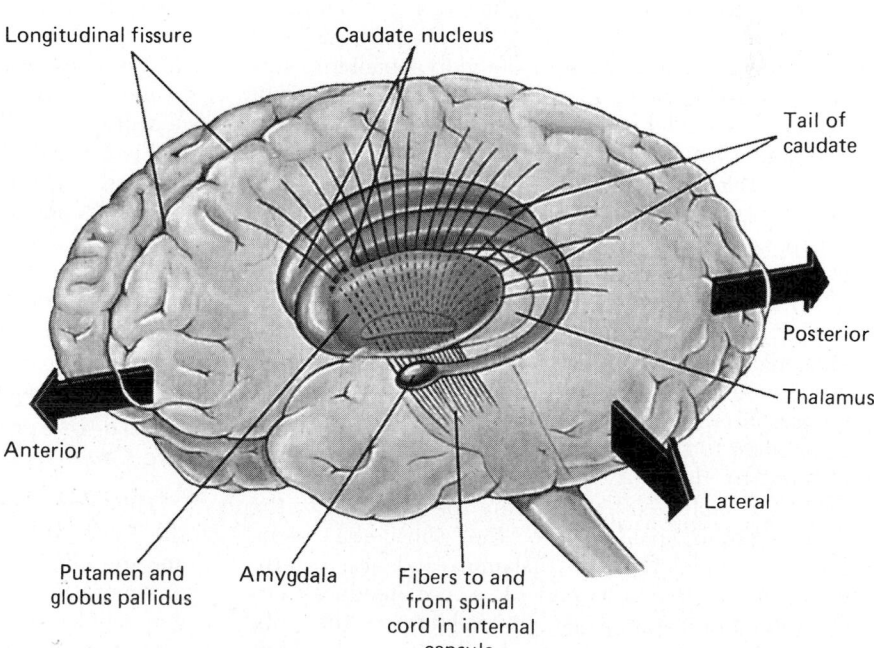

Figure 56–9. Anatomical relations of the basal ganglia to the cerebral cortex and thalamus, shown in three-dimensional view. (From Guyton: Basic Neuroscience: Anatomy and Physiology. Philadelphia, W. B. Saunders Company, 1992.)

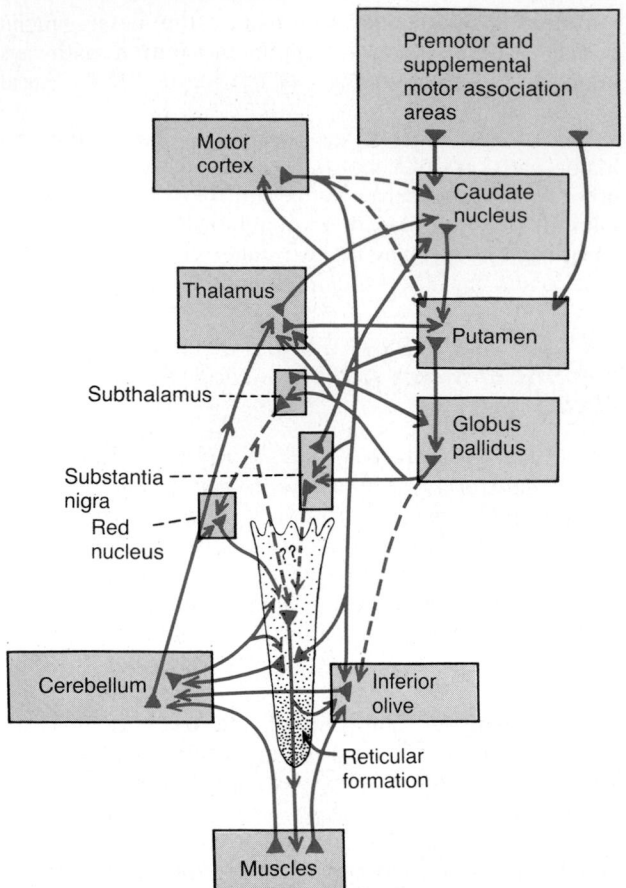

Figure 56–10. Relation of the basal ganglial circuitry to the corticospinal-cerebellar system for movement control.

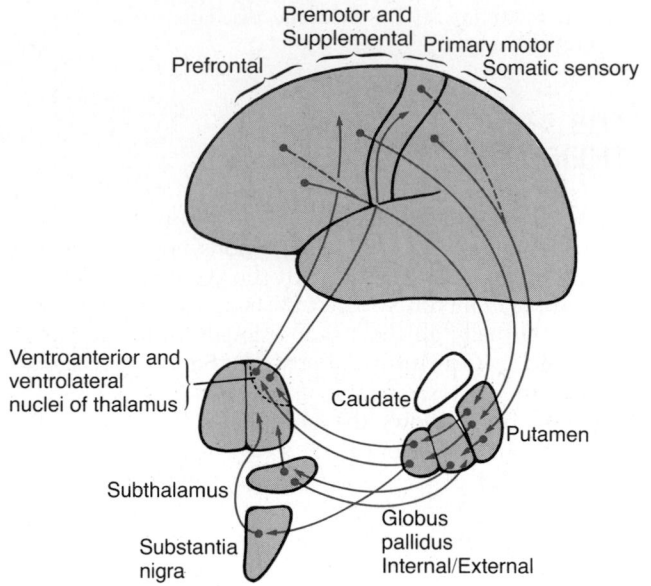

The Putamen Circuit

Figure 56–11. Putamen circuit through the basal ganglia for subconscious execution of learned patterns of movement.

Next they pass, as shown in bright red in the figure, to the putamen (mainly bypassing the caudate nucleus), then to the internal portion of the globus pallidus, next to the ventroanterior and ventrolateral relay nuclei of the thalamus, and finally return to the primary motor cortex and portions of the premotor and supplementary areas closely associated with the primary motor cortex. Thus, the putamen circuit has its inputs mainly from those parts of the brain adjacent to the primary motor cortex but not much from the primary motor cortex itself. Then its outputs do go mainly back to the *primary* motor cortex or closely associated *premotor* and *supplementary* cortex.

Functioning in close association with this primary putamen circuit are three ancillary circuits: (1) from the putamen to the external globus pallidus, to the subthalamus, to the relay nuclei of the thalamus, and back to the motor cortex; (2) from the putamen to the internal globus pallidus, to the substantia nigra, to the relay nuclei of the thalamus, and also returning to the motor cortex; and (3) a local feedback circuit from the external globus pallidus to the subthalamus and returning again to the external globus pallidus.

ABNORMALITIES OF FUNCTION IN THE PUTAMEN CIRCUIT: ATHETOSIS, HEMIBALLISMUS, AND CHOREA. How does the putamen circuit function in the execution of patterns of movement? The answer is only poorly known. When any portion of the circuit is damaged or blocked, certain patterns of movement become severely abnormal. For instance, lesions in the *globus pallidus* frequently lead to spontaneous and often continuous *writhing movements* of a hand, an arm, the neck, or the face, movements called *athetosis*.

A lesion in the *subthalamus* often leads to sudden *flailing movements* of an entire limb, a condition called *hemiballismus*.

Multiple small lesions in the *putamen* lead to *flicking movements* in the hands, face, and other parts of the body, called *chorea*.

And lesions of the *substantia nigra* lead to the common and extremely severe disease of rigidity, akinesia, and tremors known as *Parkinson's disease* which we discuss in more detail later.

Role of the Basal Ganglia for Cognitive Control of Sequences of Motor Patterns— The Caudate Circuit

The term cognition means the thinking processes of the brain, using both the sensory input to the brain and the information already stored in memory. Most of our motor actions occur as a consequence of thoughts generated in the mind, a process called *cognitive control of motor activity*. The caudate nucleus plays a major role in this cognitive control of motor activity.

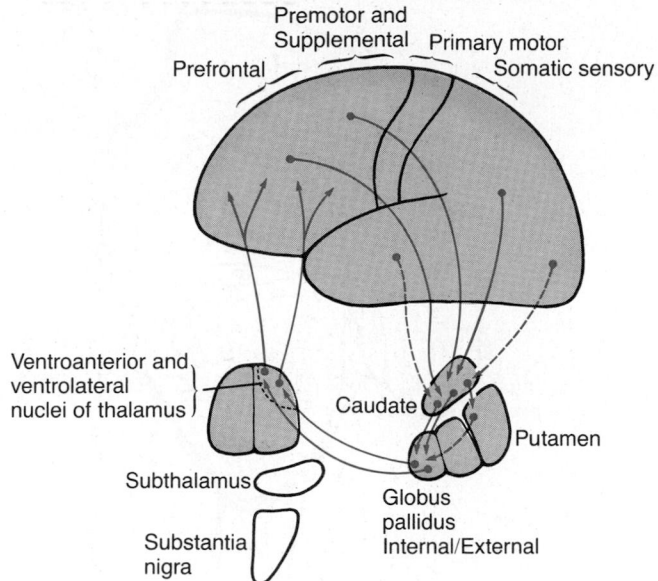

Figure 56–12. Caudate circuit through the basal ganglia for cognitive planning of sequential and parallel motor patterns to achieve specific conscious goals.

The Caudate Circuit

The neural connections between the corticospinal motor control system and the caudate nucleus, shown in Figure 56–12, are somewhat different from those of the putamen circuit. Part of the reason for this is that the caudate nucleus extends into all lobes of the cerebrum, beginning anteriorly in the frontal lobes, then passing posteriorly through the parietal and occipital lobes, and finally curving forward again like the letter "C" into the temporal lobes, as shown in Figure 56–9. Furthermore, the caudate nucleus receives large amounts of its input from the *association areas* of the cerebral cortex, the areas that integrate the different types of sensory and motor information into usable thought patterns.

After the signals pass from the cerebral cortex to the caudate nucleus, they are transmitted to the internal globus pallidus, then to the relay nuclei of the ventroanterior and ventrolateral thalamus, and finally back to the prefrontal, premotor, and supplementary motor areas of the cerebral cortex, but with almost none of the returning signals passing directly to the primary motor cortex. Instead, the returning signals go to those accessory motor regions that are concerned with putting together sequential patterns of movement instead of exciting individual muscle movements.

A good example of this would be for a person to see a lion approach and then respond instantaneously and automatically by (1) turning away from the lion, (2) beginning to run, and (3) even attempting to climb a tree. Without the cognitive functions, the person might not have the instinctive knowledge, without thinking for too long a time, to respond quickly and appropriately. Thus, cognitive control of motor activity determines subconsciously which patterns of movement will be used together and in what sequence to achieve a complex goal.

Function of the Basal Ganglia to Change the Timing and to Scale the Intensity of Movements

Two important capabilities of the brain in controlling movement are (1) to determine how rapidly the movement is to be performed and (2) to control how large the movement will be. For instance, one may write the letter "a" slowly or rapidly. Also, he may write a small "a" on a piece of paper or a large letter "a" on a chalkboard. Regardless of his choices, the proportional characteristics of the letter remain the same. This is also true even though the person might use the fingers for writing the letter in one instance or the whole arm at another time.

In patients with severe lesions of the basal ganglia, these timing and scaling functions are poor; in fact, sometimes they are nonexistent. Here again, the basal ganglia do not function alone; they function in close association with the cerebral cortex. One especially important cortical area is the posterior parietal cortex, which is the locus of the spatial coordinates for all parts of the body as well as for the relation of the body and its parts to all surroundings. Figure 56–13 shows the way in which a person lacking a left posterior parietal cortex might draw the face of another human being, providing proper proportions for the right side of the face but almost ignoring the left side (which is in his right field of vision). Also, such a person will try always to avoid using his right arm, right hand, or other portions of his right body for the performance of tasks, almost not knowing that these parts of his body exist.

Because it is the caudate circuit of the basal ganglial system that functions mainly with the association areas of the cortex, such as the posterior parietal cortex,

Figure 56–13. Typical drawing that might be made by a person who has severe damage in his or her left parietal cortex, where the spatial coordinates of the right field of vision are stored.

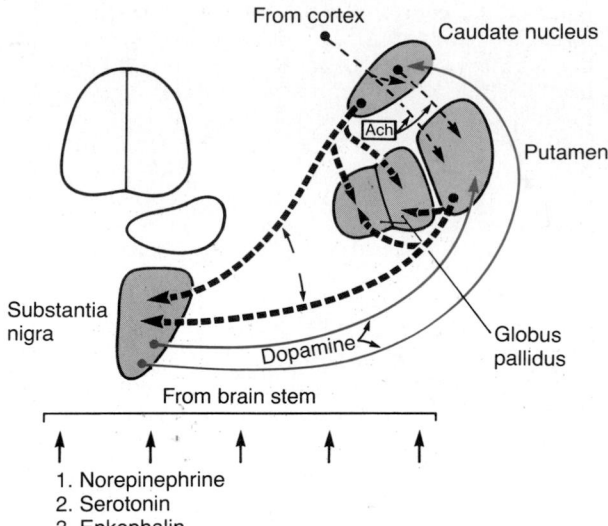

1. Norepinephrine
2. Serotonin
3. Enkephalin

Figure 56–14. Neuronal pathways that secrete different types of neurotransmitter substances in the basal ganglia.

presumably the timing and scaling of movements are functions of this caudate cognitive motor control circuit.

Functions of Specific Neurotransmitters in the Basal Ganglial System

Figure 56–14 demonstrates the interplay of some specific neurotransmitters that are known to function within the basal ganglia, showing (1) a *dopamine* pathway from the substantia nigra to the caudate nucleus and putamen; (2) a *gamma-aminobutyric acid (GABA)* pathway from the caudate nucleus and putamen to the globus pallidus and substantia nigra; (3) *acetylcholine* pathways from the cortex to the caudate nucleus and putamen; and (4) multiple general pathways from the brain stem that secrete *norepinephrine, serotonin, enkephalin,* and several other neurotransmitters in the basal ganglia as well as in other parts of the cerebrum. In addition to all these are multiple glutamate pathways that provide most of the excitatory signals (not shown in the figure) that balance out the large numbers of inhibitory signals transmitted, especially by the dopamine, GABA, and serotonin transmitters. We have more to say about some of these hormonal systems in the next sections when we discuss diseases of the basal ganglia as well as in subsequent chapters when we discuss behavior, sleep, wakefulness, and functions of the autonomic nervous system.

For the present, it should be remembered that the neurotransmitter GABA always functions as an inhibitory agent. Therefore, the GABA neurons in the feedback loops from the cortex through the basal ganglia and then back to the cortex make virtually all these loops *negative feedback loops,* rather than positive feedback loops, thus lending stability to the motor control systems. Dopamine also functions as an inhibitory neurotransmitter in most parts of the brain, so that it, too, may function as a stabilizer.

Clinical Syndromes Resulting from Damage to the Basal Ganglia

Aside from athetosis and hemiballismus, which have already been mentioned in relation to lesions in the globus pallidus and the subthalamus, two other major diseases result from damage in the basal ganglia. They are Parkinson's disease and Huntington's disease.

Parkinson's Disease

Parkinson's disease, also known as *paralysis agitans,* results from *widespread destruction of that portion of the substantia nigra, the pars compacta, that sends dopamine-secreting nerve fibers to the caudate nucleus and putamen.* The disease is characterized by (1) *rigidity* of much, if not most, of the musculature of the body, (2) *involuntary tremor* of the involved areas even when the person is resting and always at a fixed rate of 3 to 6 cycles per second, and (3) a serious difficulty in initiating movement, called *akinesia.*

The causes of these abnormal motor effects are almost entirely unknown. However, if the dopamine secreted in the caudate nucleus and putamen functions as an inhibitory transmitter, then destruction of the dopaminergic neurons in the substantia nigra theoretically would allow the caudate and putamen to become overly active and possibly cause continuous output of excitatory signals to the corticospinal motor control system. These signals could certainly overly excite many or all of the muscles of the body, thus leading to *rigidity*. Some of the feedback circuits might easily *oscillate* because of high feedback gains after loss of their inhibition, leading to the *tremor* of Parkinson's disease. This tremor is quite different from that of cerebellar disease because it occurs during all waking hours and is therefore called an *involuntary tremor,* in contradistinction to cerebellar tremor, which occurs only when the person performs intentionally initiated movements and therefore is called *intention tremor.*

The *akinesia* that occurs in Parkinson's disease is often much more distressing to the patient than are the symptoms of muscle rigidity and tremor, because to perform even the simplest movement in severe parkinsonism, the person must exert the highest degree of concentration. The mental effort, even mental anguish, that is necessary to make the movement occur at all is often at the limit of the patient's will power. Then, when the movement does occur, it is usually stiff and staccato in character instead of occurring smoothly. The cause of this akinesia is still speculative. However, dopamine secretion in the limbic system, especially in the *nucleus accumbens,* is often decreased along with its decrease in the basal ganglia. It has been suggested that this might reduce the psychic drive for motor activity so greatly that akinesia results. Or another possibility is the following. Because *patterns of movement* require sequential changes between excitation and inhibition, any effect that would lock basal ganglia activity always in one direction, such as loss of the inhibitory effects of dopamine, would prevent the initiation of and progression through sequential patterns, which require excitatory steps in addition to inhibitory steps. This is exactly what happens in akinesia.

TREATMENT WITH L-DOPA. Administration of the drug L-DOPA to patients with Parkinson's disease usually ameliorates many of the symptoms, especially the rigidity and akinesia. The reason for this is believed to be that L-DOPA is converted in the brain into dopamine, and the dopamine then restores the normal balance between inhibition and excitation in the caudate nucleus and putamen. Administration of dopamine itself does not have the same effect because dopamine has a chemical structure that will not allow it to pass through the blood-brain barrier, even though the slightly different structure of L-DOPA does allow it to pass.

TREATMENT WITH L-Deprenyl. Another treatment for Parkinson's disease is the drug L-deprenyl. This drug inhibits monoamine oxidase, which is responsible for destruction of most of the dopamine after it has been secreted. Therefore, any dopamine that is released remains in the basal ganglial tissues for a longer time. In addition, for reasons not understood, this treatment helps to slow destruction of the dopamine-secreting neurons in the substantia nigra. Therefore, appropriate combinations of L-DOPA therapy along with L-deprenyl therapy usually provide much better treatment than use of one of these drugs alone.

TREATMENT WITH TRANSPLANTED FETAL DOPAMINE CELLS. Transplantation of dopamine-secreting cells into the caudate nuclei and putamen (cells obtained from the brains of aborted fetuses) has been used with some success to treat Parkinson's disease. However, the cells do not persist for more than a few months. If persistence could be achieved, perhaps this would become the treatment of the future.

TREATMENT BY DESTROYING PART OF THE FEEDBACK CIRCUITRY IN THE BASAL GANGLIA. Because abnormal signals from the basal ganglia to the motor cortex cause most of the abnormalities in Parkinson's disease, multiple attempts have been made to treat the patients by blocking these signals. For a number of years, lesions were made in the ventrolateral and ventroanterior nuclei of the thalamus, which blocked the feedback circuit from the basal ganglia to the cortex; variable degrees of success—as well as sometimes serious neurological damage—was achieved. In monkeys with Parkinson's disease, lesions placed in the subthalamus have been used with surprisingly good results.

Huntington's Disease (Huntington's Chorea)

Huntington's disease is a hereditary disorder that usually begins to cause symptoms in the fourth or fifth decade of life. It is characterized at first by flicking movements at individual joints and then progressive severe distortional movements of the entire body. In addition, severe dementia develops along with the motor dysfunctions.

The abnormal movements of Huntington's disease are *believed to be caused by loss of most of the cell bodies of the GABA-secreting neurons in the caudate nucleus and putamen and acetylcholine-secreting neurons in many parts of the brain.* The axon terminals of the GABA neurons normally cause inhibition in the globus pallidus and substantia nigra. This loss of inhibition is believed to allow spontaneous outbursts of globus pallidus and substantia nigra activity that cause the distortional movements.

The dementia in Huntington's disease probably does not result from the loss of GABA neurons but from loss of acetylcholine-secreting neurons, perhaps especially in the thinking areas of the cerebral cortex.

The abnormal gene that causes Huntington's disease has been found; it has a many-times repeating codon, CAG, that codes for multiple glutamine amino acids in an abnormal neuronal cell protein that causes the disease. How this protein causes the disease effects is now the big question for a major research effort.

INTEGRATION OF ALL PARTS OF THE TOTAL MOTOR CONTROL SYSTEM

Finally, we need to summarize as best we can what is known about overall control of movement. To do this, let us first give a synopsis of the different levels of control.

Spinal Level

Programmed in the spinal cord are local patterns of movement for all muscle areas of the body—for instance, programmed withdrawal reflexes that pull any part of the body away from a source of pain. And the cord is the locus also of complex patterns of rhythmical motions such as to-and-fro movement of the limbs for walking plus reciprocal activity of opposite sides of the body or of the hindlimbs versus the forelimbs.

All these programs of the cord can be commanded

into action by the higher levels of motor control, or they can be inhibited while the higher levels take over control.

Hindbrain Level

The hindbrain provides two major functions for general motor control of the body: (1) maintenance of axial tone of the body for the purpose of standing and (2) continuous modification of the different degrees of tone in the different muscles in response to continuous information from the vestibular apparatuses for the purpose of maintaining equilibrium.

Corticospinal Level

The corticospinal system provides most of the motor signals to the spinal cord. It functions partly by issuing commands that set into motion various cord patterns of motor control. It can also change the intensity of the different patterns or modify their timing or other characteristics. When needed, the corticospinal system can bypass the cord patterns, replacing them with higher-level patterns from the brain stem or the cerebral cortex. The cortical patterns usually are more complex; also, they can be learned by practice, whereas the cord patterns are mainly set by heredity and are said to be "hard wired."

ASSOCIATED FUNCTION OF THE CEREBELLUM. The cerebellum functions with all levels of muscle control. It functions with the spinal cord especially to enhance the stretch reflex, so that when a contracting muscle meets an unexpectedly heavy load, a long stretch reflex signal transmitted all the way through the cerebellum and back again to the cord strongly facilitates the load-resisting effect of the basic stretch reflex.

At the brain stem level, the cerebellum functions to make the postural movements of the body, especially the rapid movements required by the equilibrium system, smooth and continuous and without abnormal oscillations.

At the cerebral cortex level, the cerebellum functions to provide many accessory motor commands, especially to provide extra motor force to turn on muscle contraction rapidly and forcefully at the start of movements. And near the end of each movement, the cerebellum turns on antagonist muscles at exactly the right time and with proper force to stop the movement at the intended point. Furthermore, there is good physiological evidence that all aspects of this turn-on/turn-off patterning by the cerebellum can be learned with experience.

In addition, the cerebellum functions with the cerebral cortex at still another level of motor planning: it helps to program in advance the muscle contractions that are required for smooth progression from the present rapid movement in one direction to the next rapid movement in another direction, all this occurring in a fraction of a second. The neural circuit for this passes from the cerebral cortex to the large lateral hemispheres of the cerebellum and then back to the cortex.

The cerebellum functions mainly with rapid movements. Without the cerebellum, slow and calculated movements can still occur, but it is difficult for the corticospinal system to achieve well-controlled rapid intended movements to a particular goal or especially to progress smoothly from one rapid movement to the next.

ASSOCIATED FUNCTIONS OF THE BASAL GANGLIA. The basal ganglia are essential to motor control in ways entirely different from those of the cerebellum. Their most important functions are (1) to help the cortex execute subconscious but *learned* patterns of movement and (2) to help plan multiple parallel and sequential patterns of movement that the mind must put together to accomplish a purposeful task.

The types of motor patterns that require the basal ganglia include those for writing all the different letters of the alphabet, for throwing a ball, for typing, and so forth. Also, the basal ganglia are required to modify these patterns for slow execution or rapid execution or to write small or write very large, thus controlling both timing and dimensions of the patterns.

At still a higher level of control is another cerebral cortex–basal ganglia circuit, beginning in the thinking processes of the brain and providing the overall sequence of action for responding to each new situation, such as planning one's immediate motor response to an assailant who hits the person in the face or one's sequential response to an unexpectedly fond embrace.

An important part of all these basal ganglial planning processes is the somatosensory cortex, especially the posterior and middle portions of the parietal lobe where the instantaneous spatial coordinates of all parts of one's body are continuously calculated and even the spatial coordinates of the relations of the body parts to the physical surroundings. If one of the two parietal cortices is severely damaged, then the person simply ignores the opposite side of his or her body and even ignores objects on the opposite side; then the movements are planned around use of only the consciously recognized side of the body.

What Drives Us to Action?

What is it that arouses us from inactivity and sets into play our trains of movement? We are beginning to learn about the motivational systems of the brain. Basically, the brain has an older core located beneath, anterior, and lateral to the thalamus—including the hypothalamus, amygdala, hippocampus, septal region anterior to the hypothalamus and thalamus, and even old regions of the thalamus and cerebral cortex themselves—all of which function together to motivate most of the motor and other functional activities of the brain. These areas are collectively called the *limbic system* of the brain. We discuss this system in detail in Chapter 58.

REFERENCES

Adams, R. D., and Victor, M.: Principles of Neurology. Hightstown, NJ, McGraw-Hill, 1994.

Bloedel, J. R., and Courville, J.: Cerebellar afferent systems. In Brooks, V. B. (ed.): Handbook of Physiology. Sec. 1, Vol. II. Bethesda, Md., American Physiological Society, 1981, p. 735.

Brodal, P.: The Central Nervous System: Structure and Function. New York, Oxford University Press, 1992.

Brooks, V. B.: The Neural Basis of Motor Control. New York, Oxford University Press, 1986.

Brooks, V. B., and Thach, W. T.: Cerebellar control of posture and movement. In Handbook of Physiology. Sec. 1, Vol. II. Bethesda, Md., American Physiological Society, 1981, p. 877.

Buneo, C. A., et al.: Muscle activation patterns for reaching: the representation of distance and time. J. Neurophysiol., 71:1546, 1994.

Burt, A. M.: Textbook of Neuroanatomy. Philadelphia, W. B. Saunders Co., 1993.

Carpenter, M. B.: Anatomy of the corpus striatum and brain stem integrating system. In Handbook of Physiology. Sec. 1, Vol. II. Bethesda, Md., American Physiological Society, 1981, p. 947.

Cordo, P., and Harnad, S.: Movement Control. New York, Cambridge University Press, 1994.

Cowan, N.: Attention and Memory: An Integrated Framework. New York, Oxford University Press, 1995.

DeLong, M., and Georgopoulos, A. P.: Motor functions of the basal ganglia. In Handbook of Physiology. Sec. 1, Vol. II. Bethesda, Md., American Physiological Society, 1981, p. 1017.

De Schutter, E., and Bower, J. M.: An active membrane model of the cerebellar purkinje cell. II. Simulation of synaptic responses. J. Neurophysiol., 71:401, 1994.

Eckmiller, R.: Neural control of pursuit eye movements. Physiol. Rev., 67:797, 1987.

Evarts, E. V., et al. (eds.): Motor System in Neurobiology. New York, Elsevier Science Publishing Co., 1986.

Forssberg, H., and Hirschfeld, H.: Movement Disorders in Children. Farmington, CT, S. Karger Publishers, Inc., 1992.

Fuster, J. M.: Prefrontal cortex in motor control. In Handbook of Physiology. Sec. 1, Vol. II. Bethesda, Md., American Physiological Society, 1981, p. 1149.

Georgopoulos, A. P.: Neural integration of movement: role of motor cortex in reaching. FASEB J., 1:2849, 1988.

Glickstein, M., and Yeo, C. (eds.): Cerebellum and Neuronal Plasticity. New York, Plenum Publishing Corp., 1987.

Harding, E. E., and Deufel, T.: Inherited Ataxias. New York, Raven Press, 1993.

Ito, M.: The Cerebellum and Neural Control. New York, Raven Press. 1984.

Jankovic, J., and Tolosa, E.: Parkinson's Disease and Movement Disorders. Baltimore, Williams & Wilkins, 1993.

Jones, E. G., and Peters, A. (eds.): Sensory-Motor Areas and Aspects of Cortical Connectivity. New York, Plenum Publishing Corp., 1986.

Joynt, R. J.: Clinical Neurology: Clinical Text in Four Looseleaf Volumes. Philadelphia, J. B. Lippincott. (Annual updates)

Keifer, J., and Houk, J. C.: Motor function of the cerebellorubrospinal system. Physiol. Rev., 74:509, 1994.

Kurlan, R.: Treatment of Movement Disorders. Philadelphia, J. B. Lippincott, 1994.

Kuypers, H. G. J. M.: Anatomy of the descending pathways. In Handbook of Physiology. Sec. 1, Vol. II. Bethesda, Md., American Physiological Society, 1981, p. 597.

Llinas, R.: Electrophysiology of the cerebellar networks. In Handbook of Physiology. Sec. 1, Vol. II. Bethesda, Md., American Physiological Society, 1981, p. 831.

Mizuno, Y.: Treatment of Parkinson's Disease. Farmington, CT, S. Karger Publishers, Inc., 1993.

Narabayashi, H., et al.: Parkinson's Disease: From Basic Research to Treatment. New York, Raven Press, 1993.

Palay, S. L., and Chan-Palay, V.: The Cerebellum—New Vistas. New York, Springer-Verlag, 1982.

Peterson, B. W., and Richmond, F. J. (eds.): Control of Head Movement. New York, Oxford University Press, 1988.

Poulton, E. C.: Human manual control. In Handbook of Physiology. Sec. 1, Vol. II. Bethesda, American Physiological Society, 1981, p. 1337.

Price, R. W., and Perry, S. W. III: HIV, AIDS, and the Brain. New York, Raven Press, 1994.

Robinson, D. A.: The windfalls of technology in the oculomotor system. Inv. Ophthal. Vis. Sci., 28:1912, 1987.

Sandler, M., et al. (eds.): Neurotransmitter Interactions in the Basal Ganglia. New York, Raven Press, 1987.

Stein, R. B., and Lee, R. G.: Tremor and clonus. In Handbook of Physiology. Sec. 1, Vol. II. Bethesda, Md., American Physiological Society, 1981, p. 325.

Trouillas, P., and Fuxe, K.: Serotonin, the Cerebellum, and Ataxia. New York, Raven Press, 1993.

Vitek, J. L., et al.: Physiological properties and somatotopic organization of the primate motor thalamus. J. Neurophysiol., 71:1498, 1994.

Weiner, W. J.: Emergent and Urgent Neurology. Philadelphia, J. B. Lippincott, 1992.

Wiesendanger, M., and Miles, T. S.: Ascending pathway of low-threshold muscle afferents to the cerebral cortex and its possible role in motor control. Physiol. Rev., 62:1234, 1982.

The Cerebral Cortex; Intellectual Functions of the Brain; and Learning and Memory

CHAPTER 57

It is ironic that of all the parts of the brain, we know least about the mechanisms of the cerebral cortex, even though it is by far the largest portion of the nervous system. But we do know the effects of destruction or specific stimulation of various portions of the cortex. In the early part of this chapter, the facts known about cortical functions are discussed; then basic theories of the neuronal mechanisms involved in thought processes, memory, analysis of sensory information, and so forth are presented briefly.

Physiologic Anatomy of the Cerebral Cortex

The functional part of the cerebral cortex comprises mainly a thin layer of neurons 2 to 5 millimeters in thickness covering the surface of all the convolutions of the cerebrum and having a total area of about one quarter square meter. The total cerebral cortex contains about 100 billion neurons.

Figure 57–1 shows the typical histologic structure of the cerebral cortex, with its successive layers of different types of cells. Most of the cells are of three types: (1) *granular* (also called *stellate*), (2) *fusiform,* and (3) *pyramidal,* the latter named for their characteristic pyramidal shape. The *granule cells,* in general, have short axons and, therefore, function mainly as intracortical interneurons. Some are excitatory, releasing mainly the excitatory neurotransmitter *glutamate;* others are inhibitory and release mainly the inhibitory neurotransmitter *gamma-aminobutyric acid (GABA).* The sensory areas of the cortex as well as the association areas between sensory and motor have large concentrations of these granule cells, suggesting a high degree of intracortical processing of the incoming sensory signals in the sensory areas and of the cognitive analytical signals in the association areas.

The *pyramidal* and *fusiform cells,* on the other hand, give rise to almost all the output fibers from the cortex. The pyramidal cells are the larger of the two and are more numerous than the fusiform cells. They are the source of the long, large nerve fibers that go all the way to the spinal cord. They also give rise to most of the large subcortical association fiber bundles that pass from one major part of the brain to the other.

To the right in Figure 57–1 is shown the typical organization of nerve fibers within the different layers of the cortex. Note particularly the large number of *horizontal fibers* that extend between adjacent areas of the cortex, but note also the *vertical fibers* that extend to and from the cortex to lower areas of the brain and to the spinal cord or to distant regions of the cerebral cortex through the long association bundles.

The functions of the specific layers of the cerebral cortex are discussed briefly in Chapters 47 and 51. By way of review, let us recall that most incoming specific sensory signals terminate in cortical layer IV. Most of the output signals leave the cortex from neurons located in layers V and VI; the very large fibers to the brain stem and cord arise generally in layer V; and the tremendous numbers of fibers to the thalamus arise in layer VI. Layers I, II, and III perform most of the intracortical association functions, with especially large numbers of neurons in layers II and III making short horizontal connections with adjacent cortical areas.

ANATOMICAL AND FUNCTIONAL RELATIONS OF THE CEREBRAL CORTEX TO THE THALAMUS AND OTHER LOWER CENTERS. All areas of the cerebral cortex have extensive to-and-fro efferent and afferent connections with deeper structures of the brain. It is especially important to emphasize the relation between the cerebral cortex and the thalamus. When the thalamus is damaged along with the cortex, the loss of cerebral

733

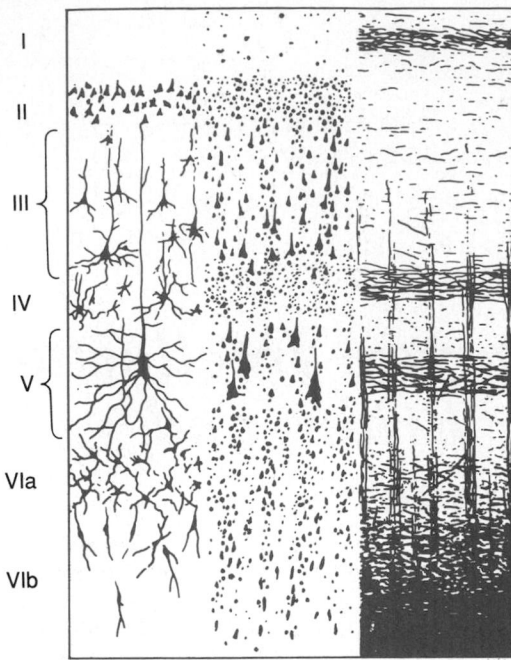

Figure 57–1. Structure of the cerebral cortex, showing *I*, molecular layer; *II*, external granular layer; *III*, layer of pyramidal cells; *IV*, internal granular layer; *V*, large pyramidal cell layer; and *VI*, layer of fusiform or polymorphic cells. (From Ranson and Clark (after Brodmann): Anatomy of the Nervous System. Philadelphia, W. B. Saunders Company, 1959.)

function is far greater than when the cortex alone is damaged because thalamic excitation of the cortex is necessary for almost all cortical activity.

Figure 57–2 shows the areas of the cerebral cortex that are connected with specific parts of the thalamus. These connections act in *two* directions, both from the thalamus to the cortex and then from the cortex back to essentially the same area of the thalamus. Furthermore, when the thalamic connections are cut, the functions of the corresponding cortical area become entirely abrogated. Therefore, the cortex operates in close association with the thalamus and can almost be

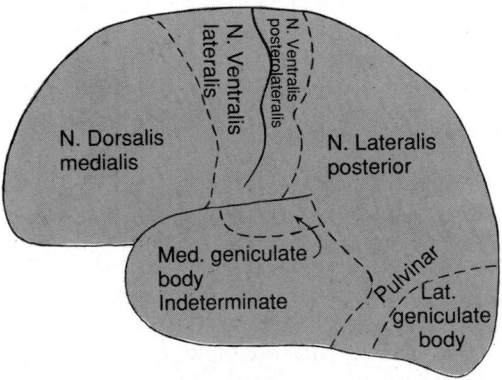

Figure 57–2. Areas of the cerebral cortex that connect with specific portions of the thalamus. (Modified from Elliott: Textbook of the Nervous System. Philadelphia, J. B. Lippincott Company.)

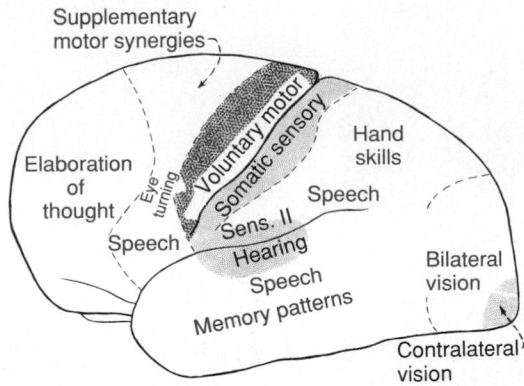

Figure 57–3. Functional areas of the human cerebral cortex as determined by electrical stimulation of the cortex during neurosurgical operations and by neurological examinations of patients with destroyed cortical regions. (From Penfield and Rasmussen: The Cerebral Cortex of Man: A Clinical Study of Localization of Function. New York, Macmillan Company, 1968.)

considered both anatomically and functionally a unit with the thalamus; for this reason, the thalamus and the cortex together are sometimes called the *thalamocortical system.* Also, all pathways from the sensory organs to the cortex pass through the thalamus, with the single exception of most of the sensory pathways of the olfactory system.

FUNCTIONS OF SPECIFIC CORTICAL AREAS

Studies in human beings by neurosurgeons, neurologists, and neuropathologists have shown that different cortical areas have separate functions. Figure 57–3 is a map of some of these functions as determined by Penfield and Rasmussen from electrical stimulation of the cortex in awake patients or during neurological examination of patients after portions of the cortex had been removed. The electrically stimulated patients told the surgeons their thoughts evoked by the stimulation, and sometimes they experienced movements. Or occasionally they spontaneously emitted sound or even a word or gave some other evidence of the stimulation. In patients in whom portions of the cortex had been removed, subsequent neurological examinations demonstrated different deficits of brain function.

Putting large amounts of information together from many different sources gives a more general map, as shown in Figure 57–4. This figure shows the major primary and secondary (both premotor and supplemental) motor areas of the cortex as well as the major primary and secondary sensory areas for somatic sensation, vision, and hearing, all of which are discussed in earlier chapters. The primary motor areas have direct connections with specific muscles for causing discrete muscle movements. The primary sensory areas detect specific sensations—visual, auditory, or somatic—transmitted to the brain from the peripheral sensory

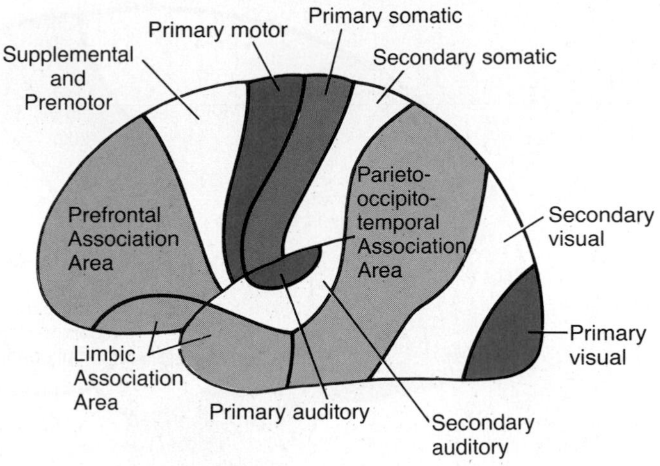

Figure 57–4. Locations of the major association areas of the cerebral cortex, shown in light pink color adjacent to the primary and secondary motor and sensory areas.

organs. The secondary areas, on the other hand, make sense out of the functions of the primary areas. For instance, the supplementary and premotor areas function along with the primary motor cortex and basal ganglia to provide "patterns" of motor activity. On the sensory side, the secondary sensory areas, located within a few centimeters of the primary areas, begin to make, sense out of the specific sensory signals, such as interpreting the shape or texture of an object in one's hand; the color, light intensity, directions of lines and angles, and other aspects of vision; and the combination of sound tones, sequence of tones, and beginning interpretation of the meanings of these auditory signals.

Association Areas

Figure 57–4 also shows several large areas of the cerebral cortex that do not fit into the rigid categories of primary and secondary motor and sensory areas. These areas are called *association areas* because they receive and analyze signals from multiple regions of both the motor and the sensory cortex as well as from subcortical structures. Yet even the association areas have their own specializations. The most important association areas are (1) the *parieto-occipitotemporal association area,* (2) the *prefrontal association area,* and (3) the *limbic association area.* The functions of these are the following.

PARIETO-OCCIPITOTEMPORAL ASSOCIATION AREA. This association area lies in the large cortical space bounded by the somatosensory cortex anteriorly, the visual cortex posteriorly, and the auditory cortex laterally. As would be expected, it provides a high level of interpretive meaning for the signals from all the surrounding sensory areas. However, even the parieto-occipitotemporal association area has its own functional subareas, which are shown in Figure 57–5.

1. Analysis of the Spatial Coordinates of the Body. An area beginning in the *posterior parietal cortex and extending into the superior occipital cortex provides continuous analysis of the spatial coordinates* of all parts of the body as well as of the surroundings of the body. This area receives visual information from the posterior occipital cortex and simultaneous somatic information from the anterior parietal cortex; from this it computes the coordinates. But why does a person need to know these spatial coordinates? The answer is that to control the body movements, the brain must know at all times where each part of the body is located as well as its relation to the surroundings. It also needs this information to analyze incoming somatosensory signals. In fact, as shown in Figure 56–13, a person missing this area of the brain loses recognition of the fact that he or she has an opposite side of the body and, as a consequence, will fail to consider the existence of the opposite side either for perceiving sensory experiences or for planning voluntary movements.

2. Area for Language Comprehension. The major area for language comprehension, called *Wernicke's area,* lies behind *the primary auditory cortex in the posterior part of the superior gyrus of the temporal lobe.* We discuss this area much more fully later; it is the most important region of the entire brain for higher intellectual function because almost all intellectual functions are language-based.

3. Area for Initial Processing of Visual Language (Reading). Posterior to the language comprehension area, lying mainly in the angular gyrus region of the occipital lobe, is a visual association area that feeds the visual information conveyed by words read from a page into Wernicke's area, the language comprehension area. This angular gyrus area is needed to make meaning out of the visually perceived words. In its absence, a person can still have excellent language comprehension through hearing but not through reading.

4. Area for Naming Objects. In the most lateral portions of both the anterior occipital lobe and the posterior temporal lobe is an area for naming objects. The names are learned mainly through auditory input, whereas the physical natures of the objects are learned mainly through visual input. In turn, the names are essential for language comprehension and intelligence,

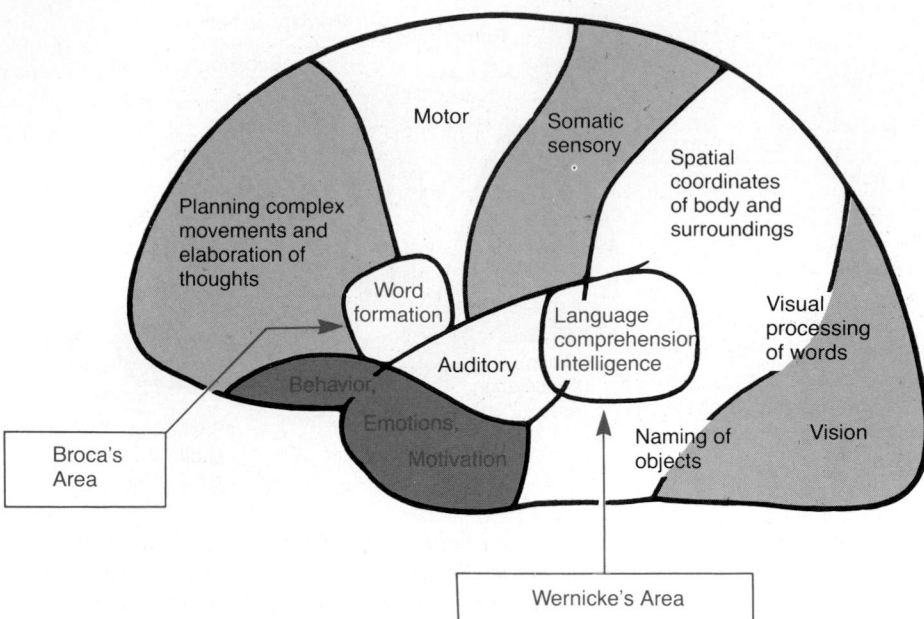

Figure 57–5. Map of specific functional areas in the cerebral cortex, showing especially Wernicke's and Broca's areas for language comprehension and speech production, which in 95 per cent of all people are located in the left hemisphere.

functions performed in Wernicke's area, located immediately superior to the "names" region.

PREFRONTAL ASSOCIATION AREA. In Chapter 56, we learn that the prefrontal association area functions in close association with the motor cortex to plan complex patterns and sequences of motor movements. To aid in this function, it receives strong input through a massive subcortical bundle of fibers connecting the parieto-occipitotemporal association area with the prefrontal association area. Through this bundle, the prefrontal cortex receives much preanalyzed sensory information, especially information on the spatial coordinates of the body that is necessary in the planning of effective movements. Much of the output from the prefrontal area into the motor control system passes through the caudate portion of the basal ganglia–thalamic feedback circuit for motor planning, which provides many of the sequential and parallel components of movement stimulation.

The prefrontal association area is also essential to carrying out prolonged thought processes in the mind. This presumably results from some of the same capabilities of the prefrontal cortex that allow it to plan motor activities. That is, it seems to be capable of processing nonmotor information from widespread areas of the brain and therefore to achieve nonmotor types of thinking as well as motor types. In fact, the prefrontal association area is frequently described simply as important for the *elaboration of thoughts.*

Broca's Area. A special region in the frontal cortex, called *Broca's area, provides the neural circuitry for word formation.* This area, shown in Figure 57–5, is located partly in the posterior lateral prefrontal cortex and partly in the premotor area. It is here that the plans and motor patterns for the expression of individual words or even short phrases are initiated and executed. This area also works in close association with

Wernicke's language comprehension center in the temporal association cortex, as we discuss more fully later in the chapter.

LIMBIC ASSOCIATION AREA. Figure 57–4 shows still another association area called the *limbic association area.* This area is found in the anterior pole of the temporal lobe, in the ventral portion of the frontal lobe, and in the cingulate gyrus on the midsurface of the cerebral hemisphere. It is concerned primarily with *behavior, emotions,* and *motivation,* as shown in Figure 57–5. We learn in Chapter 58 that the limbic cortex is part of a much more extensive system, the *limbic system,* that includes a complex set of neuronal structures in the midbasal regions of the brain. This limbic system provides most of the emotional drives for setting the other areas of the brain into action and even provides the motivational drive for the process of learning itself.

An Area for Recognition of Faces

An interesting type of brain abnormality called *prosophenosia* is the inability to recognize faces. This occurs in people who have extensive damage on the medial undersides of both occipital lobes and along the medioventral surfaces of the temporal lobes, as shown in Figure 57–6. Loss of these face recognition areas, strangely enough, results in little other abnormality of brain function.

One wonders why so much of the cerebral cortex should be reserved for the simple task of face recognition. However, when it is remembered that most of our daily tasks involve associations with other people, one can see the importance of this intellectual function.

The occipital portion of this area is contiguous with the visual cortex, and the temporal portion is closely

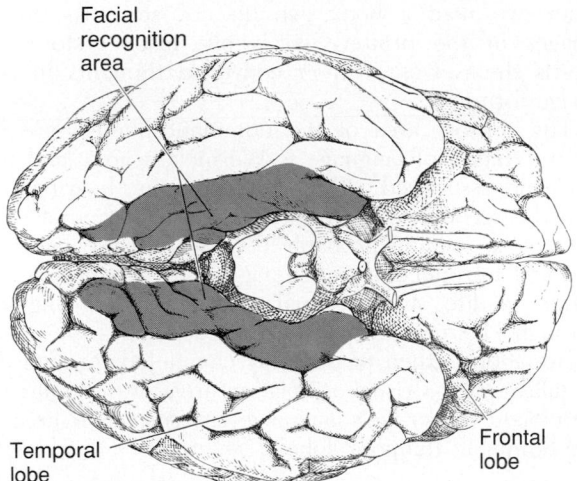

Figure 57–6. Facial recognition areas located on the underside of the brain in the medial occipital and temporal lobes. (From Geschwind: *Sci. Am., 241:*180, 1979. © 1979 by Scientific American, Inc. All rights reserved.)

associated with the limbic system that has to do with emotions, brain activation, and control of one's behavioral response to the environment, as we see in Chapter 58.

Interpretative Function of the Posterior Superior Temporal Lobe—"Wernicke's Area" (A General Interpretative Area)

The somatic, visual, and auditory association areas all meet one another in the posterior part of the superior temporal lobe, as shown in Figure 57–7, where the temporal, parietal, and occipital lobes all come together. This area of confluence of the different sensory interpretative areas is especially highly developed in the *dominant* side of the brain—the *left side* in

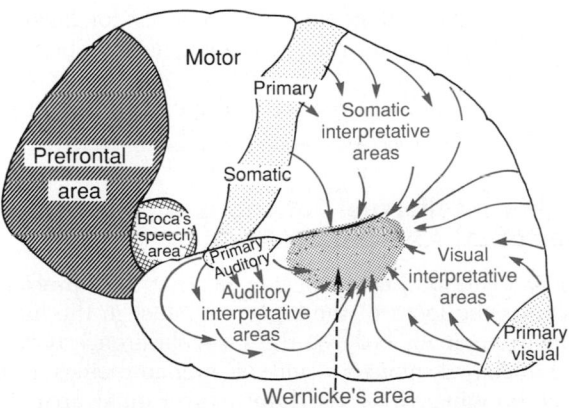

Figure 57–7. Organization of the somatic, auditory, and visual association areas into a general mechanism for interpretation of sensory experience. All these feed also into *Wernicke's area,* located in the posterosuperior portion of the temporal lobe. Note also the prefrontal area and Broca's speech area in the frontal lobe.

almost all right-handed people—and it plays the greatest single role of any part of the cerebral cortex in the higher levels of brain function that we call *intelligence.* Therefore, this region has been called by different names suggestive of an area that has almost global importance: the *general interpretative area,* the *gnostic area,* the *knowing area,* the *tertiary association area,* and so forth. It is best known as *Wernicke's area* in honor of the neurologist who first described its special significance in intellectual processes.

After severe damage in Wernicke's area, a person might hear perfectly well and even recognize different words but still be unable to arrange these words into a coherent thought. Likewise, the person may be able to read words from the printed page but be unable to recognize the thought that is conveyed.

Electrical stimulation in Wernicke's area of the conscious person occasionally causes a highly complex thought. This is particularly true when the stimulatory electrode is passed deep enough into the brain to approach the corresponding connecting areas of the thalamus. The types of thoughts that might be experienced include complicated visual scenes that one might remember from childhood, auditory hallucinations such as a specific musical piece, or even a statement by a specific person. For this reason, it is believed that activation of Wernicke's area can call forth complicated memory patterns that involve more than one sensory modality even though most of the memory patterns may be stored elsewhere. This belief is in accord with the importance of Wernicke's area in interpretation of the complicated meanings of different sensory experiences.

ANGULAR GYRUS—INTERPRETATION OF VISUAL INFORMATION. The angular gyrus is the most anteroinferior portion of the posterior parietal lobe, lying immediately behind Wernicke's area and fusing posteriorly into the visual areas of the occipital lobe as well. If this region is destroyed while Wernicke's area in the temporal lobe is still intact, the person can still interpret auditory experiences as usual, but the stream of visual experiences passing into Wernicke's area from the visual cortex is mainly blocked. Therefore, the person may be able to see words and even know that they are words but not be able to interpret their meanings. This is the condition called *dyslexia,* or *word blindness.*

Let us again emphasize the global importance of Wernicke's area for processing most intellectual functions of the brain. Loss of this area in an adult usually leads thereafter to a lifetime of almost demented existence.

Concept of the Dominant Hemisphere

The general interpretative functions of Wernicke's area and the angular gyrus as well as the functions of the speech and motor control areas are usually much more highly developed in one cerebral hemisphere than in the other. Therefore, this hemisphere is called the *dominant hemisphere.* In about 95 per cent of all people, the left hemisphere is the dominant one. Even at birth, the area of the cortex that will eventually become Wernicke's area is as much as 50 per cent larger in the left hemisphere than in the right in more than one half of neonates. Therefore, it is easy to understand why the left side of the brain might become dominant over the right side. However, if for some reason this left side area is damaged or removed in early childhood, the opposite side of the brain can develop full dominant characteristics.

A theory that can explain the capability of one hemisphere to dominate the other hemisphere is the following.

The attention of the "mind" seems to be directed to one principal thought at a time. Presumably, because the left posterior temporal lobe at birth is usually larger than the right, the left side normally begins to be used to a greater extent than the right. Thenceforth, because of the tendency to direct one's attention to the better developed region, the rate of learning in the cerebral hemisphere that gains the first start increases rapidly, whereas that in the opposite side remains slight. Therefore, the left side normally becomes dominant over the right.

In about 95 per cent of all people, the left temporal lobe and angular gyrus become dominant, and in the remaining 5 per cent, either both sides develop simultaneously to have dual dominance or, more rarely, the right side alone becomes highly developed.

As discussed later in the chapter, the premotor speech area (Broca's area), located far laterally in the intermediate frontal lobe, is also almost always dominant on the left side of the brain. This speech area causes the formation of words by exciting simultaneously the laryngeal muscles, respiratory muscles, and muscles of the mouth.

The motor areas for controlling the hands are also dominant on the left side of the brain in about 9 of 10 persons, thus causing right-handedness in most people.

Although the interpretative areas of the temporal lobe and angular gyrus as well as many of the motor areas are usually highly developed in only the left hemisphere, these areas receive sensory information from both hemispheres and are capable also of controlling motor activities in both hemispheres. For this purpose, they use mainly fiber pathways in the *corpus callosum* for communication between the two hemispheres. This unitary, cross-feeding organization prevents interference between the functions of the two sides of the brain; such interference could create havoc with both thoughts and motor responses.

Role of Language in the Function of Wernicke's Area and in Intellectual Functions

A major share of our sensory experience is converted into its language equivalent before being stored in the memory areas of the brain and before being processed for other intellectual purposes. For instance,

when we read a book, we do not store the visual images of the printed words but instead store the words themselves or their conveyed thoughts in language form.

The sensory area of the dominant hemisphere for interpretation of language is Wernicke's area, and this is closely associated with both the primary hearing area and the secondary hearing areas of the temporal lobe. This close relation probably results from the fact that the first introduction to language is by way of hearing. Later in life, when visual perception of language through the medium of reading develops, the visual information is then presumably channeled through the angular gyrus, a visual association area, into the already developed Wernicke's language interpretative area of the dominant temporal lobe.

Functions of the Parieto-occipitotemporal Cortex in the Nondominant Hemisphere

When Wernicke's area in the dominant hemisphere of a fully grown person is destroyed, the person normally loses almost all intellectual functions associated with language, or manipulation of these, verbal symbolism, such as ability to read, ability to perform mathematical operations, and even the ability to think through logical problems. However, many other types of interpretative capabilities, some of which use the temporal lobe and angular gyrus regions of the opposite hemisphere, are retained. Psychological studies in patients with damage to the nondominant hemisphere have suggested that this hemisphere may be especially important for understanding and interpreting music, nonverbal visual experiences (especially visual patterns), spatial relations between the person and the surroundings, the significance of "body language" and intonations of people's voices, and probably many somatic experiences related to use of the limbs and hands.

Thus, even though we speak of the "dominant" hemisphere, this dominance is primarily for language- or verbal symbolism–related intellectual functions; the so-called nondominant hemisphere is actually dominant for some other types of intelligence.

Higher Intellectual Functions of the Prefrontal Association Area

For years it has been taught that the prefrontal cortex is the locus of the higher intellect in the human being, principally because the main difference between the brains of monkeys and of human beings is the great prominence of the human prefrontal areas. Yet efforts to show that the prefrontal cortex is more important in higher intellectual functions than other portions of the brain have not been successful. Indeed, destruction of the language comprehension area in the posterior superior temporal lobe (Wernicke's area) and the adjacent angular gyrus region in the dominant hemisphere causes infinitely more harm to the intellect than does destruction of the prefrontal areas. The prefrontal areas do, however, have less definable but nevertheless important intellectual functions of their own. They can best be explained by describing what happens to patients in whom the prefrontal areas have become nonfunctional as follows.

Several decades ago, before the advent of modern drugs for treating psychiatric conditions, it was found that some patients could receive significant relief from severe psychotic depression by severing the neuronal connections between the prefrontal areas of the brain and the remainder of the brain, that is, by a procedure called *prefrontal lobotomy*. This was done by inserting a blunt, thin-bladed knife through small openings in the lateral frontal skull on both sides and slicing the brain from top to bottom. Subsequent studies in these patients showed the following mental changes:

1. The patients lost their ability to solve complex problems.
2. They became unable to string together sequential tasks to reach specific goals.
3. They became unable to learn to do several parallel tasks at the same time.
4. Their level of aggressiveness was decreased, sometimes markedly, and in general they lost all ambition.
5. Their social responses were often inappropriate for the occasion, including loss of morals and little reticence in relation to sex and excretion.
6. The patients could still talk and comprehend language, but they were unable to carry through any long trains of thought, and their moods changed rapidly from sweetness to rage to exhilaration to madness.
7. The patients could also still perform most of the usual patterns of motor function that they had performed throughout life, but often without purpose.

From this information, let us try to piece together a coherent understanding of the function of the prefrontal association areas.

DECREASED AGGRESSIVENESS AND INAPPROPRIATE SOCIAL RESPONSES. These two characteristics probably result from loss of the ventral parts of the frontal lobes on the underside of the brain. As explained earlier and shown in Figure 57–4, this area is considered part of the limbic association cortex, rather than the prefrontal association cortex. This limbic area helps to control behavior, which is discussed in detail in Chapter 58.

INABILITY TO PROGRESS TOWARD GOALS OR TO CARRY THROUGH WITH SEQUENTIAL THOUGHTS. We learned earlier in the chapter that the prefrontal association areas have the capability of calling forth information from widespread areas of the brain and then using it in deeper thought patterns for attaining goals. If these goals include motor action, so be it. If they do not, then the thought processes attain intellectual analytical goals. Although people without prefrontal cortices can still think, they show little concerted thinking in logical sequence for longer than a few seconds or a minute or so at most. One of the results is that people without prefrontal cortices are *easily distracted from*

the central theme of thought, whereas people with functioning prefrontal cortices can drive themselves to completion of their thought goals irrespective of distractions.

ELABORATION OF THOUGHT, PROGNOSTICATION, AND PERFORMANCE OF HIGHER INTELLECTUAL FUNCTIONS BY THE PREFRONTAL AREAS—THE CONCEPT OF A "WORKING MEMORY." Another function that has been ascribed to the prefrontal areas by psychologists and neurologists is *elaboration of thought*. This means simply an increase in depth and abstractness of the different thoughts put together from multiple sources of information. Psychological tests have shown that prefrontal lobectomized lower animals presented with successive bits of sensory information fail to keep track of these bits even in temporary memory, probably because they are distracted so easily that they cannot hold thoughts long enough for memory storage to take place.

This ability of the prefrontal areas to keep track of many bits of information simultaneously and then to cause recall of this information instantaneously as it is needed for subsequent thoughts is called the brain's "working memory." This could well explain the many functions of the brain that we associate with higher intelligence. In fact, studies have shown that the prefrontal areas are divided into separate segments for storing different types of temporary memory, such as one area for storing shape and form of an object or a part of the body and another for storing movement. By combining all these temporary bits of working memory, we have the abilities to (1) prognosticate, (2) plan for the future, (3) delay action in response to incoming sensory signals so that the sensory information can be weighed until the best course of response is decided, (4) consider the consequences of motor actions even before they are performed, (5) solve complicated mathematical, legal, or philosophical problems, (6) correlate all avenues of information in diagnosing rare diseases; and (7) control our activities in accord with moral laws.

FUNCTION OF THE BRAIN IN COMMUNICATION—LANGUAGE INPUT AND LANGUAGE OUTPUT

One of the most important differences between human beings and lower animals is the facility with which human beings can communicate with one another. Furthermore, because neurological tests can easily assess the ability of a person to communicate with others, we know more about the sensory and motor systems related to communication than about any other segment of cortical function. Therefore, we will review, with the help of anatomical maps of neural pathways in Figure 57–8, the function of the cortex in communication. From this one can see immediately how the principles of sensory analysis and motor control apply to this art.

There are two aspects to communication: first, the *sensory aspect* (language input), involving the ears and

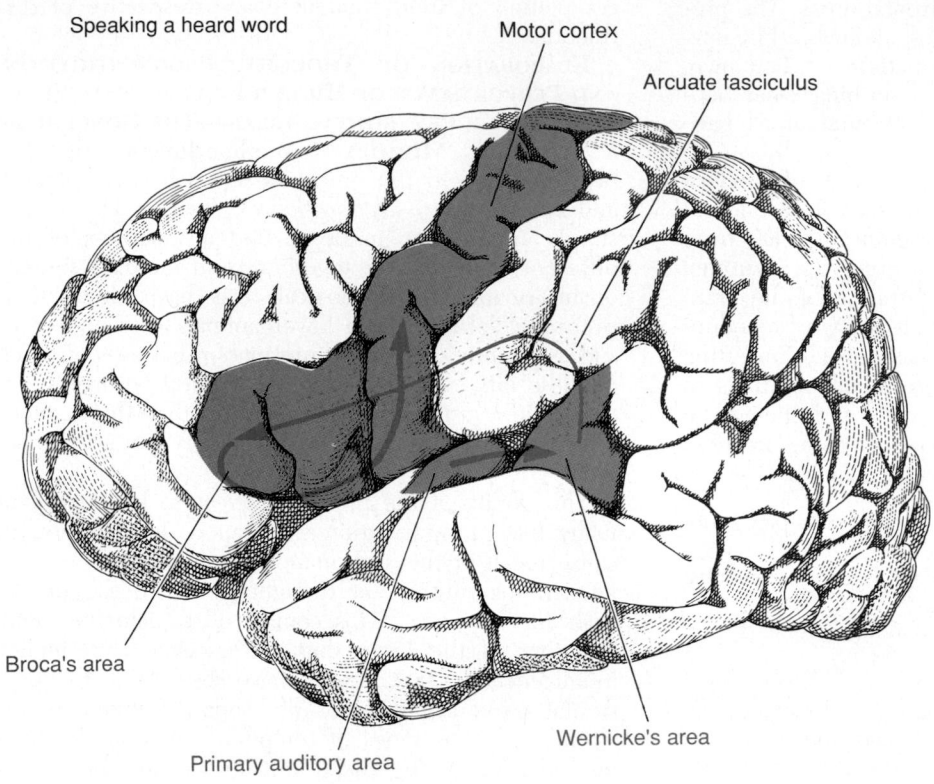

Speaking a heard word

Motor cortex

Arcuate fasciculus

Broca's area

Primary auditory area

Wernicke's area

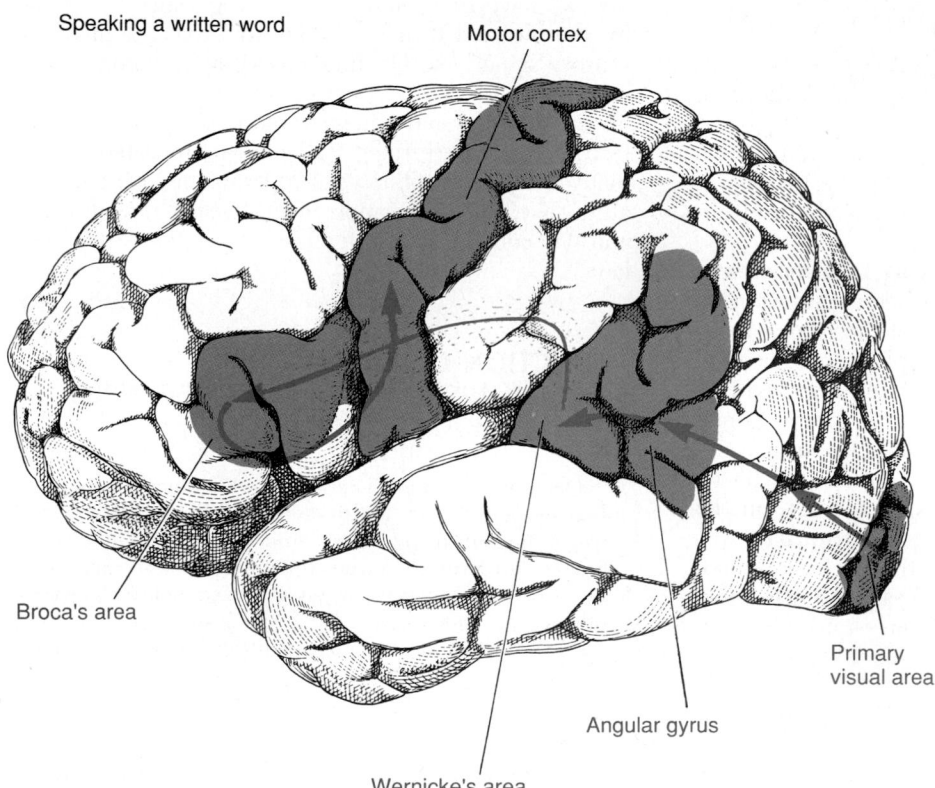

Speaking a written word

Motor cortex

Broca's area

Primary visual area

Angular gyrus

Wernicke's area

Figure 57–8. Brain pathways for *(top)* perception of the heard word and then speaking the same word and *(bottom)* perception of the written word and then speaking the same word. (From Geschwind: *Sci. Am., 241:*180, 1979. © 1979 by Scientific American, Inc. All rights reserved.)

eyes, and, second, the *motor aspect* (language output), involving vocalization and its control.

Sensory Aspects of Communication

We noted earlier in the chapter that destruction of portions of the *auditory* and *visual association areas* of the cortex can result in inability to understand the spoken word or the written word. These effects are called, respectively, *auditory receptive aphasia* and *visual receptive aphasia* or, more commonly, *word deafness* and *word blindness* (also called *dyslexia*).

WERNICKE'S APHASIA AND GLOBAL APHASIA. Some people are capable of understanding either the spoken word or the written word but are *unable to interpret the thought* that is expressed. This results most frequently when *Wernicke's area* in the *posterior portion of the dominant hemisphere superior temporal gyrus* is damaged or destroyed. Therefore, this type of aphasia is called *Wernicke's aphasia.*

When the lesion in Wernicke's area is widespread and extends (1) backward into the angular gyrus region, (2) inferiorly into the lower areas of the temporal lobe, and (3) superiorly into the superior border of the sylvian fissure, the person is likely to be almost totally demented for language understanding or communication and therefore is said to have *global aphasia.*

Motor Aspects of Communication

The process of speech involves two principal stages of mentation: (1) formation in the mind of thoughts to be expressed and choice of words to be used and then (2) motor control of vocalization and the actual act of vocalization itself. The formation of thoughts and even most choices of words are the function of sensory association areas of the brain. Again, it is Wernicke's area in the posterior part of the superior temporal gyrus that is most important for this ability. Therefore, a person with either Wernicke's aphasia or global aphasia is unable to formulate the thoughts that are to be communicated. Or, if the lesion is less severe, the person may be able to formulate thoughts but be unable to put together the appropriate sequence of words to express the thought. The person often is fluent in words but the words are jumbled.

LOSS OF BROCA'S AREA CAUSES MOTOR APHASIA. Sometimes a person is capable of deciding what he wants to say and vocalizing but cannot make the vocal system emit words instead of noises. This effect, called *motor aphasia*, results from damage to *Broca's speech area*, which lies in the *prefrontal* and *premotor* facial region of the cortex—about 95 per cent of the time in the left hemisphere, as shown in Figures 57–5 and 57–8. Therefore, the *skilled motor patterns* for control of the larynx, lips, mouth, respiratory system, and other accessory muscles of speech are all initiated from this area.

ARTICULATION. Finally, we have the act of articulation, which means the muscular movements of the mouth, tongue, larynx, vocal cords, and so forth that are responsible for the intonations, timing, and rapid changes in intensities of the sequential sounds. The *facial and laryngeal regions of the motor cortex* activate these muscles, and the *cerebellum, basal ganglia,* and *sensory cortex* all help control the sequences and intensities of muscle contractions, making liberal use of basal ganglial and cerebellar feedback mechanisms described in Chapters 55 and 56. Destruction of any of these regions can cause either total or partial inability to speak distinctly.

Summary

Figure 57–8 shows two principal pathways for communication. The upper half of the figure shows the pathway involved in hearing and speaking. This sequence is the following: (1) reception in the primary auditory area of the sound signals that encode the words; (2) interpretation of the words in Wernicke's area; (3) determination, also in Wernicke's area, of the thoughts and the words to be spoken; (4) transmission of signals from Wernicke's area to Broca's area by way of the *arcuate fasciculus;* (5) activation of the skilled motor programs in Broca's area for control of word formation; and (6) transmission of appropriate signals into the motor cortex to control the speech muscles.

The lower figure illustrates the comparable steps in reading and then speaking in response. The initial receptive area for the words is in the primary visual area rather than in the primary auditory area. Then the information passes through early stages of interpretation in the *angular gyrus region* and finally reaches its full level of recognition in Wernicke's area. From here, the sequence is the same as for speaking in response to the spoken word.

FUNCTION OF THE CORPUS CALLOSUM AND ANTERIOR COMMISSURE TO TRANSFER THOUGHTS, MEMORIES, TRAINING, AND OTHER INFORMATION BETWEEN THE TWO CEREBRAL HEMISPHERES

Fibers in the *corpus callosum* provide abundant bidirectional neural connections between most of the respective cortical areas of the two hemispheres except for the anterior portions of the temporal lobes; these temporal areas, including especially the *amygdala*, are interconnected by fibers that pass through the *anterior commissure.*

Because of the tremendous number of fibers in the corpus callosum, it was assumed from the beginning that this massive structure must have some important function to correlate activities of the two cerebral hemispheres. However, when the corpus callosum was destroyed in laboratory animals, it was at first difficult to discern deficits in brain function. Therefore, for a long time, the function of the corpus callosum was a mystery.

Properly designed psychological experiments have now demonstrated extremely important functions for the corpus callosum and anterior commissure. These functions can best be explained by describing one of the experiments. A monkey is first prepared by cutting the corpus callosum and splitting the optic chiasm longitudinally, so that signals from each eye can go only to the cerebral hemisphere on the side of the eye. Then the monkey is taught to recognize different objects with its right eye while its left eye is covered. Next, the right eye is covered and the monkey is tested to determine whether or not its left eye can recognize the same objects. The answer to this is that the left eye *cannot* recognize the objects. Yet, on repeating the same experiment in another monkey with the optic chiasm split but the corpus callosum intact, it is found invariably that recognition in one hemisphere of the brain creates recognition in the opposite hemisphere.

Thus, one of the functions of the corpus callosum and the anterior commissure is to make information stored in the cortex of one hemisphere available to corresponding cortical areas of the opposite hemisphere. Important examples of such cooperation between the two hemispheres are the following.

1. Cutting the corpus callosum blocks transfer of information from Wernicke's area of the dominant hemisphere to the motor cortex on the opposite side of the brain. Therefore, the intellectual functions of Wernicke's area, located in the left hemisphere, lose control over the right motor cortex and therefore the voluntary motor functions of the left hand and arm, even though the usual subconscious movements of the left hand and arm are normal.

2. Cutting the corpus callosum prevents transfer of somatic and visual information from the right hemisphere into Wernicke's area of the dominant hemisphere. Therefore, somatic and visual information from the left side of the body frequently fails to reach this general interpretative area of the brain and therefore cannot be used for decision making.

3. Finally, people whose corpus callosum is completely sectioned have two entirely separate conscious portions of the brain. For example, in a teenage boy with a sectioned corpus callosum, only the left half of his brain could understand both the written word and the spoken word because it was the dominant hemisphere. On the other hand, the right side of the brain could understand the written word but not the spoken word. Furthermore, the right cortex could elicit a motor response to the written word without the left cortex ever knowing why the response was performed. The effect was quite different when an emotional response was evoked in the right side of the brain: in this case, a subconscious emotional response occurred in the left side of the brain as well. This undoubtedly occurred because the areas of the two sides of the brain for emotions, the anterior temporal cortices and adjacent areas, were still communicating with each other through the anterior commissure that was not sectioned. For instance, when the command "kiss" was written for the right half of his brain to see, the boy immediately and with full emotion said, "No way!" This response required function of Wernicke's area and the motor areas for speech in the left hemisphere because these left-side areas were necessary to speak the words "No way!" But when questioned why he said this, the boy could not explain. Thus, the two halves of the brain have independent capabilities for consciousness, memory storage, communication, and control of motor activities. The corpus callosum is required for the two sides to operate cooperatively, and the anterior commis-

sure plays an important additional role in unifying the emotional responses of the two sides of the brain.

THOUGHTS, CONSCIOUSNESS, AND MEMORY

Our most difficult problem in discussing consciousness, thoughts, memory, and learning is that we do not know the neural mechanisms of a thought and we know little about the mechanisms of memory. We know that destruction of large portions of the cerebral cortex does not prevent a person from having thoughts, but it does reduce the *degree* of awareness of the surroundings.

Each thought certainly involves simultaneous signals in many portions of the cerebral cortex, thalamus, limbic system, and reticular formation of the brain stem. Some crude thoughts probably depend almost entirely on lower centers; the thought of pain is probably a good example because electrical stimulation of the human cortex seldom elicits anything more than the mildest degree of pain, whereas stimulation of certain areas of the hypothalamus and mesencephalon can cause excruciating pain. On the other hand, a type of thought pattern that requires mainly the cerebral cortex is that of vision because loss of the visual cortex causes complete inability to perceive visual form or color.

Therefore, we might formulate a provisional definition of a thought in terms of neural activity as follows: A thought results from a "pattern" of stimulation of many parts of the nervous system at the same time and in definite sequence, probably involving most importantly the cerebral cortex, thalamus, limbic system, and upper reticular formation of the brain stem. This is called the *holistic theory* of thoughts. The stimulated areas of the limbic system, thalamus, and reticular formation are believed to determine the general nature of the thought, giving it such qualities as pleasure, displeasure, pain, comfort, crude modalities of sensation, localization to gross areas of the body, and other general characteristics. On the other hand, the stimulated areas of the cerebral cortex determine the discrete characteristics of the thought, such as specific localization of sensations on the body and of objects in the fields of vision, the feeling of the texture of silk, visual recognition of the rectangular pattern of a concrete block wall, and other individual characteristics that enter into the overall awareness of a particular instant.

Consciousness can perhaps be described as our continuing stream of awareness of either our surroundings or our sequential thoughts.

Memory—Roles of Synaptic Facilitation and Synaptic Inhibition

Physiologically, memories are caused by changes in the capability of synaptic transmission from one neuron to the next as a result of previous neural activity. These changes in turn cause new pathways or facilitated pathways to develop for transmission of signals through the neural circuits of the brain. The new or facilitated pathways are called *memory traces.* They are important because once they are established, they can be activated by the thinking mind to reproduce the memories.

Experiments in lower animals have demonstrated that memory traces can occur at all levels of the nervous system. Even spinal cord reflexes can change at least slightly in response to repetitive cord activation, which is part of the memory process. Also, some long-term memories result from changed synaptic conduction in the lower brain centers.

There is much reason to believe that most of the memory that we associate with intellectual processes is based on memory traces mainly in the cerebral cortex.

POSITIVE AND NEGATIVE MEMORY—"SENSITIZATION" OR "HABITUATION" OF SYNAPTIC TRANSMISSION. Although we often think of memories as being *positive* recollections of previous thoughts or experiences, probably the greater share of our memories are *negative* memories, not positive. That is, our brain is inundated with sensory information from all our senses. If our minds attempted to remember all this information, the memory capacity of the brain would be exceeded within minutes. Fortunately, the brain has the peculiar capacity to learn to ignore information that is of no consequence. This results from *inhibition* of the synaptic pathways for this type of information; the resulting effect is called *habituation.* This is, in a sense, a type of negative memory.

On the other hand, for those types of incoming information that cause important consequences, such as pain or pleasure, the brain has the automatic capability of enhancing and storing the memory traces. This is positive memory. It results from *facilitation* of the synaptic pathways, and the process is called *memory sensitization.* We learn later that special areas in the basal limbic regions of the brain determine whether information is important or unimportant and make the subconscious decision whether to store the thought as an enhanced memory trace or to suppress it.

CLASSIFICATION OF MEMORIES. We know that some memories last for only a few seconds and others last for hours, days, months, or years. For the purpose of discussing these, let us use a common classification of memories that divides memories into (1) *short-term memory,* which includes memories that last for seconds or at most minutes unless they are converted into longer-term memories; (2) *intermediate long-term memories,* which last for days to weeks but eventually are lost; and (3) *long-term memory,* which, once stored, can be recalled up to years or even a lifetime later.

Besides this general classification of memories, we also discussed earlier (in connection with the prefrontal lobes) another type of memory, called "working memory," that includes varying amounts of each of the other types of memory.

Short-Term Memory

Short-term memory is typified by one's memory of 7 to 10 numerals in a telephone number (or 7 to 10 other discrete facts) for a few seconds to a few minutes at a time but lasting only so long as the person continues to think about the numbers or facts.

Many physiologists have suggested that this short-term memory is caused by continual neural activity resulting from nerve signals that travel around and around in a temporary memory trace through a *circuit of reverberating neurons.* It has not yet been possible to prove this theory.

Another possible explanation of short-term memory is *presynaptic facilitation or inhibition.* This occurs at synapses that lie on presynaptic terminals, not on the subsequent neuron. The neurotransmitters secreted at such terminals frequently cause prolonged facilitation or inhibition (depending on the type of transmitter secreted) lasting for as long as seconds or even several minutes. Circuits of this type could lead to short-term memory.

A final possibility for explaining short-term memory is synaptic potentiation, which can enhance synaptic conduction. It can result from the accumulation of large amounts of calcium ions in the presynaptic terminals. That is, when a train of impulses passes through a presynaptic terminal, the amount of calcium ions entering the presynaptic terminal itself through the presynaptic membrane increases with each successive action potential. When the amount of calcium ions becomes greater than the mitochondria and endoplasmic reticulum can absorb, the excess calcium then causes prolonged presynaptic release of transmitter substance at the synapse. Thus, this, too, could be a mechanism for short-term memory.

Intermediate Long-Term Memory

Now we come to intermediate long-term memories that may last for many minutes or even weeks. They will eventually be lost unless the memory traces become more permanent; then they are classified as long-term memories. Experiments in primitive animals have demonstrated that memories of this type can result from temporary chemical or physical changes, or both, in either the presynaptic terminals or the postsynaptic membrane, changes that can persist for a few minutes up to several weeks. These mechanisms are so important that they deserve special description.

Memory Based on Chemical Changes in the Presynaptic Terminal or Postsynaptic Neuronal Membrane

Figure 57–9 shows a mechanism of memory studied especially by Kandel and his colleagues that can cause memories lasting from a few minutes up to 3 weeks in the large snail *Aplysia.* In this figure, there are two presynaptic terminals. One terminal is from a primary

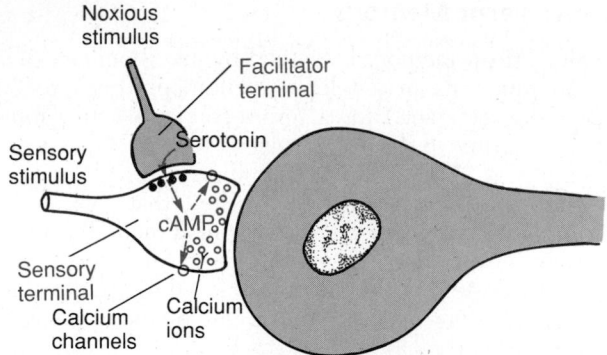

Figure 57–9. Memory system that has been discovered in the snail *Aplysia.*

input sensory neuron and terminates on the surface of the neuron that is to be stimulated; this is called the *sensory terminal.* The other terminal lies on the surface of the sensory terminal and is called the *facilitator terminal.* When the sensory terminal is stimulated repeatedly but without stimulating the facilitator terminal, signal transmission at first is great, but it becomes less and less intense with repeated stimulation until transmission almost ceases. This phenomenon is called *habituation,* as was explained earlier. It is a type of *negative* memory that causes the neuronal circuit to lose its response to repeated events that are insignificant.

On the other hand, if a noxious stimulus excites the facilitator terminal at the same time that the sensory terminal is stimulated, then instead of the transmitted signal into the postsynaptic neuron becoming progressively weaker, the ease of transmission becomes much stronger and will remain strong for minutes, hours, days, or, with more intense training, up to about 3 weeks even without further stimulation of the facilitator terminal. Thus, the noxious stimulus causes the memory pathway to become *facilitated* for days or weeks thereafter. It is especially interesting that even after habituation has occurred, the pathway can be converted to a facilitated pathway with only a few noxious stimuli.

MOLECULAR MECHANISM OF INTERMEDIATE MEMORY

Mechanism for Habituation. At the molecular level, the habituation effect in the sensory terminal results from progressive closure of calcium channels of the presynaptic terminal membrane, although the cause of this is not fully known. Nevertheless, much smaller than normal amounts of calcium then diffuse into this terminal when action potentials occur, and much less transmitter is therefore released because calcium entry is the stimulus for transmitter release (as discussed in Chapter 45).

Mechanism for Facilitation. In the case of facilitation, at least part of the molecular mechanism is believed to be the following.

1. Stimulation of the facilitator terminal at the same time that the sensory terminal is stimulated causes serotonin release at the facilitator synapse on the surface of the sensory presynaptic terminal.

2. The serotonin acts on *serotonin receptors* in the sensory terminal membrane, and these receptors activate the enzyme *adenylcyclase* inside the membrane. This causes formation of *cyclic adenosine monophosphate (cAMP)* also inside the sensory presynaptic terminal.

3. The cyclic AMP activates a *protein kinase* that causes phosphorylation of a protein that is part of the potassium channels in the sensory terminal membrane, which blocks the channels for potassium conductance. This blockage can last for minutes up to several weeks.

4. Lack of potassium conductance causes a greatly prolonged action potential in the presynaptic terminal because the flow of potassium ions out of the terminal is necessary for rapid recovery from the action potential.

5. The prolonged action potential causes prolonged activation of the calcium pores, allowing tremendous quantities of calcium ions to enter the sensory terminal. These calcium ions cause greatly increased transmitter release, thereby greatly facilitating synaptic transmission.

Thus, in a very indirect way, the associative effect of stimulating the facilitator neuron at the same time that the sensory neuron is stimulated causes a prolonged increase in excitatory sensitivity of the sensory terminal, which establishes the memory trace.

In addition, studies by Byrne and colleagues, also in the snail *Aplysia,* have suggested still another mechanism of synaptic memory: Their studies have shown that stimuli from separate sources acting on a single neuron, under appropriate conditions, can cause long-term changes in the *membrane properties of the postsynaptic neuron* instead of presynaptic membrane but leading to essentially the same memory effects. Thus, this is another possible mechanism of intermediate long-term memory.

Long-Term Memory

There is no real demarcation between the more prolonged types of intermediate long-term memory and true long-term memory. The distinction is one of degree. However, long-term memory is generally believed to result from actual *structural changes,* instead of chemical changes, at the synapses that enhance or suppress signal conduction. Again, let us recall experiments in primitive animals (where the nervous systems are much easier to study) that have aided immensely in understanding possible mechanisms of long-term memory.

Structural Changes Occur in Synapses During the Development of Long-Term Memory

Electron microscopic pictures taken from invertebrate animals have demonstrated multiple physical structural changes in the synapses during development of long-term memory traces. The structural changes will not occur if a drug is given that blocks the DNA system for replication of proteins in the presynaptic neuron, nor will the permanent memory trace develop. Therefore, it seems clear that development of true long-term memory depends on physically restructuring the synapses themselves in a way that increases their sensitivity for transmitting nervous signals.

The most important of the physical changes that occur are the following:

1. *Increase in the number of vesicle release sites for secretion of transmitter substance.* Recall from the discussion in Chapter 7 that each presynaptic terminal has specific *release sites* in its membrane that release transmitter substance into the synaptic cleft when the terminal is stimulated. Within minutes after a memory training session is begun, one can find in electron micrographs a beginning increase in the number of these release sites in the terminal. The new sites are created by development of new protein vesicle release structures on the inner surface of the presynaptic membrane. Then, when an action potential arrives in the presynaptic terminal, the extra sites cause greatly increased vesicular exocytosis of the transmitter substance into the synaptic cleft.

2. *Increase in the number of transmitter vesicles.* Not only does the number of vesicle release sites increase, but the number of transmitter vesicles in each presynaptic terminal also increases.

3. *Increase in the number of presynaptic terminals.* With intense training, even the number of presynaptic terminals also increases, sometimes to more than double normal. Along with the increase in terminals, the dendrites of the successive neuron increase in length to accommodate the extra synapses.

Thus, in several ways, the structural capability of synapses to transmit signals increases during establishment of true long-term memory traces.

The Number of Neurons and Their Connectivities Change Drastically During Early Life Learning

During the first few weeks, months, and perhaps even year or so of life, many parts of the brain produce a great excess of neurons, and the neurons send out numerous dividing axon branches to make connections with other neurons. If the new axons fail to connect with appropriate subsequent neurons, muscle cells, or gland cells, the new axons will dissolute within a few weeks. The proper connections are determined by specific but different *nerve growth factors* released by the subsequent neurons. If the type of growth factor released is of the wrong type, the growing axon will not connect and soon the axon will die. Furthermore, when insufficient connectivity occurs, the entire neuron that is sending out the axon branches will disappear.

Thus, soon after the birth of a human baby, there is a principle of "use it or lose it" that governs the final number of neurons and their connectivities in the respective parts of the brain. This is a type of learning. To give an example, if the vision of one eye of a newborn animal is blocked for multiple weeks after birth, many of the neurons in the alternate visual stripes of the visual cortex—that is, those neurons

normally connected to the covered eye—will degenerate and the eye will remain either partially or totally blinded for the remainder of life.

Therefore, in many areas of the cerebral cortex, 50 per cent or more of the original neurons are eliminated. This mechanism in itself is a type of long-term learning. Furthermore, this pliability of the nervous system in the early stages of life emphasizes the importance of giving the young child vast exposure to multiple learning experiences. The child then has the best possible foundation for a lifetime of subsequent education.

Consolidation of Memory

For short-term memory to be converted into long-term memory that can be recalled weeks or years later, it must become "consolidated." That is, the memory must in some way initiate the chemical, physical, and anatomical changes in the synapses that are responsible for the long-term type of memory. This process requires 5 to 10 minutes for minimal consolidation and 1 hour or more for strong consolidation. For instance, if a strong sensory impression is made on the brain but is then followed within a minute or so by an electrically induced brain convulsion, the sensory experience will not be remembered. Likewise, brain concussion, sudden application of deep general anesthesia, or any other effect that temporarily blocks the dynamic function of the brain can prevent consolidation.

If the strong electrical shock is delayed for more than 10 minutes, at least part of the memory trace will have become established. If the shock is delayed for 1 hour, the memory will have become much more fully *consolidated.*

Consolidation and the time required for it can probably be explained by the phenomenon of *rehearsal* of the short-term memory as follows.

Role of Rehearsal in Transference of Short-Term Memory into Long-Term Memory. Psychological studies have shown that rehearsal of the same information again and again in the mind accelerates and potentiates the degree of transfer of short-term memory into long-term memory and therefore accelerates and potentiates consolidation. The brain has a natural tendency to rehearse newfound information, especially newfound information that catches the mind's attention. Therefore, over a period of time, the important features of sensory experiences become progressively more and more fixed in the memory stores. This explains why a person can remember small amounts of information studied in depth far better than large amounts of information studied only superficially. It also explains why a person who is wide awake can consolidate memories far better than a person who is in a state of mental fatigue.

Codifying of Memories During Consolidation. One of the most important features of consolidation is that memories are codified into different classes of information. During this process, similar information is recalled from the memory storage bins and used to help process the new information. The new and old are compared for similarities and differences, and part of the storage process is to store the information about these similarities and differences, rather than to store the information unprocessed. Thus, during consolidation, the new memories are not stored randomly in the brain but are stored in direct association with other memories of the same type. This is necessary if one is to be able to "search" the memory store at a later date to find the required information.

Role of Specific Parts of the Brain in the Memory Process

THE HIPPOCAMPUS CAN PROMOTE STORAGE OF MEMORIES—ANTEROGRADE AMNESIA AFTER HIPPOCAMPAL LESIONS. The hippocampus is the most medial portion of the temporal lobe cortex where it folds first medially underneath the brain and then upward into the lower surface of the lateral ventricle. The two hippocampi have been removed for the treatment of epilepsy in a few patients. This procedure does not seriously affect the person's memory for information stored in the brain before removal of the hippocampi. However, after removal, these people have virtually no capability thereafter for storing *verbal and symbolic types* of memories in long-term memory, or even in intermediate memory lasting longer than a few minutes. Therefore, these people are unable to establish new long-term memories of those types of information that are the basis of intelligence. This is called *anterograde amnesia*.

But why is the hippocampus so important in helping the brain to store new memories? The probable answer is that the hippocampus is one of the most important output pathways from the "reward" and "punishment" areas of the limbic system. The sensory stimuli or thoughts that cause pain or aversion excite the limbic *punishment centers*, whereas the stimuli that cause pleasure, happiness, or a sense of reward excite the limbic *reward centers*. All of these together provide the background mood and motivations of the person. Among these motivations is the drive in the brain to remember those experiences and thoughts that are either pleasant or unpleasant. The hippocampus especially and to a lesser degree the dorsal medial nuclei of the thalamus, another limbic structure, have proved especially important in making the decision about which of our thoughts are important enough on a basis of reward or punishment to be worthy of memory.

RETROGRADE AMNESIA. *Retrograde amnesia* means inability to recall memories from the past—that is, from the long-term memory storage bins—even though the memories are known to be still there. When retrograde amnesia occurs, the degree of amnesia for recent events is likely to be much greater than for events of the distant past. The reason for this difference is probably that the distant memories have been rehearsed so many times that the memory traces are deeply engrained, and elements of these memories are stored in widespread areas of the brain.

In some people who have hippocampal lesions, some degree of retrograde amnesia occurs along with anterograde amnesia just discussed, which suggests that these two types of amnesia are at least partially related and that hippocampal lesions can cause both. However, it has been claimed that damage in some thalamic areas can lead specifically to retrograde amnesia without causing significant anterograde amnesia. A possible explanation of this is that the thalamus might play a role in helping the person "search" the memory storehouses and thus "read out" the memories. That is, the memory process not only requires the storing of memories but also an ability to search and find the memory at a later date. The possible function of the thalamus in this process is discussed further in Chapter 58.

THE HIPPOCAMPI ARE NOT IMPORTANT IN REFLEXIVE LEARNING. People with hippocampal lesions usually do not have difficulty in learning physical skills that do not involve verbalization or symbolic types of intelligence. For instance, these people can still learn hand and physical skills as required in many types of sports. This type of learning is called *reflexive learning;* it depends on physically repeating the required tasks over and over again, rather than on symbolical rehearsing in the mind.

REFERENCES

Adams, J. H., and Duchen, L. W.: Greenfield's Neuropathology. New York, Oxford University Press, 1992.

Avoli, M., et al. (eds.): Neurotransmitters and Cortical Function. New York, Plenum Publishing Corp., 1988.

Baddeley, A.: Working Memory. New York, Oxford University Press, 1987.

Barclay, L.: Clinical Geriatric Neurology. Baltimore, Williams & Wilkins, 1993.

Benson, D. F.: The Neurology of Thinking. New York, Oxford University Press, 1994.

Brown, T. H., et al.: Long-term synaptic potentiation. Science, 242:724, 1988.

Conn, P. M.: Neuroscience in Medicine. Philadelphia, J. B. Lippincott, 1994.

Cowan, N.: Attention and Memory: An Integrated Framework. New York, Oxford University Press, 1995.

Dagenbach, D., and Carr, T. H.: Inhibitory Processes in Attention, Memory, and Language. San Diego, CA, Academic Press, 1994.

Damasio, A. R., and Geschwind, N.: The neural basis of language. Annu. Rev. Neurosci., 7:127, 1984.

David, A. S., and Cutting, J. C.: The Neuropsychology of Schizophrenia. Hillsdale, NJ, Lawrence Erlbaum Assoc. Inc., 1994.

De Valois, R. L., and De Valois, K. K.: Spatial Vision. New York, Oxford University Press, 1988.

de Gelder, B., and Morais, J.: Speech and Reading. Hillsdale, NJ, Lawrence Erlbaum Assoc. Inc., 1995.

Elbert, T., et al.: Chaos and physiology: deterministic chaos in excitable cell assemblies. Physiol. Rev., 74:1, 1994.

Fuster, J. M.: Memory in the Cerebral Cortex: An Empirical Approach to Neural Networks in the Human and Nonhuman Primate. Cambridge, MA, MIT Press/Bradford Books, 1994.

Goldman-Rakic, P. S.: Topography of cognition: Parallel distributed networks in primate association cortex. Annu. Rev. Neurosci., 11:137, 1988.

Hundert, E. M.: Philosophy, Psychiatry and Neuroscience—Three Approaches to the Mind. New York, Oxford University Press, 1989.

Huppert, F. A., et al.: Dementia and Normal Aging. New York, Cambridge University Press, 1994.

Hyvarinen, J.: Posterior parietal lobe of the primate brain. Physiol. Rev., 62:1060, 1982.

Iaccino, J. F.: Left Brain–Right Brain Differences: Inquiries, Evidence, and New Approaches. Hillsdale, NJ, Lawrence Erlbaum Assoc. Inc., 1993.

Ito, M.: Long-term depression. Annu. Rev. Neurosci., 12:85, 1989.

Kandel, E. R.: Neuronal plasticity and the modification of behavior. In Brookhart, J. M., and Mountcastle, V. B. (eds.): Handbook of Physiology. Sec. 1, Vol. I. Baltimore, Williams & Wilkins, 1977, p. 1137.

Kapur, N.: Memory Disorders in Clinical Practice. Hillsdale, NJ, Lawrence Erlbaum Assoc. Inc., 1994.

Kosslyn, S. M.: Aspects of a cognitive neuroscience of mental imagery. Science, 240:1621, 1988.

Kosslyn, S. M.: Image and Brain: The Resolution of the Imagery Debate. Cambridge, MA, The MIT Press, 1994.

LeDoux, J. E.: Emotion, memory and the brain: the neural routes underlying the formation of memories about primitive emotional experiences, such as fear, have been traced. Sci. Am., June:50, 1994.

Levin, H. S., et al.: Frontal Lobe Function and Dysfunction. New York, Oxford University Press, 1991.

Lieke, E. E., et al.: Optical imaging of cortical activity. Annu. Rev. Physiol., 51:543, 1989.

MacNamara, J., and Reyes, G.: The Logical Foundations of Cognition. Oxford University Press, 1994.

Madden, J. IV: Neurobiology of Learning, Emotion, and Affect. New York, Raven Press, 1991.

McNeil, M. R. (ed.): The Dysarthrias, Physiology, Acoustics, and Perception Management. San Diego, College-Hill Press, 1984.

McGaugh, J. L., et al.: Brain and Memory: Modulation and Mediation of Neuroplasticity. New York, Oxford University Press, 1994.

McGaugh, J. L., et al.: Plasticity in the Central Nervous System: Learning and Memory. Hillsdale, NJ, Lawrence Erlbaum Assoc. Inc., 1995.

Naatanen, R.: Attention and Brain Function. Hillsdale, NJ, Lawrence Erlbaum Assoc. Inc., 1992.

Nappi, G., et al.: Stress and the Aging Brain: Integrative Mechanisms. New York, Raven Press, 1990.

Parks, R. W., et al.: Neuropsychology of Alzheimer's Disease and Other Dementias. New York, Oxford University Press, 1993.

Passingham, R. E.: The Frontal Lobes and Voluntary Action. New York, Oxford University Press, 1993.

Peters, A., and Jones, E. G. (eds.): Development and Maturation of Cerebral Cortex. New York, Plenum Publishing Corp., 1988.

Plum, F. (ed.): Language, Communication, and the Brain. New York, Raven Press, 1988.

Restak, R.: The Modular Brain: How New Discoveries in Neuroscience are Answering Age-old Questions About Memory, Free Will, Consciousness, and Personal Identity. New York, Lisa Drew Book/Charles Scribner's Sons, 1994.

Schulkin, J.: Hormonally Induced Changes in Mind and Brain. San Diego, CA, Academic Press, 1993.

Spear, N. E., et al.: Neurobehavioral Plasticity: Learning, Development and Response to Brain Insults. Hillsdale, NJ, Lawrence Erlbaum Assoc. Inc., 1995.

Squire, L. R.: Mechanisms of memory. Science, 232:1612, 1986.

Squire, L. R.: Memory and Brain. New York, Oxford University Press, 1987.

Stein, B. E., and Meredity, M. A.: The Merging of the Senses. Cambridge, MA, The MIT Press, 1993.

Terry, R. D., et al.: Alzheimer Disease. New York, Raven Press, 1994.

Trevarthen, C.: Hemispheric specialization. In Darian-Smith, I. (ed.): Handbook of Physiology. Sec. 1, Vol. III. Bethesda, Md., American Physiological Society, 1984, p. 1129.

Waldrop, M. M.: Soar: A unified theory of cognition? Science, 241:296, 1988.

Walters, E. T., and Byrne, J. H.: Associative conditioning of single sensory neurons suggest a cellular mechanism for learning. Science, 219:405, 1983.

Winokur, G., and Clayton, P.: The Medical Basis of Psychiatry. Philadelphia, W. B. Saunders Co., 1994.

Woody, C. D., et al.: (eds.): Cellular Mechanisms of Conditioning and Behavioral Plasticity. New York, Plenum Publishing Corp., 1988.

Zucker, R. S.: Short-term synaptic plasticity. Annu. Rev. Neurosci., 12:13, 1989.

Behavioral and Motivational Mechanisms of the Brain— The Limbic System and the Hypothalamus

CHAPTER 58

The control of behavior is a function of the entire nervous system. Even the discrete cord reflexes are an element of behavior, and the wakefulness and sleep cycle discussed in Chapter 59 is certainly one of the most important of our behavioral patterns. In this chapter, we deal first with those mechanisms that control the levels of activity in the different parts of the brain. Then we discuss the bases of motivational drives, especially the motivational control of the learning process and the feelings of pleasure and punishment. These functions of the nervous system are performed mainly by the basal regions of the brain, which together are loosely called the *limbic,* meaning "border," *system.*

ACTIVATING-DRIVING SYSTEMS OF THE BRAIN

In the absence of continuous transmission of nerve signals from the brain stem into the cerebrum, the brain becomes useless. In fact, severe compression of the brain stem at the juncture between the mesencephalon and cerebrum, often resulting from a pineal tumor, usually causes the person to go into unremitting coma lasting for the remainder of the person's life.

Nerve signals in the brain stem activate the cerebral part of the brain in two ways: (1) by directly stimulating the background level of activity in wide areas of the brain and (2) by activating neurohormonal systems that release specific facilitory or inhibitory hormone-like neurotransmitter substances into selected areas of the brain. These activating systems always function together and cannot be distinguished entirely from each other; nevertheless, let us first discuss each as a separate entity.

Control of Cerebral Activity by Continuous Excitatory Signals from the Brain Stem

Reticular Excitatory Area of the Brain Stem

Figure 58–1 shows a general system for controlling the level of activity of the brain. The central driving component of this system is an excitatory area located in the reticular substance of the pons and mesencephalon. This area is also known by the name *bulboreticular facilitory area.* We also discuss this area in Chapter 55 because it is the same brain stem reticular area that transmits facilitory signals downward to the spinal cord to maintain tone in the antigravity muscles and to control the level of activity of the spinal cord reflexes. In addition to these downward signals, this area sends a profusion of signals in the upward direction. Most of these synapse in the thalamus and are distributed from there to all regions of the cerebral cortex, although many others go to the other subcortical structures besides the thalamus.

The signals passing through the thalamus are of two types. One type is rapidly transmitted action potentials that excite the cerebrum for only a few milliseconds. These originate from large neuronal cell bodies that lie throughout the brain stem reticular area. Their nerve endings release the neurotransmitter substance *acetyl-*

749

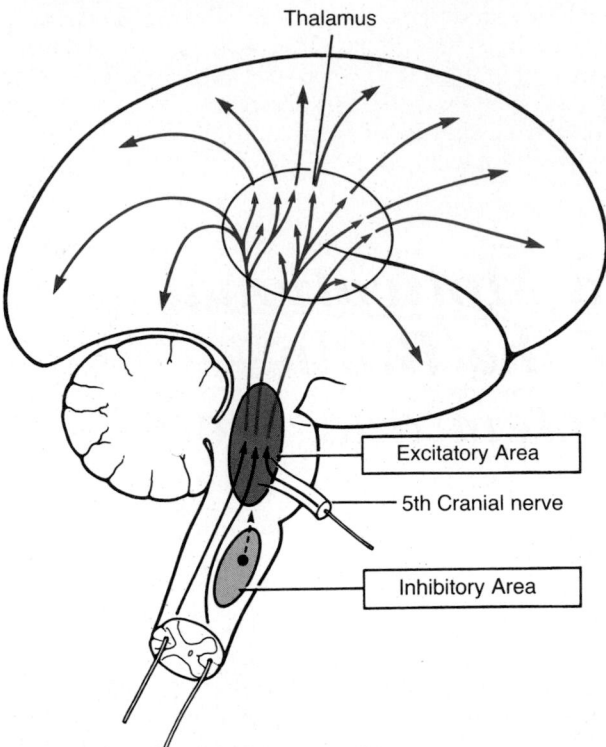

Thalamus

Excitatory Area

5th Cranial nerve

Inhibitory Area

Figure 58–1. Excitatory-activating system of the brain. Also shown is an inhibitory area in the medulla that can inhibit or depress the activating system.

choline, which serves as the excitatory agent, lasting for only a few milliseconds before it is destroyed.

The second type of excitatory signal originates from large numbers of small neurons spread throughout the brain stem reticular excitatory area. Again, most of these pass to the thalamus, but this time through small, slowly conducting fibers and synapsing mainly in the intralaminar nuclei of the thalamus and in the reticular nuclei over the surface of the thalamus. From here, additional small fibers are distributed everywhere in the cerebral cortex. The excitatory effect caused by this system of fibers can build up progressively for many seconds to a minute or more, which suggests that its signals are especially important for controlling the longer-term background excitability level of the brain.

EXCITATION OF THE BRAIN STEM EXCITATORY AREA BY PERIPHERAL SENSORY SIGNALS. The level of activity of the brain stem excitatory area, and therefore the level of activity of the entire brain, is determined to a great extent by sensory signals that enter the excitatory area from the periphery. Pain signals in particular increase the activity in this area and therefore strongly excite the brain to attention.

The importance of sensory signals in activating the excitatory area is demonstrated by the effect of cutting the brain stem above the point where the bilateral fifth nerves enter the pons. They are the highest nerves that transmit significant numbers of somatosensory sig-

nals into the brain. When all these signals are gone, the level of activity in the excitatory area diminishes abruptly and the brain proceeds instantly to greatly reduced activity, actually approaching a permanent state of coma. When the brain stem is transected below the fifth nerves, which leaves much input of the sensory signals from the facial and oral regions, the coma is averted.

INCREASED ACTIVITY OF THE BRAIN STEM EXCITATORY AREA CAUSED BY SIGNALS FROM THE CEREBRUM. Not only do excitatory signals pass to the cerebrum from the bulboreticular excitatory area of the brain stem, but signals also return from the cerebrum back to the bulbar regions. Therefore, any time the cerebral cortex becomes activated by either thought processes or motor processes, signals are sent back to the brain stem excitatory areas, which in turn send still more excitatory signals to the cerebral cortex. This helps to maintain the level of excitation of the cerebral cortex or even enhance it. This is a general mechanism of *positive feedback* that allows any beginning activity in the cerebrum to support still more activity, thus leading to an awake mind.

THE THALAMUS IS A DISTRIBUTION CENTER THAT CONTROLS ACTIVITY IN SPECIFIC REGIONS OF THE CORTEX. As pointed out in Chapter 57 and shown in Figure 57–2, almost every area of the cerebral cortex connects with its own highly specific area in the thalamus. Therefore, electrical stimulation of a specific point in the thalamus activates a specific small region of the cortex. Furthermore, signals regularly reverberate back and forth between the thalamus and the cerebral cortex, the thalamus exciting the cortex and the cortex then re-exciting the thalamus by way of return fibers. It has been suggested that the part of the thinking process that helps us to establish long-term memories might result from just such back-and-forth reverberation of signals.

Can the thalamus also function to call forth specific memories or to activate specific thought processes? The answer to this is not known. Certainly the thalamus does have the appropriate neuronal circuitry to do this.

A Reticular Inhibitory Area Located in the Lower Brain Stem

Figure 58–1 shows still another area that is important in controlling brain activity. This is the *reticular inhibitory area,* located medially and ventrally in the medulla. In Chapter 55, we see that this area can inhibit the reticular facility area of the upper brain stem and thereby reduce the tonic nerve signals transmitted by way of the spinal cord to the antigravity muscles. Likewise, this same inhibitory area, when excited, decreases activity in the superior portions of the brain as well. One of the mechanisms that it uses is to excite serotonergic neurons; these in turn secrete the inhibitory neurohormone serotonin at crucial points in the brain; we discuss this in more detail later.

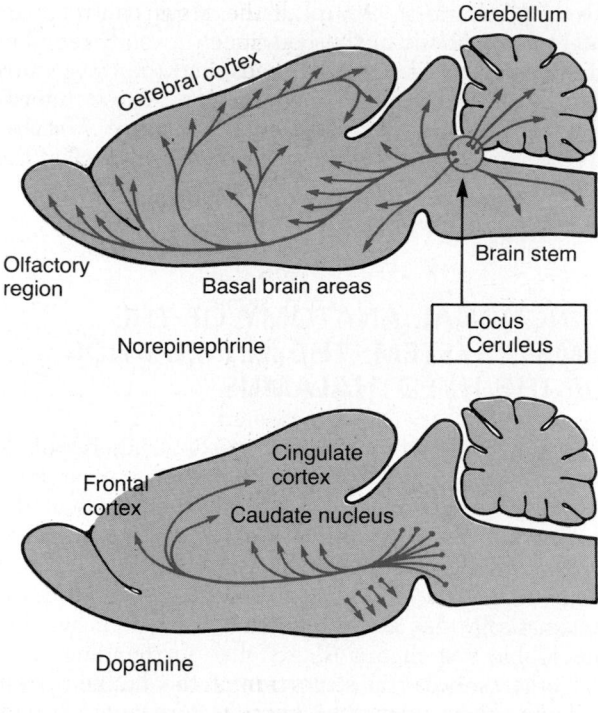

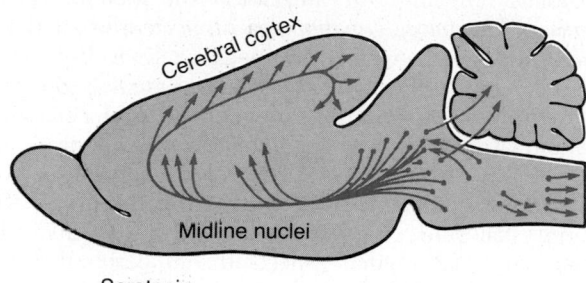

Figure 58–2. Three neurohormonal systems that have been mapped in the rat brain: a *norepinephrine system,* a *dopamine system,* and a *serotonin system.* (Adapted from Kelly, after Cooper, Bloom, and Roth, in Kandel and Schwartz: Principles of Neural Science, 2nd Ed. New York, Elsevier, 1985.)

Neurohormonal Control of Brain Activity

Aside from direct control of brain activity by specific transmission of nerve signals from the lower brain areas to the cortical regions of the brain, still another method is used to control brain activity. This is the release of excitatory or inhibitory neurotransmitter hormonal agents into the substance of the brain. These neurohormones often persist for minutes or even hours and thereby provide long periods of control, rather than instantaneous activation or inhibition.

Figure 58–2 shows three neurohormonal systems that have been mapped in detail in the rat brain: (1) a *norepinephrine system,* (2) a *dopamine system,* and (3) a *serotonin system.* Norepinephrine usually functions as an excitatory hormone, whereas serotonin usually is

inhibitory and dopamine is excitatory in some areas but inhibitory in others. Therefore, as would be expected, these three systems have different effects on the levels of excitability in different parts of the brain. The norepinephrine system spreads to virtually every area of the brain, whereas the serotonin and dopamine systems are directed to much more specific brain regions, the dopamine system mainly into the basal ganglial regions and the serotonin system more into the midline structures.

NEUROHORMONAL SYSTEMS IN THE HUMAN BRAIN. Figure 58–3 shows the brain stem areas in the human brain for activating four neurohormonal systems, the same three discussed previously for the rat and one other, the *acetylcholine system.* Some of the specific functions of these are as follows:

1. *The locus ceruleus and the norepinephrine system.* The *locus ceruleus* is a small area located bilaterally and posteriorly at the juncture between the pons and the mesencephalon. Nerve fibers from this area spread throughout the brain, the same as shown for the rat in the top frame of Figure 58–2, and they secrete *norepinephrine.* The norepinephrine generally excites the brain to increased activity. However, it has inhibitory effects in a few brain areas because of inhibitory receptors at certain neuronal synapses. In Chapter 59, we see that this system probably plays an important role in causing dreaming, thus leading to a type of sleep called rapid eye movement sleep.

2. *The substantia nigra and the dopamine system.* The *substantia nigra* is discussed in Chapter 56 in relation to the basal ganglia. It lies anteriorly in the superior mesencephalon, and its neurons send nerve endings mainly to the caudate nucleus and putamen, where they secrete *dopamine.* Other neurons located in adjacent regions also secrete dopa-

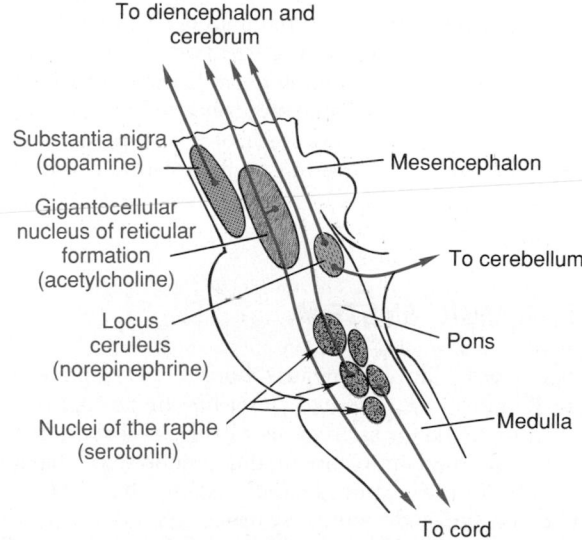

Figure 58–3. Multiple centers in the brain stem, the neurons of which secrete different transmitter substances. These neurons send control signals upward into the diencephalon and cerebrum and downward into the spinal cord.

mine, but they send their endings into more ventral areas of the brain, especially to the hypothalamus and the limbic system. The dopamine is believed to act as an inhibitory transmitter in the basal ganglia, but in some of the other areas of the brain, it is possibly excitatory. Also, remember from Chapter 56 that destruction of the dopaminergic neurons in the substantia nigra is the basic cause of Parkinson's disease.

3. *The raphe nuclei and the serotonin system.* In the midline of the pons and medulla are several thin nuclei called the *raphe nuclei.* Many of the neurons in these nuclei secrete *serotonin.* They send many fibers into the diencephalon and a few fibers to the cerebral cortex; still many others descend to the spinal cord. The cord fibers have the ability to suppress pain, which is discussed in Chapter 48. The serotonin released in the diencephalon and cerebrum almost certainly plays an essential inhibitory role to help cause normal sleep, as we discuss in Chapter 59.

4. *The gigantocellular neurons of the reticular excitatory area and the acetylcholine system.* Earlier we discussed the gigantocellular neurons (the *giant cells*) in the reticular excitatory area of the pons and mesencephalon. The fibers from these large cells divide immediately into two branches, one passing upward to the higher levels of the brain and the other passing downward through the reticulospinal tracts into the spinal cord. The neurohormone secreted at their terminals is *acetylcholine.* In most places, the acetylcholine functions as an excitatory neurotransmitter at specific synapses. Activation of this set of neurons leads to an acutely awake and excited mind.

Still other acetylcholine-secreting neurons are present in some regions of the diencephalon; some psychiatric disorders of the brain have been associated with decreased function or even destruction of some of these neurons.

OTHER NEUROTRANSMITTERS AND NEUROHORMONAL SUBSTANCES SECRETED IN THE BRAIN. Without describing their function, the following is a list of still other neurohormonal substances that function either at synapses or by release into the fluids of the brain: enkephalins, gamma-aminobutyric acid, glutamate, vasopressin, adrenocorticotropic hormone, epinephrine, endorphins, angiotensin II, neurotensin.

Thus, there are multiple neurohormonal systems in the brain, the activation of each of which plays its own role in controlling a different quality of brain function.

THE LIMBIC SYSTEM

The word "limbic" means "border." Originally, the term "limbic" was used to describe the border structures around the basal regions of the cerebrum, but as we have learned more about the functions of the limbic system, the term *"limbic system"* has been expanded to mean the entire neuronal circuitry that controls emotional behavior and motivational drives.

A major part of the limbic system is the *hypothalamus,* with its related structures. In addition to their roles in behavioral control, these areas control many internal conditions of the body, such as body temperature, osmolality of the body fluids, and the drive to eat and drink and control body weight. These internal functions are collectively called *vegetative functions* of the brain, and their control is closely related to behavior.

FUNCTIONAL ANATOMY OF THE LIMBIC SYSTEM; THE KEY POSITION OF THE HYPOTHALAMUS

Figure 58–4 shows the anatomical structures of the limbic system, demonstrating that they are an interconnected complex of basal brain elements. Located in the middle of all these is the *hypothalamus* which, from a physiological point of view, is one of the central elements of the limbic system. Figure 58–5 shows schematically this key position of the hypothalamus in the limbic system and shows that surrounding it are the other subcortical structures of the limbic system, including the *septum,* the *paraolfactory area,* the *epithalamus,* the *anterior nucleus of the thalamus,* portions of the basal ganglia, the *hippocampus,* and the *amygdala.*

Surrounding the subcortical limbic areas is the *limbic cortex,* composed of a ring of cerebral cortex (1) beginning in the *orbitofrontal area* on the ventral surface of the frontal lobes, (2) extending upward in the *subcallosal gyrus* beneath the anterior limb of the corpus callosum, (3) over the top of the corpus callosum onto the medial aspect of the cerebral hemisphere in the *cingulate gyrus,* and finally (4) passing behind the corpus callosum and downward onto the ventromedial surface of the temporal lobe to the *parahippocampal gyrus* and *uncus.* Thus, on the medial and ventral surfaces of each cerebral hemisphere is a ring mostly of *paleocortex* that surrounds a group of deep structures intimately associated with overall behavior and with emotions. In turn, this ring of limbic cortex functions as a two-way communication and association linkage between the *neocortex* and the lower limbic structures.

Many of the behavioral functions elicited from the hypothalamus and other limbic structures are mediated through the reticular nuclei in the brain stem and their associated nuclei. It is pointed out in Chapter 55 and earlier in this chapter that stimulation of the excitatory portion of this reticular formation can cause high degrees of cerebral excitability as well as increase the excitability of much of the spinal cord synapses. And in Chapter 60, we see that most of the hypothalamic signals for control of the autonomic nervous system are also transmitted through nuclei located in the brain stem.

An important route of communication between the limbic system and the brain stem is the *medial fore-*

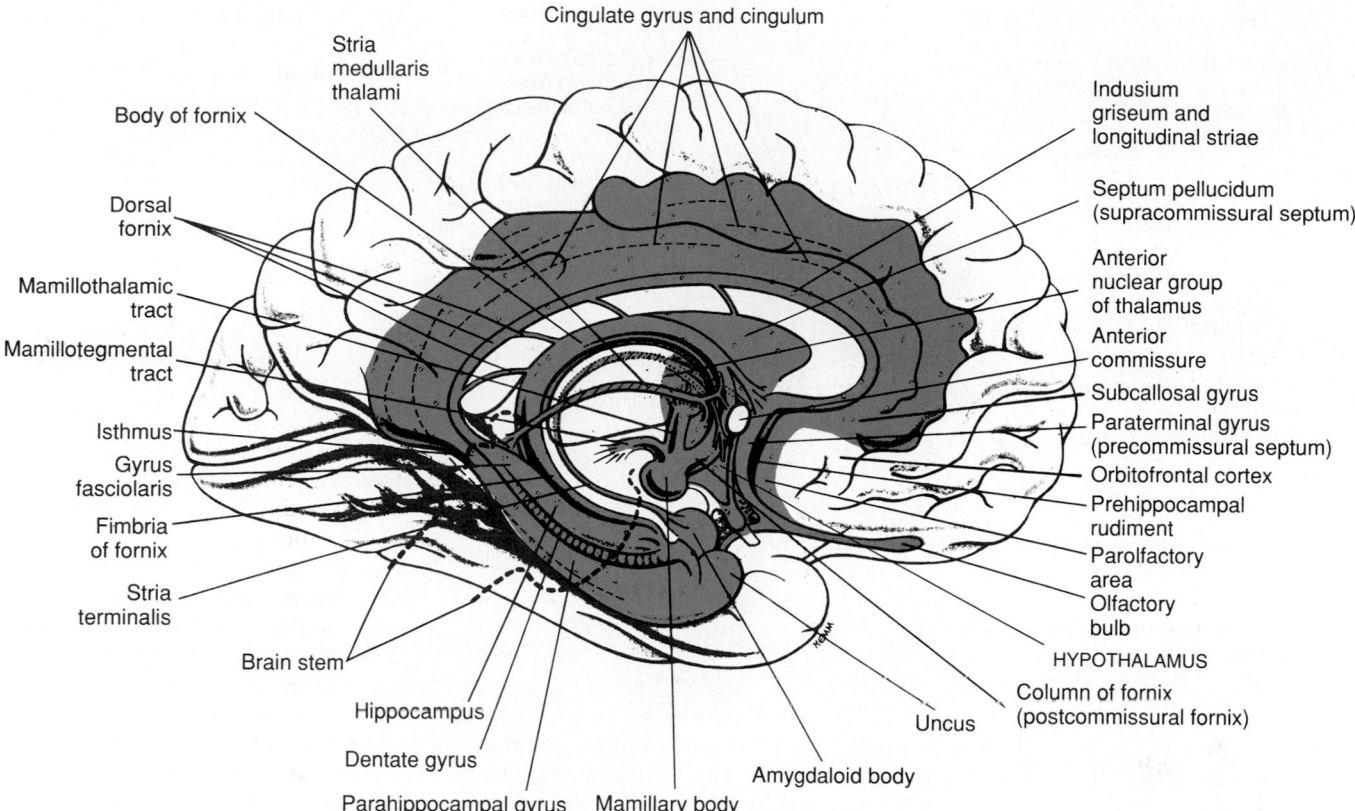

Figure 58–4. Anatomy of the limbic system shown by the colored areas of the figure. (From Warwick and Williams: Gray's Anatomy. 35th Br. Ed. London, Longman Group, Ltd, 1973.)

brain bundle, which extends from the septal and orbitofrontal cortical regions downward through the middle of the hypothalamus to the brain stem reticular formation. This bundle carries fibers in both directions, forming a trunk line communication system. A second route of communication is through short pathways among the reticular formation of the brain stem, thalamus, hypothalamus, and most of the other contiguous areas of the basal brain.

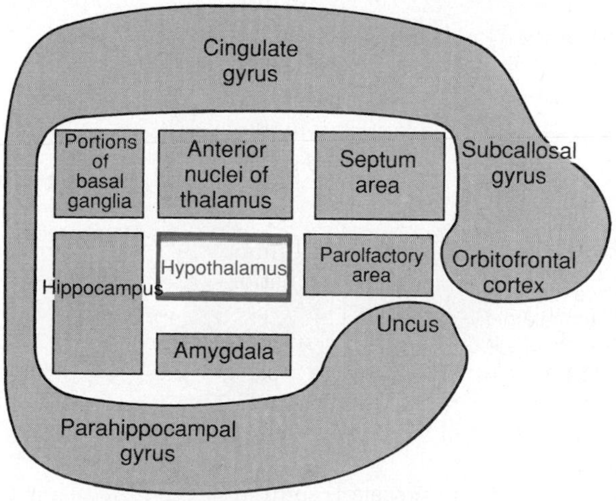

THE HYPOTHALAMUS, A MAJOR CONTROL HEADQUARTERS FOR THE LIMBIC SYSTEM

The hypothalamus has two-way communicating pathways with all levels of the limbic system. In turn, it and its closely allied structures send output signals in three directions: (1) downward to the brain stem, mainly into the reticular areas of the mesencephalon, pons, and medulla and then into the autonomic nervous system; (2) upward toward many higher areas of the diencephalon and cerebrum, especially to the anterior thalamus and limbic cortex; and (3) into the infundibulum to control most of the secretory functions of both the posterior and the anterior pituitary glands.

The hypothalamus, which represents less than 1 per cent of the brain mass, is one of the most important of the output control pathways of the limbic system. It controls most of the vegetative and endocrine functions of the body as well as many aspects of emotional behavior. Let us discuss first the vegetative and endo-

Figure 58–5. Limbic system.

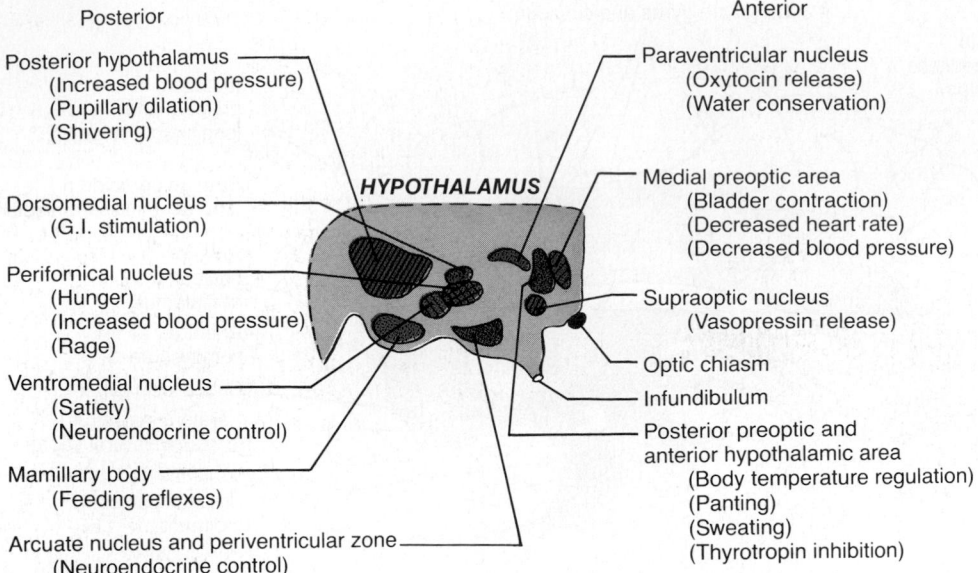

Posterior

Anterior

Posterior hypothalamus
  (Increased blood pressure)
  (Pupillary dilation)
  (Shivering)

Paraventricular nucleus
  (Oxytocin release)
  (Water conservation)

HYPOTHALAMUS

Dorsomedial nucleus
  (G.I. stimulation)

Medial preoptic area
  (Bladder contraction)
  (Decreased heart rate)
  (Decreased blood pressure)

Perifornical nucleus
  (Hunger)
  (Increased blood pressure)
  (Rage)

Supraoptic nucleus
  (Vasopressin release)

Ventromedial nucleus
  (Satiety)
  (Neuroendocrine control)

Optic chiasm

Infundibulum

Mamillary body
  (Feeding reflexes)

Posterior preoptic and
anterior hypothalamic area
  (Body temperature regulation)
  (Panting)
  (Sweating)
  (Thyrotropin inhibition)

Arcuate nucleus and periventricular zone
  (Neuroendocrine control)

Lateral hypothalamic area (not shown)
  (Thirst and hunger)

Figure 58–6. Control centers of the hypothalamus (sagittal view).

crine control functions and then return to the behavioral functions of the hypothalamus to see how all these operate together.

Vegetative and Endocrine Control Functions of the Hypothalamus

The different hypothalamic mechanisms for controlling the vegetative and endocrine functions of the body are discussed in various chapters throughout this text. For instance, the role of the hypothalamus in arterial pressure regulation is discussed in Chapter 18, thirst and water conservation in Chapter 29, temperature regulation in Chapter 73, and endocrine control in Chapter 75. To illustrate the organization of the hypothalamus as a functional unit, let us summarize the more important of its vegetative and endocrine functions here as well.

Figures 58–6 and 58–7 show enlarged sagittal and coronal views of the hypothalamus, which represents only a small area in Figure 58–4. Take a few minutes to study these diagrams, especially to read in Figure 58–6 the multiple activities that are excited or inhibited when respective hypothalamic nuclei are stimulated. In addition to those centers that are shown in Figure 58–6, a large *lateral hypothalamic* area (shown in Figure 58–7) is present on each side of the hypothalamus. The lateral areas are especially important in controlling thirst, hunger, and many of the emotional drives.

A word of caution must be issued for studying these diagrams because the areas that cause specific activities are not nearly so accurately localized as suggested in the figures. Also, it is not known whether the effects noted in the figures result from stimulation of specific control nuclei or whether they result merely from acti-

vation of fiber tracts leading from control nuclei located elsewhere. With this caution in mind, we can give the following general description of the vegetative and control functions of the hypothalamus.

CARDIOVASCULAR REGULATION. Stimulation of different areas throughout the hypothalamus can cause every known type of neurogenic effect on the cardiovascular system, including increased arterial pressure, decreased arterial pressure, increased heart rate, and decreased heart rate. In general, stimulation in the *posterior* and *lateral hypothalamus* increases the arterial pressure and heart rate, whereas stimulation in the *preoptic area* often has opposite effects, causing a decrease in both heart rate and arterial pressure. These effects are transmitted mainly through the cardiovascular control centers in the reticular regions of the medulla and pons.

REGULATION OF BODY TEMPERATURE. The anterior portion of the hypothalamus, especially the *preoptic*

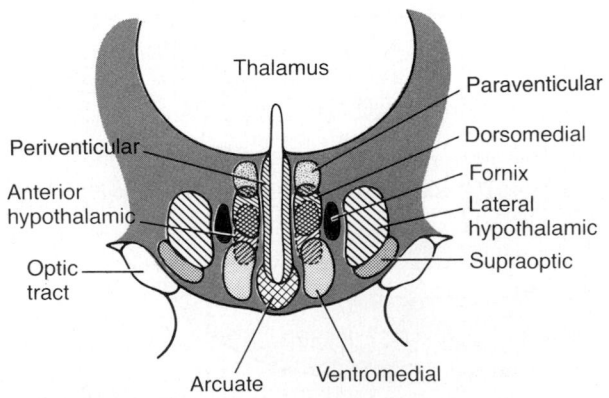

Thalamus

Paraventicular

Dorsomedial

Fornix

Periventicular

Anterior
hypothalamic

Lateral
hypothalamic

Supraoptic

Optic
tract

Arcuate

Ventromedial

Figure 58–7. Coronal view of the hypothalamus, showing the mediolateral positions of the respective hypothalamic nuclei.

area, is concerned with regulation of body temperature. An increase in the temperature of the blood flowing through this area increases the activity of temperature-sensitive neurons, whereas a decrease in temperature decreases their activity. In turn, these neurons control the mechanisms for increasing or decreasing body temperature, as discussed in Chapter 73.

REGULATION OF BODY WATER. The hypothalamus regulates body water in two ways: (1) by creating the sensation of thirst, which makes an animal drink water, and (2) by controlling the excretion of water into the urine. An area called the *thirst center* is located in the lateral hypothalamus. When the electrolytes inside the neurons either of this center or allied areas of the hypothalamus become too concentrated, the animal develops an intense desire to drink water; it will search out the nearest source of water and drink enough to return the electrolyte concentration of the thirst center neurons to normal.

Control of renal excretion of water is vested mainly in the *supraoptic* nucleus. When the body fluids become too concentrated, the neurons of this area become stimulated. The nerve fibers from these neurons project downward through the infundibulum of the hypothalamus into the posterior pituitary gland, where the nerve endings secrete a hormone called *antidiuretic hormone* (also called *vasopressin*). This hormone is then absorbed into the blood and acts on the collecting ducts of the kidneys to cause massive reabsorption of water, thereby decreasing the loss of water into the urine. These functions are presented in Chapter 28.

REGULATION OF UTERINE CONTRACTILITY AND OF MILK EJECTION BY THE BREASTS. Stimulation of the *paraventricular nucleus* causes its neuronal cells to secrete the hormone *oxytocin*. This in turn causes increased contractility of the uterus as well as contraction of the myoepithelial cells that surround the alveoli of the breasts, which then causes the alveoli to empty the milk through the nipples. At the end of pregnancy, especially large quantities of oxytocin are secreted, and this secretion helps to promote labor contractions that expel the baby. Also, when a baby suckles the mother's breast, a reflex signal from the nipple to the hypothalamus causes oxytocin release, and the oxytocin then performs the necessary function of expelling the milk through the nipples so that the baby can nourish itself. These functions are discussed in Chapter 82.

GASTROINTESTINAL AND FEEDING REGULATION. Stimulation of several areas of the hypothalamus causes an animal to experience extreme hunger, a voracious appetite, and an intense desire to search for food. The area most associated with hunger is the *lateral hypothalamic area*. On the other hand, damage to this causes the animal to lose desire for food, sometimes causing lethal starvation.

A center that opposes the desire for food, called the *satiety center*, is located in the *ventromedial nucleus*. When this center is stimulated electrically, an animal that is eating food suddenly stops eating and shows complete indifference to food. On the other hand, if this area is destroyed bilaterally, the animal cannot be satiated; instead, its hypothalamic hunger centers become overactive, so that it has a voracious appetite, resulting in tremendous obesity.

Another area of the hypothalamus that enters into the overall control of gastrointestinal activity is the *mamillary bodies;* these control at least partially the patterns of many feeding reflexes, such as licking the lips and swallowing.

Hypothalamic Control of Endocrine Hormone Secretion by the Anterior Pituitary Gland

Stimulation of certain areas of the hypothalamus also causes the *anterior* pituitary gland to secrete its endocrine hormones. This subject is discussed in detail in Chapter 74 in relation to the neural control of the endocrine glands. Briefly, the basic mechanisms are the following.

The anterior pituitary gland receives its blood supply mainly from blood that flows first through the lower part of the hypothalamus and then into the anterior pituitary vascular sinuses. As the blood courses through the hypothalamus before reaching the anterior pituitary, *releasing* and *inhibitory hormones* are secreted into the blood by various hypothalamic nuclei. These hormones are then transported in the blood to the anterior pituitary, where they act on the glandular cells to control the release of the anterior pituitary hormones.

The cell bodies of the neurons that secrete these releasing and inhibitory hormones are located mainly in the medial basal nuclei of the hypothalamus, especially in the *periventricular zone*, the *arcuate nucleus*, and part of the *ventromedial nucleus*. The axons from these nuclei then project to the *median eminence*, which is an enlarged area of the pituitary stalk (infundibulum) where it arises from the inferior border of the hypothalamus. It is here that the nerve terminals actually secrete their releasing and inhibitory hormones. These hormones are then absorbed into the blood capillaries in the median eminence and carried in the blood down along the stalk to the anterior pituitary vascular sinuses.

SUMMARY. A number of areas of the hypothalamus control specific vegetative functions. These areas are still poorly delimited, so much so that the specification of different areas for different hypothalamic functions given above is still tentative.

Behavioral Functions of the Hypothalamus and Associated Limbic Structures

EFFECTS CAUSED BY STIMULATION. Aside from the vegetative and endocrine functions of the hypothalamus, stimulation of or lesions in the hypothalamus often have profound effects on the emotional behavior of animals and human beings.

In animals, some of the behavioral effects of stimulation are the following:

1. Stimulation in the *lateral hypothalamus* not only causes thirst and eating, as discussed above, but also increases the general level of activity of the animal, sometimes leading to overt rage and fighting, as discussed subsequently.

2. Stimulation in the *ventromedial nucleus* and surrounding areas mainly causes effects opposite to those caused by lateral hypothalamic stimulation—that is, a sense of *satiety, decreased eating,* and *tranquility.*

3. Stimulation of a *thin zone of the periventricular nuclei*, located immediately adjacent to the third ventricle (or also stimulation of the central gray area of the mesencepha-

lon that is continuous with this portion of the hypothalamus), usually leads to *fear* and *punishment reactions.*

4. *Sexual drive* can be stimulated from several areas of the hypothalamus, especially the most anterior and most posterior portions of the hypothalamus.

EFFECTS CAUSED BY HYPOTHALAMIC LESIONS. Lesions in the hypothalamus, in general, cause effects opposite to those caused by stimulation. For instance:

1. Bilateral lesions in the lateral hypothalamus will decrease drinking and eating almost to zero, often leading to lethal starvation. These lesions cause extreme *passivity* of the animal as well, with loss of most of its overt drives.

2. Bilateral lesions of the ventromedial areas of the hypothalamus cause effects that are mainly opposite to those caused by lesions of the lateral hypothalamus: excessive drinking and eating as well as hyperactivity and often continuous savagery along with frequent bouts of extreme rage on the slightest provocation.

Stimulation or lesions in other regions of the limbic system, especially the amygdala, the septal area, and areas in the mesencephalon, often cause effects similar to those elicited from the hypothalamus. We discuss some of these in more detail later.

"Reward" and "Punishment" Function of the Limbic System

From the preceding discussion, it is already clear that several limbic structures are particularly concerned with the *affective* nature of sensory sensations — that is, whether the sensations are *pleasant* or *unpleasant.* These affective qualities are also called *reward* or *punishment,* or *satisfaction* or *aversion.* Electrical stimulation of certain limbic areas pleases or satisfies the animal, whereas electrical stimulation of other regions causes terror, pain, fear, defense, escape reactions, and all the other elements of punishment. The degrees of stimulation of these two oppositely responding systems greatly affect the behavior of the animal.

Reward Centers

Figure 58–8 shows a technique that has been used for localizing specific reward and punishment areas of the brain. In this figure, a lever is placed at the side of the cage and is arranged so that depressing the lever makes electrical contact with a stimulator. Electrodes are placed successively at different areas in the brain so that the animal can stimulate the area by pressing the lever. If stimulating the particular area gives the animal a sense of reward, then it will press the lever again and again, sometimes as much as thousands of times per hour. Furthermore, when offered the choice of eating some delectable food as opposed to the opportunity to stimulate the reward center, the animal often chooses the electrical stimulation.

By using this procedure, the major reward centers have been found to be located *along the course of the medial forebrain bundle,* especially in the *lateral* and

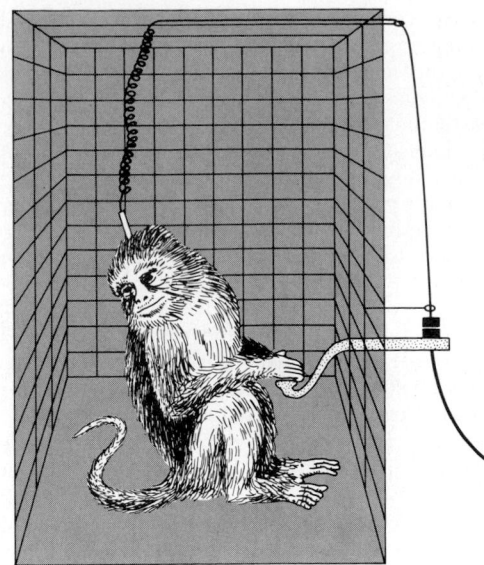

Figure 58–8. Technique for localizing reward and punishment centers in the brain of a monkey.

ventromedial nucleus of the hypothalamus. It is strange that the lateral nucleus should be included among the reward areas — indeed, it is one of the most potent of all — because even stronger stimuli in this area can cause rage. But this is true in many areas, with weaker stimuli giving a sense of reward and stronger ones a sense of punishment.

Less potent reward centers, which are perhaps secondary to the major ones in the hypothalamus, are found in the septum, the amygdala, certain areas of the thalamus and basal ganglia, and extending downward into the basal tegmentum of the mesencephalon.

Punishment Centers

The apparatus shown in Figure 58–8 can also be connected so that the stimulus to the brain continues all the time *except* when the lever is pressed. In this case, the animal will not press the lever to turn the stimulus off when the electrode is in one of the reward areas; but when it is in certain other areas, the animal immediately learns to turn it off. Stimulation in these areas causes the animal to show all the signs of displeasure, fear, terror, pain, and punishment. Furthermore, prolonged stimulation for 24 hours or more often causes the animal to become severely sick and can lead to death.

By means of this technique, the most potent areas for punishment and escape tendencies have been found in the *central gray area surrounding the aqueduct of Sylvius in the mesencephalon* and extending upward into the *periventricular zones of the hypothalamus and thalamus.* Less potent punishment areas are found in some locations in the *amygdala* and the *hippocampus.*

It is particularly interesting that stimulation in the punishment centers can frequently inhibit the reward and pleasure centers completely, demonstrating that *punishment and fear can take precedence over pleasure and reward.*

Rage—Its Association with Punishment Centers

An emotional pattern that involves the punishment centers of the hypothalamus and other limbic structures and has also been well characterized is the *rage pattern*. This can be described as follows.

Strong stimulation of the punishment centers of the brain, especially in the *periventricular zone of the hypothalamus* and in the *lateral hypothalamus*, causes the animal to (1) develop a defense posture, (2) extend its claws, (3) lift its tail, (4) hiss, (5) spit, (6) growl, and (7) develop piloerection, wide-open eyes, and dilated pupils. Furthermore, even the slightest provocation causes an immediate savage attack. This is approximately the behavior that one would expect from an animal being severely punished, and it is a pattern of behavior that is called *rage*.

Stimulation of the more rostral portions of the punishment areas—in the midline preoptic areas—causes mainly fear and anxiety, associated with a tendency for the animal to run away.

In the normal animal, the rage phenomenon is held in check mainly by counterbalancing activity of the ventromedial nuclei of the hypothalamus. In addition, the hippocampus, amygdala, and anterior portions of the limbic cortex, especially of the anterior cingulate gyrus and subcallosal gyrus, help suppress the rage phenomenon. Conversely, if these portions of the limbic system are damaged or destroyed, the animal (also the human being) becomes far more susceptible to bouts of rage.

PLACIDITY AND TAMENESS. Exactly the opposite emotional behavior patterns occur when the reward centers are stimulated: placidity and tameness.

Importance of Reward and Punishment in Behavior

Almost everything that we do is related in some way to reward and punishment. If we are doing something that is rewarding, we continue to do it; if it is punishing, we cease to do it. Therefore, the reward and punishment centers undoubtedly constitute one of the most important of all the controllers of our bodily activities, our drives, our aversions, our motivations.

EFFECT OF TRANQUILIZERS ON THE REWARD AND PUNISHMENT CENTERS. Administration of a tranquilizer, such as chlorpromazine, usually inhibits both the reward and the punishment centers, thereby greatly decreasing the affective reactivity of the animal. Therefore, it is presumed that tranquilizers function in psychotic states by suppressing many of the important behavioral areas of the hypothalamus and its associated regions of the limbic brain, a subject that we discuss more fully later.

Importance of Reward and Punishment in Learning and Memory—Habituation Versus Reinforcement

Animal experiments have shown that a sensory experience causing neither reward nor punishment is remembered hardly at all. Electrical recordings show that new types of sensory stimuli always excite wide areas in the cerebral cortex. But repetition of the stimulus over and over leads to almost complete extinction of the cortical response if the sensory experience does not elicit a sense of either reward or punishment. The animal becomes *habituated* to the sensory stimulus and thereafter ignores it.

If the stimulus causes either reward or punishment rather than indifference, the cortical response becomes progressively more and more intense during repeated stimulation instead of fading away, and the response is said to be *reinforced*. An animal builds up strong memory traces for sensations that are either rewarding or punishing but, on the other hand, develops complete habituation to indifferent sensory stimuli.

It is evident that the reward and punishment centers of the limbic system have much to do with selecting the information that we learn, usually throwing away more than 99 per cent of it and selecting less than 1 per cent for retention.

SPECIFIC FUNCTIONS OF OTHER PARTS OF THE LIMBIC SYSTEM

Functions of the Hippocampus

The hippocampus is the elongated, medial portion of the temporal cortex that folds upward and inward to form the ventral surface of the inferior horn of the lateral ventricle. One end of the hippocampus abuts the amygdaloid nuclei, and it also fuses along one of its borders with the parahippocampal gyrus, which is the cortex of the ventromedial surface of the temporal lobe.

The hippocampus (and its adjacent temporal lobe structures, all together called the *hippocampal formation*) has numerous but mainly indirect connections with many portions of the cerebral cortex as well as with the basic structures of the limbic system—the amygdala, the hypothalamus, the septum, and the mamillary bodies. Almost any type of sensory experience causes activation of at least some part of the hippocampus, and the hippocampus in turn distributes many outgoing signals to the anterior thalamus, hypothalamus, and other parts of the limbic system, especially through the *fornix*, its major output pathway. Thus, the hippocampus is an additional channel through which incoming sensory signals can lead to appropriate behavioral reactions but for different purposes, as can be seen later.

As in other limbic structures, stimulation of different areas in the hippocampus can cause almost any one of different behavioral patterns, such as rage, passivity, and excess sex drive.

Another feature of the hippocampus is that it is hyperexcitable; for instance, weak electrical stimuli can cause local epileptic seizures in the hippocampal areas themselves that persist for many seconds after the stimulation is over, suggesting that the hippocampus can perhaps give off prolonged output signals even under normal functioning conditions. During hippocampal seizures in the human being, the person experiences various psychomotor effects, including olfactory, visual, auditory, tactile, and other types of hallucinations that cannot be suppressed even though the person has not lost consciousness and knows these hallucinations to be unreal. Probably one of the reasons for this hyperexcitability of the hippocampus is that it has a different type of cortex from that elsewhere in the cerebrum, having only three nerve cell layers in some of its areas instead of the six layers found elsewhere.

Role of the Hippocampus in Learning

EFFECT OF BILATERAL REMOVAL OF THE HIPPO-CAMPI—INABILITY TO LEARN. The hippocampi have been surgically removed bilaterally in a few human beings for the treatment of epilepsy. These people can recall most previously learned memories satisfactorily. However, they can learn essentially no new information that is based on verbal symbolism. In fact, they cannot even learn the names of people with whom they come in contact every day. Yet they can remember for a moment or so what transpires during the course of their activities. Thus, they are capable of short-term memory for seconds up to a minute or two, although their ability to establish long-term memories lasting longer than a few minutes is either completely or almost completely abolished. This is the phenomenon called *anterograde amnesia*, discussed in Chapter 57.

Destruction of the hippocampi also causes some deficit in previously learned memories (retrograde amnesia), a little more so for memories of the past year or so than for memories of the distant past.

THEORETICAL FUNCTION OF THE HIPPOCAMPUS IN LEARNING. The hippocampus originated as part of the olfactory cortex. In the very lowest animals, it plays essential roles in determining whether the animal will eat a particular food, whether the smell of a particular object suggests danger, and whether the odor is sexually inviting, thus making decisions that are of life-or-death importance. Very early in the evolutionary development of the brain, the hippocampus presumably became a critical decision-making neuronal mechanism, determining the importance and type of importance of the incoming sensory signals. Once this critical decision-making capability had been established, presumably the remainder of the brain began to call on it for the same decision making. If the hippocampus says that a neuronal signal is important, it is likely to be committed to memory.

Earlier in this chapter (as well as in Chapter 57), it is pointed out that reward and punishment play a major role in determining the importance of information and especially whether or not the information will be stored in memory. A person rapidly becomes habituated to indifferent stimuli but learns assiduously any sensory experience that causes either pleasure or punishment. What is the mechanism by which this occurs? It has been suggested that the hippocampus provides the drive that causes translation of short-term memory into long-term memory—that is, the hippocampus transmits some type of signal or signals that seem to make the mind rehearse over and over the new information until permanent storage takes place.

Whatever the mechanism, without the hippocampi, *consolidation* of long-term memories of the verbal or symbolic type does not take place.

Functions of the Amygdala

The amygdala is a complex of nuclei located immediately beneath the cortex of the medial anterior pole of each temporal lobe. It has abundant bidirectional connections with the hypothalamus as well as with other adjacent areas of the limbic system.

In lower animals, the amygdala is also concerned to a great extent with olfactory stimuli and their interrelations with the limbic brain. Indeed, it is pointed out in Chapter 53 that one of the major divisions of the olfactory tract terminates in a portion of the amygdala called the *corticomedial nuclei*, which lies immediately beneath the cortex in the olfactory pyriform area of the temporal lobe. In the human being, another portion of the amygdala, the *basolateral nuclei*, has become much more highly developed than this olfactory portion and plays exceedingly important roles in many behavioral activities not generally associated with olfactory stimuli.

The amygdala receives neuronal signals from all portions of the limbic cortex as well as from the neocortex of the temporal, parietal, and occipital lobes, especially from the auditory and visual association areas. Because of these multiple connections, the amygdala has been called the "window" through which the limbic system sees the place of the person in the world. In turn, the amygdala transmits signals (1) back into these same cortical areas, (2) into the hippocampus, (3) into the septum, (4) into the thalamus, and (5) especially into the hypothalamus.

EFFECTS OF STIMULATING THE AMYGDALA. In general, stimulation in the amygdala can cause almost all the same effects as those elicited by stimulation of the hypothalamus plus other effects. The effects that are mediated through the hypothalamus include (1) increases or decreases in arterial pressure, (2) increases or decreases in heart rate, (3) increases or decreases in gastrointestinal motility and secretion, (4) defecation and micturition, (5) pupillary dilatation or, rarely, constriction, (6) piloerection, and (7) secretion of the various anterior pituitary hormones, especially the gonadotropins and adrenocorticotropic hormone.

Aside from these effects mediated through the hypothalamus, amygdala stimulation can cause different types of involuntary movement. These include (1) tonic movements, such as raising the head or bending the body; (2) circling movements; (3) occasionally clonic, rhythmical movements; and (4) different types of movements associated with olfaction and eating, such as licking, chewing, and swallowing.

In addition, stimulation of certain amygdaloid nuclei can, rarely, cause a pattern of rage, escape, punishment, and fear similar to the rage pattern elicited from the hypothalamus as described earlier. And stimulation of other nuclei can give reactions of reward and pleasure.

Finally, excitation of still other portions of the amygdala can cause sexual activities that include erection, copulatory movements, ejaculation, ovulation, uterine activity, and premature labor.

EFFECTS OF BILATERAL ABLATION OF THE AMYGDALA—THE KLÜVER-BUCY SYNDROME. When the anterior portions of both temporal lobes are destroyed in a monkey, this removes not only the temporal cortex but also the amygdalas that lie inside these parts of the temporal lobes. This causes a combination of changes in behavior called the Klüver-Bucy syndrome, which includes (1) excessive tendency to examine objects orally; (2) loss of fear; (3) decreased aggressiveness; (4) tameness; (5) changes in dietary habits, even to the extent that a herbivorous animal frequently becomes carnivorous; (6) sometimes psychic blindness; and (7) often excessive sex drive. The characteristic picture is of an animal that is not afraid of anything, has extreme curiosity about everything, forgets rapidly, has a tendency to place everything in its mouth and sometimes even tries to eat solid objects, and, finally, often has a sex drive so strong that it attempts to copulate with immature animals, animals of the wrong sex, and animals of a different species.

Although similar lesions in human beings are rare, afflicted people respond in a manner not too different from that of the monkey.

OVERALL FUNCTION OF THE AMYGDALA. The amygdala seems to be a behavioral awareness area that operates at a semiconscious level. It also seems to project into the limbic system one's current status in relation to both surroundings and thoughts. On the basis of this information, the amygdala is believed to help pattern the person's behavioral response so that it is appropriate for each occasion.

Function of the Limbic Cortex

The most poorly understood portion of the limbic system is the ring of cerebral cortex called the *limbic cortex* that surrounds the subcortical limbic structures. This cortex functions as a transitional zone through which signals are transmitted from the remainder of the cortex into the limbic system and also in the opposite direction. Therefore, the limbic cortex in effect functions as a cerebral *association area for control of behavior*.

Stimulation of the different regions of the limbic cortex has failed to give any real idea of their functions. However, as is true of so many other portions of the limbic system, essentially all the behavioral patterns that have already been described can also be elicited by stimulation in different portions of the limbic cortex. Likewise, ablation of a few limbic cortical areas can cause persistent changes in an animal's behavior, as follows.

Ablation of the Anterior Temporal Cortex. When the anterior temporal cortex is ablated bilaterally, the amygdala is almost invariably damaged as well. This was discussed earlier, and it was pointed out that the Klüver-Bucy syndrome occurs. The animal especially develops consummatory behavior, investigates any and all objects, has intense sex drives toward inappropriate animals or even inanimate objects, and loses all fear—thus develops tameness as well.

Ablation of the Posterior Orbital Frontal Cortex. Bilateral removal of the posterior portion of the orbital frontal cortex often causes an animal to develop insomnia and an intense degree of motor restlessness, becoming unable to sit still but moving about continually.

Ablation of the Anterior Cingulate Gyri and Subcallosal Gyri. The anterior cingulate gyri and the subcallosal gyri are the portions of the limbic cortex that communicate between the prefrontal cerebral cortex and the subcortical limbic structures. Destruction of these gyri bilaterally releases the rage centers of the septum and hypothalamus from any prefrontal inhibitory influence. Therefore, the animal can become vicious and much more subject to fits of rage than normally.

SUMMARY. Until further information is available, it is perhaps best to state that the cortical regions of the limbic system occupy intermediate associative positions between the functions of the remainder of the cerebral cortex and the functions of the subcortical limbic structures for control of behavioral patterns. Thus, in the anterior temporal cortex, one especially finds gustatory and olfactory associations. In the parahippocampal gyri, there is a tendency for complex auditory associations as well as complex thought associations derived from Wernicke's area of the posterior temporal lobe. In the middle and posterior cingulate cortex, there is reason to believe that sensorimotor associations occur.

REFERENCES

Andrews, G., et al: Treatment of Anxiety Disorders: Clinician's Guide and Treatment Manuals. New York, Cambridge University Press, 1994.

Aoki, C., and Siekevitz, P.: Plasticity in brain development. Sci. Am., December, 1988, p. 56.

Avoli, M., et al. (eds.): Neurotransmitters and Cortical Function. New York, Plenum Publishing Corp., 1988.

Barnes, D. M.: Neural models yield data on learning. Science 236:1628, 1987.

Benton, A. L., et al.: Contributions to Neuropsychological Assessment: A Clinical Manual. New York, Oxford University Press, 1994.

Bloom, F. E., and Kupfer, D. J.: Psychopharmacology: The Fourth Generation of Progress. New York, Raven Press, 1994.

Borbely, A. A., and Tobler, I.: Endogenous sleep-promoting substances and sleep regulation. Physiol. Rev., 69:605, 1989.

Byrne, J. H.: Cellular analysis of associative learning. Physiol. Rev., 67:329, 1987.

Carr, C.: Effectors, Behavior and Evolution. Farmington, CT, S. Karger Publishers, Inc., 1992.

Chiu, E., and Ames, D.: Functional Psychiatric Disorders of the Elderly. New York, Cambridge University Press, 1994.

Cohen, N. J.: Memory, Amnesia, and the Hippocampal System. Cambridge, MA, The MIT Press, 1993.

Conn, P. M.: Neuroscience in medicine. Philadelphia, J. B. Lippincott, 1994.

Davis, K., et al.: Foundations of Psychiatry. Philadelphia, W. B. Saunders Co., 1991.

Dietz, V.: Human neuronal control of automatic functional movements: interaction between central programs and afferent input. Physiol. Rev., 72:33, 1992.

Doane, B. K., and Livingston, K. E. (eds.): The Limbic System. New York, Raven Press, 1986.

Engel, J., et al. (eds.): Brain Reward Systems and Abuse. New York, Raven Press, 1987.

Gazzaniga, M. S.: The Cognitive Neurosciences. Cambridge, MA, The MIT Press, 1994.

Givens, J. R.: The Hypothalamus in Health and Disease. Chicago, Year Book Medical Publishers, 1984.

Jacobs, B. L., and Azmitia, E. C.: Structure and function of the brain serotonin system. Physiol. Rev., 72:165, 1992.

Jones, E. G., and Peters, A. (eds.): Further Aspects of Cortical Function, Including Hippocampus. New York, Plenum Publishing Corp., 1987.

Kandel, E. R.: Molecular Neurobiology in Neurology and Psychiatry. New York, Raven Press, 1987.

Kaplan, H. I., and Sadock, B. J.: Comprehensive Textbook of Psychiatry. Baltimore, Williams & Wilkins, 1995.

Kertesz, A.: Localization and neuroimaging in neuropsychology. San Diego, CA, Academic Press, 1994.

Kryger, M. H., et al.: Principles and Practice of Sleep Medicine. Philadelphia, W. B. Saunders Co., 1994.

Le Moal, M., and Simon, H.: Mesocorticolimbic dopaminergic network: functional and regulatory roles. Physiol. Rev., 71:155, 1991.

Levine, M. D., et al.: Developmental-Behavioral Pediatrics. Philadelphia, W. B. Saunders Co., 1992.

Lezak, M. D.: Neuropsychological Assessment. New York, Oxford University Press, 1995.

McHugh, P. R., and McKusick, V. A.: Genes, Brain, and Behavior. New York, Raven Press, 1991.

Mishkin, M., and Appenzeller, T.: The Anatomy of Memory. Sci. Am., Special Report, 1987, p. 2.

Nerozzi, D., et al. (eds.): Hypothalamic Dysfunction in Neuropsychiatric Disorders. New York, Raven Press, 1987.

Netter, P., et al.: Psychobiology: Psychophysiological and Psychohumoral Processes Combined. Farmington, CT, S. Karger Publishers, Inc., 1992.

Penrose, R.: Shadows of the Mind: A Search for the Missing Science of Consciousness. Oxford University Press, 1994.

Rescorla, R. A.: Behavioral studies of Pavlovian conditioning. Annu. Rev. Neurosci., 11:329, 1988.

Rowland, L. P.: Merritt's Textbook of Neurology. Baltimore, Williams & Wilkins, 1995.

Sahgal, A.: Behavioral Neuroscience: A Practical Approach. New York, Oxford University Press, 1993.

Schatzberg, A. F., and Nemeroff, C. B. (eds.): The Hypothalamic-Pituitary-Adrenal Axis. New York, Raven Press, 1988.

Smith, O. A., and DeVito, J. L.: Central neural integration for the control of autonomic responses associated with emotion. Annu. Rev. Neurosci., 7:43, 1984.

Spreen, O., et al.: Developmental Neuropsychology. New York, Oxford University Press, 1995.

Stoudemire, A.: Clinical psychiatry for medical students. 2nd Ed. Philadelphia, J. B. Lippincott, 1994.

Stoudemire, A.: Human Behavior: An Introduction for Medical Students. Philadelphia, J. B. Lippincott, 1994.

Stephenson, R. B.: Modification of reflex regulation of blood pressure by behavior. Annu. Rev. Physiol., 46:133, 1984.

Steriade, M., and Llinas, R. R.: The functional states of the thalamus and the associated neuronal interplay. Physiol. Rev., 68:649, 1988.

Swanson, L. W., and Sawchenko, P. E.: Hypothalamic integration: Organization of the paraventricular and supraoptic nuclei. Annu. Rev. Neurosci., 6:269, 1983.

Traub, R. D., and Miles, R.: Neuronal networks of the hippocampus. New York, Cambridge University Press, 1991.

Usdin, E.: Stress. The Role of Catecholamines and Other Neurotransmitters. New York, Gordon Press Publishers, 1984.

Weller, M. P. I., and Eysenck, M. W.: The Scientific Basis of Psychiatry. Philadelphia, W. B. Saunders Co., 1992.

Winokur, G., and Clayton, P.: The Medical Basis of Psychiatry. Philadelphia, W. B. Saunders Co., 1994.

Wolman, B. B.: Psychosomatic Disorders. New York, Plenum Publishing Corp., 1988.

Woody, C. D., et al. (eds.): Cellular Mechanisms of Conditioning and Behavioral Plasticity. New York, Plenum Publishing Corp., 1988.

States of Brain Activity— Sleep; Brain Waves; Epilepsy; Psychoses

CHAPTER 59

All of us are aware of the many different states of brain activity, including sleep, wakefulness, and extreme excitement, and even different levels of mood, such as exhilaration, depression, and fear. All these states result from different activating or inhibiting forces generated usually within the brain itself. In Chapter 58, we began a partial discussion of this subject when we described different systems that are capable of activating either large or isolated portions of the brain. In this chapter, we present brief surveys of what is known about other states of brain activity, beginning with sleep.

SLEEP

Sleep is defined as unconsciousness from which the person can be aroused by sensory or other stimuli. It is to be distinguished from *coma,* which is unconsciousness from which the person cannot be aroused. There are multiple stages of sleep, from very light sleep to very deep sleep; sleep researchers also divide sleep into two entirely different types of sleep that have different qualities, as follows.

TWO TYPES OF SLEEP. During each night, a person goes through stages of two types of sleep that alternate with each other. They are called (1) *slow-wave sleep* because in this type of sleep the brain waves are very slow, as we discuss later, and (2) *rapid-eye-movement* (*REM*) *sleep* because in this type of sleep the eyes undergo rapid movements despite the fact that the person is still asleep.

Most sleep during each night is of the slow-wave variety; this is the deep, restful type of sleep that the person experiences during the first hour of sleep after having been kept awake for many hours. Episodes of REM sleep occur periodically during sleep and occupy about 25 per cent of the sleep time of the young adult; they normally recur about every 90 minutes. This type of sleep is not so restful, and it is usually associated with vivid dreaming, as we discuss later.

Slow-Wave Sleep

Most of us can understand the characteristics of deep slow-wave sleep by remembering the last time that we were kept awake for more than 24 hours and then remembering the deep sleep that occurred during the first hour after going to sleep. This sleep is exceedingly restful and is associated with a decrease in both peripheral vascular tone and many other vegetative functions of the body. In addition, there is a 10 to 30 per cent decrease in blood pressure, respiratory rate, and basal metabolic rate.

Although slow-wave sleep is frequently called "dreamless sleep," dreams do occur during slow-wave sleep, and sometimes nightmares even occur during this type of sleep. The difference between the dreams that occur in slow-wave sleep and those that occur in REM sleep is that those of REM sleep are likely to be remembered, whereas those of slow-wave sleep usually are not remembered. That is, during slow-wave sleep, consolidation of the dreams in memory does not occur.

REM Sleep (Paradoxical Sleep, Desynchronized Sleep)

In a normal night of sleep, bouts of REM sleep lasting 5 to 30 minutes usually appear on the average every 90 minutes, the first such period occurring 80 to 100 minutes after the person falls asleep. When the person is extremely sleepy, the duration of each bout of REM sleep is short and may even be absent. On the other hand, as the person becomes more rested through the night, the duration of the REM bouts greatly increases.

There are several important characteristics of REM sleep.

1. It is usually associated with active dreaming.
2. The person is even more difficult to arouse by sensory stimuli than during deep slow-wave sleep, and yet people usually awaken in the morning during an episode of REM sleep, not from slow-wave sleep.
3. The muscle tone throughout the body is exceedingly depressed, indicating strong inhibition of the spinal projections from the excitatory areas of the brain stem.
4. The heart rate and respiratory rate usually become irregular, which is characteristic of the dream state.
5. Despite the extreme inhibition of the peripheral muscles, a few irregular muscle movements occur. These include, in particular, rapid movements of the eyes.
6. The brain is highly active in REM sleep, and the overall brain metabolism may be increased as much as 20 per cent. Also, the electroencephalogram (EEG) shows a pattern of brain waves similar to those that occur during wakefulness. This type of sleep is also called *paradoxical sleep* because it is a paradox that a person can still be asleep despite marked activity in the brain.

In summary, REM sleep is a type of sleep in which the brain is quite active. However, the brain activity is not channeled in the proper direction for people to be fully aware of their surroundings and therefore to be awake.

Basic Theories of Sleep

SLEEP IS BELIEVED TO BE CAUSED BY AN ACTIVE INHIBITORY PROCESS. An earlier theory of sleep was that the excitatory areas of the upper brain stem, which were called the *reticular activating system,* simply fatigued during the period of a waking day and, therefore, became inactive as a result. This was called the *passive theory of sleep.* An important experiment changed this view to the current belief that *sleep is probably caused by an active inhibitory process.* This was an experiment in which it was discovered that transecting the brain stem in the midpontile region leads to a brain that, based on electrical recordings, never goes to sleep. In other words, there seems to be some center or centers located below the midpontile level of the brain stem that are required to cause sleep by inhibiting other parts of the brain.

Neuronal Centers, Neurohumoral Substances, and Mechanisms That Can Cause Sleep—A Possible Specific Role for Serotonin

Stimulation of several specific areas of the brain can produce sleep with characteristics near those of natural sleep. Some of these areas are the following.

1. The most conspicuous stimulation area for causing almost natural sleep is the *raphe nuclei in the lower half of the pons and in the medulla.* They are a thin sheet of nuclei located in the midline. Nerve fibers from these nuclei spread widely in the reticular formation and upward into the thalamus, neocortex, hypothalamus, and most areas of the limbic system. In addition, they extend downward into the spinal cord, terminating in the posterior horns where they can inhibit incoming pain signals, as discussed in Chapter 48. It is also known that many of the endings of fibers from these raphe neurons secrete *serotonin.* Also, when a drug that blocks the formation of serotonin is administered to an animal, the animal often cannot sleep for the next several days. Therefore, it has been assumed that serotonin is a major transmitter substance associated with production of sleep.
2. Stimulation of some areas in the *nucleus of the tractus solitarius,* which is the sensory region of the medulla and pons for the visceral sensory signals entering the brain by way of the vagus and glossopharyngeal nerves, also promotes sleep. This will not occur if the raphe nuclei have been destroyed. Therefore, these regions perhaps act by exciting the raphe nuclei and the serotonin system.
3. Stimulation of several regions in the diencephalon can also help promote sleep, including (a) the rostral part of the hypothalamus, mainly in the suprachiasmal area, and (b) an occasional area in the diffuse nuclei of the thalamus.

(Now comes a major problem with the serotonin theory of sleep: *The blood concentrations of serotonin are lower during sleep than during wakefulness. Therefore, a new search is on for some other possible sleep-promoting substance besides serotonin but still associated with the brain stem raphe system.*)

LESIONS IN THE SLEEP-PROMOTING CENTERS CAN CAUSE INTENSE WAKEFULNESS. Discrete lesions in the raphe nuclei lead to a high state of wakefulness. This is also true of bilateral lesions in the mediorostral suprachiasmal portion of the anterior hypothalamus. In both instances, the excitatory reticular nuclei of the mesencephalon and upper pons seem to become released from inhibition. Indeed, sometimes lesions of the anterior hypothalamus can cause such intense wakefulness that the animal actually dies of exhaustion.

OTHER POSSIBLE TRANSMITTER SUBSTANCES RELATED TO SLEEP. Experiments have shown that the cerebrospinal fluid as well as the blood and urine of animals that have been kept awake for several days contain a substance or substances that will cause sleep when injected into the ventricular system of an animal. One likely substance has been identified as *muramyl peptide,* a low-molecular-weight substance that accumulates in the cerebrospinal fluid and urine in animals kept awake for several days. When only micrograms of this sleep-producing substance are injected into the third ventricle, almost natural sleep occurs within a few minutes, and the animal may then stay asleep for several hours. Another substance that has similar effects in causing sleep is a nonapeptide isolated from the blood of sleeping animals. Still a third sleep factor has been isolated from the neuronal tissues of the brain stem of animals kept awake for days. Therefore, it is possible that prolonged wakefulness causes progressive accumulation of a sleep factor or factors in the brain stem or in the cerebrospinal fluid that leads to sleep.

Possible Cause of REM Sleep

Why slow-wave sleep is broken periodically by REM sleep is not understood. However, drugs that mimic the action of acetylcholine will increase the occurrence of REM sleep. Therefore, it has been postulated that the large acetylcholine-secreting neurons in the upper brain stem reticular formation might, through their extensive fibers, activate many portions of the brain. This theoretically could cause the excess activity that occurs in certain regions of the brain in REM sleep, even though the signals are not channeled appropriately in the brain to cause normal conscious awareness that is characteristic of wakefulness.

The Cycle Between Sleep and Wakefulness

The preceding discussions have merely identified neuronal areas, transmitters, and mechanisms that are related to sleep. They have not explained the cyclic, reciprocal operation of the sleep-wakefulness cycle. There is as yet no explanation. Therefore, we can let our imaginations run wild and suggest the following possible mechanism for causing the rhythmicity of the sleep-wakefulness cycle.

When the sleep centers are not activated, the release from inhibition of the mesencephalic and upper pontile reticular nuclei allows this region to become spontaneously active. This in turn excites both the cerebral cortex and the peripheral nervous system, both of which then send numerous positive feedback signals back to the same reticular nuclei to activate them still further. Once wakefulness begins, it has a natural tendency to sustain itself because of all this positive feedback activity.

After the brain remains activated for many hours, even the neurons within the activating system presumably will become fatigued. Consequently, the positive feedback cycle between the mesencephalic reticular nuclei and the cortex will fade and the inhibitory effects of the sleep centers will take over, leading to rapid transition from the wakefulness state to the sleep state.

Then, one could postulate that during prolonged sleep, the excitatory neurons of the reticular activating system gradually become more and more excitable because of the prolonged rest, whereas the inhibitory neurons of the sleep centers become less excitable because of their overactivity, thus leading to a new cycle of wakefulness.

This overall theory can explain the rapid transitions from sleep to wakefulness and from wakefulness to sleep. It can also explain arousal, the insomnia that occurs when a person's mind becomes preoccupied with a thought, the wakefulness that is produced by bodily activity, and many other conditions that affect the person's state of sleep or wakefulness.

Physiological Effects of Sleep

Sleep causes two major types of physiological effects: first, effects on the nervous system itself and, second, effects on other structures of the body. The first of these seems to be by far the more important because any person who has a transected spinal cord in the neck shows no harmful effects in the body beneath the level of transection that can be attributed to a sleep and wakefulness cycle; that is, lack of this sleep-wakefulness cycle in the nervous system at any point below the brain causes neither harm to the bodily organs nor any deranged function. On the other hand, lack of sleep certainly does affect the functions of the central nervous system.

Prolonged wakefulness is often associated with progressive malfunction of the mind and sometimes even causes abnormal behavioral activities of the nervous system. We are all familiar with the increased sluggishness of thought that occurs toward the end of a prolonged wakeful period, but in addition, a person can become irritable or even psychotic after forced wakefulness for prolonged periods. Therefore, we can assume that sleep in some way not currently understood restores both normal levels of activity and normal "balance" among the different parts of the central nervous system. This might be likened to the "rezeroing" of electronic analog computers after prolonged use because all computers of this type gradually lose their "baseline" of operation; it is reasonable to assume that the same effect occurs in the central nervous system because overuse of some brain areas during wakefulness could easily throw these out of balance with the remainder of the nervous system. Therefore, in the absence of any definitely demonstrated functional value of sleep, we might postulate that the principal value of sleep is to restore the natural balance among the neuronal centers.

Even though, as we pointed out earlier, neither wakefulness nor sleep has been shown to be harmful to the somatic functions of the body, the cycle of enhanced and depressed nervous excitability that follows the cycle of wakefulness and sleep does have moderate physiological effects on the peripheral body. For instance, during wakefulness, there is enhanced sympathetic activity as well as enhanced numbers of skeletal nerve impulses to the skeletal musculature to increase muscle tone. Conversely, during slow-wave sleep, sympathetic activity decreases while parasympathetic activity increases. Therefore, a "restful" sleep ensues—arterial blood pressure falls, pulse rate decreases, skin vessels dilate, activity of the gastrointestinal tract sometimes increases, muscles fall into a mainly relaxed state, and the overall basal metabolic rate of the body falls by 10 to 30 per cent.

BRAIN WAVES

Electrical recordings from the surface of the brain and from the outer surface of the head demonstrate continuous electrical activity in the brain. Both the in-

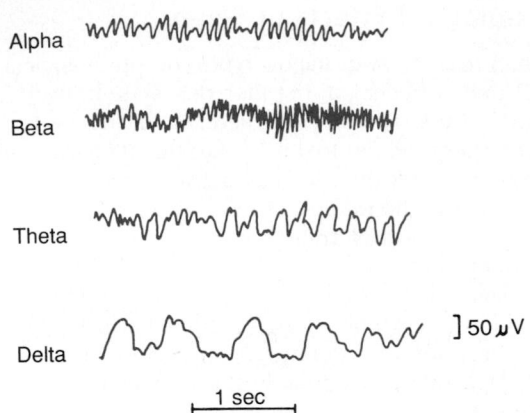

Alpha

Beta

Theta

Delta   ] 50 μV

|← 1 sec →|

Figure 59–1. Different types of normal electroencephalographic waves.

tensity and the patterns of this electrical activity are determined to a great extent by the level of excitation of the brain resulting from *sleep, wakefulness,* and brain diseases such as *epilepsy* and even some *psychoses.* The undulations in the recorded electrical potentials, shown in Figure 59–1, are called *brain waves,* and the entire record is called an EEG.

The intensities of the brain waves on the surface of the scalp range from 0 to 200 microvolts, and their frequencies range from once every few seconds to 50 or more per second. The character of the waves is highly dependent on the degree of activity of the cerebral cortex, and the waves change markedly between the states of wakefulness and sleep and coma.

Much of the time, the brain waves are irregular and no general pattern can be discerned in the EEG. At other times, distinct patterns appear. Some of these are characteristic of specific abnormalities of the brain, such as epilepsy, which is discussed later. Others occur even in normal people, and most of them can be classified as *alpha, beta, theta,* and *delta waves,* which are shown in Figure 59–1.

Alpha waves are rhythmical waves that occur at a frequency of between 8 and 13 per second and are found in the EEGs of almost all normal adult people when they are awake in a quiet, resting state of cerebration. These waves occur most intensely in the occipital region but can also be recorded from the parietal and frontal regions of the scalp. Their voltage usually is about 50 microvolts. During deep sleep, the alpha waves disappear. When the awake person's attention is directed to some specific type of mental activity, the alpha waves are replaced by asynchronous, higher-frequency but lower-voltage *beta* waves. Figure 59–2 shows the effect on the alpha waves of simply opening the eyes in bright light and then closing the eyes. Note that the visual sensations cause immediate cessation of the alpha waves and that these are replaced by low-voltage, asynchronous beta waves.

Beta waves occur at frequencies of more than 14 cycles per second and as high as 80 cycles per second. They are always recorded from the parietal and frontal regions of the scalp during extra activation of the central nervous system or during tension.

Theta waves have frequencies of between 4 and 7 cycles per second. They occur mainly in the parietal and temporal regions in children, but they also occur during emotional stress in some adults, particularly during disappointment and frustration. Theta waves also occur in many brain disorders, often in degenerative brain states.

Delta waves include all the waves of the EEG below 3.5 cycles per second. They occur in very deep sleep, in infancy, and in serious organic brain disease. They also occur in the cortex of animals that have had subcortical transections separating the cerebral cortex from the thalamus. Therefore, delta waves can occur strictly in the cortex independent of activities in lower regions of the brain.

Origin in the Brain of the Brain Waves

The discharge of a single neuron or single nerve fiber in the brain can never be recorded from the surface of the head. Instead, many thousands or even millions of neurons or fibers *must fire synchronously;* only then will the potentials from the individual neurons or fibers summate enough to be recorded all the way through the skull. Thus, the intensity of the brain waves from the scalp is determined mainly by the number of neurons and fibers that fire in synchrony with one another, not by the total level of electrical activity in the brain. In fact, strong *nonsynchronous* nerve signals often nullify one another in the recorded brain waves because they are of opposing polarities. This is demonstrated in Figure 59–2, which shows, when the eyes were closed, synchronous discharge of many neurons in the cerebral cortex at a frequency of about 12 per second, which gave the *alpha waves.* Then, when the eyes were opened, the activity of the brain increased greatly, but the synchronization of the signals became so little that the brain waves mainly nullified one another and the resultant effect was weak waves of generally higher but irregular frequency, called *beta waves.*

ORIGIN OF ALPHA WAVES. Alpha waves will *not* occur in the cortex without connections with the thalamus. Also, stimulation in the nonspecific *reticular nuclei* that surround the thalamus as well as in the "diffuse" nuclei deep inside the thalamus often sets up waves in the thalamocortical system at a frequency of between 8 and 13 per second, which is the natural frequency of the alpha waves. Therefore, it is likely that the alpha waves result from spontaneous feedback oscillation in this diffuse thalamocortical system, possibly including the brain stem activating system as well. This oscillation presumably causes both the periodicity of the alpha waves and the synchronous activation of literally millions of cortical neurons during each wave.

ORIGIN OF DELTA WAVES. Transection of the fiber tracts from the thalamus to the cortex, which blocks the thalamic activation of the cortex and eliminates the alpha waves, nevertheless does not block all delta waves in the cortex. This indicates that some synchronizing mechanism can occur in the cortical neurons themselves—mainly independently of lower structures in the brain—to cause the delta waves.

Eyes open   Eyes closed

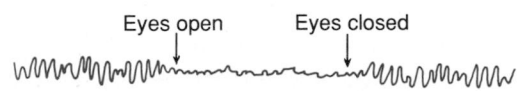

Figure 59–2. Replacement of the alpha rhythm by an asynchronous, low-voltage beta rhythm on opening the eyes.

Figure 59–3. Effect of varying degrees of cerebral activity on the basic rhythm of the electroencephalogram. (From Gibbs and Gibbs: Atlas of Electroencephalography, 2nd Ed. Vol. I. Reading, Mass., Addison-Wesley, 1974. Reprinted by permission.)

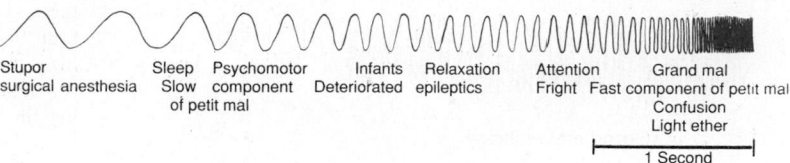

Stupor   Sleep   Psychomotor   Infants   Relaxation   Attention   Grand mal
surgical anesthesia   Slow   component   Deteriorated   epileptics   Fright   Fast component of petit mal
      of petit mal   Confusion
                                          Light ether

1 Second

Delta waves also occur in deep slow-wave sleep; this suggests that the cortex then is mainly released from the activating influences of the lower centers.

Effect of Varying Degrees of Cerebral Activity on the Basic Frequency of the EEG

There is a general relation between the degree of cerebral activity and the average frequency of the EEG rhythm, the average frequency increasing progressively with higher degrees of activity. This is demonstrated in Figure 59–3, which shows the existence of delta waves in stupor, surgical anesthesia, and sleep; theta waves in psychomotor states and in infants; alpha waves during relaxed states; and beta waves during periods of intense mental activity. *During periods of mental activity, the waves usually become asynchronous rather than synchronous, so that the voltage falls considerably, despite increased cortical activity,* as shown in Figure 59–2.

EEG Changes in the Different Stages of Wakefulness and Sleep

Figure 59–4 shows the EEG from a typical person in different stages of wakefulness and sleep. Alert wakefulness is characterized by high-frequency *beta waves,* whereas quiet wakefulness is usually associated with *alpha waves,* as demonstrated by the first two EEGs of the figure.

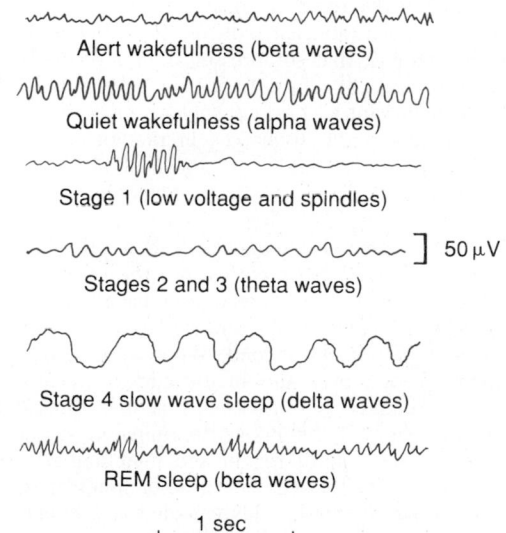

Alert wakefulness (beta waves)

Quiet wakefulness (alpha waves)

Stage 1 (low voltage and spindles)

] 50 μV

Stages 2 and 3 (theta waves)

Stage 4 slow wave sleep (delta waves)

REM sleep (beta waves)

1 sec

Figure 59–4. Progressive change in the characteristics of the brain waves during different stages of wakefulness and sleep.

Slow-wave sleep is divided into four stages. In the first stage, a stage of very light sleep, the voltage of the EEG waves becomes very low; this is broken by *"sleep spindles,"* that is, short spindle-shaped bursts of alpha waves that occur periodically. In stages 2, 3, and 4 of slow-wave sleep, the frequency of the EEG becomes progressively slower until it reaches a frequency of only 1 to 3 waves per second in stage 4; these are typical *delta waves.*

Finally, the bottom record in Figure 59–4 shows the EEG during REM sleep. It is often difficult to tell a difference between this brain wave pattern and that of an alert awake person. The waves are irregular high-frequency beta waves, which are normally suggestive of excess but desynchronized nervous activity as found in the awake state. Therefore, REM sleep is frequently called *desynchronized sleep* because there is a lack of synchrony in the firing of the neurons, despite significant brain activity.

EPILEPSY

Epilepsy (also called "seizures") is characterized by *uncontrolled* excessive activity of either a part or all of the central nervous system. A person who is predisposed to epilepsy has attacks when the basal level of excitability of the nervous system (or of the part that is susceptible to the epileptic state) rises above a certain critical threshold. As long as the degree of excitability is held below this threshold, no attack occurs.

Epilepsy can be classified into three major types: *grand mal epilepsy, petit mal epilepsy,* and *focal epilepsy.*

Grand Mal Epilepsy

Grand mal epilepsy is characterized by extreme neuronal discharges in all areas of the brain—in the cortex, in the deeper parts of the cerebrum, and even in the brain stem and thalamus. Also, discharges transmitted all the way into the spinal cord cause generalized *tonic seizures* of the entire body, followed toward the end of the attack by alternating tonic and then spasmodic muscle contractions called *tonic-clonic seizures.* Often the person bites or "swallows" the tongue and may have difficulty breathing, sometimes to the extent of developing cyanosis. Also, signals transmitted from the brain to the viscera frequently cause urination and defecation.

The grand mal seizure lasts from a few seconds to 3 to 4 minutes. It is also characterized by postseizure depression of the entire nervous system; the person remains in stupor for 1 to many minutes after the attack and then often remains severely fatigued and asleep for many hours thereafter.

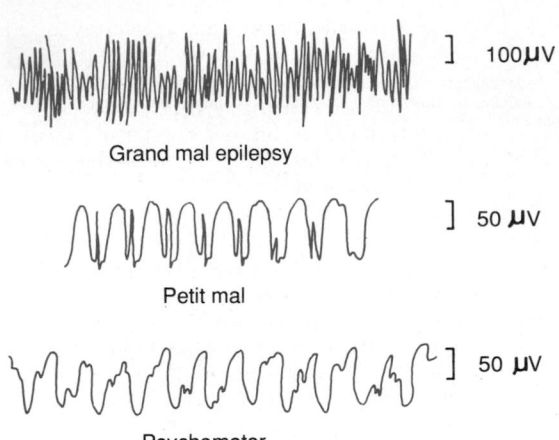

Grand mal epilepsy] 100μV

Petit mal] 50 μV

Psychomotor] 50 μV

Figure 59–5. Electroencephalograms in different types of epilepsy.

The top recording of Figure 59–5 shows a typical EEG from almost any region of the cortex during the tonic phase of a grand mal attack. This demonstrates that high-voltage, synchronous discharges occur over the entire cortex. Furthermore, the same type of discharge occurs on both sides of the brain at the same time, demonstrating that the abnormal neuronal circuitry responsible for the attack strongly involves the basal regions of the brain that drive the cortex.

In laboratory animals or even in human beings, grand mal attacks can be initiated by administering neuronal stimulants, such as the drug Metrazol, or they can be caused by insulin hypoglycemia or the passage of alternating electrical current directly through the brain. Electrical recordings from the thalamus as well as from the reticular formation of the brain stem during the grand mal attack show typical high-voltage activity in both of these areas similar to that recorded from the cerebral cortex.

Presumably, therefore, a grand mal attack involves not only abnormal activation of the thalamus and cerebral cortex but also abnormal activation in the lower portions of the brain activating system itself.

WHAT INITIATES A GRAND MAL ATTACK? Most people who have grand mal attacks have a hereditary predisposition to epilepsy, a predisposition that occurs in about 1 of every 50 to 100 persons. In such people, some of the factors that can increase the excitability of the abnormal "epileptogenic" circuitry enough to precipitate attacks are (1) strong emotional stimuli, (2) alkalosis caused by overbreathing, (3) drugs, (4) fever, and (5) loud noises or flashing lights. Also, even in people who are not genetically predisposed, traumatic lesions in almost any part of the brain can cause excess excitability of local brain areas, as we discuss shortly; these, too, sometimes transmit signals into the activating systems of the brain to elicit grand mal seizures.

WHAT STOPS THE GRAND MAL ATTACK? The cause of the extreme neuronal overactivity during a grand mal attack is presumed to be massive activation of many reverberating pathways throughout the brain. Presumably, also, the major factor, or at least one of the major factors, that stops the attack after a few minutes is the phenomenon of neuronal *fatigue*. A second factor is probably *active inhibition* by inhibitory neurons that have also been activated by the attack. The stupor and total body fatigue that occur after a grand mal seizure is over are believed to result from the intense fatigue of the neuronal synapses after their intensive activity during the grand mal attack.

Petit Mal Epilepsy

Petit mal epilepsy almost certainly involves the basic thalamocortical brain activating system. It is usually characterized by 3 to 30 seconds of unconsciousness or diminished consciousness during which the person has several twitch-like contractions of the muscles, usually in the head region, especially blinking of the eyes; this is followed by return of consciousness and resumption of previous activities. This total sequence is called the *absence syndrome* or absence epilepsy. The patient may have one such attack in many months or, in rare instances, may have a rapid series of attacks, one after the other. The usual course is for the petit mal attacks to appear in late childhood and then to disappear by the age of 30. On occasion, a petit mal epileptic attack will initiate a grand mal attack.

The brain wave pattern in petit mal epilepsy is demonstrated by the middle record of Figure 59–5, which is typified by a *spike and dome pattern*. The spike and dome can be recorded over most or all of the cerebral cortex, showing that the seizure involves much or most of the thalamocortical activating system of the brain. In fact, animal studies suggest that it results from oscillation of a system of neurons involving thalamic reticular neurons (which are *inhibitory* gamma-aminobutyric acid [GABA]-producing neurons) and other *excitatory* thalamocortical and corticothalamic neurons.

Focal Epilepsy

Focal epilepsy can involve almost any part of the brain, either localized regions of the cerebral cortex or deeper structures of both the cerebrum and brain stem. Almost always, focal epilepsy results from some localized organic lesion or functional abnormality, such as scar tissue in the brain that pulls on the adjacent neuronal tissue, a tumor that compresses an area of the brain, a destroyed area of brain tissue, or congenitally deranged local circuitry. Lesions such as these can promote extremely rapid discharges in the local neurons; when their discharge rate rises above about 1000 per second, synchronous waves begin to spread over the adjacent cortical regions. These waves presumably result from *localized reverberating circuits* that gradually recruit adjacent areas of the cortex into the epileptic discharge zone. The process spreads to adjacent areas at a rate as slow as a few millimeters a minute to as fast as several centimeters per second. When such a wave of excitation spreads over the motor cortex, it causes a progressive "march" of muscle contractions throughout the opposite side of the body, beginning most characteristically in the mouth region and marching progressively downward to the legs but at other times marching in the opposite direction. This is called *jacksonian epilepsy*.

A focal epileptic attack may remain confined to a single area of the brain, but in many instances, the strong signals from the convulsing cortex excite the mesencephalic portion of the brain activating system so greatly that a grand mal epileptic attack ensues as well.

Another type of focal epilepsy is the so-called *psychomotor seizure,* which may cause (1) a short period of amnesia; (2) an attack of abnormal rage; (3) sudden anxiety, discomfort, or fear; (4) a moment of incoherent speech or mumbling of some trite phrase; or (5) a motor act to attack someone, to rub the face with the hand, and so forth. Sometimes the person cannot remember his activities during the attack, but at other times, he will have been conscious of everything that he had been doing but unable to control it. Attacks of this type characteristically involve part of the limbic portion of the brain, such as the hippocampus, the amygdala, the septum, and the temporal cortex.

The lower tracing of Figure 59–5 demonstrates a typical EEG during a psychomotor attack, showing a low-frequency rectangular wave with a frequency of between 2 and 4 per second and with superimposed 14-per-second waves.

SURGICAL EXCISION OF EPILEPTIC FOCI CAN OFTEN PREVENT SEIZURES. The EEG can often be used to localize abnormal spiking waves originating in areas of organic brain disease that predispose to focal epileptic attacks. Once such a focal point is found, surgical excision of the focus frequently prevents future attacks.

PSYCHOTIC BEHAVIOR AND DEMENTIA—ROLES OF SPECIFIC NEUROTRANSMITTER SYSTEMS

Clinical studies of patients with different psychoses as well as some types of dementia have suggested that many of these conditions result from diminished function of neurons that secrete specific neurotransmitters. The use of appropriate drugs to counteract the loss of the respective transmitters has been successful in treating some patients.

In Chapter 56, we discuss the cause of Parkinson's disease, the loss of the neurons in the substantia nigra whose axons secrete dopamine in the caudate nucleus and putamen. Also, we point out that in Huntington's disease, the loss of GABA-secreting neurons and acetylcholine-secreting neurons is associated with the abnormal motor patterns that are observed as well as with the dementia that develops in the same patients. In this section, we discuss other abnormalities and other classes of neurons that lead to additional types of psychotic behavior or dementia.

Depression and Manic-Depressive Psychoses— Decreased Activity of the Norepinephrine and Serotonin Neurotransmitter Systems

Much evidence has accumulated suggesting that the *mental depression psychosis,* which afflicts about 8 million people in the United States at any one time, might be caused by diminished formation of either norepinephrine or serotonin, or both. (New evidence has implicated still other neurotransmitters.) Depressed patients experience symptoms of grief, unhappiness, despair, and misery. In addition, they lose their appetite and sex drive and have severe insomnia. Often associated with all these is a state of psychomotor agitation despite the depression.

Moderate numbers of *norepinephrine-secreting neu-*

rons are located in the brain stem, especially in the *locus ceruleus.* They send fibers upward to most parts of the limbic system, the thalamus, and the cerebral cortex. Also, many *serotonin-producing neurons* are located in the *midline raphe nuclei* of the lower pons and medulla and project fibers to many areas of the limbic system and to some other areas of the brain.

A principal reason for believing that depression is caused by diminished activity of the norepinephrine and serotonin systems is that drugs that block the secretion of norepinephrine and serotonin, such as reserpine, frequently cause depression. Conversely, about 70 per cent of depressive patients can be treated effectively with drugs that increase the excitatory effects of norepinephrine and serotonin at the nerve endings—for instance, (1) *monoamine oxidase inhibitors,* which block destruction of norepinephrine and serotonin once they are formed; (2) *tricyclic antidepressants,* such as *imipramine* and *amitriptyline,* which block reuptake of norepinephrine and serotonin by the nerve endings so that these transmitters remain active for longer periods after secretion; and (3) a new class of drugs that *enhance the action of serotonin* alone, often causing fewer side effects.

Mental depression can also be treated effectively by electroconvulsive therapy—commonly called "shock therapy." In this therapy, an electrical shock is used to cause a generalized seizure similar to that of an epileptic attack. This has also been shown to enhance norepinephrine transmission efficiency.

Some patients with mental depression alternate between depression and mania, which is called either *bipolar disorder* or *manic-depressive psychosis,* and a few people exhibit only mania without the depressive episodes. Drugs that diminish the formation or action of norepinephrine and serotonin, such as lithium compounds, can be effective in treating the manic condition.

Therefore, it is presumed that the norepinephrine system and the serotonin system normally provide drive to the limbic system to increase a person's sense of well-being, to create happiness, contentment, good appetite, appropriate sex drive, and psychomotor balance—although too much of a good thing can cause mania. In support of this concept is the fact that the pleasure and reward centers of the hypothalamus and surrounding areas receive large numbers of nerve endings from the norepinephrine and serotonin systems.

Schizophrenia—Possible Exaggerated Function of Part of the Dopamine System

Schizophrenia comes in many varieties. One of the most common types is seen in the person who hears voices and has delusions of grandeur, intense fear, or other types of feelings that are unreal. Many schizophrenics are highly paranoid, with a sense of persecution from outside sources; they may develop incoherent speech, dissociation of ideas, and abnormal sequences of thought; they are often withdrawn, sometimes with abnormal posture and even rigidity.

There are reasons to believe that schizophrenia results from one or more of three possibilities: (1) multiple areas in the *prefrontal lobes* where neural signals become blocked or where processing of the signals becomes dysfunctional, (2) excessive excitement of a group of neurons that secrete *dopamine* in the behavioral centers of the brain, including in the frontal lobes, and/

or (3) abnormal function of a crucial part of the brain's *limbic behavioral control system centered around the hippocampus.*

The reason for believing that the prefrontal lobes are involved in schizophrenia is that a schizophrenic-like pattern of mental activity can be induced in monkeys by making multiple minute lesions in widespread areas of the prefrontal lobes. Also, it is likely, in other instances, that dysfunction of the prefrontal lobes can result from diminished neural drive to these lobes from other areas of the brain.

Dopamine has been implicated as a possible cause of schizophrenia because many patients with Parkinson's disease develop schizophrenic-like symptoms when they are treated with the drug called L-DOPA. This drug releases dopamine in the brain, which is advantageous for treating Parkinson's disease, but at the same time it depresses various portions of the prefrontal lobes and other related areas. It has been suggested that in schizophrenia excess dopamine is secreted by a group of dopamine-secreting neurons whose cell bodies lie in the ventral tegmentum of the mesencephalon, medial and superior to the substantia nigra. These neurons give rise to the so-called *mesolimbic dopaminergic system* that projects nerve fibers to the medial and anterior portions of the limbic system, especially to the hippocampus, amygdala, anterior caudate nucleus, and portions of the prefrontal lobes. All of these are powerful behavioral control centers. An even more compelling reason for believing that schizophrenia might be caused by excess production of dopamine is that many drugs that are effective in treating schizophrenia—such as chlorpromazine, haloperidol, and thiothixene—all decrease the secretion of dopamine at the dopaminergic nerve endings or decrease the effect of dopamine on subsequent neurons.

Finally, possible involvement of the hippocampus in schizophrenia was discovered only recently when it was learned that *in schizophrenia, the hippocampus is often reduced in size*, especially in the dominant hemisphere. Also, other portions of the brain's behavioral control system connected with the hippocampus, including the prefrontal lobes, have been found to be hypofunctional.

Alzheimer's Disease—Amyloid Plaques and Loss of the Memory Process

Alzheimer's disease is defined as premature aging of the brain, usually beginning in mid-adult life and progressing rapidly to extreme loss of mental powers—similar to that seen in very, very old age. These patients usually require continuous care within a few years after the disease begins.

Pathologically, one finds accumulation of *amyloid plaques* ranging in size from 10 micrometers to several hundred micrometers in widespread areas of the brain, including in the cortex, hippocampus, basal ganglia, thalamus, and even the cerebellum. Thus, Alzheimer's disease appears to be a metabolic degenerative disease. To support this belief, research has demonstrated in most patients with Alzheimer's disease abnormality of the gene that controls production of *apoliproprotein E,* a blood protein that transports cholesterol to tissues.

One consistent finding in Alzheimer's disease is *loss of neurons in that part of the limbic pathway that drives the memory process.* Loss of this memory function is devastating.

REFERENCES

Aicardi, J.: Epilepsy in Children. New York, Raven Press, 1994.

Andreasen, N. C.: Schizophrenia: Positive and Negative Symptoms and Syndromes. Farmington, CT, S. Karger Publishers, Inc., 1990.

Barlow, J. S.: The Electroencephalogram. Cambridge, MA, The MIT Press, 1993.

Blume, W. T.: Atlas of Adult Electro-Encephalography. New York, Raven Press, 1994.

Chauvel, P., et al.: Frontal Lobe Seizures and Epilepsies. New York, Raven Press, 1992.

Chiu, E., and Ames, D.: Functional Psychiatric Disorders of the Elderly. New York, Cambridge University Press, 1994.

Daly, D. D., and Pedley, T. A.: Current Practice of Clinical Electro-Encephalography. New York, Raven Press, 1990.

DeKornfeld, T. J., and Stanford, T.: Anesthesiology. 9th Ed. Redding, MA, Appleton and Lange, 1995.

Elliott, J. M., et al.: Experimental Approaches to Anxiety and Depression. New York, John Wiley and Sons, Inc., 1992.

Georgotas, A., and Cancro, R. (eds.): Depression and Mania. New York, Elsevier Science Publishing Co., 1988.

Gillberg, C., and Coleman, M.: The Biology of the Autistic Syndromes. New York, Cambridge University Press, 1993.

Guilleminault, C., et al.: Fatal Familial Insomnia: Inherited Prion Diseases, Sleep, and the Thalamus. New York, Raven Press, 1994.

Kaplan, H. I., and Sadock, B. J.: Comprehensive Textbook of Psychiatry. Baltimore, Williams & Wilkins, 1995.

Keefe, R. S. E., and Harvey, P. D.: Understanding Schizophrenia: A Guide to the New Research on Causes and Treatment. New York, Free Press (Macmillan), 1994.

Kirby, R. R., and Gravenstein, N.: Clinical Anesthesia Practice. Philadelphia, W. B. Saunders Co., 1994.

Knight, R. G., and Longmore, B. E.: Clinical Neuropsychology of Alcoholism. Hillsdale, NJ, Lawrence Erlbaum Assoc. Inc., 1994.

Kryger, M. H., et al.: Principles and Practice of Sleep Medicine. Philadelphia, W. B. Saunders Co., 1994.

Leonard, C. S., and Llinás, R.: Serotonergic and cholinergic inhibition of mesopontine cholinergic neurons controlling REM sleep: an in vitro electrophysiological study. Neuroscience, 59:309, 1994.

Mancia, M., and Marini, G.: The Diencephalon and Sleep. New York, Raven Press, 1990.

McHugh, P. R., and McKusick, V. A.: Genes, Brain, and Behavior. New York, Raven Press, 1991.

Morgan, G. E., and Mikhail, M. S.: Clinical Anesthesiology. 2nd Ed. Redding, MA, Appleton & Lange, 1995.

Nappi, G., et al. (eds.): Neurodegenerative Disorders. New York, Raven Press, 1988.

Nerozzi, D., et al. (eds.): Hypothalamic Dysfunction in Neuropsychiatric Disorders. New York, Raven Press, 1987.

Niedermeyer, E., and da Silva, F. L.: Electroencephalography: Basic Principles, Clinical Applications, and Related Fields. Baltimore, Williams & Wilkins, 1993.

Nitsch, R. M., et al.: Alzheimer's Disease: Amyloid Precursor Proteins, Signal Transduction, and Neuronal Transplantation. New York, New York, Academy of Sciences, 1993.

Nunez, P. L.: Neocortical Dynamics and Human EEG Rhythms. New York, Oxford University Press, 1995.

Schou, M.: Lithium Treatment of Manic-Depressive Illnesses. Farmington, CT, S. Karger Publishers, Inc., 1993.

Schwartzkroin, P. A.: Epilepsy: Models, Mechanisms and Concepts. New York, Cambridge University Press, 1993.

Sheldon, S. H., et al.: Pediatric Sleep Medicine. Philadelphia, W. B. Saunders Co., 1992.

Shorvon, S.: Status Epilepticus: Its Clinical Features and Treatment in Children and Adults. New York, Cambridge University Press, 1994.

Silver, P. A.: Psychotropic Drug Use in the Medically Ill. Farmington, CT, S. Karger Publishers, Inc., 1994.

Silverman, D. G., et al.: Review of Clinical Anesthesia. Philadelphia, J. B. Lippincott, 1994.

Spitzer, M.: Hallucinations. New York, John Wiley and Sons, Inc., 1993.

Stoudemire, A.: Clinical Psychiatry for Medical Students. 2nd Ed. Philadelphia, J. B. Lippincott, 1994.

Strong, R., et al. (eds.): Central Nervous System Disorders of Aging. New York, Raven Press, 1988.

Tucek, S.: Regulation of acetylcholine synthesis in the brain. J. Neurochem., 44:11, 1985.

Terzano, M. G., et al.: Phasic Events and Dynamic Organization of Sleep. New York, Raven Press, 1991.

Wauquier, A., et al.: Slow Wave Sleep: Physiological, Pathophysiological, and Functional Aspects. New York, Raven Press, 1989.

Wolman, B. B.: Psychosomatic Disorders. New York, Plenum Publishing Corp., 1988.

Wylie, E.: The Treatment of Epilepsy: Principles and Practice. Baltimore, Williams & Wilkins, 1993.

The Autonomic Nervous System;
The Adrenal Medulla

*C*HAPTER *60*

The portion of the nervous system that controls the visceral functions of the body is called the *autonomic nervous system*. This system helps control arterial pressure, gastrointestinal motility and secretion, urinary bladder emptying, sweating, body temperature, and many other activities, some of which are controlled almost entirely and some only partially by the autonomic nervous system.

One of the most striking characteristics of the autonomic nervous system is the rapidity and intensity with which it can change visceral functions. For instance, within 3 to 5 seconds it can increase the heart rate to twice normal, and within 10 to 15 seconds the arterial pressure can be doubled; or, at the other extreme, the arterial pressure can be decreased low enough within 4 to 5 seconds to cause fainting. Sweating can begin within seconds, and the bladder may empty involuntarily, also within seconds. It is these extremely rapid changes that are measured by the lie detector polygraph, reflecting the innermost feelings of a person.

GENERAL ORGANIZATION OF THE AUTONOMIC NERVOUS SYSTEM

The autonomic nervous system is activated mainly by centers located in the *spinal cord, brain stem,* and *hypothalamus.* Also, portions of the cerebral cortex, especially of the limbic cortex, can transmit impulses to the lower centers and in this way influence autonomic control. The autonomic nervous system also often operates by means of *visceral reflexes*. That is, sensory signals entering the autonomic ganglia, cord,

brain stem, or hypothalamus can elicit appropriate reflex responses directly back to the visceral organs to control their activities.

The efferent autonomic signals are transmitted to the body through two major subdivisions called the *sympathetic nervous system* and the *parasympathetic nervous system*, the characteristics and functions of which follow.

Physiologic Anatomy of the Sympathetic Nervous System

Figure 60–1 shows the general organization of the peripheral portions of the sympathetic nervous system, including one of the two *paravertebral sympathetic chains of ganglia* that lie to the two sides of the vertebral column, two *prevertebral ganglia* (the *celiac* and *hypogastric*), and nerves extending from the ganglia to the different internal organs. The sympathetic nerves originate in the spinal cord between the segments T-1 and L-2 and pass from here first into the sympathetic chain and then to the tissues and organs that are stimulated by the sympathetic nerves.

Preganglionic and Postganglionic Sympathetic Neurons

The sympathetic nerves are different from skeletal motor nerves in the following way: Each sympathetic pathway from the cord to the stimulated tissue is composed of two neurons, a *preganglionic neuron* and a *postganglionic neuron*, in contrast to only a single neuron in the skeletal motor pathway. The cell body of each preganglionic neuron lies in the *intermediolateral horn* of the spinal cord; its fiber passes, as shown in

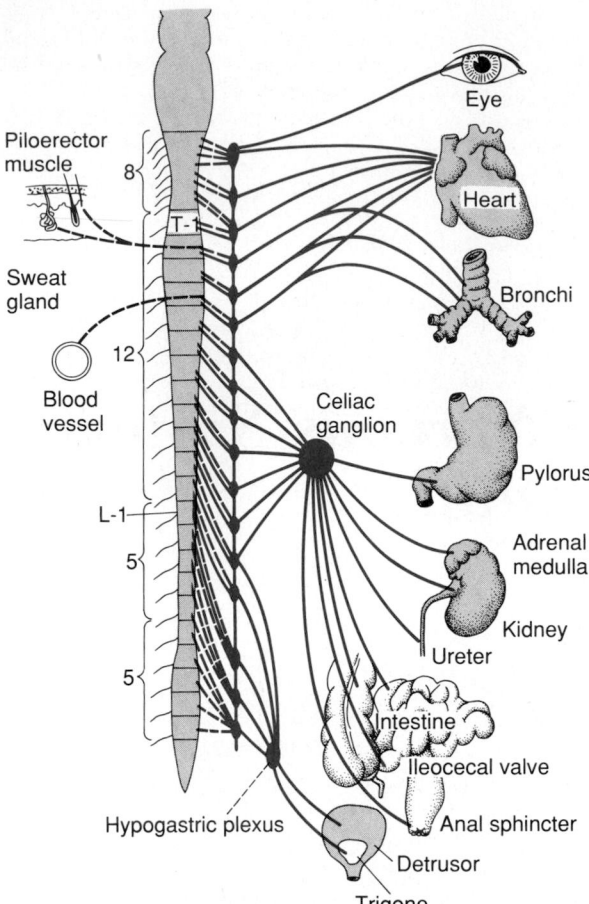

Figure 60–1. Sympathetic nervous system. Dashed lines represent postganglionic fibers in the gray rami leading back into the spinal nerves for distribution to blood vessels, sweat glands, and pilorector muscles.

parts of the body in the skeletal nerves. They control the blood vessels, sweat glands, and piloerector muscles of the hairs. About 8 per cent of the fibers in the average skeletal nerve are sympathetic fibers, a fact that indicates their great importance.

SEGMENTAL DISTRIBUTION OF THE SYMPATHETIC NERVES. The sympathetic pathways that originate in the different segments of the spinal cord are not necessarily distributed to the same part of the body as the somatic spinal nerve fibers from the same segments. Instead, the *sympathetic fibers from cord segment T-1 generally pass up the sympathetic chain to the head; from T-2 into the neck; from T-3, T-4, T-5, and T-6 into the thorax; from T-7, T-8, T-9, T-10, and T-11 into the abdomen; and from T-12, L-1, and L-2 into the legs.* This distribution is only approximate and overlaps greatly.

The distribution of sympathetic nerves to each organ is determined partly by the position in the embryo where the organ originates. For instance, the heart receives many sympathetic nerve fibers from the neck portion of the sympathetic chain because the heart originates in the neck of the embryo. Likewise, the abdominal organs receive most of their sympathetic innervation from the lower thoracic segments because most of the primitive gut originates in this area.

SPECIAL NATURE OF THE SYMPATHETIC NERVE ENDINGS IN THE ADRENAL MEDULLAE. Preganglionic sympathetic nerve fibers pass, *without synapsing,* all the way from the intermediolateral horn cells of the spinal cord, through the sympathetic chains, then through the splanchnic nerves, and finally into the adrenal medullae. There they end directly on modified neuronal cells that secrete *epinephrine* and *norepinephrine* into the blood stream. These secretory cells embryologically are derived from nervous tissue and are analogous to postganglionic neurons; indeed, they even have rudimentary

Figure 60–2, through the *anterior root* of the cord into the corresponding *spinal nerve.*

Immediately after the spinal nerve leaves the spinal canal, the preganglionic sympathetic fibers leave the nerve and pass through the *white ramus* into one of the *ganglia* of the *sympathetic chain.* Then the course of the fibers can be one of the following three: (1) It can synapse with postganglionic neurons in the ganglion that it enters. (2) It can pass upward or downward in the chain and synapse in one of the other ganglia of the chain. Or (3) it can pass for variable distances through the chain and then through one of the *sympathetic nerves* radiating outward from the chain, finally terminating in one of the *prevertebral ganglia.*

The postganglionic neuron then originates either in one of the sympathetic chain ganglia or in one of the prevertebral ganglia. From either of these two sources, the postganglionic fibers travel to their destinations in the various organs.

SYMPATHETIC NERVE FIBERS IN THE SKELETAL NERVES. Some of the postganglionic fibers pass back from the sympathetic chain into the spinal nerves through *gray rami* at all levels of the cord, shown in Figure 60–2. These spinal nerve sympathetic nerve fibers are all very small type C fibers that extend to all

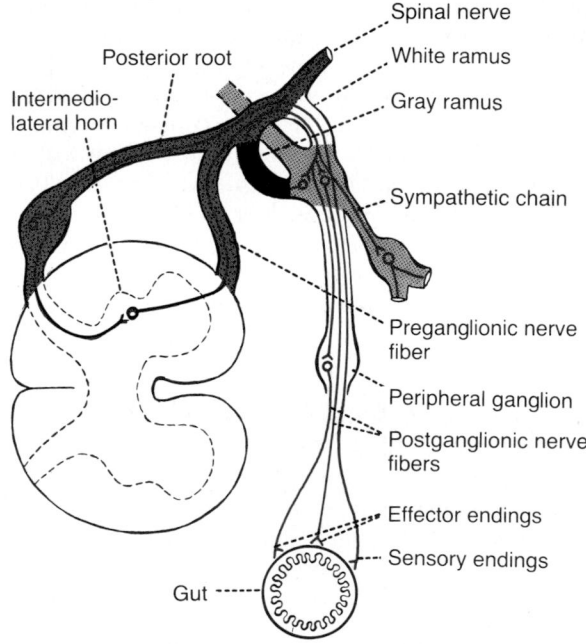

Figure 60–2. Nerve connections between the spinal cord, sympathetic chain, spinal nerves, and peripheral sympathetic nerves.

nerve fibers, and it is these fibers that secrete the hormones.

Physiologic Anatomy of the Parasympathetic Nervous System

The parasympathetic nervous system is shown in Figure 60–3, demonstrating that parasympathetic fibers leave the central nervous system through cranial nerves III, VII, IX, and X; the second and third sacral spinal nerves; and occasionally the first and fourth sacral nerves. About 75 per cent of all parasympathetic nerve fibers are in the vagus nerves (cranial nerve X), passing to the entire thoracic and abdominal regions of the body. Therefore, a physiologist speaking of the parasympathetic nervous system often thinks mainly of the two vagus nerves. The vagus nerves supply parasympathetic nerves to the heart, lungs, esophagus, stomach, entire small intestine, proximal half of the colon, liver, gallbladder, pancreas, and upper portions of the ureters.

Parasympathetic fibers in the *third nerve* flow to the pupillary sphincters and ciliary muscles of the eye. Fibers from the *seventh nerve* pass to the lacrimal, nasal, and submandibular glands, and fibers from the *ninth nerve* pass to the parotid gland.

The sacral parasympathetic fibers congregate in the *pelvic nerves*, which leave the sacral plexus on each side of the cord at the S-2 and S-3 levels and distribute their peripheral fibers to the descending colon, rectum, bladder, and lower portions of the ureters. Also, this sacral group of parasympathetics supplies nerve signals to the external genitalia to cause erection.

PREGANGLIONIC AND POSTGANGLIONIC PARASYMPATHETIC NEURONS. The parasympathetic system, like the sympathetic, has both preganglionic and postganglionic neurons. However, except in the case of a few cranial parasympathetic nerves, the *preganglionic fibers* pass uninterrupted all the way to the organ that is to be controlled. Then, in the wall of the organ are located the *postganglionic neurons*. The preganglionic fibers synapse with these, and short postganglionic fibers, 1 millimeter to several centimeters in length, leave the neurons to spread through the substance of the organ. This location of the parasympathetic postganglionic neurons in the visceral organ itself is quite different from the arrangement of the sympathetic ganglia because the cell bodies of the sympathetic postganglionic neurons are almost always located in ganglia of the sympathetic chain or in various other discrete ganglia in the abdomen, rather than in the excited organ itself.

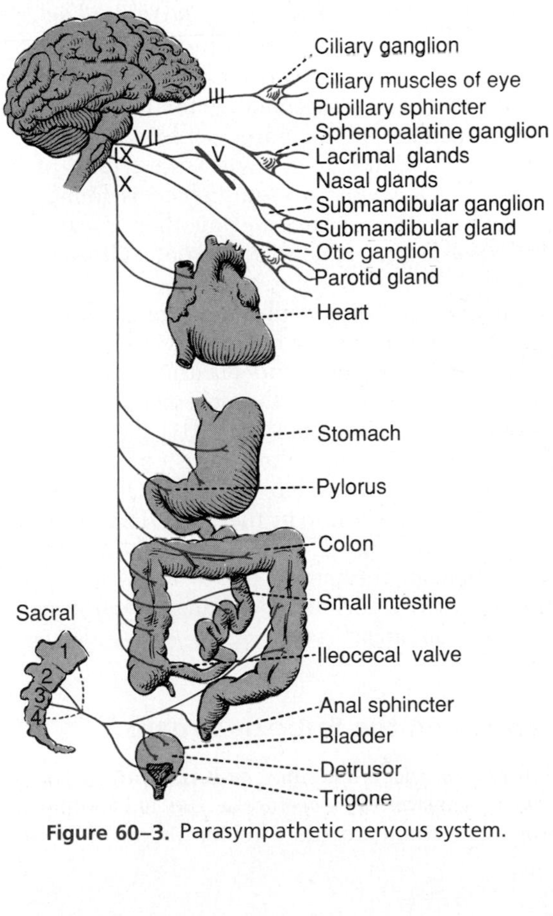

BASIC CHARACTERISTICS OF SYMPATHETIC AND PARASYMPATHETIC FUNCTION

Cholinergic and Adrenergic Fibers— Secretion of Acetylcholine or Norepinephrine

The sympathetic and parasympathetic nerve fibers secrete one or the other of two synaptic transmitter substances, *acetylcholine* or *norepinephrine*. Those that secrete acetylcholine are said to be *cholinergic*. Those that secrete norepinephrine are said to be *adrenergic*, a term derived from *adrenalin*, which is the British name for epinephrine.

All *preganglionic neurons* are *cholinergic* in both the sympathetic and the parasympathetic nervous systems. Therefore, acetylcholine or acetylcholine-like substances, when applied to the ganglia, will excite both sympathetic and parasympathetic postganglionic neurons.

Either *all or almost all of the postganglionic neurons of the parasympathetic system are also cholinergic.*

On the other hand, *most of the postganglionic sympathetic neurons are adrenergic*, although this is not entirely true because the postganglionic sympathetic nerve fibers to the sweat glands, the piloerector muscles, and a few blood vessels are cholinergic.

Thus, the terminal nerve endings of the parasympathetic system *all or virtually all* secrete *acetylcholine*, and *most* of the sympathetic nerve endings secrete *norepinephrine*. These hormones in turn act on the different organs to cause the respective parasympathetic and sympathetic effects. Therefore, acetylcholine is called the *parasympathetic transmitter* and norepinephrine is called the *sympathetic transmitter*.

The molecular structures of acetylcholine and nor-

Ciliary ganglion
Ciliary muscles of eye
Pupillary sphincter
Sphenopalatine ganglion
Lacrimal glands
Nasal glands
Submandibular ganglion
Submandibular gland
Otic ganglion
Parotid gland
Heart
Stomach
Pylorus
Colon
Small intestine
Ileocecal valve
Anal sphincter
Bladder
Detrusor
Trigone
Sacral

Figure 60–3. Parasympathetic nervous system.

epinephrine are as follows:

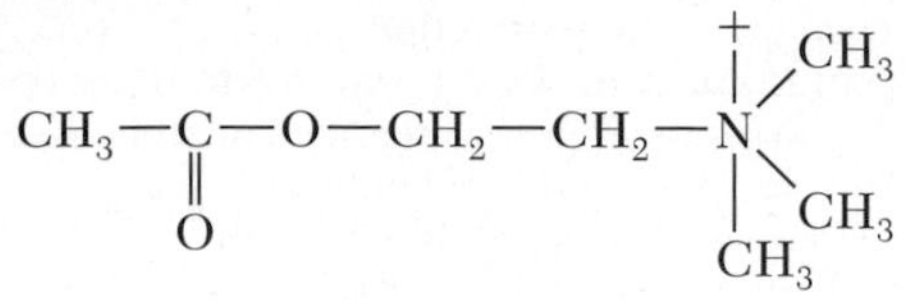

$$CH_3-\underset{\underset{O}{\|}}{C}-O-CH_2-CH_2-\overset{+}{\underset{\underset{CH_3}{|}}{\overset{|}{N}}}\overset{CH_3}{\underset{CH_3}{}}$$

Acetylcholine

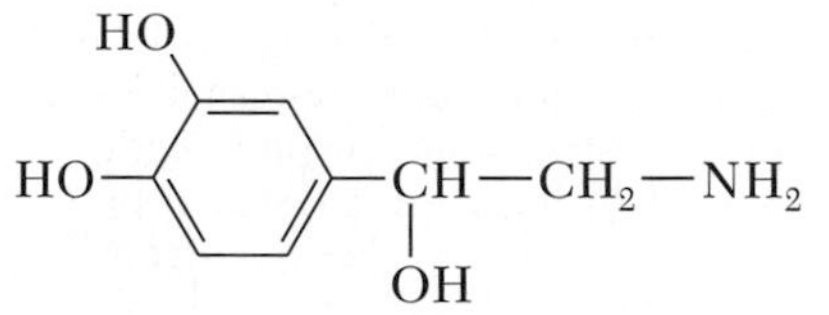

Norepinephrine

Mechanisms of Transmitter Secretion and Removal at the Postganglionic Endings

SECRETION OF ACETYLCHOLINE AND NOREPINEPHRINE BY POSTGANGLIONIC NERVE ENDINGS. A few of the postganglionic autonomic nerve endings, especially those of the parasympathetic nerves, are similar to but much smaller than those of the skeletal neuromuscular junction. However, some of the parasympathetic nerve fibers and almost all the sympathetic fibers merely touch the effector cells of the organs that they innervate as they pass by; in some instances, they terminate in connective tissue located adjacent to the cells that are to be stimulated. Where these filaments pass over or near the effector cells, they usually have bulbous enlargements called *varicosities;* it is in these varicosities that the transmitter vesicles of acetylcholine or norepinephrine are found. Also in the varicosities are large numbers of mitochondria to supply the adenosine triphosphate required to energize acetylcholine or norepinephrine synthesis.

When an action potential spreads over the terminal fibers, the depolarization process increases the permeability of the fiber membrane to calcium ions, allowing them to diffuse into the nerve terminals or nerve varicosities. There the calcium ions interact with those secretory vesicles that are adjacent to the membrane, causing them to fuse with the membrane and to empty their contents to the exterior. Thus, the transmitter substance is secreted.

SYNTHESIS OF ACETYLCHOLINE, ITS DESTRUCTION AFTER SECRETION, AND ITS DURATION OF ACTION. Acetylcholine is synthesized in the terminal endings of cholinergic nerve fibers. Most of this synthesis occurs in the axoplasm outside the vesicles. Then the acetylcholine is transported to the interior of the vesicles, where it is stored in a highly concentrated form until it is released. The basic chemical reaction of this synthesis is the following:

$$\text{Acetyl-CoA + Choline} \xrightarrow{\substack{\textit{choline acetyl-}\\ \textit{transferase}}} \text{Acetylcholine}$$

Once the acetylcholine has been secreted by the cholinergic nerve ending, it persists in the tissue for a few seconds; then most of it is split into an acetate ion and choline by the enzyme *acetylcholinesterase* that is bound with collagen and glycosaminoglycans in the local connective tissue. Thus, this is the same mechanism of acetylcholine destruction that occurs at the neuromuscular junctions of skeletal nerve fibers. The choline that is formed is in turn transported back into the terminal nerve ending, where it is used again for synthesis of new acetylcholine.

SYNTHESIS OF NOREPINEPHRINE, ITS REMOVAL, AND ITS DURATION OF ACTION. Synthesis of norepinephrine begins in the axoplasm of the terminal nerve endings of adrenergic nerve fibers but is completed inside the vesicles. The basic steps are the following:

1. Tyrosine $\xrightarrow{hydroxylation}$ DOPA
2. DOPA $\xrightarrow{decarboxylation}$ Dopamine
3. Transport of dopamine into the vesicles
4. Dopamine $\xrightarrow{hydroxylation}$ Norepinephrine

In the adrenal medulla, this reaction goes still one step further to transform about 80 per cent of the norepinephrine into epinephrine, as follows:

5. Norepinephrine $\xrightarrow{methylation}$ Epinephrine

After secretion of norepinephrine by the terminal nerve endings, it is removed from the secretory site in three ways: (1) reuptake into the adrenergic nerve endings themselves by an active transport process— accounting for removal of 50 to 80 per cent of the secreted norepinephrine; (2) diffusion away from the nerve endings into the surrounding body fluids and then into the blood—accounting for removal of most of the remainder of the norepinephrine; and (3) destruction by enzymes to a slight extent (one of these enzymes is *monoamine oxidase,* which is found in the nerve endings themselves, and another is *catechol-O-methyl transferase,* which is present diffusely in all tissues).

Ordinarily, the norepinephrine secreted directly into a tissue remains active for only a few seconds, demonstrating that its reuptake and diffusion away from the tissue are rapid. However, the norepinephrine and epinephrine secreted into the blood by the adrenal medullae remain active until they diffuse into some tissue, where they are destroyed by catechol-O-methyl transferase; this occurs mainly in the liver. Therefore, when secreted into the blood, both norepinephrine and epinephrine remain very active for 10 to 30 seconds, and then their activity declines, becoming much weaker for 1 to several minutes.

Receptors on the Effector Organs

Before the acetylcholine, norepinephrine, or epinephrine transmitter secreted at the autonomic nerve endings can stimulate the effector organ, it must first bind with highly specific *receptors* on the effector cells. The receptor is on the outside of the cell membrane, bound as a prosthetic group to a protein molecule that penetrates all the way through the cell membrane. When the transmitter binds with the receptor, this causes a conformational change in the structure of the protein molecule. In turn, the altered protein molecule excites or inhibits the cell, most often by (1) causing a change in the cell membrane permeability to one or more ions or (2) activating or inactivating an enzyme attached to the other end of the receptor protein where it protrudes into the interior of the cell.

EXCITATION OR INHIBITION OF THE EFFECTOR CELL BY CHANGING ITS MEMBRANE PERMEABILITY. Because the receptor protein is an integral part of the cell membrane, a conformational change in the structures of the receptor proteins of many organ cells opens or closes *ion channels* through the interstices of the protein molecules themselves, thus altering the permeability of the cell membrane to various ions. For instance, sodium and/or calcium ion channels frequently become opened and allow rapid influx of the respective ions into the cell, usually depolarizing the cell membrane and exciting the cell. At other times, potassium channels are opened, allowing potassium ions to diffuse out of the cell, and this usually inhibits the cell because loss of electropositive potassium ions creates hypernegativity inside the cell.

Also, in some cells, the changed intracellular ion environment will cause an internal cell action, such as the direct effect of calcium ions in promoting smooth muscle contraction.

RECEPTOR ACTION BY ALTERING INTRACELLULAR ENZYMES. Another way the receptor functions is by activating or inactivating an enzyme (or other intracellular chemical) inside the cell. The enzyme often is attached to the receptor protein where the receptor protrudes into the interior of the cell. For instance, binding of epinephrine with its receptor on the outside of many cells increases the activity of the enzyme *adenylcyclase* on the inside of the cell, and this then causes the formation of *cyclic adenosine monophosphate (cAMP)*. The cAMP in turn can initiate any one of many different intracellular actions, the exact effect depending on the chemical machinery of the effector cell.

Therefore, it is easy to understand how an autonomic transmitter substance can cause inhibition in some organs or excitation in others. This is usually determined by the nature of the receptor protein in the cell membrane and the effect of receptor binding on its conformational state. In each organ, the resulting effects are likely to be entirely different from those in other organs.

There Are Two Principal Types of Acetylcholine Receptors—Muscarinic and Nicotinic Receptors

Acetylcholine activates two types of receptors. They are called *muscarinic* and *nicotinic* receptors. The reason for these names is that muscarine, a poison from toadstools, activates only the muscarinic receptors and will not activate the nicotinic receptors, whereas nicotine will activate only the nicotinic receptors; acetylcholine activates both of them.

The muscarinic receptors are found in all effector cells stimulated by the postganglionic neurons of the parasympathetic nervous system as well as in those stimulated by the postganglionic cholinergic neurons of the sympathetic system.

The nicotinic receptors are found in the synapses between the preganglionic and postganglionic neurons of both the sympathetic and parasympathetic systems. (These receptors are also present in many nonautonomic nerve endings—for instance, in the membranes of skeletal muscle fibers at the neuromuscular junction [discussed in Chapter 7].)

An understanding of the two types of receptors is especially important because specific drugs are frequently used in medicine to stimulate or block one or the other of the two types of receptors.

Adrenergic Receptors—Alpha and Beta Receptors

Research experiments using different drugs that mimic the action of norepinephrine on sympathetic effector organs (called *sympathomimetic drugs*) have shown that there are also two major types of adrenergic receptors, *alpha receptors* and *beta receptors*. (The beta receptors in turn are divided into $beta_1$ and $beta_2$ receptors because certain drugs affect only some beta receptors. Also, there is a less distinct division of alpha receptors into $alpha_1$ and $alpha_2$ receptors.)

Norepinephrine and epinephrine, both of which are secreted into the blood by the adrenal medulla, have somewhat different effects in exciting the alpha and beta receptors. Norepinephrine excites mainly alpha receptors but excites the beta receptors to a less extent as well. On the other hand, epinephrine excites both types of receptors approximately equally. Therefore, the relative effects of norepinephrine and epinephrine on different effector organs are determined by the types of receptors in the organs. If they are all beta receptors, epinephrine will be the more effective excitant.

Table 60–1 gives the distribution of alpha and beta receptors in some of the organs and systems controlled by the sympathetics. Note that certain alpha functions are excitatory, whereas others are inhibitory. Likewise, certain beta functions are excitatory and others are inhibitory. Therefore, alpha and beta receptors are not necessarily associated with excitation or inhibition but simply with the affinity of the hormone for the receptors in the given effector organ.

A synthetic hormone chemically similar to epinephrine and norepinephrine, *isopropyl norepinephrine,* has an extremely strong action on beta receptors but essentially no action on alpha receptors.

Table 60–1 ADRENERGIC RECEPTORS AND FUNCTION

Alpha Receptor	Beta Receptor
Vasoconstriction	Vasodilatation (β_2)
Iris dilatation	Cardioacceleration (β_1)
Intestinal relaxation	Increased myocardial strength (β_1)
Intestinal sphincter contraction	Intestinal relaxation (β_2)
	Uterus relaxation (β_2)
Pilomotor contraction	Bronchodilatation (β_2)
Bladder sphincter contraction	Calorigenesis (β_2)
	Glycogenolysis β_2)
	Lipolysis (β_1)
	Bladder wall relaxation (β_2)

Excitatory and Inhibitory Actions of Sympathetic and Parasympathetic Stimulation

Table 60–2 lists the effects on different visceral functions of the body caused by stimulating either the parasympathetic nerves or the sympathetic nerves. From this table, it can be seen again that *sympathetic stimulation causes excitatory effects in some organs but inhibitory effects in others. Likewise, parasympathetic stimulation causes excitation in some but inhibition in others.* Also, when sympathetic stimulation excites a particular organ, parasympathetic stimulation sometimes inhibits it, demonstrating that the two systems occasionally act reciprocally to each other. Most organs are dominantly controlled by one or the other of the two systems.

There is no generalization one can use to explain whether sympathetic or parasympathetic stimulation will cause excitation or inhibition of a particular organ. Therefore, to understand sympathetic and parasympathetic function, one must learn all the separate functions of these two nervous systems on each organ, as listed in Table 60–2. Some of these functions need to be clarified in still greater detail.

Effects of Sympathetic and Parasympathetic Stimulation on Specific Organs

THE EYES. Two functions of the eyes are controlled by the autonomic nervous system. They are the pupillary opening and the focus of the lens. Sympathetic stimulation contracts the *meridional fibers of the iris* that dilate the pupil, whereas parasympathetic stimulation contracts the *circular muscle of the iris* to constrict the pupil. The parasympathetics that control the pupil are reflexly stimulated when excess light enters the eyes, which is explained in Chapter 51; this reflex reduces the pupillary opening and decreases the amount of light that strikes the retina. On the other hand, the sympathetics become stimulated during periods of excitement and therefore increase the pupillary opening at these times.

Focusing of the lens is controlled almost entirely by the parasympathetic nervous system. The lens is normally held in a flattened state by intrinsic elastic tension of its radial ligaments. Parasympathetic excitation con-

tracts the *ciliary muscle,* which releases this tension and allows the lens to become more convex. This causes the eye to focus on objects near at hand. The focusing mechanism is discussed in Chapters 49 and 51 in relation to the function of the eyes.

THE GLANDS OF THE BODY. The *nasal, lacrimal, salivary,* and many *gastrointestinal glands* are strongly stimulated by the parasympathetic nervous system, usually resulting in copious quantities of watery secretion. The glands of the alimentary tract most strongly stimulated by the parasympathetics are those of the upper tract, especially those of the mouth and stomach. The glands of the small and large intestines are controlled principally by local factors in the intestinal tract itself and by the intestinal enteric nervous system and much less by the autonomic nerves.

Sympathetic stimulation has a direct effect on glandular cells in causing formation of a concentrated secretion that contains extra enzymes and mucus. It also causes vasoconstriction of the blood vessels that supply the glands and in this way often reduces their rates of secretion.

The *sweat glands* secrete large quantities of sweat when the sympathetic nerves are stimulated, but no effect is caused by stimulating the parasympathetic nerves. However, the sympathetic fibers to most sweat glands are *cholinergic* (except for a few adrenergic fibers to the palms and soles) in contrast to almost all other sympathetic fibers, which are adrenergic. Furthermore, the sweat glands are stimulated primarily by centers in the hypothalamus that are usually considered to be parasympathetic centers. Therefore, sweating could be called a parasympathetic function, even though it is controlled by nerve fibers that anatomically are distributed through the sympathetic nervous system.

The *apocrine glands* in the axillae secrete a thick, odoriferous secretion as a result of sympathetic stimulation, but they do not react to parasympathetic stimulation. The apocrine glands, despite their close embryological relation to sweat glands, are controlled by adrenergic fibers rather than by cholinergic fibers and by the sympathetic centers of the central nervous system rather than by the parasympathetic centers.

THE GASTROINTESTINAL SYSTEM. The gastrointestinal system has its own intrinsic set of nerves known as the *intramural plexus* or the *intestinal enteric nervous system.* However, both parasympathetic and sympathetic stimulation can affect gastrointestinal activity, mainly by increasing or decreasing specific actions in the intramural plexus. Parasympathetic stimulation, in general, increases the overall degree of activity of the gastrointestinal tract by promoting peristalsis and relaxing the sphincters, thus allowing rapid propulsion of contents along the tract. This propulsive effect is associated with simultaneous increases in rates of secretion by many of the gastrointestinal glands, described earlier.

Normal function of the gastrointestinal tract is not very dependent on sympathetic stimulation. However, strong sympathetic stimulation inhibits peristalsis and increases the tone of the sphincters. The net result is greatly slowed propulsion of food through the tract and sometimes decreased secretion as well.

THE HEART. In general, sympathetic stimulation increases the overall activity of the heart. This is accomplished by increasing both the rate and the force of heart contraction. Parasympathetic stimulation causes mainly the opposite effects. To express these effects in

Table 60–2 AUTONOMIC EFFECTS ON VARIOUS ORGANS OF THE BODY

Organ	Effect of Sympathetic Stimulation	Effect of Parasympathetic Stimulation
Eye		
Pupil	Dilated	Constricted
Ciliary muscle	Slight relaxation (far vision)	Constricted (near vision)
Glands	Vasoconstriction and slight secretion	Stimulation of copious secretion (containing many enzymes for enzyme-secreting glands)
Nasal		
Lacrimal		
Parotid		
Submandibular		
Gastric		
Pancreatic		
Sweat glands	Copious sweating (cholinergic)	Sweating on palms of hands
Apocrine glands	Thick, odoriferous secretion	None
Blood vessels	Most often constricted	Most often little or no effect
Heart		
Muscle	Increased rate	Slowed rate
	Increased force of contraction	Decreased force of contraction (especially of atria)
Coronaries	Dilated (β_2); constricted (α)	Dilated
Lungs		
Bronchi	Dilated	Constricted
Blood vessels	Mildly constricted	? Dilated
Gut		
Lumen	Decreased peristalsis and tone	Increased peristalsis and tone
Sphincter	Increased tone (most times)	Relaxed (most times)
Liver	Glucose released	Slight glycogen synthesis
Gallbladder and bile ducts	Relaxed	Contracted
Kidney	Decreased output and renin secretion	None
Bladder		
Detrusor	Relaxed (slight)	Contracted
Trigone	Contracted	Relaxed
Penis	Ejaculation	Erection
Systemic arterioles		
Abdominal viscera	Constricted	None
Muscle	Constricted (adrenergic α)	None
	Dilated (adrenergic β_2)	
	Dilated (cholinergic)	
Skin	Constricted	None
Blood		
Coagulation	Increased	None
Glucose	Increased	None
Lipids	Increased	None
Basal metabolism	Increased up to 100%	None
Adrenal medullary secretion	Increased	None
Mental activity	Increased	None
Piloerector muscles	Contracted	None
Skeletal muscle	Increased glycogenolysis	None
	Increased strength	
Fat cells	Lipolysis	None

another way, sympathetic stimulation increases the effectiveness of the heart as a pump, as is required during heavy exercise, whereas parasympathetic stimulation decreases its pumping capability but allows the heart some degree of rest between bouts of strenuous activity.

SYSTEMIC BLOOD VESSELS. Most systemic blood vessels, especially those of the abdominal viscera and the skin of the limbs, are constricted by sympathetic stimulation. Parasympathetic stimulation has almost no effects on most blood vessels except to dilate vessels in certain restricted areas, such as in the blush area of the face. Under some conditions, the beta function of the sympathetics causes vascular dilatation instead of the usual sympathetic vascular constriction, but this occurs rarely except after drugs have paralyzed the sympathetic alpha

vasoconstrictor effects, which are usually by far dominant over the beta effects.

EFFECT OF SYMPATHETIC AND PARASYMPATHETIC STIMULATION ON ARTERIAL PRESSURE. The arterial pressure is determined by two factors, the propulsion of blood by the heart and the resistance to flow of this blood through the blood vessels. Sympathetic stimulation increases both propulsion by the heart and resistance to flow, which usually causes the arterial pressure to increase greatly.

On the other hand, parasympathetic stimulation decreases the pumping by the heart but has virtually no effect on peripheral resistance. The usual effect is a slight fall in pressure. Yet strong vagal parasympathetic stimulation can almost stop or occasionally stop the heart entirely and cause loss of all or most arterial pressure.

EFFECTS ON SYMPATHETIC AND PARASYMPATHETIC STIMULATION ON OTHER FUNCTIONS OF THE BODY. Because of the great importance of the sympathetic and parasympathetic control systems, they are discussed many times in this text in relation to myriad body functions that are not considered in detail here. In general, most of the entodermal structures, such as the ducts of the liver, gallbladder, ureter, bladder, and bronchi, are inhibited by sympathetic stimulation but excited by parasympathetic stimulation. Sympathetic stimulation also has metabolic effects, causing release of glucose from the liver, increase in blood glucose concentration, increase in glycogenolysis in both liver and muscle, increase in muscle strength, increase in basal metabolic rate, and increase in mental activity. Finally, the sympathetics and parasympathetics are involved in the execution of the male and female sexual acts, as explained in Chapters 80 and 81.

Function of the Adrenal Medullae

Stimulation of the sympathetic nerves to the adrenal medullae causes large quantities of epinephrine and norepinephrine to be released into the circulating blood, and these two hormones in turn are carried in the blood to all tissues of the body. On the average, about 80 per cent of the secretion is epinephrine and 20 per cent is norepinephrine, although the relative proportions can change considerably under different physiological conditions.

The circulating epinephrine and norepinephrine have almost the same effects on the different organs as those caused by direct sympathetic stimulation, except that *the effects last 5 to 10 times as long* because these hormones are removed from the blood slowly.

The circulating norepinephrine causes constriction of essentially all the blood vessels of the body; it causes increased activity of the heart, inhibition of the gastrointestinal tract, dilation of the pupils of the eyes, and so forth.

Epinephrine causes almost the same effects as those caused by norepinephrine, but the effects differ in the following respects: First, epinephrine, because of its greater effect in stimulating the beta receptors, has a greater effect on cardiac stimulation than does norepinephrine. Second, epinephrine causes only weak constriction of the blood vessels in the muscles, in comparison with much stronger constriction caused by norepinephrine. Because the muscle vessels represent a major segment of the vessels of the body, this difference is of special importance because norepinephrine greatly increases the total peripheral resistance and elevates arterial pressure, whereas epinephrine raises the arterial pressure to a lesser extent but increases the cardiac output considerably more because of its excitatory effect on the heart.

A third difference between the actions of epinephrine and norepinephrine relates to their effects on tissue metabolism. Epinephrine has 5 to 10 times as great a metabolic effect as norepinephrine. Indeed, the epinephrine secreted by the adrenal medullae can increase the metabolic rate of the whole body often to as much as 100 per cent above normal, in this way increasing the activity and excitability of the body. It also increases the rate of other metabolic activities, such as glycogenolysis in the liver and muscle and glucose release into the blood.

In summary, stimulation of the adrenal medullae causes the release of hormones that have almost the same effects throughout the body as direct sympathetic stimulation, except that the effects are greatly prolonged, 1 to 2 minutes after the stimulation is over. The only significant differences are caused by the beta effects of the epinephrine in the secretion, which mainly increase the rate of metabolism and cardiac output to a greater extent than is caused by direct sympathetic nerve stimulation, which releases only norepinephrine.

VALUE OF THE ADRENAL MEDULLAE TO THE FUNCTION OF THE SYMPATHETIC NERVOUS SYSTEM. Epinephrine and norepinephrine are almost always released by the adrenal medullae at the same time that the different organs are stimulated directly by generalized sympathetic activation. Therefore, the organs are actually stimulated in two ways simultaneously: directly by the sympathetic nerves and indirectly by the medullary hormones. The two means of stimulation support each other, and either can, in most instances, substitute for the other. For instance, destruction of the direct sympathetic pathways to the different body organs does not abrogate excitation of the organs because norepinephrine and epinephrine are still released into the circulating blood and indirectly cause stimulation. Likewise, loss of the two adrenal medullae usually has little effect on the operation of the sympathetic nervous system because the direct pathways can still perform almost all the necessary duties. Thus, the dual mechanism of sympathetic stimulation provides a safety factor, one mechanism substituting for the other when it is missing.

Another important value of the adrenal medullae is the capability of epinephrine and norepinephrine to stimulate structures of the body that are not innervated by direct sympathetic fibers. For instance, the metabolic rate of every cell of the body is increased by these hormones, especially by epinephrine, even though only a small proportion of all the cells in the body are innervated directly by sympathetic fibers.

Relation of Stimulus Rate to Degree of Sympathetic and Parasympathetic Effect

A special difference between the autonomic nervous system and the skeletal nervous system is that only a low frequency of stimulation is required for full activation of autonomic effectors. In general, only one nerve impulse every second or so suffices to maintain normal sympathetic or parasympathetic effect, and full activation occurs when the nerve fibers discharge 10 to 20 times per second. This compares with full activation in the skeletal nervous system at 50 to 500 or more impulses per second.

Sympathetic and Parasympathetic "Tone"

The sympathetic and parasympathetic systems are continually active, and the basal rates of activity are known, respectively, as *sympathetic tone* and *parasympathetic tone.*

The value of tone is that *it allows a single nervous system to increase or decrease the activity of a stimulated organ.* For instance, sympathetic tone normally keeps almost all the systemic arterioles constricted to about one half their maximum diameter. By increasing the degree of sympathetic stimulation, these vessels can be constricted even more; on the other hand, by inhibiting the normal tone, they can be dilated. If it were not for the continual background sympathetic tone, the sympathetic system could cause only vasoconstriction, never vasodilatation.

Another interesting example of tone is the background "tone" of the parasympathetics in the gastrointestinal tract. Surgical removal of the parasympathetic supply to most of the gut by cutting the vagus nerves can cause serious and prolonged gastric and intestinal "atony" with resulting blockage of gastrointestinal propulsion and consequent serious constipation, thus demonstrating that parasympathetic tone to the gut is normally very much required. This tone can be decreased by the brain, thereby inhibiting gastrointestinal motility, or it can be increased, thereby promoting increased gastrointestinal activity.

TONE CAUSED BY BASAL SECRETION OF EPINEPHRINE AND NOREPINEPHRINE BY THE ADRENAL MEDULLAE. The normal resting rate of secretion by the adrenal medullae is about 0.2 μg/kg/min of epinephrine and about 0.05 μg/kg/min of norepinephrine. These quantities are considerable—indeed, enough to maintain the blood pressure almost up to the normal value even if all direct sympathetic pathways to the cardiovascular system are removed. Therefore, it is obvious that much of the overall tone of the sympathetic nervous system results from basal secretion of epinephrine and norepinephrine in addition to the tone resulting from direct sympathetic stimulation.

EFFECT OF LOSS OF SYMPATHETIC OR PARASYMPATHETIC TONE AFTER DENERVATION. Immediately after a sympathetic or parasympathetic nerve is cut, the innervated organ loses its sympathetic or parasympathetic tone. In the case of the blood vessels, for instance, cutting the sympathetic nerves results immediately in almost maximal vasodilatation. However, over minutes, hours, days, or weeks, *intrinsic tone* in the smooth muscle of the vessels increases—that is, increased tone caused by increased muscle contractile

force that is *not* the result of sympathetic stimulation but of chemical adaptations in the smooth muscle fibers themselves. This intrinsic tone eventually restores almost normal vasoconstriction.

Essentially the same events occur in most effector organs whenever sympathetic or parasympathetic tone is lost. That is, intrinsic compensation soon develops to return the function of the organ almost to its normal basal level. However, in the parasympathetic system, the compensation sometimes requires many months. For instance, loss of parasympathetic tone to the heart after cardiac vagotomy increases the heart rate to 160 beats per minute in a dog, and this will still be partially elevated 6 months later.

Denervation Supersensitivity of Sympathetic and Parasympathetic Organs After Denervation

During the first week or so after a sympathetic or parasympathetic nerve is destroyed, the innervated organ becomes more and more sensitive to injected norepinephrine or acetylcholine, respectively. This effect is demonstrated in Figure 60–4, showing the blood flow in the forearm before removal of the sympathetics to be about 200 ml/min; a test dose of norepinephrine causes only a slight depression in flow. Then the stellate ganglion is removed, and normal sympathetic tone is lost. At first, the blood flow rises markedly because of the lost vascular tone, but over a period of days to weeks the blood flow returns almost to normal because of progressive increase in intrinsic tone of the vascular musculature itself, thus compensating for the loss of sympathetic tone. Another test dose of norepinephrine is then administered, and the blood flow decreases much more than before, demonstrating that the blood vessels become about two to four times as responsive to norepinephrine as previously. This phenomenon is called *denervation supersensitivity.* It occurs in both sympathetic and parasympathetic organs but to a far greater extent in some organs than in others, often increasing the response more than 10-fold.

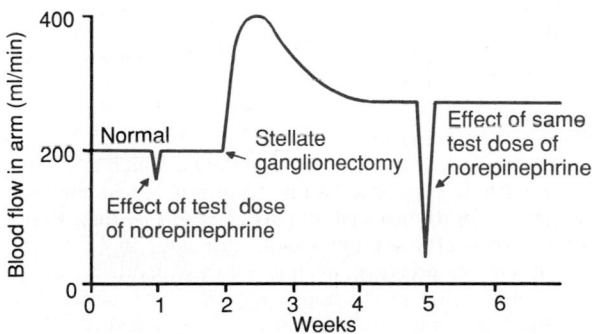

Figure 60–4. Effect of sympathectomy on blood flow in the arm and the effect of a test dose of norepinephrine before and after sympathectomy, showing sensitization of the vasculature to norepinephrine.

MECHANISM OF DENERVATION SUPERSENSITIVITY. The cause of denervation supersensitivity is only partially known. Part of the answer is that the number of receptors in the postsynaptic membranes of the effector cells increases—sometimes manyfold—when norepinephrine or acetylcholine is no longer released at the synapses, a process called "up-regulation" of the receptors. Therefore, when a dose of the hormone is now injected into the circulating blood, the effector reaction is vastly enhanced.

AUTONOMIC REFLEXES

Many of the visceral functions of the body are regulated by *autonomic reflexes.* Throughout this text the functions of these reflexes are discussed in relation to individual organ systems; to illustrate their importance, a few are presented here briefly.

CARDIOVASCULAR AUTONOMIC REFLEXES. Several reflexes in the cardiovascular system help to control especially the arterial blood pressure and the heart rate. One of these is the *baroreceptor reflex,* which is described in Chapter 18 along with other cardiovascular reflexes. Briefly, stretch receptors called *baroreceptors* are located in the walls of the major arteries, including the carotid arteries and the aorta. When these become stretched by high pressure, signals are transmitted to the brain stem, where they inhibit the sympathetic impulses to the heart and blood vessels; this allows the arterial pressure to fall back toward normal.

GASTROINTESTINAL AUTONOMIC REFLEXES. The uppermost part of the gastrointestinal tract as well as the rectum are controlled principally by autonomic reflexes. For instance, the smell of appetizing food or the presence of food in the mouth initiates signals from the nose and mouth to the vagal, glossopharyngeal, and salivatory nuclei of the brain stem. These in turn transmit signals through the parasympathetic nerves to the secretory glands of the mouth and stomach, causing secretion of digestive juices even before food enters the mouth. And when fecal matter fills the rectum at the other end of the alimentary canal, sensory impulses initiated by stretching the rectum are sent to the sacral portion of the spinal cord and a reflex signal is retransmitted through the parasympathetics to the distal parts of the colon; these result in strong peristaltic contractions that cause defecation.

OTHER AUTONOMIC REFLEXES. Emptying of the bladder is controlled in the same way as emptying the rectum; stretching of the bladder sends impulses to the sacral cord, and this in turn causes reflex contraction of the bladder and relaxation of the urinary sphincters, thereby promoting micturition.

Also important are the sexual reflexes, which are initiated both by psychic stimuli from the brain and by stimuli from the sexual organs. Impulses from these sources converge on the sacral cord and, in the man, result first in erection, mainly a parasympathetic function, and then in ejaculation, a sympathetic function.

Other autonomic reflexes include reflex contributions to the regulation of pancreatic secretion, gallbladder emptying, kidney excretion of urine, sweating, blood glucose concentration, and many other visceral functions, all of which are discussed in detail at other points in this text.

STIMULATION OF DISCRETE ORGANS IN SOME INSTANCES AND MASS STIMULATION IN OTHER INSTANCES BY THE SYMPATHETIC AND PARASYMPATHETIC SYSTEMS

THE SYMPATHETIC SYSTEM OFTEN RESPONDS BY MASS DISCHARGE. In many instances, the sympathetic nervous system discharges almost as a complete unit, a phenomenon called *mass discharge.* This frequently occurs when the hypothalamus is activated by fright or fear or severe pain. The result is a widespread reaction throughout the body called the *alarm* or *stress response,* which we shall discuss shortly.

At other times, sympathetic activation occurs in isolated portions of the system, mainly in response to reflexes that involve the spinal cord but not the brain. The most important of these are the following: (1) In the process of heat regulation, the sympathetics control sweating and blood flow in the skin without affecting other organs innervated by the sympathetics. (2) During muscular activity in some animals, specific cholinergic vasodilator fibers of the skeletal muscles are stimulated independently, apart from the remainder of the sympathetic system. (3) Many "local reflexes" involving sensory afferent fibers that travel centrally in the sympathetic nerves to the sympathetic ganglia and spinal cord cause highly localized reflex responses. For instance, heating a local skin area causes local vasodilatation and enhanced local sweating, whereas cooling causes the opposite effects. (4) Many of the sympathetic reflexes that control gastrointestinal functions are discrete, operating sometimes by way of nerve pathways that do not even enter the spinal cord, merely passing from the gut to the sympathetic ganglia, mainly the prevertebral ganglia, and then back to the gut through the sympathetic nerves to control motor or secretory activity.

THE PARASYMPATHETIC SYSTEM USUALLY CAUSES SPECIFIC LOCALIZED RESPONSES. In contrast to the common mass discharge response of the sympathetic system, control functions of the parasympathetic system are much more likely to be highly specific. For instance, parasympathetic cardiovascular reflexes usually act only on the heart to increase or decrease its rate of beating. Likewise, other parasympathetic reflexes cause secretion mainly in the mouth glands, whereas in other instances secretion is mainly in the stomach glands. Finally, the rectal emptying reflex does not affect other parts of the bowel to a major extent.

Yet there is often association between closely allied parasympathetic functions. For instance, although salivary secretion can occur independently of gastric secretion, these two also often occur together, and pancreatic secretion frequently occurs at the same time. Also, the rectal emptying reflex often initiates a bladder emptying reflex, resulting in simultaneous emptying of both the bladder and the rectum. Conversely, the bladder emptying reflex can help initiate rectal emptying.

"Alarm" or "Stress" Response of the Sympathetic Nervous System

When large portions of the sympathetic nervous system discharge at the same time—that is, a *mass discharge*—this increases in many ways the ability of the body to perform vigorous muscle activity. Let us summarize these ways:

1. Increased arterial pressure
2. Increased blood flow to active muscles concurrent with decreased blood flow to organs such as the gastrointestinal tract and the kidneys that are not needed for rapid motor activity
3. Increased rates of cellular metabolism throughout the body
4. Increased blood glucose concentration
5. Increased glycolysis in the liver and in muscle
6. Increased muscle strength
7. Increased mental activity
8. Increased rate of blood coagulation

The sum of these effects permits a person to perform far more strenuous physical activity than would otherwise be possible. Because it is mental or physical *stress* that usually excites the sympathetic system, it is frequently said that the purpose of the sympathetic system is to provide extra activation of the body in states of stress: this is called the sympathetic *stress response.*

The sympathetic system is especially strongly activated in many emotional states. For instance, in the state of *rage,* which is elicited mainly by stimulating the hypothalamus, signals are transmitted downward through the reticular formation of the brain stem and into the spinal cord to cause massive sympathetic discharge, and all the sympathetic events listed above ensue immediately. This is called the sympathetic *alarm reaction.* It is also called the *fight or flight reaction* because an animal in this state decides almost instantly whether to stand and fight or to run. In either event, the sympathetic alarm reaction makes the animal's subsequent activities vigorous.

Medullary, Pontine, and Mesencephalic Control of the Autonomic Nervous System

Many areas in the reticular substance and tractus solitarius of the medulla, pons, and mesencephalon as well as many special nuclei (Fig. 60–5) control different autonomic functions, such as arterial pressure, heart rate, glandular secretion in the gastrointestinal tract, gastrointestinal peristalsis, and degree of contraction of the urinary bladder. The control of each of these is discussed at appropriate points in this text. Suffice it to point out here that the most important factors controlled in the brain stem are arterial pressure, heart rate, and respiratory rate. Indeed, transection of the brain stem above the midpontine level allows normal basal control of arterial pressure to continue as before but prevents its modulation by higher nervous centers, particularly the hypothalamus. On the other hand, transection immediately below the medulla causes the arterial pressure to fall to less than one-half normal for several hours or several days after the transection.

Closely associated with the cardiovascular regulatory centers in the medulla are the medullary and pontine centers for regulation of respiration, which are discussed in Chapter 41. Although this is not considered to be an autonomic function, it is one of the *involuntary* functions of the body.

CONTROL OF BRAIN STEM AUTONOMIC CENTERS BY HIGHER AREAS. Signals from the hypothalamus and even from the cerebrum can affect the activities of almost all the brain stem autonomic control centers. For instance, stimulation in appropriate areas of the hypothalamus can activate the medullary cardiovascular control centers strongly enough to increase the arterial pressure to more than double normal. Likewise, other hypothalamic centers can control body temperature, increase or decrease salivation and gastrointestinal activity, or cause bladder emptying. To some extent, therefore, the autonomic centers in the brain stem act as relay stations for control activities initiated at higher levels of the brain.

In Chapters 58 and 59, it is pointed out that many of our behavioral responses are mediated through the hypothalamus, the reticular areas of the brain stem,

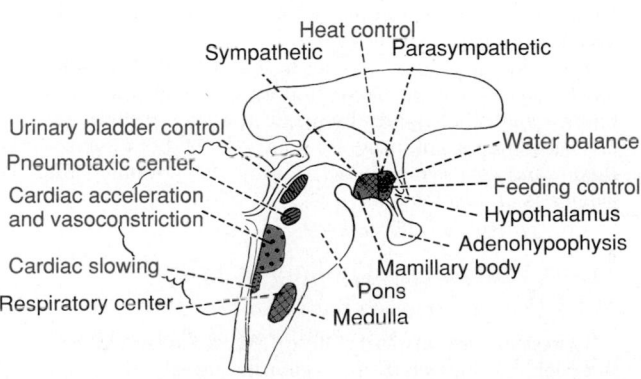

Figure 60–5. Autonomic control areas of the brain stem and hypothalamus.

and the autonomic nervous system. Indeed, the higher areas of the brain can alter the function of the whole autonomic nervous system or of portions of it strongly enough to cause severe autonomic-induced disease, such as peptic ulcer, constipation, heart palpitation, and even heart attacks.

PHARMACOLOGY OF THE AUTONOMIC NERVOUS SYSTEM

Drugs That Act on Adrenergic Effector Organs—The Sympathomimetic Drugs

From the foregoing discussion, it is obvious that intravenous injection of norepinephrine causes essentially the same effects throughout the body as sympathetic stimulation. Therefore, norepinephrine is called a *sympathomimetic*, or *adrenergic*, *drug*. Epinephrine and *methoxamine* are also sympathomimetic drugs, and there are many others. They differ from one another in the degree to which they stimulate different sympathetic effector organs and in their duration of action. Norepinephrine and epinephrine have actions as short as 1 to 2 minutes, whereas the actions of most other commonly used sympathomimetic drugs last for 30 minutes to 2 hours.

Important drugs that stimulate specific adrenergic receptors but not the others are *phenylephrine*—alpha receptors; *isoproterenol*—beta receptors; and *albuterol*—only beta$_2$ receptors.

Drugs That Cause Release of Norepinephrine from Nerve Endings. Certain drugs have an indirect sympathomimetic action rather than directly exciting adrenergic effector organs. These drugs include *ephedrine*, *tyramine*, and *amphetamine*. Their effect is to cause release of norepinephrine from its storage vesicles in the sympathetic nerve endings. The released norepinephrine in turn causes the sympathetic effects.

Drugs That Block Adrenergic Activity. Adrenergic activity can be blocked at several points in the stimulatory process as follows:

1. The synthesis and storage of norepinephrine in the sympathetic nerve endings can be prevented. The best known drug that causes this effect is *reserpine*.

2. Release of norepinephrine from the sympathetic endings can be blocked. This is caused by *guanethidine*.

3. The *alpha* receptors can be blocked. Two drugs that cause this effect are *phenoxybenzamine* and *phentolamine*.

4. The beta receptors can be blocked. A drug that blocks all beta receptors is *propranolol*. One that blocks only beta$_1$ receptors is *metoprolol*.

5. Sympathetic activity can be blocked by drugs that block transmission of nerve impulses through the autonomic ganglia. They are discussed in a later section, but the most important drug for blockade of both sympathetic and parasympathetic transmission through the ganglia is *hexamethonium*.

Drugs That Act on Cholinergic Effector Organs

Parasympathomimetic Drugs (Muscarinic Drugs). Acetylcholine injected intravenously usually does not cause exactly the same effects throughout the body as

parasympathetic stimulation because the acetylcholine is destroyed by cholinesterase in the blood and body fluids before it can reach all the effector organs. Yet a number of other drugs that are not so rapidly destroyed can produce typical parasympathetic effects, and they are called *parasympathomimetic drugs*.

Two commonly used parasympathomimetic drugs are *pilocarpine* and *methacholine*. They act directly on the muscarinic type of cholinergic receptors.

Parasympathomimetic drugs act on the effector organs of cholinergic *sympathetic* fibers also. For instance, these drugs cause profuse sweating. Also, they cause vascular dilatation in some organs, this effect occurring even in some vessels not innervated by cholinergic fibers.

Drugs That Have a Parasympathetic Potentiating Effect—Anticholinesterase Drugs. Some drugs do not have a direct effect on parasympathetic effector organs but do potentiate the effects of the naturally secreted acetylcholine at the parasympathetic endings. They are the same drugs as those discussed in Chapter 7 that potentiate the effect of acetylcholine at the neuromuscular junction. They include *neostigmine*, *pyridostigmine*, and *ambenonium*. These drugs inhibit acetylcholinesterase, thus preventing rapid destruction of the acetylcholine liberated by the parasympathetic nerve endings. As a consequence, the quantity of acetylcholine acting on the effector organs progressively increases with successive stimuli and the degree of action also increases.

Drugs That Block Cholinergic Activity at Effector Organs—Antimuscarinic Drugs. Atropine and similar drugs, such as *homatropine* and *scopolamine*, block the action of acetylcholine on the muscarinic type of cholinergic effector organs. These drugs do not affect the nicotinic action of acetylcholine on the postganglionic neurons or on skeletal muscle.

Drugs That Stimulate or Block Sympathetic and Parasympathetic Postganglionic Neurons

Drugs That Stimulate Autonomic Postganglionic Neurons. The preganglionic neurons of both the parasympathetic and the sympathetic nervous systems secrete acetylcholine at their endings, and this acetylcholine in turn stimulates the postganglionic neurons. Furthermore, injected acetylcholine can also stimulate the postganglionic neurons of both systems, thereby causing at the same time both sympathetic and parasympathetic effects throughout the body. *Nicotine* is a drug that can also stimulate postganglionic neurons in the same manner as acetylcholine because the membranes of these neurons all contain the *nicotinic type of acetylcholine receptor*. Therefore, drugs that cause autonomic effects by stimulating the postganglionic neurons are called *nicotinic drugs*. Some drugs, such as *acetylcholine* itself and *methacholine*, have both nicotinic and muscarinic actions, whereas pilocarpine has only muscarinic actions.

Nicotine excites both the sympathetic and the parasympathetic postganglionic neurons at the same time, resulting in strong sympathetic vasoconstriction in the abdominal organs and limbs but at the same time resulting in parasympathetic effects, such as increased gastrointestinal activity and, sometimes, slowing of the heart.

GANGLIONIC BLOCKING DRUGS. Many important drugs block impulse transmission from the preganglionic neurons to the postganglionic neurons, including *tetraethyl ammonium ion, hexamethonium ion,* and *pentolinium.* These drugs block acetylcholine stimulation of the postganglionic neurons in both the sympathetic and the parasympathetic systems simultaneously. They are often used for blocking sympathetic activity but seldom for blocking parasympathetic activity because the effects of sympathetic blockade usually far overshadow the effects of parasympathetic blockade. The ganglionic blocking drugs can especially reduce the arterial pressure in patients with hypertension, but these drugs are not very useful for this purpose because their effects are difficult to control.

REFERENCES

Andresen, M. C., and Kunze, D. L.: Nucleus tractus solitarius—gateway to neural circulatory control. Annu. Rev. Physiol., 56:93, 1994.

Bannister, Sir R. (ed.): Autonomic Failure. New York, Oxford University Press, 1988.

Buckley, J. P., et al. (eds.): Brain Peptides and Catecholamines in Cardiovascular Regulation. New York, Raven Press, 1987.

Burchfield, S. R. (ed.): Stress. Physiological and Psychological Interactions. Washington, D.C., Hemisphere Publishing Corp., 1985.

Burt, A. M.: Textbook of Neuroanatomy. Philadelphia, W. B. Saunders Co., 1993.

Carter, L. P., et al.: Neurovascular surgery. Hightstown, NJ, McGraw-Hill, 1994.

Christensen, N. J., and Galbo, H.: Sympathetic nervous activity during exercise. Annu. Rev. Physiol., 45:139, 1983.

Christensen, N. J., et al.: The Sympathoadrenal System: Physiology and Pathophysiology. New York, Raven Press, 1986.

Conn, P. M.: Neuroscience in Medicine. Philadelphia, J. B. Lippincott, 1994.

Cotman, C. W., et al. (eds.): The Neuro-Immune-Endocrine Connection. New York, Raven Press, 1987.

Dampney, R. A. L.: Functional organization of central pathways regulating the cardiovascular system. Physiol Rev., 74:323, 1994.

Fedida, D., et al.: α_1-Adrenoceptors in myocardium: functional aspects and transmembrane signaling mechanisms. Physiol. Rev., 73:469, 1993.

Fillenz, M.: Noradrenergic neurons. New York, Cambridge University Press, 1990.

Givens, J. R.: The Hypothalamus in Health and Disease. Chicago, Year Book Medical Publishers, 1984.

Greenspan, F. S., and Baxter, J. D.: Basic and Clinical Endocrinology. 4th Ed. Redding, MA, Appleton & Lange, 1994.

Goldstein, D. S.: Stress, Catecholamines, and Cardiovascular Disease. New York, Oxford University Press, 1995.

Guyton, A. C., and Gillespie, W. M., Jr.: Constant infusion of epinephrine: Rate of epinephrine secretion and destruction in the body. Am. J. Physiol., 164:319, 1951.

Guyton, A. C., and Reeder, R. C.: Quantitative studies on the autonomic actions of curare. J. Pharmacol. Exp. Ther., 98:188, 1950.

Herd, J. A.: Cardiovascular response to stress. Physiol. Rev., 71:305, 1991.

Hirst, G. D. S., and Edwards, F. R.: Sympathetic neuroeffector transmission in arteries and arterioles. Physiol. Rev., 69:546, 1989.

Janig, W.: Pre- and postganglionic vasoconstrictor neurons: Differentiation, types, and discharge properties. Annu. Rev. Physiol., 50:525, 1988.

Nijkamp, F. P., et al.: Mechanisms of β-adrenergic receptor regulation in lungs and its implications for physiological responses. Physiol. Rev., 72:323, 1992.

Perkins, J. D.: The Beta-Adrenergic Receptors. Totowa, NJ, Humana Press, Inc., 1991.

Rowell, L. B.: Reflex control of regional circulations in humans. J. Auton. Nerv. Syst., 11:101, 1984.

Stella, A., and Zanchetti, A.: Functional role of renal afferents. Physiol. Rev., 71:659, 1991.

Stiles, G. L., et al.: β-Adrenergic receptors: Biochemical mechanisms of physiological regulation. Physiol. Rev., 64:661, 1984.

Strange, P. G.: Brain Biochemistry and Brain Disorders. New York, Oxford University Press, 1993.

Ungar, A., and Phillips, J. H.: Regulation of the adrenal medulla. Physiol. Rev., 63:787, 1983.

Usdin, E.: Stress. The Role of Catecholamines and Other Neurotransmitters. New York, Gordon Press Publishers, 1984.

van Giersbergen, P. L. M., et al.: Involvement of neurotransmitters in the nucleus tractus solitarii in cardiovascular regulation. Physiol. Rev., 72:789, 1992.

Vanhoutte, P. M.: Vasodilatation: Vascular Smooth Muscle, Peptides, Autonomic Nerves, and Endothelium. New York, Raven Press, 1988.

Youmans, J. R. (ed.): Neurological Surgery. Philadelphia, W. B. Saunders Co., 1989.

Cerebral Blood Flow, the Cerebrospinal Fluid, and Brain Metabolism

CHAPTER 61

Thus far, we have discussed the function of the brain as if it were independent of its blood flow, its metabolism, and its fluids. However, this is far from the truth because abnormalities of any of these can profoundly affect brain function. For instance, total cessation of blood flow to the brain causes unconsciousness within 5 to 10 seconds. This is true because lack of oxygen delivery to the brain cells shuts down most of their metabolism. Also, on a longer time scale, abnormalities of the cerebrospinal fluid, either in its composition or in its fluid pressure, can have equally severe effects on brain function.

CEREBRAL BLOOD FLOW

Normal Rate of Cerebral Blood Flow

The normal blood flow through the brain tissue of the adult averages 50 to 65 milliliters per 100 grams of brain per minute. For the entire brain, this is 750 to 900 ml/min, or 15 per cent of the total resting cardiac output.

Regulation of Cerebral Blood Flow

Metabolic Control of Flow

As in most other vascular areas of the body, cerebral blood flow is highly related to the metabolism of the cerebral tissue. At least three metabolic factors have potent effects in controlling cerebral blood flow. They are carbon dioxide concentration, hydrogen ion concentration, and oxygen concentration. An *increase* in either the carbon dioxide or the hydrogen ion concentration increases cerebral blood flow, whereas a *decrease* in oxygen concentration increases the flow.

REGULATION OF CEREBRAL BLOOD FLOW IN RESPONSE TO EXCESS CARBON DIOXIDE OR HYDROGEN ION CONCENTRATION. An increase in carbon dioxide concentration in the arterial blood perfusing the brain greatly increases cerebral blood flow. This is demonstrated in Figure 61-1, which shows that a 70 per cent increase in arterial P_{CO_2} approximately doubles the blood flow.

Carbon dioxide is believed to increase cerebral blood flow almost entirely by combining first with water in the body fluids to form carbonic acid, with subsequent dissociation to form hydrogen ions. The hydrogen ions then cause vasodilatation of the cerebral vessels—the dilatation being almost directly proportional to the increase in hydrogen ion concentration up to a blood flow limit about twice normal.

Any other substance that increases the acidity of the brain tissue, and therefore also increases the hydrogen ion concentration, will increase blood flow as well. Such substances include lactic acid, pyruvic acid, and any other acidic material formed during the course of metabolism.

Importance of Cerebral Blood Flow Control by Carbon Dioxide and Hydrogen Ions. Increased hydrogen ion concentration greatly depresses neuronal activity. Therefore, it is fortunate that an increase in hydrogen ion concentration causes an increase in blood flow, which in turn carries both carbon dioxide and other acidic substances away from the brain tissues. Loss of the carbon dioxide removes carbonic acid from the tissues; this, along with the removal of other acids, reduces the hydrogen ion concentration back toward normal. Thus, this mechanism helps maintain a constant hydrogen ion concentration in the cerebral fluids and thereby helps to maintain the normal level of neuronal activity.

OXYGEN DEFICIENCY AS A REGULATOR OF CEREBRAL BLOOD FLOW. Except during periods of intense brain activity, the utilization of oxygen by the brain tissue remains within narrow limits—within a few percentage points of 3.5 milliliters of oxygen per 100 grams of brain tissue per min-

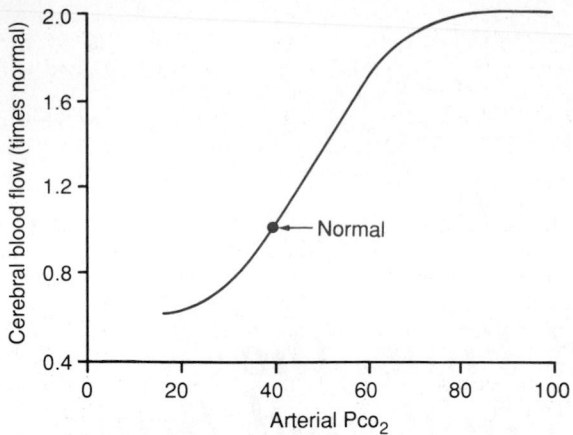

Figure 61–1. Relationship between arterial PCO₂ and cerebral blood flow.

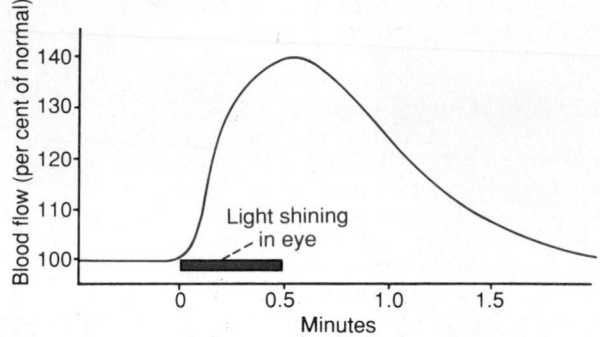

Figure 61–2. Increase in blood flow to the occipital regions of the brain when a light is shone in the eyes of an animal.

ute. If the blood flow to the brain ever becomes insufficient and cannot supply this needed amount of oxygen, the oxygen deficiency mechanism for causing vasodilatation, discussed in Chapter 17, which functions in essentially all tissues of the body, immediately causes vasodilatation, returning the blood flow and transport of oxygen to the cerebral tissues to near normal. Thus, this local blood flow regulatory mechanism is much the same in the brain as in the coronary and skeletal muscle circulation and in many other circulatory areas of the body.

Experiments have shown that a decrease in cerebral *tissue* Po₂ below about 30 mm Hg (normal value is 35 to 40 mm Hg) will immediately begin to increase cerebral blood flow. This is fortuitous because brain function becomes deranged at not much lower values of Po₂, especially so at Po₂ levels below 20 mm Hg. Even coma can result at these low levels. Thus, the oxygen mechanism for local regulation of cerebral blood flow is also an important protective response against diminished cerebral neuronal activity and, therefore, against derangement of mental capability.

Measurement of Cerebral Blood Flow and Effect of Cerebral Activity on the Flow. A method has been developed to record blood flow in as many as 256 isolated segments of the human cerebral cortex simultaneously. A radioactive substance, usually radioactive xenon, is injected into the carotid artery; then the radioactivity of each segment of the cortex is recorded as the radioactive substance passes through the brain tissue. For this to be done, 256 small radioactive scintillation detectors are focused on the same number of separate parts of the cortex; the rapidity of decay of the radioactivity after it once peaks in each tissue segment is a direct measure of the rate of blood flow through the segment.

With this technique, it has become clear that the blood flow in each individual segment of the brain changes within seconds in response to changes in local neuronal activity. For instance, simply making a fist of the hand causes an immediate increase in blood flow in the motor cortex of the opposite side of the brain. Or, reading a book increases the blood flow in multiple areas of the brain, especially in the occipital cortex and in the language perception areas of the temporal cortex. This measuring procedure can also be used for localizing the origin of epileptic attacks because the blood flow increases acutely and markedly at the focal point of the attack at its onset.

Demonstrating the effect of local neuronal activity on cerebral blood flow, Figure 61–2 shows a typical increase in occipital blood flow recorded in a cat when intense light is shone into its eyes for 0.5 minute.

Autoregulation of Cerebral Blood Flow When the Arterial Pressure Is Changed. Cerebral blood flow is autoregulated extremely well between the arterial pressure limits of 60 and 140 mm Hg. That is, the arterial pressure can be decreased acutely to as low as 60 mm Hg or increased to as high as 140 mm Hg without a significant change in cerebral blood flow. In people who have hypertension, this autoregulatory range shifts to even higher pressure levels, up to 180 to 200 mm Hg. This effect is demonstrated in Figure 61–3, which shows cerebral blood flows measured both in normal human beings and in hypertensive patients. Note the extreme constancy of cerebral blood flow between the limits of 60 and 180 mm Hg mean arterial pressure. On the other hand, if the arterial pressure does fall below 60 mm Hg, cerebral blood flow then becomes severely compromised, and if the pressure rises above the upper limit of autoregulation, the blood flow rises rapidly and can cause severe overstretching or rupture of the cerebral blood vessels, sometimes resulting in serious brain edema or cerebral hemorrhage.

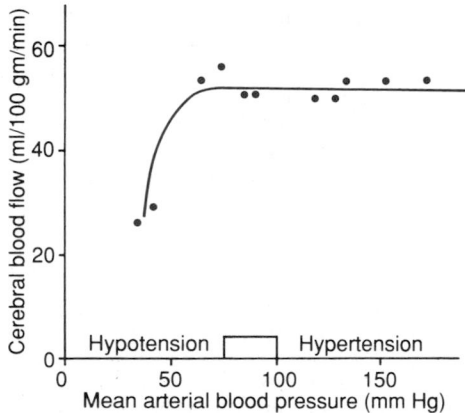

Figure 61–3. Relation of mean arterial pressure to cerebral blood flow in normotensive, hypotensive, and hypertensive people. (Modified from Lassen: *Physiol. Rev., 39*:183, 1959.)

Role of the Sympathetic Nervous System in Regulating Cerebral Blood Flow

The cerebral circulatory system has a strong sympathetic innervation that passes upward from the superior cervical sympathetic ganglia along with the cerebral arteries. This innervation supplies both the large superficial arteries and the small arteries that penetrate into the substance of the brain. Neither transection of these sympathetic nerves nor mild to moderate stimulation of them usually causes significant change in the cerebral blood flow because the blood flow autoregulation mechanism can override the nervous effects. Therefore, it has long been stated that the sympathetic nerves play essentially no role in regulating cerebral blood flow.

However, experiments have now shown that cerebral sympathetic stimulation can, under some conditions, become activated strongly enough to constrict the cerebral arteries markedly. For instance, when the arterial pressure rises to a high level during strenuous exercise and during other states of excessive circulatory activity, the sympathetic nervous system constricts the large- and intermediate-sized arteries enough to prevent the high pressure from reaching the smaller blood vessels. Experiments have shown that this is important in preventing the occurrence of vascular hemorrhages into the brain—that is, for preventing the occurrence of cerebral stroke.

Also, sympathetic reflexes are believed to cause vasospasm in the intermediate-sized and large arteries in some instances of brain damage, such as after a cerebral stroke has occurred or in patients with a subdural hematoma or brain tumor.

Cerebral Microcirculation

As in almost all other tissues of the body, the density of the blood capillaries in the brain is greatest where the metabolic needs are greatest. The overall metabolic rate of the brain gray matter, where the neuronal cell bodies lie, is about four times as great as that of white matter; correspondingly, the number of capillaries and rate of blood flow are also about four times as great in the gray matter.

Another important structural characteristic of the brain capillaries is that they are much less "leaky" than the capillaries in almost any other tissue of the body. Most important, the capillaries are supported on all sides by "glial feet," which are small projections from the surrounding glia that abut against all surfaces of the capillaries and provide physical support to prevent overstretching of the capillaries in case of high pressure. In addition, the walls of the small arterioles leading to the brain capillaries become greatly thickened in people who develop high blood pressure, and these arterioles remain significantly constricted all the time to prevent transmission of the high pressure to the capillaries. We shall see later in the chapter that whenever these systems for protecting against transudation of fluid into the brain break down, serious brain edema ensues, which can lead rapidly to coma and death.

A Cerebral "Stroke" Occurs When Cerebral Blood Vessels Are Blocked

Almost all old people have at least some blockage of the arterial blood supply to the brain, and as many as 10 per cent eventually have enough blockage to cause a disturbance in function, a condition called a "stroke."

Most strokes are caused by arteriosclerotic plaques that occur in one or more of the feeder arteries to the brain. The plaque usually activates the clotting mechanism of the blood, causing a clot to develop and block the artery, thereby leading to acute loss of brain function in a localized area. Or, in about one quarter of people who develop strokes, the cause is high blood pressure that makes one of the blood vessels burst; hemorrhage then occurs, compressing the local brain tissue.

The neurological effects of a stroke are determined by the brain area affected. One of the most common types of stroke is blockage of one of the middle cerebral arteries that supplies the midportion of one brain hemisphere. For instance, if the middle cerebral artery is blocked on the left side of the brain, the person is likely to become almost totally demented because of lost function in Wernicke's speech comprehension area; he or she also becomes unable to speak words because of loss of Broca's motor area for word formation. In addition, lost function in other neural motor control areas of the left hemisphere can create spastic paralysis of all or most muscles on the opposite side of the body.

In a similar manner, blockage of a posterior cerebral artery will cause infarction of the occipital pole of the hemisphere on the same side and loss of vision in both eyes in the half of the retina on the same side as the stroke lesion. Especially devastating are strokes that involve the blood supply to the hindbrain and midbrain because they can block conduction in major pathways between the brain and spinal cord, causing totally incapacitating sensory and motor abnormalities.

CEREBROSPINAL FLUID SYSTEM

The entire cavity enclosing the brain and spinal cord has a volume of about 1600 to 1700 milliliters; about 150 milliliters of this volume is occupied by cerebrospinal fluid and the remainder, by the brain and cord. This fluid, as shown in Figure 61–4, is found in the *ventricles of the brain,* in the *cisterns around the brain,* and in the *subarachnoid space around both the brain and the spinal cord.* All these chambers are connected with one another, and the pressure of the fluid is regulated at a constant level.

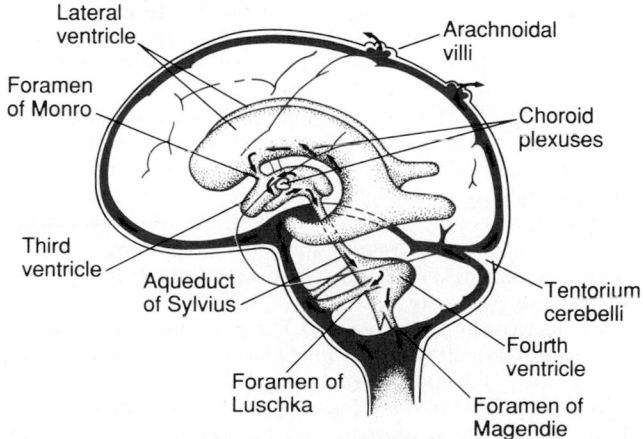

Figure 61–4. Pathway of cerebrospinal fluid flow from the choroid plexuses in the lateral ventricles to the arachnoidal villi protruding into the dural sinuses.

Cushioning Function of the Cerebrospinal Fluid

A major function of the cerebrospinal fluid is to cushion the brain within its solid vault. The brain and the cerebrospinal fluid have about the same specific gravity (only about 4 per cent different), so that the brain simply floats in the fluid. Therefore, a blow to the head moves the entire brain simultaneously with the skull, causing no one portion of the brain to be momentarily contorted by the blow.

CONTRECOUP. When a blow to the head is extremely severe, it usually does not damage the brain on the side of the head where the blow is struck but on the opposite side. This phenomenon is known as "contrecoup," and the reason for this effect is the following: When the blow is struck, the fluid on the struck side is so incompressible that as the skull moves, the fluid pushes the brain at the same time. On the opposite side, the sudden movement of the skull causes it to pull away from the brain momentarily because of the brain's inertia, creating for a split second a vacuum space in the cranial vault at this point. Then, when the skull is no longer being accelerated by the blow, the vacuum suddenly collapses and the brain strikes the inner surface of the skull. Because of this effect, the damage to the brain of a boxer usually does not occur in the frontal regions where he is struck most often but in the occipital regions.

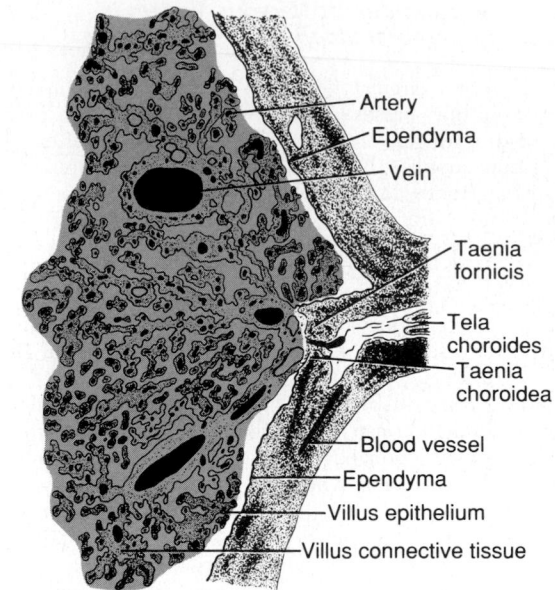

Figure 61–5. Choroid plexus. (Modified from Clara: Das Nervensystem des Menschen. Barth.)

Formation, Flow, and Absorption of Cerebrospinal Fluid

Cerebrospinal fluid is formed at a rate of about 500 milliliters each day, which is three to four times as much as the total volume of fluid in the entire cerebrospinal fluid system. Probably two thirds or more of this fluid originates as a secretion from the choroid plexuses in the four ventricles, mainly in the two lateral ventricles. Additional amounts of fluid are secreted by all the ependymal surfaces of the ventricles and the arachnoidal membranes, and a small amount comes from the brain itself through the perivascular spaces that surround the blood vessels entering the brain.

The arrows in Figure 61–4 show the main channels of fluid flow from the choroid plexuses and then through the cerebrospinal fluid system. The fluid secreted in the lateral ventricles and the third ventricle passes along the *aqueduct of Sylvius* into the fourth ventricle, where a small amount of additional fluid is added. It then passes out of the fourth ventricle through three small openings, two lateral *foramina of Luschka* and a midline *foramen of Magendie,* entering the *cisterna magna,* a large fluid space that lies behind the medulla and beneath the cerebellum. The cisterna magna is continuous with the *subarachnoid space* that surrounds the entire brain and spinal cord. Almost all the cerebrospinal fluid then flows upward from the cisterna magna through the subarachnoid space surrounding the cerebrum. From here, the fluid flows into multiple *arachnoidal villi* that project into the large sagittal venous sinus and other venous sinuses of the cerebrum. Finally, the fluid empties into the venous blood through the surfaces of these villi.

SECRETION BY THE CHOROID PLEXUS. The choroid plexus, a section of which is shown in Figure 61–5, is a cauliflower-like growth of blood vessels covered by a thin layer of epithelial cells. This plexus projects into (1 and 2) the temporal horn of each lateral ventricle, (3) the posterior portion of the third ventricle, and (4) the roof of the fourth ventricle.

The secretion of fluid by the choroid plexus depends mainly on active transport of sodium ions through the epithelial cells that line the outsides of the plexus. The sodium ions in turn pull along large amounts of chloride ions as well because the positive charge of the sodium ion attracts the chloride ion's negative charge. The two of these together increase the quantity of osmotically active substances in the cerebrospinal fluid, which then causes almost immediate osmosis of water through the membrane, thus providing the fluid of the secretion. Less important transport processes move *small amounts of glucose into the cerebrospinal fluid* and *both potassium and bicarbonate ions out of the cerebrospinal fluid into the capillaries.* Therefore, the resulting characteristics of the cerebrospinal fluid become the following: osmotic pressure, approximately equal to that of plasma; sodium ion concentration, also approximately equal to that of plasma; chloride, about 15 per cent greater than in plasma; potassium, approximately 40 per cent less; and glucose, about 30 per cent less.

ABSORPTION OF CEREBROSPINAL FLUID THROUGH THE ARACHNOIDAL VILLI. The *arachnoidal villi* are microscopic fingerlike projections of the arachnoidal membrane through the walls of the venous sinuses. Large conglomerates of these villi are usually found together and form macroscopic structures called *arachnoidal granulations* that can be seen protruding into the sinuses. The endothelial cells covering the villi have been shown by electron microscopy to have large vesicular holes directly through the bodies of the cells. It has been proposed that these are large enough to allow relatively free flow of cerebrospinal fluid, protein molecules, and even particles as large as red and white blood cells into the venous blood.

PERIVASCULAR SPACES AND CEREBROSPINAL FLUID. The blood vessels that supply the brain pass first along the surface of the brain and then penetrate inward, carrying a layer of *pia mater,* the membrane that covers the brain, with them, as shown in Figure 61–6. The pia is only loosely adherent to the vessels, so that a space, the *perivascular space,* exists between it and each vessel. Therefore, perivas-

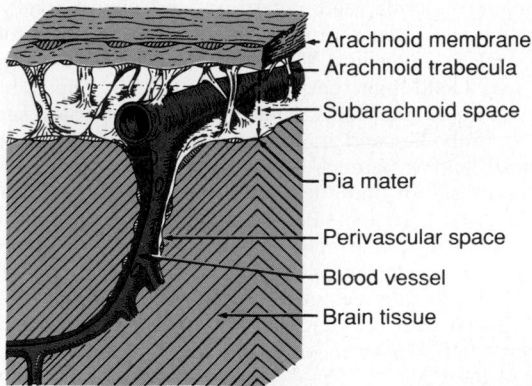

— Arachnoid membrane
— Arachnoid trabecula
— Subarachnoid space
— Pia mater
— Perivascular space
— Blood vessel
— Brain tissue

Figure 61–6. Drainage of the perivascular spaces into the subarachnoid space. (From Ranson and Clark: Anatomy of the Nervous System. Philadelphia, W. B. Saunders Company, 1959.)

cular spaces follow both the arteries and the veins into the brain as far as the arterioles and venules go but not to the capillaries.

Lymphatic Function of the Perivascular Spaces. As is true elsewhere in the body, a small amount of protein leaks out of the parenchymal capillaries into the interstitial spaces of the brain; because no true lymphatics are present in brain tissue, this protein leaves the tissue mainly flowing with fluid through the perivascular spaces into the subarachnoid spaces. On reaching the subarachnoid spaces, the protein then flows with the cerebrospinal fluid to be absorbed through the *arachnoidal villi* into the cerebral veins. The perivascular spaces in effect are a specialized lymphatic system for the brain.

In addition to transporting fluid and proteins, the perivascular spaces transport extraneous particulate matter from the brain into the subarachnoid space. For instance, whenever infection occurs in the brain, dead white blood cells and other infectious debris are carried away through the perivascular spaces.

Cerebrospinal Fluid Pressure

The normal pressure in the cerebrospinal fluid system when one is lying in a horizontal position averages 130 millimeters of water (10 mm Hg), although this may be as low as 65 millimeters of water or as high as 195 millimeters of water even in the normal person.

REGULATION OF CEREBROSPINAL FLUID PRESSURE BY THE ARACHNOIDAL VILLI. The cerebrospinal fluid pressure normally is regulated almost entirely by absorption of the fluid through the arachnoidal villi. The reason for this is that the normal rate of cerebrospinal fluid formation is constant, so that changes in fluid formation are seldom a factor in pressure control. On the other hand, the villi function like "valves" that allow the fluid and its contents to flow readily into the blood of the venous sinuses while not allowing blood to flow backward in the opposite direction. Normally, this valve action of the villi allows cerebrospinal fluid to begin to flow into the blood when its pressure is about 1.5 mm Hg greater than the pressure of the blood in the venous sinuses.

Then, as the cerebrospinal fluid pressure rises still higher, the valves open widely, so that under normal conditions, the pressure almost never rises more than a few millimeters of mercury higher than the pressure in the venous sinuses.

On the other hand, in disease states the villi sometimes become blocked by large particulate matter, by fibrosis, or even by excesses of plasma protein molecules that have leaked into the cerebrospinal fluid in brain diseases. Such blockage can cause high cerebrospinal fluid pressure, as we discuss later.

CEREBROSPINAL FLUID PRESSURE IN PATHOLOGICAL CONDITIONS OF THE BRAIN. Often a large *brain tumor* elevates the cerebrospinal fluid pressure by decreasing the rate of absorption of fluid. For instance, if the tumor is above the tentorium and becomes so large that it compresses the brain downward, the upward flow of fluid through the subarachnoid space around the brain stem where it passes through the tentorial opening may become blocked and the absorption of fluid by the cerebral arachnoidal villi greatly curtailed. As a result, the cerebrospinal fluid pressure below the tentorium can rise to 500 millimeters of water (37 mm Hg) or more.

The pressure also rises considerably when *hemorrhage* or *infection* occurs in the cranial vault. In both these conditions, large numbers of cells suddenly appear in the cerebrospinal fluid, and they can cause serious blockage of the small channels for absorption through the arachnoidal villi. This sometimes elevates the cerebrospinal fluid pressure to 400 to 600 millimeters of water (about four times normal).

Some babies are born with high cerebrospinal fluid pressure. This is often caused by abnormally high resistance to fluid reabsorption through the arachnoidal villi, resulting either from too few arachnoidal villi or from villi with abnormal absorptive properties. This is discussed later in connection with *hydrocephalus.*

MEASUREMENT OF CEREBROSPINAL FLUID PRESSURE. The usual procedure for measuring cerebrospinal fluid pressure is the following: First, the subject lies exactly horizontally on the side, so that the spinal fluid pressure is equal to the pressure in the cranial vault. A spinal needle is then inserted into the lumbar spinal canal below the lower end of the cord and connected to a glass tube. The spinal fluid is allowed to rise in the tube as high as it will. If it rises to a level 136 millimeters above the level of the needle, the pressure is said to be 136 millimeters of water pressure or, dividing this by 13.6, which is the specific gravity of mercury, about 10 mm Hg pressure.

HIGH CEREBROSPINAL FLUID PRESSURE CAUSES EDEMA OF THE OPTIC DISC—PAPILLEDEMA. Anatomically, the dura of the brain extends as a sheath around the optic nerve and then connects with the sclera of the eye. When the pressure rises in the cerebrospinal fluid system, it also rises in the optic nerve inside the optic nerve sheath. The retinal artery and vein pierce this sheath a few millimeters behind the eye and then pass with the optic nerve into the eye itself. The high pressure in the optic sheath pushes fluid along the spaces between the optic nerve fibers to the interior of the eyeball. Also, the pressure in the sheath impedes the flow of blood in the retinal vein, thereby also increasing the retinal capillary pressure throughout the eye, which results in additional retinal edema. The tissues of the optic disc are much more distensible than those of the remainder of the retina, so that the disc becomes far more edematous than the remainder of the retina and swells into the cavity of the eye. The swelling of the disc, which can be observed with an ophthalmoscope, is called *papilledema.* Neurologists can estimate the cerebrospinal fluid pressure level by assessing the extent to which the optic disc protrudes into the eyeball.

Obstruction to the Flow of Cerebrospinal Fluid Causes Hydrocephalus

"Hydrocephalus" means excess water in the cranial vault. This condition is frequently divided into *communicating hydrocephalus* and *noncommunicating hydrocephalus*. In communicating hydrocephalus fluid flows readily from the ventricular system into the subarachnoid space, whereas in noncommunicating hydrocephalus fluid flow out of one or more of the ventricles is blocked.

Usually the *noncommunicating* type of hydrocephalus is caused by a block in the aqueduct of Sylvius, resulting from *atresia* (closure) before birth in many babies or from a brain tumor at any age. As fluid is formed by the choroid plexuses in the two lateral and the third ventricles, the volumes of these three ventricles increase greatly. This flattens the brain into a thin shell against the skull. In neonates, the increased pressure also causes the whole head to swell because the skull bones have not fused.

The *communicating* type of hydrocephalus is usually caused by blockage of fluid flow in the subarachnoid space around the basal regions of the brain or blockage of the arachnoidal villi themselves. Fluid therefore collects both inside the ventricles and on the outside of the brain, causing the head to swell tremendously if it occurs in infants when the skull is still pliable and can be stretched and usually damaging the brain severely at any age.

The most effective therapy for hydrocephalus is surgical institution of a silicone rubber tube shunt all the way from one of the ventricles to the peritoneal cavity, into an intestine or elsewhere in the abdominal cavity, where the fluid can then be absorbed or excreted.

Blood–Cerebrospinal Fluid and Blood-Brain Barriers

It has already been pointed out that the constituents of the cerebrospinal fluid are not exactly the same as those of the extracellular fluid elsewhere in the body. Furthermore, many large molecular substances hardly pass at all from the blood into the cerebrospinal fluid or into the interstitial fluids of the brain, even though these same substances pass readily into the usual interstitial fluids of the body. Therefore, it is said that barriers, called the *blood–cerebrospinal fluid barrier* and the *blood-brain barrier,* exist between the blood and the cerebrospinal fluid and brain fluid, respectively. These barriers exist both at the choroid plexus and at the tissue capillary membranes in essentially all areas of the brain parenchyma *except in some areas of the hypothalamus, pineal gland,* and *area postrema,* where substances diffuse with ease into the tissue spaces. The ease of diffusion in these areas is important because they have sensory receptors that respond to different changes in the body fluids, such as changes in osmolality and glucose concentration; these responses provide the signals for feedback regulation of each of the factors.

In general, the blood–cerebrospinal fluid and blood-brain barriers are highly permeable to water, carbon dioxide, oxygen, and most lipid-soluble substances, such as alcohol and most anesthetics; slightly permeable to the electrolytes, such as sodium, chloride, and potassium; and almost totally impermeable to plasma proteins and most non–lipid-soluble large organic molecules. Therefore, the blood–cerebrospinal fluid and blood-brain barriers often make it impossible to achieve effective concentrations of therapeutic drugs, such as protein antibodies and non–lipid-soluble drugs, in the cerebrospinal fluid or parenchyma of the brain.

The cause of the low permeability of the blood–cerebrospinal fluid and blood-brain barriers is the manner in which the endothelial cells of the capillaries are joined to one another. They are joined by so-called *tight junctions.* That is, the membranes of the adjacent endothelial cells are tightly fused with one another rather than having extensive slit-pores between them, as is the case in most other capillaries of the body.

DIFFUSION BETWEEN THE CEREBROSPINAL FLUID AND THE BRAIN INTERSTITIAL FLUID. The surfaces of the ventricles are lined with a thin cuboidal epithelium called the *ependyma* and the cerebrospinal fluid on the outer surfaces of the brain is separated from the brain tissue by a thin membrane called the *pia mater.* Both the ependyma and the pia mater are quite permeable, so that almost all substances that enter the cerebrospinal fluid can then diffuse readily into the interstitial fluid of the brain tissues. Therefore, some drugs that have no effect on the brain when introduced into the blood stream can have important effects on the brain when injected into the cerebrospinal fluid.

Brain Edema

One of the most serious complications of abnormal cerebral hemodynamics and fluid dynamics is the development of brain edema. Because the brain is encased in a solid vault, the accumulation of edema fluid compresses the blood vessels, often causing seriously decreased blood flow and destruction of brain tissue.

The usual cause of brain edema is either greatly increased capillary pressure or damage to the capillary wall. One cause of excessively high capillary pressure is a sudden increase in cerebral arterial blood pressure to levels too high for the brain's autoregulatory mechanism to cope with. However, the most common cause is brain concussion, in which the brain tissues and capillaries are traumatized and capillary fluid leaks into the traumatized tissues. Once brain edema begins, it often initiates two vicious cycles because of the following positive feedbacks: (1) Edema compresses the vasculature. This in turn decreases the blood flow and causes brain ischemia. The ischemia causes arteriolar dilatation with still further increase in capillary pressure. The increased capillary pressure then causes more edema fluid, so that the edema becomes progressively worse. (2) The decreased blood flow also decreases oxygen delivery. This increases the permeability of the capillaries, allowing still more fluid leakage. It also turns off the sodium pumps of the tissue cells, thus allowing them to swell.

Once these two vicious cycles have begun, heroic measures must be used to prevent total destruction of the brain. One such measure is to infuse intravenously a concentrated osmotic substance, such as a very concentrated mannitol solution. This pulls fluid by osmosis from the brain tissue and breaks up the vicious cycle. Another procedure is to remove fluid quickly from the lateral ventricles of the brain by means of ventricular puncture, thereby relieving the intracerebral pressure.

BRAIN METABOLISM

Like other tissues, the brain requires oxygen and solid nutrients to supply its metabolic needs. However, there are special peculiarities of brain metabolism that need to be mentioned.

TOTAL BRAIN METABOLIC RATE AND METABOLIC RATE OF NEURONS. Under resting conditions, the metabolism of the brain accounts for about 15 per cent of the total metabolism in the body, even though the mass of the brain is only 2 per cent of the total body mass. Therefore, under resting conditions, brain metabolism is about 7.5 times the average metabolism in the rest of the body.

Most of this excess metabolism of the brain occurs in the neurons, not in the glial supportive tissues. The major need for metabolism in the neurons is to pump ions through their membranes, mainly to transport sodium and calcium ions to the outside of the neuronal membrane and potassium and chloride ions to the interior. Each time a neuron conducts an action potential, these ions move through the membranes, increasing the need for membrane transport to restore the proper ionic concentrations. Therefore, during excessive brain activity, neuronal metabolism can increase severalfold.

SPECIAL REQUIREMENT OF THE BRAIN FOR OXYGEN—LACK OF SIGNIFICANT ANAEROBIC METABOLISM. Most tissues of the body can go without oxygen for several minutes and some, as long as 30 minutes. During this time, the tissue cells obtain their energy through processes of anaerobic metabolism, which means the release of energy by partial breakdown of glucose and glycogen but without combining with oxygen. This delivers energy only at the expense of consuming tremendous amounts of glucose and glycogen. However, it does keep the tissues functioning.

The brain is not capable of much anaerobic metabolism. One of the reasons for this is the high metabolic rate of the neurons, so that far more energy is required by each brain cell than is required in most tissues. An additional reason is that the amount of glycogen stored in the neurons is slight, so that anaerobic breakdown of glycogen cannot supply much energy. The stores of oxygen in brain tissues are also slight. Therefore, most neuronal activity depends on second-by-second delivery of glucose and oxygen from the blood.

Putting these various factors together, one can understand why a sudden cessation of blood flow to the brain or a sudden lack of oxygen in the blood can cause unconsciousness within 5 to 10 seconds.

UNDER NORMAL CONDITIONS, MOST BRAIN ENERGY IS SUPPLIED BY GLUCOSE. Under normal conditions, almost all the energy used by the brain cells is supplied by glucose derived from the blood. As is true for oxygen, most of this is derived minute by minute and second by second from the capillary blood, with a total of only about a 2-minute supply of glucose normally stored as glycogen in the neurons at any given time.

A special feature of glucose delivery to the neurons is that its transport into the neurons through the cell membrane is not dependent on insulin, even though insulin is needed for most other body cells. Therefore, even in patients who have serious diabetes with essentially zero secretion of insulin, glucose still diffuses readily into the neurons—which is most fortunate in preventing loss of mental function in diabetic patients. When a diabetic patient is overtreated with insulin, the blood glucose concentration can fall extremely low because the excess insulin causes almost all the glucose in the blood to be transported rapidly into the insulin-sensitive non-neural cells throughout the body, especially muscle and liver cells. When this happens, not enough glucose is left in the blood to supply the neurons, and mental function does then become seriously deranged, leading sometimes to coma but more often to mental imbalances and psychotic disturbances.

REFERENCES

Angerson, W. J., et al. (eds.): Blood Flow in the Brain. New York, Oxford University Press, 1989.

Bannister, C. M., and Tew, B.: Current Concepts in Spina Bifida and Hydrocephalus. New York, Cambridge University Press, 1992.

Bevan, J. A., et al.: Sympathetic control of cerebral arteries: specialization in receptor type, reserve, affinity, and distribution. FASEB J., 1:193, 1987.

Bevan, R. D., and Bevan, J. A.: The Human Brain Circulation: Functional Changes in Disease. Totowa, NJ, Humana Press, Inc., 1994.

Dorndorf, W., and Marx, P.: Stroke Prevention. Farmington, CT, S. Karger Publishers, Inc., 1992.

Edvinsson, L., et al.: Cerebral Blood Flow and Metabolism. New York, Raven Press, 1993.

Ermisch, A., et al.: Peptides and blood-brain barrier transport. Physiol. Rev., 73:489, 1993.

Fenstermacher, J. D., and Rapoport, S. I.: Blood-brain barrier. In Renkin, E. M., and Michel, C. C. (eds.): Handbook of Physiology. Sec. 2, Vol. IV. Bethesda, Md., American Physiological Society, 1984, p. 969.

Finger, S., et al. (eds.): Brain Injury and Recovery. New York, Plenum Publishing Corp., 1988.

Guyton, A. C., et al.: Circulatory Physiology. II. Dynamics and Control of the Body Fluids. Philadelphia, W. B. Saunders Co., 1975.

Hibbard, L. S., et al.: Three-dimensional representation and analysis of brain energy metabolism. Science, 236:1641, 1987.

Hochwald, G. M.: Animal models of hydrocephalus: Recent developments. Proc. Soc. Exp. Biol. Med., 178:1, 1985.

Kazemi, H., and Johnson, D. C.: Regulation of cerebrospinal fluid acid-base balance. Physiol. Rev., 66:953, 1986.

Kelly, P. A. T., et al.: Cerebrovascular autoregulation in response to hypertension induced by N^G-nitro-L-arginine methyl ester. Neuroscience, 59:13, 1994.

Kogure, K., and Yoshimoto, T.: Proceedings of the Sixteenth International Symposium on Cerebral Blood Flow and Metabolism. New York, Raven Press, 1993.

Long, D. M.: Brain Edema: Pathogenesis, Imaging, and Therapy. New York, Raven Press, 1990.

Mayhan, W. G., et al.: Cerebral microcirculation. News Physiol. Sci., 3:164, 1988.

Minns, R. A.: Problems of Intracranial Pressure in Childhood. New York, Cambridge University Press, 1991.

Neuwelt, E. A. (ed.): Implications of the Blood-Brain Barrier and Its Manipulation. New York, Plenum Publishing Corp., 1989.

Orgogozo, J. M., and Dyken, M.: Advances in Stroke Prevention. Farmington, CT, S. Karger Publishers, Inc., 1993.

Pardridge, W. M.: The Blood-Brain Barrier: Cellular and Molecular Biology. New York, Raven Press, 1993.

Pullicino, P. M., et al.: Cerebral Small Artery Disease. New York, Raven Press, 1993.

Rescigno, A., and Boicelli, A.: Cerebral Blood Flow. New York, Plenum Publishing Corp., 1988.

Sánchez-Armass, S., et al.: Regulation of pH in rat brain snyaptosomes. I. Role of sodium, bicarbonate, and potassium. J. Neurophysiol., 71:2236, 1994.

Siesjo, B. K.: Cerebral circulation and metabolism. J. Neurosurg., 60:883, 1984.

Willard, F. H., and Perl, D. B.: Medical Neuroanatomy: A Problem-Oriented Manual with Annotated Atlas. Philadelphia, J. B. Lippincott, 1993.

Wood, J. H. (ed.): Cerebral Blood Flow: Physiologic and Clinical Aspects. New York, McGraw-Hill Book Co., 1987.

Yonas, H.: Cerebral Blood Flow Measurement with Stable Xenon-Enhanced Computed Tomography. New York, Raven Press, 1992.

Zierler, R. E.: Surgical Management of Cerebrovascular Disease. Hightstown, NJ, McGraw-Hill, 1994.

GASTROINTESTINAL PHYSIOLOGY

UNIT XII

General Principles of Gastrointestinal Function— Motility, Nervous Control, and Blood Circulation

CHAPTER 62

The alimentary tract provides the body with a continual supply of water, electrolytes, and nutrients. To achieve this requires (1) movement of food through the alimentary tract; (2) secretion of digestive juices and digestion of the food; (3) absorption of the digestive products, water, and the various electrolytes; (4) circulation of blood through the gastrointestinal organs to carry away the absorbed substances; and (5) control of all these functions by the nervous and hormonal systems.

Figure 62–1 shows the entire alimentary tract. Each part is adapted to its specific functions: some to simple passage of food, such as the esophagus; others to storage of food, such as the stomach; and others to digestion and absorption, such as the small intestine. In this chapter, we discuss the basic principles of function in the entire alimentary tract; in the following chapters, the specific functions of the different segments of the tract are discussed.

GENERAL PRINCIPLES OF GASTROINTESTINAL MOTILITY

Characteristics of the Gastrointestinal Wall

Figure 62–2 shows a typical section of the intestinal wall, including the following layers from the outer surface inward: (1) the *serosa*, (2) a *longitudinal muscle layer,* (3) a *circular muscle layer,* (4) the *submucosa,* and (5) the *mucosa.* In addition, a sparse layer of smooth muscle fibers, the *muscularis mucosae,* lies in the deeper layers of the mucosa. The motor functions of the gut are performed by the different layers of smooth muscle.

The general characteristics of smooth muscle and its function are discussed in Chapter 8, which should be reviewed as a background for the following sections of this chapter. The specific characteristics of smooth muscle in the gut are the following.

GASTROINTESTINAL SMOOTH MUSCLE—ITS FUNCTION AS A SYNCYTIUM. The individual smooth muscle fibers in the gastrointestinal tract are between 200 and 500 micrometers in length and 2 and 10 micrometers in diameter, and they are arranged in bundles of as many as 1000 parallel fibers. In the longitudinal muscle layer, these bundles extend longitudinally down the intestinal tract; in the circular muscle layer they extend around the gut. Within each bundle the muscle fibers are electrically connected with one another through large numbers of *gap junctions* that allow low-resistance movement of ions from one cell to the next. Therefore, electrical signals can travel readily from one fiber to the next within each bundle but more rapidly along the length of the bundle than sideways.

Each bundle of smooth muscle fibers is separated from the next by loose connective tissue, but the bundles fuse with one another at many points, so that in reality each muscle layer represents a branching latticework of smooth muscle bundles. Therefore, each muscle layer functions as a *syncytium;* that is, when an action potential is elicited anywhere within the muscle mass, it generally travels in all directions in the muscle. The distance that it travels depends on the excitability of the muscle; sometimes it stops after only a few millimeters and at other times it travels many centimeters or even the entire length of the intestinal

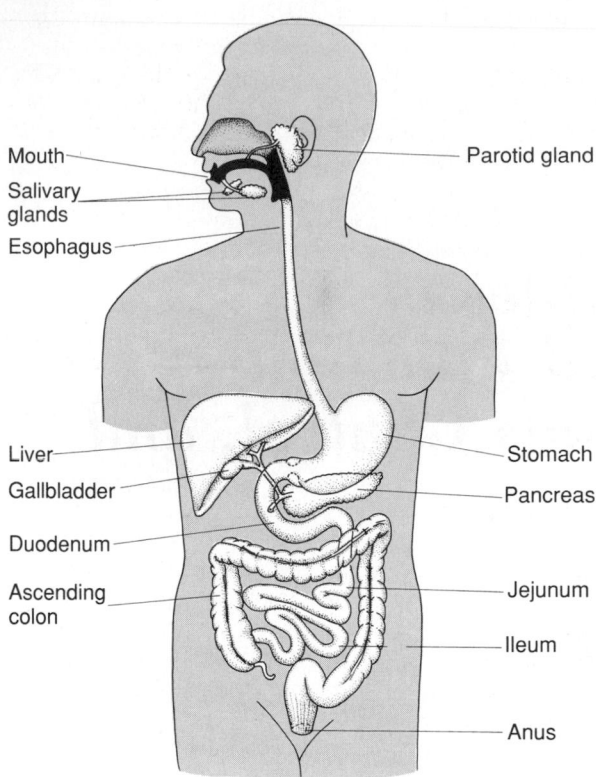

Figure 62–1. Alimentary tract.

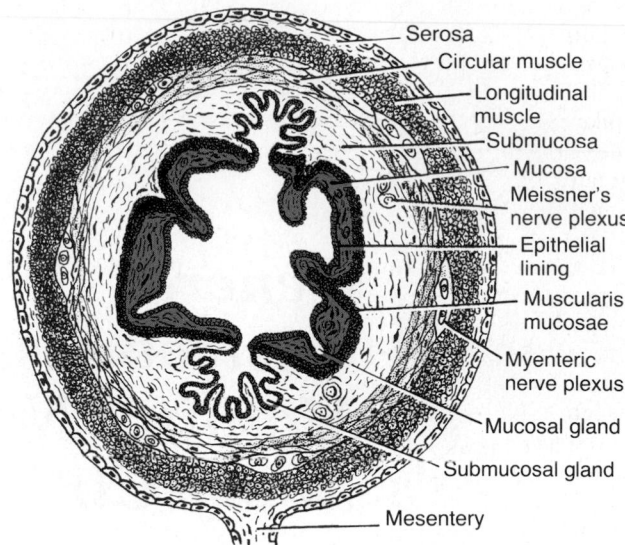

Figure 62–2. Typical cross section of the gut.

tract. Also, a few connections exist between the longitudinal and circular muscle layers, so that excitation of one of these layers often excites the other as well.

Electrical Activity of Gastrointestinal Smooth Muscle

The smooth muscle of the gastrointestinal tract is subject to almost continual but slow electrical activity. This activity tends to have two basic types of electrical waves: (1) *slow waves* and (2) *spikes,* both of which are shown in Figure 62–3. In addition, the voltage of the resting membrane potential of the gastrointestinal smooth muscle can change to different levels, and this too can have important effects in controlling motor activity of the gastrointestinal tract.

SLOW WAVES. Most gastrointestinal contractions occur rhythmically, and this rhythm is determined mainly by the frequency of the so-called slow waves in the smooth muscle membrane potential. These waves, shown in Figure 62–3, are not action potentials. Instead, they are slow, undulating changes in the resting membrane potential. Their intensity usually varies between 5 and 15 millivolts, and their frequency ranges in different parts of the human gastrointestinal tract between 3 and 12 per minute: about 3 in the body of the stomach, as much as 12 in the duodenum, and changing to about 8 or 9 in the terminal ileum. Therefore, the rhythm of contraction of the body of the stomach usually is about 3 per minute, of the duode-

num about 12 per minute, and of the ileum 8 to 9 per minute.

The cause of the slow waves is not known. However, it is believed that they might be caused by a slow undulation of the pumping activity of the sodium-potassium pump.

The slow waves themselves usually do not cause muscle contraction in most parts of the gastrointestinal tract *except in the stomach.* Instead, they mainly control the appearance of intermittent spike potentials, and the spike potentials in turn actually cause most of the muscle contraction.

SPIKE POTENTIALS. The spike potentials are true action potentials. They occur automatically when the resting membrane potential of the gastrointestinal smooth muscle becomes more positive than about − 40 millivolts (the normal resting membrane potential

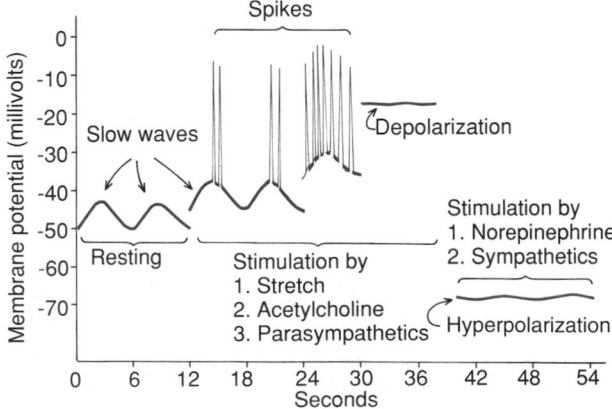

Figure 62–3. Membrane potentials in intestinal smooth muscle. Note the slow waves, the spike potentials, total depolarization, and hyperpolarization, all of which occur under different physiological conditions of the intestine.

is between −50 and −60 millivolts). Thus, note in Figure 62–3 that each time the peaks of the slow waves rise temporarily above the −40 millivolt level—that is, become less negative than −40 millivolts—spike potentials appear on these peaks. And the higher the slow wave potential rises above this level, the greater the frequency of the spike potentials, usually ranging between 1 and 10 spikes per second.

The spike potentials last 10 to 40 times as long in the gastrointestinal muscle as the action potentials in large nerve fibers, each lasting as long as 10 to 20 milliseconds. Another important difference between the action potentials of the gastrointestinal smooth muscle and those of nerve fibers is the manner in which they are generated. In nerve fibers, the action potentials are caused almost entirely by rapid entry of sodium ions through sodium channels to the interior of the fibers. In the gastrointestinal smooth muscle, the channels responsible for the action potentials are quite different; they allow especially large numbers of calcium ions to enter along with smaller numbers of sodium ions and therefore are called *calcium-sodium channels*. These channels are much slower to open and close than are the rapid sodium channels, which accounts for the longer duration of the action potentials. Also, the movement of large amounts of calcium ions to the interior of the muscle fiber during the action potential plays a special role in causing the intestinal smooth muscle to contract, as we discuss shortly.

CHANGES IN THE VOLTAGE OF THE RESTING MEMBRANE POTENTIAL. In addition to the slow waves and spike potentials, the voltage level of the resting membrane potential can change. Under normal conditions, the resting membrane potential averages about −56 millivolts, but multiple factors can change this level. When the potential becomes more positive, which is called *depolarization* of the membrane, the muscle fiber becomes more excitable. When the potential becomes more negative, which is called *hyperpolarization*, the fiber becomes less excitable.

Factors that depolarize the membrane—that is, make it more excitable—are (1) stretching of the muscle, (2) stimulation by acetylcholine, (3) stimulation by the parasympathetic nerves that secrete acetylcholine at their endings, and (4) stimulation by several specific gastrointestinal hormones.

Important factors that make the membrane potential more negative—that is, hyperpolarize the membrane and make the muscle fiber less excitable—are (1) the effect of norepinephrine or epinephrine on the muscle membrane and (2) stimulation of the sympathetic nerves that secrete norepinephrine at their endings.

CALCIUM IONS AND MUSCLE CONTRACTION. Muscle contraction occurs in response to the entry of calcium into the muscle fiber. As explained in Chapter 8, the calcium ions, acting through a calmodulin control mechanism, activate the myosin filaments in the fiber, causing attractive forces between these and the actin filaments and thereby causing the muscle to contract.

The slow waves do not cause calcium ions to enter

the smooth muscle fiber but only sodium ions. Therefore, the slow waves by themselves usually cause no contraction. Instead, it is during the spike potentials, generated at the peaks of the slow waves, that large quantities of calcium ions do enter the fibers and cause most of the contraction.

TONIC CONTRACTION OF SOME GASTROINTESTINAL SMOOTH MUSCLE. Some smooth muscle of the gastrointestinal tract exhibits *tonic contraction* as well as or instead of rhythmical contractions. Tonic contraction is continuous, not associated with the basic electrical rhythm of the slow waves but often lasting several minutes or even several hours. The tonic contraction often increases or decreases in intensity but continues. Tonic contraction is sometimes caused by repetitive spike potentials—the greater the frequency, the greater the degree of contraction. At other times, tonic contraction is caused by hormones or other factors that bring about continuous depolarization of the smooth muscle membrane without causing action potentials. And still a third cause of tonic contraction is continuous entry of calcium into the interior of the cell, brought about in ways not associated with changes in the membrane potential. The details of these mechanisms are still unclear.

NEURAL CONTROL OF GASTROINTESTINAL FUNCTION

The gastrointestinal tract has a nervous system all its own called the *enteric nervous system.* It lies entirely in the wall of the gut, beginning in the esophagus and extending all the way to the anus. The number of neurons in this enteric system is about 100 million, almost exactly equal to the number in the entire spinal cord; this demonstrates the importance of the enteric system for controlling gastrointestinal function. It especially controls gastrointestinal movements and secretion.

The enteric system is composed mainly of two plexuses, as shown in Figures 62–2 and 62–4: (1) an outer plexus lying between the longitudinal and circular muscle layers, called the *myenteric plexus* or *Auerbach's plexus,* and (2) an inner plexus, called the *submucosal plexus* or *Meissner's plexus,* that lies in the submucosa. The nervous connections within and between these two plexuses are shown in Figure 62–4. The myenteric plexus controls mainly the gastrointestinal movements, and the submucosal plexus controls mainly gastrointestinal secretion and local blood flow.

Note in Figure 62–4 the sympathetic and parasympathetic fibers that connect with both the myenteric and the submucosal plexuses. Although the enteric nervous system can function on its own, independently of these extrinsic nerves, stimulation by the parasympathetic and sympathetic systems can further activate or inhibit gastrointestinal functions, as we discuss later.

Also shown in Figure 62–4 are sensory nerve end-

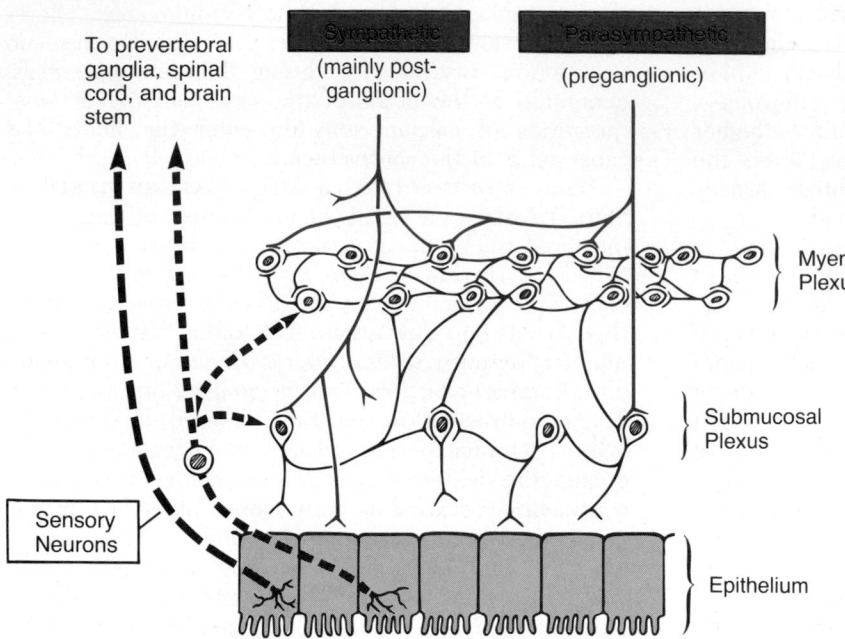

Figure 62–4. Neural control of the gut wall, showing (1) the myenteric and submucosal plexuses; (2) extrinsic control of these plexuses by the sympathetic and parasympathetic nervous systems; and (3) sensory fibers passing from the luminal epithelium and gut wall to the enteric plexuses and from there to the prevertebral ganglia, spinal cord, and brain stem.

ings that originate in the gastrointestinal epithelium or gut wall and then send afferent fibers to both plexuses of the enteric system as well as to the prevertebral ganglia of the sympathetic nervous system, some traveling through the sympathetic nerves to the spinal cord and others traveling in the vagus nerves all the way to the brain stem. These sensory nerves can elicit local reflexes within the gut itself and still other reflexes that are relayed back to the gut from either the prevertebral ganglia or the basal regions of the central nervous system.

Differences Between the Myenteric and Submucosal Plexuses

The myenteric plexus consists mostly of linear chains of many interconnecting neurons that extend the entire length of the gastrointestinal tract. One of these linear chains is shown in Figure 62–4; similar chains are located parallel to one another a few millimeters apart all the way around the gut wall. The separate chains also have lateral fiber connections from one to the other as well as connections with the deeper-lying submucosal plexus.

Because the myenteric plexus is a linear plexus extending all the way down the intestinal wall and because it lies between the longitudinal and circular masses of intestinal smooth muscle, it is concerned mainly with controlling motor activity along the length of the gut. When it is stimulated, its principal effects are (1) increased tonic contraction, or "tone," of the gut wall, (2) increased intensity of the rhythmical contractions, (3) slightly increased rate of the rhythm of contraction, and (4) increased velocity of conduction of excitatory waves along the gut wall, causing more rapid movement of the peristaltic waves.

The myenteric plexus must not be considered en-

tirely excitatory because some of the neurons are inhibitory; their fiber endings secrete an inhibitory transmitter, possibly vasoactive intestinal polypeptide or some other peptide. The resulting inhibitory signals are especially useful for inhibiting some of the intestinal sphincter muscles that impede movement of food between successive segments of the gastrointestinal tract, such as the *pyloric sphincter,* which controls the emptying of the stomach, and the *sphincter of the ileocecal valve,* which controls the emptying of the small intestine into the cecum.

The submucosal plexus, in contrast to the myenteric plexus, is mainly concerned with controlling function within the inner wall of each minute segment of the intestine. For instance, many sensory signals originate from the gastrointestinal epithelium and are then integrated in the submucosal plexus to help control local intestinal secretion, local absorption, and local contraction of the submucosal muscle that causes various degrees of infolding of the stomach mucosa.

Types of Neurotransmitters Secreted by the Enteric Neurons

In an attempt to better understand the multiple functions of the enteric nervous system, research workers the world over have identified a dozen or more different neurotransmitter substances that are released by the nerve endings of different types of enteric neurons. Two of them with which we are already familiar are (1) *acetylcholine* and (2) *norepinephrine.* Others are (3) *adenosine triphosphate,* (4) *serotonin,* (5) *dopamine,* (6) *cholecystokinin,* (7) *substance P,* (8) *vasoactive intestinal polypeptide,* (9) *somatostatin,* (10) *leu-enkephalin,* (11) *met-enkephalin,* and (12) *bombesin.* The specific functions of most of these are not well enough known to justify extensive discussion, other than to point out the following: Acetylcholine most often excites gastrointestinal activity. Norepinephrine, on the other hand, almost always inhibits gastrointestinal activity. This is also true of epinephrine, which reaches the gastrointestinal tract by way of the blood after it is secreted by the adrenal medullae into the circulation. The other transmitter substances listed above are a mixture of excitatory and inhibitory agents, but their importance and even their functions are still mainly to be determined.

Autonomic Control of the Gastrointestinal Tract

PARASYMPATHETIC INNERVATION. The parasympathetic supply to the gut is divided into *cranial* and *sacral divisions,* which are discussed in Chapter 60. Except for a few parasympathetic fibers to the mouth and pharyngeal regions of the alimentary tract, the cranial parasympathetics are transmitted almost entirely in the *vagus nerves.* These fibers provide extensive innervation to the esophagus, stomach, and pancreas and somewhat less to the intestines down through the first half of the large intestine. The sacral parasympathetics originate in the second, third, and fourth sacral segments of the spinal cord and pass through the *pelvic nerves* to the distal half of the large intestine. The sigmoidal, rectal, and anal regions of the large intestine are considerably better supplied with parasympathetic fibers than are the other intestinal areas. These fibers function especially in the defecation reflexes, which are discussed in Chapter 63.

The *postganglionic neurons* of the parasympathetic system are located in the myenteric and submucosal plexuses, and stimulation of the parasympathetic nerves causes a general increase in activity of the entire enteric nervous system. This in turn enhances the activity of most gastrointestinal functions but not all because some of the enteric neurons are inhibitory and therefore inhibit certain of the functions.

SYMPATHETIC INNERVATION. The sympathetic fibers to the gastrointestinal tract originate in the spinal cord between the segments T-5 and L-2. Most of the preganglionic fibers that innervate the gut, after leaving the cord, enter the sympathetic chains and pass through the chains to outlying ganglia, such as the *celiac ganglion* and various *mesenteric ganglia.* Here, most of the *postganglionic neuron bodies* are located, and postganglionic fibers spread from them through postganglionic sympathetic nerves to all parts of the gut, terminating principally on neurons in the enteric nervous system. The sympathetics innervate essentially all portions of the gastrointestinal tract, rather than being more extensively supplied to the portions nearest the oral cavity and anus, as is true of the parasympathetics. The sympathetic nerve endings secrete *norepinephrine.*

In general, stimulation of the sympathetic nervous system inhibits activity in the gastrointestinal tract, causing many effects opposite to those of the parasympathetic system. It exerts its effects in two ways: (1) to a slight extent by direct effect of the norepinephrine on the smooth muscle to inhibit this (except the muscularis mucosae, which it excites) and (2) to a major extent by an inhibitory effect of the norepinephrine on the neurons of the enteric nervous system. Thus, strong stimulation of the sympathetic system can block movement of food through the gastrointestinal tract.

Afferent Sensory Nerve Fibers from the Gut

Many afferent sensory nerve fibers arise in the gut. Some of them have their cell bodies in the enteric nervous system itself. These nerves can be stimulated by (1) irritation of the gut mucosa, (2) excessive distention of the gut, or (3) the presence of specific chemical substances in the gut. Signals transmitted through these fibers can cause excitation or, under other conditions, inhibition of intestinal movements or intestinal secretion.

In addition to the afferent fibers that terminate in the enteric nervous system, two other types of afferent fibers are associated with this system. One of these has its cell bodies in the enteric nervous system but sends its axons through the autonomic nerves to terminate in the *prevertebral sympathetic ganglia,* that is, in the *celiac, mesenteric,* and *hypogastric ganglia.* The other type of afferent fibers has its cell bodies in the dorsal root ganglia of the spinal cord or in the cranial nerve ganglia; these fibers transmit their signals directly into the spinal cord or brain stem, traveling in the same nerve trunks along with the sympathetic or parasympathetic nerve fibers. For example, 80 per cent of the nerve fibers in the vagus nerves are afferent rather than efferent. These fibers transmit afferent signals into the medulla, which in turn initiates many vagal reflex signals that return to the gastrointestinal tract to control many of its functions.

Gastrointestinal Reflexes

The anatomical arrangement of the enteric nervous system and its connections with the sympathetic and parasympathetic systems supports three types of gastrointestinal reflexes that are essential to gastrointestinal control. They are the following:

1. *Reflexes that occur entirely within the enteric nervous system.* They include reflexes that control gastrointestinal secretion, peristalsis, mixing contractions, local inhibitory effects, and so forth.
2. *Reflexes from the gut to the prevertebral sympathetic ganglia and then back to the gastrointestinal tract.* These reflexes transmit signals for long distances in the gastrointestinal tract, such as signals from the stomach to cause evacuation of the colon (the *gastrocolic reflex*), signals from the colon and small intestine to inhibit stomach motility and stomach secretion (the *enterogastric reflexes*), and reflexes from the colon to inhibit emptying of ileal contents into the colon (the *colonoileal reflex*).
3. *Reflexes from the gut to the spinal cord or brain stem and then back to the gastrointestinal tract.* They include especially (a) reflexes from the stomach and duodenum to the brain stem and back to the stomach—by way of the

vagus nerves—to control gastric motor and secretory activity; (b) pain reflexes that cause general inhibition of the entire gastrointestinal tract; and (c) defecation reflexes that travel to the spinal cord and back again to produce the powerful colonic, rectal, and abdominal contractions required for defecation (the defecation reflexes).

Hormonal Control of Gastrointestinal Motility

In Chapter 64, we discuss the extreme importance of several hormones for controlling gastrointestinal secretion. Most of these same hormones also affect the motility of some parts of the gastrointestinal tract. Although the motility effects are less important than the secretory effects of the hormones, some of the more important of them for motility control are the following.

Cholecystokinin is secreted by "I" cells in the mucosa of the duodenum and jejunum mainly in response to the presence of breakdown products of fat, fatty acids, and monoglycerides in the intestinal contents. It has a potent effect in increasing contractility of the gallbladder, thus expelling bile into the small intestine, where the bile then plays important roles in emulsifying fatty substances, allowing them to be digested and absorbed. Cholecystokinin also inhibits stomach motility moderately. Therefore, at the same time that this hormone causes emptying of the gallbladder, it also slows the emptying of food from the stomach to give adequate time for digestion of the fats in the upper intestinal tract.

Secretin is secreted by "S" cells in the mucosa of the duodenum in response to acidic gastric juice emptied from the stomach through the pylorus. It has a mild inhibitory effect on the motility of most of the gastrointestinal tract.

Gastric inhibitory peptide is secreted by the mucosa of the upper small intestine, mainly in response to fatty acids and amino acids but to a lesser extent in response to carbohydrate. It has a mild effect in decreasing motor activity of the stomach and therefore slows the emptying of gastric contents into the duodenum when the upper small intestine is already oversupplied with food products.

FUNCTIONAL TYPES OF MOVEMENTS IN THE GASTROINTESTINAL TRACT

Two types of movements occur in the gastrointestinal tract: (1) *propulsive movements,* which cause food to move forward along the tract at an appropriate rate for digestion and absorption, and (2) *mixing movements,* which keep the intestinal contents thoroughly mixed at all times.

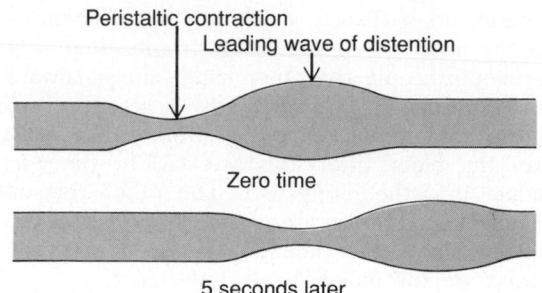

Figure 62–5. Peristalsis.

Propulsive Movements—Peristalsis

The basic propulsive movement of the gastrointestinal tract is *peristalsis,* which is shown in Figure 62–5. A contractile ring appears around the gut and then moves forward; this is analogous to putting one's fingers around a thin distended tube and then constricting the fingers and sliding them forward along the tube. Any material in front of the contractile ring is moved forward.

Peristalsis is an inherent property of many syncytial smooth muscle tubes; stimulation at any point can cause a contractile ring to appear in the circular muscle of the gut, and this ring then spreads along the tube. Thus, peristalsis occurs in the gastrointestinal tract, the bile ducts, other glandular ducts throughout the body, the ureters, and many other smooth muscle tubes of the body.

The usual stimulus for peristalsis is *distention of the gut.* That is, if a large amount of food collects at any point in the gut, the stretching of the gut wall stimulates the enteric nervous system to contract the gut 2 to 3 centimeters above this point, and a contractile ring appears that initiates a peristaltic movement. Other stimuli that can initiate peristalsis include irritation of the epithelium lining the gut and extrinsic nervous signals, particularly parasympathetic, that excite the gut.

FUNCTION OF THE MYENTERIC PLEXUS IN PERISTALSIS. Peristalsis occurs only weakly, if at all, in any portion of the gastrointestinal tract that has congenital absence of the myenteric plexus. Also, it is greatly depressed or completely blocked in the entire gut when the person is treated with atropine to paralyze the cholinergic nerve endings of the myenteric plexus. Therefore, *effectual* peristalsis requires an active myenteric plexus.

DIRECTIONAL MOVEMENT OF PERISTALTIC WAVES TOWARD THE ANUS. Peristalsis, theoretically, can occur in either direction from a stimulated point, but it normally dies out rapidly in the orad direction while continuing for a considerable distance analward. The exact cause of this directional transmission of peristalsis has never been ascertained, although it probably results mainly from the fact that the myenteric plexus itself is "polarized" in the anal direction, which can be explained as follows.

Peristaltic Reflex and the "Law of the Gut." When a segment of the intestinal tract is excited by distention and thereby initiates peristalsis, the contractile ring causing the peristalsis normally begins to move slightly on the orad side of the distended segment; it then moves toward the distended segment, thus pushing the intestinal contents in the anal direction for 5 to 10 centimeters before dying out. At the same time, the gut sometimes relaxes several centimeters downstream toward the anus, which is called "receptive relaxation," thus allowing the food to be propelled more easily analward than in the orad direction.

This complex pattern does not occur in the absence of the myenteric plexus. Therefore, the complex is called the *myenteric reflex* or the *peristaltic reflex*. The peristaltic reflex plus the analward direction of movement of the peristalsis is called the "law of the gut."

Mixing Movements

The mixing movements are quite different in different parts of the alimentary tract. In some areas, the peristaltic contractions themselves cause most of the mixing. This is especially true when forward progession of the intestinal contents is blocked by a sphincter, so that a peristaltic wave can then only churn the intestinal contents, rather than propelling them forward. At other times, *local constrictive contractions* occur every few centimeters in the gut wall. These constrictions usually last for only a few seconds; then new constrictions occur at other points in the gut, thus "chopping" the contents first here and then there. These peristaltic and constrictive movements are modified in different parts of the gastrointestinal tract for proper propulsion and mixing, as are discussed for each portion of the tract in Chapter 63.

GASTROINTESTINAL BLOOD FLOW

The blood vessels of the gastrointestinal system are part of a more extensive system called the *splanchnic circulation*, shown in Figure 62–6. It includes the blood flow through the gut itself plus the blood flow through the spleen, pancreas, and liver. The design of this system is such that all the blood that courses through the gut, spleen, and pancreas then flows immediately into the liver by way of the *portal vein*. In the liver, the blood passes through millions of fine *liver sinusoids* and finally leaves the liver by way of the hepatic veins that empty into the vena cava of the general circulation. This secondary flow of blood through the liver allows the *reticuloendothelial cells* that line the liver sinusoids to remove bacteria and other particulate matter that might enter the general blood stream from the gastrointestinal tract, thus preventing direct access of potentially harmful agents into the remainder of the body.

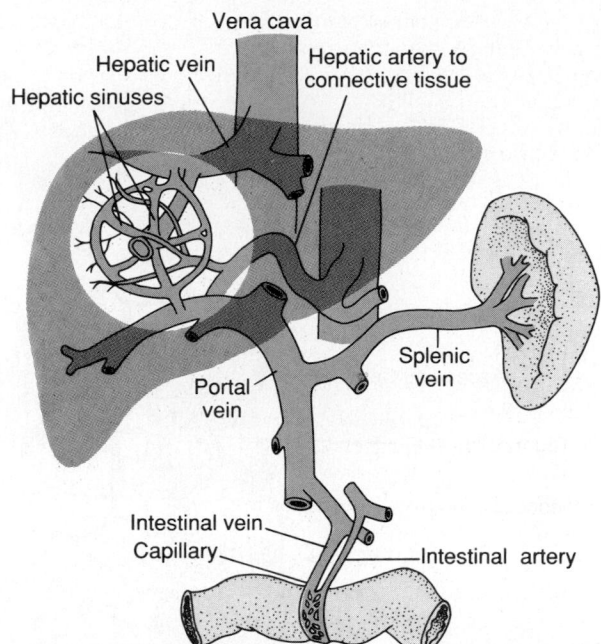

Figure 62–6. Splanchnic circulation.

Most of the nonfat, water-soluble nutrients absorbed from the gut are also transported in the portal venous blood to the same liver sinusoids. Here, both the reticuloendothelial cells and the principal parenchymal cells of the liver, the *hepatic cells,* absorb from the blood and store temporarily from one half to three quarters of all the absorbed nutrients. Much intermediary processing of these nutrients occurs in the liver as well. We discuss these nutritional functions of the liver in Chapters 67 through 71.

The non–water-soluble, fat-based nutrients are almost all absorbed into the intestinal lymphatics and then conducted to the blood by way of the thoracic duct.

Anatomy of the Gastrointestinal Blood Supply

Figure 62–7 shows the general plan of the blood supply to the gut, including the superior mesenteric and inferior mesenteric arteries supplying the walls of the small and large intestines by way of an arching arterial system. Not shown in the figure is the celiac artery, which provides a similar blood supply to the stomach.

On entering the wall of the gut, the arteries branch and send smaller arteries circling in both directions around the gut, with the tips of these meeting on the side of the gut wall opposite the mesenteric attachment. From the circling arteries, still much smaller arteries penetrate into the intestinal wall and spread (1) along the muscle bundles, (2) into the villi, and (3) into submucosal vessels beneath the epithelium to serve the secretory and absorptive functions of the gut.

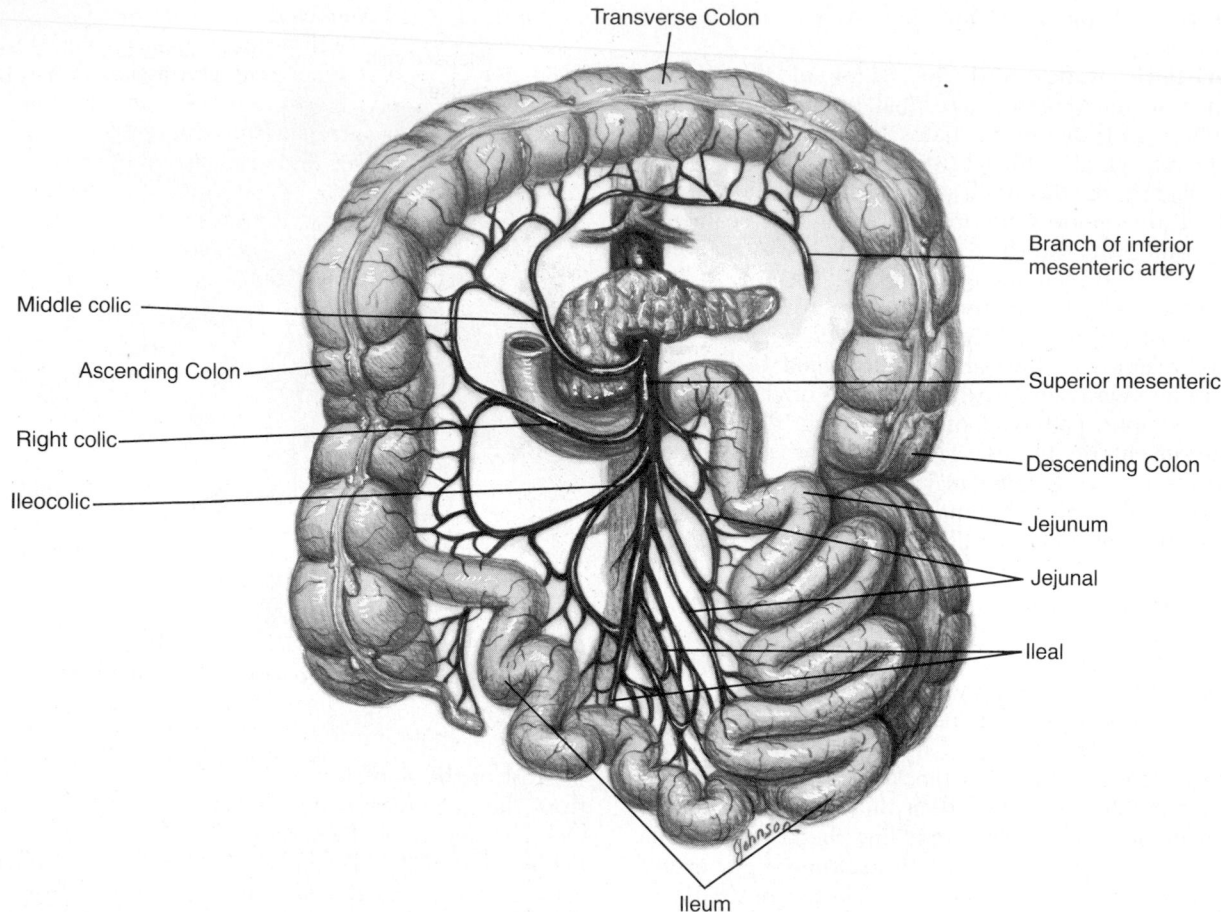

Figure 62–7. Blood supply to the intestines through the mesenteric web.

Figure 62–8 shows the special organization of the blood flow through an intestinal villus, including a small arteriole and venule that interconnect with a system of multiple looping capillaries. The walls of the arterioles are highly muscular and are highly active in controlling villus blood flow.

Effect of Gut Activity and Metabolic Factors on Gastrointestinal Blood Flow

Under normal conditions, the blood flow in each area of the gastrointestinal tract as well as in each layer of the gut wall is directly related to the level of local activity. For instance, during active absorption of nutrients, blood flow in the villi and adjacent regions of the submucosa is greatly increased, as much as eightfold or more at times. Likewise, blood flow in the muscle layers of the intestinal wall increases with increased motor activity in the gut. For instance, after a meal, the motor activity, secretory activity, and absorptive activity all increase; likewise, the blood flow increases greatly during the next hour or so and then decreases back to the resting level over another 2 to 4 hours.

POSSIBLE CAUSES OF THE INCREASED BLOOD FLOW DURING ACTIVITY. Although the precise cause or causes of the increased blood flow during increased gastrointestinal activity are still unclear, some facts are known.

First, several vasodilator substances are released from the mucosa of the intestinal tract during the digestive process. Most of these are peptide hormones, including *cholecystokinin, vasoactive intestinal peptide, gastrin,* and *secretin.* These same hormones are also important in controlling certain specific motor and secretory activities of the gut, as we see in Chapters 63 and 64.

Second, some of the gastrointestinal glands also release into the gut wall two kinins, *kallidin* and *bradykinin,* at the same time that they secrete their secretions into the lumen. These kinins are powerful vasodilators that are believed by some researchers to cause much of the increased mucosal vasodilatation that occurs along with secretion.

Third, *decreased oxygen concentration* in the gut wall can increase intestinal blood flow at least 50 per cent; therefore, the increased metabolic rate during gut activity probably lowers the oxygen concentration enough to cause much of the vasodilatation. The lack

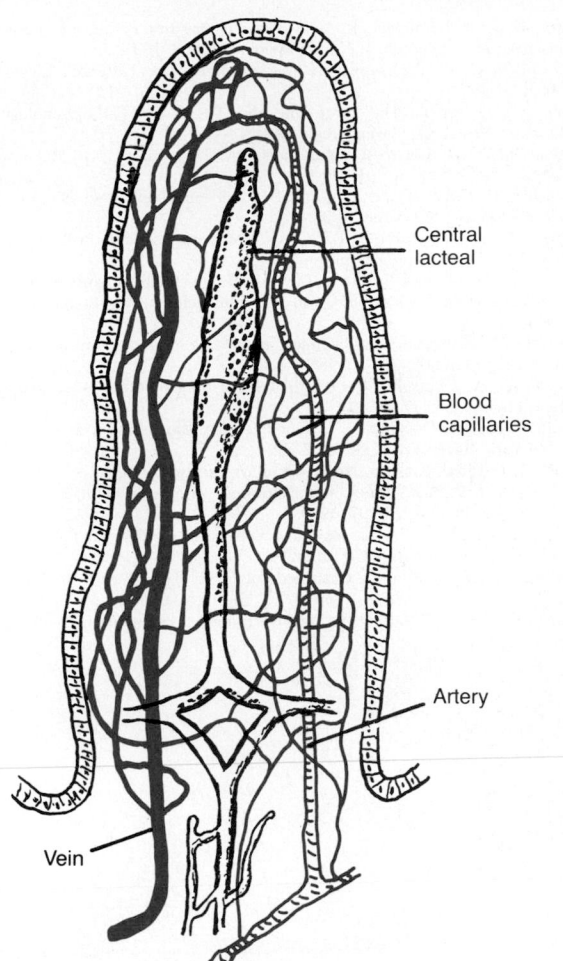

Central
lacteal

Blood
capillaries

Artery

Vein

Figure 62–8. Microvasculature of the villus, showing a countercurrent arrangement of blood flow in the arterioles and venules.

of oxygen can also lead to the release by as much as fourfold of *adenosine,* a well-known vasodilator that could be responsible for much of the increased flow.

The reason for the increased blood flow during increased gastrointestinal activity is not clear. It is probably a combination of all or many of the factors above plus still others yet undiscovered.

COUNTERCURRENT BLOOD FLOW MECHANISM IN THE VILLI. Note in Figure 62–8 that the arterial flow into the villus and the venous flow out of the villus are in directions opposite to each other, and the vessels lie in close apposition to each other. Because of this vascular arrangement, much of the blood oxygen diffuses out of the arterioles directly into the adjacent venules without ever being carried in the blood to the tips of the villi. As much as 80 per cent of the oxygen may take this short-circuit route and therefore not be available for the local metabolic functions of the villi. The reader will recognize that this type of countercurrent mechanism in the villi is analogous to the countercurrent mechanism in the vasa recta of the kidney medulla, discussed in detail in Chapter 28.

Under normal conditions, this shunting of oxygen from the arterioles to the venules is not harmful to the villi, but in disease conditions in which blood flow to the gut becomes greatly curtailed, such as in circulatory shock, the oxygen deficit in the tips of the villi can become so great that the villus tip or even the whole villus suffers ischemic death and can disintegrate. Therefore, for this reason and others, in many gastrointestinal diseases the villi become seriously blunted, leading to greatly diminished gastrointestinal absorptive capacity.

Nervous Control of Gastrointestinal Blood Flow

Stimulation of the parasympathetic nerves to the *stomach* and *lower colon* increases local blood flow at the same time that it increases glandular secretion. This increased flow probably results secondarily from the increased glandular activity and not as a direct effect of nervous stimulation.

Sympathetic stimulation, in contrast, has a direct effect on essentially all the gastrointestinal tract by causing intense vasoconstriction of the arterioles with greatly decreased blood flow. After a few minutes of this vasoconstriction, the flow often returns almost to normal by means of a mechanism called "autoregulatory escape." That is, the local metabolic vasodilator mechanisms that are elicited by ischemia become prepotent over the sympathetic vasoconstriction and, therefore, redilate the arterioles, thus causing return of necessary nutrient blood flow to the gastrointestinal glands and muscle.

IMPORTANCE OF NERVOUS CONTROL WHEN OTHER PARTS OF THE BODY NEED EXTRA BLOOD FLOW. A major value of sympathetic vasoconstriction in the gut is that it allows the shutting off of gastrointestinal and other splanchnic blood flow for short periods during heavy exercise when increased flow is needed by the skeletal muscle and heart. Also, in circulatory shock, when all the vital tissues are in danger of cellular death for lack of blood flow—especially the brain and the heart—sympathetic stimulation can block splanchnic blood flow almost entirely for as long as 1 hour.

Sympathetic stimulation also causes especially strong vasoconstriction of the intestinal and mesenteric *veins.* Furthermore, this venous vasoconstriction does not "escape." Instead, it decreases the volume of these veins and thereby displaces large amounts of blood into other parts of the circulation. In hemorrhagic shock or other states of low blood volume, this mechanism can provide probably 200 to 300 milliliters of extra blood to sustain the general circulation.

REFERENCES

Bouchier, I. A. D., et al.: Gastroenterology: Clinical Science & Practice. Philadelphia, W. B. Saunders Co., 1994.
Chou, C. C.: Relationship between intestinal blood flow and motility. Annu. Rev. Physiol., 44:29, 1982.

Costa, M., et al.: Histochemistry of the enteric nervous system. In: Johnson, L. R. (ed.) Physiology of the Gastrointestinal Tract, 2nd Ed. New York, Raven Press, 1987.

Donald, D. E.: Splanchnic circulation. In Shephard, J. T., and Abboud, F. M. (eds.): Handbook of Physiology. Sec. 2, Vol. III. Bethesda, Md., American Physiological Society, 1983, p. 219.

Evans, G. S., et al.: Primary cultures for studies of cell regulation and physiology in intestinal epithelium. Annu. Rev. Physiol., 56:399, 1994.

Gonella, J., et al.: Extrinsic nervous control of motility of small and large intestines and related sphincters. Physiol. Rev. 67:902, 1987.

Haubrich, W. S., et al.: Bockus Gastroenterology. Philadelphia, W. B. Saunders Co., 1994.

Heinz-Erian, P., et al.: Regulatory Gut Peptides in Paediatric Gastroenterology and Nutrition. Farmington, CT, S. Karger Publishers, Inc., 1992.

Hunt, J. N.: Mechanisms and disorders of gastric emptying. Annu. Rev. Med., 34:219, 1983.

Jacobson, E. D., and Levine, J. S.: Clinical GI Physiology for the Exam Taker. Philadelphia, W. B. Saunders Co., 1994.

Janig, W., and McLachlan E. M.: Organization of lumbar spinal outflow to distal colon and pelvic organs. Physiol. Rev., 67:1332, 1987.

Johnson, L. R., et al.: Physiology of the Gastrointestinal Tract. New York, Raven Press, 1994.

Kirsner, J. B., and Shorter R. G. (eds.): Diseases of the Colon, Rectum, and Anal Canal. Baltimore, Williams & Wilkins, 1988.

Kumar, D., and Wingate, D. L.: An Illustrated Guide to Gastrointestinal Motility. 2nd Ed. New York, Churchill Livingstone, 1994.

Legg, C. R., and Booth, D. A.: Appetite; Neural and Behavioural Bases. New York, Oxford University Press, 1994.

Loewenstein, W. R.: Junctional intercellular communication: The cell-to-cell membrane channel. Physiol. Rev., 61:829, 1981.

Lynch, M. A., et al.: Burket's Oral Medicine. Philadelphia, J. B. Lippincott, 1994.

Ostry, D. J., and Munhall, K. G.: Control of jaw orientation and position in mastication and speech. J. Neurophysiol., 71:1528, 1994.

Pearson, F. G., et al.: Esophageal Surgery. New York, Churchill Livingstone, 1994.

Phillips, S. F., et al.: The Large Intestine: Physiology, Pathophysiology, and Disease. New York, Raven Press, 1991.

Plavsic, B., et al. Gastrointestinal Radiology. Hightstown, NJ, McGraw-Hill, 1992.

Schuster, M.: Atlas of Gastrointestinal Motility in Health and Disease. Baltimore, Williams & Wilkins, 1993.

Shaffer, E., and Thomson, A. B. R. (eds.): Modern Concepts in Gastroenterology. New York, Plenum Publishing Corp., 1989.

Sleisenger, M. H., and Fordtran, J. S.: Gastrointestinal Disease: Pathophysiology, Diagnosis, Management, 5th Ed. Philadelphia, W. B. Saunders Co., 1993.

Sternini, C.: Structural and chemical organization of the myenteric plexus. Annu. Rev. Physiol., 50:81, 1988.

Surprenant, A.: Control of the gastrointestinal tract by enteric neurons. Annu. Rev. Physiol., 56:117, 1994.

Walsh, J. H., and Dockray, G. J.: Gut Peptides: Biochemistry and Physiology. New York, Raven Press, 1994.

Walsh, J. H., et al.: Gastrin. New York, Raven Press, 1993.

Williamson, J. R., and Monck, J. R.: Hormone effects on cellular Ca^{2+} fluxes. Annu. Rev. Physiol., 51:107, 1989.

Wong, M. E. K., et al.: Atlas of Major Oral and Maxillofacial Surgery. Philadelphia, W. B. Saunders Co., 1994.

Yamada, T., et al.: Atlas of Gastroenterology. Philadelphia, J. B. Lippincott, 1992.

Yamada, T., et al.: Textbook of Gastroenterology. Philadelphia, J. B. Lippincott, 1991.

Transport and Mixing of Food in the Alimentary Tract

*C*HAPTER 63

For food to be processed optimally in the alimentary tract, the time that it remains in each part of the tract is critical. Also, appropriate mixing must be provided. Yet because the requirements for mixing and propulsion are quite different at each stage of processing, multiple automatic nervous and hormonal feedback mechanisms control each aspect of these so that they will both occur optimally, not too rapidly, not too slowly.

The purpose of this chapter is to discuss these movements and especially the automatic mechanisms that control them.

INGESTION OF FOOD

The amount of food that a person ingests is determined principally by the intrinsic desire for food called *hunger.* The type of food that a person preferentially seeks is determined by *appetite.* These mechanisms in themselves are extremely important automatic regulatory systems for maintaining an adequate nutritional supply for the body, and they are discussed in Chapter 71 in relation to nutrition of the body. The current discussion of food ingestion is confined to the mechanical aspects of food ingestion, especially *mastication* and *swallowing.*

Mastication (Chewing)

The teeth are admirably designed for chewing, the anterior teeth (incisors) providing a strong cutting action and the posterior teeth (molars), a grinding action. All the jaw muscles working together can close the teeth with a force as great as 55 pounds on the incisors and 200 pounds on the molars.

Most of the muscles of chewing are innervated by the motor branch of the fifth cranial nerve, and the chewing process is controlled by nuclei in the brain stem. Stimulation of the reticular formation near the brain stem centers for taste can cause continual rhythmical chewing movements. Also, stimulation of areas in the hypothalamus, amygdala, and even the cerebral cortex near the sensory areas for taste and smell can often cause chewing.

Much of the chewing process is caused by the *chewing reflex,* which may be explained as follows: The presence of a bolus of food in the mouth at first initiates reflex inhibition of the muscles of mastication, which allows the lower jaw to drop. The drop in turn initiates a stretch reflex of the jaw muscles that leads to *rebound* contraction. This automatically raises the jaw to cause closure of the teeth, but it also compresses the bolus again against the linings of the mouth, which inhibits the jaw muscles once again, allowing the jaw to drop and rebound another time; this is repeated again and again.

Chewing of the food is important for digestion of all foods, but it is especially important for most fruits and raw vegetables because these have undigestible cellulose membranes around their nutrient portions that must be broken before the food can be used. Also, chewing aids in the digestion of food for still another simple reason: Because the *digestive enzymes act only on the surfaces of food particles,* the rate of digestion is highly dependent on the total surface area exposed to the intestinal secretions. In addition, grinding the food to a very fine particulate consistency prevents

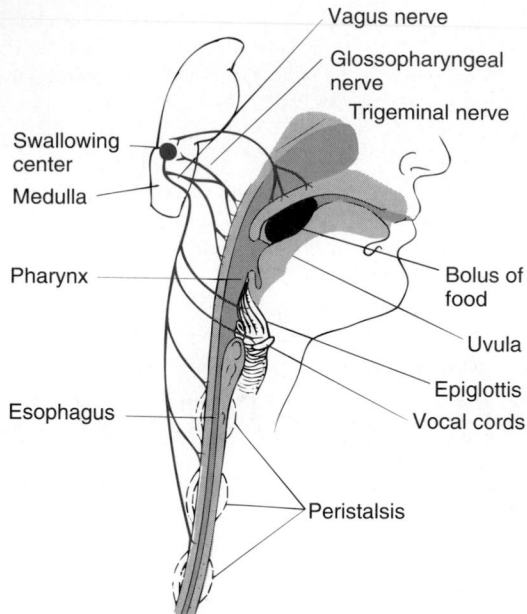

Figure 63–1. Swallowing mechanism.

excoriation of the gastrointestinal tract and increases the ease with which food is emptied from the stomach into the small intestine and then into all succeeding segments of the gut.

Swallowing (Deglutition)

Swallowing is a complicated mechanism, principally because the pharynx most of the time subserves several other functions besides swallowing and is converted for only a few seconds at a time into a tract for propulsion of food. It is especially important that respiration not be compromised because of swallowing.

In general, swallowing can be divided into (1) a *voluntary stage,* which initiates the swallowing process; (2) a *pharyngeal stage,* which is involuntary and constitutes the passage of food through the pharynx into the esophagus; and (3) an *esophageal stage,* another involuntary phase that promotes passage of food from the pharynx to the stomach.

VOLUNTARY STAGE OF SWALLOWING. When the food is ready for swallowing, it is "voluntarily" squeezed or rolled posteriorly into the pharynx by pressure of the tongue upward and backward against the palate, as shown in Figure 63–1. From here on, swallowing becomes entirely—or almost entirely—automatic and ordinarily cannot be stopped.

PHARYNGEAL STAGE OF SWALLOWING. As the bolus of food enters the posterior mouth and pharynx, it stimulates *swallowing receptor areas* all around the opening of the pharynx, especially on the tonsillar pillars, and impulses from these pass to the brain stem to initiate a series of automatic pharyngeal muscle contractions as follows:

1. The soft palate is pulled upward to close the posterior nares, in this way preventing reflux of food into the nasal cavities.

2. The palatopharyngeal folds on either side of the pharynx are pulled medially to approximate each other. In this way, these folds form a sagittal slit through which the food must pass into the posterior pharynx. This slit performs a selective action, allowing food that has been masticated sufficiently to pass with ease while impeding the passage of large objects. Because this stage of swallowing lasts for less than 1 second, any large object is usually impeded too much to pass through the pharynx into the esophagus.

3. The vocal cords of the larynx are strongly approximated, and the larynx is pulled upward and anteriorly by the neck muscles. This action, combined with the presence of ligaments that prevent upward movement of the epiglottis, causes the epiglottis to swing backward over the opening of the larynx. Both effects prevent passage of food into the trachea. Most essential is the tight approximation of the vocal cords, but the epiglottis helps to prevent food from ever getting as far as the vocal cords. Destruction of the vocal cords or of the muscles that approximate them can cause strangulation. On the other hand, removal of the epiglottis usually does not cause serious debility in swallowing.

4. The upward movement of the larynx also pulls up and enlarges the opening of the esophagus. At the same time, the upper 3 to 4 centimeters of the esophageal muscular wall, an area called the *upper esophageal sphincter* or the *pharyngoesophageal sphincter,* relaxes, thus allowing food to move easily and freely from the posterior pharynx into the upper esophagus. This sphincter, between swallows, remains strongly contracted (as much as 60 mm Hg pressure in the esophageal lumen), thereby preventing air from going into the esophagus during respiration. The upward movement of the larynx also lifts the glottis out of the main stream of food flow, so that the food usually passes on either side of the epiglottis rather than over its surface; this adds still another protection against entry of food into the trachea.

5. At the same time that the larynx is raised and the pharyngoesophageal sphincter is relaxed, the entire muscular wall of the pharynx contracts, beginning in the superior part of the pharynx and spreading downward as a rapid peristaltic wave over the middle and inferior pharyngeal areas and then into the esophagus, which propels the food into the esophagus.

To summarize the mechanics of the pharyngeal stage of swallowing: the trachea is closed, the esophagus is opened, and a fast peristaltic wave originating in the pharynx forces the bolus of food into the upper esophagus, the entire process occurring in less than 2 seconds.

Nervous Control of the Pharyngeal Stage of Swallowing. The most sensitive tactile areas of the posterior mouth and pharynx for initiation of the pharyngeal stage of swallowing lie in a ring around the pharyngeal opening, with greatest sensitivity in the tonsillar pillars. Impulses are transmitted from these areas through the sensory portions of the trigeminal and glossopharyngeal nerves into a region of the medulla oblongata either in or closely associated with the *tractus solitarius,* which receives essentially all sensory impulses from the mouth.

The successive stages of the swallowing process are

then automatically controlled in orderly sequence by neuronal areas of the brain stem distributed throughout the reticular substance of the medulla and lower portion of the pons. The sequence of the swallowing reflex is the same from one swallow to the next, and the timing of the entire cycle also remains constant from one swallow to the next. The areas in the medulla and lower pons that control swallowing are collectively called the *deglutition* or *swallowing center.*

The motor impulses from the swallowing center to the pharynx and upper esophagus that cause swallowing are transmitted by the 5th, 9th, 10th, and 12th cranial nerves and even a few of the superior cervical nerves.

In summary, the pharyngeal stage of swallowing is principally a reflex act. It is almost never initiated by direct stimuli to the swallowing center from higher regions of the central nervous system. Instead, it is almost always initiated by voluntary movement of food into the back of the mouth, which in turn excites the sensory receptors that elicit the swallowing reflex.

Effect of the Pharyngeal Stage of Swallowing on Respiration. The entire pharyngeal stage of swallowing occurs in less than 2 seconds, thereby interrupting respiration for only a fraction of a usual respiratory cycle. The swallowing center specifically inhibits the respiratory center of the medulla during this time, halting respiration at any point in its cycle to allow swallowing to proceed. Yet even while a person is talking, swallowing interrupts respiration for such a short time that it is hardly noticeable.

Esophageal Stage of Swallowing

The esophagus functions primarily to conduct food from the pharynx to the stomach, and its movements are organized specifically for this function.

The esophagus normally exhibits two types of peristaltic movements: *primary peristalsis* and *secondary peristalsis.* Primary peristalsis is simply a continuation of the peristaltic wave that begins in the pharynx and spreads into the esophagus during the pharyngeal stage of swallowing. This wave passes all the way from the pharynx to the stomach in about 8 to 10 seconds. Food swallowed by a person who is in the upright position is usually transmitted to the lower end of the esophagus even more rapidly than the peristaltic wave itself, in about 5 to 8 seconds, because of the additional effect of gravity pulling the food downward. If the primary peristaltic wave fails to move all the food that has entered the esophagus into the stomach, *secondary peristaltic waves* result from distention of the esophagus by the retained food, and they continue until all the food has emptied into the stomach. These secondary waves are initiated partly by intrinsic neural circuits in the esophageal myenteric nervous system and partly by reflexes that are transmitted through *vagal afferent fibers* from the esophagus to the medulla and then back again to the esophagus through *vagal efferent fibers.*

The musculature of the pharynx and the upper third of the esophagus is *striated muscle.* Therefore, the peristaltic waves in these regions are controlled only by skeletal nerve impulses in the glossopharyngeal and vagus nerves. In the lower two thirds of the esophagus, the musculature is smooth, but this portion of the esophagus is also strongly controlled by the vagus nerves acting through their connections with the myenteric nervous system. When the vagus nerves to the esophagus are sectioned, the myenteric nerve plexus of the esophagus becomes excitable enough after several days to cause strong secondary peristaltic waves even without support from the vagal reflexes. Therefore, after paralysis of the swallowing reflex, food forced in some other way into the lower esophagus still passes readily into the stomach.

Receptive Relaxation of the Stomach. As the esophageal peristaltic wave passes toward the stomach, a wave of relaxation, transmitted through myenteric inhibitory neurons, precedes the peristalsis. Furthermore, the entire stomach and, to a lesser extent, even the duodenum become relaxed as this wave reaches the lower end of the esophagus and thus are prepared ahead of time to receive the food propelled down the esophagus during the swallowing act.

Function of the Lower Esophageal Sphincter (Gastroesophageal Sphincter)

At the lower end of the esophagus, extending from about 2 to 5 centimeters above its juncture with the stomach, the esophageal circular muscle functions as a *lower esophageal sphincter* or *gastroesophageal sphincter.* Anatomically, this sphincter is no different from the remainder of the esophagus. Physiologically, it normally remains tonically constricted (with an intraluminal pressure at this point in the esophagus of about 30 mm Hg), in contrast to the midportion of the esophagus between the upper and lower sphincters, which normally remains relaxed. When a peristaltic swallowing wave passes down the esophagus, "receptive relaxation" relaxes the lower esophageal sphincter ahead of the peristaltic wave and allows easy propulsion of the swallowed food into the stomach. Rarely, the sphincter does not relax satisfactorily, resulting in a condition called *achalasia,* which is discussed in Chapter 66.

The stomach contents are highly acidic and contain many proteolytic enzymes. The esophageal mucosa, except in the lower one eighth of the esophagus, is not capable of resisting for long the digestive action of gastric secretions. The tonic constriction of the lower esophageal sphincter helps to prevent significant reflux of stomach contents into the esophagus except under abnormal conditions.

ADDITIONAL PREVENTION OF REFLUX BY VALVE-LIKE CLOSURE OF THE DISTAL END OF THE ESOPHAGUS. Another factor that prevents reflux is a valve-like mechanism of that short portion of the esophagus that lies immediately beneath the diaphragm before reaching the stomach. Increased intra-abdominal pressure caves the esophagus inward at this point at the same time that this pressure increases the intragastric pressure. Thus, this valve-like closure of the lower esophagus prevents high abdominal pressure from forcing stomach contents into the esophagus. Otherwise, every time we walked, coughed, or breathed hard, we might expel acid into the esophagus.

MOTOR FUNCTIONS OF THE STOMACH

The motor functions of the stomach are threefold: (1) storage of large quantities of food until the food can be processed in the duodenum, (2) mixing of this food with gastric secretions until it forms a semifluid mixture called *chyme,* and (3) slow emptying of the food from the stomach into the small intestine at a rate suitable for proper digestion and absorption by the small intestine.

Figure 63–2 shows the basic anatomy of the stomach. Anatomically, the stomach can be divided into two major parts: (1) the *body* and (2) the *antrum*. Physiologically, it is more appropriately divided into (1) the "orad" portion, comprising about the first two thirds of the body, and (2) the "caudad" portion, comprising the remainder of the body plus the antrum.

Storage Function of the Stomach

As food enters the stomach, it forms concentric circles in the orad portion of the stomach, the newest food lying closest to the esophageal opening and the oldest food lying nearest the wall of the stomach. Normally, when food enters the stomach, a "vagovagal reflex" from the stomach to the brain stem and then back to the stomach reduces the tone in the muscular wall of the body of the stomach so that the wall can

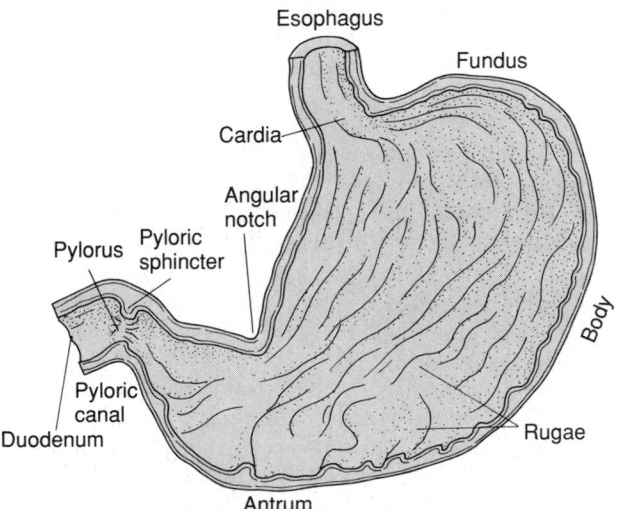

Figure 63–2. Physiologic anatomy of the stomach.

bulge progressively outward, accommodating greater and greater quantities of food up to a limit in the completely relaxed stomach of about 1.5 liters. The pressure in the stomach remains low until this limit is approached.

Mixing and Propulsion of Food in the Stomach—The Basic Electrical Rhythm of the Stomach

The digestive juices of the stomach are secreted by the *gastric glands,* which cover almost the entire wall of the body of the stomach except along a strip on the lesser curvature of the stomach. These secretions come immediately into contact with that portion of the stored food lying against the mucosal surface of the stomach; when the stomach contains food, weak peristaltic *constrictor waves,* also called *mixing waves,* begin in the midportion of the stomach wall and move toward the antrum along the stomach wall about once every 15 to 20 seconds. These waves are initiated by the *basic electrical rhythm* discussed in Chapter 62 and consists of electrical "slow waves" that occur spontaneously in the stomach wall. In most parts of the gastrointestinal tract, these waves are not strong enough to cause contractions unless they elicit superimposed action potentials, but in the stomach their positive peaks often do rise above the threshold for excitation even without action potentials.

As the constrictor waves progress from the body of the stomach into the antrum, they become more intense, some becoming extremely intense and providing powerful *peristaltic constrictor rings* that force the antral contents under high pressure toward the pylorus. These constrictor rings also play an exceedingly important role in mixing the stomach contents in the following way: Each time a peristaltic wave passes over the antrum toward the pylorus, it digs deeply into the contents of the antrum. Yet the opening of the pylorus is small enough that only a few milliliters or even much less of antral contents are expelled into the duodenum with each peristaltic wave. Also, as each peristaltic wave approaches the pylorus, the pyloric muscle itself contracts, which further impedes emptying through the pylorus. Therefore, most of the antral contents are squirted backward through the peristaltic ring toward the body of the stomach. Thus, the moving peristaltic constrictive ring, combined with this squirting action, called "retropulsion," is an exceedingly important mixing mechanism of the stomach.

CHYME. After the food has become mixed with the stomach secretions, the resulting mixture that passes down the gut is called *chyme.* The degree of fluidity of chyme depends on the relative amounts of food and stomach secretions and on the degree of digestion that has occurred. The appearance of chyme is that of a murky, milky semifluid or paste.

HUNGER CONTRACTIONS. Besides the peristaltic contractions that occur when food is present in the stomach, another type of intense contractions, called *hunger contractions,* often occurs when the stomach has been empty for several hours or more. They are rhythmical peristaltic contractions in the *body* of the stomach. When they become extremely strong, they often fuse together to cause a continuing tetanic contraction that lasts for 2 to 3 minutes.

Hunger contractions are most intense in young, healthy people with a high degree of gastrointestinal tonus; they are also greatly increased by a low level of blood sugar.

When hunger contractions occur in the stomach, the person sometimes experiences pain in the pit of the stomach, called *hunger pangs.* Hunger pangs usually do not begin until 12 to 24 hours after the last ingestion of food; in starvation, they reach their greatest intensity in 3 to 4 days and then gradually weaken in succeeding days.

Emptying of the Stomach

Stomach emptying is promoted by the intense peristaltic contractions of the stomach antrum. At the same time, emptying is opposed by varying degrees of resistance to the passage of chyme at the pylorus.

INTENSE ANTRAL PERISTALTIC CONTRACTIONS DURING STOMACH EMPTYING—THE "PYLORIC PUMP." Most of the time, the rhythmical stomach contractions are weak and function mainly to cause mixing of the food and gastric secretions. However, about 20 per cent of the time while food is in the stomach, the contractions become intense, beginning in midstomach and spreading through the caudad stomach no longer as weak mixing contractions but as strong peristaltic, very tight ring-like constrictions. As the stomach becomes progressively more and more empty, these constrictions begin farther and farther up the body of the stomach, gradually pinching off the lowermost portions of the food stored in the body of the stomach and adding this food to the chyme in the antrum. These intense peristaltic contractions often create 50 to 70 centimeters of water pressure, which is about six times as powerful as the usual mixing type of peristaltic waves. The intensity of this antral peristalsis is the principal factor that determines the rate of stomach emptying.

When pyloric tone is normal, each strong peristaltic wave forces up to several milliliters of chyme into the duodenum. Thus, the peristaltic waves provide a pumping action that is called the "pyloric pump."

ROLE OF THE PYLORUS IN CONTROLLING STOMACH EMPTYING. The distal opening of the stomach is the *pylorus.* Here the thickness of the circular muscle becomes 50 to 100 per cent greater than in the earlier portions of the stomach antrum, and it remains slightly tonically contracted almost all the time. Therefore, the pyloric circular muscle is called the *pyloric sphincter.*

Despite the tonic contraction of the pyloric sphincter, the pylorus usually opens enough for water and other fluids to empty from the stomach with ease. On the other hand, the constriction usually prevents passage of most food particles until they have become mixed in the chyme to an almost fluid consistency.

The degree of constriction of the pylorus can be increased or decreased under the influence of nervous and humoral signals from both the stomach and the duodenum, as discussed shortly.

Regulation of Stomach Emptying

The rate at which the stomach empties is regulated by signals from both the stomach and the duodenum. However, the duodenum provides by far the more potent of the signals, always controlling the emptying of chyme into the duodenum at a rate no greater than it can be digested and absorbed in the small intestine.

The Weak Gastric Factors That Promote Emptying

EFFECT OF GASTRIC FOOD VOLUME ON RATE OF EMPTYING. Increased food volume in the stomach promotes increased emptying from the stomach. This increased emptying does not occur for the reasons that one would expect. It is not increased storage pressure of the food in the stomach that causes the increased emptying because in the usual normal range of volume, the increase in volume does not increase the pressure significantly. On the other hand, stretching of the stomach wall does elicit mainly local myenteric reflexes in the wall that greatly excite the activity of the pyloric pump and at the same time slightly inhibit the pylorus.

EFFECT OF THE HORMONE GASTRIN ON STOMACH EMPTYING. In Chapter 64, we see that stretch as well as the presence of certain types of foods in the stomach—particularly digestive products of meat—elicit release of a hormone called *gastrin* from the antral mucosa. This has potent effects to cause secretion of highly acidic gastric juice by the stomach glands. Gastrin has mild to moderate stimulatory effects on motor functions of the stomach as well. Most important, it seems to enhance the activity of the pyloric pump. Thus, it, too, probably aids at least to some degree in promoting stomach emptying.

The Powerful Duodenal Factors That Inhibit Emptying

Inhibitory Effect of Enterogastric Nervous Reflexes from the Duodenum. When food enters the duodenum, multiple nervous reflexes are initiated from the duodenal wall that pass back to the stomach and slow or even stop stomach emptying as the volume of chyme in the duodenum becomes too much. These reflexes are mediated by three routes: (1) directly from the duodenum to the stomach through the enteric nervous system in the gut wall, (2) through extrinsic nerves that go to the prevertebral sympathetic ganglia and then back through inhibitory sympathetic nerve fibers to the stomach, and (3) probably to a slight extent through the vagus nerves all the way to the brain stem, where they inhibit the normal excitatory signals transmitted to the stomach through the vagi.

All these parallel reflexes have two effects on stomach emptying: first, they strongly inhibit the antral propulsive contraction, and second, they probably increase slightly to moderately the tone of the pyloric sphincter.

The types of factors that are continually monitored in the duodenum and that can excite the enterogastric reflexes include the following:

1. The degree of distention of the duodenum
2. The presence of any degree of irritation of the duodenal mucosa
3. The degree of acidity of the duodenal chyme
4. The degree of osmolality of the chyme
5. The presence of certain breakdown products in the chyme, especially breakdown products of proteins and perhaps to a lesser extent of fats

The enterogastric reflexes are especially sensitive to the presence of irritants and acids in the duodenal chyme, often becoming strongly activated in as little as 30 seconds. For instance, whenever the pH of the chyme in the duodenum falls below about 3.5 to 4, the reflexes frequently block further release of acidic stomach contents into the duodenum until the duodenal chyme can be neutralized by pancreatic and other secretions.

Breakdown products of protein digestion also elicit these reflexes; by slowing the rate of stomach emptying, sufficient time is ensured for adequate protein digestion in the upper portion of the small intestine.

Finally, either hypotonic or hypertonic fluids (especially hypertonic) elicit the reflexes. Thus, too rapid flow of nonisotonic fluids into the small intestine is prevented, thereby also preventing rapid changes in electrolyte concentrations in the extracellular fluid during absorption of the intestinal contents.

HORMONAL FEEDBACK FROM THE DUODENUM INHIBITS GASTRIC EMPTYING—ROLE OF FATS AND THE HORMONE CHOLECYSTOKININ. Not only do nervous reflexes from the duodenum to the stomach inhibit stomach emptying, but hormones released from the upper intestine do so as well. The stimulus for producing the hormones is mainly fats entering the duodenum, although other types of foods can increase the hormones to a lesser degree.

On entering the duodenum, the fats extract several different hormones from the duodenal and jejunal epithelium, either by binding with "receptors" in the epithelial cells or in some other way. In turn, the hormones are carried by way of the blood to the stomach, where they inhibit the activity of the pyloric pump and at the same time slightly increase the strength of contraction of the pyloric sphincter. These effects are important because fats are much slower to be digested than most other foods.

Precisely which hormones cause the hormonal feedback inhibition of the stomach is not fully clear. The most potent appears to be *cholecystokinin* (CCK), which is released from the mucosa of the jejunum in response to fatty substances in the chyme. This hormone acts as a competitive inhibitor to block the increased stomach motility caused by gastrin.

Other possible inhibitors of stomach emptying are the hormones *secretin* and *gastric inhibitory peptide* (GIP). Secretin is released mainly from the duodenal mucosa in response to gastric acid released from the stomach through the pylorus. This hormone has a general but only weak effect of decreasing gastrointestinal motility. GIP is released from the upper small intestine in response mainly to fat in the chyme but to carbohydrates as well. Although it is known to inhibit gastric motility under some conditions, its effect at physiological concentrations is probably mainly to stimulate the secretion of insulin by the pancreas. These hormones are discussed at greater length elsewhere in this text, especially in Chapter 64, in relation to the control of gallbladder emptying and pancreatic secretion.

In summary, several hormones are known that could serve as mechanisms for inhibiting gastric emptying when excess quantities of chyme, especially acidic or fatty chyme, enter the duodenum from the stomach. CCK is probably the most important.

Summary of the Control of Stomach Emptying

Emptying of the stomach is controlled only to a moderate degree by stomach factors, such as the degree of filling in the stomach and the excitatory effect of gastrin on stomach peristalsis. Probably the more important control of stomach emptying resides in feedback signals from the duodenum, including both the enterogastric nervous system feedback reflexes and hormonal feedback. These two feedback inhibitory mechanisms work together to slow the rate of emptying when (1) too much chyme is already in the small intestine or (2) the chyme is excessively acid, contains too much unprocessed protein or fat, is hypotonic or hypertonic, or is irritating. In this way, the rate of stomach emptying is limited to that amount of chyme that the small intestine can process.

MOVEMENTS OF THE SMALL INTESTINE

The movements of the small intestine, as elsewhere in the gastrointestinal tract, can be divided into *mixing contractions* and *propulsive contractions*. To a great extent, this separation is artificial because essentially all movements of the small intestine cause at least some degree of both mixing and propulsion. The usual classification of these processes is the following.

Mixing Contractions (Segmentation Contractions)

When a portion of the small intestine becomes distended with chyme, the stretch of the intestinal wall elicits localized concentric contractions spaced at intervals along the intestine. The longitudinal length of each one of the contractions is only about 1 centimeter, so that each set of contractions causes "segmentation" of the small intestine, as shown in Figure 63–3. That is, they divide the intestine into spaced segments that have the appearance of a chain of sausages. As one set of segmentation contractions relaxes, a new set begins, but the contractions this time occur mainly at new points between the previous contractions. These segmentation contractions usually "chop" the chyme about two to three times a minute, in this way promoting progressive mixing of the solid food particles with the secretions of the small intestine.

The maximum frequency of the segmentation contractions in the small intestine is determined by the frequency of the *slow waves* in the intestinal wall, which is the basic electrical rhythm, as explained in Chapter 62. Because this frequency is about 12 per minute in the duodenum and proximal jejunum, the maximum frequency of the segmentation contractions in these areas is also about 12 per minute, but this occurs only under extreme conditions of stimulation. In the terminal ileum, the maximum frequency is usually eight to nine contractions per minute.

The segmentation contractions become exceedingly weak when the excitatory activity of the enteric nervous system is blocked by atropine. Therefore, even though it is the slow waves in the smooth muscle itself that control the segmentation contractions, these contractions are not effective without background excitation by the enteric nervous system, especially by the myenteric plexus.

Propulsive Movements

PERISTALSIS IN THE SMALL INTESTINE. Chyme is propelled through the small intestine by *peristaltic waves*. These can occur in any part of the small intestine, and they move analward at a velocity of 0.5 to 2.0 cm/sec, much faster in the proximal intestine and much slower in the terminal intestine. They normally

are very weak and usually die out after traveling only 3 to 5 centimeters, very rarely farther than 10 centimeters, so that movement of the chyme is also very poor, so poor in fact that the *net* movement of the chyme along the small intestine averages only 1 cm/min. This means that 3 to 5 hours are required for passage of chyme from the pylorus to the ileocecal valve.

CONTROL OF PERISTALSIS BY NERVOUS AND HORMONAL SIGNALS. Peristaltic activity of the small intestine is greatly increased after a meal. This is caused partly by the beginning entry of chyme into the duodenum but also by a so-called *gastroenteric reflex* that is initiated by distention of the stomach and conducted principally through the myenteric plexus from the stomach down along the wall of the small intestine.

In addition to the nervous signals that affect small intestinal peristalsis, several hormonal factors affect peristalsis. They include *gastrin, CCK, insulin,* and *serotonin,* all of which enhance intestinal motility and are secreted during various phases of food processing. On the other hand, *secretin* and *glucagon* inhibit small intestinal motility. The quantitative importance of each of these hormonal factors for controlling motility is still questionable.

The function of the peristaltic waves in the small intestine is not only to cause progression of the chyme toward the ileocecal valve but also to spread out the chyme along the intestinal mucosa. As the chyme enters the intestine from the stomach and causes initial distention of the proximal intestine, the elicited peristaltic waves begin immediately to spread the chyme along the intestine; this process intensifies as additional chyme enters the duodenum. On reaching the ileocecal valve, the chyme is sometimes blocked for several hours until the person eats another meal, when a new *gastroenteric* (also called *gastroileal*) reflex intensifies the peristalsis in the ileum and forces the remaining chyme through the ileocecal valve into the cecum.

PROPULSIVE EFFECT OF THE SEGMENTATION MOVEMENTS. The segmentation movements, although they last for only a few seconds, often also travel in the anal direction and help propel the food down the intestine. Therefore, the difference between the segmentation and the peristaltic movements is not as great as might be implied by their separation into these two classifications.

PERISTALTIC RUSH. Although peristalsis in the small intestine is normally very weak, intense irritation of the intestinal mucosa, as occurs in some severe cases of infectious diarrhea, can cause both powerful and rapid peristalsis, called the *peristaltic rush.* This is initiated partly by extrinsic nervous reflexes to the autonomic nervous system ganglia and the brain stem and then back again to the gut and partly by direct enhancement of the myenteric plexus reflexes. The powerful peristaltic contractions then travel long distances in the small intestine within minutes, sweeping the contents of the intestine into the colon and thereby relieving the small intestine of either irritative chyme or excessive distention.

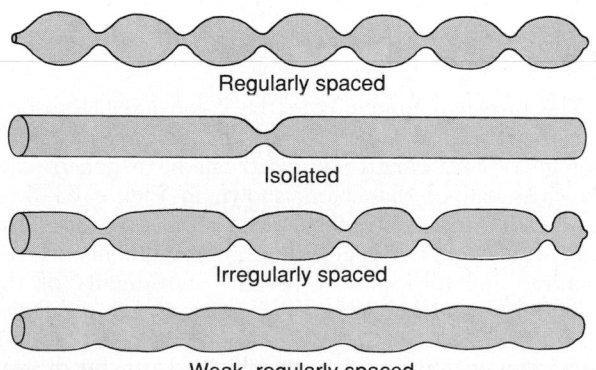

Regularly spaced

Isolated

Irregularly spaced

Weak, regularly spaced

Figure 63–3. Segmentation movements of the small intestine.

Peristalsis in the Fasting Human Being—The "Migrating Motor Complex"

After a person or an animal eats a meal, the nature of the gastrointestinal motor functions is determined mainly by the stimulating effects of the food in the gastrointestinal tract itself. However, many hours after a meal or at other times when the human being is fasting, a special pattern of activity recurs approximately every 90 minutes in the stomach and small intestine, called the *migrating motor complex.* This migrating complex causes moderately active peristaltic waves to sweep slowly downward along the stomach and small intestine, sweeping any excess digestive secretions or other intra-intestinal debris into the colon and thereby preventing their accumulation in the upper gastrointestinal tract.

The migrating complex begins in the body of the stomach and spreads all the way through the ileum. At any one time, only about 40 centimeters of the intestinal tract is actively engaged in the peristaltic waves, and this lasts for only 6 to 10 minutes; this 40-centimeter area moves slowly along the intestinal tract at a velocity of 6 to 12 cm/min. Then approximately at the time that one migrating complex reaches the end of the ileum, a new one begins in the stomach.

Movements Caused by the Muscularis Mucosae and Muscle Fibers of the Villi

The muscularis mucosae can cause short or long folds to appear in the intestinal mucosa; it can also cause the folds to move to progressively newer areas of mucosa. In addition, individual fibers from this muscle extend into the intestinal villi and cause them to contract intermittently. The mucosal folds increase the surface area exposed to the chyme, thereby increasing the rate of absorption. The contractions of the villi—shortening, elongating, and shortening again—"milk" the villi, so that lymph flows freely from the central lacteals into the lymphatic system. Both of these types of contraction also agitate the fluids surrounding the villi, so that progressively new areas of fluid become exposed to absorption.

These mucosal and villous contractions are initiated by local nervous reflexes in the submucosal plexus that occur in response to chyme in the small intestine.

Function of the Ileocecal Valve

A principal function of the ileocecal valve is to prevent backflow of fecal contents from the colon into the small intestine. As shown in Figure 63–4, the lips of the ileocecal valve protrude into the lumen of the cecum and therefore are forcefully closed when excess pressure builds up in the cecum and tries to push the cecal contents backward against the lips. The valve usually can resist reverse pressure of 50 to 60 centimeters of water.

In addition, the wall of the ileum for several centimeters immediately preceding the ileocecal valve has a thickened muscular coat called the *ileocecal sphincter.* This sphincter normally remains mildly constricted and slows the emptying of ileal contents into the cecum except immediately after a meal, when a gastroileal reflex (described earlier) intensifies the peristalsis in the ileum.

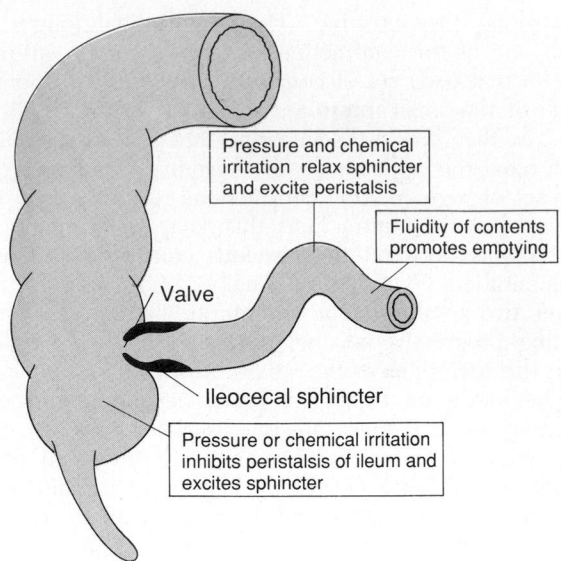

Figure 63–4. Emptying at the ileocecal valve.

The resistance to emptying at the ileocecal valve prolongs the stay of chyme in the ileum and thereby facilitates absorption. Only about 1500 milliliters of chyme empty into the cecum each day.

FEEDBACK CONTROL OF THE ILEOCECAL SPHINCTER. The degree of contraction of the ileocecal sphincter as well as the intensity of peristalsis in the terminal ileum is also controlled significantly by reflexes from the cecum. Whenever the cecum is distended, the contraction of the ileocecal sphincter is intensified and ileal peristalsis inhibited, which greatly delays emptying of additional chyme from the ileum. Also, any irritant in the cecum delays emptying. For instance, when a person has an inflamed appendix, the irritation of this vestigial remnant of the cecum can cause such intense spasm of the ileocecal sphincter and paralysis of the ileum that they block emptying of the ileum. These reflexes from the cecum to the ileocecal sphincter and ileum are mediated both by way of the myenteric plexus in the gut wall itself and through extrinsic nerves, especially reflexes by way of the prevertebral sympathetic ganglia.

MOVEMENTS OF THE COLON

The principal functions of the colon are (1) absorption of water and electrolytes from the chyme and (2) storage of fecal matter until it can be expelled. The proximal half of the colon, shown in Figure 63–5, is concerned principally with absorption and the distal half, with storage. Because intense movements are not required for these functions, the movements of the colon are normally very, very sluggish. Yet in a sluggish manner, the movements still have characteristics similar to those of the small intestine and can be divided once again into mixing movements and propulsive movements.

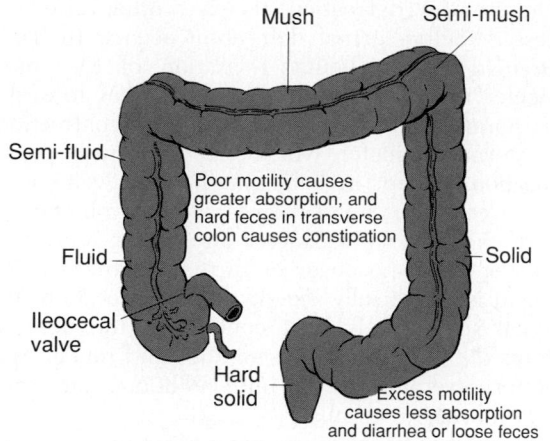

Figure 63–5. Absorptive and storage functions of the large intestine.

MIXING MOVEMENTS—HAUSTRATIONS. In the same manner that segmentation movements occur in the small intestine, large circular constrictions occur in the large intestine. At each of these constriction points, about 2.5 centimeters of the circular muscle contracts, sometimes constricting the lumen of the colon to almost occlusion. At the same time, the longitudinal muscle of the colon, which is aggregated into three longitudinal strips called the *teniae coli,* contract. These combined contractions of the circular and longitudinal strips of muscle cause the unstimulated portion of the large intestine to bulge outward into bag-like sacs called *haustrations.* The haustral contractions, once initiated, usually reach peak intensity in about 30 seconds and then disappear during the next 60 seconds. They also at times move slowly analward during their period of contraction, especially in the cecum and ascending colon, and thereby provide a minor amount of forward propulsion of the colonic contents. After another few minutes, new haustral contractions occur in other areas nearby. Therefore, the fecal material in the large intestine is slowly *dug into and rolled over* in much the same manner that one spades the earth. In this way, all the fecal material is gradually exposed to the surface of the large intestine, and fluid and dissolved substances are progressively absorbed until only 80 to 200 milliliters of the daily load of chyme is lost in the feces.

PROPULSIVE MOVEMENTS—"MASS MOVEMENTS." Peristaltic waves of the type seen in the small intestine only seldom occur in most parts of the colon. Instead, most propulsion occurs by (1) the slow analward movement of the *haustral contractions* just discussed and (2) *mass movements.*

Much of the propulsion in the cecum and ascending colon results from the slow but persistent haustral contractions, requiring as many as 8 to 15 hours to move the chyme only from the ileocecal valve through the transverse colon, while the chyme itself becomes fecal in quality and becomes a semisolid slush instead of a semifluid.

From the beginning of the transverse colon to the sigmoid, mass movements mainly take over the propulsive role. These movements usually occur only one to three times each day, most abundantly for about 15 minutes during the first hour after eating breakfast.

A mass movement is a modified type of peristalsis characterized by the following sequence of events: First, a constrictive ring occurs at a distended or irritated point in the colon, usually in the transverse colon, and then rapidly thereafter the 20 or more centimeters of colon *distal* to the constriction lose their haustrations and instead contract as a unit, forcing the fecal material in this segment en masse down the colon. The contraction develops progressively more force for about 30 seconds, and relaxation then occurs during the next 2 to 3 minutes before another mass movement occurs, this time perhaps farther along the colon. The whole series of mass movements usually persists for only 10 to 30 minutes, and they then return perhaps a half day or even a day later. When they have forced a mass of feces into the rectum, the desire for defecation is felt.

Initiation of Mass Movements by the Gastrocolic and Duodenocolic Reflexes. The appearance of mass movements after meals is facilitated by *gastrocolic* and *duodenocolic reflexes.* These reflexes result from distention of the stomach and duodenum. They occur either not at all or hardly at all when the extrinsic nerves are removed; therefore, these reflexes almost certainly are conducted through the extrinsic nerves of the autonomic nervous system.

Irritation in the colon can also initiate intense mass movements. For instance, a person who has an ulcerated condition of the colon (*ulcerative colitis*) frequently has mass movements that persist almost all the time.

Mass movements can also be initiated by intense stimulation of the parasympathetic nervous system or by overdistention of a segment of the colon.

Defecation

Most of the time, the rectum is empty of feces. This results partly from the fact that a weak functional sphincter exists about 20 centimeters from the anus at the juncture between the sigmoid and the rectum. There is also a sharp angulation here that contributes additional resistance to filling of the rectum. When a mass movement forces feces into the rectum, the desire for defecation is normally initiated, including reflex contraction of the rectum and relaxation of the anal sphincters.

Continual dribble of fecal matter through the anus is prevented by tonic constriction of (1) the *internal anal sphincter,* a several-centimeters-long thickening of the circular smooth muscle that lies immediately inside the anus, and (2) the *external anal sphincter,* composed of striated voluntary muscle that both surrounds the internal sphincter and extends distal to it. The external sphincter is controlled by nerve fibers in the pudendal nerve, which is part of the somatic nervous system and therefore is under *voluntary, conscious control;* subconsciously, it is usually kept continuously constricted unless conscious signals inhibit the constriction.

DEFECATION REFLEXES. Ordinarily, defecation is initiated by *defecation reflexes.* One of these reflexes is an *intrinsic reflex* mediated by the local enteric nervous system. This can be described as follows: When feces enter the rectum, distention of the rectal wall initiates afferent signals that spread through the *myenteric plexus* to initiate peristaltic waves in the descending colon, sigmoid, and rectum, forcing feces toward the anus. As the peristaltic wave approaches the anus, the internal anal sphincter is relaxed by inhibitory signals from the myenteric plexus; if the external anal sphincter is consciously, voluntarily relaxed at the same time, defecation will occur.

However, the intrinsic defecation reflex functioning by itself is relatively weak. To be effective in causing defecation, it usually must be fortified by another type of defecation reflex, a *parasympathetic defecation reflex* that involves the sacral segments of the spinal cord, as shown in Figure 63–6. When the nerve endings in the rectum are stimulated, signals are transmitted first into the spinal cord and then reflexly back to the descending colon, sigmoid, rectum, and anus by way of parasympathetic nerve fibers in the *pelvic nerves*. These parasympathetic signals greatly intensify the peristaltic waves as well as relaxing the internal anal sphincter and thus convert the intrinsic defecation reflex from a weak movement into a powerful process of defecation that is sometimes effective in emptying the large bowel all at once all the way from the splenic flexure of the colon to the anus.

Also, the afferent signals entering the spinal cord initiate other effects, such as taking a deep breath, closure of the glottis, and contraction of the abdominal wall muscles to force the fecal contents of the colon downward and at the same time cause the pelvic floor to extend downward and pull outward on the anal ring to evaginate the feces.

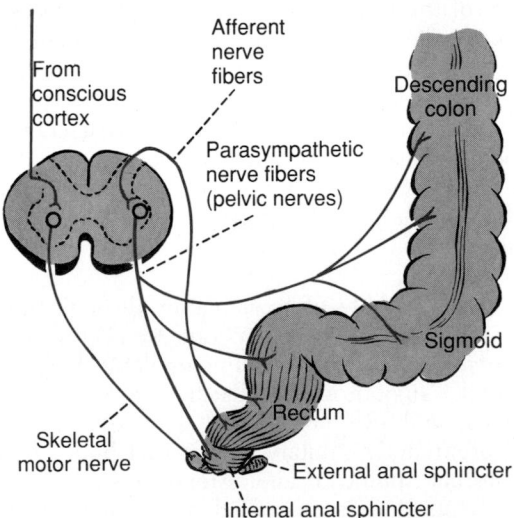

Figure 63–6. Afferent and efferent pathways of the parasympathetic mechanism for enhancing the defecation reflex.

Despite the defecation reflexes, other effects are necessary before actual defecation occurs. In the toilet-trained human being, relaxation of the internal sphincter and forward movement of feces toward the anus normally initiate an instantaneous contraction of the external sphincter, which still temporarily prevents defecation. Except in babies and mentally inept people, the conscious mind then takes over voluntary control of the external sphincter and either relaxes it to allow defecation to occur or further contracts it if the moment is not socially acceptable for defecation. If the external sphincter is kept contracted, the defecation reflexes die out after a few minutes and remain quiescent for several hours or until additional amounts of feces enter the rectum.

When it becomes convenient for the person to defecate, the defecation reflexes can sometimes be excited by taking a deep breath to move the diaphragm downward and then contracting the abdominal muscles to increase the pressure in the abdomen, thus forcing fecal contents into the rectum to elicit new reflexes. Reflexes initiated in this way are almost never as effective as those that arise naturally, for which reason people who too often inhibit their natural reflexes are likely to become severely constipated.

In neonates and in some people with transected spinal cords, the defecation reflexes cause automatic emptying of the lower bowel at inconvenient times during the day because of lack of conscious control exercised through voluntary contraction of the external anal sphincter.

OTHER AUTONOMIC REFLEXES THAT AFFECT BOWEL ACTIVITY

Aside from the duodenocolic, gastrocolic, gastroileal, enterogastric, and defecation reflexes that have been discussed in this chapter, several other important nervous reflexes can affect the overall degree of bowel activity. They are the peritoneointestinal reflex, renointestinal reflex, vesicointestinal reflex, and somatointestinal reflex. All these reflexes are initiated by sensory signals that pass to the prevertebral sympathetic ganglia or to the spinal cord and then are transmitted through the sympathetic nervous system back to the gut. And they all *inhibit* gastrointestinal activity, often severely blocking movement of food through the intestines.

The *peritoneointestinal reflex* results from irritation of the peritoneum; it strongly inhibits the excitatory enteric nerves and thereby can cause intestinal paralysis, especially so in patients with peritonitis. The *renointestinal* and *vesicointestinal reflexes* inhibit intestinal activity as a result of kidney and bladder irritation, respectively. Finally, the *somatointestinal reflex* causes intestinal inhibition when the skin over the abdomen is irritatingly stimulated.

REFERENCES

Bouchier, I. A. D., et al.: Gastroenterology: Clinical Science & Practice. Philadelphia, W. B. Saunders Co., 1994.
Chou, C. C.: Relationship between intestinal blood flow and motility. Annu. Rev. Physiol., 44:29, 1982.

Cohen, S., et al.: Gastrointestinal motility. In Crane, R. K. (ed.): International Review of Physiology: Gastrointestinal Physiology III. Vol. 19. Baltimore, University Park Press, 1979, p. 107.

Davenport, H. W.: A Digest of Digestion, 2nd Ed. Chicago, Year Book Medical Publishers, 1978.

Haubrich, W. S., et al.: Bockus Gastroenterology. Philadelphia, W. B. Saunders Co., 1994.

Hunt, J. N.: Mechanisms and disorders of gastric emptying. Annu. Rev. Med., 34:219, 1983.

Johnson, L. R., et al.: Physiology of the Gastrointestinal Tract, 2nd Ed. New York, Raven Press, 1987.

Keighley, M. R. B., and Williams, N. S.: Surgery of the Anus, Rectum and Colon. Philadelphia, W. B. Saunders Co., 1994.

Klimov, P. K.: Behavior of the organs of the digestive system. Neurosci. Behav. Physiol., 14:333, 1984.

Kumar, D., and Wingate, D. L.: An Illustrated Guide to Gastrointestinal Motility. 2nd Ed. New York, Churchill Livingstone, 1994.

Lewis, J. H.: A Pharmacological Approach to Gastrointestinal Diseases. Baltimore, Williams & Wilkins, 1994.

Luschei, E. S., and Goldberg, L. J.: Neural mechanisms of mandibular control: Mastication and voluntary biting. In Brooks, V. B. (ed.): Handbook of Physiology. Sec. 1, Vol. II. Bethesda, Md., American Physiological Society, 1981, p. 1237.

Magee, D. F.: Interdigestive activity in the gastrointestinal tract. News in Physiol. Sci., 2:101, 1987.

Mei, N.: Intestinal chemosensitivity. Physiol. Rev., 65:211, 1985.

Miller, A. J.: Deglutition. Physiol. Rev., 62:129, 1982.

Murphy, R. A.: Muscle cells of hollow organs. News Physiol. Sci., 3:124, 1988.

Philips, S. F., and Devroede, G. J.: Functions of the large intestine. In Crane, R. K. (ed.): International Review of Physiology: Gastrointestinal Physiology III. Vol. 19, Baltimore, University Park Press, 1979, p. 263.

Rehfeld, J. F.: Gastrointestinal hormones, In Crane, R. K. (ed.): International Review of Physiology: Gastrointestinal Physiology III. Vol. 19., Baltimore, University Park Press, 1979, p. 291.

Sant'Ambrogio, G., and Mathew, O. P.: Control of upper airway muscles. News Physiol. Sci., 3:167, 1988.

Scarpignato, C., and Galmiche, J.-P.: Clinical investigation in Esophageal Disease. Farmington, CT, S. Karger Publishers, Inc., 1994.

Schuster, M.: Atlas of Gastrointestinal Motility in Health and Disease. Baltimore, Williams & Wilkins, 1993.

Sternini, C.: Structural and chemical organization of the myenteric plexus. Annu. Rev. Physiol., 50:81, 1988.

Targan, S. R., and Shanahan, F.: Inflammatory Bowel Disease: From Bench to Bedside. Baltimore, Williams & Wilkins, 1994.

Thompson, J. C., et al. (eds.): Gastrointestinal Endocrinology, New York, McGraw-Hill, 1987.

Wong, M. E. K., et al.: Atlas of Major Oral and Maxillofacial Surgery. Philadelphia, W. B. Saunders Co., 1994.

Worthington, P., and Evans, J. R.: Controversies in Oral and Maxillofacial Surgery. Philadelphia, W. B. Saunders Co., 1994.

Secretory Functions of the Alimentary Tract

CHAPTER 64

Throughout the gastrointestinal tract, secretory glands subserve two primary functions: First, *digestive enzymes* are secreted in most areas from the mouth to the distal end of the ileum. Second, mucous glands, from the mouth to the anus, provide *mucus* for lubrication and protection of all parts of the alimentary tract.

Most digestive secretions are formed only in response to the presence of food in the alimentary tract, and the quantity secreted in each segment of the tract is almost exactly the amount needed for proper digestion. Furthermore, in some portions of the gastrointestinal tract, even the types of enzymes and other constituents of the secretions are varied in accordance with the types of food present. The purpose of this chapter, therefore, is to describe the different alimentary secretions, their functions, and regulation of their production. (Before proceeding, it is especially important that the reader understand or re-review the general principles of cell transport mechanisms and intracellular formation of substances that are to be secreted. These principles are discussed in detail in Chapters 2 through 4.)

GENERAL PRINCIPLES OF ALIMENTARY TRACT SECRETION

Anatomical Types of Glands

Several types of glands provide the different types of secretions in the alimentary tract. First, on the surface of the epithelium in most parts of the gastrointestinal tract are literally billions of *single-cell mucous glands* called simply *mucous cells* or sometimes *goblet cells*. They function mainly by themselves in response to local stimulation of the epithelium, and they simply extrude their mucus directly onto the epithelial surface to act as a lubricant and to protect the surfaces from excoriation and digestion.

Second, many surface areas of the gastrointestinal tract are lined by pits that represent invaginations of the epithelium into the submucosa. In the small intestine, these pits, called *crypts of Lieberkühn*, are deep and contain specialized secretory cells. One of these cells is shown in Figure 64–1.

Third, in the stomach and upper duodenum are found large numbers of deep *tubular glands*. A typical tubular gland can be seen in Figure 64–4, which shows an acid- and pepsinogen-secreting gland of the stomach.

Fourth, also associated with the alimentary tract are several complex glands—the *salivary glands, pancreas,* and *liver*—which provide secretions for digestion or emulsification of food. The liver has a highly specialized structure that is discussed in Chapter 70. The salivary glands and the pancreas are compound acinous glands of the type shown in Figure 64–2. These glands lie outside the walls of the alimentary tract and, in this, differ from all other alimentary glands. They contain millions of *acini* lined with secreting glandular cells; these acini feed into a system of ducts that finally empty into the alimentary tract itself.

Basic Mechanisms of Stimulation of the Alimentary Tract Glands

EFFECT OF CONTACT OF FOOD WITH THE EPITHELIUM: FUNCTION OF ENTERIC NERVOUS STIMULI. The mechanical presence of food in a particular segment of the gastrointestinal tract usually causes the glands of

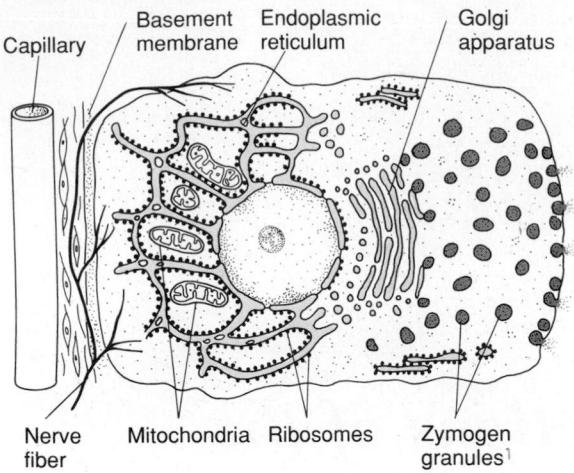

Figure 64–1. Typical function of a glandular cell in formation and secretion of enzymes and other secretory substances.

that region and often of adjacent regions to secrete moderate to large quantities of digestive juices. Part of this local effect results from direct contact stimulation of the surface glandular cells by contact with the food. In addition, epithelial stimulation also activates the *enteric nervous system* of the gut wall. The types of stimuli that do this are (1) tactile stimulation, (2) chemical irritation, and (3) distention of the gut wall. The resulting nervous reflexes stimulate both the mucous cells on the epithelial surface and the deeper glands in the mucosa to increase their secretion.

Autonomic Stimulation of Secretion

Parasympathetic Stimulation. Stimulation of the parasympathetic nerves to the alimentary tract almost invariably increases the rates of glandular secretion. This is especially true of the glands in the upper portion of the tract, innervated by the vagus and other cranial parasympathetic nerves, including the salivary glands, esophageal glands, gastric glands, pancreas, and Brunner's glands in the duodenum. It is also true of the glands of the distal portion of the large intestine, innervated by the pelvic parasympathetic nerves. Secretion in the remainder of the small intestine and in the first two thirds of the large intestine occurs mainly in response to local neural and hormonal stimuli in each segment of the gut.

Sympathetic Stimulation. Stimulation of sympathetic nerves in some parts of the gastrointestinal tract causes a slight to moderate increase in secretion by the respective glands. On the other hand, sympathetic stimulation also results in constriction of the blood vessels that supply the glands. Therefore, sympathetic stimulation can have a dual effect: First, sympathetic stimulation alone usually slightly increases secretion. But second, if parasympathetic or hormonal stimulation is causing copious secretion by the glands, superimposed sympathetic stimulation usually reduces the

secretion, sometimes significantly, mainly because of reduced blood supply.

Regulation of Glandular Secretion by Hormones. In the stomach and intestine, several different *gastrointestinal hormones* help regulate the volume and character of the secretions. These hormones are liberated from the gastrointestinal mucosa in response to the presence of foods in the lumen of the gut. The hormones then are absorbed into the blood and carried to the glands, where they stimulate secretion. This type of stimulation is particularly valuable in increasing the output of gastric juice and pancreatic juice when food enters the stomach or duodenum. Also, hormonal stimulation of the gallbladder wall causes it to empty its stored bile into the duodenum. Other hormones that are still of doubtful effectiveness have been postulated to stimulate secretion by the glands of the small intestine.

Chemically, the gastrointestinal hormones are polypeptides or polypeptide derivatives.

Basic Mechanism of Secretion by Glandular Cells

Secretion of Organic Substances. Although all the basic mechanisms by which glandular cells function are not known, experimental evidence points to the following basic principles of secretion by glandular cells, as shown in Figure 64–1.

1. The nutrient material needed for formation of the secretion must diffuse or be actively transported from the capillary into the base of the glandular cell.

2. Many *mitochondria* located inside the cell near its base use oxidative energy for formation of adenosine triphosphate (ATP).

3. Energy from the ATP, along with appropriate substrates provided by the nutrients, is then used for synthesis of the organic substances; this synthesis occurs almost entirely in the *endoplasmic reticulum* and *Golgi complex*. The

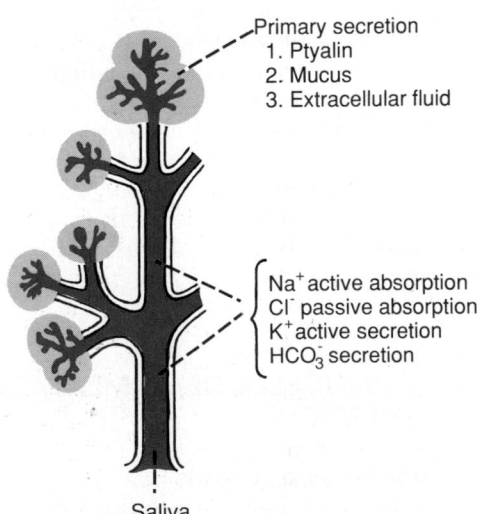

Figure 64–2. Formation and secretion of saliva by a salivary gland.

ribosomes adherent to this reticulum are specifically responsible for formation of the proteins that are to be secreted.

4. The secretory materials are transported through the tubules of the endoplasmic reticulum, passing in about 20 minutes all the way to the vesicles of the Golgi complex, which lies near the secretory ends of the cells.

5. In the Golgi complex, the materials are modified, added to, concentrated, and discharged into the cytoplasm in the form of *secretory vesicles,* which are stored in the apical ends of the secretory cells.

6. These vesicles remain stored until nervous or hormonal control signals cause the cells to extrude the vesicular contents through the cell's surface. This probably occurs in the following way: The control signal first increases the cell membrane permeability to calcium, and calcium enters the cell. The calcium in turn causes many of the vesicles to fuse with the cell membrane and then to break open on their outer surfaces, thus emptying their contents to the exterior; this process is called *exocytosis.*

WATER AND ELECTROLYTE SECRETION. A second necessity for glandular secretion is secretion of sufficient water and electrolytes along with the organic substances. The following is a postulated method by which nervous stimulation causes water and salts to pass through the glandular cells in great profusion, which washes the organic substances through the secretory border of the cells at the same time:

1. Nerve stimulation has a specific effect on the *basal* portion of the cell membrane to cause active transport of chloride ions to the interior.

2. The resulting increase in electronegativity induced inside the cell by excess chloride ions then causes positive ions also to move to the interior of the cell.

3. The excess of both the negative and the positive ions inside the cell creates an osmotic force that pulls water to the interior, thereby increasing the hydrostatic pressure inside the cell and causing the cell itself to swell.

4. The pressure in the cell then results in minute ruptures of the secretory border of the cell and causes flushing of water, electrolytes, and organic materials out of the secretory end of the glandular cell and into the lumen of the gland.

In support of this theory have been the following findings: First, the nerve endings on glandular cells are principally on the bases of the cells. Second, microelectrode studies show that the normal electrical potential across the membrane at the base of the cell is between 30 and 40 millivolts, with negativity on the interior and positivity on the exterior. Parasympathetic stimulation increases this polarization voltage to values some 10 and 20 millivolts more negative than normal. This increase in polarization occurs a second or more after the nerve signal has arrived, indicating that it is caused by movement of negative ions (presumably chloride ions) through the membrane to the interior of the cell.

Although this mechanism for secretion is still partly theoretical, it does explain how it would be possible for nerve impulses to regulate secretion. Hormones acting on the cell membrane could cause similar results to those caused by nervous stimulation.

Lubricating and Protective Properties of Mucus and Its Importance in the Gastrointestinal Tract

Mucus is a thick secretion composed mainly of water, electrolytes, and a mixture of several glycoproteins, which themselves are composed of large polysaccharides bound with much smaller quantities of protein. Mucus is slightly different in different parts of the gastrointestinal tract, but everywhere it has several important characteristics that make it both an excellent lubricant and a protectant for the wall of the gut. *First,* mucus has adherent qualities that make it adhere tightly to the food or other particles and spread as a thin film over the surfaces. *Second,* it has sufficient *body* that it coats the wall of the gut and prevents actual contact of the food particles with the mucosa. *Third,* mucus has a low resistance for slippage, so that the particles can slide along the epithelium with great ease. *Fourth,* mucus causes fecal particles to adhere to one another to form the fecal masses that are expelled during a bowel movement. *Fifth,* mucus is strongly resistant to digestion by the gastrointestinal enzymes. And, *sixth,* the glycoproteins of mucus have amphoteric properties, which means that they are capable of buffering small amounts of either acids or alkalics; also, mucus often contains moderate quantities of bicarbonate ions, which specifically neutralize acids.

In summary, mucus has the ability to allow easy slippage of food along the gastrointestinal tract and to prevent excoriative or chemical damage to the epithelium. A person becomes acutely aware of the lubricating qualities of mucus when the salivary glands fail to secrete saliva because under these circumstances, it is extremely difficult to swallow solid food even when it is taken with large amounts of water.

SECRETION OF SALIVA

SALIVARY GLANDS; CHARACTERISTICS OF SALIVA. The principal glands of salivation are the *parotid, submandibular,* and *sublingual* glands; in addition, there are many small *buccal* glands. The daily secretion of saliva normally ranges between about 800 and 1500 milliliters, as shown in Table 64–1.

Saliva contains two major types of protein secretion: (1) a *serous secretion* that contains *ptyalin* (an α-amylase), which is an enzyme for digesting starches, and

Table 64–1 DAILY SECRETION OF INTESTINAL JUICES

	Daily Volume (ml)	pH
Saliva	1000	6.0–7.0
Gastric secretion	1500	1.0–3.5
Pancreatic secretion	1000	8.0–8.3
Bile	1000	7.8
Small intestine secretion	1800	7.5–8.0
Brunner's gland secretion	200	8.0–8.9
Large intestinal secretion	200	7.5–8.0
Total	6700	

(2) *mucous secretion* that contains *mucin* for lubricating and for surface protective purposes. The parotid glands secrete entirely the serous type, and the submandibular and sublingual glands secrete both the serous type and mucus. The buccal glands secrete only mucus. Saliva has a pH between 6.0 and 7.0, a favorable range for the digestive action of ptyalin.

SECRETION OF IONS IN THE SALIVA. Saliva contains especially large quantities of potassium and bicarbonate ions. On the other hand, the concentrations of both sodium and chloride ions are several times less in saliva than in plasma. One can understand these special concentrations of ions in the saliva from the following description of the mechanism for secretion of saliva.

Figure 64–2 shows secretion by the submaxillary gland, a typical compound gland that contains *acini* and *salivary ducts*. Salivary secretion is a two-stage operation: the first stage involves the acini and the second, the salivary ducts. The acini secrete a *primary secretion* that contains ptyalin and/or mucin in a solution of ions in concentrations not greatly different from those of typical extracellular fluid. As the primary secretion flows through the ducts, two major active transport processes take place that markedly modify the ionic composition of the saliva.

First, *sodium ions* are actively reabsorbed from all the salivary ducts and *potassium ions* are actively secreted in exchange for the sodium. Therefore, the sodium concentration of the saliva becomes greatly reduced, whereas the potassium ion concentration becomes increased. However, there is excess of sodium reabsorption over potassium secretion, and this creates negativity of about − 70 millivolts in the salivary ducts; this in turn causes chloride ions to be reabsorbed passively; therefore, the chloride ion concentration falls to a very low level, matching the decrease in sodium ion concentration.

Second, *bicarbonate ions* are secreted by the ductal epithelium into the lumen of the duct. This is at least partly caused by exchange of bicarbonate for chloride ions, but it may also result partly from an active secretory process.

The net result of these transport processes is that *under resting conditions,* the concentrations of sodium and chloride ions in the saliva are only about 15 mEq/liter each, about one seventh to one tenth their concentrations in plasma. On the other hand, the concentration of potassium ions is about 30 mEq/liter, seven times as great as its concentration in plasma; and the concentration of bicarbonate ions is 50 to 70 mEq/liter, about two to three times that of plasma.

During maximal salivation, the salivary ionic concentrations change considerably because the rate of formation of primary secretion by the acini can increase as much as 20-fold. As a result, this acinar secretion flows through the ducts so rapidly that the ductal reconditioning of the secretion is considerably reduced. Therefore, when copious quantities of saliva are secreted, the sodium chloride concentration rises to about one-half to two-thirds that of plasma, whereas the potassium concentration falls to only four times that of plasma.

In the presence of excess aldosterone secretion, the sodium and chloride reabsorption and the potassium secretion become greatly increased, so that the sodium chloride concentration in the saliva is sometimes reduced almost to zero while the potassium concentration increases even higher than the normal seven times the plasma potassium level.

Because of the high potassium ion concentration of saliva, in any abnormal state in which the saliva is lost to the exterior of the body for long periods, a person can develop serious depletion of potassium ions in the body, leading in occasional circumstances to serious hypokalemia and paralysis.

FUNCTION OF SALIVA FOR ORAL HYGIENE. Under basal conditions, about 0.5 milliliter of saliva, almost entirely of the mucous type, is secreted each minute all the time except during sleep, when the secretion becomes very little. This secretion plays an exceedingly important role in maintaining healthy oral tissues. The mouth is loaded with pathogenic bacteria that can easily destroy tissues and cause dental caries. Saliva helps prevent the deteriorative processes in several ways: *First,* the flow of saliva itself helps wash away the pathogenic bacteria as well as the food particles that provide their metabolic support. *Second,* the saliva contains several factors that destroy bacteria. One of these is *thiocyanate ions* and another is several *proteolytic enzymes*—most important, *lysozyme*—that (1) attack the bacteria; (2) aid the thiocyanate ions in entering the bacteria, where they in turn become bactericidal; and (3) digest food particles, thus helping further to remove the bacterial metabolic support. And, *third,* saliva often contains significant amounts of protein antibodies that can destroy oral bacteria, including those that cause dental caries.

Therefore, in the absence of salivation, the oral tissues become ulcerated and otherwise infected, and caries of the teeth becomes rampant.

NERVOUS REGULATION OF SALIVARY SECRETION. Figure 64–3 shows the parasympathetic nervous pathways for regulation of salivation, demonstrating that the salivary glands are controlled mainly by *parasympathetic nervous signals* from the *superior* and *inferior salivatory nuclei* in the brain stem. The salivatory nuclei are located approximately at the juncture of the medulla and pons and are excited by both taste and tactile stimuli from the tongue and other areas of the mouth and pharynx. Many taste stimuli, especially the sour taste, elicit copious secretion of saliva—often 8 to 20 times the basal rate of secretion. Also, certain tactile stimuli, such as the presence of smooth objects in the mouth (a pebble, for instance), cause marked salivation, whereas rough objects cause less salivation and occasionally even inhibit salivation.

Salivation can also be stimulated or inhibited by nervous signals arriving in the salivatory nuclei from higher centers of the central nervous system. For instance, when a person smells or eats favorite foods, salivation is greater than when disliked food is smelled or eaten. The *appetite area* of the brain, which partially regulates these effects, is located in proximity to

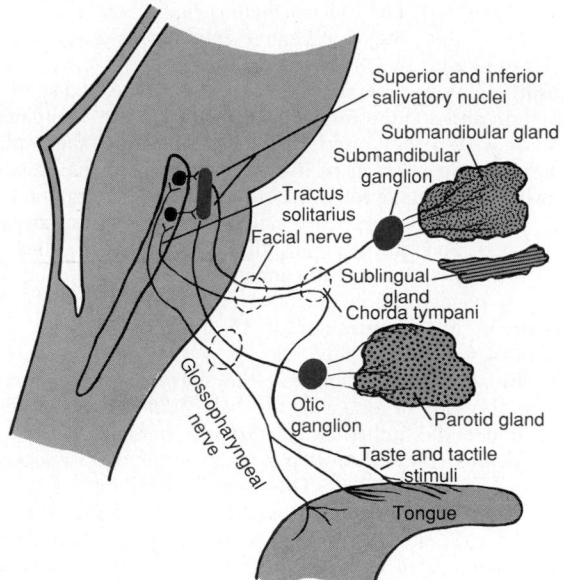

Figure 64–3. Parasympathetic nervous regulation of salivary secretion.

the parasympathetic centers of the anterior hypothalamus, and it functions to a great extent in response to signals from the taste and smell areas of the cerebral cortex or amygdala.

Salivation also occurs in response to reflexes originating in the stomach and upper intestines—particularly when irritating foods are swallowed or when a person is nauseated because of some gastrointestinal abnormality. The swallowed saliva presumably helps to remove the irritating factor in the gastrointestinal tract by diluting or neutralizing the irritating substances.

Sympathetic stimulation can also increase salivation a moderate amount, but much less so than does parasympathetic stimulation. The sympathetic nerves originate from the superior cervical ganglia and then travel along the blood vessels to the salivary glands.

A secondary factor that also affects secretion is the *blood supply to the glands* because secretion always requires adequate nutrition. The parasympathetic nerve signals that induce copious salivation at the same time dilate the blood vessels. But, in addition, salivation itself directly dilates the blood vessels, thus providing increased nutrition as it is needed. Part of this additional vasodilator effect is caused by *kallikrien* secreted by the activated salivary cells, which in turn acts as an enzyme to split one of the blood proteins, an alpha$_2$-globulin, to form *bradykinin*, a strong vasodilator.

ESOPHAGEAL SECRETION

The esophageal secretions are entirely mucoid in character and principally provide lubrication for swallowing. The main body of the esophagus is lined with many *simple mucous glands;* at the gastric end and to a lesser extent in the initial portion of the esophagus, there are many *compound mucous glands.* The mucus secreted by the compound glands in the upper esophagus prevents mucosal excoriation by the newly entering food, whereas the compound glands near the esophagogastric junction protect the esophageal wall from digestion by gastric juices that often reflux from the stomach back into the lower esophagus. Despite this protection, a peptic ulcer at times may occur at the gastric end of the esophagus.

GASTRIC SECRETION

Characteristics of the Gastric Secretions

In addition to mucus-secreting cells that line the entire surface of the stomach, the stomach mucosa has two important types of tubular glands: the *oxyntic* (or *gastric*) *glands* and the *pyloric glands.* The oxyntic (acid-forming) glands secrete *hydrochloric acid, pepsinogen, intrinsic factor,* and *mucus.* The pyloric glands secrete mainly *mucus* for protection of the pyloric mucosa but also some *pepsinogen* and, very important, the hormone *gastrin.* The oxyntic glands are located on the inside surfaces of the body and fundus of the stomach, constituting the proximal 80 per cent of the stomach. The pyloric glands are located in the antral portion of the stomach.

Secretions from the Oxyntic Glands

A typical oxyntic gland is shown in Figure 64–4. It is composed of three types of cells: (1) the *mucous neck cells,* which secrete mainly mucus but also some pepsinogen; (2) the *peptic* (or *chief*) *cells,* which secrete large quantities of pepsinogen; and (3) the *parietal* (or *oxyntic*) *cells,* which secrete hydrochloric acid and *intrinsic factor.* A postulated mechanism for secretion of mucus and pepsinogen by the mucous neck cells and the peptic cells was given earlier in the chapter and is shown in Figure 64–1. Secretion of

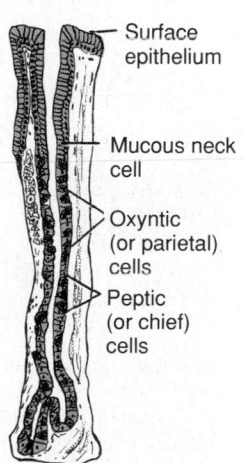

Figure 64–4. Oxyntic gland from the body of the stomach.

hydrochloric acid by the parietal cells involves special mechanisms as follows.

BASIC MECHANISM OF HYDROCHLORIC ACID SECRETION. When stimulated, the parietal cells secrete an acid solution that contains about 160 millimoles of hydrochloric acid per liter, which is almost exactly isotonic with the body fluids. The pH of this acid is about 0.8, demonstrating its extreme acidity. At this pH, the hydrogen ion concentration is about 3 million times that of the arterial blood. And to concentrate the hydrogen ions, this tremendous amount requires more than 1500 calories of energy per liter of gastric juice, as discussed in Chapter 4 in relation to membrane transport mechanisms.

Figure 64–5 shows schematically the functional structure of a parietal cell, demonstrating that it contains many large branching *intracellular canaliculi.* When these cells secrete their acid juice, the membranes of the canaliculi open widely to empty their secretion directly into the lumen of the oxyntic gland. The hydrochloric acid is formed at the villus-like membranes of these canaliculi and then conducted to the exterior.

Different suggestions for the precise mechanism of hydrochloric acid formation have been offered. One of these is shown in Figure 64–6 and consists of the following steps.

1. Chloride ion is actively transported from the cytoplasm of the parietal cell into the lumen of the canaliculus, and sodium ions are actively transported out of the lumen. These two effects together create a negative potential of -40 to -70 millivolts in the canaliculus, which in turn causes passive diffusion of positively charged potassium ions and a small number of sodium ions from the cell cytoplasm also into the canaliculus. Thus, in effect, mainly potassium chloride but also much smaller amounts of sodium chloride enter the canaliculus.

2. Water becomes dissociated into hydrogen ions and hydroxyl ions in the cell cytoplasm. The hydrogen ions are then actively secreted into the canaliculus in exchange for potassium ions; this active exchange process is catalyzed by H^+,K^+-ATPase. In addition, the sodium ions are actively reabsorbed by a separate sodium pump. Thus, most of the potassium and sodium ions that had diffused into the canaliculus are reabsorbed, and hydrogen ions take their place, giving a strong solution of hydrochloric acid in the canaliculus, which is then secreted into the lumen of the gland.

3. Water passes into the canaliculus by osmosis because of the secretion of the ions into the canaliculus. Thus, the final secretion entering the canaliculus contains hydrochloric acid in a concentration of 150 to 160 mEq/liter, potassium chloride in a concentration of 15 mEq/liter, and a small amount of sodium chloride.

4. Finally, carbon dioxide, either formed during metabolism in the cell or entering the cell from the blood, combines under the influence of *carbonic anhydrase* with the hydroxyl ions (formed in step 2 when water was dissociated) to form bicarbonate ions. This then diffuses out of the cell into the extracellular fluid in exchange for chloride ions that enter the cell and later will be secreted into the canaliculus. The importance of carbon dioxide in the chemical reactions for formation of hydrochloric acid is demonstrated by the fact that carbonic anhydrase inhibition by the drug *acetazolamide* diminishes the formation of hydrochloric acid.

SECRETION AND ACTIVATION OF PEPSINOGEN. Several slightly different types of pepsinogen are secreted by the peptic and mucous cells of the gastric glands. Even so, all the pepsinogens perform the same functions. When the pepsinogens are first secreted, they have no digestive activity. However, as soon as they come in contact with hydrochloric acid, and especially when they come in contact with previously formed pepsin plus the hydrochloric acid, they are activated to form active *pepsin.* In this process, the pepsinogen molecule, having a molecular weight of about 42,500, is split to form the pepsin molecule, having a molecular weight of about 35,000.

Pepsin is an active proteolytic enzyme in a highly acid medium (optimum pH 1.8 to 3.5), but above a pH of about 5 it has little proteolytic activity and actually becomes completely inactivated in a short time. Therefore, hydrochloric acid is as necessary as pepsin for protein digestion in the stomach; this is discussed in Chapter 65.

Secretion of Other Enzymes. Small quantities of other enzymes are also secreted in the stomach juices, including *gastric lipase, gastric amylase,* and a *gelatinase.* Gastric lipase is of little quantitative importance and is actually a *tributyrase* because its principal activity is on tributyrin, which is butterfat; it has almost no lipolytic activity on the other fats. Gastric amylase plays a minor role in digestion of starches, and the gelatinase helps to liquefy some of the proteoglycans in meats.

Secretion of Intrinsic Factor. The substance *intrinsic factor,* essential for absorption of vitamin B_{12} in the ileum, is secreted by the *parietal cells* along with the secretion of hydrochloric acid. When the acid-producing cells of the stomach are destroyed, which frequently occurs in chronic gastritis, the person develops not only *achlorhydria* but often also *pernicious anemia* because of failure of maturation of the red blood cells in the absence of vitamin B_{12} stimulation of the bone marrow. This is discussed in detail in Chapter 32.

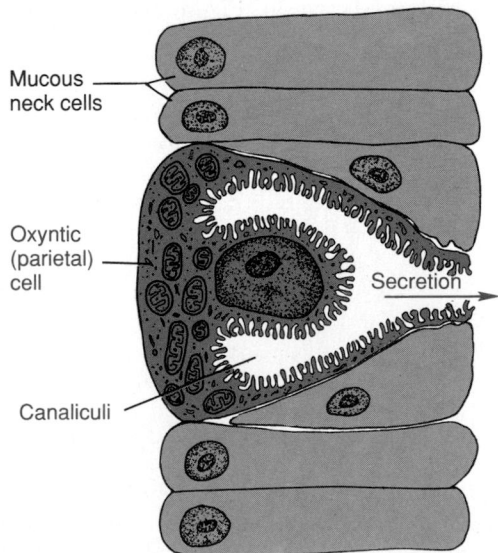

Figure 64–5. Schematic anatomy of the canaliculi in a parietal (oxyntic) cell.

Mucous
neck cells

Oxyntic
(parietal)
cell

Secretion

Canaliculi

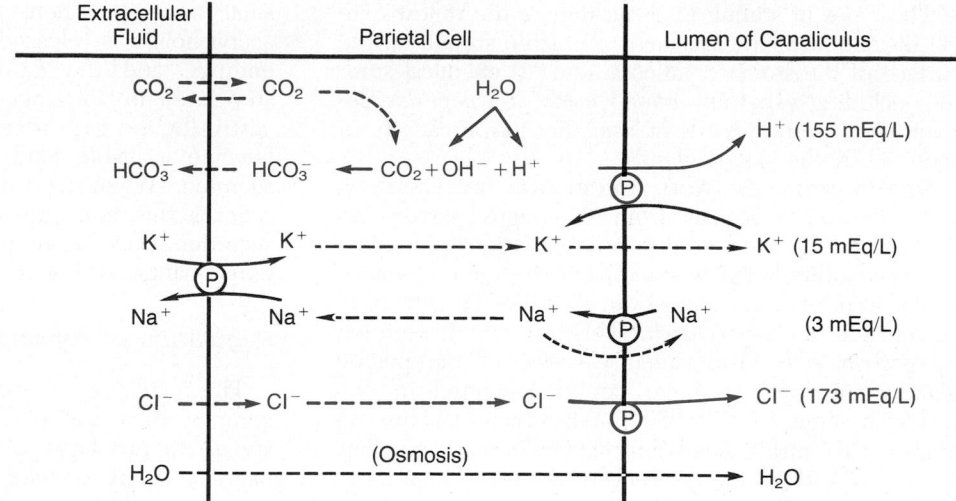

Figure 64–6. Postulated mechanism for the secretion of hydrochloric acid.

PYLORIC GLANDS—SECRETION OF MUCUS AND GASTRIN. The pyloric glands are structurally similar to the oxyntic glands but contain few peptic cells and almost no parietal cells. Instead, they contain mostly mucous cells that are identical with the mucous neck cells of the gastric glands. These cells secrete a small amount of pepsinogen, as discussed earlier, and an especially large amount of thin mucus that helps to lubricate food movement as well as to protect the stomach wall from digestion by the gastric enzymes.

The pyloric glands also secrete the hormone *gastrin,* which plays a key role in controlling gastric secretion, as we discuss shortly.

SURFACE MUCOUS CELLS. The entire surface of the stomach mucosa between glands has a continuous layer of a different type of mucous cells called simply "surface mucous cells." They secrete large quantities of a far more *viscid mucus* that is mainly insoluble and coats the mucosa with a gel layer of mucus often more than 1 millimeter thick, thus providing a major shell of protection for the stomach wall as well as contributing to lubrication of food transport.

Another characteristic of this mucus is that *it is alkaline.* Therefore, the *normal* underlying stomach wall is never directly exposed to the highly acidic, proteolytic stomach secretion. Even the slightest contact with food or especially any irritation of the mucosa directly stimulates the mucous cells to secrete copious quantities of this thick, alkaline viscid mucus.

Regulation of Gastric Secretion by Nervous and Hormonal Mechanisms

Basic Factors That Stimulate Gastric Secretion: Acetylcholine, Gastrin, and Histamine

The basic neurotransmitters or hormones that directly stimulate secretion by the gastric glands are *acetylcholine, gastrin,* and *histamine.* All these function by binding first with specific receptors for each on the secretory cells. Then the receptors activate the secre-tory processes. Acetylcholine excites secretion by all the secretory cell types in the gastric glands, including secretion of *pepsinogen* by the *peptic cells, hydrochloric acid* by the *parietal cells,* and *mucus* by the *mucus cells.* On the other hand, both *gastrin* and *histamine* stimulate strongly the *secretion of acid* by the parietal cells but have little effect in stimulating the other cells.

A few other substances also stimulate the gastric secretory cells, such as circulating amino acids, caffeine, and alcohol. The stimulatory effects of these are slight in comparison with acetylcholine, gastrin, and histamine.

Stimulation of Acid Secretion

NERVOUS STIMULATION. About one half of the nerve signals to the stomach that cause gastric secretion originate in the *dorsal motor nuclei of the vagi* and pass by way of the *vagus nerves* first to the *enteric nervous system* of the stomach wall and then to the gastric glands. The other one half of the nervous secretory signals are generated by local reflexes that occur entirely within the wall of the stomach itself in the enteric nervous system. All the secretory nerves release acetylcholine as the neurotransmitter at their endings on the glandular cells, with one exception: for those signals that go to the gastrin-secreting cells in the pyloric glands, an intermediate neuron serves as the final path and secretes *gastrin-releasing peptide,* which is probably the peptide *bombesin,* as the neurotransmitter.

Nerve stimulation of gastric secretion can be initiated by signals that originate either in the brain, especially in the limbic system, or in the stomach itself. The stomach-initiated signals can activate two types of reflexes: (1) *long vagovagal reflexes* that are transmitted from the stomach mucosa all the way to the brain stem and then back to the stomach through the vagus nerves and (2) *short reflexes* that originate locally and are transmitted entirely through the local enteric nervous system.

The types of stimuli that can initiate the reflexes are (1) distention of the stomach, (2) tactile stimuli on the surface of the stomach mucosa, and (3) chemical stimuli, including especially *amino acids* and *peptides* derived from protein foods or *acid* that has already been secreted by the gastric glands.

STIMULATION OF ACID SECRETION BY GASTRIN. Both the nerve signals from the vagus nerves and those from the local enteric reflexes, aside from causing direct stimulation of glandular secretion of stomach juices, also cause the mucosa in the stomach antrum to secrete the hormone *gastrin*. This hormone is secreted by *gastrin cells,* also called *G cells,* in the pyloric glands. Gastrin is a large peptide secreted in two forms, a large form called G-34, which contains 34 amino acids, and a smaller form, G-17, which contains 17 amino acids. Although both of these are important, the smaller is more abundant.

Gastrin is absorbed into the blood and carried to the *oxyntic glands* in the body of the stomach; there it stimulates the *parietal cells* strongly and the peptic cells as well, but to a much less extent. Thus, the important effect is to increase the rate of parietal cell hydrochloric acid secretion, often as much as eight times. In turn, the hydrochloric acid excites still additional enteric reflex activity that not only further increases hydrochloric acid secretion but also stimulates secondarily the secretion of enzymes by the peptic cells to increase two to four times.

ROLE OF HISTAMINE IN CONTROLLING GASTRIC SECRETION. *Histamine,* an amino acid derivative, also stimulates acid secretion by the *parietal cells.* A small amount of histamine is formed continually in the gastric mucosa, either in response to acid in the stomach or for other reasons. This amount, acting by itself, causes little acid secretion. However, whenever acetylcholine and gastrin stimulate the parietal cells at the same time, then even the small normal amounts of histamine greatly enhance acid secretion. We know this to be true because when the action of histamine is blocked by an appropriate antihistaminic drug such as *cimetidine,* neither acetylcholine nor gastrin can then cause significant amounts of acid secretion. Thus, histamine is a necessary *cofactor* for exciting significant acid secretion.

The histamine receptors on the parietal cells are of the H_2 type, not of the H_1 type. Therefore, only antihistaminic drugs that block the action of histamine H_2 receptors are effective in blocking stomach acid secretion. The first important drug of this type was *cimetidine,* but others are now available.

MULTIPLICATIVE EFFECT OF ACETYLCHOLINE, GASTRIN, AND HISTAMINE IN STIMULATING ACID SECRETION. Because no one of the primary stimulators of the acid-secreting parietal cells—acetylcholine, gastrin, or histamine—is effective in causing secretion of more than slight amounts of acid when functioning alone, it has been postulated that all three of the receptors to these separate transmitter-hormonal substances must be activated simultaneously to give a truly effective stimulus for gastric acid secretion. The histamine seems to be always present under normal conditions in

small amounts. Then, when the vagi are stimulated, acetylcholine is released at the parasympathetic nerve endings, and the gastrin-releasing peptide neurons stimulated by the vagi cause simultaneous release of gastrin by the gastrin cells. Therefore, all three stimuli become available, and copious amounts of acid are secreted. When food in the stomach elicits enteric reflexes, this, too, causes both gastrin and acetylcholine secretion, once again promoting the flow of tremendous quantities of acid.

Regulation of Pepsinogen Secretion

The regulation of *pepsinogen* secretion is much less complex than that of acid secretion; it occurs in response to two types of signals: (1) stimulation of the *peptic cells* by *acetylcholine* released from the *vagus nerves* or other enteric nerves and (2) stimulation of peptic secretion in response to acid in the stomach. The acid probably does not stimulate the peptic cells directly but elicits additional enteric reflexes, thus further supporting the original nervous signals to the peptic cells. It is possible, too, that the gastrin released during acid secretion has an additional slight direct effect in stimulating the peptic cells, although this is not clear. Nevertheless, the rate of secretion of *pepsinogen*, the precursor of pepsin that causes protein digestion, is strongly influenced by the amount of acid in the stomach. In people who have lost the ability to secrete normal amounts of acid, the secretion of pepsinogen is very little, even though the peptic cells may still be normal.

FEEDBACK INHIBITION OF GASTRIC SECRETION BY EXCESS ACID. When the acidity of the gastric juices increases to a pH below 3.0, the gastrin mechanism for stimulating gastric secretion becomes blocked. This effect results from two factors. First, greatly enhanced acidity depresses or blocks the secretion of gastrin itself by the G cells. Second, the acid seems to cause an inhibitory nervous reflex that inhibits gastric secretion.

This feedback inhibition of the gastric glands plays an important role in protecting the stomach against excessive acidity, which would promote peptic ulceration. In addition to this protective effect, the feedback mechanism is important in maintaining optimal pH for function of the peptic enzymes in the digestive process, which is a pH of about 3.0.

Phases of Gastric Secretion

Gastric secretion is said to occur in three phases (as shown in Fig. 64–7): a *cephalic phase,* a *gastric phase,* and an *intestinal phase.* As will be apparent in the following discussion, these three phases fuse together.

CEPHALIC PHASE. The cephalic phase of gastric secretion occurs even before food enters the stomach or while it is being eaten. It results from the sight, smell, thought, or taste of food, and the greater the appetite, the more intense is the stimulation. Neurogenic signals that cause the cephalic phase of secretion can originate in the cerebral cortex or in the appetite centers of the amygdala or hypothalamus. They are transmitted through the dorsal motor nuclei of the vagi to the stom-

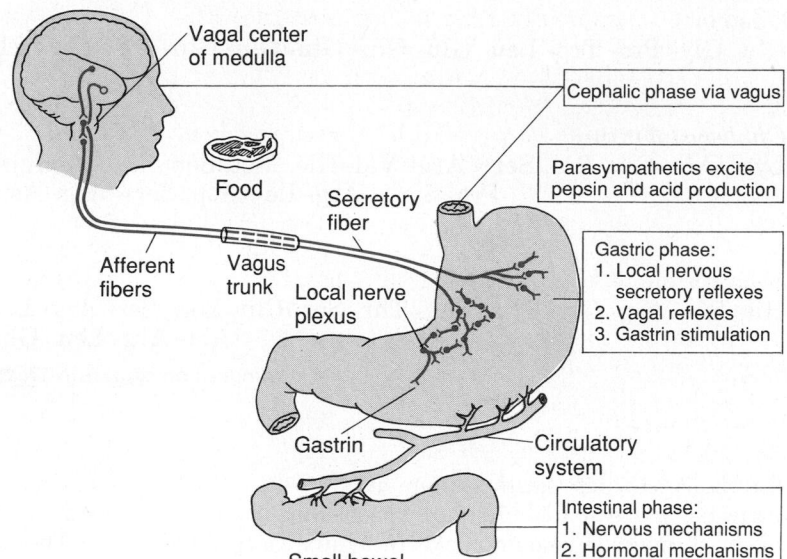

Figure 64–7. Phases of gastric secretion and their regulation.

ach. This phase of secretion normally accounts for about 20 per cent of the gastric secretion associated with eating a meal.

GASTRIC PHASE. Once the food enters the stomach, it excites the long vagovagal reflexes, the local enteric reflexes, and the gastrin mechanism, which in turn cause secretion of gastric juice that continues throughout the several hours that the food remains in the stomach.

The gastric phase of secretion accounts for about 70 per cent of the total gastric secretion associated with eating a meal and therefore accounts for most of the total daily gastric secretion of about 1500 milliliters.

INTESTINAL PHASE. The presence of food in the upper portion of the small intestine, particularly in the duodenum, can cause the stomach to secrete small amounts of gastric juice, probably partly because of the small amounts of gastrin that are also released by the duodenal mucosa in response to distention or chemical stimuli of the same type as those that stimulate the stomach gastrin mechanism. In addition, amino acids absorbed into the blood as well as several other hormones or reflexes play minor roles in causing secretion of gastric juice.

Inhibition of Gastric Secretion by Intestinal Factors

Although chyme stimulates gastric secretion during the intestinal phase of secretion, it paradoxically often inhibits secretion during the gastric phase. This inhibition results from at least two influences.

1. The presence of food in the small intestine initiates an *enterogastric reflex,* transmitted through the enteric nervous system as well as through the extrinsic sympathetic and vagus nerves, that inhibits stomach secretion. This reflex can be initiated by distention of the small bowel, the presence of acid in the upper intestine, the presence of protein breakdown products, or irritation of the mucosa. This is part of the complex mechanism discussed in Chapter 63 for slowing down stomach emptying when the intestines are already filled.

2. The presence of acid, fat, protein breakdown products, hyperosmotic or hypo-osmotic fluids, or any irritating factor in the upper small intestine causes the release of several intestinal hormones. One of these is *secretin,* which is especially important for control of pancreatic secretion. In addition to having this effect, secretin opposes stomach secretion. Three other hormones—*gastric inhibitory peptide, vasoactive intestinal polypeptide,* and *somatostatin*—have slight to moderate effects in inhibiting gastric secretion.

The functional purpose of the inhibition of gastric secretion by intestinal factors is probably to slow the release of chyme from the stomach when the small intestine is already filled. In fact, the enterogastric reflex and these inhibitory hormones usually reduce stomach motility at the same time that they reduce gastric secretion, as discussed in Chapter 63.

SECRETION DURING THE INTERDIGESTIVE PERIOD. The stomach secretes a few milliliters of gastric juice per hour during the "interdigestive period," when little or no digestion is occurring anywhere in the gut. The secretion that does occur is almost entirely of the so-called nonoxyntic type, meaning that it is composed mainly of mucus that contains little pepsin and almost no acid. Strong emotional stimuli frequently increase the interdigestive secretion to 50 milliliters or more of highly peptic and acidic gastric juice per hour, in very much the same manner that the cephalic phase of gastric secretion excites secretion at the onset of a meal. This increase of secretion during the presence of emotional stimuli is believed to be one of the factors in the development of peptic ulcers, as discussed in Chapter 66.

Chemical Composition of Gastrin and Other Gastrointestinal Hormones

Figure 64–8 shows the amino acid compositions of *gastrin-17* as well as of *cholecystokinin* and *secretin,* which are discussed later in the chapter. Note that all are polypeptides and that the terminal five amino acids in the gastrin and cholecystokinin molecular chains are the same. The activity of gastrin resides in the terminal four amino acids and in the terminal eight amino acids

Gastrin:
Glu- Gly- Pro- Trp- Leu- Glu- Glu- Glu- Glu- Glu- Ala- Tyr- Gly- Trp- Met- Asp- Phe- NH_2
$|$
HSO_3

Cholecystokinin:
Lys- (Ala, Gly, Pro, Ser)- Arg- Val- (Ile, Met, Ser)- Lys- Asn- (Asn, Gln, His, Leu_2,
Pro, Ser_2)- Arg- Ile- (Asp, Ser)- Arg- Asp- Tyr- Met- Gly- Trp- Met- Asp- Phe- NH_2
$|$
HSO_3

Secretin:
His- Ser- Asp- Gly- Thr- Phe- Thr- Ser- Glu- Leu- Ser- Arg- Leu- Arg- Asp- Ser-
Ala- Arg- Leu- Gln- Arg- Leu- Leu- Gln- Gly- Leu- Val- NH_2

Figure 64–8. Amino acid composition of gastrin-17, cholecystokinin, and secretin.

for cholecystokinin; all the amino acids in the secretin molecule are essential. A synthetic gastrin, composed of the terminal four amino acids of natural gastrin plus the amino acid alanine, has all the same physiological properties as the natural gastrin. This synthetic product is called *pentagastrin*.

PANCREATIC SECRETION

The pancreas, which lies parallel to and beneath the stomach, is a large compound gland with an internal structure similar to that of the salivary glands, shown in Figure 64–2. In addition to secreting insulin by the islets of Langerhans in the pancreas, digestive enzymes are secreted by the pancreatic acini, and large volumes of sodium bicarbonate solution are secreted by both the small ductules and the larger ducts leading from the acini. The combined product then flows through a long pancreatic duct that usually joins the hepatic duct immediately before it empties into the duodenum through the papilla of Vater that is surrounded by the sphincter of Oddi. Pancreatic juice is secreted most abundantly in response to the presence of chyme in the upper portions of the small intestine, and the characteristics of the pancreatic juice are determined to some extent by the types of food in the chyme.

Secretion of the Pancreatic Enzymes

Pancreatic secretion contains enzymes for digesting all three major types of food: proteins, carbohydrates, and fats. It also contains large quantities of bicarbonate ions, which play an important role in neutralizing the acid chyme emptied by the stomach into the duodenum.

The more important of the proteolytic enzymes are *trypsin, chymotrypsin,* and *carboxypolypeptidase*. Less important are several *elastases* and *nucleases*. By far the most abundant of these is trypsin. The trypsin and chymotrypsin split whole and partially digested proteins into peptides of various sizes but do not cause the release of individual amino acids. On the other hand, carboxypolypeptidase splits some peptides into individual amino acids, thus completing the digestion of much of the proteins all the way to the amino acid state.

The pancreatic digestive enzyme for carbohydrates is *pancreatic amylase,* which hydrolyzes starches, glycogen, and most other carbohydrates (except cellulose) to form disaccharides and a few trisaccharides.

The main enzymes for fat digestion are *pancreatic lipase,* which is capable of hydrolyzing neutral fat into fatty acids and monoglycerides; *cholesterol esterase,* which causes hydrolysis of cholesterol esters; and *phospholipase,* which splits fatty acids from phospholipids.

When synthesized in the pancreatic cells, the proteolytic enzymes are in the inactive forms *trypsinogen, chymotrypsinogen,* and *procarboxypolypeptidase,* which are all enzymatically inactive. They become activated only after they are secreted into the intestinal tract. Trypsinogen is activated by an enzyme called *enterokinase,* which is secreted by the intestinal mucosa when chyme comes in contact with the mucosa. Also, trypsinogen can be autocatalytically activated by trypsin that has already been formed from trypsinogen. Chymotrypsinogen is activated by trypsin to form chymotrypsin, and procarboxypolypeptidase is activated in a similar manner.

SECRETION OF TRYPSIN INHIBITOR PREVENTS DIGESTION OF THE PANCREAS. It is important that the proteolytic enzymes of the pancreatic juice not become activated until they have been secreted into the intestine because the trypsin and other enzymes would digest the pancreas itself. The same cells that secrete the proteolytic enzymes into the acini of the pancreas secrete simultaneously another substance called *trypsin inhibitor.* This substance is stored in the cytoplasm of the glandular cells that surround the enzyme granules, and it prevents activation of trypsin both inside the secretory cells and in the acini and ducts of the pancreas. Because it is trypsin that activates the other pancreatic proteolytic enzymes, trypsin inhibitor prevents the subsequent activation of the others as well.

When the pancreas becomes severely damaged or when a duct becomes blocked, large quantities of pancreatic secretion become pooled in the damaged areas of the pancreas. Under these conditions, the effect of trypsin inhibitor is sometimes overwhelmed, in which case the pancreatic secretions rapidly become activated and literally digest the entire pancreas within a few hours, giving rise to the condition called *acute pancreatitis*. This often is lethal because of accompanying shock; even if not lethal, it usually leads to a lifetime of pancreatic insufficiency.

Secretion of Bicarbonate Ions

Although the enzymes of the pancreatic juice are secreted entirely by the acini of the pancreatic glands, the other two important components of pancreatic juice, bicarbonate ions and water, are secreted in large amounts mainly by the epithelial cells of the ductules and ducts that lead from the acini. We shall see later that the stimulatory mechanisms for enzyme production and production of bicarbonate ions and water are also quite different. When the pancreas is stimulated to secrete copious quantities of pancreatic juice, the bicarbonate ion concentration can rise to as high as 145 mEq/liter, a value about five times that of bicarbonate ions in the plasma. This provides a large quantity of alkali in the pancreatic juice that serves to neutralize acid emptied into the duodenum from the stomach.

The basic steps in the cellular mechanism for secreting sodium bicarbonate solution into the pancreatic ductules and ducts are shown in Figure 64–9. They are the following.

1. Carbon dioxide diffuses to the interior of the cell from the blood and combines with water under the influence of carbonic anhydrase to form carbonic acid. The carbonic acid in turn dissociates into bicarbonate ions and hydrogen ions. Then the bicarbonate ions are actively transported (by secondary active transport in exchange for chloride ions) through the *luminal* border of the cell into the lumen of the duct.

2. The hydrogen ions formed by dissociation of carbonic acid inside the cell are exchanged for sodium ions through

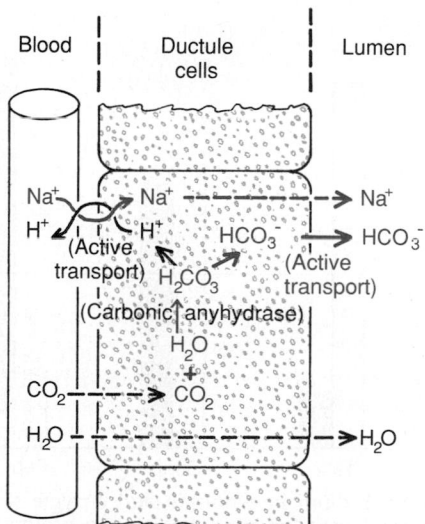

Figure 64–9. Secretion of isosmotic sodium bicarbonate solution by the pancreatic ductules and ducts.

the *blood* border of the cell, also by a secondary active transport process. The sodium ions in turn either diffuse through or are actively transported through the *luminal* border into the pancreatic duct to provide electrical neutrality for the secreted bicarbonate ions.

3. Any excess of sodium ions still in the cell is actively transported through the base of the cell into the blood.

4. The movement of sodium and bicarbonate ions from the blood to the lumen creates an osmotic gradient that causes osmosis of water also into the pancreatic duct, thus forming an almost completely isosmotic bicarbonate solution.

Regulation of Pancreatic Secretion
Basic Stimuli of Pancreatic Secretion

Three basic stimuli are important in causing pancreatic secretion:

1. *Acetylcholine*, which is released from the parasympathetic vagus nerve endings as well as from other cholinergic nerves in the enteric nervous system
2. *Cholecystokinin*, which is secreted by the duodenal and upper jejunal mucosa when food enters the small intestine
3. *Secretin*, which is secreted by the same duodenal and jejunal mucosa when highly acid food enters the small intestine

The first two stimuli, acetylcholine and cholecystokinin, stimulate the acinar cells of the pancreas much more than the ductal cells. Therefore, they cause production of large quantities of digestive enzymes but relatively small quantities of fluid to go with the enzymes. Without the fluid, most of the enzymes remain temporarily stored in the acini and ducts until more fluid secretion comes along to wash them into the duodenum.

Secretin, in contrast to the other two basic stimuli, mainly stimulates the secretion of large quantities of sodium bicarbonate solution by the ductal epithelium but is responsible for almost no stimulation of enzyme secretion.

MULTIPLICATIVE EFFECTS OF THE DIFFERENT STIMULI. When all the different stimuli of pancreatic secretion occur at once, the secretion is far greater than the sum of the secretions caused by each one separately. Therefore, the various stimuli are said to "multiply," or "potentiate," one another. Thus, pancreatic secretion normally results from the combined effects of multiple basic stimuli, not from one alone.

Phases of Pancreatic Secretion

Pancreatic secretion occurs in three phases, the same as for gastric secretion: the *cephalic* phase, the *gastric* phase, and the *intestinal phase*. Their characteristics are as follows.

CEPHALIC AND GASTRIC PHASES. During the cephalic phase of pancreatic secretion, the same nervous signals that cause secretion in the stomach also cause acetylcholine release by the vagal nerve endings in the pancreas. This causes moderate amounts of enzymes to be secreted into the pancreatic acini and ducts, accounting for about 20 per cent of the total secretion of pancreatic enzymes after a meal. Little of the secretion flows out the pancreatic ducts into the intestine because only small amounts of water and electrolytes are secreted along with the enzymes.

During the gastric phase, the nervous stimulation of enzyme secretion continues, accounting for another 5 to 10 per cent of the enzymes secreted after a meal. Still only small amounts reach the duodenal lumen because of continued lack of significant quantities of fluid secretion.

INTESTINAL PHASE. After chyme enters the small intestine, pancreatic secretion becomes copious, mainly in response to the hormone *secretin.* In addition, *cholecystokinin* causes still much more increase in the secretion of enzymes.

Secretin Stimulates Secretion of Copious Quantities of Bicarbonate—Neutralization of the Acidic Chyme. Secretin is a polypeptide containing 27 amino acids (molecular weight about 3400) that is present in so-called *S cells* in the mucosa of the upper small intestine (duodenum and jejunum) in an inactive form, *prosecretin.* When acid chyme with pH less than 4.5 to 5.0 enters the duodenum from the stomach, it causes the release and activation of secretin, which is subsequently absorbed into the blood. The one truly potent constituent of chyme that causes secretin release is hydrochloric acid, although several other constituents, such as fatty acids, also contribute slightly to its release.

Secretin in turn causes the pancreas to secrete large quantities of fluid that contains a high concentration of bicarbonate ion (up to 145 mEq/liter) but a low concentration of chloride ion. However, this fluid contains few enzymes when the pancreas is stimulated only by secretin because secretin has little effect by itself to stimulate the acinar cells.

The secretin mechanism is especially important for two reasons: *First,* secretin begins to be released from the mucosa of the small intestine when the pH of the duodenal contents falls below 4.5 to 5.0, and its release increases greatly as the pH falls to 3.0 and as more and more acid reaches deeper into the duodenum and jejunum. This immediately causes copious quantities of pancreatic juice containing abundant amounts of sodium bicarbonate to be secreted, which results in the following reaction in the contents of the duodenum:

$$HCl + NaHCO_3 \longrightarrow NaCl + H_2CO_3.$$

The carbonic acid immediately dissociates into carbon dioxide and water and the carbon dioxide is absorbed into the blood and expired through the lungs, thus leaving a neutral solution of sodium chloride in the duodenum. In this way, the acid contents emptied into the duodenum from the stomach become neu-

tralized, and the peptic activity of the gastric juices is immediately blocked. Because the mucosa of the small intestine cannot withstand the digestive action of acid gastric juice, this is a highly important and even essential protective mechanism against the development of duodenal ulcers, discussed in further detail in Chapter 66.

Second, bicarbonate secretion by the pancreas provides an appropriate pH for action of the pancreatic enzymes. They function optimally in a slightly alkaline or neutral medium. The pH of the sodium bicarbonate secretion averages 8.0.

Cholecystokinin—Control of Enzyme Secretion by the Pancreas. The presence of food in the upper small intestine also causes a second hormone, cholecystokinin, a polypeptide containing 33 amino acids, to be released from a still different group of cells, called I cells, in the mucosa of the duodenum and upper jejunum. This results especially from the presence of *proteoses* and *peptones* (which are products of partial protein digestion) and of *long-chain fatty acids;* hydrochloric acid from the stomach juices also causes its release in smaller quantities.

Cholecystokinin, like secretin, passes by way of the blood to the pancreas but instead of causing sodium bicarbonate secretion causes mainly secretion of large quantities of digestive enzymes by the acinar cells. This effect is similar to that of vagal stimulation but even more pronounced, accounting for 70 to 80 per cent of the total secretion of the pancreatic enzymes after a meal.

The differences between the stimulatory effects of secretin and cholecystokinin are shown in Figure 64–10, which demonstrates (1) intense sodium bicarbonate secretion in response to acid in the duodenum, stimulated by secretin, (2) a dual effect in response to soap (a fat), and (3) intense enzyme secretion in response to peptones, stimulated by cholecystokinin.

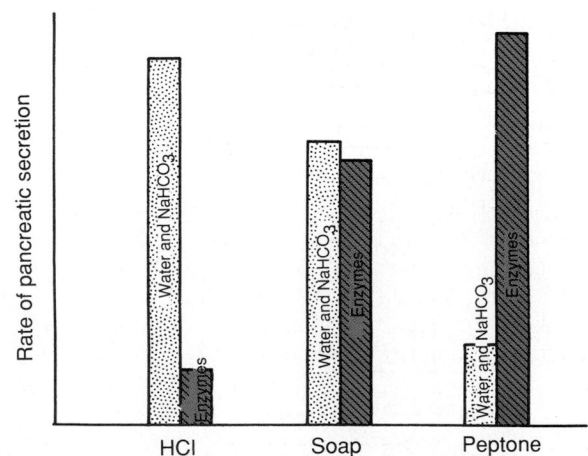

Figure 64–10. Sodium bicarbonate and enzyme secretion by the pancreas, caused, respectively, by the presence of acid, fat (soap), or peptone solutions in the duodenum.

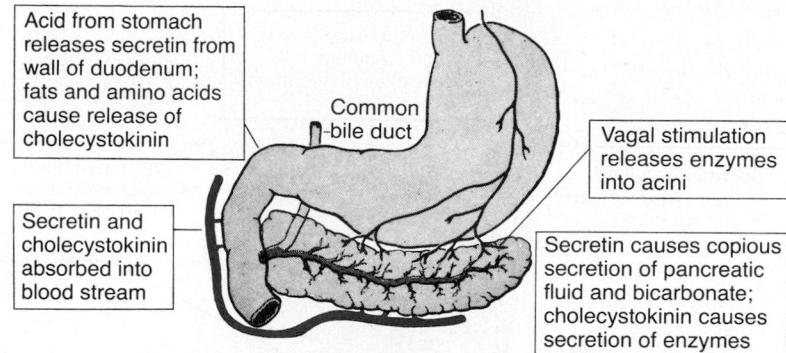

Acid from stomach releases secretin from wall of duodenum; fats and amino acids cause release of cholecystokinin

Common bile duct

Vagal stimulation releases enzymes into acini

Secretin and cholecystokinin absorbed into blood stream

Secretin causes copious secretion of pancreatic fluid and bicarbonate; cholecystokinin causes secretion of enzymes

Figure 64–11. Regulation of pancreatic secretion.

Figure 64–11 summarizes the more important factors in the regulation of pancreatic secretion. The total amount secreted each day is about 1000 milliliters.

SECRETION OF BILE BY THE LIVER; FUNCTIONS OF THE BILIARY TREE

One of the many functions of the liver is to secrete bile, normally between 600 and 1200 ml/day. Bile serves two important functions: *First,* it plays an important role in fat digestion and absorption, not because of any enzymes in the bile that cause fat digestion but because of *bile acids* in the bile that do two things: (1) they help to emulsify the large fat particles of the food into many minute particles that can be attacked by the lipase enzyme secreted in pancreatic juice and (2) they aid in the transport and absorption of the digested fat end products to and through the intestinal mucosal membrane. *Second,* bile serves as a means for excretion of several important waste products from the blood. These include especially *bilirubin,* an end product of hemoglobin destruction, and excesses of *cholesterol* synthesized by the liver cells.

Physiologic Anatomy of Biliary Secretion

Bile is secreted in two stages by the liver: (1) The initial portion is secreted by the liver hepatocytes; this initial secretion contains large amounts of bile acids, cholesterol, and other organic constituents. It is secreted into the minute *bile canaliculi* that lie between the hepatic cells in the hepatic plates. (2) Next, the bile flows peripherally toward the interlobular septa, where the canaliculi empty into *terminal bile ducts* and then into progressively larger ducts, finally reaching the *hepatic duct* and *common bile duct,* from which the bile either empties directly into the duodenum or is diverted through the *cystic duct* into the gallbladder, shown in Figure 64–12. In its course through these bile ducts, a second portion of secretion is added to the initial bile. This additional secretion is a watery solution of sodium and bicarbonate ions secreted by secretory epithelial cells that line the ductules and ducts. This second secretion sometimes increases the total quantity of bile by as much as an additional 100 per cent. The second secretion is stimulated by *secretin,* thus causing increased quantities of bicarbonate ions that supplement the pancreatic secretions in neutralizing acid from the stomach.

STORAGE AND CONCENTRATION OF BILE IN THE GALLBLADDER. The bile secreted continually by the liver cells is normally stored in the gallbladder until needed in the duodenum. The maximum volume of the gallbladder is only 30 to 60 milliliters. Nevertheless, as much as 12 hours of bile secretion (usually about 450 milliliters) can be stored in the gallbladder because water, sodium, chloride, and most other small electrolytes are continually absorbed by the gallbladder mucosa, concentrating the other bile constituents, including the bile salts, cholesterol, lecithin, and bilirubin. Most of this absorption is caused by active transport of sodium through the gallbladder epithelium, and this is followed by secondary absorption of chloride ions, water, and most other soluble constituents. Bile is normally concentrated in this way about 5-fold, but it can be concentrated up to a maximum of 20-fold.

COMPOSITION OF BILE. Table 64–2 gives the composition of bile when it is first secreted by the liver and then after it has been concentrated in the gallbladder. This table shows that by far the most abundant substance secreted in the bile is the *bile salts,* accounting for about one half of the total solutes of bile; also secreted or excreted in large concentrations are *bilirubin, cholesterol, lecithin,* and the usual *electrolytes* of plasma. In the concentrating process in the gallbladder, water and large portions of the electrolytes (except calcium ions) are reabsorbed by the gallbladder mucosa; essentially all the other constituents, including especially in the bile salts and the lipid substances cholesterol and lecithin, are not reabsorbed and, therefore, become highly concentrated in the gallbladder bile.

EMPTYING OF THE GALLBLADDER—ROLE OF CHOLECYSTOKININ. When food begins to be digested in the upper gastrointestinal tract, the gallbladder begins to empty, especially as fatty foods enter the duodenum about 30 minutes after a meal. The basic cause of the emptying is rhythmical contractions of the wall of the gallbladder, but effective emptying also requires simul-

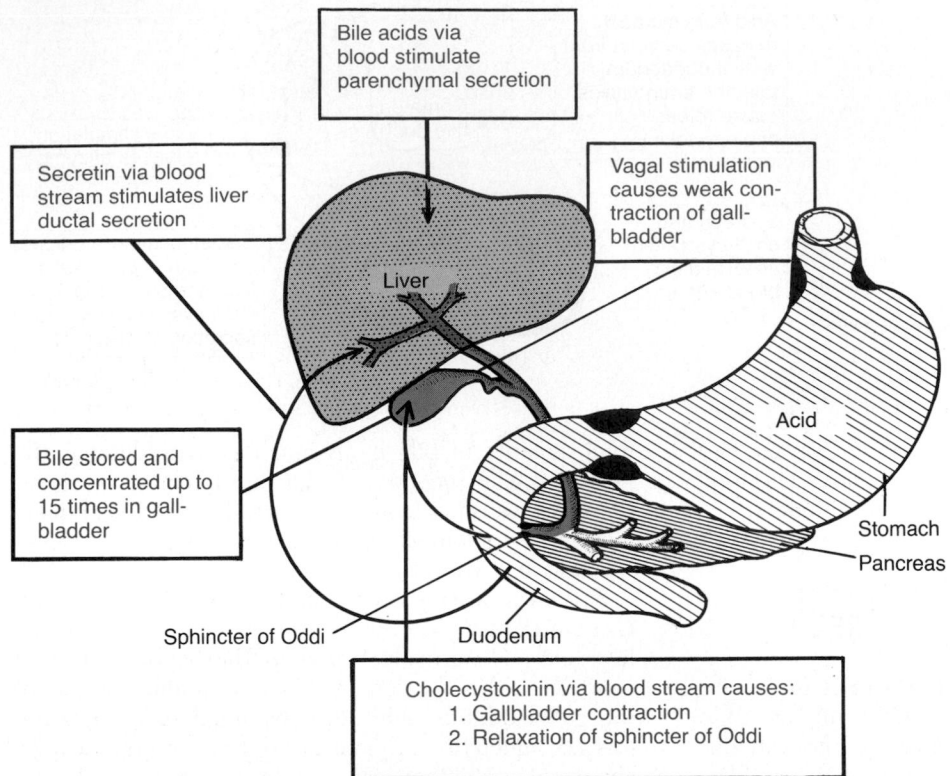

Figure 64–12. Liver secretion and gallbladder emptying.

taneous relaxation of the *sphincter of Oddi* that guards the exit of the common bile duct into the duodenum.

By far the most potent stimulus for causing the gallbladder contractions is the hormone *cholecystokinin*. This is the same cholecystokinin that causes increased secretion of enzymes by the acinar cells of the pancreas. The stimulus for its release into the blood from the duodenal mucosa is mainly the fatty foods themselves that enter the duodenum.

In addition to cholecystokinin, the gallbladder is stimulated less strongly by acetylcholine-secreting nerve fibers from both the vagi and the enteric nervous system. They are the same nerves that promote motility and secretion in other parts of the upper gastrointestinal tract.

Even with relatively strong contractions of the gallbladder, emptying can be difficult because the sphincter of Oddi normally remains tonically contracted. Therefore, before emptying of the gallbladder will occur, the sphincter of Oddi, too, must be relaxed. At least three factors help in this: First, cholecystokinin, instead of stimulating the sphincter of Oddi, has a relaxing effect, but this effect is usually not sufficient alone to allow significant emptying. Second, the rhythmical contractions of the gallbladder transmit peristaltic waves down the common bile duct to the sphincter of Oddi, causing a leading wave of relaxation that partially inhibits the sphincter in advance of the peristaltic wave. But this, too, is usually not enough to allow large amounts of emptying. Third, when intestinal peristaltic waves travel over the wall of the duodenum itself, the relaxation phase of each of these waves strongly relaxes the sphincter of Oddi along with the relaxation of the muscle of the gut wall. This seems to be by far the most potent of all the relaxant effects on the sphincter of Oddi. As a result, bile usually enters the duodenum in the form of squirts that are synchronized with the relaxation phase of the duodenal peristaltic waves.

In summary, the gallbladder empties its store of concentrated bile into the duodenum mainly in response to the cholecystokinin stimulus. When fat is not in the meal, the gallbladder empties poorly, but when adequate quantities of fat are present, the gallbladder normally empties completely in about 1 hour.

Figure 64–12 summarizes the secretion of bile, its

Table 64–2 COMPOSITION OF BILE

	Liver Bile		Gallbladder Bile	
Water	97.5	gm/dl	92	gm/dl
Bile salts	1.1	gm/dl	6	gm/dl
Bilirubin	0.04	gm/dl	0.3	gm/dl
Cholesterol	0.1	gm/dl	0.3 to 0.9	gm/dl
Fatty acids	0.12	gm/dl	0.3 to 1.2	gm/dl
Lecithin	0.04	gm/dl	0.3	gm/dl
Na^+	145	mEq/liter	130	mEq/liter
K^+	5	mEq/liter	12	mEq/liter
Ca^+	5	mEq/liter	23	mEq/liter
Cl^-	100	mEq/liter	25	mEq/liter
HCO_3^-	28	mEq/liter	10	mEq/liter

storage in the gallbladder, and its release from the bladder to the gut.

Bile Salts and Their Function

The liver cells synthesize about 0.6 gram of *bile salts* daily. The precursor of the bile salts is *cholesterol,* which is either supplied in the diet or synthesized in the liver cells during the course of fat metabolism and then converted to *cholic acid* or *chenodeoxycholic acid* in about equal quantities. These acids then combine principally with glycine and to a lesser extent with taurine to form *glyco-* and *tauro-conjugated bile acids.* The salts of these acids are secreted in the bile.

The bile salts have two important actions in the intestinal tract.

First, they have a detergent action on the fat particles in the food, which decreases the surface tension of the particles and allows the agitation in the intestinal tract to break the fat globules into minute sizes. This is called the *emulsifying* or *detergent function* of bile salts.

Second, and even more important than the emulsifying function, bile salts help in the absorption of fatty acids, monoglycerides, cholesterol, and other lipids from the intestinal tract. They do this by forming minute complexes with these lipids; the complexes are called *micelles,* and they are highly soluble because of the electrical charges of the bile salts. The lipids are "ferried" in this form to the mucosa, where they are then absorbed; this mechanism is described in detail in Chapter 65. Without the presence of bile salts in the intestinal tract, up to 40 per cent of the ingested lipids are lost into the stools, and the person often develops a metabolic deficit because of this nutrient loss.

ENTEROHEPATIC CIRCULATION OF BILE SALTS. About 94 per cent of the bile salts are reabsorbed by the small intestine, about one half of this by *diffusion* through the mucosa in the early portions of the small intestine and the remainder by an *active transport* process through the intestinal mucosa in the distal ileum. They enter the portal blood and pass to the liver. On reaching the liver, these salts are absorbed almost totally on the first passage through the venous sinusoids into the hepatic cells and then resecreted into the bile. In this way, about 94 per cent of all the bile salts are recirculated into the bile, so that on the average these salts make the entire circuit some 18 times before being carried out in the feces. The small quantities of bile salts lost into the feces are replaced by new amounts formed continually by the liver cells. This recirculation of the bile salts is called the *enterohepatic circulation.*

The quantity of bile secreted by the liver each day is highly dependent on the availability of bile salts—the greater the quantity of bile salts in the enterohepatic circulation (usually a total of about 2.5 grams), the greater the rate of bile secretion. Indeed, ingestion of an excess of bile salts can increase bile secretion by several hundred milliliters per day.

If a bile fistula empties the bile salts to the exterior for several days to several weeks so that they cannot be reabsorbed from the ileum, the liver increases its production of bile salts 6- to 10-fold, which increases the rate of bile secretion most of the way back to normal. This demonstrates that the daily rate of bile salt secretion is actively controlled by the availability (or lack of availability) of bile salts in the enterohepatic circulation.

ROLE OF SECRETIN IN CONTROLLING BILE SECRETION. In addition to the strong stimulating effect of bile acids on bile secretion, the hormone *secretin* increases bile secretion, sometimes more than doubling the secretion rate for several hours after a meal. This increase in secretion represents almost entirely secretion of a bicarbonate-rich watery solution by the epithelial cells of the bile ductules and ducts and not increased secretion by the liver parenchymal cells themselves. The bicarbonate in turn passes into the small intestine and joins the bicarbonate from the pancreas in neutralizing the acid from the stomach. Thus, the secretin feedback mechanism for neutralizing duodenal acid operates not only through its effects on pancreatic secretion but also through its effect on the secretion by the liver ductules and ducts.

Secretion of Cholesterol; Gallstone Formation

Bile salts are formed in the hepatic cells from cholesterol, and in the process of secreting the bile salts about 1 to 2 gm/day cholesterol is also secreted into the bile. No specific function is known for the cholesterol in the bile, and it is presumed that it is simply a byproduct of bile salt formation and secretion.

Cholesterol is almost completely insoluble in pure water, but the bile salts and lecithin in bile combine physically with the cholesterol to form ultramicroscopic *micelles* that are soluble, as is explained in more detail in Chapter 65. When the bile becomes concentrated in the gallbladder, the bile salts and lecithin become concentrated along with the cholesterol, which keeps the cholesterol in solution.

Under abnormal conditions, the cholesterol may precipitate, resulting in the formation of *cholesterol gallstones,* as shown in Figure 64–13. The different conditions that can cause cholesterol precipitation are (1) too much absorption of water from the bile, (2) too much absorption of bile salts and lecithin from the bile, (3) too much secretion of cholesterol in the bile, and (4) inflammation of the epithelium of the gallbladder. The latter two require special explanation.

The amount of cholesterol in the bile is determined partly by the quantity of fat that the person eats because the hepatic cells synthesize cholesterol as one of the products of fat metabolism in the body. For this reason, people on a high-fat diet over a period of many years are prone to the development of gallstones.

Inflammation of the gallbladder epithelium often results from low-grade chronic infection; this changes the absorptive characteristics of the gallbladder mucosa, sometimes allowing excessive absorption of water, bile salts, or other substances that are necessary to keep the cholesterol in solution. As a result, cholesterol begins to precipitate, usually forming many small crystals of cholesterol on the surface of the inflamed mucosa or on small precipitated particles of bilirubin that themselves are the result of deconjugation of soluble bilirubin glu-

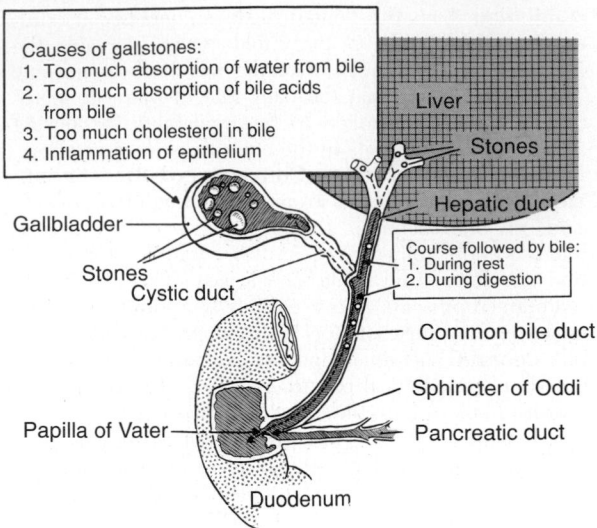

Causes of gallstones:
1. Too much absorption of water from bile
2. Too much absorption of bile acids from bile
3. Too much cholesterol in bile
4. Inflammation of epithelium

Course followed by bile:
1. During rest
2. During digestion

Figure 64–13. Formation of gallstones.

curonide by bacterial enzymes. The bilirubin particles in turn act as nidi for further precipitation of cholesterol, and the crystals grow larger. Occasionally tremendous numbers of sand-like stones develop, but much more frequently they coalesce to form a few large gallstones or even a single stone that fills the entire gallbladder. Also, calcium ions, which are usually concentrated fivefold or more in the gallbladder, often precipitate in the gallstones, making the stones x-ray–opaque, so that they can be seen in radiographs of the abdomen.

MEDICAL THERAPY FOR DISSOLVING GALLSTONES. Simple cholesterol gallstones in many patients can be dissolved over a period of 1 to 2 years by feeding the patient 1 to 1.5 grams of *chenodeoxycholic acid* daily. This is one of the naturally secreted bile acids, and its exogenous administration adds greatly to the enterohepatic pool of bile acids. This can cause dissolution and reabsorption of the gallstones in the following ways: (1) The increased quantity of bile acids increases the volume of bile formed and therefore decreases the concentration of cholesterol in the bile. (2) The increase of bile acids in the bile makes the cholesterol that is present more soluble. (3) The exogenous administration of bile acids decreases the formation of bile acids by the liver, which at the same time reduces the secretion of cholesterol.

SECRETIONS OF THE SMALL INTESTINE

Secretion of Mucus by Brunner's Glands

An extensive array of compound mucous glands, called *Brunner's glands,* is located in the first few centimeters of the duodenum, mainly between the pylorus and the papilla of Vater, where the pancreatic juices and bile empty into the duodenum. These glands secrete an alkaline mucus in response to (1) tactile stimuli or irritating stimuli of the overlying mucosa; (2) vagal stimulation, which causes secretion con-

currently with increase in stomach secretion; and (3) gastrointestinal hormones, especially secretin.

The function of the mucus secreted by Brunner's glands is to protect the duodenal wall from digestion by the gastric juice. Their rapid and intense response to irritating stimuli is especially geared to this purpose. In addition, the secretin-stimulated secretion by the glands contains a large excess of bicarbonate ions, which add to the bicarbonate ions from pancreatic secretion and liver bile in neutralizing acid entering the duodenum from the stomach.

Brunner's glands are inhibited by sympathetic stimulation; therefore, such stimulation is likely to leave the duodenal bulb unprotected and is perhaps one of the factors that cause this area of the gastrointestinal tract to be the site of peptic ulcers in about 50 per cent of cases.

Secretion of the Intestinal Digestive Juices by the Crypts of Lieberkühn

Located on the entire surface of the small intestine are small pits called *crypts of Lieberkühn,* one of which is shown in Figure 64–14. These crypts lie between the intestinal villi, and the intestinal surfaces of both the crypts and the villi are covered by an epithelium composed of two types of cells: (1) a moderate number of *goblet cells,* which secrete mucus that lubricates and protects the intestinal surfaces, and (2) a large number of *enterocytes,* which, in the crypts, secrete large quantities of water and electrolytes and, over the surfaces of the villi, reabsorb the water and electrolytes along with the end products of digestion. The intestinal secretions are formed by the enterocytes of the crypts at a rate of about 1800 ml/day. The secretions are almost pure extracellular fluid and have a slightly alkaline pH in the range of 7.5 to 8.0. They are rapidly reabsorbed by the villi. This circulation of fluid from the crypts to the villi supplies a watery vehicle for absorption of substances from the chyme as

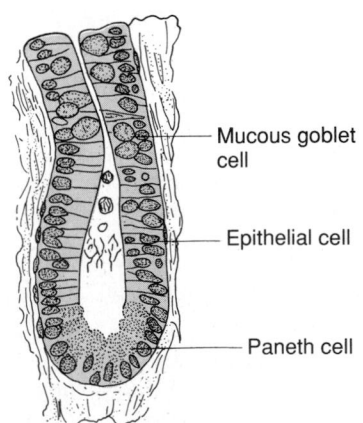

— Mucous goblet cell

— Epithelial cell

— Paneth cell

Figure 64–14. Crypt of Lieberkühn, found in all parts of the small intestine between the villi, which secretes almost pure extracellular fluid.

it comes in contact with the villi, which is the primary function of the small intestine.

In addition to the watery secretion, interspersed goblet cells in the epithelium secrete moderate amounts of mucus, which provides its usual functions of lubrication and protection of the intestinal mucosa.

MECHANISM OF SECRETION OF THE WATERY FLUID. The exact mechanism that causes the marked secretion of watery fluid by the crypts of Lieberkühn is not known. It is believed to involve at least two active secretory processes: (1) active secretion of chloride ions into the crypts and (2) active secretion of bicarbonate ions. The secretion of these ions, especially the chloride ions, causes electrical drag of sodium ions through the membrane as well. Finally, all these ions together cause osmotic movement of water.

ENZYMES IN THE SMALL INTESTINAL SECRETION. When secretions of the small intestine are collected without cellular debris, they have almost no enzymes. However, the enterocytes of the mucosa, especially those that cover the villi, do contain digestive enzymes that digest specific food substances *while* they are being absorbed through the epithelium. These enzymes are the following: (1) several *peptidases* for splitting small peptides into amino acids, (2) four enzymes for splitting disaccharides into monosaccharides —*sucrase, maltase, isomaltase,* and *lactase*—and (3) small amounts of *intestinal lipase* for splitting neutral fats into glycerol and fatty acids. Most if not all of these enzymes are mainly in the brush border of the enterocytes. They are believed to catalyze hydrolysis of the foods on the outside surfaces of the microvilli before absorption of the end products.

The epithelial cells deep in the crypts of Lieberkühn continually undergo mitosis, and the new cells gradually migrate along the basement membrane upward out of the crypts toward the tips of the villi, thus continually replacing the villus epithelium. As the villus cells age, they are finally shed into the intestinal secretions. The life cycle of an intestinal epithelial cell is about 5 days. This rapid growth of new cells also allows rapid repair of excoriations that occur in the mucosa.

Regulation of Small Intestinal Secretion

LOCAL STIMULI. By far the most important means for regulating small intestinal secretion are various local nervous reflexes, especially reflexes initiated by tactile or irritative stimuli and by the increase in enteric nervous activity associated with the gastrointestinal movements. Therefore, for the most part, secretion in the small intestine occurs simply in response to the presence of chyme in the intestine—the greater the amount of chyme, the greater the secretion.

HORMONAL REGULATION. Some of the same hormones that promote secretion elsewhere in the gastrointestinal tract also increase small intestinal secretion, especially secretin and cholecystokinin. Also, some experiments suggest that other hormonal substances extracted from the small intestinal mucosa by the chyme might help to control secretion. In general, the local enteric reflex mechanisms almost certainly play by far the dominant role.

SECRETIONS OF THE LARGE INTESTINE

MUCUS SECRETION. The mucosa of the large intestine, like that of the small intestine, has many crypts of Lieberkühn, but in this mucosa, unlike that of the small intestine, there are no villi. Also, the epithelial cells contain almost no enzymes. Instead, they consist mainly of mucous cells that secrete only mucus.

Therefore, the great preponderance of secretion in the large intestine is mucus. This mucus contains large amounts of bicarbonate ions caused by active transport through other epithelial cells that lie between the mucus-secreting epithelial cells. The rate of secretion of mucus is regulated principally by direct, tactile stimulation of the mucous cells on the surface of the mucosa and by local nervous reflexes to the mucous cells in the crypts of Lieberkühn. Stimulation of the pelvic nerves, which carry the parasympathetic innervation to the distal one half to two thirds of the large intestine, also causes marked increase in the secretion of mucus. This occurs along with an increase in motility, which is discussed in Chapter 63. Therefore, during extreme parasympathetic stimulation, often caused by emotional disturbances, so much mucus may be secreted into the large intestine that the person has a bowel movement of ropy mucus as often as every 30 minutes; this mucus contains little or no fecal material.

Mucus in the large intestine protects the wall against excoriation, but in addition, it provides the adherent medium for holding fecal matter together. Furthermore, it protects the intestinal wall from the great amount of bacterial activity that takes place inside the feces, and it plus the alkalinity of the secretion (pH of 8.0 caused by large amounts of sodium bicarbonate) provides a barrier to keep acids formed deep in the feces from attacking the intestinal wall.

SECRETION OF WATER AND ELECTROLYTES IN RESPONSE TO IRRITATION. Whenever a segment of the large intestine becomes intensely irritated, as occurs when bacterial infection becomes rampant during *enteritis,* the mucosa secretes large quantities of water and electrolytes in addition to the normal viscid solution of alkaline mucus. This acts to dilute the irritating factors and to cause rapid movement of the feces toward the anus. The usual result is *diarrhea,* with loss of large quantities of water and electrolytes. But the diarrhea also washes away the irritant factor, which promotes earlier recovery from the disease than would otherwise occur.

REFERENCES

Allen, A., et al.: Gastroduodenal mucosal protection. Physiol. Rev., 73:823, 1993.

Berglindh, T.: The mammalian gastric parietal cell in vitro. Annu. Rev. Physiol., 46:377, 1984.

Bouchier, I. A. D., et al.: Gastroenterology: Clinical Science & Practice. Philadelphia, W. B. Saunders Co., 1994.

Burns, G. P., and Bank, S.: Disorders of the Pancreas. Hightstown, NJ, McGraw-Hill, 1992.

Cheli, R., et al.: Gastric Protection. New York, Raven Press, 1988.

Chew, C. S.: Parietal cell culture: new models and directions. Annu. Rev. Physiol., 56:445, 1994.

Cooke, H. J.: Role of the "little brain" in the gut in water and electrolyte homeostasis. FASEB J., 3:127, 1989.

Daughtery, D., and Yamada, T.: Posttranslational processing of gastrin. Physiol. Rev., 69:482, 1989.

Evans, G. S., et al.: Primary cultures for studies of cell regulation and physiology in intestinal epithelium. Annu. Rev. Physiol. 56:399, 1994.

Fushiki, T., and Iwai, K.: Two hypotheses on the feedback regulation of pancreatic enzyme secretion. FASEB J., 3:121, 1989.

Go, V. L. W., et al.: The Pancreas: Biology, Pathobiology, and Disease. New York, Raven Press, 1993.

Hersey, S. J., et al.: Cellular control of pepsinogen secretion. Annu. Rev. Physiol. 46:393, 1984.

Hopfer, U., and Liedtke, C. M.: Proton and bicarbonate transport mechanisms in the intestine. Annu. Rev. Physiol., 49:51, 1987.

Johnson, L. R., et al.: Physiology of the Gastrointestinal Tract, 2nd Ed. New York, Raven Press, 1987.

Lauger, P.: Dynamics of ion transport systems in membranes. Physiol. Rev., 67:1296, 1987.

Lipkin, M.: Growth and development of the stomach. Annu. Rev. Physiol., 47:175, 1985.

Lundgren, O.: Microcirculation of the gastrointestinal tract and pancreas. In Renkin, E. M., and Michel, C. C. (eds.): Handbook of Physiology. Sec. 2, Vol. IV. Bethesda, Md., American Physiological Society, 1984, p. 799.

Machen, T. E., and Paradiso, A. M.: Regulation of intracellular pH in the stomach. Annu. Rev. Physiol., 49:19, 1987.

Morris, A. P., and Frizzell, R. A.: Vesicle targeting and ion secretion in epithelial cells: implications for cystic fibrosis. Annu. Rev. Physiol., 56:371, 1994.

Muallem, S.: Calcium transport pathways of pancreatic acinar cells. Annu. Rev. Physiol., 51:83, 1989.

Petersen, O. H., and Gallacher, D. V.: Electrophysiology of pancreatic and salivary acinar cells. Annu. Rev. Physiol., 50:65, 1988.

Putney, J. W., Jr.: Identification of cellular activation mechanisms associated with salivary secretion. Annu. Rev. Physiol., 48:75, 1986.

Reuss, L., and Stoddard, J. S.: Role of H^+ and HCO_3^- in salt transport in gallbladder epithelium. Annu. Rev. Physiol., 49:35, 1987.

Schultz, I., and Stolze, H. H.: The exocrine pancreas: The role of secretagogues, cyclic nucleotides and calcium in enzyme secretion. Annu. Rev. Physiol., 42:127, 1980.

Strange, R. C.: Hepatic bile flow. Physiol. Rev., 64:1055, 1984.

Streebny, L. M.: The Salivary System. Boca Raton, CRC Press Inc., 1987.

Surprenant, A.: Control of the gastrointestinal tract by enteric neurons. Annu. Rev. Physiol., 56:117, 1994.

Szurszewski, J. H.: Physiology of mammalian prevertebral ganglia. Annu. Rev. Physiol., 43:53, 1981.

Tache, Y.: CNS peptides and regulation of gastric acid secretion. Annu. Rev. Physiol., 50:19, 1988.

Tavoloni, N., and Berk, P. D.: Hepatic Transport and Bile Secretion: Physiology and Pathophysiology. New York, Raven Press, 1993.

Thompson, J. C., et al.: Gastrointestinal Endocrinology. New York, McGraw-Hill Book Co., 1987.

Trede, M., and Carter, D. C.: Surgery of the Pancreas. New York, Churchill Livingstone, 1993.

Walsh, J. H., and Dockray, G. J.: Gut Peptides: Biochemistry and Physiology. New York, Raven Press, 1994.

Walsh, J. H.: Peptides as regulators of gastric acid secretion. Annu. Rev. Physiol., 50:41, 1988.

Yamada, T., et al.: Textbook of Gastroenterology. Philadelphia, J. B. Lippincott, 1991.

Williams, J. A., and Blevins, G. T. Jr.: Cholecystokinin and regulation of pancreatic acinar cell function. Physiol. Rev., 73:701, 1993.

Digestion and Absorption in the Gastrointestinal Tract

CHAPTER 65

The foods on which the body lives, with the exception of small quantities of substances such as vitamins and minerals, can be classified as *carbohydrates, fats,* and *proteins.* They generally cannot be absorbed in their natural forms through the gastrointestinal mucosa and, for this reason, are useless as nutrients without the preliminary process of digestion. Therefore, this chapter discusses, first, the processes by which carbohydrates, fats, and proteins are digested into small enough compounds for absorption and, second, the mechanisms by which the digestive end products as well as water, electrolytes, and other substances are absorbed.

DIGESTION OF THE VARIOUS FOODS

HYDROLYSIS AS THE BASIC PROCESS OF DIGESTION. Almost all the carbohydrates of the diet are large *polysaccharides* or *disaccharides,* which are combinations of *monosaccharides* bound to one another by *condensation.* This means that a hydrogen ion has been removed from one of the monosaccharides, while a hydroxyl ion has been removed from the next one. The two monosaccharides then are combined with each other at these sites of removal, and the hydrogen and hydroxyl ions combine to form water. When the carbohydrates are digested back into monosaccharides, specific enzymes return the hydrogen and hydroxyl ions to the polysaccharides and thereby separate the monosaccharides from each other. This process, called

hydrolysis, is the following (in which R″-R′ is a disaccharide):

$$R''\text{-}R' + H_2O \xrightarrow[enzyme]{digestive} R''OH + R'H$$

Almost the entire fat portion of the diet consists of triglycerides (neutral fats), which are combinations of three *fatty acid* molecules condensed with a single *glycerol* molecule. In condensation, three molecules of water have been removed. Digestion of the triglycerides consists of the reverse process, the fat-digesting enzymes returning molecules of water to the triglyceride molecule and thereby splitting the fatty acid molecules away from the glycerol. Here again, the digestive process is one of *hydrolysis.*

Finally, proteins are formed from *amino acids* that are bound together by *peptide linkages.* In this linkage, a hydroxyl ion is removed from one amino acid and a hydrogen ion is removed from the succeeding one; thus, the amino acids in the protein chain are bound together by condensation, and digestion occurs by the reverse effect of hydrolysis, the proteolytic enzymes returning the water to the protein molecules to split them into their constituent amino acids.

Therefore, the chemistry of digestion is simple because in the case of all three major types of food, the same basic process of *hydrolysis* is involved. The only difference lies in the enzymes required to promote the reactions for each type of food.

833

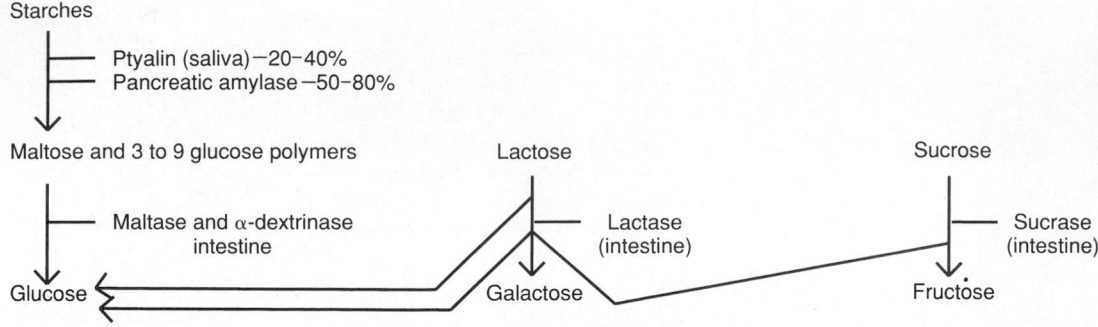

Figure 65–1. Digestion of carbohydrates.

All the digestive enzymes are proteins. Their secretion by the different gastrointestinal glands is discussed in Chapter 64.

Digestion of Carbohydrates

CARBOHYDRATE FOODS OF THE DIET. Only three major sources of carbohydrates exist in the normal human diet. They are *sucrose,* which is the disaccharide known popularly as cane sugar; *lactose,* which is a disaccharide in milk; and *starches,* which are large polysaccharides present in almost all nonanimal foods and particularly in the grains. Other carbohydrates ingested to a slight extent are amylose, glycogen, alcohol, lactic acid, pyruvic acid, pectins, dextrins, and minor quantities of carbohydrate derivatives in meats. The diet also contains a large amount of cellulose, which is a carbohydrate. However, no enzymes capable of hydrolyzing cellulose are secreted in the human digestive tract. Consequently, cellulose cannot be considered a food for the human being.

DIGESTION OF CARBOHYDRATES IN THE MOUTH AND STOMACH. When food is chewed, it is mixed with the saliva, which contains the enzyme *ptyalin* (an α-amylase) secreted mainly by the parotid glands. This enzyme hydrolyzes starch into the disaccharide *maltose* and other small polymers of glucose that contain three to nine glucose molecules (such as *maltotriose* and α *limit dextrins* that are the branch points of the starch molecule), as shown in Figure 65–1. But the food remains in the mouth only a short time, and probably not more than 5 per cent of all the starches that are eaten will have become hydrolyzed by the time the food is swallowed. Digestion continues in the body and fundus of the stomach for as long as 1 hour before the food becomes mixed with the stomach secretions. Then the activity of the salivary amylase is blocked by the acid of the gastric secretions because it is essentially nonactive as an enzyme once the pH of the medium falls below about 4.0. Nevertheless, on the average, before the food becomes completely mixed with the gastric secretions, as much as 30 to 40 per cent of the starches will have been hydrolyzed mainly to maltose.

DIGESTION OF CARBOHYDRATES IN THE SMALL INTESTINE

Digestion by Pancreatic Amylase. Pancreatic secretion, like saliva, contains a large quantity of α-amylase that is almost identical in its function with the α-amylase of saliva but is several times as powerful. Therefore, within 15 to 30 minutes after the chyme empties from the stomach into the duodenum and mixes with pancreatic juice, virtually all the starches are digested. In general, the starches are almost totally converted into *maltose* and *other very small glucose polymers* before they have passed beyond the duodenum or upper jejunum.

HYDROLYSIS OF DISACCHARIDES AND SMALL GLUCOSE POLYMERS INTO MONOSACCHARIDES BY THE INTESTINAL EPITHELIAL ENZYMES. The enterocytes lining the villi of the small intestine contain the four enzymes *lactase, sucrase, maltase,* and α-*dextrinase,* which are capable of splitting the disaccharides lactose, sucrose, and maltose as well as the other small glucose polymers into their constituent monosaccharides. These enzymes are located *in the membranes of the microvilli brush border* of the enterocytes, and the disaccharides are digested as they come in contact with these membranes. Lactose splits into a molecule of *galactose* and a molecule of *glucose.* Sucrose splits into a molecule of *fructose* and a molecule of *glucose.* Maltose and the other small glucose polymers all split into *molecules of glucose.* Thus, the final products of carbohydrate digestion are all monosaccharides, and they are absorbed immediately into the portal blood.

In the ordinary diet, which contains far more starches than all other carbohydrates combined, glucose represents more than 80 per cent of the final products of carbohydrate digestion, and galactose and fructose each seldom represent more than 10 per cent of the products of carbohydrate digestion.

The major steps in carbohydrate digestion are summarized in Figure 65–1.

Digestion of Proteins

PROTEINS OF THE DIET. The dietary proteins are formed of long chains of amino acids bound together by *peptide linkages.* A typical linkage is the following:

$$R-CH-\underset{\underset{O}{\parallel}}{C}-\overset{\overset{NH_2}{\mid}}{\underset{}{}}\!\!\! OH + H-N-CH-COOH \longrightarrow$$

$$R-CH-\underset{\underset{O}{\parallel}}{C}-N-CH-COOH + H_2O$$

The characteristics of each type of protein are determined by the types of amino acids in the protein molecule and by the arrangement of these amino acids. The physical and chemical characteristics of the different proteins are discussed in Chapter 69.

DIGESTION OF PROTEINS IN THE STOMACH. *Pepsin*, the important peptic enzyme of the stomach, is most active at a pH of 2.0 to 3.0 and is inactive at a pH above about 5.0. Consequently, for this enzyme to cause any digestive action on protein, the stomach juices must be acidic. As explained in Chapter 64, the gastric glands secrete a large quantity of hydrochloric acid. This hydrochloric acid is secreted by the parietal cells at a pH of about 0.8, but by the time it is mixed with the stomach contents and with the secretions from the nonparietal glandular cells of the stomach, the pH ranges around 2.0 to 3.0, a highly favorable range of acidity for pepsin activity.

One of the important features of pepsin digestion is its ability to digest collagen, an albuminoid that is affected little by other digestive enzymes. Collagen is a major constituent of the intercellular connective tissue of meats; therefore, for the digestive enzymes of the digestive tract to penetrate meats and digest the cellular proteins, it is first necessary that the collagen fibers be digested. Consequently, in people who lack peptic activity in the stomach, the ingested meats are less well penetrated by the digestive enzymes and, therefore, may be poorly digested.

As shown in Figure 65–2, pepsin only begins the process of protein digestion, usually providing only 10 to 20 per cent of the total protein digestion. This splitting of proteins is a process of hydrolysis that occurs at the peptide linkages between the amino acids.

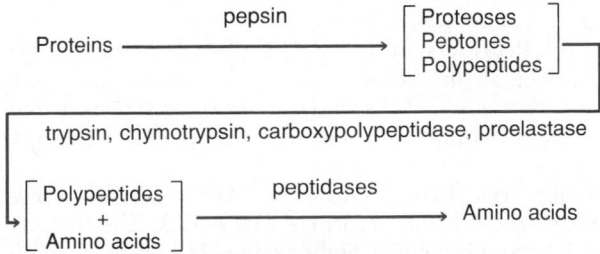

Figure 65–2. Digestion of proteins.

DIGESTION OF PROTEINS BY PANCREATIC SECRETIONS. Most protein digestion occurs principally in the upper small intestine, in the duodenum and jejunum, under the influence of the proteolytic enzymes of the pancreatic secretion. When the proteins leave the stomach, they ordinarily are mainly in the form of proteoses, peptones, and large polypeptides. Immediately on entering the small intestine, the partial breakdown products are attacked by the major proteolytic pancreatic enzymes *trypsin, chymotrypsin, carboxypolypeptidase,* and *proelastase,* as shown in Figure 65–2. Both trypsin and chymotrypsin can split protein molecules into small polypeptides; carboxypolypeptidase then cleaves individual amino acids from the carboxyl ends of the polypeptides. Proelastase gives rise to *elastase* that in turn digests the elastin fibers that hold meats together. Only a small percentage of the proteins are digested all the way to their constituent amino acids by the pancreatic juices. Most remain as dipeptides, tripeptides, and some even larger.

DIGESTION OF PEPTIDES BY PEPTIDASES IN THE ENTEROCYTES THAT LINE THE SMALL INTESTINAL VILLI. The last digestion of the proteins in the intestinal lumen is achieved by the enterocytes that line the villi of the small intestine, mainly in the duodenum and jejunum. These cells have a *brush border* that consists literally of hundreds of microvilli projecting from the surface of each cell. In the cell membrane of each of these microvilli are multiple *peptidases* that protrude through the membranes to the exterior, where they come in contact with the intestinal fluids. Two types of peptidase enzymes are especially important, *aminopolypeptidase* and several *dipeptidases*. They succeed in splitting the remaining larger polypeptides into tripeptides and dipeptides and a few all the way to amino acids. Both the amino acids and the dipeptides and tripeptides are easily transported through the microvillar membrane to the interior of the enterocyte.

Finally, inside the cytosol of the enterocyte are multiple other peptidases that are specific for the remaining types of linkages between the amino acids. Within minutes, virtually all the last dipeptides and tripeptides are digested to the final stage of single amino acids; they then pass on through the opposite side of the enterocyte into the blood. More than 99 per cent of the final protein digestive products that are absorbed are individual amino acids, with only rare absorption of peptides, and very, very rare absorption of whole protein molecules. Even these very few molecules of protein can sometimes cause serious allergic or immunological disturbances, as discussed in Chapter 34.

Digestion of Fats

FATS OF THE DIET. By far the most abundant fats of the diet are the neutral fats, also known as *triglycerides*, each molecule of which is composed of a glycerol nucleus and three fatty acids, as shown in Figure 65–3. Neutral fat is a major constituent in food of animal origin and much less so in food of plant origin. In the usual diet are also small quantities of phos-

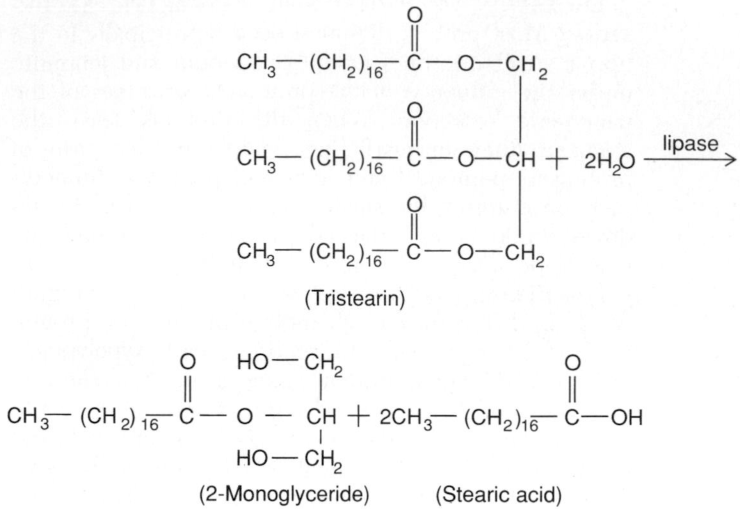

Figure 65-3. Hydrolysis of neutral fat catalyzed by lipase.

pholipids, cholesterol, and cholesterol esters. The phospholipids and cholesterol esters contain fatty acid and therefore can be considered fats themselves. Cholesterol, on the other hand, is a sterol compound that contains no fatty acid, but it does exhibit some of the physical and chemical characteristics of fats; it is derived from fats, and it is metabolized similarly to fats. Therefore, cholesterol is considered, from a dietary point of view, a fat.

DIGESTION OF FATS IN THE INTESTINE. A small amount of triglycerides is digested in the stomach by *lingual lipase* that is secreted by lingual glands in the mouth and swallowed with the saliva. The amount of digestion is less than 10 per cent and generally unimportant. Instead, essentially all fat digestion occurs in the small intestine as follows.

Emulsification of Fat by Bile Acids and Lecithin. The first step in fat digestion is to break the fat globules into small sizes so that the water-soluble digestive enzymes can act on the globule surfaces. This process is called emulsification of the fat, and it is achieved partly by agitation in the stomach along with the products of stomach digestion but mainly under the influence of *bile*, the secretion of the liver that does not contain any digestive enzymes. However, bile does contain a large quantity of *bile salts* as well as the phospholipid *lecithin*, both of which, *but especially the lecithin*, are extremely important for the emulsification of fat. The polar parts (the points where ionization occurs in water) of the bile salt and lecithin molecules are highly soluble in water, whereas most of the remaining portions of their molecules are highly soluble in fat. Therefore, the fat-soluble portions dissolve in the surface layer of the fat globule with the polar portions projecting outward and soluble in the surrounding fluids; this effect greatly decreases the interfacial tension of the fat.

When the interfacial tension of a globule of nonmiscible fluid is low, this nonmiscible fluid, on agitation, can be broken up into many minute particles far more easily than it can when the interfacial tension is great. Consequently, a major function of the bile salts and lecithin, especially the lecithin, in the bile is to make the fat globules readily fragmentable by agitation in the small bowel. This action is the same as that of many detergents that are widely used in household cleaners for removing grease.

Each time the diameters of the fat globules are decreased by a factor of 2 as a result of agitation in the small intestine, the total surface area of the fat increases two times. In other words, the total surface area of the fat particles in the intestinal contents is inversely proportional to the diameters of the particles. Because the average size of the emulsified fat particles in the intestine is less than 1 micrometer, this represents an increase of as much as 1000-fold in the total surface area of the fats caused by the emulsification process.

The lipases are water-soluble compounds and can attack the fat globules only on their surfaces. Consequently, it can be readily understood how important this detergent function of bile salts is for the digestion of fats.

DIGESTION OF TRIGLYCERIDES BY PANCREATIC LIPASE. By far the most important enzyme for the digestion of triglycerides is *pancreatic lipase* in the pancreatic juice. This is present in enormous quantities in pancreatic juice, enough to digest all triglycerides that it can reach within a few minutes. In addition, the enterocytes of the small intestine contain a minute quantity of lipase known as *enteric lipase*, but this is usually unimportant.

End Products of Fat Digestion. Most of the triglycerides of the diet are split by pancreatic lipase into *free fatty acids* and *2-monoglycerides*, as shown in Figure 65-4. Minute portions remain in the diglyceride state.

ROLE OF BILE SALTS IN ACCELERATING FAT DIGESTION—FORMATION OF MICELLES. The hydrolysis of triglycerides is a highly reversible process; there-

Fat $\xrightarrow{\text{(Bile + Agitation)}}$ Emulsified fat

Emulsified fat $\xrightarrow{\text{Pancreatic lipase}}$ Fatty acids and 2-monoglycerides

Figure 65–4. Digestion of fats.

fore, accumulation of monoglycerides and free fatty acids in the vicinity of digesting fats quickly blocks further digestion. The bile salts play an important role in removing the monoglycerides and free fatty acids from the vicinity of the digesting fat globules almost as rapidly as these end products of digestion are formed. This occurs in the following way.

Bile salts, when in high enough concentration, have the propensity to form *micelles,* which are small spherical, cylindrical globules 3 to 6 nanometers in diameter composed of 20 to 40 molecules of bile salt. They develop because each bile salt molecule is composed of a sterol nucleus, most of which is highly fat-soluble, and a polar group that is highly water-soluble. The sterol nuclei of the 20 to 40 bile salt molecules of the micelle aggregate, together with the fat digestates, to form a small fat globule in the middle of the micelle with the polar groups of the bile salts projecting outward to cover the surface of the micelle. Because these polar groups are negatively charged, they allow the entire micelle globule to become dissolved in the water of the digestive fluids and to remain in stable solution despite the large size of the micelle.

During triglyceride digestion, as rapidly as the monoglycerides and free fatty acids are formed, the fatty portions of them become dissolved in the central fatty portion of the micelles, which immediately reduces the concentrations of these end products of digestion in the vicinity of the digesting fat globules. Consequently, the digestive process can proceed unabated.

The bile salt micelles also act as a transport medium to carry the monoglycerides and the free fatty acids, both of which would otherwise be relatively insoluble, to the brush borders of the intestinal epithelial cells. There the monoglycerides and free fatty acids are absorbed, as discussed later. On delivery of these substances to the brush border, the bile salts are again released back into the chyme to be used again and again for this "ferrying" process.

DIGESTION OF CHOLESTEROL ESTERS AND PHOSPHOLIPIDS. Most of the cholesterol in the diet is in the form of cholesterol esters, which are combinations of free cholesterol and one molecule of fatty acid. Phospholipids also contain fatty acid chains within their molecules. Both the cholesterol esters and the phospholipids are hydrolyzed by two other lipases in the pancreatic secretion that free the fatty acids—the enzyme *cholesterol ester hydrolase* to hydrolyze the cholesterol ester and *phospholipase* A_2 to hydrolyze the phospholipid.

The bile salt micelles play the same role in "ferrying" free cholesterol and the remaining portions of the digested phospholipid molecules as they play in "ferrying" monoglycerides and free fatty acids. Indeed, this role of the micelles is essential to the absorption of cholesterol because essentially no cholesterol can be absorbed without the function of the micelles. On the other hand, as much as 60 per cent of the triglycerides can be digested and absorbed even in the absence of the bile salt micelles.

BASIC PRINCIPLES OF GASTROINTESTINAL ABSORPTION

It is suggested that the reader review the basic principles of transport through cells that are discussed in detail in Chapter 4. The following sections discuss specialized applications of the transport processes during gastrointestinal absorption.

Anatomical Basis of Absorption

The total quantity of fluid that must be absorbed each day is equal to the ingested fluid (about 1.5 liters) plus that secreted in the various gastrointestinal secretions (about 7 liters). This comes to a total of about 8 to 9 liters. All but about 1.5 liters of this is absorbed in the small intestine, leaving only this 1.5 liters to pass through the ileocecal valve into the colon each day.

The stomach is a poor absorptive area of the gastrointestinal tract because it lacks the typical villus type of absorptive membrane and because the junctions between the epithelial cells are tight junctions. Only a few highly lipid-soluble substances, such as alcohol and some drugs like aspirin, can be absorbed in small quantities.

ABSORPTIVE SURFACE OF THE INTESTINAL MUCOSA —THE VILLI. Figure 65–5 demonstrates the absorptive surface of the intestinal mucosa, showing many folds called *valvulae conniventes* (or *folds of Kerckring*), which increase the surface area of the absorptive mucosa about threefold. These folds extend circularly most of the way around the intestine and are especially well developed in the duodenum and jejunum, where they often protrude as much as 8 millimeters into the lumen.

Located over the entire surface of the small intestine, from about the point at which the common bile duct empties into the duodenum down to the ileocecal valve, are literally millions of small *villi,* which project about 1 millimeter from the surface of the mucosa, as shown on the surfaces of the valvulae conniventes in Figure 65–5 and in more detail in Figure 65–6. These villi lie so close to one another in the upper small intestine that they touch in most areas, but their distribution is less profuse in the distal small intestine. The presence of villi on the mucosal surface enhances the absorptive area another 10-fold.

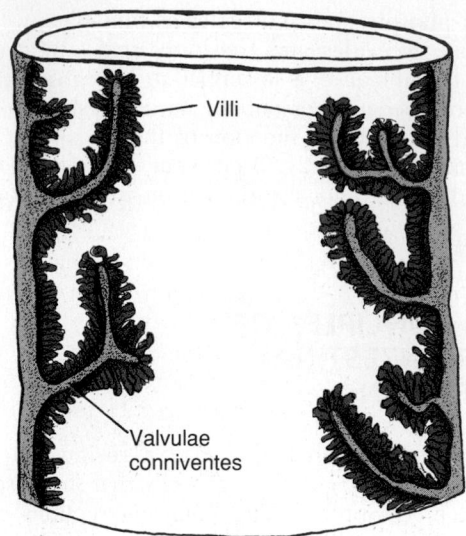

Figure 65–5. Longitudinal section of the small intestine, showing the valvulae conniventes covered by villi.

Finally, each intestinal epithelial cell is characterized by a brush border, consisting of as many as 1000 *microvilli* 1 micrometer in length and 0.1 micrometer in diameter protruding into the intestinal chyme; these microvilli are shown in the electron micrograph in Figure 65–7. This increases the surface area exposed to the intestinal materials at least another 20-fold. Thus, the combination of the folds of Kerckring, the villi, and the microvilli increases the absorptive area of the mucosa perhaps 1000-fold, making a tremendous total area of 250 or more square meters for the entire small intestine—about the surface area of a tennis court.

Figure 65–6A shows the general organization of the villus, emphasizing especially the advantageous arrangement of the vascular system for absorption of fluid and dissolved material into the portal blood and the arrangement of the *central lacteal* for absorption into the lymph. Figure 65–6B shows the cross section of the villus, and Figure 65–7 shows many small *pinocytic* vesicles, which are pinched-off portions of infolded enterocyte membrane that contain inside the vesicles extracellular materials that have been entrapped. Minute amounts of substances are absorbed by this physical process of *pinocytosis*, although the amounts are very, very small in proportion to total absorption. Also, extending linearly into each microvillus of the brush border are multiple actin filaments that contract intermittently and cause continual movements of the microvilli, keeping them constantly exposed to new quantities of intestinal fluid.

Basic Mechanisms of Absorption

Absorption through the gastrointestinal mucosa occurs by *active transport*, by *diffusion*, and, possibly, by *solvent drag*. The basic physical principles of these processes are explained in Chapter 4.

Briefly, active transport imparts energy to the substance as it is being transported for the purpose of concentrating it on the other side of the membrane or moving it against an electrical potential. On the other hand, transport by "diffusion" means simply transport of substances through the membrane as a result of molecular movement *along*, rather than against, an electrochemical gradient. Transport by solvent drag means that anytime a solvent is absorbed because of physical absorptive forces, the movement of the solvent will "drag" dissolved substances along at the same time.

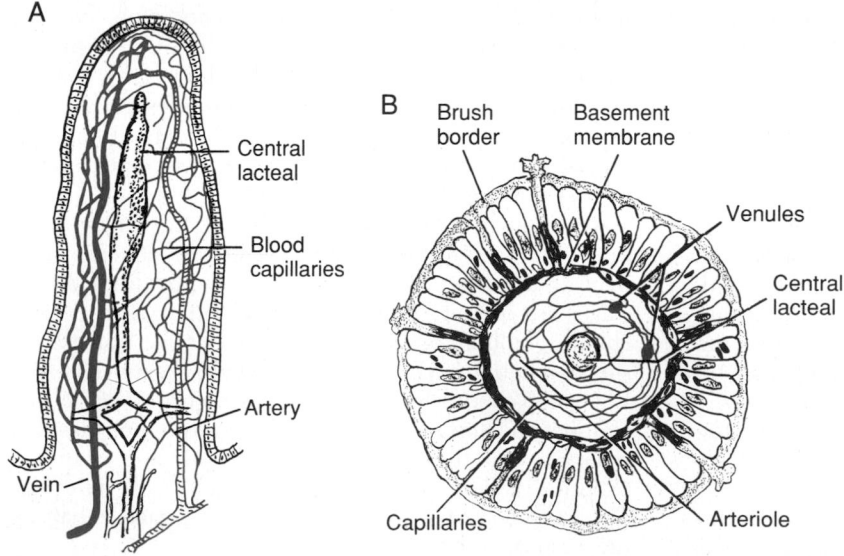

Figure 65–6. Functional organization of the villus. *A,* Longitudinal section. *B,* Cross section showing the epithelial cells and basement membrane.

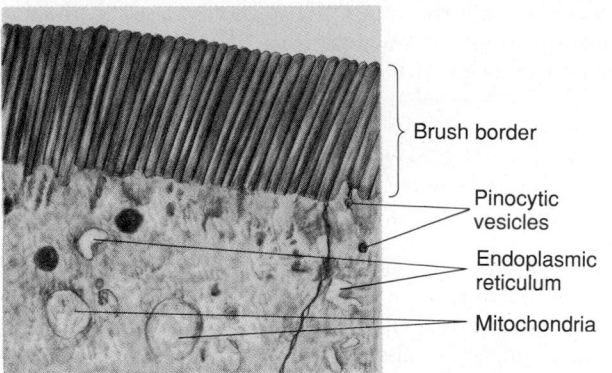

Figure 65–7. Brush border of the gastrointestinal epithelial cell, showing also pinocytic vesicles, mitochondria, and endoplasmic reticulum lying immediately beneath the brush border. (Courtesy of Dr. William Lockwood.)

ABSORPTION IN THE SMALL INTESTINE

Absorption from the small intestine each day consists of several hundred grams of carbohydrates, 100 or more grams of fat, 50 to 100 grams of amino acids, 50 to 100 grams of ions, and 7 to 8 liters of water. The absorptive *capacity* of the normal small intestine is far greater than this: as much as several kilograms of carbohydrates per day, 500 grams of fat per day, 500 to 700 grams of proteins per day, and 20 or more liters of water per day. The *large* intestine can absorb still more water and ions, although almost no nutrients.

Absorption of Water

ISOSMOTIC ABSORPTION. Water is transported through the intestinal membrane entirely by *diffusion.* Furthermore, this diffusion obeys the usual laws of osmosis. Therefore, when the chyme is dilute, water is absorbed through the intestinal mucosa into the blood of the villi by osmosis.

On the other hand, water can also be transported in the opposite direction, from the plasma into the chyme. This occurs especially when hyperosmotic solutions are discharged from the stomach into the duodenum. Within minutes, sufficient water usually is transferred by osmosis to make the chyme isosmotic with the plasma.

As dissolved substances are absorbed from the lumen of the gut into the blood, the absorption tends to decrease the osmotic pressure of the chyme, but water diffuses so readily through the intestinal membrane (because of large 0.7- to 1.5-nanometer paracellular pores through the so-called tight junctions between the epithelial cells) that it almost instantaneously "follows" the absorbed substances into the blood. Therefore, as ions and nutrients are ab-

sorbed, so also is an isosmotic equivalent of water absorbed.

Absorption of Ions

ACTIVE TRANSPORT OF SODIUM. Twenty to 30 grams of sodium are secreted into the intestinal secretions each day. In addition, the normal person eats 5 to 8 grams of sodium each day. Combining these, the small intestine must absorb 25 to 35 grams of sodium each day, which is equal to about one seventh of all the sodium present in the body. Therefore, one can well understand that whenever intestinal secretions are lost to the exterior, as in extreme diarrhea, the sodium reserves of the body can be depleted to a lethal level within hours. Normally, however, less than 0.5 per cent of the intestinal sodium is lost in the feces each day because of its rapid absorption through the intestinal mucosa. The sodium also plays an important role in the absorption of sugars and amino acids, as we shall see in subsequent discussions.

The basic mechanism of sodium absorption from the intestine is shown in Figure 65–8. The principles of this mechanism, which are discussed in Chapter 4, are essentially the same as those for absorption of sodium from the gallbladder and the renal tubules as discussed in Chapter 27. The motive power for the sodium absorption is provided by active transport of sodium from inside the epithelial cells through the basal and side walls of these cells into the paracellular spaces. This is demonstrated by the heavy black arrows in Figure 65–8. This active transport obeys the usual laws of active transport: it requires energy, and the energy process is catalyzed by appropriate adenosine triphosphatase enzymes in the cell membrane (see Chap. 4). Part of the sodium is absorbed simultaneously with chloride ions; the chloride ions are passively "dragged" along by the positive electrical charges of the sodium ion. Other sodium ions are absorbed while either potassium or hydrogen ions are transported in the opposite direction in exchange for the sodium ions.

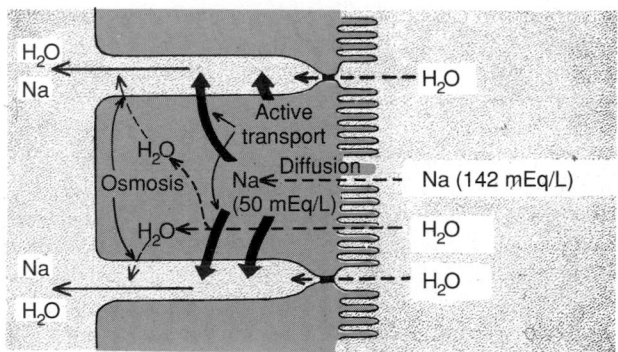

Figure 65–8. Absorption of sodium through the intestinal epithelium. Note also the osmotic absorption of water—that is, the water "follows" the sodium through the epithelial membrane.

The active transport of sodium through the basolateral membranes of the cell reduces the sodium concentration inside the cell to a low value (about 50 mEq/liter), as also shown in Figure 65–8. Because the sodium concentration in the chyme is normally about 142 mEq/liter (that is, about equal to that in the plasma), sodium moves down a steep electrochemical gradient from the chyme through the brush border of the epithelial cell into the epithelial cell cytoplasm. This replaces the sodium that is actively transported out of the epithelial cells into the paracellular spaces.

The next step in the transport process is osmosis of water into the paracellular spaces. This is caused by the osmotic gradient created by the elevated concentration of ions in the paracellular space. Most of this osmosis occurs through the tight junctions between the apical borders of the epithelial cells, as discussed earlier, but a smaller proportion occurs through the cells themselves. The osmotic movement of water creates a flow of fluid into and through the paracellular space and, finally, into the circulating blood of the villus.

Aldosterone Greatly Enhances Sodium Absorption. When a person becomes dehydrated, large amounts of aldosterone almost always are secreted by the adrenal glands. Within 1 to 3 hours the excess aldosterone greatly enhances all the enzyme and transport mechanisms for all aspects of sodium absorption by the intestinal epithelial cells. The increased sodium absorption then causes secondary increases in absorption of chloride ions, water, and some other substances. This effect of aldosterone is especially important in the colon because it allows virtually no loss of sodium chloride in the feces and little water loss. This effect of aldosterone in the intestinal tract is the same as that activated by aldosterone in the renal tubules, which also serves to conserve salt and water in the body when a person becomes dehydrated.

Absorption of Chloride Ions in the Duodenum and Jejunum. In the upper part of the small intestine, chloride absorption is rapid and mainly by passive diffusion. The absorption of sodium ions through the epithelium creates slight electronegativity in the chyme and electropositivity on the basal side of the epithelial cells. Then chloride ions move along this electrical gradient to "follow" the sodium ions.

Absorption of Bicarbonate Ions in the Duodenum and Jejunum. Often large quantities of bicarbonate ions must be reabsorbed from the upper small intestine because of the large amounts of bicarbonate ions in both the pancreatic secretion and the bile. The bicarbonate ion is absorbed in an indirect way as follows: When sodium ions are absorbed, moderate amounts of hydrogen ions are secreted into the lumen of the gut in exchange for some of the sodium, as explained earlier. These hydrogen ions in turn combine with the bicarbonate ions to form carbonic acid (H_2CO_3), which then dissociates to form water and carbon dioxide. The water remains as part of the chyme in the intestines, but the carbon dioxide is readily absorbed into the blood and subsequently expired through the lungs. Thus, this is a so-called active absorption of bicarbonate ions. It is the same mechanism that occurs in some of the tubules of the kidneys.

Secretion of Bicarbonate Ions in the Ileum and Large Intestine—Simultaneous Absorption of Chloride Ions

The epithelial cells on the surfaces of the villi in the ileum as well as on all surfaces of the large intestine have a special capability of secreting bicarbonate ions in exchange for absorption of chloride ions. This is important because it provides alkaline bicarbonate ions that are used to neutralize acid products formed by bacteria, especially so in the large intestine. The entire mechanism of this exchange is not clear, but basically, it depends on an exchange protein in the luminal membrane of the epithelial cell that forcibly exchanges bicarbonate ions formed inside the cell for chloride ions in the intestinal lumen. Then the excess chloride in the cell is transported by facilitated diffusion through the basolateral membrane of the epithelial cell, thus completing the chloride absorption.

Extreme Secretion of Chloride Ions, Sodium Ions, and Water from the Crypts of Lieberkühn in Cholera and Some Other Types of Diarrhea. Deep in the crypts of Lieberkühn are immature epithelial cells that continually divide to form new epithelial cells that then spread outward over the luminal surfaces of the intestines. These new cells, while still in the crypts, have properties different from those of the mature cells already on the outer luminal surfaces. Even normally, they secrete small quantities of sodium chloride and water into the intestinal lumen, but this secretion is immediately reabsorbed by the older epithelial cells outside the crypts, thus providing a watery solution for absorbing intestinal digestates. The toxins of cholera and some other types of diarrheal bacteria can stimulate the crypt secretion so greatly that this secretion inundates the reabsorption, thus often causing loss of 5 to 10 liters of water and salt as diarrhea each day. Within 1 to 5 days, many severely affected patients die of this loss of fluid alone.

This extreme secretion is initiated by entry of a subunit of the cholera toxin into the cell. This stimulates the formation of excess cyclic adenosine monophosphate, which then opens tremendous numbers of chloride channels, allowing chloride ions to flow rapidly from inside the cell into the crypts. In turn, this is believed to activate a sodium pump that pumps sodium ions into the crypts to go along with the chloride ions. Finally, all this extra sodium chloride causes extreme osmosis of water into the crypts as well, thus providing the rapid flow of fluid along with the salt. Initially, all this excess fluid washes away the bacteria and is of value in combating the disease, but too much of a good thing can be lethal because of the serious dehydration of the body that may ensue.

In most instances, the life of the cholera victim can be saved by simple administration of tremendous amounts of sodium chloride solution to make up for the loss.

Absorption of Other Ions. Calcium ions are actively absorbed, especially from the duodenum, and calcium ion absorption is exactly controlled in relation to the need of the body for calcium. One important factor controlling calcium absorption is parathyroid hormone secreted by the parathyroid glands and another is vitamin D. The parathyroid hormone activates vitamin D in the kidneys, and the activated vitamin D in turn greatly enhances calcium absorption. These effects are discussed in Chapter 79.

Iron ions are also actively absorbed from the small intestine. The principles of iron absorption and the regulation of its absorption in proportion to the body's need for iron, especially for the formation of hemoglobin, are discussed in Chapter 32.

Potassium, magnesium, phosphate, and probably still other ions can also be actively absorbed through the mucosa. In general, the monovalent ions are absorbed with ease and in great quantities. On the other hand, bivalent ions are normally absorbed in only small amounts; for instance, the maximum absorption of calcium ions is only 1/50 as great as the normal absorption of sodium ions. Fortunately, only small quantities of the bivalent ions are normally needed by the body.

Absorption of Nutrients

Absorption of Carbohydrates

Essentially all food carbohydrates are absorbed in the form of monosaccharides; only a small fraction are absorbed as disaccharides and almost none as larger carbohydrate compounds. By far the most abundant of the absorbed monosaccharides is glucose, usually accounting for more than 80 per cent of carbohydrate calories absorbed. The reason for this is that glucose is the final digestion product of our most abundant carbohydrate food, the starches. The remaining 20 per cent of absorbed monosaccharides are composed almost entirely of galactose and fructose, the galactose derived from milk and the fructose as one of the monosaccharides in cane sugar.

Virtually all the monosaccharides are absorbed by an active transport process or secondarily to active transport. Let us first discuss the absorption of glucose.

Glucose Is Basically Transported by a Sodium Co-Transport Mechanism. In the absence of sodium transport through the intestinal membrane, virtually no glucose can be absorbed. The reason is that glucose absorption occurs in a co-transport mode with the active transport of sodium. There are two stages in the transport of sodium through the intestinal enterocyte. First is active transport of sodium through the basolateral membranes into the paracellular spaces, thereby depleting the sodium inside the cells. This decrease of sodium inside the cells then causes sodium in the intestinal lumen to diffuse through the brush border of the enterocyte to its interior by facilitated diffusion. The sodium first combines with a transport protein, but the transport protein will not transport the sodium to the interior of the cell until it also combines with some other appropriate substance such as glucose. Therefore, intestinal glucose also combines simultaneously with the same transport protein, and then both the sodium and the glucose are transported together to the interior of the cell. It is the low concentration of sodium inside the cell that literally "drags" sodium to the interior cell and along with it the glucose at the same time. Once inside the enterocyte, other transport proteins and enzymes cause facilitated diffusion of the glucose outward through the enterocyte's basolateral membrane into the paracellular space. To summarize, it is the initial active transport of sodium through the basolateral membranes of the enterocyte that provides the eventual motive force for moving glucose through the enterocyte to the paracellular space.

The digestion of disaccharides and trisaccharides at the brush border enhances glucose transport. When disaccharides and trisaccharides composed of glucose come in contact with the brush border, digestive enzymes attached to the membranes of the brush microvilli cause hydrolysis of the disaccharides and trisaccharides into glucose. This greatly increases the concentration of the glucose in this immediate area adjacent to the absorptive membrane of the enterocyte. The high concentration of glucose then plays an additional special role in enhancing the rate at which glucose is transported to the interior of the enterocyte, and then all the way through the enterocyte into the paracellular space.

Additional transport of glucose occurs by "solvent drag" through the cell junctions into the paracellular spaces. This is probably important at high concentrations. When glucose is transported through the enterocyte and finally into the paracellular space, as discussed above, this causes greatly increased glucose concentration in the paracellular space. The high glucose concentration in turn causes a high osmotic pressure in the paracellular space, which in turn causes water to be absorbed osmotically from the intestinal lumen through the cell junctions directly into the paracellular space without going through the interior of the intestinal enterocyte. At low concentrations of carbohydrates in the intestinal lumen, this is not an important pathway for absorption. At high concentrations, however, the active transport of glucose through the enterocyte becomes limited because there are limited amounts of the required enzymes and carrier proteins in the enterocyte membranes. Yet the osmotic flow of fluid through the cell junctions continues to increase and carries with it anything dissolved in the fluid, ending up in the paracellular space. This is called the "solvent drag" mechanism. At high concentrations of glucose in the intestine, the quantity of glucose absorbed by this solvent drag mechanism might be as great as two to three times the sodium co-transport of glucose through the enterocyte itself. The solvent drag mechanism is even further enhanced by the fact that actomyosin molecules inside the enterocyte walls adjacent to the cell junctions contract under conditions of high glucose and literally open the cell junctions to larger sizes.

ABSORPTION OF OTHER MONOSACCHARIDES. Galactose is transported by almost exactly the same mechanism as glucose. On the other hand, fructose transport does not occur by the sodium co-transport mechanism. Instead, fructose is transported by facilitated diffusion all the way through the enterocyte but not coupled with sodium transport. Also, much of the fructose is converted into glucose on its way through the enterocyte. That is, on entering the enterocyte, much of the fructose becomes phosphorylated in the cell, then converted to glucose, and finally transported in the form of glucose the rest of the way into the paracellular space. Because fructose is not co-transported with sodium, its overall rate of transport is only about one half that of glucose or galactose.

Absorption of Proteins

As explained earlier in the chapter, most proteins are absorbed through the luminal membranes of the intestinal epithelial cells (the small intestinal enterocytes) in the form of dipeptides, tripeptides, and a few free amino acids. The energy for most of this transport is supplied by a sodium co-transport mechanism in the same way that sodium co-transport of glucose occurs. That is, most peptide or amino acid molecules bind in the enterocyte microvillus membrane with a specific transport protein that also requires sodium binding before transport can occur. The sodium ion then moves down its electrochemical gradient to the interior of the cell and pulls the amino acid or peptide along with it. This is called *co-transport* or *secondary active transport of the amino acids or peptides.*

A few amino acids do not require this sodium co-transport mechanism but instead are transported by special membrane transport proteins in the same way that fructose is transported, by facilitated diffusion.

At least five types of amino acid and peptide transport proteins have been characterized in the luminal membrane of intestinal epithelial cells. This multiplicity of transport proteins is required because of the diverse binding properties of the different amino acids and peptides.

Absorption of Fats

Earlier in this chapter, it was pointed out that as fats are digested to form monoglycerides and free fatty acids, both of these digestive end products become dissolved in the central lipid portion of the bile acid micelles. Because of the molecular dimensions of these micelles, only 3 to 6 nanometers in diameter, and because of their highly charged exterior, they are soluble in the chyme. In this form, the monoglycerides and the fatty acids are carried to the surfaces of the microvilli in the brush border, even penetrating into the recesses among the moving, agitating microvilli. Here, both the monoglycerides and the fatty acids diffuse immediately through the enterocyte cell membrane to the interior of the enterocyte; this is possible because these lipids are as soluble in the enterocyte membrane as in the micelles. This leaves the bile acid micelles still in the chyme. The micelles then diffuse back through the chyme and absorb still more monoglycerides and fatty acids and similarly carry these also to the epithelial cells. Thus, the micelles perform a "ferrying" function, which is highly important for fat absorption. In the presence of an abundance of bile acid micelles, about 97 per cent of the fat is absorbed; in the absence of bile acids, only 40 to 50 per cent is normally absorbed.

The undigested triglycerides as well as the diglycerides are highly soluble in the lipid membrane of the intestinal enterocyte. Even so, only small quantities of them are normally absorbed because the bile acid micelles will not dissolve either triglycerides or diglycerides and therefore will not ferry them to the enterocyte membrane.

After entering the enterocyte, the fatty acids and monoglycerides are taken up by the smooth endoplasmic reticulum, and here they are mainly recombined to form new triglycerides. A few of the monoglycerides are further digested into glycerol and fatty acids by an intracellular lipase. Then these free fatty acids, too, are reconstituted by the smooth endoplasmic reticulum into triglycerides, using for this purpose new glycerol that is synthesized de novo from alpha-glycerophosphate, this synthesis requiring both energy from adenosine triphosphate and a complex of enzymes to catalyze the reactions.

FORMATION OF CHYLOMICRONS. Once formed, the reconstituted triglycerides aggregate first within the endoplasmic reticulum and then in the Golgi apparatus into globules that contain absorbed cholesterol, absorbed phospholipids, and small amounts of newly synthesized cholesterol and phospholipids. The phospholipids arrange themselves in these globules with the fatty portion of the phospholipid toward the center and the polar portions located on the surface. This provides an electrically charged surface that makes these globules miscible with the fluids of the cell. In addition, small amounts of several types of *apoprotein,* also synthesized by the endoplasmic reticulum, coat part of the surface of each globule. In this form, the globules are released from the Golgi apparatus and excreted by cellular *exocytosis* into the basolateral spaces around the cell; from there, they pass into the lymph in the central lacteal of the villus. These globules are then called *chylomicrons.*

The apoproteins are essential for cellular exocytosis of the chylomicrons to occur, especially apoprotein B, because they provide a means for attaching the fatty globule to the cell membrane before it is extruded. In people who have a genetic inability to form apoprotein B, the enterocytes become engorged with fatty products that cannot proceed the rest of the way to be absorbed.

TRANSPORT OF THE CHYLOMICRONS IN THE LYMPH. From the basolateral surfaces of the enterocytes, the chylomicrons wend their way into the central lacteals of the villi and from here are propelled, along with the lymph, by the lymphatic pump upward through the thoracic duct to be emptied into the great veins of the neck. Between 80 and 90 per cent of all fat absorbed from the gut is absorbed in this manner and transported to the blood by way of the thoracic lymph in the form of chylomicrons.

DIRECT ABSORPTION OF FATTY ACIDS INTO THE PORTAL BLOOD. Small quantities of short- and medium-chain fatty acids, such as those from butterfat, are absorbed directly into the portal blood rather than being converted into triglycerides and absorbed into the lymphatics. The cause of this difference between shorter- and longer-chain fatty acid absorption is that the shorter-chain fatty acids are more water-soluble and mostly are not reconverted into triglycerides by the endoplasmic reticulum. This allows direct diffusion of these fatty acids from the epithelial cells into the capillary blood of the villi.

ABSORPTION IN THE LARGE INTESTINE: FORMATION OF THE FECES.

About 1500 milliliters of chyme normally pass through the ileocecal valve into the large intestine each day. Most of the water and electrolytes in the chyme are absorbed in the colon, usually leaving less than 100 milliliters of fluid to be excreted in the feces. Also, essentially all the ions are absorbed, leaving only 1 to 5 milliequivalents each of sodium and chloride ions to be lost in the feces.

Most of the absorption in the large intestine occurs in the proximal one half of the colon, giving this portion the name *absorbing colon,* whereas the distal colon functions principally for storage and is therefore called the *storage colon.*

ABSORPTION AND SECRETION OF ELECTROLYTES AND WATER. The mucosa of the large intestine, like that of the small intestine, has a high capability for active absorption of sodium, and the electrical potential created by the absorption of the sodium causes chloride absorption as well. The tight junctions between the epithelial cells of the large intestinal epithelium are much tighter than those of the small intestine. This prevents significant amounts of back-diffusion of ions through these junctions, thus allowing the large intestinal mucosa to absorb sodium ions far more completely—that is, against a much higher concentration gradient—than can occur in the small intestine. This is especially true when large quantities of aldosterone are available because aldosterone greatly enhances the sodium transport capability.

In addition, as in the distal portion of the small intestine, the mucosa of the large intestine secretes *bicarbonate ions* while it simultaneously absorbs an equal number of chloride ions in an exchange transport process that has already been described. The bicarbonate helps neutralize the acidic end products of bacterial action in the colon.

The absorption of sodium and chloride ions creates an osmotic gradient across the large intestinal mucosa, which in turn causes absorption of water.

MAXIMUM ABSORPTION CAPACITY OF THE LARGE INTESTINE. The large intestine can absorb a maximum of about 5 to 7 liters of fluid and electrolytes each day. When the total quantity entering the large intestine through the ileocecal valve or by way of large intestinal secretion exceeds this amount, the excess appears in the feces as diarrhea. As noted earlier in the chapter, the toxins from cholera or certain other bacterial infections often cause the crypts of Lieberkühn in the terminal ileum and the large intestine to secrete 10 or more liters of fluid each day, leading to severe and sometimes lethal diarrhea.

BACTERIAL ACTION IN THE COLON. Numerous bacteria, especially colon bacilli, are present even normally in the absorbing colon. They are capable of digesting small amounts of cellulose, in this way providing a few calories of nutrition to the body each day. In herbivorous animals, this source of energy is significant, although it is of negligible importance in human beings. Other substances formed as a result of bacterial activity are vitamin K, vitamin B_{12}, thiamin, riboflavin, and various gases that contribute to *flatus* in the colon, especially carbon dioxide, hydrogen gas, and methane. Vitamin K is especially important because the amount of this vitamin in the ingested foods is normally insufficient to maintain adequate blood coagulation.

COMPOSITION OF THE FECES. The feces normally are about three-fourths water and one-fourth solid matter that itself is composed of about 30 per cent dead bacteria, 10 to 20 per cent fat, 10 to 20 per cent inorganic matter, 2 to 3 per cent protein, and 30 per cent undigested roughage of the food and dried constituents of digestive juices, such as bile pigment and sloughed epithelial cells. The large amount of fat derives mainly from fat formed by bacteria and fat in the sloughed epithelial cells.

The brown color of feces is caused by *stercobilin* and *urobilin,* derivatives of bilirubin. The odor is caused principally by the products of bacterial action; these products vary from one person to another, depending on each person's colonic bacterial flora and on the type of food eaten. The actual odoriferous products include *indole, skatole, mercaptans,* and *hydrogen sulfide.*

REFERENCES

Anderson, P. O.: Vascular control in the colon and rectum. Scand. J. Gastroenterol., 93:65, 1984.

Bickel, H. (ed.): Digestion and Absorption of Nutrients. Ft. Lee, N. J., J. K. Burgess, 1983.

Buddington, R. K., and Diamond, J. M.: Ontogenetic development of intestinal nutrient transporters. Annu. Rev. Physiol., 51:601, 1989.

Case, R. M.: The role of Ca^{2+} stores in secretion. Cell-Calcium, 5:89, 1984.

Christensen, H. N.: The regulation of amino acid and sugar absorption by diet. Nutr. Rev., 42:237, 1984.

Donowitz, M., et al.: Cytosol free Ca^{++} in the regulation of active intestinal Na and Cl transport, KROC Found. Ser., 17:171, 1984.

Donowitz, M., and Welsh, M. J.: Ca^{2+} and cyclic AMP in regulation of intestinal Na, K, and Cl transport. Annu. Rev. Physiol., 48:135, 1986.

Ferraris, R. P., and Diamond, J. M.: Specific regulation of intestinal nutrient transporters by their dietary substrates. Annu. Rev. Physiol., 51:125, 1989.

Flemstrom, G., and Garner, A.: Some characteristics of duodenal epithelium. Ciba Found. Symp., 109:94, 1984.

Gardner, M. L.: Intestinal assimilation of intact peptides and proteins from the diet—a neglected field? Biol. Rev., 59:289, 1984.

Hopfer, U., and Liedtke, C. M.: Proton and bicarbonate transport mechanisms in the intestine. Annu. Rev. Physiol., 49:51, 1987.

Horl, W. H., and Heidland, A. (eds.): Proteases. New York, Plenum Publishing Corp., 1988.

Johnson, L. R.: Regulation of gastrointestinal mucosal growth. Physiol. Rev., 68:456, 1988.

Johnson, L., et al.: Physiology of the Gastrointestinal Tract, 2nd Ed. New York, Raven Press, 1987.

Kenny, A. J., and Maroux, S.: Topology of microvillar membrane hydrolases of kidney and intestine. Physiol. Rev., 62:91, 1982.

Liedtke, C. M.: Regulation of chloride transport in epithelia. Annu. Rev. Physiol., 51:143, 1989.

Mailman, D.: Relationships between intestinal absorption and hemodynamics. Annu. Rev. Physiol., 44:43, 1982.

Norum, K. R., et al.: Transport of cholesterol. Physiol. Rev., 63:1343, 1983.

Ockner, R. K., and Isselbacher, K. J.: Recent concepts of intestinal fat absorption. Rev. Physiol. Biochem. Pharmacol., 71:107, 1984.

Pappenheimer, J. R., et al.: Intestinal absorption and excretion of octapeptides composed of D amino acids. Proc. Natl. Acad., Sci. USA, 91:1942, 1994.

Schultz, S. G.: A cellular model for active sodium absorption by mammalian colon. Annu. Rev. Physiol., 46:435, 1984.

Schultz, S. G., and Hudson, R. L.: How do sodium-absorbing cells do their job and survive? News in Physiol. Sci., 1:185, 1986.

Setchell, K. D. R., et al. (eds.): The Bile Acids. New York, Plenum Publishing Corp., 1988.

Sjovall, H., et al.: Sympathetic control of intestinal fluid and electrolyte transport. News Physiol. Sci., 2:214, 1987.

Smith, M. W.: Expression of digestive and absorptive function in differentiating enterocytes. Annu. Rev. Physiol., 47:247, 1985.

Smith, P. L., and McCabe, R. D.: Mechanisms of regulation of transcellular potassium transport by the colon. Am. J. Physiol., 247:G445, 1984.

Stead, R. H., et al.: Neuro-immuno-physiology of Gastrointestinal Mucosa: Implications for Inflammatory Diseases. New York Academy of Sciences, 1992.

Stevens, B. R., et al.: Intestinal transport of amino acids and sugars: Advances using membrane vesicles. Annu. Rev. Physiol., 46:417, 1984.

Thompson, J. C., et al. (eds.): Gastrointestinal Endocrinology. New York, McGraw-Hill Book Co., 1987.

Ugolev, A. M.: Membrane transport and hydrolytic enzymes under physiological vs. acute experimental conditions. News Physiol. Sci., 2:186, 1987.

Watson, D. W., and Sodeman, W. A., Jr.: The small intestine. In Sodeman, W. A., Jr., and Sodeman, T. M. (eds.): Pathologic Physiology: Mechanisms of Disease, 6th Ed. Philadelphia, W. B. Saunders Co., 1979, p. 824.

Wheeler, T. J., and Hinkle, P. C.: The glucose transporter of mammalian cells. Annu. Rev. Physiol., 1985.

Physiology of Gastrointestinal Disorders

CHAPTER 66

The logical therapy of most gastrointestinal disorders depends on a basic knowledge of gastrointestinal physiology. The purpose of this chapter, therefore, is to discuss a few representative types of malfunction that have special physiological bases or consequences.

DISORDERS OF SWALLOWING AND OF THE ESOPHAGUS

PARALYSIS OF THE SWALLOWING MECHANISM. Damage to the 5th, 9th, or 10th nerve can cause paralysis of significant portions of the swallowing mechanism. Also, a few diseases, such as poliomyelitis and encephalitis, can prevent normal swallowing by damaging the swallowing center in the brain stem. Finally, paralysis of the swallowing muscles, as occurs in *muscle dystrophy* or in failure of neuromuscular transmission in *myasthenia gravis* or *botulism,* can also prevent normal swallowing.

When the swallowing mechanism is partially or totally paralyzed, the abnormalities that can occur include (1) complete abrogation of the swallowing act so that swallowing cannot occur, (2) failure of the glottis to close so that food passes into the lungs instead of the esophagus, and (3) failure of the soft palate and uvula to close the posterior nares so that food refluxes into the nose during swallowing.

One of the most serious instances of paralysis of the swallowing mechanism occurs when patients are under deep anesthesia. Often, while on the operating table, they vomit large quantities of materials from the stomach into the pharynx; then, instead of swallowing the materials again, they simply suck them into the trachea because the anesthetic has blocked the reflex mechanism of swallowing. As a result, such patients occasionally choke to death on their own vomitus.

ACHALASIA AND MEGAESOPHAGUS. Achalasia is a condition in which the lower esophageal sphincter fails to relax during the swallowing mechanism. As a result, food transmission from the esophagus into the stomach is impeded or prevented. Pathological studies have shown damage in the neural network of the myenteric plexus in the lower two thirds of the esophagus. The musculature of the lower esophagus remains spastically contracted, and the myenteric plexus has lost the ability to transmit a signal to cause "receptive relaxation" of the gastroesophageal sphincter as food approaches this area during the swallowing process.

When achalasia becomes severe, the esophagus may not empty the swallowed food into the stomach for many hours, instead of the few seconds that is the normal time. Over months and years, the esophagus becomes tremendously enlarged until it often can hold as much as 1 liter of food, which becomes putridly infected during the long periods of esophageal stasis. The infection may also cause ulceration of the esophageal mucosa, sometimes leading to severe substernal pain or even rupture and death. Considerable benefit can be achieved by stretching the lower end of the esophagus by means of a balloon inflated on the end of a swallowed esophageal tube. Antispasmotic drugs (drugs that relax smooth muscle), such as isosorbide dinitrate and nifedipine, can also be helpful.

DISORDERS OF THE STOMACH

GASTRITIS. Gastritis means inflammation of the gastric mucosa. Mild to moderate chronic gastritis is exceedingly common in the population as a whole, especially in the later years of adult life.

The inflammation of gastritis can be only superficial and therefore not very harmful, or it can penetrate deeply into

845

the gastric mucosa and in many long-standing cases cause almost complete atrophy of the gastric mucosa. In a few cases, gastritis can be acute and severe, with ulcerative excoriation of the stomach mucosa by the stomach's own peptic secretions.

Research suggests that much gastritis is caused by chronic bacterial infection of the gastric mucosa. This often can be treated successfully by an intensive regimen of antibacterial therapy.

In addition, certain ingested irritant substances can be damaging to the protective gastric mucosal barrier—that is, to the mucous glands and the tight epithelial junctions between the gastric lining cells—often leading to severe acute or chronic gastritis. Two of the most common of these substances are *alcohol* and *aspirin.*

The Gastric Barrier and Its Penetration in Gastritis. Absorption from the stomach is normally slight. This low level of absorption is mainly caused by two specific features of the gastric mucosa: (1) it is lined with highly resistant mucous cells that secrete a viscid and adherent mucus and (2) it has tight junctions between the adjacent epithelial cells. These two together plus other impediments to gastric absorption are called the "gastric barrier." This barrier normally is so resistant to diffusion that even the highly concentrated hydrogen ions of the gastric juice, averaging about 100,000 times the concentration of the hydrogen ions in the plasma, seldom diffuse even to the slightest extent through the lining mucus as far as the epithelial membrane itself. In gastritis, this barrier permeability is greatly increased. The hydrogen ions do then diffuse into the stomach epithelium, creating additional havoc and leading to a vicious cycle of progressive stomach mucosal damage and atrophy. It also makes the mucosa susceptible to peptic digestion, thus frequently resulting in gastric ulcer.

GASTRIC ATROPHY. In many people who have chronic gastritis, the mucosa gradually becomes atrophic until little or no gastric gland activity remains. It is also believed that some people develop autoimmunity against the gastric mucosa, this also leading eventually to gastric atrophy. Loss of the stomach secretions in gastric atrophy leads to achlorhydria and, occasionally, *pernicious anemia.*

Achlorhydria (and Hypochlorhydria). Achlorhydria means simply that the stomach fails to secrete hydrochloric acid, and it is diagnosed when the pH of the gastric secretions fails to decrease below 6.5 after maximal stimulation. Hypochlorhydria means diminished acid secretion. When acid is not secreted, pepsin also usually is not secreted; even when it is, the lack of acid prevents it from functioning because pepsin requires an acid medium for activity.

Although achlorhydria is associated with depressed or even no digestive capability by the stomach, the overall digestion of food in the entire gastrointestinal tract is still almost normal. This is because trypsin and other enzymes secreted by the pancreas are capable of digesting virtually all the protein in the diet.

Pernicious Anemia in Gastric Atrophy. Pernicious anemia is a common accompaniment of achlorhydria and gastric atrophy. The normal gastric secretions contain a glycoprotein called *intrinsic factor,* which is secreted by the same parietal cells that secrete the hydrochloric acid; intrinsic factor must be present for adequate absorption of vitamin B_{12} from the ileum. The intrinsic factor combines with vitamin B_{12} in the stomach and thereafter protects it from being digested and destroyed as it passes through the gastrointestinal tract. Then, when the intrinsic factor–vitamin B_{12} complex reaches the terminal ileum, the intrinsic factor binds with receptors on the ileal epithelial surface. This in turn makes it possible

for the vitamin B_{12} to be absorbed. In the absence of intrinsic factor, only about $\frac{1}{50}$ of the vitamin B_{12} is absorbed. Therefore, an adequate amount of vitamin B_{12} is not made available from the foods. As a result, maturation failure occurs in the bone marrow, resulting in pernicious anemia. This subject is discussed in more detail in Chapter 32.

Pernicious anemia also occurs frequently when most of the stomach has been removed for treatment of either stomach ulcer or gastric cancer or when the terminal ileum, where vitamin B_{12} is almost entirely absorbed, is removed.

Peptic Ulcer

A peptic ulcer is an excoriated area of the mucosa caused by the digestive action of gastric juice. Figure 66–1 shows the points in the gastrointestinal tract at which peptic ulcers most frequently occur, demonstrating that the most frequent site is in the first few centimeters of the duodenum. In addition, peptic ulcers frequently occur along the lesser curvature of the antral end of the stomach or, more rarely, in the lower end of the esophagus where stomach juices frequently reflux. A peptic ulcer called a *marginal ulcer* also frequently occurs wherever a surgical opening, such as a gastrojejunostomy, is made between the stomach and some portion of the small intestine.

BASIC CAUSE OF PEPTIC ULCERATION. The usual cause of peptic ulceration is an *imbalance* between the rate of secretion of gastric juice and the degree of protection afforded by the gastroduodenal mucosal barrier as well as the neutralization of the gastric acid by duodenal juices. It will be recalled that all areas normally exposed to gastric juice are well supplied with mucous glands, beginning with the compound mucous glands of the lower esophagus and including the mucous cell coating of the stomach mucosa, the mucous neck cells of the gastric glands, the deep pyloric glands that secrete mainly mucus, and, finally, the glands of Brunner of the upper duodenum, which secrete a highly alkaline mucus.

In addition to the mucus protection of the mucosa, the duodenum is protected by the alkalinity of the small intestinal secretions. Especially important is pancreatic secretion, which contains large quantities of sodium bicarbonate that neutralize the hydrochloric acid of the gastric juice, thus inactivating the pepsin to prevent digestion of the mucosa. In addition, large amounts of bicarbonate ions are provided in the secretions of the large Brunner's glands in the first few inches of the duodenal wall and in the bile coming from the liver.

Finally, two feedback control mechanisms ensure that this neutralization of gastric juices is complete.

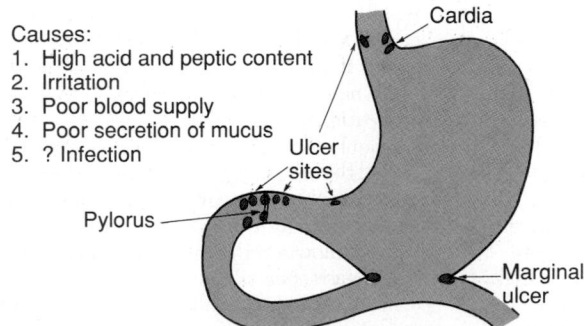

Causes:
1. High acid and peptic content
2. Irritation
3. Poor blood supply
4. Poor secretion of mucus
5. ? Infection

Figure 66–1. Peptic ulcer.

1. When excess acid enters the duodenum, it reflexly inhibits gastric secretion and peristalsis in the stomach, both nervously and hormonally, thereby decreasing the rate of gastric emptying.

2. The presence of acid in the small intestine liberates secretin from the intestinal mucosa, which then passes by way of the blood to the pancreas to promote rapid secretion of pancreatic juice that contains a high concentration of sodium bicarbonate, thus making more sodium bicarbonate available for neutralization of the acid.

Therefore, a peptic ulcer can be caused in either of two ways: (1) *excess secretion of acid and pepsin* by the gastric mucosa or (2) *diminished ability of the gastroduodenal mucosal barrier to protect against the digestive properties of the acid-pepsin complex.*

Specific Causes of Peptic Ulcer in the Human Being

BACTERIAL INFECTION BY *HELICOBACTER PYLORI* BREAKS DOWN THE GASTRODUODENAL MUCOSAL BARRIER. Within the past 5 years, at least 75 per cent of peptic ulcer patients have been found to have chronic infection of the terminal portions of the gastric mucosa and initial portions of the duodenal mucosa by the bacterium *Helicobacter pylori*. Once this infection begins, it can last for a lifetime unless it is eradicated by antibacterial therapy. Furthermore, the bacterium is capable of penetrating the mucosal barrier both by virtue of its physical capability to burrow through the barrier and by releasing digestive enzymes that liquefy the barrier. As a result, the strong acidic digestive juices of the stomach secretions can penetrate into the underlying epithelium and literally digest the epithelial cells—in fact, in severe cases, even the sublying tissues. This leads to peptic ulceration.

INCREASED SECRETION OF ACID–PEPTIC JUICES CAN CONTRIBUTE TO ULCERATION. In most people who have peptic ulcer in the initial portion of the duodenum, the rate of gastric acid secretion is greater than normal, sometimes as much as twice normal. Although part of this increased secretion may be stimulated by the bacterial infection, experiments in animals plus evidence of excess neural stimulation of gastric acid secretion in human beings with peptic ulcer suggest that excess secretion of gastric juices for any reason (for instance, in psychic disturbances) is often a primary cause of peptic ulceration.

In addition to bacterial infection and excess secretion of gastric juices, other factors that predispose to ulcers include (1) smoking, presumably because of increased nervous stimulation of the stomach secretory glands; (2) alcohol, because it tends to break down the mucosal barrier; and (3) aspirin, which also has a strong propensity for breaking down this barrier.

PHYSIOLOGY OF TREATMENT. Since discovery of the infectious basis for much if not most of peptic ulceration, therapy has changed immensely. Initial reports are that almost all patients with peptic ulceration can be treated effectively by two measures: (1) the use of *antibiotics* such as tetracycline along with other agents to kill the infectious bacteria and (2) the administration of an acid-suppressant drug, especially *ranitidine,* an antihistaminic that blocks the stimulatory effect of histamine on gastric gland histamine$_2$ receptors, thus reducing gastric acid secretion by 70 to 80 per cent.

In the past, before these approaches to peptic ulcer therapy were developed, it was often necessary to remove as much as four fifths of the stomach, thus reducing stomach juices enough to cure most patients. Another therapy was to cut the two vagus nerves that supply parasympathetic stimulation to the gastric glands. This temporarily blocks almost all secretion of acid and pepsin and often cures the ulcer or ulcers within 1 week after the operation is performed. However, much of the basal stomach secretion returns after a few months, and in many patients the ulcer also returns.

It is clear that the newer physiological approaches to therapy have proved to be miraculous. Even so, in a few instances, the patient's condition is so severe—including massive bleeding from the ulcer—that heroic operative procedures must still be used.

DISORDERS OF THE SMALL INTESTINE

Abnormal Digestion of Food in the Small Intestine—Pancreatic Failure

A serious cause of abnormal digestion is failure of the pancreas to secrete its juice into the small intestine. Lack of pancreatic secretion frequently occurs (1) in *pancreatitis* (which is discussed later), (2) when the *pancreatic duct is blocked* by a gallstone at the papilla of Vater, or (3) after the *head of the pancreas has been removed* because of malignancy. Loss of pancreatic juice means loss of trypsin, chymotrypsin, carboxypolypeptidase, pancreatic amylase, pancreatic lipase, and still a few other digestive enzymes. Without these enzymes, as much as 60 per cent of the fat entering the small intestine may go unabsorbed as well as one third to one half of the proteins and carbohydrates. As a result, large portions of the ingested food are not used for nutrition, and copious, fatty feces are excreted.

PANCREATITIS. Pancreatitis means inflammation of the pancreas, and this can occur in the form of either *acute pancreatitis* or *chronic pancreatitis*. The most common cause of pancreatitis is alcohol and the second most common cause is blockage of the papilla of Vater by a gallstone; the two together account for more than 90 per cent of all cases. When a gallstone blocks the papilla of Vater, this blocks the main secretory duct from the pancreas as well as the common bile duct. The pancreatic enzymes are then dammed up in the ducts and acini of the pancreas. Eventually, so much trypsinogen accumulates that it *overcomes the trypsin inhibitor* in the secretions, and a small quantity of trypsinogen becomes activated to form trypsin. Once this happens, the trypsin activates still more trypsinogen as well as chymotrypsinogen and carboxypolypeptidase, resulting in a vicious cycle until most of the proteolytic enzymes in the pancreatic ducts and acini become activated. These enzymes rapidly digest large portions of the pancreas itself, sometimes completely and permanently destroying the ability of the pancreas to secrete digestive enzymes. The process may become so severe that it leads to death within days or even hours.

Malabsorption by the Small Intestinal Mucosa— Sprue

Occasionally, nutrients are not adequately absorbed from the small intestine even though the food is well digested. Several diseases can cause decreased absorption by the mucosa; they are often classified together under the general heading of *"sprue."* Malabsorption also can occur when large portions of the small intestine have been removed.

Nontropical Sprue. One type of sprue, called variously *idiopathic sprue, celiac disease* (in children), and *gluten enteropathy,* results from the toxic effects of *gluten* present in certain types of grains, especially wheat and rye. Only some people are susceptible to this effect, but in these people, gluten has a direct destructive effect on the intestinal enterocytes or perhaps as the result of an immunological or allergic reaction. In milder forms of the disease, the microvilli of the absorbing enterocytes on the villi are destroyed, thus decreasing the absorptive surface area as much as twofold. In the more severe forms, the villi themselves become blunted or disappear altogether, thus still further reducing the absorptive area of the gut. Removal of wheat and rye flour from the diet frequently results in a miraculous cure within weeks, especially in children with this disease.

Tropical Sprue. A different type of sprue called *tropical sprue* frequently occurs in the tropics and can often be treated with antibacterial agents. Even though no specific bacterium has been implicated as the cause, it is believed that this variety of sprue is usually caused by inflammation of the intestinal mucosa resulting from unidentified infectious agents.

Malabsorption in Sprue. In the early stages of sprue, the absorption of fats is more impaired than the absorption of other digestive products. The fat that appears in the stools is almost entirely in the form of salts of fatty acids rather than undigested neutral fat, demonstrating that the problem is one of absorption and not of digestion. In this stage of sprue, the condition is frequently called *idiopathic steatorrhea,* which means simply excess fats in the stools as a result of unknown causes.

In more severe cases of sprue, the absorption of proteins, carbohydrates, calcium, vitamin K, folic acid, and vitamin B_{12} as well as many other important substances becomes greatly impaired. As a result, the person suffers (1) severe nutritional deficiency, often developing severe wasting of the body; (2) osteomalacia (demineralization of the bones because of calcium lack); (3) inadequate blood coagulation caused by lack of vitamin K; and (4) macrocytic anemia of the pernicious anemia type, owing to diminished vitamin B_{12} and folic acid absorption.

DISORDERS OF THE LARGE INTESTINE

Constipation

Constipation means slow movement of feces through the large intestine; it is often associated with large quantities of dry, hard feces in the descending colon that accumulate because of the long time available for absorption of fluid.

Any pathology of the intestines that obstructs movement of intestinal contents, such as tumors, adhesions that constrict the intestines, and ulcers, can cause constipation. A frequent functional cause of constipation is irregular bowel habits that have developed through a lifetime of inhibition of the normal defecation reflexes. Infants are seldom constipated, but part of their training in the early years of life requires that they learn to control defecation, and this control is effected by inhibiting the natural defecation reflexes. Clinical experience shows that if one fails to allow defecation to occur when the defecation reflexes are excited or if one overuses laxatives to take the place of natural bowel function, the reflexes themselves become progressively less strong over a period of time and the colon becomes *atonic.* For this reason, if a person establishes regular bowel habits early in life, usually defecating in the morning after breakfast when the gastrocolic and duodenocolic reflexes cause mass movements in the large intestine, the development of constipation in later life can generally be prevented.

Constipation can also result from spasm of a small segment of the sigmoid colon. It should be recalled that motility, even normally, is weak in the large intestine, so that even a slight degree of spasm is often capable of causing serious constipation. After the constipation has continued for several days and excessive feces have accumulated above the spastic sigmoid colon, excessive colonic secretions often lead to a day or so of diarrhea. After this, the cycle begins again, with repeated bouts of alternating constipation and diarrhea.

Megacolon

Occasionally, constipation is so severe that bowel movements occur only once every week or so. This allows tremendous quantities of fecal matter to accumulate in the colon, causing the colon sometimes to distend to a diameter of 3 to 4 inches. The condition is called megacolon, or *Hirschsprung's disease.*

The most frequent cause of megacolon is a lack or deficiency of ganglion cells in the myenteric plexus in a segment of the sigmoid colon. As a consequence, neither defecation reflexes nor strong peristaltic motility can occur through this area of the large intestine. The sigmoid itself becomes small and almost spastic while feces accumulate proximal to this area, causing megacolon. Thus, the diseased portion of the large intestine where motility is absent appears normal by radiographic examination, whereas the originally normal portion of the large intestine appears greatly enlarged.

Diarrhea

Diarrhea results from rapid movement of fecal matter through the large intestine. Several causes of diarrhea with important physiological overtones are the following.

Enteritis. Enteritis means infection caused either by a virus or by bacteria in the intestinal tract. In the usual infectious diarrhea, the infection is most extensive in the entire large intestine and the distal end of the ileum. Everywhere the infection is present, the mucosa becomes extensively irritated, and its rate of secretion becomes greatly enhanced. In addition, the motility of the intestinal wall usually increases manyfold. As a result, large quantities of fluid are made available for washing the infectious agent toward the anus, and at the same time strong propulsive movements propel this fluid forward. This is an important mechanism for ridding the intestinal tract of the debilitating infection.

Of special interest is the diarrhea caused by *cholera* (and sometimes by other bacteria such as some pathogenic colon bacilli). As explained in Chapter 66, cholera toxin directly stimulates excessive secretion of electrolytes and fluid from the crypts of Lieberkühn in the distal ileum and colon. The amount can be 10 to 12 liters per day, and the colon can usually reabsorb a maximum of only 6 liters per day. Therefore, the loss of fluid and electrolytes can be so debilitating within a day or so that death ensues. Therefore, the most important physiological basis of therapy is to replace the fluid and electrolytes as rapidly as they are lost, mainly by giving the patient intravenous saline and glucose solutions. With proper therapy of this type along with the use of antibiotics, almost no cholera patients die, but without therapy, as many as 50 per cent do.

PSYCHOGENIC DIARRHEA. Everyone is familiar with the diarrhea that accompanies periods of nervous tension, such as during examination time or when a soldier is about to go into battle. This type of diarrhea, called *psychogenic* emotional diarrhea, is caused by excessive stimulation of the parasympathetic nervous system, which greatly excites both motility and secretion of mucus in the distal colon. These two effects added together can cause marked diarrhea.

ULCERATIVE COLITIS. Ulcerative colitis is a disease in which extensive areas of the walls of the large intestine become inflamed and ulcerated. The motility of the ulcerated colon is often so great that mass movements occur most of the time, rather than the usual 10 to 30 minutes per day. Also, the colon's secretions are greatly enhanced. As a result, the patient has repeated diarrheal bowel movements.

The cause of ulcerative colitis is unknown. Some clinicians believe that it results from an allergic or immune destructive effect, but it also could result from a chronic bacterial infection not yet understood. Whatever the cause, there is a strong hereditary tendency for susceptibility to ulcerative colitis. Once the condition has progressed very far, the ulcers are perpetuated by superimposed bacterial infection and seldom will heal until an ileostomy is performed to allow the intestinal contents to drain to the exterior rather than to flow through the colon. Even then the ulcers sometimes fail to heal, and the only solution is removal of the entire colon.

Paralysis of Defecation in Spinal Cord Injuries

From Chapter 63 it will be recalled that defecation is normally initiated by the movement of feces into the rectum, which causes a cord-mediated defecation reflex passing from the rectum to the spinal cord and then back to the descending colon, sigmoid, rectum, and anus. This reflex adds greatly to the activity of the intrinsic defecation reflex mediated through the myenteric plexus of the sigmoid and colon wall itself.

Frequently, the *cord defecation reflex* is blocked or altered by spinal cord injuries. For instance, compressive destruction of the *conus medullaris* of the spinal cord destroys the sacral segments of the cord where the cord reflex is integrated and therefore almost paralyzes defecation. In such instances, defecation requires extensive supportive measures, such as cathartics and large enemas.

But more frequently, the spinal cord is injured somewhere between the conus medullaris and the brain, in which case the voluntary portion of the defecation act is blocked while the basic cord reflex for defecation is still intact. Nevertheless, loss of the voluntary aid to defecation—that is, loss of the increased abdominal pressure and the relaxation of the voluntary anal sphincter—often makes defecation a difficult process in the person with this type of cord injury. Because the cord defecation reflex can still occur, a small enema to excite the action of this reflex, usually given in the morning shortly after a meal, can often cause adequate defecation. In this way, people with spinal cord injuries that do not destroy the conus medullaris can usually control their bowel movements each day.

GENERAL DISORDERS OF THE GASTROINTESTINAL TRACT

Vomiting

Vomiting is the means by which the upper gastrointestinal tract rids itself of its contents when almost any part of the upper tract becomes excessively irritated, overdistended, or even overexcitable. Excessive distention or irritation of the duodenum provides an especially strong stimulus for vomiting. Impulses are transmitted, as shown in Figure 66–2, by both vagal and sympathetic afferents to the bilateral *vomiting center* of the medulla, which lies near the tractus solitarius at the level of the dorsal motor nucleus of the vagus. Appropriate automatic motor reactions are then instituted to cause the vomiting act. The motor impulses that cause the actual vomiting are transmitted from the vomiting center through the 5th, 7th, 9th, 10th, and 12th cranial nerves to the upper gastrointestinal tract and through the spinal nerves to the diaphragm and abdominal muscles.

ANTIPERISTALSIS, THE PRELUDE TO VOMITING. In the early stages of excessive gastrointestinal irritation or overdistention, *antiperistalsis* begins to occur, often many minutes before vomiting appears. The antiperistalsis may begin as far down in the intestinal tract as the ileum, and the antiperistaltic wave travels backward up the intestine at a rate of 2 to 3 cm/sec; this process can actually push a large share of the intestinal contents all the way back to the duodenum and stomach within 3 to 5 minutes. Then, as these upper por-

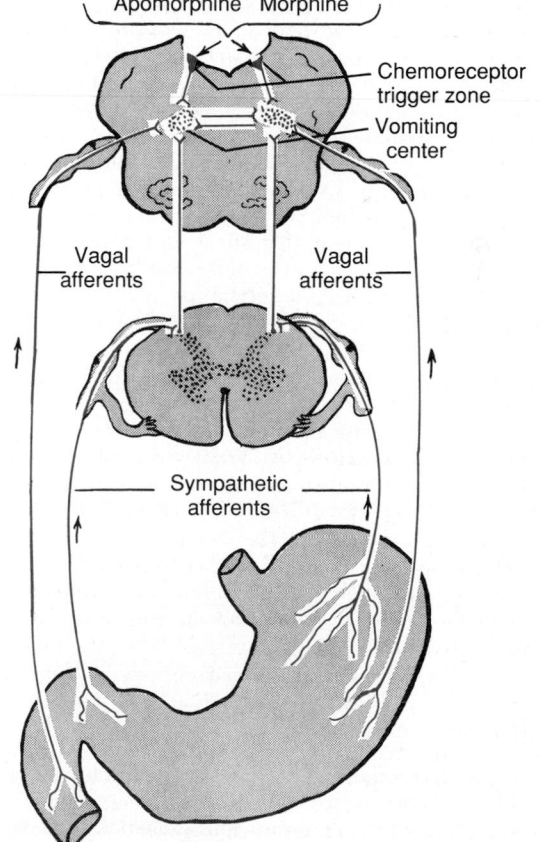

Figure 66–2. Different connections of the vomiting center.

tions of the gastrointestinal tract, especially the duodenum, become overly distended, this distention becomes the exciting factor that initiates the actual vomiting act. At the onset of vomiting, strong intrinsic contractions occur in both the duodenum and the stomach, along with partial relaxation of the lower esophageal sphincter, thus allowing the vomitus to begin moving into the esophagus. From here, a specific vomiting act involving the abdominal muscles takes over and expels the vomitus to the exterior, as explained in the next paragraph.

THE VOMITING ACT. Once the vomiting center has been sufficiently stimulated and the vomiting act instituted, the first effects are (1) a deep breath, (2) raising of the hyoid bone and the larynx to pull the upper esophageal sphincter open, (3) closing of the glottis, and (4) lifting of the soft palate to close the posterior nares. Next comes a strong downward contraction of the diaphragm along with simultaneous contraction of all the abdominal wall muscles. This squeezes the stomach between the diaphragm and the abdominal muscles, building the intragastric pressure to a high level. Finally, the lower esophageal sphincter relaxes completely, allowing expulsion of the gastric contents upward through the esophagus.

Thus, the vomiting act results from a squeezing action of the muscles of the abdomen associated with sudden opening of the esophageal sphincters so that the gastric contents can be expelled.

CHEMORECEPTOR TRIGGER ZONE OF THE MEDULLA FOR INITIATION OF VOMITING BY DRUGS OR BY MOTION SICKNESS. Aside from the vomiting initiated by irritative stimuli in the gastrointestinal tract itself, vomiting can be caused by nervous signals arising in areas of the brain outside the vomiting center. This is particularly true of a small area located bilaterally on the floor of the fourth ventricle near the area postrema and called the *chemoreceptor trigger zone.* Electrical stimulation of this area also initiates vomiting; more important, administration of certain drugs, including apomorphine, morphine, and some of the digitalis derivatives, can directly stimulate the chemoreceptor trigger zone and initiate vomiting. Destruction of this area blocks this type of vomiting but does not block vomiting resulting from irritative stimuli in the gastrointestinal tract itself.

Also, it is well known that rapidly changing directions of motion of the body cause certain people to vomit. The mechanism for this is the following: The motion stimulates the receptors of the labyrinth, and impulses are transmitted mainly by way of the vestibular nuclei into the cerebellum, then to the chemoreceptor trigger zone, and finally to the vomiting center to cause vomiting.

CEREBRAL EXCITATION OF VOMITING. Various psychic stimuli, including disquieting scenes, noisome odors, and other similar psychological factors, can cause vomiting. Stimulation of certain areas of the hypothalamus also causes vomiting. The precise neuronal connections for these effects are not known, although it is probable that the impulses pass directly to the vomiting center and do not involve the chemoreceptor trigger zone.

Nausea

Everyone has experienced the sensation of nausea and knows that it is often a prodrome of vomiting. Nausea is the conscious recognition of subconscious excitation in an area of the medulla closely associated with or part of the vomiting center, and it can be caused by irritative impulses coming from the gastrointestinal tract, impulses that originate in the

lower brain associated with motion sickness, or impulses from the cerebral cortex to initiate vomiting. Vomiting occasionally occurs without the prodromal sensation of nausea, indicating that only certain portions of the vomiting centers are associated with the sensation of nausea.

Gastrointestinal Obstruction

The gastrointestinal tract can become obstructed at almost any point along its course, as shown in Figure 66–3; some common causes of obstruction are *cancer, fibrotic constriction resulting from ulceration or from peritoneal adhesions, spasm of a segment of the gut,* or *paralysis of a segment of the gut.*

The abnormal consequences of obstruction depend on the point in the gastrointestinal tract that becomes obstructed. If the obstruction occurs at the pylorus, which results often from fibrotic constriction after peptic ulceration, persistent vomiting of stomach contents occurs. This depresses bodily nutrition; it also causes excessive loss of hydrogen ions from the body and can result in various degrees of alkalosis.

If the obstruction is beyond the stomach, antiperistaltic reflux from the small intestine causes intestinal juices to flow backward into the stomach, and these juices are vomited along with the stomach secretions. In this instance, the person loses large amounts of water and electrolytes, so that he becomes severely dehydrated, but the loss of acids and bases may be approximately equal, so that little change in acid-base balance occurs. If the obstruction is near the lower end of the small intestine, then it is possible to vomit more basic than acidic substances; in this case, acidosis may result. In addition, the vomitus, after a few days of obstruction, becomes fecal in character.

Also important in obstruction of the small intestine is marked distention of the intestine proximal to the obstructed point. Large quantities of fluid and electrolytes continue to be secreted into the lumen of the small intestine; even large amounts of proteins are lost from the blood stream, partly into the intestinal lumen and partly into the gut wall, which becomes edematous as a result of the excessive distention. The plasma volume diminishes because of the protein loss, and severe circulatory shock often ensues. One might immediately ask: Why does the small intestine not reabsorb these fluids and electrolytes? The answer is that distention of the gut usually stimulates the secretory activity of the gut but does not correspondingly increase the rate of absorption.

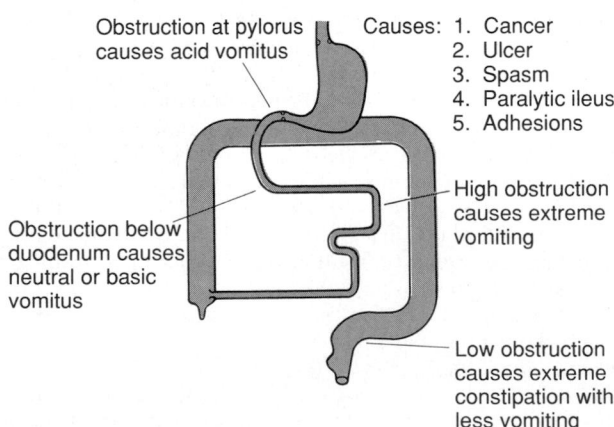

Figure 66–3. Obstruction in different parts of the gastrointestinal tract.

Normally this would act to flush the chyme further down the small intestine and therefore relieve the distention. But if obstruction is present, this normal mechanism cannot occur; instead, a vicious cycle of more and more distention occurs.

If the obstruction is near the distal end of the large intestine, feces can accumulate in the colon for several weeks. The patient develops an intense feeling of constipation, but in the first stages of the obstruction vomiting is not severe. After the large intestine has become completely filled and it finally becomes impossible for additional chyme to move from the small intestine into the large intestine, vomiting does then begin severely. Prolonged obstruction of the large intestine will finally cause rupture of the intestine itself or dehydration and circulatory shock resulting from the severe vomiting.

Gases in the Gastrointestinal Tract; "Flatus"

Gases can enter the gastrointestinal tract from three sources: (1) swallowed air, (2) gases formed as a result of bacterial action, and (3) gases that diffuse from the blood into the gastrointestinal tract.

Most gases in the stomach are nitrogen and oxygen derived from swallowed air, and in the normal person most of these gases are expelled by belching.

Only small amounts of gas are usually present in the small intestine, and much of this gas is air that passes from the stomach into the intestinal tract. In addition, considerable carbon dioxide often occurs because the reaction between acidic gastric juice and bicarbonate in pancreatic juice sometimes is too rapid for the liberated carbon dioxide to be absorbed.

In the large intestine, a greater proportion of the gases is derived from bacterial action, including especially *carbon dioxide, methane,* and *hydrogen.* They occur along with varying amounts of oxygen and nitrogen from swallowed air. When the methane and hydrogen become suitably mixed with oxygen from swallowed air, an actual explosive mixture is occasionally formed; the use of the electric cautery during sigmoidoscopy has, on rare occasions, caused colon explosions.

Certain foods are known to cause greater expulsion of flatus through the anus than others—beans, cabbage, onion, cauliflower, corn, and certain irritant foods such as vinegar. Some of these foods serve as a suitable medium for gas-forming bacteria, especially because of unabsorbed fermentable types of carbohydrates (for instance, beans contain an undigestible sugar that passes into the colon and becomes a superior food for the colonic bacteria), but in other instances, excess expulsion of gas results from irritation of the large intestine, which promotes rapid peristaltic expulsion of the gases before they can be absorbed.

The amount of gases entering or forming in the large intestine each day averages 7 to 10 liters, whereas the average amount expelled through the anus is usually only about 0.6 liter. The remainder is absorbed through the intestinal mucosa.

REFERENCES

Allen, A., et al.: Gastroduodenal mucosal protection. Physiol. Rev., 73:823, 1993.

Anuras, S.: Motility Disorders of the Gastrointestinal Tract: Principles and Practice. New York, Raven Press, 1992.

Bongiovanni, G. L. (ed.): Essentials of Clinical Gastroenterology, 2nd Ed. New York, McGraw-Hill Book Co., 1988.

Bouchier, I. A. D., et al.: Gastroenterology: Clinical Science & Practice. Philadelphia, W. B. Saunders Co., 1994.

Bradley, E. L., III: Acute Pancreatitis: Principles and Practice. New York, Raven Press, 1993.

Castell, D. O.: Calcium-channel blocking agents for gastrointestinal disorders. Am J. Cardiol., 55:210B, 1985.

Cervero, F.: Sensory innervation of the viscera: peripheral basis of visceral pain. Physiol. Rev., 74:95, 1994.

Cotran, R. S., et al.: Robbins Pathologic Basis of Disease. Philadelphia, W. B. Saunders Co., 1994.

Danzi, J. T., and Landman, S.: Case Atlas of Gastroenterology. Baltimore, Williams & Wilkins, 1994.

Gitnick, G. (ed.): Handbook of Gastrointestinal Emergencies, 2nd Ed. New York Elsevier Science Publishing Co., 1988.

Gitnick, G., et al. Principles and Practice of Gastroenterology and Hepatology, 2nd Ed. Redding, MA, Appleton & Lange, 1994.

Grendell, J. J., et al.: Current Gastroenterology Diagnosis and Treatment. Redding, MA, Appleton & Lange, 1995.

Haubrich, W. S., et al.: Bockus Gastroenterology. Philadelphia, W. B. Saunders Co., 1994.

Jacobson, E. D., and Levine, J. S.: Clinical GI Physiology for the Exam Taker. Philadelphia, W. B. Saunders Co., 1994.

Keighley, M. R. B., and Williams, N. S.: Surgery of the Anus, Rectum and Colon. Philadelphia, W. B. Saunders Co., 1994.

Kirsner, J. B.: Inflammatory Bowel Disease. Baltimore, Williams & Wilkins, 1995.

Lebenthal, E., and Duffey, M. E.: Textbook of Secretory Diarrhea. New York, Raven Press, 1990.

Lewis, J. H.: A Pharmacological Approach to Gastrointestinal Diseases. Baltimore, Williams & Wilkins, 1994.

Misiewicz, J. J., et al.: Atlas of Clinical Gastroenterology, Philadelphia, Lea & Febiger, 1987.

Nelson, R. L., and Nyhus, L. M.: Surgery of the Small Intestine. East Norwalk, CT, Appleton & Lange, 1987.

Quigley, E. M., and Sorrell, M. F.: The Gastrointestinal Surgical Patient. Baltimore, Williams & Wilkins, 1994.

Rigas, B.: Clinical Gastroenterology. Hightstown, NJ, McGraw-Hill, 1994.

Rodolfo, C., et al. (eds.): Gastric Protection. New York, Raven Press, 1988.

Shaffer, E., and Thomson, A. B. R.: Modern Concepts in Gastroenterology. New York, Plenum Publishing Corp., 1989.

Sleisenger, M. H., and Fordtran, J. S.: Gastrointestinal Disease: Pathophysiology, Diagnosis, Management. Philadelphia, W. B. Saunders Co., 1993.

Snape, W. J., Jr. (ed.): Pathogenesis of Functional Bowel Disease. New York, Plenum Publishing Corp., 1989.

Spiro, H. M.: Clinical gastroenterology. Hightstown, NJ, McGraw-Hill, 1993.

Targan, S. R., and Shanahan, F.: Inflammatory Bowel Disease: From Bench to Bedside. Baltimore, Williams & Wilkins, 1994.

Yamada, T., et al.: Atlas of Gastroenterology. Philadelphia, J. B. Lippincott, 1992.

Yamada, T., et al.: Textbook of Gastroenterology. Philadelphia, J. B. Lippincott, 1991.

METABOLISM AND TEMPERATURE REGULATION

UNIT XIII

Metabolism of Carbohydrates and Formation of Adenosine Triphosphate

CHAPTER 67

The next few chapters deal with metabolism in the body, which means the chemical processes that make it possible for the cells to continue living. It is not the purpose of this textbook to present the chemical details of all the various cellular reactions because this lies in the discipline of biochemistry. Instead, these chapters are devoted to (1) a review of the principal chemical processes of the cell and (2) an analysis of their physiological implications, especially in relation to the manner in which they fit into the overall concept of homeostasis.

Release of Energy from Foods and the Concept of "Free Energy"

A great proportion of the chemical reactions in the cells is concerned with making the energy in foods available to the various physiological systems of the cell. For instance, energy is required for muscle activity, secretion by the glands, maintenance of membrane potentials by the nerve and muscle fibers, synthesis of substances in the cells, absorption of foods from the gastrointestinal tract, and many other functions.

COUPLED REACTIONS. All the energy foods— carbohydrates, fats, and proteins—can be oxidized in the cells, and in this process, large amounts of energy are released. These same foods can also be burned with pure oxygen outside the body in an actual fire, again releasing large amounts of energy. This time the energy is released suddenly, all in the form of heat. The energy needed by the physiological processes of the cells is not heat, but energy to cause mechanical movement in the case of muscle function, to concentrate solutes in the case of glandular secretion, and to effect other functions. To provide this energy, the chemical reactions must be "coupled" with the systems responsible for these physiological functions. This coupling is accomplished by special cellular enzyme and energy transfer systems, some of which are explained in this and subsequent chapters.

"Free Energy." The amount of energy liberated by complete oxidation of a food is called *the free energy of oxidation of the food,* and this is generally represented by the symbol ΔG. Free energy is usually expressed in terms of calories per mole of substance. For instance, the amount of free energy liberated by oxidation of 1 mole of glucose (180 grams of glucose) is 686,000 calories.

Role of Adenosine Triphosphate in Metabolism

Adenosine triphosphate (ATP) is a labile chemical compound that is present in all cells and has the chemical structure shown in Figure 67–1.

From this formula it can be seen that ATP is a combination of adenine, ribose, and three phosphate radicals. The last two phosphate radicals are connected with the remainder of the molecule by so-called *high-energy bonds,* which are indicated by the symbol $\sim$. The amount of free energy in each of these high-energy bonds per mole of ATP is about 7300 calories under standard conditions and about 12,000 calories under the conditions of temperature and concentrations of the reactants in the body. Therefore, removal of each of the last two phosphate radicals liberates 12,000 calories of energy. After loss of one phosphate radical from ATP, the compound becomes *adenosine diphosphate* (ADP), and after loss of the second phosphate radical, it becomes *adenosine monophosphate* (AMP). The interconversions between ATP, ADP, and AMP are the following:

$$\text{ATP} \underset{+\ 12,000\ \text{cal}}{\overset{-\ 12,000\ \text{cal}}{\rightleftharpoons}} \left\{ \begin{array}{c} \text{ADP} \\ + \\ \text{PO}_3 \end{array} \right\} \underset{+\ 12,000\ \text{cal}}{\overset{-\ 12,000\ \text{cal}}{\rightleftharpoons}} \left\{ \begin{array}{c} \text{AMP} \\ + \\ 2\text{PO}_3 \end{array} \right\}$$

855

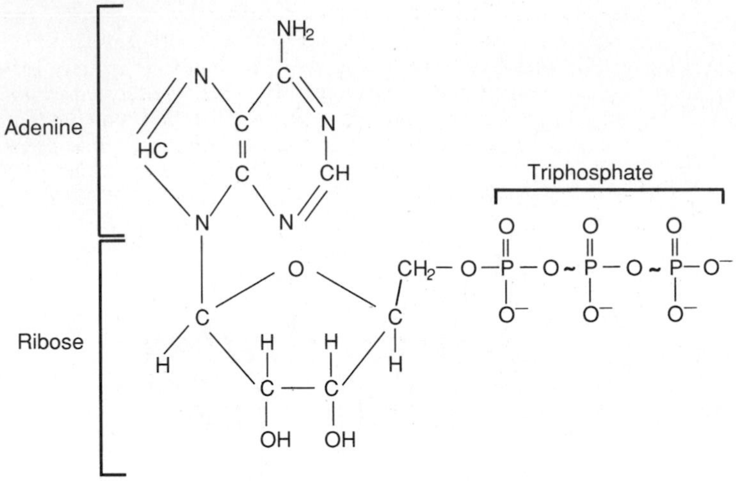

Figure 67–1. Chemical structure of adenosine triphosphate.

ATP is present everywhere in the cytoplasm and nucleoplasm of all cells, and essentially all the physiological mechanisms that require energy for operation obtain this directly from the ATP (or another similar high-energy compound—guanosine triphosphate). In turn, the food in the cells is gradually oxidized, and the released energy is used to re-form the ATP, thus always maintaining a supply of this substance; all these energy transfers take place by means of coupled reactions.

In summary, ATP is an intermediary compound that has the peculiar ability to enter into many coupled reactions—reactions with the food to extract energy and reactions in many physiological mechanisms to provide energy for their operation. For this reason, ATP has been called the *energy currency* of the body that can be gained and spent again and again.

The principal purpose of this chapter is to explain how the energy from carbohydrates can be used to form ATP in the cells. Normally, 90 per cent or more of all the carbohydrates utilized by the body are used for this purpose.

CENTRAL ROLE OF GLUCOSE IN CARBOHYDRATE METABOLISM

From Chapter 65 it will be recalled that the final products of carbohydrate digestion in the alimentary tract are almost entirely glucose, fructose, and galactose—with glucose representing, on the average, about 80 per cent of these. After absorption from the intestinal tract, much of the fructose and almost all the galactose are also rapidly converted in the liver into glucose. Therefore, little fructose and galactose are present in the circulating blood. Glucose thus becomes the final common pathway for transport of almost all carbohydrates to the tissue cells.

In liver cells, appropriate enzymes are available to promote interconversions among the monosaccharides, as shown in Figure 67–2. Furthermore, the dynamics of the reactions are such that when the liver releases the monosaccharides back into the blood, the final product is almost entirely

glucose. The reason for this is that the liver cells contain large amounts of *glucose phosphatase*. Therefore, glucose-6-phosphate can be degraded back to glucose and phosphate, and the glucose can be transported through the liver cell membrane back into the blood.

Once again it should be emphasized that usually more than 95 per cent of all the monosaccharides that circulate in the blood are the final conversion product, glucose.

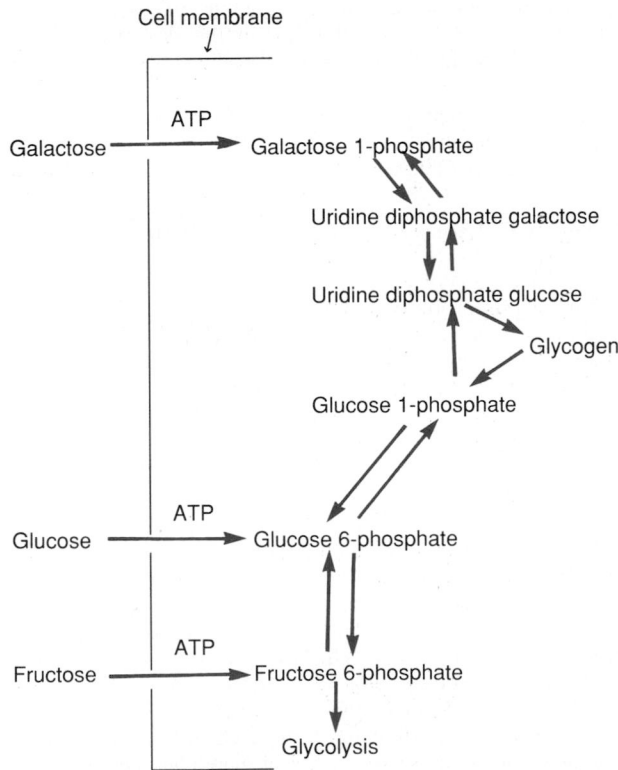

Figure 67–2. Interconversions of the three major monosaccharides—glucose, fructose, and galactose—in liver cells.

TRANSPORT OF GLUCOSE THROUGH THE CELL MEMBRANE

Before glucose can be used by the body's tissue cells, it must be transported through the cell membrane into the cellular cytoplasm. However, glucose *cannot diffuse through the pores* of the cell membrane because the maximum molecular weight of particles that can do this is about 100, whereas glucose has a molecular weight of 180. Yet glucose does pass to the interior of the cells with a reasonable degree of freedom by the mechanism of *facilitated diffusion.* The principles of this type of transport are discussed in Chapter 4. Basically, they are the following: Penetrating through the lipid matrix of the cell membrane are large numbers of protein *carrier* molecules that can bind with glucose. In this bound form, the glucose can be transported by the carrier from one side of the membrane to the other side and then released. Therefore, if the concentration of glucose is greater on one side of the membrane than on the other side, more glucose will be transported from the high concentration area than in the opposite direction.

The transport of glucose through the membranes of most tissue cells is quite different from that which occurs through the gastrointestinal membrane or through the epithelium of the renal tubules. In both these latter cases, the glucose is transported by the mechanism of *active sodium-glucose co-transport,* in which active transport of sodium provides energy for absorbing glucose *against a concentration difference.* This sodium co-transport mechanism functions only in certain special epithelial cells that are specifically adapted for active absorption of glucose. At all other cell membranes, glucose is transported only from higher concentration toward lower concentration by facilitated diffusion, made possible by the special binding properties of the membrane glucose carrier protein. The details of *facilitated diffusion* for cell membrane transport are presented in Chapter 4.

Effect of Insulin to Increase Facilitated Diffusion of Glucose

The rate of glucose transport as well as the transport of some other monosaccharides is greatly increased by insulin. When large amounts of insulin are secreted by the pancreas, the rate of glucose transport into most cells increases to 10 or more times the rate of transport when no insulin is secreted. Conversely, the amounts of glucose that can diffuse to the insides of most cells of the body in the absence of insulin, with the exceptions of the liver and brain cells, are far too little to supply anywhere near the amount of glucose normally required for energy metabolism. Therefore, in effect, the rate of carbohydrate utilization by most cells is controlled by the rate of insulin secretion from the pancreas. The functions of insulin and its control of carbohydrate metabolism are discussed in detail in Chapter 78.

Phosphorylation of Glucose

Immediately on entry into the cells, glucose combines with a phosphate radical in accordance with the following reaction:

$$\text{Glucose} \xrightarrow[\text{+ ATP}]{\text{glucokinase or hexokinase}} \text{Glucose-6-phosphate}$$

This phosphorylation is promoted mainly by the enzyme *glucokinase* in the liver or hexokinase in most other cells.

The phosphorylation of glucose is almost completely irreversible *except in the liver cells,* the *renal tubular epithelium,* and the *intestinal epithelial cells;* in these cells, another enzyme, *glucose phosphatase,* is also available, and when this is activated, it can reverse the reaction. Therefore, in most tissues of the body, phosphorylation serves to *capture* the glucose in the cell. That is, because of its almost instantaneous binding with phosphate, the glucose will not diffuse back out except from those special cells, especially liver cells, that have the phosphatase.

STORAGE OF GLYCOGEN IN LIVER AND MUSCLE

After absorption into the cells, glucose can either be used immediately for release of energy to the cells or be stored in the form of *glycogen,* which is a large polymer of glucose.

All cells of the body are capable of storing at least some glycogen, but certain cells can store large amounts, especially the liver cells, which can store up to 5 to 8 per cent of their weight as glycogen, and muscle cells, which can store up to 1 to 3 per cent glycogen. The glycogen molecules can be polymerized to almost any molecular weight, the average molecular weight being 5 million or greater; most of the glycogen precipitates in the form of solid granules. This conversion of the monosaccharides into a high-molecular-weight precipitated compound makes it possible to store large quantities of carbohydrates without significantly altering the osmotic pressure of the intracellular fluids. High concentrations of low-molecular-weight soluble monosaccharides would play havoc with the osmotic relations between intracellular and extracellular fluids.

Glycogenesis

Glycogenesis is the process of glycogen formation, the chemical reactions for which are shown in Figure 67–3. From this figure, it can be seen that *glucose-6-phosphate* first becomes *glucose-1-phosphate;* then this is converted to *uridine diphosphate glucose,* which is then converted into glycogen. Several specific enzymes are required to cause these conversions, and any monosaccharide that can be converted into glucose can enter into the reactions. Certain smaller compounds, including *lactic acid, glycerol, pyruvic acid,* and *some deaminated amino acids,* can also be converted into glucose or closely allied compounds and then into glycogen.

Removal of Stored Glycogen— Glycogenolysis

Glycogenolysis means the breakdown of the cell's stored glycogen to reform glucose in the cells. The glucose can then be used to provide energy. Glycogenolysis does not occur by reversal of the same chemical reactions that serve to form glycogen; instead, each succeeding glucose molecule on each branch of the glycogen polymer is split away by *phosphorylation,* catalyzed by the enzyme *phosphorylase.*

Under resting conditions, the phosphorylase is in an inactive form, so that glycogen can be stored but not reconverted into glucose. When it is necessary to re-form glucose from glycogen, therefore, the phosphorylase must first be

Cell membrane

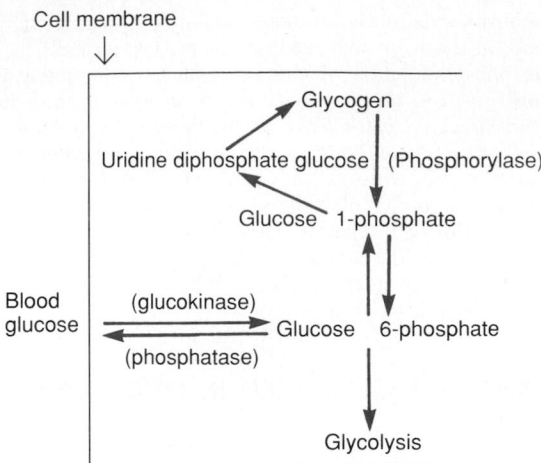

Figure 67–3. Chemical reactions of glycogenesis and glycogenolysis, showing also the interconversions between blood glucose and liver glycogen. (The phosphatase required for release of glucose from the cell is present in liver cells but not in most other cells.)

activated. This can be accomplished in several ways, including the following two.

ACTIVATION OF PHOSPHORYLASE BY EPINEPHRINE AND GLUCAGON. Two hormones, *epinephrine* and *glucagon,* can specifically activate phosphorylase and thereby cause rapid glycogenolysis. The initial effect of each of these hormones is to cause the formation of *cyclic AMP* in the cells. This substance then initiates a cascade of chemical reactions that activates the phosphorylase. This is discussed in more detail in Chapter 78.

Epinephrine is released by the adrenal medullae when the sympathetic nervous system is stimulated. Therefore, one of the functions of the sympathetic nervous system is to increase the availability of glucose for rapid energy metabolism. This function of epinephrine occurs markedly both in liver cells and in muscle, thereby contributing, along with other effects of sympathetic stimulation, to preparing the body for action, as discussed more fully in Chapter 60.

Glucagon is a hormone secreted by the *alpha cells* of the pancreas when the blood glucose concentration falls too low. It stimulates the formation of cyclic AMP mainly in the liver cells. Therefore, its effect is primarily to convert liver glycogen into glucose and to release this into the blood, thereby elevating blood glucose concentration. The function of glucagon in blood glucose regulation is discussed more fully in Chapter 78.

RELEASE OF ENERGY FROM THE GLUCOSE MOLECULE BY THE GLYCOLYTIC PATHWAY

Because complete oxidation of 1 gram-molecule of glucose releases 686,000 calories of energy and only 12,000 calories of energy are required to form 1 gram-molecule of ATP, it would be wasteful of energy if glucose should be decomposed all the way into water and carbon dioxide at once while forming only a single ATP molecule. All cells contain an extensive series of different protein enzymes that cause the glucose molecule to split a little at a time in many successive steps, with its energy released in small packets to form one molecule of ATP at a time, forming a total of 38 moles of ATP for each mole of glucose utilized by the cells.

The purpose of the next sections is to describe the basic principles of the processes by which the glucose molecule is progressively dissected and its energy released to form ATP.

Glycolysis; The Formation of Pyruvic Acid

By far the most important means by which energy release from the glucose molecule is initiated is by *glycolysis.* Then the end products of glycolysis are mainly oxidized to provide the energy. Glycolysis means splitting of the glucose molecule to form *two molecules of pyruvic acid.* This occurs by 10 successive steps of chemical reactions, as shown in Figure 67–4. Each step is catalyzed by at least one specific protein enzyme. Note that glucose is first converted into fructose 1,6-phosphate and then split into two three-carbon atom molecules, each of which is then converted through five successive steps into pyruvic acid.

FORMATION OF ATP DURING GLYCOLYSIS. Despite the many chemical reactions in the glycolytic series, only a small portion of the free energy in the glucose molecule is released at each step. However, between the 1,3-diphosphoglyceric acid and the 3-phosphoglyceric acid stages and again between the phosphoenolpyruvic acid and the pyruvic acid stages, the packets of energy released are greater than 12,000 calories per mole, the amount required to form ATP, and the reactions are coupled in such a way that ATP is formed. Thus, a total of 4 moles of ATP are formed for each mole of fructose 1,6-phosphate that is split into pyruvic acid.

Yet 2 moles of ATP had been required to phosphorylate the original glucose to form fructose 1,6-phosphate before glycolysis could begin. Therefore, the net gain in ATP molecules by the entire glycolytic process is only 2 moles for each mole of glucose utilized. This amounts to 24,000 calories of energy stored in the form of ATP, but during glycolysis, a total of 56,000 calories of energy is lost from the original

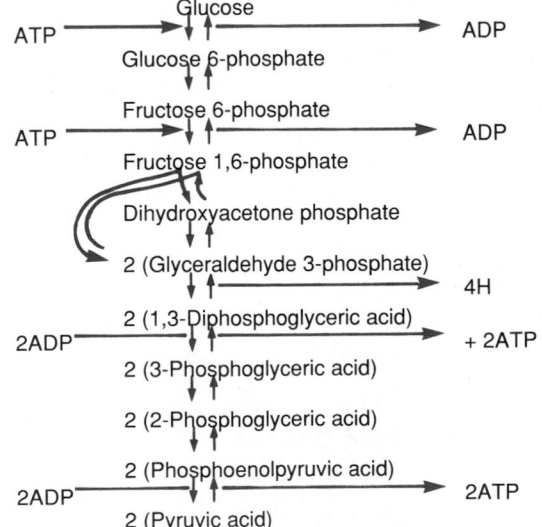

NET REACTION PER MOLECULE OF GLUCOSE:

Glucose + 2ADP + 2PO₄⁻⁻⁻ ⟶ 2 Pyruvic acid + 2ATP + 4H

Figure 67–4. Sequence of chemical reactions responsible for glycolysis.

glucose, giving an overall *efficiency* for ATP formation of 43 per cent. The remaining 57 per cent of the energy is lost in the form of heat.

Conversion of Pyruvic Acid to Acetylcoenzyme A

The next stage in the degradation of glucose is (1) facilitated transport of the two derivative pyruvic acid molecules into the matrix of the mitochondria and then (2) conversion of these into two molecules of *acetylcoenzyme A* (acetyl-CoA) in accordance with the following reaction:

$$2CH_3-\overset{\overset{\displaystyle O}{\|}}{C}-COOH + 2CoA-SH \rightarrow$$
$$\text{(Pyruvic Acid)} \qquad \text{(Coenzyme A)}$$

$$2CH_3-\overset{\overset{\displaystyle O}{\|}}{C}-S-CoA + 2CO_2 + 4H$$
$$\text{(Acetyl-CoA)}$$

From this reaction, it can be seen that two carbon dioxide molecules and four hydrogen atoms are released, while the remaining portions of the two pyruvic acid molecules combine with coenzyme A, a derivative of the vitamin pantothenic acid, to form two molecules of acetyl-CoA. In this conversion, no ATP is formed, but up to six molecules of ATP are formed when the four released hydrogen atoms are later oxidized, as discussed in a later section.

Citric Acid Cycle

The next stage in the degradation of the glucose molecule is called the *citric acid cycle* (also called the *tricarboxylic acid cycle* or *Krebs' cycle*). This is a sequence of chemical reactions in which the acetyl portion of acetyl-CoA is degraded to carbon dioxide and hydrogen atoms. These reactions all occur *in the matrix of the mitochondrion.* The released hydrogen atoms are subsequently oxidized, as discussed later, releasing tremendous amounts of energy to form ATP.

Figure 67–5 shows the different stages of the chemical reactions in the citric acid cycle. The substances to the left are added during the chemical reactions, and the products of the chemical reactions are shown to the right. Note at the top of the column that the cycle begins with *oxaloacetic acid,* and then at the bottom of the chain of reactions *oxaloacetic acid* is formed once again. Thus, the cycle can continue over and over.

In the initial stage of the citric acid cycle, *acetyl-CoA* combines with *oxaloacetic acid to form citric acid.* The coenzyme A portion of the acetyl-CoA is released and can be used again and again for the formation of still more quantities of acetyl-CoA from pyruvic acid. The acetyl portion, however, becomes an integral part of the citric acid molecule. During the successive stages of the citric acid cycle, several molecules of water are added, and *carbon dioxide* and *hydrogen atoms* are released at various stages in the cycle, as shown on the right in the figure.

The net results of the entire citric acid cycle are given in the legend at the bottom of Figure 67–5, demonstrating that for each molecule of glucose originally metabolized, two acetyl-CoA molecules enter into the citric acid cycle along with six molecules of water. These then are degraded into 4 carbon dioxide molecules, 16 hydrogen atoms, and 2 molecules

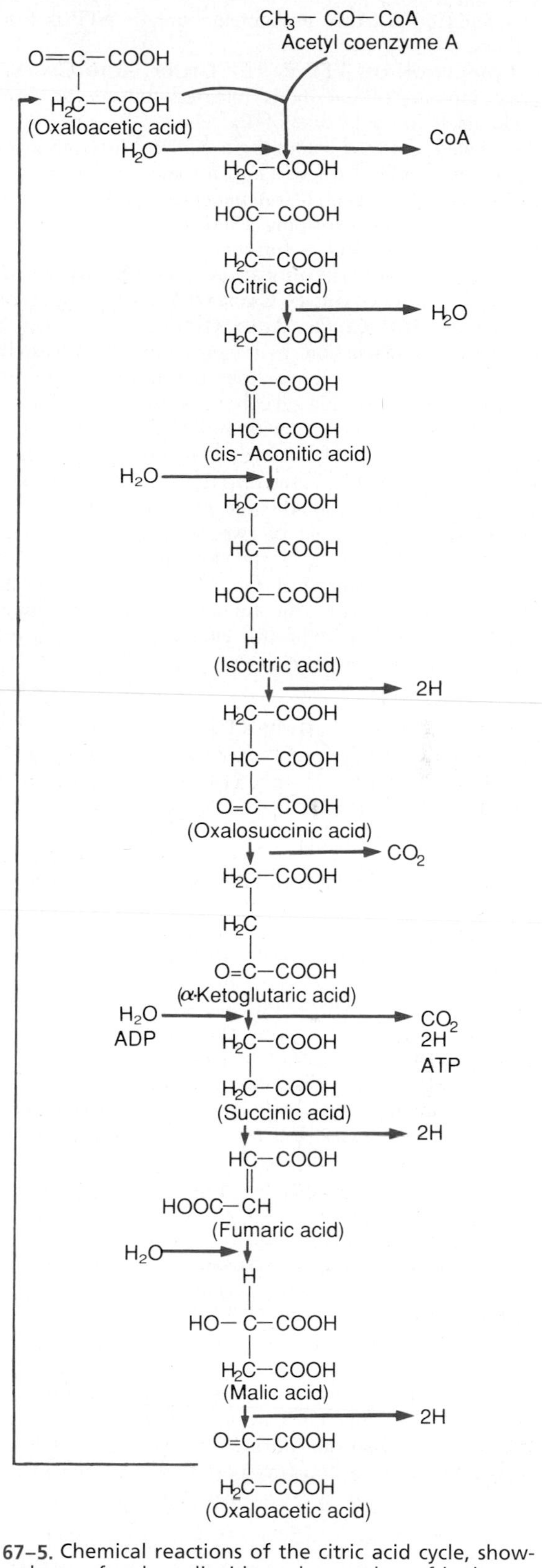

Figure 67–5. Chemical reactions of the citric acid cycle, showing the release of carbon dioxide and a number of hydrogen atoms during the cycle.

Net reaction per molecule of glucose:

2 Acetyl-CoA + 6H$_2$O + 2ADP

$$\longrightarrow 4CO_2 + 16H + 2CoA + 2ATP$$

of coenzyme A. Also, a slight amount of ATP is formed as follows.

FORMATION OF ATP IN THE CITRIC ACID CYCLE. Not a great amount of energy is released during the citric acid cycle itself; in only one of the chemical reactions—during the change from α-ketoglutaric acid to succinic acid—is a molecule of ATP formed. Thus, for each molecule of glucose metabolized, two acetyl-CoA molecules pass through the citric acid cycle, each forming a molecule of ATP; or a total of two molecules of ATP is formed.

FUNCTION OF DEHYDROGENASES AND NICOTINAMIDE ADENINE DINUCLEOTIDE IN CAUSING RELEASE OF HYDROGEN ATOMS IN THE CITRIC ACID CYCLE. As noted at several points in this discussion, hydrogen atoms are released during the different chemical reactions of the citric acid cycle—4 hydrogen atoms during glycolysis, 4 during the formation of acetyl-CoA from pyruvic acid, and 16 in the citric acid cycle; this makes a total of 24 hydrogen atoms for each original molecule of glucose. However, the hydrogen atoms are not simply turned loose in the intracellular fluid. Instead, they are released in packets of two, and in each instance, the release is catalyzed by a specific protein enzyme called a *dehydrogenase*. Twenty of the 24 hydrogen atoms immediately combine with nicotinamide adenine dinucleotide (NAD^+), a derivative of the vitamin niacin, in accordance with the following reaction:

$$\text{Substrate} \underset{H}{\overset{H}{\diagup}} + NAD^+ \xrightarrow{\quad \textit{dehydrogenase} \quad}$$

$$NADH + H^+ + \text{Substrate}$$

This reaction will not occur without the intermediation of the specific dehydrogenase or without the availability of NAD^+ to act as a hydrogen carrier. Both the free hydrogen ion and the hydrogen bound with NAD^+ subsequently enter into the oxidative chemical reactions that form tremendous quantities of ATP, as discussed below.

The remaining four hydrogen atoms released during the breakdown of glucose—the four released during the citric acid cycle between the succinic and fumaric acid stages—combine with a specific dehydrogenase but are not then subsequently released to NAD^+. Instead, they pass directly from the dehydrogenase into the oxidative process.

FUNCTION OF DECARBOXYLASES IN CAUSING RELEASE OF CARBON DIOXIDE. Referring again to the chemical reactions of the citric acid cycle as well as to those for formation of acetyl-CoA from pyruvic acid, we find that there are three stages in which carbon dioxide is released. To cause the carbon dioxide release, other specific protein enzymes, called *decarboxylases*, split the carbon dioxide away from the substrate. The carbon dioxide in turn becomes dissolved in the body fluids and is then transported to the lungs, where it is expired from the body (see Chap. 40).

Formation of Large Quantities of ATP by Oxidation of Hydrogen (The Process of Oxidative Phosphorylation)

Despite all the complexities of (1) glycolysis, (2) the citric acid cycle, (3) dehydrogenation, and (4) decarboxylation, piti-

fully small amounts of ATP are formed during all these processes—only two ATP molecules in the glycolysis scheme and another two in the citric acid cycle for each molecule of glucose metabolized. Instead, almost 90 per cent of the total ATP formed by glucose metabolism is formed during subsequent oxidation of the hydrogen atoms that were released during these earlier stages of glucose degradation. Indeed, the principal function of all these earlier stages is to make the hydrogen of the glucose molecule available in forms that can be used for oxidation.

Oxidation of hydrogen is accomplished by a series of enzymatically catalyzed reactions *in the mitochondria* that (1) split each hydrogen atom into a hydrogen ion and an electron and (2) use the electrons eventually to combine the dissolved oxygen of the fluids with water molecules to form hydroxyl ions. Then the hydrogen and hydroxyl ions combine with each other to form water. During the sequence of oxidative reactions, tremendous quantities of energy are released to form ATP. Formation of ATP in this manner is called *oxidative phosphorylation*. This occurs entirely in the mitochondria by a highly specialized process called the *chemiosmotic mechanism*.

Chemiosmotic Mechanism of the Mitochondria to Form ATP

IONIZATION OF HYDROGEN, THE ELECTRON TRANSPORT CHAIN, AND FORMATION OF WATER. The first step in oxidative phosphorylation in the mitochondria is to ionize the hydrogen atoms that have been removed from the food substrates. As described earlier, these hydrogen atoms are removed in pairs: one immediately becomes a hydrogen ion, H^+; the other combines with NAD^+ to form NADH. The upper portion of Figure 67–6 shows in color the subsequent fate of the NADH and H^+. The initial effect is to release the other hydrogen atom from the NADH to form another

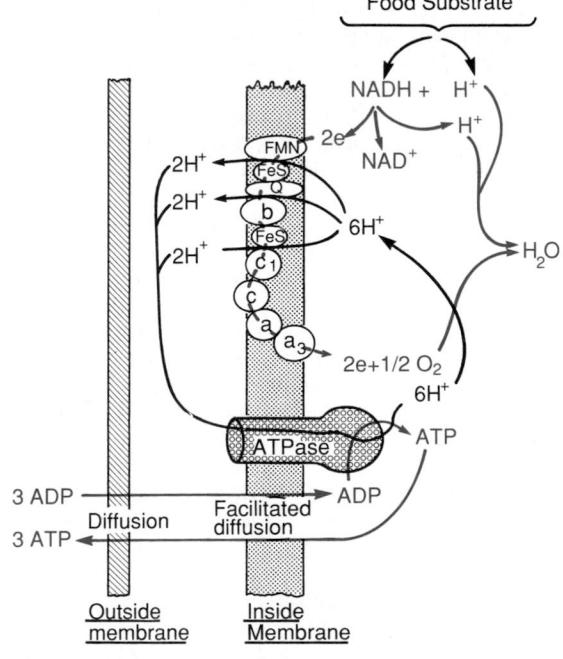

Figure 67–6. Chemiosmotic mechanism of oxidative phosphorylation for forming great quantities of ATP.

hydrogen ion, H $^+$; this process also reconstitutes NAD$^+$ that will be reused again and again.

The electrons that are removed from the hydrogen atoms to cause the ionization immediately enter an *electron transport chain of electron acceptors* that are an integral part of the inner membrane (the shelf membrane) of the mitochondrion. The electron acceptors can be reversibly reduced or oxidized by accepting or giving up electrons. The important members of this electron transport chain include *flavoprotein, several iron sulfide proteins, ubiquinone,* and *cytochromes B, C$_1$, C, A, and A$_3$*. Each electron is shuttled from one of these acceptors to the next until it finally reaches cytochrome A$_3$, which is called *cytochrome oxidase* because it is capable, by giving up two electrons, of reducing elemental oxygen to form ionic oxygen, which then combines with hydrogen ions to form water.

Thus, Figure 67–6 shows transport of electrons through the electron chain and then ultimate use of these by cytochrome oxidase to cause the formation of water molecules. During the transport of these electrons through the electron transport chain, energy is released that is used to cause synthesis of ATP, as follows.

PUMPING OF HYDROGEN IONS INTO THE OUTER CHAMBER OF THE MITOCHONDRION, CAUSED BY THE ELECTRON TRANSPORT CHAIN. As the electrons pass through the electron transport chain, large amounts of energy are released. This energy is used to pump hydrogen ions from the inner matrix of the mitochondrion (to the right side in Fig. 67–6) into the outer chamber between the inner and outer mitochondrial membranes. This creates a high concentration of positively charged hydrogen ions in this chamber; it also creates a strong negative electrical potential in the inner matrix.

FORMATION OF ATP. The next step in oxidative phosphorylation is to convert ADP into ATP. This occurs in conjunction with a large protein molecule that protrudes all the way through the inner mitochondrial membrane and projects with a knob-like head into the inner matrix. This molecule is an ATPase, the physical nature of which is shown in Figure 67–6. It is called *ATP synthetase.* The high concentration of the positively charged hydrogen ions in the outer chamber and the large electrical potential difference across the inner membrane cause the hydrogen ions to flow into the mitochondrial matrix *through the substance of the ATPase molecule.* In doing so, energy derived from this hydrogen ion flow is used by the ATPase to convert ADP into ATP by combining ADP with a free ionic phosphate radical (Pi), thus adding an additional high-energy phosphate bond to the molecule.

The final step in the process is transfer of the ATP from the inside of the mitochondrion back to the cytoplasm. This occurs by facilitated diffusion outward through the inner membrane and then by simple diffusion through the permeable outer mitochondrial membrane. In turn, ADP is continually transferred in the other direction for continual conversion into ATP.

For each two electrons that pass through the entire electron transport chain (representing the ionization of two hydrogen atoms), up to three ATP molecules are synthesized.

Summary of ATP Formation During the Breakdown of Glucose

We can now determine the total number of ATP molecules that, under optimal conditions, can be formed by the energy from one molecule of glucose.

1. During glycolysis, four molecules of ATP are formed, whereas two are expended to cause the initial phosphorylation of glucose to start the process going. This gives a net gain of *two molecules of ATP.*

2. During each revolution of the citric acid cycle, one molecule of ATP is formed. However, because each glucose molecule splits into two pyruvic acid molecules, there are two revolutions of the cycle for each molecule of glucose metabolized, giving a net production of *two more molecules of ATP.*

3. During the entire schema of glucose breakdown, a total of 24 hydrogen atoms are released during glycolysis and during the citric acid cycle. Twenty of these atoms are oxidized in conjunction with the chemiosmotic mechanism shown in Figure 67–6, with the release of up to three ATP molecules per two atoms of hydrogen metabolized. This gives an additional *30 ATP molecules.*

4. The remaining four hydrogen atoms are released by their dehydrogenase into the chemiosmotic oxidative schema in the mitochondrion beyond the first stage of Figure 67–6, so that for these four hydrogen atoms, only two ATP molecules are usually released for each two hydrogen atoms oxidized, giving a total of *four more ATP molecules.*

Now, adding all the ATP molecules formed, we find a maximum of *38 ATP molecules* formed for each molecule of glucose degraded to carbon dioxide and water. Thus, 456,000 calories of energy can be stored in the form of ATP, whereas 686,000 calories are released during the complete oxidation of each gram-molecule of glucose. This represents an overall maximum *efficiency* of energy transfer of 66 per cent. The remaining 34 per cent of the energy becomes heat and, therefore, cannot be used by the cells to perform specific functions.

Control of Energy Release from Stored Glycogen When the Body Needs the Energy: Effect of ATP and ADP Concentrations in the Cell to Control the Rate of Glycolysis

Continual release of energy from glucose when the energy is not needed by the cells would be an extremely wasteful process. Glycolysis and the subsequent oxidation of hydrogen atoms are continually controlled in accordance with the needs of the cells for ATP. This control is accomplished by multiple feedback control mechanisms within the chemical schemata. Among the more important of these are the effects of the cell concentrations of both ADP and ATP in controlling the rates of chemical reactions in the energy metabolism sequence.

One important way in which ATP helps control energy metabolism is its effect in causing an *allosteric inhibition of the enzyme phosphofructokinase.* Because this enzyme promotes the formation of fructose 1,6-phosphate, one of the initial steps in the glycolytic series of reactions, the net effect of excess cellular ATP, therefore, is to stop glycolysis, which in turn stops most carbohydrate metabolism. Conversely, ADP (and AMP as well) causes the opposite allosteric change in this enzyme, greatly increasing its activity. Therefore, whenever ATP is used by the tissues for energy, this reduces the ATP inhibition of the enzyme but at the same time increases its activity as a result of the ADP that is formed, and the glycolytic process is set in motion. Soon the total cellular store of ATP is replenished.

Another control linkage is the *citrate ion* formed in the

citric acid cycle. An excess of this ion also *strongly inhibits phosphofructokinase,* thus preventing the glycolytic process from getting ahead of the ability of the citric acid cycle to use the pyruvic acid formed during glycolysis.

A third way by which the ATP-ADP-AMP system controls carbohydrate metabolism as well as controlling energy release from fats and proteins is the following: Referring back to the various chemical reactions for energy release, we see that if all the ADP in the cell has already been converted into ATP, then additional ATP simply cannot occur. As a result, the entire sequence for use of foodstuffs—glucose, fats, and proteins—to form ATP is stopped. Then, when ATP is used to energize the different physiological functions in the cell, the resulting newly formed ADP and AMP turn on the energy processes again and the ADP and AMP are almost instantly returned to the ATP state. In this way, essentially a full store of ATP is automatically maintained all the time, except during extreme cellular activity as might occur in heavy exercise.

Anaerobic Release of Energy—"Anaerobic Glycolysis"

Occasionally, oxygen becomes either unavailable or insufficient, so that oxidative phosphorylation cannot take place. Yet even under these conditions, a small amount of energy can still be released to the cells by the glycolysis stage of carbohydrate degradation because the chemical reactions in the glycolytic breakdown of glucose to pyruvic acid do not require oxygen. This process is extremely wasteful of glucose because only 24,000 calories of energy are used to form ATP for each molecule of glucose used, which represents only a little over 3 per cent of the total energy in the glucose molecule. Nevertheless, this release of glycolytic energy to the cells, which is called *anaerobic energy,* can be a life-saving measure for up to a few minutes' time when oxygen becomes unavailable.

FORMATION OF LACTIC ACID DURING ANAEROBIC GLYCOLYSIS ALLOWS RELEASE OF EXTRA ANAEROBIC ENERGY. The *law of mass action* states that as the end products of a chemical reaction build up in a reacting medium, the rate of the reaction approaches zero. The two end products of the glycolytic reactions (see Fig. 67–4) are (1) pyruvic acid and (2) hydrogen atoms combined with NAD^+ to form NADH and H^+. The buildup of either or both of these would stop the glycolytic process and prevent further formation of ATP. When their quantities begin to be excessive, these two end products react with each other to form lactic acid in accordance with the following equation:

$$CH_3-\overset{\overset{O}{\|}}{C}-COOH + NADH + H^+ \underset{}{\overset{lactic\ dehydrogenase}{\rightleftharpoons}}$$
(Pyruvic Acid)

$$CH_3-\overset{\overset{OH}{|}}{\underset{\overset{|}{H}}{C}}-COOH + NAD^+$$
(Lactic acid)

Thus, under anaerobic conditions, by far the major portion of the pyruvic acid is converted into lactic acid, which diffuses readily out of the cells into the extracellular fluids and even into the intracellular fluids of other less active cells. Therefore, lactic acid represents a type of "sinkhole" into which the glycolytic end products can disappear, thus allowing glycolysis to proceed far longer than would otherwise be possible. Indeed, glycolysis could proceed for only a few seconds without this conversion. Instead, it can proceed for several minutes, supplying the body with considerable quantities of ATP even in the absence of respiratory oxygen.

RECONVERSION OF LACTIC ACID WHEN OXYGEN BECOMES AVAILABLE AGAIN. When a person begins to breathe oxygen again after a period of anaerobic metabolism, the lactic acid is rapidly reconverted to pyruvic acid and NADH plus H^+. Large portions of these are immediately oxidized to form large quantities of ATP. This excess ATP then causes as much as three fourths of the remaining excess pyruvic acid to be converted back into glucose.

Thus, the great amount of lactic acid that forms during anaerobic glycolysis does not become lost from the body because when oxygen is again available, the lactic acid either can be reconverted to glucose or can be used directly for energy. By far the greater portion of this reconversion occurs in the liver, but a small amount can also occur in other tissues.

Use of Lactic Acid by the Heart for Energy. Heart muscle is especially capable of converting lactic acid to pyruvic acid and then using it for energy. This occurs to a great extent in heavy exercise, during which large amounts of lactic acid are released into the blood from the skeletal muscles and then are consumed as an extra energy source by the heart.

RELEASE OF ENERGY FROM GLUCOSE BY THE PENTOSE PHOSPHATE PATHWAY

Though, in muscle, essentially all the carbohydrates utilized for energy are first degraded to pyruvic acid by glycolysis and then oxidized, this glycolytic scheme is not the only means by which glucose can be degraded and then used to provide energy. A second important mechanism for breakdown and oxidation of glucose is called the *pentose phosphate pathway* (or *phosphogluconate pathway*), which is responsible for as much as 30 per cent of the glucose breakdown *in the liver and even more than this in fat cells.* It is especially important because it can provide energy independently of all the enzymes of the citric acid cycle and therefore is an alternate pathway for energy metabolism when some enzymatic abnormalities occur in cells. Also, it has a special capacity for providing energy to multiple cellular synthetic processes, as we shall see.

RELEASE OF CARBON DIOXIDE AND HYDROGEN BY MEANS OF THE PENTOSE PHOSPHATE PATHWAY. Figure 67–7 shows most of the basic chemical reactions in the pentose phosphate pathway. It demonstrates that glucose, during several stages of conversion, releases one molecule of carbon dioxide and four atoms of hydrogen, with resultant formation of a five-carbon sugar, D-ribulose. This substance in turn can change progressively into several other five-, four-, seven-, and three-carbon sugars. Finally, various combinations of these sugars can resynthesize glucose. However, *only five molecules of glucose are resynthesized for every six molecules of glucose that initially enter into the reactions.* That is, the pentose phosphate pathway is a cyclic process in which one molecule of glucose is metabolized for each revolution of the cycle. Thus, by revolution of the cycle again and again, all the glucose can eventually be converted into carbon dioxide and hydrogen, and the hydrogen in turn can enter the oxida-

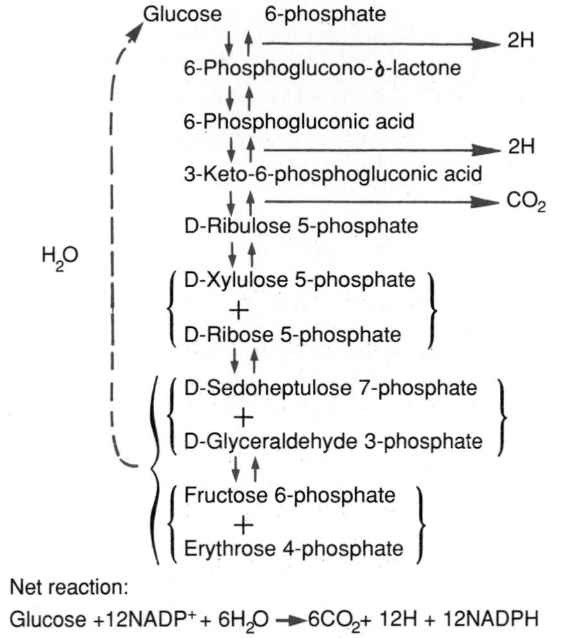

Glucose   6-phosphate

6-Phosphoglucono-δ-lactone → 2H

6-Phosphogluconic acid

3-Keto-6-phosphogluconic acid → 2H

D-Ribulose 5-phosphate → CO₂

{ D-Xylulose 5-phosphate + D-Ribose 5-phosphate }

{ D-Sedoheptulose 7-phosphate + D-Glyceraldehyde 3-phosphate }

{ Fructose 6-phosphate + Erythrose 4-phosphate }

H₂O

Net reaction:

Glucose + 12NADP⁺ + 6H₂O → 6CO₂ + 12H + 12NADPH

Figure 67–7. Pentose phosphate pathway for glucose metabolism.

tive phosphorylation pathway to form ATP, or more often, it is used for synthesis of fat or other substances as follows.

USE OF HYDROGEN TO SYNTHESIZE FAT; THE FUNCTION OF NICOTINAMIDE ADENINE DINUCLEOTIDE PHOSPHATE. The hydrogen released during the pentose phosphate cycle does not combine with NAD⁺ as in the glycolytic pathway but combines with nicotinamide adenine dinucleotide phosphate (NADP⁺), which is almost identical to NAD⁺ except for an extra phosphate radical, P. This difference is extremely significant because only hydrogen bound with NADP⁺ in the form of NADPH can be used for synthesis of fats from carbohydrates, which is discussed in Chapter 68, as well as synthesis of some other substances. When the glycolytic pathway for using glucose becomes slowed because of cellular inactivity, the pentose phosphate pathway remains operative (mainly in the liver) to break down any excess glucose that continues to be transported into the cells, and NADPH becomes abundant to help convert acetyl-CoA, also derived from glucose, into long fatty acid chains. This is another way in which the energy in the glucose molecule is used besides the formation of ATP, in this instance for the formation and storage of fat in the body.

Glucose Conversion to Glycogen or Fat

When glucose is not immediately required for energy, the extra glucose that continually enters the cells is either stored as glycogen or converted into fat. Glucose is preferentially stored as glycogen until the cells have stored as much glycogen as they can—an amount sufficient to supply the energy needs of the body for only 12 to 24 hours. When the cells (primarily liver and muscle cells) approach saturation with glycogen, the additional glucose is converted into fat in the liver and in fat cells and is stored in the fat cells. Other steps in the chemistry of this conversion are discussed in Chapter 68.

FORMATION OF CARBOHYDRATES FROM PROTEINS AND FATS— "GLUCONEOGENESIS"

When the body's stores of carbohydrates decrease below normal, moderate quantities of glucose can be formed from *amino acids* and the *glycerol* portion of fat. This process is called *gluconeogenesis*. About 60 per cent of the amino acids in the body proteins can be converted easily into carbohydrates; the remaining 40 per cent have chemical configurations that make this difficult. Each amino acid is converted into glucose by a slightly different chemical process. For instance, alanine can be converted directly into pyruvic acid simply by deamination; the pyruvic acid then is converted into glucose or stored glycogen. Several of the more complicated amino acids can be converted into different sugars that contain three-, four-, five-, or seven-carbon atoms; they can then enter the phosphogluconate pathway and eventually form glucose. Thus, by means of deamination plus several simple interconversions, many of the amino acids become glucose. Similar interconversions can change glycerol into glucose or glycogen.

REGULATION OF GLUCONEOGENESIS. Diminished carbohydrates in the cells and decreased blood sugar are the basic stimuli that set off an increase in the rate of gluconeogenesis. The diminished carbohydrates can directly cause reversal of many of the glycolytic and phosphogluconate reactions, thus allowing conversion of deaminated amino acids and glycerol into carbohydrates. In addition, the hormone cortisol is especially important in this regulation, as follows.

Effect of Corticotropin and Glucocorticoids on Gluconeogenesis. When normal quantities of carbohydrates are not available to the cells, the adenohypophysis, for reasons not completely understood, begins to secrete increased quantities of *corticotropin*. This stimulates the adrenal cortex to produce large quantities of *glucocorticoid hormones*, especially *cortisol*. In turn, cortisol mobilizes proteins from essentially all cells of the body, making these available in the form of amino acids in the body fluids. A high proportion of these immediately become deaminated in the liver and provide ideal substrates for conversion into glucose. Thus, one of the most important means by which gluconeogenesis is promoted is through the release of glucocorticoids from the adrenal cortex.

BLOOD GLUCOSE

The normal blood glucose concentration in a person who has not eaten a meal within the past 3 to 4 hours is about 90 mg/dl. After a meal containing large amounts of carbohydrates, this level seldom rises above 140 mg/dl unless the person has diabetes mellitus, which is discussed in Chapter 78.

The regulation of blood glucose concentration is intimately related to the hormones insulin and glucagon; this subject is discussed in detail in Chapter 78 in relation to the functions of these hormones.

REFERENCES

Becker, K. L., et al.: Principles and Practice of Endocrinology and Metabolism. Philadelphia, J. B. Lippincott, 1990.
Cornish-Bowden, A.: Fundamentals of Enzyme Kinetics. Boston, Butterworths, 1979.

Dickerson, R. E.: Cytochrome C and the evolution of energy metabolism. Sci. Am., 242(3):136, 1980.

Edvinsson, L., et al.: Cerebral Blood Flow and Metabolism. New York, Raven Press, 1993.

Felig, P.: Disorders of carbohydrate metabolism. In Bondy, P. K., and Rosenberg, L. E. (eds.): Metabolic Control and Disease, 8th Ed. Philadelphia, W. B. Saunders Co., 1980, p. 276.

Friedmann, H. C. (ed.): Enzymes. Stroudsburg, Pa., Dowden, Hutchinson & Ross, 1980.

Frohman, L. A.: CNS peptides and glucoregulation. Annu. Rev. Physiol., 45:95, 1983.

Golinick, P. D.: Metabolism of substrates: Energy substrate metabolism during exercise and as modified by training. Fed. Proc., 44:353, 1985.

Gracey, M., et al.: Sugars in Nutrition. New York, Raven Press, 1991.

Hediger, M. A., and Rhoads, D. B.: Molecular physiology of sodium-glucose cotransporters. Physiol. Rev., 74:993, 1994.

Hems, D. A., and Whitton, P. D.: Control of hepatic glycogenolysis. Physiol. Rev., 60:1, 1980.

Hetenyi, G., Jr., et al.: Turnover and precursor-product relationships of non-lipid metabolites. Physiol. Rev., 63:606, 1983.

Howell, S. L., and Tyhurst, M.: Insulin secretion: The effector system. Experimentia. 40:1098, 1984.

Jackson, R. L., et al.: Glycosaminoglycans: molecular properties, protein interactions, and role in physiological processes. Physiol. Rev., 71:481, 1991.

Jacquez, J. A.: Red blood cell as glucose carrier: Significance for placental and cerebral glucose transfer. Am. J. Physiol., 246:R289, 1984.

Jequier, E., and Flatt, J.-P.: Recent advances in human energetics. News Physiol. Sci., 1:112, 1986.

Jungas, R. L., et al.: Quantitative analysis of amino acid oxidation and related gluconeogenesis in humans. Physiol. Rev., 72:419, 1992.

Kraus-Friedmann, N.: Hormonal regulation of hepatic gluconeogenesis. Physiol. Rev., 64:170, 1984.

Krebs, H. A.: The tricarboxylic acid cycle. Harvey Lectures, 44:165, 1948–1949.

Lemasters, J. J., et al. (eds.): Integration of Mitochondrial Function. New York, Plenum Publishing Corp., 1988.

Murad, F.: Cyclic GMP: Synthesis, Metabolism, and Function. San Diego, CA, Academic Press, 1994.

Nestler, J. E.: Assessment of Insulin Resistance. New York, Scientific American Science and Medicine, September/October, 1994.

Oomura, Y., and Yoshimatsu, H.: Neural network of glucose monitoring system. J. Auton. Nerv. Syst., 10:359, 1984.

Rombeau, J. L., and Caldwell, M. D.: Clinical Nutrition: Parenteral Nutrition. Philadelphia, W. B. Saunders Co., 1993.

Ruderman, N., et al.: Hyperglycemia, Diabetes, and Vascular Disease. New York, Oxford University Press, 1992.

Sairam, M. R.: Role of carbohydrates in glycoprotein hormone signal transduction. FASEB J., 3:1915, 1989.

Senior, A. E.: ATP synthesis by oxidative phosphorylation. Physiol. Rev., 68:177, 1988.

Storlien, L. H.: The role of the ventromedial hypothalamic area in periprandial glucoregulation. Life-Sci., 36:505, 1985.

Wang, J. H.: Coupling of proton flux to the hydrolysis and synthesis of ATP. Annu. Rev. Biophys. Bioeng., 12:21, 1983.

Lipid Metabolism

CHAPTER 68

A number of chemical compounds in the food and in the body are classified as *lipids*. They include (1) *neutral fat,* known also as *triglycerides;* (2) the *phospholipids;* (3) *cholesterol;* and (4) a few others of less importance. Chemically, the basic lipid moiety of the triglycerides and the phospholipids is *fatty acids,* which are simply long-chain hydrocarbon organic acids. A typical fatty acid, palmitic acid, is the following:

$$CH_3(CH_2)_{14}COOH$$

Although cholesterol does not contain fatty acid, its sterol nucleus is synthesized from degradation products of fatty acid molecules, thus giving it many of the physical and chemical properties of other lipid substances.

The triglycerides are used in the body mainly to provide energy for the different metabolic processes; this function they share almost equally with the carbohydrates. However, some lipids, especially cholesterol, the phospholipids, and small amounts of triglycerides, are used throughout the body to form the membranes of all cells of the body and to perform other cellular functions.

BASIC CHEMICAL STRUCTURE OF TRIGLYCERIDES (NEUTRAL FAT). Because most of this chapter deals with the utilization of triglycerides for energy, the following basic structure of the triglyceride molecule must be understood:

$$CH_3-(CH_2)_{16}-COO-CH_2$$
$$CH_3-(CH_2)_{16}-COO-CH$$
$$CH_3-(CH_2)_{16}-COO-CH_2$$

Tristearin

Note that three long-chain fatty acid molecules are bound with one molecule of glycerol. In the human body, the three fatty acids most commonly present in the triglycerides are (1) *stearic acid* (shown above), which has an 18-carbon chain and is fully saturated with hydrogen atoms; (2) *oleic acid,* which also has an 18-carbon chain but has one double bond in the middle of the chain; and (3) *palmitic acid,* which has 16 carbon atoms and is fully saturated.

TRANSPORT OF LIPIDS IN THE BODY FLUIDS

Transport of Triglycerides and Other Lipids by the Lymph from the Gastrointestinal Tract—The Chylomicrons

As explained in Chapter 65, almost all the fats of the diet, with the principal exception of the short-chain fatty acids, are absorbed from the intestines into the lymph. During digestion, most triglycerides are split into monoglycerides and fatty acids. Then, while passing through the intestinal epithelial cells, they are resynthesized into new molecules of triglycerides that aggregate and enter the lymph as minute, dispersed droplets called *chylomicrons,* having sizes between 0.08 and 0.6 micrometer. A small amount of the protein apoprotein B adsorbs to the outer surfaces of the chylomicrons; this increases their suspension stability in the fluid of the lymph and prevents their adherence to the lymphatic vessel walls.

Most of the cholesterol and phospholipids absorbed from the gastrointestinal tract as well as small amounts of phospholipids that are continually synthesized by the intestinal mucosa also enter the chylomicrons. Thus, chylomicrons are composed principally of triglycerides, but they contain about 9 per cent phospholipids, 3 per cent cholesterol, and 1 per

865

cent apoprotein B as well. The chylomicrons are then transported up the thoracic duct and emptied into the venous blood at the juncture of the jugular and subclavian veins.

Removal of the Chylomicrons from the Blood

About 1 hour after a meal that contains large quantities of fat, the chylomicron concentration in the plasma may rise to 1 to 2 per cent, and because of the large sizes of the chylomicrons, the plasma appears turbid and sometimes yellow. However, the chylomicrons have a half-life of less than 1 hour, so that the plasma becomes clear again within a few hours. The fat of the chylomicrons is removed mainly in the following way.

HYDROLYSIS OF THE CHYLOMICRON TRIGLYCERIDES BY LIPOPROTEIN LIPASE; FAT STORAGE IN THE FAT AND LIVER CELLS. Most of the chylomicrons are removed from the circulating blood as they pass through the capillaries of adipose tissue and the liver. Both adipose tissue and the liver contain large quantities of the enzyme called *lipoprotein lipase*. This enzyme is active in the capillary endothelium, where it hydrolyzes the triglycerides of chylomicrons that stick to the endothelial wall, releasing fatty acids and glycerol. The fatty acids, being highly miscible with the membranes of the cells, immediately diffuse into the fat cells of adipose tissue and the liver cells. Once within these cells, the fatty acids are resynthesized into triglycerides, new glycerol being supplied by the metabolic processes of the cells, as discussed later in the chapter. The lipase also causes hydrolysis of phospholipids, this, too, releasing fatty acids to be stored in the cells in the same way. Thus, most of the mass of the chylomicrons is removed from the circulating blood; then the remnants are taken up mainly by the liver.

Transport of Fatty Acids in the Blood in Combination with Albumin—"Free Fatty Acid"

When the fat that has been stored in the adipose tissue is to be used elsewhere in the body, usually to provide energy, it must first be transported to the other tissues. It is transported mainly in the form of *free fatty acid*. This is achieved by hydrolysis of the triglycerides back into fatty acids and glycerol. At least two classes of stimuli play important roles in promoting this hydrolysis. First, when the availability of glucose to the fat cell falls too low, one of its breakdown products, *α-glycerophosphate*, also becomes too low. This substance is required to form the glycerol portion of newly synthesized triglycerides, and in its absence, the equilibrium shifts in favor of hydrolysis. Second, a *hormone-sensitive cellular lipase* can be activated by several hormones, and this promotes rapid hydrolysis of the triglyceride. This also is discussed later in the chapter.

On leaving the fat cells, the fatty acids ionize strongly in the plasma and immediately combine with albumin molecules of the plasma proteins. The fatty acid bound in this manner is called *free fatty acid* or *nonesterified fatty acid* to distinguish it from other fatty acids in the plasma that exist in the form of esters of glycerol, cholesterol, or other substances.

The concentration of free fatty acid in the plasma under resting conditions is about 15 mg/dl, which is a total of only 0.45 gram of fatty acids in the entire circulatory system. Strangely enough, even this small amount accounts for almost all the transport of fatty acids from one part of the body to another for the following reasons:

1. Despite the minute amount of free fatty acid in the blood, its rate of "turnover" is extremely rapid; *one half the plasma fatty acid is replaced by new fatty acid every 2 to 3 minutes.* One can calculate that at this rate, almost all the normal energy requirements of the body can be provided by oxidation of the transported free fatty acid without using any carbohydrates or proteins for energy.

2. All conditions that increase the rate of utilization of fat for cellular energy also increase the free fatty acid concentration in the blood; this concentration sometimes increases fivefold to eightfold. This increase occurs especially in starvation and diabetes, in both of which conditions the person derives little or no energy from carbohydrates.

Under normal conditions, about 3 molecules of fatty acid combine with each molecule of albumin, but as many as 30 fatty acid molecules can combine with a single albumin molecule when the need for fatty acid transport is extreme. This shows how variable the rate of lipid transport can be with different physiological needs.

Lipoproteins—Their Special Function in Transporting Cholesterol and Phospholipids

In the postabsorptive state—that is, after all the chylomicrons have been removed from the blood—more than 95 per cent of all the lipids in the plasma are in the form of *lipoproteins,* which are small particles much smaller than chylomicrons but qualitatively similar in composition, containing *triglycerides, cholesterol, phospholipids,* and *protein.* The protein averages about one fourth to one third of the total constituents and the remainder is lipids. The total concentration of lipoproteins in the plasma averages about 700 mg/dl, and this can be broken down into the following average concentrations of the individual constituents:

	mg/dl of plasma
Cholesterol	180
Phospholipids	160
Triglycerides	160
Protein	200

TYPES OF LIPOPROTEINS. Aside from the chylomicrons, which are themselves large lipoproteins, there are four major classes of lipoproteins classified by their densities as measured in the ultracentrifuge: (1) *very low density lipoproteins,* which contain high concentrations of triglycerides and moderate concentrations of both cholesterol and phospholipids; (2) *intermediate-density lipoproteins,* which are very low density lipoproteins from which a large share of the triglycerides have been removed, so that the concentrations of cholesterol and phospholipids are increased; (3) *low-density lipoproteins,* which are intermediate-density lipoproteins from which almost all the triglycerides have been removed, leaving an especially high concentration of cholesterol and a moderately high concentration of phospholipids; and (4) *high-density lipoproteins,* which contain a high concentration of protein, about 50 per cent, but smaller concentrations of cholesterol and phospholipids.

FORMATION AND FUNCTION OF THE LIPOPROTEINS. Almost all the lipoproteins are formed in the liver, which is where most of the plasma cholesterol, phospholipids, and triglycerides (except those absorbed from the intestines in the chylomicrons) are synthesized. Small quantities of high-density lipoproteins are also synthesized in the intestinal epithelium during the absorption of fatty acids from the intestines.

The primary function of the lipoproteins is to transport their lipid components in the blood. The very low density lipoproteins transport triglycerides synthesized in the liver mainly to the adipose tissue, whereas the other lipoproteins are especially important in the different stages of phospholipid and cholesterol transport from the liver to the peripheral tissues or from the periphery back to the liver. Later in the chapter, we discuss in more detail special problems of cholesterol transport in relation to the disease *atherosclerosis*.

FAT DEPOSITS

Adipose Tissue

Large quantities of fat are stored in two major tissues of the body, the adipose tissue and the liver. The adipose tissue is usually called the *fat deposits*, or simply the *fat depots*.

The major function of adipose tissue is storage of triglycerides until they are needed to provide energy elsewhere in the body. A subsidiary function is to provide heat insulation for the body, as discussed in Chapter 73.

FAT CELLS. The fat cells of adipose tissue are modified fibroblasts that are capable of storing almost pure triglycerides in quantities equal to 80 to 95 per cent of their volume. The triglycerides are generally in a liquid form, and when the tissues of the skin are exposed to prolonged cold, the fatty acid chains of the triglycerides, over a period of weeks, become either shorter or more unsaturated to decrease their melting point, thereby always allowing the fat to remain in a liquid state. This is particularly important because only liquid fat can be hydrolyzed and transported from the cells.

Fat cells can synthesize very, very small amounts of fatty acids and triglycerides from carbohydrates, this function supplementing the synthesis of fat in the liver, as discussed later in the chapter.

EXCHANGE OF FAT BETWEEN THE ADIPOSE TISSUE AND THE BLOOD—TISSUE LIPASES. As discussed above, large quantities of lipases are present in adipose tissue. Some of these enzymes catalyze the deposition of triglycerides from the chylomicrons and very low density lipoproteins. Others, when activated by hormones, cause splitting of the triglycerides of the fat cells to release free fatty acids. Because of rapid exchanges of the fatty acids, the triglycerides in the fat cells are renewed about once every 2 to 3 weeks, which means that the fat stored in the tissues today is not the same fat that was stored last month, thus emphasizing the dynamic state of the storage fat.

Liver Lipids

The principal functions of the liver in lipid metabolism are (1) to degrade fatty acids into small compounds that can be used for energy; (2) to synthesize triglycerides, mainly from carbohydrates but to a lesser extent from proteins as well; and (3) to synthesize other lipids from fatty acids, especially cholesterol and phospholipids.

Large quantities of triglycerides appear in the liver (1) during starvation, (2) in diabetes mellitus, and (3) in any other condition in which fat is being used rapidly for energy. In these conditions, large quantities of triglycerides are mo-

bilized from the adipose tissue, transported as free fatty acids in the blood, and then redeposited as triglycerides in the liver, where the initial stages of much of the fat degradation begin. Thus, under normal physiological conditions, the total amount of triglycerides in the liver is determined to a great extent by the overall rate at which lipids are being used for energy.

The liver cells, in addition to containing triglycerides, contain large quantities of phospholipids and cholesterol, which are continually synthesized by the liver. Also, the liver cells are much more capable than other tissues of desaturating fatty acids, so that the liver triglycerides normally are much more unsaturated than the triglycerides of the adipose tissue. This capability of the liver to desaturate fatty acids is functionally important to all the tissues of the body because many of the structural members of all cells contain reasonable quantities of unsaturated fats, and their principal source is the liver. This desaturation is accomplished by a dehydrogenase in the liver cells.

USE OF TRIGLYCERIDES FOR ENERGY: FORMATION OF ADENOSINE TRIPHOSPHATE

About 40 per cent of the calories in the normal American diet are derived from fats, which is almost equal to the calories derived from carbohydrates. Therefore, the use of fats by the body for energy is as important as the use of carbohydrates. In addition, many of the carbohydrates ingested with each meal are converted into triglycerides, then stored, and later used as fatty acids released from the triglycerides for energy.

HYDROLYSIS OF THE TRIGLYCERIDES. The first stage in the use of triglycerides for energy is hydrolysis of the triglycerides into fatty acids and glycerol. Then, both the fatty acids and the glycerol are transported to the active tissues where they will be oxidized to give energy. Almost all cells, with some degree of exception for brain tissue, can use fatty acids almost interchangeably with glucose for energy.

The glycerol, on entering the active tissue, is immediately changed by intracellular enzymes into *glycerol 3-phosphate*, which enters the glycolytic pathway for glucose breakdown and in this way is used for energy. Before the fatty acids can be used for energy, they must be processed further in the following way.

ENTRY OF FATTY ACIDS INTO THE MITOCHONDRIA. The degradation and oxidation of fatty acids occur only in the mitochondria. Therefore, the first step for use of the fatty acids is their transport into the mitochondria. This is a carrier-mediated process that uses *carnitine* as the carrier substance. Once inside the mitochondria, the fatty acid splits away from the carnitine and is then degraded and oxidized.

DEGRADATION OF FATTY ACID TO ACETYLCOENZYME A BY BETA OXIDATION. The fatty acid molecule is degraded in the mitochondria by progressive release of two-carbon segments in the form of acetylcoenzyme A (acetyl-CoA). This process, which is shown in Figure 68–1, is called the *beta oxidation* process for degradation of fatty acids.

To understand the essential steps in the beta oxidation process, note in Equation 1 that the first step is combination of the fatty acid molecule with coenzyme A (CoA) to form fatty acyl-CoA.

Then, in Equations 2, 3, and 4, through several chemical

Thiokinase

(1) $RCH_2CH_2CH_2COOH + CoA + ATP \rightleftharpoons RCH_2CH_2CH_2COCoA + AMP + Pyrophosphate$
(Fatty acid)   (Fatty acyl CoA)

Acyl dehydrogenase

(2) $RCH_2CH_2CH_2COCoA + FAD \longrightarrow RCH_2CH=CHCOCoA + FADH_2$
(Fatty acyl CoA)

Enoyl hydrase

(3) $RCH_2CH=CHCOCoA + H_2O \rightleftharpoons RCH_2CHOHCH_2COCoA$

β -Hydroxyacyl

(4) $RCH_2CHOHCH_2COCoA + NAD^+ \rightleftharpoons RCH_2COCH_2COCoA + NADH + H^+$

Dehydrogenase

Thiolase

(5) $RCH_2COCH_2COCoA + CoA \rightleftharpoons RCH_2COCoA + CH_3COCoA$
                                          (Fatty acyl CoA)(Acetyl CoA)

Figure 68–1. Beta oxidation of fatty acids to yield acetylcoenzyme A.

steps, the *beta carbon* (the second carbon from the right) of the fatty acyl-CoA binds with an oxygen molecule—that is, the beta carbon is oxidized.

Then, in Equation 5, the right-hand two-carbon portion of the molecule is split off to release acetyl-CoA into the cell fluid. At the same time, another coenzyme A (CoA) molecule binds at the end of the remaining portion of the fatty acid molecule, and this forms a new fatty acyl-CoA molecule, but this time two carbon atoms shorter than before because of loss of the acetyl-CoA from its terminal end.

Next, this shorter fatty acyl-CoA enters into Equation 2 and progresses through Equations 3, 4, and 5 to release still another acetyl-CoA molecule, thus shortening the original fatty acid molecule still another two carbons.

In this manner, by multiple revolutions of this set of equations, the entire original fatty acid molecule is dissected away to form multiple acetyl-CoA molecules. For each acetyl-CoA molecule split from the fatty acid, a total of four hydrogen atoms are released. They are later oxidized in the mitochondria to form large amounts of adenosine triphosphate (ATP), as we discuss shortly.

OXIDATION OF ACETYL-CoA. The acetyl-CoA molecules formed by beta oxidation of fatty acids in the mitochondria enter immediately into the *citric acid cycle*, as explained in Chapter 67, combining first with oxaloacetic acid to form citric acid, which then is degraded into carbon dioxide and hydrogen atoms. The hydrogen is subsequently oxidized by the *chemiosmotic oxidative system of the mitochondria*, which is explained in Chapter 67. The net reaction in the citric acid cycle for each molecule of acetyl-CoA is the following:

$$CH_3COCo\text{-}A + \text{Oxaloacetic acid} + 3H_2O + ADP$$

$$\xrightarrow{\textit{Citric acid cycle}}$$

$$2CO_2 + 8H + HCo\text{-}A + ATP + \text{Oxaloacetic acid}$$

Thus, after the initial degradation of fatty acids to acetyl-CoA, their final breakdown is precisely the same as that of the acetyl-CoA formed from pyruvic acid during the metabolism of glucose. Then the hydrogen is oxidized by the same *chemiosmotic oxidative system of the mitochondria* that is used in carbohydrate oxidation, liberating tremendous amounts of ATP.

TREMENDOUS AMOUNTS OF ATP ARE FORMED BY OXIDATION OF FATTY ACID. In Figure 68–1, note also that four

hydrogen atoms are released in the forms of $FADH_2$, NADH, and H^+ each time a molecule of acetyl-CoA is formed from the fatty acid chain. Therefore, for every stearic acid molecule that is split, a total of 32 hydrogen atoms are removed. In addition, for each of the 9 molecules of acetyl-CoA degraded by the citric acid cycle, 8 hydrogen atoms are removed, making an additional 72 hydrogens for each molecule of stearic acid metabolized. This, added to the 32 hydrogen atoms above, makes a total of 104 hydrogen atoms. Of this group, 34 are removed from the degrading fatty acid by flavoproteins and 70 are removed by nicotinamide adenine dinucleotide (NAD^+) as NADH and H^+. These two groups of hydrogen atoms are oxidized in the mitochondria, as discussed in Chapter 67, but they enter the oxidative system at different points, so that up to 1 molecule of ATP is synthesized for each of the 34 flavoprotein hydrogens and up to 1.5 molecules of ATP are synthesized for each of the 70 NADH and H^+ hydrogens. This makes 34 plus 105, or a total of 139 molecules of ATP formed by the oxidation of hydrogen derived from each molecule of stearic acid. Another nine molecules of ATP are formed in the citric acid cycle, one for each of the nine acetyl-CoA molecules metabolized. Thus, 148 molecules of ATP are formed during the complete oxidation of one molecule of stearic acid. However, two high-energy bonds are consumed in the initial combination of coenzyme A with the fatty acid molecule, making a net gain of 146 molecules of ATP.

Formation of Acetoacetic Acid in the Liver and Its Transport in the Blood

A large share of the initial degradation of fatty acids occurs in the liver, especially when excessive amounts of lipids are being mobilized for energy. However, the liver uses only a small proportion of the fatty acids for its own intrinsic metabolic processes. Instead, when the fatty acid chains have been split into acetyl-CoA, two molecules of acetyl-CoA condense to form one molecule of acetoacetic acid, which is then transported in the blood to the other cells throughout the body, where it is used for energy. The chemical processes are the following:

$$2CH_3COCo\text{-}A + H_2O \underset{\text{other cells}}{\overset{\text{liver cells}}{\rightleftharpoons}}$$
Acetyl-CoA

$$CH_3COCH_2COOH + 2HCo\text{-}A$$
Acetoacetic acid

Part of the acetoacetic acid is also converted into *β-hydroxybutyric acid,* and minute quantities into *acetone* in accord with the following reactions:

$$CH_3-\overset{\overset{\displaystyle O}{\|}}{C}-CH_2-\overset{\overset{\displaystyle O}{\|}}{C}-OH$$

Acetoacetic acid

+ 2H

$$CH_3-\overset{\overset{\displaystyle OH}{|}}{CH}-CH_2-\overset{\overset{\displaystyle O}{\|}}{C}-OH \qquad \xrightarrow{-CO_2} \qquad CH_3-\overset{\overset{\displaystyle O}{\|}}{C}-CH_3$$

β-Hydroxybutyric acid   Acetone

The acetoacetic acid, β-hydroxybutyric acid, and acetone diffuse freely through the liver cell membranes and are transported by the blood to the peripheral tissues. Here they again diffuse into the cells, where reverse reactions occur and acetyl-CoA molecules are formed. These in turn enter the citric acid cycle and are oxidized for energy, as already explained.

Normally, the acetoacetic acid and β-hydroxybutyric acid that enter the blood are transported so rapidly to the tissues that their combined concentration in the plasma seldom rises above 3 mg/dl. Yet despite the small quantities in the blood, large amounts are actually transported; this is analogous to the high rate of free fatty acid transport. The rapid transport of both these substances results from their high solubility in the membranes of the target cells, which allows almost instantaneous diffusion into the cells.

Ketosis and Its Occurrence in Starvation, Diabetes, and Other Diseases

The concentrations of acetoacetic acid, β-hydroxybutyric acid, and acetone occasionally rise high in the blood and interstitial fluids; this condition is called *ketosis* because acetoacetic acid is a keto acid, and the three compounds are called *ketone bodies.* Ketosis occurs especially in starvation, in diabetes mellitus, and sometimes even when a person's diet is composed almost entirely of fat. In all these states, essentially no carbohydrates are metabolized—in starvation and after a high-fat diet because carbohydrates are not available and in diabetes because insulin is not available to cause glucose transport into the cells.

When carbohydrates are not used for energy, almost all the energy of the body must come from metabolism of fats. We see later in the chapter that lack of availability of carbohydrates automatically increases the rate of removal of fatty acids from adipose tissues; in addition, several hormonal factors—such as increased secretion of glucocorticoids by the adrenal cortex, increased secretion of glucagon by the pancreas, and decreased secretion of insulin by the pancreas —further enhance the removal of fatty acids from the fat tissues. As a result, tremendous quantities of fatty acids become available to the peripheral tissue cells to be used for energy and to the liver cells, where much of the fatty acids is converted to ketone bodies.

The ketone bodies pour out of the liver to be carried to the cells. Yet the cells are limited in the amount of ketone bodies that can be oxidized for several reasons, the most important of which is the following: One of the products of carbohydrate metabolism is the *oxaloacetate* that is required to bind with acetyl-CoA before it can be processed in the citric acid cycle. Therefore, deficiency of oxaloacetate derived from carbohydrates limits the entry of acetyl-CoA into the citric acid cycle, and because of the simultaneous outpouring of tremendous quantities of acetoacetic acid and the other ketone bodies from the liver, the blood concentrations of acetoacetic acid and β-hydroxybutyric acid sometimes rise to as high as 20 times normal, thus leading to extreme acidosis, as explained in Chapter 30.

The acetone that is formed during ketosis is a volatile substance, some of which is blown off in small quantities in the expired air of the lungs, often giving the breath an acetone smell. This smell is frequently used as a diagnostic criterion of ketosis.

Adaptation to a High-Fat Diet. On changing slowly from a carbohydrate diet to an almost completely fat diet, a person's body adapts to the use of far more acetoacetic acid than usual, and in this instance, ketosis normally does not occur. For instance, the Eskimos, who sometimes live almost entirely on a fat diet, do not develop ketosis. Undoubtedly several factors enhance the rate of acetoacetic acid metabolism by the cells. Even the brain cells, which normally derive almost all their energy from glucose, after a few weeks can derive 50 to 75 per cent of their energy from fats.

Synthesis of Triglycerides from Carbohydrates

Whenever a greater quantity of carbohydrates enters the body than can be used immediately for energy or stored in the form of glycogen, the excess is rapidly converted into triglycerides and stored in this form in the adipose tissue. In human beings, most triglyceride synthesis occurs in the liver, but minute quantities are also synthesized in the adipose tissue. The triglycerides formed in the liver are transported mainly by the very low density lipoproteins to the adipose tissue.

CONVERSION OF ACETYL-CoA INTO FATTY ACIDS. The first step in the synthesis of triglycerides is conversion of carbohydrates into acetyl-CoA. It will be recalled from Chapter 67 that this occurs during the normal degradation of glucose by the glycolytic system. It will also be remembered from earlier in this chapter that fatty acids are actually large polymers of acetic acid. Therefore, it is easy to understand how acetyl-CoA could be converted into fatty acids.

However, synthesis of the fatty acids from acetyl-CoA is not achieved by simply reversing the oxidative degradation described above. Instead, this occurs by the two-step process shown in Figure 68–2, using *malonyl CoA* and NADPH as the principal intermediates in the polymerization process.

COMBINATION OF FATTY ACIDS WITH α-GLYCEROPHOSPHATE TO FORM TRIGLYCERIDES. Once the synthesized fatty acid chains have grown to contain 14 to 18 carbon atoms, they then become bound to glycerol to form triglycerides. The enzymes that cause this conversion are highly specific for fatty acids with chain lengths of 14 carbon atoms or greater, a factor that controls the physical quality of the triglycerides stored in the body.

As shown in Figure 68–3, the glycerol portion of the triglyceride is furnished by α-glycerophosphate, which is a product derived from the glycolytic scheme of glucose degradation. The mechanism of this is discussed in Chapter 67.

Step 1:

$$CH_3COCoA + CO_2 + ATP$$

$\Uparrow$ (Acetyl CoA carboxylase)

COOH
|
CH_2 $+ ADP + PO_4^{---}$
|
$O = C - CoA$
Malonyl CoA

Step 2:

$$1 \ Acetyl \ CoA + 8 \ Malonyl \ CoA + 16NADPH + 16H^+ \longrightarrow$$

$$1 \ Stearic \ acid + 8CO_2 + 9CoA + 16NADP^+ + 7H_2O$$ Figure 68–2. Synthesis of fatty acids.

EFFICIENCY OF CARBOHYDRATE CONVERSION INTO FAT. During triglyceride synthesis, only about 15 per cent of the original energy in the glucose is lost in the form of heat, whereas the remaining 85 per cent is transferred to the stored triglycerides.

IMPORTANCE OF FAT SYNTHESIS AND STORAGE. Fat synthesis from carbohydrates is especially important for two reasons: (1) The ability of the different cells of the body to store carbohydrates in the form of glycogen is generally slight; a maximum of only a few hundred grams of glycogen can be stored in the liver, the skeletal muscles, and all other tissues of the body put together. In contrast, many kilograms of fat can be stored. Therefore, fat synthesis provides a means by which the energy of excess ingested carbohydrates (and proteins) can be stored for later use. Indeed, the average person has almost 150 times as much energy stored in the form of fat as stored in the form of carbohydrate. (2) Each gram of fat contains almost two and one-half times as many calories of energy as each gram of glycogen. Therefore, for a given weight gain, a person can store several times as much energy in the form of fat as in the form of carbohydrate, which is exceedingly important when an animal must be highly motile to survive.

FAILURE TO SYNTHESIZE FATS FROM CARBOHYDRATES IN THE ABSENCE OF INSULIN. When no insulin is available, as occurs in serious diabetes mellitus, fats are poorly, if at all, synthesized for the following reasons: First, when insulin is not available, glucose does not enter the fat and liver cells satisfactorily, so that little of the acetyl-CoA and NADPH needed for fat synthesis can be derived from glucose. Second, lack of glucose in the fat cells greatly reduces the availability of α-glycerophosphate, which also makes it difficult for the tissues to form triglycerides.

Synthesis of Triglycerides from Proteins

Many amino acids can be converted into acetyl-CoA, as is discussed in Chapter 69. This can then be synthesized into triglycerides. Therefore, when people have more proteins in their diets than their tissues can use as proteins, a large share of the excess is stored as fat.

REGULATION OF ENERGY RELEASE FROM TRIGLYCERIDES

CARBOHYDRATES ARE USED IN PREFERENCE TO FAT WHEN EXCESS CARBOHYDRATE IS AVAILABLE. When excess quantities of carbohydrates are available in the body, carbohydrates are used preferentially over triglycerides for energy. There are several reasons for this "fat-sparing" effect of carbohydrate. One of the most important is the following: The fats in adipose tissue cells are present in two forms, the stored triglycerides and small quantities of free fatty acids. They are in constant equilibrium with each other. When excess quantities of α-*glycerophosphate* are present (which occurs when excess carbohydrate is available), the excess α-glycerophosphate binds the free fatty acids in the form of stored triglycerides. As a result, the equilibrium between free fatty acids and triglycerides shifts toward the stored triglycerides; consequently, only minute quantities of fatty acids are then available to be used for energy. Because α-glycerophosphate is an important product of glucose metabolism, the availability of large amounts of glucose automatically inhibits the use of fatty acids for energy.

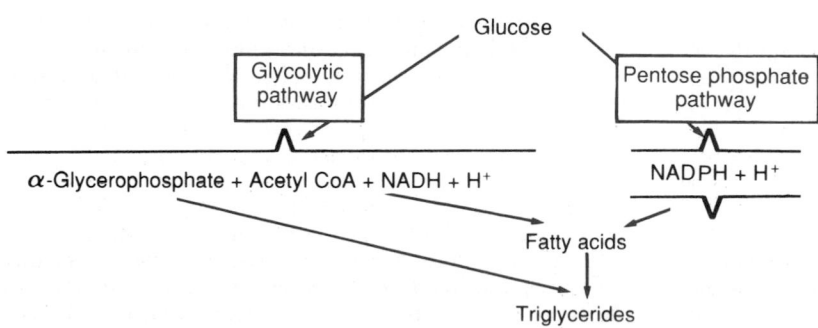

Figure 68–3. Overall schema for synthesis of triglycerides from glucose.

Second, when carbohydrates are available in excess, fatty acids are synthesized more rapidly than they are degraded. This effect is caused partially by the large quantities of acetyl-CoA formed from the carbohydrates and by the low concentration of free fatty acids in the adipose tissue, thus creating conditions appropriate for conversion of acetyl-CoA into fatty acids. An even more important effect that promotes conversion of carbohydrates to fats is the following: The first step, and the rate-limiting step, in the synthesis of fatty acids is carboxylation of acetyl-CoA to form malonyl-CoA, as discussed previously. The rate of this reaction is controlled primarily by the enzyme *acetyl-CoA carboxylase*, the activity of which is accelerated in the presence of the intermediates of the citric acid cycle. When excess carbohydrates are being used, these intermediates increase, thus automatically causing increased synthesis of fatty acids. Thus, an excess of carbohydrates in the diet not only acts as a fat-sparer but also increases the fat in the fat stores. In fact, all the excess carbohydrate not used for energy or stored in the small glycogen deposits of the body is converted to fat and stored as such.

ACCELERATION OF FAT UTILIZATION FOR ENERGY IN THE ABSENCE OF CARBOHYDRATES. All the fat-sparing effects of carbohydrates are lost and actually reversed when carbohydrates are not available or are available in short supply. Therefore, the equilibrium now shifts in the opposite direction, and fat is mobilized from the adipose cells and used for energy in place of the absent carbohydrates.

But also important are several hormonal changes that take place to promote rapid fatty acid mobilization from adipose tissue. Among the most important of these is a marked decrease in pancreatic secretion of insulin caused by the absence of carbohydrates. This not only reduces the rate of glucose utilization by the tissues but also decreases fat storage, which further shifts the equilibrium in favor of fat metabolism in place of carbohydrates.

HORMONAL REGULATION OF FAT UTILIZATION. At least seven of the hormones secreted by the endocrine glands have significant or even large effects on fat utilization. Some important hormonal effects on fat metabolism—in addition to the important effect of *insulin lack* discussed in the previous paragraph—are listed here.

Probably the most dramatic increase that occurs in fat utilization is that observed during heavy exercise. This results almost entirely from release of *epinephrine* and *norepinephrine* by the adrenal medullae during exercise, as a result of sympathetic stimulation. These two hormones directly activate *hormone-sensitive triglyceride lipase* that is present in abundance in the fat cells, and this causes rapid breakdown of triglycerides and mobilization of fatty acids. Sometimes the free fatty acid concentration in the blood rises as much as eightfold, and the use of these fatty acids by the muscles for energy is correspondingly increased. Other types of stress that activate the sympathetic nervous system can also increase fatty acid mobilization and utilization in a similar manner.

Stress also causes large quantities of *corticotropin* to be released by the anterior pituitary gland, and this in turn causes the adrenal cortex to secrete excessive quantities of *glucocorticoids*. Both the corticotropin and glucocorticoids activate either the same hormone-sensitive triglyceride lipase as that activated by epinephrine and norepinephrine or a similar lipase. Therefore, this is still another mechanism for increasing the release of fatty acids from fat tissue. When corticotropin and glucocorticoids are secreted in excessive amounts for long periods, as occurs in the endocrine disease called Cushing's disease, fats are frequently

mobilized to such a great extent that ketosis results. Corticotropin and glucocorticoids are then said to have a *ketogenic effect*.

Growth hormone has an effect similar to but less than that of corticotropin and glucocorticoids in activating the hormone-sensitive lipase. Therefore, growth hormone can also have a mild ketogenic effect.

Finally, *thyroid hormone* causes rapid mobilization of fat, which is believed to result indirectly from an increased overall rate of energy metabolism in all cells of the body under the influence of this hormone. The resulting reduction in acetyl-CoA and other intermediates of both fat and carbohydrate metabolism in the cells would then be a stimulus to cause fat mobilization.

The effects of the different hormones on metabolism are discussed further in the chapters dealing with each of them.

Obesity

Obesity means deposition of excess fat in the body. This subject is discussed in relation to dietary balances in Chapter 71, but briefly, it is caused by ingestion of greater amounts of food than can be used by the body for energy. The excess food, whether fats, carbohydrates, or proteins, is then stored as fat in the adipose tissue to be used later for energy.

Strains of rats have been found in which *hereditary obesity* occurs. In at least one of these, the obesity is caused by ineffective mobilization of fat from the adipose tissue by tissue lipase while synthesis and storage of fat continue normally. Such a one-way process causes progressive enhancement of the fat stores, resulting in severe obesity.

PHOSPHOLIPIDS AND CHOLESTEROL

Phospholipids

The major types of body phospholipids are *lecithins, cephalins,* and *sphingomyelins*, typical chemical formulas for which are shown in Figure 68–4.

Phospholipids always contain one or more fatty acid molecules and one phosphoric acid radical, and they usually contain a nitrogenous base. Although the chemical structures of phospholipids are somewhat variant, their physical properties are similar because they are all lipid-soluble, transported in lipoproteins, and used throughout the body for various structural purposes, such as for use in cell membranes and intracellular membranes.

FORMATION OF PHOSPHOLIPIDS. Phospholipids are synthesized in essentially all cells of the body, although certain cells have a special ability to form great quantities of them. Probably 90 per cent are formed in the liver cells; reasonable quantities are also formed by the intestinal epithelial cells during lipid absorption from the gut.

The rate of phospholipid formation is governed to some extent by the usual factors that control the rate of fat metabolism because, when triglycerides are deposited in the liver, the rate of phospholipid formation increases. Also, certain specific chemical substances are needed for formation of some phospholipids. For instance, *choline*, either in the diet or synthesized in the body, is needed for the formation of lecithin because choline is the nitrogenous base of the leci-

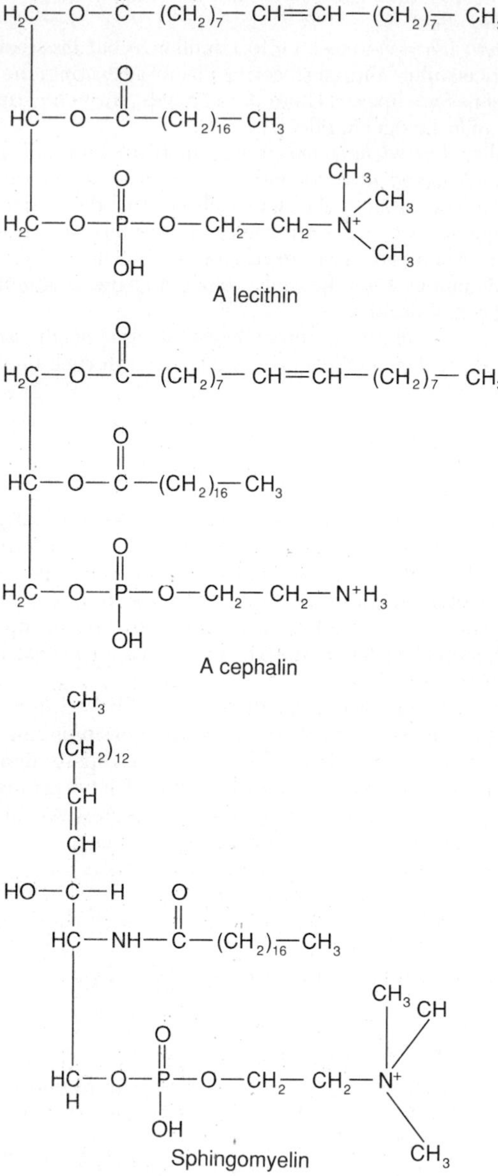

A lecithin

A cephalin

Sphingomyelin

Figure 68–4. Typical phospholipids.

thin molecule. Also, *inositol* is needed for the formation of some cephalins.

SPECIFIC USE OF PHOSPHOLIPIDS. Several isolated functions of the phospholipids are the following: (1) Phospholipids are an important constituent of lipoproteins in the blood and are essential for the formation and function of most of these; in their absence, serious abnormalities of transport of cholesterol and other lipids can occur. (2) Thromboplastin, which is needed to initiate the clotting process, is composed mainly of one of the cephalins. (3) Large quantities of sphingomyelin are present in the nervous system; this substance acts as an insulator in the myelin sheath around nerve fibers. (4) Phospholipids are donors of phosphate radicals when they are needed for different chem-

ical reactions in the tissues. (5) Perhaps the most important of all the functions of phospholipids is participation in the formation of structural elements—mainly membranes—within the cells throughout the body, as discussed later in connection with a similar function for cholesterol.

Cholesterol

Cholesterol, the formula of which is shown in Figure 68–5, is present in the diet of all people, and it can be absorbed slowly from the gastrointestinal tract into the intestinal lymph. It is highly fat-soluble but only slightly soluble in water and is capable of forming esters with fatty acids. Indeed, about 70 per cent of the cholesterol in the lipoproteins of the plasma is in the form of cholesterol esters.

FORMATION OF CHOLESTEROL. Besides the cholesterol absorbed each day from the gastrointestinal tract, which is called *exogenous cholesterol,* an even greater quantity is formed in the cells of the body, called *endogenous cholesterol.* Essentially all the endogenous cholesterol that circulates in the lipoproteins of the plasma is formed by the liver, but all the other cells of the body form at least some cholesterol, which is consistent with the fact that many of the membranous structures of all cells are partially composed of this substance.

As demonstrated by the formula of cholesterol, its basic structure is a sterol nucleus. This is synthesized entirely from multiple molecules of acetyl-CoA. In turn, the sterol nucleus can be modified by means of various side chains to form (a) cholesterol; (b) cholic acid, which is the basis of the bile acids formed in the liver; and (c) many important steroid hormones secreted by the adrenal cortex, the ovaries, and the testes (these hormones are discussed in later chapters).

FACTORS THAT AFFECT PLASMA CHOLESTEROL CONCENTRATION—FEEDBACK CONTROL OF BODY CHOLESTEROL. Among the important factors that affect plasma cholesterol concentration are the following:

1. An increase in the *amount of cholesterol ingested each day* increases the plasma concentration slightly. However, when cholesterol is ingested, the rising concentration of cholesterol inhibits the most essential enzyme for endogenous synthesis of cholesterol, 3-hydroxy-3-methylglutaryl CoA reductase, thus providing an intrinsic feedback control system to prevent excessive increase in plasma cholesterol concen-

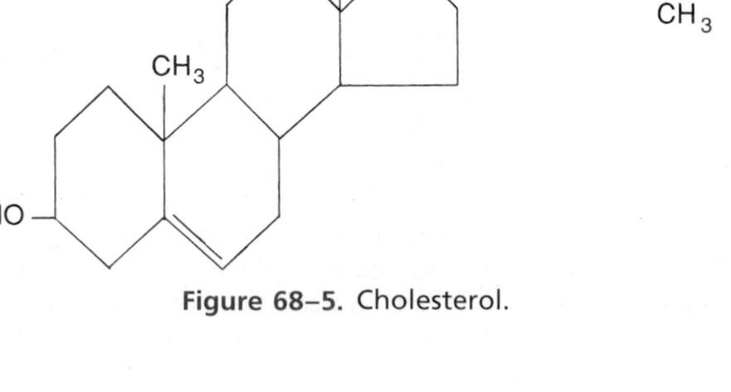

Figure 68–5. Cholesterol.

tration. As a result, plasma cholesterol concentration *usually* is not changed upward or downward more than ± 15 per cent by altering the amount of cholesterol in the diet, although the response of individuals differs markedly.

2. A *highly saturated* fat diet increases blood cholesterol concentration 15 to 25 per cent. This results from increased fat deposition in the liver, which then provides increased quantities of acetyl-CoA in the liver cells for production of cholesterol. Therefore, to decrease the blood cholesterol concentration, it is usually equally as important, if not even more important, to maintain a diet low in saturated fat than to maintain a diet low in cholesterol.

3. Ingestion of fat containing highly *unsaturated fatty acids* usually depresses the blood cholesterol concentration a slight to moderate amount. The mechanism of this effect is unknown, despite the fact that this observation is the basis of much present-day dietary strategy.

4. *Lack of insulin* or *thyroid hormone* increases the blood cholesterol concentration, whereas excess thyroid hormone decreases the concentration. These effects are probably caused mainly by changes in the activities of specific enzymes responsible for the metabolism of lipid substances.

SPECIFIC USES OF CHOLESTEROL. By far the most abundant nonmembranous use of cholesterol in the body is to form cholic acid in the liver. As much as 80 per cent of the cholesterol is converted into cholic acid. As explained in Chapter 70, this is conjugated with other substances to form bile salts, which promote digestion and absorption of fats.

A small quantity of cholesterol is used by (a) the adrenal glands to form *adrenocortical hormones,* (b) the ovaries to form *progesterone* and *estrogen,* and (c) the testes to form *testosterone.* These glands can also synthesize their own sterols and then form their hormones from these, as discussed in the chapters on endocrinology later in the text.

A large amount of cholesterol is precipitated in the corneum of the skin. This along with other lipids make the skin highly resistant to the absorption of water-soluble substances and to the action of many chemical agents because cholesterol and the other lipids are highly inert to acids and many solvents that might otherwise easily penetrate the body. Also, these lipid substances help to prevent water evaporation from the skin; without this protection, the amount of evaporation (as occurs in burn patients who have lost their skin) can be 5 to 10 liters per day instead of the usual 300 to 400 milliliters.

Cellular Structural Functions of Phospholipids and Cholesterol—Especially for Membranes

The above-listed uses of phospholipids and cholesterol are of only minor importance in comparison with their importance for forming specialized structures in all cells of the body, mainly for formation of membranes.

In Chapter 2, it is pointed out that large quantities of phospholipids and cholesterol are present in both the cell membrane and the membranes of the internal organelles of all cells. It is also known that the *ratio* of membrane cholesterol to phospholipids is especially important in determining the fluidity of the cell membranes.

For membranes to be formed, substances that are not soluble in water must be available. In general, the only substances in the body that are not soluble in water (besides the inorganic substances of bone) are the lipids and some proteins. Thus, the physical integrity of cells everywhere in the body is based mainly on phospholipids, cholesterol, and certain insoluble proteins. The polar charges on the phospholipids also reduce the interfacial tension between the cell membranes and the surrounding fluids.

Another fact that indicates the importance of phospholipids and cholesterol for the formation of structural elements of the cells is the slow turnover rates of these substances in most nonhepatic tissues—turnover rates measured in months or years. For instance, their function in the cells of the brain is mainly related to their indestructible physical properties.

ATHEROSCLEROSIS

Atherosclerosis is a disease of the large and intermediate-sized arteries in which fatty lesions called *atheromatous plaques* develop on the inside surfaces of the arterial walls. These plaques begin by deposition of minute crystals of cholesterol in the intima and sublying smooth muscle. With time, the crystals grow larger and coalesce to form large mat-like beds of crystals. In addition, the surrounding fibrous and smooth muscle tissues proliferate to form larger and larger plaques. The cholesterol deposits plus the cellular proliferation can become so large that the plaque bulges deeply into the lumen and greatly reduces blood flow, sometimes even to complete vessel occlusion. Even without occlusion, the fibroblasts of the plaque eventually deposit such extensive amounts of dense connective tissue that *sclerosis* (fibrosis) becomes so great that the arteries become stiff and unyielding. Still later, calcium salts often precipitate with the cholesterol and other lipids of the plaques, leading to bony-hard calcifications that make the arteries at times rigid tubes. Both of these latter stages of the disease are called "hardening of the arteries."

Arteriosclerotic arteries lose most of their distensibility, and because of the degenerative areas in their walls, they are easily ruptured. Also, where the plaques protrude into the flowing blood, the roughness of their surfaces causes blood clots to develop, with resultant thrombus or embolus formation (see Chapter 36), thus blocking all blood flow in the artery suddenly. Almost one half of all people in the United States and Europe die of arteriosclerosis. About two thirds of these deaths are caused by thrombosis of one or more coronary arteries. The remaining one third are caused by thrombosis or hemorrhage of vessels in other organs of the body, especially the brain, causing strokes, but also in the kidneys, liver, gastrointestinal tract, limbs, and so forth.

Basic Causes of Atherosclerosis—The Roles of Cholesterol and Lipoproteins

The factor most important in causing atherosclerosis is a high blood plasma concentration of cholesterol in the form of low-density lipoproteins. As explained earlier in this chapter, the plasma concentration of these high-cholesterol low-density lipoproteins is directly increased by increased saturated fat in the daily diet. To a lesser extent, it is also increased by increased cholesterol in the diet. Therefore, both or either of these dietary factors can contribute to the development of atherosclerosis. An interesting example of this occurs in rabbits that normally have low plasma cholesterol concentrations because of their vegetarian diet. Simply feeding these animals large quantities of cholesterol as part of their daily nutrition will lead to serious atherosclerotic plaques all through their arterial systems.

THE LIPOPROTEIN SYSTEM CONTROLS CHOLESTEROL DEPOSITION IN ALL THE BODY'S TISSUES. Three of the lipoproteins are especially concerned with controlling cholesterol transport to the deposition sites in the tissue cells. They are the *very low density lipoproteins*, the *intermediate-density lipoproteins*, and the *low-density lipoproteins*, all of which work together in a common system as follows: Of these three lipoproteins, only one is initially formed in the liver, the *very low density lipoproteins*. They contain large quantities of liver-formed triglycerides in addition to less cholesterol and phospholipids. However, as these lipoproteins circulate in the blood, the *lipoprotein lipase* in the walls of the tissue capillaries (especially in the adipose tissue) hydrolyzes a large share of the triglycerides into glycerol and fatty acids, releasing them to be stored in the fat tissue or to be used for energy, as explained earlier.

After removal from the very low density lipoproteins of much of the triglycerides, the density of the remnant lipoproteins becomes slightly greater, and they are then called *intermediate-density lipoproteins*. At this stage, many of these intermediate-density lipoproteins are attracted back to the liver cells because of receptors on the liver cell membranes for a surface protein on the lipoprotein, called *apoprotein B-100*. Normally, the liver removes about one half of these intermediate-density lipoproteins. Those that remain in the blood continue to lose almost all their remaining triglycerides by further hydrolysis in the capillaries under the influence of capillary lipoprotein lipase. As a result, the density of the lipoproteins becomes still greater, and their cholesterol and phospholipid reach their greatest concentrations. The lipoproteins are then called *low-density lipoproteins*.

The structure of the low-density lipoproteins is shown in Figure 68–6. The center of this lipoprotein is composed almost entirely of *fat-soluble esterified cholesterol*. Most of the surface, on the other hand, is composed of *phospholipids* and *nonesterified cholesterol*. Both of these surface sub-

stances have electrically charged prosthetic groups that protrude outward and provide a negative electrical charge over the surface of the lipoprotein; this negative charge allows the lipoprotein to remain soluble in the plasma. At one pole of the low-density lipoprotein is a single large molecule of apoprotein B-100 (molecular weight 400,000), a protein that provides a recognition site for the low-density lipoprotein receptors on the cell membranes of virtually all cells of the body. Attachment of this protein to the receptors causes the entire lipoprotein to be transported to the inside of the cell by the process of pinocytosis; then it is digested internally to release the lipoprotein constituents. In this way, the low-density lipoproteins deliver cholesterol and phospholipids to virtually all cells of the body to be used for cellular structural purposes.

CONTROL OF INTRACELLULAR CHOLESTEROL CONCENTRATION. When the concentration of cholesterol inside a cell becomes too great, decreased production of that cell's low-density lipoprotein receptors occurs; this now reduces the absorption of additional low-density lipoproteins. In this way, each cell controls its own internal concentration of cholesterol.

CONTROL OF CHOLESTEROL SYNTHESIS BY THE LIVER. The liver cells, like all other cells of the body, participate in the attraction and ingestion of the low-density lipoproteins. In addition, the liver cells also ingest about one half of the intermediate-density lipoproteins, as already discussed. This return of both of these types of lipoproteins to the liver cells maintains a high concentration of cholesterol in the liver cells; the greater the return, the greater the concentration. In turn, this excess cholesterol inhibits the liver enzyme system for formation of new cholesterol. Therefore, when the tissue cells throughout the body do not use cholesterol, the remainder returns to the liver and further synthesis of cholesterol is diminished.

FAMILIAL HYPERCHOLESTEROLEMIA. *Familial hypercholesterolemia* is a hereditary disease in which the person inherits defective genes for the formation of low-density lipoprotein receptors on the membrane surfaces of the body's cells. In the absence of these receptors, the liver cannot reabsorb either the intermediate-density lipoproteins or the low-density lipoproteins. Without this normal redelivery of cholesterol to the liver cells, the cholesterol machinery of the liver cells goes on a rampage of producing new cholesterol, no longer responsive to the feedback inhibition of too much plasma cholesterol. As a result, the number of very low density lipoproteins released by the liver into the plasma increases immensely, and the patient with full-blown familial hypercholesterolemia will have a blood cholesterol concentration of 600 to 1000 mg/dl, a level four to six times normal. Almost all these people die before the age of 20, some as early as 4 to 6, because of myocardial infarction or other sequelae of atherosclerotic blockage of blood vessels throughout the body.

ROLE OF HIGH-DENSITY LIPOPROTEINS IN HELPING TO PREVENT ATHEROSCLEROSIS. Much less is known about the function of high-density lipoproteins than about that of low-density lipoproteins. They are formed mainly in the liver and to a lesser extent in the intestinal epithelium during the absorption of fat from the gut. They do not contain apoprotein B-100 and, therefore, have an entirely different life history than do the very low density, intermediate-density, and low-density lipoproteins, all of which do have apoprotein B-100. An especially large portion of the high-density lipoproteins contain one of two other types of apoproteins, apoprotein A-I or apoprotein A-II, which lie on the outer sur-

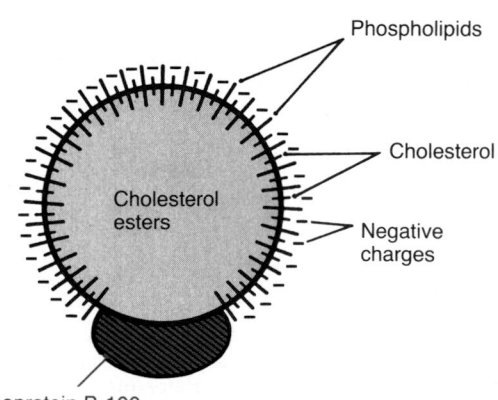

Figure 68–6. Structure and composition of a low-density lipoprotein.

face of the lipoprotein. These apoproteins have affinities for entirely different receptors on the tissue cells than those for the apoprotein B of the low-density lipoproteins, and the high-density lipoproteins also have entirely different functions. It is believed that the high-density lipoproteins can actually absorb cholesterol crystals that are beginning to deposit in the arterial walls. Whether this mechanism is true or not, the high-density lipoproteins do help to protect against the development of atherosclerosis. Consequently, when a person has a high *ratio* of high-density to low-density lipoproteins, the likelihood of developing atherosclerosis is considerably reduced.

Other Factors That Lead to Atherosclerosis

In some people with perfectly normal levels of cholesterol and lipoproteins, atherosclerosis will still develop. In most instances, the reasons are not known. Special factors that do predispose to atherosclerosis are (1) diabetes mellitus, (2) hypothyroidism, and (3) smoking. In addition, the human male in his early and mid-adult years is several times more likely to have atherosclerosis than the female, suggesting that the male sex hormones might be atherogenic or, conversely, that the female sex hormones might be protective. Some of these factors cause atherosclerosis by increasing the concentration of low-density lipoproteins in the plasma. Others lead to atherosclerosis by causing local changes in the vascular tissues that in turn predispose to cholesterol deposition.

To add to the complexity of atherosclerosis, experimental studies suggest that excess blood levels of iron can lead to atherosclerosis, perhaps by forming free radicals in the blood that damage the vessel walls. And about one quarter of all people have a special type of low-density lipoprotein called lipoprotein(a), containing in addition to apoprotein B-100 an additional protein, *apoprotein(a)*, that almost doubles the incidence of atherosclerosis. The mechanisms of these atherogenic effects are still to be discovered.

Prevention of Atherosclerosis

By far the most important preventive measure against the development of atherosclerosis is to eat a low-fat diet that contains mainly unsaturated fat with low cholesterol content.

Two types of drug therapy have also proved to be valuable. First, most of the cholesterol formed in the liver is converted into bile acids and secreted in this form into the duodenum; then more than 90 per cent of these same bile acids are reabsorbed in the terminal ileum and used over and over again in the bile. Therefore, any agent that will combine with the bile acids in the gastrointestinal tract and prevent their return to the circulation can decrease the total bile acid pool in the circulating blood. This in turn causes far more of the liver cholesterol to be converted into new bile acids. Thus, eating *oat bran*, which binds bile acids and is a constituent of many breakfast cereals, will increase the proportion of the liver cholesterol that forms new bile acids, rather than forming new low-density lipoproteins.

Second, new drugs, including *mevinolin*, have been developed that inhibit the liver enzyme system for cholesterol synthesis. They sometimes reduce the person's low-density lipoprotein cholesterol level 25 to 45 per cent.

In general, the preliminary studies show that for each 1 mg/dl decrease in low-density lipoprotein cholesterol in the plasma there is about a 2 per cent decrease in mortality from atherosclerotic heart disease. Therefore, preventive measures will probably prove to be valuable in decreasing the incidence of heart attacks.

REFERENCES

Birdi, K. S.: Lipid and Biopolymer Monolayers at Liquid Interfaces. New York, Plenum Publishing Corp., 1989.

Bjorntorp, P., and Brodoff, B. N. Obesity. Philadelphia, J. B. Lippincott, 1994.

Bracco, U., and Deckelbaum, R. J.: Polyunsaturated Fatty Acids in Human Nutrition. New York, Raven Press, 1992.

Breslow, J. L.: Apoliproprotein genetic variation and human disease. Physiol. Rev., 68:85, 1988.

Breslow, J. L.: Insights into lipoprotein metabolism from studies in transgenic mice. Annu. Rev. Physiol., 56:797, 1994.

Brown, M. S., and Goldstein, J. L.: A receptor-mediated pathway for cholesterol homeostasis. Science, 232:34, 1986.

Brownell, K. D., et al.: Eating, Body Weight and Performance in Athletes: Disorders of Modern Society. Williams & Wilkins, 1992.

Campbell, J. H., and Campbell, G. R.: Potential role of heparinase in atherosclerosis. News Physiol. Sci., 4:9, 1989.

Carey, M. C., et al.: Lipid digestion and absorption. Annu. Rev. Physiol., 45:651, 1983.

Catapano, A. L., et al.: High-Density Lipoproteins: Physiopathology and Clinical Relevance. New York, Raven Press, 1993.

Cummings, J. H., et al.: Physiological and Clinical Aspects of Short Chain Fatty Acids. New York, Cambridge University Press, 1994.

Disalvo, E. A., and Simon, S. A.: Permeability and Stability of Lipid Bilayers. Boca Raton, FL, CRC Press, Inc., 1994.

Gaber, B. P., and Schnur, J. M.: Biotechnological Applications of Lipid Microstructures. New York, Plenum Publishing Corp., 1989.

Giorgi, P. L., et al.: The Obese Child. Farmington, CT, S. Karger Publishers, Inc., 1992.

Glatz, J. F. C., et al.: Fatty acid-binding proteins and their physiological significance. News Physiol. Sci., 3:41, 1988.

Golinick, P. D.: Metabolism of substrates: Energy substrate metabolism during exercise and as modified by training. Fed. Proc., 44:353, 1985.

Gotto, A. M. Jr., and Paoletti, R.: Triglycerides: Their Role in Diabetes and Atheroslerosis. New York, Raven Press, 1991.

Guthrie, H. A.: Introductory Nutrition, 7th Ed. St. Louis, C. V. Mosby Co., 1988.

Havel, R. J.: Functional activities of hepatic lipoprotein receptors. Annu. Rev. Physiol., 48:119, 1986.

Hidalgo, C. (ed.): Physical Properties of Biological Membranes and Their Functional Implications. New York, Plenum Publishing Corp., 1988.

Hilderson, H. J. (ed.): Fluorescence Studies on Biological Membranes. New York, Plenum Publishing Corp., 1988.

Holtzman, E.: Lysosomes. New York, Plenum Publishing Corp., 1989.

Kissebah, A. H., and Krakower, G. R.: Regional adiposity and morbidity. Physiol. Rev., 74:761, 1994.

Levy, R. I., et al.: Lipoproteins and Atherosclerosis. New York, Raven Press, 1988.

Mahley, R. W.: Apolipoprotein E: cholesterol transport protein with expanding role in cell biology. Science, 240:611, 1988.

Oberman, A., et al.: Principles and Management of Lipid Disorders. Baltimore, Williams & Wilkins, 1992.

Paige, D. M.: Clinical Nutrition, St. Louis, C. V. Mosby Co., 1988.

Redgrave, T. G.: A new approach to the physiology of lipid transport. News Physiol. Sci., 3:10, 1988.

Rosell, S.: Microcirculation and transport in adipose tissue. In Renkin, E. M., and Michel, C. C. (eds.): Handbook of Physiology. Sec. 2, Vol. IV. Bethesda, Md. American Physiological Society, 1984, p. 949.

Rosenberg, R. D.: Role of heparin and heparin-like molecules in thrombosis and atherosclerosis, Fed. Proc., 44:404, 1985.

Samuel, P., et al.: The role of diet in the etiology and treatment of atherosclerosis. Annu. Rev. Med., 34:179, 1983.

Stallones, R. A.: Ischemic heart disease and lipids in blood and diet. Annu. Rev. Nutr., 3:155, 1983.

Steiner, G., and Shafrir, E.: Primary Hyperlipoproteinemias. Hightstown, NJ, McGraw-Hill, 1991.

Stokes, J. I., and Mancini, M.: Hypercholesterolemia: Clinical and Therapeutic Implications. New York, Raven Press, 1988.

Stunkard, A. J., and Wadden, T. A.: Obesity: Theory and Therapy. New York, Raven Press, 1993.

Sugano, M., and Beynen, A. C.: Dietary Proteins, Cholesterol Metabolism and Atherosclerosis. Farmington, CT, S. Karger Publishers, Inc., 1990.

Weber, P. C., and Leaf, A.: Atherosclerosis: Cellular Interactions, Growth Factors, and Lipids. New York, Raven Press, 1994.

Wight, T. N.: Proteoglycans in pathological conditions: Atherosclerosis. Fed. Proc., 44:381, 1985.

Williams, S. R.: Nutrition and Diet Therapy, 6th Ed. St. Louis, C. V. Mosby Co., 1989.

Wong, P. Y.-K., and Serhan, C. N. (eds.): Lipoxins. New York, Plenum Publishing Corp., 1988.

Woolf, N., and Davies, M. J.: Arterial plaque and thrombus formation. New York, Scientific American Science and Medicine, September/October, 1994.

Protein Metabolism

CHAPTER 69

About three quarters of the body solids are proteins. These include *structural proteins, enzymes, nucleoproteins, proteins that transport oxygen, proteins of the muscle that cause contraction,* and many other types that perform specific functions both intracellularly and extracellularly throughout the body.

The basic chemical properties of proteins that explain their diverse functions are so extensive that they constitute a major portion of the entire discipline of biochemistry. For this reason, the current discussion is confined to a few specific aspects of protein metabolism that are important as background for other discussions in this text.

BASIC PROPERTIES

Amino Acids

The principal constituents of proteins are amino acids, 20 of which are present in the body proteins in significant quantities. Figure 69–1 shows the chemical formulas of these 20 amino acids, demonstrating that they all have two features in common: Each amino acid has an acidic group (—COOH) and a nitrogen radical attached to the molecule near the acidic radical, usually represented by the amino group (—NH_2).

PEPTIDE LINKAGES AND PEPTIDE CHAINS. In proteins, the amino acids are aggregated into long chains by means of *peptide linkages.* The chemical nature of this linkage is demonstrated by the following reaction:

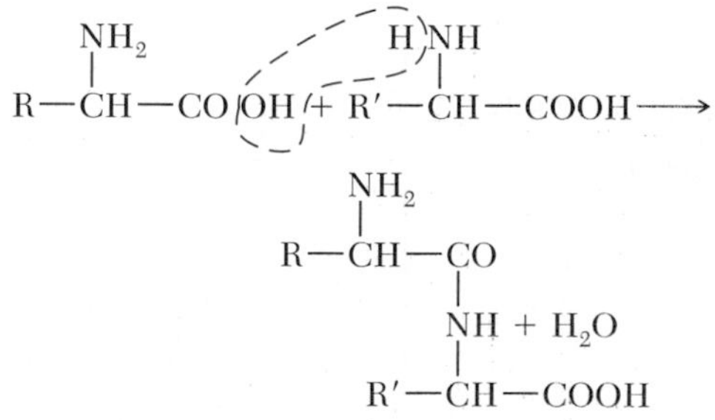

$$R-CH-CO|OH + R'-CH-COOH \longrightarrow$$

Note in this reaction that the nitrogen of the amino radical of one amino acid bonds with the carbon of the carboxyl radical of the other amino acid. A hydrogen ion is released from the amino radical and a hydroxyl ion is released from the carboxyl radical; these two combine to form a molecule of water. After the peptide linkage has been formed, an amino radical and a carboxyl radical are still at opposite ends of the new molecule, both of which are capable of combining with additional amino acids to form a *peptide chain.* Some complicated protein molecules have many thousand amino acids combined together by peptide linkages, and even the smallest protein usually has more than 20 amino acids combined together by peptide linkages. The average is about 400 amino acids.

OTHER LINKAGES IN PROTEIN MOLECULES. Some protein molecules are composed of several peptide chains rather than a single chain, and these in turn are bound with one another by other linkages, often by *hydrogen bonding* between the CO and NH radicals of the peptides as follows:

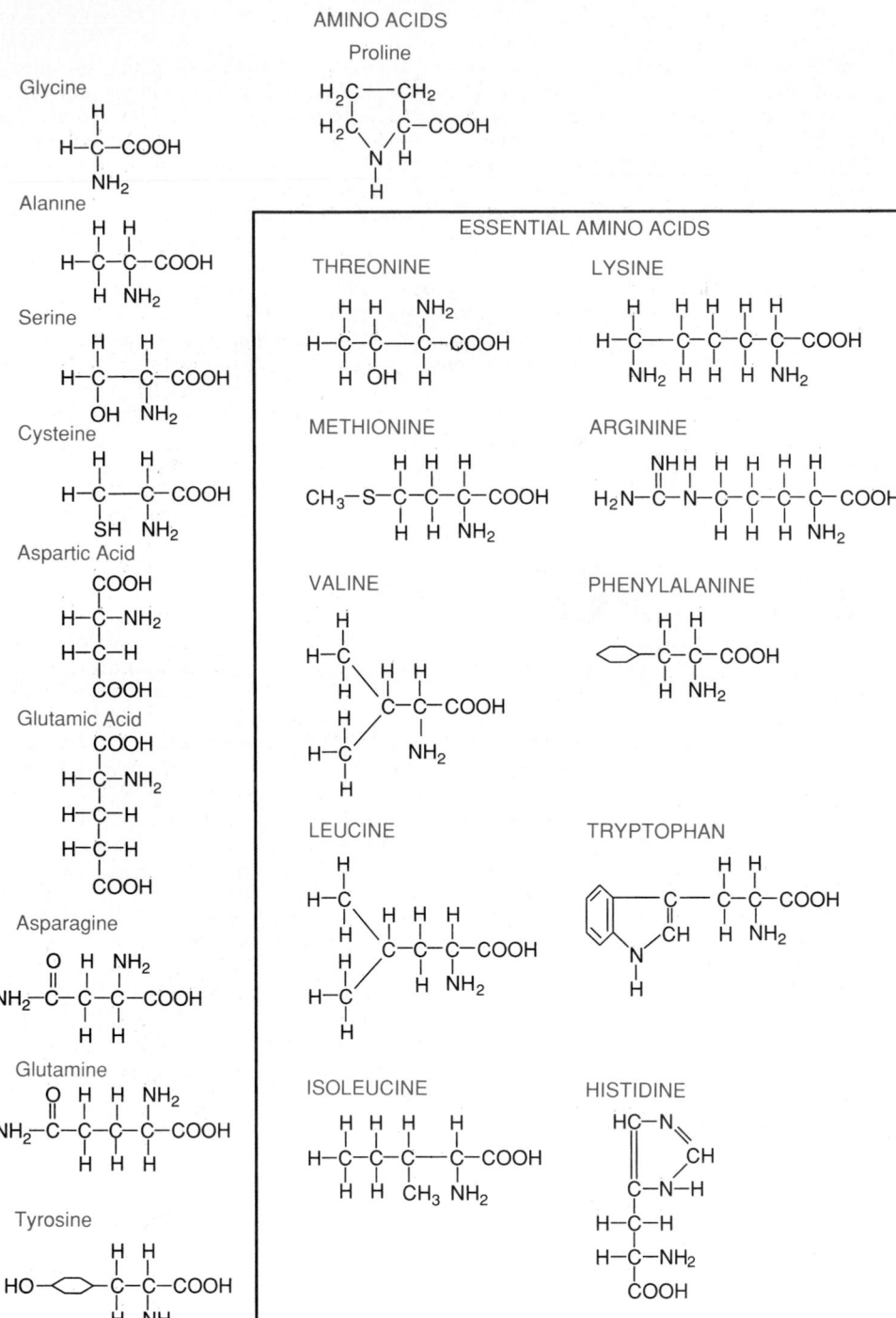

Figure 69–1. Amino acids, showing the 10 essential amino acids, which cannot be synthesized either at all or in sufficient quantity in the body.

Also, many peptide chains are coiled or folded, and the successive coils or folds are held in a tight spiral or in other shapes by similar hydrogen bonding and other forces. In addition to hydrogen bonds, separate peptide chains can be held together by hydrophobic bonds, electrostatic forces, and sulfhydryl, phenolic, and salt linkages as well as by others.

TRANSPORT AND STORAGE OF AMINO ACIDS

Blood Amino Acids

The normal concentration of amino acids in the blood is between 35 and 65 mg/dl. This is an average of about 2 mg/dl for each of the 20 amino acids, although some are present in far greater concentrations than others. Because the amino acids are relatively strong acids, they exist in the blood principally in the ionized state and account for 2 to 3 milliequivalents of the negative ions in the blood. The precise distribution of the different amino acids in the blood depends to some extent on the types of proteins ingested, but the concentrations of at least some individual amino acids are regulated by selective synthesis in the different cells.

FATE OF AMINO ACIDS ABSORBED FROM THE GASTROINTESTINAL TRACT. It will be recalled from Chapter 65 that the products of protein digestion and absorption in the gastrointestinal tract are almost entirely amino acids and that only rare polypeptide or whole protein molecules are absorbed from the digestive tract into the blood. Immediately after a meal, the amino acid concentration in the blood rises, but the rise is usually only a few milligrams per deciliter for two reasons: First, protein digestion and absorption are usually extended over 2 to 3 hours, which allows only small quantities of amino acids to be absorbed at a time. Second, after entering the blood, the excess amino acids are absorbed within 5 to 10 minutes by cells throughout the entire body, especially by the liver. Therefore, almost never do large concentrations of amino acids accumulate in the blood. Nevertheless, the turnover rate of the amino acids is so rapid that many grams of proteins can be carried from one part of the body to another in the form of amino acids each hour.

ACTIVE TRANSPORT OF AMINO ACIDS INTO THE CELLS. The molecules of essentially all the amino acids are much too large to diffuse through the pores of the cell membranes. Therefore, significant quantities of amino acids can be transported through the membrane only by facilitated or active transport using carrier mechanisms. The nature of some of the carrier mechanisms is still poorly understood, but some are discussed in Chapter 4.

Renal Threshold for Amino Acids. One of the special functions of carrier transport of amino acids is to prevent loss of amino acids in the urine. All the different amino acids can be *actively transported* through the proximal tubular epithelium, which removes them from the glomerular filtrate and returns them to the blood. However, as is true of other active transport mechanisms in the renal tubules, there is an upper limit to the rate at which each type of amino acid can be transported. For this reason, when a particular type of amino acid rises to too high a concentration in the plasma and glomerular filtrate, the excess above that which can be actively reabsorbed is lost into the urine.

In Chapter 31, it is pointed out that appropriate carrier systems for active reabsorption of certain amino acids are often deficient or lacking in the renal tubular epithelium. Under these conditions, the plasma threshold at which each affected amino acid begins to be lost in the urine is greatly reduced.

Storage of Amino Acids as Proteins in the Cells

Almost immediately after entry into the cells, amino acids are combined by peptide linkages, under the direction of the messenger RNA and ribosomal system, to form cellular proteins. Therefore, the concentrations of free amino acids inside the cells usually remain low. Thus, storage of large quantities of amino acids as such does not occur in the cells; instead, they are mainly stored in the form of actual proteins. Yet many intracellular proteins can be rapidly decomposed again into amino acids under the influence of intracellular lysosomal digestive enzymes, and these amino acids in turn can be transported back out of the cell into the blood. Special exceptions to this are the proteins in the chromosomes of the nucleus and the structural proteins such as collagen and muscle contractile proteins; these proteins do not participate significantly in this reversible storage of amino acids.

Some tissues of the body participate in the storage of amino acids to a greater extent than others. For instance, the liver, which is a large organ and has special systems for processing amino acids, can store large quantities of rapidly exchangeable proteins; this is also true to a lesser extent of the kidneys and the intestinal mucosa.

RELEASE OF AMINO ACIDS FROM THE CELLS AND REGULATION OF PLASMA AMINO ACID CONCENTRATION. Whenever the plasma amino acid concentrations fall below their normal levels, amino acids are transported out of the cells to replenish the supply in the plasma. In this way, the plasma concentration of each type of amino acid is maintained at a reasonably constant value. Later, it is noted that some of the hormones secreted by the endocrine glands are able to alter the balance between tissue proteins and circulating amino acids. Growth hormone and insulin increase the formation of tissue proteins, and the adrenocortical glucocorticoid hormones increase the concentration of circulating amino acids.

REVERSIBLE EQUILIBRIUM BETWEEN THE PROTEINS OF DIFFERENT PARTS OF THE BODY. Because cellular proteins in the liver (and to a much less extent in other tissues) can be synthesized rapidly from plasma amino acids and many of these in turn can be degraded and returned to the plasma almost equally as rapidly, there is constant equilibrium between the plasma amino acids and the labile proteins in the cells of the body. Therefore, it follows that there is also equilibrium between the proteins from one type of cell and another. For instance, if any particular tissue requires proteins, it can synthesize new proteins from the amino acids of the blood; in turn, these are replenished by degradation of proteins from other cells of the body, especially from the liver cells. These effects are particularly noticeable in relation to protein synthesis in cancer cells. Cancer cells are prolific users of amino acids; therefore, the proteins of the other cells can become markedly depleted.

UPPER LIMIT TO THE STORAGE OF PROTEINS. Each particular type of cell has an upper limit to the amount of proteins it can store. After all the cells have reached their limits, the excess amino acids in the circulation are then degraded into other products and used for energy, as discussed subsequently, or they are converted to fat or glycogen and stored in these forms.

FUNCTIONAL ROLES OF THE PLASMA PROTEINS

The major types of protein present in the plasma are *albumin, globulin,* and *fibrinogen.* The principal function of albumin is to provide *colloid osmotic pressure* in the plasma, which in turn prevents plasma loss from the capillaries, as discussed in Chapter 16. The globulins perform a number of enzymatic functions in the plasma, but equally important, they are principally responsible for both the natural and the acquired immunity that a person has against invading organisms, discussed in Chapter 34. The fibrinogen polymerizes into long fibrin threads during blood coagulation, thereby forming blood clots that help to repair leaks in the circulatory system, discussed in Chapter 36.

FORMATION OF THE PLASMA PROTEINS. Essentially all the albumin and fibrinogen of the plasma proteins as well as 50 to 80 per cent of the globulins are formed in the liver. The remainder of the globulins are formed in the lymphoid tissues. They are mainly the gamma globulins that constitute the antibodies.

The rate of plasma protein formation by the liver can be extremely high, as much as 30 gm/day. Certain disease conditions often cause rapid loss of plasma proteins; severe burns that denude large surface areas of the skin can cause loss of many liters of plasma through the denuded areas each day. The rapid production of plasma proteins by the liver is valuable in preventing death in such states. Furthermore, occasionally, a person with severe renal disease loses as much as 20 grams of plasma protein in the urine each day for months, and it is continually replaced.

USE OF PLASMA PROTEINS AS A SOURCE OF AMINO ACIDS FOR THE TISSUES. When the tissues become depleted of proteins, the plasma proteins can act as a source for rapid replacement of the tissue proteins. Indeed, in addition to use of blood amino acids to make new tissue protein, whole plasma proteins can be imbibed in toto by the tissue macrophages by pinocytosis; then, once in the cells, they are split into amino acids that are transported back into the blood and used throughout the body to build cellular proteins wherever needed. In this way, the plasma proteins function as a labile protein storage medium and represent a rapidly available source of amino acids whenever a particular tissue requires them.

REVERSIBLE EQUILIBRIUM BETWEEN THE PLASMA PROTEINS AND THE TISSUE PROTEINS. There is a constant state of equilibrium, as shown in Figure 69–2, between the plasma proteins, the amino acids of the blood, and the tissue proteins. It has been estimated from radioactive tracer studies that about 400 grams of body protein are synthesized and degraded each day as part of the continual state of flux of amino acids. This demonstrates once again the general principle of reversible exchange of amino acids among the different proteins of the body. Even during starvation or severe debilitating diseases, the ratio of total tissue proteins to total plasma proteins in the body remains relatively constant at about 33 to 1.

Because of this reversible equilibrium between plasma proteins and the other proteins of the body, one of the most effective of all therapies for severe acute protein deficiency is intravenous administration of plasma protein. Within a few days or sometimes within hours, the amino acids of the administered protein become distributed throughout the cells of the body to form proteins where they are needed.

Essential and Nonessential Amino Acids

Ten of the amino acids normally present in animal proteins can be synthesized in the cells, whereas the other 10 either cannot be synthesized or are synthesized in quantities too small to supply the body's needs. This second group of amino acids that cannot be synthesized are called *essential amino acids*. The use of the word "essential" does not mean that the other 10 "nonessential" amino acids are not equally as essential for formation of the proteins but only that the

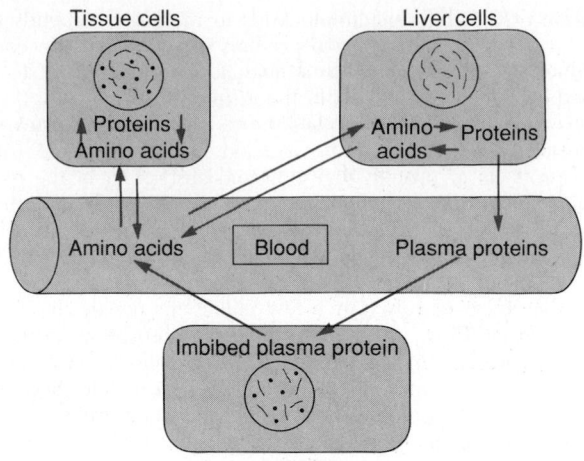

Figure 69–2. Reversible equilibrium among the tissue proteins, plasma proteins, and plasma amino acids.

others are *not essential in the diet* because they can be synthesized in the body.

Synthesis of the nonessential amino acids depends mainly on the formation first of appropriate α-keto acids, the precursors of the respective amino acids. For instance, *pyruvic acid*, which is formed in large quantities during the glycolytic breakdown of glucose, is the keto acid precursor of the amino acid *alanine*. Then, by the process of *transamination*, an amino radical is transferred to the α-keto acid while the keto oxygen is transferred to the donor of the amino radical. This reaction is shown in Figure 69–3. Note in this reaction that the amino radical is transferred to the pyruvic acid from another chemical that is closely allied to the amino acids, *glutamine*. Glutamine is present in the tissues in large quantities, and one of its principal functions is to serve as an amino radical storehouse. In addition, amino radicals can be transferred from *asparagine, glutamic acid,* and *aspartic acid.*

Transamination is promoted by several enzymes among which are the *aminotransferases*, which are derivatives of pyridoxine, one of the B vitamins (B_6). Without this vitamin, the nonessential amino acids are synthesized only poorly and, therefore, protein formation cannot proceed normally.

Use of Proteins for Energy

Once the cells are filled to their limits for protein storage, any additional amino acids in the body fluids are degraded and used for energy or stored mainly as fat or slightly as glycogen. This degradation occurs almost entirely in the liver, and it begins with *deamination*. To initiate this process, the excess amino acids in the cells, especially in the liver, induce the activation of large quantities of *aminotransferases*, the enzymes responsible for initiating most deamination.

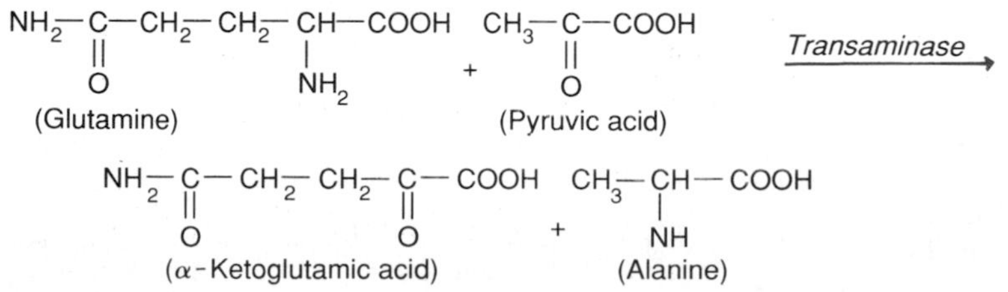

$$NH_2-\overset{\overset{\displaystyle O}{\|}}{C}-CH_2-CH_2-\underset{\underset{\displaystyle NH_2}{|}}{CH}-COOH \quad CH_3-\overset{\overset{\displaystyle O}{\|}}{C}-COOH$$
(Glutamine)   +   (Pyruvic acid) $\xrightarrow{\text{\textit{Transaminase}}}$

$$NH_2-\overset{\overset{\displaystyle O}{\|}}{C}-CH_2-CH_2-\overset{\overset{\displaystyle O}{\|}}{C}-COOH \quad CH_3-\underset{\underset{\displaystyle NH}{|}}{CH}-COOH$$
(α-Ketoglutamic acid)   +   (Alanine)

Figure 69–3. Synthesis of alanine from pyruvic acid by transamination.

DEAMINATION. Deamination means removal of the amino groups from the amino acids. This occurs mainly by transamination, which means transfer of the amino group to some acceptor substance, which is the reverse of the transamination explained above in relation to the synthesis of amino acids.

The greatest amount of deamination occurs by the following transamination schema:

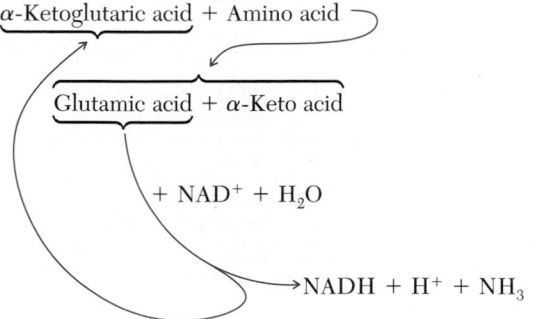

Note from this schema that the amino group from the amino acid is transferred to α-ketoglutaric acid, which then becomes glutamic acid. The glutamic acid can then transfer the amino group to still other substances or can release it in the form of ammonia (NH_3). In the process of losing the amino group, the glutamic acid once again becomes α-ketoglutaric acid, so that the cycle can be repeated again and again.

Urea Formation by the Liver. The ammonia released during deamination is removed from the blood almost entirely by conversion into urea; two molecules of ammonia and one molecule of carbon dioxide combine in accordance with the following net reaction:

$$2NH_3 + CO_2 \longrightarrow H_2N-\underset{\underset{O}{\|}}{C}-NH_2 + H_2O$$

Essentially all urea formed in the human body is synthesized in the liver. In the absence of the liver or in serious liver disease, ammonia accumulates in the blood. This in turn is extremely toxic, especially to the brain, often leading to a state called *hepatic coma.*

The stages in the formation of urea are essentially the following:

Ornithine $\xrightarrow[-H_2O]{+ CO_2 + NH_3}$ Citrulline

$\xrightarrow[-H_2O]{+NH_3}$ Arginine

Arginine $\xrightarrow[+ H_2O]{(Arginase)}$ Urea

After its formation, the urea diffuses from the liver cells into the body fluids and is excreted by the kidneys.

OXIDATION OF DEAMINATED AMINO ACIDS. Once amino acids have been deaminated, the resulting keto acid products can, in most instances, be oxidized to release energy for metabolic purposes. This usually involves two successive processes: (1) The keto acid is changed into an appropriate chemical substance that can enter the citric acid cycle; and (2) this substance is then degraded by this cycle and used for energy in the same manner that acetyl coenzyme A (acetyl-CoA) derived from carbohydrate and lipid metabolism is used.

In general, the amount of adenosine triphosphate formed for each gram of protein that is oxidized is slightly less than that formed for each gram of glucose oxidized, about 4 Calories per gram of protein.

GLUCONEOGENESIS AND KETOGENESIS. Certain deaminated amino acids are similar to the substrates normally used by the cells, mainly the liver cells, to synthesize glucose or fatty acids. For instance, deaminated alanine is pyruvic acid. This can be converted into glucose or glycogen. Or it can be converted into acetyl-CoA, which can then be polymerized into fatty acids. Also, two molecules of acetyl-CoA can condense to form acetoacetic acid, which is one of the ketone bodies, as explained in Chapter 68.

The conversion of amino acids into glucose or glycogen is called *gluconeogenesis,* and the conversion of amino acids into keto acids or fatty acids is called *ketogenesis.* Eighteen of 20 of the deaminated amino acids have chemical structures that allow them to be converted into glucose, and 19 can be converted into fatty acids.

Obligatory Degradation of Proteins

When a person eats no proteins, a certain proportion of that person's own body proteins continues to be degraded into amino acids and then deaminated and oxidized. This involves 20 to 30 grams of protein each day, which is called the *obligatory loss* of proteins. Therefore, to prevent a net loss of protein from the body, one must ingest a minimum of 20 to 30 grams of protein each day, and to be on the safe side, a minimum of 60 to 75 grams is usually recommended.

The ratios of the different amino acids in the dietary protein must be about the same as the ratios in the body tissues if the entire protein is to be fully usable. If one particular type of essential amino acid is low in concentration, the others become unusable as well because cells form either whole proteins or none at all, as explained in Chapter 3 in relation to protein synthesis. The unusable amino acids are then deaminated and oxidized. A protein that has a ratio of amino acids different from that of the average body protein is called a *partial protein* or *incomplete protein,* and such a protein is less valuable for nutrition than is the *complete protein.*

EFFECT OF STARVATION ON PROTEIN DEGRADATION. Except for the excess protein in the diet or the 20 to 30 grams of obligatory protein degradation each day, the body uses almost entirely carbohydrates or fats for energy as long as they are available. However, after several weeks of starvation, when the quantity of stored fats begins to run out, the amino acids of the blood begin to be rapidly deaminated and oxidized for energy. From this point on, the proteins of the tissues degrade rapidly—as much as 125 grams daily—and the cellular functions deteriorate precipitously.

Because carbohydrate and fat utilization for energy normally occurs in preference to protein utilization, carbohydrates and fats are called *protein sparers.*

HORMONAL REGULATION OF PROTEIN METABOLISM

GROWTH HORMONE. Growth hormone increases the rate of synthesis of cellular proteins, causing the tissue proteins to increase. The precise mechanism by which growth hormone increases the rate of protein synthesis is not known, but it is believed mainly to enhance the transport of amino acids through the cell membranes and/or to accelerate the DNA and RNA transcription and translation processes for protein synthesis. Part of the action might also result from the effect of growth hormone on fat metabolism because this hormone causes increased rate of fat liberation from the fat depots, making this available for energy. This in turn reduces the rate of oxidation of amino acids and, consequently, makes increased quantities of amino acids available to the tissues to be synthesized into proteins.

INSULIN. Lack of insulin reduces protein synthesis almost to zero. The mechanism by which this hormone affects protein metabolism is also unknown, but insulin does accelerate transport of some amino acids into the cells, which could be the stimulus to protein synthesis. Also, insulin increases the availability of glucose to the cells, so that the use of amino acids for energy becomes correspondingly reduced.

GLUCOCORTICOIDS. The glucocorticoids secreted by the adrenal cortex *decrease* the quantity of protein in *most* tissues while increasing the amino acid concentration in the plasma as well as *both the liver proteins and the plasma proteins.* It is believed that the glucocorticoids act by increasing the rate of breakdown of extrahepatic proteins, thereby making increased quantities of amino acids available in the body fluids. This in turn supposedly allows the liver to synthesize increased quantities of hepatic cellular proteins and plasma proteins.

The effects of glucocorticoids on protein metabolism are especially important in promoting ketogenesis and gluconeogenesis; in the absence of the glucocorticoids, insufficient quantities of amino acids are usually available in the plasma to allow either of these processes to occur significantly.

TESTOSTERONE. Testosterone, the male sex hormone, causes increased deposition of protein in the tissues throughout the body, including especially an increase in the contractile proteins of the muscles (30 to 50 per cent increase). The mechanism by which this effect comes about is unknown, but it is definitely different from the effect of growth hormone in the following way: Growth hormone causes tissues to continue growing almost indefinitely, whereas testosterone causes the muscles and other protein tissues to enlarge only for several months. Beyond that time, despite continued administration of testosterone, further protein deposition ceases.

Estrogen, the principal female sex hormone, also causes slight deposition of protein, but its effect is relatively insignificant in comparison with that of testosterone.

THYROXINE. Thyroxine increases the rate of metabolism of all cells and, as a result, indirectly affects protein metabolism. If insufficient carbohydrates and fats are available for energy, thyroxine causes rapid degradation of proteins and uses them for energy. On the other hand, if adequate quantities of carbohydrates and fats are available and excesses of amino acids are also available in the extracellular fluid, thyroxine can increase the rate of protein synthesis.

Conversely, in growing animals, deficiency of thyroxine causes growth to be greatly inhibited because of lack of protein synthesis. In essence, it is believed that thyroxine has little specific direct effect on protein metabolism but does have an important general effect in increasing the rates of both normal anabolic and normal catabolic protein reactions.

REFERENCES

Benevenga, N. J., and Steele, R. D.: Adverse effects of excessive consumption of amino acids. Annu. Rev. Nutr., 4:157, 1984.

Christensen, H. N.: Interorgan amino acid nutrition. Physiol. Rev., 62:1193, 1982.

DeMartino, G. N., and Croall, D. E.: Calcium-dependent proteases: A prevalent proteolytic system of uncertain function. News Physiol. Sci., 2:82, 1987.

Dohm, G. L., et al.: Protein metabolism during endurance exercise. Fed. Proc., 44:348, 1985.

Fasman, G. D. (ed.): Prediction of Protein Structure and the Principles of Protein Conformation. New York, Plenum Publishing Corp., 1989.

Francis, S. H., and Corbin, J. D.: Structure and function of cyclic nucleotide-dependent protein kinases. Annu. Rev. Physiol., 56:237, 1994.

Friedmann, H. C. (ed.): Enzymes. Stroudsburg, Pa., Dowden, Hutchinson & Ross, 1980.

Gevers, W.: Protein metabolism of the heart. J. Mol. Cell Cardiol., 16:3, 1984.

Guthrie, H. A.: Introductory Nutrition, 7th Ed. St. Louis, C. V. Mosby Co., 1988.

Harper, A. E., et al.: Branched-chain amino acid metabolism. Annu. Rev. Nutr., 4:409, 1984.

Jackson, R. L., et al.: Glycosaminoglycans: Molecular properties, protein interactions, and role in physiological processes. Physiol. Rev., 71:481, 1991.

Jungas, R. L., et al.: Quantitative analysis of amino acid oxidation and related gluconeogenesis in humans. Physiol. Rev., 72:419, 1992.

Kimball, S. R., et al.: Regulation of protein synthesis by insulin. Annu. Rev. Phyiol., 56:321, 1994.

Kinney, J. M., and Elwyn, D. H.: Protein metabolism and injury. Annu. Rev. Nutr., 3:433, 1983.

Mignatti, P., and Rifkin, D. B.: Biology and biochemistry of proteinases in tumor invasion. Physiol. Rev., 73:161, 1993.

Mumby, M. C., and Walter, G.: Protein serine/threonine phosphatases: Structure, regulation, and functions in cell growth. Physiol. Rev., 73:673, 1993.

Passonneau, J. V., and O. H. Lowry: Enzymatic Analysis: A Practical Guide. Totowa, NJ, Humana Press, Inc., 1993.

Rombeau, J. L., and Caldwell, M. D.: Clinical Nutrition: Parenteral Nutrition. Philadelphia, W. B. Saunders Co., 1993.

Rosenberg, L. E., and Scriver, C. R.: Disorders of amino acid metabolism. In Bondy, P. K., and Rosenberg, L. E. (eds.): Metabolic Control and Disease, 8th Ed. Philadelphia, W. B. Saunders Co., 1980, p. 583.

Rothschild, M. A.: Albumin synthesis. In Javitt, N. B. (ed.): International Review of Physiology: Liver and Biliary Tract Physiology I. Vol. 21. Baltimore, University Park Press, 1980, p. 249.

Schaub, J., et al.: Inborn Errors of Metabolism. New York, Raven Press, 1991.

Seegmiller, J. E.: Diseases of purine and pyrimidine metabolism. In Bondy, P. K., and Rosenberg, L. E. (eds.): Metabolic Control and Disease, 8th Ed. Philadelphia, W. B. Saunders Co., 1980, p. 777.

Stryer, L.: Biochemistry. New York, W. H. Freeman Co., 1988.

Sugano, M., and Beynen, A. C.: Dietary Proteins, Cholesterol Metabolism and Atherosclerosis. Farmington, CT, S. Karger Publishers, Inc., 1990.

Sugden, P. H.: The effects of hormonal factors on cardiac protein turnover. Adv. Myocardial, 5:105, 1985.

Walsh, D. A., et al.: Motifs of protein phosphorylation and mechanisms of reversible covalent regulation. Physiol. Rev., 71:285, 1991.

Weinsier, R. L., et al.: Handbook of Clinical Nutrition: Clinician's Manual for the Diagnosis and Management of Nutritional Problems. St. Louis, C. V. Mosby Co., 1988.

White, S. H.: Vol. 1: Membrane Protein Structure Experimental Approaches. Bethesda, MD, Am. Physiol. Soc., 1994.

Williams, S. R.: Basic Nutrition and Diet Therapy, 8th Ed. St. Louis, C. V. Mosby Co., 1989.

Zamyatnin, A. A.: Amino acid, peptide, and protein volume in solution. Annu. Rev. Biophys. Bioeng., 13:145, 1984.

The Liver as an Organ

CHAPTER 70

In many previous chapters we have encountered the liver, and each time it is performing a different function. Yet we often forget that the liver is a discrete organ and that many of its various functions interrelate to one another. This becomes particularly evident in clinical abnormalities of the liver because many of its functions are disturbed simultaneously in different combinations, depending on the nature of the disorder. Therefore, the purpose of this chapter is to summarize the different functions of the liver and to show how the liver operates as an individual organ.

The basic functions of the liver can be divided into (1) its vascular functions for storage and filtration of blood, (2) its metabolic functions concerned with the majority of the metabolic systems of the body, and (3) its secretory and excretory functions that are responsible for forming the bile that flows through the bile ducts into the gastrointestinal tract.

Physiologic Anatomy of the Liver

The basic functional unit of the liver is the *liver lobule*, which is a cylindrical structure several millimeters in length and 0.8 to 2 millimeters in diameter. The human liver contains 50,000 to 100,000 individual lobules.

The liver lobule, shown in cut-away format in Figure 70–1, is constructed around a *central vein* that empties into the hepatic veins and then into the vena cava. The lobule itself is composed principally of many *hepatic cellular plates* (two of which are shown in Fig. 70–1) that radiate centrifugally from the central vein like spokes in a wheel. Each hepatic plate is usually two cells thick, and between the adjacent cells lie small *bile canaliculi* that empty into *bile ducts* in the fibrous septa separating the adjacent liver lobules (shown at the bottom of the figure).

Also in the septa are small *portal venules* that receive their blood mainly from the venous outflow of the gastrointestinal tract by way of the portal veins. From these venules blood flows into flat, branching *hepatic sinusoids* that lie between the hepatic plates and then into the central vein. Thus, the hepatic cells are exposed continuously to portal venous blood.

In addition to the portal venules, *hepatic arterioles* are present in the interlobular septa. These arterioles supply arterial blood to the septal tissues between the adjacent lobules, and many of the small arterioles also empty directly into the hepatic sinusoids, most frequently emptying into these about one third of the distance away from the interlobular septa, as shown in Figure 70–1.

In addition to the hepatic cells, the venous sinusoids are lined by two other types of cell: (1) typical *endothelial cells* and (2) large *Kupffer cells* (also called reticuloendothelial cells), which are a type of macrophage capable of phagocytizing bacteria and other foreign matter in the hepatic sinus blood. The endothelial lining of the sinusoids has extremely large pores, some of which are almost 1 micrometer in diameter. Beneath this lining, lying between the endothelial cells and the hepatic cells, is a narrow tissue space called the *space of Disse*. The millions of spaces of Disse in turn connect with lymphatic vessels in the interlobular septa. Therefore, excess fluid in these spaces is removed through the lymphatics. Because of the large pores in the endothelium, substances in the plasma move freely into the space of Disse. Even large portions of the plasma proteins diffuse freely into this space.

FUNCTION OF THE HEPATIC VASCULAR SYSTEM

The function of the hepatic vascular system is discussed in Chapter 15 in connection with the portal veins. Briefly, it can be summarized as follows.

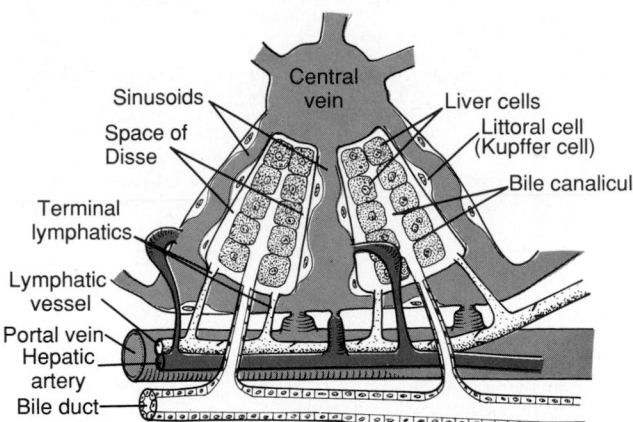

Figure 70–1. Basic structure of a liver lobule, showing the hepatic cellular plates, the blood vessels, the bile-collecting system, and the lymph flow system composed of the spaces of Disse and the interlobular lymphatics. (Reprinted from Guyton, Taylor, and Granger [as modified from Elias]: Circulatory Physiology, Vol. 2: Dynamics and Control of the Body Fluids. Philadelphia, W. B. Saunders Company, 1975.)

Blood Flow Through the Liver

About 1100 milliliters of blood flows from the portal vein into the liver sinusoids each minute, and about an additional 350 milliliters flows into the sinusoids from the hepatic artery, the total averaging about 1450 ml/min. This amounts to about 29 per cent of the resting cardiac output, almost one third of the total body blood flow.

MEASUREMENT OF HEPATIC BLOOD FLOW. Total hepatic blood flow per minute can be measured by a modified Fick procedure in which the dye *indocyanine green*, a dye that is removed from the blood only by the liver, is infused continuously into the circulatory system until its concentration in the arterial blood remains constant. When this is achieved, one knows the rate of excretion of the dye because it is then exactly equal to the rate of infusion. Next, the concentration of the dye is measured in the arterial blood as well as in venous blood collected from the hepatic vein by means of a catheter. From these measurements, one can calculate the arteriovenous difference (A-V difference) of the dye. The hepatic blood flow is then calculated by the usual Fick formula, as follows:

$$\frac{\text{Hepatic blood}}{\text{flow (ml/min)}} = \frac{\text{Rate of dye excretion (mg/min)}}{\text{A-V difference in dye (mg/ml)}}$$

Pressures and Resistance in the Hepatic Vessels

The pressure in the portal vein leading into the liver averages about 9 mm Hg, and the pressure in the hepatic vein leading from the liver into the vena cava normally averages almost exactly 0 mm Hg. This small pressure difference, only 9 mm Hg, shows that the resistance to blood flow through the hepatic sinusoids is normally very low, especially when one considers that about 1.45 liters of blood flows by this route each minute.

BLOCKAGE OF THE PORTAL SYSTEM—"CIRRHOSIS" OF THE LIVER. Frequently, extreme amounts of fibrous tissue develop within the liver structure, destroying many of the parenchymal cells and eventually contracting around the blood vessels, thereby greatly impeding the flow of portal blood through the liver. This disease process is known as *cirrhosis of the liver*. It results most commonly from alcoholism, but it can also follow ingestion of poisons such as carbon tetrachloride, virus diseases such as infectious hepatitis, and infectious processes in the bile ducts.

The portal system is also occasionally blocked by a large clot that develops in the portal vein or its major branches. When the portal system is suddenly blocked, the return of blood from the intestines and spleen through the liver portal blood flow system to the systemic circulation is tremendously impeded, the capillary pressure in the intestinal wall rising to 15 to 20 mm Hg above normal. The patient often dies within a few hours because of excessive loss of fluid from the capillaries into the lumens and walls of the intestines.

Reservoir Function of the Liver

Because the liver is an expandable organ, large quantities of blood can be stored in its blood vessels. Its normal blood volume, including both that in the hepatic veins and that in the hepatic sinuses, is 450 milliliters, or almost 10 per cent of the body's total blood volume. When high pressure in the right atrium causes back pressure in the liver, the liver expands and 0.5 to 1 liter of extra blood occasionally is thereby stored in the hepatic veins and sinuses. This occurs especially in cardiac failure with peripheral congestion, which is discussed in Chapter 22.

Thus, in effect, the liver is a large, expandable, venous organ capable of acting as a valuable blood reservoir in times of excess blood volume and capable of supplying extra blood in times of diminished blood volume.

Very High Lymph Flow from the Liver

Because the pores in the hepatic sinusoids are very permeable and allow ready passage of both fluid and proteins into the spaces of Disse, the lymph draining from the liver usually has a protein concentration of about 6 gm/dl, which is only slightly less than the protein concentration of plasma. Also, the extreme permeability of the liver sinusoid epithelium allows large quantities of lymph to form. Therefore, about one half of all the lymph formed in the body under resting conditions arises in the liver.

EFFECTS OF HIGH HEPATIC VASCULAR PRESSURES IN CAUSING FLUID TRANSUDATION INTO THE ABDOMINAL CAVITY FROM THE LIVER AND PORTAL CAPILLARIES—ASCITES. When the pressure in the hepatic veins where they empty into the vena cava rises only 3 to 7 mm Hg above normal, excessive amounts of fluid already begin to transude into the lymph and to leak through the outer surface of the liver capsule directly into the abdominal cavity. This fluid is almost pure plasma, containing 80 to 90 per cent as much protein as normal plasma. At still higher vena caval pressures, at 10 to 15 mm Hg, hepatic lymph flow increases to as much as 20 times normal, and the "sweating" from the surface of the liver can be so great that it causes large amounts of free fluid in the abdominal cavity, which is called *ascites*.

Blockage of portal flow through the liver also causes high capillary pressures in the entire portal vascular system of the gastrointestinal tract, resulting in edema of the gut wall and transudation of fluid through the serosa of the gut into the abdominal cavity. This, too, can cause ascites but is less likely to do so than is sweating from the liver surface because collateral vascular channels develop rapidly from the portal veins to the systemic veins, decreasing the intestinal capillary pressure back to a safe value.

Hepatic Macrophage System—The Blood-Cleansing Function of the Liver

Blood flowing through the intestinal capillaries picks up many bacteria from the intestines. Indeed, a sample of blood from the portal veins before entering the liver almost always grows colon bacilli when cultured, whereas growth of colon bacilli from blood in the systemic circulation is extremely rare. Special high-speed motion pictures of the action of Kupffer cells, the large phagocytic macrophages that line the hepatic venous sinuses, have demonstrated that these cells can cleanse blood extremely efficiently as it passes through the sinuses; when a bacterium comes into momentary contact with a Kupffer cell, in less than 0.01 second the bacterium passes inward through the wall of the Kupffer cell to become permanently lodged therein until it is digested. Probably not more than 1 per cent of the bacteria entering the portal blood from the intestines succeeds in passing through the liver into the systemic circulation.

METABOLIC FUNCTIONS OF THE LIVER

The hepatic cells all together are a large chemically reactant pool that have a high rate of metabolism, sharing substrates and energy from one metabolic system to another, processing and synthesizing multiple substances that are transported to other areas of the body, and performing myriad other metabolic functions. For all of these reasons, a major share of the entire discipline of biochemistry is devoted to the metabolic reactions in the liver. But here, let us summarize those metabolic functions that are especially important in understanding the integrated physiology of the body.

Carbohydrate Metabolism

In carbohydrate metabolism, the liver performs the following specific functions, as summarized from Chapter 67: (1) storage of glycogen, (2) conversion of galactose and fructose to glucose, (3) gluconeogenesis, and (4) formation of many important chemical compounds from the intermediate products of carbohydrate metabolism.

The liver is especially important for maintaining a normal blood glucose concentration. For instance, storage of glycogen allows the liver to remove excess glucose from the blood, store it, and then return it to the blood when the blood glucose concentration begins to fall too low. This is called the *glucose buffer function* of the liver. As an example, immediately after a meal that contains large amounts of carbohydrates, the blood glucose concentration rises about three times as much in a person with a nonfunctional liver as in a person with a normal liver.

Gluconeogenesis in the liver is also concerned with maintaining a normal blood glucose concentration because gluconeogenesis occurs to a significant extent only when the glucose concentration falls below normal. In such a case, large amounts of amino acids and glycerol from the triglycerides are converted into glucose, thereby contributing still another means for maintaining a relatively normal blood glucose concentration.

Fat Metabolism

Although some fat metabolism can take place in all cells of the body, certain aspects of fat metabolism occur mainly in the liver. Specific functions of the liver in fat metabolism, summarized from Chapter 68 on the lipids, are (1) a high rate of oxidation of fatty acids to supply energy for other bodily functions, (2) formation of most of the lipoproteins, (3) synthesis of large quantities of cholesterol and phospholipids, and (4) conversion of large quantities of carbohydrates and proteins into fat.

To derive energy from neutral fats, the fat is first split into glycerol and fatty acids and then the fatty acids are split by *beta oxidation* into two-carbon acetyl radicals that then form *acetylcoenzyme A* (acetyl-CoA). This in turn can enter the citric acid cycle and be oxidized to liberate tremendous amounts of energy. Beta oxidation can take place in all cells of the body, but it occurs especially rapidly in the hepatic cells. The liver itself cannot use all the acetyl-CoA that is formed; instead, it is converted by condensation of two molecules of acetyl-CoA into *acetoacetic acid*, a highly soluble acid that passes from the hepatic cells into the extracellular fluids and then is transported throughout the body to be absorbed by the other tissues. These tissues in turn reconvert the acetoacetic acid into acetyl-CoA and then oxidize this in the usual manner. In these ways, the liver is responsible for a major part of the metabolism of fats.

About 80 per cent of the cholesterol synthesized in the liver is converted into bile salts, which in turn are secreted into the bile; the remainder is transported in the lipoproteins, carried by the blood to the tissue cells everywhere in the body. Phospholipids are likewise synthesized in the liver and transported principally in the lipoproteins. Both the cholesterol and the phospholipids are used by the cells to form membranes, intracellular structures, and multiple derived chemical substances that are important to cellular function.

Almost all the fat synthesis in the body from carbohydrates and proteins also occurs in the liver. After fat is synthesized in the liver, it is transported in the lipoproteins to the adipose tissue to be stored.

Protein Metabolism

Even though a large proportion of the processes for carbohydrate and fat metabolism occurs in the liver, the body could probably dispense with many of these functions of the liver and still survive. On the other hand, the body cannot dispense with the services of the liver in protein metabolism for more than a few days without death ensuing. The most important functions of the liver in protein metabolism, as summarized from Chapter 69 on proteins, are (1) deamination of amino acids, (2) formation of urea for removal of ammonia from the body fluids, (3) formation of plasma proteins, and (4) interconversions among the different amino acids as well as between the amino acids and other compounds important to the metabolic processes of the body.

Deamination of the amino acids is required before they can be used for energy or before they can be converted into carbohydrates or fats. A small amount of deamination can occur in the other tissues of the body, especially in the kidneys, but the percentage of deamination that occurs extrahepatically is so small that it is almost unimportant.

Formation of urea by the liver removes ammonia from the body fluids. Large amounts of ammonia are formed by the deamination process, and still additional amounts are continually formed in the gut by bacteria and are then absorbed into the blood. Therefore, in the absence of this function of the liver to form urea, the plasma ammonia concentration rises rapidly and results in *hepatic coma* and death. Indeed, even greatly decreased blood flow through the liver—as occurs occasionally when a shunt develops between the portal vein and the vena cava—can cause excessive ammonia in the blood, an exceedingly toxic condition.

Essentially all the plasma proteins, with the exception of part of the gamma globulins, are formed by the hepatic cells. This accounts for about 90 per cent of all the plasma proteins. The remaining gamma globulins are the antibodies formed mainly by plasma cells in the lymph tissue of the body. The liver can form plasma proteins at a maximum rate of 15 to 50 gm/day. Therefore, after loss of as much as one half the plasma proteins from the body, they can be replenished in 1 or 2 weeks. It is particularly interesting that plasma protein depletion causes rapid mitosis of the hepatic cells and growth of the liver to a larger size; these effects are coupled with rapid output of plasma proteins until the plasma concentration returns to normal.

Among the most important functions of the liver is its ability to synthesize certain amino acids and to synthesize other important chemical compounds from amino acids. For instance, the so-called nonessential amino acids can all be synthesized in the liver. To do this, a keto acid having the same chemical composition (except at the keto oxygen) as that of the amino acid to be formed is first synthesized. Then an amino radical is transferred through several stages of *transamination* from an available amino acid to the keto acid to take the place of the keto oxygen.

Miscellaneous Metabolic Functions of the Liver

STORAGE OF VITAMINS. The liver has a particular propensity for storing vitamins and has long been known as an excellent source of certain vitamins in treating patients. The single vitamin stored in greatest quantity in the liver is vitamin A, but large quantities of vitamin D and vitamin B_{12} are normally stored as well. Sufficient quantities of vitamin A can be stored to prevent vitamin A deficiency for as long as 10 months. Sufficient vitamin D can be stored to prevent deficiency for 3 to 4 months, and enough vitamin B_{12} can be stored to last for at least 1 year and maybe several years.

RELATION OF THE LIVER TO BLOOD COAGULATION. The liver forms a large proportion of the blood substances used in the coagulation process. They are *fibrinogen, prothrombin, accelerator globulin, Factor VII,* and several other important coagulation factors. Vitamin K is required by the metabolic processes of the liver for the formation of several of these substances, especially prothrombin and Factors VII, IX, and X. In the absence of vitamin K, the concentrations of all these fall low, and this almost prevents blood coagulation.

STORAGE OF IRON. Except for the iron in the hemoglobin of the blood, by far the greater proportion of the iron in the body is usually stored in the liver in the form of *ferritin.* The hepatic cells contain large amounts of a protein called *apoferritin,* which is capable of combining with either small or large quantities of iron. Therefore, when iron is available in the body fluids in extra quantities, it combines with the apoferritin to form ferritin and is stored in this form in the hepatic cells until needed elsewhere. When the iron in the circulating body fluids reaches a low level, the ferritin releases the iron. Thus, the apoferritin-ferritin system of the liver acts as a *blood iron buffer* as well as an iron storage

medium. Other functions of the liver in relation to iron metabolism and red blood cell formation are considered in Chapter 32.

REMOVAL OR EXCRETION OF DRUGS, HORMONES, AND OTHER SUBSTANCES BY THE LIVER. The active chemical medium of the liver is well known for its ability to detoxify or excrete into the bile many drugs, including sulfonamides, penicillin, ampicillin, and erythromycin. In a similar manner, several of the hormones secreted by the endocrine glands are either chemically altered or excreted by the liver, including thyroxine and essentially all the steroid hormones such as estrogen, cortisol, and aldosterone. Liver damage can often lead to excess accumulation of one or more of these hormones in the body fluids and therefore can cause overactivity of the hormonal systems. Finally, one of the major routes for excreting calcium from the body is first secretion by the liver into the bile and then passage into the gut and loss in the feces.

EXCRETION OF BILIRUBIN IN THE BILE— USE OF THIS AS A CLINICAL DIAGNOSTIC TOOL

The formation of bile by the liver and the function of the bile salts in the digestive and absorptive processes of the intestinal tract are discussed in Chapters 64 and 65. In addition, many substances are excreted in the bile and then eliminated in the feces. One of these is the greenish yellow pigment *bilirubin.* This is a major end product of hemoglobin degradation, as pointed out in Chapter 32. But even more important, it provides *an exceedingly valuable tool for the clinician in diagnosing both hemolytic blood diseases and various types of liver diseases.* Therefore, while referring constantly to Figure 70–2, let us explain this.

Briefly, when the red blood cells have lived out their life span, averaging 120 days, and have become too fragile to exist longer in the circulatory system, their cell membranes rupture and the released hemoglobin is phagocytized by tissue macrophages (also called the "reticuloendothelial system") throughout the body. Here, the hemoglobin is first split into *globin* and *heme,* and the heme ring is opened to give (1) free iron that is transported in the blood by transferrin and (2) a straight chain of four pyrrole nuclei that is the substrate from which bilirubin will eventually be formed. The first substance formed is *biliverdin,* but this is rapidly reduced to *free bilirubin,* which is gradually released from the macrophages into the plasma. The free bilirubin immediately combines strongly with the plasma albumin and is transported in this combination throughout the blood and interstitial fluids. Even when bound with the plasma protein, this bilirubin is still called "free bilirubin" to distinguish it from "conjugated bilirubin," which is discussed later.

Within hours, the free bilirubin is absorbed through the hepatic cell membrane. In passing to the insides of the hepatic cells, it is released from the plasma albumin and soon thereafter conjugated about 80 per cent with glucuronic acid to form *bilirubin glucuronide,* about 10 percent with sulfate to form *bilirubin sulfate,* and the final 10 per cent with a multitude of other substances. In these forms, the bilirubin is excreted from the hepatocytes by an active transport process into the bile canaliculi and then into the intestines.

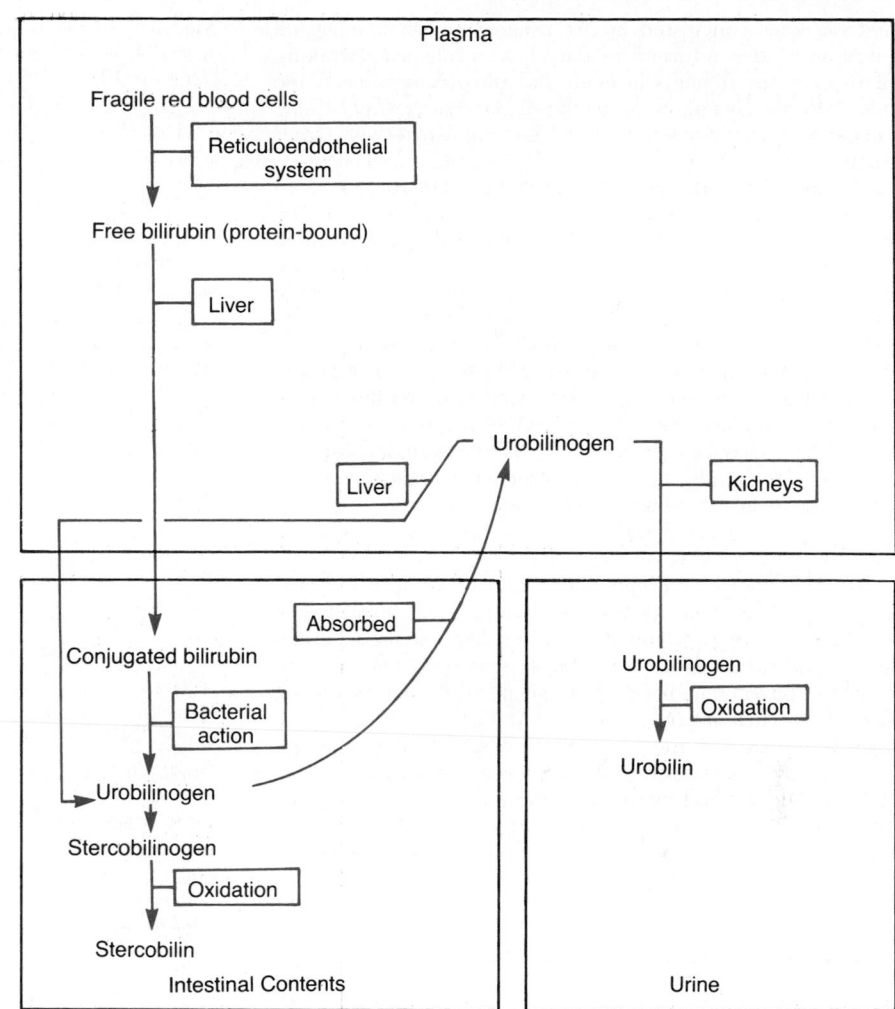

Figure 70–2. Bilirubin formation and excretion.

FORMATION AND FATE OF UROBILINOGEN. Once in the intestine, about one half of the "conjugated" bilirubin is converted by bacterial action into the substance *urobilinogen,* which is highly soluble. Some of the urobilinogen is reabsorbed through the intestinal mucosa back into the blood. Most of this is once again re-excreted by the liver back into the gut, but about 5 per cent is excreted by the kidneys into the urine. After exposure to air in the urine, the urobilinogen becomes oxidized to *urobilin,* or in the feces, it becomes altered and oxidized to form *stercobilin.* These interrelations of bilirubin and the other bilirubin products are shown in Figure 70–2.

Jaundice

The word "jaundice" means a yellowish tint to the body tissues, including yellowness of the skin as well as the deep tissues. The usual cause of jaundice is large quantities of bilirubin in the extracellular fluids, either free bilirubin or conjugated bilirubin. The normal plasma concentration of bilirubin, which is almost entirely the free form, averages 0.5 mg/dl of plasma. In certain abnormal conditions, this can rise to as high as 40 mg/dl, and much of it can become of the conjugated type. The skin usually begins to appear jaundiced when the concentration rises to about three times normal—that is, above 1.5 mg/dl.

The common causes of jaundice are (1) increased destruction of red blood cells with rapid release of bilirubin into the blood and (2) obstruction of the bile ducts or damage to the liver cells so that even the usual amounts of bilirubin cannot be excreted into the gastrointestinal tract. These two types of jaundice are called, respectively, *hemolytic jaundice* and *obstructive jaundice.* They differ from each other in the following ways.

HEMOLYTIC JAUNDICE. In hemolytic jaundice, the excretory function of the liver is not impaired in the least, but red blood cells are hemolyzed rapidly and the hepatic cells simply cannot excrete the bilirubin as rapidly as it is formed. Therefore, the plasma concentration of free bilirubin rises to levels much above normal. Likewise, the rate of formation of *urobilinogen* in the intestine is greatly increased, and much of this is absorbed into the blood and later excreted in the urine.

OBSTRUCTIVE JAUNDICE. In obstructive jaundice, caused either by obstruction of the bile ducts (which most often occurs when a gallstone or cancer blocks the common bile duct) or by damage to the hepatic cells (which occurs in *hepatitis*), the rate of bilirubin formation is normal but the bilirubin formed cannot pass from the blood into the intestines. The free bilirubin usually does still enter the liver cells and becomes conjugated in the usual way. This conjugated bilirubin is then returned to the blood, probably by rupture of the congested bile canaliculi and direct emptying of the bile into the lymph leaving the liver. Thus, *most of the bilirubin in the plasma becomes the conjugated type* rather than the free type.

DIAGNOSTIC DIFFERENCES BETWEEN HEMOLYTIC AND OBSTRUCTIVE JAUNDICE. Chemical laboratory tests can be used to differentiate between free and conjugated bilirubin in the plasma. In hemolytic jaundice, almost all the bilirubin is in the "free" form. In obstructive jaundice, it is mainly in the "conjugated" form; a test called the *van den Bergh reaction* can be used to differentiate between these two.

When there is total obstruction of bile flow, no bilirubin can reach the intestines to be converted into urobilinogen by bacteria. Therefore, no urobilinogen is reabsorbed into the blood and none can be excreted by the kidneys into the urine. Consequently, in *total* obstructive jaundice, tests for urobilinogen in the urine are completely negative. Also, the stools become clay-colored for lack of stercobilin and other bile pigments.

Another major difference between free and conjugated bilirubin is that the kidneys can excrete small quantities of the highly soluble conjugated bilirubin but not the albumin-bound "free" bilirubin. Therefore, in severe obstructive jaundice, significant quantities of conjugated bilirubin appear in the urine. This can be demonstrated simply by shaking the urine and observing the foam, which turns an intense yellow.

Thus, by understanding the physiology of bilirubin excretion by the liver and by use of a few simple tests, it is often possible to differentiate among multiple types of both hemolytic diseases and liver diseases as well as to determine the severity of the disease.

REFERENCES

Alison, M. R.: Regulation of hepatic growth. Physiol. Rev., 66:499, 1986.

Arias, I. M., et al.: The Liver: Biology and Pathobiology. New York, Raven Press, 1994.

Blumgart, L. H.: Surgery of the liver and biliary tract. 2nd Ed. New York, Churchill Livingstone, 1994.

Cousins, R. J.: Absorption, transport, and hepatic metabolism of copper and zinc: Special reference to metallothionine and ceruloplasmin. Physiol. Rev., 65:238, 1985.

El Maghrabi, M. R., and Pilkis, S. J.: Rat liver 6-phosphofructo-2-kinase/fructose 2,6-biophosphatase: A review of relationships between the two activities of the enzyme. J. Cell. Biochem., 26:1, 1984.

Farrell, G. C.: Drug-Induced Liver Disease. New York, Churchill Livingstone, 1994.

Gitnick, G. (ed.): Modern Concepts of Acute and Chronic Hepatitis. New York, Plenum Publishing Corp., 1989.

Gitnick, G., et al. Principles and Practice of Gastroenterology and Hepatology, 2nd Ed. Redding, MA, Appleton & Lange, 1994.

Goresky, C. A., and Groom, A. C.: Microcirculatory events in the liver and the spleen. In Renkin, E. M., and Michel, C. C. (eds.): Handbook of Physiology. Sec. 2, Vol. IV. Bethesda, Md., American Physiological Society, 1984, p. 689.

Gustafsson, J. A.: Sex steroid induced changes in hepatic enzymes. Annu. Rev. Physiol., 45:51, 1983.

Havel, R. J.: Functional activities of hepatic lipoprotein receptors. Annu. Rev. Physiol., 48:119, 1986.

Henderson, J. R., and Daniel, P. M.: Capillary beds and portal circulations. In Renkin, E. M., and Michel, C. C. (eds.): Handbook of Physiology. Sec. 2, Vol. IV. Bethesda, Md., American Physiological Society, 1984. p. 1035.

Heyworth, M. F., and Jones, A. L.: Immunology of the Gastrointestinal Tract and Liver. New York, Raven Press, 1988.

Hollinger, F. B., et al.: Viral Hepatitis. New York, Raven Press, 1991.

Huebers, H. A., and Finch, C. A.: The physiology of transferrin and transferrin receptors. Physiol. Rev., 67:520, 1987.

Johnson, L. R., et al.: Physiology of the Gastrointestinal Tract, 2nd Ed. New York, Raven Press, 1987.

Jungermann, K., and Katz, N.: Functional specialization of different hepatocyte populations. Physiol. Rev., 69:708, 1989.

Kaplowitz, N.: Liver and Biliary Diseases. Baltimore, Williams & Wilkins, 1992.

Kraus-Friedmann, N.: Hormonal regulation of hepatic gluconeogenesis. Physiol. Rev., 64:170, 1984.

Krawitt, E. L., and Wiesner, R. H.: Autoimmune Liver Diseases. New York, Raven Press, 1991.

Messmer, K., and Menger, M. D.: Liver Microcirculation and Hepatobiliary Function. Farmington, CT, S. Karger Publishers, Inc., 1993.

Pappas, S. C.: Fulminant Hepatic Failure. New York, Raven Press, 1994.

Rappaport, A. M.: Hepatic blood flow: Morphologic aspects and physiologic regulation. In Javitt, N. B. (ed.): International Review of Physiology: Liver and Biliary Tract Physiology I. Vol. 21. Baltimore, University Park Press, 1980, p. 1.

Reichen, J., and Paumgartner, G.: Excretory function of the liver. In Javitt, N. B. (ed.): International Review of Physiology: Liver and Biliary Tract Physiology I. Vol. 21. Baltimore, University Park Press, 1980, p. 103.

Richardson, P. D. I., and Withrington, P. G.: Physiological regulation of the hepatic circulation. Annu. Rev. Physiol., 44:57, 1982.

Rustgi, V. K., and Van Thiel, D. H.: The Liver in Systemic Disease. New York, Raven Press, 1993.

Schachter, D.: Fluidity and function of hepatocyte plasma membranes. Hepatology, 4:140, 1984.

Scharschmidt, B. F., and Van Dyke, R. W.: Proton transport by hepatocyte organelles and isolated membrane vesicles. Annu. Rev. Physiol., 49:69, 1987.

Schiff, L., and Schiff, E. R.: Diseases of the Liver. Philadelphia, J. B. Lippincott, 1993.

Seeff, L. B., and Lewis, J. H. (eds.): Current Perspectives in Hepatology. New York, Plenum Publishing Corp., 1989.

Setchell, K. D. R., et al. (eds.): The Bile Acids. New York, Plenum Publishing Corp., 1988.

Soloway, R. D. (ed.): Chronic Active Liver Disease. New York, Churchill Livingstone, 1983.

Strange, R. C.: Hepatic bile flow. Physiol. Rev., 64:1055, 1984.

Tavoloni, N., and Berk, P. D.: Hepatic Transport and Bile Secretion: Physiology and Pathophysiology. New York, Raven Press, 1993.

Williams, J. A., and Blevins, G. T. Jr.: Cholecystokinin and Regulation of Pancreatic Acinar Cell Function. Physiol. Rev., 73:701, 1993.

Zuckerman, A., and Thomas, H. C.: Viral Hepatitis: Scientific Basis and Clinical Management. New York, Churchill Livingstone, 1994.

Dietary Balances; Regulation of Feeding; Obesity and Starvation; Vitamins and Minerals

CHAPTER *71*

The intake of food must always be sufficient to supply the metabolic needs of the body and yet not so much as to cause obesity. Also, because different foods contain different proportions of proteins, carbohydrates, fats, minerals, and vitamins, appropriate balance must be maintained among these so that all segments of the body's metabolic systems can be supplied with the requisite materials. This chapter, therefore, discusses the problems of balance among the types of food as well as the mechanisms by which the intake of food is regulated in accordance with the metabolic needs of the body.

DIETARY BALANCES

Energy Available in Foods

The energy liberated from each gram of carbohydrate as it is oxidized to carbon dioxide and water is 4.1 Calories and that liberated from fat is 9.3 Calories. The energy liberated from metabolism of the average protein of the diet as each gram is oxidized to carbon dioxide, water, and urea is 4.35 Calories. Also, these substances vary in the average percentages that are absorbed from the gastrointestinal tract: about 98 per cent of the carbohydrate, 95 per cent of the fat, and 92 per cent of the protein. Therefore, in round figures, the average *physiologically available energy* in each gram of the three foodstuffs in the diet is as follows:

	Calories
Carbohydrates	4.0
Fat	9.0
Protein	4.0

Average Americans receive about 15 per cent of their energy from protein, about 40 per cent from fat, and 45 per cent from carbohydrates. In most other parts of the world, the quantity of energy derived from carbohydrates far exceeds that derived from both proteins and fats. Indeed, in some parts of the world where meat is scarce, the energy received from fats and proteins combined is said to be no greater than 15 to 20 per cent.

Table 71–1 gives the compositions of selected foods, demonstrating especially the high proportions of fats and proteins in meat products and the high proportions of carbohydrates in most vegetable and grain products. Fat is deceptive in the diet because it usually exists as 100 per cent fat, whereas both proteins and carbohydrates are mixed in watery media so that each of these normally represents less than 25 per cent of the weight. Therefore, the fat of one pat of butter mixed with an entire helping of potato sometimes contains as much energy as the potato itself.

DAILY REQUIREMENT FOR PROTEIN. Twenty to 30 grams of the body proteins are degraded and used for producing other body chemicals daily. Therefore, all cells must continue to form new proteins to take the place of those that are being destroyed, and a supply of protein is needed in the diet for this purpose. An average person can maintain normal stores of protein, provided that the *daily intake is above 30 to 55 grams*.

However, as discussed in Chapter 69, some proteins have inadequate quantities of certain essential amino acids and therefore cannot be used to form the body proteins. Such proteins are called *partial proteins*, and when they are present in large quantities in the diet, the daily requirement of protein is much greater than normal. In general, proteins derived from animal foodstuffs are more nearly complete than are proteins derived from vegetable and grain sources. A special example of dietary deficiency resulting from partial proteins

889

Table 71–1 PROTEIN, FAT, AND CARBOHYDRATE CONTENT OF DIFFERENT FOODS

Food	% Protein	% Fat	% Carbohydrate	Fuel Value per 100 Grams, Calories
Apples	0.3	0.4	14.9	64
Asparagus	2.2	0.2	3.9	26
Bacon, fat	6.2	76.0	0.7	712
broiled	25.0	55.0	1.0	599
Beef, medium	17.5	22.0	1.0	268
Beets, fresh	1.6	0.1	9.6	46
Bread, white	9.0	3.6	49.8	268
Butter	0.6	81.0	0.4	733
Cabbage	1.4	0.2	5.3	29
Carrots	1.2	0.3	9.3	45
Cashew nuts	19.6	47.2	26.4	609
Cheese, Cheddar, American	23.9	32.3	1.7	393
Chicken, total edible	21.6	2.7	1.0	111
Chocolate	(5.5)	52.9	(18.0)	570
Corn (maize), entire	10.0	4.3	73.4	372
Haddock	17.2	0.3	0.5	72
Lamb, leg, intermediate	18.0	17.5	1.0	230
Milk, fresh whole	3.5	3.9	4.9	69
Molasses, medium	0.0	0.0	(60.0)	240
Oatmeal, dry, uncooked	14.2	7.4	68.2	396
Oranges	0.9	0.2	11.2	50
Peanuts	26.9	44.2	23.6	600
Peas, fresh	6.7	0.4	17.7	101
Pork, ham, medium	15.2	31.0	1.0	340
Potatoes	2.0	0.1	19.1	85
Spinach	2.3	0.3	3.2	25
Strawberries	0.8	0.6	8.1	41
Tomatoes	1.0	0.3	4.0	23
Tuna, canned	24.2	10.8	0.5	194
Walnuts, English	15.0	64.4	15.6	702

occurs in the diet of many African natives who subsist primarily on corn meal. The protein of corn has almost no tryptophan, one of the essential amino acids; therefore, for practical purposes, the entire diet of these natives is almost completely protein deficient. As a result, these natives, especially the children, develop the protein deficiency syndrome called *kwashiorkor*, which consists of failure to grow, lethargy, depressed mentality, and hypoprotein edema.

CARBOHYDRATES AND FATS AS "PROTEIN SPARERS." It is also noted in Chapters 67 and 68 that when the diet contains an abundance of carbohydrates and fats, almost all the body's energy is derived from these two substances and little is derived from proteins. Therefore, both carbohydrates and fats are said to be *protein sparers*. On the other hand, in starvation, after the carbohydrates and fats have been depleted, the body's protein stores are consumed rapidly for energy, sometimes at rates approaching several hundred grams

per day rather than at the normal daily rate of 30 to 55 grams.

Clinical and Experimental Methods for Determining Metabolic Utilization of Proteins, Carbohydrates, and Fats

In animal experiments and in human beings who have metabolic problems, it is often important to determine the relative rates of utilization of proteins, carbohydrates, and fats. Two principal methods are used for this purpose.

DETERMINATION OF THE RATE OF PROTEIN METABOLISM IN THE BODY. The average protein of the diet contains about 16 per cent nitrogen. During metabolism of the protein, about 90 per cent of this nitrogen is excreted in the urine in the form of urea, uric acid, creatinine, and other, less important nitrogen products. The remaining 10 per cent is excreted in the feces. Therefore, it is a simple matter to accurately estimate the rate of protein breakdown in the body by simply measuring the amount of nitrogen in the urine, then adding 11 per cent for the nitrogen excreted in the feces, and, finally, multiplying by 6.25 to determine the total amount of protein metabolism in grams per day. Thus, excretion of 8 grams of nitrogen in the urine each day means that there has been about 55 grams of protein breakdown.

If the daily intake of protein is less than the daily breakdown of protein, the person is said to have a *negative nitrogen balance*, which means that his or her body stores of protein are decreasing daily.

RELATIVE UTILIZATION OF FAT AND CARBOHYDRATES—THE "RESPIRATORY QUOTIENT." When carbohydrates are metabolized with oxygen, for each molecule of oxygen consumed exactly one carbon dioxide molecule is formed. This ratio of carbon dioxide output to oxygen usage is called the *respiratory quotient*, and therefore the respiratory quotient for carbohydrates is 1.0.

On the other hand, when fat is oxidized in the body's cells, on average only 70 carbon dioxide molecules are formed for each 100 molecules of oxygen consumed. Therefore, the respiratory quotient for the metabolism of fat averages 0.70. In a similar manner, when proteins are oxidized by the cells, the average respiratory quotient is 0.80. The reason for these lower respiratory quotients for fat and protein than for carbohydrates is that a large share of the oxygen metabolized with these foods is required to combine with the excess hydrogen atoms present in their molecules so that less carbon dioxide is formed in relation to the oxygen used.

Now let us see how one can make use of this respiratory quotient to determine the relative utilization of different foods by the body. First, it will be recalled from Chapter 39 that the output of carbon dioxide by the lungs divided by the uptake of oxygen during the same period is called the *respiratory exchange ratio*. Over a period of 1 hour or more, the respiratory exchange ratio exactly equals the average respiratory quotient of the metabolic reactions throughout the body. If a person has a respiratory quotient of 1.0, he or she will be metabolizing almost entirely carbohydrates because the respiratory quotients for both fat and protein metabolism are considerably less than 1.0. Likewise, when the respiratory quotient is about 0.70, the body is metabolizing fat almost entirely to the exclusion of carbohydrates and proteins. And, finally, if we ignore the normally small amount of protein metabolism, respiratory quotients between 0.70 and 1.0 describe the approximate ratios of carbohydrate to fat metabolism. To be more exact, one can first determine the protein utilization by measuring nitrogen excretion and then, using appropriate mathematical formulation, calculate almost exactly the utilization of the three foodstuffs.

Some of the important findings from studies of respiratory quotient are the following:

1.   Immediately after a meal almost all the food that is metabolized is carbohydrates, so that the respiratory quotient at that time approaches 1.0.

2.   About 8 to 10 hours after a meal, the body has already used up most of its readily available carbohydrate, and the respiratory quotient approaches that for fat metabolism, about 0.70.

3.   In untreated diabetes mellitus, little carbohydrate can be used by the body's cells under any conditions because insulin is required for this. Therefore, when diabetes is severe, the respiratory quotient remains most of the time near to that for fat metabolism, 0.70.

REGULATION OF FOOD INTAKE

HUNGER. The term "hunger" means a craving for food, and it is associated with a number of objective sensations. For instance, in a person who has not had food for many hours, the stomach undergoes intense rhythmical contractions called *hunger contractions.* These contractions cause a tight or gnawing feeling in the pit of the stomach and sometimes cause pain called *hunger pangs.* Even after the stomach is removed, the psychic sensations of hunger still occur, and craving for food still makes the person search for an adequate food supply.

APPETITE. The term "appetite" is often used in the same sense as hunger except that it usually implies desire for specific types of food instead of food in general. Therefore, appetite helps a person choose the quality of food to eat.

SATIETY. Satiety is the opposite of hunger. It means a feeling of fulfillment in the quest for food. Satiety usually results from a filling meal, particularly when the person's nutritional storage depots, the adipose tissue and glycogen stores, are already filled.

Neural Centers for Regulation of Food Intake

HUNGER AND SATIETY CENTERS. Stimulation of the *lateral hypothalamus* causes an animal to eat voraciously, which is called *hyperphagia.* On the other hand, stimulation of the *ventromedial nuclei of the hypothalamus* causes complete satiety; even in the presence of highly appetizing food, the animal will refuse to eat, which is *aphagia.* Conversely, destructive

lesions of the two areas cause results opposite to those caused by stimulation. That is, ventromedial lesions cause voracious and continued eating until the animal becomes extremely obese, sometimes as large as four times normal size. And lesions of the lateral nuclei on the two sides of the hypothalamus cause lack of desire for food and often progressive inanition of the animal. Therefore, we label the lateral nuclei of the hypothalamus as a *feeding center* and the ventromedial nuclei of the hypothalamus as a *satiety center.*

The lateral hypothalamic feeding center operates by exciting the motor drives of the animal for all activity and specifically the emotional drive to search for food (while stimulating other emotional drives as well—see Chapters 58 and 60). On the other hand, it is believed that the satiety center operates to give the animal a sense of nutritional satisfaction, and this secondarily inhibits the feeding center.

OTHER NEURAL CENTERS THAT ENTER INTO FEEDING. Many other areas of the brain have also been implicated as sensors of the nutritional status of the body or as neural centers that drive an animal to search for and devour food. For instance, lesions in the *paraventricular nuclei* often cause excess eating and have been claimed specifically to cause excess eating of carbohydrates. Conversely, lesions in the *dorsomedial nuclei* of the hypothalamus usually depress eating. In addition, lesions in or stimulation of areas of the *lower brain stem,* such as in the *area postrema,* the *caudal medial nucleus of the solitary tract,* or the *vagus nerve,* can affect the degree of eating. How all these brain areas are used in the overall pattern of feeding regulation is still unclear.

Another aspect of feeding is the mechanical act of the feeding process itself. If the brain is sectioned below the hypothalamus but above the mesencephalon, the animal can still perform the basic mechanical features of the feeding process. It can salivate, lick its lips, chew food, and swallow. Therefore, *the actual mechanics of feeding are controlled by centers in the brain stem.* The function of the other centers in feeding, then, is to control the quantity of food intake and to excite these feeding-mechanics centers to activity.

Higher centers than the hypothalamus also play important roles in the control of feeding, particularly in the control of appetite. These centers include especially the *amygdala* and the *prefrontal cortex,* which are closely coupled with the hypothalamus. It will be recalled from the discussion of the sense of smell in Chapter 53 that portions of the amygdala are a major part of the olfactory nervous system. Destructive lesions in the amygdala have demonstrated that some of its areas increase feeding, whereas others inhibit feeding. In addition, stimulation of some areas of the amygdala elicits the mechanical act of feeding. The most important effect of destruction of the amygdala on both sides of the brain is a "psychic blindness" in the choice of foods. In other words, the animal (and presumably the human being as well) loses or at least partially loses the appetite control of type and quality of food that it eats.

Factors That Regulate Quantity of Food Intake

We can divide the regulation of quantity of food intake into (1) *energy regulation* (or *long-term regulation*), which is concerned primarily with long-term maintenance of normal quantities of energy stores in the body, and (2) *alimentary regulation* (or *short-term regulation*), which is concerned primarily with preventing overeating at the time of each meal.

Energy Regulation (Long-Term Regulation)

An animal that has been starved for a long time and is then presented with unlimited food eats a far greater quantity than does an animal that has been on a regular diet. Conversely, an animal that has been force-fed for several weeks eats very little when allowed to eat according to its own desires. Thus, the feeding control mechanism of the body is geared to the nutritional status of the body. Some of the nutritional factors that control the degree of activity of the feeding center are the following.

EFFECT OF BLOOD CONCENTRATIONS OF GLUCOSE, AMINO ACIDS, AND LIPIDS ON HUNGER AND FEEDING—THE GLUCOSTATIC, AMINOSTATIC, AND LIPOSTATIC THEORIES. It has long been known that a decrease in blood glucose concentration causes hunger, which has led to the so-called *glucostatic theory of hunger and feeding regulation.* Similar studies have more recently demonstrated the same effect for blood amino acid concentration and blood concentration of breakdown products of lipids such as the keto acids and some fatty acids, leading to the *aminostatic* and *lipostatic* theories of regulation. That is, when the availability of any of the three major types of food decreases, the animal automatically increases its feeding, which eventually returns the blood metabolite concentrations back toward normal.

Neurophysiological studies of function in certain specific areas of the brain have also substantiated the glucostatic, aminostatic, and lipostatic theories by the following observations: (1) an increase in blood glucose level *increases* the rate of firing of *glucoreceptor neurons* in the *satiety center in the ventromedial nucleus of the hypothalamus.* (2) The same increase in blood glucose level simultaneously *decreases* the firing of neurons called *glucosensitive neurons* in the *hunger center of the lateral hypothalamus.* In addition, some amino acids and lipid substances affect the rates of firing of these same neurons or other closely associated neurons.

Still other neurons, found in the *dorsomedial nuclei of the hypothalamus,* respond to the rate of utilization of all the foodstuffs that provide energy for the cells. This has led to a more global theory of hunger and feeding regulation based on *energy availability* for *power generation* within these cells.

INTERRELATION BETWEEN BODY TEMPERATURE AND FOOD INTAKE. When an animal is exposed to cold, it tends to overeat; when it is exposed to heat, it tends to undereat. This is caused by interaction within the hypothalamus between the temperature-regulating system (see Chapter 73) and the food intake–regulating system. It is important because increased food intake in the cold animal (1) increases its metabolic rate and (2) provides increased fat for insulation, both of which tend to correct the cold state.

SUMMARY OF LONG-TERM REGULATION. Even though our information on the different feedback factors in long-term feeding regulation is imprecise, we can make the following general statement: When the energy stores of the body fall below normal, the feeding centers of the hypothalamus and other areas of the brain become highly active, and the person exhibits increased hunger as well as searching for food; on the other hand, when the energy stores (mainly the fat stores) are already abundant, the person loses the hunger and develops a state of satiety.

Alimentary Regulation of Feeding (Short-Term Regulation)

When a person is driven by hunger to eat voraciously and rapidly, what turns off the eating when he or she has eaten enough? It is not the energy feedback mechanisms that we have discussed above because all of them take hours before enough quantities of the nutritional factors are absorbed into the blood to cause the necessary inhibition of eating. Yet it is important that the person not overeat and even that he or she eat an amount of food that approximates nutritional needs. The following are several types of rapid feedback signals that are important for this purpose.

Gastrointestinal Filling. When the gastrointestinal tract becomes distended, especially the stomach and the duodenum, stretch inhibitory signals are transmitted mainly by way of the vagi to suppress the feeding center, thereby reducing the desire for food.

Humoral and Hormonal Factors That Suppress Feeding—Cholecystokinin, Glucagon, and Insulin. The gastrointestinal hormone *cholecystokinin,* released mainly in response to fat entering the duodenum, has a strong direct effect on the feeding centers to reduce further eating. In addition, for reasons that are not entirely understood, the presence of food in the stomach and duodenum causes the pancreas to secrete significant quantities of *glucagon* and *insulin,* both of which also suppress the neurogenic feeding signals from the brain.

Metering of Food by Oral Receptors. When a person with an esophageal fistula is fed large quantities of food, even though this food is immediately lost again to the exterior, the degree of hunger is decreased after a reasonable quantity of food has passed through the mouth. This effect occurs despite the fact that the gastrointestinal tract does not become the least bit filled. Therefore, it is postulated that various "oral factors" related to feeding, such as chewing, salivation, swallowing, and tasting, "meter" the food as it passes through the mouth, and after a certain amount has passed, the hypothalamic feeding center becomes inhibited. However, the inhibition caused by this metering mechanism is considerably less intense and less lasting, usually lasting for only 20 to 40 minutes, than is the inhibition caused by gastrointestinal filling.

Importance of Having Both Long- and Short-Term Regulatory Systems for Feeding

The long-term regulatory system for feeding, which includes all the nutritional, energy feedback mechanisms, helps to maintain constant stores of nutrients in the tissues, preventing them from becoming too low or becoming too high. The short-term regulatory stimuli, on the other hand, serve two other purposes. First, they make the person or animal eat smaller quantities at each eating session, thus allowing food to pass through the gastrointestinal tract at a steadier pace, so that its digestive and absorptive mechanisms can work at optimal rates rather than becoming periodically overburdened. Second, they prevent the person or animal from eating amounts of food at each meal that would be too much for the metabolic storage systems once the food has all been absorbed.

OBESITY

ENERGY INPUT VERSUS ENERGY OUTPUT. When greater quantities of energy (in the form of food) enter the body than are expended, the body weight increases. Therefore, obesity is caused by excess energy input over energy output. For each 9.3 Calories of excess energy that enters the body, 1 gram of fat is stored.

Excess energy input occurs *only during the developing phase of obesity*. Once a person has become obese, all that is required to remain obese is that the energy input equal the energy output. For the person to reduce in weight, the input must be *less* than the output. Indeed, studies of obese people have shown that the intake of food of most of them in the static stage of obesity (after the obesity has already been attained) is about the same as that for normal people.

EFFECT OF MUSCULAR ACTIVITY ON ENERGY OUTPUT. About one third of the energy used each day by the normal person goes into muscular activity, and in the laborer, as much as two thirds or occasionally three fourths is used in this way. Because muscular activity is by far the most important means by which energy is expended in the body, it is frequently said that obesity in the otherwise normal person results from *too high a ratio of food intake to daily exercise.*

Abnormal Feeding Regulation as a Pathological Cause of Obesity

We have already emphasized that the rate of feeding is normally regulated in proportion to the energy stores in the body. When these stores exceed the optimal level for a normal person, feeding is automatically reduced to prevent overstorage. However, in many obese people, this is not true, because feeding in these persons does not slacken until body weight is far above normal. Therefore, in effect, obesity is usually caused by an abnormality of the feeding regulatory mechanism. This

can result from either psychogenic factors that affect this regulation or actual abnormalities of the regulatory system itself.

PSYCHOGENIC OBESITY. Studies of obese patients show that a large proportion of obesity results from psychogenic factors. Perhaps the most common psychogenic factor contributing to obesity is the prevalent idea that healthy eating habits require three meals a day and that each meal must be filling. Many young children are forced into this habit by overly solicitous parents, and the children continue to practice it throughout life. In addition, people are known often to gain large amounts of weight during or after stressful situations, such as the death of a parent, a severe illness, or even mental depression. It seems that eating is often a means of release from tension.

NEUROGENIC ABNORMALITIES AS A CAUSE OF OBESITY. In the preceding discussion of feeding regulation, it is pointed out that lesions in the ventromedial nuclei of the hypothalamus cause an animal to eat excessively and become obese. Such lesions also cause excess insulin production, which in turn increases fat deposition. In addition, many people with hypophysial tumors that encroach on the hypothalamus develop progressive obesity, demonstrating that obesity in the human being, too, can definitely result from damage to the hypothalamus.

Yet in the normal obese person, hypothalamic damage is almost never found. Nevertheless, it is possible that the functional organization of the hypothalamic or other neurogenic feeding centers is different in the obese person from that of the nonobese person. In support of this, a normally obese person who has reduced to normal weight by strict dietary measures usually develops intense hunger that is demonstrably far greater than that of the normal person. This indicates that the "setpoint" of the obese person's feeding control system is at a much higher level of nutrient storage than that of the normal person.

GENETIC FACTORS IN OBESITY. Obesity definitely runs in families. Furthermore, identical twins usually maintain weight levels within 2 pounds of each other throughout life if they live under similar conditions, or within 5 pounds of each other if their conditions of life differ markedly. This might result partly from eating habits engendered during childhood, but it is generally believed that this close similarity between twins is genetically controlled.

The genes can direct the degree of feeding in several ways, including (1) a genetic abnormality of the feeding centers to set the level of energy storage high or low and (2) abnormal hereditary psychic factors that either whet the appetite or cause the person to eat as a "release" mechanism.

Genetic abnormalities in the *chemistry of fat storage* are also known to cause obesity in certain strains of rats and mice. In one strain of rats, fat is easily stored in the adipose tissue, but the quantity of hormone-sensitive lipase in the adipose tissue is greatly reduced, so that little of the fat can be removed. This results in a one-way path, the fat continually being deposited but never released. In a strain of obese mice, an excess of fatty acid synthetase has been found that causes excess synthesis of fatty acids. Therefore, similar genetic mechanisms are possible causes of obesity in human beings.

CHILDHOOD OVERNUTRITION AS A POSSIBLE CAUSE OF OBESITY. The rate of formation of new fat cells is especially rapid in the first few years of life, and the greater the rate of fat storage, the greater also becomes the number of fat cells. In obese children, the number of fat cells is often as much as three times that in normal children. However, after adolescence, the number of fat cells remains almost identically the same throughout the remainder of life. Therefore, it has been suggested that overnutrition of children, especially in infancy and to a lesser extent during the later years of childhood, can lead to a lifetime of obesity. The person who has excess fat cells is thought to have a higher setting of fat storage by the neurogenic feedback autoregulatory mechanism for control of adipose tissues.

In people who become obese in middle or old age, most of the obesity results from hypertrophy of already existing fat cells without developing additional cells. This type of obesity is far more susceptible to treatment than is the lifelong type.

Treatment of Obesity

Treatment of obesity depends simply on decreasing energy input below energy expenditure. In other words, this means partial starvation. For this purpose, most reducing diets are designed to contain large quantities of "bulk," which, in general, are made up of non-nutritive cellulose substances. This bulk distends the stomach and thereby partially appeases the hunger. In most lower animals, such a procedure simply makes the animal increase its food intake still further, but human beings can often fool themselves because their food intake is sometimes controlled as much by habit as by hunger. As pointed out later in connection with starvation, it is important to prevent vitamin deficiencies during the dieting period.

Various *drugs for decreasing the degree of hunger* have been used in the treatment of obesity. The most important of these is *amphetamine* (or amphetamine derivatives), which directly inhibits the feeding centers in the brain. There is danger in using this drug because it simultaneously overexcites the central nervous system, making the person nervous and elevating the blood pressure. Also, a person soon adapts to the drug, so that weight reduction is usually no greater than 5 to 10 per cent.

Finally, the more exercise one takes, the greater is the daily energy expenditure and the more rapidly the obesity disappears. Therefore, forced exercise is often an essential part of the treatment for obesity.

INANITION

Inanition is the exact opposite of obesity. In addition to inanition caused by inadequate availability of food, both psychogenic and hypothalamic abnormalities can, on occasion, cause greatly decreased feeding. One such condition, *anorexia nervosa,* is an abnormal psychic state in which a person loses all desire for food and even becomes nauseated by food; as a result, severe inanition occurs. Also, destructive lesions of the hypothalamus, often caused by vascular thrombosis in some people, frequently cause a condition called *cachexia;* the term simply means severe inanition.

STARVATION

Depletion of Food Stores in the Body Tissues During Starvation. Even though the tissues preferentially use carbohydrate for energy over both fat and protein, the quantity of carbohydrate normally stored in the entire body is only a few hundred grams (mainly glycogen in the liver and muscles), and it can supply the energy required for body function for perhaps half a day. Therefore, except for the first few hours of starvation, the major effects are progressive depletion of tissue fat and protein. Because fat is the prime source of energy (100 times as much fat-energy is stored in the normal person as carbohydrate-energy), the rate of fat depletion continues unabated, as shown in Figure 71–1, until most of the fat stores in the body are gone.

Protein undergoes three phases of depletion: rapid depletion at first, then greatly slowed depletion, and, finally, rapid depletion again shortly before death. The initial rapid depletion is caused by use of easily mobilizable protein for direct metabolism or for conversion to glucose and then metabolism of the glucose mainly by the brain. After the readily mobilizable protein stores have been depleted during the early phase of starvation, the remaining protein is not so easily removed. At this time, the rate of gluconeogenesis decreases to one third to one fifth its previous rate, and the rate of depletion of protein becomes greatly decreased. The lessened availability of glucose then initiates a series of events that leads to excessive fat utilization and conversion of some of the fat breakdown products to ketone bodies, producing the state of *ketosis,* which is discussed in Chapter 68. The ketone bodies, like glucose, can cross the blood-brain barrier and can be used by the brain cells for energy. Therefore, about two thirds of the brain's energy now is derived from these ketone bodies, principally from beta-hydroxybutyrate. This sequence of events thus leads to at least partial preservation of the protein stores of the body.

There finally comes a time when the fat stores also are almost depleted, and the only remaining source of energy is proteins. At that time, the protein stores once again enter a stage of rapid depletion. Because proteins are also essential for maintenance of cellular function,

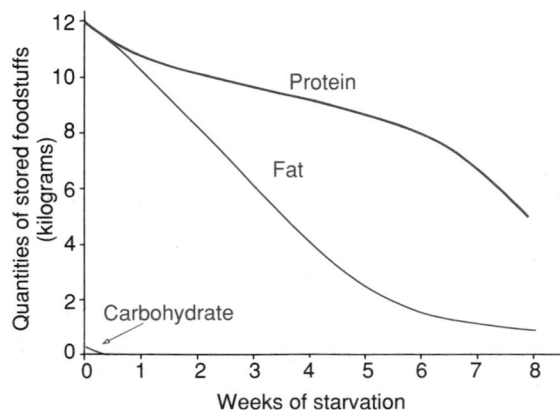

Figure 71–1. Effect of starvation on the food stores of the body.

death ordinarily ensues when the proteins of the body have been depleted to about one half their normal level.

VITAMIN DEFICIENCIES IN STARVATION. The stores of some of the vitamins, especially the water-soluble vitamins—the vitamin B group and vitamin C—do not last long during starvation. Consequently, after a week or more of starvation, mild vitamin deficiencies usually begin to appear, and after several weeks severe vitamin deficiencies can occur. These deficiencies can add to the debility that leads to death.

VITAMINS

DAILY REQUIREMENTS OF VITAMINS. A vitamin is an organic compound needed in small quantities for normal bodily metabolism and that cannot be manufactured in the cells of the body. When lacking in the diet, vitamins can cause specific metabolic deficits.

Table 71-2 lists the amounts of important vitamins required daily by the average man. These requirements vary considerably, depending on such factors as body size, rate of growth, amount of exercise, and pregnancy.

STORAGE OF VITAMINS IN THE BODY. Vitamins are stored to a slight extent in all cells. Some vitamins are stored to a major extent in the liver. For instance, the quantity of vitamin A stored in the liver may be sufficient to maintain a person without any intake of vitamin A for 5 to 10 months, and ordinarily the quantity of vitamin D stored in the liver is sufficient to maintain a person for 2 to 4 months without any additional intake of vitamin D.

The storage of most water-soluble vitamins is relatively slight. This applies especially to most vitamin B compounds because when a person's diet is deficient in vitamin B compounds, clinical symptoms of the deficiency can sometimes be recognized within a few days (except for vitamin B_{12}, which can last in the liver in a bound form for a year or longer). Absence of vitamin C, another water-soluble vitamin, can cause symptoms within a few weeks and can cause death from scurvy in 20 to 30 weeks.

Vitamin A

Vitamin A occurs in animal tissues as *retinol,* the formula of which is shown below. This vitamin does not occur in foods of vegetable origin, but *provitamins* for the formation of vitamin A do occur in abundance in many vegetable foods. They are the yellow and red *carotenoid pigments,* which, because they have chemical structures similar to that of vitamin A, can be changed into vitamin A in the liver.

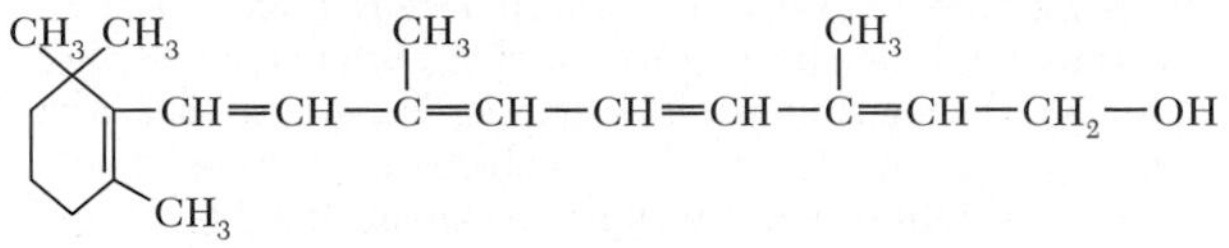

Vitamin A$_1$

The basic function of vitamin A in the metabolism of the body is not clear except in relation to its use in the formation of the retinal pigments of the eye, which is discussed in Chapter 50. Nevertheless, some of the effects of vitamin A lack have been well documented. In addition to the need for vitamin A to form the visual pigments and, therefore, to prevent night blindness, it is necessary for normal growth of most cells of the body and especially for normal growth and proliferation of the different types of epithelial cells. When vitamin A is lacking, the epithelial structures of the body tend to become stratified and keratinized. Vitamin A deficiency manifests itself by (1) scaliness of the skin and sometimes acne; (2) failure of growth of young animals, including cessation of skeletal growth; (3) failure of reproduction, associated especially with atrophy of the germinal epithelium of the testes and sometimes with interruption of the female sexual cycle; and (4) keratinization of the cornea with resultant corneal opacity and blindness.

Also, in vitamin A deficiency, the damaged epithelial structures often become infected, for example, the conjunctivas of the eyes, the linings of the urinary tract, and the respiratory passages. Therefore, vitamin A has been called an "anti-infection" vitamin.

Thiamine (Vitamin B$_1$)

Thiamine operates in the metabolic systems of the body principally as *thiamine pyrophosphate;* this compound functions as a *cocarboxylase,* operating mainly in conjunction with a protein decarboxylase for decarboxylation of pyruvic acid and other α-keto acids, as discussed in Chapter 67.

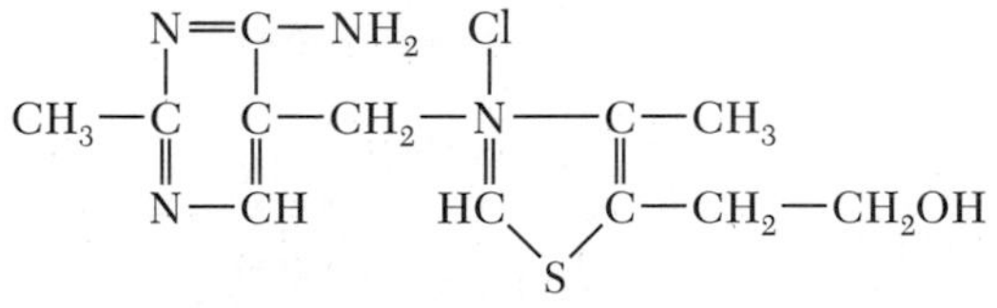

Thiamine chloride

Thiamine deficiency causes decreased utilization of pyruvic acid and some amino acids by the tissues but increased utilization of fats. Thus, thiamine is specifically needed for final metabolism of carbohydrates and many amino acids. The decreased utilization of these nutrients is the responsible factor for many debilities associated with thiamine deficiency.

Table 71-2 REQUIRED DAILY AMOUNTS OF THE VITAMINS

A	5000 IU
Thiamine	1.5 mg
Riboflavin	1.8 mg
Niacin	20 mg
Ascorbic acid	45 mg
D	400 IU
E	15 IU
K	70 μg
Folic acid	0.4 mg
B$_{12}$	3 μg
Pyridoxine	2 mg
Pantothenic acid	Unknown

THIAMINE DEFICIENCY AND THE NERVOUS SYSTEM.
The central nervous system normally depends almost entirely on the metabolism of carbohydrates for its energy. In thiamine deficiency, the utilization of glucose by nervous tissue may be decreased 50 to 60 per cent, which is replaced by utilization of ketone bodies derived from fat metabolism. The neuronal cells of the central nervous system frequently show chromatolysis and swelling during thiamine deficiency, changes that are characteristic of neuronal cells with poor nutrition. Such changes as these can disrupt communication in many portions of the central nervous system.

Also, thiamine deficiency can cause *degeneration of myelin sheaths* of nerve fibers both in the peripheral nerves and in the central nervous system. The lesions in the peripheral nerves frequently cause these nerves to become extremely irritable, resulting in "polyneuritis" characterized by pain radiating along the course of one or many peripheral nerves. Also, fiber tracts in the cord can degenerate to such an extent that *paralysis* occasionally results; even in the absence of paralysis, the muscles atrophy, with resultant severe weakness.

THIAMINE DEFICIENCY AND THE CARDIOVASCULAR SYSTEM. Thiamine deficiency also weakens the heart muscle, so that a person with severe thiamine deficiency eventually develops *cardiac failure.* Furthermore, the venous return of blood to the heart may be increased to as much as two times normal. This occurs because thiamine deficiency causes *peripheral vasodilation* throughout the circulatory system, presumably as a result of decreased release of metabolic energy in the tissues and this leading to local vascular dilation. Therefore, the cardiac effects of thiamine deficiency are due partly to excessive load of blood flow into the heart and partly to primary weakness of the cardiac muscle. *Peripheral edema* and *ascites* also occur to a major extent in some people with thiamine deficiency mainly because of the cardiac failure.

THIAMINE DEFICIENCY AND THE GASTROINTESTINAL TRACT. Among the gastrointestinal symptoms of thiamine deficiency are indigestion, severe constipation, anorexia, gastric atony, and hypochlorhydria. All these effects presumably result from failure of the smooth muscle and glands of the gastrointestinal tract to derive sufficient energy from carbohydrate metabolism.

The overall picture of thiamine deficiency, including polyneuritis, cardiovascular symptoms, and gastrointestinal disorders, is frequently referred to as "beriberi"— especially when the cardiovascular symptoms predominate.

Niacin

Niacin, also called *nicotinic acid,* functions in the body as coenzymes in the forms of nicotinamide adenine dinucleotide (NAD) and nicotinamide adenine dinucleotide phosphate (NADP). These coenzymes are hydrogen acceptors; they combine with hydrogen atoms as they are removed from food substrates by many types of dehydrogenases. The typical operation of both of them is presented in Chapter 67. When a deficiency of niacin exists, the normal rate of dehydrogenation cannot be maintained; therefore, oxidative delivery of energy from the foodstuffs to the

functioning elements of all cells cannot occur at normal rates.

Niacin

In the early stages of niacin deficiency, simple physiological changes such as muscle weakness and poor glandular secretion may occur, but in severe niacin deficiency, actual death of tissues ensues. Pathological lesions appear in many parts of the central nervous system, and permanent dementia or any of many types of psychoses may result. Also, the skin develops a cracked, pigmented scaliness in areas that are exposed to mechanical irritation or sun irradiation; thus, it seems as though the skin were unable to repair the different types of irritative damage.

Niacin deficiency causes intense irritation and inflammation of the mucous membranes of the mouth and other portions of the gastrointestinal tract, thus instituting many digestive abnormalities, leading in severe cases to widespread gastrointestinal hemorrhage. It is possible that this results from generalized depression of metabolism in the gastrointestinal epithelium and failure of appropriate epithelial repair.

The clinical entity called *pellagra* and the canine disease called *black tongue* are caused mainly by niacin deficiency. Pellagra is greatly exacerbated in people on a corn diet (such as many of the natives of Africa) because corn is deficient in the amino acid tryptophan, which can be converted in limited quantities to niacin in the body.

Riboflavin (Vitamin B$_2$)

Riboflavin normally combines in the tissues with phosphoric acid to form two coenzymes, *flavin mononucleotide (FMN)* and *flavin adenine dinucleotide (FAD).* They in turn operate as hydrogen carriers in important oxidative systems of the mitochondria. NAD, operating in association with specific dehydrogenases, usually accepts hydrogen removed from various food substrates and then passes the hydrogen to FMN or FAD; finally, the hydrogen is released as an ion into the mitochondrial matrix to become oxidized by oxygen, the system for which is described in Chapter 67.

Riboflavin

Deficiency of riboflavin in lower animals causes severe *dermatitis, vomiting, diarrhea, muscle spasticity* that finally becomes muscle weakness, and then *death* preceded by coma and decline in body temperature. Thus, severe riboflavin deficiency can cause many of the same effects as a lack of niacin in the diet; presumably, the debilities that result in each instance are due to generally depressed oxidative processes within the cells.

In the human being, riboflavin deficiency has never been known to be severe enough to cause the marked debilities noted in animal experiments, but mild riboflavin deficiency is probably common. Such deficiency causes digestive disturbances, burning sensations of the skin and eyes, cracking at the corners of the mouth, headaches, mental depression, forgetfulness, and so on.

Although the manifestations of riboflavin deficiency are usually relatively mild, this deficiency frequently occurs in association with deficiency of thiamine and/or niacin. Therefore, many deficiency syndromes, including *pellagra, beriberi, sprue,* and *kwashiorkor,* are probably due to a combined deficiency of a number of the vitamins as well as to other aspects of malnutrition.

Vitamin B$_{12}$

Several *cobalamin* compounds that possess the common prosthetic group shown below exhibit so-called "vitamin B$_{12}$" activity.

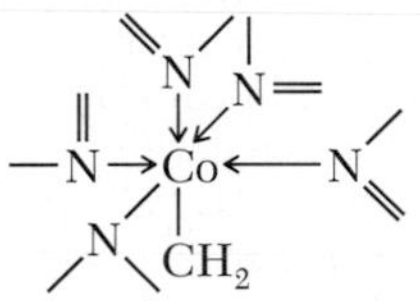

Note that this prosthetic group contains cobalt, which has coordination bonds similar to those of iron in the hemoglobin molecule. It is likely that the cobalt atom functions in much the same way that the iron atom functions to combine reversibly with other substances.

Vitamin B$_{12}$ performs several metabolic functions, acting as a hydrogen acceptor coenzyme. Its most important function is to act as a coenzyme for reducing ribonucleotides to deoxyribonucleotides, a step that is necessary in the replication of genes. This could explain the major functions of vitamin B$_{12}$: (1) promotion of growth and (2) promotion of red blood cell formation and maturation. This red cell function is described in detail in Chapter 32 in relation to pernicious anemia, a type of anemia caused by failure of red blood cell maturation when vitamin B$_{12}$ is deficient.

A special effect of vitamin B$_{12}$ deficiency is often demyelination of the large nerve fibers of the spinal cord, especially of the posterior columns and occasionally of the lateral columns. As a result, many people with pernicious anemia have much loss of peripheral sensation and, in severe cases, even become paralyzed.

The usual cause of vitamin B$_{12}$ deficiency is not lack of this vitamin in the food but deficiency of formation of *intrinsic factor,* which is normally secreted by the parietal cells of the gastric glands and is essential for absorption of vitamin B$_{12}$ by the ileal mucosa. This is discussed in Chapters 32 and 66.

Folic Acid (Pteroylglutamic Acid)

Several pteroylglutamic acids, one of which is shown below, exhibit the "folic acid effect." Folic acid functions as a carrier of hydroxymethyl and formyl groups. Perhaps its most important use in the body is in the synthesis of purines and thymine, which are required for formation of DNA. Therefore, folic acid, like vitamin B$_{12}$, is required for replication of the cellular genes. This perhaps explains one of the most important functions of folic acid—that is, to promote growth. Indeed, when it is absent from the diet, an animal will grow very little.

Folic acid is an even more potent growth promoter than vitamin B$_{12}$ and, like vitamin B$_{12}$, is important for the maturation of red blood cells, as discussed in Chapter 32. However, vitamin B$_{12}$ and folic acid each perform specific and different chemical functions in promoting growth and maturation of red blood cells. One of the significant effects of folic acid deficiency is development of macrocytic anemia almost identical to that which occurs in pernicious anemia. This often can be treated effectively with folic acid alone. (The structural formula for folic acid is shown at the bottom of this page.)

Pyridoxine (Vitamin B$_6$)

Pyridoxine exists in the form of *pyridoxal phosphate* in the cells and functions as a coenzyme for many chemical reactions related to amino acid and protein metabolism. Its most important role is that of coenzyme in the transamination process for the synthesis of amino acids. As a result, pyridoxine plays many key roles in metabolism, especially protein metabolism. Also, it is believed to act in the transport of some amino acids across cell membranes.

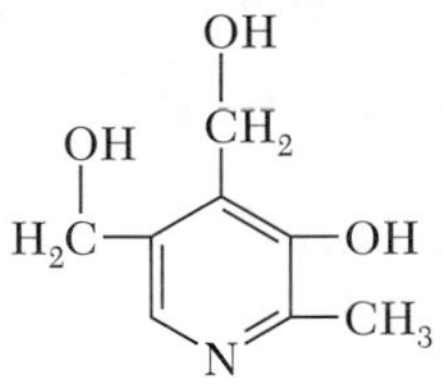

Pyridoxine

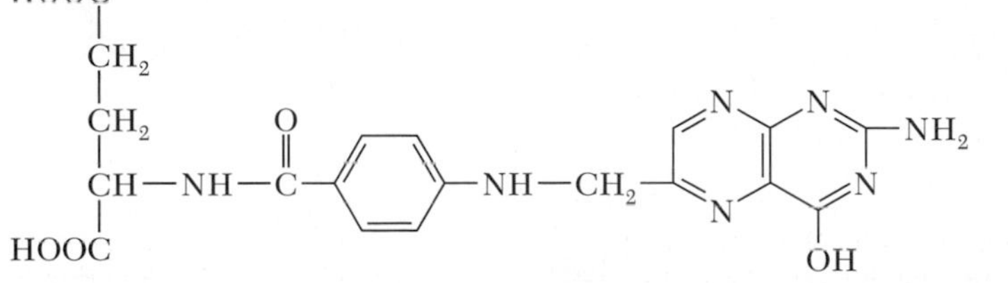

Folic acid (pteroylglutamic acid)

Dietary lack of pyridoxine in lower animals can cause dermatitis, decreased rate of growth, development of fatty liver, anemia, and evidence of mental deterioration. Rarely, in children, pyridoxine deficiency has been known to cause seizures, dermatitis, and gastrointestinal disturbances such as nausea and vomiting.

Pantothenic Acid

Pantothenic acid mainly is incorporated in the body into *coenzyme A* (CoA), which has many metabolic roles in the cells. Two of these discussed at length in Chapters 67 and 68 are (1) conversion of decarboxylated pyruvic acid into acetyl-CoA before its entry into the citric acid cycle and (2) degradation of fatty acid molecules into multiple molecules of acetyl-CoA. Thus, lack of pantothenic acid can lead to depressed metabolism of both carbohydrates and fats.

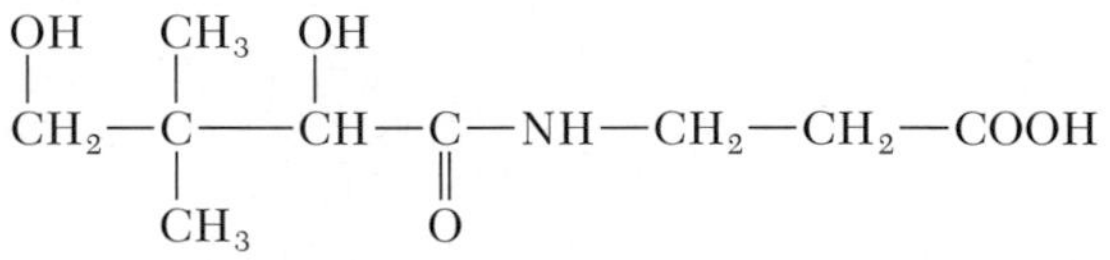

Pantothenic acid

Deficiency of pantothenic acid in lower animals can cause retarded growth, failure of reproduction, graying of the hair, dermatitis, fatty liver, and hemorrhagic adrenal cortical necrosis. In the human being, no definite deficiency syndrome has been proved, presumably because of the wide occurrence of this vitamin in almost all foods and because small amounts of the vitamin can probably be synthesized in the body. This does not mean that pantothenic acid is not of value in the metabolic systems of the body; indeed, it is perhaps as necessary as any other vitamin.

Ascorbic Acid (Vitamin C)

Ascorbic acid is essential for activating the enzyme *prolyl hydroxylase*, which promotes the hydroxylation step in the formation of hydroxyproline, an integral constituent of collagen. Without ascorbic acid, the collagen fibers that are formed in virtually all tissues of the body are defective and weak. Therefore, this vitamin is essential for the growth and strength of the fibers in subcutaneous tissue, cartilage, bone, and teeth.

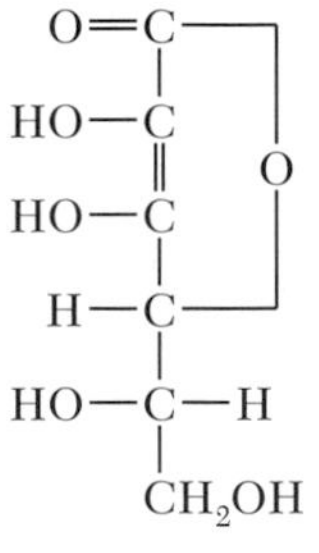

Ascorbic acid
(vitamin C)

Deficiency of ascorbic acid for 20 to 30 weeks, which occurred frequently during long sailing ship voyages in the past, causes *scurvy*, some effects of which are the following.

One of the most important effects of scurvy is *failure of wounds to heal*. This is caused by failure of the cells to deposit collagen fibrils and intercellular cement substances. As a result, healing of a wound may require several months instead of the several days ordinarily necessary.

Lack of ascorbic acid causes *cessation of bone growth*. The cells of the growing epiphyses continue to proliferate, but no new collagen is laid down between the cells, and the bones fracture easily at the point of growth because of failure to ossify. Also, when an already ossified bone fractures in a person with ascorbic acid deficiency, the osteoblasts cannot form new bone matrix. Consequently, the fractured bone does not heal.

The *blood vessel walls become extremely fragile* in scurvy because of (a) failure of the endothelial cells to be cemented together properly and (b) failure to form the collagen fibrils normally present in vessel walls. The capillaries especially are likely to rupture, and as a result, many small petechial hemorrhages occur throughout the body. The hemorrhages beneath the skin cause purpuric blotches, sometimes over the entire body. To test for ascorbic acid deficiency, one can produce such petechial hemorrhages by inflating a blood pressure cuff over the upper arm; this occludes the venous return of blood, the capillary pressure rises, and red blotches occur in the skin of the forearm if there is a sufficiently severe ascorbic acid deficiency.

In extreme scurvy, the muscle cells sometimes fragment; lesions of the gums with loosening of the teeth occur; infections of the mouth develop; vomiting of blood, bloody stools, and cerebral hemorrhage can all occur; and, finally, high fever often develops before death.

Vitamin D

Vitamin D increases calcium absorption from the gastrointestinal tract and helps control calcium deposition in the bone. The mechanism by which vitamin D increases calcium absorption is mainly to promote active transport of calcium through the epithelium of the ileum. It especially increases the formation of a calcium-binding protein in the intestinal epithelial cells that aids in calcium absorption. The specific functions of vitamin D in relation to overall body calcium metabolism and to bone formation are presented in Chapter 79.

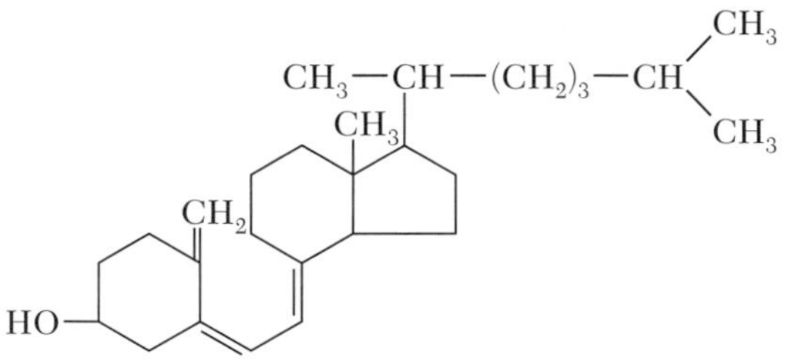

Vitamin D_3 (cholecalciferol)

Vitamin E

Several related compounds, one of which is shown below, exhibit so-called "vitamin E activity." Only rare instances of proved vitamin E deficiency have occurred in human beings. In lower animals, lack of vitamin E can cause degeneration of the germinal epithelium in the testis and, therefore, can cause male sterility. Lack of vitamin E can also cause resorption of a fetus after conception in the female. Because of these effects of vitamin E deficiency, vitamin E is sometimes called the "antisterility vitamin."

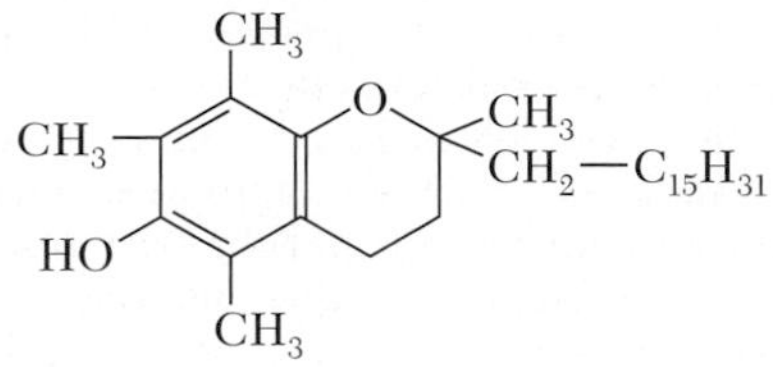

Vitamin E (alpha-tocopherol)

Also, as is true of almost all the vitamins, deficiency of vitamin E prevents normal growth and sometimes causes degeneration of the renal tubular cells and the muscle cells.

Vitamin E is believed to function mainly in relation to unsaturated fatty acids, providing a protective role to prevent oxidation of the unsaturated fats. In the absence of vitamin E, the quantity of unsaturated fats in the cells becomes diminished, causing abnormal structure and function of such cellular organelles as the mitochondria, the lysosomes, and even the cell membrane.

Vitamin K

Vitamin K is necessary for the formation by the liver of prothrombin, Factor VII (proconvertin), Factor IX, and Factor X, all of which are important in blood coagulation. Therefore, when vitamin K deficiency occurs, blood clotting is retarded. The function of this vitamin and its relations with some of the anticoagulants, such as dicumarol, are presented in greater detail in Chapter 35. (The structural formula is shown at the bottom of the page.)

Several compounds, both natural and synthetic, exhibit vitamin K activity. The chemical formula for one of the natural vitamin K compounds is shown below. Because vitamin K is synthesized by bacteria in the colon, it is rare when a person has a bleeding tendency because of vitamin K deficiency in the diet. However, when the bacteria of the colon are destroyed by administration of large quantities of antibiotic drugs, vitamin K deficiency occurs rapidly because of the paucity of this compound in the normal diet.

MINERAL METABOLISM

The functions of many of the minerals, such as sodium, potassium, and chloride, are presented at appropriate points in the text. Therefore, only specific functions of minerals not covered elsewhere are mentioned here.

The body content of the most important minerals is listed in Table 71–3, and the daily requirements of these are given in Table 71–4.

MAGNESIUM. Magnesium is about one sixth as plentiful in cells as potassium. Magnesium is especially required as a catalyst for many intracellular enzymatic reactions, particularly those related to carbohydrate metabolism.

The extracellular fluid magnesium concentration is slight, only 1.8 to 2.5 mEq/liter. Increased extracellular concentration of magnesium depresses activity in the nervous system as well as skeletal muscle contraction. This latter effect can be blocked by administration of calcium. Low magnesium concentration causes increased irritability of the nervous system, peripheral vasodilatation, and cardiac arrhythmias, especially after acute myocardial infarction.

CALCIUM. Calcium is present in the body mainly in the form of calcium phosphate in the bone. This subject is discussed in detail in Chapter 79, as is the calcium content of extracellular fluids.

Excess quantities of calcium ions in extracellular fluids can cause the heart to stop in systole and can act as a mental depressant. At the other extreme, low levels of calcium can cause spontaneous discharge of nerve fibers, resulting in tetany. This, too, is discussed in Chapter 79.

PHOSPHORUS. Phosphate is the major anion of intracellular fluids. Phosphates have the ability to combine

Table 71–3 CONTENT IN GRAMS OF A 70-KILOGRAM MAN

Water	41,400
Fat	12,600
Protein	12,600
Carbohydrate	300
Na	63
K	150
Ca	1,160
Mg	21
Cl	85
P	670
S	112
Fe	3
I	0.014

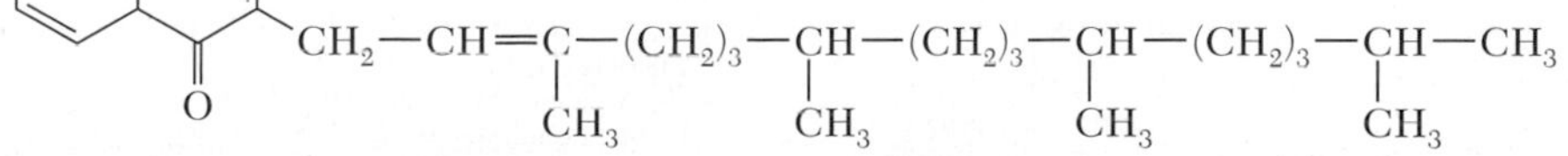

Vitamin K$_1$ (2-methyl-3-phytyl-1,4-naphthoquinone)

Table 71–4 REQUIRED DAILY AMOUNTS OF MINERALS

Na	3.0 gm
K	1.0 gm
Cl	3.5 gm
Ca	1.2 gm
P	1.2 gm
Fe	18.0 mg
I	150.0 μg
Mg	0.4 gm
Co	Unknown
Cu	Unknown
Mn	Unknown
Zn	15 mg

reversibly with many coenzyme systems and with multiple other compounds that are necessary for operation of the metabolic processes. Many important reactions of phosphates have been catalogued at other points in this text, especially in relation to the functions of adenosine triphosphate, adenosine diphosphate, phosphocreatine, and so forth. Also, bone contains a tremendous amount of calcium phosphate, which is discussed in Chapter 79.

IRON. The function of iron in the body, especially in relation to the formation of hemoglobin, is discussed in Chapter 32. Two thirds of the iron in the body is in the form of hemoglobin, although smaller quantities are present in other forms, especially in the liver and the bone marrow. Electron carriers containing iron (especially the cytochromes) are present in the mitochondria of all cells of the body and are essential for most of the oxidation that occurs in the cells. Therefore, iron is absolutely essential both for transport of oxygen to the tissues and for operation of oxidative systems within the tissue cells, without which life would cease within a few seconds.

IMPORTANT TRACE ELEMENTS IN THE BODY. A few elements are present in the body in such small quantities that they are called *trace elements*. The amounts of these elements in foods usually are also minute. Yet without any one of them, a specific deficiency syndrome is likely to develop. Three of the most important are iodine, zinc, and fluorine.

Iodine. The best known of the trace elements is iodine. This element is discussed in Chapter 76 in connection with the formation and function of thyroid hormone; as shown in Table 71–3, the entire body contains an average of only 14 milligrams. Iodine is essential for the formation of *thyroxine* and *triiodothyronine,* the two thyroid hormones that are essential for maintenance of normal metabolic rates in all cells of the body.

Zinc. Zinc is an integral part of many enzymes, one of the most important of which is *carbonic anhydrase,* present in especially high concentration in the red blood cells. This enzyme is responsible for rapid combination of carbon dioxide with water in the red blood cells of the peripheral capillary blood and for rapid release of carbon dioxide from the pulmonary capillary blood into the alveoli. Carbonic anhydrase is also present to a major extent in the gastrointestinal mucosa, in the tubules of the kidney, and in the epithelial cells of many glands of the body. Consequently, zinc in small quantities is essential for the performance of many reactions related to carbon dioxide metabolism.

Zinc is also a component of *lactic dehydrogenase* and,

therefore, is important for the interconversions between pyruvic acid and lactic acid. Finally, zinc is a component of some *peptidases* and, therefore, is important for digestion of proteins in the gastrointestinal tract.

Fluorine. Fluorine does not seem to be a necessary element for metabolism, but the presence of a small quantity of fluorine in the body during the period of life when the teeth are being formed subsequently protects against carious teeth. Fluorine does not make the teeth themselves stronger but has a poorly understood effect in suppressing the cariogenic process. It has been suggested that fluorine is deposited in the hydroxyapatite crystals of the tooth enamel and combines with and, therefore, blocks the functions of various trace metals that are necessary for activation of the bacterial enzymes that cause caries. Therefore, when fluorine is present, the enzymes remain inactive and cause no caries.

Excessive intake of fluorine causes *fluorosis,* which is manifest in its mild state by mottled teeth and in a more severe state by enlarged bones. It has been postulated that in this condition, fluorine combines with trace metals in some of the metabolic enzymes, including the phosphatases, so that various metabolic systems become partially inactivated. According to this theory, the mottled teeth and enlarged bones are due to abnormal enzyme systems in the odontoblasts and osteoblasts. Even though mottled teeth are highly resistant to the development of caries, the structural strength of these teeth is considerably lessened by the mottling process.

REFERENCES

Alderton, P.: The Vitamin, Mineral Connection. Louisville, Booksworld, 1985.

Anderson, K. E., et al.: Nutrient regulation of chemical metabolism in humans. Fed. Proc., 44:130, 1985.

Bendich, A., and Olson, J. A.: Biological actions of carotenoids. FASEB J., 3:1927, 1989.

Berger, H.: Vitamins and Minerals in Pregnancy and Lactation. New York, Raven Press, 1988.

Blomhoff, R., et al.: Vitamin A metabolism: new perspectives on absorption, transport, and storage. Physiol. Rev., 71:951, 1991.

Bjorntorp, P., and Brodoff, B. N. Obesity. Philadelphia, J. B. Lippincott, 1994.

Burk, R. F.: Biological activity of selenium. Annu. Rev. Nutr., 3:53, 1983.

Chandra, R. K.: Trace Elements in Nutrition of Children—II. New York, Raven Press, 1991.

Christensen, H. N.: The regulation of amino acid and sugar absorption by diet. Nutr. Rev., 42:237, 1984.

Cohen, R. D., et al.: The Metabolic and Molecular Basis of Acquired Disease. Philadelphia, W. B. Saunders Co., 1990.

Cousins, R. J.: Absorption, transport, and hepatic metabolism of copper and zinc with special reference to metallothionein and ceruloplasmin. Physiol. Rev., 65:238, 1985.

Cunningham, J. J.: Introduction to Nutritional Physiology. Philadelphia, G. F. Stickley, 1983.

de Rouffignac, C., and Quamme, G.: Renal magnesium handling and its hormonal control. Physiol. Rev., 74:305, 1994.

Forbes, R. M., and Erdman, J. W., Jr.: Bioavailability of trace mineral elements. Annu. Rev. Nutr., 3:213, 1983.

Gershwin, M. E.: Nutrition and Immunity. New York, Academic Press, 1985.

Guthrie, H. A.: Introductory Nutrition, 7th Ed. St. Louis, C. V. Mosby Co., 1988.

Hanson, L. A.: Biology of Human Milk. New York, Raven Press, 1988.

Hausinger, R. P.: Biochemistry of Nickel. New York, Plenum Publishing Corp., 1993.

Haust, M. D., et al.: Genetic Metabolic Diseases. Farmington, CT, S. Karger Publishers, Inc., 1993.

Henderson, L. M.: Niacin. Annu. Rev. Nutr., 3:289, 1983.

Huxtable, R. J.: Physiological actions of taurine. Physiol. Rev., 72:101, 1992.

Katch, F., and McArdle, W. D.: Nutrition, Weight Control, and Exercise. Philadelphia, Lea & Febiger, 1988.

Katch, F., and McArdle, W. D.: Introduction to Nutrition, Exercise, and Health. Baltimore, Williams & Wilkins, 1994.

Kinney, J. M., et al.: Nutrition and Metabolism in Patient Care. Philadelphia, W. B. Saunders Co., 1988.

Klurfeld, D. M.: Human nutrition: A Comprehensive Treatise. Vol. 8. Nutrition and Immunology. New York, Plenum Publishing Corp., 1993.

Krishnamurti, D.: Vitamin Receptors: Vitamins as Ligands in Cell Communication-Metabolic Indicators. New York, Cambridge University Press, 1993.

Linder, M. C.: Biochemistry of copper. New York, Plenum Publishing Corp., 1991.

Luigi, B., et al.: Nutrition in Gastrointestinal Disease. New York, Raven Press, 1987.

Magnen, J. L.: Body energy balance and food intake: A neuroendocrine regulatory mechanism. Physiol. Rev., 63:314, 1983.

Maho, Y. L., et al.: Starvation as a treatment for obesity: The need to conserve body protein. News Physiol. Sci., 3:21, 1988.

Mino, M., et al.: Vitamin E. Farmington, CT, S. Karger Publishers, Inc., 1993.

Morley, J. E., et al.: Geriatric Nutrition: A Comprehensive Review. New York, Raven Press, 1994.

Muno, H. N., and Danford, D. E.: Nutrition, Aging, and the Elderly. New York, Plenum Publishing Corp., 1989.

Nicolaidis, S.: What determines food intake? The ischymetric theory. News Physiol. Sci., 2:104, 1987.

Nielsen, F. H.: Ultratrace elements in nutrition. Annu. Rev. Nutr., 4:21, 1984.

Olson, R. E.: The function and metabolism of vitamin K. Annu. Rev. Nutr., 4:281, 1984.

Oomura, Y., and Yoshimatsu, H.: Neural network of glucose monitoring system. J. Auton. Nerv. Syst., 10:359, 1984.

Oomura, Y.: Regulation of feeding by neural responses to endogenous factors. News Physiol. Sci., 2:199, 1987.

Paige, D. M.: Clinical Nutrition. St. Louis, C. V. Mosby Co., 1988.

Prasad, A. S.: Biochemistry of Zinc. New York, Plenum Publishing Corp., 1993.

Prohaska, J. R.: Functions of trace elements in brain metabolism. Physiol. Rev., 67:858, 1987.

Rombeau, J. L., and Caldwell, M. D.: Clinical Nutrition: Parenteral Nutrition. Philadelphia, W. B. Saunders Co., 1993.

Schrier, R. W.: Geriatric Medicine. Philadelphia, W. B. Saunders Co., 1990.

Shils, M. E., and Young, V. R.: Modern Nutrition in Health and Disease, 7th Ed. Philadelphia, Lea & Febiger, 1988.

Storlien, L. H.: The role of the ventromedial hypothalamic area in periprandial glucoregulation. Life. Sci., 360:505, 1985.

Sturman, J. A.: Taurine in development. Physiol. Rev., 73:119, 1993.

Sullivan, A. C., and Gruen, R. K.: Mechanisms of appetite modulation by drugs. Fed. Proc., 44:139, 1985.

Tielsch, J. M., and Sommer, A.: The epidemiology of vitamin A deficiency and xerophthalmia. Annu. Rev. Nutr., 4:183, 1984.

Tresguerres, J. A. F., et al.: Dietary essential polyunsaturated fatty acids and blood pressure. News Physiol. Sci., 4:64, 1989.

Vallee, B. L., and Falchuk, K. H.: The biochemical basis of zinc physiology. Physiol. Rev., 73:79, 1993.

Weinsier, R. L., et al.: Handbook of Clinical Nutrition: Clinician's Manual for the Diagnosis and Management of Nutritional Problems. St. Louis, C. V. Mosby Co., 1988.

Williams, S. R.: Nutrition and Diet Therapy, 6th Ed. St. Louis, C. V. Mosby Co., 1989.

Wolf, G.: Multiple functions of vitamin A. Physiol. Rev., 64:873, 1984.

Energetics and Metabolic Rate

C *HAPTER* 72

ADENOSINE TRIPHOSPHATE FUNCTIONS AS AN "ENERGY CURRENCY" IN METABOLISM

In the past few chapters, it has been pointed out that carbohydrates, fats, and proteins can all be used by cells to synthesize large quantities of adenosine triphosphate (ATP) and that in turn the ATP can be used as an energy source for almost all other cellular functions. For these reasons, ATP has been called an energy "currency" that can be created and expended. Indeed, the cells can transfer energy from the different foodstuffs to most functional systems of the cells only through this medium of ATP (or the similar nucleotide guanosine triphosphate, GTP). Many of the attributes of ATP are presented in Chapter 2, but others require discussion at this point.

An attribute of ATP that makes it highly valuable as a means of energy currency is the large quantity of free energy (about 7300 Calories per mole under standard conditions but as much as 12,000 Calories under physiological conditions) vested in each of its two high-energy phosphate bonds. The amount of energy in each bond, when liberated by decomposition of ATP, is enough to cause almost any step of any chemical reaction in the body to take place if appropriate transfer of the energy is achieved. Some chemical reactions that require ATP energy use only a few hundred of the available 12,000 Calories, and the remainder of this energy is then lost in the form of heat. Even this inefficiency in the utilization of energy is better than not being able to energize the necessary chemical reactions.

ATP ENERGIZES THE SYNTHESIS OF MOST IMPORTANT CELLULAR COMPONENTS. Among the most important of the intracellular processes that require ATP energy is the formation of peptide linkages between amino acids during the synthesis of proteins. The different peptide linkages, depending on which types of amino acids are linked together,

require from 500 to 5000 calories of energy per mole. It will be recalled from the discussion of protein synthesis in Chapter 3 that four high-energy phosphate bonds are expended during the cascade of reactions required for forming each peptide linkage. This provides a total of 48,000 calories of energy, which is far more than the 500 to 5000 calories eventually stored in each of the peptide linkages.

Also, it will be recalled from the preceding chapters that ATP energy is used in the synthesis of glucose from lactic acid and in the synthesis of fatty acids from acetyl coenzyme A. In addition, ATP energy is used for synthesis of cholesterol, phospholipids, the hormones, and almost all other substances of the body. Even the urea excreted by the kidneys requires ATP to cause its formation from ammonia. One might wonder at the advisability of expending energy to form urea, which then is simply thrown away from the body. However, remembering the extreme toxicity of ammonia in the body fluids, one can see the value of this reaction, which keeps the ammonia concentration of the body fluids always at a low level.

ATP ENERGIZES MUSCLE CONTRACTION. Muscle contraction will not occur without energy from ATP. Myosin, one of the important contractile proteins of the muscle fiber, acts as an enzyme to cause breakdown of ATP into adenosine diphosphate (ADP), thus causing release of the energy required to cause contraction. Only a small amount of ATP is normally degraded in muscles when muscle contraction is not occurring, but this rate of ATP usage can rise to at least 150 times the resting level during short bursts of maximal contraction. The postulated mechanism by which ATP energy is used to cause muscle contraction is discussed in Chapter 6.

ATP ENERGIZES ACTIVE TRANSPORT ACROSS MEMBRANES. In Chapters 4, 27, and 65, active transport of electrolytes and various nutrients across cell membranes and from the renal tubules and gastrointestinal tract into the

903

blood is discussed. In each instance, it is noted that active transport of most electrolytes and substances such as glucose, amino acids, and acetoacetate can occur against an electrochemical gradient, even though the natural diffusion of the substances would be in the opposite direction. To oppose the electrochemical gradient requires energy, as discussed in Chapter 4. This energy is provided by ATP.

Energy for Glandular Secretion. The same principles apply to glandular secretion as to the absorption of substances against concentration gradients because energy is required to concentrate substances as they are secreted by the glandular cell. In addition, energy is required to synthesize the organic compounds to be secreted.

Energy for Nerve Conduction. The energy used during propagation of a nerve impulse is derived from the potential energy stored in the form of concentration differences of ions across the membranes. That is, a high concentration of potassium inside the fiber and a low concentration outside the fiber constitute a type of energy storage. Likewise, a high concentration of sodium on the outside of the membrane and a low concentration on the inside represent another store of energy. The energy needed to pass each action potential along the fiber membrane is derived from this energy storage, with small amounts of potassium transferring out of the cell and sodium into the cell during each of the action potentials. However, active transport systems energized by ATP then retransport the ions back through the membrane to their former positions.

Phosphocreatine as an Accessory Storage Depot for Energy and for Buffering the Concentration of ATP

Despite the paramount importance of ATP as a coupling agent for energy transfer, this substance is not the most abundant store of high-energy phosphate bonds in the cells. On the contrary, *phosphocreatine*, which also contains high-energy phosphate bonds, is three to eight times as abundant. Also, the high-energy bond (~) of phosphocreatine contains about 8500 calories per mole under standard conditions and as much as 13,000 calories per mole under conditions in the body (37°C and low concentrations of the reactants). This is somewhat greater than the 12,000 calories per mole in each of the two high-energy phosphate bonds of ATP. The formula for creatine phosphate is the following:

$$\underset{\substack{| \\ \text{H}}}{\text{HOOC}-\text{CH}_2-\text{N}-\overset{\substack{\text{CH}_3 \quad \text{NH} \quad \text{H} \quad \text{O} \\ | \qquad \| \quad\;\; | \qquad \|}}{\underset{\substack{| \\ \text{O} \\ | \\ \text{H}}}{\text{N}-\text{C}-\text{N}\sim\text{P}-\text{OH}}}}$$

Phosphocreatine cannot act in the same manner as ATP as a direct coupling agent for transfer of energy between the foods and the functional cellular systems, but it can transfer energy interchangeably with ATP. When extra amounts of ATP are available in the cell, much of its energy is used to synthesize phosphocreatine, thus building up this storehouse of energy. Then, when the ATP begins to be used up, the energy in the phosphocreatine is transferred rapidly back to ATP and from this to the functional systems of the cells. This reversible interrelation between ATP and phosphocrea-

tine is demonstrated by the following equation:

$$\text{Phosphocreatine} + \text{ADP}$$
$$\Updownarrow$$
$$\text{ATP} + \text{Creatine}$$

Note particularly that the higher energy level of the high-energy phosphate bond in phosphocreatine (1000 to 1500 calories per mole greater than that in ATP) causes the reaction between phosphocreatine and ADP to proceed rapidly toward the formation of new ATP every time even the slightest amount of ATP expends its energy elsewhere. Therefore, the slightest usage of ATP by the cells calls forth the energy from the phosphocreatine to synthesize new ATP. This effect keeps the concentration of ATP at an almost constant high level as long as any phosphocreatine remains. For this reason, we can also call the ATP-phosphocreatine system an ATP "buffer" system. One can readily understand the importance of keeping the concentration of ATP nearly constant because the rates of almost all the metabolic reactions in the body depend on this constancy.

Anaerobic Versus Aerobic Energy

Anaerobic energy means energy that can be derived from foods without the simultaneous utilization of oxygen; *aerobic energy* means energy that can be derived from foods only by oxidative metabolism. In the discussions in Chapters 67 through 69, it is noted that carbohydrates, fats, and proteins can all be oxidized to cause synthesis of ATP. However, carbohydrates are the only significant foods that can be used to provide energy without utilization of oxygen; this energy release occurs during glycolytic breakdown of glucose or glycogen to pyruvic acid. For each mole of glucose that is split into pyruvic acid, 2 moles of ATP are formed. However, when stored glycogen in a cell is split to pyruvic acid, each mole of glucose in the glycogen gives rise to 3 moles of ATP. The reason for this difference is that free glucose entering the cell must be phosphorylated by using 1 mole of ATP before it can begin to be split, whereas this is not true of glucose derived from glycogen because it comes from the glycogen already in the phosphorylated state without additional expenditure of ATP. Thus, the best source of energy under anaerobic conditions is the stored glycogen of the cells.

ANAEROBIC ENERGY DURING HYPOXIA. One of the prime examples of anaerobic energy utilization occurs in acute hypoxia. When a person stops breathing, there is already a small amount of oxygen stored in the lungs and an additional amount stored in the hemoglobin of the blood. This oxygen is sufficient to keep the metabolic processes functioning only for about 2 minutes. Continued life beyond this time requires an additional source of energy. This can be derived for another minute or so from glycolysis, that is, the glycogen of the cells splitting into pyruvic acid and the pyruvic acid in turn becoming lactic acid, which diffuses out of the cells, as described in Chapter 67.

ANAEROBIC ENERGY USAGE DURING STRENUOUS BURSTS OF ACTIVITY. It is common knowledge that muscles can perform extreme feats of strength for a few seconds but are much less capable during prolonged activity. Most of the extra energy required during these bursts of activity cannot come from the oxidative processes because they are too slow to respond. Instead, the extra energy comes from anaerobic sources: (1) ATP already present in the muscle cells,

(2) phosphocreatine in the cells, and (3) anaerobic energy released by glycolytic breakdown of glycogen to lactic acid.

The maximum amount of ATP in muscle is only about 5 mmol/liter of intracellular fluid, and this amount can maintain maximum muscle contraction for no more than a second or so. The amount of phosphocreatine in the cells is three to eight times this amount, but even by utilization of all the phosphocreatine, the amount of time that maximum contraction can be maintained is still only 5 to 10 seconds. Release of energy by glycolysis can occur much more rapidly than can oxidative release of energy. Consequently, most of the extra energy required during strenuous activity that lasts for more than 5 to 10 seconds but less than 1 to 2 minutes is derived from anaerobic glycolysis. As a result, the glycogen content of muscles during strenuous bouts of exercise becomes reduced, whereas the lactic acid concentration of the blood rises. Then, after the exercise is over, oxidative metabolism is used to reconvert about four fifths of the lactic acid into glucose; the remainder becomes pyruvic acid and is degraded and oxidized in the citric acid cycle. The reconversion to glucose occurs principally in the liver cells, and the glucose is then transported in the blood back to the muscles, where it is stored once more in the form of glycogen.

OXYGEN DEBT. After a period of strenuous exercise, a person continues to breathe hard and to consume excessive amounts of oxygen for at least a few minutes and sometimes for as long as 1 hour thereafter. This excess oxygen is used (1) to reconvert the lactic acid that has accumulated during exercise back into glucose, (2) to reconvert adenosine monophosphate and ADP to ATP, (3) to reconvert creatine and phosphate to phosphocreatine, (4) to re-establish normal concentrations of oxygen bound with hemoglobin and myoglobin, and (5) to raise the concentration of oxygen in the lungs up to its normal level. This excess consumption of oxygen after the exercise is over is called the *oxygen debt.*

The principle of oxygen debt is discussed further in Chapter 84 in relation to sports physiology; the ability of a person to build up an oxygen debt is especially important in many types of athletics.

Summary of Energy Utilization by the Cells

With the background of the past few chapters and of the preceding discussion, we can now synthesize a composite picture of overall energy utilization by the cells, as shown in Figure 72–1. This figure demonstrates the anaerobic utilization of glycogen and glucose to form ATP and the aerobic utilization of compounds derived from carbohydrates, fats, proteins, and other substances for the formation of still additional ATP. In turn, ATP is in reversible equilibrium with phosphocreatine in the cells, and because larger quantities of phosphocreatine are present in the cell than ATP, much of the stored energy of the cell is in this energy storehouse.

Energy from ATP can be used by the different functioning systems of the cells to provide for synthesis and growth, muscle contraction, glandular secretion, nerve impulse conduction, active absorption, and other cellular activities. If greater amounts of energy are called forth for cellular activities than can be provided by oxidative metabolism, the phosphocreatine storehouse is first used and then anaerobic breakdown of glycogen follows rapidly. Thus, oxidative metabolism cannot deliver bursts of extreme energy to the cells nearly so rapidly as can the anaerobic processes, but in contrast, at slower rates of usage, it is quantitatively almost inexhaustible because the oxidative processes can continue indefinitely.

CONTROL OF ENERGY RELEASE IN THE CELL

RATE CONTROL OF ENZYME-CATALYZED REACTIONS. Before it is possible to discuss the control of energy release in the cell, it is necessary to consider the basic principles of *rate control* of enzymatically catalyzed chemical reactions, which are the types of reactions that occur almost universally throughout the body.

The mechanism by which an enzyme catalyzes a chemical

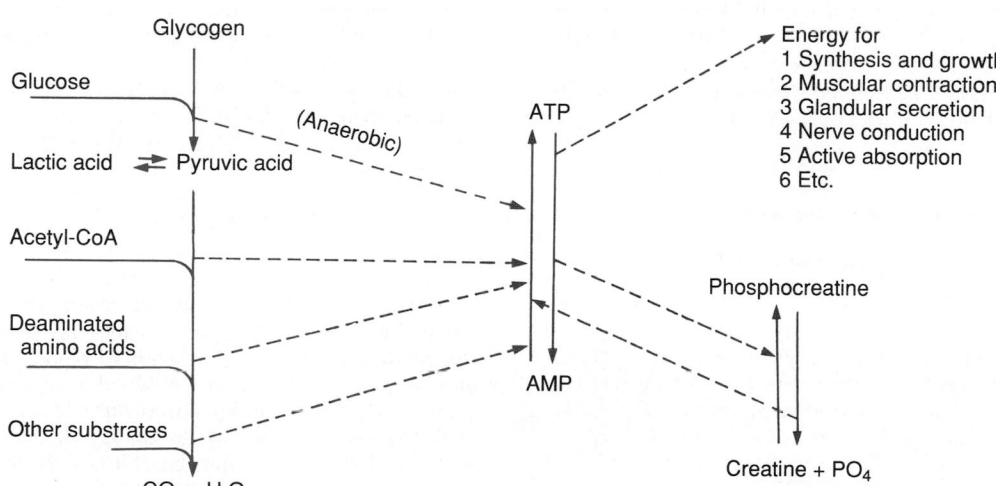

Figure 72–1. Overall schema of energy transfer from foods to the adenylic acid system and then to the functional elements of the cells. (Modified from Soskin and Levine: Carbohydrate Metabolism. Chicago, University of Chicago Press. ©1946, 1952 by The University of Chicago. All rights reserved.)

reaction is for the enzyme first to combine loosely with one of the substrates of the reaction. This alters the bonding forces on the substrate sufficiently that it can then react with other substances. Therefore, the rate of the overall chemical reaction is determined by both the concentration of the enzyme and the concentration of the substrate that binds with the enzyme. The basic equation expressing this concept is as follows:

$$\text{Rate of reaction} = \frac{K_1 \cdot \text{Enzyme} \cdot \text{Substrate}}{K_2 + \text{Substrate}}$$

This is called the *Michaelis-Menten equation*. Figure 72–2 shows the application of this equation.

Role of Enzyme Concentration in the Regulation of Metabolic Reactions. Figure 72–2 shows that *when the substrate is present in reasonably high concentration*, as shown in the right half of the figure, the rate of a chemical reaction is then determined almost entirely by the concentration of the enzyme. Thus, as the enzyme concentration increases from an arbitrary value of 1 up to 2, 4, or 8, the rate of the reaction increases proportionately, as demonstrated by the rising levels of the curves. As an example, when large quantities of glucose enter the renal tubules in diabetes mellitus—that is, the substrate glucose is then in great excess in the tubules—the rate of reabsorption of the glucose is determined almost entirely by the concentration of the transport enzymes in the proximal tubular cells, not by the concentration of the glucose itself.

Role of Substrate Concentration in Regulation of Metabolic Reactions. Note also in Figure 72–2 that when the substrate concentration becomes low enough that only a small portion of the enzyme is required in the reaction, the rate of the reaction then becomes directly proportional to the substrate concentration as well as the enzyme concentration. This is the effect seen in the absorption of substances from the intestinal tract and renal tubules when their concentrations are low.

Rate Limitation in a Series of Reactions. Almost all chemical reactions of the body occur in series, the product of one reaction acting as a substrate for the next reaction and so on. Therefore, the overall rate of a complex series of chemical reactions is determined mainly by the rate of reaction of the slowest step in the series. This is called the *rate-limiting step* in the entire series.

ADP Concentration as a Rate-Controlling Factor in Energy Release. Under *resting* conditions, the concentration of ADP in the cells is extremely slight, so that the chemical reactions that depend on ADP as one of the substrates likewise are quite slow. They include all the oxidative metabolic pathways that release energy from the food, as well as essentially all other pathways for release of energy in the body. Thus, *ADP is a major rate-limiting factor* for almost all energy metabolism of the body.

When the cells become active, regardless of the type of activity, ATP is converted into ADP, increasing the concentration of ADP in direct proportion to the degree of activity of the cell. This ADP then automatically increases the rates of all the reactions for metabolic release of energy from food. Thus, by this simple process, the amount of energy released in the cell is controlled by the degree of activity of the cell. In the absence of cellular activity, the release of energy stops because all the ADP soon becomes ATP.

METABOLIC RATE

The *metabolism* of the body means simply all the chemical reactions in all the cells of the body, and the *metabolic rate* is normally expressed in terms of the rate of heat liberation during the chemical reactions.

Heat Is the End Product of Almost All the Energy Released in the Body. In discussing many of the metabolic reactions of the preceding chapters, we have noted that not all the energy in foods is transferred to ATP; instead, a large portion of this energy becomes heat. On the average, 35 per cent of the energy in foods becomes heat during ATP formation. Then, still more energy becomes heat as it is transferred from ATP to the functional systems of the cells, so that even under the best of conditions, not more than 27 per cent of all the energy from food is finally used by the functional systems.

Even when 27 per cent of the energy reaches the functional systems of the cells, almost all of this also eventually becomes heat for the following reasons: We might first consider the synthesis of protein and other growing elements of the body. When proteins are synthesized, large portions of ATP are used to form the peptide linkages, and this stores energy in these linkages. But we also noted in our discussions of proteins in Chapter 69 that there is continuous turnover of proteins, some being degraded while others are being formed. When the proteins are degraded, the energy stored in the peptide linkages is released in the form of heat into the body.

Now let us consider the energy used for muscle activity. Much of this energy simply overcomes the viscosity of the muscles themselves or of the tissues so that the limbs can move. The viscous movement in turn causes friction within the tissues, which generates heat.

We might also consider the energy expended by the heart in pumping blood. The blood distends the arterial system, the distention in itself representing a reservoir of potential energy. As the blood flows through the peripheral vessels, the friction of the different layers of blood flowing over one another and the friction of the blood against the walls of the vessels turn all this energy into heat.

Therefore, essentially all the energy expended by the body is eventually converted into heat. The only significant exception to this occurs when the muscles are used to perform some form of work outside the body. For instance, when the muscles elevate an object to a height or carry the person's body up steps, a type of potential energy is thus created by raising a mass against gravity. But when external expenditure of energy is not taking place, it is safe to consider that all

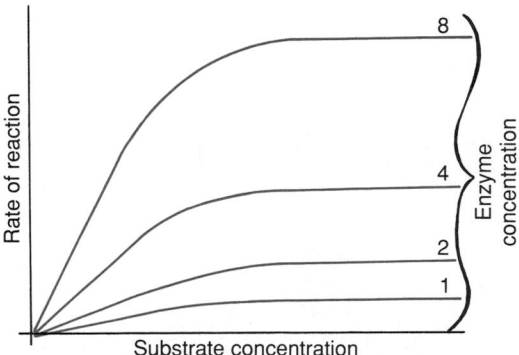

Figure 72–2. Effect of substrate and enzyme concentrations on the rate of enzyme-catalyzed reaction.

the energy released by the metabolic processes eventually becomes body heat.

THE CALORIE. To discuss the metabolic rate of the body and related subjects intelligently, it is necessary to use some unit for expressing the quantity of energy released from the different foods or expended by the different functional processes of the body. Most often, the *Calorie* is the unit used for this purpose. It will be recalled that 1 *calorie,* spelled with a small "c" and often called a *gram calorie,* is the quantity of heat required to raise the temperature of 1 gram of water 1°C. The calorie is much too small a unit for ease of expression in speaking of energy in the body. Consequently, the large Calorie, sometimes spelled with a capital "C" and often called a *kilocalorie,* which is equivalent to 1000 calories, is the unit ordinarily used in discussing energy metabolism.

Measurement of the Whole-Body Metabolic Rate

DIRECT CALORIMETRY. Because a person ordinarily is not performing any external work, the whole-body metabolic rate can be determined by simply measuring the total quantity of heat liberated from the body in a given time. This method is called *direct calorimetry.*

In determining the metabolic rate by direct calorimetry, one measures the quantity of heat liberated from the body in a large, specially constructed *calorimeter.* The subject is placed in an air chamber that is so well insulated that no heat can leak through the walls of the chamber. Heat formed by the subject's body warms the air of the chamber. However, the air temperature within the chamber is maintained at a constant level by forcing the air through pipes in a cool water bath. The rate of heat gain by the water bath, which can be measured with an accurate thermometer, is equal to the rate at which heat is liberated by the subject's body.

Direct calorimetry is physically difficult to perform and, therefore, is used only for research purposes.

INDIRECT CALORIMETRY—THE "ENERGY EQUIVALENT" OF OXYGEN. Because more than 95 per cent of the energy expended in the body is derived from reactions of oxygen with the different foods, the whole-body metabolic rate can also be calculated with a high degree of accuracy from the rate of oxygen utilization. When 1 liter of oxygen is metabo-

lized with glucose, 5.01 Calories of energy are released; when metabolized with starches, 5.06 Calories are released; with fat, 4.70 Calories; and with protein, 4.60 Calories.

From these figures, it is striking how nearly equivalent are the quantities of energy liberated per liter of oxygen, regardless of the type of food that is being metabolized. For the average diet, the *quantity of energy liberated per liter of oxygen used in the body averages about 4.825 Calories.* This is called the *energy equivalent* of oxygen; using this energy equivalent, one can calculate with a high degree of precision the rate of heat liberation in the body from the quantity of oxygen used in a given period of time.

If a person should metabolize only carbohydrates during the period of the metabolic rate determination, the calculated quantity of energy liberated, based on the value for the average energy equivalent of oxygen (4.825 Calories per liter), would be about 4 per cent too little. On the other hand, if the person were obtaining most energy from fat, the calculated value would be about 4 per cent too great.

The Metabolator. Figure 72–3 shows the metabolator usually used for indirect calorimetry. This apparatus contains a floating drum that holds an oxygen chamber connected to a mouthpiece through two rubber tubes. A valve in one of these rubber tubes allows air to pass from the oxygen chamber into the mouth, while air passing from the mouth back to the chamber is directed by means of another valve through the second tube. Before the expired air from the mouth enters the upper portion of the oxygen chamber, it flows through a lower chamber that contains pellets of soda lime, which combine chemically with the carbon dioxide in the expired air. Therefore, as oxygen is used by the person's body and the carbon dioxide is absorbed by the soda lime, the floating oxygen chamber, which is precisely balanced by a weight, gradually sinks in the water owing to the oxygen loss. This chamber is coupled to a pen that records on a moving drum the rate at which the chamber sinks in the water and thereby records the rate at which the body uses oxygen.

Factors That Affect the Metabolic Rate

Factors that increase the chemical activity in the cells also increase the metabolic rate. Some of these are the following.

EXERCISE. The factor that causes by far the most dramatic effect on metabolic rate is strenuous exercise. Short bursts of

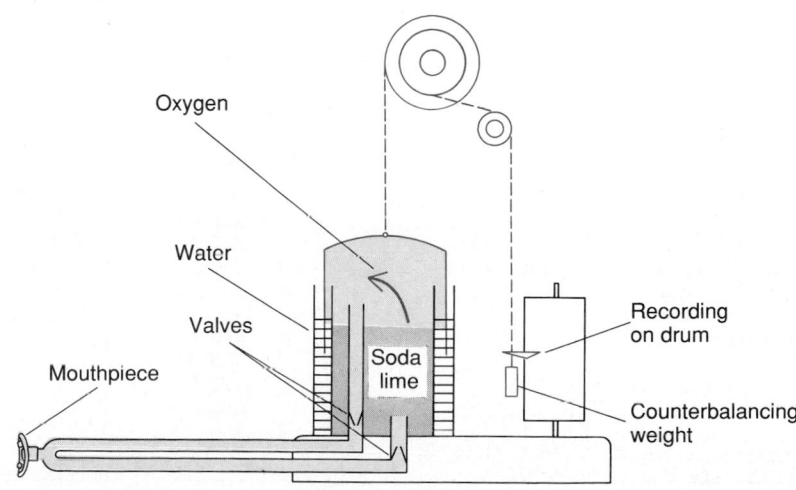

Figure 72–3. Metabolator.

maximal muscle contraction in any single muscle can liberate as much as 100 times its normal resting amount of heat for a few seconds at a time. In considering the entire body, maximal muscle exercise can increase the overall heat production of the body for a few seconds to about 50 times normal or sustained for several minutes to about 20 times normal in the well-trained athlete, which is an increase in metabolic rate to 2000 per cent of normal.

Energy Requirements for Daily Activities. When an average man of 70 kilograms lies in bed all day, he uses about 1650 Calories of energy. The process of eating and digesting food increases the amount of energy used each day by an additional 200 or more Calories, so that the same man lying in bed and eating a reasonable diet requires a dietary intake of about 1850 Calories per day. If he sits in a chair all day, his total energy requirement reaches 2000 to 2250 Calories. Therefore, in round figures, it can be assumed that the daily energy requirement simply for existing (that is, performing essential functions only) is about 2000 Calories.

Effects of Different Types of Work on Daily Energy Requirements. Table 72–1 lists the rates of energy utilization while one performs different types of activities. Note that walking up stairs requires about 17 times as much energy as lying in bed asleep. In general, over a 24-hour period, a laborer can achieve a maximum rate of energy utilization as great at 6000 to 7000 Calories—in other words, as much as 3.5 times the basal rate of metabolism.

SPECIFIC DYNAMIC ACTION OF PROTEIN. After a meal is ingested, the metabolic rate increases. This is believed to result to a slight extent from the different chemical reactions associated with digestion, absorption, and storage of food in the body. However, it mainly results from a stimulatory effect on the cellular chemical processes by certain of the amino acids derived from the proteins of the ingested food.

After a meal that contains a large quantity of carbohydrates or fats, the metabolic rate usually increases only about 4 per cent. However, after a meal that contains large quantities of protein, the metabolic rate usually begins rising within 1 hour, reaching a maximum about 30 per cent above normal, and this lasts for 3 to 12 hours. This effect of protein on the metabolic rate is called the *specific dynamic action* of protein.

AGE. The metabolic rate of a young child in relation to the child's size is almost twice that of an old person. This is demonstrated in Figure 72–4, which shows the declining metabolic rates of both males and females from birth until old age. The high metabolic rate of young children results from high rates of cellular reactions, including partly the rapid synthesis of cellular materials and growth of the body, which require moderate quantities of energy.

THYROID HORMONE. When the thyroid gland secretes maximal quantities of thyroxine, the metabolic rate sometimes rises to 50 to 100 per cent above normal. On the other hand, total loss of thyroid secretion decreases the metabolic rate to 40 to 60 per cent of normal. These effects can readily be explained by the basic function of thyroxine of increasing the rates of activity of almost all the chemical reactions in all cells of the body. This relation between thyroxine and metabolic rate is discussed in much greater detail in Chapter 76 in connection with thyroid function.

SYMPATHETIC STIMULATION—"NONSHIVERING THERMOGENESIS." Stimulating the sympathetic nervous system, with liberation of norepinephrine and epinephrine, increases the metabolic rate of many tissues of the body. These hormones have a direct effect on muscle and liver cells in causing glycogenolysis; this, along with other intracellular effects, increases cellular activity. Even more important is the effect of sympathetic stimulation on a certain type of fat tissue called *brown fat* in causing marked liberation of heat. This type of fat contains large numbers of mitochondria and many small globules of fat instead of one large fat globule. In these cells, the process of oxidative phosphorylation in the mitochondria is mainly "uncoupled." That is, when the cells are stimulated by the sympathetic nerves, their mitochondria produce a large amount of heat but almost no ATP, so that almost all the released oxidative energy immediately becomes heat. The neonate has a considerable number of such fat cells, and maximal sympathetic stimulation can increase the child's metabolism more than 100 per cent. This is called *nonshivering thermogenesis*. The magnitude of this type of thermogenesis in the adult human being, who has virtually no brown fat, is in question—probably less than 15 per cent—although this might increase after cold adaptation.

MALE SEX HORMONE. The male sex hormone can increase the basal metabolic rate about 10 to 15 per cent and the female sex hormone, perhaps a small amount but usually not enough to be of significance. The difference in metabolic rates of males and females is shown in Figure 72–4.

GROWTH HORMONE. Growth hormone can increase the basal metabolic rate 15 to 20 per cent as a result of direct stimulation of cellular metabolism.

FEVER. Fever, regardless of its cause, increases the metabolic rate. This is because all chemical reactions, either in the body or in the test tube, increase their rates of reaction an average of 120 per cent for every 10°C rise in temperature. The body's temperature control system diminishes this effect somewhat, as we discuss in Chapter 73.

CLIMATE. Studies of metabolic rates of people living in the different geographical zones have shown metabolic rates 10 to 20 per cent lower in tropical regions than in arctic regions. This difference is caused at least in part by adaptation of the thyroid gland, with increased secretion in cold climates and decreased secretion in hot climates. Indeed, far more people develop hyperthyroidism in cold regions of the earth than in tropical regions.

SLEEP. The metabolic rate falls 10 to 15 per cent below normal during sleep. This fall is presumably due to two principal factors: (1) decreased tone of the skeletal muscula-

Table 72–1 ENERGY EXPENDITURE DURING DIFFERENT TYPES OF ACTIVITY FOR A 70-KILOGRAM MAN

Form of Activity	Calories per Hour
Sleeping	65
Awake lying still	77
Sitting at rest	100
Standing relaxed	105
Dressing and undressing	118
Tailoring	135
Typewriting rapidly	140
"Light" exercise	170
Walking slowly (2.6 miles per hour)	200
Carpentry, metal working, industrial painting	240
"Active" exercise	290
"Severe" exercise	450
Sawing wood	480
Swimming	500
Running (5.3 miles per hour)	570
"Very severe" exercise	600
Walking very fast (5.3 miles per hour)	650
Walking up stairs	1100

Extracted from data compiled by Professor M. S. Rose.

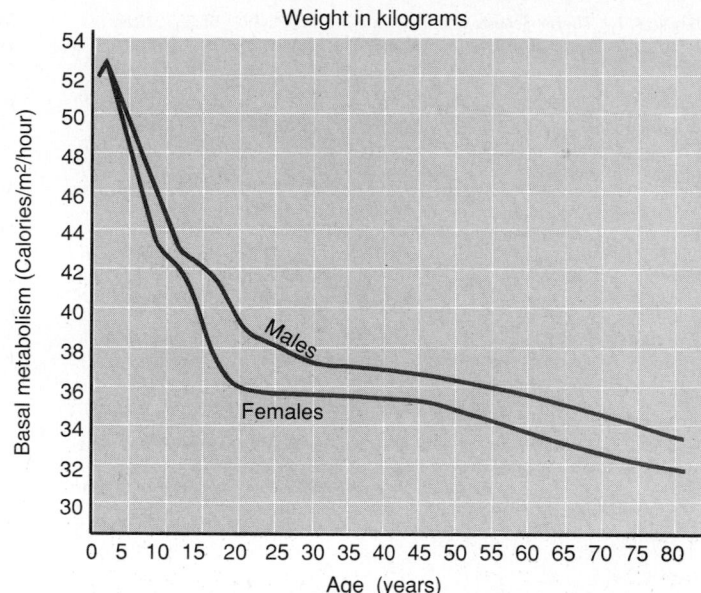

Figure 72–4. Normal basal metabolic rates at different ages for each sex.

ture during sleep and (2) decreased activity of the sympathetic nervous system.

MALNUTRITION. Prolonged malnutrition can decrease the metabolic rate 20 to 30 per cent; this decrease is presumably caused by paucity of the necessary food substances in the cells.

In the final stages of many disease conditions, the inanition that accompanies the disease causes a marked premortem decrease in metabolic rate, even to the extent that the body temperature may fall a number of degrees shortly before death.

Basal Metabolic Rate

THE BASAL METABOLIC RATE IS A METHOD FOR COMPARING METABOLIC RATES BETWEEN INDIVIDUALS. It has been important to establish a procedure that will measure the inherent metabolic rate of the tissues independently of exercise and other extraneous factors that would make it impossible to compare one person's metabolic rate with that of another person. To do this, the metabolic rate is usually measured under so-called *basal conditions;* the metabolic rate then measured is called the *basal metabolic rate* (BMR).

BASAL CONDITIONS. The BMR means the rate of energy utilization in the body during absolute rest but while the person is awake. The following basal conditions are necessary for measuring the BMR:

1. The person must not have eaten any food for at least 12 hours.
2. The BMR is determined after a night of restful sleep.
3. No strenuous exercise is performed during the preceding hour or more.
4. All psychic and physical factors that cause excitement must be eliminated.
5. The temperature of the air must be comfortable and be somewhere between the limits of 68° and 80°F.

USUAL TECHNIQUE FOR DETERMINING THE BMR. The usual method for determining the BMR is to measure the rate of oxygen utilization using a metabolator of the type shown in Figure 72–3. Then the BMR is calculated in terms of *Calories per hour.* This normally averages about 60 Calories per hour in young men and about 53 Calories per hour in young women.

EXPRESSING THE BMR IN TERMS OF SURFACE AREA. If one subject is much larger than another, the total amount of energy used by the two subjects will be considerably different simply because of differences in body size. Experimentally, among normal people, the average BMR varies approximately *in proportion to the body surface area,* and the surface area can be determined from height and weight tables.

Therefore, to compare BMRs between people, they are expressed as *Calories per hour per square meter.* The normal values for both males and females at different ages are shown in Figure 72–4.

EXPRESSING THE BMR IN PERCENTAGE ABOVE OR BELOW NORMAL. Finally, because of the changing BMRs at different ages, it has become the practice to compare the actual rate with the normal rate shown in Figure 72–4. Then, the rate is expressed as a percentage above or below normal. Thus, the BMR is expressed as *plus 25* when it is 25 per cent too high or as *minus 15* when it is 15 per cent too low.

CONSTANCY OF THE METABOLIC RATE IN THE SAME PERSON. BMRs have been measured in many subjects at repeated intervals for 20 or more years. As long as a subject remains healthy, almost invariably the BMR, when expressed as a percentage of normal, does not vary more than 5 to 10 per cent except for the age-related changes.

CONSTANCY OF THE BMR FROM PERSON TO PERSON. When the BMR is measured in a wide variety of people and comparisons are made within single age, weight, and sex groups, 85 per cent of normal people have been found to have BMRs within 10 per cent of the mean. Thus, measurements of metabolic rates performed under basal conditions offer a reasonable means for comparing the rate of metabolism from one person to another.

REFERENCES

Baldwin, R. L., and Bywater, A. C.: Nutritional energetics of animals. Annu. Rev. Nutr., 4:101, 1984.

Becker, D. J.: The endocrine responses to protein-calorie malnutrition. Annu. Rev. Nutr., 3:187, 1983.

Block, B. A.: Thermogenesis in muscle. Annu. Rev. Physiol., 56:535, 1994.

Bray, G. A.: Regulation of energy balance. Physiologist, 28:186, 1985.

Burrow, G. N., et al.: Thyroid Function and Disease. Philadelphia, W. B. Saunders Co., 1989.

Calder, W. A. III: Scaling energetics of homeothermic vertebrates: An operational allometry. Annu. Rev. Physiol., 49:107, 1987.

Clausen, T., et al.: Significance of cation transport in control of energy metabolism and thermogenesis. Physiol. Rev., 71:733, 1991.

Cohen, R. D., et al.: The Metabolic and Molecular Basis of Acquired Disease. Philadelphia, W. B. Saunders Co., 1990.

Durnin, J. V.: Energy balance in childhood and adolescence. Proc. Nutr. Soc., 43:271, 1984.

Driedzic, W. R., and Gesser, H.: Energy metabolism and contractility in ectothermic vertebrate hearts: hypoxia, acidosis, and low temperature. Physiol. Rev., 74:221, 1994.

Edmond, J., and Clark, J. B.: Functional Aspects of Energy Metabolism in Neural Tissue. Farmington, CT, S. Karger Publishers, Inc., 1993.

Fabris, F., et al., Sedentary Life and Nutrition. New York, Raven Press, 1991.

Felig, P., et al.: Endocrinology and Metabolism. Hightstown, NJ, McGraw-Hill, 1994.

Golinick, P. D.: Metabolism of substrates: Energy substrate metabolism during exercise and as modified by training. Fed. Proc., 44:353, 1985.

Guyton, A. C., and Farish, C. A.: A rapidly responding continuous oxygen consumption recorder. J. Appl. Physiol., 14:143, 1959.

Jequier, E., and Flatt, J.-P.: Recent advances in human energetics. News Physiol. Sci., 1:112, 1986.

Kim, C. H., et al. (eds.): Advances in Membrane Biochemistry and Bioenergetics. New York, Plenum Publishing Corp., 1989.

Kinney, J. M., and Tucker, H. N.: Energy Metabolism: Tissue Determinants and Cellular Corollaries. New York, Raven Press, 1992.

Kinney, J. M., and Tucker, H. N.: Organ Metabolism and Nutrition: Ideas for Future Critical Care. New York, Raven Press, 1994.

Magnen, J. L.: Body energy balance and food intake: A neuroendocrine regulatory mechanism. Physiol. Rev., 63:314, 1983.

McArdle, W. D., et al.: Exercise Physiology: Energy, Nutrition, and Human Performance. Baltimore, Williams & Wilkins, 1991.

Nicholls, D. G., and Locke, R. M.: Thermogenic mechanisms in brown fat. Physiol. Rev., 64:1, 1984.

Oppenheimer, J. H.: Thyroid hormone action at the nuclear level. Ann. Intern. Med., 102:374, 1985.

Plowman, P. N.: Endocrinology and Metabolic Diseases. New York, Elsevier Science Publishing Co., 1987.

Storlein, L. H.: The role of the ventromedial hypothalamic area in periprandial glucoregulation. Life Sci., 36:505, 1985.

Van der Laarse, W. J., and Woledge, R. C.: Energetics at the single cell level. News Physiol. Sci., 4:91, 1989.

Wyndham, C. H., and Loots, H.: Responses to cold during a year in Antarctica. J. Appl. Physiol., 27:696, 1969.

Body Temperature, Temperature Regulation, and Fever

CHAPTER 73

Normal Body Temperatures

CORE TEMPERATURE AND SKIN TEMPERATURE. The temperature of the deep tissues of the body—the "core"—remains almost exactly constant, within ± 1°F (± 0.6°C), day in and day out except when a person develops a febrile illness. Indeed, a nude person can be exposed to temperatures as low as 55°F or as high as 130°F in *dry* air and still maintain an almost constant internal body temperature. The mechanisms for control of body temperature represent a beautifully designed control system. The purpose of this chapter is to discuss this system as it operates in health and in disease.

The *skin temperature,* in contrast to the *core temperature,* rises and falls with the temperature of the surroundings. This is the temperature that is important when we refer to the ability of the skin to lose heat to the surroundings.

NORMAL CORE TEMPERATURE. No single temperature level can be considered to be normal because measurements on many normal people have shown a *range* of normal temperatures measured orally, as shown in Figure 73–1, from less than 97°F (36°C) to over 99.5°F (37.5°C). The average normal temperature is generally considered to be between 98.0°F and 98.6°F when measured orally and about 1°F higher when measured rectally.

The body temperature varies still more with exercise and with extremes of temperature of the surroundings because the temperature regulatory mechanisms are not 100 per cent perfect. When excessive heat is produced in the body by strenuous exercise, the tempera-

ture can rise temporarily to as high as 101° to 104°F. On the other hand, when the body is exposed to cold, the temperature can often fall to values below 96°F.

BODY TEMPERATURE IS CONTROLLED BY BALANCING HEAT PRODUCTION AGAINST HEAT LOSS

When the rate of heat production in the body is greater than the rate at which heat is being lost, heat builds up in the body and the body temperature rises. Conversely, when heat loss is greater, both body heat and body temperature decrease. Therefore, most of the remainder of this chapter is concerned with this balance between heat production and heat loss and the mechanisms by which the body controls each of these.

Heat Production

Heat production is a principal by-product of metabolism. In Chapter 72, which summarizes body energetics, we discuss the different factors that determine the rate of heat production, called the *metabolic rate of the body.* The most important of these factors are listed again here: (1) basal rate of metabolism of all the cells of the body; (2) extra rate of metabolism caused by muscle activity, including muscle contractions caused by shivering; (3) extra metabolism caused by the effect of thyroxine (and to a less extent other hormones, such as growth hormone and testosterone) on the cells; (4) extra metabolism caused by the effect

911

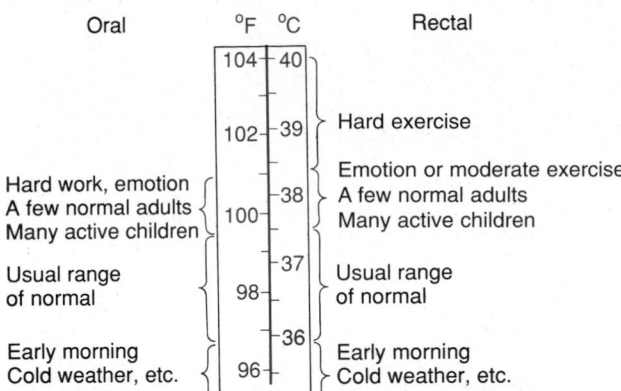

Figure 73–1. Estimated range of body temperature in normal people. (From E. F. DuBois: Fever. Springfield, Ill., Charles C Thomas, 1948.)

of epinephrine, norepinephrine, and sympathetic stimulation on the cells; and (5) extra metabolism caused by increased chemical activity in the cells themselves, especially when the cell temperature increases.

Heat Loss

Most of the heat produced in the body is generated in the deep organs, especially in the liver, brain, heart, and the skeletal muscles during exercise. Then this heat is transferred from the deeper organs and tissues to the skin, where it is lost to the air and other surroundings. Therefore, the rate at which heat is lost is determined almost entirely by two factors: (1) how rapidly heat can be conducted from where it is produced in the body core to the skin and (2) how rapidly heat can then be transferred from the skin to the surroundings. Let us begin by discussing the insulator system that insulates the core from the skin surface.

Insulator System of the Body

The skin, the subcutaneous tissues, and especially the fat of the subcutaneous tissues are a heat insulator for the body. The fat is important because it conducts heat only *one third* as readily as other tissues. When no blood is flowing from the heated internal organs to the skin, the insulating properties of the normal male body are about equal to three quarters the insulating properties of a usual suit of clothes. In women, this insulation is still better.

The insulation beneath the skin is an effective means of maintaining normal internal core temperature, even though it allows the temperature of the skin to approach the temperature of the surroundings.

Blood Flow to the Skin from the Body Core Provides Heat Transfer

Blood vessels penetrate the fatty subcutaneous insulator tissues and are distributed profusely immediately beneath the skin. Especially important is a continuous venous plexus that is supplied by inflow of blood from the skin capillaries, shown in Figure 73–2. In the most exposed areas of the body–the hands, feet, and ears–blood is also supplied to the plexus directly from the small arteries through highly muscular *arteriovenous anastomoses.*

The rate of blood flow into the venous plexus can vary tremendously—from barely above zero to as great as 30 per cent of the total cardiac output. A high rate of blood flow causes heat to be conducted from the core of the body to the skin with great efficiency, whereas reduction in the rate of blood flow decreases the heat conduction from the core. Figure 73–3 shows quantitatively the effect of skin blood flow on conductance of heat from the body core to the skin surface, demonstrating an approximate eightfold increase in heat conductance between the fully vasoconstricted state and the fully vasodilated state.

Therefore, the skin is an effective controlled "heat radiator" system, and the flow of blood to the skin is a most effective mechanism of heat transfer from the body core to the skin.

CONTROL OF HEAT CONDUCTION TO THE SKIN BY THE SYMPATHETIC NERVOUS SYSTEM. Heat conduction to the skin by the blood is controlled by the degree of vasoconstriction of the arterioles and arteriovenous anastomoses that supply blood to the venous plexus of the skin. This vasoconstriction in turn is controlled almost entirely by the sympathetic nervous system in response to changes in the body core temperature and changes in the environmental temperature. This is discussed later in the chapter in connection with control of body temperature by the hypothalamus.

Basic Physics of How Heat Is Lost from the Skin Surface

The various methods by which heat is lost from the skin to the surroundings are shown in Figure 73–4. They include *radiation, conduction,* and *evaporation* and may be explained as follows.

RADIATION. As shown in Figure 73–4, a nude person in a room at normal room temperature loses about 60 per cent of the total heat loss (about 15%) by radiation.

Loss of heat by radiation means loss in the form of infrared heat rays, a type of electromagnetic wave. Most infrared heat rays that radiate from the body have wavelengths of 5 to 20 micrometers, 10 to 30 times the wavelengths of light rays. All objects that are not at absolute zero temperature radiate such rays. The human body radiates heat rays in all directions. Heat rays are also being radiated from the walls and other objects toward the body. If the temperature of the body is greater than the temperature of the surroundings, a greater quantity of heat is radiated from the body than is radiated to the body.

CONDUCTION. As shown in Figure 73–4, only minute quantities of heat are normally lost from the body by direct conduction from the surface of the body *to other objects,* such as a chair or a bed. On the other

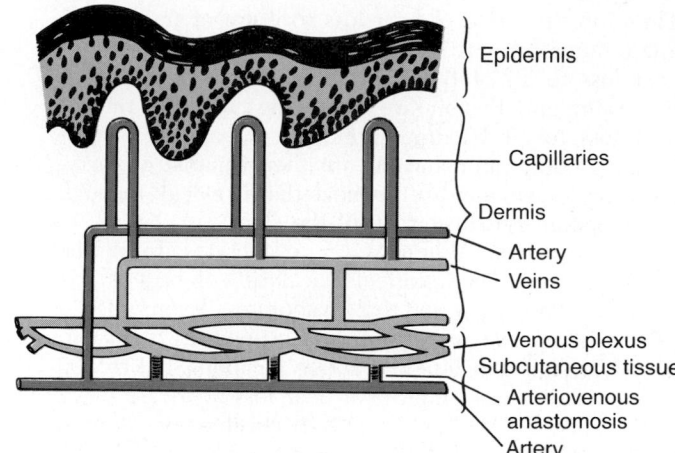

Figure 73–2. Skin circulation.

hand, loss of heat by *conduction to air* does represent a sizable proportion of the body's heat loss (about 15 per cent) even under normal conditions. It will be recalled that heat is actually the kinetic energy of molecular motion, and the molecules of the skin are continually undergoing vibratory motion. Much of the energy of this motion can be transferred to the air if the air is colder than the skin, thus increasing the velocity of motion of the air molecules. Once the temperature of the air immediately adjacent to the skin equals the temperature of the skin, no further loss of heat occurs because now an equal amount of heat is conducted from the air to the body. Therefore, conduction of heat from the body to the air is self-limited *unless the heated air moves away from the skin,* so that new, unheated air is continually brought in contact with the skin, a phenomenon called air convection.

Convection. The removal of heat from the body by convection air currents is commonly called heat loss by convection. Actually, the heat must first be *conducted* to the air and then carried away by the convection currents.

A small amount of convection almost always occurs around the body because of the tendency for the air adjacent to the skin to rise as it becomes heated. Therefore, a nude person seated in a comfortable room without gross air movement still loses about 15 per cent of his or her heat by conduction to the air and then by air convection away from the body.

Cooling Effect of Wind. When the body is exposed to wind, the layer of air immediately adjacent to the skin is replaced by new air much more rapidly than normally and heat loss by convection increases accordingly. The cooling effect of wind at low velocities is about proportional to the *square root of the wind velocity.* For instance, a wind of 4 miles per hour is about twice as effective for cooling as a wind of 1 mile per hour.

Conduction and Convection of Heat from a Person Exposed to Water. Water has a specific heat several thousand times as great as that of air, so that each unit portion of water adjacent to the skin can absorb far greater quantities of heat than can air. Also, the conductivity of water for heat is marked in comparison with that of air. Consequently, it is impossible for the body to heat a thin layer of water next to the body to form an "insulator zone" as occurs in air.

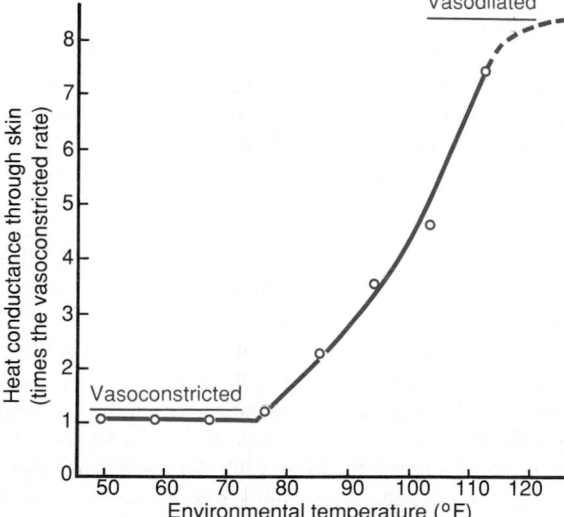

Figure 73–3. Effect of changes in the environmental temperature on heat conductance from the body core to the skin surface. (Modified from Benzinger: Heat and Temperature: Fundamentals of Medical Physiology. New York, Dowden, Hutchinson and Ross, 1980.)

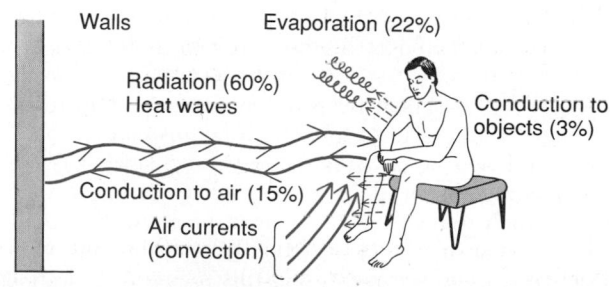

Figure 73–4. Mechanisms of heat loss from the body.

Therefore, the rate of heat loss to water at moderately low temperatures is many times as great as the rate of heat loss to air of the same temperature. When both the water and the air are extremely cold, the rate of heat loss to air becomes almost as great as to water because both water and air are then capable of carrying away essentially all the heat that can leak through the subcutaneous insulation of the skin.

EVAPORATION. When water evaporates from the body surface, 0.58 Calorie (kilocalorie) of heat is lost for each gram of water that evaporates. Even when a person is not sweating, water still evaporates *insensibly* from the skin and lungs at a rate of about 450 to 600 ml/day. This causes continual heat loss at a rate of 12 to 16 Calories per hour. This insensible evaporation through the skin and lungs cannot be controlled for purposes of temperature regulation because it results from continual diffusion of water molecules through the skin and respiratory surfaces. However, loss of heat by *evaporation of sweat* can be controlled by regulating the rate of sweating, which is discussed later in the chapter.

Evaporation Is a Necessary Cooling Mechanism at Very High Air Temperatures. As long as skin temperature is greater than the temperature of the surroundings, heat can be lost by radiation and conduction. But when the temperature of the surroundings is greater than that of the skin, instead of losing heat, the body gains heat by both radiation and conduction. Under these conditions, *the only means by which the body can rid itself of heat is evaporation.* Therefore, anything that prevents adequate evaporation when the surrounding temperatures are higher than skin temperature will cause the body temperature to rise. This occurs occasionally in human beings who are born with congenital absence of sweat glands. These people can stand cold temperatures as well as can normal people, but they are likely to die of heatstroke in tropical zones because without the evaporative refrigeration system, they cannot prevent a rise in body temperature when the air temperature is above that of the body.

EFFECT OF CLOTHING ON CONDUCTIVE HEAT LOSS. Clothing entraps air next to the skin and in the weave of the cloth, thereby increasing the thickness of the so-called private zone of air adjacent to the skin and also decreasing the flow of convection air currents. Consequently, the rate of heat loss from the body by conduction and convection is greatly depressed. A usual suit of clothes decreases the rate of heat loss to about half that from a nude body, whereas arctic-type clothing can decrease this heat loss to as little as one sixth.

About half the heat transmitted from the skin to the clothing is radiated to the clothing instead of being conducted across the small intervening space. Therefore, coating the inside of clothing with a thin layer of gold, which reflects radiant heat back to the body, makes the insulating properties of clothing far more effective than otherwise. Using this technique, clothing for use in the arctic can be decreased in weight by about half.

The effectiveness of clothing in maintaining body temperature is almost completely lost when it becomes wet because the high conductivity of water increases the rate of heat transmission through cloth 20-fold or more. Therefore, one of the most important factors for protecting the body against cold in arctic regions is extreme caution against allowing the clothing to become wet. Indeed, one must be careful not to overheat oneself even temporarily because sweating in one's clothes makes them much less effective thereafter as an insulator.

Sweating and Its Regulation by the Autonomic Nervous System

Stimulation of the anterior hypothalamus–preoptic area either electrically or by excess heat causes sweating. The impulses from this area that cause sweating are transmitted in the autonomic pathways to the cord and then through the sympathetic outflow to the skin everywhere in the body.

It should be recalled from the discussion of the autonomic nervous system in Chapter 60 that the sweat glands are innervated by sympathetic *cholinergic* nerve fibers (fibers that secrete acetylcholine). These glands can also be stimulated by epinephrine or norepinephrine circulating in the blood, even though the glands themselves do not have adrenergic innervation. This is important during exercise, when these hormones are secreted by the adrenal medullae and the body needs to lose the extra heat produced by the active muscles.

MECHANISM OF SWEAT SECRETION. In Figure 73–5, the sweat gland is shown to be a tubular structure consisting of two parts: (1) a deep subdermal *coiled portion* that secretes the sweat and (2) a *duct portion* that passes outward through the dermis and epidermis of the skin. As is true of so many other glands, the secretory portion of the sweat gland secretes a fluid called the *primary secretion* or *precursor secretion;* then the concentrations of the constituents in the fluid are modified as the fluid flows through the duct.

The precursor secretion is an active secretory product of the epithelial cells lining the coiled portion of the sweat gland. Cholinergic sympathetic nerve fibers ending on or near the glandular cells elicit the secretion.

The composition of the precursor secretion is similar to that of plasma except that it does not contain the plasma proteins. The concentration of sodium is about 142 mEq/liter and chloride, about 104 mEq/liter, with much smaller concentrations of the other solutes of plasma. As this precursor solution flows through the duct portion of the gland, it is modified by reabsorption of most of the sodium and chloride ions. The degree of this reabsorption depends on the rate of sweating as follows.

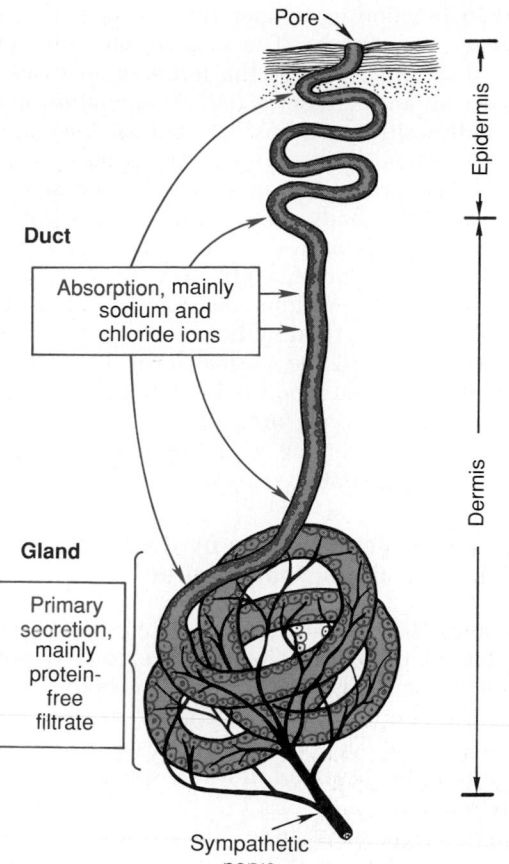

Figure 73–5. Sweat gland innervated by an acetylcholine-secreting sympathetic nerve. A *primary secretion* is formed by the glandular portion, but much, if not most, of the electrolytes are reabsorbed in the duct, leaving a dilute, watery secretion.

When the sweat glands are stimulated only slightly, the precursor fluid passes through the duct slowly. In this instance, essentially all the sodium and chloride ions are reabsorbed and the concentration of each of these falls to as low as 5 mEq/liter. This reduces the osmotic pressure of the sweat fluid to such a low level that most of the water is then also reabsorbed, which concentrates most of the other constituents. Therefore, at low rates of sweating, such constituents as urea, lactic acid, and potassium ions are usually very concentrated.

On the other hand, when the sweat glands are strongly stimulated by the sympathetic nervous system, large amounts of precursor secretion are formed, and the duct may now reabsorb only slightly more than one half of the sodium chloride; the concentrations of the sodium and chloride ions are then (in the *unacclimatized* person) a maximum of about 50 to 60 mEq/liter, slightly less than one half the concentrations in plasma. Furthermore, the sweat flows through the glandular tubules so rapidly that little of the water is reabsorbed. Therefore, the other dissolved constituents of the sweat are only moderately increased in concentration: urea is then about twice that in the plasma; lactic acid, about four times; and potassium, about 1.2 times.

Note especially the large loss of sodium chloride in the sweat when the person is unacclimatized. This is quite different once the person has become acclimatized to heat as follows.

ACCLIMATIZATION OF THE SWEATING MECHANISM— ROLE OF ALDOSTERONE. Although a normal, unacclimatized person can seldom produce more than about 1 liter of sweat per hour, when exposed to hot weather for 1 to 6 weeks, the person sweats progressively more profusely, often increasing the maximum sweat production to as much as 2 to 3 liters/hr. Evaporation of this much sweat can remove heat from the body at a rate *more than 10 times* the normal basal rate of heat production. This increased effectiveness of the sweating mechanism is caused by a direct increase in the sweating capability of the sweat glands themselves.

Also associated with acclimatization is further decrease in the concentration of sodium chloride in the sweat, which allows progressively better conservation of body salt. Most of this effect is caused by *increased secretion of aldosterone,* which in turn results from a slight decrease in the level of sodium chloride in the extracellular fluid and plasma. An *unacclimatized* person who sweats profusely often loses 15 to 30 grams of salt each day for the first few days. After 4 to 6 weeks of acclimatization, the loss is usually 3 to 5 gm/day.

Loss of Heat by Panting

Many lower animals have little ability to lose heat from the surfaces of their bodies for two reasons: (1) the surfaces are usually covered with fur and (2) the skin of most lower animals is not supplied with sweat glands, which prevents most of the evaporative loss of heat from the skin. A substitute mechanism, the *panting* mechanism, is used by many lower animals as a means of dissipating heat.

The phenomenon of panting is "turned on" by the thermoregulator centers of the brain. That is, when the blood becomes overheated, the hypothalamus initiates neurogenic signals to decrease the body temperature. One of these signals initiates panting. The actual panting process is then controlled by a *panting center* that is closely related to the pneumotaxic respiratory center in the pons.

When an animal pants, it breathes in and out rapidly, so that large quantities of new air from the exterior come in contact with the upper portions of the respiratory passages; this cools the blood in the mucosa as a result of water evaporation from the mucosal surfaces, especially evaporation of saliva from the tongue. Yet panting does not increase the alveolar ventilation more than is required for proper control of the blood gases because each breath is extremely shallow; therefore, most of the air that enters the alveoli is dead space air.

REGULATION OF BODY TEMPERATURE—ROLE OF THE HYPOTHALAMUS

Figure 73–6 shows approximately what happens to the temperature of a nude body after a few hours' exposure to *dry* air ranging from 30° to 160°F. The precise dimensions of this curve depend on the movement of the air, the amount of moisture in the air, and even the nature of the surroundings. In general, a nude body in dry air between 55° and 130°F is capable of maintaining a normal body core temperature somewhere between 97° and 100°F.

The temperature of the body is regulated almost entirely by nervous feedback mechanisms, and almost all these operate through *temperature-regulating centers* located in the *hypothalamus.* For these feedback mechanisms to operate, there must also exist temperature detectors to determine when the body temperature becomes either too hot or too cold.

Thermostatic Detection of Temperature in the Hypothalamus—Role of the Anterior Hypothalamus–Preoptic Area

Experiments have been performed in which minute areas in the brain have been either heated or cooled by use of a so-called *thermode.* This small, needle-like device is heated by electrical means or by passing hot water through it, or it is cooled by cold water. The principal area in the brain in which heat from a thermode affects body temperature control consists of the preoptic and anterior hypothalamic nuclei of the hypothalamus.

Using the thermode, the anterior hypothalamic–preoptic area has been found to contain large numbers of heat-sensitive neurons as well as about one third as many cold-sensitive neurons. These neurons are be-

lieved to function as temperature sensors for controlling body temperature. The heat-sensitive neurons increase their firing rate as the temperature rises, 2- to 10-fold with an increase in body temperature of 10°C. The cold-sensitive neurons, by contrast, increase their firing rate when the body temperature falls.

When the preoptic area is heated, the skin everywhere over the body immediately breaks out into a profuse sweat while at the same time the skin blood vessels over the entire body become greatly vasodilated. Thus, this is an immediate reaction to cause the body to lose heat, thereby helping to return the body temperature toward the normal level. In addition, excess body heat production is inhibited. Therefore, it is clear that the preoptic area of the hypothalamus has the capability of serving as a thermostatic body temperature control center.

Detection of Temperature by Receptors in the Skin and Deep Body Tissues

Although the signals generated by the temperature receptors of the hypothalamus are extremely powerful in controlling body temperature, receptors in other parts of the body also play important roles in temperature regulation. This is especially true of temperature receptors in the skin and in a few specific deep tissues of the body.

It will be recalled from the discussion of sensory receptors in Chapter 48 that the skin is endowed with both *cold* and *warmth* receptors. There are far more cold receptors than warmth receptors; in fact, 10 times as many in many parts of the skin. Therefore, peripheral detection of temperature mainly concerns detecting cool and cold instead of warm temperatures.

When the skin is chilled over the entire body, immediate reflex effects are invoked to increase the temperature of the body in several ways: (1) by providing a strong stimulus to cause shivering, with resultant increase in the rate of body heat production; (2) by inhibiting the process of sweating if this should be occurring; and (3) by promoting skin vasoconstriction to diminish the transfer of body heat to the skin.

Deep body temperature receptors are also found in certain parts of the body, mainly in the *spinal cord,* in the *abdominal viscera,* and in or around the *great veins.* These deep receptors function differently from the skin receptors because they are exposed to the body core temperature rather than the body surface temperature. Yet, like the skin temperature receptors, they mainly detect cold rather than warmth. It is probable that both the skin and the deep body receptors are concerned with preventing hypothermia—that is, preventing low body temperatures.

The Posterior Hypothalamus Summates the Central and Peripheral Temperature Sensory Signals

Even though many temperature sensory signals arise in peripheral receptors, these signals contribute to

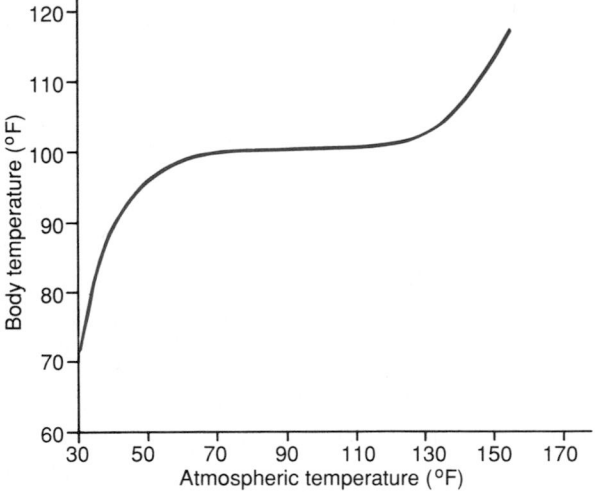

Figure 73–6. Effect of high and low atmospheric temperature for several hours' duration on the internal body temperature, showing that the internal body temperature remains stable despite wide changes in atmospheric temperature.

body temperature control mainly through the hypothalamus. The area of the hypothalamus that they stimulate is an area located bilaterally in the posterior hypothalamus approximately at the level of the mammary bodies. The temperature sensory signals from the anterior hypothalamus–preoptic area are also transmitted into this posterior hypothalamus area. Here the signals from the preoptic area and the signals from the body periphery are combined to control the heat-producing and heat-conserving reactions of the body.

Neuronal Effector Mechanisms That Decrease or Increase Body Temperature

When the hypothalamic temperature centers detect that the body temperature is either too hot or too cold, they institute appropriate temperature-decreasing or temperature-increasing procedures. The student is familiar with most of them from personal experience, but special features are the following.

Temperature-Decreasing Mechanisms When the Body Is Too Hot

The temperature control system uses three important mechanisms to reduce body heat when the body temperature becomes too great:

1. *Vasodilatation.* In almost all areas of the body, the skin blood vessels become intensely dilated. This is caused by *inhibition of the sympathetic centers in the posterior hypothalamus that cause vasoconstriction.* Full vasodilatation can increase the rate of heat transfer to the skin as much as eightfold.

2. *Sweating.* The effect of increased temperature on causing sweating is demonstrated by the solid curve in Figure 73–7, which shows a sharp increase in the rate of evaporative heat loss resulting from sweating when the body core temperature rises above the critical temperature level of 37°C (98.6°F). An additional 1°C increase in body temperature causes enough sweating to remove 10 times the basal rate of body heat production.

3. *Decrease in heat production.* The mechanisms that cause excess heat production, such as shivering and chemical thermogenesis, are strongly inhibited.

Temperature-Increasing Mechanisms When the Body Is Too Cold

When the body is too cold, the temperature control system institutes exactly opposite procedures. They are:

1. *Skin vasoconstriction throughout the body.* This is caused by stimulation of the posterior hypothalamic sympathetic centers.

2. *Piloerection.* Piloerection means hairs "standing on end." Sympathetic stimulation causes the arrector pili muscles attached to the hair follicles to contract, which brings the hairs to an upright stance. This is not important in the human being, but in lower animals, upright projection of the hairs allows them to entrap a thick layer of "insulator air" next to the skin, so that the transfer of heat to the surroundings is greatly depressed.

3. *Increase in heat production.* Heat production by the metabolic systems is increased by promoting (a) shivering,

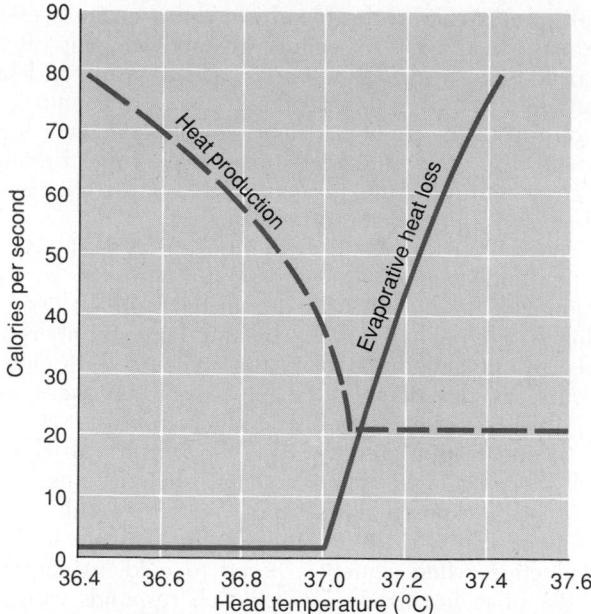

Figure 73–7. Effect of hypothalamic temperature on (1) evaporative heat loss from the body and (2) heat production caused primarily by muscle activity and shivering. This figure demonstrates the extremely critical temperature level at which increased heat loss begins and heat production reaches a minimum stable level. (Drawn from data in Benzinger, Kitzinger, and Pratt, in Hardy [ed.]: Temperature, Part 3, p. 637. Reinhold Publishing Corp.)

(b) sympathetic excitation of heat production, and (c) thyroxine secretion. These methods of increasing heat require additional explanation, as follows.

Hypothalamic Stimulation of Shivering. Located in the dorsomedial portion of the posterior hypothalamus near the wall of the third ventricle is an area called the *primary motor center for shivering.* This area is normally inhibited by signals from the heat center in the anterior hypothalamic–preoptic area but is excited by cold signals from the skin and spinal cord. Therefore, as shown by the sudden increase in the "heat production" (see dashed curve in Figure 73–7), this center becomes activated when the body temperature falls even a fraction of a degree below a critical temperature level. It then transmits signals that cause shivering through bilateral tracts down the brain stem, into the lateral columns of the spinal cord, and, finally, to the anterior motor neurons. These signals are nonrhythmical and do not cause the actual muscle shaking. Instead, they increase the tone of the skeletal muscles throughout the body by facilitating the activity of the anterior motor neurons. When the tone rises above a certain critical level, shivering begins. This probably results from feedback oscillation of the muscle spindle stretch reflex mechanism, which is discussed in Chapter 54. During maximum shivering, body heat production can rise to four to five times normal.

Sympathetic "Chemical" Excitation of Heat Production. As pointed out in Chapter 72, either sympathetic stimulation or circulating norepinephrine and epinephrine in the blood can cause an immediate increase in the rate of cellular metabolism; this effect is called *chemical thermogenesis,* and it results at least partially from the ability of norepinephrine and epinephrine to *uncouple* oxidative phosphorylation, which means that excess foodstuffs are oxidized and thereby release energy in the form of heat but do not cause adenosine triphosphate to be formed.

The degree of chemical thermogenesis that occurs in an animal is almost directly proportional to the amount of *brown* fat that exists in the animal's tissues. This is a type of fat that contains large numbers of special mitochondria where the uncoupled oxidation occurs, as described in Chapter 72; these cells are supplied by strong sympathetic innervation.

Acclimatization greatly affects the intensity of chemical thermogenesis; some animals, such as rats, that have been exposed for several weeks to a cold environment exhibit a 100 to 500 per cent increase in heat production when acutely exposed to cold, in contrast to the unacclimatized animal, which responds with an increase of perhaps one third as much.

In adult human beings, who have almost no brown fat, it is rare that chemical thermogenesis increases the rate of heat production more than 10 to 15 per cent. However, in infants, who *do* have a small amount of brown fat in the interscapular space, chemical thermogenesis can increase the rate of heat production 100 per cent, which is probably an important factor in maintaining normal body temperature in the neonate.

Increased Thyroxine Output as a Long-Term Cause of Increased Heat Production. Cooling the anterior hypothalamic–preoptic area of the hypothalamus also increases the production of the neurosecretory hormone *thyrotropin-releasing hormone* by the hypothalamus. This hormone is carried by way of the hypothalamic portal veins to the anterior pituitary gland, where it stimulates the secretion of *thyroid-stimulating hormone.* Thyroid-stimulating hormone in turn stimulates increased output of *thyroxine* by the thyroid gland, as explained in Chapter 76. The increased thyroxine increases the rate of cellular metabolism throughout the body, which is yet another mechanism of *chemical thermogenesis.* This increase in metabolism does not occur immediately but requires several weeks for the thyroid gland to hypertrophy before it reaches its new level of thyroxine secretion.

Exposure of animals to extreme cold for several weeks can cause their thyroid glands to increase in size 20 to 40 per cent. However, human beings seldom allow themselves to be exposed to the same degree of cold as that to which animals are often subjected. Therefore, we still do not know, quantitatively, how important the thyroid method of adaptation to cold is in the human being. Isolated measurements have shown that military personnel residing for several months in the Arctic develop increased metabolic rates; Eskimos as well have abnormally high basal metabolic rates. Also, the continuous stimulatory effect of cold on the thyroid gland can probably explain the much higher incidence of toxic thyroid goiters in people who live in colder climates than in those who live in warmer climates.

Concept of a "Set-Point" for Temperature Control

In the example of Figure 73–7, it is clear that at a critical body core temperature, at a level of almost exactly 37.1°C, drastic changes occur in the rates of both heat loss and heat production. At temperatures above this level, the rate of heat loss is greater than that of heat production, so that the body temperature falls and reapproaches the 37.1°C level. At temperatures below this level, the rate of heat production is greater than heat loss, so that now the body temperature rises and again approaches the 37.1°C level. This crucial temperature level is called the "set-point" of the temperature control mechanism. That is, all the temperature control mechanisms continually attempt to bring the body temperature back to this set-point level.

FEEDBACK GAIN FOR BODY TEMPERATURE CONTROL. Let us recall for a moment the discussion of feedback gain of control systems presented in Chapter 1. Feedback gain is a measure of the effectiveness of a control system. In the case of body temperature control, it is important for the internal core temperature to change as little as possible even though the environmental temperature changes greatly. The *feedback gain* of the temperature control system is equal to the ratio of the change in environmental temperature to the change in body temperature minus 1.0 (see Chapter 1 for this formula). Experiments have shown that the body temperature of humans changes about 1°C for each 25° to 30°C change in environmental temperature. Therefore, the feedback gain of the total mechanism for control of body temperature averages about 27 (28/1.0 − 1.0 = 27), which is an extremely high gain for a biological control system (the baroreceptor arterial pressure control system, for instance, has a gain of less than 2).

The Skin Temperature Can Alter Slightly the Set-Point Level for Core Temperature Control

The critical temperature set-point in the hypothalamus above which sweating begins and below which shivering begins is determined mainly by the degree of activity of the heat temperature receptors in the anterior hypothalamic–preoptic area of the hypothalamus. However, temperature signals from the peripheral areas of the body, especially from the skin and certain deep body tissues (the spinal cord and the abdominal viscera), also contribute slightly to body temperature regulation. But how do they contribute? The answer is that they alter the set-point of the hypothalamic temperature control center. This effect is shown in Figures 73–8 and 73–9.

Figure 73–8 demonstrates the effect of different skin temperatures on the set-point, showing that it

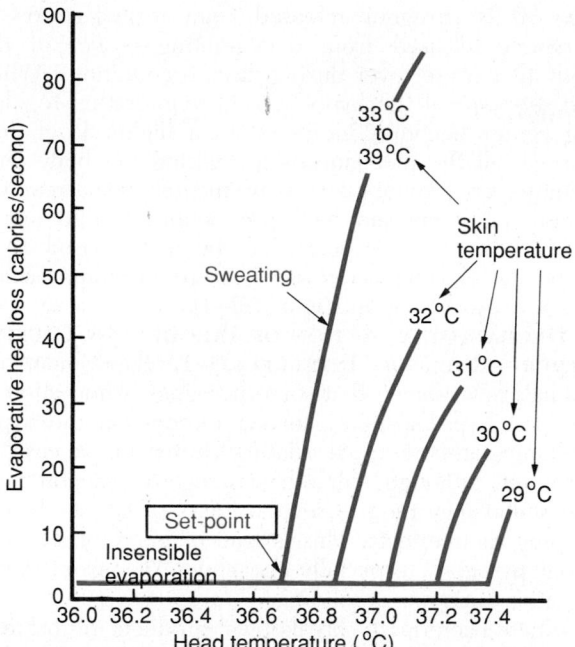

Figure 73–8. Effect of changes in the internal head temperature on the rate of evaporative heat loss from the body. Note also that the skin temperature determines the set-point level at which sweating begins. (Courtesy of Dr. T. H. Benzinger.)

increases as the skin temperature decreases. Thus, for the person represented in this figure, the hypothalamic set-point increased from 36.7°C when the skin temperature was higher than 33°C to a set-point of 37.4°C when the skin temperature had fallen to 29°C. Therefore, when the skin temperature was high, sweating began at a much lower hypothalamic temperature than when the skin temperature was low. One can readily understand the value of a system such as this because it is important that sweating be inhibited when the skin temperature is low; otherwise, the combined effect of a low skin temperature and sweating could cause far too much loss of body heat.

A similar effect occurs in shivering, as shown in Figure 73–9. That is, when the skin becomes cold, it drives the hypothalamic centers to the shivering threshold even when the hypothalamic temperature itself is still quite hot. Here again, one can well understand the value of the control system because a cold skin temperature will soon lead to deeply depressed body temperature unless heat production is increased. Thus, this effect of a cold skin temperature on increasing heat production actually "anticipates" a fall in internal body temperature and prevents its occurrence.

Behavioral Control of Body Temperature

Aside from the subconscious mechanisms for body temperature control, the body has still another temperature-controlling mechanism that is even more potent. This is *behavioral control of temperature*, which

can be explained as follows: Whenever the internal body temperature becomes too high, signals from the brain temperature-controlling areas give the person a psychic sensation of being overheated. Conversely, whenever the body becomes too cold, signals from the skin and probably from the deep body receptors elicit the feeling of cold discomfort. Therefore, the person makes appropriate environmental adjustments to re-establish comfort—such as moving into a heated room in freezing weather. This is a much more powerful system of body temperature control than most physiologists have recognized in the past. Indeed, this is the only really effective mechanism for body heat control in severely cold environs.

Local Skin Temperature Reflexes

When a person places a foot under a hot lamp and leaves it there for a short time, *local vasodilatation* and mild *local sweating* occur. Conversely, placing the foot in cold water causes local vasoconstriction and local cessation of sweating. These reactions are caused by local effects of temperature directly on the blood vessels and by local cord reflexes conducted from the skin receptors to the spinal cord and back to the same skin area and to sweat glands. The *intensity* of these local effects is controlled by the central brain temperature controller, so that the overall effect is about proportional to the hypothalamic heat control signal *times* the local signal. Such reflexes can help prevent excessive heat exchange from locally cooled or heated portions of the body.

REGULATION OF INTERNAL BODY TEMPERATURE AFTER CUTTING THE SPINAL CORD. After cutting the spinal cord in the neck above the sympathetic outflow from the cord, regulation of body temperature becomes

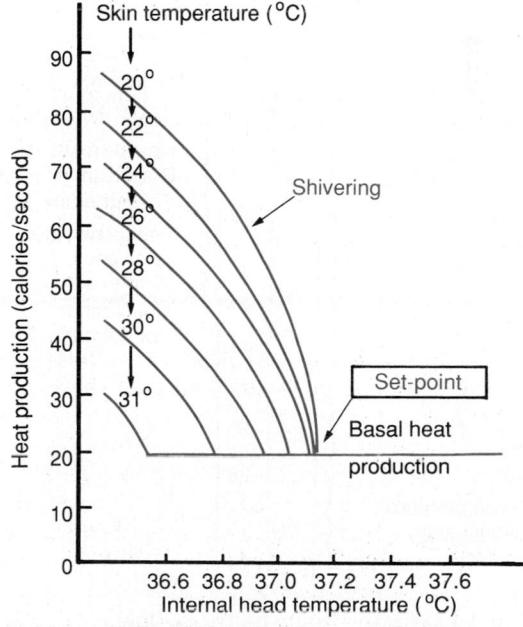

Figure 73–9. Effect of changes in the internal head temperature on the rate of heat production by the body. Note also that the skin temperature determines the set-point level at which shivering begins. (Courtesy of Dr. T. H. Benzinger.)

extremely poor because the hypothalamus can then no longer control either skin blood flow or the degree of sweating anywhere in the body. This is true even though the local temperature reflexes originating in the skin, spinal cord, and intra-abdominal receptors still exist. These reflexes are weak in comparison with hypothalamic control of body temperature. In people with this condition, body temperature must be regulated principally by the patients' psychic response to cold and hot sensations in the head region—that is, by behavioral control of clothing and by moving into the appropriate heat or cold environment.

ABNORMALITIES OF BODY TEMPERATURE REGULATION

Fever

Fever, which means a body temperature above the usual range of normal, can be caused by abnormalities in the brain itself or by toxic substances that affect the temperature-regulating centers. Some causes of fever are presented in Figure 73–10. They include bacterial diseases, brain tumors, and environmental conditions that may terminate in heatstroke.

Resetting the Hypothalamic Temperature-Regulating Center in Febrile Diseases—Effect of Pyrogens

Many proteins, breakdown products of proteins, and certain other substances, especially lipopolysaccharide toxins released from bacterial cell membranes, can cause the set-point of the hypothalamic thermostat to rise. Substances that cause this effect are called *pyro-*

gens. It is pyrogens released from toxic bacteria or pyrogens released from degenerating tissues of the body that cause fever during disease conditions. When the set-point of the hypothalamic temperature-regulating center becomes increased to a higher level than normal, all the mechanisms for raising the body temperature are brought into play, including heat conservation and increased heat production. Within a few hours after the set-point has been increased to a higher level, the body temperature also approaches this level, as shown in Figure 73–11.

MECHANISM OF ACTION OF PYROGENS IN CAUSING FEVER—ROLE OF INTERLEUKIN-1. Experiments in animals have shown that some pyrogens, when injected into the hypothalamus, can act directly on the hypothalamic temperature-regulating center to increase its set-point, although still other pyrogens function indirectly and may require several hours of latency before causing their effects. This is true of many of the bacterial pyrogens, especially the *endotoxins* from gram-negative bacteria, as follows.

When bacteria or breakdown products of bacteria are present in the tissues or the blood, they are *phagocytized by the blood leukocytes, tissue macrophages,* and *large granular killer lymphocytes.* All these cells in turn digest the bacterial products and then release into the body fluids the *substance interleukin-1,* also called *leukocyte pyrogen* or *endogenous pyrogen.* The interleukin-1, on reaching the hypothalamus, immediately produces fever, increasing the body temperature in 8 to 10 minutes. As little as one ten-millionth of a gram of endotoxin lipopolysaccharide from the bacteria, acting in this manner in concert with the blood leukocytes, tissue macrophages, and killer lymphocytes, can cause fever. The amount of interleukin-1 that is formed in response to the lipopolysaccharide to cause the fever is only a few nanograms.

Several experiments have suggested that interleukin-1 causes fever by first inducing the formation of one of the prostaglandins, mainly prostaglandin E_2, or a similar substance and this in turn acting in the hypothala-

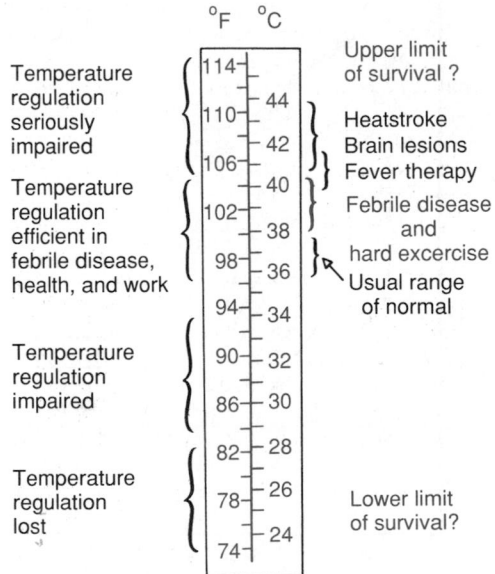

Figure 73–10. Body temperatures under different conditions. (From E. F. DuBois: Fever. Springfield, Ill., Charles C Thomas, 1948.)

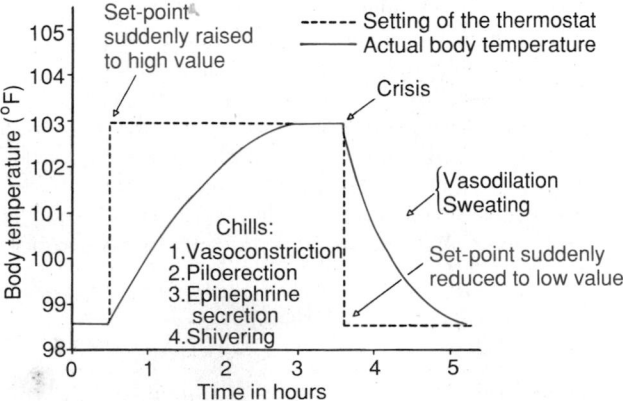

Figure 73–11. Effects of changing the set-point of the hypothalamic temperature controller.

mus to elicit the fever reaction. When prostaglandin formation is blocked by drugs, the fever is either completely abrogated or at least reduced. In fact, this may be the explanation for the manner in which aspirin reduces the degree of fever because aspirin impedes the formation of prostaglandins from arachidonic acid. It also would explain why aspirin does not lower the body temperature in a normal person because a normal person does not have any interleukin-1. Drugs such as aspirin that reduce the level of fever are called *antipyretics.*

FEVER CAUSED BY BRAIN LESIONS. When a brain surgeon operates in the region of the hypothalamus, severe fever almost always occurs; rarely, however, the opposite effect occurs, thus demonstrating both the potency of the hypothalamic mechanisms for body temperature control and the ease with which abnormalities of the hypothalamus can alter the set-point of temperature control. Another condition that frequently causes prolonged high temperature is compression of the hypothalamus by a brain tumor.

Characteristics of Febrile Conditions

CHILLS. When the set-point of the hypothalamic temperature-control center is suddenly changed from the normal level to higher than normal as a result of tissue destruction, pyrogenic substances, or dehydration, the body temperature usually takes several hours to reach the new temperature set-point. Figure 73–11 shows this, demonstrating the effect of suddenly increasing the set-point to a level of 103°F. Because the blood temperature is now less than the set-point of the hypothalamic temperature controller, the usual responses that cause elevation of body temperature occur. During this period, the person experiences chills and feels extremely cold, even though his or her body temperature may already be above normal. Also, the skin becomes cold because of vasoconstriction, and the person shivers. Chills continue until the body temperature reaches the hypothalamic setting of 103°F. Then the person no longer experiences chills but instead feels neither cold nor hot. As long as the factor that is causing the hypothalamic temperature controller to be set at this high value continues, the body temperature is regulated more or less in the normal manner but at the high temperature set-point level.

THE CRISIS, OR "FLUSH." If the factor that is causing the high temperature is suddenly removed, the set-point of the hypothalamic temperature controller is suddenly reduced to a lower value—perhaps even back to the normal level, as shown in Figure 73–11. In this instance, the body temperature is still 103°F, but the hypothalamus is attempting to regulate the temperature to 98.6°F. This situation is analogous to excessive heating of the anterior hypothalamic–preoptic area, which causes intense sweating and sudden development of a hot skin because of vasodilatation everywhere. This sudden change of events in a febrile disease is known as the "crisis" or, more appropriately, the "flush." In the days before the advent of antibiotics, the crisis was always awaited because once this occurred, the doctor knew immediately that the patient's temperature would soon be falling.

Heatstroke

The limits of extreme heat that one can stand depend almost entirely on whether the heat is dry or wet. If the air is dry and sufficient convection air currents are flowing to promote rapid evaporation from the body, a person can withstand several hours of air temperature at 130°F. On the other hand, if the air is 100 per cent humidified or if the body is in water, the body temperature begins to rise whenever the environmental temperature rises above about 94°F. If the person is performing heavy work, this critical temperature level may be as low as 85° to 90°F.

There is a limit to the rate at which the body can lose heat even with maximal sweating. Furthermore, when the hypothalamus becomes excessively heated, above 105 to 108°F, its heat-regulating ability becomes greatly depressed and sweating diminishes or stops. As a result, a high body temperature tends to perpetuate itself unless measures are taken specifically to decrease body heat.

When the body temperature rises beyond a critical temperature, into the range of 105° to 108°F, the person is likely to develop *heatstroke.* The symptoms include dizziness, abdominal distress sometimes with vomiting, sometimes delirium, and eventually loss of consciousness if the body temperature is not soon decreased. These symptoms are often exacerbated by a degree of *circulatory shock* brought on by excessive loss of fluid and electrolytes in the sweat. The hyperpyrexia itself is also exceedingly damaging to the body tissues, especially to the brain, and therefore is responsible for many of the effects. In fact, even a few minutes of high body temperature can sometimes be fatal. For this reason, many authorities recommend immediate treatment of heatstroke by placing the person in an ice-water bath. Because this often induces uncontrollable shivering with considerable increase in rate of heat production, others have suggested that sponge or spray cooling of the skin is likely to be more effective for rapidly decreasing the body core temperature.

HARMFUL EFFECTS OF THE HIGH TEMPERATURE. The pathological findings in a person who dies of hyperpyrexia are local hemorrhages and parenchymatous degeneration of cells throughout the entire body, but especially in the brain. Once neuronal cells are destroyed, they can never be replaced. Damage to the liver, kidneys, and other body organs can often be great enough that failure of one or more of these eventually causes death, sometimes not until several days after the heatstroke.

ACCLIMATIZATION TO HEAT. It is often extremely important to acclimatize people to extreme heat; some examples are (1) acclimatization of soldiers for tropical duty and (2) acclimatization of miners for work in the 2-mile-deep gold mines of South Africa, where the temperature approaches body temperature and the humidity approaches 100 per cent. Exposure of a person to heat for several hours each day while working a reasonably heavy work load will develop increased tolerance to hot and humid conditions in 1 to 3 weeks. Among the most important physiological changes that occur during this acclimatization process are about a twofold increase in the maximum rate of sweating, an increase in the plasma volume, and diminished loss of salt in the sweat and urine to almost none; the latter two of these effects result from increased secretion of aldosterone by the adrenal glands.

Exposure of the Body to Extreme Cold

Unless treated immediately, a person exposed to ice water for about 20 to 30 minutes ordinarily dies because of heart standstill or heart fibrillation. By that time, the internal body temperature will have fallen to about 77°F. If warmed rapidly by application of external heat, the person's life can often be saved.

LOSS OF TEMPERATURE REGULATION AT LOW TEMPERATURES. As noted in Figure 73–10, once the body temperature has fallen below about 85°F, the ability of the hypothalamus to regulate temperature is lost; it is greatly impaired even when the body temperature falls below about 94°F. Part of the reason for this loss of temperature regulation is that the rate of chemical heat production in each cell is depressed almost twofold for each 10°F decrease in body temperature. Also, sleepiness at first and later even coma develop, which depress the activity of the central nervous system heat control mechanisms and prevent shivering.

FROSTBITE. When the body is exposed to extremely low temperatures, surface areas can freeze; the freezing is called *frostbite*. This occurs especially in the lobes of the ears and in the digits of the hands and feet. If the freeze has been sufficient to cause extensive formation of ice crystals in the cells, permanent damage usually results, such as permanent circulatory impairment as well as local tissue damage. Often gangrene follows thawing, and the frostbitten areas must be removed surgically.

Cold-Induced Vasodilation Is a Final Protection Against Frostbite at Almost Freezing Temperature. When the temperature of tissues falls almost to freezing, the smooth muscle in the vascular wall becomes paralyzed because of the cold itself, and sudden vasodilation occurs, often manifested by a flush of the skin. This mechanism helps prevent frostbite by delivering warm blood to the skin. This mechanism is far less well-developed in humans than in most lower animals that live in the cold all the time.

ARTIFICIAL HYPOTHERMIA. It is easy to decrease the temperature of a person by first administering a strong sedative to depress the reactivity of the hypothalamic temperature controller and then cooling the person with ice, with cooling blankets, or otherwise until the temperature falls. The temperature can then be maintained below 90°F for several days to a week or more by continual sprinkling of cool water or alcohol on the body. Such artificial cooling is often used during heart surgery so that the heart can be stopped artificially for many minutes at a time. Cooling to this extent does not cause severe physiological results. It does slow the heart and greatly depresses cell metabolism, so that the body's cells can survive 30 minutes to more than 1 hour without blood flow during the surgical procedure.

REFERENCES

Benzinger, T. H.: Heat regulation: Homeostasis of central temperature in man. Physiol. Rev., 49:671, 1969.

Block, B. A.: Thermogenesis in muscle. Annu. Rev. Physiol., 56:535, 1994.

Boulant, J. A., and Dean, J. B.: Temperature receptors in the central nervous system. Annu. Rev. Physiol., 48:639, 1986.

Brengelmann, G. L.: Circulatory adjustments to exercise and heat stress. Annu. Rev. Physiol., 45:191, 1983.

Bukowiecki, L. J.: Mechanisms of stimulus-calorigenesis coupling in brown adipose tissue. Can. J. Biochem. Cell. Biol., 62:623, 1984.

Calder, W. A., III: Scaling of physiological processes in homeothermic animals. Annu. Rev. Physiol., 43:301, 1981.

Clausen, T., et al.: Significance of cation transport in control of energy metabolism and thermogenesis. Physiol. Rev., 71:733, 1991.

Cohen, R. D., et al.: The Metabolic and Molecular Basis of Acquired Disease. Philadelphia, W. B. Saunders Co., 1990.

Crawshaw, L. I.: Temperature regulation in vertebrates. Annu. Rev. Physiol., 42:473, 1980.

Eisenberg, M., et al.: Emergency Medical Therapy. Philadelphia, W. B. Saunders Co., 1994.

Felig, P., et al. (eds.): Endocrinology and Metabolism, 2nd Ed. New York, McGraw-Hill, 1987.

Galanter, E.: Detection and discrimination and environmental change. In Darian-Smith, I. (ed.): Handbook of Physiology, Sec. 1, Vol. III. Bethesda, Md., American Physiological Society, 1984, p. 103.

Gilly, F. N., et al.: Clinical Hyperthermia. Farmington, CT, S. Karger Publishers, Inc., 1993.

Gordon, C. J., and Heath, J. E.: Integration and central processing in temperature regulation. Annu. Rev. Physiol., 48:595, 1986.

Gordon, C. J.: Temperature Regulation in Laboratory Rodents. New York, Cambridge University Press, 1993.

Hales, J. E. (ed.): Thermal Physiology. New York, Raven Press, 1984.

Hardy, J. D.: Physiology of temperature regulation. Physiol. Rev., 41:521, 1961.

Harrison, M. H.: Effects of thermal stress and exercise on blood volume in humans. Physiol. Rev., 65:149, 1985.

Hellon, R.: Thermoreceptors. In Shepherd, J. T., and Abboud, F. M. (eds.): Handbook of Physiology. Sec. 1, Vol. III. Bethesda, Md., American Physiological Society, 1983, p. 659.

Hensel, H.: Neutral processes in thermoregulation. Physiol. Rev., 53:948, 1973.

Hensel, H.: Thermoreceptors. Annu. Rev. Physiol., 36:233, 1974.

Hong, S. K., et al.: Humans can acclimatize to cold: A lesson from Korean women divers. News Physiol. Sci., 2:79, 1987.

Kelso, S. R., et al.: Thermosensitive single-unit activity of in vitro hypothalamic slices. Am. J. Physiol., 242:R77, 1982.

Kluger, M. J.: Temperature regulation, fever, and disease. In Robertshaw, D. (ed.): International Reveiw of Physiology: Environmental Physiology III. Vol. 20. Baltimore, University Park Press, 1979, p. 209.

Kluger, M. J.: Fever: Role of pyrogens and cryogens. Physiol. Rev., 71:93, 1991.

Lipton, J. M., and Clark, W. G.: Neurotransmitters in temperature control. Annu. Rev. Physiol., 48:613, 1986.

Mitchell, D., and Laburn, H. P.: Pathophysiology of temperature regulation. Physiologist, 28:507, 1985.

Myers, R. D.: Neurochemistry of thermoregulation. Physiologist, 27:41, 1984.

Nicholls, D. G., and Locke, R. M.: Thermogenic mechanisms in brown fat. Physiol. Rev., 64:1, 1984.

Niederman, M. S., et al.: Respiratory Infections: A Scientific Basis for Management. Philadelphia, W. B. Saunders Co., 1994.

Prosser, C. L., and Nelson, D. O.: The role of nervous systems in temperature adaptation of poikilotherms. Annu. Rev. Physiol., 43:281, 1981.

Quinton, P. M: Sweating and its disorders. Annu. Rev. Med., 34:453, 1983.

Robertshaw, D.: Role of the adrenal medulla in thermoregulation. Int. Rev. Physiol., 15:189, 1977.

Schmidt-Nielsen, K.: Animal Physiology: Adaptation and Environment. London, Cambridge University Press, 1975.

Simon, E., et al.: Central and peripheral thermal control of effectors in homeothermic temperature regulation. Physiol. Rev., 66:235, 1986.

Spray, D. C.: Cutaneous temperature receptors. Annu. Rev. Physiol., 48:625, 1986.

Wolkomir, R.: Chilling out for science. Discover, February, 1988, p. 44.

Wyndham, C. H.: The physiology of exercise under heat stress. Annu. Rev. Physiol., 35:193, 1973.

ENDOCRINOLOGY AND REPRODUCTION
UNIT XIV

Introduction to Endocrinology

CHAPTER 74

The functions of the body are regulated by two major control systems: (1) the nervous system, which has already been discussed, and (2) the hormonal, or endocrine, system. In general, the hormonal system is concerned principally with control of the different metabolic functions of the body, such as the rates of chemical reactions in the cells and the transport of substances through cell membranes or other aspects of cellular metabolism like growth and secretion. Some hormonal effects occur in seconds, whereas others require several days simply to start but then continue for weeks or even months.

Many interrelations exist between the hormonal and nervous systems. For instance, at least two glands secrete their hormones almost entirely in response to appropriate neural stimuli, the *adrenal medullae* and the *pituitary gland*. In turn, the different pituitary hormones control secretion by the majority of the other endocrine glands, as we see in the following chapters.

Nature of a Hormone

A hormone is a chemical substance that is secreted into the internal body fluids by one cell or a group of cells and has a physiological *control* effect on other cells of the body.

At many points in this text, we discuss different hormones. Some are *local hormones* and others are *general hormones.* Examples of local hormones are *acetylcholine* released at the parasympathetic and skeletal nerve endings; *secretin,* released by the duodenal

wall and transported in the blood to the pancreas to cause a watery pancreatic secretion; and *cholecystokinin,* released in the small intestine and transported to the gallbladder to cause it to contract and to the pancreas to cause digestive enzyme secretion. These hormones have specific local effects, from whence comes the name local hormones.

Most of the general hormones are secreted by specific *endocrine glands.* Two examples with which we are already familiar are *epinephrine* and *norepinephrine,* both of which are secreted by the *adrenal medullae* in response to sympathetic stimulation. These hormones are transported in the blood to all parts of the body and cause many different reactions, especially constriction of the blood vessels and elevation of the arterial pressure.

A few of the general hormones affect all or almost all the cells of the body; examples are *growth hormone* from the anterior pituitary gland, which causes growth in all or most parts of the body, and *thyroid hormone* from the thyroid gland, which increases the rates of most chemical reactions in almost all the body's cells.

Other hormones affect only specific tissues, called *target tissues* because only these tissues have the specific target cell *receptors* that will bind the hormones to initiate their actions. For instance, *adrenocorticotropin* from the anterior pituitary gland specifically stimulates the adrenal cortex, causing it to secrete adrenocortical hormones, and the *ovarian hormones* have specific effects on the female sex organs as well as on the secondary sexual characteristics of the female body. Many examples of target tissues will become apparent in the following chapters.

Overview of the Important Endocrine Glands and Their Hormones

Figure 74–1 shows the anatomical loci of the important endocrine glands of the body except for the testes and the placenta, which are additional important sources of sex hormones. Let us give a preview of the important hormones secreted by these glands and their most important actions.

ANTERIOR PITUITARY HORMONES

1. *Growth hormone:* causes growth of almost all cells and tissues of the body
2. *Adrenocorticotropin:* causes the adrenal cortex to secrete adrenocortical hormones
3. *Thyroid-stimulating hormone:* causes the thyroid gland to secrete thyroxine and triiodothyronine
4. *Follicle-stimulating hormone:* causes growth of follicles in the ovaries before ovulation; promotes the formation of sperm in the testes
5. *Luteinizing hormone:* plays an important role in causing ovulation; also causes secretion of female sex hormones by the ovaries and testosterone by the testes
6. *Prolactin:* promotes development of the breasts and secretion of milk

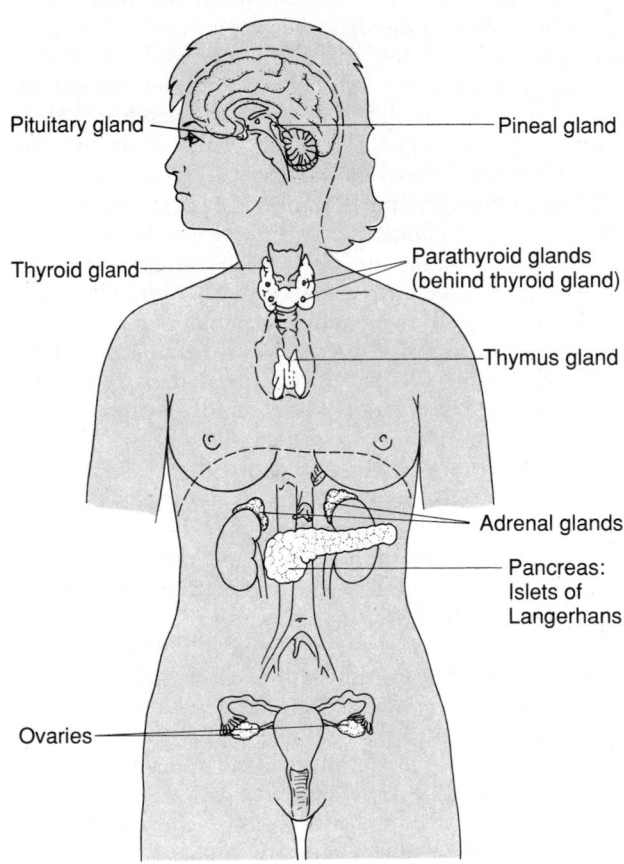

Figure 74–1. Anatomical loci of the principal endocrine glands of the body.

POSTERIOR PITUITARY HORMONES

1. *Antidiuretic hormone* (also called *vasopressin*): causes the kidneys to retain water, thus increasing the water content of the body; also, in high concentrations, causes constriction of the blood vessels throughout the body and elevates the blood pressure
2. *Oxytocin:* contracts the uterus during the birthing process, thus helping expel the baby; also contracts myoepithelial cells in the breasts, thereby expressing milk from the breasts when the baby suckles

ADRENAL CORTEX

1. *Cortisol:* has multiple metabolic functions for control of the metabolism of proteins, carbohydrates, and fats
2. *Aldosterone:* reduces sodium excretion by the kidneys and increases potassium excretion, thus increasing sodium in the body while decreasing the amount of potassium

THYROID GLAND

1 and 2. *Thyroxine and triidothyronine:* increase the rates of chemical reactions in almost all cells of the body, thus increasing the general level of body metabolism
3. *Calcitonin:* promotes the deposition of calcium in the bones and thereby decreases calcium concentration in the extracellular fluid

ISLETS OF LANGERHANS IN THE PANCREAS

1. *Insulin:* promotes glucose entry into most cells of the body, in this way controlling the rate of metabolism of most carbohydrates
2. *Glucagon:* increases the synthesis and release of glucose from the liver into the circulating body fluids

OVARIES

1. *Estrogens:* stimulate the development of the female sex organs, the breasts, and various secondary sexual characteristics
2. *Progesterone:* stimulates secretion of "uterine milk" by the uterine endometrial glands; also helps promote development of the secretory apparatus of the breasts

TESTES

1. *Testosterone:* stimulates growth of the male sex organs; also promotes the development of male secondary sex characteristics

PARATHYROID GLAND

1. *Parathormone:* controls the calcium ion concentration in the extracellular fluid by controlling (a) absorption of calcium from the gut, (b) excretion of calcium by the kidneys, and (c) release of calcium from the bones

PLACENTA

1. *Human chorionic gonadotropin:* promotes growth of the corpus luteum and secretion of estrogens and progesterone by the corpus luteum
2. *Estrogens:* promote growth of the mother's sex organs and some of the tissues of the fetus

3. *Progesterone:* promotes special development of the uterine endometrium in advance of implantation of the fertilized ovum; probably promotes development of some of the fetal tissues and organs; helps promote development of the secretory apparatus of the mother's breasts

4. *Human somatomammotropin:* probably promotes growth of some fetal tissues as well as aiding in the development of the mother's breasts

From this overview of the endocrine system, it is clear that most of the metabolic functions of the body are controlled in one way or another by the endocrine glands. For instance, without growth hormone, a person remains a dwarf. Without thyroxine and triidothyronine from the thyroid gland, almost all the chemical reactions of the body become sluggish and the person becomes sluggish as well. Without insulin from the pancreas, the body's cells can use little of the food carbohydrates for energy. And without the sex hormones, sexual development and sexual functions are absent.

Chemistry of the Hormones

Chemically, the hormones are of three types:

1. *Steroid hormones:* These hormones all have a chemical structure based on the steroid nucleus, similar to that of cholesterol and in most instances are derived from cholesterol itself. Different steroid hormones are secreted by (a) the adrenal cortex (*cortisol* and *aldosterone*), (b) the ovaries (*estrogen* and *progesterone*), (c) the testes (*testosterone*), and (d) the placenta (*estrogen* and *progesterone*).

2. *Derivatives of the amino acid tyrosine:* Two groups of hormones are derivatives of the amino acid tyrosine. The two metabolic thyroid hormones, *thyroxine* and *triiodothyronine,* are iodinated forms of tyrosine derivatives. And the two principal hormones of the adrenal medullae, *epinephrine* and *norepinephrine,* are both catecholamines, also derived from tyrosine.

3. *Proteins or peptides:* All the remaining important endocrine hormones are either proteins, peptides, or immediate derivatives of these. The anterior pituitary hormones are either proteins or large polypeptides; the posterior pituitary hormones, antidiuretic hormone and oxytocin, are peptides, each containing only nine amino acids. Insulin, glucagon, and parathormone are all large polypeptides.

Storage and Secretion of Hormones

As discussed in the following chapters, there is no single way in which all the endocrine glands store and secrete their hormones. However, several general patterns occur for most of the hormones.

For instance, all the protein hormones are formed by the granular endoplasmic reticulum of the glandular cells in the same manner that other secretory proteins are formed, as described in Chapter 3. However, the initial protein formed by the endoplasmic reticulum almost never is the final hormone itself. Instead, it is larger than the active hormone and is called a *preprohormone.* Then this large protein is further cleaved, usually while still in the endoplasmic reticulum, to form a smaller protein called the *prohormone.* This in turn is transported in the endoplasmic reticulum transport vesicles to the Golgi apparatus, where still another section of the protein is cleaved; in this way, the final active protein hormone is formed. The Golgi apparatus usually also compacts the hormone molecules into small membrane-encapsulated vesicles called *secretory vesicles* or *secretory granules.* These vesicles then remain stored in the cytoplasmic compartment of the endocrine cell until a specific signal, such as a nerve signal, another hormonal signal, or a local chemical or physical signal, comes along to cause secretion.

The two groups of hormones derived from tyrosine, the thyroid and the adrenal medullary hormones, are both formed by the actions of enzymes in the cytoplasmic compartments of the glandular cells. In the case of the adrenal medullary hormones norepinephrine and epinephrine, they are then absorbed into preformed vesicles and stored until they are to be secreted. The thyroid metabolic hormones, thyroxine and triiodothyronine, on the other hand, are first formed as component parts of a large protein molecule called thyroglobulin, and this is then stored in large follicles within the thyroid gland. When the thyroid hormones are to be secreted, specific enzyme systems within the thyroid glandular cells cleave the thyroglobulin molecule, thereby allowing released thyroid hormones to be secreted into the blood.

For the steroid hormones formed in the adrenal cortex, ovaries, or testes, the amounts stored in the glandular cells are usually quite small, but large amounts of precursor molecules, especially cholesterol and various intermediates between cholesterol and the final hormones, are present in the cells. On appropriate stimulation, enzymes in these cells can within minutes cause the necessary chemical conversions to the final hormones, followed almost immediately by secretion.

Onset of Hormone Secretion After a Stimulus, and Durations of Action of Different Hormones

Some hormones, such as norepinephrine and epinephrine, are secreted within seconds after the gland is stimulated, and they may develop full action within another few seconds to minutes; the actions of other hormones, such as thyroxine or growth hormone, may require months for full effect.

Thus, each of the different hormones has its own characteristic onset and duration of action—each tailored to perform its specific control function.

CONCENTRATIONS OF HORMONES IN THE CIRCULATING BLOOD, AND HORMONAL SECRETION RATES. The quantitative amounts of hormones that are required to control most metabolic and endocrine functions are incredibly small. Their concentrations in the blood range from as little as 1 picogram (which is one millionth of a millionth of a gram) in each milliliter of blood up to at most a few micrograms (a few millionths of a gram) per milliliter of blood. Similarly, the rates of secretion of the various hormones are extremely small, usually measured in micrograms or milligrams per day. We shall see later in this chapter that highly specialized mechanisms are available in the target tissues that allow even these minute quantities of hormones to exert powerful control over the physiological systems.

Control of Hormone Secretion Rate—The Role of Negative Feedback

Without exception, the rate of secretion of every hormone that has ever been studied is itself controlled very exactly by some internal control system. In most instances, this control generally is exerted through a negative feedback mechanism as follows:

1. The endocrine gland has a natural tendency to *oversecrete* its hormone.
2. Because of this tendency, the hormone exerts more and more of its control effect on the target organ.
3. The target organ in turn performs its function.
4. But when too much function occurs, usually some factor about the function then *feeds back* to the endocrine gland and causes a *negative* effect on the gland to decrease its secretory rate. Thus, the function of the hormone is monitored by the control mechanism, and this information in turn provides negative feedback control of the secretory rate by the gland.

Careful consideration of each feedback mechanism will show that the important factor to be controlled usually is not the secretory rate of the hormone itself but the degree of activity of the target organ. Therefore, only when the target organ activity rises to an appropriate level will the feedback to the gland become powerful enough to slow further secretion of the hormone. If the target organ responds poorly to the hormone, the endocrine gland will almost always secrete still more and more of its hormone until the target organ eventually reaches the appropriate level of activity, but at the expense of excessive secretion of the controlling hormone.

Hormone Receptors and Their Activation

The endocrine hormones most often do not act directly on the intracellular machinery to control the final cellular chemical reactions; instead, they usually first combine with *hormone receptors* on the surfaces of the cells or inside the cells. The combination of hormone and receptor then usually initiates a *cascade of reactions* in the cell, with each stage of reaction in the cascade becoming more powerfully activated than the previous stage, so that even a small initiating hormonal stimulus leads to a large final effect.

Either all or almost all hormonal receptors are large proteins, and each cell that is to be stimulated usually has some 2000 to 100,000 receptors.

Also, each receptor is usually highly specific for a single hormone; this determines the type of hormone that will act on a particular tissue. The target tissues that are affected by a hormone are those that contain its specific receptors.

The locations of the receptors for the different types of hormones are generally the following:

1. *In or on the surface of the cell membrane.* The membrane receptors are specific mostly to the protein, peptide, and catecholamine (epinephrine and norepinephrine) hormones.
2. *In the cell cytoplasm.* The receptors for the different steroid hormones are found almost entirely in the cytoplasm.
3. *In the cell nucleus.* The receptors for the metabolic thyroid hormones (thyroxine and triiodothyronine) are found in the nucleus, believed to be located in direct association with one or more of the chromosomes.

REGULATION OF THE NUMBER OF RECEPTORS. The number of receptors in a target cell usually does not remain constant from day to day, or even from minute to minute, because the receptor proteins themselves are often inactivated or destroyed during the course of their function, and at other times, they are either reactivated or new ones are manufactured by the protein-manufacturing mechanism of the cell. For instance, binding of a hormone with its target cell receptors often, if not usually, causes the number of active receptors to decrease, either because of inactivation of some of the receptor molecules or because of decreased production of the molecules. In either event, this is called *down-regulation* of the receptors. This decreases the responsiveness of the target tissue to the hormone as the number of active receptors decreases.

In a few instances, hormones cause *up-regulation* of receptors; that is, the stimulating hormone induces the formation of more receptor molecules than normal by the protein-manufacturing machinery of the target cell. In this instance, the target tissue becomes progressively more sensitive to the stimulating effects of the hormone.

MECHANISMS OF HORMONAL ACTION

Hormonal Receptors Play a Major Role in Hormonal Action

Almost without exception, a hormone affects its target tissues by first activating target receptors in the tissue cells. This alters the function of the receptor itself, and this receptor is then the direct cause of the hormonal effects. To explain this, let us give a few examples.

1. CHANGE IN MEMBRANE PERMEABILITY. Virtually all the neurotransmitter substances, which are themselves local hormones, combine with receptors in the postsynaptic membrane. Almost always this causes a conformational change in the protein structure of the receptor, usually opening or closing a channel for one or more ions. Some receptors provide open (or closed) channels for sodium ions, others for potassium ions, others for calcium ions, and so forth. Then, it is the altered movement of these ions through the channels that causes the subsequent effects on the postsynaptic cells.

A few of the general hormones also have similar effects in opening or closing membrane ion channels. This is most notably true for many of the actions of the adrenal medullary secretions, norepinephrine and epinephrine. For instance, norepinephrine and epinephrine have especially powerful effects in opening or closing membrane ion channels for sodium, potassium, or both, thus changing the membrane potentials of specific smooth muscle cells and causing excitation in some instances or inhibition in others.

2. ACTIVATION OF AN INTRACELLULAR ENZYME WHEN A HORMONE COMBINES WITH A MEMBRANE RECEPTOR. Another common effect of membrane receptor binding is activation (or occasionally inactivation) of an enzyme immediately inside the cell membrane. A good example of this is the effect of insulin. Insulin binds with the portion of its membrane receptor that protrudes to the exterior of the cell. This produces a structural change in the receptor molecule itself, causing the portion of the molecule that protrudes to the inside to become an activated kinase. This kinase then promotes phosphorylation of several different substances inside the cell. Most of the actions of insulin on the cell then result secondarily from these phosphorylation processes.

A second example, one widely used in hormonal control of cell function, is for the hormone to bind with a special transmembrane receptor that then becomes the activated enzyme *adenyl cyclase* at its end that protrudes to the interior of the cell. This cyclase in turn causes the formation of the substance *cyclic adenosine monophosphate (cAMP)*. And cAMP has a multitude of effects inside the cell to control cell activity, as discussed in greater detail later. The cAMP is called a *second messenger* because it is not the hormone itself that directly institutes the intracellular changes; instead, it is the cAMP that serves as a "second messenger" to cause these effects.

In a few instances, *cyclic guanosine monophosphate (cGMP)*, which is only slightly different from cAMP, serves in a similar manner as a "second messenger."

3. ACTIVATION OF GENES BY BINDING WITH INTRACELLULAR RECEPTORS. Several hormones, especially the steroid hormones and thyroid hormones, bind with protein receptors inside the cell, not in the cell membrane. The activated hormone–receptor complex then binds with or activates specific portions of the DNA strands of the cell nucleus, which in turn initiates transcription of specific genes to form messenger RNA. Therefore, minutes, hours, or even days after the hormone has entered the cell, newly formed proteins appear in the cell and become the controllers of new or increased cellular functions.

Other hormones enhance the translation of messenger RNA in the cytoplasm. This is believed to be especially true of one of the functions of growth hormone and perhaps of insulin.

Second Messenger Mechanisms for Mediating Intracellular Hormonal Functions

We noted previously that one of the means by which hormones exert intracellular actions is to cause the "second messenger" cAMP to be formed inside the cell membrane. Then the cAMP in turn causes all or most of the intracellular effects of the hormone. Thus, the only direct effect that the hormone has on the cell is to activate a single type of membrane receptor. The second messenger does the rest.

Cyclic AMP is not the only second messenger used by the different hormones. Two other especially important ones are (a) calcium ions and associated *calmodulin* and (b) products of membrane phospholipid breakdown.

Intracellular Mechanism of the cAMP "Second Messenger" System

The cAMP mechanism has been shown to be the way in which all the following hormones (and many more) mainly stimulate their target tissues:

1. Adrenocorticotropin
2. Thyroid-stimulating hormone
3. Luteinizing hormone
4. Follicle-stimulating hormone
5. Vasopressin
6. Parathyroid hormone
7. Glucagon
8. Catecholamines
9. Secretin
10. Most hypothalamic releasing hormones

Figure 74–2 shows the function of the cAMP mechanism in more detail. The stimulating hormone first binds with a specific "receptor" for that hormone on the membrane surface of the target cell. The specificity of the receptor determines which hormone will affect the target cell. After binding with the membrane receptor, the portion of the receptor that protrudes to the interior of the cell membrane is activated to become the protein enzyme *adenyl cyclase*. This enzyme in turn causes immediate *conversion of a small amount of the cytoplasmic adenosine triphosphate into cAMP*, which is the compound *cyclic 3',5'-adenosine monophosphate*.

Once cAMP is formed inside the cell, it activates still other enzymes. In fact, it usually activates a *cascade of enzymes*. That is, a first enzyme is activated, which then activates another enzyme, which activates still a third, and so forth. The importance of this mechanism is that only a few molecules of activated adenyl cyclase immediately inside the cell membrane can cause many more molecules of the next enzyme to be activated, which can cause still many times that many molecules of the third enzyme to be activated, and so forth. In this way, even the slightest amount of hormone acting on the cell surface can initiate a powerful cascading activating force for the entire cell.

The specific action that occurs in response to cAMP

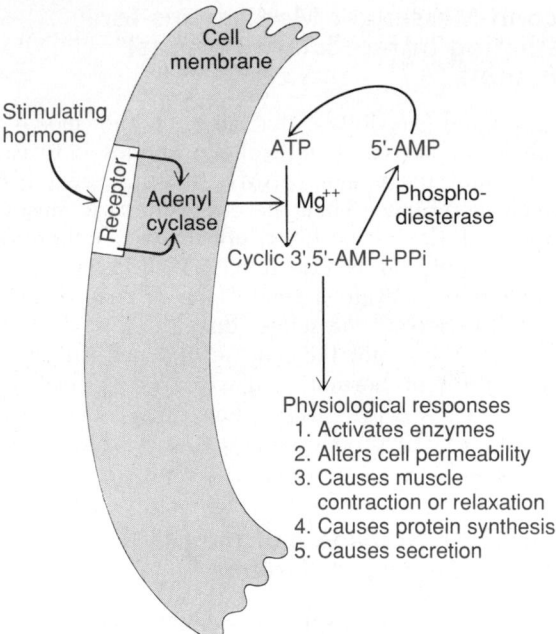

Figure 74–2. Cyclic 3′,5′-adenosine monophosphate (cAMP) mechanism by which many hormones exert their control of cell function.

in each type of target cell depends on the nature of the intracellular machinery, some cells having one set of enzymes and other cells having other enzymes. Therefore, different functions are elicited in different target cells—such functions as initiating synthesis of specific intracellular chemicals, causing muscle contraction or relaxation, initiating secretion by the cells, and altering the cell permeability.

Thus, a thyroid cell stimulated by cAMP forms the metabolic hormones thyroxine and triiodothyronine, whereas the same cAMP in an adrenocortical cell causes secretion of the adrenocortical steroid hormones. On the other hand, cAMP affects epithelial cells of the renal tubules by increasing their permeability to water.

Role of Calcium Ions and "Calmodulin" as a Second Messenger System

Another second messenger system operates in response to the entry of calcium ions into cells. The calcium entry may be initiated by a change in the membrane electrical potential that opens membrane calcium channels or by hormones interacting with membrane receptors that similarly open calcium channels.

On entering the cell, the calcium ions bind with a protein called *calmodulin.* This protein has four calcium sites; when as many as three or four of these sites have bound with calcium, a conformational change occurs that activates the calmodulin, causing multiple effects inside the cell in the same way that cAMP functions. However, it activates a different set of enzymes from those activated by cAMP, thus causing a different set of intracellular reactions. For in-

stance, one of the specific functions of calmodulin is to activate myosin kinase that then acts directly on the myosin of smooth muscle to cause smooth muscle contraction.

The normal calcium ion concentration in most cells of the body is about 10^{-7} to 10^{-8} mol/liter, which is not enough to activate the calmodulin system. But when the calcium ion concentration rises to 10^{-6} to 10^{-5} mol/liter, enough binding then occurs to cause all the intracellular actions of calmodulin. This is almost exactly the same amount of calcium ion change that is required in skeletal muscle to activate troponin C, which in turn causes skeletal muscle contraction, as explained in Chapter 7. It is interesting that troponin C is similar to calmodulin both in function and in protein structure.

Membrane Phospholipid Breakdown Products as Second Messengers—The Phosphatidylinositol System

Some hormones activate transmembrane receptors that then activate the enzyme *phospholipase C* attached to the inside projections of the receptors. This enzyme in turn causes some phospholipids in the cell membrane itself to split into smaller substances that have widespread "second messenger" intracellular effects. The types of hormones that cause this effect are mainly local hormones, most notably hormonal factors released by tissue immune and allergic reactions, as explained in Chapter 34.

The most important membrane phospholipid broken down in this way is *phosphatidyl inositol bisphosphate.* And the most important products that serve as second messengers are *inositol trisphosphate* and *diacylglycerol.* The inositol trisphosphate especially mobilizes calcium ions from both the mitochondria and the endoplasmic reticulum, and the calcium ions then promote all their own second messenger effects such as smooth muscle contraction, changes in secretion by secreting cells, and changes in ciliary action.

The diacylglycerol, the other lipid second messenger, activates the enzyme *protein kinase C.* This activation is further enhanced by the increased calcium ions that have been released in response to inositol trisphosphate. In turn, the activated protein kinase C plays an especially important role in promoting cell division and cell proliferation.

In addition, the lipid portion of diacylglycerol is *arachidonic acid,* which is the precursor for the *prostaglandins* and other local hormones that cause a wide variety of local effects in tissues throughout the body.

Hormones That Act Mainly on the Genetic Machinery of the Cell

Action of Steroid Hormones on the Genes to Cause Protein Synthesis

Another means by which hormones act—specifically the steroid hormones secreted by the adrenal cortex, ovaries, and testes—is to cause synthesis of proteins in the target cells; these proteins then function as enzymes, transport proteins, or structural proteins that in turn provide other functions of the cells.

The sequence of events in steroid function is essentially the following:

1. The steroid hormone enters the cytoplasm of the cell, where it binds with a specific *receptor protein.*
2. The combined receptor protein/hormone then diffuses into or is transported into the nucleus.
3. The combination now binds at specific points on the DNA strands in the chromosomes, which activates the transcription process of specific genes to form messenger RNA.
4. The messenger RNA diffuses into the cytoplasm, where it promotes the translation process at the ribosomes to form new proteins.

To give an example, *aldosterone*, one of the hormones secreted by the adrenal cortex, enters the cytoplasm of renal tubular cells, which contain its specific receptor protein. Therefore, in these cells, the sequence of events cited above ensues. After about 45 minutes, proteins begin to appear in the renal tubular cells that promote sodium reabsorption from the tubules and potassium secretion into the tubules. Thus, there is a characteristic delay in the beginning action of the steroid hormone of at least 45 minutes and up to several hours or even days for full action, which is in marked contrast to the almost instantaneous action of some of the peptide and amino acid–derived hormones, such as vasopressin and norepinephrine.

Action of the Thyroid Hormones to Cause Gene Transcription in the Cell Nucleus

The thyroid hormones *thyroxine* and *triiodothyronine* cause increased transcription by certain genes in the nucleus. To accomplish this, these hormones first bind directly with receptor proteins in the nucleus itself; these receptors are probably protein molecules located within the chromosomal complex, and they probably control the function of the genetic promotors or operators, as explained in Chapter 3.

Two important features of thyroid hormone function in the nucleus are the following:

1. They activate the genetic mechanisms for formation of many types of intracellular proteins—probably 100 or more. Many of these are enzymes that promote enhanced intracellular metabolic activity in virtually all cells of the body.
2. Once bound to the intranuclear receptors, the thyroid hormones can continue to express their control functions for days or even weeks.

MEASUREMENT OF HORMONE CONCENTRATIONS IN THE BLOOD

Most hormones are present in the blood in extremely minute quantities, some in concentrations as low as one billionth of a milligram (1 picogram) per milliliter. Therefore, except in a few instances, it has been almost impossible to measure these concentrations by the usual chemical means. An extremely sensitive method, however, was developed about 30 years ago that revolutionized the measurement of hormones, their precursors, and their metabolic end products. This is the method of *radioimmunoassay.*

Radioimmunoassay

The principle of radioimmunoassay is as follows.

First, an antibody is made in large quantities in some lower animal that is highly specific for the hormone to be measured.

Second, a small quantity of this antibody is (1) mixed with a quantity of fluid from the animal containing the hormone to be measured and (2) mixed simultaneously with an appropriate amount of purified standard hormone that has been tagged with a radioactive isotope. However, one specific condition must be met: there must be too little antibody to bind completely both the radioactively tagged hormone and the hormone in the fluid to be assayed. Therefore, the natural hormone in the assay fluid and the radioactive standard hormone *compete for the binding sites* of the antibody. In the process of competing, the quantity of each of the two hormones, the natural and the radioactive, that binds is proportional to its concentration.

Third, after binding has reached equilibrium, the antibody-hormone complex is separated from the remainder of the solution, and the quantity of radioactive hormone bound in this complex is measured by radioactive counting techniques. If a *large amount of radioactive hormone* has bound with the antibody, then it is clear that there was only a *small amount of natural hormone* to compete with the radioactive hormone, and therefore, the concentration of the natural hormone in the assayed fluid was small. Conversely, if only a small amount of radioactive hormone has bound, it is clear that there was a large amount of natural hormone to compete for the binding sites.

Fourth, to make the assay highly quantitative, the radioimmunoassay procedure is performed also for "standard" solutions of untagged hormone at several concentration levels. Then, a "standard curve" is plotted, as shown in Figure 74–3. By comparing the radioactive counts recorded from the "unknown" assay procedures with the standard curve, one can determine within an error of ± 10 to 15 per cent the concentration of the hormone in the "unknown" assayed fluid. As little as billionths or even trillionths of a gram of hormone can often be assayed in this way.

OTHER COMPETITIVE BINDING ASSAY PROCEDURES. Several other competitive binding techniques for assay of minute quantities of hormones have also been used. One of these is to use in place of the antibody a specific carrier globulin of the plasma that is a natural binding agent for some specific hormone. For instance, the plasma-binding globulin for the adrenocortical hormone cortisol is highly specific. Therefore, this globulin is substituted for the antibody in the assay process for cortisol; the assay is then carried out in exactly the same way as the radioimmunoassay procedure.

Measurement of the Metabolic Clearance Rate of Hormones

Two factors can increase or decrease the concentration of a hormone in the blood. One of these is the rate of secretion of the hormone. The second is the rate of removal of the hormone from the blood, which is called

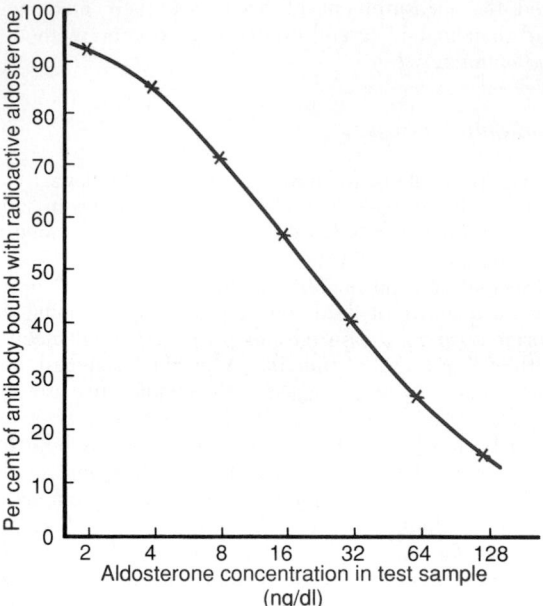

Figure 74–3. "Standard curve" for radioimmunoassay of aldosterone. (Courtesy of Dr. Manis Smith.)

the *metabolic clearance rate*. This is expressed in terms of the number of milliliters of plasma cleared of the hormone per minute. To calculate this clearance rate, one makes the following two measurements:

1. The rate of disappearance of the hormone from the plasma per minute
2. The concentration of the hormone in each milliliter of plasma

Then, the metabolic clearance rate is calculated by the following formula:

$$\text{Metabolic clearance rate} = \frac{\left\{\begin{array}{c}\text{Rate of disappearance of}\\\text{hormone from the plasma}\end{array}\right\}}{\left\{\begin{array}{c}\text{Concentration of hormone in}\\\text{each milliliter of plasma}\end{array}\right\}}$$

The usual procedure for making this measurement is the following: A purified solution of the hormone to be measured is tagged with a radioactive substance. Then the radioactive hormone is infused at a constant rate into the blood stream until the radioactive concentration in the plasma becomes steady. At this time, the rate of disappearance of the radioactive hormone from the plasma equals the rate at which it is infused, which gives one the rate of disappearance. At the same time, the plasma concentration of the radioactive hormone is measured with the use of a standard radioactive counting procedure. Then, with the formula just cited, the metabolic clearance rate is calculated.

Hormones are "cleared" from the plasma in several ways, including (1) metabolic destruction by the tissues, (2) binding with the tissues, (3) excretion by the liver into the bile, and (4) excretion by the kidneys into the urine. For certain hormones, a decreased metabolic clearance rate frequently causes an excessively high concentration of the hormone in the circulating body fluids. For instance, this occurs for several of the steroid hormones when the liver is diseased because these hormones are mainly conjugated in the liver and then "cleared" into the bile.

Measurement of the Rate of Hormone Secretion

A simple method for estimating the rate of hormone secretion is, first, to measure the concentration of the natural hormone in the plasma by means of a radioimmunoassay procedure. Next, the metabolic clearance rate is measured. Then, by multiplying the concentration of the natural hormone times the metabolic clearance rate, one derives a value that is equal to the steady-state rate of hormone production.

However, hormone production often increases and decreases rapidly. In such cases, one can measure the changing rates of secretion only by collecting samples of both the arterial blood entering the gland and the venous blood leaving the gland and simultaneously measuring the rate of blood flow through the gland. Then, by multiplying the rate of blood flow times the venous increase in hormone concentration over that in the arterial blood, one can derive the instantaneous secretion rate. With methods such as this, it has been possible to show that many endocrine glands can be stimulated to extremely high secretory rates within minutes, and many glands also secrete hormones intermittently, rather than steadily. This is especially true for the pituitary hormones as well as for cortisol from the adrenal gland.

REFERENCES

Arky, R. A., and Kettyle, W. M.: Endocrine Pathophysiology: A Problem-Oriented Approach. Philadelphia, J. B. Lippincott, 1994.
Attanasio, A.: Growth and Metabolism: Obesity, Insulin Action and Use of Anthropometry. Farmington, CT, S. Karger Publishers, Inc., 1993.
Barrow, D. L., and Selman, W.: Neuroendocrinology. Baltimore, Williams & Wilkins, 1992.
Becker, K. L., et al.: Principles and Practice of Endocrinology and Metabolism. Philadelphia, J. B. Lippincott, 1990.
Blalock, J. E.: A Molecular basis for bidirectional communication between the immune and neuroendocrine systems. Physiol. Rev., 69:1, 1989.
Brown, E. M.: Extracellular CA^{2+} sensing, regulation of parathyroid cell function, and role of CA^{2+} and other ions as extracellular (first) messengers. Physiol. Rev., 71:371, 1991.
Bertrand, J., et al.: Pediatric Endocrinology: Physiology, Pathophysiology, Clinical Aspects. Baltimore, Williams & Wilkins, 1993.
Conn, P. M. (ed.): Neuroendocrine Peptide Methodology. San Diego, Calif., Academic Press, 1988.
Degroot, L. J., et al.: Endocrinology. Philadelphia, W. B. Saunders Co., 1994.
Evans, R. M.: The steroid and thyroid hormone receptor superfamily. Science, 240:889, 1988.
Goldbeter, A.: Periodic signaling as an optimal mode of intercellular communication. News Physiol. Sci. 3:103, 1988.
Goodman, H. M.: Basic Medical Endocrinology. New York, Raven Press, 1994.
Greengard, P., and Alan R. G.: Advances in Second Messenger and Phosphoprotein Research. New York, Raven Press, 1988.
Greenspan, F. S., and Baxter, J. D.: Basic and Clinical Endocrinology. 4th Ed. Redding, MA, Appleton & Lange, 1994.
Martini, L., and Ganong, W. F., (eds.): Frontiers in Neuroendocrinology. New York, Raven Press, 1988.
Motta, M.: Brain Endocrinology. New York, Raven Press, 1991.
Shenoklkar, S.: Protein phosphorylation: hormones, drugs, and bioregulation. FASEB J., 2:2753, 1988.
Sowers, J. R., and Felicetta, J. V.: Endocrinology of Aging. New York, Raven Press, 1988.
Rubanyi, G. M.: Endothelin. New York, Oxford University Press, 1992.
Swartz, D. P.: Hormone Replacement Therapy. Baltimore, Williams & Wilkins, 1992.
Weiss, E. R., et al.: Receptor activation of G proteins. FASEB J., 2:2841, 1988.
Wilson, J. D., and Foster, D. W.: Williams Textbook of Endocrinology. Philadelphia, W. B. Saunders Co, 1992.

The Pituitary Hormones and Their Control by the Hypothalamus

CHAPTER 75

THE PITUITARY GLAND AND ITS RELATION TO THE HYPOTHALAMUS

The *pituitary gland* (Fig. 75–1), also called the *hypophysis*, is a small gland—about 1 centimeter in diameter and 0.5 to 1 gram in weight—that lies in the *sella turcica,* a bony cavity at the base of the brain, and is connected to the hypothalamus by the *pituitary* (or *hypophysial*) stalk. Physiologically, the pituitary gland is divisible into two distinct portions: the *anterior pituitary*, also known as the *adenohypophysis*, and the *posterior pituitary,* also known as the *neurohypophysis*. Between these is a small, relatively avascular zone called the *pars intermedia*, which is almost absent in the human being while much larger and much more functional in some lower animals.

Embryologically, the two portions of the pituitary originate from different sources, the anterior pituitary from *Rathke's pouch,* which is an embryonic invagination of the pharyngeal epithelium, and the posterior pituitary from an outgrowth of the hypothalamus. The origin of the anterior pituitary from the pharyngeal epithelium explains the epithelioid nature of its cells, whereas the origin of the posterior pituitary from neural tissue explains the presence of large numbers of glial-type cells in this gland.

Six important hormones plus several less important ones are secreted by the *anterior* pituitary, and two important hormones are secreted by the *posterior* pituitary. The hormones of the anterior pituitary play major roles in the control of metabolic functions throughout the body, as shown in Figure 75–2. (1) *Growth hormone* promotes growth of the entire body by affecting protein formation, cell multiplication, and cell differentiation. (2) *Adrenocorticotropin (corticotropin)* controls the secretion of some of the adrenocortical hormones, which in turn affect the metabolism of glucose, proteins, and fats. (3) *Thyroid-stimulating hormone (thyrotropin)* controls the rate of secretion of thyroxine and triiodothyronine by the thyroid gland, and these hormones in turn control the rates of most intracellular chemical reactions of the entire body. (4) *Prolactin* promotes mammary gland development and milk production. And two separate gonadotropic hormones, (5) *follicle-stimulating hormone* and (6) *luteinizing hormone,* control growth of the gonads as well as their hormonal and reproductive activities.

The two hormones secreted by the posterior pituitary play other roles. (1) *Antidiuretic hormone* (also called *vasopressin*) controls the rate of water excretion into the urine and in this way helps control the concentration of water in the body fluids. (2) *Oxytocin* (a) helps deliver milk from the glands of the breast to the nipples during suckling and (b) possibly helps in the delivery of the baby at the end of gestation.

Cell Types in the Anterior Pituitary Gland

The anterior pituitary gland contains many types of secretory cells, as shown in Figure 75–3. Usually, there is one cell type for each major hormone formed in this gland. With special stains attached to high-affinity antibodies that bind with the distinctive hormones, at least five cell types can be differentiated one from another, as follows:

933

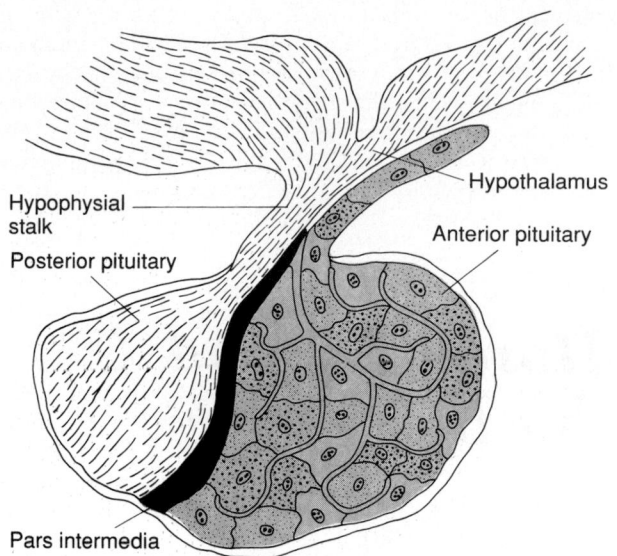

Figure 75–1. Pituitary gland.

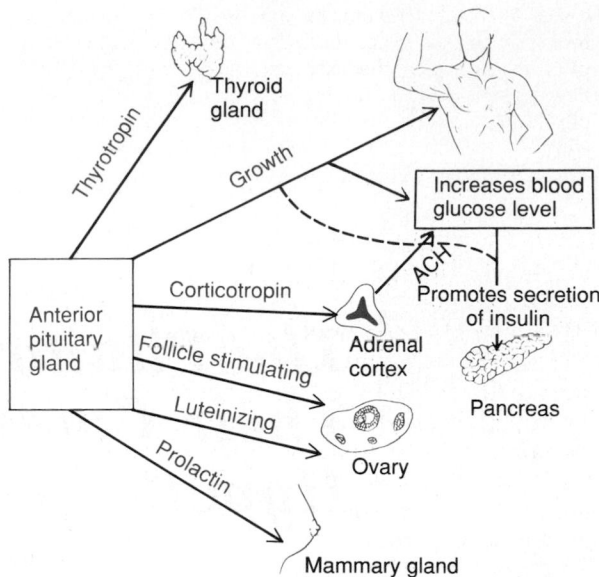

Figure 75–2. Metabolic functions of the anterior pituitary hormones.

1.   *Somatotropes*—human growth hormone (hGH)
2.   *Corticotropes*—adrenocorticotropin (ACTH)
3.   *Thyrotropes*—thyroid-stimulating hormone (TSH)
4.   *Gonadotropes*—gonadotropic hormones, which include both luteinizing hormone (LH) and follicle-stimulating hormone (FSH)
5.   *Lactotropes*—prolactin (PRL)

About 30 to 40 per cent of the anterior pituitary cells are somatotropes that secrete growth hormone, and about 20 per cent are corticotropes that secrete ACTH. The other cell types each number only 3 to 5 per cent

of the total; nevertheless, they secrete the extremely powerful hormones for controlling thyroid function, sexual functions, and milk secretion by the breasts.

Somatotropes stain strongly with acid dyes and, therefore, are called *acidophils.* Thus, pituitary tumors that secrete large quantities of human growth hormone are called *acidophilic tumors.*

The cell bodies of the cells that secrete the *posterior* pituitary hormones are not located in the posterior pituitary gland itself but are large neurons located in the

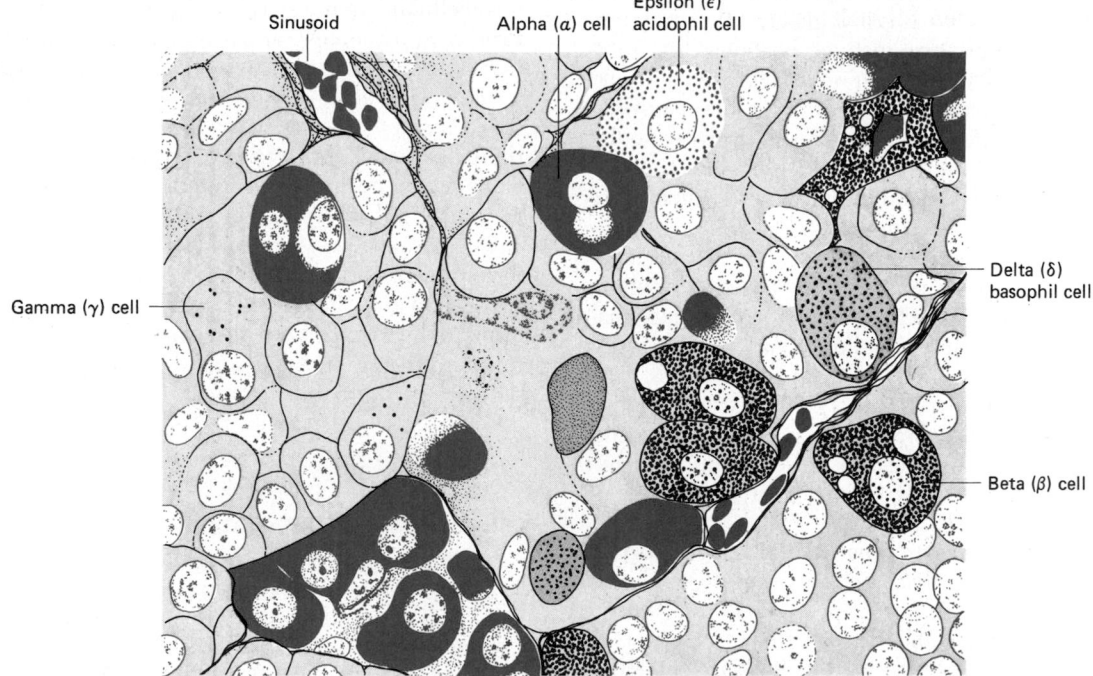

Figure 75–3. Cellular structure of the anterior pituitary gland. (From Guyton: Physiology of the Human Body. 6th Ed. Philadelphia, Saunders College Publishing, 1984.)

supraoptic and *paraventricular nuclei* of the hypothalamus; the hormones are then transported to the posterior pituitary glands in the axoplasm of the neurons' nerve fibers passing from the hypothalamus to the posterior pituitary gland. This is discussed more fully later in the chapter.

CONTROL OF PITUITARY SECRETION BY THE HYPOTHALAMUS

Almost all secretion by the pituitary is controlled by either hormonal or nervous signals from the hypothalamus. Indeed, when the pituitary gland is removed from its normal position beneath the hypothalamus and transplanted to some other part of the body, its rates of secretion of the different hormones (except for prolactin) fall to low levels—in the case of some of the hormones, to zero.

Secretion from the posterior pituitary is controlled by nerve signals that originate in the hypothalamus and terminate in the posterior pituitary. In contrast, secretion by the anterior pituitary is controlled by hormones called *hypothalamic releasing* and *inhibitory hormones* (or *factors*) secreted within the hypothalamus itself and then conducted, as shown in Figure 75–4, to the anterior pituitary through minute blood vessels called *hypothalamic-hypophysial portal vessels*. In the anterior pituitary, these releasing and inhibitory hormones act on the glandular cells to control their secretion. This system of control is discussed in detail later in the chapter.

The hypothalamus in turn receives signals from almost all possible sources in the nervous system. Thus, when a person is exposed to pain, a portion of the pain signal is transmitted into the hypothalamus. Likewise, when a person experiences some powerful depressing or exciting thought, a portion of the signal is transmitted into the hypothalamus. Olfactory stimuli denoting pleasant or unpleasant smells transmit strong signal components directly and through the amygdaloid nu-

clei into the hypothalamus. *Even the concentrations of nutrients, electrolytes, water, and various hormones* in the blood excite or inhibit various portions of the hypothalamus. Thus, the hypothalamus is a collecting center for information concerned with the internal well-being of the body, and in turn, much of this information is used to control secretions of the many globally important pituitary hormones.

Hypothalamic-Hypophysial Portal System

The anterior pituitary is a highly vascular gland with extensive capillary sinuses among the glandular cells. Almost all the blood that enters these sinuses passes first through another capillary bed in the lower hypothalamus. The blood then flows through small *hypothalamic-hypophysial portal vessels* into the anterior pituitary sinuses. Thus, Figure 75–4 shows the lowermost portion of the hypothalamus called the *median eminence* that connects inferiorly with the pituitary stalk. Small arteries penetrate into the substance of the median eminence and then additional small vessels return to its surface, coalescing to form the hypothalamic-hypophysial portal vessels. These in turn pass downward along the pituitary stalk to supply blood to the anterior pituitary sinuses.

SECRETION OF HYPOTHALAMIC RELEASING AND INHIBITORY HORMONES INTO THE MEDIAN EMINENCE. Special neurons in the hypothalamus synthesize and secrete the *hypothalamic releasing* and *inhibitory hormones* that control the secretion of the anterior pituitary hormones. These neurons originate in various parts of the hypothalamus and send their nerve fibers into the median eminence and the tuber cinereum, an extension of hypothalamic tissue that extends into the pituitary stalk. The endings of these fibers are different from most endings in the central nervous system in that their function is not to transmit signals from one neuron to another but merely to secrete the hypothalamic releasing and inhibitory hormones into the tissue fluids. These hormones are immediately absorbed into the hypothalamic-hypophysial portal system and carried directly to the sinuses of the anterior pituitary gland.

FUNCTION OF THE RELEASING AND INHIBITORY HORMONES IN THE ANTERIOR PITUITARY. The function of the releasing and inhibitory hormones is to control the secretion of the anterior pituitary hormones. For most of the anterior pituitary hormones, it is the releasing hormones that are important, but for prolactin, an inhibitory hormone probably exerts most control. The important hypothalamic releasing and inhibitory hormones are the following:

1. *Thyrotropin-releasing hormone* (TRH), which causes release of thyroid-stimulating hormone
2. *Corticotropin-releasing hormone* (CRH), which causes release of adrenocorticotropin
3. *Growth hormone releasing hormone* (GHRH), which causes release of growth hormone, and *growth hormone inhibitory hormone* (GHIH), which is the same as the hor-

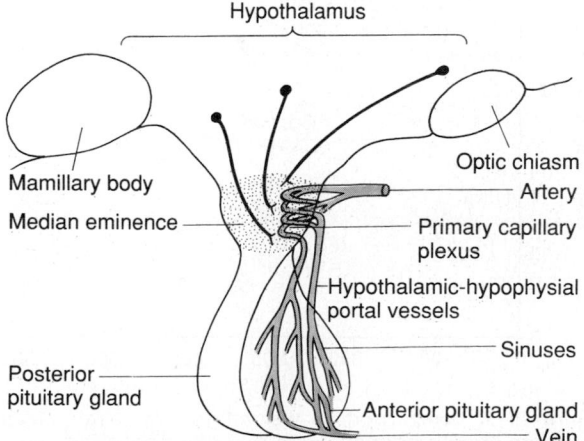

Figure 75–4. Hypothalamic-hypophysial portal system.

Hypothalamus

Mamillary body
Median eminence

Posterior pituitary gland

Optic chiasm
Artery
Primary capillary plexus
Hypothalamic-hypophysial portal vessels
Sinuses
Anterior pituitary gland
Vein

mone *somatostatin* and which inhibits the release of growth hormone

4. *Gonadotropin-releasing hormone* (GnRH), which causes release of the two gonadotropic hormones, luteinizing hormone and follicle-stimulating hormone

5. *Prolactin inhibitory hormone* (PIH), which causes inhibition of prolactin secretion

In addition to these more important hypothalamic hormones, still another probably excites the secretion of prolactin and several possibly inhibit some of the other anterior pituitary hormones. Each of the more important hypothalamic hormones are discussed in detail at the time that the specific hormonal system controlled by them is presented in this and subsequent chapters.

Specific Areas in the Hypothalamus That Control Secretion of Specific Hypothalamic Releasing and Inhibitory Hormones. All or most of the hypothalamic hormones are secreted at nerve endings in the median eminence before being transported to the anterior pituitary gland. Electrical stimulation of this region excites these nerve endings and, therefore, causes release of essentially all the hypothalamic hormones. However, the neuronal cell bodies that give rise to these median eminence nerve endings are located in other discrete areas of the hypothalamus or in closely related areas of the basal brain. The specific loci of the neuronal cell bodies that form the different hypothalamic releasing or inhibitory hormones are still poorly known, so that it would be misleading to attempt a delineation here.

PHYSIOLOGICAL FUNCTIONS OF GROWTH HORMONE

All the major anterior pituitary hormones besides growth hormone exert their principal effects by stimulating target glands, including the thyroid gland, the adrenal cortex, the ovaries, the testicles, and the mammary glands. The functions of each of these pituitary hormones are so intimately concerned with the functions of the respective target glands that, except for growth hormone, their functions are discussed in subsequent chapters along with the target glands. Growth hormone, in contrast to other hormones, does not function through a target gland but exerts its effects on all or almost all tissues of the body.

Effect of Growth Hormone to Cause Growth

Growth hormone, also called *somatotropic hormone* or *somatotropin,* is a small protein molecule that contains 191 amino acids in a single chain and has a molecular weight of 22,005. It causes growth of almost all tissues of the body that are capable of growing. It promotes increased sizes of the cells and increased mitosis with development of increased numbers of cells and specific differentiation of certain types of cells such as bone growth cells and early muscle cells.

Figure 75–5 shows typical weight charts of two growing rats, one of which received daily injections of growth hormone, compared with a littermate that did not receive growth hormone. This figure shows marked exacerbation of growth by growth hormone—both in the early days of life and even after the two rats had reached adulthood. In the early stages of development, all organs of the treated rat increased proportionately in size, but after adulthood was reached, most of the bones ceased lengthening while many of the soft tissues continued to grow. This results from the fact that once the epiphyses of the long bones have united with the shafts, further growth of bone length cannot occur even though most other tissues of the body can continue to grow throughout life.

Metabolic Effects of Growth Hormone

Aside from its general effect in causing growth, GH has many specific metabolic effects as well, including especially (1) increased rate of protein synthesis in all cells of the body; (2) increased mobilization of fatty acids from adipose tissue, increased free fatty acids in the blood, and increased use of the fatty acids for energy; and (3) decreased rate of glucose utilization throughout the body.

Thus, in effect, growth hormone enhances the body protein, uses up the fat stores, and conserves carbohydrate.

Role of Growth Hormone in Promoting Protein Deposition

Although the most important mechanism by which growth hormone increases protein deposition is not known, a series of different effects are known, all of which could lead to enhanced protein.

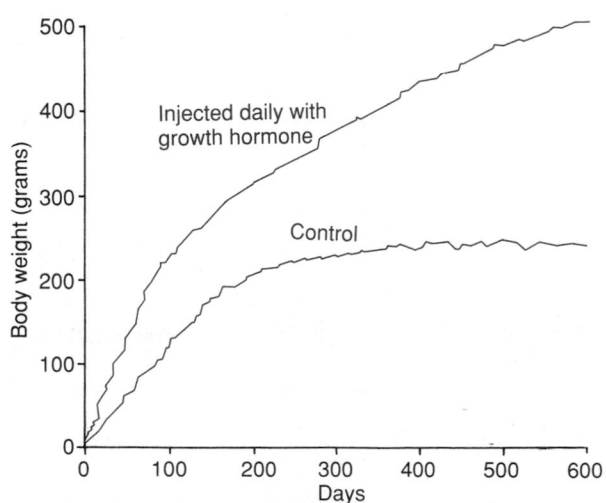

Figure 75–5. Comparison of weight gain of a rat injected daily with growth hormone with that of a normal rat.

1. ENHANCEMENT OF AMINO ACID TRANSPORT THROUGH THE CELL MEMBRANES. Growth hormone directly enhances transport of at least some and perhaps most amino acids through the cell membranes to the interior of the cells. This increases the concentrations of the amino acids in the cells and is presumed to be at least partly responsible for the increased protein synthesis. This control of amino acid transport is similar to the effect of insulin in controlling glucose transport through the membrane, as discussed in Chapters 67 and 78.

2. ENHANCEMENT OF RNA TRANSLATION TO CAUSE PROTEIN SYNTHESIS BY THE RIBOSOMES. Even when the amino acids are not increased in the cells, growth hormone still excites increased RNA translation, causing protein to be synthesized in increased amounts by the ribosomes in the cytoplasm.

3. INCREASED NUCLEAR TRANSCRIPTION OF DNA TO FORM RNA. Over more prolonged periods (24 to 48 hours), growth hormone also stimulates the transcription of DNA in the nucleus, causing formation of increased quantities of RNA. This in turn promotes more protein synthesis and promotes growth if sufficient energy, amino acids, vitamins, and other necessities for growth are available. In the long run, this perhaps is the most important of all the functions of growth hormone.

4. DECREASED CATABOLISM OF PROTEIN AND AMINO ACIDS. In addition to the increase in protein synthesis, there is a decrease in the breakdown of cell protein. A probable reason for this is that growth hormone also mobilizes large quantities of free fatty acids from the adipose tissue, and these in turn are used to supply most of the energy for the body cells, thus acting as a potent "protein sparer."

SUMMARY. Growth hormone enhances almost all facets of amino acid uptake and protein synthesis by cells, while at the same time reducing the breakdown of proteins.

Effect of Growth Hormone in Enhancing Fat Utilization for Energy

Growth hormone has a specific effect in causing release of fatty acids from adipose tissue and, therefore, increasing the concentration of fatty acids in the body fluids. In addition, in the tissues throughout the body, it enhances the conversion of fatty acids to acetylcoenzyme A (acetyl-CoA) with subsequent utilization of this for energy. Therefore, under the influence of growth hormone, fat is used for energy in preference to both carbohydrates and proteins.

Some research workers have considered this fat mobilization effect of growth hormone one of its most important functions and have considered the protein-sparing effect of excess fat usage the most important factor that promotes protein deposition and growth. However, growth hormone mobilization of fat requires hours to occur, whereas enhancement of cellular protein synthesis can begin in minutes under the influence of growth hormone.

Ketogenic Effect of Growth Hormone. Under the influence of excessive amounts of growth hormone, fat mobilization from adipose tissue sometimes becomes so great that large quantities of acetoacetic acid are formed by the liver and released into the body fluids, thus causing *ketosis*. This excessive mobilization of fat from the adipose tissue also frequently causes a fatty liver.

Effect of Growth Hormone on Carbohydrate Metabolism

Growth hormone has four major effects on cellular metabolism of glucose: (1) decreased use of glucose for energy, (2) enhancement of glycogen deposition in the cells, (3) diminished uptake of glucose by the cells, and (4) increased insulin secretion and decreased sensitivity to insulin.

1. DECREASED USE OF GLUCOSE FOR ENERGY. We do not know the precise mechanism by which growth hormone decreases glucose utilization by the cells. However, the decrease probably results partially from the increased mobilization and utilization of fatty acids for energy caused by growth hormone. That is, the fatty acids form large quantities of acetyl-CoA that in turn initiate feedback effects to block the glycolytic breakdown of glucose and glycogen.

2. ENHANCEMENT OF GLYCOGEN DEPOSITION IN THE CELLS. Because, in the presence of excess growth hormone, glucose and glycogen cannot easily be used for energy, the glucose that does enter the cells is rapidly polymerized into glycogen and deposited. Therefore, the cells rapidly become saturated with glycogen and can store no more.

3. DIMINISHED UPTAKE OF GLUCOSE BY THE CELLS AND INCREASED BLOOD GLUCOSE CONCENTRATION—"PITUITARY DIABETES." When growth hormone is first administered to an animal, the cellular uptake of glucose is enhanced and the blood glucose concentration falls slightly. This effect lasts for only 30 minutes to 1 hour or so and is then followed by exactly the opposite effect—decreased transport of glucose into the cell. This probably results from the fact that the cells have already taken up an excess of glucose that they are having difficulty using. Without normal cellular uptake and use, the blood concentration of glucose often increases to 50 per cent or more above normal, and the condition is called *pituitary diabetes.* When this diabetes is treated with insulin, it is "insulin insensitive," requiring excessive amounts of insulin to cause even a modest decrease in the blood glucose level.

4. INCREASED SECRETION OF INSULIN—DIABETOGENIC EFFECT OF GROWTH HORMONE. The increase in blood glucose concentration caused by growth hormone stimulates the beta cells of the islets of Langerhans to secrete extra insulin. In addition, the growth hormone has a direct stimulatory effect on the beta cells. The combination of these two effects sometimes so greatly overstimulates insulin secretion by the beta cells that they literally "burn out." When this occurs, diabetes mellitus develops, a disease that is discussed in detail in Chapter 78. Therefore, growth hormone is said to have a *diabetogenic effect.*

NECESSITY OF INSULIN AND CARBOHYDRATE FOR THE GROWTH-PROMOTING ACTION OF GROWTH HORMONE. Growth hormone fails to cause growth in an animal that lacks a pancreas; it also fails to cause growth if carbohydrates are excluded from the diet. This shows that adequate insulin activity as well as adequate availability of carbohydrates are necessary for growth hormone to be effective. Part of this requirement for carbohydrates and insulin is to provide the energy needed for the metabolism of growth. But there seem to be other effects as well. Especially important is the specific effect of insulin in enhancing the transport of some amino acids into cells in the same way that it enhances glucose transport.

Stimulation of Cartilage and Bone Growth

Although growth hormone stimulates increased deposition of proteins and increased growth in almost all tissues of the body, its most obvious effect is to increase growth of the skeletal frame. This results from multiple effects of growth hormone on bone, including (1) increased deposition of protein by the chondrocytic and osteogenic cells that cause bone growth, (2) increased rate of reproduction of these cells as well, and (3) the specific effect of converting chondrocytes into osteogenic cells, thus causing specific deposition of new bone.

There are two principal mechanisms of bone growth: In one of these, the long bones grow in length at the epiphyseal cartilages, where the epiphyses at the ends of the bone are separated from the shaft. This growth first causes deposition of new cartilage, followed by conversion of this into new bone, thus elongating the shaft and pushing the epiphyses farther and farther apart. At the same time, the epiphyseal cartilage itself is progressively used up, so that by late adolescence no additional epiphyseal cartilage remains to provide for further growth. At this time, bony fusion occurs between the shaft and the epiphysis at each end, so that no further lengthening of the long bone can occur. Growth hormone stimulates all these processes of epiphyseal cartilage growth and growth of the long bones. However, once the epiphyses have united with the shafts, growth hormone has no further ability to lengthen the bones.

For the second mechanism of bone growth, *osteoblasts* in the bone periosteum and in some bone cavities deposit new bone on the surfaces of older bone. Simultaneously, *osteoclasts* in the bone (discussed in detail in Chapter 79) remove old bone. When the rate of deposition is greater than that of resorption, the thickness of the bone increases. Growth hormone strongly stimulates the osteoblasts. Therefore, the

bones can continue to enlarge throughout life under the influence of growth hormone; this is especially true for the membranous bones. For instance, the jaw bones can be stimulated to grow even after adolescence, causing forward protrusion of the chin and lower teeth. Likewise, the bones of the skull grow in thickness and give rise to bony protrusions over the eyes.

Growth Hormone Exerts Much of Its Effects Through Intermediate Substances Called "Somatomedins," Also Called "Insulin-Like Growth Factors"

When growth hormone is supplied directly to cartilage chondrocytes cultured outside the body, proliferation or enlargement of the chondrocytes usually fails to occur. Yet growth hormone injected into the intact animal does cause proliferation and growth of the same cells.

In brief, it has been found that growth hormone causes the liver (and to much less extent other tissues) to form several small proteins called *somatomedins* that in turn have the potent effect of increasing all aspects of bone growth. Many of the somatomedin effects on growth are similar to the effects of insulin on growth. Therefore, the somatomedins are called insulin-like growth factors (IGF) as well as the name somatomedins.

At least four somatomedins have been isolated, but by far the most important of these is *somatomedin C* (also called insulin-like growth factor I [IGF-I]). The molecular weight of somatomedin C is about 7500, and its concentration in the plasma normally follows closely the rate of secretion of growth hormone. The pygmies of Africa have a congenital inability to synthesize significant amounts of somatomedin C. Therefore, even though their plasma concentration of growth hormone is either normal or high, there remain diminished amounts of somatomedin C in the plasma, thus apparently accounting for the small stature of these people. Some other dwarfs (the Levi-Lorain dwarf) also have this problem.

So it has been postulated that most, if not nearly all, of the growth effects of growth hormone result from somatomedin C and other somatomedins, rather than from direct effects of growth hormone on the bones and other peripheral tissues. Even so, experiments have demonstrated that injection of growth hormone directly into the epiphyseal cartilages of bones of living animals will cause specific growth of these cartilage areas alone and that the amount of growth hormone required for this is minute. Therefore, some aspects of the somatomedin hypothesis are still questionable. One possibility is that growth hormone can also cause formation of enough somatomedin C in the local tissue to cause the local growth. It is also possible that growth hormone itself is directly responsible for increased growth in some tissues and that the somatomedin mechanism is an alternative means of increasing growth but not always a necessary one.

SHORT DURATION OF ACTION OF GROWTH HORMONE BUT PROLONGED ACTION OF SOMATOMEDIN C. Growth hormone is only weakly attached to the plasma proteins in the blood. Therefore, it is released from the blood into the tissues rapidly, having a half-time in the blood of less than 20 minutes. By contrast, somatomedin C attaches strongly to a carrier protein in the blood that, like somatomedin C, is itself produced in response to growth hormone. As a result, somatomedin C is released only slowly from the blood to the tissues with a half-time of about 20 hours. This greatly prolongs the growth-promoting effects of the bursts of growth hormone secretion that we see in Figure 75–6.

Regulation of Growth Hormone Secretion

For many years it was believed that growth hormone was secreted primarily during the period of growth but then disappeared from the blood at adolescence. This has proved not to be true because after adolescence, secretion decreases only slowly with aging, finally falling to about 25 per cent of the adolescent level in very old age.

The rate of growth hormone secretion increases and decreases within minutes, as shown in Figure 75–6, sometimes for reasons that are not understood but at other times definitely in relation to the person's state of nutrition or stress, such as during (1) *starvation*, especially with severe *protein deficiency;* (2) *hypoglycemia* or *low concentration of fatty acids in the blood;* (3) *exercise;* (4) *excitement;* and (5) *trauma.* And it characteristically increases during the first 2 hours of *deep sleep,* as also is shown in Figure 75–6.

The normal concentration of growth hormone in the plasma of an adult is between 1.6 and 3 ng/ml and in a child or adolescent about 6 ng/ml. These values often increase to as high as 50 ng/ml after depletion of the body stores of proteins or carbohydrates during prolonged starvation.

Under acute conditions, hypoglycemia is a far more potent stimulator of growth hormone secretion than is an acute decrease in protein intake. On the other hand, in chronic conditions, growth hormone secretion seems to correlate more with the degree of cellular protein depletion than with the degree of glucose insufficiency. For instance, the extremely high levels of growth hormone that occur during starvation are closely related to the amount of protein depletion. Figure 75–7 demonstrates this relation: the first column shows growth hormone levels in children with extreme protein deficiency during the malnutrition disease called *kwashiorkor;* the second column shows the levels in the same children after 3 days of treatment with more than adequate quantities of carbohydrates in their diets, demonstrating that the carbohydrates did not lower the plasma growth hormone concentration; the third and fourth columns show the levels after treatment with protein supplement in their diet for 3 and 25 days, with concomitant decrease in the hormone. These results demonstrate that under severe conditions of protein malnutrition, adequate calories alone are not sufficient to correct the excess production of growth hormone. Instead, the protein deficiency must also be corrected before the growth hormone concentration will return to normal.

Role of the Hypothalamus, Growth Hormone Releasing Hormone, and Somatostatin in the Control of Growth Hormone Secretion

From the preceding description of the many factors that can affect growth hormone secretion, one can readily understand the perplexity that has faced physiologists in their attempt to unravel the mysteries of the regulation of growth hormone secretion. It is known

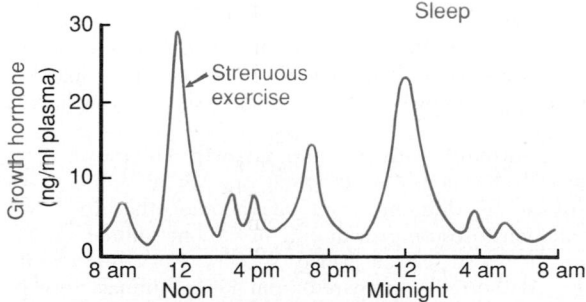

Figure 75–6. Typical variations in growth hormone secretion throughout the day, demonstrating the especially powerful effect on secretion caused by strenuous exercise and the high rate of growth hormone secretion that occurs during the first few hours of deep sleep.

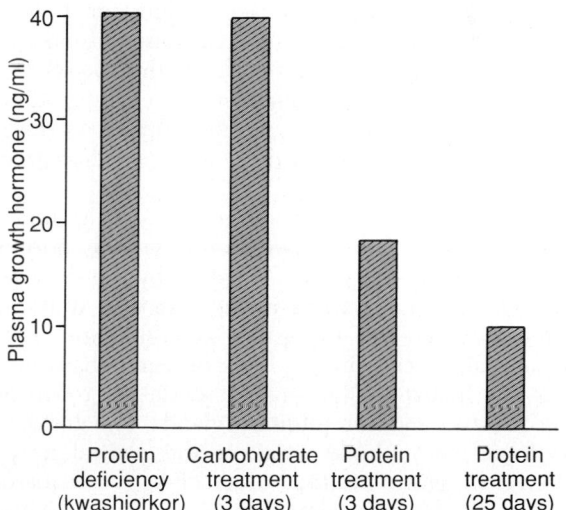

Figure 75–7. Effect of extreme protein deficiency on the concentration of growth hormone in the plasma in the disease kwashiorkor. Also shown is the failure of carbohydrate treatment but the effectiveness of protein treatment in lowering growth hormone concentration. (Drawn from data in Pimstone: *Am. J. Clin. Nutr.,* 21:482, 1968.)

that growth hormone secretion is controlled almost entirely in response to two factors secreted in the hypothalamus and then transported to the anterior pituitary gland through the hypothalamic-hypophysial portal vessels. They are *growth hormone releasing hormone* and *growth hormone inhibitory hormone,* also called *somatostatin.* Both of them are polypeptides; growth hormone releasing hormone is composed of 44 amino acids, and somatostatin, 14 amino acids.

The hypothalamic nucleus that causes secretion of growth hormone releasing hormone is the ventromedial nucleus, the same area of the hypothalamus that is known to be sensitive to hypoglycemia and to cause hunger in hypoglycemic states. The secretion of somatostatin is controlled by other nearby areas of the hypothalamus. Therefore, it is reasonable to believe that some of the same signals that modify a person's behavioral feeding instincts also alter the rate of growth hormone secretion. In a similar manner, hypothalamic signals depicting emotions, stress, and trauma can all affect hypothalamic control of growth hormone secretion. In fact, definitive experiments have shown that catecholamines, dopamine, and serotonin, each of which is released by a different neuronal system in the hypothalamus, all increase the rate of growth hormone secretion.

Most of the control of growth hormone secretion is probably mediated through growth hormone releasing hormone rather than through the inhibitory hormone somatostatin. Growth hormone releasing hormone stimulates growth hormone secretion by attaching to cell membrane receptors specific for growth hormone releasing hormone on the outer surfaces of the growth hormone cells in the pituitary gland. The growth hormone releasing hormone activates the adenyl cyclase system inside the cell, increasing the level of cyclic adenosine monophosphate (cAMP). This in turn has both a short-term effect and a long-term effect. The short-term effect is to increase calcium ion transport into the cell, which within minutes causes fusion of the growth hormone secretory vesicles with the cell membrane and release of the hormone into the blood. The long-term effect is to increase transcription in the nucleus by the genes that cause new growth hormone synthesis.

Somatostatin plays many other roles in the body in addition to its inhibitory effect on growth hormone secretion. For instance, it is secreted by the delta cells of the islets of Langerhans in the pancreas, it also can inhibit the secretion of insulin and glucagon by the beta and alpha cells in the islets of Langerhans in the same way that it inhibits the secretion of growth hormone by the anterior pituitary gland. Somatostatin is also found in multiple areas of the central nervous system and gastrointestinal tract. For these reasons, somatostatin has widespread roles in modulating the functions of multiple hormonal and other physiological systems.

When growth hormone is administered to an animal over a period of hours, the rate of endogenous growth hormone secretion decreases. This demonstrates that growth hormone secretion, as is true for essentially all other hormones, is subject to typical negative feedback control. However, the nature of this feedback mechanism and whether it is mediated mainly through inhibition of growth hormone releasing hormone or enhancement of somatostatin are uncertain.

In summary, our knowledge of the regulation of growth hormone secretion is not sufficient to describe a composite picture. Yet, because of the extreme secretion of growth hormone during starvation and its undoubted long-term effect on promoting protein synthesis and growth of tissues, one would like to propose the following: that the major long-term controller of growth hormone secretion is the long-term state of nutrition of the tissues themselves, especially their level of protein nutrition. That is, nutritional deficiency or excess tissue need for cellular proteins—for instance, after a severe bout of exercise when the muscles' nutritional status has been taxed—would in some way increase the rate of growth hormone secretion. The growth hormone in turn would promote synthesis of new proteins while at the same time conserving the proteins already present in the cells.

Abnormalities of Growth Hormone Secretion

PANHYPOPITUITARISM. This term means decreased secretion of all the anterior pituitary hormones. The decrease in secretion may be congenital (present from birth), or it may occur suddenly or slowly at any time during the life of the individual.

Dwarfism. Most instances of dwarfism result from generalized deficiency of anterior pituitary secretion (panhypopituitarism) during childhood. In general, the features of the body develop in appropriate proportion to one another, but the rate of development is greatly decreased. A child who has reached the age of 10 years may have the bodily development of a child of 4 to 5 years, and the same person on reaching the age of 20 years may have the bodily development of a child of 7 to 10 years.

The panhypopituitary dwarf does not pass through puberty and never secretes a sufficient quantity of gonadotropic hormones to develop adult sexual functions. In one third of the dwarfs, the deficiency is of growth hormone alone; these individuals do mature sexually and occasionally reproduce.

In one type of dwarfism (the African pygmy and the Levi-Lorain dwarf), the rate of growth hormone secretion is normal or high, but there is a hereditary inability to form somatomedin C in response to the growth hormone.

Treatment with Human Growth Hormone. The growth hormones of different species of animals are sufficiently different from one another that they will function to cause growth only in the one animal species or, at most, closely allied species. For this reason, growth hormone prepared from lower animals (except to some extent from primates) is not effective in human beings. Therefore, the growth hormone of the human being is called *human growth hormone* to distinguish it from the others.

In the past, it has been difficult to obtain sufficient

quantities of human growth hormone to treat patients with growth hormone deficiency except on an experimental basis because it has had to be prepared from human pituitary glands. Human growth hormone can now be synthesized by *Escherichia coli* bacteria as a result of successful application of recombinant DNA technology. Therefore, this hormone is now beginning to become available in sufficient quantities for treatment purposes. Dwarfs who have pure growth hormone deficiency can be completely cured. Also, human growth hormone might prove to be beneficial in other metabolic disorders because of its widespread metabolic functions.

Panhypopituitarism in the Adult. Panhypopituitarism occurring in adulthood frequently results from one of three common abnormalities: Two tumorous conditions, craniopharyngiomas and chromophobe tumors, may compress the pituitary gland until the functioning anterior pituitary cells are totally or almost totally destroyed. The third cause is thrombosis of the pituitary blood vessels. Specifically, this abnormality occasionally occurs when a mother develops circulatory shock after the birth of a baby.

The effects of adult panhypopituitarism in general are (1) hypothyroidism, (2) depressed production of glucocorticoids by the adrenal glands, and (3) suppressed secretion of the gonadotropic hormones so that sexual functions are lost. Thus, the picture is that of a lethargic person (from lack of thyroid hormones) who is gaining weight because of lack of fat mobilization by growth, adrenocorticotropic, adrenocortical, and thyroid hormones, and who has lost all sexual functions. Except for the abnormal sexual functions, the patient can usually be treated satisfactorily by administration of adrenocortical and thyroid hormones.

GIGANTISM. Occasionally, the acidophilic, growth hormone–producing cells of the anterior pituitary gland become excessively active, and sometimes even acidophilic tumors occur in the gland. As a result, large quantities of growth hormone are produced. All body tissues grow rapidly, including the bones, and if the condition occurs before adolescence, that is, before the epiphyses of the long bones have become fused with the shafts, height increases so that the person becomes a giant as tall as 8 feet.

The giant ordinarily has *hyperglycemia,* and the beta cells of the islets of Langerhans in the pancreas are prone to degenerate, partially because they become overactive owing to the hyperglycemia and partially because of a direct overstimulating effect of growth hormone on the islet cells. Consequently, in about 10 per cent of giants, *full-blown diabetes mellitus* finally develops.

In most giants, panhypopituitarism eventually develops if they remain untreated because the gigantism is usually caused by a tumor of the pituitary gland that grows until the gland itself is destroyed. This eventual general deficiency of pituitary hormones usually causes death in early adulthood. However, once gigantism is diagnosed, further development can often be blocked by microsurgical removal of the tumor from the pituitary gland or by irradiation of the gland.

ACROMEGALY. If an acidophilic tumor occurs after adolescence—that is, after the epiphyses of the long bones have fused with the shafts—the person cannot grow taller, but the soft tissues can continue to grow and the bones can grow in thickness. This condition, shown in Figure 75–8, is known as *acromegaly.* Enlargement is especially marked in the bones of the hands and feet and in the *membranous bones,* including the cranium, nose, bosses on the forehead, supraorbital ridges, lower jawbone, and portions of the vertebrae because their growth does not cease at adolescence. Consequently, the lower jaw protrudes forward, some-

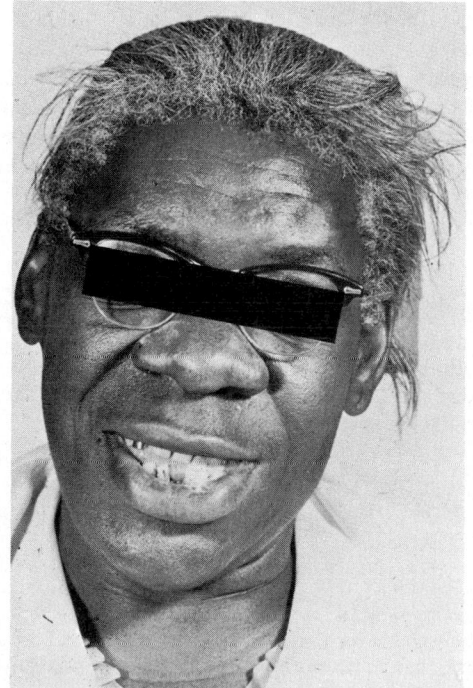
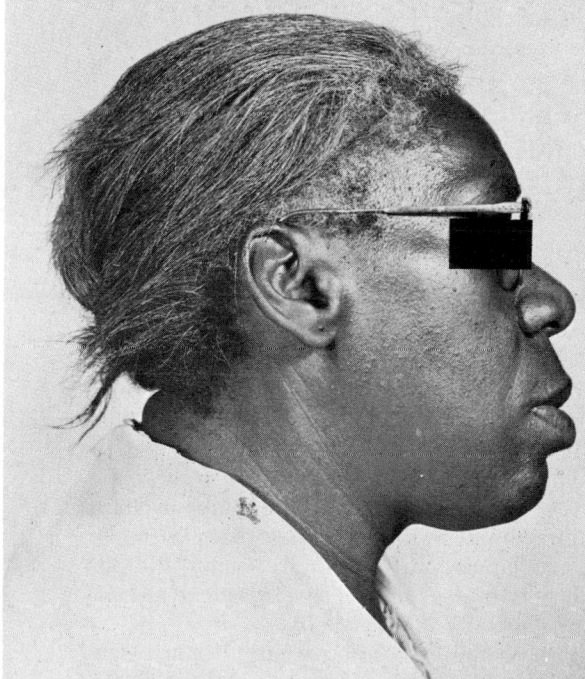

Figure 75–8. Acromegalic patient.

times as much as one-half inch, the forehead slants forward because of excess development of the supraorbital ridges, the nose increases to as much as twice normal size, the feet require size 14 or larger shoes, and the fingers become extremely thickened so that the hands develop a size almost twice normal. In addition to these effects, changes in the vertebrae ordinarily cause a hunched back, which is known clinically as *kyphosis.* Finally, many soft tissue organs, such as the tongue, liver, and especially the kidneys, become greatly enlarged.

Possible Role of Decreased Growth Hormone Secretion in Causing Aging

In people who have lost their ability to secrete growth hormone, the aging process accelerates. For instance, a person at age 50 who has been without growth hormone for many years will probably have the appearance of a person aged 65. The aging seems to result mainly from decreased protein deposition in most tissues of the body and in its place increased deposition of fat. The physical and physiological effects are increased wrinkling of the skin, diminished rates of function of some of the organs, and diminished muscle mass and strength.

As one ages, the average plasma concentration of growth hormone changes approximately as follows:

	ng/ml
5 to 20 years	6
20 to 40 years	3
40 to 70 years	1.6

Thus, it is highly possible that part of the aging effects in normal older life result from diminished growth hormone secretion. In fact, multiple tests of growth hormone therapy in older people have demonstrated three important effects that suggest antiaging actions: (1) increased protein deposition in the body, especially in the muscles; (2) decreased fat deposits; and (3) a feeling of increased energy and physical drive.

THE POSTERIOR PITUITARY GLAND AND ITS RELATION TO THE HYPOTHALAMUS

The *posterior pituitary gland,* also called the *neurohypophysis,* is composed mainly of glial-like cells called *pituicytes.* The pituicytes do not secrete hormones; they act simply as a supporting structure for large numbers of *terminal nerve fibers* and *terminal nerve endings* from nerve tracts that originate in the *supraoptic* and *paraventricular nuclei* of the hypothalamus, as shown in Figure 75–9. These tracts pass to the neurohypophysis through the *pituitary stalk* (hypophysial stalk). The nerve endings are bulbous knobs that contain many secretory granules that lie on the surfaces of capillaries onto which they secrete the posterior pituitary hormones *antidiuretic hormone* (ADH), also called *vasopressin,* and *oxytocin.*

If the pituitary stalk is cut above the pituitary gland but the entire hypothalamus is left intact, the posterior

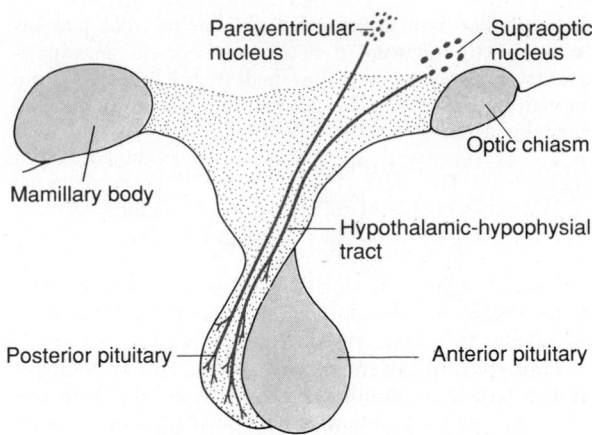

Figure 75–9. Hypothalamic control of the posterior pituitary.

pituitary hormones continue, after a transient decrease for a few days, to be secreted almost normally; they are then secreted by the cut ends of the fibers within the hypothalamus and not by the nerve endings in the posterior pituitary. The reason for this is that the hormones are initially synthesized in the cell bodies of the supraoptic and paraventricular nuclei and are then transported in combination with "carrier" proteins called *neurophysins* down to the nerve endings in the posterior pituitary gland, requiring several days to reach the gland.

ADH is formed primarily in the supraoptic nuclei, whereas oxytocin is formed primarily in the paraventricular nuclei. Each of these nuclei can synthesize about one sixth as much of the second hormone as of its primary hormone.

When nerve impulses are transmitted downward along the fibers from the supraoptic or paraventricular nuclei, the hormone is immediately released from the secretory granules in the nerve endings by the usual secretory mechanism of *exocytosis* and is absorbed into adjacent capillaries. Both the neurophysin and the hormone are secreted together, but because they are only loosely bound to each other, the hormone separates almost immediately. The neurophysin has no known function after leaving the nerve terminals.

Chemical Nature of ADH and Oxytocin

Both oxytocin and ADH (vasopressin) are polypeptides, each containing nine amino acids. Their amino acid sequences are the following:

Vasopressin:
    Cys-Tyr-Phe-Gln-Asn-Cys-Pro-Arg-GlyNH$_2$
Oxytocin:
    Cys-Tyr-Ile-Gln-Asn-Cys-Pro-Leu-GlyNH$_2$

Note that these two hormones are almost identical except that in vasopressin, phenylalanine and arginine replace isoleucine and leucine of the oxytocin molecule.

The similarity of the molecules explains their partial functional similarities.

Physiological Functions of ADH

Extremely minute quantities of ADH—as small as 2 nanograms—when injected into a person can cause antidiuresis, that is, decreased excretion of water by the kidneys. This antidiuretic effect is discussed in detail in Chapter 28. Briefly, in the absence of ADH, the collecting tubules and ducts are almost impermeable to water, which prevents significant reabsorption of water and therefore allows extreme loss of water into the urine, also causing extreme dilution of the urine. On the other hand, in the presence of ADH, the permeability of the collecting ducts and tubules to water increases greatly and allows most of the water to be reabsorbed as the tubular fluid passes through these ducts, thereby conserving water in the body and producing very concentrated urine.

The precise mechanism by which ADH acts on the ducts to increase their permeability is partially known. Without ADH, the luminal membranes of the tubular cells are almost impermeable to water. However, immediately inside the cell membrane are a large number of special vesicles that have highly water-permeable pores. When ADH acts on the cell, it first combines with membrane receptors that cause the formation of cAMP. This in turn causes phosphorylation of elements in the special vesicles, which then causes the vesicles to insert into the apical cell membrane, thus providing many areas of high water permeability. All of this occurs within 5 to 10 minutes. Then, in the absence of ADH, the entire process reverses in another 5 to 10 minutes. Thus, this process temporarily provides many new pores that allow free diffusion of water from the tubular to the peritubular fluids. Water is then absorbed from the collecting tubules and ducts by osmosis, as explained in Chapter 28 in relation to the concentrating mechanism of the kidneys.

Regulation of ADH Production

OSMOTIC REGULATION. When a concentrated electrolyte solution is injected into the artery that supplies the hypothalamus, the ADH neurons in the supraoptic and paraventricular nuclei immediately transmit impulses into the posterior pituitary to release large quantities of ADH into the circulating blood, sometimes increasing its secretion to as high as 20 times normal. Conversely, injection of a dilute solution into this artery causes cessation of the impulses and essentially cessation of ADH secretion. The ADH that is already in the tissues is destroyed at a rate of about one half every 15 to 20 minutes. Thus, the concentration of ADH in the body fluids can change from small amounts to large amounts, or vice versa, in only a few minutes.

The precise way that the osmotic concentration of the extracellular fluids controls ADH secretion is not clear. Yet somewhat in or near the hypothalamus are modified neuron receptors called *osmoreceptors*. When the extracellular fluid becomes too concentrated, fluid is pulled by osmosis out of the osmoreceptor cell, decreasing its size and initiating appropriate nerve signals in the hypothalamus to cause additional ADH secretion. Conversely, when the extracellular fluid becomes too dilute, water moves by osmosis in the opposite direction into the cell, and this decreases the signal for ADH secretion. Although some research workers place these osmoreceptors in the hypothalamus itself (possibly even in the supraoptic nuclei themselves), others believe that they are located in the *organum vasculosum*, a highly vascular structure located in the anteroventral wall of the third ventricle. Regardless of the mechanism, concentrated body fluids do stimulate the supraoptic nuclei, whereas dilute body fluids inhibit them. Therefore, a feedback control system is available to control the total osmotic pressure of the body fluids, operating as follows.

When the body fluids become highly concentrated, the supraoptic nuclei become excited, impulses are transmitted to the posterior pituitary, and ADH is secreted. This passes by way of the blood to the kidneys, where it increases the permeability of the collecting ducts to water. As a result, most of the water is then reabsorbed from the tubular fluid, whereas electrolytes continue to be lost into the urine. This dilutes the extracellular fluid, returning it to a reasonably normal osmotic composition.

Further details on the function of ADH to control renal function and body fluid osmolality are presented in Chapter 28.

Vasoconstrictor and Pressor Effects of ADH, and Increased ADH Secretion Caused by Low Blood Volume

Aside from the effect of minute concentrations of ADH in causing increased water conservation by the kidneys, higher concentrations of ADH have a potent effect of constricting the arterioles everywhere in the body and therefore of increasing the arterial pressure. For this reason, ADH has another name, *vasopressin*.

One of the stimuli for causing intense ADH (vasopressin) secretion is decreased blood volume. This occurs especially strongly when the blood volume decreases 15 to 25 per cent, with the secretory rate then rising sometimes to as high as 50 times normal. The cause of this is the following.

The atria, especially the right atrium, have stretch receptors that are excited by overfilling. When excited, they send signals to the brain to inhibit ADH secretion. Conversely, when unexcited as a result of underfilling, the opposite occurs, with greatly increased ADH secretion. Furthermore, aside from the atrial stretch receptors, decreased stretch of the baroreceptors of the carotid, aortic, and pulmonary regions participates in increasing ADH secretion.

For further details about this blood volume-pressure

feedback mechanism, please refer to Chapter 28, in which the ADH-vasopressin mechanisms are discussed more fully.

Oxytocic Hormone

EFFECT ON THE UTERUS AND ON BIRTH. An oxytocic substance is one that causes contraction of the pregnant uterus. The hormone *oxytocin,* in accordance with its name, powerfully stimulates the pregnant uterus, especially so toward the end of gestation. Therefore, many obstetricians believe that this hormone is at least partially responsible for effecting birth. This is supported by the following facts: (1) In a hypophysectomized animal, the duration of labor is prolonged, thus indicating a possible effect of oxytocin during delivery. (2) The amount of oxytocin in the plasma increases during labor, especially during the last stage. (3) Stimulation of the cervix in a pregnant animal elicits nervous signals that pass to the hypothalamus and cause increased secretion of oxytocin. These effects and this possible mechanism for aiding the birth of the baby are discussed in much more detail in Chapter 82.

EFFECT OF OXYTOCIN ON MILK EJECTION. Oxytocin plays an especially important role in lactation, a role that is far better understood than its role in delivery of the baby. In lactation, it causes milk to be expressed from the alveoli into the ducts so that the baby can obtain it by suckling. This mechanism works as follows: The suckling stimuli on the nipple of the breast cause signals to be transmitted through the sensory nerves to the brain. The signals finally reach the oxytocin neurons in the paraventricular and supraoptic nuclei in the hypothalamus, to cause release of oxytocin by the posterior pituitary gland. The oxytocin then is carried by the blood to the breasts, where it causes contraction of *myoepithelial cells* that lie outside of and form a latticework surrounding the alveoli of the mammary glands. In less than a minute after the beginning of suckling, milk begins to flow. This mechanism is called *milk letdown* or *milk ejection*. This process is discussed further in Chapter 82 in relation to lactation.

REFERENCES

Arky, R. A., and Kettyle, W. M.: Endocrine Pathophysiology: A Problem-Oriented Approach. Philadelphia, J. B. Lippincott, 1994.

Attanasio, A.: Multiple Endocrine Diseases: Growth Hormone Action: Intersexuality. Farmington, CT, S. Karger Publishers, Inc., 1992.

Barrow, D. L., and Selman W.: Neuroendocrinology. Baltimore, Williams & Wilkins, 1992.

Bercu, B. B. (ed.): Basic and Clinical Aspects of Growth Hormone. New York, Plenum Publishing Corp., 1988.

Brown, R. E.: An Introduction to Neuroendocrinology. New York, Cambridge University Press, 1993.

Becker, J. B., et al.: Behavioral Endocrinology. Cambridge, MA, The MIT Press, 1992.

Becker, K. L., et al.: Principles and Practice of Endocrinology and Metabolism. Philadelphia, J. B. Lippincott, 1990.

Bertrand, J., et al.: Pediatric Endocrinology: Physiology, Pathophysiology, Clinical Aspects. Baltimore, Williams & Wilkins, 1993.

Campion, D. R., et al. (eds.): Animal Growth Regulation. New York, Plenum Publishing Corp., 1989.

Collu, R., et al.: Pediatric Endocrinology, 2nd Ed. New York, Raven Press, 1989.

DeGroot, L. J., et al.: Endocrinology. Philadelphia, W. B. Saunders Co., 1994.

Doris, P. A.: Vasopressin and central integrative processes. Neuroendocrinology, 38:75, 1984.

Dubocovish, M. L.: Pharmacology and function of melatonin receptors. FASEB J., 2:2765, 1988.

Eberle, A. N.: The Melanotropins. Farmington, CT, S. Karger Publishers, Inc., 1988.

Felig, P., et al. (eds.): Endocrinology and Metabolism, 2nd Ed. New York, McGraw-Hill, 1987.

Girard, J., and Christiansen, J. S.: hGH Symposium. Farmington, CT, S. Karger Publishers, Inc., 1992.

Greenspan, F. S., and Baxter, J. D.: Basic and Clinical Endocrinology, 4th Ed. Redding, MA, Appleton & Lange, 1994.

Gash, D. M., and Boer, G. J. (eds.): Vasopressin. New York, Plenum Publishing Corp., 1987.

Goodman, H. M.: Basic Medical Endocrinology. New York, Raven Press, 1988.

Harvey, S., et al.: Growth Hormone. Boca Raton, FL, CRC Press, Inc., 1994.

Hughes, J. P., and Friesen, H. G.: The nature and regulation of the receptors for pituitary growth hormone. Annu. Rev. Physiol., 47:469, 1985.

Isaksson, O. G. P., et al.: Mode of action of pituitary growth hormone on target cells. Annu. Rev. Physiol., 47:483, 1985.

Kannan, C. R.: The Pituitary Gland. New York, Plenum Publishing Corp., 1987.

Kolata, G.: New growth industry in human growth hormone? Science, 234:22, 1986.

Kudlow, J. E., et al. (eds.): Biology of Growth Factors. New York, Plenum Publishing Corp., 1988.

Lloyd, R. V.: Surgical Pathology of the Pituitary Gland. Philadelphia, W. B. Saunders Co., 1993.

Menninger, R. P.: Current concepts of volume regulation of vasopressin release. Fed. Proc., 4:55, 1985.

Muller, E. E.: Neural control of somatotrophic function. Physiol. Rev., 67:962, 1987.

North, W. G., et al.: The Neurohypophysis: A Window on Brain Function. New York Academy of Sciences, 1993.

Pierpaoli, W., et al.: The Aging Clock: The Pineal Gland and Other Pacemakers in the Progression of Aging and Carcinogenesis, New York Academy of Sciences, 1994.

Richard, P., et al.: Central effects of oxytocin. Physiol. Rev., 71:331, 1991.

Robbins, R. J., and Melmed, S. (eds.): Acromegaly. New York, Plenum Publishing Corp., 1987.

Saez, J. M.: Growth Factors in Endocrinology: Recent Advances, Therapeutic Prospects. Farmington, CT, S. Karger Publishers, Inc., 1994.

Sara, V. R., et al.: Growth Factors—From Genes to Clinical Application. New York, Raven Press, 1990.

Shiverick, K. T., and Rosenbloom, A. L. Human Growth Hormone Pharmacology: Basic and Clinical Aspects. Boca Raton, FL, CRC Press, Inc. 1995.

Sklar, A. H., and Schrier, R. W.: Central nervous system mediators of vasopressin release. Physiol. Rev., 63:1243, 1983.

Stahnke, N., and Zachmann, M.: Mammalian Cell-Derived Recombinant Human Growth Hormone: Pharmacology, Metabolism, and Clinical Results. Farmington, CT, S. Karger Publishers, Inc., 1992.

Thompson, R. G.: Growth Hormone Therapy in Turner Syndrome. Farmington, CT, S. Karger Publishers, Inc., 1993.

Wilson, J. D., and Foster, D. W.: Williams Textbook of Endocrinology. Philadelphia, W. B. Saunders Co., 1992.

The Thyroid Metabolic Hormones

CHAPTER 76

The thyroid gland, which is located immediately below the larynx on either side of and anterior to the trachea, secretes two significant hormones, *thyroxine* and *triiodothyronine*, commonly called T_4 and T_3, that have the profound effect of increasing the metabolic rate of the body. It also secretes *calcitonin*, an important hormone for calcium metabolism that is considered in detail in Chapter 79. Complete lack of thyroid secretion usually causes the basal metabolic rate to fall 40 to 50 per cent below normal, and extreme excesses of thyroid secretion can cause the basal metabolic rate to rise 60 to 100 per cent above normal. Thyroid secretion is controlled primarily by thyroid-stimulating hormone (TSH) secreted by the anterior pituitary gland.

The purpose of this chapter is to discuss the formation and secretion of the thyroid hormones, their functions in the metabolic scheme of the body, and regulation of their secretion.

FORMATION AND SECRETION OF THE THYROID HORMONES

About 93 per cent of the metabolically active hormones secreted by the thyroid gland is *thyroxine* and 7 per cent is *triiodothyronine*. However, almost all the thyroxine is eventually converted to triiodothyronine in the tissues, so that both are important functionally. The functions of these two hormones are qualitatively the same, but they differ in rapidity and intensity of action. Triiodothyronine is about four times as potent as thyroxine, but it is present in the blood in much

smaller quantities and persists for a much shorter time than does thyroxine.

PHYSIOLOGIC ANATOMY OF THE THYROID GLAND. The thyroid gland is composed, as shown in Figure 76–1, of large numbers of closed *follicles* (100 to 300 micrometers in diameter) filled with a secretory substance called *colloid* and lined with *cuboidal epithelioid cells* that secrete into the interior of the follicles. The major constituent of colloid is the large glycoprotein *thyroglobulin*, which contains the thyroid hormones within its molecule. Once the secretion has entered the follicles, it must be absorbed back through the follicular epithelium into the blood before it can function in the body. The thyroid gland has a blood flow about five times the weight of the gland each minute, which is a blood supply as rich as that of any other area of the body, with the possible exception of the adrenal cortex.

Iodine Requirements for Formation of Thyroxine

To form normal quantities of thyroxine, about 50 milligrams of ingested iodine in the form of iodides are required *each year*, or about *1 mg/week*. To prevent iodine deficiency, common table salt is iodized with about 1 part sodium iodide to every 100,000 parts sodium chloride.

FATE OF INGESTED IODIDES. Iodides ingested orally are absorbed from the gastrointestinal tract into the blood in about the same manner as chlorides. Normally, most of them are rapidly excreted by the kidneys, but only after about one fifth are selectively removed from the circulating blood by the cells of the

945

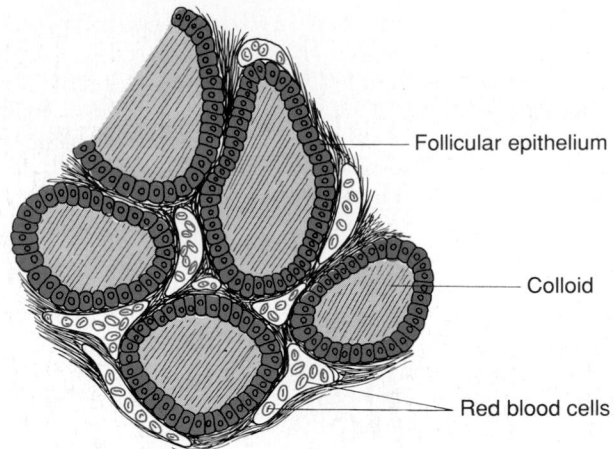

Figure 76–1. Microscopic appearance of the thyroid gland, showing the secretion of the thyroglobulin into the follicles.

Labels: Follicular epithelium; Colloid; Red blood cells

thyroid gland and used for synthesis of the thyroid hormones.

Iodide Pump (Iodide Trapping)

The first stage in the formation of thyroid hormones, shown in Figure 76–2, is transport of iodides from the blood into the thyroid glandular cells and follicles. The basal membrane of the thyroid cell has the specific ability to pump the iodide actively to the interior of the cell. This is called *iodide trapping*. In a normal gland, the iodide pump concentrates the iodide to about 30 times its concentration in the blood. When the thyroid gland becomes maximally active, the concentration ratio can rise to as high as 250 times.

Thyroglobulin and Chemistry of Thyroxine and Triiodothyronine Formation

FORMATION AND SECRETION OF THYROGLOBULIN BY THE THYROID CELLS. The thyroid cells are typical

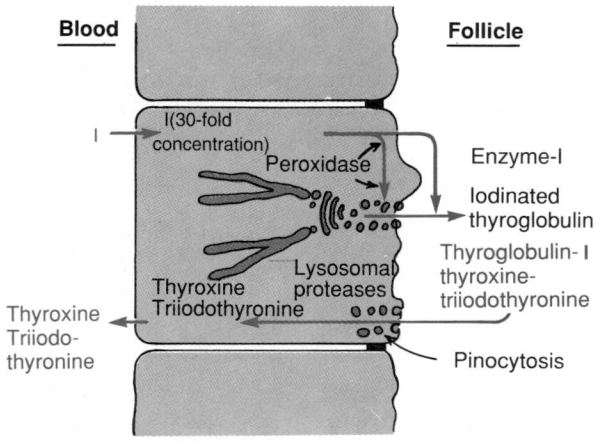

Figure 76–2. Thyroid cellular mechanisms for iodine transport, thyroxine and triiodothyronine formation, and thyroxine and triiodothyronine release into the blood.

protein-secreting glandular cells, as shown in Figure 76–2. The endoplasmic reticulum and Golgi apparatus synthesize and secrete into the follicles a large glycoprotein molecule called *thyroglobulin*, with a molecular weight of about 335,000.

Each molecule of thyroglobulin contains 70 tyrosine amino acids, and they are the major substrates that combine with iodine to form the thyroid hormones, which form *within* the thyroglobulin molecule. That is, the thyroxine and triiodothyronine hormones formed from the tyrosine amino acids remain part of the thyroglobulin molecule during synthesis of the thyroid hormones and even afterward as stored hormones in the follicular colloid.

In addition to secreting the thyroglobulin, the glandular cells process the iodine and provide the enzymes and other substances necessary for thyroid hormone synthesis.

OXIDATION OF THE IODIDE ION. The first essential step in the formation of the thyroid hormones is conversion of the iodide ions to an *oxidized form of iodine*, either nascent iodine (I^0) or I_3^-, that is then capable of combining directly with the amino acid tyrosine.

This oxidation of iodine is promoted by the enzyme *peroxidase* and its accompanying *hydrogen peroxide*, which provide a potent system capable of oxidizing iodides. The peroxidase is located either in the apical membrane of the cell or attached to it, thus providing the oxidized iodine at exactly the point in the cell where the thyroglobulin molecule issues forth from the Golgi apparatus and then through the membrane into the stored colloid. When the peroxidase system is blocked or when it is hereditarily absent from the cells, the rate of formation of thyroid hormones falls to zero.

IODINATION OF TYROSINE AND FORMATION OF THE THYROID HORMONES—"ORGANIFICATION" OF THYROGLOBULIN. The binding of iodine with the thyroglobulin molecule is called *organification* of the thyroglobulin. Oxidized iodine even in the molecular form will bind directly but slowly with the amino acid tyrosine, but in the thyroid cells the oxidized iodine is associated with an *iodinase* enzyme (enzyme-I in Figure 76–2) that causes the process to occur within seconds or minutes. Therefore, almost as rapidly as the thyroglobulin molecule is released from the Golgi apparatus or as it is secreted through the apical cell membrane into the follicle, iodine binds with about one sixth of the tyrosine amino acids within the thyroglobulin molecule.

Figure 76–3 shows the successive stages of iodination of tyrosine and the final formation of the two important thyroid hormones, thyroxine and triiodothyronine. Tyrosine is first iodized to *monoiodotyrosine* and then to *diiodotyrosine*. Then, during the next few minutes, hours, and even days, more and more of the diiodotyrosine residues become *coupled* with one another. The mechanism of the coupling is not understood, but this possibly results from coupling between two adjacent thyroglobulin molecules because the final stored follicular thyroglobulin has a molecular weight

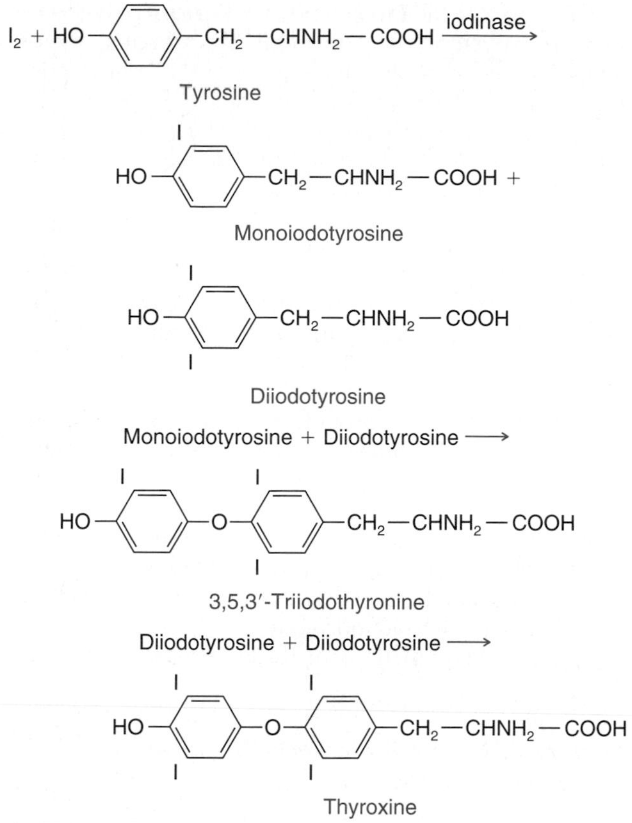

Figure 76–3. Chemistry of thyroxine and triiodothyronine formation.

of about 670,000, twice that of the originally secreted thyroglobulin.

The major hormonal product of the coupling reaction is the molecule *thyroxine* that remains part of the thyroglobulin molecule. Or one molecule of monoiodotyrosine couples with one molecule of diiodotyrosine to form *triiodothyronine*, which represents about one fifteenth of the stored hormone.

STORAGE OF THYROGLOBULIN. After synthesis of the thyroid hormones has run its course, each thyroglobulin molecule contains 1 to 3 thyroxine molecules and an average of 1 triiodothyronine molecule for every 14 molecules of thyroxine. In this form, the thyroid hormones are stored in the follicles in an amount sufficient to supply the body with its normal requirements of thyroid hormones for 2 to 3 months. Therefore, when synthesis of thyroid hormone ceases, the effects of deficiency are not observed for several months.

Release of Thyroxine and Triiodothyronine from the Thyroid Gland

Thyroglobulin itself is not released into the circulating blood in measurable amounts; instead, thyroxine and triiodothyronine are first cleaved from the thyroglobulin molecule, and then these free hormones are released. This process occurs as follows: The apical surface of the thyroid cells sends out pseudopod extensions that close around small portions of the colloid to form *pinocytic vesicles* that enter the apex of the thyroid cell. Then *lysosomes* immediately fuse with these vesicles to form digestive vesicles that contain the digestive enzymes from the lysosomes mixed with the colloid. The *proteinases* among these enzymes digest the thyroglobulin molecules and release the thyroxine and triiodothyronine, which then diffuse through the base of the thyroid cell into the surrounding capillaries. Thus, the thyroid hormones are released into the blood.

About three quarters of the iodinated tyrosine in the thyroglobulin never becomes thyroid hormones but remains monoiodotyrosine and diiodotyrosine. During the digestion of the thyroglobulin molecule to cause release of thyroxine and triiodothyronine, these iodinated tyrosines also are freed from the thyroglobulin molecules. However, they are not secreted into the blood. Instead, their iodine is cleaved from them by a *deiodinase enzyme* that makes virtually all of this iodine available again for recycling within the gland for forming additional thyroid hormones. In the congenital absence of this deiodinase enzyme, many people become iodine-deficient because of failure of this recycling process.

DAILY RATE OF SECRETION OF THYROXINE AND TRIIODOTHYRONINE. About 93 per cent of the thyroid hormone released from the thyroid gland is normally thyroxine and only 7 per cent is triiodothyronine. However, during the ensuing few days, most of the thyroxine is slowly deiodinated to form additional triiodothyronine. Therefore, the hormone finally delivered to and used by the tissues is mainly triiodothyronine, a total of about 35 micrograms of triiodothyronine per day. (Another 35 micrograms of so-called reverse triiodothyronine is formed each day by removal of one of the iodines of thyroxine from the wrong point on the molecule, that is, from near the carboxyl end instead of the hydroxyl end. This reverse triiodothyronine is almost totally inactive, and it is eventually destroyed.)

Transport of Thyroxine and Triiodothyronine to the Tissues

BINDING OF THYROXINE AND TRIIODOTHYRONINE WITH PLASMA PROTEINS. On entering the blood, all but a fraction of 1 per cent of the thyroxine and triiodothyronine combine immediately with several of the plasma proteins. They combine mainly with *thyroxine-binding globulin,* but much less so with *thyroxine-binding prealbumin* and *albumin.*

SLOW RELEASE OF THYROXINE AND TRIIODOTHYRONINE TO THE TISSUE CELLS. Because of the high affinity of the plasma-binding proteins for the thyroid hormones, these substances—in particular, thyroxine—are released to the tissue cells slowly. Half the thyroxine in the blood is released to the tissue cells about every 6 days, whereas half the triiodothyronine—because of its lower affinity—is released to the cells in about 1 day.

On entering the tissue cells, both of these hormones

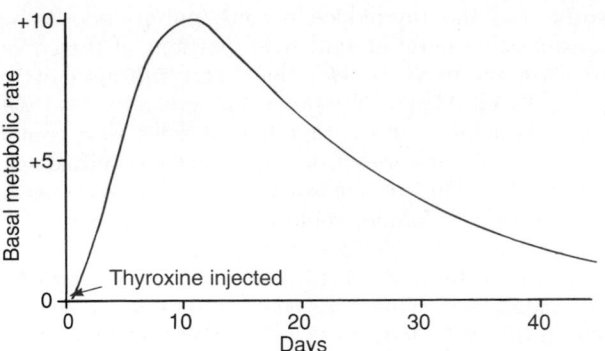

Figure 76–4. Approximate prolonged effect on the basal metabolic rate caused by administering a single large dose of thyroxine.

again bind with intracellular proteins, the thyroxine once again binding more strongly than the triiodothyronine. Therefore, they are again stored, but this time in the functional cells themselves, and they are used slowly over a period of days or weeks.

LATENCY AND DURATION OF ACTION OF THE THYROID HORMONES. After injection of a large quantity of thyroxine into a human being, essentially no effect on the metabolic rate can be discerned for 2 to 3 days, thereby demonstrating that there is a *long latent* period before thyroxine activity begins. Once activity does begin, it increases progressively and reaches a maximum in 10 to 12 days, as shown in Figure 76–4. Thereafter, it decreases with a half-life of about 15 days. Some of the activity persists as long as 6 weeks to 2 months later.

The actions of triiodothyronine occur about four times as rapidly as those of thyroxine, with the latent period as short as 6 to 12 hours and maximum cellular activity occurring within 2 to 3 days.

Most of the latency and prolonged period of action of these hormones is probably caused by their binding with proteins both in the plasma and in the tissue cells, followed by their slow release. However, we shall see in subsequent discussions that part of the latent period also results from the manner in which these hormones perform their functions in the cells themselves.

FUNCTIONS OF THE THYROID HORMONES IN THE TISSUES

The Thyroid Hormones Increase the Transcription of Large Numbers of Genes

The general effect of thyroid hormone is to cause nuclear transcription of large numbers of genes. Therefore, in virtually all cells of the body, great numbers of protein enzymes, structural proteins, transport proteins, and other substances increase. The net result of all this is a generalized increase in functional activity throughout the body.

CONVERSION OF THYROXINE TO TRIIODOTHYRONINE AND ACTIVATION OF NUCLEAR RECEPTORS. Before acting on the genes to increase genetic transcription, almost all the thyroxine is deiodinated by one iodide ion, thus forming triiodothryonine. This in turn has a very high binding affinity to the intracellular thyroid hormone receptors. Consequently, about 90 per cent of the thyroid hormone molecules that bind with the receptors is triiodothryonine, and only 10 per cent, thyroxine.

The thyroid hormone receptors are either attached to the DNA genetic strands or in proximity to them. On binding with thyroid hormone, the receptors become activated and initiate the transcription process. Then large numbers of different types of messenger RNA are formed, followed within another few minutes and hours by RNA translation on the cytoplasmic ribosomes to form hundreds of new types of proteins. However, not all proteins are increased by similar percentages—some only slightly and others at least as much as sixfold. It is believed that most, if not all, of the actions of thyroid hormone result from the enzymatic and other functions of these new proteins.

Important Types of Increased Cellular Metabolic Activity

The thyroid hormones increase the metabolic activities of all or almost all the tissues of the body. The basal metabolic rate can increase to 60 to 100 per cent above normal when large quantities of the hormones are secreted. The rate of utilization of foods for energy is greatly accelerated. Although the rate of protein synthesis is increased, at the same time the rate of protein catabolism is also increased. The growth rate of young people is greatly accelerated. The mental processes are excited, and the activity of most of the endocrine glands is increased.

EFFECT OF THYROID HORMONES ON MITOCHONDRIA. When thyroxine or triiodothyronine is given to an animal, the mitochondria in most cells of the body increase in size as well as number. Furthermore, the total membrane surface area of the mitochondria increases almost directly in proportion to the increased metabolic rate of the whole animal. Therefore, it seems almost to be an obvious deduction that one of the principal functions of thyroxine might be simply to increase the number and activity of mitochondria, and they in turn increase the rate of formation of adenosine triphosphate (ATP) to energize cellular function. However, the increase in the number and activity of mitochondria could be the *result* of increased activity of the cells as well as the cause of the increase.

When *extremely* high concentrations of thyroid hormone are administered, the mitochondria swell inordinately, and there is then uncoupling of the oxidative phosphorylation process with production of large amounts of heat but little ATP. Under natural conditions, it is questionable whether the concentration of thyroid hormones ever becomes high enough to cause this effect even in people who have thyrotoxicosis.

EFFECT OF THYROID HORMONE IN INCREASING ACTIVE TRANSPORT OF IONS THROUGH CELL MEMBRANES. One of the enzymes that becomes increased in response to thyroid hormone is *Na,K-ATPase.* This in turn increases the rate of transport of both sodium and potassium through the cell membranes of some tissues. Because this process uses energy and increases the amount of heat produced in the body, it has been suggested that this might be one of the mechanisms by which thyroid hormone increases the body's metabolic rate. In fact, thyroid hormone also causes the cell membranes of most cells to become leaky to sodium ions, which further activates the sodium pump and further increases heat production.

Effect of Thyroid Hormone on Growth

Thyroid hormone has both general and specific effects on growth. For instance, it has long been known that thyroid hormone is essential for the metamorphic change of the tadpole into the frog. In the human being, the effect of thyroid hormone on growth is manifest mainly in growing children. In those who are hypothyroid, the rate of growth is greatly retarded. In those who are hyperthyroid, excessive skeletal growth often occurs, causing the child to become considerably taller at an earlier age. However, the bones also mature more rapidly and the epiphyses close at an early age, so that the duration of growth and the eventual height of the adult may actually be shortened.

An important effect of thyroid hormone is to promote growth and development of the brain during fetal life and for the first few years of postnatal life. If the fetus does not secrete sufficient quantities of thyroid hormone, growth and maturation of the brain both before birth and afterward are greatly retarded and the brain remains smaller than normal. Without specific thyroid therapy within days or weeks after birth, the child without a thyroid gland will remain mentally deficient throughout life. This is discussed more fully later in the chapter.

Effects of Thyroid Hormone on Specific Bodily Mechanisms

EFFECT ON CARBOHYDRATE METABOLISM. Thyroid hormone stimulates almost all aspects of carbohydrate metabolism, including rapid uptake of glucose by the cells, enhanced glycolysis, enhanced gluconeogenesis, increased rate of absorption from the gastrointestinal tract, and even increased insulin secretion with its resultant secondary effects on carbohydrate metabolism. All these effects probably result from the overall increase in cellular metabolic enzymes caused by thyroid hormone.

EFFECT ON FAT METABOLISM. Essentially all aspects of fat metabolism are also enhanced under the influence of thyroid hormone. Because fats are the major source of long-term energy supplies, the fat stores of the body are depleted to a greater extent than are most of the other tissue elements. In particular, lipids are mobilized from the fat tissue, which increases the free fatty acid concentration in the plasma; thyroid hormone also greatly accelerates the oxidation of free fatty acids by the cells.

Effect on Plasma and Liver Fats. Increased thyroid hormone *decreases* the quantity of cholesterol, phospholipids, and triglycerides in the plasma, even though it *increases* the free fatty acids. Conversely, *decreased* thyroid secretion greatly increases the plasma concentrations of cholesterol, phospholipids, and triglycerides and almost always causes excessive deposition of fat in the liver as well. The large increase in circulating plasma cholesterol in prolonged hypothyroidism is often associated with severe arteriosclerosis, discussed in Chapter 68.

One of the mechanisms by which thyroid hormone decreases the plasma cholesterol concentration is to increase significantly the rate of cholesterol secretion in the bile and consequent loss in the feces. A possible mechanism for the increased cholesterol secretion is that thyroid hormone induces increased numbers of low-density lipoprotein receptors on the liver cells, leading to rapid removal of low-density lipoproteins from the plasma and subsequent secretion of the lipoprotein cholesterol by the liver cells.

EFFECT ON VITAMIN METABOLISM. Because thyroid hormone increases the quantities of many of the enzymes and because vitamins are essential parts of some of the enzymes or coenzymes, thyroid hormone causes increased need for vitamins. Therefore, a relative vitamin deficiency can occur when excess thyroid hormone is secreted, unless at the same time increased quantities of vitamins are available.

EFFECT ON BASAL METABOLIC RATE. Because thyroid hormone increases metabolism in almost all cells of the body, excessive quantities of the hormone can occasionally increase the basal metabolic rate to 60 to 100 per cent above normal. On the other hand, when no thyroid hormone is produced, the basal metabolic rate falls almost to one half normal; that is, the basal metabolic rate becomes -30 to -50, as discussed in Chapter 72. Figure 76–5 shows the approximate relation between the daily supply of thyroid hormones and the basal metabolic rate. Extreme amounts of the hormones are required to cause high basal metabolic rates.

EFFECT ON BODY WEIGHT. Greatly increased thyroid hormone almost always decreases the body weight, and greatly decreased hormone almost always increases the body weight; these effects do not always occur because thyroid hormone increases the appetite, and this may overbalance the change in the metabolic rate.

EFFECT ON THE CARDIOVASCULAR SYSTEM

Blood Flow and Cardiac Output. Increased metabolism in the tissues causes more rapid utilization of

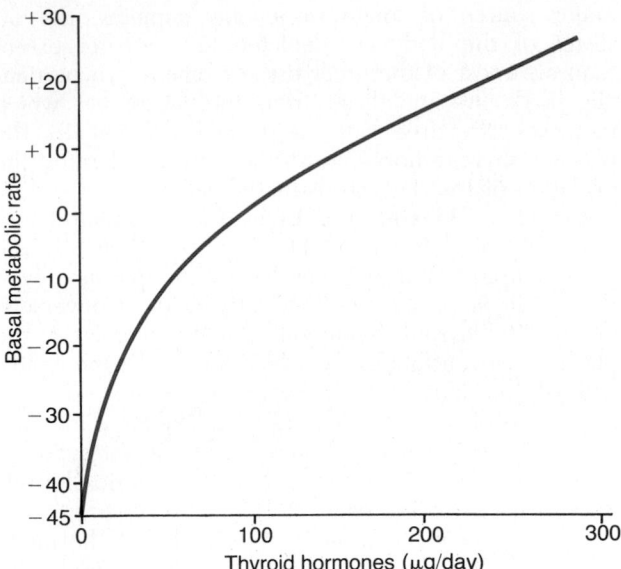

Figure 76-5. Approximate relation of thyroid hormone (T$_4$ and T$_3$) daily rate of secretion to the basal metabolic rate.

oxygen than normal and greater than normal quantities of metabolic end products to be released from the tissues. These effects cause vasodilatation in most of the body tissues, thus increasing blood flow. The rate of blood flow in the skin especially increases because of the increased need for heat elimination.

As a consequence of the increased blood flow, cardiac output also increases, sometimes rising to 60 per cent or more above normal when excessive thyroid hormone is present and falling to only 50 per cent of normal in severe hypothyroidism.

Heart Rate. The heart rate increases considerably more under the influence of thyroid hormone than would be expected from the increase in cardiac output. Therefore, thyroid hormone probably has a direct effect on the excitability of the heart, which in turn increases the heart rate. This effect is of particular importance because the heart rate is one of the sensitive physical signs that the clinician uses in determining whether a patient has excessive or diminished thyroid hormone production.

Strength of Heartbeat. The increased enzymatic activity caused by increased thyroid hormone production apparently increases the strength of the heart when only a slight excess of thyroid hormone is secreted. This is analogous to the increase in strength of heartbeat that occurs in mild fevers and during exercise. However, when thyroid hormone is increased markedly, the heart muscle strength becomes depressed because of excessive protein catabolism. Indeed, some severely thyrotoxic patients die of cardiac decompensation secondary to myocardial failure and increased cardiac load imposed by the increased output.

Blood Volume. Thyroid hormone causes the blood volume to increase slightly. The effect probably results at least partly from vasodilatation, which allows increased quantities of blood to collect in the circulatory system.

Arterial Pressure. The *mean* arterial pressure usually is unchanged. However, because of the increased blood flow through the tissues between heartbeats, the pulse pressure is often increased, with the systolic pressure elevated in hyperthyroidism 10 to 15 mm Hg and the diastolic pressure reduced in a corresponding manner.

EFFECT ON RESPIRATION. The increased rate of metabolism increases the utilization of oxygen and the formation of carbon dioxide; these effects activate all the mechanisms that increase the rate and depth of respiration.

EFFECT ON THE GASTROINTESTINAL TRACT. In addition to increased appetite and food intake, which has been discussed, thyroid hormone increases both the rate of secretion of the digestive juices and the motility of the gastrointestinal tract. Diarrhea often results. Lack of thyroid hormone can cause constipation.

EFFECT ON THE CENTRAL NERVOUS SYSTEM. In general, thyroid hormone increases the rapidity of cerebration but also often dissociates this; on the other hand, lack of thyroid hormone decreases this function. The hyperthyroid individual is likely to have extreme nervousness and many psychoneurotic tendencies, such as anxiety complexes, extreme worry, and paranoia.

EFFECT ON THE FUNCTION OF THE MUSCLES. Slight increase in thyroid hormone usually makes the muscles react with vigor, but when the quantity of hormone becomes excessive, the muscles become weakened because of excess protein catabolism. On the other hand, lack of thyroid hormone causes the muscles to become sluggish, and they relax slowly after a contraction.

Muscle Tremor. One of the most characteristic signs of hyperthyroidism is a fine muscle tremor. This is not the coarse tremor that occurs in Parkinson's disease or in shivering because it occurs at the rapid frequency of 10 to 15 times per second. The tremor can be observed easily by placing a sheet of paper on the extended fingers and noting the degree of vibration of the paper. This tremor is believed to be caused by increased reactivity of the neuronal synapses in the areas of the cord that control muscle tone. The tremor is an important means for assessing the degree of thyroid hormone effect on the central nervous system.

EFFECT ON SLEEP. Because of the exhausting effect of thyroid hormone on the musculature and on the central nervous system, the hyperthyroid subject often has a feeling of constant tiredness, but because of the excitable effects of thyroid hormone on the synapses, it is difficult to sleep. On the other hand, extreme somnolence is characteristic of hypothyroidism, with sleep sometimes lasting 12 to 14 hours a day.

EFFECT ON OTHER ENDOCRINE GLANDS. Increased thyroid hormone increases the rates of secretion of most other endocrine glands, but it also increases the need of the tissues for the hormones. For instance, increased thyroxine secretion increases the rate of glucose metabolism everywhere in the body and therefore causes a corresponding need for increased insulin secretion by the pancreas. Also, thyroid hormone increases many metabolic activities related to bone formation and, as a consequence, increases the need for parathyroid hormone. Finally, thyroid hormone increases the rate at which adrenal glucocorticoids are inactivated by the liver. This leads to feedback increase in adrenocorticotropic hormone production by the anterior pituitary and, therefore, increased rate of glucocorticoid secretion by the adrenal glands.

EFFECT OF THYROID HORMONE ON SEXUAL FUNCTION. For normal sexual function, thyroid secretion needs to be approximately normal. In men, lack of thyroid hormone is likely to cause loss of libido; on the other hand, great excesses of the hormone frequently cause impotence. In women, lack of thyroid hormone often causes *menorrhagia* and *polymenorrhea,* that is, respectively, excessive and frequent menstrual bleeding. Yet, strangely enough, in other women thyroid lack may cause irregular periods and occasionally even *amenorrhea.* A hypothyroid woman, like a man, is likely to have greatly decreased libido. To make the picture still more confusing, in the hyperthyroid woman, *oligomenorrhea,* which means greatly reduced bleeding, is common, and occasionally amenorrhea results.

The action of thyroid hormone on the gonads cannot be pinpointed to a specific function but probably results from a combination of direct metabolic effects on the gonads and excitatory and inhibitory feedback effects operating through the anterior pituitary hormones that control the sexual functions.

REGULATION OF THYROID HORMONE SECRETION

To maintain normal levels of metabolic activity in the body, precisely the right amount of thyroid hormone must be secreted at all times; to provide this, specific feedback mechanisms operate through the hypothalamus and anterior pituitary gland to control the rate of thyroid secretion. These mechanisms can be explained as follows:

EFFECTS OF TSH (FROM THE ANTERIOR PITUITARY GLAND) ON THYROID SECRETION. TSH, also known as *thyrotropin,* is an anterior pituitary hormone, a glycoprotein with a molecular weight of about 28,000, discussed in Chapter 74; it increases the secretion of thyroxine and triiodothyronine by the thyroid gland. Its specific effects on the thyroid gland are as follows:

1. Increased proteolysis of the thyroglobulin that has already been stored in the follicles, with the resultant release of the thyroid hormones into the circulating blood and diminishment of the follicular substance itself
2. Increased activity of the iodide pump, which increases the rate of "iodide trapping" in the glandular cells, sometimes increasing the ratio of intracellular to extracellular iodide concentration to as much as eight times normal
3. Increased iodination of tyrosine and increased coupling to form the thyroid hormones

4. Increased size and increased secretory activity of the thyroid cells
5. Increased number of thyroid cells plus a change from cuboidal to columnar cells and much infolding of the thyroid epithelium into the follicles

In summary, TSH *increases all the known secretory activities of the thyroid glandular cells.*

The most important early effect after administration of TSH is almost immediate beginning proteolysis of the thyroglobulin, which causes release of thyroxine and triiodothyronine into the blood within 30 minutes. The other effects require hours or even days and weeks to develop fully.

ROLE OF CYCLIC ADENOSINE MONOPHOSPHATE IN THE STIMULATORY EFFECT OF TSH. In the past, it was difficult to explain the many and varied effects of TSH on the thyroid cell. It is now clear that most, if not all, of these effects result from activation of the "second messenger" *cyclic adenosine monophosphate (cAMP)* system of the cell. The first event in this activation is binding of TSH with specific TSH receptors on the basal membrane surfaces of the thyroid cell. This then activates *adenylcyclase* in the membrane, which increases the formation of cAMP in the cell. Finally, the cAMP acts as a *second messenger* to activate protein kinase, which causes multiple phosphorylations throughout the cell. The result is both an immediate increase in secretion of thyroid hormones and prolonged growth of the thyroid glandular tissue itself. This method for control of thyroid cell activity is similar to the function of cAMP in many other target tissues of the body.

TSH Secretion Is Regulated by Thyrotropin-Releasing Hormone from the Hypothalamus

Anterior pituitary secretion of TSH is controlled by a hypothalamic hormone, *thyrotropin-releasing hormone* (TRH), which is secreted by nerve endings in the median eminence of the hypothalamus and then transported from there to the anterior pituitary in the hypothalamic-hypophysial portal blood, as explained in Chapter 74. The precise nuclei of the hypothalamus that are responsible for causing TRH secretion in the median eminence are not known. However, injection of radioactive antibodies that attach specifically to TRH have shown this hormone to be present in many different hypothalamic loci, including the (1) *dorsomedial nucleus,* (2) *suprachiasmatic nucleus,* (3) *ventromedial nucleus,* (4) *anterior hypothalamus,* (5) *preoptic area,* and (6) *paraventricular nucleus.*

TRH has been obtained in pure form. It is a simple substance, a tripeptide amide—*pyroglutamyl-histidyl-proline-amide.* TRH directly affects the anterior pituitary gland cells to increase their output of TSH. When the portal system from the hypothalamus to the anterior pituitary gland is blocked, the rate of secretion of TSH by the anterior pituitary is greatly decreased but not reduced to zero.

The molecular mechanism by which TRH causes the

TSH-secreting cells of the anterior pituitary to produce TSH is first to bind with TRH receptors in the pituitary cell membrane. This in turn *activates the phospholipase second messenger* system to produce large amounts of phospholipase C, followed by many other second messenger products, including calcium ions and diacyl glycerol, which eventually leads to TSH release.

EFFECTS OF COLD AND OTHER NEUROGENIC STIMULI ON TRH AND TSH SECRETION. One of the best-known stimuli for increasing the rate of TRH secretion by the hypothalamus, and therefore TSH secretion by the anterior pituitary, is exposure of an animal to cold. This effect almost certainly results from excitation of the hypothalamic centers for body temperature control. Exposure of rats for several weeks to severe cold increases the output of thyroid hormones sometimes more than 100 per cent of normal and can increase the basal metabolic rate as much as 50 per cent. Indeed, people moving to arctic regions have been known to develop basal metabolic rates 15 to 20 per cent above normal; however, the behavioral propensity of human beings to protect themselves from cold usually prevents much of this effect.

Various emotional reactions can also affect the output of TRH and TSH and therefore indirectly affect the secretion of thyroid hormones. On the other hand, excitement and anxiety—conditions that greatly stimulate the sympathetic nervous system—cause acute decrease in secretion of TSH, perhaps because these states increase the metabolic rate and the body heat and exert an inverse effect on the heat control center.

Neither these emotional effects nor the effect of cold is observed after the hypophysial stalk has been cut, demonstrating that both these effects are mediated by way of the hypothalamus.

Feedback Effect of Thyroid Hormone in Decreasing Anterior Pituitary Secretion of TSH—Feedback Regulation of Thyroid Secretion

Increased thyroid hormone in the body fluids decreases the secretion of TSH by the anterior pituitary. When the rate of thyroid hormone secretion rises to about 1.75 times normal, the rate of TSH secretion falls essentially to zero. Almost all this feedback depressant effect occurs even when the anterior pituitary has been separated from the hypothalamus. Therefore, as shown in Figure 76–6, it is probable that increased thyroid hormone inhibits anterior pituitary secretion of TSH mainly by a direct effect on the anterior pituitary itself, although perhaps secondarily by much weaker effects acting through the hypothalamus.

Regardless of the mechanism of the feedback, its effect is to maintain an almost constant concentration of free thyroid hormones in the circulating body fluids.

If there is a feedback effect through the hypothalamus in addition to the direct feedback to the pituitary gland itself, this probably operates slowly and might be caused at least partly by changes in body heat and its effect on the body temperature–controlling centers of

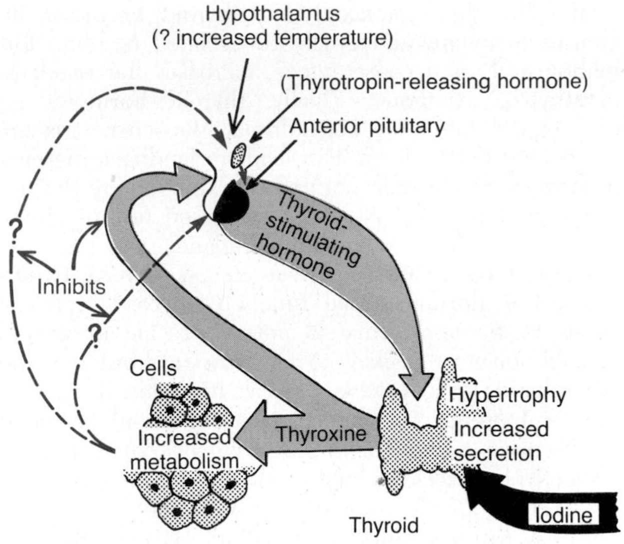

Figure 76–6. Regulation of thyroid secretion.

the hypothalamus, which are known to have a significant effect in controlling the thyroid hormone system.

Antithyroid Substances

Drugs that suppress thyroid secretion are called antithyroid substances. The best known of these substances are thiocyanate, propylthiouracil, and high concentrations of inorganic iodides. The mechanism by which each of these blocks thyroid secretion is different from the others, and they can be explained as follows.

DECREASED IODIDE TRAPPING CAUSED BY THIOCYANATE IONS. The same active pump that pumps iodide ions into the thyroid cells can also pump thiocyanate ions, perchlorate ions, and nitrate ions. Therefore, the administration of thiocyanate (or one of the other ions as well) in high enough concentration can cause competitive inhibition of iodide transport into the cell, that is, inhibition of the iodide-trapping mechanism.

The decreased availability of iodide in the glandular cells does not stop the formation of thyroglobulin, it merely prevents the thyroglobulin that is formed from becoming iodinated and therefore forming the thyroid hormones. This deficiency of the thyroid hormones in turn leads to increased secretion of TSH by the anterior pituitary gland, which causes overgrowth of the thyroid gland even though the gland still does not form adequate quantities of thyroid hormones. Therefore, the use of thiocyanates and some other ions to block thyroid secretion can lead to the development of an enlarged thyroid gland, which is called a *goiter.*

DEPRESSION OF THYROID HORMONE FORMATION BY PROPYLTHIOURACIL. Propylthiouracil (and other, similar compounds, such as methimazole and carbimazole) prevents formation of thyroid hormone from iodides and tyrosine. The mechanism of this is partly to block the peroxidase enzyme that is required for iodination of tyrosine and partly to block the coupling of two iodinated tyrosines to form thyroxine or triiodothyronine.

Propylthiouracil, like thiocyanate, does not prevent formation of thyroglobulin. Therefore, the absence of thyroxine and triiodothyronine in the thyroglobulin can lead to tremendous feedback enhancement of TSH secretion by the anterior pituitary gland, thus promoting growth of the glandular tissue and forming a goiter.

DECREASE IN THYROID ACTIVITY AND THYROID GLAND SIZE CAUSED BY IODIDES. When iodides are present in the blood *in high concentration* (100 times the normal plasma level), most activities of the thyroid gland are decreased, but often they remain decreased for only a few weeks. The effect is to reduce the rate of iodide trapping, and the rate of iodination of tyrosine to form thyroid hormones is also decreased. Even more important, the normal endocytosis of colloid from the follicles by the thyroid glandular cells is paralyzed by the high iodide concentrations. Because this is the first step in release of the thyroid hormones from the storage colloid, there is almost immediate shutdown of thyroid hormone secretion into the blood.

Because iodides in high concentrations decrease all phases of thyroid activity, they slightly decrease the size of the thyroid gland and especially decrease its blood supply, in contradistinction to the opposite effects caused by most of the other antithyroid agents. For this reason, iodides are frequently administered to patients for 2 to 3 weeks before surgical removal of the thyroid gland to decrease the necessary amount of surgery, especially to decrease the amount of bleeding.

DISEASES OF THE THYROID

Hyperthyroidism

Most effects of hyperthyroidism are obvious from the preceding discussion of the various physiological effects of thyroid hormone. However, some specific effects should be mentioned in connection especially with the development, diagnosis, and treatment of hyperthyroidism.

CAUSES OF HYPERTHYROIDISM (TOXIC GOITER, THYROTOXICOSIS, GRAVES' DISEASE). In most patients with hyperthyroidism, the thyroid gland is increased to two to three times normal size, with tremendous hyperplasia and infolding of the follicular cell lining into the follicles, so that the number of cells is increased several more times than the size of the gland is increased. Also, each cell increases its rate of secretion severalfold; radioactive iodine uptake studies indicate that some of these hyperplastic glands secrete thyroid hormone at rates 5 to 15 times normal.

The changes in the thyroid gland are similar to those caused by excessive TSH. However, radioimmunoassay studies have shown the plasma TSH concentrations to be less than normal rather than enhanced in almost all patients and often to be essentially zero. On the other hand, other substances that have actions similar to those of TSH are found in the blood of almost all these patients. These substances are immunoglobulin antibodies that bind with the same membrane receptors that bind TSH. They induce continual activation of the cAMP system of the cells, with resultant development of hyperthyroidism. These antibodies are called *thyroid-stimulating immunoglobulin* and are designated "TSI." They have a prolonged stimulating effect on the thyroid gland, lasting for as long as 12 hours in contrast to a little over 1 hour for TSH. The high level of thyroid hormone secretion caused by TSI in turn suppresses anterior pituitary formation of TSH.

The antibodies that cause hyperthyroidism almost certainly develop as the result of autoimmunity that has developed against thyroid tissue. Presumably, at some time in the history of the person, an excess of thyroid cell antigens has been released from the thyroid cells, and this has resulted in the formation of antibodies against the thyroid gland itself.

Thyroid Adenoma. Hyperthyroidism occasionally results from a localized adenoma (a tumor) that develops in the thyroid tissue and secretes large quantities of thyroid hormone. This is different from the more usual type of hyperthyroidism in that it usually is not associated with evidence of autoimmune disease. An interesting effect of the adenoma is that as long as it continues to secrete large quantities of thyroid hormone, function in the remainder of the thyroid gland is almost totally inhibited because the thyroid hormone from the adenoma depresses the production of TSH by the pituitary gland.

Symptoms of Hyperthyroidism

The symptoms of hyperthyroidism are obvious from the preceding discussion of the physiology of the thyroid hormones: a high state of excitability, intolerance to heat, increased sweating, mild to extreme weight loss (sometimes as much as 100 pounds), varying degrees of diarrhea, muscle weakness, nervousness or other psychic disorders, extreme fatigue but inability to sleep, and tremor of the hands.

EXOPHTHALMOS. Most people with hyperthyroidism develop some degree of protrusion of the eyeballs, as shown in Figure 76–7. This condition is called *exophthalmos*. A major degree of exophthalmos occurs in about one third of hyperthyroid patients, and the condition sometimes becomes severe enough that the eyeball protrusion stretches the optic nerve enough to damage vision. Much more often the eyes are damaged because the eyelids do not close completely when the person blinks or is asleep. As a result, the epithelial surfaces of the eyes become dry and irritated and often infected, resulting in ulceration of the cornea.

The cause of the protruding eyes is edematous swelling of the retro-orbital tissues and degenerative changes

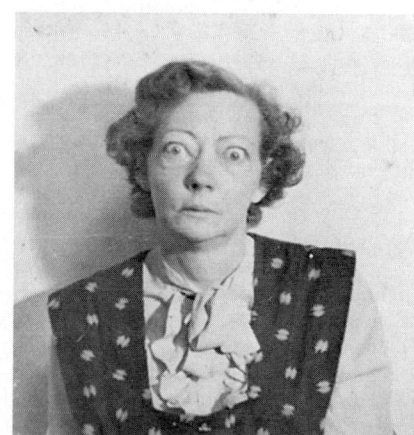

Figure 76–7. Patient with exophthalmic hyperthyroidism. Note protrusion of the eyes and retraction of the superior eyelids. The basal metabolic rate was +40. (Courtesy of Dr. Leonard Posey.)

in the extraocular muscles. The factor or factors that initiate these changes are in serious dispute. In most patients, immunoglobulins can be found in the blood that react with the eye muscles. Furthermore, the concentration of these immunoglobulins is usually highest in patients who have high concentrations of TSIs. Therefore, there is much reason to believe that exophthalmos, like hyperthyroidism itself, is an autoimmune process. The exophthalmos usually is greatly ameliorated with treatment of the hyperthyroidism.

DIAGNOSTIC TESTS FOR HYPERTHYROIDISM. For the usual case of hyperthyroidism, the most accurate diagnostic test is direct measurement of the concentration of "free" thyroxine (and sometimes triiodothyronine) in the plasma, using appropriate radioimmunoassay procedures.

Other tests that are sometimes used are as follows:

1. The basal metabolic rate is usually increased to $+30$ to $+60$ in severe hyperthyroidism.

2. The concentration of TSH in the plasma is measured by radioimmunoassay. In the usual type of thyrotoxicosis, the anterior pituitary secretion of TSH is so completely suppressed by the large amounts of circulating thyroxine and triiodothyronine that there is almost no plasma TSH.

3. The concentration of TSI is measured by radioimmunoassay. This is normally high in the usual type of thyrotoxicosis but low in thyroid adenoma.

PHYSIOLOGY OF TREATMENT IN HYPERTHYROIDISM. The most direct treatment for hyperthyroidism is surgical removal of most of the thyroid gland. In general, it is desirable to prepare the patient for surgical removal of the gland before the operation. This is done by administering propylthiouracil, usually for several weeks, until the basal metabolic rate of the patient has returned to normal. Then, the administration of high concentrations of iodides for 1 to 2 weeks immediately before operation causes the gland itself to recede in size and its blood supply to diminish. By using these preoperative procedures, the operative mortality is less than 1 in 1000 in the better hospitals, whereas before development of modern procedures, the operative mortality was 1 in 25.

Treatment of the Hyperplastic Thyroid Gland with Radioactive Iodine. Eighty to 90 per cent of an injected dose of iodide is absorbed by the hyperplastic, toxic thyroid gland within 1 day after injection. If this injected iodine is radioactive, it can destroy internally the secretory cells of the thyroid gland. Usually 5 millicuries of radioactive iodine is given to the patient, whose condition is reassessed several weeks later. If the patient is still hyperthyroid, additional doses are administered until normal thyroid status is reached.

Hypothyroidism

The effects of hypothyroidism in general are opposite to those of hyperthyroidism, but here again, a few physiological mechanisms peculiar to hypothyroidism alone are involved.

Hypothyroidism, like hyperthyroidism, probably also results in most instances from autoimmunity against the thyroid gland, but immunity that destroys the gland rather than stimulates it. The thyroid glands of most of these patients first have "thyroiditis," which means thyroid inflammation. This causes progressive deterioration

and finally fibrosis of the gland, with resultant diminished or absent secretion of thyroid hormone. Several other types of hypothyroidism also occur, often associated with development of enlarged thyroid glands, called thyroid goiter, as follows:

ENDEMIC COLLOID GOITER. The term "goiter" means a greatly enlarged thyroid gland. As pointed out in the discussion of iodine metabolism, about 50 milligrams of iodine is necessary each year for the formation of adequate quantities of thyroid hormone. In certain areas of the world, notably in the Swiss Alps, the Andes, and the Great Lakes region of the United States, insufficient iodine is present in the soil for the foodstuffs to contain even this minute quantity of iodine. Therefore, in the days before iodized table salt, many people who lived in these areas developed extremely large thyroid glands, called *endemic goiters*.

The mechanism for development of the large endemic goiters is the following: Lack of iodine prevents production of both thyroxine and triiodothyronine but does not stop the formation of thyroglobulin. As a result, no hormone is available to inhibit production of TSH by the anterior pituitary; this allows the pituitary to secrete excessively large quantities of TSH. The TSH then causes the thyroid cells to secrete tremendous amounts of thyroglobulin (colloid) into the follicles, and the gland grows larger and larger. But because of lack of iodine, thyroxine and triiodothyronine production still does not occur within the thyroglobulin and therefore does not cause the normal suppression of TSH production by the anterior pituitary. The follicles become tremendous in size, and the thyroid gland may increase to 10 to 20 times normal size.

IDIOPATHIC NONTOXIC COLLOID GOITER. Enlarged thyroid glands similar to those of endemic colloid goiter also frequently occur in people who do not have iodine deficiency. These goitrous glands may secrete normal quantities of thyroid hormones, but more frequently, the secretion of hormone is depressed, as in endemic colloid goiter.

The exact cause of the enlarged thyroid gland in patients with idiopathic colloid goiter is not known, but most of these patients show signs of mild thyroiditis; therefore, it has been suggested that the thyroiditis causes slight hypothyroidism, which then leads to increased TSH secretion and progressive growth of the noninflamed portions of the gland. This could explain why these glands usually are nodular, with some portions of the gland growing while other portions are being destroyed by thyroiditis.

In some people with colloid goiter, the thyroid gland has an abnormality of the enzyme system required for the formation of the thyroid hormones. Among the abnormalities often encountered are the following:

1. Deficient iodide-trapping mechanism, in which iodine is not pumped adequately into the thyroid cells

2. Deficient peroxidase system, in which the iodides are not oxidized to the iodine state

3. Deficient coupling of iodinated tyrosines in the thyroglobulin molecule, so that the final thyroid hormones cannot be formed

4. Deficiency of the deiodinase enzyme, which prevents recovery of iodine from the iodinated tyrosines that are not coupled to form the thyroid hormones (this is about two thirds of the iodine), thus leading to iodine deficiency

Finally, some foods contain *goitrogenic substances* that have a propylthiouracil-type of antithyroid activity, thus also leading to TSH-stimulated enlargement of the thyroid gland. Such goitrogenic substances are found especially in some varieties of turnips and cabbages.

CHARACTERISTICS OF HYPOTHYROIDISM. Whether hypothyroidism is due to thyroiditis, endemic colloid goiter, idiopathic colloid goiter, destruction of the thyroid gland by irradiation, or surgical removal of the thyroid gland, the physiological effects are the same. They include fatigue and extreme somnolence with sleeping up to 12 to 14 hours a day, extreme muscular sluggishness, slowed heart rate, decreased cardiac output, decreased blood volume, sometimes increased weight, constipation, mental sluggishness, failure of many trophic functions in the body evidenced by depressed growth of hair and scaliness of the skin, development of a frog-like husky voice, and, in severe cases, development of an edematous appearance throughout the body called myxedema.

Myxedema. *Myxedema* develops in the patient with almost total lack of thyroid function. Figure 76–8 shows such a patient, demonstrating bagginess under the eyes and swelling of the face. In this condition, for reasons not explained, greatly increased quantities of hyaluronic acid and chondroitin sulfate bound with protein form excessive tissue gel in the interstitial spaces, and this causes the total quantity of interstitial fluid to increase. Because of the gel nature of the excess fluid, it is relatively immobile, and the edema is nonpitting in type.

Arteriosclerosis in Hypothyroidism. As pointed out earlier, lack of thyroid hormone increases the quantity of blood cholesterol because of diminished liver excretion of cholesterol in the bile. The increase in blood cholesterol is usually associated with increased atherosclerosis and arteriosclerosis. Therefore, in many hypothyroid patients, particularly those with myxedema, arteriosclerosis develops, which results in peripheral vascular disease, deafness, and often extreme coronary sclerosis with consequent early death.

DIAGNOSTIC TESTS IN HYPOTHYROIDISM. The tests already described for diagnosis of hyperthyroidism give the opposite results in hypothyroidism. The free thyroxine in the blood is low. The basal metabolic rate in myxedema ranges between − 30 and − 50. And the secretion of TSH by the anterior pituitary when a test dose of TRH is administered is usually greatly increased (except in those rare instances of hypothyroidism caused by depressed response of the pituitary gland to the TRH).

TREATMENT OF HYPOTHYROIDISM. Figure 76–4 shows the effect of thyroxine on the basal metabolic rate, demonstrating that the hormone normally has a duration of action of more than 1 month. Consequently, it is easy to maintain a steady level of thyroid hormone activity in the body by daily oral ingestion of a tablet or more containing thyroxine. Furthermore, proper treatment of the hypothyroid patient results in such complete normality that formerly myxedematous patients have lived into their 90s after treatment for more than 50 years.

Cretinism

Cretinism is caused by extreme hypothyroidism during fetal life, infancy, and childhood. This condition is characterized especially by failure of growth and by mental retardation. It results from congenital lack of a thyroid gland *(congenital cretinism)*, from failure of the thyroid gland to produce thyroid hormone because of a genetic defect of the gland, or from iodine lack in the diet *(endemic cretinism)*. The severity of endemic cretinism varies greatly, depending on the amount of iodine in the diet, and whole populaces of an endemic area have been known to have cretinoid tendencies.

A neonate without a thyroid gland may have normal appearance and function because it had been supplied with some (but usually not enough) thyroid hormone by the mother while in utero, but a few weeks after birth, its movements become sluggish and both its physical and mental growth are greatly retarded. Treatment of the cretin at any time usually causes normal return of physical growth, but unless the cretin is treated within a few weeks after birth, its mental growth is permanently retarded. This results from retardation of the growth, branching, and myelination of the neuronal cells of the central nervous system at this critical time in the normal development of the mental powers.

Skeletal growth in the cretin is characteristically more inhibited than is soft tissue growth. As a result of this disproportionate rate of growth, the soft tissues are likely to enlarge excessively, giving the cretin the appearance of an obese and stocky short child. Occasionally the tongue becomes so large in relation to the skeletal growth that it obstructs swallowing and breathing, inducing a characteristic guttural breathing that sometimes chokes the baby.

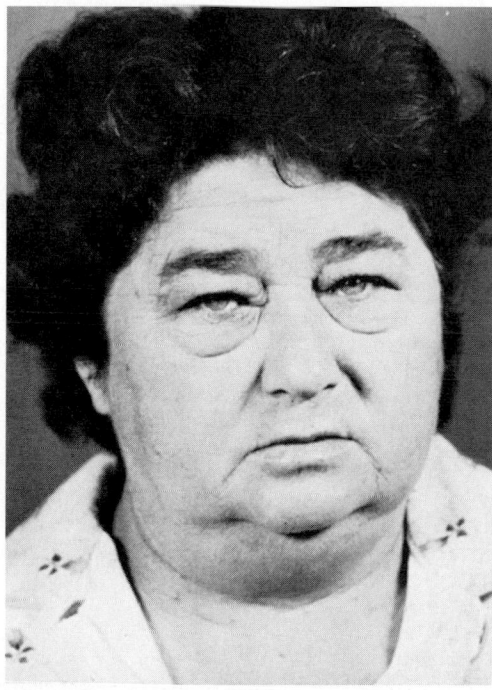

Figure 76–8. Patient with myxedema. (Courtesy of Dr. Herbert Langford.)

REFERENCES

Bayliss, R. I.: Thyroid Disease. New York. Oxford University Press, 1982.
Becker, J. B., et al.: Behavioral Endocrinology. Cambridge, MA, The MIT Press, 1992.

Becker, K. L., et al.: Principles and Practice of Endocrinology and Metabolism. Philadelphia, J. B. Lippincott, 1990.

Braverman, L. E., and Utiger, R. D.: Werner and Ingbar's The Thyroid. Philadelphia, J. B. Lippincott, 1992.

Burrow, G. N., et al.: Thyroid Function and Disease. Philadelphia, W. B. Saunders Co., 1989.

Cady, B., and Rossi, R. L.: Surgery of the Thyroid and Parathyroid Glands. Philadelphia, W. B. Saunders Co., 1991.

DeGroot, L. J.: Endocrinology. Philadelphia, W. B. Saunders Co., 1994.

Delang, F., et al.: Iodine Deficiency in Europe. New York, Plenum Publishing Corp., 1993.

Delang, F., et al. (eds.): Research in Congenital Hypothyroidism. New York, Plenum Publishing Corp., 1989.

DeLong, G. R., et al. (eds.): Iodine and the Brain. New York, Plenum Publishing Corp., 1989.

Dumont, J. E., et al.: Physiological and pathological regulation of thyroid cell proliferation and differentiation by thyrotropin and other factors. Physiol. Rev., 72:667, 1992.

Dussault, J. H., and Ruel, J.: Thyroid hormones and brain development. Annu. Rev. Physiol., 49:321, 1987.

Felig, P., et al. (eds.): Endocrinology and Metabolism, 2nd Ed. New York, McGraw-Hill Book Co., 1987.

Gershengorn, M. C.: Mechanism of thyrotropin releasing hormone stimulation of pituitary hormone secretion. Annu. Rev. Physiol., 48:515, 1986.

Greer, M.: The Thyroid Gland. New York, Raven Press, 1990.

King, D. B., and May, J. D.: Thyroidal influence on body growth. J. Exp. Zool., 232:453, 1984.

Kouridos, I. A., et al.: The regulation and organization of thyroid stimulating hormone genes. Recent Prog. Horm. Res., 40:79, 1984.

Kreiger, D. T. (ed.): Current Therapy in Endocrinology, 1983–1984. St. Louis, C. V. Mosby, 1984.

Lenzen, S., and Bailey, C. J.: Thyroid hormones, gonadal and adrenocortical steroids and the function of the islets of Langerhans. Endocrinol. Rev., 5:411, 1984.

LiVolsi, V. A., and DeLellis, R. A.: Pathology of the Parathyroid and Thyroid Glands. Baltimore, Williams & Wilkins, 1993.

LiVolsi, V. A.: Pathology of the Thyroid. Philadelphia, W. B. Saunders Co., 1989.

Matovinovic, J.: Endemic goiter and cretinism at the dawn of the third millennium. Annu. Rev. Nutr., 3:341, 1983.

Mederiros-Neto, G. A., and Gaitan, E. (eds.): Frontiers in Thyroidology. New York, Plenum Publishing Corp., 1987.

Oppenheimer, J. H.: Thyroid hormone action at the nuclear level. Ann. Intern. Med., 102:374, 1985.

Pinchera, A., et al. (eds.): Thyroid Autoimmunity. New York, Plenum Publishing Corp., 1987.

Samuels, H. H., et al.: Regulation of gene expression by thyroid hormone. Annu. Rev. Physiol., 51:623, 1989.

Wilson, J. D., and Foster, D. W.: Williams Textbook of Endocrinology. Philadelphia, W. B. Saunders Co., 1992.

The Adrenocortical Hormones

CHAPTER 77

The two *adrenal glands,* each of which weighs about 4 grams, lie at the superior poles of the two kidneys. As shown in Figure 77–1, each gland is composed of two distinct parts, the *adrenal medulla* and the *adrenal cortex.* The adrenal medulla, the central 20 per cent of the gland, is functionally related to the sympathetic nervous system; it secretes the hormones *epinephrine* and *norepinephrine* in response to sympathetic stimulation. In turn, these hormones cause almost the same effects as direct stimulation of the sympathetic nerves in all parts of the body. These hormones and their effects are discussed in detail in Chapter 60 in relation to the sympathetic nervous system.

The adrenal cortex secretes an entirely different group of hormones, called *corticosteroids.* These hormones are all synthesized from the steroid cholesterol, and they all have similar chemical formulas. However, slight differences in their molecular structures give them several very different but very important functions.

THE CORTICOSTEROIDS—MINERALOCORTICOIDS, GLUCOCORTICOIDS, AND ANDROGENS. Two major types of adrenocortical hormones, the *mineralocorticoids* and the *glucocorticoids,* are secreted by the adrenal cortex. In addition to these, small amounts of sex hormones are secreted, especially *androgenic hormones,* which exhibit about the same effects in the body as the male sex hormone testosterone. They are normally of only slight importance, although in certain abnormalities of the adrenal cortices extreme quantities can be secreted (which is discussed later in the chapter) and can then result in masculinizing effects.

The *mineralocorticoids* have gained this name because they especially affect the electrolytes (the "minerals") of the extracellular fluids—sodium and potassium, in particular. The *glucocorticoids* have gained their name because they exhibit an important effect in increasing blood glucose concentration. They have additional effects on both protein and fat metabolism that are as important to body function as their effects on carbohydrate metabolism, if not more so.

More than 30 steroids have been isolated from the adrenal cortex, but only two are of exceptional importance to the normal endocrine function of the human body: *aldosterone,* which is the principal mineralocorticoid, and *cortisol,* which is the principal glucocorticoid.

CHEMISTRY OF ADRENOCORTICAL SECRETION

LAYERS OF THE ADRENAL CORTEX, AND HORMONE FORMATION. Figure 77–1 shows that the adrenal cortex is composed of three relatively distinct layers. *Aldosterone* is secreted by the *zona glomerulosa,* the outermost and very thin layer of cells on the surface. *Cortisol* and several other glucocorticoids are secreted by both the *zona fasciculata,* the middle layer, and the *zona reticularis,* the deep layer. The *adrenal androgens* are also secreted by both these layers.

The conditions that increase the output of aldosterone cause hypertrophy of the zona glomerulosa while having no effect on the other two zones. On the other hand, the factors that cause increased secretion of cortisol and adrenal androgens cause hypertrophy of the zona fasciculata and zona reticularis while having little or no effect on the zona glomerulosa; this is especially true of stimulation of the gland by adrenocorticotropic hormone

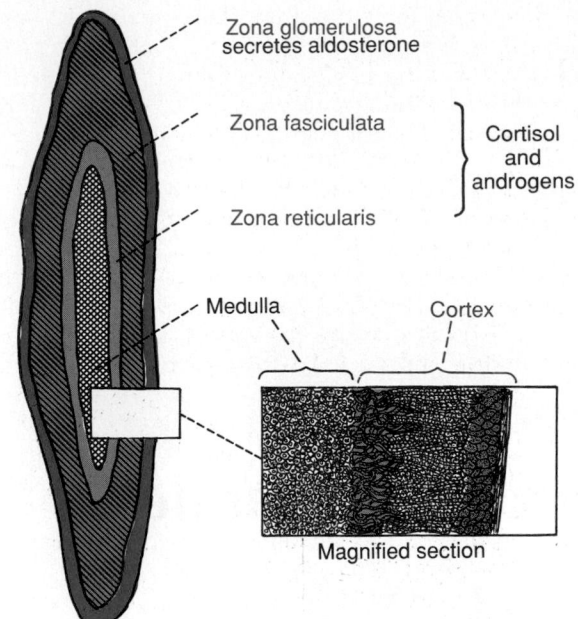

Figure 77–1. Secretion of adrenocortical hormones by the different zones of the adrenal cortex.

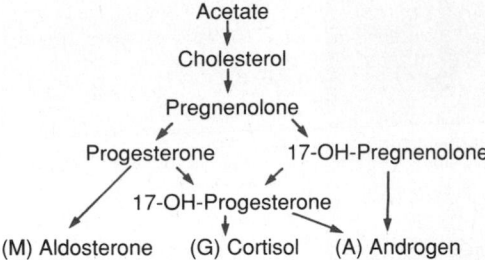

Figure 77–2. Major steps in the synthesis of the principal adrenal steroids. The physiological characteristics are designated (M) mineralocorticoid effect, (G) glucocorticoid effect, and (A) androgenic effect.

(ACTH) from the anterior pituitary gland, as discussed later in this chapter.

CHEMISTRY OF THE ADRENOCORTICAL HORMONES. All the adrenocortical hormones are steroid compounds. They are formed mainly from cholesterol absorbed directly from the circulating blood by endocytosis through the cell membrane. This membrane has specific receptors for the low-density lipoproteins that contain high concentrations of cholesterol, and attachment of these lipoproteins to the membrane promotes the endocytotic process. Small amounts of cholesterol are also synthesized within the cortical cells from acetyl coenzyme A; this, too, can be used for forming the adrenocortical hormones.

Figure 77–2 gives the principal steps in the formation of the important steroid products of the adrenal cortex: aldosterone, cortisol, and the androgens. Essentially all these steps occur in two of the organelles of the cell, the *mitochondria* and the *endoplasmic reticulum,* some steps occurring in one of these organelles and some in the other. Each step is catalyzed by a specific enzyme

system. A change in even a single enzyme in the schema can cause vastly different types and relative proportions of hormones to be formed, such as especially large quantities of masculinizing or, rarely, feminizing sex hormones or other steroid compounds not normally present in the blood but that have either mineralocorticoid or glucocorticoid actions or a combination of both.

Figure 77–3 shows the chemical formulas of aldosterone and cortisol. It is the oxygen atom bound at the number 18 carbon of the cholesterol nucleus that is most important in providing the mineralocorticoid activity of aldosterone. The glucocorticoid activity of cortisol is caused principally by the presence of the keto-oxygen on carbon number 3 and the hydroxylation of carbon numbers 11 and 21.

In addition to aldosterone and cortisol, which are, respectively, the principal mineralocorticoid and glucocorticoid hormones, other steroids having one or both of these activities are secreted in small amounts even normally by the adrenal cortex. And several additional potent steroid hormones not normally formed in the adrenal glands have been synthesized and are used in various forms of therapy. The more important of the adrenocortical hormones, including the synthetic ones, are the following:

Mineralocorticoids

Aldosterone (very potent, accounts for about 90 per cent of all mineralocorticoid activity)
Desoxycorticosterone (one fifteenth as potent as aldosterone but very small quantities secreted)

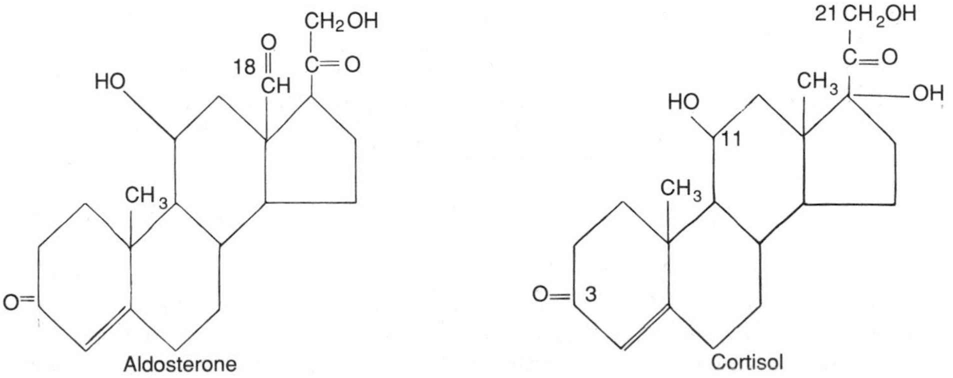

Figure 77–3. Two important corticosteroids.

Corticosterone (slight mineralocorticoid activity)

9α-Fluorocortisol (synthetic, slightly more potent than aldosterone)

Cortisol (very slight mineralocorticoid activity but large quantity secreted)

Cortisone (synthetic, slight mineralocorticoid activity)

Glucocorticoids

Cortisol (very potent, accounts for about 95 per cent of all glucocorticoid activity)

Corticosterone (provides about 4 per cent of total glucocorticoid activity but much less potent than cortisol)

Cortisone (synthetic, almost as potent as cortisol)

Prednisone (synthetic, four times as potent as cortisol)

Methylprednisone (synthetic, five times as potent as cortisol)

Dexamethasone (synthetic, 30 times as potent as cortisol)

It is clear from this list that some of these hormones have both glucocorticoid and mineralocorticoid activities. It is especially significant that cortisol has a small amount of mineralocorticoid activity because some syndromes of excess cortisol secretion can cause significant mineralocorticoid effects, along with its much more potent glucocorticoid effects.

The intense glucocorticoid activity of the synthetic hormone dexamethasone, which has almost zero mineralocorticoid activity, makes this an especially important drug for stimulating specific glucocorticoid activity.

TRANSPORT AND FATE OF THE ADRENAL HORMONES. Cortisol combines in the blood mainly with a globulin called *cortisol-binding globulin* or *transcortin* and to a lesser extent with albumin—about 94 per cent is normally transported in the bound form and about 6 per cent is free. On the other hand, aldosterone combines only loosely with the plasma proteins so that about 50 per cent is in the free form. In both the combined and free forms, the hormones are transported throughout the extracellular fluid compartment. In general, the hormones become fixed in the target tissues or destroyed within 1 or 2 hours for cortisol and within about 30 minutes for aldosterone.

The adrenal steroids are degraded mainly in the liver and conjugated especially to form glucuronides and to a lesser extent sulfates. About 25 per cent of these are excreted in the bile and then in the feces and the remaining 75 per cent, in the urine. The conjugated forms of these hormones are inactive.

The normal concentration of aldosterone in blood is about 6 nanograms (six billionths of a gram) per deciliter, and the secretory rate is 150 to 250 µg/day.

The concentration of cortisol in the blood averages 12 µg/dl, and the secretory rate averages 15 to 20 mg/day.

FUNCTIONS OF THE MINERALOCORTICOIDS— ALDOSTERONE

Total loss of adrenocortical secretion usually causes death within 3 days to 2 weeks unless the person receives extensive salt therapy or injection of mineralocorticoids. Without mineralocorticoids, the potassium ion concentration of the extracellular fluid rises markedly, the sodium and chloride concentrations decrease, and the total extracellular fluid volume and blood volume become greatly reduced. The person soon develops diminished cardiac output, which proceeds to a shock-like state followed by death. This entire sequence can be prevented by the administration of aldosterone or some other mineralocorticoid. Therefore, the mineralocorticoids are said to be the acute "life-saving" portion of the adrenocortical hormones; the glucocorticoids are equally necessary, allowing the person to resist the destructive effects of life's intermittent physical and mental "stresses," as discussed later in the chapter.

Aldosterone exerts nearly 90 per cent of the mineralocorticoid activity of the adrenocortical secretion, but cortisol, the major glucocorticoid secreted by the adrenal cortex, also provides a significant amount of mineralocorticoid activity—its mineralocorticoid activity is only 1/400 that of aldosterone, but about 80 times as much is secreted! Other adrenal steroids secreted in small amounts that have mineralocorticoid effects are *corticosterone*, which exerts mainly glucocorticoid effects but some mineralocorticoid effects as well, and *deoxycorticosterone*, which has almost the same effects as aldosterone but with a potency 1/30 that of aldosterone.

Renal and Circulatory Effects of Aldosterone

By far the most important function of aldosterone is to promote transport of sodium and potassium through some portions of the renal tubular walls; a less important function is to promote the transport of hydrogen ions. The mechanisms of these effects are discussed in detail in Chapters 27 through 30. Let us summarize the renal and body fluid effects of aldosterone.

EFFECT ON TUBULAR REABSORPTION OF SODIUM AND TUBULAR SECRETION OF POTASSIUM. It will be recalled from Chapter 27 that aldosterone causes increased exchange transport of sodium and potassium—that is, absorption of sodium and simultaneous excretion of potassium by the tubular epithelial cells—especially in the collecting tubule and to a lesser extent in the distal tubule and collecting duct. Therefore, aldosterone causes sodium to be conserved in the extracellular fluid while more potassium is excreted into the urine.

A high concentration of aldosterone in the plasma can decrease the sodium loss into the urine to as little as a few milliequivalents a day. At the same time, potassium loss into the urine increases manyfold.

Conversely, total lack of aldosterone secretion can cause loss of 10 to 20 grams of sodium in the urine a day, an amount equal to one tenth to one fifth of all the sodium in the body. At the same time, potassium is conserved tenaciously in the extracellular fluid.

Therefore, the net effect of excess aldosterone in the plasma is to increase the total quantity of sodium in the extracellular fluid while decreasing the potassium.

EFFECT ON EXTRACELLULAR FLUID VOLUME AND ARTERIAL PRESSURE. Even though aldosterone has a potent effect in decreasing the rate of sodium ion excretion by the kidneys, the concentration of sodium in the extracellular fluid rises very little. The reason for this is that when sodium is reabsorbed by the tubules, there is simultaneous osmotic absorption of almost equivalent amounts of water. Therefore, the extracellular fluid volume increases almost as much as the retained sodium but without much change in sodium concentration.

An increase in extracellular fluid volume lasting more than 1 to 2 days leads to an increase in arterial pressure, as explained in Chapter 19. The increase in arterial pressure then causes greatly increased kidney excretion of both water and salt, which is the phenomenon called *pressure diuresis*. Thus, in a roundabout manner, after the extracellular fluid volume has increased to about 5 to 15 per cent above normal in response to the excesses of aldosterone, the arterial pressure also has increased some 15 to 25 mm Hg, and this hypertension (high blood pressure) returns the renal output of water and salt to normal despite the excess aldosterone. This secondary increase in water and salt excretion by the kidneys as a result of pressure diuresis is called *aldosterone escape* because the rate of gain of salt and water by the body thereafter is zero. But in the meantime the person has developed hypertension, which lasts as long as the person remains exposed to the excess aldosterone.

Conversely, when aldosterone secretion becomes zero, large amounts of salt are lost in the urine, not only diminishing the amount of sodium chloride in the extracellular fluid but also decreasing the extracellular fluid volume. The result is severe extracellular fluid dehydration and low blood volume, leading to *circulatory shock*. Without immediate therapy, this usually causes death within a few days after the adrenal glands suddenly stop secreting aldosterone.

EXCESS ALDOSTERONE CAUSES *HYPOKALEMIA* AND MUSCLE WEAKNESS; TOO LITTLE ALDOSTERONE CAUSES *HYPERKALEMIA* AND CARDIAC TOXICITY. The excessive loss of potassium ions from the extracellular fluid into the urine under the influence of excess aldosterone causes a serious decrease in the plasma potassium concentration, often from the normal value of 4.5 mEq/liter to 1 to 2 mEq/liter. This condition is called *hypokalemia*. When the potassium ion concentration falls below about one half to one third normal, severe muscle weakness often develops. This is caused by alteration of the electrical excitability of the nerve and muscle fiber membranes (see Chapter 5), which prevents transmission of normal action potentials.

On the other hand, when aldosterone is deficient, the extracellular fluid potassium ion concentration can rise far above normal. When it rises to 60 to 100 per cent above normal, serious cardiac toxicity, including weakness of heart contraction and development of arrhythmia, becomes evident; progressively higher concentrations of potassium lead inevitably to heart failure.

EFFECT OF ALDOSTERONE ON INCREASING TUBULAR HYDROGEN ION SECRETION, WITH RESULTANT MILD ALKALOSIS. Although aldosterone mainly causes potassium to be secreted into the tubules in exchange for sodium reabsorption, to a much smaller extent it causes instead tubular secretion of hydrogen ions in exchange for sodium. The obvious effect of this is to decrease the hydrogen ion concentration in the extracellular fluid. This effect is not a strong one, usually causing only a mild degree of alkalosis.

Effects of Aldosterone on Sweat Glands, Salivary Glands, and Intestinal Absorption

Aldosterone has almost the same effects on sweat glands and salivary glands that it has on the renal tubules. Both these glands form a primary secretion that contains large quantities of sodium chloride, but much of the sodium chloride, on passing through the excretory ducts, is reabsorbed, whereas potassium and bicarbonate ions are secreted. Aldosterone greatly increases the reabsorption of sodium chloride and the secretion of potassium by the ducts. The effect on the sweat glands is important to conserve body salt in hot environments, and the effect on the salivary glands is necessary to conserve salt when excessive quantities of saliva are lost.

Aldosterone also greatly enhances sodium absorption by the intestines, especially in the colon, which prevents loss of sodium in the stools. On the other hand, in the absence of aldosterone, sodium absorption can be poor, leading to failure to absorb chloride and other anions and water as well. The unabsorbed sodium chloride and water then lead to diarrhea, with further loss of salt from the body.

Cellular Mechanism of Aldosterone Action

Although for many years we have known the overall effects of mineralocorticoids on the body, the basic action of aldosterone on the tubular cells to increase transport of sodium is still only partly understood. However, the cellular sequence of events that leads to increased sodium reabsorption seems to be the following.

First, because of its lipid solubility in the cellular membranes, aldosterone diffuses readily to the interior of the tubular epithelial cells.

Second, in the cytoplasm of the tubular cells, aldosterone combines with a highly specific cytoplasmic *receptor protein,* a protein that has a stereomolecular configuration that will allow only aldosterone or extremely similar compounds to combine.

Third, the aldosterone-receptor complex or a product of this complex diffuses into the nucleus, where it may undergo further alterations, finally inducing one or more specific portions of the DNA to form one or more types of messenger RNA related to the process of sodium and potassium transport.

Fourth, the messenger RNA diffuses back into the cytoplasm, where, operating in conjunction with the ribosomes, it causes protein formation. The proteins formed are a mixture of (1) one or more enzymes and (2) membrane transport proteins that all acting together are required for sodium, potassium, and hydrogen transport through the cell membrane. One of the enzymes especially increased is *sodium-potassium adenosine triphosphatase,* which serves as the principal part of the pump for sodium and potassium exchange at the *basolateral membranes* of the renal tubular cells. Another protein, perhaps equally important, is a channel protein inserted into the *luminal membrane* of the same tubular cells that allows rapid diffusion of sodium ions from the tubular lumen into the cell; then the sodium is pumped the rest of the way by the sodium-potassium pump located in the basolateral membranes of the cell.

Thus, aldosterone does not have an immediate effect on sodium transport but must await the sequence of events that leads to the formation of the specific intracellular substance or substances required for sodium transport. About 30 minutes is required before new RNA appears in the cells, and about 45 minutes is required before the rate of sodium transport begins to increase; the effect reaches maximum only after several hours.

Regulation of Aldosterone Secretion

The regulation of aldosterone secretion is so deeply intertwined with the regulation of extracellular fluid electrolyte concentrations, extracellular fluid volume, blood volume, arterial pressure, and many special aspects of renal function that it is not possible to discuss the regulation of aldosterone secretion independently of all these other factors. This subject is presented in detail in Chapters 28 and 29, to which the reader is referred. However, it is important to list here as well the more important points of aldosterone secretion control.

Let us note first that aldosterone is secreted by the *zona glomerulosa,* a thin zone of cells located on the surface of the adrenal cortex immediately beneath the capsule. These cells function almost entirely independently of the deeper cells in the zona reticularis and zona fasciculata, which secrete cortisol and the androgens. The regulation of aldosterone secretion is almost entirely independent of the regulation of these other hormones.

Four factors are known to play essential roles in the regulation of aldosterone. In the probable order of their importance, they are as follows:

1. Increased potassium ion concentration in the extracellular fluid greatly *increases* aldosterone secretion.

2. Increased activity of the renin-angiotensin system also greatly *increases* aldosterone secretion.

3. Increased sodium ion concentration in the extracellular fluid *very slightly decreases* aldosterone secretion.

4. Adrenocorticotropic hormone (ACTH) from the anterior pituitary gland is necessary for aldosterone secretion but has little effect in controlling the rate of secretion.

Of the factors above, *potassium ion concentration* and the *renin-angiotensin system* are by far the most potent in regulating aldosterone secretion. A low percentage increase in potassium concentration can cause a severalfold increase in aldosterone secretion. Likewise, activation of the renin-angiotensin system, usually in response to diminished blood flow to the kidneys, can cause a severalfold increase in aldosterone secretion. In turn, the aldosterone acts on the kidneys (a) to help them excrete the excess potassium ions and (b) to increase the blood volume and arterial pressure, thus returning the renin-angiotensin system also back toward its normal level of activity. These feedback control mechanisms are essential for maintaining life, and the reader is referred again to Chapters 28 and 29 for a full understanding of their functions.

Figure 77–4 shows the effects on plasma aldosterone concentration caused by continuous infusion of angiotensin II at two different infusion rates for 2 weeks. Note especially the large acute effect, which allows the renin-angiotensin-aldosterone control system to correct abnormalities of extracellular fluid electrolytes as rapidly as possible.

By contrast, the effects of sodium ion concentration and of ACTH in controlling aldosterone secretion are usually minor. Nevertheless, a 10 to 20 per cent decrease in extracellular fluid sodium ion concentration, which occurs on rare occasions, can perhaps double aldosterone secretion. In the case of ACTH, if there is even a small amount of ACTH secreted by the anterior pituitary gland, it is usually enough to permit the adrenal glands to secrete whatever amount of aldosterone is required, but total absence of ACTH can significantly reduce aldosterone secretion.

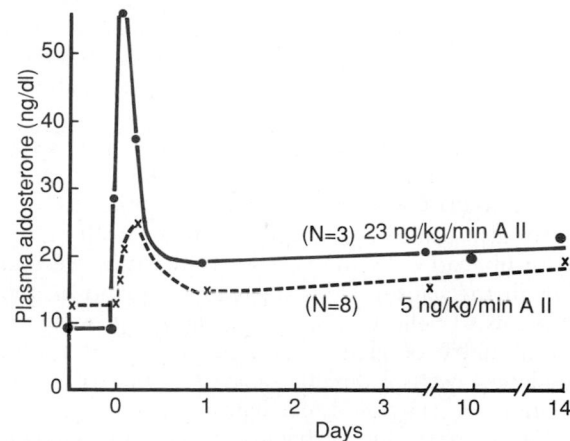

Figure 77–4. Effects on plasma aldosterone concentration caused by continuous infusion of angiotensin II into dogs at two different infusion rates for 2 weeks. Note the marked acute effect but the much weaker chronic effect. (Drawn from data in Cowley and McCaa, *Circ. Res., 39:*788, 1976.)

FUNCTIONS OF THE GLUCOCORTICOIDS

Even though mineralocorticoids can save the life of an acutely adrenalectomized animal, the animal still is far from normal. Instead, its metabolic systems for utilization of proteins, carbohydrates, and fats remain considerably deranged. Furthermore, the animal cannot resist different types of physical or even mental stress, and minor illnesses such as respiratory tract infections can lead to death. Therefore, the glucocorticoids have functions just as important to the long-continued life of the animal as those of the mineralocorticoids. They are explained in the following sections.

At least 95 per cent of the glucocorticoid activity of the adrenocortical secretions results from the secretion of *cortisol*, known also as *hydrocortisone*. In addition to this, a small but significant amount of glucocorticoid activity is provided by *corticosterone*.

Effects of Cortisol on Carbohydrate Metabolism

STIMULATION OF GLUCONEOGENESIS. By far the best-known metabolic effect of cortisol and other glucocorticoids on metabolism is their ability to stimulate gluconeogenesis (formation of carbohydrate from proteins and some other substances) by the liver, often increasing the rate of gluconeogenesis as much as 6- to 10-fold. This results mainly from two effects of cortisol.

First, cortisol increases all the enzymes required to convert amino acids into glucose in the liver cells. This results from the effect of the glucocorticoids to activate DNA transcription in the liver cell nuclei in the same way that aldosterone functions in the renal tubular cells, with formation of messenger RNAs that in turn lead to the array of enzymes required for gluconeogenesis.

Second, cortisol causes mobilization of amino acids from the extrahepatic tissues, mainly from muscle. As a result, more amino acids become available in the plasma to enter into the gluconeogenesis process of the liver and thereby to promote the formation of glucose.

One of the effects of increased gluconeogenesis is a marked increase in glycogen storage in the liver cells.

DECREASED GLUCOSE UTILIZATION BY THE CELLS. Cortisol also causes a moderate decrease in the rate of glucose utilization by the cells everywhere in the body. Although the cause of this decrease is unknown, most physiologists believe that somewhere between the point of entry of glucose into the cells and its final degradation cortisol directly delays the rate of glucose utilization. A suggested mechanism is based on the observation that glucocorticoids depress the oxidation of nicotinamide-adenine dinucleotide (NADH) to form NAD⁺. Because NADH must be oxidized to allow glycolysis, this effect could account for the diminished utilization of glucose by the cells.

ELEVATED BLOOD GLUCOSE CONCENTRATION AND ADRENAL DIABETES. Both the increased rate of gluconeogenesis and the moderate reduction in the rate of glucose utilization by the cells cause the blood glucose concentrations to rise. The increase in concentration is occasionally great enough (50 per cent or more above normal) that the condition is called *adrenal diabetes*. It has many similarities to pituitary diabetes, discussed in Chapter 75. Administration of insulin lowers the blood glucose concentration only a moderate amount in adrenal diabetes, not nearly so much as it does in pancreatic diabetes. On the other hand, insulin causes greater decrease in blood glucose concentration in adrenal diabetes than in pituitary diabetes. Therefore, *pituitary diabetes is said to be weakly insulin sensitive, adrenal diabetes moderately insulin sensitive, and pancreatic diabetes strongly insulin sensitive.*

Effects of Cortisol on Protein Metabolism

REDUCTION IN CELLULAR PROTEIN. One of the principal effects of cortisol on the metabolic systems of the body is reduction of the protein stores in essentially all body cells except those of the liver. This is caused by both decreased protein synthesis and increased catabolism of protein already in the cells. Both these effects may result from decreased amino acid transport into extrahepatic tissues, as discussed later; this probably is not the major cause because cortisol also depresses the formation of RNA and subsequent protein synthesis in many extrahepatic tissues, especially in muscle and lymphoid tissue.

In the presence of great excesses of cortisol, the muscles can become so weak that the person cannot rise from the squatting position. And the immunity functions of the lymphoid tissue can be decreased to a small fraction of normal.

Increased Liver Protein and Plasma Proteins Caused by Cortisol. Coincidentally with the reduced proteins elsewhere in the body, the liver proteins become enhanced. Furthermore, the plasma proteins (which are produced by the liver and then released into the blood) are also increased. These increases are exceptions to the protein depletion that occurs elsewhere in the body. It is believed that this difference results from a possible effect of cortisol in enhancing amino acid transport into liver cells (but not into most other cells) and of enhancement of the liver enzymes required for protein synthesis.

INCREASED BLOOD AMINO ACIDS, DIMINISHED TRANSPORT OF AMINO ACIDS INTO EXTRAHEPATIC CELLS, AND ENHANCED TRANSPORT INTO HEPATIC CELLS. Studies in isolated tissues have demonstrated that cortisol depresses amino acid transport into muscle cells and perhaps into other extrahepatic cells.

The decreased transport of amino acids into extrahepatic cells decreases their intracellular amino acid concentrations and consequently decreases the synthesis of protein. Yet catabolism of proteins in the cells continues to release amino acids from the already existing proteins, and these diffuse out of the cells to increase the plasma amino acid concentration. Therefore, *cortisol mobilizes amino acids from the nonhepatic tissues* and in doing so diminishes the tissue stores of protein.

The increased plasma concentration of amino acids plus the fact that cortisol enhances transport of amino acids into the hepatic cells could also account for enhanced utilization of amino acids by the liver to cause such effects as (1) increased rate of deamination of amino acids by the liver, (2) increased protein synthesis in the liver, (3) increased formation of plasma proteins by the liver, and (4) increased conversion of amino acids to glucose—that is, enhanced gluconeogenesis.

Thus, it is possible that many of the effects of cortisol on the metabolic systems of the body result mainly from this ability of cortisol to mobilize amino acids from the peripheral tissues while at the same time increasing the liver enzymes required for the hepatic effects.

Effects of Cortisol on Fat Metabolism

MOBILIZATION OF FATTY ACIDS. In much the same manner that cortisol promotes amino acid mobilization from muscle, it promotes mobilization of fatty acids from adipose tissue. This increases the concentration of free fatty acids in the plasma, which also increases their utilization for energy. Cortisol seems also to have a direct effect to enhance the oxidation of fatty acids in the cells.

The mechanism by which cortisol promotes fatty acid mobilization is not understood. However, part of the effect probably results from diminished transport of glucose into the fat cells. It will be remembered that α-glycerophosphate, which is derived from glucose, is required for both deposition and maintenance of triglycerides in these cells, and in its absence the fat cells begin to release fatty acids.

The increased mobilization of fats by cortisol combined with increased oxidation of fatty acids in the cells help shift the metabolic systems of the cells in times of starvation or other stresses from utilization of glucose for energy to utilization of fatty acids. This cortisol mechanism, however, requires several hours to become fully developed—not nearly so rapid or so powerful an effect as a similar shift elicited by a decrease in insulin, as we discuss in Chapter 78. Nevertheless, the increased use of fatty acids for metabolic energy is an important factor for long-term conservation of body glucose and glycogen.

OBESITY CAUSED BY CORTISOL. Despite the fact that cortisol can cause a moderate degree of fatty acid mobilization from adipose tissue, many people with excess cortisol secretion develop a peculiar type of obesity, with excess deposition of fat in the chest and head regions of the body, giving a buffalo-like torso and a rounded face, a "moon face." Although the cause is unknown, it has been suggested that this obesity results from excess stimulation of food intake, so

that fat is generated in some tissues of the body more rapidly than it is mobilized and oxidized.

Function of Cortisol in Stress and Inflammation

It is amazing that almost any type of stress, whether physical or neurogenic, will cause an immediate and marked increase in ACTH secretion by the anterior pituitary gland, followed within minutes by greatly increased adrenocortical secretion of cortisol. This is demonstrated dramatically by the experiment shown in Figure 77–5, in which corticosteroid formation and secretion increased sixfold in a rat within 4 to 20 minutes after fracture of two leg bones.

Some of the different types of stress that increase cortisol release are the following:

1. Trauma of almost any type
2. Infection
3. Intense heat or cold
4. Injection of norepinephrine and other sympathomimetic drugs
5. Surgery
6. Injection of necrotizing substances beneath the skin
7. Restraining an animal so that it cannot move
8. Almost any debilitating disease

Thus, a wide variety of nonspecific stimuli can cause marked increase in the rate of cortisol secretion by the adrenal cortex.

Even though we know that cortisol secretion often increases greatly in stressful situations, we are not sure

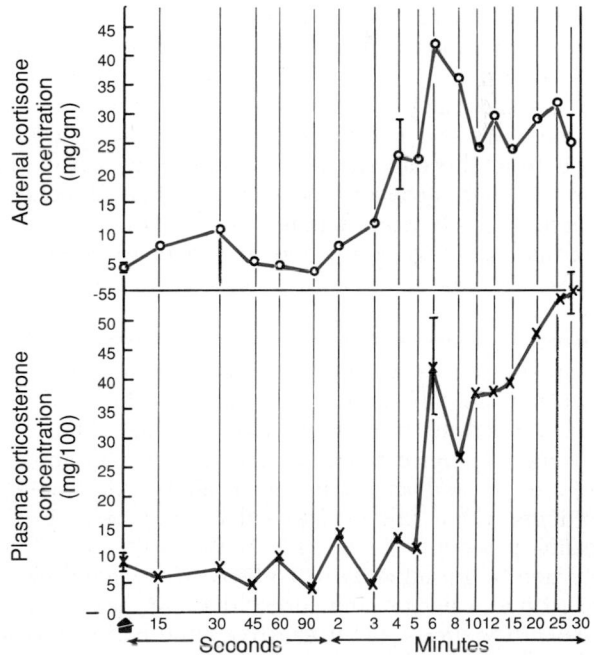

Figure 77–5. Rapid reaction of the adrenal cortex of a rat to stress caused by fracture of the tibia and fibula. (In the rat, corticosterone is secreted in place of cortisol.) (Courtesy of Drs. Guillemin, Dear, and Lipscomb.)

why this is of significant benefit to the animal. One guess, which is probably as good as any other, is that the glucocorticoids cause rapid mobilization of amino acids and fats from their cellular stores, making them immediately available both for energy and for synthesis of other compounds, including glucose, needed by the different tissues of the body. Indeed, it has been shown in a few instances that damaged tissues that are momentarily severely depleted of proteins can use the newly available amino acids to form new proteins that are essential to the lives of the cells. Also, the amino acids are perhaps used to synthesize such other essential intracellular substances as purines, pyrimidines, and creatine phosphate, which are necessary for maintenance of cellular life and reproduction of new cells.

But all this is mainly supposition. It is supported only by the fact that cortisol usually does not mobilize the basic functional proteins of the cells, such as the muscle contractile proteins and the proteins of neurons, until almost all other proteins have been released. This preferential effect of cortisol in mobilizing labile proteins could make amino acids available to needy cells to synthesize substances essential to life.

Anti-inflammatory Effects of Cortisol

When tissues are damaged by trauma, by infection with bacteria, or in almost any other way, they almost always become "inflamed." In some conditions, the inflammation is more damaging than the trauma or disease itself, such as in rheumatoid arthritis. The administration of large amounts of cortisol can usually block this inflammation or even reverse many of its effects once it has begun. Before attempting to explain the way in which cortisol functions to block inflammation, let us review the basic steps in the inflammation process, discussed in more detail in Chapter 33.

There are five main stages of inflammation: (1) release from the damaged tissue cells of chemical substances that activate the inflammation process—chemicals such as histamine, bradykinin, proteolytic enzymes, prostaglandins, and leukotrienes; (2) an increase in blood flow in the inflamed area caused by some of the released products from the tissues, an effect called *erythema;* (3) leakage of large quantities of almost pure plasma out of the capillaries into the damaged areas because of increased capillary permeability, followed by clotting of the tissue fluid, thus causing a *nonpitting type of edema;* (4) infiltration of the area by leukocytes; and then, after days or weeks, (5) ingrowth of fibrous tissue that often helps in the healing process.

When large amounts of cortisol are secreted or injected into a person, the cortisol has two basic *anti-inflammatory effects:* (1) it can block the early stages of the inflammation process before inflammation even begins or (2) if inflammation has already begun, it causes rapid resolution of the inflammation and increased rapidity of healing. These effects are explained further as follows.

PREVENTION OF THE DEVELOPMENT OF INFLAMMATION—LYSOSOME STABILIZATION AND OTHER EFFECTS. Cortisol has the following effects in preventing inflammation.

1. One of the most important anti-inflammatory effects of cortisol is its ability to cause *stabilization of the lysosomal membranes.* That is, cortisol makes it much more difficult than is normal for the membranes of the intracellular lysosomes to rupture. Therefore, most of the proteolytic enzymes that are released by damaged cells to cause inflammation, which are mainly stored in the lysosomes, are released in greatly decreased quantity.

2. Cortisol decreases the permeability of the capillaries, probably as a secondary effect of the reduced release of proteolytic enzymes. This prevents loss of plasma into the tissues.

3. Cortisol decreases both migration of white blood cells into the inflamed area and phagocytosis of the damaged cells. These effects probably result from the fact that cortisol diminishes the formation of prostaglandins and leukotrienes that otherwise would increase vasodilation, capillary permeability, and mobility of white blood cells.

4. Cortisol suppresses the immune system, causing lymphocyte reproduction to decrease markedly. The T lymphocytes are especially suppressed. In turn, reduced amounts of T cells and antibodies in the inflamed area lessen the tissue reactions that would otherwise promote further the inflammation process.

5. Cortisol lowers fever mainly because it reduces the release of interleukin-1 from the white blood cells, which is one of the principal excitants to the hypothalamic temperature control system. The decreased temperature in turn reduces the degree of vasodilatation.

Thus, cortisol has an almost global effect in reducing all aspects of the inflammatory process. How much of this results from the simple effect of cortisol in stabilizing lysosomal and cell membranes versus its effect to reduce the formation of prostaglandins and leukotrienes from arachidonic acid in damaged cell membranes and versus still other effects of cortisol is unknown.

EFFECT OF CORTISOL IN CAUSING RESOLUTION OF INFLAMMATION. Even after inflammation has become well established, the administration of cortisol can often reduce inflammation within hours to a few days. The immediate effect is to block most of the factors that are promoting the inflammation. But in addition, the rate of healing is enhanced. This probably results from the same, mainly undefined factors that allow the body to resist many other types of physical stress when large quantities of cortisol are secreted: perhaps this results from the mobilization of amino acids and use of these to repair the damaged tissues; perhaps it results from the increased glucogenesis that makes extra glucose available in critical metabolic systems; perhaps it results from increased amounts of fatty acids available for cellular energy; or perhaps it depends on some effect of cortisol for inactivating or removing inflammatory products.

Regardless of the precise mechanisms by which the anti-inflammatory effect occurs, this effect of cortisol plays a major role in combating certain types of diseases, such as rheumatoid arthritis, rheumatic fever, and acute glomerulonephritis. All these diseases are characterized by severe local inflammation, and the harmful effects on the body are caused mainly by the inflammation itself and not by other aspects of the disease. When cortisol or other glucocorticoids are administered to patients with these diseases, almost invariably the inflammation begins to subside within 24 hours. And even though the cortisol does not correct the basic disease condition, merely preventing the damaging effects of the inflammatory response, this alone can often be a lifesaving measure.

Other Effects of Cortisol

Effect on Allergy

Cortisol blocks the inflammatory response to allergic reactions in the same way that it blocks other types of inflammatory responses. The basic allergic reaction between antigen and antibody is not affected by cortisol, and even some of the secondary effects of the allergic reaction still occur. However, because the inflammatory response is responsible for many of the serious and sometimes lethal effects of allergic reactions, administration of cortisol, followed by its effect in reducing inflammation and the release of inflammatory products, can be lifesaving. For instance, cortisol effectively prevents shock or death in anaphylaxis, which otherwise kills many people, as explained in Chapter 34.

Effect on Blood Cells and on Immunity in Infectious Diseases

Cortisol decreases the number of eosinophils and lymphocytes in the blood; this effect begins within a few minutes after the injection of cortisol and becomes marked within a few hours. Indeed, a finding of lymphocytopenia or eosinopenia is an important diagnostic criterion for overproduction of cortisol by the adrenal gland.

Likewise, the administration of large doses of cortisol causes significant atrophy of all the lymphoid tissue throughout the body, which in turn decreases the output of both T cells and antibodies from the lymphoid tissue. As a result, the level of immunity for almost all foreign invaders of the body is decreased. This occasionally can lead to fulminating infection and death from diseases that would otherwise not be lethal, such as fulminating tuberculosis in a person whose disease had previously been arrested. On the other hand, this ability of cortisol and other glucocorticoids to suppress immunity makes them among the most useful of all drugs to prevent immunological rejection of transplanted hearts, kidneys, and other tissues.

Cortisol increases the production of red blood cells, the cause of which is unknown. When excess cortisol is secreted by the adrenal glands, polycythemia often results, and conversely, when the adrenal glands secrete no cortisol, anemia often results.

Regulation of Cortisol Secretion—Adrenocorticotropic Hormone from the Pituitary Gland

CONTROL OF CORTISOL SECRETION BY ACTH. Unlike aldosterone secretion by the zona glomerulosa, which is controlled mainly by potassium and angiotensin acting directly on the adrenocortical cells, *almost no stimuli have direct control effects* on the adrenal cells that secrete cortisol. Instead, *secretion of cortisol is controlled almost entirely by ACTH secreted by the anterior pituitary gland*. This hormone, also called *corticotropin* or *adrenocorticotropin*, also enhances the production of adrenal androgens.

CHEMISTRY OF ACTH. ACTH has been isolated in pure form from the anterior pituitary. It is a large polypeptide, having a chain length of 39 amino acids. A smaller polypeptide, a digested product of ACTH having a chain length of 24 amino acids, has all the effects of the total molecule.

CONTROL OF ACTH SECRETION BY THE HYPOTHALAMUS—CORTICOTROPIN-RELEASING FACTOR. In the same way that other pituitary hormones are controlled by releasing hormones or factors from the hypothalamus, so also does an important releasing factor control ACTH secretion. This is called *corticotropin-releasing factor* (CRF). It is secreted into the primary capillary plexus of the hypophysial portal system in the median eminence of the hypothalamus and then carried to the anterior pituitary gland, where it induces ACTH secretion. CRF is a peptide composed of 41 amino acids. The cell bodies of the neurons that secrete CRF are located mainly in the paraventricular nucleus of the hypothalamus. This nucleus in turn receives many nervous connections from the limbic system and lower brain stem.

The anterior pituitary gland can secrete only minute quantities of ACTH in the absence of CRF. Instead, most conditions that cause high ACTH secretory rates initiate this secretion by signals that begin in the basal regions of the brain, including the hypothalamus, and are then transmitted by CRF to the anterior pituitary gland.

MECHANISM BY WHICH ACTH ACTIVATES ADRENOCORTICAL CELLS TO PRODUCE STEROIDS—FUNCTION OF CYCLIC ADENOSINE MONOPHOSPHATE (cAMP). The principal effect of ACTH on the adrenocortical cells is to activate *adenyl cyclase* in the cell membrane. This then induces the formation of *cAMP* in the cell cytoplasm, reaching its maximum effect in about 3 minutes. The cAMP in turn activates the intracellular enzymes that cause formation of the adrenocortical hormones. This is another example of cAMP as a *second messenger* hormone.

The most important of all the ACTH-stimulated steps for controlling adrenocortical secretion is activation of the enzyme *protein kinase A*, which *causes initial conversion of cholesterol to pregnenolone.* This initial conversion is the "rate-limiting" step for all the adrenocortical hormones, which explains why ACTH normally is necessary for any adrenocortical hormones to be formed. Long-term stimulation of the adrenal cortex by ACTH not only increases secretory activity but also causes hypertrophy and proliferation of the adrenocortical cells, especially in the zona fasciculata and zona reticularis, where cortisol and the androgens are secreted.

Effect of Physiological Stress on ACTH Secretion and Adrenocortical Secretion

As pointed out earlier in the chapter, almost any type of physical or mental stress can lead within minutes to greatly enhanced secretion of ACTH and consequently cortisol as well, often increasing cortisol secretion as much as 20-fold. This effect was demonstrated forcefully by the rapid and strong adrenocortical secretory responses after trauma shown in Figure 77–5.

Pain stimuli caused by any type of physical stress or tissue damage are transmitted first upward through the brain stem and eventually to the median eminence of the hypothalamus, as shown in Figure 77–6. Here CRF is secreted into the hypophysial portal system. Within minutes the entire control sequence leads to large quantities of cortisol in the blood.

Mental stress can cause equally as rapid increase in ACTH secretion. This is believed to result from increased activity in the limbic system, especially in the region of the amygdala and hippocampus, both of these then transmitting signals to the posterior medial hypothalamus.

INHIBITORY EFFECT OF CORTISOL ON THE HYPOTHALAMUS AND ON THE ANTERIOR PITUITARY TO CAUSE DECREASED ACTH SECRETION. Cortisol has direct negative feedback effects on (1) the hypothalamus to decrease the formation of CRF and (2) the anterior pituitary gland to decrease the formation of ACTH. Both of these feedbacks help regulate the plasma concentration of cortisol. That is, whenever the concentration becomes too great, the feedbacks automatically reduce the ACTH toward a normal control level.

Summary of the Control System

Figure 77–6 shows the overall system for control of cortisol secretion. The central key to this control is the excitation of the hypothalamus by different types of stress. They activate the entire system to cause rapid release of cortisol, and the cortisol in turn initiates a series of metabolic effects directed toward relieving the damaging nature of the stressful state. In addition, there is direct feedback of the cortisol to both the hypothalamus and the anterior pituitary gland to decrease the concentration of cortisol in the plasma at times when the body is not experiencing stress. However, the stress stimuli are the prepotent ones; they can always break through this direct inhibitory feedback of cortisol, causing either periodic exacerbations of cortisol secretion at multiple times during the day (Fig. 77–7) or prolonged cortisol secretion in times of chronic stress.

CIRCADIAN RHYTHM OF GLUCOCORTICOID SECRETION. The secretory rates of CRF, ACTH, and cortisol are high in the early morning but low in the late evening, as shown in Figure 77–7; the plasma cortisol level ranges between a high of about 20 μg/dl an hour before arising in the morning and a low of about 5 μg/dl around midnight. This effect results from a 24-hour cyclic alteration in the signals from the hypothalamus that cause cortisol secretion. When a person changes daily sleeping habits, the cycle changes correspondingly. One of the reasons the cycle is so important is that measurements of blood cortisol levels are meaningful only when

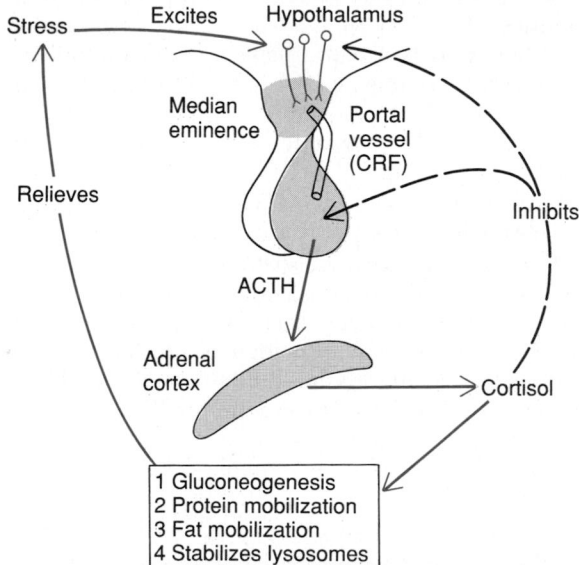

Figure 77–6. Mechanism for regulation of glucocorticoid secretion.

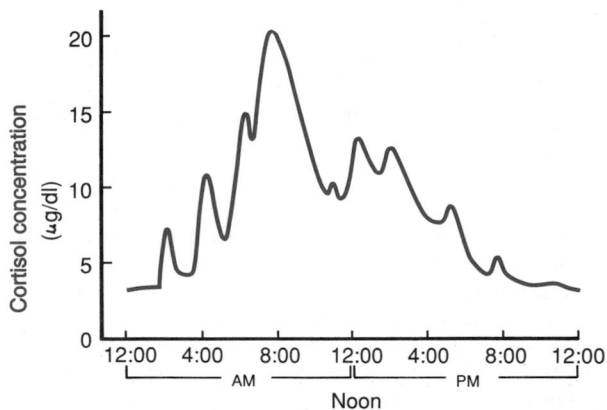

Figure 77–7. Typical pattern of cortisol concentration during the 24-hour day. Note the oscillations in secretion as well as a daily secretory surge an hour or so after awaking in the morning.

expressed in terms of the time in the cycle at which the measurements are made.

Secretion of Melanocyte-Stimulating Hormone, Lipotropin, and Endorphin in Association with ACTH

When ACTH is secreted by the anterior pituitary gland, several other hormones that have similar chemical structures are secreted simultaneously. The reason for this is that the RNA molecule that causes ACTH formation, initially causes the formation of a considerably larger protein molecule, a preprohormone, that contains ACTH as only one of its subunits. This same preprohormone also contains *melanocyte-stimulating hormone (MSH), beta-lipotropin,* and *beta-endorphin.* Under normal conditions, none of these hormones is secreted in enough quantity to have a significant effect on the human body, but this is not true, especially for MSH, when the rate of secretion of ACTH is high, as occurs in some types of Addison's disease, which is discussed later.

MSH causes the *melanocytes,* which are located in abundance between the dermis and epidermis of the skin, to form the black pigment *melanin* and to disperse this in the epidermis. Injection of MSH into a person over 8 to 10 days can greatly increase darkening of the skin. The effect is much greater in people who have genetically dark skins than in light-skinned people.

In some lower animals, an intermediate "lobe" of the pituitary gland, called the *pars intermedia,* is highly developed, lying between the anterior and posterior pituitary lobes. This lobe secretes an especially large amount of MSH. Furthermore, this secretion is independently controlled by the hypothalamus in response to the amount of light to which the animal is exposed or in response to other environmental factors. For instance, some arctic animals develop darkened fur in the summer and yet have entirely white fur in the winter.

ACTH, because of its similarity to MSH, has about $\frac{1}{30}$ as much melanocyte-stimulating effect as MSH. Furthermore, because the quantities of pure MSH secreted in the human being are extremely small, whereas those of ACTH are large, it is likely that ACTH normally is more important than MSH in determining the amount of melanin in the skin.

ADRENAL ANDROGENS

Several moderately active male sex hormones called *adrenal androgens* (the most important of which is *dehydroepiandrosterone*) are continually secreted by the adrenal cortex, especially during fetal life, as discussed more fully in Chapter 83. Also, progesterone and estrogens, which are female sex hormones, are secreted in minute quantities.

Normally, in the human being, the adrenal androgens

have only weak effects. It is possible that part of the early development of the male sex organs results from childhood secretion of adrenal androgens. The adrenal androgens also exert mild effects in the female, not only before puberty but also throughout life. Much of the growth of the pubic and axillary hair in the female results from the action of these hormones. Some of the adrenal androgens are converted to testosterone, the major male sex hormone, in the extra-adrenal tissues, which probably accounts for much of their androgenic activity. The physiological effects of androgens are discussed in Chapter 80 in relation to male sexual function.

ABNORMALITIES OF ADRENOCORTICAL SECRETION

Hypoadrenalism—Addison's Disease

Addison's disease results from failure of the adrenal cortices to produce adrenocortical hormones, and this in turn is most frequently caused by *primary atrophy* of the adrenal cortices. In about 80 per cent of the cases, the atrophy is caused by autoimmunity against the cortices. Adrenal gland hypofunction is also frequently caused by tuberculous destruction of the adrenal glands or invasion of the adrenal cortices by cancer. The disturbances in Addison's disease are as follows.

MINERALOCORTICOID DEFICIENCY. Lack of aldosterone secretion greatly decreases renal tubular sodium reabsorption and consequently allows sodium ions, chloride ions, and water to be lost into urine in great profusion. The net result is a greatly decreased extracellular fluid volume. Furthermore, hyponatremia, hyperkalemia, and mild acidosis develop because of failure of potassium and hydrogen ions to be secreted in exchange for sodium reabsorption.

As the extracellular fluid becomes depleted, plasma volume falls, red blood cell concentration rises markedly, cardiac output decreases, and the patient dies in shock, death usually occurring in the untreated patient 4 days to 2 weeks after cessation of mineralocorticoid secretion.

GLUCOCORTICOID DEFICIENCY. Loss of cortisol secretion makes it impossible for a person with Addison's disease to maintain normal blood glucose concentration between meals because he cannot synthesize significant quantities of glucose by gluconeogenesis. Furthermore, lack of cortisol reduces the mobilization of both proteins and fats from the tissues, thereby depressing many other metabolic functions of the body. This sluggishness of energy mobilization when cortisol is not available is one of the major detrimental effects of glucocorticoid lack. Even when excess quantities of glucose and other nutrients are available, the person's muscles are weak, indicating that glucocorticoids are needed to maintain other metabolic functions of the tissues in addition to energy metabolism.

Lack of adequate glucocorticoid secretion also makes a person with Addison's disease highly susceptible to the deteriorating effects of different types of stress, and even a mild respiratory infection can cause death.

MELANIN PIGMENTATION. Another characteristic of most people with Addison's disease is melanin pigmentation of the mucous membranes and skin. This melanin is not always deposited evenly but occasionally in blotches, and it is deposited especially in the thin skin areas, such as the mucous membranes of the lips and the thin skin of the nipples.

The cause of the melanin deposition is believed to be the following: When cortisol secretion is depressed, the normal negative feedback to the hypothalamus and anterior pituitary gland is also depressed, therefore allowing tremendous rates of ACTH secretion as well as simultaneous secretion of increased amounts of MSH. Probably the tremendous amounts of ACTH cause most of the pigmenting effect because they can stimulate formation of melanin by the melanocytes in the same way that MSH does. Even though MSH has 30 times as much melanocyte-stimulating effect as ACTH, the amounts secreted by the human being are extremely small.

TREATMENT OF PEOPLE WITH ADDISON'S DISEASE. An untreated person with total adrenal destruction dies within a few days to a few weeks because of consuming weakness and usually circulatory shock. Yet such a person can live for years if small quantities of mineralocorticoids and glucocorticoids are administered daily.

ADDISONIAN CRISIS. As noted earlier in the chapter, great quantities of glucocorticoids are occasionally secreted in response to different types of physical or mental stress. In a person with Addison's disease, the output of glucocorticoids does not increase during stress. Yet whenever different types of trauma, disease, or other stresses, such as surgical operations, supervene, a person is likely to have an acute need for excessive amounts of glucocorticoids and often must be given 10 or more times the normal quantities of glucocorticoids to prevent death.

This critical need for extra glucocorticoids and the associated severe debility in times of stress is called an *addisonian crisis.*

Hyperadrenalism—Cushing's Syndrome

Hypersecretion by the adrenal cortex often causes a complex of hormonal effects called *Cushing's disease,* and this results from either a cortisol-secreting tumor of one adrenal cortex or general hyperplasia of both adrenal cortices. The hyperplasia in turn is usually caused by increased secretion of ACTH by the anterior pituitary or by "ectopic secretion" of ACTH by a tumor elsewhere in the body, such as an abdominal carcinoma. Most abnormalities of Cushing's syndrome are ascribable to abnormal amounts of cortisol, but secretion of androgens is also of significance.

A special characteristic of Cushing's disease is mobilization of fat from the lower part of the body, with concomitant extra deposition of fat in the thoracic and upper abdominal regions, giving rise to a buffalo torso. The excess secretion of steroids also leads to an edematous appearance of the face, and the androgenic potency of some of the hormones sometimes causes acne and hirsutism (excess growth of facial hair). The total appearance of the face is frequently described as a moon face, as demonstrated in the patient with Cushing's syndrome before treatment shown to the left in Figure 77–8. About 80 per cent of the patients have hypertension, presumably because of the slight mineralocorticoid effects of cortisol.

EFFECTS ON CARBOHYDRATE AND PROTEIN METABOLISM. The abundance of cortisol secreted in Cushing's syndrome can cause increased blood glucose concentration, sometimes to values as high as 200 mg/dl after

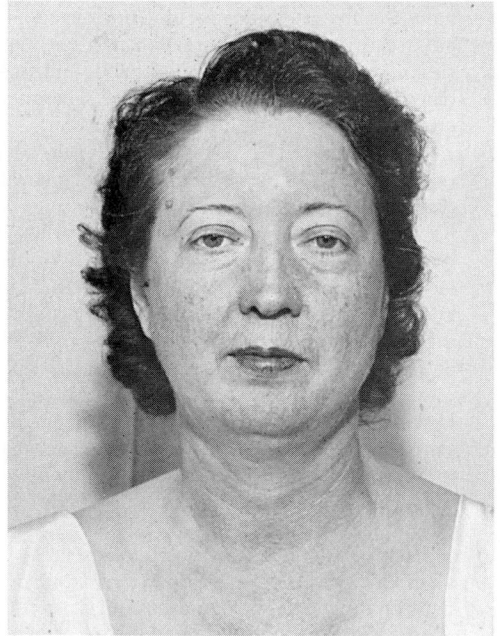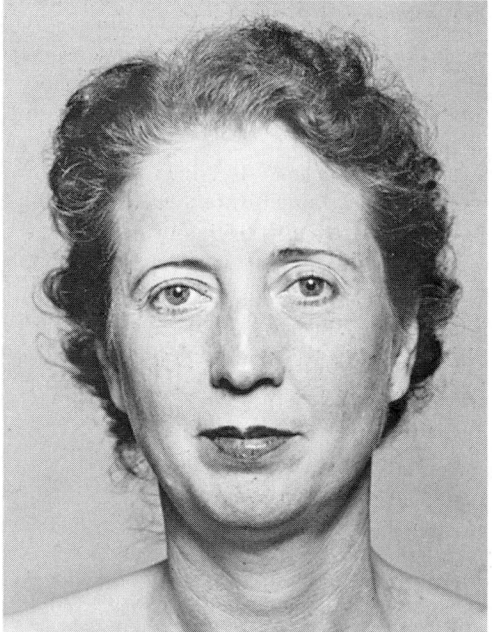

Figure 77–8. A person with Cushing's disease before subtotal adrenalectomy *(left)* and after subtotal adrenalectomy *(right).* (Courtesy of Dr. Leonard Posey.)

meals, as much as twice normal. This results mainly from enhanced gluconeogenesis.

The effects of glucocorticoids on protein catabolism are often profound in Cushing's syndrome, causing greatly decreased tissue proteins almost everywhere in the body with the exception of the liver and the plasma proteins. The loss of protein from the muscles in particular causes severe weakness. The loss of protein synthesis in the lymphoid tissues leads to a suppressed immune system, so that many of these patients die of infections. Even the protein collagen fibers in the subcutaneous tissue are diminished so that the subcutaneous tissues tear easily, resulting in development of large *purplish striae* where the subcutaneous tissues have torn apart. In addition, severely diminished protein deposition in the bones often causes severe *osteoporosis* with consequent weakness of the bones.

TREATMENT OF CUSHING'S DISEASE. Treatment in Cushing's disease consists of removing an adrenal tumor if this is the cause or decreasing the secretion of ACTH, if this is possible. Hypertrophied pituitary glands or even small tumors in the pituitary that oversecrete ACTH can be surgically or microsurgically partially or totally removed or be destroyed by radiation. If ACTH secretion cannot easily be decreased, the only satisfactory treatment is usually bilateral partial (or even total) adrenalectomy, followed by administration of adrenal steroids to make up for any insufficiency that develops.

Primary Aldosteronism

Occasionally a small tumor of the zona glomerulosa cells occurs and secretes large amounts of aldosterone, which is called "primary aldosteronism." Also, in a few instances, hyperplastic adrenal cortices secrete aldosterone rather than cortisol. The effects of the excess aldosterone are discussed in detail earlier in the chapter. The most important effects are hypokalemia, slight increase in extracellular fluid volume and blood volume, very slight increase in plasma sodium concentration (usually not over a 2 to 3 per cent increase), and, almost always, hypertension. Especially interesting in primary aldosteronism are occasional periods of muscle paralysis caused by the hypokalemia. The paralysis is caused by a depressant effect of low extracellular potassium concentration on action potential transmission by the nerve fibers, as explained in Chapter 5.

One of the diagnostic criteria of primary aldosteronism is a decreased plasma renin concentration. This results from feedback suppression of renin secretion caused by the excess aldosterone or by the excess extracellular fluid volume and arterial pressure resulting from the aldosteronism.

Treatment of primary aldosteronism is usually surgical removal of the tumor or of most of the adrenal tissue when hyperplasia is the cause.

Adrenogenital Syndrome

An occasional adrenocortical tumor secretes excessive quantities of androgens that cause intense masculinizing effects throughout the body. If this occurs in a female, she develops virile characteristics, including growth of a beard, a much deeper voice, occasionally baldness if she also has the genetic inheritance for baldness, masculine distribution of hair on the body and the pubis, growth

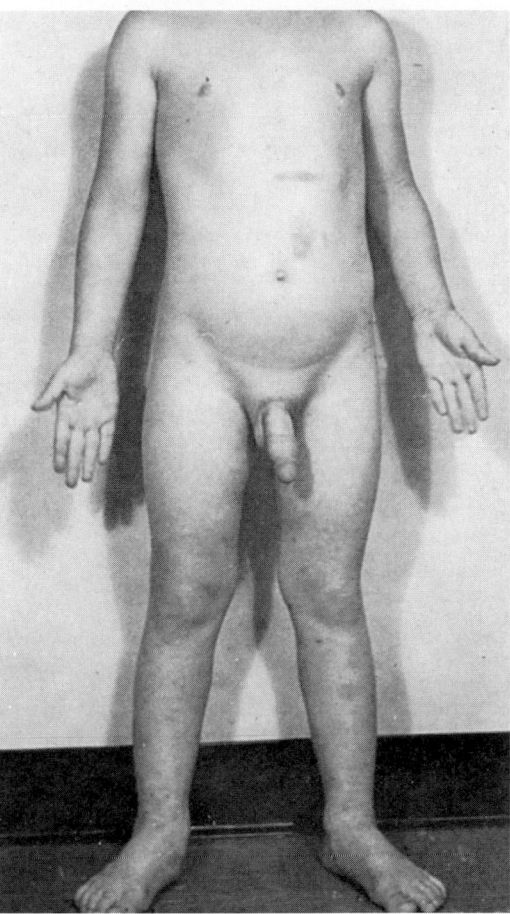

Figure 77–9. Adrenogenital syndrome in a 4-year-old boy. (Courtesy of Dr. Leonard Posey.)

of the clitoris to resemble a penis, and deposition of proteins in the skin and especially in the muscles to give typical masculine characteristics.

In the prepubertal male, a virilizing adrenal tumor causes the same characteristics as in the female plus rapid development of the male sexual organs and creation of male sexual desires. Typical development of the male sexual organs in a 4-year-old boy with adrenogenital syndrome is shown in Figure 77–9.

In the adult male, the virilizing characteristics of adrenogenital syndrome are usually obscured by the normal virilizing characteristics of the testosterone secreted by the testes. Therefore, it is often difficult to make a diagnosis of adrenogenital syndrome in the adult male.

In adrenogenital syndrome, the excretion of 17-ketosteroids (which are derived from androgens) in the urine may be 10 to 15 times normal. This finding can be used in diagnosing the disease.

REFERENCES

Burnstein, K. L., and Cidlowski, J. A.: Regulation of gene expression by glucocorticoids. Annu. Rev. Physiol., 51:683, 1989.
Burtis, C. A., and Ashwood, E. R.: Tietz Textbook of Clinical Chemistry. Philadelphia, W. B. Saunders Co., 1994.

D'Agata, R., and Chrousos, G. P.: Recent Advances in Adrenal Regulation and Function. New York, Raven Press, 1987.

DeGroot, L. J. (ed.): Endocrinology, 2nd Ed. Philadelphia, W. B. Saunders Co., 1989.

de Kloet, E. R., and Sutanto, W.: Neurobiology of Steroids. San Diego, CA, Academic Press, 1994.

Felig, P., et al. (eds.): Endocrinology and Metabolism, 2nd Ed. New York, McGraw-Hill Book Co., 1987.

Funder, J. W., and Sheppard, K.: Adrenocortical steroids and the brain. Annu. Rev. Physiol., 49:397, 1987.

Funder, J. W.: Adrenal steroids: New answers, new questions. Science, 237:236, 1987.

Goodman, H. M.: Basic Medical Endocrinology. New York, Raven Press, 1994.

Hall, J. E., et al.: Control of arterial pressure and renal function during glucocorticoid excess in dogs. Hypertension, 2:139, 1980.

James, V. H. T.: The Adrenal Gland. New York, Raven Press, 1992.

Jones, M. T., and Gilham, B.: Factors involved in the regulation of adrenocorticotropic hormone/β-lipotropic hormone. Physiol. Rev., 68:743, 1988.

Kannan, C. R.: The Adrenal Gland. New York, Plenum Publishing Corp., 1988.

Lack, E. E.: Pathology of Adrenal and Extra-Adrenal Paraganglia. Philadelphia, W. B. Saunders Co., 1994.

Ludecke, D. K., et al.: ACTH, Cushing's Syndrome, and Other Hypercortisolemic States. New York, Raven Press, 1990.

McDougall, J. G.: The physiology of aldosterone secretion. News Physiol. Sci., 2:126, 1987.

McEwen, B. S., et al.: Adrenal steroid receptors and actions in the nervous system. Physiol. Rev., 66:1121, 1986.

Meyer, J. S.: Biochemical effects of corticosteroids on neural tissues. Physiol. Rev., 65:946, 1985.

Miller, M.: Assessment of hormonal disorders of water metabolism. Clin. Lab. Med., 4:729, 1984.

Morel, F., and Doucet, A.: Hormonal control of kidney functions at the cell level. Physiol. Rev., 66:377, 1986.

Moudgil, V. K. (ed.): Steroid Receptors in Health and Disease. New York, Plenum Publishing Corp., 1988.

Mulrow, P. J. (ed.): The Adrenal Gland. New York, Elsevier Science Publishing Co., 1986.

Parker, L.: Adrenal Androgens in Clinical Medicine. San Diego, CA, Academic Press, 1988.

Quinn, S. J., and Williams, G. H.: Regulation of aldosterone secretion. Annu. Rev. Physiol., 50:409, 1988.

Schatzberg, A. F., and Nemeroff, C. B. (eds.): The Hypothalamic-Pituitary-Adrenal Axis. New York, Raven Press, 1988.

Schneider, E. G., et al.: Effect of osmolality on aldosterone secretion. Endocrinology 116:1621, 1985.

Seldin, D. W., and Giebisch, G.: The Regulation of Potassium Balance. New York, Raven Press, 1989.

Seldin, D. W., and Giebisch, G.: The Regulation of Sodium and Chloride Balance. New York, Raven Press, 1989.

Smith, P. L., and McCabe, R. D.: Mechanism and regulation of transcellular potassium transport by the colon. Am. J. Physiol., 247:G445, 1984.

Wehling, M.: Genomic and Non-Genomic Effects of Aldosterone. Boca Raton, FL, CRC Press, Inc., 1994.

Young, D. B., and Guyton, A. C.: Steady state aldosterone dose response relationships. Circ. Res., 40(2):138, 1977.

Insulin, Glucagon, and Diabetes Mellitus

CHAPTER 78

The pancreas, in addition to its digestive functions, secretes two important hormones, *insulin* and *glucagon*. The purpose of this chapter is to discuss the functions of these hormones in regulating glucose, lipid, and protein metabolism as well as to discuss briefly the two diseases—*diabetes mellitus* and *hyperinsulinism*—caused, respectively, by hyposecretion of insulin and excess secretion of insulin.

PHYSIOLOGIC ANATOMY OF THE PANCREAS. The pancreas is composed of two major types of tissues, as shown in Figure 78–1: (1) the *acini*, which secrete digestive juices into the duodenum, and (2) the *islets of Langerhans*, which do not have any means for emptying their secretions externally but instead secrete insulin and glucagon directly into the blood. The digestive secretions of the pancreas are discussed in Chapter 64.

The pancreas of the human being has 1 to 2 million islets of Langerhans, each only about 0.3 millimeter in diameter and organized around small capillaries into which its cells secrete their hormones. The islets contain three major types of cells, *alpha, beta,* and *delta* cells, which are distinguished from one another by their morphologic and staining characteristics. The beta cells, constituting about 60 per cent of all the cells, lie mainly in the middle of each islet and secrete *insulin.* The alpha cells, about 25 per cent of the total, secrete *glucagon.* And the delta cells, about 10 per cent of the total, secrete *somatostatin.* In addition, at least one other type of cell, the PP cell, is present in small numbers in the islets and secretes a hormone of uncertain function called *pancreatic polypeptide.*

The close interrelations among these cell types in the islets of Langerhans allow direct control of secretion of some of the hormones by the other hormones. For instance, insulin inhibits glucagon secretion, and somato-statin inhibits the secretion of both insulin and glucagon.

INSULIN AND ITS METABOLIC EFFECTS

Insulin was first isolated from the pancreas in 1922 by Banting and Best, and almost overnight the outlook for the severely diabetic patient changed from one of rapid decline and death to that of a nearly normal person.

Historically, insulin has been associated with "blood sugar," and true enough, insulin has profound effects on carbohydrate metabolism. Yet it is abnormalities of fat metabolism, causing such conditions as acidosis and arteriosclerosis, that are the usual causes of death of a diabetic patient. Also, in patients with prolonged diabetes, diminished ability to synthesize proteins leads to wasting of the tissues as well as many cellular functional disorders. Therefore, it is clear that insulin affects fat and protein metabolism almost as much as it does carbohydrate metabolism.

Insulin Is a Hormone Associated with Energy Abundance

As we discuss insulin in the next few pages, it will become apparent that insulin secretion is associated with energy abundance. That is, when there is great abundance of energy-giving foods in the diet, especially excess amounts of carbohydrates and proteins, insulin is secreted in great quantity. This is especially

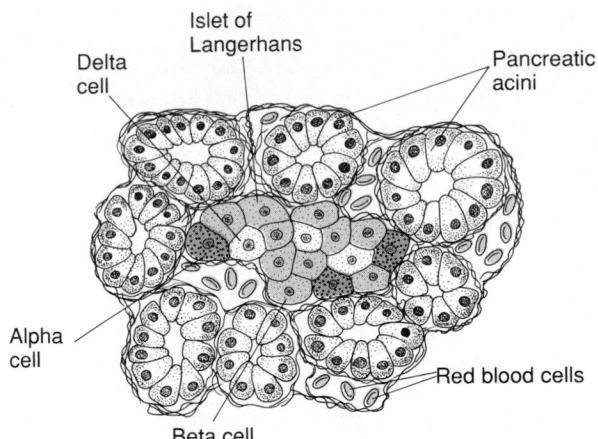

Delta cell
Islet of Langerhans
Pancreatic acini
Alpha cell
Red blood cells
Beta cell

Figure 78–1. Physiologic anatomy of an islet of Langerhans in the pancreas.

true for excess carbohydrates, less for excess proteins, but only slightly even for fats.

In turn, the insulin plays an important role in storing the excess energy substances. In the case of excess carbohydrates, it causes them to be stored as glycogen mainly in the liver and muscles. It causes fat storage in the adipose tissue. Also, all the excess carbohydrates that cannot be stored as glycogen are converted under the stimulus of insulin into fats and also stored in the adipose tissue. In the case of proteins, insulin has a direct effect in promoting amino acid uptake by cells and conversion of these amino acids into protein. In addition, it inhibits the breakdown of the proteins that are already in the cells.

Chemistry of Insulin

Insulin is a small protein; human insulin has a molecular weight of 5808. It is composed of two amino acid chains, shown in Figure 78–2, connected to each other by disulfide linkages. When the two amino acid chains are split apart, the functional activity of the insulin molecule is lost.

Insulin is synthesized in the beta cells by the usual cell machinery for protein synthesis, as explained in Chapter 3, beginning with translation of the insulin RNA by ribosomes attached to the endoplasmic reticulum to form an *insulin preprohormone*. This initial preprohormone has a molecular weight of about 11,500, but it is then cleaved in the endoplasmic reticulum to form a *proinsulin* with a molecular weight of

about 9000; most of this is further cleaved in the Golgi apparatus to form insulin before being packaged in the secretory granules. However, about one sixth of the final secreted product is still in the form of proinsulin. The proinsulin has virtually no insulin activity.

When insulin is secreted into the blood, it circulates almost entirely in an unbound form; it has a plasma half-life that averages only about 6 minutes, so that it is mainly cleared from the circulation within 10 to 15 minutes. Except for that portion of the insulin that combines with receptors in the target cells, the remainder is degraded by the enzyme insulinase mainly in the liver, to a lesser extent in the kidneys and muscle, and slightly in most other tissues. This rapid removal from the plasma is important because at times, it is equally as important to turn off rapidly as to turn on the control functions of insulin.

Activation of Target Cell Receptors by Insulin and the Resulting Cellular Effects

To initiate its effects on target cells, insulin first binds with and activates a membrane receptor protein that has a molecular weight of about 300,000. It is the activated receptor, not the insulin, that causes the subsequent effects.

The insulin receptor is a combination of four subunits held together by disulfide linkages, *two alpha subunits* that lie entirely outside the cell membrane and *two beta subunits* that penetrate through the membrane, protruding into the cell cytoplasm. The insulin binds with the alpha subunits on the outside of the cell, but because of the linkages with the beta subunits, the portions of the beta subunits protruding into the cell become autophosphorylated. This makes them become an activated enzyme, a local *protein kinase*, which in turn causes phosphorylation of multiple other intracellular enzymes. The net effect is to activate some of these enzymes while inactivating others. Thus, in this roundabout way, insulin directs the intracellular metabolic machinery to produce the desired effects. From this point on, the molecular mechanisms are almost entirely unknown.

The end effects of insulin stimulation are clear and are the following.

1. Within seconds after insulin binds with its membrane receptors, the membranes of about 80 per cent of the body's cells become highly permeable to glucose. This is especially true of muscle cells and adipose cells but *is not true of most neurons in the brain.* The increased permeability to glucose in turn allows rapid entry of glucose into the cells. Inside the

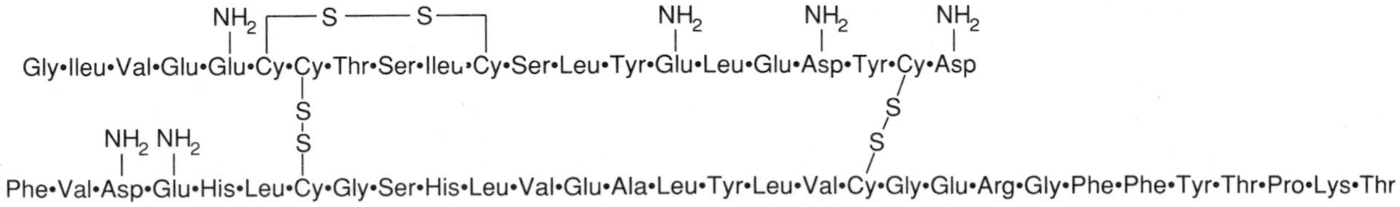

Gly·Ileu·Val·Glu·Glu·Cy·Cy·Thr·Ser·Ileu·Cy·Ser·Leu·Tyr·Glu·Leu·Glu·Asp·Tyr·Cy·Asp

Phe·Val·Asp·Glu·His·Leu·Cy·Gly·Ser·His·Leu·Val·Glu·Ala·Leu·Tyr·Leu·Val·Cy·Gly·Glu·Arg·Gly·Phe·Phe·Tyr·Thr·Pro·Lys·Thr

Figure 78–2. Human insulin molecule.

cell, the glucose is immediately phosphorylated and becomes a substrate for all the usual carbohydrate metabolic functions. The increased glucose transport is believed to result from fusion of multiple intracellular vesicles with the cell membrane, these vesicles carrying in their own membranes multiple molecules of glucose transport protein, a membrane protein with a molecular weight of about 55,000. When insulin is no longer available, these vesicles separate from the cell membrane within about 3 to 5 minutes and move back to the cell interior to be used again and again as needed.

2. In addition to increased membrane permeability for glucose, the cell membrane becomes more permeable for many of the amino acids, potassium ions, and phosphate ions.

3. Slower effects occur during the next 10 to 15 minutes to change the activity levels of many more intracellular metabolic enzymes. These effects result mainly from the changed states of phosphorylation of the enzymes.

4. Much slower effects continue to occur for hours and even several days. They result from changed rates of translation of messenger RNAs at the ribosomes to form new proteins and still slower effects from changed rates of transcription of DNA in the cell nucleus. In this way, insulin remolds much of the cellular enzymatic machinery to achieve its metabolic goals.

Effect of Insulin on Carbohydrate Metabolism

Immediately after a high-carbohydrate meal, the glucose that is absorbed into the blood causes rapid secretion of insulin, which is discussed in detail later in the chapter. The insulin in turn causes rapid uptake, storage, and use of glucose by almost all tissues of the body, but especially by the muscles, adipose tissue, and liver.

Effect of Insulin in Promoting Glucose Metabolism in Muscle

During much of the day, muscle tissue depends not on glucose for its energy but on fatty acids. The principal reason for this is that the normal *resting muscle* membrane is only slightly permeable to glucose except when the muscle fiber is stimulated by insulin; between meals, the amount of insulin that is secreted is too small to promote significant amounts of glucose entry into the muscle cells.

However, under two conditions the muscles do use large amounts of glucose. One of these is during moderate or heavy exercise. This usage of glucose does not require large amounts of insulin because exercising muscle fibers, for reasons not understood, become permeable to glucose even in the absence of insulin because of the contraction process itself.

The second condition for muscle usage of large amounts of glucose is during the few hours after a meal. At this time the blood glucose concentration is high; the pancreas is secreting large quantities of insulin. The extra insulin causes rapid transport of glucose into the muscle cells. This causes the muscle cell during this period to use glucose preferentially over fatty acids, as we discuss later.

STORAGE OF GLYCOGEN IN MUSCLE. If the muscles are not exercising after a meal and yet glucose is transported into the muscle cells in abundance, then most of the glucose is stored in the form of muscle glycogen instead of being used for energy, up to a limit of 2 to 3 per cent concentration. The glycogen can later be used for energy by the muscle. It is especially useful for short periods of extreme energy use by the muscles and even to provide spurts of anaerobic energy for a few minutes at a time by glycolytic breakdown of the glycogen to lactic acid, which can occur even in the absence of oxygen.

QUANTITATIVE DEGREE OF INSULIN FACILITATION OF GLUCOSE TRANSPORT THROUGH THE MUSCLE CELL MEMBRANE. The quantitative effect of insulin to facilitate glucose transport through the muscle cell membrane is demonstrated by the experimental results shown in Figure 78–3. The lower curve labeled "control" shows the concentration of free glucose measured inside the cell, demonstrating that the glucose concentration remained almost zero despite increased extracellular glucose concentration up to as high as 750 mg/dl. In contrast, the curve labeled "insulin" demonstrates that the intracellular glucose concentration rose to as high as 400 mg/dl when insulin was added. Thus, it is clear that insulin can increase the rate of transport of glucose into the resting muscle cell by at least 15-fold.

Effect of Insulin on Promoting Liver Uptake, Storage, and Use of Glucose

One of the most important of all the effects of insulin is to cause most of the glucose absorbed after a meal to be stored almost immediately in the liver in the form of glycogen. Then, between meals, when food is not available and the blood glucose concentration begins to fall, insulin secretion decreases rapidly and the liver glycogen is split back into glucose, which is

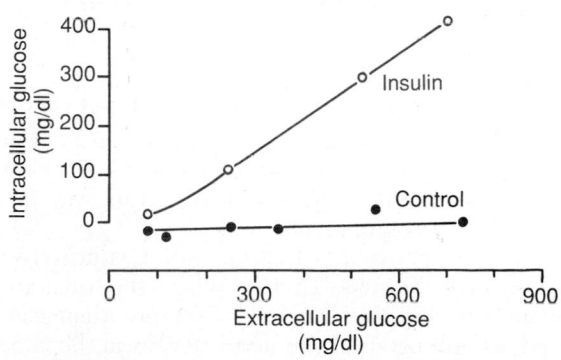

Figure 78–3. Effect of insulin in enhancing the concentration of glucose inside muscle cells. Note that in the absence of insulin (control), the intracellular glucose concentration remained near zero despite high extracellular glucose concentrations. (From Park, Morgan, Kaji, and Smith, *in* Eisenstein [ed.]: The Biochemical Aspects of Hormone Action. Boston, Little, Brown & Co.)

released back into the blood to keep the blood glucose concentration from falling too low.

The mechanism by which insulin causes glucose uptake and storage in the liver includes several almost simultaneous steps.

1. Insulin *inactivates liver phosphorylase,* the principal enzyme that causes liver glycogen to split into glucose. This prevents breakdown of the glycogen that has been stored in the liver cells.

2. Insulin causes *enhanced uptake of glucose* from the blood by the liver cells. It does this by *increasing the activity of the enzyme glucokinase,* which is one of the enzymes that causes the initial phosphorylation of glucose after it diffuses into the liver cells. Once phosphorylated, the glucose is *temporarily* trapped inside the liver cells because phosphorylated glucose cannot diffuse back through the cell membrane.

3. Insulin also increases the activities of the enzymes that promote glycogen synthesis, including especially *glycogen synthase,* which is responsible for polymerization of the monosaccharide units to form the glycogen molecules.

The net effect of all these actions is to increase the amount of glycogen in the liver. The glycogen can increase to a total of about 5 to 6 per cent of the liver mass, which is equivalent to almost 100 grams of stored glycogen in the whole liver.

RELEASE OF GLUCOSE FROM THE LIVER BETWEEN MEALS. After the meal is over and the blood glucose level begins to fall to a low level, several events now transpire that cause the liver to release glucose back into the circulating blood.

1. The decreasing blood glucose causes the pancreas to decrease its insulin secretion.

2. The lack of insulin then reverses all the effects listed above for glycogen storage, essentially stopping further synthesis of glycogen in the liver and preventing further uptake of glucose by the liver from the blood.

3. The lack of insulin (along with increase of glucagon, which is discussed later) activates the enzyme *phosphorylase,* which causes the splitting of glycogen into *glucose phosphate.*

4. The enzyme *glucose phosphatase,* which had been inhibited by insulin, now becomes activated by the insulin lack and causes the phosphate radical to split away from the glucose; this allows the free glucose to diffuse back into the blood.

Thus, the liver removes glucose from the blood when it is present in excess after a meal and returns it to the blood when the blood glucose concentration falls between meals. Ordinarily, about 60 per cent of the glucose in the meal is stored in this way in the liver and then returned later.

OTHER EFFECTS OF INSULIN ON CARBOHYDRATE METABOLISM IN THE LIVER. When the quantity of glucose entering the liver cells is more than can be stored as glycogen or be used for local hepatocyte metabolism, *insulin promotes the conversion of all this excess glucose into fatty acids.* These fatty acids are subsequently packaged as triglycerides in very low density lipoproteins and transported in this form by way of the blood to the adipose tissue and deposited as fat.

Insulin also *inhibits gluconeogenesis.* It does this mainly by decreasing the quantities and activities of the liver enzymes required for gluconeogenesis. However, part of the effect is caused by an action of insulin that decreases the release of amino acids from muscle and other extrahepatic tissues and in turn the availability of these necessary precursors required for gluconeogenesis. This is discussed further in relation to the effect of insulin on protein metabolism.

Lack of Effect of Insulin on Glucose Uptake and Usage by the Brain

The brain is quite different from most other tissues of the body in that insulin has little or no effect on uptake or use of glucose. Instead, *the brain cells are even normally permeable to glucose and can use glucose without the intermediation of insulin.*

The brain cells are also quite different from most other cells of the body in that they normally use only glucose for energy and can use other energy substrates, such as fats, only with difficulty. Therefore, it is essential that the blood glucose level be maintained always above a critical level, which is one of the most important functions of the blood glucose control system. When the blood glucose does fall too low, into the range of 20 to 50 mg/dl, symptoms of *hypoglycemic shock* develop, characterized by progressive nervous irritability that leads to fainting, seizures, and even coma.

Effect of Insulin on Carbohydrate Metabolism in Other Cells

Insulin increases glucose transport into and glucose usage by most other cells of the body (with the exception of the brain cells, as noted) in the same way that it affects glucose transport and usage in the muscle cell. The transport of glucose into adipose cells mainly provides the glycerol portion of the fat molecule. Therefore, in this indirect way, insulin promotes deposition of fat in these cells.

Effect of Insulin on Fat Metabolism

Although not quite as visible as the acute effects of insulin on carbohydrate metabolism, insulin also affects fat metabolism in ways that, in the long run, are equally as important. Especially dramatic is the long-term effect of *insulin lack* in causing extreme atherosclerosis, often leading to heart attacks, cerebral strokes, and other vascular accidents. But first, let us discuss the acute effects of insulin on fat metabolism.

Effect of Insulin Excess on Fat Synthesis and Storage

Insulin has several effects that lead to fat storage in adipose tissue. First, insulin increases the utilization of glucose by most of the body's tissues, which automatically decreases the utilization of fat, thus functioning as a "fat sparer." However, insulin also promotes fatty acid synthesis. This is especially true when more carbohydrates are ingested than can be used for immediate energy, thus providing the substrate for fat synthesis. Almost all this synthesis occurs in the liver cells, and the fatty acids are then transported from the liver by way of the blood lipoproteins to the adipose cells to be stored. The different factors that lead to increased fatty acid synthesis in the liver include the following.

1. Insulin increases the transport of glucose into the liver cells. After the liver glycogen concentration reaches 5 to 6 per cent, this in itself inhibits further glycogen synthesis. Then all the additional glucose entering the liver cells becomes available to form fat. The glucose is first split to pyruvate in the glycolytic pathway, and the pyruvate subsequently is converted to acetyl coenzyme A (acetyl-CoA), the substrate from which fatty acids are synthesized.

2. An excess of *citrate* and *isocitrate ions* is formed by the citric acid cycle when excess amounts of glucose are being used for energy. These ions then have a direct effect in activating *acetyl-CoA carboxylase,* the enzyme required to carboxylate acetyl-CoA to form malonyl-CoA, the first stage of fatty acid synthesis.

3. Most of the fatty acids are then synthesized within the liver itself and used to form triglycerides, the usual form of storage fat. They are released from the liver cells to the blood in the lipoproteins. Insulin activates *lipoprotein lipase* in the capillary walls of the adipose tissue, which splits the triglycerides again into fatty acids, a requirement for them to be absorbed into the adipose cells, where they are again converted to triglycerides and stored.

STORAGE OF FAT IN THE ADIPOSE CELLS. Insulin has two other essential effects that are required for fat storage in adipose cells.

1. Insulin *inhibits the action of hormone-sensitive lipase.* This is the enzyme that causes hydrolysis of the triglycerides already stored in the fat cells. Therefore, the release of fatty acids from the adipose tissue into the circulating blood is inhibited.

2. Insulin *promotes glucose transport through the cell membrane into the fat cells* in exactly the same way that it promotes glucose transport into muscle cells. Some of this glucose is then used to synthesize minute amounts of fatty acids, but more important, it also forms large quantities of *α-glycerol phosphate.* This substance supplies the *glycerol* that combines with fatty acids to form the triglycerides that are the storage form of fat in adipose cells. Therefore, when insulin is not available, even storage of the large amounts of fatty acids transported from the liver in the lipoproteins is almost blocked.

Increased Metabolic Use of Fat Caused by Insulin Lack

All aspects of fat breakdown and use for providing energy are greatly *enhanced in the absence of insulin.* This occurs even normally between meals when secretion of insulin is minimal, but it becomes extreme in diabetes mellitus when secretion of insulin is almost zero. The resulting effects are as follows.

LIPOLYSIS OF STORAGE FAT AND RELEASE OF FREE FATTY ACIDS DURING INSULIN LACK. In the absence of insulin, all the effects of insulin noted above that

cause storage of fat are reversed. The most important effect is that the enzyme *hormone-sensitive lipase* in the fat cells becomes strongly activated. This causes hydrolysis of the stored triglycerides, releasing large quantities of fatty acids and glycerol into the circulating blood. Consequently, the plasma concentration of free fatty acids begins to rise within minutes. This free fatty acid then becomes the main energy substrate used by essentially all tissues of the body besides the brain. Figure 78–4 shows the effect of insulin lack on the plasma concentration of free fatty acids, glucose, and acetoacetic acid. Note that immediately after removal of the pancreas, the free fatty acid concentration in the plasma begins to rise, rising considerably more rapidly even than the concentration of glucose.

EFFECT OF INSULIN LACK ON PLASMA CHOLESTEROL AND PHOSPHOLIPID CONCENTRATIONS. The excess of fatty acids in the plasma also promotes liver conversion of some of the fatty acids into phospholipids and cholesterol, two of the major products of fat metabolism. These two substances, along with excess triglycerides formed at the same time in the liver, are then discharged into the blood in the lipoproteins. Occasionally the plasma lipoproteins increase as much as threefold in the absence of insulin, giving a total concentration of plasma lipids of several per cent rather than the normal 0.6 per cent. This high lipid concentration—especially the high concentration of cholesterol—leads to rapid development of atherosclerosis in people with serious diabetes.

THE EXCESS USAGE OF FATS DURING INSULIN LACK CAUSES KETOSIS AND ACIDOSIS. Insulin lack also causes excessive amounts of *acetoacetic acid* to be formed in the liver cells. This results from the following effect: In the absence of insulin but in the presence of excess fatty acids in the liver cells, the carnitine transport mechanism for transporting fatty acids

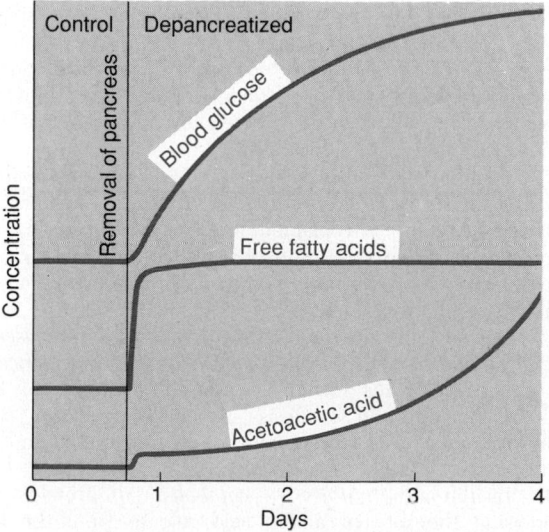

Figure 78–4. Effect of removing the pancreas on the approximate concentrations of blood glucose, plasma free fatty acids, and acetoacetic acid.

into the mitochondria becomes increasingly activated. In the mitochondria, beta oxidation of the fatty acids then proceeds exceedingly rapidly, releasing extreme amounts of acetyl-CoA. A large part of this excess acetyl-CoA is then condensed to form acetoacetic acid, which in turn is released into the circulating blood. Most of this passes to the peripheral cells, where it is again converted into acetyl-CoA and used for energy in the usual manner.

At the same time, the absence of insulin also depresses the utilization of acetoacetic acid in the peripheral tissues. Thus, so much acetoacetic acid is released from the liver that it cannot all be metabolized by the tissues. Therefore, as shown in Figure 78–4, its concentration rises during the days after cessation of insulin secretion, sometimes reaching concentrations of 10 mEq/liter or more, which is a severe state of body fluid acidosis. As explained in Chapter 68, some of the acetoacetic acid is also converted into β-hydroxybutyric acid and *acetone*. These two substances, along with the acetoacetic acid, are called *ketone bodies,* and their presence in large quantities in the body fluids is called *ketosis.* We see later that in severe diabetes the acetoacetic acid and the β-hydroxybutyric acid can cause severe *acidosis* and *coma,* which often leads to death.

Effect of Insulin on Protein Metabolism and on Growth

EFFECT OF INSULIN ON PROTEIN SYNTHESIS AND STORAGE. During the few hours after a meal when excess quantities of nutrients are available in the circulating blood, not only carbohydrates and fats but proteins as well are stored in the tissues; insulin is required for this to occur. The manner in which insulin causes protein storage is not as well understood as the mechanisms for both glucose and fat storage. Some of the facts follow.

1.   Insulin causes active transport of many of the amino acids into the cells. Among the amino acids most strongly transported are *valine, leucine, isoleucine, tyrosine,* and *phenylalanine.* Thus, insulin shares with growth hormone the capability of increasing the uptake of amino acids into cells. However, the amino acids affected are not necessarily the same ones.

2.   Insulin has a direct effect on the ribosomes in *increasing the translation of messenger RNA,* thus forming new proteins. In some unexplained way, insulin "turns on" the ribosomal machinery. In the absence of insulin, the ribosomes simply stop working, almost as if insulin operates an "on-off" mechanism.

3.   Over a longer period of time, insulin also *increases the rate of transcription of selected DNA genetic sequences* in the cell nuclei, thus forming increased quantities of RNA and still more protein synthesis—especially promoting a vast array of enzymes for storage of carbohydrates, fats, and proteins.

4.   Insulin also *inhibits the catabolism of proteins,* thus decreasing the rate of amino acid release from the cells, especially from the muscle cells. Presumably this results from some ability of the insulin to diminish the normal degradation of proteins by the cellular lysosomes.

5.   In the liver, insulin *depresses the rate of gluconeogenesis.* It does this by decreasing the activity of the enzymes that promote gluconeogenesis. Because the substrates most used for synthesis of glucose by gluconeogenesis are the plasma amino acids, this suppression of gluconeogenesis conserves the amino acids in the protein stores of the body.

In summary, insulin promotes protein formation and prevents the degradation of proteins.

INSULIN LACK CAUSES PROTEIN DEPLETION AND INCREASED PLASMA AMINO ACIDS. Virtually all protein storage comes to a halt when insulin is not available. The catabolism of proteins increases, protein synthesis stops, and large quantities of amino acids are dumped into the plasma. The plasma amino acid concentration rises considerably, and most of the excess amino acids are used either directly for energy or as substrates for gluconeogenesis. This degradation of the amino acids also leads to enhanced urea excretion in the urine.

The resulting protein wasting is one of the most serious of all the effects of severe diabetes mellitus. It can lead to extreme weakness as well as many deranged functions of the organs.

EFFECT OF INSULIN ON GROWTH—ITS SYNERGISTIC EFFECT WITH GROWTH HORMONE. Because insulin is required for the synthesis of proteins, it is equally as essential for growth of an animal as is growth hormone. This is demonstrated in Figure 78–5, which shows that a depancreatized, hypophysectomized rat without therapy hardly grows at all. Furthermore, the administration of either growth hormone or insulin one at a time causes almost no growth. Yet a combination of these hormones does cause dramatic growth. Thus, it appears that the two hormones function synergistically to promote growth, each performing a specific function that is separate from that of the other. Perhaps a small part of this necessity for both hormones results from the fact that each promotes cellular uptake of a different selection of amino acids, all of which are required if growth is to be achieved.

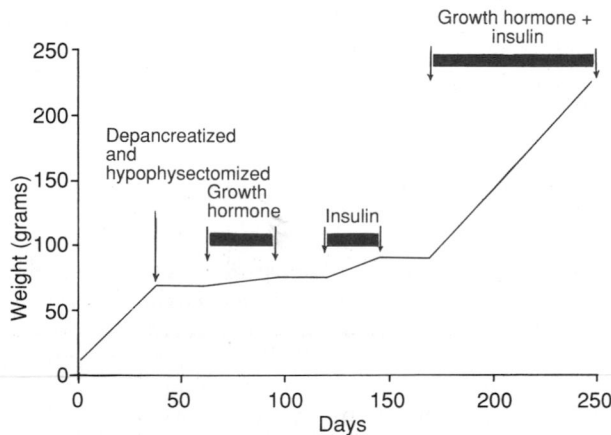

Figure 78–5. Effect of growth hormone, insulin, and growth hormone plus insulin on growth in a depancreatized and hypophysectomized rat.

Control of Insulin Secretion

Formerly, it was believed that insulin secretion was controlled almost entirely by the blood glucose concentration. However, as more has been learned about the metabolic functions of insulin for protein and fat metabolism, it has been learned that blood amino acids and other factors also play important roles in controlling insulin secretion.

STIMULATION OF INSULIN SECRETION BY BLOOD GLUCOSE. At the normal *fasting* level of blood glucose of 80 to 90 mg/dl, the rate of insulin secretion is minimal—on the order of 25 ng/min/kg of body weight, a level that has only slight physiological activity. If the blood glucose concentration is suddenly increased to a level two to three times normal and kept at this high level thereafter, insulin secretion increases markedly in two stages, as shown by the changes in plasma insulin concentration seen in Figure 78–6.

1. Plasma insulin concentration increases almost 10-fold within 3 to 5 minutes after the acute elevation of the blood glucose; this results from immediate dumping of preformed insulin from the beta cells of the islets of Langerhans. However, the initial high rate of secretion is not maintained; instead, the insulin concentration decreases about halfway back toward normal in another 5 to 10 minutes.

2. Beginning at about 15 minutes, insulin secretion rises a second time and reaches a new plateau in 2 to 3 hours, this time usually at a rate of secretion even greater than that in the initial phase. This secretion results both from additional release of preformed insulin and from activation of the enzyme system that synthesizes and releases new insulin from the cells.

FEEDBACK RELATION BETWEEN BLOOD GLUCOSE CONCENTRATION AND INSULIN SECRETION RATE. As the concentration of blood glucose rises above 100 mg/dl of blood, the rate of insulin secretion rises rapidly, reaching a peak some 10 to 25 times the basal level at blood glucose concentrations between 400 and 600 mg/dl, as shown in Figure 78–7. Thus, the increase in insulin secretion under a glucose stimulus is

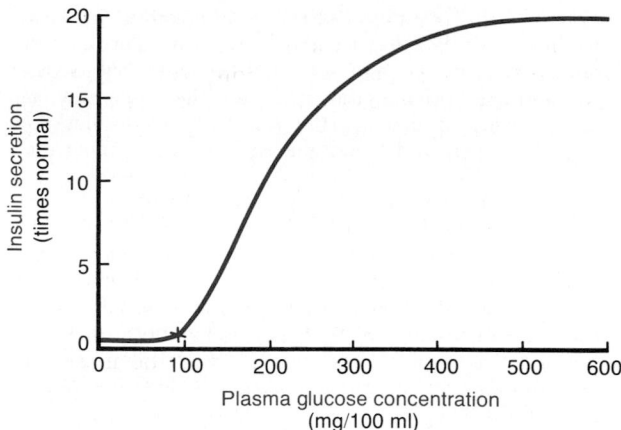

Figure 78–7. Approximate insulin secretion at different plasma glucose levels.

dramatic both in its rapidity and in the tremendous level of secretion achieved. Furthermore, the turn-off of insulin secretion is almost equally as rapid, occurring within 3 to 5 minutes after reduction in blood glucose concentration back to the fasting level.

This response of insulin secretion to an elevated blood glucose concentration provides an extremely important feedback mechanism for regulating blood glucose concentration. That is, any rise in blood glucose increases insulin secretion, and the insulin in turn increases transport of glucose into liver, muscle, and other cells, thereby reducing the blood glucose concentration back toward the normal value.

Other Factors That Stimulate Insulin Secretion

AMINO ACIDS. In addition to the stimulating of insulin secretion by excess blood glucose, some of the amino acids have a similar effect. The most potent of these are *arginine* and *lysine*. This effect differs from glucose stimulating of insulin secretion in the following way: Amino acids administered in the absence of a rise in blood glucose cause only a small increase in insulin secretion. However, when administered at the same time that the blood glucose concentration is elevated, the glucose-induced secretion of insulin may be as much as doubled in the presence of the excess amino acids. Thus, *the amino acids strongly potentiate the glucose stimulus* for insulin secretion.

The stimulation of insulin secretion by amino acids seems to be a purposeful response because the insulin in turn promotes transport of amino acids into the tissue cells as well as intracellular formation of protein. That is, insulin is important for proper utilization of excess amino acids in the same way that it is important for the utilization of carbohydrates.

GASTROINTESTINAL HORMONES. A mixture of several important gastrointestinal hormones—*gastrin, secretin, cholecystokinin,* and *gastric inhibitory peptide* (which seems to be the most potent of them all)—will cause a moderate increase in insulin secretion. These hormones are released in the gastrointestinal tract after a person

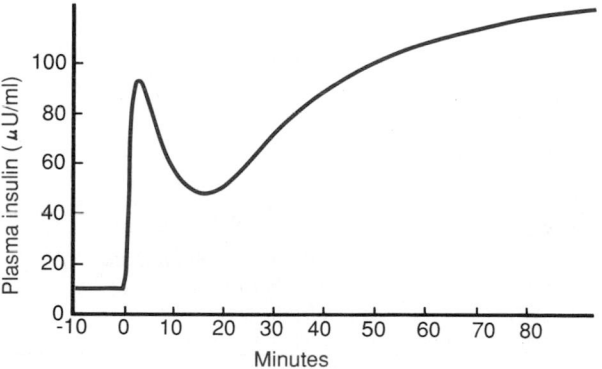

Figure 78–6. Increase in plasma insulin concentration after a sudden increase in blood glucose to two to three times the normal range. Note an initial rapid surge in insulin concentration and then a delayed but higher and continuing increase in concentration beginning 15 to 20 minutes later.

eats a meal. They then cause an "anticipatory" increase in blood insulin in preparation for the glucose and amino acids to be absorbed from the meal. These gastrointestinal hormones generally act the same way as amino acids to increase the sensitivity of insulin response to increased blood glucose, almost doubling the rate of insulin secretion as the blood glucose level rises.

OTHER HORMONES AND THE AUTONOMIC NERVOUS SYSTEM. Other hormones that either directly increase insulin secretion or potentiate the glucose stimulus for insulin secretion include *glucagon, growth hormone, cortisol*, and, to a lesser extent, *progesterone* and *estrogen*. The importance of the stimulatory effects of these hormones is that prolonged secretion of any one of them in large quantities can occasionally lead to exhaustion of the beta cells of the islets of Langerhans and thereby cause diabetes mellitus. Indeed, diabetes often occurs in people who are maintained on high pharmacological doses of some of these hormones. Diabetes is particularly common in giants or acromegalic people with growth hormone–secreting tumors or in people whose adrenal glands or adrenal gland tumors secrete excess glucocorticoids.

Under some conditions, stimulation of either the parasympathetic or the sympathetic nerves to the pancreas can increase insulin secretion. However, it is doubtful that either of these effects is of physiological significance for regulating insulin secretion.

Role of Insulin (and Other Hormones) in "Switching" Between Carbohydrate and Lipid Metabolism

From the preceding discussions, it should be clear that insulin promotes the utilization of carbohydrates for energy, whereas it depresses the utilization of fats. Conversely, lack of insulin causes fat utilization mainly to the exclusion of glucose utilization, except by brain tissue. Furthermore, the signal that controls this switching mechanism is principally the blood glucose concentration. When the glucose concentration is low, insulin secretion is suppressed and fat is used almost exclusively for energy everywhere except in the brain. When the glucose concentration is high, insulin secretion is stimulated and carbohydrate is used instead of fat, and the excess blood glucose is stored in the form of liver glycogen, liver fat, and muscle glycogen. Therefore, one of the most important functional roles of insulin in the body is to control which of these two foods from moment to moment will be used by the cells for energy.

At least four other known hormones also play important roles in this switching mechanism: *growth hormone* from the anterior pituitary gland, *cortisol* from the adrenal cortex, *epinephrine* from the adrenal medulla, and *glucagon* from the alpha cells of the islets of Langerhans in the pancreas. Glucagon is discussed in the next section of this chapter. Both growth hormone and cortisol are secreted in response to hypoglycemia, and both inhibit cellular utilization of glucose while promoting fat utilization. However, the effects of both of these hormones develop slowly, usually requiring many hours to obtain maximal levels.

Epinephrine is especially important in increasing plasma glucose concentration during periods of stress when the sympathetic nervous system is excited. However, epinephrine acts differently from the other hormones in that it increases the plasma fatty acid concentration at the same time. The reasons for these effects are as follows: (1) epinephrine has the potent effect of causing glycogenolysis in the liver, thus releasing within minutes large quantities of glucose into the blood; (2) it also has a direct lipolytic effect on the adipose cells because it activates adipose tissue hormone-sensitive lipase, thus greatly enhancing the blood concentration of fatty acids as well. Quantitatively, the enhancement of fatty acids is far greater than the enhancement of blood glucose. Therefore, epinephrine especially enhances the utilization of fat in such stressful states as exercise, circulatory shock, and anxiety.

GLUCAGON AND ITS FUNCTIONS

Glucagon, a hormone secreted by the *alpha cells* of the islets of Langerhans when the blood glucose concentration falls, has several functions that are diametrically opposed to those of insulin. Most important of these functions is to increase the blood glucose concentration, an effect that is exactly the opposite of that of insulin.

Like insulin, glucagon is a large polypeptide. It has a molecular weight of 3485 and is composed of a chain of 29 amino acids. On injection of purified glucagon into an animal, a profound *hyper*glycemic effect occurs. Only 1 μg/kg of glucagon can elevate the blood glucose concentration about 20 mg/dl of blood (a 25 per cent increase) in about 20 minutes. For this reason, glucagon is also called the *hyperglycemic hormone*.

Effects on Glucose Metabolism

The major effects of glucagon on glucose metabolism are (1) breakdown of liver glycogen (*glycogenolysis*) and (2) increased *gluconeogenesis* in the liver. Both of these effects greatly enhance the availability of glucose to the other organs of the body.

GLYCOGENOLYSIS AND INCREASED BLOOD GLUCOSE CONCENTRATION CAUSED BY GLUCAGON. The most dramatic effect of glucagon is its ability to cause glycogenolysis in the liver, which in turn increases the blood glucose concentration within minutes.

It does this by the following complex cascade of events.

1. Glucagon activates *adenyl cyclase* in the hepatic cell membrane,
2. Which causes the formation of *cyclic adenosine monophosphate*,
3. Which activates *protein kinase regulator protein*,
4. Which activates *protein kinase*,
5. Which activates *phosphorylase b kinase*,
6. Which converts *phosphorylase b* into *phosphorylase a*,
7. Which promotes the degradation of glycogen into glucose-1-phosphate,
8. Which then is dephosphorylated; and the glucose is released from the liver cells.

This sequence of events is exceedingly important for several reasons. First, it is one of the most thoroughly studied of all the *second messenger* functions of cyclic adenosine monophosphate. Second, it demonstrates a cascading system in which *each succeeding product is produced in greater quantity than the preceding product.* Therefore, it represents a potent *amplifying* mechanism; this type of amplifying mechanism is widely used throughout our body for controlling many, if not most, cellular metabolic systems, often causing as much as a millionfold amplification in response. This explains how *only a few micrograms of glucagon can cause the blood glucose level to double or more within a few minutes.*

Infusion of glucagon for about 4 hours can cause such intensive liver glycogenolysis that all the liver stores of glycogen become depleted.

INCREASED GLUCONEOGENESIS CAUSED BY GLUCAGON. Even after all the glycogen in the liver has been exhausted under the influence of glucagon, continued infusion of this hormone still causes continued hyperglycemia. This results from the effect of glucagon to increase the rate of amino acid uptake by the liver cells and then the conversion of many of them to glucose by gluconeogenesis. This is achieved by activating multiple enzymes that are required for amino acid transport and gluconeogenesis, especially activation of the enzyme system for converting pyruvate to phosphoenolpyruvate, a rate-limiting step in gluconeogenesis.

Other Effects of Glucagon

Most other effects of glucagon occur only when its concentration rises well above the maximum normally found in the blood. Perhaps the most important effect is that glucagon *activates adipose cell lipase,* making increased quantities of fatty acids available to the energy systems of the body. It also inhibits the storage of triglycerides in the liver, which prevents the liver from removing fatty acids from the blood; this also helps make additional amounts of fatty acids available for the other tissues of the body.

Glucagon in very abnormally large concentrations also (1) enhances the strength of the heart, (2) enhances bile secretion, and (3) inhibits gastric acid secretion. All these effects are probably unimportant in the normal function of the body.

Regulation of Glucagon Secretion

INCREASED BLOOD GLUCOSE INHIBITS GLUCAGON SECRETION. The blood glucose concentration is by far the most potent factor that controls glucagon secretion. Note specifically, however, that *the effect of blood glucose concentration on glucagon secretion is in exactly the opposite direction from the effect of glucose on insulin secretion.* This is demonstrated in Figure 78–8, showing that a *decrease* in the blood glucose concentration from its normal fasting level of about 90 mg/dl of blood down to hypoglycemic levels can increase the plasma concentration of glucagon severalfold. On the other hand, increasing the blood glucose to hyperglycemic levels decreases plasma glucagon. Thus, in hy-

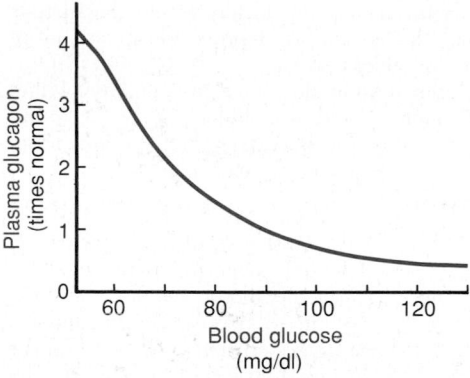

Figure 78–8. Approximate plasma glucagon concentration at different blood glucose levels.

poglycemia, glucagon is secreted in large amounts; it then greatly increases the output of glucose from the liver and thereby serves the important function of correcting the hypoglycemia.

EXCITATORY EFFECT OF AMINO ACIDS. High concentrations of amino acids, as occur in the blood after a protein meal (especially the amino acids *alanine* and *arginine*), *stimulate* the secretion of glucagon. This is the same effect that amino acids have in stimulating insulin secretion. *Thus, in this instance, the glucagon and insulin responses are* not *opposites.*

The importance of amino acid stimulation of glucagon secretion is that the glucagon then promotes rapid conversion of the amino acids to glucose, thus making even more glucose available to the tissues.

EXCITATORY EFFECT OF EXERCISE. In exhaustive exercise, the blood concentration of glucagon often increases fourfold to fivefold. What causes this is not understood because the blood glucose concentration does not necessarily fall. A beneficial effect of the glucagon is that it prevents a decrease in blood glucose. One of the factors that might increase glucagon secretion in exercise is increased circulating amino acids. Other factors, such as autonomic nervous stimulation of the islets of Langerhans, also play a role.

SOMATOSTATIN—ITS EFFECT ON INHIBITION OF GLUCAGON AND INSULIN SECRETION

The *delta cells* of the islets of Langerhans secrete the hormone *somatostatin,* a polypeptide containing only 14 amino acids that has an extremely short half-life in the circulating blood of only 3 minutes. Almost all factors related to the ingestion of food stimulate somatostatin secretion. They include (1) increased blood glucose, (2) increased amino acids, (3) increased fatty acids, and (4) increased concentrations of several of the gastrointestinal hormones released from the upper gastrointestinal tract in response to food intake.

In turn, somatostatin has multiple inhibitory effects as follows:

1. Somatostatin acts locally within the islets of Langerhans themselves to depress the secretion of both insulin and glucagon.

2. Somatostatin decreases the motility of the stomach, duodenum, and gallbladder.

3. Somatostatin decreases both secretion and absorption in the gastrointestinal tract.

Putting all this information together, it has been suggested that the principal role of somatostatin is to extend the period of time over which the food nutrients are assimilated into the blood. At the same time, the effect of somatostatin to depress insulin and glucagon secretion decreases the utilization of the absorbed nutrients by the tissues, thus preventing rapid exhaustion of the food and therefore making it available over a longer period of time.

It should also be recalled that somatostatin is the same chemical substance as *growth hormone inhibitory hormone,* which is secreted in the hypothalamus and suppresses anterior pituitary gland growth hormone secretion.

SUMMARY OF BLOOD GLUCOSE REGULATION

In a normal person, the blood glucose concentration is narrowly controlled, usually between 80 and 90 mg/dl of blood in the fasting person each morning before breakfast. This concentration increases to 120 to 140 mg/dl during the first hour or so after a meal, but the feedback systems for control of blood glucose return the glucose concentration rapidly back to the control level, usually within 2 hours after the last absorption of carbohydrates. Conversely, in starvation, the gluconeogenesis function of the liver provides the glucose that is required to maintain the fasting blood glucose level.

The mechanisms for achieving this high degree of control have been presented in this chapter. Let us summarize them.

1. The liver functions as an important *blood glucose buffer system.* That is, when the blood glucose rises to a high concentration after a meal and the rate of insulin secretion also increases, as much as two thirds of the glucose absorbed from the gut is almost immediately stored in the liver in the form of glycogen. Then, during the succeeding hours, when both the blood glucose concentration and the rate of insulin secretion fall, the liver releases the glucose back into the blood. In this way, the liver decreases the fluctuations in blood glucose concentration to about one third what they would otherwise be. In fact, in patients with severe liver disease, it becomes almost impossible to maintain a narrow range of blood glucose concentration.

2. It is clear that both insulin and glucagon function as important feedback control systems for maintaining a normal blood glucose concentration. When the glucose concentration rises too high, insulin is secreted; the insulin in turn causes the blood glucose concentration to decrease toward normal. Conversely, a decrease in blood glucose stimulates glucagon secretion; the glucagon then functions in the opposite direction to increase the glucose up toward normal. Under most normal conditions, the insulin feedback mechanism is many times as important as the glucagon mechanism, but in in-

stances of starvation or excessive utilization of glucose during exercise and other stressful situations, the glucagon mechanism also becomes valuable.

3. Also, in severe hypoglycemia, a direct effect of low blood glucose on the hypothalamus stimulates the sympathetic nervous system. In turn, the epinephrine secreted by the adrenal glands causes still further release of glucose from the liver. This, too, helps protect against severe hypoglycemia.

4. And finally, over a period of hours and days, both growth hormone and cortisol are secreted in response to prolonged hypoglycemia, and they both decrease the rate of glucose utilization by most cells of the body, converting instead to greater amounts of fat utilization. This, too, helps return the blood glucose concentration toward normal.

IMPORTANCE OF BLOOD GLUCOSE REGULATION. One might ask the question: Why is it so important to maintain a constant blood glucose concentration, particularly since most tissues can shift to utilization of fats and proteins for energy in the absence of glucose? The answer is that glucose is the only nutrient that normally can be used by the *brain, retina,* and *germinal epithelium of the gonads* in sufficient quantities to supply them optimally with their required energy. Therefore, it is important to maintain the blood glucose concentration at a sufficiently high level to provide this necessary nutrition.

Most of the glucose formed by gluconeogenesis during the interdigestive period is used for metabolism in the brain. Indeed, it is important that the pancreas not secrete any insulin during this time; otherwise, the scant supplies of glucose that are available would all go into the muscles and other peripheral tissues, leaving the brain without a nutritive source.

On the other hand, it is also important that the blood glucose concentration not rise too high for three reasons: First, glucose exerts a large amount of osmotic pressure in the extracellular fluid, and if the glucose concentration rises to excessive values, this can cause considerable cellular dehydration. Second, an excessively high level of blood glucose concentration causes loss of glucose in the urine. Third, this causes osmotic diuresis by the kidneys, which can deplete the body of its fluids and electrolytes.

DIABETES MELLITUS

In most instances, diabetes mellitus results from diminished secretion of insulin by the beta cells of the islets of Langerhans. Heredity usually plays a major role in determining in whom diabetes will develop and in whom it will not. Sometimes it does this by increasing the susceptibility of the beta cells to destruction by viruses or by favoring the development of autoimmune antibodies against the beta cells, thus also leading to their destruction. In other instances, there appears to be a simple hereditary tendency for beta cell degeneration.

Obesity also plays a role in the development of clinical diabetes. One reason is that obesity decreases the number of insulin receptors in the insulin target cells throughout the body, thus making the amount of insulin that is available less effective in promoting its usual metabolic effects.

Pathological Physiology of Diabetes Mellitus

Most of the pathological features of diabetes mellitus can be attributed to one of the following major effects of insulin lack: (1) decreased utilization of glucose by the body cells, with resultant increase in blood glucose concentration to 300 to 1200 mg/dl; (2) markedly increased mobilization of fats from the fat storage areas, causing abnormal fat metabolism as well as deposition of cholesterol in arterial walls, causing atherosclerosis; and (3) depletion of protein in the tissues of the body.

LOSS OF GLUCOSE IN THE URINE IN DIABETES. Whenever the quantity of glucose entering the kidney tubules in the glomerular filtrate rises above a critical level, a significant proportion of the excess glucose cannot be reabsorbed and instead spills into the urine. This normally occurs when the blood glucose concentration rises above 180 mg/dl, a level that is called the blood "threshold" for the appearance of glucose in the urine. When the blood glucose level rises to 300 to 500 mg/dl—common values in people with severe untreated diabetes—100 or more grams of glucose can be lost into the urine each day.

DEHYDRATING EFFECT OF ELEVATED BLOOD GLUCOSE LEVELS IN DIABETES. Blood glucose levels as high as 1200 mg/dl, 12 times normal, have been known to occur in extreme untreated diabetes, and levels of 300 to 500 mg/dl are common. Yet the only significant effect of the elevated glucose is dehydration of the tissue cells. This occurs partly because glucose does not diffuse easily through the pores of the cell membrane, and the increased osmotic pressure in the extracellular fluids causes osmotic transfer of water out of the cells.

In addition to the direct cellular dehydrating effect of excessive glucose, the loss of glucose in the urine causes *osmotic diuresis*. That is, the osmotic effect of glucose in the renal tubules greatly decreases tubular reabsorption of fluid. The overall effect is massive loss of fluid in the urine, causing dehydration of the extracellular fluid, which in turn causes compensatory dehydration of the intracellular fluid for reasons discussed in Chapter 26. Thus, one of the important features of diabetes is a tendency for both extracellular and intracellular dehydration to develop, and this can contribute to the development of *circulatory shock*.

ACIDOSIS AND COMA IN DIABETES. The shift from carbohydrate to fat metabolism in diabetes has already been discussed. When the body depends almost entirely on fat for energy, the level of the keto acids acetoacetic acid and β-hydroxybutyric acid in the body fluids may rise from 1 mEq/liter to 10 mEq/liter or greater. All this extra acid is likely to result in acidosis.

A second effect that is usually even more important in causing acidosis than is the direct increase in blood keto acids is a decrease in sodium concentration caused by the following: Keto acids have a low threshold for excretion by the kidneys; therefore, when the keto acid concentration rises in diabetes, 100 to 200 grams of keto acids can be excreted in the urine each day. Because these acids are strong, having a pK averaging 4.0 or less, little of them can be excreted in the acidic form; instead, they are excreted combined with sodium derived from the extracellular fluid. As a result, the sodium concentration in the extracellular fluid usually decreases, and part of the sodium is replaced by increased quantities of hydrogen ions, thus adding greatly to the acidosis.

All the usual physiological reactions that occur in metabolic acidosis take place in diabetic acidosis. They include *rapid and deep breathing* called "Kussmaul respiration," which causes excessive expiration of carbon dioxide, and *marked decrease in bicarbonate content of the extracellular fluids*. Although these extreme effects occur only in the most severe instances of uncontrolled diabetes, they can lead to *acidotic coma* and death within hours when the pH of the blood falls below about 7.0. The overall changes in the electrolytes of the blood as a result of severe diabetic acidosis are shown in Figure 78–9.

RELATION OF OTHER DIABETIC SYMPTOMS TO THE PATHOLOGICAL PHYSIOLOGY OF INSULIN LACK. *Polyuria* (excessive elimination of urine), *polydipsia* (excessive drinking of water), *polyphagia* (excessive eating), *loss of weight*, and *asthenia* (lack of energy) are the earliest symptoms of diabetes. As explained, the polyuria is due to the osmotic diuretic effect of glucose in the kidney tubules. In turn, the polydipsia is due to dehydration resulting from polyuria. The failure of glucose (and protein) metabolism by the body causes loss of weight and a tendency toward polyphagia. The asthenia apparently is caused mainly by loss of body protein but also by diminished utilization of carbohydrates for energy.

Physiology of Diagnosis

The usual methods for diagnosing diabetes are based on various chemical tests of the urine and the blood.

URINARY SUGAR. Simple office tests or more complicated quantitative laboratory tests may be used for determining the quantity of glucose lost in the urine. In general, a normal person loses undetectable amounts of glucose, whereas a person with diabetes loses glucose in small to large amounts, in proportion to the severity of disease and the intake of carbohydrates.

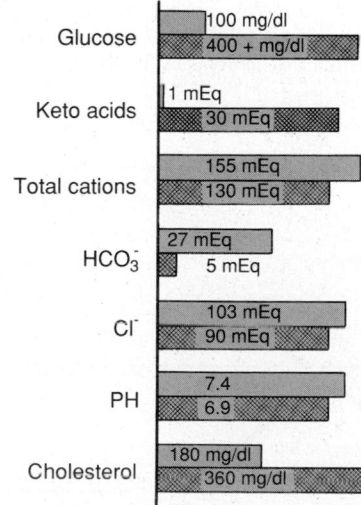

Figure 78–9. Changes in blood constituents in diabetic coma, showing normal values (light bars) and diabetic coma values (dark bars).

FASTING BLOOD GLUCOSE LEVEL. The fasting blood sugar level in the early morning is normally 80 to 90 mg/dl, and 110 mg/dl is considered to be the upper limit of normal. A fasting blood sugar level above this value often indicates diabetes mellitus or, much less commonly, either pituitary diabetes or adrenal diabetes.

GLUCOSE TOLERANCE TEST. As demonstrated by the bottom curve in Figure 78–10, called a "glucose tolerance curve," when a normal, fasting person ingests 1 gram of glucose per kilogram of body weight, the blood glucose level rises from about 90 mg/dl to 120 to 140 mg/dl and falls back to below normal in about 2 hours.

In a person with diabetes, the fasting blood glucose concentration is almost always above 110 mg/dl and often above 140 mg/dl. Also, the glucose tolerance test is almost always abnormal. On ingestion of glucose, these people exhibit a much greater than normal rise in blood glucose level, as demonstrated by the upper curve, and the glucose level falls back to the control value only after 4 to 6 hours; furthermore, it fails to fall *below* the control level. This slow fall of the curve and its failure to fall below the control level demonstrate that the normal increase in insulin secretion after glucose ingestion does not occur in a person with diabetes, and a diagnosis of diabetes mellitus can usually be definitely established on the basis of such a curve.

ACETONE BREATH. As pointed out in Chapter 68, small quantities of acetoacetic acid in the blood, which increase greatly in severe diabetes, are converted to acetone. This is volatile and vaporized into the expired air. Consequently, one frequently can make a diagnosis of diabetes mellitus simply by smelling acetone on the breath of a patient. Also, keto acids can be detected by chemical means in the urine, and their quantitation aids in determining the severity of the diabetes.

Treatment of Diabetes

The theory of treatment of diabetes mellitus is to administer enough insulin so that the patient will have as nearly normal carbohydrate, fat, and protein metabolism as possible. Insulin is available in several forms. "Regular" insulin has a duration of action that lasts from 3 to 8 hours, whereas other forms of insulin (precipitated with zinc or with various protein derivatives) are absorbed slowly from the injectate site and, therefore, have effects that last as long as 10 to 48 hours. Ordinar-

ily, a patient with severe diabetes is given a single dose of one of the longer-acting insulins each day to increase overall carbohydrate metabolism throughout the day. Then additional quantities of regular insulin are given at those times of the day when the blood glucose level tends to rise too high, such as at mealtimes. Thus, each patient is established on an individualized pattern of treatment.

In the past, the insulin used for treatment has been derived from animal pancreases. Recently, however, human insulin produced by the recombinant DNA process has come into use because many patients develop immunity and sensitization against animal insulin, thus limiting its effectiveness.

RELATION OF TREATMENT TO ARTERIOSCLEROSIS. Diabetic patients, mainly because of their high levels of circulating cholesterol and other lipids, develop atherosclerosis, arteriosclerosis, severe coronary heart disease, and multiple microcirculatory lesions far more easily than do normal people. Indeed, those who have relatively poorly controlled diabetes throughout childhood are likely to die of heart disease in early adulthood.

In the early days of treating diabetes, the tendency was to reduce severely the carbohydrates in the diet so that the insulin requirements would be minimized. This procedure kept the blood sugar level down to normal values and prevented loss of glucose in the urine, but it did not prevent many of the abnormalities of fat metabolism. Consequently, the current tendency is to allow the patient an almost normal carbohydrate diet and to give simultaneously large enough quantities of insulin to metabolize the carbohydrates. This decreases the rate of fat metabolism and depresses the high level of blood cholesterol.

Because the complications of diabetes—such as atherosclerosis, greatly increased susceptibility to infection, diabetic retinopathy, cataracts, hypertension, and chronic renal disease—are more closely associated with the level of the blood lipids than with the level of blood glucose, it is the object of some clinics treating diabetes to administer sufficient glucose and insulin so that the quantity of blood lipids becomes normal.

HYPERINSULINISM

Although much rarer than diabetes, increased insulin production, known as *hyperinsulinism*, does occasionally occur. This usually results from an adenoma of an islet of Langerhans. About 10 to 15 per cent of these adenomas are malignant, and occasionally metastases from the islets of Langerhans spread throughout the body, causing tremendous production of insulin by both the primary and the metastatic cancers. Indeed, to prevent hypoglycemia, in some of these patients, more than 1000 grams of glucose have had to be administered every 24 hours.

Diagnosis of hyperinsulinism is made with assurance by measuring high levels of plasma insulin by means of radioimmunoassay, especially when the insulin remains high constantly throughout the day without rising significantly with increased carbohydrate intake.

INSULIN SHOCK AND HYPOGLYCEMIA. As already emphasized, the central nervous system normally derives essentially all its energy from glucose metabolism, and insulin is not necessary for this use of glucose. However, if insulin causes the level of blood glucose to fall to low

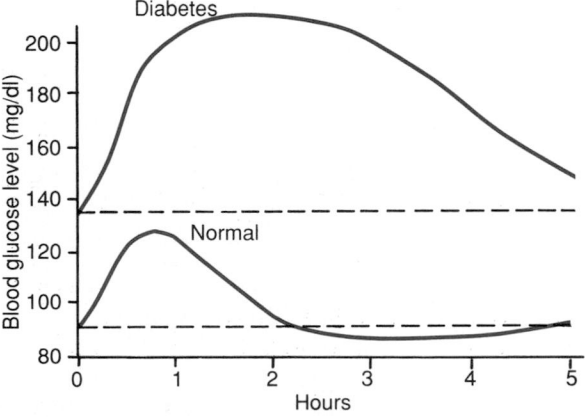

Figure 78–10. Glucose tolerance curve in a normal person and in a person with diabetes.

values, the metabolism of the central nervous system becomes depressed. Consequently, in patients with hyperinsulinism or in patients with diabetes who administer too much insulin to themselves, the syndrome called *insulin shock* may occur as follows.

As the blood sugar level falls into the range of 50 to 70 mg/dl, the central nervous system usually becomes quite excitable because this degree of hypoglycemia sensitizes neuronal activity. Sometimes various forms of hallucinations result, but more often the patient simply experiences extreme nervousness, trembles all over, and breaks out in a sweat. As the blood glucose level falls to 20 to 50 mg/dl, clonic seizures and loss of consciousness are likely to occur. As the glucose level falls still lower, the seizures cease and only a state of coma remains. Indeed, at times it is difficult by simple clinical observation to distinguish between diabetic coma as a result of insulin lack acidosis and coma due to hypoglycemia caused by excess insulin. The acetone breath and the rapid, deep breathing of diabetic coma are not present in hypoglycemic coma.

Proper treatment for a patient who has hypoglycemic shock or coma is immediate intravenous administration of large quantities of glucose. This usually brings the patient out of shock within a minute or more. Also, the administration of glucagon (or, less effectively, epinephrine) can cause glycogenolysis in the liver and thereby increase the blood glucose level extremely rapidly.

If treatment is not effected immediately, permanent damage to the neuronal cells of the central nervous system often occurs; this happens especially in prolonged hyperinsulinism caused by pancreatic tumors.

REFERENCES

Baskin, D. G., et al.: Insulin in the brain. Annu. Rev. Physiol., 49:335, 1987.

Beger, H. G., et al.: The Role of Enzyme Treatment in Pancreatic Disease. Farmington, CT, S. Karger Publishers, Inc., 1993.

Belifiore, F., et al.: Current Topics in Diabetes Research. Farmington, CT, S. Karger Publishers, Inc., 1993.

Bonner-Weir, S., and Weir, G. C.: The islets of Langerhans and diabetes mellitus. Current Concepts. Upjohn, November, 1986.

Büchler, M. W., et al.: The Role of Somatostatin and Octreotide in Pancreatic Disease. Farmington, CT, S. Karger Publishers, Inc., 1994.

Cheng, K., and Larner, J.: Intracellular mediators of insulin action. Annu. Rev. Physiol., 47:405, 1985.

Czech, M. P.: New perspectives on the mechanism of insulin action. Recent Prog. Horm. Res., 40:347, 1984.

Czech, M. P.: The nature and regulation of the insulin receptor: Structure and function. Annu. Rev. Physiol., 47:357, 1985.

Czernichow, P., and Levy-Marshal, C.: Epidemiology and Etiology of Insulin-Dependent Diabetes in the Young. Farmington, CT, S. Karger Publishers, Inc., 1991.

DeGroot, L. J. (ed.): Endocrinology, 2nd Ed. Philadelphia, W. B. Saunders Co., 1989.

Fain, J. N.: Insulin secretion and action. Metabolism, 33:672, 1984.

Felig, P., et al. (eds.): Endocrinology and Metabolism, 2nd Ed. New York, McGraw-Hill, 1987.

Githens, S.: Pancreatic duct cell cultures. Annu. Rev. Phyiol., 56:419, 1994.

Goren, H. J., et al.: Insulin Action and Diabetes. New York, Raven Press, 1988.

Go, V. L. W., et al.: The Pancreas: Biology, Pathobiology, and Disease. New York, Raven Press, 1993.

Guyton, J. R., et al.: A model of glucose-insulin homeostasis in man that incorporates the heterogeneous fast pool theory of pancreatic insulin release. Diabetes, 27:1027, 1978.

Hales, C. N.: Fetal Nutrition and Adult Diabetes. New York, Scientific American Science and Medicine, July/August, 1994.

Hediger, M. A., and Rhoads, D. B.: Molecular physiology of sodium-glucose cotransporters. Physiol. Rev., 74:993, 1994.

Howell, S. L., and Tyhurst, M.: Insulin secretion: The effector system. Experientia, 40:1098, 1984.

Kahn, C. R., and Weir, G. C.: Joslin's Diabetes Mellitus. Baltimore, Williams & Wilkins, 1994.

Kimball, S. R., et al.: Regulation of protein synthesis by insulin. Annu. Rev. Phyiol., 56:321, 1994.

Kerstein, M. D.: Diabetes and Vascular Disease. Philadelphia, J. B. Lippincott, 1989.

Krall, L. P., and Beaser, R.: Joslin diabetes manual, 12th Ed. Philadelphia, Lea & Febiger, 1989.

Leslie, R. D. G., and Robbins, D. C.: Diabetes: Clinical Science in Practice. New York, Cambridge University Press, 1994.

Makino, H., et al.: Role of ATP in insulin actions. Annu. Rev. Physiol., 56:273, 1994.

Meisler, M. H., and Howard, G.: Effects of insulin on gene transcription. Annu. Rev. Physiol., 51:701, 1989.

Mignon, M., and Jensen, R. T.: Endocrine Tumors of the Pancreas. Farmington, CT, S. Karger Publishers, Inc., 1995.

Nagano, M., and Dhalla, N. S.: The Diabetic Heart. New York, Raven Press, 1991.

Nestler, J. E.: Assessment of insulin resistance. New York, Scientific American Science and Medicine, September/October, 1994.

Nishimoto, I., Kojima, I.: Calcium signalling system triggered by insulin-like growth factor II. News Physiol. Sci., 4:94, 1989.

Nujima, A., and Mei, N.: Glucose sensors in viscera and control of blood glucose level. News Physiol. Sci., 2:164, 1987.

Pipeleers, D.: Islet cell interactions with pancreatic β-cells. Experientia, 40:1114, 1984.

Post, R. L., et al.: Regulation of glucose uptake in muscle III. The interaction of membrane transport and phosphorylation in the control of glucose uptake. J. Biol. Chem., 236:269, 1961.

Prentki, M., and Matschinsky, F. M.: Ca^{2+}-cAMP, and phospholipid-derived messengers in coupling mechanisms of insulin secretion. Physiol. Rev., 67:1185, 1987.

Raizada, M. K., and LeRoth, D.: The Role of Insulin-like Growth Factors in the Nervous System. New York, New York Academy of Sciences, 1993.

Rosen, O. M.: After insulin binds. Science, 237:1452, 1987.

Ruderman, N., et al.: Hyperglycemia, Diabetes and Vascular Disease. New York, Oxford University Press, 1992.

Samols, E.: The Endocrine Pancreas. New York, Raven Press, 1991.

Sonne, O.: Receptor-mediated endocytosis and degradation of insulin. Physiol. Rev., 68:1129, 1988.

Squifflet, J. P.: Pancreas Transplantation: Experimental and Clinical Studies. Farmington, CT, S. Karger Publishers, Inc., 1990.

Standaert, M. L., and Pollet, R. J.: Insulin-glycerolipid mediators and gene expression. FASEB J., 2:2453, 1988.

Trede, M., and Carter, D. C.: Surgery of the pancreas. New York, Churchill Livingstone, 1993.

Parathyroid Hormone, Calcitonin, Calcium and Phosphate Metabolism, Vitamin D, Bone, and Teeth

CHAPTER 79

The physiology of calcium and phosphate metabolism, the function of vitamin D, and the formation of bone and teeth are all tied together in a common system along with the two regulatory hormones *parathyroid hormone* and *calcitonin*. Therefore, they are discussed together in this chapter.

CALCIUM AND PHOSPHATE IN THE EXTRACELLULAR FLUID AND PLASMA—FUNCTION OF VITAMIN D

Absorption and Excretion of Calcium and Phosphate

INTESTINAL ABSORPTION OF CALCIUM AND PHOSPHATE. Normally, divalent cations such as calcium ions are poorly absorbed from the intestines. However, as discussed later, vitamin D promotes an active calcium transport system, allowing the necessary calcium absorption.

For phosphate ion absorption, this normally occurs very easily, as is true for most anions. The usual rates of intake are about 1 gm/day each for calcium and phosphorous or slightly greater, about the amounts in 1 liter of milk.

EXCRETION OF CALCIUM IN FECES AND URINE; NET RATE OF INTESTINAL ABSORPTION. About nine tenths of the daily intake of calcium is excreted in the feces; the remaining one tenth is excreted in the urine. The approximate daily turnover rates for calcium in the adult are as follows:

	mg
Intake	1000
Intestinal absorption	350
Secretion in gastrointestinal juices	250
Net absorption over secretion	100
Loss in the feces	900
Excretion in the urine	100

Excretion of calcium in the urine conforms to much the same principles as sodium excretion. About 90 per cent of the calcium in the glomerular filtrate is reabsorbed in the proximal tubules, loops of Henle, and early distal tubules. Then in the late distal tubules and early collecting ducts, reabsorption of the remaining 10 per cent calcium is very selective, depending on the calcium ion concentration in the blood. When low, this reabsorption is great, so that almost no calcium is lost in the urine. On the other hand, even a minute increase in blood calcium ion concentration above normal increases calcium excretion markedly. We shall see later in the chapter that the most important factor controlling this reabsorption of calcium in the distal portions of the nephron, and therefore controlling the rate of calcium excretion, is parathyroid hormone.

INTESTINAL AND URINARY EXCRETION OF PHOSPHATE. Except for the portion of phosphate that is excreted in the feces in combination with nonabsorbed calcium, almost all the dietary phosphate is absorbed into the blood from the gut and later excreted in the urine.

Phosphate is a so-called *renal threshold substance*. That is, when its concentration in the plasma is below the critical value of about 1 mmol/liter, no phosphate is lost into the urine because all that is in the glomeru-

985

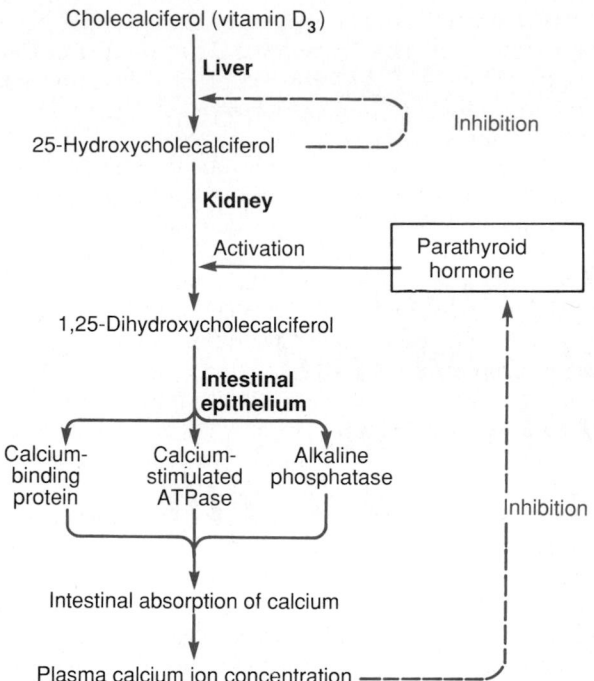

Figure 79–1. Activation of vitamin D$_3$ to form *1,25-dihydroxy-cholecalciferol;* and the role of vitamin D in controlling the plasma calcium concentration.

lar filtrate is reabsorbed. But above this critical concentration, the rate of phosphate loss is directly proportional to the additional increase. Thus, the kidneys regulate the phosphate concentration in the extracellular fluid by altering the rate of phosphate excretion in accordance with the plasma phosphate concentration.

However, as discussed later in the chapter, phosphate excretion by the kidneys is greatly increased by parathyroid hormone, thereby playing an important role in the control of plasma phosphate concentration as well as calcium concentration.

Vitamin D and Its Role in Calcium and Phosphate Absorption

Vitamin D has a potent effect on increasing calcium absorption from the intestinal tract; it also has important effects on both bone deposition and bone reabsorption, as discussed later in the chapter. However, vitamin D itself is not the active substance that actually causes these effects. Instead, the vitamin D must first be converted through a succession of reactions in the liver and the kidneys to the final active product, *1,25-dihydroxycholecalciferol,* also called 1,25-(OH)$_2$-D$_3$. Figure 79–1 shows the succession of steps that lead to the formation of this substance from vitamin D. Let us discuss these steps.

VITAMIN D COMPOUNDS. Several compounds derived from sterols belong to the vitamin D family, and they all perform more or less the same functions. The most important of these, called vitamin D$_3$, is *cholecalciferol.* Most of this substance is formed in the skin as a result of irradiation of *7-dehydrocholesterol,* a

substance even normally in the skin, by ultraviolet rays from the sun. Consequently, appropriate exposure to the sun prevents vitamin D deficiency. The additional vitamin D compounds that we ingest in our food are identical to the cholecalciferol formed in our skin except for the substitution of one or more atoms that do not affect their function.

CONVERSION OF CHOLECALCIFEROL TO 25-HYDROXYCHOLECALCIFEROL IN THE LIVER AND ITS FEEDBACK CONTROL. The first step in the activation of cholecalciferol is to convert it to 25-hydroxycholecalciferol; this occurs in the liver. The process is itself a limited one because the 25-hydroxycholecalciferol has a feedback inhibitory effect on the conversion reactions. This feedback effect is extremely important for two reasons.

First, the feedback mechanism regulates precisely the concentration of 25-hydroxycholecalciferol in the plasma, an effect that is shown in Figure 79–2. Note that the intake of vitamin D$_3$ can change many times, and yet the concentration of 25-hydroxycholecalciferol remains within a few percentage points of its normal mean value. This high degree of feedback control prevents excessive action of vitamin D when it is present in too great a quantity.

Second, this controlled conversion of vitamin D$_3$ to 25-hydroxycholecalciferol conserves the vitamin D stored in the liver for future use because once it is converted, it persists in the body for only a few weeks, whereas in the vitamin D form, it can be stored in the liver for many months.

FORMATION OF 1,25-DIHYDROXYCHOLECALCIFEROL IN THE KIDNEYS AND ITS CONTROL BY PARATHYROID HORMONE. Figure 79–1 also shows the conversion in the proximal tubules of the kidneys of 25-hydroxycholecalciferol to *1,25-dihydroxycholecalciferol.* This latter substance is by far the most active form of vitamin D because the previous products in the scheme of Figure 79–1 have less than 1/1000 as much vitamin D effect.

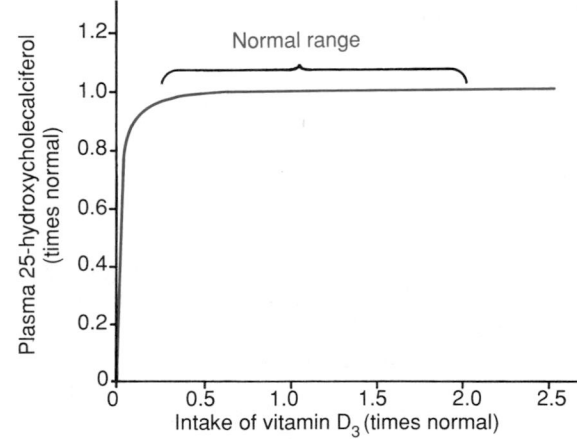

Figure 79–2. Effect of increasing vitamin D$_3$ intake on the plasma concentration of 25-hydroxycholecalciferol. This figure shows that tremendous changes in vitamin D intake have little effect on the final quantity of activated vitamin D that is formed.

Therefore, in the absence of the kidneys, vitamin D loses almost all its effectiveness.

Note also in Figure 79–1 that the conversion of 25-hydroxycholecalciferol to 1,25-dihydroxycholecalciferol requires parathyroid hormone. In the absence of this hormone, either none or almost none of the 1,25-dihydroxycholecalciferol is formed. Therefore, parathyroid hormone exerts a potent effect in determining the functional effects of vitamin D in the body.

Effect of Calcium Ion Concentration on Control of the Formation of 1,25-Dihydroxycholecalciferol. Figure 79–3 demonstrates that the plasma concentration of 1,25-dihydroxycholecalciferol is inversely affected by the concentration of calcium in the plasma. There are two reasons for this. First, the calcium ion itself has a slight effect in preventing the conversion of 25-hydroxycholecalciferol to 1,25-dihydroxycholecalciferol. Second, and even more important, as we shall see later in the chapter, the rate of secretion of parathyroid hormone is greatly suppressed when the plasma calcium ion concentration rises above 9 to 10 mg/dl. Therefore, at calcium concentrations below 9 to 10 mg/dl, parathyroid hormone promotes the conversion of 25-hydroxycholecalciferol to 1,25-dihydroxycholecalciferol in the kidneys. At higher calcium concentrations, when parathyroid hormone is suppressed, the 25-hydroxycholecalciferol is converted to a different compound—24,25-dihydroxycholecalciferol—that has almost no vitamin D effect.

Therefore, when the plasma calcium concentration is already too high, the formation of 1,25-dihydroxycholecalciferol is greatly depressed. Lack of this in turn decreases the absorption of calcium from the intestines, the bones, and the renal tubules, thus causing the calcium ion concentration to fall back toward its normal level.

"Hormonal" Effect of 1,25-Dihydroxychole- calciferol on the Intestinal Epithelium in Promoting Calcium Absorption. 1,25-Dihydroxychole- calciferol itself functions as a type of "hormone" to promote intestinal absorption of calcium. It does this principally by increasing over a period of about 2 days formation of a *calcium-binding protein* in the intestinal epithelial cells. This protein functions in the brush border of these cells to transport calcium into the cell cytoplasm, and the calcium then moves through the basolateral membrane of the cell by facilitated diffusion. The rate of calcium absorption seems to be directly proportional to the quantity of this calcium-binding protein. Furthermore, this protein remains in the cells for several weeks after the 1,25-dihydroxycholecalciferol has been removed from the body, thus causing a prolonged effect on calcium absorption.

Other effects of this "hormone," 1,25-dihydroxycholecalciferol, that might play a role in promoting calcium absorption are (1) the formation of a calcium-stimulated ATPase in the brush border of the epithelial cells and (2) the formation of an alkaline phosphatase in the epithelial cells. The precise details of all these effects are unknown.

Effect of Vitamin D on Phosphate Absorption. Much less is known about the effect of vitamin D on phosphate absorption than on calcium absorption. Also, this is much less important because phosphate is usually absorbed easily anyway. However, phosphate flux through the gastrointestinal epithelium is enhanced by vitamin D. It is believed that this results from a direct effect of 1,25-dihydroxycholecalciferol, but it is possible that it results secondarily from this hormone's action on calcium absorption, the calcium in turn acting as a transport mediator for the phosphate.

Calcium in the Plasma and Interstitial Fluid

The concentration of calcium in the plasma averages about 9.4 mg/dl, normally varying between 9 and 10 mg/dl. This is equivalent to about 2.4 mmol of calcium per liter. It is apparent that the calcium level in the plasma is regulated within narrow limits—and mainly by parathyroid hormone, as discussed later in the chapter.

The calcium in the plasma is present in three forms, as shown in Figure 79–4. (1) About 40 per cent (1 mmol/liter) of the calcium is combined with the plasma proteins and in this form is nondiffusible through the capillary membrane. (2) About 10 per cent of the calcium (0.2 mmol/liter) is diffusible through the capillary membrane but is combined with other substances of the plasma and interstitial fluids (citrate and phosphate, for instance) in such a manner that it is not ionized. (3) The remaining 50 per cent of the calcium in the plasma is both diffusible through the capillary membrane and ionized. Thus, the plasma and interstitial fluids have a normal *calcium ion concentration of about 1.2 mmol/liter (or 2.4 mEq/liter because*

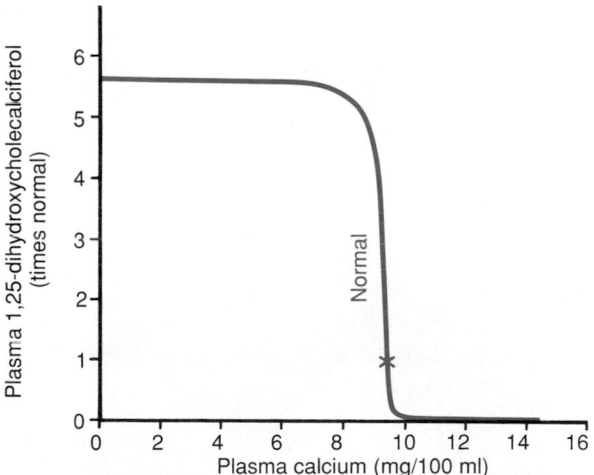

Figure 79–3. Effect of plasma calcium concentration on the plasma concentration of 1,25-dihydroxycholecalciferol. This figure shows that a slight decrease in calcium concentration below normal causes marked formation of activated vitamin D, which in turn leads to greatly increased absorption of calcium from the intestine.

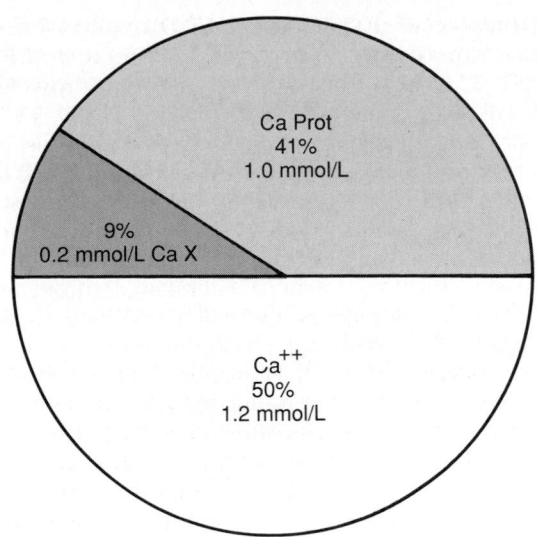

Figure 79–4. Distribution of ionic calcium (Ca^{++}), diffusible but un-ionized calcium (Ca X), and nondiffusible calcium proteinate (Ca Prot) in blood plasma.

it is a divalent ion), a level only one half the total plasma calcium concentration. This ionic calcium is the calcium form that is important for most functions of calcium in the body, including the effect of calcium on the heart, the nervous system, and bone formation.

Inorganic Phosphate in the Extracellular Fluids

Inorganic phosphate in the plasma is mainly in two forms: HPO_4^{--} and $H_2PO_4^{-}$. The concentration of HPO_4^{--} is about 1.05 mmol/liter and the concentration of $H_2PO_4^{-}$, about 0.26 mmol/liter. When the total quantity of phosphate in the extracellular fluid rises, so does the quantity of each of these two types of phosphate ions. Furthermore, when the pH of the extracellular fluid becomes more acidic, there is a relative increase in $H_2PO_4^{-}$ and decrease in HPO_4^{--}, whereas the opposite occurs when the extracellular fluid becomes alkaline. These relations were presented in the discussion of acid-base balance in Chapter 30.

Because it is difficult to determine chemically the exact quantities of HPO_4^{--} and $H_2PO_4^{-}$ in the blood, ordinarily the total quantity of phosphate is expressed in terms of milligrams of *phosphorus* per deciliter of blood. The average total quantity of inorganic phosphorus represented by both phosphate ions is about 4 mg/dl, varying between normal limits of 3 to 4 mg/dl in adults and 4 to 5 mg/dl in children.

Non-Bone Physiological Effects of Altered Calcium and Phosphate Concentrations in the Body Fluids

Changing the level of phosphate in the extracellular fluid from far below normal to two to three times normal does not cause significant immediate effects on the body.

On the other hand, even slight elevations or decreases of calcium ion in the extracellular fluid often cause extreme immediate physiologic effects. In addition, chronic hypocalcemia or hypophosphatemia greatly decrease bone mineralization, as explained later in the chapter.

TETANY RESULTING FROM HYPOCALCEMIA. When the extracellular fluid concentration of calcium ions falls below normal, the nervous system becomes progressively more and more excitable because this causes increased neuronal membrane permeability to sodium ions, allowing easy initiation of action potentials. At plasma calcium ion concentrations about 50 per cent below normal, the peripheral nerve fibers become so excitable that they begin to discharge spontaneously, initiating trains of nerve impulses that pass to the peripheral skeletal muscles to elicit tetanic muscle contraction. Consequently, hypocalcemia causes tetany. It also occasionally causes seizures because of its action of increasing excitability in the brain.

Figure 79–5 shows tetany in the hand, which usually occurs before tetany develops in most other parts of the body. This is called "carpopedal spasm."

Tetany ordinarily occurs when the blood concentration of calcium falls from its normal level of 9.4 mg/dl to about 6 mg/dl, which is only 35 per cent below the normal calcium concentration, and it is usually lethal at about 4 mg/dl.

In laboratory animals, in which the level of calcium can be reduced beyond the normal lethal stage, extreme hypocalcemia can cause other effects that seldom are evident in patients; these effects are marked dilatation of the heart, changes in cellular enzyme activities, increased cell membrane permeability in some cells (in addition to nerve cells), and impaired blood clotting.

HYPERCALCEMIA. When the level of calcium in the body fluids rises above normal, the nervous system is depressed, and reflex activities of the central nervous system can become sluggish. Also, increased calcium ion concentration decreases the QT interval of the heart and causes constipation and lack of appetite, probably because of depressed contractility of the muscle walls of the gastrointestinal tract.

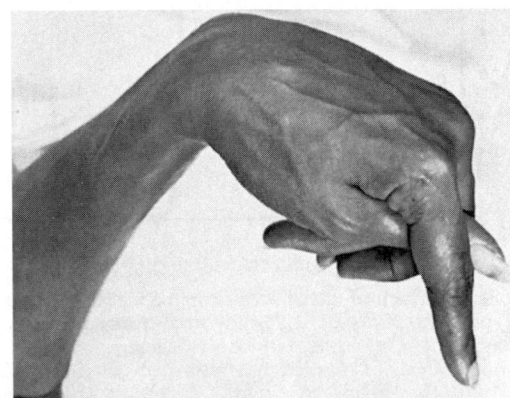

Figure 79–5. Hypocalcemic tetany in the hand, called "carpopedal spasm." (Courtesy of Dr. Herbert Langford.)

The depressive effects of increased calcium level begin to appear when the blood level of calcium rises above about 12 mg/dl, and they can become marked as the calcium level rises above 15 mg/dl. When the level of calcium rises above about 17 mg/dl in the blood, calcium phosphate crystals are likely to precipitate throughout the body; this condition is discussed shortly in connection with parathyroid poisoning.

BONE AND ITS RELATIONS TO EXTRACELLULAR CALCIUM AND PHOSPHATES

Bone is composed of a tough *organic matrix* that is greatly strengthened by deposits of *calcium salts.* Average *compact bone* contains by weight about 30 per cent matrix and 70 per cent salts. *Newly formed bone* may have a considerably higher percentage of matrix in relation to salts.

ORGANIC MATRIX OF BONE. The organic matrix of bone is 90 to 95 per cent *collagen fibers,* and the remainder is a homogeneous gelatinous medium called *ground substance.* The collagen fibers extend primarily along the lines of tensional force. These fibers give bone its powerful tensile strength.

The ground substance is composed of extracellular fluid plus *proteoglycans,* especially *chondroitin sulfate* and *hyaluronic acid.* The precise function of each of these is not known, although they do help to control the deposition of calcium salts.

BONE SALTS. The crystalline salts deposited in the organic matrix of bone are composed principally of *calcium* and *phosphate.* The formula for the major crystalline salt, known as *hydroxyapatite,* is the following:

$$Ca_{10}(PO_4)_6(OH)_2$$

Each crystal—about 400 angstroms long, 10 to 30 angstroms thick, and 100 angstroms wide—is shaped like a long, flat plate. The relative ratio of calcium to phosphorus can vary markedly under different nutritional conditions, the Ca/P ratio on a weight basis varying between 1.3 and 2.0.

Magnesium, sodium, potassium, and *carbonate* ions are also present among the bone salts, although x-ray diffraction studies fail to show definite crystals formed by them. Therefore, they are believed to be conjugated to the hydroxyapatite crystals rather than organized into distinct crystals of their own. This ability of many types of ions to conjugate to bone crystals extends to many ions normally foreign to bone, such as *strontium, uranium, plutonium, the other transuranic elements, lead, gold, other heavy metals,* and *at least 9 of 14 of the major radioactive products released by explosion of the hydrogen bomb.* Deposition of radioactive substances in the bone can cause prolonged irradiation of the bone tissues, and if a sufficient amount is deposited, an osteogenic sarcoma (bone cancer) almost invariably eventually develops.

TENSILE AND COMPRESSIONAL STRENGTH OF BONE. Each collagen fiber of *compact* bone is composed of repeating periodic segments every 640 angstroms along its length; hydroxyapatite crystals lie adjacent to each segment of the fiber, bound tightly to it. This intimate bonding prevents "shear" in the bone; that is, it prevents the crystals and collagen fibers from slipping out of place, which is essential in providing strength to the bone. In addition, the segments of adjacent collagen fibers overlap one another, also causing hydroxyapatite crystals to be overlapped like bricks keyed to one another in a brick wall.

The collagen fibers of bone, like those of tendons, have great tensile strength, whereas the calcium salts, which are similar in physical properties to marble, have great compressional strength. These combined properties plus the degree of bondage between the collagen fibers and the crystals provide a bony structure that has both extreme tensile strength and extreme compressional strength. Thus, bones are constructed in the same way that reinforced concrete is constructed. The steel of reinforced concrete provides the tensile strength, while the cement, sand, and rock provide the compressional strength. Indeed, the compressional strength of bone is greater than that of even the best reinforced concrete, and the tensile strength approaches that of reinforced concrete.

Precipitation and Absorption of Calcium and Phosphate in Bone—Equilibrium with the Extracellular Fluids

SUPERSATURATED STATE OF CALCIUM AND PHOSPHATE IONS IN THE EXTRACELLULAR FLUIDS WITH RESPECT TO HYDROXYAPATITE. The concentrations of calcium and phosphate ions in extracellular fluid are considerably greater than those required to cause precipitation of hydroxyapatite. However, inhibitors are present in almost all tissues of the body as well as in plasma to prevent such precipitation; one such inhibitor is pyrophosphate. Therefore, hydroxyapatite crystals fail to precipitate in normal tissues except in bone despite the state of supersaturation of the ions.

MECHANISM OF BONE CALCIFICATION. The initial stage in bone production is the secretion of *collagen molecules* (called collagen monomers) and *ground substance* (mainly proteoglycans) by *osteoblasts.* The collagen monomers polymerize rapidly to form collagen fibers; the resultant tissue becomes *osteoid,* a cartilage-like material but differing from cartilage in that calcium salts will soon precipitate in it. As the osteoid is formed, some of the osteoblasts become entrapped in the osteoid and then are called *osteocytes.*

Within a few days after the osteoid is formed, calcium salts begin to precipitate on the surfaces of the collagen fibers. The precipitates first appear at intervals along each collagen fiber, forming minute nidi that rapidly multiply and grow over a period of days and weeks into the finished product, *hydroxyapatite crystals.*

The initial calcium salts to be deposited are not hydroxyapatite crystals but amorphous compounds (noncrystalline), a probable mixture of such salts as $CaHPO_4 \cdot 2H_2O$, $Ca_3(PO_4)_2 \cdot 3H_2O$, and others. Then by a process of substitution and addition of atoms, or reabsorption and reprecipitation, these salts are converted into the hydroxyapatite crystals over a period of weeks or months. Yet as much as a few per cent may remain permanently in the amorphous form. This is important because these amorphous salts can be absorbed rapidly when there is need for extra calcium in the extracellular fluid.

It is not known what causes calcium salts to be deposited in osteoid. One theory holds that at the time of formation, the collagen fibers are specially constituted in advance for causing precipitation of calcium salts. The osteoblasts supposedly also secrete a substance into the osteoid to neutralize an inhibitor (believed to be *pyrophosphate*) that normally prevents hydroxyapatite crystallization. Once the pyrophosphate has been neutralized, then the natural affinity of the collagen fibers for calcium salts supposedly causes the precipitation. In support of this theory is the fact that properly prepared collagen fibers from other tissues of the body besides bone will also cause precipitation of hydroxyapatite crystals from plasma.

Precipitation of Calcium in Nonosseous Tissues Under Abnormal Conditions. Although calcium salts almost never precipitate in normal tissues besides bone, under abnormal conditions, they do precipitate. For instance, they precipitate in arterial walls in the condition called arteriosclerosis and cause the arteries to become bone-like tubes. Likewise, calcium salts frequently deposit in degenerating tissues or in old blood clots. Presumably, in these instances, the inhibitor factors that normally prevent deposition of calcium salts disappear from the tissues, thereby allowing precipitation.

Exchangeable Calcium

If soluble calcium salts are injected intravenously, the calcium ion concentration can be made to increase immediately to high levels. However, within 30 minutes to 1 hour or more, the calcium ion concentration returns to normal. Likewise, if large quantities of calcium ions are removed from the circulating body fluids, the calcium ion concentration again returns to normal within 30 minutes to 1 hour or so. These effects result in great part from the fact that the bone contains a type of *exchangeable* calcium that is always in equilibrium with the calcium ions in the extracellular fluids. A small portion of this exchangeable calcium is also that calcium found in all tissue cells, especially in highly permeable types of cells such as those of the liver and the gastrointestinal tract. However, most of the exchangeable calcium, as shown by studies using radioactively tagged calcium, is in the bone. It normally amounts to about 0.4 to 1.0 per cent of the total bone calcium. This calcium is deposited in the bones in a form of readily mobilizable salt such as $CaHPO_4$ and other amorphous calcium salts.

The importance of exchangeable calcium is that it

provides a rapid *buffering* mechanism to keep the calcium ion concentration in the extracellular fluids from rising to excessive levels or falling to very low levels under transient conditions of excess or hypoavailability of calcium.

Deposition and Absorption of Bone—Remodeling of Bone

DEPOSITION OF BONE BY THE OSTEOBLASTS. Bone is continually being deposited by *osteoblasts,* and it is continually being absorbed where *osteoclasts* are active. Osteoblasts are found on the outer surfaces of the bones and in the bone cavities. A small amount of osteoblastic activity occurs continually in all living bones (on about 4 per cent of all surfaces at any given time in an adult, as shown in Figure 79–6), so that at least some new bone is being formed constantly.

ABSORPTION OF BONE—FUNCTION OF THE OSTEOCLASTS. Bone is also being continually absorbed in the presence of osteoclasts, which are large phagocytic, multinucleated cells (as many as 50 nuclei), derivatives of monocytes or monocyte-like cells formed in the bone marrow. The osteoclasts are normally active on less than 1 per cent of the bone surfaces of an adult. Later in the chapter we see that parathyroid hormone controls the bone absorptive activity of osteoclasts.

Histologically, bone absorption occurs immediately adjacent to the osteoclasts, as shown in Figure 79–6. The mechanism of this absorption is believed to be the following: The osteoclasts send out villus-like projections toward the bone, forming a so-called ruffled border adjacent to the bone. The villi secrete two types of substances: (1) proteolytic enzymes, released from the lyso-

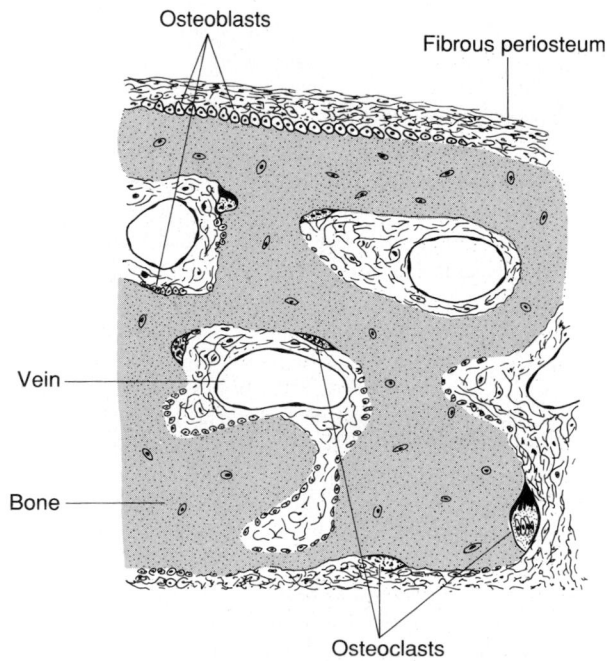

Figure 79–6. Osteoblastic and osteoclastic activity in the same bone.

somes of the osteoclasts, and (2) several acids, including citric acid and lactic acid, released from the mitochondria and secretory vesicles. The enzymes digest or dissolve the organic matrix of the bone, and the acids cause solution of the bone salts. The osteoclastic cells also embibe by phagocytosis minute particles of bone matrix and crystals, eventually also dissoluting these and releasing the products into the blood.

EQUILIBRIUM BETWEEN BONE DEPOSITION AND ABSORPTION. Normally, except in growing bones, the rates of bone deposition and absorption are equal to each other, so that the total mass of bone remains constant. Osteoclasts usually exist in small but concentrated masses, and once a mass of osteoclasts begins to develop, it usually eats away at the bone for about 3 weeks, eating out a tunnel that ranges in diameter from 0.2 to 1 millimeter and is several millimeters long. At the end of this time, the osteoclasts disappear and the tunnel is invaded by osteoblasts instead; then new bone begins to develop. Bone deposition then continues for several months, the new bone being laid down in successive layers of concentric circles (*lamellae*) on the inner surfaces of the cavity until the tunnel is filled. Deposition of new bone ceases when the bone begins to encroach on the blood vessels supplying the area. The canal through which these vessels run, called the *haversian canal,* therefore, is all that remains of the original cavity. Each new area of bone deposited in this way is called an *osteon,* as shown in Figure 79–7.

Value of Continual Remodeling of Bone. The continual deposition and absorption of bone has a number of physiologically important functions. First, bone ordinarily adjusts its strength in proportion to the degree of bone stress. Consequently, bones thicken when subjected to heavy loads. Second, even the shape of the bone can be rearranged for proper support of mechanical forces by deposition and absorption of bone in accordance with stress patterns. Third, because old bone becomes relatively brittle and weak, new organic matrix is needed as the old organic matrix degenerates. In this manner, the normal toughness of bone is maintained. Indeed, the bones of children, in whom the rates of deposition and absorption are rapid, show little brittleness in comparison with the bones of old age, at which time the rates of deposition and absorption are slow.

CONTROL OF THE RATE OF BONE DEPOSITION BY BONE "STRESS." Bone is deposited in proportion to the compressional load that the bone must carry. For instance, the bones of athletes become considerably heavier than those of nonathletes. Also, if a person has one leg in a cast but continues to walk on the opposite leg, the bone of the leg in the cast becomes thin and as much as 30 per cent decalcified within a few weeks, whereas the opposite bone remains thick and normally calcified. Therefore, continual physical stress stimulates osteoblastic deposition and calcification of bone.

Bone stress also determines the shape of bones under certain circumstances. For instance, if a long bone of the leg breaks in its center and then heals at an angle, the compression stress on the inside of the angle causes increased deposition of bone, and increased absorption occurs on the outer side of the angle where the bone is not compressed. After many years of increased deposition on the inner side of the angulated bone and absorption on the outer side, the bone can become almost straight, especially in children because of the rapid remodeling of bone at younger ages.

The deposition of bone at points of compressional stress has been suggested to be caused by a *piezoelectric* effect, as follows: Compression of bone causes a negative electrical potential in the compressed areas and a positive potential elsewhere in the bone. It has been shown that minute quantities of electric current flowing in bone cause osteoblastic activity at the negative end of the current flow, which could explain the increased bone deposition at compression sites. On the other hand, usual osteoclastic activity could account for reabsorption of bone elsewhere.

REPAIR OF A FRACTURE. Fracture of a bone in some way maximally activates all the periosteal and intraosseous osteoblasts involved in the break. Also, immense numbers of new osteoblasts are formed almost immediately from so-called *osteoprogenitor cells,* which are bone stem cells in the surface tissue lining bone, called the "bone membrane." Therefore, within a short time, a large bulge of osteoblastic tissue and new organic bone matrix, followed shortly by the deposition of calcium salts, develops between the two broken ends of the bone. This is called a *callus.*

Many bone surgeons use the phenomenon of bone stress to accelerate the rate of fracture healing. This is done by use of special mechanical fixation apparatuses for holding the ends of the broken bone together so that the patient can continue to use the bone immediately. This causes stress on the opposed ends of the broken bones, which accelerates osteoblastic activity at the break and often shortens convalescence.

BLOOD ALKALINE PHOSPHATASE AS AN INDICATION OF THE RATE OF BONE DEPOSITION. The osteoblasts secrete large quantities of alkaline phosphatase when they are actively depositing bone matrix. This phosphatase is believed either to increase the local concentration of inorganic phosphate or to activate the collagen

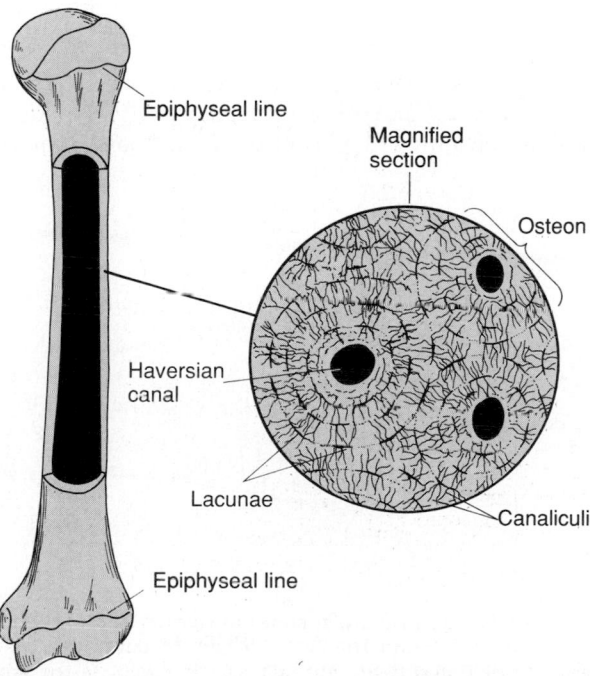

Figure 79–7. Structure of bone.

Epiphyseal line

Magnified section

Osteon

Haversian canal

Lacunae

Canaliculi

Epiphyseal line

fibers in such a way that they cause the deposition of calcium salts. Because some alkaline phosphatase diffuses into the blood, the blood level of alkaline phosphatase is usually a good indicator of the rate of bone formation.

PARATHYROID HORMONE

For many years it has been known that increased activity of the parathyroid gland causes rapid absorption of calcium salts from the bones, with resultant *hypercalcemia* in the extracellular fluid; conversely, hypofunction of the parathyroid glands causes *hypocalcemia,* often with resultant tetany. Also, parathyroid hormone is important in phosphate metabolism as well as in calcium metabolism.

PHYSIOLOGIC ANATOMY OF THE PARATHYROID GLANDS. Normally there are four parathyroid glands in the human being; they are located immediately behind the thyroid gland—one behind each of the upper and each of the lower poles of the thyroid. Each parathyroid gland is about 6 millimeters long, 3 millimeters wide, and 2 millimeters thick and has a macroscopic appearance of dark brown fat. The parathyroid glands are difficult to locate during thyroid operations because they often look like just another lobule of the thyroid gland. For this reason, before the importance of these glands was generally recognized, total or subtotal thyroidectomy frequently resulted in total removal of the parathyroid glands as well.

Removal of half the parathyroid glands usually causes little physiological abnormality. However, removal of three of the four normal glands usually causes transient hypoparathyroidism. But even a small quantity of remaining parathyroid tissue is usually capable of hypertrophying satisfactorily to perform the function of all the glands.

The parathyroid gland of the adult human being, shown in Figure 79–8, contains mainly *chief cells* and a small to moderate number of *oxyphil cells,* but oxyphil cells are absent in many animals and in young human beings. The chief cells are believed to secrete most, if not all, of the parathyroid hormone. The function of the oxyphil cells is not certain; they are believed to be modified or depleted chief cells that no longer secrete hormone.

CHEMISTRY OF PARATHYROID HORMONE. Parathyroid hormone has been isolated in a pure form. It is first synthesized on the ribosomes in the form of a preprohormone, a polypeptide chain of 110 amino acids. This is cleaved first to a prohormone with 90 amino acids, then to the hormone itself with 84 amino acids by the endoplasmic reticulum and Golgi apparatus, and finally packaged in secretory granules in the cytoplasm of the cells. The final hormone has a molecular weight of about 9500. Smaller compounds with as few as 34 amino acids adjacent to the N terminus of the molecule have also been isolated from the parathyroid glands that exhibit full parathyroid hormone activity. In fact, because the kidneys rapidly remove the whole 84–amino acid hormone within minutes but fail to remove many of the fragments for hours, a large share of the hormonal activity is caused by the fragments.

Effect of Parathyroid Hormone on Calcium and Phosphate Concentrations in the Extracellular Fluid

Figure 79–9 shows the approximate effects on the blood calcium and phosphate concentrations caused by suddenly beginning to infuse parathyroid hormone into an animal and continuing this for a number of hours. Note that at the onset of infusion the calcium ion concentration begins to rise and reaches a plateau level in about 4 hours. On the other hand, the phosphate concentration falls more rapidly than the calcium rises and reaches a depressed level within 1 or 2 hours. The rise in calcium concentration is caused principally by two effects: (1) an effect of parathyroid hormone in causing calcium and phosphate absorption from the

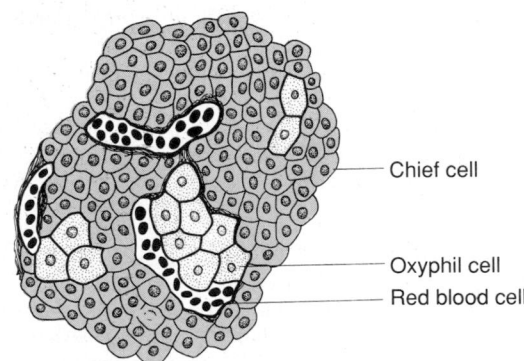

Figure 79–8. Histological structure of a parathyroid gland.

Chief cell

Oxyphil cell
Red blood cell

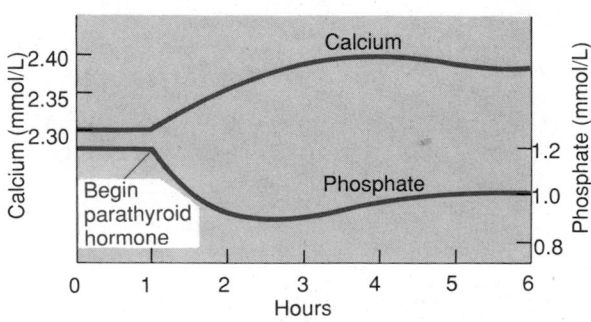

Figure 79–9. Approximate changes in calcium and phosphate concentrations during the first 5 hours of parathyroid hormone infusion at a moderate rate.

bone and (2) a rapid effect of parathyroid hormone in decreasing the excretion of calcium by the kidneys. The decline in phosphate concentration, on the other hand, is caused by a strong effect of parathyroid hormone on the kidneys in causing excessive renal phosphate excretion, an effect that is usually great enough to override increased phosphate absorption from the bone.

Calcium and Phosphate Absorption from the Bone Caused by Parathyroid Hormone

Parathyroid hormone seems to have two effects on bone in causing absorption of calcium and phosphate. One is a rapid phase that begins in minutes and increases progressively for several hours. This phase is believed to result from activation of the already existing bone cells (mainly the osteocytes) to promote calcium and phosphate absorption. The second phase is a much slower one, requiring several days or even weeks to become fully developed; it results from proliferation of the osteoclasts, followed by greatly increased osteoclastic reabsorption of the bone itself, not merely absorption of the calcium phosphate salts from the bone.

RAPID PHASE OF CALCIUM AND PHOSPHATE ABSORPTION—OSTEOLYSIS. When large quantities of parathyroid hormone are injected, the calcium ion concentration in the blood begins to rise within minutes, long before any new bone cells can be developed. Histological and physiological studies have shown that the parathyroid hormone causes removal of bone salts from two areas in the bone: (1) from the bone matrix in the vicinity of the osteocytes lying within the bone itself, and (2) in the vicinity of the osteoblasts along the bone surface. Yet, strangely enough, one does not usually think of either osteoblasts or osteocytes functioning to cause bone salt absorption because both these types of cells are osteoblastic in nature and normally associated with bone deposition and its calcification. However, more recent studies have shown that the osteoblasts and osteocytes form a system of interconnected cells that spreads all through the bone and over all the bone surfaces except the small surface areas adjacent to the osteoclasts. In fact, long, filmy processes extend from osteocyte to osteocyte throughout the bone structure, and these processes also connect with the surface osteocytes and osteoblasts. This extensive system is called the *osteocytic membrane system,* and it is believed to provide a membrane that separates the bone itself from the extracellular fluid. Between the osteocytic membrane and the bone is a small amount of fluid called simply *bone fluid.* Experiments suggest that the osteocytic membrane pumps calcium ions from the bone fluid into the extracellular fluid, creating a calcium ion concentration in the bone fluid only one-third that in the extracellular fluid. When the osteocytic pump becomes excessively activated, the bone fluid calcium concentration falls even lower and calcium phosphate salts are then absorbed from the bone. This effect is called *osteolysis,* and it occurs without absorption of the bone's fibrous and gel matrix. When the pump is inactivated, the bone fluid calcium concentration rises to a higher level and calcium phosphate salts are redeposited in the matrix.

But where does parathyroid hormone fit into this picture? First, the cell membranes of both the osteoblasts and the osteocytes have receptor proteins for binding parathyroid hormone. It seems that parathyroid hormone can activate the calcium pump strongly, thereby causing rapid removal of calcium phosphate salts from those amorphous bone crystals that lie near the cells. The parathyroid hormone is believed to stimulate this pump by increasing the calcium permeability of the bone fluid side of the osteocytic membrane, thus allowing calcium ions to diffuse into the membrane cells from the bone fluid. Then the calcium pump on the other side of the cell membrane transfers the calcium ions the rest of the way into the extracellular fluid.

SLOW PHASE OF BONE ABSORPTION AND CALCIUM PHOSPHATE RELEASE—ACTIVATION OF THE OSTEOCLASTS. A much better known effect of parathyroid hormone and one for which the evidence is much clearer is its activation of the osteoclasts. Yet the osteoclasts do not themselves have membrane receptor proteins for parathyroid hormone. Instead, it is believed that the activated osteoblasts and osteocytes send a secondary but unknown "signal" to the osteoclasts, causing them to set about their usual task of gobbling up the bone over a period of weeks or months.

Activation of the osteoclastic system occurs in two stages: (1) immediate activation of the osteoclasts that are already formed and (2) formation of new osteoclasts. Several days of excess parathyroid hormone usually cause the osteoclastic system to become well developed, but it can continue to grow for literally months under the influence of strong parathyroid hormone stimulation.

After a few months, osteoclastic resorption of bone can lead to weakened bones and secondary stimulation of the osteoblasts that attempt to correct the weakened state. Therefore, the late effect is actually to enhance both osteoblastic and osteoclastic activity. Still, even in the late stages, there is more bone absorption than bone deposition in the presence of continued excess parathyroid hormone.

Bone contains such great amounts of calcium in comparison with the total amount in all the extracellular fluids (about 1000 times as much) that even when parathyroid hormone causes a great rise in calcium concentration in the fluids, it is impossible to discern any immediate effect on the bones. Prolonged administration or secretion of parathyroid hormone—over a period of many months or years—finally results in very evident absorption in all the bones and even development of large cavities filled with large, multinucleated osteoclasts.

Effect of Parathyroid Hormone on Phosphate and Calcium Excretion by the Kidneys

Administration of parathyroid hormone causes immediate and rapid loss of phosphate in the urine owing to the effect of the hormone to diminish proximal tubular reabsorption of phosphate ions.

Parathyroid hormone also increases tubular reabsorption of calcium at the same time that it diminishes phosphate reabsorption. Moreover, it also increases the rate of reabsorption of magnesium ions and hydrogen ions while it decreases the reabsorption of sodium, potassium, and amino acid ions in much the same way that it affects phosphate. The increased calcium absorption occurs mainly in the *late distal tubules,* the *collecting tubules,* and the early collecting ducts.

Were it not for the effect of parathyroid hormone on the kidneys to increase calcium reabsorption, continual loss of calcium into the urine would eventually deplete both the extracellular fluid and the bones of this mineral.

Effect of Parathyroid Hormone on Intestinal Absorption of Calcium and Phosphate

At this point, we should be reminded again that parathyroid hormone greatly enhances both calcium and phosphate absorption from the intestines by increasing the formation in the kidneys of 1,25-dihydroxycholecalciferol from vitamin D, as discussed earlier in the chapter.

Effect of Vitamin D on Bone and Its Relation to Parathyroid Hormone Activity

Vitamin D plays important roles in both bone absorption and bone deposition. The administration of *extreme quantities* of vitamin D causes absorption of bone in much the same way that administration of parathyroid hormone does. Also, in the absence of vitamin D, the effect of parathyroid hormone in causing bone absorption is greatly reduced or even prevented. The mechanism of this action of vitamin D is not known, but it is believed to result from the effect of 1,25-dihydroxycholecalciferol (the principal active product of vitamin D) in increasing calcium transport through cellular membranes.

Vitamin D in smaller quantities promotes bone calcification. One of the ways in which it does this is to increase calcium and phosphate absorption from the intestines. However, even in the absence of such increase, it enhances the mineralization of bone. Here again, the mechanism of the effect is unknown, but it probably also results from the ability of 1,25-dihydroxycholecalciferol to cause transport of calcium ions through cell membranes—but in this instance perhaps in the opposite direction through the osteoblastic or osteocytic cell membranes.

Role of Cyclic Adenosine Monophosphate as a Mediator of Parathyroid Stimulation. A large share of the effect of parathyroid hormone on its target organs is mediated by the cyclic adenosine monophosphate (cAMP) *second messenger* mechanism. Within a few minutes after parathyroid hormone administration, the concentration of cAMP increases in the osteocytes, osteoclasts, and other target cells. This cAMP in turn is probably responsible for such functions as osteoclastic secretion of enzymes and acids to cause bone reabsorption, formation of 1,25-dihydroxycholecalciferol in the kidneys, and so forth. There are probably other direct effects of parathyroid hormone that function independently of the second messenger mechanism.

Control of Parathyroid Secretion by the Calcium Ion Concentration

Even the slightest decrease in calcium ion concentration in the extracellular fluid causes the parathyroid glands to increase their rate of secretion within minutes; if the decreased calcium concentration persists, the glands will hypertrophy, sometimes fivefold or more. For instance, the parathyroid glands become greatly enlarged in *rickets,* in which the level of calcium is usually depressed only a small amount; also, they become greatly enlarged in *pregnancy,* even though the decrease in calcium ion concentration in the mother's extracellular fluid is hardly measurable; and they are greatly enlarged during *lactation* because calcium is used for milk formation.

On the other hand, any condition that increases the calcium ion concentration above normal causes decreased activity and reduced size of the parathyroid glands. Such conditions include (1) excess quantities of calcium in the diet, (2) increased vitamin D in the diet, and (3) bone absorption caused by factors other than parathyroid hormone (for example, bone absorption caused by disuse of the bones).

Figure 79–10 shows quantitatively the approximate relation between plasma calcium concentration and plasma parathyroid hormone concentration. The solid red curve shows the acute relation when the calcium concentration is changed over a period of a few hours. This shows that even small decreases in calcium concentration from the normal value can double or triple the plasma parathyroid hormone. On the other hand, the approximate chronic relation that one finds when the calcium ion concentration changes over a period of many weeks, thus allowing time for the glands to hypertrophy greatly, is shown by the dashed red line; this demonstrates that a decrease of only a fraction of a milligram per deciliter in plasma calcium concentration can double parathyroid hormone secretion. This is the basis of the body's extremely potent feedback system for long-term control of plasma calcium ion concentration.

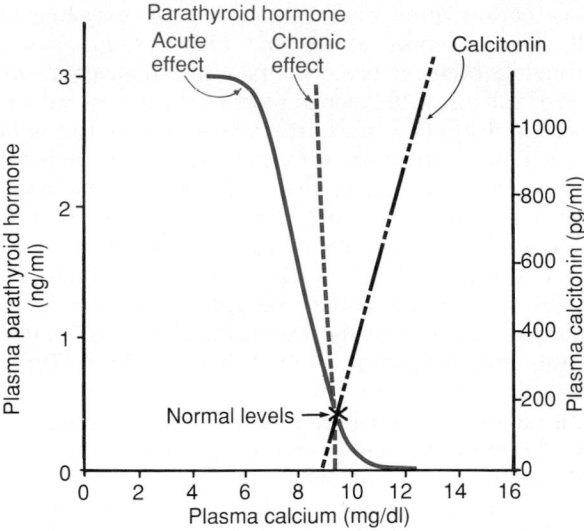

Figure 79–10. Approximate effect of plasma calcium concentration on the plasma concentrations of parathyroid hormone and calcitonin. Note especially that long-term, chronic changes in calcium concentration can cause as much as a 100 per cent change in parathyroid hormone concentration for only a few per cent change in calcium concentration.

CALCITONIN

In the early 1960s, a new hormone that has *weak* effects on blood calcium but opposite those of parathyroid hormone was discovered. This hormone, named *calcitonin* because it reduces the blood calcium ion concentration, is secreted in the human being not by the parathyroid glands but by the thyroid gland. In fish, amphibians, reptiles, and birds, calcitonin is secreted by the *ultimobranchial glands;* it plays an especially important role in helping control the blood calcium ion concentration when these animals change their habitat from freshwater to seawater, where there are great excesses of calcium. Furthermore, its concentration in these ultimobranchial glands is extremely great. In the human being, ultimobranchial glands do not exist as such but have become incorporated into the thyroid gland. The so-called *parafollicular cells*, or C cells, in the interstitial tissue between the follicles of the human thyroid gland are remnants of the ultimobranchial glands of lower animals, constituting only 0.1 per cent of the thyroid gland; it is these cells that secrete the calcitonin.

Calcitonin is a large polypeptide with a molecular weight of about 3400 and a chain of 32 amino acids.

EFFECT OF CALCITONIN IN DECREASING PLASMA CALCIUM CONCENTRATION. In some young animals but little in the human being, calcitonin decreases blood calcium ion concentration rapidly, beginning within minutes after injection of the calcitonin. Thus, the effect of calcitonin on blood calcium ion concentration is opposite that of parathyroid hormone, and it occurs several times as rapidly.

Calcitonin reduces plasma calcium concentration in at least two ways.

1. The immediate effect is to decrease the absorptive activities of the osteoclasts and possibly the osteolytic effect of the osteocytic membrane throughout the bone, thus shifting the balance in favor of deposition of calcium in the exchangeable bone calcium salts. This effect is especially significant in young animals because of the rapid interchange of absorbed and deposited calcium.

2. The second and more prolonged effect of calcitonin is to decrease the formation of new osteoclasts. Also, because osteoclastic resorption of bone leads secondarily to osteoblastic activity, the depressed numbers of osteoclasts are followed by depressed numbers of osteoblasts. Therefore, over a long period of time, the net result is simply greatly reduced osteoclastic and osteoblastic activity; consequently, there is not a significant prolonged effect on plasma calcium ion concentration. That is, the effect on plasma calcium is mainly a transient one, lasting for a few hours to a few days at most.

Calcitonin also has minor effects on calcium handling in the kidney tubules and intestinal tract. Again, the effects are opposite those of parathyroid hormone, but quantitatively, they appear to be of such little import that they are seldom considered.

IMPORTANCE OF THE CALCITONIN EFFECT ON PLASMA CALCIUM CONCENTRATION. Calcitonin has only a weak effect on plasma calcium concentration in the adult human being. The reason for this is twofold. First, any initial reduction of the calcium ion concentration caused by calcitonin leads within hours to a powerful stimulation of parathyroid hormone secretion, which almost overrides the calcitonin effect. Second, in the adult, the daily rates of absorption and deposition of calcium are small, and even after the rate of absorption is slowed by calcitonin, this still has only a small effect on plasma calcium ion concentration. The effect in children is much more marked because bone remodeling occurs rapidly in children, with absorption and deposition of calcium as great as 5 or more grams per day—equal to 5 to 10 times the total calcium in all the extracellular fluid. Also, in certain bone diseases, such as Paget's disease, in which osteoclastic activity is greatly accelerated, calcitonin has a much more potent effect of reducing the calcium absorption.

Effect of Plasma Calcium Concentration on the Secretion of Calcitonin

In young animals but much less so in older animals and in human beings, an increase in plasma calcium concentration of about 10 per cent causes an immediate twofold or more increase in the rate of secretion of calcitonin, which is shown by the dot-dash line of Figure 79–10. This provides a second hormonal feedback mechanism for controlling the plasma calcium ion concentration, but one that works in a way opposite that of the parathyroid hormone system.

There are two major differences between the calcitonin and the parathyroid feedback systems. First, the calcitonin mechanism operates more rapidly, reaching peak activity in less than 1 hour, in contrast to the 3 to 4 hours required for peak activity to be attained after the onset of parathyroid secretion.

The second difference is that the calcitonin mechanism acts only weakly and only as a short-term regulator of calcium ion concentration because it is rapidly overridden by the much more powerful parathyroid control mechanism. Also, the calcitonin receptors on the osteoclasts seem to down-regulate within minutes to hours in response to a calcitonin stimulus. Therefore, over a prolonged period of time, it is almost entirely the parathyroid system that sets the long-term level of calcium ions in the extracellular fluid.

When the thyroid gland is removed and calcitonin is no longer secreted, the long-term blood calcium ion concentration is not measurably altered, which again demonstrates the overriding effect of the parathyroid hormonal system of control.

OVERALL CONTROL OF CALCIUM ION CONCENTRATION

At times, the amount of calcium absorbed into or lost from the body fluids is as much as 0.3 gram in 1 hour. For instance, in cases of diarrhea, several grams of calcium can be secreted in the intestinal juices, passed into the intestinal tract, and lost into the feces each day. Conversely, after ingestion of large quantities of calcium, particularly when there is also an excess of vitamin D activity, a person may absorb as much as 0.3 gram in 1 hour. This figure compares with a *total quantity of calcium in all the extracellular fluid of about 1 gram.* The addition or subtraction of 0.3 gram to or from such a small amount of calcium in the extracellular fluid would cause serious hypercalcemia or hypocalcemia. However, there is a first line of defense to prevent this from occurring even before the parathyroid and calcitonin hormone feedback systems have a chance to act.

BUFFER FUNCTION OF THE EXCHANGEABLE CALCIUM IN BONES. The exchangeable calcium salts in the bones, discussed earlier in this chapter, are amorphous calcium phosphate compounds, probably mainly $CaHPO_4$ or some similar compound loosely bound in the bone and in reversible equilibrium with the calcium and phosphate ions in the extracellular fluid. The quantity of these salts that is available for exchange is about 0.5 to 1 per cent of the total calcium salts of the bone, a total of 5 to 10 grams of calcium. Because of the ease of deposition of these exchangeable salts and their ease of resolubility, an increase in the concentrations of extracellular fluid calcium and phosphate ions above normal causes immediate deposition of exchangeable salt. Conversely, a decrease in these concentrations causes immediate absorption of exchangeable salt. This reaction is so rapid that a single passage through a bone of blood containing a high concentration of calcium will remove most of the excess calcium. This rapid effect results from the fact that the amorphous bone crystals are extremely small, and their total surface area exposed to the fluids of the bone is perhaps 1 acre or more. Also, about 5 per cent of all the blood flows through the bones each minute—that is, about 1 per cent of all the extracellular fluid each minute. Therefore, about one half of any excess calcium that appears in the extracellular fluid is removed by this buffer function of the bones in about 70 minutes.

In addition to the buffer function of the bones, the mitochondria of many of the tissues of the body, especially of the liver and intestine, contain a reasonable amount of exchangeable calcium (a total of about 10 grams in the whole body) that provides an additional buffer system for helping to maintain constancy of the extracellular fluid calcium ion concentration.

HORMONAL CONTROL OF CALCIUM ION CONCENTRATION, THE SECOND LINE OF DEFENSE. At the same time that the exchangeable calcium mechanism in the bones is "buffering" the calcium in the extracellular fluid, both the parathyroid and the calcitonin hormonal systems are beginning to act. Within 3 to 5 minutes after an acute increase in the calcium ion concentration, the rate of parathyroid hormone secretion decreases. As already explained, this sets into play multiple mechanisms for reducing the calcium ion concentration back toward normal. This is a slow process. On the other hand, at the same time that parathyroid hormone decreases, calcitonin increases. In young animals and possibly in young human children (but probably little in adults), the calcitonin causes rapid deposition of calcium in the bones, and perhaps in some cells of other tissues. Therefore, in very young animals, excess calcitonin can cause a high calcium ion concentration to return to normal perhaps considerably more rapidly than can be achieved by the exchangeable calcium-buffering mechanism alone.

In prolonged calcium excess or prolonged calcium deficiency, only the parathyroid hormonal mechanism seems to be really important in maintaining a normal plasma calcium ion concentration. When a person has a continuing deficiency of calcium in the diet, parathyroid hormone often can stimulate enough calcium absorption from the bones to maintain a normal plasma calcium ion concentration for 1 year or more, but eventually, even the bones will run out of calcium. Thus, in effect, the bones are a large buffer-reservoir of calcium that can be manipulated by parathyroid hormone. Yet, when the bone reservoir either runs out of calcium or, oppositely, becomes saturated with calcium, then the long-term control of extracellular calcium ion concentration resides almost entirely in the roles of parathyroid hormone and vitamin D in controlling calcium absorption from the gut and calcium excretion in the urine.

PHYSIOLOGY OF PARATHYROID AND BONE DISEASES

Hypoparathyroidism

When the parathyroid glands do not secrete sufficient parathyroid hormone, the osteocytic reabsorption of exchangeable calcium decreases and the osteoclasts become almost totally inactive. As a result, calcium reabsorption from the bones is so depressed that the level of calcium in the body fluids decreases. Yet, because calcium and phosphates are not being absorbed from the bone, the bone usually remains strong.

When the parathyroid glands are suddenly removed, the calcium level in the blood falls from the normal of 9.4 to 6 to 7 mg/dl within 2 to 3 days and the blood phosphate concentration may double. When this low calcium level is reached, the usual signs of tetany develop. Among the muscles of the body especially sensitive to tetanic spasm are the laryngeal muscles. Spasm of these muscles obstructs respiration, which is the usual cause of death in tetany unless appropriate treatment is applied.

Treatment of Hypoparathyroidism

Parathyroid Hormone (Parathormone). Parathyroid hormone is occasionally used for treating hypoparathyroidism. However, because of the expense of this hormone, because its effect lasts for a few hours at most, and because the tendency of the body to develop immune bodies against it makes it progressively less and less effective, treatment of hypoparathyroidism with parathyroid hormone is rare in present-day therapy.

Vitamin D and Calcium Therapy. In most patients, the administration of extremely large quantities of vitamin D, to as high as 100,000 units per day, along with intake of 1 to 2 grams of calcium will suffice to keep the calcium ion concentration in a normal range. At times, it might be necessary to administer 1,25-dihydroxycholecalciferol instead of the nonactivated form of vitamin D because of its much more potent and much more rapid action. This can also cause unwanted effects because it is sometimes difficult to prevent overactivity by this activated form of vitamin D.

Hyperparathyroidism

The cause of hyperparathyroidism ordinarily is a tumor of one of the parathyroid glands; such tumors occur much more frequently in women than in men or children, mainly because pregnancy and lactation stimulate the parathyroid glands and therefore predispose to the development of such a tumor.

Hyperparathyroidism causes extreme osteoclastic activity in the bones. This elevates the calcium ion concentration in the extracellular fluid while usually depressing the concentration of phosphate ions because of increased renal excretion of phosphate.

Bone Disease in Hyperparathyroidism. Although in mild hyperparathyroidism new bone can be deposited rapidly enough to compensate for the increased osteoclastic reabsorption of bone, in severe hyperparathyroidism, the osteoclastic absorption soon far outstrips osteoblastic deposition and the bone may be eaten away almost entirely. Indeed, the reason a hyperparathyroid person seeks medical attention is often a broken bone. Radiographs of the bone show extensive decalcification and, occasionally, large punched-out cystic areas of the bone that are filled with osteoclasts in the form of so-called giant cell osteoclast "tumors." Multiple fractures of the weakened bones can result from only slight trauma, especially where cysts develop. The cystic bone disease of hyperparathyroidism is called *osteitis fibrosa cystica.*

Osteoblastic activity in the bones also increases greatly in a vain attempt to form enough new bone to make up for the old bone absorbed by the osteoclastic activity. As explained earlier in the chapter, when the osteoblasts become active, they secrete large quantities of alkaline phosphatase. Therefore, one of the important diagnostic findings in hyperparathyroidism is a high level of plasma alkaline phosphatase.

Effects of Hypercalcemia in Hyperparathyroidism. Hyperparathyroidism can at times cause the plasma calcium level to rise to 12 to 15 mg/dl and, rarely, even higher. The effects of such elevated calcium levels, as detailed earlier in the chapter, are depression of the central and peripheral nervous systems, muscle weakness, constipation, abdominal pain, peptic ulcer, lack of appetite, and depressed relaxation of the heart during diastole.

Parathyroid Poisoning and Metastatic Calcification. When, on rare occasions, extreme quantities of parathyroid hormones are secreted, the level of calcium in the body fluids rises rapidly to high values. Even the extracellular fluid phosphate concentration often rises markedly instead of falling, as is usually the case, probably because the kidneys cannot excrete rapidly enough all the phosphate being absorbed from the bone. Therefore, the calcium and phosphate in the body fluids become greatly supersaturated, so that calcium phosphate ($CaHPO_4$) crystals begin to deposit in the alveoli of the lungs, the tubules of the kidneys, the thyroid gland, the acid-producing area of the stomach mucosa, and the walls of the arteries throughout the body. This extensive *metastatic* deposition of calcium phosphate can develop within a few days.

Ordinarily, the level of calcium in the blood must rise above 17 mg/dl before there is danger of parathyroid poisoning, but once such elevation develops along with concurrent elevation of phosphate, death can occur in only a few days.

Formation of Kidney Stones in Hyperparathyroidism. Most patients with mild hyperparathyroidism show few signs of bone disease and few general abnormalities as a result of elevated calcium, but they do have an extreme tendency to form kidney stones. The reason is that the excess calcium and phosphate absorbed from the intestines or mobilized from the bones in hyperparathyroidism must eventually be excreted by the kidneys, causing a proportionate increase in the concentrations of these substances in the urine. As a result, crystals of calcium phosphate tend to precipitate in the kidney, forming calcium phosphate stones. Also, calcium oxalate stones develop because even normal levels of oxalate will cause calcium precipitation at high calcium levels. Because the solubility of most renal stones is slight in alkaline media, the tendency for formation of renal calculi is considerably greater in alkaline urine than in acid urine. For this reason, acidotic diets and acidic drugs are frequently used for treating renal calculi.

Rickets

Rickets occurs mainly in children. It results from calcium or phosphate deficiency in the extracellular fluid, usually caused by lack of vitamin D. If the child is properly exposed to sunlight, the 7-dehydrocholesterol in the skin becomes activated by the ultraviolet rays and forms vitamin D_3, which prevents rickets by promoting calcium and phosphate absorption from the intestines, as discussed earlier in the chapter.

Children who remain indoors through the winter in general do not receive adequate quantities of vitamin D without some supplementary therapy in the diet. Rickets tends to occur especially in the spring months because vitamin D formed during the preceding summer is stored in the liver and available for use during the early winter months. Also, calcium and phosphate absorption from the bones can prevent clinical signs of rickets for the first few months of vitamin D deficiency.

Plasma Concentrations of Calcium and Phosphate in Rickets. Ordinarily, the plasma calcium concentration in rickets is only slightly depressed, but the level of phosphate is greatly depressed. This is because the parathyroid glands prevent the calcium level from falling by promoting bone absorption every time the calcium level begins to fall. On the other hand, there is no good regulatory system for preventing a falling level of phosphate, and the increased parathyroid activity actually increases the excretion of phosphates in the urine.

Effect of Rickets on the Bones. During prolonged rickets, the marked compensatory increase in parathyroid hormone secretion causes extreme osteoclastic absorption of the bone; this in turn causes the bone to become progressively weaker and imposes marked physical stress on the bone, resulting in rapid osteoblastic activity as well. The osteoblasts lay down large quantities of osteoid, which does not become calcified because of insufficient calcium and phosphate ions. Consequently, the newly formed, uncalcified, and weak osteoid gradually takes the place of the older bone that is being reabsorbed.

Tetany in Rickets. In the early stages of rickets, tetany almost never occurs because the parathyroid glands continually stimulate osteoclastic absorption of bone and, therefore, maintain an almost normal level of calcium in the extracellular fluid. However, when the bones finally become exhausted of calcium, the level of calcium may fall rapidly. As the blood level of calcium falls below 7 mg/dl, the usual signs of tetany develop and the child may die of tetanic respiratory spasm unless intravenous calcium is administered, which relieves the tetany immediately.

TREATMENT. The treatment of rickets depends on supplying adequate calcium and phosphate in the diet and, equally important, on administering large amounts of vitamin D. If vitamin D is not administered, little calcium and phosphate are absorbed from the gut.

Osteomalacia

Osteomalacia is rickets in adults and is frequently called "adult rickets."

Normal adults seldom have a serious *dietary* deficiency of vitamin D or calcium because large quantities of calcium are not needed for bone growth as in children. However, serious deficiencies of both vitamin D and calcium occasionally occur as a result of steatorrhea (failure to absorb fat) because vitamin D is fat-soluble and calcium tends to form insoluble soaps with fat; consequently, in steatorrhea, both vitamin D and calcium tend to pass into the feces. Under these conditions, an adult occasionally has such poor calcium and phosphate absorption that adult rickets can occur, although this almost never proceeds to the stage of tetany but often is a cause of severe bone disability.

OSTEOMALACIA AND RICKETS CAUSED BY RENAL DISEASE. "Renal rickets" is a type of osteomalacia that results from prolonged kidney damage. The cause of this condition is mainly failure of the damaged kidneys to form 1,25-dihydroxycholecalciferol, the active form of vitamin D. In patients whose kidneys have been removed or destroyed and who are being treated by hemodialysis, the problem of renal rickets is often a severe one.

Another type of renal disease that leads to rickets and osteomalacia is *congenital hypophosphatemia* resulting from congenitally reduced reabsorption of phosphates by the renal tubules. This type of rickets must be treated with phosphate compounds instead of calcium and vitamin D, and it is called *vitamin D–resistant rickets.*

Osteoporosis

Osteoporosis is the most common of all bone diseases in adults, especially in old age. It is a different disease from osteomalacia and rickets because it results from diminished organic bone matrix rather than from poor bone calcification. In osteoporosis the osteoblastic activity in the bone usually is less than normal, and consequently the rate of bone osteoid deposition is depressed. But occasionally, as in hyperparathyroidism, the cause of the diminished bone is excess osteoclastic activity.

The many common causes of osteoporosis are (1) lack of physical stress on the bones because of inactivity; (2) malnutrition to the extent that sufficient protein matrix cannot be formed; (3) lack of vitamin C, which is necessary for the secretion of intercellular substances by all cells, including formation of osteoid by the osteoblasts; (4) postmenopausal lack of estrogen secretion because estrogens have an osteoblast-stimulating activity; (5) old age, in which growth hormone and other growth factors diminish greatly plus the fact that many of the protein anabolic functions are poor anyway, so that bone matrix cannot be deposited satisfactorily; and (6) Cushing's disease because massive quantities of glucocorticoids secreted in this disease cause decreased deposition of protein throughout the body and increased catabolism of protein and have the specific effect of depressing osteoblastic activity. Thus, many diseases of deficiencies of protein metabolism can cause osteoporosis.

PHYSIOLOGY OF THE TEETH

The teeth cut, grind, and mix the food eaten. To perform these functions, the jaws have powerful muscles capable of providing an occlusive force between the front teeth of 50 to 100 pounds and for the jaw teeth, 150 to 200 pounds. Also, the upper and lower teeth are provided with projections and facets that interdigitate, so that the upper set of teeth fits with the lower. This fitting is called *occlusion,* and it allows even small particles of food to be caught and ground between the tooth surfaces.

Function of the Different Parts of the Teeth

Figure 79–11 shows a sagittal section of a tooth, demonstrating its major functional parts: the *enamel, dentine, cementum,* and *pulp.* The tooth can also be divided into the *crown,* which is the portion that protrudes out from the gum into the mouth, and the *root,* which is the portion that protrudes into the bony socket of the jaw. The collar between the crown and the root where the tooth is surrounded by the gum is called the *neck.*

DENTINE. The main body of the tooth is composed of dentine, which has a strong, bony structure. Dentine is made up principally of hydroxyapatite crystals similar to those in bone but much more dense. These crystals are embedded in a strong meshwork of collagen fibers. In other words, the principal constituents of dentine are very much the same as those of bone. The major difference is its histological organization because dentine does not contain any osteoblasts, osteocytes, osteoclasts, or spaces for blood vessels or nerves. Instead, it is deposited and nourished by a layer of cells called *odontoblasts,* which line its inner surface along the wall of the pulp cavity.

The calcium salts in dentine make it extremely resistant to compressional forces, while the collagen fibers make it tough and resistant to tensional forces

that might result when the teeth are struck by solid objects.

ENAMEL. The outer surface of the tooth is covered by a layer of enamel that is formed before eruption of the tooth by special epithelial cells called *ameloblasts.* Once the tooth has erupted, no more enamel is formed. Enamel is composed of very large and very dense crystals of hydroxyapatite with adsorbed carbonate, magnesium, sodium, potassium, and other ions embedded in a fine meshwork of strong and almost insoluble protein fibers that are similar in physical characteristics (but not chemically identical) to the keratin of hair. The crystalline structure of the salts makes the enamel extremely hard—much, much harder than the dentine. Also, the special protein fiber meshwork, although constituting only about 1 per cent of the enamel mass, nevertheless makes enamel resistant to acids, enzymes, and other corrosive agents because this protein is one of the most insoluble and resistant proteins known.

CEMENTUM. Cementum is a bony substance secreted by cells of the *periodontal membrane,* which lines the tooth socket. Many collagen fibers pass directly from the bone of the jaw, through the periodontal membrane, and then into the cementum. These collagen fibers and the cementum hold the tooth in place. When the teeth are exposed to excessive strain, the layer of cementum becomes thicker and stronger. Also, it increases in thickness and strength with age, causing the teeth to become progressively more firmly seated in the jaws as one reaches adulthood and older.

PULP. The pulp cavity of each tooth is filled with *pulp,* which is composed of connective tissue with an abundant supply of nerve fibers, blood vessels, and lymphatics. The cells lining the surface of the pulp cavity are the odontoblasts, which, during the formative years of the tooth, lay down the dentine but at the same time encroach more and more on the pulp cavity, making it smaller. In later life, the dentine stops growing and the pulp cavity remains essentially constant in size. However, the odontoblasts are still viable and send projections into small *dentinal tubules* that penetrate all the way through the dentine; they are of importance for exchange of calcium, phosphate, and other minerals with the dentine.

Dentition

Each human being and most other mammals develop two sets of teeth during a lifetime. The first teeth are called the *deciduous teeth,* or *milk teeth,* and they number 20 in the human being. They erupt between the 7th month and the 2nd year of life, and they last until the 6th to the 13th year. After each deciduous tooth is lost, a permanent tooth replaces it, and an additional 8 to 12 molars appear posteriorly in the jaws, making the total number of permanent teeth 28 to 32, depending on whether the four *wisdom teeth* finally appear, which does not occur in everyone.

FORMATION OF THE TEETH. Figure 79–12 shows the formation and eruption of teeth. Figure 79–12A

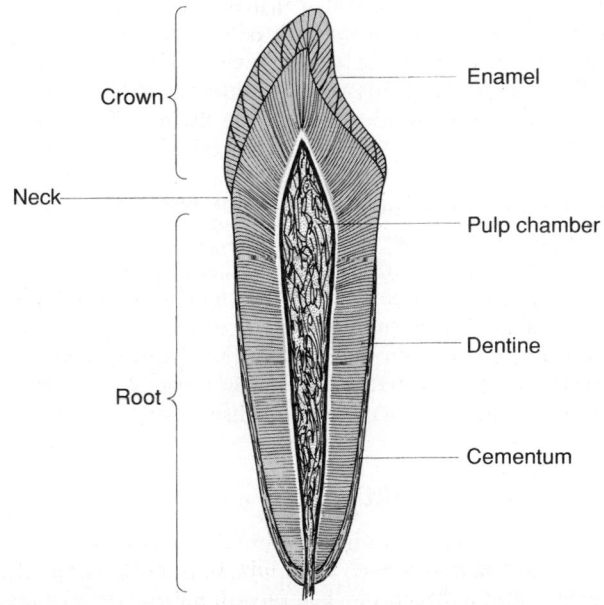

Figure 79–11. Functional parts of a tooth.

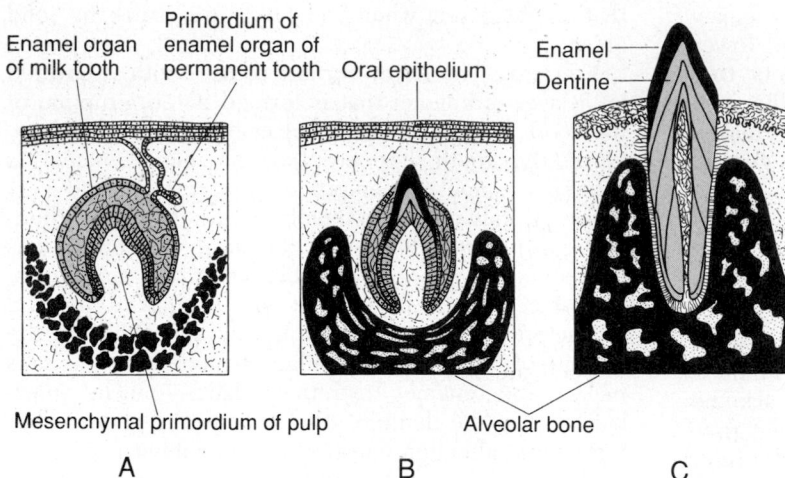

Enamel organ of milk tooth
Primordium of enamel organ of permanent tooth
Oral epithelium
Enamel
Dentine
Mesenchymal primordium of pulp
Alveolar bone
A   B   C

Figure 79–12. *A,* Primordial tooth organ. *B,* Developing tooth. *C,* Erupting tooth.

shows invagination of the oral epithelium into the *dental lamina;* this is followed by the development of a tooth-producing organ. The epithelial cells above form ameloblasts, which form the enamel on the outside of the tooth. The epithelial cells below invaginate upward into the middle of the tooth to form the pulp cavity and the odontoblasts that secrete dentine. Thus, enamel is formed on the outside of the tooth and dentine, on the inside, giving rise to an early tooth, as shown in Figure 79–12B.

Eruption of Teeth. During early childhood, the teeth begin to protrude outward from the bone through the oral epithelium into the mouth. The cause of "eruption" is unknown, although several theories have been offered in an attempt to explain this phenomenon. The most likely theory is that growth of the tooth root as well as of the bone underneath the tooth progressively shoves the tooth forward.

Development of the Permanent Teeth. During embryonic life, a tooth-forming organ also develops in the deeper dental lamina for each permanent tooth that will be needed after the deciduous teeth are gone. These tooth-producing organs slowly form the permanent teeth throughout the first 6 to 20 years of life. When each permanent tooth becomes fully formed, it, like the deciduous tooth, pushes outward through the bone. In so doing, it erodes the root of the deciduous tooth and eventually causes it to loosen and fall out. Soon thereafter, the permanent tooth erupts to take the place of the original one.

Metabolic Factors in Development of the Teeth. The rate of development and the speed of eruption of teeth can be accelerated by both thyroid and growth hormones. Also, the deposition of salts in the early-forming teeth is affected considerably by various factors of metabolism, such as the availability of calcium and phosphate in the diet, the amount of vitamin D present, and the rate of parathyroid hormone secretion. When all these factors are normal, the dentine and enamel will be correspondingly healthy, but when they are deficient, the calcification of the teeth also may be defective, so that the teeth will be abnormal throughout life.

Mineral Exchange in Teeth

The salts of teeth, like those of bone, are composed of hydroxyapatite with adsorbed carbonates and various cations bound together in a hard crystalline substance. Also, new salts are constantly being deposited while old salts are being reabsorbed from the teeth, as occurs in bone. Experiments indicate that deposition and reabsorption occur mainly in the dentine and cementum and little occurs in the enamel. Most of that which does occur in the enamel occurs by diffusive exchange of minerals with the saliva instead of with the fluids of the pulp cavity.

The rate of absorption and deposition of minerals in the cementum is about equal to that in the surrounding bone of the jaw, whereas the rate of deposition and absorption of minerals in the dentine is only one-third that of bone. The cementum has characteristics almost identical to those of usual bone, including the presence of osteoblasts and osteoclasts, whereas dentine does not have these characteristics, as explained above. This difference undoubtedly explains the different rates of mineral exchange.

The mechanism by which minerals are deposited and reabsorbed from the dentine is not clear. It is probable that the thin, tubular processes of the odontoblasts that protrude into the tubules of the dentine are capable of absorbing salts and then of providing new salts to take the place of the old.

In summary, continual mineral exchange occurs in the dentine and cementum of teeth, although the mechanism of this exchange in dentine is unclear. On the other hand, enamel exhibits extremely slow mineral exchange, so that it maintains most of its original mineral complement throughout life.

Dental Abnormalities

The two most common dental abnormalities are *caries* and *malocclusion.* Caries means erosions of the teeth, whereas malocclusion means failure of the projections of the upper and lower teeth to interdigitate properly.

CARIES AND THE ROLE OF FLUORINE. It is generally agreed by all research investigators of dental caries that caries results from the action of bacteria on the teeth, the most common of which is *Streptococcus mutans*. The first event in the development of caries is the deposit of *plaque*, a film of precipitated products of saliva and food, on the teeth. Large numbers of bacteria inhabit this plaque and are readily available to cause caries. These bacteria depend to a great extent on carbohydrates for their food. When carbohydrates are available, their metabolic systems are strongly activated and they multiply. In addition, they form acids (particularly lactic acid) and proteolytic enzymes. The acids are the major culprit in the causation of caries because the calcium salts of teeth are slowly dissolved in a highly acidic medium. And once the salts have become absorbed, the remaining organic matrix is rapidly digested by the proteolytic enzymes.

Enamel is far more resistant to demineralization by acids than is dentine, primarily because the crystals of enamel are dense but also because each enamel crystal is about 200 times as large in volume as each dentine crystal. Therefore, the enamel of the tooth is the primary barrier to the development of caries. Once the carious process has penetrated through the enamel to the dentine, it then proceeds many times as rapidly because of the high degree of solubility of the dentine salts.

Because of the dependence of the caries bacteria on carbohydrates for their nutrition, it has frequently been taught that eating a diet high in carbohydrate content will lead to excessive development of caries. However, it is not the quantity of carbohydrate ingested but the frequency with which it is eaten that is important. If eaten in many small parcels throughout the day, such as in the form of candy, the bacteria are supplied with their preferential metabolic substrate for many hours of the day and the development of caries is extreme. If eaten in large amounts only at mealtimes, the extensiveness of the caries is greatly reduced.

Some teeth are more resistant to caries than others. Studies show that teeth formed in children who drink water that contains small amounts of fluorine develop enamel that is more resistant to caries than the enamel in children who drink water that does not contain fluorine. Fluorine does not make the enamel harder than usual but fluorine ions replace many of the hydroxyl ions in the hydroxyapatite crystals, which in turn makes the enamel several times less soluble. It is also believed that the fluorine is toxic to the bacteria. Finally, when small pits do develop in the enamel, fluorine is believed to promote deposition of calcium phosphate to "heal" the enamel surface. Regardless of the precise means by which fluorine protects the teeth, it is known that small amounts of fluorine deposited in enamel make teeth about three times as resistant to caries as teeth without fluorine.

MALOCCLUSION. Malocclusion is usually caused by a hereditary abnormality that causes the teeth of one jaw to grow to abnormal positions. In malocclusion, the teeth do not interdigitate properly and, therefore, cannot perform their normal grinding or cutting action adequately. Malocclusion occasionally also results in abnormal displacement of the lower jaw in relation to the upper jaw, causing such undesirable effects as pain in the mandibular joint and deterioration of the teeth.

The orthodontist can usually correct malocclusion by applying prolonged gentle pressure against the teeth with appropriate braces. The gentle pressure causes absorption of alveolar jaw bone on the compressed side of the tooth and deposition of new bone on the tensional side of the tooth. In this way, the tooth gradually moves to a new position as directed by the applied pressure.

REFERENCES

Ash, M. M.: Wheeler's Dental Anatomy, Physiology and Occlusion. Philadelphia, W. B. Saunders Co., 1993.

Avioli, L. V., and Krane, S. M.: Metabolic Bone Disease and Clinically Related Disorders. Philadelphia, W. B. Saunders Co., 1990.

Baum, L.: Textbook of Operative Dentistry. Philadelphia, W. B. Saunders Co., 1994.

Barnes, D. M.: Close encounters with an osteoclast. Science, 236:914, 1987.

Bilezikian, J. P., et al.: The Parathyroids: Basic and Clinical Concepts. New York, Raven Press, 1994.

Bröll, H., and Dambacher, M.: Osteoporosis. Farmington, CT, S. Karger Publishers, Inc., 1994.

Brown, E. M.: Extracellular CA^{2+} sensing, regulation of parathyroid cell function, and role of CA^{2+} and other ions as extracellular (first) messengers. Physiol. Rev., 71:371, 1991.

Burstein, A. H., and Wright, T. M.: Fundamentals of Orthopaedic Biomechanics. Baltimore, Williams & Wilkins, 1994.

Cady, B., and Rossi, R. L.: Surgery of the Thyroid and Parathyroid Glands. Philadelphia, W. B. Saunders Co., 1991.

Canalis, E.: Bone-related growth factors. Triangle, 27:11, 1988.

Carney, S. L., and Muir, H.: The structure and function of cartilage proteoglycans. Physiol. Rev., 68:858, 1988.

Chambers, T. J.: The effect of calcitonin on the osteoclast. Triangle, 27:53, 1988.

Carafoli, E.: Calcium pump of the plasma membrane. Physiol. Rev., 71:129, 1991.

Chapman, M. W., and Madison, M.: Operative Orthopaedics. Philadelphia, J. B. Lippincott, 1993.

Coe, F. L., and Favus, M. J.: Disorders of Bone and Mineral Metabolism. New York, Raven Press, 1992.

Croall, D. E., and DeMartino, G. N.: Calcium-activated neutral protease (calpain) system: structure, function, and regulation. Physiol. Rev., 71:813, 1991.

Curzon, M. E. J., and ten Cate, J. M.: Efficacy of Caries Preventive Strategies. Farmington, CT, S. Karger Publishers, Inc., 1993.

DeGroot, L. J.: Endocrinology. Philadelphia, W. B. Saunders Co., 1994.

DeLuca, H. F.: The vitamin D story: a collaborative effort of basic science and clinical medicine. FASEB J., 2:224, 1988.

Einhorn, T. A.: Biomechanical properties of bone. Triangle, 27:27, 1988.

Epps, C. H. Jr.: Complications in Orthopaedic Surgery. Philadelphia, J. B. Lippincott, 1994.

Favus, M. J., et al.: Primer on the Metabolic Bone Diseases and Disorders of Mineral Metabolism. New York, Raven Press, 1993.

Fiskum, G. (ed.): Cell Calcium Metabolism. New York, Plenum Publishing Corp., 1989.

Garel, J.-M.: Hormonal control of calcium metabolism during the reproductive cycle in mammals. Physiol. Rev., 67:1, 1987.

Glorieux, F. H.: Rickets. New York, Raven Press, 1991.

Habener, J. F., et al.: Parathyroid hormone: Biochemical aspects of biosynthesis, secretion, action, and metabolism. Physiol. Rev., 64:985, 1984.

Harris, N. O., and Christen, A. G.: Primary Preventive Dentistry. 4th Ed. Redding, MA, Appleton & Lange, 1994.

Hauschka, P. V., et al.: Osteocalcin and matrix Gla protein: Vitamin K-dependent proteins in bone. Physiol. Rev., 69:990, 1989.

Ingle, J. I., and Bakland, L. K.: Endodontics. Baltimore, Williams & Wilkins, 1994.

Kasle, M. J.: An Atlas of Dental Radiographic Anatomy. Philadelphia, W. B. Saunders Co., 1994.

Kohama, K.: Calcium Inhibition: A New Mode For CA2+ Regulation. Boca Raton, FL, CRC Press, Inc., 1993.

Lindsay, R.: Osteoporosis: A Guide to Diagnosis, Prevention, and Treatment. New York, Raven Press, 1992.

LiVolsi, V. A., and DeLellis, R. A.: Pathology of the Parathyroid and Thyroid Glands. Baltimore, Williams & Wilkins, 1993.

Malcolm, A. J.: Diagnostic Histopathology of Bones and Joints. New York, Churchill Livingstone, 1994.

Marcove, R. C., and Arlen, M.: Atlas of Bone Pathology: With Clinical and Radiographic Correlations. Philadelphia, J. B. Lippincott Co., 1992.

Martin, R. B., and Burr, D. B.: Structure, Function, and Adaptation of Compact Bone. New York, Raven Press, 1989.

Minghetti, P. P., and Norman, A. W.: 1.25(OH)$_2$-vitamin D$_3$ receptors: gene regulation and genetic circuitry. FASEB J., 2:3043, 1988.

Newman, H. N.: Dental Plaque: The Ecology of the Flora on Human Teeth. Springfield, Ill., Charles C. Thomas, 1980.

Nuccitelli, R.: A Practical Guide to the Study of Calcium in Living Cells. San Diego, CA, Academic Press, 1994.

Petersen, O. H., et al.: Calcium and hormone action. Annu. Rev. Physiol., 56:297, 1994.

Pitkin, R. M.: Calcium metabolism in pregnancy and the perinatal period: A review. Am. J. Obstet. Gynecol., 151:99, 1985.

Pozzan, T., et al.: Molecular and cellular physiology of intracellular calcium stores. Physiol. Rev., 74:595, 1994.

Provenza, D. V.: Fundamentals of Oral Histology and Embryology, 2nd Ed. Philadelphia, Lea & Febiger, 1988.

Quamme, G. A.: Magnesium Homeostasis. Farmington, CT, S. Karger Publishers, Inc., 1993.

Raisz, L. G.: Bone metabolism and its hormonal regulation: An update. Triangle, 27:5, 1988.

Rao, G. S.: Dietary intake and bioavailability of fluoride. Annu. Rev. Nutr., 4:115, 1984.

Richmond, V. L.: Thirty years of fluoridation: A review. Am. J. Clin. Nutr., 41:129, 1985.

Schoutens, A., et al.: Bone Circulation and Vascularization in Normal and Pathological Conditions. New York, Plenum Publishing Corp., 1993.

Sonis, S. T., et al.: Principles and Practice of Oral Medicine. Philadelphia, W. B. Saunders Co., 1994.

Stroller, D. W.: Magnetic Resonance Imaging in Orthopaedics and Sports Medicine. Philadelphia, J. B. Lippincott, 1992.

Stevenson, J. C.: Pathophysiology of osteoporosis. Triangle, 27:47, 1988.

Tam, C. S., et al. (eds.): Metabolic Bone Disease: Cellular and Tissue Mechanisms. Boca Raton, CRC Press, Inc., 1988.

Thaller, S. R., and Montgomery, W. W. (eds.): Guide to Dental Problems for Physicians and Surgeons. Baltimore, Williams & Wilkins, 1988.

Tsang, R. C., and Mimouni, F.: Calcium Nutriture for Mothers and Children. New York, Raven Press, 1992.

Uhthoff, H. K., and Wiley, J. J.: Behavior of the Growth Plate. New York, Raven Press, 1988.

Reproductive and Hormonal Functions of the Male (and the Pineal Gland)

C HAPTER 80

The reproductive functions of the male can be divided into three major subdivisions: first, spermatogenesis, which means simply the formation of sperm; second, performance of the male sexual act; and third, regulation of male reproductive functions by the various hormones. Associated with these reproductive functions are the effects of the male sex hormones on the accessory sexual organs, cellular metabolism, growth, and other functions of the body.

PHYSIOLOGIC ANATOMY OF THE MALE SEXUAL ORGANS. Figure 80–1A shows the various portions of the male reproductive system, and Figure 80–1B gives a more detailed structure of the testis and epididymis. The testis is composed of up to 900 coiled *seminiferous tubules,* each averaging more than .5 meter long, in which the sperm are formed. The sperm then empty into the *epididymis,* another coiled tube about 6 meters long. The epididymis leads into the *vas deferens,* which enlarges into the *ampulla of the vas deferens* immediately before the vas enters the body of the *prostate gland.* A *seminal vesicle,* one located on each side of the prostate, empties into the prostatic end of the ampulla, and the contents from both the ampulla and the seminal vesicle pass into an *ejaculatory duct* leading through the body of the prostate gland to empty into the *internal urethra. Prostatic ducts* in turn empty from the prostate gland into the ejaculatory duct. Finally, the *urethra* is the last connecting link from the testis to the exterior. The urethra is supplied with mucus derived from a large number of minute *urethral glands* located along its entire extent and even more so from bilateral *bulbourethral glands* (Cowper's glands) located near the origin of the urethra.

SPERMATOGENESIS

Spermatogenesis occurs in all the seminiferous tubules during active sexual life as the result of stimulation by anterior pituitary gonadotropic hormones, beginning at an average age of 13 years and continuing throughout the remainder of life.

Steps of Spermatogenesis

The seminiferous tubules, a cross section of one of which is shown in Figure 80–2A, contain large numbers of germinal epithelial cells called *spermatogonia,* located in two to three layers along the outer border of the tubular structure. They continually proliferate to replenish themselves, and a portion of them differentiate through definite stages of development to form sperm, as shown in Figure 80–2B.

In the first stage of spermatogenesis, primordial spermatogonia located immediately adjacent to the basement membrane of the germinal epithelium, called *type A spermatogonia,* divide four times to form 16 slightly more differentiated cells, the *type B spermatogonia.*

At this stage, the spermatogonia migrate centrally among the *Sertoli cells.* These Sertoli cells are very large, with overflowing cytoplasmic envelopes that extend from the spermatogonial cell layers all the way to the central lumen of the tubule. The membranes of the Sertoli cells are tightly adherent to one another at

1003

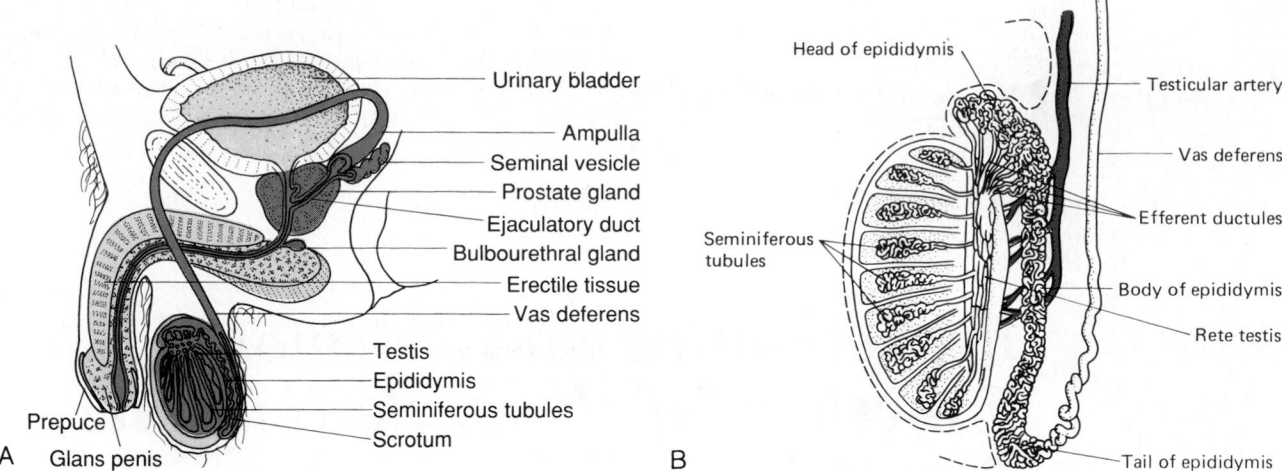

Figure 80–1. *A,* Male reproductive system. (Modified from Bloom and Fawcett: A Textbook of Histology. 10th Ed. Philadelphia, W. B. Saunders Company, 1975.) *B,* Internal structure of the testis and relation of the testis to the epididymis. (From Guyton: Anatomy and Physiology. Philadelphia, Saunders College Publishing, 1985.)

their bases and sides, forming a barrier that prevents penetration from the capillaries that surround the tubules of large protein molecules such as immunoglobulins that might interfere with further development of the spermatogonia into spermatozoa. Yet the spermatogonia that are destined to become spermatozoa do penetrate this barrier and become enveloped within the enfolding cytoplasmic processes of the Sertoli cells. This close relation with the Sertoli cells continues throughout the remainder of spermatozoon development.

MEIOSIS. For a period averaging 24 days, each spermatogonium that crosses the barrier into the Sertoli cell layer becomes progressively modified and enlarged to form a large *primary spermatocyte.* At the end of the 24 days, each primary spermatocyte divides to form two *secondary spermatocytes.* This division is not a normal division. Instead, it is called the *first meiotic division.* In the initial stage of this division, all the DNA in the 46 chromosomes is replicated. In this process, each of the 46 chromosomes becomes two *chromatids* that remain bound together at the centromeres, the two chromatids having duplicate genes of that chromosome. At this time, the primary spermatocyte divides into two secondary spermatocytes, each pair of chromosomes separating so that 23 chromosomes, each containing two chromatids, go to one of the secondary spermatocytes while the other 23 chromosomes go to the other secondary spermatocyte. Within 2 to 3 days, a *second meiotic division* occurs in which the two chromatids in each of the 23 chromosomes splits apart at the centromeres, forming two sets of 23 chromosomes, one set passing into one daughter spermatid and the other set passing into the second daughter spermatid.

The importance of these two meiotic divisions is that each spermatid that is finally formed carries only 23 chromosomes, having only one half the genes of the original spermatogonium. Therefore, the eventual spermatozoon that fertilizes the female ovum will provide one half of the genetic material to the fertilized ovum and the ovum will provide the other one half.

DEVELOPMENT OF THE SPERM AFTER MEIOSIS. During the next few weeks after meiosis, each spermatid is nursed and physically reshaped by its enveloping Sertoli cell, changing it slowly into a spermatozoon (a sperm) by (1) losing some of its cytoplasm, (2) reorganizing the chromatin material of its nucleus to form a compact head, and (3) collecting the remaining cytoplasm and cell membranes at one end of the cell to form a tail.

All the stages of the final conversion of the spermatocytes into sperm occur with the spermatocytes and spermatids actually embedded in the Sertoli cells. The Sertoli cells nurture and control the spermatogenesis process. The entire period of spermatogenesis, from germinal cell to sperm, takes about 64 days.

SEX CHROMOSOMES. In each spermatogonium, one of the 23 pairs of chromosomes carries the genetic information that determines the sex of the eventual offspring. This pair is composed of one X chromosome, which is called the *female chromosome,* and one Y chromosome, the *male chromosome.* During meiotic division, the sex-determining chromosomes divide among the secondary spermatocytes so that one half the sperm become *male sperm* that contain the Y chromosome and the other one half become *female sperm* that contain the X chromosome. The sex of the offspring is determined by which of these two types of sperm fertilizes the ovum. This is discussed further in Chapter 82.

FORMATION OF SPERM. When the spermatids are first formed, they still have the usual characteristics of epithelioid cells, but soon each spermatid begins to

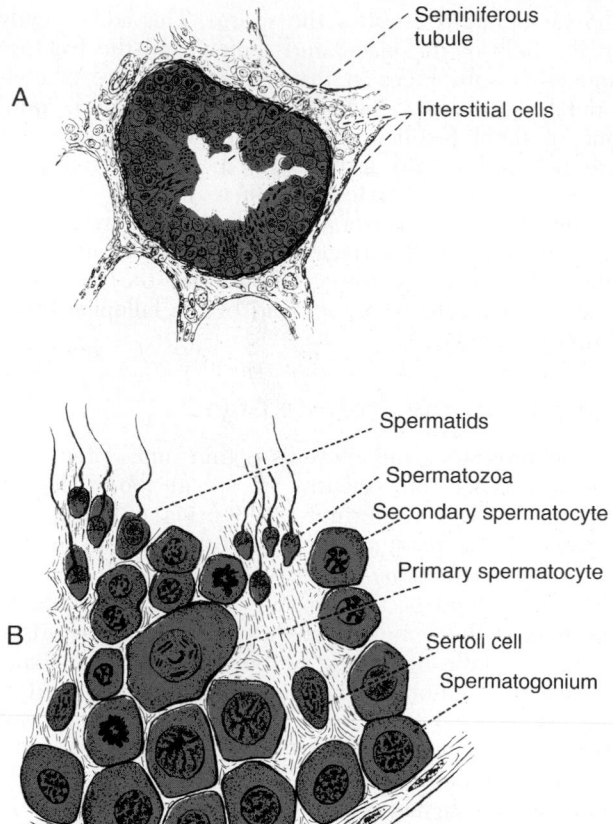

Figure 80–2. *A,* Cross section of a seminiferous tubule and, *B,* the stages in the development of sperm from spermatogonia.

elongate into a spermatozoon, shown in Figure 80–3, composed of a *head* and a *tail.* The head is composed of the condensed nucleus of the cell with only a thin cytoplasmic and cell membrane layer around its surface. On the outside of the anterior two thirds of the head is a thick cap called the *acrosome* that has been formed mainly from the Golgi apparatus. This contains a number of enzymes similar to those found in lysosomes of the typical cell, including *hyaluronidase,* which can digest proteoglycan filaments of tissues, and powerful *proteolytic enzymes,* which can digest proteins. These enzymes play important roles in allowing the sperm to fertilize the ovum.

The tail of the sperm, called the *flagellum,* has three major components: (1) a central skeleton constructed of 11 microtubules, collectively called the *axoneme—* the structure of this is similar to that of cilia described in Chapter 2; (2) a thin cell membrane covering the axoneme; and (3) a collection of mitochondria surrounding the axoneme in the proximal portion of the tail (called the *body of the tail*).

To and fro movement of the tail (flagellar movement) provides motility for the sperm. This movement results from a rhythmical longitudinal sliding motion between the anterior and posterior tubules that make

up the axoneme. The energy for this process is supplied in the form of adenosine triphosphate synthesized by the mitochondria in the body of the tail. Normal sperm move in a straight line at a velocity of 1 to 4 mm/min. This allows them to move through the female genital tract in quest of the ovum.

Hormonal Factors That Stimulate Spermatogenesis

We shall speak more about the role of hormones in reproduction later, but at this point, we need to note that several hormones play essential roles in spermatogenesis. Some of these are as follows:

1. *Testosterone,* secreted by the Leydig cells located in the interstitium of the testis, is essential for growth and division of the germinal cells in forming sperm.

2. *Luteinizing hormone,* secreted by the anterior pituitary gland, stimulates the Leydig cells to secrete testosterone.

3. *Follicle-stimulating hormone,* also secreted by the anterior pituitary gland, stimulates the Sertoli cells; without this stimulation, the conversion of the spermatids to sperm (the process of spermiogenesis) will not occur.

4. *Estrogens,* formed from testosterone by the Sertoli cells when they are stimulated by follicle-stimulating hormone, are probably also essential for spermiogenesis. The Sertoli cells also secrete an *androgen-binding protein* that binds both testosterone and estrogens and carries these into the fluid in the seminiferous tubular lumen, making both these hormones available to the maturing sperm.

5. *Growth hormone* (as well as most of the other hormones) is necessary for controlling background metabolic functions of the testes. Growth hormone specifically promotes early division of the spermatogonia themselves; in its absence, as in pituitary dwarfs, spermatogenesis is severely deficient or absent.

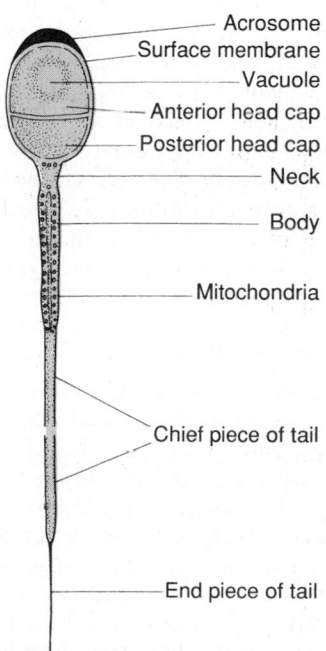

Figure 80–3. Structure of the human spermatozoon.

Maturation of Sperm in the Epididymis

After formation in the seminiferous tubules, the sperm require several days to pass through the 6-meter long *epididymis.* Sperm removed from the seminiferous tubules and from the early portions of the epididymis are nonmotile, and they cannot fertilize an ovum. However, after the sperm have been in the epididymis for some 18 to 24 hours, they develop the capability of motility, even though several inhibitory proteins in the epididymal fluid still prevent actual motility until after ejaculation.

After ejaculation, the sperm do become motile, and they also become capable of fertilizing the ovum, a process called *maturation.* The Sertoli cells and the epithelium of the epididymis secrete a special nutrient fluid that is ejaculated along with the sperm. This fluid contains hormones (both testosterone and estrogens), enzymes, and special nutrients that may be important or even essential for sperm maturation.

Storage of Sperm. The two testes of the young adult form about 120 million sperm each day. A small quantity of them can be stored in the epididymis, but most of them are stored in the vas deferens and the ampulla of the vas deferens. They can remain stored, maintaining their fertility, in the genital ducts for at least a month. During all this time, they are kept in a deeply suppressed inactive state because of multiple inhibitory substances in the secretions of the ducts. On the other hand, with excessive sexual activity, storage may be no longer than a few days at most.

Physiology of the Mature Sperm. The usual motile and fertile sperm are capable of flagellated movement through the fluid media at a speed of about 1 to 4 mm/min. Furthermore, *normal* sperm tend to travel in a straight line, rather than with a circuitous movement. The activity of sperm is greatly enhanced in neutral and slightly alkaline media as exists in the ejaculated semen, but it is greatly depressed in mildly acidic media. Strong acidic media can cause rapid death of sperm. The activity of sperm increases markedly with increasing temperature, but so does the rate of metabolism, causing the life of the sperm to be considerably shortened. Although sperm can live for many weeks in the genital ducts of the testes, the life of sperm in the female genital tract is only 1 to 2 days.

Function of the Seminal Vesicles

From early anatomical studies of the seminal vesicles, it was erroneously believed that sperm were stored in them, whence came the name "seminal vesicles." However, these structures are only secretory glands, not sperm storage areas.

Each seminal vesicle is a tortuous, loculated tube lined with a secretory epithelium that secretes a mucoid material that contains an abundance of *fructose, citric acid,* and other nutrient substances as well as large quantities of *prostaglandins* and *fibrinogen.* During the process of emission, each seminal vesicle empties its contents into the ejaculatory duct shortly after the vas deferens empties the sperm. This adds greatly to the bulk of the ejaculated semen, and the fructose and other substances in the seminal fluid are of considerable nutrient value for the ejaculated sperm until one of them fertilizes the ovum. The prostaglandins are believed to aid fertilization in two ways: (1) by reacting with the cervical mucus to make it more receptive to sperm movement and (2) possibly causing reverse peristaltic contractions in the uterus and fallopian tubes to move the sperm toward the ovaries (a few sperm reach the upper end of the fallopian tubes within 5 minutes).

Function of the Prostate Gland

The prostate gland secretes a thin, milky fluid that contains citrate ion, calcium, phosphate ion, a clotting enzyme, and a profibrinolysin. During emission, the capsule of the prostate gland contracts simultaneously with the contractions of the vas deferens so that the thin, milky fluid of the prostate gland adds further to the bulk of the semen. A slightly alkaline characteristic of the prostatic fluid may be quite important for successful fertilization of the ovum because the fluid of the vas deferens is relatively acidic owing to the presence of citric acid and metabolic end products of the sperm and, consequently, helps to inhibit sperm fertility. Also, the vaginal secretions of the female are acidic (pH of 3.5 to 4.0). Sperm do not become optimally motile until the pH of the surrounding fluids rises to about 6.0 to 6.5. Consequently, it is probable that prostatic fluid helps to neutralize the acidity of these other fluids after ejaculation and thus enhances the motility and fertility of the sperm.

Semen

Semen, which is ejaculated during the male sexual act, is composed of the fluid and sperm from the vas deferens (about 10 per cent of the total), fluid from the seminal vesicles (about 60 per cent), fluid from the prostate gland (about 30 per cent), and small amounts from the mucous glands, especially the bulbourethral glands. Thus, the bulk of the semen is seminal vesicle fluid, which is the last to be ejaculated and serves to wash the sperm out of the ejaculatory duct and urethra. The average pH of the combined semen is about 7.5, the alkaline prostatic fluid having neutralized the mild acidity of the other portions of the semen. The prostatic fluid gives the semen a milky appearance, and fluid from the seminal vesicles and the mucous glands gives the semen a mucoid consistency. Also, a clotting enzyme of the prostatic fluid causes the fibrinogen of the seminal vesicle fluid to form a weak coagulum that holds the semen in the deeper regions of the vagina where the uterine cervix lies. The coagulum then dissolves during the next 15 to 30 minutes because of lysis by fibrinolysin formed from the prostatic profibrinolysin. In the early minutes after ejaculation, the sperm remain relatively immobile, possibly because of the viscosity of the coagulum. As the coagulum dissolves, the sperm simultaneously become highly motile.

Although sperm can live for many weeks in the male genital ducts, once they are ejaculated in the semen, their maximal life span is only 24 to 48 hours at body temperature. At lowered temperatures, however, semen may be stored for several weeks, and when frozen at temperatures below −100°C, sperm have been preserved for years.

Capacitation of the Spermatozoa—Making It Possible for Them to Penetrate the Ovum

Although the spermatozoa are said to be "mature" when they leave the epididymis, their activity is held in check by multiple inhibitory factors secreted by the genital duct epithelia. Therefore, when they are first expelled in the semen, they are unable to perform their duties in fertilizing the ovum. However, on coming in contact with the fluids of the female genital tract, multiple changes occur that activate the sperm for the final processes of fertilization. These collective changes are called *capacitation of the spermatozoa*. This normally requires from 1 to 10 hours. Some changes that are believed to occur are the following.

1. The uterine and fallopian tube fluids wash away the various inhibitory factors that had suppressed sperm activity in the male genital ducts.

2. While the spermatozoa remained in the fluid of the male genital ducts, they were continually exposed to many floating vesicles from the seminiferous tubules containing large amounts of cholesterol. This cholesterol was continually donated to the cellular membrane covering the sperm acrosome, toughening this membrane and preventing release of its enzymes. After ejaculation, the sperm that are deposited in the vagina swim away from the cholesterol vesicles upward into the uterine fluid, and they gradually lose much of their excess cholesterol over the next few hours. In so doing, the membrane at the head of the sperm becomes much weaker.

3. The membrane of the sperm head also becomes much more permeable to calcium ions, so that calcium now enters the sperm in abundance and changes the activity of the flagellum, giving it a powerful whiplash motion in contrast to its previously weak undulating motion. In addition, the calcium ions probably cause changes in the intracellular membrane that covers the leading edge of the acrosome, making it possible for the acrosome to release its enzymes rapidly and easily as the sperm penetrates the granulosa cell mass surrounding the ovum and even more so as it attempts to penetrate the zona pellucida of the ovum itself.

Thus, multiple changes occur during capacitation. Without them, the sperm cannot make its way to the interior of the ovum to cause fertilization.

Acrosome Enzymes, the "Acrosome Reaction," and Penetration of the Ovum

Stored in the acrosomes of the sperm are large quantities of *hyaluronidase* and *proteolytic enzymes*. Hyaluronidase depolymerizes the hyaluronic acid polymers in the intercellular cement that holds the granu-losa cells together. The proteolytic enzymes digest proteins in the structural elements of tissues still adherent to the ovum.

When the ovum is expelled from the ovarian follicle into the abdominal cavity and fallopian tube, it carries with it multiple layers of granulosa cells. Before a sperm can fertilize the ovum, it must pass through the granulosa cell layer, and then it must penetrate through the thick covering of the ovum itself, the *zona pellucida*. During capacitation of the sperm, the anterior membrane of the acrosome fuses with the cell membrane at the head of the sperm. This allows beginning release of small amounts of the acrosome enzymes. It is believed that the hyaluronidase among these enzymes is especially important in opening pathways between the granulosa cells so that the sperm can reach the ovum.

On reaching the zona pellucida of the ovum, the anterior membrane of the sperm binds specifically with a receptor protein in the zona pellucida. Then, rapidly, the entire anterior membrane of the acrosome dissolves and all the acrosomal enzymes are immediately released. Within minutes, they open a penetrating pathway for passage of the sperm head through the zona pellucida. The head at first enters the *perivitelline space* lying beneath the zona pellucida but outside the membrane of the underlying oocyte. Within 30 minutes, the membranes of the sperm head and the oocyte fuse, and the sperm genetic material enters the oocyte to cause fertilization; then the embryo begins to develop, as discussed in Chapter 82.

Why Does Only One Sperm Enter the Oocyte? With as many sperm as there are, why does only one enter the oocyte? The reason is not entirely known, but some of the facts are the following.

First, only a few sperm ever get as far as the zona pellucida, so that it might be 10, 20, or even 30 minutes before the second sperm arrives.

Second, within a few minutes after the first sperm penetrates the zona pellucida, calcium ions diffuse through the oocyte membrane and cause multiple cortical granules to be released by exocytosis from the oocyte into the perivitelline space. These granules contain substances that permeate all portions of the zona pellucida and prevent binding of additional sperm and even cause the sperm that have already bound to fall off.

And, third, changes in the oocyte membrane after its fusion with the sperm are believed to cause electrical depolarization; this, too, may play a role in fending off subsequent sperm.

At any rate, almost never does more than one sperm enter the oocyte during fertilization.

Abnormal Spermatogenesis and Male Fertility

The seminiferous tubular epithelium can be destroyed by a number of diseases. For instance, bilateral orchitis resulting from mumps causes sterility in a large percentage of afflicted males. Also, many male infants are born with degenerate tubular epithelium as a result of strictures in the genital ducts or genetic abnormalities. Finally, another cause of sterility, usually temporary, is excessive temperature of the testes.

EFFECT OF TEMPERATURE ON SPERMATOGENESIS. Increasing the temperature of the testes can prevent spermatogenesis by causing degeneration of most cells of the seminiferous tubules besides the spermatogonia.

It has often been stated that the reason the testes are located in the dangling scrotum is to maintain the temperature of these glands below the internal temperature of the body, although usually only about 2°C below the internal temperature. On cold days, scrotal reflexes cause the musculature of the scrotum to contract, pulling the testes close to the body to maintain this 2° differential. Thus, the scrotum theoretically acts as a cooling mechanism for the testes (but a controlled cooling), without which spermatogenesis is said to be deficient during hot weather.

Cryptorchidism. Cryptorchidism means failure of a testis to descend from the abdomen into the scrotum. During the development of the male fetus, the testes are derived from the genital ridges in the abdomen. However, at about 3 weeks to 1 month before the birth of the baby, the testes normally descend through the inguinal canals into the scrotum. Occasionally this descent does not occur or occurs incompletely, so that one or both testes remain in the abdomen, in the inguinal canal, or elsewhere along the route of descent.

A testis that remains throughout life in the abdominal cavity is incapable of forming sperm. The tubular epithelium is degenerate, leaving only the interstitial structures of the testis. It has been claimed that even the few degrees' higher temperature in the abdomen than in the scrotum is sufficient to cause degeneration of the tubular epithelium and, consequently, to cause sterility, although this is not certain. Nevertheless, for this reason, operations to relocate the cryptorchid testes from the abdominal cavity into the scrotum before the beginning of adult sexual life are frequently performed on boys who have undescended testes.

Testosterone secretion by the fetal testes themselves is the normal stimulus that causes the testes to descend into the scrotum from the abdomen. Therefore, many, if not most, instances of cryptorchidism are caused by abnormally formed testes that are unable to secrete enough testosterone. The surgical operation for cryptorchidism in these patients is unlikely to be successful.

EFFECT OF SPERM COUNT ON FERTILITY. The usual quantity of semen ejaculated at each coitus averages about 3.5 milliliters, and in each milliliter of semen is an average of about 120 million sperm, although even in "normal" people this can vary from 35 to 200 million. This means an average total of 400 million sperm are usually present in each ejaculate. When the number of sperm in each milliliter falls below about 20 million, the person is likely to be infertile. Thus, even though only a single sperm is necessary to fertilize the ovum, for reasons not completely understood, the ejaculate usually must contain a tremendous number of sperm for one sperm to fertilize the ovum.

EFFECT OF SPERM MORPHOLOGY AND MOTILITY ON FERTILITY. Occasionally a man has a normal number of sperm but is still infertile. When this occurs, sometimes as many as one half the sperm are found to be abnormal physically, having two heads, abnormally shaped heads, or abnormal tails, as shown in Figure 80–4; at

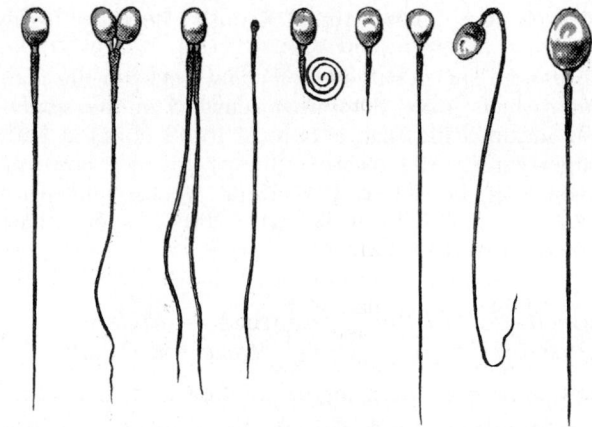

Figure 80–4. Abnormal infertile sperm, compared with a normal sperm on the right.

other times, the sperm appear to be structurally normal, but for reasons not understood, they are either entirely nonmotile or relatively nonmotile. Whenever the majority of the sperm are morphologically abnormal or are found to be nonmotile, the person is likely to be infertile, even though the remainder of the sperm appear to be normal.

THE MALE SEXUAL ACT

Neuronal Stimulus for Performance of the Male Sexual Act

The most important source of sensory nerve signals for initiating the male sexual act is the glans penis. The glans contains an especially sensitive sensory end-organ system that transmits into the central nervous system that special modality of sensation called *sexual sensation.* The massaging action of intercourse on the glans stimulates the sensory end-organs, and the sexual signals in turn pass through the pudendal nerve, then through the sacral plexus into the sacral portion of the spinal cord, and finally up the cord to undefined areas of the brain. Impulses may also enter the spinal cord from areas adjacent to the penis to aid in stimulating the sexual act. For instance, stimulation of the anal epithelium, the scrotum, and perineal structures in general can send signals into the cord that add to the sexual sensation. Sexual sensations can even originate in internal structures, such as irritated areas of the urethra, bladder, prostate, seminal vesicles, testes, and vas deferens. Indeed, one of the causes of "sexual drive" is filling of the sexual organs with secretions. Infection and inflammation of these sexual organs sometimes cause almost continual sexual desire, and "aphrodisiac" drugs, such as cantharides, increase the sexual desire by irritating the bladder and urethral mucosa.

PSYCHIC ELEMENT OF MALE SEXUAL STIMULATION. Appropriate psychic stimuli can greatly enhance the ability of a person to perform the sexual act. Simply thinking sexual thoughts or even dreaming that the act of intercourse is being performed can cause the male sexual act to occur and to culminate in ejaculation. Indeed, *nocturnal emissions* during dreams occur in many males during some stages of sexual life, especially during the teens.

INTEGRATION OF THE MALE SEXUAL ACT IN THE SPINAL CORD. Although psychic factors usually play an important part in the male sexual act and can initiate or inhibit it, brain function is probably not necessary for its performance because appropriate genital stimulation can cause ejaculation in some animals and occasionally in a human being after their spinal cords have been cut above the lumbar region. Therefore, the male sexual act results from inherent reflex mechanisms integrated in the sacral and lumbar spinal cord, and these mechanisms can be initiated by either psychic stimulation from the brain or actual sexual stimulation but usually it is a combination of both.

Stages of the Male Sexual Act

ERECTION; ROLE OF THE PARASYMPATHETIC NERVES. Erection is the first effect of male sexual stimulation, and the degree of erection is proportional to the degree of stimulation, whether psychic or physical.

Erection is caused by parasympathetic impulses that pass from the sacral portion of the spinal cord through the pelvic nerves to the penis. These parasympathetic nerve fibers, in contrast to most other parasympathetic fibers, are believed to secrete *nitric oxide* instead of acetylcholine. The nitric oxide relaxes the arteries of the penis as well as the trabecular meshwork of smooth muscle fibers in the *erectile tissue* of the *corpora cavernosa* and *corpus spongiosum* in the shaft of the penis, shown in Figure 80–5. This erectile tissue is nothing more than large, cavernous sinusoids, which are normally relatively empty but which become dilated tremendously when arterial blood flows rapidly into them under pressure while the venous outflow is partially occluded. Also, the erectile bodies, especially the two corpora cavernosa, are surrounded by strong fibrous coats; therefore, high pressure within the sinusoids causes ballooning of the erectile tissue to such an extent that the penis becomes hard and elongated.

LUBRICATION, A PARASYMPATHETIC FUNCTION. During sexual stimulation, the parasympathetic impulses in addition to promoting erection cause the urethral glands and the bulbourethral glands to secrete mucus. This mucus flows through the urethra during intercourse to aid in the lubrication of coitus. However, most of the lubrication of coitus is provided by the female sexual organs rather than by the male. Without satisfactory lubrication, the male sexual act is seldom successful because unlubricated intercourse causes grating, painful sensations that inhibit rather than excite sexual sensations.

EMISSION AND EJACULATION; FUNCTION OF THE SYMPATHETIC NERVES. Emission and ejaculation are the culmination of the male sexual act. When the sexual stimulus becomes extremely intense, the reflex centers of the spinal cord begin to emit *sympathetic impulses* that leave the cord at L-1 and L-2 and pass to the genital organs through the hypogastric and pelvic sympathetic plexuses to initiate emission, the forerunner of ejaculation.

Emission begins with contraction of the vas deferens and the ampulla to cause expulsion of sperm into the internal urethra. Then, contractions of the muscular coat of the prostate gland followed finally by contraction of the seminal vesicles expel prostatic and seminal fluid, forcing the sperm forward. All these fluids mix in the internal urethra with the mucus already secreted by the bulbourethral glands to form the semen. The process to this point is *emission*.

The filling of the internal urethra simultaneously elicits sensory signals that are transmitted through the pudendal nerves to the sacral regions of the cord, giving the feeling of sudden fullness in the internal genital organs. Also, these sensory signals further excite the rhythmical contraction of the internal genital organs and cause contraction of the ischiocavernosus and bulbocavernosus muscles that compress the bases of the penile erectile tissue. These effects together cause rhythmical, wave-like increases in pressure in the genital ducts and urethra, which "ejaculate" the semen from the urethra to the exterior. The process is called *ejaculation*. At the same time, rhythmical contractions of the pelvic muscles and even of some of the muscles of the body trunk cause thrusting movements of the pelvis and penis, which also help propel the semen into the deepest recesses of the vagina and perhaps even slightly into the cervix of the uterus.

This entire period of emission and ejaculation is called the *male orgasm*. At its termination, the male sexual excitement disappears almost entirely within 1 to 2 minutes and erection ceases, a process called *resolution*.

TESTOSTERONE AND OTHER MALE SEX HORMONES

Secretion, Metabolism, and Chemistry of the Male Sex Hormone

SECRETION OF TESTOSTERONE BY THE INTERSTITIAL CELLS OF LEYDIG IN THE TESTES. The testes secrete several male sex hormones, which are collectively

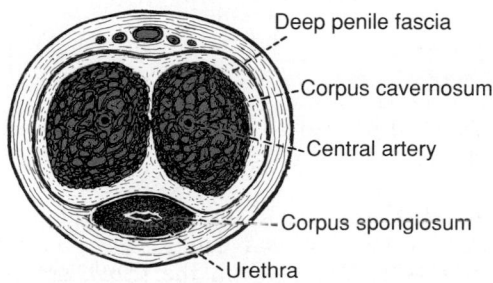

Deep penile fascia

Corpus cavernosum

Central artery

Corpus spongiosum

Urethra

Figure 80–5. Erectile tissue of the penis.

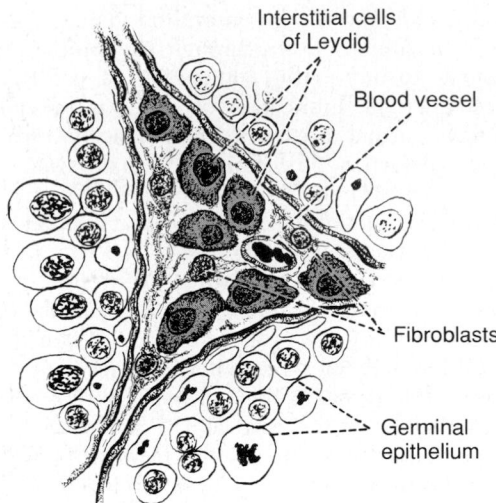

Figure 80–6. Interstitial cells of Leydig, the cells that secrete testosterone, located in the interstices between the seminiferous tubules.

called *androgens*, including *testosterone, dihydrotestosterone*, and *androstenedione*. Testosterone is so much more abundant than the others that one can consider it the significant testicular hormone, although as we shall see, much, if not most, of the testosterone is converted into the more active hormone dihydrotestosterone in the target tissues.

Testosterone is formed by the *interstitial cells of Leydig*, which lie in the interstices between the seminiferous tubules and constitute about 20 per cent of the mass of the adult testes, as shown in Figure 80–6. Leydig cells are almost nonexistent in the testes during childhood, when the testes secrete almost no testosterone, but they *are* numerous in the newborn male infant for the first few months of life and in the adult male any time after puberty; at both these times the testes secrete large quantities of testosterone. Furthermore, when tumors develop from the interstitial cells of Leydig, great quantities of testosterone are secreted. Finally, when the germinal epithelium of the testes is destroyed by x-ray treatment or excessive heat, the Leydig cells, which are less easily destroyed, continue to produce testosterone.

SECRETION OF ANDROGENS ELSEWHERE IN THE BODY. The term "androgen" means any steroid hormone that has masculinizing effects, including testosterone itself; it also includes male sex hormones produced elsewhere in the body besides the testes. For instance, the adrenal glands secrete at least five androgens, although the total masculinizing activity of all these is normally so slight (less than 5 per cent of the total in the adult male) that they do not cause significant masculine characteristics even in women, except for causing growth of pubic and axillary hair. But when an adrenal tumor of the adrenal androgen–producing cells occurs, the quantity of androgenic hormones may then become great enough to cause all the usual male secondary sexual characteristics. These effects are described in connection with the adrenogenital syndrome in Chapter 77.

Rarely, embryonic rest cells in the ovary can develop into a tumor that produces excessive quantities of androgens in women; one such tumor is the *arrhenoblastoma.* The normal ovary also produces minute quantities of androgens, but they are not significant.

CHEMISTRY OF THE ANDROGENS. All androgens are steroid compounds, as shown by the formulas in Figure 80–7 for *testosterone* and *dihydrotestosterone.* Both in the testes and in the adrenals, the androgens can be synthesized either from cholesterol or directly from acetyl coenzyme A.

METABOLISM OF TESTOSTERONE. After secretion by the testes, about 97 per cent of the testosterone becomes either loosely bound with plasma albumin or more tightly bound with a beta globulin called *sex hormone–binding globulin* and circulates in the blood in these states for 30 minutes to 1 hour or so. By that time, the testosterone either becomes fixed to the tissues or is degraded into inactive products that are subsequently excreted.

Much of the testosterone that becomes fixed to the tissues is converted within the cells to *dihydrotestosterone*, especially in certain target organs such as the prostate gland in the adult and in the external genitalia of the fetal male. Some actions of testosterone are dependent on this conversion, whereas other actions are not. The intracellular functions are discussed later in the chapter.

Degradation and Excretion of Testosterone. The testosterone that does not become fixed to the tissues is rapidly converted, mainly by the liver, into *androsterone* and *dehydroepiandrosterone* and simultaneously conjugated as either glucuronides or sulfates (glucuronides, particularly). These are excreted either into the gut in the liver bile or into the urine through the kidneys.

PRODUCTION OF ESTROGEN IN THE MALE. In addition to testosterone, small amounts of estrogens are formed in the male (about one-fifth the amount in the nonpregnant female), and a reasonable quantity of them can be recovered from a man's urine.

The exact source of the estrogens in the male is doubtful, but the following are known: (1) The concentration of estrogens in the fluid of the seminiferous tubules is quite high and probably plays an important role in spermiogenesis. This estrogen is believed to be formed by the Sertoli cells by converting some of the testosterone to estradiol. (2) Estrogens are formed from testosterone and androstanediol in other tissues of the body, especially the liver, probably accounting for as

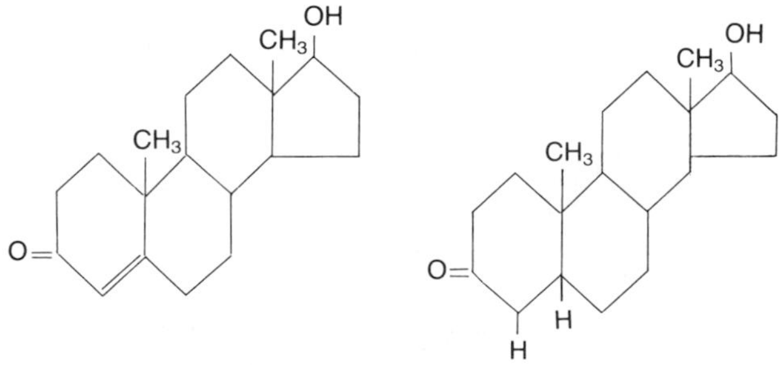

Testosterone   Dihydrotestosterone

Figure 80–7. Testosterone and dihydrotestosterone.

much as 80 per cent of the total male estrogen production.

Functions of Testosterone

In general, testosterone is responsible for the distinguishing characteristics of the masculine body. Even during fetal life, the testes are stimulated by chorionic gonadotropin from the placenta to produce moderate quantities of testosterone throughout the entire period of fetal development and for 10 or more weeks after birth; thereafter, essentially no testosterone is produced during childhood until about the ages of 10 to 13 years. Then testosterone production increases rapidly under the stimulus of anterior pituitary gonadotropic hormones at the onset of puberty and lasts throughout most of the remainder of life, as shown in Figure 80–8, dwindling rapidly beyond age 50 to become 20 to 50 per cent of the peak value by age 80.

Functions of Testosterone During Fetal Development

Testosterone begins to be elaborated by the male fetal testes at about the 7th week of embryonic life. Indeed, one of the major functional differences between the female and the male sex chromosome is that the male chromosome causes the newly developing genital ridge to secrete testosterone, whereas the female chromosome causes this ridge to secrete estrogens. Injection of large quantities of male sex hormone into pregnant animals causes development of male sexual organs even though the fetus is female. Also, removal of the testes in the early male fetus causes development of female sexual organs. Therefore, testosterone secreted first by the genital ridges and later by the fetal testes is responsible for the development of the male body characteristics, including the formation of a penis and a scrotum rather than the formation of a clitoris and a vagina. Also, it causes formation of the prostate gland, seminal vesicles, and male genital ducts, while at the same time suppressing the formation of female genital organs.

EFFECT OF TESTOSTERONE TO CAUSE DESCENT OF THE TESTES. The testes usually descend into the scrotum during the last 2 to 3 months of gestation, when the testes are secreting reasonable quantities of testosterone. If a male child is born with undescended but otherwise normal testes, the administration of testosterone can cause the testes to descend in the usual manner if the inguinal canals are large enough to allow the testes to pass. The administration of gonadotropic hormones, which stimulate the Leydig cells of the newborn child's testes to produce testosterone, can also cause the testes to descend. Thus, the stimulus for descent of the testes is testosterone, indicating again that testosterone is an important hormone for male sexual development during fetal life.

Effect of Testosterone on Development of Adult Primary and Secondary Sexual Characteristics

After puberty, the reinitiation of testosterone secretion causes the penis, scrotum, and testes to enlarge about eightfold before the age of 20 years. In addition, testosterone causes the "secondary sexual characteristics" of the male to develop at the same time, beginning at puberty and ending at maturity. These secondary sexual characteristics, in addition to the sexual organs themselves, distinguish the male from the female as follows.

EFFECT ON THE DISTRIBUTION OF BODY HAIR. Testosterone causes growth of hair (1) over the pubis, (2) upward along the linea alba sometimes to the umbilicus and above, (3) on the face, (4) usually on the chest, and (5) less often on other regions of the body, such as the back. It also causes the hair on most other portions of the body to become more prolific.

BALDNESS. Testosterone decreases the growth of hair on the top of the head; a man who does not have functional testes does not become bald. However, many virile men never become bald because baldness is a result of two factors: first, a *genetic background* for the development of baldness and, second, superimposed on this genetic background, *large quantities of androgenic hormones.* A woman who has the appropriate genetic background and who develops a long-sustained androgenic tumor becomes bald in the same manner as does a man.

EFFECT ON THE VOICE. Testosterone secreted by the testes or injected into the body causes hypertrophy of the laryngeal mucosa and enlargement of the larynx. The effects cause at first a relatively discordant, "cracking" voice, but this gradually changes into the typical adult masculine bass voice.

EFFECT ON THE SKIN, AND DEVELOPMENT OF ACNE. Testosterone increases the thickness of the skin over the entire body and increases the ruggedness of the subcutaneous tissues.

Testosterone increases the rate of secretion by some

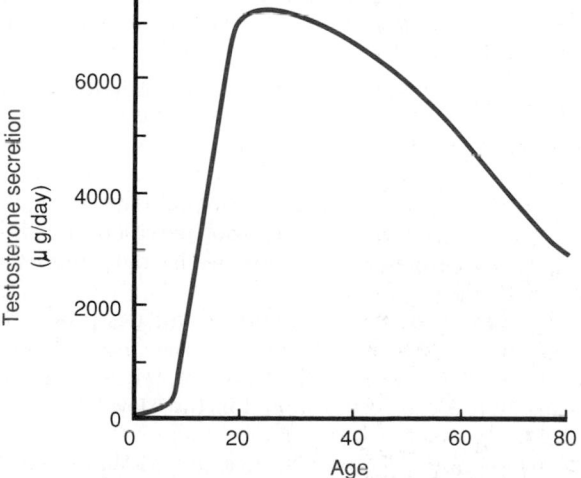

Figure 80–8. Approximate rates of testosterone secretion at different ages.

or perhaps all the sebaceous glands. Especially important is the excessive secretion by the sebaceous glands of the face because oversecretion of these glands can result in *acne.* Therefore, acne is one of the most common features of adolescence when the male body is first becoming introduced to increased testosterone. After several years of testosterone secretion, the skin normally adapts itself to the testosterone in some way that allows it to overcome the acne.

EFFECT ON PROTEIN FORMATION AND MUSCLE DEVELOPMENT. One of the most important male characteristics is the development of increasing musculature after puberty, averaging about a 50 per cent increase in muscle mass over that in the female. This is associated with increased protein in the nonmuscle parts of the body as well. Many of the changes in the skin are due to deposition of proteins in the skin, and the changes in the voice probably also result mainly from this protein anabolic function of testosterone.

Because of the great effect that testosterone has on the body musculature, it (or more usually a synthetic androgen) is widely used by athletes to improve their muscular performance. This practice is to be severely deprecated because of prolonged harmful effects of excess testosterone, as we discuss in Chapter 84 in relation to sports physiology. Testosterone is also used in old age as a "youth hormone" to improve muscle strength and vigor.

EFFECT ON BONE GROWTH AND CALCIUM RETENTION. After the great increase in circulating testosterone at puberty or after prolonged injection of testosterone, the bones grow considerably in thickness and deposit considerable additional calcium salts. Thus, testosterone increases the total quantity of bone matrix and causes calcium retention. The increase in bone matrix is believed to result from the general protein anabolic function of testosterone and the deposition of calcium salts, to result secondarily to the increased bone matrix.

Finally, testosterone has a specific effect on the pelvis to (1) narrow the pelvic outlet, (2) lengthen it, (3) cause a funnel-like shape instead of the broad ovoid shape of the female pelvis, and (4) greatly increase the strength of the entire pelvis for load-bearing. In the absence of testosterone, the male pelvis develops into a pelvis that is similar to that of the female.

Because of the ability of testosterone to increase the size and strength of bones, it is often used in old age in men to treat osteoporosis.

When great quantities of testosterone (or any other androgen) are secreted abnormally in the still-growing child, the rate of bone growth increases markedly, causing a spurt in total body height as well. However, the testosterone also causes the epiphyses of the long bones to unite with the shafts of the bones at an early age. Therefore, despite the rapidity of growth, this early uniting of the epiphyses prevents the person from growing as tall as he would have grown had testosterone not been secreted at all. Even in normal men, the final adult height is slightly less than that which would have been attained had the person been castrated before puberty.

EFFECT ON BASAL METABOLISM. Injection of large quantities of testosterone can increase the basal metabolic rate by as much as 15 per cent. Also, even the usual quantity of testosterone secreted by the testes during adolescence and early adult life increases the rate of metabolism some 5 to 10 per cent above the value that it would be were the testes not active. This increased rate of metabolism is possibly an indirect result of the effect of testosterone on protein anabolism, the increased quantity of proteins—the enzymes especially—increasing the activities of all cells.

EFFECT ON RED BLOOD CELLS. When normal quantities of testosterone are injected into a castrated adult, the number of red blood cells per cubic millimeter of blood increases 15 to 20 per cent. Also, the average man has about 700,000 more red blood cells per cubic millimeter than the average woman. This difference may be due partly to the increased metabolic rate after testosterone administration rather than to a direct effect of testosterone on red blood cell production.

EFFECT ON ELECTROLYTE AND WATER BALANCE. As pointed out in Chapter 77, many steroid hormones can increase the reabsorption of sodium in the distal tubules of the kidneys. Testosterone has such an effect but only to a minor degree in comparison with the adrenal mineralocorticoids. Nevertheless, after puberty, the blood and extracellular fluid volumes of the male in relation to his weight increase to a slight extent.

Basic Intracellular Mechanism of Action of Testosterone

Probably all or almost all the effects just listed result from increased rate of protein formation in the target cells. This has been studied extensively in the prostate gland, one of the organs that is most affected by testosterone. In this gland, testosterone enters the cells within a few minutes after secretion, is there converted, under the influence of the intracellular enzyme *5α-reductase,* to *dihydrotestosterone,* and binds with a cytoplasmic "receptor protein." This combination then migrates to the nucleus where it binds with a nuclear protein and induces the DNA-RNA transcription process. Within 30 minutes, RNA polymerase has become activated and the concentration of RNA begins to increase in the cells; this is followed by progressive increase in cellular protein. After several days, the quantity of DNA in the gland has also increased and there has been a simultaneous increase in the number of prostatic cells.

Therefore, testosterone greatly stimulates production of proteins virtually everywhere in the body, although increasing more specifically those proteins in "target" organs or tissues responsible for the development of secondary sexual characteristics.

Some important target tissues do not have the reductase enzyme in their cells to convert testosterone into dihydrotestosterone. In these tissues, testosterone functions directly, although usually with only about one half the potency, to induce the formation of cellular proteins. For instance, this direct effect of testosterone is essential in the male fetus for the development of the epididymis, vas deferens, and seminal vesicles. The direct effect is probably also responsible for much of the effects of testosterone to promote spermatogenesis.

Control of Male Sexual Functions by Hormones from the Hypothalamus and Anterior Pituitary Gland

A major share of the control of sexual functions in both the male and the female begins with the secretion of *gonadotropin-releasing hormone* (GnRH) by the hypothalamus. This hormone in turn stimulates the anterior pituitary gland to secrete two other hormones called gonadotropic hormones: (1) *luteinizing hormone (LH)* and (2) *follicle-stimulating hormone (FSH)*. In turn, LH is the primary stimulus for the secretion of testosterone by the testes, and FSH mainly stimulates spermatogenesis.

GnRH and Its Effect in Increasing the Secretion of LH and FSH

GnRH is a 10–amino acid peptide secreted by neurons whose cell bodies are located in the *arcuate nuclei of the hypothalamus*. The endings of these neurons terminate mainly in the median eminence of the hypothalamus, where they release GnRH into the hypothalamic-hypophysial portal vascular system. Then the GnRH is transported to the anterior pituitary gland in the portal blood and stimulates the release of the two gonadotropins, LH and FSH.

GnRH is secreted intermittently a few minutes at a time once every 1 to 3 hours. The intensity of this hormone's stimulus is determined in two ways: (1) by the frequency of these cycles of secretion and (2) by the quantity of GnRH released with each cycle. The secretion of LH by the anterior pituitary gland is also cyclic, with LH following fairly faithfully the pulsatile release of GnRH. On the other hand, FSH secretion increases and decreases only slightly with the fluctuating GnRH; instead, it changes more slowly over a period of many hours in response to longer-term changes in GnRH. Because of the much closer relation between GnRH secretion and LH secretion, GnRH has also been widely known as *LH-releasing hormone*.

Gonadotropic Hormones: LH and FSH

Both of the gonadotropic hormones, LH and FSH, are secreted by the same cells, called *gonadotropes,* in the anterior pituitary gland. In the absence of GnRH from the hypothalamus, the gonadotropes in the pituitary gland secrete almost no LH or FSH.

LH and FSH are *glycoproteins;* however, the quantity of carbohydrate bound with the protein in the molecules varies considerably under different conditions, which may change the potencies of activity.

Both LH and FSH exert their effects on their target tissues in the testes mainly by *activating the cyclic adenosine monophosphate second messenger system,* which in turn activates specific enzyme systems in the respective target cells.

TESTOSTERONE—REGULATION OF ITS PRODUCTION BY LH. *Testosterone* is secreted by the *interstitial cells of Leydig* in the testes but only when they are stimulated by LH from the pituitary gland. Furthermore, the quantity of testosterone secreted increases approximately in direct proportion to the amount of LH available.

Mature Leydig cells are not normally found in the child's testes (except for a few weeks after birth) until after the age of about 10 years. However, either injection of purified LH into a child at any age or secretion of LH at puberty causes cells that look like fibroblasts in the interstitial areas of the testes to evolve into interstitial cells of Leydig.

RECIPROCAL INHIBITION OF ANTERIOR PITUITARY SECRETION OF LH AND FSH BY TESTOSTERONE— NEGATIVE FEEDBACK CONTROL OF TESTOSTERONE SECRETION. The testosterone secreted by the testes in response to LH has the reciprocal effect of turning off anterior pituitary secretion of LH. It does this in two ways.

1. By far the greater part of the inhibition results from the direct effect of testosterone on the hypothalamus of decreasing the secretion of GnRH. This in turn causes a corresponding decrease in the secretion of both LH and FSH by the anterior pituitary, and the decrease in LH decreases the secretion of testosterone by the testes. Thus, whenever the secretion of testosterone becomes too great, this automatic negative feedback effect, operating through the hypothalamus and anterior pituitary gland, reduces the testosterone secretion back toward its normal operating level. Conversely, too little testosterone allows the hypothalamus to secrete large amounts of GnRH, with a corresponding increase in anterior pituitary LH and FSH secretion and an increase in testicular testosterone secretion.

2. Testosterone probably also has a weak negative feedback effect, acting directly on the anterior pituitary gland in addition to its feedback effect on the hypothalamus. This pituitary feedback is believed specifically to diminish LH secretion. Consequently, a small degree of regulation of testosterone secretion is believed to occur in this way as well.

Regulation of Spermatogenesis by FSH and Testosterone

FSH binds with specific FSH receptors attached to the Sertoli cells in the seminiferous tubules. This causes these cells to grow and secrete various spermatogenic substances. Simultaneously, testosterone diffusing into the tubules from the Leydig cells in the interstitial spaces also has a strong tropic effect on spermatogenesis. To initiate spermatogenesis, both FSH and testosterone are necessary, although once the initial stimulation has occurred, testosterone alone can maintain spermatogenesis for a long time thereafter.

NEGATIVE FEEDBACK CONTROL OF SEMINIFEROUS TUBULE ACTIVITY—ROLE OF THE HORMONE INHIBIN. When the seminiferous tubules fail to produce sperm, the secretion of FSH by the anterior pituitary gland increases markedly. Conversely, when spermatogenesis proceeds too rapidly, the secretion of FSH diminishes. The cause of this negative feedback effect on the anterior pituitary is believed to be secretion by the Sertoli cells of still another hormone called *inhibin*. This hormone has a strong direct effect on the anterior pituitary gland in inhibiting the secretion of FSH and possibly a slight effect on the hypothalamus in inhibiting secretion of GnRH.

Inhibin is a glycoprotein, like both LH and FSH, having a molecular weight between 10,000 and 30,000. It has been isolated from cultured Sertoli cells. Its potent inhibitory feedback effect on the anterior pituitary gland provides an important negative feedback mechanism for control of spermatogenesis, operating simultaneously with and in parallel to the negative feedback mechanism for control of testosterone secretion.

Psychic Factors That Affect Gonadotropin Secretion and Sexual Activity

Many psychic factors, feeding especially from the limbic system of the brain into the hypothalamus, can affect the rate of secretion of GnRH by the hypothalamus and therefore most other aspects of sexual and reproductive functions in both the male and the female. For instance, transporting a prize bull in a rough truck is said to inhibit fertility—and the human male is hardly different.

Human Chorionic Gonadotropin and Its Effect on the Fetal Testes

During pregnancy, still another hormone, *human chorionic gonadotropin* (hCG), is secreted by the placenta and circulates both in the mother and in the fetus. This hormone has almost the same effects on the sexual organs as LH.

During pregnancy, if the fetus is a male, hCG from the placenta causes the testes to secrete testosterone. This testosterone is critical for promoting formation of the male sexual organs, as pointed out earlier. We discuss hCG and its functions during pregnancy in greater detail in Chapter 82.

Puberty and Regulation of Its Onset

Initiation of the onset of puberty has long been a mystery. In the earliest history of humanity, the belief was simply that the testes "ripened" at this time. With the discovery of the gonadotropins, ripening of the anterior pituitary gland was then considered responsible. But now it is known from experiments in which both testicular and pituitary tissues have been transplanted from infant animals into adult animals that both the testes and the anterior pituitary of the infant are capable of performing adult functions if appropriately stimulated. Therefore, it is now certain that *during childhood the hypothalamus simply does not secrete significant amounts of GnRH*. One of the reasons for this is that during childhood, the slightest secretion of sex steroid hormones exerts a strong inhibitory effect on hypothalamic secretion of GnRH.

THE MALE ADULT SEXUAL LIFE AND THE MALE CLIMACTERIC. After puberty, gonadotropic hormones are produced by the male pituitary gland for the remainder of life, and at least some spermatogenesis usually continues until death. Most men, however, begin to exhibit slowly decreasing sexual functions in their late 40s or 50s, and one study showed that the average age for terminating intersexual relations was 68, although the variation was great. This decline in sexual function is related to decrease in testosterone secretion, as shown in Figure 80–8. The decrease in male sexual function is called the *male climacteric*. Occasionally the male climacteric is associated with symptoms of hot flashes, suffocation, and psychic disorders similar to the menopausal symptoms of the female. These symptoms can be abrogated by administration of testosterone, synthetic androgens, or even estrogens, which are used for treatment of menopausal symptoms in the female.

ABNORMALITIES OF MALE SEXUAL FUNCTION

The Prostate Gland and Its Abnormalities

The prostate gland remains relatively small throughout childhood and begins to grow at puberty under the stimulus of testosterone. This gland reaches an almost stationary size by the age of about 20 years and remains at this size up to the age of about 50 years. At that time in some men it begins to involute, along with the decreased production of testosterone by the testes. A benign prostatic fibroadenoma frequently develops in the prostate in many older men and can cause urinary obstruction. This hypertrophy is not caused by testosterone.

Cancer of the prostate gland is a common cause of death, resulting in about 2 to 3 per cent of all male deaths.

Once cancer of the prostate gland does occur, the cancerous cells are usually stimulated to more rapid growth by testosterone and are inhibited by removal of the testes, so that testosterone cannot be formed. Also, prostatic cancer usually can be inhibited by the administration of estrogens. Even some patients who have prostatic cancer that has already metastasized to almost all the bones of the body can be successfully treated for a few months to years by removal of the testes, by estrogen therapy, or by both; after this therapy the metastases usually diminish in size and the bones partially heal. This treatment does not stop the cancer but does slow it and sometimes greatly diminishes the severe bone pain.

Hypogonadism in the Male

When the testes are nonfunctional during fetal life or when there is genetic absence of androgen receptors in the target cells, none of the male sexual organs develop. Instead, normal female organs are formed. The reason for this is that the basic genetic characteristic of the fetus, whether male or female, is to form female sexual organs if there are no sex hormones. But in the presence of testosterone, formation of the female sexual organs is suppressed, and instead, male organs are induced.

When a boy loses his testes before puberty, a state of *eunuchism* ensues in which he continues to have infantile sexual characteristics throughout life. The height of the adult eunuch is slightly greater than that of the normal man, although the bones are quite thin, the muscles are considerably weaker than those of the normal man, and the sexual organs and secondary sexual characteristics remain those of a child rather than those of an adult. The voice is child-like, there is no loss of hair on the head, and the normal masculine hair distribution on the face and elsewhere does not occur.

When a man is castrated after puberty, some male secondary sexual characteristics revert to those of a child and others remain of adult masculine character. The sexual organs regress slightly in size but not to a child-like state and the voice regresses from the bass quality only slightly. On the other hand, there is loss of masculine hair production, loss of the thick masculine bones, and loss of the musculature of the virile male.

Also in the castrated adult male, sexual desires are decreased but not lost, provided sexual activities have been practiced previously. Erection can still occur as before, although with less ease, but it is rare that ejaculation can take place, primarily because the semen-forming organs become degenerate and there has been a loss of the testosterone-driven psychic desire.

Some instances of hypogonadism are caused by a genetic inability of the hypothalamus to secrete normal amounts of GnRH. This often is associated with a simultaneous abnormality of the feeding center of the hypothalamus, causing the person to greatly overeat. Consequently, obesity occurs along with eunuchism. A patient with this condition is shown in Figure 80–9; the condition is called *adiposogenital syndrome*, *Fröhlich's syndrome*, or *hypothalamic eunuchism*.

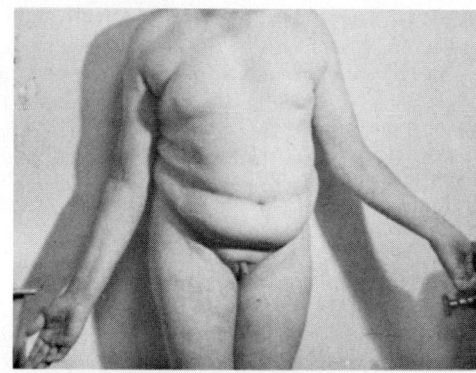

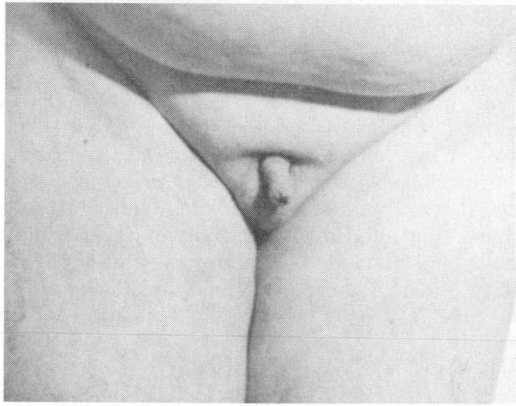

Figure 80–9. Adiposogenital syndrome in an adolescent male. Note the obesity and the child-like sexual organs. (Courtesy of Dr. Leonard Posey.)

and cause the condition called *gynecomastia*, overgrowth of the breasts.

Testicular Tumors and Hypergonadism in the Male

Interstitial Leydig cell tumors develop very rarely in the testes, but when they do develop, they sometimes produce as much as 100 times the normal quantities of testosterone. When such tumors develop in young children, they cause rapid growth of the musculature and bones but also cause early uniting of the epiphyses, so that the eventual adult height actually is less than that which would have been achieved otherwise. Such interstitial cell tumors cause excessive development of the male sexual organs in males, the clitoris in females, all muscles, and other secondary male sexual characteristics. In the adult male, small interstitial cell tumors are difficult to diagnose because masculine features are already present.

Much more common than the interstitial Leydig cell tumors are tumors of the germinal epithelium. Because germinal cells are capable of differentiating into almost any type of cell, many of these tumors contain multiple tissues, such as placental tissue, hair, teeth, bone, skin, and so forth, all found together in the same tumorous mass called a *teratoma*. These tumors often secrete no hormones, but if a significant quantity of placental tissue develops in the tumor, it may secrete large quantities of hCG with functions similar to those of LH. Also, estrogenic hormones are frequently secreted by these tumors

THE PINEAL GLAND—ITS FUNCTION IN CONTROLLING SEASONAL FERTILITY IN SOME ANIMALS

For as long as the pineal gland has been known to exist, to it has been ascribed myriad functions, including being the seat of the soul. It is known from comparative anatomy that the pineal gland is a vestigial remnant of what was a third eye high in the back of the head in lower animals. Many physiologists have been content with the idea that this gland is a nonfunctional remnant, but others have claimed for many years that it plays important roles in the control of sexual activities and reproduction, functions that still others said were nothing more than the fanciful imaginings of physiologists preoccupied with sexual delusions.

But now, after years of dispute, it looks as though the sex advocates have won and that the pineal gland does indeed play an important regulatory role in sexual and reproductive function. In lower animals that bear their young at certain seasons of the year and in which the pineal gland has been removed or the nervous circuits to the pineal gland have been sectioned, the normal periods of seasonal fertility are lost. To these animals such seasonal fertility is important because it allows birth of the offspring at the time of year when survival is most likely. The mechanism of this effect is not entirely clear, but it seems to be the following.

First, the pineal gland is controlled by the amount of light or "time pattern" of light seen by the eyes each day. For instance, in the hamster, greater than 13 hours of *darkness* each day activates the pineal gland, whereas less than that amount of darkness fails to activate it, with a critical balance between activation and nonactivation. The nervous pathway involved is this: passage of light signals from the eyes to the suprachiasmal nucleus of the hypothalamus and then to the pineal gland activates pineal secretion.

Second, the pineal gland secretes *melatonin* and several other, similar substances. Either melatonin or one of the other substances then is believed to pass either by way of the blood or through the fluid of the third ventricle to the anterior pituitary gland to *decrease* gonadotropic hormone secretion.

Thus, in the presence of pineal gland secretion, gonadotropic hormone secretion is suppressed in some species of animals and the gonads become inhibited and even partly involuted. This is what presumably occurs during the early winter months when there is increasing darkness. But after about 4 months of dysfunction, the gonadotropic hormone secretion breaks through the inhibitory effect of the pineal gland and the gonads become functional once more, ready for a full springtime of activity.

But does the pineal gland have a similar function for control of reproduction in humans? The answer to this is far from known. However, tumors often occur in the region of the pineal gland. Some of them secrete excessive quantities of pineal hormones, whereas others are tumors of surrounding tissue and press on the pineal gland to destroy it. Both types of tumors are often associated with serious hypogonadal or hypergonadal function. So perhaps the pineal gland does play at least some role in controlling sexual drive and reproduction in humans.

REFERENCES

Ackland, J. F., et al.: Nonsteroidal signals originating in the gonads. Physiol. Rev., 72:731, 1992.

Acosta, A. A., et al.: Human Spermatozoa in Assisted Reproduction. Baltimore, Williams & Wilkins, 1990.

Adashi, E. Y., et al.: Reproductive Endocrinology, Surgery, and Technology. New York, Raven Press, 1995.

Barratt, C., and Cooke, I.: Donor Insemination. New York, Cambridge University Press, 1993.

Bennett, A. H.: Impotence: Diagnosis and Management of Erectile Dysfunction. Philadelphia, W. B. Saunders Co., 1994.

Berger, F. G., and Watson, G: Androgen-regulated gene expression. Annu. Rev. Physiol., 51:51, 1989.

Beyer, C., and Feder, H. H.: Sex steroids and afferent input: Their roles in brain sexual differentiation. Annu. Rev. Physiol., 49:349, 1987.

Burger, H., and de Kretsner, D.: The Testis. New York, Raven Press, 1989.

Byskov, A. G.: Differentiation of mammalian embryonic gonad. Physiol. Rev., 66:71, 1986.

Colpi, G. M., and Balerna, M.: Treating Male Infertility. Farmington, CT, S. Karger Publishers, Inc., 1994.

Conn, P. M., et al.: Mechanism of action of gonadotropin releasing hormone. Annu. Rev. Physiol., 48:495, 1986.

Cooke, B. A., and Sharpe, R. M.: The Molecular and Cellular Endocrinology of the Testis. New York, Raven Press, 1988.

Cowan, B.: Reproductive Endocrinology. Philadelphia, J. B. Lippincott, 1994.

Diczfalusy, E., and Bygdeman, M.: Fertility Regulation Today and Tomorrow. New York, Raven Press, 1987.

Hashmat, A. I., and Das, S.: The Penis. Baltimore, Williams & Wilkins, 1993.

Holmes, K. K., et al. Sexually transmitted diseases. Hightstown, NJ, McGraw-Hill, 1990.

Inster, V., and Lunenfeld, B.: Infertility: Male and Female. 2nd Ed. New York, Churchill Livingstone, 1993.

Jaffee, R., et al.: Diagnostic imaging in infertility and reproductive endocrinology. Philadelphia, J. B. Lippincott, 1994.

Keye, W. R., et al.: Infertility: Evaluation and Treatment. Philadelphia, W. B. Saunders Co., 1994.

Knobil, E.: A hypothalamic pulse generator governs mammalian reproduction. News Physiol. Sci., 2:42, 1987.

Knobil, E., and Neill, J. D.: The Physiology of Reproduction. New York, Raven Press, 1994.

Lechtenberg, R., and Ohl, D. A.: Sexual Dysfunction. Baltimore, Williams & Wilkins, 1994.

Lepor, H., and Lawson, R. K.: Prostate Diseases. Philadelphia, W. B. Saunders Co., 1993.

Leung P. C. K., et al. (eds): Endocrinology and Physiology of Reproduction. New York, Plenum Publishing Corp., 1987.

LeVay, S.: The Sexual Brain. Cambridge, MA, The MIT Press, 1993.

Marx, J. L.: Sexual responses are—almost—all in the brain. Science, 241:903, 1988.

Menkveld, R., et al.: Atlas of Human Sperm Morphology. Baltimore, Williams & Wilkins, 1991.

Ochiai, K. (ed.): Endocrine Correlates of Reproduction. New York, Springer-Verlag, 1984.

Reppert, S. M., et al.: Putative melatonin receptors in a human biological clock. Science, 242:78, 1988.

Riley, A. J., et al.: Sexual Pharmacology. New York, Oxford University Press, 1994.

Scialli, A. R., and Zinaman, M. J.: Reproductive toxicology and infertility. Hightstown, NJ, McGraw-Hill, 1993.

Soules, M. R.: Problems in Reproductive Endocrinology and Infertility. New York, Elsevier Science Publishing Co., 1989.

Spark, R. F.: The Infertile Male. New York, Plenum Publishing Corp., 1988.

Taketani Y., and Kawagoe, S.: Aging of Reproductive Organs. Farmington, CT, S. Karger Publishers, Inc., 1993.

Wassarman, P. M.: Eggs, sperm, and sugar: A recipe for fertilization. News Physiol. Sci., 3:120, 1988.

Whitehead, E. D., and Nagler, H. M.: Management of Impotence and Infertility. Philadelphia, J. B. Lippincott, 1994.

Yen, S. S. C., and Jaffe, R. B.: Reproductive Endocrinology: Physiology, Pathophysiology, and Clinical Management. Philadelphia, W. B. Saunders Co., 1991.

Female Physiology Before Pregnancy; and the Female Hormones

CHAPTER 81

Female reproductive functions can be divided into two major phases: first, preparation of the female body for conception and gestation and, second, the period of gestation itself. This chapter is concerned with the preparation of the female body for gestation and Chapter 82 presents the physiology of pregnancy.

PHYSIOLOGIC ANATOMY OF THE FEMALE SEXUAL ORGANS

Figures 81–1 and 81–2 show the principal organs of the human female reproductive tract, the most important of which are the *ovaries, fallopian tubes, uterus,* and *vagina.* Reproduction begins with the development of ova in the ovaries. A single ovum is expelled from an ovarian follicle into the abdominal cavity in the middle of each monthly sexual cycle. This ovum then passes through one of the fallopian tubes into the uterus; if it has been fertilized by a sperm, it implants in the uterus, where it develops into a fetus, a placenta, and fetal membranes.

During fetal life, the outer surface of the ovary is covered by a *germinal epithelium,* which embryologically is derived directly from the epithelium of the germinal ridges. As the fetus develops, *primordial ova* differentiate from the germinal epithelium and migrate into the substance of the ovarian cortex. Each ovum then collects around it a layer of spindle cells from the ovarian *stroma* (the supporting tissue of the ovary) and causes them to take on epithelioid characteristics; they are the *granulosa cells.* The ovum surrounded by a single layer of granulosa cells is called a *primordial follicle.* The ovum itself at this stage is still immature, requiring two more cell divisions for maturity, and is called a *primary oocyte.*

At the 30th week of gestation, the number of ova reaches about 6 million; most of them soon degenerate, so that only about 2 million are present in the two ovaries at birth and only 300,000 to 400,000 at puberty. Then, during all the reproductive years of women, between about 13 and 46 years of age, about 400 of these follicles develop enough to expel their ova—one each month; the remainder degenerate (become *atretic*). At the end of reproductive capability, at the *menopause,* only a few primordial follicles remain in the ovaries, and even these degenerate soon thereafter.

FEMALE HORMONAL SYSTEM

The female hormonal system, like that of the male, consists of three hierarchies of hormones, as follows:

1. A hypothalamic releasing hormone, *gonadotropin-releasing hormone (GnRH),* previously also called *luteinizing hormone–releasing hormone*
2. The anterior pituitary hormones, *follicle-stimulating hormone (FSH)* and *luteinizing hormone (LH),* both of which are secreted in response to the releasing hormone GnRH from the hypothalamus
3. The ovarian hormones, *estrogen* and *progesterone,* which are secreted by the ovaries in response to the two hormones from the anterior pituitary gland

These various hormones are not secreted in constant amounts throughout the female monthly sexual cycle but are secreted at drastically differing rates during different parts of the cycle. Figure 81–3 shows the approximate changing concentrations of the anterior pituitary gonadotropic hormones, FSH and LH, and of

1017

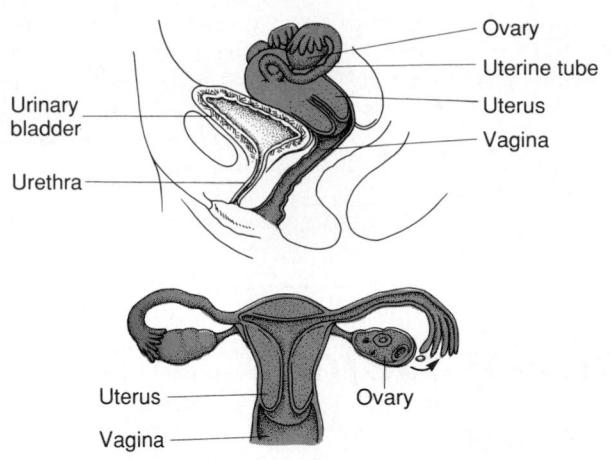

Figure 81–1. Female reproductive organs.

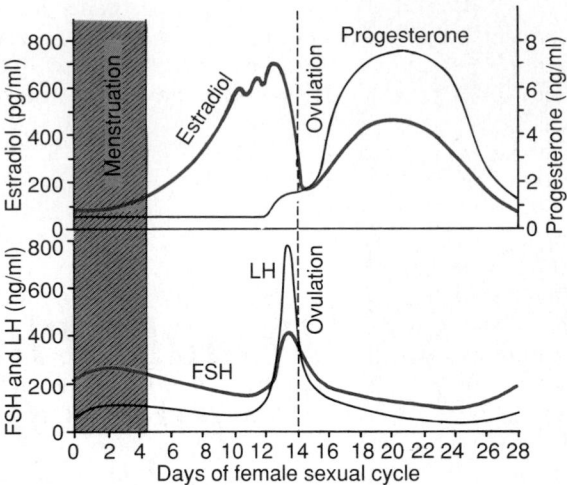

Figure 81–3. Approximate plasma concentrations of the gonadotropins and ovarian hormones during the normal female sexual cycle.

the ovarian hormones, estradiol (estrogen) and progesterone.

The GnRH from the hypothalamus increases and decreases much, much less during the monthly sexual cycle, but perhaps slightly. It is secreted in short pulses averaging once every 1 to 3 hours, as occurs in the male.

THE MONTHLY OVARIAN CYCLE AND FUNCTION OF THE GONADOTROPIC HORMONES

The normal reproductive years of the female are characterized by monthly rhythmical changes in the rates of secretion of the female hormones and corresponding changes in the ovaries and sexual organs. This rhythmical pattern is called the *female sexual cycle* (or, less accurately, the *menstrual cycle*). The duration of the cycle averages 28 days. It may be as short as 20 days or as long as 45 days even in normal women, although abnormal cycle length is frequently associated with decreased fertility.

There are two significant results of the female sexual cycle. First, only a *single* ovum is normally released from the ovaries each month, so that normally only a single fetus can begin to grow at a time. Second, the

uterine endometrium is prepared for implantation of the fertilized ovum at the required time of the month.

Gonadotropic Hormones and Their Effects on the Ovaries

The ovarian changes during the sexual cycle depend completely on the gonadotropic hormones, *FSH* and *LH*, secreted by the anterior pituitary gland. Ovaries that are not stimulated by these hormones remain inactive, which is the case throughout childhood, when almost no gonadotropic hormones are secreted. At the ages of 9 to 10 years, the pituitary begins to secrete progressively more FSH and LH, which culminates in the initiation of monthly sexual cycles between the ages of 11 and 16 years. This period of change is called *puberty,* and the first menstrual cycle is called *menarche.*

(The ovaries also function during fetal life because of stimulation by another gonadotropic hormone, *chorionic gonadotropin,* secreted by the placenta, as we discuss in Chapter 83. But within a few weeks after birth this stimulus is lost, and the ovaries become almost dormant until the prepubertal period.)

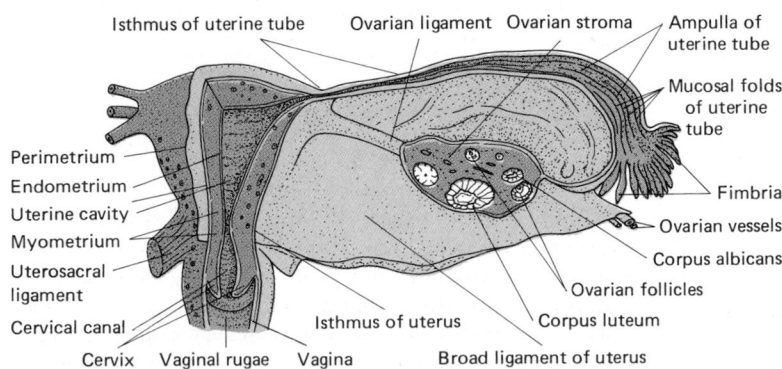

Figure 81–2. Internal structures of the uterus, ovary, and a uterine tube. (From Guyton: Physiology of the Human Body. 6th ed. Philadelphia, Saunders College Publishing, 1984.)

Both FSH and LH are small glycoproteins having molecular weights of about 30,000. The only significant effects of FSH and LH are on the testes in the male, as explained in Chapter 80, and the ovaries in the female.

During each month of the female sexual cycle, there is a cyclic increase and decrease of both FSH and LH, as shown in Figure 81–3. These cyclic variations in turn cause the cyclic ovarian changes, which are explained in the following sections.

Both FSH and LH stimulate their ovarian target cells by combining with highly specific FSH and LH receptors in the cell membranes. The activated receptors in turn increase both the rates of secretion of these cells and the growth and proliferation of the cells. Almost all these stimulatory effects result from *activation of the cyclic adenosine monophosphate second messenger system* in the cell cytoplasm, which in turn causes the formation of protein kinase and then multiple phosphorylations of key enzymes that excite many intracellular functions, as explained in Chapter 74.

Ovarian Follicle Growth—The "Follicular" Phase of the Ovarian Cycle

Figure 81–4 shows the various stages of follicular growth in the ovaries. At birth in the female child, each of the ova is surrounded by a single layer of granulosa cells, and the ovum, with its granulosa cell sheath, is called a *primordial follicle,* as shown in the figure. Throughout childhood, the granulosa cells are believed to provide nourishment for the ovum and to secrete an *oocyte maturation-inhibiting factor* that keeps the ovum in its primordial state, suspended during this entire time in the prophase stage of meiotic division. Then, after puberty, when FSH and LH from the anterior pituitary gland begin to be secreted in large quantity, the entire ovaries, together with some of the follicles within them, begin to grow.

The first stage of follicular growth is moderate enlargement of the ovum itself, which increases in diameter twofold to threefold. Then follows growth of additional layers of granulosa cells, and the follicle becomes known as a *primary follicle.* At least some of the development to this stage can occur even in the absence of FSH and LH, but development beyond this point is not possible without these two hormones.

DEVELOPMENT OF ANTRAL AND VESICULAR FOLLICLES. During the first few days after the beginning of menstruation, the concentrations of FSH and LH increase slightly to moderately, with the increase in FSH slightly greater than and preceding that of LH by a few days. These hormones, especially the FSH, cause accelerated growth of 6 to 12 primary follicles each month. The initial effect is rapid proliferation of the granulosa cells, giving rise to many more layers of granulosa cells. In addition, many spindle cells derived from the ovarian interstitium collect in several layers outside the granulosa cells, giving rise to a second class of cells called the *theca.* This is divided into two sublayers: in the *theca interna,* the cells take on epithelioid characteristics similar to those of the granulosa cells and develop an ability to secrete steroid hormones, similar to the ability of the granulosa cells to secrete slightly different hormones. The outer sublayer, the *theca externa,* is a highly vascular connective tissue capsule. This becomes the capsule of the developing follicle.

After the early proliferative phase of growth, lasting for a few days, the mass of granulosa cells secretes a *follicular fluid* that contains a high concentration of estrogen, one of the important female sex hormones that is discussed later. The accumulation of this fluid causes an *antrum* to appear within the mass of granulosa cells, as shown in Figure 81–4. Once the antrum is formed, the granulosa and theca cells proliferate even more rapidly, the rate of secretion accelerates, and each of the growing follicles becomes an *antral follicle.*

The early growth of the primary follicle up to the antral stage is stimulated mainly by FSH alone. Then greatly accelerated growth occurs in the antral follicles, leading to much larger follicles called *vesicular follicles.* This accelerated growth is caused by the following: (1) Estrogen is secreted into the follicle and causes the granulosa cells to form increasing numbers of FSH receptors; this causes a positive feedback effect because it makes the granulosa cells more sensitive than ever to the FSH from the anterior pituitary gland. (2) The pituitary FSH and the estrogens combine to promote LH receptors as well on the original granulosa cells, thus allowing LH stimulation of these cells in addition to the FSH stimulation and creating a rapid increase in follicular secretion. (3) The increasing

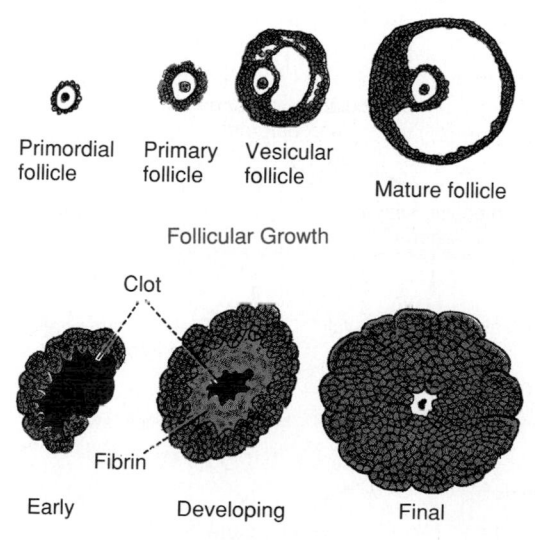

Primordial follicle   Primary follicle   Vesicular follicle   Mature follicle

Follicular Growth

Clot

Fibrin

Early   Developing   Final

Corpus Luteum

Figure 81–4. Stages of follicular growth in the ovary, showing also formation of the corpus luteum. (Modified from Arey: Developmental Anatomy. 7th ed. Philadelphia, W. B. Saunders Company, 1974.)

estrogens from the follicle plus the increasing LH from the anterior pituitary gland act together to cause proliferation of the follicular thecal cells and increase their secretion as well. Therefore, once the antral follicles begin to grow, their further growth occurs rapidly. The ovum itself also enlarges in diameter still another threefold to fourfold, giving a total diameter increase from the beginning of up to 10-fold, or a mass increase of 1000-fold.

As the follicle enlarges, the ovum itself remains embedded in a mass of granulosa cells located at one pole of the follicle, as shown in Figure 81–4. The ovum (still in the primary oocyte stage), together with its surrounding granulosa cells, is called the *cumulus oophorus.*

FULL MATURATION OF ONLY ONE FOLLICLE, ATRESIA OF THE REMAINDER. After a week or more of growth—but before ovulation occurs—one of the follicles begins to outgrow all the others; the remainder begin to involute (a process called *atresia*), and these follicles are said to become *atretic*. The cause of the atresia is unknown, but it has been postulated to be the following: The one follicle that becomes more highly developed than the others also secretes more estrogen. Furthermore, this estrogen causes a *positive* feedback effect in that single local follicle because the FSH (1) enhances both granulosa cell and thecal cell proliferation, which leads to further estrogen production and still a new cycle of cellular proliferation— thus the positive feedback, and (2) the combination of FSH and estrogens causes increasing numbers of both FSH and LH receptors on the granulosa cells and to a lesser extent on the thecal cells, thus promoting another positive feedback cycle. These effects together cause an explosive increase in the rate of secretion of fluid and hormones in this one rapidly developing follicle. At the same time, the large amounts of estrogen from this follicle act on the hypothalamus to depress further enhancement of secretion of FSH by the anterior pituitary gland, believed in this way to block further growth of the less well developed follicles that have not yet initiated their own intrinsic positive feedback stimulation. Therefore, the largest follicle continues to grow because of its intrinsic positive feedback effects while all the other follicles stop growing and, indeed, involute.

This process of atresia is important in that it allows only one of the follicles to grow large enough to ovulate. The single follicle reaches a size of 1 to 1.5 centimeters at the time of ovulation and is called the *mature follicle.*

Ovulation

Ovulation in a woman who has a normal 28-day female sexual cycle occurs 14 days after the onset of menstruation.

Shortly before ovulation, the protruding outer wall of the follicle swells rapidly, and a small area in the center of the capsule, called the *stigma,* protrudes like a nipple. In another 30 minutes or so, fluid begins to ooze from the follicle through the stigma. About 2

minutes later, as the follicle becomes smaller because of loss of fluid, the stigma ruptures widely and a more viscous fluid that has occupied that central portion of the follicle is evaginated outward into the abdomen. This viscous fluid carries with it the ovum surrounded by several thousand small granulosa cells called the *corona radiata.*

NEED FOR LH IN OVULATION—OVULATORY SURGE OF LH. LH is necessary for final follicular growth and ovulation. Without this hormone, even though large quantities of FSH are available, the follicle will not progress to the stage of ovulation.

About 2 days before ovulation, for reasons that are not completely known but that are discussed in more detail later in the chapter, the rate of secretion of LH by the anterior pituitary gland increases markedly, rising 6- to 10-fold and peaking about 16 hours before ovulation. FSH also increases about twofold to threefold at the same time, and the two hormones act synergistically to cause rapid swelling of the follicle during the last few days before ovulation. The LH also has the specific effect on the granulosa and theca cells of converting them more to progesterone-secreting cells and less estrogen secretion. Therefore, the rate of secretion of estrogen begins to fall about 1 day before ovulation, while small amounts of progesterone begin to be secreted.

It is in this environment of (1) rapid growth of the follicle, (2) diminishing estrogen secretion after a prolonged phase of excessive estrogen secretion, and (3) beginning secretion of progesterone that ovulation occurs. Without the initial preovulatory surge of LH, ovulation will not take place.

INITIATION OF OVULATION. Figure 81–5 gives a schema for the initiation of ovulation. It shows especially the role of the large quantity of LH secreted by

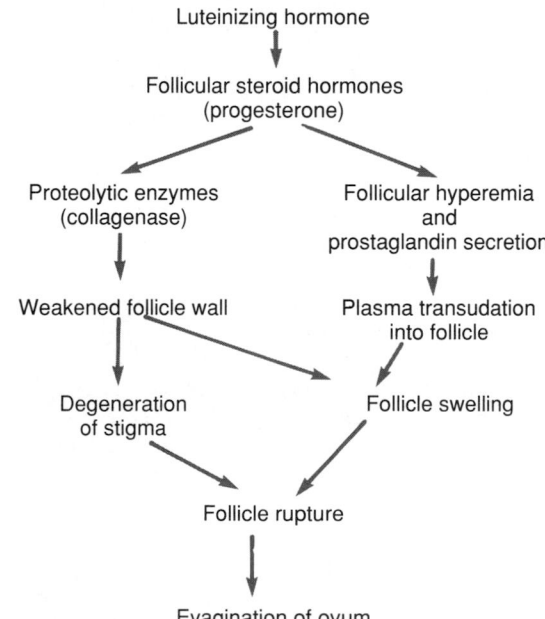

Figure 81–5. Postulated mechanism of ovulation. (Based primarily on the research studies of H. Lipner.)

the anterior pituitary gland. The LH in turn causes rapid secretion of follicular steroid hormones that contain a small amount of progesterone for the first time. Within a few hours two events occur, both of which are necessary for ovulation: (1) The theca externa (the capsule of the follicle) begins to release proteolytic enzymes from lysosomes that cause dissolution of the capsular wall and consequent weakening of the wall, resulting in further swelling of the entire follicle and degeneration of the stigma. (2) Simultaneously, there is rapid growth of new blood vessels into the follicle wall, and at the same time prostaglandins (local hormones that cause vasodilation) are secreted in the follicular tissues. These two effects in turn cause plasma transudation into the follicle, which also contributes to follicle swelling. Finally, the combined follicle swelling and simultaneous degeneration of the stigma cause follicle rupture with discharge of the ovum.

The Corpus Luteum—The "Luteal" Phase of the Ovarian Cycle

During the first few hours after expulsion of the ovum from the follicle, the remaining granulosa and theca interna cells change rapidly into *lutein cells.* They enlarge in diameter two or more times and become filled with lipid inclusions that give them a yellowish appearance. This process is called *luteinization,* and the total mass of cells together is called the *corpus luteum,* which is shown in the bottom of Figure 81–4. A well-developed vascular supply also grows into the corpus luteum.

The granulosa cells in the corpus luteum develop an extensive smooth endoplasmic reticulum that forms large amounts of the female sex hormones *progesterone* and *estrogen* but more so of progesterone. The theca cells form mainly the androgens *androstenedione* and *testosterone* rather than female sex hormones. However, most of them are then converted by the granulosa cells into the female hormones.

In the normal female, the corpus luteum grows to about 1.5 centimeters in diameter, reaching this stage of development about 7 to 8 days after ovulation. Then it begins to involute and eventually loses its secretory function as well as its yellowish, lipid characteristic about 12 days after ovulation, becoming the so-called *corpus albicans;* during the ensuing few weeks this is replaced by connective tissue.

LUTEINIZING FUNCTION OF LH. The change of granulosa and theca interna cells into lutein cells is mainly dependent on the LH secreted by the anterior pituitary gland. In fact, this function gave LH its name "luteinizing" for "yellowing." Luteinization of the granulosa cells also depends on extrusion of the ovum from the follicle. A yet undiscovered local hormone in the follicular fluid, called *luteinization-inhibiting factor,* seems to hold the luteinization process mainly in check until after ovulation. Also for this reason, a corpus luteum only seldom develops in a follicle that does not ovulate.

SECRETION BY THE CORPUS LUTEUM: FUNCTION OF LH. The corpus luteum is a highly secretory organ,

secreting large amounts of both *progesterone* and *estrogen.* Once LH (mainly that which has been secreted during the ovulatory surge) has acted on the granulosa and theca cells to cause luteinization, the newly formed lutein cells seem to be programmed to go through a preordained sequence of (1) proliferation, (2) enlargement, and (3) secretion, then to be followed by (4) degeneration. Even in the absence of further secretion of LH by the anterior pituitary gland, this process will still occur, but it will last only 4 to 8 days. On the other hand, in the presence of LH the degree of growth of the corpus luteum is enhanced, its secretion is greater, and its life is extended, usually to about 12 days. We shall see in the discussion of pregnancy in Chapter 82 that another hormone that has almost exactly the same properties as LH, *chorionic gonadotropin,* which is secreted by the placenta, can also act on the corpus luteum to prolong its life—usually maintaining it for at least the first 2 to 4 months of pregnancy.

INVOLUTION OF THE CORPUS LUTEUM AND ONSET OF THE NEXT OVARIAN CYCLE. Estrogen in particular and progesterone to a lesser extent secreted by the corpus luteum during the luteal phase of the ovarian cycle have the strong feedback effect on the anterior pituitary gland of maintaining low secretory rates of both FSH and LH. In addition, the luteal cells secrete small amounts of the hormone *inhibin,* the same as the inhibin secreted by the Sertoli cells of the male testes. This hormone inhibits secretion by the anterior pituitary gland, especially FSH secretion. As a result, both FSH and LH in the blood fall to low concentrations, and loss of these hormones causes the corpus luteum to degenerate completely, a process called *involution* of the corpus luteum. Final involution occurs at the end of almost exactly 12 days of corpus luteum life, which is about on the 26th day of the normal female sexual cycle, 2 days before menstruation begins. Now, the lack of secretion of the estrogen, progesterone, and inhibin by the corpus luteum removes the feedback inhibition of the anterior pituitary gland, allowing it again to begin secreting increasing amounts of FSH and a few days later slightly increased quantities of LH. The FSH and LH initiate growth of new follicles to begin a new ovarian cycle. But before these follicles can progress significantly, the paucity of secretion of progesterone and estrogen leads to menstruation by the uterus, as explained later.

Summary

About every 28 days, gonadotropic hormones from the anterior pituitary gland cause new follicles to begin to grow in the ovaries. One of the follicles finally becomes "mature" and ovulates on the 14th day of the cycle. During growth of the follicles, mainly estrogen is secreted.

After ovulation, the secretory cells of the follicle develop into a corpus luteum that secretes large quantities of the female hormones progesterone and estrogen. After another 2 weeks, the corpus luteum degenerates, whereupon the ovarian hormones estrogen and progesterone decrease greatly and menstruation begins. A new ovarian cycle then follows.

FUNCTIONS OF THE OVARIAN HORMONES—ESTRADIOL AND PROGESTERONE

The two types of ovarian sex hormones are the *estrogens* and the *progestins*. By far the most important of the estrogens is the hormone *estradiol*, and by far the most important progestin is *progesterone*. The estrogens mainly promote proliferation and growth of specific cells in the body and are responsible for development of most secondary sexual characteristics of the female. On the other hand, the progestins are concerned almost entirely with final preparation of the uterus for pregnancy and the breasts for lactation.

Chemistry of the Sex Hormones

THE ESTROGENS. In the normal nonpregnant female, estrogens are secreted in major quantities only by the ovaries, although minute amounts are also se-

creted by the adrenal cortices. In pregnancy, tremendous quantities are also secreted by the placenta, as we discuss in Chapter 82.

Only three estrogens are present in significant quantities in the plasma of the human female: *β-estradiol, estrone,* and *estriol,* the formulas for which are shown in Figure 81–6. The principal estrogen secreted by the ovaries is *β*-estradiol. Small amounts of estrone are also secreted, but most of this is formed in the peripheral tissues from androgens secreted by the adrenal cortices and by the ovarian thecal cells. Estriol is a weak estrogen; it is an oxidative product derived from both estradiol and estrone, the conversion occurring mainly in the liver.

The estrogenic potency of *β*-estradiol is 12 times that of estrone and 80 times that of estriol. Considering these relative potencies, one can see that the total estrogenic effect of *β*-estradiol is usually many times that of the other two together. For this reason, *β*-estradiol is considered the major estrogen, although the estrogenic effects of estrone are far from negligible.

THE PROGESTINS. By far the most important of the progestins is progesterone. However, small amounts of another progestin, 17-α-hydroxyprogesterone, also are secreted along with progesterone and have essentially

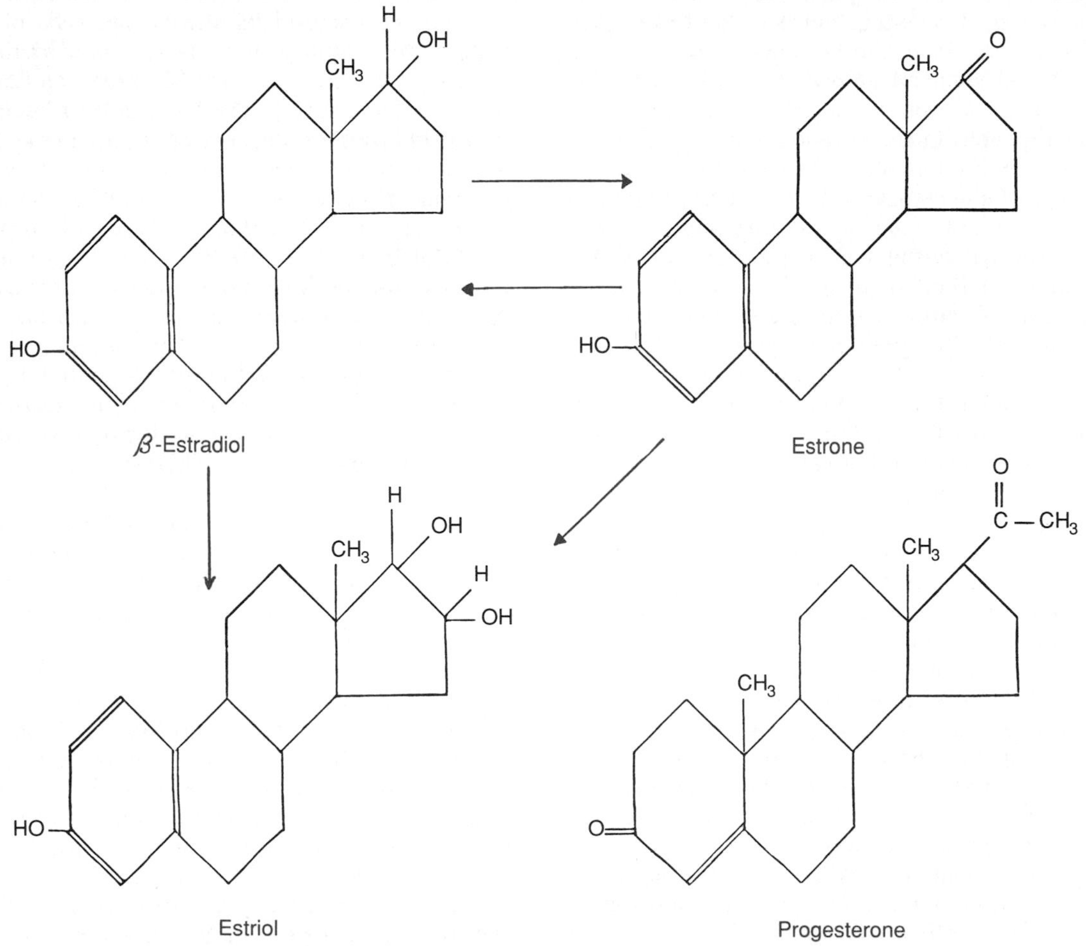

Figure 81–6. Chemical formulas of the principal female hormones.

the same effects. Yet for practical purposes it is usually reasonable to consider progesterone the single important progestin.

In the normal nonpregnant female, progesterone is secreted in significant amounts only during the latter half of each ovarian cycle, when it is secreted by the corpus luteum. Only minute amounts of progesterone appear in the plasma during the first half of the ovarian cycle, secreted approximately equally by the ovaries and the adrenal cortices. Yet, as we see in Chapter 82, large amounts of progesterone are also secreted by the placenta during pregnancy, especially after the 4th month of gestation.

SYNTHESIS OF THE ESTROGENS AND PROGESTINS. Note from the chemical formulas of the estrogens and progesterone in Figure 81–6 that they are all steroids. They are synthesized in the ovaries mainly from cholesterol derived from the blood but to a slight extent also from acetyl coenzyme A, multiple molecules of which can combine to form the appropriate steroid nucleus. During synthesis, progesterone and the male sex hormone testosterone are mainly synthesized first; then, during the follicular phase of the ovarian cycle, before these initial two hormones can leave the ovaries, almost all the testosterone and much of the progesterone are converted into estrogens by the granulosa cells. During the luteal phase of the cycle, far too much progesterone is formed for all of it to be converted, which accounts for the large progesterone secretion at this time. Also, about one fifteenth as much testosterone is secreted into the plasma of the female by the ovaries as is secreted into the plasma of the males by the testes.

TRANSPORT OF ESTROGENS AND PROGESTERONE IN THE BLOOD. Both estrogens and progesterone are transported in the blood bound mainly with plasma albumin and specific estrogen- and progesterone-binding globulins. The binding between these hormones and the plasma proteins is loose enough that they are rapidly released to the tissues over a period of 30 minutes or so.

FATE OF ESTROGENS; FUNCTION OF THE LIVER IN ESTROGEN DEGRADATION. The liver conjugates the estrogens to form glucuronides and sulfates, and about one fifth of these conjugated products is excreted in the bile while most of the remainder is excreted in the urine. Also, the liver converts the potent estrogens estradiol and estrone into the almost totally impotent estrogen estriol. Therefore, diminished liver function actually *increases* the activity of estrogens in the body, sometimes causing *hyperestrinism.*

FATE OF PROGESTERONE. Within a few minutes after secretion, almost all the progesterone is degraded to other steroids that have no progesteronic effect. Here, as is also true with the estrogens, the liver is especially important for this metabolic degradation.

The major end product of progesterone degradation is *pregnanediol.* About 10 per cent of the original progesterone is excreted in the urine in this form. Therefore, one can estimate the rate of progesterone formation in the body from the rate of this excretion.

Functions of the Estrogens—Their Effects on the Primary and Secondary Female Sex Characteristics

The principal function of the estrogens is to cause cellular proliferation and growth of the tissues of the sex organs and other tissues related to reproduction.

EFFECT ON THE UTERUS AND EXTERNAL FEMALE SEX ORGANS. During childhood, estrogens are secreted only in minute quantities, but at puberty, the quantity of estrogens secreted under the influence of the pituitary gonadotropic hormones increases 20-fold or more. At this time, the female sex organs change from those of a child to those of an adult. The ovaries, fallopian tubes, uterus, and vagina all increase several times in size. Also, the external genitalia enlarge, with deposition of fat in the mons pubis and labia majora and with enlargement of the labia minora.

In addition, estrogens change the vaginal epithelium from a cuboidal into a stratified type, which is considerably more resistant to trauma and infection than is the prepubertal epithelium. Infections in children, such as gonorrheal vaginitis, can be cured by the administration of estrogens simply because of the resulting increased resistance of the vaginal epithelium.

During the few years after puberty, the size of the uterus increases twofold to threefold. More important than increases in size are the changes that take place in the endometrium under the influence of estrogens because estrogens cause marked proliferation of the endometrial stroma and greatly increased development of the endometrial glands that will later be used to aid in nutrition of the implanting ovum. These effects are discussed later in the chapter in connection with the endometrial cycle.

EFFECT OF ESTROGENS ON THE FALLOPIAN TUBES. The estrogens have an effect on the mucosal lining of the fallopian tubes similar to that on the uterine endometrium: They cause the glandular tissues to proliferate, and especially important, they cause the number of ciliated epithelial cells that line the fallopian tubes to increase. Also, the activity of the cilia is considerably enhanced, these always beating toward the uterus. This helps propel the fertilized ovum toward the uterus.

EFFECT OF ESTROGENS ON THE BREASTS. The primordial breasts of both female and male are exactly alike, and under the influence of appropriate hormones, the masculine breast, at least during the first two decades of life, can develop sufficiently to produce milk in the same manner as the female breast.

Estrogens cause (1) development of the stromal tissues of the breasts, (2) growth of an extensive ductile system, and (3) deposition of fat in the breasts. The lobules and alveoli of the breast develop to a slight extent under the influence of estrogens alone, but it is progesterone and prolactin that cause the ultimate determinative growth and function of these structures. In summary, the estrogens initiate growth of the breasts and of the milk-producing apparatus. They are also responsible for the characteristic growth and external appearance of the mature female breast. However, they do not complete the job of converting the breasts into milk-producing organs.

EFFECT OF ESTROGENS ON THE SKELETON. Estrogens cause increased osteoblastic activity. Therefore, at puberty, when the female enters her reproductive years, her growth rate becomes rapid for several years. However, estrogens have another potent effect on skeletal growth—that is, they cause early uniting of the epiphyses with the shafts of the long bones. This effect is much stronger in the female than is a similar effect of testosterone in the male. As a result, growth of the female usually ceases several years earlier than growth of the male. The female eunuch who is devoid of estrogen production usually grows several inches taller than the normal mature female because her epiphyses do not unite early.

Osteoporosis of the Bones Caused by Estrogen Deficiency in Old Age. After the menopause, almost no estrogens are secreted by the ovaries. This estrogen deficiency leads to (1) diminished osteoblastic activity in the bones, (2) decreased bone matrix, and (3) decreased deposition of bone calcium and phosphate. In some women, this effect is extremely severe, and the resulting condition is *osteoporosis,* described in Chapter 79. Because this can greatly weaken the bones and lead to bone fracture, especially fracture of the vertebrae, a large share of postmenopausal women are treated prophylactically with substitutive estrogens.

EFFECT OF ESTROGENS ON PROTEIN DEPOSITION. Estrogens cause a slight increase in total body protein, which is evidenced by a slight positive nitrogen balance when estrogens are administered. This probably results from the growth-promoting effect of estrogen on the sexual organs, the bones, and a few other tissues of the body. The enhanced protein deposition caused by testosterone is much more general and many times as powerful as that caused by estrogens.

EFFECT OF ESTROGENS ON METABOLISM AND FAT DEPOSITION. Estrogens increase the metabolic rate slightly but only about one third as much as does the male sex hormone testosterone. They also cause deposition of increased quantities of fat in the subcutaneous tissues. As a result, the overall specific gravity of the female body, as judged by flotation in water, is considerably less than that of the male body, which contains more protein and less fat. In addition to deposition of fat in the breasts and subcutaneous tissues, estrogens cause the deposition of fat in the buttocks and thighs that is characteristic of the feminine figure.

EFFECT OF ESTROGENS ON HAIR DISTRIBUTION. Estrogens do not greatly affect hair distribution. However, hair does develop in the pubic region and in the axillae after puberty. Androgens formed in increased quantities by the female adrenal glands after puberty are mainly responsible for this.

EFFECT OF ESTROGENS ON THE SKIN. Estrogens cause the skin to develop a texture that is soft and usually smooth, but nevertheless, the skin is thicker than that of the child or the female castrate. Also, estrogens cause the skin to become more vascular than normal; this is often associated with increased warmth of the skin and often results in greater bleeding of cut surfaces than is observed in men.

The adrenal androgens, which are secreted in increased quantities after puberty, cause increased secretion by the axillary sweat glands and often cause acne.

EFFECT OF ESTROGENS ON ELECTROLYTE BALANCE. The chemical similarity of estrogenic hormones to adrenocortical hormones has been pointed out, and estrogens, like aldosterone and some other adrenocortical hormones, cause sodium and water retention by the kidney tubules. This effect of estrogens is slight and rarely of significance except in pregnancy, as discussed in Chapter 82.

INTRACELLULAR FUNCTIONS OF ESTROGENS. Thus far, we have discussed the gross effects of estrogens on the female body. The cellular mechanism behind these effects is as follows: Estrogens circulate in the blood for only a few minutes before they are delivered to the target cells. On entry into these cells, they combine within 10 to 15 seconds with a "receptor" protein in the cytoplasm and then, in combination with this protein, activate specific portions of the chromosomal DNA. This immediately initiates the process of transcription; therefore, RNA begins to be produced within a few minutes. In addition, over many hours, new DNA may also be produced, resulting eventually in division of the cell. The RNA diffuses to the cytoplasm, where it causes greatly increased protein formation and subsequently altered cellular function.

One of the principal differences between the protein anabolic effects of the estrogens and those of testosterone is that estrogens cause their effects almost exclusively in a few specific target organs, such as the uterus, the breasts, the skeleton, and certain fatty areas of the body, whereas testosterone has a more generalized protein anabolic effect throughout the body.

Functions of Progesterone

EFFECT OF PROGESTERONE ON THE UTERUS. By far the most important function of progesterone is *to promote secretory changes in the uterine endometrium* during the latter half of the monthly female sexual cycle, thus preparing the uterus for implantation of the fertilized ovum. This function is discussed later in connection with the endometrial cycle of the uterus.

In addition to this effect on the endometrium, progesterone decreases the frequency and intensity of uterine contractions, thereby helping to prevent expulsion of the implanted ovum.

EFFECT OF PROGESTERONE ON THE FALLOPIAN TUBES. Progesterone also promotes secretory changes in the mucosal lining of the fallopian tubes. These secretions are necessary for nutrition of the fertilized, dividing ovum as it traverses the fallopian tube before implantation.

EFFECT OF PROGESTERONE ON THE BREASTS. Progesterone promotes development of the lobules and alveoli of the breasts, causing the alveolar cells to proliferate, enlarge, and become secretory in nature. However, progesterone does not cause the alveoli actually to secrete milk because, as discussed in Chapter 82, milk is secreted only after the prepared breast is further stimulated by prolactin from the anterior pituitary gland.

Progesterone also causes the breasts to swell. Part of this swelling is due to the secretory development in the lobules and alveoli, but part also results from increased fluid in the subcutaneous tissue.

EFFECT OF PROGESTERONE ON ELECTROLYTE BALANCE. Progesterone in large quantity, like estrogens, testosterone, and adrenocortical hormones, can enhance sodium, chloride, and water reabsorption from the distal tubules of the kidney. Yet, strangely enough, progesterone more often causes increased sodium and water excretion. The cause of this is competition between progesterone and aldosterone, which probably occurs as follows: These two substances combine with the same receptor proteins that cause transport of sodium ions through the epithelial cells of the tubules. When progesterone combines with these proteins, aldosterone cannot combine. Yet progesterone exerts many times less sodium transport effect than does aldosterone. Therefore, despite the fact that under appropriate conditions, progesterone can weakly promote sodium and water retention by the renal tubules, it blocks the far more potent effect of aldosterone, thus usually resulting in net loss of sodium and water from the body.

The Monthly Endometrial Cycle and Menstruation

Associated with the monthly cyclic production of estrogens and progesterone by the ovaries is an endometrial cycle that operates through the following stages: first, proliferation of the uterine endometrium; second, development of secretory changes in the endometrium; and third, desquamation of the endometrium, which is known as *menstruation*. The various phases of this endometrial cycle are shown in Figure 81–7.

PROLIFERATIVE PHASE (ESTROGEN PHASE) OF THE ENDOMETRIAL CYCLE, OCCURRING BEFORE OVULATION. At the beginning of each monthly cycle, most of the endometrium is desquamated by menstruation. After menstruation, only a thin layer of endometrial stroma remains at the base of the original endometrium and the only epithelial cells that are left are those located in the remaining deep portions of the glands and crypts of the endometrium. *Under the influence of estrogens,* secreted in increasing quantities by the ovary during the first part of the monthly ovarian cycle, the stromal cells and the epithelial cells

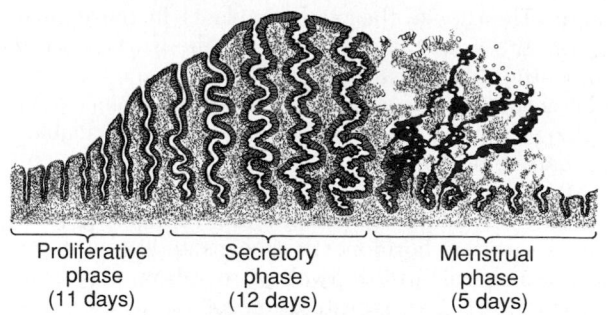

Figure 81–7. Phases of endometrial growth and menstruation during each monthly female sexual cycle.

proliferate rapidly. The endometrial surface is re-epithelialized within 4 to 7 days after the beginning of menstruation. Then, during the next week and a half —that is, before ovulation occurs—the endometrium increases greatly in thickness, owing to increasing numbers of stromal cells and to progressive growth of the endometrial glands and new blood vessels into the endometrium. At the time of ovulation, the endometrium is about 3 to 4 millimeters thick. The endometrial glands, especially those of the cervical region, secrete a thin, stringy mucus. The mucous strings actually align themselves along the length of the cervical canal, forming channels that help guide the sperm in the proper direction into the uterus.

SECRETORY PHASE (PROGESTATIONAL PHASE) OF THE ENDOMETRIAL CYCLE, OCCURRING AFTER OVULATION. During most of the latter half of the monthly cycle, after ovulation has occurred, progesterone and estrogen are secreted in large quantity by the corpus luteum. The estrogens cause slight additional cellular proliferation in the endometrium during this phase of the endometrial cycle, whereas progesterone causes marked swelling and secretory development of the endometrium. The glands increase in tortuosity; an excess of secretory substances accumulates in the glandular epithelial cells. Also, the cytoplasm of the stromal cells increases; lipid and glycogen deposits increase greatly in the stromal cells; and the blood supply to the endometrium further increases in proportion to the developing secretory activity, the blood vessels becoming highly tortuous. At the peak of the secretory phase, about 1 week after ovulation, the endometrium has a thickness of 5 to 6 millimeters.

The whole purpose of all these endometrial changes is to produce a highly secretory endometrium that contains large amounts of stored nutrients that can provide appropriate conditions for implantation of a fertilized ovum during the latter half of the monthly cycle. From the time a fertilized ovum enters the uterine cavity from the fallopian tube (3 to 4 days after ovulation) until the time the ovum implants (7 to 9 days after ovulation), the uterine secretions, called "uterine milk," provide nutrition for the early dividing ovum. Then, once the ovum implants in the endometrium, the trophoblastic cells on the surface of the implanting blastocyst begin to digest the endometrium and absorb the endometrial stored substances, thus making far greater quantities of nutrients available to the early embryo.

MENSTRUATION. About 2 days before the end of the monthly cycle, the corpus luteum suddenly involutes and the ovarian hormones, estrogens and progesterone, decrease sharply to low levels of secretion, as shown in Figure 81–3. Then menstruation follows.

Menstruation is caused by this sudden reduction of the estrogens and progesterone, especially the progesterone, at the end of the monthly ovarian cycle. The first effect is decreased stimulation of the endometrial cells by these two hormones, followed rapidly by involution of the endometrium itself to about 65 per cent of its previous thickness. Then, during the 24 hours preceding the onset of menstruation, the tortuous blood vessels leading to the mucosal layers of the endometrium become vasospastic, presumably because of some effect of the involution, such as release of a vasoconstrictor material, possibly one of the vasoconstrictor types of prostaglandins that are present in abundance at that time. The vasospasm and loss of hormonal stimulation cause beginning necrosis in the endometrium, especially of blood vessels. As a result, blood at first seeps into the vascular layer of the endometrium, and the hemorrhagic areas grow more rapidly over a period of 24 to 36 hours. Gradually, the necrotic outer layers of the endometrium separate from the uterus at the site of the hemorrhages, until, about 48 hours after the onset of menstruation, all the superficial layers of the endometrium have desquamated. The mass of desquamated tissue and blood in the uterine cavity, possibly plus contractile effects of the prostaglandins, initiate uterine contractions that expel the uterine contents.

During normal menstruation, 40 milliliters of blood and an additional 35 milliliters of serous fluid are lost. The menstrual fluid is normally nonclotting because a *fibrinolysin* is released along with the necrotic endometrial material. If excessive bleeding occurs from the uterine surface, the quantity of fibrinolysin may not be sufficient to prevent clotting. The presence of clots during menstruation ordinarily is clinical evidence of uterine pathology.

Within 4 to 7 days after menstruation starts, the loss of blood ceases because by this time, the endometrium has become re-epithelialized.

Leukorrhea During Menstruation. During menstruation, tremendous numbers of leukocytes are released along with the necrotic material and blood. It is probable that some substance liberated by the endometrial necrosis causes this outflow of leukocytes. As a result of these many leukocytes and possible other factors, the uterus is highly resistant to infection during menstruation even though the endometrial surfaces are denuded. This is of extreme protective value.

REGULATION OF THE FEMALE MONTHLY RHYTHM—INTERPLAY BETWEEN THE OVARIAN AND HYPOTHALAMIC-PITUITARY HORMONES

Now that we have presented the major cyclic changes that occur during the monthly female sexual cycle, we can attempt to explain the basic rhythmical mechanism that causes these cyclic variations.

The Hypothalamus Secretes GnRH; This Causes the Hypothalamus to Secrete LH and FSH

As pointed out in Chapter 74, secretion of most of the anterior pituitary hormones is controlled by "releasing hormones" formed in the hypothalamus and then transported to the anterior pituitary gland by way of the hypothalamic-hypophysial portal system. In the case of the gonadotropins, one releasing hormone, *GnRH*, is important. This hormone has been purified and found to be a decapeptide having the following formula:

$$\text{Glu-His-Trp-Ser-Tyr-Gly-Leu-Arg-Pro-Gly-NH}_2$$

INTERMITTENT, PULSATILE SECRETION OF GnRH BY THE HYPOTHALAMUS–AND PULSATILE RELEASE OF LH FROM THE ANTERIOR PITUITARY. Experiments have demonstrated that the hypothalamus does not secrete GnRH continuously but instead secretes it in pulses lasting for several minutes that occur every 1 to 3 hours. The lower curve in Figure 81–8 shows the electrical pulses in the hypothalamus that cause the hypothalamic pulsatile output of GnRH.

It is intriguing that when GnRH is infused continuously so that it is available all the time rather than in these pulses, its effects in causing release of LH and FSH by the anterior pituitary gland are lost. Therefore, for reasons unknown, the pulsatile nature of GnRH release is essential to its function.

The pulsatile release of GnRH also causes pulsatile output of LH. This is shown by the upper curve in Figure 81–8.

To a slight extent, the secretion of FSH is also increased upward and downward by the hypothalamic pulses of GnRH, but there is a more important prolonged effect on FSH secretion that persists for many hours rather than changing greatly from one pulse to the next.

HYPOTHALAMIC CENTERS FOR RELEASE OF GnRH. The neuronal activity that causes pulsatile release of GnRH occurs primarily in the mediobasal hypothalamus, especially in the arcuate nuclei of this area. Therefore, it is believed that these arcuate nuclei control most female sexual activity, although other neurons located in the preoptic area of the anterior hypothalamus also secrete GnRH in moderate amounts, the function of which is unclear. Multiple neuronal centers

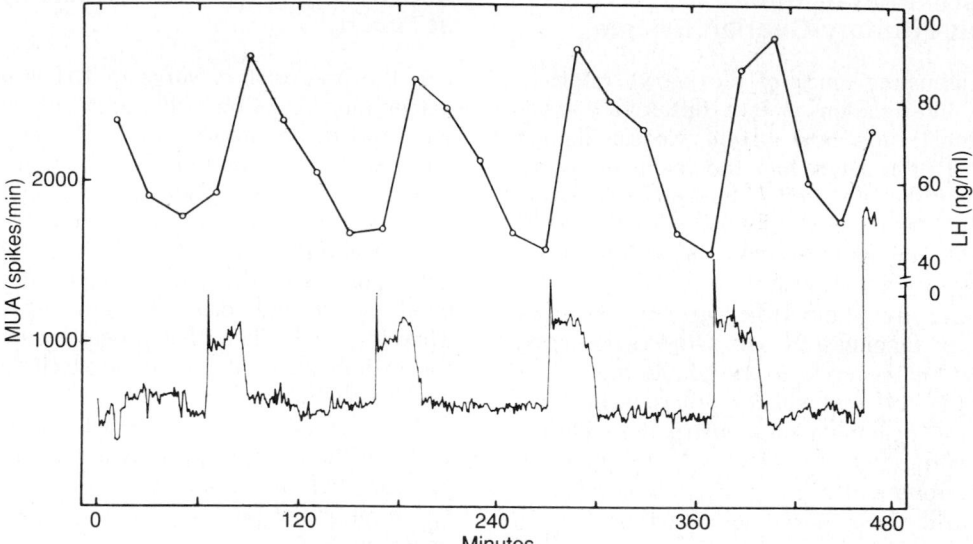

Figure 81–8. *Lower curve:* Changes in multiunit electrical activity (MUA) recorded from medio-basal hypothalamus. *Upper curve:* Luteinizing hormone (LH) pulses in peripheral circulation in a pentobarbital-anesthetized ovariectomized rhesus monkey. (From Wilson, Kesner, Kaufman, Uemura, Akema, and Knobil: *Neuroendocrinology* 39:256, 1984.)

in the brain's limbic system transmit signals into the arcuate nuclei to modify both the intensity of GnRH release and the frequency of the pulses, thus providing a possible explanation of why psychic factors often modify female sexual function.

Negative Feedback Effect of Estrogen and Progesterone in Decreasing Both LH and FSH Secretion

Estrogen in small amounts has a strong effect in inhibiting the production of both LH and FSH. Also, when progesterone is available, the inhibitory effect of estrogen is multiplied, even though progesterone by itself has little effect.

These feedback effects seem to operate mainly directly on the anterior pituitary gland but to a lesser extent on the hypothalamus to decrease secretion of GnRH, especially by altering the frequency of the GnRH pulses.

THE HORMONE INHIBIN FROM THE CORPUS LUTEUM INHIBITS FSH AND LH SECRETION. In addition to the feedback effects of estrogen and progesterone, still another hormone seems to be involved. This is *inhibin,* which is secreted along with the steroid sex hormones by the granulosa cells of the corpus luteum in the same way that Sertoli cells secrete the same hormone in the male testes. This inhibin has the same effect in the female as in the male of inhibiting the secretion of FSH by the anterior pituitary gland and LH to a lesser extent as well. Therefore, it is believed that inhibin might be especially important in causing the decrease in secretion of FSH and LH at the end of the female sexual month.

Positive Feedback Effect of Estrogen Before Ovulation—The Preovulatory LH Surge

For reasons not completely understood, the anterior pituitary gland secretes greatly increased amounts of LH for 1 to 2 days beginning 24 to 48 hours before ovulation. This effect is demonstrated in Figure 81–3, and the figure shows a much smaller preovulatory surge of FSH as well.

Experiments have shown that infusion of estrogen into a female above a critical rate for 2 to 3 days during the latter part of the first half of the ovarian cycle will cause rapidly accelerating growth of the follicles as well as rapidly accelerating secretion of ovarian estrogens. During this period, the secretion of both FSH and LH by the anterior pituitary gland is at first suppressed slightly. Then abruptly the secretion of LH increases sixfold to eightfold and the secretion of FSH increases about twofold. This greatly increased secretion of LH causes ovulation to occur.

The cause of this abrupt surge in LH secretion is not known. However, several possible explanations are as follows: (1) It has been suggested that estrogen at this point in the cycle has a peculiar *positive feedback effect* of stimulating pituitary secretion of LH and less so of FSH; this is in sharp contrast to its normal negative feedback effect that occurs during the remainder of the female monthly cycle. (2) The granulosa cells of the follicles begin to secrete small but increasing quantities of progesterone a day or so before the preovulatory LH surge, and it has been suggested that this might be the factor that sensitizes the excess LH secretion.

Without this normal preovulatory surge of LH, ovulation will not occur.

Feedback Oscillation of the Hypothalamic-Pituitary-Ovarian System

Now, after discussing much of the known information about the interrelations of the different components of the female hormonal system, we can digress from the area of proven fact into the realm of speculation and attempt to explain the feedback oscillation that controls the rhythm of the female sexual cycle. It seems to operate in approximately the following sequence of three successive events.

1. Postovulatory Secretion of the Ovarian Hormones and Depression of Gonadotropins. The easiest part of the cycle to explain is the events that occur during the postovulatory phase—between ovulation and the beginning of menstruation. During this time, the corpus luteum secretes large quantities of both progesterone and estrogen as well as the hormone inhibin. All these hormones together have a combined negative feedback effect on the anterior pituitary gland and hypothalamus to cause suppression of both FSH and LH, decreasing them to their lowest levels at about 3 to 4 days before the onset of menstruation. These effects are shown in Figure 81–3.

2. Follicular Growth Phase. Two to 3 days before menstruation, the corpus luteum has regressed to almost total involution and the corpus luteum secretion of estrogen, progesterone, and inhibin decreases to a low ebb. This releases the hypothalamus and anterior pituitary from the feedback effect of these hormones. A day or so later, at about the time that menstruation begins, pituitary secretion of FSH begins to increase again, increasing as much as twofold; then, several days after menstruation begins, LH secretion increases slightly as well. These hormones initiate new follicular growth and progressive increase in the secretion of estrogen, reaching a peak estrogen secretion at about 12.5 to 13 days after the onset of menstruation. During the first 11 to 12 days of this follicular growth, the rates of pituitary secretion of the gonadotropins FSH and LH decrease slightly because of the negative feedback effect mainly of estrogen on the anterior pituitary gland. Then, strangely, there comes a sudden increase in secretion of LH and less so FSH. This is the preovulatory surge of LH and FSH, which is followed by ovulation.

3. Preovulatory Surge of LH and FSH; Ovulation. At about 11.5 to 12 days after the onset of menstruation, the decline in secretion of FSH and LH comes to an abrupt halt. It is believed that the high level of estrogens at this time (or the beginning secretion of progesterone by the follicles) causes a positive feedback stimulatory effect on the anterior pituitary, as explained earlier, which leads to a terrific surge of secretion of LH and to a lesser extent of FSH. Whatever the cause of this preovulatory LH and FSH surge, the LH leads to both ovulation and subsequent development of and secretion by the corpus luteum. Thus, the hormonal system begins a new round of the cycle until the next ovulation.

Anovulatory Cycles—The Sexual Cycles at Puberty

If the preovulatory surge of LH is not of sufficient magnitude, ovulation will not occur, and the cycle is then said to be "anovulatory." The cyclic variations of the sexual cycle continue, but they are altered in the following ways: First, lack of ovulation causes failure of development of the corpus luteum, and consequently, there is almost no secretion of progesterone during the latter portion of the cycle. Second, the cycle is shortened by several days, but the rhythm continues. Therefore, it is likely that progesterone is not required for maintenance of the cycle itself, although it can alter its rhythm.

Anovulatory cycles are usual during the first few cycles at the onset of puberty as well as several months to years before menopause, presumably because the LH surge is not potent enough at these times to cause ovulation.

Puberty and Menarche

Puberty means the onset of adult sexual life, and menarche means the onset of menstruation. The period of puberty is caused by a gradual increase in gonadotropic hormone secretion by the pituitary, beginning about in the 8th year of life, as shown in Figure 81–9, and usually culminating in the onset of menstruation between ages 11 and 16 years (average, 13 years).

In the female, as in the male, the infantile pituitary gland and ovaries are capable of full function if appropriately stimulated. However, as is also true in the male and for reasons not understood, the hypothalamus does not secrete significant quantities of GnRH during childhood. Experiments have shown that the hypothalamus itself is capable of secreting this hormone, but there is lack of the appropriate signal from

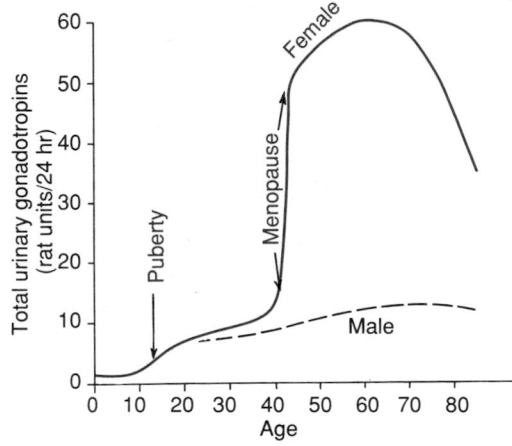

Figure 81–9. Total rates of secretion of gonadotropic hormones throughout the sexual lives of females and males, showing an especially abrupt increase in gonadotropic hormones at the menopause in the female.

some other brain area to cause the secretion. Therefore, it is now believed that the onset of puberty is initiated by some maturation process that occurs elsewhere in the brain, perhaps somewhere in the limbic system.

Figure 81–10 shows (1) the increasing levels of estrogen secretion at puberty, (2) the cyclic variation during the monthly sexual cycles, (3) the further increase in estrogen secretion during the first few years of reproductive life, (4) then progressive decrease in estrogen secretion toward the end of reproductive life, and, finally, (5) almost no estrogen or progesterone secretion beyond the menopause.

Menopause

At ages 40 to 50 years, the sexual cycles usually become irregular and ovulation fails to occur during many of the cycles. After a few months to a few years, the cycles cease altogether, as shown in Figure 81–10. This period during which the cycles cease and the female sex hormones diminish to almost none is called the *menopause.*

The cause of the menopause is "burning out" of the ovaries. Throughout a woman's reproductive life, about 400 of the primordial follicles grow into vesicular follicles and ovulate, while literally hundreds of thousands of the ova degenerate. At the age of about 45 years, only a few primordial follicles remain to be stimulated by FSH and LH and the production of estrogens by the ovary decreases as the number of primordial follicles approaches zero (also shown in Figure 81–10). When estrogen production falls below a critical value, the estrogens can no longer inhibit the production of FSH and LH; nor can they cause an ovulatory surge of LH and FSH to cause ovulatory cycles. Instead, as shown in Figure 81–9, FSH and LH (mainly FSH) are produced after menopause in large and continuous quantities. Estrogens are produced in subcritical quantities for a short time after the menopause, but over a few years, as the remaining primordial follicles become atretic, the production of estrogens by the ovaries falls almost to zero.

At the time of the menopause, a woman must readjust her life from one that has been physiologically stimulated by estrogen and progesterone production to one devoid of these hormones. The loss of the estrogens often causes marked physiological changes in the function of the body, including (1) "hot flushes" characterized by extreme flushing of the skin, (2) psychic sensations of dyspnea, (3) irritability, (4) fatigue, (5) anxiety, (6) occasionally various psychotic states, and (7) decreased strength and calcification of bones throughout the body. These symptoms are of sufficient magnitude in about 15 per cent of women to warrant treatment. If counseling fails, daily administration of an estrogen in small quantities will reverse the symptoms, and by gradually decreasing the dose, the postmenopausal woman is likely to avoid severe symptoms.

ABNORMALITIES OF SECRETION BY THE OVARIES

HYPOGONADISM. Less than normal secretion by the ovaries can result from poorly formed ovaries, lack of ovaries, or genetically abnormal ovaries that secrete the wrong hormones because of missing enzymes in the secretory cells. When ovaries are absent from birth or when they become nonfunctional before puberty, *female eunuchism* occurs. In this condition, the usual secondary sexual characteristics do not appear and the sexual organs remain infantile. Especially characteristic of this condition is prolonged growth of the long bones because the epiphyses do not unite with the shafts of these bones at as early an age as in the normal adolescent woman. Consequently, the female eunuch is essentially as tall as or perhaps even slightly taller than her male counterpart of similar genetic background.

When the ovaries of a fully developed woman are removed, the sexual organs regress to some extent so that the uterus becomes almost infantile in size, the vagina becomes smaller, and the vaginal epithelium becomes thin and easily damaged. The breasts atrophy and become pendulous and the pubic hair becomes thinner. These same changes occur in women after menopause.

Irregularity of Menses, and Amenorrhea Caused by Hypogonadism. As pointed out in the preceding discussion of the menopause, the quantity of estrogens produced by the ovaries must rise above a critical value if they are to be able to cause rhythmical sexual cycles. Consequently, in hypogonadism or when the gonads are secreting small quantities of estrogens as a result of other factors, such as *hypothyroidism,* the ovarian cycle probably will not occur normally. Instead, several months may elapse between menstrual periods or menstruation may cease altogether (amenorrhea). Prolonged ovarian cycles are frequently associated with failure of ovulation, presumably because of insufficient secretion of LH at the time of the preovulatory surge of LH, which is necessary for ovulation.

HYPERSECRETION BY THE OVARIES. Extreme hypersecretion of ovarian hormones by the ovaries is a rare clinical entity because excessive secretion of estrogens

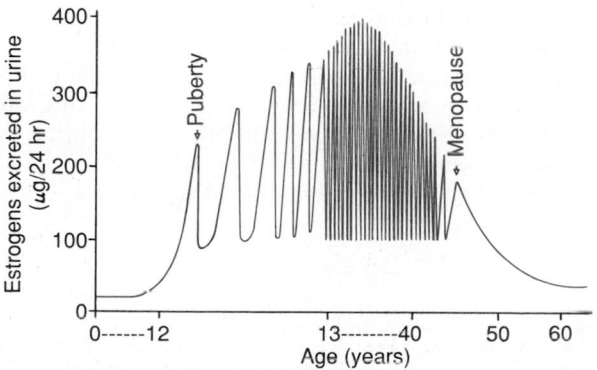

Figure 81–10. Estrogen secretion throughout sexual life.

automatically decreases the production of gonadotropins by the pituitary, and this in turn limits the production of the ovarian hormones. Consequently, hypersecretion of feminizing hormones is usually recognized clinically only when a feminizing tumor develops.

A rare *granulosa cell tumor* can develop in an ovary, occurring more often after menopause than before. These tumors secrete large quantities of estrogens, which exert the usual estrogenic effects, including hypertrophy of the uterine endometrium and irregular bleeding from this endometrium. In fact, bleeding is often the first and only indication that such a tumor exists.

FEMALE SEXUAL ACT

STIMULATION OF THE FEMALE SEXUAL ACT. As is true in the male sexual act, successful performance of the female sexual act depends on both psychic stimulation and local sexual stimulation.

Also, as in men, the thinking of erotic thoughts can lead to female sexual desire; this aids greatly in the performance of the female sexual act. Such desire is based very much on a woman's background training as well as on her physiological drive, although sexual desire does increase in proportion to the level of secretion of the sex hormones. Desire also changes during the sexual month, reaching a peak near the time of ovulation, probably because of the high levels of estrogen secretion during the preovulatory period.

Local sexual stimulation in women occurs in more or less the same manner as in men because massage and other types of stimulation of the vulva, vagina, and other perineal regions and even of the urinary tract create sexual sensations. The glans of the *clitoris* is especially sensitive for initiating sexual sensations. As in the male, the sexual sensory signals are transmitted to the sacral segments of the spinal cord through the pudendal nerve and sacral plexus. Once these signals have entered the spinal cord, they are transmitted to the cerebrum. Also, local reflexes integrated in the sacral and lumbar spinal cord are at least partly responsible for the female sexual reactions.

FEMALE ERECTION AND LUBRICATION. Located around the introitus and extending into the clitoris is erectile tissue almost identical with the erectile tissue of the penis. This erectile tissue, like that of the penis, is controlled by the parasympathetic nerves that pass through the nervi erigentes from the sacral plexus to the external genitalia. In the early phases of sexual stimulation, parasympathetic signals dilate the arteries of the erectile tissues, probably resulting from release of nitric oxide at the nerve endings. This allows rapid accumulation of blood in the erectile tissue so that the introitus tightens around the penis; this in turn aids the male greatly in his attainment of sufficient sexual stimulation for ejaculation to occur.

Parasympathetic signals also pass to the bilateral Bartholin's glands located beneath the labia minora to cause secretion of mucus immediately inside the introitus. This mucus is responsible for much of the lubrication during sexual intercourse, although much is also provided by mucus secreted by the vaginal epithelium and a small amount from the male urethral glands. The lubrication in turn is necessary for establishing during intercourse a satisfactory massaging sensation rather than an irritative sensation, which may be provoked by a dry vagina. A massaging sensation constitutes the optimal stimulus for evoking the appropriate reflexes that culminate in both the male and female climaxes.

FEMALE ORGASM. When local sexual stimulation reaches maximum intensity, and especially when the local sensations are supported by appropriate psychic conditioning signals from the cerebrum, reflexes are initiated that cause the female orgasm, also called the *female climax.* The female orgasm is analogous to emission and ejaculation in the male, and it perhaps helps promote fertilization of the ovum. Indeed, the human female is known to be somewhat more fertile when inseminated by normal sexual intercourse rather than by artificial methods, thus indicating an important function of the female orgasm. Possible reasons for this are as follows.

First, during the orgasm, the perineal muscles of the female contract rhythmically, which results from spinal cord reflexes similar to those that cause ejaculation in the male. It is possible that these same reflexes increase uterine and fallopian tube motility during the orgasm, thus helping transport the sperm upward through the uterus toward the ovum, but information on this subject is scanty. Also, the orgasm seems to cause dilation of the cervical canal for up to 30 minutes, thus allowing easy transport of the sperm.

Second, in many lower animals, copulation causes the posterior pituitary gland to secrete oxytocin; this effect is probably mediated through the amygdaloid nuclei and then through the hypothalamus to the pituitary. The oxytocin in turn causes increased rhythmical contractions of the uterus, which have been postulated to cause rapid transport of the sperm. A few sperm have been shown to traverse the entire length of the fallopian tube in the cow in about 5 minutes, a rate at least 10 times as fast as that which the swimming motions of the sperm themselves could possibly achieve. Whether or not this occurs in the human female is unknown.

In addition to the possible effects of the orgasm on fertilization, the intense sexual sensations that develop during the orgasm also pass to the cerebrum and cause intense muscle tension throughout the body. But after culmination of the sexual act, this gives way during the succeeding minutes to a sense of satisfaction characterized by relaxed peacefulness, an effect called *resolution.*

FEMALE FERTILITY

THE FERTILE PERIOD OF EACH SEXUAL CYCLE. The ovum remains viable and capable of being fertilized after it is expelled from the ovary probably no longer than 24 hours. Therefore, sperm must be available soon after ovulation if fertilization is to take place. On the other hand, a few sperm can remain fertile in the female reproductive tract for up to 72 hours, although most of them for not more than 24 hours. Therefore, for fertilization to take place, intercourse usually must occur some time between 1 to 2 days before ovulation and up to 1 day after ovulation. Thus, the period of female fertility during each sexual cycle is short.

RHYTHM METHOD OF CONTRACEPTION. One of the often practiced methods of contraception is to avoid intercourse near the time of ovulation. The difficulty with this method of contraception is the impossibility of predicting the exact time of ovulation. Yet the interval from ovulation until the next succeeding onset of menstruation is almost always between 13 and 15 days. Therefore, if the menstrual cycle is regular, with a periodicity of 28 days, ovulation usually occurs within 1 day of the 14th day of the cycle. If, on the other hand, the periodicity of the cycle is 40 days, ovulation usually occurs within 1 day of the 26th day of the cycle. Finally, if the periodicity of the cycle is 21 days, ovulation usually occurs within 1 day of the 7th day of the cycle. Therefore, it is usually stated that avoidance of intercourse for 4 days before the calculated day of ovulation and 3 days afterward prevents conception. But such a method of contraception can be used only when the periodicity of the menstrual cycle is regular.

HORMONAL SUPPRESSION OF FERTILITY—"THE PILL." It has long been known that the administration of either estrogen or progesterone, if given in appropriate quantity during the first half of the monthly female cycle, can inhibit ovulation. The reason for this is that appropriate administration of either of these hormones can prevent the preovulatory surge of LH secretion by the pituitary gland, which, it will be recalled, is essential in causing ovulation.

The reason the administration of estrogen or progesterone prevents the preovulatory surge of LH secretion is not fully understood. However, experimental work has suggested that immediately before the surge occurs, there is probably a sudden depression of estrogen secretion by the ovarian follicles, and that this might be the necessary signal for causing the subsequent feedback effect on the anterior pituitary that leads to the surge. The administration of the sex hormones could prevent the initial ovarian hormonal depression that might be the initiating signal for ovulation.

The problem in devising methods for hormonal suppression of ovulation has been in developing appropriate combinations of estrogens and progestins that will suppress ovulation but not cause unwanted effects of these two hormones. For instance, too much of either of the hormones can cause abnormal menstrual bleeding patterns. However, the use of certain synthetic progestins in place of progesterone, especially the 19-norsteroids, along with small amounts of estrogens will usually prevent ovulation and yet allow almost a normal pattern of menstruation. Therefore, almost all "pills" used for control of fertility consist of some combination of synthetic estrogens and synthetic progestins. The main reason for using synthetic estrogens and synthetic progestins is that the *natural* hormones are almost entirely destroyed by the liver within a short time after they are absorbed from the gastrointestinal tract into the portal circulation. However, many of the *synthetic* hormones can resist

this destructive propensity of the liver, thus allowing oral administration.

Two of the most commonly used synthetic estrogens are *ethynyl estradiol* and *mestranol*. Among the most commonly used progestins are *norethindrone, norethynodrel, ethynodiol,* and *norgestrel*. The medication is usually begun in the early stages of the monthly cycle and continued beyond the time that ovulation normally would have occurred. Then the medication is stopped, allowing menstruation to occur and a new cycle to begin.

ABNORMAL CONDITIONS THAT CAUSE FEMALE STERILITY. About one of every six to eight marriages is infertile; in about 60 per cent of these marriages, the infertility is due to female sterility.

Occasionally, no abnormality can be discovered in the female genital organs, in which case it must be assumed that the infertility is due either to abnormal physiological function of the genital system or to abnormal genetic development of the ova themselves.

Probably by far the most common cause of female sterility is failure to ovulate. This can result from hyposecretion of gonadotropic hormones, in which case the intensity of the hormonal stimuli simply is not sufficient to cause ovulation; or it can result from abnormal ovaries that will not allow ovulation. For instance, thick capsules occasionally exist on the outside of the ovaries that make ovulation difficult.

Because of the high incidence of anovulation in sterile women, special methods are often used to determine whether or not ovulation occurs. These methods are all based on the effects of progesterone on the body because the normal increase in progesterone secretion usually does not occur during the latter half of anovulatory cycles. In the absence of progesteronic effects, the cycle can be assumed to be anovulatory. One of these tests is simply to analyze the urine for a surge in pregnanediol, the end product of progesterone metabolism, during the latter half of the sexual cycle, the lack of which indicates failure of ovulation. Another common test is for the woman to chart her body temperature throughout the cycle. Secretion of progesterone during the latter half of the cycle raises the body temperature about 0.5°F, the temperature rise coming abruptly at the time of ovulation. Such a temperature chart, showing the point of ovulation, is shown in Figure 81–11.

Lack of ovulation caused by hyposecretion of the pituitary gonadotropic hormones can sometimes be treated by appropriately timed administration of *human chorionic gonadotropin*, a hormone that is discussed in Chapter 82 and that is extracted from the human placenta. This hormone, although secreted by the placenta, has almost the same effects as LH and, therefore, is a powerful stimulator of ovulation. However, excess use of this hormone can cause ovulation from many follicles simultaneously; this results in multiple births, an effect that has caused as many as seven babies to be born to mothers treated for infertility with this hormone.

One of the most common causes of female sterility is *endometriosis*, a common condition in which endometrial tissue almost identical to that of the uterine endometrium grows and even menstruates in the pelvic cavity surrounding the uterus, fallopian tubes, and ovaries. Endometriosis causes fibrosis throughout the pelvis, and this fibrosis sometimes so enshrouds the ovaries that an ovum cannot be released into the abdominal cavity. Often, also, endometriosis occludes the fallopian tubes,

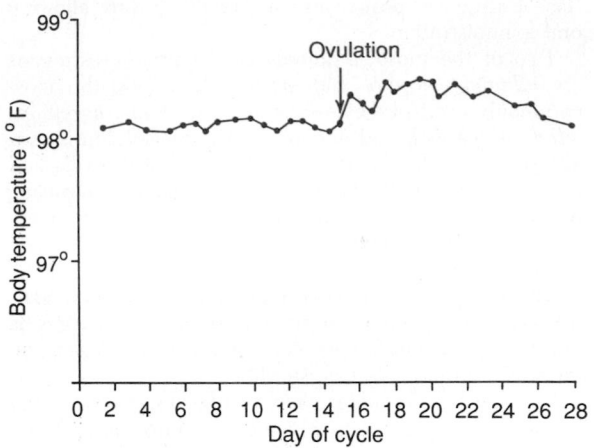

Figure 81–11. Elevation in body temperature shortly after ovulation.

either at the fimbriated ends or elsewhere along their extent.

Another common cause of female infertility is salpingitis, that is, inflammation of the fallopian tubes; this causes fibrosis in the tubes, thereby occluding them. In past years, such inflammation was extremely common as a result of gonococcal infection, but with modern therapy, this is becoming a less prevalent cause of female infertility.

Still another cause of infertility is secretion of abnormal mucus by the uterine cervix. Ordinarily, at the time of ovulation, the hormonal environment of estrogen causes secretion of a thin mucus with special characteristics that will allow rapid mobility of sperm into the uterus and actually guide the sperm up along mucous "threads." Abnormalities of the cervix itself, such as low-grade infection or inflammation, or abnormal hormonal stimulation of the cervix can lead to a viscous mucous plug that will prevent fertilization.

REFERENCES

Ackland, J. F., et al.: Nonsteroidal signals originating in the gonads. Physiol. Rev., 72:731, 1992.

Adashi, E. Y., and Leung, P. C. K.: The Ovary. New York, Raven Press, 1993.

Bland, K. I., and Copeland, III, E. M.: The Breast: Comprehensive Management of Benign and Malignant Diseases. Philadelphia, W. B. Saunders Co., 1991.

Chard, T., and Grudzinskas, J. G.: The Uterus. New York, Cambridge University Press, 1994.

Chen, C., et al.: Recent Advances in the Management of Infertility. Hightstown, NJ, McGraw-Hill, 1990.

Compel, C., and Silverberg, S. G.: Pathology in Gynecology and Obstetrics. Philadelphia, J. B. Lippincott, 1994.

Copeland, L. J.: Textbook of Gynecology. Philadelphia, W. B. Saunders Co., 1993.

Cowan, B.: Reproductive Endocrinology. Philadelphia, J. B. Lippincott, 1994.

DeJong, F. H.: Inhibin. Physiol. Rev., 68:555, 1988.

Dufau, M. L.: Endocrine regulation and communicating functions of the Leydig cell. Annu. Rev. Physiol., 50:483, 1988.

Dunnihoo, D. R.: Fundamentals of Gynecology and Obstetrics. 2nd Ed. Philadelphia, J. B. Lippincott, 1992.

Eskin, B. A.: The Menopause: Comprehensive Management. Hightstown, NJ, McGraw-Hill, 1994.

Garel, J.-M.: Hormonal control of calcium metabolism during the reproductive cycle in mammals. Physiol. Rev., 67:1, 1987.

Girard, J.: Endocrinology of Puberty. Farmington, CT, S. Karger Publishers, Inc., 1991.

Goldzieher, J. W., and Fotherby, K.: Pharmacology of the Contraceptive Steroids. New York, Raven Press, 1994.

Grave, G. D., and Cutler, G. B., Jr.: Sexual Precocity: Etiology, Diagnosis, and Management. New York, Raven Press, 1993.

Gruhn, J. G., and Kazer, R. R.: Hormonal Regulation of the Menstrual Cycle. New York, Plenum Publishing Corp., 1989.

Hacker, N. F., and Moore, J. G.: Essentials of Obstetrics and Gynecology. Philadelphia, W. B. Saunders Co., 1992.

Harris, J. R., et al.: Breast Diseases. 2nd Ed. Philadelphia, J. B. Lippincott, 1991.

Holmes, K. K., et al. Sexually Transmitted Diseases. Hightstown, NJ, McGraw-Hill, 1990.

Inster, V., and Lunenfeld, B.: Infertility: Male and Female. 2nd Ed. New York, Churchill Livingstone, 1993.

Jones, H. W., III, et al. (eds.): Novak's Textbook of Gynecology, 11th Ed. Baltimore, Williams & Wilkins, 1988.

Karsch, F. J.: Central actions of ovarian steroids in the feedback regulation of pulsatile secretion of luteinizing hormone. Annu. Rev. Physiol., 49:365, 1987.

Keyes, P. L., and Wiltbank, M. C.: Endocrine regulation of the corpus luteum. Annu. Rev. Physiol., 50:465, 1988.

Keye, W. R., et al.: Infertility: Evaluation and Treatment. Philadelphia, W. B. Saunders Co., 1994.

Knobil, E., et al. (eds.): The Physiology of Reproduction. New York, Raven Press, 1988.

Knobil, E.: A hypothalamic pulse generator governs mammalian reproduction. News Physiol. Sci., 2:42, 1987.

Lechtenberg, R., and Ohl, D. A.: Sexual Dysfunction. Baltimore, Williams & Wilkins, 1994.

Leung, P. C. K., et al. (eds.): Endocrinology and Physiology of Reproduction. New York, Plenum Publishing Corp., 1987.

Lobo, R. A.: Treatment of the Postmenopausal Woman: Basic and Clinical Aspects. New York, Raven Press, 1994.

Mackay, E. V., et al.: Illustrated Textbook of Gynecology. Philadelphia, W. B. Saunders Co., 1992.

Mahesh, V. B., et al. (eds.): Regulation of Ovarian and Testicular Function. New York, Plenum Publishing Corp., 1987.

Millar, R. P., and King, J. A.: Evolution of gonadotropin-releasing hormone: Multiple usage of a peptide. News Physiol. Sci., 3:49, 1988.

Muske, L. E.: The Neurobiology of Reproductive Behavior. Farmington, CT, S. Karger Publishers, Inc., 1993.

Newton, M., and Newton, E. R.: Complications of Gynecologic and Obstetric Management. Philadelphia, W. B. Saunders Co., 1988.

Scott, J., et al.: Danforth's Obstetrics and Gynecology. Philadelphia, J. B. Lippincott, 1994.

Seibel, M. M., and Blackwell, R. E.: Ovulation Induction. New York, Raven Press, 1994.

Soules, M. F.: Problems in Reproductive Endocrinology and Infertility. New York, Elsevier Science Publishing Co., 1989.

Speroff L., and Darney, P. D.: A Clinical Guide for Contraception. Baltimore, Williams & Wilkins, 1992.

Speroff, L., et al.: Clinical Gynecologic Endocrinology and Infertility. Baltimore, Williams & Wilkins, 1994.

Stouffer, R. L. (ed.): The Primate Ovary. New York, Plenum Publishing Corp., 1987.

Taketani Y., and Kawagoe, S.: Aging of Reproductive Organs. Farmington, CT, S. Karger Publishers, Inc., 1993.

Whitehead, E. D., and Nagler, H. M.: Management of Impotence and Infertility. Philadelphia, J. B. Lippincott, 1994.

Wynn, R. M., and Jollie, W. (eds.): The Biology of the Uterus, 2nd Ed. New York, Plenum Publishing Corp., 1989.

Yen, S. S. C., and Jaffe, R. B.: Reproductive Endocrinology: Physiology, Pathophysiology, and Clinical Management. Philadelphia, W. B. Saunders Co., 1991.

Pregnancy and Lactation
*C*HAPTER *82*

In Chapters 80 and 81, the sexual functions of the male and female are described to the point of fertilization of the ovum. If the ovum becomes fertilized, a new sequence of events called *gestation,* or *pregnancy,* takes place, and the fertilized ovum eventually develops into a full-term fetus. The purpose of this chapter is to discuss the early stages of ovum development after fertilization and then to discuss the physiology of pregnancy. In Chapter 83, some special problems of fetal and early childhood physiology are discussed.

Maturation of the Ovum

Shortly before the *primary oocyte* (the stage of the ovum in the ovary) is released from the follicle, its nucleus divides by meiosis and a *first polar body* is expelled from the nucleus of the oocyte. The primary oocyte then becomes the *secondary oocyte.* In this process, each of the 23 pairs of chromosomes loses one of its partners that becomes incorporated in the polar body that is expelled, so that 23 *unpaired* chromosomes remain in the secondary oocyte. A few hours after the sperm enters the oocyte, the nucleus divides again and a *second polar body* is expelled, thus forming the *mature ovum,* which still contains 23 unpaired chromosomes.

One of the 23 chromosomes in the ovum is a female chromosome, called an *X chromosome.* When this combines with a sperm that also carries an X chromosome, giving an XX combination, a female child is formed, as explained in Chapter 80. But when the X chromosome of the ovum is paired with a sperm that carries a Y chromosome, giving an XY combination, a male child is formed.

Transport, Fertilization, and Implantation of the Developing Ovum

ENTRY OF THE OVUM INTO THE FALLOPIAN TUBE (THE OVIDUCT). When ovulation occurs, the ovum, along with a hundred or more attached granulosa cells that constitute the *corona radiata,* is expelled directly into the peritoneal cavity and must then enter one of the fallopian tubes to reach the cavity of the uterus. The fimbriated ends of each fallopian tube fall naturally around the ovaries. The inner surfaces of the fimbriated tentacles are lined with ciliated epithelium, and the *cilia,* which are activated by estrogen from the ovaries, continually beat toward the opening, the *ostium,* of the fallopian tube. One can actually see a slow fluid current flowing toward the ostium. By this means the ovum enters one or the other fallopian tube.

It would seem likely that many ova might fail to enter the fallopian tubes. However, on the basis of conception studies, it is probable that as many as 98 per cent succeed in this task. Indeed, cases are on record in which women with one ovary removed and the opposite fallopian tube removed have had multiple children with relative ease of conception, thus demonstrating that ova can even enter the opposite fallopian tube.

FERTILIZATION OF THE OVUM. After ejaculation, a few sperm are transported within 5 to 10 minutes through the uterus to the *ampullae* in the ovarian ends of the fallopian tubes, aided by contractions of the uterus and fallopian tubes stimulated by prostaglandins in the seminal fluid and oxytocin released from the posterior pituitary gland during the female orgasm. Of the almost half billion sperm deposited in the vagina, only a few thousand succeed in reaching the ampulla.

1033

Fertilization of the ovum normally takes place soon after the ovum enters the ampulla. Before a sperm can enter the ovum, it must penetrate the multiple layers of granulosa cells attached to the outside of the ovum, called the *corona radiata,* and must bind to and penetrate the *zona pellucida* surrounding the ovum itself. The mechanisms used by the sperm for these purposes are presented in Chapter 80.

Once a sperm has entered the ovum, its head swells rapidly to form a *male pronucleus,* which is shown in Figure 82–1. Later, the 23 unpaired chromosomes of the male pronucleus and the 23 unpaired chromosomes of the *female pronucleus* align themselves to reform a complete complement of 46 chromosomes (23 pairs) in the fertilized ovum.

Transport in the Fallopian Tubes

After fertilization has occurred, an additional 3 to 4 days is normally required for transport of the ovum through the fallopian tube into the cavity of the uterus. This transport is effected mainly by a feeble fluid current in the tube resulting from epithelial secretion plus action of the ciliated epithelium that lines the tube, the cilia always beating toward the uterus. It is possible also that weak contractions of the fallopian tube aid the ovum passage.

The fallopian tubes are lined with a rugged, cryptoid surface that impedes the passage of the ovum despite the fluid current. Also, the *isthmus* of the fallopian tube (the last 2 centimeters before the uterus is entered) remains spastically contracted for the first 3 days after ovulation. After this time, the rapidly increasing progesterone secreted by the ovarian corpus luteum first promotes increasing progesterone receptors on the fallopian tube smooth muscle cells and then activates them, exerting a relaxing effect that allows entry of the ovum into the uterus.

The delayed transport of the ovum through the fallopian tube allows several stages of division to occur before the dividing ovum, now called a *blastocyst* with about 100 cells, enters the uterus. During this time, large quantities of secretions are formed by secretory cells that alternate with ciliated cells lining the fallopian tube. These secretions are for nutrition of the blastocyst.

Implantation of the Blastocyst in the Uterus

After reaching the uterus, the developing blastocyst usually remains in the uterine cavity an additional 1 to 3 days before it implants in the endometrium; thus, implantation ordinarily occurs on about the 5th to 7th day after ovulation. Before implantation, the blastocyst obtains its nutrition from the endometrial secretions called "uterine milk."

Implantation results from the action of *trophoblast cells* that develop over the surface of the blastocyst. These cells secrete proteolytic enzymes that digest and liquefy the adjacent cells of the endometrium. The fluid and nutrients thus released are actively transported by the same trophoblast cells into the blastocyst, adding still further sustenance for growth. Figure 82–2 shows an early implanted human blastocyst, with a small embryo shown in color.

Once implantation has taken place, the trophoblast cells and other adjacent cells both from the blastocyst and from the uterine endometrium proliferate rapidly, forming the placenta and the various membranes of pregnancy.

EARLY INTRAUTERINE NUTRITION OF THE EMBRYO

In Chapter 81, it is pointed out that the progesterone secreted by the ovarian corpus luteum during the latter half of each sexual cycle has a special effect on

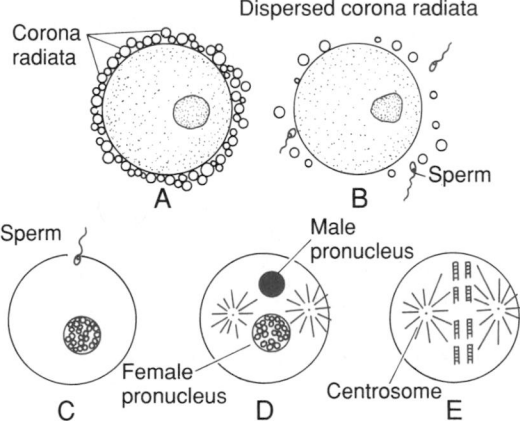

Figure 82–1. Fertilization of the ovum, showing *A,* the mature ovum surrounded by the corona radiata; *B,* dispersal of the corona radiata; *C,* entry of the sperm; *D,* formation of the male and female pronuclei; and *E,* reorganization of a full complement of chromosomes and beginning division of the ovum. (Modified from Arey: Developmental Anatomy. 7th ed. Philadelphia, W. B. Saunders Company, 1974.)

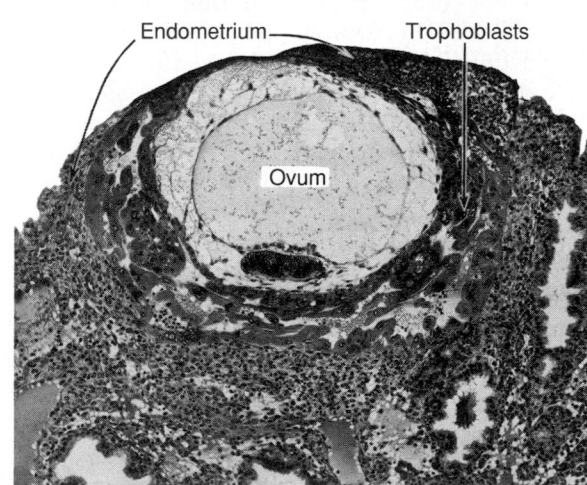

Figure 82–2. Implantation of the early human embryo, showing trophoblastic digestion and invasion of the endometrium. (Courtesy of Dr. Arthur Hertig.)

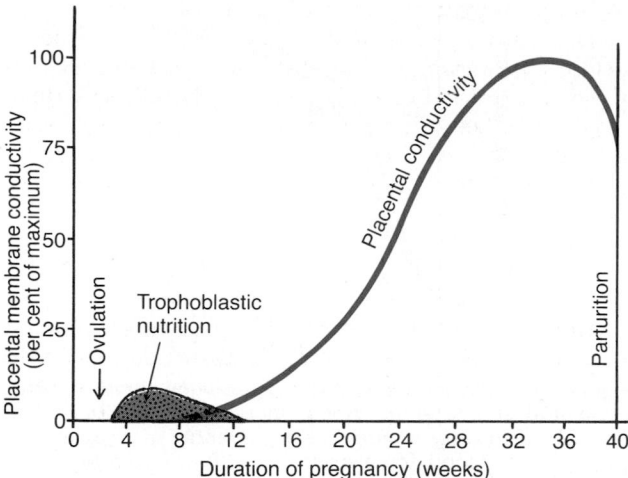

Figure 82–3. Nutrition of the fetus, demonstrating that most of the early nutrition is due to trophoblastic digestion and absorption of nutrients from the endometrial decidua and that essentially all the later nutrition results from diffusion through the placental membrane.

The final structure of the placenta is shown in Figure 82–4. Note that the fetus's blood flows through two *umbilical arteries,* then to the capillaries of the villi, and then back through a single *umbilical vein* into the fetus. At the same time, the mother's blood flows from her *uterine arteries* into large *maternal sinuses* that surround the villi and then back into the *uterine veins* of the mother.

The lower part of Figure 82–4 shows the relation between the fetal blood of each fetal placental villus and the blood of the mother in the fully developed placenta. The capillaries of the villus are lined with an extremely thin endothelium and surrounded by a layer of connective tissue that is covered on the outside of the villus by a layer of *syncytial trophoblast* cells.

The total surface area of all the villi of the mature placenta is only a few square meters—many times less than the area of the pulmonary membrane. Also, remember that even at full maturity, the placental membrane is still several cell layers thick and the minimum distance between the maternal blood and the fetal blood is 3.5 micrometers, almost 10 times the distance across

the endometrium to convert the endometrial stromal cells into large swollen cells that contain extra quantities of glycogen, proteins, lipids, and even some necessary minerals for development of the conceptus. Then, when the conceptus implants in the endometrium, the continued secretion of progesterone causes the endometrial cells to swell still more and to store even more nutrients. These cells are now called *decidual cells,* and the total mass of cells is called the *decidua.*

As the trophoblast cells invade the decidua, digesting and imbibing it, the stored nutrients in the decidua are used by the embryo for growth and development. During the 1st week after implantation, this is the only means by which the embryo can obtain nutrients, and the embryo continues to obtain much of its nutrition in this way for up to 8 weeks, although the placenta also begins to provide nutrition after about the 16th day beyond fertilization (a little more than 1 week after implantation). Figure 82–3 shows this trophoblastic period of nutrition, which gradually gives way to placental nutrition.

FUNCTION OF THE PLACENTA

Developmental and Physiologic Anatomy of the Placenta

While the trophoblastic cords from the blastocyst are attaching to the uterus, blood capillaries grow into the cords from the vascular system of the embryo, and by the 16th day of fertilization, blood begins to flow. Simultaneously, *blood sinuses* supplied with blood from the mother develop around the trophoblastic cords. The trophoblast cells send out more and more projections, which become the *placental villi* into which fetal capillaries grow. Thus, the villi, carrying fetal blood, are surrounded by sinuses that contain maternal blood.

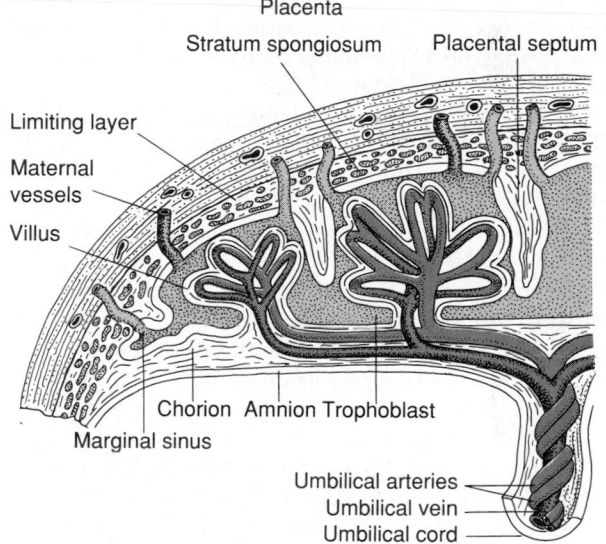

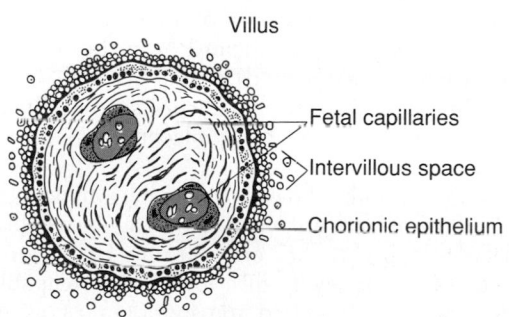

Figure 82–4. *Above,* Organization of the mature placenta. *Below,* Relation of the fetal blood in the villus capillaries to the mother's blood in the intervillous spaces. (Modified from Gray and Goss: Anatomy of the Human Body. 25th ed. Philadelphia, Lea & Febiger, 1948; and from Arey: Developmental Anatomy. 7th ed. Philadelphia, W. B. Saunders Company, 1974.)

the alveolar membranes of the lung. Nevertheless, nutrients and other substances pass through this placental membrane mainly by diffusion in much the same manner that diffusion occurs through the alveolar membranes of the lungs and the capillary membranes elsewhere in the body.

Placental Permeability and Membrane Diffusion Conductance

The major function of the placenta is to provide for *diffusion of foodstuffs* from the mother's blood into the fetus's blood and *diffusion of excretory products* from the fetus back into the mother. Therefore, it is important to know the permeability and total membrane diffusion conductance of the placental membrane.

In the early months of pregnancy, the placental membrane is still thick because it is not fully developed. Therefore, its permeability is low. Furthermore, the surface area is slight because the placenta has not grown significantly. Therefore, the total diffusion conductance is minuscule at first. On the other hand, in later pregnancy, the permeability increases because of thinning of the membrane diffusion layers, and the surface area becomes enormous because of growth, thus giving the tremendous increase in placental conductance shown in Figure 82–3.

Rarely, "breaks" occur in the placental membrane, which allow fetal blood cells to pass into the mother or, more rarely, the mother's cells to pass into the fetus. Indeed, there are instances in which the fetus bleeds severely into the mother's circulation because of a ruptured placental membrane.

DIFFUSION OF OXYGEN THROUGH THE PLACENTAL MEMBRANE. Almost the same principles are applicable for the diffusion of oxygen through the placental membrane as for the diffusion of oxygen through the pulmonary membrane; these principles are discussed in detail in Chapter 39. The dissolved oxygen in the blood of the large maternal sinuses passes into the fetal blood by *simple diffusion,* driven by an oxygen pressure gradient from the mother's blood to the fetus's blood. The mean P_{O_2} in the mother's blood in the maternal sinuses is about 50 mm Hg toward the end of pregnancy, and the mean P_{O_2} in the fetal blood after it becomes oxygenated in the placenta is about 30 mm Hg. Therefore, the mean pressure gradient for diffusion of oxygen through the placental membrane is about 20 mm Hg.

One might wonder how it is possible for a fetus to obtain sufficient oxygen when the fetal blood leaving the placenta has a P_{O_2} of only 30 mm Hg. There are three reasons why even this low P_{O_2} is capable of allowing the fetal blood to transport almost as much oxygen to the fetal tissues as is transported by the mother's blood to her tissues.

First, the hemoglobin of the fetus is mainly *fetal hemoglobin,* a type of hemoglobin synthesized in the fetus before birth. Figure 82–5 shows the comparative

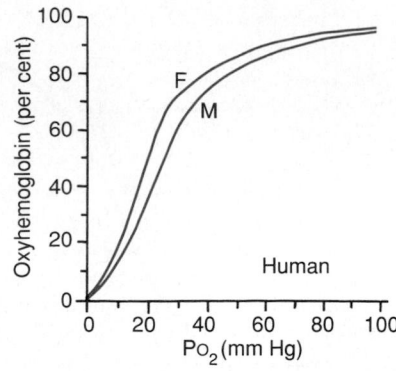

Figure 82–5. Oxygen-hemoglobin dissociation curves for maternal *(M)* and fetal *(F)* blood, showing the ability of the fetal blood to carry a much greater quantity of oxygen than can maternal blood for a given blood P_{O_2}. (From Metcalfe, Moli, and Bartels: *Fed Proc., 23*:775, 1964.)

oxygen dissociation curves of maternal hemoglobin and fetal hemoglobin, demonstrating that the curve for fetal hemoglobin is shifted to the left of that for maternal hemoglobin. This means that at the low P_{O_2} levels in fetal blood, the fetal hemoglobin can carry 20 to 50 per cent more oxygen than can maternal hemoglobin.

Second, the *hemoglobin concentration of fetal blood is about 50 per cent greater than that of the mother;* therefore, this is an even more important factor in enhancing the amount of oxygen transported to the fetal tissues.

Third, the *Bohr effect,* which is explained in relation to the exchange of carbon dioxide and oxygen in the lung in Chapter 40, provides another factor that enhances the transport of oxygen by the fetal blood. That is, hemoglobin can carry more oxygen at a low P_{CO_2} than it can at a high P_{CO_2}. The fetal blood entering the placenta carries large amounts of carbon dioxide, but much of this carbon dioxide diffuses from the fetal blood into the maternal blood. Loss of the carbon dioxide makes the fetal blood more alkaline, whereas the increased carbon dioxide in the maternal blood makes this more acidic. These changes cause the combining capacity of fetal blood for oxygen to become increased and that of the maternal blood to become decreased. This forces more oxygen from the maternal blood while enhancing oxygen uptake by the fetal blood. Thus, the Bohr shift operates in one direction in the maternal blood and in the other in the fetal blood, these two effects making the Bohr shift twice as important here as it is for oxygen exchange in the lungs; therefore, it is called the *double Bohr effect.*

By these three means, the fetus is capable of receiving more than adequate oxygen through the placenta despite the fact that the fetal blood leaving the placenta has a P_{O_2} of only 30 mm Hg.

The total *diffusing capacity* of the entire placenta for oxygen at term is about 1.2 milliliters of oxygen per minute per millimeter of oxygen difference across the membrane. This compares favorably with that of the lungs of the neonate.

DIFFUSION OF CARBON DIOXIDE THROUGH THE PLACENTAL MEMBRANE. Carbon dioxide is continually formed in the tissues of the fetus in the same way that it is formed in maternal tissues. And the only means for excreting the carbon dioxide from the fetus is through the placenta into the mother's blood. The P_{CO_2} of the fetal blood is 2 to 3 mm Hg higher than that of the maternal blood. This small pressure gradient for carbon dioxide across the membrane is sufficient to allow adequate diffusion of carbon dioxide because the extreme solubility of carbon dioxide in the placental membrane allows carbon dioxide to diffuse about 20 times as rapidly as oxygen.

The P_{CO_2} of the mother's blood in the placental sinuses is usually less than the normal value of 40 mm Hg found in the blood of the nonpregnant woman because the estrogens and progesterone of pregnancy overdrive the mother's breathing, thus causing her to blow off excess carbon dioxide from her lungs. This helps keep the P_{CO_2} in the fetus's blood at a low level as well, usually near to the normal value of 40 mm Hg.

DIFFUSION OF FOODSTUFFS THROUGH THE PLACENTAL MEMBRANE. Other metabolic substrates needed by the fetus diffuse into the fetal blood in the same manner as oxygen does. For instance, in the late stages of pregnancy, the fetus often uses as much glucose as the entire body of the mother. To provide this much glucose, the trophoblast cells lining the placental villi provide for *facilitated diffusion* of glucose through the placental membrane. That is, the glucose is transported by carrier molecules in the trophoblast cell membrane. Even so, the glucose level in the fetal blood is still 20 to 30 per cent lower than that in the maternal blood.

Because of the high solubility of fatty acids in cell membranes, these also diffuse from the maternal blood into the fetal blood but more slowly than glucose, so that glucose is preferentially used by the fetus for nutrition. Also, such substances as ketone bodies and potassium, sodium, and chloride ions diffuse from the maternal blood into the fetal blood.

EXCRETION OF WASTE PRODUCTS THROUGH THE PLACENTAL MEMBRANE. In the same manner that carbon dioxide diffuses from the fetal blood into the maternal blood, other excretory products formed in the fetus also diffuse into the maternal blood and then are excreted along with the excretory products of the mother. These include especially the *nonprotein nitrogens* such as *urea, uric acid,* and *creatinine.* The level of urea in the fetal blood is only slightly greater than that in maternal blood because urea diffuses through the placental membrane with considerable ease. On the other hand, creatinine, which does not diffuse as easily, has a concentration gradient of a considerably higher percentage. Therefore, excretion from the fetus occurs mainly, if not entirely, as a result of diffusion

gradients across the placental membrane—that is, because of higher concentrations of the excretory products in the fetal blood than in the maternal blood.

HORMONAL FACTORS IN PREGNANCY

In pregnancy, the placenta forms especially large quantities of *human chorionic gonadotropin, estrogens, progesterone,* and *human chorionic somatomammotropin,* the first three of which, and probably the fourth as well, are all essential to a normal pregnancy.

Human Chorionic Gonadotropin and Its Effect to Cause Persistence of the Corpus Luteum and in Preventing Menstruation

Menstruation normally occurs about 14 days after ovulation, at which time most of the endometrium of the uterus sloughs away from the uterine wall and is expelled to the exterior. If this should happen after an ovum has implanted, the pregnancy would terminate. However, this is prevented by the secretion of human chorionic gonadotropin by the newly developing tissues in the following manner.

Coincidentally with the development of the trophoblast cells from an early fertilized ovum, the hormone *human chorionic gonadotropin* is secreted by the syncytial trophoblast cells into the fluids of the mother, as shown in Figure 82–6. The secretion of this hormone can first be measured in the blood 8 to 9 days after ovulation, shortly after the blastocyst implants in the endometrium. Then the rate of secretion rises rapidly to reach a maximum about 10 to 12 weeks after ovulation and decreases to a much lower value by 16 to 20 weeks after ovulation. It continues at this level for the remainder of pregnancy.

FUNCTION OF HUMAN CHORIONIC GONADOTROPIN. Human chorionic gonadotropin is a glycoprotein having a molecular weight of about 39,000 and much the

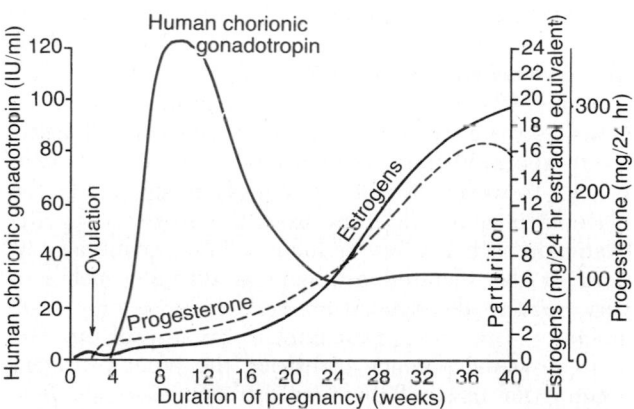

Figure 82–6. Rates of secretion of estrogens, progesterone, and chorionic gonadotropin at different stages of pregnancy.

same molecular structure and function as luteinizing hormone secreted by the pituitary. By far, its most important function is to prevent the normal involution of the corpus luteum at the end of the female sexual cycle. Instead, it causes the corpus luteum to secrete even larger quantities of its usual sex hormones, progesterone and estrogens, for the next few months. These sex hormones prevent menstruation and cause the endometrium to continue growing and to store large amounts of nutrients rather than being shed in the menstruum. As a result, the *decidua-like cells* that develop in the endometrium during the normal female sexual cycle become actual, greatly swollen, and nutritious *decidual cells* at about the time that the blastocyst implants.

Under the influence of human chorionic gonadotropin, the corpus luteum grows to about twice its initial size by a month or so after pregnancy begins, and its continued secretion of estrogens and progesterone maintains the decidual nature of the uterine endometrium, which is necessary for the early development of the fetus. If the corpus luteum is removed before approximately the 7th week of pregnancy, spontaneous abortion almost always occurs, sometimes even up to the 12th week. After that time, the placenta itself secretes sufficient quantities of progesterone and estrogens to maintain pregnancy for the remainder of the gestation period. The corpus luteum involutes slowly after the 13th to 17th week of gestation.

Effect of Human Chorionic Gonadotropin on the Fetal Testes. Human chorionic gonadotropin also exerts an *interstitial cell–stimulating effect* on the testes, thus resulting in the production of testosterone in male fetuses up until the time of birth. This small secretion of testosterone during gestation is what causes the fetus to grow male sex organs instead of female organs. Near the end of pregnancy, the testosterone secreted by the fetal testes also causes the testes to descend into the scrotum.

Secretion of Estrogens by the Placenta

The placenta, like the corpus luteum, secretes both estrogens and progesterone. Both histochemical and physiological studies show that these two hormones, like most other placental hormones, are secreted by the *syncytial trophoblast cells.*

Figure 82–6 shows that the daily production of placental estrogens increases, in terms of estrogenic activity, to about 30 times normal toward the end of pregnancy. However, the secretion of estrogens by the placenta is quite different from the secretion by the ovaries in several ways, as follows: First, quantitatively, most of the secreted estrogens is estriol, which is a very, very weak estrogen and is formed in only small amounts in the nongravid female. Because of the very low estrogenic potency of estriol, the other estrogens account for most of the total estrogenic activity. Second, the estrogens secreted by the placenta are not synthesized de novo from basic substrates in the pla-

centa. Instead, they are formed almost entirely from androgenic steroid compounds, *dehydroepiandrosterone* and *16-hydroxydehydroepiandrosterone,* which are formed both in the mother's adrenal glands and in the adrenal glands of the fetus. These weak androgens are transported by the blood to the placenta and converted by the trophoblast cells into estradiol, estrone, and estriol. (The cortices of the fetal adrenal glands are extremely large, composed about 80 per cent by a so-called *fetal zone,* the primary function of which seems to be to secrete the dehydroepiandrosterone.)

FUNCTION OF ESTROGEN IN PREGNANCY. In the discussions of estrogens in Chapter 81, it is pointed out that these hormones exert mainly a proliferative function on most reproductive and associated organs of the mother. During pregnancy, the extreme quantities of estrogens cause (1) enlargement of the mother's uterus, (2) enlargement of the mother's breasts and growth of the breast ductal structure, and (3) enlargement of the mother's female external genitalia.

The estrogens also relax the various pelvic ligaments of the mother, so that the sacroiliac joints become relatively limber and the symphysis pubis becomes elastic. These changes make for easier passage of the fetus through the birth canal.

There is much reason to believe that estrogens also affect many general aspects of the development of the fetus during pregnancy, for example, by affecting the rate of cell reproduction in the early embryo.

Secretion of Progesterone by the Placenta

Progesterone is also a hormone essential for pregnancy—in fact equally as important as estrogen. In addition to being secreted in moderate quantities by the corpus luteum at the beginning of pregnancy, it is secreted in tremendous quantities by the placenta, averaging about 0.25 gm/day toward the end of pregnancy. Indeed, the rate of progesterone secretion increases by about 10-fold during the course of pregnancy, as shown in Figure 82–6.

The special effects of progesterone that are essential for normal progression of pregnancy and even for its continuance are the following:

1. As pointed out earlier, progesterone causes decidual cells to develop in the uterine endometrium, and these cells then play an important role in the nutrition of the early embryo.

2. Progesterone has a special effect on decreasing the contractility of the pregnant uterus, thus preventing uterine contractions from causing spontaneous abortion.

3. Progesterone contributes to the development of the conceptus even before implantation because it specifically increases the secretions of the fallopian tubes and uterus to provide appropriate nutritive matter for the developing *morula* and *blastocyst.* There is reason to believe, too, that progesterone even affects cell cleavage in the early-developing embryo.

4. The progesterone secreted during pregnancy helps the estrogen prepare the mother's breasts for lactation, which is discussed later in the chapter.

Human Chorionic Somatomammotropin

A newly discovered placental hormone is called *human chorionic somatomammotropin.* This is a protein, having a molecular weight of about 38,000, that begins to be secreted by the placenta at about the 5th week of pregnancy. Secretion of this hormone increases progressively throughout the remainder of pregnancy in direct proportion to the weight of the placenta. Although the functions of chorionic somatomammotropin are uncertain, this hormone is secreted in quantities several times as great as that of all other pregnancy hormones combined. It has several possible important effects.

First, when administered to several types of lower animals, human chorionic somatomammotropin causes at least partial development of the breasts and in some instances causes lactation. Because this was the first function of the hormone discovered, it was at first named *human placental lactogen* and was believed to have functions similar to those of prolactin. However, attempts to promote lactation in the human being with its use have not been successful.

Second, this hormone has weak actions similar to those of growth hormone, causing deposition of protein tissues in the same way as growth hormone does. It also has a chemical structure similar to that of growth hormone, but 100 times as much human chorionic somatomammotropin as growth hormone is required to promote growth.

Third, human chorionic somatomammotropin causes decreased insulin sensitivity and decreased utilization of glucose in the mother, thereby making larger quantities of glucose available to the fetus. Because glucose is the major substrate used by the fetus to energize its growth, the possible importance of such a hormonal effect as this is obvious. Furthermore, the hormone promotes release of free fatty acids from the fat stores of the mother, thus providing this alternative source of energy for her metabolism.

Therefore, it is beginning to appear that human chorionic somatomammotropin is a general metabolic hormone that has specific nutritional implications for both the mother and the fetus.

Other Hormonal Factors in Pregnancy

Almost all the nonsexual endocrine glands of the mother react markedly to pregnancy. This results mainly from the increased metabolic load on the mother but also to some extent from effects of placental hormones on the pituitary and other glands. Some of the most notable effects are the following.

PITUITARY SECRETION. The anterior pituitary gland enlarges at least 50 per cent during pregnancy and increases its production of *corticotropin, thyrotropin,* and *prolactin.* On the other hand, follicle-stimulating hormone and luteinizing hormone secretions are almost suppressed as a result of the inhibitory effects of estrogens and progesterone from the placenta.

CORTICOSTEROID SECRETION. The rate of adrenocortical secretion of the *glucocorticoids* is moderately increased throughout pregnancy. It is possible that the glucocorticoids help mobilize amino acids from the mother's tissues so that these can be used for synthesis of tissues in the fetus.

Pregnant women also usually have about a twofold increase in secretion of *aldosterone,* reaching the peak at the end of gestation. This, along with the actions of the estrogens, causes a tendency for even the normal pregnant woman to reabsorb excess sodium from the renal tubules and, therefore, to retain fluid, occasionally leading to hypertension.

SECRETIONS BY THE THYROID GLAND. The mother's thyroid gland ordinarily enlarges up to 50 per cent during pregnancy and increases its production of thyroxine a corresponding amount. The increased thyroxine production is caused at least partly by a thyrotropic effect of *human chorionic gonadotropin* and by small quantities of a specific thyroid-stimulating hormone, *human chorionic thyrotropin,* secreted by the placenta.

SECRETION BY THE PARATHYROID GLANDS. The mother's parathyroid glands usually enlarge during pregnancy; this is especially true if the mother is on a calcium-deficient diet. Enlargement of these glands causes calcium absorption from the mother's bones, thereby maintaining normal calcium ion concentration in the mother's extracellular fluids as the fetus removes calcium for ossifying its own bones. This secretion of parathyroid hormone is even more intensified during lactation after the birth of the baby because the baby then requires many times more calcium than the fetus.

SECRETION OF "RELAXIN" BY THE OVARIES AND PLACENTA. An additional substance besides the estrogens and progesterone, a hormone called relaxin, is secreted by the corpus luteum of the ovary and by the placenta. Its secretion by the corpus luteum is increased by human chorionic gonadotropin at the same time that the corpus luteum secretes large quantities of estrogen and progesterone.

Relaxin is a polypeptide having a molecular weight of about 9000. This hormone, when injected, causes relaxation of the ligaments of the symphysis pubis in the estrous rat and guinea pig. This effect is poor or possibly even absent in the pregnant woman. Instead, this role is probably played mainly by the estrogens, which also cause relaxation of the pelvic ligaments.

It has also been claimed that relaxin softens the cervix of the pregnant woman at the time of delivery.

RESPONSE OF THE MOTHER'S BODY TO PREGNANCY

Most apparent among the many reactions of the mother to the fetus and to the excessive hormones of pregnancy is the increased size of the various sexual organs. For instance, the uterus increases from about 50 grams to about 1100 grams, and the breasts approximately double in size. At the same time the vagina enlarges and the introitus opens more widely. Also, the various hormones can cause marked changes in the appearance of the woman, sometimes resulting in the development of edema, acne, and masculine or acromegalic features.

Weight Gain in the Pregnant Woman

The average weight gain during pregnancy is about 24 pounds, most of this gain occurring during the last two trimesters. Of this, about 7 pounds is fetus and 4 pounds is amniotic fluid, placenta, and fetal membranes. The uterus increases about 2 pounds and the breasts another 2 pounds, still leaving an average increase in weight of the woman's body of 9 pounds. About 6 pounds of this is extra fluid in the blood and extracellular fluid, and the remaining 3 pounds in general is fat accumulation. The extra fluid is excreted in the urine during the first few days after birth, that is, after loss of the fluid-retaining hormones of the placenta.

Often during pregnancy a woman has a greatly increased desire for food, partly as a result of removal of food substrates from the mother's blood by the fetus and partly because of hormonal factors. Without appropriate prenatal care, the mother's weight gain can be as great as 75 pounds.

Metabolism During Pregnancy

As a consequence of the increased secretion of many hormones during pregnancy, including thyroxine, adrenocortical hormones, and the sex hormones, the basal metabolic rate of the pregnant woman increases about 15 per cent during the latter half of the pregnancy. As a result, she frequently has sensations of becoming overheated. Also, owing to the extra load that she is carrying, greater amounts of energy than normal must be expended for muscle activity.

NUTRITION DURING PREGNANCY. By far the greatest growth of the fetus occurs during the last trimester of pregnancy; its weight almost doubles during the last 2 months of pregnancy. Ordinarily, from her diet, the mother does not absorb sufficient protein, calcium, phosphates, and iron from the gastrointestinal tract during the last months of pregnancy to supply the fetus. However, anticipating these extra needs toward the end of pregnancy, the mother's body has been storing these substances, some in the placenta but most in the normal storage depots of the mother.

If appropriate nutritional elements are not present in the pregnant woman's diet, a number of maternal deficiencies can occur. Deficiencies often occur in calcium, phosphates, iron, and the vitamins. For example, about 375 milligrams of iron is needed by the fetus to form its blood, and an additional 600 milligrams is needed by the mother to form her own extra blood. The normal store of nonhemoglobin iron in the mother at the outset of pregnancy is often only 100 or so milligrams and almost never more than 700 milligrams. Therefore, without sufficient iron in her food, a pregnant woman usually develops anemia. Also, it is especially important that she receive vitamin D because although the total quantity of calcium used by the fetus is small, calcium even normally is poorly absorbed by the gastrointestinal tract. Finally, shortly before birth of the baby, vitamin K is often added to the mother's diet so that the baby will have sufficient prothrombin to prevent hemorrhage, particularly brain hemorrhage, caused by the birth process.

Changes in the Maternal Circulatory System During Pregnancy

BLOOD FLOW THROUGH THE PLACENTA AND CARDIAC OUTPUT DURING PREGNANCY. About 625 milliliters of blood flows through the maternal circulation of the placenta each minute during the latter phases of pregnancy. This, plus the general increase in metabolism, increases the mother's cardiac output to 30 to 40 per cent above normal by the 27th week of pregnancy, but then, for reasons unexplained, the cardiac output falls to only a little above normal during the last 8 weeks of pregnancy, despite the high uterine blood flow.

BLOOD VOLUME DURING PREGNANCY. The maternal blood volume shortly before term is about 30 per cent above normal. This increase occurs mainly during the latter half of pregnancy, as shown by the curve of Figure 82-7. The cause of the increased volume is mainly hormonal because both aldosterone and estrogens, which are greatly increased in pregnancy, cause increased fluid retention by the kidneys. Also, the bone marrow becomes increasingly active and produces extra red blood cells to go with the excess fluid volume. Therefore, at the time of birth of the baby, the mother has about 1 to 2 liters of extra blood in her circulatory system. Only about one fourth of this amount is normally lost through bleeding during delivery of the baby, thereby allowing a considerable safety factor for the mother.

Respiration During Pregnancy

Because of the increased basal metabolic rate of the pregnant woman and because of her increase in size, the total amount of oxygen used by the mother shortly before birth of the baby is about 20 per cent above normal, and a commensurate amount of carbon dioxide is formed. These effects cause the mother's minute ventilation to increase. It is also believed that the high levels of progesterone during pregnancy increase the minute ventilation still more because progesterone increases the sensitivity of the respiratory center to carbon dioxide. The net result is an increase in minute ventilation of about 50 per cent and a decrease in arterial P_{CO_2} to several millimeters of mercury below that of the normal woman. Simultaneously, the growing uterus presses upward against the abdominal contents, and these in turn press upward against the diaphragm, so that the total excursion of the diaphragm is decreased. Consequently, the respiratory rate is increased to maintain the extra ventilation.

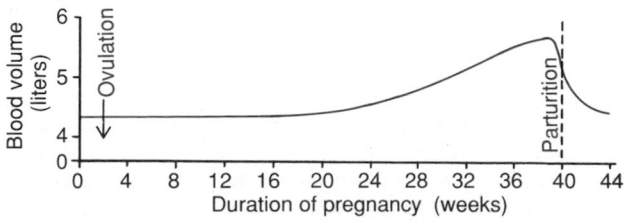

Figure 82-7. Effect of pregnancy on blood volume.

Function of the Maternal Urinary System During Pregnancy

The rate of urine formation by the pregnant woman is usually slightly increased because of increased fluid intake and increased load of excretory products. But in addition, several special alterations of urinary function occur.

First, reabsorptive capacity for sodium, chloride, and water by the renal tubules is increased as much as 50 per cent as a consequence of increased production of steroid hormones by the placenta and adrenal cortex.

Second, the glomerular filtration rate also increases as much as 50 per cent during pregnancy, which tends to increase the rate of water and electrolyte loss in the urine. This factor normally almost balances the first, so that the mother ordinarily accumulates only about 6 pounds of extra water and salt except when she develops *preeclampsia;* this condition is discussed later in the chapter.

Amniotic Fluid and Its Formation

Normally, the volume of amniotic fluid is between 500 milliliters and 1 liter, but it can be only a few milliliters or as much as several liters. Studies with isotopes on the rate of formation of amniotic fluid show that on the average, the water in amniotic fluid is replaced once every 3 hours, and the electrolytes sodium and potassium are replaced an average of once every 15 hours. A portion of the fluid is derived from renal excretion by the fetus. Likewise, a certain amount of absorption occurs by way of the gastrointestinal tract and lungs of the fetus. However, even after in utero death of a fetus, the rate of turnover of the amniotic fluid is still one-half as great as it is when the fetus is normal, which indicates that much of the fluid is formed and absorbed directly through the amniotic membranes. The total volume of amniotic fluid is probably regulated mainly by the amniotic membranes, but the volume also increases when the fetal excretion of urine increases and decreases when there is no urine output.

Preeclampsia and Eclampsia

About 4 per cent of all pregnant women experience a rapid rise in arterial blood pressure associated with loss of large amounts of protein in the urine during the latter few months of pregnancy. This condition is called *preeclampsia* or *toxemia of pregnancy.* It is often also characterized by salt and water retention by the mother's kidneys, weight gain, and development of edema. In addition, arterial spasm occurs in many parts of the mother's body, most significantly in the kidneys, brain, and liver. Both the renal blood flow and the glomerular filtration rate are decreased, which is exactly opposite to the changes that occur in the normal pregnant woman. The renal effects are caused by thickened glomerular tufts that contain a protein deposit in the basement membranes.

Various attempts have been made to prove that preeclampsia is caused by excessive secretion of placental or adrenal hormones, but proof of a hormonal basis is still lacking. Indeed, another plausible theory is that preeclampsia results from some type of autoimmunity or allergy resulting from the presence of the fetus. In support of this, the acute symptoms usually disappear within a few days after birth of the baby.

The severity of preeclampsia symptoms is closely associated with an accompanying increase in arterial pressure. In fact, an increasing pressure seems to set off a vicious cycle that intensifies the arterial spasm and other pathological effects of preeclampsia.

Eclampsia is an extreme degree of the same effects as those observed in preeclampsia, characterized by extreme vascular spasm throughout the body, clonic seizures sometimes followed by coma, greatly decreased kidney output, malfunction of the liver, often extreme hypertension, and a generalized toxic condition of the body. It usually occurs shortly before parturition. Without treatment, a high percentage of eclamptic patients die. However, with optimal and immediate use of rapidly acting vasodilating drugs to reduce the arterial pressure to normal, followed by immediate termination of pregnancy—by cesarean section if necessary—the mortality has been reduced to 1 per cent or less.

PARTURITION

Increased Uterine Excitability Near Term

Parturition means simply the process by which the baby is born. Toward the end of pregnancy, the uterus becomes progressively more excitable until finally it begins strong rhythmical contractions with such force that the baby is expelled. The exact cause of the increased activity of the uterus is not known, but at least two major categories of effects lead up to the culminating contractions responsible for parturition: first, progressive hormonal changes that cause increased excitability of the uterine musculature and, second, progressive mechanical changes.

HORMONAL FACTORS THAT CAUSE INCREASED UTERINE CONTRACTILITY

Ratio of Estrogens to Progesterone. Progesterone inhibits uterine contractility during pregnancy, thereby helping to prevent expulsion of the fetus. On the other hand, estrogens have a definite tendency to increase the degree of uterine contractility, at least partly because estrogens increase the number of gap junctions between the adjacent uterine smooth muscle cells. Both progesterone and estrogen are secreted in progressively greater quantities throughout most of pregnancy, but from the 7th month onward estrogen secretion continues to increase while progesterone secretion remains constant or perhaps even decreases slightly. Therefore, it has been postulated that the *estrogen-to-progesterone ratio* increases sufficiently toward the end of pregnancy to be at least partly responsible for the increased contractility of the uterus.

Effect of Oxytocin on the Uterus. Oxytocin is a hormone secreted by the neurohypophysis that specifically causes uterine contraction (see Chap. 75). There are four reasons for believing that oxytocin might be important in increasing the contractility of the uterus near term: (1) The uterine muscle increases its oxytocin receptors and, therefore, increases its responsiveness to a given dose of oxytocin during the latter few months of pregnancy. (2) The rate of oxytocin secretion by the neurohypophysis is considerably increased at the time of labor. (3) Although hypophysectomized animals and human beings can still deliver their young at term, labor is prolonged. (4) Experiments in animals indicate that irritation or stretching of the uterine cervix, as occurs during labor, can cause a neurogenic reflex through the paraventricular and supraoptic nuclei of the hypothalamus that causes the posterior pituitary gland (the neurohypophysis) to increase its secretion of oxytocin.

Effect of Fetal Hormones on the Uterus. The fetus's pituitary gland also secretes increasing quantities of oxytocin that could possibly play a role in exciting the uterus, and its adrenal gland secretes large quantities of cortisol that are also a possible uterine stimulant. In addition, the fetal membranes release prostaglandins in high concentration at the time of labor. These, too, can increase the intensity of the uterine contractions.

MECHANICAL FACTORS THAT INCREASE THE CONTRACTILITY OF THE UTERUS

Stretch of the Uterine Musculature. Simply stretching smooth muscle organs usually increases their contractility. Furthermore, intermittent stretch, as occurs repetitively in the uterus because of movements of the fetus, can also elicit smooth muscle contraction.

Note especially that twins are born on the average *19 days* earlier than a single child, which emphasizes the probable importance of mechanical stretch in eliciting uterine contractions.

Stretch or Irritation of the Cervix. There is much reason to believe that stretching or irritation of the uterine cervix is particularly important in eliciting uterine contractions. For instance, the obstetrician frequently induces labor by rupturing the membranes so that the head of the baby stretches the cervix more forcefully than usual or irritates it in other ways.

The mechanism by which cervical irritation excites the body of the uterus is not known. It has been suggested that stretching or irritation of nerves in the cervix initiates reflexes to the body of the uterus, but the effect could also result simply from myogenic transmission of signals from the cervix to the body of the uterus.

Onset of Labor—A Positive Feedback Theory for Its Initiation

During most of the months of pregnancy, the uterus undergoes periodic episodes of weak and slow rhythmical contractions called *Braxton Hicks contractions*. These contractions become progressively stronger toward the end of pregnancy; then they change rather suddenly, within hours, to become exceptionally strong contractions that start stretching the cervix and later force the baby through the birth canal, thereby causing

parturition. This process is called *labor*, and the strong contractions that result in final parturition are called *labor contractions.*

Yet, strangely enough, we do not know what suddenly changes the slow and weak rhythmicity of the uterus into the strong labor contractions. However, on the basis of experience during the past few years with other types of physiological control systems, a theory has been proposed for explaining the onset of labor based on positive feedback. This theory suggests that stretching of the cervix by the fetus's head finally becomes great enough to elicit strong reflex increase in contractility of the uterine body. This pushes the baby forward, which stretches the cervix still more and initiates still more positive feedback to the uterine body. Thus, the process continues again and again until the baby is expelled. This theory is shown in Figure 82–8, and the data supporting it are the following.

First, labor contractions obey all the principles of positive feedback. That is, once the strength of uterine contraction becomes greater than a critical value, each contraction leads to subsequent contractions that become stronger and stronger until maximum effect is achieved. Referring to the discussion in Chapter 1 of positive feedback in control systems, one will see that this is the precise nature of all positive feedback mechanisms when the feedback gain becomes greater than a critical value.

Second, two known types of positive feedback increase uterine contractions during labor: (1) Stretching of the cervix causes the entire body of the uterus to contract, and this contraction stretches the cervix still more because of the downward thrust of the baby's head. (2) Cervical stretching also causes the pituitary gland to secrete oxytocin, which is still another means for increasing uterine contractility.

To summarize the theory, we can assume that multiple factors increase the contractility of the uterus toward the end of pregnancy. Eventually a uterine

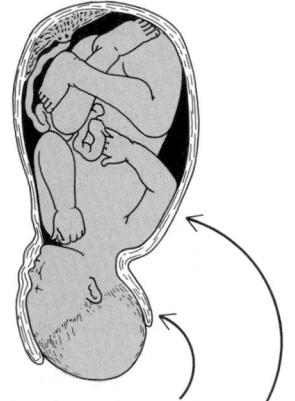

1. Baby's head stretches cervix...
2. Cervical stretch excites fundic contraction
3. Fundic contraction pushes baby down and stretches cervix some more...
4. Cycle repeats over and over again...

Figure 82–8. Theory for the onset of intensely strong contractions during labor.

contraction becomes strong enough to irritate the uterus, especially at the cervix, and this increases uterine contractility still more because of positive feedback, resulting in a second uterine contraction stronger than the first, a third stronger than the second, and so forth. Once these contractions become strong enough to cause this type of feedback, with each succeeding contraction greater than the one preceding, the process proceeds to completion—all simply *because positive feedback initiates a vicious circle when the gain of the feedback is greater than a critical level.*

One might immediately ask about the many instances of false labor in which the contractions become stronger and stronger and then fade away. Remember that for a vicious cycle to continue, *each* new cycle of the positive feedback must be stronger than the previous one. If at any time after labor starts some contractions fail to re-excite the uterus sufficiently, the positive feedback could go into a retrograde decline and the labor contractions would fade away.

Abdominal Muscle Contractions During Labor

Once the uterine contractions become strong during labor, pain signals then originate both from the uterus itself and from the birth canal. These signals, in addition to causing suffering, elicit neurogenic reflexes from the spinal cord to the abdominal muscles, causing intense contractions of these muscles. The abdominal contractions in turn add greatly to the force that causes expulsion of the baby.

Mechanics of Parturition

The uterine contractions during labor begin at the top of the uterine fundus and spread downward over the body of the uterus. Also, the intensity of contraction is great in the top and body of the uterus but weak in the lower segment of the uterus adjacent to the cervix. Therefore, each uterine contraction tends to force the baby downward toward the cervix.

In the early part of labor, the contractions might occur only once every 30 minutes. As labor progresses, the contractions finally appear as often as once every 1 to 3 minutes, and the intensity of contraction increases greatly, with only a short period of relaxation between contractions.

The combined contractions of the uterine and abdominal musculature during delivery of the baby cause downward force on the fetus of about 25 pounds during each strong contraction.

It is fortunate that the contractions of labor occur intermittently because strong contractions impede or sometimes even stop blood flow through the placenta and would cause death of the fetus were the contractions continuous. Indeed, in clinical use of various uterine stimulants, such as oxytocin, overuse of the drugs can cause uterine spasm rather than rhythmical contractions and can lead to death of the fetus.

In 19 of 20 births, the head is the first part of the baby to be expelled, and in most of the remaining instances, the buttocks are presented first. The head acts as a wedge to open the structures of the birth canal as the fetus is forced downward.

The first major obstruction to expulsion of the fetus is the uterine cervix. Toward the end of pregnancy, the cervix becomes soft, which allows it to stretch when labor contractions begin in the uterus. The so-called *first stage of labor* is the period of progressive cervical dilatation, lasting until the opening is as large as the head of the fetus. This stage usually lasts for 8 to 24 hours in the first pregnancy but often only a few minutes after many pregnancies.

Once the cervix has dilated fully, the fetal membranes usually rupture and the amniotic fluid is lost suddenly through the vagina. Then the fetus's head moves rapidly into the birth canal, and with additional force from above, it continues to wedge its way through the canal until delivery is effected. This is called the *second stage of labor,* and it may last from as little as 1 minute after many pregnancies to 30 minutes or more in the first pregnancy.

Separation and Delivery of the Placenta

During the succeeding 10 to 45 minutes after birth of the baby, the uterus contracts to a small size, which causes a *shearing* effect between the walls of the uterus and the placenta, thus separating the placenta from its implantation site. Separation of the placenta opens the placental sinuses and causes bleeding. The amount of bleeding is limited to an average of 350 milliliters by the following mechanism: The smooth muscle fibers of the uterine musculature are arranged in figures of 8 around the blood vessels as the vessels pass through the uterine wall. Therefore, contraction of the uterus after delivery of the baby constricts the vessels that had previously supplied blood to the placenta. In addition, it is believed that vasoconstrictor prostaglandins formed at the placental separation site cause additional blood vessel spasm.

Labor Pains

With each uterine contraction, the mother experiences considerable pain. The cramping pain in early labor is probably caused mainly by hypoxia of the uterine muscle resulting from compression of the blood vessels to the uterus. This pain is not felt when the visceral sensory *hypogastric nerves,* which carry the visceral sensory fibers leading from the uterus, have been sectioned. However, during the second stage of labor, when the fetus is being expelled through the birth canal, much more severe pain is caused by cervical stretching, perineal stretching, and stretching or tearing of structures in the vaginal canal itself. This pain is conducted by somatic nerves instead of by the visceral sensory nerves.

Involution of the Uterus

During the first 4 to 5 weeks after parturition, the uterus involutes. Its weight becomes less than one-half its immediate postpartum weight within 1 week, and in 4 weeks, if the mother lactates, the uterus may be as small as it had been before pregnancy. This effect of lactation results from the suppression of pituitary gonadotropin and ovarian hormone secretion during the first few months of lactation, as discussed later. During early involution of the uterus, the placental site on the endometrial surface autolyzes, causing a vaginal discharge known as "lochia," which is first bloody and then serous in nature, continuing in all for about 10 days. After this time, the endometrial surface will have become re-epithelialized and ready for normal, nongravid sex life again.

LACTATION

Development of the Breasts

The breasts, shown in Figure 82–9, begin to develop at puberty; this development is stimulated by the same estrogens of the monthly sexual cycles; they stimulate growth of the breast's *mammary gland* plus deposition of fat to give mass to the breasts. In addition, far greater growth occurs during pregnancy, and the glandular tissue only then becomes completely developed for production of milk.

GROWTH OF THE DUCTAL SYSTEM—ROLE OF THE ESTROGENS. All through pregnancy, the tremendous quantities of estrogens secreted by the placenta cause the ductal system of the breasts to grow and branch. Simultaneously, the stroma of the breasts increases in quantity, and large quantities of fat are laid down in the stroma.

Also important in growth of the ductal system are at least four other hormones: *growth hormone, prolactin,* the *adrenal glucocorticoids,* and *insulin.* Each of these is known to play at least some role in protein metabolism, which presumably explains their function in the development of the breasts.

DEVELOPMENT OF THE LOBULE-ALVEOLAR SYSTEM — ROLE OF PROGESTERONE. Final development of the breasts into milk-secreting organs also requires progesterone. Once the ductal system has developed, progesterone, acting synergistically especially with estrogen but also with all the other hormones just mentioned, causes additional growth of the lobules, budding of alveoli, and development of secretory characteristics in the cells of the alveoli. These changes are analogous to

the secretory effects of progesterone on the endometrium of the uterus during the latter half of the female menstrual cycle.

Initiation of Lactation—Function of Prolactin

Although estrogen and progesterone are essential for the physical development of the breasts during pregnancy, a specific effect of both these hormones is to inhibit *the actual secretion of milk.* On the other hand, the hormone *prolactin* has exactly the opposite effect, promotion of milk secretion. This hormone is secreted by the mother's pituitary gland, and its concentration in her blood rises steadily from the 5th week of pregnancy until birth of the baby, at which time it has risen to 10 to 20 times the normal nonpregnant level. This high level of prolactin at the end of pregnancy is shown in Figure 82–10. In addition, the placenta secretes large quantities of *human chorionic somatomammotropin,* which probably also has mild lactogenic properties, thus supporting the prolactin from the mother's pituitary. Even so, because of the suppressive effects on the breasts of estrogen and progesterone, never more than a few milliliters of fluid are secreted each day until after the baby is born. The fluid that is secreted the last few days or weeks before parturition is called *colostrum;* it contains essentially the same concentrations of proteins and lactose as milk but almost no fat, and its maximum rate of production is about 1/100 the subsequent rate of milk production.

Immediately after the baby is born, the sudden loss of both estrogen and progesterone secretion by the placenta now allows the lactogenic effect of the prolactin from the mother's pituitary gland to assume its natural milk-promoting role, and over the next 1 to 7 days, the breasts progressively begin to secrete copious quantities of milk instead of colostrum. This secretion of milk requires an adequate background secretion of

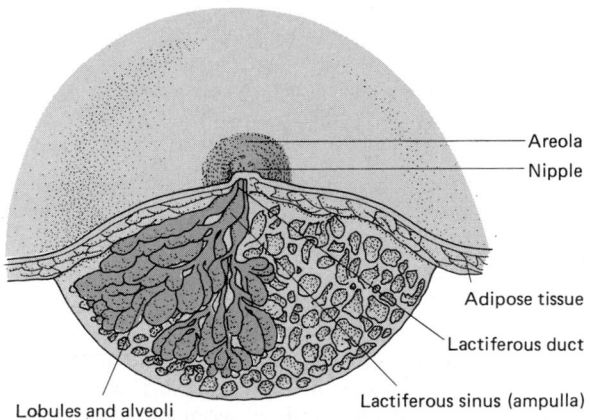

Figure 82–9. The breast and its mammary gland.

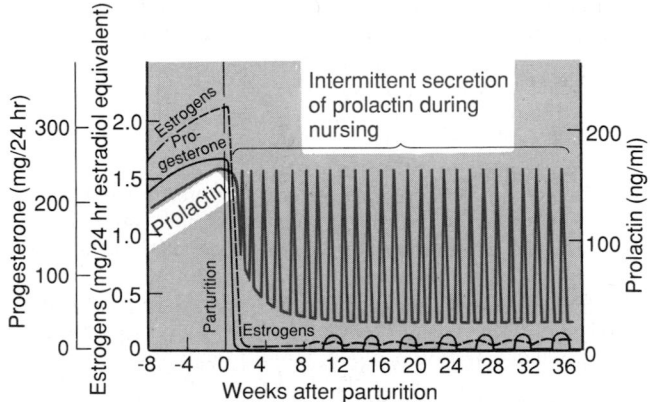

Figure 82–10. Changes in rates of secretion of estrogens, progesterone, and prolactin for 8 weeks before parturition and for 36 weeks thereafter. Note especially the decrease of prolactin secretion back to basal levels within a few weeks but also the intermittent periods of marked prolactin secretion (for about 1 hour at a time) during and after periods of nursing.

most of the mother's other hormones as well, but most important of all are *growth hormone, cortisol, parathyroid hormone,* and *insulin.* These hormones are necessary to provide the amino acids, fatty acids, glucose, and calcium that are required for milk formation.

After birth of the baby, the *basal level* of prolactin secretion returns during the next few weeks to the nonpregnant level, as shown in Figure 82–10. However, each time the mother nurses her baby, nervous signals from the nipples to the hypothalamus cause a 10- to 20-fold surge in prolactin secretion that lasts for about 1 hour, which is also shown in the figure. This prolactin in turn acts on the breasts to keep the mammary glands secreting milk into the alveoli for the subsequent nursing periods. If this prolactin surge is absent or blocked as a result of hypothalamic or pituitary damage or if nursing does not continue, the breasts lose their ability to produce milk within 1 week or so. However, milk production can continue for several years if the child continues to suckle, although the rate of milk formation normally decreases considerably after 7 to 9 months.

HYPOTHALAMIC CONTROL OF PROLACTIN SECRETION. The hypothalamus plays an essential role in controlling prolactin secretion, as it does also for the control of secretion of almost all the other anterior pituitary hormones. However, this control is different in one aspect: the hypothalamus mainly *stimulates* the production of all the other hormones, but it mainly *inhibits* prolactin production. Consequently, damage to the hypothalamus or blockage of the hypothalamic-hypophysial portal system increases prolactin secretion while it depresses secretion of the other anterior pituitary hormones.

Therefore, it is believed that anterior pituitary secretion of prolactin is controlled either entirely or almost entirely by an inhibitory factor formed in the hypothalamus and transported through the hypothalamic-hypophysial portal system to the anterior pituitary gland. This factor is called *prolactin inhibitory hormone.* It is almost certainly the catecholamine *dopamine,* which is known to be secreted in the arcuate nuclei of the hypothalamus and which can decrease prolactin secretion as much as 10-fold.

SUPPRESSION OF THE FEMALE OVARIAN CYCLES IN NURSING MOTHERS FOR MANY MONTHS AFTER DELIVERY. In most nursing mothers, the ovarian cycle and ovulation do not resume until a few weeks after cessation of nursing the baby. The cause of this seems to be that the same nervous signals from the breasts to the hypothalamus that cause prolactin secretion during suckling, either because of the nervous signals themselves or because of a subsequent effect of the increased prolactin, inhibit secretion of gonadotropin-releasing hormone by the hypothalamus, which in turn suppresses formation of the pituitary gonadotropic hormones, luteinizing hormone and follicle-stimulating hormone. Yet after several months of lactation, in some mothers, especially in those who nurse their babies only part of the time, the pituitary begins again to secrete sufficient gonadotropic hormones to rein-

state the monthly sexual cycle even though nursing continues.

Ejection (or "Let-Down") Process in Milk Secretion—Function of Oxytocin

Milk is secreted continuously into the alveoli of the breasts, but milk does not flow easily from the alveoli into the ductal system and, therefore, does not continually leak from the breast nipples. Instead, the milk must be ejected from the alveoli and into the ducts before the baby can obtain it. This process is called "let-down" of the milk. It is caused by a combined neurogenic and hormonal reflex that involves the posterior pituitary hormone *oxytocin.*

When the baby first suckles, it receives virtually no milk. Instead, sensory impulses must first be transmitted through somatic nerves from the nipples to the spinal cord and then to the hypothalamus, there causing *oxytocin* secretion at the same time that they cause prolactin secretion. The oxytocin is carried in the blood to the breasts, where it causes the *myoepithelial cells* that surround the outer walls of the alveoli to contract, thereby expressing the milk from the alveoli into the ducts at a pressure of plus 10 to 20 mm Hg. Then the baby's suckling does become effective in removing the milk. Thus, within 30 seconds to 1 minute after a baby begins to suckle, milk begins to flow. This process is called *milk ejection* or *milk let-down.*

Suckling on one breast causes milk flow not only in that breast but also in the opposite breast. It is especially interesting that fondling of the baby by the mother or hearing the baby crying also often gives enough of an emotional signal to her hypothalamus to cause milk ejection.

INHIBITION OF MILK EJECTION. A particular problem in nursing the baby comes from the fact that many psychogenic factors or generalized sympathetic stimulation throughout the body can inhibit oxytocin secretion and consequently depress milk ejection. For this reason, many mothers must have an undisturbed puerperium if they are to be successful in nursing their babies.

Milk Composition and the Metabolic Drain on the Mother Caused by Lactation

Table 82–1 lists the contents of human milk and cow's milk. The concentration of lactose in human milk is about 50 per cent greater than in cow's milk, but on the other hand, the concentration of protein in cow's milk is ordinarily two or more times as great as in human milk. Finally, the ash, which contains the calcium and other minerals, is only one-third as much in human milk as in cow's milk.

At the height of lactation, 1.5 liters of milk may be formed each day (and even more if the mother has twins). With this degree of lactation, great quantities of metabolic substrates are drained from the mother. For instance, about 50 grams of fat then enter the milk

Table 82–1 PERCENTAGE COMPOSITION OF MILK

	Human Milk	Cow's Milk
Water	88.5	87
Fat	3.3	3.5
Lactose	6.8	4.8
Casein	0.9	2.7
Lactalbumin and other protein	0.4	0.7
Ash	0.2	0.7

each day, and about 100 grams of lactose, which must be derived from glucose, are lost from the mother each day. Also, 2 to 3 grams of calcium phosphate may be lost each day; unless the mother is drinking large quantities of milk and has an adequate intake of vitamin D, the output of calcium and phosphate by the lactating mammae will often be much greater than the intake of these substances. To supply the needed calcium and phosphate, the parathyroid glands enlarge greatly, and the bones become progressively decalcified. The problem of bone decalcification is usually not great during pregnancy, but it can be a distinct problem during lactation.

REFERENCES

Ada, G. L., and Griffin, P. D.: Vaccines for Fertility Regulation: The Assessment of Their Safety and Efficacy. New York, Cambridge University Press, 1991.

Ben-Jonathan, N., et al.: Suckling-induced rise in prolactin: Mediation by prolactin-releasing factor from posterior pituitary. News Physiol. Sci., 3:172, 1988.

Benrubi, G. I.: Obstetric and Gynecologic Emergencies. Philadelphia, J. B. Lippincott, 1994.

Blackburn, S. T., and Loper, D. L.: Maternal, Fetal, and Neonatal Physiology: A Clinical Perspective. Philadelphia, W. B. Saunders Co., 1992.

Bland, K. I., and Copeland, III, E. M.: The Breast: Comprehensive Management of Benign and Malignant Diseases. Philadelphia, W. B. Saunders Co., 1991.

Briggs, G. G.: Drugs in Pregnancy and Lactation: A Guide to Fetal and Neonatal Risk. Baltimore, Williams & Wilkins, 1994.

Cherry, S. H., and Merkatz, I. R.: Complications of Pregnancy: Medical, Surgical, Gynecologic, Psychosocial, and Perinatal. Baltimore, Williams & Wilkins, 1991.

Compel, C., and Silverberg, S. G.: Pathology in Gynecology and Obstetrics. Philadelphia, J. B. Lippincott, 1994.

Conn, P. M., et al.: Gonadotropin-releasing hormone: Molecular and cell biology, physiology, and clinical applications. Fed. Proc., 43:2351, 1984.

Cowan, B.: Reproductive Endocrinology. Philadelphia, J. B. Lippincott, 1994.

Creasy, R. K., and Resnik, R.: Maternal-Fetal Medicine: Principles and Practice. Philadelphia, W. B. Saunders Co., 1994.

Dimmick, J. E., and Kalousek, D. K.: Developmental Pathology of the Embryo and Fetus. Philadelphia, J. B. Lippincott, 1992.

Hacker, N. F., and Moore, J. G.: Essentials of Obstetrics and Gynecology. Philadelphia, W. B. Saunders Co., 1992.

Hanson, L. A.: Biology of Human Milk. New York, Raven Press, 1988.

Harris, J. R., et al.: Breast Diseases. Philadelphia, J. B. Lippincott, 1991.

Jaffe, L. A., and Cross, N. L.: Electrical regulation of sperm-egg fusion. Annu. Rev. Physiol., 48:191, 1986.

Jones, H. W. Jr., et al. (eds.): In Vitro Fertilization. Baltimore, Williams & Wilkins, 1986.

Kaufmann, P., and Miller, R. K., (eds.): Placental Vascularization and Blood Flow. New York, Plenum Publishing Corp., 1988.

Keyes, P. L., and Wiltbank, M. C.: Endocrine regulation of the corpus luteum. Annu. Rev. Physiol., 50:465, 1988.

Knobil, E., and Neill, J. D.: The Physiology of Reproduction. New York, Raven Press, 1994.

Leung, P. C. K., et al. (eds.): Endocrinology and Physiology of Reproduction. New York, Plenum Publishing Corp., 1987.

Lotgering, F. K., et al.: Maternal and fetal responses to exercise during pregnancy. Physiol. Rev., 65:1, 1985.

Mackay, E. V., et al.: Illustrated Textbook of Gynaecology. Philadelphia, W. B. Saunders Co., 1992.

Mitchell, Jr., G. W., and Bassett, L.: The Female Breast and Its Disorders. Baltimore, Williams & Wilkins, 1990.

Moore, T. R., et al.: Gynecology and Obstetrics: A Longitudinal Approach. New York, Churchill Livingstone, 1993.

Neville, M. C., and Daniel, C. W. (eds.): The Mammary Gland. New York, Plenum Publishing Corp., 1987.

Olsson, K.: Pregnancy—A challenge to water balance. News Physiol. Sci., 1:131, 1986.

Peaker, M., and Wilde, C. J.: Milk secretion: Autocrine control. News Physiol. Sci., 2:124, 1987.

Pitkin, R. M.: Calcium metabolism in pregnancy and the perinatal period: A review. Am J. Obstet. Gynecol., 151:99, 1985.

Sampson, D. A., and Jansen, G. R.: Protein and energy nutrition during lactation. Annu. Rev. Nutr., 4:43, 1984.

Scott, J., et al.: Danforth's Obstetrics and Gynecology. Philadelphia, J. B. Lippincott, 1994.

Seibel, M. M., and Blackwell, R. E.: Ovulation Induction. New York, Raven Press, 1994.

Speroff, L., et al.: Clinical Gynecologic Endocrinology and Infertility. Baltimore, Williams & Wilkins, 1994.

Stern, L., et al.: Physiologic Foundations of Perinatal Care. New York, Elsevier Science Publishing Co., 1989.

Thorburn, G. D., and Challis, J. R. G.: Endocrine control of parturition. Physiol. Rev., 59:863, 1979.

Tulchinsky, D., and Little, A. B.: Maternal-Fetal Endocrinology. Philadelphia, W. B. Saunders Co., 1994.

Vonderhaar, B. K., and Ziska, S. E.: Hormonal regulation of milk protein gene expression. Annu. Rev. Physiol., 51:641, 1989.

Wintour, E. M.: Amniotic fluid—Our first environment. News Physiol. Sci., 1:95, 1986.

Wolf, D. P., et al. (eds.): In Vitro Fertilization and Embryo Transfer. New York, Plenum Publishing Corp., 1988.

Yen, S. C., and Jaffe, R. B.: Reproductive Endocrinology: Physiology, Pathophysiology, and Clinical Management. Philadelphia, W. B. Saunders Co., 1991.

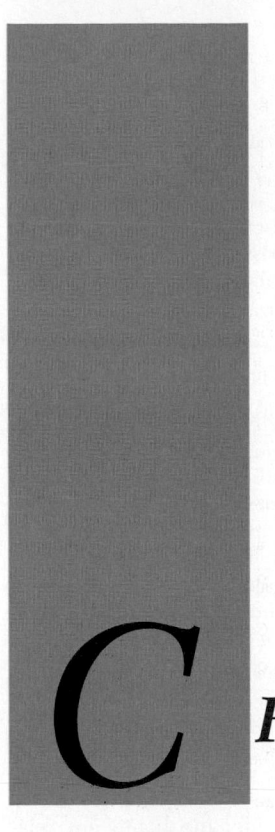

Fetal and Neonatal Physiology

*C*HAPTER 83

A complete discussion of fetal development, functioning of the child immediately after birth, and growth and development through the early years of life lies within the province of formal courses in obstetrics and pediatrics. However, many aspects of these are strictly physiological problems, some of which relate to the physiological principles that we have discussed for the adult and some of which are peculiar to the infant itself. This chapter discusses some of the more important of these special problems.

GROWTH AND FUNCTIONAL DEVELOPMENT OF THE FETUS

Initial development of the placenta and fetal membranes occurs far more rapidly than development of the fetus itself. In fact, during the first 2 to 3 weeks after implantation of the blastocyst, the fetus remains almost microscopic in size, but thereafter, as shown in Figure 83–1, the length of the fetus increases almost in proportion to age. At 12 weeks, the length is about 10 centimeters; at 20 weeks, 25 centimeters; and at term (40 weeks), 53 centimeters (about 21 inches). Because the weight of the fetus is proportional to the cube of the length, the weight increases approximately in proportion to the cube of the age of the fetus. Note in Figure 83–1 that the weight remains minuscule during the first 12 weeks and reaches 1 pound only at 23 weeks (5½ months) of gestation. Then, during the last trimester of pregnancy, the fetus gains tremendously, so that 2 months before birth, the weight averages 3 pounds, 1 month before birth 4.5 pounds, and at birth 7 pounds—the birth weight varying from as low as 4.5 pounds to as high as 11 pounds in normal infants with normal gestational periods.

Development of the Organ Systems

Within 1 month after fertilization of the ovum, the gross characteristics of all the different organs of the fetus have already begun to develop, and during the next 2 to 3 months, most of the details of the different organs are established. Beyond the 4th month, the organs of the fetus are mainly the same as those of the neonate. However, cellular development in each organ is usually far from complete at this time and requires the full remaining 5 months of pregnancy for complete development. Even at birth, certain structures, particularly the nervous system, the kidneys, and the liver, lack full development, which is discussed in more detail later in the chapter.

Circulatory System

The human heart begins beating during the 4th week after fertilization, contracting at the rate of about 65 beats per minute. This increases to a rate of about 140 beats per minute immediately before birth.

FORMATION OF BLOOD CELLS. Nucleated red blood cells begin to be formed in the yolk sac and mesothelial layers of the placenta at about the 3rd week of fetal development. This is followed 1 week later (at 4 to 5 weeks) by the formation of non-nucleated red blood cells by the fetal mesenchyme and by the endothelium of the fetal blood vessels. Then, at about 6 weeks, the liver begins to form blood cells, and in the 3rd month, the spleen and other lymphoid tissues of the body also begin forming blood cells. Finally, from about the 3rd month on, the bone marrow also forms red and white blood cells. During the midportion of fetal life, the extramarrow areas are the major sources of the fetus's blood cells; during the latter 3 months of fetal life, the bone marrow gradually takes over while the other blood cell–

1047

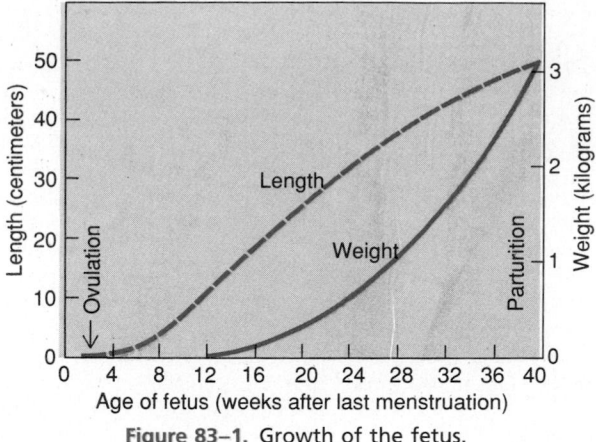

Figure 83–1. Growth of the fetus.

forming areas lose their ability to form blood cells, except for lymphocytes and plasma cells produced in lymphoid tissue.

Respiratory System

Respiration cannot occur during fetal life. However, respiratory movements do take place beginning at the end of the first trimester of pregnancy. Tactile stimuli or fetal asphyxia especially cause respiratory movements.

During the latter 3 to 4 months of pregnancy, the respiratory movements of the fetus are mainly inhibited, for reasons unknown, and the lungs remain almost completely deflated, so that the alveoli contain little fluid. This could possibly result from (1) special chemical conditions in the body fluids of the fetus, (2) the presence of fluid in the fetal lungs, or (3) other possible unexplored stimuli.

The inhibition of respiration during the latter months of fetal life prevents filling of the lungs with debris from the meconium excreted by the fetus's gastrointestinal tract into the amniotic fluid. Also, fluid is secreted into the lungs by the alveolar epithelium up until the moment of birth, thus keeping only clean fluid in the lungs.

Nervous System

Most of the skin reflexes of the fetus are present by the 3rd to 4th month of pregnancy. However, those functions of the central nervous system that involve the cerebral cortex are still mainly undeveloped even at birth. Indeed, myelinization of some major tracts of the central nervous system becomes complete only after about 1 year of postnatal life.

Gastrointestinal Tract

By midpregnancy, the fetus ingests and absorbs large quantities of amniotic fluid, and during the last 2 to 3 months, gastrointestinal function approaches that of the normal neonate. Small quantities of *meconium* are continually formed in the gastrointestinal tract and excreted from the bowels into the amniotic fluid. Meconium is composed partly of residue from amniotic fluid and partly of excretory products from the gastrointestinal mucosa and glands.

The Kidneys

The fetal kidneys are capable of excreting urine during at least the latter half of pregnancy, and urination occurs normally in utero. However, the renal control systems for regulating extracellular fluid electrolyte balances and especially acid-base balance are almost nonexistent until after midfetal life and do not reach full development until a few months after birth.

Fetal Metabolism

The fetus uses mainly glucose for energy, and it has a high rate of storage of fat and protein, much if not most of the fat being synthesized from glucose, rather than being absorbed from the mother's blood. Aside from these generalities, there are some special problems of fetal metabolism in relation to calcium, phosphate, iron, and some vitamins.

METABOLISM OF CALCIUM AND PHOSPHATE. Figure 83–2 shows the rates of calcium and phosphate accumulation in the fetus, demonstrating that about 22.5 grams of calcium and 13.5 grams of phosphorus are accumulated in the average fetus during gestation. About one half of this accumulates during the last 4 weeks of gestation, which is also coincident with the period of rapid ossification of the fetal bones as well as with the period of rapid weight gain of the fetus.

During the earlier part of fetal life, the bones are relatively unossified and have mainly a cartilaginous matrix. Indeed x-ray films ordinarily will not show any ossification until about the 4th month of pregnancy.

Note especially that the total amounts of calcium and phosphate needed by the fetus during gestation represent only about 1/50 the quantities of these substances in the mother's bones. Therefore, this is a minimal drain from the mother. A much greater drain occurs after birth during lactation.

ACCUMULATION OF IRON. Figure 83–2 also shows that iron accumulates in the fetus even more rapidly than calcium and phosphates. Most of the iron is in the form of hemoglobin, which begins to be formed as early as the 3rd week after fertilization of the ovum.

Small amounts of iron are concentrated in the mother's uterine progestational endometrium even before implantation

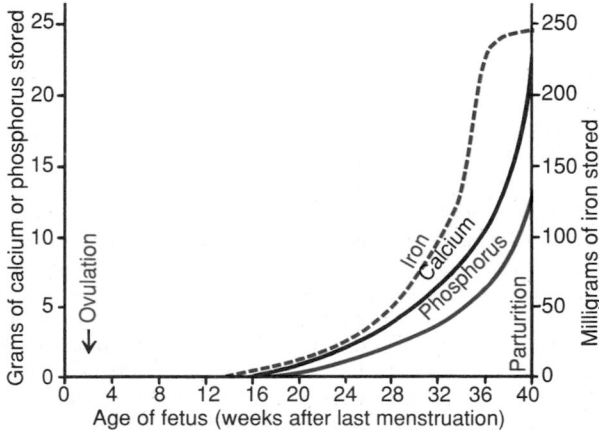

Figure 83–2. Calcium, phosphorus, and iron storage in the fetus at different stages of gestation.

of the ovum; this iron is ingested into the embryo by the trophoblastic cells for early formation of the red blood cells.

About one third of the iron in a fully developed fetus is normally stored in the liver. This iron can then be used for several months after birth by the neonate for formation of additional hemoglobin.

UTILIZATION AND STORAGE OF VITAMINS. The fetus needs vitamins equally as much as the adult and in some instances to a far greater extent. In general, the vitamins function the same in the fetus as in the adult, as discussed in Chapter 71. Special functions of several vitamins should be mentioned, however.

The B vitamins, especially vitamin B_{12} and folic acid, are necessary for formation of red blood cells and nervous tissue as well as for overall growth of the fetus.

Vitamin C is necessary for appropriate formation of intercellular substances, especially the bone matrix and fibers of connective tissue.

Vitamin D probably is needed for normal bone growth in the fetus, but even more important, the mother needs it for adequate absorption of calcium from her gastrointestinal tract. If the mother has plenty of this vitamin in her body fluids, large quantities of the vitamin will be stored by the fetal liver to be used by the neonate for several months after birth.

Vitamin E, although the mechanisms of its functions are not clear, is necessary for normal development of the early embryo. In its absence in laboratory animals, spontaneous abortion usually occurs at an early age.

Vitamin K is used by the fetal liver for formation of Factor VII, prothrombin, and several other blood coagulation factors. When vitamin K is insufficient in the mother, Factor VII and prothrombin become deficient in the fetus as well as in the mother. Because most vitamin K is formed by bacterial action in the colon, the neonate has no adequate source of vitamin K for the 1st week or so of life until normal colonic bacterial flora become established. Therefore, prenatal storage in the fetal liver of at least small amounts of vitamin K derived from the mother is helpful in preventing hemorrhage, particularly hemorrhage in the brain when the head is traumatized by squeezing through the birth canal.

ADJUSTMENTS OF THE INFANT TO EXTRAUTERINE LIFE

Onset of Breathing

The most obvious effect of birth on the baby is loss of the placental connection with the mother and, therefore, loss of this means of metabolic support. By far the most important immediate adjustment required of the infant is to begin breathing.

CAUSE OF BREATHING AT BIRTH. After normal delivery from a mother who has not been depressed by anesthetics, the child ordinarily begins to breathe within seconds and has a normal respiratory rhythm within less than 1 minute after birth. The promptness with which the fetus begins to breathe indicates that breathing is initiated by sudden exposure to the exterior world, probably resulting from a slightly asphyxiated state incident to the birth process but also from sensory impulses that originate in the suddenly cooled skin. In an infant who does not breathe immediately, the body becomes progressively more hypoxic and hypercapnic, which provides additional stimulus to the respiratory center and

usually causes breathing within an additional minute after birth.

DELAYED AND ABNORMAL BREATHING AT BIRTH—DANGER OF HYPOXIA. If the mother has been depressed by a general anesthetic during delivery, which at least partially anesthetizes the child as well, the initiation of respiration is likely to be delayed for several minutes, thus demonstrating the importance of using as little obstetrical anesthesia as feasible. Also, many infants who have had head trauma during delivery or who undergo prolonged delivery are slow to breathe or sometimes will not breathe at all. This can result from two possible effects: First, in a few infants, intracranial hemorrhage or brain contusion causes a concussion syndrome with a greatly depressed respiratory center. Second, and probably much more important, prolonged fetal hypoxia during delivery can cause serious depression of the respiratory center. Hypoxia frequently occurs during delivery because of (1) compression of the umbilical cord; (2) premature separation of the placenta; (3) excessive contraction of the uterus, which cuts off the mother's blood flow to the placenta; or (4) excessive anesthesia of the mother, which depresses the oxygenation even of her blood.

Degree of Hypoxia That an Infant Can Tolerate. In the adult, failure to breathe for only 4 minutes often causes death, but a neonate often survives as long as 10 minutes of failure to breathe after birth. Permanent and very evident brain impairment often ensues if breathing is delayed more than 8 to 10 minutes. Indeed, actual lesions develop mainly in the thalamus, the inferior colliculi, and in other brain stem areas, thus permanently affecting many of the motor functions of the body.

Expansion of the Lungs at Birth

At birth, the walls of the alveoli are at first kept collapsed by the surface tension of the viscid fluid that fills them. More than 25 mm Hg of negative inspiratory pressure is required to oppose the effects of this surface tension when opening the alveoli for the first time. But once the alveoli are open, further respiration can be effected with relatively weak respiratory movements. The first inspirations of the neonate are extremely powerful, usually capable of creating as much as 60 mm Hg negative pressure in the intrapleural space.

Figure 83–3 shows the tremendous negative intrapleural pressures required to open the lungs at the onset of breathing. To the left is shown the pressure-volume curve (compliance curve) for the first breath after birth. Observe, first, the lower part of the curve *beginning at the zero pressure point* and moving to the right. The curve shows that the

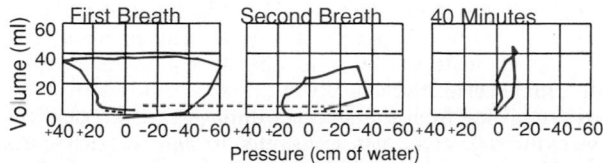

Figure 83–3. Pressure-volume curves of the lungs (compliance curves) of a neonate immediately after birth, showing *(a)* the extreme forces required for breathing during the first two breaths of life and *(b)* development of nearly normal compliance curves within 40 minutes after birth. (From Smith: The first breath. *Sci. Amer., 209:*32, 1963, © 1963 by Scientific American, Inc. All rights reserved.)

volume of air in the lungs remains almost exactly zero until the negative pressure has reached −40 centimeters water (−30 mm Hg). Then, as the negative pressure increases to −60 centimeters of water, about 40 milliliters of air enters the lungs. To deflate the lungs, considerable positive pressure, about +40 centimeters of water, is required because of the viscous resistance offered by the fluid in the bronchioles.

Note that the second breath is much easier, with far less both negative and positive pressures required. Breathing does not become completely normal until about 40 minutes after birth, as shown by the third compliance curve, the shape of which compares favorably with that for the normal adult, as shown in Chapter 38.

RESPIRATORY DISTRESS SYNDROME CAUSED BY LACK OF SURFACTANT SECRETION. A small number of infants, especially premature infants and infants born of diabetic mothers, develop severe respiratory distress in the early hours to several days after birth, and many succumb within the next day or so. The alveoli of these infants at death contain large quantities of proteinaceous fluid, almost as if pure plasma had leaked out of the capillaries into the alveoli. The fluid also contains desquamated alveolar epithelial cells. This condition is called *hyaline membrane disease* because microscopic slides of the lung show the material filling the alveoli to look like a hyaline membrane.

One of the most characteristic findings in respiratory distress syndrome is failure to secrete adequate quantities of *surfactant,* a substance normally secreted into the alveoli that decreases the surface tension of the alveolar fluid, therefore allowing the alveoli to open easily during inspiration. The surfactant-secreting cells (the type II alveolar epithelial cells) do not begin to secrete surfactant until the last 1 to 3 months of gestation. Therefore, many premature babies and some full-term babies are born without the capability of secreting surfactant, which therefore causes both a collapse tendency of the alveoli and development of pulmonary edema. The role of surfactant in preventing these effects is discussed in Chapter 37.

Circulatory Readjustments at Birth

Equally as essential as the onset of breathing at birth are the immediate circulatory adjustments that allow adequate blood flow through the lungs. Also, circulatory adjustments during the first few hours of life shunt more and more blood through the liver. To describe these readjustments, we must first consider the anatomical structure of the fetal circulation.

SPECIFIC ANATOMICAL STRUCTURE OF THE FETAL CIRCULATION. Because the lungs are mainly nonfunctional during fetal life and because the liver is only partially functional, it is not necessary for the fetal heart to pump much blood through either the lungs or the liver. On the other hand, the fetal heart must pump large quantities of blood through the placenta. Therefore, special anatomical arrangements cause the fetal circulatory system to operate considerably differently from that of the adult. First, as shown in Figure 83–4, blood returning from the placenta through the umbilical vein passes through the *ductus venosus,* mainly bypassing the liver. Then most of the blood entering the right atrium from the inferior vena cava is directed in a straight pathway across the posterior aspect of the right atrium and then through the *foramen ovale* directly into the left atrium. Thus, this well-oxygenated blood from the placenta enters mainly the left ventricle of the heart, rather than the right ventricle, and is pumped by the left ventricle mainly into the vessels of the head and forelimbs.

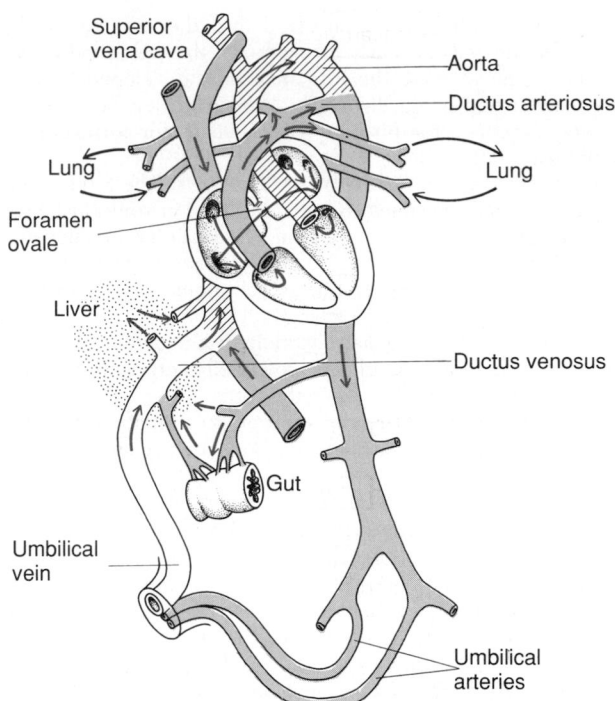

Figure 83–4. Organization of the fetal circulation. (Modified from Arey: Developmental Anatomy. 7th ed. Philadelphia, W. B. Saunders Company, 1974.)

The blood entering the right atrium from the superior vena cava is directed downward through the tricuspid valve into the right ventricle. This blood is mainly deoxygenated blood from the head region of the fetus, and it is pumped by the right ventricle into the pulmonary artery and then mainly through the *ductus arteriosus* into the descending aorta and through the two umbilical arteries into the placenta, where the deoxygenated blood becomes oxygenated.

Figure 83–5 gives the relative percentages of the total blood pumped by the heart that passes through the different vascular circuits of the fetus. This figure shows that 55 per cent of all the blood goes through the placenta, leaving only 45 per cent to pass through all the tissues of the fetus. Furthermore, during fetal life, only 12 per cent of the blood flows through the lungs; immediately after birth, virtually all the blood flows through the lungs, a manyfold increase at birth.

CHANGES IN THE FETAL CIRCULATION AT BIRTH. The basic changes in the fetal circulation at birth are discussed in Chapter 23 in relation to congenital anomalies of the ductus arteriosus and foramen ovale that persist throughout life in a few persons. Briefly, these changes are the following.

Primary Changes in Pulmonary and Systemic Vascular Resistance at Birth. The primary changes in the circulation at birth are, first, loss of the tremendous blood flow through the placenta, which *approximately doubles the systemic vascular resistance at birth.* This *increases the aortic pressure* as well as the pressures in the left ventricle and left atrium.

Second, the *pulmonary vascular resistance greatly decreases* as a result of expansion of the lungs. In the unexpanded fetal lungs, the blood vessels had been compressed because of the small volume of the lungs. Immediately on

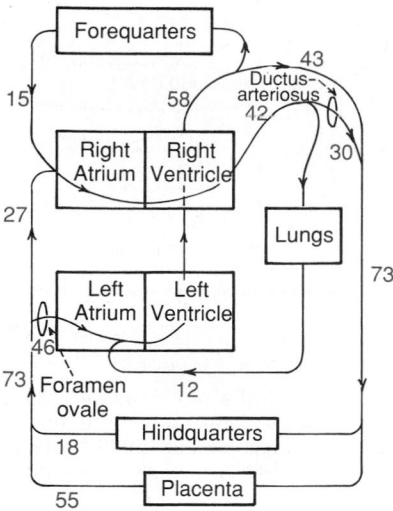

Figure 83–5. Diagram of the fetal circulatory system, showing relative distribution of blood flow to the different vascular areas. The numerals represent the percentage of the total cardiac output flowing through the particular area.

expansion, these vessels are no longer compressed and the resistance to blood flow decreases severalfold. Also, in fetal life, the hypoxia of the lungs causes considerable tonic vasoconstriction of the lung blood vessels, but vasodilation takes place when aeration of the lungs eliminates the hypoxia. All these changes together reduce the resistance to blood flow through the lungs as much as fivefold, which *reduces the pulmonary arterial pressure,* right ventricular pressure, and right atrial pressure.

Closure of the Foramen Ovale. The *low right atrial pressure* and the *high left atrial pressure* that occur secondarily to the changes in pulmonary and systemic resistances at birth cause blood now to attempt to flow backward through the foramen ovale, that is, from the left atrium into the right atrium, rather than in the other direction, as occurred during fetal life. Consequently, the small valve that lies over the foramen ovale on the left side of the atrial septum closes over this opening, thereby preventing further flow through the foramen ovale. In two thirds of all people, the valve becomes adherent over the foramen ovale within a few months to a few years and forms a permanent closure. But even if permanent closure does not occur, the left atrial pressure throughout life remains 2 to 4 mm Hg greater than the right atrial pressure, and the back pressure keeps the valve closed.

Closure of the Ductus Arteriosus. The ductus arteriosus also closes but for different reasons. First, the increased systemic resistance *elevates the aortic pressure* while the decreased pulmonary resistance *reduces the pulmonary arterial pressure.* As a consequence, after birth, blood begins to flow backward from the aorta into the pulmonary artery through the ductus arteriosus, rather than in the other direction as in fetal life. However, after only a few hours the muscle wall of the ductus arteriosus constricts markedly, and within 1 to 8 days, the constriction is usually sufficient to stop all blood flow. This is called *functional closure* of the ductus arteriosus. Then, during the next 1 to 4 months the ductus arteriosus ordinarily becomes anatomically *occluded* by growth of fibrous tissue into its lumen.

The cause of ductus closure relates to the increased oxygenation of the blood flowing through the ductus. In fetal life the Po_2 of the ductus blood is only 15 to 20 mm Hg, but it increases to about 100 mm Hg within a few hours after birth. Furthermore, many experiments have shown that the degree of contraction of the smooth muscle in the ductus wall is highly related to the availability of oxygen.

In one of several thousand infants, the ductus fails to close, resulting in a *patent ductus arteriosus,* the consequences of which are discussed in Chapter 23. In some instances at least, the failure of closure might be excessive ductus dilation caused by vasodilating prostaglandins in the ductus wall. The administration of the drug indomethacin, which blocks the prostaglandin dilating effect, often leads to closure.

Closure of the Ductus Venosus. In fetal life, the portal blood from the fetus's abdomen joins the blood from the umbilical vein, and these together pass through the *ductus venosus* directly into the vena cava, thus bypassing the liver. Immediately after birth, blood flow through the umbilical vein ceases, but most of the portal blood still flows through the ductus venosus, with only a small amount passing through the channels of the liver. However, within 1 to 3 hours the muscle wall of the ductus venosus contracts strongly and closes this avenue of flow. As a consequence, the portal venous pressure rises from near 0 to 6 to 10 mm Hg, which is enough to force blood flow through the liver sinuses. Although the ductus venosus almost never fails to close, we know almost nothing about what causes the closure.

Nutrition of the Neonate

The fetus obtains almost all its energy from glucose obtained from the mother's blood. Immediately after birth, the amount of glucose stored in the infant's body in the form of liver and muscle glycogen is sufficient to supply the infant's needs for only a few hours, and the liver of the neonate is still far from functionally adequate at birth, which prevents significant gluconeogenesis. Therefore, the infant's blood glucose concentration frequently falls the 1st day to as low as 30 to 40 mg/dl of plasma, less than one half the normal value. Appropriate mechanisms are available for the infant to use its stored fats and proteins for metabolism until mother's milk can be provided 2 to 3 days later.

Special problems are also frequently associated with getting an adequate fluid supply to the neonate because the infant's rate of body fluid turnover averages seven times that of an adult, and the mother's milk supply requires several days to develop. Ordinarily, the infant's weight decreases 5 to 10 per cent and sometimes as much as 20 per cent within the first 2 to 3 days of life. Most of this weight loss is loss of fluid rather than of body solids.

SPECIAL FUNCTIONAL PROBLEMS IN THE NEONATE

A most important characteristic of the neonate is instability of the various hormonal and neurogenic control systems. This results partly from immature development of the different organs of the body and partly from the fact that the control systems simply have not become adjusted to the new way of life.

Respiratory System

The normal rate of respiration in the neonate is about 40 breaths per minute, and tidal air with each breath averages 16 milliliters. This gives a total minute respiratory volume of 640 ml/min, which is about twice as great in relation to the body weight as that of an adult. *The functional residual capacity of the infant is only one half that of an adult in relation to body weight.* This difference causes excessive cyclic changes in blood gas concentrations when the respiratory rate becomes slowed because it is the residual air in the lungs that smooths out the blood gas variations in the adult.

Circulation

BLOOD VOLUME. The blood volume of a neonate immediately after birth averages about 300 milliliters, but if the infant is left attached to the placenta for a few minutes after birth or if the umbilical cord is stripped to force blood out of its vessels into the baby, an additional 75 milliliters of blood enters the infant, to make a total of 375 milliliters. Then, during the ensuing few hours, fluid is lost into the neonate's tissue spaces from this blood, which increases the hematocrit but returns the blood volume once again to the normal value of about 300 milliliters. Some pediatricians believe that this extra blood volume in some instances causes mild pulmonary edema with some degree of respiratory distress.

CARDIAC OUTPUT. The cardiac output of the neonate averages 500 ml/min, which, like respiration and body metabolism, is about twice as much in relation to body weight as in the adult. Occasionally a child is born with an especially low cardiac output caused by hemorrhage of much of its blood volume through the placental membrane into the mother's body before birth.

ARTERIAL PRESSURE. The arterial pressure during the 1st day after birth averages about 70/50; this increases slowly during the next several months to about 90/60. Then there is a much slower rise during the subsequent years until the adult pressure of 115/70 is attained at adolescence.

BLOOD CHARACTERISTICS. The red blood cell count in the neonate averages about 4 million per microliter. If blood is stripped from the cord into the infant, the red blood cell count rises an additional 0.5 to 0.75 million during the first few hours of life, giving a red blood cell count of about 4.75 million per microliter, as shown in Figure 83–6. Subsequent to this, however, few new red blood cells are formed in the infant during the first few weeks of life, presumably because the hypoxic stimulus of fetal life is no longer present to stimulate red cell production. Thus, as shown in Figure 83–6, the average red blood cell count falls to less than 4 million per microliter by about 6 to 8 weeks of age. From that time on, increasing activity by the baby provides the appropriate stimulus for returning the red blood cell count to normal within another 2 to 3 months.

Immediately after birth, the white blood cell count of the neonate is about 45,000 per microliter, which is about five times as great as that of the normal adult.

Neonatal Jaundice and Erythroblastosis Fetalis. Bilirubin formed in the fetus can cross the placenta into the mother and be excreted through the liver of the mother, but immediately after birth, the only means for ridding the neonate of bilirubin is through the neonate's own liver, which for the 1st week or so of life functions poorly and is incapable of conjugating significant quantities of bilirubin with glucuronic acid for excretion into the bile. Consequently, the

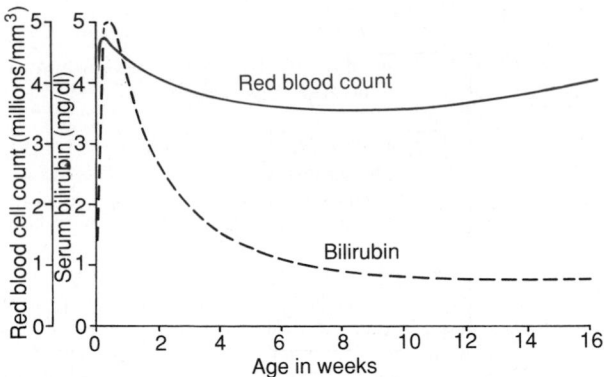

Figure 83–6. Changes in the red blood cell count and in the serum bilirubin concentration during the first 16 weeks of life, showing "physiologic anemia" at 6 to 12 weeks of life and "physiologic hyperbilirubinemia" during the first 2 weeks of life.

plasma bilirubin concentration rises from a normal value of less than 1 mg/dl to an average of 5 mg/dl during the first 3 days of life and then gradually falls back to normal as the liver becomes functional. This condition, called *physiologic hyperbilirubinemia*, is shown in Figure 83–6, and it is associated with a mild *jaundice* (yellowness) of the infant's skin and especially of the sclerae of its eyes.

However, by far the most important abnormal cause of serious neonatal jaundice is *erythroblastosis fetalis*, which is discussed in detail in Chapter 32 in relation to Rh factor incompatibility between the infant and mother. Briefly, the erythroblastotic baby usually inherits Rh-positive red cells from the father while the mother is Rh negative. The mother then becomes immunized against the Rh positive factor (a protein) in the fetus's blood cells, and her antibodies in turn destroy the fetal red cells, releasing extreme quantities of bilirubin into the fetus's plasma and often causing death for lack of adequate red cells. Before the advent of modern obstetrics, this condition occurred either mildly or seriously in 1 of every 50 to 100 neonates.

Fluid Balance, Acid-Base Balance, and Renal Function

The rate of fluid intake and fluid excretion in the infant is seven times as great in relation to weight as in the adult, which means that even a slight percentage alteration of fluid intake or fluid output can cause rapidly developing abnormalities. Second, the rate of metabolism in the infant is twice as great in relation to body mass as in the adult, which means that twice as much acid is normally formed, which leads to a tendency toward acidosis in the infant. Third, functional development of the kidneys is not complete until the end of about the 1st month of life. For instance, the kidneys of the neonate can concentrate urine to only 1.5 times the osmolality of the plasma instead of the normal three to four times in the adult.

Therefore, considering the immaturity of the kidneys, together with the marked fluid turnover in the infant and rapid formation of acid, one can readily understand that among the most important problems of infancy are acidosis, dehydration, and, more rarely, overhydration.

Liver Function

During the first few days of life, liver function may be quite deficient, as evidenced by the following effects:

1. The liver of the neonate conjugates bilirubin with glucuronic acid poorly and therefore excretes bilirubin only slightly during the first few days of life.

2. The liver of the neonate is deficient in forming plasma proteins, so that the plasma protein concentration falls during the first weeks of life to 15 to 20 per cent less than that for older children. Occasionally the protein concentration falls so low that the infant develops hypoproteinemic edema.

3. The gluconeogenesis function of the liver is particularly deficient. As a result, the blood glucose level of the unfed neonate falls to about 30 to 40 mg/dl, and the infant must depend on its stored fats for energy until sufficient feeding can occur.

4. The liver of the neonate usually also forms too little of the factors needed for normal blood coagulation.

Digestion, Absorption, Metabolism of Energy Foods, and Nutrition

In general, the ability of the neonate to digest, absorb, and metabolize foods is not different from that of the older child, with the following three exceptions.

First, secretion of pancreatic amylase in the neonate is deficient, so that the neonate uses starches less adequately than do older children.

Second, absorption of fats from the gastrointestinal tract is somewhat less than that in the older child. Consequently, milk with a high fat content, such as cow's milk, is frequently inadequately absorbed.

Third, because the liver functions imperfectly during at least the 1st week of life, the glucose concentration in the blood is unstable and low.

The neonate is especially capable of synthesizing and storing proteins. Indeed, with an adequate diet, as much as 90 per cent of the ingested amino acids is used for formation of body proteins. This is a much higher percentage than in adults.

METABOLIC RATE AND BODY TEMPERATURE. The normal metabolic rate of the neonate in relation to body weight is about twice that of the adult, which accounts also for the twice as great cardiac output and twice as great minute respiratory volume in relation to body weight in the infant.

Because the body surface area is large in relation to body mass, heat is readily lost from the body. As a result, the body temperature of the neonate, particularly of premature infants, falls easily. Figure 83–7 shows that the body temperature of even the normal infant often falls several degrees during the first few hours after birth but returns to normal in 7 to 10 hours. Still, the body temperature regulatory mechanisms remain poor during the early days of life, allowing marked deviations in temperature, which are also shown in Figure 83–7.

NUTRITIONAL NEEDS DURING THE EARLY WEEKS OF LIFE. At birth, a neonate is usually in complete nutritional balance, provided the mother has had an adequate diet. Furthermore, function of the gastrointestinal system is usually more than adequate to digest and assimilate all the nutritional needs of the infant if they are provided in the diet. However, three specific problems do occur in the early nutrition of the infant.

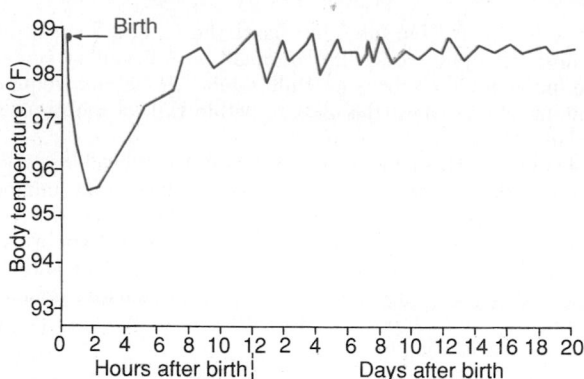

Figure 83–7. Fall in body temperature of the neonate immediately after birth, and instability of body temperature during the first few days of life.

Need for Calcium and Vitamin D. The neonate is in a stage of rapid ossification of its bones at birth, so that a ready supply of calcium throughout infancy is needed. This is ordinarily supplied adequately by the usual diet of milk. Yet absorption of calcium by the gastrointestinal tract is poor in the absence of vitamin D. Therefore, the vitamin D–deficient infant can develop severe rickets in only a few weeks. This is particularly true in premature babies because their gastrointestinal tracts absorb calcium even less effectively than those of normal infants.

Necessity for Iron in the Diet. If the mother has had adequate amounts of iron in her diet, the liver of the infant usually has stored enough iron to keep forming blood cells for 4 to 6 months after birth. But if the mother has had insufficient iron in her diet, severe anemia is likely to occur in the infant after about 3 months of life. To prevent this possibility, early feeding of egg yolk, which contains reasonably large quantities of iron, or the administration of iron in some other form is desirable by the 2nd or 3rd month of life.

Vitamin C Deficiency in Infants. Ascorbic acid (vitamin C) is not stored in significant quantities in the fetal tissues; yet it is required for proper formation of cartilage, bone, and other intercellular structures of the infant. Furthermore, milk has poor supplies of ascorbic acid, especially cow's milk, which has only one fourth as much as mother's milk. For this reason, orange juice or other sources of ascorbic acid are often prescribed by the 3rd week of life.

Immunity

The neonate inherits much immunity from the mother because many antibodies diffuse from the mother's blood through the placenta into the fetus. However, the neonate does not form antibodies of its own to a significant extent. By the end of the 1st month, the baby's gamma globulins, which contain the antibodies, have decreased to less than one half the original level, with corresponding decrease in immunity. Thereafter, the baby's own immunity system begins to form antibodies, and the gamma globulin concentration returns essentially to normal by the age of 12 to 20 months.

Despite the decrease in gamma globulins soon after birth, the antibodies inherited from the mother protect the infant for about 6 months against most major childhood infectious diseases, including diphtheria, measles, and polio. Therefore, immunization against these diseases before 6 months is usually unnecessary. On the other hand, the inherited antibodies against whooping cough are normally insufficient to protect the neonate; therefore, for full safety, the infant requires immunization against this disease within the 1st month or so of life.

ALLERGY. The neonate is seldom subject to allergy. Several months later, however, when the neonate's own antibodies first begin to form, extreme allergic states can develop, often resulting in serious eczema, gastrointestinal abnormalities, and even anaphylaxis. As the child grows older and still higher degrees of immunity develop, these allergic manifestations usually disappear. This relation of immunity to allergy is discussed in Chapter 34.

Endocrine Problems

Ordinarily the endocrine system of the infant is highly developed at birth, and the infant seldom exhibits any immediate endocrine abnormalities. However, there are special instances in which the endocrinology of infancy is important.

1.   If a pregnant mother bearing a female child is treated with an androgenic hormone or if an androgenic tumor develops during pregnancy, the child will be born with a high degree of masculinization of her sexual organs, thus resulting in a type of *hermaphroditism.*

2.   The sex hormones secreted by the placenta and by the mother's glands during pregnancy occasionally cause the neonate's breasts to form milk during the first days of life. Sometimes the breasts then become inflamed or infectious mastitis develops.

3.   An infant born of an untreated diabetic mother will have considerable hypertrophy and hyperfunction of the islets of Langerhans. As a consequence, the infant's blood glucose concentration may fall to lower than 20 mg/dl shortly after birth. In the neonate, unlike the adult, insulin shock or coma from this low level of blood glucose concentration only rarely develops.

Because of metabolic deficits in the diabetic mother, the fetus is often stunted in growth, and growth and tissue maturation of the neonate are often impaired. Also, there is a high rate of intrauterine mortality, and among those fetuses that do come to term, there is still a high mortality rate. Two thirds of the infants who die succumb to the respiratory distress syndrome, described earlier in the chapter.

4.   Occasionally a child is born with hypofunctional adrenal cortices, often resulting from *agenesis* of the adrenal glands or *exhaustion atrophy,* which can occur when the adrenal glands have been vastly overstimulated.

5.   If a pregnant woman has hyperthyroidism or is treated with excess thyroid hormone, the infant is likely to be born with a temporarily hyposecreting thyroid gland. On the other hand, if before pregnancy a woman had had her thyroid gland removed, her pituitary may secrete great quantities of thyrotropin during gestation, and the child might be born with temporary hyperthyroidism.

6.   In a fetus lacking thyroid hormone secretion, the bones grow poorly and there is mental retardation. This causes the condition called *cretin dwarfism,* discussed in Chapter 76.

SPECIAL PROBLEMS OF PREMATURITY

All the problems just noted for neonatal life are severely exacerbated in prematurity. They can be categorized under the following two headings: (1) immaturity of certain organ systems and (2) instability of the different homeostatic control systems. Because of these effects, a premature baby seldom lives if it is born more than 2.5 to 3 months before term.

Immature Development of the Premature Infant

Almost all the organ systems of the body are immature in the premature infant, but some require particular attention if the life of the premature baby is to be saved.

RESPIRATION. The respiratory system is especially likely to be underdeveloped in the premature infant. The vital capacity and the functional residual capacity of the lungs are especially small in relation to the size of the infant. Also, surfactant secretion is depressed or absent. As a consequence, the respiratory distress syndrome is a common cause of death. Also, the low functional residual capacity in the premature infant is often associated with periodic breathing of the Cheyne-Stokes type.

GASTROINTESTINAL FUNCTION. Another major problem of the premature infant is to ingest and absorb adequate food. If the infant is more than 2 months premature, the digestive and absorptive systems are almost always inadequate. The absorption of fat is so poor that the premature infant must have a low-fat diet. Furthermore, the premature infant has unusual difficulty in absorbing calcium and, therefore, can develop severe rickets before the difficulty is recognized. For this reason, special attention must be paid to adequate calcium and vitamin D intake.

FUNCTION OF OTHER ORGANS. Immaturity of other organ systems that frequently causes serious difficulties in the premature infant includes (1) immaturity of the liver, which results in poor intermediary metabolism and often a bleeding tendency as a result of poor formation of coagulation factors; (2) immaturity of the kidneys, which are particularly deficient in their ability to rid the body of acids, thereby predisposing to acidosis as well as to serious fluid balance abnormalities; (3) immaturity of the blood-forming mechanism of the bone marrow, which allows rapid development of anemia; and (4) depressed formation of gamma globulin by the lymphoid system, which often leads to serious infection.

Instability of the Control Systems in the Premature Infant

Immaturity of the different organ systems in the premature infant creates a high degree of instability in the homeostatic mechanisms of the body. For instance, the acid-base balance can vary tremendously, particularly when the food intake varies from time to time. Likewise, the blood protein concentration is usually somewhat low because of immature liver development, often leading to *hypoproteinemic edema.* And inability of the infant to regulate its calcium ion concentration frequently brings on hypocalcemic tetany. Also, the blood glucose concentration can vary between the extremely wide limits of 20 to more than 100 mg/dl, depending principally on the regularity of feeding. It is no wonder, then, with these extreme variations in the internal environment of the premature infant, that mortality is high.

INSTABILITY OF BODY TEMPERATURE. One of the particular problems of the premature infant is inability to maintain normal body temperature. Its temperature tends to approach that of its surroundings. At normal room temperature, the infant's temperature may stabilize in the low 90s or even in the 80s. Statistical studies show that a body temperature maintained below 96°F (35.5°C) is associated with a particularly high incidence of death, which explains the common use of the incubator in the treatment of prematurity.

Danger of Blindness Caused by Oxygen Therapy in the Premature Infant

Because premature infants frequently develop respiratory distress, oxygen therapy has often been used in treating prematurity. However, it has been discovered that use of high oxygen concentrations in treating premature infants, especially in early prematurity, can lead to blindness. The reason is that too much oxygen stops the growth of new blood vessels in the retina. Then when oxygen therapy is stopped, the blood vessels try to make up for lost time and burst forth with a great mass of vessels growing all through the vitreous humor, blocking light from the pupil to the retina. And still later, the vessels are replaced with a mass of fibrous tissue where the eye's clear vitreous humor should be. This condition, known as *retrolental fibroplasia,* causes permanent blindness. For this reason, it is particularly important to avoid treatment of premature infants with high concentrations of respiratory oxygen. Physiological studies indicate that the premature infant is usually safe with up to 40 per cent oxygen, but some child physiologists believe that complete safety can be achieved only at normal oxygen concentration.

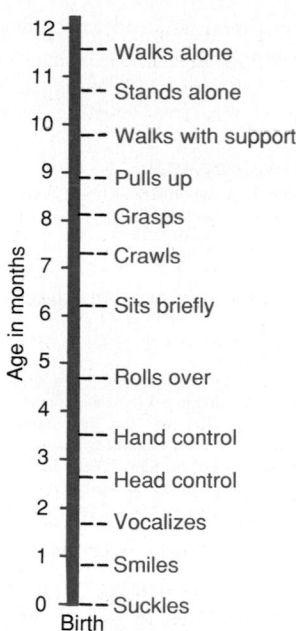

Figure 83–9. Behavioral development of the infant during the 1st year of life.

GROWTH AND DEVELOPMENT OF THE CHILD

The major physiological problems of the child beyond the neonatal period are related to special metabolic needs for growth, which have been fully covered in the sections on metabolism and endocrinology.

Figure 83–8 shows the changes in heights of boys and girls from the time of birth until the age of 20 years. Note especially that these parallel each other almost exactly until the end of the first decade of life. Between the ages of 11 and 13 years, the female estrogens cause rapid growth but early uniting of the epiphyses of the long bones at about the 14th to 16th year of life, so that growth in height ceases. This contrasts with the effect of testosterone in the male, which causes growth at a slightly later age—mainly between ages 13 and 17 years. The male, however, undergoes much more prolonged growth because of much delayed uniting of the epiphyses, so that his final height is considerably greater than that of the female.

Behavioral Growth

Behavioral growth is principally a problem of maturity of the nervous system. Here, it is extremely difficult to dissociate maturity of the anatomical structures of the nervous system from maturity caused by training. Anatomical studies show that certain major tracts in the central nervous system are not completely myelinated until the end of the 1st year of life. For this reason, it is frequently stated that the nervous system is not fully functional at birth. The brain cortex and its associated mechanisms such as vision seem to require several months after birth for most rapid functional development.

At birth, the infant brain mass is only 26 per cent of the adult brain mass and 55 per cent at 1 year but reaches almost adult proportions by the end of the 2nd year. This is also associated with closure of the fontanels and sutures of the skull, which allows only 20 per cent additional growth of the brain beyond the first 2 years of life.

Figure 83–9 shows a normal progress chart for the infant during the 1st year of life. Comparison of this chart with the

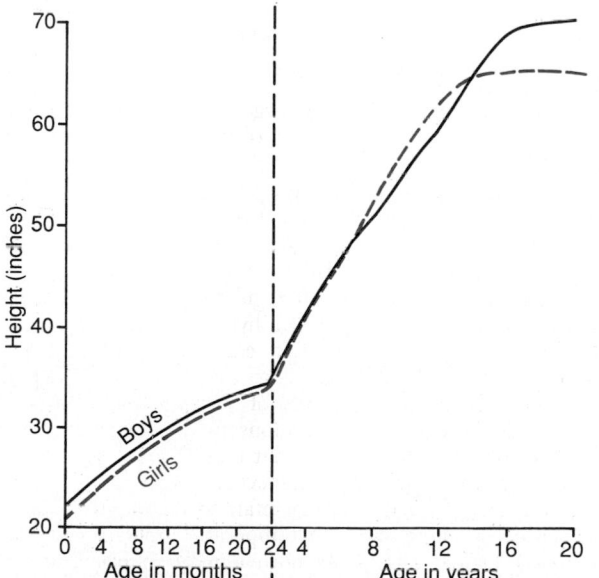

Figure 83–8. Height of boys and girls from infancy to 20 years of age.

baby's actual development is used for clinical assessment of mental and behavioral growth.

REFERENCES

Avery, G. B., et al.: Neonatology: Pathophysiology and Management of the Newborn. Philadelphia, J. B. Lippincott, 1994.

Beckerman, R. C., et al.: Respiratory Control Disorders in Infants and Children. Baltimore, Williams & Wilkins, 1992.

Benrubi, G. I.: Obstetric and Gynecologic Emergencies. Philadelphia, J. B. Lippincott, 1994.

Bertrand, J., et al.: Pediatric Endocrinology: Physiology, Pathophysiology, Clinical Aspects. Baltimore, Williams & Wilkins, 1993.

Bissonnette, J. M., et al.: Regulation of cerebral blood flow in the fetus. J. Dev. Physiol., 6:275, 1984.

Blackburn, S. T., and Loper, D. L.: Maternal, Fetal, and Neonatal Physiology: A Clinical Perspective. Philadelphia, W. B. Saunders Co., 1992.

Boynton, B. R., et al.: New Therapies for Neonatal Respiratory Failure: A Physiological Approach. New York, Cambridge University Press, 1994.

Byard, R. W., and Cohle, S. D.: Sudden Death in Infancy, Childhood and Adolescence. New York, Cambridge University Press, 1994.

Byskov, A. G.: Differentiation of mammalian embryonic gonad. Physiol. Rev., 66:71, 1986.

Cabaniss, M.: Fetal Monitoring Interpretation. Philadelphia, J. B. Lippincott, 1993.

Compel, C., and Silverberg, S. G.: Pathology in Gynecology and Obstetrics. Philadelphia, J. B. Lippincott, 1994.

Castaneda, A. R., et al.: Cardiac Surgery of the Neonate and Infant. Philadelphia, W. B. Saunders Co., 1994.

Cooke, J.: The early embryo and the formation of body pattern. Am. Sci., 76:35, 1988.

Creasy, R. K., and Resnik, R.: Maternal-Fetal Medicine: Principles and Practice. Philadelphia, W. B. Saunders Co., 1994.

Dimmick, J. E., and Kalousek, D. K.: Developmental Pathology of the Embryo and Fetus. Philadelphia, J. B. Lippincott, 1992.

Fletcher, M. A., and MacDonald, M. G.: Atlas of Procedures in Neonatology. Philadelphia, J. B. Lippincott, 1993.

Garland, T. Jr., and Carter, P. A.: Evolutionary physiology. Annu. Rev. Physiol., 56:579, 1994.

Girard, J., et al.: Adaptations of glucose and fatty acid metabolism during perinatal period and suckling-weaning transition. Physiol. Rev., 72:507, 1992.

Gomella, T. L., and Cunningham, M. D.: Neonatology. East Norwalk, Conn., Appleton & Lange, 1988.

Gootman, P. M., et al.: Postnatal maturation of the respiratory rhythm generator. In The Physiological Development of the Fetus and Newborn. London, Academic Press, 1985, p. 223.

Hacker, N. F., and Moore, J. G.: Essentials of Obstetrics and Gynecology. Philadelphia, W. B. Saunders Co., 1992.

Hanson, M. A., et al.: Fetus and Neonate: Physiology and Clinical Applications, Volume 1: The Circulation. New York, Cambridge University Press, 1993.

Jones, C. T., and Rolph, T. P.: Metabolism during fetal life: A functional assessment of metabolic development. Physiol. Rev., 65:357, 1985.

Lebenthal, E.: Textbook of Gastroenterology and Nutrition in Infancy. New York, Raven Press, 1989.

Long, W. A.: Fetal and Neonatal Cardiology. Philadelphia, W. B. Saunders Co., 1990.

Lumbers, E. R.: Renal function during intrauterine life. News Physiol. Sci., 2:220, 1987.

Moore, K. L., et al.: Color Atlas of Clinical Embryology. Philadelphia, W. B. Saunders Co., 1994.

Mortola, J. P.: Dynamics of breathing in newborn mammals. Physiol. Rev., 67:187, 1987.

Perelman, R. H., et al.: Developmental aspects of lung lipids. Annu. Rev. Physiol., 47:803, 1985.

Perloff, J. K.: The Clinical Recognition of Congenital Heart Disease. Philadelphia, W. B. Saunders Co., 1994.

Polin, R. A., and Fox, W. W.: Fetal and Neonatal Physiology. Philadelphia, W. B. Saunders Co., 1992.

Read, D. J. and Henderson-Smart, D. J.: Regulation of breathing in the newborn during different behavioral states. Annu. Rev. Physiol., 46:675, 1984.

Reece, A. E., et al.: Medicine of the Fetus and Mother. Philadelphia, J. B. Lippincott, 1992.

Rigatto, H.: Control of ventilation in the newborn. Annu. Rev. Physiol., 46:661, 1984.

Rories, C., and Spelsberg, T. C.: Ovarian steroid action on gene expression: Mechanisms and models. Annu. Rev. Physiol., 51:653, 1989.

Saliba, E.: International Symposium on Fetal and Neonatal Neurology. Farmington, CT, S. Karger Publishers, Inc., 1992.

Saling, E.: Perinatology. New York, Raven Press, 1992.

Salle, B. L., and Swyer, P. R.: Nutrition of the Low Birthweight Infant. New York, Raven Press, 1993.

Schreiner, R. L., and Bradbum, N. C.: Care of the Newborn, 2nd Ed. New York, Raven Press, 1987.

Scott, J., et al.: Danforth's Obstetrics and Gynecology. 7th Ed. Philadelphia, J. B. Lippincott, 1994.

Stark, J., and de Leval, M.: Surgery for Congenital Heart Defects. Philadelphia, W. B. Saunders Co., 1993.

Stocker, J. T., and Dehner, L. P.: Pediatric Pathology. Philadelphia, J. B. Lippincott, 1992.

Strang, L. B.: Fetal lung liquid: Secretion and reabsorption. Physiol. Rev., 71:991, 1991.

Tsang, R. C., et al.: Growth Factors in Perinatal Development. New York, Raven Press, 1993.

Tulchinsky, D., and Little A. B.: Maternal-Fetal Endocrinology. Philadelphia, W. B. Saunders Co., 1994.

Walker, D. W.: Peripheral and central chemoreceptors in the fetus and newborn. Annu. Rev. Physiol., 46:687, 1984.

SPORTS PHYSIOLOGY
UNIT XV

Sports Physiology

CHAPTER 84

There are no other normal stresses to which the body is exposed that even nearly approach the extreme stresses of heavy exercise. In fact, if some of the extremes of exercise were continued for even slightly prolonged periods, they might easily be lethal. Therefore, in the main, sports physiology is a discussion of the ultimate limits to which most of the bodily mechanisms can be stressed. To give one simple example: In a person who has extremely high fever, approaching the level of lethality, the body metabolism increases to about 100 per cent above normal. By comparison, the metabolism of the body during a marathon race increases to 2000 per cent above normal.

The Female and the Male Athlete

Most of the quantitative data that are given in this chapter are for the young man, not because it is desirable to know only these values but because it is only in this class of athletes that relatively complete measurements have been made. However, for those measurements that have been made in women, almost identically the same basic physiological principles apply equally as to men except for quantitative differences caused by differences in body size, body composition, and the presence or absence of the male sex hormone testosterone. In general, most quantitative values for women —such as muscle strength, pulmonary ventilation, and cardiac output, all of which are related mainly to the muscle mass—will vary between two thirds and three quarters of the values recorded in men. On the other hand, when measured in terms of strength per square centimeter of cross-sectional area, the female muscle can achieve almost exactly the same maximum force of contraction as that of the male —between 3 and 4 kg/cm². Therefore, much of the difference in total muscle performance lies in the extra percentage of the male body that is muscle, caused by endocrine differences that we discuss later.

A good indication of the relative performance capabilities of the female versus the male athlete comes from the relative times required for running the marathon race. In a recent comparison, the top female performer had a running time 12 per cent greater than that of the top male performer. On the other hand, for some events, women have at times held records over men—for instance, for the two-way swim across the English channel, where the availability of extra fat seems to be of advantage for heat insulation, buoyancy, and extra long-term energy.

The hormonal differences between women and men certainly account for a large part, if not most, of the differences in athletic performance. Testosterone secreted by the male testes has a powerful anabolic effect in causing greatly increased deposition of protein everywhere in the body, especially in the muscles. In fact, even the man who participates in very little sports activity but who nevertheless is well endowed with testosterone will have muscles that grow to sizes 40 per cent or more greater than those of his female counterpart and with a corresponding increase in strength.

The female sex hormone estrogen probably also accounts for some of the difference between female and male performance, although not nearly so much as testosterone. Estrogen is known to increase the deposition of fat in the female, especially in certain tissues such as the breasts, hips, and subcutaneous tissue. At least partly for this reason, the average nonathletic female has about 27 per cent body fat composition in contrast to the nonathletic male, who has about 15 per cent. This is a detriment to the highest levels of athletic performance in those events in which performance depends on speed or on ratio of total body muscle strength to body weight.

Finally, one cannot neglect the effect of the sex hormones on temperament. There is no doubt that testosterone promotes aggressiveness and that estrogen is associated with a milder temperament. Certainly a large part of competitive

1059

sports is the aggressive spirit that drives a person to maximum effort, often at the expense of judicious restraint.

THE MUSCLES IN EXERCISE

Strength, Power, and Endurance of Muscles

The final common denominator in athletic events is what the muscles can do for you—what strength they can give when it is needed, what power they can achieve in the performance of work, and how long they can continue in their activity.

The strength of a muscle is determined mainly by its size, with a *maximum contractile force between 3 and 4 kg/cm²* of muscle cross-sectional area. Thus, the man who is well laced with testosterone and therefore has correspondingly enlarged muscles will be much stronger than those people without the testosterone advantage. Also, the athlete who has enlarged his muscles through an exercise training program likewise will have increased muscle strength.

To give an example of muscle strength, a world-class weight lifter might have a quadriceps muscle with a cross-sectional area as great as 150 square centimeters. This would translate into a maximum contractile strength of 525 kilograms (or 1155 pounds), with all this force applied to the patellar tendon. Therefore, one can readily understand how it is possible for this tendon at times to be ruptured or actually to be avulsed from its insertion into the tibia below the knee. Also, when such forces occur in tendons that span a joint, similar forces are applied as well to the surfaces of the joints or sometimes to ligaments spanning the joints, thus accounting for such happenings as displaced cartilages, compression fractures about the joint, and torn ligaments.

The *holding strength* of muscles is about 40 per cent greater than the contractile strength. That is, if a muscle is already contracted and a force then attempts to stretch out the muscle, as occurs when landing after a jump, this requires about 40 per cent more force than can be achieved by a shortening contraction. Therefore, the force of 525 kilograms calculated previously for the patellar tendon during muscle contraction becomes 735 kilograms (1617 pounds). This further compounds the problems of the tendons, joints, and ligaments. It can also lead to internal tearing in the muscle itself. In fact, stretching out of a maximally contracted muscle is one of the best ways to ensure the highest degree of muscle soreness.

The *power* of muscle contraction is different from muscle strength because power is a measure of the total amount of work that the muscle performs in a unit of time. This is determined not only by the strength of muscle contraction but also by its *distance of contraction* and the *number of times that it contracts each minute*. Muscle power is generally measured in *kilogram meters (kg-m) per minute*. That is, a muscle that can lift a kilogram weight to a height of 1 meter or that can move some object laterally against a force of 1 kilogram for a distance of 1 meter in 1 minute is said to have a power of 1 kg-m/min. The maximum power achievable by all the muscles in the body of a highly trained athlete with all the muscles working together is approximately the following:

	kg-m/min
First 8 to 10 seconds	7000
Next 1 minute	4000
Next 30 minutes	1700

Thus, it is clear that a person has the capability of an extreme power surge for a short period of time, such as during a 100-meter dash that can be completed entirely within the first 10 seconds, whereas for long-term endurance events, the power output of the muscles is only one fourth as great as during the initial power surge.

This does not mean that one's athletic performance is four times as great during the initial power surge as it is for the next 30 minutes because the efficiency for translation of muscle power output into athletic performance is often much less during rapid activity than during less rapid but sustained activity. Thus, the velocity of the 100-meter dash is only 1.75 times as great as the velocity of the 30-minute race despite the fourfold difference in short-term versus long-term muscle power capability.

The final measure of muscle performance is *endurance*. This, to a great extent, depends on the nutritive support for the muscle—more than anything else on the amount of glycogen that has been stored in the muscle before the period of exercise. A person on a high-carbohydrate diet stores far more glycogen in muscles than a person on either a mixed diet or a high-fat diet. Therefore, endurance is greatly enhanced by a high-carbohydrate diet. When athletes run at speeds typical for the marathon race, their endurance as measured by the time that they can sustain the race until complete exhaustion is approximately the following:

	Minutes
High-carbohydrate diet	240
Mixed diet	120
High-fat diet	85

The corresponding amounts of glycogen stored in the muscle before the race started explain these differences. The amounts stored are approximately the following:

	gm/kg Muscle
High-carbohydrate diet	40
Mixed diet	20
High-fat diet	6

Muscle Metabolic Systems in Exercise

The same basic metabolic systems are present in muscle as in all other parts of the body; these were discussed in detail in Chapters 67 through 73. However, special quantitative measures of the activities of three metabolic systems are exceedingly important in understanding the limits of physical activity. These systems are (1) the *phosphagen system*, (2) the *glycogen–lactic acid system*, and (3) the *aerobic system*.

Phosphagen System

ADENOSINE TRIPHOSPHATE. The basic source of energy for muscle contraction is adenosine triphosphate (ATP), which has the following basic formula:

$$\text{Adenosine} - PO_3 \sim PO_3 \sim PO_3^-$$

The bonds attaching the last two phosphate radicals to the

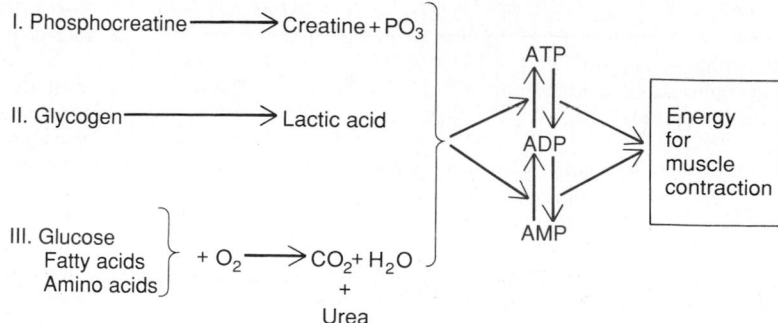

Figure 84–1. Important metabolic systems that supply energy for muscle contraction.

molecule, designated by the symbol ~, are *high-energy phosphate bonds.* Each of these bonds stores 7300 Calories of energy per mole of ATP under standard conditions (and even more than this under the physical conditions in the body, which is discussed in detail in Chapter 67). Therefore, when one phosphate radical is removed from the molecule, 7300 Calories of energy that can be used to energize the muscle contractile process is released. Then, when the second phosphate radical is removed, still another 7300 Calories becomes available. Removal of the first phosphate converts the ATP into *adenosine diphosphate* (ADP), and removal of the second converts this ADP into *adenosine monophosphate* (AMP).

The amount of ATP present in the muscles, even in the well-trained athlete, is sufficient to sustain maximal muscle power for only about 3 seconds, maybe enough for one half of a 50-meter dash. Therefore, except for a few seconds at a time, it is essential that new ATP be formed continuously, even during the performance of short athletic events. Figure 84–1 shows the overall metabolic system, demonstrating the breakdown of ATP first to ADP and then to AMP, with the release of energy to the muscles for contraction. To the left-hand side of the figure are shown the three metabolic systems that are responsible for reconstituting a continuous supply of ATP in the muscle fibers. They are the following.

RELEASE OF ENERGY FROM PHOSPHOCREATINE. Phosphocreatine (also called *creatine phosphate*) is another chemical compound that has a high-energy phosphate bond, with the following formula:

$$Creatine \sim PO_3^-$$

This can decompose to *creatine* and *phosphate ion,* as shown to the left in Figure 84–1, and in doing so releases large amounts of energy. In fact, the high-energy phosphate bond of phosphocreatine has more energy than the bond of ATP —10,300 Calories per mole, in comparison with 7300. Therefore, the phosphocreatine can easily provide enough energy to reconstitute the high-energy bond of ATP. Furthermore, most muscle cells have two to four times as much phosphocreatine as ATP.

A special characteristic of energy transfer from phosphocreatine to ATP is that it occurs within a small fraction of a second. Therefore, in effect, all the energy stored in the muscle phosphocreatine is instantaneously available for muscle contraction, just as is the energy stored in ATP.

The combined amounts of cell ATP and cell phosphocreatine are called the *phosphagen energy system.* These together can provide maximal muscle power for 8 to 10 seconds, almost enough for the 100-meter run. Thus, the

energy from the phosphagen system is used for maximal short bursts of muscle power.

Glycogen–Lactic Acid System

The stored glycogen in muscle can be split into glucose and the glucose then used for energy. The initial stage of this process, called *glycolysis,* occurs without use of oxygen and, therefore, is said to be *anaerobic metabolism* (see Chapter 67). During glycolysis, each glucose molecule is split into two *pyruvic acid molecules,* and energy is released to form four ATP molecules for each original glucose molecule, as explained in Chapter 67. Ordinarily the pyruvic acid then enters the mitochondria of the muscle cells and reacts with oxygen to form still many more ATP molecules. However, when there is insufficient oxygen for this second stage (the oxidative stage) of glucose metabolism to occur, most of the pyruvic acid is converted into *lactic acid,* which then diffuses out of the muscle cells into the interstitial fluid and blood. Therefore, in effect, much of the muscle glycogen becomes lactic acid, but in doing so, considerable amounts of ATP are formed entirely without the consumption of oxygen.

Another characteristic of the glycogen–lactic acid system is that it can form ATP molecules about 2.5 times as rapidly as can the oxidative mechanism of the mitochondria. Therefore, when large amounts of ATP are required for short to moderate periods of muscle contraction, this anaerobic glycolysis mechanism can be used as a rapid source of energy. It is not as rapid as the phosphagen system but about one half as rapid.

Under optimal conditions, the glycogen–lactic acid system can provide 1.3 to 1.6 minutes of maximal muscle activity in addition to the 8 to 10 seconds provided by the phosphagen system, although at somewhat reduced muscle power.

Aerobic System

The aerobic system means the oxidation of foodstuffs in the mitochondria to provide energy. That is, as shown to the left in Figure 84–1, glucose, fatty acids, and amino acids from the foods—after some intermediate processing—combine with oxygen to release tremendous amounts of energy that are used to convert AMP and ADP into ATP, as discussed in Chapter 67.

In comparing this aerobic mechanism of energy supply with the glycogen–lactic acid system and the phosphagen system, the relative *maximum rates of power generation* in terms of ATP generation per minute are the following:

	M of ATP/min
Phosphagen system	4
Glycogen–lactic acid system	2.5
Aerobic system	1

On the other hand, when comparing systems for endurance, the relative values are the following:

	Time
Phosphagen system	8 to 10 seconds
Glycogen–lactic acid system	1.3 to 1.6 minutes
Aerobic system	Unlimited time (as long as nutrients last)

Thus, one can readily see that the phosphagen system is the one used by the muscle for power surges of a few seconds, and the aerobic system is required for prolonged athletic activity. In between is the glycogen–lactic acid system, which is especially important for giving extra power during such intermediate races as the 200- to 800-meter runs.

What Types of Sports Use Which Energy Systems?

By considering the vigor of a sports activity and its duration, one can estimate closely which of the energy systems are used for each activity. Various approximations are presented in Table 84–1.

Table 84–1 ENERGY SYSTEMS USED IN VARIOUS SPORTS

Phosphagen system, almost entirely
100 meter dash
Jumping
Weight lifting
Diving
Football dashes
Phosphagen and glycogen–lactic acid systems
200 meter dash
Basketball
Baseball home run
Ice hockey dashes
Glycogen–lactic acid system, mainly
400 meter dash
100 meter swim
Tennis
Soccer
Glycogen–lactic acid and aerobic systems
800 meter dash
200 meter swim
1500 meter skating
Boxing
2000 meter rowing
1500 meter run
1 mile run
400 meter swim
Aerobic system
10,000 meter skating
Cross-country skiing
Marathon run (26.2 miles, 42.2 km)
Jogging

Recovery of the Muscle Metabolic Systems After Exercise

In the same way that the energy from phosphocreatine can be used to reconstitute ATP, so also can energy from the glycogen–lactic acid system be used to reconstitute both phosphocreatine and ATP. And then energy from the oxidative metabolism of the aerobic system can be used to reconstitute all the other systems—the ATP, the phosphocreatine, and the glycogen–lactic acid system.

Reconstitution of the lactic acid system means mainly the removal of the excess lactic acid that has accumulated in all the fluids of the body. This is especially important because *lactic acid causes extreme fatigue.* When adequate amounts of energy are available from oxidative metabolism, removal of lactic acid is achieved in two ways: First, a small portion of it is converted back into pyruvic acid and then metabolized oxidatively by all the body tissues. Second, the remaining lactic acid is reconverted into glucose mainly in the liver, and the glucose in turn is used to replenish the glycogen stores of the muscles.

RECOVERY OF THE AEROBIC SYSTEM AFTER EXERCISE. Even during the early stages of heavy exercise, a portion of one's aerobic energy capability is depleted. This results from two effects: (1) the so-called *oxygen debt* and (2) *depletion of the glycogen stores* of the muscles.

Oxygen Debt. The body normally contains about 2 liters of stored oxygen that can be used for aerobic metabolism even without breathing any new oxygen. This stored oxygen consists of the following: (1) 0.5 liter in the air of the lungs, (2) 0.25 liter dissolved in the body fluids, (3) 1 liter combined with the hemoglobin of the blood, and (4) 0.3 liter stored in the muscle fibers themselves, combined with myoglobin, an oxygen-binding chemical similar to hemoglobin.

In heavy exercise, almost all this stored oxygen is used within a minute or so for aerobic metabolism. Then, after the exercise is over, this stored oxygen must be replenished by breathing extra amounts of oxygen over and above the normal requirements. In addition, about 9 liters more oxygen must be consumed to provide for reconstituting both the phosphagen system and the lactic acid system. All this extra oxygen that must be "repaid," about 11.5 liters, is called the oxygen debt.

Figure 84–2 shows this principle of oxygen debt. During

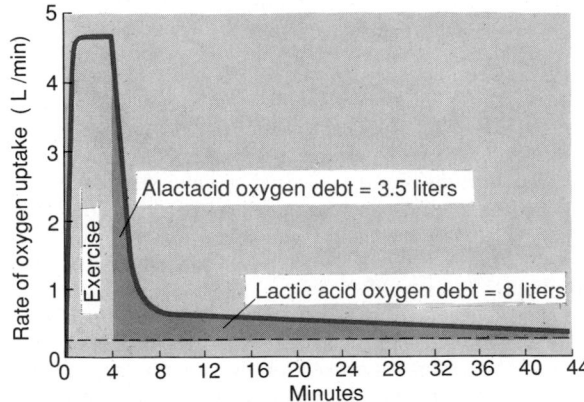

Figure 84–2. Rate of oxygen uptake by the lungs during maximal exercise for 4 minutes and then for almost 1 hour after the exercise is over. This figure demonstrates the principle of *oxygen debt.*

the first 4 minutes of the figure, the person exercises heavily, and the rate of oxygen uptake increases more than 15-fold. Then, even after the exercise is over, the oxygen uptake still remains above normal, at first very high while the body is reconstituting the phosphagen system and repaying the stored oxygen portion of the oxygen debt and then for another hour at a lower level while the lactic acid is removed. The early portion of the oxygen debt is called the *alactacid oxygen debt* and amounts to about 3.5 liters. The latter portion is called the *lactic acid oxygen debt* and amounts to about 8 liters.

Recovery of Muscle Glycogen. Recovery from exhaustive muscle glycogen depletion is not a simple matter. This often requires days, rather than the seconds, minutes, or hours required for recovery of the phosphagen and lactic acid metabolic systems. Figure 84–3 shows this recovery process under three conditions: first in people on a high-carbohydrate diet; second, in people on a high-fat/high-protein diet; and third, in people with no food. Note that on a high-carbohydrate diet, full recovery occurs in about 2 days. On the other hand, people on a high-fat/high-protein diet or on no food at all show extremely little recovery even after as long as 5 days. The messages of this comparison are (1) that it is important for an athlete to have a high-carbohydrate diet before a grueling athletic event and (2) not to participate in exhaustive exercise during the 48 hours preceding the event.

Nutrients Used During Muscle Activity

Although we have emphasized the importance of a high-carbohydrate diet and large stores of muscle glycogen for maximal athletic performance, this does not mean that only carbohydrates are used for muscle energy—it means simply that carbohydrates are used by preference. Actually, the muscles use large amounts of fat for energy in the form of *fatty acids* and *acetoacetic acid* (see Chapter 68) and use to a much less extent proteins in the form of *amino acids*. In fact, even under the best conditions, in those endurance athletic events that last longer than 4 to 5 hours, the glycogen stores of the muscle become almost totally depleted and

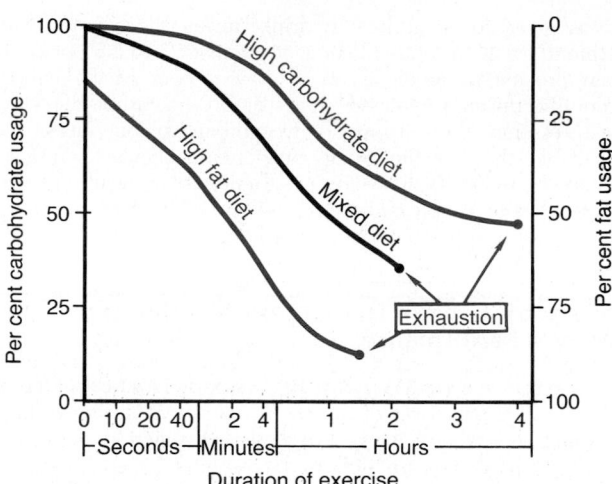

Figure 84–4. Effect of duration of exercise as well as type of diet on relative percentages of carbohydrate or fat used for energy by muscles. (Based partly on data in Fox: Sports Physiology. Philadelphia, Saunders College Publishing, 1979.)

are then of little further use for energizing muscle contraction. Instead, the muscle now depends on energy from other sources, mainly from fats.

Figure 84–4 shows the approximate relative usage of carbohydrates and fat for energy during prolonged exhaustive exercise under three dietary conditions: high-carbohydrate diet, mixed diet, and high-fat diet. Note that most of the energy is derived from carbohydrate during the first few seconds or minutes of the exercise, but at the time of exhaustion, as much as 60 to 85 per cent of the energy is being derived from fats, rather than carbohydrates.

Not all the energy from carbohydrates comes from the stored muscle glycogen. In fact, almost as much glycogen is stored in the liver as in the muscles, and this can be released into the blood in the form of glucose and then taken up by the muscles as an energy source. In addition, glucose solu-

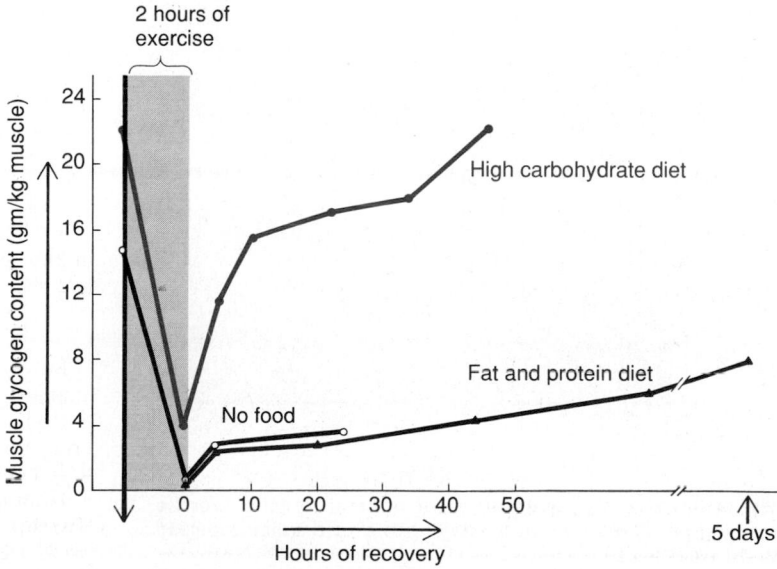

Figure 84–3. Effect of diet on the rate of muscle glycogen replenishment after prolonged exercise. (Reprinted from Fox: Sports Physiology. Philadelphia, Saunders College Publishing, 1979.)

tions given to an athlete to drink during the course of an athletic event (in optimal concentrations of 2 to 2.5 per cent) can provide as much as 30 to 40 per cent of the energy required during prolonged events such as marathon races.

In essence, then, if muscle glycogen and blood glucose are available, they are the energy nutrients of choice for intense muscle activity. Even so, for a real endurance event, one can expect fat to supply more than 50 per cent of the required energy after about the first 3 to 4 hours.

Effect of Athletic Training on Muscles and Muscle Performance

IMPORTANCE OF MAXIMUM RESISTANCE TRAINING. One of the cardinal principles of muscle development during athletic training is the following: Muscles that function under no load, even if they are exercised for hours on end, increase little in strength. At the other extreme, muscles that contract at more than 50 per cent maximal force of contraction will develop strength rapidly even if the contractions are performed only a few times each day. Using this principle, experiments on muscle building have shown that six nearly maximal muscle contractions performed in three sets 3 days a week give approximately optimal increase in muscle strength and without producing chronic muscle fatigue. The upper curve in Figure 84–5 shows the approximate percentage increase in strength that can be achieved in the previously untrained young person by this resistive training program, demonstrating that the muscle strength increases about 30 per cent during the first 6 to 8 weeks but almost plateaus after that time. Along with this increase in strength is an approximately equal percentage increase in muscle mass, which is called *muscle hypertrophy.*

In old age, many people become so sedentary that their muscles atrophy tremendously. In these instances, muscle training often increases muscle strength more than 100 per cent.

MUSCLE HYPERTROPHY. The basic size of a person's muscles is determined mainly by heredity plus the level of testosterone secretion, which, in men, causes considerably larger muscles than in women. However, with training, the muscles can become hypertrophied perhaps an additional 30 to 60 per cent. Most of this hypertrophy results from in-

creased diameter of the muscle fibers rather than increased numbers of fibers, but this probably is not entirely true because a very few greatly enlarged muscle fibers are believed to split down the middle along their entire length to form entirely new fibers, thus increasing the number of fibers slightly.

The changes that occur inside the hypertrophied muscle fibers themselves include (1) increased numbers of myofibrils, proportionate to the degree of hypertrophy; (2) up to 120 per cent increase in mitochondrial enzymes; (3) as much as 60 to 80 per cent increase in the components of the phosphagen metabolic system, including both ATP and phosphocreatine; (4) as much as 50 per cent increase in stored glycogen; and (5) as much as 75 to 100 per cent increase in stored triglyceride (fat). Because of all these changes, the capabilities of both the anaerobic and the aerobic metabolic systems are increased, increasing especially the maximum oxidation rate and efficiency of the oxidative metabolic system as much as 45 per cent.

Fast-Twitch and Slow-Twitch Muscle Fibers

In the human being, all muscles have varying percentages of *fast-twitch* and *slow-twitch muscle fibers.* For instance, the gastrocnemius muscle has a higher preponderance of fast-twitch fibers, which gives it the capability of forceful and rapid contraction of the type used in jumping. On the other hand, the soleus muscle has a higher preponderance of slow-twitch muscle fibers and therefore is used to a greater extent for prolonged lower leg muscle activity.

The basic differences between the fast-twitch and the slow-twitch fibers are the following:

1. Fast-twitch fibers are about twice as large in diameter.

2. The enzymes that promote rapid release of energy from the phosphagen and glycogen–lactic acid energy systems are two to three times as active in fast-twitch fibers as in slow-twitch fibers, thus making the maximal power that can be achieved by fast-twitch fibers as great as twice that of slow-twitch fibers.

3. Slow-twitch fibers are mainly organized for endurance, especially for generation of aerobic energy. They have far more mitochondria than the fast-twitch fibers. In addition, they contain considerably more myoglobin, a hemoglobin-like protein that combines with oxygen within the muscle fiber; the extra myoglobin increases the rate of diffusion of oxygen throughout the fiber by shuttling oxygen from one molecule of myoglobin to the next. In addition, the enzymes of the aerobic metabolic system are considerably more active in slow-twitch fibers than in fast-twitch fibers.

4. The number of capillaries per mass of fibers is greater in the vicinity of slow-twitch fibers than in the vicinity of fast-twitch fibers.

In summary, fast-twitch fibers can deliver extreme amounts of power for a few seconds to a minute or so. On the other hand, slow-twitch fibers provide endurance, delivering prolonged strength of contraction over many minutes to hours.

HEREDITARY DIFFERENCES AMONG ATHLETES FOR FAST-TWITCH VERSUS SLOW-TWITCH MUSCLE FIBERS. Some people have considerably more fast-twitch than slow-twitch

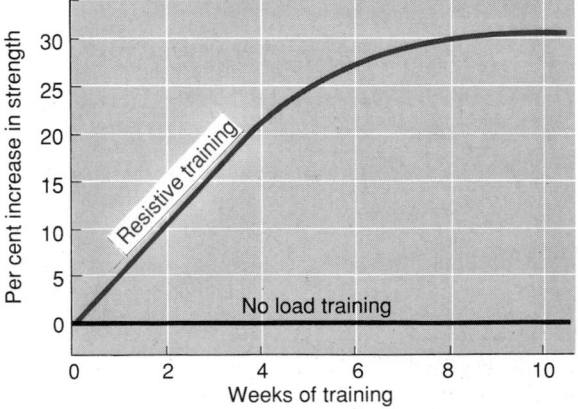

Figure 84–5. Approximate effect of optimal resistive exercise training on increase in muscle strength over a training period of 10 weeks.

fibers, and others have more slow-twitch fibers; this could determine to some extent the athletic capabilities of different individuals. Athletic training has not been shown to change the relative proportions of fast-twitch and slow-twitch fibers however much an athlete might want to develop one type of athletic prowess over another. Instead, this seems to be determined almost entirely by genetic inheritance, and this in turn helps determine which area of athletics is most suited to each person: some people appear to be born to be marathoners; others are born to be sprinters and jumpers. For example, the following are recorded percentages of fast-twitch versus slow-twitch fiber in the quadriceps muscles of different types of athletes:

	Fast-Twitch	*Slow-Twitch*
Marathoners	18	82
Swimmers	26	74
Average man	55	45
Weight lifters	55	45
Sprinters	63	37
Jumpers	63	37

RESPIRATION IN EXERCISE

Although one's respiratory ability is of relatively little concern in the performance of sprint types of athletics, it is critical for maximal performance in endurance athletics.

OXYGEN CONSUMPTION AND PULMONARY VENTILATION IN EXERCISE. Normal oxygen consumption for a young man at rest is about 250 ml/min. However, under maximal conditions this can be increased to approximately the following average levels:

	ml/min
Untrained average male	3600
Athletically trained average male	4000
Male marathon runners	5100

Figure 84–6 shows the relation between *oxygen consumption* and *total pulmonary ventilation* at different levels of exercise. It is clear from this figure, as would be expected,

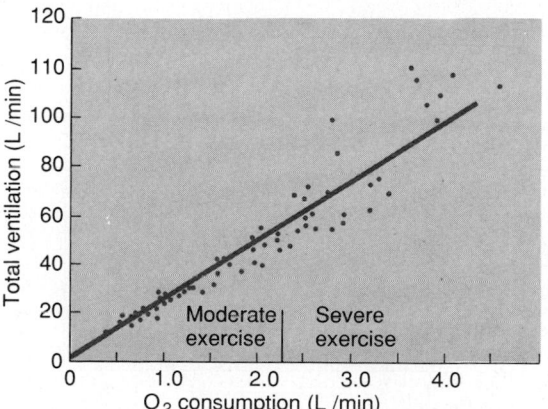

Figure 84–6. Effect of exercise on oxygen consumption and ventilatory rate. (From J.S. Gray: Pulmonary Ventilation and Its Physiological Regulation. Springfield, Ill., Charles C Thomas, 1950.)

that there is a linear relation. In round numbers, both oxygen consumption and total pulmonary ventilation increase about 20-fold between the resting state and maximum intensity of exercise *in the well-trained athlete.*

LIMITS OF PULMONARY VENTILATION. How severely do we stress our respiratory systems during exercise? This can be answered by the following comparison for the normal young man:

	Liters/min
Pulmonary ventilation at maximal exercise	100 to 110
Maximal breathing capacity	150 to 170

Thus, the maximal breathing capacity is about 50 per cent greater than the actual pulmonary ventilation during maximal exercise. This provides an element of safety for athletes, giving them extra ventilation that can be called on in such conditions as (1) exercise at high altitudes, (2) exercise under very hot conditions, and (3) abnormalities in the respiratory system.

The important point is that the respiratory system is not normally the most limiting factor in the delivery of oxygen to the muscles during maximal muscle aerobic metabolism. We shall see shortly that the ability of the heart to pump blood to the muscles is usually a greater limiting factor.

EFFECT OF TRAINING ON $\dot{V}O_2$ MAX. The abbreviation for the rate of oxygen usage under maximal aerobic metabolism is $\dot{V}O_2$ Max. Figure 84–7 shows the progressive effect of athletic training on $\dot{V}O_2$ Max recorded in a group of subjects beginning at the level of no training and then pursuing the training program for 7 to 13 weeks. In this study, it is surprising that the $\dot{V}O_2$ Max increased only about 10 per cent. Furthermore, the frequency of training, whether two times or five times per week, had little effect on the increase in $\dot{V}O_2$ Max. Yet, as pointed out earlier, the $\dot{V}O_2$ Max of marathoners is about 45 per cent greater than that of the untrained person. Part of this greater $\dot{V}O_2$ Max of the marathoner probably is genetically determined; that is, those people who have greater chest sizes in relation to body size and stronger respiratory muscles select themselves to become marathoners. However, it is also likely that the prolonged training of the marathoner does increase the $\dot{V}O_2$ Max by values considerably greater than the 10 per cent that has been recorded in short-term experiments such as that in Figure 84–7.

OXYGEN DIFFUSING CAPACITY OF ATHLETES. The *oxygen diffusing capacity* is a measure of the rate at which oxygen can diffuse from the alveoli into the blood. This is expressed in terms of *milliliters of oxygen that will diffuse each minute for each millimeter of mercury difference between alveolar partial pressure of oxygen and pulmonary blood oxygen pressure.* That is, if the partial pressure of oxygen in the alveoli is 91 mm Hg while the oxygen pressure in the blood is 90 mm Hg, the amount of oxygen that diffuses through the respiratory membrane each minute is equal to the diffusing capacity. The following are measured values for different diffusing capacities:

	ml/min
Nonathlete at rest	23
Nonathlete during maximum exercise	48
Speed skaters during maximum exercise	64
Swimmers during maximum exercise	71
Oarsmen during maximum exercise	80

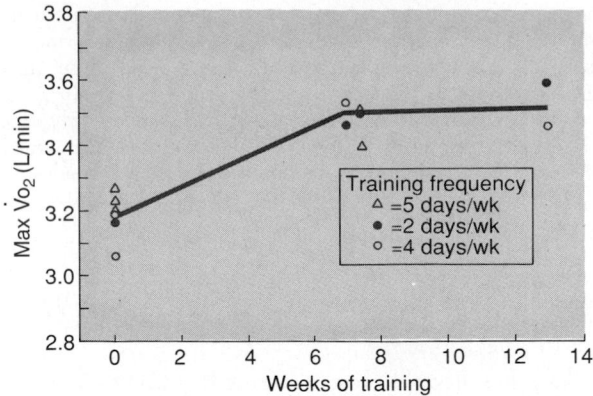

Figure 84–7. Increase in V̇O₂ Max over a period of 7 to 13 weeks of athletic training. (Reprinted from Fox: Sports Physiology. Philadelphia, Saunders College Publishing, 1979.)

The most startling fact about these results is the almost threefold increase in diffusing capacity between the resting state and the state of maximum exercise. This results mainly from the fact that blood flow through many of the pulmonary capillaries is sluggish or even dormant in the resting state, whereas in exercise, increased blood flow through the lungs causes all the pulmonary capillaries to be perfused at their maximum level, thus providing far greater surface area through which oxygen can diffuse into the pulmonary capillary blood.

It is also clear from these values that those athletes who require greater amounts of oxygen per minute have higher diffusing capacities. Is this because people with naturally greater diffusing capacities choose these types of sports, or is it because something about the training procedures increases the diffusing capacity? The answer is not known, but one must believe that training does play a role, particularly in endurance training.

BLOOD GASES DURING EXERCISE. Because of the great usage of oxygen by the muscles in exercise, one would expect the oxygen pressure of the arterial blood to decrease markedly during strenuous athletics and the carbon dioxide pressure of the venous blood to increase far above normal. However, this normally is not the case. Both of these values remain nearly normal, demonstrating the extreme ability of the respiratory system to provide adequate aeration of the blood even in heavy exercise, which demonstrates another important point: *The blood gases do not have to become abnormal for respiration to be stimulated in exercise.* Instead, respiration is stimulated mainly by neurogenic mechanisms in exercise, as discussed in Chapter 41. Part of this stimulation results from direct stimulation of the respiratory center by the same nervous signals that are transmitted from the brain to the muscles to cause the exercise. An additional part is believed to result from sensory signals transmitted into the respiratory center from the contracting muscles and moving joints. All this nervous stimulation of respiration is normally sufficient to provide almost exactly the proper increase in pulmonary ventilation to keep the blood respiratory gases—the oxygen and the carbon dioxide—almost normal.

EFFECT OF SMOKING ON PULMONARY VENTILATION IN EXERCISE. It is widely known that smoking can decrease an athlete's "wind." This is true for many reasons. First, one effect of nicotine is constriction of the terminal bronchioles of the lungs, which increases the resistance of air flow into

and out of the lungs. Second, the irritating effects of smoke cause increased fluid secretion into the bronchial tree as well as some swelling of the epithelial linings. Third, nicotine paralyzes the cilia on the surfaces of the respiratory epithelial cells that normally beat continuously to remove excess fluids and foreign particles from the respiratory tract. As a result, much debris accumulates in the respiratory passageways and adds further to the difficulty of breathing. Putting all these factors together, even the light smoker will feel respiratory strain during maximal exercise, and the level of performance may be reduced.

Much more severe are the effects of chronic smoking. There are few chronic smokers in whom some degree of emphysema does not develop. In this disease, the following occur: (1) chronic bronchitis, (2) obstruction of many of the terminal bronchioles, and (3) destruction of many alveolar walls. In severe emphysema, as much as four fifths of the respiratory membrane can be destroyed; then even the slightest exercise can cause respiratory distress. In fact, many such patients cannot even perform the athletic feat of walking across the floor of a single room without gasping for breath. Such is the indictment of smoking.

THE CARDIOVASCULAR SYSTEM IN EXERCISE

MUSCLE BLOOD FLOW. The final common denominator of cardiovascular function in exercise is to deliver oxygen and other nutrients to the muscles. For this purpose, the muscle blood flow increases drastically during exercise. Figure 84–8 shows a recording of muscle blood flow in the calf of a person for a period of 6 minutes during moderately strong intermittent contraction. Note not only the great increase in flow—about 13-fold—but also that the flow decreased during each muscle contraction. Two points can be made from this study: (1) The actual contractile process itself temporarily decreases muscle blood flow because the contracting muscle compresses the intramuscular blood vessels; therefore, strong *tonic* muscle contractions can cause rapid muscle fatigue because of lack of delivery of enough oxygen and nutrients during the continuous contraction. (2) The blood

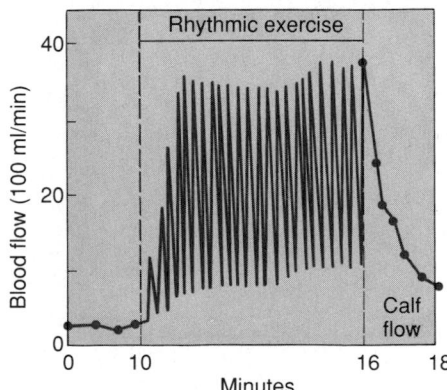

Figure 84–8. Effects of muscle exercise on blood flow in the calf of a leg during strong rhythmical contraction. The blood flow was much less during contraction than between contractions. (From Barcroft and Dornhorst: *J. Physiol.*, 109:402, 1949.)

flow to muscles during exercise can increase markedly. The following comparison shows the maximum increase in blood flow that can occur in the well-trained athlete.

ml/100 gm Muscle/min

Resting blood flow	3.6
Blood flow during maximal exercise	90

Thus, muscle blood flow can increase a maximum of about 25-fold during the most strenuous exercise. Almost one half this increase in flow results from intramuscular vasodilation caused by the direct effects of increased muscle metabolism, as explained in Chapter 21. The remaining increase results from multiple factors, the most important of which is probably the moderate increase in arterial blood pressure that occurs in exercise, usually about a 30 per cent increase. The increase in pressure not only forces more blood through the blood vessels but also stretches the walls of the arterioles and further reduces the vascular resistance. Therefore, a 30 per cent increase in blood pressure can often more than double the blood flow; this multiplies the great increase in flow already caused by the metabolic vasodilation at least another twofold.

WORK OUTPUT, OXYGEN CONSUMPTION, AND CARDIAC OUTPUT DURING EXERCISE. Figure 84–9 shows the interrelations among work output, oxygen consumption, and cardiac output during exercise. It is not surprising that all these are directly related to one another, as shown by the linear functions, because the muscle work output increases oxygen consumption, and oxygen consumption in turn dilates the muscle blood vessels, thus increasing venous return and cardiac output. Typical cardiac outputs at several levels of exercise are the following:

Liters/min

Average young man at rest	5.5
Maximum output during exercise in young untrained man	23
Maximum average output during exercise in male marathoner	30

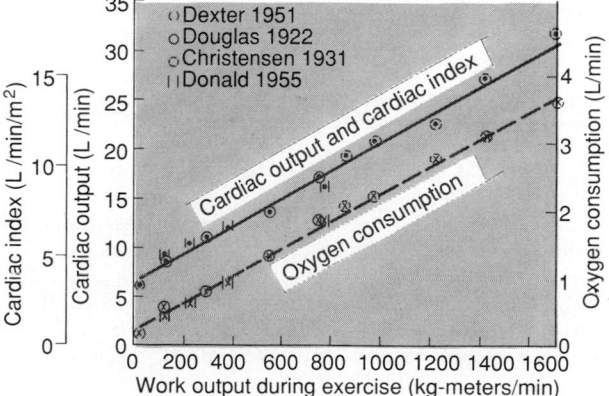

Figure 84–9. Relation between cardiac output and work output (solid line) and between oxygen consumption and work output (dashed line) during different levels of exercise. (From Guyton, Jones, and Coleman: Circulatory Physiology: Cardiac Output and Its Regulation. Philadelphia, W. B. Saunders Company, 1973.)

Table 84–2 COMPARISON OF CARDIAC OUTPUT BETWEEN MARATHONER AND NONATHLETE

	Stroke Volume (ml)	Heart Rate (beats/min)
Resting		
Nonathlete	75	75
Marathoner	105	50
Maximum		
Nonathlete	110	195
Marathoner	162	185

Thus, the normal untrained person can increase cardiac output a little over fourfold, and the well-trained athlete can increase output about sixfold. Individual marathoners have been clocked at cardiac outputs as great as 35 to 40 liters/min, seven to eight times normal resting output.

EFFECT OF TRAINING ON HEART HYPERTROPHY AND ON CARDIAC OUTPUT. From the foregoing data, it is clear that marathoners can achieve maximum cardiac outputs about 40 per cent greater than that achieved by the untrained person. This results mainly from the fact that the heart chambers of marathoners enlarge about 40 per cent; along with enlargement of the chambers, the heart mass increases 40 per cent or more. Therefore, not only do the skeletal muscles hypertrophy during athletic training but the heart does also. However, heart enlargement and increased pumping capacity occur only in the endurance types, not in the sprint types, of athletic training.

Even though the heart of the marathoner is considerably larger than that of the normal person, resting cardiac output is almost exactly the same as that in the normal person. However, this normal cardiac output is achieved by a large stroke volume at a reduced heart rate. Table 84–2 compares stroke volume and heart rate in the untrained person and the marathoner.

Thus, the heart-pumping effectiveness of each heartbeat is 40 to 50 per cent greater in the highly trained athlete than in the untrained person, but there is a corresponding decrease in heart rate at rest.

ROLE OF STROKE VOLUME AND HEART RATE IN INCREASING THE CARDIAC OUTPUT. Figure 84–10 shows the approximate changes in stroke volume and heart rate as the cardiac output increases from its resting level of about 5.5 liters/min to 30 liters/min in the marathon runner. The *stroke volume* increases from 105 to 162 milliliters, an increase of about 50 per cent, whereas the heart rate increases from 50 to 185 beats per minute, an increase of 270 per cent. Therefore, the heart rate increase accounts by far for a greater proportion of the increase in cardiac output than does the increase in stroke volume during strenuous exercise. The stroke volume normally reaches its maximum by the time the cardiac output has increased only halfway to its maximum. Any further increase in cardiac output must occur by increasing the heart rate.

RELATION OF CARDIOVASCULAR PERFORMANCE TO V̇O₂ Max. During maximal exercise, both the heart rate and the stroke volume are increased to about 95 per cent of their maximal levels. Because the cardiac output is equal to stroke volume *times* heart rate, one finds that the cardiac output is about 90 per cent of the maximum that the person can achieve. This is in contrast to about 65 per cent of maximum for pulmonary ventilation. Therefore, one can readily see

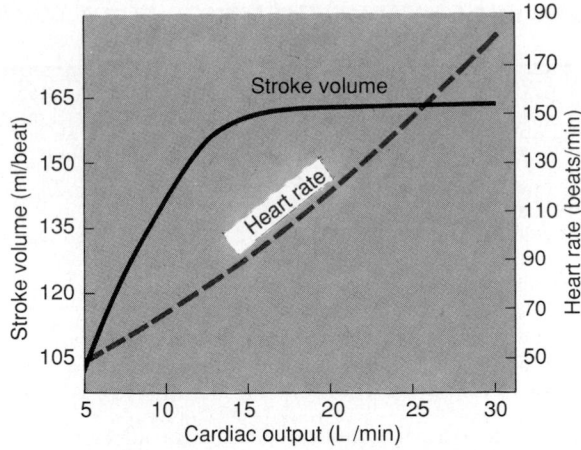

Figure 84–10. Approximate stroke volume output and heart rate at different levels of cardiac output in a marathon athlete.

that the cardiovascular system is normally much more limiting on $\dot{V}O_2$ Max than is the respiratory system because oxygen utilization by the body can never be more than the rate at which the cardiovascular system can transport oxygen to the tissues. For this reason, it is frequently stated that the performance that can be achieved by the marathoner mainly depends on his or her heart because this is the most limiting link in the delivery of adequate oxygen to the exercising muscles. Therefore, the 40 per cent greater cardiac output that the marathoner can achieve over the average untrained male is probably the single most important physiological benefit of the marathoner's training program.

EFFECT OF HEART DISEASE AND OLD AGE ON ATHLETIC PERFORMANCE. Because of the critical limitation that the cardiovascular system places on maximal performance in endurance athletics, one can readily understand that any type of heart disease that reduces the maximum cardiac output will cause an almost corresponding decrease in achievable total body muscle power. Therefore, a person with congestive heart failure frequently has difficulty achieving even the muscle power required to climb out of bed, much less to walk across the floor.

The maximum cardiac output of older people also decreases considerably—there is as much as a 50 per cent decrease between the teens and the age of 80. Also, there is even more decrease in maximal breathing capacity. For these reasons as well as reduced muscle mass, the maximum achievable muscle power is tremendously reduced.

BODY HEAT IN EXERCISE

Almost all the energy released by the internal metabolism of nutrients is eventually converted into body heat. This applies even to the energy that causes muscle contraction, for the following reasons: First, the maximum efficiency for conversion of nutrient energy into muscle work, even under the best of conditions, is only 20 to 25 per cent; the remainder of the nutrient energy is converted into heat during the course of the intracellular chemical reactions. Second,

almost all the energy that does go into creating muscle work still becomes body heat because all but a small portion of this energy is used for (1) overcoming viscous resistance to the movement of the muscles and joints, (2) overcoming the friction of the blood flowing through the blood vessels, and (3) other, similar effects—all of which convert the muscle contractile energy into heat.

Now, recognizing that the oxygen consumption by the body can increase as much as 20-fold in the well-trained athlete and that the amount of heat liberated in the body is almost exactly proportional to the oxygen consumption (as discussed in Chapter 72), one quickly realizes that tremendous amounts of heat are injected into the internal body tissues during endurance athletic events.

Next, with a vast rate of heat flow into the body, on a very hot and humid day so that the sweating mechanism cannot eliminate the heat, an intolerable and even lethal condition called *heatstroke* can easily develop in the athlete.

HEATSTROKE. During endurance athletics, even under normal environmental conditions, the body temperature often rises from its normal level of 98.6°F to 102°F to 103°F (37°C to 40°C). With very hot and humid conditions or excess clothing, the body temperature can easily rise to 106°F to 108°F (41°C to 42°C). At this level, the elevated temperature itself becomes destructive to tissue cells, especially destructive to brain cells. When this happens, multiple symptoms begin to appear, including extreme weakness, exhaustion, headache, dizziness, nausea, profuse sweating, confusion, staggering gait, collapse, and unconsciousness.

This whole complex is called heatstroke, and failure to treat it immediately can lead to death. In fact, even though the person has stopped exercising, the temperature does not easily decrease by itself. One of the reasons for this is that at these high temperatures, the temperature-regulating mechanism itself often fails (see Chapter 73). A second reason is that in heatstroke, the very high body temperature itself approximately doubles the rates of all intracellular chemical reactions, thus liberating still far more heat.

The treatment of heatstroke is to reduce the body temperature as rapidly as possible. The most practical way to do this is to remove all clothing, maintain a spray of water on all surfaces of the body or continually sponge the body, and blow air over the body with a strong fan. Experiments have shown that this treatment can reduce the temperature either as rapidly or almost as rapidly as any other procedure, though some physicians prefer total immersion of the body in ice water containing a mush of crushed ice if available.

BODY FLUIDS AND SALT IN EXERCISE

As much as a 5- to 10-pound weight loss has been recorded in athletes in a period of 1 hour during endurance athletic events under hot and humid conditions. Essentially all this weight loss results from loss of sweat. Loss of enough sweat to decrease body weight only 3 per cent can significantly diminish a person's performance, and a 5 to 10 per cent rapid decrease in weight can often be serious, leading to muscle cramps, nausea, and other effects. Therefore, it is essential to replace fluid as it is lost.

REPLACEMENT OF SALT AND POTASSIUM. Sweat contains a large amount of salt, for which reason it has long been stated that all athletes should take salt (sodium chloride) tablets when performing exercise on hot and humid days. Overuse of salt tablets has often done as much harm as good. Furthermore, if an athlete becomes acclimatized to the heat by progressive increase in athletic exposure over a period of 1 to 2 weeks rather than performing maximal athletic feats on the first day, the sweat glands also become acclimatized, so that the amount of salt lost in the sweat is only a small fraction of that before acclimatization. This sweat gland acclimatization results mainly from increased aldosterone secretion by the adrenal cortex. The aldosterone in turn has a direct effect on the sweat glands, increasing the reabsorption of sodium chloride from the sweat before it issues forth from the sweat gland tubules onto the surface of the skin. Once the athlete is acclimatized, only rarely do salt supplements need to be considered during athletic events.

On the other hand, experience by military units exposed to heavy exercise in the desert has demonstrated still another electrolyte problem—the problem of potassium loss. Potassium loss results partly from the increased secretion of aldosterone during heat acclimatization, which increases the loss of potassium in the urine as well as in the sweat. As a consequence of these new findings, some of the newer supplemental fluids for athletics are beginning to contain properly proportioned amounts of potassium along with sodium, usually in the form of fruit juices.

DRUGS AND ATHLETES

Without belaboring this issue, let us list some of the effects of drugs in athletics.

First, *caffeine* is believed by some to increase athletic performance. In one experiment on a marathon runner, his running time for the marathon was reduced by 7 per cent by judicious use of caffeine in amounts similar to those found in one to three cups of coffee. Yet experiments by others have failed to confirm any advantage, thus leaving this issue in doubt.

Second, use of *male sex hormones (androgens)* or other anabolic steroids to increase muscle strength undoubtedly can increase athletic performance under some conditions, especially in women and even in men who are poorly endowed with normal testosterone secretion. These anabolic steroids also greatly increase the risk of cardiovascular disease because they often cause hypertension, decreased high-density blood lipoproteins, and increased low-density lipoproteins, all of which promote heart attacks and strokes. In men, any type of male sex hormone preparation also leads to decreased testicular function, including both decreased formation of sperm and decreased secretion of the person's own natural testosterone, with residual effects lasting at least for many months and perhaps indefinitely. In a woman, even more dire effects can occur because she is not normally adapted to the male sex hormone—hair on the face, a bass voice, ruddy skin, cessation of menses.

Other drugs, such as *amphetamines* and *cocaine,* have been reputed to increase one's athletic performance. It is equally true that overuse of these drugs can lead to deterioration of performance. Furthermore, experiments have failed to prove the value of these drugs except as a psychic stimulant. Some athletes have been known to die during athletic events because of interaction between such drugs and the norepinephrine and epinephrine released by the sympathetic nervous system during exercise. One of the causes of death under these conditions is overexcitability of the heart, leading to ventricular fibrillation, which is lethal within seconds.

BODY FITNESS PROLONGS LIFE

Multiple studies have now shown that people who maintain appropriate body fitness, using judicious regimens of exercise and weight control, have the additional benefit of prolonged life. Especially between the ages of 50 and 70, studies have shown mortality to be three times less in the most fit people than in the least fit.

But why does body fitness prolong life? The following are the two most evident reasons.

First, body fitness and weight control greatly reduce cardiovascular disease. This results from (a) maintenance of moderately lower blood pressure and (b) reduced blood cholesterol and low-density lipoprotein along with increased high-density lipoprotein. As pointed out earlier, these changes all work together to reduce the number of heart attacks and brain strokes.

Second, and perhaps equally important, the athletically fit person has more bodily reserves to call on when he or she does become sick. For instance, an 80-year-old non-fit person may have a respiratory system that limits oxygen usage to no more than 1 liter/min; this means a *respiratory reserve of no more than threefold to fourfold.* On the other hand, the athletically fit old person may have twice as much reserve. This is especially important in preserving life when the older person develops conditions such as pneumonia that can rapidly require all available respiratory reserve. In addition, the ability to increase cardiac output in times of need (the "cardiac reserve") is often 50 per cent greater in the athletically fit old person than in the non-fit person.

REFERENCES

Australian Sports Medicine Foundation: The Textbook of Sports Nutrition. Hightstown, NJ, McGraw-Hill, 1994.
Barrow, H. M., and Brown, J. D.: Man and Movement: Principles of Physical Education, 4th Ed. Philadelphia, Lea & Febiger, 1988.
Booth, F. W., and Thomason, D. B.: Molecular and cellular adaptation of muscle in response to exercise: Perspectives of various models. Physiol. Rev., 71:541, 1991.
Brukner, P., and Miran-Khan, K.: Clinical Sports Medicine. Hightstown, NJ, McGraw-Hill, 1993.
Burstein, A. H., and Wright, T. M.: Fundamentals of Orthopaedic Biomechanics. Baltimore, Williams & Wilkins, 1994.
Chapman, M. W., and Madison, M.: Operative Orthopaedics. Philadelphia, J. B. Lippincott, 1993.
Daniels, L., and Worthingham, C.: Muscle Testing: Techniques of Manual Examination, 5th Ed. Philadelphia, W. B. Saunders Co., 1986.
DeLee, J. C., and Drez, Jr., D.: Orthopaedic Sports Medicine: Principles and Practice. Philadelphia, W. B. Saunders Co., 1994.
Engel, A. G., and Franzini-Armstrong, C. Myology. 2nd Ed. Blue Ridge Summit, PA, McGraw-Hill, 1994.
Fisher, A. G., and Jensen, C. R.: Scientific Basis of Athletic Conditioning, 3rd Ed. Philadelphia, Lea & Febiger, 1989.
Franklin, B. A., et al. (eds.): Exercise in Modern Medicine. Baltimore, Williams & Wilkins, 1988.
Fu, F. H., and Stone, D. A.: Sports Injuries: Mechanisms, Prevention, and Treatment. Baltimore, Williams & Wilkins, 1994.
Gowitzke, B. A., and Milner, M.: Scientific Bases of Human Movement, 3rd Ed. Baltimore, Williams & Wilkins, 1988.
Grana, W. A., and Kalenak, A.: Clinical Sports Medicine. Philadelphia, W. B. Saunders Co., 1991.
Guyton, A. C., et al.: Circulatory Physiology: Cardiac Output and Its Regulation. 2nd Ed. Philadelphia, W. B. Saunders Co., 1973.
Karvonen, J., et al.: Medicine in Sports Training and Coaching. Farmington, CT, S. Karger Publishers, Inc., 1992.
Kendall, F. P., et al.: Muscles: Testing and Function. Baltimore, Williams & Wilkins, 1993.
Kuettner, K. E., et al.: Articular Cartilage and Osteoarthritis. New York, Raven Press, 1992.
LeVeau, B. F.: Williams & Lissner's Biomechanics of Human Motion. Philadelphia, W. B. Saunders Co., 1992.
Marconnet, P., et al.: Muscle Fatigue Mechanisms in Exercise and Training. Farmington, CT, S. Karger Publishers, Inc., 1992.

McArdle, W. D., et al.: Essentials of Exercise Physiology. Baltimore, Williams & Wilkins, 1994.

McArdle, W. D., et al.: Exercise Physiology: Energy, Nutrition, and Human Performance. Baltimore, Williams & Wilkins, 1991.

McMeeken, J., et al.: Sports Physiotherapy. New York, Churchill Livingstone, 1994.

Miyashita, M., et al.: Medicine and Science in Aquatic Sports. Farmington, CT, S. Karger Publishers, Inc., 1994.

Nadel, E. R.: Physiological adaptations to aerobic training. Am. Sci., 73:334, 1985.

Nordin, M., and Frankel, V. H. (eds.): Basic Biomechanics of the Musculo-skeletal System, 2nd Ed. Philadelphia, Lea & Febiger, 1989.

Partridge, L. D., and Benton, L. A.: Muscle, the motor. In Brooks, V. B. (ed.): Handbook of Physiology. Sec. 1, Vol. II. Bethesda, MD., American Physiological Society, 1981, p. 43.

Pollock, M. L., and Wilmore, J. H.: Exercise in Health and Disease: Evaluation and Prescription for Prevention and Rehabilitation. Philadelphia, W. B. Saunders Co., 1990.

Poortmans, J. R.: Principles of Exercise Biochemistry. Farmington, CT, S. Karger Publishers, Inc., 1993.

Pritchard, W. G., and Pritchard, J. K.: Mathematical models of running. Am. Sci., 82:546, 1994.

Rasch, P. J. (ed.): Kinesiology and Applied Anatomy, 7th Ed. Philadelphia, Lea & Febiger, 1989.

Rash, J.: Neuromuscular Atlas. New York, Praeger Publishers, 1984.

Reider, B.: Sports Medicine: The School-Age Athlete. Philadelphia, W. B. Saunders Co., 1991.

Rose, J., and Gamble, J. G.: Human Walking. Baltimore, Williams & Wilkins, 1994.

Sanders, B.: Sports Physical Therapy. Redding, MA, Appleton & Lange, 1990.

Stanitski, C. L., et al: Pediatric and Adolescent Sports Medicine. Philadelphia, W. B. Saunders Co., 1994.

Stone, H. L.: Control of the coronary circulation during exercise. Annu. Rev. Physiol., 45:213, 1983.

Strauss, R. H.: Sports Medicine. Philadelphia, W. B. Saunders Co., 1991.

Taylor, C. R.: Carrying loads: The cost of generating muscular force. News Physiol. Sci., 1:153, 1986.

The American College of Sports Medicine: Resource Manual for Guidelines for Exercise Testing and Prescription. Baltimore, Williams & Wilkins, 1993.

Thomas, J. A. (ed.): Drugs, Athletes, and Physical Performance. New York, Plenum Publishing Corp., 1988.

Viru, A.: Adaptation in Sports Training. Boca Raton, FL, CRC Press, Inc., 1994.

Wagner, P. D.: The lungs during exercise. News Physiol. Sci., 2:6, 1987.

Walsh, E. G.: Muscles, Masses and Motion: The Physiology of Normality, Hypotonicity, Spasticity and Rigidity. New York, Cambridge University Press, 1993.

Wasserman, K., et al.: Principles of Exercise Testing and Interpretation. Baltimore, Williams & Wilkins, 1994.

Wood, S. C., and Roach, R. C.: Sports and Exercise Medicine. New York, Marcel Dekker, Inc., 1994.

Index

Note: Pages in *italics* indicate illustrations; those followed by t refer to tables.

A

A bands, in muscle fibers, 73, *75*
Abdomen, effusion fluid in, 312
  fluid leakage into, from liver, 884–885
  pain from, parietal transmission of, 616, *616*
  sympathetic innervation of, 770
  temperature receptors in, 916
  veins of, blood reservoir function of, 179–180
    compression of, 176–177, *177*
Abdominal muscles, compression of, for centrifugal acceleratory forces, 553
  compression reflex of, 217–218
  contraction of, during labor, 1043
  in breathing, 477, *478*
  spasm of, peritonitis and, 695
Abdominal pressure, lower extremity venous pressure effects of, 177
Abducens nucleus, *657*
Abortion, spontaneous, 1038
Absence syndrome, 766
Absorption, gastrointestinal, 837–838, *838*, *839*. See specific part of gastrointestinal tract.
  renal. See *Renal tubules, reabsorption in.*
Acceleration, angular, rate of, 710
  semicircular duct in, 709–710, *710*
  body effects of, 551–554, *553*, *554*
  centrifugal, 551–552
    negative G, 553
    positive G, 553
      circulatory effects of, 553, *553*
      vertebral effects of, 553
    protection against, 553
  linear, 553–554, *554*
  maculae in, 709
  position in, 554
Acclimatization, to high altitude, cellular, 551
  circulation in, 551
  diffusing capacity in, 551
  increased capillarity in, 551
  natural, 551, *552*
  oxygen levels in, 531, 550–551
  pulmonary ventilation in, 551

Acclimatization *(Continued)*
  red blood cells in, 551
  work capacity in, 551–552
Accommodation, 627
  autonomic control of, 660
  by crystalline lens, 627, *627*–628
  control of, 660
  in astigmatism, 630
  of cell membrane, 65
  of sensory receptors, 586–587
  oscillation of, 660
  parasympathetic control of, 628, 774
  pupillary reaction to, 661
Acetate, in vasodilation, 207
Acetazolamide, carbonic anhydrase inhibition by, 410
Acetoacetic acid, formation of, in liver, 868–869, 885
  insulin lack and, 975–976
  in high-fat diet, 869
  in ketosis, 869
  muscle usage of, 1063
  plasma concentration of, insulin lack and, 975, 976
  transport of, 868–869
Acetone, acetoacetic acid conversion into, 869
  breath smell of, 982
  in ketosis, 869
Acetylcholine, 925
  as parasympathetic transmitter, 771
  as small-molecule transmitter, 573
  as synaptic transmitter, 573
  cardiac effects of, 126
  destruction of, 89, 772
  D-tubocurarine effect on, 91
  duration of action of, 772
  exocytosis of, 90
  gating action of, curare and, 89, *90*
  in action potential transmission, 570
  in coronary blood flow control, 258
  in gastric secretion, 821, 822
  in muscle spasm, 91
  in nitric oxide release, 204
  in pancreatic secretion, 825
  in potassium ion leakage, 126

Acetylcholine *(Continued)*
  in skeletal muscle contraction, 75–76
  in smooth muscle contraction, 99, 101
  in vagus nerve, 126
  molecular structure of, 772
  muscle fiber effect of, 90
  pain stimulation by, 610
  pathways for, in basal ganglia, 728, *728*
  postganglionic neuron stimulation by, 780
  receptors for, 88, *88*, 773
  secretion of, at neuromuscular junction, 87–91
    by autonomic nervous system, 99
    by enteric neurons, 796
    by gigantocellular neurons, 752
    by parasympathetic nervous system, 771
    by postganglionic nerve endings, 772
    by sympathetic nervous system, 771
    calcium ions in, 88
    in REM sleep, 763
    in smooth muscle contraction, 101
  sensitivity to, after denervation, 777
  synthesis of, 90, 772
    at neuromuscular junction, 88
    botulinum toxin and, 89, *90*
    in thalamus, 749–750
    in vasodilator fibers, 212
    retinal, 646
Acetylcholine channels, 88–89, *89*
  curare and, 89, *90*
  gating of, 46
  in myasthenia gravis, 91
  in skeletal muscle contraction, 76
Acetylcholine receptors, nicotinic type of, 780
Acetylcholinesterase, action of, 87, 89, 90
  in acetylcholine destruction, 772
  inactivation of, 91
Acetyl-CoA. See *Acetylcoenzyme A.*
Acetylcoenzyme A, in citric acid cycle, 23, 859, *859*
  in diabetes mellitus, 870
  in fatty acid synthesis, 869, *870*, 871
  oxidation of, 868
  synthesis of, 23
    fatty acid beta oxidation in, 867–868, *868*

Fibrinogen (*Continued*)
function of, 879
in seminal vesicles, 1006
polymerization of, into fibrin, *465*, 465–466
structure of, 465
thrombin action on, 465
Fibrinolysin, in menstrual fluid, 1026
Fibrin-stabilizing factor, 463, 465
Fibroblast growth factor, 205
Fibroblasts, ameboid locomotion in, 24
in coagulation, 464
Fibrogen, colloid osmotic pressure effect of, 191
Fibronectin, in ameboid locomotion, 24, *24*
Fibroplasia, retrolental, 205
Fibrosis, after myocardial infarction, 262
in cardiac hypertrophy, 282
Fick principle, in cardiac output measurement, 250–251, *251*
Fight or flight reaction, 779
Filaments, in cell, 17, *17*
in muscle fibers. See *Actin filaments; Myosin filaments.*
Filtration, glomerular. See *Glomerular filtration.*
of intraocular fluid, 632, *633*
Filtration coefficient, in capillary membrane, 192–193
Filtration fraction, calculation of, 347
in peritubular colloid osmotic pressure, 343
Filtration pressure, at pulmonary capillary membrane, 496
in capillaries, capillary pressure and, 193
Flatus, 843, 851
Flavin adenine dinucleotide, 896
Flavin mononucleotide, 896
Flavoproteins, in hydrogen synthesis, 868
in oxidative phosphorylation, 861
Flexion, nucleus precommissuralis in, 712
Flexor reflex, afterdischarge in, 693, *693*
decremental contraction in, 593, *594*
fatigue of, 693, *693*
myogram of, 692–693, *693*
neuronal mechanism of, 692, 692–693
reciprocal inhibition in, 693–694, *694*
withdrawal pattern in, 693
Flocculonodular lobe, 716, 716–717, *717*
in equilibrium, 711–712
semicircular duct function and, 710
Flowmetry, Doppler, 164, *165*
electromagnetic, 164, *165*
in cardiac output measurement, 250, *250*
Fluids. See also *Blood; Water.*
blood volume and, 369, 369–370
cell volume effect of, 304, *304*
corrected osmolar activities of, 301t, 304
diffusion in, 503
distribution of, among compartments, 298, 298–299
volume of, 300–302, *301*
extracellular. See *Extracellular fluid.*
hypertonic, 304, *304*
hypotonic, 304, *304*
in homeostasis, 351
in potential spaces, 312–313
insensible loss of, 297–298, 298t
interstitial. See *Interstitial fluid.*
intracellular. See *Intracellular fluid.*
isotonic, 304, *304*
loss of, in exercise, 1068–1069
in sweat, 298
volume reflex in, 216
nitrogen bubbles in, in decompression sickness, 560, *560*

Fluids (*Continued*)
osmolarity of, 301t, 303–304
osmotic pressure of, 301t, 304
oxygen diffusion into, *514*, 514–515, *515*
regulation of, *298*
renal failure effect on, *417*, 417–418
retention of, aldosterone in, 229
angiotensin in, 228, 229, *229*
in cardiac failure, 266–268, *267*, 270
in myocardial infarction, 266–267
in peripheral edema, 270
translocation of, in weightlessness, 555
volume of, abnormalities of, 308
balance of, 297–298, 298t
in neonate, 1052
measurement of, 300–302, *301*
Fluorine, 900
caries and, 1000
Fluorosis, 900
Flush, in fever, *920*, 921
Folds of Kerckring, 837, *838*
Folic acid, 897–898
deficiency of, 429
in red blood cell maturation, 428–429
Follicle(s), antral, *1019*, 1019–1020
atresia of, 1020
development of, *1019*, 1019–1020
estrogen in, 1019, 1020
follicle-stimulating hormone in, 1019, 1020, 1028
luteinizing hormone in, 1019, 1020, 1028
maturation of, 1020
primary, 1019
primordial, 1019, *1019*
stigma of, 1020
vesicular, *1019*, 1019–1020
Follicle-stimulating hormone, 926, 1017
cell type of, 934
cyclic variation of, *1018*, 1019
function of, 933
in follicular growth, 1019, 1020, 1028
in ovulation, 1020
in spermatogenesis, 1005, 1013–1014
ovarian stimulation by, 1018, 1019
preovulatory surge of, 1028
secretion of, after menopause, *1028*, 1029
by anterior pituitary gland, 1013, 1028
corporal luteal involution and, 1021
during ovarian cycle, *1018*
estrogen effect on, 1027
gonadotropin-releasing hormone in, 1013
inhibin effect on, 1027
testosterone inhibition of, 1013
suppression of, 1028
Follicular fluid, 1019
Food, carbohydrate content of, 890t
energy availability in, 22, 889–890, 890t
energy release from, 855
fat content of, 890t
goitrogenic substances in, 955
in gastric secretion, 815–816
ingestion of, 803–806
glucose usage after, 973
regulation of, 891–893
aminostatic theory of, 891
energy usage in, 892
glucostatic theory of, 891
lipostatic theory of, 891
long-term, 892, 893
neural centers for, 891–892
short-term, 892–893
temperature and, 891
oxidation of, free energy of, 855
protein content of, 890t

Food (*Continued*)
storage of, in starvation, *894*, 894–895
volume of, in stomach emptying, 807
Foramen of Magendie, in cerebrospinal fluid flow, 785, *786*
Foramen ovale, closure of, 1051
in fetal circulation, 1050, *1050*
Foramina of Luschka, in cerebrospinal fluid flow, 785, *786*
Forced expiratory vital capacity, 539, *539*
Forced expiratory volume, 539, *539*
Fornix, 753
Fovea, cones in, 631, 648
diameter of, 631
flicking movements of, 657–658, *658*
focus clarity in, 660
in acute vision, 637, *638*
visual acuity of, 653
Fracture, muscle spasm from, 695
repair of, 991
Frank-Starling mechanism, 240
in cardiac output control, 115–116, *116*, 240
in cardiac regulation, 115–116, *116*
ventricular function curves of, 116, *116*
Free-water clearance, 358
Frequency principle, 668
Fröhlich's syndrome, 1015, *1015*
Frontal lobe, in vasomotor center control, 212
orbitofrontal area of, 752, *753*
Frontotectal tract, *657*
Frostbite, 922
Fructose, absorption of, 842
hydrolysis of, 834
in glucose synthesis, liver in, 885
in liver cells, 856, *856*
in seminal vesicles, 1006
FSH. See *Follicle-stimulating hormone.*
Functional residual capacity, 482, *483*, 504
determination of, 484
neonatal, 1052
Furosemide, action of, 409, 409t
in pulmonary edema, 270
Fusiform cells, of cerebral cortex, 733, *734*

G

G cells, gastrin secretion from, 822
GABA, as synaptic transmitter, 573
pathways for, in basal ganglia, 728, *728*
release of, by granular cells, 733
in presynaptic inhibition, 577–578
retinal, 646
G-actin, in actin filament, 77, *77*
Galactose, absorption of, 842
hydrolysis of, 834
in glucose synthesis, 885
in liver cells, 856, *856*
synthesis of, in Golgi apparatus, 21
Gallbladder, 827, *828*
bile storage in, 827
cholecystokinin in, 798, 827–829, *828*
contractility of, 798
emptying of, 827–829, *828*
pain referred from, *616*
parasympathetic effects on, 775t
sympathetic effects on, 775t
Galloping reflexes, 695
Gallstones, dissolution of, 830
formation of, 829–830, *830*
pancreatitis from, 847
Gamma globulin, formation of, in premature infant, 1054

P